Nester

Microbiology: A Human Perspective

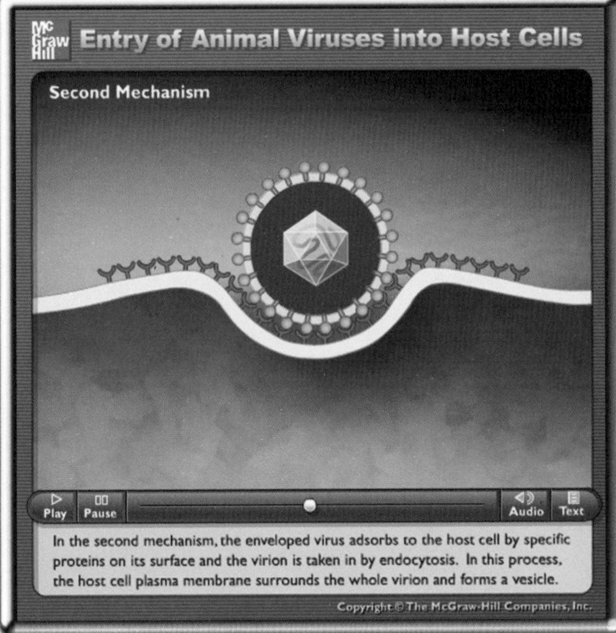
Mc Graw Hill
Entry of Animal Viruses into Host Cells
Second Mechanism
Play
Pause
Audio
Text
In the second mechanism, the enveloped virus adsorbs to the host cell by specific proteins on its surface and the virion is taken in by endocytosis. In this process, the host cell plasma membrane surrounds the whole virion and forms a vesicle.
Copyright © The McGraw-Hill Companies, Inc.

Question 1

A(n) __________ recognizes and cleaves DNA at the site of a specific palindromic sequence.
- **A)** restriction enzyme
- **B)** plasmid
- **C)** ligase
- **D)** electrophoresis

Question 2

When a restriction enzyme makes a straight cut across a strand of DNA, this is known as a
- **A)** sticky end.
- **B)** blunt end.
- **C)** ligase.
- **D)** genetic fingerprint.

Question 3

Gel electrophoresis utilizes
- **A)** ribose gel.
- **B)** an electric current.
- **C)** gene probes.
- **D)** a hybridization test.

Question 4

A hybridization test
- **A)** utilizes a nitrocellulose filter.
- **B)** lyses red blood cells.
- **C)** adds fluorescently tagged probes.
- **D)** is exposed to ultraviolet light.

ON TRACK . . .

These interactive activities and quizzing keep you motivated and on track in mastering key concepts:

- **Animations** Access to over 100 animations of key microbial processes will help you visualize and comprehend important concepts depicted in the text. The animations even include quiz questions to help ensure that you are retaining the information.
- **Test Yourself** Take a chapter quiz at the ARIS website to gauge your mastery of chapter content. Each quiz is specially constructed to test your comprehension of key concepts. Immediate feedback explains incorrect responses. You can even e-mail your quiz results to your professor!
- **Learning Activities** Helpful and engaging learning experiences await you at the *Microbiology, A Human Perspective* ARIS site. In addition to interactive online quizzing and animations, each chapter offers relevant case study presentations, digital images for creating PowerPoints®, vocabulary flash cards, and other activities designed to reinforce learning.

Mc Graw Hill
Higher Education

MICROBIOLOGY

A HUMAN PERSPECTIVE

uKnown #21

sixth edition

MICROBIOLOGY

A HUMAN PERSPECTIVE

Eugene W. Nester
University of Washington

Denise G. Anderson
University of Washington

C. Evans Roberts, Jr.
University of Washington

Martha T. Nester

McGraw Hill **Higher Education**

Boston Burr Ridge, IL Dubuque, IA New York San Francisco St. Louis
Bangkok Bogotá Caracas Kuala Lumpur Lisbon London Madrid Mexico City
Milan Montreal New Delhi Santiago Seoul Singapore Sydney Taipei Toronto

MICROBIOLOGY: A HUMAN PERSPECTIVE, SIXTH EDITION

Published by McGraw-Hill, a business unit of The McGraw-Hill Companies, Inc., 1221 Avenue of the Americas, New York, NY 10020.

 This book is printed on recycled, acid-free paper containing 10% postconsumer waste.

2 3 4 5 6 7 8 9 0 QPD/QPD 0 9

ISBN 978-0-07-299543-5
MHID 0-07-299543-2

Publisher: *Michelle Watnick*
Senior Sponsoring Editor: *James F. Connely*
Director of Development: *Kristine Tibbetts*
Senior Developmental Editor: *Lisa A. Bruflodt*
Project Coordinator: *Mary Jane Lampe*
Senior Production Supervisor: *Laura Fuller*
Senior Media Project Manager: *Tammy Juran*
Senior Designer: *David W. Hash*
Cover/Interior Designer: *Jamie E. O'Neal*
(USE) Cover Image: *color enhanced photomicrograph of Salmonella Enteritidis, ©Dennis Kunkel Microscopy, Inc.*
Senior Photo Research Coordinator: *John C. Leland*
Photo Research: *David Tietz/Editorial Image, LLC*
Compositor: *Electronic Publishing Services Inc., NY*
Typeface: *10/12 Times*
Printer: *Quebecor World Dubuque, IA*

The credits section for this book begins on page C-1 and is considered an extension of the copyright page.

Library of Congress Cataloging-in-Publication Data

Microbiology : a human perspective / Eugene W. Nester ... [et al.]. — 6th ed.
p. cm.
Includes index.
ISBN 978-0-07-299543-5 — ISBN 0-07-299543-2 (hard copy : alk. paper) 1. Microbiology.
I. Nester, Eugene W.
[DNLM: 1. Microbiological Techniques. 2. Communicable Diseases—microbiology. QW 4 M62555 2009]
QR41.2.M485 2009 2008019596
616.9'041—dc22

www.mhhe.com

We dedicate this book to our students;
we hope it helps to enrich their lives and to make them
better informed citizens,

to our families
whose patience and endurance
made completion of this project a reality,

to Anne Nongthanat Panarak Roberts
in recognition of her invaluable help,
patience, and understanding,

to our colleagues
for continuing encouragement
and advice.

BRIEF CONTENTS

CONTENTS

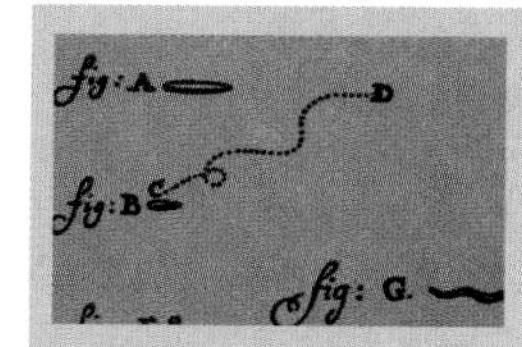

PART I
LIFE AND DEATH OF MICROORGANISMS

CHAPTER ONE
Humans and the Microbial World 1

CHAPTER TWO
The Molecules of Life 18

CHAPTER THREE
Microscopy and Cell Structure 40

CHAPTER FOUR
Dynamics of Prokaryotic Growth 83

CHAPTER FIVE
Control of Microbial Growth 107

CHAPTER SIX
Metabolism: Fueling Cell Growth 126

CHAPTER NINE

Biotechnology and Recombinant DNA 212

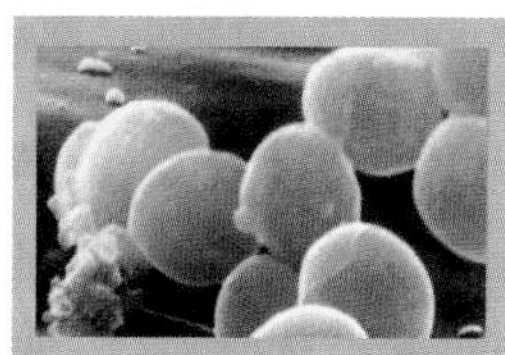

PART II
THE MICROBIAL WORLD

CHAPTER TEN

Identification and Classification of Prokaryotic Organisms 232

CHAPTER ELEVEN

The Diversity of Prokaryotic Organisms 251

CHAPTER TWELVE

The Eukaryotic Members of the Microbial World 280

CHAPTER THIRTEEN

Viruses of Bacteria 302

CHAPTER FOURTEEN

Viruses, Prions, and Viroids: Infectious Agents of Animals and Plants 320

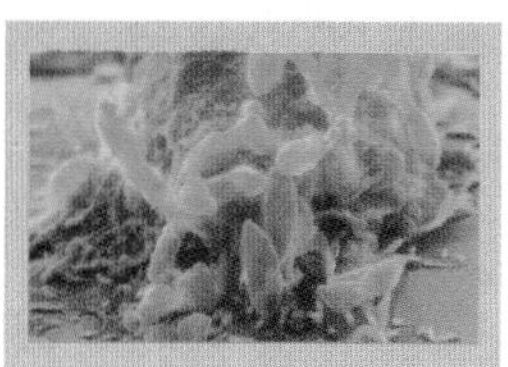

PART III
MICROORGANISMS AND HUMANS

CHAPTER FIFTEEN

The Innate Immune Response 346

CHAPTER SIXTEEN

The Adaptive Immune Response 366

CHAPTER SEVENTEEN

Host-Microbe Interactions 391

CHAPTER EIGHTEEN

Immunologic Disorders 414

CHAPTER NINETEEN

Applications of Immune Responses 431

IMMUNIZATION

IMMUNOLOGICAL TESTING

CHAPTER TWENTY

Epidemiology 450

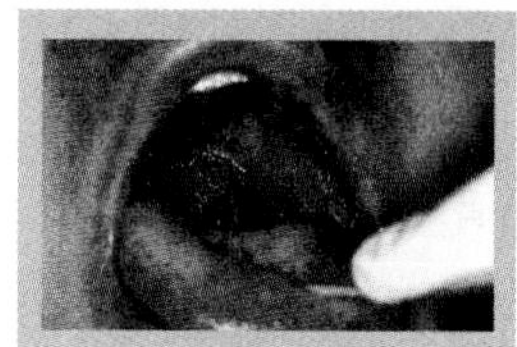

PART IV
INFECTIOUS DISEASES

CHAPTER TWENTY TWO
Respiratory System Infections 495

CHAPTER TWENTY THREE
Skin Infections 531

CHAPTER TWENTY FOUR
Wound Infections 559

PART V
APPLIED MICROBIOLOGY

CHAPTER THIRTY

CHAPTER THIRTY ONE

CHAPTER THIRTY TWO

ABOUT THE AUTHORS

Eugene Nester

Eugene (Gene) Nester performed his undergraduate work at Cornell University and received his Ph.D. in Microbiology from Case Western University. He then pursued postdoctoral work in the Department of Genetics at Stanford University with Joshua Lederberg. Since 1962, Gene has been a faculty member in the Department of Microbiology at the University of Washington. Gene's research has focused on gene transfer systems in bacteria. His laboratory demonstrated that *Agrobacterium* transfers DNA into plant cells, the basis for the disease crown gall. He continues to study this unique system of gene transfer which has become a cornerstone of plant biotechnology.

In 1990, Gene Nester was awarded the inaugural Australia Prize along with an Australian and a German scientist for their work on *Agrobacterium* transformation of plants. In 1991, he was awarded the Cetus Prize in Biotechnology by the American Society of Microbiology. He has been elected to Fellowship in the National Academy of Sciences, the American Academy for the Advancement of Science, the American Academy of Microbiology, and the National Academy of Sciences in India. Throughout his career, Gene has been actively involved with the American Society for Microbiology in several leadership positions.

In addition to his research activities, Gene has taught an introductory microbiology course for students in the allied health sciences for many years. He wrote the original version of the present text, *Microbiology: Molecules, Microbes and Man,* with C. Evans Roberts, Brian McCarthy, and Nancy Pearsall more than 30 years ago because they felt no suitable text was available for this group of students. The original text pioneered the organ system approach to the study of infectious disease.

Gene enjoys traveling, museum hopping, and the study and collecting of Northwest Coast Indian Art. He and his wife, Martha, live on Lake Washington with their labradoodle, Twana, and a well-used kayak. Their two children and four grandchildren live in the Seattle area.

Denise Anderson

Denise Anderson is a Senior Lecturer in the Department of Microbiology at the University of Washington, where she teaches a variety of courses including general microbiology, recombinant DNA techniques, medical bacteriology laboratory, and medical mycology/parasitology laboratory. Equipped with a diverse educational background, including undergraduate work in nutrition and graduate work in food science and in microbiology, she first discovered a passion for teaching when she taught microbiology laboratory courses as part of her graduate training. Her enthusiastic teaching style, fueled by regular doses of Seattle's famous caffeine, receives high reviews by her students.

Outside of academic life, Denise relaxes in the Phinney Ridge neighborhood of Seattle, where she lives with her husband, Richard Moore, and dog, Dudley (neither of whom are well trained). When not planning lectures, grading papers, or writing textbook chapters, she can usually be found chatting with the neighbors, fighting the weeds in her garden, or enjoying a fermented beverage at the local pub.

C. Evans Roberts, Jr.

Evans Roberts was a mathematics student at Haverford College when a chance encounter landed him a summer job at the Marine Biological Laboratory in Woods Hole, Massachusetts. There, interactions with leading scientists awakened an interest in biology and medicine. After finishing his degree at Haverford, he went on to get a M.D. degree at Columbia University College of Physicians and Surgeons, complete an intership at University of Rochester School of Medicine and Dentistry, and a residency in medicine at University of Washington School of Medicine where he also completed a fellowship in Infectious Diseases under Dr. William M. M. Kirby, and a traineeship in Diagnostic Microbiology under Dr. John Sherris.

Subsequently, Dr. Roberts taught microbiology at University of Washington, University of Oregon, and Chiang Mai University, in Chiagmai, Thailand, returning to University of Washington thereafter. He has directed diagnostic medical microbiology laboratories, served on hospital infection control committees, and taught infectious diseases to nurse practitioners in a camp for Karen refugees in Northern Thailand. He has had extensive experience in the practice of medicine as it relates to infectious diseases. He is certified both by the American Board of Microbiology and the American Board of Internal Medicine.

Evans Roberts worked with Gene Nester in the early development of *Microbiology: A Human Perspective*. His professional publications concern susceptibility testing as a guide to treatment of infectious diseases, etiology of Whipple's disease, group A streptococcal epidemiology, use of fluorescent antibody in diagnosis, bacteriocin typing, antimicrobial resistance in gonorrhea and tuberculosis, Japanese B encephalitis, and rabies. For relaxation, he enjoys hiking, bird watching and traveling worldwide.

Martha Nester

Martha Nester received an undergraduate degree in biology from Oberlin College and a Master's degree in education from Stanford University. She has worked in university research laboratories and has taught elementary school. She currently works in an environmental education program at the Seattle Audubon Society. Martha has worked with her husband, Gene, for more than 40 years on microbiology textbook projects, at first informally as an editor and sounding board, and then as one of the authors of *Microbiology: A Human Perspective.* Martha's favorite activities include spending time with their four grandchildren, all of whom live in the Seattle area. She also enjoys playing the cello with a number of musical groups in the Seattle area.

PREFACE

This is an exciting yet challenging time to be teaching and learning about microbiology. The need to provide accurate and current information about the good and bad microbes seems greater than ever. Almost every day a newspaper article describes illness arising from a contaminated food, the discovery of microbes in an environment once considered impossible to sustain life, the sequencing of another microbial genome, or the death of an individual from a rare infectious disease. Anyone glancing at the front page cannot help but realize the impact that microorganisms have in our daily lives. The announcements of the many scientific advances being made about the microbial world often bring with them vehement arguments related to the science. Are plants that contain genes of microorganisms safe to eat? Is it wise to put antimicrobial agents in soaps and animal feed? What agents of biological warfare might endanger the citizens of the world? Are we facing another flu pandemic? This book presents what we believe are the most important facts and concepts about the microbial world and the important role its members play in our daily lives. With the information presented, students should be able to form reasoned opinions and discuss intelligently their views on these questions.

An important consideration in revising this textbook is the diverse interests among students who take an introductory microbiology course. As always, many students take microbiology as a prerequisite for nursing, pharmacy, and dental programs. A suitable textbook must provide a solid foundation in health-related aspects of microbiology, including coverage of medically important bacteria, antimicrobial medications, and immunization. An increasing number of students take microbiology as a step in the pursuit of other fields, including biotechnology, food science, and ecology. For these students, topics such as recombinant DNA technologies, fermentation processes, and microbial diversity are essential. With the recent outbreaks of foodborne illnesses traced to products that had been distributed widely, the subject of microbial identification becomes more relevant. Microbiology is also popular as an elective for biology students, who are particularly interested in topics that highlight the relevance of microorganisms in the biological world. Because of the wide range of career goals and interests of students, we have made a particular effort to maintain a broad scope, providing a balanced approach, yet retaining our strength in medical microbiology.

Diversity in the student population is manifested not only in the range of career goals, but also in educational backgrounds. For some, microbiology may be their first college-level science course; for others, microbiology builds on an already strong background in biology and chemistry. To address this broad range of student backgrounds, we have incorporated numerous learning aids that will facilitate review for some advanced students, and will be a tremendous support to those who are seeing this material for the first time.

Preparing a textbook that satisfies such a broad range of needs and interests is a daunting task, but also extremely rewarding. We hope you will find that the approach and structure of this edition presents a modern and balanced view of microbiology in our world, acknowledging the profound and essential impact that microbes have on our lives today and their possible roles in our lives tomorrow.

Features of the Sixth Edition

Completely updated and including the most current topics in microbiology today, *Microbiology: A Human Perspective,* sixth edition, continues to be a classic. It has always been our goal to present sound scientific content that students can understand and rely upon for accuracy and currency, and thereby succeed in their preparation for meaningful careers. We have used constructive comments from numerous microbiology instructors and their students to continue to enhance the robust features of this proven text.

Expert Approach to Writing

We, as a strong and diverse team of scientists and teachers, solidly present the connection between microorganisms and humans. Because of our individual specializations and our research and educational backgrounds, we remain in the hub of the scientific community and can provide accurate and modern coverage spanning the breadth of microbiology. More importantly, as teachers, we constantly strive to present material that easily speaks to the students reading it.

We recognize that a textbook, no matter how exciting the subject matter, is not a novel. Few students will read the text from cover to cover and few instructors will include all of the topics covered in their course. We have used judicious redundancy to help present each major topic as a complete unit. We have avoided the chatty, superficial style of writing in favor of clarity and conciseness. The text is not "watered down" but rather provides students the depth of coverage needed to fully understand and appreciate the role of microorganisms in the biological sciences and human affairs.

> *"Without a doubt Microbiology: A Human Perspective is one of the most readable science texts I have ever had the pleasure of reading. The text is not scary or overly weighty in its approach to microbiology."* (Robyn Senter, Lamar State College–Orange)

"I like the simple, straight-forward wording. An introductory student with no or little background in biology should have no problems understanding these concepts." (Karen Nakaoka, Weber State University)

"Students can relate to the examples/analogies and apply them to their daily lives. The text clearly demonstrates the connection between microbes and humans!" (Michelle Fisher, Three Rivers Community College)

Instructive Art Program that Speaks a Thousand Words

Microorganisms, by definition, are invisible to the naked eye. It becomes ever more important to allow students to visualize organisms as well as processes to reinforce learning. The art program continues as a key element of the learning process. Each figure in *Microbiology: A Human Perspective* was developed as the narrative was written and is referenced in bold in the supporting text. Colors and symbols are used consistently throughout the text. Legends are short, clear, and descriptive. Various types of art styles are used as needed to bring concepts to life.

Overview Figures simplify complex interactions and provide a sound study tool. **Image Pathways** help students follow the progression of a discussion over several pages by highlighting and visualizing in detail each step of an overview figure.

Process Figures include step-by-step descriptions and supporting text so that the figure walks through a compact summary of important concepts.

Combination Figures tie together the features that can be illustrated by an artist with the appearance of organisms in the real world.

Stunning Micrographs used generously throughout the text bring the microbial world to life. In the chapters presenting infectious diseases (chapters 22 to 29), micrographs are often combined with photographs showing the symptoms that the organisms cause.

Unmatched Clinical Coverage

Evans Roberts, Jr.—a member of the author team who is licensed and board certified in internal medicine by the American Board of Internal Medicine, and in public health and medical laboratory microbiology by the American Board of Microbiology—ensures that clinical coverage is accurate, modern, and instructive to those planning to enter health careers. The incomparable treatment of infectious diseases, which are organized by human body systems, is supported with generous photographs, summary tables, case histories, and critical thinking questions. Elements of the unparalleled clinical coverage include:

- Consistent coverage of all diseases, including individual sections that describe the symptoms, pathogenesis, causative agent, epidemiology, prevention, and treatment.
- Disease summaries that feature a drawing of a human showing symptoms, portals of entry and exit, location of pathology, and a step-by-step description of the infection process for each major disease.
- Case presentations of realistic clinical situations.
- Modern coverage of topics such as emerging diseases, new vaccines, and nosocomial infections.
- Dedicated chapters covering wound infections and HIV.

Learning System that Actively Involves Students

In today's classroom, it is important to pursue active learning by students. This edition of *Microbiology* challenges students to think critically by providing several avenues of practice in analyzing data, drawing conclusions, synthesizing information, interpreting graphs, and applying concepts to practical situations. These learning tools, developed by critical thinking expert Robert Allen, will benefit students pursuing any discipline.

What's New In This Edition?

We moved the chapter on host-microbe interactions so that it now immediately follows the chapters on innate and adaptive immunity. This makes it easier for instructors to present a trilogy of topics: Part I, "The Immune Wars" (innate and adaptive immunity); Part II, "The Microbes Fight Back" (pathogenesis); and Part III, "The Return of the Humans" (vaccination, epidemiology, and antimicrobial medications). We also moved the chapter on respiratory infections forward. This puts the major discussion of *Streptococcus pyogenes* early in the infectious disease section, providing students with a solid framework to help them understand the additional coverage in subsequent chapters. The following are new features in each chapter. Other changes and updates include:

Chapter Highlights

Chapter 1

Humans and the Microbial World

- New figure showing advances in microbiology in the context of other historical events

Chapter 2

The Molecules of Life

- New section on molarity
- New table summarizing the characteristics of various sugars and their importance

Chapter 3

Microscopy and Cell Structure

- Description of the bacterial cytoskeleton has been added
- Lipid rafts in eukaryotic membranes are described
- New figure of a model bacterium emphasizing the layers that envelop the cell

Chapter 4

Dynamics of Prokaryotic Growth

- New table highlighting the impact of exponential growth
- The concept of limiting nutrients is described
- Updated figure and description of an anaerobe container

Chapter 5

Control of Microbial Growth

- New figure on membrane filtration

Chapter 6

Metabolism: Fueling Cell Growth

- The importance of microbial metabolism in the production of biofuels is discussed
- The description of the steps of glycolysis has been simplified by grouping them into two phases: investment and payoff

Chapter 7

The Blueprint of Life, from DNA to Protein

- New section describing the role of RNA interference in eukaryotic gene expression
- Alternative sigma factors are now discussed in the section on mechanisms to control transpiration
- Figures showing quorum sensing and two component regulatory systems have been added

Chapter 8

Bacterial Genetics

- Reorganized to create a new section on mobile genetic elements, highlighting the importance of horizontal gene transfer
- New table that lists mobile genetic elements

Chapter 9

Biotechnology and Recombinant DNA

- Reorganized so that methods immediately follow applications
- In recognition of the fact that many of the applications of Southern Blotting have been replaced by PCR, information on the technique has been moved to the web
- Updated information and explanatory figure on DNA sequencing
- The Human Microbiome Project is described
- Discussion of metagenomics has been added

Chapter 10

Identification and Classification of Prokaryotic Organisms

- Updated boxed story on tracing an *E. coli* O157:H7 outbreak
- Updated example of the importance of distinguishing different strains of a species

Chapter 11

The Diversity of the Prokaryotic Organisms

- New description of *Epulopiscium*
- New description of *Wolbachia*
- New equations that emphasize the energy sources and terminal electron acceptors used by the microbes covered in the section on metabolic diversity

Chapter 12

The Eukaryotic Members of the Microbial World

- Revised figure on the anatomy of the mosquito
- New Future Challenge

Chapter 13

Viruses of Bacteria

- Expanded discussion on the importance of phage
- New figure on restriction-modification
- New Perspective on mimiviruses
- Revised figure on restriction-modification

Chapter 15

The Innate Immune Response

- New figure that illustrates how lymph is formed
- Neutrophil extracellular traps (NETs) are described

Chapter 16

The Adaptive Immune Response

- The importance of regulatory T cells in preventing autoimmune disease is included
- Information on the recently discovered T_H17 cells is included in the subsets of effector helper T cells

Chapter 17

Host-Microbe Interactions

- Moved chapter forward so that it directly follows the information about innate and adaptive immunity, emphasizing the importance of evading the immune response in pathogenesis
- Added description of the hygiene hypothesis
- New Future Challenge on probiotics

Chapter 18

Immunologic Disorders

- Added information on childhood allergies and bone marrow transplantation
- Revised sections on immunotherapy, transfusion reactions, and erythroblastosis fetalis

Chapter 19

Applications of the Immune Response

- New information about the HPV vaccine
- Mention of a lipid A derivative as a new adjuvant has been added
- New application question that directs student to the vaccine schedule on the CDC website

Chapter 20

Epidemiology

- Expanded Future Challenge on bioterrorism to include category A, B, and C agents
- Expanded and renamed section on nosocomial infections so that it now reflects the general concerns regarding healthcare-associated infections
- Updated coverage of Universal Precautions (Perspective 20.1)

Chapter 21

Antimicrobial Medications

- Information about entry inhibitors and integrase inhibitors in the section on antiviral medications has been added
- Added new information about glycylcyclines

Chapter 22

Respiratory Infections

- Moved this chapter topic to the beginning of the coverage of infectious diseases so that the complete description of *Streptococcus pyogenes* is now consolidated in the section on strep throat
- Consolidated material on *Streptococcus pyogenes* from other chapters
- Information on avian influenza has been added

Chapter 23

Skin Infections

- Consolidated material on *Staphylococcus aureus* from other chapters
- Added information on MRSA
- Added a photograph of individual with erythema infectiosum

Chapter 24

Wound Infections

- Added a new case presentation on gangrene

Chapter 25

Digestive System Infections

- Revised figure on *Helicobacter pylori* infection
- Photograph of individual with herpes simplex labialis has been added
- Revised figure on cholera mode of action
- Updated figures on mumps and hepatitis A

Chapter 26

Genitourinary Infections

- Updated information on herpes simplex latency, prevention of papilloma virus infection, and changes in the HIV/AIDS pandemic

Chapter 27

Nervous System Infections

- Added a new table on the causes of meningitis; updated illustrations on West Nile and invasive *Haemophilus influenzae*

Chapter 28

Blood and Lymphatic Infections

- Updated illustrations on tularemia, yellow fever, and malaria incidence

Chapter 29

HIV Disease and Complications of Immunodeficiency

- Updated information on HIV/AIDS distribution, deaths, impact on women
- Updated nomenclature for the causative agent of pneumocystosis
- Added normal comparison figure for CMV eye involvement

Chapter 30

Microbial Ecology

- Figure illustrating how dead zones develop has been added

Chapter 32

Food Microbiology

- Updated example of an *E. coli* O157:H7 outbreak

-Teaching and Learning Supplements-

ARIS

The ARIS (Assessment, Review, and Instruction System) website that accompanies this textbook includes self-quizzing with immediate feedback, animations of key processes with self-quizzing, electronic flashcards to review key vocabulary, additional clinical case presentations and more—a whole semester's worth of study help for students. Instructors will find an instructor's manual, PowerPoint lecture outlines, and test questions that are directly tied to *Microbiology, 6/e* as well as a complete electronic homework management system where they can create and share course materials and assignments with colleagues in just a few clicks of the mouse. Instructors can also edit questions, import their own content, and create announcements and/or due dates for assignments. ARIS offers automatic grading and reporting of easy-to-assign homework, quizzing, and testing. Check out www.aris.mhhe.com, select your subject and textbook, and start benefiting today!

Presentation Center

Part of the ARIS website, the Presentation Center, contains assets such as photos, artwork, animations, PowerPoints, and other media resources that can be used to create customized lectures, visually enhance tests and quizzes, and design compelling course websites or attractive, printed support materials. All assets are copyrighted by McGraw-Hill Higher Education but can be used by instructors for classroom purposes. The visual resources in this collection include:

Art—Full-color digital files of all illustrations in the book can be readily imported into lecture presentations, exams, or custom-made classroom materials. In addition, all files are pre-inserted into blank PowerPoint slides for ease of lecture preparation.

Photos—The photos collection contains digital files of photographs from the text that can be reproduced for multiple classroom uses.

Tables—Every table that appears in the text has been saved in electronic form for use in classroom presentation and/or quizzes.

Animations—More than 50 full-color animations are available to harness the visual impact of processes in motion. Import these dynamic files into classroom presentations or online course materials.

Lecture Outlines—Specially prepared custom outlines for each chapter are offered in easy-to-use PowerPoint slides.

Online Computerized Test Bank

A comprehensive bank of test questions is provided within a computerized test bank powered by McGraw-Hill's flexible electronic testing program, EZ Test Online. EZ Test Online allows instructors to create and access paper or online tests and quizzes in an easy-to-use program anywhere, at any time, without installing the testing software. Now, with EZ Test Online, instructors can select questions from multiple McGraw-Hill test banks or author their own, and then either print the test for paper distribution or give it online.

Laboratory Manual

The sixth edition of *Microbiology Experiments: A Health Science Perspective*, by the late John Kleyn and by Mary Bicknell, has been prepared to directly support the text (although it may also be used with other microbiology textbooks). The laboratory manual features health-oriented experiments and endeavors that also reflect the goals and safety regulation guidelines of the American Society for Microbiology. Engaging student projects introduce some more intriguing members of the microbial world and expand the breadth of the manual beyond health-related topics. New experiments introduce modern techniques in biotechnology such as the use of restriction enzymes and use of a computer database to identify sequence information.

McGraw-Hill publishes additional microbiology laboratory manuals. Please contact your McGraw-Hill sales representative for more information.

Preparator's Manual for the Laboratory Manual

This invaluable guide includes answers to exercises, tips for successful experiments, lists of microbial cultures with sources and storage information, formulae and sources for stains and reagents, directions and recipes for preparing culture media, and sources of supplies. The *Preparator's Manual* is available to instructors through ARIS.

Transparencies

A set of acetate transparencies can be customized for your course. Please contact your McGraw-Hill sales representative for details.

Electronic Books

CourseSmart is a new way for faculty to find and review eTextbooks. It's also a great option for students who are interested in saving money by accessing their course materials digitally. CourseSmart offers thousands of the most commonly adopted textbooks across hundreds of courses from a wide variety of higher education publishers. It is the only place for faculty to review and compare the full text of a textbook online, providing immediate access without the environmental impact of requesting a print exam copy. At CourseSmart, students can save up to 50% off the cost of a print book, reduce their impact on the environment, and gain access to powerful web tools for learning including full text search, notes and highlighting, and email tools for sharing notes between classmates. **www.CourseSmart.com**

McGraw-Hill: Biology Digitized Video Clips

McGraw-Hill is pleased to offer adopting instructors an outstanding presentation tool—digitized biology video clips on DVD! Licensed from some of the highest-quality science video producers in the world, these brief segments range from about 5 seconds to just under 3 minutes in length and cover all areas of general biology from cells to ecosystems. Engaging and informative, McGraw-Hill's digitized videos will help capture students' interest while illustrating key biological concepts and processes such as Virus Lytic Cycle, Osmotic Effects on Blood Cells, and Anti-Immune Responses.

Course Delivery Systems

In addition to McGraw-Hill's ARIS course management options, instructors can also design and control their course content with help from our partners, WebCT, Blackboard, Top-Class, and eCollege. Course cartridges containing website content, online testing, and powerful student tracking features are readily available for use within these or any other HTML-based course management platforms.

— Reviewers of the Sixth Edition —

Gene Nester, Evans Roberts, Brian McCarthy, and Nancy Pearsall shared a vision many years ago to write a new breed of microbiology textbook especially for students planning to enter nursing and other health-related careers. Today there are other books of this type, but we were extremely gratified to learn that a majority of the students we surveyed intend to keep their copies of *Microbiology: A Human Perspective* because they feel it will benefit them greatly as they pursue their studies in these fields. Special thanks to the many students who used *Microbiology: A Human Perspective* over the years and who shared their thoughts with us about how to improve the presentation for the students who will use this edition of the text.

We offer our sincere appreciation to the many gracious and expert professionals who helped us with this revision by offering helpful suggestions. In addition to thanking those individuals listed here who carefully reviewed chapters, we also thank those who responded to our information surveys, those who participated in regional focus groups, and those participants who chose not to be identified. All of you have contributed significantly to this work and we thank you.

Cynthia Anderson, *Mt. San Antonio College*
James Barbaree, *Auburn University*
Morris Blaylock, *Darton College*
Alfred Brown, *Auburn University*
George Bullerjahn, *Bowling Green State University*
Thomas Danford, *West Virginia Northern Community College*
Charlie Dick, *Pasco Hernando Community College*
James Dickson, *Iowa State University*
Matthew Dodge, *Simmons College*
Fahd Z. Eissa, *Voorhees College*
Melissa Elliott, *Butler Community College*
Noel Espina, *Schenectady County Community College*
Michelle Fisher, *Three Rivers Community College*
Joe Gauthier, *University of AL–Birmingham*
Virginia Gutierrez-Osborne, *Fresno City College*
Katina Harris, *Tidewater Community College*
Daniel Herman, *University of WI–Eau Claire*
Chike Igboechi, *Medgar Evers College of CUNY*
Judith Krey, *Waubonsee Community College*
Ruhul Kuddus, *Utah Valley State University*
William Lorowitz, *Weber State University*
Shannon Meadows, *Roane State Community College*
Catherine Murphy, *Ocean County College*
Karen Nakaoka, *Weber State University*
Joseph Newhouse, *Lock Haven University*
Marcia Pierce, *Eastern Kentucky University*
Madhura Pradhan, *Ohio State University*
Carmen Rexach, *Mt. San Antonio College*
Susan Roman, *Georgia State University*
Barbara Rundell, *College of DuPage*
Pushpa Samkutty, *Southern University–Baton Rouge*
Robyn Senter, *Lamar State College–Orange*
Sasha Showsh, *Univ. of WI–Eau Claire*
Christina Strickland, *Clackamas Community College*
Renato Tameta, *Schenectady Community College*
Steve Thurlow, *Jackson Community College*
Michael Troyan, *Penn State University*
Richard Van Enk, *Western Michigan University*
Roger Wainwright, *University of Central Arkansas*
Winfred Watkins, *McLennan Community College*
Alan Wilson, *Darton College*
Carola Wright, *Mt. San Antonio College*

Acknowledgments

We thank our colleagues in the Department of Microbiology at the University of Washington who have lent their support of this project over many years. Our special thanks go to John Leigh, Mary Bicknell, Mark Chandler, Kendall Gray, Jimmie Lara, Sharon Schultz, Michael Lagunoff, and James Staley for their general suggestions and encouragement.

We would also like to thank Denise's husband, Richard Moore, who was "forced" to proofread and critique many of the chapters. Although he has no formal scientific education, or perhaps because of that fact, his suggestions have been instrumental in making the text more "reader-friendly." Much to his own surprise, Richard has learned enough about the fundamentals of microbiology to actually become intrigued with the subject.

Special thanks to the reviewers and other instructors who helped guide us in this revision. Deciding what to eliminate, what to add, and what to rearrange is always difficult, so we appreciate your input.

Thanks also to Deborah Allen and David Hurley, who helped shape the book through their work on earlier editions. Deborah taught us the true meaning of excellence, both by example and through gentle guidance. David was instrumental in helping us navigate the murky waters during a substantial revision that updated the coverage of innate and adaptive immunity.

A list of acknowledgments is not complete without thanking the people from McGraw-Hill—Jim Connely, Lisa Bruflodt, Tami Petsche, and Peggy Lucas—who gave inspiration and sound advice throughout this revision. Jayne Klein, Mary Jane Lampe, and our copyeditor, Sue Dillon, were instrumental in making sure the correct words actually made it onto paper.

Additionally, we would like to thank Joseph Gauthier, Elizabeth McPherson, and Donald Rubbelke for producing new media resources to support us and other instructors who lecture from our text.

We hope very much that this text will be interesting, educational for students, a help to their instructors, and will convey the excitement that we all feel for the subject. We would appreciate any comments and suggestions from our readers.

Eugene Nester
Denise Anderson
C. Evans Roberts, Jr.
Martha Nester

GUIDED TOUR

Instructive Artwork Makes the Difference

A picture is worth a thousand words, especially in microbiology. *Microbiology: A Human Perspective* employs a combination of art styles to bring concepts to life and to provide concrete, visual reinforcement of the topics discussed throughout the text.

Overview Figures

Overview figures simplify complex interactions and provide a sound study tool.

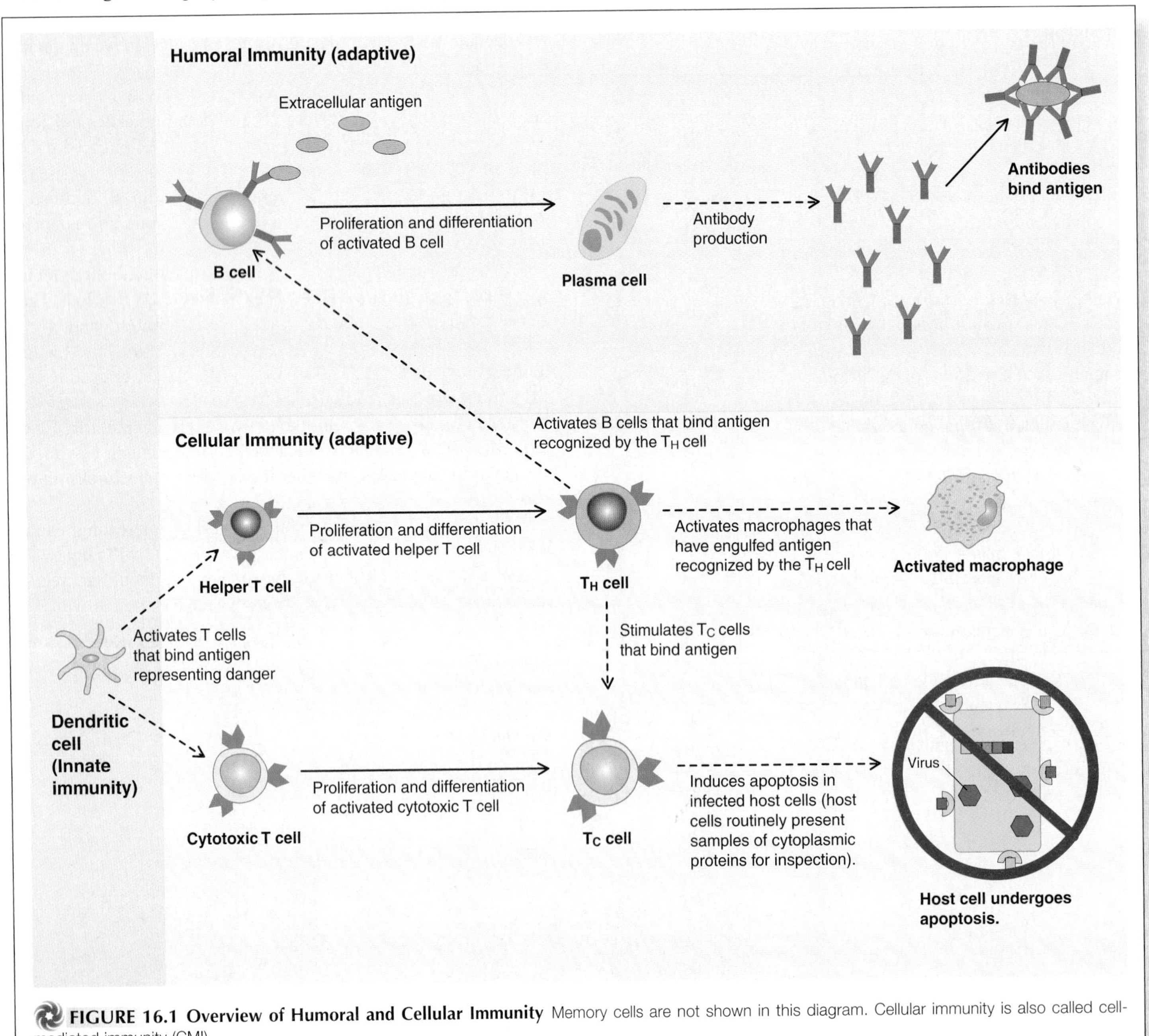

FIGURE 16.1 Overview of Humoral and Cellular Immunity Memory cells are not shown in this diagram. Cellular immunity is also called cell-mediated immunity (CMI).

Image Pathways

Image pathways help students follow the progression of a discussion over several pages by highlighting and illustrating each step of an overview figure.

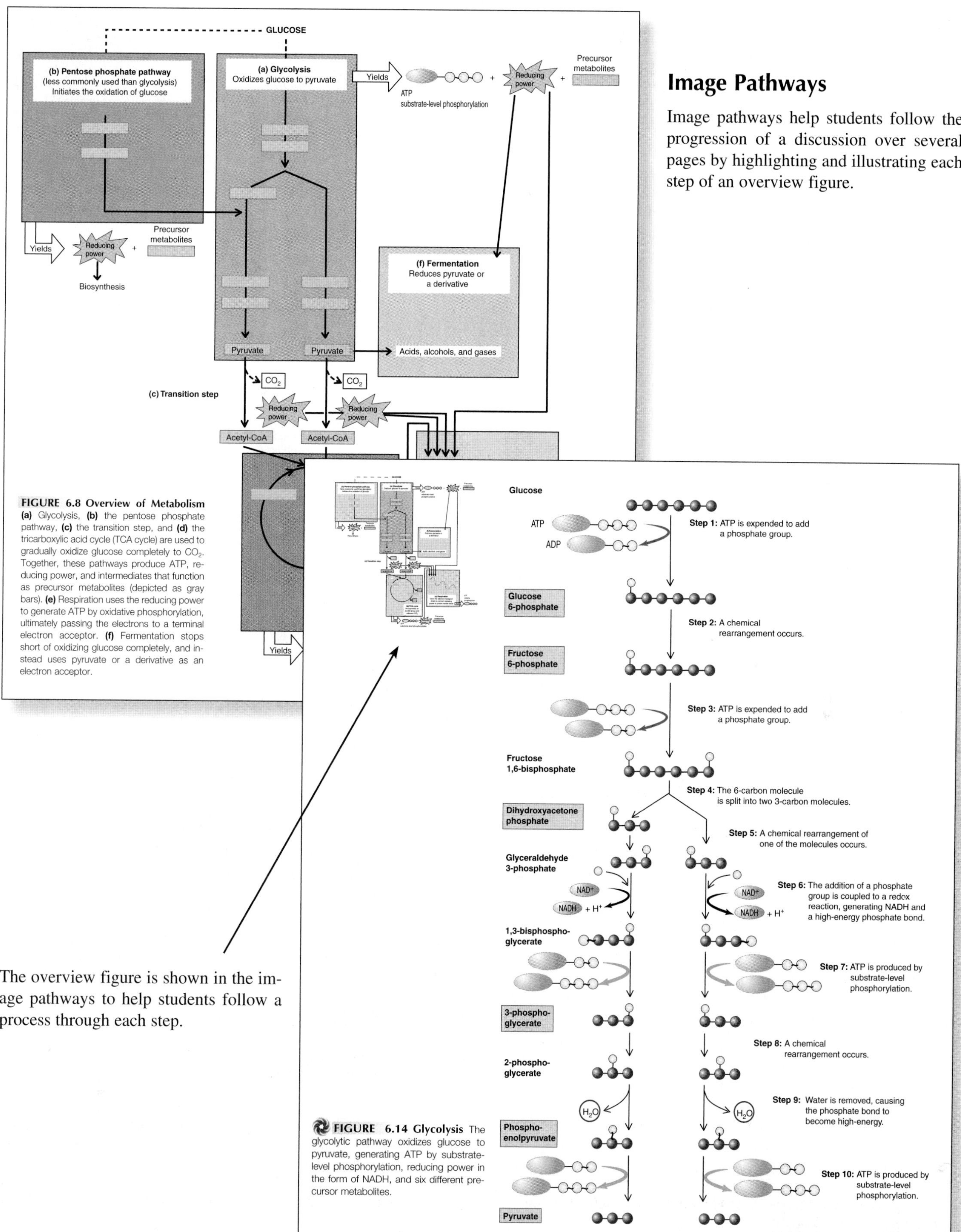

FIGURE 6.8 Overview of Metabolism **(a)** Glycolysis, **(b)** the pentose phosphate pathway, **(c)** the transition step, and **(d)** the tricarboxylic acid cycle (TCA cycle) are used to gradually oxidize glucose completely to CO_2. Together, these pathways produce ATP, reducing power, and intermediates that function as precursor metabolites (depicted as gray bars). **(e)** Respiration uses the reducing power to generate ATP by oxidative phosphorylation, ultimately passing the electrons to a terminal electron acceptor. **(f)** Fermentation stops short of oxidizing glucose completely, and instead uses pyruvate or a derivative as an electron acceptor.

FIGURE 6.14 Glycolysis The glycolytic pathway oxidizes glucose to pyruvate, generating ATP by substrate-level phosphorylation, reducing power in the form of NADH, and six different precursor metabolites.

The overview figure is shown in the image pathways to help students follow a process through each step.

Process Figures

Process figures include step-by-step descriptions to walk the student through a compact summary of important concepts.

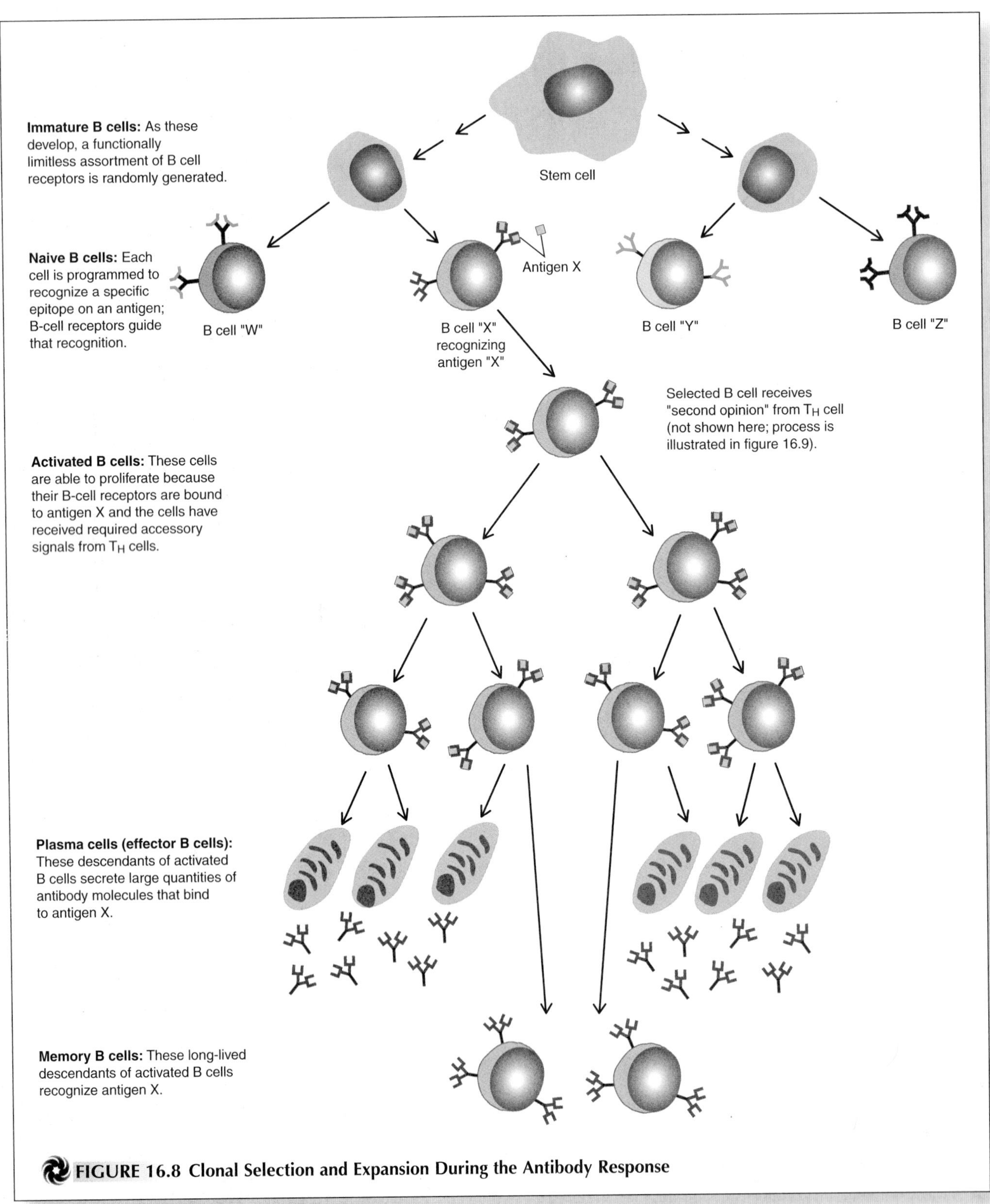

Immature B cells: As these develop, a functionally limitless assortment of B cell receptors is randomly generated.
Stem cell
Naive B cells: Each cell is programmed to recognize a specific epitope on an antigen; B-cell receptors guide that recognition.
Antigen X
B cell "W"
B cell "X" recognizing antigen "X"
B cell "Y"
B cell "Z"
Selected B cell receives "second opinion" from T_H cell (not shown here; process is illustrated in figure 16.9).
Activated B cells: These cells are able to proliferate because their B-cell receptors are bound to antigen X and the cells have received required accessory signals from T_H cells.
Plasma cells (effector B cells): These descendants of activated B cells secrete large quantities of antibody molecules that bind to antigen X.
Memory B cells: These long-lived descendants of activated B cells recognize antigen X.

FIGURE 16.8 Clonal Selection and Expansion During the Antibody Response

Combination Figures

Combination figures tie together the appearance of organisms in the real world with features that can be illustrated by an artist.

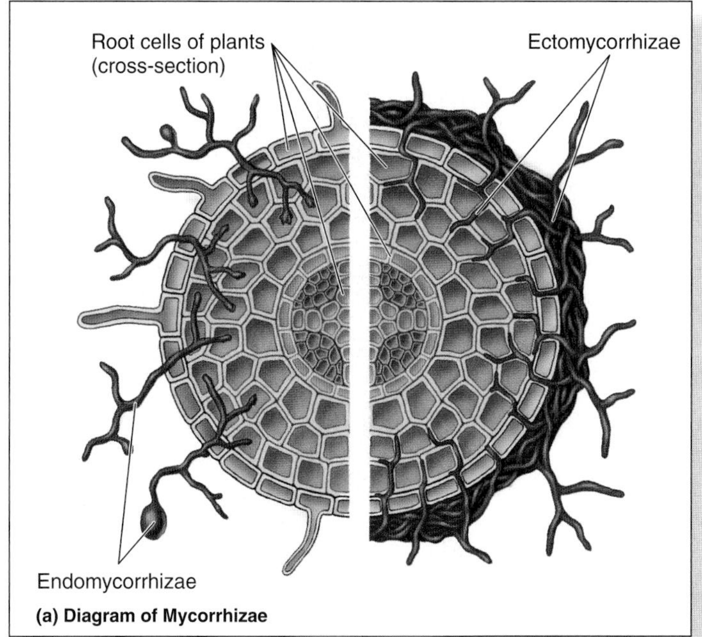

(a) Diagram of Mycorrhizae

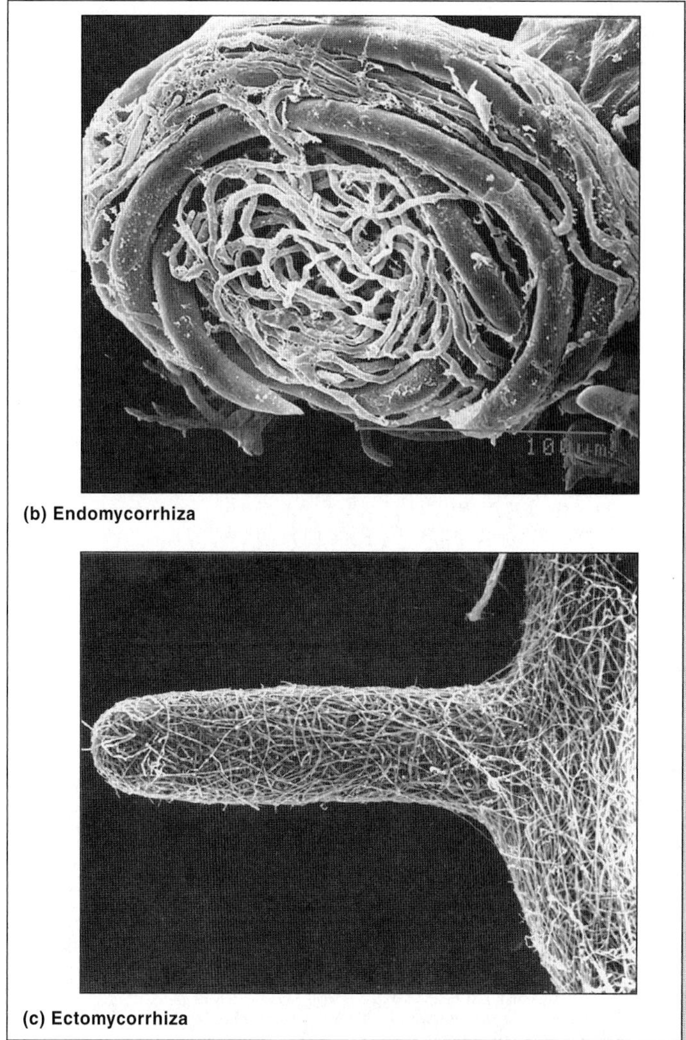

(b) Endomycorrhiza

(c) Ectomycorrhiza

Micrographs

Stunning micrographs used generously throughout the text bring the microbial world to life.

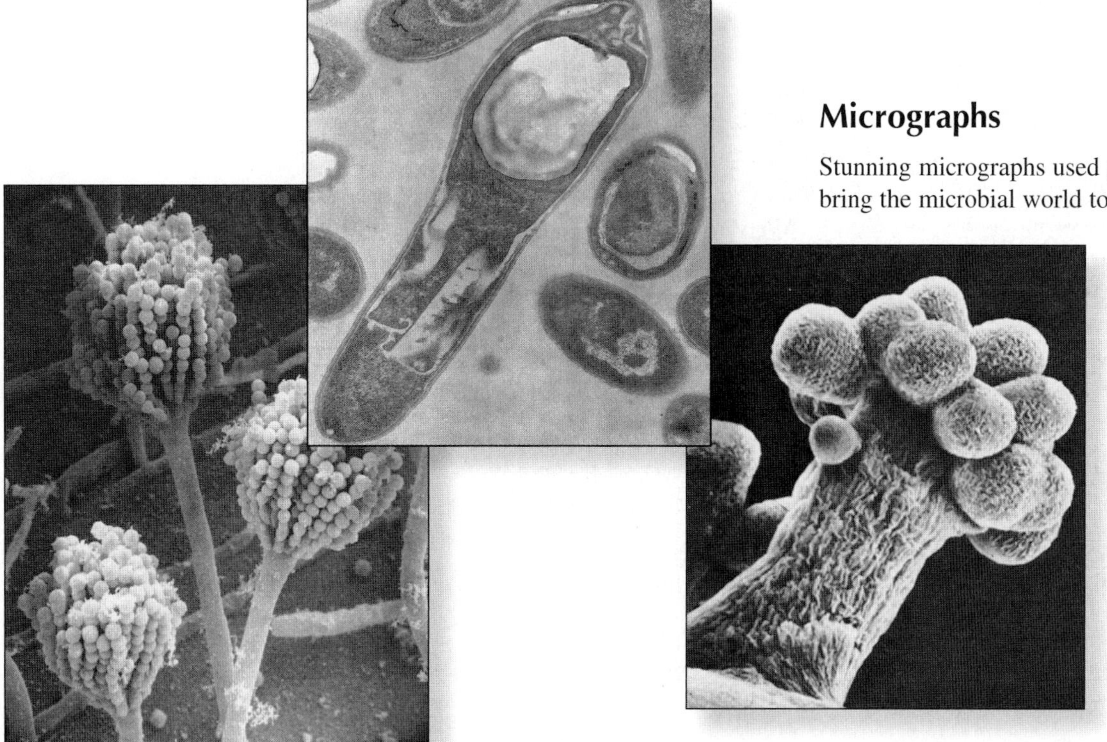

Unmatched Clinical Coverage

Organized by human body systems, the infectious disease chapters (chapters 22 to 29) are highlighted with yellow shading in the top corner of the page for easy reference. Additional case presentations and clinical reference material are available through ARIS (aris.mhhe.com).

gens. Farther inside the nasal passages, the microbial population increasingly resembles that of the **nasopharynx** (the part of pharynx behind the nose). The nasopharynx contains mostly α-hemolytic viridans streptococci, non-hemolytic streptococci, *Moraxella catarrhalis,* and diphtheroids. Anaerobic Gram-negative bacteria, including species of *Bacteroides,* are also present in large numbers in the nasopharynx. In addition, commonly pathogenic bacteria such as *Streptococcus pneumoniae, Haemophilus influenzae,* and *Neisseria meningitidis* are often found, especially during the cooler seasons of the year. ■ viridans streptococci, p. 97

MICROCHECK 22.2

Except in parts of the upper respiratory tract, the respiratory system is free of a normal microbiota. The upper respiratory tract microbiota is highly diverse, including aerobes, anaerobes, facultative anaerobes, and aerotolerant bacteria. Although most of them are of low virulence, these organisms can sometimes cause disease.

✓ What are some possible advantages to the body of providing a niche for normal flora in the upper respiratory tract?

✓ How can strict anaerobes exist in the upper respiratory tract?

INFECTIONS OF THE UPPER RESPIRATORY SYSTEM

22.3 Bacterial Infections of the Upper Respiratory System

Focus Points

- Compare the distinctive characteristics of strep throat and diphtheria.
- List the parts of upper respiratory system commonly infected by *Streptococcus pneumoniae* and *Haemophilus influenzae*.

A number of different species of bacteria can infect the upper respiratory system. Some, such as *Haemophilus influenzae* and β-hemolytic streptococci of Lancefield group C, can cause sore throats but generally do not require treatment because the bacteria are quickly eliminated by the immune system. Other infections require treatment because they are not so easily eliminated and can cause serious complications.

Strep Throat (Streptococcal Pharyngitis)

Sore throat is one of the most common reasons that people in the United States seek medical care, resulting in about 27 million doctor visits per year. Many of these visits are due to a justifiable fear of streptococcal **pharyngitis,** commonly known as strep throat.

Symptoms

Streptococcal pharyngitis typically is characterized by pain, difficulty swallowing, and fever. The throat is red, with patches of adhering pus and scattered tiny hemorrhages. The lymph nodes in the neck are enlarged and tender. Abdominal pain or headache may be prominent in older children and young adults. Not usually present are red, weepy eyes, cough, or runny nose, common symptoms with viral pharyngitis. Most patients with streptococcal sore throat recover spontaneously after about a week. In fact, many infected people have only mild symptoms or no symptoms at all.

Causative Agent

Streptococcus pyogenes, the cause of strep throat, is a Gram-positive coccus that grows in chains of varying lengths **(figure 22.2).** It can be differentiated from other streptococci that normally inhabit the throat by its characteristic colonial morphology when grown on blood agar. *Streptococcus pyogenes* produces hemolysins, enzymes that lyse red blood cells, which result in the colonies being surrounded by a zone of β-hemolysis **(figure 22.3).** Because of their characteristic hemolysis, *S. pyogenes* and other streptococci that show a similar phenotype are called β-hemolytic streptococci. In contrast, species of *Streptococcus* that are typically part of the normal throat microbiota are either non-hemolytic, or they produce α-hemolysis, characterized by a zone of incomplete, often greenish clearing around colonies grown on blood agar. ■ streptococcal hemolysis, p. 97

Streptococcus pyogenes is commonly referred to as the group A streptococcus. The group A carbohydrate in the cell wall of *S. pyogenes* differs antigenically from that of most other streptococci and serves as a convenient basis for identification (see figure 19.9). Lancefield grouping uses antibodies to differentiate the various

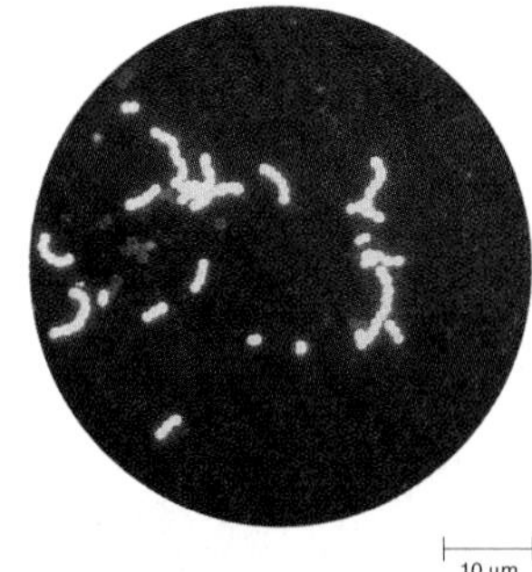

FIGURE 22.2 ***Streptococcus pyogenes*** Chain formation in fluid as revealed by fluorescence microscopy.

TABLE 22.2 Virulence Factors of *Streptococcus pyogenes*

Product	Effect
C5a peptidase	Inhibits attraction of phagocytes by destroying C5a
Hyaluronic acid capsule	Inhibits phagocytosis; aids penetration of epithelium
M protein	Interferes with phagocytosis by causing breakdown of C3b opsonin
Protein F	Responsible for attachment to host cells
Protein G	Interferes with phagocytosis by binding Fc segment of IgG
Streptococcal pyrogenic exotoxins (SPEs)	Superantigens responsible for scarlet fever, toxic shock, "flesh-eating" fasciitis
Streptolysins O and S	Lyse leukocytes and erythrocytes
Tissue degrading enzymes	Enhance spread of bacteria by breaking down DNA, proteins, blood clots, tissue hyaluronic acid

Incomparable Treatment of Diseases

Each disease is presented systematically and predictably. Individual sections describe the disease's symptoms, causative agents, pathogenesis, epidemiology, and prevention and treatment.

Disease Summaries

Major diseases are represented with a summary table that includes an outline of pathogenesis keyed to a human figure showing the entry and exit of the pathogen.

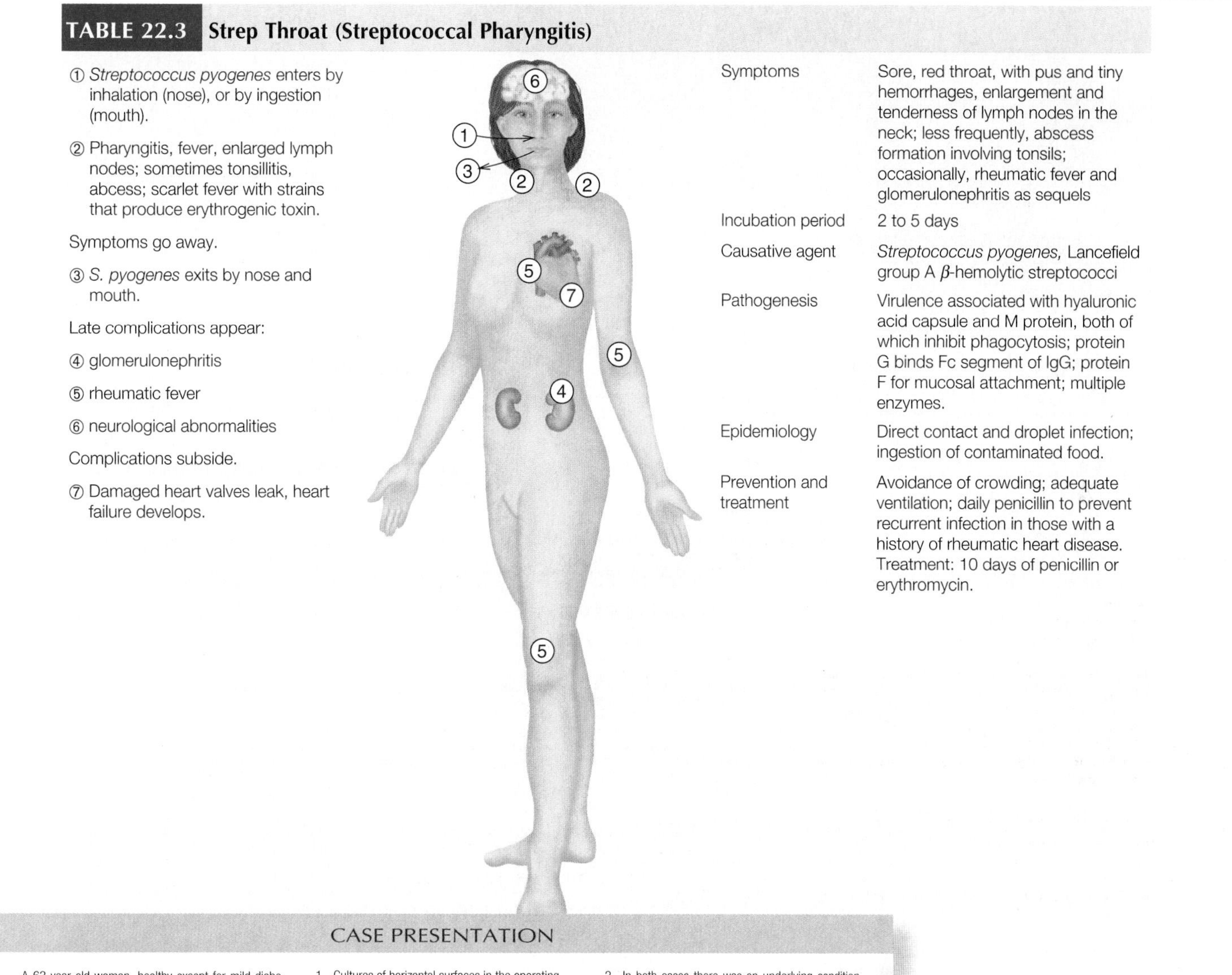

TABLE 22.3 Strep Throat (Streptococcal Pharyngitis)

① *Streptococcus pyogenes* enters by inhalation (nose), or by ingestion (mouth).

② Pharyngitis, fever, enlarged lymph nodes; sometimes tonsillitis, abcess; scarlet fever with strains that produce erythrogenic toxin.

Symptoms go away.

③ *S. pyogenes* exits by nose and mouth.

Late complications appear:

④ glomerulonephritis

⑤ rheumatic fever

⑥ neurological abnormalities

Complications subside.

⑦ Damaged heart valves leak, heart failure develops.

Symptoms	Sore, red throat, with pus and tiny hemorrhages, enlargement and tenderness of lymph nodes in the neck; less frequently, abscess formation involving tonsils; occasionally, rheumatic fever and glomerulonephritis as sequels
Incubation period	2 to 5 days
Causative agent	*Streptococcus pyogenes,* Lancefield group A β-hemolytic streptococci
Pathogenesis	Virulence associated with hyaluronic acid capsule and M protein, both of which inhibit phagocytosis; protein G binds Fc segment of IgG; protein F for mucosal attachment; multiple enzymes.
Epidemiology	Direct contact and droplet infection; ingestion of contaminated food.
Prevention and treatment	Avoidance of crowding; adequate ventilation; daily penicillin to prevent recurrent infection in those with a history of rheumatic heart disease. Treatment: 10 days of penicillin or erythromycin.

CASE PRESENTATION

A 63-year-old woman, healthy except for mild diabetes, underwent surgery for a diseased gallbladder. The surgery went well, but within 72 hours the repaired surgical incision became swollen and pale. Within hours the swollen area widened and developed a bluish discoloration. The woman's surgeon suspected gangrene. Antibiotic therapy was started and she was rushed back to the operating room where the entire swollen area, including the repaired operative incision, was surgically removed. After that, the wound healed normally, although she required a skin graft to close the large skin deficit. Large numbers of *Clostridium perfringens* grew from the wound culture.

Six days later, a 58-year-old woman underwent surgery in the same operating room for a malignant tumor of the colon. The surgery was performed without difficulty, but 48 hours later she developed rapidly advancing swelling and bluish discoloration of her surgical wound. As with the first case, gangrene was suspected and she was treated with antibiotics and surgical removal of the affected tissue. She also required skin grafting. Her wound culture also showed a heavy growth of *Clostridium perfringens.*

Because the surgery department had never had any of its patients develop surgical wound infections with *Clostridium perfringens,* much less two cases so close together, the hospital epidemiologist was asked to do an investigation. Among the findings of the investigation:

1. Cultures of horizontal surfaces in the operating room grew large numbers of *Clostridium perfringens:*
2. Unknown to the medical staff, a workman had recently serviced a fan in the ventilation system of the operating room, and for a time air was allowed to flow into the operating room, rather than out of it.
3. Heavy machinery was doing grading outside the hospital, creating clouds of dust.

As a result of these findings, the operating room and its ventilating system were cleaned and upgraded. No further cases of surgical wound gangrene developed.

1. Was the surgeon's diagnosis correct?
2. Many other patients had surgery in the same operation room. Why did only these two patients develop wound gangrene?
3. What could be done to help identify the source of the patients' infections?

Discussion

1. *Clostridium perfringens* is commonly cultivated from wounds without any evidence of infection. However, in these cases, there was not only a heavy growth of the organism but a clinical picture compatible with gangrene. The surgeon's diagnosis was undoubtedly correct.
2. In both cases there was an underlying condition that increased the two patients' susceptibility to infection—cancer in one, and diabetes in the other. Moreover, both had a recognized source for the organism. Cultures of as many as 20% of diseased gallbladders are positive for *Clostridium perfringens,* while the organism is commonly found in large numbers in the human intestine—a potential source in the case involving removal of the bowel malignancy.
3. The surgeon favored the idea that the infecting organism came from the patients themselves because such strains tend to be much more virulent than strains that live and sporulate in the soil. One the other hand, the gross contamination of the operating room as revealed by the cultures of its surfaces could indicate a very large infecting dose at the operative site, possibly compensating for lesser virulence. Moreover, no further cases occurred after cleaning the operating room and fixing the ventilation system. Unfortunately, in this case, no cultures of the excised gallbladder or bowel tumor were done, nor were the strains isolated from the wounds and the environment compared. Comparing the antibiotic susceptibility, toxin production, and other characteristics of the different isolates could have helped identify the source of the infections.

Case Presentations

Each infectious disease chapter includes a case presentation of a realistic clinical situation.

Applications Promote Further Interest

Applications throughout *Microbiology: A Human Perspective* not only help students understand microbiology's history but also how microbiology influences their daily lives and their futures.

Wine—a beverage produced using microbial metabolism.

Metabolism: Fueling Cell Growth

A Glimpse of History

In the 1850s, Louis Pasteur, a chemist, accepted the challenge of studying how alcohol arises from grape juice. Biologists had already observed that when grape juice is held in large vats, alcohol and carbon dioxide are produced and the number of yeast cells increases. They argued that the multiplying yeast cells convert the sugar in the juice to alcohol and carbon dioxide. Pasteur agreed, but could not convince two very powerful and influential German chemists, Justus von Liebig and Friedrich Wöhler, who refused to believe that microorganisms caused the breakdown of sugar. Both men lampooned the hypothesis and tried to discredit it by publishing pictures of yeast cells looking like
grape juice through one orifice and rel
alcohol through the other.

Pasteur studied the relationship b
production using a strategy common
today—that is, simplifying the experimen
ships can be more easily identified. First,
of sugar, ammonia, mineral salts, and tra
a few yeast cells. As the yeast grew, the
the alcohol level increased, indicating tha
verted to alcohol as the cells multiplied. T
living cells caused the chemical transform
would not believe the process was actua
ganisms. To convince him, Pasteur tried
inside the yeast cells that would conve
many others before him.

In 1897, Eduard Buchner, a Germ
crushed yeast cells could convert suga
now know that enzymes of the crushed
formation. For these pioneering studies,

Nobel Prize in 1907. He was the first of many investigators who received Nobel Prizes for studies on the processes by which cells degrade sugars.

To grow, all cells must accomplish two fundamental tasks. They must continually synthesize new components including cell walls, membranes, ribosomes, nucleic acids, and surface structures such as flagella. These allow the cell to enlarge and eventually divide. In addition, cells need to harvest energy and

126

Glimpse of History

Each chapter opens with an engaging story about the men and women who pioneered the field of microbiology.

Perspective Boxes

Perspective boxes introduce a "human" perspective by showing how microorganisms and their products influence our lives in a myriad of different ways.

PERSPECTIVE 1.1

The Long and the Short of It

We might assume that because prokaryotes have been so intensively studied over the past hundred years, no major surprises are left to be discovered. This, however is far from the truth. In the mid-1990s, a large, peculiar-looking organism was seen when the intestinal tracts of certain fish from both the Red Sea in the Middle East and the Great Barrier Reef in Australia were examined. This organism, named *Epulopisicium* cannot be cultured in the laboratory **(figure 1).** Its large size, 600 μm long and 80 μm wide, makes it clearly visible without any magnification, and suggested that this organism was a eukaryote. It did not, however have a membrane bound nucleus. A chemical analysis of the cell confirmed that it was a prokaryote and a member of the domain *Bacteria*. This very long, slender
ism is an exception to the rule that prokaryote
always smaller than eukaryotes.

In 1999, an even larger prokaryote in v
was isolated from the sulfurous muck of the

floor off the coast of Namibia in Africa. It is a spherical organism 70 times larger in volume than *Epulopisicium*. Since it grows on sulfur compounds and contains glistening globules of sulfur, it was named *Thiomargarita namibiensis*, which means "sulfur pearl of Namibia" **(figure 2).** Although scientists were initially skeptical that prokaryotes could be so large, there is no question in their minds now. In contrast to these large bacteria, a cell was isolated in the Mediterranean Sea that is 1 μm in width. It is a eukaryote because it contains a nucleus even though it is about the size of a typical bacterium.

How small can an organism be? An answer to this question may be at hand as a result of a new

member of the *Archaea* **(figure 3).** These tiny organisms, also members of the *Archaea*, have been named *Nanoarchaeum equitans*, which means "riding the fire sphere." The organism to which *N. equitans* is attached is *Ignicoccus*, which means "fire ball." *Ignicoccus* grows very well without its rider. *N. equitans* is spherical and only about 400 nanometers in diameter, about a quarter the diameter of *Ignicoccus*. Also, the amount of genetic information (DNA) contained in *N. equitans* is less than in any known organism, and only about one-tenth the amount found in the common gut organism, *Escherichia coli*. This sets the record for the smallest amount of DNA in any organism. Thus, this organism may contain only the essential DNA required for life. Further analysis of these cells

FIGURE 1 Longest Known Bacterium *Epulopisicium* Mixed With Paramecia Note how large this prokaryote is compared with the four eukaryotic paramecia.

FUTURE CHALLENGES

Maintaining Vigilance Against Bioterrorism

Today, an unfortunate challenge in epidemiology is to maintain vigilance against **bioterrorism**—the deliberate release of infectious agents or their toxins as a means to cause harm. Even as we work to control, and seek to eradicate, some diseases, we must be aware that microbes pose a threat as agents of bioterrorism. Hopefully, future attacks will never occur, but it is crucial to be prepared for the possibility. Prompt recognition of such an event, followed by rapid and appropriate isolation and treatment procedures, can help to minimize the consequences. The CDC, in cooperation with the Association for Professionals in Infection Control and Epidemiology (APIC), has prepared a bioterrorism readiness plan to be used as a template by healthcare facilities. Many of the recommendations are based on the Standard Precautions already employed by hospitals to prevent the spread of infectious agents (see Perspective 20.1).

The CDC separates bioterrorism agents into three categories based on the ease of spread and severity of disease. **Category A agents** pose the highest risk because they are easily spread or transmitted from person to person and result in high mortality. These agents include:

- ***Bacillus anthracis.*** Endospores of this bacterium were used in the bioterrorism events of 2001. The most severe outcome, **inhalational anthrax,** results when an individual breathes in the airborne spores. It can lead to a rapidly fatal systemic illness. **Cutaneous anthrax,** which occurs when the organism enters the skin, manifests as a blister that develops into a skin ulcer with a black center. Although this usually heals without treatment, it can also progress to a fatal bloodstream infection. **Gastrointestinal anthrax** results from consuming contaminated food, leading to vomiting of blood and severe diarrhea; it is not common but has a high mortality rate. Anthrax can be prevented by vaccination, but that option is not widely available. Prophylaxis with antimicrobial medications is possible for those who might have been exposed, but this requires prompt recognition of exposure. Fortunately, person-to-person transmission of the agent is not likely.
- **Botulism.** Botulism is caused naturally by the ingestion of botulinum toxin, produced by *Clostridium botulinum*. Any mucous membrane can absorb the toxin, so aerosolized toxin could be used as a weapon. Botulism can be prevented by vaccination, but that option is not widely available. An antitoxin is also available in limited supplies. Botulism is not contagious.
- ***Yersinia pestis.*** Pneumonic plague, caused by inhalation of *Yersinia pestis*, is the most likely form of plague to result from a biological weapon. Although no effective vaccine is available, post-exposure prophylaxis with antimicrobial medications is possible. Special isolation precautions must be used for patients who have pneumonic plague because the disease is easily transmitted by respiratory droplets.
- **Smallpox.** Although a vaccine is available to prevent infection with this virus, routine immunization was stopped over 30 years ago because the natural disease has been eradicated. As is the case with nearly all infections caused by viruses, effective drug therapy is not available. Special isolation precautions must be used for smallpox patients because the virus can be acquired through droplet, airborne, or contact transmission.
- ***Francisella tularensis.*** This bacterium, naturally found in animals such as rodents and rabbits, causes the disease tularemia. Inhalation of the bacterium results in severe pneumonia, which is incapacitating but would probably have a lower mortality rate than inhalational anthrax or plague. A vaccine is not available, but post-exposure prophylaxis with antimicrobial medications is possible. Fortunately, person-to-person transmission of the agent is not likely.
- **Viruses that cause hemorrhagic fevers.** These include various viruses such as Ebola and Marburg. Symptoms vary depending on the virus, but severe cases show signs of bleeding from many sites. There are no vaccines against these viruses, and generally no treatment. Some, but not all, of these viruses can be transmitted from person to person, so patient isolation in these cases is important.

Category B agents pose moderate risk because they are relatively easy to spread and cause moderate morbidity. These agents include organisms that cause food- and waterborne illness, various biological toxins, *Brucella* species, *Burkholderia mallei* and *pseudomallei*, *Coxiella burnetii*, and *Chlamydophila* (*Chlamydia*) *psittaci*. **Category C agents** are emerging pathogens that could be engineered for easy dissemination. These include Nipah virus, which was first recognized in 1999, and hantavirus, first recognized in 1993.

Future Challenges

Many chapters end with a pending challenge facing microbiologists and future microbiologists.

An Active Learning System

In today's classroom, it is important to pursue active learning by students. Carefully devised question and problem sets have been provided throughout the text and at the end of each chapter, allowing students to build their working knowledge of microbiology while also developing reasoning and analytical skills.

Microchecks

Major sections end with a short "Microcheck" that summarizes the major concepts in that section and offers both review questions and critical thinking questions (in blue) to assess understanding of the preceding section.

MICROCHECK 3.4

The cytoplasmic membrane is a phospholipid bilayer embedded with a variety of different proteins. It serves as a barrier between the cell and the surrounding environment, allowing relatively few types of molecules to pass through freely. The electron transport chain within the membrane expels protons, generating a proton motive force.

✓ Explain the fluid mosaic model.

✓ Name three molecules that can pass freely through the lipid bilayer.

✓ Why is the word "fluid" in fluid mosaic model an appropriate term?

160 CHAPTER SIX Metabolism: Fueling Cell Growth

6.9 Carbon Fixation

Calvin Cycle (figure 6.26)

The most common pathway used to incorporate CO_2 into an organic form is the **Calvin cycle.**

6.10 Anabolic Pathways—Synthesizing Subunits from Precursor Molecules (figure 6.27)

Lipid Synthesis

The fatty acid components of fat are synthesized by progressively adding 2-carbon units to an acetyl group. The glycerol component is synthesized from dihydroxyacetone phosphate.

Amino Acid Synthesis

Synthesis of glutamate from α-ketoglutarate and ammonia provides a mechanism for cells to incorporate nitrogen into organic molecules (figure 6.28). Synthesis of aromatic amino acids requires a multistep branching pathway. Allosteric enzymes regulate key steps of the pathway (figure 6.29).

Nucleotide Synthesis

Purine nucleotides are synthesized on the sugar-phosphate component; the pyrimidine ring is made first and then attached to the sugar-phosphate (figure 6.30).

REVIEW QUESTIONS

Short Answer

1. Explain the difference between catabolism and anabolism.
2. How does ATP serve as a carrier of free energy?
3. How do enzymes catalyze chemical reactions?
4. Explain how precursor molecules serve as junctions between catabolic and anabolic pathways.
5. How do cells regulate enzyme activity?
6. Why do the electrons carried by $FADH_2$ result in less ATP production than those carried by NADH?
7. Name three food products produced with the aid of microorganisms.
8. In photosynthesis, what is encompassed by the term "light-independent reactions?"
9. Unlike the cyanobacteria, the anoxygenic photosynthetic bacteria do not evolve oxygen (O_2). Why not?
10. What is the role of transamination in amino acid biosynthesis?

Multiple Choice

1. Which of these factors does not affect enzyme activity?
   a) temperature b) inhibitors c) coenzymes
   d) humidity e) pH
2. Which of the following statements is false? Enzymes
   a) bind to substrates.
   b) lower the energy of activation.
   c) convert coenzymes to products.
   d) speed up biochemical reactions.
   e) can be named after the kinds of reaction they catalyze.
3. Which of these is not a coenzyme?
   a) FAD b) coenzyme A c) NAD^+ d) ATP e) $NADP^+$
4. What is the end product of glycolysis?
   a) glucose b) citrate c) oxaloacetate
   d) α-ketoglutarate e) pyruvate
5. The major pathway(s) of central metabolism are
   a) glycolysis and the TCA cycle only.
   b) glycolysis, the TCA cycle, and the pentose phosphate pathway.
   c) glycolysis only.
   d) glycolysis and the pentose phosphate pathway only.
   e) the TCA cycle only.
6. Which of these pathways gives a cell the potential to produce the most ATP?
   a) TCA cycle
   b) pentose phosphate pathway
   c) lactic acid fermentation
   d) glycolysis
7. In fermentation, the terminal electron acceptor is
   a) oxygen (O_2). b) hydrogen (H_2).
   c) carbon dioxide (CO_2). d) an organic compound.
8. In the process of oxidative phosphorylation, the energy of proton motive force is used to generate
   a) NADH. b) ADP. c) ethanol. d) ATP. e) glucose.
9. In the TCA cycle, the carbon atoms contained in acetate are converted into
   a) lactic acid. b) glucose. c) glycerol.
   d) CO_2. e) all of these.
10. Degradation of fats as an energy source involves all of the following, *except*
    a) β-oxidation. b) acetyl-CoA. c) glycerol.
    d) lipase. e) transamination.

Applications

1. A worker in a cheese-making facility argues that whey, a nutrient-rich by-product of cheese, should be dumped in a nearby pond where it could serve as fish food. Explain why this proposed action could actually kill the fish by depleting the oxygen in the pond.
2. Scientists working with DNA *in vitro* often store it in solutions that contain EDTA, a chelating agent that binds magnesium (Mg^{2+}). This is done to prevent enzymes called DNases from degrading the DNA. Explain why EDTA would interfere with enzyme activity.

Critical Thinking

1. A student argued that aerobic and anaerobic respiration should produce the same amount of ATP. He reasoned that they both use basically the same process; only the terminal electron acceptor is different. What is the primary error in this student's argument?
2. Chemolithotrophs near hydrothermal vents support a variety of other life forms there. Explain how their role there is analogous to that of photosynthetic organisms in terrestrial environments.

End-of-Chapter Review

Short Answer questions review major chapter concepts. **Multiple Choice** questions allow self-testing; answers are provided in Appendix V. **Applications** provide an opportunity to use knowledge of microbiology to solve real-world problems. **Critical Thinking** questions, written by leading critical thinking expert, Robert Allen, encourage practice in analysis and problem solving that can be used in the study of any subject.

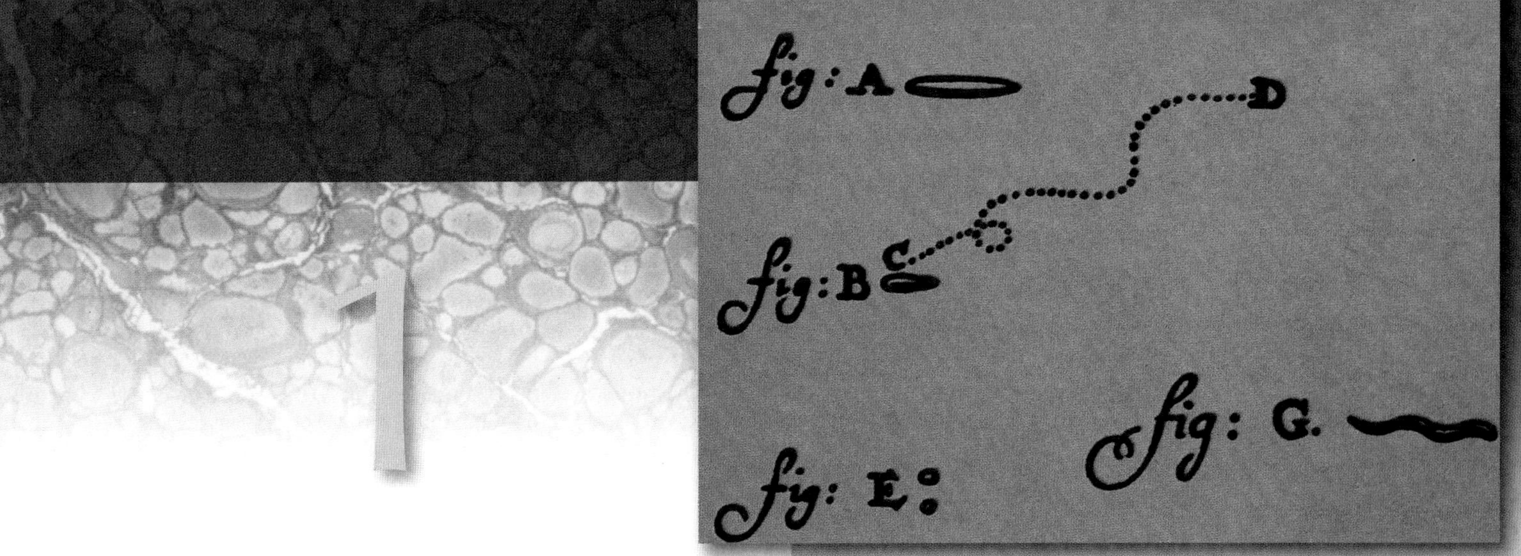

Van Leeuwenhoek's engravings (1.5×), 1695. Drawings that van Leeuwenhoek made in 1695 of the shapes of microorganisms he saw through his single lens microscope. He also observed the movement of organism B moving from C to D.

Humans and the Microbial World

A Glimpse of History

Microbiology as a science was born in 1674 when Antony van Leeuwenhoek (1632–1723), an inquisitive Dutch drapery merchant, peered at a drop of lake water through a glass lens he had carefully ground. For several centuries it was known that curved glass would magnify objects, but it took the skillful hands of a craftsman and his questioning mind to revolutionize the understanding of the world in which we live. What he observed through this simple magnifying glass was undoubtedly one of the most startling and amazing sights that humans have ever beheld—the first glimpse of the world of microbes. As van Leeuwenhoek wrote in a letter to the Royal Society of London, he saw

> "Very many little animalcules, whereof some were roundish, while others a bit bigger consisted of an oval. On these last, I saw two little legs near the head, and two little fins at the hind most end of the body. Others were somewhat longer than an oval, and these were very slow a-moving, and few in number. These animalcules had diverse colours, some being whitish and transparent; others with green and very glittering little scales, others again were green in the middle, and before and behind white; others yet were ashed grey. And the motion of most of these animalcules in the water was so swift, and so various, upwards, downwards, and round about, that 'twas wonderful to see."

Although van Leeuwenhoek was the first to observe bacteria, Robert Hooke, an English microscopist was the first to observe a microorganism. In 1665, he published a description of a microfungus, which he called a "microscopical mushroom." His drawing was so accurate that his specimen could later be identified as the common bread mold. Hooke also described how to make the kind of microscope that van Leeuwenhoek made almost 10 years later. In light of their almost simultaneous discovery of the microbial world, both men should be given equal credit for first describing the organisms you are about to study.

Microorganisms are the foundation for all life on earth. It has been said that the twentieth century was the age of physics. Now we can say that the twenty-first century will be the age of biology and biotechnology, with microbiology as the most important branch.

1.1 The Origin of Microorganisms

Focus Points

- Describe the key experiments that disproved spontaneous generation. Name the scientists who carried them out.
- Explain why endospores confused the studies on spontaneous generation.

Microorganisms have existed on earth for about 3.5 billion years, and over this time, plants and animals have evolved from these microscopic forms.

The discovery of microorganisms raised an intriguing question: "Where did these microscopic forms originate?" The theory of **spontaneous generation** suggested that organisms, such as tiny worms, can arise spontaneously from non-living material. It was completely debunked by Francesco Redi, an Italian biologist and physician, at the end of the seventeenth century. By a simple experiment, he demonstrated conclusively that worms found on rotting meat originated from the eggs of flies, not directly from the decaying meat as proponents of spontaneous generation believed. To prove this, he simply covered the meat with gauze fine enough to prevent flies from depositing their eggs. No worms appeared.

KEY TERMS

Biodiversity The variety of species inhabiting a particular environment.

Bioremediation The degradation of environmental pollutants by living organisms.

Domain The highest level in classification above the level of kingdom. All organisms can be assigned to one of three domains: *Bacteria*, *Archaea*, and *Eucarya*.

Emerging Diseases Diseases that have increased in incidence in the past 20 years.

Eukaryote Organism composed of one or more eukaryotic cells; members of the domain *Eucarya* are eukaryotes.

Eukaryotic Cell Cell type characterized by a membrane-bound nucleus.

Normal Microbiota The population of microorganisms that normally grow on the healthy human body or other specified environment.

Obligate Intracellular Parasite An organism or other agent that can only multiply inside living cells.

Pathogen An organism or virus able to cause disease.

Prokaryote Single-celled organism consisting of a prokaryotic cell; members of the domains *Bacteria* and *Archaea* are prokaryotes.

Prokaryotic Cell Cell type characterized by the lack of a membrane-bound nucleus.

Spontaneous Generation Living organisms arising from non-living material.

Theory of Spontaneous Generation Revisited

Despite Redi's work that explained the origin of worms on decaying meat, the theory of spontaneous generation of microorganisms was difficult to disprove. In fact, it took 200 more years to conclusively refute this idea. One reason for the delay was that various experiments carried out in different laboratories yielded conflicting results. For example, in 1749, John Needham, a scientist and Catholic priest, showed that various infusions (solutions obtained by soaking hay, chicken, or other nutrient source in water) gave rise to microorganisms even when the solutions had been boiled and sealed with a cork. Because even brief boiling was thought to kill all organisms, this suggested that microbes did indeed arise spontaneously. In 1776, however, the animal physiologist and priest, Father Spallanzani, observed quite the opposite; no bacteria appeared in his infusions after boiling. His experiments differed from Needham's in two significant ways: Spallanzani boiled the infusions for longer periods and he sealed the flasks by melting their glass necks. Using these techniques, he repeatedly demonstrated that infusions remained sterile (free of microorganisms). However, if the neck of the flask cracked, the infusion rapidly became cloudy. Spallanzani concluded that the microbes must have entered the broth with the air, and that the corks used by many investigators did not keep out air. However, others argued that heating the contents of the flask destroyed a "vital force" necessary for spontaneous generation. The controversy continued to rage and further experiments, with proper controls, were needed to settle the matter.

Experiments of Pasteur

One giant in science who did much to disprove the theory of spontaneous generation was the French chemist Louis Pasteur, considered by many to be the father of modern microbiology. In 1861, Pasteur refuted spontaneous generation by a series of clever experiments. First, he demonstrated that air is filled with microorganisms. He did this by filtering air through a cotton plug, trapping organisms that he then examined with a microscope. Many of these trapped organisms looked identical microscopically to those that had previously been observed by others in many infusions. Pasteur further showed that if the cotton plug was then dropped into a sterilized infusion, it became cloudy because the organisms quickly multiplied.

Most importantly, Pasteur's experiment demonstrated that sterile infusions would remain sterile in specially constructed flasks even when they were left open to the air. Organisms from the air settled in the bends and sides of these swan-necked flasks, never reaching the fluid in the bottom of the flask **(figure 1.1).** Only when the flasks were tipped would bacteria be able to enter the broth and grow. These simple and elegant experiments ended the arguments that unheated air or the infusions themselves contained a "vital force" necessary for spontaneous generation.

Experiments of Tyndall

Although most scientists were convinced by Pasteur's experiments, others were not. This skepticism in part stemmed from the fact that some reputable scientists could not reproduce Pasteur's results. One of these was an English physicist, John Tyndall. It was Tyndall who finally explained differences in experimental results obtained in different laboratories and, in turn, proved Pasteur correct. Tyndall concluded that different infusions required different boiling times to be sterilized. Thus, boiling for 5 minutes would sterilize some materials, whereas others, most notably hay infusions, could be boiled for 5 hours and they still contained living organisms! Furthermore, if hay was in the laboratory, it became almost impossible to sterilize the infusions that had previously been sterilized by boiling for 5 minutes. What did hay contain that caused this effect? Tyndall finally realized that heat-resistant forms of life were being brought into his laboratory on the hay. These heat-resistant life forms must then have been transferred to all other infusions on dust particles, thereby making everything difficult to sterilize. Tyndall concluded that some microorganisms could exist in two forms: a cell that is readily killed by boiling, and one that is heat resistant. In the same year (1876), a German botanist, Ferdinand Cohn, also discovered the heat-resistant forms of bacteria, now termed **endospores.** The following year, Robert Koch demonstrated that anthrax was caused by *Bacillus anthracis* and that the usual means of transmission in animals involved resistant spores. In 2001, the deliberate transmission of anthrax spores to humans was instigated by bioterrorists in the United States.

■ endospores, p. 69

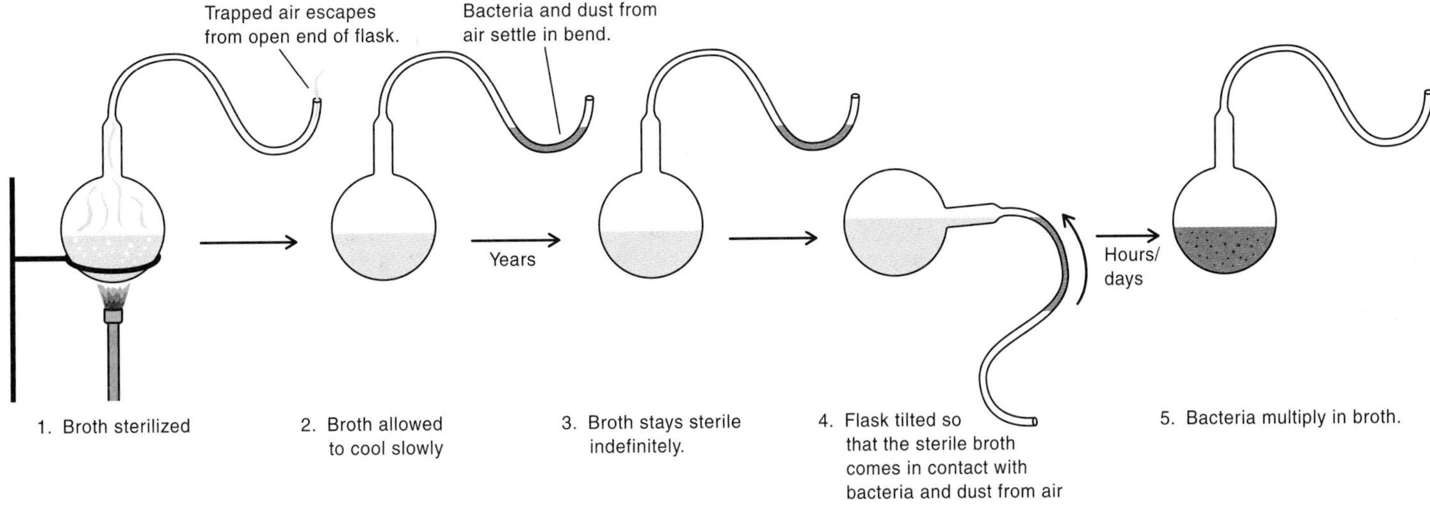
Trapped air escapes from open end of flask.
Bacteria and dust from air settle in bend.
Years
Hours/ days
1. Broth sterilized
2. Broth allowed to cool slowly
3. Broth stays sterile indefinitely.
4. Flask tilted so that the sterile broth comes in contact with bacteria and dust from air
5. Bacteria multiply in broth.

FIGURE 1.1 Pasteur's Experiment with the Swan-Necked Flask If the flask remains upright, no microbial growth occurs. **(1–3)** If the flask is tipped, the microorganisms trapped in the neck reach the sterile liquid and grow. **(4, 5)** Why did bacteria grow in the flask only after the flask was tipped?

The extreme heat resistance of endospores explains the differences between Pasteur's results and those of other investigators. Organisms that produce endospores are commonly found in the soil and most likely were present in hay infusions. Because Pasteur used only infusions prepared from sugar or yeast extract, his broth most likely did not contain endospores. At the time of these experiments on spontaneous generation, scientists did not appreciate the importance of the source of the infusion. In hindsight, the source was critical to the results observed and conclusions drawn.

These experiments on spontaneous generation point out an important lesson for all scientists. In repeating an experiment and comparing results with previous experiments, it is absolutely essential to reproduce all conditions of an experiment as closely as possible.

It may seem surprising that the concept of spontaneous generation was disproved less than a century and a half ago in light of the remarkable progress that has been made since that time. **Figure 1.2** lists some of the more important advances made over the centuries in the context of other historical events. Rather than cover more history of microbiology at this time, we will return to many of these milestones in more detail in subsequent chapters in vignettes which open each chapter. How far the science of microbiology and all biological sciences have advanced over the last 150 years!

MICROCHECK 1.1

Antony van Leeuwenhoek first observed bacteria about 300 years ago. Pasteur and Tyndall finally refuted the theory of spontaneous generation less than 150 years ago.

- ✓ Give two reasons why it took so long to disprove the theory of spontaneous generation.
- ✓ What experiment disproved the notion that a "vital force" in air was responsible for spontaneous generation?
- ✓ If Pasteur's swan-necked flasks had contained endospores, what results would have been observed?

SOMETIMES I WISH I'D GONE INTO MICRO-BIOLOGY.
4-1
THAVES
E-mail:BobThaves@aol.com
©2003 Thaves / Dist. by NEA, Inc.
www.comics.com

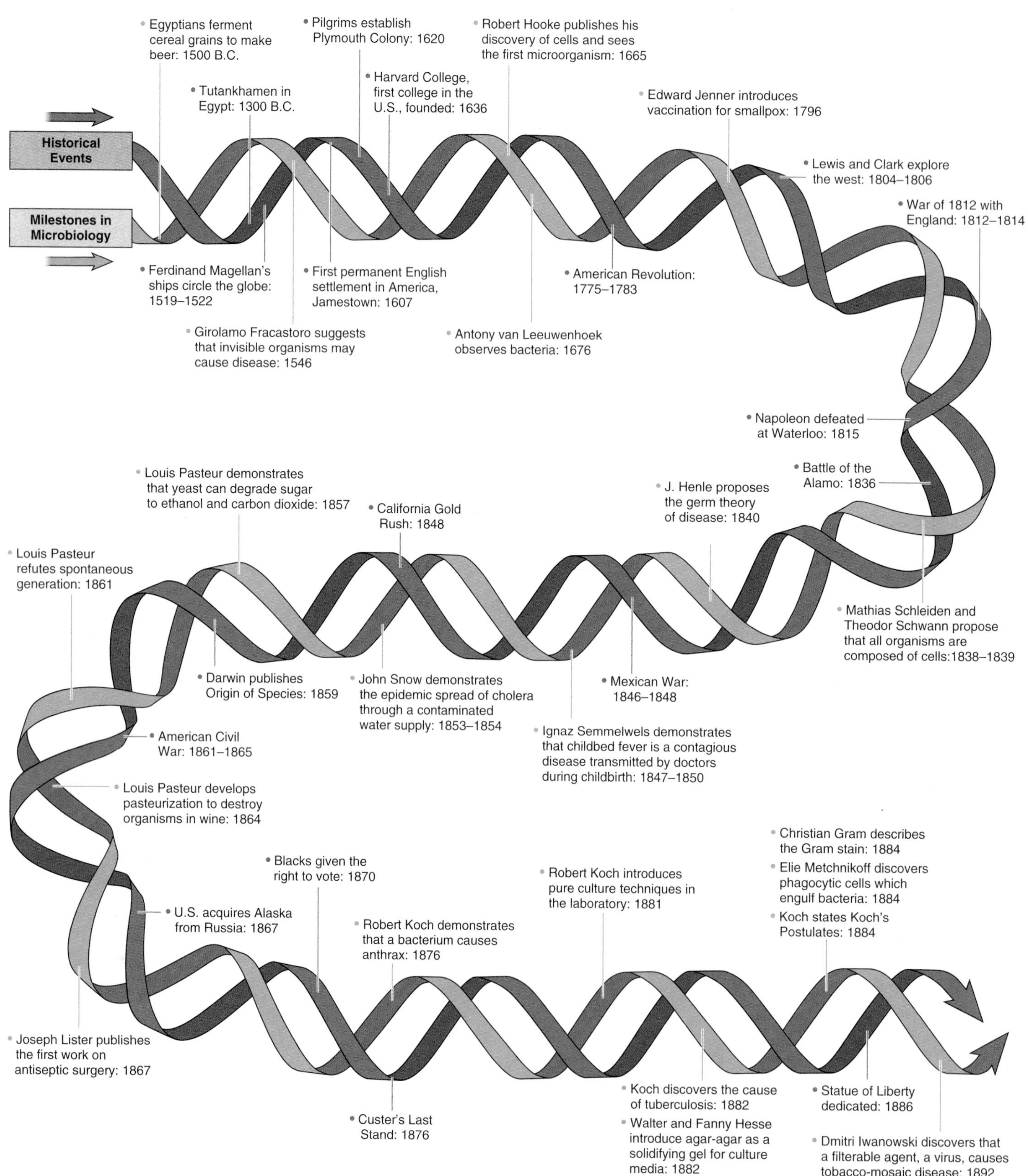

FIGURE 1.2 Some major milestones in microbiology—and their timeline in relation to other historical events.

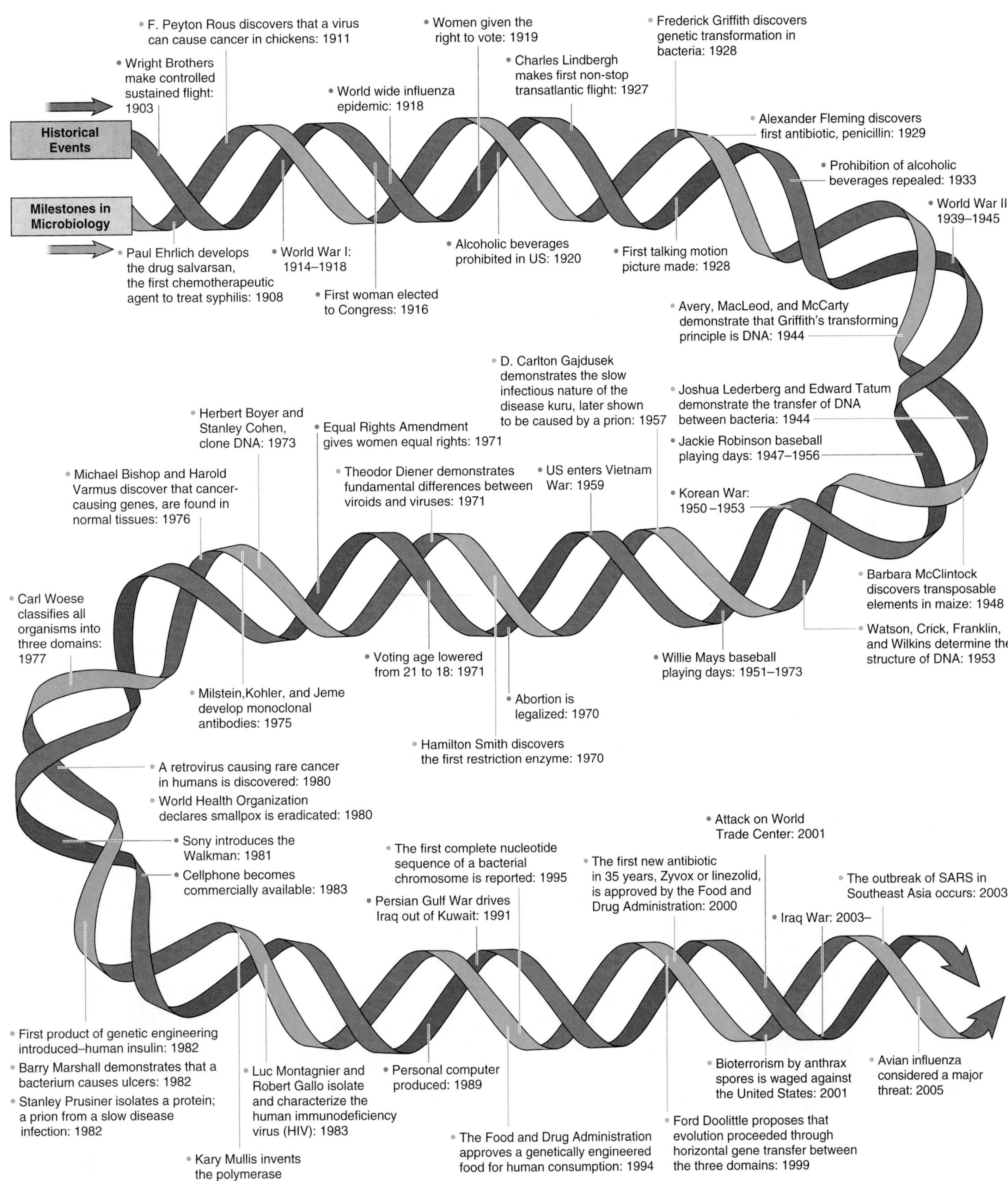

FIGURE 1.2 *(Continued)*

1.2

Microbiology: A Human Perspective

Focus Points

- List the reasons why life could not exist without microorganisms.
- Describe five applications of microbiology.
- Discuss why emerging diseases are appearing in industrialized countries.

Microoganisms have had, and continue to have, an enormous impact on all living things. On one hand, microorganisms and other infectious agents, the viruses, have killed far more people than have ever been killed in war. Yet, without microorganisms, life as we know it could not exist. They are responsible for continually recycling the carbon, oxygen, and nitrogen that all living beings require.

Features of the Microbial World

Members of the microbial world are incredibly diverse. They include bacteria, archaea, protozoa, algae, fungi, some multicellular parasites, and non-living agents such as viruses, viroids, and prions. Because they include both living and non-living forms, as a group, they are called **microbes.**

The most common feature of most members of the microbial world is that they can be seen only with the aid of a microscope. Other than their small size, microorganisms share few other properties. They are extremely diverse in their appearance, metabolism, physiology, and genetics. In fact, plants are more closely related to animals than certain bacteria are to one another. There is tremendous **biodiversity** in the microbial world, biodiversity being the variety of species present in a particular environment. The visible forms, the plants and animals, by which biodiversity is usually measured, represent only a tiny fraction of the organisms that contribute to biodiversity. Microorganisms not only represent the most abundant forms of life on earth in terms of weight, or **biomass,** but they are also the oldest and therefore have had the longest time to evolve. The most important and underappreciated forms of life are those that cannot be seen. It has been estimated that 500 to 1,000 species of bacteria live on the healthy human body and for every human cell, there are 10 bacterial cells. Yet, the true contribution and biological role of microorganisms is vastly underestimated because less than 1% of the total number of microorganisms in any environment can be cultivated in the laboratory.

Let us now consider some of the roles that microorganisms play in our lives, both beneficial and harmful. In large part this section and the remainder of this chapter will introduce you to what will be covered in more detail in later chapters.

Vital Activities of Microorganisms

The activities of microorganisms are responsible for the survival of all other organisms, including humans on this planet. A few examples readily prove this point. Nitrogen is an essential part of most of the important molecules in our bodies, such as nucleic acids and proteins. Nitrogen is also the most common gas in the atmosphere. Neither plants nor animals, however, can use nitrogen gas. Without certain bacteria that are able to convert the nitrogen in air into a form that plants can use, life as we know it would not exist.

All animals including humans require oxygen (O_2) to breathe. The supply of O_2 in the atmosphere, however, would be depleted in about 20 years, were it not replenished. On land, plants are important producers of O_2, but when all land and aquatic environments are considered, microorganisms are primarily responsible for continually replenishing the supply of O_2.

Microorganisms can also break down a wide variety of materials that no other forms of life can degrade. For example, the bulk of the carbohydrate in terrestrial (land) plants is in the form of cellulose, which humans and most animals cannot digest. Certain microorganisms can, however. As a result, leaves and downed trees do not pile up in the environment. Cellulose is also degraded by billions of microorganisms in the digestive tracts of cattle, sheep, deer, and other ruminants. The digestion products are used by the cattle for energy. Without these bacteria, ruminants would not survive. Microorganisms also play an indispensable role in degrading a wide variety of materials in sewage and wastewater.

Applications of Microbiology

In addition to the crucial roles that microorganisms play in maintaining all life, they also have made life more comfortable for humans over the centuries.

Food Production

By taking advantage of what microorganisms do naturally, Egyptian bakers as early as 2100 B.C. used yeast to make bread. Today, bakeries use essentially the same technology. ■ **breadmaking, p. 761**

The excavation of early tombs in Egypt revealed that by 1500 B.C., Egyptians employed a highly complex procedure for fermenting cereal grains to produce beer. Today, brewers use the same fundamental techniques to make beer and other fermented drinks. ■ **beer, p. 759**

Virtually every human population that has domesticated milk-producing animals such as cows and goats also has developed the technology to ferment milk to produce foods such as yogurt, cheeses, and buttermilk. Today, the bacteria added to some fermented milk products are being touted as protecting against intestinal infections and bowel cancer, the field of **probiotics.** ■ **milk products, p. 756** ■ **probiotics, p. 411**

Bioremediation

The use of living organisms to degrade environmental pollutants is termed **bioremediation.** Bacteria are being used to destroy such dangerous chemical pollutants as polychlorinated biphenyls (PCBs), dichlorodiphenyltrichloroethane (DDT), and trichloroethylene, a highly toxic solvent used in dry cleaning. All three organic compounds and many more have been detected in contaminated soil and water. Bacteria are also being used to degrade oil, assist in the cleanup of oil spills, and treat radioactive wastes. A bacterium was discovered that can live on trinitrotoluene (TNT). ■ **bioremediation, p. 86**

Useful Products from Bacteria

Bacteria can synthesize a wide variety of different products in the course of their metabolism, some of which have great commercial value. Although these same products can be synthesized in factories, bacteria often can do it faster and cheaper. For example, different bacteria produce:

- Cellulose used in stereo headsets
- Hydroxybutyric acid used in the manufacture of disposable diapers and plastics
- Ethanol, as a substitute for gasoline—a "biofuel"
- Chemicals poisonous to insects
- Antibiotics used in the treatment of disease
- Amino acids, used as dietary supplements

Medical Microbiology

In addition to the useful roles that many microbes play in our daily lives, some also play a sinister role. For example, more Americans died of influenza in 1918–1919 than were killed in World War I, World War II, the Korean War, and the Vietnam and Iraq wars combined **(figure 1.3).** Modern sanitation, vaccination, and effective antibiotic treatments have reduced the incidence of some of the worst diseases, such as smallpox, bubonic plague, and influenza, to a small fraction of their former numbers. To maintain this decrease, however, we must educate future generations to continue their vigilance. Meanwhile, another disease, acquired immunodeficiency syndrome (AIDS), has risen as a modern-day plague for which no vaccine is effective.

Past Triumphs

About the time that spontaneous generation was finally disproved to everyone's satisfaction, the Golden Age of medical microbiology was born. Between the years 1875 and 1918, most disease-causing bacteria were identified, and early work on viruses had begun. Once people realized that some of these invisible agents could cause disease, they tried to prevent their spread from sick to healthy people. The great successes in the area of human health in the last 100 years have resulted from the prevention of infectious diseases with vaccines and treatment of these diseases with antibiotics. The results have been astounding!

FIGURE 1.3 Students wearing gauze masks to protect themselves against infection with the influenza virus in 1918.

The viral disease smallpox was one of the greatest killers the world has ever known. Approximately 10 million people have died from this disease over the past 4,000 years. It was brought to the New World by the Spaniards and made it possible for Hernando Cortez, with fewer than 600 soldiers, to conquer the Aztec Empire, whose subjects numbered in the millions. During a crucial battle in Mexico City, an epidemic of smallpox raged, killing mainly the Aztecs who had never been exposed to the disease before. In recent times, an active worldwide vaccination program has resulted in no cases being reported since 1977. Although the disease will probably never reappear on its own, its potential use as an agent in bioterrorist attacks is raising great concern.

Plague has been another great killer. One-third of the entire population of Europe, approximately 25 million people, died of this bacterial disease between 1346 and 1350. Now, generally less than 100 people in the entire world die each year from this disease. In large part, this dramatic decrease is a result of controlling the population of rodents that harbor the bacterium. Further, the discovery of antibiotics in the early twentieth century made the isolated outbreaks treatable and the disease is no longer the scourge it once was.

Epidemics are not limited to human populations. In 2001, a catastrophic outbreak of foot-and-mouth disease of animals ran out of control in England. To contain this disease, one of the most contagious diseases known, almost 4 million pigs, sheep, and cattle were destroyed. Epidemic spread of diseases of food plants has led to starvation in human populations.

Present and Future Challenges

Although progress has been very impressive against bacterial diseases, a great deal still remains to be done, especially in the treatment of viral diseases and diseases that are prevalent in developing countries. Even in wealthy developed countries with their sophisticated health care systems, infectious diseases remain a serious threat. For example, about 750 million cases of infectious diseases occur in the United States each year, leading to 200,000 deaths and costing tens of billions of health care dollars. Respiratory infections and diarrheal diseases cause most illness and deaths in the world today.

Emerging Diseases In addition to the well-recognized diseases, seemingly "new" **emerging diseases** continue to arise. In the last several decades, they have included:

- Legionnaires' disease, p. 517
- Toxic shock syndrome, p. 626
- Lyme disease, p. 542
- Acquired immunodeficiency syndrome (AIDS), p. 698
- Hantavirus pulmonary syndrome, p. 522
- Hemolytic uremic syndrome, p. 598
- Mad cow disease (Bovine spongiform encephalopathy), p. 320
- West Nile virus disease, p. 660
- Severe acute respiratory syndrome (SARS), p. 528

Few of these diseases are really new, but an increased occurrence and wider distribution have brought them to the attention of health workers. Using the latest techniques, biomedical scientists have isolated, characterized, and identified the agents causing the diseases. Now, better methods need to be developed to prevent them.

A number of factors account for these emerging diseases arising even in industrially advanced countries. One reason is that changing lifestyles bring new opportunities for infectious agents to cause disease. For example, the vaginal tampons used by women provide an environment in which the organism causing toxic shock syndrome can grow and produce a toxin. In another example, the suburbs of cities are expanding into rural areas, bringing people into closer contact with animals previously isolated from humans. Consequently, people become exposed to viruses and infectious organisms that had been far removed from their environment. A good example is the hantavirus. This virus infects rodents, usually without causing disease. The infected animals, however, shed virus in urine, feces, and saliva; from there, it can be inhaled by humans as an aerosol. This disease, as well as Lyme disease, are only two of many emerging human diseases associated with small-animal reservoirs.

Some emerging diseases arise because the infectious agents change abruptly and gain the ability to infect new hosts. It appears that HIV (Human Immunodeficiency Virus), the cause of AIDS, arose from a virus that once could infect only chimpanzees. The virus causing SARS is related to viruses found in animals and may have been transmitted from animals to humans. Some bacterial **pathogens,** organisms capable of causing disease, differ from their non-pathogenic relatives in that the pathogens contain large pieces of DNA that confer on the organism the ability to cause disease. These pieces of DNA may have originated in unrelated organisms.

Figure 1.4 shows the countries in the world where, since 1976, new infectious diseases of humans and animals have first appeared. Are there other agents out there that may cause "new" diseases in the future? The answer is undoubtedly yes!

Resurgence of Old Diseases Not only are "new" diseases emerging, but many infectious diseases once on the wane in the United States have begun to increase again. One reason for this resurgence is that thousands of foreign visitors and U.S. citizens returning from travel abroad enter this country daily. About one in five comes from a country where such diseases as malaria, cholera, plague, and yellow fever still exist. In developed countries these diseases have been eliminated largely through sanitation, vaccination, and quarantine. An international traveler incubating a disease in his or her body, however, could theoretically circle the globe, touch down in several countries, and expose many people before he or she became ill. As a result these diseases are recurring in countries where they had been virtually eliminated. Further, many of these diseases are more serious today because the causative agents resist the antibiotics once used to treat them.

A second reason that certain diseases are on the rise is that in both developed and developing countries many childhood diseases have been so effectively controlled by childhood vaccinations that some parents have become lax about having their children vaccinated. The unvaccinated children are highly susceptible, and the number of those infected has increased dramatically. These diseases include measles, polio, mumps, whooping cough, and diphtheria.

A third reason for the rise in infectious diseases is that the population contains an increasing proportion of elderly

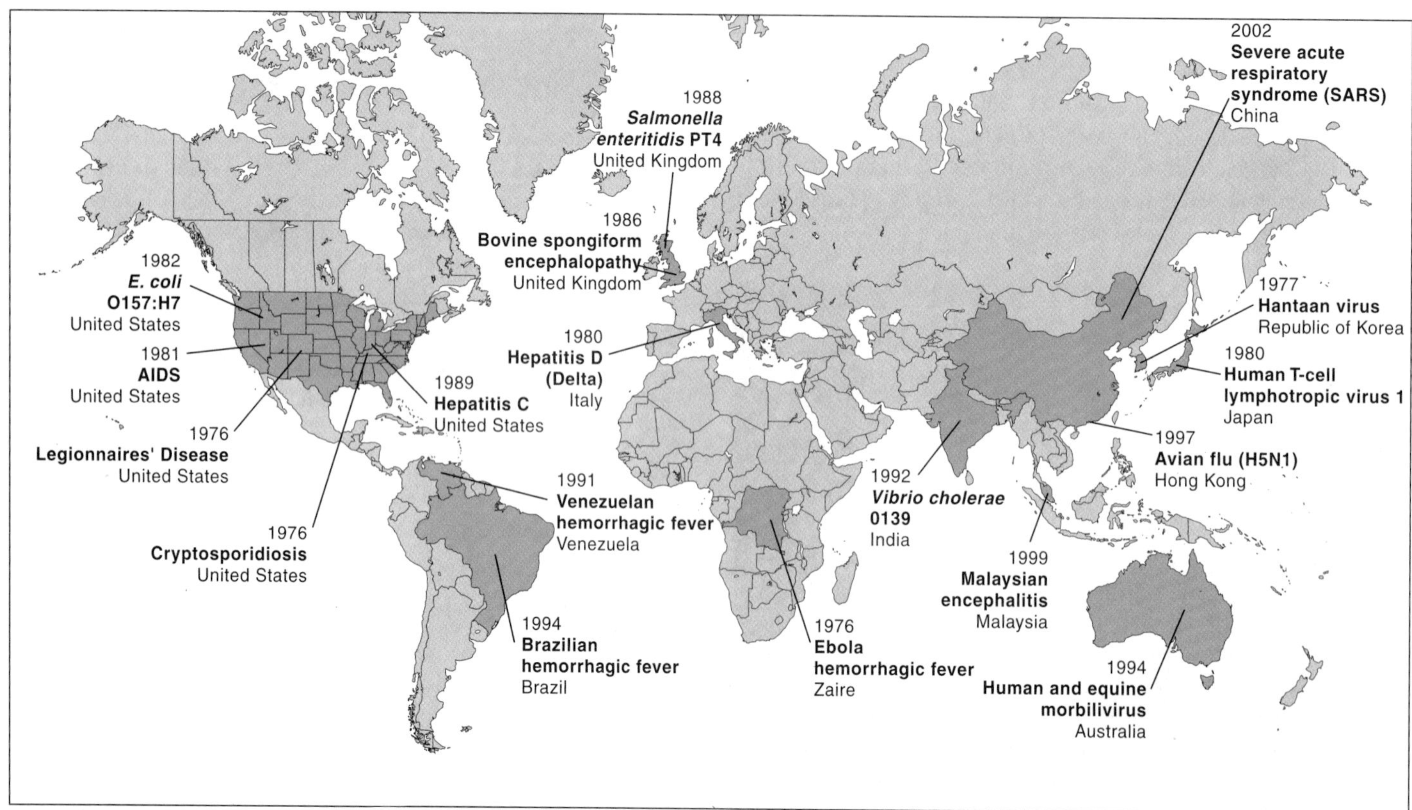

FIGURE 1.4 "New" Infectious Diseases in Humans and Animals Since 1976 Countries where cases first appeared or were identified appear in a darker shade. Why are the United States and Western European countries so prominent?

people, who have weakened immune systems and are susceptible to diseases that younger people readily resist. In addition, individuals infected with HIV are especially susceptible to a wide variety of diseases, such as tuberculosis and Kaposi's sarcoma.

Chronic Diseases Caused by Bacteria In addition to the diseases long recognized as being caused by microorganisms or viruses, some illnesses once attributed to other causes may in fact be caused by bacteria. The best-known example is peptic ulcers. This common affliction has been shown to be caused by a bacterium, *Helicobacter pylori,* and is treatable with antibiotics. Chronic indigestion, which affects 25% to 40% of the people in the Western world, may also be caused by the same bacterium. Some scientists have also implicated a bacterium in Crohn's Disease, an inflammation of the gastrointestinal tract.

In 2002, it was shown that the worm responsible for the tropical disease river blindness must contain a specific bacterium that apparently causes the disease. Infectious agents likely play roles in other diseases of unknown origin.

Host-Bacterial Interactions

All surfaces of the human body are populated with bacteria, most of which protect against disease. These bacteria are termed **normal microbiota** or normal flora. These bacteria play a number of indispensable roles in the life of the body. They successfully compete with occasional disease-causing bacteria and keep them from breaching host defenses that prevent disease. Further, they play important roles in the development of the intestine. Bacteria also degrade foodstuffs in the intestine that the body cannot digest.

Pathogenic microbes, including bacteria, fungi, protozoa, and viruses, can damage tissues of the body, leading to symptoms of disease. They use the human body as a habitat for multiplication, persistence, and transmission to other hosts. The disease symptoms are often offshoots of the body's defense mechanisms, which may damage the host as well as the pathogen.

Microorganisms As Model Organisms

Microorganisms are wonderful model organisms to study because they display the same fundamental metabolic and genetic properties found in higher forms of life. For example, all cells are composed of the same elements and synthesize their cell structures by the same basic mechanisms. They all duplicate their DNA by similar processes, and they degrade food materials to harvest energy via the same metabolic pathways. To paraphrase a Nobel Prize-winning microbiologist, Dr. Jacques Monod, what is true of elephants is also true of bacteria. The study of bacteria has many advantages. They are easy to study and results can be obtained very quickly because they grow rapidly and form billions of cells per milliliter on simple inexpensive media. Thus, most of the major advances made in the last century toward understanding life have come through the study of microorganisms. The number of Nobel Prizes that have been awarded to microbiologists, and especially the ones awarded in 2001, proves this point (see inside cover). Such studies constitute basic research, and they continue today.

MICROCHECK 1.2

Microorganisms are essential to all life and affect the life of humans in both beneficial and harmful ways. Microorganisms have been used for food production for thousands of years using essentially the same techniques that are used today. They are now being used to degrade toxic pollutants and produce a variety of compounds more cheaply than can be done in the chemical laboratory. Enormous progress has been made in preventing and curing most infectious diseases, but new ones continue to emerge around the world. Microbes represent wonderful model organisms, and many principles of biochemistry and genetics have been discovered from studying them.

- ✓ Discuss activities that microbes carry out that are essential to life on earth.
- ✓ Discuss several reasons for the reemergence of old diseases.
- ✓ Why would it seem logical, even inevitable, that at least some bacteria would attack the human body and cause disease?

1.3

Members of The Microbial World

Focus Points

- Name the three domains of life and their distinguishing properties.
- Compare and contrast the three eukaryotic groups of the microbial world.

The microbial world includes the kinds of cells that van Leeuwenhoek observed looking through his simple microscope **(figure 1.5).** Although he could not realize it at the time, members

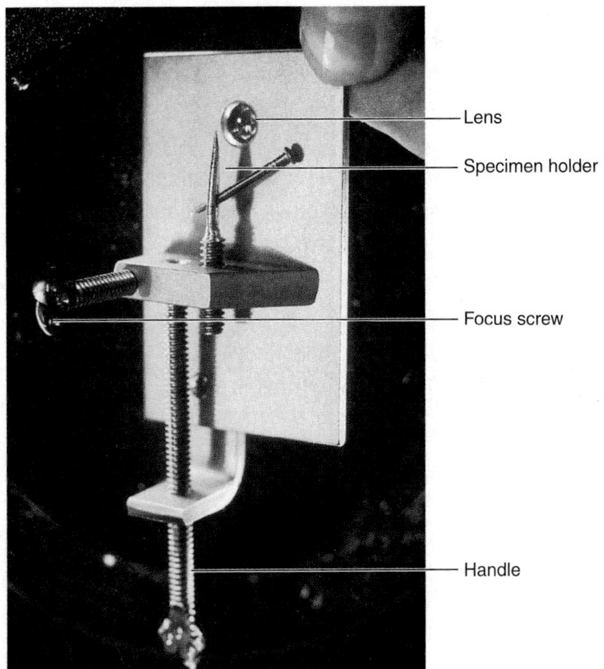

FIGURE 1.5 Model of van Leeuwenhoek's Microscope The original made in 1673 could magnify the object being viewed almost 300 times. The object being viewed is brought into focus with the adjusting screws. Note the small size.

of the microbial world, in fact all living organisms, can be separated into three distinct groups called **domains.** Organisms in each domain consist of cells with the same properties that distinguish them from members of the other domains. Many characteristics, however, are common among members of different domains.

The three domains are the ***Bacteria*** (formerly called Eubacteria), ***Archaea*** (formerly called Archaebacteria, meaning ancient bacteria), and ***Eucarya.*** Microscopically, members of the *Bacteria* and *Archaea* look indentical. Both are **prokaryotes,** meaning they are single-celled organisms consisting of a **prokaryotic cell** (meaning "prenucleus"). This type of cell does not contain a membrane-bound nucleus nor any other intracellular lipid-bound organelles. Their genetic information is stored in deoxyribonucleic acid (DNA) in a region called the **nucleoid.** Although all members of *Bacteria* and *Archaea* are prokaryotes, they are genetically quite different. In fact, the *Archaea* are as closely related to humans as they are to the *Bacteria.*

Members of the *Eucarya* are **eukaryotes,** meaning they are composed of one or more **eukaryotic cells** (meaning "true nucleus"). These cells always contain a membrane-bound nucleus and other organelles, making them far more complex than the simple prokaryotes. All algae, fungi, protozoa, and multicellular parasites are eukaryotes.

Bacteria

Most of the prokaryotes covered in this text are members of the domain *Bacteria.* Even within this group, much diversity is seen in the shape and properties of the organisms. Their most prominent features are:

- They are all single-celled prokaryotes.
- Most have specific shapes, most commonly cylindrical (rod-shaped), spherical (round), or spiral. ■ bacterial shapes, p. 52
- Most have rigid cell walls, which are responsible for the shape of the organism. The walls contain an unusual chemical compound called **peptidoglycan,** which is not found in organisms in the other domains (see figure 3.32).
- They multiply by **binary fission** in which one cell divides into two cells, each generally identical to the original cell. ■ binary fission, p. 53
- Many can move using appendages extending from the cell, called **flagella** (sing: **flagellum**). ■ flagella, p. 65

Archaea

Archaea have the same shape, size, and appearance as the *Bacteria.* Like the *Bacteria,* the *Archaea* multiply by binary fission and move primarily by means of flagella. They also have rigid cell walls. The chemical composition of their cell wall, however, differs from that in the *Bacteria.* The *Archaea* do not have peptidoglycan as part of their cell walls. Other chemical differences also exist between these two groups.

An interesting feature of many members of the *Archaea* is their ability to grow in extreme environments in which most organisms cannot survive. For example, some archaea can grow in salt concentrations 10 times higher than that found in seawater. These organisms grow in such habitats as the Great Salt Lake and the Dead Sea. Other archaea grow best at extremely high temperatures. One member can grow at a temperature of 121°C. (100°C is the temperature at which water boils at sea level). Some archaea can be found in the boiling hot springs at Yellowstone National Park. Members of the *Archaea,* however, are spread far beyond extreme environments. They are widely distributed in the oceans, and they are found in the cold surface waters of Antarctica and Alaska.

Eucarya

The microbial members of the domain *Eucarya* comprise single-celled and multicellular organisms that have a eukaryotic cell structure. These members include **algae** (sing: **alga**), **fungi** (sing: **fungus**), and **protozoa** (sing: **protozoan**). Algae and protozoa are also referred to as **protists.** In addition, some multicellular organisms are considered in this text because they kill millions of people around the world, especially in developing nations. They are given the general name of **helminths** and include organisms such as roundworms and tapeworms. Since they derive nutrients from the host organism they are termed **parasites.**

The *Bacteria, Archaea,* and *Eucarya* are compared in **table 1.1.**

Algae

The algae are a diverse group of eukaryotes; some are single-celled and others, multicellular. Many different shapes and sizes are represented, but they all share some fundamental characteristics **(figure 1.6).** They all contain chloroplasts, some of which have a green pigment, **chlorophyll.** Some also contain other pigments that give them characteristic colors. The pigments absorb the energy of light, which is used in **photosynthesis.** Algae are usually found

TABLE 1.1 Comparison of *Bacteria, Archaea,* and *Eucarya*

	Bacteria	*Archaea*	*Eucarya*
Typical Size	0.3–2 μm	0.3–2 μm	5–50 μm
Nuclear Membrane	No	No	Yes
Cell Wall	Peptidoglycan present	No peptidoglycan	No peptidoglycan
Membrane-bound Organelles	No	No	Yes
Where Found	In all environments	In all environments	In environments that are not extreme

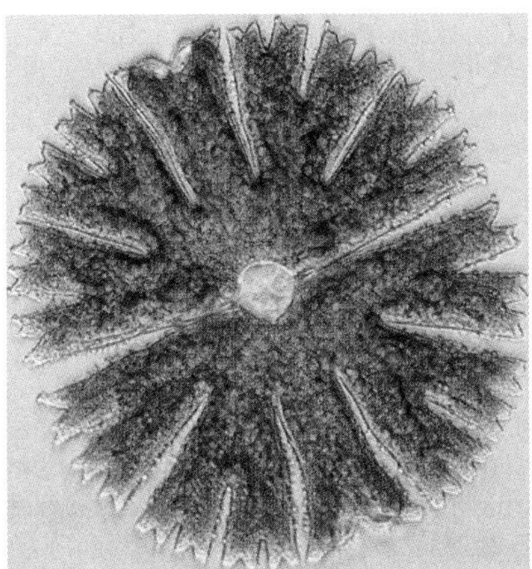

FIGURE 1.6 Alga *Micrasterias*, a green alga composed of two symmetrical halves (100×).

near the surface of either salt or fresh water. Their cell walls are rigid, but their chemical composition is quite distinct from that of the *Bacteria* and the *Archaea*. Many algae move by means of flagella, which are structurally more complex and unrelated to flagella in prokaryotes.

Fungi

Fungi are also a diverse group of eukaryotes. Some are single-celled yeasts, but many are large multicellular organisms such as molds and mushrooms **(figure 1.7)**. In contrast to algae, fungi gain their energy from degrading organic materials and are found wherever organic materials are present. Unlike algae, which live primarily in water, fungi live mostly on land.

Protozoa

Protozoa are a diverse group of microscopic, single-celled organisms that live in both aquatic and terrestrial environments. Although microscopic, they are very complex organisms and much larger than prokaryotes **(figure 1.8)**. Unlike algae and fungi, protozoa do not have a rigid cell wall. However, many do have a specific shape based on a gelatinous region just beneath the plasma membrane of the cell. Most protozoa require organic compounds as sources of food, which they ingest as particles. Most groups of protozoa are motile, and a major feature of their classification is their means of locomotion.

The eukaryotic members of the microbial world are compared in **table 1.2.**

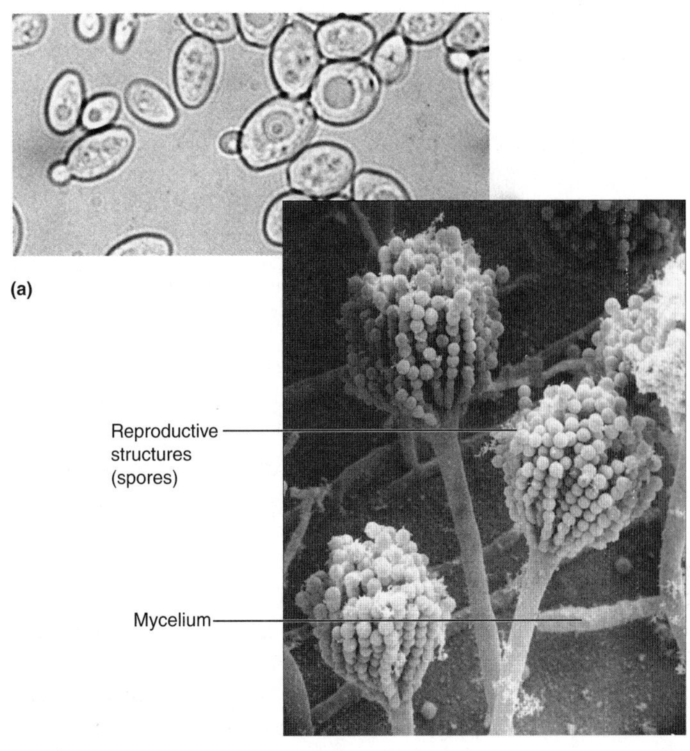

FIGURE 1.7 Two Forms of Fungi (a) Living cells of yeast form, *Cryptococcus neoformans*. **(b)** *Aspergillus*, a typical mold form whose reproductive structures rise above the mycelium.

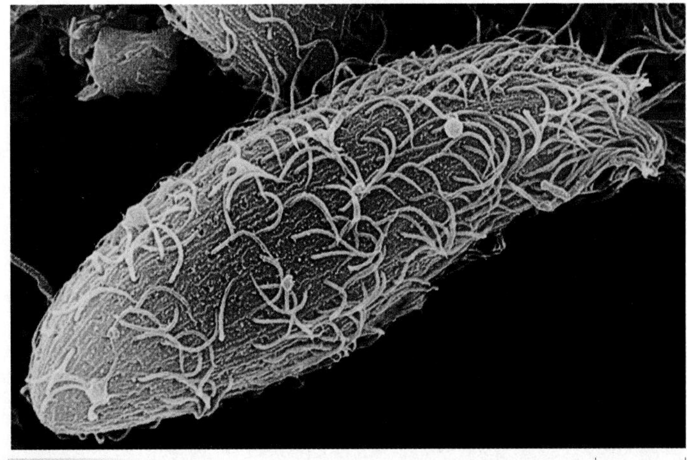

FIGURE 1.8 Protozoan A paramecium moves with the aid of cilia on the cell surface. 20 μm

TABLE 1.2 Comparison of Eukaryotic Members of the Microbial World

	Algae	Fungi	Protozoa
Cell Organization	Single- or multicellular	Single- or multicellular	Single-celled
Source of Energy	Sunlight	Organic compounds	Organic compounds
Size	Microscopic or macroscopic	Microscopic or macroscopic	Microscopic

Nomenclature

In biology, the Binomial System of Nomenclature devised by Carl Linnaeus refers to a two word naming system. The first word in the name indicates the **genus,** with the first letter always capitalized; the second indicates the **species** and is not capitalized. Both words are always italicized or underlined, for example, *Escherichia coli*. The genus name is commonly abbreviated, with the first letter capitalized: that is, *E. coli*. A number of different species are included in the same genus. Members of the same species may vary from one another in minor ways, but not enough to give the organisms different species names. These differences, however, may result in the organism being given different **strain designations,** for example, *E. coli* strain *B* or *E. coli* strain *K12*.

Bacteria are often referred to informally by names resembling genus names but are not italicized. For example, species of *Staphylococcus* are often called staphylococci.

MICROCHECK 1.3

All organisms fall into one of three large groups based on their cell structure and chemical composition: the *Bacteria*, the *Archaea*, or the *Eucarya*. The *Bacteria* and the *Archaea* are prokaryotes. Both are identical in appearance but distinctly different in many aspects of their chemical composition. The *Eucarya* are eukaryotes. The algae, fungi, protozoa, and multicellular parasites belong to this group. Bacteria, like all organisms, are classified according to the Binomial System of Nomenclature.

- ✓ Name one feature that distinguishes the domain *Bacteria* from the domain *Archaea*.
- ✓ List two features that distinguish prokaryotes from eukaryotes.
- ✓ The binomial system of classification uses both a genus and a species name. Why bother with two names? Wouldn't it be easier to use a single, unique name for each different kind of microorganism?

1.4

Viruses, Viroids, and Prions

Focus Points

- Distinguish among viruses, viroids, and prions.
- Discuss the reasons why viruses, viroids, and prions are not organisms.

The organisms discussed so far are living members of the microbial world. In order to be alive, an organism must be composed of one or more cells. Viruses, viroids, and prions are not living, are **acellular** and are termed **agents.** ■ viroids, p. 342 ■ prions, p. 341

Viruses consist of a piece of nucleic acid surrounded by a protein coat. They come in a variety of shapes **(figure 1.9).** Viruses need to produce copies of themselves, otherwise they would not exist in nature. Viruses can only multiply inside living host cells, whose multiplication machinery and nutrients they use for reproduction. Outside the hosts, they are inactive. Thus, viruses are **obligate intracellular parasites.** All forms of life including members of the *Bacteria*, *Archaea*, and *Eucarya* can be infected by viruses but of different types. Although viruses frequently kill the cells in which they multiply, some types exist harmoniously within the host cell without causing obvious ill effects.

Viroids are simpler than viruses, consisting of a single, short piece of ribonucleic acid (RNA), without a protective coat. They are much smaller than viruses **(figure 1.10),** and, like viruses, can reproduce only inside cells. Viroids cause a number of plant diseases, and some scientists speculate that they may cause diseases in humans.

Prions consist of only protein, without any nucleic acid **(figure 1.11).** They are very unusual agents consisting of an abnormal form of a cellular protein and are responsible for at least seven neurodegenerative diseases in humans and animals; these are always fatal.

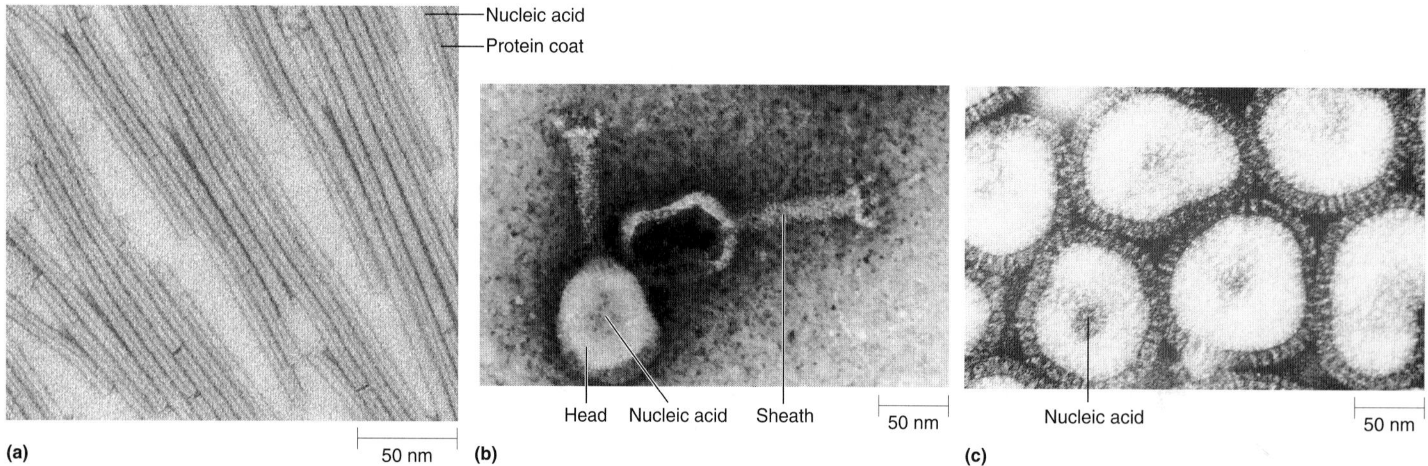

FIGURE 1.9 Viruses That Infect Three Kinds of Organisms (a) Tobacco mosaic virus that infects tobacco plants. A long hollow protein coat surrounds a molecule of RNA. **(b)** A bacterial virus (bacteriophage), which infects bacteria. Nucleic acid is surrounded by a protein coat (head). **(c)** Influenza virus, thin section. This virus infects humans and causes flu.

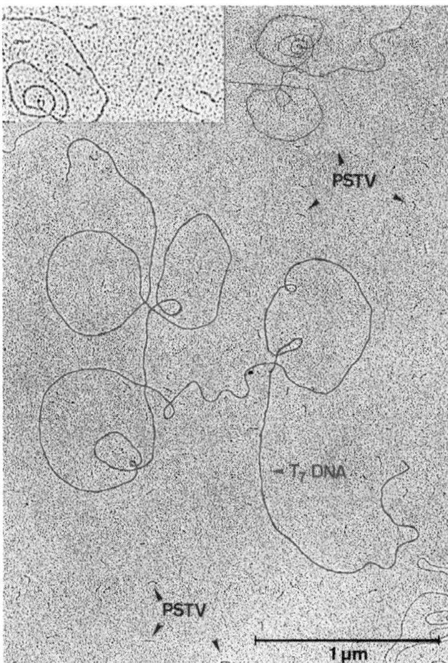

FIGURE 1.10 The Size of a Viroid Compared with a Molecule of DNA from a Virus That Infects Bacteria (T_7) The red arrows point to the potato spindle tuber viroids (PSTV); the other arrow points to the bacterial virus T7 DNA.

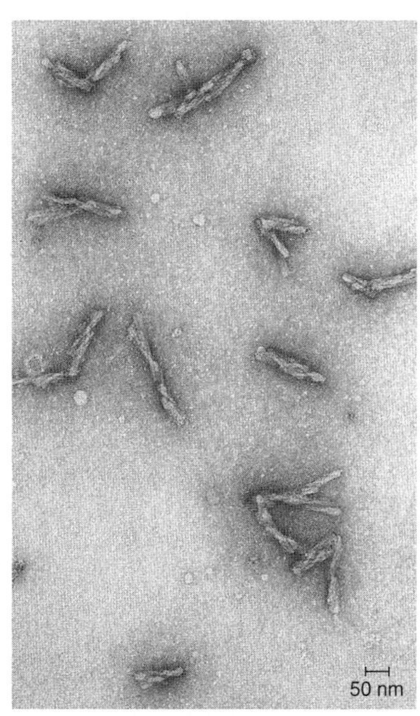

FIGURE 1.11 Prion Prions isolated from the brain of a scrapie-infected hamster. This neurodegenerative disease is caused by a prion.

The distinguishing features of the non-living members of the microbial world are given in **table 1.3.** The relationships of the major groups of the microbial world to one another are presented in **figure 1.12.**

MICROCHECK 1.4

The acellular agents are viruses, viroids, and prions.

- ✓ Compare the chemical composition of viruses, viroids, and prions.
- ✓ What groups of organisms are infected by each of the following: viruses, viroids, prions?
- ✓ How might one argue that viruses are actually living organisms?

TABLE 1.3 Distinguishing Characteristics of Viruses, Viroids, and Prions

Viruses	Viroids	Prions
Obligate intracellular agents	Obligate intracellular agents	Abnormal form of a cellular protein
Consist of either DNA or RNA, surrounded by a protein coat	Consist only of RNA; no protein coat	Consist only of protein; no DNA or RNA

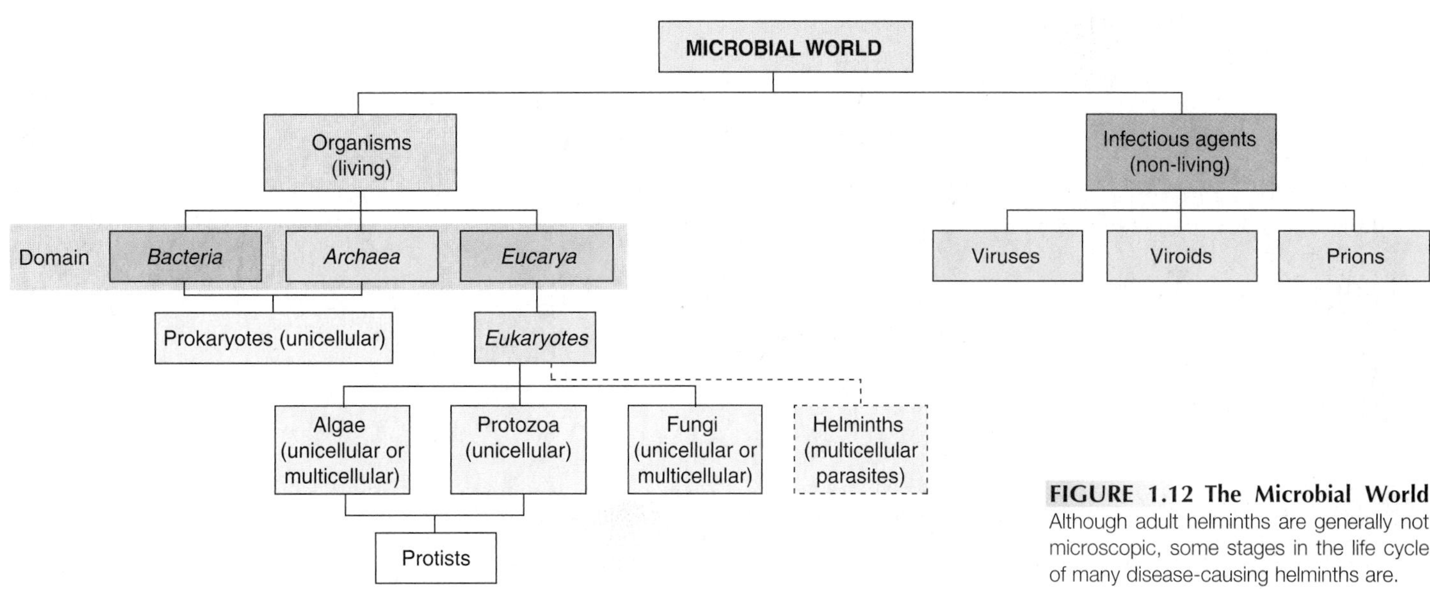

FIGURE 1.12 The Microbial World Although adult helminths are generally not microscopic, some stages in the life cycle of many disease-causing helminths are.

1.5

Size in the Microbial World

Focus Point

- Compare the differences in sizes among members of the microbial world.

Members of the microbial world cover a tremendous range of sizes, as seen in **figure 1.13.** The smallest viruses are about 1 million times smaller than the largest eukaryotic cells. Even within a single group, wide variations exist. For example, *Bacillus megaterium* and *Mycoplasma pneumoniae* are both bacteria, but they differ enormously in size (see figure 1.13). The variation in size of bacteria was recently expanded when a bacterium longer than 0.5 mm was discovered (see **Perspective 1.1**). In fact, it is so big that it is visible to the naked eye. More recently, an even larger bacterium, round in shape, was discovered. Its volume is 70 times larger than the previous record holder. Likewise, a eukaryotic cell was recently discovered that is not much larger than a typical bacterium. These, however, are rare exceptions to the rule that eukaryotes are larger than prokaryotes, which in turn are larger than viruses.

As you might expect, the small size and broad size range of some members of the microbial world have required the use of measurements not commonly used in everyday life. The use of logarithms has proved to be enormously helpful, especially in designating the sizes of prokaryotes and viruses. A brief discussion of measurements and logarithms is given in Appendix I.

MICROCHECK 1.5

The range in size of the members of the microbial world is tremendous. As a general rule, the obligate intracellular parasites are the smallest and the eukaryotes the largest.

✓ Place in order with respect to typical size (arrange from smallest to largest) bacteria, eukaryotic cells, and viruses.

✓ What factor limits the size of free-living cells?

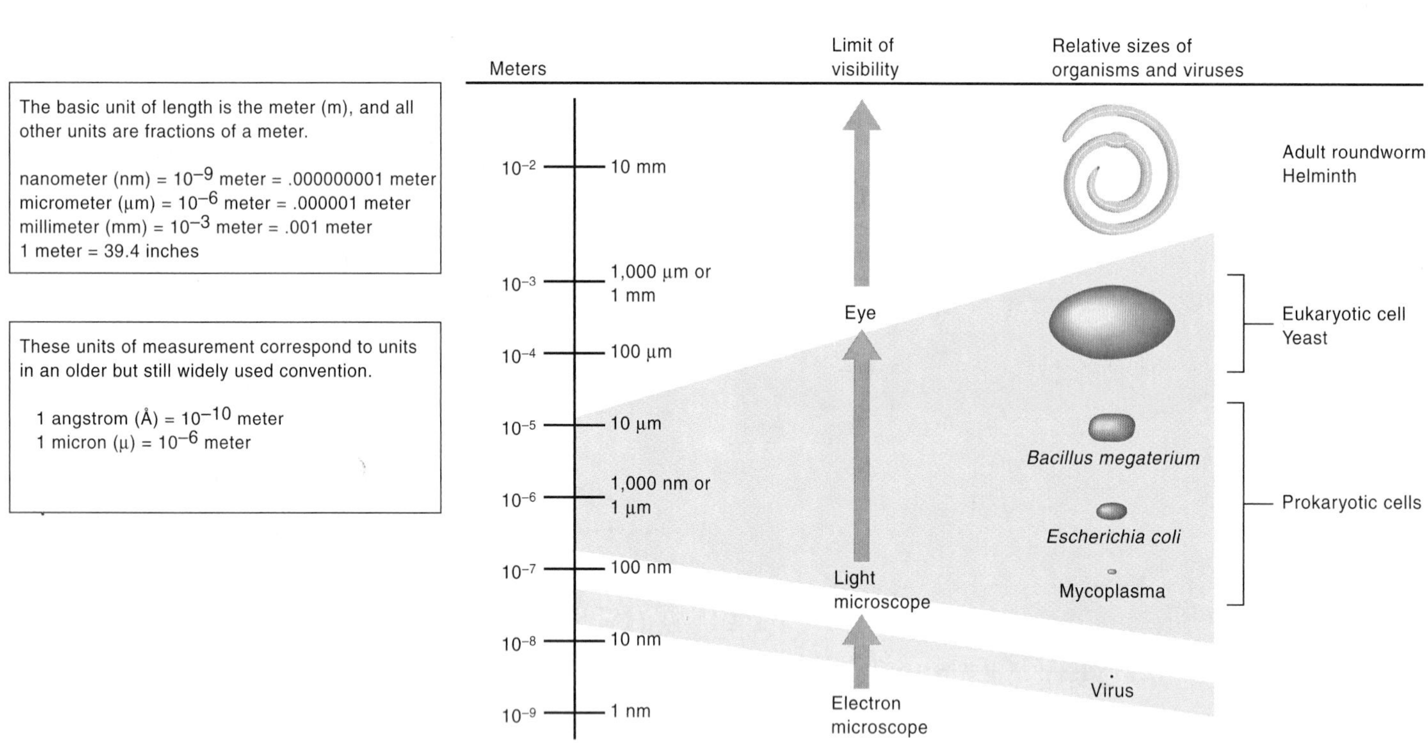

FIGURE 1.13 Sizes of Organisms and Viruses

PERSPECTIVE 1.1

The Long and the Short of It

We might assume that because prokaryotes have been so intensively studied over the past hundred years, no major surprises are left to be discovered. This, however is far from the truth. In the mid-1990s, a large, peculiar-looking organism was seen when the intestinal tracts of certain fish from both the Red Sea in the Middle East and the Great Barrier Reef in Australia were examined. This organism, named *Epulopisicium* cannot be cultured in the laboratory **(figure 1).** Its large size, 600 μm long and 80 μm wide, makes it clearly visible without any magnification, and suggested that this organism was a eukaryote. It did not, however have a membrane bound nucleus. A chemical analysis of the cell confirmed that it was a prokaryote and a member of the domain *Bacteria*. This very long, slender organism is an exception to the rule that prokaryotes are always smaller than eukaryotes.

In 1999, an even larger prokaryote in volume was isolated from the sulfurous muck of the ocean floor off the coast of Namibia in Africa. It is a spherical organism 70 times larger in volume than *Epulopisicium*. Since it grows on sulfur compounds and contains glistening globules of sulfur, it was named *Thiomargarita namibiensis*, which means "sulfur pearl of Namibia" **(figure 2).** Although scientists were initially skeptical that prokaryotes could be so large, there is no question in their minds now. In contrast to these large bacteria, a cell was isolated in the Mediterranean Sea that is 1 μm in width. It is a eukaryote because it contains a nucleus even though it is about the size of a typical bacterium.

How small can an organism be? An answer to this question may be at hand as a result of a new microorganism discovered off the coast of Iceland. The organism, found in an ocean vent where the temperature was close to the boiling point of water, cannot be grown in the laboratory by itself, but only grows when it is attached to another much larger member of the *Archaea* **(figure 3).** These tiny organisms, also members of the *Archaea*, have been named *Nanoarchaeum equitans*, which means "riding the fire sphere." The organism to which *N. equitans* is attached is *Ignicoccus*, which means "fire ball." *Ignicoccus* grows very well without its rider. *N. equitans* is spherical and only about 400 nanometers in diameter, about a quarter the diameter of *Ignicoccus*. Also, the amount of genetic information (DNA) contained in *N. equitans* is less than in any known organism, and only about one-tenth the amount found in the common gut organism, *Escherichia coli*. This sets the record for the smallest amount of DNA in any organism. Thus, this organism may contain only the essential DNA required for life. Further analysis of these cells suggests that they may resemble the earliest cells and therefore the ancestor of all life. The scientists who discovered *N. equitans* suggest that many more unusual organisms related to *N. equitans* will be discovered. They are probably right!

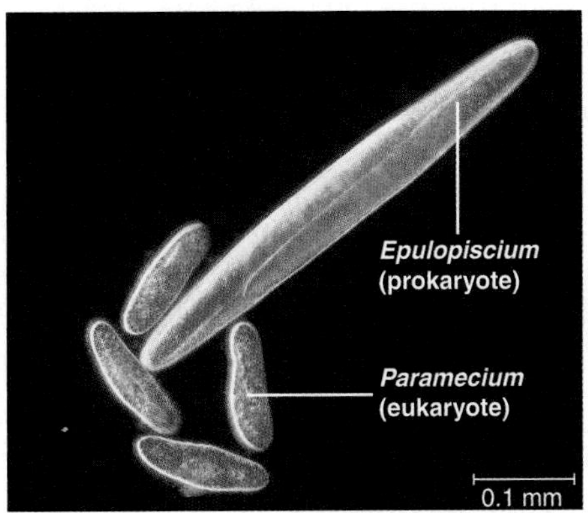

FIGURE 1 Longest Known Bacterium, *Epulopisicium* Mixed With Paramecia Note how large this prokaryote is compared with the four eukaryotic paramecia.

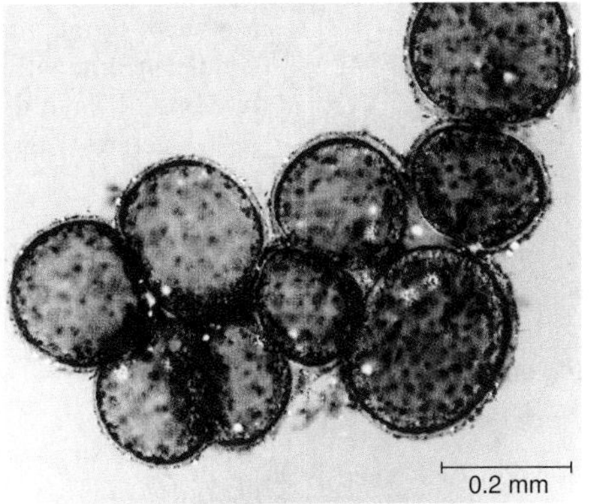

FIGURE 2 *Thiomargarita namibiensis* The average *Thiomargarita namibiensis* is two-tenths of a millimeter, but some reach three times that size. ■ *Thiomargarita*, p. 263

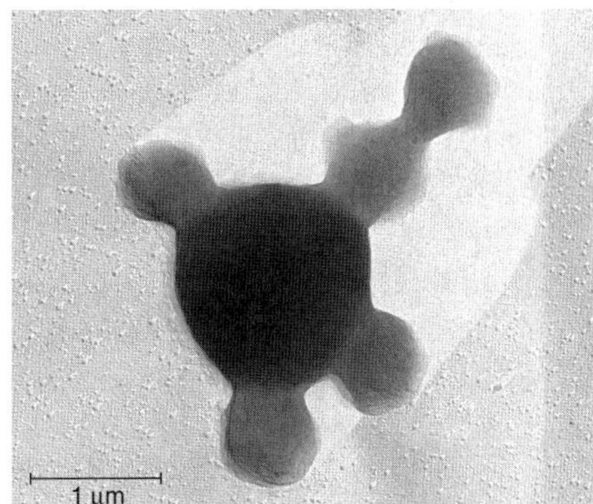

FIGURE 3 Five Cells of "*N. equitans*," Attached on the Surface of the (Central) *Ignicoccus* Cell Platinum shadowed.

FUTURE CHALLENGES

Entering a New Golden Age

For all the information that has been gathered about the microbial world, it is remarkable how little we know about its prokaryotic members. This is not surprising in view of the fact that less than 1% of the prokaryotes have ever been studied. In large part, this is because only one in a hundred of the prokaryotes in the environment can be cultured in the laboratory. Part of the current revolution in microbiology, however, will allow us to inventory the millions of species that are out there waiting to be discovered. This is now being done. Using techniques that helped decipher the human genome, scientists have begun to analyze the biological content of the oceans. In a small volume of water from the Sargasso Sea, an area of the ocean that contains few nutrients and therefore presumably few organisms, scientists found 1,800 species of bacteria that were previously unknown. The biodiversity of the microbial world is astounding!

Exploring the unknowns in the microbial world is a major challenge and should answer many intriguing questions fundamental to understanding the biological world. What are the extremes of temperature, salt, pH, radioactivity, and pressure in which prokaryotes can live? Are there organisms growing in even more extreme environments? If life can exist on this planet under such extreme conditions, what does this mean about the possibility of finding living organisms on other planets? Although considered highly unlikely, is it possible that living organisms exist whose chemical structure is not based on the carbon atom? Will living organisms be found whose genetic information is coded in a chemical other than deoxyribonucleic acid? What new metabolic pathways remain to be discovered? As extreme environments are mined for their living biological diversity, there seems little doubt that many surprises will be found. In many cases these surprises will be translated into new biotechnology products on this planet, and they will help shape the way we look for life on other planets.

One hundred years ago we were in the Golden Age of medical microbiology. We are now entering the Golden Age of microbial biodiversity.

SUMMARY

1.1 The Origin of Microorganisms

Theory of Spontaneous Generation Revisited

The experiments of Pasteur refuted the theory of spontaneous generation (figure 1.1). Tyndall and Cohn demonstrated the existence of heat-resistant forms of bacteria that could account for the growth of bacteria in heated infusions.

1.2 Microbiology: A Human Perspective

Features of the Microbial World

Microorganisms represent the most diverse forms of life on earth.

Vital Activities of Microorganisms

The activities of microorganisms are vital for the survival of all other organisms, including humans. Bacteria are necessary to convert the nitrogen gas in air into a form that plants and other organisms can use. Microorganisms replenish the oxygen on earth and degrade waste materials.

Applications of Microbiology

For thousands of years, bread, wine, beer, and cheeses have been made by using technology still applied today. Bacteria are being used to degrade dangerous toxic pollutants. Bacteria are used to synthesize a variety of different products, such as cellulose, hydroxybutyric acid, ethanol, antibiotics, and amino acids.

Medical Microbiology

Many devastating diseases such as smallpox, plague, and influenza have determined the course of history (figure 1.3). "New" emerging diseases are arising, partly, because people are engaging in different lifestyles and living in regions where formerly only animals lived (figure 1.4). "Old" diseases once on the wane have begun to reemerge. Many are brought to this country by people visiting foreign lands. Several chronic diseases such as ulcers are caused by bacteria.

Microorganisms As Model Organisms

Microorganisms are excellent model organisms to study because they grow rapidly on simple, inexpensive media, but follow the same genetic, metabolic, and biochemical principles as higher organisms.

1.3 Members of the Microbial World (figure 1.12)

Members of the microbial world consist of two major cell types: the simple **prokaryotic** and the complex **eukaryotic**. All organisms fall into one of three **domains**, based on their chemical composition and cell structure. These are the ***Bacteria***, the ***Archaea***, and the ***Eucarya*** (table 1.1).

Bacteria

Bacteria are single-celled prokaryotes that have peptidoglycan in their cell wall.

Archaea

Archaea are single-celled prokaryotes that are identical in appearance to the *Bacteria*. They do not have peptidoglycan in their cell walls and are unrelated to the *Bacteria*. Many of the *Archaea* grow in extreme environments such as hot springs and salt flats.

Eucarya

Eucarya have eukaryotic cell structures and may be single-celled or multicellular. Microbial members of the *Eucarya* are the **algae, fungi,** and **protozoa.** Algae can be single-celled or multicellular, and use sunlight as a source of energy (figure 1.6, table 1.2). Fungi are either single-celled yeasts or multicellular molds and mushrooms and use organic compounds as food (figure 1.7, table 1.2). Protozoa are motile single-celled organisms and use organic compounds as food (figure 1.8, table 1.2).

Nomenclature

Organisms are named according to a binomial system.

Each organism has a **genus** and a **species** name, written in italics or underlined.

1.4 Viruses, Viroids, and Prions

The non-living members of the microbial world are not composed of cells but are **obligate intracellular parasites** and include **viruses** and **viroids.** Viruses are a piece of nucleic acid surrounded by a protein coat (figure 1.9, table 1.3). Viroids are composed of a single, short RNA molecule (figure 1.10, table 1.3). Prions consist only of protein, without any nucleic acid and are an abnormal cellular protein (figure 1.11, table 1.3).

1.5 Size in the Microbial World

Sizes of members of the microbial world vary enormously (figure 1.13).

REVIEW QUESTIONS

Short Answer

1. Name the prokaryotic groups in the microbial world.
2. List five beneficial applications of bacteria.
3. Name three non-living groups in the microbial world and describe their major properties.
4. In the designation *Escherichia coli B,* what is the genus? What is the species? What is the strain?
5. Where would you go to isolate members of the *Archaea*?
6. How might you distinguish a prokaryotic cell from a eukaryotic cell?
7. Give three reasons why life could not exist without the activities of microorganisms.
8. Why are viruses not microorganisms?
9. Name two diseases that have been especially destructive in the past. What is the status of those diseases today?
10. State three reasons why there is a resurgence of infectious diseases today.

Multiple Choice

1. The prokaryotic members of the microbial world include
   1. algae. 2. fungi. 3. prions. 4. bacteria. 5. archaea.
   a) 1, 2 b) 2, 3 c) 3, 4 d) 4, 5 e) 1, 5
2. The *Archaea*
   1. are microscopic.
   2. are commonly found in extreme environments.
   3. contain peptidoglycan.
   4. contain mitochondria.
   5. are most commonly found in the soil.
   a) 1, 2 b) 2, 3 c) 3, 4 d) 4, 5 e) 1, 5
3. The most fundamental division of cell types is between the
   a) algae, fungi, and protozoa.
   b) eukaryotes and prokaryotes.
   c) viruses and viroids.
   d) *bacteria* and *archaea.*
   e) *Eucarya*, *Bacteria*, and *Archaea*.
4. The number of bacteria in the human body compared to the number of non-bacterial cells is estimated to be
   a) about 10 times more non-bacterial cells than bacteria.
   b) about equal numbers of bacteria and non-bacterial cells.
   c) about 10 times more bacteria than non-bacteria.
5. An organism isolated from a hot spring in an acidic environment is most likely a member of the
   a) *Bacteria.* b) *Archaea.* c) *Eucarya.*
   d) virus family. e) Fungi.
6. The agent that contains no nucleic acid is a
   a) virus. b) prion. c) viroid.
   d) bacterium. e) fungus.
7. Prokaryotes do not have
   a) cell walls. b) flagella. c) a nuclear membrane.
   d) specific shapes. e) genetic information.
8. Nucleoids are associated with
   1. genetic information. 2. prokaryotes.
   3. eukaryotes. 4. viruses. 5. prions.
   a) 1, 2 b) 2, 3 c) 3, 4 d) 4, 5 e) 1, 5
9. Which of the following are eukaryotes?
   1. Algae 2. Viruses 3. Bacteria
   4. Prions 5. Protozoa
   a) 1, 2 b) 2, 3 c) 3, 4 d) 4, 5 e) 1, 5
10. The person best known for his microscopy of microorganisms is
    a) Antony van Leeuwenhoek.
    b) Louis Pasteur.
    c) John Tyndall.
    d) Ferdinand Cohn.

Applications

1. The American Society of Microbiology is preparing a "Microbe-Free" banquet to emphasize the importance of microorganisms in the diet. What foods would not be on the menu if microorganisms were not available for our use?
2. If you were asked to nominate one of the individuals mentioned in this chapter for the Nobel Prize, who would it be? Make a statement supporting your choice.

Critical Thinking

1. A microbiologist obtained two pure isolated biological samples: one of a virus, and the other of a viroid. Unfortunately, the labels had been lost from the two samples. The microbiologist felt she could distinguish the two by analyzing for the presence or absence of a single chemical element. What element would she search for and why?
2. Why are the spores of *Bacillus anthracis* such an effective agent of bioterrorism?

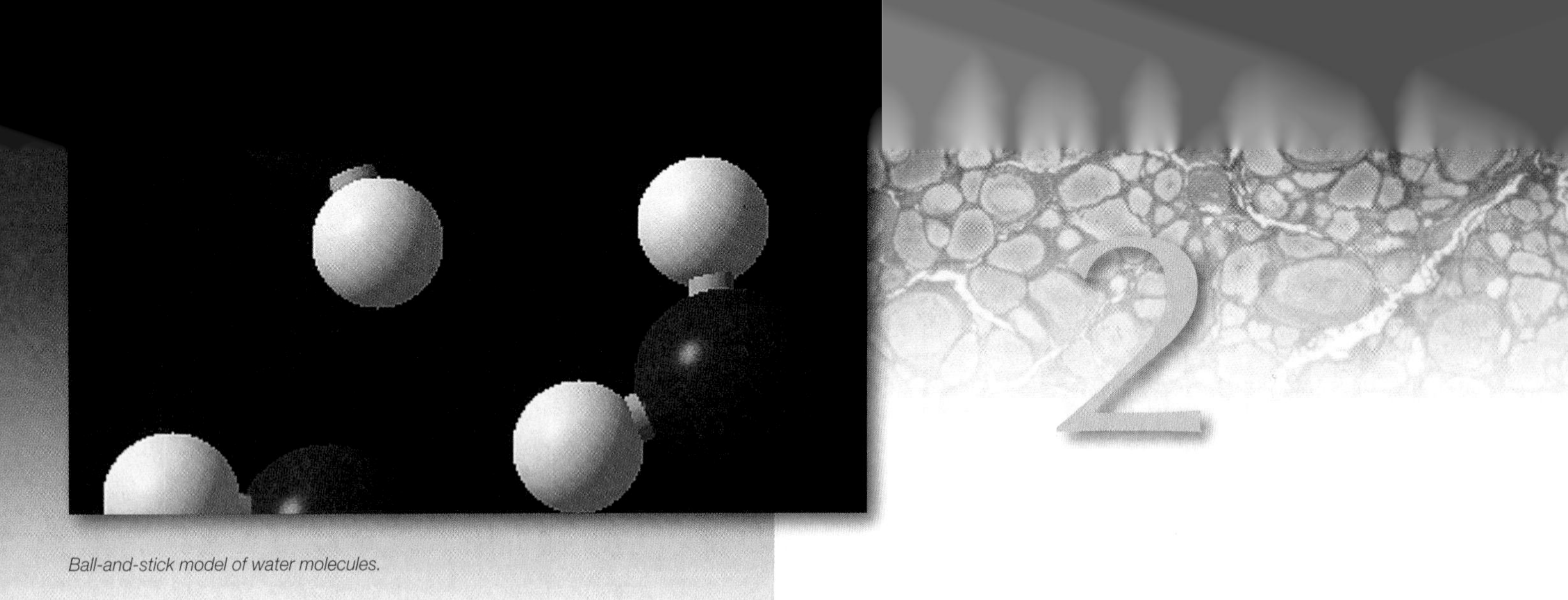

Ball-and-stick model of water molecules.

The Molecules of Life

A Glimpse of History

Louis Pasteur (1822–1895) is often considered the father of bacteriology. His contributions to this science, especially in its early formative years, were enormous and are discussed in many of the succeeding chapters. Pasteur started his scientific career as a chemist, initially working in the science of crystallography.

He first studied two compounds, tartaric and paratartaric acids, which form thick crusts within wine barrels. These two substances form crystals that have the same number and arrangement of atoms, yet they rotate (twist) polarized light differently when that light passes through the crystal. Tartaric acid rotates the light; paratartaric acid does not. Therefore, the two molecules must differ in some way, even though they are chemically identical. Pasteur was determined to find out how the crystals differed. Under a microscope, he saw that the crystals of tartaric acid all looked identical but paratartaric acid consisted of two different kinds of crystals. Using tweezers, he carefully separated the two kinds into separate piles and dissolved each in a separate flask of water. When he shone polarized light through each solution, one rotated the light to the left and the other rotated it to the right. When he mixed equal numbers of each kind of crystal into water and shone polarized light through the solution, the light was not rotated. Apparently, the two components counteracted each other, and as a result, the mixture did not rotate the light. Pasteur concluded that paratartaric acid is a mixture of two compounds, each being the mirror image, or **stereoisomer,** of the other. This mixture of two stereoisomers can be viewed as a mixture of right- and left-handed molecules, represented as a right and left hand facing each other (see figure 2.14). They cannot be superimposed on each other, much as a right-handed glove cannot fit the left hand.

Stereoisomers of the same molecule have greatly different properties. For example, the amino acid phenylalanine, one of the key ingredients in the artificial sweetener aspartame, makes aspartame sweet when it is in one form but bitter when in the other form. Thus, what Pasteur studied as a straightforward problem in chemistry has implications far beyond what he ever imagined. It is often difficult to predict where research will lead or the significance of interesting but seemingly unimportant observations.

To understand how cells live and interact with one another and with their environment, we must be familiar with the molecules that comprise all living matter. For some, this information may serve as a review of material already studied in other courses. For others, it may be a first encounter with the chemistry of biological molecules. In this case, you likely will return to this chapter frequently. The discussion proceeds from the lowest level of chemical organization, the atoms and elements, to the highly complex associations between small molecules that often form large molecules, the macromolecules.

2.1 Atoms and Elements

Focus Point

- Name the three major components of atoms and describe their properties.

Atoms, the basic units of all matter, are made up of three major components: the negatively charged **electrons;** positively charged **protons;** and uncharged **neutrons (figure 2.1).** The protons and neutrons, the heaviest components, are found in the heaviest part of

KEY TERMS

ATP An abbreviation for adenosine triphosphate, the form in which chemical energy is stored in the cell.

Carbohydrate A compound characterized by a large number of —OH groups and containing principally carbon, hydrogen, and oxygen in a ratio of 1:2:1.

Covalent Bond A strong chemical bond formed by the sharing of electrons between atoms.

Dehydration Synthesis A chemical reaction that joins two molecules to form a larger molecule by removing water.

Hydrogen Bond A weak bond resulting from the attraction between a positively charged hydrogen atom in one compound and a negatively charged atom in another compound.

Inorganic Compound A chemical that has no carbon-hydrogen bonds.

Lipid A heterogenous group of organic molecules characterized by being insoluble in water, but soluble in organic solvents.

Macromolecule A very large molecule usually consisting of repeating subunits.

Nucleic Acid A macromolecule consisting of chains of nucleotide subunits to form either DNA or RNA, the two types of nucleic acid.

Organic Compound A chemical in which a carbon atom is covalently bonded to a hydrogen atom.

Peptide Bond A covalent bond formed between the —COOH group of one amino acid and the —NH_2 group of another amino acid; their formation is an important reaction in the synthesis of a protein.

pH The abbreviation for potential hydrogen, a measure on a scale of 0 to 14 of the acidity of a solution.

Protein A macromolecule consisting of one or more chains of amino acids.

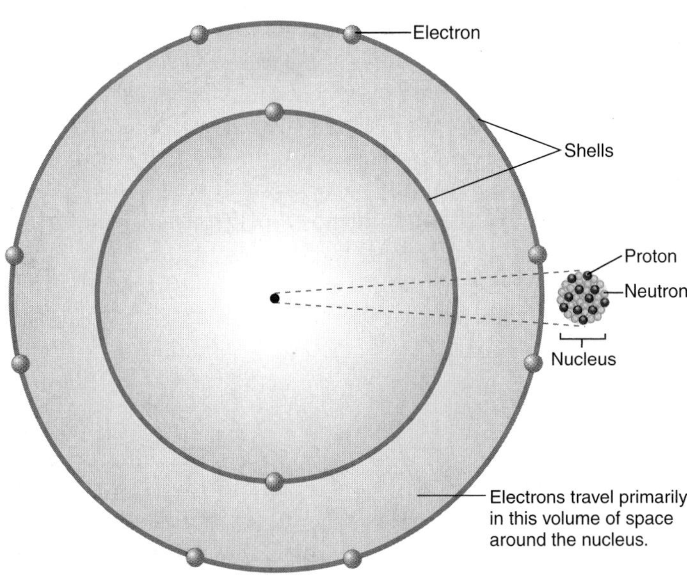

FIGURE 2.1 Atom The proton has a positive charge, the neutron has a neutral charge, and the electron has a negative charge. The electrons that orbit the nucleus are arranged in shells of different energy levels.

the atom, the nucleus. The very light electrons orbit the nucleus. The number of protons normally equals the number of electrons, and so the atom as a whole has no charge. The relative sizes and motion of the parts of an atom can be illustrated by the following analogy. If a single atom were enlarged to the size of a football stadium, the nucleus would be the size of a marble and it would be positioned somewhere above the 50-yard line. The electrons would resemble fruit flies zipping around the stands. Their orbits would be mostly inside the stadium, but on occasion they would travel outside it.

An **element** is a substance that consists of a single type of atom. Although 92 naturally occurring elements exist, four elements make up over 99% of all living material by weight. These are carbon (abbreviated C), hydrogen (H), oxygen (O), and nitrogen (N). Two other elements, phosphorus (P), and sulfur (S), together make up an additional 0.5% of the elements in living systems **(table 2.1).** All of the remaining elements together account for less than 0.5% of living material. In general, the basic chemical composition of all living cells is remarkably similar.

Each element is identified by two numbers: its atomic number and its atomic weight or mass. The **atomic number** is the number of protons, which equals the number of electrons. For example, hydrogen has 1 proton, and thus its atomic number is 1; oxygen

TABLE 2.1 Atomic Structure of Elements Commonly Found in the Living World

Element	Symbol	Atomic Number (Total Number of Protons)	Atomic Weight (Protons + Neutrons)	Number of Possible Covalent Bonds*	Approximate % of Atoms in Cells
Hydrogen	H	1	1	1	49
Carbon	C	6	12	4	25
Nitrogen	N	7	14	3	0.5
Oxygen	O	8	16	2	25
Phosphorus	P	15	31	3	0.1
Sulfur	S	16	32	2	0.4

*The number of electrons required to fill the outer shell equals the number of possible covalent bonds—its valence. The number of electrons in a completed outer shell varies depending on the distance of the shell from the nucleus.

has 8 protons, and its atomic number is 8. The **atomic weight** is the sum of the number of protons and neutrons (electrons are too light to contribute to the weight). The atomic weight of hydrogen is 1, which is abbreviated ^{1}H, reflecting 1 proton and no neutrons. It is the lightest element known. The atomic weight of oxygen is 16, consisting of 8 protons and 8 neutrons, and is abbreviated ^{16}O.

Electrons are arranged in **shells** of differing energy levels. The electrons farthest from the nucleus with its positive charge travel the fastest and have the highest energy level. Electrons can move from one shell to another as they gain or lose energy (see figure 2.1). Each shell can contain only a certain number of electrons with the first shell (closest to the nucleus) containing a maximum of 2 electrons, the next 8, and the next also 8. Other atoms, which have little biological importance, have additional electrons. Each shell must be filled, starting with the one closest to the nucleus, before electrons can occupy the next outer shell.

MICROCHECK 2.1

All living organisms contain the same elements. The four most important are carbon, hydrogen, oxygen, and nitrogen. The basic unit of all matter, the atom, is composed of protons, electrons, and neutrons.

- ✓ Of all the elements found in cells, which element is found most frequently?
- ✓ Why is the energy level of an electron higher the farther it is from the nucleus?

2.2

Chemical Bonds and the Formation of Molecules

Focus Points

- Name the strongest bond and two weak bonds.
- Explain the difference between polar and non-polar covalent bonds and explain why polar bonds are important in biology.
- Describe the properties of the carbon atom that make it the most important atom in all organisms.

For an atom to be stable, its outer shell must contain the maximum number of electrons. Most atoms do not have that number, however, and therefore to achieve it they tend to gain, lose, or share electrons with other atoms. This is the basis for chemical bonds. When two or more atoms are joined by chemical bonds, the substance is called a **molecule.** The atoms that make up a molecule may be of the same or different elements. For example, H_2 is a molecule of hydrogen gas formed from two atoms of hydrogen; water (H_2O) is an association of two hydrogen atoms with one oxygen atom. Water is a **compound,** a molecule that consists of two or more different elements.

Three general types of chemical bonds join atoms—ionic, covalent, and hydrogen. Covalent bonds are generally the strongest of the three. Ionic bonds are typically weaker than covalent bonds, and hydrogen bonds are weaker still. Even relatively weak bonds are important in biological systems, however, because they allow molecules to recognize one another. This recognition depends on large numbers of atoms on the surfaces of molecules matching each other precisely. In addition, large numbers of bonds can hold molecules tightly together. An analogy would be the hook and loop fasteners of Velcro™. A single hook and loop attachment does not provide much strength, but many such attachments result in a strong connection. Like hook and loop attachments, weak bonds can be formed and broken quickly and easily, allowing the molecules to separate.

Ionic Bonds

Ionic bonds join charged atoms termed **ions (figure 2.2).** As already mentioned, an atom can fill its outer shell by gaining, losing, or sharing electrons. If electrons from one atom are attracted very strongly by another nearby atom, the electrons completely leave the first atom and become a part of the outer electron shell of the second, without any sharing. The loss or gain of electrons leads to an electrically charged atom. An atom that gains electrons becomes negatively charged, whereas an atom that gives up electrons becomes positively charged. The type and amount of charge, which is the difference between the

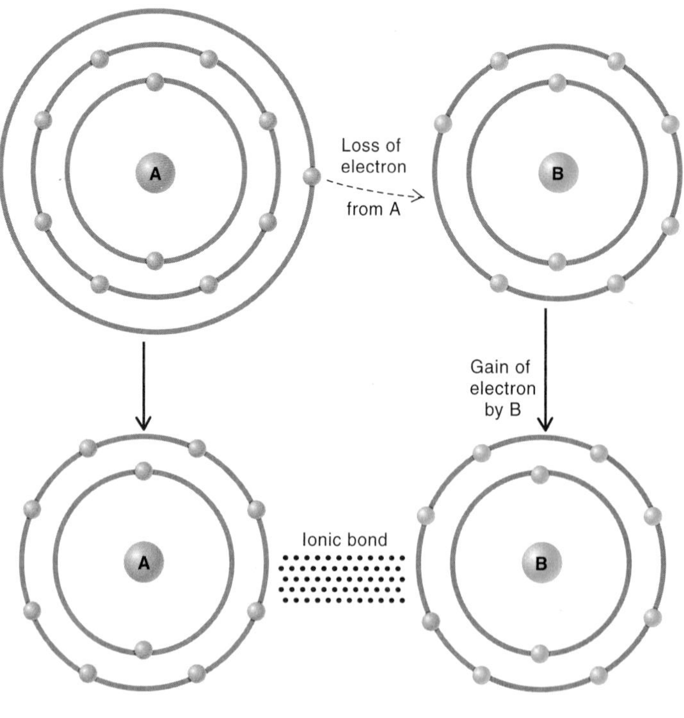

FIGURE 2.2 Ionic Bond Atom A gives up an electron to atom B; hence, atom A acquires a positive charge and atom B a negative charge. Both atoms then have their outer shells filled with the maximum number of electrons leading to maximum stability. The attraction of the positively charged A^+ ion to the negatively charged B^- ion forms the bond.

number of protons and electrons in the ion, is indicated by a superscript number. If only a $^+$ or $^-$ is indicated, then the charge is 1. For example, Na^+ indicates a Na ion with one positive charge. Positively charged ions are called **cations;** negatively charged ones are **anions.** The attraction of a cation to an anion forms the ionic bond.

In water (aqueous solutions), ionic bonds are about 100 times weaker than strong covalent bonds because water molecules tend to move between the ions, thereby greatly reducing their attraction to one another. Thus, in aqueous solution, which is common in all biological systems, weak ionic bonds are readily broken at room temperature.

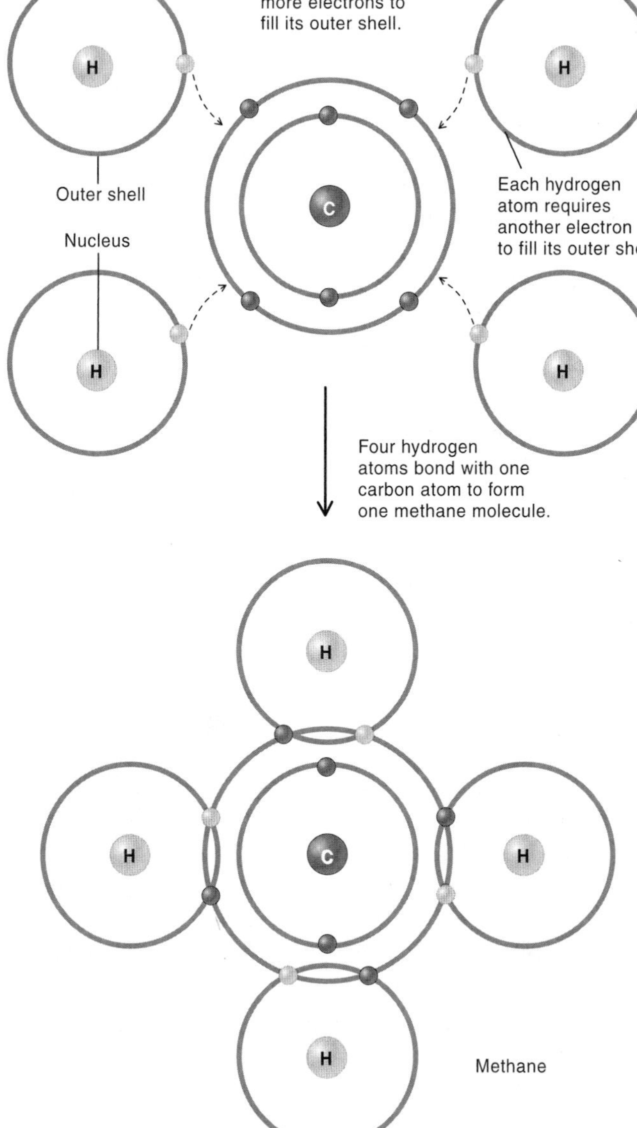
Carbon needs four more electrons to fill its outer shell.
H
H
Outer shell
C
Each hydrogen atom requires another electron to fill its outer shell.
Nucleus
H
H
Four hydrogen atoms bond with one carbon atom to form one methane molecule.
H
H
C
H
H
Methane

FIGURE 2.3 Covalent Bonds The carbon atom fills its outer electron shell by sharing a total of 8 electrons. Four belong to the four H atoms and four belong to the one carbon atom. The outer shells of each of these atoms are then filled—2 in the case of the H atom and 8 in the case of the C atom.

Covalent Bonds

Atoms often achieve stability by sharing electrons with other atoms, thereby filling their outer shells. This sharing creates strong **covalent bonds.** The number of covalent bonds an atom can form—its **valence**—is the number of electrons the atom must gain or lose to fill its outer shell (see table 2.1).

Carbon (C), the most important single atom in biology, is frequently involved in covalent bonding. This element has four electrons but requires a total of eight to fill its outer shell. Thus, its valence is 4. A hydrogen (H) atom has one electron and requires an additional one to fill its outer shell. Accordingly, the C atom can fill its outer shell by sharing electrons with four H atoms, creating methane (CH_4) (**figure 2.3;** see also table 2.1). Because a C atom can bond with four other atoms, it can build up a large number of different molecules, which explains why it is the key atom in all cells. When C forms covalent bonds with H atoms, an **organic** compound is formed. **Inorganic** compounds do not contain C to H bonds.

A single covalent bond is designated by a dash between the two atoms sharing the electrons and is written as C—H. Sometimes two pairs of electrons are shared between atoms in order to fill their outer shells. This forms a double covalent bond indicated by two lines between the atoms—for example O═C═O (CO_2).

Covalent bonds are strong. The stronger the bond, the more difficult it is to break. Consequently, covalent bonds do not break unless exposed to strong chemicals or large amounts of energy, generally as heat. Molecules formed by covalent bonds do not break apart spontaneously at temperatures compatible with life. Because most biological systems cannot tolerate the high temperatures required to break these bonds, cells utilize protein catalysts called **enzymes,** which can break covalent bonds at the temperatures found in living systems. Enzymes and their functions are covered in chapter 6. ■ **enzymes, pp. 129, 134**

Non-Polar and Polar Covalent Bonds

Two atoms connected by a covalent bond may have the same or different attractions for the electrons. In covalent bonds between identical atoms, such as H—H, the electrons are shared equally. Equal sharing also occurs between different atoms, such as C—H, if both have a similar attraction for electrons **(table 2.2).**

TABLE 2.2 Non-Polar and Polar Covalent Bonds

Type of Covalent Bond	Atoms Involved and Charge Distribution	
Non-polar	C—C C—H H—H	C and H have equal attractions for electrons, so there is an equivalent charge on each atom.
Polar	O—H N—H O—C N—C	The O and N atoms have a stronger attraction for electrons than do C and H, so the O and N have a slight negative charge; the C and H have a slight positive charge.

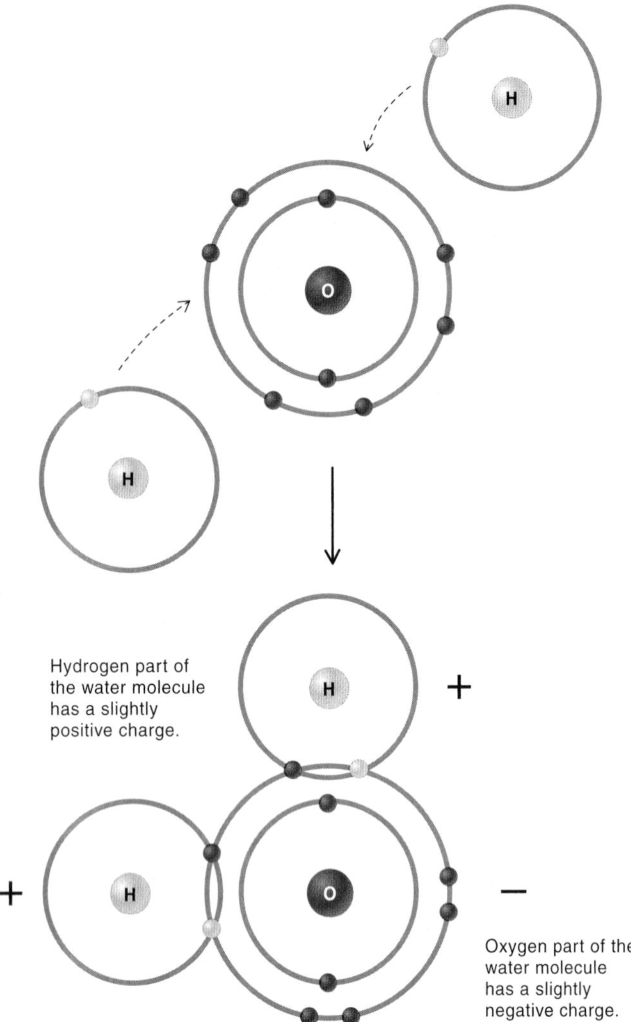

FIGURE 2.4 Formation of Polar Covalent Bonds in a Water Molecule In a water molecule the oxygen atom has a greater attraction for the shared electrons than do the hydrogen atoms. Hence, the electron is closer to the oxygen and confers a slight negative charge on a portion of this atom. Each of the hydrogen atoms has a slight positive charge on a portion of the atom. Because of these charges, water is a polar molecule.

This results in a **non-polar covalent bond.** If, however, one atom has a much greater attraction for electrons than the other, the electrons are shared unequally, and **polar covalent bonds** result. One part of the molecule has a slight positive charge and another, a slight negative charge. An example is water, in which the oxygen atom attracts the shared electrons more strongly than do the hydrogen atoms **(figure 2.4).** Consequently, the oxygen atom has a slight negative charge and the two hydrogen atoms, a slight positive charge. Polar covalent bonds play a key role in biological systems because they result in the formation of hydrogen bonds.

Hydrogen Bonds

Hydrogen bonds are weak bonds formed when a positively charged hydrogen atom in a polar molecule is attracted to a negatively charged atom, frequently oxygen (O) or nitrogen (N) in another polar molecule **(figure 2.5;** see also table 2.2). Such bonds can form between atoms in the same molecule or between two different molecules. Molecules that contain nitrogen or oxygen atoms bonded to hydrogen atoms are common in biological systems, thereby creating the possibility for many hydrogen bonds. Like other weak bonds, these are important in recognizing matching surfaces and holding molecules on these surfaces together **(figure 2.6).** For example, in order for an enzyme to break covalent bonds of a compound (the substrate), the enzyme binds to the substrate through many weak non-covalent bonds.

Hydrogen bonds between water molecules are constantly being formed and broken at room temperature because the energy produced by the movement of water is enough to break these bonds. The average lifetime of a single hydrogen bond is only a fraction of a second at room temperature, so enzymes are not needed to form or break these bonds. Although a single hydrogen bond is too weak to keep molecules together, a large number can hold them together firmly. A good example is the double-stranded DNA molecule. The two strands are held together by many hydrogen bonds up and down the length of the molecule. The two strands will come apart only if energy is supplied, usually in the form of heat approaching temperatures of 100°C.

Molarity

A chemical reaction can be likened to a recipe you follow in the kitchen—it uses relative quantities of different substances. But while chefs work with measures such as a dozen, chemists work with moles; one **mole** is 6.022×10^{23} molecules. That number is not important from a practical standpoint, but the concept is essential in chemistry—a mole of one compound has the same number of molecules as a mole of any other.

A more practical definition of a mole is that it is the amount (in grams) equal to the sum of the atomic weights of the atoms that make up the molecule. That sum is the **molecular weight** of the

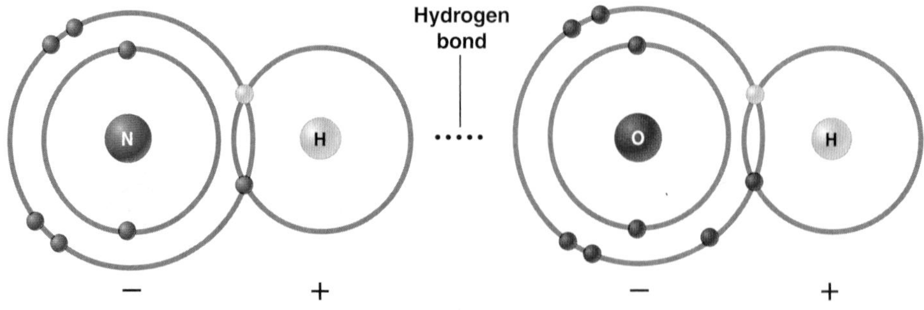

FIGURE 2.5 Hydrogen Bond Formation The N atom has a greater attraction for electrons than the H atom, thereby conferring a slight positive charge on the H atom. The O atom has a greater attraction for electrons than the H atom to which it is covalently bonded, thereby gaining a slight negative charge. The bonds between N—H and O—H are polar covalent bonds. The positively charged H atom weakly bonds to the negatively charged O atom, thereby forming a hydrogen bond.

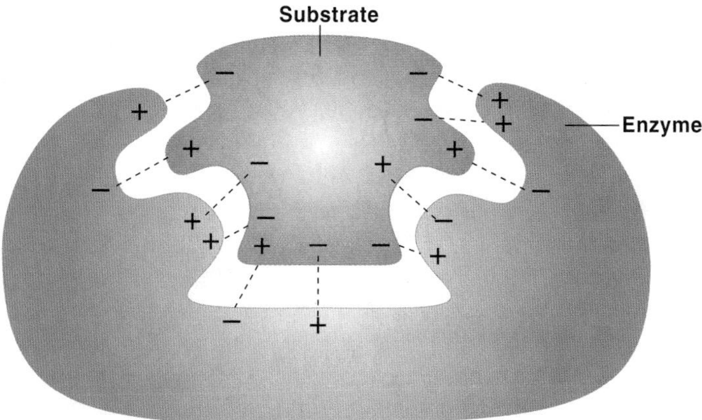

FIGURE 2.6 Weak Ionic Bonds and Molecular Recognition Weak bonds, such as ionic and hydrogen bonds, are important for molecules to recognize each other. Many weak bonds are required to hold the two molecules together, in this case, the substrate binding to an enzyme.

compound. Sodium chloride (NaCl), for example, has a molecular weight of 58.4, meaning that 58.4 grams of NaCl has approximately 6.022×10^{23} molecules. The **molarity** of a solution is defined as the number of moles of a compound dissolved in water to make 1 liter of solution. Therefore, a 1-molar solution of NaCl has 58.4 grams of that chemical dissolved in 1 liter of solution. Note that a 1-molar solution of two different compounds will contain different numbers of grams but the same number of molecules.

MICROCHECK 2.2

Molecules are formed by bonding between atoms. Bonds are formed when electrons from one atom interact with another atom. The bonds between the atoms that make up a molecule are strong covalent bonds; bonds between molecules are generally weak bonds such as ionic and hydrogen bonds.

✓ Compare the relative strengths of covalent, hydrogen, and ionic bonds.

✓ Which type of bond requires an enzyme to break it?

✓ Why does an atom that gives up electrons become positively charged? What causes the positive charge?

2.3 Chemical Components of the Cell

Focus Points

- Describe the bonding properties of a water molecule and explain why they are important in biology.
- Define pH, and state what the pH numbers tell you about the acidity of a solution.
- Name the four macromolecules found in all cells.

The most important molecule in the cell is water and the life of all organisms depends on its special properties.

Water

Unquestionably, water is the most important molecule in the world. It makes up over 70% of all living organisms by weight. The importance of water in large part depends on its unusual properties.

Bonding Properties of Water

Hydrogen bonding plays a very important role in giving water the properties required for life **(figure 2.7a).** Since water is a polar molecule, the positive H portion of the molecule is attracted to the negative O portion of other water molecules, thereby creating hydrogen bonds. The extent of hydrogen bonding between water molecules depends on the temperature. At room temperature, the weak bonds continually form and break. As the temperature is lowered, the breakage and formation decreases, and in ice, a crystalline structure forms. Each water molecule bonds to four other molecules to create a rigid lattice structure (figure 2.7b). When ice melts, the water molecules move closer together. Consequently, liquid water is denser than ice, which explains

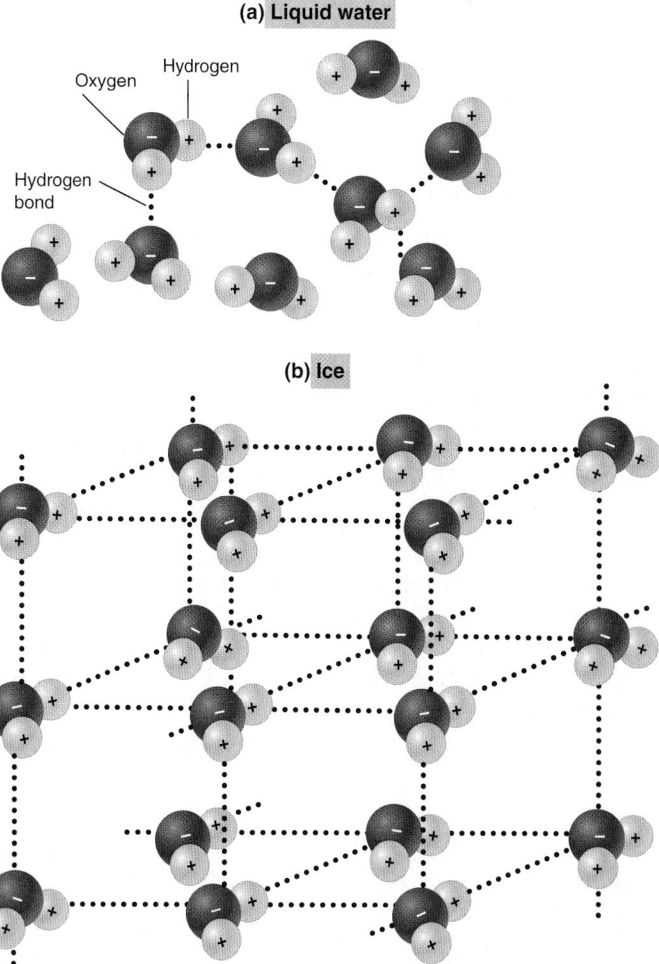

FIGURE 2.7 Water (a) In liquid water, each H_2O molecule hydrogen bonds to one or more H_2O molecules. These bonds continuously break and re-form. **(b)** In ice, each H_2O molecule hydrogen bonds to four other H_2O molecules, forming a rigid crystalline structure. The bonds do not break continuously.

why ice floats. This explains how fish can live in apparently frozen bodies of water. They actually live in the water, which remains liquid below the ice.

The polar nature of water also accounts for its ability to dissolve a large number of compounds. Water has been referred to as *the universal solvent of life* because it dissolves so many compounds. To dissolve in water, compounds must consist of atoms with positive or negative charges. In water, they ionize or split into their component charged atoms. For example, NaCl dissolves in water to form Na^+ and Cl^- ions. In solution, these ions tend to be surrounded by water molecules. The OH^- of HOH forms weak bonds with Na^+, whereas the H^+ forms weak bonds with Cl^- **(figure 2.8).** The Na^+ and the Cl^- cannot associate, and this accounts for the solubility of NaCl in water.

Water containing dissolved substances freezes at a lower temperature than pure water and so in nature, most water does not freeze unless the temperature drops below 0°C. Consequently, microorganisms can usually multiply in liquids below 0°C, because the water remains liquid.

pH

An important property of every aqueous solution is how acidic it is. This property is measured as the **pH** of the solution (an abbreviation for **p**otential **H**ydrogen), defined as the concentration of H^+ in moles per liter.

pH is measured on a logarithmic scale of 0 to 14 in which the lower the number, the more acidic the solution. Water has a slight tendency to split (ionize) into hydrogen ions H^+ (protons), which are acidic, and OH^- ions (hydroxyl), which are basic or alkaline. In pure water, the number of H^+ and OH^- ions is equal, and the concentration of each is 10^{-7} molar (10^{-7} M). The product of the concentration of H^+ and OH^- must always be 10^{-14} M ($10^{-7} \times 10^{-7}$). (Exponents are added when numbers are multiplied.) Thus, if H^+ ions are added to an aqueous solution such that the concentration of H^+ increases tenfold to 10^{-6} M, then the concentration of OH^- must decrease by a factor of 10 (to 10^{-8} M).

The pH scale ranges from 0 to 14 because the concentrations of H^+ and OH^- ions vary within these limits **(figure 2.9).** When the concentrations of H^+ and OH^- are equal, the pH of the solution is 7 and is neutral. However, for every unit on the log scale, the concentration of H^+ ions changes by a factor of 10. Most bacteria can live within only a narrow pH range, near neutrality **(neutrophiles).** Some, however, can live under very acidic conditions **(acidophiles)** and a few under alkaline conditions **(alkalophiles).** ■ acidophiles, p. 93 ■ alkalophiles, p. 93

Compounds called **buffers** stabilize the pH of solutions. These are sometimes added to bacterial growth media because bacteria often produce acids and, less commonly, bases when they degrade compounds to harvest energy. Buffers prevent drastic shifts in pH, which would be deleterious to growth. A common buffer is a mixture of two salts of phosphoric acid, Na_2HPO_4 and NaH_2PO_4. These salts can combine chemically with the H^+ ions of acids and the OH^- of bases to produce neutral compounds, thereby maintaining the pH of the solution near neutrality.

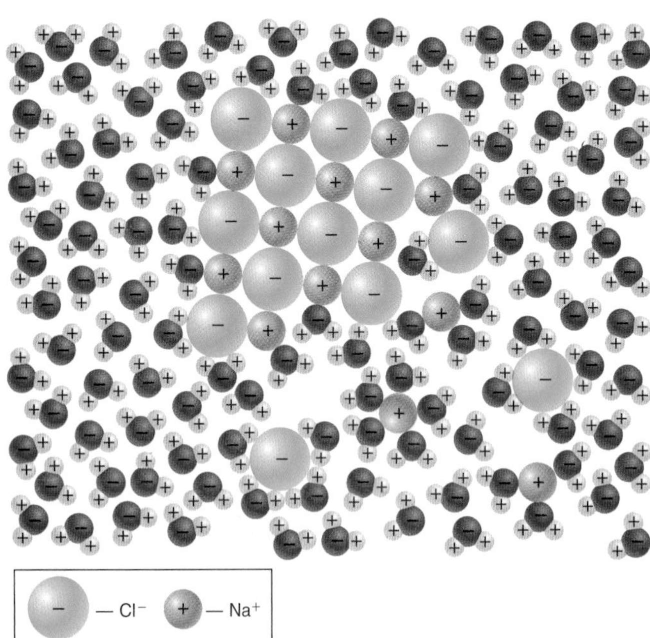

FIGURE 2.8 Salt (NaCl) Dissolving in Water In water, the Na^+ and Cl^- are separated by H_2O molecules. The Na^+ hydrogen bonds to the slightly negatively charged O^- and the Cl^- hydrogen bonds to the slightly positively charged H^+ portion of the water molecules. In the absence of water, the salt is highly structured because of ionic bond formation between Na^+ and Cl^- ions.

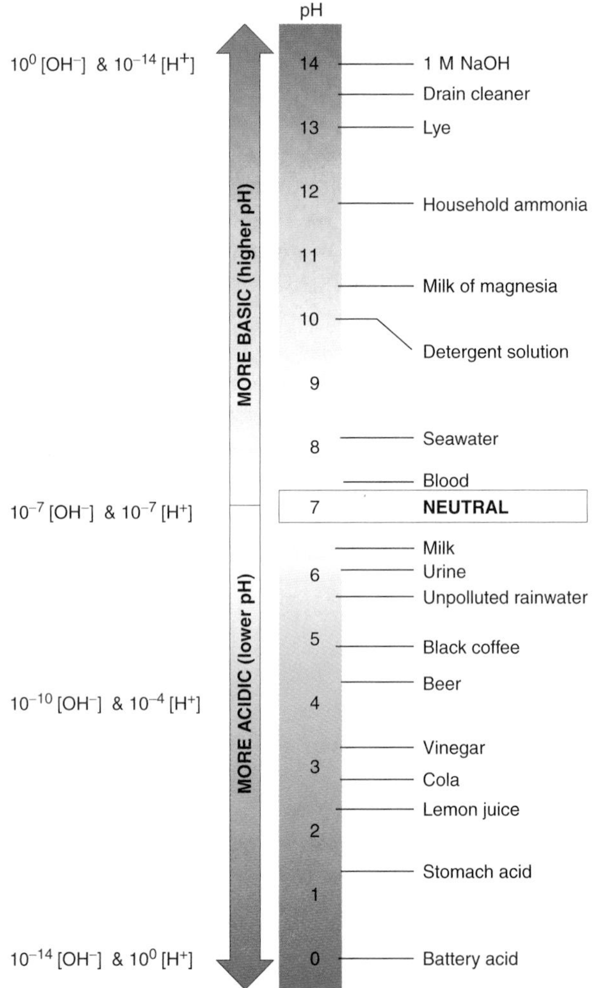

FIGURE 2.9 pH Scale The concentration of H^+ ions varies by a factor of 10 between each pH number since the scale is logarithmic.

Small Molecules in the Cell

All cells contain a variety of small organic and inorganic molecules, many of which occur in the form of ions. About 1% of the weight of a bacterial cell, once the water is removed (dry weight), is composed of inorganic ions, principally Na^+ (sodium), K^+ (potassium), Mg^{2+} (magnesium), Ca^{2+} (calcium), Fe^{2+} (iron), Cl^- (chloride), PO_4^{3-} (phosphate), and SO_4^{2-} (sulfate). Certain enzymes require positively charged ions in minute amounts to function.

The organic small molecules include the building blocks of large molecules, the **macromolecules,** which will be considered in the next section. The building blocks include amino acids, purines and pyrimidines, and various sugars.

An especially important small organic molecule is **adenosine triphosphate (ATP),** the storage form of energy in the cell. The molecule is composed of the sugar ribose, the purine adenine, and three phosphate groups, arranged in tandem **(figure 2.10).** This is an energy-rich molecule because two of the bonds which join the three phosphate molecules are readily broken with the release of energy. The breakage of the terminal high-energy bond of ATP results in the formation of **adenosine diphosphate (ADP),** inorganic phosphate, and the release of energy. The role of ATP in energy metabolism is covered more fully in chapter 6.

Macromolecules and Their Component Parts

Macromolecules are very large molecules (*macro* means "large"). The four major classes of biologically important macromolecules are **proteins, polysaccharides, nucleic acids,** and **lipids.** These four groups differ from each other in their chemical structure. However, other aspects of their structure, as well as how they are synthesized, have features in common.

Most macromolecules are **polymers** (*poly* means "many"), large molecules formed by joining together small molecules, the **subunits.** Each different class of macromolecules is composed of different subunits, each with a different structure. The synthesis of macromolecules involves subunits being joined, one by one, generally forming a chain. This involves chemical reactions in which H_2O is removed, termed **dehydration synthesis (figure 2.11a).** The reverse reactions, breaking a macromolecule down to its subunits, use H_2O and are called **hydrolytic reactions** or **hydrolysis** (figure 2.11b). Both reactions require specific enzymes.

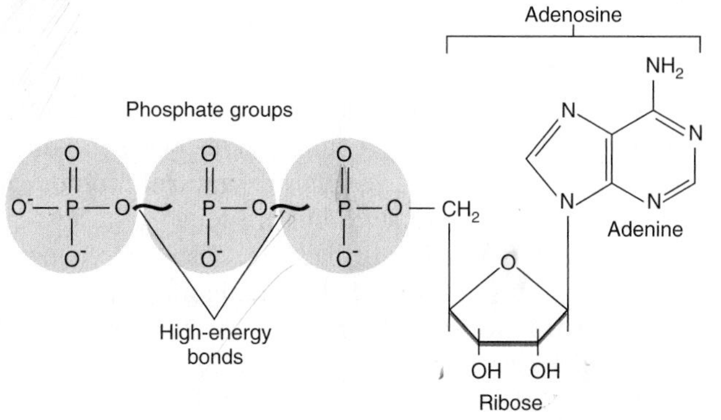

FIGURE 2.10 ATP Adenosine triphosphate (ATP) serves as the energy currency of a cell. The bonds are high energy because of the tandem arrangement of the negatively charged phosphate groups.

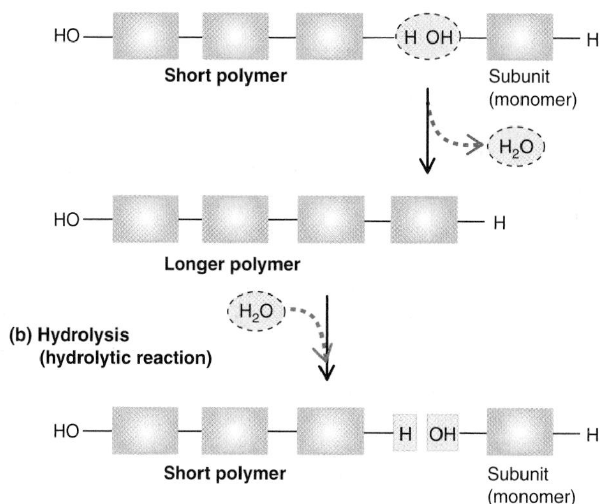

FIGURE 2.11 The Synthesis and Breakdown of Polymers (a) Subunits are joined (polymerized) by removal of water, a dehydration reaction. **(b)** In the reverse reaction, hydrolysis, the addition of water breaks bonds between the subunits. These reactions take place in the formation of many different polymers (macromolecules).

MICROCHECK 2.3

The weak polar bonds of water molecules are responsible for the many properties of water required for life. The degree of acidity of an aqueous solution is expressed as pH. Macromolecules consist of many repeating subunits, each subunit being similar or identical to the other subunits.

- ✓ Why is water a polar molecule? Give three examples of why this property is important in microbiology.
- ✓ Name the four important classes of large molecules in cells.
- ✓ In pure water, what must be done to decrease the OH^- concentration? To decrease the H^+ concentration?

2.4

Proteins and Their Functions

Focus Points

- Name the subunits of proteins and the bonds that join them.
- Name the four levels of protein structure and what distinguishes each level.

Proteins constitute more than 50% of the dry weight of cells and a typical bacterial cell contains 600 to 800 different kinds of proteins at any one time. Of all the cellular macromolecules, they are the most versatile.

In the microbial world, proteins are responsible for:

- Catalyzing all enzymatic reactions of the cell. ■ enzymes, pp. 129, 134
- The structure and shape of certain structures such as ribosomes, the protein-building machinery. ■ ribosomes, p. 68
- Cell movement by flagella. ■ flagella, p. 65

PERSPECTIVE 2.1

Isotopes: Valuable Tools for the Study of Biological Systems

One important tool in the analysis of living cells is the use of **isotopes,** variant forms of the same element that have different atomic weights. The nuclei of certain elements can have greater or fewer neutrons than usual and thereby be heavier or lighter than is typical. For example, the most common form of the hydrogen atom contains one proton and zero neutrons and has an atomic weight of 1 (^{1}H). Another form, however, also exists in nature in very low amounts. This isotope, ^{2}H (deuterium), contains one neutron. A third, even heavier isotope, ^{3}H (tritium), is not found in nature but can be made by a nuclear reaction in which stable atoms are bombarded with high-energy particles. This latter isotope is unstable and gives off radiation (decays) in the form of rays or electrons, which can be very sensitively measured by a radioactivity counter. Once the atom has finished disintegrating, it no longer gives off radiation and is stable.

The other properties of isotopes are very similar to their non-radioactive counterparts. For example, tritium combines with oxygen to form water and with carbon to form hydrocarbons, and both molecules have biological properties similar to those of their non-radioactive counterparts. The only difference is that the molecules containing tritium can be detected by the radiation they emit.

Isotopes are used in numerous ways in biological research. They are frequently added to growing cells in order to label particular molecules, thereby making them detectable. For example, tritiated thymidine (a component of DNA) added to growing bacteria will specifically label DNA and no other molecules. Tritiated uridine, a component of RNA, will label RNA. Isotopes are also used in medical diagnosis. For example, to evaluate proper functioning of the human thyroid gland, which produces the iodine-containing hormone thyroxin, doctors often administer radioactive iodine and then scan the gland later to locate the gland and determine if the amount and distribution of the iodine in the gland is normal **(figure 1).**

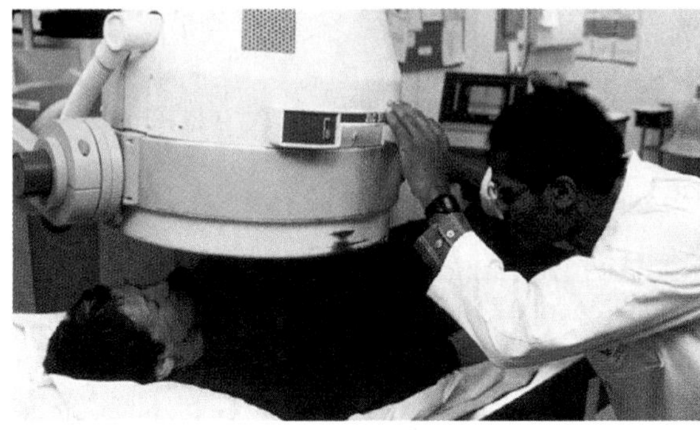
(a)

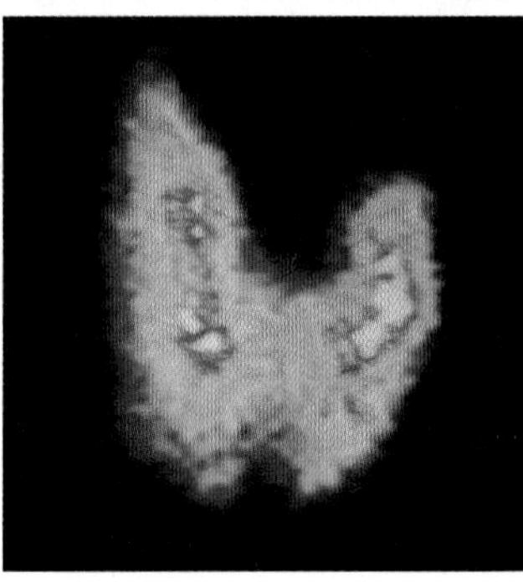
(b)

FIGURE 1 Radioactive Isotopes (a) Physicians use scintillation counters such as this to detect radioactive isotopes. **(b)** A scan of the thyroid gland 24 hours after the patient received radioactive iodine.

- Taking nutrients into the cell. ■ **transport proteins, p. 58**
- Turning genes on and off. ■ **gene regulation, p. 176**
- Determining certain properties of various membranes in the cell. ■ **cytoplasmic membrane, p. 55** ■ **outer membrane, p. 61**

Amino Acid Subunits

Proteins are unbranched macromolecules composed of numerous combinations of 20 major amino acids. The properties of a protein depend mainly on its shape, which in turn depends on the arrangement of the **amino acids** that make up the protein.

All amino acids have at one end a carbon atom to which a carboxyl group and an amino group are bonded **(figure 2.12).** This carbon atom also is bonded to a side chain or backbone (labeled R), which gives each amino acid its characteristic properties. In solution at pH 7, both the amino and carboxyl groups are ionized. The $—NH_2$ group becomes $—NH_3^+$ and the —COOH group becomes $—COO^-$, with the overall charge being zero. The amino acids are subdivided into several different groups based on properties of their side chains **(figure 2.13).** The types and positions of the various side chains in the amino acids that make up a protein govern the solubility of that protein, its shape, and how it interacts with other proteins inside the cell. Amino acids that contain many methyl (CH_3) groups are non-polar and therefore do not interact with water molecules. Thus, they are poorly soluble in water and are termed **hydrophobic** ("water-fearing"). These amino acids tend to be on the inside of protein molecules away from water molecules. Other amino acids contain polar side chains, which make them more soluble in water. They are termed **hydrophilic** ("water-loving") and usually occur on the surface of protein molecules.

All amino acids except glycine can exist in two stereoisomeric forms, a D (right-handed) or L (left-handed) form. Each is a mirror image or stereoisomer of the other **(figure 2.14;** see **A Glimpse of History).** Only L-amino acids occur in proteins, and accordingly, they are designated the **natural amino acids.** D-amino acids are rare in nature but are found in a few compounds mostly associated with bacteria. They are found primarily in the cell walls and in certain antibiotics, antimicrobial medications that many bacteria produce. The bacterium *Bacillus anthracis*, which causes the disease anthrax and has been used as an agent of bioterrorism in the United States, has an outer coat of D-glutamic acid.

```
   H        H      O
   |        |      ||
H— N  —  —  C —    C — OH
            |
            R
Amino   Side chain  Carboxyl
group               group
```

FIGURE 2.12 Generalized Amino Acid This figure illustrates the three groups that all amino acids possess. The R side chain differs with each amino acid and determines the properties of the amino acid.

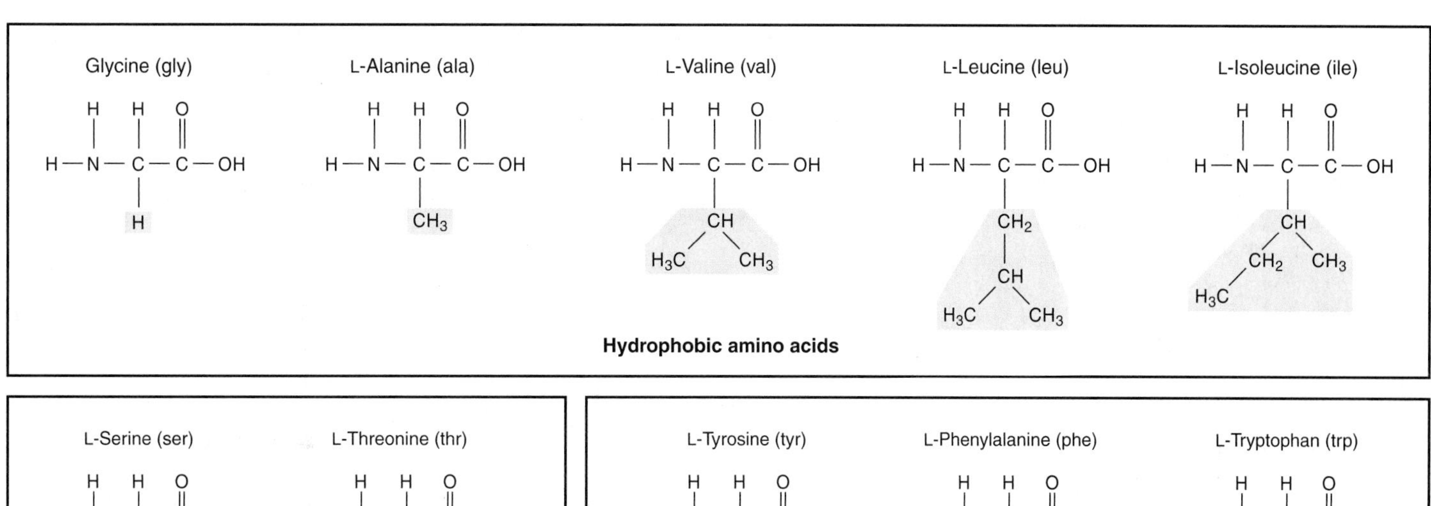

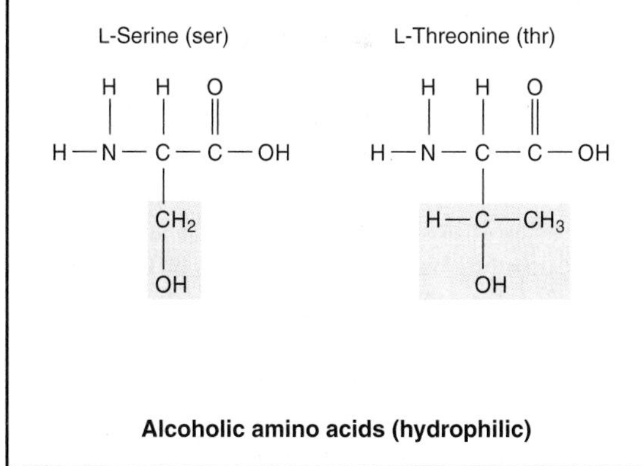

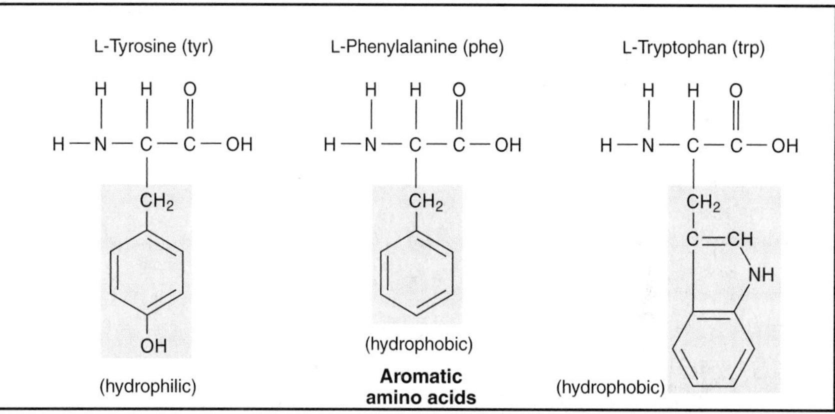

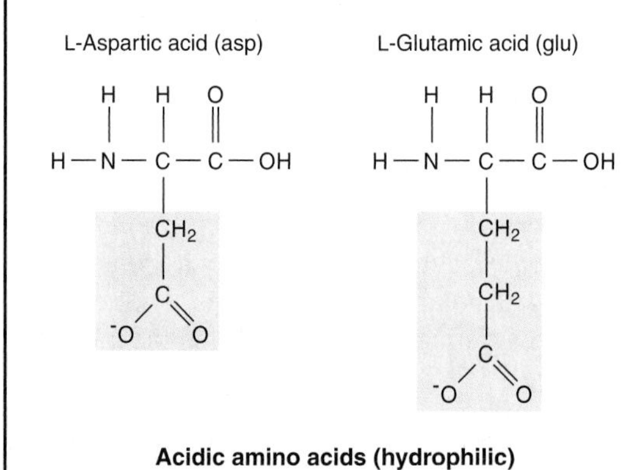

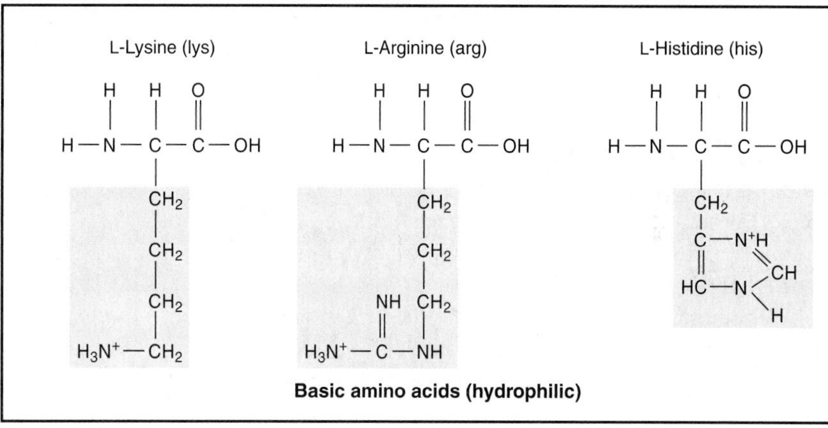

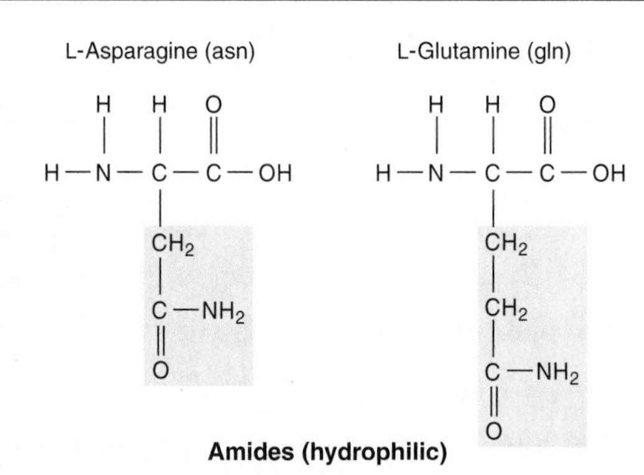

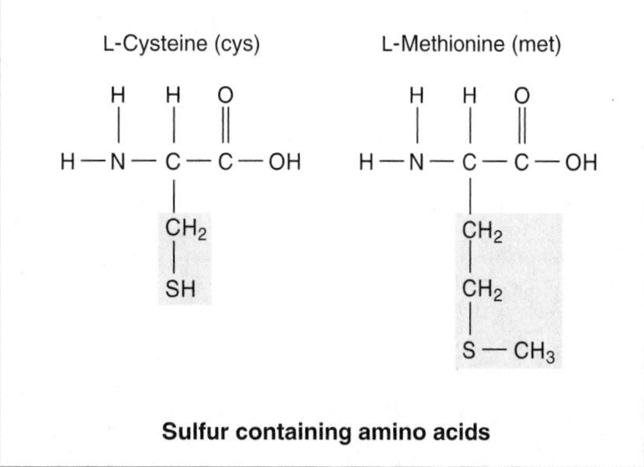

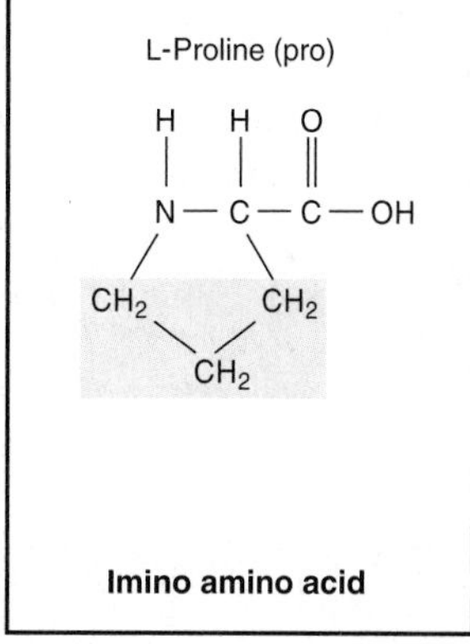

FIGURE 2.13 Common Amino Acids All amino acids have one feature in common—a carboxyl group and an amino group bonded to the same carbon atom. This carbon atom is also bonded to a side chain (shaded). In solution, the —COOH group is ionized to $—COO^-$ and the $—NH_2$ group to $—NH_3$ giving a net charge of zero to the amino acid. The basic and acidic amino acids have a net positive or negative charge, respectively. The three-letter code name for each amino acid is given.

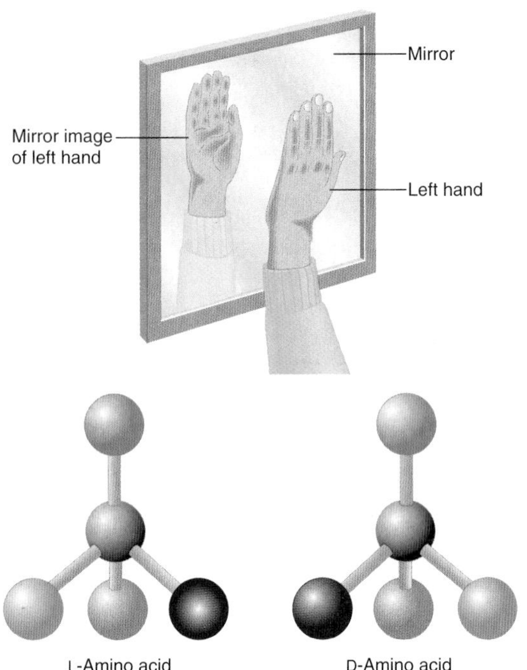

FIGURE 2.14 Mirror Images (Stereoisomers) of an Amino Acid The joining of a carbon atom to four different groups leads to asymmetry in the molecule. The molecule can exist in either the L - or D - form, each being the mirror image of the other. There is no way that the two molecules can be rotated in space to give two identical molecules.

Peptide Bonds and Their Synthesis

Proteins are made up of amino acids held together by **peptide bonds,** a unique type of covalent linkage formed when the carboxyl group of one amino acid reacts with the amino group of another, with the release of water (dehydration synthesis) **(figure 2.15).**

A chain of amino acids joined by peptide bonds is called a **polypeptide chain.** A protein molecule is a long polypeptide chain. One end of the chain has a free amino (—NH_2) group, termed the **N terminal,** or **amino terminal,** end; the other has a free carboxyl (—COOH) group, **the C terminal,** or **carboxyl terminal,** end. Some proteins consist of a single polypeptide chain, whereas others consist of one or more chains joined together by weak bonds. Sometimes, the chains are identical; in other cases, they are different. In proteins that consist of several chains, the individual polypeptide chains generally do not have biological activity by themselves. Proteins vary greatly in size, but an average-size protein consists of a single polypeptide chain of about 400 amino acids.

■ protein synthesis, p. 170

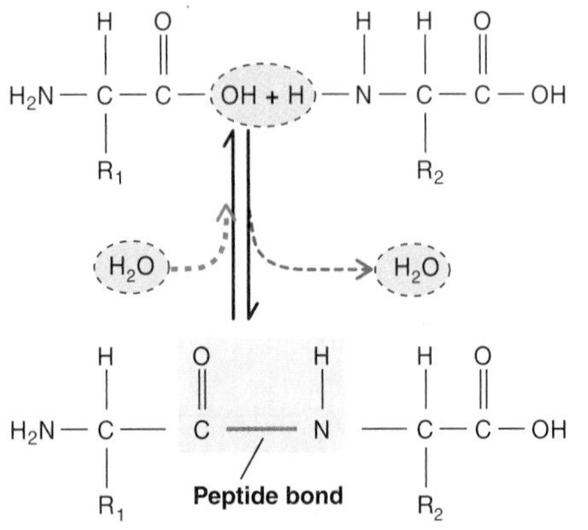

$$H_2N-\underset{R_1}{\overset{H}{C}}-\overset{O}{\overset{\|}{C}}-OH + H-\overset{H}{N}-\underset{R_2}{\overset{H}{C}}-\overset{O}{\overset{\|}{C}}-OH \rightleftharpoons H_2N-\underset{R_1}{\overset{H}{C}}-\overset{O}{\overset{\|}{C}}-\overset{H}{N}-\underset{R_2}{\overset{H}{C}}-\overset{O}{\overset{\|}{C}}-OH + H_2O$$

Peptide bond

FIGURE 2.15 Peptide Bond Formation by Dehydration Synthesis

Protein Structure

Proteins have four levels of structure: primary, secondary, tertiary, and quaternary. The number and arrangement or sequence of amino acids in the polypeptide chain determines its **primary structure (figure 2.16a).** The primary structure in large part determines the other features of the protein.

Parts of the polypeptide chain can form helixes, or folds. This is the protein's **secondary structure** (figure 2.16b). A helical structure is termed an **alpha (α) helix** and a **pleated structure** is called a **beta (β) sheet** (figure 2.16b). These structures result from the amino acids forming weak bonds, such as hydrogen bonds, with other amino acids. This explains why certain sequences of amino acids lead to distinctive secondary structures in various parts of the molecule.

The entire protein next folds into its distinctive three-dimensional shape, its **tertiary structure** (figure 2.16c). Two major shapes exist: **globular,** which tends to be spherical; and **fibrous,** which has an elongated structure (figure 2.16c). The shape is determined in large part by the sequence of amino acids and whether or not they interact with water. Hydrophilic amino acids are located on the outside of the protein molecule, where they can interact with charged polar water molecules. Hydrophobic amino acids are pushed together and cluster inside the molecule to avoid water molecules. The non-polar amino acids form weak interactions with each other, termed **hydrophobic interactions.** In addition to these weak bonds, some amino acids can form strong covalent bonds with other amino acids. One example is the formation of bonds between sulfur atoms (S—S bonds) in different cysteine molecules. The combination of strong and weak bonds between the various amino acids results in the proteins' tertiary structure.

Proteins often consist of more than one polypeptide chain, either identical or different, held together by many weak bonds. The specific shape is termed the **quaternary structure** of the protein (figure 2.16d). Of course, only proteins that consist of more than one polypeptide chain have a quaternary structure.

Sometimes different proteins, each having different functions, associate with one another to make even larger structures termed **multiprotein complexes.** For example, sometimes enzymes involved in the pathway of synthesis of the same amino acid are joined in a **multi-enzyme complex.** On occasion, enzymes involved in the degradation of a particular compound form a multi-enzyme complex.

(a) Primary structure

Amino terminal end

H_2N

Peptide (covalent) bonds

Amino acids

Carboxyl terminal end

O ‖ C — OH

Direction of chain growth

(b) Secondary structure

Amino acids

Hydrogen bond

Helical structure (α-helix)

Hydrogen bond

Amino acids

Pleated structure (β-sheet)

(c) Tertiary structure

Pleated sheet

Helix

Globular protein

Fibrous protein

α chains

β chains

(d) Quaternary structure

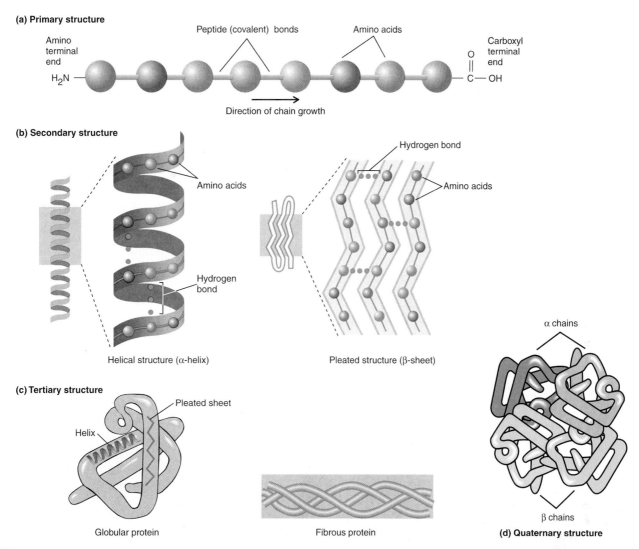

FIGURE 2.16 Protein Structures (a) The primary structure is determined by the amino acid composition. **(b)** The secondary structure results from folding of the various parts of the protein into two major patterns—helices and sheets. **(c)** The tertiary structure is the overall shape of the molecule, globular and fibrous. **(d)** Quaternary structure results from several polypeptide chains interacting to form the protein. This protein is hemoglobin and consists of two pairs of identical chains, *α* and *β*.

Proteins form extremely rapidly. Within seconds, cellular processes join amino acids together to yield a polypeptide chain. How this occurs will be discussed in chapter 7. The polypeptide chain then folds into its correct shape. Although many shapes are possible, only one is functional. Most proteins fold spontaneously into their most stable state correctly. To help certain proteins assume their proper shape, however, cells have proteins called **chaperones** that aid proteins to fold correctly. Incorrectly folded proteins are degraded into their amino acid subunits, which are then used to make more proteins.

Protein Denaturation

A protein must have its proper shape to function. When a protein encounters different conditions such as high temperature, high or low pH, or certain solvents, bonds within the protein are broken and its shape changes **(figure 2.17)**. The protein becomes **denatured** and no longer functions. This explains why most bacteria cannot grow at very high temperatures. Denaturation may be

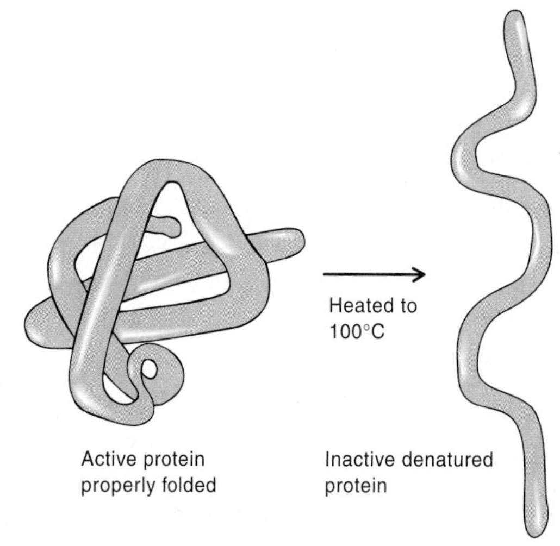

FIGURE 2.17 Denaturation of a Protein

reversible in some cases; in other cases, it is irreversible. For example, boiling an egg denatures the egg white protein, an irreversible process since cooling the egg does not restore the protein to its original appearance. If the denaturating agent is a chemical and is removed, the protein may refold spontaneously into its original shape.

Substituted Proteins

The proteins that play important roles in certain structures of the cell often have other molecules covalently bonded to the side chains of amino acids and are called **substituted proteins.** The proteins are named after the molecules that are covalently joined to the amino acids. If sugar molecules are bonded, the protein is termed a **glycoprotein;** if lipids, the protein is termed a **lipoprotein.** Sugars and lipids are covered later in this chapter.

MICROCHECK 2.4

The side chains of amino acids determine their properties. The sequence of amino acids in a protein determines how the protein folds into its three-dimensional shape.

✓ What type of bond joins amino acids to form proteins?

✓ Name two groups of amino acids that are hydrophilic.

✓ What elements must all amino acids contain? What elements will only some amino acids contain?

2.5 Carbohydrates

Focus Points

- Distinguish among the various carbohydrates based on the number of their subunits.
- Name the most characteristic feature of carbohydrates in terms of their chemical composition.

Carbohydrates comprise a heterogeneous group of compounds of various sizes that play important roles in the life of all organisms:

- Carbohydrates are a common food source from which organisms can harvest energy and make cellular material. ■ metabolism, p. 126
- Two sugars form a part of the nucleic acids, DNA and RNA. ■ nucleic acids, p. 32
- Certain carbohydrates can be stored as a reserve source of nutrients in bacteria. ■ storage granules, p. 68
- Sugars form a part of the bacterial cell wall. ■ cell wall structure, p. 59

All carbohydrates (which means a "hydrate of carbon") contain carbon, hydrogen, and oxygen atoms in an approximate ratio of 1:2:1 respectively. This is because they contain a large number of alcohol groups (—OH) in which the C is also bonded to an H atom to form H—C—OH. **Polysaccharides** are high molecular weight compounds that are linear or branched polymers of their subunits. **Oligosaccharides** are short chains. The term **sugar** is often applied to both **monosaccharides** (*mono* means "one"), a single subunit, and **disaccharides** (*di* means "two"), which are two monosaccharides joined together by covalent bonds.

Carbohydrates also usually contain an aldehyde group

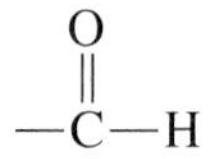

$$\overset{\displaystyle O}{\overset{\|}{—C—H}}$$

and less commonly, a keto group

$$\overset{\displaystyle O}{\overset{\|}{—C—}}$$

The —OH groups on sugars can be replaced by carboxyl, amino, acetyl, or other groups to form molecules important in the structures of the cell. For example, acetyl glucosamine is an important component of the cell wall of bacteria. ■ bacterial cell wall, p. 60

Monosaccharides

Monosaccharides are classified by the number of carbon atoms they contain. The most common monosaccharides have 5- or 6-carbon atoms **(table 2.3).** The 5-carbon sugars, **ribose** and **deoxyribose,** are the sugars in nucleic acids **(figure 2.18).** Note that these monosaccharides are identical except that deoxyribose has one less molecule of oxygen than does ribose (*de* means "away from"). Thus, deoxyribose is ribose "away from" oxygen. Common 6-carbon sugars include **glucose, galactose,** and **fructose.** The carbon atoms are numbered, with carbon atom 1 being closest to the aldehyde or keto group.

Sugars occur in two interconvertible forms: a linear and a ring form (figure 2.18). Both naturally occur in the cell, but most molecules are in the ring form. In diagrams, the lower portion of the ring form is thickened to suggest a three-dimensional structure.

Sugars can exist in two different forms termed alpha (α) and beta (β) based on whether the —OH group on the carbon atom that carries the aldehyde or ketone is above or below the plane of the ring **(figure 2.19).** The α and β forms are interchangeable but once the carbon atom is joined to another sugar molecule, the α or β form is frozen.

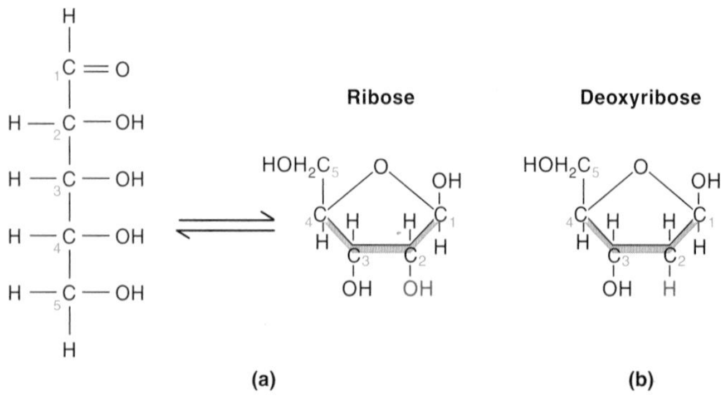

FIGURE 2.18 Ribose and Deoxyribose with the Carbon Atoms Numbered (a) Ribose in linear and ring form. **(b)** Deoxyribose in ring form. Although both structures occur in the cell, the ring form predominates. The plane of the ring is perpendicular to the plane of the paper with the shaded line on the ring closest to the reader.

TABLE 2.3 Common Monosaccharides, Disaccharides, and Polysaccharides

Name of Sugar	Components	Comments
Monosaccharides		
(5-carbon)		
Ribose		Component of RNA
Deoxyribose		Component of DNA
(6-carbon)		
Glucose		Common subunit of disaccharides
Galactose		Component of milk sugar (see below)
Fructose		Fruit sugar
Mannose		Found on the surface of some microbes
Disaccharides		
Lactose	Glucose + galactose	Milk sugar
Maltose	Glucose + glucose	Breakdown product of starch
Sucrose	Glucose + fructose	Table sugar from sugar canes and beets
Polysaccharides		
Agar	Polymer of galactose	Hardening agent in bacteriological media; extracted from the cell walls of some algae
Cellulose	Polymer of glucose, in a β 1, 4 linkage; no branching	Main structural polysaccharide in plant cell walls
Chitin	Polymer of N-acetyl-glucosamine	Major organic component in exoskeleton of insects and crustaceans
Dextran	Polymer of glucose in an α 1, 6 linkage; branching	Storage product in some bacterial cells
Glycogen	Polymer of glucose in an α 1, 4 linkage; branching	Main storage polysaccharide in animal and bacterial cells
Starch	Polymer of glucose	Main storage product in plants

Sugars also form **structural isomers,** molecules that contain the same number of the same elements but in different arrangements that are not mirror images. They are different sugars with different names. For example, common hexoses of biological importance include glucose, galactose, and mannose. They all contain the same atoms but differ in the arrangements of the —H and —OH groups relative to the carbon atoms. Glucose and galactose are identical except for the arrangement of the —H and —OH groups attached to carbon 4. Mannose and glucose differ in the arrangement of the —H and —OH groups joined to carbon 2 **(figure 2.20).** Structural isomers result in three distinct sugars with different properties and different names. For example, glucose has a sweet taste as does mannose, but mannose has a bitter aftertaste.

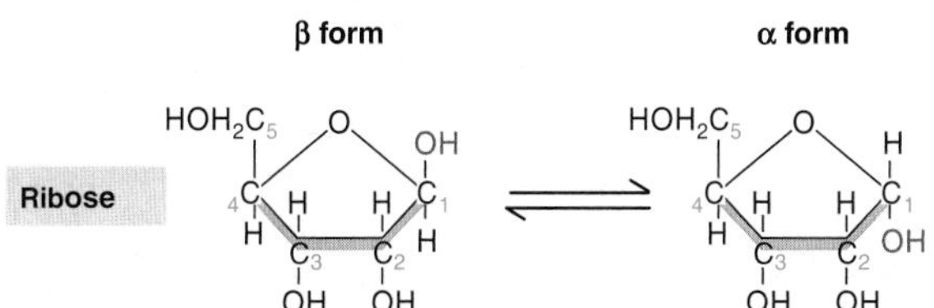

FIGURE 2.19 α and β Links The α and β forms of ribose are interconvertible and only differ in whether the OH group on carbon 1 is above or below the plane of the ring.

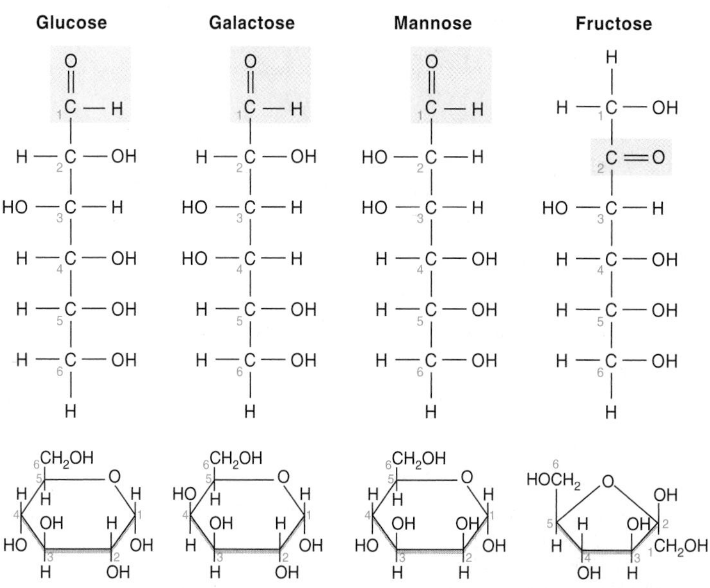

FIGURE 2.20 Formulas of Some Common Sugars Represented in Their Linear and Ring Forms Note that glucose, galactose, and mannose all have an aldehyde group, involving C atom 1 (shaded), whereas fructose has a keto group, involving C atom 2 (shaded).

Disaccharides

The two most common disaccharides in nature are the milk sugar, **lactose,** and the common table sugar, **sucrose** (table 2.3). Lactose consists of glucose and galactose, whereas sucrose, which comes from sugar cane or sugar beets, is composed of glucose and fructose. The monosaccharides are joined together by a dehydration reaction between hydroxyl groups of two monosaccharides, with the loss of water. Note that this reaction is similar to that used to join two amino acids. The reaction is reversible, so that the addition of a water molecule, the process of hydrolysis, yields the two original molecules. Great diversity is possible in molecules formed by joining monosaccharides. The carbon atoms involved in the joining together of monosaccharides may differ and the position of the —OH groups, α and β, involved in the bonding may also differ.

Polysaccharides

Polysaccharides, which are found in many different places in nature, serve different functions (see table 2.3). **Cellulose,** the most abundant organic molecule on earth, is a polymer of glucose subunits and is the principal constituent of plant cell walls. Some bacteria synthesize cellulose in the form of fibrils that attach bacteria to various surfaces. **Glycogen,** a carbohydrate storage product of animals and some bacteria, and **dextran,** which is also synthesized by bacteria as a storage product for carbon and energy, resemble cellulose in some ways.

Cellulose, glycogen, and dextran are composed of glucose subunits, but they differ from one another in many important ways. These include (1) the size of the polymer; (2) the degree of chain branching (the side chains of monosaccharides can branch from the main chain); (3) the particular carbon atoms of the two sugar molecules involved in covalent bond formation, such as a 1, 4 linkage when carbon atom number 1 of one sugar is joined to number 4 carbon atom of the adjacent sugar; and (4) the orientation of the covalent bond between the sugar molecules. Thus, the same subunits can yield a large variety of polysaccharides that have different properties. How these various features of the structure of a polysaccharide fit into the structures of cellulose, glycogen, and dextran are shown in **figure 2.21.**

Polysaccharides and oligosaccharides can also contain different monosaccharide subunits in the same molecule. For example, the cell walls of the domain *Bacteria* contain a polysaccharide consisting of alternating subunit molecules of two different amino sugars.

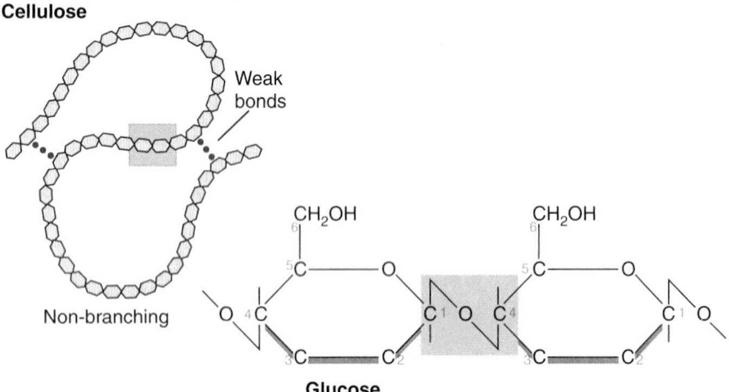

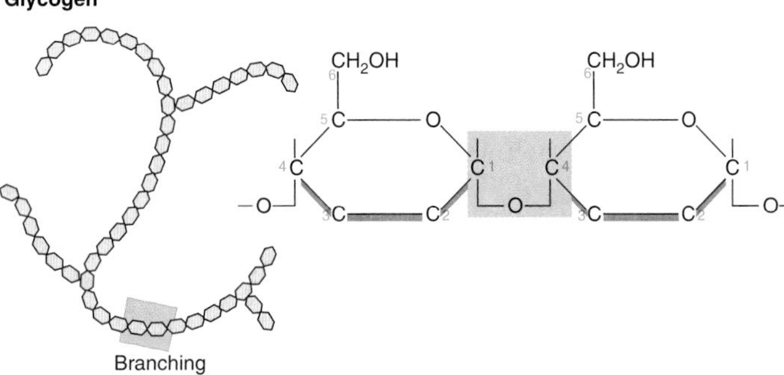

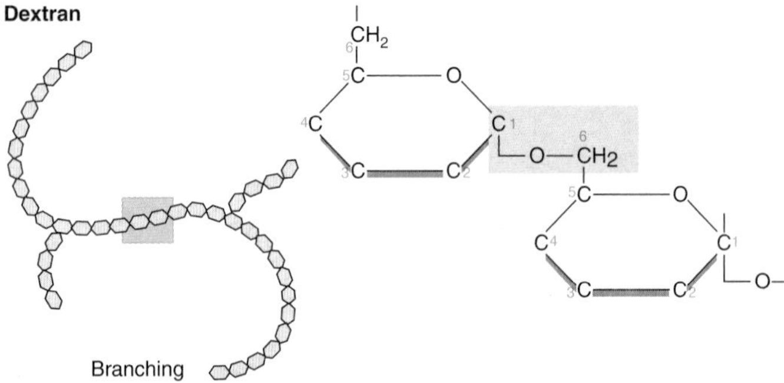

FIGURE 2.21 Structures of Three Important Polysaccharides The three molecules shown consist of the same subunit, glucose, yet they are distinctly different molecules because of differences in linkage that join the molecules (α and β; 1,4 or 1,6), the degree of branching, and the bonds involved in branching (not shown). Weak hydrogen bonds are also involved.

MICROCHECK 2.5

Carbohydrates perform a variety of functions in cells, including serving as a source of energy and forming part of the cells' structures. Carbohydrates with the same subunit composition can have distinct properties because of different arrangements of the atoms in the molecules.

- ✓ Distinguish between structural isomers and stereoisomers.
- ✓ What is the general name given to a single sugar?
- ✓ How could you distinguish sucrose and lactose from a protein molecule by analyzing the elements in the molecules?

2.6

Nucleic Acids

Focus Points

- Compare and contrast the chemical compositions of RNA and DNA.
- Describe the major functions of RNA and DNA.

Nucleic acids carry the genetic information that is then decoded into the sequence of amino acids in protein molecules. There are two types of nucleic acids: **deoxyribonucleic acid (DNA)** and **ribonucleic acid (RNA),** and their subunits are **nucleotides.**

DNA

DNA is the master molecule of the cell—all of the cell's properties are determined by its DNA. This information is coded in the sequence of nucleotides. The code is then converted into a specific sequence of amino acids that make up protein molecules. The details of this process are covered in chapter 7.

In addition to their role in the structure of DNA, nucleotides play other roles in the cell.

- They carry chemical energy in their bonds. ■ adenosine triphosphate, p. 130

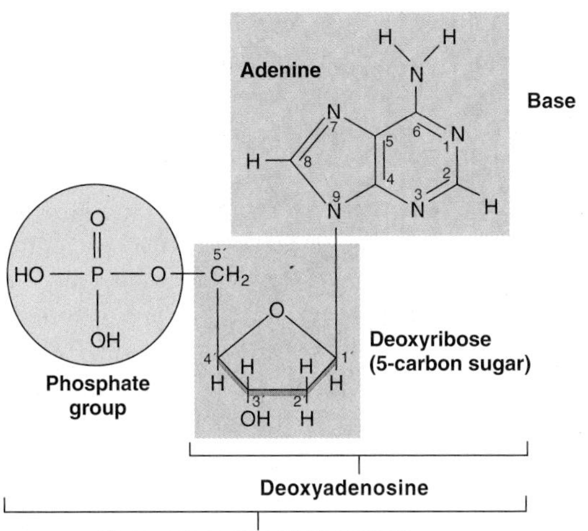

FIGURE 2.22 A Nucleotide This is one subunit of DNA. This subunit is called adenylic acid or deoxyadenosine-5′-phosphate because the base is adenine. If the base is thymine, the nucleotide is thymidylic acid; if guanine, guanylic acid; and if cytosine, cytidylic acid. If the nucleotide lacks the phosphate molecule, it is called a nucleoside, in this case, deoxyadenosine.

- They are part of certain enzymes. ■ CoA, p. 136
- They serve as specific signaling molecules. ■ cyclic AMP, p. 179

The nucleotides of DNA are composed of three different parts: a nitrogen-containing ring compound, called a **base;** which is covalently bonded to a 5-carbon sugar molecule, **deoxyribose;** which in turn is bonded to a **phosphate** molecule **(figure 2.22).** The four different bases in nucleic acids can be divided into two groups according to their

Purines

Adenine

Guanine

Pyrimidines

Cytosine

Thymine

Uracil

FIGURE 2.23 Formulas of Purines and Pyrimidines Both N and C atoms are numbered consecutively.

ring structures: two **purines** (adenine and guanine), which consist of two fused rings; and two **pyrimidines** (cytosine and thymine), which consist of a single ring **(figure 2.23).**

To form nucleic acid chains, the nucleotide subunits are joined by a covalent bond between the phosphate of one nucleotide and the sugar of the adjacent nucleotide **(figure 2.24a).** Thus, the phosphate is a bridge that joins the number 3 carbon atom (termed

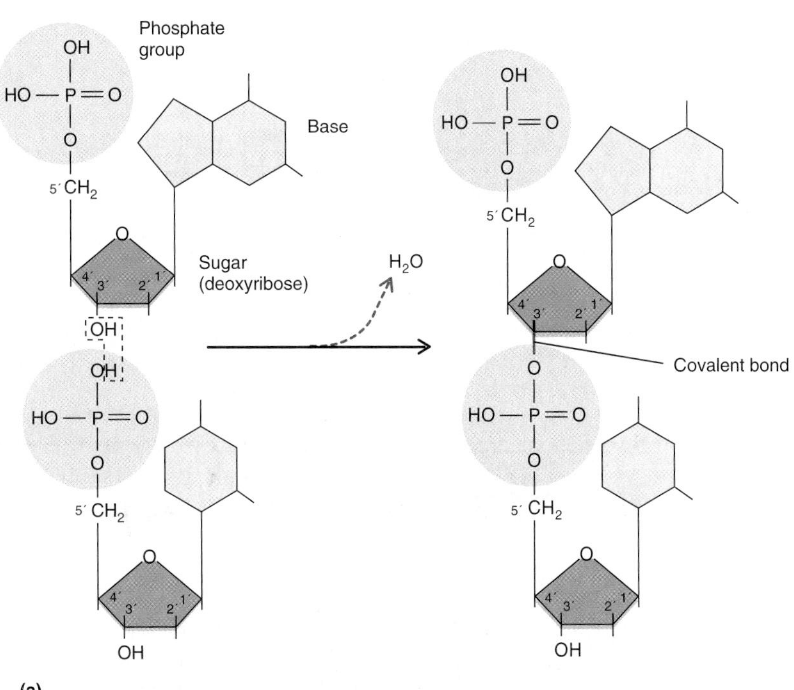

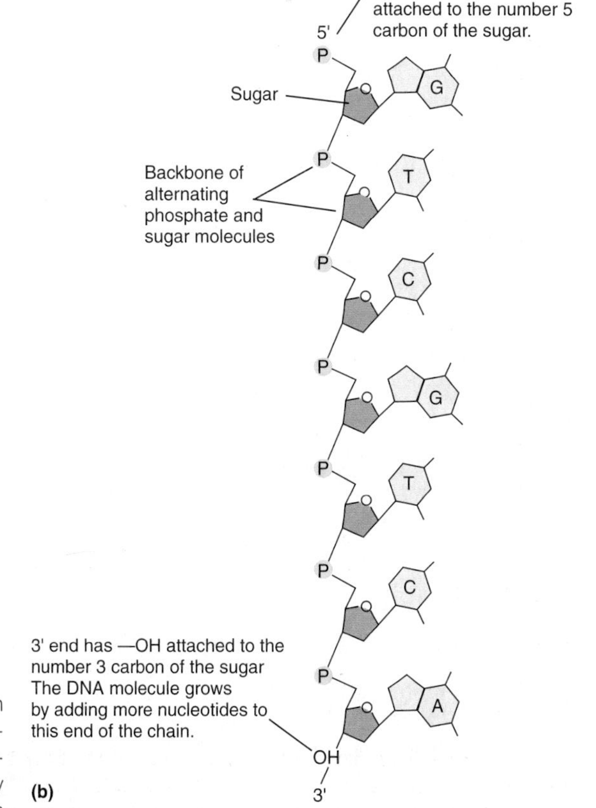

FIGURE 2.24 Joining Nucleotide Subunits (a) Formation of covalent bond between nucleotides by dehydration synthesis. The nucleotide that is added comes from a nucleoside triphosphate and not a nucleotide as illustrated. The two terminal phosphate groups of the nucleoside triphosphate are released as the covalent bond is formed between the nucleotides by dehydration synthesis. This release provides the energy for the joining together of the nucleotides by dehydration synthesis. **(b)** Chain of nucleotides showing the differences between 5′ end and 3′ end. The chain always is extended at the 3′ end, which has the unbonded —OH hydroxyl group.

3 prime and written 3′) of one sugar to the number 5 carbon atom (termed 5′) of the other. This results in a molecule with a backbone of alternating sugar and phosphate molecules, with two different ends. The 5′ end has a phosphate molecule attached to the sugar; the 3′ end has a hydroxyl group (figure 2.24b). During DNA synthesis, the chain is elongated by adding more nucleotides to the 3′ end. This topic is covered in chapter 7.

The DNA of a typical bacterium is a single molecule composed of nucleotides joined together and arranged in a double-stranded helix, with about 4 million nucleotides in each strand **(figure 2.25).** This molecule can be pictured as a spiral staircase with two railings. The railings represent the sugar-phosphate backbone of the molecule, and the stairs are a pair of bases attached to the railings. Each pair of stairs (bases) is held together by weak hydrogen bonds. A specificity exists in the bonding between bases, however, in that adenine (A) can only hydrogen bond to thymine (T), and guanine (G) to cytosine (C). The pair of bases that bond are **complementary** to each other. Thus, G is complementary to C, and A to T. As a result, one entire strand of DNA is complementary to the other strand. This explains why in all DNA molecules, the number of adenine molecules equals the number of thymine molecules and the number of guanines equals the number of cytosines.

Three hydrogen bonds join each G to C, but only two join A to T. Each of the hydrogen bonds is weak, but their large number in a DNA molecule holds the two strands together. In addition to the differences in their sequence of bases, the two complementary strands differ from each other in orientation. The two strands are arranged in opposite directions. One goes in the 3′ to the 5′ direction; the other in the 5′ to 3′. Consequently, the two ends of the strands opposite each other differ; one is a 5′ end, the other, a 3′ end (see figure 2.25).

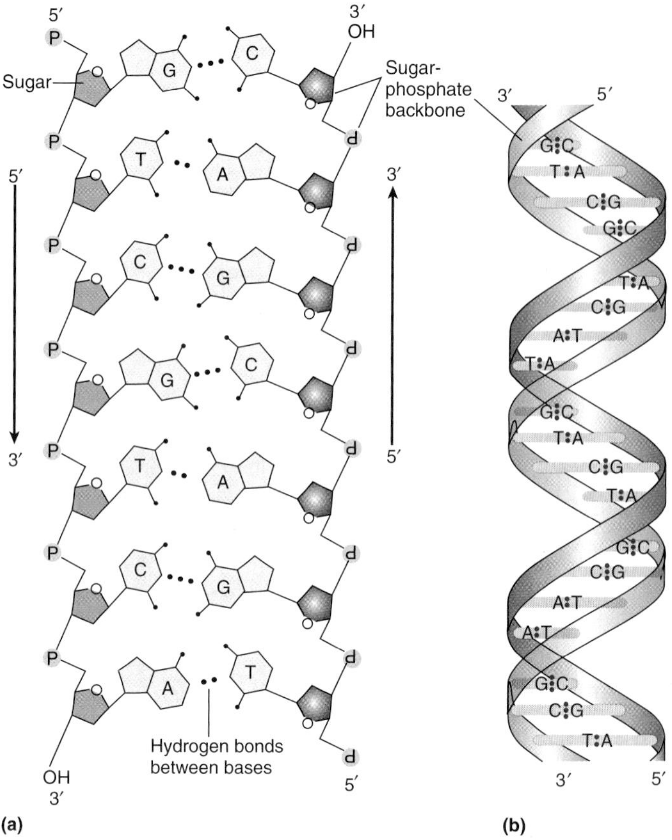

FIGURE 2.25 DNA Double-Stranded Helix (a) The sugar-phosphate backbone and the hydrogen bonding between bases. There are two hydrogen bonds between adenine and thymine and three between guanine and cytosine. **(b)** The spiral staircase of the sugar-phosphate backbone with the bases on the inside. The railings go in opposite directions.

RNA

RNA is involved in decoding the information in the DNA into a sequence of amino acids in proteins. This complex multi-step process will be examined in chapter 7.

Although the structure of RNA resembles that of DNA, it differs in several ways. First, RNA contains the pyrimidine **uracil** in place of thymine and the sugar **ribose** in place of deoxyribose (see figures 2.23 and 2.18). Also, whereas DNA is a long, double-stranded helix, RNA is considerably shorter and exists as a single chain of nucleotides. Although single-stranded, it may form short, double-stranded stretches as a result of hydrogen bonding between complementary bases in the single strand.

MICROCHECK 2.6

DNA carries the genetic code in the sequence of purine and pyrimidine bases in its double-helical structure. The information is transferred to RNA and then into a sequence of amino acids in proteins.

- ✓ What are the two types of nucleic acids?
- ✓ If the DNA molecule were placed in boiling water, how would the molecule change?

FRANK & ERNEST: © Thaves/Dist. By Newspaper Enterprise Association, Inc.

2.7 Lipids

Focus Points

- Name the one property common to all lipids.
- Explain how the chemical structure of a phospholipid prevents the entry and exit of substances into and out of the cell.

Lipids play an indispensable role in all living cells. They are critically important in the structure of all membranes, which act as gatekeepers of cells. They keep a cell's internal contents inside the cell and keep many molecules from entering the cell. ■ cytoplasmic membrane, p. 55

Lipids are a very heterogeneous group of molecules. Their defining feature is their slight solubility in water contrasted with their great solubility in most organic solvents such as ether, benzene, and chloroform. These solubility properties result from their non-polar, hydrophobic nature. Lipids have molecular weights of no more than a few thousand and so are the smallest of the macromolecules we have discussed. Further, unlike the other macromolecules, they are not composed of similar subunits; rather, they consist of a wide variety of substances that differ in their chemical structure. Lipids can be divided into two general classes: the **simple** and the **compound,** which differ in important aspects of their chemical composition.

Simple Lipids

Simple lipids contain only carbon, hydrogen, and oxygen. The most common are **fats,** a combination of **fatty acids** and **glycerol** that are solid at room temperature **(figure 2.26).** Fatty acids are molecules with long chains of C atoms bonded to H atoms with an acidic group (—COOH) on one end **(figure 2.27).** Since glycerol has three hydroxyl groups, a maximum of three fatty acid molecules, either the same or different, can be linked through covalent bonds between the —OH group of glycerol and the —COOH group of the fatty acid. If only one fatty acid is bound to glycerol, the fat is called a **monoglyceride;** when two are joined, it is a **diglyceride;** when three are bound, a **triglyceride** is formed. Fatty acids are stored in the body as an energy reserve in the form of triglycerides.

Although hundreds of different fatty acids exist, they can be divided into two groups based on whether or not any double bonds are present in the portion of the molecule containing only carbon and hydrogen atoms. If there are no double bonds, the fatty acid is termed **saturated** with H atoms. If it contains one or more double bonds, it is **unsaturated.** Unsaturated fats tend to be liquid and are then called **oils.** Oils are liquid because these unsaturated fatty acids develop kinks in their long tails that prevent tight packing. The saturated fats can pack their straight, long tails tightly together and therefore are solid (figure 2.27a). Oleic acid, with its one double bond, is a common **monounsaturated** fatty acid (figure 2.27b). Other fatty acids containing numerous double bonds are **polyunsaturated.** Different lipids are called highly saturated or highly unsaturated when they contain mostly saturated or unsaturated fatty acids.

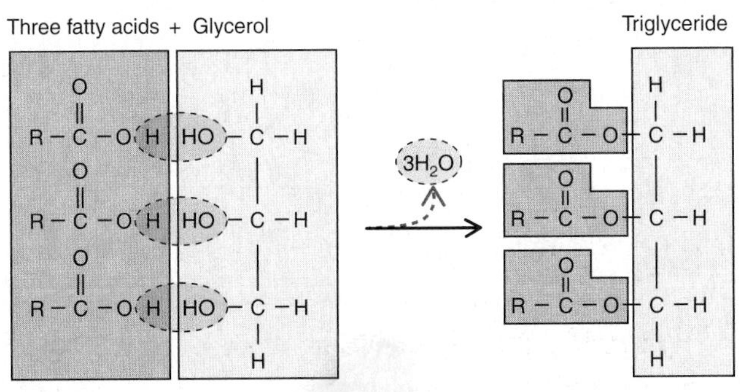

FIGURE 2.26 Formation of a Fat The R group of the fatty acids commonly contains 16 or 18 carbon atoms bonded to hydrogen atoms.

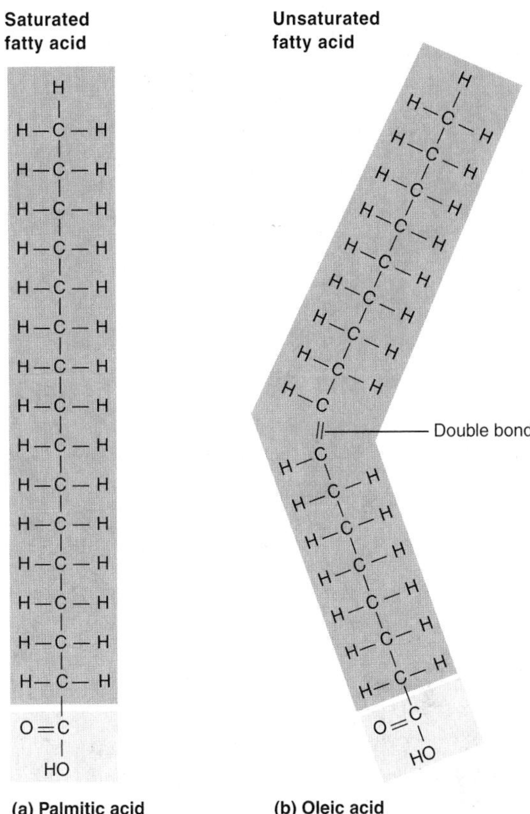

FIGURE 2.27 Fatty Acids The saturated fatty acids **(a)** are solids, and the unsaturated fatty acids **(b)** are liquids.

Another very important group of simple lipids is the **steroids.** All members of this group have the four-membered ring structure shown in **(figure 2.28a).** These compounds differ from the fats in

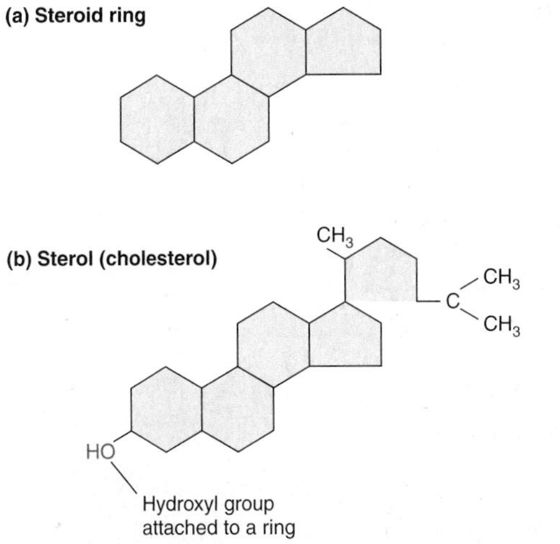

FIGURE 2.28 Steroid (a) General formula showing the four-membered ring and **(b)** the —OH group that make the molecule a sterol. The sterol shown here is cholesterol. The carbon atoms in the ring structures and the attached hydrogen atoms are not shown.

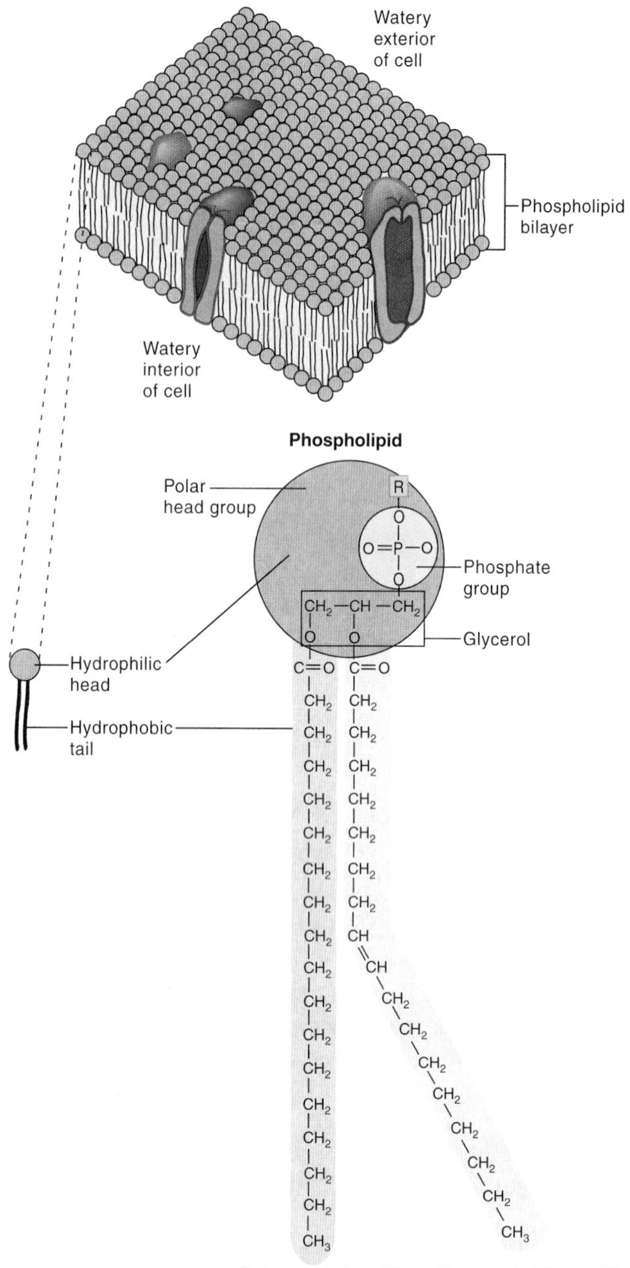

FIGURE 2.29 Phospholipid and the Bilayer That Phospholipids Form in the Membrane of Cells In phospholipids, two of the —OH groups of glycerol are linked to fatty acids and the third —OH group is linked to a hydrophilic head group, which contains a phosphate ion and a polar molecule, labeled R.

chemical structure, but both are classified as lipids because they are insoluble in water. If a hydroxyl group is attached to one of the rings, the steroid is called a **sterol,** an example being **cholesterol** (figure 2.28b). Other important compounds in this group of lipids are certain hormones such as cortisone, progesterone, and testosterone.

Compound Lipids

Compound lipids contain fatty acids and glycerol as well as elements other than carbon, hydrogen, and oxygen, Biologically, some of the most important members of this group are the **phospholipids,** which contain a phosphate molecule in addition to the fatty acids and glycerol **(figure 2.29).** The phosphate is further linked to a variety of other polar molecules, such as an alcohol, a sugar, or one of certain amino acids. This entire group is referred to as a **polar head** and is soluble in water (hydrophilic). In contrast, the fatty acid portion is insoluble in water (hydrophobic).

Phospholipids are an integral part of cytoplasmic membranes, the structure that separates the internal contents of a cell from the outside environment (see figure 2.29). The phospholipid molecules orient themselves in the membrane as opposing layers, forming a **bilayer.** In other words, the hydrophilic polar heads face outward, toward either the external, in the case of one of the bilayers, or internal (cytoplasmic) environment, in the case of the other. The fatty acids orient themselves inward, interacting hydrophobically with the fatty acids of the phospholipid molecules in the opposing bilayer. Water-soluble substances, which are the most common and most important in the cell's environment, cannot pass through the hydrophobic portion. Therefore, the cell has special mechanisms to bring these molecules into the cell. These will be discussed in chapter 3.

Other compound lipids are found in the outer covering of bacterial cells and will also be discussed in chapter 3. These include the **lipoproteins,** covalent associations of proteins and lipids, and the **lipopolysaccharides,** molecules of lipid linked with polysaccharides through covalent bonds.

Some of the most important properties of macromolecules of biological importance are summarized in **table 2.4.**

MICROCHECK 2.7

Phospholipids, with one end hydrophilic and the other, hydrophobic, form a major part of cell membranes where they exist as a bilayer. They limit the entry and exit of molecules into and out of cells.

- ✓ What are the two main types of lipids and how do they differ from one another?
- ✓ What are the main functions of lipids in cells?
- ✓ Some molecules such as many alcohols are soluble in both water and hydrophobic liquids such as oils. How easily do you think these molecules would cross the cell membrane?

TABLE 2.4 Structure and Function of Macromolecules

Name	Subunit	Some Functions of Macromolecules
Protein	Amino acid	Catalysts; structural portion of many cell components
Nucleic acids	Nucleotide	RNA—Various roles in protein synthesis; DNA—Carrier of genetic information
Polysaccharide	Monosaccharide	Structural component of plant cell wall; storage products
Lipids	Varies—Subunits are not similar	Important in structure of cell membranes

FUTURE CHALLENGES

Fold Properly: Do Not Bend or Mutilate

The properties of all organisms depend on the proteins they contain. These include the structural proteins as well as enzymes. Even though a cell may be able to synthesize a protein, unless that protein is folded correctly and achieves its correct shape, it will not function properly. A major challenge is to understand how proteins fold correctly—the **protein-folding problem.** Not only is this an important problem from a purely scientific point of view, but a number of serious neurodegenerative diseases result from protein misfolding. These include Alzheimers disease and the neurodegenerative diseases caused by prions. If we could understand why proteins fold incorrectly, we might be able to prevent such diseases.

The information that determines how a protein folds into its three-dimensional shape is contained in the sequence of its amino acids. It is not yet possible, however, to predict accurately how a protein will fold from its amino acid sequence. The folding occurs in a matter of seconds after the protein is synthesized. The protein folds rapidly into its secondary structure and then more slowly into its tertiary structure. These slower reactions are still poorly understood, but various attractive and repulsive interactions between the amino acid side chains allow the flexible molecule to "find its way" to the correct tertiary structure. Proteins called **chaperones** can assist the process by preventing detrimental interactions. Mistakes still occur, but improperly folded proteins can be recognized and degraded by enzymes called **proteases.** The protein-folding problem has such important implications for medicine and is so challenging a scientific question, that a super-computer with a huge memory is now being used to help predict the three-dimensional structure of a protein from its amino acid sequence.

■ **prions, pp. 12, 341**

SUMMARY

2.1 Atoms and Elements

Atoms are composed of **electrons, protons,** and **neutrons** (figure 2.1). An element consists of a single type of atom.

2.2 Chemical Bonds and the Formation of Molecules

For maximum stability, the outer **shell** of electrons of an atom must be filled. The electrons in different shells have different energy levels. **Bonds** form between atoms to fill their outer shells with electrons.

Ionic Bonds

When electrons leave the shells of one atom and enter the shells of another atom, an **ionic bond** forms between the atoms (figure 2.2).

Covalent Bonds

Covalent bonds are strong bonds formed by atoms sharing electrons (figure 2.3). When atoms have an equal attraction for electrons, a **non-polar covalent bond** is formed between them (table 2.2). When one atom has a greater attraction for electrons than another atom, **polar covalent bonds** are formed between them (figure 2.4).

Hydrogen Bonds

Hydrogen bonds are weak bonds that result from the attraction of a positively charged hydrogen atom in a polar molecule to a negatively charged atom in another polar molecule (figure 2.5). Hydrogen bonds are important in the weak association of enzymes with their substrate (figure 2.6).

2.3 Chemical Components of the Cell

Water

Water is the most important molecule in the cell. Water makes up over 70% of all living organisms by weight. Hydrogen bonding plays a very important role in the properties of water (figures 2.7, 2.8).

pH

pH is the degree of acidity of a solution; it is measured on a scale of 0 to 14. Buffers prevent the rise or fall of pH (figure 2.9).

Small Molecules in the Cell

All cells contain a variety of small organic and inorganic molecules. A key element in all cells is carbon; it occurs in all organic molecules. ATP, the energy currency of the cell, stores energy in two high-energy phosphate bonds which, when broken, release energy (figure 2.10).

Macromolecules and Their Component Parts

Macromolecules are large molecules usually composed of **subunits** with similar properties. Synthesis of macromolecules occurs by **dehydration synthesis,** the removal of water, and their degradation occurs by **hydrolysis,** the addition of water.

2.4 Proteins and Their Functions

Proteins are the most versatile of the macromolecules in what they do. Activities of proteins include catalyzing reactions, being a component of cell structures, moving cells, taking nutrients into the cell, turning genes on and off, and being a part of cell membranes.

Amino Acid Subunits

Proteins are composed of 20 major **amino acids** (figure 2.13). All amino acids consists of a **carboxyl group** at one end and an **amino group** bonded to the same carbon atom as the carboxyl group and a side chain which confers unique properties on the amino acid (figure 2.12).

Peptide Bonds and Their Synthesis

Amino acids are joined through **peptide bonds,** joining an amino with a carboxyl group and splitting out water (figure 2.15).

Protein Structure (figure 2.16)

The **primary structure** of a protein is its amino acid sequence. The **secondary structure** of a protein is determined by intramolecular bonding between amino acids to form **helices** and **sheets.** The **tertiary structure** of a protein describes the three-dimensional shape of the protein, either **globular** or **fibrous.** The **quaternary structure** describes the structure resulting from the interaction of several **polypeptide** chains. When the intramolecular bonds within the protein are broken, the protein changes shape and no longer functions; the proteins are **denatured** (figure 2.17).

Substituted Proteins

Substituted proteins contain other molecules such as **sugars** and **lipids,** bonded to the side chains of amino acids in the protein.

2.5 Carbohydrates

Carbohydrates comprise a heterogeneous group of compounds that perform a variety of functions in the cell. Carbohydrates have carbon, hydrogen, and oxygen atoms in a ratio of approximately 1:2:1.

Monosaccharides

Monosaccharides are classified by the number of carbon atoms they contain, most commonly 5 or 6 (table 2.3, figure 2.20). Sugars can exist in two interchangeable forms: α and β, depending on whether the —OH group on carbon atom 1 is above or below the plane of the ring (figure 2.19). Sugars can exist as structural isomers—molecules that have the same number of the same elements but are arranged differently (figure 2.20, table 2.3).

Disaccharides

Disaccharides consist of two monosaccharides joined by a covalent bond between their hydroxyl groups (table 2.3).

Polysaccharides

Polysaccharides are macromolecules consisting of monosaccharide subunits, sometimes identical, other times not (figure 2.21, table 2.3).

2.6 Nucleic Acids

Nucleic acids are macromolecules whose subunits are **nucleotides** (figure 2.22). There are two types of nucleic acids: **deoxyribonucleic acid (DNA)** and **ribonucleic acid (RNA).**

DNA

DNA is the master molecule of the cell and carries all of the cell's genetic information in its sequence of nucleotides.

DNA is a double-stranded helical molecule with a backbone composed of covalently bonded sugar and phosphate groups. The **purine** and **pyrimidine** bases extend into the center of the helix (figure 2.23, figure 2.25a). The two strands of DNA are **complementary** and are held together by hydrogen bonds between the bases (figure 2.25b).

RNA

RNA is involved in decoding the genetic information contained in DNA. RNA is a single-stranded molecule and contains **uracil** in place of **thymine** in DNA (figure 2.23).

2.7 Lipids

Lipids are a heterogeneous group of molecules that are slightly soluble in water and very soluble in most organic solvents. They comprise two groups: **simple** and **compound lipids.**

Simple Lipids

Simple lipids contain carbon, hydrogen, and oxygen and may be liquid or solid at room temperature. **Fats** are common simple lipids and consist of **glycerol** bound to **fatty acids** (figure 2.26). **Fatty acids** may be **saturated,** in which the fatty acid contains no double-bonds between carbon atoms, or **unsaturated,** in which one or more double bonds exist (figure 2.27). Some simple lipids consist of a four-membered ring, and include **steroids** and **sterols** (figure 2.28).

Compound Lipids

Compound lipids contain elements other than carbon, hydrogen, and oxygen. **Phospholipids** are common and important examples of compound lipids. They are essential components of bilayer membranes in cells (figure 2.29).

REVIEW QUESTIONS

Short Answer

1. Differentiate between an atom, an element, an ion, and a molecule.
2. Which solution is more acidic, one with a pH of 4 or a pH of 5? What is the concentration of H^+ ions in each? The concentration of OH^- ions?
3. How do the two types of nucleic acids differ from one another in (a) composition, (b) size, and (c) function?
4. Name the subunits of proteins, polysaccharides, and nucleic acids.
5. What are the two major groups of lipids? Give an example of each group. What feature is common to all lipids?
6. How does the primary structure of a protein determine its overall structure?
7. Why is water a good solvent?
8. Give an example of a dehydration synthesis reaction. Give an example of a hydrolysis reaction. How are these types of reactions related?
9. List four functions of proteins.
10. What is a steroid?

Multiple Choice

1. Choose the list that goes from the lightest to the heaviest:
   a) Proton, atom, molecule, compound, electron
   b) Atom, proton, compound, molecule, electron
   c) Electron, proton, atom, molecule, compound
   d) Atom, electron, proton, molecule, compound
   e) Proton, atom, electron, molecule, compound
2. The strongest chemical bonds between two atoms in solution are
   a) covalent. b) ionic.
   c) hydrogen bonds. d) hydrophobic interactions.
3. Dehydration synthesis is involved in the synthesis of all of the following, *except*
   a) DNA b) proteins c) polysaccharides
   d) lipids e) monosaccharides
4. The primary structure of a protein relates to its
   a) sequence of amino acids b) length c) shape
   d) solubility e) bonds between amino acids
5. Pure water has all of the following properties, *except*
   a) polarity. b) ability to dissolve lipids. c) pH of 7.
   d) covalent joining of its atoms. e) ability to form hydrogen bonds.
6. The macromolecules that are composed of carbon, hydrogen, and oxygen in an approximate ratio of 1 : 2 : 1 are
   a) proteins. b) lipids. c) polysaccharides.
   d) DNA. e) RNA.
7. In proteins, α helices and β pleated structures are associated with the
   a) primary structure. b) secondary structure.
   c) tertiary structure. d) quaternary structure.
   e) multiprotein complexes.
8. Complementarity plays a major role in the structure of
   a) proteins. b) lipids. c) polysaccharides.
   d) DNA. e) RNA.

9. A bilayer is associated with
   a) proteins. b) DNA. c) RNA.
   d) complex polysaccharides. e) phospholipids.
10. Isomers are associated with
   1. carbohydates. 2. amino acids. 3. nucleotides.
   4. RNA. 5. fatty acids.
   a) 1, 2 b) 2, 3 c) 3, 4 d) 4, 5 e) 1, 5

Applications

1. A group of bacteria known as thermophiles thrive at high temperatures that would normally destroy other bacteria. Yet these thermophiles cannot survive well at the lower temperatures normally found on the earth. Propose a plausible explanation for this observation.
2. Microorganisms use hydrogen bonds to attach themselves to the surfaces that they live upon. Many of them lose hold of the surface because of the weak nature of these bonds and end up dying. Contrast the benefits and disadvantages of using covalent bonds as a means of attaching to surfaces.

Critical Thinking

1. What properties of the carbon atom make it ideal as the key atom for all molecules in organisms?
2. A biologist determined the amounts of several amino acids in two separate samples of pure protein. His data are shown here:

Amino Acid	Leucine	Alanine	Histidine	Cysteine	Glycine
Protein A	7%	12%	4%	2%	5%
Protein B	7%	12%	4%	2%	5%

He concluded that protein A and protein B were the same protein. Do you agree with this conclusion? Justify your answer.

3. This table indicates the freezing and boiling points of several molecules:

Molecule	Freezing Point (°C)	Boiling Point (°C)
Water	0	100
Carbon tetrachloride (CCl_4)	– 23	77
Methane (CH_4)	– 182	– 164

Carbon tetrachloride and methane are non-polar molecules. How does the polarity and non-polarity of these molecules explain why the freezing and boiling points for methane and carbon tetrachloride are so much lower than those for water?

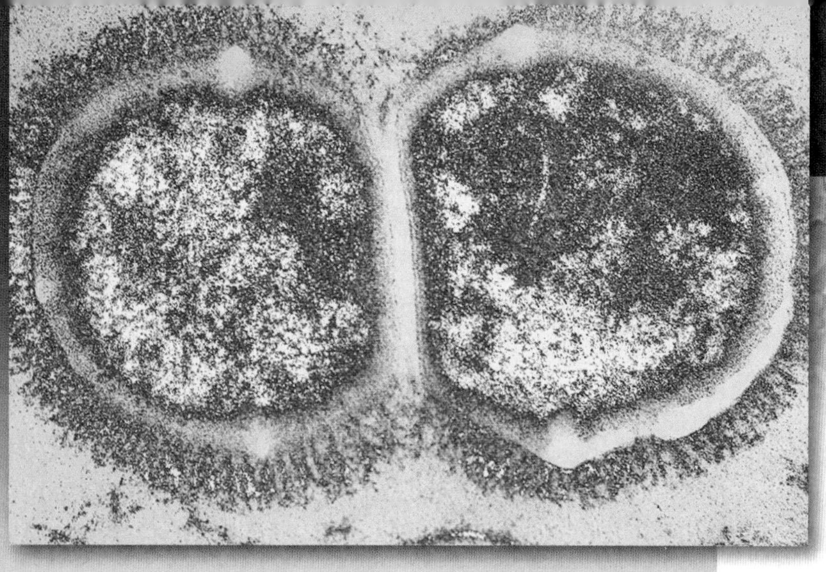

Color-enhanced TEM of bacterial cells.

3

Microscopy and Cell Structure

A Glimpse of History

Hans Christian Joachim Gram (1853–1938) was a Danish physician working in a laboratory at the morgue of the City Hospital in Berlin, microscopically examining the lungs of patients who had died of pneumonia. He was working under the direction of Dr. Carl Friedlander, who was trying to identify the cause of pneumonia by studying patients who had died of it. Gram's task was to stain the infected lung tissue to make the bacteria easier to see under the microscope. Strangely, one of the methods he developed did not stain all bacteria equally; some types retained the first dye applied in this multistep procedure, whereas others did not. Gram's staining method revealed that two different kinds of bacteria were causing pneumonia, and that these types retained the dye differently. We now recognize that this important staining method, called the Gram stain, efficiently identifies two large, distinct groups of bacteria: Gram-positive and Gram-negative. The variation in the staining outcome of these two groups reflects a fundamental difference in the structure and chemistry of their cell walls.

For a long time, historians thought that Gram did not appreciate the significance of his discovery. In more recent years, however, several letters show that Gram did not want to offend the famous Dr. Friedlander under whom he worked; therefore, he played down the importance of his staining method. In fact, the Gram stain has been used as a key test in the initial identification of bacterial species ever since the late 1880s. ■

Imagine the astonishment Antony van Leeuwenhoek must have felt in the 1600s when he first observed microorganisms with his handcrafted microscopes, instruments that could magnify images approximately 300-fold (300×). Even today, observing diverse microbes interacting in a sample of stagnant pond water can provide enormous education and entertainment.

Microscopic study of cells has revealed two fundamental types: prokaryotic and eukaryotic. The cells of all members of the Domains *Bacteria* and *Archaea* are prokaryotic. In contrast, cells of all animals, plants, protozoa, fungi, and algae are eukaryotic. The similarities and differences between these two basic cell types are important from a scientific standpoint and also have significant consequences to human health. For example, chemicals that interfere with processes unique to prokaryotic cells can be used to selectively destroy bacteria without harming humans. ■ **prokaryotic cells, p. 10**

Prokaryotic cells are generally much smaller than most eukaryotic cells—a trait that carries with it certain advantages as well as disadvantages. On one hand, their high surface area relative to their low volume makes it easier for these cells to take in nutrients and excrete waste products. Because of this, they can multiply much more rapidly than can their eukaryotic counterparts. On the other hand, their small size makes them vulnerable to an array of threats. Predators, parasites, and competitors constantly surround them. Prokaryotic cells, although simple in structure, have developed many unique attributes that enhance their evolutionary success.

Eukaryotic cells are considerably more complex than prokaryotic cells. Not only are they larger, but many of their cellular processes take place within membrane-bound compartments. Eukaryotic cells are defined by the presence of a membrane-bound nucleus, which contains the chromosomes. Although eukaryotic cells share many of the same characteristics as prokaryotic cells, many of their structures and cellular processes are fundamentally different. ■ **eukaryotic cells, p. 10**

KEY TERMS

Capsule A distinct, thick gelatinous material that surrounds some microorganisms.

Chemotaxis Directed movement of an organism toward or away from a certain chemical in the environment.

Cytoplasmic Membrane A phospholipid bilayer embedded with proteins that surrounds the cytoplasm and defines the boundary of the cell.

Endospore A type of dormant cell that is extraordinarily resistant to damaging conditions including heat, desiccation, ultraviolet light, and toxic chemicals.

Flagellum A structure that provides a mechanism for motility.

Gram-Negative Bacteria Bacteria that have a cell wall composed of a thin layer of peptidoglycan surrounded by an outer membrane; when Gram stained, these cells are pink.

Gram-Positive Bacteria Bacteria that have a cell wall composed of a thick layer of peptidoglycan; when Gram stained, these cells are purple.

Lipopolysaccharide (LPS) Molecule that makes up the outer layer of the outer membrane of Gram-negative bacteria.

Peptidoglycan A macromolecule that provides rigidity to the cell wall; it is found only in bacteria.

Periplasm The gel-like material that fills the region between the cytoplasmic membrane and the outer membrane of Gram-negative bacteria.

Pili Cell surface structures that generally enable cells to adhere to certain surfaces; some types are involved in a mechanism of DNA transfer.

Plasmid Extrachromosomal DNA molecule that replicates independently of the chromosome.

Ribosome Structure intimately involved in protein synthesis.

Transport Systems Mechanisms used to transport nutrients and other small molecules across the cytoplasmic membrane.

MICROSCOPY AND CELL MORPHOLOGY

3.1 Microscopic Techniques: The Instruments

Focus Points

- Describe the importance of magnification, resolution, and contrast in microscopy.
- Compare and contrast light microscopes, electron microscopes, and atomic force microscopes.

One of the most important tools for studying microorganisms is the **light microscope,** which uses visible light for observing objects. These instruments can magnify images approximately 1,000×, making it relatively easy to observe the size, shape, and motility of prokaryotic cells. The **electron microscope,** introduced in 1931, can magnify images in excess of 100,000×, revealing many fine details of cell structure. A major advancement came in the 1980s with the development of the **atomic force microscope,** which allows scientists to produce images of individual atoms on a surface.

Principles of Light Microscopy: The Bright-Field Microscope

In light microscopy, light typically passes through a specimen and then through a series of magnifying lenses. The most common type of light microscope, and the easiest to use, is the **bright-field microscope,** which evenly illuminates the field of view.

Magnification

The modern light microscope has two magnifying lenses—an **objective lens** and an **ocular lens**—and is called a **compound microscope (figure 3.1).** These lenses in combination visually enlarge an object by a factor equal to the product of each lens' magnification. For example, an object is magnified 1,000-fold when it is viewed through a 10× ocular lens in conjunction with a 100× objective lens. Most compound microscopes have a selection of objective lenses that are of different powers—typically 4×, 10×, 40×, and 100×. This makes a choice of different magnifications possible with the same instrument.

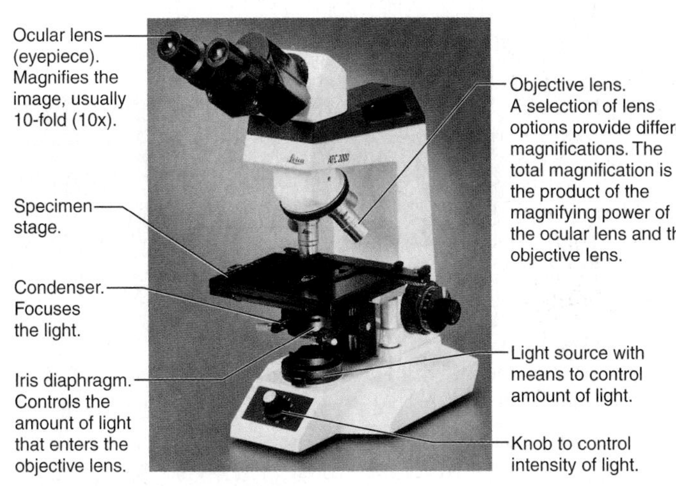

Ocular lens (eyepiece). Magnifies the image, usually 10-fold (10x).
Specimen stage.
Condenser. Focuses the light.
Iris diaphragm. Controls the amount of light that enters the objective lens.
Objective lens. A selection of lens options provide different magnifications. The total magnification is the product of the magnifying power of the ocular lens and the objective lens.
Light source with means to control amount of light.
Knob to control intensity of light.

FIGURE 3.1 A Modern Light Microscope The compound microscope employs a series of magnifying lenses.

FIGURE 3.2 Resolving Power These images of an onion root tip magnified 450x illustrate the difference in resolving power between a light microscope and an electron microscope. Note the difference in the degree of detail that can be seen at the same magnification.

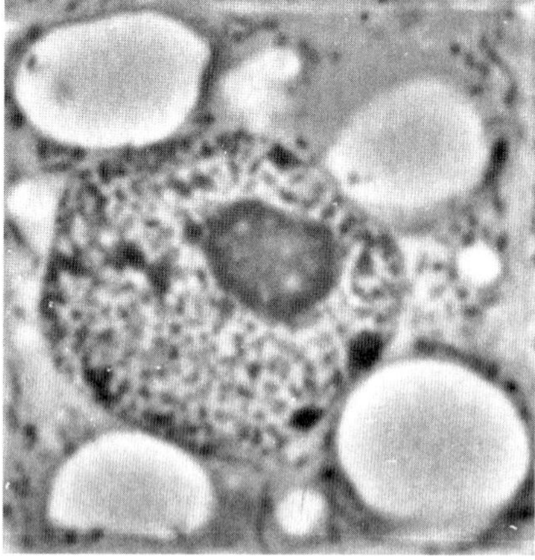

Light microscope (450x)

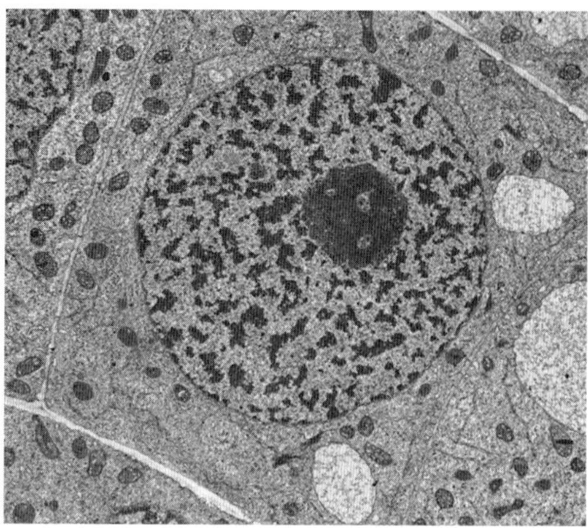

Electron microscope (450x)

The **condenser lens** does not affect the magnification but, positioned between the light source and the specimen, is used to focus the light on the specimen.

Resolution

The usefulness of a microscope depends both on its degree of magnification, and its ability to clearly separate, or **resolve,** two objects that are very close together. The **resolving power** is defined as the minimum distance existing between two objects when those objects can still be observed as separate entities. The resolving power therefore determines how much detail actually can be seen **(figure 3.2).**

The resolving power of a microscope depends on the quality and type of lens, wavelength of the light, magnification, and how the specimen under observation has been prepared. The maximum resolving power of the best light microscope is 0.2 μm. This is sufficient to observe the general morphology of a prokaryotic cell but too low to distinguish a particle the size of most viruses.

To obtain maximum resolution when using certain high-power objectives such as the 100× lens, oil must be used to displace the air between the lens and the specimen. This avoids the bending of light rays, or **refraction,** that occurs when light passes from glass to air **(figure 3.3).** Refraction can prevent those rays from entering the relatively small openings of higher-power objective lenses. The oil has nearly the same **refractive index** as glass. Refractive index is a measure of the relative velocity of light as it passes through a medium. As light travels from a medium of one refractive index to another, those rays are bent. When oil displaces air at the interface of the glass slide and glass lens, light rays pass with little refraction occurring.

Contrast

Contrast reflects the number of visible shades in a specimen—high contrast being just two shades, black and white. Different specimens require various degrees of contrast to reveal the most information. One example is bacteria, which are essentially transparent against

FIGURE 3.3 Refraction As light passes from one medium to another, the light rays may bend, depending on the refractive index of the two media. **(a)** The pencil in water appears bent because the refractive index of water is different from that of air. **(b)** Light rays bend as they pass from air to glass because of the different refractive indexes of these media; some rays are lost to the objective. Oil and glass have the same refractive index, and therefore the light rays are not bent.

(a)

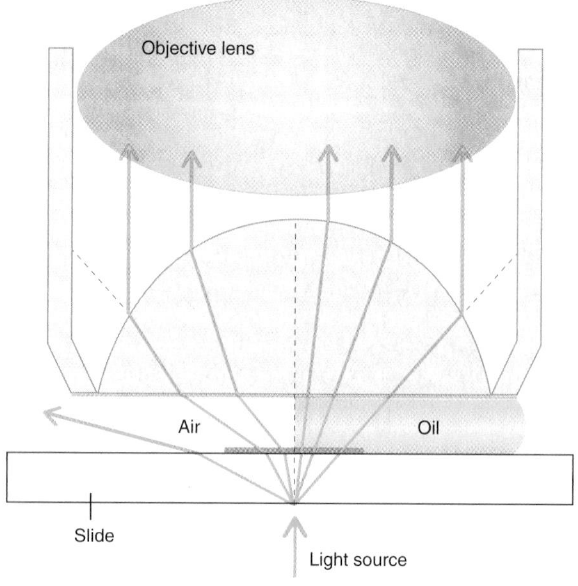

(b)

a bright colorless background. The lack of contrast presents a problem when viewing objects (see figure 11.18). One way to overcome this problem is to stain the bacteria with any one of a number of dyes. The types and characteristics of these stains will be discussed shortly.

Light Microscopes That Increase Contrast

Special light microscopes that increase the contrast between microorganisms and their surroundings overcome some of the difficulties of observing unstained bacteria. Staining kills microbes; therefore, some of these microscopes are invaluable when the goal is to examine characteristics of living organisms such as motility. Characteristics of these and other microscopes are summarized in **table 3.1.**

The Phase-Contrast Microscope

The **phase-contrast microscope** amplifies the slight difference between the refractive index of cells and the surrounding medium, resulting in a darker appearance of the denser material **(figure 3.4).** As light passes through cells, it is refracted slightly differently than when it passes through its surroundings. Special optical devices boost those differences, thereby increasing the contrast.

The Interference Microscope

The **interference microscope** causes the specimen to appear as a three-dimensional image **(figure 3.5).** This microscope, like the phase-contrast microscope, depends on differences in refractive index as light passes through different materials. The most frequently used microscope of this type is the **Nomarski differential interference contrast (DIC) microscope,** which has a device for separating light into two beams that pass through the specimen and then recombine. The light waves are out of phase when they recombine, thereby yielding the three-dimensional appearance of the specimen.

The Dark-Field Microscope

Organisms viewed through a **dark-field microscope** stand out as bright objects against a dark background **(figure 3.6).** The microscope operates on the same principle that makes dust visible when a beam of bright light shines into a dark room. A special mechanism directs light toward the specimen at an angle, so that only light scattered by the specimen enters the objective lens. Dark-field microscopy can detect *Treponema pallidum*, the causative agent of syphilis. These thin, spiral-shaped organisms stain poorly and are difficult to see via bright-field microscopy (see figure 11.26).

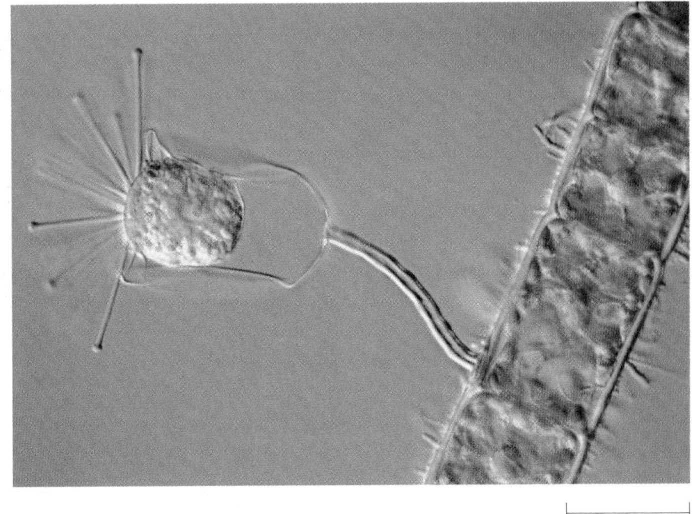

FIGURE 3.5 Nomarski Differential Interference Contrast (DIC) Microscopy Protozoan (*Paracineta*) attached to a green alga (*Spongomorpha*).

The Fluorescence Microscope

The **fluorescence microscope** is used to observe cells or other materials that are either naturally fluorescent or have been stained or tagged with fluorescent dyes. A **fluorescent** molecule absorbs light at one wavelength (usually ultraviolet light) and then emits light of a longer wavelength.

The fluorescence microscope projects ultraviolet light through a specimen, but then captures only the light emitted by the fluorescent molecules to form the image. This allows

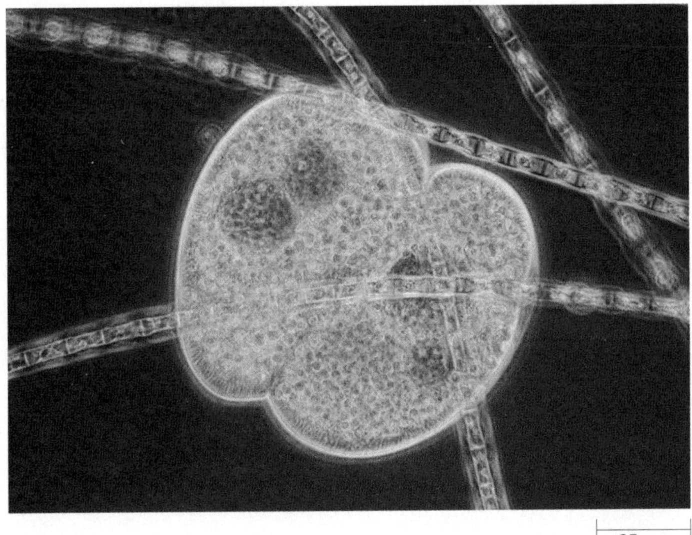

FIGURE 3.4 Phase-Contrast Photomicrograph *Paramecium bursaria* containing endosymbiotic *Chlorella* (a green alga).

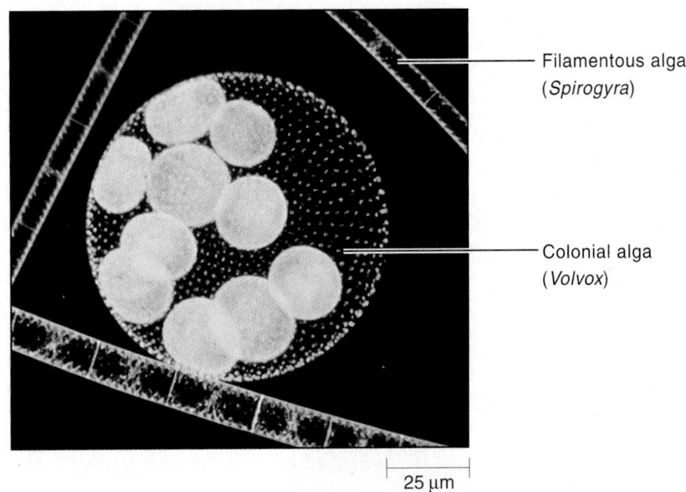

FIGURE 3.6 Dark-Field Photomicrograph *Volvox* (sphere) and *Spirogyra* (filaments), both of which are eukaryotes.

TABLE 3.1 A Summary of Microscopic Instruments and Their Characteristics

Instrument	Mechanism	Uses/Comment
Light Microscopes	Visible light passes through a series of lenses to produce a magnified image.	Relatively easy to use; considerably less expensive than confocal and electron microscopes.
Bright-field	Illuminates the field of view evenly.	Most common type of microscope.
Phase-contrast 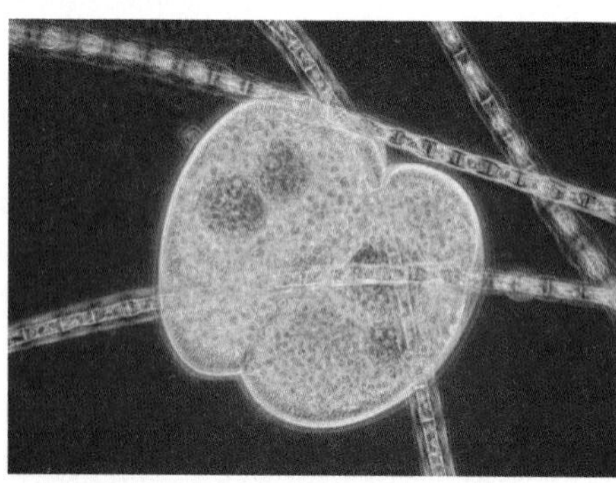	Amplifies differences in refractive index to create contrast.	Makes unstained cells more readily visible.
Interference 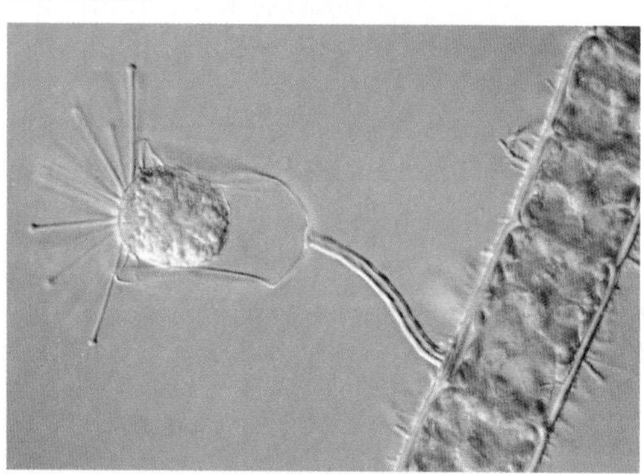	Two light beams pass through the specimen and then recombine.	Causes the specimen to appear as a three-dimensional image.
Dark-field 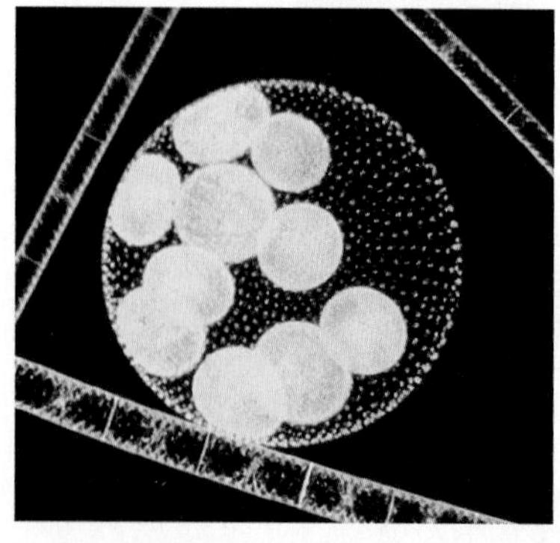	Light is directed toward the specimen at an angle.	Makes unstained cells more readily visible; organisms stand out as bright objects against a dark background.

TABLE 3.1 A Summary of Microscopic Instruments and Their Characteristics (*continued*)

Instrument	Mechanism	Uses/Comment
Fluorescence 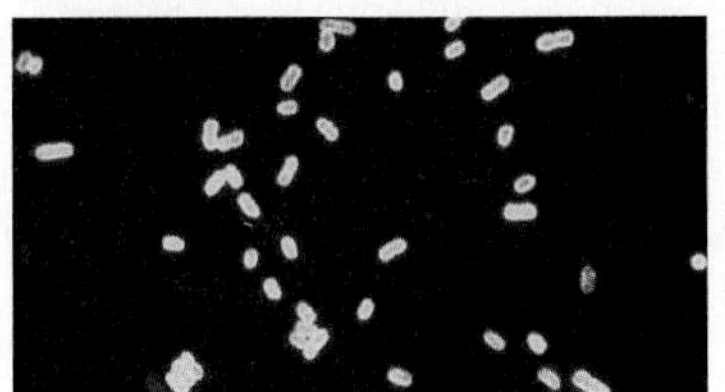	Projects ultraviolet light, causing fluorescent molecules in the specimen to emit longer wavelength light.	Used to observe cells that have been stained or tagged with a fluorescent dye.
Confocal 	Mirrors scan a laser beam across successive regions and planes of a specimen. From that data, a computer constructs an image.	Used to construct a three-dimensional image of a structure; provides detailed sectional views of intact cells.
Electron Microscopes	Electron beams are used in place of visible light to produce the magnified image.	Can clearly magnify images 100,000×.
Transmission  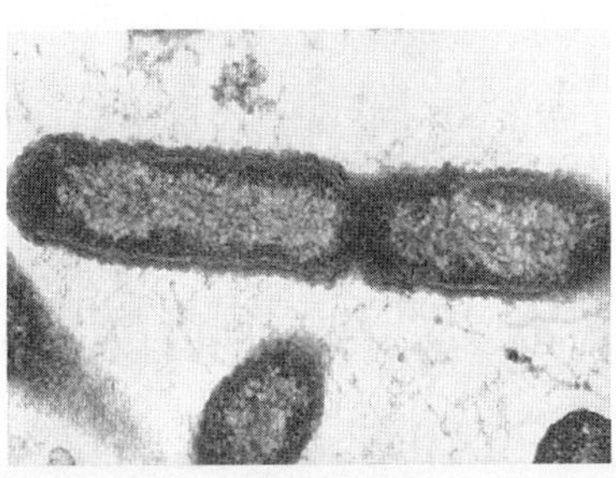	Transmits a beam of electrons through a specimen.	Elaborate specimen preparation is required.
Scanning 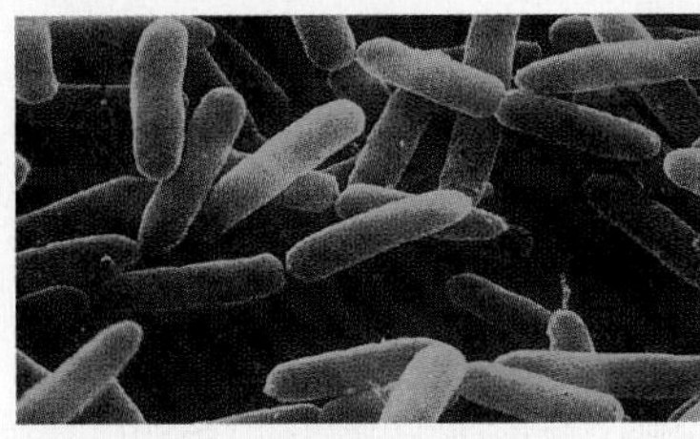	A beam of electrons scans back and forth over the surface of a specimen.	Used for observing surface details; produces a three-dimensional effect.
Atomic Force Microscope 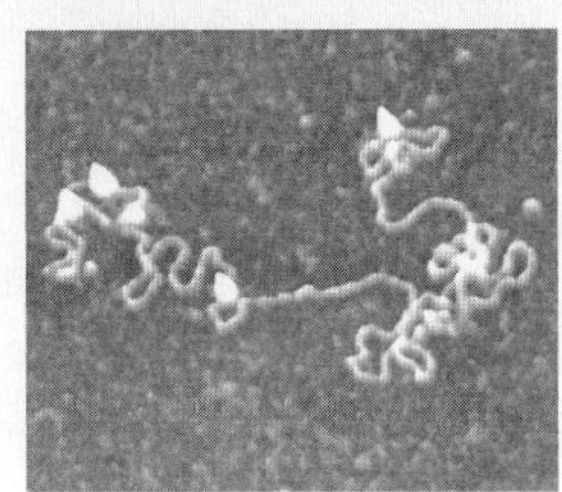	A probe moves in response to even the slightest force between it and the sample.	Produces a map showing the bumps and valleys of the atoms on the surface of the sample.

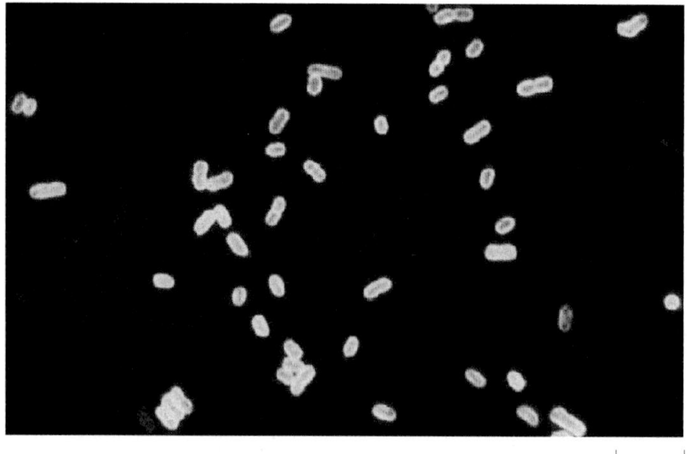

FIGURE 3.7 Fluorescence Photomicrograph A rod-shaped bacterium tagged with fluorescent marker.

fluorescent cells to stand out as illuminated objects against a dark background **(figure 3.7).** The types and characteristics of fluorescent dyes and tags will be discussed shortly. ■ **fluorescent dyes and tags, p. 51**

A common variation of the standard fluorescence microscope is the **epifluorescence microscope,** which projects the ultraviolet light through the objective lens and onto the specimen. Because the light is not transmitted through the specimen, cells attached to soil particles or other opaque materials can be observed.

The Confocal Scanning Laser Microscope

The **confocal scanning laser microscope** is used to construct a three-dimensional image of a thick structure such as a community of microorganisms **(figure 3.8).** The instrument can also provide detailed sectional views of the interior of an intact cell. In confocal microscopy, lenses focus a laser beam to illuminate a given point on one vertical plane of a specimen. Mirrors then scan the laser beam across the specimen, illuminating successive regions and planes until the entire specimen has been scanned. Each plane corresponds to an image of one fine slice of the specimen. A computer then assembles the data and constructs a three-dimensional image, which is displayed on a screen. In effect, this microscope is a miniature CAT scan for cells.

Frequently, the specimens are first stained or tagged with a fluorescent dye. By using certain fluorescent tags that bind specifically to a given protein or other compound, the precise cellular location of that compound can be determined. In some cases, multiple different tags that bind to specific molecules are used, each having a distinct color.

Electron Microscopes

Electron microscopy is in some ways comparable to light microscopy. Rather than using glass lenses, visible light, and the eye to observe the specimen, the electron microscope uses electromagnetic lenses, electrons, and a fluorescent screen to produce the magnified image **(figure 3.9).** That image can be captured on photographic film to create an **electron photomicrograph.** Sometimes, the black and white images are artificially enhanced with color to add visual clarity. ■ **electrons, p. 18**

Since the electrons have a wavelength about 1,000 times shorter than visible light, the resolving power increases about 1,000-fold, to about 0.3 nanometers (nm) or 0.3×10^{-3} μm (see figure 1.13). Consequently, considerably more detail can be observed due to the much higher resolution. These instruments can clearly magnify an image 100,000×. One of the biggest drawbacks of the microscope is that the lenses and specimen must all be in a vacuum. Otherwise, the molecules composing air would interfere with the path of the electrons. This results in an expensive, bulky unit and requires substantial and complex specimen preparation.

The Transmission Electron Microscope

The **transmission electron microscope (TEM)** is used to observe fine details of cell structure, such as the number of layers that envelop a cell. The instrument directs a beam of electrons at a

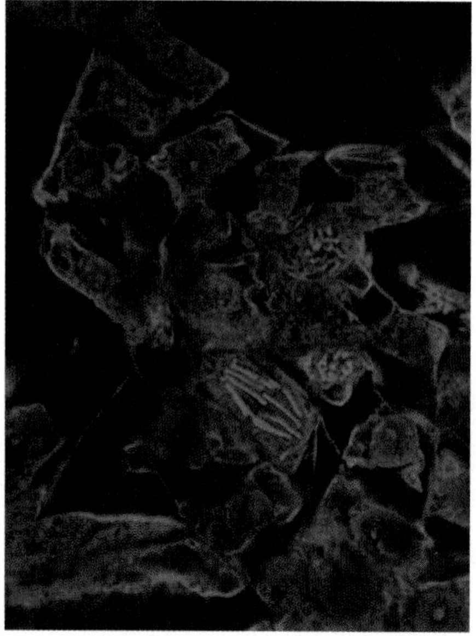

(a)

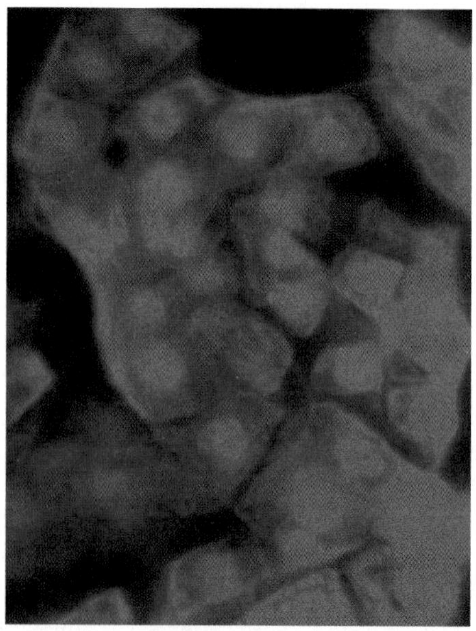

(b)

FIGURE 3.8 Confocal Microscopy This can be used to produce a clear image of a single plane in a thick structure. **(a)** Confocal photomicrography of fava bean mitosis. **(b)** Regular photomicrograph.

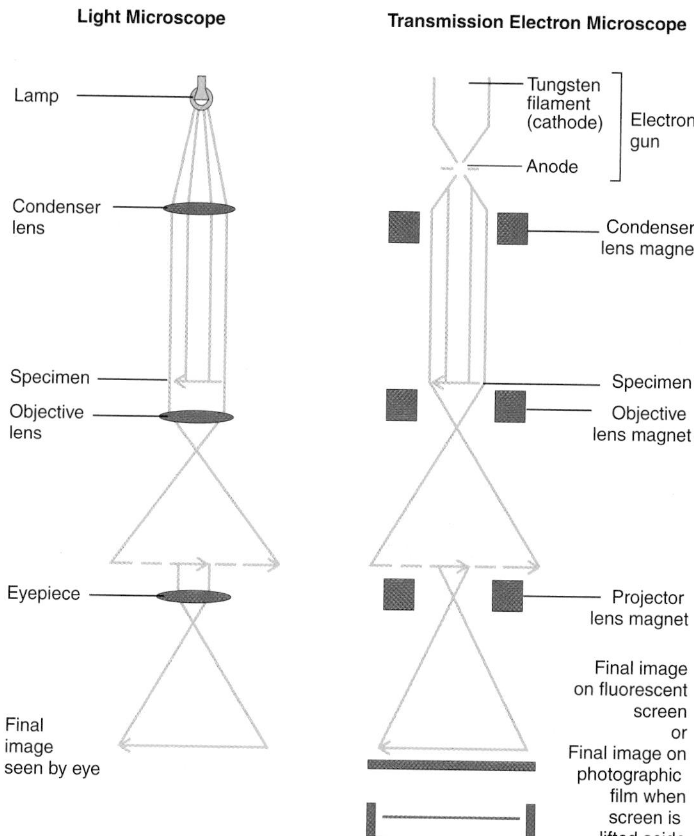

FIGURE 3.9 Comparison of the Principles of Light and the Electron Microscopy For the sake of comparison, the light source for the light microscope has been inverted (the light is shown at the top and the eyepiece, or ocular lens, at the bottom).

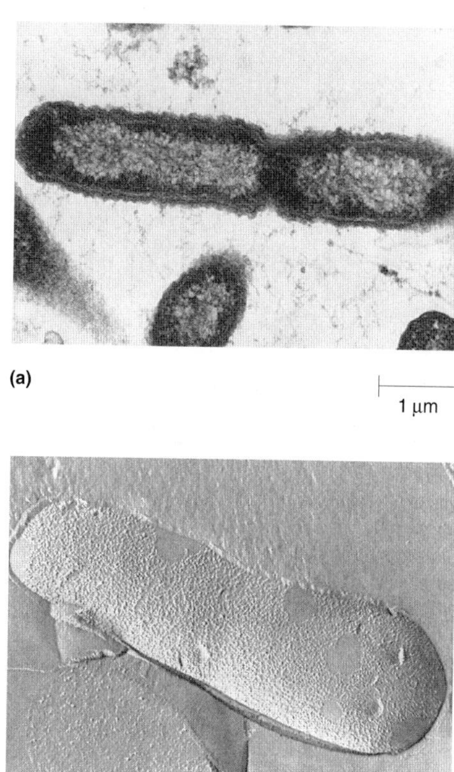

FIGURE 3.10 Transmission Electron Photomicrograph A rod-shaped bacterium prepared by **(a)** thin section; **(b)** freeze etching.

specimen. Depending on the density of a particular region in the specimen, electrons will either pass through or be scattered to varying degrees. The darker areas of the resulting image correspond to the denser portions of the specimen **(figure 3.10).**

Transmission electron microscopy requires elaborate and painstaking specimen preparation. To view details of internal structure, a process called **thin sectioning** is used. Cells are carefully treated with a preservative and dehydrated in an organic solvent before being embedded in a plastic resin. Once embedded, they can be cut into exceptionally thin slices with a diamond or glass knife and then stained with heavy metals. Even a single bacterial cell must be cut into slices this way to be viewed via TEM. Unfortunately, the procedure can severely distort the cells. Consequently, a major concern in using TEM is distinguishing actual cell components from artifacts occurring as a result of specimen preparation.

A process called **freeze fracturing** is used to observe the shape of structures within the cell. The specimen is rapidly frozen and then fractured by striking it with a knife blade. The cells break open, usually along the middle of internal membranes. Next, the surface of the section is coated with a thin layer of carbon to create a replica of the surface. This replica is then examined in the electron microscope. A variation of freeze fracturing is **freeze etching.** In this process, the frozen surface exposed by fracturing is dried slightly under vacuum, which allows underlying regions to be exposed.

The Scanning Electron Microscope

The **scanning electron microscope (SEM)** is used for observing surface details of cells. A beam of electrons scans back and forth over the surface of a specimen coated with a thin film of metal. As those beams move, electrons are released from the specimen and reflected back into the viewing chamber. This reflected radiation is observed with the microscope. Relatively large specimens can be viewed, and a dramatic three-dimensional effect is observed with the SEM **(figure 3.11).**

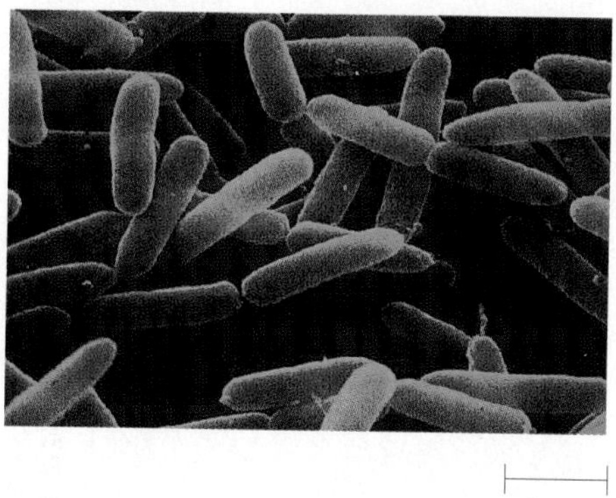

FIGURE 3.11 Scanning Electron Photomicrograph A rod-shaped bacterium.

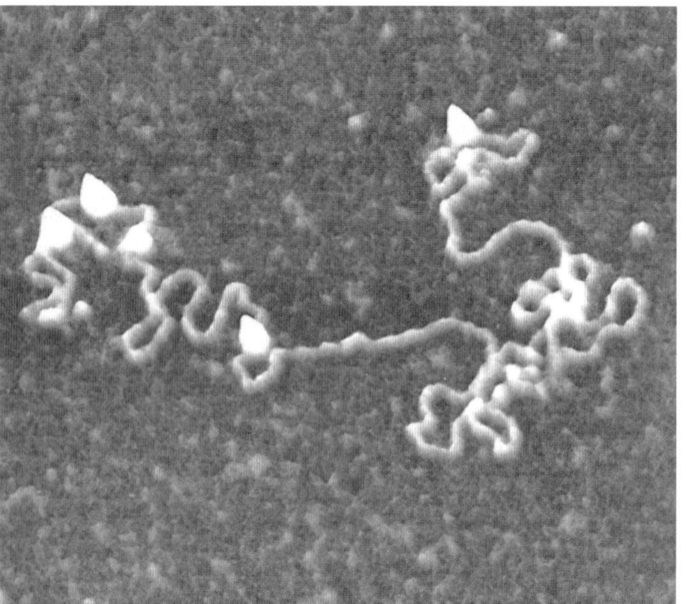

FIGURE 3.12 Atomic Force Microscopy Micrograph of a fragment of DNA. The bright peaks are enzymes attached to the DNA.

Atomic Force Microscopy

The **atomic force microscope** (AFM) produces detailed images of surfaces **(figure 3.12).** The resolving power is much greater than that of an electron microscope, and the samples do not need the special preparation required for electron microscopy. In fact, the instrument can inspect samples either in air or submerged in liquid.

The mechanics of AFM can be compared to that of a stylus mounted on the arm of a record player. A very sharp probe (stylus) moves across the surface of the sample, "feeling" the bumps and valleys of the atoms on that surface. As the probe scans the sample, a laser measures its motion, and a computer produces a surface map of the sample. ■ atom, p. 18

MICROCHECK 3.1

The usefulness of a microscope depends on its resolving power. The most common type of microscope is the bright-field microscope. Variations of light microscopes are designed to increase contrast between a microorganism and its surroundings. The fluorescence microscope is used to observe microbes stained with special dyes. The confocal scanning laser microscope is used to construct a three-dimensional image of a thick structure. Electron microscopes can magnify images 100,000x. The atomic force microscope produces detailed images of surfaces.

- ✓ Why must oil be employed when using the 100x lens?
- ✓ Why are microscopes that enhance contrast used to view live rather than stained specimens?
- ✓ If an object being viewed under the phase-contrast microscope has the same refractive index as the background material, how would it appear?

3.2 Microscopic Techniques: Dyes and Staining

Focus Points

- Describe the principles of the Gram stain and the acid-fast stain.
- Describe the techniques used to observe capsules, endospores, and flagella.
- Describe the benefits of using fluorescent dyes and tags.

It can be difficult to observe living microorganisms with the bright-field microscope. Most microorganisms are nearly transparent and often move rapidly about the slide. To remedy this problem, cells are frequently immobilized and stained with dyes. Many different dyes and staining procedures can be used; each has specific applications **(table 3.2).**

TABLE 3.2 A Summary of Stains and Their Characteristics

Stain	Characteristics
Simple Stains	Employ a basic dye to impart a color to a cell. Easy way to increase the contrast between otherwise colorless cells and a transparent background.
Differential Stains	Distinguish one group of microorganisms from another.
Gram stain	Used to separate bacteria into two major groups, Gram-positive and Gram-negative. The staining characteristics of these groups reflect a fundamental difference in the chemical structure of their cell walls. This is by far the most widely used staining procedure.
Acid-fast stain	Used to detect organisms that do not readily take up stains, such as members of the genus *Mycobacterium*.
Special Stains	Stain specific structures inside or outside of a cell.
Capsule stain	Capsule stains exploit the fact that viscous capsules do not readily take up certain stains; the capsules stand out against a stained background. This is an example of a negative stain.
Endospore stain	Stains endospores, a type of dormant cell that does not readily take up stains. These are produced by *Bacillus* and *Clostridium* species.
Flagella stain	The staining agent adheres to and coats the otherwise thin flagella, making them visible with the light microscope.
Fluorescent Dyes and Tags	Fluorescent dyes and tags absorb ultraviolet light and then emit light of a longer wavelength. They are used in conjunction with a fluorescence microscope.
Fluorescent dyes	Some fluorescent dyes bind to compounds found in all cells; others bind to compounds specific to only certain types of cells.
Fluorescent tags	Antibodies to which a fluorescent molecule has been attached are used to tag specific molecules.

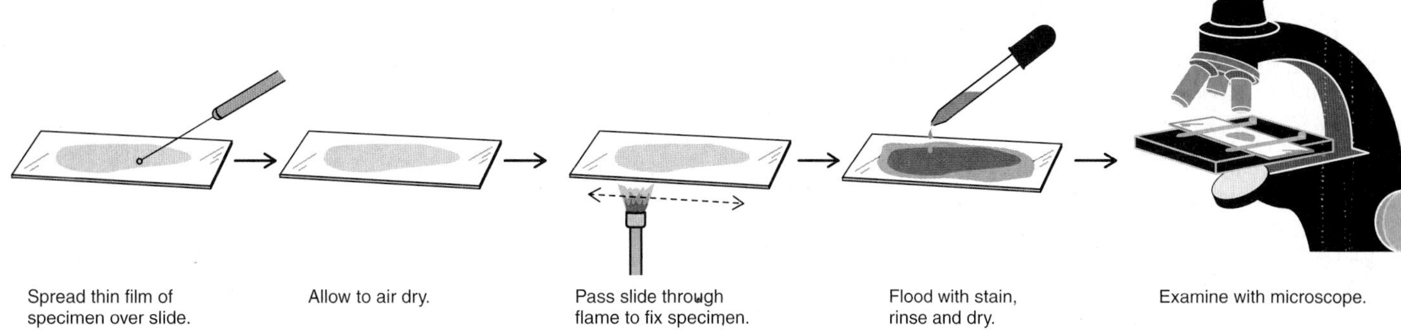
Spread thin film of specimen over slide.
Allow to air dry.
Pass slide through flame to fix specimen.
Flood with stain, rinse and dry.
Examine with microscope.

FIGURE 3.13 Staining Bacteria for Microscopic Observation

Basic dyes, which carry a positive charge, are more commonly used for staining than are negatively charged **acidic dyes.** Because opposite charges attract, basic dyes stain the negatively charged components of cells, including nucleic acid and many proteins, whereas acidic dyes are repelled. Common basic dyes include methylene blue, crystal violet, safranin, and malachite green. **Simple staining** employs one of these basic dyes to stain the cells. Acidic dyes are sometimes used to stain backgrounds against which colorless cells can be seen, a technique called **negative staining.**

To stain microorganisms, a drop of liquid containing the microbe is placed on a microscope slide and allowed to dry. The resulting specimen forms a film, or **smear.** The organisms are then attached, or **fixed,** to the slide, usually by passing the slide over a flame **(figure 3.13).** Dye is then applied and the excess washed off with water. Heat fixing and subsequent staining steps kill the microorganisms and may distort their shape.

Differential Stains

Differential staining procedures are used to distinguish one group of bacteria from another. The two most frequently used differential staining techniques are the Gram stain and the acid-fast stain.

Gram Stain

The **Gram stain** is by far the most widely used procedure for staining bacteria. The basis for it was developed over a century ago by Dr. Hans Christian Gram (see **A Glimpse of History**). He showed that bacteria can be separated into two major groups: **Gram-positive** and **Gram-negative.** We now know that the difference in the staining properties of these two groups reflects a fundamental difference in the structure of their cell walls.

Gram staining involves four basic steps **(figure 3.14).**

1. The smear is first flooded with the **primary stain,** crystal violet in this case. The primary stain is the first dye applied in any multistep staining procedure and generally stains all cells.
2. The smear is rinsed to remove excess crystal violet and then flooded with a dilute solution of iodine, called Gram's iodine. Iodine is a mordant, a substance that increases the affinity of cellular components for a dye. The iodine combines with the crystal violet to form a dye-iodine complex, thereby decreasing the solubility of the dye within the cell.

	Steps in Staining	State of Bacteria
	Step 1: Crystal violet (primary stain)	Cells stain purple.
	Step 2: Iodine (mordant)	Cells remain purple.
	Step 3: Alcohol (decolorizer)	Gram-positive cells remain purple; Gram-negative cells become colorless.
	Step 4: Safranin (counterstain)	Gram-positive cells remain purple; Gram-negative cells appear red.

(a)

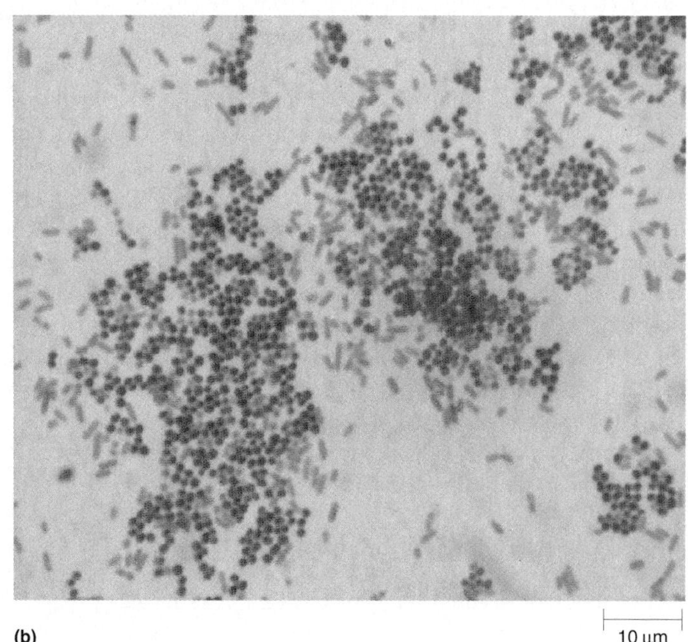
10 µm

(b)

FIGURE 3.14 Gram Stain (a) Steps in the Gram stain procedure. **(b)** Results of a Gram stain. The Gram-positive cells (purple) are *Staphylococcus aureus;* the Gram-negative cells (reddish-pink) are *Escherichia coli.*

3. The stained smear is rinsed again, and then 95% alcohol or a mixture of alcohol and acetone is briefly added. These solvents act as **decolorizing agents** and readily remove the dye-iodine complex from Gram-negative, but not Gram-positive, bacteria.
4. A **counterstain** is then applied to impart a contrasting color to the now colorless Gram-negative bacteria. For this purpose, the red dye safranin is used. This dye stains Gram-negative as well as Gram-positive bacteria, but because the latter are already stained purple, it imparts little difference to those cells.

To obtain reliable results, the Gram stain must be done properly. One of the most common mistakes is to decolorize a smear for too long a time period. Even Gram-positive cells can lose the crystal violet-iodine complex during prolonged decolorization. An over-decolorized Gram-positive cell will appear pink after counterstaining. Another important consideration is the age of the culture. As bacterial cells age, they lose their ability to retain the crystal violet–iodine dye complex, presumably because of changes in their cell wall. As a result, cells from old cultures may appear pink. Thus, the Gram stain results of fresh cultures (less than 24 hours old) are more reliable.

Acid-Fast Stain

The **acid-fast stain** is a procedure used to stain a small group of organisms that do not readily take up stains. Among these are members of the genus *Mycobacterium*, including a species that causes tuberculosis and one that causes Hansen's disease (leprosy). The cell wall of these bacteria contains high concentrations of lipid, preventing the uptake of dyes, including those used in the Gram stain. Therefore, harsh methods are needed to stain these organisms. Once stained, however, these same cells are very resistant to decolorization. Because mycobacteria are among the few organisms that retain the dye in this procedure, the acid-fast stain can be used to presumptively identify them in clinical specimens that might contain a variety of different bacteria. ■ tuberculosis, p. 514 ■ Hansen's disease, p. 655

The acid-fast stain, like the Gram stain, requires multiple steps. The primary stain in this procedure is carbol fuchsin, a red dye. In the classic procedure, the stain-flooded slide is heated, which facilitates the staining. A current variation does not employ heat, instead it uses a prolonged application of a more concentrated solution of dye. The slide is then rinsed briefly to remove the residual stain before being flooded with acid-alcohol, a potent decolorizing agent. This step removes the carbol fuchsin from tissue cells and most bacteria. Those few species that retain the dye are called **acid-fast.** Methylene blue is then used as a counter-stain, imparting a blue color to non-acid-fast cells. Acid-fast organisms, which do not take up the methylene blue, appear a bright reddish-pink **(figure 3.15).**

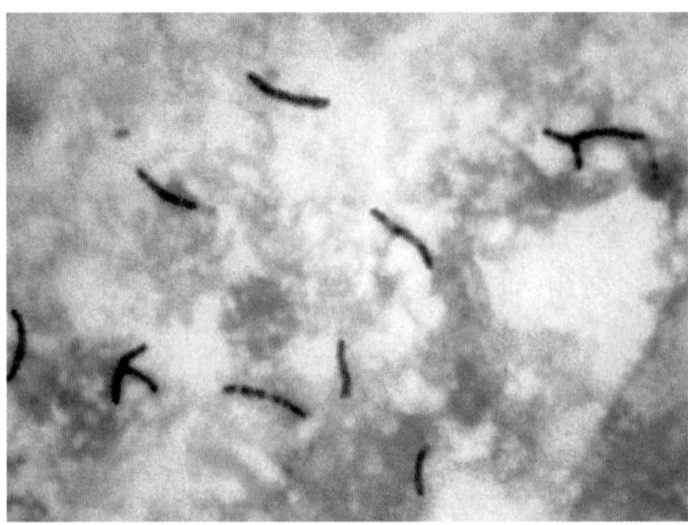

FIGURE 3.15 Acid-Fast Stain *Mycobacterium* species retain the red primary stain, carbol fuchsin. Counterstaining with methylene blue imparts a blue color to cells that are not acid-fast.

Special Stains to Observe Cell Structures

Dyes can also be used to stain specific structures inside or outside the cell. The staining procedure for each component of the cell is different, being geared to the chemical composition and properties of that structure. The function of each of these structures will be discussed in more depth later in the chapter.

Capsule Stain

A capsule is a viscous layer that envelops a cell and is sometimes correlated with an organism's ability to cause disease. Capsules stain poorly, a characteristic exploited with a **capsule stain,** an example of a negative stain. It colors the background, allowing the capsule to stand out as a halo around an organism **(figure 3.16).** ■ capsule, p. 64

In one method to observe capsules, a liquid specimen is placed on a slide next to a drop of India ink. A thin glass coverslip is then placed over the two drops, causing them to flow together. This creates a gradient of India ink concentration across the specimen. Unlike the stains discussed previously, this capsule stain is done as a **wet mount**—a drop of liquid on which a coverslip has been placed—rather than as a smear. At the optimum concentration of

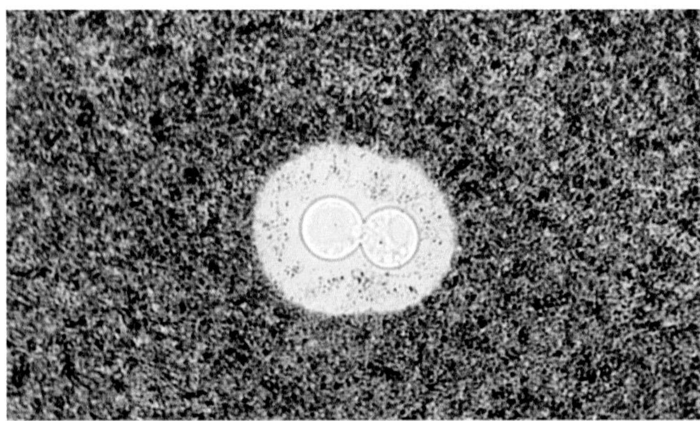

FIGURE 3.16 Capsule Stain Capsules stain poorly, and so they stand out against the India ink-stained background as a halo around the organism. This photomicrograph shows *Cryptococcus neoformans,* an encapsulated yeast.

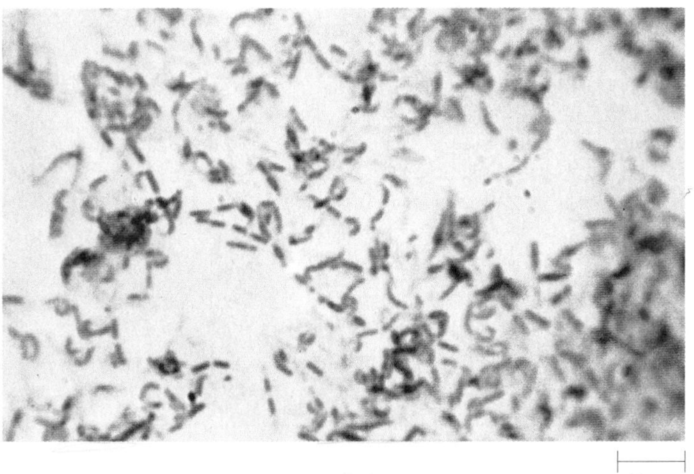

FIGURE 3.17 Endospore Stain Endospores retain the green primary stain, malachite green. Counterstaining with safranin imparts a red color to other cells.

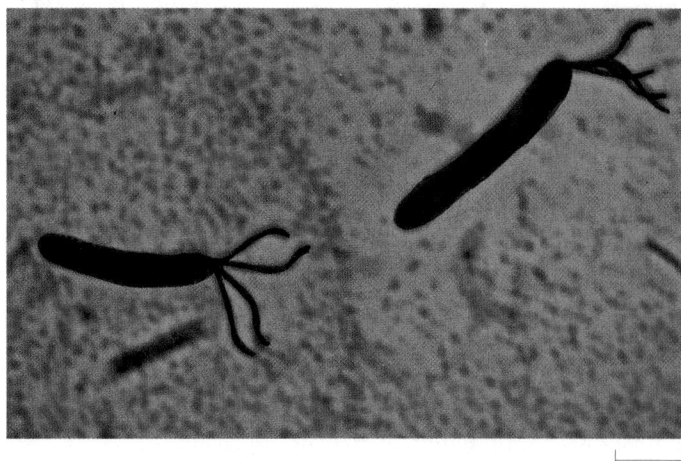

FIGURE 3.18 Flagella Stain The staining agent adheres to and coats the flagella. This increases their diameter so they can be seen with the light microscope.

India ink, the fine particles of the stain darken the background enough to allow the capsule to be visible.

Endospore Stain

Members of certain Gram-positive genera, including *Bacillus* and *Clostridium,* form a special type of dormant cell, an endospore, that is resistant to destruction and to staining. Although these structures do not stain with the Gram stain, they can often be seen as clear, smooth objects within otherwise purple-stained cells. To make endospores more readily noticeable, a **spore stain** is used. ■ endospore, p. 69

The endospore stain is a multistep procedure that employs a primary stain as well as a counterstain. Generally, malachite green is used as a primary stain. Its uptake by endospores is facilitated by gentle heat. When water is then used to rinse the smear, only endospores retain the malachite green. The smear is then counterstained, most often with the red dye safranin. The endospores appear green amid a background of pink cells **(figure 3.17).**

Flagella Stain

Flagella are appendages that provide the most common mechanism of motility for prokaryotic cells, but they are ordinarily too thin to be seen with the light microscope. The **flagella stain** employs a mordant that allows the staining agent to adhere to and coat the thin flagella, effectively increasing their diameter—which makes them visible using light microscopy. Not all bacteria have flagella, but those that do can have them in different arrangements around a cell, so that the presence and distribution of these appendages can be used to identify bacteria **(figure 3.18).** Unfortunately, this staining procedure is difficult and requires patience and expertise. ■ flagella, p. 65

Fluorescent Dyes and Tags

Depending on the procedure employed, fluorescence can be used to observe total cells, a subset of cells, or cells with certain proteins on their surface **(figure 3.19).**

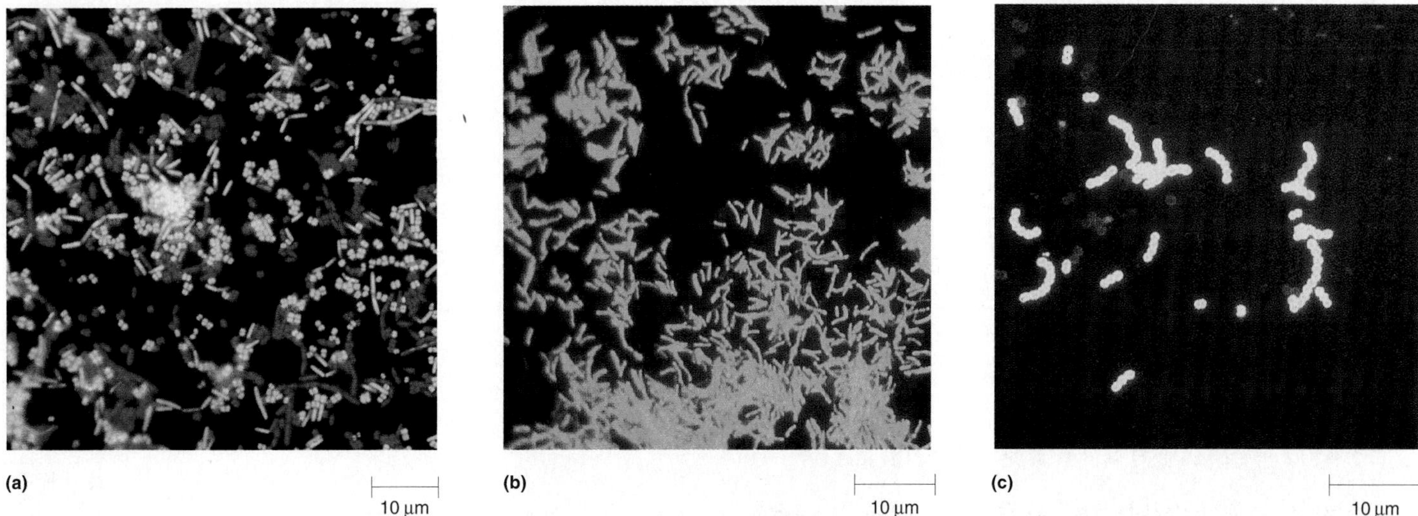

FIGURE 3.19 Fluorescent Dyes and Tags (a) Dyes that cause live cells to fluoresce green and dead ones red. **(b)** Auramine is used to stain *Mycobacterium* species in a modification of the acid-fast technique. **(c)** Fluorescent antibodies tag specific molecules—in this case, the antibody binds to a molecule unique to *Streptococcus pyogenes.*

Fluorescent Dyes

Some fluorescent dyes bind to compounds found in all cells. For example, acridine orange binds DNA, making it useful for determining the total number of microorganisms in a sample. Other fluorescent dyes are changed by cellular processes of living cells, enabling microbiologists to distinguish between cells that are alive and those that are dead. For example, the dye CTC is made fluorescent by cellular proteins involved in respiration. Consequently, CTC only fluoresces when bound to live cells. There are also fluorescent dyes that bind to compounds primarily found in certain types of cells. **Calcofluor white** binds to a component of the cell walls of fungi and certain bacteria, causing those cells to fluoresce bright blue. The fluorescent dyes **auramine** and **rhodamine** bind to a compound found in the cell walls of members of the genus *Mycobacterium*. These two dyes can be used in a staining procedure analogous to the acid-fast stain; cells of *Mycobacterium* will emit a bright yellow or orange fluorescence. ■ respiration, p. 132

Immunofluorescence

Immunofluorescence is a technique used to tag specific proteins with a fluorescent compound. By tagging a protein unique to a given microbe, immunofluorescence can be used to detect that specific organism in a sample containing a mixture of cells. Immunofluorescence uses an antibody to deliver the fluorescent tag (see figure 19.10). An antibody is produced by the immune system in response to a foreign compound, usually a protein; it binds specifically to that compound. ■ antibody, pp. 367, 371

MICROCHECK 3.2

Dyes can be used to stain cells so they stand out against the unstained background. The Gram stain is by far the most widely used differential stain. The acid-fast stain detects species of *Mycobacterium*. Specific dyes and techniques can be used to observe cell structures such as capsules, endospores, and flagella. Fluorescent dyes and tags can be used to observe total cells, a subset of cells, or cells that have certain proteins on their surface.

- ✓ What are the functions of a primary stain and a counterstain?
- ✓ Describe one error in the staining procedure that would result in a Gram-positive bacterium appearing pink.
- ✓ What color would a Gram-negative bacterium be in an acid-fast stain?

3.3 Morphology of Prokaryotic Cells

Focus Points

- Describe the common shapes and groupings of bacteria.
- Describe two multicellular associations of bacteria.

Prokaryotic cells come in a variety of simple shapes and often form characteristic groupings. Some aggregate, living as multicellular associations.

Shapes

Most common bacteria are one of two shapes: spherical, called a **coccus** (plural: cocci); and cylindrical, called a **rod (figure 3.20).**

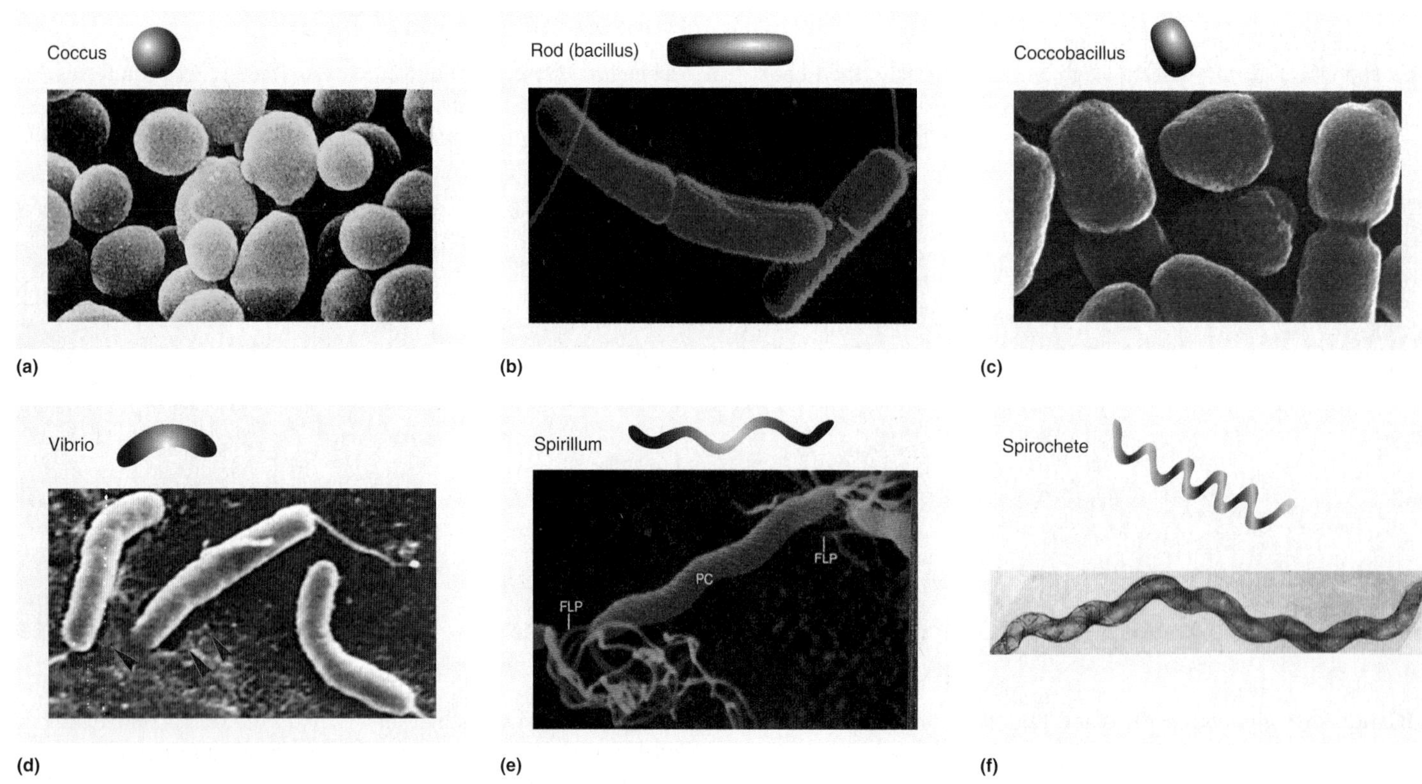

FIGURE 3.20 Typical Shapes of Common Bacteria **(a)** Coccus; **(b)** rod; **(c)** coccobacillus; **(d)** vibrio; **(e)** spirillum; **(f)** spirochete. Electron micrographs.

A rod-shaped bacterium is sometimes called a **bacillus** (plural: bacilli). The descriptive term "bacillus" should not be confused with *Bacillus*, the name of a genus. While members of the genus *Bacillus* are rod-shaped, so are many other bacteria, including *Escherichia coli*.

Cells have a variety of other shapes. A rod-shaped bacterium so short that it can easily be mistaken for a coccus is often called a **coccobacillus.** A short, curved rod is called a **vibrio** (plural: vibrios), whereas a curved rod long enough to form spirals is called a **spirillum** (plural: spirilla). A long, helical cell with a flexible cell wall and a unique mechanism of motility is a **spirochete.** Bacteria that characteristically vary in their shape are called **pleomorphic** (*pleo* meaning "many" and *morphic* referring to shape).

Perhaps the greatest diversity in cell shapes is found in aquatic environments, where maximizing their surface area helps microbes absorb dilute nutrients **(figure 3.21).** Some aquatic bacteria have cytoplasmic extensions, giving them a starlike appearance. Square, tilelike archaeal cells have been found in the salty pools of the Sinai Peninsula in Egypt.

Groupings

Most prokaryotes divide by **binary fission,** a process in which one cell divides into two. Cells adhering to one another following division form a characteristic arrangement that depends on the planes in which the organisms divide. This is seen especially in the cocci because they may divide in more than one plane **(figure 3.22).** Cells that divide in one plane may form chains of varying length. Cocci that typically occur in pairs are routinely called **diplococci.** An important clue in the identification of *Neisseria gonorrhoeae* is its characteristic diplococcus arrangement. Some cocci form long chains; this characteristic is typical of some, but not all, members of the genus *Streptococcus*.

Cocci that divide in two or three planes perpendicular to one another form cubical packets. Members of the genus *Sarcina* form such packets. Cocci that divide in several planes at random may form clusters. Species of *Staphylococcus* typically form characteristic grapelike clusters.

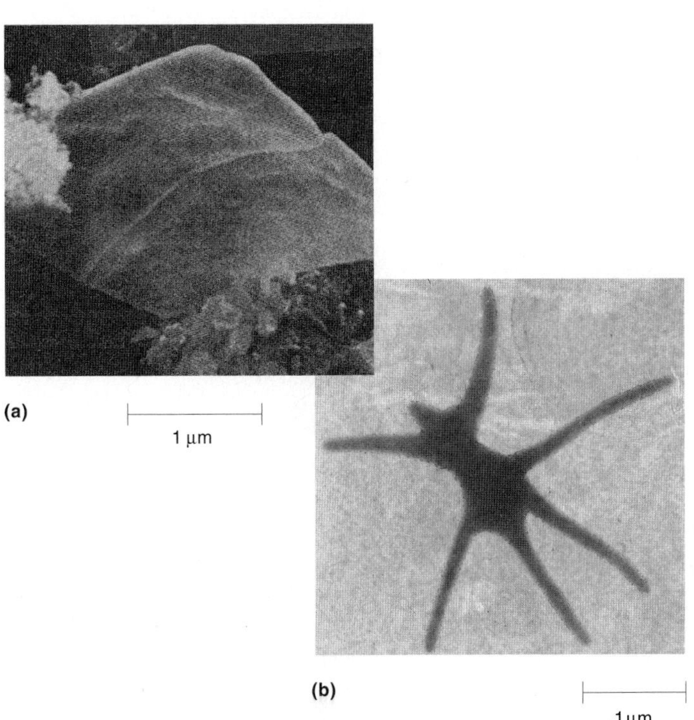

FIGURE 3.21 Diverse Shapes of Aquatic Prokaryotes (a) Square, tilelike archaeal cell. **(b)** *Ancalomicrobium adetum*, a bacterium that has a starlike appearance.

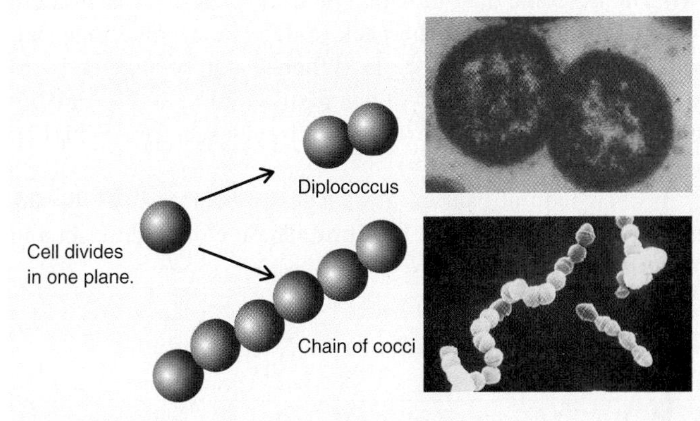

(a) Chains

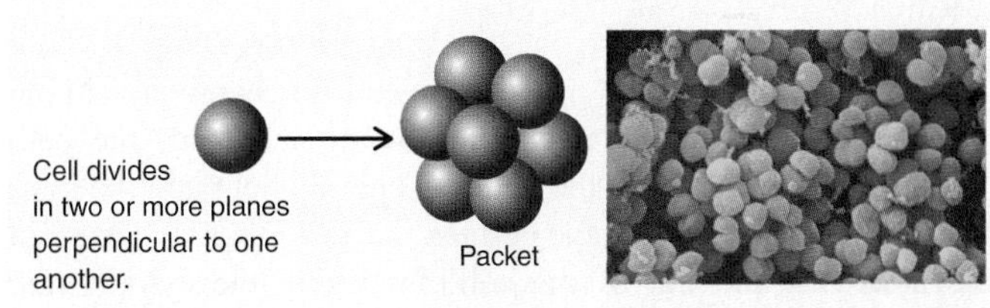

(b) Packets

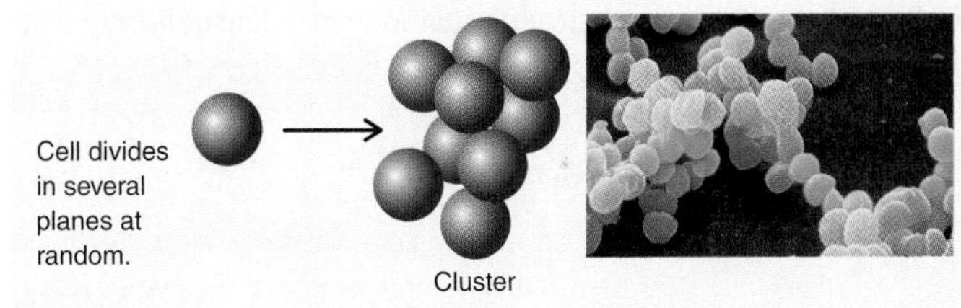

(c) Clusters

FIGURE 3.22 Typical Cell Groupings The planes in which cells divide determine the arrangement of the cells. These characteristic arrangements can provide important clues in the identification of certain bacteria: **(a)** chains; **(b)** packets; **(c)** clusters.

Multicellular Associations

Some types of bacteria typically live as multicellular associations. For example, members of a group of bacteria called myxobacteria glide over moist surfaces together, forming a swarm of cells that move as a pack. These cells release enzymes, which enables the pack to degrade organic material, including other bacterial cells. When water or nutrients are depleted, cells aggregate to form a structure called a **fruiting body,** which is visible to the naked eye (see figure 11.17). ■ myxobacteria, p. 264

In their natural habitat, most types of bacteria live on surfaces in associations called **biofilms.** These will be described in chapter 4. ■ biofilms, p. 85

MICROCHECK 3.3

Most common prokaryotes are cocci or rods, but as a group, they come in a variety of shapes and sizes. Cells may form characteristic arrangements such as chains or clusters. Some form multicellular associations.

✓ Which environmental habitat has the greatest diversity of bacterial shapes?

✓ What causes some bacteria to form characteristic cell arrangements?

THE STRUCTURE OF THE PROKARYOTIC CELL

The overall structure of the prokaryotic cell is deceptively simple **(figure 3.23).** The **cytoplasmic membrane** surrounds the cell, acting as a barrier between the external environment and the interior of the cell. This membrane permits the passage of only certain molecules into and out of the cell. Enclosing the cytoplasmic membrane is the **cell wall,** a rigid barrier that functions as a tight corset to keep the cell contents from bursting out. Cloaking the wall may be additional layers, some of which serve to protect the cell from predators and environmental assaults. The cell may also have appendages, giving it useful traits, including motility and the ability to adhere to certain surfaces.

The capsule (if present), cell wall, and cytoplasmic membrane together make up the **cell envelope.** Enclosed within this envelope are the contents of the cell—the cytoplasm and nucleoid. The **cytoplasm** is a viscous fluid composed of a variety of substances including water, enzymes and other proteins, carbohydrates, lipids, and various inorganic molecules. Structures within the cytoplasm include ribosomes and various storage granules. The **nucleoid** is the gel-

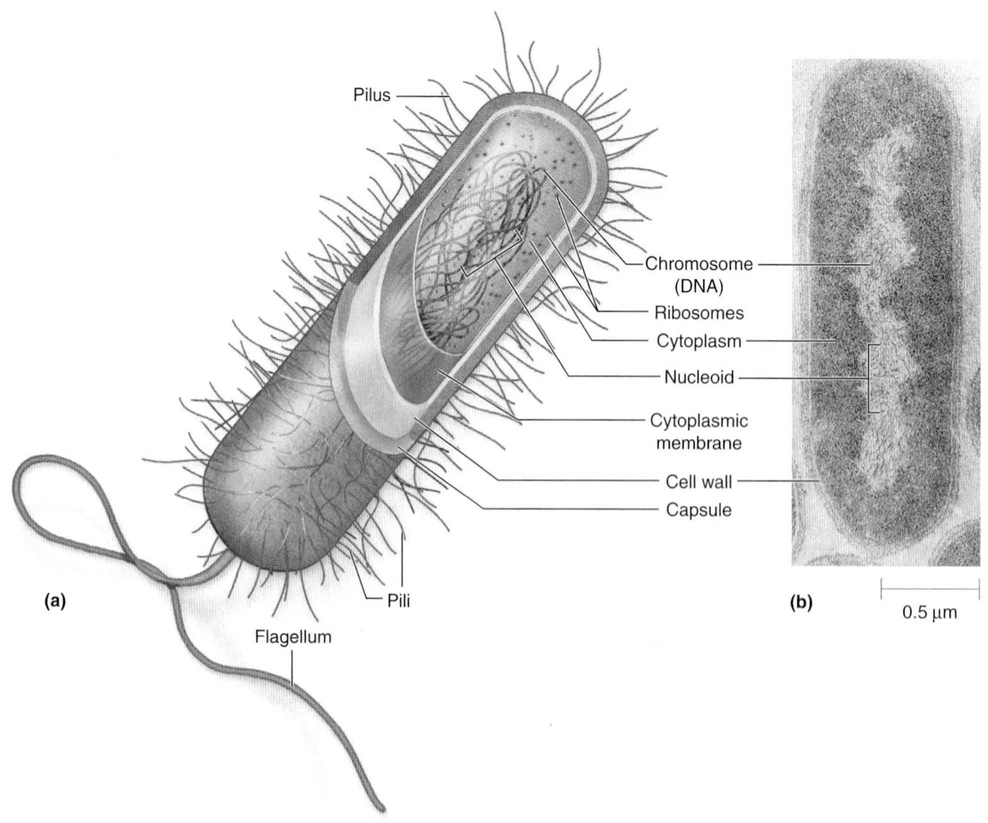

FIGURE 3.23 Typical Prokaryotic Cell A representation of typical structures within and outside a bacterial cell. **(a)** Diagrammatic representation. **(b)** Electron micrograph.

like region where the chromosome resides. Unlike the nucleus that characterizes eukaryotic cells, the nucleoid is not enclosed within a membrane.

The components of a prokaryote, together, enable the cell to survive and multiply in a given environment. Some structures are essential for survival and, as such, are common among all prokaryotic cells; others might be considered optional. Without these "optional" components, the cell may grow in the protected confines of a laboratory, but might not survive in the competitive surroundings of the outside world. The characteristics of typical structures of prokaryotic cells are summarized in **table 3.3.**

Understanding the components of prokaryotic cells is essential for recognizing how these microbes function, but the information is relevant for other reasons as well. For example, certain features are characteristic of select groups of microbes and can be used as identifying markers, allowing scientists to distinguish different groups of bacteria. In addition, structures and processes unique to bacteria are potential targets for selective toxicity. By interfering with these, we can kill bacteria or inhibit their growth without harming the human host.

3.4 The Cytoplasmic Membrane

Focus Points

- Describe the structure and chemistry of the cytoplasmic membrane, focusing on how it relates to membrane permeability.
- Briefly describe how the electron transport chain generates a proton motive force, focusing on how it relates to membrane permeability.

The cytoplasmic membrane is a delicate, thin, fluid structure that surrounds the cytoplasm and defines the boundary of the cell. It serves as an important semipermeable barrier between the cell and its external environment. Although the membrane primarily allows only water, gases, and some small hydrophobic molecules to pass through freely, specific proteins embedded within it act as selective gates. These permit nutrients to enter the cell, and waste products to exit. Other proteins within the membrane serve as sensors of environmental conditions. Thus, while the cytoplasmic

TABLE 3.3 A Summary of Prokaryotic Cell Structures

Structure	Characteristics
Extracellular	
Filamentous appendages	Composed of protein subunits that form a helical chain.
Flagella	Provide the most common mechanism of motility.
Pili	Different types of pili have different functions. The common types, often called fimbriae, enable cells to adhere to surfaces. A few types mediate twitching or gliding motility. Sex pili join cells as a prelude to DNA transfer.
Capsules and slime layers	Layers outside the cell wall, usually made of polysaccharide.
Capsule	Distinct and gelatinous. Enables bacteria to adhere to specific surfaces; allows some organisms to thwart innate defense systems and thus cause disease.
Slime layer	Diffuse and irregular. Enables bacteria to adhere to specific surfaces.
Cell wall	Peptidoglycan provides rigidity to bacterial cell walls, preventing the cells from lysing.
Gram-positive	Thick layer of peptidoglycan that contains teichoic adds and lipoteichoic acids.
Gram-negative	Thin layer of peptidoglycan surrounded by an outer membrane. The outer layer of the outer membrane is lipopolysaccharide.
Cell Boundary	
Cytoplasmic membrane	Phospholipid bilayer embedded with proteins. Surrounds the cytoplasm, separating it from the outside environment. Also functions as a discriminating conduit between the cell and its surroundings.
Intracellular	
DNA	Contains the genetic information of the cell.
Chromosomal	Carries the genetic information essential to a cell, Typically a single, circular, double-stranded DNA molecule.
Plasmid	Extrachromosomal DNA molecule. Generally carries only genetic information that may be advantageous to a cell in certain situations.
Endospore	A type of dormant cell. Generally extraordinarily resistant to heat, desiccation, ultraviolet light, and toxic chemicals.
Cytoskeleton	Involved in cell division and controls cell shape.
Gas vesicles	Small, rigid structures that provides buoyancy to a cell.
Granules	Accumulations of high molecular weight polymers, synthesized from a nutrient available in relative excess.
Ribosomes	Intimately involved in protein synthesis. Two subunits, 30S and 50S, join to form the 70S ribosome.

membrane acts as a barrier, it also functions as an effective and highly discriminating conduit between the cell and its surroundings. ■ hydrophobic, p. 26

Structure and Chemistry of the Cytoplasmic Membrane

The structure of the bacterial cytoplasmic membrane is typical of other biological membranes—a lipid bilayer embedded with proteins **(figure 3.24).** The bilayer consists of two opposing layers composed of phospholipids. At one end of each phospholipid molecule are two fatty acid chains, which act as **hydrophobic tails.** The other end, containing glycerol, a phosphate group, and other polar molecules, functions as a **hydrophilic head.** The phospholipid molecules are arranged in each layer of the bilayer so that their hydrophobic tails face in, toward the other layer. Their hydrophilic heads face outward. As a consequence, the inside of the bilayer is water insoluble whereas the two surfaces interact freely with aqueous solutions. ■ phospholipids, p. 36 ■ hydrophilic, p. 26

More than 200 different **membrane proteins** have been found in *E. coli*. Many function as **receptors,** binding to specific molecules in the environment. This provides a mechanism for the cell to sense and adjust to its surroundings. Proteins are not stationary within the fluid bilayer; rather, they constantly change position. Such movement is necessary for the functions the membrane performs. This structure, with its resulting dynamic nature, is called the **fluid mosaic model.**

Members of the *Bacteria* and *Archaea* have the same general structure of their cytoplasmic membranes, but the lipid compositions are distinctly different. The side chains of the membrane lipids of *Archaea* are connected to glycerol by a different type of chemical linkage. In addition, the side chains are hydrocarbons rather than fatty acids. These differences represent important distinguishing characteristics between these two domains of prokaryotes. ■ fatty acid, p. 35

Permeability of the Cytoplasmic Membrane

The cytoplasmic membrane is **selectively permeable;** relatively few types of molecules can pass through freely. Those that do, move through by a process called simple diffusion. Other molecules must be transported across the membrane by specific transport mechanisms that will be discussed later.

Simple Diffusion

Simple diffusion is the process by which some molecules move freely into and out of the cell. Water, small hydrophobic molecules, and gases such as oxygen and carbon dioxide are among the few substances that move through the cytoplasmic membrane by simple diffusion. The speed and direction of diffusion depend on the relative concentration of molecules on each side of the membrane. The greater the difference in concentration, the higher the rate of diffusion (from the higher to the lower concentration). The molecules continue to pass through at a diminishing rate until their concentration is the same on both sides of the membrane.

The ability of water to move freely through the membrane has important biological consequences. The cytoplasm of a cell is a concentrated solution of inorganic salts, sugars, amino acids, and various other molecules. However, the environments in which prokaryotes normally grow contain only small amounts of some salts and other small molecules. Since the concentration of dissolved molecules, or **solute,** tends to equalize inside and outside the cell, water flows from the surrounding medium into the cell, thereby reducing the concentration of solute inside the cell **(figure 3.25).** This is the process of **osmosis.** This

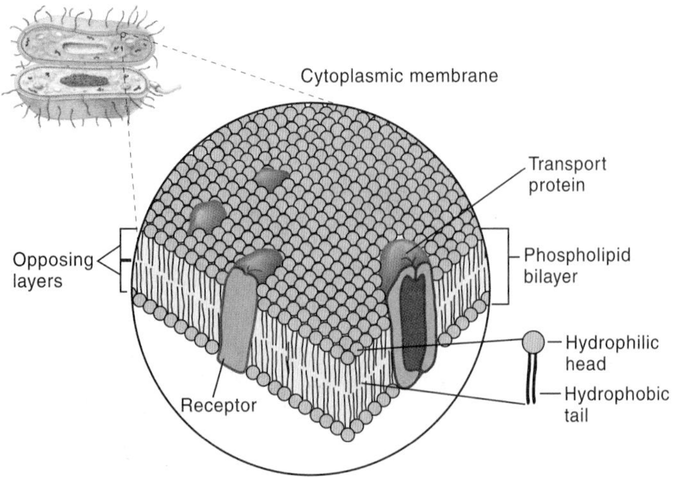

FIGURE 3.24 The Structure of the Cytoplasmic Membrane Two opposing leaflets make up the phospholipid bilayer. Embedded within the bilayer are a variety of different proteins, some of which span the membrane.

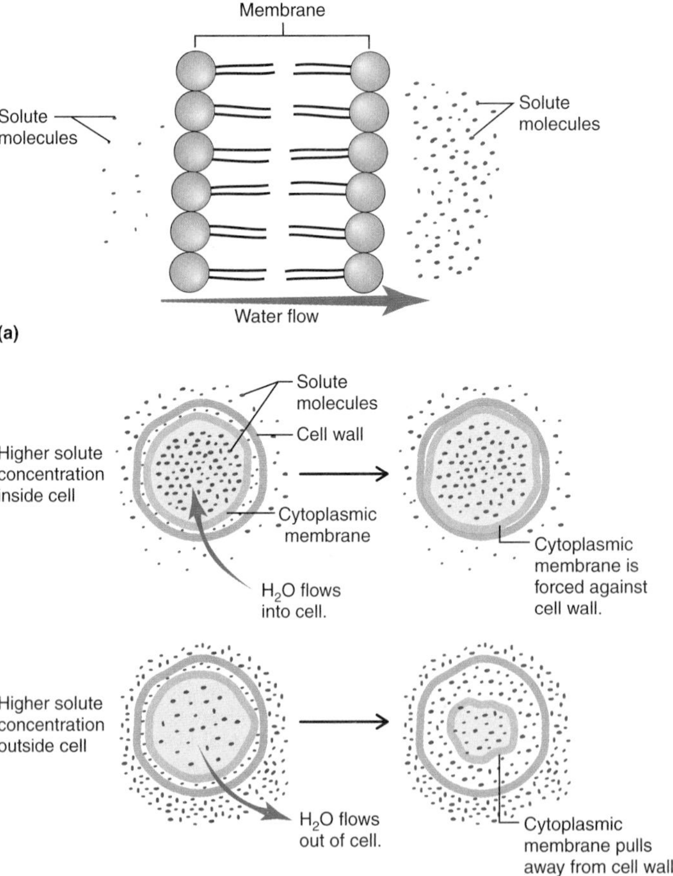

FIGURE 3.25 Osmosis (a) Water flows across a membrane toward the side that has the highest concentration of molecules and ions, thereby equalizing the concentrations on both sides. **(b)** The effect of osmosis on cells.

inflow of water exerts tremendous **osmotic pressure** on the cytoplasmic membrane, much more than it generally can resist. However, the rigid cell wall surrounding the membrane generally withstands such high pressure. The cytoplasmic membrane is forced up against the wall but cannot balloon further. Damage to the cell wall weakens the structure, and consequently, cells may burst or **lyse.**

The Role of the Cytoplasmic Membrane in Energy Transformation

The cytoplasmic membrane of prokaryotic cells plays an indispensable role in converting energy to a usable form. This is an important distinction between prokaryotic and eukaryotic cells; in eukaryotic cells energy is transformed in membrane-bound organelles, which will be discussed later in this chapter.

As part of their energy-harvesting processes, most prokaryotes have a series of protein complexes, the **electron transport chain,** embedded in their membrane. These sequentially transfer electrons and, in the process, eject protons from the cell. The details of these processes will be explained in chapter 6. The expulsion of protons by the electron transport chain results in the formation of a proton gradient across the cell membrane. Positively charged protons are concentrated immediately outside the membrane, whereas negatively charged hydroxyl ions accumulate directly inside the membrane **(figure 3.26).** This separation of charged ions creates an **electrochemical gradient** across the membrane; inherent in it is a form of energy, called **proton motive force.** This is analogous to the energy stored in a battery. ■ electron transport chain, p. 142

Energy of the proton motive force can be harvested when protons are allowed to move back into the cell. This is used directly to drive certain cellular processes, including some transport mechanisms that carry small molecules across the membrane and some forms of motility. It is also used to synthesize ATP. ■ ATP, p. 25

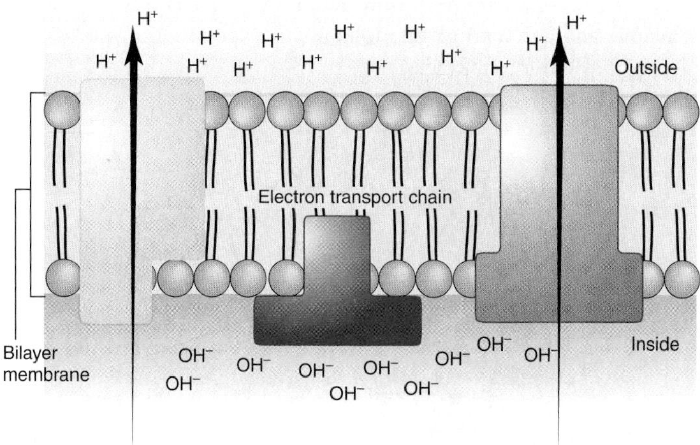

FIGURE 3.26 Proton Motive Force The electron transport chain, a series of protein complexes within the membrane, ejects protons from the cell. This creates an electrochemical gradient, a form of energy called proton motive force.

MICROCHECK 3.4

The cytoplasmic membrane is a phospholipid bilayer embedded with a variety of different proteins. It serves as a barrier between the cell and the surrounding environment, allowing relatively few types of molecules to pass through freely. The electron transport chain within the membrane expels protons, generating a proton motive force.

- ✓ Explain the fluid mosaic model.
- ✓ Name three molecules that can pass freely through the lipid bilayer.
- ✓ Why is the word "fluid" in fluid mosaic model an appropriate term?

3.5

Directed Movement of Molecules Across the Cytoplasmic Membrane

Focus Points

- Compare and contrast facilitated diffusion and active transport.
- Describe the role of signal sequences in secretion.

Nearly all molecules that enter or exit a cell must cross the otherwise impermeable cytoplasmic membrane through proteins that function as selective gates. Mechanisms allowing nutrients and other small molecules to enter the cell are called **transport systems.** These systems are also used to expel wastes and compounds such as antibiotics and disinfectants that are otherwise deleterious to the cell.

Cells actively move certain proteins they synthesize out of the cell—a process called **secretion.** Some of these secreted proteins make up structures such as flagella, which are appendages used for motility. Others are enzymes secreted to break down substances that would otherwise be too large to transport into the cell.

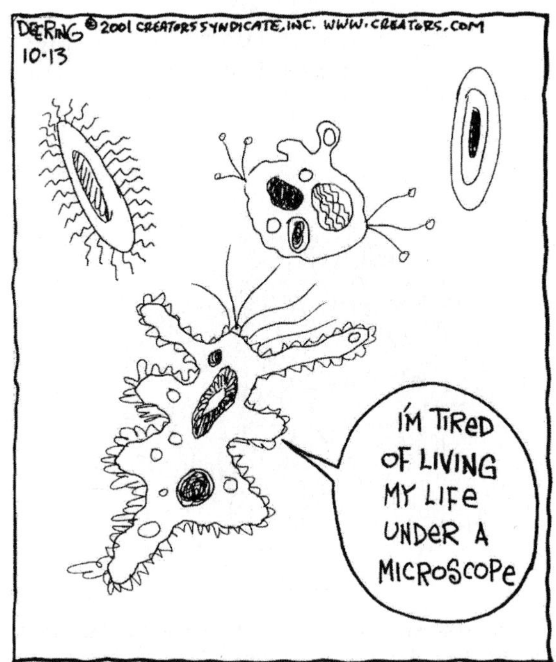

By permission of John Deering and Creators Syndicate, Inc.

Transport Systems

Mechanisms used to transport molecules across the membrane employ highly specific proteins called **transport proteins, permeases,** or **carriers.** These proteins span the membrane, so that one end projects into the surrounding environment and the other into the cell. The interaction between the transport protein and the molecule it carries is highly specific. Consequently, a single carrier generally transports only a specific type of molecule. As a carrier transports a molecule, its shape changes, facilitating the passage of the molecule **(figure 3.27).** The mechanisms of transport are summarized in **table 3.4.**

Facilitated Diffusion

Facilitated diffusion, or **passive transport,** moves substances from one side of the membrane to the other by exploiting a concentration gradient. Molecules are transported across until their concentration is the same on both sides of the membrane. This mechanism can only eliminate a difference in concentration, it cannot create one. As prokaryotes typically grow in relatively nutrient-poor environments, they generally cannot rely on facilitated diffusion to take in nutrients.

Active Transport

Active transport moves compounds against a concentration gradient. This requires an expenditure of energy. There are two primary mechanisms of active transport, each utilizing a different form of energy.

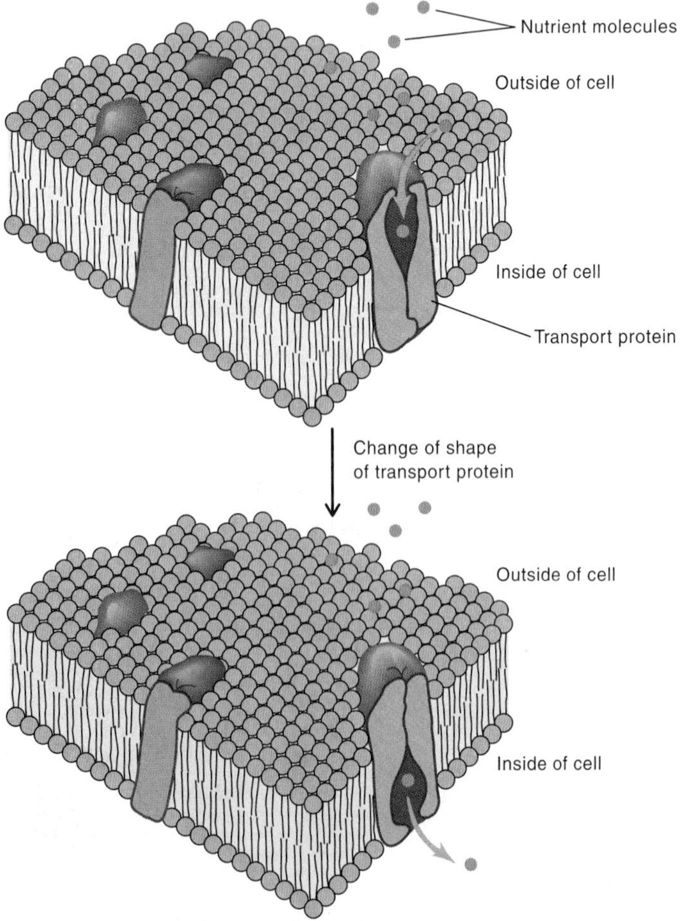

FIGURE 3.27 Transport Protein A transport protein changes its shape to facilitate passage of a compound across the cytoplasmic membrane.

TABLE 3.4 A Summary of Transport Mechanisms Used by Prokaryotic Cells

Transport Mechanism	Characteristics
Facilitated Diffusion	Rarely used by prokaryotes. Exploits a concentration gradient to move molecules; can only eliminate a gradient, not create one. No energy is expended.
Active Transport	Energy is expended to accumulate molecules against a concentration gradient.
Major facilitator superfamily	In bacteria, the proton motive force drives these transporters. As a proton is allowed into the cell another substance is either brought along or expelled.
ABC transporters	ATP is used as an energy source. Extracellular binding proteins deliver a molecule to the transporter.
Group Translocation	The transported molecule is chemically altered as it passes into the cell.

Transport Systems That Use Proton Motive Force Many bacterial transport systems can accumulate or extrude small molecules and ions using the energy of a proton motive force. Transporters of this type allow a proton into the cell and simultaneously either bring along or expel another substance **(figure 3.28).** For example, the permease that transports lactose brings the sugar into the cell along with a proton. Expulsion of waste products, on the other hand, relies on transporters that eject the compound as a proton passes in. **Efflux pumps,** which are used by some bacteria to oust antimicrobial drugs, use this latter mechanism. These systems are part of a large group of transporters, collectively known as the **major facilitator superfamily (MFS),** found in prokaryotes as well as eukaryotes.

Transport Systems That Use ATP Transport mechanisms called **ABC transport systems** require ATP as an energy source (ABC stands for ATP Binding-Cassette). These systems are relatively elaborate, involving multiple protein components **(figure 3.29).** ABC transport systems use **binding proteins** that reside immediately outside of the cytoplasmic membrane. These proteins each scavenge and deliver a given molecule to a specific transport complex within the membrane.

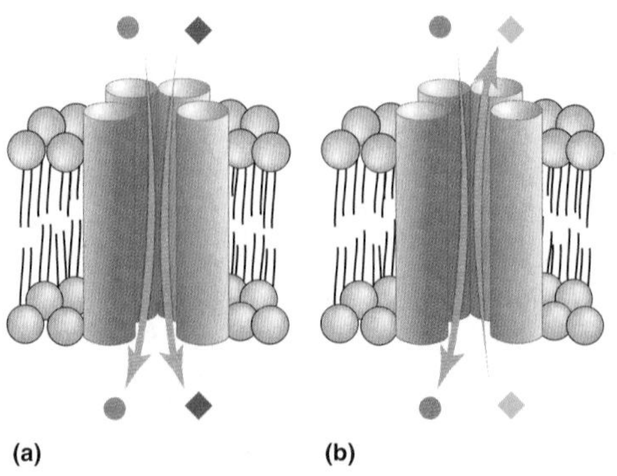

FIGURE 3.28 Active Transport Systems That Use Proton Motive Force Transporters of this type allow a proton into cell and simultaneously either **(a)** bring along another substance or **(b)** expel a substance.

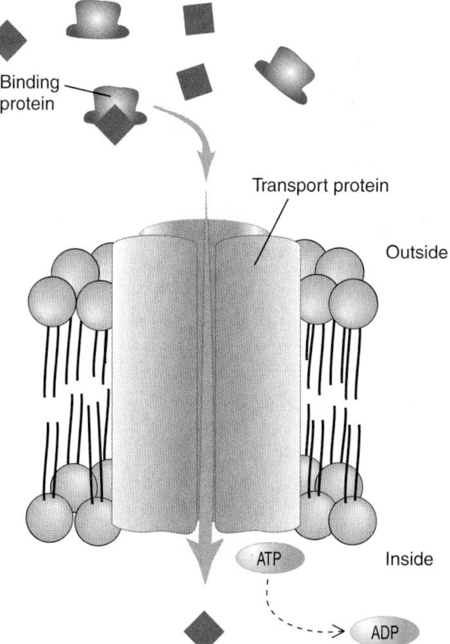

FIGURE 3.29 Active Transport Systems That Use ATP ABC transport systems require energy in the form of ATP. A binding protein that resides outside of the cytoplasmic membrane delivers a given molecule to a specific transport protein.

Group Translocation

Group translocation is a transport process that chemically alters a molecule during its passage through the cytoplasmic membrane **(figure 3.30).** Glucose and several other sugars are phosphorylated during their transport into the cell by the **phosphotransferase system.** The energy expended to phosphorylate the sugar can be regained when that sugar is later broken down to provide energy.

Secretion

The **general secretory pathway** is the primary mechanism used to secrete proteins synthesized by the cell. It recognizes proteins destined for secretion by their **signal sequence,** a characteristic sequence of amino acids that make up one end.

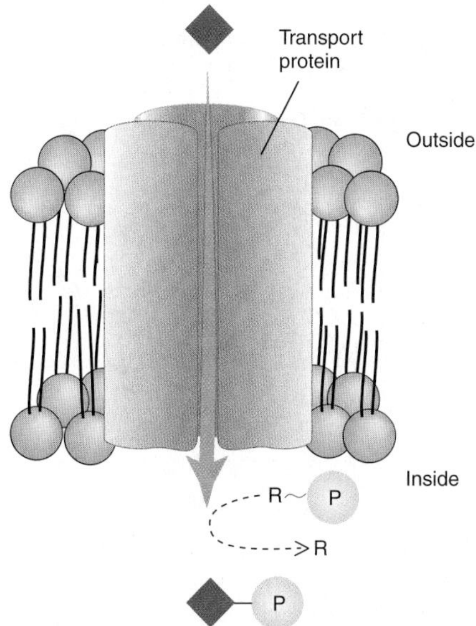

FIGURE 3.30 Group Translocation This process chemically alters a molecule during its passage through the cytoplasmic membrane.

The general secretory pathway requires at least 11 different proteins and uses ATP to drive the process. However, the precise mechanism by which the proteins move across the membrane is still poorly understood.

MICROCHECK 3.5

Facilitated transport does not require energy but is rarely used by bacteria. Active transport via the major facilitator superfamily uses proton motive force. Active transport via an ABC transporter uses ATP as an energy source. Group translocation chemically modifies a molecule as it enters the cell. Proteins that have a signal sequence are secreted from a cell by the general secretory pathway.

✓ Transport proteins may be referred to as what two other terms?

✓ Describe the role of binding proteins in an ABC transport system.

✓ Can you argue that group translocation is a form of active transport?

3.6 Cell Wall

Focus Points

- Describe the chemistry and structure of peptidoglycan.
- Compare and contrast the structure and chemistry of the Gram-positive and Gram-negative cell walls, and describe how the differences account for the Gram staining characteristics.
- Explain why Gram-negative bacteria are typically less susceptible than Gram-positive bacteria to penicillin and lysozyme, and why *Mycoplasma* species are not affected by these agents.
- Describe the cell walls of members of the *Archaea*.

The cell wall of most common prokaryotes is a rigid structure that determines the shape of the organism. A primary function of the wall is to hold the cell together and prevent it from bursting. If the cell wall is somehow breached, undamaged parts maintain their original shape **(figure 3.31).**

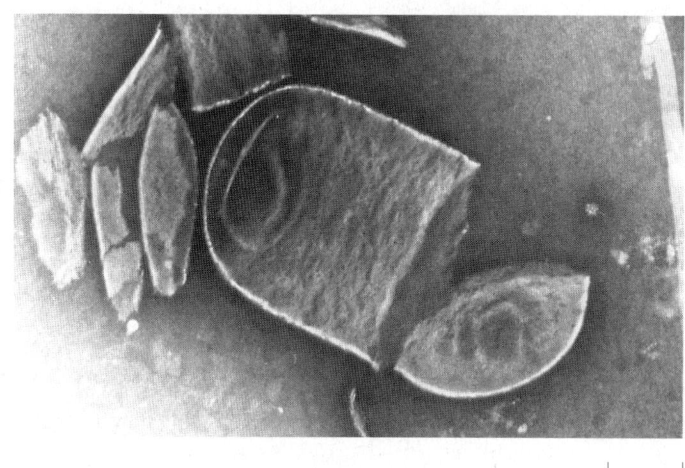

FIGURE 3.31 The Rigid Cell Wall Determines the Shape of the Bacterium Even though the cell has split apart, the cell wall maintains its original shape.

The type of cell wall distinguishes two main groups of bacteria—Gram-positive and Gram-negative. A comparison of the features of these groups is presented in **table 3.5.**

Peptidoglycan

Although the structure varies in Gram-positive and Gram-negative cells, the rigidity of bacterial cell walls is due to a layer of **peptidoglycan,** a macromolecule found only in bacteria. The basic structure of peptidoglycan is an alternating series of two major subunits, **N-acetylmuramic acid (NAM)** and **N-acetylglucosamine (NAG).** These subunits, which are related to glucose, are covalently joined to one another to form a **glycan chain (figure 3.32).**

TABLE 3.5 Comparison of Features of Gram-Positive and Gram-Negative Bacteria

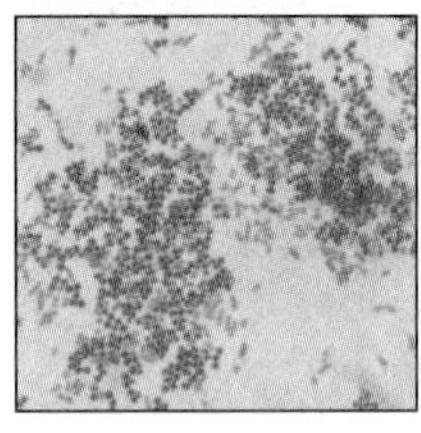

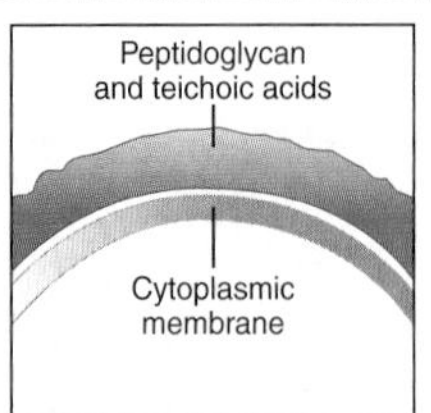

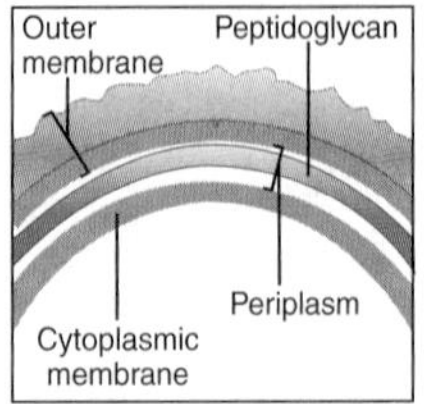

	Gram-Positive	Gram-Negative
Color of Gram Stained Cell	Purple	Reddish-pink
Representative Genera	*Bacillus, Staphylococcus, Streptococcus*	*Escherichia, Neisseria, Pseudomonas*
Distinguishing Structures/Components		
Peptidoglycan	Thick layer	Thin layer
Teichoic acids	Present	Absent
Outer membrane	Absent	Present
Lipopolysaccharide (endotoxin)	Absent	Present
Porin proteins	Absent (unnecessary because there is no outer membrane)	Present; allow passage of molecules through outer membrane
Periplasm	Absent	Present
General Characteristics		
Sensitivity to penicillin	Generally more susceptible (with notable exceptions)	Generally less susceptible (with notable exceptions)
Sensitivity to lysozyme	Yes	No

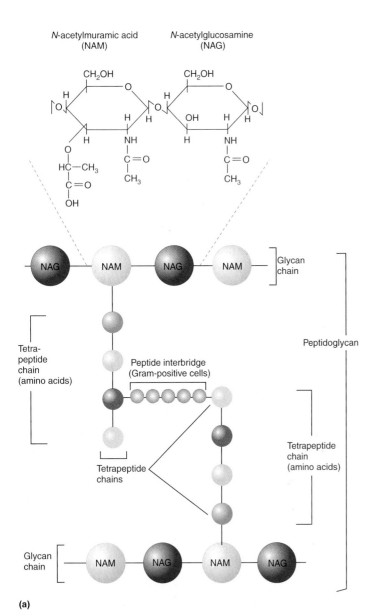

(a)

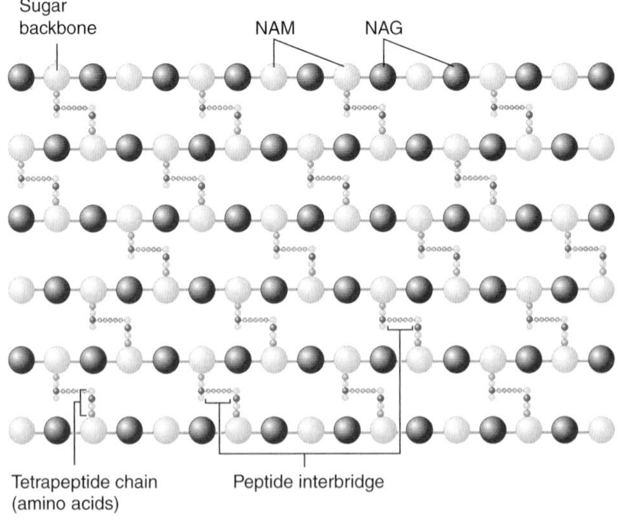

(b)

FIGURE 3.32 Components and Structure of Peptidoglycan **(a)** Chemical structure of *N*-acetylglucosamine (NAG) and *N*-acetylmuramic acid (NAM); the ring structures of the two molecules are glucose. Glycan chains are composed of alternating subunits of NAG and NAM joined by covalent bonds. Adjacent glycan chains are cross-linked via their tetrapeptide chains to create peptidoglycan. **(b)** Interconnected glycan chains form a very large three-dimensional molecule of peptidoglycan.

This high molecular weight linear polymer serves as the backbone of the peptidoglycan molecule.

Attached to each of the NAM molecules is a string of four amino acids, a tetrapeptide chain, that plays an important role in the structure of the peptidoglycan molecule. Cross-linkages can form between tetrapeptide chains, thus joining adjacent glycan chains to form a single, very large three-dimensional molecule. In Gram-negative bacteria, tetrapeptides are joined directly. In Gram-positive bacteria, they are usually joined indirectly by a series of amino acids, a peptide interbridge, the composition of which may vary among species.

An assortment of only a few different amino acids make up the tetrapeptide chain. One of these, diaminopimelic acid, which is related to the amino acid lysine, is not found in any other place in nature. Some of the others are D-isomers, a form not found in proteins. ■ D-Isomer, p. 26 ■ lysine, p. 27

The Gram-Positive Cell Wall

A relatively thick layer of peptidoglycan characterizes the cell wall of Gram-positive bacteria **(figure 3.33).** As many as 30 layers, or sheets, of interconnected glycan chains make up the polymer. Regardless of its thickness, peptidoglycan is fully permeable to many substances including sugars and amino acids.

A prominent component of the Gram-positive cell wall is a group of molecules called **teichoic acids** (from the Greek word *teichos*, meaning wall). These are chains of a common subunit, either ribitol-phosphate or glycerol-phosphate, to which various sugars and D-alanine are usually attached. Teichoic acids are joined to the peptidoglycan molecule through covalent bonds to *N*-acetylmuramic acid. Some, which are called **lipoteichoic acids,** are linked to the cytoplasmic membrane. Teichoic acids and lipoteichoic acids both stick out above the peptidoglycan layer and, because they are negatively charged, give the cell its negative polarity.

The Gram-Negative Cell Wall

The cell wall of Gram-negative bacteria is far more complex than that of Gram-positive organisms **(figure 3.34).** It contains only a thin layer of peptidoglycan. Outside of that layer is the **outer membrane,** a unique lipid bilayer embedded with proteins. The peptidoglycan layer is sandwiched between the cytoplasmic membrane and the outer membrane.

The Outer Membrane

Like the cytoplasmic membrane, which in Gram-negative bacteria is sometimes called the **inner membrane,** the outer membrane serves as a barrier to the passage of most molecules. Thus, it serves as a protective barrier, excluding many compounds that are deleterious to the cell, including certain antimicrobial medications. This is one reason why Gram-negative bacteria are generally less sensitive to many such medications. Small molecules and ions can cross the membrane through **porins,** specialized channel-forming proteins that span the outer membrane. Some porins are specific for certain molecules; others allow many different molecules to pass. Proteins produced by the cell that are destined for secretion are moved across the outer membrane by mechanisms known as **secretion systems.**

The outer membrane is unlike any other membrane in nature. Its lipid bilayer structure is typical of other membranes, but the outside layer is made up of **lipopolysaccharides** rather than phospholipids. For this reason, the outer membrane is also called the **lipopolysaccharide layer,** or **LPS.** The outer membrane is joined to peptidoglycan by means of lipoprotein molecules. ■ lipoproteins, p. 30

The Lipopolysaccharide Molecule The lipopolysaccharide molecule is extremely important from a medical standpoint. When purified lipopolysaccharide is injected into an animal, it elicits symptoms characteristic of infections caused by live bacteria. The same symptoms occur regardless of the bacterial species. To reflect the fact that the molecule that elicits the symptoms is an inherent part of the cell wall, it is called **endotoxin.** ■ endotoxin, p. 407

Two parts of the LPS molecule are notable for their medical significance (see figure 3.34c):

- **Lipid A** is the portion that anchors the LPS molecule in the lipid bilayer. Its chemical make-up plays a significant role in the body's ability to recognize the presence of invading bacteria. When lipid A is introduced into the body in small amounts, such as when microbes contaminate a small lesion, the defense system responds and effectively eliminates the invader. If, however, large amounts of lipid A are present, such as when Gram-negative bacteria are actively growing in the bloodstream, the magnitude of the defense system's response damages even our own cells. This response to lipid A causes the symptoms associated with endotoxin.
- The **O-specific polysaccharide** side chain is the portion of LPS directed away from the membrane, at the end opposite that of lipid A. It is made up of a chain of sugar molecules, the number and composition of which varies among different species of bacteria. The differences can be exploited to identify certain species or strains. For example, the "O157" in *E. coli* O157:H7 refers to the characteristic O-side chain of the strains.

Periplasm

The region between the cytoplasmic membrane and the outer membrane of Gram-negative bacteria is filled with a gel-like fluid called **periplasm.** All secreted proteins in these bacteria are contained within the periplasm unless they are specifically moved across the outer membrane as well. Thus, the periplasm is filled with proteins involved in a variety of cellular activities, including nutrient degradation and transport. For example, the enzymes that cells secrete to break down peptides and other molecules are found in the periplasm. Similarly, the binding proteins of the ABC transport systems are found there.

Antibacterial Substances that Target Peptidoglycan

Compounds that interfere with the synthesis of peptidoglycan or alter its structural integrity weaken the rigid molecule to a point where it cannot prevent the cell from bursting. These substances have no effect on eukaryotic cells because peptidoglycan is unique to bacteria. Examples of compounds that target peptidoglycan include the antibiotic penicillin and the enzyme lysozyme.

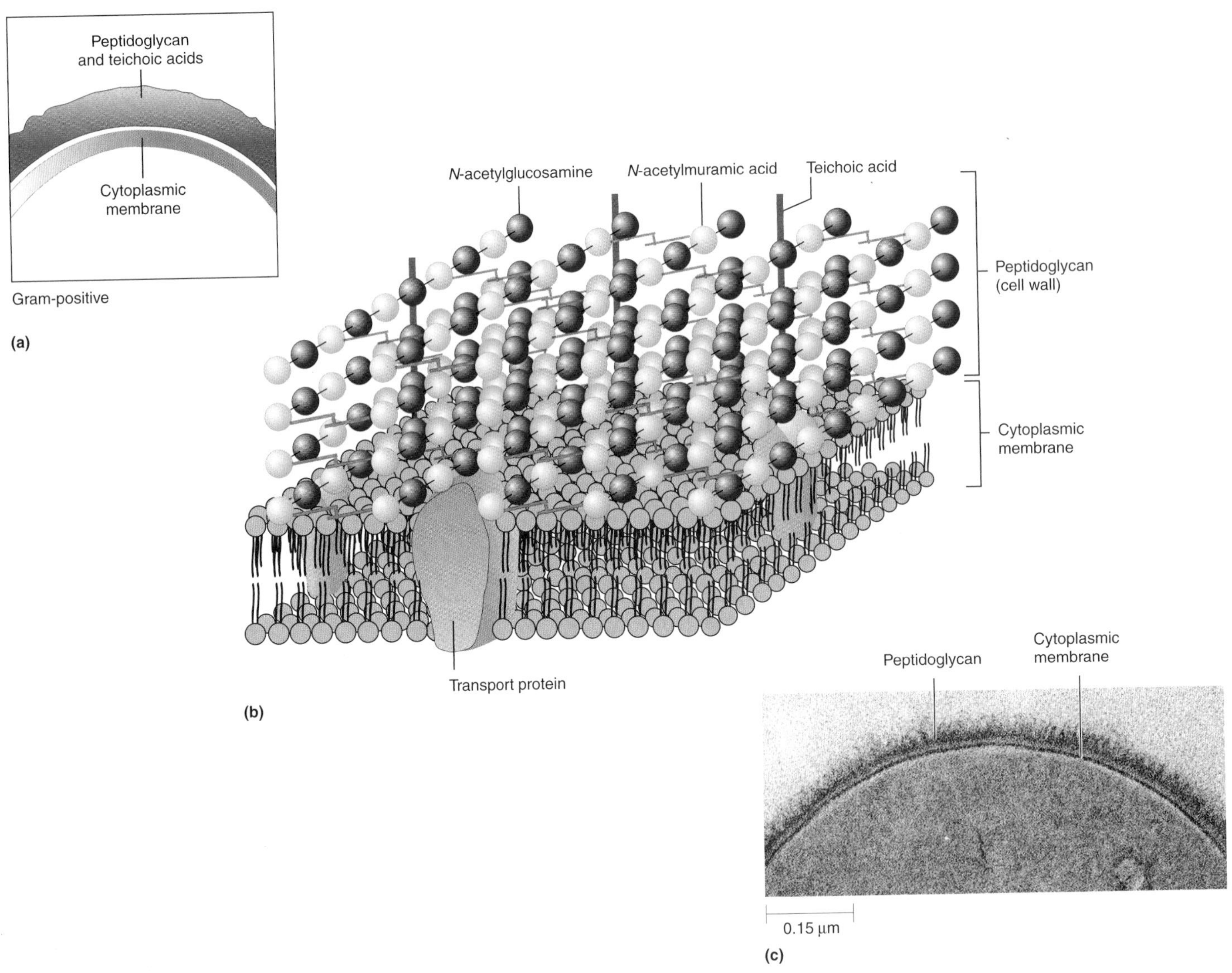

FIGURE 3.33 Gram-Positive Cell Wall (a) The Gram-positive cell wall is characterized by a relatively thick layer of peptidoglycan. **(b)** It is made up of many sheets of interconnected glycan chains. **(c)** Transmission electron photomicrograph of a typical Gram-positive cell (*Bacillus subtilis*).

Penicillin

Penicillin is the most thoroughly studied of a group of antibiotics that interfere with peptidoglycan synthesis. Penicillin binds proteins involved in cell wall synthesis and, subsequently, prevents the cross-linking of adjacent glycan chains. ■ penicillin, p. 475

Generally, but with notable exceptions, penicillin is far more effective against Gram-positive cells than Gram-negative cells. This is because the outer membrane of Gram-negative cells prevents the medication from reaching its site of action, the peptidoglycan layer.

Lysozyme

Lysozyme, an enzyme found in many body fluids including tears and saliva, breaks the bond that links the alternating *N*-acetylglucosamine and *N*-acetylmuramic acid molecules of peptidoglycan. This destroys the structural integrity of the glycan chain, the backbone of the peptidoglycan molecule. Lysozyme is sometimes used in the laboratory to remove the peptidoglycan layer from bacteria for experimental purposes.

Differences in Cell Wall Composition and the Gram Stain

Differences in the cell wall composition of Gram-positive and Gram-negative bacteria account for their staining characteristics. It is not the cell wall, however, but the inside of the cell that is stained by the crystal violet-iodine complex. The Gram-positive cell wall somehow retains the crystal violet-iodine complex within the cell even when subjected to acetone-alcohol treatment, whereas the Gram-negative cell wall cannot.

The precise mechanism that accounts for the differential aspect of the Gram stain is not entirely understood.

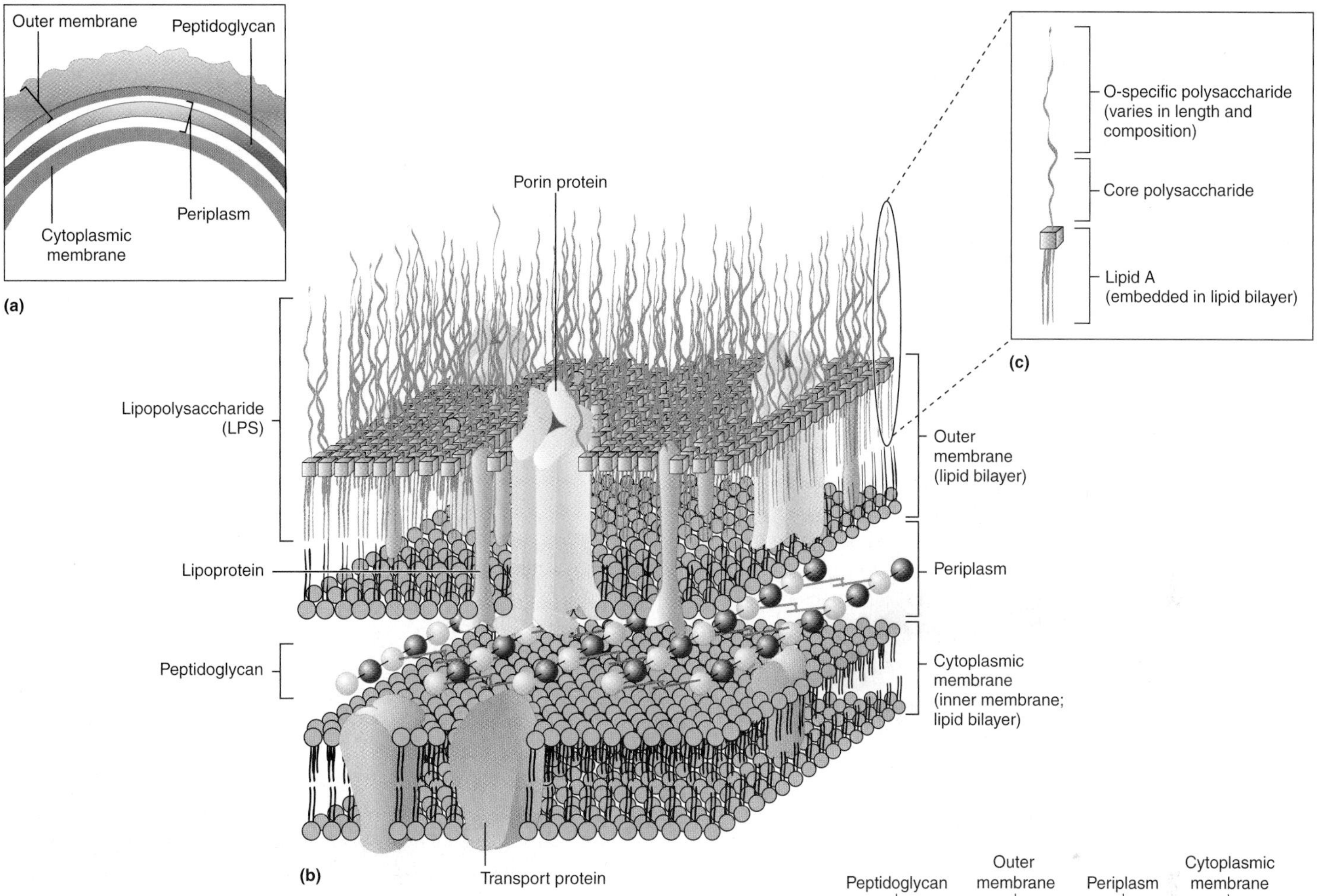

Outer membrane
Peptidoglycan
Periplasm
Cytoplasmic membrane
(a)
Porin protein
Lipopolysaccharide (LPS)
Lipoprotein
Peptidoglycan
Outer membrane (lipid bilayer)
Periplasm
Cytoplasmic membrane (inner membrane; lipid bilayer)
(b)
Transport protein
O-specific polysaccharide (varies in length and composition)
Core polysaccharide
Lipid A (embedded in lipid bilayer)
(c)
Peptidoglycan
Outer membrane
Periplasm
Cytoplasmic membrane
(d)
0.15 μm

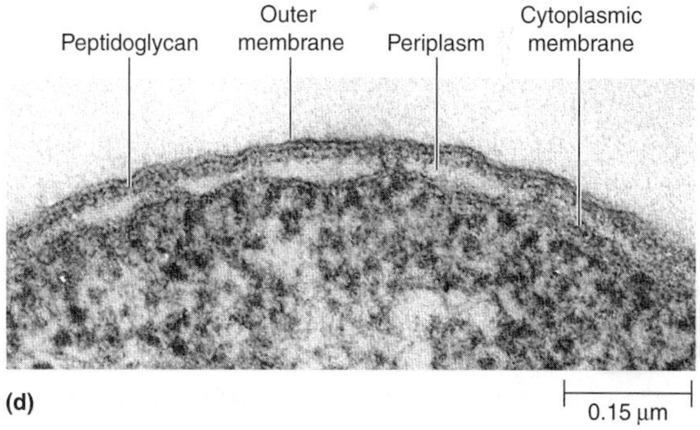

FIGURE 3.34 Gram-Negative Cell Wall **(a)** The Gram-negative cell wall is characterized by a very thin layer of peptidoglycan surrounded by an outer membrane. **(b)** The peptidoglycan layer is made up of only one or two sheets of interconnected glycan chains. The outer membrane is a typical phospholipid bilayer, except the outer leaflet contains lipolysaccharide. Porins span the membrane to allow specific molecules to pass. Periplasm fills the region between the cytoplasmic and outer membranes. **(c)** Structure of lipopolysaccharide. The Lipid A portion, which anchors the LPS molecule in the lipid bilayer, is responsible for the symptoms associated with endotoxin. The composition and length of the O-specific polysaccharide side chain varies among different species of bacteria. **(d)** A transmission electron micrograph of a typical Gram-negative cell wall *(Pseudomonas aeruginosa).*

Presumably, the decolorizing agent dehydrates the thick layer of peptidoglycan; in this dehydrated state the wall acts as a permeability barrier, holding the dye within the cell. In contrast, the solvent action of acetone-alcohol easily damages the outer membrane of Gram-negative bacteria, and the relatively thin layer of peptidoglycan cannot retain the dye complex. These bacteria lose the dye complex more readily than their Gram-positive counterparts.

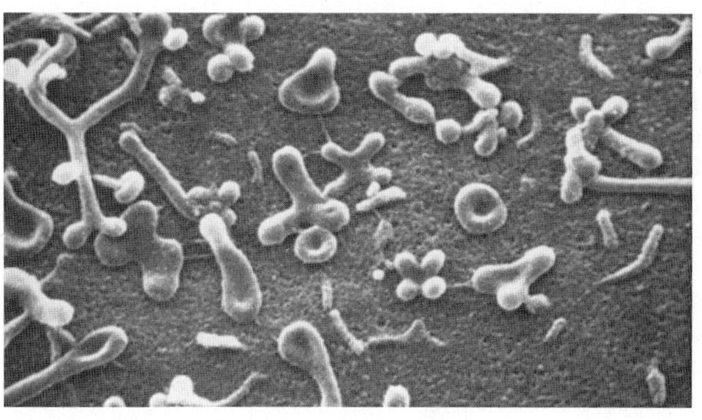

2 μm

FIGURE 3.35 ***Mycoplasma pneumoniae*** These cells vary in shape because they lack a cell wall.

Characteristics of Bacteria that Lack a Cell Wall

Some bacteria naturally lack a cell wall. Species of *Mycoplasma*, one of which causes a mild form of pneumonia, have an extremely variable shape because they lack a rigid cell wall **(figure 3.35).** As expected, neither penicillin nor lysozyme affects these organisms. *Mycoplasma* and related bacteria can survive without a cell wall because their cytoplasmic membrane is stronger than that of most other bacteria. They have **sterols** in their membrane; these rigid, planar molecules stabilize membranes, making them stronger.

Cell Walls of the Domain *Archaea*

As a group, members of the *Archaea* inhabit a wide range of extreme environments, and so it is not surprising they contain a greater variety of cell wall types than do members of the *Bacteria*. However, because most of these organisms have not been studied as extensively as the *Bacteria*, less is known about the structure of their walls. None contain peptidoglycan, but some do have a similar molecule, **pseudopeptidoglycan.**

MICROCHECK 3.6

Peptidoglycan is a molecule unique to bacteria that provides rigidity to the cell wall. The Gram-positive cell wall is composed of a relatively thick layer of peptidoglycan as well as teichoic acids. The Gram-negative cell wall has a thin layer of peptidoglycan and an outer membrane, which contains lipopolysaccharide. The outer membrane excludes molecules with the exception of those that pass through porins; proteins are secreted via special mechanisms. Penicillin and lysozyme interfere with the structural integrity of peptidoglycan. *Mycoplasma* species lack a cell wall. Members of the *Archaea* have a variety of cell wall types.

✓ What is the significance of lipid A?

✓ How does the action of penicillin differ from that of lysozyme?

✓ Explain why penicillin will kill only actively multiplying cells, whereas lysozyme will kill cells in any stage of growth.

3.7 Capsules and Slime Layers

Focus Point

- Compare and contrast the structure and function of capsules and slime layers.

Many bacteria envelop themselves with a gel-like layer that generally functions as a mechanism of either protection or attachment **(figure 3.36).** If the layer is distinct and gelatinous, it is called a **capsule.** If, instead, the layer is diffuse and irregular, it is called a **slime layer.** Colonies that form either of these often appear moist and glistening.

Capsules and slime layers vary in their chemical composition depending on the species of bacteria. Most are composed of polysaccharides, and are commonly referred to as a **glycocalyx** (*glyco* means "sugar" and *calyx* means "shell"). A few capsules consist of polypeptides made up of repeating subunits of only one or two amino acids. Interestingly, the amino acids are generally of the D-stereoisomeric form, one of the few places D-amino acids are found in nature. ■ **polysaccharide, pp. 30, 32 ■ D-amino acid, p. 26**

Some types of capsules and slime layers enable bacteria to adhere to specific surfaces, including teeth, rocks, and other bacteria. These often enable microorganisms to grow as a **biofilm,**

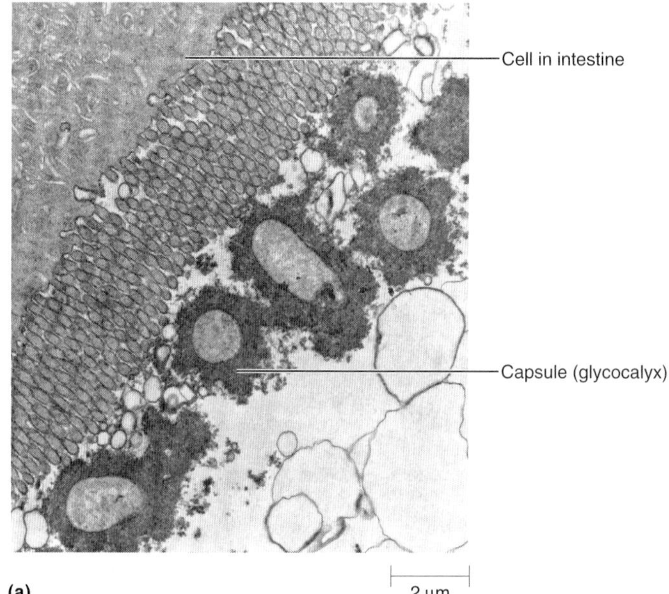

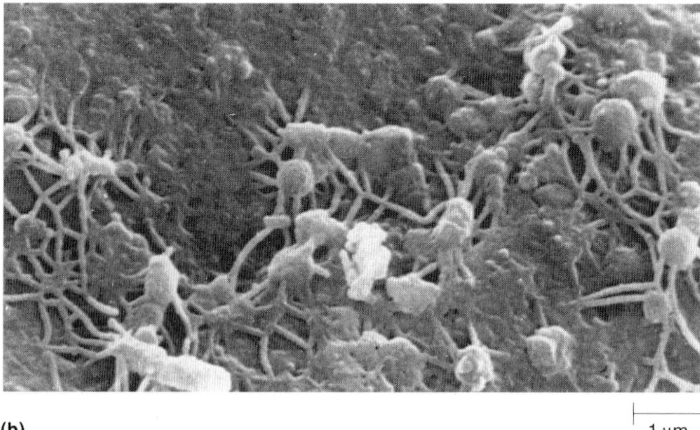

FIGURE 3.36 Capsules and Slime Layers These layers enable bacteria to attach to specific surfaces. **(a)** Capsules facilitating the attachment of bacteria to cells in the intestine (EM). **(b)** Masses of cells of *Eikenella corrodens* adhering in a layer of slime (SEM).

a polysaccharide-encased mass of bacteria coating a surface. One example is **dental plaque,** a biofilm on teeth. *Streptococcus mutans* uses sucrose to synthesize a capsule, which enables it to adhere to and grow in the crevices of the tooth. Other bacteria can then adhere to the layer created by *S. mutans.* Acid production by bacteria in the biofilm damages the tooth surface. ■ ***Streptococcus mutans*, p. 587 ■ sucrose, p. 32 ■ dental caries, p. 586**

Some capsules enable bacteria to thwart innate defense systems that otherwise protect against infection. *Streptococcus pneumoniae*, an organism that causes bacterial pneumonia, can only cause disease if it has a capsule. Unencapsulated cells are quickly engulfed and killed by phagocytes, an important cell of the body's defense system. ■ ***Streptococcus pneumoniae*, p. 509 ■ phagocytes, p. 358**

MICROCHECK 3.7

Capsules and slime layers enable organisms to adhere to surfaces and sometimes protect bacteria from our innate defense system.

✓ How do capsules differ from slime layers?

✓ What is dental plaque?

✓ Explain why a sugary diet can lead to tooth decay.

3.8

Filamentous Protein Appendages

Focus Points

- Describe the structure and function of flagella, and explain how the direction of their rotation is involved in chemotaxis.
- Compare and contrast the structure and function of fimbriae and sex pili.

Many bacteria have protein appendages that are anchored in the cytoplasmic membrane and protrude out from the surface. These structures are not essential to the life of the cell, but they do allow some bacteria to exist in certain environments in which they otherwise might not survive.

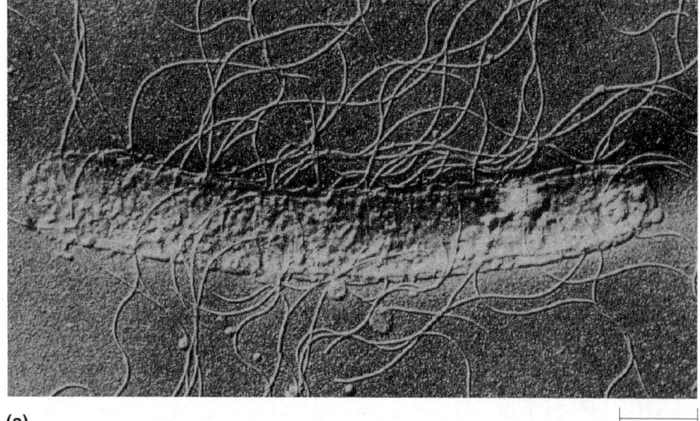

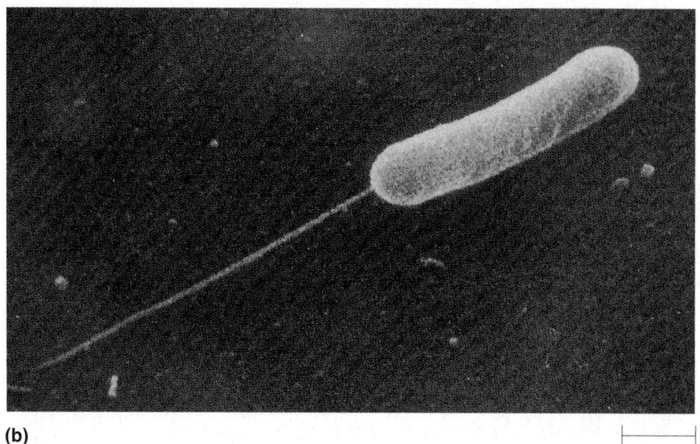

FIGURE 3.37 Flagella (a) Peritrichous flagella (SEM); **(b)** polar flagellum (SEM).

Flagella

The **flagellum** is a long protein structure responsible for most types of bacterial motility **(figure 3.37).** By spinning like a propeller, using proton motive force as energy, the flagellum pushes the bacterium through liquid much as a ship is driven through water. Flagella must work very hard to move a cell, since water has the same relative viscosity to bacteria as molasses has to humans. Nevertheless, their speed is quite phenomenal; flagella can rotate more than 100,000 revolutions per minute (rpm), propelling the cell at a rate of 20 body lengths per second. This is the equivalent of a 6-foot man running 82 miles per hour!

In some cases, flagella are important in the ability of an organism to cause disease. For example, *Helicobacter pylori*, the bacterium that causes gastric ulcers, has powerful multiple flagella at one end of its spiral-shaped cell. These flagella allow *H. pylori* to penetrate the viscous mucous gel that coats the stomach epithelium. ■ *Helicobacter pylori*, p. 589

Structure and Arrangement of Flagella

Flagella are composed of three basic parts **(figure 3.38).** The **filament** is the portion extending into the exterior environment. It is composed of identical subunits of a protein called **flagellin.** These

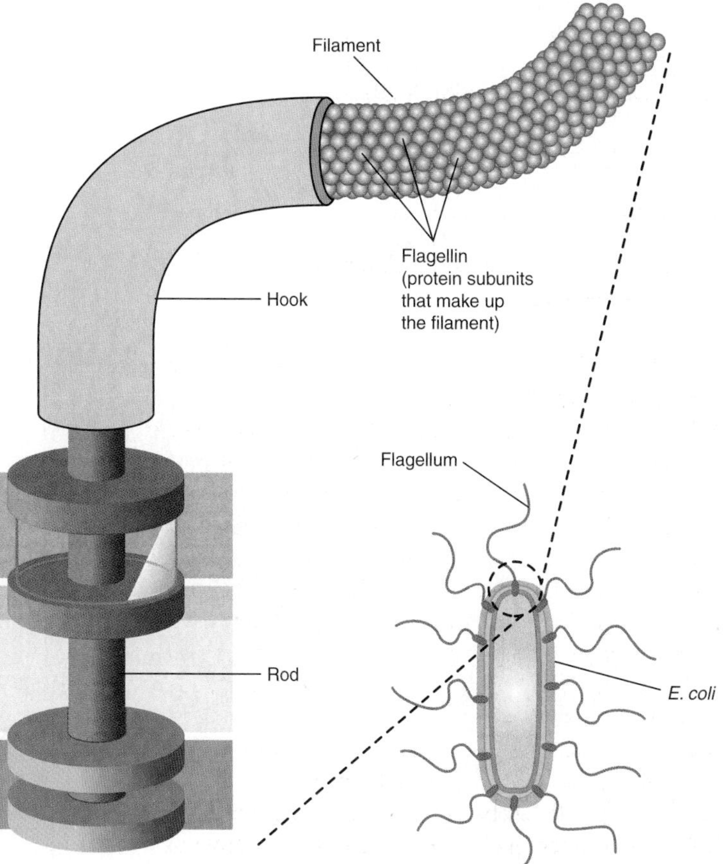

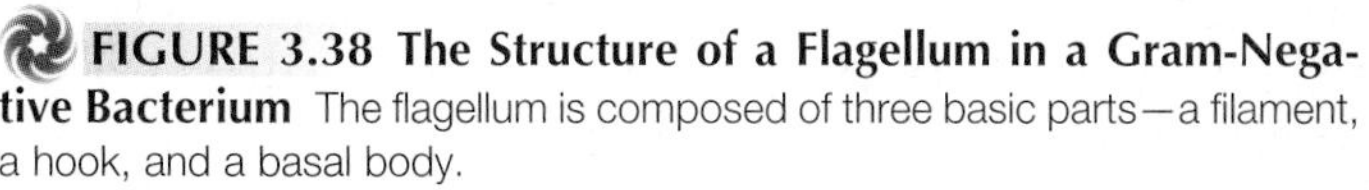

FIGURE 3.38 The Structure of a Flagellum in a Gram-Negative Bacterium The flagellum is composed of three basic parts—a filament, a hook, and a basal body.

subunits form a chain that twists into a helical structure with a hollow core. Connecting the filament to the cell surface is a curved structure, the **hook.** The **basal body** anchors the flagellum to the cell wall and cytoplasmic membrane.

The numbers and arrangement of flagella can be used to characterize flagellated bacteria. For example, *E. coli* have flagella distributed over the entire surface, an arrangement called **peritrichous** (*peri* means "around"). Other common bacteria have a **polar flagellum,** a single flagellum at one end of the cell. Other arrangements include a tuft of flagella at one or both ends of a cell (see figures 3.18 and 3.20e).

Chemotaxis

Motile bacteria sense the presence of chemicals and respond by moving in a certain direction—a phenomenon called **chemotaxis.** If a compound is a nutrient, it may serve as an **attractant,** enticing cells to move toward it. On the other hand, if the compound is toxic, it may act as a **repellent,** causing cells to move away.

The movement of a bacterium toward an attractant is anything but direct **(figure 3.39).** When *E. coli* travels, it progresses in a given direction for a short time, then stops and tumbles for a fraction of a second, and then moves again in a relatively straight line. But after rolling around, the cell is often oriented in a completely different direction. The seemingly odd pattern of movement is due to the rotation of the flagella. When the flagella rotate counterclockwise, the bacterium is propelled in a forward movement called a **run.** The flagella of *E. coli* and other peritrichously flagellated bacteria rotate in a coordinated fashion, forming a tight propelling bundle. After a brief period, the direction of rotation of the flagella is reversed. This abrupt change causes the cell to stop and roll, called a **tumble.** Movement toward an attractant is due to runs of longer duration that occur when cells are going in the right direction; this occurs because cells tumble less frequently when they sense they are moving closer to an attractant. In contrast, they tumble more frequently when they sense they are moving closer to a repellent.

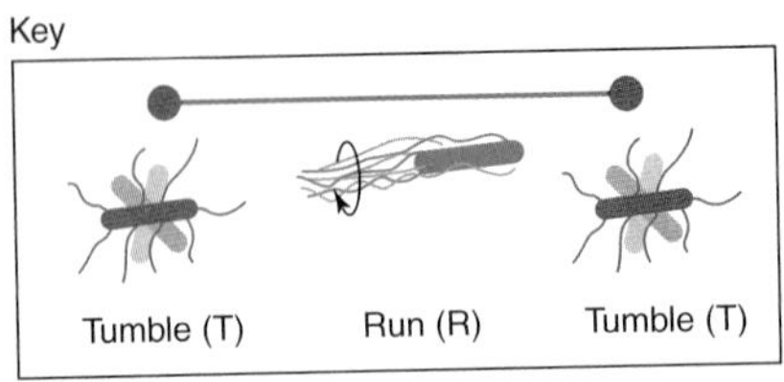

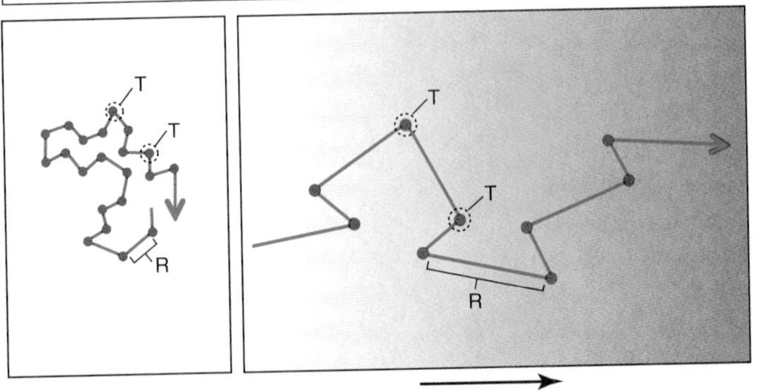

(a) No attractant or repellent

(b) Gradient of attractant concentration

FIGURE 3.39 Chemotaxis (a) A cell moves via a random series of short runs and tumbles when the attractant or repellent is uniformly distributed. **(b)** The cell tumbles less frequently resulting in longer runs when it senses that it is moving closer to the attractant.

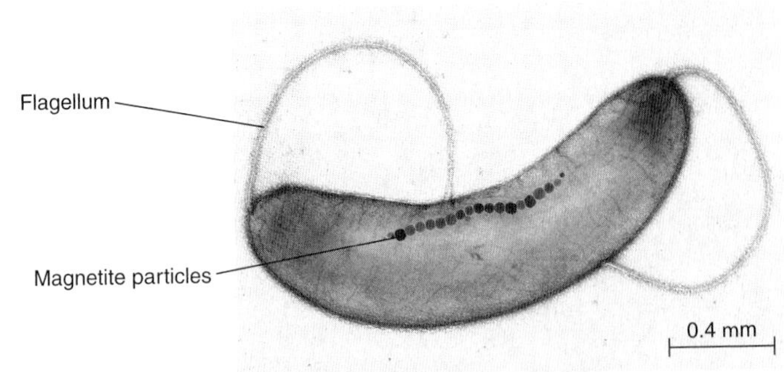

FIGURE 3.40 Magnetic Particles Within a Magnetotactic Bacterium The chain of particles of magnetite (Fe_3O_4) within the spirillum *Magnetospirillum magnetotacticum* serve to align the cell along geomagnetic lines (TEM).

In addition to reacting to chemicals, some bacteria can respond to variations in light **(phototaxis).** Other bacteria can respond to the concentration of oxygen **(aerotaxis).** Organisms that require oxygen for growth will move toward it, whereas bacteria that grow only in its absence tend to be repelled by it. Certain motile bacteria can react to the earth's magnetic field by the process of **magnetotaxis.** They actually contain a row of magnetic particles that cause the cells to line up in a north-to-south direction much as a compass does **(figure 3.40).** The magnetic forces of the earth attract the organisms so that they move downward and into sediments where the concentration of oxygen is low, which is the environment best suited for their growth. Some bacteria can also move toward certain temperatures **(thermotaxis).**

Pili

Pili are considerably shorter and thinner than flagella, but they have a similar structural theme to the filament of flagella—a string of protein subunits arranged helically to form a long cylindrical molecule with a hollow core **(figure 3.41).** The functions of pili, however, are distinctly different from those of flagella.

Many types of pili enable attachment of cells to specific surfaces; these pili are also called **fimbriae.** At the tip or along the length of the molecule is located another protein, an **adhesin,** that adheres by binding to a very specific molecule. For example, certain strains of *E. coli* that cause a severe watery diarrhea can attach to the cells that line the small intestine. They do this through specific interactions between adhesins on their pili and the intestinal cell surface. Without the ability to attach, these cells would simply be propelled through the small intestine along with the other intestinal contents. ■ enterotoxigenic *E. coli*, p. 600

Pili also appear to play a role in the movement of populations of cells on solid media. **Twitching motility,** characterized by short, jerking movements, and some types of **gliding motility,** characterized by smooth sliding motion, involve pili.

Another type of pilus, called a **sex pilus,** is used to join one bacterium to another as a prelude to a specific type of DNA transfer. This and other mechanisms of DNA transfer will be described in chapter 8. ■ DNA transfer, p. 199

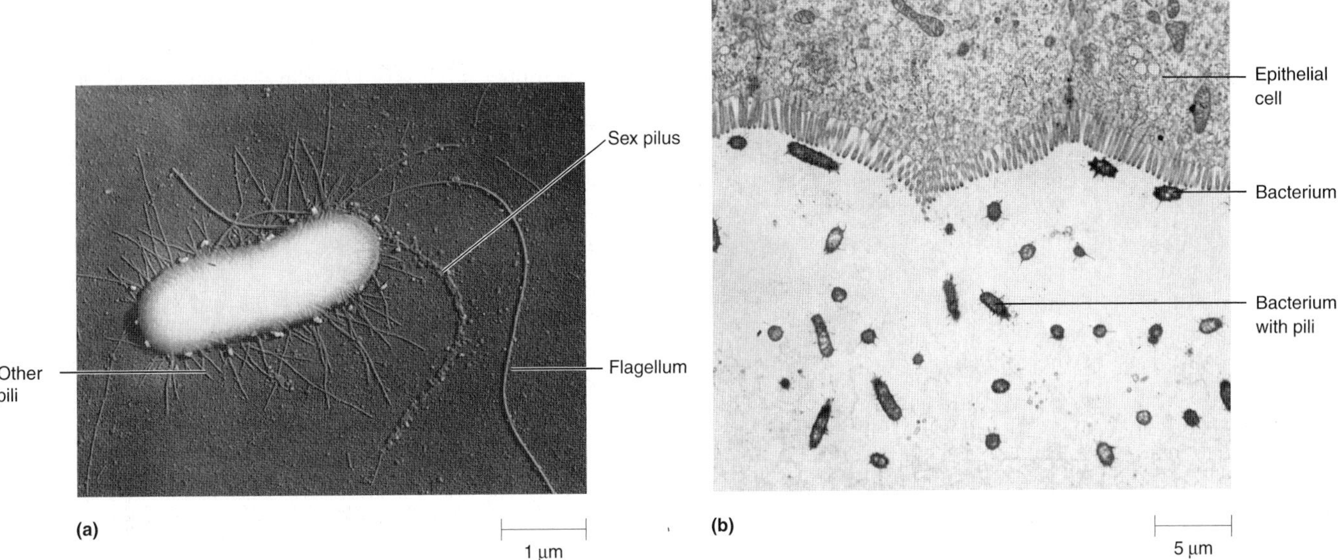

FIGURE 3.41 Pili **(a)** Pili on an *Escherichia coli* cell. The short pili (fimbriae) mediate adherence; the sex pilus is involved in DNA transfer. **(b)** *Escherichia coli* attaching to epithelial cells in the small intestine of a pig.

MICROCHECK 3.8

Flagella are the most common mechanism for bacterial motility. Chemotaxis is the directed movement of cells toward an attractant or away from a repellent. Pili provide a mechanism for attachment to specific surfaces and, in some cases, a type of motility. Sex pili join cells as a prelude to DNA transfer.

✓ What role does a series of runs and tumbles play in chemotaxis?

✓ *E. coli* cells have peritrichous flagella. What does this mean?

3.9 Internal Structures

Focus Points

- Describe the structure and function of the chromosomes, plasmids, ribosomes, storage granules, gas vesicles, and endospores.
- Describe the processes of sporulation and germination.

Prokaryotic cells have a variety of structures within the cell. Some, such as the chromosome and ribosomes, are essential for the life of all cells, whereas others confer certain selective advantages.

The Chromosome

The chromosome of prokaryotes resides as an irregular mass within the cytoplasm, forming a gel-like region called the **nucleoid.** Typically, it is a single, circular double-stranded DNA molecule that contains all the genetic information required by a cell.

Chromosomal DNA is tightly packed into about 10% of the total volume of the cell **(figure 3.42).** Rather than being a loose circle it is typically in a twisted form called **supercoiled,** which

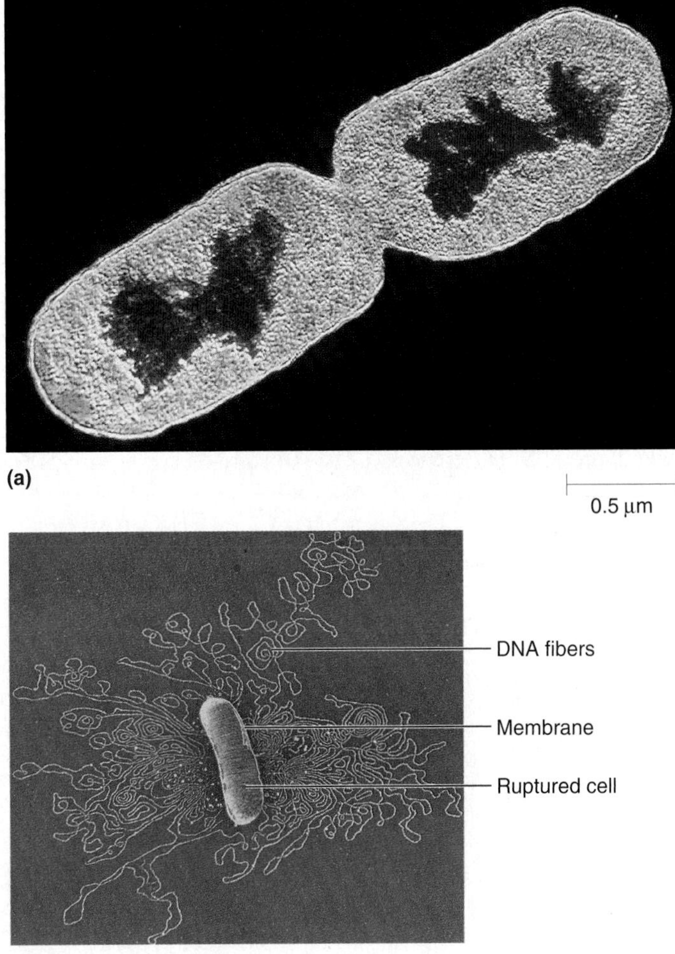

FIGURE 3.42 The Chromosome **(a)** Color-enhanced transmission electron micrograph of a thin section of *Escherichia coli*, with the DNA shown in red. **(b)** Chromosome released from a gently lysed cell of *E. coli*. Note how tightly packed the DNA must be inside the bacterium.

appears to be stabilized by the binding of positively charged proteins. Supercoiling can be visualized by cutting a rubber band and twisting one end several times before rejoining the ends. The resulting circle will twist and coil in response.

Plasmids

Most **plasmids** are circular, supercoiled, double-stranded DNA molecules. They are generally 0.1% to 10% of the size of the chromosome and carry from a few to several hundred genes. A single cell can harbor multiple types of plasmids.

A cell generally does not require the genetic information carried by a plasmid. However, the encoded characteristics may be advantageous in certain situations. For example, many plasmids code for the production of one or more enzymes that destroy certain antibiotics, enabling the organism to resist the otherwise lethal effect of these medications. Because a bacterium can sometimes transfer a copy of a plasmid to another bacterial cell, this accessory genetic information can spread, which accounts in large part for the increasing frequency of antibiotic-resistant organisms worldwide. ■ mobile gene pool, p. 205

Ribosomes

Ribosomes are intimately involved in protein synthesis, where they serve as the structures that facilitate the joining of amino acids. Each ribosome is composed of a large and a small subunit, which are made up of **ribosomal proteins** and **ribosomal RNAs.** ■ function of ribosomes, p. 171

The relative size and density of ribosomes and their subunits are expressed as a distinct unit, **S** (for Svedberg), that reflects how fast they settle when spun at very high speeds in an ultracentrifuge. The faster they move toward the bottom, the higher the S value and the greater the density. Prokaryotic ribosomes are 70S ribosomes. Note that S units are not strictly arithmetic; the 70S ribosome is composed of a 30S and a 50S subunit **(figure 3.43).**

Prokaryotic ribosomes differ from eukaryotic ribosomes, which are 80S. Differences in the structures serve as targets for certain antibiotics.

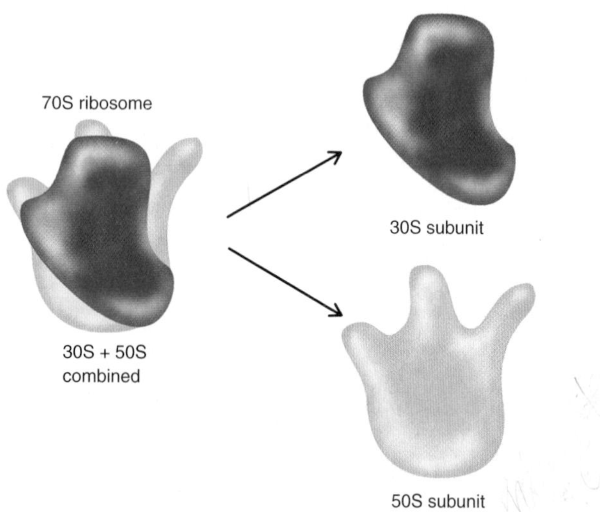

FIGURE 3.43 The Ribosome The 70S ribosome is composed of 50S and 30S subunits.

Cytoskeleton

It was once thought that bacteria lacked a **cytoskeleton,** an interior protein framework. Several bacterial proteins that have similarities to those of the eukaryotic cytoskeleton have now been characterized, and these appear to be intimately involved in cell division and controlling cell shape.

Storage Granules

Storage granules are accumulations of high molecular weight polymers synthesized from a nutrient that a cell has in relative excess. For example, if nitrogen and/or phosphorus are lacking, *E. coli* cannot multiply even if a carbon and energy source such as glucose is plentiful. Rather than waste the carbon/energy source, cells use it to produce **glycogen,** a glucose polymer. Later, when conditions are appropriate, cells degrade and use the glycogen. Other bacterial species store carbon and energy as **poly-β-hydroxybutyrate (figure 3.44).** This microbial compound is now being employed to produce a biodegradable polymer, which can be used in place of petroleum-based plastics.

Some types of granules can be readily detected by light microscopy. **Volutin** granules, a storage form of phosphate, stain red with blue dyes such as methylene blue, whereas the surrounding cellular material stains blue. Because of this, they are often called **metachromatic granules** (*meta* means "change" and *chromatic* means "color"). Recent evidence suggests that the role of these granules is more complex than originally thought. The volutin granules of some bacteria are membrane-bound and appear to resemble eukaryotic organelles that are thought to be involved with energy storage and pH balance. Regardless of the precise function of the granules, bacteria that produce them are beneficial in wastewater treatment because they scavenge phosphate, an environmental pollutant.

Gas Vesicles

Some aquatic bacteria produce **gas vesicles**—small, rigid, protein-bound compartments that provide buoyancy to the cell. Gases, but not water, flow freely into the vesicles, thereby decreasing the

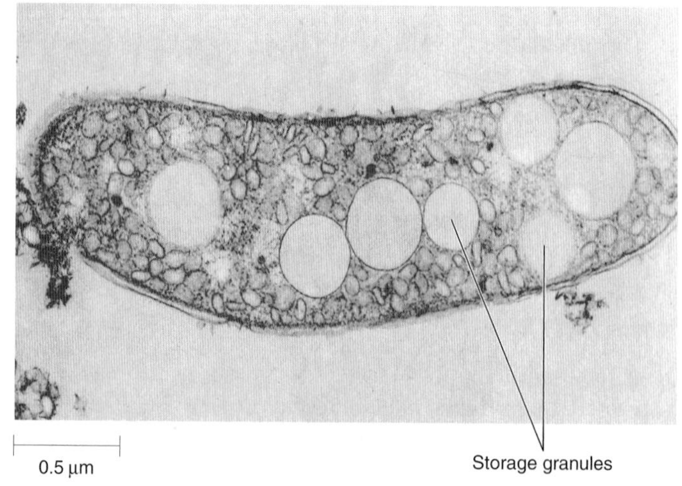

FIGURE 3.44 Storage Granules The large unstained areas in the photosynthetic bacterium *Rhodospirillum rubrum* are granules of *poly-β-hydroxybutyrate.*

density of the cell. By regulating the number of gas vesicles within the cell, an organism can float or sink to its ideal position in the water column. For example, bacteria that use sunlight as a source of energy float closer to the surface, where light is available.

Endospores

An **endospore** is a unique type of dormant cell produced by a process called **sporulation** within cells of certain bacterial species, such as members of the genera *Bacillus* and *Clostridium* **(figure 3.45).** The structures may remain dormant for perhaps 100 years, or even longer, and are extraordinarily resistant to damaging conditions including heat, desiccation, toxic chemicals, and ultraviolet irradiation. Immersion in boiling water for hours may not kill them. Endospores that survive these treatments can **germinate,** or exit the dormant stage, to become a typical multiplying cell, called a **vegetative cell.**

The consequences of these resistant dormant forms are far-reaching. Because endospores can survive so long in a variety of conditions, they can be found virtually anywhere. They are common in soil, which can make its way into environments such as laboratories and hospitals and onto products such as food, media used to cultivate microbes, and medical devices. Because the exclusion of microbes in these environments and on these products is of paramount importance, special precautions must be taken to destroy these resistant structures.

Endospores are sometimes called **spores.** However, this latter term is also used to refer to the structures produced by unrelated microorganisms such as fungi. Bacterial endospores are much more resistant to environmental conditions than are other types of spores.

Several species of endospore-formers can cause disease. For example, botulism results from the ingestion of a deadly toxin produced by vegetative cells of *Clostridium botulinum.* Other disease-causing species of endospore-formers include *Clostridium tetani,* which causes tetanus; *Clostridium perfringens,* which causes gas gangrene; and *Bacillus anthracis,* which causes anthrax. ■ **botulism, p. 657** ■ **tetanus, p. 566** ■ **gas gangrene, p. 568** ■ **anthrax, p. 512**

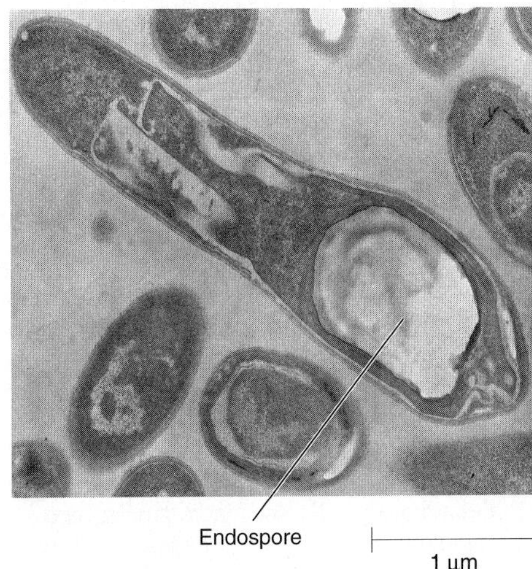
Endospore
1 µm

FIGURE 3.45 Endospores Endospore inside a vegetative cell of a *Clostridium* species (TEM).

Sporulation

Endospore formation is a complex, highly ordered sequence of changes that initiates when sporeforming bacteria are grown in low amounts of carbon or nitrogen **(figure 3.46).** Apparently the cells sense starvation conditions and therefore begin the 8-hour process that prepares them for rough times ahead.

After vegetative growth stops, DNA is duplicated and then a septum forms between the two chromosomes, dividing the cell asymmetrically. The larger compartment then engulfs the smaller compartment, forming a **forespore** within a **mother cell.** These two portions take on different roles in synthesizing the components that will make up the endospore. The forespore, which is enclosed by two membranes, will ultimately become the **core** of the endospore. Peptidoglycan-containing material is laid down between these two membranes, forming the **core wall** and the **cortex.** Meanwhile, the mother cell makes proteins that will form the **spore coat.** Ultimately, the mother cell is degraded and the endospore released.

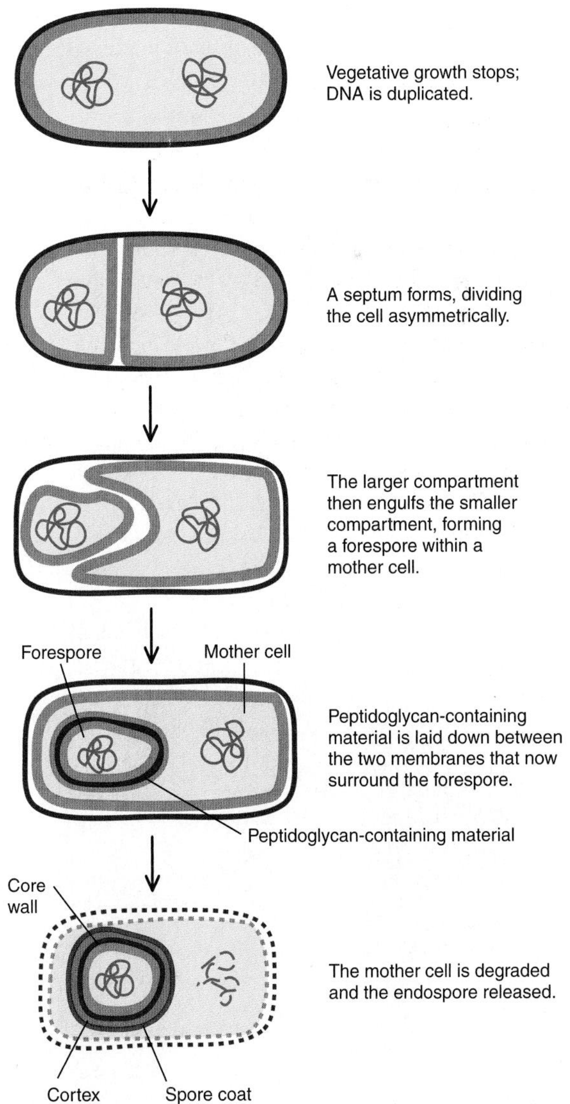
Vegetative growth stops; DNA is duplicated.
A septum forms, dividing the cell asymmetrically.
The larger compartment then engulfs the smaller compartment, forming a forespore within a mother cell.
Forespore
Mother cell
Peptidoglycan-containing material is laid down between the two membranes that now surround the forespore.
Peptidoglycan-containing material
Core wall
The mother cell is degraded and the endospore released.
Cortex
Spore coat

FIGURE 3.46 The Process of Sporulation

The layers of the endospore shield it from damage. The spore coat is thought to function as a sieve, excluding molecules such as lysozyme. The cortex helps maintain the core in a dehydrated state, protecting it from the effects of heat. In addition, the core has small, acid-soluble proteins that bind the DNA, thereby protecting it from damage. The core is rich in an unusual compound called dipicolinic acid, which combines with calcium ions. This complex appears to also play an important role in spore resistance.

Germination

Germination can be triggered by a brief exposure to heat or certain chemicals. Following such exposure, the endospore takes on water and swells. The spore coat and cortex then crack open, and a vegetative cell grows out. Since one vegetative cell gives rise to one endospore, sporulation is not a means of cell reproduction.

MICROCHECK 3.9

The prokaryotic chromosome is usually a circular, double-stranded DNA molecule that contains all of the genetic information required by a cell. Plasmids generally only encode information that is advantageous to a cell in certain conditions. Ribosomes are the structures that facilitate the joining of amino acids to form a protein. The cytoskeleton is an interior framework involved in cell division and controlling cell shape.Storage granules are polymers synthesized from a nutrient a cell has in relative excess. Gas vesicles provide buoyancy to a cell. An endospore is a highly resistant dormant stage produced by certain bacterial species.

✓ Explain how glycogen granules benefit a cell.

✓ Explain why endospores are an important consideration for the canning industry.

✓ Why are the processes of sporulation and germination not considered a mechanism of multiplication?

THE EUKARYOTIC CELL

Eukaryotic cells are generally much larger than prokaryotic cells, and their internal structures are far more complex **(figure 3.47).** One of their distinguishing characteristics is the abundance of membrane-enclosed compartments or **organelles.** The most important of these is the nucleus, which contains the DNA. The organelles, which can take up half the total cell volume, enable the cell to perform complex functions in separated regions. For example, degradative enzymes contained within an organelle digest food and other material without posing a threat to the integrity of the cell itself.

Each organelle contains a variety of proteins and other molecules, many of which are synthesized at other locations. To deliver these to the **lumen,** or interior, of another organelle, an elaborate transportation system is required. To transfer material,

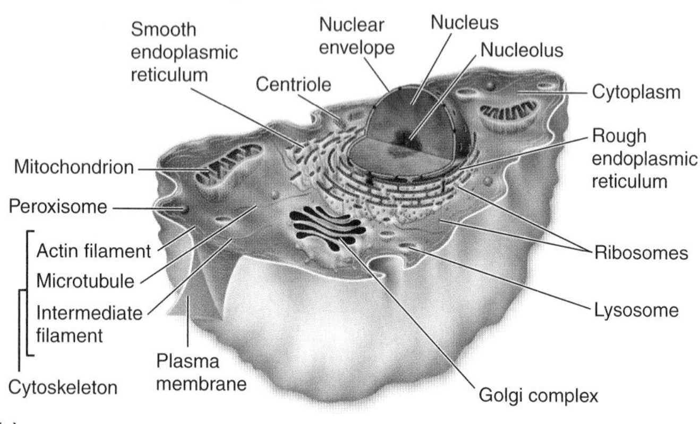

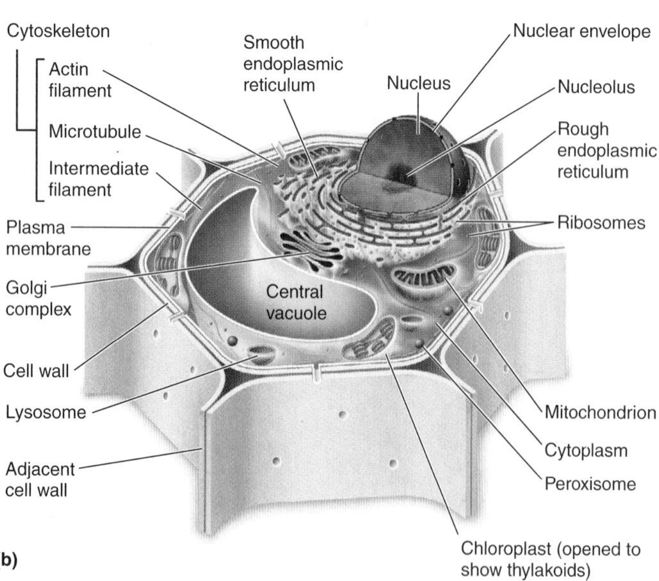

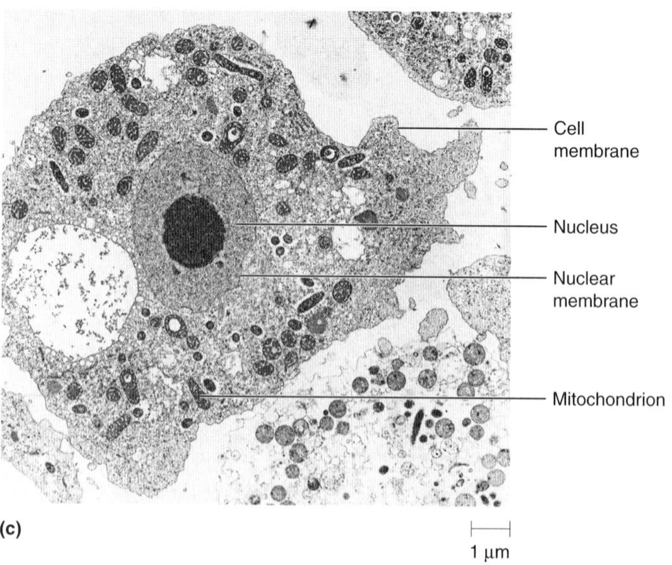

FIGURE 3.47 Eukaryotic Cells (a) Diagrammatic representation of an animal cell. **(b)** Diagrammatic representation of a plant cell. **(c)** Micrograph of an animal cell shows several membrane-bound structures including mitochondria and a nucleus.

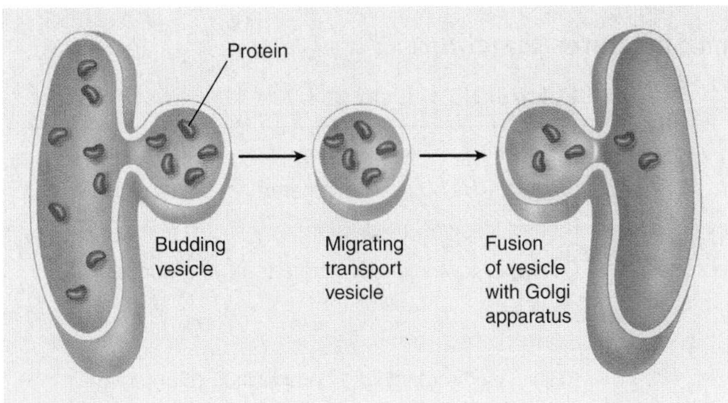

FIGURE 3.48 Vesicle Formation and Fusion A vesicle forms when a section of an organelle buds off. The mobile vesicle can then move to other parts of the cell, ultimately fusing with the membrane of another organelle.

a section of an organelle will bud or pinch off, forming a small membrane-enclosed **vesicle (figure 3.48).** This mobile vesicle, containing a sampling of the contents of the organelle, can move to other parts of the cell. When that vesicle encounters the lipid membrane of another organelle, the two membranes fuse to become one contiguous unit. By doing so, the vesicle introduces its contents to the lumen of the organelle. A similar process is used to export molecules synthesized within an organelle to the external environment.

As a group, eukaryotic cells are highly variable. For example, protozoa, which are single-celled organisms, must function exclusively as self-contained units that seek and ingest food. These cells must be mobile and flexible to take in food particles. Consequently, they lack cell walls that would otherwise provide rigidity. Animal cells also lack a cell wall, because they too must be flexible to accommodate movement. Fungi, on the other hand, are stationary and benefit from the protection provided by a rigid cell wall. Compounds that make up fungal cell walls include glucan and mannan, which are polysaccharides, and chitin, a polymer of *N*-acetylglucosamine that is also found in crustaceans and insects. Plant cells, which are also stationary, have cell walls composed of cellulose, a polymer of glucose. ■ polysaccharides, pp. 30, 32

The individual cells of a multicellular organism can be distinctly different from one another. Mammals, for example, are composed of several hundred different types of cells, and it is obvious that a liver cell is quite different from a bone cell. Cells of plants and animals function in cooperative associations called **tissues.** The tissues in your body include muscle, connective, nerve, epithelial, blood, and lymphoid. Each of these provides a unique function. Combinations of various tissues function together to make up larger units, **organs.** These include skin, heart, and liver. Organs and the tissues that constitute them will be covered in more detail in chapters 22 through 29 on infectious diseases.

A comprehensive coverage of all aspects of eukaryotic cells is beyond the scope of this textbook. Instead, this section will focus on key characteristics, particularly those that directly affect the interaction of a microbe with a human host. These characteristics are summarized in **table 3.6.** A comparison of

TABLE 3.6 A Summary of Eukaryotic Cell Structures

	Characteristics
Plasma Membrane	Asymmetric lipid bilayer embedded with proteins. Selective permeability, conduit to external environment.
Internal Protein Structures	
Cilia	Appear to project out of a cell. Beat in synchrony to provide movement. Composed of microtubules in a 9 + 2 arrangement.
Cytoskeleton	Dynamic filamentous network that provides structure to the cell.
Flagella	Appear to project out of a cell. Propel or push the cell with a whiplike or thrashing motion. Composed of microtubules in a 9 + 2 arrangement.
Ribosomes	Two subunits, 60S and 40S, join to form the 80S ribosome.
Membrane-Bound Organelles	
Chloroplasts	Site of photosynthesis; the organelle harvests the energy of sunlight to generate ATP, which is then used to convert CO_2 to carbohydrates. Within the stroma are chlorophyll-containing, disclike thylakoids. The membranes of these contain the components of the electron transport chain and the proteins that use proton motive force to synthesize ATP.
Endoplasmic reticulum	Site of synthesis of macromolecules destined for other organelles or the external environment.
Rough	Attached ribosomes thread proteins they are synthesizing into the lumen of the organelle.
Smooth	Site of lipid synthesis and degradation, and Ca^{2+} storage.
Golgi apparatus	Site where macromolecules synthesized in the endoplasmic reticulum are modified before they are transported in vesicles to other destinations.
Lysosome	Digestion of macromolecules.
Mitochondria	Harvest the energy released during the degradation of organic compounds to generate ATP. Within the highly folded inner membrane are the components of the electron transport chain and the proteins that use proton motive force to synthesize ATP.
Nucleus	Contains the DNA.
Peroxisome	Oxidation of lipids and toxic chemicals occurs.

TABLE 3.7 Comparison of Prokaryotic and Eukaryotic Cell Structures/Functions

	Prokaryotic	Eukaryotic
General Characteristics		
Size	Generally 0.3– 2 μm in diameter.	Generally 5– 50 μm in diameter.
Cell Division	Chromosome replication followed by binary fission.	Mitosis followed by division.
Chromosome location	Located in the nucleoid, which is not membrane-bound.	Contained within the membrane-bound nucleus.
Structures		
Cell membrane	Relatively symmetric with respect to the lipid content of the bilayers.	Highly asymmetric; lipid composition of outer layer differs significantly from that of inner layer.
Cell wall	Composed of peptidoglycan (*Bacteria*); Gram-negative bacteria have an outer membrane as well.	Absent in animal cells; composition in other cell types may include: chitin, glucans and mannans (fungi), and cellulose (plants).
Chromosome	Single, circular DNA molecule is typical.	Multiple, linear DNA molecules. DNA is wrapped around histones.
Flagella	Composed of protein subunits.	Made up of a 9 + 2 arrangement of microtubules.
Membrane-bound organelles	Absent.	Present; includes the nucleus, mitochondria, chloroplasts (only in plant cells), endoplasmic reticulum, Golgi apparatus, lysosomes, and peroxisomes.
Nucleus	Absent; DNA resides as an irregular mass forming the nucleoid region.	Present.
Ribosomes	70S ribosomes, which are made up of 50S and 30S subunits.	80S ribosomes, which are made up of 60S and 40S subunits. Mitochondria and chloroplasts have 70S ribosomes.
Functions		
Degradation of extracellular substances	Enzymes are secreted that degrade macromolecules outside of the cell. The resulting small molecules are transported into the cell.	Macromolecules are brought into the cell by endocytosis. Lysosomes carry digestive enzymes.
Motility	Generally involves flagella, which are composed of protein subunits. Flagella rotate like propellers, using proton motive force for energy.	Involves cilia and flagella, which are made up of a 9 + 2 arrangement of microtubules, Cilia move in synchrony; flagella propel a cell with a whiplike motion or thrash back and forth to pull a cell forward. Both use ATP for energy.
Protein secretion	A characteristic signal sequence marks proteins for secretion by the general secretory pathway.	Secreted proteins are moved to the lumen of the rough endoplasmic reticulum as they are being synthesized. From there, they are transported to the Golgi apparatus for processing and packaging.
Strength and rigidity	Peptidoglycan-containing cell wall (*Bacteria*).	Cytoskeleton composed of microtubules, intermediate filaments, and microfilaments. Some have a cell wall; some have sterols in the membrane.
Transport	Primarily active transport. Group translocation.	Facilitated diffusion and active transport. Ion channels.

functional aspects of prokaryotic and eukaryotic cells is presented in **table 3.7.**

3.10 The Plasma Membrane

Focus Point

- Describe the structure of the eukaryotic plasma membrane, comparing and contrasting it with the bacterial cytoplasmic membrane.

All eukaryotic cells have a cytoplasmic membrane, or **plasma membrane,** which is similar in chemical structure and function to that of prokaryotic cells. It is a typical phospholipid bilayer embedded with proteins. The lipid and protein composition of the layer that faces the cytoplasm, however, differs significantly from that facing the outside of the cell. The same is true for membranes that surround the organelles. The layer facing the lumen of the organelle is similar to its counterpart facing the cell exterior. This lack of symmetry reflects the important role these membranes play in the complex processes occuring within the eukaryotic cell.

The proteins in the lipid bilayer perform a variety of functions. Some are involved in transport and others are attached to internal structures, helping to maintain cell integrity. Those in the outer layer often function as receptors. Typically, these receptors are **glycoproteins,** proteins that have various sugars attached. A given receptor binds a specific molecule, which is referred to as its **ligand.** These receptor-ligand interactions are extremely important in multicellular organisms because they allow cells to communicate with each other—a process called **signaling.** For example, in our bodies, some cells secrete a specific protein when they encounter certain compounds perceived as dangerous. Other cells of the immune system have receptors for that protein on their surface. When the protein binds its receptor, those cells recognize the signal as a call for help and respond accordingly. This cell-to-cell communication enables a multicellular organism to function as a cohesive unit. ■ glycoproteins, p. 30

The membranes of many eukaryotic cells contain sterols, which provide strength to the otherwise fluid structure. Recall that *Mycoplasma,* a group of bacteria lacking a cell wall, also have sterols in their membrances. The sterol found in animal cell membranes is **cholesterol,** whereas fungal membranes contain **ergosterol.** This difference is exploited by antifungal medications that act by interfering with ergosterol synthesis or function. ■ antifungal medications, p. 488 ■ oligosaccharides, p. 30

Within the fluid lipid layer of the plasma membrane are cholesterol-rich regions called **lipid rafts.** The role of these regions is still being elucidated, but they appear to be important in allowing the cell to detect and respond to signals in the external environment. From a microbiologist's perspective, they are also important because many viruses appear to use these regions when they exit a cell.

The plasma membrane plays no role in ATP synthesis; instead, that task is performed by mitochondria (discussed later). Although proton motive force is not generated across the membrane, an electrochemical gradient is maintained by energy-consuming mechanisms that expel either sodium ions or protons. ■ electrochemical gradient, p. 57

MICROCHECK 3.10

The plasma membrane is an asymmetric lipid bilayer embedded with proteins. Specific receptors on the outer layer mediate cell-to-cell signaling. Sterols provide strength to the fluid membrane.

- ✓ What is the medical significance of ergosterol in the fungal membranes?
- ✓ Describe why signaling is important in animal cells.
- ✓ How could one argue that the lumen of an organelle is "outside" of a cell?

3.11

Transfer of Molecules Across the Plasma Membrane

Focus Points

- Compare and contrast the roles of channels and carriers in transport.
- Describe the processes of endocytosis and exocytosis.
- Describe the role of the endoplasmic reticulum in secretion.

Nutrients, signaling molecules, and waste products pass through the plasma membrane. Some of these enter and exit the cell via transport proteins; others are taken in through a process called **endocytosis. Exocytosis,** the reverse of endocytosis, can be used to expel material.

Transport Proteins

The transport proteins of eukaryotic cells function as either carriers or channels. **Carriers** are analogous to proteins in prokaryotic cells that mediate facilitated diffusion and active transport. **Channels** are pores in the membrane. These pores are so small that only specific ions can diffuse through. They allow ions to move with the concentration gradient; they do not create such a gradient. To control ion passage, the channel has a gate, which can be either opened or closed, depending on environmental conditions.

Cells of multicellular organisms can often take up nutrients by facilitated diffusion, because the nutrient concentration of surrounding environments can be controlled. For example, glucose levels in the blood are maintained at a concentration higher than in most tissues. Consequently, animal cells generally do not need to expend energy transporting glucose.

The active transport mechanisms of eukaryotic cells are structurally analogous to those of prokaryotic cells. Some are of medical interest because they can eject drugs from the cell. For example, some human cancer cells use an ABC transporter that ejects therapeutic drugs intended to kill those cells.

Endocytosis and Exocytosis

Endocytosis is the process by which eukaryotic cells take up material from the surrounding environment (**figure 3.49).** The type of endocytosis common to most animal cells is **pinocytosis.** In this process a cell internalizes and pinches off small pieces of its own membrane, bringing along a small volume of liquid and any material attached to the membrane. This **endocytic vesicle** becomes a

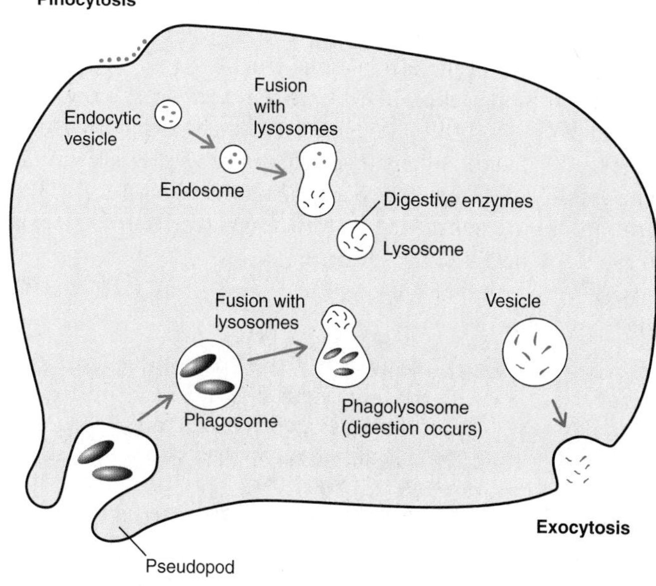

FIGURE 3.49 Endocytosis and Exocytosis Endocytosis includes pinocytosis and phagocytosis.

membrane-enclosed, low-pH compartment called an **endosome.** This then fuses with digestive organelles called lysosomes. The characteristics of lysosomes will be discussed shortly. ■ lysosome, p. 79

Animal cells often take up material by **receptor-mediated endocytosis,** which can be viewed as a variation of pinocytosis that allows cells to internalize extracellular ligands that bind to receptors on the cell's surface. When the receptors bind their ligand, the region is internalized to form an endocytic vesicle that contains the receptors along with their bound ligands. The low pH of the endosome frees the ligands from the receptors, which are often recycled. The endosome then fuses with lysosomes. Many viruses, including those that cause influenza and rabies, exploit receptor-mediated endocytosis to enter animal cells. By binding to a specific receptor, they too are taken up. ■ influenza, p. 519 ■ rabies, p. 663

Protozoa and phagocytes, both of which ingest bacteria and large debris, use a specific type of endocytosis called **phagocytosis.** Phagocytes are important cells of the body's defense system. The cells send out armlike extensions, **pseudopods,** which surround and enclose extracellular material, including bacteria. This action envelops the material, bringing it into the cell in an enclosed compartment called a **phagosome.** These ultimately fuse with lysosomes to form a **phagolysosome.** Phagocytes have a greater abundance of lysosomes than do other animal cells, which reflects their specialized function. In addition, their lysosomes contain a wider array of powerful digestive enzymes. Thus, most microbes are readily dispatched within the phagolysosome. Those that resist the killing effects are able to cause disease. ■ phagocytes, p. 358 ■ survival within a phagocyte, p. 403

The process of exocytosis is the reverse of endocytosis. Membrane-bound vesicles inside the cell fuse with the plasma membrane and release their contents into the external medium. The processes of endocytosis and exocytosis result in the exchange of material between the inside and outside of the cell.

Secretion

Proteins destined for a non-cytoplasmic region, either outside of the cell or the lumen of an organelle, must be moved across a membrane. Ribosomes synthesizing a protein that will be secreted attach to the membrane of the endoplasmic reticulum (ER). The characteristics of this organelle will be described shortly. As the protein is being made, it is threaded through the membrane and into the lumen of the ER. The lumen of any organelle can be viewed as equivalent to an extracellular space. Once a protein or any substance is there, it can readily be transported by vesicles to the lumen of another organelle, or to the exterior of the cell. ■ endoplasmic reticulum, p. 77

MICROCHECK 3.11

Gated channels allow specific ions to pass across the membrane. Carriers facilitate the passage of molecules across the membrane and often use energy. Pinocytosis allows cells to internalize small molecules. Protozoa and phagocytes internalize bacteria and debris by phagocytosis. Exocytosis is used to expel material. Secreted proteins are moved across the membrane of the endoplasmic reticulum as they are being made.

✓ How does a cell bring in ligands?

✓ How is the formation of an endocytic vesicle different from that of a phagosome?

✓ How might a bacterium resist the killing effects of a phagolysosome?

3.12 Protein Structures Within the Cell

Focus Point

- Describe the structure and function of the eukaryotic cytoskeleton, flagella, and cilia.

Eukaryotic cells have a variety of important protein structures within the cell. These include ribosomes, the cytoskeleton, flagella, and cilia.

Ribosomes

The eukaryotic ribosome is 80S, which is made up of a 60S and a 40S subunit. Recall that the prokaryotic ribosomes are 70S. Like prokaryotic ribosomes, eukaryotic ribosomes are composed of ribosomal RNA and protein.

Cytoskeleton

The threadlike proteins that make up the cytoskeleton continually reconstruct to adapt to the cell's constantly changing needs. The network is composed of three elements: microtubules, actin filaments, and intermediate filaments **(figure 3.50).**

Microtubules, the thickest of the cytoskeleton structures, are long hollow cylinders composed of protein subunits called **tubulin.** Microtubules form the mitotic spindles, the machinery that partitions chromosomes between two cells in the process of cell division. Without mitotic spindles, cells could not reproduce. Microtubules also are the main structures that make up the cilia and flagella, the mechanisms of locomotion in certain eukaryotic cells. In addition, microtubules also function as the framework along which organelles and vesicles move within a cell. Organelles called centrioles are involved in the assembly of microtubules. The antifungal drug griseofulvin is thought to interfere with the structural integrity of the microtubules of some fungi. ■ mitosis, p. 284

Actin filaments enable the cell cytoplasm to move. They are composed of a polymer of **actin,** which can rapidly assemble and subsequently disassemble, causing motion. For example, pseudopod formation relies on actin polymerization in one part of the cell and depolymerization in another. Some intracellular pathogens exploit the process and trigger a rapid polymerization of actin, propelling them within that cell. This can move the pathogens with enough force to be ejected into an adjacent cell.

Intermediate filaments function like ropes, strengthening the cell mechanically. They enable cells to resist physical stresses.

Flagella and Cilia

Flagella and **cilia** are flexible structures that appear to project out of a cell yet are covered by an extension of the plasma membrane **(figure 3.51).** Both are composed of long microtubules grouped in what is called a 9 + 2 arrangement: nine pairs of microtubules surrounding two individual ones. They originate from a basal body within the cell; the basal body has a slightly different arrangement of microtubules.

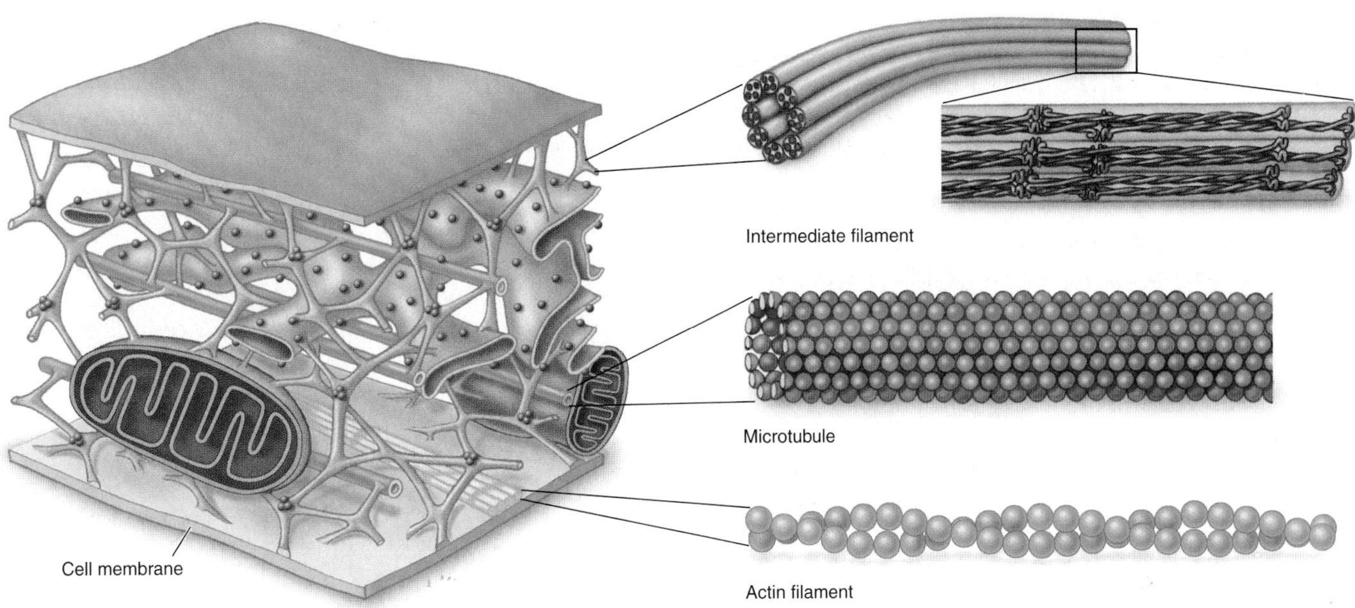

FIGURE 3.50 Cytoskeleton Diagrammatic representation of the dynamic filamentous network that provides structure to the cell; the cytoskeleton is composed of three elements—microtubules, actin filaments, and intermediate filaments.

Although eukaryotic flagella provide cells with motility, they are structurally very different from their prokaryotic counterparts. Using ATP as a source of energy, they either propel the cell with a whiplike motion or thrash back and forth to pull the cell forward.

Cilia are shorter than flagella, often covering a cell and moving in synchrony (see figure 1.8). This motion can move a cell forward in an aqueous solution, or propel surrounding material along a stationary cell. For example, epithelial cells that line the respiratory tract have cilia that beat together in a directed fashion. This moves the mucus film that covers those cells, directing it upward toward the mouth, where it can be swallowed. This action removes microorganisms that have been inhaled before they can enter the lungs.

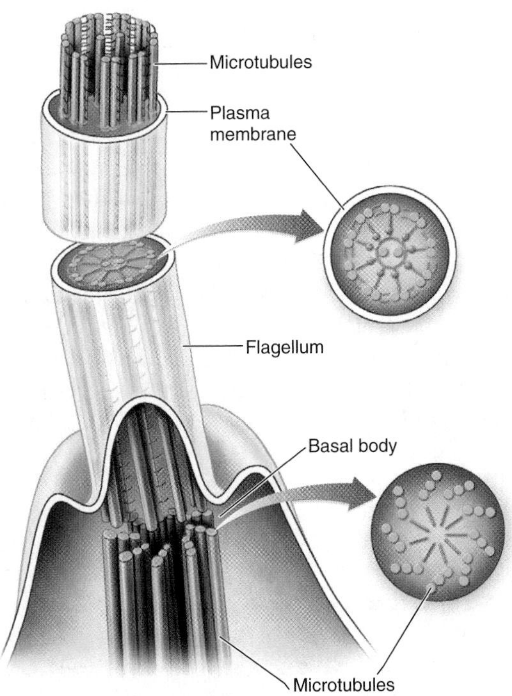

FIGURE 3.51 Flagella Flexible structures involved in movement.

MICROCHECK 3.12

The 80S eukaryotic ribosome is composed of 60S and 40S subunits. The cytoskeleton is a dynamic filamentous network that provides structure to the cell; it is composed of microtubules, actin filaments, and intermediate filaments. Flagella function in motility. Cilia either propel a cell or move material along a stationary cell.

- ✓ Explain how actin filaments are related to phagocytosis.
- ✓ Explain what is meant by the 9 + 2 structure of cilia and flagella.

3.13 Membrane-Bound Organelles

Focus Point

- Describe the function of the nucleus, mitochondria, chloroplasts, endoplasmic reticulum, Golgi apparatus, lysosomes, and peroxisomes.

The presence of membrane-bound organelles is an important feature that sets eukaryotic cells apart from their prokaryotic counterparts.

The Nucleus

The predominant distinguishing feature of the eukaryotic cell is the **nucleus,** which contains the DNA. The boundary of this structure is the **nuclear envelope,** which is composed of two lipid bilayer

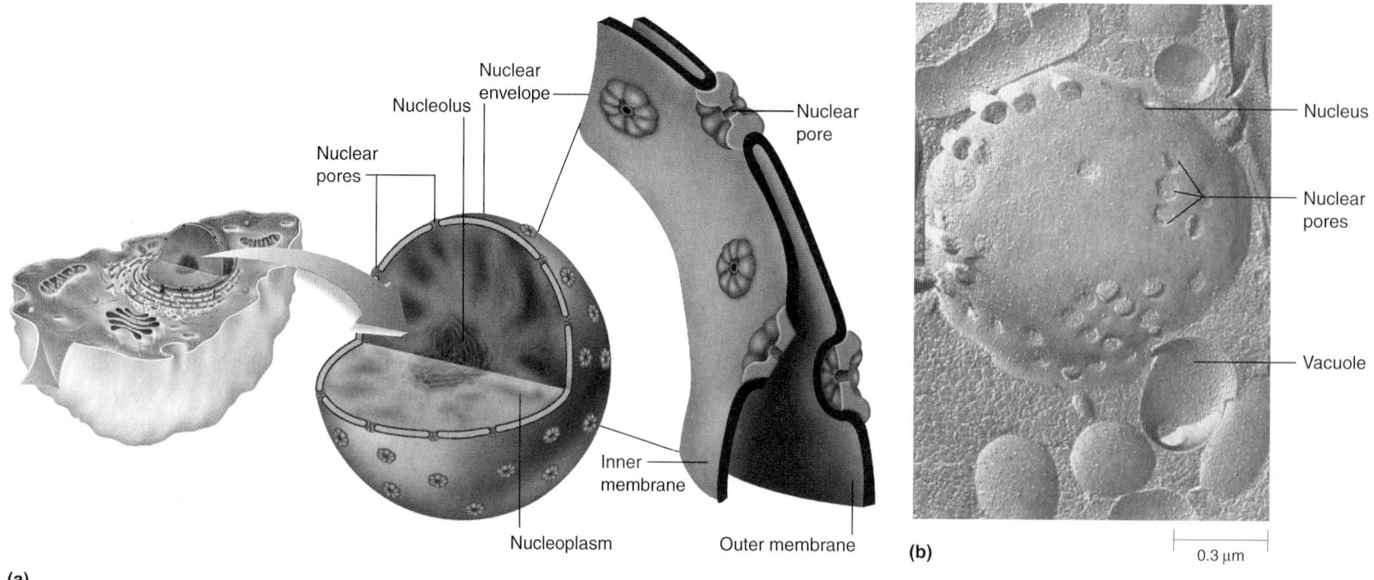

FIGURE 3.52 Nucleus Organelle that contains the DNA. **(a)** Diagrammatic representation. **(b)** Electron micrograph of a yeast cell (*Geotrichum candidium*) by freeze-fracture technique.

membranes: the inner membrane and the outer membrane. Spanning the envelope are complex protein structures that form nuclear pores, allowing large molecules such as ribosomal subunits and proteins to be transported into and out of the nucleus **(figure 3.52).** The **nucleolus** is a region within the nucleus where ribosomal RNAs are synthesized.

The nucleus contains multiple chromosomes, each one encoding different genetic information. Unlike the situation in most prokaryotic cells, double-stranded chromosomal DNA is linear. To add structure and order to the long DNA molecule, it is packed by winding it around positively charged proteins called **histones.** These bind tightly to the negatively charged DNA molecule. One packing unit, called a nucleosome, consists of a complex of histones around which the linear DNA wraps twice. The complex of DNA and proteins that together form the chromosomes is called **chromatin.**

Events that take place in the nucleus during cell division distinguish eukaryotes from prokaryotes. In eukaryotic cells, after DNA is replicated, chromosomes go through a nuclear division process called **mitosis,** which ensures the daughter cells receive the same number of chromosomes as the original parent. Through mitosis, a cell that is **diploid,** or has two copies of each chromosome, will generate two diploid daughter cells. A different process, **meiosis,** generates haploid daughter cells, which each have a single copy of each chromosome.

Mitochondria

Mitochondria function as ATP-generating powerhouses, and are found in nearly all eukaryotic cells. They are highly complex structures about the size of a bacterial cell, and are bounded by two lipid bilayer membranes **(figure 3.53).** These are referred to as the outer and inner membranes. The outer membrane is smooth, but the inner membrane is highly folded, forming invaginations called **cristae.** These folds increase the surface area of the membrane, maximizing the ATP-generating capabilities of the organelle (the processes will be discussed in chapter 6).

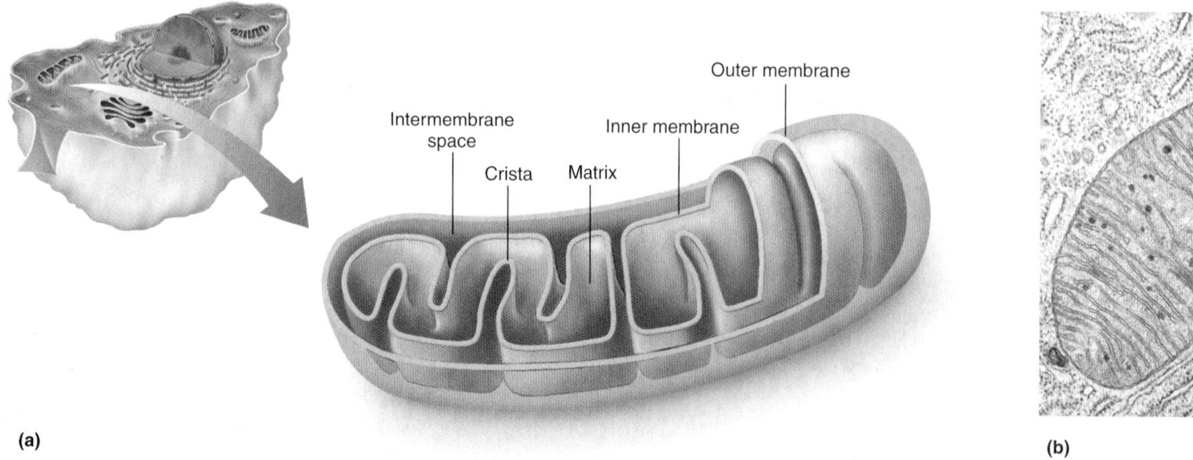

FIGURE 3.53 Mitochondria These harvest the energy released during the degradation of organic compounds to synthesize ATP. **(a)** Diagrammatic representation. **(b)** Electron micrograph.

PERSPECTIVE 3.1

The Origins of Mitochondria and Chloroplasts

Mitochondria and chloroplasts bear such a striking similarity to prokaryotic cells that it is no wonder scientists speculated for many decades that these organelles evolved from bacteria. The **endosymbiont theory** states that the ancestors of mitochondria and chloroplasts were bacteria residing within other cells in a mutually beneficial partnership. The intracellular bacterium in such a partnership is called an **endosymbiont.** As time went on each partner became indispensable to the other, and the endosymbiont eventually lost key features such as a cell wall and the ability to replicate independently.

Several early observations have supported the endosymbiont theory. Mitochondria and chloroplasts, unlike other eukaryotic organelles, both carry some of the genetic information necessary for their function. These include genes for some of the ribosomal proteins and ribosomal RNAs that make up their 70S ribosomes. These ribosomes contrast with the typical 80S ribosomes that characterize eukaryotic cells and, in fact, are equivalent to the prokaryotic 70S ribosomes. Interestingly, nuclear DNA encodes some of the components that make up these ribosomes. Another characteristic that supports the theory that mitochondria and chloroplasts were once intracellular bacteria is the double membrane that surrounds these organelles. Present-day endosymbionts retain their cytoplasmic membranes and live within membrane-bound compartments in their eukaryotic host cell.

Evidence in favor of the endosymbiont theory continues to accumulate. Recent technology enables scientists to readily determine the precise order or sequence, of nucleotides that make up DNA. This allows comparison of the nucleotide sequences of organelle DNA with genomes of different bacteria. It has become apparent that some mitochondrial DNA sequences bear a striking resemblance to DNA sequences of members of a group of obligate intracellular parasites, the rickettsias. These are probably relatives of modern-day mitochondria.

A tremendous effort is now under way to determine the nucleotide sequence of mitochondria from a wide variety of eukaryotes, including plants, animals, and protists. While the size of mitochondrial DNA varies a great deal among these different eukaryotic organisms, common sequence themes are emerging. Today, researchers are no longer discussing "if" but "when" these organelles evolved from intracellular prokaryotes.

Enclosed by the inner membrane is the **matrix,** which contains DNA, ribosomes, and other molecules necessary for protein synthesis. Notably, the ribosomes are 70S rather than the 80S ribosome found in the cytoplasm of eukaryotic cells. These observations, along with the fact that mitochondria elongate and divide in a fashion similar to that of bacteria, were among the first pieces of evidence that led scientists to conclude that mitochondria evolved from bacterial cells (see **Perspective 3.1**).

Chloroplasts

Chloroplasts, found exclusively in plants and algae, are the site of photosynthesis in eukaryotic cells. They harvest the energy of sunlight to generate ATP, which is then used to convert CO_2 to organic compounds like sugar and starch. Like mitochondria, chloroplasts are bounded by two membranes **(figure 3.54).** Within the chloroplast's **stroma,** the region analogous to the mitochondrial matrix, are membrane-bound, disclike structures called **thylakoids.** Chlorophyll and other pigments that capture radiant energy are embedded in the thylakoid membranes.

Like mitochondria, chloroplasts appear to have evolved from bacterial cells (see Perspective 3.1). They contain DNA and 70S ribosomes, elongate and divide, and have photosynthetic mechanisms similar to a group of bacteria called cyanobacteria.

Endoplasmic Reticulum (ER)

The **endoplasmic reticulum (ER)** is a complex, three-dimensional internal membrane system of flattened sheets, sacs, and tubes

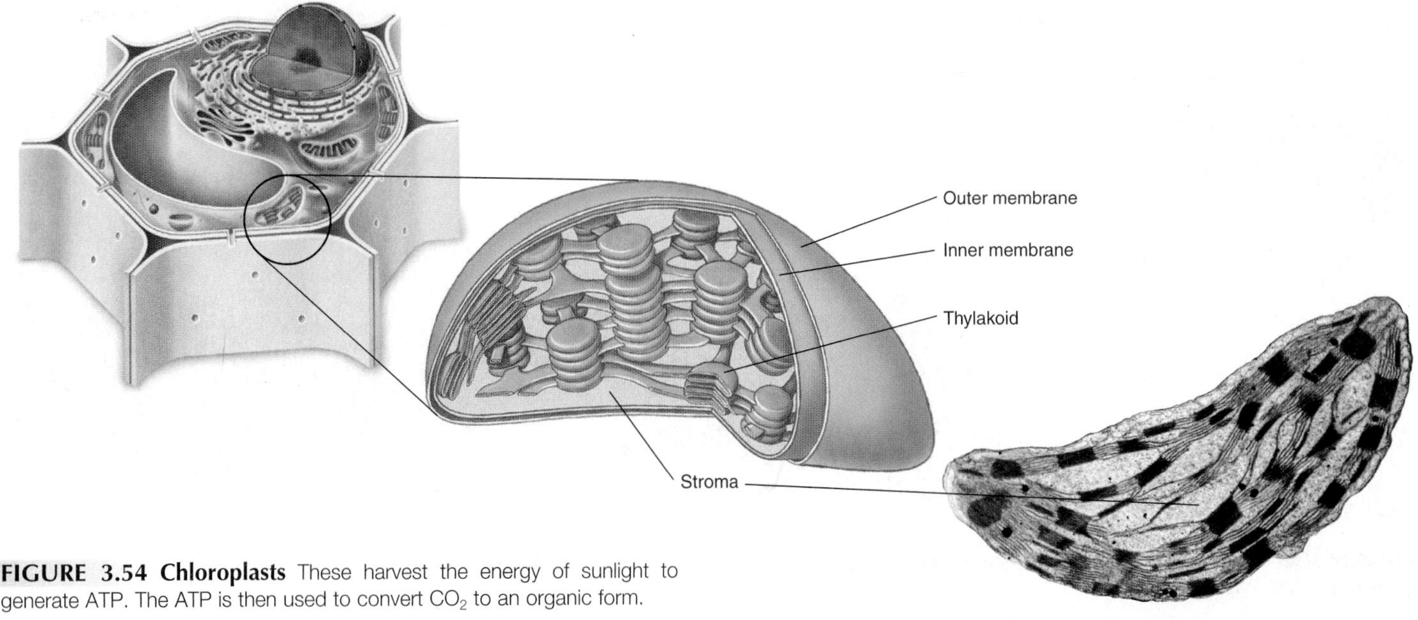

FIGURE 3.54 Chloroplasts These harvest the energy of sunlight to generate ATP. The ATP is then used to convert CO_2 to an organic form.

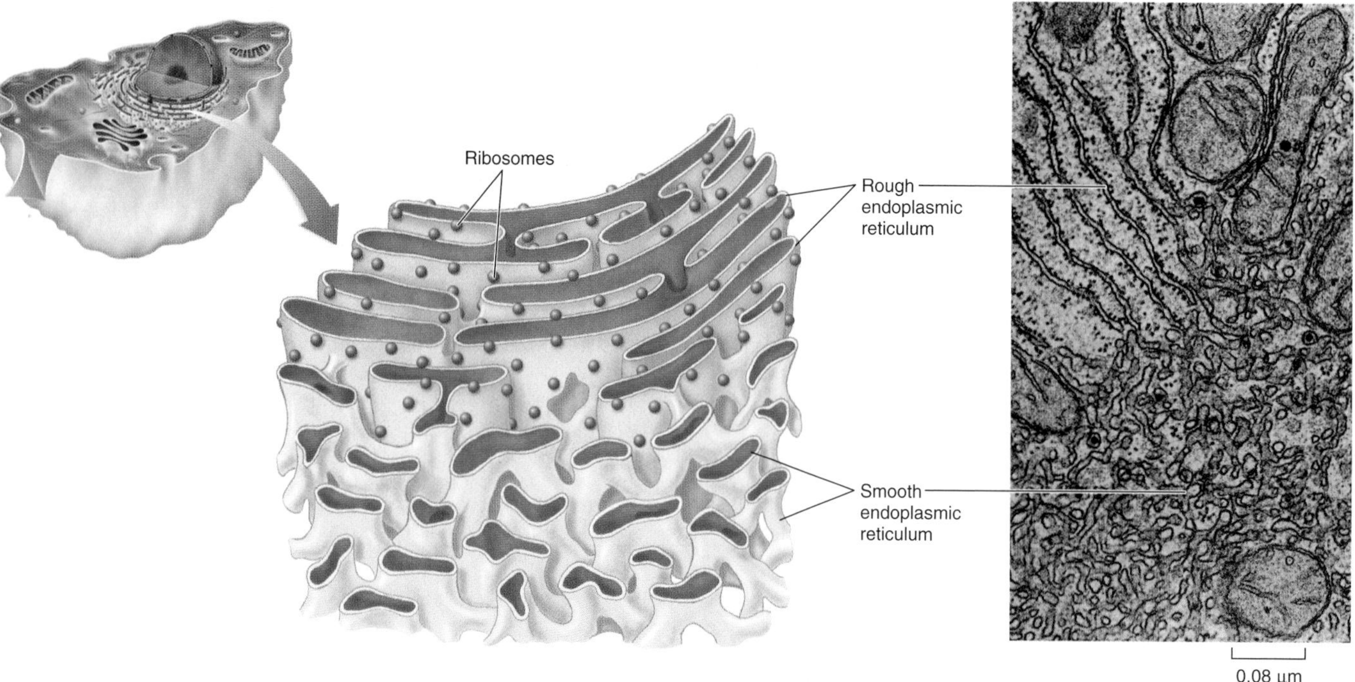

FIGURE 3.55 Endoplasmic Reticulum Site of synthesis of macromolecules destined for other organelles or the external environment.

(figure 3.55). The **rough endoplasmic reticulum** has a characteristic bumpy appearance due to the multitude of ribosomes coating it. It is the site where proteins not destined for the cytoplasm are synthesized. These include proteins targeted for the lumen of an organelle or for secretion outside the cell. Membrane proteins such as receptors are also synthesized on the rough ER. The ribosomes making these proteins attach to the ER surface. As the ribosomes synthesize the proteins, they thread them through gated pores in the membrane, delivering the proteins to the lumen of the ER. There, the proteins fold to assume their three-dimensional shapes. Vesicles that bud off from the ER transfer the newly synthesized molecules to the Golgi apparatus for further modification and sorting.

Some regions of the ER are smooth. This **smooth endoplasmic reticulum** functions in lipid synthesis and degradation, and calcium ion storage. As with material made in the rough ER, vesicles transfer compounds from the smooth ER to the Golgi apparatus.

The Golgi Apparatus

The **Golgi apparatus** consists of a series of membrane-bound flattened sacs **(figure 3.56).** It is the site where macromolecules synthesized in the endoplasmic reticulum are modified before they are transported to other destinations. These modifications, such as the addition of carbohydrate and phosphate groups, take place in

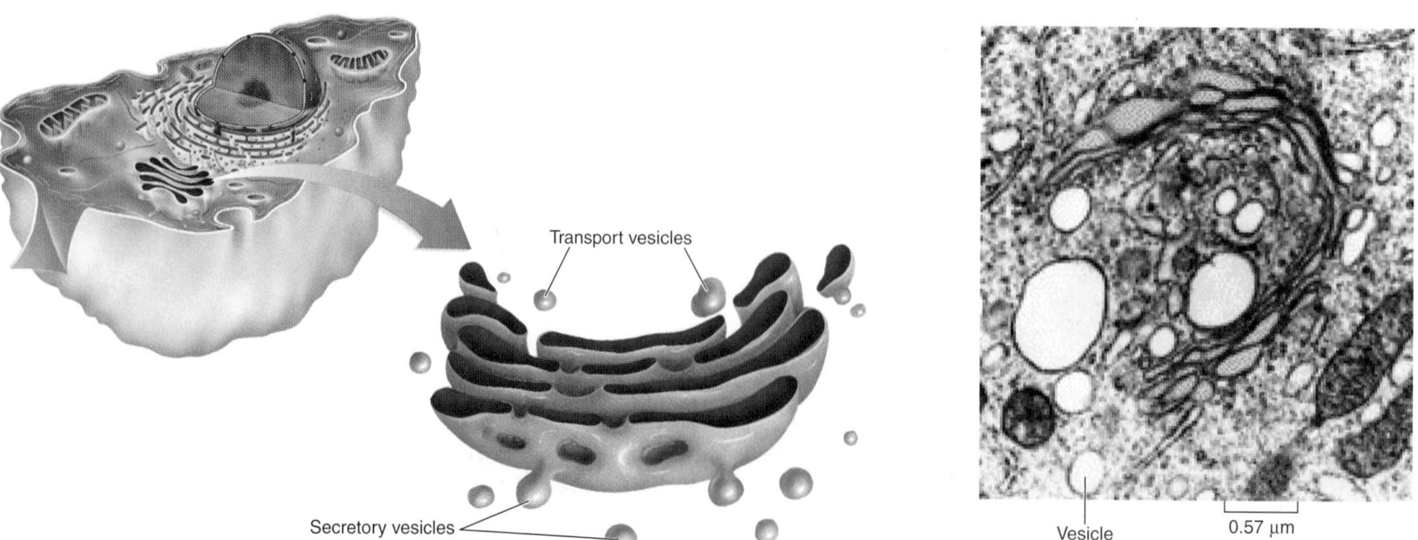

FIGURE 3.56 Golgi Apparatus Site where macromolecules synthesized in the endoplasmic reticulum are modified before being transported to other destinations in vesicles.

a sequential order in different Golgi sacs. Much like an assembly line, the molecules are transferred in vesicles from one Golgi sac to another. These various molecules are then sorted and delivered in vesicles destined either for specific cellular compartments or to the outside of the cell.

Lysosomes and Peroxisomes

Lysosomes are organelles that contain a number of powerful degradative enzymes that could destroy the cell if not contained within the organelle. Endosomes and phagosomes fuse with lysosomes, allowing digestion of material taken up by the cell. In a similar manner, exhausted organelles can fuse with lysosomes so that their contents are digested.

Peroxisomes are the organelles in which oxygen is used to oxidize substances, breaking down lipids and detoxifying certain chemicals. As a consequence, their enzymes generate highly reactive molecules such as hydrogen peroxide and superoxide. The peroxisome contains these molecules and ultimately degrades them, protecting the cell from their toxic effects. ■ **hydrogen peroxide, p. 92** ■ **superoxide, p. 92**

MICROCHECK 3.13

The nucleus, which contains DNA, is the predominant distinguishing feature of eukaryotes. Mitochondria are ATP-generating powerhouses. Chloroplasts are the site of photosynthesis. The rough endoplasmic reticulum is the site where proteins not destined for the cytoplasm are synthesized. The smooth endoplasmic reticulum functions in lipid synthesis and degradation, and calcium ion storage. The Golgi apparatus modifies and sorts molecules synthesized in the rough ER. Lysosomes are the structures within which digestion takes place; peroxisomes are the organelles in which oxygen is used to oxidize substances.

✓ Describe the structure of the nucleus.
✓ How does the function of the rough endoplasmic reticulum differ from that of the smooth endoplasmic reticulum?
✓ If enzymes contained in a peroxisome are to act on a substrate, what must first occur?

FUTURE CHALLENGES

A Case of Breaking and Entering

Unraveling the complex mechanisms that prokaryotic and eukaryotic cells use to transport materials across their membranes can potentially aid in the development of new antimicrobial medications. Armed with a precise model of the structure and function of bacterial transporter proteins, scientists might be able to design new drugs that exploit these systems. One strategy would be to design compounds that irreversibly bind to transporter molecules and jam the mechanism. If the microbes can be prevented from bringing in nutrients and removing wastes, their growth would cease. Another strategy would be to enhance the uptake or decrease the efflux of a specific compound that interferes with intracellular processes. This is already being done to some extent as new derivatives of current antibiotics are being produced, but more precise understanding of the processes by which bacteria take up or remove compounds could expedite drug development.

A more thorough understanding of eukaryotic uptake systems could be used to develop better antiviral drugs. Viruses exploit the process of receptor-mediated endocytosis to gain entry into the cell. Once they are enclosed within the endosome, their protective protein coat is removed, releasing their genetic material. New drugs can potentially be developed that block these steps, preventing the entry or uncoating of infectious viral particles.

SUMMARY

Microscopy and Cell Morphology

3.1 Microscopic Techniques: The Instruments (table 3.1)

Principles of Light Microscopy: The Bright-Field Microscope

The most commonly used type of microscope is the **bright-field** microscope (figure 3.1). The **objective lens** and the **ocular lens** in combination magnify an object by a factor equal to the product of the magnification of each of the individual lenses. The usefulness of a microscope depends largely on its **resolving power** (figure 3.2).

Light Microscopes That Increase Contrast

The **phase-contrast** microscope amplifies differences in refraction (figure 3.4). The **interference microscope** causes the specimen to appear as a three-dimensional image (figure 3.5). The **dark-field microscope** makes organisms stand out as bright objects against a dark background (figure 3.6). The **fluorescence microscope** is used to observe cells that have been stained with fluorescent dyes (figure 3.7). The **confocal scanning laser microscope** is used to construct a three-dimensional image of a thick structure and to provide detailed sectional views of the interior of an intact cell (figure 3.8).

Electron Microscopes

Transmission electron microscopes (TEMs) transmit electrons through a specimen that has been prepared by **thin sectioning, freeze fracturing,** or **freeze etching** (figure 3.10). **Scanning electron microscopes (SEM)** scan a beam of electrons back and forth over the surface of a specimen, producing a three-dimensional effect (figure 3.11).

Atomic Force Microscopes

Atomic force microscopes map the bumps and valleys of a surface on an atomic scale (figure 3.12).

3.2 Microscopic Techniques: Dyes and Staining (table 3.2)

Differential Stains

The **Gram stain** is widely used for staining bacteria; Gram-positive bacteria stain purple and Gram-negative bacteria stain pink (figure 3.14). The **acid-fast stain** is used to stain organisms such as *Mycobacterium* species; **acid-fast** organisms stain pink and all other organisms stain blue (figure 3.15).

Special Stains to Observe Cell Structures

The **capsule stain** allows the capsule to stand out as a halo around an organism (figure 3.16). The **spore stain** uses heat to facilitate the staining of endospores (figure 3.17). The **flagella stain** employs a mordant that enables the stain to adhere to and coat the otherwise thin flagella (figure 3.18).

Fluorescent Dyes and Tags

Some fluorescent dyes bind compounds that characterize all cells; others bind compounds specific to certain cell types (figure 3.19). **Immunofluorescence** is used to tag a specific protein of interest with a fluorescent compound.

3.3 Morphology of Prokaryotic Cells

Shapes

Most prokaryotes are **cocci** or **rods;** other shapes include **coccobacilli, vibrios, spirilla,** and **spirochetes. Pleomorphic** bacteria have variable shapes (figure 3.20).

Groupings

Cells adhering to one another following division form characteristic arrangements such as chains, packets, and clusters (figure 3.22).

Multicellular Associations

Cells within **biofilms** often alter their activities when a critical number of cells are present.

The Structure of the Prokaryotic Cell (figure 3.23, table 3.3)

3.4 The Cytoplasmic Membrane

Structure and Chemistry of the Cytoplasmic Membrane (figure 3.24)

The cytoplasmic membrane is a **phospholipid bilayer** embedded with a variety of different proteins including transport proteins and receptors.

Permeability of the Cytoplasmic Membrane

The cytoplasmic membrane is selectively permeable; water, gases, and small hydrophobic molecules are among the few compounds that can pass through by **simple diffusion.**

The Role of the Cytoplasmic Membrane in Energy Transformation

The **electron transport chain** expels protons, generating an **electrochemical gradient,** a source of energy called **proton motive force** (figure 3.26).

3.5 Directed Movement of Molecules Across the Cytoplasmic Membrane

Transport Systems (table 3.4)

Facilitated diffusion, or **passive transport,** moves compounds by exploiting a concentration gradient (figure 3.27). **Active transport** uses energy, either proton motive force or ATP, to accumulate compounds against a concentration gradient. **Group translocation** chemically modifies a molecule during its transport (figure 3.30).

Secretion

The presence of a characteristic **signal sequence** targets proteins for secretion.

3.6 Cell Wall

Peptidoglycan (figure 3.32)

Peptidoglycan is found only in the domain *Bacteria* and provides rigidity to the cell wall. Peptidoglycan is composed of **glycan strands,** which are alternating subunits of ***N*-acetylmuramic acid (NAM)** and ***N*-acetylglucosamine (NAG),** interconnected via the tetrapeptide chains on NAM.

The Gram-Positive Cell Wall (figure 3.33)

The Gram-positive cell wall contains a relatively thick layer of peptidoglycan. **Teichoic acids** project out of the peptidoglycan layer.

The Gram-Negative Cell Wall (figure 3.34)

The Gram-negative cell wall has a relatively thin layer of peptidoglycan sandwiched between the cytoplasmic membrane and an **outer membrane.** The outer membrane contains **lipopolysaccharides.** The **lipid A** portion of the lipopolysaccharide molecule is responsible for the toxic effects, which is why LPS is called **endotoxin. Porins** form small channels that permit small molecules to pass through the outer membrane. **Periplasm** contains a variety of proteins, including those involved in nutrient degradation and transport.

Antibacterial Substances That Target Peptidoglycan

Penicillin prevents peptidoglycan synthesis. **Lysozyme** destroys the structural integrity of peptidoglycan.

Differences in Cell Wall Composition and the Gram Stain

The Gram-positive, but not the Gram-negative, cell wall retains the crystal violet-iodine dye complex within the cell even when subjected to acetone-alcohol treatment.

Characteristics of Bacteria That Lack a Cell Wall

Mycoplasma species are extremely variable in shape and are not affected by lysozyme or penicillin (figure 3.35).

Cell Walls of the Domain Archaea

Archaea have a greater variety of cell wall types than do the *Bacteria* and they all lack peptidoglycan.

3.7 Capsules and Slime Layers

Capsules and **slime layers** enable bacteria to adhere to surfaces. Some capsules allow disease-causing microorganisms to thwart the innate defense system (figure 3.36).

3.8 Filamentous Protein Appendages

Flagella (figure 3.37)

The **flagellum** is a long protein structure commonly responsible for bacterial motility (figure 3.38). **Chemotaxis** is the directed movement toward an attractant or away from a repellent (figure 3.39). **Phototaxis, aerotaxis, magnetotaxis,** and **thermotaxis** are directed movements toward light, oxygen, a magnetic field, and temperature, respectively.

Pili (figure 3.41)

Many types of **pili (fimbriae)** enable specific attachment of cells to surfaces. **Sex pili** are involved in a form of DNA transfer.

3.9 Internal Structure

The Chromosome (figure 3.42)

The **chromosome** of prokaryotes resides in the **nucleoid** rather than within a membrane bound nucleus; it contains all the genetic information required by a cell.

Plasmids

Plasmids only encode genetic information that may be advantageous, but not required by the cell.

Ribosomes (figure 3.43)

Ribosomes facilitate the joining of amino acids. The 70S bacterial ribosome is composed of a 50S and a 30S subunit.

Cytoskeleton

The **cytoskeleton** is an interior framework involved in cell division and regulation of cell shape.

Storage Granules (figure 3.44)

Storage granules are synthesized from a nutrient that a cell has in relative excess.

Gas Vesicles

Gas vesicles provide buoyancy to aquatic cells.

Endospores

Endospores are extraordinarily resistant to heat, desiccation, toxic chemicals, and ultraviolet irradiation; they can **germinate** to become **vegetative cells** (figures 3.45, 3.46).

The Eukaryotic Cell (figure 3.47, table 3.6)

3.10 The Plasma Membrane

The **plasma membrane** is a phospholipid bilayer embedded with proteins. Proteins in the membrane are involved in transport, structural integrity, and **signaling.**

3.11 Transfer of Molecules Across the Plasma Membrane

Transport Proteins

Carriers mediate facilitated diffusion and active transport. **Channels** are pores in the membrane that are so small that only specific ions can pass through. These channels are gated.

Endocytosis and Exocytosis (figure 3.49)

Pinocytosis is the most common form of endocytosis in animal cells. The **endocytic vesicle** fuses with an **endosome,** which then fuses with a **lysosome.** Protozoa and phagocytes take up bacteria and debris through the process of **phagocytosis.** The **phagosome** fuses with the **lysosome,** where the material is digested. **Exocytosis** expels material.

Secretion

Proteins destined for a non-cytoplasmic region are made by ribosomes bound to the **endoplasmic reticulum.**

3.12 Protein Structures Within the Cell

Ribosomes

The **80S** ribosome is composed of **60S** and **40S subunits.**

Cytoskeleton (figure 3.50)

The cytoskeleton is composed of **microtubules, actin filaments, and intermediate filaments.**

Flagella and Cilia (figure 3.51)

Flagella propel a cell or pull the cell forward. **Cilia** move in synchrony to either propel a cell or move material along a stationary cell.

3.13 Membrane-Bound Organelles

The Nucleus (figure 3.52)

The **nucleus** is the predominant distinguishing feature of eukaryotic cells.

Mitochondria

Mitochondria use the energy released during the degradation of organic compounds to generate ATP (figure 3.53).

Chloroplasts

Chloroplasts capture the energy of sunlight; this is then used to synthesize ATP that is expended to convert CO_2 to an organic form (figure 3.54).

Endoplasmic Reticulum (ER) (figure 3.55)

The **rough endoplasmic reticulum** is the site where proteins not located in the cytoplasm are synthesized. Within the **smooth endoplasmic reticulum,** lipids are synthesized and degraded, and calcium is stored.

The Golgi Apparatus (figure 3.56)

The **Golgi apparatus** modifies and sorts molecules synthesized in the endoplasmic reticulum.

Lysosomes and Peroxisomes

Lysosomes carry digestive enzymes. **Peroxisomes** are the organelles in which oxygen is used to oxidize certain substances.

REVIEW QUESTIONS

Short Answer

1. Explain why resolving power is important in microscopy.
2. Explain why basic dyes are used more frequently than acidic dyes in staining.
3. Describe what happens at each step in the Gram stain.
4. Compare and contrast ABC transport systems with group translocation.
5. Give two reasons that the outer membrane of Gram-negative bacteria is significant medically.
6. Compare and contrast penicillin and lysozyme.
7. Describe how a plasmid can help a cell.
8. How is an organ different from tissue?
9. How is receptor-mediated endocytosis different from phagocytosis?
10. Explain how the Golgi apparatus cooperatively functions with the endoplasmic reticulum.

Multiple Choice

1. Which of the following is most likely to be used in a typical microbiology laboratory?
   a) Bright-field microscope
   b) Confocal scanning microscope
   c) Phase-contrast microscope
   d) Scanning electron microscope
   e) Transmission electron microscope

2. Which of the following stains is used to detect *Mycobacterium* species?
   a) Acid-fast stain b) Capsule stain c) Endospore stain
   d) Gram stain e) Simple stain
3. Penicillin
   1. is generally effective against Gram-positive bacteria.
   2. is generally effective against Gram-negative bacteria.
   3. functions in the cytoplasm of the cell.
   4. is effective against mycoplasma.
   5. kills only growing cells.
   a) 1,2 b) 2,3 c) 3,4 d) 4,5 e) 1,5
4. Endotoxin is associated with
   a) Gram-positive bacteria. b) Gram-negative bacteria.
   c) the cytoplasmic membrane. d) the endospore.
5. In prokaryotes, lipid bilayers are associated with the
   1. Gram-positive cell wall. 2. Gram-negative cell wall.
   3. cytoplasmic membrane. 4. capsule.
   5. nuclear membrane.
   a) 1,2 b) 2,3 c) 3,4 d) 4,5 e) 1,5
6. In bacteria, the cytoplasmic membrane functions in
   1. protein synthesis. 2. ribosome synthesis.
   3. generation of ATP. 4. transport of molecules.
   5. attachment.
   a) 1,2 b) 2,3 c) 3,4 d) 4,5 e) 1,5
7. Attachment is mediated by the
   1. capsule. 2. cell wall. 3. cytoplasmic membrane.
   4. periplasm. 5. pilus.
   a) 1,2 b) 2,3 c) 3,4 d) 4,5 e) 1,5
8. Endocytosis is associated with
   a) mitochondria. b) prokaryotic cells. c) eukaryotic cells.
   d) chloroplasts. e) ribosomes.
9. Protein synthesis is associated with
   1. lysosomes. 2. the cytoplasmic membrane.
   3. the Golgi apparatus. 4. rough endoplasmic reticulum.
   5. ribosomes.
   a) 1,2 b) 2,3 c) 3,4 d) 4,5 e) 1,5
10. All of the following are composed of tubulin, *except*:
    a) actin b) cilia c) eukaryotic flagella
    d) microtubules e) more than one of these

Applications

1. You are working in a laboratory producing new antibiotics for human and veterinary use. One compound with potential value inhibits the action of prokaryotic ribosomes. The compound, however, was shown to inhibit the growth of animal cells in culture. What is one possible explanation for its effect on animal cells?
2. A research laboratory is investigating environmental factors that would inhibit the growth of *Archaea*. One question they have is if adding the antibiotic penicillin would be effective in controlling their growth. Explain the probable results of an experiment in which penicillin is added to a culture of *Archaea*.

Critical Thinking

1. This graph shows facilitated diffusion of a compound across a cytoplasmic membrane and into a cell. As the external concentration of the compound is increased, the rate of uptake increases until it reaches a point where it slows and then begins to plateau. This is not the case with passive diffusion, where the rate of uptake continually increases. Why does the rate of uptake slow and then eventually plateau with facilitated diffusion?

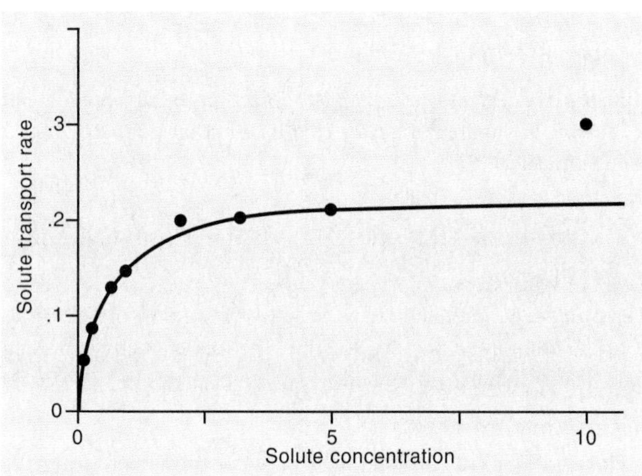

2. Most medically useful antibiotics interfere with either peptidoglycan synthesis or ribosome function. Why would the cytoplasmic membrane be a poor target for antibacterial medications?

Bacteria on an agar plate.

Dynamics of Prokaryotic Growth

A Glimpse of History

The greatest contributor to methods of cultivating bacteria was Robert Koch (1843–1910), a German physician who combined a medical practice with a productive research career for which he received a Nobel Prize in 1905. Koch was primarily interested in identifying disease-causing bacteria. To do this, however, he soon realized it was necessary to have simple methods to isolate and grow these particular species. He recognized that a single bacterial cell could multiply on a solid medium in a limited area and form a distinct visible mass of descendants.

Koch initially experimented with growing bacteria on the cut surfaces of potatoes, but he found that a lack of nutrients in the potatoes prevented growth of some species. To overcome this difficulty, Koch realized it would be advantageous to be able to solidify any liquid nutrient medium. Gelatin was used initially, but there were two major drawbacks—it melts at the temperature preferred by many medically important organisms and some bacteria can digest it. In 1882, Fannie Hess, the wife of an associate of Koch, suggested using agar. This solidifying agent was used to harden jelly at the time and proved to be the perfect answer.

Today, we take pure culture techniques for granted because of their relative ease and simplicity. Their development in the late 1800s, however, had a major impact on microbiology. Within 20 years, the agents causing most of the major bacterial diseases of humans were isolated and characterized.

Prokaryotes can be found growing even in the harshest climates and the most severe conditions. Environments that no unprotected human could survive, such as the ocean depths, volcanic vents, and the polar regions, have thriving species of prokaryotes. Indeed many scientists believe that if life exists on other planets, it may resemble these microorganisms. Each species, however, has a limited set of environmental conditions in which it can grow; even then, it will grow only if specific nutrients are available. Some prokaryotes can grow at temperatures above the boiling point of water but not at room temperature. Many species can only grow within an animal host, and then only in specific areas of that host.

Because of the medical significance of some bacteria, as well as the nutritional and industrial use of microbial by-products, microbiologists must be able to identify, isolate, and cultivate many species. To do this, one needs to understand the basic principles involved in prokaryotic growth while recognizing that a vast sea of information is yet to be discovered.

KEY TERMS

Biofilm Polysaccharide-encased community of microorganisms.

Chemically Defined Medium Bacteriological growth medium composed of precise mixtures of known pure chemicals; generally used for specific experiments when nutrients must be precisely controlled.

Complex Medium Bacteriological medium that contains protein digests, extracts, or other ingredients that vary in their chemical composition.

Differential Medium Bacteriological medium that contains an ingredient that can be changed by certain bacteria in a recognizable way; used to differentiate organisms based on their metabolic traits.

Exponential (Log) Phase Stage of growth in which cells divide at a constant rate; generation time is measured during this period of active multiplication.

Facultative Anaerobe Organism that grows best if O_2 is available, but can also grow without it.

Generation Time The time it takes for a population to double in number.

Obligate Aerobe Organism that requires molecular oxygen (O_2).

Obligate Anaerobe Organism that cannot multiply, and is often killed, in the presence of O_2.

Plate Count Method to measure the concentration of viable cells by determining the number of colonies that arise from a sample added to an agar plate.

Pure Culture A population of organisms descended from a single cell and therefore separated from all other species.

Selective Medium Bacteriological medium to which additional ingredients have been added that inhibit the growth of many organisms other than the one being sought.

4.1

Principles of Prokaryotic Growth

Focus Point

- Describe binary fission and explain how it relates to generation time.

Prokaryotes generally multiply by the process of **binary fission (figure 4.1).** After a cell has increased in size and doubled its components, it divides. One cell divides into two, those two divide to become four, those four become eight, and so on. In other words, the increase in cell numbers is exponential. Because it is neither practical, nor particularly meaningful, to determine the relative size of the cells in a given population, **microbial growth** is defined as an increase in the number of cells in a population.

The time it takes for a population to double in number is the **generation,** or **doubling, time.** This varies greatly depending on the species of the organism and the conditions in which it is grown. Some common organisms, such as *Escherichia coli*, can double in approximately 20 minutes; others, such as the causative agent of tuberculosis, *Mycobacterium tuberculosis*, require at least 12 hours to double even under the most favorable conditions. The environmental and nutritional factors that affect the rate of growth will be discussed shortly.

The exponential multiplication of bacteria has important health consequences. For example, a mere 10 cells of a food-borne pathogen in a potato salad, sitting for 4 hours in the warm sun at a picnic, may multiply to more than 40,000 cells. A simple equation expresses the relationship between the number of cells in a population at a given time (N_t), the original number of cells in the population (N_0), and the number of divisions those cells have undergone during that time (n). If any two values are known, the third can be easily calculated from the equation:

$$N_t = N_0 \times 2^n$$

In this example, let us assume that we know that 10 cells of a disease-causing organism were initially added to the potato salad and we also know that the organism has a generation time of 20 minutes. The first step is to determine the number of cell divisions that will occur in a given time. Because the organism divides every 20 minutes, 3 times every hour, we know that in 4 hours it will divide 12 times. Now that we know the original number of cells and the number of divisions, we can solve for N_t:

$$10 \times 2^{12} = N_t = 40{,}960$$

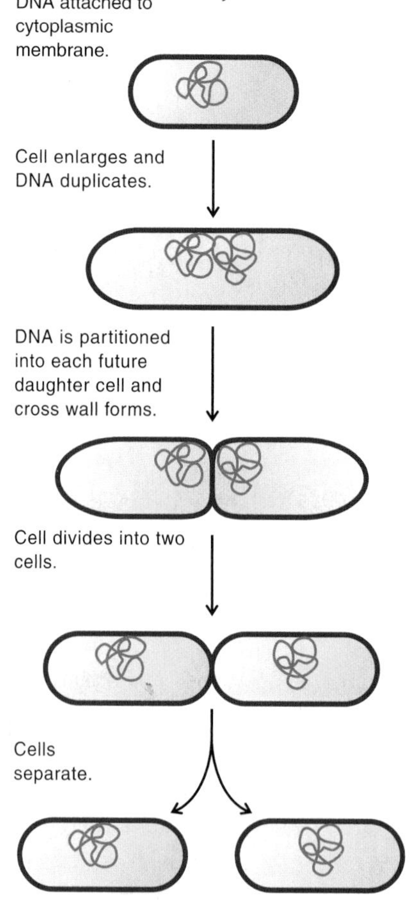

DNA attached to cytoplasmic membrane.
Cell enlarges and DNA duplicates.
DNA is partitioned into each future daughter cell and cross wall forms.
Cell divides into two cells.
Cells separate.
Daughter cells

FIGURE 4.1 Binary Fission The chromosomal DNA is attached to the cytoplasmic membrane. As the cell increases in length, the DNA is replicated and then partitioned into each of the two daughter cells.

TABLE 4.1 Example of Exponential Growth

Time in Minutes (t)	Initial Population (N_0)	Number of Generations (n)	2^n	Population $N_0 \times 2^n$
0	10	0		10
20	10	1	$2^1(=2)$	20
40	10	2	$2^2(=2 \times 2)$	40
60 (1 hour)	10	3	$2^3(=2 \times 2 \times 2)$	80
80	10	4	$2^4(=2 \times 2 \times 2 \times 2)$	160
100	10	5	$2^5(=2 \times 2 \times 2 \times 2 \times 2)$	320
120 (2 hours)	10	6	2^6	640
140	10	7	2^7	1,280
160	10	8	2^8	2,560
180 (3 hours)	10	9	2^9	5,120
200	10	10	2^{10}	10,240
220	10	11	2^{11}	20,480
240 (4 hours)	10	12	2^{12}	40,960

Thus, after 4 hours the potato salad in the example will have 40,960 cells of our pathogen **(table 4.1).** Keep this in mind, and your potato salad in a cooler, the next time you go to a picnic!

MICROCHECK 4.1

Most prokaryotes multiply by binary fission. Microbial growth is an increase in the number of cells in a population. The time required for a population to double in number is the generation time.

- ✓ Explain why microbial growth refers to a population rather than a cell size.
- ✓ If a bacterium has a generation time of 30 minutes, and you start with 100 cells at time 0, how many cells will you have in 30, 60, 90, and 120 minutes?

Bacterial Growth in Nature

Focus Points

- Describe a biofilm and give one positive and one negative impact that biofilms have on humans.
- Explain why bacteria that grow naturally in mixed communities sometimes cannot be grown in pure culture.

Historically, microorganisms have been studied by growing them in the laboratory, but scientists now recognize that the dynamic and complex conditions of the natural environment, which differ greatly from the conditions in the laboratory, have profound effects on microbial growth and behavior. When growing in a running stream, for example, prokaryotes frequently synthesize slime layers or other structures that allow them to attach to rocks or other solid surfaces. They may not produce these adherent structures when growing in the laboratory. In fact, microbial cells actually sense various surrounding chemicals and then respond by synthesizing compounds useful for growth in that particular environment. Cells often grow in multicellular associations that function cooperatively to increase the chance of survival of the population as a whole. ■ slime layer, p. 64

Biofilms

In nature, prokaryotes can live suspended in an aqueous environment, but many attach to surfaces and live in polysaccharide-encased communities called **biofilms (figure 4.2).** Biofilms cause the slipperiness of rocks in a stream bed, the slimy "gunk" that

FIGURE 4.2 Biofilm on a Stainless Steel Surface A biofilm is a polysaccharide-encased community of microorganisms.

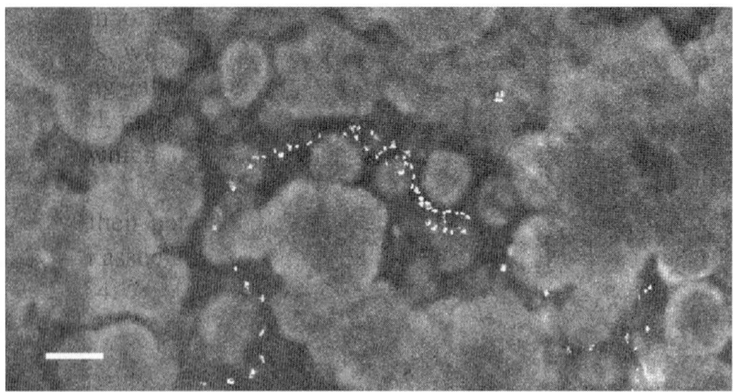

FIGURE 4.3 Architecture of a Biofilm Superimposed time sequence image shows a single latex bead moving through a biofilm water channel. The large light gray shapes are clusters of bacteria (scale bar = 50μm).

coats kitchen drains, the scum that gradually accumulates in toilet bowls, and the dental plaque that forms on teeth. Biofilm formation begins when planktonic, or free-floating, bacteria adhere to a surface where they multiply and synthesize slime layers to which unrelated cells can attach and grow.

Surprisingly, biofilms are not generally haphazard mixtures of microbes in a layer of slime, but instead have characteristic architectures with open channels through which nutrients and waste materials can pass **(figure 4.3).** Cells communicate with one another by synthesizing and responding to chemical signals, an exchange that appears to be important in establishing structure.

Biofilms are more than just an unsightly annoyance. Dental plaque leads to tooth decay and gum disease. Even troublesome, persistent ear infections and the complications of cystic fibrosis are thought to be due to bacteria that grow as a biofilm. In fact, it is estimated that 65% of human bacterial infections involve biofilms. Treatment of these infections is difficult because microorganisms growing within the protective slime are often able to resist the effects of antibiotics, as well as the body's defenses. Biofilms are also important in industry, where their growth in pipes, drains, and cooling water towers can interfere with processes and damage equipment. Again, the structure of the biofilm shields the microbes growing within it, and bacteria in a biofilm may be hundreds of times more resistant to disinfectants than are their planktonic counterparts. ■ **disinfectants, p. 108**

While biofilms can be damaging, they also can be beneficial. Many bioremediation efforts, which use bacteria to degrade harmful chemicals, are enhanced by biofilms. Thus, as some industries are exploring ways to destroy biofilms, others, such as wastewater treatment facilities, are looking for ways to foster their development. ■ **bioremediation, p. 749** ■ **wastewater treatment, p. 739**

Interactions of Mixed Microbial Communities

Prokaryotes in the environment regularly grow in close associations with many different species. Sometimes the interactions are cooperative, even fostering the growth of members that otherwise could not survive. For example, organisms that cannot multiply in the presence of O_2 can grow in the mouth. This is because other microbes found there consume O_2 during their metabolism, creating microenvironments that lack O_2. In addition, the metabolic wastes of one species may serve as a nutrient for another. Often, however, cells in these communities compete for nutrients, and some even resort to a type of biological warfare, synthesizing toxic compounds that inhibit competitors. Understandably, the conditions in these close associations are exceedingly difficult to reproduce in the laboratory. ■ **microbial competition and antagonism, p. 723**

MICROCHECK 4.2

Biofilms have a characteristic architecture with open channels through which nutrients and waste products can pass. In nature, prokaryotes often grow in close associations with many different species.

- ✓ Give three examples of biofilms.
- ✓ Describe a situation in which the activities of one species benefit another.
- ✓ Why would bacteria in a biofilm be more resistant to harmful chemicals?

4.3

Obtaining a Pure Culture

Focus Point

- Describe how the streak-plate method is used to obtain a pure culture, and how the resulting culture can be stored.

In the laboratory, prokaryotes are generally isolated and grown in **pure culture** in order to identify them and study the activities of a particular species. A pure culture is defined as a population of organisms descended from a single cell and therefore separated from all other species. Results obtained using pure cultures are much easier to interpret, but as discussed earlier, the organisms sometimes behave differently than they do in their natural environment. Another complicating issue is that only an estimated 1% of all prokaryotes can currently be cultivated successfully. This makes it exceedingly difficult to study the vast majority of environmental microorganisms. Fortunately for humanity, most known medically significant bacteria can be grown in pure culture.

Pure cultures are obtained using a variety of special techniques. All glassware, media, and instruments must be **sterile,** or free of microbes, prior to use. These are then handled using **aseptic techniques,** procedures that minimize the chance of other organisms being accidentally introduced. The medium that the cells are grown in, or on, is a mixture of nutrients dissolved in water and may be in a liquid broth or a solidified gel-like form. The medium is called a **culture medium.** ■ **aseptic techniques, p. 109** ■ **sterilization, p. 108**

Cultivating Bacteria on a Solid Culture Medium

The basic requirements for obtaining a pure culture are a solid culture medium, a media container that can be maintained in an aseptic condition, and a method to separate individual bacterial cells. A single bacterium, supplied with the right nutrients and conditions, will multiply on the solid medium in a limited area to form a **colony,** a mass of cells descended from the original one **(figure 4.4).** About 1 million cells are required for a colony to be easily visible to the naked eye.

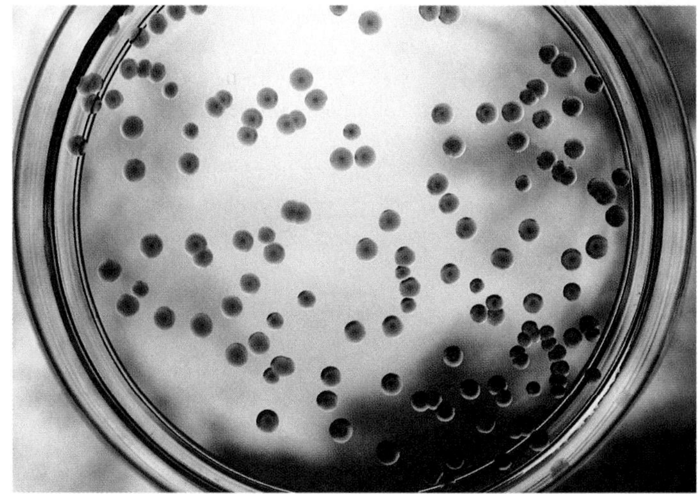

FIGURE 4.4 Colonies Growing on Agar Medium

Agar, a polysaccharide extracted from marine algae, is used to solidify a liquid culture medium. Unlike other gelling agents such as gelatin, very few bacteria can degrade agar. It is not destroyed at high temperatures and can therefore be sterilized by heating, a process that also liquefies it. Melted agar will stay liquid until cooled to a temperature below 45°C. Therefore, nutrients that would be destroyed at high temperatures can be added at lower temperatures before the agar hardens. Once solidified, an agar medium will remain so until it is heated above 95°C. Thus, unlike gelatin, which is liquid at 37°C, agar remains solid over the entire temperature range at which the majority of bacteria grow. ■ polysaccharide, pp. 30, 32

The culture medium is contained in a **Petri dish**—a two-part, covered container made of glass or plastic. While not airtight, the Petri dish does exclude airborne microbial contaminants. A Petri dish containing a medium is commonly referred to as a plate of that medium type—for example, a nutrient agar plate or, more simply, an **agar plate.**

The Streak-Plate Method

The **streak-plate** method is the simplest and most commonly used technique for isolating bacteria **(figure 4.5).** A sterilized inoculating loop is dipped into a solution containing the organism of interest and then lightly drawn several times across an agar plate, creating a set of parallel streaks covering approximately one-third of the plate. The loop is then sterilized and

(1) Loop is sterilized.
(2) Loop is inoculated.
(3) First set of streaks made.
Agar containing nutrients
(4) Loop is sterilized.
Starting point
(5) Second set of streaks made.
(6) Loop is sterilized.
(7) Final set of streaks made.
(8) Isolated colonies develop after incubation.

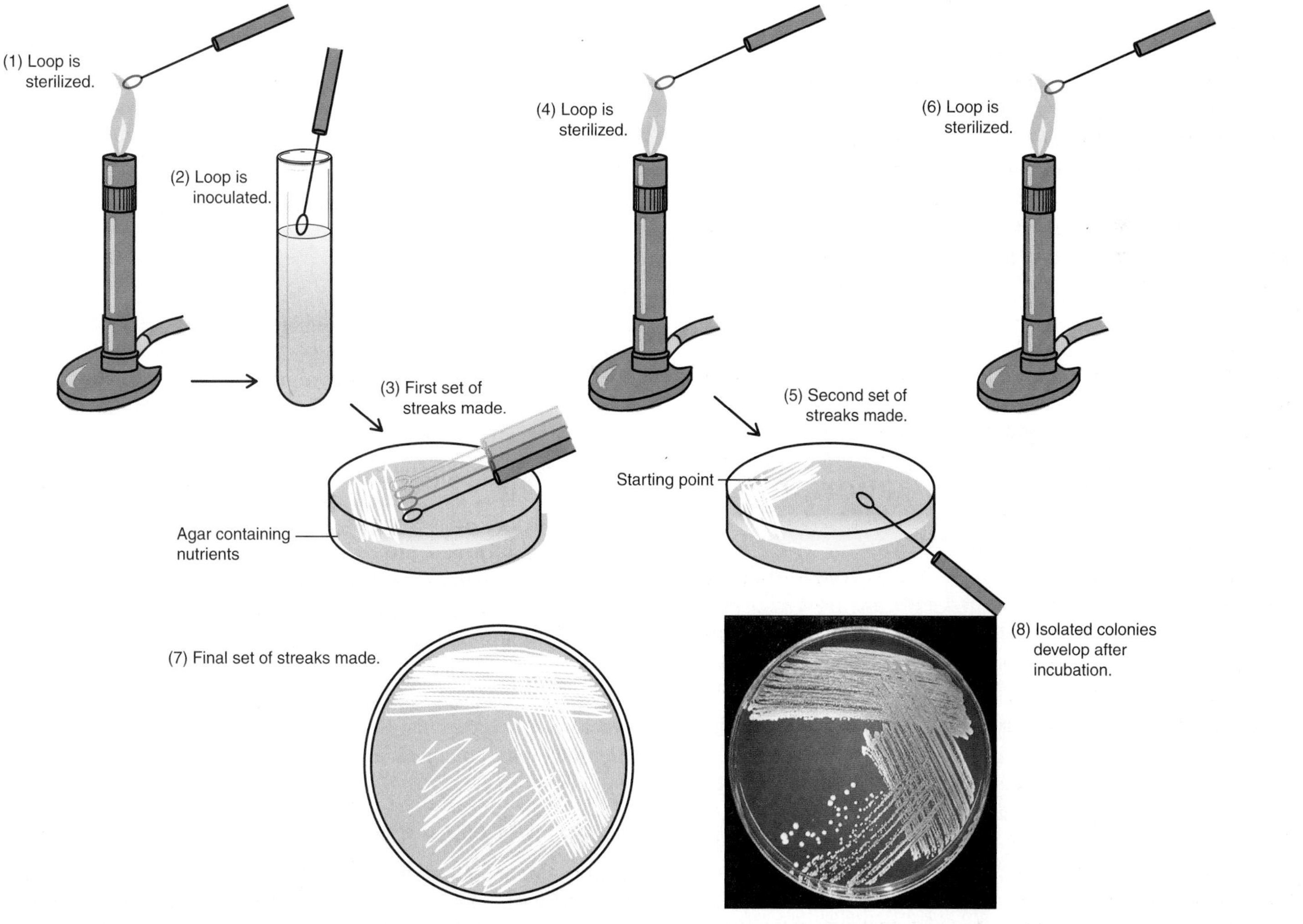

FIGURE 4.5 The Streak-Plate Method A sterilized inoculating loop **(1)** is dipped into a culture **(2)** and is then lightly drawn several times across an agar plate **(3).** The loop is sterilized again **(4),** and a new series of streaks is made at an angle to the first set **(5).** The loop is sterilized again **(6),** and another set of parallel streaks is made **(7).** The successive streaks dilute the concentration of cells. By the third set of streaks, cells should be separated enough so that isolated colonies develop after incubation **(8).**

a new series of parallel streaks is made across and at an angle to the previous ones, covering another third of the plate. This drags some of those cells streaked onto the first portion of the plate over to a previously uninoculated portion, creating a region containing a more dilute inoculum. The loop is sterilized again, and another set of parallel streaks is made, dragging into a third area some of the organisms that had been moved into the second section. The object is to reduce the number of cells being spread with each successive series of streaks, effectively diluting the sample. By the third set of streaks, cells should be separated enough so that distinct, well-isolated colonies will form.

Maintaining Stock Cultures

Once a pure culture has been obtained, it can be maintained as a **stock culture,** a culture stored for use as an inoculum in later procedures. Often, a stock culture is stored in the refrigerator as growth on an **agar slant.** This is agar medium in a tube that was held at a shallow angle as the medium solidified, creating a larger surface area. For long-term storage, stock cultures can be frozen at –70°C in a solution that prevents ice crystals from forming and damaging cells. Alternatively, cells can be **lyophilized,** or freeze-dried. ■ lyophilization, p. 122

MICROCHECK 4.3

Only an estimated 1% of prokaryotes can be cultivated in the laboratory. Agar is used to solidify nutrient-containing broth. The streak-plate method is used to obtain a pure culture.

- ✓ What properties of agar make it ideal for use in bacteriological media?
- ✓ How does the streak-plate method separate individual cells?
- ✓ What might be a reason that medically significant bacteria can be grown in pure culture more often than environmental organisms?

Bacterial Growth in Laboratory Conditions

Focus Point

- Describe the five distinct stages of a growth curve, and compare this closed system to colony growth and continuous culture.

In the laboratory, bacteria are typically grown in broth contained in a tube or flask, or on an agar plate. These are considered **closed** or **batch systems** because nutrients are not renewed, nor are waste products removed. Under these conditions, the cell population increases in number in a predictable fashion and then eventually declines. As the population in a closed system grows, it follows a pattern of stages, called a **growth curve.** This growth pattern is most distinct in a shaken broth culture, because all cells are exposed to the same environment. In a colony, cells on the outer edge of a colony experience very different conditions from those at the center.

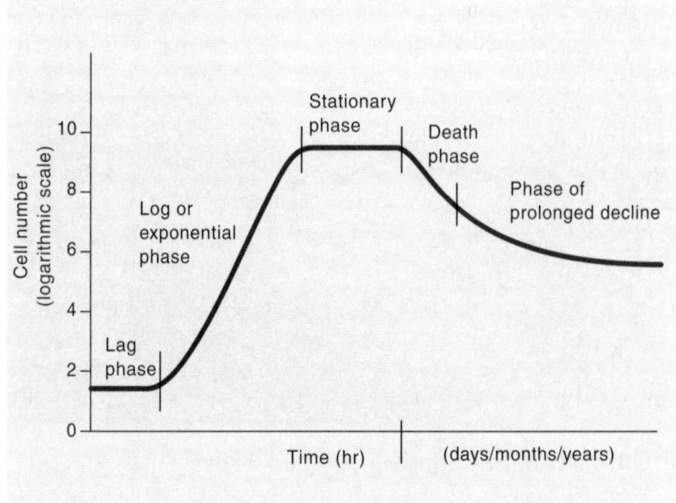

FIGURE 4.6 Growth Curve The growth curve is characterized by five distinct stages: lag phase, exponential or log phase, stationary phase, death phase, and phase of prolonged decline.

To maintain cells in a state of continuous growth, nutrients must be continuously added and waste products removed. This is called an **open system,** or **continuous culture.**

The Growth Curve

A growth curve is characterized by five distinct stages—the lag phase, the exponential or log phase, the stationary phase, the death phase, and the phase of prolonged decline **(figure 4.6).**

Lag Phase

When a bacterial culture is diluted and then transferred into a different medium, the number of viable cells does not immediately increase. They go through a "tooling up" or **lag phase** prior to active multiplication. During this time they synthesize macromolecules required for multiplication, including enzymes, ribosomes, and nucleic acids, and they generate energy in the form of ATP.

The length of the lag phase depends on conditions in the original culture and the medium into which the bacteria are transferred. If cells are transferred from a nutrient-rich medium to one containing fewer nutrients, the lag time tends to be longer. This is because cells must begin making enzymes to synthesize components missing in the new medium. A similar situation occurs when a stock culture stored in the refrigerator for several weeks is inoculated into fresh medium. In contrast, if young cells are transferred to a medium similar in composition, the lag time is quite short.

Exponential Phase (Log Phase)

During the **exponential** or **log phase,** cells divide at a constant rate and their numbers increase by the same percentage during each time interval. The generation time is measured during this period of active multiplication. Because bacteria are most susceptible to antibiotics and other chemicals during this time, the log phase is important medically.

During the initial phase of exponential growth, the cells' activities are directed toward increasing cell mass. Cells produce compounds such as amino acids and nucleotides, the respective building blocks of proteins and nucleic acids. Cells are remarkably precise in their ability to regulate the synthesis of these compounds, ensuring that each is made in the appropriate relative amount for efficient assembly into macromolecules. Compounds synthesized during this period of active multiplication are called **primary metabolites.** A metabolite is any product of a chemical reaction in a cell and includes compounds required for growth, as well as waste materials. Some primary metabolites are commercially valuable as flavoring agents and food supplements. Understandably, industries that harvest these compounds are working to develop methods to manipulate bacteria to overproduce certain primary metabolites. ■ regulation of gene expression, p. 176

Cells' activities shift as they enter a stage called **late log phase,** which marks the transition to stationary phase. This change occurs in response to multiple factors inevitable in a closed system, such as depletion of nutrients and buildup of waste products. If the cells are able to form endospores, they initiate the process of sporulation. If they cannot, they still "hunker down" in preparation for the starvation conditions ahead. The cells become rounder in shape and more resistant to harmful chemicals and radiation. Changes in the composition of their cell walls and cytoplasmic membranes also occur. As their surrounding environment changes, cells begin synthesizing a new group of metabolites, termed **secondary metabolites (figure 4.7).** Commercially, the most important of these are antibiotics, which inhibit the growth of or kill other organisms.

Stationary Phase

Cells enter the **stationary phase** when they no longer have supplies of energy and nutrients adequate for sustained growth. The total number of viable cells in the overall population remains relatively constant, but some cells are dying while others are multiplying. How can cells multiply when they have exhausted their supply of nutrients? Cells that die release their contents, providing a source of nutrients and energy to fuel the growth of other cells.

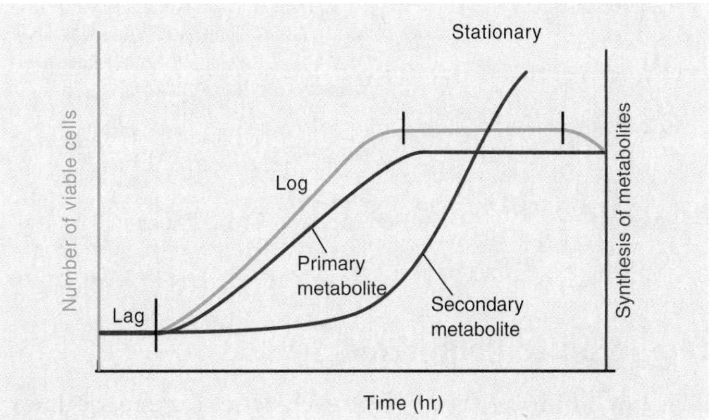

FIGURE 4.7 Primary and Secondary Metabolite Production Primary metabolites are synthesized during the period of active multiplication. Late in the log phase, cells begin synthesizing secondary metabolites. These compounds, which continue to be synthesized in stationary phase, appear to make the cells more resistant to environmental conditions.

During the stationary phase, the viable cells continue to synthesize secondary metabolites and maintain the altered properties they demonstrated in late log phase.

The length of time cells remain in the stationary phase varies depending on the species and on environmental conditions. Some populations remain in the stationary phase for only a few hours, whereas others remain for days.

Death Phase

The **death phase** is the period when the total number of viable cells in the population decreases as cells die off at a constant rate. Like bacterial growth, death is exponential. However, the cell population usually dies off much more slowly than it multiplies during the log phase. Once about 99% of the cells have died off, the remaining members of the population enter a different phase.

Phase of Prolonged Decline

The **phase of prolonged decline** is marked by a very gradual decrease in the number of viable cells in the population, lasting for days to years. Superficially, it might seem like a gradual march towards death of the population, but dynamic changes are actually occurring. Many members of the population are dying and releasing their nutrients, while a few "fitter" cells more able to cope with the deteriorating environmental conditions are multiplying. This dynamic process generates successive waves of slightly modified populations, each more fit to survive than the previous ones **(figure 4.8).** Thus, the statement "survival of the fittest" even holds true for closed cultures of bacteria.

Colony Growth

Growth of a bacterial colony on a solid medium involves many of the same features as bacteria growing in liquid, but it is marked by some important differences. After a lag phase, cells multiply exponentially and eventually compete with one another for available nutrients and become very crowded. Unlike a liquid culture, the position of a single cell within a colony markedly determines its environment. Cells multiplying on the edge of the colony face relatively little competition and can use O_2 in the air and obtain nutrients from the agar medium. In contrast, in the center of the colony the high density of cells rapidly depletes available O_2 and

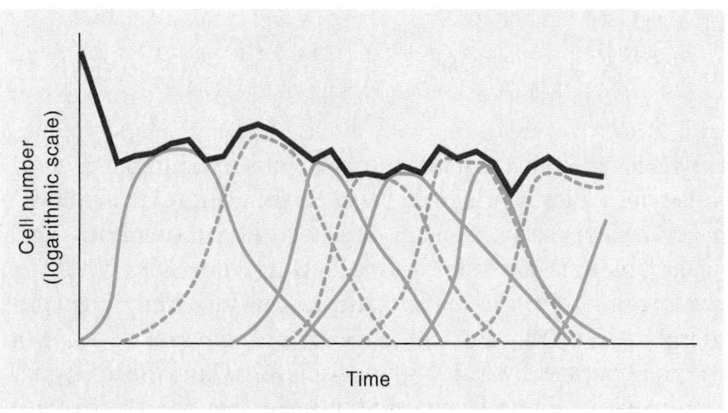

FIGURE 4.8 Dynamic Population Changes in the Phase of Prolonged Decline Many members of the population are dying and releasing their nutrients, while a few "fitter" cells are actively multiplying.

nutrients. Toxic metabolic wastes such as acids accumulate. As a consequence, cells at the edge of the colony may be growing exponentially, whereas those in the center may be in the death phase. Cells in locations between these two extremes may be in stationary phase.

Continuous Culture

Bacteria can be maintained in a state of continuous exponential growth by using a **chemostat.** This device continually drips fresh medium into a liquid culture contained in a growth chamber. With each drop that enters, an equivalent volume—containing cells, wastes, and spent medium—leaves through an outlet. By manipulating the concentration of nutrients in the medium and the rate at which it enters the growth chamber, a constant cell density and generation time of log phase cells can be maintained. This makes it possible to study a uniform population of log phase cells over a long period of time. The effect of adding various supplements to the medium or altering the cellular environment on long-term cell growth can be determined.

MICROCHECK 4.4

When grown in a closed system, a bacterial population goes through five distinct phases: lag, log, stationary, death, and prolonged decline. Cells within a colony may be in any one of the growth phases, depending on their relative location.

- ✓ Explain the difference between the lag phase and the log phase.
- ✓ Describe how a chemostat keeps a culture in a continuous stage of growth.
- ✓ Why would bacteria be more susceptible to antibiotics during the log phase?

4.5

Environmental Factors That Influence Microbial Growth

Focus Point

- List the descriptive terms that express a prokaryote's requirements for temperature, oxygen, pH, and water availability.

As a group, prokaryotes inhabit nearly every environment on earth. Those we associate with disease and rapid food spoilage live in habitats that humans consider quite comfortable. Some prokaryotes, however, live in harsh environments that would kill most other organisms. Most of these, called **extremophiles** (*phile* means "loving"), are members of the Domain *Archaea.*

Recognizing the environmental factors that influence microbial growth—such as temperature, amount of oxygen, pH, and water availability—helps scientists study microorganisms in the laboratory and aids in understanding their role in the complex ecology of the planet. The major environmental conditions that influence the growth of microorganisms are summarized in **table 4.2.**

TABLE 4.2 Environmental Factors that Influence Microbial Growth

Environmental Factor/ Descriptive Terms	Characteristics
Temperature	Thermostability appears to be due to protein structure.
Psychrophile	Optimum temperature between –5°C and 15°C.
Psychrotroph	Optimum temperature between 20°C and 30°C, but grows well at refrigeration temperatures.
Mesophile	Optimum temperature between 25°C and 45°C.
Thermophile	Optimum temperature between 45°C and 70°C.
Hyperthermophile	Optimum temperature of 70°C or greater.
Oxygen (O_2) Availability	Oxygen (O_2) requirement/tolerance reflects the organism's energy-converting mechanisms (aerobic respiration, anaerobic respiration, and fermentation) and its ability to detoxify O_2 derivatives.
Obligate aerobe	Requires O_2.
Obligate anaerobe	Cannot multiply in the presence of O_2.
Facultative anaerobe	Grows best if O_2 is present, but can also grow without it.
Microaerophile	Requires small amounts of O_2, but higher concentrations are inhibitory.
Aerotolerant anaerobe (obligate fermenter)	Indifferent to O_2.
pH	Prokaryotes that live in pH extremes appear to maintain a near neutral internal pH by pumping protons out of or into the cell.
Neutrophile	Multiplies in the range of pH 5 to 8.
Acidophile	Grows optimally at a pH below 5.5.
Alkalophile	Grows optimally at a pH above 8.5.
Water Availability	Prokaryotes that can grow in high solute solutions maintain the availability of water in the cell by increasing their internal solute concentration.
Halotolerant	Can grow in relatively high salt solutions, up to approximately 10% NaCl.
Halophile	Requires high levels of sodium chloride.

Temperature Requirements

Each species of prokaryote has a well-defined upper and lower temperature limit within which it grows. Within this range lies the **optimum growth temperature,** the temperature at which the organism multiplies most rapidly. As a general rule, this optimum temperature is close to the upper limit of the organism's range. This is because the speed of enzymatic reactions in the

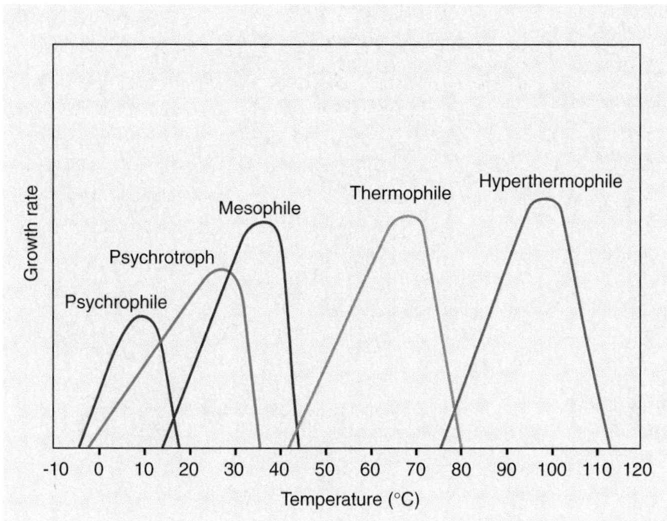

FIGURE 4.9 Temperature Requirements for Growth Prokaryotes are commonly divided into five groups based on their optimum growth temperatures. This graph depicts a typical example of each group. Note that the optimum temperature, the point at which the growth rate is highest, is near the upper limit of the range.

cell approximately doubles for each 10°C rise in temperature. At a critical point, however, the temperature becomes too high and enzymes required for growth are denatured and can no longer function. As a result, the cells die.

Prokaryotes are commonly divided into five groups based on their optimum growth temperatures **(figure 4.9).** Note, however, that this merely represents a convenient organization scheme. In reality, no sharp dividing line exists between each group. Furthermore, not every organism in a group can grow in the entire temperature range typical for its group.

- **Psychrophiles** have their optimum between –5°C and 15°C. These organisms are usually found in such environments as the Arctic and Antarctic regions and in lakes fed by glaciers.
- **Psychrotrophs** have a temperature optimum between 20°C and 30°C, but grow well at lower temperatures. They are an important cause of food spoilage. ■ food spoilage, p. 762
- **Mesophiles** which include *E. coli* and most other common bacteria, have their optimum temperature between 25°C and about 45°C. Disease-causing bacteria, which are adapted to growth in the human body, typically have an optimum between 35°C and 40°C. Mesophiles that inhabit soil, a colder environment, generally have a lower optimum, close to 30°C.
- **Thermophiles** have an optimum temperature between 45°C and 70°C. These organisms commonly occur in hot springs and compost heaps. They also are found in artificially created thermal environments such as water heaters. ■ composting, p. 747
- **Hyperthermophiles** have an optimum growth temperature of 70°C or greater. These are usually members of the *Archaea.* One member, isolated from the wall of a hydrothermal vent deep in the ocean, has a maximum growth temperature of 121°C, the highest yet recorded.

Why can some prokaryotes withstand very high temperatures but most cannot? As a general rule, proteins from thermophiles are not denatured at high temperatures. This thermostability is due to the sequence of the amino acids in the protein. This controls the number and position of the bonds that form within the protein, which in turn determines its three-dimensional structure. For example, the formation of many covalent bonds, as well as many hydrogen and other weak bonds, prevents denaturation of proteins. Heat-stable enzymes that degrade fats and other proteins are being used in high-temperature detergents. ■ protein denaturation, p. 29

Temperature and Food Preservation

Storage of fruits, vegetables, and cheeses at refrigeration temperatures (approximately 4°C) retards food spoilage because it limits the growth of otherwise fast-growing mesophiles. Psychrophiles and psychrotrophs, however, can still multiply and consequently spoilage will still occur, albeit more slowly. Because of this, foods and other perishable products that can withstand below-freezing temperatures should be frozen for long-term storage. Microorganisms, which require liquid water to grow, cannot multiply under these conditions. It is important to recognize, however, that freezing is not an effective means of destroying microbes. Recall that freezing is routinely used to preserve stock cultures. ■ low-temperature storage, p. 122 ■ food spoilage, p. 762

Temperature and Disease

Significant variations exist in the temperature of various parts of the human body. Although the heart, brain, and gastrointestinal tract are near 37°C, the temperature of the extremities may be much lower. For these reasons, some microorganisms can cause disease more readily in certain body parts but not in others. For example, Hansen's disease (leprosy) typically involves the coolest regions of the body (ears, hands, feet, and fingers) because the causative organism, *Mycobacterium leprae*, grows best at these lower temperatures. The same situation applies to syphilis, in which lesions appear on the genitalia and then on the lips, tongue, and throat. Indeed, for more than 30 years the major treatment of syphilis was to induce fever by deliberately introducing the agent that causes malaria, which results in very high fevers. ■ Hansen's disease, p. 655 ■ syphilis, p. 633

Oxygen (O_2) Requirements

The oxygen (O_2) level in different environments varies greatly, providing many different habitats with respect to its availability. Gaseous oxygen accounts for about 20% of the earth's atmosphere. Beneath the surface of soil and in swamps, however, very limited amounts, if any, may be available. The human body alone provides many different habitats. While the surface of the skin is exposed to the atmosphere, the stomach and intestines are relatively **anaerobic,** meaning they contain little or no O_2.

Like humans, some bacteria have an absolute requirement for O_2. Others thrive in anaerobic environments, and many of these are killed if O_2 is present. The O_2 requirements of some organisms can be determined by growing them in **shake tubes.** To prepare a shake tube, a tube of nutrient agar is boiled, which both melts the agar and drives off the O_2. The agar is then allowed to cool

TABLE 4.3 Oxygen (O_2) Requirements of Prokaryotes

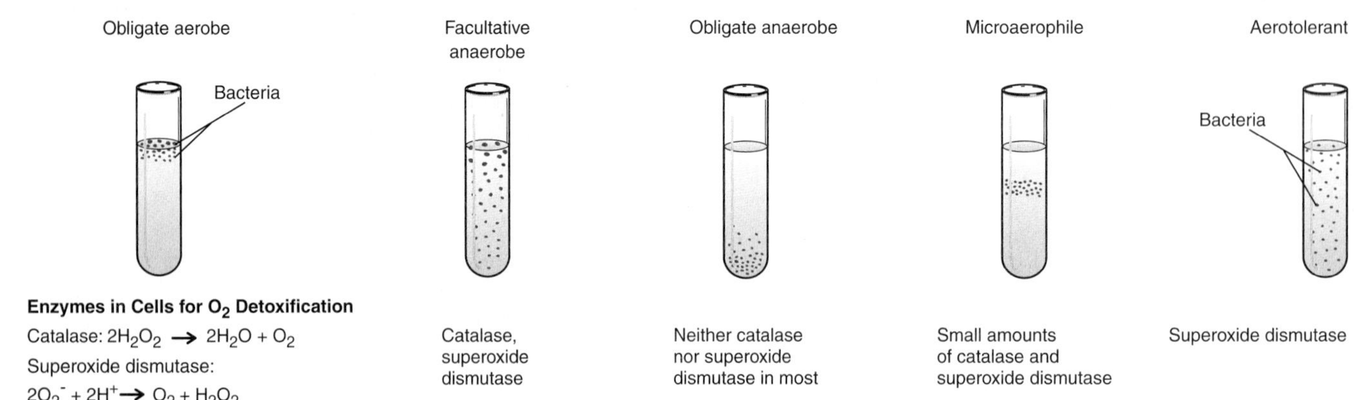

Obligate aerobe	Facultative anaerobe	Obligate anaerobe	Microaerophile	Aerotolerant
Enzymes in Cells for O_2 Detoxification Catalase: $2H_2O_2 \rightarrow 2H_2O + O_2$ Superoxide dismutase: $2O_2^- + 2H^+ \rightarrow O_2 + H_2O_2$	Catalase, superoxide dismutase	Neither catalase nor superoxide dismutase in most	Small amounts of catalase and superoxide dismutase	Superoxide dismutase

to 50°C. Next, the test organism is added and dispersed by gentle shaking or swirling. The agar is allowed to harden and the tube is incubated at an appropriate temperature. Because the solidified agar impedes the diffusion of O_2, the level of O_2 in the tube is high at the top, whereas the bottom portion is anaerobic. The bacteria grow in the region that has the level of O_2 that suits their requirements **(table 4.3).**

Based on their O_2 requirements, prokaryotes can be separated into these groups:

- **Obligate aerobes** have an absolute requirement for oxygen (O_2). They use it to transform energy in the process of aerobic respiration. This and other ATP-generating pathways will be discussed in detail in chapter 6. Obligate aerobes include *Micrococcus* species, which are common in the environment. ■ aerobic respiration, pp. 133, 144 ■ ATP, p. 25
- **Obligate anaerobes** cannot multiply if any O_2 is present; in fact, they are often killed in environments that have even traces of O_2 because of its toxic derivatives, which will be discussed shortly. Obligate anaerobes transform energy by fermentation or anaerobic respiration; the details of these processes will be discussed in chapter 6. Obligate anaerobes include members of the genus *Bacteroides* (the major inhabitants of the large intestine), *Clostridium botulinum* (the causative agent of botulism), and many others. In fact, it is estimated that one-half of all the cytoplasm on earth is in anaerobic bacteria! ■ fermentation, pp. 133, 147 ■ anaerobic respiration, pp. 133, 145
- **Facultative anaerobes** grow better if O_2 is present, but can also grow without it. The term "facultative" means that the organism is flexible, in this case in its requirements for O_2. Facultative anaerobes use aerobic respiration if oxygen is available, but use fermentation or anaerobic respiration in its absence. Growth is more rapid when oxygen is present because aerobic respiration yields the most ATP of all these processes. An example is *E. coli,* a common inhabitant of the large intestine.
- **Microaerophiles** require small amounts of O_2 (2% to 10%) for aerobic respiration; higher concentrations are inhibitory. An example is *Helicobacter pylori,* which causes gastric and duodenal ulcers.
- **Aerotolerant anaerobes** are indifferent to O_2. They can grow in its presence, but they do not use it to transform energy. Because they do not use aerobic or anaerobic respiration, they are also called **obligate fermenters.** An example is *Streptococcus pyogenes,* which causes strep throat.

Toxic Derivatives of Oxygen (O_2)

Although not toxic itself, O_2 can be converted into a number of compounds that are highly toxic. Some of these, such as **superoxide** (O_2^-), are produced both as a part of normal metabolic processes and as chemical reactions involving oxygen and light. Others, such as **hydrogen peroxide** (H_2O_2), result from metabolic processes involving oxygen. To survive in an environment containing O_2, cells must have enzymes that can convert these toxic derivatives to non-toxic forms. The enzyme **superoxide dismutase** degrades superoxide to produce hydrogen peroxide. **Catalase** breaks down hydrogen peroxide to H_2O and O_2. Together, these two enzymes detoxify these reactive products of O_2.

Although most strict anaerobes do not have superoxide dismutase, some do, while a few aerobes lack it. Therefore, other unknown factors must also be playing a role in protecting organisms from the toxic forms of oxygen.

pH

Each bacterial species can survive within a range of pH values; within this range is its pH optimum. Despite the pH of the external environment, cells maintain a constant internal pH, typically near neutral. ■ pH, p. 24

Most bacteria can live and multiply within the range of pH 5 (acidic) to pH 8 (basic) and have a pH optimum near neutral (pH 7). These bacteria are called **neutrophiles.** Preservation methods that acidify foods, such as pickling, are intended to inhibit the growth of these organisms. Surprisingly, some neutrophiles have adapted special mechanisms that enable them to grow at a very low pH. For example, *Helicobacter pylori* grows in the stomach, where it can cause ulcers. To maintain the pH close to neutral in its immediate surroundings, *H. pylori* produces the enzyme **urease,** which splits urea in the stomach into carbon dioxide and ammonia. The ammonia neutralizes the stomach acid in the bacterium's immediate surroundings. ■ pickling, p. 758

Acidophiles grow optimally at a pH below 5.5. For example, *Acidothiobacillus ferroxidans,* grows best at a pH of approximately 2. This bacterium obtains its energy by oxidizing sulfur compounds, producing sulfuric acid in the process. It maintains its internal pH near neutral by pumping out protons (H^+) as quickly as they enter the cell. *Picrophilus oshimae*, a member of the *Archaea*, has an optimum pH of less than 1! This prokaryote, which was isolated from the dry, acid soils of a gas-emitting volcanic fissure in Japan, has an unusual cytoplasmic membrane that is unstable at a pH above 4.0.

Alkalophiles grow optimally at a pH above 8.5. For instance, the bacterium *Bacillus alcalophilus* grows best at pH 10.5. It appears that alkalophiles maintain a relatively neutral internal pH by exchanging internal sodium ions for external protons. Alkalophiles often live in alkaline lakes and soils.

Water Availability

All microorganisms require water for growth. Even if water is present, however, it may not be available in certain environments. For example, dissolved substances such as salt (NaCl) and sugars interact with water molecules and make the water unavailable to the cell. In many environments, particularly in certain natural habitats such as salt marshes, prokaryotes are faced with this situation. If the solute concentration is higher in the medium than in the cell, water diffuses out of the cell due to osmosis. This causes the cytoplasm to dehydrate and shrink from the cell wall, a phenomenon called **plasmolysis (figure 4.10).** ■ solute p. 56 ■ osmosis, p. 56

Prokaryotes able to live in high-salt environments maintain the availability of water in the cell by increasing their internal solute concentration. Some bacteria do this by synthesizing certain small organic compounds, such as the amino acid proline, that have no detrimental effect on normal cellular activity. ■ proline, p. 27

Bacteria that can tolerate high concentrations of salt, up to approximately 10% NaCl, are called **halotolerant.** *Staphylococcus* species, which reside on the dry salty environment of the skin, are an example. Some organisms actually require high levels of sodium chloride to grow and are called **halophiles** (*halo* means "salt"). Many marine bacteria are mildly halophilic, requiring concentrations of approximately 3% sodium chloride. Certain members of the *Archaea* are **extreme halophiles,** requiring 9% sodium chloride or more. Extreme halophiles are found in environments such as the salt flats of Utah and the Dead Sea.

The growth-inhibiting effect of high concentrations of salt and sugars is used in food preservation. High levels of salt are added to preserve such foods as bacon, salt pork, and anchovies. High concentrations of sugars can also inhibit the growth of bacteria. Many foods with a high sugar content, such as jams, jellies, and honey, are naturally preserved. ■ food preservation, p. 765

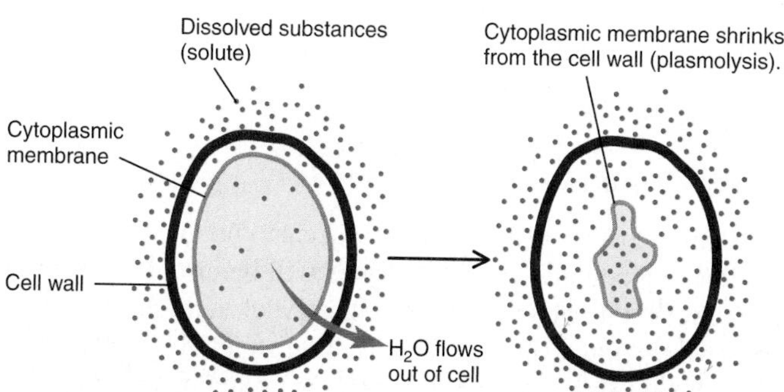

FIGURE 4.10 Effects of Solute Concentration on Cells The cytoplasmic membrane allows water molecules to pass through freely. If the solute concentration is higher outside of the cell, water moves out. The dehydrated cytoplasm shrinks from the cell wall, a process called plasmolysis.

MICROCHECK 4.5

A prokaryotic species can be categorized according to its optimum growth temperature. A species can also be grouped according to its oxygen requirements. Most species grow best near neutral pH, although some prefer acidic conditions and others grow best in alkaline conditions. Halophiles require high-salt conditions.

✓ List four environmental factors that influence bacterial growth.
✓ List the categories into which bacteria can be classified according to their requirements for oxygen.
✓ Why would small organic compounds affect the water content of cells?

4.6

Nutritional Factors That Influence Microbial Growth

Focus Points

- Give an example of a bacterium that is fastidious.
- Define the terms *photoautotroph, chemolithoautotroph, photoheterotroph,* and *chemoorganoheterotroph.*

Growth of any prokaryote depends not only on a suitable physical environment, but also on the availability of nutrients. From these, the cell must synthesize all of the cell components discussed in chapter 3, including lipid membranes, cell walls, proteins, and nucleic acids. These components are made from building blocks such as fatty acids, sugars, amino acids, and nucleotides. In turn, each of these building blocks is composed of a variety of elements, including carbon and nitrogen. What sets the prokaryotic world apart from all other forms of life is their remarkable ability to use diverse sources of these elements. For example, prokaryotes are the only organisms able to use atmospheric nitrogen (N_2) as a nitrogen source.

Required Elements

Elements that make up cell constituents are called **major elements.** These include carbon, oxygen, hydrogen, nitrogen, sulfur, phosphorus, potassium, magnesium, calcium, and iron. They are

PERSPECTIVE 4.1

Can Prokaryotes Live on Only Rocks and Water?

Prokaryotes have been isolated from diverse environments that previously were thought to be incapable of sustaining life. For example, members of the *Archaea* have been isolated from environments 10 times more acidic than that of lemon juice. Other *Archaea* have been isolated from oil wells a mile below the surface of the earth at temperatures of 70°C and pressures of 160 atmospheres (at sea level, the pressure is 1 atmosphere). The isolation of these organisms suggests that thermophiles may be widespread in the earth's crust.

Perhaps the most unusual environment from which prokaryotes have been isolated are the volcanic rocks 1 mile below the earth's surface near the Columbia River in Washington State. What do these organisms use for food? They apparently get their energy from hydrogen gas that is produced chemically in a reaction between the iron-rich minerals in the rock and the groundwater. The groundwater also contains dissolved CO_2, which the bacteria use as a source of carbon. Thus, these bacteria apparently exist on nothing more than rocks and water.

the essential components of proteins, carbohydrates, lipids, and nucleic acids **(table 4.4).** ■ elements, p. 19

The source of carbon distinguishes different groups of prokaryotes. Those that use organic carbon are called **heterotrophs** (*hetero* means "different" and *troph* means "nourishment"). Medically important bacteria are typically heterotrophs, using organic carbon sources such as glucose. **Autotrophs** (*auto* means "self") use inorganic carbon in the form of carbon dioxide. They play a critical role in the cycling of carbon in the environment because they can convert inorganic carbon (CO_2) to an organic form, the process of **carbon fixation.** Without carbon fixation, the earth would quickly run out of organic carbon, which is essential to humans and other animals. ■ carbon cycle, p. 728

Some prokaryotes are able to use nitrogen gas (N_2), converting it to ammonia, which can be incorporated into cellular material.This process, called **nitrogen fixation,** is unique to prokaryotes. Like carbon fixation, it is essential to life on this planet. ■ nitrogen cycle, p. 730

Phosphorus and iron are important because they are often **limiting nutrients,** meaning they are present at the lowest concentration relative to need. Just as the quantity of chocolate chips in your kitchen would limit the number of chocolate chip cookie batches you can make (assuming the other ingredients are on hand), a limiting nutrient dictates the maximum level of microbial growth. Algal blooms in a small Seattle lake were curtailed by chemical treatment that removed excess phosphate in the lake.

Some elements, termed **trace elements,** are required in very minute amounts by all cells. They include cobalt, zinc, copper, molybdenum, and manganese, which are required for enzyme function. Very small amounts of these trace elements are found in most natural environments, including water.

TABLE 4.4 Representative Functions of the Major Elements

Chemical	Function
Carbon, oxygen, and hydrogen	Component of cellular constituents including amino acids, lipids, nucleic acids, and sugars.
Nitrogen	Component of amino acids and nucleic acids.
Sulfur	Component of some amino acids.
Phosphorus	Component of nucleic acids, membrane lipids, and ATP.
Potassium, magnesium, and calcium	Required for the functioning of certain enzymes; additional functions as well.
Iron	Part of certain enzymes.

Growth Factors

Some bacteria cannot synthesize some of their cell constitutents, such as amino acids, vitamins, purines, and pyrimidines. Consequently, these organisms can only grow in environments that contain these compounds. Low molecular weight compounds required by a particular bacterium are called **growth factors.** ■ purines, p. 33 ■ pyrimidines, p. 33

Microorganisms display a wide spectrum in their growth factor requirements, reflecting differences in their biosynthetic capabilities. For example, *E. coli* is quite versatile and does not require any growth factors. It grows in a medium containing only glucose and six different inorganic salts. In contrast, species of *Neisseria* require at least 40 additional ingredients, including 7 vitamins and all of the 20 amino acids. Bacteria such as *Neisseria* that require many growth factors are called **fastidious.**

Fastidious bacteria are exploited to determine the quantity of specific vitamins in food products. To do this, a well-characterized species that requires a specific vitamin to grow is inoculated into a medium that lacks the vitamin but is supplemented with a measured amount of the food product. The amount of growth of the bacterium is related to the amount of test vitamin in the product.

Energy Sources

Organisms derive energy either from sunlight or by oxidizing chemical compounds. These processes will be discussed in chapter 6. Organisms that harvest the energy of sunlight are called **phototrophs** (*photo* means "light"). These include plants, algae, and photosynthetic bacteria. Organisms that obtain energy by oxidizing chemical compounds are called **chemotrophs** (*chemo* means "chemical"). Mammalian cells, fungi, and many types of bacteria oxidize organic compounds such as sugars, amino acids,

TABLE 4.5 Energy and Carbon Sources Used by Different Groups of Prokaryotes

Type	Energy Source	Carbon Source
Photoautotroph	Sunlight	CO_2
Photoheterotroph	Sunlight	Organic compounds
Chemolithoautotroph	Inorganic chemicals (H_2, NH_3, NO_2^-, Fe^{2+}, H_2S)	CO_2
Chemoorganoheterotroph	Organic compounds (sugars, amino acids, etc.)	Organic compounds

and fatty acids. Some prokaryotes can extract energy from seemingly unlikely sources such as hydrogen sulfide, hydrogen gas, and other inorganic compounds, an ability that distinguishes them from eukaryotes.

Nutritional Diversity

Microbiologists often group prokaryotes according to the energy and carbon sources they utilize **(table 4.5):**

- **Photoautotrophs** use the energy of sunlight and the CO_2 in the atmosphere to make organic compounds. These are eventually consumed by other organisms, including humans. Because of this, photoautotrophs are called **primary producers.** Cyanobacteria are important examples that inhabit soil and both freshwater and saltwater environments. Many can fix nitrogen, providing another indispensable role in the biosphere; they are the only organisms that can fix both N_2 and CO_2.
- **Chemolithoautotrophs** (*lith* means "stone"), commonly referred to simply as chemoautotrophs or chemolithotrophs, use inorganic compounds for energy and derive their carbon from CO_2. These prokaryotes live in seemingly inhospitable environments such as sulfur hot springs, which are rich in reduced inorganic compounds such as hydrogen sulfide. In some regions of the ocean depths, near hydrothermal vents, chemoautotrophs serve as the primary producers, supporting rich communities of life in these habitats utterly devoid of sunlight (see figure 30.11). ■ **hydrothermal vents, p. 732**
- **Photoheterotrophs** use the energy of sunlight and derive their carbon from organic compounds. Some are facultative in their nutritional capabilities. For example, some members of a group of bacteria called the purple nonsulfur bacteria can grow anaerobically using light as an energy source and organic compounds as a carbon source (photoheterotrophs). They can also grow aerobically in the dark using organic sources of carbon and energy (chemoheterotrophs). ■ **purple nonsulfur bacteria, p. 257**
- **Chemoorganoheterotrophs,** also referred to as chemoheterotrophs or chemoorganotrophs use organic compounds for energy and as a carbon source. They are by far the most common group associated with humans and other animals. Individual species of chemoheterotrophs differ in the number of organic compounds they can use. For example, certain members of the genus *Pseudomonas* can derive carbon and/or energy from more than 80 different organic compounds, including such unusual compounds as naphthalene (the ingredient associated with the smell of mothballs). At the other extreme, some organisms can degrade only a few compounds. For example, *Bacillus fastidiosus* can use only urea and certain of its derivatives as a source of both carbon and energy.

MICROCHECK 4.6

Organisms require a source of major and trace elements. Heterotrophs use an organic carbon source, and autotrophs use CO_2. Phototrophs harvest the energy of sunlight, and chemotrophs obtain energy by oxidizing chemicals.

- ✓ List the major elements required for growth of bacteria.
- ✓ What is the carbon source of a photoautotroph? Of a chemoautotroph?
- ✓ Why would human-made materials (such as many plastics) be degraded only slowly or not at all?

4.7 Cultivating Prokaryotes in the Laboratory

Focus Points

- Compare and contrast complex, chemically defined, selective, and differential media.
- Explain how the correct atmospheric conditions are provided to cultivate obligate aerobes, capnophiles, microaerophiles, and obligate anaerobes.
- Describe the purpose of an enrichment culture.

By knowing the environmental and nutritional factors that influence growth of specific prokaryotes, it is often possible to provide appropriate conditions for their cultivation. These include a medium on which to grow the organisms and a suitable atmosphere.

General Categories of Culture Media

Considering the diversity of bacteria, it is not surprising that a wide variety of media is used to cultivate them. For routine purposes, one of the many types of **complex media** is used; **chemically defined media** are generally used only for specific research

TABLE 4.6 Characteristics of Representative Media Used to Cultivate Bacteria

Medium	Characteristic
Blood agar	Complex medium used routinely in clinical labs. Differential because colonies of hemolytic organisms are surrounded by a zone of clearing of the red blood cells. Not selective.
Chocolate agar	Complex medium used to culture fastidious bacteria, particularly those found in clinical specimens. Not selective or differential.
Glucose-salts	Chemically defined medium. Used in laboratory experiments to study nutritional requirements of bacteria. Not selective or differential.
MacConkey agar	Complex medium used to isolate Gram-negative rods that typically reside in the intestine. Selective because bile salts and dyes inhibit Gram-positive organisms and Gram-negative cocci. Differential because the pH indicator turns pink-red when the sugar in the medium, lactose, is fermented.
Nutrient agar	Complex medium used for routine laboratory work. Supports the growth of a variety of nonfastidious bacteria. Not selective or differential.
Thayer-Martin	Complex medium used to isolate *Neisseria* species, which are fastidious. Selective because it contains antibiotics that inhibit most organisms except *Neisseria* species. Not differential.

experiments when the type and quantity of nutrients must be precisely controlled. **Table 4.6** summarizes the characteristics of various types of media.

Complex Media

A **complex medium** contains a variety of ingredients such as meat juices and digested proteins, making what might be viewed as a tasty soup for microbes. Although a specific amount of each ingredient is in the medium, the exact chemical composition of these can be highly variable. One common ingredient is **peptone.** This is a mixture of amino acids and short peptides produced by digesting protein from any of a variety of different sources with enzymes, acids or alkali. **Extracts,** which are the water-soluble components of a substance, are also common ingredients. For example, beef extract is a water extract of lean meat and provides vitamins, minerals, and other nutrients. A commonly used complex medium, **nutrient broth,** consists of peptone and beef extract in distilled water. If agar is added, then **nutrient agar** results.

Many medically important bacteria are fastidious, requiring a medium even richer than nutrient agar. One rich medium commonly used in clinical laboratories is **blood agar.** This contains red blood cells, which supply a variety of nutrients including hemin, in addition to other ingredients. A medium used to cultivate even more fastidious bacteria is **chocolate agar,** named for its brownish appearance rather than its ingredients. Chocolate agar contains lysed red blood cells and additional nutrients.

Additional ingredients are often incorporated into complex media to counteract compounds that may be toxic to some exquisitely sensitive bacteria. For example, cornstarch sometimes included because it binds fatty acids, which can be toxic to *Neisseria* species.

Hundreds of different types of media are manufactured. Even with the availability of all of these, however, some medically important organisms and most environmental ones have not yet been grown on culture media.

Chemically Defined Media

Chemically defined media are composed of precise amounts of pure chemicals. This type of medium is invaluable when studying nutritional requirements of bacteria. **Glucose-salts,** which supports the growth of *E. coli,* contains only those chemicals listed in **table 4.7.** More elaborate recipes containing as many as 46 different ingredients can be used to make chemically defined media that support the growth of fastidious bacteria such as *Neisseria gonorrhoeae*, the organism that causes gonorrhea. ■ gonorrhea, p. 629

To maintain the pH near neutrality, buffers are often added to the medium. They are especially important in a defined medium because some bacteria produce so much acid as a by-product of metabolism that they inhibit their own growth. This typically is not as much of a problem in complex media because the amino acids and other natural components provide at least some buffering function. ■ buffer, p. 24

Special Types of Culture Media

To detect or isolate a bacterium that is part of a mixed population, it is often necessary to make it more prevalent or more obvious. For these purposes selective and differential media are used. These can be either complex or chemically defined, depending on the needs of the microbiologist.

Selective Media

Selective media inhibit the growth of organisms other than the one being sought. For example, **Thayer-Martin agar** is used to isolate *Neisseria gonorrhoeae,* the cause of gonorrhea, from clinical specimens. This is chocolate agar to which three or more antimicrobial drugs have been added. The antimicrobials inhibit fungi, Gram-positive bacteria, and Gram-negative rods. Because these drugs do not inhibit most strains of *N. gonorrhoeae*, they allow those strains to grow with little competition from other organisms.

TABLE 4.7 Ingredients in Two Representative Types of Media that Support the Growth of *E. coli*

Nutrient Broth (complex medium)	Glucose-Salts (defined medium)
Peptone	Glucose
Meat extract	Dipotassium phosphate
Water	Monopotassium phosphate
	Magnesium sulfate
	Ammonium sulfate
	Calcium chloride
	Iron sulfate
	Water

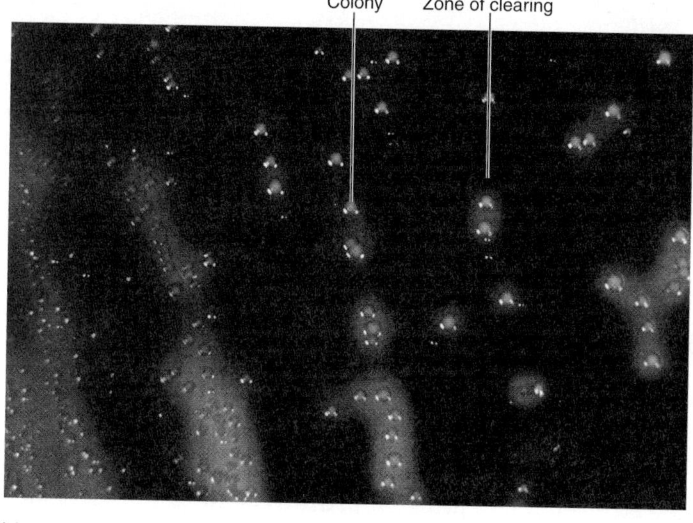

(a)

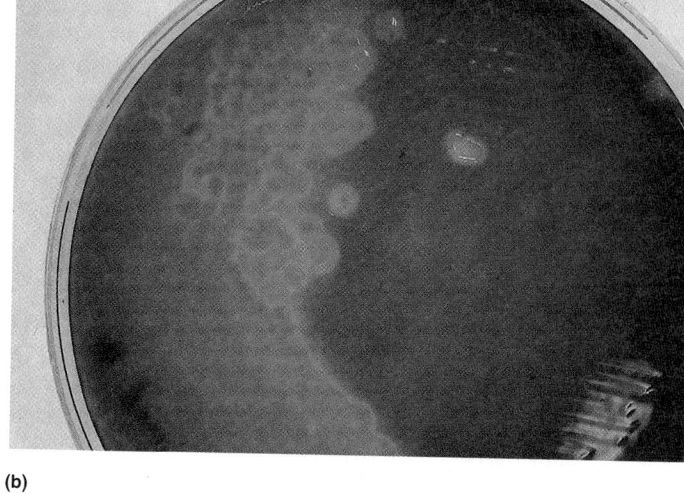

(b)

FIGURE 4.11 Blood Agar This complex medium is differential for hemolysis. **(a)** A zone of complete clearing around a colony growing on blood agar is called beta hemolysis. **(b)** A zone of greenish clearing is called alpha hemolysis.

MacConkey agar is used to isolate Gram-negative rods from various clinical specimens such as urine. This complex medium contains, in addition to peptones and other nutrients, two inhibitory compounds: crystal violet, a dye, inhibits Gram-positive bacteria, and bile salts inhibit most non-intestinal bacteria.

Differential Media

Differential media contain a substance that certain bacteria change in a recognizable way. For example, blood agar, in addition to being nutritious, is differential; it is used to detect bacteria that produce a **hemolysin,** which lyses red blood cells **(figure 4.11).** The lysis appears as a zone of clearing around the colony growing on the blood agar plate. The type of hemolysis is used as an identifying characteristic. For example, species of *Streptococcus* that reside harmlessly in the throat often cause a type of hemolysis called **alpha hemolysis,** characterized by a zone of greenish partial clearing around the colonies. In contrast, *Streptococcus pyogenes*, which causes strep throat, causes **beta hemolysis,** characterized by a clear zone of hemolysis. Still other bacteria have no effect on red blood cells. ■ ***Streptococcus pyogenes*, p. 499**

MacConkey agar, which is selective, is also differential **(figure 4.12).** In addition to containing peptones and other nutrients, it has lactose and a pH indicator. Bacteria that ferment the sugar produce acid, which turns the pH indicator pink. Thus, *E. coli* and other lactose-fermenting bacteria growing on MacConkey agar form pink colonies. Lactose-negative bacteria form tan or colorless colonies. ■ **lactose, p. 32**

Providing Appropriate Atmospheric Conditions

To cultivate bacteria in the laboratory, appropriate atmospheric conditions must be provided. For instance, broth cultures of obligate aerobes grow best when tubes or flasks containing the media are shaken, providing maximum aeration. Special methods create atmospheric environments such as increased CO_2, microaerophilic, and anaerobic conditions.

Increased CO_2

Providing an environment with increased levels of CO_2 enhances the growth of many medically important bacteria, including species of *Neisseria* and *Haemophilus*. Organisms requiring increased CO_2, along with approximately 15% oxygen, are called **capnophiles.** One of the simplest ways to provide this atmosphere is to incubate the bacteria in a closed **candle jar.** A lit candle in the jar consumes some of the O_2 in the air, generating CO_2 and H_2O; the flame soon extinguishes because of insufficient oxygen. Although a candle jar atmosphere contains about 3.5% CO_2, enough O_2 remains to support the growth of obligate aerobes and prevent the growth of obligate anaerobes. Special incubators are also available that maintain CO_2 at prescribed levels.

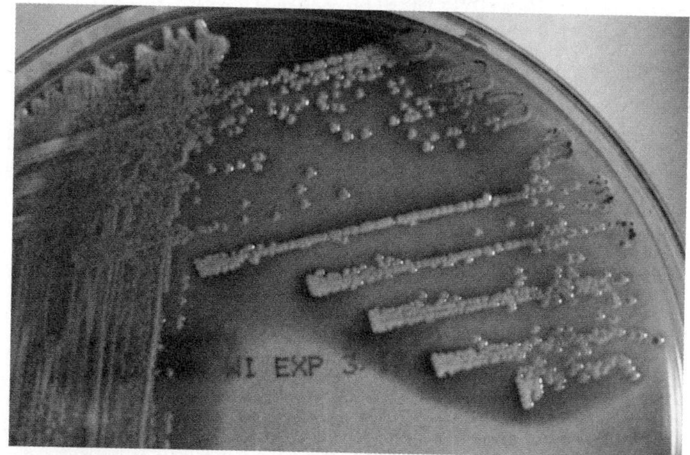

FIGURE 4.12 MacConkey Agar This complex medium is differential for lactose fermentation and selective for Gram-negative rods. Bacteria that ferment the sugar produce acid, which turns the pH indicator pink, resulting in pink colonies. Lactose-negative colonies are tan or colorless. The bile salts and dyes in the media inhibit all but certain Gram-negative rods.

Microaerophilic

Microaerophilic bacteria typically require O_2 concentrations less than what is achieved in a candle jar. Therefore, these bacteria are often incubated in a gastight container with a special disposable packet; the packet holds chemicals that react with O_2, reducing its concentration to approximately 5–15%.

Anaerobic

Cultivation of obligate anaerobes presents a great challenge to the microbiologist, because the cells may be killed if they are exposed to O_2 for even a short time. Obviously, special techniques to exclude O_2 are required.

Anaerobes that can tolerate a brief exposure to O_2 are cultivated in an **anaerobe container (figure 4.13).** This is the same type of container used to incubate microaerophiles, but the chemical composition of the disposable packet produces an anaerobic environment.

Another method to cultivate anaerobes incorporates **reducing agents** into the culture medium. These react with O_2 and thus eliminate dissolved O_2; they include sodium thioglycollate, cysteine, and ascorbic acid. In some cases, immediately before the bacteria are inoculated, the medium is boiled to drive out dissolved O_2. Media that employ reducing agents frequently contain an O_2-indicating dye such as methylene blue.

A more stringent method for working with anaerobes is to use an **anaerobic chamber,** an enclosed compartment that can be maintained as an anaerobic environment **(figure 4.14).** A special port, which can be filled with an inert gas, is used to add or remove items. Airtight gloves enable researchers to handle items within the chamber.

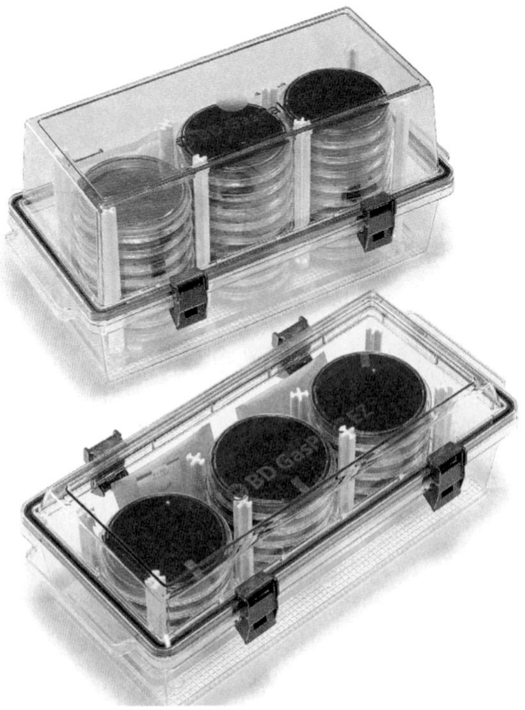

FIGURE 4.13 Anaerobe Container A disposable packet contains chemicals that react with O_2, thereby producing an anaerobic environment.

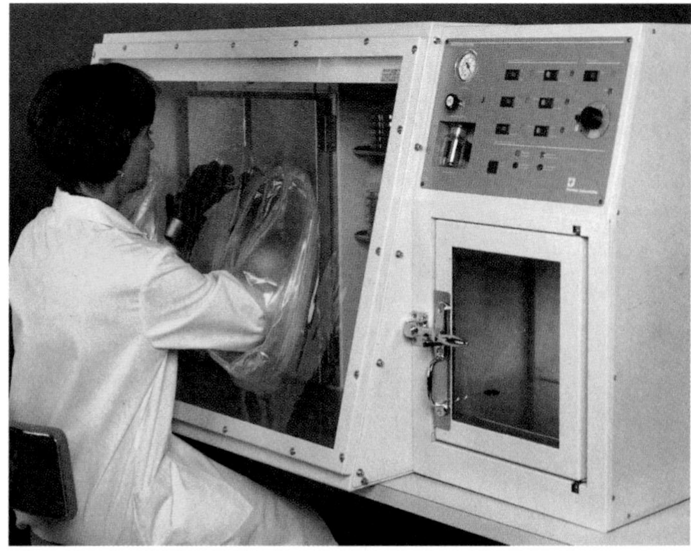

FIGURE 4.14 Anaerobic Chamber The enclosed compartment can be maintained as an anaerobic environment. A special port (visible on the right side of this device), which can be filled with inert gas, is used to add or remove items. The airtight gloves enable the researcher to handle items within the chamber.

Enrichment Cultures

An **enrichment culture** provides conditions in a broth that preferentially enhance the growth of one particular species in a mixed population **(figure 4.15).** This is helpful in isolating an organism from natural sources when the bacterium of interest is present in relatively small numbers. For example, if an organism is present at a concentration of only 1 cell/ml and it is outnumbered 10,000-fold by other organisms, isolating it using the streak-plate method would be difficult, even if a selective medium were used.

To enrich for a species, a sample such as pond water is placed into a liquid medium that favors the growth of the desired organism over others. For example, if the target organism can grow using atmospheric nitrogen as a source of nitrogen, then nitrogen is left out of the medium. If it can use an unusual carbon source such as phenol, then that is added as the only carbon source. In some cases selective agents such as bile are added: the procedure is then referred to as a **selective enrichment.** The culture is incubated under temperature and atmospheric conditions that preferentially promote the growth of the desired organism. During this time, the relative concentration of a microorganism that initially made up only a minor fraction of the population can increase dramatically. A pure culture can then be obtained by streaking the enrichment onto an appropriate agar medium and selecting a single colony.

MICROCHECK 4.7

Culture media can be either complex or chemically defined. Some media contain additional ingredients that make them selective or differential. Appropriate atmospheric conditions must be provided to isolate microaerophiles and anaerobes. An enrichment culture increases the relative concentration of an organism growing in a broth.

- ✓ Distinguish between a complex and a chemically defined medium.
- ✓ Describe two methods to create anaerobic conditions.
- ✓ Would bacteria that cannot utilize lactose be able to grow on MacConkey agar?

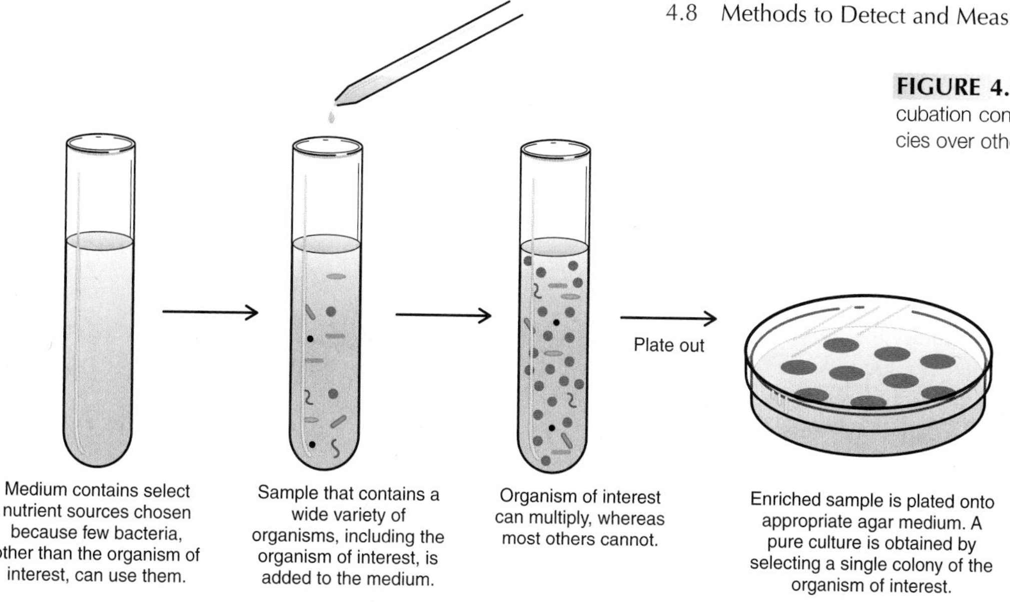

Plate out
Medium contains select nutrient sources chosen because few bacteria, other than the organism of interest, can use them.
Sample that contains a wide variety of organisms, including the organism of interest, is added to the medium.
Organism of interest can multiply, whereas most others cannot.
Enriched sample is plated onto appropriate agar medium. A pure culture is obtained by selecting a single colony of the organism of interest.

FIGURE 4.15 Enrichment Culture Medium and incubation conditions favor the growth of the desired species over other bacteria in the same sample.

4.8

Methods to Detect and Measure Bacterial Growth

Focus Point

- Compare and contrast direct cell counts, viable cell counts, measuring biomass, and detecting cell products to measure bacterial growth.

A variety of techniques are available to monitor bacterial growth. The choice depends on various characteristics of the sample and the goals of the measurements. Characteristics of the common methods for measuring bacterial growth are summarized in **table 4.8.**

Direct Cell Counts

Direct cell counts are particularly useful for determining the total numbers of bacteria in a specimen, including those that

TABLE 4.8 Methods Used to Measure Bacterial Growth

Method	Characteristics and Limitations
Direct Cell Counts	Used to determine total number of cells; counts include living and dead cells.
Direct microscopic count	Rapid, but at least 10^7 cells/ml must be present to be effectively counted.
Cell-counting instruments	Coulter counters and flow cytometers count total cells in dilute solutions. Flow cytometers can also be used to count organisms to which fluorescent dyes or tags have been attached.
Viable Cell Counts	Used to determine the number of viable bacteria in a sample, but that number only includes those that can grow in given conditions. Requires an incubation period of approximately 24 hours or longer. Selective and differential media can be used to enumerate specific species of bacteria.
Plate count	Time-consuming but technically simple method that does not require sophisticated equipment. Generally used only if the sample has at least 10^2 cells/ml.
Membrane filtration	Concentrates bacteria by filtration before they are plated; thus can be used to count cells in dilute environments.
Most probable number	Statistical estimation of likely cell number; it is not a precise measurement. Can be used to estimate numbers of bacteria in relatively dilute solutions.
Measuring Biomass	Biomass can be correlated to cell number.
Turbidity	Very rapid method; used routinely. A one-time correlation with plate counts is required in order to use turbidity for determining cell number.
Total weight	Tedious and time-consuming; however, it is one of the best methods for measuring the growth of filamentous microorganisms.
Measuring Cell Products	Methods are rapid but results must be correlated to cell number. Frequently used to detect growth, but not routinely used for quantitation.
Acid	Titration can be used to quantify acid production. A pH indicator is often used to detect growth.
Gases	Carbon dioxide can be detected by using a molecule that fluoresces when the medium becomes slightly more acidic. Gases can be trapped in an inverted Durham tube in a tube of broth.
ATP	Firefly luciferase catalyzes light-emitting reaction when ATP is present.

cannot be grown in culture. Unfortunately, they generally do not distinguish between living and dead cells.

Direct Microscopic Count

One of the most rapid methods of determining the number of cells in a suspension is the direct microscopic count. The number of cells in a measured volume of liquid is counted using special glass slides—**counting chambers**—that hold a known volume of liquid **(figure 4.16).** These can be viewed under the light microscope, and the number of cells can be counted precisely. At least 10 million bacteria (10^7) per milliliter are required to gain an accurate estimate. Otherwise, few, if any, cells will be seen in the microscope field.

Cell-Counting Instruments

A **Coulter counter** is an electronic instrument that counts cells in a suspension as they pass single file through a minute aperture **(figure 4.17).** The suspending liquid must be an electrically conducting fluid, because the machine counts the brief changes in resistance that occur when non-conducting particles such as bacteria pass by.

A **flow cytometer** is similar in principle to a Coulter counter except it measures the scattering of light by cells as they pass by a laser. The instrument can be used to count either total cells or, by using special techniques, a specific population. This is done by first staining cells with a fluorescent dye or tag that binds only to the cells of interest; the flow cytometer then counts those cells that carry the fluorescent marker. ■ **fluorescent dyes and tags, p. 51**

Viable Cell Counts

Viable cell counts are used to quantify the number of cells capable of multiplying. These methods require knowledge of appropriate growth conditions for a particular microorganism as well as the time to allow growth to occur. By using selective and differential media, a particular species of bacteria can often be enumerated.

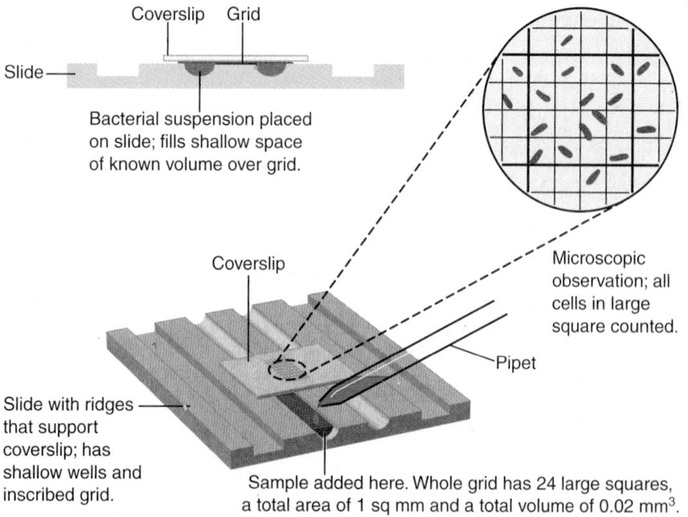

FIGURE 4.16 A Counting Chamber This special glass slide holds a known volume of liquid. The number of bacteria in that volume can be counted precisely.

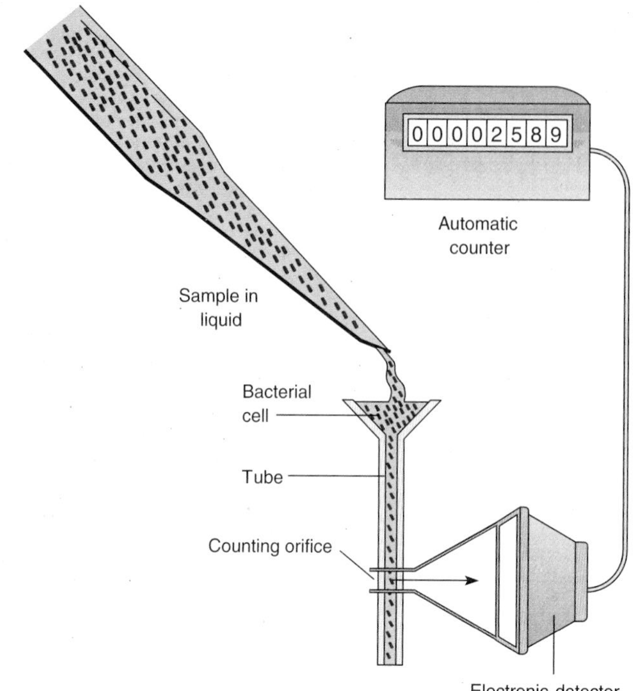

FIGURE 4.17 A Coulter Counter This instrument counts cells as they pass through a minute aperture. The bacteria, which are suspended in an electrically conducting liquid, cause a brief change in resistance as they pass by the counter.

Viable cell counts are invaluable for monitoring bacterial growth in samples such as food and water that often contain numbers too low to be seen using a direct microscopic count.

Plate Counts

Plate counts measure the number of viable cells in a sample by exploiting the fact that an isolated cell on a nutrient agar plate will give rise to one colony. A simple count of the colonies determines how many cells were in the initial sample **(figure 4.18).** As the ideal number of colonies to count is between 30 and 300, and samples frequently contain many more bacteria than this, it is usually necessary to dilute the samples before plating the cells. Samples are normally diluted in 10-fold increments, making the resulting math relatively simple. The **diluent,** or sterile solution used to make the dilutions, is generally physiological saline (0.85% NaCl in water).

In the **pour-plate** method, 0.1 to 1.0 ml of the final dilution is transferred into a sterile Petri dish and then overlaid with melted nutrient agar that has been cooled to 50°C. At this temperature, agar is still liquid. The dish is then gently swirled to mix the bacteria with the liquid agar. When the agar hardens, the individual cells are fixed in place and, after incubation, form distinguishable colonies.

In the **spread-plate** method, 0.1 to 0.2 ml of the final dilution is transferred directly onto a plate of solidified nutrient agar. This solution is then spread over the surface of the agar with a sterilized bent glass rod, which resembles a miniature hockey stick.

In both methods the plates are then incubated for a specific time period to allow the colonies to form, which can then be counted. By knowing how much the sample was diluted prior to being plated, along with the amount of the dilution used in

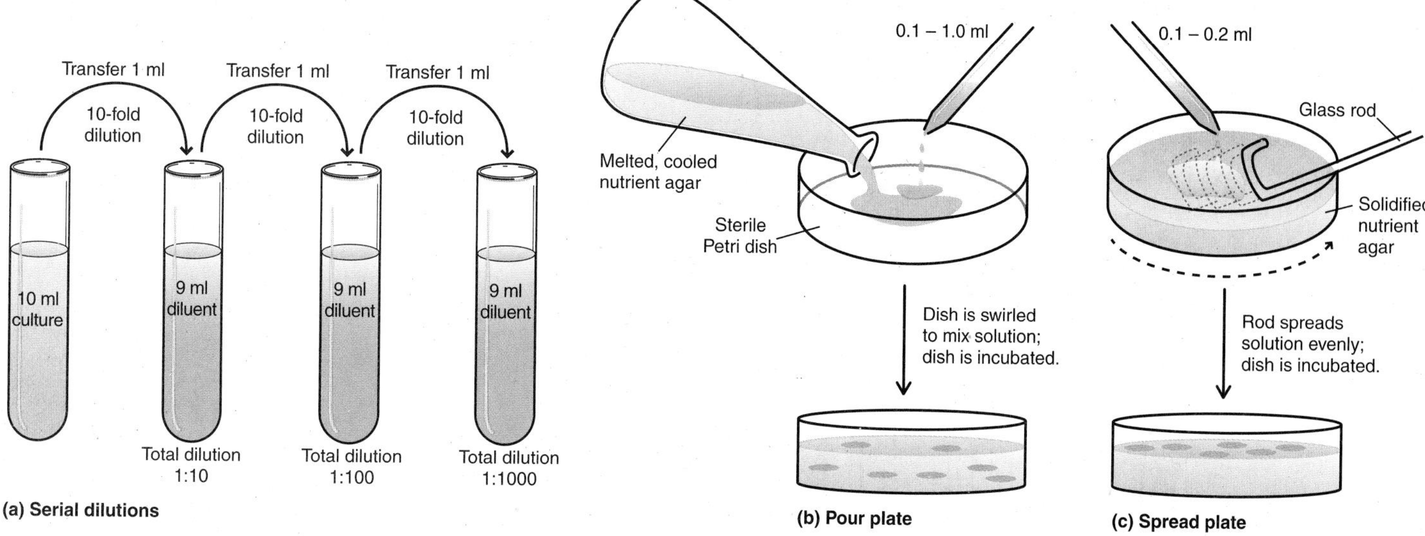

FIGURE 4.18 Plate Counts **(a)** A sample is first diluted in 10-fold increments. **(b)** In the pour-plate method, 0.1–1.0 ml of a dilution is transferred to a sterile Petri dish and mixed with melted, cooled nutrient agar. When the agar hardens, the plate is incubated and colonies form on the surface and within the agar. **(c)** In the spread-plate method, 0.1–0.2 ml of a dilution is spread on a hardened agar plate with a sterile glass rod. After incubation, colonies form only on the surface of the agar.

plating, the concentration of viable cells in the original sample can then be calculated. Because bacterial cells often attach to one another and then grow to form a single colony, counts are expressed as **colony-forming units.**

Pour plates and spread plates are generally only used if a sample contains more than 100 organisms/ml. Otherwise, few if any cells will be transferred to the plates. In these situations, alternative methods give more reliable results.

Membrane Filtration

Membrane filtration is used when the numbers of organisms in a sample are relatively low, as might occur in dilute environments such as natural waters. This method concentrates the bacteria by filtration before they are plated. A known volume of liquid is passed through a sterile membrane filter, which has a pore size that retains bacteria **(figure 4.19).** The filter is subsequently placed on an appropriate agar medium and then incubated. The number of colonies that grow on the filter indicates the number of bacteria in the volume filtered.

Most Probable Number (MPN)

The **most probable number (MPN)** method is a statistical assay of cell numbers based on the theory of probability. The goal is to successively dilute a sample and determine the point at which subsequent dilutions receive no cells.

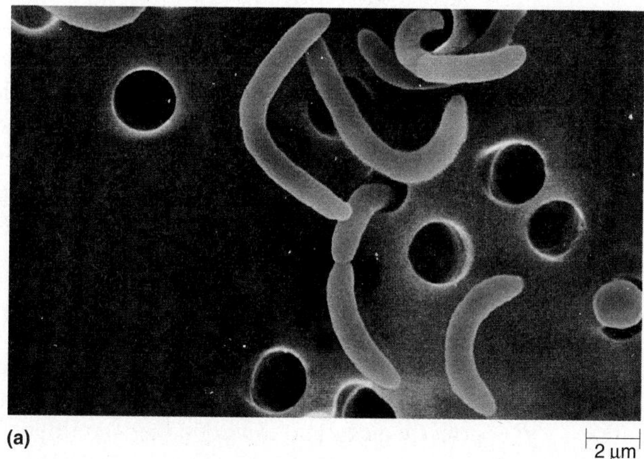

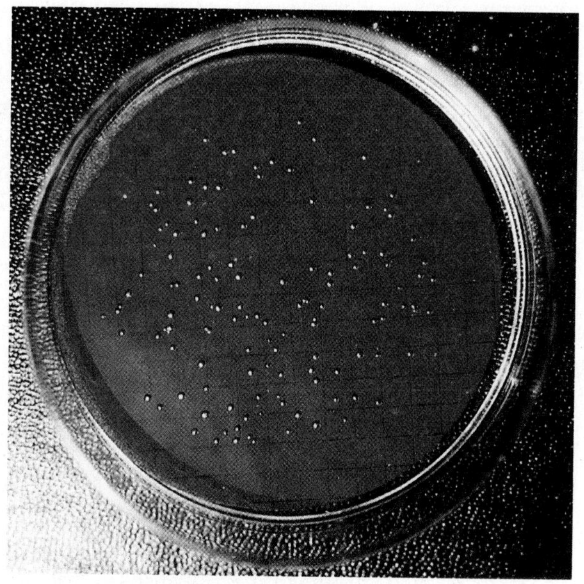

FIGURE 4.19 Membrane Filtration This technique concentrates bacteria before they are plated. **(a)** A known volume of liquid is passed through a sterile membrane filter, which has a pore size that retains bacteria. **(b)** The filter is then placed on an appropriate agar medium and incubated. The number of colonies that grow on the filter indicates the number of bacteria that were in the volume filtered.

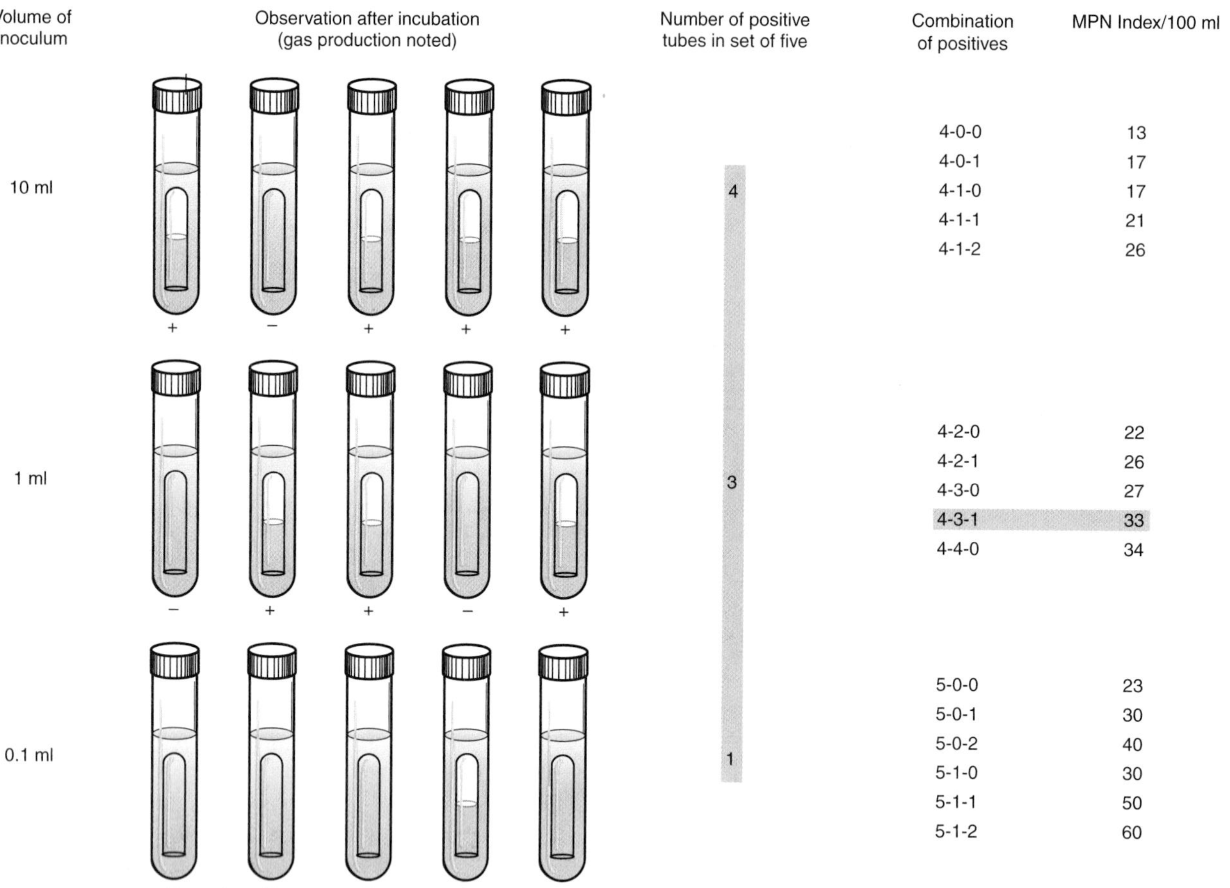

Volume of inoculum
Observation after incubation (gas production noted)
Number of positive tubes in set of five
Combination of positives
MPN Index/100 ml
10 ml
+ − + + +
4
4-0-0 13
4-0-1 17
4-1-0 17
4-1-1 21
4-1-2 26
1 ml
− + + − +
3
4-2-0 22
4-2-1 26
4-3-0 27
4-3-1 33
4-4-0 34
0.1 ml
− − − + −
1
5-0-0 23
5-0-1 30
5-0-2 40
5-1-0 30
5-1-1 50
5-1-2 60

FIGURE 4.20 The Most Probable Number (MPN) Method In this example, three sets of five tubes containing the same growth medium were prepared. Each set received the indicated amount of inoculum. After incubation the presence or absence of gas in each tube was noted. The results were then compared to an MPN table to get a statistical estimate of the concentration of gas-producing bacteria.

To determine the MPN, three sets of three or five tubes containing the same growth medium are prepared **(figure 4.20).** Each set receives a measured amount of a sample such as water, soil, or food. The amount added is determined, in part, by the expected bacterial concentration in that sample. What is important is that the second set receives 10-fold less than the first, and the third set 100-fold less. In other words, each set is inoculated with an amount 10-fold less than the previous set. After incubation, the presence or absence of turbidity or other indication of growth is noted; the results are then compared against an MPN table, which gives a statistical estimate of the cell concentration. The MPN method is most commonly used to determine the approximate number of **coliforms** in a water sample. Coliforms are lactose-fermenting, Gram-negative rods that typically reside in the intestine and thus serve as a bacterial indicator of fecal contamination. ■ coliforms, p. 746

Measuring Biomass

Instead of measuring the number of cells, the cell mass can be determined. This can be done by measuring the turbidity or the total weight.

Turbidity

Cloudiness or **turbidity** of a bacterial suspension such as a broth culture is due to the scattering of light passing through the liquid by cells **(figure 4.21).** The amount scattered is proportional to the concentration of cells. To measure turbidity, a **spectrophotometer** is used. This instrument transmits light through a specimen and measures the percentage that reaches a light detector. That number is inversely proportional to the optical density. To use turbidity to estimate cell numbers, a one-time correlation between optical density and cell concentration for the specific organism and conditions under study must be made. Once this correlation has been determined, the turbidity measurement becomes a rapid and relatively accurate assay.

One limitation of assaying turbidity is that a medium must contain relatively high numbers of bacteria in order to be cloudy. A solution containing 1 million bacteria (10^6) per ml is still perfectly clear, and if it contains 10 million cells (10^7) per ml, it is barely turbid. Thus, although a turbid culture indicates that bacteria are present, a clear solution does not guarantee their absence. Not recognizing these facts can have serious consequences in the laboratory as well as outside. Experienced hikers, for example,

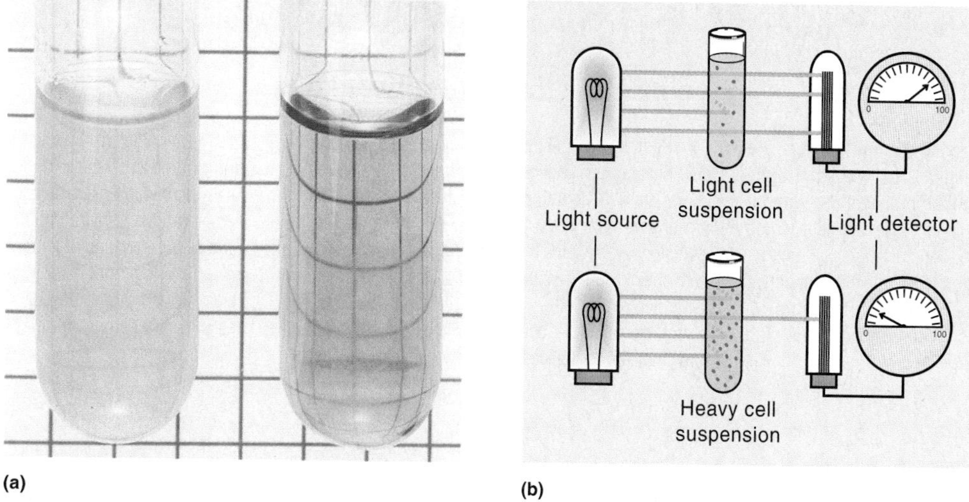

FIGURE 4.21 Measuring Turbidity with a Spectrophotometer **(a)** The cloudiness, or turbidity, of the liquid in the tube on the left is proportional to the concentration of cells. **(b)** The percentage of light that reaches the detector is inversely proportional to the optical density. To use turbidity to estimate cell number, a one-time experiment must be done to determine the correlation between cell concentration and optical density of a culture.

know that the clarity of mountain streams does not necessarily mean that the water is free of *Giardia* or other harmful organisms.

■ giardiasis, p. 609

Total Weight

Determining the total weight of a culture is a tedious and time-consuming method that can be used to measure growth of filamentous organisms. These do not readily separate into the individual cells necessary for a valid plate count. To measure the **wet weight,** cells growing in liquid culture are centrifuged and the liquid supernate removed. The weight of the resulting packed cell mass is proportional to the number of cells in the culture. The **dry weight** can be determined by drying the centrifuged cells at approximately 100°C for 8 to 12 hours before weighing them. About 70% of the weight of a cell is water.

Detecting Cell Products

Products of microbial growth can be used to estimate the number of microorganisms or, more commonly, to confirm their presence. These products include acids, gases, and ATP.

Acid Production

As a consequence of the metabolic breakdown of sugars, which are used as an energy source, microorganisms produce a variety of acids. The precise amount of acid can be measured using chemical means. Most commonly, however, acid production is used to detect growth by incorporating a **pH indicator** into a medium. A pH indicator changes from one color to another as the pH of a medium changes. Several pH indicators are available, and they differ in the pH value at which their color changes.

Gases

Production of gases such as CO_2 can be monitored in several ways. A method used in clinical labs employs a fluorescent molecule to detect bacteria growing in blood taken from patients who are suspected of having a bloodstream infection. The slight decrease in pH that accompanies the production of CO_2 increases the fluorescence.

ATP

The presence of ATP can be detected by adding the firefly enzyme luciferase. The enzyme catalyzes a chemical reaction that uses ATP as an energy source to produce light. This method is sometimes used to assess the effectiveness of chemical agents formulated to kill bacteria. Light is produced only if viable organisms remain.

MICROCHECK 4.8

Direct microscopic counts and cell-counting instruments generally do not distinguish between living and dead cells. Plate counts determine the number of cells capable of multiplying; membrane filtration can be used to concentrate the sample. The most probable number is a statistical assay based on probability. Turbidity of a culture is a rapid measurement that can be correlated to cell number. The total weight of a culture can be correlated to the number of cells present. Microbial growth can be detected by the presence of cell products such as acid, gas, and ATP.

✓ Why is an MPN an estimate rather than an accurate number?

✓ Why would a direct microscopic count yield a higher number than a pour plate if a sample of seawater was examined by both methods?

FUTURE CHALLENGES

Seeing How the Other 99% Lives

One of the biggest challenges for the future is the development of methodologies to cultivate and study a wider array of environmental prokaryotes. Without these microbes, humans and other animals would not be able to exist. Yet, considering their importance, we still know very little about most species, including the relative contributions of each to such fundamental processes as O_2 generation and N_2 and CO_2 fixation.

Studying environmental microorganisms can be difficult. Much of our understanding of prokaryotic processes comes from work with pure cultures. Yet, over 99% of prokaryotes have never been successfully grown in the laboratory. At the same time, when organisms are removed from their natural habital, and especially when they are separated from other organisms, their environment changes drastically. Consequently, the study of pure cultures may not be the ideal for studying natural situations.

Technological advances such as flow cytometry and fluorescent labeling, along with the recombinant DNA techniques discussed in chapter 9, may make it easier to study environmental bacteria. This may well lead to a better understanding of the diversity and the roles of microorganisms in our ecosystem. Scientists have learned a great deal about microorganisms since the days of Pasteur, but most of the microbial world is still a mystery.

SUMMARY

4.1 Principles of Prokaryotic Growth

Most prokaryotes multiply by **binary fission** (figure 4.1). **Microbial growth** is an increase in the number of cells in a population. The time required for a population to double in number is the **generation time** (table 4.1).

4.2 Bacterial Growth in Nature

Biofilms (figure 4.2)

Prokaryotes often live in a **biofilm,** a polysaccharide-encased community.

Interactions of Mixed Microbial Communities

Prokaryotes often grow in close associations containing multiple different species: the metabolic activities of one organism often affects the growth of another.

4.3 Obtaining a Pure Culture

Only an estimated 1% of prokaryotes have been cultivated in the laboratory.

Cultivating Bacteria on a Solid Culture Medium

A single bacterial cell deposited on a solid medium will multiply to form a visible colony (figure 4.4).

The Streak-Plate Method (figure 4.5)

The **streak-plate** method is used to isolate bacteria in order to obtain a **pure culture.**

Maintaining Stock Cultures

Stock cultures can be stored on an **agar slant** in the refrigerator, frozen, or **lyophilized.**

4.4 Bacterial Growth in Laboratory Conditions

The Growth Curve (figure 4.6)

When grown in a **closed system,** a population of bacterial cells goes through five phases: **lag, log, stationary, death,** and **prolonged decline.**

Colony Growth

The position of a single cell within a colony markedly determines its environment.

Continuous Culture

Bacteria can be maintained in a state of continuous exponential growth by using a **chemostat.**

4.5 Environmental Factors That Influence Microbial Growth (table 4.2)

Temperature Requirements (figure 4.9)

Organisms can be grouped as **psychrophiles, psychrotrophs, mesophiles, thermophiles,** or **hyperthermophiles** based on their optimum growth temperatures.

Oxygen (O_2) Requirements (table 4.3)

Organisms can be grouped as **obligate aerobes, obligate anaerobes, facultative anaerobes, microaerophiles** or **aerotolerant anaerobes** based on their oxygen (O_2) requirements. Although O_2 itself is not toxic, it can be converted to **superoxide** and **hydrogen peroxide,** both of which are toxic. **Superoxide dismutase** and **catalase** can break these down.

pH

Organisms can be grouped as **neutrophiles, acidophiles,** or **alkalophiles** based on their optimum pH.

Water Availability

Halophiles are adapted to live in high salt environments.

4.6 Nutritional Factors That Influence Microbial Growth

Required Elements (table 4.4)

The **major elements** make up cell constituents and include carbon, nitrogen, sulfur, and phosphorus. **Trace elements** are required in very minute amounts.

Growth Factors

Bacteria that cannot synthesize cell constituents such as amino acids and vitamins require these as **growth factors.**

Energy Sources

Organisms derive energy either from sunlight or from the oxidation of chemical compounds.

Nutritional Diversity (table 4.5)

Photoautotrophs use the energy of sunlight and the carbon in the atmosphere to make organic compounds. **Chemolithoautotrophs** use inorganic compounds for energy and derive their carbon from CO_2. **Photoheterotrophs** use the energy of sunlight and derive their carbon from organic compounds. **Chemoorganoheterotrophs** use organic compounds for energy and as a carbon source.

4.7 Cultivating Prokaryotes in the Laboratory

General Categories of Culture Media (table 4.6)

A **complex medium** contains a variety of ingredients such as peptones and extracts. A **chemically defined** medium is composed of precise mixtures of pure chemicals.

Special Types of Culture Media

A **selective medium** inhibits organisms other than the one being sought. A **differential medium** contains a substance that certain bacteria change in a recognizable way.

Providing Appropriate Atmospheric Conditions

A **candle jar** provides increased CO_2, which enhances the growth of many medically important bacteria. Microaerophilic bacteria are incubated in a gastight container along with a packet that generates low O_2 conditions. Anaerobes may be incubated in an **anaerobe container** or an **anaerobic chamber** (figures 4.13, 4.14).

Enrichment Cultures (figure 4.15)

An **enrichment culture** provides conditions in a broth that enhance the growth of one particular organism in a mixed population.

4.8 Methods to Detect and Measure Bacterial Growth

(table 4.8)

Direct Cell Counts

Direct cell counts do not distinguish between living and dead cells. One of the most rapid methods of determining the number of cells is the **direct microscopic count** (figure 4.16). Both a **Coulter counter** and a **flow cytometer** count cells as they pass through a minute aperture (figure 4.17).

Viable Cell Counts

Plate counts measure the number of viable cells by exploiting the fact that an isolated cell will form a single colony (figure 4.18). **Membrane filtration** concentrates bacteria by filtration; the filter is then incubated on an agar plate (figure 4.19). The **most probable number (MPN)** method is a statistical assay used to estimate cell numbers (figure 4.20).

Measuring Biomass

Turbidity of a culture can be correlated with the number of cells; a **spectrophotometer** is used to measure turbidity (figure 4.21). **Wet weight** and **dry weight** are proportional to the number of cells in a culture.

Detecting Cell Products

Products including acid, gas, and ATP can indicate growth.

REVIEW QUESTIONS

Short Answer

1. Define a *pure culture*.
2. If the number of bacteria in lake water were determined using both a direct microscopic count and a plate count, which method would most likely give a higher number? Why?
3. List the five categories of optimum temperature, and describe a corresponding environment in which a representative might thrive.
4. Explain why obligate anaerobes are significant to the canning industry.
5. Explain why O_2-containing atmospheres kill some bacteria.
6. Explain why photoautotrophs are the primary producers.
7. Distinguish between a selective medium and a differential medium.
8. Explain what occurs during each of the five phases of growth.
9. Explain how the environment of a colony differs from that of cells growing in a liquid broth.
10. Describe a detrimental and a beneficial effect of biofilms.

Multiple Choice

1. *E. coli* is present in a liquid sample at a concentration of between 10^4 and 10^6 bacteria per ml. To determine the precise number of living bacteria in the sample, it would be best to
   a) use a counting chamber.
   b) plate out an appropriate dilution of the sample on nutrient agar.
   c) determine cell number by using a spectrophotometer.
   d) Any of these three methods would be satisfactory.
   e) None of these three methods would be satisfactory.
2. *E. coli*, a facultative anaerobe, is grown for 24 hours on the same solid medium, but under two different conditions: one aerobic, the other anaerobic. The size of the colonies would be
   a) the same under both conditions.
   b) larger when grown under aerobic conditions.
   c) larger when grown under anaerobic conditions.
3. A soil sample is placed in liquid and the number of bacteria in the sample determined in two ways: (1) by colony count and (2) by counting the cells in a counting chamber (slide). How would the results compare?
   a) Methods 1 and 2 would give approximately the same number of bacteria.
   b) Many more bacteria would be estimated by method 1.
   c) Many more bacteria would be estimated by method 2.
   d) Depending on the soil sample, sometimes method 1 would be higher and sometimes method 2 would be higher.
4. Nutrient broth is an example of a
   a) synthetic medium. b) complex medium. c) selective medium.
   d) indicator medium. e) defined medium.
5. *E. coli* does not use vitamins in the medium in which it grows. This is because *E. coli*
   a) does not use vitamins for growth.
   b) gets vitamins from its host.
   c) is a chemoheterotroph.
   d) can synthesize vitamins from the simple compounds provided in the medium.
6. Cells are most sensitive to penicillin during which phase of the growth curve?
   a) lag b) log c) stationary
   d) death e) more than one of these.
7. *Streptomyces* cells would most likely synthesize antibiotics during which phase of the growth curve?
   a) lag b) log c) stationary
   d) death e) more than one of these.
8. Compared with their growth in the laboratory, bacteria in nature generally grow
   a) more slowly.
   b) faster.
   c) at the same rate.

9. If there are 10^3 cells per ml at the middle of log phase, and the generation time of the cells is 30 minutes, how many cells will there be 2 hours later?
   a) 2×10^3 b) 4×10^3 c) 8×10^3
   d) 1.6×10^4 e) 1×10^7
10. The major effect of a temperature of 60°C on a mesophile is to
    a) destroy the cell wall.
    b) denature proteins.
    c) destroy nucleic acids.
    d) destroy the cytoplasmic membrane.
    e) cause the formation of endospores.

Applications

1. You are a microbiologist working for a pharmaceutical company and discover a new metabolite that can serve as a medication. Your company asked you to oversee the production of the metabolite. What are some factors you must consider if you need to grow 5,000-liter cultures of bacteria?
2. High-performance boat manufacturers know that bacteria can collect on a boat, ruining the boat's hydrodynamic properties. Periodic cleaning of the boat's surface and repainting eventually ruin that surface and do not solve the problem. A boat-manufacturing facility recently hired you to help with this problem because of your microbiology background. What strategies can you use to come up with a long-term remedy for the problem?

Critical Thinking

1. This figure shows a growth curve plotted on a non-logarithmic, or linear, scale. Compare this with figure 4.6. In both figures, the number of cells increases dramatically during the log or exponential phase. In this phase, the cell number increases more and more rapidly (this effect is more apparent in the accompanying figure). Why should the increase be speeding up?

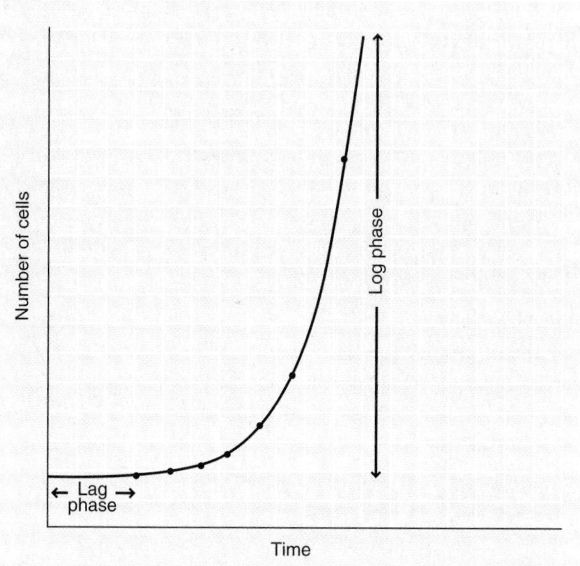

2. In question 1, how would the curve appear if the availability of nutrients were increased?

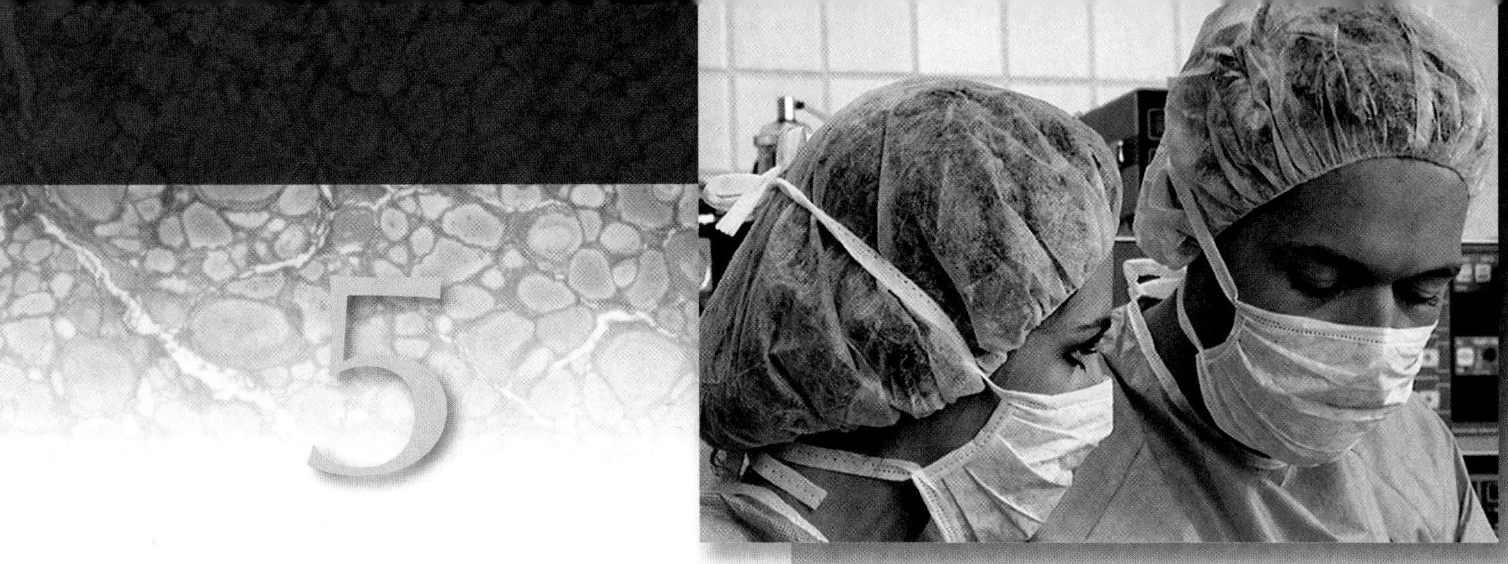

Medical settings warrant a high level of microbial control.

5 Control of Microbial Growth

A Glimpse of History

The *British Medical Journal* stated that the British physician Joseph Lister (1827–1912) "saved more lives by the introduction of his system than all the wars of the 19th century together had sacrificed." He revolutionized surgery by developing effective methods that prevent surgical wounds from becoming infected. Impressed with Pasteur's work on fermentation (said to be caused by "minute organisms suspended in the air"), Lister wondered if "minute organisms" might be responsible also for the pus that formed in surgical wounds. He then experimented with a phenolic compound, carbolic acid, introducing it at full strength into wounds by means of a saturated rag. Lister was particularly proud of the fact that, after carbolic acid wound dressings became standard in his practice, his patients no longer developed gangrene. Lister's work provided impressive evidence for the germ theory of disease, even though microorganisms specific for various diseases were not identified for another decade.

Later, Lister improved his methods by introducing surgical procedures that excluded bacteria from wounds by maintaining a clean environment in the operating room and by sterilizing instruments. These procedures were preferable to killing the bacteria after they had entered wounds because they avoided the toxic effects of the disinfectant on the wound.

Lister was knighted in 1883 and subsequently became a baron and a member of the House of Lords.

Until the late nineteenth century, patients undergoing even minor surgery were at great risk of developing fatal infections due to unsanitary medical practices and hospital conditions. Physicians did not know that their hands could pass diseases from one patient to the next. Nor did they understand that airborne microscopic organisms could infect open wounds. Fortunately, today's modern hospitals use rigorous procedures to avoid microbial contamination, allowing surgical operations to be performed with relative safety.

The growth of microorganisms affects more than our health. Producers of a wide variety of goods recognize that unless microbial growth is controlled, the quality of their products can be compromised. This ranges from undesirable changes in the safety, appearance, taste, or odor of food products to the decay of untreated lumber.

This chapter covers methods to destroy, remove, and inhibit the growth of microorganisms on inanimate objects and some body surfaces. Most of these approaches are non-selective in that they can adversely impact all forms of life. Antibiotics and other antimicrobial medications will be discussed in chapter 21. These medications are particularly valuable in combating infectious diseases because their toxicity is specifically targeted to microbes.

5.1 Approaches to Control

Focus Points

- Define the terms *sterile, disinfection, disinfectant, biocide, germicide, antiseptic, degerming, pasteurization, decontamination, sanitize,* and *preservation*.
- Compare and contrast the rigor of the methods used to control microbial growth in daily life, and in hospitals, microbiology laboratories, food and food production facilities, water treatment facilities, and other industries.

The processes used to control microorganisms are either physical or chemical, though a combination of both can be used. **Physical methods** include heat treatment, irradiation, filtration, and mechanical removal (washing). **Chemical**

KEY TERMS

Antiseptic A disinfectant that is non-toxic enough to be used on skin.

Aseptic Technique Procedures that minimize the chance of unwanted microbes being accidentally introduced.

Bactericidal Kills bacteria.

Bacteriostatic Prevents the growth of, but does not kill, bacteria.

Degerm Treatment used to decrease the number of microbes in an area, usually skin.

Disinfectant A chemical used to destroy many microorganisms and viruses.

Germicide Kills microorganisms and viruses.

Pasteurization A treatment, usually brief heating, used to reduce the number of spoilage organisms and to kill disease-causing microbes.

Preservation The process of inhibiting the growth of microorganisms in products to delay spoilage.

Sterilant A chemical used to destroy all microorganisms and viruses in a product, rendering it sterile.

Sterile Completely free of all viable microbes; an absolute term.

Sterilization The process of destroying or removing all microorganisms and viruses, through physical or chemical means.

methods use any of a variety of antimicrobial chemicals. The method chosen depends on the circumstances and degree of control required.

Principles of Control

The process of removing or destroying all microorganisms and viruses on or in a product is called **sterilization.** These procedures include removing microbes by filtration, or destroying them using heat, certain chemicals, or irradiation. Destruction of microorganisms means they cannot be "revived" to multiply even when transferred from the sterilized product to an ideal growth medium. A **sterile** item is one that is absolutely free of microbes, including endospores and viruses. It is important to note, however, that the term *sterile* does not consider prions. These infectious protein particles are not destroyed by standard sterilization procedures. ■ endospores, p. 69 ■ prions, p. 341

Disinfection is the process that eliminates most or all pathogens on or in a material. Unlike sterilization, disinfection suggests that some viable microbes may persist. In practice, the term *disinfection* generally implies the use of antimicrobial chemicals. Those used for disinfecting inanimate objects are called **disinfectants.** Disinfectants are **biocides** (*bio* means "life," and *cida* means "to kill"). Although they are at least somewhat toxic to many forms of life, they are typically used in a manner that targets microscopic organisms, including bacteria and their endospores, fungi, and viruses. Thus, they are often called **germicides.** They are also described as **bactericidal,** meaning they kill bacteria. When disinfectants are formulated for use on skin they are called **antiseptics.** Antiseptics are routinely used to decrease the number of bacteria on skin to prepare for invasive procedures such as surgery. ■ pathogen, p. 394

Pasteurization uses a brief heat treatment to reduce the number of spoilage organisms and kill pathogens. Foods and inanimate objects can be pasteurized.

Decontamination is a treatment used to reduce the number of pathogens to a level considered safe to handle. The treatment can be as simple as thorough washing, or it may involve the use of heat or disinfectants.

Degerming is a treatment used to decrease the number of microbes in an area, particularly the skin. In other words, antiseptics are degerming agents.

Sanitized generally implies a substantially reduced microbial population that meets accepted health standards. Most people also expect a sanitized object to be clean in appearance. Note that this term does not denote any specific level of control.

Preservation is the process of delaying spoilage of foods or other perishable products. This is done by adding growth-inhibiting ingredients or adjusting storage conditions to impede growth of microorganisms.

Situational Considerations

Methods used to control microbial growth vary greatly depending on the situation and degree of control required **(figure 5.1).** Control measures adequate for routine circumstances of daily life might not be sufficient for situations such as hospitals, microbiology laboratories, foods and food production facilities, water treatment facilities, and other industries.

Daily Life

Washing and scrubbing with soaps and detergents achieves routine control of undesirable microorganisms and viruses. In fact, simple handwashing with plain soap and water is considered the single most important step in preventing the spread of many infectious diseases. Plain soap itself generally does not destroy many organisms; it simply aids in the mechanical removal of transient microbes, including most pathogens, as well as dirt, organic material, and cells of the outermost layer of skin. Regular handwashing and bathing does not adversely impact the beneficial normal skin microbiota, which reside more deeply on underlying layers of skin cells and in hair follicles. ■ normal microbiota of the skin, p. 533

Other methods used to control microorganisms in daily life include cooking foods, cleaning surfaces, and refrigeration.

Hospitals

Minimizing the numbers of microorganisms in a hospital is particularly important because of the danger of hospital-acquired, or **nosocomial,** infections. Hospitalized patients are often more susceptible to infectious agents because of their weakened condition. In addition, patients may be subject to invasive procedures such as surgery, which breaches the intact skin that would otherwise help prevent infection. Finally, pathogens are more likely to be found in hospitals because of the high concentration of patients with infectious disease. These patients may shed pathogens in their feces, urine, respiratory droplets, or other body secretions. Thus, hospitals must be scrupulous in their control of microorganisms. Nowhere is this more important than in the operating rooms, where instruments used in invasive procedures must be sterile to avoid introducing even normally benign microbes into deep body tissue where they could easily establish infection. ■ nosocomial infections, p. 462

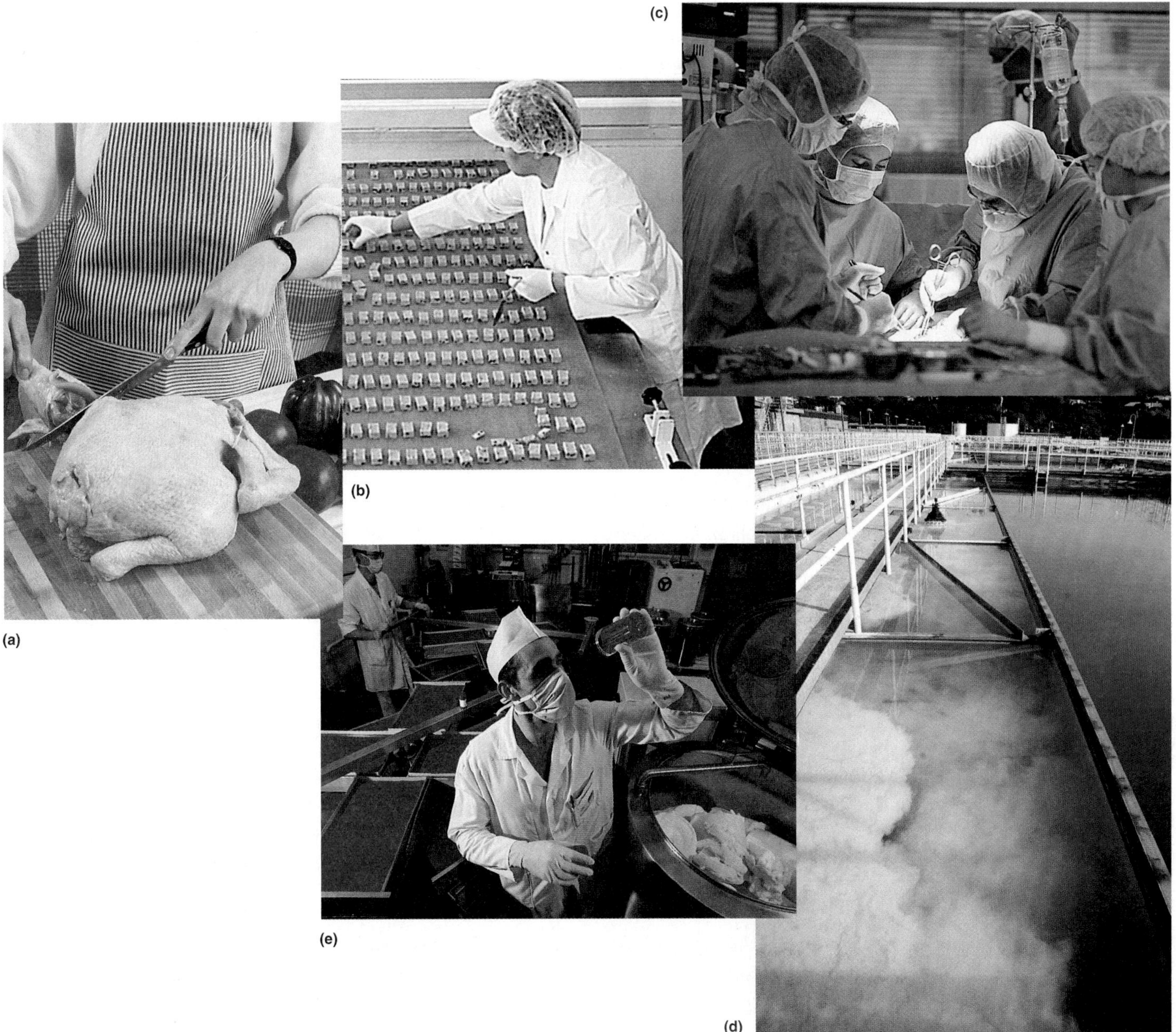

FIGURE 5.1 Situations that Warrant Different Levels of Microbial Control **(a)** Daily home life; **(b)** foods and food production facilities; **(c)** hospitals; **(d)** water treatment facilities; **(e)** other industries.

Prions are a relatively new concern for hospitals. Fortunately, disease caused by prions is thought to be exceedingly rare, less than 1 case per 1 million persons per year. Hospitals, however, must take special precautions when handling tissue that may be contaminated with prions, because these infectious particles are very difficult to destroy.

Microbiology Laboratories

Microbiology laboratories routinely work with microbial cultures and consequently must use rigorous methods of controlling microorganisms. To work with pure cultures, all media and instruments that contact the culture must first be rendered sterile to avoid contaminating the culture with environmental microbes. All materials used to grow microorganisms must again be treated before disposal to avoid contamination of workers and the environment. The use of specific methods to exclude contaminating microorganisms from an environment is called **aseptic technique.** Although all microbiology laboratory personnel must use these prudent measures, those who work with known disease-causing microbes must be even more diligent.

Foods and Food Production Facilities

Foods and other perishable products retain their quality longer when the growth of contaminating microorganisms is prevented. This can be accomplished by physically removing or destroying microorganisms or by adding chemicals that impede their growth.

Heat treatment is the most common and reliable method used to kill microbes, but heating can alter the flavor and appearance of food. Irradiation can destroy microbes without causing perceptible changes in food, but the Food and Drug Administration (FDA) has approved this technology to treat only certain foods. Chemicals can prevent the growth of microorganisms, but the risk of toxicity must always be a concern. Because of this, the FDA regulates chemical additives used in food and must deem them safe for consumption.

Food-processing facilities need to keep surfaces relatively free of microorganisms to avoid contamination. If machinery used to grind meat, for example, is not cleaned properly, it can create an environment in which bacteria multiply, eventually contaminating large quantities of product.

Water Treatment Facilities

Water treatment facilities need to ensure that drinking water is free of pathogenic bacteria, protozoa, and viruses. Chlorine has traditionally been used to disinfect water, saving hundreds of thousands of lives by preventing transmission of waterborne illnesses such as cholera. Disinfectants including chlorine, however, can react with naturally occurring chemicals in the water to form compounds called **disinfection by-products (DBPs).** Some of these have been linked to long-term health risks. In addition, certain pathogens, particularly the oocysts of *Cryptosporidium parvum*, are resistant to traditional chemical disinfection procedures. To address these problems, water treatment regulations have been amended to require that facilities minimize the level of both DBPs and *C. parvum* oocysts in treated water. ■ *Cryptosporidium parvum*, p. 611

Other Industries

Many diverse industries have specialized concerns regarding microbial growth. Manufacturers of cosmetics, deodorants, or any other product that will be applied to the skin must avoid microbial contamination that could affect the product's quality or safety.

MICROCHECK 5.1

The methods used to control microbial growth depend on the situation and the degree of control required.

✓ How is sterilization different from disinfection?
✓ What is an antiseptic?
✓ Why would the term *sterilization* not necessarily encompass prions?

5.2

Selection of an Antimicrobial Procedure

Focus Point

- Explain why the type of microbe, number of microbes initially present, environmental conditions, potential risk of infection, and composition of the item influence the selection of an antimicrobial procedure.

Selection of an effective antimicrobial procedure is complicated by the fact that every procedure has drawbacks that limit its use. An ideal, multipurpose, non-toxic method simply does not exist. The ultimate choice depends on many factors including the type of microbes present, extent of contamination, environmental conditions, potential risk of infection associated with use of the item, and the composition of the item.

Type of Microorganism

One of the most critical considerations in selecting an antimicrobial procedure is the type of microbial population present on or in the product. Products contaminated with microorganisms highly resistant to killing require a more rigorous heat or chemical treatment. Some of the highly resistant microbes include:

- Bacterial endospores. The endospores of *Bacillus* and *Clostridium* (and related genera) are the most resistant form of life typically encountered. Only extreme heat or chemical treatment ensures their complete destruction. Chemical treatments that kill vegetative bacteria in 30 minutes may require 10 hours to destroy their endospores. ■ endospores, p. 69
- Protozoan cysts and oocysts. Cysts and oocysts are stages in the life cycle of certain intestinal protozoan pathogens such as *Giardia lamblia* and *Cryptosporidium parvum*. These disinfectant-resistant forms are excreted in the feces of infected animals, including humans, and can cause diarrheal disease if ingested. They are of particular concern in water treatment. Unlike endospores, they are readily destroyed by boiling. ■ *Cryptosporidium parvum*, p. 611 ■ *Giardia lamblia*, p. 609
- *Mycobacterium* species. The waxy cell walls of mycobacteria make them resistant to many chemical treatments. Thus, stronger, more toxic disinfectants must be used to disinfect environments that may contain *Mycobacterium tuberculosis*, the causative agent of tuberculosis. ■ tuberculosis, p. 514
- *Pseudomonas* species. These common environmental organisms are not only resistant to some chemical disinfectants, but in some cases can actually grow in them. *Pseudomonas* species are of particular importance in hospitals, where they are a common cause of infections. ■ Pseudomonas infections, p. 564
- Naked viruses. Viruses such as poliovirus that lack a lipid envelope are more resistant to disinfectants. Conversely, enveloped viruses, such as HIV, tend to be very sensitive to heat and chemical disinfectants. ■ naked viruses, p. 303 ■ enveloped viruses, p. 303

Numbers of Microorganisms Initially Present

The time it takes for heat or chemicals to kill a population of microorganisms is dictated in part by the number of cells initially present. It takes more time to kill a large population than it does to kill a small population, because only a fraction of organisms die during a given time interval. For example, if 90% of a bacterial population is killed during the first 3 minutes, then approximately 90% of those remaining will be killed during the next 3 minutes, and so on.

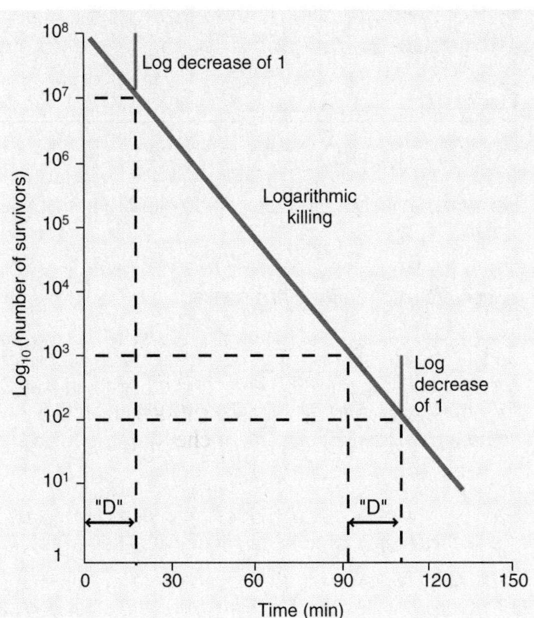

FIGURE 5.2 D Value The D value is the time it takes to reduce the population by 90%.

In the commercial canning industry, the **decimal reduction time,** or **D value,** is the time required for killing 90% of a population of bacteria under specific conditions **(figure 5.2).** The temperature of the process may be indicated by a subscript, for example, D_{121}. A one D process reduces the number of cells by one exponent. Thus, if the D value for an organism is 2 minutes, then it would take 4 minutes (2 D values) to reduce a population of 100 (10^2) cells to only one (10^0) survivor. It would take 20 minutes (10 D values) to reduce a population of 10^{10} cells to only one survivor. Removing organisms by washing or scrubbing can minimize the time necessary to sterilize or disinfect a product.

Environmental Conditions

Factors such as pH, temperature, and presence of fats and other organic materials strongly influence microbial death rates. A solution of sodium hypochlorite (household bleach) can kill a suspension of *M. tuberculosis* in 150 seconds at a temperature of 50°C; whereas it takes only 60 seconds to kill the same suspension with bleach if the temperature is increased to 55°C. The hypochlorite solution is even more effective at a low pH.

The presence of dirt, grease, and organic compounds such as blood and other body fluids can interfere with heat penetration and the action of chemical disinfectants. This is another reason why it is important to thoroughly clean items before disinfection or sterilization.

Potential Risk of Infection

To guide medical biosafety personnel in their selection of germicidal procedures, medical items such as surgical instruments, endoscopes, and stethoscopes are categorized according to their potential risk of transmitting infectious agents. Those that pose the greatest threat of transmitting disease must be subject to more rigorous germicidal procedures.

- **Critical instruments** come into direct contact with body tissues. These items, including needles, scalpels, and biopsy forceps, must be sterilized to avoid transmission of all infectious agents.
- **Semicritical instruments** come into contact with mucous membranes, but do not penetrate body tissue. These items, including gastrointestinal endoscopes and endotracheal tubes, must be free of all viruses and vegetative bacteria including mycobacteria. Low numbers of endospores that may remain on semicritical instruments pose little risk of infection because mucous membranes are effective barriers against their entry into deeper tissue.
- **Non-critical instruments** and surfaces pose little risk of infection because they only come into contact with unbroken skin. Countertops, stethoscopes, and blood pressure cuffs are examples of non-critical items.

Composition of the Item

Some sterilization and disinfection procedures are inappropriate for certain types of material. For example, although heat treatment is generally the method of choice because it is so dependable and relatively inexpensive, many plastics and other materials are heat-sensitive. In addition, moist heat corrodes metals, dulling some instruments. Heat-sensitive material can be irradiated, but the process damages some types of plastics. Moisture-sensitive material cannot be treated with liquid chemical disinfectants, which can also damage metals and rubber.

MICROCHECK 5.2

The types and numbers of microorganisms initially present, environmental conditions, the potential risks associated with use of the item, and the composition of the item must all be considered when determining which sterilization or disinfection procedure to employ.

- ✓ Describe three groups of microorganisms that are resistant to certain chemical treatments.
- ✓ Define the term *D value*.
- ✓ Would it be safe to say that if all bacterial endospores had been killed, then all other medically important microorganisms had also been killed?

5.3 Using Heat to Destroy Microorganisms and Viruses

Focus Points

- Compare and contrast pasteurization, sterilization using pressurized steam, and the commercial canning process.
- Explain the drawbacks and benefits of using dry heat rather than moist heat to kill microorganisms.

TABLE 5.1 Physical Methods Used to Destroy Microorganisms and Viruses

	Characteristics	Uses
Moist Heat	Denatures proteins. Relatively fast, reliable, safe, and inexpensive.	Widely used.
Boiling	Boiling for 5 minutes destroys most microorganisms and viruses; a notable exception is endospores.	Boiling for at least 5 minutes can be used to treat drinking water.
Pasteurization	Significantly decreases the numbers of heat-sensitive microorganisms, including spoilage microbes and pathogens (except sporeformers).	Milk is pasteurized by heating it to 72°C for 15 seconds. Juices are also routinely pasteurized.
Pressurized steam (autoclaving)	Typical treatment is 121°C/15 psi for 15 minutes or longer, a process that destroys endospores.	Widely used to sterilize microbiological media, laboratory glassware, surgical instruments, and other items that can be penetrated by steam. The canning process renders foods commercially sterile.
Dry Heat		
Incineration	Oxidizes cell components to ashes.	Flaming of wire inoculating loops. Also used to destroy medical wastes and contaminated animal carcasses.
Dry heat ovens	Oxidizes cell components and denatures proteins. Less efficient than moist heat, requiring longer times and higher temperatures.	Laboratory glassware is sterilized by heating it to 160°C to 170°C for 2 to 3 hours. Powders, oils, and other anhydrous materials are also sterilized in ovens.
Filtration	Filter retains microbes while letting the suspending fluid or air pass through small holes.	
Filtration of fluids	Various pore sizes are available; 0.2 μm is commonly used to remove bacteria.	Used to produce beer and wine, and to sterilize some heat-sensitive medications.
Filtration of air	HEPA filters are used to remove microbes that have a diameter greater than 0.3 μm.	Used in biological safety cabinets, specialized hospital rooms, and airplanes. Also used in some vacuum cleaners and home air purification units.
Radiation	Type of cell damage depends on the wavelength of the radiation.	
Ionizing radiation	Destroys DNA and possibly damages cytoplasmic membranes. Produces reactive molecules that damage other cell components. Items can be sterilized even after packaging.	Used to sterilize heat-sensitive materials including medical equipment, disposable surgical supplies, and drugs such as penicillin. Also used to destroy microbes in spices, herbs, and approved types of produce and meats.
Ultraviolet radiation	Damages DNA. Penetrates poorly.	Used to destroy microbes in the air and drinking water, and to disinfect surfaces.
High Pressure	Treatments of 130,000 psi are thought to denature proteins and alter the permeability of the cell. Products retain color and flavor.	Used to extend the shelf life of certain commercial food products such as guacamole.

Heat treatment is one of the most useful methods of microbial control because it is reliable, safe, relatively fast and inexpensive, and it does not introduce potentially toxic substances into the material being treated. Some heat-based methods sterilize the product, whereas others decrease the numbers of microorganisms and viruses. **Table 5.1** summarizes the characteristics of heat treatment and other physical methods of control.

Moist Heat

Moist heat destroys microorganisms by irreversibly coagulating their proteins. Examples of moist heat treatment include boiling, pasteurization, and pressurized steam.

Boiling

Boiling (100°C at sea level) easily destroys most microorganisms and viruses. Because of this, drinking water that has potentially been contaminated because of floods or other emergency situations should be boiled for at least 5 minutes. Boiling is not an effective means of sterilization, however, because endospores can survive many hours of the treatment.

Pasteurization

Louis Pasteur developed the brief heat treatment we now call pasteurization as a way of avoiding spoilage of wine. The process does not sterilize substances but significantly reduces the numbers of heat-sensitive organisms, including pathogens. Today, pasteurization is still used to destroy spoilage organisms in wine, vinegar, and a few other foods, but it is most widely used for killing pathogens in milk and juices. It increases the shelf life of foods and protects consumers by killing organisms that cause diseases such as tuberculosis, brucellosis, salmonellosis, and typhoid fever, without significantly altering the quality of the food. ■ food spoilage, p. 762

Today, most pasteurization protocols employ the **high-temperature-short-time (HTST) method.** Using this method, milk is heated to 72°C and held for 15 seconds. The parameters

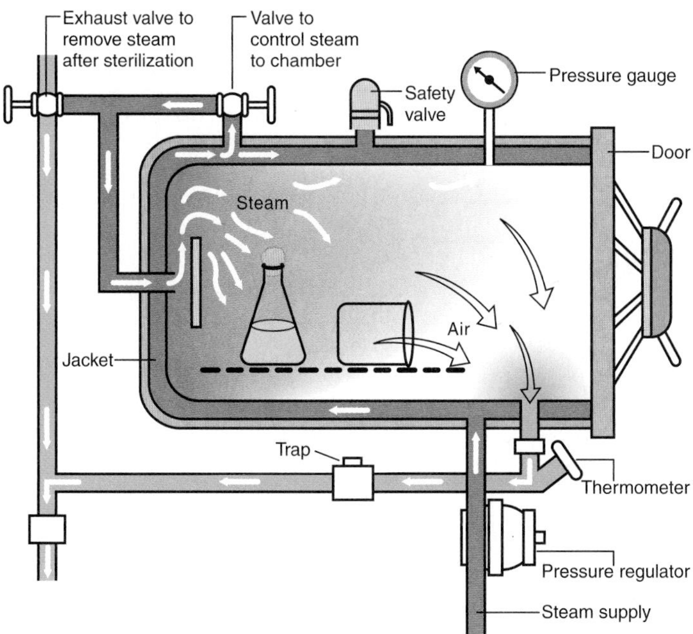

FIGURE 5.3 Autoclave Steam first travels in an enclosed layer, or jacket, surrounding the chamber. It then enters the autoclave, displacing the air downward and out through a port in the bottom of the chamber.

must be adjusted to the individual food product. For example, ice cream, which is richer in fats than is milk, requires a pasteurization process of 82°C for about 20 seconds.

The single-serving containers of cream served in restaurants are processed using the **ultra-high-temperature (UHT) method.** Because this process is designed to render the product free of all microorganisms that can grow under normal storage conditions, it is technically not a type of pasteurization. The milk is rapidly heated to a temperature of 140°C to 150°C, held for several seconds, then rapidly cooled. The product is then aseptically packaged in containers that have been treated with the chemical germicide hydrogen peroxide. Shelf-stable boxed juices and milk are processed and packaged in a similar manner.

Items such as cloth and rubber can be pasteurized by regulating the temperature of the water in a washing machine. For example, hospital anesthesia masks can be pasteurized at 80°C for 15 minutes. The temperatures and times used vary according to the organisms present and the heat stability of the material.

Sterilization Using Pressurized Steam

Pressure cookers and their commercial counterpart, the **autoclave,** heat water in an enclosed vessel that achieves temperatures above 100°C **(figure 5.3).** As heated water in the vessel forms steam, the steam causes the pressure in the vessel to increase beyond atmospheric pressure. The higher pressure, in turn, increases the temperature at which steam forms. Whereas steam produced at atmospheric pressure never exceeds 100°C, steam produced at an additional 15 psi (pounds/square inch) is 121°C, a temperature that kills endospores. Note that the pressure itself plays no direct role in the killing.

Autoclaving is generally the preferred method to sterilize heat- and moisture-tolerant items that steam can penetrate. Examples include surgical instruments, most microbiological media, reusable glassware and other supplies. Autoclaving is also used to sterilize microbial cultures and other biohazards before disposal.

Typical conditions used for sterilization are 15 psi and 121°C for 15 minutes. Longer time periods are necessary when sterilizing large volumes because it takes longer for heat to completely penetrate the liquid. For example, it takes longer to sterilize 4 liters of liquid in a flask than it would if the same volume were distributed into small tubes. When rapid sterilization is important, such as in operating rooms when sterile instruments must always be available, **flash autoclaving** at higher temperature can be used. By increasing the temperature to 135°C, sterilization is achieved in only 3 minutes. Autoclaving at a temperature of 132°C for 4.5 hours is thought to destroy prions.

Autoclaving is a consistently effective means of sterilizing most objects, provided the process is done correctly. The temperature and pressure gauge should both be monitored to ensure proper operating conditions. It is also critical that steam enter items and displace the air. Long, thin containers should be placed on their sides. Likewise, containers and bags should never be closed tightly.

To provide a visual signal that an item has been heated, tape that contains a heat-sensitive indicator can be attached to the item before it is autoclaved. The indicator turns black during autoclaving **(figure 5.4a).** A changed indicator, however, does not always mean that the object is sterile, because heating may not have been uniform.

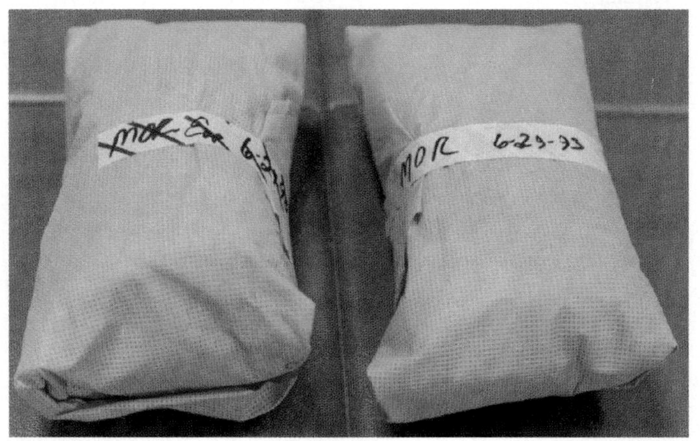

(a)

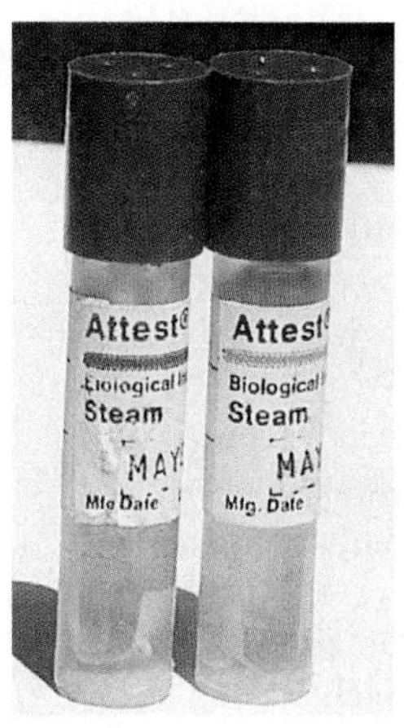

(b)

FIGURE 5.4 Indicator Used in Autoclaving **(a)** Chemical indicators. The pack on the left has been autoclaved. Diagonal marks on the tape have turned black, indicating that the object was exposed to heat. **(b)** Biological indicators. Following incubation, a change of color to yellow indicates growth of endospore-forming organisms. Why would a biological indicator be better than other indicators to determine if sterilization had been completely effective?

Biological indicators are used to ensure that the autoclave is working properly (figure 5.4b). A tube containing the heat-resistant endospores of *Geobacillus (Bacillus) stearothermophilus* is placed near the center of an item or package being autoclaved. After autoclaving, the endospores are mixed with a growth medium by crushing a container within the tube. Following incubation, a change of color of the medium indicates growth of the organisms and thus faulty autoclaving.

The Commercial Canning Process

The commercial canning process uses pressurized steam in an industrial-sized autoclave called a **retort.** Conditions of the process are designed to ensure that endospores of *Clostridium botulinum* are destroyed. This is critical because surviving spores can germinate and the resulting vegetative cells can grow in the anaerobic conditions of low-acid canned foods, such as vegetables and meats, and produce botulinum toxin, one of the most potent toxins known. In killing all endospores of *C. botulinum*, the process also kills all other organisms capable of growing under normal storage conditions. Endospores of some thermophilic bacteria may survive the canning process, but these are usually of no concern because they only can grow at temperatures well above those of normal storage. Because of this, canned foods are called **commercially sterile** to reflect the fact that the endospores of some thermophiles may survive. **Figure 5.5** shows the steps involved in the commercial canning of foods. ■ botulism, p. 657

Several factors dictate the time and temperature of the canning process. First, as discussed earlier, the higher the temperature, the shorter the time needed to kill all organisms. Second, the higher the concentration of bacteria, the longer the heat treatment required to kill all organisms. To provide a wide margin of safety, the commercial canning process is designed to reduce a population of 10^{12} *C. botulinum* endospores to only one. In other words, it is a 12 D process. It is virtually impossible for a food to have this high a level of initial concentration of endospores, and so the process has a wide safety margin.

Dry Heat

Dry heat is not as efficient as wet heat in killing microbes, requiring longer times and higher temperatures. For example, 200°C for 90 minutes of dry heat is the killing equivalent of 121°C for 15 minutes of moist heat.

Incineration oxidizes the cell components to ashes. In microbiology laboratories, for example, the wire loops continually reused to transfer bacterial cultures are sterilized by **flaming**—heating them in a flame until they are red hot. Alternatively, they can be heated to the same point in a benchtop incinerator designed for this purpose. Incineration is also used to destroy medical wastes and contaminated animal carcasses.

Temperatures achieved in hot air ovens oxidize cell components and irreversibly denature proteins. Glass Petri dishes and glass pipets are sterilized in ovens with non-circulating air at temperatures of 160°C to 170°C for 2 to 3 hours. Ovens with a fan that circulates the hot air can sterilize in a shorter time because of the more efficient transfer of heat. Powders, oils, and other anhydrous material are also sterilized in hot ovens.

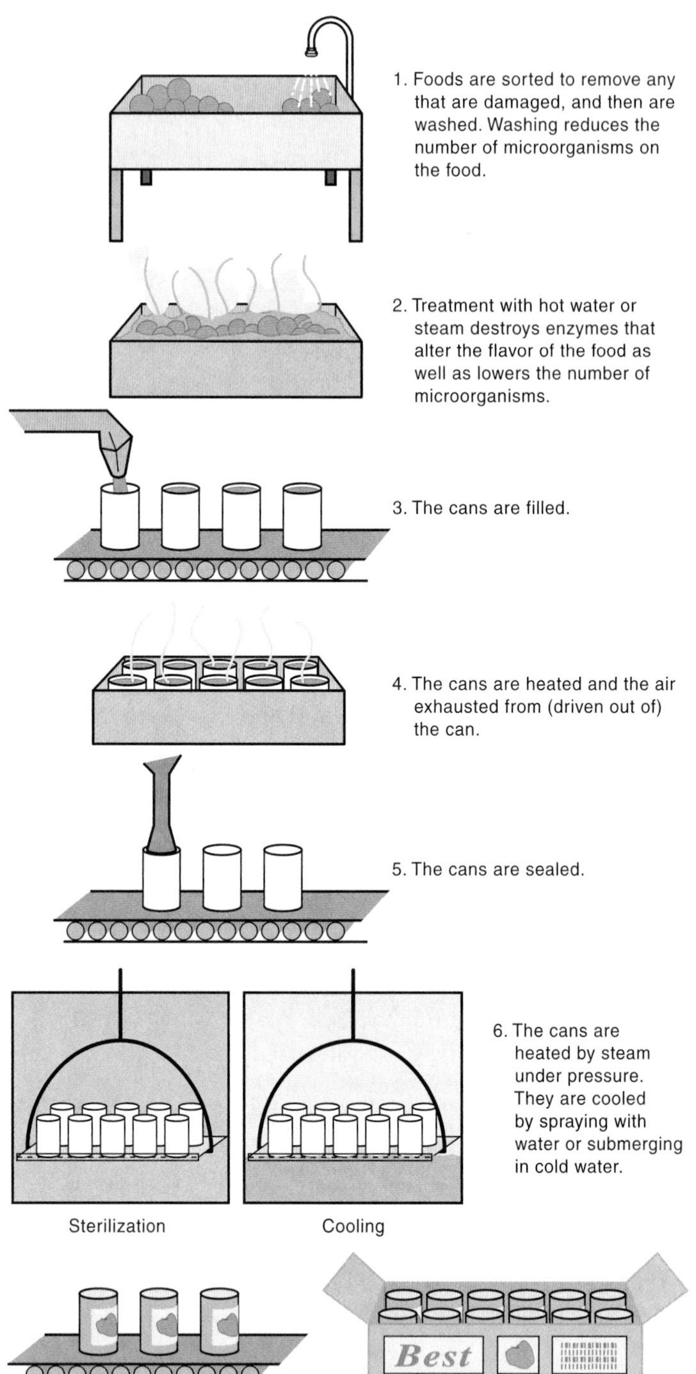

1. Foods are sorted to remove any that are damaged, and then are washed. Washing reduces the number of microorganisms on the food.
2. Treatment with hot water or steam destroys enzymes that alter the flavor of the food as well as lowers the number of microorganisms.
3. The cans are filled.
4. The cans are heated and the air exhausted from (driven out of) the can.
5. The cans are sealed.
6. The cans are heated by steam under pressure. They are cooled by spraying with water or submerging in cold water.
Sterilization
Cooling
Best
7. Cans are labeled, packaged, and shipped.

FIGURE 5.5 Steps in the Commercial Canning of Foods

MICROCHECK 5.3

Moist heat such as boiling water destroys most microorganisms and viruses. Pasteurization significantly reduces the numbers of heat-sensitive organisms. Autoclaves use pressurized steam to achieve high temperatures that kill microbes, including endospores. The commercial canning process is designed to destroy the endospores of *Clostridium botulinum*. Dry heat takes longer than moist heat to kill microbes.

✓ Why is it important that the commercial canning process destroys the endospores of *Clostridium botulinum*?

✓ What are two purposes of pasteurization?

✓ Would endospores be destroyed in the pasteurization process?

5.4

Using Other Physical Methods to Remove or Destroy Microbes

Focus Points

- Describe how depth filters, membrane filters, and HEPA filters are used to remove microorganisms.
- Describe how gamma irradiation, ultraviolet irradiation, and microwaves destroy microorganisms.

Some materials are either heat-sensitive or impractical to treat using heat. For these items, other physical methods including filtration, irradiation, and high-pressure treatment can be used to either destroy or remove microorganisms.

Filtration

Recall that membrane filtration, which is used to determine the number of bacteria in a liquid medium, retains bacteria while allowing the fluid to pass through. That same principle can be employed to physically remove microbes from liquids or air.

■ membrane filtration, p. 101

Filtration of Fluids

Filtration is used extensively to remove organisms from heat-sensitive fluids. Examples include production of unpasteurized beer, sterilization of sugar solutions, and clarification of wine. Specially designed filtration units are also used by backpackers and campers to remove *Giardia* cysts and bacteria from water.

Paper-thin **membrane filters** have microscopic pores that allow liquid to flow through while trapping particles that are too large to pass through the pores **(figure 5.6).** A vacuum is commonly used to help pull the liquid through the filter; alternatively, pressure may be applied to push the liquid through. Membrane filters are available in a variety of different pore sizes, extending below the dimensions of the smallest known viruses. Pore sizes smaller than necessary should be avoided, however, because they slow the flow. Filters with a pore size of 0.2 micrometers (μm) are commonly used to remove bacteria. The filters are made of compounds such as polycarbonate or cellulose nitrate that are relatively inert chemically and absorb very little of the fluid or its biologically important constituents such as enzymes.

Depth filters trap material within thick filtration material such as cellulose fibers or diatomaceous earth. They have complex, torturous passages that retain microorganisms while letting the suspending fluid pass through the small holes. The diameter of the passages is often considerably larger than that of the microorganisms they retain, and trapping of microbes partly results from electrical charges on the walls of the filter passages.

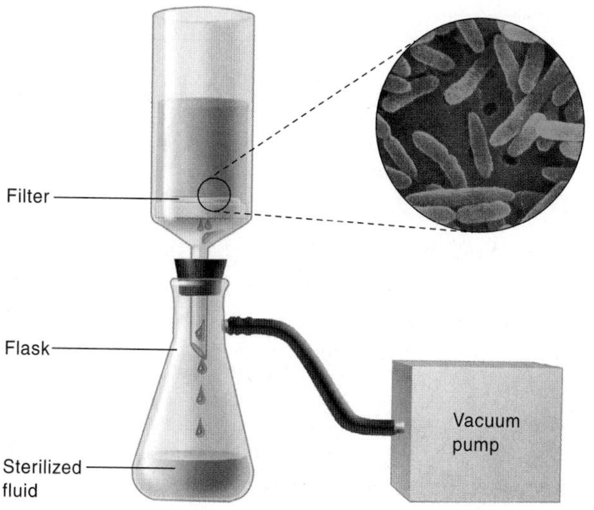

FIGURE 5.6 Filtration of Fluids Using a Membrane Filter The liquid to be sterilized flows through the filter on top of the flask in response to a vacuum produced in the flask by means of a pump. Scanning electron micrograph (5,000×) shows a membrane filter retaining cells of *Pseudomonas*.

Filtration of Air

Special filters called **high-efficiency particulate air (HEPA) filters** remove nearly all microorganisms that have a diameter greater than 0.3 μm from air. These filters are employed for keeping microorganisms out of specialized hospital rooms designed for patients who are extremely susceptible to infection. The filters are also used in biological safety cabinets, **laminar flow hoods,** in which laboratory personnel work with dangerous airborne pathogens such as *Mycobacterium tuberculosis*. A continuous flow of incoming and outgoing air is filtered through the HEPA filters to contain microorganisms within the cabinet. Biological safety cabinets are used not only to protect the worker from contamination by the sample, but also to protect the sample from environmental contamination.

Radiation

Radio waves, microwaves, visible and ultraviolet light rays, X rays, and gamma rays are all examples of a form of energy called **electromagnetic radiation.** This energy travels at the speed of light in waves and has no mass. The amount of energy in electromagnetic radiation is related to its **wavelength,** which is the distance from crest to crest (or trough to trough) of a wave, and **frequency,** which is the number of waves per second. Radiation that has short waves, and therefore high frequency, has more energy than that which has long waves and low frequency. The full range of wavelengths is called the **electromagnetic spectrum (figure 5.7).**

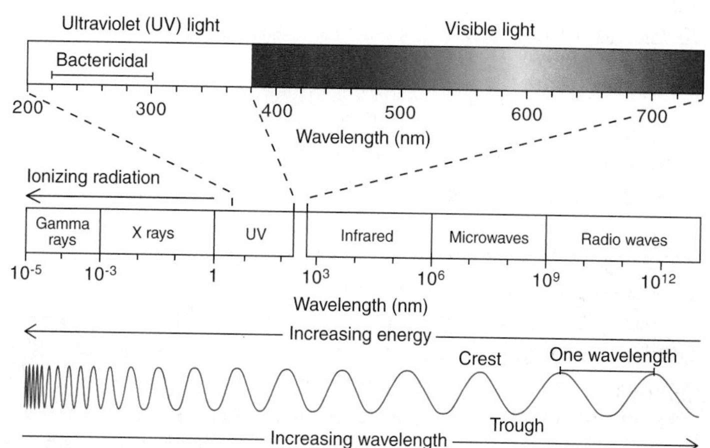

FIGURE 5.7 The Electromagnetic Spectrum Visible wavelengths include the colors of the rainbow.

Electromagnetic radiation can be either ionizing, meaning it can strip electrons off of atoms, or non-ionizing. Both types can be used to destroy microbes, but the mechanisms of destruction differ.

Ionizing Radiation

There are three sources of ionizing radiation: gamma rays, which are emitted from decaying radioisotopes (such as cobalt-60), X rays, and electron accelerators. The radiation causes biological harm both directly, by destroying DNA and possibly damaging cytoplasmic membranes, and indirectly, by producing reactive molecules such as superoxide and hydroxyl free radicals. The latter is a highly unstable molecule because it has one unpaired electron. Bacterial endospores are among the most radiation-resistant microbial forms, whereas Gram-negative bacteria such as *Salmonella* and *Pseudomonas* species are among the most susceptible. ■ superoxide, p. 92

Radiation is used extensively to sterilize heat-sensitive materials including medical equipment, disposable surgical supplies, and drugs such as penicillin. Radiation can generally be carried out after packaging.

Foods can be either sterilized or pasteurized using radiation, depending on the doses employed. Treatments designed to sterilize food can cause undesirable flavor changes, however, which limits their usefulness. More commonly, food is irradiated as a method of pasteurization, eliminating pathogens and decreasing the numbers of spoilage organisms. For example, it can be used to kill pathogens such as *Salmonella* species in poultry with little or no change in taste of the product.

In the United States, irradiation has been used for many years to control microorganisms on spices and herbs. The Food and Drug Administration (FDA) has also approved irradiation of fruits, vegetables, and grains to control insects; pork to control the trichina parasite; and most recently, meats including poultry, beef, lamb, and pork to control pathogens such as *Salmonella* species and *E. coli* O157:H7.

Many consumers refuse to accept irradiated products, even though the FDA and officials of the World Health and the United Nations Food and Agriculture Organizations have endorsed the technique. Some people erroneously believe that irradiated products are radioactive. Others think that irradiation-induced toxins or carcinogens are present in food, even though available scientific evidence indicates that consumption of irradiated food is safe. Another argument raised against irradiation is that it will cause a relaxation of other prudent food-handling practices. Irradiation, however, is intended to complement, not replace, proper food-handling procedures by producers, processors, and consumers.

Ultraviolet Radiation

Ultraviolet light in wavelengths of approximately 220 to 300 nm destroys microorganisms by damaging their DNA. Actively multiplying organisms are the most easily killed, whereas bacterial endospores are the most UV-resistant.

Ultraviolet light is used extensively to destroy microbes in the air and drinking water and to disinfect surfaces. It penetrates poorly, however, so even a thin film of grease on the UV bulb or extraneous material covering microorganisms can markedly reduce its effective microbial killing. It is not useful for destroying microbes in solid substances or turbid liquids. Since most types of glass and plastic screen out ultraviolet radiation, UV light is most effective when used at close range against exposed microorganisms. It must be used carefully because UV rays can also damage the skin and eyes and promote the development of skin cancers.

Microwaves

Microwaves do not affect microorganisms directly, but they can kill microbes by the heat they generate in an item. Organisms often survive microwave cooking, however, because the food heats unevenly.

High Pressure

High-pressure processing is used to pasteurize commercial food products such as guacamole without the use of high temperatures. The process, which employs pressures of up to 130,000 psi (pounds per square inch), is thought to destroy microorganisms by denaturing proteins and altering the permeability of the cell. Products treated with high-pressure processes retain the color and flavor associated with fresh foods.

MICROCHECK 5.4

Filters can be used to remove microorganisms and viruses from liquids and air. Gamma irradiation can be used to sterilize products and to decrease the number of microorganisms in foods. Ultraviolet light can be used to disinfect surfaces and air. Microwaves do not kill microbes directly, but by the heat they generate. Extreme pressure can kill microorganisms.

- ✓ What is the difference between the mechanism of a depth filter and that of a membrane filter?
- ✓ How does ultraviolet light kill microorganisms?
- ✓ Why could sterilization by gamma irradiation be carried out even after packaging?

5.5

Using Chemicals to Destroy Microorganisms and Viruses

Focus Points

- Describe the difference between sterilants, high-level disinfectants, intermediate-level disinfectants, and low-level disinfectants.
- Describe five important factors to consider when selecting an appropriate germicidal chemical.
- Compare and contrast the characteristics and use of alcohols, aldehydes, biguanides, ethylene oxide gas, halogens, metals, ozone, peroxygens, phenolic compounds, and quaternary ammonium compounds as germicidal chemicals.

Germicidal chemicals can be used to disinfect and, in some cases, sterilize. Most chemical germicides react irreversibly with vital proteins, DNA, cytoplasmic membranes or viral envelopes

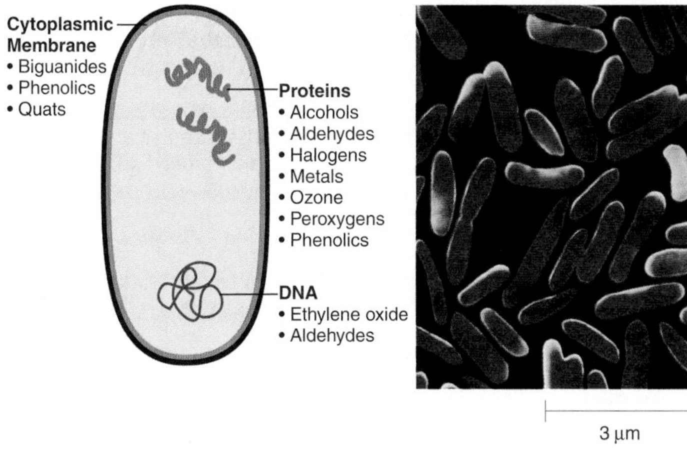

FIGURE 5.8 Sites of Action Germicidal Chemicals

(figure 5.8). Their precise mechanisms of action, however, are often not completely understood. Although generally less reliable than heat, these chemicals are suitable for treating large surfaces and many heat-sensitive items. Some are sufficiently non-toxic to be used as antiseptics. Those that have a **bacteriostatic** action, meaning they prevent the growth of (but do not kill) bacteria, can be used as preservatives.

Potency of Germicidal Chemical Formulations

Numerous different germicidal chemicals are marketed for medical and industrial use under a variety of trade names. Frequently, they contain more than one antimicrobial chemical as well as other chemicals such as buffers that can influence their antimicrobial activity. In the United States, the Food and Drug Administration (FDA) is responsible for regulating chemicals that can be used to process medical devices in order to ensure they perform as claimed. Most chemical disinfectants are considered pesticides and, as such, are regulated by the Environmental Protection Agency (EPA). To be registered with either the FDA or EPA, manufacturers of germicidal chemicals must document the potency of their products using testing procedures originally defined by the EPA. Germicides are grouped according to their potency:

- **Sterilants** can destroy all microorganisms, including endospores and viruses. Destruction of endospores usually requires a 6- to 10-hour treatment. Sterilants are used to treat heat-sensitive critical instruments such as scalpels.
- **High-level disinfectants** destroy all viruses and vegetative microorganisms, but they do not reliably kill endospores. Most are simply sterilants used for time periods as short as 30 minutes, not long enough to ensure endospore destruction. They can be used to treat semicritical instruments such as gastrointestinal endoscopes.
- **Intermediate-level disinfectants** destroy all vegetative bacteria including mycobacteria, fungi, and most, but not all, viruses. They do not kill endospores even with prolonged exposure. They are used to disinfect non-critical instruments such as stethoscopes.
- **Low-level disinfectants** destroy fungi, vegetative bacteria except mycobacteria, and enveloped viruses. They do not kill endospores, nor do they reliably destroy naked viruses.

  Intermediate-level and low-level disinfectants are also called **general-purpose disinfectants.** They are used in hospitals for disinfecting furniture, floors, and walls.

To perform properly, germicides must be used strictly according to the manufacturer's directions, especially as they relate to dilution, temperature, and the amount of time they must be in contact with the object being treated. It is extremely important that the object be thoroughly cleaned and free of organic material before the germicidal procedure is begun.

Selecting the Appropriate Germicidal Chemical

Selecting the appropriate germicide is a complex decision. Some points to consider include:

- **Toxicity.** Germicides are at least somewhat toxic to humans and the environment. Therefore, the benefit of disinfecting or sterilizing an item or surface must be weighed against the risks associated with using the germicidal procedure. For example, the risk of being exposed to a pathogenic microorganism in a hospital environment warrants using the most effective chemical germicides, even considering the potential risks of their use. The microbiological risks associated with typical household and office situations, however, may not justify the use of many of those same germicides.
- **Activity in the presence of organic matter.** Many germicidal chemicals, such as hypochlorite, are readily inactivated by organic matter and are not appropriate to use in situations where organic material is present. Chemicals such as phenolics, however, tolerate the presence of some organic matter.
- **Compatibility with the material being treated.** Items such as electrical equipment often cannot tolerate liquid chemical germicides, and so gaseous alternatives must be employed. Likewise, corrosive germicides such as hypochlorite often damage some metals and rubber.
- **Residue.** Many chemical germicides leave a residue that is toxic or corrosive. If a germicide that leaves a residue is used to sterilize or disinfect an item, the item must be thoroughly rinsed with sterile water to entirely remove the residue.
- **Cost and availability.** Some germicides are less expensive and more readily available than others. For example, hypochlorite can easily be purchased in the form of household bleach. On the other hand, ethylene oxide gas is not only more expensive, but it must be used in a special chamber, which influences the cost and practicality of the procedure.
- **Storage and stability.** Some germicides are available in concentrated stock solutions, decreasing the required storage space. The stock solutions are simply diluted according to the manufacturer's instructions before use. Others, such as chlorine dioxide, come in two-component systems that have a limited shelf life once mixed.
- **Environmental risk.** Germicides that retain their antimicrobial activity after use can interfere with sewage treatment systems that utilize microorganisms to degrade sewage. The activity of those germicides must be neutralized before disposal.

TABLE 5.2 Chemicals Used in Sterilization, and Disinfection, and Preservation of Non-Food Substances

Chemical (examples)	Characteristics	Uses
Alcohols (ethanol and isopropanol)	Easy to obtain and inexpensive. Rapid evaporation limits their contact time.	Aqueous solutions of alcohol are used as antiseptics to degerm skin in preparation for procedures that break intact skin, and as disinfectants for treating instruments.
Aldehydes (glutaraldehyde, orthophthalaldehyde, and formaldehyde)	Capable of destroying all forms of microbial life. Irritating to the respiratory tract, skin, and eyes.	Glutaraldehyde and orthophthalaldehyde are used to sterilize medical instruments. Formalin is used in vaccine production and to preserve biological specimens.
Biguanides (chlorhexidine)	Relatively low toxicity, destroys a wide range of microbes, adheres to and persists on skin and mucous membranes.	Chlorhexidine is widely used as an antiseptic in soaps and lotions, and impregnated into catheters and surgical mesh.
Ethylene Oxide Gas	Easily penetrates hard-to-reach places and fabrics and does not damage moisture-sensitive material. It is toxic, explosive, and potentially carcinogenic.	Commonly used to sterilize medical devices.
Halogens (chlorine and iodine)	Chlorine solutions are inexpensive and readily available; however, organic compounds and other impurities neutralize the activity. Some forms of chlorine may react with organic compounds to form toxic chlorinated products. Iodine is more expensive than chlorine and does not reliably kill endospores.	Solutions of chlorine are widely used to disinfect inanimate objects, surfaces, drinking water, and wastewater. Tincture of iodine and iodophores can be used as disinfectants or antiseptics.
Metals (silver)	Most metal compounds are too toxic to be used medically.	Silver sulfadiazine is used in topical dressings to prevent infection of burns. Silver nitrate drops can be used to prevent eye infections caused by *Neisseria gonorrhoeae* in newborns. Some metal compounds are used to prevent microbial growth in industrial processes.
Ozone	This unstable form of molecular oxygen readily breaks down.	Used to disinfect drinking water and wastewater.
Peroxygens (hydrogen peroxide and peracetic acid)	Readily biodegradable and less toxic than traditional alternatives. The effectiveness of hydrogen peroxide as an antiseptic is limited because the enzyme catalase breaks it down. Peracetic acid is a more potent germicide than is hydrogen peroxide.	Hydrogen peroxide is used to sterilize containers for aseptically packaged juices and milk. Peracetic acid is widely used to disinfect and sterilize medical devices.
Phenolic Compounds (triclosan and hexachlorophene)	Wide range of activity, reasonable cost, remains effective in the presence of detergents and organic contaminants, leaves an active antimicrobial residue.	Triclosan is used in a variety of personal care products, including toothpastes, lotions, and deodorant soaps. Hexachlorophene is highly effective against *Staphylococcus aureus,* but its use is limited because it can cause neurological damage.
Quaternary Ammonium Compounds (benzalkonium chloride and cetylpyridinium chloride)	Non-toxic enough to be used on food preparation surfaces. Inactivated by anionic soaps and detergents.	Widely used to disinfect inanimate objects and to preserve non-food substances.

Classes of Germicidal Chemicals

Germicides are represented in a number of chemical families. Each type has characteristics that make it more or less appropriate for specific uses **(table 5.2).**

Alcohols

Aqueous solutions of 60% to 80% ethyl or isopropyl alcohol rapidly kill vegetative bacteria and fungi. They do not, however, reliably destroy bacterial endospores and some naked viruses. Alcohol probably acts by coagulating enzymes and other essential proteins and by damaging lipid membranes. Proteins are more soluble and denature more easily in alcohol mixed with water, which is why the aqueous solutions are more effective than pure alcohol.

Alcohol solutions are commonly used as antiseptics to degerm skin before procedures such as injections that break intact skin. In addition, the Centers for Disease Control recently recommended that alcohol-based hand sanitizers be used routinely by healthcare personnel as a means to protect patients.

Alcohol solutions are also used as disinfectants for treating instruments and surfaces. They are relatively non-toxic and inexpensive, and do not leave a residue, but they evaporate quickly, which limits their effective contact time and, consequently, their germicidal effectiveness. In addition, they may damage some materials, such as rubber and some plastics.

Other antimicrobial chemicals are sometimes dissolved in alcohol. These alcohol-based solutions, called **tinctures,** can be more effective than the corresponding aqueous solutions.

Aldehydes

The aldehydes **glutaraldehyde, orthophthalaldehyde (OPA),** and **formaldehyde** destroy microorganisms and viruses by inactivating proteins and nucleic acids. A 2% solution of alkaline glutaraldehyde is one of the most widely used liquid chemical sterilants for treating heat-sensitive medical items. Immersion in this solution for 10 to 12 hours destroys all forms of microbial life, including endospores, and viruses. Soaking times as short as 10 minutes can be used to destroy vegetative bacteria. Glutaraldehyde is toxic, however, so treated items must be thoroughly rinsed with sterile water before use.

Orthophthalaldehyde is a relatively new type of disinfectant that provides an alternative to glutaraldehyde. It requires shorter processing times and is less irritating to eyes and nasal passages, but it stains proteins grey, including those of the skin.

Formaldehyde is used as a gas or an aqueous 37% solution called **formalin.** It is an extremely effective germicide that kills most forms of microbial life within minutes. Formalin is used to kill bacteria and to inactivate viruses for use as vaccines. It has also been used to preserve biological specimens. Formaldehyde's irritating vapors and suspected carcinogenicity, however, now limit its use.

Biguanides

Chlorhexidine, the most effective of a group of chemicals called **biguanides,** is extensively used in antiseptic products. It adheres to and persists on skin and mucous membranes, is of relatively low toxicity, and destroys a wide range of microbes, including vegetative bacteria, fungi, and some enveloped viruses. Chlorhexidine is an ingredient in many products including antiseptic skin creams, disinfectants, and mouthwashes. Chlorhexidine-impregnated catheters and implanted surgical mesh are used in medical procedures. Even tiny chips have been developed that can be inserted into periodontal pockets, where they slowly release chlorhexidine to treat periodontal gum disease. Adverse side effects of chlorhexidine are rare, but severe allergic reactions have been reported.

Ethylene Oxide

Ethylene oxide is an extremely useful gaseous sterilizing agent that destroys all microbes, including endospores and viruses, by reacting with proteins. As a gas, it penetrates well into fabrics, equipment, and implantable devices such as pacemakers and artificial hips. It is particularly useful for sterilizing heat- or moisture-sensitive items such as electrical equipment, pillows, and mattresses. Many disposable laboratory items, including plastic Petri dishes and pipets, are also sterilized with ethylene oxide.

A special chamber that resembles an autoclave is used to sterilize items with ethylene oxide. This allows careful control of factors such as temperature, relative humidity, and ethylene oxide concentration, all of which influence the effectiveness of the gas. Because ethylene oxide is explosive, it is generally mixed with a non-flammable gas such as carbon dioxide. Under these carefully controlled conditions, objects can be sterilized in 3 to 12 hours. The toxic ethylene oxide must then be eliminated from the treated material using heated forced air for 8 to 12 hours. Absorbed ethylene oxide must be allowed to dissipate because of its irritating effects on tissues and persistent antimicrobial effect, which, in the case of Petri dishes and other items used for culturing bacteria, is unacceptable.

Ethylene oxide is mutagenic and therefore potentially carcinogenic. Indeed, studies have shown a slightly increased risk of malignancies in long-term users of the gas.

Halogens

Chlorine and iodine are common disinfectants that are thought to act by oxidizing proteins and other essential cell components.

Chlorine Chlorine destroys all types of microorganisms and viruses but is too irritating to skin and mucous membranes to be used as an antiseptic. Chlorine-releasing compounds such as sodium hypochlorite can be used to disinfect waste liquids, swimming pool water, instruments, and surfaces, and at much lower concentrations, to disinfect drinking water.

Chlorine solutions are inexpensive, readily available disinfectants. An effective disinfection solution can easily be made by diluting liquid household bleach (5.25% sodium hypochlorite) 1:100 in water, resulting in a solution of 500 ppm (parts per million) chlorine. This concentration is several hundred times the amount required to kill most pathogenic microorganisms and viruses, but it is usually necessary for fast, reliable killing. In situations when excessive organic material is present, a 1:10 dilution of bleach may be required. This is because chlorine readily reacts with organic compounds and other impurities in water, disrupting its germicidal activity. The use of high concentrations, however, should be avoided when possible, because chlorine is both corrosive and toxic. Diluted solutions of liquid bleach deteriorate over time; thus, fresh solutions need to be prepared regularly. More stable forms of chlorine, including sodium dichloroisocyanurate and chloramines, are often used in hospitals.

Properly chlorinated drinking water contains approximately 0.5 ppm chlorine, much less than that used for disinfectant solutions. The exact amount of chlorine that must be added depends on the amount of organic material in the water. The presence of organic compounds is also a problem because chlorine can react with some organic compounds to form trihalomethanes, which are potential carcinogens. Note also that the concentrations of chlorine typically used to disinfect drinking water are not effective against *Cryptosporidium parvum* oocysts and *Giardia lamblia* cysts.

Chlorine dioxide (ClO_2) is a strong oxidizing agent that is increasingly being used as a disinfectant and sterilant. It has an advantage over chlorine-releasing compounds in that it does not react with organic compounds to form trihalomethanes or other toxic chlorinated products. Compressed chlorine dioxide gas, however, is explosive and liquid solutions decompose readily, so that it must be generated on-site. It is used to treat drinking water, wastewater, and swimming pools.

Iodine Iodine, unlike chlorine, does not reliably kill endospores, but it can be used as a disinfectant. It is used as a tincture in which the iodine is dissolved in alcohol, or more commonly as an **iodophore,** in which the iodine is linked to a carrier molecule that releases free (unbound) iodine slowly. Iodophores are not as irritating to the skin as tincture of iodine nor are they as likely to stain. Iodophores used as disinfectants contain more free iodine (30 to 50 ppm) than do those used as antiseptics (1 to 2 ppm). Stock solutions must be strictly diluted according to the manufacturer's instructions because dilution affects the amount of free iodine available.

Surprisingly, some *Pseudomonas* species survive in the concentrated stock solutions of iodophores. The reasons are unclear, but it could be due to inadequate levels of free iodine in concentrated solutions, because iodine may be released from the carrier only with dilution. *Pseudomonas* species also can form biofilms, which are less permeable to chemicals. Nosocomial infections can result if a *Pseudomonas*-contaminated iodophore is unknowingly used to disinfect instruments. ■ biofilm, p. 85

Metal Compounds

Metal compounds kill microorganisms by combining with sulfhydryl groups (—SH) of enzymes and other proteins, thereby interfering with their function. Unfortunately, most metals at high concentrations are too toxic to human tissue to be used medically.

Silver is one of the few metals still used as a disinfectant. Creams containing silver sulfadiazine, a combination of silver and a sulfa drug, are applied topically to prevent infection of second- and third-degree burns. Commercially available bandages with silver-containing pads can be used on minor scalds, cuts, and scrapes. For many years, doctors were required by law to instill drops of another silver compound, 1% silver nitrate, into the eyes of newborns to prevent ophthalmia neonatorum, an eye infection caused by *Neisseria gonorrhoeae*, which is acquired from infected mothers during the birth process. Drops of antibiotics have now largely replaced use of silver nitrate because they are less irritating to the eye and more effective against another genitally acquired pathogen, *Chlamydia trachomatis*. ■ *Neisseria gonorrhoeae*, p. 629 ■ *Chlamydia trachomatis*, p. 631

Compounds of mercury, tin, arsenic, copper, and other metals were once widely used as preservatives in industrial products and to prevent microbial growth in recirculating cooling water. Their extensive use resulted in serious pollution of natural waters, which has prompted strict controls.

Ozone

Ozone (O_3) is an unstable form of oxygen that is a powerful oxidizing agent. It decomposes quickly, however, so it must be generated on-site, usually by passing air or oxygen between two electrodes. Ozone is used as an alternative to chlorine for disinfecting drinking water and wastewater.

Peroxygens

Hydrogen peroxide and peracetic acid are powerful oxidizing agents that under controlled conditions can be used as sterilants. They are readily biodegradable and, in normal concentrations of use, appear to be less toxic than the traditional alternatives, ethylene oxide and glutaraldehyde.

Hydrogen Peroxide The effectiveness of hydrogen peroxide (H_2O_2) as a germicide depends in part on whether it is used on living tissue, such as a wound, or on an inanimate object. This is because all cells that use aerobic metabolism, including the body's cells, produce the enzyme catalase, which inactivates hydrogen peroxide by breaking it down to water and oxygen gas. Thus, when a solution of 3% hydrogen peroxide is applied to a wound, our cellular enzymes quickly break it down. When the same solution is used on an inanimate surface, however, it overwhelms the relatively low concentration of catalase produced by microscopic organisms. ■ catalase, p. 92

Hydrogen peroxide is particularly useful as a disinfectant because it leaves no residue and does not damage stainless steel, rubber, plastic, or glass. Hot solutions are commonly used in the food industry to yield commercially sterile containers for aseptically packaged juices and milk. Vapor-phase hydrogen peroxide is more effective than liquid solutions and can be used as a sterilant.

Peracetic Acid Peracetic acid is an even more potent germicide than hydrogen peroxide. A 0.2% solution of peracetic acid, or a combination of peracetic acid and hydrogen peroxide, can be used to sterilize items in less than 1 hour. It is effective in the presence of organic compounds, leaves no residue, and can be used on a wide range of materials. It has a sharp, pungent odor, however, and like other oxidizing agents, it is irritating to the skin and eyes.

Phenolic Compounds (Phenolics)

Phenol (carbolic acid) is important historically because it was one of the earliest disinfectants, but its use is now limited because it has an unpleasant odor and irritates the skin. Derivatives of phenol, called **phenolics,** have greater germicidal activity, which enables effective use of more dilute and therefore less irritating solutions. Phenolic compounds are the active ingredients in Lysol™.

Phenolics destroy cytoplasmic membranes of microorganisms and denature proteins. They kill most vegetative bacteria and, in high concentrations (from 5% to 19%), many can kill *Mycobacterium tuberculosis*. They do not, however, reliably inactivate all groups of viruses. The major advantages of phenolic compounds include their wide range of activity, reasonable cost, and ability to remain effective in the presence of detergents and organic contaminants. They also leave an active antimicrobial residue, which in some cases is desirable.

Some phenolics, such as triclosan and hexachlorophene, are sufficiently non-toxic to be used in soaps and lotions. Triclosan is widely used as an ingredient in a variety of personal care products such as deodorant soaps, lotions, and toothpaste. Hexachlorophene has substantial activity against *Staphylococcus aureus*, the leading cause of wound infections, but high levels have been associated with symptoms of neurotoxicity. Although once widely used in over-the-counter products, antiseptic skin cleansers containing hexachlorophene are now available only with a prescription.

PERSPECTIVE 5.1

Contamination of an Operating Room by a Bacterial Pathogen

A patient with burns infected with *Pseudomonas aeruginosa* was taken to the operating room for cleaning of the wounds and removal of dead tissue. After the procedure was completed, samples of various surfaces in the room were cultured to determine the extent of contamination. *P. aeruginosa* was recovered from all parts of the room. **Figure 1** shows how readily and extensively an operating room can become contaminated by an infected patient. Operating rooms and other patient care rooms must be thoroughly cleaned after use, in a process known as terminal cleaning.

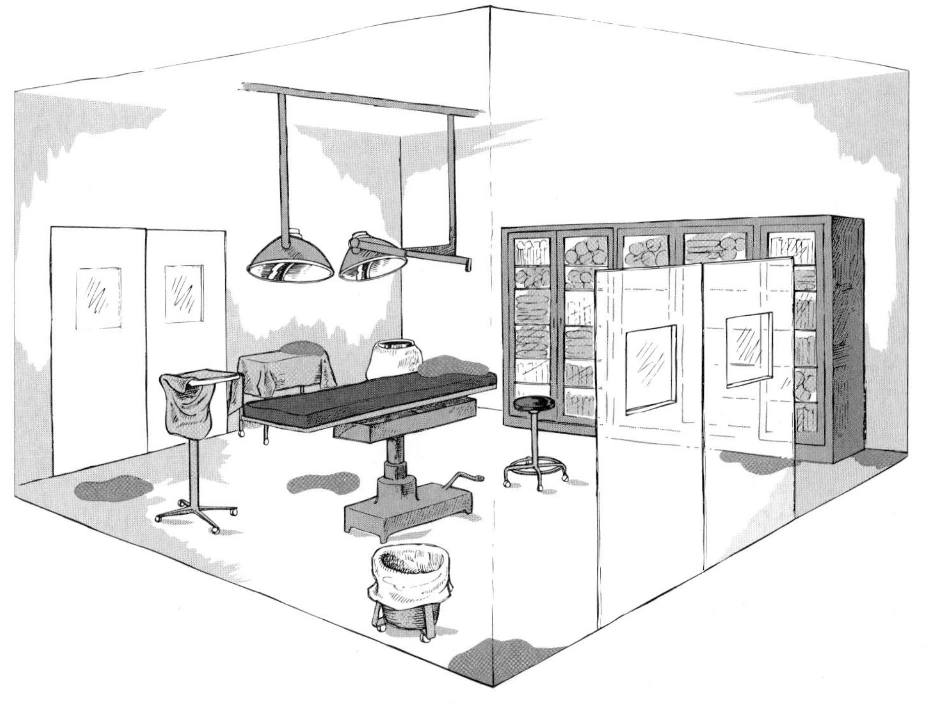

FIGURE 1 Diagram of an operating room in which dead tissue infected with *Pseudomonas aeruginosa* was removed from a patient with burns. Reddish areas indicate places where *P. aeruginosa* was recovered following the surgical procedure.

Quaternary Ammonium Compounds (Quats)

Quaternary ammonium compounds, also commonly called quats, are cationic (positively charged) detergents that are nontoxic enough to be used to disinfect food preparation surfaces. Like all detergents, quats have both a charged hydrophilic region and an uncharged hydrophobic region. This enables them to reduce the surface tension of liquids and help wash away dirt and organic material, facilitating the mechanical removal of microorganisms from surfaces. Unlike most common household soaps and detergents, however, which are anionic (negatively charged) and repelled by the negatively charged microbial cell surface, quats are attracted to the cell surface. They react with membranes, destroying many vegetative bacteria and enveloped viruses. They are not effective, however, against endospores, mycobacteria, or naked viruses.

Quaternary ammonium compounds are economical and effective agents that are widely used to disinfect clean inanimate objects and to preserve non-food substances. The ingredients of many personal care products include quats such as benzalkonium chloride or cetylpyridinium chloride. They also enhance the effectiveness of some other disinfectants. Cationic soaps and organic material such as gauze, however, can neutralize their effectiveness. In addition, *Pseudomonas*, a troublesome cause of nosocomial infections, resists the effects of quats and can even grow in solutions preserved with them.

MICROCHECK 5.5

Germicidal chemicals can be used to disinfect and, in some cases, sterilize, but they are less reliable than heat. They are especially useful for destroying microorganisms and viruses on heat-sensitive items and large surfaces.

✓ Describe four factors that must be considered when selecting a germicidal chemical.

✓ Explain why it is essential to dilute iodophores properly.

✓ Why would a heavy metal be a more serious pollutant than most organic compounds?

5.6 Preservation of Perishable Products

Focus Point

- Explain how chemical preservatives, low-temperature storage, adding salt or sugar, and drying food can all be used to preserve perishable products.

Preventing or slowing the growth of microorganisms extends the shelf life of products such as food, soaps, medicines, deodorants, cosmetics, and contact lens solutions. Preservative chemicals are often added to these products to prevent or slow the growth of microbes that are inevitably introduced from the environment. Other common methods of decreasing the growth rate of microbes include low-temperature storage such as refrigeration or freezing, and reducing available water. These methods are particularly important in preserving foods. ■ food spoilage, p. 762 ■ factors influencing the growth of microorganisms in foods, p. 754

Chemical Preservatives

Some of the germicidal chemicals previously described can be used to preserve non-food items. For example, mouthwash may contain a quaternary ammonium compound, nasal sprays may contain thimerosal, and leather belts may be treated with one or more phenol derivatives. Food preservatives, however, must be non-toxic for repeated safe ingestion.

Benzoic, sorbic, and propionic acids are weak organic acids that are sometimes added to foods such as bread, cheese, and juice to prevent microbial growth. At a low pH, these weak acids alter cell membrane functions and interfere with energy transformation. The low pH at which they are most effective is itself sufficient to prevent the growth of most bacteria, so these preservatives are primarily added to acidic foods to prevent the growth of fungi, which otherwise grow well at acidic pH. These organic acids also occur naturally in some foods such as cranberries and Swiss cheese.

Another preservative, nitrate, and its reduced form, nitrite, serve a dual purpose in processed meats. From a microbiological viewpoint, their most important function is to inhibit the germination of endospores and subsequent growth of *Clostridium botulinum*. Without the addition of low levels of nitrate or nitrite to cured meats such as bologna, ham, bacon, and smoked fish, *C. botulinum* may grow and produce deadly botulinum toxin. At higher concentrations than are required for preservation, nitrate and nitrite react with myoglobin in the meat to form a stable pigment that gives a desirable pink color associated with fresh meat. Nitrates and nitrites also pose a potential hazard, however, because they can be converted to nitrosamines during the frying of meats in hot oil or by the metabolic activities of intestinal bacteria. Nitrosamines are potent carcinogens, which has caused concern regarding the use of nitrate and nitrite as preservatives.

Low-Temperature Storage

The growth of many pathogens and spoilage microorganisms is inhibited by refrigeration because their critical enzyme reactions are slowed or stopped. Thus, low temperature storage is extremely useful in preservation. Psychrotrophic and some psychrophilic organisms, however, can grow at normal refrigeration temperatures. ■ psychrophiles, p. 91

Freezing is also an important means of preserving foods and other products. Freezing essentially stops all microbial growth. While the formation of ice crystals can kill some of the microbial cells, the remaining organisms can grow and spoil foods once they are thawed.

Reducing the Available Water

For many years, salting and drying have been used to preserve food. Both processes decrease the availability of water in food below the limits required for growth of most microorganisms. The high-solute environment causes plasmolysis, which damages microbial cells (see figure 4.10). ■ plasmolysis, p. 93 ■ water availability, p. 93

Adding Salt or Sugar

Sugar and salt draw water out of cells, dehydrating them. High concentrations of sugars or salts are added to many foods as preservatives. For example, fruit is made into jams and jellies by adding sugar, and fish and meats are cured by soaking them in salty water, or **brine.** Some caution should be exercised when using salt as a preservative, however, because the food-poisoning bacterium *Staphylococcus aureus* can grow under quite high salt conditions. ■ *Staphylococcus aureus*, p. 763

Drying Food

Removing water, or **desiccating,** food is often supplemented by salting or adding high concentrations of sugar or small amounts of chemical preservatives. For example, meat jerkies usually have added salt and sometimes sugar.

Lyophilization (freeze-drying) is widely used for preserving foods such as coffee, milk, meats, and vegetables. In the process of freeze-drying, the food is first frozen and then dried in a vacuum. When water is added to the lyophilized material, it reconstitutes. The quality of the reconstituted product is often much better than that of products treated with ordinary drying methods. The light weight and stability without refrigeration of freeze-dried foods make them popular with hikers.

Although drying stops microbial growth, it does not reliably kill bacteria and fungi in or on foods. For example, numerous cases of salmonellosis have been traced to dried eggs. Eggshells and even egg yolks may be heavily contaminated with *Salmonella* species from the gastrointestinal tract of the hen. To prevent the transmission of such pathogens, some states have laws requiring dried eggs to be pasteurized before they are sold.

MICROCHECK 5.6

Preservation techniques slow or halt the growth of microorganisms to delay spoilage.

- ✓ What organism that causes food poisoning is able to grow under high-salt conditions?
- ✓ What is the risk of consuming nitrate- or nitrite-free cured meats?
- ✓ Preservation by freezing is sometimes compared to drying. Why would this be so?

FUTURE CHALLENGES

Too Much of a Good Thing?

In our complex world, the solution to one challenge may inadvertently lead to the creation of another. Scientists have long been pursuing less toxic alternatives to many traditional biocidal chemicals. For example, glutaraldehyde has now largely replaced the more toxic formaldehyde, chlorhexidine is generally used in place of hexachlorophene, and gaseous alternatives to ethylene oxide are now being sought. Meanwhile, ozone and hydrogen peroxide, which are both readily biodegradable, may eventually replace glutaraldehyde. While these less toxic alternatives are better for human health and the environment, their widespread acceptance and use may be unwittingly contributing to an additional problem—the overuse and misuse of germicidal chemicals. Many products, including soaps, toothbrushes, and even clothing and toys are marketed with the claim of containing antimicrobial ingredients. Already there are reports of bacterial resistance to some of the chemicals included in these products.

The issues surrounding the excessive use of antimicrobial chemicals are complicated. On the one hand, there is no question that some microorganisms cause disease. Even those that are not harmful to human health can be troublesome because they produce metabolic end products that ruin the quality of perishable products. Based on that information, it seems prudent to destroy or inhibit the growth of microorganisms whenever possible. The role of microorganisms in our life, however, is not that simple. Our bodies actually harbor a greater number of microbial cells than human cells, and this normal microbiota plays an important role in maintaining our health. Excessive use of antiseptics or other antimicrobials may actually predispose a person to infection by damaging the normal microbiota.

An even more worrisome concern is that overuse of disinfectants and other germicidal chemicals will select for microorganisms that are more resistant to those chemicals, a situation analogous to our current problems with antibiotic resistance. By using antimicrobial chemicals indiscriminately, we may eventually make these useful tools obsolete. Excessive use of disinfectants may even be contributing to the problems of antibiotic resistance. Disinfectant-resistant bacteria sometimes over-produce efflux pumps that expel otherwise damaging chemicals, including antibiotics, from the cell. Thus, by overusing disinfectants, we may be inadvertently increasing antibiotic resistance.

Another concern is over the misguided belief that "non-toxic" or "biodegradable" chemicals cause no harm, and the common notion that "if a little is good, more is even better." For example, concentrated solutions of hydrogen peroxide, though biodegradable, can cause serious damage, even death, when used improperly. Other chemicals, such as chlorhexidine, can elicit severe allergic reactions in some people.

As less toxic germicidal chemicals are developed, people must be educated on the appropriate use of these alternatives.

SUMMARY

5.1 Approaches to Control

The methods used to destroy or remove microorganisms and viruses can be **physical,** such as heat treatment, irradiation, and filtration, or **chemical.**

Principles of Control

A variety of terms are used to describe antimicrobial agents and processes.

Situational Considerations (figure 5.1)

Situations encountered in daily life, hospitals, microbiology laboratories, food production facilities, water treatment facilities, and other industries warrant different degrees of microbial control.

5.2 Selection of an Antimicrobial Procedure

Type of Microorganism

One of the most critical considerations in selecting a method of destroying microorganisms and viruses is the type of microbial population thought to be present on or in the product.

Numbers of Microorganisms Initially Present

The amount of time it takes for heat or chemicals to kill a population of microorganisms is dictated in part by the number of cells initially present. Microbial death generally occurs at a constant rate. The **D value,** or **decimal reduction time,** is the time it takes to kill 90% of a population of bacteria under specific conditions (figure 5.2).

Environmental Conditions

Factors such as pH and presence of organic materials influence microbial death rates.

Potential Risk of Infection

To guide medical biosafety personnel in their selection of germicidal procedures, instruments are categorized as **critical, semicritical,** and **non-critical** according to their potential risk of transmitting infectious agents.

Composition of the Item

Some sterilization and disinfection procedures are inappropriate for certain types of material.

5.3 Using Heat to Destroy Microorganisms and Viruses (figure 5.1)

Moist Heat

Moist heat destroys microorganisms by causing irreversible coagulation of their proteins. **Pasteurization** utilizes a brief heat treatment to destroy spoilage and disease-causing organisms. Pressure cookers and **autoclaves** heat water in an enclosed vessel that causes the pressure in the vessel to increase beyond atmospheric pressure, increasing the temperature of steam, which kills endospores (figure 5.3). The most important aspect of the commercial canning process is to ensure that endospores of *Clostridium botulinum* are destroyed (figure 5.5).

Dry Heat

Incineration oxidizes cell components to ashes. Temperatures achieved in hot air ovens oxidize cell components and irreversibly denature proteins.

5.4 Using Other Physical Methods to Remove or Destroy Microbes

Filtration (figure 5.6)

Membrane filters and **depth filters** retain microorganisms while letting the suspending fluid pass through. **High-efficiency particulate air (HEPA)** filters remove nearly all microorganisms.

Radiation (figure 5.7)

Gamma rays cause biological damage by producing superoxide and hydroxyl free radicals. Ultraviolet light damages the structure and function of nucleic acids. Microwaves do not affect microorganisms directly, but kill microorganisms by the heat they generate in a product.

High Pressure

High pressure is thought to destroy microorganisms by denaturing proteins and altering the permeability of the cell.

5.5 Using Chemicals to Destroy Microorganisms and Viruses

Potency of Germicidal Chemical Formulations

Germicides are grouped according to their potency as **sterilants, high-level disinfectants, intermediate-level disinfectants,** or **low- level disinfectants.**

Selecting the Appropriate Germicidal Chemical

Factors that must be included in the selection of an appropriate germicidal chemical include toxicity, residue, activity in the presence of organic matter, compatibility with the material being treated, cost and availability, storage and stability, and ease of disposal.

Classes of Germicidal Chemicals (table 5.2)

Solutions of 60% to 80% ethyl or isopropyl alcohol in water rapidly kill vegetative bacteria and fungi by coagulating enzymes and other essential proteins, and by damaging lipid membranes. **Glutaraldehyde, orthophthalaldehyde** and **formaldehyde** destroy microorganisms and viruses by inactivating proteins and nucleic acids. **Chlorhexidine** is a **biguanide** extensively used in antiseptic products. **Ethylene oxide** is a gaseous sterilizing agent that penetrates well and destroys microorganisms and viruses by reacting with proteins. Sodium hypochlorite (liquid bleach) is one of the least expensive and most readily available forms of chlorine. **Chlorine dioxide** is used as a sterilant and disinfectant. **Iodophores** are iodine-releasing compounds used as antiseptics. Metals interfere with protein function. Silver-containing compounds are used to prevent wound infections. Ozone is used as an alternative to chlorine in the disinfection of drinking water and wastewater. Peroxide and peracetic acid are both strong oxidizing agents that can be used alone or in combination as sterilants. **Phenolic compounds** destroy cytoplasmic membranes and denature proteins. Triclosan is used in lotions and deodorant soaps. **Quaternary ammonium compounds** are cationic detergents; they are nontoxic enough to be used to disinfect food preparation surfaces.

5.6 Preservation of Perishable Products

Chemical Preservatives

Benzoic, sorbic, and propionic acids are sometimes added to foods to prevent microbial growth. Nitrate and nitrite are added to some foods to inhibit the germination and subsequent growth of *Clostridium botulinum* endospores.

Low-Temperature Storage

Low temperatures above freezing inhibit microbial growth. Freezing essentially stops all microbial growth.

Reducing the Available Water

Sugar and salt draw water out of cells, preventing the growth of microorganisms. **Lyophilization** is used for preserving food. The food is first frozen and then dried in a vacuum.

REVIEW QUESTIONS

Short Answer

1. What is the primary reason that milk is pasteurized?
2. What is the primary reason that wine is pasteurized?
3. What is the most chemically resistant non-spore-forming bacterial pathogen?
4. Why are low acid foods processed at higher temperatures than high acid foods?
5. Explain why it takes longer to kill a population of 10^9 cells than it does to kill a population of 10^3 cells.
6. How is an iodophore different from a tincture of iodine?
7. How does microwaving a food product kill bacteria?
8. How is preservation different from pasteurization?
9. How are heat-sensitive liquids sterilized?
10. Name two products commonly sterilized using ethylene oxide gas.

Multiple Choice

1. Unlike a disinfectant, an antiseptic
   a) sanitizes objects rather than sterilizes them.
   b) destroys all microorganisms.
   c) is nontoxic enough to be used on human skin.
   d) requires heat to be effective.
   e) can be used in food products.
2. The D value is defined as the time it takes to kill
   a) all bacteria in a population.
   b) all pathogens in a population.
   c) 99.9% of bacteria in a population.
   d) 90% of bacteria in a population.
   e) 10% of bacteria in a population.
3. Which of the following is the most resistant to destruction by chemicals and heat?
   a) Bacterial endospores
   b) Fungal spores
   c) Mycobacterium tuberculosis
   d) *E. coli*
   e) HIV
4. Ultraviolet light kills bacteria by
   a) generating heat.
   b) damaging DNA.
   c) inhibiting protein synthesis.
   d) damaging cell walls.
   e) damaging cytoplasmic membranes.

5. Which concentration of ethyl alcohol is the most effective germicide?
   a) 100% b) 75% c) 50% d) 25% e) 5%
6. Which of the following chemical agents can most reliably be used to sterilize objects?
   a) Alcohol b) Phenolic compounds c) Ethylene oxide gas
   d) Iodine
7. All of the following are routinely used to preserve foods, except
   a) high concentrations of sugar.
   b) high concentrations of salt.
   c) benzoic acid.
   d) freezing.
   e) ethylene oxide.
8. Aseptically boxed juices and cream containers are processed using which of the following heating methods?
   a) Canning
   b) High-temperature-short-time (HTST) method
   c) Low-temperature-long-time (LTLT) method
   d) Ultra-high-temperature (UHT) method
9. Commercial canning processes are designed to ensure destruction of which of the following?
   a) All vegetative bacteria
   b) All vegetative bacteria and their endospores
   c) Endospores of *Clostridium botulinum*
   d) *E. coli*
   e) *Mycobacterium tuberculosis*
10. Which of the following is false?
    a) A chemical that is a high-level disinfectant cannot be used as a sterilant.
    b) Critical items must be sterilized before use.
    c) Low numbers of endospores may remain on semicritical items.
    d) Standard sterilization procedures do not destroy prions.
    e) Quaternary ammonium compounds can be used to disinfect food preparation surfaces.

Applications

1. An agriculture extension agent is preparing pamphlets on preventing the spread of disease. In the pamphlet, he must explain the appropriate situations for using disinfectants around the house. What situations should the agent discuss?
2. As a microbiologist representing a food corporation, you have been asked to serve on a health food panel to debate the need for chemical preservatives in foods. Your role is to prepare a statement that compares the benefits of chemical preservatives and the risks. What points must you bring up that indicate the benefits of chemical preservatives?

Critical Thinking

1. This graph shows the time it takes to kill populations of the same microorganism under different conditions. What conditions would explain the differences in lines a, b, and c?

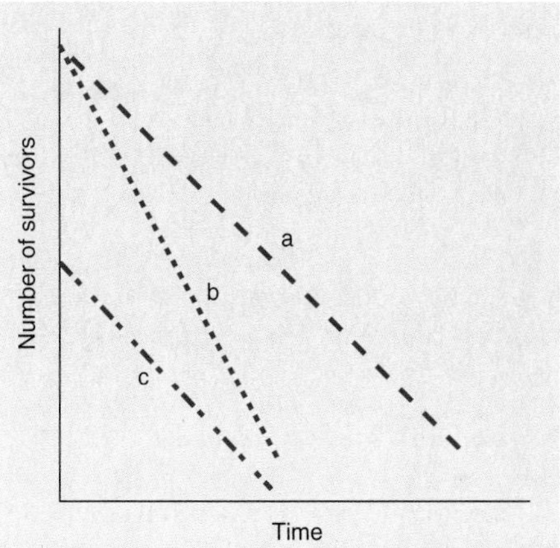

2. This diagram shows the filter paper method used to evaluate the inhibitory effect of chemical agents, heavy metals, and antibiotics on bacterial growth. A culture of test bacteria is spread uniformly over the surface of an agar plate. Small filter paper discs containing the material to be tested are then placed on the surface of the medium. A disc that has been soaked in sterile distilled water is sometimes added as a control. After incubation, a film of growth will cover the plate, but a clear zone will surround those discs that contain an inhibitory compound. The size of the zone reflects several factors, one of which is the effectiveness of the inhibitory agent. What are two other factors that might affect the size of the zone of inhibition? What is the purpose of the control disc? If a clear area were apparent around the control disc, how would you interpret the observation?

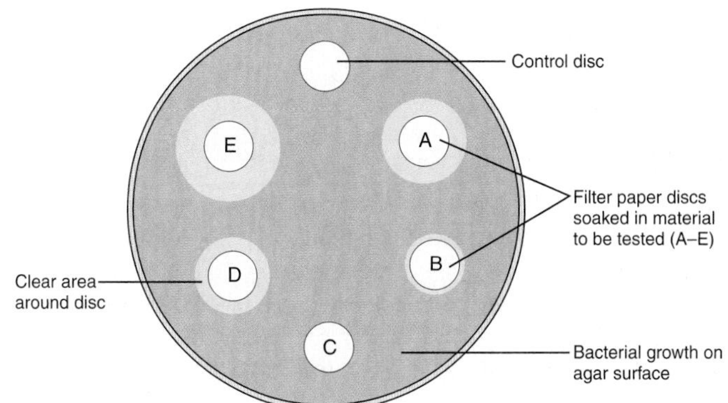

Wine—a beverage produced using microbial metabolism.

Metabolism: Fueling Cell Growth

A Glimpse of History

In the 1850s, Louis Pasteur, a chemist, accepted the challenge of studying how alcohol arises from grape juice. Biologists had already observed that when grape juice is held in large vats, alcohol and carbon dioxide are produced and the number of yeast cells increases. They argued that the multiplying yeast cells convert the sugar in the juice to alcohol and carbon dioxide. Pasteur agreed, but could not convince two very powerful and influential German chemists, Justus von Liebig and Friedrich Wöhler, who refused to believe that microorganisms caused the breakdown of sugar. Both men lampooned the hypothesis and tried to discredit it by publishing pictures of yeast cells looking like miniature animals taking in grape juice through one orifice and releasing carbon dioxide and alcohol through the other.

Pasteur studied the relationship between yeast and alcohol production using a strategy commonly employed by scientists today—that is, simplifying the experimental system so that relationships can be more easily identified. First, he prepared a clear solution of sugar, ammonia, mineral salts, and trace elements. He then added a few yeast cells. As the yeast grew, the sugar level decreased and the alcohol level increased, indicating that the sugar was being converted to alcohol as the cells multiplied. This strongly suggested that living cells caused the chemical transformation. Liebig, however, still would not believe the process was actually occurring inside microorganisms. To convince him, Pasteur tried to extract something from inside the yeast cells that would convert the sugar. He failed, like many others before him.

In 1897, Eduard Buchner, a German chemist, showed that crushed yeast cells could convert sugar to ethanol and CO_2. We now know that enzymes of the crushed cells carried out this transformation. For these pioneering studies, Buchner was awarded the Nobel Prize in 1907. He was the first of many investigators who received Nobel Prizes for studies on the processes by which cells degrade sugars.

To grow, all cells must accomplish two fundamental tasks. They must continually synthesize new components including cell walls, membranes, ribosomes, nucleic acids, and surface structures such as flagella. These allow the cell to enlarge and eventually divide. In addition, cells need to harvest energy and convert it to a form that is usable to power biosynthetic reactions, transport nutrients and other molecules, and in some cases, move. The sum total of chemical reactions used for biosynthetic and energy-harvesting processes is called **metabolism.**

Bacterial metabolism is important to humans for a number of reasons. Many bacterial products are commercially or medically important. For example, as scientists look for new supplies of energy, some are investigating **biofuels,** which are fuels made from a renewable biological source such as plants and organic waste products. Microorganisms or their enzymes are currently producing these fuels, breaking down solid materials such as corn stalks, sugar cane, and wood to a fuel such as ethanol. As another example, cheese-makers intentionally add *Lactococcus* and *Lactobacillus* species to milk because the metabolic wastes of these bacteria contribute to the flavor and texture of various cheeses. Yet some of these same products contribute to tooth decay when related bacteria are growing on teeth. Microbial metabolism is also important in the laboratory, because products that are characteristic of a specific group of

KEY TERMS

Adenosine Triphosphate (ATP) The energy currency of cells. Hydrolysis of its unstable phosphate bonds can be used to power endergonic (energy-consuming) reactions.

Anabolism Processes that utilize energy stored in ATP to synthesize and assemble the subunits (building blocks) of macromolecules that make up the cell; biosynthesis.

Catabolism Processes that harvest energy released during the breakdown of compounds such as glucose, using it to synthesize ATP.

Electron Transport Chain Group of membrane-embedded electron carriers that pass electrons from one to another, and, in the process, move protons across the membrane to create a proton motive force.

Enzyme A protein that functions as a catalyst, speeding up a biological reaction.

Fermentation Metabolic process that stops short of oxidizing glucose or other organic compounds completely, using an organic intermediate such as pyruvate or a derivative as a terminal electron acceptor.

Oxidative Phosphorylation Synthesis of ATP using the energy of a proton motive force created by harvesting chemical energy.

Photophosphorylation Synthesis of ATP using the energy of a proton motive force created by harvesting radiant energy.

Precursor Metabolites Metabolic intermediates that can either be used to make the subunits of macromolecules, or be oxidized to generate ATP.

Proton Motive Force Form of energy generated as an electron transport chain moves protons across a membrane, creating a chemiosmotic gradient.

Respiration Process that involves transfer of electrons stripped from a chemical energy source to an electron transport chain, generating a proton motive force that is then used to synthesize ATP.

Substrate-Level Phosphorylation Synthesis of ATP using the energy released in an exergonic (energy-releasing) chemical reaction.

Terminal Electron Acceptor Chemical such as O_2 that is ultimately reduced as a consequence of fermentation or respiration.

microorganisms can be used as identifying markers. In addition, the metabolic pathways of organisms such as *E. coli* have served as an invaluable model for studying analogous processes in eukaryotic cells, including those of humans. Metabolic processes unique to prokaryotes are potential targets for antimicrobial drugs.

6.1 Principles of Metabolism

Focus Points

- Compare and contrast catabolism and anabolism.
- Describe the energy sources used by photosynthetic organisms and chemoorganoheterotrophs.
- Describe the components of metabolic pathways (enzymes, ATP, chemical energy source, redox reactions, electron carriers, and precursor metabolites).
- List the three central metabolic pathways.
- Distinguish between respiration and fermentation.

Metabolism can be viewed as having two components—**catabolism** and **anabolism (figure 6.1).** Catabolism encompasses processes that harvest energy released during the disassembly or breakdown of compounds such as glucose, using that energy to synthesize **ATP,** the energy currency of all cells. In contrast, anabolism, or **biosynthesis,** includes processes that utilize energy stored in ATP to synthesize and assemble subunits (building blocks) of macromolecules that make up the cell. These subunits include amino acids, nucleotides, and lipids. ■ ATP, p. 25

Although catabolism and anabolism are often discussed separately, they are intimately linked. As mentioned, ATP generated during catabolism is used in anabolism. In addition, some of the compounds produced in steps of the catabolic processes can be diverted by the cell and used as precursors of subunits employed in anabolic processes.

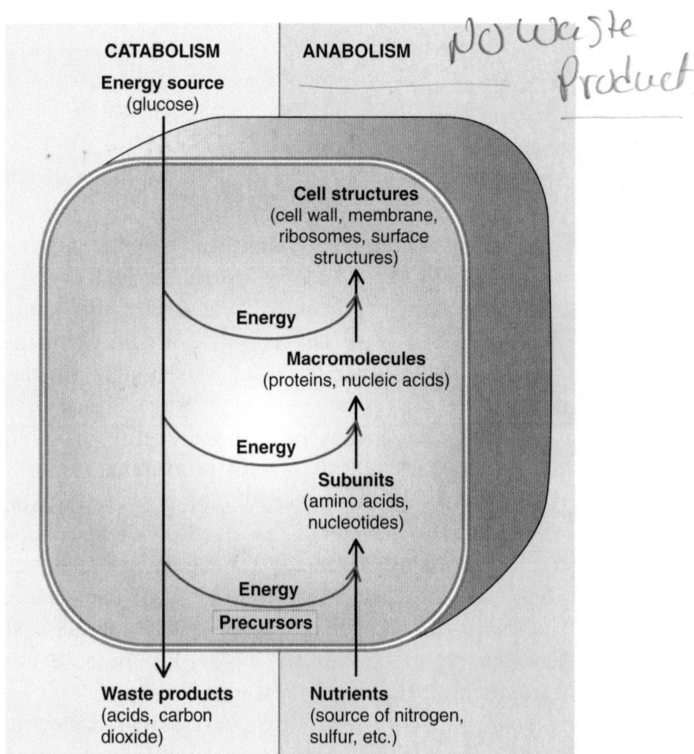

FIGURE 6.1 The Relationship Between Catabolism and Anabolism Catabolism encompasses processes that harvest energy released during disassembly of compounds, using it to synthesize ATP; it also provides precursor metabolites used in biosynthesis. Anabolism, or biosynthesis, includes processes that utilize ATP and precursor metabolites to synthesize and assemble subunits of macromolecules that make up the cell.

FIGURE 6.2 Forms of Energy Potential energy is stored energy, such as water held behind a dam. Kinetic energy is the energy of motion, such as movement of water from behind the dam.

Harvesting Energy

Energy is defined as the capacity to do work. It can exist as **potential energy,** which is stored energy, and **kinetic energy,** which is energy of motion **(figure 6.2).** Potential energy can be stored in various forms including chemical bonds, a rock on a hill, or water behind a dam.

Energy in the universe can never be created or destroyed; however, it can be changed from one form to another. In other words, while energy cannot be created, potential energy can be converted to kinetic energy and vice versa, and one form of potential energy can be converted to another. For example, hydroelectric dams unleash the potential energy of water stored behind a dam, creating the kinetic energy of moving water; this can then be used to generate an electrical current, which can then be used to charge a battery.

Photosynthetic organisms harvest the energy of sunlight, using it to power the synthesis of organic compounds such as glucose **(figure 6.3).** In other words, they convert the kinetic energy of photons to the potential energy of chemical bonds. **Chemoorganotrophs** obtain energy by degrading organic compounds such as glucose, releasing the energy of their chemical bonds. Thus, most chemoorganotrophs ultimately depend on solar energy harvested by photosynthetic organisms, because this is what is used to power the synthesis of glucose.

The amount of energy available for harvest by breaking down a compound can be explained by the concept of **free energy.** This is the energy available to do work; from a biological perspective, it is the energy that can be released when a chemical bond is broken. In a chemical reaction, some bonds are broken and others are formed. If the **reactants,** or starting compounds, have more free energy than the **products,** or final compounds, energy is released in the reaction. The reaction is said to be **exergonic.** In contrast, if

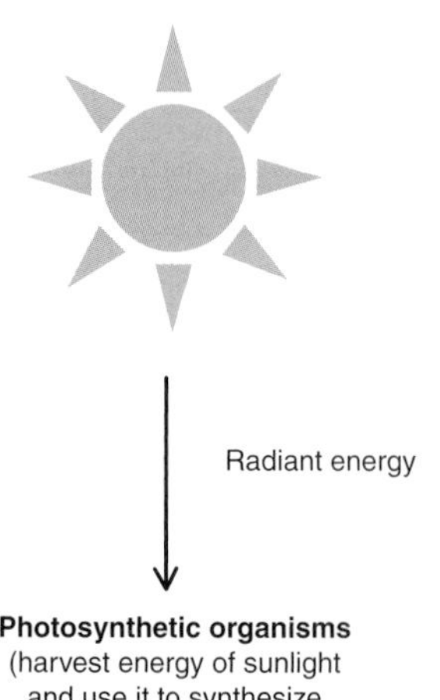

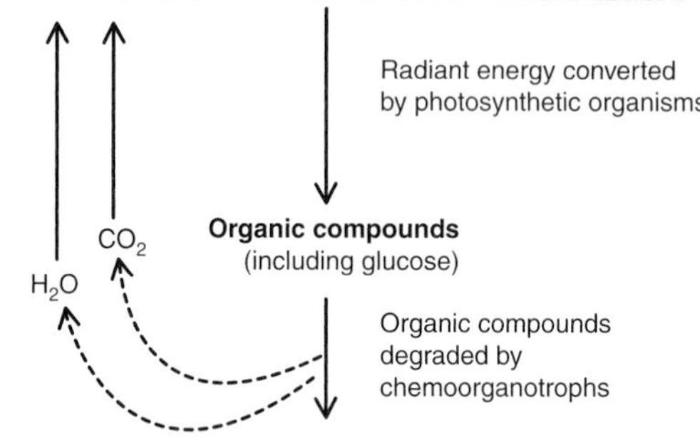

FIGURE 6.3 Most Chemoorganotrophs Depend on the Radiant Energy Harvested by Photosynthetic Organisms Photosynthetic organisms use the energy of sunlight to power the synthesis of organic compounds; chemoorganotrophs can then use those organic compounds as an energy source.

the products have more free energy than the reactants, the reaction requires an input of energy and is termed **endergonic.**

The change in free energy for a given reaction is the same regardless of the number of steps involved. For example, converting glucose to carbon dioxide and water in a single step by combustion releases the same amount of energy as degrading it in a series of steps. Cells exploit this fact to slowly release free energy from compounds, harvesting the energy released at each step. A specific energy-releasing reaction is used to power an energy-utilizing reaction.

Components of Metabolic Pathways

Metabolic processes often occur as a series of sequential chemical reactions, which constitute a **metabolic pathway (figure 6.4).** A series of **intermediates** are produced as the starting compound is gradually converted into the final product, or **end product.** A metabolic pathway can be linear, branched, or cyclical, and, like the flow of a river controlled by dams, its activity can be modulated at certain points. In this way, a cell can regulate certain processes, ensuring that specific molecules are produced in precise quantities when needed. If a metabolic step is blocked, all products "downstream" of that blockage will be affected.

The intermediates and end products of metabolic pathways are sometimes organic acids, which are weak acids. Depending on the pH, these may exist primarily as either the undissociated form or the dissociated (ionized) form. Biologists often use the names of the two forms interchangeably—for example, pyruvic acid and pyruvate. Note, however, that at the near-neutral pH inside the cell, the ionized form predominates, whereas outside of the cell, the acid may predominate. ■ pH, p. 24

To recognize what metabolic pathways accomplish, it is helpful to first understand the critical components—enzymes, ATP, the chemical energy source, electron carriers, and precursor metabolites.

The Role of Enzymes

A specific **enzyme** facilitates each step of a metabolic pathway. Enzymes are proteins that function as biological catalysts, accelerating the conversion of one substance, the **substrate,** into another, the **product.** Without enzymes, energy-yielding reactions would still occur, but at rates so slow they would be imperceptible.

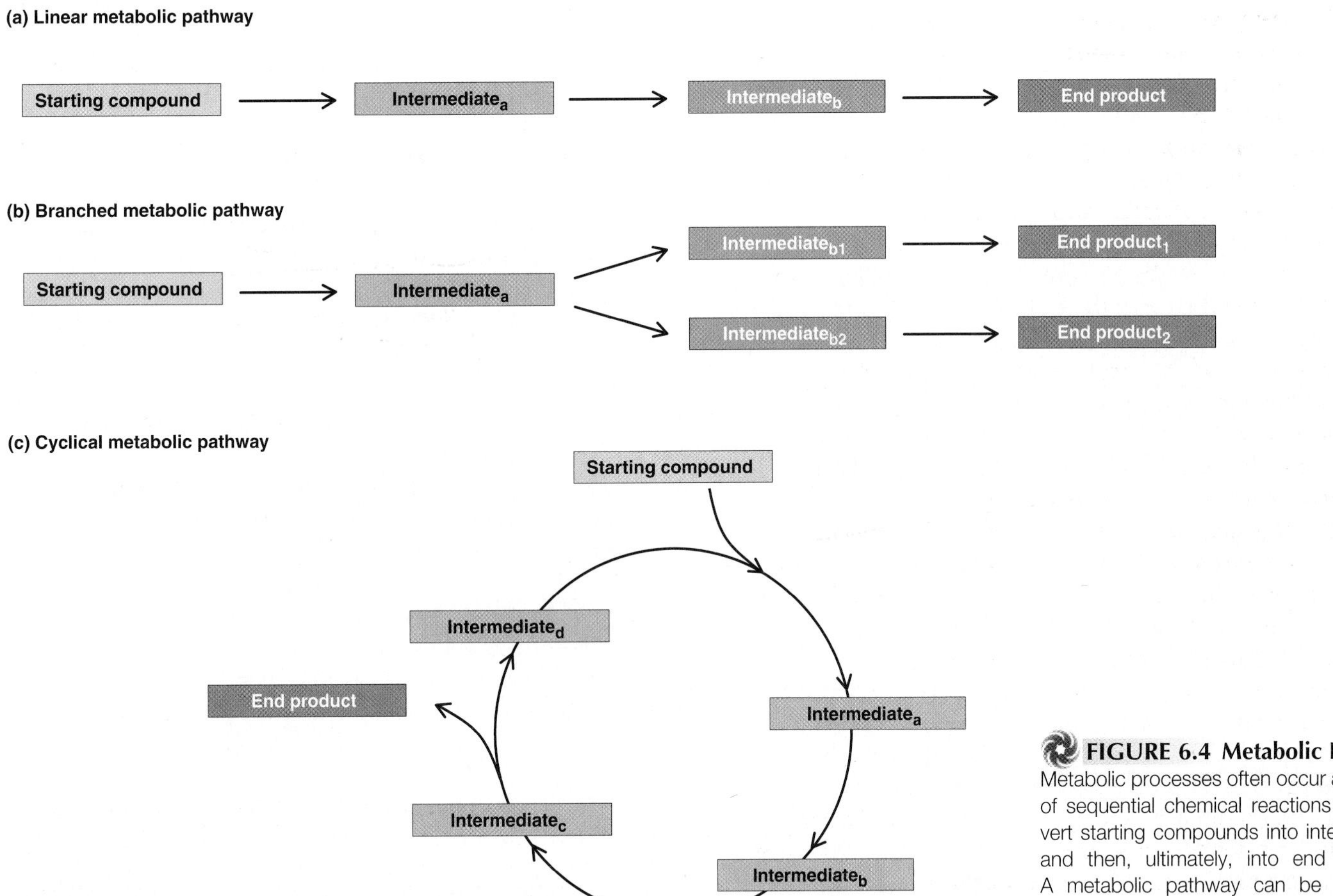

FIGURE 6.4 Metabolic Pathways Metabolic processes often occur as a series of sequential chemical reactions that convert starting compounds into intermediates and then, ultimately, into end products. A metabolic pathway can be **(a)** linear, **(b)** branched, or **(c)** cyclical.

An enzyme catalyzes a chemical reaction by lowering the **activation energy** of that reaction **(figure 6.5).** This is the energy it takes to initiate a chemical reaction; even exergonic chemical reactions have an activation energy. By lowering the activation energy barrier, enzymes allow chemicals to undergo rearrangements. Enzymes will be described in more detail later in the chapter. ■ enzymes, p. 134

The Role of ATP

Adenosine triphosphate (ATP) is the energy currency of a cell, serving as the ready and immediate donor of free energy. It is composed of the sugar ribose, the nitrogenous base adenine, and three phosphate groups (see figure 2.10). Its counterpart, **adenosine diphosphate (ADP),** can be viewed as an acceptor of free energy. An input of energy is required to add an inorganic phosphate group (P_i) to ADP, forming ATP; energy is released when that group is removed from ATP, yielding ADP **(figure 6.6).** ■ ATP, p. 25

The phosphate groups of ATP are arranged in tandem (see figure 2.10). Their negative charges repel each other, making the bonds that join them unstable. The bonds are readily hydrolyzed, releasing the phosphate group and a sufficient amount of energy to power an endergonic reaction. Because of the relatively high amount of free energy released when the bonds between the phosphate groups are hydrolyzed, they are called **high-energy phosphate bonds,** denoted by the symbol ~. ■ hydrolysis, p. 25

Cells constantly turn over ATP, powering biosynthetic reactions by hydrolyzing the high-energy phosphate bond, and then exploiting energy-releasing reactions to form it again. Two different processes are used by chemoorganotrophs to provide the energy necessary to form the high-energy phosphate bond. **Substrate-level phosphorylation** uses the chemical energy released in an exergonic reaction to add P_i to ADP; **oxidative phosphorylation** harvests the energy of proton motive force to do the same thing. Recall from chapter 3 that **proton motive force** is the form of energy that results from the electrochemical gradient established as protons are expelled from the cell (see figure 3.26). The electron transport chain that generates this type of energy will be discussed later in this chapter. Photosynthetic organisms can generate ATP using the process of **photophosphorylation,** utilizing radiant energy of the sun to drive the formation of a proton motive force. The mechanisms they use to do this will be discussed later. ■ proton motive force, p. 57

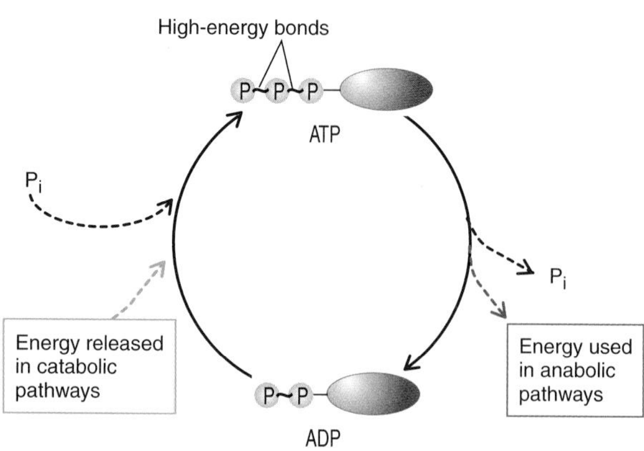

FIGURE 6.6 ATP Energy is released when unstable ("high-energy") phosphate bonds are broken; an input of energy is required to convert ADP back to ATP.

The Role of the Chemical Energy Source

The compound broken down by a cell to release energy is called the **energy source.** As a group, prokaryotes show remarkable diversity in the variety of energy sources they can use. Many use organic compounds such as glucose. Others use inorganic compounds including hydrogen sulfide and ammonia. Harvesting energy from a compound involves a series of coupled oxidation-reduction reactions.

Oxidation-Reduction Reactions In **oxidation-reduction reactions,** or **redox reactions,** one or more electrons are transferred from one substance to another **(figure 6.7).** The molecule that loses electrons becomes **oxidized;** the one that gains those electrons becomes **reduced.** ■ electrons, p. 18

When electrons are removed from a molecule in a biochemical reaction, protons (H^+) often follow. In other words, an electron-proton pair, or hydrogen atom, is often removed. Thus, the removal of a hydrogen atom is an oxidation; correspondingly, the addition of a hydrogen atom is a reduction. An oxidation reaction

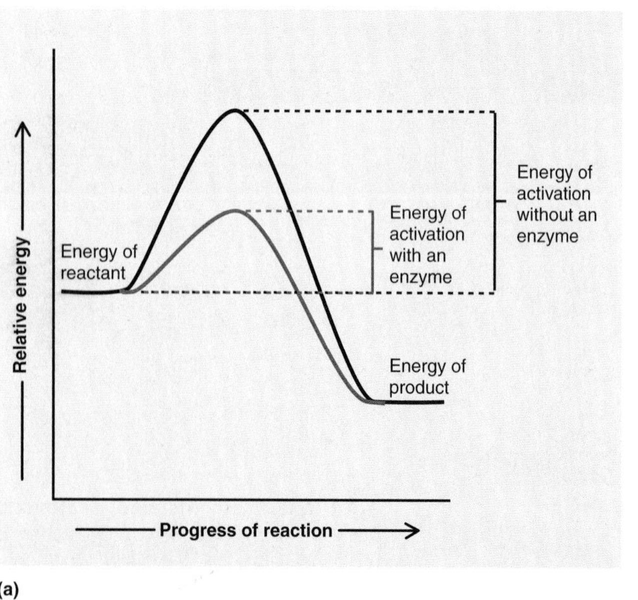

(a)

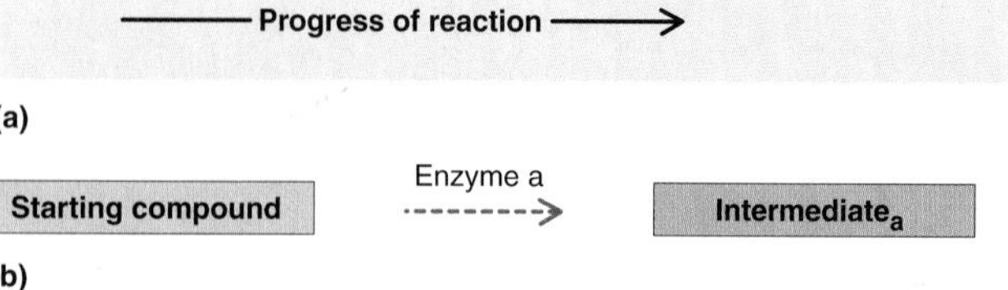

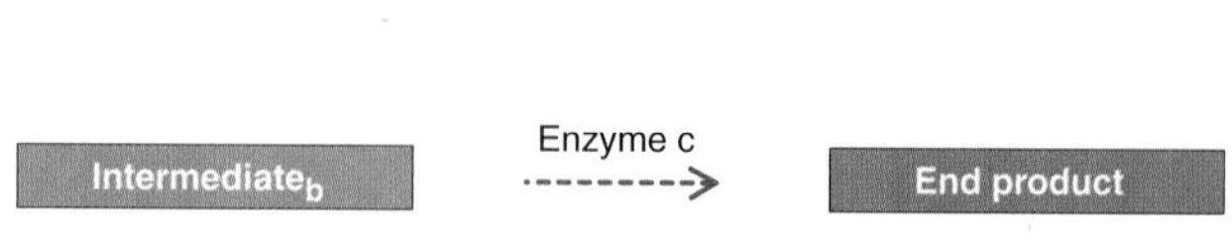

(b)

FIGURE 6.5 The Role of Enzymes Enzymes function as biological catalysts. **(a)** An enzyme catalyzes a chemical reaction by lowering the activation energy of the reaction. **(b)** A specific enzyme facilitates each step of a metabolic pathway.

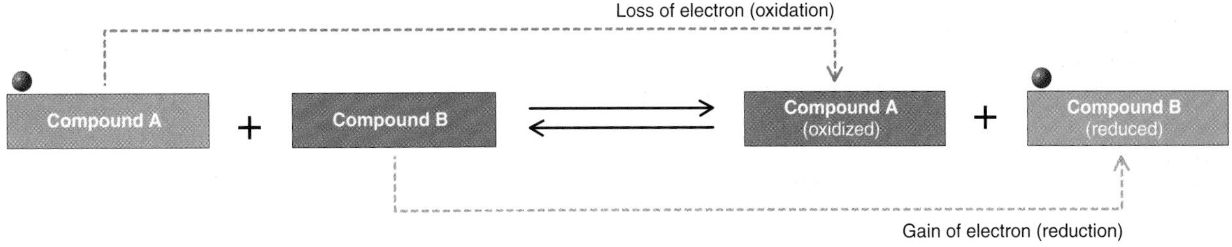

FIGURE 6.7 Oxidation-Reduction Reactions The compound that loses one or more electrons becomes oxidized; the compound that gains those electrons become reduced.

in which an electron and an accompanying proton are removed is called a **dehydrogenation.** A reduction reaction in which an electron and an accompanying proton are added is called a **hydrogenation.**

When electrons are removed from the energy source, or **electron donor,** they are temporarily transferred to a specific molecule that serves as an **electron carrier.** That carrier can also be viewed as a **hydrogen carrier** if a proton accompanies the electron. Protons, however, unlike electrons, do not require carriers when in an aqueous solution. Because of this, the whereabouts of protons in biological reactions are often ignored.

The Role of Electron Carriers

Just as cells use ATP as a carrier of free energy, they use designated molecules as carriers of electrons. Cells have several different types of electron carriers, and each serves a different function.

Three types of electron carriers directly participate in reactions that oxidize the energy source **(table 6.1).** They are **NAD^+** (nicotinamide adenine dinucleotide), **FAD** (flavin adenine dinucleotide), and **$NADP^+$** (NAD phosphate). The reduced forms of these carriers are **NADH, $FADH_2$,** and **NADPH,** respectively. These electron carriers can also be considered hydrogen carriers because along with electrons, they carry protons. NAD^+ and $NADP^+$ can each carry a hydride ion, which consists of two electrons and one proton; $FADH_2$ carries two electrons and two protons.

Reduced electron carriers represent **reducing power** because their bonds contain a form of usable energy. The reducing power of NADH and $FADH_2$ is used to generate the proton motive force, which drives the synthesis of ATP in the process of oxidative phosphorylation. Ultimately the electrons are transferred to a molecule such as O_2 that functions as a **terminal electron acceptor.** The reducing power of NADPH has an entirely different fate; it is used in biosynthetic reactions when a reduction is required. Note, however, that many microbial cells have a membrane-associated enzyme that is able to use proton motive force to reduce $NADP^+$. This allows them to convert reducing power in the form of NADH to NADPH.

Precursor Metabolites

Precursor metabolites are metabolic intermediates produced at specific steps in catabolic pathways that can be used in anabolic pathways. In anabolism, they serve as raw material used to make the subunits of macromolecules (see figure 6.1). For example, the precursor metabolite pyruvate can be converted to the amino acid alanine.

Many organisms, including *Escherichia coli,* can make all of their cell components, including proteins, lipids, carbohydrates, and nucleic acids, using only a dozen or so precursor metabolites. Recall from chapter 4 that *E. coli* can grow in glucose-salts medium, which contains only glucose and a few inorganic salts. The glucose in the medium not only serves as the energy source,

TABLE 6.1 Electron Carriers

Carrier	Oxidized Form	Reduced Form	Typical Fate of Electrons Carried
		electron; electron carrier	
Nicotinamide adenine dinucleotide (carries 2 electrons and 1 proton)	$NAD^+ + 2e^- + 2H^+ \rightleftharpoons$	$NADH + H^+$	Used to generate a proton motive force that can drive ATP synthesis
Flavin adenine dinucleotide (carries 2 electrons and 2 protons; i.e., 2 hydrogen atoms)	$FAD^+ + 2e^- + 2H^+ \rightleftharpoons$	$FADH_2$	Used to generate a proton motive force that can drive ATP synthesis
Nicotinamide adenine dinucleotide phosphate (carries 2 electrons and 1 proton)	$NADP^+ + 2e^- + 2H^+ \rightleftharpoons$	$NADPH + H^+$	Biosynthesis

TABLE 6.2 Precursor Metabolites

Precursor Metabolite	Pathway Generated	Biosynthetic Role
Glucose 6-phosphate	Glycolysis	Lipopolysaccharide
Fructose 6-phosphate	Glycolysis	Peptidoglycan
Dihydroxyacetone phosphate	Glycolysis	Lipids (glycerol component)
3-phosphoglycerate	Glycolysis	Protein (the amino acids cysteine, glycine, and serine)
Phosphoenolpyruvate	Glycolysis	Protein (the amino acids phenylalanine, tryptophan, and tyrosine)
Pyruvate	Glycolysis	Proteins (the amino acids alanine, leucine, and valine)
Ribose 5-phosphate	Pentose phosphate cycle	Nucleic acids and proteins (the amino acid histidine)
Erythrose 4-phosphate	Pentose phosphate cycle	Protein (the amino acids phenylalanine, tryptophan, and tyrosine)
Acetyl-CoA	Transition step	Lipids (fatty acids)
α-ketoglutarate	TCA cycle	Protein (the amino acids arginine, glutamate, glutamine, and proline)
Oxaloacetate	TCA cycle	Protein (the amino acids aspartate, asparagine, isoleucine, lysine, methionine, and threonine)

Some organisms use succinyl-coA as a precursor in heme biosynthesis; *E. coli* uses glutamate.

but also the source of precursor metabolites from which all other cell components are made **(table 6.2).** Some organisms, however, are not as versatile as *E. coli* with respect to their biosynthetic capabilities. Any essential compounds that a cell cannot synthesize from the appropriate precursor metabolite must be provided from an external source. ■ glucose-salts medium, p. 96

Overview of Metabolism

Three key metabolic pathways, called the **central metabolic pathways,** are used to gradually oxidize glucose, the preferred energy source of many cells, completely to carbon dioxide **(figure 6.8).** The central metabolic pathways include:

- Glycolysis
- Pentose phosphate pathway
- Tricarboxylic acid cycle (TCA cycle)

In a step-wise process, the central metabolic pathways provide cells with energy in the form of ATP, reducing power, and the precursor metabolites needed to synthesize the cells' building blocks. The pathways are catabolic, but the precursor metabolites and reducing power they generate can also be diverted for use in biosynthesis. To reflect the dual role of these pathways, they are sometimes called **amphibolic pathways** (*amphi* meaning "both kinds").

The most common pathway that initiates the breakdown of sugars is **glycolysis** (*glycos* means "sugar" and *lysis* means "dissolution") (figure 6.8a). This pathway is also called the **Embden-Meyerhof-Parnas** pathway (to honor the scientists who described it). This multistep pathway gradually oxidizes the 6-carbon sugar glucose to form two molecules of pyruvate, a 3-carbon compound. At about midpoint in the pathway, a 6-carbon derivative of glucose is split into two 3-carbon molecules. Both of these latter molecules then undergo the same series of transformations to produce pyruvate molecules. Glycolysis provides the cell with a small amount of energy in the form of ATP, some reducing power in the form of NADH, and a number of different precursor metabolites. Some bacteria have a different pathway called the **Entner-Doudoroff pathway** (named after the scientists who described it) instead of or in addition to the glycolytic pathway; some archaea have a slightly modified version of the Entner-Doudoroff pathway. Like glycolysis, the Entner-Doudoroff pathway generates pyruvate, but it uses different enzymes, generates reducing power in the form of NADPH, and yields less ATP.

The **pentose phosphate pathway** also breaks down glucose, but its primary role in metabolism is the production of compounds used in biosynthesis, including reducing power in the form of NADPH and precursor metabolites. It operates in conjunction with other glucose-degrading pathways (glycolysis and the Entner-Doudoroff pathway) (figure 6.8b). Most intermediates it generates are drawn off for use in biosynthesis, but one compound is directed to a mid-point step of glycolysis for further breakdown.

Pyruvate generated in any of the preceding pathways must then be converted into a specific 2-carbon fragment. This is accomplished in a complex reaction called the **transition step,** which removes CO_2, generates reducing power, and joins the resulting acetyl group to a compound called coenzyme A, forming acetyl-CoA (figure 6.8c). Note that the transition step is repeated twice for each molecule of glucose broken down.

The 2-carbon acetyl group of acetyl-CoA enters the **tricarboxylic acid cycle (TCA cycle),** also called the **Krebs cycle** (in honor of the scientist who first described it), or the **citric acid cycle** (figure 6.8d). This initiates a series of oxidations that result in the release of two molecules of CO_2. For every acetyl-CoA that enters the TCA cycle, the cyclic pathway "turns" once. Therefore, it must "turn" twice to complete the oxidation of one molecule of glucose. The TCA cycle generates precursor metabolites, a great deal of reducing power, and ATP.

Respiration uses the reducing power accumulated in glycolysis, the transition step, and the TCA cycle to generate ATP by oxidative phosphorylation (figure 6.8e). The electron carriers NADH and $FADH_2$ transfer their electrons to the electron transport chain, which ejects protons from the cell (or the matrix of a

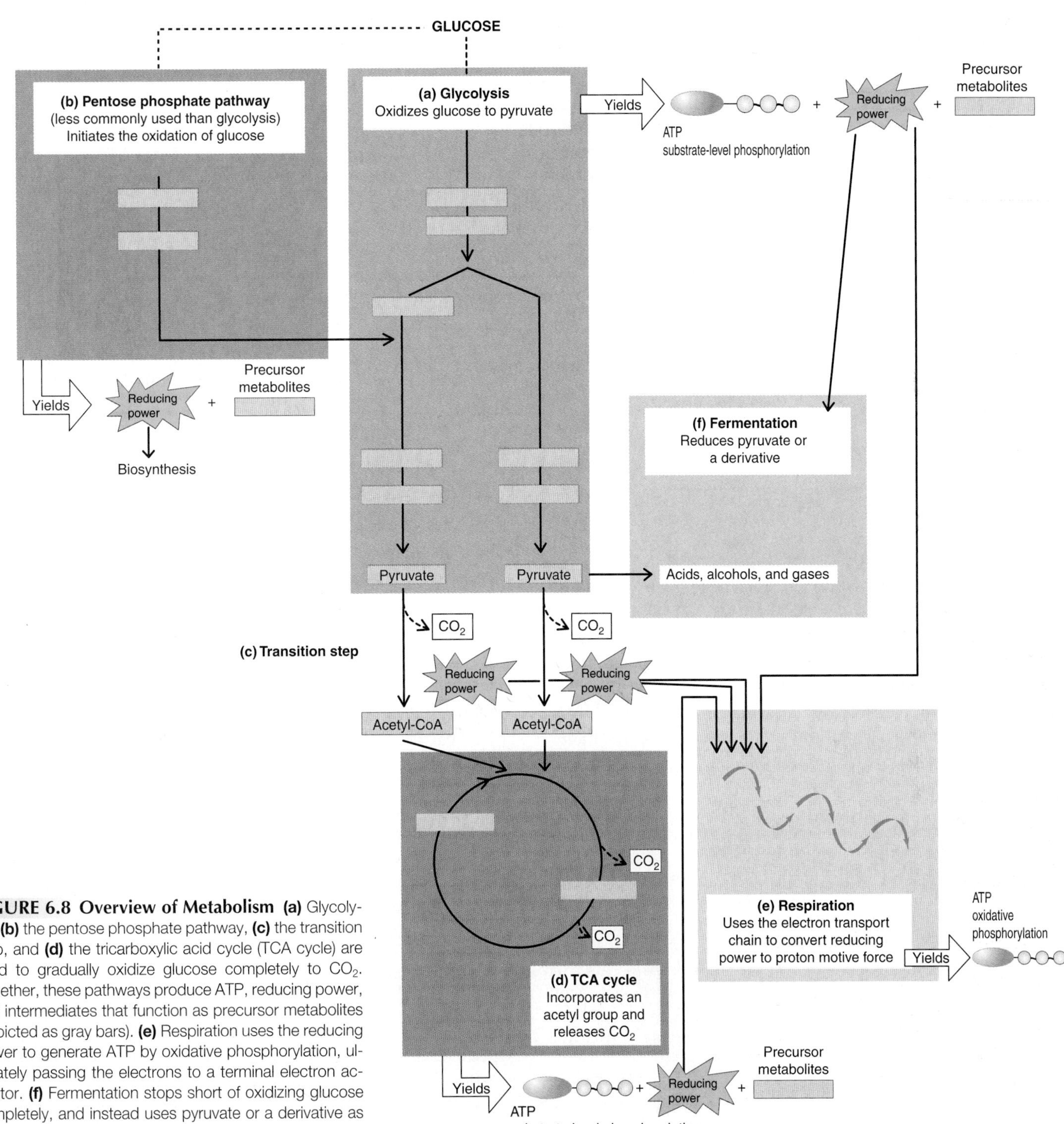

FIGURE 6.8 Overview of Metabolism **(a)** Glycolysis, **(b)** the pentose phosphate pathway, **(c)** the transition step, and **(d)** the tricarboxylic acid cycle (TCA cycle) are used to gradually oxidize glucose completely to CO_2. Together, these pathways produce ATP, reducing power, and intermediates that function as precursor metabolites (depicted as gray bars). **(e)** Respiration uses the reducing power to generate ATP by oxidative phosphorylation, ultimately passing the electrons to a terminal electron acceptor. **(f)** Fermentation stops short of oxidizing glucose completely, and instead uses pyruvate or a derivative as an electron acceptor.

mitochondrion) to generate a proton motive force. This transfer of electrons also serves to recycle the carriers so they can once again accept electrons during catabolic reactions. In **aerobic respiration,** electrons are ultimately passed to molecular oxygen (O_2), the terminal electron acceptor, producing water. **Anaerobic respiration** is similar to aerobic respiration, but uses a molecule other than O_2 as a terminal electron acceptor. In addition, modified versions of the TCA cycle that generate less reducing power are used during anaerobic respiration. Organisms that use respiration, either aerobic or anaerobic, are said to **respire.**

Cells that cannot respire are limited by their relative inability to recycle reduced electron carriers. A cell only has a limited number of carrier molecules; if electrons are not removed from the reduced carriers, none will be available to accept electrons. As a consequence, subsequent catabolic processes cannot occur. **Fermentation** provides a solution to this problem, but it results in only the partial oxidation of glucose (figure 6.8f). Thus, compared with respiration, fermentation produces relatively little ATP. It is used by facultative anaerobes when a suitable inorganic terminal electron acceptor is not available and by organisms that

TABLE 6.3 ATP-Generating Processes of Prokaryotic Chemoorganoheterotrophs

Metabolic Process	Uses an Electron Transport Chain	Terminal Electron Acceptor	ATP Generated by Substrate-Level Phosphorylation (Theoretical Maximum)	ATP Generated by Oxidative Phosphorylation (Theoretical Maximum)	Total ATP Generated (Theoretical Maximum)
Aerobic respiration	Yes	O_2	2 in glycolysis (net) 2 in the TCA cycle 4 total	34	38
Anaerobic respiration	Yes	Molecule other than O_2 such as nitrate (NO_3^-), nitrite (NO_2^-), sulfate (SO_4^{2-})	Number varies; however, the ATP yield of anaerobic respiration is less than that of aerobic respiration but more than that of fermentation.		
Fermentation	No	Organic molecule (pyruvate or a derivative)	2 in glycolysis (net) 2 total	0	2

lack an electron transport chain. These cells stop short of oxidizing glucose completely, thereby limiting the amount of reducing power generated. Instead of oxidizing pyruvate in the TCA cycle, they use pyruvate or a derivative as a terminal electron acceptor. By transferring the electrons carried by NADH to pyruvate or a derivative, NAD^+ is regenerated so it can once again accept electrons in the steps of glycolysis. Note that fermentation always uses an organic molecule as the terminal electron acceptor. Although fermentation does not use the TCA cycle, organisms that ferment still employ certain key steps of it to generate the precursor molecules required for biosynthesis. ■ facultative anaerobes, p. 92

A comparison of aerobic respiration, anaerobic respiration, and fermentation is summarized in **table 6.3.**

MICROCHECK 6.1

Catabolic pathways gradually oxidize an energy source to harvest energy. A specific enzyme catalyzes each step. Substrate-level phosphorylation uses chemical energy to synthesize ATP; oxidative phosphorylation employs a proton motive force to do the same. Reducing power in the form of NADH and $FADH_2$ is used to generate the proton motive force; the reducing power of NADPH is utilized in biosynthesis. Precursor metabolites are metabolic intermediates that can be used in biosynthesis. The central metabolic pathways generate ATP, reducing power, and precursor metabolites.

✓ How does the fate of electrons carried by NADPH differ from those carried by NADH?

✓ Why are the central metabolic pathways called amphibolic pathways?

✓ Why does fermentation release less energy than respiration?

6.2 Enzymes

Focus Points

- Describe the active site of an enzyme and explain how it relates to the enzyme-substrate complex.
- Compare and contrast cofactors and coenzymes.
- List two environmental factors that influence enzyme activity.
- Describe allosteric regulation.
- Compare and contrast non-competitive enzyme inhibition and competitive enzyme inhibition.

Recall that enzymes are proteins that act as biological catalysts, facilitating the conversion of a substrate into a product (see figure 6.5). They do this with extraordinary specificity and speed, usually acting on only one, or a very limited number of, substrates. They are neither consumed nor permanently changed during a reaction, allowing a single enzyme molecule to be rapidly used over and over again. In only one second, the fastest enzymes can transform more than 10^4 substrate molecules to products. More than a thousand different enzymes exist in a cell; most are given a common name that reflects their function and ends with the suffix -ase. For example, those that degrade proteins are collectively called proteases. ■ enzymes, p. 129

Mechanisms and Consequences of Enzyme Action

An enzyme has on its surface an **active,** or **catalytic, site,** typically a relatively small crevice **(figure 6.9).** This is the critical site to which a substrate binds by weak forces. The binding of the substrate to the active site causes the shape of the flexible enzyme to change slightly. This mutual interaction, or **induced fit,** results in a temporary intermediate called an **enzyme-substrate complex.** The substrate is held within this complex in a specific orientation so that the activation energy for a given reaction is lowered, allowing the products to be formed. The products are then released, leaving the enzyme unchanged and free to combine with new substrate molecules. Note that enzymes may also catalyze reactions that join two substrates to create one product. Theoretically, all enzyme-catalyzed reactions are reversible. The free energy

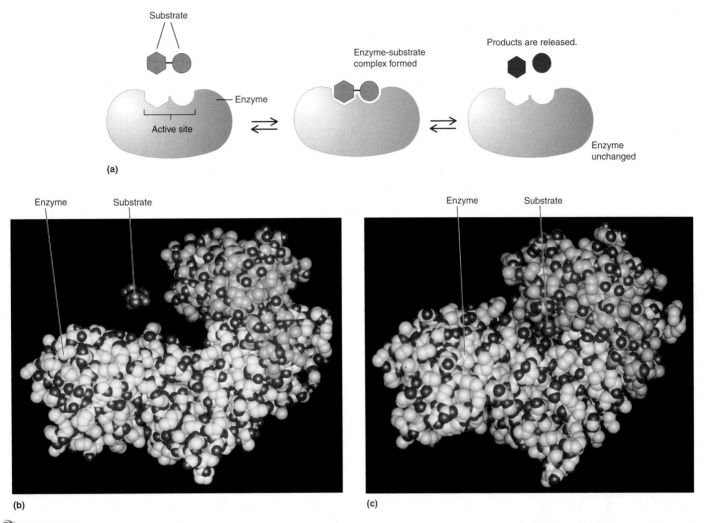

FIGURE 6.9 Mechanism of Enzyme Action **(a)** The substrate binds to the active site, forming an enzyme-substrate complex. The products are then released, leaving the enzyme unchanged and free to combine with new substrate molecules. **(b)** A model showing an enzyme and its substrate. **(c)** The binding of the substrate to the active site causes the shape of the flexible enzyme to change slightly.

change of certain reactions, however, makes them effectively non-reversible.

The interaction of an enzyme with its substrate is very specific. The substrate fits into the active site like a hand into a glove. Not only must it fit spatially, but appropriate chemical interactions such as hydrogen and ionic bonding need to occur to induce the fit. This requirement for a precise fit and interaction explains why, with minor exceptions, a different enzyme is required to catalyze every reaction in a cell. Very few molecules of any particular enzyme are needed, however, as each is swiftly reused again and again. ■ **hydrogen bonds, p. 22** ■ **ionic bonds, p. 20**

Cofactors and Coenzymes

Some enzymes act with the assistance of a non-protein component called a **cofactor (figure 6.10). Coenzymes** are organic cofactors that act as loosely bound carriers of molecules or

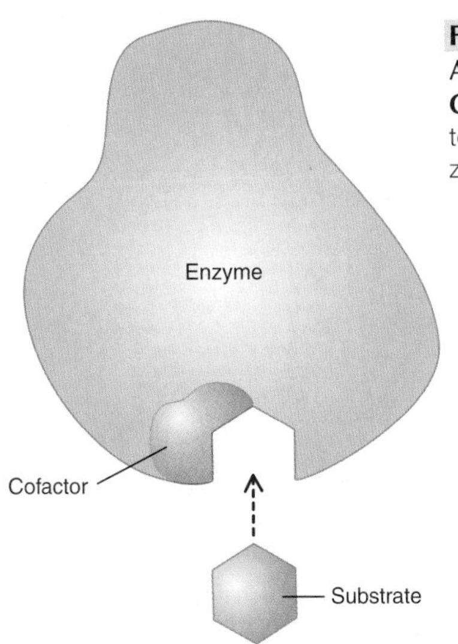

FIGURE 6.10 Some Enzymes Act in Conjunction with a Cofactor Cofactors are non-protein components, either coenzymes or trace elements.

TABLE 6.4 Some Coenzymes and Their Function

Coenzyme	Vitamin from Which It Is Derived	Substance Transferred	Example of Use
Nicotinamide adenine dinucleotide (NAD^+)	Niacin	Hydride ions (2 electrons and 1 proton)	Carrier of reducing power
Flavin adenine dinucleotide (FAD)	Riboflavin	Hydrogen atoms (2 electrons and 2 protons)	Carrier of reducing power
Coenzyme A	Pantothenic acid	Acyl groups	Carries the acetyl group that enters the TCA cycle
Thiamin pyrophosphate	Thiamine	Aldehydes	Facilitates the removal of CO_2 from pyruvate in the transition step
Pyridoxal phosphate	Pyridoxine	Amino groups	Transfers amino groups in amino acid synthesis
Tetrahydrofolate	Folic acid	1-carbon molecules	Used in nucleotide synthesis

electrons **(table 6.4).** They include the electron carriers FAD, NAD^+, and $NADP^+$. Other cofactors attach tightly to enzymes. For example, magnesium, zinc, copper, and other trace elements required for growth often function as cofactors. ■ **trace elements, p. 94**

All coenzymes transfer substances from one compound to another, but they function in different ways. Some remain bound to the enzyme during the transfer process, whereas others separate from the enzyme, carrying the substance being transferred along with them. The same coenzyme can assist different enzymes. Because of this, far fewer different coenzymes are required than enzymes. Like enzymes, coenzymes are recycled as they function and, consequently, are needed only in minute quantities.

Most coenzymes are derived from vitamins (see table 6.4). Some bacteria, such as *E. coli,* can synthesize vitamins and convert them to the necessary coenzymes. In contrast, humans and other animals must be provided with vitamins from external sources. Most often they must be supplied in the diet, but in some cases vitamins synthesized by bacteria residing in the intestine can be absorbed. If an animal lacks a vitamin, the functions of all the different enzymes whose activity requires the corresponding coenzyme are impaired. Thus, a single vitamin deficiency has serious consequences.

Environmental Factors That Influence Enzyme Activity

Several environmental factors influence how well enzymes function and in this way determine how rapidly microorganisms multiply **(figure 6.11).** Each enzyme has a narrow range of factors—including temperature, pH, and salt concentration—at which it operates optimally. A 10°C rise in temperature approximately doubles the speed of enzymatic reactions, until optimal activity is reached; this explains why bacteria tend to grow more rapidly at higher temperatures. If the temperature gets too high, however, proteins will denature and no longer function. Most enzymes operate best at low salt concentrations and at pH values slightly above 7. Not surprisingly then, most microbes grow fastest under these same conditions. Some prokaryotes, however, particularly certain members of the *Archaea,* are found in environ-

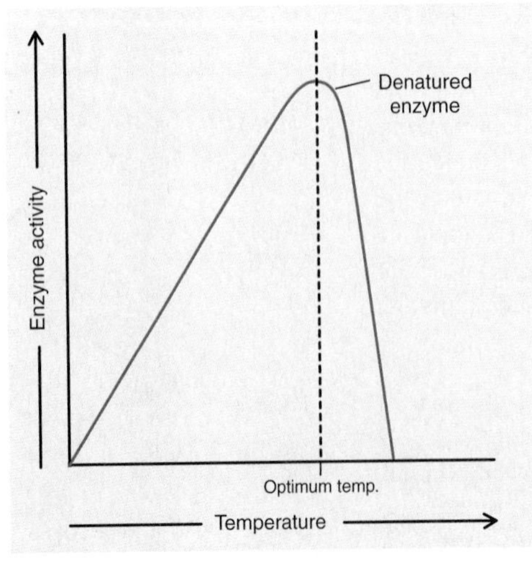

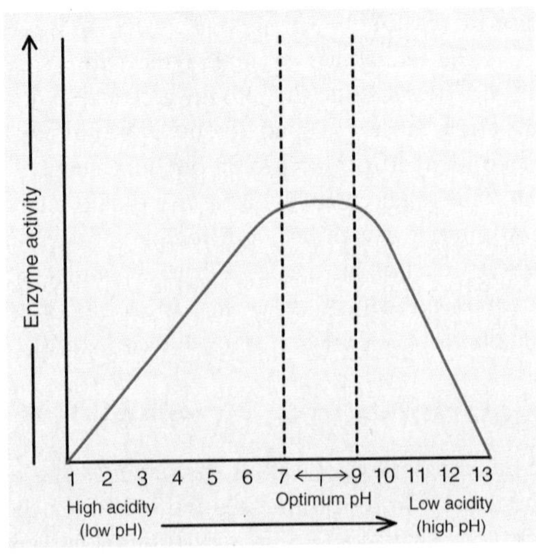

FIGURE 6.11 Environmental Factors That Influence Enzyme Activity **(a)** A rise in temperature increases the speed of enzymatic activity until the optimum temperature is reached. If the temperature gets too high, the enzyme denatures and no longer functions. **(b)** Most enzymes function best at pH values slightly above 7.

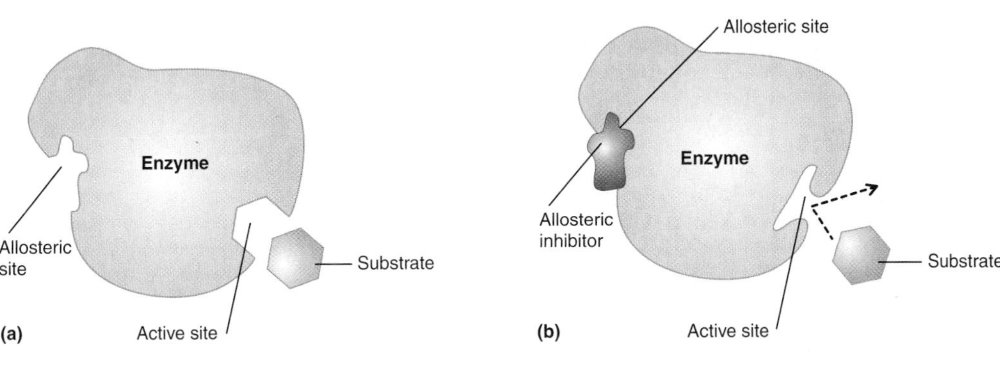

FIGURE 6.12 Regulation of Allosteric Enzymes **(a)** Allosteric enzymes have, in addition to the active site, an allosteric site. **(b)** The binding of regulatory molecule to the allosteric site causes the shape of the enzyme to change, altering the relative affinity of the enzyme for its substrate. **(c)** The end product of a given biosynthetic pathway generally acts as an allosteric inhibitor of the first enzyme of that pathway.

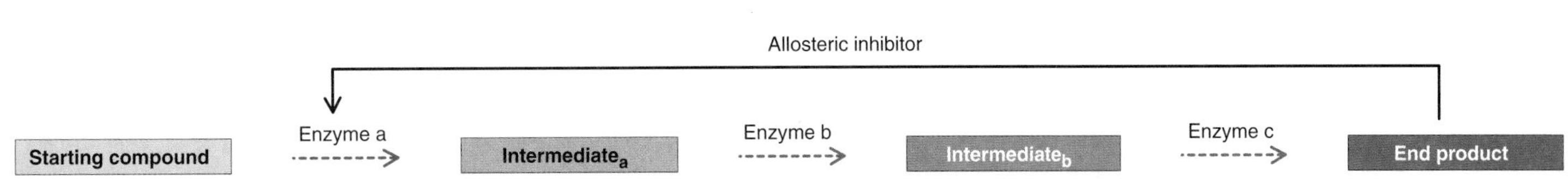

ments where conditions are extreme. They may require high salt concentrations, grow under very acidic conditions, or be found where temperatures are near boiling. ■ pH, p. 24 ■ temperature and growth requirements, p. 90

Allosteric Regulation

Cells can rapidly fine-tune or regulate the activity of certain key enzymes using other molecules that reversibly bind to and distort them **(figure 6.12).** This has the effect of regulating the activity of metabolic pathways. These enzymes can be controlled because they are **allosteric enzymes** (*allo* means "other"), which have a binding site called an **allosteric site** that is separate from their active site. When a regulatory molecule binds to the allosteric site, the shape of the enzyme changes. This distortion alters the relative **affinity,** or chemical attraction, of the enzyme for its substrate. In some cases the binding of the regulatory molecule enhances the affinity for the substrate, but in other cases it decreases it.

Allosteric enzymes generally catalyze the step that either initiates or commits to a given pathway. Because their activity can be controlled, they provide the cell with a means to modulate the pace of metabolic processes, turning off some pathways and activating others. Cells can also control the amount of enzyme they synthesize; this control mechanism, which will be discussed in chapter 7, also involves allosteric proteins. ■ regulation, p. 176

The end product of a given biosynthetic pathway generally acts as an allosteric inhibitor of the first enzyme of that pathway—a mechanism called **feedback inhibition** (figure 6.12c). This mechanism allows the product of the pathway to modulate its own synthesis. For example, the first enzyme of the multistep pathway used to convert the amino acid threonine to isoleucine is an allosteric enzyme that is inhibited by the binding of isoleucine. This amino acid must be present at a relatively high concentration, however, to bind and inhibit the enzyme. Thus, the pathway will only be shut down when a cell accumulates sufficient isoleucine to fill its immediate needs. Because the binding of the inhibitor is reversible, the enzyme can again become active when isoleucine levels decrease.

Compounds that reflect a cell's relative energy stores often regulate allosteric enzymes of catabolic pathways, enabling cells to modulate the flow of these pathways in response to changing energy needs. High levels of ATP inhibit certain enzymes and, as a consequence, slow down catabolic processes. In contrast, high levels of ADP warn that a cell's energy stores are low, and they function to stimulate the activity of some enzymes.

Enzyme Inhibition

Enzymes can be inhibited by a variety of compounds other than the regulatory molecules normally used by the cell **(table 6.5).**

TABLE 6.5 Characteristics of Enzyme Inhibitors

Type	Characteristics
Non-competitive inhibition (by regulatory molecules)	Inhibitor temporarily changes the enzyme, altering the enzyme's relative affinity for the substrate. This mechanism provides cells with a means to control the activity of allosteric enzymes.
Non-competitive inhibition (by enzyme poisons)	Inhibitor permanently changes the enzyme, rendering the enzyme non-functional. Enzyme poisons such as mercury are used in certain antimicrobial compounds.
Competitive inhibition	Inhibitor binds to the active site of the enzyme, obstructing the access of the substrate. Competitive inhibitors such as sulfa drugs are used as antibacterial medications.

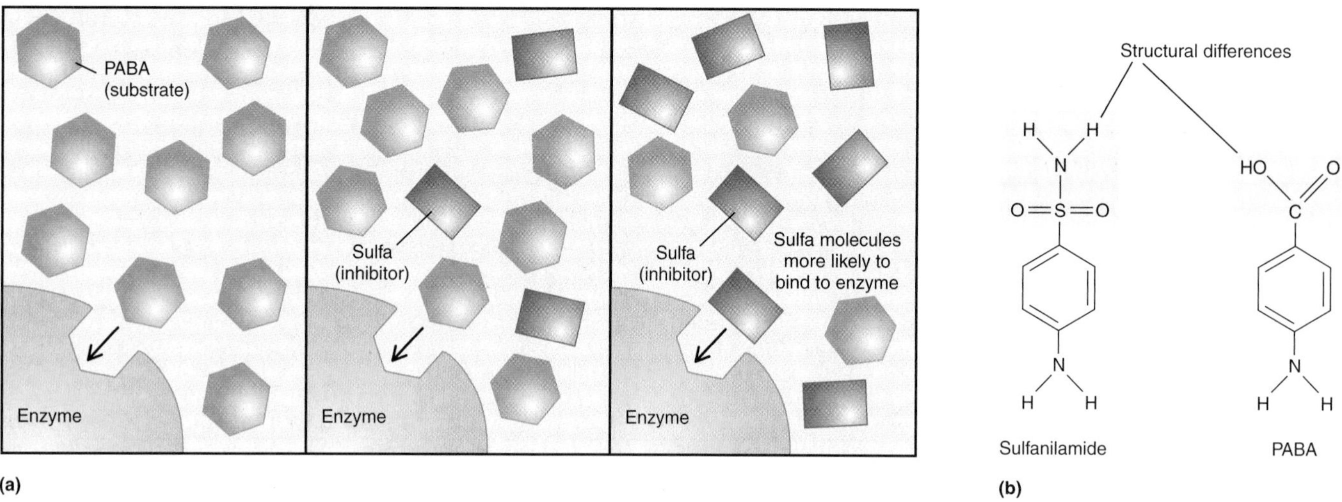

FIGURE 6.13 Competitive Inhibition of Enzymes (a) The inhibitor competes with the normal substrate for binding to the active site. The greater the proportion of inhibitor relative to substrate, the more likely the active site of the enzyme will be occupied by an inhibitor. **(b)** A competitive inhibitor generally has a chemical structure similar to the normal substrate.

These inhibitory compounds can be exploited to prevent microbial growth. The site on the enzyme to which the molecules bind determines whether they function as competitive or non-competitive inhibitors.

Non-Competitive Inhibition

Non-competitive inhibition occurs when the inhibitor and the substrate act at different sites on the enzyme. Allosteric inhibition, discussed previously, is an example of non-competitive reversible inhibition and is used by the cell to modulate its processes (see figure 6.12). Non-competitive, non-reversible inhibitors damage the enzyme permanently so that it can no longer function; the inhibitor acts as an enzyme poison. For example, mercury in the antibacterial compound mercurochrome inhibits growth because it oxidizes the S—H groups of the amino acid cysteine in proteins. This converts cysteine to cystine, which cannot form the important covalent disulfide bond (S—S). As a result, the protein cannot achieve its proper shape.

Competitive Inhibition

In **competitive inhibition,** the inhibitor binds to the active site of the enzyme, obstructing access of the substrate to that site **(figure 6.13).** Generally this occurs because the inhibitor has a chemical structure similar to the normal substrate.

A good example of competitive inhibition is the action of sulfanilamide, one of the sulfa drugs used as an antimicrobial medication. Sulfa drugs inhibit an enzyme in the pathway that bacteria use to synthesize the vitamin folic acid by binding to the active site of the enzyme. The drug does not affect human metabolism because humans cannot synthesize folic acid; it must be provided in the diet. Sulfa drugs have a structure similar to para-aminobenzoic acid (PABA), an intermediate in the bacterial pathway for folic acid synthesis. Because of this, they fit into the active site of the enzyme that normally uses PABA as a substrate, preventing the attachment of PABA. The greater the proportion of sulfa molecules relative to PABA molecules, the more likely the active site of the enzyme will be occupied by a sulfa molecule. Once the sulfa is removed, the enzyme functions normally with PABA as the substrate. ■

sulfa drugs, p. 479

MICROCHECK 6.2

Enzymes facilitate the conversion of a substrate into a product with extraordinary speed and specificity. They are neither consumed nor permanently changed in the reaction. Some enzymes act with the assistance of a cofactor. Environmental factors influence enzyme activity and, by doing so, determine how rapidly microorganisms multiply. The activity of allosteric enzymes can be regulated. A variety of different compounds adversely affect enzyme activity.

✓ Explain why sulfa drugs inhibit the growth of bacteria without harming the human host.

✓ Explain the function of a coenzyme.

✓ Why is it important for a cell that allosteric inhibition be reversible?

6.3 The Central Metabolic Pathways

Focus Point

- List the amount of ATP and reducing power and the number of different precursor molecules generated by each of the central metabolic pathways.

The three central metabolic pathways—glycolysis, the pentose phosphate pathway, and the tricarboxylic acid cycle—modify organic molecules in a step-wise fashion to form:

- Intermediates with high-energy bonds that can be used to synthesize ATP by substrate-level phosphorylation
- Intermediates that can be oxidized to generate reducing power
- Intermediates and end products that function as precursor metabolites

This section describes how a molecule of glucose is broken down in the central metabolic pathways, but bear in mind that a cell has many millions of molecules of glucose, and different molecules can have different fates. For example, a cell might oxidize one glucose molecule completely to CO_2, thereby producing the maximum amount of ATP. Another glucose molecule might enter glycolysis, or perhaps the pentose phosphate pathway, only to be siphoned off as a precursor metabolite for use in biosynthesis. The step and rate at which the various intermediates are removed for biosynthesis will dramatically affect the overall energy gain of catabolism. This is generally overlooked in descriptions of the ATP-generating functions of these pathways for the sake of simplicity. However, because these pathways serve more than one function, the energy yields are only theoretical.

The pathways of central metabolism are compared in **table 6.6.** The entire pathways with chemical formulas and enzyme names are illustrated in Appendix IV.

Glycolysis

Glycolysis is the primary pathway used by many organisms to convert glucose to pyruvate **(figure 6.14).** In the 10-step pathway, one molecule of glucose is converted into two molecules of pyruvate. This generates a net gain of two molecules of ATP and two molecules of NADH. The overall process can be summarized as:

$$\text{glucose (6 C)} + 2\ NAD^+ + 2\ ADP + 2\ P_i \longrightarrow 2\ \text{pyruvate (3 C)} + 2\ NADH + 2\ H^+ + 2\ ATP$$

In addition to generating ATP and reducing power (NADH), the pathway produces six different precursor molecules needed by *E. coli* (see table 6.2).

The 10-step pathway can be viewed as having two phases:

- **Investment** or **preparatory phase** (figure 6.14, steps 1 through 5)—This consumes energy because two different steps transfer a high-energy phosphate group to the 6-carbon sugar. In eukaryotic cells, both of the high-energy phosphates come from ATP, as shown in figure 6.14. In bacteria, the first high-energy phosphate is added as glucose is transported into the cell via group translocation and the other comes from ATP. The 6-carbon sugar is then split to yield two 3-carbon molecules.
- **Pay-off phase** (figure 6.14, steps 6 through 10)—This oxidizes and rearranges the 3-carbon molecules to form pyruvate, generating 1 NADH and 2 ATP. Note that the steps of this phase occur twice for each molecule of glucose that entered glycolysis because the 6-carbon sugar was split into two 3-carbon molecules in the previous phase.

Yield of Glycolysis

For every glucose molecule degraded, the steps of glycolysis produce:

- **ATP**—The maximum possible energy gain as ATP in glycolysis is:

Energy expended	2 ATP molecules (investment phase)
Energy harvested	4 ATP molecules (pay-off phase)
Net gain	2 ATP molecules

- **Reducing power**—The payoff phase converts 2 NAD^+ to 2 NADH + 2 H^+.
- **Precursor metabolites**—Five intermediates of glycolysis as well as the end product, pyruvate, are precursor metabolites used by *E. coli.*

Pentose Phosphate Pathway

The other central metabolic pathway used by cells to break down glucose is the pentose phosphate pathway. This complex pathway generates 5- and 7-carbon sugars. In addition, glyceraldehyde 3-phosphate (G3P) is produced, and can be directed to a step in glycolysis for further breakdown. The greatest importance of the pentose phosphate pathway is its contribution to biosynthesis. The reducing power it generates is in the form of NADPH, which is used in biosynthetic reactions when a reduction is required. In addition, two of its intermediates, ribose 5-phosphate and erythrose 4-phosphate, are important precursor metabolites.

TABLE 6.6 Comparison of the Central Metabolic Pathways

Pathway	Characteristics
Glycolysis	Glycolysis generates: • 2 ATP (net) by substrate-level phosphorylation • 2 NADH + 2 H^+ • six different precursor metabolites
Pentose phosphate cycle	The pentose phosphate cycle generates: • NADPH + H^+ (amount varies) • two different precursor metabolites
Transition step	The transition step, repeated twice to oxidize two molecules of pyruvate to acetyl-CoA, generates: • 2 NADH + 2 H^+ • one precursor metabolite
TCA cycle	The TCA cycle, repeated twice to incorporate two acetyl groups, generates: • 2 ATP by substrate-level phosphorylation (may involve conversion of GTP) • 6 NADH + 6 H^+ • 2 $FADH_2$ • two different precursor metabolites

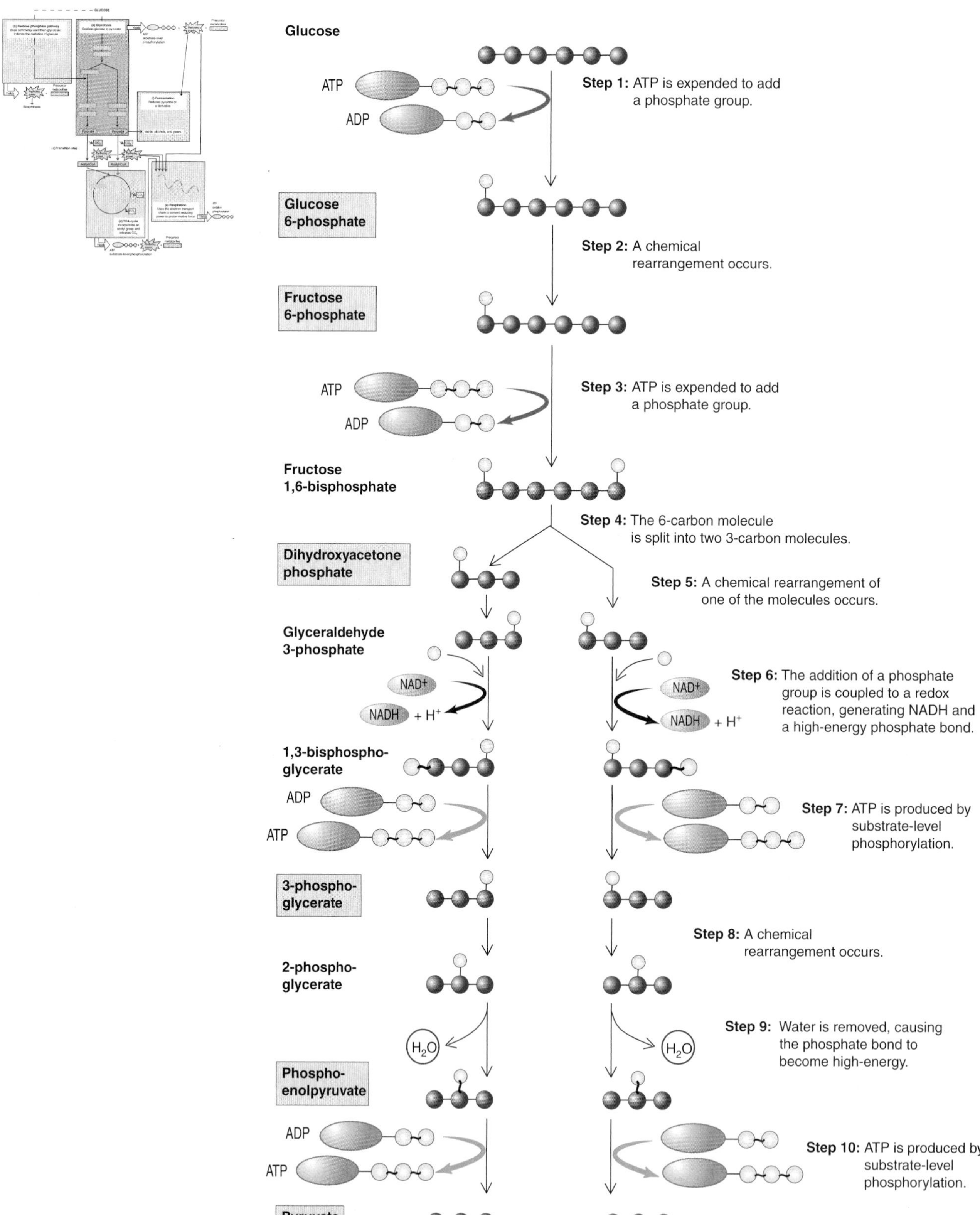

Glucose
ATP
ADP
Step 1: ATP is expended to add a phosphate group.
Glucose 6-phosphate
Step 2: A chemical rearrangement occurs.
Fructose 6-phosphate
ATP
ADP
Step 3: ATP is expended to add a phosphate group.
Fructose 1,6-bisphosphate
Step 4: The 6-carbon molecule is split into two 3-carbon molecules.
Dihydroxyacetone phosphate
Step 5: A chemical rearrangement of one of the molecules occurs.
Glyceraldehyde 3-phosphate
NAD+
NADH + H+
Step 6: The addition of a phosphate group is coupled to a redox reaction, generating NADH and a high-energy phosphate bond.
1,3-bisphospho-glycerate
ADP
ATP
Step 7: ATP is produced by substrate-level phosphorylation.
3-phospho-glycerate
Step 8: A chemical rearrangement occurs.
2-phospho-glycerate
H2O
Step 9: Water is removed, causing the phosphate bond to become high-energy.
Phospho-enolpyruvate
ADP
ATP
Step 10: ATP is produced by substrate-level phosphorylation.
Pyruvate

FIGURE 6.14 Glycolysis The glycolytic pathway oxidizes glucose to pyruvate, generating ATP by substrate-level phosphorylation, reducing power in the form of NADH, and six different precursor metabolites.

Yield of the Pentose Phosphate Pathway

The yield of the pentose phosphate pathway varies, depending on which of several possible alternatives are taken. It can produce:

- **Reducing power**—A variable amount of reducing power in the form of NADPH is produced.
- **Precursor metabolites**—Two intermediates of the pentose phosphate pathway are precursor metabolites.

Transition Step

The transition step links glycolysis to the TCA cycle **(figure 6.15).** In prokaryotic cells, the entire oxidation process takes place in the cytoplasm. In eukaryotic cells, however, pyruvate must first enter the mitochondria since the enzymes of the glycolytic pathway are located in the cytoplasm of the cell, whereas those of the TCA cycle are found only within the matrix of the mitochondria. ■ **mitochondria, p. 76**

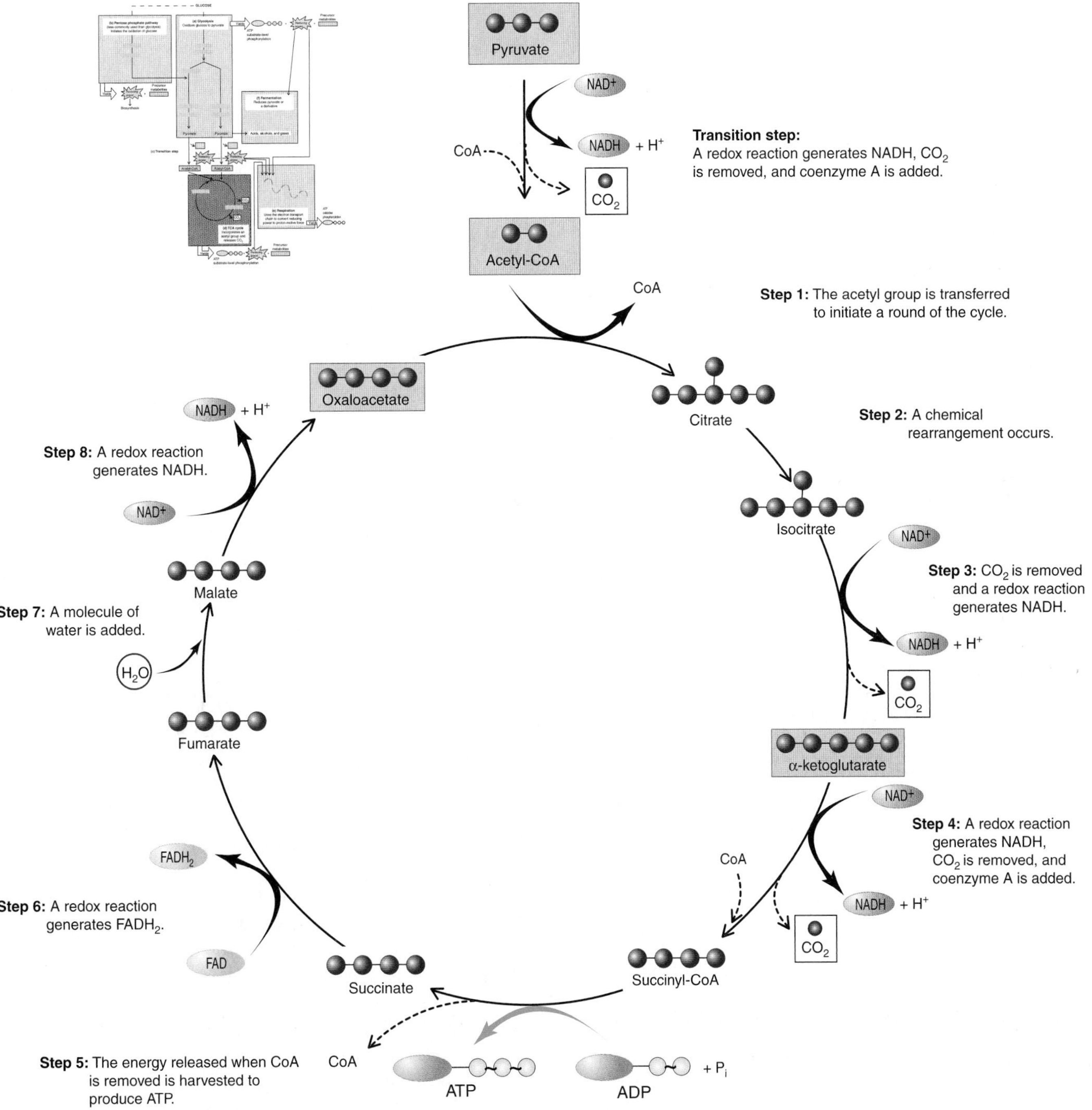

FIGURE 6.15 The Transition Step and the Tricarboxylic Acid Cycle The transition step links glycolysis and the TCA cycle, converting pyruvate to acetyl-CoA; it generates reducing power and one precursor metabolite. The TCA cycle incorporates the acetyl group of acetyl-CoA and, using a series of steps, releases CO_2; it generates ATP, reducing power in the form of both NADH and $FADH_2$, and two different precursor metabolites.

The transition step involves several integrated reactions catalyzed by a large multi-enzyme complex. In the concerted series of reactions, carbon dioxide is first removed from the pyruvate, a process called decarboxylation. Then, an oxidation occurs, reducing NAD^+ to form NADH + H^+. Finally, the remaining 2-carbon acetyl group is joined to the coenzyme A to form acetyl-CoA.

Yield of the Transition Step

- **Reducing power**—The transition step, which occurs twice for every molecule of glucose that enters glycolysis, oxidizes pyruvate. This reduces 2 NAD^+ to form 2 NADH + 2 H^+.
- **Precursor metabolites**—The end product of the transition step, acetyl-CoA, is a precursor metabolite.

Tricarboxylic Acid (TCA) Cycle

The eight steps of the tricarboxylic acid (TCA) cycle complete the oxidation of glucose (see figure 6.15). The cycle incorporates the acetyl groups from the transition step, releasing CO_2 in this net reaction:

$$2 \text{ acetyl groups (2 C)} + 6\,NAD^+ + 2\,FAD + 2\,ADP + 2\,P_i \longrightarrow 4\,CO_2 + 6\,NADH + 6\,H^+ + 2\,FADH_2 + 2\,ATP$$

In addition to generating ATP and reducing power, the steps of the TCA cycle form two more precursor metabolites used by *E. coli* (see table 6.2).

Step 1 The cycle begins when CoA transfers its acetyl group to the 4-carbon compound oxaloacetate, thereby forming the 6-carbon compound citrate.

Step 2 Citrate is chemically rearranged to form a structural isomer, isocitrate. ■ **structural isomer, p. 31**

Step 3 Isocitrate is oxidized and a molecule of CO_2 is removed, forming the 5-carbon compound α-ketoglutarate. During the oxidation, NAD^+ is reduced to form NADH + H^+.

Step 4 Like the transition step that converts pyruvate to acetyl-CoA, this involves a group of reactions catalyzed by a complex of enzymes. In this step, α-ketoglutarate is oxidized, CO_2 is removed, and CoA is added, producing the 4-carbon compound succinyl-CoA. During the oxidation, NAD^+ is reduced to form NADH + H^+.

Step 5 This removes CoA from succinyl-CoA, harvesting the energy to make ATP. The reaction forms succinate. Note that some types of cells make guanosine triphosphate (GTP) rather than ATP at this step. This compound, however, can be converted to ATP.

Step 6 Succinate is oxidized to form fumarate. During the oxidation, FAD is reduced to form $FADH_2$.

Step 7 A molecule of water is added to fumarate, forming malate.

Step 8 Malate is oxidized to form oxaloacetate; note that oxaloacetate is the starting compound to which acetyl-CoA is added to initiate the cycle. During the oxidation, NAD^+ is reduced to form NADH + H^+.

Yield of the TCA Cycle

The tricarboxylic acid cycle "turns" once for each acetyl-CoA that enters. Because two molecules of acetyl-CoA are generated for each glucose molecule that enters glycolysis, the breakdown of one molecule of glucose causes the cycle to "turn" twice. These two "turns" generate:

- **ATP**—2 ATP produced in step 5.
- **Reducing power**—Redox reactions at steps 3, 4, 6 and 8 produce a total of 6 NADH + 6 H^+ and 2 $FADH_2$.
- **Precursor metabolites**—Two precursor metabolites used by *E. coli* are formed as a result of steps 3 and 8.

MICROCHECK 6.3

Glycolysis oxidizes glucose to pyruvate, yielding some ATP and NADH and six different precursor metabolites. The pentose phosphate pathway initiates the breakdown of glucose; its greatest significance is its contribution of two different precursor metabolites and NADPH for biosynthesis. The transition step and the TCA cycle, repeated twice, complete the oxidation of glucose, yielding some ATP, a great deal of reducing power, and three different precursor metabolites.

✓ What is the product of the transition step?
✓ Explain why the TCA cycle ultimately results in a greater ATP gain than glycolysis.
✓ Which compound contains more free energy—pyruvate or oxaloacetate? Why?

6.4 Respiration

Focus Points

- Describe how the electron transport chain generates a proton motive force.
- Compare and contrast the electron transport chains of eukaryotes and prokaryotes.
- Describe how proton motive force is used to synthesize ATP.

As mentioned earlier, respiration uses the NADH and $FADH_2$ generated in gycolysis, the transition step, and the TCA cycle to synthesize ATP. The process, called **oxidative phosphorylation,** occurs through the combined action of two mechanisms—the **electron transport chain,** which generates proton motive force, and an enzyme called **ATP synthase,** which harvests the energy of the proton motive force to drive the synthesis of ATP. In 1961, the British scientist Peter Mitchell originally proposed the **chemiosmotic theory,** which describes the remarkable mechanism by which ATP synthesis is linked to electron transport, but his hypothesis was widely dismissed. Only through years of self-funded research was he finally able to convince others of its validity, and he was awarded the Nobel Prize in 1978. ■ **proton motive force, p. 57**

The Electron Transport Chain—Generating Proton Motive Force

The electron transport chain is a group of membrane-embedded electron carriers that pass electrons sequentially from one to another. In prokaryotes, it is located in the cytoplasmic membrane, whereas in eukaryotic cells it is in the inner membrane of mitochondria (see figure 3.53). Because of the asymmetrical arrangement of the electron carriers, the sequential oxidation/reduction reactions result in the ejection of protons to the outside of the cell or, in the case of mitochondria, to the space between the inner and outer membranes. This expulsion of protons creates a proton gradient, or electrochemical gradient, across the membrane. Energy of this gradient, proton motive force, can be harvested by cells and used to fuel the synthesis of ATP. Recall from chapter 3 that prokaryotes can also use proton motive force as a source of energy to transport substances into or out of the cell, and to power the rotation of flagella.

Four types of electron carriers participate in the electron transport chain: flavoproteins, iron-sulfur proteins, quinones, and cytochromes. **Flavoproteins** are proteins to which an organic molecule called a flavin is attached. FAD is an example of a flavin. **Iron-sulfur proteins** are proteins that contain iron and sulfur molecules arranged in a cluster. **Quinones** are lipid-soluble molecules that move freely in the membrane and can therefore transfer electrons between different enzyme structures in the membrane. Several types of quinones exist, one of the most common being ubiquinone (meaning ubiquitous quinone). **Cytochromes** are proteins that contain heme, a chemical structure that holds an iron atom in the center. Several different cytochromes exist, each distinguished with a letter after the term, for example, cytochrome *c*.

Because of the order of the carriers in the electron transport chain, energy is gradually released as the electrons are passed from one carrier to another, much like a ball falling down a flight of stairs **(figure 6.16).** Energy release is coupled to the ejection of protons to establish a proton gradient.

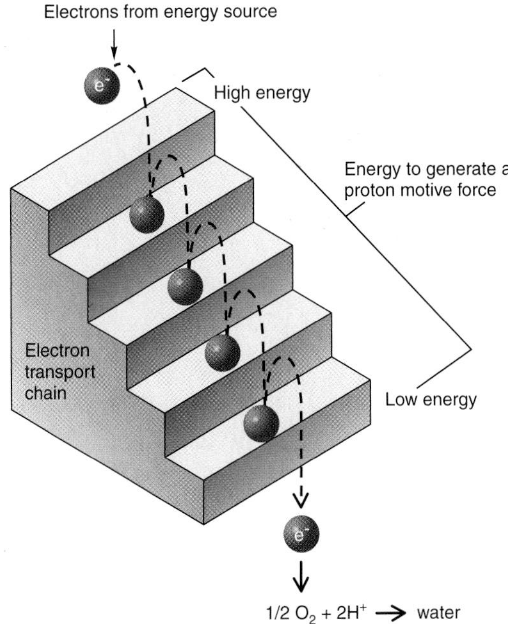

FIGURE 6.16 Electron Transport Energy released as electrons are passed along carriers of the electron transport chain is used to establish a proton gradient.

General Mechanisms of Proton Ejection

An important characteristic of the electron carriers is that some accept only hydrogen atoms (proton-electron pairs), whereas others accept only electrons. The spatial arrangement of these two types of carriers in the membrane causes protons to be shuttled from the inside of the membrane to the outside. This occurs because a hydrogen carrier that receives electrons from an electron carrier must pick up protons; because of the hydrogen carrier's relative location in the membrane, those protons come from inside the cell (or matrix of the mitochondrion). Conversely, when a hydrogen carrier passes electrons to a carrier that accepts electrons but not protons, the protons are released to the outside of the cell (or intermembrane space of the mitochondrion). The net effect of these processes is that protons are pumped from one side of the membrane to the other, establishing the concentration gradient across the membrane.

Most carriers of the electron transport chain are grouped into several large protein complexes that function as **proton pumps;** other carriers shuttle electrons from one complex to the next.

The Electron Transport Chain of Mitochondria

Mitochondria have four different protein complexes, three of which function as proton pumps (complexes I, III, and IV). In addition, two electron carriers (coenzyme Q and cytochrome *c*) shuttle electrons between the complexes. The electron transport chain of mitochondria consists of these components **(figure 6.17):**

- **Complex I** (also called NADH dehydrogenase complex). This accepts electrons from NADH, ultimately transferring them to coenzyme Q; in the process, four protons are pumped across the membrane.
- **Complex II** (also called succinate dehydrogenase complex). This accepts electrons from the TCA cycle, when $FADH_2$ is formed during the oxidation of succinate (see figure 6.15, step 6). Electrons are then transferred to coenzyme Q.
- **Coenzyme Q** (also called ubiquinone). This lipid soluble carrier accepts electrons from either complex I or complex II and then shuttles them to complex III. Note that the electrons carried by $FADH_2$ have entered the electron transport chain "downstream" of those carried by NADH. Because of this, a pair of electrons carried by NADH result in more protons being expelled than does a pair carried by $FADH_2$.
- **Complex III** (also called cytochrome bc_1 complex). This accepts electrons from coenzyme Q, ultimately transferring them to cytochrome *c;* in the process, four protons are pumped across the membrane.
- **Cytochrome *c*.** This accepts electrons from complex III and then shuttles them to complex IV.

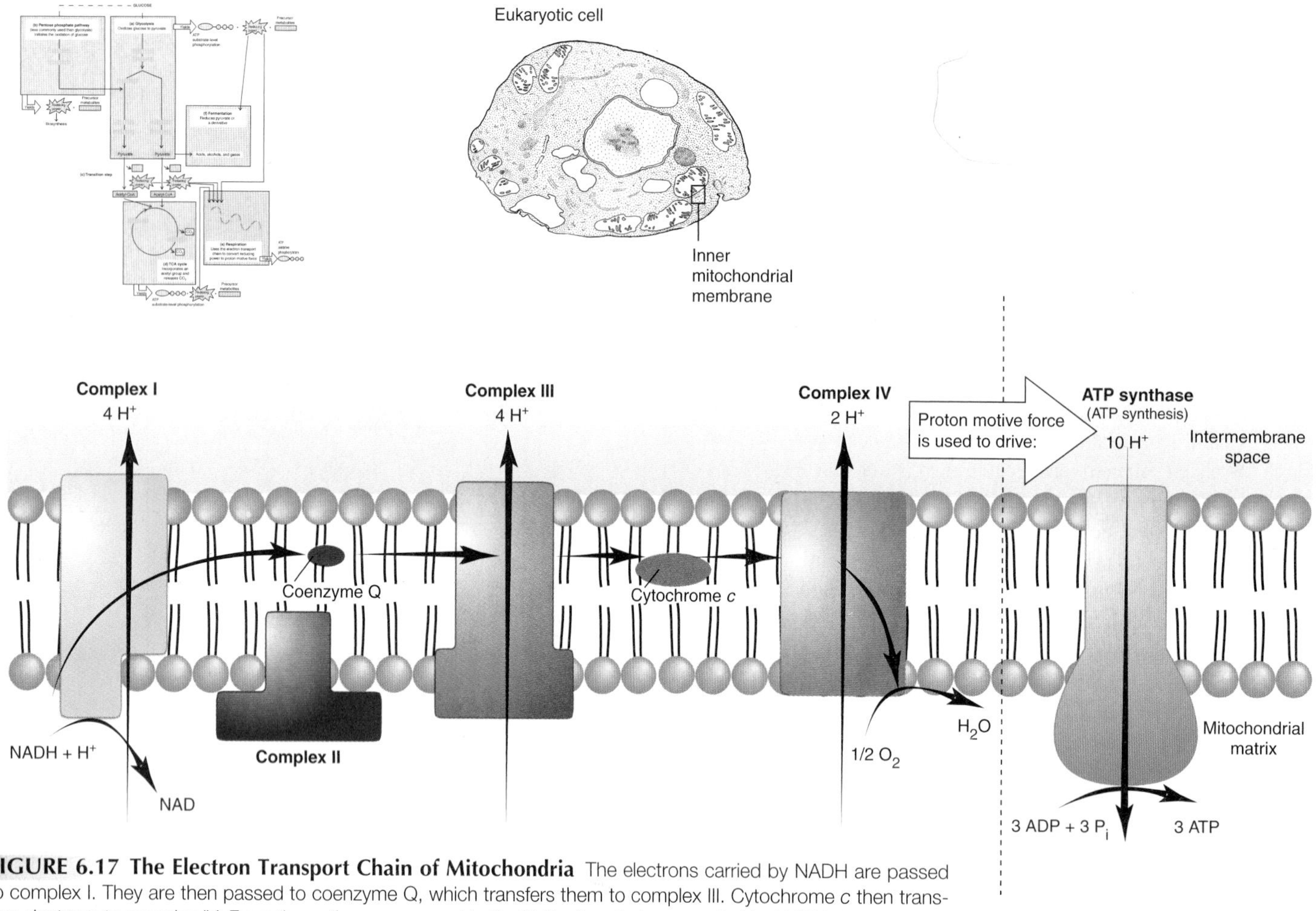

FIGURE 6.17 The Electron Transport Chain of Mitochondria The electrons carried by NADH are passed to complex I. They are then passed to coenzyme Q, which transfers them to complex III. Cytochrome *c* then transfers electrons to complex IV. From there, they are passed to O_2. Unlike the electrons carried by NADH, those carried by $FADH_2$ are passed to complex II, which then passes them to coenzyme Q; from there, the electrons follow the same path as the ones donated by NADH. Protons are shuttled from the mitochondrial matrix to the intermembrane space by complex I, III and IV, creating the proton motive force. ATP synthase allows protons to reenter the mitochondrial matrix, using the energy released to drive ATP synthesis.

- **Complex IV** (also called cytochrome *c* oxidase complex). This accepts electrons from cytochrome *c*, ultimately transferring them to oxygen (O_2), forming H_2O. In the process, two protons are pumped across the membrane. Complex IV is a terminal oxidoreductase, meaning that it transfers the electrons to the terminal electron acceptor, which, in this case, is O_2.

The Electron Transport Chains of Prokaryotes

Considering the flexibility and diversity of prokaryotes, it is not surprising that they vary with respect to the types and arrangement of their electron transport components. In fact, a single species may have several alternative carriers so that the system as a whole can function optimally under changeable growth conditions. In the laboratory, the different electron transport components provide a mechanism to distinguish between certain types of bacteria. For example, the activity of cytochrome *c* oxidase, which is found in species of *Neisseria, Pseudomonas, Campylobacter,* and certain other genera, is detected using the rapid biochemical test called the **oxidase test** and is important in the identification scheme of these organisms (see table 10.5).

The electron transport chain of *E. coli* provides an excellent example of the diversity found even in a single organism. This organism preferentially uses aerobic respiration, but when molecular oxygen is not available, it can switch to anaerobic respiration provided that a suitable terminal electron acceptor such as nitrate is available. The *E. coli* electron transport chain serves as a model for both aerobic and anaerobic respiration.

Aerobic Respiration When growing aerobically in a glucose-containing medium, *E. coli* can use two different NADH dehydrogenases **(figure 6.18).** One is a proton pump functionally equivalent to complex I of the mitochondrion. *E. coli* also has a succinate dehydrogenase that is functionally equivalent to complex II of the mitochondrion. In addition to these enzyme complexes, *E. coli* can produce several alternatives, enabling the organism to optimally use a variety of different energy sources, including hydrogen gas. *E. coli* does not have the equivalent of complex III or cytochrome *c*; instead quinones, including ubiquinone, shuttle the electrons

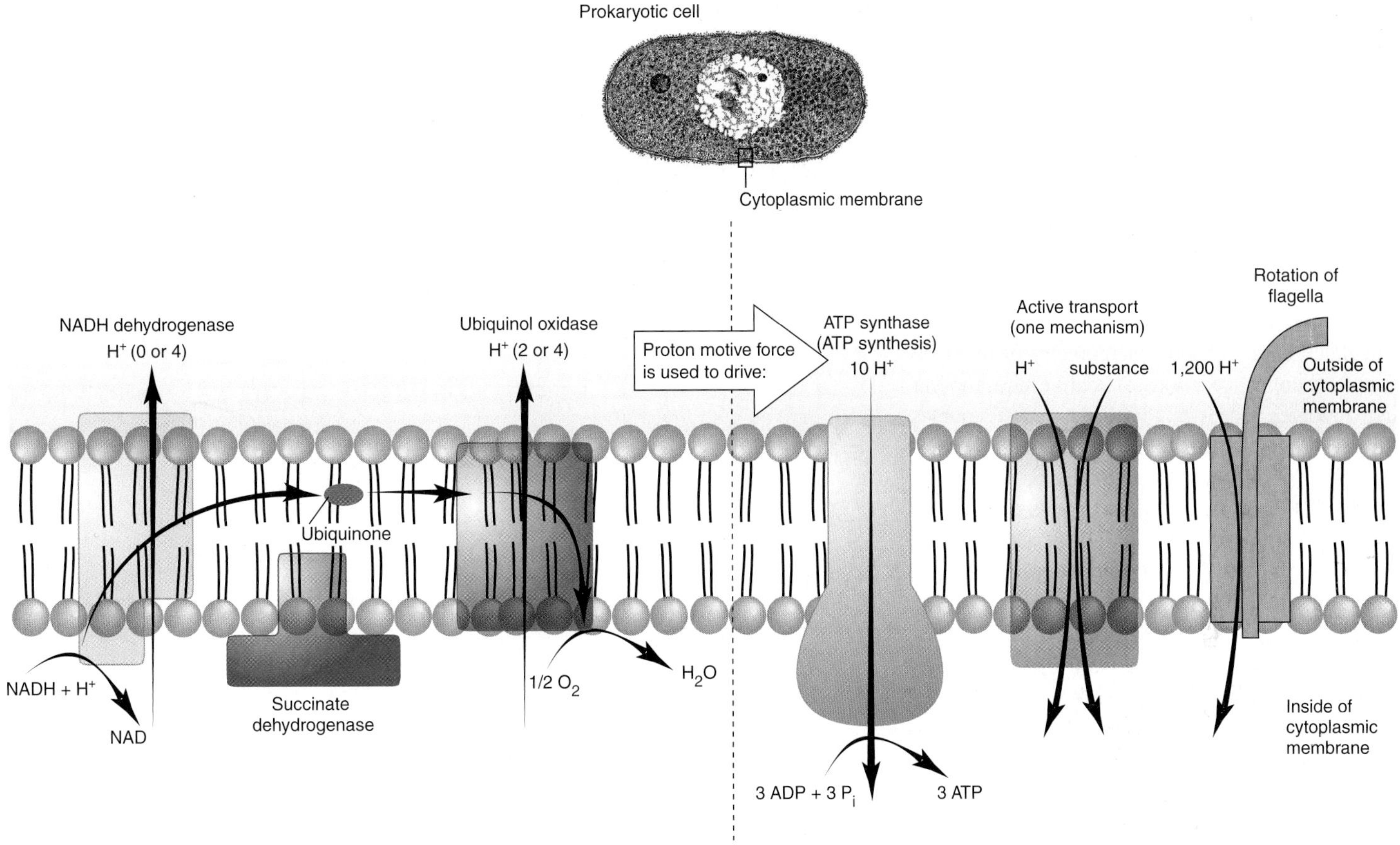

FIGURE 6.18 The Electron Transport Chain of *E. coli* Growing Aerobically in a Glucose-Containing Medium The electrons carried by NADH are passed to one of two different NADH dehydrogenases. They are then passed to ubiquinone, which transfers them to one of two ubiquinol oxidases. From there they are passed to O_2. Unlike the electrons carried by NADH, those carried by $FADH_2$ are passed to succinate dehydrogenase, which then transfers them to ubiquinone; from there, the electrons follow the same path as the ones donated by NADH. Protons are ejected by one of the two NADH dehydrogenases and both ubiquinol oxidases, creating the proton motive force. ATP synthase allows protons to reenter the cell, using the energy released to drive ATP synthesis. The proton motive force is also used to drive one form of active transport and to power the rotation of flagella. *E. coli* has other components of the electron transport chain that function under different growth conditions.

directly to a terminal oxidoreductase. When O_2 is available to serve as a terminal electron acceptor, one of two variations of a terminal oxidoreductase called ubiquinol oxidase is used. One form functions optimally only in high O_2 conditions and results in the expulsion of 4 protons. The other results in the ejection of only 2 protons, but it can more effectively scavenge O_2 and thus is particularly useful when the supply of O_2 is limited.

Anaerobic Respiration Anaerobic respiration is a less efficient form of energy transformation than is aerobic respiration. This is partly due to the lesser amount of energy released in reactions that involve the reduction of chemicals other than molecular oxygen. Alternative electron carriers are used in the electron transport chain during anaerobic respiration.

When oxygen is absent and nitrate is available, *E. coli* responds by synthesizing a terminal oxidoreductase that uses nitrate as a terminal electron acceptor, producing nitrite. The organism then converts nitrite to ammonia, avoiding the toxic effects of nitrite. Other bacteria can reduce nitrate further than *E. coli* can, forming compounds such as nitrous oxide (N_2O), and nitrogen gas (N_2).

The quinone that bacteria use during anaerobic respiration, menaquinone, provides humans and other mammals with a source of the nutrient called vitamin K. This vitamin is required for the proper coagulation of blood, and mammals are able to obtain at least part of their requirement by absorbing menaquinone produced by bacteria growing in the intestinal tract.

A group of obligate anaerobes called the **sulfate-reducers** use sulfate (SO_4^{2-}) as a terminal electron acceptor, producing hydrogen sulfide as an end product. The diversity and ecology of sulfate-reducing bacteria will be discussed in chapter 11. ■ **sulfate-reducing bacteria, p. 254**

ATP Synthase—Harvesting the Proton Motive Force to Synthesize ATP

Just as energy is required to establish a concentration gradient, energy is released when a gradient is eased. The enzyme ATP synthase uses that energy to synthesize ATP. It permits protons to flow back into the bacterial cell (or matrix of the mitochondrion) in a controlled manner, harvesting the energy released to fuel the addition of a phosphate group to ADP. One molecule of ATP is formed from the entry of approximately three protons. The precise mechanism of how this occurs is not well understood.

Theoretical ATP Yield of Oxidative Phosphorylation

The complexity of oxidative phosphorylation makes it exceedingly difficult to determine the actual maximum yield of ATP. Unlike the yield of substrate-level phosphorylation, which can be calculated based on the stoichiometry of relatively simple chemical reactions, oxidative phosphorylation involves processes that have many variables. This is particularly true for prokaryotic cells because they use proton motive force to drive activities other than ATP synthesis, including flagella rotation and membrane transport. In addition, as a group, they use different carriers in their electron transport chain, and these may vary in the number of protons ejected per pair of electrons passed.

For each pair of electrons transferred to the electron transport chain by NADH, between 2 and 3 ATP may be generated; for each pair transferred by $FADH_2$, the yield is between 1 and 2 ATP. Although experimental studies using rat mitochondria indicate that the yield is approximately 2.5 ATP/NADH and 1.5 ATP/$FADH_2$, for simplicity we will use whole numbers (3 ATP/NADH and 2 ATP/$FADH_2$) to calculate the maximum ATP gain of oxidative phosphorylation in a prokaryotic cell. Note, however, that these numbers are only theoretical and serve primarily as a means of comparing the relative energy gains of respiration and fermentation.

The ATP gain as a result of oxidative phosphorylation will be at least slightly different in eukaryotic cells than in prokaryotic cells because of the fate of the reducing power (NADH) generated during glycolysis. Recall that in eukaryotic cells, glycolysis takes place in the cytoplasm, whereas the electron transport chain is located in the mitochondria. Consequently, the electrons carried by cytoplasmic NADH must be translocated across the mitochondrial membrane before they can enter the electron transport chain. This requires an expenditure of approximately 2 ATP.

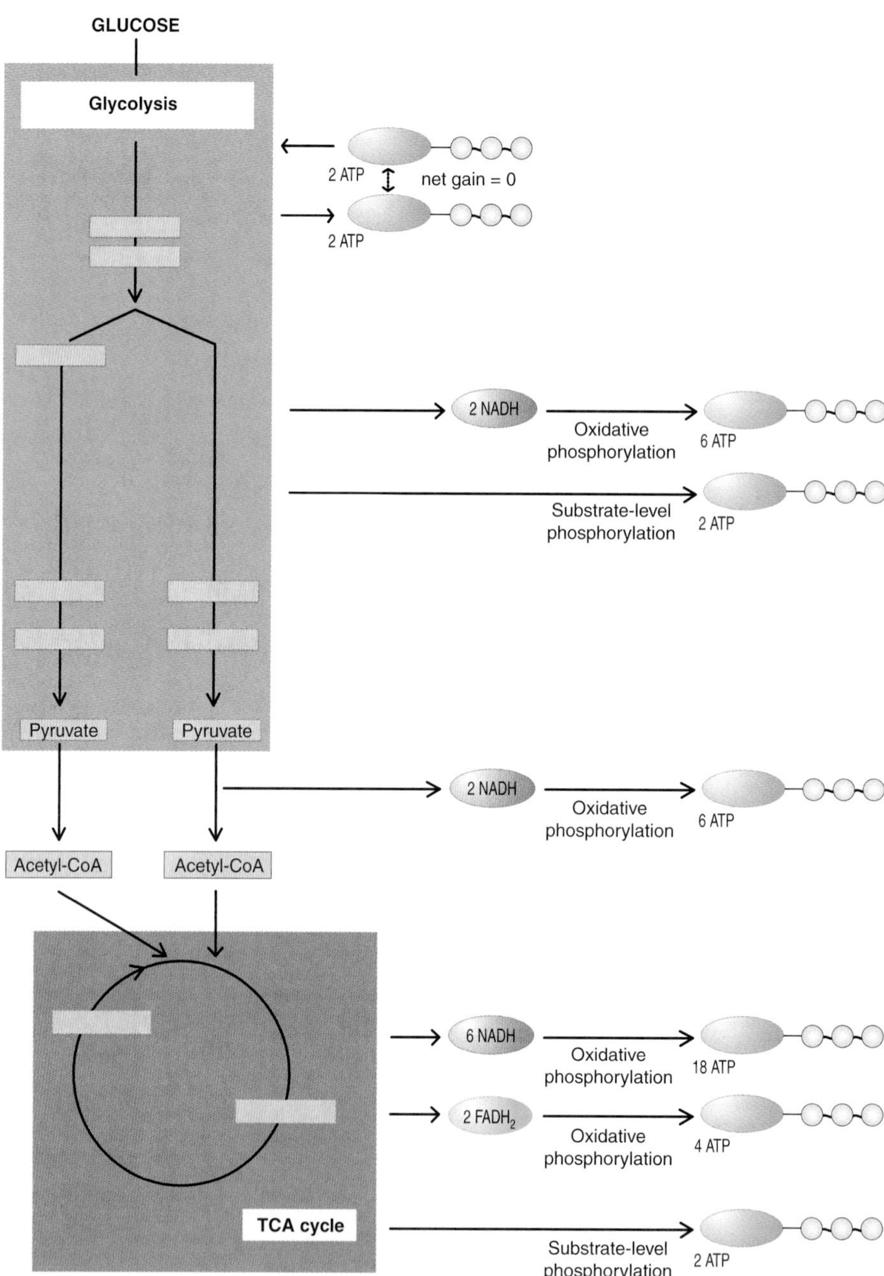

FIGURE 6.19 Maximum Theoretical Energy Yield from Aerobic Respiration in a Prokaryotic Cell This maximum energy yield calculation assumes that for every pair of electrons transferred to the electron transport chain, 3 ATP are synthesized; and for every pair of electrons donated by $FADH_2$, 2 ATP are synthesized. Note that these values are theoretical; a variety of factors, including the electron carriers employed and use of the proton motive force to drive other processes, affects the yield.

The maximum theoretical energy yield for oxidative phosphorylation in a prokaryotic cell that uses an electron transport chain similar to that of mitochondria is:

- **From glycolysis:**
  2 NADH ⟶ 6 ATP (assuming 3 for each NADH)
- **From the transition step:**
  2 NADH ⟶ 6 ATP (assuming 3 for each NADH)
- **From the TCA cycle:**
  6 NADH ⟶ 18 ATP (assuming 3 for each NADH)
  2 $FADH_2$ ⟶ 4 ATP (assuming 2 for each $FADH_2$)
- **Total maximum ATP yield from oxidative, phosphorylation = 34**

ATP Yield of Aerobic Respiration in Prokaryotes

Now that the ATP-yielding components of the central metabolic pathways have been considered, we can calculate the theoretical maximum ATP yield of aerobic respiration in prokaryotes. This yield is illustrated in **figure 6.19.**

- **Substrate-level phosphorylation:**
  2 ATP (from glycolysis; net gain)
  2 ATP (from the TCA cycle)
  4 ATP (total; substrate-level phosphorylation)
- **Oxidative phosphorylation:**
  6 ATP (from the reducing power gained in glycolysis)
  6 ATP (from the reducing power gained in the transition step)
  22 ATP (from the reducing power gained in the TCA cycle)
  34 (total; oxidative phosphorylation)
- **Total ATP gain (theoretical maximum) = 38**

MICROCHECK 6.4

Respiration uses the NADH and $FADH_2$ generated in glycolysis, the transition step, and the TCA cycle to synthesize ATP. The electron transport chain is used to convert reducing power into a proton motive force. ATP synthase then harvests that energy to synthesize ATP. The overall process is called oxidative phosphorylation. In aerobic respiration, O_2 serves as the terminal electron acceptor; anaerobic respiration employs a molecule other than O_2.

✓ Why is the overall ATP yield in aerobic respiration only a theoretical number?

✓ In bacteria, what is the role of the molecule that serves as a source of vitamin K for humans?

✓ Why could an oxidase also be called a reductase?

6.5 Fermentation

Focus Point

- Describe six common end products of fermentation, and explain the importance of each.

Fermentation is used by organisms that cannot respire, either because a suitable inorganic terminal electron acceptor is not available or because they lack an electron transport chain. *Escherichia coli* is a facultative anaerobe that has the ability to use any of three ATP-generating options—aerobic respiration, anaerobic respiration, and fermentation; the choice depends in part on the availability of terminal electron acceptors. In contrast, members of a group of aerotolerant anaerobes called the **lactic acid bacteria** lack the ability to respire; they only ferment, regardless of the presence of oxygen (O_2). Because they can grow in the presence of oxygen but never use it as a terminal electron acceptor, they are sometimes called **obligate fermenters.** The situation is different for obligate anaerobes that use fermentation pathways; they cannot even grow in the presence of O_2, and many are rapidly killed in its presence.

In general, the only ATP-yielding reactions of fermentation are those of glycolysis, and involve substrate-level phosphorylation. The other steps function primarily to consume excess reducing power, thereby providing a mechanism for recycling NADH **(figure 6.20).** If this reduced carrier were not recycled, no NAD^+ would be available to accept electrons in subsequent rounds of glycolysis, blocking that ATP-generating pathway. To consume reducing power, fermentation pathways use an organic intermediate such as pyruvate or a derivative as a terminal electron acceptor.

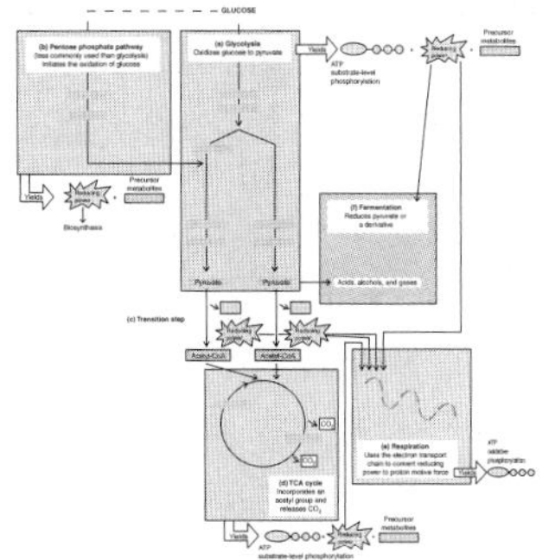

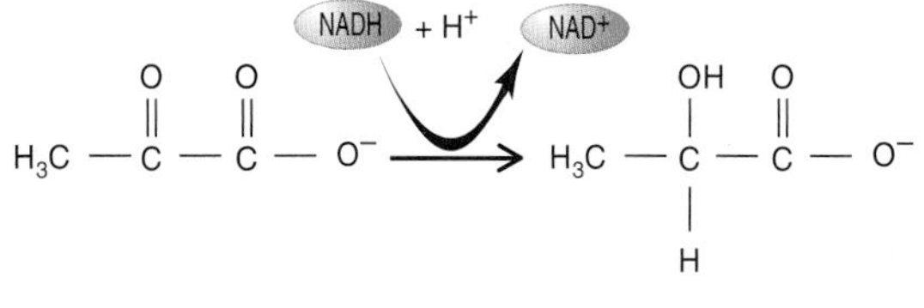

(a)

CO2
NADH
+ H+
NAD+
Pyruvate
Acetaldehyde
Ethanol

(b)

FIGURE 6.20 Fermentation Pathways Use Pyruvate or a Derivative As a Terminal Electron Acceptor **(a)** In lactic acid fermentation, pyruvate serves directly as a terminal electron acceptor, producing lactate. **(b)** In ethanol fermentation, pyruvate is first converted to acetaldehyde, which then serves as the terminal electron acceptor, producing ethanol.

The end products of fermentation are significant for a number of reasons **(figure 6.21).** Because a given type of organism uses a characteristic fermentation pathway, end products can sometimes be used as a marker to aid in identification. In addition, some end products are commercially valuable. In fact, much of chapter 32 is devoted to the fermentations used to produce certain beverages and food products. Important end products of fermentation pathways include the following (note that organic acids produced during fermentation are traditionally referred to by the name of their undissociated form):

- **Lactic acid.** Lactic acid (the ionized form is lactate) is produced when pyruvate itself serves as the terminal electron acceptor. The end products of a group of Gram-positive organisms called the lactic acid bacteria are instrumental in creating the flavor and texture of cheese, yogurt, pickles, cured sausages, and other foods. On the other hand, lactic acid causes tooth decay and spoilage of some foods. Some animal

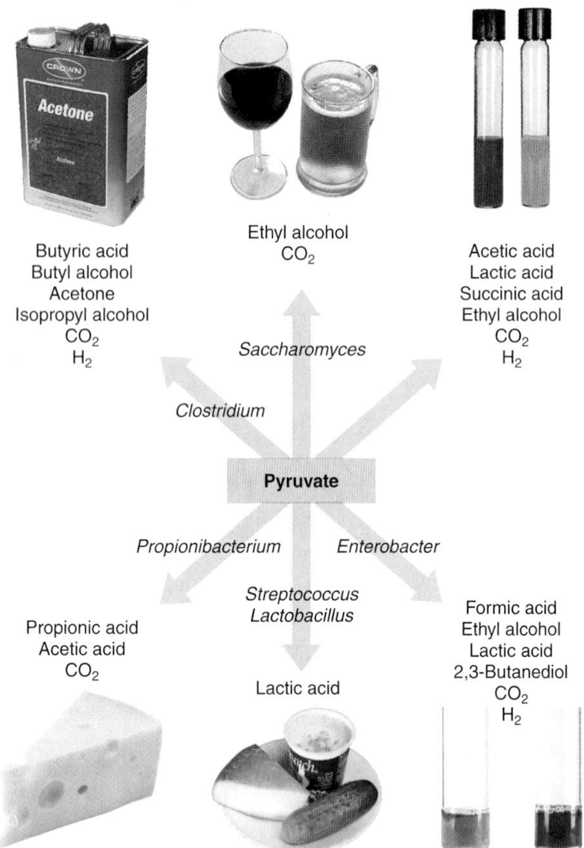

FIGURE 6.21 End Products of Fermentation Pathways Because a given type of organism uses a characteristic fermentation pathway, the end products can be used as an identifying marker. Some end products are commercially valuable.

cells use this fermentation pathway on a temporary basis when molecular oxygen is in short supply; the accumulation of lactic acid in muscle tissue causes the pain and fatigue sometimes associated with strenuous exercise. ■ lactic acid bacteria, p. 255 ■ cheese, yogurt and other fermented milk products, p. 756 ■ pickled vegetables, p. 758 ■ fermented meat products, p. 758

- **Ethanol.** Ethanol is produced in a pathway that first removes CO_2 from pyruvate, generating acetaldehyde, which then serves as the terminal electron acceptor. The end products of these sequential reactions are ethanol and CO_2, which are used to make wine, beer, spirits, and bread (see figures 32.4, 32.5, and 32.6). Ethanol is also an important biofuel. Members of *Saccharomyces* (yeast) and *Zymomonas* (bacteria) use this pathway. ■ wine, p. 758 ■ beer, p. 759 ■ distilled spirits, p. 760 ■ bread, p. 761
- **Butyric acid.** Butyric acid (the ionized form is butyrate) and a variety of other end products are produced in a complex multistep pathway used by species of *Clostridium,* which are obligate anaerobes. Under certain conditions, some species use a variation of this pathway to produce the organic solvents butanol and acetone.
- **Propionic acid.** Propionic acid (the ionized form is propionate) is generated in a multistep pathway that first removes CO_2 from pyruvate, generating a compound that then serves as a terminal electron acceptor. After NADH reduces this, it is further modified to form propionate. Members of the genus *Propionibacterium* use this pathway; their growth is encouraged in the production of Swiss cheese. The CO_2 they form makes the holes, and propionic acid gives the cheese its characteristic flavor. ■ cheese, p. 757
- **2, 3-Butanediol.** This is produced in a multistep pathway that uses two molecules of pyruvate to generate acetoin and two molecules of CO_2. Acetoin is then used as the terminal electron acceptor. The primary significance of this pathway is that it serves to differentiate certain members of the family *Enterobacteriaceae;* the **Voges-Proskauer test** detects acetoin, distinguishing members that use this pathway, such as *Klebsiella* and *Enterobacter,* from those that do not, such as *E. coli* (see table 10.4). ■ Voges-Proskauer test, p. 240
- **Mixed acids.** These are produced in a multistep branching pathway, generating a variety of different fermentation end products including lactic acid, succinic acid (the ionized form is succinate), ethanol, acetic acid (the ionized form is acetate), CO_2, and gases. This is another pathway used to differentiate certain members of the family *Enterobacteriaceae;* the **methyl-red test** detects its end products, distinguishing members that use this pathway, such as *E. coli,* from those that do not, such as *Klebsiella* and *Enterobacter* (see table 10.4). ■ methyl-red test, p. 240

MICROCHECK 6.5

Fermentation stops short of the TCA cycle, using pyruvate or a derivative of it as a terminal electron acceptor. Many end products of fermentation are commercially valuable.

- ✓ How do the Voges-Proskauer and methyl-red tests differentiate between certain members of the *Enterobacteriaceae?*
- ✓ Why do cells use fermentation rather than having pyruvate as the end product?
- ✓ Fermentation is used as a means of preserving foods. Why would it slow spoilage?

6.6

Catabolism of Organic Compounds Other Than Glucose

Focus Point

- Briefly describe how polysaccharides and disaccharides, lipids, and proteins are degraded and utilized by a cell.

Microbes can use a variety of organic compounds other than glucose as energy sources, including macromolecules such as polysaccharides, lipids, and proteins. To break these down into their respective sugar, amino acid, and lipid subunits, cells synthesize **hydrolytic enzymes,** which break bonds by adding water. To use a macromolecule in the surrounding medium, a cell must secrete the appropriate hydrolytic enzyme and then transport the resulting subunits into the cell. Inside the cell, the subunits are further degraded to form appropriate precursor metabolites **(figure 6.22).** Recall that precursor metabolites can be either

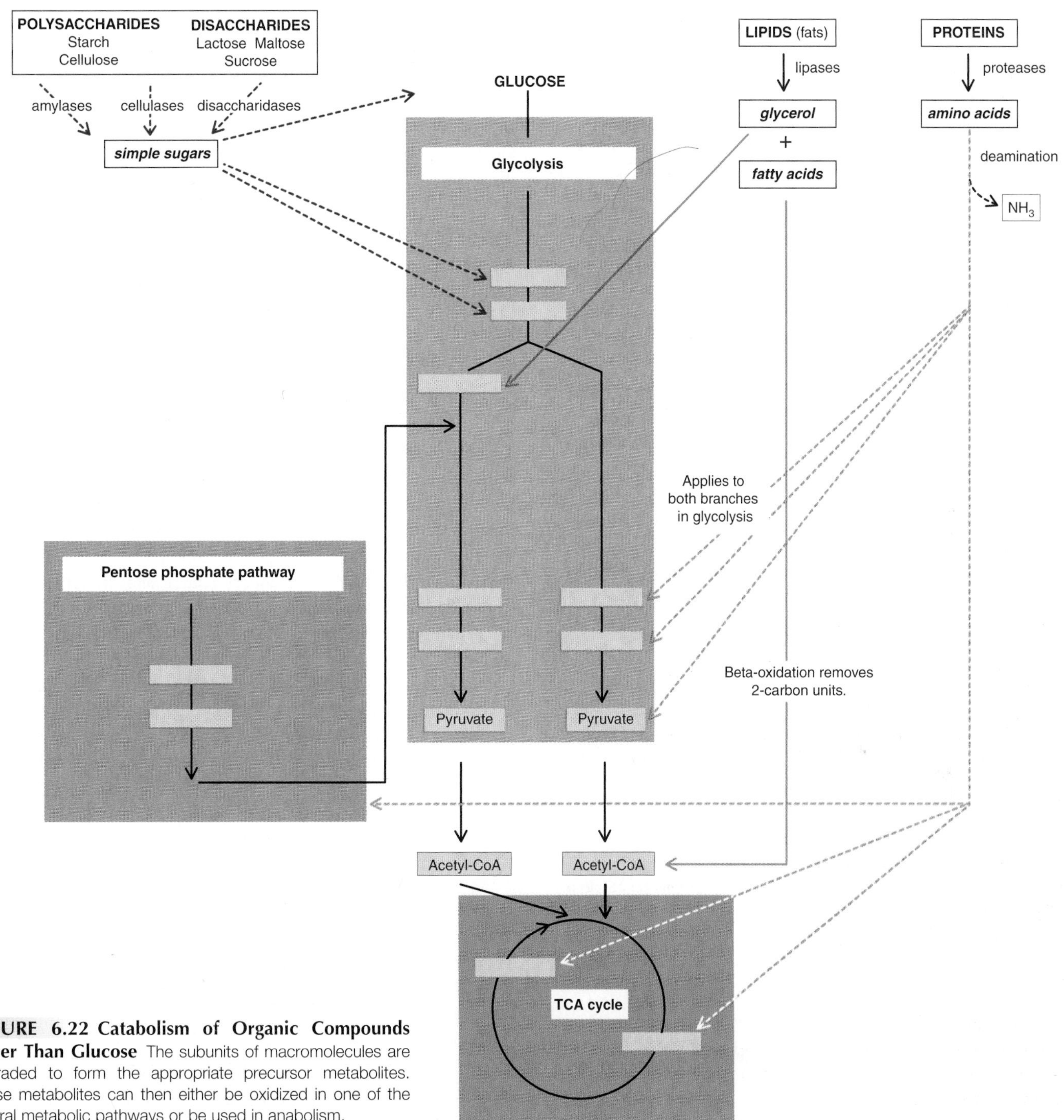

FIGURE 6.22 Catabolism of Organic Compounds Other Than Glucose The subunits of macromolecules are degraded to form the appropriate precursor metabolites. These metabolites can then either be oxidized in one of the central metabolic pathways or be used in anabolism.

oxidized in one of the central metabolic pathways or used in biosynthesis. ■ **hydrolysis, p. 25**

Polysaccharides and Disaccharides

Starch and cellulose are both polymers of glucose, but different types of chemical bonds join their subunits. The nature of this difference profoundly affects the mechanisms by which they are degraded. Enzymes called **amylases** are produced by a wide variety of organisms to digest starches. In contrast, cellulose is digested by enzymes called **cellulases,** which are produced by relatively few organisms. Among the organisms that can degrade cellulose are bacteria that reside in the rumen of animals, and many types of fungi. Considering that cellulose is the most abundant organic compound on earth, it is not surprising that fungi are important decomposers in terrestrial habitats. The glucose subunits released when polysaccharides are hydrolyzed can then enter glycolysis to be oxidized to pyruvate. ■ **polysaccharides, p. 32** ■ **cellulose, p. 32** ■ **rumen, p. 735**

Disaccharides including lactose, maltose, and sucrose are hydrolyzed by specific **disaccharidases.** For example, the enzyme β-galactosidase breaks down lactose, forming glucose and galactose. Glucose can enter glycolysis directly, but the other monosaccharides must first be modified. ■ **disaccharides, p. 32**

Lipids

The most common simple lipids are fats, which are a combination of fatty acids joined to glycerol. Fats are hydrolyzed by enzymes called **lipases.** The glycerol component is then converted to the precursor metabolite dihydroxyacetone phosphate, which then enters the glycolytic pathway. The fatty acids are degraded using a series of reactions collectively called **β-oxidation.** Each sequential reaction transfers a 2-carbon unit from the end of the fatty acid to coenzyme A, forming acetyl-CoA; this can enter the TCA cycle. Each reaction is a redox reaction, generating one NADH + H^+ and one $FADH_2$. ■ simple lipids, p. 35

Proteins

Proteins are hydrolyzed by enzymes called **proteases,** which break peptide bonds that join amino acid subunits. The amino group of the resulting amino acids is removed by a reaction called a **deamination.** The remaining carbon skeletons are then converted into the appropriate precursor molecules. ■ protein, p. 25

MICROCHECK 6.6

In order for polysaccharides, lipids, and proteins to be used as energy sources, they are first hydrolyzed to release their respective subunits. These are then converted to the appropriate precursor metabolites so they can enter a central metabolic pathway.

✓ Why do cells secrete hydrolytic enzymes?
✓ Explain the process used to degrade fatty acids.
✓ How would cellulose-degrading bacteria in the rumen of a cow benefit the animal?

6.7 Chemolithotrophs

Focus Point

- Explain how chemolithotrophs obtain energy.

Prokaryotes as a group are unique in their ability to use reduced inorganic chemicals such as hydrogen sulfide (H_2S) and ammonia (NH_3) as a source of energy. Note that these are the very compounds produced as a result of anaerobic respiration, when inorganic molecules such as sulfate and nitrate serve as terminal electron acceptors. This is one important example of how nutrients are cycled; the waste products of one organism serve as an energy source for another. ■ biogeochemical cycling and energy flow, p. 728

Chemolithotrophs fall into four general groups **(table 6.7):**

- **Hydrogen bacteria** oxidize hydrogen gas.
- **Sulfur bacteria** oxidize hydrogen sulfide.
- **Iron bacteria** oxidize reduced forms of iron.
- **Nitrifying bacteria** include two groups of bacteria—one oxidizes ammonia, forming nitrite, and the other oxidizes nitrite, forming nitrate.

The chemolithotrophs extract electrons from inorganic energy sources and then use the electrons to generate ATP by oxidative phosphorylation. The electrons are passed along an electron transport chain to generate a proton motive force, analogous to the processes described earlier. The amount of energy gained in metabolism depends on the energy source and the terminal electron acceptor; **figure 6.23** illustrates this relationship.

Chemolithotrophs generally thrive in very specific environments where reduced inorganic compounds are found. For example, *Acidithiobacillus ferrooxidans* is found in certain acidic environments that are rich in sulfides. Because these organisms oxidize metal sulfides, they can be used to enhance the recovery of metals (see **Perspective 6.1**). Thermophilic chemolithotrophs thrive near hydrothermal vents of the deep ocean, harvesting the energy of reduced inorganic compounds that spew from the vents. The diversity and ecology of some of these organisms will be discussed in chapter 11.

Unlike organisms that use organic molecules to fill both their energy and carbon needs, chemolithotrophs incorporate inorganic carbon, CO_2, into an organic form. This process, called **carbon fixation,** will be described later.

TABLE 6.7 Metabolism of Chemolithotrophs

Common Name of Organism	Source of Energy	Oxidation Reaction (Energy Yielding)	Important Features of Group	Common Genera In Group
Hydrogen bacteria	H_2 gas	$H_2 + \frac{1}{2} O_2 \longrightarrow H_2O$	Can also use simple organic compounds for energy	*Hydrogenomonas*
Sulfur bacteria (non-photosynthetic)	H_2S	$H_2S + \frac{1}{2} O_2 \longrightarrow H_2O + S$ $S + 1\frac{1}{2} O_2 + H_2O \longrightarrow H_2SO_4$	Some members of this group can live at a pH of less than 1.	*Acidithiobacillus* *Thiobacillus* *Beggiatoa* *Thiothrix*
Iron bacteria	Reduced Iron (Fe^{2+})	$2\,Fe^{2+} + \frac{1}{2} O_2 + H_2O \longrightarrow 2\,Fe^{3+} + 2\,OH^-$	Iron oxide present in the sheaths of these bacteria	*Sphaerotilus* *Gallionella*
Nitrifying bacteria	NH_3	$NH_3 + 1\frac{1}{2} O_2 \longrightarrow HNO_2 + H_2O$	Important in nitrogen cycle	*Nitrosomonas*
	HNO_2	$HNO_2 + \frac{1}{2} O_2 \longrightarrow HNO_3$	Important in nitrogen cycle	*Nitrobacter*

PERSPECTIVE 6.1

Mining with Microbes

Microorganisms have been used for thousands of years in the production of bread and wine. It is only in the past several decades, however, that they are being used with increasing frequency in another area—the mining industry. The mining process traditionally consists of digging crude ores from the earth, crushing them, and then extracting the desired minerals from the contaminants. The extraction process of such minerals as copper and gold frequently involves harsh conditions, such as smelting, and burning off the contaminants before extracting the metal with cyanide. Such activities are expensive and deleterious to the environment. With the development of biomining, some of these problems are being solved.

In the process of biomining copper, the low grade ore is dumped outside the mine and then treated with sulfuric acid. The acidic conditions encourage the growth of *Acidithiobacillus* species present naturally in the ore. These acidophilic bacteria use CO_2 as a source of carbon and gain energy by oxidizing sulfides of iron first to sulfur and then to sulfuric acid. The sulfuric acid dissolves the insoluble copper and gold from the ore. Currently about 25% of all copper produced in the world comes from the process of biomining. Similar processes are being applied to gold mining.

The current process of biomining employs microbes indigenous to the ore. Many improvements should be possible. For example, the oxidation of the minerals generates heat to the point that the bacteria may be killed. The use of thermophiles should overcome this problem. Further, many ores contain heavy metals, such as mercury, cadmium, and arsenic, which are toxic to the bacteria. It should be possible to isolate bacteria that are resistant to these metals. Biomining is still in its infancy.

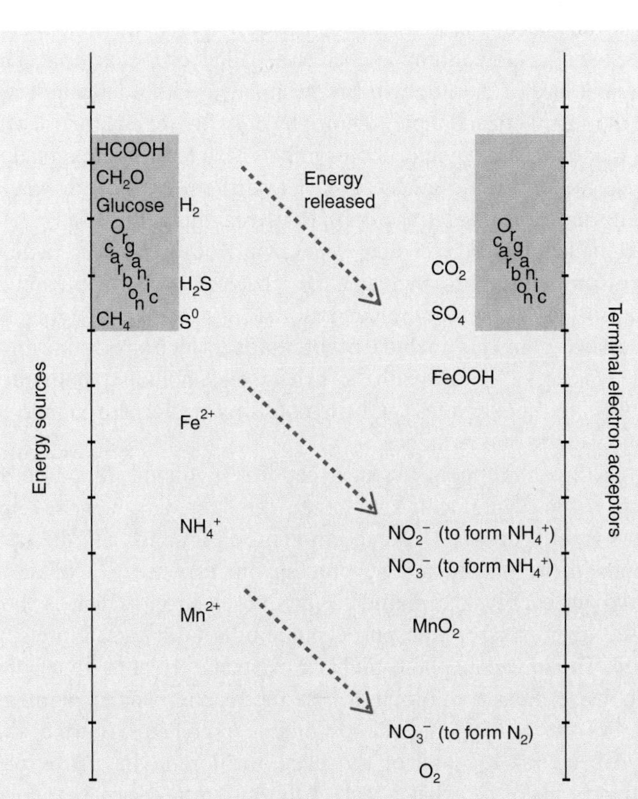

FIGURE 6.23 Relative Energy Gain of Different Types of Metabolism The left axis shows potential energy sources, ordered according to their relative tendency to give up electrons; those at the top lose electrons most easily. The right axis shows potential terminal electron acceptors, ordered according to their relative tendency to gain electrons; those at the bottom accept electrons most readily. Energy is released only when electrons are transferred from an energy source to a terminal electron acceptor that is lower on the chart; the greater the downward slope, the more energy that can be harvested to make ATP.

MICROCHECK 6.7

Chemolithotrophs use reduced inorganic compounds as an energy source. They use carbon dioxide as a carbon source.

✓ Describe the roles of hydrogen sulfide and carbon dioxide in chemolithoautotrophic metabolism.

✓ Which energy source, Fe^{2+} or H_2S, would result in the greatest energy yield when O_2 is used as a terminal electron acceptor (hint: refer to figure 6.23)?

6.8 Photosynthesis

Focus Points

- Describe the role of chlorophylls, bacteriochlorophylls, accessory pigments, reaction-center pigments, and antennae pigments in capturing radiant energy.
- Compare and contrast the tandem photosystems of cyanobacteria and photosynthetic eukaryotes with the single photosystems of purple and green bacteria.

Plants, algae, and several groups of bacteria harvest the radiant energy of sunlight, and then use it to power the synthesis of organic compounds from CO_2. This capture and subsequent conversion of light energy into chemical energy is called **photosynthesis.** The general reaction of photosynthesis can be summarized as:

$$6\,CO_2 + 12\,H_2X \xrightarrow{\text{Light Energy}} C_6H_{12}O_6 + 12\,X + 6\,H_2O$$

Photosynthetic processes are generally considered in two distinct stages. The **light-dependent reactions,** often simply called the **light reactions,** are used to capture the energy from light and convert it to chemical energy in the form of ATP. The **light-independent reactions,** also termed the **dark reactions,** use that energy to synthesize organic compounds. The process that converts carbon

TABLE 6.8 Comparison of the Photosynthetic Mechanisms Used by Different Organisms

	Oxygenic Photosynthesis		Anoxygenic Photosynthesis	
	Plants, Algae	**Cyanobacteria**	**Purple Bacteria**	**Green Bacteria**
Location of the photosystem	In membranes of thylakoids, which are within the stroma of chloroplasts	In membranes of thylakoids, located within the cell	Within the cytoplasmic membrane; extensive invaginations in that membrane effectively increase the surface area.	Primarily within the cytoplasmic membrane; chlorosomes attached to the inner surface of the membrane contain the accessory pigments.
Type of photosystem	Photosystem I and photosystem II		Similar to photosystem II	Similar to photosystem I
Primary light harvesting pigment	Chlorophyll *a*		Bacteriochlorophylls	
Mechanism for generating reducing power	Non-cyclic photophosphorylation using both photosystems		Reversed electron transport	Non-cyclic use of the photosystem
Source of electrons for reducing power	H_2O		Varies among the organisms in the group; may include H_2S, H_2, or organic compounds.	
CO_2 fixation	Calvin cycle		Calvin cycle	Reversed TCA cycle
Accessory pigments	Carotenoids	Carotenoids, phycobilins	Carotenoids	Carotenoids

dioxide into organic compounds is called **carbon fixation.** We will describe the steps of carbon fixation in a separate section (see section 6.9) because, as mentioned earlier, a variety of prokaryotes other than photosynthetic ones use the process. Characteristics of various photosynthetic mechanisms are summarized in **table 6.8.**

Capturing Radiant Energy

Photosynthetic organisms are highly visible in their natural habitats because they possess various colored pigments that capture light energy. The color we observe is due to the wavelengths reflected by the pigment; for example, pigments that absorb only blue and red light will appear green (see figure 5.7). Multiple pigments are involved in photosynthesis, increasing the range of wavelengths of light that can be absorbed by a cell. The pigments are located together in protein complexes called **photosystems,** which specialize in capturing and using light **(figure 6.24).**

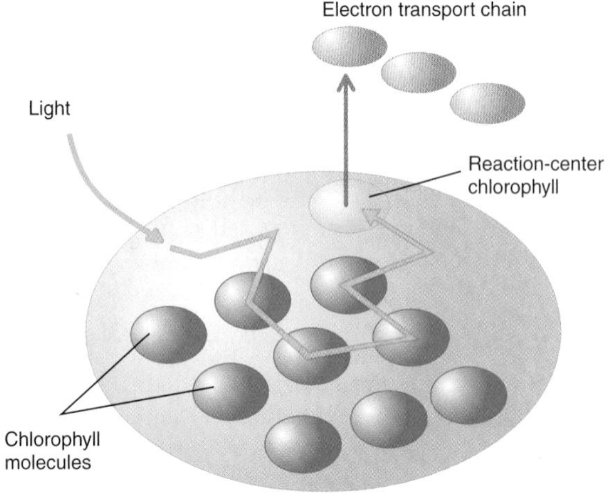

FIGURE 6.24 Photosystem Chlorophyll and other pigments capture the energy of light and then transfer it to reaction-center chlorophyll, which emits an electron that is then passed to an electron transport chain.

Photosynthetic pigments include chlorophylls, bacteriochlorophylls, and accessory pigments. **Chlorophylls** are found in plants, algae, and a large group of bacteria called cyanobacteria. The various types of chlorophylls are designated with a letter following the term; for example, chlorophyll *a*. **Bacteriochlorophylls** are found in anoxygenic photosynthetic bacteria ("anoxygenic" means they do not generate O_2). These pigments absorb wavelengths not absorbed by chlorophylls, enabling the bacteria to grow in habitats where other photosynthetic organisms cannot. **Accessory pigments** increase the efficiency of light capture by absorbing wavelengths not absorbed by the other pigments. Accessory pigments include **carotenoids,** which are found in a wide variety of photosynthetic organisms, including both prokaryotes and eukaryotes, and **phycobillins,** which are unique to cyanobacteria and red algae.

Within the photosystems, certain pigments function as reaction-center pigments and others function as antennae pigments (see figure 6.24). **Reaction-center pigments** are electron donors in the photosynthetic process. In response to excitation by radiant energy, the pigment emits an electron, which is then passed to an electron transport chain similar to that used in respiration. The oxygenic photosynthetic organisms (plants, algae, and cyanobacteria) use chlorophyll *a* as the reaction-center pigment, whereas the anoxygenic photosynthetic organisms (purple and green bacteria) use one of the bacteriochlorophylls. **Antennae pigments** make up what is called the **antenna complex,** which acts as a funnel, capturing the energy of light and then transferring it to the reaction-center pigment.

The photosystems of cyanobacteria are embedded in the membranes of stacked structures called **thylakoids** located within the cells. Plants and algae also have thylakoids, in the stroma of the chloroplast (see figure 3.54). The similarity between the structure of chloroplasts and cyanobacteria is not surprising considering that genetic evidence indicates that the organelle descended from an ancestor of a cyanobacterium (see Perspective 3.1).

The photosystems of the purple and green bacteria are embedded in the cytoplasmic membrane. Purple bacteria have extensive invaginations in the membrane that maximize the surface area. Green bacteria have specialized structures called **chlorosomes** attached to the inner surface of the cytoplasmic membrane. These structures contain the accessory pigments.

Converting Radiant Energy into Chemical Energy

Photosynthetic organisms use the light-dependent reactions to accomplish two tasks. First, they must use radiant energy to fuel the synthesis of ATP, the process of **photophosphorylation.** They also need to generate reducing power so they can fix CO_2. Depending on the method used to fix CO_2, the type of reducing power required may be either NADPH or NADH.

Light-Dependent Reactions in Cyanobacteria and Photosynthetic Eukaryotic Cells

Cyanobacteria and chloroplasts have two distinct photosystems that work in tandem **(figure 6.25).** The sequential absorption of energy by the two photosystems allows the process to raise the energy level of electrons stripped from water high enough to be used to generate a proton motive force as well as produce reducing power. The process is **oxygenic**—that is, it generates O_2.

First we will consider the simplest situation, which occurs when the cell needs to synthesize ATP but not reducing power (NADPH). To accomplish this, only photosystem I is used. Radiant energy is absorbed by this photosystem, exciting the reaction-center chlorophylls, which causes them to emit high-energy electrons. The electrons are then passed to an electron carrier, which transports them to a proton pump; this pump is analogous to complex III in the respiratory chain of mitochondria. After being used to pump protons across the membrane, thus generating a proton motive force, the electrons are returned to photosystem I. As occurs in oxidative phosphorylation, ATP synthase harvests the energy of the proton motive force to synthesize ATP. This overall process is called **cyclic photophosphorylation** because the molecule that serves as the electron donor, reaction-center chlorophyll, is also the terminal electron acceptor; the electrons have followed a cyclical path.

When cells must produce both ATP and reducing power, **noncyclic photophosphorylation** is used. In this process, the electrons emitted by photosystem I are not passed to the proton pump, but instead are donated to $NADP^+$ to produce NADPH. While this action provides reducing power, the cell must now replenish the electrons emitted by reaction center chlorophyll from another source. In addition, the cell must still generate a proton motive force in order to synthesize ATP. Photosystem II plays a pivotal role in this process. When photosystem II absorbs radiant energy, the reaction-center chlorophylls emit high-energy electrons that can be donated to photosystem I. First, however, the electrons are passed to the proton pump, which uses some of their energy to establish the proton motive force. In order to replenish the electrons emitted from photosystem II, an enzyme within that complex extracts the electrons from water, donating them to the reaction-center chlorophyll. Removal of electrons from two molecules of water generates O_2. In essence, photosystem II captures the energy of light and then uses it to raise the energy level of electrons stripped

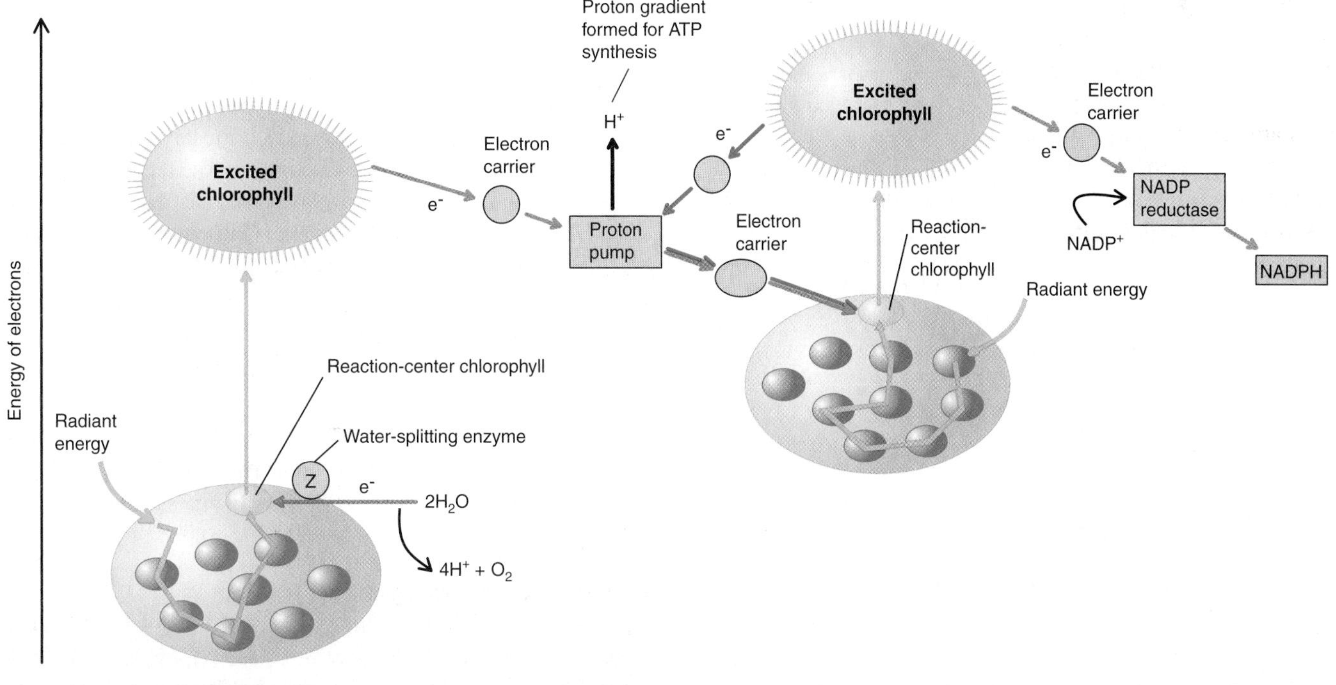

FIGURE 6.25 The Tandem Photosystems of Cyanobacteria and Chloroplasts Radiant energy captured by photosynthetic pigments excites the reaction-center chlorophyll, causing it to emit a high-energy electron, which is then passed to an electron transport chain. In cyclic photophosphorylation, electrons emitted by photosystem I are returned to that photosystem; the path of the electrons is shown in green arrows. In non-cyclic photophosphorylation, the electrons used to replenish photosystem I are donated by radiant energy-excited photosystem II; the path of these electrons is shown in orange arrows. In turn, photosystem II replenishes its own electrons by stripping them from water, producing O_2.

from water molecules to a high enough level that they can be used to power photophosphorylation. Photosystem I then accepts those electrons, which still retain some residual energy, and again captures the energy of light to boost the energy of the electrons to an even higher level so they can be used to reduce NADPH.

Light-Dependent Reactions in Anoxygenic Photosynthetic Bacteria

Anoxygenic photosynthetic bacteria employ only a single photosystem and are unable to use water as an electron donor for reducing power. This is why they are **anoxygenic,** or do not evolve O_2. Molecules used as electron donors by these bacteria include hydrogen gas (H_2), hydrogen sulfide (H_2S), and organic compounds. There are two general groups of anoxygenic photosynthetic bacteria—the purple bacteria and the green bacteria.

Purple bacteria use a photosystem similar to the photosystem II of cyanobacteria and eukaryotes to synthesize ATP. However, the photosystem does not raise the electrons to a high enough energy level to reduce NAD^+ (or $NADP^+$), so the purple bacteria must use an alternative mechanism to generate reducing power. To do this they employ a process called reversed electron transport, which uses ATP to run the electron transport chain in the reverse direction, or "uphill."

Green bacteria employ a photosystem similar to photosystem I. The electrons emitted from this photosystem can be used to either generate a proton motive force or reduce NAD^+.

MICROCHECK 6.8

Photosynthetic organisms harvest the energy of sunlight and use it to power the synthesis of organic compounds from CO_2. Various pigments are used to capture radiant energy. These pigments are arranged in complexes called photosystems. When reaction-center chlorophyll absorbs the energy of light, a high-energy electron is emitted. This is then passed along an electron transport chain to generate a proton motive force, which is used to synthesize ATP. Plants and cyanobacteria use water as a source of electrons for reducing power, generating oxygen. Anoxygenic photosynthetic bacteria obtain electrons from a reduced compound other than water, and therefore do not evolve oxygen.

- ✓ β-carotene is a carotenoid that mammals can use as a source of vitamin A. What is the function of carotenoids in photosynthetic organisms?
- ✓ What is the advantage of having tandem photosystems?
- ✓ It requires energy to reverse the flow of the electron transport chain. Why would this be so?

6.9 Carbon Fixation

Focus Point

- Describe the three stages of the Calvin cycle.

Chemolithoautotrophs and photoautotrophs use carbon dioxide to synthesize organic compounds, the process of **carbon fixation.** In photosynthetic organisms, the process occurs in the light-independent reactions. Carbon fixation consumes a great deal of ATP and reducing power, which should not be surprising considering that the reverse process—oxidizing those same compounds to CO_2—liberates a great deal of energy. The Calvin cycle is the most common pathway used to fix carbon, but some prokaryotes incorporate CO_2 using other mechanisms. For example, the green bacteria and some members of the *Archaea* use a pathway that effectively reverses the steps of the TCA cycle.

Calvin Cycle

The **Calvin cycle,** or Calvin-Benson cycle, named in honor of the scientists who described much of it, is a complex cycle that can be viewed as having three essential stages—incorporation of CO_2 into an organic compound, reduction of the resulting molecule, and regeneration of the starting compound **(figure 6.26).** Because of the complexities of the cycle, it is easiest to consider the process as consisting of six "turns" of the cycle. Together, these six "turns" generate a net gain of two molecules of glyceraldehyde 3-phosphate, which can be converted into one molecule of fructose 6-phosphate. The Calvin cycle consists of three stages:

- **Stage 1** Carbon dioxide enters the cycle when the enzyme ribulose bisphosphate carboxylase, commonly called **rubisco,** joins it to a 5-carbon compound, ribulose 1, 5-bisphosphate. The resulting compound spontaneously hydrolyzes to produce two molecules of a 3-carbon compound, 3-phosphoglycerate (3PG). Interestingly, although rubisco is unique to autotrophs, it is thought to be the most abundant enzyme on earth!
- **Stage 2** A sequential input of energy (ATP) and reducing power (NADPH) is used in steps that, together, convert 3PG to glyceraldehyde 3-phosphate (G3P). This compound is identical to the precursor metabolite formed as an intermediate in glycolysis. It can be converted to a number of different compounds used in biosynthesis, oxidized to make other precursor compounds, or converted to a 6-carbon sugar. A critical aspect of the pathway of CO_2 fixation, however, stems from the fact that it operates as a cycle—ribulose 1, 5-bisphosphate must be regenerated from G3P for the process to continue. Consequently, in six cycles, a maximum of 2 G3P can be converted to a 6-carbon sugar, the rest is used to regenerate ribulose 1, 5-bisphosphate.
- **Stage 3** Many of the steps used to regenerate ribulose 1, 5-bisphosphate involve reactions of the pentose phosphate cycle.

Yield of the Calvin Cycle

One molecule of the 6-carbon sugar fructose can be generated for every six "turns" of the cycle. These six "turns" consume 18 ATP and 12 NADPH + H^+.

MICROCHECK 6.9

The process of carbon dioxide fixation consumes a great deal of ATP and reducing power. The Calvin cycle is the most common pathway used to incorporate inorganic carbon into an organic form.

- ✓ What is the role of rubisco?
- ✓ What would happen if ribulose 1, 5-bisphosphate were depleted in a cell?

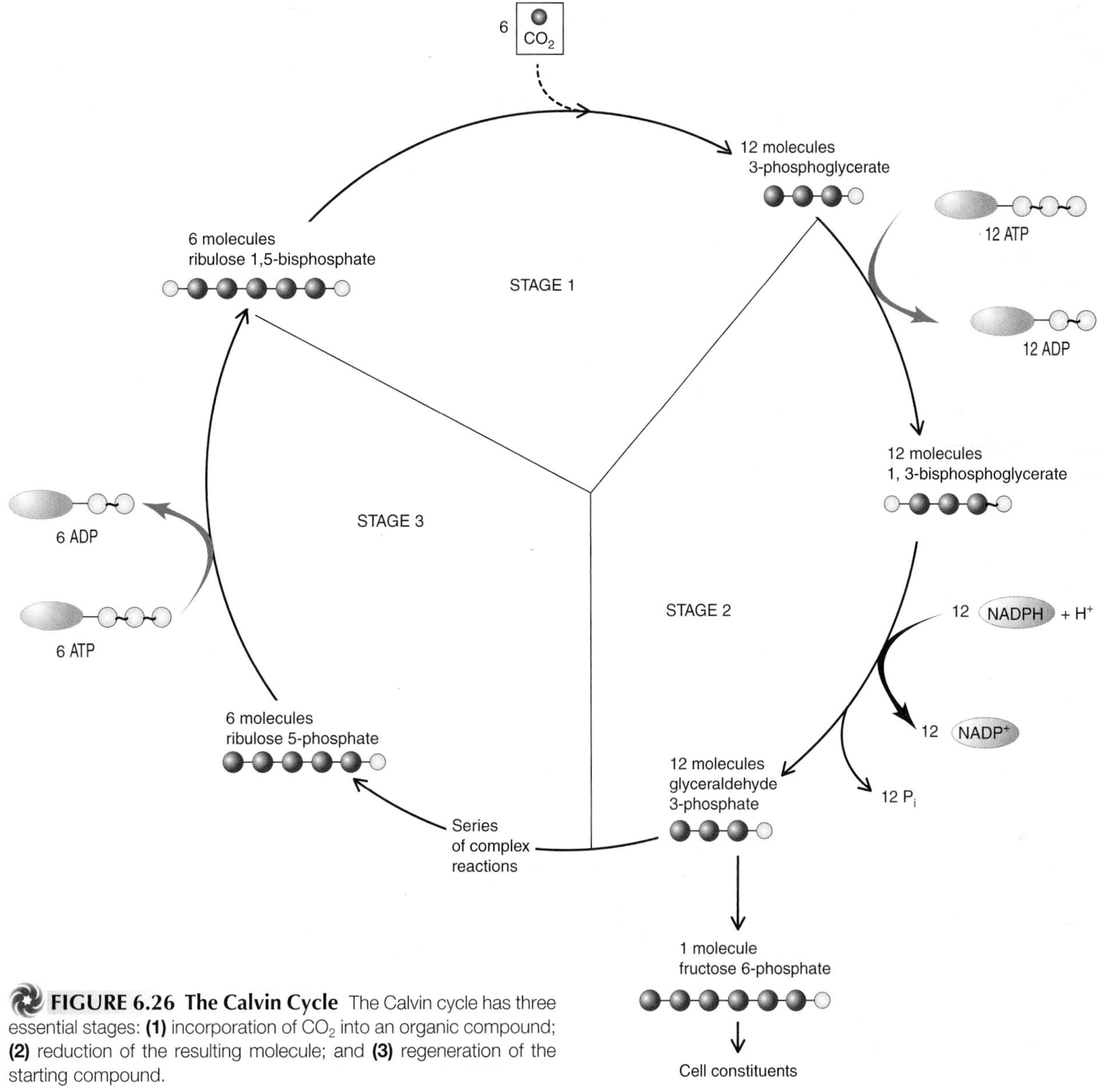

FIGURE 6.26 The Calvin Cycle The Calvin cycle has three essential stages: **(1)** incorporation of CO_2 into an organic compound; **(2)** reduction of the resulting molecule; and **(3)** regeneration of the starting compound.

6.10

Anabolic Pathways—Synthesizing Subunits from Precursor Molecules

Focus Point

- Describe the synthesis of lipids, amino acids, and nucleotides.

Prokaryotes, as a group, are highly diverse with respect to the compounds they use for energy, but they are remarkably similar in their biosynthetic processes. They synthesize the necessary subunits, employing specific anabolic pathways that use ATP, reducing power in the form of NADPH, and the precursor metabolites formed in the central metabolic pathways **(figure 6.27).** Organisms lacking one or more enzymes in a given biosynthetic pathway must have the end product provided from an external source. This is why fastidious bacteria, such as lactic acid bacteria, require many different growth factors. Once the subunits are synthesized or taken up, they can be assembled to make macromolecules. Various different macromolecules can then be joined to form the structures that make up the cell. ■ fastidious, p. 94

Lipid Synthesis

Synthesis of most lipids in microorganisms can be viewed as having two essential components—fatty acid synthesis and glycerol synthesis. Synthesis starts with transfer of the acetyl group of the precursor metabolite produced in the transition step, acetyl-CoA, to a carrier protein called **acyl carrier protein (ACP).** This carrier holds the fatty acid chain as 2-carbon units are progressively added. When the newly synthesized fatty acid reaches its required length, usually 14, 16, or 18 carbons long, it is released from ACP.

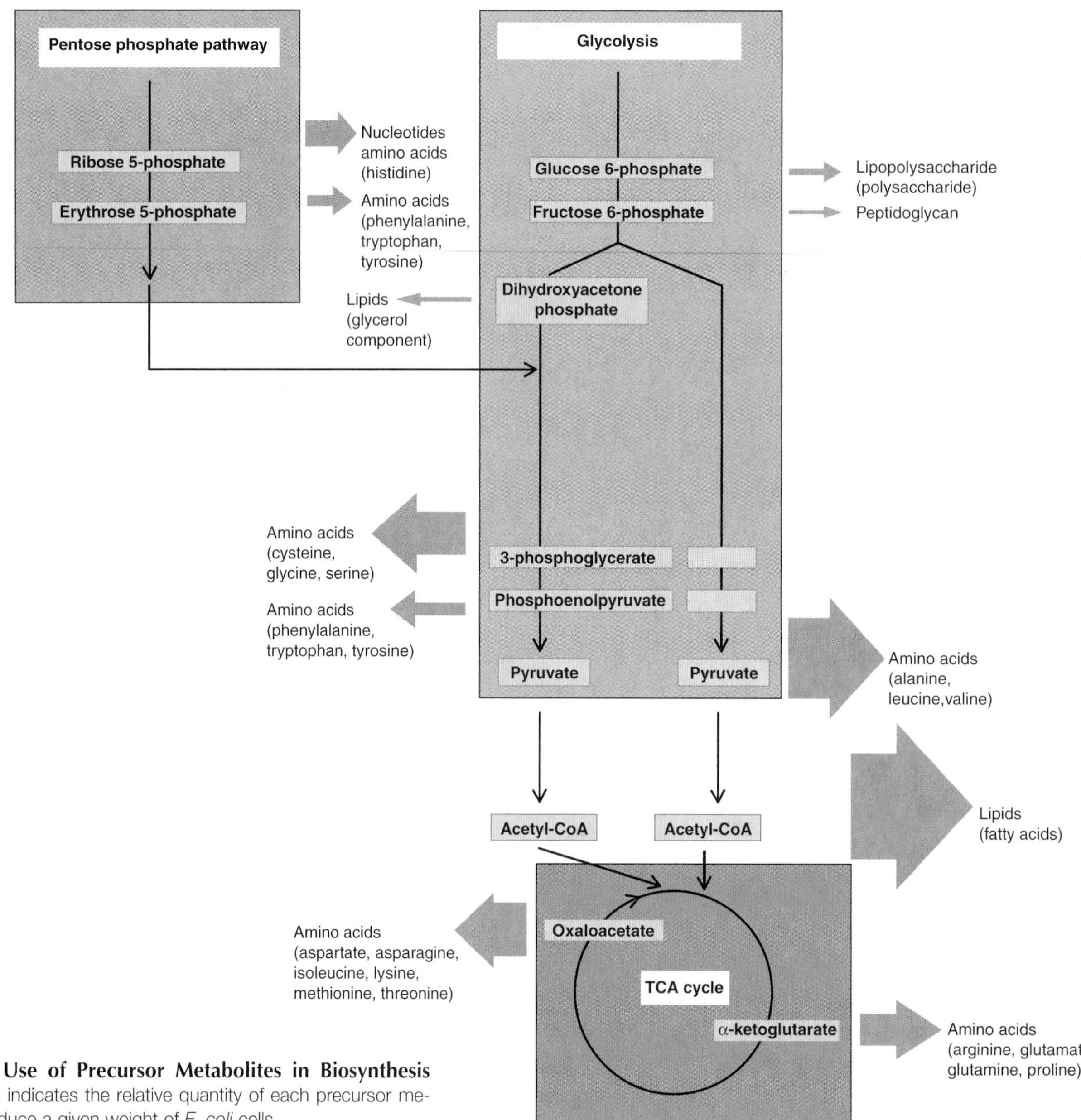

FIGURE 6.27 The Use of Precursor Metabolites in Biosynthesis The size of the arrows indicates the relative quantity of each precursor metabolite needed to produce a given weight of *E. coli* cells.

The glycerol component of the fat is synthesized from the precursor metabolite dihydroxyacetone phosphate, which is generated in glycolysis.

Amino Acid Synthesis

Proteins are composed of various combinations of 20 different amino acids. Amino acids can be grouped into structurally related families that share common pathways of biosynthesis. Some are synthesized from precursor metabolites formed during glycolysis, while others are derived from compounds of the TCA cycle (see table 6.2).

Glutamate

Amino acids are necessary for protein synthesis, but glutamate is especially important because its synthesis provides a mechanism for bacteria to incorporate nitrogen into an organic material. Recall from chapter 4 that many bacteria utilize ammonium (NH_4^+) provided in the medium as their source of nitrogen; it is primarily through the synthesis of glutamate that they do this.

Glutamate is synthesized in a single-step reaction that adds ammonia to the precursor metabolite α-ketoglutarate, produced in the TCA cycle **(figure 6.28a).** Once glutamate has been formed, its amino group can be transferred to other carbon compounds to produce amino acids such as aspartate (figure 6.28b). This transfer of the amino group, a **transamination,** regenerates α-ketoglutarate from glutamate. The α-ketoglutarate can then be used again to incorporate more ammonia. ■ **amino group, p. 26**

Aromatic Amino Acids

Synthesis of aromatic amino acids such as tyrosine, phenylalanine, and tryptophan requires a multistep, branching pathway **(figure 6.29).**

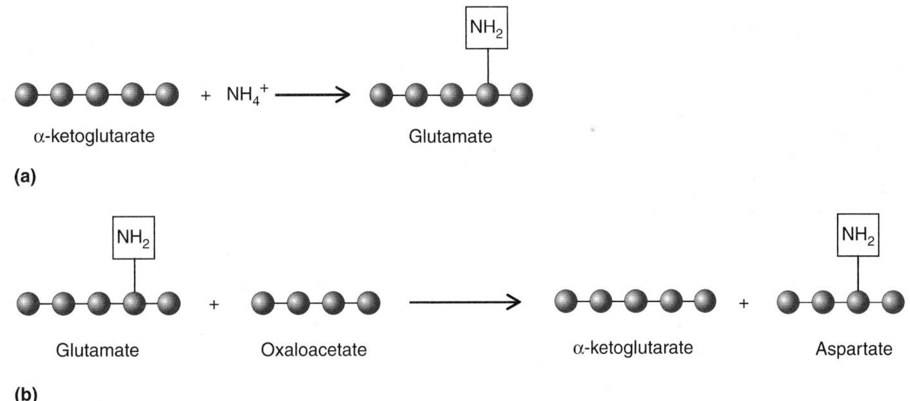

FIGURE 6.28 Glutamate (a) Glutamate is synthesized in a single-step reaction that adds ammonia to the precursor metabolite α-ketoglutarate. **(b)** The amino group of glutamate can be transferred to other carbon compounds in order to produce other amino acids. For example, transferring it to oxaloacetate produces aspartate.

This serves as an excellent illustration of many important features of the regulation of amino acid synthesis.

The pathway begins with the formation of a 7-carbon compound, resulting from the joining of two precursor metabolites, erythrose 4-phosphate (4-carbon) and phosphoenolpyruvate (3-carbon). These precursors originate in the pentose phosphate pathway and glycolysis, respectively. The 7-carbon compound is modified through a series of steps until a branch point is reached. At this juncture, two options are possible. If synthesis proceeds in one direction, tryptophan is produced. In the other direction, another branch point is reached; from there, either tyrosine or phenylalanine can be made.

When a given amino acid is provided to a cell, it would be a waste of carbon, energy, and reducing power for that cell to continue synthesizing it. But when only one product of a branched pathway is present, how does the cell control synthesis? In the pathway for aromatic acid biosynthesis, this partly occurs by regulating the enzymes at the branch points. Tryptophan acts as a feedback inhibitor of the enzyme that directs the branch to its synthesis; this sends the pathway to the steps leading to the synthesis of the other amino acids—tyrosine and phenylalanine. Likewise, these two amino acids each inhibit the first enzyme of the branch leading to their synthesis.

In addition, the three amino acids each control the first step of the full pathway, the formation of the 7-carbon compound. Three different enzymes can catalyze this step; each has the same active site, but they have different allosteric sites. Each aromatic amino acid acts as a feedback inhibitor for one of the enzymes. If all three amino acids are present in the environment, then very little of the 7-carbon compound will be synthesized. If only one or two of those amino acids are present, then proportionally more of the compound will be synthesized. ■ **allosteric enzymes, p. 137**

Nucleotide Synthesis

Nucleotide subunits of DNA and RNA are composed of three units: a 5-carbon sugar, a phosphate group, and a nitrogenous base, either a purine or a pyrimidine. They are synthesized as ribonucleotides, but these can then be converted to deoxyribonucleotides by replacing the hydroxyl group on the 2′ carbon of the sugar with a hydrogen atom. ■ **nucleotides, p. 32** ■ **purine, p. 33** ■ **pyrimidine, p. 33**

The purine nucleotides are synthesized in a distinctly different manner from the pyrimidine nucleotides. Purine nucleotides are synthesized on the sugar phosphate component in a very complex process. In fact, nearly every carbon and nitrogen of

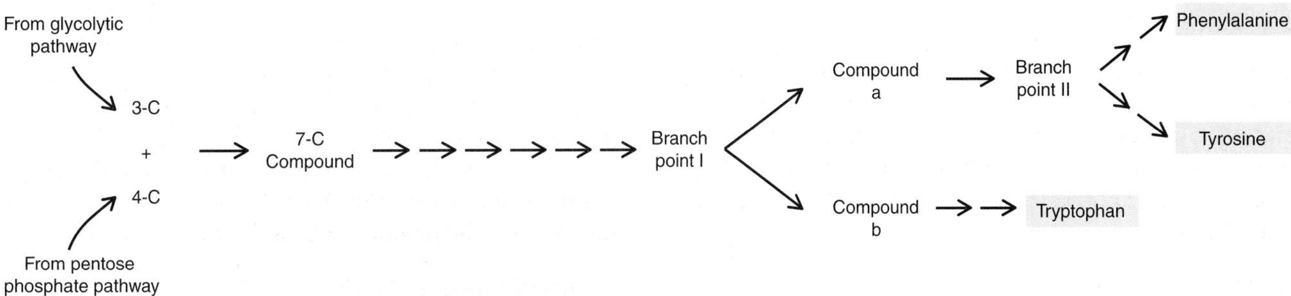

FIGURE 6.29 Synthesis of Aromatic Amino Acids A multistep branching pathway is used to synthesize aromatic amino acids. The end product of a branch inhibits the first enzyme of that branch; in addition, the end product inhibits one of the three enzymes that catalyze the first step of the pathway.

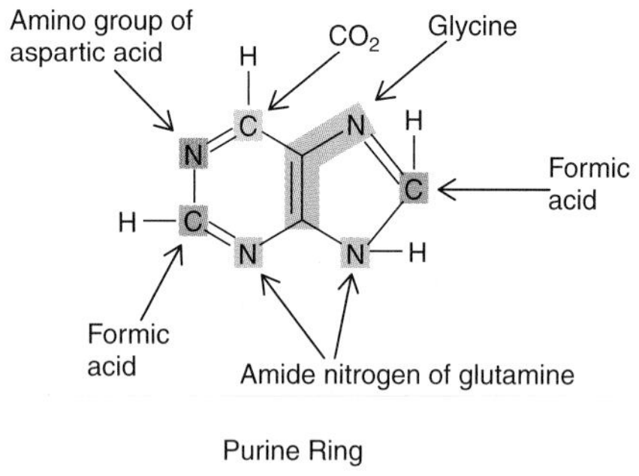

Purine Ring

FIGURE 6.30 Source of the Carbons and Nitrogen Atoms in Purine Rings Nearly every carbon and nitrogen of the purine ring comes from a different source.

the purine ring comes from a different source **(figure 6.30).** The starting compound is ribose 5-phosphate, a precursor metabolite generated in the pentose phosphate pathway. Then, in a highly ordered sequence, atoms from the other sources are added. Once this purine is formed, it is converted to adenylic or guanylic acid, which are components of the nucleic acids. To synthesize pyrimidine nucleotides, the pyrimidine ring is made first, and then attached to ribose 5-phosphate. After one pyrimidine nucleotide is formed, the base component can be converted into one of the other pyrimidines. ■ **adenylic acid, p. 33** ■ **guanylic acid, p. 33**

MICROCHECK 6.10

Biosynthetic processes of different organisms are remarkably similar, using precursor metabolites, NADPH, and ATP to form subunits. Synthesis of the amino acid glutamate provides a mechanism for bacteria to incorporate nitrogen in the form of ammonia into organic material. Allosteric enzymes are used to regulate certain biosynthetic pathways. The purine nucleotides are synthesized in a very different manner from the pyrimidine nucleotides.

- ✓ Explain why the synthesis of glutamate is particularly important for a cell.
- ✓ What are three general products of the central metabolic pathways that a cell requires in order to carry out biosynthesis?
- ✓ With a branched biochemical pathway, why would it be important for a cell to shut down the first step as well as branching steps?

FUTURE CHALLENGES

Going to Extremes

The remarkable speed and precision of enzyme activity are already exploited in a number of different processes. For example, the enzyme glucose isomerase is used to modify corn syrup, converting some of the glucose into fructose, which is much sweeter than glucose. The resulting high-fructose corn syrup is used in the commercial production of a variety of beverages and food products. Other enzymes, including proteases, amylases, and lipases, are used in certain laundry detergents to facilitate stain removal. These enzymes break down proteins, starches, and fats, respectively, which otherwise adhere strongly to fabrics. Similar enzymes are being added to some dishwashing detergents, decreasing the reliance on chlorine bleaching agents and phosphates that can otherwise pollute the environment. Enzymes are also used by the pulp and paper industry to facilitate the bleaching process.

Even with the current successes of enzyme technology, however, only a small fraction of enzymes in nature have been characterized. Recognizing that the field is still in its infancy, some companies are actively searching diverse environments for microorganisms that produce novel enzymes, hoping that some may be commercially valuable. Among the most promising are those produced by the extremophiles, microbes that preferentially live in conditions inhospitable to other forms of life. Because these organisms live in severe environments, their enzymes likely withstand the harsh conditions that characterize certain processes. For example, the enzymes of the extreme thermophiles should withstand temperatures that would quickly inactivate enzymes of mesophiles.

With the aid of enzymes, many of which are still to be discovered, scientists may eventually be able to precisely control a greater variety of commercially important chemical processes. This will hopefully result in fewer unwanted by-products and a decreased reliance on chemicals that damage the environment.

SUMMARY

6.1 Principles of Metabolism

Catabolism encompasses processes that capture and store energy by breaking down complex molecules. **Anabolism** includes processes that use energy to synthesize and assemble the building blocks of a cell (figure 6.1).

Harvesting Energy

Photosynthetic organisms harvest the energy of sunlight, using it to power the synthesis of organic compounds. **Chemoorganotrophs** harvest energy contained in organic compounds (figure 6.3). **Exergonic** reactions release energy; **endergonic** reactions utilize energy.

Components of Metabolic Pathways

A specific **enzyme** facilitates each step of a metabolic pathway (figure 6.5). **ATP** is the energy currency of the cell. The **energy source** is **oxidized** to release its energy; the **redox** reactions **reduce** an electron carrier (figure 6.7). **NAD$^+$, NADP$^+$,** and **FAD** are electron carriers (table 6.1).

Precursor Metabolites

Precursor metabolites are used to make the subunits of macromolecules, and they can also be oxidized to generate energy in the form of ATP (table 6.2).

Overview of Metabolism (figure 6.8)

The **central metabolic pathways** are **glycolysis,** the **pentose phosphate pathway,** and the **tricarboxylic acid cycle (TCA cycle). Respiration** uses the reducing power accumulated in the central metabolic pathways to generate ATP by oxidative phosphorylation. **Aerobic respiration** uses O_2 as a terminal electron acceptor; **anaerobic respiration** uses a molecule other than O_2 as a terminal electron acceptor (table 6.3). **Fermentation** uses pyruvate or a derivative as a terminal electron acceptor; this recycles the reduced electron carrier NADH.

6.2 Enzymes

Enzymes function as biological catalysts; they are neither consumed nor permanently changed during a reaction.

Mechanisms and Consequences of Enzyme Action (figure 6.9)

The substrate binds to the **active site** or **catalytic site** to form a temporary intermediate called an **enzyme-substrate complex.**

Cofactors and Coenzymes (figure 6.10, table 6.4)

Enzymes sometimes act in conjunction with **cofactors** such as **coenzymes** and trace elements.

Environmental Factors That Influence Enzyme Activity (figure 6.11)

The factors most important in influencing enzyme activities are temperature, pH, and salt concentration.

Allosteric Regulation (figure 6.12)

Cells can fine-tune the activity of an **allosteric enzyme** by using a regulatory molecule that binds to the **allosteric site** of the enzyme.

Enzyme Inhibition

Non-competitive inhibition occurs when the inhibitor and the substrate act at different sites on the enzyme. **Competitive inhibition** occurs when the inhibitor competes with the normal substrate for the active binding site (figure 6.13).

6.3 The Central Metabolic Pathways (table 6.6)

Glycolysis (figure 6.14)

Glycolysis converts one molecule of glucose into two molecules of pyruvate; the theoretical net yield is 2 ATP, 2 NADH + H^+, and six different precursor metabolites.

Pentose Phosphate Pathway

The pentose phosphate pathway forms $NADPH^+$ and two different precursor metabolites.

Transition Step (figure 6.15)

The transition step converts pyruvate to acetyl-CoA. Repeated twice, this produces 2 NADH + 2 H^+ and 1 precursor metabolite.

Tricarboxylic Acid (TCA) Cycle (figure 6.15)

The TCA cycle completes the oxidation of glucose; the theoretical yield of two "turns" is 6 NADH + 6 H^+, 2 $FADH_2$, 2 ATP, and two different precursor metabolites.

6.4 Respiration

The Electron Transport Chain—Generating Proton Motive Force

The electron transport chain sequentially passes electrons, and, as a result, ejects protons. The mitochondrial electron transport chain has three different complexes (complexes I, III and IV) that function as proton pumps (figure 6.17). Prokaryotes vary with respect to the types and arrangements of their electron transport components (figure 6.18). Some prokaryotes can use molecules other than O_2 as terminal electron acceptors. This process of anaerobic respiration harvests less energy than aerobic respiration.

ATP Synthase—Harvesting the Proton Motive Force to Synthesize ATP

ATP synthase permits protons to flow back across the membrane, harvesting the energy released to fuel the synthesis of ATP.

ATP Yield of Aerobic Respiration in Prokaryotes (figure 6.19)

The theoretical maximum yield of ATP of aerobic respiration is 38 ATP.

6.5 Fermentation

In general, the only ATP-yielding reactions of fermentations are those of the glycolytic pathway; the other steps provide a mechanism for recycling NADH (figure 6.20). Some end products of fermentation are commercially valuable (figure 6.21). Because a given type of organism uses a specific fermentation pathway, the end products can be used as markers that aid in identification.

6.6 Catabolism of Organic Compounds Other Than Glucose (figure 6.22)

Hydrolytic enzymes break down macromolecules into their respective subunits.

Polysaccharides and Disaccharides

Amylases digest starch, releasing glucose subunits, and are produced by many organisms. **Cellulases** degrade cellulose. Sugar subunits released when polysaccharides are broken down can then enter glycolysis to be oxidized to pyruvate.

Lipids

Fats are hydrolyzed by **lipase,** releasing glycerol and fatty acids. Glycerol is converted to dihydroxyacetone phosphate; fatty acids are degraded by **β-oxidation,** generating reducing power and the precursor metabolite acetyl-CoA.

Proteins

Proteins are hydrolyzed by **proteases. Deamination** removes the amino group; the remaining carbon skeleton is then converted into the appropriate precursor molecule.

6.7 Chemolithotrophs

Prokaryotes, as a group, are unique in their ability to use reduced inorganic compounds such as hydrogen sulfide (H_2S) and ammonia (NH_3) as a source of energy (figure 6.23). Chemolithotrophs are autotrophs.

6.8 Photosynthesis

The **light-dependent reactions** capture energy from light and convert it to chemical energy in the form of ATP. The **light-independent reactions** use that energy to synthesize organic carbon compounds.

Capturing Radiant Energy

Various pigments such as **chlorophylls, bacteriochlorophylls, carotenoids,** and **phycobilins** are used to capture radiant energy. **Reaction center pigments** function as the electron donor in the photosynthetic process; **antennae pigments** funnel radiant energy to the reaction center pigment.

Converting Radiant Energy into Chemical Energy

The high-energy electrons emitted by reaction center chlorophylls are passed to an electron transport chain, which uses them to generate a proton motive force. The energy of proton motive force is harvested by ATP synthase to fuel the synthesis of ATP (figure 6.24). Photosystems I and II of cyanobacteria and chloroplasts raise the energy level of electrons stripped from water to a high enough level to be used to generate a proton motive force and produce reducing power; this process evolves oxygen (figure 6.25). Purple and green bacteria employ only a single photosystem; they must obtain electrons from a reduced compound other than water and therefore do not evolve oxygen.

6.9 Carbon Fixation

Calvin Cycle (figure 6.26)

The most common pathway used to incorporate CO_2 into an organic form is the **Calvin cycle.**

6.10 Anabolic Pathways—Synthesizing Subunits from Precursor Molecules (figure 6.27)

Lipid Synthesis

The fatty acid components of fat are synthesized by progressively adding 2-carbon units to an acetyl group. The glycerol component is synthesized from dihydroxyacetone phosphate.

Amino Acid Synthesis

Synthesis of glutamate from α-ketoglutarate and ammonia provides a mechanism for cells to incorporate nitrogen into organic molecules (figure 6.28). Synthesis of aromatic amino acids requires a multistep branching pathway. Allosteric enzymes regulate key steps of the pathway (figure 6.29).

Nucleotide Synthesis

Purine nucleotides are synthesized on the sugar-phosphate component; the pyrimidine ring is made first and then attached to the sugar-phosphate (figure 6.30).

REVIEW QUESTIONS

Short Answer

1. Explain the difference between catabolism and anabolism.
2. How does ATP serve as a carrier of free energy?
3. How do enzymes catalyze chemical reactions?
4. Explain how precursor molecules serve as junctions between catabolic and anabolic pathways.
5. How do cells regulate enzyme activity?
6. Why do the electrons carried by $FADH_2$ result in less ATP production than those carried by NADH?
7. Name three food products produced with the aid of microorganisms.
8. In photosynthesis, what is encompassed by the term "light-independent reactions?"
9. Unlike the cyanobacteria, the anoxygenic photosynthetic bacteria do not evolve oxygen (O_2). Why not?
10. What is the role of transamination in amino acid biosynthesis?

Multiple Choice

1. Which of these factors does not affect enzyme activity?
   a) temperature b) inhibitors c) coenzymes
   d) humidity e) pH
2. Which of the following statements is false? Enzymes
   a) bind to substrates.
   b) lower the energy of activation.
   c) convert coenzymes to products.
   d) speed up biochemical reactions.
   e) can be named after the kinds of reaction they catalyze.
3. Which of these is not a coenzyme?
   a) FAD b) coenzyme A c) NAD^+ d) ATP e) $NADP^+$
4. What is the end product of glycolysis?
   a) glucose b) citrate c) oxaloacetate
   d) α-ketoglutarate e) pyruvate
5. The major pathway(s) of central metabolism are
   a) glycolysis and the TCA cycle only.
   b) glycolysis, the TCA cycle, and the pentose phosphate pathway.
   c) glycolysis only.
   d) glycolysis and the pentose phosphate pathway only.
   e) the TCA cycle only.
6. Which of these pathways gives a cell the potential to produce the most ATP?
   a) TCA cycle
   b) pentose phosphate pathway
   c) lactic acid fermentation
   d) glycolysis
7. In fermentation, the terminal electron acceptor is
   a) oxygen (O_2). b) hydrogen (H_2).
   c) carbon dioxide (CO_2). d) an organic compound.
8. In the process of oxidative phosphorylation, the energy of proton motive force is used to generate
   a) NADH. b) ADP. c) ethanol. d) ATP. e) glucose.
9. In the TCA cycle, the carbon atoms contained in acetate are converted into
   a) lactic acid. b) glucose. c) glycerol.
   d) CO_2. e) all of these.
10. Degradation of fats as an energy source involves all of the following, *except*
    a) β-oxidation. b) acetyl-CoA. c) glycerol.
    d) lipase. e) transamination.

Applications

1. A worker in a cheese-making facility argues that whey, a nutrient-rich by-product of cheese, should be dumped in a nearby pond where it could serve as fish food. Explain why this proposed action could actually kill the fish by depleting the oxygen in the pond.
2. Scientists working with DNA *in vitro* often store it in solutions that contain EDTA, a chelating agent that binds magnesium (Mg^{2+}). This is done to prevent enzymes called DNases from degrading the DNA. Explain why EDTA would interfere with enzyme activity.

Critical Thinking

1. A student argued that aerobic and anaerobic respiration should produce the same amount of ATP. He reasoned that they both use basically the same process; only the terminal electron acceptor is different. What is the primary error in this student's argument?
2. Chemolithotrophs near hydrothermal vents support a variety of other life forms there. Explain how their role there is analogous to that of photosynthetic organisms in terrestrial environments.

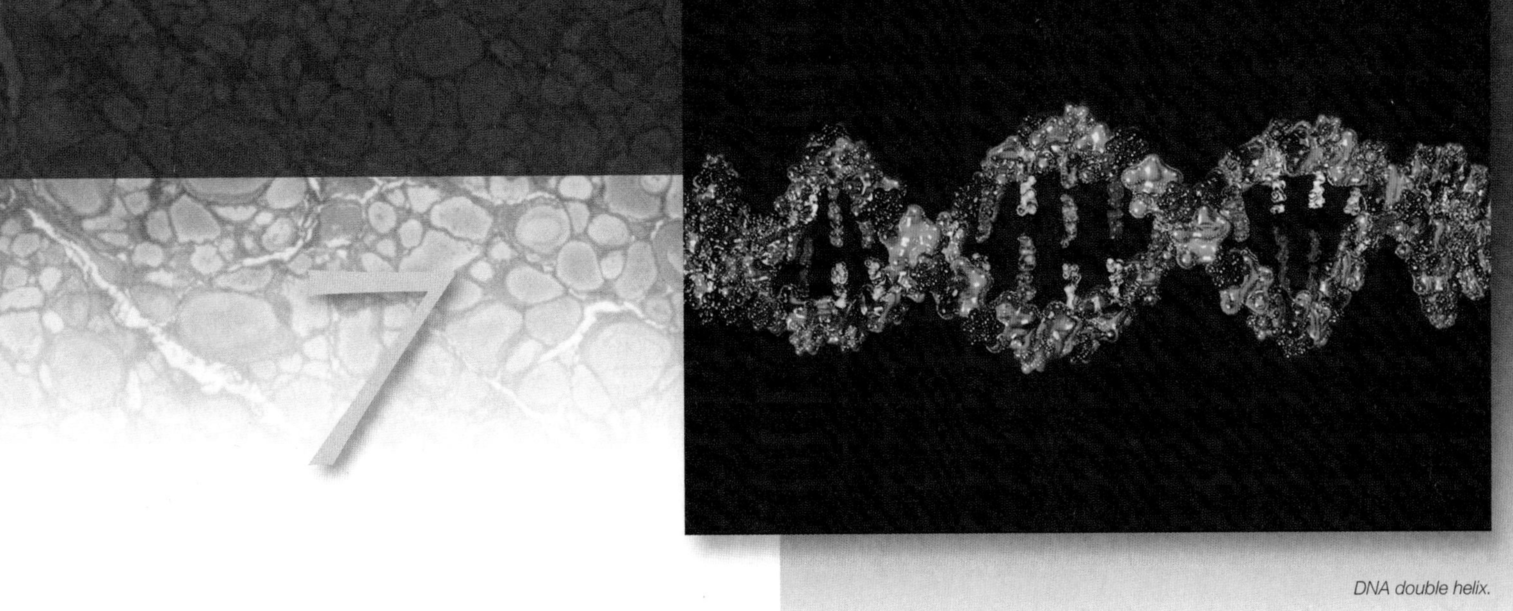

DNA double helix.

The Blueprint of Life, from DNA to Protein

A Glimpse of History

In 1866, the Czech monk Gregor Mendel showed that traits are inherited by means of physical units, which we now call genes. It was not until 1941, however, that the precise function of genes was revealed when George Beadle, a geneticist, and Edward Tatum, a chemist, published a scientific paper reporting that genes determine the structure of enzymes. Biochemists had already shown that enzymes catalyze the conversion of one compound into another in a biochemical pathway.

Beadle and Tatum studied *Neurospora crassa,* a common bread mold that grows on a very simple medium containing sugar and simple inorganic salts. Beadle and Tatum created *N. crassa* strains with altered properties, mutants, by treating cells with X rays, which were known to alter genes. Some of these mutants could no longer grow on the glucose-salts medium unless growth factors such as vitamins were added to the medium. To isolate these, Beadle and Tatum had to laboriously screen thousands of progeny to find the relatively few that required the growth factors. Each mutant presumably contained a defective gene.

The next task for Beadle and Tatum was to identify the specific biochemical defect of each mutant. To do this, they added different growth factors, one at a time, to each mutant culture. The one that allowed a particular mutant to grow had presumably bypassed the function of a defective enzyme. In this manner, they were able to pinpoint in each mutant the specific step in the biochemical pathway that was defective. Then, using these same mutants, Beadle and Tatum showed that the requirement for each growth factor was inherited as a single gene, ultimately leading to their conclusion that a single gene determines the production of one enzyme. Their conclusion has been modified somewhat, because we now know that some enzymes are made up of more than one protein. A single gene usually determines the production of one protein. In 1958, Beadle and Tatum shared the Nobel Prize in Medicine, largely for these pioneering studies that ushered in the era of modern biology.

Consider for a moment the vast diversity of cellular life forms that exist. Our world contains a remarkable variety of microorganisms and specialized cells that make up plants and animals. Every characteristic of each of these cells, from its shape to its function, is dictated by information contained in its deoxyribonucleic acid (DNA). DNA encodes the master plan, the blueprint, for all cell structures and processes. Yet for all the complexity this would seem to require, DNA is a string composed of only four different nucleotides, each containing a particular nitrogenous base: adenine (A), thymine (T), cytosine (C), or guanine (G). ■ **nucleotides, p. 32**

While it might seem improbable that the vast array of life forms can be encoded by a molecule consisting of only four different units, think about how much information can be transmitted by binary code, the language of all computers, which has a base of only two. A simple series of ones and zeros can code for each letter of the alphabet. String enough of these together in the right sequence and the letters become words, and the words can become complete sentences, chapters, books, or even whole libraries.

The four nucleotides of a DNA molecule convey information in a similar fashion. A set of three nucleotides encodes a specific amino acid. In turn, a string of amino acids makes up a protein, the function of which is dictated by the order of the amino acid subunits. Some proteins serve as structural components of a cell. Others, such as enzymes, mediate cellular activities including biosynthesis and energy conversion. Together, proteins synthesized by a cell are responsible for every aspect of that cell. Thus, the sequential order of nucleotide bases in a cell's DNA ultimately dictates the characteristics of that cell. ■ **amino acids, p. 26** ■ **protein structure, p. 28** ■ **enzymes, pp. 129, 134**

This chapter will focus on the processes used to replicate DNA and convert the information encoded within it into proteins, concentrating primarily on the mechanisms used by bacterial cells. The eukaryotic processes have many similarities, but are considerably

KEY TERMS

Codon A series of three nucleotides that code for a specific amino acid.

DNA Polymerase Enzyme that synthesizes DNA, using an existing strand as a template to synthesize the complementary strand.

DNA Replication Duplication of a DNA molecule.

Gene The functional unit of the genome; it encodes a product, most often a protein.

Genome Complete set of genetic information in a cell or a virus.

Messenger RNA (mRNA) Type of RNA molecule that is translated during protein synthesis.

Primer Fragment of nucleic acid to which DNA polymerase can add nucleotides.

Promoter Nucleotide sequence to which RNA polymerase binds to initiate transcription.

Ribosomal RNA (rRNA) Type of RNA molecule present in ribosomes.

Ribosome Structure that facilitates the joining of amino acids during the process of translation; it is composed of ribosomal RNA (rRNA) and protein.

RNA Polymerase Enzyme that synthesizes RNA using one strand of DNA as a template.

Transcription The process that copies the information encoded by DNA into RNA.

Transfer RNA (tRNA) Type of RNA molecule that acts as a key, interpreting the genetic code; each tRNA molecule carries a specific amino acid.

Translation The process that interprets the information carried by mRNA to synthesize the encoded protein.

more complicated and will only be discussed briefly. The processes in archaea are often similar to those of bacteria, but sometimes resemble those of eukaryotic cells.

7.1 Overview

Focus Points

- Compare and contrast the characteristics of DNA and RNA.
- Explain why it is important that a cell be able to regulate the expression of certain genes.

The complete set of genetic information for a cell is referred to as its **genome.** Technically, this includes plasmids as well as the chromosome; however, the term "genome" is often used interchangeably with chromosome. The genome of all cells is composed of DNA, but some viruses have an RNA genome. The functional unit of the genome is a **gene.** A gene encodes a product, the **gene product,** most commonly a protein. The study of the function and transfer of genes is called **genetics,** whereas the study and analysis of the nucleotide sequence of DNA is called **genomics.** ■ chromosome, p. 67 ■ plasmid, p. 68

All living cells must accomplish two general tasks in order to multiply. The double-stranded DNA must be duplicated before cell division so that its encoded information can be passed on to the next generation. This is the process of **DNA replication.** In addition, the information encoded by the DNA must be deciphered, or **expressed,** so that the cell can synthesize the necessary gene products at the appropriate time. Gene expression involves two interrelated processes—transcription and translation. **Transcription** copies the information encoded in DNA into a slightly different molecule, RNA. The RNA serves as a transitional, temporary form of the genetic information and is the one actually deciphered. **Translation** interprets information carried by RNA to synthesize the encoded protein.

The flow of information from DNA to RNA to protein is often referred to as the **central dogma of molecular biology (figure 7.1).** It was once believed that information flow proceeded only in this direction. Although this direction is by far the most

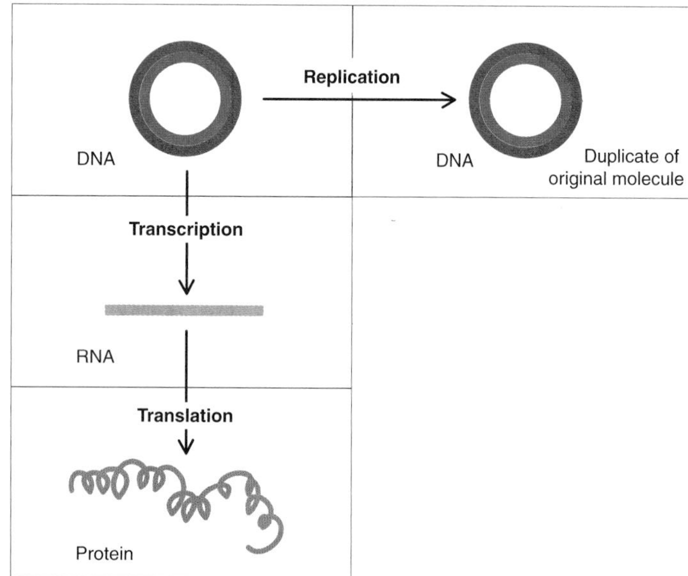

FIGURE 7.1 Overview of Replication, Transcription, and Translation DNA replication is the process that duplicates DNA so that its encoded information can be passed on to future generations. Transcription is the process that copies the genetic information into a transitional form, RNA. Translation is the process that deciphers the encoded information to synthesize a specific protein.

common, certain viruses, **retroviruses,** have an RNA genome but copy that information into the form of DNA. HIV is a retrovirus.

Characteristics of DNA

A single strand of DNA is composed of a series of deoxyribonucleotide subunits, more commonly called nucleotides. These are joined in a chain by a covalent bond between the $5'PO_4$ (5 prime phosphate) of one nucleotide and the 3′OH (3 prime hydroxyl) of the next. Note that the designations 5′ and 3′ refer to the numbered carbon atoms of the pentose sugar of the nucleotide (see figure 2.22). Joining of the nucleotides in this manner creates a series of alternating sugar and phosphate units, called the **sugar-phosphate backbone.** Connected to each sugar is one of the nitrogenous bases—an adenine (A), thymine (T), guanine (G), or cytosine (C). Because of the chemical structure of the nucleotides and how

A highly coiled line is used to depict genomic DNA.

Red and blue lines placed in a helical arrangement depict the two complementary strands and highlight the three-dimensional structure of DNA.

A circular arrangement of the red and blue lines is used as the simplified form of prokaryotic DNA.

Red and blue lines separated by a thin black line are used as a simple representation of the double-stranded DNA molecule.

Two parallel lines are used to emphasize the base-pairing interactions and nucleotide sequence characteristics of the two complementary strands. The "tracks" between the lines are not intended to depict a specific number of base pairs, only the general interaction between complementary strands.

Either a red or a blue line can be depicted as the "top" strand, since DNA is a three-dimensional structure.

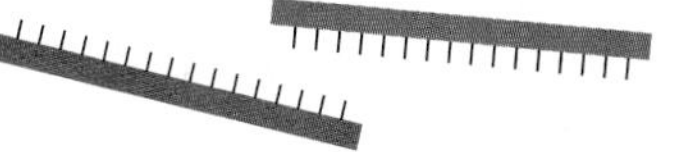

Denatured DNA is depicted as separate red and blue lines to emphasize its single-stranded nature.

FIGURE 7.2 Diagrammatic Representations of the Structure of DNA Although DNA is a double-stranded helical structure, explanatory diagrams may depict it in a number of different ways. In chapter 7 we will use these representations.

they are joined, a single strand of DNA will always have a $5'PO_4$ at one end and a 3′OH at the other. These ends, often referred to as the **5′ end** and the **3′ end,** have important implications in DNA and RNA synthesis that will be discussed later. ■ **deoxyribonucleic acid (DNA), p. 32,** ■ **nucleotides, p. 32**

DNA in a cell usually occurs as a double-stranded, helical structure **(figure 7.2).** The two strands are held together by weak hydrogen bonds between the nitrogenous bases of the opposing strands. While individual hydrogen bonds are readily broken, the duplex structure of double-stranded DNA is generally quite stable because of the sheer number of bonds that occurs along its length. Short fragments of DNA have correspondingly fewer hydrogen bonds, so they are readily separated into single-stranded pieces. Separating the two strands is called **melting,** or **denaturing.** ■ **hydrogen bonds, p. 22**

The two strands of double-stranded DNA are complementary **(figure 7.3).** Wherever an adenine is in one strand, a

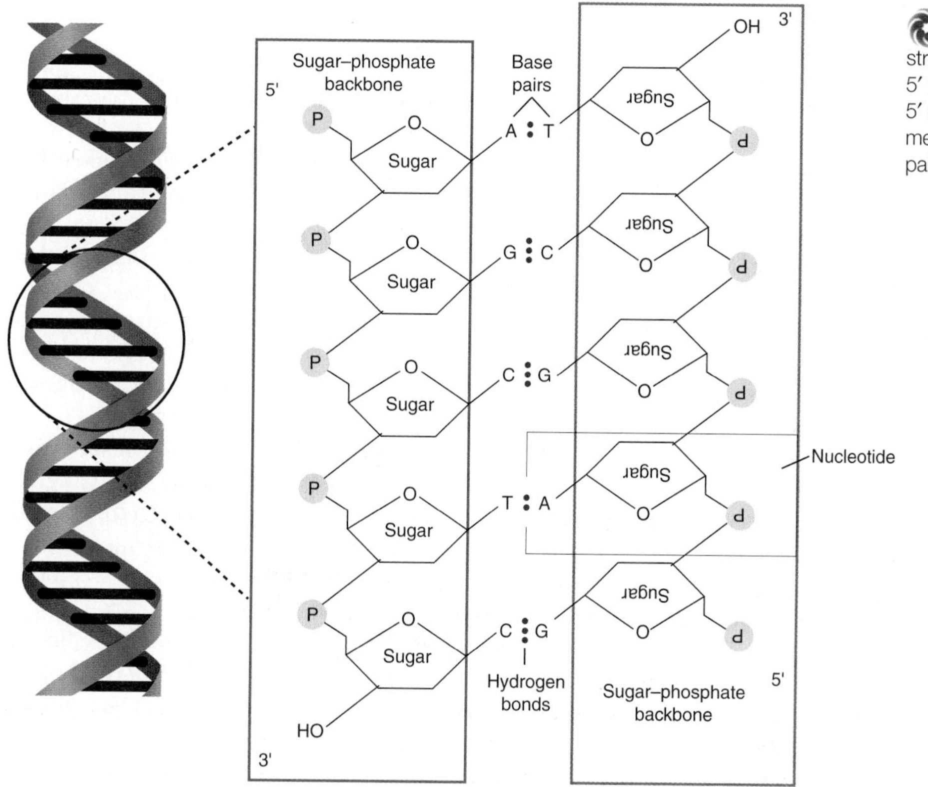

FIGURE 7.3 The Double Helix of DNA The two strands of DNA are antiparallel; one strand is oriented in the 5′ to 3′ direction, and its complement is oriented in the 3′ to 5′ direction. Hydrogen bonding occurs between the complementary base pairs; three bonds form between a G—C base pair, and two bonds form between an A—T base pair.

thymine is in the other; these opposing bases are held together by two hydrogen bonds between them. Similarly, wherever a cytosine is in one strand, a guanine is in the other. These are held together by three hydrogen bonds, a slightly stronger attraction than that of an A-T pair. The characteristic bonding of A to T and G to C is called **base-pairing** and is fundamental to the remarkable functionality of DNA. Because of the rules of base-pairing, one strand can always be used as a **template** for the synthesis of the complementary opposing strand. ■ complementary, p. 34

While the two strands of DNA in the double helix are complementary, they are also **antiparallel.** That is, they are oriented in opposite directions. One strand is oriented in the 5′ to 3′ direction and its complement is oriented in the 3′ to 5′ direction. This also has important implications in the function and synthesis of nucleic acids.

Characteristics of RNA

RNA is in many ways comparable to DNA, but with some important exceptions. One difference is that RNA is made up of ribonucleotides rather than deoxynucleotides, although in both cases these are usually referred to simply as nucleotides. Another distinction is that RNA contains the nitrogenous base uracil in place of the thymine found in DNA. Like DNA, RNA consists of a sequence of nucleotides, but RNA usually exists as a single-stranded linear molecule that is much shorter than DNA. ■ ribonucleic acid (RNA), p. 34 ■ nucleotide, p. 32

A fragment of RNA, a **transcript,** is synthesized using a region of one of the two strands of DNA as a template. In making the RNA transcript, the same base-pairing rules of DNA apply except that uracil, rather than thymine, base-pairs with adenine. This base-pairing is only transient, however, and the molecule quickly leaves the DNA template. Numerous different RNA transcripts can be generated from a single chromosome using specific regions as templates. Either strand may serve as the template. In a region the size of a single gene, however, only one of the two strands is generally transcribed. As a result, two complementary strands of RNA are not normally generated.

There are three different functional groups of RNA molecules, each transcribed from different genes. Most genes encode proteins and are transcribed into **messenger RNA (mRNA).** These molecules are translated during protein synthesis. Encrypted information in mRNA is deciphered according to the **genetic code,** which correlates each set of three nucleotides, called a **codon,** to a particular amino acid. Some genes are never translated into proteins; instead the RNAs themselves are the ultimate products. These genes encode either **ribosomal RNA (rRNA)** or **transfer RNA (tRNA),** each of which plays a different but critical role in protein synthesis.

Regulating the Expression of Genes

Although the basic structure of DNA and RNA is relatively simple, the information encoded is extensive and complex. The nucleotide sequence of the genes codes for the amino acid sequence of proteins, but because not all proteins are required by a cell in the same quantity and at all times, mechanisms that determine the extent and duration of their synthesis are needed. In other words, DNA must also code for mechanisms to regulate expression of genes.

One of the key mechanisms a cell uses to control protein synthesis is to regulate the synthesis of mRNA molecules. Unless a gene is transcribed into mRNA, the encoded protein cannot be synthesized. The number of mRNA copies of the gene also influences the level of expression. If transcription of a gene ceases, the level of gene expression rapidly declines. This is because mRNA is generally short-lived, often lasting only a few minutes, due to the activity of enzymes called **RNases** that rapidly degrade it. Eukaryotic cells have mechanisms to modulate the stability of RNA, providing an additional level of control.

MICROCHECK 7.1

Replication is the process of duplicating double-stranded DNA. Transcription is the process of copying the information encoded in DNA into RNA. Translation is the process of interpreting the information carried by messenger RNA in order to synthesize the encoded protein.

✓ How does the 5′ end of DNA differ from the 3′ end?

✓ If the nucleotide sequence of one strand of DNA is 5′ ACGTTGCA 3′, what is the sequence of the complementary strand?

✓ Why is a short-lived RNA important in cell control mechanisms?

7.2 DNA Replication

Focus Point

- Describe the process of replication, focusing on initiation of replication and the events that occur at the replication fork.

DNA is replicated in order to create a second DNA molecule, identical to the original. Each of the two cells generated during binary fission then receives one complete copy. ■ binary fission, p. 84

DNA replication is generally **bidirectional.** From a distinct starting point in circular DNA, replication proceeds in opposite directions, creating an ever-expanding "bubble" of two identical replicated portions of the chromosome **(figure 7.4).** Bidirectional replication allows an entire chromosome to be replicated in half the time it would take if replication were unidirectional.

Replication of double-stranded DNA is **semiconservative.** Each of the two molecules generated contains one of the original strands (the template strand) and one newly synthesized strand. Thus, the two cells produced as a result of division each have one of the original strands of DNA paired with a new complementary strand.

The process of DNA replication requires the coordinated action of many different enzymes and other proteins **(table 7.1).** The most critical of these exist together as a complex that appears

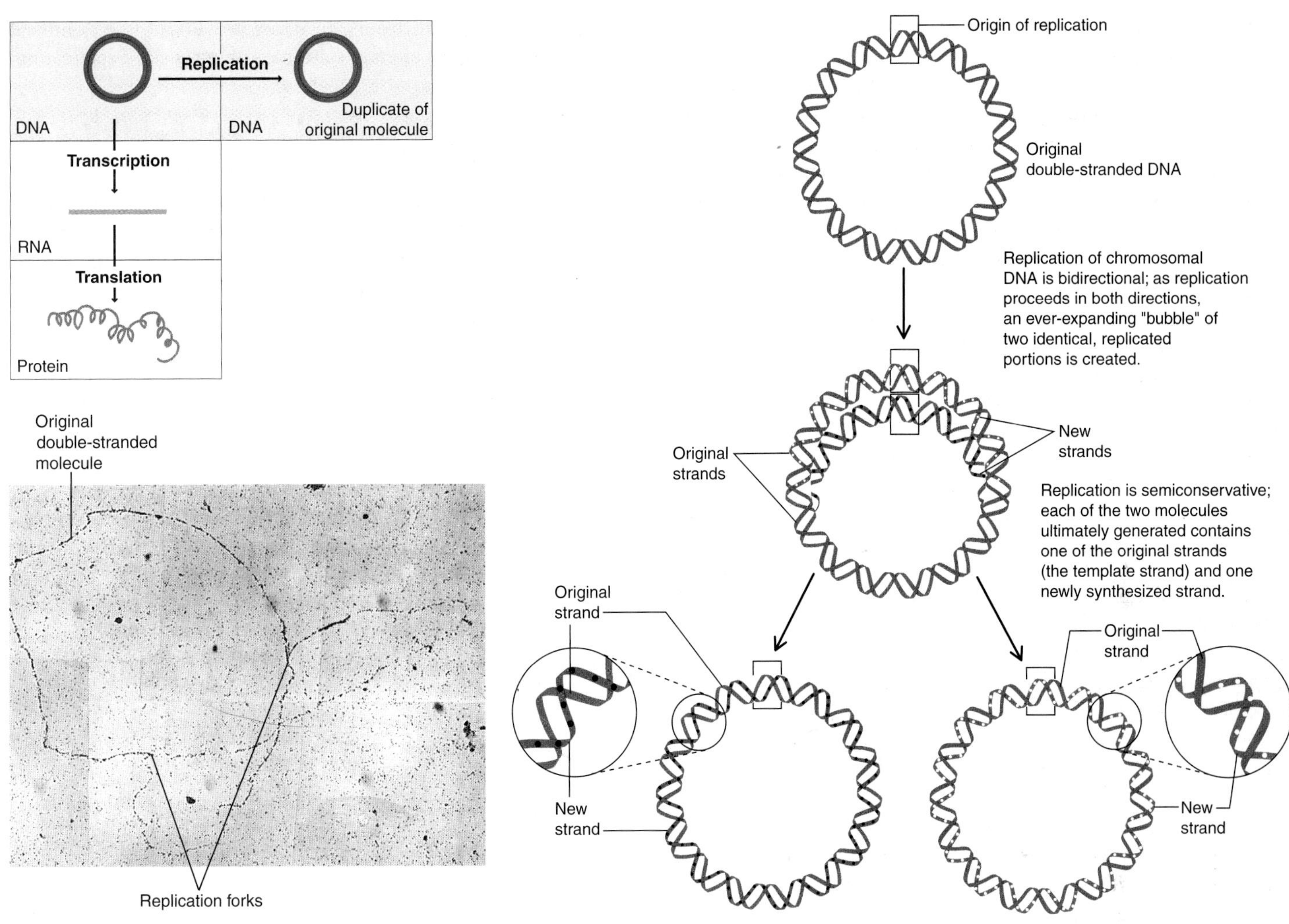

FIGURE 7.4 Replication of Chromosomal DNA of Prokaryotes

TABLE 7.1 Components of DNA Replication in Bacteria

Component	Comments
DNA gyrase	Enzyme that temporarily breaks the strands of DNA, relieving the tension caused by unwinding the two strands of the DNA helix.
DNA ligase	Enzyme that joins two DNA fragments by forming a covalent bond between the sugar-phosphate residues of adjacent nucleotides.
DNA polymerases	Enzymes that synthesize DNA; they use one strand of DNA as a template to generate the complementary strand. Nucleotides can only be added to the 3′ end of an existing fragment, therefore synthesis always occurs in the 5′ to 3′ direction.
Helicases	Enzymes that unwind the DNA helix ahead of the replication fork.
Okazaki fragment	Nucleic acid fragment generated during discontinuous replication of the lagging strand of DNA.
Origin of replication	Distinct region of a DNA molecule at which replication is initiated.
Primase	Enzyme that synthesizes small fragments of RNA to serve as primers for DNA synthesis.
Primer	Fragment of nucleic acid to which DNA polymerase can add nucleotides (the enzyme can only add nucleotides to an existing fragment).

to act as a fixed DNA-synthesizing factory, reeling in the DNA to be replicated. **DNA polymerases** are enzymes that synthesize DNA, using one strand as a template to generate the complementary strand. These enzymes can only add nucleotides onto a preexisting fragment of nucleic acid, either DNA or RNA. Thus, the fragment serves as a **primer** from which synthesis can continue.

DNA is synthesized one nucleotide at a time as a subunit (dATP, dGTP, dCTP, or dTTP) is covalently joined to the nucleotide at the 3′ end of the growing strand. Hydrolysis of a phosphate bond in the incoming molecule provides energy for the reaction. DNA polymerase always adds the nucleotide to the 3′ end of the chain, elongating the strand in the 5′ to 3′ direction **(figure 7.5).** The base-pairing rules determine the specific nucleotides added.

The replication process is very accurate, resulting in only one mistake approximately every billion nucleotides. Part of the reason for this remarkable precision is the proofreading ability of some DNA polymerases. If an incorrect nucleotide is incorporated into the growing chain, the enzyme can edit the mistake by replacing that nucleotide before moving on.

It takes approximately 40 minutes for the chromosome of *E. coli* to be replicated. How, then, can *E. coli* sometimes multiply with a generation time of only 20 minutes? Under favorable growing conditions, a cell initiates replication before the preceding round of replication is completed. In this way, each of the two progeny resulting from cell division will get one complete chromosome that has already started another round of replication.
■ generation time, p. 84

Initiation of DNA Replication

To begin the process of DNA replication, specific proteins must recognize and bind to a distinct region of the DNA, an **origin of replication.** All molecules of DNA, including chromosomes and plasmids, must have this region of approximately 250 nucleotides for replication to be initiated. The binding of the proteins causes localized melting of a specific region within the origin. Using the exposed single strands as templates, a **primase** synthesizes small fragments of RNA to serve as primers for DNA synthesis. The enzymes that synthesize RNA do not require a primer.

The Replication Fork

The bidirectional progression of replication around a circular DNA molecule creates two advancing Y-shaped regions where active replication is occurring. Each of these is called a **replication fork.** The template strands continue to "unzip" at each fork due to the activity of enzymes called **helicases.** Synthesis of one new strand proceeds continuously in the 5′ to

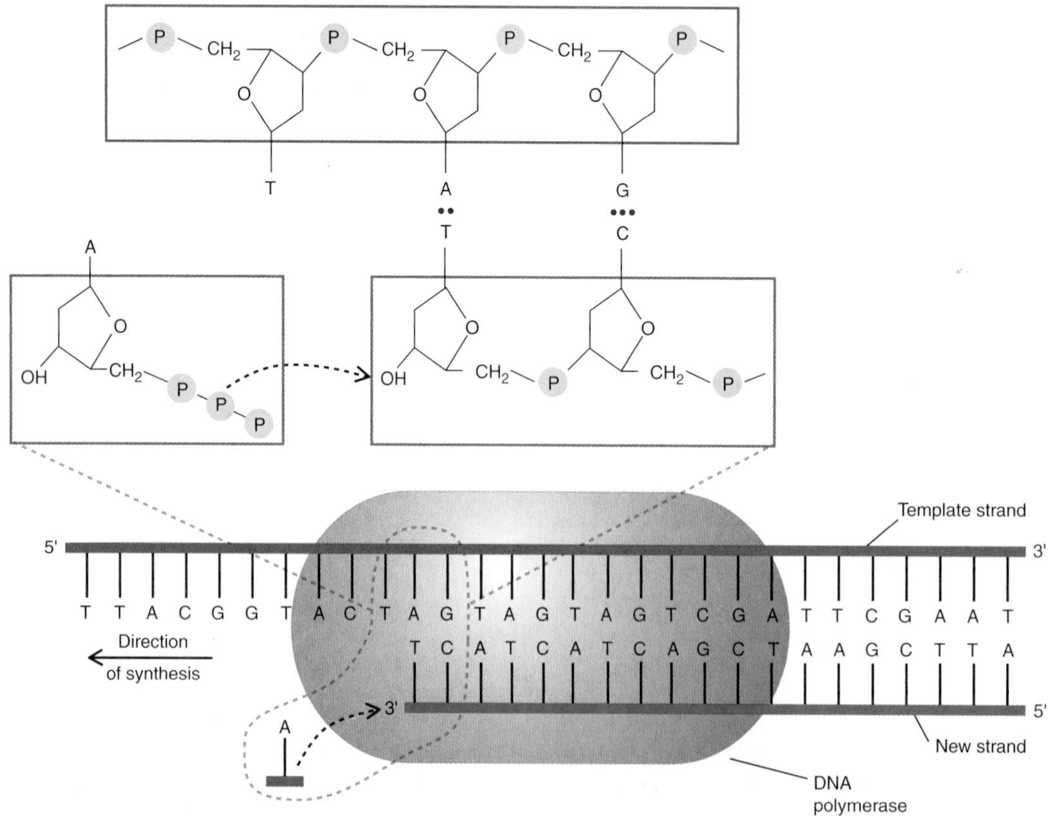

FIGURE 7.5 The Process of DNA Synthesis DNA polymerase synthesizes a new strand by adding one nucleotide at a time to the 3′ end of the elongating strand. The base-pairing rules determine the specific nucleotides that are added.

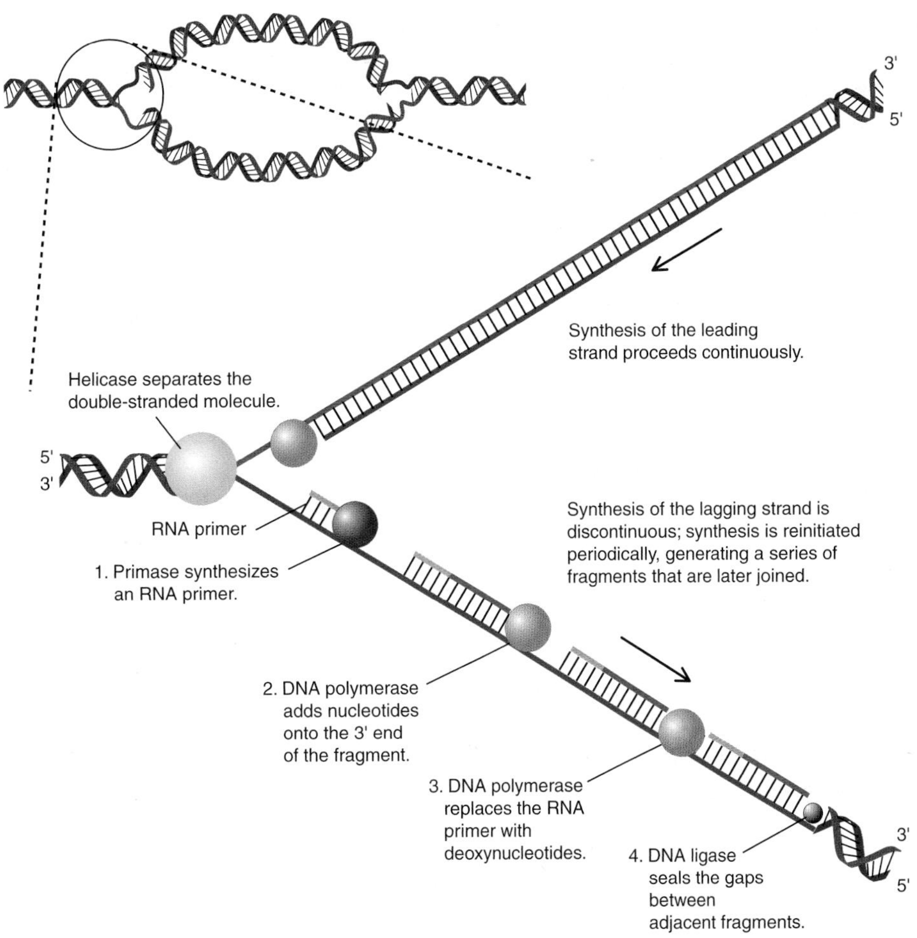
3'
5'
Synthesis of the leading strand proceeds continuously.
Helicase separates the double-stranded molecule.
5'
3'
RNA primer
1. Primase synthesizes an RNA primer.
Synthesis of the lagging strand is discontinuous; synthesis is reinitiated periodically, generating a series of fragments that are later joined.
2. DNA polymerase adds nucleotides onto the 3' end of the fragment.
3. DNA polymerase replaces the RNA primer with deoxynucleotides.
4. DNA ligase seals the gaps between adjacent fragments.
3'
5'

FIGURE 7.6 The Replication Fork This simplified diagram of the replication fork highlights the key steps in the synthesis of the lagging strand.

3′ direction, as fresh single-stranded template DNA is exposed **(figure 7.6).** This strand is called the **leading strand.** Synthesis of the opposing strand, the **lagging strand,** is considerably more complicated because the DNA polymerase cannot add nucleotides to the 5′ end of DNA. Instead, synthesis must be reinitiated periodically as advancement of the replication fork exposes more of the template DNA. Each initiation event must be preceded by the synthesis of an RNA primer by a primase. The result is the synthesis of a series of fragments, called **Okazaki fragments,** each of which begins with a short stretch of RNA. As DNA polymerase adds nucleotides to the 3′ end of an Okazaki fragment, it eventually reaches the initiating point of the previous fragment. A different type of DNA polymerase then removes those RNA primer nucleotides and simultaneously replaces them with deoxynucleotides. The enzyme **DNA ligase** seals the gaps between fragments by catalyzing the formation of a covalent bond between the adjacent nucleotides.

Several other proteins are also involved in DNA replication. Among them is **DNA gyrase,** an enzyme that temporarily breaks the strands of DNA, relieving the tension caused by the unwinding of the two strands of the DNA helix. This enzyme is a target of ciprofloxacin and other members of a class of antibacterial drugs called fluoroquinolones. By inhibiting the function of gyrase, the fluoroquinolones interfere with bacterial DNA replication and prevent the growth of bacteria.

■ **fluoroquinolones, p. 478**

MICROCHECK 7.2

DNA polymerases synthesize DNA in the 5′ to 3′ direction, using one strand as a template to generate the complementary strand. Replication of DNA begins at a specific sequence called the origin of replication, and then proceeds bidirectionally, creating two replication forks.

- ✓ Why is a primer required for DNA synthesis?
- ✓ How does synthesis of the lagging strand differ from that of the leading strand?
- ✓ If DNA replication were shown to be "conservative," what would this mean?

7.3

Gene Expression in Bacteria

Focus Points

- Describe the process of transcription, focusing on the role of RNA polymerase, sigma (σ) factor, promoters, and terminators.
- Describe the process of translation, focusing on the role of mRNA, ribosomes, ribosome-binding sites, rRNAs, tRNAs, and codons.

Gene expression involves two separate but interrelated processes, transcription and translation. Transcription is the process of synthesizing RNA from a DNA template. During translation, information encoded on an mRNA transcript is deciphered to synthesize a protein.

Transcription

The enzyme RNA polymerase catalyzes the process of transcription, producing a single-stranded RNA molecule complementary and antiparallel to the DNA template **(figure 7.7).** To describe the two strands of DNA in a region that is transcribed into RNA, the terms **minus (–) strand** and **plus (+) strand** are sometimes used **(table 7.2).** The strand that serves as the template for RNA synthesis is called the minus (–) strand, whereas its complement is called the plus (+) strand. Recall that the base-pairing rules of DNA and RNA are the same, except that RNA contains uracil in place of thymine. Therefore, because the RNA is complementary to the (–) strand, its nucleotide sequence is the same as the (+) strand, except it has uracil rather than thymine. Likewise, the RNA transcript has the same 5′ to 3′ direction, or **polarity,** as the (+) strand.

In prokaryotes, an mRNA molecule can carry the information for one or multiple genes. A transcript that carries one gene is called **monocistronic** (a cistron is synonymous with a gene). Those that carry multiple genes are called **polycistronic.** Generally, the proteins encoded on a polycistronic message are all involved in a single biochemical pathway. This enables the cell to express related genes in a coordinated manner.

Transcription begins when **RNA polymerase** recognizes a nucleotide sequence on DNA called a **promoter.** The promoter identifies the region of the DNA molecule that will be transcribed into RNA. In addition, the promoter orients the RNA polymerase in one of the two possible directions. This dictates which of the two DNA strands is used as a template **(figure 7.8).** Like DNA polymerase, RNA polymerase can only add nucleotides to the 3′ end of a chain, and therefore synthesizes nucleic acid in the 5′ to 3′ direction. Unlike DNA polymerase, however, RNA polymerase can initiate synthesis without a primer.

The RNA molecule can be used as a reference point to describe direction on the analogous DNA. **Upstream** implies the direction toward the 5′ end of the (+) strand of DNA, whereas **downstream** implies the direction toward the 3′ end. Thus, a promoter is upstream of a gene.

TABLE 7.2 Components of Transcription in Bacteria

Component	Comments
(–) strand	Strand of DNA that serves as the template for RNA synthesis; the resulting RNA molecule is complementary to this strand.
(+) strand	Strand of DNA complementary to the one that serves as the template for RNA synthesis; the sequence of the resulting RNA molecule is analogous to this strand.
Promoter	Nucleotide sequence to which RNA polymerase binds to initiate transcription.
RNA polymerase	Enzyme that synthesizes RNA using single-stranded DNA as a template; synthesis always occurs in the 5′ to 3′ direction.
Sigma (σ) factor	Component of RNA polymerase that recognizes the promoter regions. A cell may have different types of σ factors that recognize different promoters. These may be expressed at different stages of cell growth, enabling the cell to transcribe specialized sets of genes as needed.
Terminator	Sequence at which RNA synthesis stops; the RNA polymerase falls off the DNA template and releases the newly synthesized RNA.

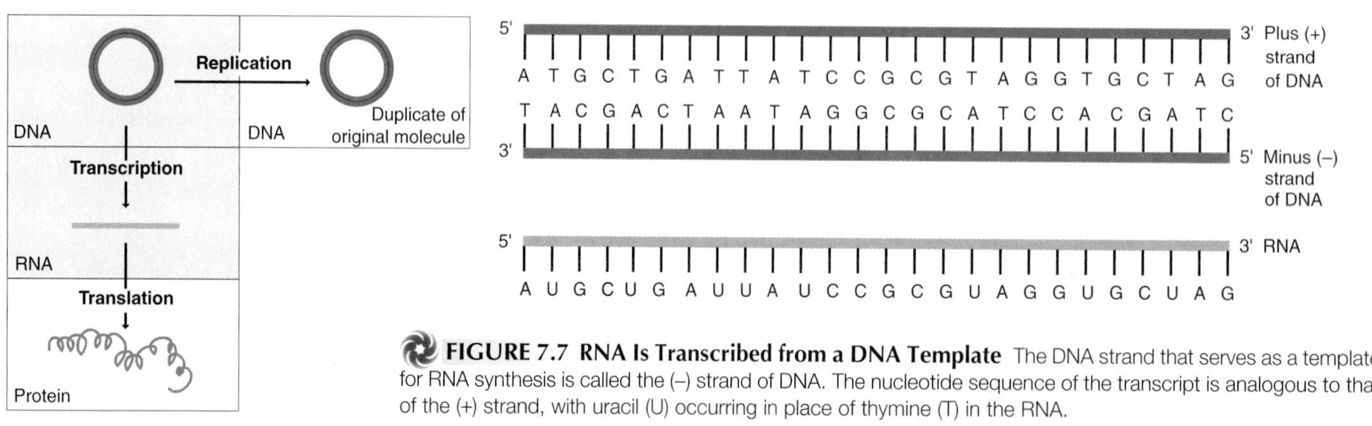

FIGURE 7.7 RNA Is Transcribed from a DNA Template The DNA strand that serves as a template for RNA synthesis is called the (–) strand of DNA. The nucleotide sequence of the transcript is analogous to that of the (+) strand, with uracil (U) occurring in place of thymine (T) in the RNA.

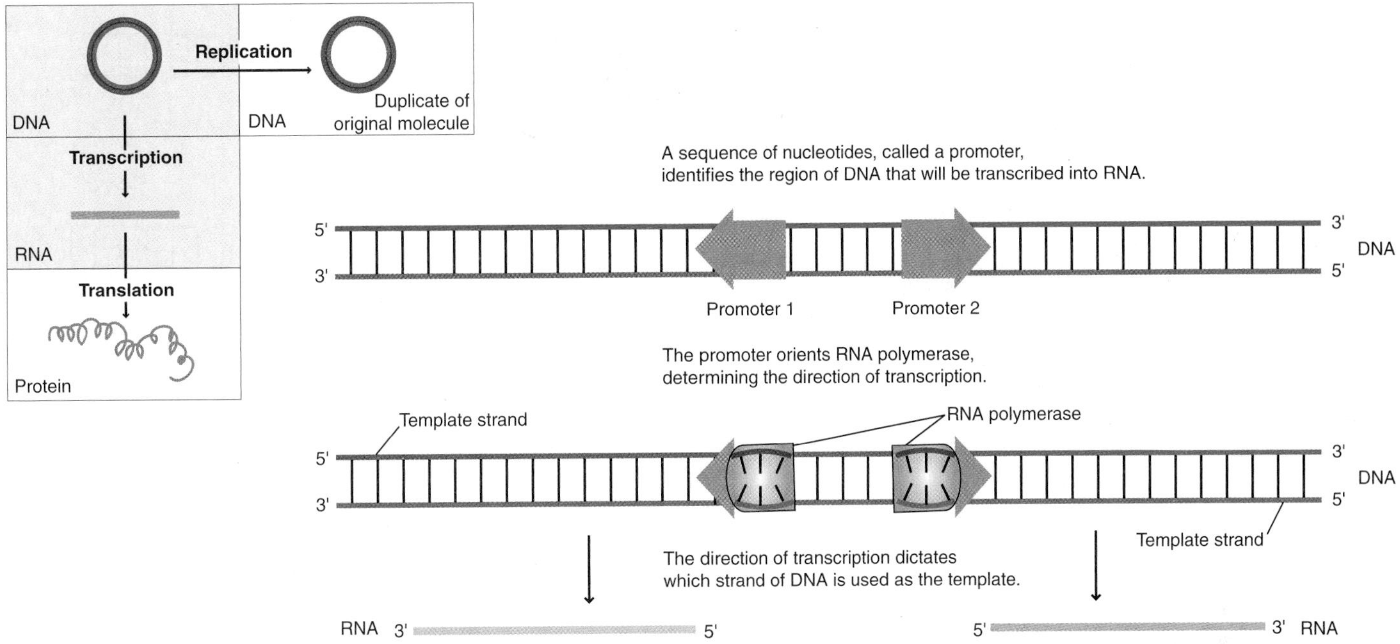

FIGURE 7.8 Promoters Direct Transcription A promoter not only identifies the region of DNA that will be transcribed into RNA, its orientation determines which strand will be used as the template. Note that the color depiction of each RNA molecule indicates which strand of DNA was used as a template. The light blue RNA was transcribed from the red DNA strand (and is therefore analogous in sequence to the blue DNA strand), whereas the pink RNA was transcribed from the blue DNA strand (and is therefore analogous in sequence to the red DNA strand).

Initiation of RNA Synthesis

Transcription begins after RNA polymerase recognizes and binds to a promoter on the double-stranded DNA molecule. The binding melts a short stretch of DNA, creating a region of exposed nucleotides that serves as a template for RNA synthesis.

In bacteria, a particular subunit of RNA polymerase recognizes the promoter region prior to the initiation of transcription. This subunit, **sigma (σ) factor,** can dissociate from the enzyme shortly after transcription is initiated. This leaves the remaining portion of RNA polymerase, called the **core enzyme,** to complete transcription. A cell can have different types of *σ* factors that recognize different promoters. These may be expressed at different stages of cell growth, enabling the cell to transcribe specialized sets of genes as needed. The RNA polymerases of eukaryotic cells and archaea use **transcription factors** to recognize promoters.

Elongation

In the elongation phase, the RNA polymerase moves along the template strand of DNA, synthesizing the complementary single-stranded RNA molecule. The RNA molecule is synthesized in the 5′ to 3′ direction as the enzyme adds nucleotides to the 3′ end of the growing chain. The core RNA polymerase advances along the DNA, melting a new stretch and allowing the previous stretch to close **(figure 7.9).** This exposes a new region of the template, permitting the elongation process to continue.

Once elongation has proceeded far enough for RNA polymerase to clear the promoter, another molecule of RNA polymerase can bind, initiating a new round of transcription. Thus, a single gene can be transcribed multiple times in a very short time interval.

FRANK & ERNEST: © Thaves/Dist. By Newspaper Enterprise Association, Inc.

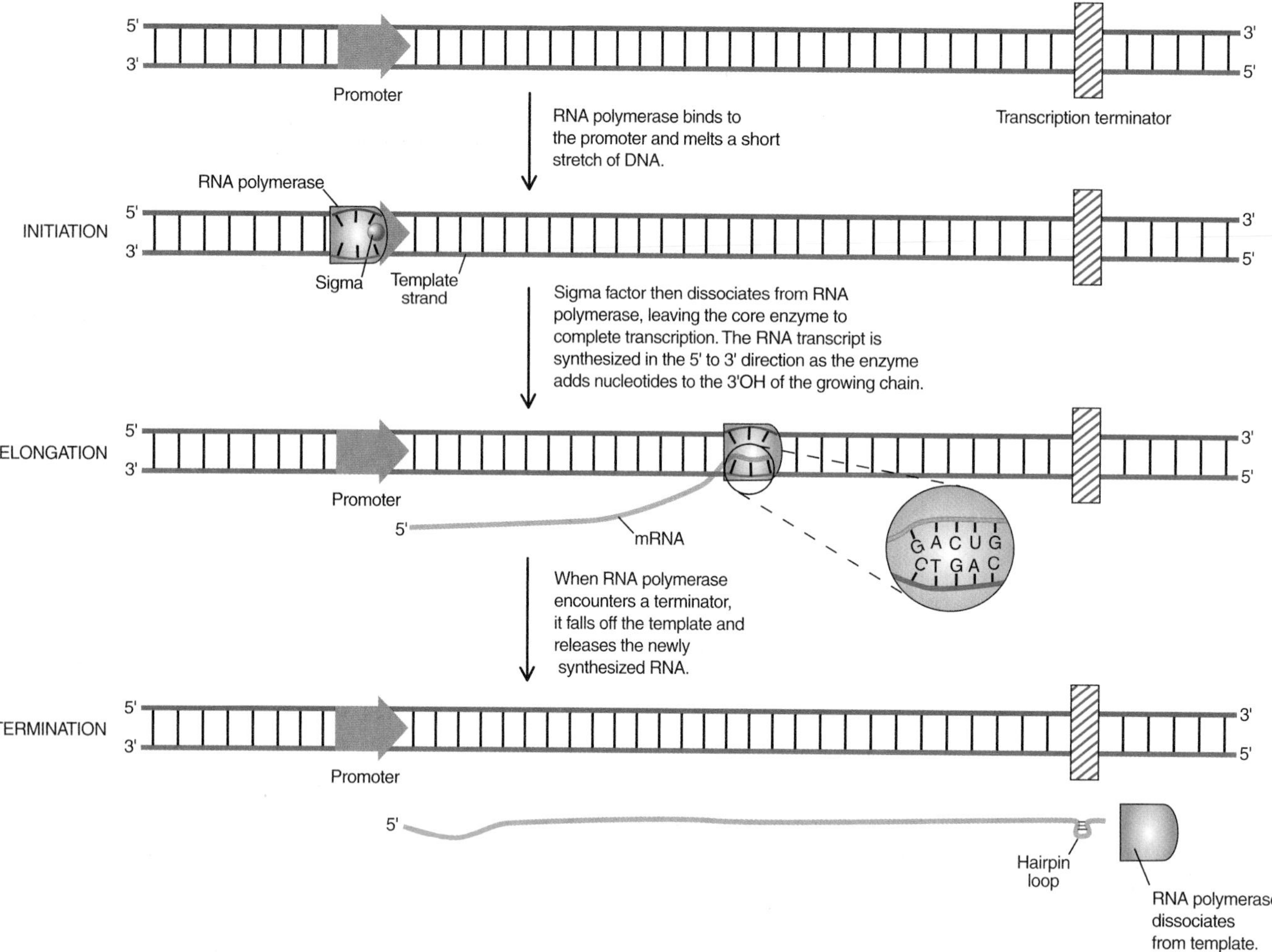

FIGURE 7.9 The Process of RNA Synthesis Bacterial RNA polymerases include a sigma subunit (as illustrated); the RNA polymerases of eukaryotic cells and archaea use transcription factors to recognize promoters.

Termination

Just as an initiation of transcription occurs at a distinct site on the DNA, so does termination. When RNA polymerase encounters a **terminator,** it falls off the DNA template and releases the newly synthesized RNA. The terminator is a sequence of nucleotides in the DNA that, when transcribed, permits two complementary regions of the resulting RNA to base-pair, forming a hairpin loop structure. For reasons that are not yet understood, this causes the RNA polymerase to stall, resulting in its dissociation from the DNA template and release of the RNA.

Translation

Translation is the process of decoding the information carried on the mRNA to synthesize the specified protein. The process requires three major components—mRNA, ribosomes, and tRNAs—in addition to various accessory proteins **(table 7.3).**

The Role of mRNA

The mRNA is a temporary copy of genetic information; it carries encoded instructions for synthesis of a specific polypeptide, or in the case of a polycistronic message, a specific group of polypeptides. That information is deciphered using the **genetic code,** which correlates each series of three nucleotides, a **codon,** with one amino acid **(figure 7.10).** The genetic code is practically universal, meaning that it is used in nearly its entirety by all living things.

Because a codon is a sequence of any combination of the four nucleotides, there are 64 different codons (4^3). Three are stop codons, which will be discussed later. The remaining 61 translate to the 20 different amino acids. This means that more than one codon can code for a specific amino acid. For example, both ACA and ACG encode the amino acid threonine. Because of this redundancy, the genetic code is said to be **degenerate.** Note, however, that different amino acids are never coded for by the same codon.

An equally important aspect of mRNA is that it carries the information that indicates where the coding region actually begins. This is critical because the genetic code is read as groups of three nucleotides. Thus, any given sequence has three possible **reading frames,** or ways in which triplets can be grouped **(figure 7.11).** If translation occurs in the wrong reading frame, a

TABLE 7.3 Components of Translation in Bacteria

Component	Comments
Anticodon	Sequence of three nucleotides in a tRNA molecule that is complementary to a particular codon in mRNA. The anticodon allows the tRNA to recognize and bind to the appropriate codon.
mRNA	Type of RNA molecule that contains the genetic information deciphered during translation.
Polyribosome (polysome)	Multiple ribosomes attached to a single mRNA molecule.
Reading frame	Grouping of a stretch of nucleotides into sequential triplets; an mRNA molecule has three reading frames, but only one is typically used in translation.
Ribosome	Structure that facilitates the joining of amino acids during the process of translation; composed of protein and ribosomal RNA. The prokaryotic ribosome (70S) consists of a 30S and 50S subunit.
Ribosome-binding site	Sequence of nucleotides in mRNA to which a ribosome binds; the first time the codon for methionine (AUG) appears after that site, translation generally begins.
rRNA	Type of RNA molecule present in ribosomes.
Start codon	Codon at which translation is initiated; it is typically the first AUG after a ribosome-binding site.
Stop codon	Codon that terminates translation, signaling the end of the protein; there are three stop codons.
tRNA	Type of RNA molecule that act as keys that interpret the genetic code; each tRNA molecule carries a specific amino acid.

very different, and generally non-functional, polypeptide would be synthesized.

The Role of Ribosomes

Ribosomes serve as the sites of translation; their structure facilitates the joining of one amino acid to another. A ribosome brings each amino acid into a favorable position so that an enzyme can catalyze the formation of a peptide bond between them. It also helps to identify key punctuation sequences on the mRNA molecule, such as the point at which protein synthesis should be initiated. The ribosome moves along the mRNA in the 5′ to 3′ direction, "presenting" each codon in a sequential order for deciphering, while maintaining the correct reading frame.

A prokaryotic ribosome is composed of a 30S subunit and a 50S subunit, each made up of protein and rRNA **(figure 7.12)**; the "S" stands for Svedberg unit, which is a measure of size. Some of the ribosomal components are important in other

Middle Letter

First Letter	U (Reading frame 5' 3')	C (Reading frame 5' 3')	A (Reading frame 5' 3')	G (Reading frame 5' 3')	Last Letter
U	UUU Phenylalanine	UCU Serine	UAU Tyrosine	UGU Cysteine	U
U	UUC Phenylalanine	UCC Serine	UAC Tyrosine	UGC Cysteine	C
U	UUA Leucine	UCA Serine	UAA (Stop)	UGA (Stop)	A
U	UUG Leucine	UCG Serine	UAG (Stop)	UGG Tryptophan	G
C	CUU Leucine	CCU Proline	CAU Histidine	CGU Arginine	U
C	CUC Leucine	CCC Proline	CAC Histidine	CGC Arginine	C
C	CUA Leucine	CCA Proline	CAA Glutamine	CGA Arginine	A
C	CUG Leucine	CCG Proline	CAG Glutamine	CGG Arginine	G
A	AUU Isoleucine	ACU Threonine	AAU Asparagine	AGU Serine	U
A	AUC Isoleucine	ACC Threonine	AAC Asparagine	AGC Serine	C
A	AUA Isoleucine	ACA Threonine	AAA Lysine	AGA Arginine	A
A	AUG Methionine (Start)	ACG Threonine	AAG Lysine	AGG Arginine	G
G	GUU Valine	GCU Alanine	GAU Aspartate	GGU Glycine	U
G	GUC Valine	GCC Alanine	GAC Aspartate	GGC Glycine	C
G	GUA Valine	GCA Alanine	GAA Glutamate	GGA Glycine	A
G	GUG Valine	GCG Alanine	GAG Glutamate	GGG Glycine	G

FIGURE 7.10 The Genetic Code The genetic code correlates each series of three nucleotides, a codon, with one amino acid. Three of the codons do not code for an amino acid and instead serve as a stop codon, terminating translation. AUG functions as a start codon.

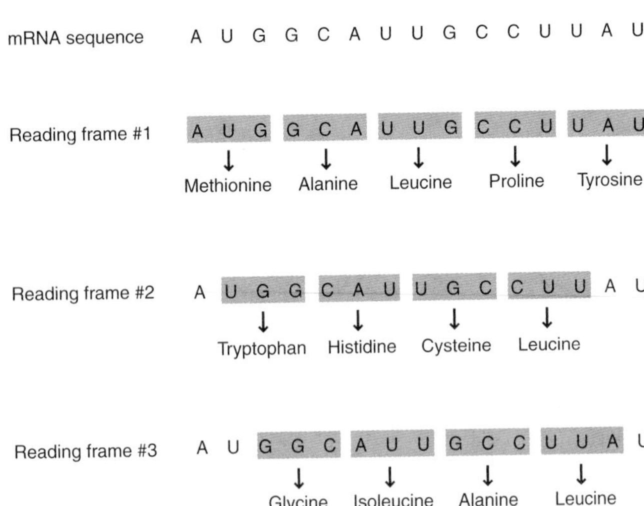

FIGURE 7.11 Reading Frames A nucleotide sequence has three potential reading frames. Because each reading frame encodes a very different order of amino acids, translation of the correct reading frame is important.

aspects of microbiology as well. For example, comparison of the nucleotide sequences of rRNA molecules is playing an increasingly prominent role in the establishment of the genetic relatedness of various organisms. Medically, ribosomal proteins and rRNA are significant because they are the targets of several groups of antimicrobial drugs. ■ **ribosomal subunits, p. 68** ■ **sequencing ribosomal RNA genes, p. 242**

The Role of Transfer RNA

The tRNAs are segments of RNA able to carry specific amino acids, and act as keys that interpret the genetic code. Each recognizes and base-pairs with a specific codon and in the process delivers the appropriate amino acid to that site. This recognition is made possible because each tRNA has an **anticodon**—three nucleotides complementary to a particular codon in the mRNA. The amino acid each tRNA carries is dictated by its anticodon and the genetic code **(figure 7.13).**

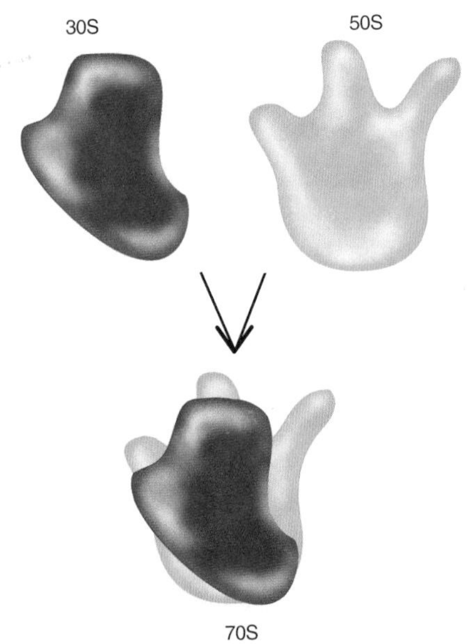

FIGURE 7.12 The Structure of the 70S Ribosome The 70S ribosome is composed of a 30S subunit and a 50S subunit.

Initiation of Translation

In prokaryotes, translation begins as the mRNA is still being synthesized **(figure 7.14).** The 30S subunit of the ribosome binds to a sequence in mRNA called the **ribosome-binding site.** The first time the codon for methionine (AUG) appears after that site, translation generally starts. Note that AUG functions as a **start**

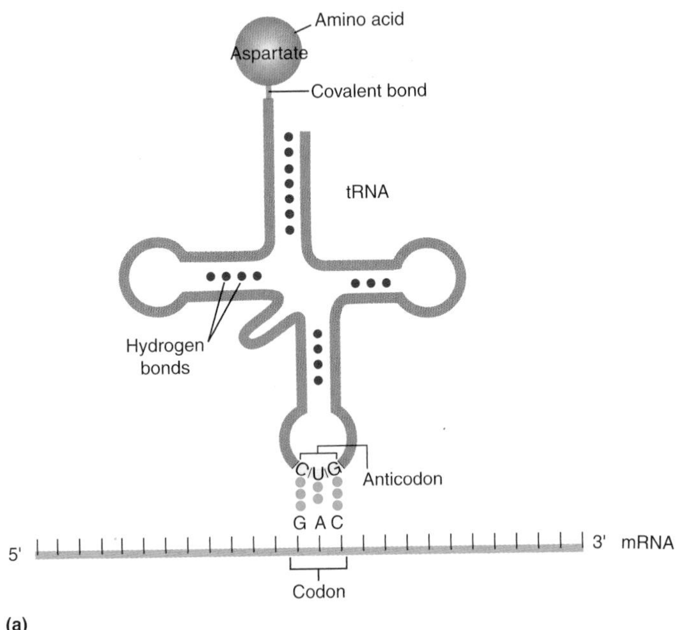

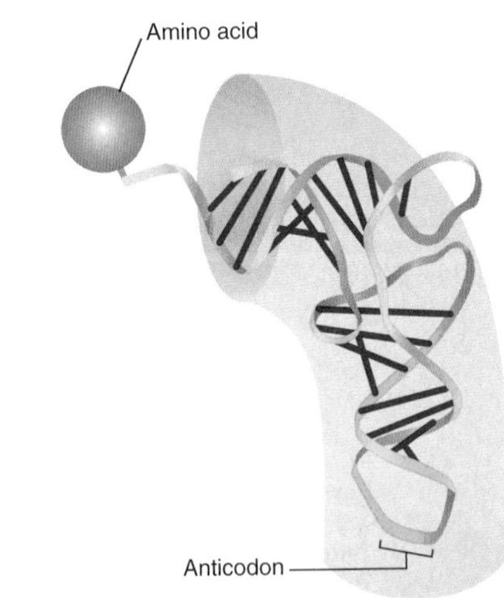

FIGURE 7.13 The Structure of Transfer RNA (tRNA) **(a)** Two dimensional illustration of tRNA. The anticodon of the tRNA base-pairs with a specific codon in the mRNA; by doing so, the appropriate amino acid is delivered to the site. The amino acid that the tRNA carries is dictated by the genetic code. The tRNA that recognizes the codon GAC carries the amino acid aspartate. **(b)** Three-dimensional illustration of tRNA.

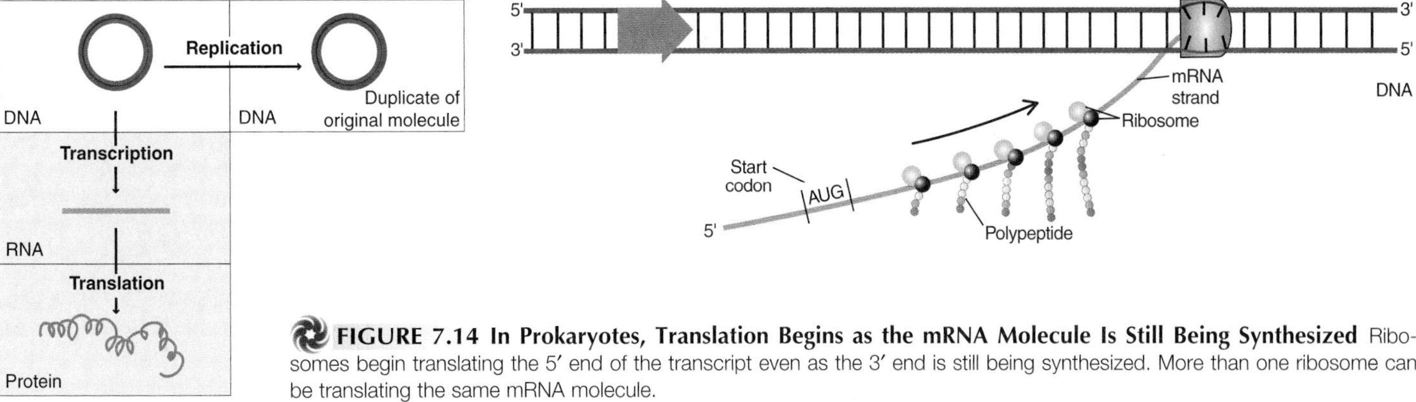

FIGURE 7.14 In Prokaryotes, Translation Begins as the mRNA Molecule Is Still Being Synthesized Ribosomes begin translating the 5′ end of the transcript even as the 3′ end is still being synthesized. More than one ribosome can be translating the same mRNA molecule.

codon only when preceded by a ribosome-binding site; at other sites, it simply encodes methionine. The position of the first AUG is critical, as it determines the reading frame used for translation of the remainder of that protein.

At that first AUG, the ribosome begins to assemble. First, an **initiation complex** forms. This consists of the 30S ribosomal subunit, a tRNA that carries a chemically altered form of the amino acid methionine, *N*-formylmethionine or **f-Met,** and proteins called **initiation factors.** Shortly thereafter, the 50S subunit of the ribosome joins that complex and the initiation factors leave, forming the 70S ribosome. The elongation phase then begins.

Elongation

The 70S ribosome has two sites to which tRNA-carrying amino acids can bind **(figure 7.15).** One is called the **P-site** (peptidyl site), and the other is called the **A-site** (aminoacyl site, commonly referred to as the acceptor site). The initiating tRNA, carrying the f-Met, binds to the P-site. A tRNA that recognizes the next codon on the mRNA then fills the unoccupied A-site. An enzyme then creates a peptide bond between the carboxyl group of the f-Met carried by the tRNA in the P-site and the amino group of the amino acid carried by the tRNA that just entered the A-site. This transfers the amino acid from the initiating tRNA to the amino acid carried by the incoming tRNA. ■ **peptide bond, p. 28**

The ribosome then advances, or **translocates,** a distance of one codon, and the tRNA that carried the f-Met is released through an adjacent site called the **E-site** (exit site). Translocation requires several different proteins, called **elongation factors.** As a result of translocation, the remaining tRNA, which now carries the two-amino-acid chain, occupies the P-site; the A-site is transiently vacant. A tRNA that recognizes the next codon then quickly fills the empty A-site, and the process repeats.

Once translation has progressed far enough for the ribosome to clear the ribosome-binding site and the first AUG, another ribosome can bind, beginning another round of synthesis of the encoded polypeptide. Thus, at any one time, multiple ribosomes can be translating a single mRNA molecule. This allows the maximal expression of protein from a single mRNA template. The assembly of multiple ribosomes attached to a single mRNA molecule is called a **polyribosome** or a **polysome.**

Termination

Elongation of the polypeptide terminates when the ribosome reaches a **stop codon,** a codon that does not code for an amino acid and is not recognized by a tRNA. At this point, enzymes called **release factors** free the polypeptide by breaking the covalent bond that joins it to the tRNA. The ribosome falls off the mRNA and dissociates into its two component subunits, 30S and 50S. These can then be reused to initiate translation at other sites.

Post-Translational Modification

Polypeptides must often be modified after they are synthesized in order to attain their functional properties. For example, some must be folded into their final functional shape, a process that requires the assistance of a protein called a **chaperone.** Polypeptides destined for transport outside of the cytoplasmic membrane also must be modified. These have a characteristic series of hydrophobic amino acids, a **signal sequence,** at their amino terminal end, which "tags" them for transport through the membrane. The signal sequence is removed during transport. ■ **chaperones, p. 29** ■ **hydrophobic amino acids, p. 26**

MICROCHECK 7.3

RNA polymerase initiates RNA synthesis after it binds to a promoter on DNA. Using one strand of DNA as a template, RNA is synthesized in the 5′ to 3′ direction. Synthesis stops when RNA polymerase encounters a terminator. Translation occurs as ribosomes move along mRNA in the 5′ to 3′ direction, with the ribosomes serving as the structure that facilitates the joining of one amino acid to another. tRNAs carry specific amino acids, thus acting to decode the genetic code.

- ✓ How does the orientation of the promoter dictate which strand is used as a template for RNA synthesis?
- ✓ Explain why it is important for the translation machinery to recognize the correct reading frame.
- ✓ Could two mRNAs have different nucleotide sequences and yet code for the same protein?

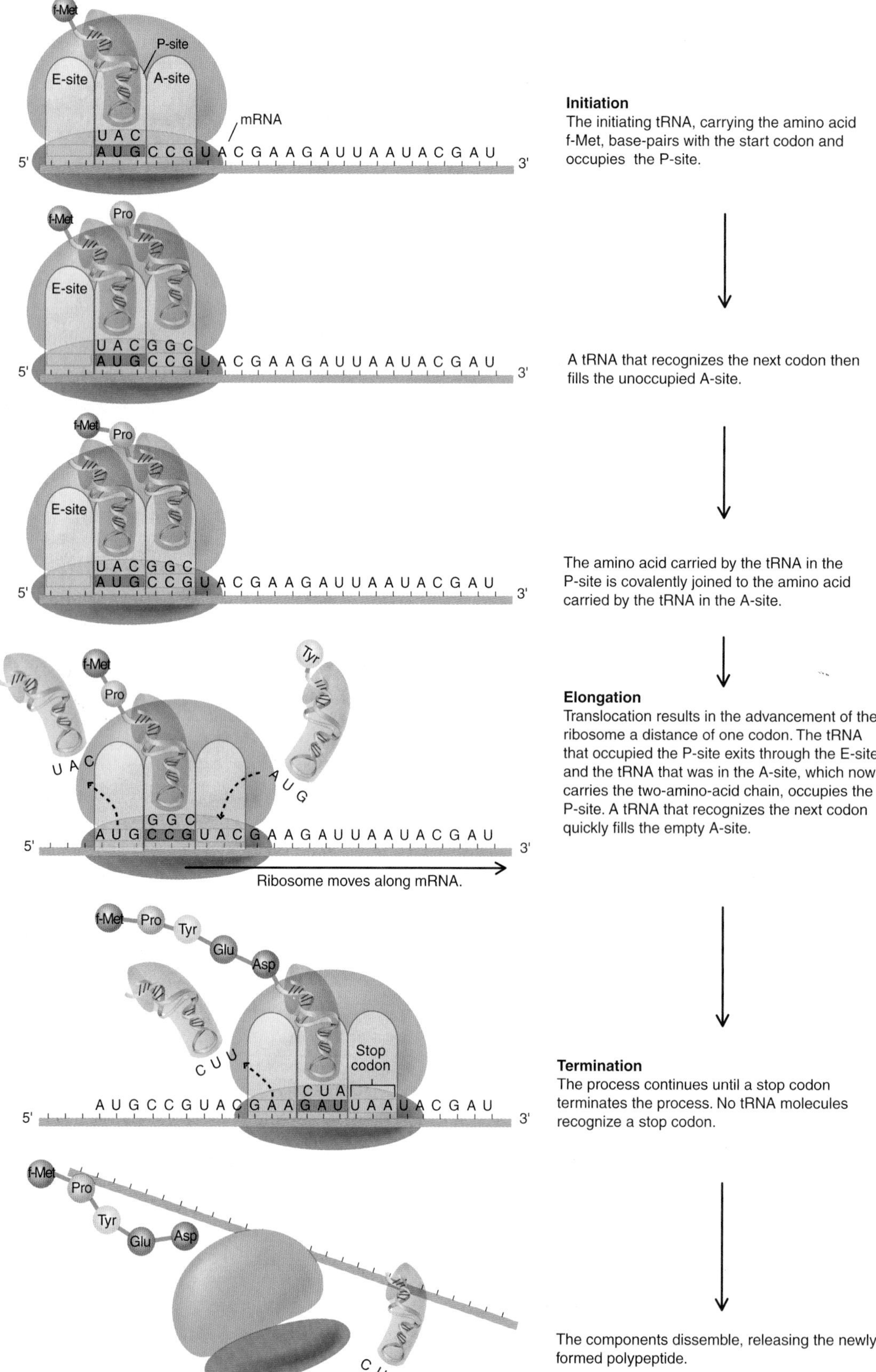

f-Met
P-site
E-site
A-site
mRNA
U A C
A U G C C G U A C G A A G A U U A A U A C G A U
5'
3'
Initiation
The initiating tRNA, carrying the amino acid f-Met, base-pairs with the start codon and occupies the P-site.
Pro
U A C G G C
A tRNA that recognizes the next codon then fills the unoccupied A-site.
The amino acid carried by the tRNA in the P-site is covalently joined to the amino acid carried by the tRNA in the A-site.
Tyr
U A C
G G C
A U G
Ribosome moves along mRNA.
Elongation
Translocation results in the advancement of the ribosome a distance of one codon. The tRNA that occupied the P-site exits through the E-site and the tRNA that was in the A-site, which now carries the two-amino-acid chain, occupies the P-site. A tRNA that recognizes the next codon quickly fills the empty A-site.
Glu
Asp
Stop codon
C U U
C U A
Termination
The process continues until a stop codon terminates the process. No tRNA molecules recognize a stop codon.
C U A
The components dissemble, releasing the newly formed polypeptide.

FIGURE 7.15 The Process of Translation For simplicity, this diagram shows a polypeptide only five amino acids long being synthesized. Note, however, that most polypeptides are over 100 amino acids long.

PERSPECTIVE 7.1

RNA: The First Macromolecule?

The 1989 Nobel Prize in Chemistry was awarded to two Americans. Sidney Altman of Yale University and Thomas Cech of the University of Colorado, who independently made the surprising and completely unexpected observation that RNA molecules can act as enzymes. Before their studies, it was believed that only proteins had enzymatic activity. Cech made the key observation in 1982 when he was trying to understand how introns were removed from precursor ribosomal RNA in the eukaryotic protozoan *Tetrahymena.* Since he was convinced that proteins were responsible for cutting out these introns, he added all of the protein in the cells' nuclei to the RNA that still contained the introns.

As expected, the introns were cut out. As a control, Cech looked at the ribosomal RNA to which no nuclear proteins had been added, fully expecting that nothing would happen. Much to his surprise, the introns were also removed. It did not make any difference whether the protein was present—the introns were removed regardless. Thus, Cech could only conclude that the RNA acted on itself to cut out pieces of RNA.

The question remained of how widespread this phenomenon was. Did RNA have catalytic properties other than that of cutting out introns from rRNA? The studies of Altman and his colleagues, carried out simultaneously to and independently of Cech's, provided answers to these further questions. Altman's group found that RNA could convert a tRNA molecule from a precursor form to its final functional state. Additional studies have shown that enzymatic reactions in which catalytic RNAs, termed **ribozymes,** play a role are very widespread. Ribozymes have been found in the mitochondria of eukaryotic cells and shown to catalyze other reactions that resemble the polymerization of RNA.

These observations have profound implications for evolution: which came first, proteins or nucleic acids? The answer seems to be that nucleic acids came first, specifically RNA, which acted both as a carrier of genetic information as well as an enzyme. Billions of years ago, before the present universe in which DNA, RNA, and protein are found, probably the only macromolecule that existed was RNA. Once tRNA became available, these adapters could carry amino acids present in the environment to specific nucleotide sequences on a strand of RNA. In this scenario, the RNA functions as the genes as well as the mRNA.

7.4 Differences Between Eukaryotic and Prokaryotic Gene Expression

Focus Point

- Describe four differences between prokaryotic and eukaryotic gene expression.

Eukaryotes differ significantly from prokaryotes in several aspects of transcription and translation **(table 7.4).** In eukaryotic cells for example, most mRNA molecules are extensively modified, or **processed,** in the nucleus during and after transcription. Shortly after transcription begins, the 5′ end of the transcript is modified, or **capped,** by the addition of a methylated guanine derivative, creating what is called a **cap.** The cap binds specific proteins that stabilize the transcript and enhance translation. The 3′ end of the molecule is also modified, even before transcription has been terminated. This process, called **polyadenylation,** involves cleaving the transcript at a specific sequence of nucleotides and then adding approximately 200 adenine derivatives to the newly exposed 3′ end. This creates what is called a **poly A tail,** which is thought to stabilize the transcript as well as enhance translation. Another important modification is **splicing,** a process that removes specific segments of the transcript **(figure 7.16).** Splicing is necessary because eukaryotic genes are not always contiguous; they are often interrupted by non-coding nucleotide sequences. These intervening sequences, or **introns,** are transcribed along with the expressed regions, or **exons,** generating what is called **precursor mRNA.** The introns must be removed from precursor mRNA to form the mature mRNA that is then translated.

The mRNA in eukaryotic cells must be transported out of the nucleus before it can be translated in the cytoplasm. Thus, the same mRNA molecule cannot be transcribed and translated at the same time or even in the same cellular location. Unlike in prokaryotes, the mRNA of eukaryotes is generally monocistronic. Translation of the message generally begins at the first occurrence of AUG in the molecule.

The ribosomes of eukaryotes are different from those of prokaryotes. Whereas the prokaryotic ribosome is 70S, made up of

TABLE 7.4 Major Differences Between Prokaryotic and Eukaryotic Transcription and Translation

Prokaryotes	Eukaryotes
mRNA is not processed.	A cap is added to the 5′ end of mRNA, and a poly A tail is added to the 3′ end.
mRNA does not contain introns.	mRNA contains introns, which are removed by splicing.
Translation of mRNA begins as it is being transcribed.	The mRNA transcript is transported out of the nucleus so that it can be translated in the cytoplasm.
mRNA is often polycistronic; translation usually begins at the first AUG that follows a ribosome-binding site.	mRNA is monocistronic; translation begins at the first AUG.

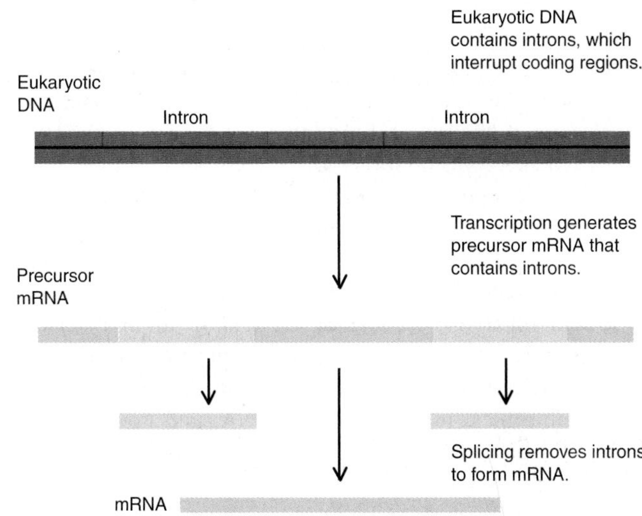

FIGURE 7.16 Splicing of Eukaryotic RNA

30S and 50S subunits, the eukaryotic ribosome is 80S, made up of 40S and 60S subunits. The differences in ribosome structure account for the ability of certain types of antibiotics to kill bacteria without causing significant harm to mammalian cells.

Some of the proteins that play essential roles in translation differ between eukaryotic and prokaryotic cells. Diphtheria toxin, which selectively kills eukaryotic but not prokaryotic cells, illustrates this difference. This toxin is produced by *Corynebacterium diphtheriae;* it binds to and inactivates one of the elongation factors of eukaryotes. Since this protein is required for translocation of the ribosome, translation ceases and the eukaryotic cell dies, resulting in the typical symptoms of diphtheria. ■ diphtheria toxin, p. 503

MICROCHECK 7.4

Eukaryotic mRNA must be processed, which involves capping, polyadenylation, and splicing. In eukaryotic cells, the mRNA must be transported out of the nucleus before it can be translated in the cytoplasm. Eukaryotic mRNA is monocistronic.

✓ What is an intron?
✓ Explain the mechanism of action of diphtheria toxin.
✓ Would a deletion of two base pairs have a greater consequence if it occurred in an intron or in an exon?

7.5 Regulation of Bacterial Gene Expression

Focus Points

- Give a functional example of a constitutive enzyme, an inducible enzyme, and a repressible enzyme.
- Using the *lac* operon as a model, explain the role of inducers and repressors.

To cope with changing conditions in their environment, microorganisms have evolved elaborate control mechanisms to synthesize the maximum amount of cell material from a limited supply of energy. This is critical, because generally a microorganism must reproduce more rapidly than its competitors to be successful.

Consider the situation of *Escherichia coli.* For over 100 million years, it has successfully inhabited the gut of mammals, where it reaches concentrations of 10^6 cells per milliliter. In this habitat, it must cope with alternating periods of feast and famine. For a limited time after a mammal eats, *E. coli* in the large intestine prosper, wallowing in the milieu of amino acids, vitamins, and other nutrients. The cells actively take up these compounds they would otherwise synthesize, expending minimal energy. Simultaneously, the cells shut down their biosynthetic pathways, channeling the conserved energy into the rapid synthesis of macromolecules, including DNA, RNA, and protein. Under these conditions, the cells divide at their most rapid rate. Famine, however, follows the feast. Between meals—a period of time that might be many days in the case of some mammals—the rich source of nutrients is depleted. Now the cells' biosynthetic pathways must be activated, using energy and markedly slowing cell division. Cells dividing several times an hour in a nutrient-rich environment might divide only once every 24 hours in a famished mammalian gut.

A cell controls its metabolic pathways by two general mechanisms. The most immediate of these is the allosteric inhibition of enzymes. The most energy-efficient strategy, however, is to control the actual synthesis of the enzymes, making only what is required. To do this, cells control expression of certain genes. ■ allosteric regulation, p. 137

Principles of Regulation

Not all genes are subjected to the same type of regulation. Many are routinely expressed, whereas others are either turned on or off by certain conditions. Enzymes are often described according to characteristics of the regulation that governs their synthesis:

- **Constitutive enzymes** are synthesized constantly; the genes that encode these enzymes are always active. Constitutive enzymes usually play indispensable roles in the central metabolic pathways. For example, the enzymes of glycolysis are constitutive. ■ central metabolic pathways, pp. 132, 138
- **Inducible enzymes** are not produced regularly; instead, their synthesis is turned on by certain conditions. Inducible enzymes are often involved in the utilization of specific energy sources. A cell would waste precious resources if it synthesized the enzyme when the energy source is not present. An example of an inducible enzyme is **β-galactosidase,** whose sole function is to break down the disaccharide lactose into its two component monosaccharides, glucose and galactose. The mechanisms by which the cell controls β-galactosidase synthesis serve as an important model for regulation and will be described shortly.
- **Repressible enzymes** are synthesized routinely, but they can be turned off by certain conditions. Repressible enzymes are generally involved in biosynthetic (anabolic) pathways, such as those that produce amino acids. Cells require a sufficient amount of a given amino acid to multiply; thus, the amino acid must be either synthesized or available as a component of the growth medium. If a certain amino acid is not present in the medium, then the cell must synthesize the enzymes involved in its manufacture. When the amino acid is supplied, however, synthesis of the enzymes would waste energy.

Mechanisms to Control Transcription

The mechanisms a cell uses to prevent or facilitate transcription must be readily reversible, allowing cells to effectively control the relative number of transcripts made. In some cases, the control mechanisms affect the transcription of only a limited number of genes; in other cases, a wide array of genes is controlled coordinately. For example, in *E. coli,* the expression of more than 300 different genes is affected by the availability of glucose as an energy source. The simultaneous regulation of numerous genes is called **global control.** Two of the most common methods of regulation in bacterial cells are alternative sigma factors and DNA-binding proteins.

Alternative Sigma Factors

Recall that sigma factor is a loose component of RNA polymerase and functions in recognizing specific promoters. *E. coli*, for example, has a standard sigma factor (sigma 70) that recognizes promoters for genes that need to be expressed during routine growth conditions. A cell can also produce other types of sigma factors, called **alternative sigma factors.** Each of these recognizes a different set of promoters, thereby controlling the expression of specific groups of genes. One alternative sigma factor (sigma S) is produced when the cell experiences a general stress such as starvation. This sigma factor directs RNA polymerase to transcribe the multitude of genes required to enter stationary phase. Another (sigma F) directs the expression of genes required for flagella synthesis. The cell can also express anti-sigma factors, which inhibit the function of specific sigma factors.

DNA-Binding Proteins

Transcription of genes is often controlled by means of a regulatory region near the promoter to which a specific protein can bind, acting as a sophisticated on/off switch. When a regulatory protein binds DNA, it can either act as a repressor, which blocks transcription, or an activator, which facilitates transcription. A set of genes coordinately controlled by a regulatory protein and transcribed as a single polycistronic message is called an **operon.**

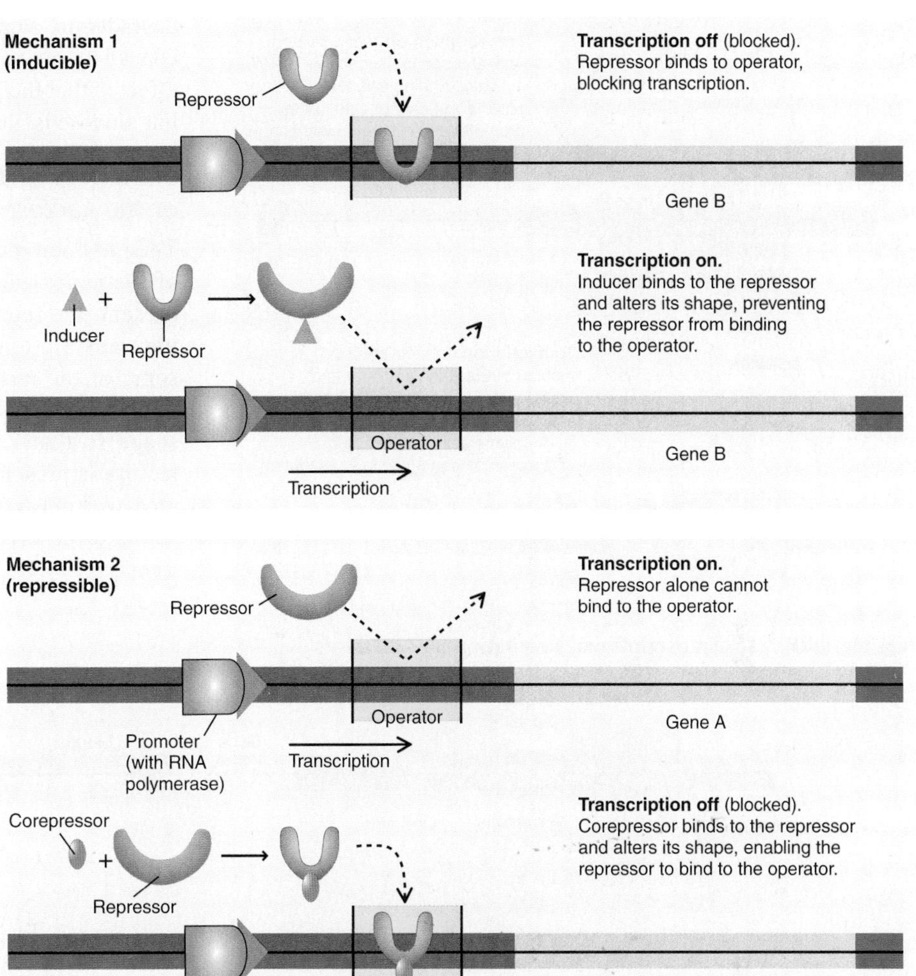

FIGURE 7.17 Transcriptional Regulation by Repressors

Repressors A **repressor** is a regulatory protein that blocks transcription. It does this by binding to a sequence of DNA called an **operator,** located immediately downstream of a promoter. When a repressor is bound to an operator, RNA polymerase cannot progress past that region. Regulation involving a repressor is called **negative control.** Specific molecules can bind to the repressor and thereby alter the ability of the repressor to bind DNA. This can occur because a repressor is an allosteric protein, meaning that the binding of a specific molecule causes the protein's shape to change. The altered shape of the repressor affects its ability to bind DNA. As shown in **figure 7.17,** there are two general mechanisms by which different repressors can function:

1. The repressor is synthesized as a form that effectively binds to the operator, blocking transcription. When a molecule called an **inducer** binds to the repressor, however, the shape of the repressor is altered so that it no longer binds to the operator. Consequently, the gene can then be transcribed.
2. The repressor is synthesized as a form that alone cannot bind to the operator. When a molecule termed a **corepressor** binds to the repressor, however, the shape of the repressor is altered so that it can bind to the operator, blocking transcription.

Activators An **activator** is a regulatory protein that facilitates transcription. Genes controlled by an activator have an ineffective promoter preceded by an **activator-binding site.** The binding of the activator to the DNA enhances the ability of RNA polymerase to initiate transcription at that promoter. Regulation involving an activator is sometimes called **positive control.**

Like repressors, activators are allosteric proteins whose function can be modulated by the binding of other molecules. When a molecule called an inducer binds to an activator, the shape of the the activator is altered so that it can effectively bind to the activator-binding site **(figure 7.18).** Thus, the term "inducer" applies to a molecule that turns on transcription, either by stimulating the function of an activator or interfering with the function of a repressor.

The *lac* Operon As a Model for Control of Metabolic Pathways

Originally elucidated in the early 1960s by Francois Jacob and Jacques Monod, the ***lac* operon** has served as an important model for understanding the control of gene expression in bacteria. The operon, which consists of three genes involved with lactose

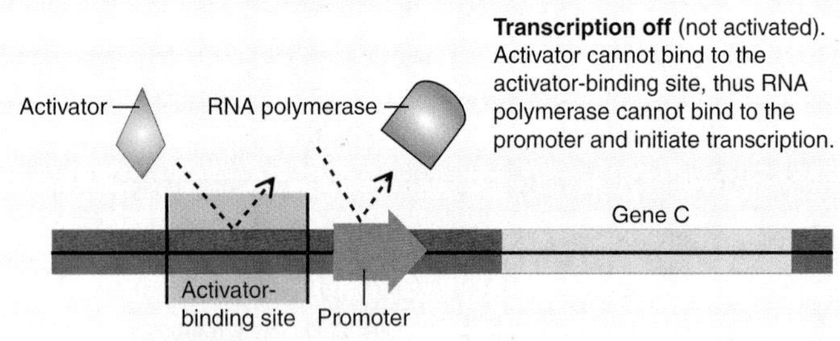

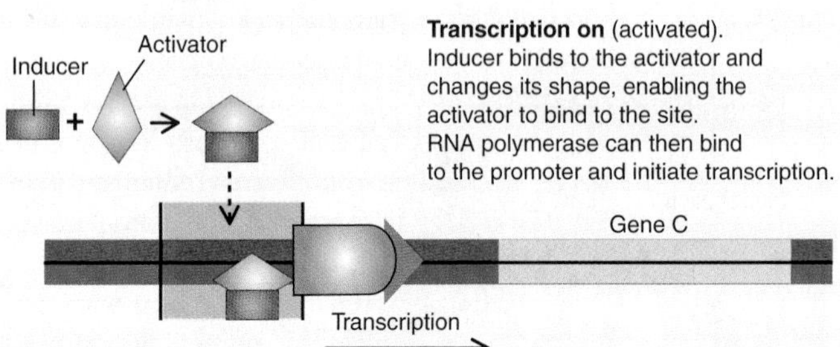

FIGURE 7.18 Transcriptional Regulation by Activators

degradation, along with regulatory components, is subject to dual control by both a repressor and an activator **(figure 7.19).** The net effect is that the genes are expressed only when lactose is present but glucose is absent.

The Effect of Lactose on the Control of the Lactose Operon

The *lac* operon employs a repressor that prevents transcription of the genes when lactose is unavailable. When lactose is not present, the repressor binds to the operator, effectively blocking transcription. When lactose is present in the cell, however, some of the molecules are converted into a compound called allolactose. This compound binds to the repressor, altering its shape so that it no longer binds to the operator. Thus, when lactose is present, the repressor no longer prevents RNA polymerase from transcribing the operon. Note, however, that the activator described in the next section is needed for successful transcription.

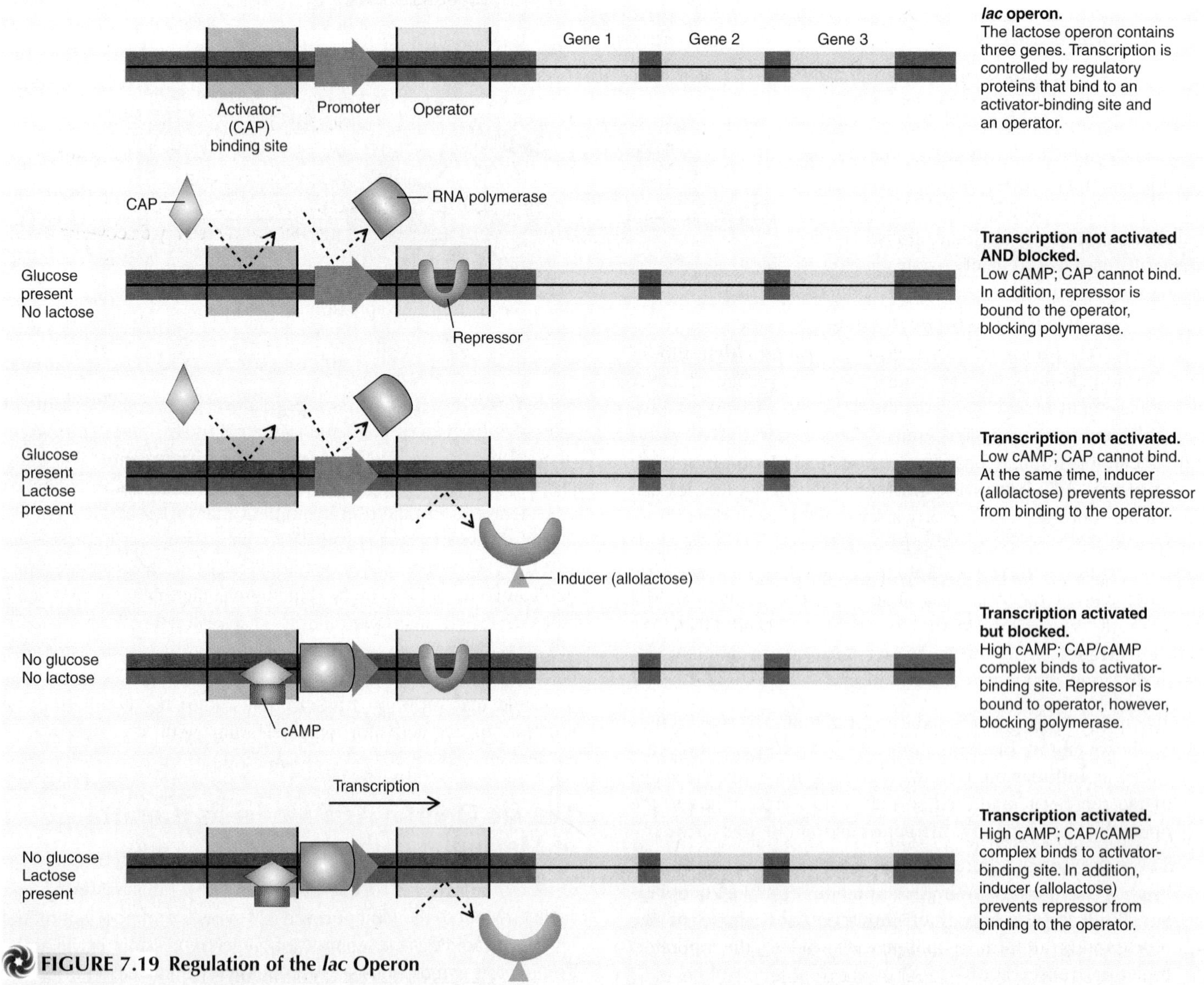

Gene 1
Gene 2
Gene 3
Activator-(CAP) binding site
Promoter
Operator

***lac* operon.**
The lactose operon contains three genes. Transcription is controlled by regulatory proteins that bind to an activator-binding site and an operator.

CAP
RNA polymerase
Glucose present No lactose
Repressor

Transcription not activated AND blocked.
Low cAMP; CAP cannot bind. In addition, repressor is bound to the operator, blocking polymerase.

Glucose present Lactose present
Inducer (allolactose)

Transcription not activated.
Low cAMP; CAP cannot bind. At the same time, inducer (allolactose) prevents repressor from binding to the operator.

No glucose No lactose
cAMP

Transcription activated but blocked.
High cAMP; CAP/cAMP complex binds to activator-binding site. Repressor is bound to operator, however, blocking polymerase.

Transcription
No glucose Lactose present

Transcription activated.
High cAMP; CAP/cAMP complex binds to activator-binding site. In addition, inducer (allolactose) prevents repressor from binding to the operator.

FIGURE 7.19 Regulation of the *lac* Operon

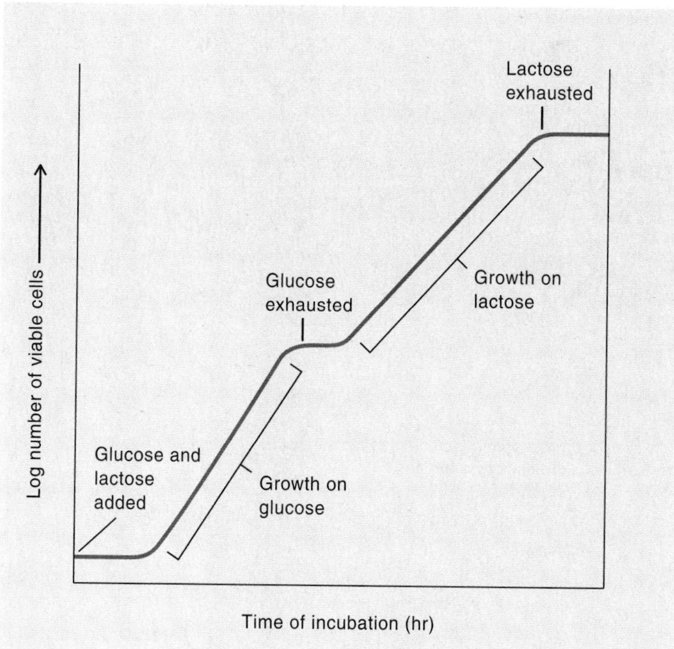

FIGURE 7.20 Diauxic Growth Curve of *E. coli* Growing in a Medium Containing Glucose and Lactose Cells preferentially use glucose. Only when the supply of glucose is exhausted do cells start metabolizing lactose. Note that the growth on lactose is slower than it is on glucose.

The Effect of Glucose on the Control of the Lactose Operon

Escherichia coli preferentially uses glucose over other sugars such as lactose. This can be demonstrated by observing growth and sugar utilization of *E. coli* in a medium containing glucose and lactose. Cells grow, metabolizing only glucose until its supply is exhausted **(figure 7.20).** Growth then ceases for a short period until the cells begin utilizing lactose. At this point, the cells start multiplying again. This two-step growth response, called **diauxic growth,** represents the ability of glucose to repress the enzymes of lactose degradation—a phenomenon called **catabolite repression.**

The regulatory mechanism of catabolite repression does not directly sense glucose in a cell. Instead, it recognizes the concentration of a nucleotide derivative, cyclic AMP (cAMP), which is low when glucose is being transported into the cell and high when it is not. cAMP is an inducer of the operon; it binds to an activator that facilitates transcription of the *lac* operon. This activator, called CAP (catabolite activator protein), is only able to bind to the *lac* promoter when cAMP is bound to it. The higher the concentration of cAMP, the more likely it is to bind to CAP. Thus, when glucose concentrations in the medium are low (and therefore cAMP levels are high), the *lac* operon can be transcribed. Note, however, that even in the presence of a functional activator, the repressor prevents transcription unless lactose is present.

Catabolite repression is significant biologically because it forces cells to first use the carbon source that is most easily metabolized. Only when the supply of glucose is exhausted do cells begin degrading lactose, a carbon source that requires additional enzymatic steps to metabolize.

MICROCHECK 7.5

Enzymes can be constitutive, inducible, or repressible. A repressor blocks transcription when it binds to an operator. An activator enhances transcription when it binds to an activator-binding site. The functioning of specific activators and repressors may require or be blocked by other molecules.

- ✓ Explain the difference between a constitutive enzyme and an inducible enzyme.
- ✓ Explain how glucose represses the lactose operon.
- ✓ Why would it be advantageous for a cell to control the activity of an enzyme as well as its synthesis?

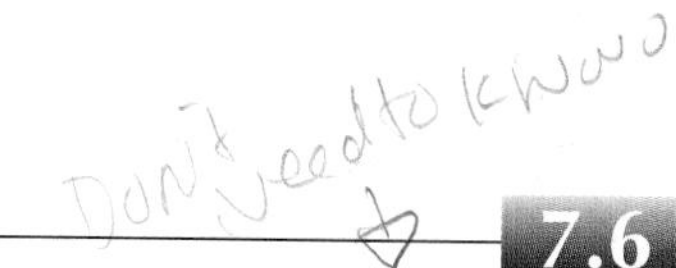

7.6 Regulation of Eukaryotic Gene Expression

Focus Point

- Describe how RNA interference silences genes.

Considering the complexity of eukaryotic cells and the diversity of cell types found in multicellular organisms, it is not surprising that gene regulation in eukaryotic cells is much more complicated than that in prokaryotic cells. Eukaryotic cells use a variety of control methods, including modifying the structure of the chromosome, regulating the initiation of transcription, and altering transcript processing and modification. We will focus only on a recently discovered mechanism called *RNA interference* (RNAi) that now makes it possible for scientists to silence select genes. Andrew Fire and Craig Mello received a Noble Prize in 2006 for their discovery of this process.

RNA interference (RNAi) uses short pieces of single-stranded RNA to direct the degradation of specific RNA transcripts. After a short RNA molecule (approximately 20 to 26 nucleotides in length) is produced by the cell, it is loaded into a multi-protein complex called an **RNA-induced silencing complex (RISC).** The single-stranded RNA molecule within the complex binds to complementary sequences on mRNA, effectively tagging that transcript for destruction by enzymes that make up the RISC **(figure 7.21).** Because the components of the RISC are not destroyed in the process, the complex is catalytic, providing a rapid and effective means of silencing genes that have already been transcribed. Two different types of RNA molecules are used in RNAi—**microRNA**

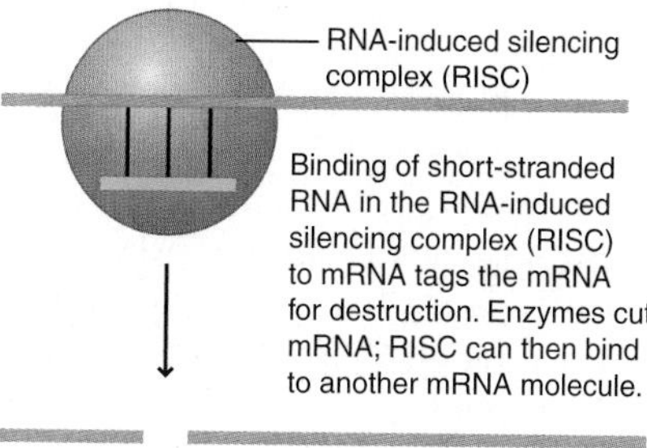

FIGURE 7.21 RNA Interference (RNAi)

(miRNA) and **short interfering RNA (siRNA).** These are functionally equivalent, but differ in how they are produced.

The discovery of RNAi has revolutionized current views on gene regulation. In addition, it provides a new mechanism to alter gene expression. By using RNAi to turn off selected genes *in vitro,* scientists are able to more precisely identify the function of those genes. An ultimate hope is that RNA interference could be used as a form of gene therapy to silence abnormal genes.

MICROCHECK 7.6

RNA interference uses short, single strands of RNA to direct the destruction of specific RNA transcripts.

✓ What is the role of miRNA and siRNA in regulation of gene expression?

7.7

Sensing and Responding to Environmental Fluctuations

Focus Points

- Describe how two-component regulatory systems and quorum sensing allow cells to adapt to fluctuating environmental conditions.
- Compare and contrast antigenic variation and phase variation.

Microorganisms adapt to fluctuating conditions by altering the level of expression of certain genes. For example, certain pathogenic bacteria have mechanisms to sense when they are within the tissues of an animal; in response they can activate certain genes that facilitate their survival against the impending onslaught of host defenses.

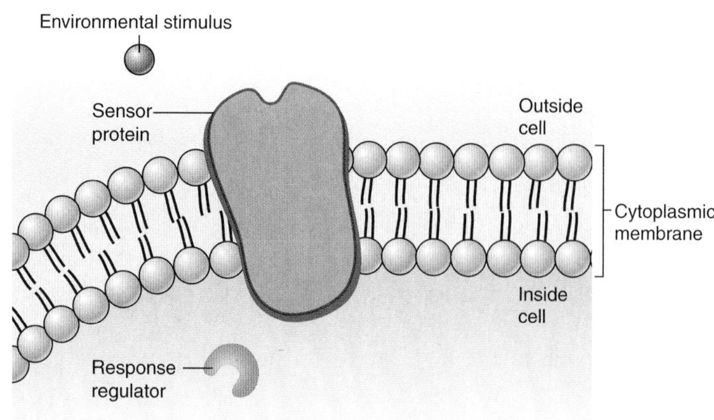

The sensor protein spans the cytoplasmic membrane. The response regulator is a protein inside the cell.

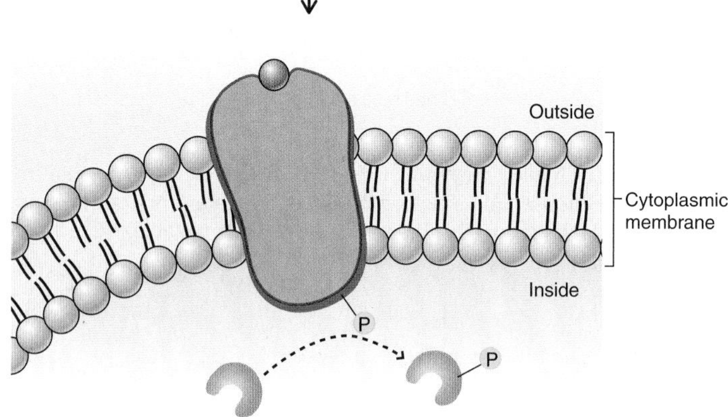

In response to a specific change in the environment, the sensor phosphorylates a region on its internal portion. The phosphoryl group is transferred to the response regulator, which can then act as an activator or a repressor, depending on the system.

FIGURE 7.22 Two-Component Regulatory System

Signal Transduction

Signal transduction is a process that transmits information from outside a cell to the inside, allowing that cell to respond to changing environmental conditions. For example, cells turn on or off certain genes in response to variations in such factors as osmotic pressure, cell concentration, and nitrogen availability.

Two-Component Regulatory Systems

An important mechanism that cells use to relay information about the external environment is a **two-component regulatory system (figure 7.22).** This relies on the coordinated activities of two different proteins, a sensor and a response regulator. The **sensor** spans the cytoplasmic membrane so that the part recognizing changes in the environment is positioned outside the cell. In response to specific external variations, the sensor chemically modifies a region on its internal portion, usually by phosphorylating a specific amino acid. The phosphoryl group is then transferred to a **response regulator.** The modified response regulator can act as either an activator or a repressor, turning on or off genes, depending on the system.

Bacteria use different two-component regulatory systems to detect and respond to a wide variety of environmental cues. *E. coli,* for example, uses such systems to control the expression of genes for its alternative types of metabolism. When nitrate is present in anaerobic conditions, cells activate genes required to use nitrate as a terminal electron acceptor. Some pathogens use two-component regulatory systems to sense environmental magnesium concentrations, and then activate specific genes in response. Because the magnesium concentration within certain tissue cells is generally lower than that of extracellular sites, these pathogens are able to recognize whether or not they are within a cell. In turn, they can activate appropriate genes that help them evade the host defenses intended to protect that relative site.

Quorum Sensing

Some organisms can "sense" the density of cells within their own population—a phenomenon called **quorum sensing.** This enables them to activate genes that are only beneficial when expressed by a critical mass of cells. The cooperative activities leading to biofilm formation are controlled by quorum sensing. ■ **biofilms, p. 85**

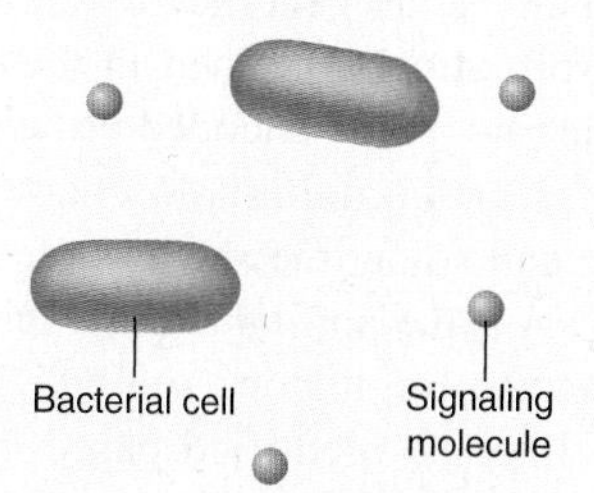

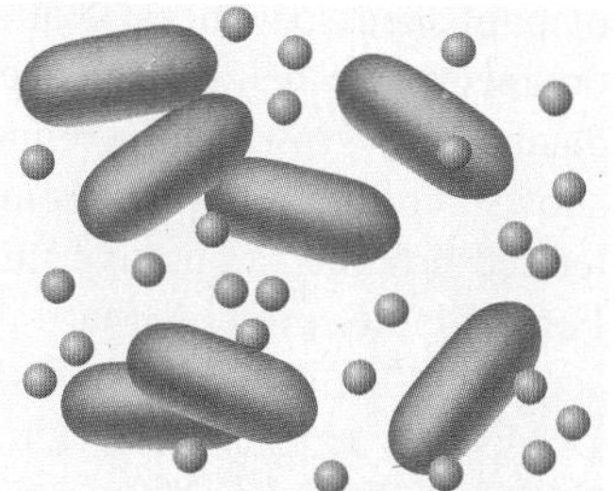

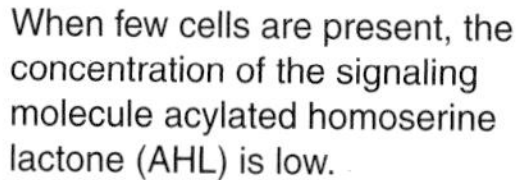
When few cells are present, the concentration of the signaling molecule acylated homoserine lactone (AHL) is low.

When many cells are present, the concentration of the AHL is high. High concentrations of AHL induce expression of specific genes.

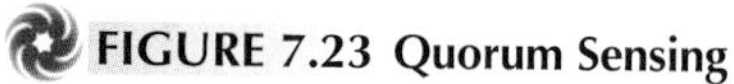
FIGURE 7.23 Quorum Sensing

In the most thoroughly studied quorum sensing systems, the bacteria synthesize one or more varieties of an **acylated homoserine lactone (AHL)** (or HSL for **homoserine lactone**). When few cells are present, the concentration of a given AHL is very low. As the cells multiply in a confined area, however, the concentration of that AHL increases proportionally. Only when it reaches a critical level does it induce the expression of specific genes **(figure 7.23).**

Natural Selection

Natural selection can also play a role in gene expression. The expression of some genes changes randomly, presumably enhancing the chances of survival of at least a part of a population under certain environmental conditions.

The role of natural selection is readily apparent in bacteria that undergo **antigenic variation,** an alteration in the characteristics of certain surface proteins such as flagella, pili, and outer membrane proteins. Pathogens able to change these proteins can stay one step ahead of the body's defenses by altering the very molecules our immune systems must learn to recognize. One of the most well characterized examples is *Neisseria gonorrhoeae,* a bacterium that successfully disguises itself from the immune system by changing several of its surface proteins. *N. gonorrhoeae* has many different genes for pilin, the protein subunit that makes up pili, yet most are silent. The only one that is expressed resides in a particular chromosomal location called an expression locus. *N. gonorrhoeae* cells have a mechanism to shuffle the pilin genes, randomly moving different ones in and out of the expression locus. In a population of 10^4 cells, at least one is expressing a different type of pilin. It appears that expression of different pilin genes is not regulated in any controlled manner but occurs randomly. Only some of the changes, however, are advantageous to a cell's survival. When the body's immune system eventually begins to respond to a specific pilin type, those cells that have already "switched" to produce a different type will survive and then multiply. Eventually, the immune system learns to recognize those, but by that time, another subpopulation will have "switched" its pilin type. ■ **pili, p. 66** ■ ***Neisseria gonorrhoeae*, p. 630**

Another mechanism of randomly altering gene expression is **phase variation,** the routine switching on and off of certain genes. Presumably, phase variation helps an organism adapt to selective pressures. By altering the expression of certain critical genes, at least a part of the population is poised for change and thus able to survive and multiply. For example, phase variation of genes that encode fimbriae may allow some members of a population to attach to a surface, while permitting others to detach and colonize surfaces elsewhere. ■ **fimbriae, p. 66**

MICROCHECK 7.7

Signal transduction allows a cell to respond to changing conditions outside of that cell. The expression of some genes changes randomly, presumably enhancing the chances of survival of at least a subset of a population of cells under varying environmental conditions.

- ✓ Explain the mechanism by which certain bacteria can "sense" the density of cells.
- ✓ Why would it be advantageous for a bacterium to synthesize more than one type of homoserine lactone?

7.8 Genomics

Focus Point

- Explain how protein-encoding regions are found when analyzing a DNA sequence.

Increasingly rapid methods of determining the nucleotide sequence of DNA have led to exciting advancements in genomics. In 1995, the sequence of the chromosome of *Haemophilus influenzae* was published, marking the first complete genomic sequence ever determined. Since then, microbial genome sequencing has become almost commonplace. A table describing some of the representative prokaryotes that have been sequenced is available at the Online Learning Center (www.mhhe.com/nester6).

Although sequencing methodologies are becoming more rapid, analyzing the resulting data and extracting the pertinent information is far more complex than it might initially seem. One of the most difficult steps is to locate and characterize the potential protein-encoding regions. Imagine trying to determine the amino acid sequence of a protein encoded by a 1,000-base-pair (bp) stretch of DNA, without knowing anything about the orientation of the promoter or the reading frame of the transcribed mRNA. Since either strand of the double-stranded DNA molecule could be the template strand, two entirely different mRNA molecules could potentially code for the protein. In turn, each of those two molecules has three reading frames, for a total of six reading frames. Yet only one of these actually codes for the protein. Understandably, computers are an invaluable aid and are used extensively in deciphering the meaning of the raw sequence data. In turn, this has resulted in the emergence of a new field, **bioinformatics,** which creates the computer technology to store, retrieve, and analyze nucleotide sequence data.

Analyzing a Prokaryotic DNA Sequence

When analyzing a DNA sequence, the nucleotide sequence of the (+) strand is used to infer information contained in the corresponding RNA transcript. Because of this, terms like start codon, which actually refers to a sequence in mRNA, are used to describe sequences in DNA. In other words, to locate the start codon AUG, which would be found in mRNA, one would look for the analogous sequence, ATG, in the (+) strand of DNA. In most cases it is not initially known which of the two strands is actually used as a template for RNA synthesis. Only after a promoter is located can this be determined.

To locate protein-encoding regions, computers are used to search for **open reading frames (ORFs),** stretches of DNA, generally longer than 300 bp, that begin with a start codon and end with a stop codon. An ORF potentially encodes a protein. Other characteristics, such as the presence of an upstream sequence that can serve as a ribosome-binding site, also indicate that an ORF encodes a protein.

The nucleotide sequence of the ORF or deduced amino acid sequence of the encoded protein can be compared with other known sequences by searching computerized databases of published sequences. Not surprisingly, as genomes of more organisms are being sequenced, information contained in these databases is growing at a remarkable rate. If the encoded protein shows certain amino acid similarities, or **homology,** to characterized proteins, a putative function can sometimes be assigned. For example, proteins that bind DNA have similar amino acid sequences in certain regions. Likewise, regulatory regions in DNA such as promoters can sometimes be identified based on similarities to known sequences.

MICROCHECK 7.8

Sequencing methodologies are quickly becoming more rapid, but analyzing the data and extracting the pertinent information is difficult.

- ✓ What is an open reading frame?
- ✓ Describe two things that you can learn by searching a computerized database for sequences that have homologies to a newly sequenced gene.
- ✓ There are characteristic differences in the nucleotide sequences of the leading and lagging strands. Why might this be so?

FUTURE CHALLENGES

Gems in the Genomes?

From a medical standpoint, one of the most exciting challenges will be to capitalize on the rapidly accruing genomic information and use that knowledge to develop new drugs and therapies. The potential gains are tremendous, particularly in the face of increasing resistance to current antimicrobial drugs. For example, by studying the genomes of pathogenic microorganisms, scientists can learn more about specific genes that enable an organism to cause disease. By learning more about the signals and mechanisms that turn these genes on and off, scientists may be able to one day design a drug that prevents the synthesis of critical bacterial proteins. Such a drug could interfere with that pathogen's ability to survive within our body and thereby render it harmless. ■ **resistance to antimicrobial drugs, p. 483**

Learning more about the human genome provides another means of developing drug therapies. Already, companies are searching genomic databases, a process called **genome mining,** to locate ORFs that may encode proteins of medical value. What they generally look for are previously uncharacterized proteins that have certain sequence similarities to proteins of proven therapeutic value. Some of their discoveries are now in clinical trials to test their efficacy. Genes encoding many other medically useful proteins are probably still hidden, waiting to be discovered.

SUMMARY

7.1 Overview (figure 7.1)

Characteristics of DNA (figure 7.3)

A single strand of DNA has a 5′ end and a 3′ end; the two strands of DNA in the double helix are **antiparallel.**

Characteristics of RNA

A single-stranded RNA fragment is transcribed from one of the two strands of DNA. There are three different functional groups of RNA molecules: **messenger RNA (mRNA), ribosomal RNA (rRNA),** and **transfer RNA (tRNA).**

Regulating the Expression of Genes

Protein synthesis is generally controlled by regulating the synthesis of mRNA.

7.2 DNA Replication

DNA replication is generally **bidirectional** and **semiconservative** (figure 7.4). The DNA chain always elongates in the 5′ to 3′ direction (figure 7.5).

Initiation of DNA Replication

DNA replication begins at the **origin of replication. DNA polymerase** synthesizes DNA in the 5′ to 3′ direction, using one strand as a **template** to generate the complementary strand.

The Replication Fork (figure 7.6)

The bidirectional progression of replication around a circular DNA molecule creates two replication forks; numerous enzymes and other proteins are involved.

7.3 Gene Expression in Bacteria

Transcription

RNA polymerase catalyzes the process of transcription, producing a single-stranded RNA molecule that is complementary and antiparallel to the DNA template (figure 7.7). **Transcription** begins after RNA polymerase recognizes and binds to a **promoter** (figure 7.8). RNA is synthesized in the 5′ to 3′ direction (figure 7.9). When **RNA polymerase** encounters a **terminator,** it falls off the DNA template and releases the newly synthesized RNA.

Translation

The information encoded by mRNA is deciphered using the genetic code (figure 7.10). A nucleotide sequence has three potential **reading frames** (figure 7.11). **Ribosomes** function as the site of translation (figure 7.12). **tRNAs** carry specific amino acids and act as keys that interpret the genetic code (figure 7.13). In prokaryotes, initiation of translation begins when the ribosome binds to the **ribosome-binding site** of the mRNA molecule. Translation starts at the first AUG downstream of that site (figure 7.14). The ribosome moves along mRNA in the 5′ to 3′ direction; translation terminates when the ribosome reaches a **stop codon** (figure 7.15). Polypeptides are often modified after they are synthesized.

7.4 Differences Between Eukaryotic and Prokaryotic Gene Expression (table 7.4)

Eukaryotic mRNA is **processed;** a **cap** and a **poly A tail** are added. Eukaryotic genes often contain **introns** which are removed from **precursor mRNA** by a process called **splicing** (figure 7.16). In eukaryotic cells, the mRNA must be transported out of the nucleus before it can be translated in the cytoplasm.

7.5 Regulation of Bacterial Gene Expression

Principles of Regulation

Constitutive enzymes are constantly synthesized. The synthesis of **inducible enzymes** can be turned on by certain conditions. The synthesis of **repressible enzymes** can be turned off by certain conditions.

Mechanisms to Control Transcription

Repressors block transcription (figure 7.17). **Activators** enhance transcription (figure 7.18).

The lac Operon As a Model for Control of Metabolic Pathways (figure 7.19)

The *lac* operon employs a repressor that prevents transcription of the genes when lactose is not available. Catabolite repression prevents transcription of the *lac* operon when glucose is available.

7.6 Regulation of Eukaryotic Gene Expression

Regulation in eukaryotic cells is much more complicated than that in prokaryotic cells. **RNA interference (RNAi)** is a recently discovered mechanism that uses either **microRNA (miRNA)** or **short interfering RNA (siRNA)** to direct the degradation of RNA transcripts.

7.7 Sensing and Responding to Environmental Fluctuations

Signal Transduction

Two-component regulatory systems utilize a sensor that recognizes changes outside the cell and then transmits that information to a **response regulator.** Bacteria that utilize **quorum sensing** synthesize a compound that activates specific genes when it reaches a critical concentration.

Natural Selection

The expression of some genes changes randomly, enhancing the chances of survival of at least a subset of a population under varying environmental conditions. **Antigenic variation** is a routine change in the expression of surface proteins. **Phase variation** is the routine switching on and off of certain genes.

7.8 Genomics

Analyzing a Prokaryotic DNA Sequence

When analyzing a DNA sequence, the nucleotide sequence of the (+) strand is used to infer information carried by the corresponding RNA transcript; computers are used to search for **open reading frames (ORFs).**

REVIEW QUESTIONS

Short Answer

1. Explain what the term *semiconservative* means with respect to DNA replication.
2. How can *E. coli* have a generation time of only 20 minutes when it takes 40 minutes to replicate its chromosome?
3. What is the function of primase in DNA replication? Why is this enzyme necessary?
4. What is polycistronic mRNA?
5. Explain why knowing the orientation of a promoter is critical when determining the amino acid sequence of an encoded protein.
6. What is the function of a sigma factor?
7. What is the fate of a protein that has a signal sequence?
8. Compare and contrast regulation by a repressor and an activator.
9. Explain how some bacteria sense the density of cells in their own population.
10. Explain why it is sometimes difficult to locate genomic regions that encode a protein.

Multiple Choice

1. All of the following are involved in transcription, *except*
   a) polymerase. b) primer. c) promoter.
   d) sigma factor. e) uracil.
2. All of the following are involved in DNA replication, *except*
   a) elongation factors. b) gyrase. c) polymerase.
   d) primase. e) primer.

3. All of the following are directly involved in translation, *except*
   a) promoter. b) ribosome. c) start codon.
   d) stop codon. e) tRNA.
4. Using the DNA strand depicted here as a template, what will be the sequence of the RNA transcript?

   promoter →
   5′ GCGTTAACGTAGGC 3′
   3′ CGCAATTGCATCCG 5′

   a) 5′ GCGUUAACGUAGGC 3′ b) 5′ CGGAUGCAAUUGCG 3′
   c) 5′ CGCAAUUGCAUCCG 3′ d) 5′ GCCUACGUUAACGC 3′
5. A ribosome binds to the following mRNA at the site indicated by the dark box. At which codon will translation likely begin?

   5′ ■ GCCGGAAUGCUGCUGGC

   a) GCC b) GGC
   c) AUG d) AAU
6. Allolactose induces the *lac* regulon by binding to a(n)
   a) operator. b) repressor.
   c) activator. d) CAP protein.
7. Under which of the following conditions will transcription of the *lac* operon occur?
   a) Lactose present/glucose present
   b) Lactose present/glucose absent
   c) Lactose absent/glucose present
   d) Lactose absent/glucose absent
   e) A and B
8. Which of the following statements about gene expression is *false*?
   a) More than one RNA polymerase can be transcribing a specific gene at a given time.
   b) More than one ribosome can be translating a specific transcript at a given time.
   c) Translation begins at a site called a promoter.
   d) Transcription stops at a site called a terminator.
   e) Some amino acids are coded for by more than one codon.
9. Which of the following is not characteristic of eukaryotic gene expression?
   a) 5′ cap is added to the mRNA.
   b) A poly A tail is added to the 3′ end of mRNA.
   c) Introns must be removed to create the mRNA that is translated.
   d) The mRNA is often polycistronic.
   e) Translation begins at the first AUG.
10. Which of the following statements is *false?*
    a) A derivative of lactose serves as an inducer of the *lac* operon.
    b) Signal transduction provides a mechanism for a cell to sense the conditions of its external environment.
    c) The function of an acylated homoserine lactone is to enable a cell to sense the density of like cells.
    d) An example of a two-component regulatory system is the lactose operon, which is controlled by a repressor and an activator.
    e) An ORF is a stretch of DNA that may encode a protein.

Applications

1. A graduate student is trying to identity the gene coding for an enzyme found in a bacterial species that degrades trinitrotoluene (TNT). The student is frustrated to find that the organism does not produce the enzyme when grown in nutrient broth, making it is difficult to collect the mRNA needed to help identify the gene. What could the student do to potentially increase the amount of the desired enzyme?
2. A student wants to remove the introns from a segment of DNA coding for protein X. Devise a strategy to do this.

Critical Thinking

1. The study of protein synthesis often uses a cell-free system where cells are ground with an abrasive to release the cell contents and then filtered to remove the abrasive. These materials are added to the system, generating the indicated results:

Materials Added	Results
Radioactive amino acids	Radioactive protein produced
Radioactive amino acids *and* RNase (an RNA-digesting enzyme)	No radioactive protein produced

   What is the best interpretation of these observations?
2. In a variation of the experiment in the previous question, the following materials were added to three separate cell-free systems, generating the indicated results:

Materials Added	Results
Radioactive amino acids	Radioactive protein produced
Radioactive amino acids *and* DNase (a DNA-digesting enzyme)	Radioactive protein produced
Several hours after grinding:	
Radioactive amino acids *and* DNase	No radioactive protein produced

   What is the best interpretation of these observations?

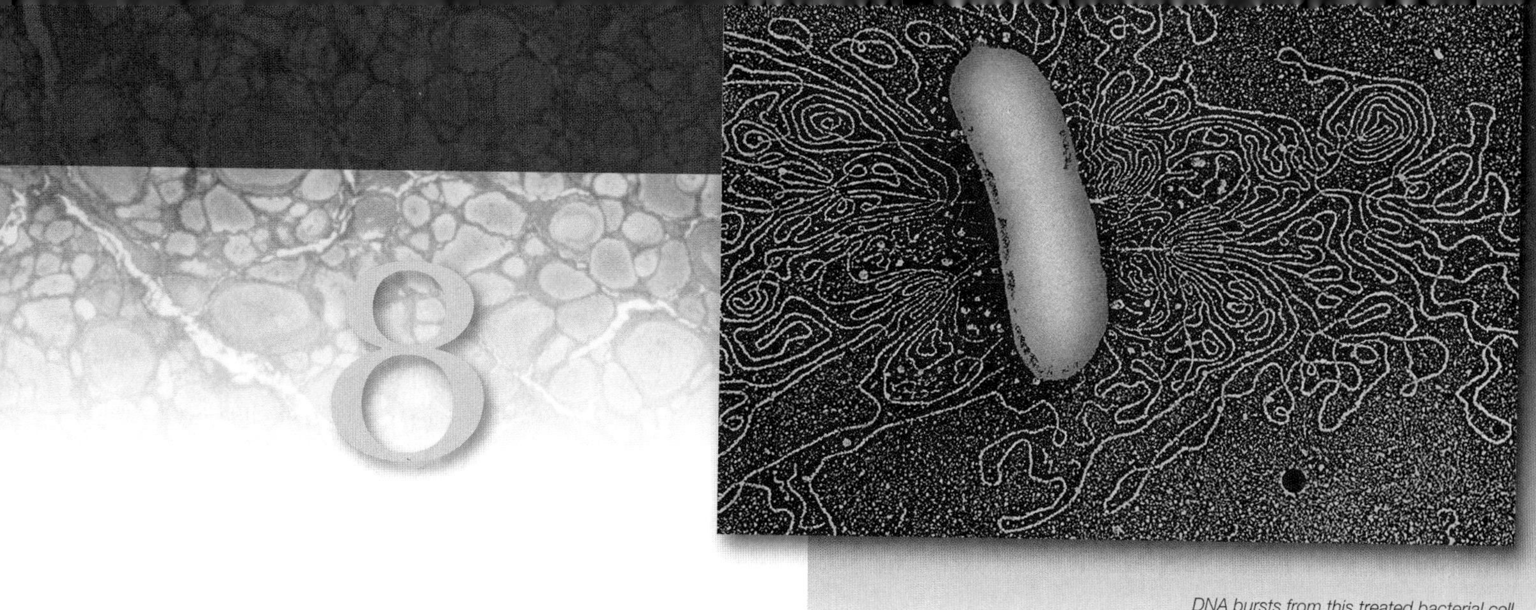

DNA bursts from this treated bacterial cell.

8 Bacterial Genetics

A Glimpse of History

Barbara McClintock (1902–1992) was a remarkable scientist who made several very important discoveries in genetics. She carried out her studies before the age of large interdisciplinary research teams and before the sophisticated tools of molecular genetics were available. Her tools consisted of a clear mind and a consuming curiosity that could make sense of confusing and revolutionary observations. She worked 12-hour days, 6 days a week in a small laboratory at Cold Spring Harbor on Long Island, New York. In 1983, at age 81, McClintock received the Nobel Prize in Medicine or Physiology largely for her discovery 40 years earlier of transposable elements, or transposons, popularly called "jumping genes." Her experimental system consisted of kernels of corn. She observed kernels of various colors produced by different enzymes (see figure 8.6). If the gene coding for an enzyme responsible for a particular color was inactivated, the kernel was not pigmented. If the enzyme was only partially inactivated, the kernel was partially pigmented. Thus, by looking at kernel colors, McClintock could detect changes in gene function. From her genetic analysis, she concluded that something, most likely pieces of DNA, must be moving into and out of genes to account for the differences in kernel color. When a piece of DNA, called a transposable element, moved into a gene, the gene could no longer function. When the transposable element left the gene, it was restored to its original state, and would function normally again.

When McClintock published her results, most scientists believed that chromosomal DNA was very stable and unchanging. Consequently, most geneticists were very skeptical of McClintock's heretical ideas. As a result, she stopped publishing many of her observations and it was not until the late 1970s that her ideas began to be accepted. By that time, transposable elements had been discovered in many organisms, including bacteria. Although transposons were first discovered in plants, once they were found in bacteria, the field moved ahead very quickly. The techniques of molecular biology, biochemistry, and genetics made the understanding of "jumping genes" possible.

Staphylococcus aureus, the Gram-positive coccus commonly called Staph, is a frequent cause of skin infections, such as boils and pimples. Since the 1970s the usual treatment for these infections has been penicillin-like antibiotics, such as methicillin. Today, however, this treatment is likely to fail. In 2003, well over 60% of the *S. aureus* strains isolated in hospitals were resistant to this antibiotic. Unfortunately, methicillin-resistant Staph is also resistant to a variety of other antibiotics. These resistant organisms now are commonly treated with a less effective antimicrobial, vancomycin, often considered the drug of last resort. However, in 2002 the situation became more worrisome—Staph isolated from foot ulcers on a diabetes patient in Detroit was vancomycin-resistant. This organism was also resistant to most common antibiotics, including penicillin, methicillin, and ciprofloxacin. How do multiple resistant strains arise and evolve? How are these resistance traits transferred so readily to other bacteria?

The answer to these and many other questions important to human health requires a basic understanding of bacterial genetics. This subject encompasses the study of heredity—how genes function (chapter 7), how they can change, and how they are transferred to other cells in the population. With this knowledge you will understand why antibiotics are no longer miracle drugs against infectious diseases.

KEY TERMS

Auxotroph A microorganism that requires an organic growth factor.

Conjugation Mechanism of horizontal gene transfer in which the donor cell must physically contact the recipient cell.

DNA-Mediated Transformation Mechanism of horizontal gene transfer between bacteria in which the bacterial DNA is transferred as "naked" DNA.

Extrachromosomal DNA that is not part of a chromosome.

Genomic Island Large segment of DNA that has been acquired from another species through horizontal gene transfer. Examples include pathogenicity islands and antibiotic resistant islands.

Genotype The sequence of nucleotides in the DNA of an organism.

Haploid Containing only a single set of genes.

Homologous Recombination Genetic exchange between stretches of similar or identical nucleotide sequences. Involves a type of breakage and rejoining of DNA into new combinations and replacement of DNA.

Horizontal Gene Transfer Transmission of DNA from one bacterium to another by conjugation, DNA-mediated transformation, or transduction. Also called lateral gene transfer.

Mutation A change in the nucleotide sequence of a cell's DNA, which is then passed on to daughter cells. A mutation can alter the protein which it encodes.

Non-Homologous Recombination Genetic recombination that does not require the two DNAs to have similar sequences in the region of recombination. Frequently the process involves addition and not replacement of genes.

Phenotype The observed characteristics of a cell resulting from expression of the genotype.

Plasmid A small extrachromosomal DNA molecule that replicates independently of the chromosome and generally encodes information not essential to the life of the cell.

Prototroph A microorganism that has no requirements for organic growth factors because it can synthesize them.

Reactive Oxygen Toxic forms of oxygen that modify and damage DNA.

Replicon A piece of DNA that has an origin of replication and is therefore capable of replicating.

Transduction Mechanism of horizontal gene transfer between bacteria in which bacterial DNA is transferred inside a phage coat.

Transposable Element (Transposon) Genes that move from one replicon to another site on the same replicon, or to another replicon in the same cell. Involves non-homologous recombination.

8.1

Genetic Change in Bacteria

Focus Points

- Name the two genetic changes that can alter the properties of bacteria.
- Distinguish between the genotype and the phenotype of a cell.

In the ever-changing conditions that characterize most environments, all organisms need to adapt in order to succeed. If they fail, competing organisms more "fit" to multiply in the new setting will soon predominate. This is the process of **natural selection.** Bacteria have two general means by which they routinely adjust to new circumstances: regulating gene expression (discussed in the previous chapter), and genetic change, which is the focus of this chapter. ■ regulation of gene expression, p. 176

A genetic change alters an organism's **genotype,** the sequence of nucleotides in its DNA. This can have a profound impact on bacteria because they are **haploid,** meaning they contain only a single set of genes. There is no "backup copy" of a gene in a haploid organism. Because of this, a change in genotype can easily alter the observable characteristics of an organism, its **phenotype.** Note, however, that the phenotype involves more than just the genetic makeup of an organism; it is also influenced by environmental conditions. For example, *Serratia marcescens* colonies are typically red when incubated at 22°C, but white when incubated at 37°C. This reversible change in the phenotype is governed by the organism's environment—temperature in this case. Altering the genotype of *S. marcescens* by removing the genes that direct production of the red pigment will also change the phenotype of the organism.

Genetic change in an organism can occur by two mechanisms—mutation and gene transfer. **Mutation** is a change in the existing nucleotide sequence of a cell's DNA, which is then passed on to daughter cells. The modified organism is referred to as a **mutant,** and the progeny (offspring) will be mutants as well. Because a mutation arises in a single cell and is then passed to the progeny, this adaptation is referred to as **vertical gene transfer (figure 8.1a).** The other mechanism of genetic change, **gene transfer,** is the acquisition of genes from another organism. It is commonly referred to as either **horizontal gene transfer** or **lateral gene transfer,** to emphasize that the cell acquires DNA from a different source (figure 8.1b). Like mutations, the changes are then passed to the progeny of the altered organism.

MICROCHECK 8.1

The properties of bacteria can change either through mutations, or by acquiring genetic information from other sources.

✓ Contrast genotype and phenotype.

✓ Which has a longer-lived effect on a cell—a change in the genotype or a change in the phenotype?

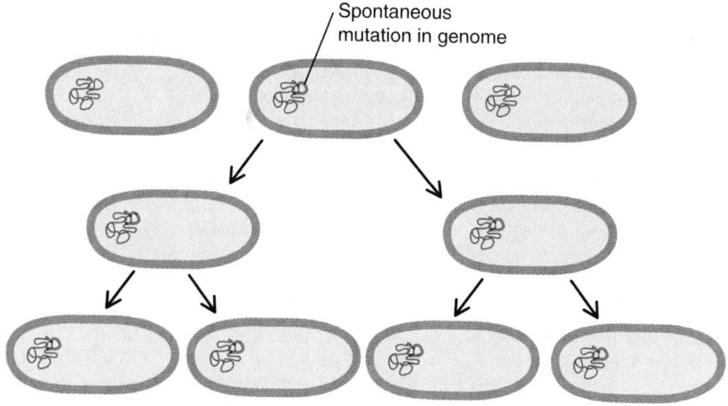

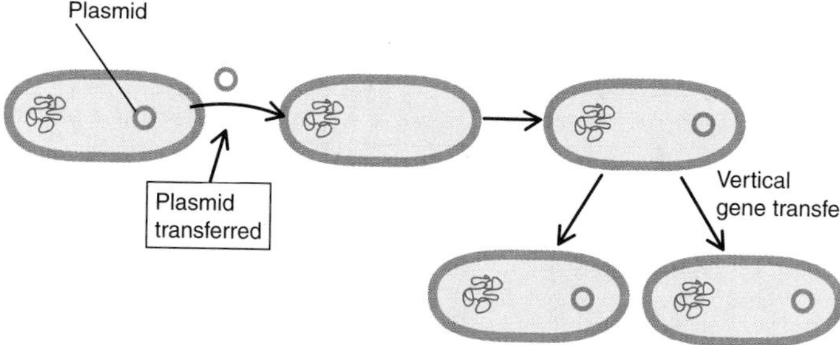

FIGURE 8.1 The Acquisition and Transfer of Genetic Information to Progeny (a) Spontaneous mutation. All progeny of mutant cells will be mutant. **(b)** Gene transfer or horizontal acquisition of genetic information. All progeny of the cell acquiring the plasmid will also carry the plasmid by vertical gene transfer. A copy of the plasmid remains in the cell that transfers the plasmid. Genes can be transferred by several different mechanisms.

GENE MUTATION AS A MECHANISM OF GENETIC CHANGE

Mutation changes the DNA base sequence so that it differs from that of the **wild-type organism,** a strain whose properties are similar to the organism first isolated from nature. The change in a nucleotide sequence may lead to an altered phenotype. A change in phenotype results when the protein coded by the gene does not function properly. The substitution of even one amino acid for another in a critical location in the protein such as in the catalytic site may cause the protein to be dysfunctional, thereby changing the properties of the cell. For example, if any gene of the tryptophan operon is altered so that the encoded enzyme no longer functions properly, the cells will grow well only if this amino acid is in its environment. A mutant that requires a growth factor is called an **auxotroph** (auxo means "increase," as an increase in requirements). Cells that grow in the absence of any added growth factors are termed **prototrophs.** ■ enzymes, pp. 129, 134 ■ protein structure, p. 28 ■ operon, p. 177 ■ growth factor, p. 94

By convention, the auxotrophic requirements of a cell are designated by three-letter abbreviations. For example, a cell that cannot make tryptophan is designated Trp^-. The first letter is in caps. Each growth factor has its own three-letter designation. If a cell grows without the addition of tryptophan, it is Trp^+. However, for convenience, only growth factors required are indicated. Likewise, only if a cell is resistant to an antimicrobial agent such as streptomycin is this indicated, such as Str^R. Otherwise, the cells are assumed to be sensitive, Str^S, and this is not indicated when the phenotype of the cell is described.

8.2 Spontaneous Mutations

Focus Points

- Name three types of mutations that can occur spontaneously.
- Name three types of mutations that can result from base substitutions.
- Explain how mutations relate to natural selection.

Spontaneous mutations are those that occur in the cell's natural environment. Mutations occur randomly, and each gene mutates spontaneously and infrequently at a characteristic rate. The rate of mutation is defined as the probability that a mutation will occur in a given gene each time a cell divides; this rate is generally expressed as a negative exponent. The mutation rate of different genes usually varies between 10^{-4} and 10^{-12} per cell division. In other words, the chances that any single gene will undergo a mutation when one cell divides into two are between one in 10,000 (10^{-4}) and one in a trillion (10^{-12}). ■ exponents, Appendix 1, p. A-1

Genes mutate independently of one another. Consequently, the chance that two given mutations will occur within the same cell is very low. Indeed, the actual occurrence is the product of the individual rates of mutation of the two genes (calculated by taking the sum of the exponents). For example, if the mutation rate to streptomycin resistance is 10^{-6} per cell division and the mutation rate to penicillin resistance is 10^{-8} per cell division, the probability that both mutations will occur within the same cell is $10^{-6} \times 10^{-8}$, or 10^{-14}. For this reason, two or more drugs may be administered simultaneously in the treatment of some diseases such as tuberculosis and AIDS. Any mutant cell or virus resistant to one antimicrobial medication is likely to still be sensitive to the other and therefore will be killed by the combination of the two antimicrobials.

Mutations are stable so that the progeny of a mutant will retain the genotype. On rare occasions, however, the mutation will change back to its original, non-mutant state. This change in a mutated gene is termed **reversion** and, like the original mutation, it occurs spontaneously at low frequencies.

Because of mutations, the concept that all cells arising from a single cell are identical is not strictly true, since every large population contains mutants. Even in a single colony that contains about 1 million cells, all cells are not completely identical because of spontaneous random mutations. These mutations provide a mechanism by which organisms, with their altered characteristics, can respond to a changing environment. This is the process of natural selection. The environment does not cause the mutation but rather selects those cells that can grow under its conditions. Thus, a spontaneous mutation to antimicrobial resistance, though rare, will result in the mutant becoming the dominant organism in a hospital environment where the antimicrobial medication is present. The antimicrobial kills the sensitive cells and thereby allows the resistant cells to take over the population.

Base Substitution

The most common type of mutation occurs during DNA synthesis, when an incorrect base is incorporated into DNA, an event called **base substitution (figure 8.2).** If only one base pair is changed, the mutation is called a **point mutation.**

Three outcomes are possible from base substitutions: a silent mutation, missense mutation, or nonsense mutation **(figure 8.3).** A **silent mutation** results from a nucleotide change that generates a codon that still specifies the wild type amino acid. This is possible because the genetic code is degenerate, meaning that different codons can encode the same amino acid. A **missense mutation** results when the new codon specifies a different amino acid. The effect of this type of mutation depends on the position of the change and the difference between the original and new amino acid. In many cases, cells with a missense mutation in a gene of tryptophan synthesis can grow slowly in the absence of tryptophan because the encoded protein is partially functional. Such a mutation is termed **leaky.** A **nonsense mutation** occurs when the new codon is a stop codon, resulting in a shortened, or **truncated,** protein. The site of the nonsense mutation dictates the length of the protein; in most cases, the truncated protein is non-functional. Any mutation that totally inactivates the gene is termed a **null** or **knockout mutation.** ■ genetic code, p. 170 ■ codon, p. 170 ■ stop codon, p. 173

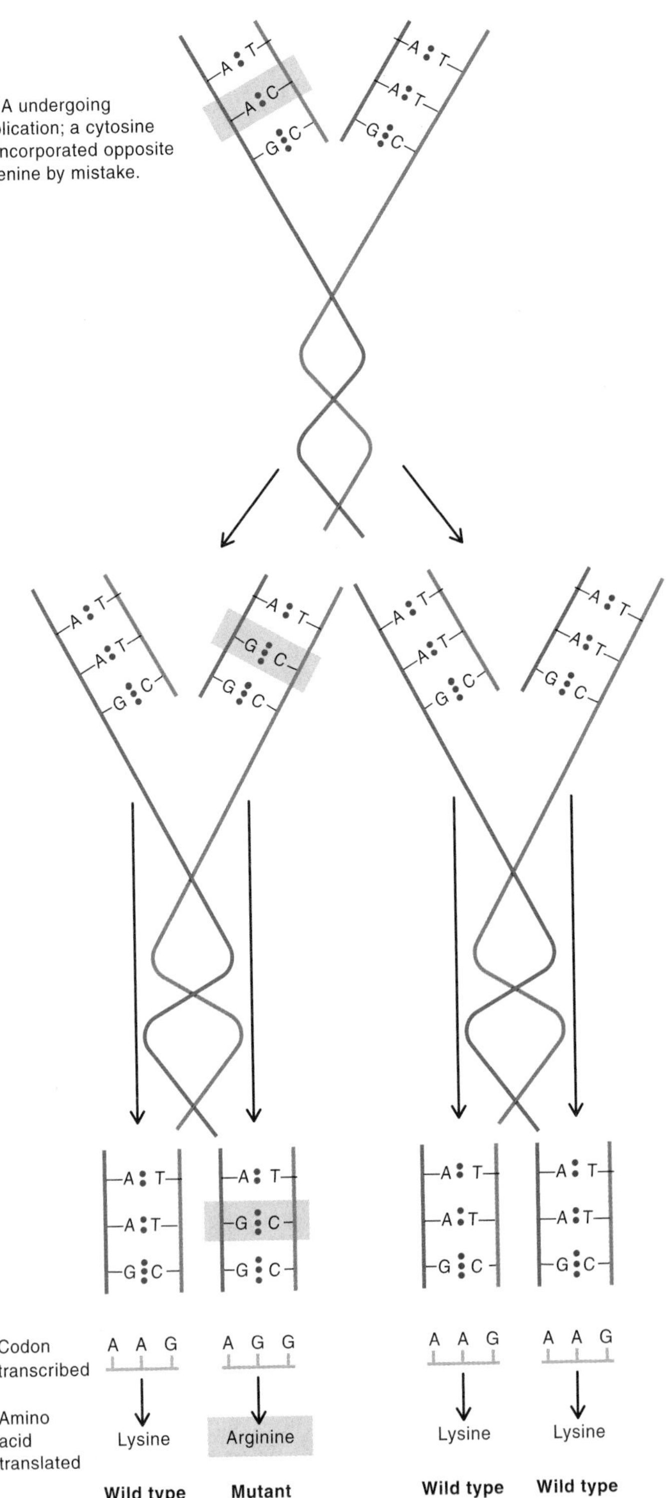

FIGURE 8.2 Base Substitution Shown here is the generation of a mutant organism as a result of the incorporation of a pyrimidine base (cytosine) in place of thymine in DNA replication. The mutation is a missense mutation.

Note that geneticists sometimes use the term "silent mutation" to indicate a mutation that does not alter the function of the protein. With this broad definition, any base substitution that does not affect protein function would be a silent mutation.

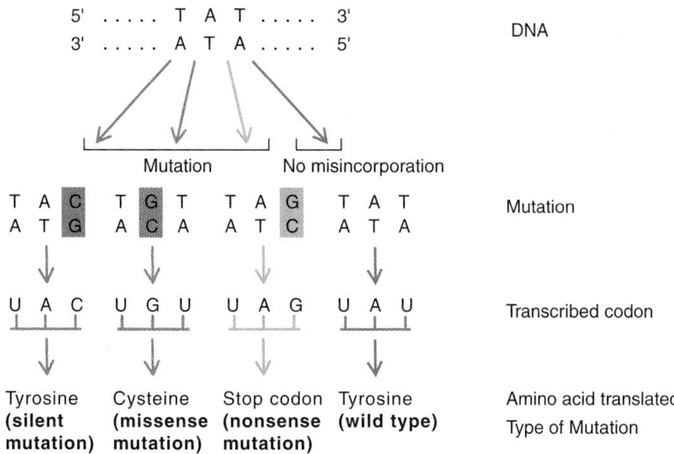

FIGURE 8.3 Three Types of Mutations Resulting from Base Substitutions The cells remain wild type if there is no misincorporation.

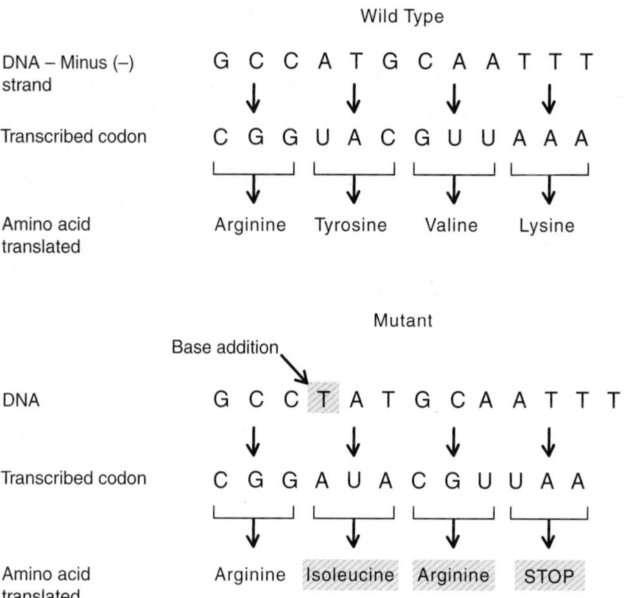

FIGURE 8.4 Frameshift Mutation As a Result of Base Addition The addition of a nucleotide (T) to the DNA results in a frameshift when the DNA is transcribed into mRNA and a new triplet code word is translated as a new amino acid. The deletion of a nucleotide would have essentially the same effect. The protein chain terminates when a stop codon appears in the DNA.

Oxygen in the environment can increase the frequency of base substitutions. As discussed in chapter 4, O_2 can be readily converted through cellular metabolism or by environmental factors into forms highly toxic to cells, such as superoxide (O_2^-) and hydrogen peroxide (H_2O_2). In part, this toxicity results from these **reactive forms of oxygen** damaging cellular DNA and causing mutations. These reactive forms of oxygen can oxidize guanine before or after it has been incorporated into DNA, and DNA polymerase often mispairs oxidized guanine with adenine rather than cytosine, thereby resulting in a base substitution and a point mutation.

Removal or Addition of Nucleotides

The deletion or addition of nucleotides, which may occur in the course of DNA replication, is another type of spontaneous mutation. The consequence of this depends on how many nucleotides are deleted or added.

If three nucleotides are deleted (or added), this effectively removes (or adds) one codon from the DNA. When the gene is expressed, one additional (or one fewer) amino acid will be in the resulting protein. How serious this change is depends on the location of the change in the encoded protein.

Adding or subtracting one or two nucleotides is more significant than adding or subtracting three because it causes a **frameshift mutation.** This results because translation of a gene begins at a specific codon and proceeds one codon at a time. Thus, the deletion or addition of a nucleotide shifts the codons of the DNA when it is transcribed into mRNA **(figure 8.4).** Consequently, a frameshift mutation changes the reading frame, so that an entirely different set of codons is used. Frequently, one of the resulting downstream codons will be a stop codon. As a result, a frameshift mutation is likely to result in a protein that is truncated and probably non-functional—a knockout mutation. ■ **reading frame, p. 171** ■ **downstream, p. 170**

Transposable Elements (Jumping Genes)

Transposable elements, also called **transposons** or jumping genes, are distinct segments of DNA that can direct their own movement in a process called **transposition.** Inside a single cell, a transposon can "jump" to a different location within the chromosome, or to a plasmid, or vice versa **(figure 8.5).** The gene into which a transposon inserts no longer encodes a functional protein because the insertion disrupts the gene; this is an example of **insertional inactivation.** Because most transposons contain transcriptional terminators that stop mRNA synthesis, the expression of genes downstream of the insertion in the same operon will also be affected. The structure and biology of transposons will be described later in this chapter. ■ **transcriptional terminators, p. 170** ■ **operon, p. 177**

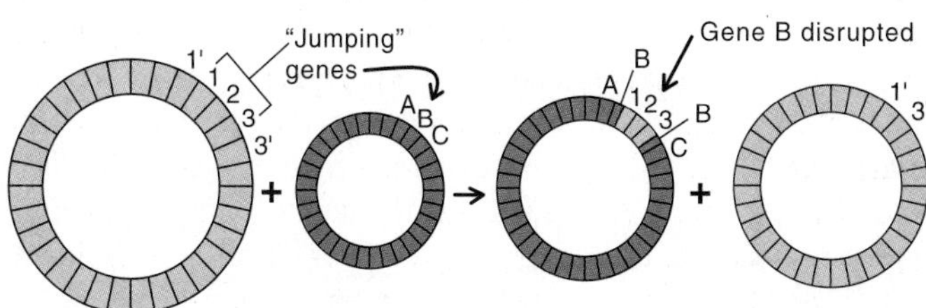

FIGURE 8.5 Transposition The transposon consisting of (genes 1, 2, 3), has the ability to "jump" from one piece of DNA to another, where it becomes integrated. In this case, the transposon has jumped into gene B, thereby mutating it.

FIGURE 8.6 Transposition Detected by Color Changes Variegation in color observed in the kernels of corn is caused by the insertion of transposable elements into genes involved in the synthesis of different pigments, thereby altering the synthesis of the pigments.

The classic studies of transposition were carried out by Dr. Barbara McClintock (see **A Glimpse of History**). She observed variation in the colors of corn kernels as a result of transposons moving into and out of genes concerned with pigment synthesis **(figure 8.6).**

MICROCHECK 8.2

Mutations, changes in the nucleotide sequences of DNA, may result in proteins that are dysfunctional, thereby altering the properties of the cell. A leaky mutation results in a partially functional protein; a knockout mutation results in a non-functional protein. Mutations most commonly occur spontaneously as a result of mistakes in DNA replication, in some cases because of reactive oxygen molecules that have modified the guanine in DNA.

- ✓ How would the growth requirements change in a cell that has a silent mutation in a gene for histidine synthesis? How would it change if the mutation were a knockout?
- ✓ If the rate of mutation to streptomycin resistance is 10^{-6} and that of penicillin resistance is 10^{-4}, what is the rate of mutation to simultaneous resistance to both antibiotics?
- ✓ Is it as effective to take two antibiotics sequentially as it is to take them simultaneously, as long as the total length of time that they are both taken is the same? Explain.

8.3

Induced Mutations

Focus Points

- Name four mechanisms by which mutagens act on DNA.
- Describe the effect of UV light on DNA.

Mutants in nature are important because they are the raw material on which natural selection operates. They are also essential for studying and understanding most aspects of genetics. Consequently, geneticists spend much time isolating mutants. One reason why bacteria represent an excellent experimental system for genetic studies and why more is known about *E. coli* than any other organism in the world is because bacterial mutants with a broad range of properties are easier to isolate than mutants in any other system. Bacteria grow rapidly to enormous numbers in very small volumes of inexpensive media. Thus, rare mutations will be represented in a small volume of medium. Further, since bacteria are haploid, with only one copy of each gene generally, the mutation will not be obscured by a wild-type gene.

Because the frequency of spontaneous mutations is so low, investigators trying to isolate certain mutants must resort to using **mutagens**—chemicals or radiation that can increase the frequency of mutations at least 1,000-fold. Such mutations are said to be **induced** by the mutagen.

Chemical Mutagens

Any chemical treatment that alters the hydrogen-bonding properties of a purine or pyrimidine base in the DNA will increase the frequency of mutations as the DNA replicates.

Chemical Modification of Purines and Pyrimidines

A large number of chemicals can alter the structure of purines and pyrimidines. The biggest group of chemical mutagens consists of **alkylating agents,** highly reactive chemicals that add alkyl groups (short chains of carbon atoms) onto purines and pyrimidines, thereby altering their hydrogen-bonding properties. A common alkylating agent used in research laboratories is nitrosoguanidine. Many compounds formerly used in cancer therapy are in this group. These compounds kill rapidly dividing cancer cells, but they also damage DNA in normal cells. As a result, these agents have caused cancers that appear more than 10 years after they were used to treat the original cancer. ■ **purines and pyrimidines, p. 33**

Another example of a powerful mutagen is nitrous acid (HNO_2). This chemical converts amino ($-NH_2$) to keto ($-C{=}O$) groups—for example, converting cytosine to uracil, which pairs with adenine rather than guanine when the DNA is replicated. Nitrous acid also removes amino groups from adenine and guanine.

Base Analogs

Base analogs are compounds that structurally resemble purine or pyrimidine bases closely enough that they can be mistakenly incorporated in place of the natural bases as nucleotides are synthesized. These can then be incorporated into DNA in place of the natural nucleotides. Base analogs such as 5-bromouracil and 2-amino purine, however, do not have the same hydrogen-bonding properties as the natural bases, thymine and adenine respectively. This difference increases the probability that, once incorporated into DNA, the base analog will pair with the wrong base as the complementary strand is being synthesized **(figure 8.7).** ■ **DNA replication, p. 164**

Intercalating Agents

A number of chemical mutagens, termed **intercalating agents,** increase the frequency of frameshift mutations. They are planar (flat) molecules of about the same size as a pair of nucleotides in DNA. These molecules do not alter hydrogen-bonding properties of the bases; rather, they insert, or **intercalate,** between adjacent base pairs

TABLE 8.1 Common Mutagens

Agent	Action	Result
Chemical Agent		
Chemical modification of bases		
Examples: nitrous acid	Converts amino group to keto group in adenine and cytosine	Base substitution
alkylating agents	Adds alkyl groups (CH_3 and others) to nitrogenous bases such as guanine	Base substitution
Base analogs Example: 5-bromo-uracil	Incorporates in place of normal nucleotide in DNA	Base substitution
Intercalating agents Example: ethidium bromide	Inserts between base pairs in either the template or new strand	Addition or subtraction of base pairs
Transposons	Random insertion of transposon into any gene	Insertional inactivation
Radiation		
Ultraviolet (UV)	Intrastrand thymine dimer formation	Base substitution
X rays	Single- and double-strand breaks in DNA	Deletion of bases

in one of the strands of DNA. This pushes the nucleotides apart, producing enough space between bases that errors are made during replication. If the intercalating agent inserts into the old strand of DNA, a base pair will be added as the new strand is synthesized. If it intercalates into the strand being synthesized, a deletion of a base pair will occur. In either case, the result is a frameshift mutation **(figure 8.8).** As in spontaneous frameshift mutants, the addition or subtraction of a nucleotide often results in a stop codon being generated prematurely in the mRNA transcribed from the altered DNA, and a shortened protein being synthesized. An intercalating agent commonly used in the laboratory to stain DNA is **ethidium bromide.** The manufacturer now warns users that ethidium bromide should be handled with great care because it likely is a **carcinogen**—a cancer-causing agent. Another intercalating agent is chloroquine, which has been used for many years to treat malaria. ■ DNA replication, p. 164

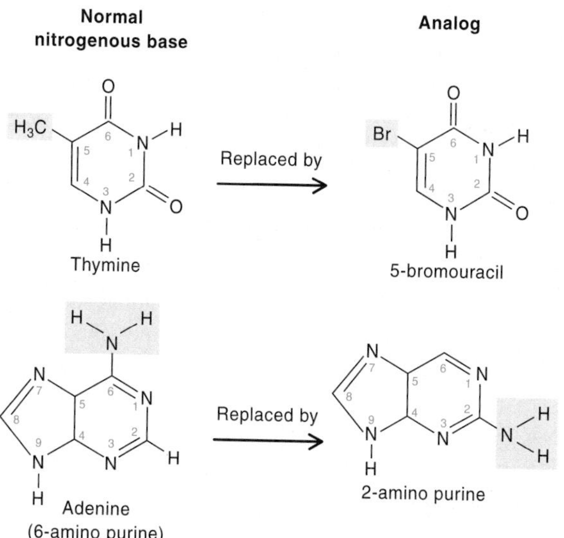

FIGURE 8.7 Common Base Analogs and the Normal Bases They Replace in DNA The important differences between the normal bases and the analogs are in boxes. Like the natural bases, they are incorporated as nucleotides into replicating DNA.

Transposition

A common procedure to generate mutants in research laboratories is to introduce a transposon into a cell. The transposon, which cannot replicate on its own because it lacks an origin of replication,

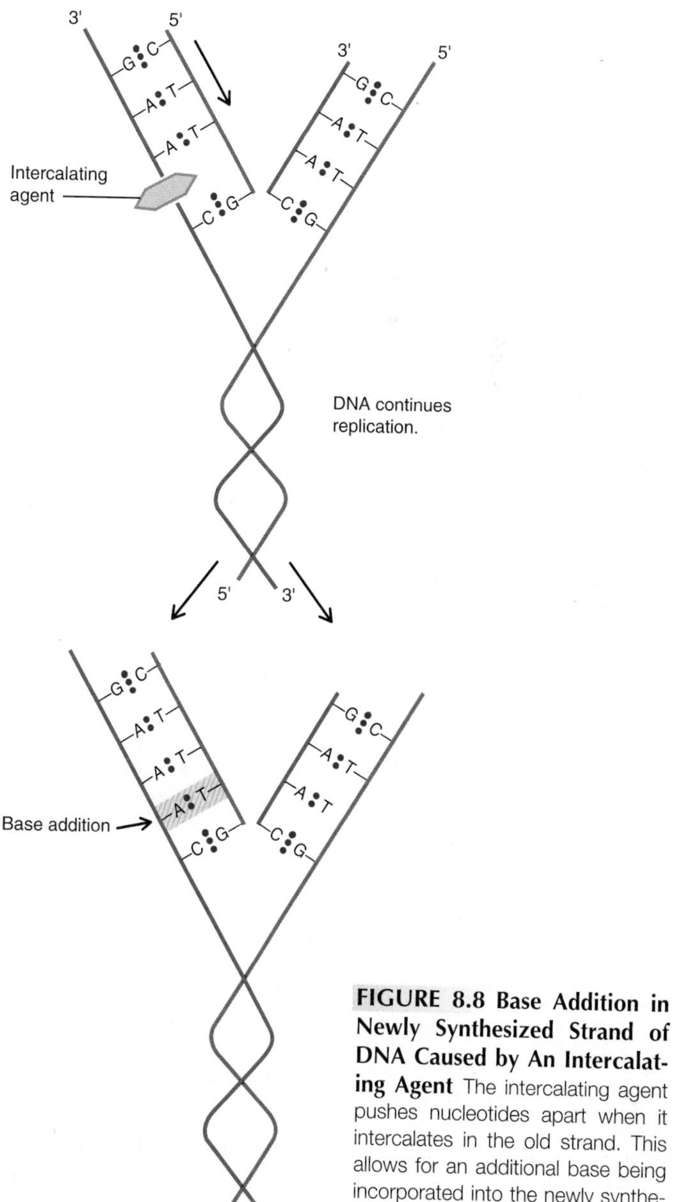

FIGURE 8.8 Base Addition in Newly Synthesized Strand of DNA Caused by An Intercalating Agent The intercalating agent pushes nucleotides apart when it intercalates in the old strand. This allows for an additional base being incorporated into the newly synthesized strand.

must integrate into the cell's genome in order to be replicated. The gene into which the transposon has inserted will usually be inactivated as a result of the insertion and usually results in a knockout mutation.

Radiation

Two kinds of radiation are mutagens: ultraviolet (UV) light and X rays. ■ wavelengths of radiation, p. 115

Ultraviolet Irradiation

Irradiation of cells with ultraviolet light causes covalent bond formation between adjacent thymine molecules on the same strand of DNA (intrastrand bonding), resulting in the formation of **thymine dimers (figure 8.9).** The covalent bonding distorts the DNA strand so much that the dimer cannot fit properly into the double helix, resulting in badly damaged DNA. The DNA molecule cannot be replicated, nor genes transcribed, beyond this site of damage and, as a result, the cells should die. How then can UV light be mutagenic? The major mutagenic action of UV light results from the cells repairing the damage by a mechanism termed **SOS repair** (see the next section). ■ gene transcription, p. 168

X Rays

X rays cause several types of damage: single- and double-strand breaks in DNA, and alterations to the bases. Double-strand breaks often result in deletions that are lethal. **Table 8.1** summarizes information on the common mutagens.

MICROCHECK 8.3

The frequency of spontaneous mutations can be increased significantly by treating cells with chemicals and radiation. These treatments induce mutations, which most frequently result from the alteration of hydrogen-bonding properties of the nitrogenous bases, or the addition or deletion of bases in DNA.

✓ How does UV light affect cells?

✓ Do you think that mutations caused by reactive oxygen should be considered spontaneous or induced? Justify your answer.

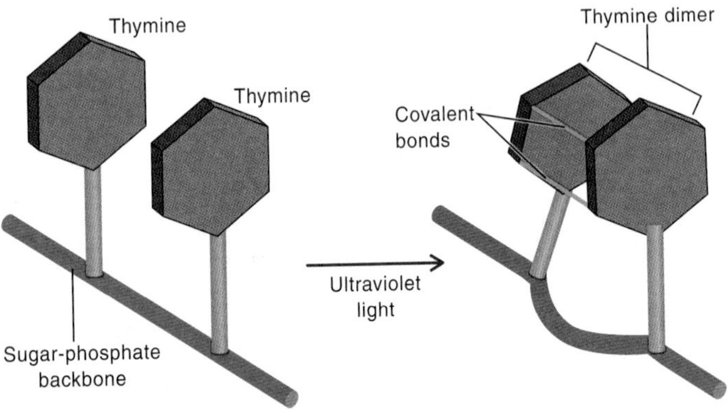

FIGURE 8.9 Thymine Dimer Formation Convalent bonds form between adjacent thymine molecules on the same strand of DNA when DNA is exposed to UV light. This distorts the shape of the DNA and prevents replication past the dimer.

8.4 Repair of Damaged DNA

Focus Points

- Explain how DNA polymerase can correct base substitutions and prevent misincorporation of nucleotides.
- Explain three mechanisms by which UV light damage can be repaired.
- Explain how a mutation caused by reactive oxygen is repaired.

Probably no function is more important to a cell than being able to repair damaged DNA because no molecule is more critical to the proper functioning of the cell. The amount of spontaneous and mutagen-induced damage to DNA in cells is enormous. Every 24 hours, the DNA in every cell in the human body is damaged spontaneously more than 10,000 times. This damage, if not repaired, can lead to cell death and, in animals, cancer. In humans, two breast cancer susceptibility genes code for enzymes that repair damaged DNA. Mutations in either one result in a high (80%) probability of breast cancer.

A major reason why observed mutations are so rare is that alterations in DNA are repaired shortly after they occur and before they can be passed on to progeny and change their properties. It is not surprising that, in the course of many million of years of evolution, all cells, both prokaryotic and eukaryotic, have developed several different mechanisms for repairing any damage that their DNA might suffer either spontaneously or by mutagens in the environment.

Repair of Errors in Base Incorporation

A common cause of spontaneous mutations is the incorporation of the wrong base by DNA polymerase as it replicates DNA. The resulting mispairing of bases results in a slight distortion in the DNA helix, which is recognized by enzymes within the cell that repair such mistakes. By quickly repairing the error before the DNA is replicated, the cell prevents the mutation. Two mechanisms exist for repairing errors in base incorporation: proofreading by DNA polymerase and mismatch repair. ■ DNA polymerase, p. 166

Proofreading by DNA Polymerase

DNA polymerases are complex enzymes that not only synthesize DNA, but also verify the accuracy of their actions—a characteristic called **proofreading** or **editing.** The enzymes can back up and excise (remove) any nucleotides that are not correctly hydrogen bonded to the base in the template strand. Following excision, the DNA polymerase then incorporates the correct nucleotide. Although the proofreading function of DNA polymerases is very efficient, it is not infallible.

Mismatch Repair

Mismatch repair fixes errors missed by the proofreading of DNA polymerase. A specific protein binds to the site of

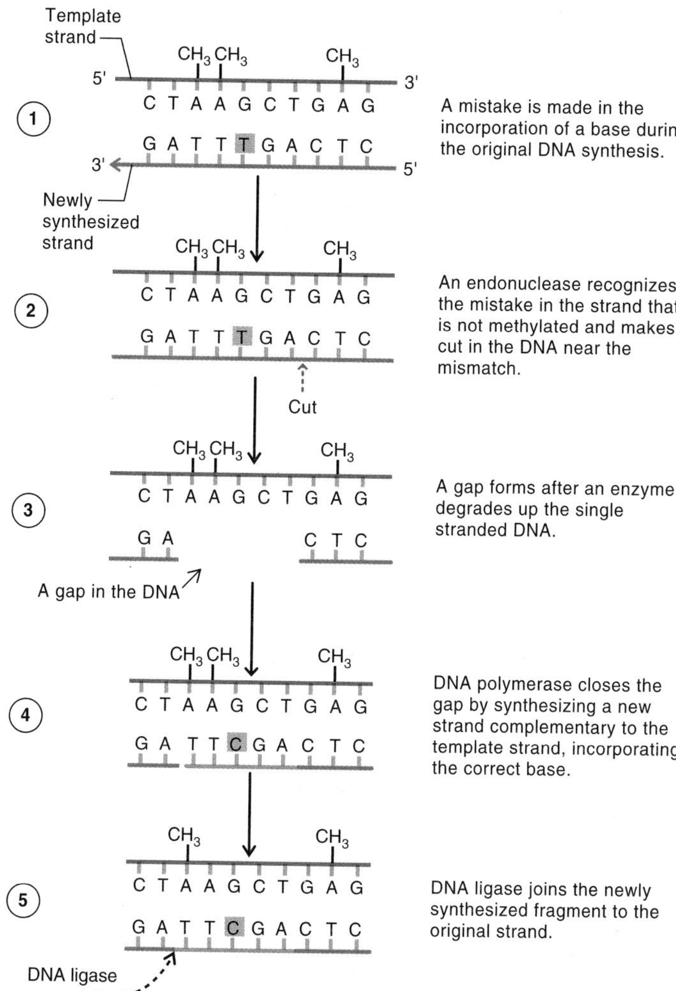

FIGURE 8.10 Mismatch Repair The endonuclease cuts the DNA near the misincorporated base and another enzyme chews up the single-stranded DNA, creating a gap in the backbone. A new complementary strand is then synthesized and joined to the original strand by DNA ligase.

the mismatched base, directing an enzyme to cut the DNA backbone of one strand **(figure 8.10).** Another enzyme then degrades a short region of DNA of that strand, thereby removing the misincorporated nucleotide. How does the cell know which strand to excise? If the enzyme cuts the original strand and not the one being synthesized, the wrong base would remain. The key lies in methylation of DNA bases. Soon after a strand of DNA is synthesized, an enzyme adds methyl groups to certain nucleotide bases. However, immediately after the new strand is synthesized, it is still unmethylated. This difference in methylation distinguishes old and recently synthesized strands of DNA and ensures that the repair enzyme cleaves the correct strand. The combined actions of DNA polymerase and DNA ligase then fill in and seal the gap left by the removal of the DNA segment. ■ DNA ligase, p. 167

Mismatch repair also occurs in humans. Defects in this repair system lead to an increased incidence of colorectal cancers.

Repair of Thymine Dimers

Because UV light is a part of sunlight, cells are frequently exposed to this mutagenic agent. Bacteria have developed several mechanisms to combat the harmful effects of these rays. In one mechanism, an enzyme uses the energy of visible light to break the covalent bond of the thymine dimer, restoring the DNA to its original state **(figure 8.11a).** Because light is required for this mechanism, it is called **photoreactivation,** or **light repair.** This enzyme is found only in prokaryotes.

Some bacteria have an enzyme that recognizes the major distortions in DNA that result from thymine dimer formation. In this process, **excision repair,** or **dark repair,** the enzyme makes single-stranded cuts that flank both sides of the damaged region, resulting in excision of the region (figure 8.11b). The actions of DNA polymerase and DNA ligase then fill in and seal the gap left by the removal of the segment.

Repair of Modified Bases in DNA

Modified bases such as oxidized guanine can result in base substitutions if they are not repaired before the DNA is replicated. An important mechanism for repairing this defect uses an enzyme called a **glycosylase,** which removes the oxidized base from the sugar-phosphate backbone **(figure 8.12).** Another enzyme recognizes that a base is missing and cuts the DNA at this site. DNA polymerase degrades a short section of this strand to remove the damage. This same enzyme then synthesizes another strand with the proper bases. DNA ligase seals the gap in the single-stranded DNA.

Humans have repair enzymes analogous to glycosylases in bacteria. Mutations in the genes coding for these repair enzymes result in an increased rate of colon cancer.

SOS Repair

If DNA is heavily damaged by UV light such that it contains many thymine dimers, photoreactivation and excision repair may not be able to correct all of the dimers and the cells will die. Therefore, bacteria have a mechanism, termed **SOS repair,** that bypasses the damaged DNA and allows replication to continue. The damaged DNA activates the expression of over 30 genes which encode the SOS system. One of the most important of these genes codes for a DNA polymerase that is able to synthesize DNA at the site of the damaged DNA. However, unlike the standard DNA polymerase, which is relatively error-free because of its proofreading ability but cannot copy at the site of the lesion, this newly synthesized polymerase makes many mistakes and incorporates the wrong bases in the DNA strand it is synthesizing. Further, it cannot correct these mistakes because it has no proofreading ability. As a result, mutations arise. This process is called **SOS mutagenesis.**

Table 8.2 summarizes the key features of the major DNA repair systems in bacteria. Note that the cell cannot repair all types of mutations, such as insertional inactivation caused by transposition. However, if the transposon jumps to a new location, the gene that it leaves may regain normal function.

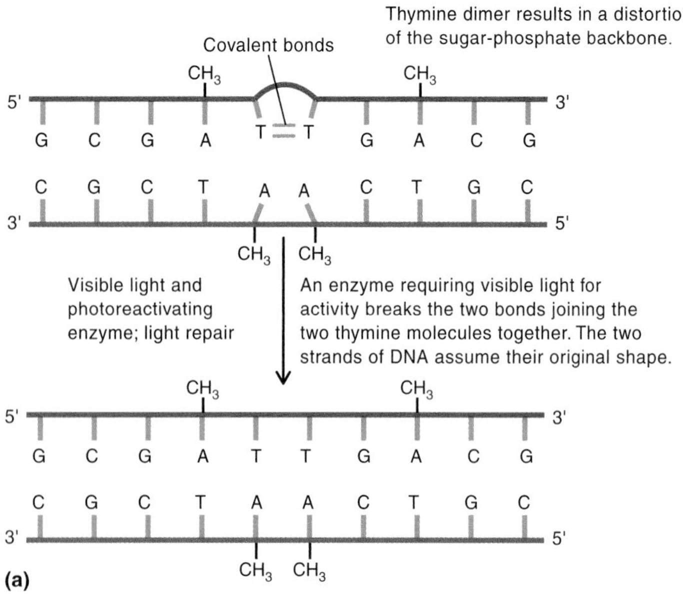
Thymine dimer results in a distortion of the sugar-phosphate backbone.
Covalent bonds
Visible light and photoreactivating enzyme; light repair
An enzyme requiring visible light for activity breaks the two bonds joining the two thymine molecules together. The two strands of DNA assume their original shape.
(a)

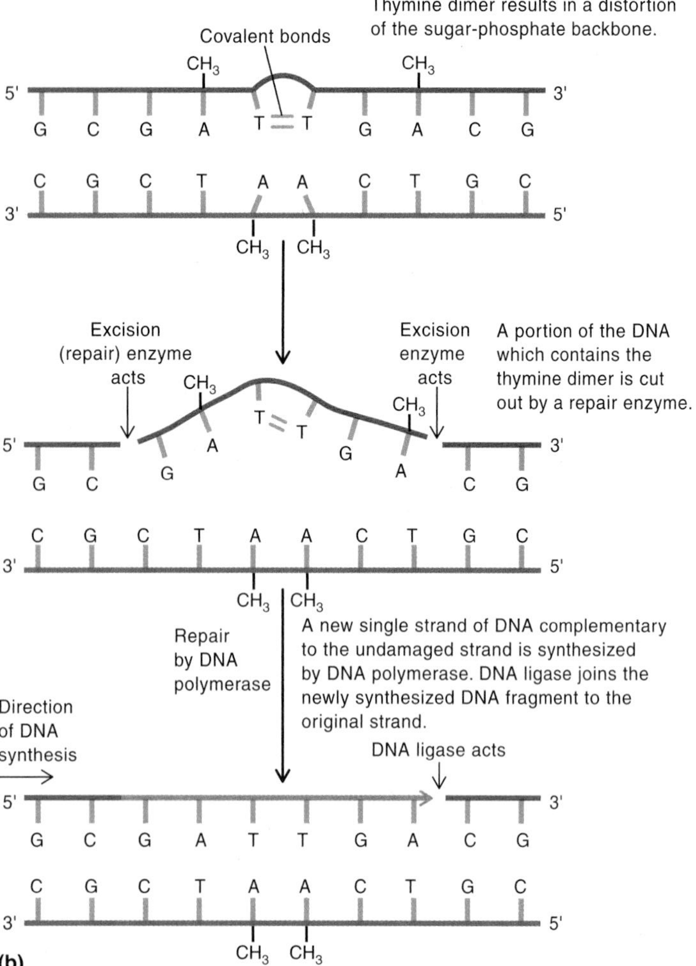
Thymine dimer results in a distortion of the sugar-phosphate backbone.
Covalent bonds
Excision (repair) enzyme acts
Excision enzyme acts
A portion of the DNA which contains the thymine dimer is cut out by a repair enzyme.
Repair by DNA polymerase
A new single strand of DNA complementary to the undamaged strand is synthesized by DNA polymerase. DNA ligase joins the newly synthesized DNA fragment to the original strand.
Direction of DNA synthesis
DNA ligase acts
(b)

FIGURE 8.11 Repair of Thymine Dimers **(a)** In photoreactivation, a light-requiring enzyme breaks the two covalent bonds (light repair). **(b)** In excision, or dark repair, the single strand of DNA containing the thymine dimer is removed and destroyed. The newly synthesized strand is joined to the end of the original strand by the enzyme DNA ligase. Light is not required.

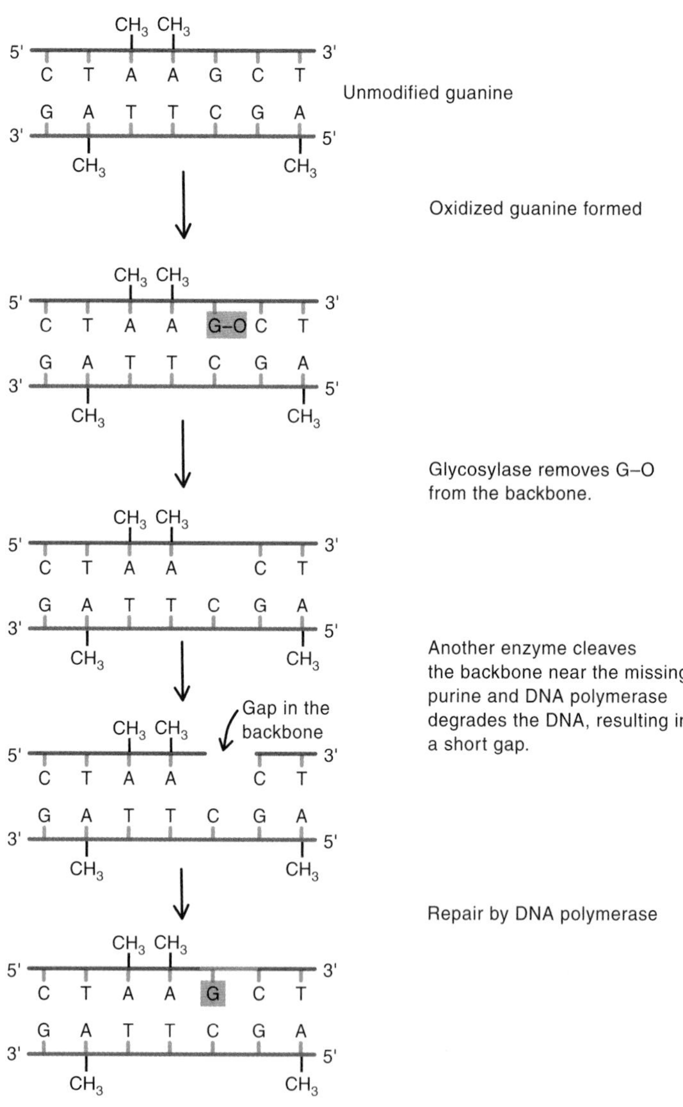
Unmodified guanine
Oxidized guanine formed
Glycosylase removes G–O from the backbone.
Another enzyme cleaves the backbone near the missing purine and DNA polymerase degrades the DNA, resulting in a short gap.
Gap in the backbone
Repair by DNA polymerase

FIGURE 8.12 Repair of Oxidized Guanine in DNA In this diagram the correct base cytosine is based-paired with oxidized guanine, G–O.

MICROCHECK 8.4

Bacteria use a variety of mechanisms to repair damaged DNA that contains errors resulting from the incorporation of wrong nucleotides. These include proofreading by DNA polymerase, and excising the nucleotide errors by mismatch repair. Specific glycosylases can remove modified bases. Thymine dimers can be repaired through light and dark repair mechanisms; severe damage can be overcome by the SOS repair system.

- ✓ How does UV light cause mutations?
- ✓ Distinguish between light and dark repair of thymine dimers.
- ✓ If you wish to maximize the number of mutations following UV irradiation, should you incubate the irradiated cells in the light or in the dark, or does it make any difference? Explain your answer.

TABLE 8.2 Repair of Damaged DNA

	Type of Defect	Repair Mechanism	Biochemical Mechanism	Result
Spontaneous	Wrong base incorporated during DNA replication	Proofreading by DNA polymerase	Removal of mispaired base by DNA polymerase	Potential mutation eliminated
		Mismatch repair	Cleavage and degradation of short stretch of single-stranded DNA and synthesis of new strand by DNA polymerase	Potential mutation in non-methylated DNA strand eliminated
	Reactive oxygen forms oxidized guanine in DNA.	Action of glycosylase	Glycosylase removes the oxidized guanine. Short piece of DNA degraded and guanine incorporated.	Potential mutation eliminated
Mutagen—Induced				
Chemical	Wrong base incorporated during DNA replication	Same as for spontaneous mutations	Same as for spontaneous mutations	Same as for spontaneous mutations
UV light	Thymine dimer formation	Photoreactivation (light repair)	Breaking of covalent bond forming thymine molecules	Original DNA molecule restored
		Excision repair (dark repair)	Excision of a short stretch of single-stranded DNA containing thymine dimer and synthesis of a new strand by DNA polymerase	Mutation eliminated; original DNA molecule restored
		SOS repair	DNA synthesis by a new DNA polymerase bypasses damaged DNA	Cell survives but numerous mutations are generated

8.5

Mutant Selection

Focus Point

- Distinguish between direct and indirect selection, and describe how mutants of each type are selected.

Even when mutagens are used, mutations rarely appear in the population. This presents a major challenge to the investigator who wants to isolate a desired mutant. As discussed in chapter 4, bacteria can multiply on simple media and produce several billion cells per milliliter of medium in less than 24 hours. In such a large population, every gene should have a mutation in at least one cell in the population. The major problem becomes how to find and identify the rare cells containing the desired mutation. Depending on the type of mutant being sought, one of two simple techniques can be used—direct or indirect selection. A major reason why the field of microbial genetics has advanced so rapidly is because the process of detection and isolation of mutants is so simple and fast.

Direct Selection

Direct selection involves inoculating cells onto a medium on which the mutant, but not the parent, can grow. For example, mutants resistant to the antibiotic streptomycin can be easily selected directly by inoculating cells onto a medium containing streptomycin. Only the rare resistant cells in the population will form a colony **(figure 8.13).** Mutants that can grow under conditions in which the parent cells cannot are usually easy to isolate by direct selection.

Indirect Selection

Indirect selection is required to isolate an auxotroph, such as a Trp^- mutant, from a prototrophic parent strain. This process is more cumbersome and takes a longer time than direct selection because there is no medium on which the desired mutant will grow and the prototroph will not. Trp^- cells can grow only on an enriched complex medium because this supplies the tryptophan they require, but Trp^+ cells readily grow on this same medium. To overcome this problem, replica plating is used, sometimes preceded by penicillin enrichment of mutants. Bacteria are treated routinely with a mutagen prior to the selection process.

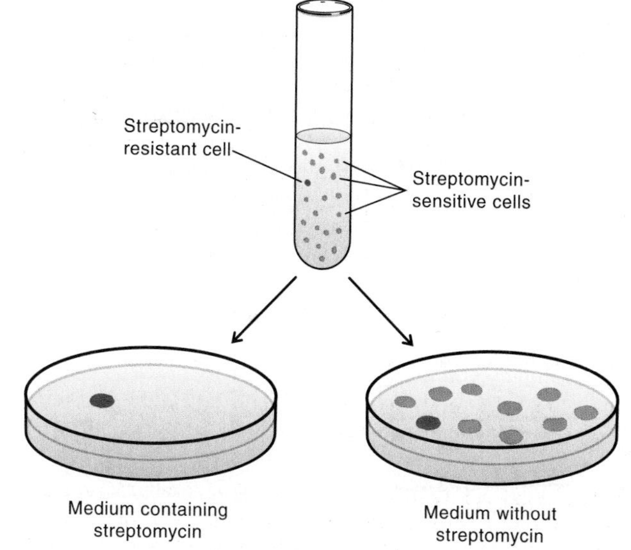

FIGURE 8.13 Direct Selection of Mutants Only the streptomycin-resistant cells will grow on the streptomycin-containing medium. All cells will grow on media without streptomycin and have the same appearance.

Replica Plating

An ingenious technique for indirect selection of auxotrophic mutants, **replica plating,** was devised by the husband-and-wife team of Joshua and Esther Lederberg in the early 1950s **(figure 8.14).** In this technique, the mutagenized bacterial culture is plated on an enriched medium on which both mutant and non-mutant cells grow as individual colonies. This serves as a master plate. The master plate is pressed onto sterile velvet, a fabric with tiny threads that stand on end like tiny bristles (figure 8.14, step 1). This operation transfers some cells of every bacterial colony onto the velvet. Next, two sterile plates—one containing a glucose-salts (minimal) medium and the second an enriched, complex medium—are pressed in succession and in the same orientation onto the same velvet (step 2). This procedure transfers cells imprinted on the velvet from the master plate to both the glucose-salts medium and the enriched medium. Following incubation, all prototrophs will form colonies on both the enriched and the glucose-salts medium, but auxotrophs will only form colonies on the enriched medium (step 3). ■ glucose-salts medium, p. 96

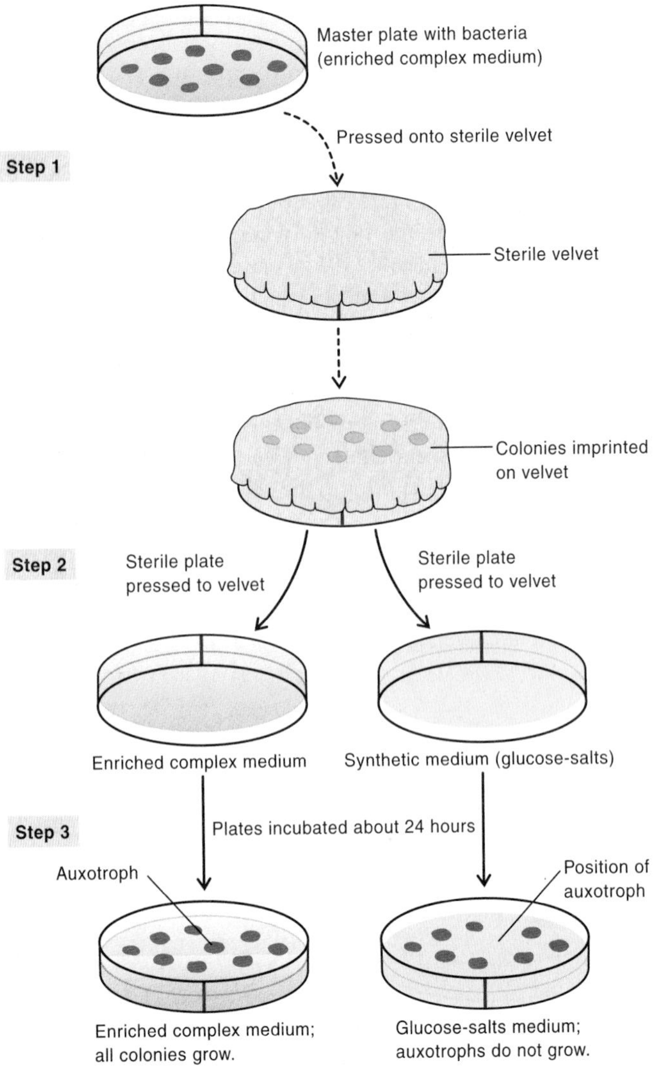

FIGURE 8.14 Indirect Selection of Mutants by Replica Plating The procedure shown is the one first used by the Lederbergs and continues to be used today in many laboratories.

By keeping the orientation of the two plates the same as they touch the velvet, any colony on the master plate that can grow on the enriched medium but not on the glucose-salts medium can be identified. The particular growth factor required can then be determined by adding the various factors individually to the glucose-salts medium and determining which one promotes cell growth.

Penicillin Enrichment of Mutants

Even using mutagenic agents, the frequency of mutation in a particular gene is low, ranging perhaps from less than one in 1,000 to one in 100 million cells. In cases where the parent cell is sensitive to penicillin, the proportion of auxotrophic mutants in the population can be increased significantly by a technique called **penicillin enrichment.** Following treatment with a mutagen, the cells are grown in a glucose-salts medium containing penicillin. Since penicillin kills only growing cells, the prototrophs will grow and be killed, while the non-multiplying auxotrophs will survive **(figure 8.15).** The enzyme **penicillinase** is then added to destroy the penicillin, and the cells are plated on an enriched medium. This plate can then be replica plated onto a glucose-salts medium. Any colonies that grow on the enriched medium but not on the glucose-salts medium must be auxotrophs. ■ action of penicillin, p. 62

Testing of Chemicals for Their Cancer-Causing Ability

Strong evidence exists that a substantial proportion of all cancers are caused by chemicals in the environment called **carcinogens.** How can the thousands of chemicals released into the environment, such as pesticides, herbicides, hair dyes, cosmetics, food additives, and the by-products of manufacturing processes, be

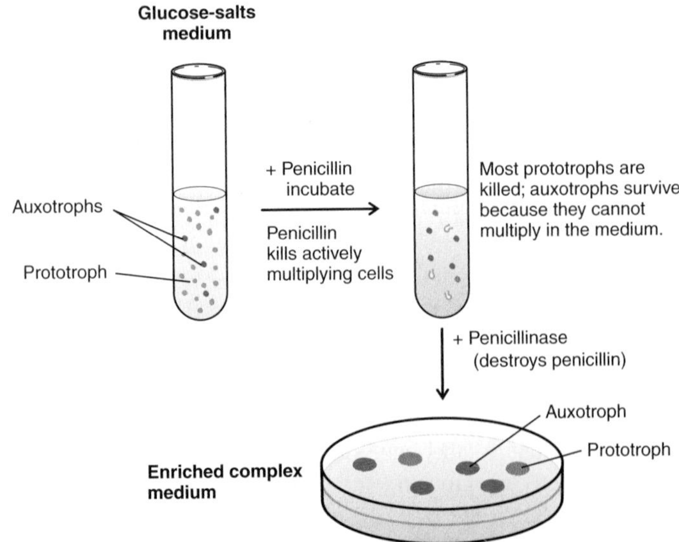

FIGURE 8.15 Penicillin Enrichment of Mutants Since auxotrophs require a growth factor to multiply, they are not killed.

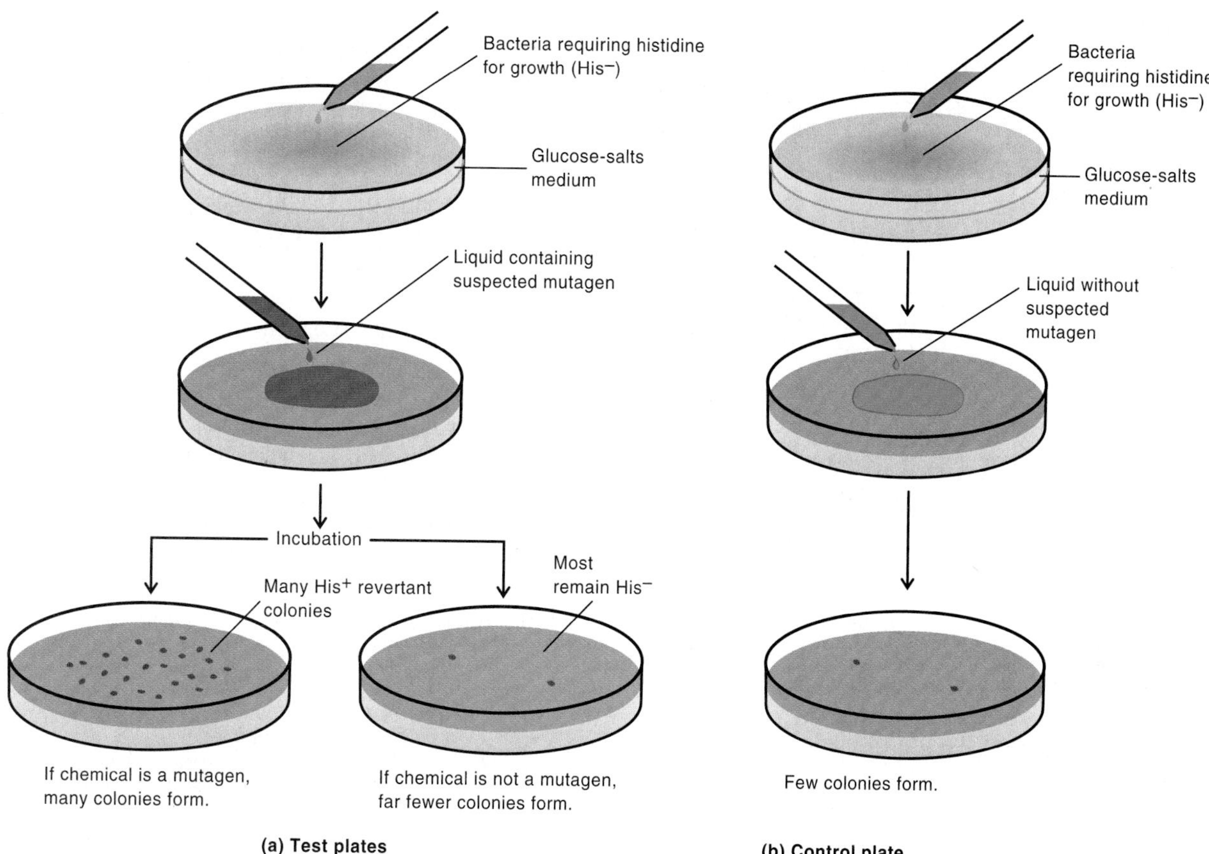

FIGURE 8.16 Ames Test to Screen for Mutagens **(a)** The chemical will increase the frequency of reversion of His⁻ to His⁺ cells if it is a mutagen and, therefore, a potential carcinogen. **(b)** The control plate contains only the liquid in which the suspected mutagen is dissolved.

tested for their carcinogenic activity? Testing in animals for tumor formation takes 2 to 3 years and may cost $100,000 or more to test a single compound. Today, a number of much less expensive, more rapid, and simpler tests have been devised. All are based on assaying the effect of the potential carcinogen on DNA in a microbiological system. The first one was devised by Bruce Ames and his colleagues in the 1960s and illustrates the concept of such tests. The **Ames test,** takes only a few days to get results and is based on three observations: (1) the reversion frequency of a mutant gene in a biosynthetic pathway, such as histidine biosynthesis, can be readily measured; (2) the low frequency of spontaneous reversions is increased by mutagens; and (3) most carcinogens affect DNA and therefore are mutagens. Specifically, the Ames test compares the effect of a test chemical on the rate of reversion of a histidine-requiring auxotroph of *Salmonella,* to the reversion frequency when no chemical is added **(figure 8.16).** If the chemical is mutagenic, it will increase the reversion rate of the strain relative to that observed when no chemical is added (the control). The test also gives some idea about how powerful the mutagen is, and therefore how potentially hazardous the chemical is, by the number of revertants that arise.

The Ames test fails to detect many carcinogens because some substances are not carcinogenic themselves but can be converted to active carcinogens by enzymatic reactions that occur in animals but not in bacteria. Therefore, an extract of ground-up rat liver, which has the enzymes to carry out these conversions, is added to the Petri plates containing the suspected mutagen (carcinogen) being analyzed by the Ames test. To increase the sensitivity of the test, a mutant strain that lacks repair enzymes is often used. As a result, more revertants will be observed because the damaged DNA cannot be repaired.

Additional testing must be done on any mutagenic agent identified in the Ames test to confirm that it is actually carcinogenic in animals. Although data are not available on the percentage of mutagens that are carcinogens, it is clear that the Ames test is useful as a rapid screening test to identify those compounds that have a high probability of being carcinogenic. Thus far, no compound with a negative Ames test has been shown to be carcinogenic in animals.

MICROCHECK 8.5

Mutants can be selected using either direct techniques or indirect techniques such as replica plating. Penicillin enrichment can often help in the isolation of auxotrophic mutants by killing multiplying cells.

- ✓ Distinguish between the kinds of mutants that can be isolated by direct and indirect selection.
- ✓ When does penicillin enrichment not work?
- ✓ How could you demonstrate by replica plating that the environment selects but does not mutate genes in bacteria?

GENE TRANSFER AS A MECHANISM OF GENETIC CHANGE

In addition to mutation, the genetic information in a cell can be altered if the cell gains genes from other cells, a common occurrence in the microbial world. The movement of DNA from one cell, the **donor,** to another, the **recipient,** is called horizontal, or lateral, gene transfer. This mechanism of genetic change is largely responsible for the rapid spread of antibiotic resistance, such as described for *S. aureus* earlier in this chapter.

Gene transfer can only be studied if genetic differences exist between the donor and recipient cells, and this is one reason why bacterial geneticists frequently isolate mutants. These differences make it possible to determine whether **genetic recombination,** the combining of DNA or genes from two different cells, has occurred. Genetic recombination can be readily recognized because the resulting cells, termed **recombinants,** have a combination of properties of each of the original strains. **Figure 8.17** shows how this can be demonstrated. Two bacterial strains, neither of which can grow on a glucose-salts medium because of multiple growth factor requirements, are mixed. Strain A requires histidine (His$^-$) and tryptophan (Trp$^-$), and is resistant to streptomycin (Str^R). Strain B does not require histidine or tryptophan, but does require leucine (Leu$^-$) and threonine (Thr$^-$), and is killed by streptomycin (Str^S). Neither population is likely to give rise to a spontaneous mutant that can grow on the glucose-salts medium because multiple simultaneous mutations in the same cell would be required. After the strains are mixed, they are plated on a glucose-salts medium that contains streptomycin. In order for cells to grow and form colonies on this medium, they must be prototrophic and resistant to streptomycin. Therefore they must acquire genes from the other strain. Following its transfer, the transferred DNA must replicate in order to be passed on to daughter cells and confer on them new genetic information. Therefore, the transferred DNA must have an origin of replication. If it does not, it must become part of a DNA molecule such as a chromosome or a plasmid that can replicate and be passed on to all daughter cells. Such a molecule is termed a **replicon (figure 8.18).**

■ origin of replication, p. 166

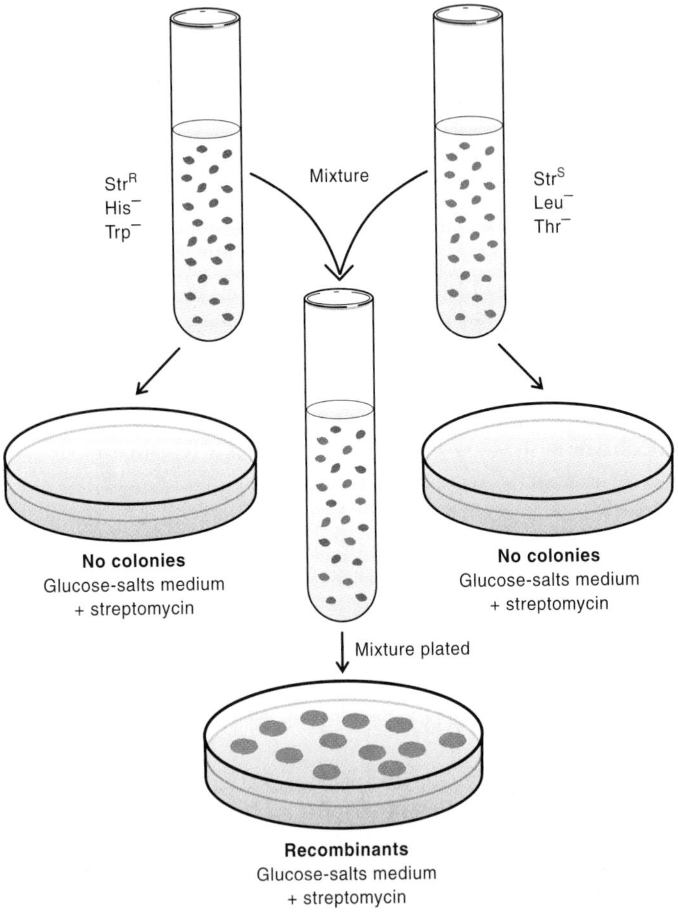

Str^R
His$^-$
Trp$^-$
Mixture
Str^S
Leu$^-$
Thr$^-$
No colonies
Glucose-salts medium
+ streptomycin
No colonies
Glucose-salts medium
+ streptomycin
Mixture plated
Recombinants
Glucose-salts medium
+ streptomycin

FIGURE 8.17 General Experimental Approach for Detecting Gene Transfer in Bacteria Recombinants will only arise if the genes transferred are part of a replicon or integrate into a replicon.

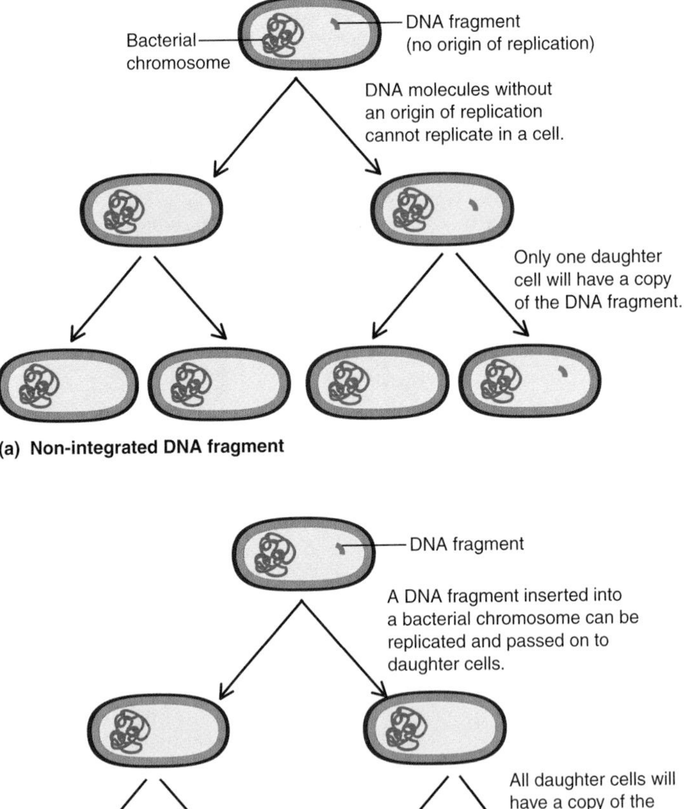

FIGURE 8.18 DNA Without an Origin of Replication Must Become Part of a Replicon in Order to Be Maintained in a Population of Cells (a) Without an origin of replication, the DNA will not be passed on to daughter cells and thus will not confer new properties on the population. **(b)** Transferred DNA becomes integrated into a replicon in the recipient cell and is passed on to all daughter cells.

Genes in nature are transferred between bacteria by three different mechanisms:

1. DNA-mediated transformation, in which DNA is transferred as "naked" DNA.
2. **Transduction,** in which bacterial DNA is transferred by a virus that infects bacterial cells.
3. **Conjugation,** in which DNA is transferred directly from one bacterium to another when the cells are in contact with one another.

To detect gene transfer, it is most convenient to select directly for recombinants by inoculating the mixture of cells on a medium on which only the recombinants will form colonies. Since several billion bacteria can be plated on agar in a single Petri dish, a few colony-forming recombinants can be detected readily and very rare events observed.

8.6 DNA-Mediated Transformation

Focus Point

- Describe the process of DNA-mediated transformation.

DNA-mediated transformation, commonly referred to as **transformation,** involves the uptake of "naked" DNA by recipient cells **(Perspective 8.1). Naked DNA** is simply DNA that is free in the surroundings; it is not contained within a cell or a virus. The fact that the DNA is naked can be demonstrated by adding **DNAse,** an enzyme that degrades DNA outside the cell. Since DNAse prevents DNA-mediated transformation, this process must involve naked DNA transfer.

PERSPECTIVE 8.1

The Biological Function of DNA: A Discovery Ahead of Its Time

In the 1930s, it was well known that DNA occurred in all cells, including bacteria. Its function, however, was a mystery. Since DNA consists of only four repeating subunits, most scientists believed that it could not be a very important molecule. Its important biological role was discovered through a series of experiments conducted during a 20-year period by scientists in England and the United States.

In the 1920s, Frederick Griffith, an English bacteriologist, was studying pneumococci, the bacteria that cause pneumonia. It was known that pneumococci could cause this disease only if they made a polysaccharide capsule. In trying to understand the role of this capsule in causing disease, Griffith killed encapsulated pneumococci and mixed them with living mutant pneumococci that could not synthesize a capsule. When he inoculated this mixture of organisms into mice, much to his surprise, they developed pneumonia and died **(figure 1).** Griffith isolated living encapsulated pneumococci from the dead mice. When he injected the dead encapsulated cells and living non-encapsulated cells into separate mice, they did not develop pneumonia.

Two years after Griffith reported these findings, another investigator, M. H. Dawson, lysed heat-killed encapsulated pneumococci and passed the suspension of ruptured cells through a very fine filter, through which only the cytoplasmic contents of the bacteria could pass. When he mixed the filtrate (the material passing through the filter) with living bacteria unable to make a capsule, some bacteria were able to make a capsule. Moreover, the progeny of these bacteria could also make a capsule. Something in the filtrate was "transforming" the harmless unencapsulated bacteria into ones that could make a capsule.

What was this transforming principle? In 1944, after years of painstaking chemical analysis of lysates capable of transforming pneumococci, three investigators from the Rockefeller Institute, Oswald T. Avery, Colin MacLeod, and Maclyn McCarty, purified the active compound and then wrote one of the most important papers ever published in biology. In it, they reported that the transforming molecule was DNA.

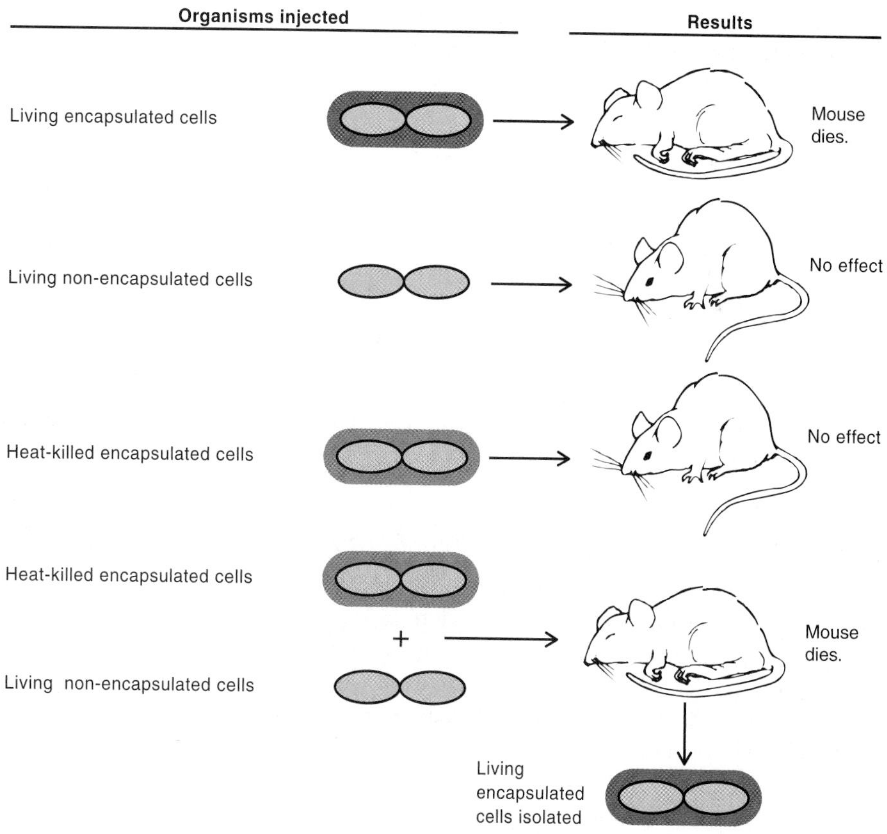

FIGURE 1 Demonstration of the Transforming Principle

The significance of their discovery was not appreciated at the time. Perhaps the discovery was premature, and scientists were slow to recognize its significance. None of the three investigators received a Nobel Prize, although many scientists believe that they deserved it. Their studies pointed out that DNA is a key molecule in the scheme of life and led to James Watson and Francis Crick's determination of its structure, which they published in 1953. The understanding of the structure and function of DNA revolutionized the study of biology and ushered in the era of molecular biology. Microbial genetics serves as its foundation.

One source of naked DNA in some genera is lysed cells in a population of bacteria. As cells burst open, the long chromosomal DNA molecules that are tightly jammed into the cells typically break up into hundreds of pieces as they explode through the broken cell walls. Other genera of bacteria secrete small segments of DNA, presumably as a means of promoting transformation.

In order to take up naked DNA, the recipient cell must be **competent**—a specific physiological state that allows DNA to enter a cell. Once inside the cell, the DNA can integrate into the recipient's genome. Over 40 species are naturally competent, but cells can also take up DNA if they are treated with certain chemicals or electric currents that alter the permeability of their cell walls.

Natural Competence

Among the species that can become naturally competent, the ability to take up DNA is a tightly controlled physiological state, and the mechanism of control varies. Some species are always competent, whereas others become so only under specific conditions, as when the population reaches a critical density or under certain nutritional conditions. In the case of *Bacillus subtilis,* a two-component regulatory system recognizes a limiting supply of nitrogen or carbon in the environment and activates a set of genes required for the competent state. Competence also requires that the bacterial concentration be high, which is a function of a quorum sensing system. Presumably, the high concentration of cells ensures that the DNA in the medium will contact the competent bacteria. However, even under optimal conditions, only 10% of the population ever becomes competent. Perhaps non-competent cells are releasing DNA that can be taken up by the competent cells. This means that presumably identical cells in a population can differ in their physiological properties. The fact that some species of bacteria become competent only under precise environmental conditions highlights the remarkable ability of these seemingly simple cells to sense their surroundings and adjust their behavior accordingly. ■ two-component regulatory system, p. 180 ■ quorum sensing, p. 180

Entry of DNA

Double-stranded DNA molecules bind to specific receptors on the surface of competent cells **(figure 8.19).** However, only one strand enters the cell; nucleases at the cell surface degrade the other. Most competent bacteria take up DNA regardless of its origin, but some only accept DNA from closely related species. The cells recognize closely related DNA by characteristic nucleotide sequences found throughout the genome.

Integration of Donor DNA

Once the donor DNA is inside the recipient cell, it integrates into the genome by the process of **homologous recombination,** which can only occur if the donor DNA is similar in sequence, or is **homologous,** to a region in the recipient cell's genome. Thus, transformation occurs only between closely related species. The single-stranded donor DNA becomes positioned next to the complementary region of the recipient DNA. A nuclease then cleaves one strand of the recipient cell's DNA on either side of where the donor DNA is aligned. This fragment of DNA is released and will be degraded by nucleases. The donor DNA then replaces precisely a single strand of the recipient DNA.

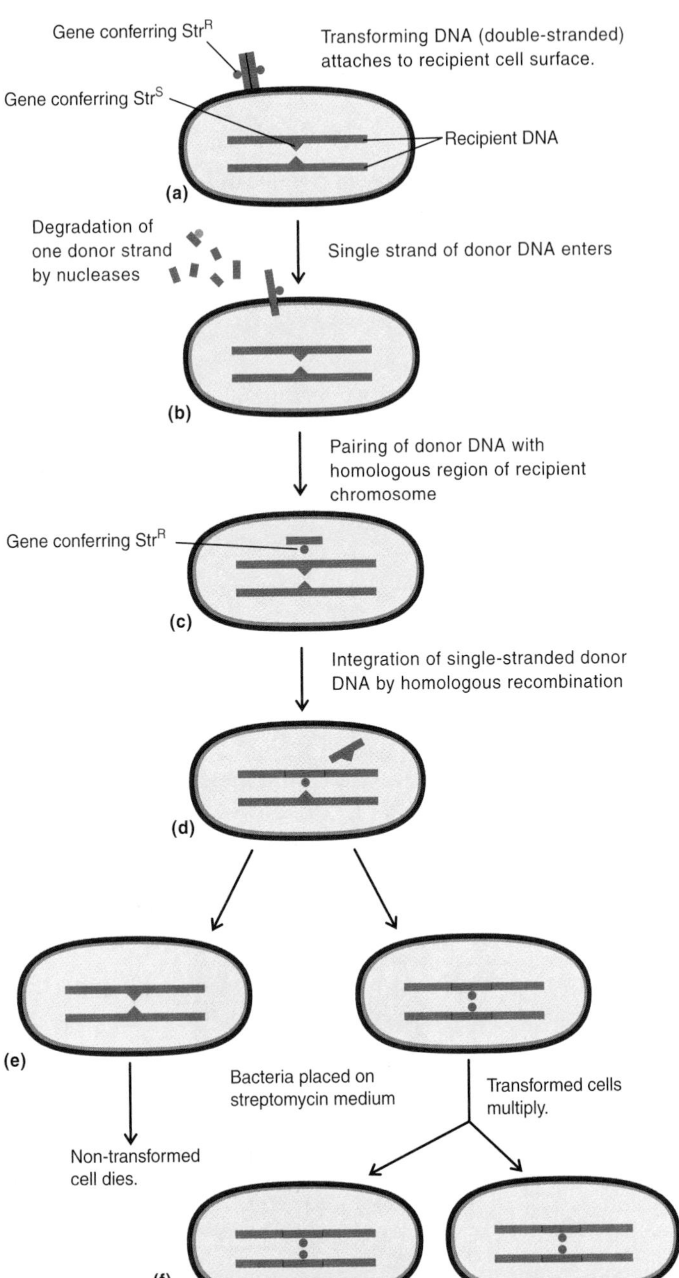

FIGURE 8.19 DNA-Mediated Transformation The donor DNA comes from a cell that is streptomycin resistant (Str^R). The recipient cell is streptomycin sensitive (Str^S). The genes for resistance and sensitivity to streptomycin may differ by only a single nucleotide.

Multiplication of Transformed Cells

In the laboratory, DNA transformation is most easily detected if the transformed cells can multiply under selective conditions in which the non-transformed cells cannot grow and form colonies. For example, if the donor cells are Str^R and the recipient cells are Str^S, then cells transformed to Str^R will grow on medium that contains streptomycin. Since only one strand of the recipient cell's DNA is transformed initially to streptomycin resistance, only half

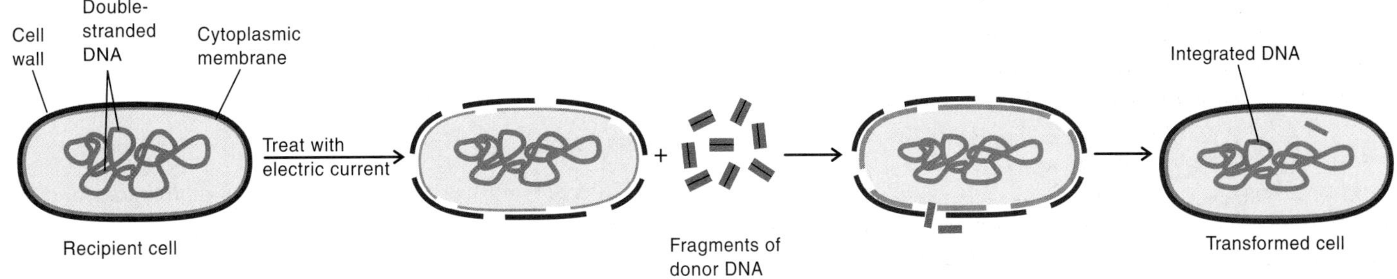

FIGURE 8.20 Electroporation The electric current makes holes in both the cell wall and the cytoplasmic membrane through which the DNA can pass. These holes are then repaired by the cell, and the DNA becomes incorporated into the chromosome of the cell.

of the daughter cells will be streptomycin resistant. The other half will be streptomycin sensitive and will die on streptomycin-containing medium (see figure 8.19d and 8.19e). Although many other donor genes besides Str^R will be transferred and integrated into the chromosome of the recipient cells, these transformants will go undetected because the donor and recipient cells are identical in these other genes.

Artificial Competence

Although not all bacteria become naturally competent, double- and single-stranded DNA can be introduced into most cells, including those of bacteria, animals, and plants, through a special treatment of the recipient cells. In one technique called **electroporation,** bacteria and DNA are mixed together and the mixture is subjected to an electric current **(figure 8.20).** The current apparently makes holes in the bacterial cell wall and cytoplasmic membrane through which the DNA enters.

MICROCHECK 8.6

Gene transfer can occur in many Gram-positive and Gram-negative bacteria by DNA-mediated transformation in which DNA is released from some cells and is taken up by other competent cells. Competent cells have undergone a number of changes that allow them to bind DNA, take DNA into the cell in a single-stranded form, and integrate the DNA. Artificial means such as electroporation can be used to get DNA into cells that do not become competent naturally.

- ✓ What effect would adding deoxyribonuclease to the culture have on transformation?
- ✓ If cells do not become competent naturally, can they still be made to take up DNA? Explain.
- ✓ Can you devise a test using DNA-mediated transformation that could test chemicals for their mutagenic activity?

8.7 Transduction

Focus Points

- Describe the process of bacterial gene transfer by transduction.
- Distinguish between generalized and specialized transduction.

Bacterial viruses, called **bacteriophages** or simply **phages,** can transfer bacterial genes from a donor to a recipient by a process called **transduction.** To understand this process, you need to know something about phages and how they infect bacterial cells. Phages consist of genetic material, either DNA or RNA, surrounded by a protein coat. They infect bacteria by attaching to a cell and then injecting their nucleic acid into that cell. Enzymes encoded by the phage genome then degrade the bacterial DNA. Next, the cell's enzymes replicate the phage nucleic acid and synthesize proteins that make up the empty phage coat. The phage nucleic acid then enters the phage coat and the various components of the phage assemble to produce complete phage particles, which are released, usually as a result of host cell lysis. The phage particles then attach to other bacterial cells, beginning new cycles of infection. Transduction results from rare errors that occur during the infection cycle, giving rise to phage progeny that carry bacterial genes in place of phage genes inside the coat. When these progeny then infect other bacteria, they inadvertently transfer bacterial genes to another bacterium.

There are two types of transduction: generalized and specialized. In **generalized transduction,** any genes of the donor cell can be transferred. It results from of an error during construction of the phage inside the infected cell; a fragment of bacterial DNA is mistakenly substituted for phage DNA within the protein coat **(figure 8.21).** The product is referred to as a **transducing particle.** Like a phage, however, the transducing particle will attach to a bacterium and inject the nucleic acid into that cell. Inside the cell, the injected DNA must integrate into the chromosome by homologous recombination if it is to be maintained by the cell. In **specialized transduction,** only a few specific genes can be transferred. This process will be described in chapter 13, after more details of the infection cycles of bacteriophages are discussed.

MICROCHECK 8.7

Transduction results from an error that occurs during the infection cycle of bacteriophages, leading to the transfer of bacterial genes from one bacterial cell to another.

- ✓ How is generalized transduction different from specialized transduction?
- ✓ Two genes are transduced simultaneously. What does this suggest about the location of the two genes relative to each other? Explain.

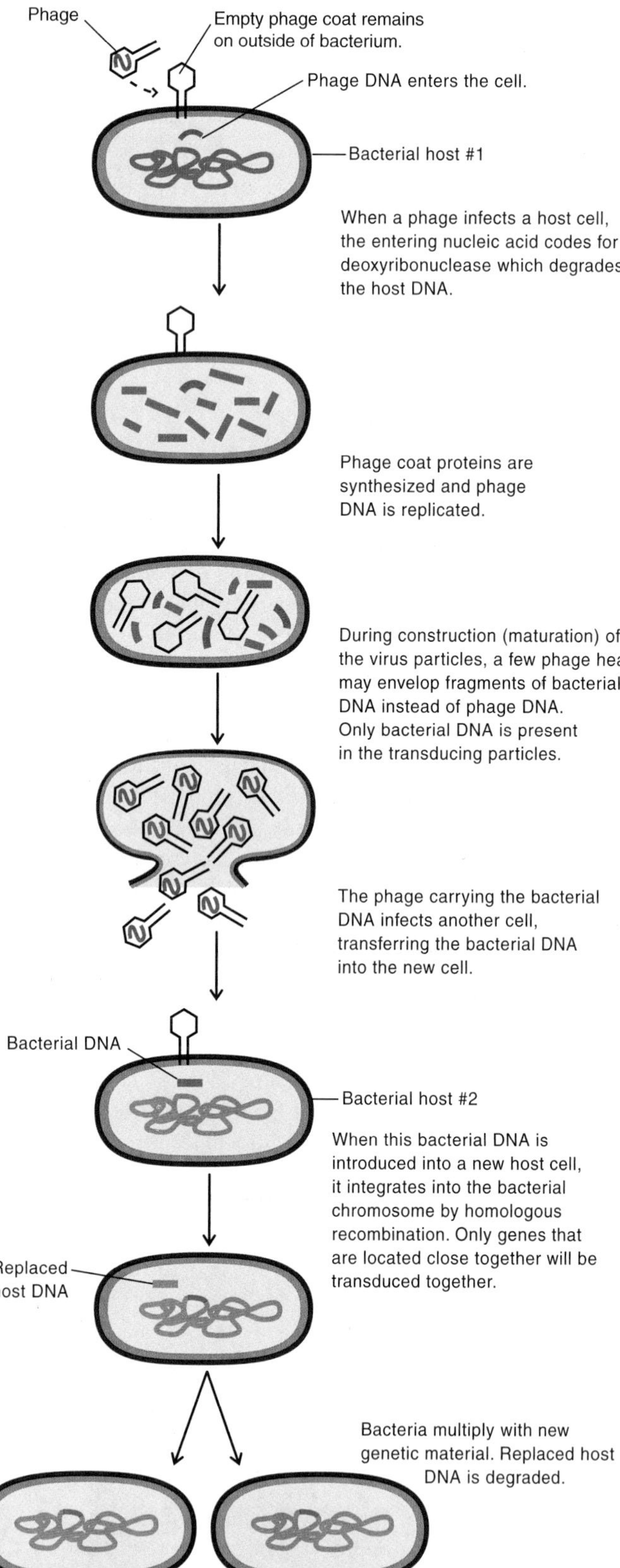

Phage
Empty phage coat remains on outside of bacterium.
Phage DNA enters the cell.
Bacterial host #1
When a phage infects a host cell, the entering nucleic acid codes for deoxyribonuclease which degrades the host DNA.
Phage coat proteins are synthesized and phage DNA is replicated.
During construction (maturation) of the virus particles, a few phage heads may envelop fragments of bacterial DNA instead of phage DNA.
Only bacterial DNA is present in the transducing particles.
The phage carrying the bacterial DNA infects another cell, transferring the bacterial DNA into the new cell.
Bacterial DNA
Bacterial host #2
When this bacterial DNA is introduced into a new host cell, it integrates into the bacterial chromosome by homologous recombination. Only genes that are located close together will be transduced together.
Replaced host DNA
Bacteria multiply with new genetic material. Replaced host DNA is degraded.

FIGURE 8.21 Transduction (Generalized) Any piece of the chromosomal DNA of the donor cell can be transferred in this process. All of the DNA molecules of the bacterial virus and the bacteria are double-stranded.

8.8 Conjugation

Focus Points

- Compare an F^+ to an F^- cell in terms of morphology and gene content.
- Compare and contrast the state of the F plasmid in a cell (1) that transfers only the plasmid, with a cell that (2) can transfer chromosomal DNA, and a cell that (3) transfers both plasmid and chromosomal DNA.

An important and common mechanism of gene transfer in both Gram-positive and Gram-negative bacteria is **conjugation.** The process is quite different in the two groups but we will only consider conjugation in Gram-negative bacteria. Conjugation requires contact between donor and recipient cells. This can be shown through the following experiment. If two different auxotrophic mutants are placed on either side of a filter through which fluids, but not bacteria, can pass, genetic recombination does not occur. If the filter is removed, however, allowing cell-to-cell contact, genetic recombination takes place. Both plasmids and chromosomal DNA can be transferred by conjugation. The process is complex and many aspects are not understood even though it was first observed in *E. coli* more than 50 years ago. ■ plasmid, p. 68

Plasmid Transfer

Transfer of **plasmids** to other cells is most frequently mediated by conjugation. **Conjugative plasmids** direct their own transfer from donor to recipient cells. Since plasmids are replicons with an origin of replication, they can replicate inside cells independent of the chromosome. The most thoroughly studied example is the **F (fertility) plasmid** of *E. coli*. Although this plasmid does not encode any notable characteristics other than those required for transfer, other conjugative plasmids encode resistance to certain antibiotics, which explains how such resistance can easily spread among a population of cells.

E. coli cells that harbor the F plasmid are designated $\mathbf{F^+}$, whereas those that do not are $\mathbf{F^-}$. The F plasmid encodes several proteins required for conjugation, including the **F pilus,** also referred to as the **sex pilus.** This pilus attaches to the recipient cell **(figure 8.22).** ■ F pilus, p. 66

Plasmid transfer can be divided into four steps **(figure 8.23):**

Step 1 Contact between donor and recipient cells. The F pilus of the donor cell recognizes and binds to a specific receptor on the cell wall of the recipient cell. After attachment, the F pilus acts as a grappling hook, pulling the two cells together.

Step 2 Mobilization or activation of DNA transfer. The plasmid becomes mobilized for transfer when a plasmid-encoded enzyme, an endonuclease, cleaves one strand of the plasmid at a specific nucleotide sequence, the **origin of transfer.** This results in the formation of a single-stranded DNA molecule with the endonuclease attached to the end.

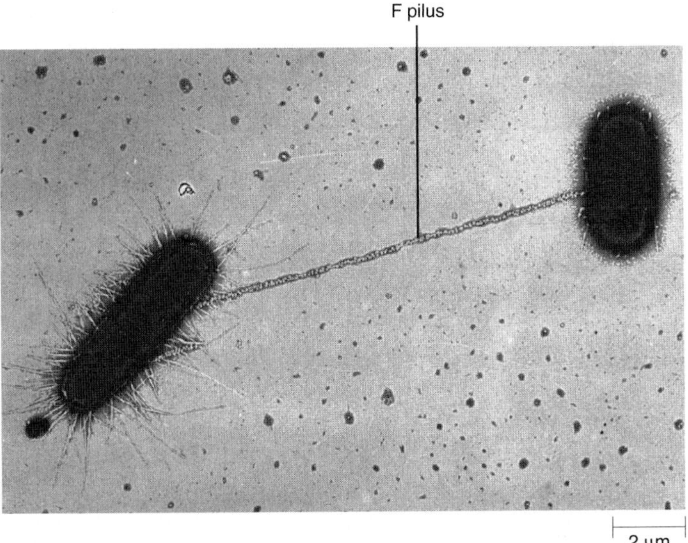

FIGURE 8.22 F or Sex Pilus Holding Together Donor and Recipient Cells of *E. coli* During DNA Transfer During the actual transfer of DNA, the pilus becomes much shorter as it pulls the cells together. The DNA passes through the pilus from donor to recipient cells.

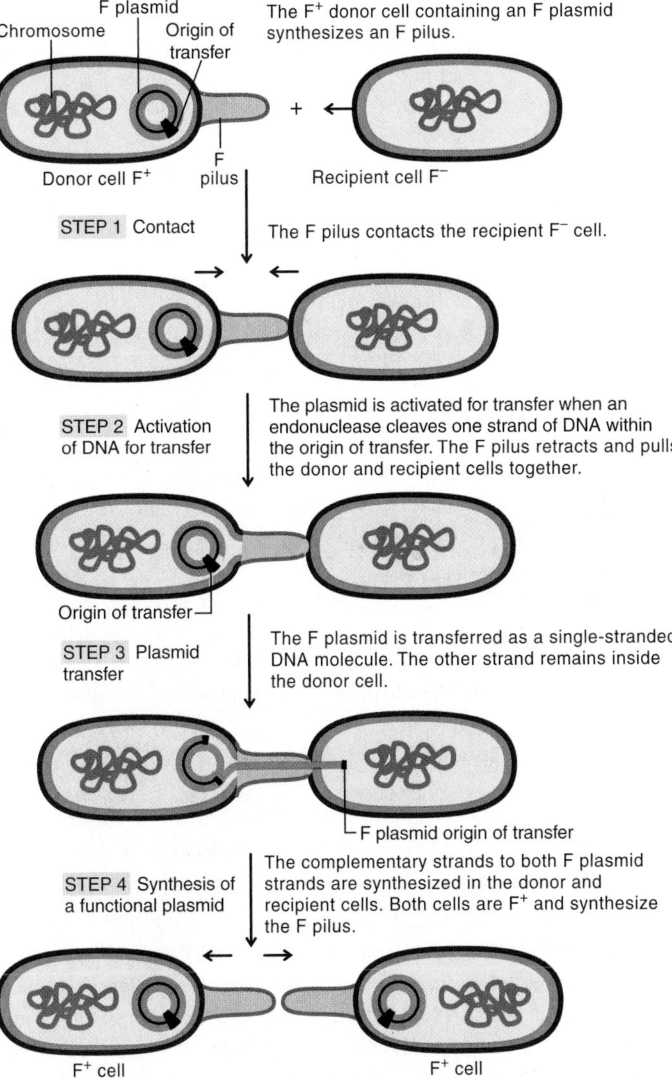

FIGURE 8.23 Conjugation—Transfer of the F Plasmid

Step 3 Plasmid transfer. Within minutes of the F^+ cell contacting the F^- cell, a single strand of the F plasmid with the attached endonuclease enters the F^- cell. This transfer takes about 2 minutes. Recent research indicates that the DNA passes through the F pilus.

Step 4 Synthesis of a functional plasmid inside the recipient and donor cells. Once inside the recipient cell, a strand of DNA complementary to the single-stranded transferred DNA is synthesized. Likewise, a strand complementary to the single-stranded plasmid DNA remaining in the donor is synthesized. Thus, both the donor and recipient cells are now F^+ and can act as donors of the F plasmid.

Chromosome Transfer

Chromosomal DNA transfer is less common than plasmid transfer and involves **Hfr strains** (meaning high frequency of recombination). These are strains in which the F plasmid has integrated into the chromosome at specific sites, which happens occasionally **(figure 8.24).** Like F^+ cells, Hfr cells produce an F pilus, and the F plasmid DNA directs its transfer to the recipient cell. However, because the F plasmid DNA is integrated into the chromosome, chromosomal DNA is also transferred as a single stranded DNA molecule **(figure 8.25).** The entire chromosome is generally not transferred because it would take approximately 100 minutes for this to occur, an unlikely event because the connection between the two cells is likely to break before this time. Unlike the F plasmid, the transferred chromosomal DNA is not a replicon, and so it must integrate into the chromosome of the recipient cell through homologous recombination if it is to be maintained.

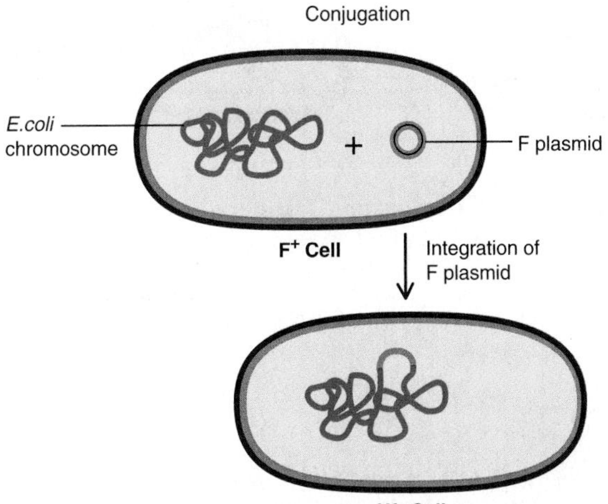

FIGURE 8.24 Hfr Formation Integration of the F plasmid into the bacterial chromosome to form Hfr. Homologous sites, termed insertion sequences, on the F plasmid and the chromosome allow the integration to occur. There is no replacement of DNA in the chromosome; the integration of the F plasmid increases the size of the chromosome. ■ **insertion sequence, p. 207**

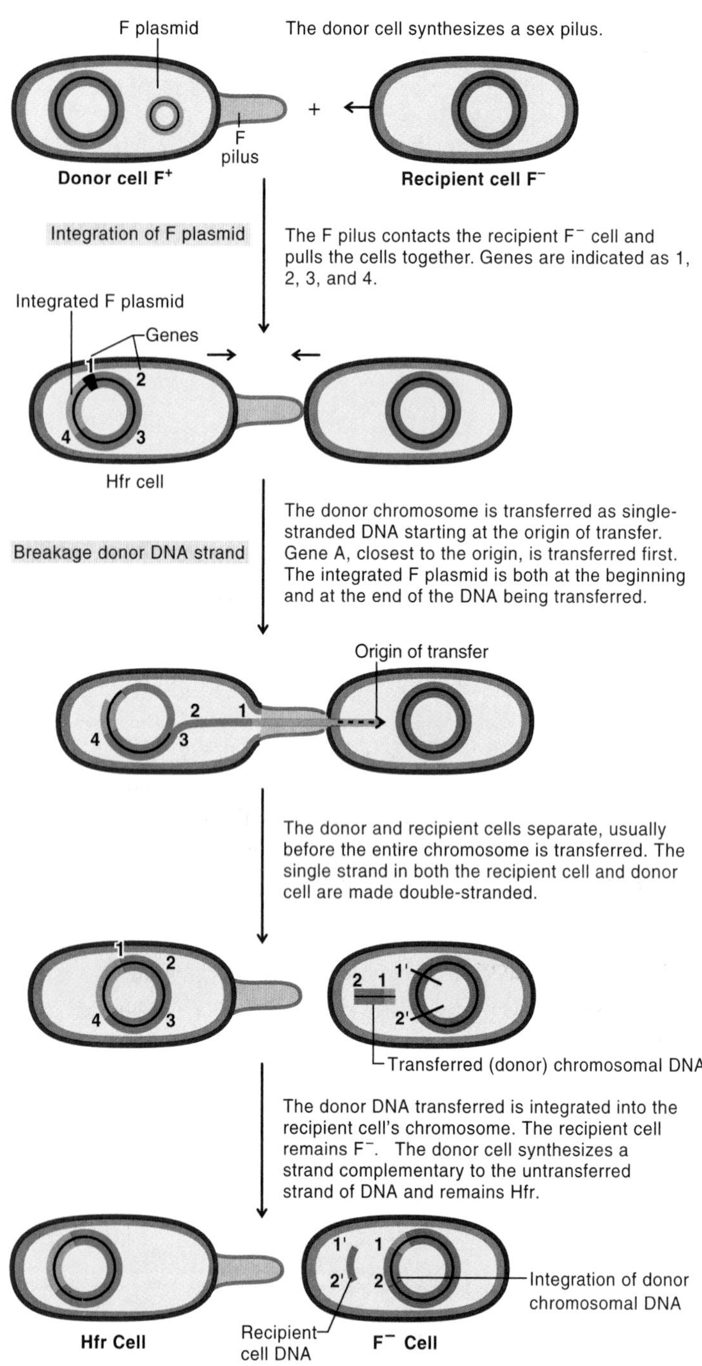

FIGURE 8.25 Conjugation-Transfer of Chromosomal DNA The DNA is transferred as a single-stranded DNA molecule. The recipient genes are designated with a prime (′). The corresponding DNA in the donor lacks the prime (′).

F′ Donors

The F plasmid in Hfr strains can be excised from the chromosome because the process of F plasmid integration into the chromosome is reversible. In some instances, an error occurs in the process of excision, and a small piece of the bacterial chromosome becomes incorporated into the F plasmid **(figure 8.26).** This F plasmid with its incorporated chromosome is called **F′** (F prime), and like the F plasmid it is rapidly and efficiently transferred to F− cells in the

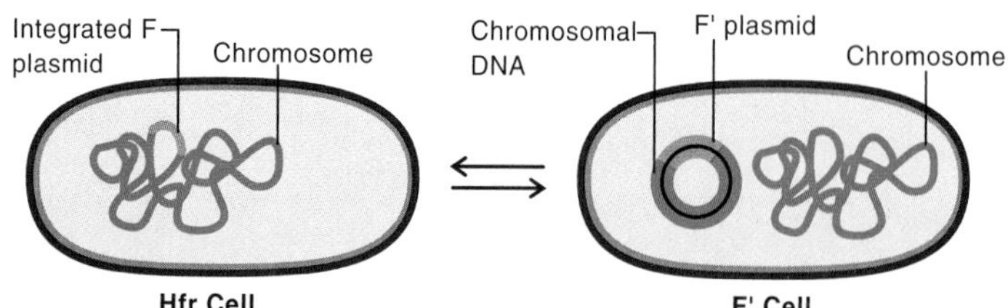

FIGURE 8.26 Formation of F′ Plasmid This F′ plasmid has the transfer properties of the F plasmid but carries chromosomal DNA. This process is reversible.

population. Consequently, any chromosomal genes incorporated into the F plasmid are also transferred. The F′ plasmid usually remains **extrachromosomal**—that is, it remains independent of the recipient cell's chromosome. On rare occasions, however, it can integrate into the chromosome of recipient cells, which then become Hfr because they contain the F plasmid integrated into the chromosome.

The three mechanisms of DNA transfer are compared in **table 8.3.**

MICROCHECK 8.8

Conjugation requires contact between donor and recipient cells. Transfer is from a donor cell that synthesizes a F pilus to one that does not. Both plasmid and chromosomal DNA can be transferred. Following transfer, plasmids replicate but chromosomal DNA must be integrated into a replicon.

- ✓ For which two characteristics essential for conjugation does the F plasmid encode?
- ✓ Describe the outcomes of the three types of matings ($F^+ \times F^-$, Hfr × F^-, and $F' \times F^-$).
- ✓ Would you expect transfer of chromosomal DNA by conjugation to be more efficient if cells were plated together on solid medium (agar) or mixed together in a liquid in a shaking flask? Explain.

TABLE 8.3 Comparison of Mechanisms of DNA Transfer

Mechanism	Main Features	Size of DNA Transferred	Sensitivity to DNase addition*
Transformation	Naked DNA transferred	About 20 genes	Yes
Transduction	DNA enclosed in a bacteriophage coat	Small fraction of the chromosome	No
Conjugation			
Plasmid transfer	Cell-to-cell contact required	Entire plasmid	No
Chromosome transfer	Cell-to-cell contact required; only certain cells can be donors (Hfr)	Variable fraction of chromosome	No

*DNase is an abbreviation of deoxyribonuclease, an enzyme that degrades DNA.

8.9

The Mobile Gene Pool

Focus Points

- Describe how plasmids differ from bacterial chromosomes.
- Compare and contrast transposons and genomic islands.

Advances in genomics have uncovered surprising variation in the gene pool (the sum of all genes) of even a single species. For example, nucleotide sequence analysis of many *E. coli* strains indicates that only about 75% of a strain's genes are found in all strains of that species. These make up the **conserved,** or **core genome** of the species. The remaining genes, which are not conserved, vary considerably among different strains, and are associated with various plasmids, transposons, and regions of DNA called genomic islands. Phage DNA, which we will discuss in chapter 13, is also in this group. Surprisingly, when all the non-conserved genome components of *E. coli* strains are considered, these sequences vastly outnumber sequences in the core genome. Many of the components of the non-conserved genome are **mobile genetic elements,** meaning they encode their own transfer, either within the genome of a cell or between cells.

Plasmids

Plasmids are common in the microbial world and are found in most members of the *Bacteria* and *Archaea*, as well as *Eucarya*. Like chromosomes, most plasmids are double-stranded DNA molecules that have an origin of replication and therefore can be replicated by the cell before it divides. The machinery required for replication is provided by the cell in which the plasmid resides. Plasmids, however, generally do not encode any information essential to the growth of cells. ■ plasmid, p. 68 ■ origin of replication, p. 166

Plasmids vary in many of their properties; some carry only a few genes, others carry a thousand. Plasmids also vary in the number of copies present in a cell. **Low-copy-number plasmids** occur in only one or a few copies per cell, whereas **high-copy-number plasmids** are present in numerous copies, perhaps 500. Most plasmids, termed **narrow host range,** can replicate in only one species. A few, however, called **broad host range,** can replicate in many different species. Plasmids are divided into different compatibility groups and only members of a different compatibility group can co-exist in the same cell.

Many bacterial plasmids are readily transferred by conjugation. **Conjugative,** or **self-transmissible,** plasmids carry all of the genetic information needed for transfer, including an origin of transfer. In contrast, **mobilizable plasmids** encode an origin of transfer but lack other genetic information required for their own transfer. A conjugative plasmid can help transfer a mobilizable plasmid present in the same cell. Some plasmids, termed **promiscuous,** can transfer between unrelated species and even between Gram-positive and Gram-negative bacteria. Certain plasmids can even be transferred into plant cells. Some of the traits encoded by plasmids are listed in **table 8.4.**

TABLE 8.4 Some Plasmid-Coded Traits

Trait	Organisms in Which Trait is Found
Antibiotic resistance	*Escherichia coli, Salmonella* sp., *Neisseria* sp., *Staphylococcus* sp., *Shigella* sp., and many other organisms
Pilus synthesis	*E. coli, Pseudomonas* sp.
Tumor formation in plants	*Agrobacterium* sp. **(see Perspective 8.2)**
Nitrogen fixation	*Rhizobium* sp.
Oil degradation	*Pseudomonas* sp.
Gas vacuole production	*Halobacterium* sp.
Insect toxin synthesis	*Bacillus thuringiensis*
Plant hormone synthesis	*Pseudomonas* sp.
Antibiotic synthesis	*Streptomyces* sp.
Increased virulence	*Yersinia enterocolitica*
Toxin production	*Bacillus anthracis*

Not only can plasmids be acquired by a cell, but they can also be lost from a cell. Occasionally, a cell will divide without distributing its plasmids to a daughter cell. If a cell can multiply faster as a consequence of losing the plasmid, the progeny of the cell will eventually predominate. In the laboratory, cells can be grown under conditions that increase the likelihood that a plasmid will be lost. The resulting population has been **cured** of the plasmid.

Resistance Plasmids

Resistance, or **R, plasmids** confer resistance to many different widely used antimicrobial medications and heavy metals, such as mercury and arsenic. Many of these plasmids are conjugative and are composed of two parts: the **resistance,** or **R, genes,** which encode the resistance traits, and a **resistance transfer factor,** or **RTF,** which encodes the properties required for conjugation **(figure 8.27).** Perhaps the most important feature of conjugative

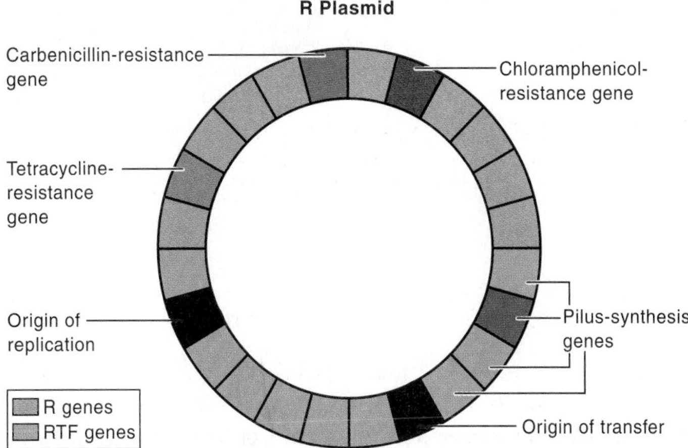

FIGURE 8.27 Two Regions of an R Plasmid The R (resistance) genes code for resistance to various antimicrobials; the RTF (resistance transfer factor) region codes for plasmid replication and the transfer of the plasmid to other bacteria.

PERSPECTIVE 8.2

Bacteria Can Conjugate with Plants: A Natural Case of Genetic Engineering

For more than 50 years, scientists have known that DNA can be transferred between bacteria. Thirty years ago, it was shown that a bacterium can even transfer its genes into plant cells, such as tobacco, carrots, and cedar trees, through a process analogous to conjugation. What led to this discovery started about 100 years ago in the laboratory of a plant pathologist, Dr. Erwin Smith. He showed that the causative agent of a common plant disease, termed **crown gall,** is a bacterium, *Agrobacterium tumefaciens.* This disease is characterized by large galls or swellings that occur on the plant at the site of infection, usually near the soil line, the crown of the plant. When other investigators cultured the diseased plant tissue, it had properties that differed from normal plant tissue. Whereas normal tissue requires several plant hormones for growth, crown gall tissue grows in the absence of these added hormones. In addition, crown gall tissue synthesizes large amounts of a compound, an **opine,** which neither normal plant tissue nor *Agrobacterium* synthesizes. The most surprising observation was that the plant cells maintained their altered nutritional requirements and the ability to synthesize opine even after the bacteria were killed by penicillin. Investigators concluded that the crown gall plant cells are permanently transformed. Although *Agrobacterium* is required to start the infection, they are not necessary to maintain the altered nutritional requirements and biosynthetic capabilities of the plant cells.

The explanation of the process by which *Agrobacterium* causes crown gall tumors and transforms plant cells was established in 1977 following a report that all strains of *Agrobacterium* capable of causing crown gall tumors contained a large plasmid termed the **tumor-inducing,** or **Ti, plasmid.** A group of microbiologists then showed that a specific piece of the Ti plasmid, termed the **transferred DNA,** or **T-DNA,** is transferred from the bacterial cell to the plant cell, where it becomes incorporated into the plant chromosome **(figure 1).** Since no regions of DNA in the plant are similar to those of bacteria, integration occurs through **non-homologous recombination.** Like conjugation between bacteria, a pilus is required for DNA transfer.

The transferred DNA acts like plant DNA because its promoters resemble those of plants rather than those of bacteria. Therefore, the genetic information in the T-DNA is expressed in plants but not in *Agrobacterium.* This DNA encodes enzymes for the synthesis of the plant hormones as well as for the opine. The expression of these genes supplies the plant cells with the plant hormones, explaining why the transformed plant cells can grow in the absence of added hormones and are able to synthesize the opine. Thus, once incorporated into the plant chromosome, the DNA provides the transformed cell with additional genetic information that confers new properties on the plant cell. ■ **promoter, p. 168**

Why does *Agrobacterium* transform plants? This bacterium has the ability to use the opine as a source of carbon and energy, whereas most other bacteria in the soil, as well as plants, cannot. Therefore, *Agrobacterium* subverts the metabolism of the plant to produce food that only *Agrobacterium* can use. Thus, *Agrobacterium* is a natural genetic engineer of plants.

The *Agrobacterium*-crown gall system is of great interest for several reasons. First, it shows that DNA can be transferred from prokaryotes to eukaryotes. Many people believed that such transfer would be impossible in nature and could only occur in the laboratory. Second, this system has spawned an industry of plant biotechnology dedicated to improving the quality of higher plants. Thus, it is possible to replace the genes of hormone and opine synthesis in the Ti plasmid with any other genes, which will then be transferred and incorporated into the plant. With this technology, genes conferring resistance to bacteria, viruses, insects, and different herbicides have been incorporated into a wide variety of plants. Rice has been transformed to synthesize high levels of β-carotene, the precursor of vitamin A.

Genetic engineering of plants became a reality once scientists learned how a common soil bacterium caused a well-recognized and serious plant disease. This system serves as a beautiful example of how solving a riddle in basic science can lead to major industrial applications.

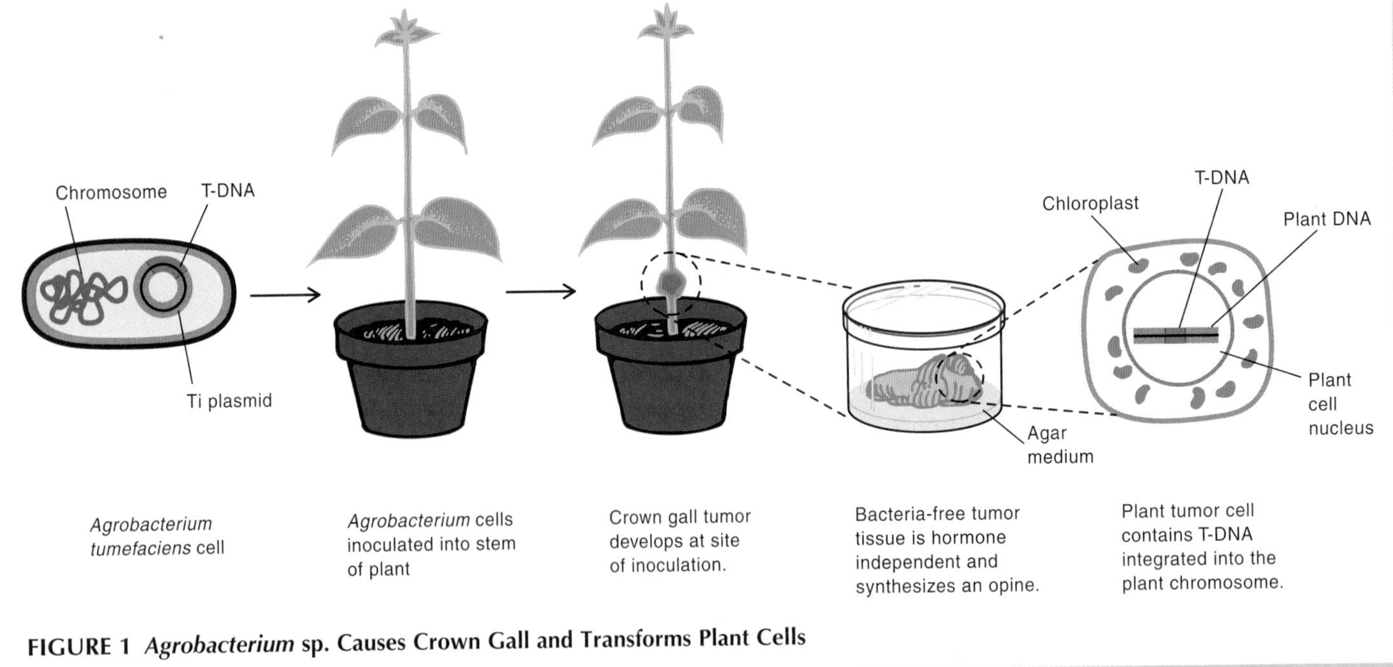

FIGURE 1 ***Agrobacterium* sp. Causes Crown Gall and Transforms Plant Cells**

R plasmids is that their transfer confers simultaneous resistance to numerous antimicrobials encoded by the R genes. Further, many R plasmids have a broad host range and can multiply in a wide variety of different Gram-negative genera, including *Shigella, Salmonella, Escherichia, Yersinia, Klebsiella, Vibrio,* and *Pseudomonas*. Transfer of promiscous plasmids can give rise to a wide array of organisms resistant to many different antimicrobials. Members of the normal microbiota, such as *E. coli,* can serve as a reservoir for R plasmids, which then can be transferred to disease-causing organisms. This is one reason why so many

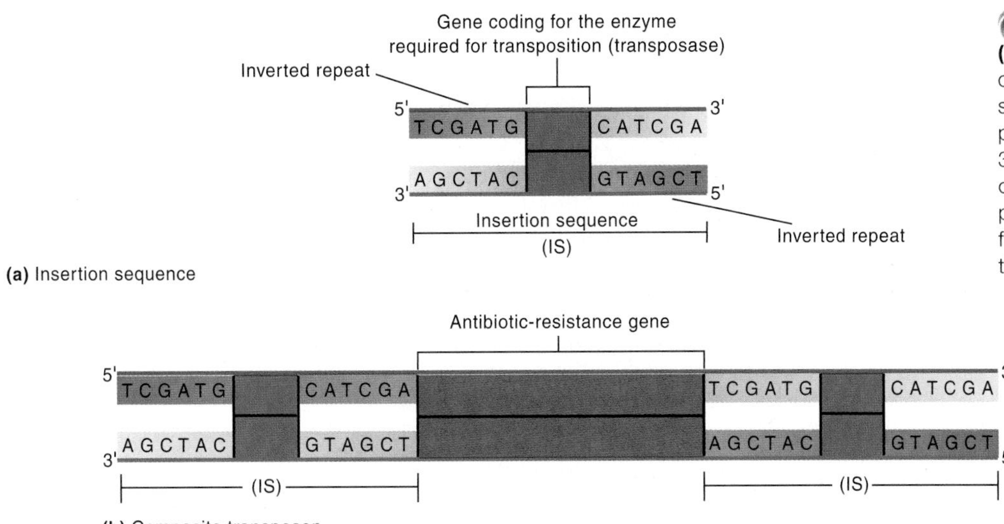

FIGURE 8.28 Transposable Elements **(a)** Insertion sequence. Note that the sequence of bases at the end of the IS element in one strand is the same as the sequence in the opposite strand reading in the same direction (5′ to 3′). This is the inverted repeat sequence. Only 6 of the 15 to 25 base pairs are shown. **(b)** Composite transposon. Antibiotic-resistance locus flanked by two insertion elements. The sizes of the various parts are not drawn to scale.

different organisms in a hospital environment are resistant to a variety of antimicrobials. ■ microbiota, p. 348

Transposons

In addition to causing mutations, transposons can provide a mechanism for mobilizing genes for transfer. Transposons can move into other replicons in the same cell without any specificity. Several types of transposons exist, varying in the complexity of their structure. The simplest, an **insertion sequence (IS),** encodes only the enzyme **transposase,** which is responsible for transposition **(figure 8.28a).** Flanking the gene are short sequences, usually 15 to 25 base pairs in length, that are identical and usually oriented in opposite directions. These are called **inverted repeats.** Two regions of DNA are said to be inverted repeats when the sequence of nucleotides in one strand in one region, when read in the 5′ to 3′ direction, is the same as the sequence in the other strand, also read in the 5′ to 3′ direction.

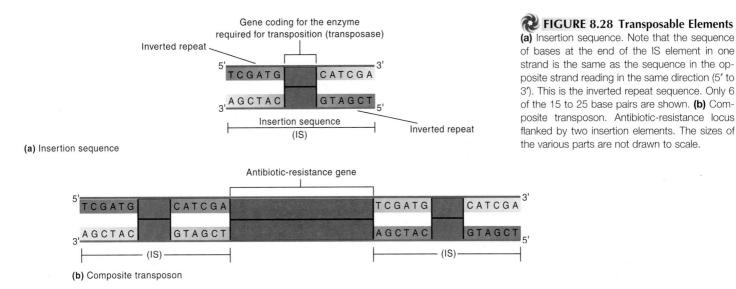

Composite transposons (named Tn1, Tn2, and so on) consist of at least one gene whose product often is easily recognized, such as a gene coding for antimicrobial resistance, flanked by ISs (figure 8.28b). Composite transposons, like insertion sequences, can move in the same replicon or from one to another in the cell. Their movement is readily followed by the easily recognized gene product they encode. They integrate into their new location through **non-homologous recombination,** a process that does not require a similar sequence in the region of recombination. If a composite transposon inserts into a conjugative plasmid, it can be transferred to other cells. In theory, any gene or group of genes can move to another site if they are bounded by ISs, but those that carry genes for antibiotic resistance are particularly important medically.

The introduction in this chapter presented a real-life example of how a transposon enabled a strain of *Staphylococcus aureus* to become resistant to vancomycin. The patient initially was infected with a strain of *S. aureus* that was susceptible to vancomycin, but resistant to many common antibiotics. After treatment with vancomycin, a strain of *S. aureus* resistant to that drug was isolated from the patient. The only obvious difference between the two isolates was that the latter had a gene that encoded resistance to vancomycin inserted into a plasmid harbored by both strains. But where did the gene come from? Further analysis suggested that it was acquired from a vancomycin-resistant strain of *Enterococcus faecalis* isolated from the same patient. The resistance gene of that bacterium was identical to the one of the *S. aureus* isolate, and was also part of a transposon integrated into a plasmid. It appears that *E. faecalis* transferred this plasmid containing the transposon to the sensitive *S. aureus* by conjugation **(figure 8.29).** This entering plasmid was apparently destroyed by enzymes in *S. aureus,* but before it was

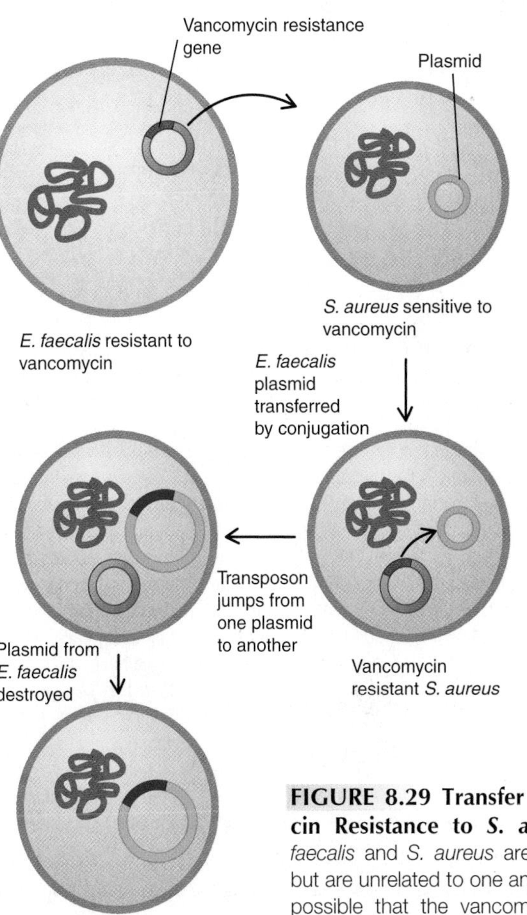

FIGURE 8.29 Transfer of Vancomycin Resistance to *S. aureus* Both *E. faecalis* and *S. aureus* are Gram + cocci but are unrelated to one another. It is quite possible that the vancomycin resistance gene was originally in the chromosome of *E. faecalis* and jumped to a plasmid.

destroyed, the transposon jumped to the plasmid that was already in the *S. aureus* cell, thereby creating the vancomycin-resistant *S. aureus*. Considering the ease at which DNA can move among a bacterial population, is it any wonder that antibiotic resistance is such a serious problem in treating infectious diseases today?

Genomic Islands

Genomic Islands are other members of the mobile gene pool. They are large DNA segments in a cell's genome that are thought to have originated in other species. This assumption is based on the fact that their base composition is quite different from the rest of the cell's genomic sequences. In general, each bacterial species has a characteristic proportion of G:C base pairs, so a large segment of DNA that has a very different G:C ratio suggests that the segment originated from a foreign source and was transferred to the cell through horizontal gene transfer. ■ **base composition, p. 248**

The characteristics encoded by genomic islands include utilization of specific energy sources, acid tolerance, development of symbiosis, and ability to cause disease. Genomic islands that encode the latter are called **pathogenicity islands.**

Some members of the mobile gene pool are summarized in **table 8.5.**

MICROCHECK 8.9

Plasmids vary in size, copy number, host range, genetic composition, compatibility to co-exist with other plasmids, and their ability to be transferred to other cells. One of the most important plasmids is the R plasmid, which codes for resistance to various antimicrobial medications and heavy metals. Transposons can move from one location to another in the same replicon or to other replicons. Transposition requires that the gene being transposed has inverted repeat sequences on its ends and the enzyme transposase as part of the insertion sequence. Genomic islands are large DNA segments that are thought to have originated in other species.

- ✓ What functions must a plasmid code for in order to be self-transmissible?
- ✓ How does transposition promote gene transfer between bacteria?
- ✓ What does the phrase, "reservoir for R plasmids" mean when referring to plasmids carried by non-disease-causing bacteria?

TABLE 8.5 Mobile Genetic Elements

Name	Composition	Properties
Transposon		Moves to different locations in DNA in same cell
Insertion Sequence (IS)	Transposase gene flanked by short repeat sequences	
Composite Transposon	Recognizable gene flanked by ISs	Same as insertion sequence
Plasmid	Circular double-stranded DNA replicon; smaller than chromosomes	Generally codes only for non-essential genetic information
Genomic Island	A large fragment of DNA in a chromosome or plasmid	Codes for genes which allow cell to occupy specific environmental locations

FUTURE CHALLENGES

Hunting for Magic Bullets

Because of the increasing resistance of microorganisms to current antimicrobial medications, the demand for new antimicrobials that will kill these resistant organisms is rapidly increasing. It is surprising and sobering to realize that only two new classes of antimicrobials have been introduced into clinics in the past 40 years. The great challenge is to develop new antimicrobial agents that strike at targets different from those attacked by current antimicrobials.

With the revolution that has been occurring in biology over the past 15 years, the development of new antimicrobial agents may become a reality. Promising new strategies are based on knowing the genomic sequence of microbial pathogens. Many of the microbial genomes which have now been sequenced are human pathogens. The study and analysis of the nucleotide sequence of DNA is called **genomics.** The next step is to identify genes necessary for the survival of the microorganism or required to cause disease. To gain some understanding of the function of any gene, one must compare its DNA sequence with the sequence of all other genes that have been put into a database, called GenBank. If the gene that has been sequenced is similar in sequence to any other gene, then it is assumed that the two genes have similar functions. Thus, if the function of a gene that has been sequenced in any organism is known, the function of all genes with a similar sequence is likely to be similar. This is the science of **bioinformatics,** which involves the analysis of the nucleotide sequence of DNA in order to understand what it codes for. Genes that are required for virulence in the pathogen but are not found in the host are potential targets. It should be possible to design a protein that inhibits the virulence protein and thereby prevents disease.

To develop an antimicrobial that inhibits an enzyme required for virulence requires a great deal of information about the enzyme, how it folds, what its three-dimensional structure is, and whether or not it interacts with other proteins.

In the past, the search for antimicrobial medications has relied on random screening. Scientists looked for growth inhibition of a pathogen by a large number of organisms isolated from soil samples collected from around the world. Today, new technologies based on microbial genomics should identify new targets and provide a rational approach to developing new antimicrobials.

SUMMARY

8.1 Genetic Change in Bacteria

The genotype of bacteria can change either through **mutations,** permanent changes in the existing nucleotide sequence of a cell's DNA, or through the acquisition of DNA from other organisms. Bacteria contain only a single set of genes (haploid) so any changes in DNA are expressed rapidly. Genes can be transferred from one organism to another, **horizontal gene transfer** (figure 8.1b), and from parent to offspring, **vertical gene transfer** (figure 8.1a).

Gene Mutation As a Source of Genetic Change

8.2 Spontaneous Mutations

Spontaneous mutations are changes in the nucleotide sequences in DNA that occur without the addition of agents known to cause mutations. These mutations are rare and occur at a characteristic frequency for each gene. They are stable but on rare occasion can undergo a change back to the non-mutant form—a **reversion.** Genes mutate independently of one another, and the chance that two mutations will occur within the same cell is the product of the individual mutation rates.

Base Substitution (figure 8.3)

Base substitutions usually occur during DNA replication; **point mutations** occur when only one base pair is changed (figure 8.2). Reactive oxygen molecules can modify guanine in DNA, leading to an increased frequency of base substitutions during DNA replication.

Removal or Addition of Nucleotides (figure 8.4)

Frameshift mutations involve the addition or deletion of nucleotides, often resulting in the formation of a **stop codon** and the synthesis of a shortened protein.

Transposable Elements (Jumping Genes) (figure 8.29)

Certain genes, called **transposons,** have the ability to move to any other location in the genome. Introduction of a transposon into another gene **insertionally inactivates** that gene (figure 8.5).

8.3 Induced Mutations (table 8.1)

Chemical Mutagens

Chemical mutagens frequently alter hydrogen-bonding properties of purines and pyrimidines, increasing the frequency of mutations. **Base analogs** with different hydrogen-bonding properties can be incorporated into DNA in place of the usual purines and pyrimidines (figure 8.7). **Intercalating agents** are flat molecules that insert into the double helix and push nucleotides apart, resulting in a frameshift mutation (figure 8.8).

Transposition

Insertional inactivation results when a transposon integrates into a new site in the cell's genome and inactivates the gene (figure 8.5).

Radiation

Ultraviolet irradiation results in **thymine dimers** due to the formation of covalent bonds between adjacent thymine molecules on the same strand of DNA (figure 8.9). X rays cause single-strand breaks, double-strand breaks, and alterations to the DNA bases.

8.4 Repair of Damaged DNA (table 8.2)

Repair of Errors in Base Incorporation

DNA polymerase has a **proofreading** function. In **mismatch repair,** an endonuclease makes a cut in one strand of DNA, and another enzyme degrades a portion of the strand, removing the misincorporated nucleotide. A new DNA strand is then synthesized (figure 8.10).

Repair of Thymine Dimers

In **light repair,** a photoreactivating enzyme breaks the bonds of the thymine dimer, thereby restoring the original molecule (figure 8.11a). In **excision,** or **dark repair,** the damaged single-stranded segment is excised by an endonuclease. A new strand is synthesized by DNA polymerase (figure 8.11b).

Repair of Modified Bases in DNA

Specific glycosylases can remove modified bases, such as oxidized guanine and alkylated bases in DNA (figure 8.12).

SOS Repair

SOS repair is a last-ditch repair mechanism in which about 20 enzymes are induced by damaged DNA, including a new DNA polymerase that can bypass the damaged DNA but does not proofread the DNA it synthesizes. Consequently, the synthesized DNA contains many mutations. SOS repair accounts for the mutagenic activity of UV irradiation.

8.5 Mutant Selection

Direct Selection

Direct selection involves inoculating cells onto a medium on which the mutant, but not the parent, can grow; these are easy mutants to isolate (figure 8.13).

Indirect Selection

Indirect selection is required when no medium supports the growth of only the desired mutant. **Replica plating** involves the simultaneous transfer of all the colonies on one plate to two other plates and the comparison of the growth of individual colonies on both plates (figure 8.14). **Penicillin enrichment** increases the proportion of mutants in a population by killing growing bacteria in a medium on which only non-mutants will grow (figure 8.15).

Testing of Chemicals for Their Cancer-Causing Ability

The Ames test measures whether a chemical, a suspected **carcinogen,** increases the frequency of reversion; a positive test indicates the subject chemical is a mutagen and is therefore a possible carcinogen (figure 8.16).

Gene Transfer As a Mechanism of Genetic Change (table 8.3)

8.6 DNA-Mediated Transformation

DNA-mediated transformation involves the transfer of "naked" DNA. DNA must be integrated into the recipient's genome to be maintained (figure 8.18). Deoxyribonuclease addition prevents the transfer (figure 8.19).

Natural Competence

Natural **competence** is the ability of a cell to take up and integrate DNA. DNA enters the cell as a single-stranded molecule and is integrated by replacing recipient cell genes via **homologous recombination.**

Artificial Competence

Cells can be made artificially competent by **electroporation;** in this process, the cells are treated with an electric current, which makes holes in the cell envelope through which DNA can pass (figure 8.20).

8.7 Transduction

Transduction involves the transfer of bacterial DNA by a **bacteriophage** (figure 8.21). DNA must be integrated to be maintained (figure 8.18).

8.8 Conjugation

Conjugation requires cell-to-cell contact (figure 8.22).

Plasmid Transfer

The **donor cells** synthesize an **F pilus** encoded on an **F plasmid,** which recipient cells do not have; the F plasmid is transferred from an F^+ to an F^- cell through the pilus (figure 8.23).

Chromosome Transfer

Chromosome transfer occurs when the F plasmid integrates into a chromosome and the resulting cell can transfer a portion of the chromosome into a recipient cell (figures 8.24, 8.25). DNA must be integrated into a replicon of the recipient in order to be maintained (figure 8.18).

F′ Donors

In an **F′** donor, the F plasmid is excised from the chromosome and carries a piece of chromosome with it. This piece is transferred together with the F plasmid (figure 8.26).

8.9 The Mobile Gene Pool (table 8.5)

Many members of the gene pool can be transferred to other locations within a cell or to other cells. These members include plasmids, transposons, and genomic islands.

Plasmids (table 8.4)

Plasmids are extrachromosomal replicons that code for non-essential information; many are readily transferred by conjugation. **Self-transmissible** or conjugative plasmids carry all the genetic information needed for their transfer. **R plasmids** code for antibiotic resistance (figure 8.27).

Transposons (figure 8.28)

Transposons provide a mechanism for mobilizing genes because they can move into other replicons in the same cell. The simplest transposon is an **insertion sequence (IS),** which only contains a gene that codes for the enzyme **transposase** (figure 8.28a). A **composite transposon** consists of one or more genes flanked by insertion sequences (figure 8.28b).

Genomic Islands

Genomic islands are large DNA segments in a cell that are thought to have originated in other species.

REVIEW QUESTIONS

Short Answer

1. What one activity must all plasmids carry out?
2. What is the term that describes a plasmid that can transfer itself from *E. coli* into *Pseudomonas,* where it can replicate?
3. What type of mutation in an operon is most likely to affect the synthesis of more than one protein?
4. What are the two necessary features of an insertion sequence?
5. What enzyme has proofreading ability? How does it function in proofreading?
6. Give an example of gene transfer that involves homologous recombination. Give one that involves non-homologous recombination.
7. Single-strand integration of DNA has been shown to be a feature of which mechanism of DNA transfer?
8. What feature of an F′ particle is similar to chromosomal DNA? To a plasmid?
9. Name three mobile genetic elements.
10. Name three ways in which plasmids differ from bacterial chromosomes.

Multiple Choice

1. A culture of *E. coli* is irradiated with ultraviolet (UV) light. Answer questions 1 and 2 based on this statement. The effect of the UV light is to specifically
   a) join the two strands of DNA together by covalent bonds.
   b) join the two strands of DNA together by hydrogen bonds.
   c) form covalent bonds between thymine molecules on the same strand of DNA.
   d) form covalent bonds between guanine and cytosine.
   e) delete bases.
2. The highest frequency of mutations would be obtained if, after irradiation, the cells were immediately
   a) placed in the dark.
   b) exposed to visible light.
   c) shaken vigorously.
   d) incubated at a temperature below their optimum for growth.
   e) The frequency would be the same no matter what the environmental conditions are after irradiation.
3. Penicillin enrichment of mutants works on the principle that
   a) only Gram-positive cells are killed.
   b) cells are most sensitive to antimicrobial medications during the lag phase of growth.
   c) most Gram-negative cells are resistant to penicillin.
   d) penicillin only kills growing cells.
   e) penicillin inhibits formation of the lipopolysaccharide layer.
4. Repair mechanisms that occur during DNA synthesis are
   1. mismatch repair.
   2. proofreading by DNA polymerase.
   3. light repair.
   4. SOS repair.
   5. excision repair.

   a) 1, 2 b) 2, 3 c) 3, 4 d) 4, 5 e) 1, 5
5. You are trying to isolate a mutant of wild-type *E. coli* that requires histidine for growth. This can best be done using
   1. direct selection.
   2. replica plating.
   3. penicillin enrichment.
   4. a procedure for isolating conditional mutants.
   5. reversion.

   a) 1, 2 b) 2, 3 c) 3, 4 d) 4, 5 e) 1, 5

6. The properties that all plasmids share are that they
   1. all carry genes for antimicrobial resistance.
   2. are self-transmissible to other bacteria.
   3. always occur in multiple copies in the cells.
   4. code for non-essential functions.
   5. replicate in the cells in which they are found.

   a) 1, 2 b) 2, 3 c) 3, 4 d) 4, 5 e) 1, 5
7. The addition of deoxyribonuclease to a mixture of donor and recipient cells will prevent gene transfer via
   a) DNA transformation.
   b) chromosome transfer by conjugation.
   c) plasmid transfer by conjugation.
   d) generalized transduction.
8. An F pilus is essential for
   1. DNA transformation.
   2. chromosome transfer by conjugation.
   3. plasmid transfer by conjugation.
   4. generalized transduction.
   5. cell movement.

   a) 1, 2 b) 2, 3 c) 3, 4 d) 4, 5 e) 1, 5
9. A plasmid that can replicate in *E. coli* and *Pseudomonas* is most likely a/an
   a) broad host range plasmid.
   b) self-transmissible plasmid.
   c) high-copy-number plasmid.
   d) essential plasmid.
   e) low-copy-number plasmid.
10. The frequency of transfer of an F′ DNA molecule by conjugation is closest to the frequency of transfer of
    a) chromosomal genes by conjugation.
    b) an F plasmid by conjugation.
    c) an F plasmid by transformation.
    d) an F plasmid by transduction.
    e) an R plasmid by DNA transformation.

Applications

1. Some bacteria are more resistant to UV light than others. Discuss two reasons why this might be the case. What experiments could you do to determine whether each of the two possibilities could be correct?
2. A pharmaceutical researcher is disturbed to discover that the major ingredient of a new drug formulation causes frameshift mutations in bacteria. What other information would the researcher want before looking for a substitute chemical?

Critical Thinking

1. You have the choice of different kinds of mutants for use in the Ames test to determine the frequency of reversion by suspected carcinogens. You can choose a deletion, a point mutation, or a frameshift mutation. Would it make any difference which one you chose? Explain.
2. You have isolated a strain of *E. coli* that is resistant to penicillin, streptomycin, chloramphenicol, and tetracycline. You also observe that when you mix this strain with cells of *E. coli* that are sensitive to the four antibiotics, they become resistant to streptomycin, penicillin, and chloramphenicol but remain sensitive to tetracycline. Explain what is going on.

Researcher in a molecular biology laboratory.

9 Biotechnology and Recombinant DNA

A Glimpse of History

In 1976, Argentinean newspapers reported a violent shootout between soldiers and the occupants of a house in suburban Buenos Aires, leaving the five extremists inside dead. Conspicuously absent from those reports was the identity of the "extremists"—a young couple and their three children, ages 6 years, 5 years, and 6 months. Over the next seven years, similar scenarios recurred as the military junta that ruled Argentina eliminated thousands of citizens it perceived as threats. This "Dirty War," as it came to be known, finally ended in 1983 with the collapse of the military junta and the election of a democratic government. The new leaders opened previously sealed records, which confirmed what many had already suspected—that more than 200 children survived the carnage and had in fact been kidnapped and placed with families in favor with the junta.

Dr. Mary-Claire King was at the University of California at Berkeley when her help was enlisted in the effort to return the children to the surviving members of their biological families. Dr. King and others recognized that DNA technology could be used for this important humanitarian cause. By analyzing certain DNA sequences, blood and tissue samples from one individual can be distinguished from those of another. These same principles can also be used to show that a particular child is the progeny of a given set of parents. Because a person has two copies of each chromosome—one inherited from each parent—one half of a child's DNA will represent maternal sequences and the other half will represent paternal traits. The case of the Argentinean children presented a great challenge, however, as most of the parents were dead or missing. Often, the only surviving relatives were aunts and grandmothers, and it is difficult to use chromosomal DNA to show genetic relatedness between a child and such relatives. Dr. King decided to investigate mitochondrial DNA (mtDNA). This organelle DNA, unlike chromosomal DNA, is inherited only from the mother. A child will have the same nucleotide sequence of mtDNA as his or her siblings, the mother and her siblings, as well as the maternal grandmother.

By comparing the nucleotide sequences of mtDNA in different individuals, Dr. King was able to locate key positions that varied extensively among unrelated people, but were similar in maternal relatives. Dr. King's technique, born out of a desire to help reunite families victimized by war, has now found many uses. Today her lab, which is now at the University of Washington, remains very active using molecular biology techniques for humanitarian efforts, identifying the remains of victims of atrocities around the world.

A revolution has occurred in molecular biology over the past several decades—the science has been transformed from a descriptive study of what cells are, to an intricate study of how they function. A driving force in that revolution was the development of simple methods to extract and manipulate DNA, the blueprint of life.

Biotechnology is the use of microbiological and biochemical techniques to solve practical problems and produce useful products. In the past, this usually meant laboriously searching for naturally occurring mutants that produced maximal product or expressed other desirable characteristics. Today, the rapid developments in **recombinant DNA techniques,** the methods scientists use to study and manipulate DNA, have made it possible to genetically alter organisms to give them more useful traits. Researchers can isolate genes from one organism, manipulate the purified DNA *in vitro,* and then transfer the genes into another organism, a process called **gene cloning.** In fact, biotechnology is now nearly synonymous with **genetic engineering,** the process of

KEY TERMS

Colony Blotting Technique used to determine which colonies on an agar plate contain a given nucleotide sequence.

DNA Microarray Technique used to study gene expression patterns in an organism.

DNA Probe Single-stranded piece of DNA tagged with a detectable marker and used to detect its complement.

DNA Sequencing Technique used to determine the sequence of nucleotides in a DNA molecule.

Fluorescence *in situ* Hybridization (FISH) Technique used to detect a given nucleotide sequence within intact cells affixed to a microscope slide.

Gel Electrophoresis A procedure used to separate DNA fragments (or other macromolecules) according to their size.

Genetic Engineering Deliberately altering an organism's genetic information using *in vitro* techniques.

Polymerase Chain Reaction (PCR) Technique used to exponentially amplify specific regions of a DNA molecule.

Recombinant DNA Molecule DNA molecule created by joining DNA fragments from two different sources.

Restriction Enzyme Type of enzyme that recognizes and cleaves a specific sequence of DNA.

Vector DNA molecule, often a plasmid, that acts as a carrier of cloned DNA.

deliberately altering an organism's genetic information using *in vitro* techniques.

Since the advent of gene cloning, a virtual toolbox of DNA technologies has been developed. These have generated information and innovations that are impacting society in innumerable ways—from agricultural practices and medical diagnoses to evidence used in the courtroom.

9.1

Fundamental Tools Used in Biotechnology

Focus Point

- Describe the role of restriction enzymes and gel electrophoresis in biotechnology.

Before exploring the application of biotechnology, it is helpful to understand some of the basic components of a molecular biologist's "tool kit." These include restriction enzymes, which cut DNA into fragments, and gel electrophoresis, which separates DNA fragments according to size. As we describe these, it is important to remember that diagrams focus on only one or a few DNA molecules in a solution to illustrate what is happening at a molecular level. In reality, scientists are generally working with DNA isolated from cultures containing well over a million cells.

Restriction Enzymes

Restriction enzymes are naturally occurring enzymes that allow scientists to easily cut DNA into fragments in a predictable and controllable manner **(figure 9.1a).** Each enzyme recognizes a specific 4 to 6 base-pair nucleotide sequence **(table 9.1).** The sequences are characteristically **palindromes,** meaning that the sequence is the same on both strands when each is read in the 5′ to 3′ direction. Wherever the palindrome recognized by a particular restriction enzyme is found, that enzyme cuts each strand within

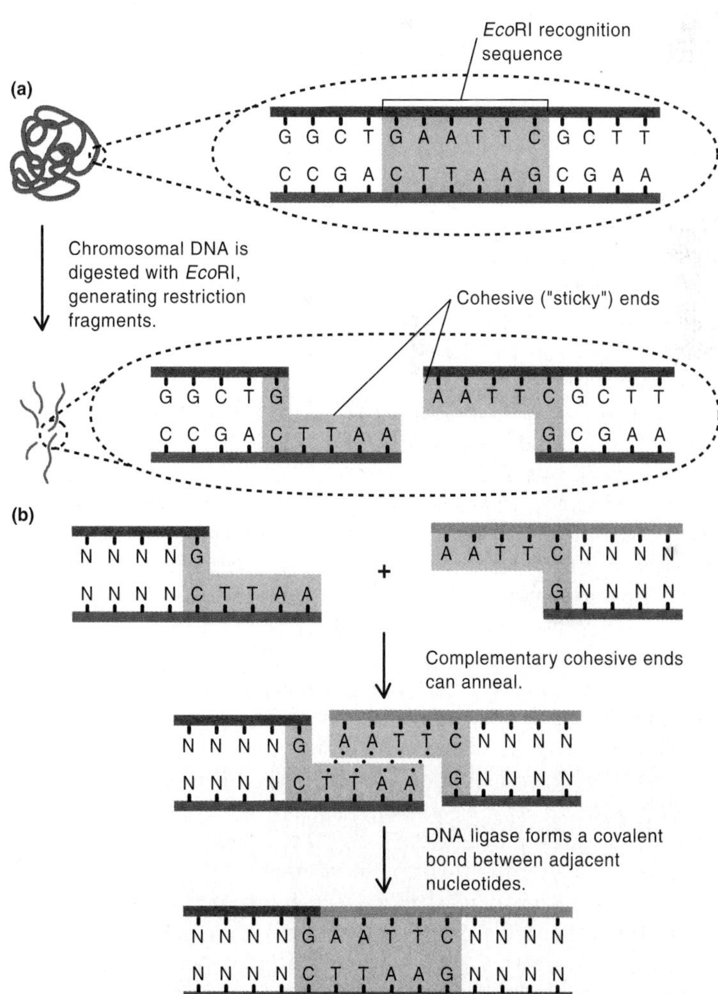

FIGURE 9.1 Action of Restriction Enzymes (a) Digesting DNA with a restriction enzyme generates restriction fragments. **(b)** Fragments that have complementary cohesive ends can anneal, regardless of their original source (N = nucleotide, meaning that it could be any of the 4 bases as long as base-pairing rules are followed).

TABLE 9.1 Examples of Common Restriction Enzymes

Enzyme	Microbial Source	Recognition Sequence (arrows indicate cleavage sites)
*Alu*I	*Arthrobacter luteus*	5′ A G↓C T 3′ 3′ T C↑G A 5′
*Bam*HI	*Bacillus amyloliquefaciens* H	5′ G↓G A T C C 3′ 3′ C C T A G↑G 5′
*Eco*RI	*Escherichia coli* RY13	5′ G↓A A T T C 3′ 3′ C T T A A↑G 5′

or near that sequence, thereby cleaving the double-stranded DNA molecules. This action digests the DNA, generating a series of pieces called **restriction fragments.** ■ restriction enzymes, p. 314

Each restriction enzyme has been given a name that represents the bacterium from which the enzyme was first isolated. The first letter represents the first letter of the genus name, and the next two letters are derived from the species name. Any other numbers or letters designate the strain and order of discovery. For example, a restriction enzyme from *E. coli* strain RY13 is called *Eco*RI.

Researchers use restriction enzymes not only to cut DNA, but also to generate fragments that can easily be joined to fragments from an entirely different source. This is possible because many restriction enzymes produce a staggered cut in the recognition sequence, generating ends with a short overhang of usually 4 bases (see figure 9.1b). The overhangs are called **sticky ends** or **cohesive ends** because they will form base pairs, or **anneal,** with one another. Any two complementary cohesive ends can anneal, even those from two different organisms. The relatively weak hydrogen bonds that hold the strands together are only temporary, however. The enzyme **DNA ligase** forms a covalent bond between the sugar-phosphate residues of adjacent nucleotides, joining the two molecules. Thus, if restriction enzymes are viewed as scissors that cut DNA into fragments, then DNA ligase is the glue that pastes the fragments together. The combined actions of restriction enzymes and DNA ligase enable researchers to join fragments of DNA from diverse sources, creating **recombinant DNA molecules.** ■ DNA ligase, p. 167

Gel Electrophoresis

Gel electrophoresis is used to separate DNA fragments according to size **(figure 9.2).** The technique uses a slab of gel that has the consistency of a very firm gelatin and is made of either **agarose,** a highly purified form of agar, or polyacrylamide. A DNA sample is put into a well in the gel; there are generally numerous wells

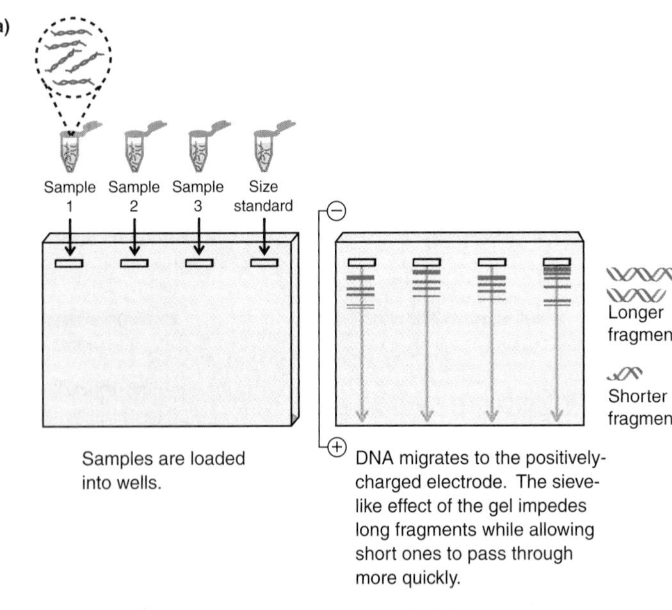
(a)
Sample 1
Sample 2
Sample 3
Size standard
Longer fragments
Shorter fragments
Samples are loaded into wells.
DNA migrates to the positively-charged electrode. The sieve-like effect of the gel impedes long fragments while allowing short ones to pass through more quickly.

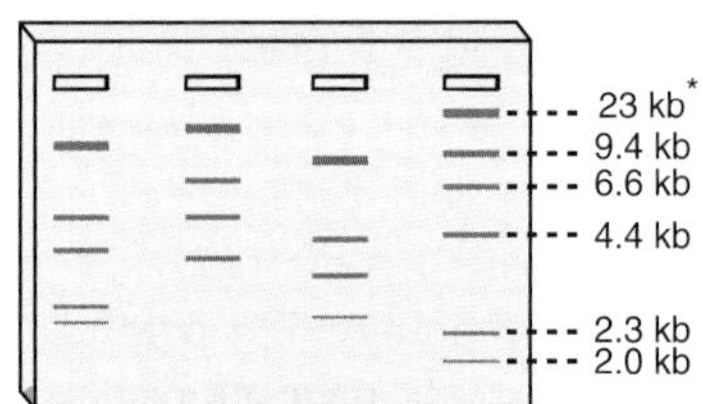
23 kb*
9.4 kb
6.6 kb
4.4 kb
2.3 kb
2.0 kb
Fragments in the samples are separated according to size.
* Fragment sizes in size standard
kb = 1,000 base pairs

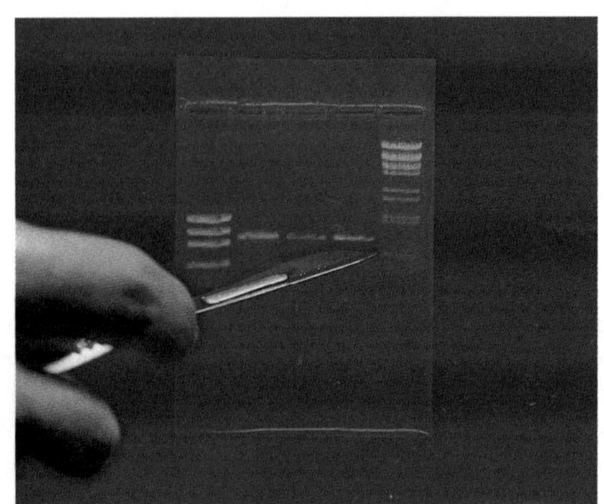
(b)

FIGURE 9.2 Gel Electrophoresis **(a)** Gel electrophoresis separates DNA fragments according to size. **(b)** DNA on the gel is visible when stained with ethidium bromide and viewed with UV light; each fluorescent band represents millions of molecules of a specific-sized fragment.

in a gel so that multiple samples can be analyzed simultaneously. As a means to eventually determine the size of the various DNA fragments in the samples, a **size standard** is routinely put into a well of the same gel. This is simply a mixture of DNA fragments of known sizes that can be used as a basis for later comparison.

The gel is then subjected to an electrical current. DNA is negatively charged, so the fragments migrate toward the positively charged electrode. As the DNA moves through the gel, however, not all fragments progress at the same rate. This is because the gel acts as a sieve, impeding the long fragments while allowing the shorter ones to pass through more quickly. Because of the sieve-like effect of the gel, the restriction fragments are separated according to their size.

The DNA is not visible in the gel unless it is stained. To do this, the gel containing the separated DNA fragments is immersed in a solution containing **ethidium bromide.** This dye binds DNA and fluoresces when viewed with UV light (see figure 9.2b). Each fluorescent band represents millions of molecules of a specific-sized fragment of DNA.

Gel electrophoresis can also be used to separate other macromolecules, specifically RNA and proteins, according to their size. The basic principles are similar to that illustrated for separating DNA, but the gel compositions differ.

MICROCHECK 9.1

Restriction enzymes recognize specific nucleotide sequences, and then cut the DNA, generating restriction fragments. Gel electrophoresis is used to separate DNA fragments according to their size.

✓ How does gel electrophoresis separate different-sized DNA fragments?

✓ What is the significance of cohesive ends?

✓ Based on the protocol for naming restriction enzymes, what is the name of one isolated from *Staphylococcus aureus* strain 3A?

TABLE 9.2 Some Applications of Genetic Engineering

Example	Use
PROTEIN PRODUCTION	
Pharmaceutical proteins	
Alpha interferon	Treating cancer and viral infections
Erythropoietin	Treating some types of anemia
Beta interferon	Treating multiple sclerosis
Deoxyribonuclease	Treating cystic fibrosis
Factor VIII	Treating hemophilia
Gamma interferon	Treating cancer
Glucocerebrosidase	Treating Gaucher disease
Growth hormone	Treating dwarfism
Insulin	Treating diabetes
Platelet derived growth factor	Treating foot ulcers in diabetics
Streptokinase	Dissolving blood clots
Tissue plasminogen activator	Dissolving blood clots
Vaccines	
Hepatitis B	Preventing hepatitis
HPV	Preventing cervical cancer
Foot-and-mouth disease	Preventing foot-and-mouth disease in animals
Other proteins	
Bovine somatotropin	Increasing milk production in cows
Chymosin	Cheese-making
Restriction enzymes	Cutting DNA into fragments
DNA PRODUCTION	
DNA for study	Determining nucleotide sequences; obtaining DNA probes
RESEARCHING GENE FUNCTION AND REGULATION	
Creating gene fusions	Studying the conditions that affect gene activity
TRANSGENIC PLANTS	
Pest-resistant plants	Insect-resistant corn, cotton, and potatoes
Herbicide-resistant plants	Biodegradable herbicide can be used to kill weeds without killing engineered plants (soybean, cotton, corn)
Plants with improved nutritional value	Rice that produces vitamin A and iron
Plants that function as edible vaccines	Enables researchers to study foods as vehicles for edible vaccines

9.2

Applications of Genetic Engineering

Focus Point

- Describe three general applications of genetically engineered bacteria, and one application of genetically engineered plants.

Genetic engineering brought biotechnology into a new era by providing a powerful tool for manipulating microorganisms for medical, industrial, and research uses **(table 9.2).** More recently, techniques that permit the genetic engineering of plants and animals have been developed.

Genetically Engineered Bacteria

Genetic engineering relies on **DNA cloning,** a process that involves isolating DNA from one organism, using restriction enzymes to cut the DNA into fragments, and then introducing those fragments into cells of another organism, most commonly *E. coli.* As part of the process, the cloned DNA must replicate in the recipient in order for it to be passed to daughter cells, generating a population of cells, each harboring a copy of the DNA fragment (see figure 8.18). Most DNA fragments, however, are unlikely to contain an origin of replication and therefore will not replicate independently in a cell, nor will they integrate into the chromosome. To overcome this problem, each fragment being cloned must be inserted into a plasmid or other independently replicating DNA molecule to from a **recombinant molecule.** The DNA molecule used as a carrier of the cloned DNA is called a **vector,** and is commonly a plasmid that has been genetically modified. The DNA that has been incorporated into the vector is called an **insert.** If a gene coding for a valuable protein is inserted into a high-copy-number vector and then introduced into a bacterium, that organism will then make increased amounts of the protein, because each gene copy can be transcribed and translated **(figure 9.3).** ■ **origin of replication, p. 166** ■ **high-copy-number plasmid, p. 205**

An approach frequently used to clone a specific gene is to clone a set of restriction fragments that together make up the entire genome of the organism being studied into a population of *E. coli* cells **(figure 9.4).** While each cell in the resulting population contains only one fragment of the genome, the entire genome is represented in the population as a whole. Because each cloned molecule can be viewed as one "book" of the total genetic information, the collection of clones is called a **DNA library.** Once a DNA library has been prepared, colony blots (which will be described later) can be used to determine which cells contain the gene of interest.

Genetically engineered bacteria have a variety of uses, including protein production, DNA production, and research. In fact, a great deal of the information described in this textbook has been revealed through research involving genetically engineered bacteria.

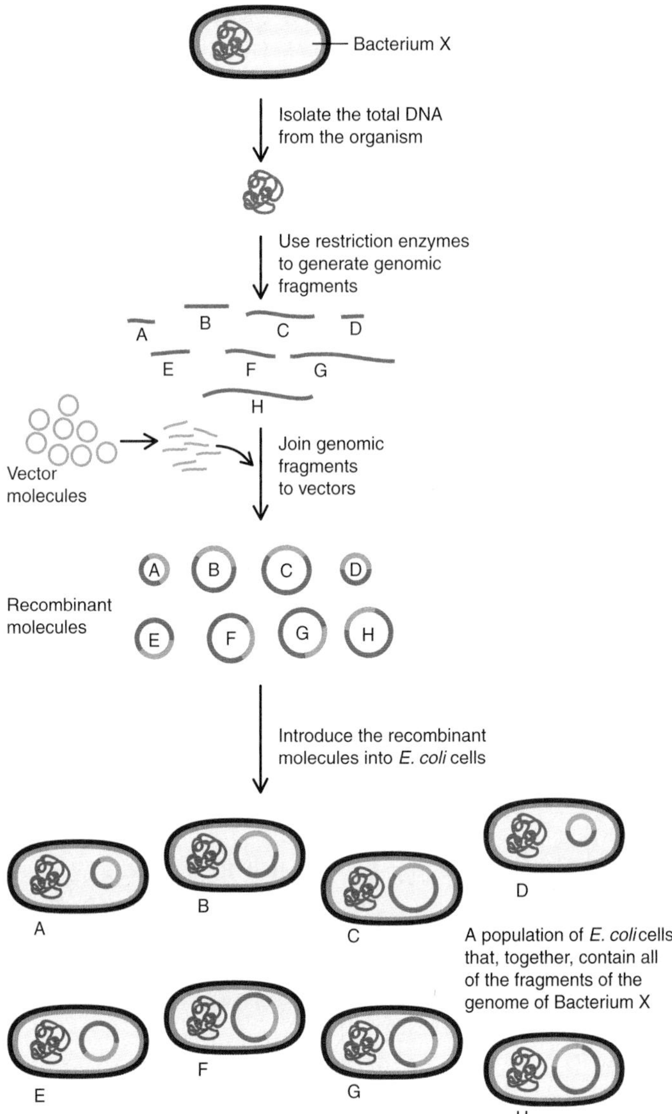

FIGURE 9.4 A DNA Library Each cell contains one fragment of a given organism's genome.

Protein Production

A number of different pharmaceutical proteins are now produced by genetically engineered microorganisms. In the past, these proteins were extracted from live animal or cadaver tissues, which made them expensive and limited in supply. Human insulin, used in treating diabetes, was one of the first important pharmaceutical proteins to be produced through genetic engineering. The original commercial product, extracted from pancreatic glands of cattle and pigs, sometimes caused allergic reactions. Once the gene for human insulin was cloned into bacteria, microorganisms became the major source of insulin sold in the United States. A 2,000-liter culture of *E. coli* that contains the insulin gene yields 100 grams of purified insulin—an amount that would require 1,600 pounds of pancreatic glands! Using microbes to produce insulin is safer and more economical than extracting it from animal tissues.

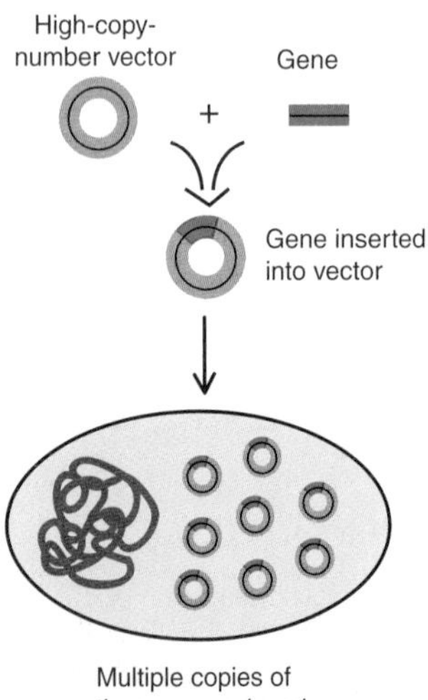

FIGURE 9.3 Cloning into a High-Copy-Number Vector When a gene is inserted into a high-copy-number vector, multiple copies of that gene will be present in a single cell, resulting in the synthesis of many more molecules of the encoded protein.

Another medically important use of genetically engineered microorganisms is vaccine production. Vaccines protect against disease by harmlessly exposing a person's immune system to killed or weakened forms of the pathogen, or to a part of the pathogen. Although vaccines are generally composed of whole bacterial cells or viral particles, only specific proteins, or parts of the proteins, are actually necessary to induce protection, or immunize, against the disease. The genes coding for these proteins can be cloned into yeast or bacteria so that these cells produce large amount of immunizing the proteins, which can then be purified. This type of vaccine is currently used to prevent hepatitis B and cervical cancer in humans and foot-and-mouth disease of domestic animals. ■ **vaccines, p. 433**

One of the most widely used proteins made by genetically engineered organisms is chymosin, a proteolytic enzyme used in cheese production. It is a natural component of rennin, a preparation from the stomach of calves. Chymosin causes milk to coagulate and produces desirable changes in the characteristics of cheeses as they ripen. Using genetically engineered bacteria to produce chymosin is preferable to isolating rennin from calves because the microbial product is less expensive and more reliably available. Other proteins produced by genetically engineered microbes include various restriction enzymes and bovine somatotropin, a growth hormone used to increase milk production in dairy cows.

DNA Production

In many cases, a researcher is interested in obtaining readily available supplies of certain DNA fragments. By cloning a segment of DNA into a well-characterized bacterium such as *E. coli,* a readily available source of that sequence is available for study and further manipulation.

Human genes are often cloned into bacteria to make them easier to study. A human cell contains an estimated 25,000 genes, whereas *E. coli* contains only 4,500 genes; thus, a human gene cloned into the bacterium on a high-copy-number vector represents a much higher percentage of the total DNA in the recipient cell than in the original cell. This makes it easier to isolate the DNA as well as the gene product.

Random samples of DNA from any environment can be cloned into *E. coli* and then the nucleotide sequence determined. By doing this, the genomic characteristics of some of the 99% of bacteria that have not been growth in culture can be studied. This "shotgun cloning" is the first step in **metagenomics,** the study of the total genomes in a sample. Metagenomics will be discussed when we describe the applications of DNA sequencing.

Researching Gene Function and Regulation

Gene function and regulation can more easily be studied in *E. coli* because systems for manipulating its DNA have been developed. For example, regulation of gene expression can be studied by creating a **gene fusion**—joining the gene being studied and a **reporter gene (figure 9.5).** The reporter gene encodes a readily observable phenotype such as fluorescence, making it possible to directly observe the expression of the gene. This, in turn, makes it possible to determine the conditions that affect gene activity. One widely used reporter gene encodes a protein called green fluorescent protein (GFP).

Genetically Engineered Eukaryotes

Yeasts can be genetically engineered to perform many of the functions described for bacteria. They serve as an important model for gene function and regulation in eukaryotic cells.

Multicellular organisms can also be genetically engineered. A plant or animal into which a cloned gene has been introduced is called a **transgenic organism.** Transgenic plants are of particular interest to microbiologists because their development was spawned as a result of basic research studying *Agrobacterium tumefaciens*; this bacterium harbors a plasmid, the Ti plasmid, that has been genetically manipulated so it can be used as a vector to deliver desirable genes to plant cells (see Perspective 8.2). Corn, cotton, and potatoes have been engineered to produce a biological insecticide called Bt-toxin, which is naturally produced by the bacterium *Bacillus thuringiensis* as it forms endospores. Unlike many chemically synthesized toxins, Bt-toxin is toxic only to insects, including their larvae. Soybeans, cotton, and corn have been engineered to resist the effects of the herbicide glyphosate (Roundup™). This enables growers to apply this biodegradable herbicide, which kills weeds and other non-engineered plants, in place of more persistent alternatives. Also, because the herbicide can be applied throughout the growing season, the soil can be tilled less frequently, preventing erosion. Plants with improved nutritional value are also being developed. For example, genes that code for the synthesis of β-carotene, a precursor of vitamin A, have been introduced into rice. The rice was also engineered to provide more dietary iron. The diet of a significant proportion of the world's population is deficient in these essential nutrients, so advances such as this could have a profound impact on world health. Plants that function as edible vaccines are also being developed. Researchers have successfully genetically engineered potatoes and rice to produce certain proteins from pathogens and have shown that the immune system responds to these, raising hopes for an edible vaccine. Whether the immune response is strong enough to protect against disease remains to be determined. ■ **endospore, p. 69**

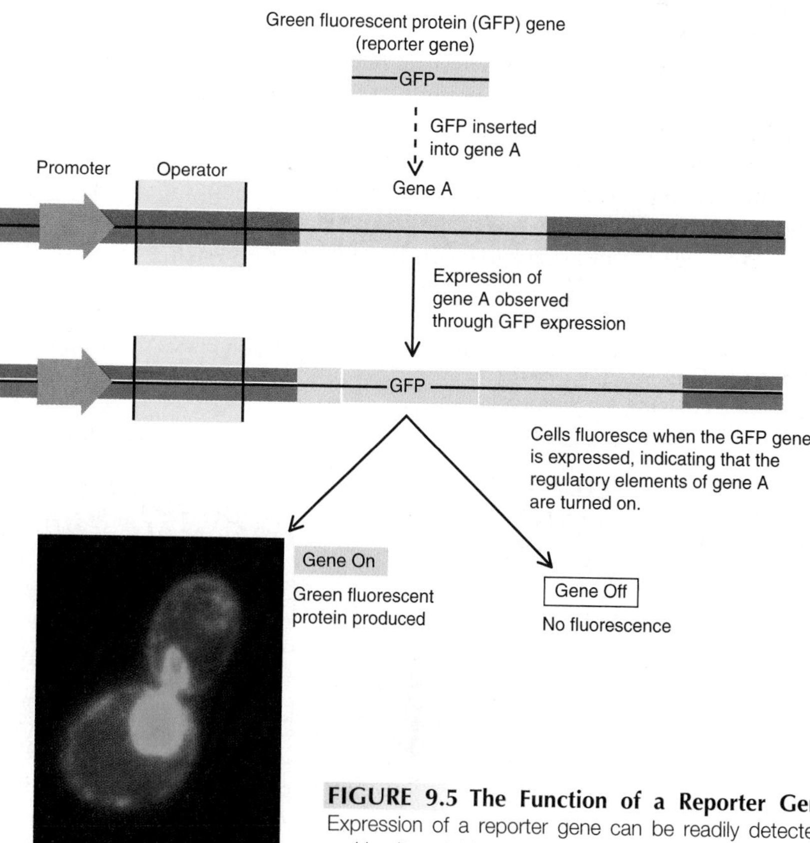

FIGURE 9.5 The Function of a Reporter Gene Expression of a reporter gene can be readily detected, making it useful in the study of gene regulation.

MICROCHECK 9.2

Genetically engineered bacteria can be used to produce a variety of products including medically important proteins, vaccines, commercially important proteins, and DNA for study. They are also used as research tools. Genetic engineering can be used to develop pest-resistant plants, herbicide-resistant plants, and plants with improved nutritional value.

✓ Name a disease that is treated by using a protein produced by genetically engineered microorganisms.

✓ Describe the attributes of two different types of transgenic plants.

Techniques Used in Genetic Engineering

Focus Points

- Explain how introns are removed from eukaryotic genes.
- Describe the characteristics of a typical vector.
- Explain how cells that harbor recombinant molecules are obtained.

This section will describe methods used to clone DNA in bacterial cells **(figure 9.6).** For information regarding the engineering of eukaryotic organisms, see Perspective 8.2 and visit the Online Learning Center (www.mhhe.com/nester6).

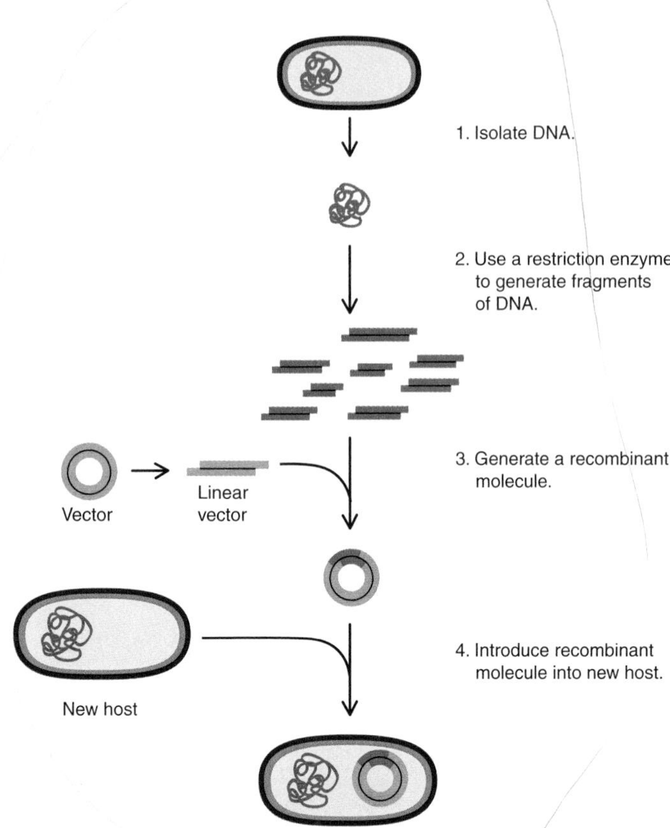

FIGURE 9.6 The Steps of a Cloning Experiment

Obtaining DNA

The first step of a cloning experiment is to obtain the DNA that will be cloned. To do this, cells in a broth culture are lysed by adding a detergent. As the cells burst open, the relatively fragile DNA is inevitably sheared into many pieces of varying lengths.

When cloning eukaryotic genes into bacteria, the introns must first be removed if the goal is protein production. To do this, mRNA from which the eukaryotic cell has already removed the introns is first isolated from the appropriate eukaryotic tissue. Then, a strand of DNA complementary to the mRNA is synthesized *in vitro* using **reverse transcriptase,** an enzyme encoded by retroviruses. That strand of DNA is then used as a template for synthesis of its complement, creating double-stranded DNA. The resulting copy of DNA, or **cDNA,** encodes the same protein as the original DNA, but lacks the introns **(figure 9.7).** ■ introns, p. 175 ■ retrovirus, p. 333

Generating a Recombinant DNA Molecule

As described earlier, restriction enzymes and DNA ligase are used to create recombinant molecules consisting of a vector and insert. The vector, usually a modified plasmid or bacteriophage, has an origin of replication and functions as a carrier of the cloned DNA **(figure 9.8).** Vectors also must have at least one restriction enzyme recognition site. This allows the circular vector to be cut, forming a linear molecule to which the insert can

FRANK & ERNEST: © Thaves/Dist. By Newspaper Enterprise Association, Inc.

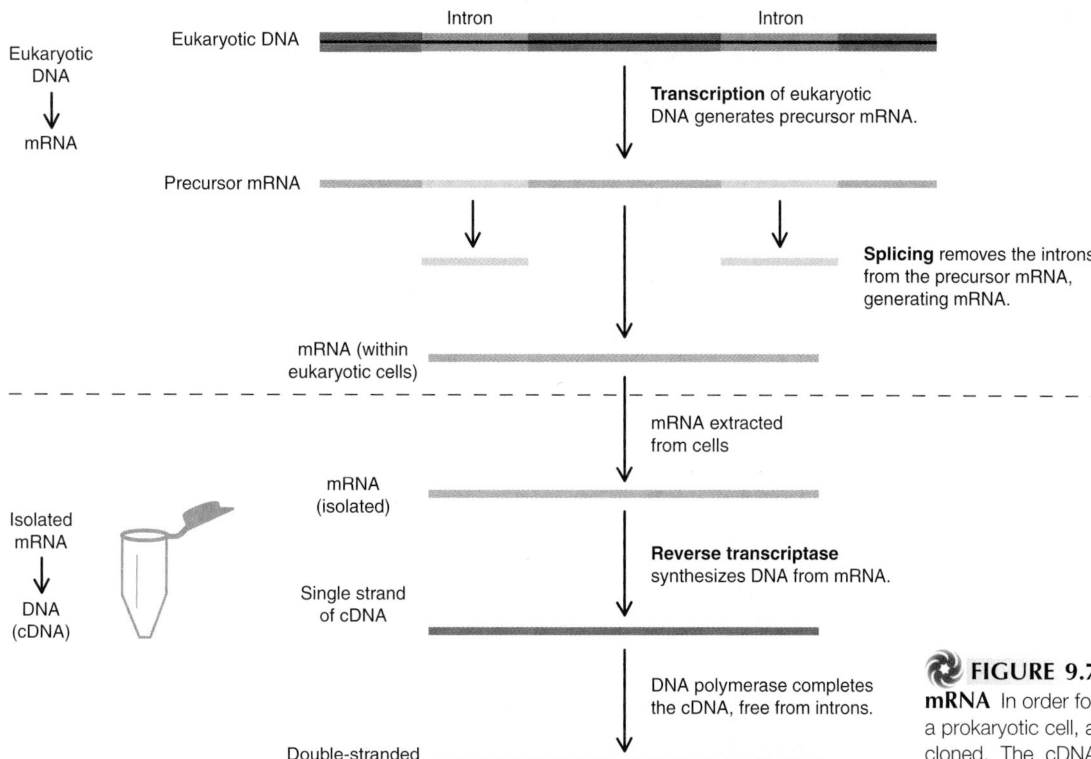

FIGURE 9.7 Making cDNA from Eukaryotic mRNA In order for eukaryotic genes to be expressed by a prokaryotic cell, a copy of DNA without introns must be cloned. The cDNA encodes the same protein as the original DNA but lacks introns.

be joined. Many vectors have been engineered to contain a short sequence called a multiple-cloning site that has the recognition sequences of several different restriction enzymes. The value of a multiple-cloning site is its versatility; a fragment obtained by digesting with any of a number of different restriction enzymes can be inserted into the site.

Vectors typically encode some type of **selectable marker,** a gene whose product allows cells to grow in conditions that would otherwise be inhibitory or lethal. A common selectable marker is a gene that codes for resistance to ampicillin or another antibiotic; cells that harbor a vector or recombinant molecule are able to grow on the antibiotic-containing medium. This is important because when DNA is added to a host, most cells do not take up that DNA. The selectable marker is used to eliminate cells that have not taken up vector sequences. ■ **antibiotic, p. 470** ■ **ampicillin, p. 476**

Most vectors have a second genetic marker, in addition to the selectable marker, used to distinguish cells that contain recombinant plasmids from those containing intact vector. This is significant because when the vector and insert DNA are both cut with the same restriction enzyme and the fragments mixed together, not all will form the desired recombinant molecules. For example, the two ends of the vector can anneal to each other, regenerating the circular vector, which can replicate when introduced into a cell. The second genetic marker is situated so that it is insertionally inactivated when the DNA to be cloned is successfully ligated into the multiple-cloning site of the vector. A good illustration of the utility of the second genetic marker is provided by the common vector,

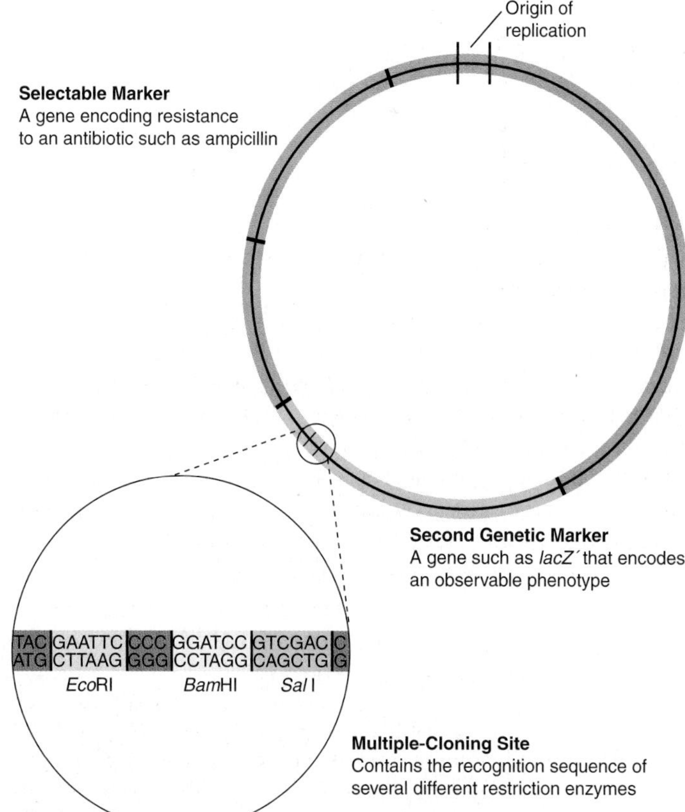

FIGURE 9.8 Typical Properties of an Ideal Vector Most vectors have an origin of replication, a selectable marker, and a multiple-cloning site. A second genetic marker, used to differentiate cells containing recombinant plasmids from those that contain intact vector, spans the multiple-cloning site.

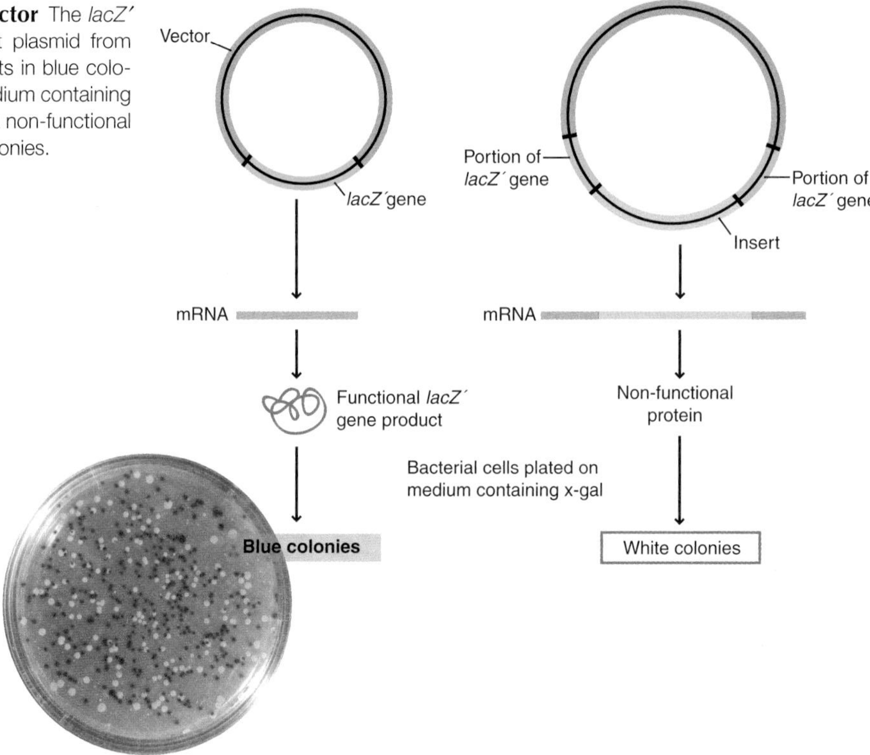

FIGURE 9.9 The Function of the *lacZ′* Gene in a Vector The *lacZ′* gene is used to differentiate cells that contain recombinant plasmid from those that contain vector alone. A functional *lacZ′* gene results in blue colonies when bacteria harboring intact vector are plated on a medium containing x-gal. Because disruption of the gene by an insert generates a non-functional product, cells harboring recombinant plasmids form white colonies.

pUC18 (**figure 9.9**). The second genetic marker of pUC18 is a gene called ***lacZ′***. The product of *lacZ′* enables cells to cleave a colorless chemical, x-gal, to form a blue compound. Because the multiple-cloning site of pUC18 is within the gene, creation of a vector-insert hybrid results in a non-functional gene product. Thus, colonies of cells that harbor intact vector have a functional *lacZ′* gene and are blue, whereas those that contain a recombinant molecule are white. ■ insertional inactivation, p. 189 ■ phenotype, p. 186

The type of vector used to clone eukaryotic DNA into a bacterial cell depends largely on the ultimate purpose of the procedure. If the goal of cloning is to produce the protein encoded by the DNA, then a vector designed to optimize transcription and translation of the insert DNA is used. To create a DNA library of a human or other eukaryotic genome, however, a vector that can carry a large insert is generally used. For more information about these vectors, visit the Online Learning Center (www.mhhe.com/nester6).

Introducing the Recombinant DNA into a New Host

Once recombinant plasmids are generated, they must be transferred into a suitable host where the molecules can replicate. For routine cloning experiments, one of the many well-characterized laboratory strains of *E. coli* is generally used. These strains are easy to grow and much is known about their genetics and biochemistry. They also have known phenotypic characteristics such as sensitivity to specific antibiotics. ■ antibiotics, p. 470

A common method of introducing DNA into a bacterial host is DNA-mediated transformation. *E. coli* cells must be specially treated to induce them to take up DNA. An alternative technique is to introduce the DNA by electroporation, a procedure that subjects the cells to an electric current. ■ DNA-mediated transformation, p. 199 ■ electroporation, p. 201

After the DNA is introduced into the new host, the transformed bacteria are cultivated on a medium that both selects for cells containing vector sequences and differentiates those carrying recombinant plasmids. The medium exploits the selective marker encoded on the vector to permit growth of only those cells that have taken up either a recombinant molecule or an intact vector. If, for example, the vector encodes resistance to the antibiotic ampicillin, then the transformed cells are grown on ampicillin-containing medium. The medium exploits the second genetic marker to differentiate the cells that took up a recombinant plasmid from those that took up intact vector. If that marker is the *lacZ′* gene, the chemical x-gal is added to the medium. Colonies that are white (rather than blue) likely harbor a vector carrying an insert; these colonies are then further characterized to determine if they harbor the gene of interest (see figure 9.9). ■ ampicillin, p. 476

MICROCHECK 9.3

Introns must first be removed if eukaryotic DNA is to be expressed in a prokaryotic cell. Vectors typically have an origin of replication, a selectable marker, a multiple-cloning site, and a second genetic marker.

✓ Explain the role of reverse transcriptase in the procedure to clone eukaryotic DNA.

✓ Explain how the *lacZ′* gene is used to distinguish colonies that harbor a recombinant plasmid.

✓ What would happen if a cell took up a molecule of circular insert DNA?

9.4

Concerns Regarding Genetic Engineering and Other DNA Technologies

Focus Point

- Describe some of the concerns regarding DNA technologies.

The advent of any new technology should also bring scrutiny about the safety and efficacy of the outcome. When recombinant DNA technologies first allowed gene cloning over two decades ago, controversies swirled about their use and possible abuse. Even the scientists who developed the technologies were concerned about potential dangers. In response, the National Institutes of Health (NIH) formed the Recombinant DNA Advisory Committee (RAC) to develop guidelines for conducting research involving recombinant DNA techniques and gene cloning. Today, we are enjoying the fruits of many of those technologies, as evidenced by the list of commercially available products in table 9.2. Despite the fact that the technologies can be used to produce life-saving products, however, there can never be a guarantee that they will not be used for malicious purposes. Today, the idea that "superbugs" are being created for the purpose of bioterrorism is a disturbing possibility.

Recent advances in genomics have generated new cause for concern, primarily involving ethical issues regarding the appropriateness and confidentiality of information gained by analyzing a person's DNA. For example, will it be in an individual's best interest to be told of a genetic life-terminating disease? Could such information be used to deny an individual certain rights and privileges? It is important that ongoing discussions about these complex issues continue as the technologies advance.

Genetically modified (GM) organisms hold many promises, but the debate over their use has raised concerns, some logical and others not. For example, some people have expressed fear over the fact that GM foods "contain DNA." Considering that DNA is consumed routinely as we eat plants and animals, this is obviously an irrational concern. Other fears include the worry that unanticipated allergens could be introduced into a food product, posing a threat to the health of some people. To address this issue, the FDA has implemented strict guidelines, including the requirement for producers to demonstrate that GM products intended for human consumption do not unduly elicit allergic reactions. Incidents such as the inadvertent use of GM corn that had not been approved for human consumption as an ingredient in tortilla chips continue to fuel apprehension about the strictness of regulatory control over GM products. Another concern about GM products is their possible unintended effects on the environment. For example, some laboratory studies have shown that pollen from plants genetically modified to produce Bt toxin can inadvertently kill monarch butterflies; other studies, however, have refuted the evidence. In addition, there are indications that herbicide-resistance genes can be transferred to weeds, decreasing the usefulness of the herbicide. As with any new technology, the impact of GM organisms will need to be carefully scrutinized to avoid any negative consequences.

MICROCHECK 9.4

Concerns about genetic engineering are varied, including ethical and moral issues associated with genomics, and potentially adverse impacts of genetically modified organisms on human health and the environment.

✓ Describe two concerns regarding information that can be gained by analyzing a person's DNA.

✓ Describe two concerns regarding the use of genetically modified organisms.

9.5

DNA Sequencing

Focus Points

- Describe two applications of DNA sequencing.
- Describe the automated dideoxy chain termination method of DNA sequencing.

The **Human Genome Project,** the completed undertaking to determine the nucleotide sequence of the human genome, resulted in highly automated and efficient **DNA sequencing** techniques. These enabled scientists to more readily determine the genomic sequence of other organisms, including both prokaryotes and eukaryotes, fueling the rapidly growing field of genomics. The resulting explosion of data spawned a new field, bioinformatics, to analyze the information. ■ genomics, p. 181 ■ bioinformatics, p. 181

By determining the DNA sequence of a genome, the amino acid sequence of encoded proteins can be determined, thereby making it possible to compare the characteristics of various proteins in different organisms. Non-coding sequences and mobile genetic elements can be compared as well. DNA sequencing also provides a mechanism for determining the evolutionary relatedness of organisms, a topic that will be discussed in the next chapter. ■ mobile genetic elements, p. 205

Sequencing technologies have become so efficient that a new project has been intiated—the **Human Microbiome Project**—which uses genomics to determine the biological diversity in the normal microbiota of the human body. This project will not only use traditional approaches by sequencing the genome of individual strains, it will also employ **metagenomics,** the study of total genomes in a sample. With a metagenomics approach, all microbes, not just cultivable entities, can be studied. Goals of the Human Microbiome Project include determining the extent of variation in the microbiota among different individuals and comparing the composition of the normal microbiota in health and disease. ■ normal microbiota, p. 9

Metagenomics is also being used to study the extent of biodiversity in the open oceans and in soils. Analyzing these sequences will probably lead to the discovery of new antibiotics and other medically useful compounds.

Techniques Used in DNA Sequencing

DNA sequencing technologies have advanced markedly in the last decades. While the Human Genome Project initiated in 1990 took 13 years to complete and cost about 440 million dollars to sequence one human genome, a company recently announced that it had sequenced an individual's genome in only a matter of months, for a cost of one million dollars. A race is now on to develop the technology to sequence 100 human genomes within 10 days for a cost of no more than $ 10,000 each!

The most widely used technique for determining the sequence of DNA is the dideoxy chain termination method. The procedure is now typically done using automated sequencing instruments, making it a rapid and efficient process. Several newer sequencing methods have been developed, and these are speeding the process even more.

Dideoxy Chain Termination Method

The fundamental aspect of the **dideoxy chain termination method** is an *in vitro* DNA synthesis reaction. This requires:

- A single-stranded piece of DNA, the **template,** from which a complementary copy is synthesized. ■ template, p. 164
- DNA polymerase, the enzyme that catalyzes DNA synthesis. ■ DNA polymerase, p. 166
- A primer that anneals to the single-stranded template. This allows the technician to dictate the site at which synthesis will initiate; recall that DNA polymerase can only add nucleotides to an existing fragment of DNA, thereby extending the fragment **(figure 9.10).** The DNA being sequenced usually has been cloned into a vector; because the nucleotide sequences of vectors are known, a primer that anneals to a portion of the vector adjacent to the DNA to be sequenced can be readily obtained. ■ primer, p. 166
- Each of the four deoxynucleotides that are used in DNA synthesis—dATP, dGTP, dCTP, and dTTP. ■ nucleotide, p. 32

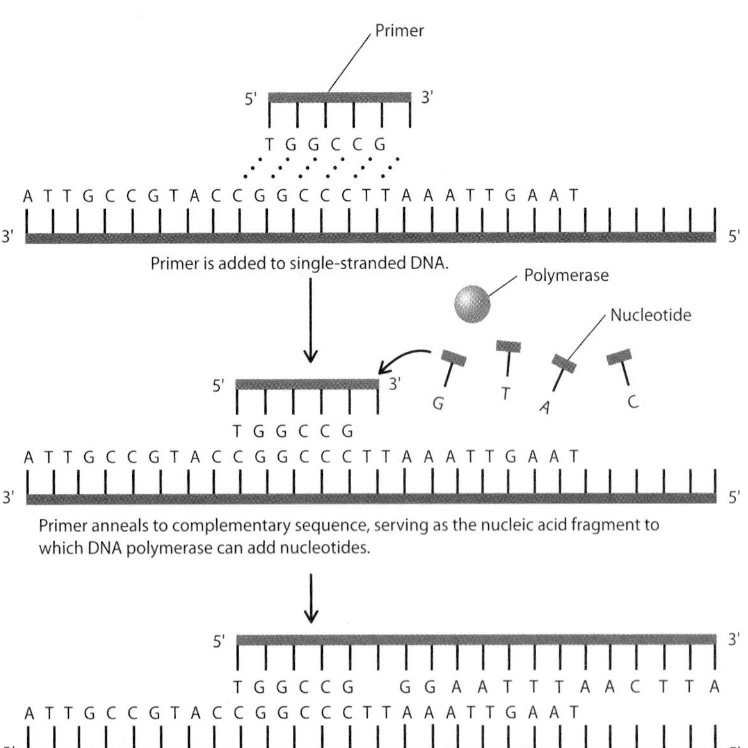

FIGURE 9.10 Primer Through appropriate primer selection, a researcher can choose the site where *in vitro* DNA synthesis will initiate.

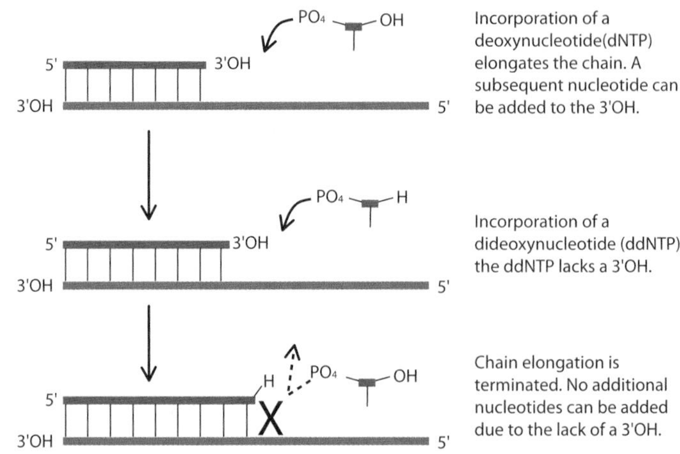

FIGURE 9.11 Chain Termination by a Dideoxynucleotide Once a dideoxynucleotide is incorporated into a growing strand, subsequent nucleotides cannot be added due to the lack of a 3′OH.

If these were the only ingredients in the reaction, full-length molecules complementary to the template DNA would be synthesized. A key additional ingredient is added, however—**dideoxynucleotides (ddNTPs).** These are identical to their deoxynucleotide (dNTP) counterparts except they lack the 3′ OH group, the portion of a nucleotide required for the addition of subsequent nucleotides during DNA synthesis **(figure 9.11).** Because they lack the 3′OH group, dideoxynucleotides are **chain terminators.** When one of these is incorporated into a growing strand of DNA, no additional nucleotides can be added, and elongation of the strand ceases. In automated sequencing reaction, each of the different dideoxynucleotides (ddATP, ddGTP, ddCTP and ddTTP) carries a distinct fluorescent marker.

In an automated sequencing reaction, DNA polymerase, template DNA, primer, the four deoxynucleotides, and a very small amount of the four labeled dideoxynucleotides are mixed together **(figure 9.12).** When the reaction is incubated at an appropriate temperature, each primer anneals to a template molecule and DNA synthesis begins. In every case, the nucleotide chain is elongated until a dideoxynucleotide is incorporated, terminating synthesis at that point. Termination, however, is an infrequent event because of the numerous template molecules and the small amount of the ddNTPs relative to the dNTPs. This generates fragments of various lengths, but in each case, the fluorescent marker of the chain terminator indicates which nucleotide was incorporated at the terminating position. The sample is then heated to denature the DNA, and electrophoresis is used to separate the DNA fragments and determine their relative size. The conditions (pH, temperature, and gel concentration) maintain the DNA in a single-stranded state and enable separation of fragments that differ by only one nucleotide in length. A laser is used to detect the colors of

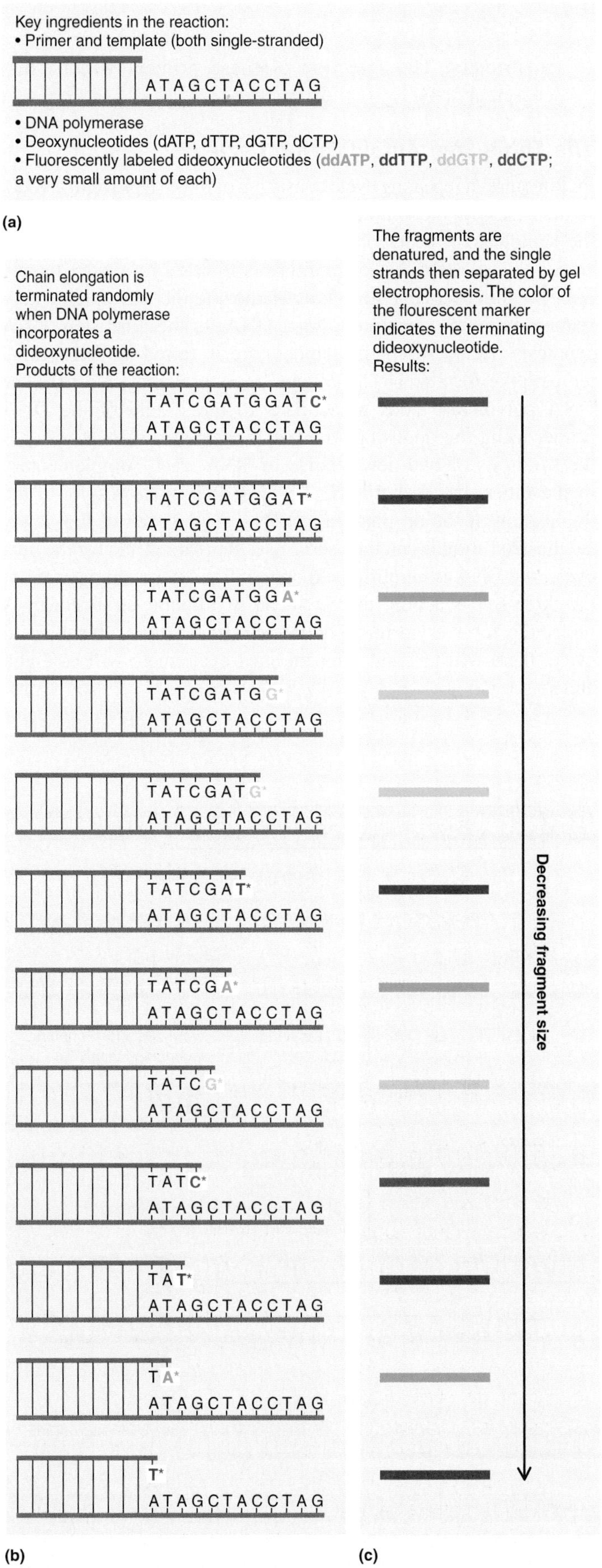

Key ingredients in the reaction:
- Primer and template (both single-stranded)

ATAGCTACCTAG

- DNA polymerase
- Deoxynucleotides (dATP, dTTP, dGTP, dCTP)
- Fluorescently labeled dideoxynucleotides (ddATP, **ddTTP**, ddGTP, **ddCTP**; a very small amount of each)

(a)

Chain elongation is terminated randomly when DNA polymerase incorporates a dideoxynucleotide.
Products of the reaction:

TATCGATGGATC*
ATAGCTACCTAG

TATCGATGGAT*
ATAGCTACCTAG

TATCGATGGA*
ATAGCTACCTAG

TATCGATGG*
ATAGCTACCTAG

TATCGATG*
ATAGCTACCTAG

TATCGAT*
ATAGCTACCTAG

TATCGA*
ATAGCTACCTAG

TATCG*
ATAGCTACCTAG

TATC*
ATAGCTACCTAG

TAT*
ATAGCTACCTAG

TA*
ATAGCTACCTAG

T*
ATAGCTACCTAG

(b)

The fragments are denatured, and the single strands then separated by gel electrophoresis. The color of the flourescent marker indicates the terminating dideoxynucleotide.
Results:

Decreasing fragment size

(c)

FIGURE 9.12 Dideoxy Chain Termination Method of DNA Sequencing This figure illustrates the principles of the automated method, which employs dideoxynucleotides that carry a fluorescent label.

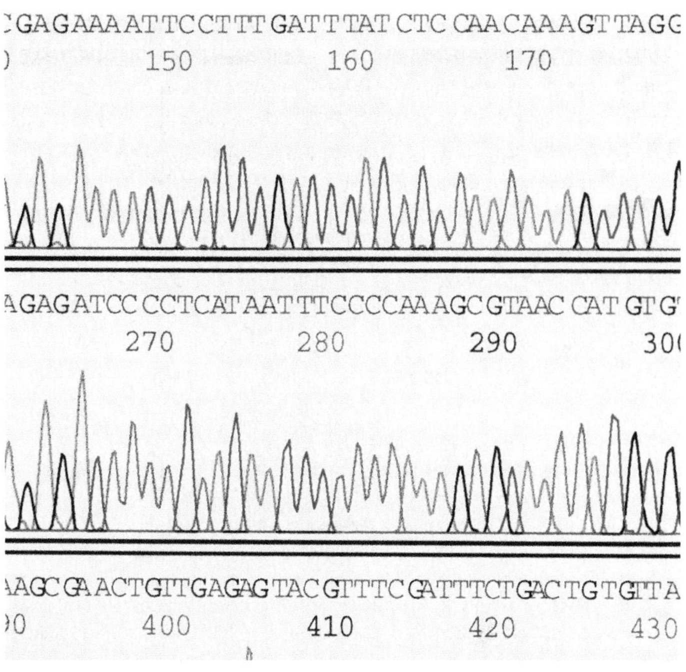

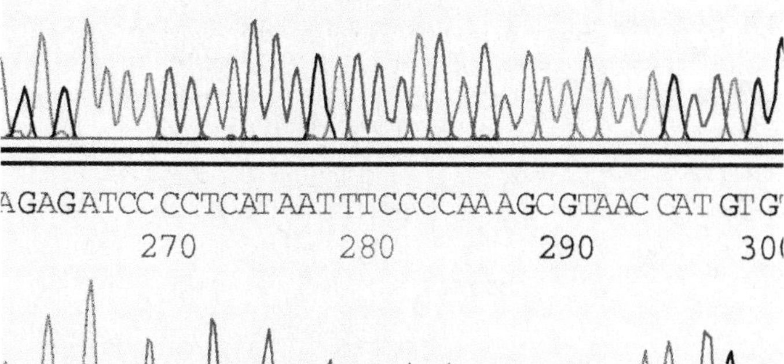

FIGURE 9.13 Results of Automated DNA Sequencing The order of the colored peaks reflects the nucleotide sequence of the DNA.

fluorescent bands as they run past, recording their intensity as a peak. The order of the colored peaks reflects the nucleotide sequence of the DNA **(figure 9.13).**

MICROCHECK 9.5

Efficient DNA sequencing methods have fueled the rapidly growing field of genomics. The automated dideoxy chain termination method is a widely used sequencing technique.

✓ What is the Human Microbiome Project?

✓ How does a ddNTP terminate DNA synthesis?

✓ What would happen in the sequencing reaction if the relative concentration of a dideoxynucleotide were increased?

9.6

Polymerase Chain Reaction (PCR)

Focus Points

- Describe one application of PCR.
- Explain how PCR can be used to exponentially amplify a select region of DNA.

Development of the **polymerase chain reaction (PCR)** revolutionized research by making it possible to create millions of copies of a given region of DNA in a matter of hours. The technique exploits the specificity of primers to selectively replicate only chosen regions, referred to as **target DNA.** As a result, starting with just a few DNA molecules, millions of molecules of target

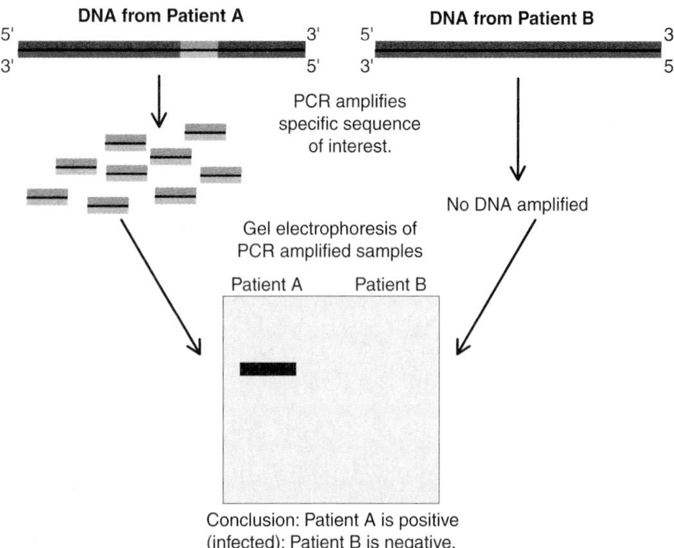

FIGURE 9.14 PCR Amplifies Selected Sequences By amplifying a chosen sequence unique to a given organism, that organism can be detected in a sample.

DNA are synthesized. In fact, target DNA can be generated in sufficient concentration to be visible to the unaided eye when fragments in the sample are separated by gel electrophoresis, stained with ethidium bromide, and illuminated with UV light **(figure 9.14).**

By choosing primers that anneal to sequences unique to a given organism, PCR can be used to detect that organism, even in sample that contains a wide assortment of different microbes. For example, if a nucleotide sequence found only in *Neisseria gonorrhoeae* can be amplified from a vaginal secretion, then *N. gonorrhoeae* must be present in that specimen, indicating that the patient has the sexually transmitted disease gonorrhea. Likewise, PCR can be used to detect HIV nucleotide sequences in a sample of blood cells, diagnosing HIV infection.

Techniques Used in PCR

The polymerase chain reaction starts with a double-stranded DNA molecule, from which millions of identical copies of a select region can be produced. The process involves a series of DNA synthesis reaction and the following key ingredients:

- Double-stranded DNA containing the region to be amplified, the **target DNA.**
- ***Taq*** polymerase, a heat-stable DNA polymerase from the thermophile *Thermus aquaticus*.
- Each of the four nucleotides (dATP, dGTP, dCTP, dTTP).
- Short, single-stranded segments of DNA, generally about 20 nucleotides in length, to serve as primers. The selection of these primers will be discussed shortly.

The Three-Step Amplification Cycle

PCR requires a repeating cycle consisting of three steps **(figure 9.15).** In the first step, the double-stranded DNA is denatured by heating the sample to a near-boiling temperature, approximately 95°C. In the second step, the temperature is lowered to approximately 50°C; within seconds, the primers anneal to their complementary sequences on the denatured target DNA. In the third step, DNA synthesis occurs when the temperature is raised to the optimal temperature for *Taq* DNA polymerase, approximately 70°C. The DNA polymerase adds nucleotides to the 3′ end of the DNA primer using the opposing strand as a template. The net result is the synthesis of two new strands of DNA, each complementary to the other. In other words, the three-step cycle results in the duplication of the original target DNA. Since each of the newly synthesized strands can then serve as a template strand for the next cycle, the DNA is amplified exponentially. After a single cycle of the three-step reaction, there will be two double-stranded DNA molecules for every original double-stranded target; after the next cycle, there will be four; after the next cycle there will be eight, and so on.

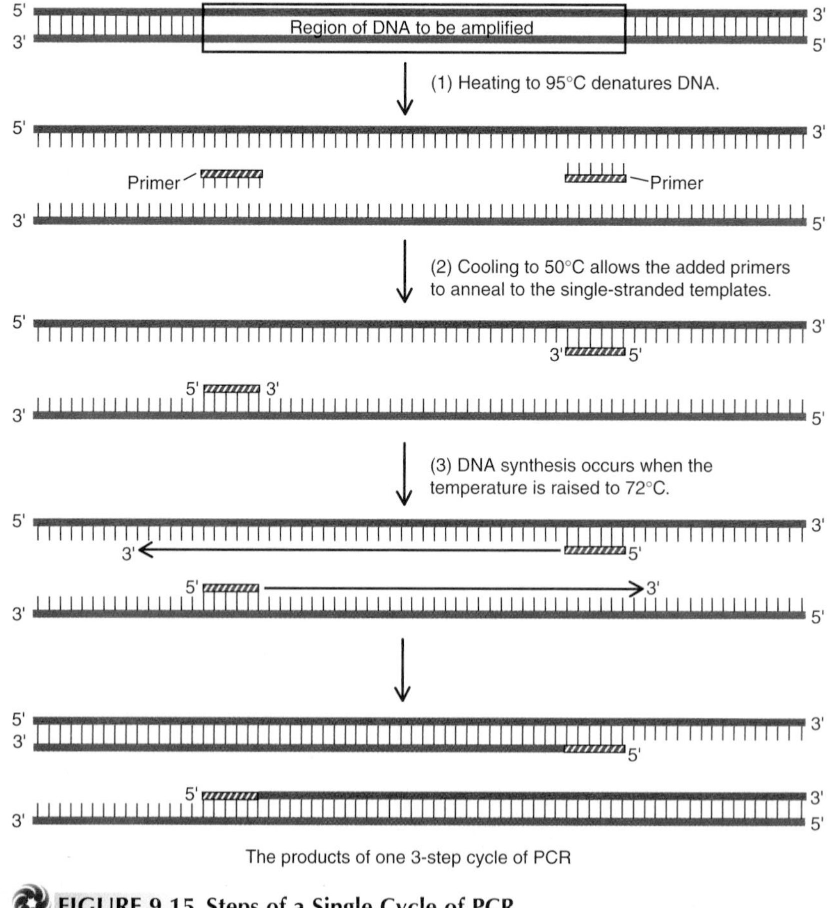

FIGURE 9.15 Steps of a Single Cycle of PCR

PERSPECTIVE 9.1

Science Takes the Witness Stand

After serving more than 10 years on a rape charge, a wrongfully convicted young man was released from prison when a new DNA typing technique exonerated him. The new technique, which used polymerase chain reaction (PCR) to amplify specific sequences, showed that the semen sample taken from the rape victim did not contain the man's DNA. Indeed, a database indicated a DNA match with a man currently in prison for an unrelated rape charge.

Stories abound about the growing power of DNA evidence for obtaining convictions and also clearing the wrongly accused, but how is DNA used in forensics? It is not practical to compare the entire nucleotide sequence of two people; instead, specific regions that vary significantly between individuals are analyzed. In the past, forensics labs have used a method that employs the probe-based technique called Southern blot hybridization to detect differences. This method provides valuable information but cannot be used on small or degraded samples and is quite time-consuming. Most forensics labs have now switched to using a PCR-based method because results can be obtained in less than 5 hours from a sample as small as a drop of blood the size of a pinhead. In fact, the FBI now catalogs PCR-based DNA profiles from unsolved crimes and convicted violent offenders, making it easier to track or link the crimes of serial offenders. The national database is called CODIS (**Co**mbined **D**NA **I**ndex **S**ystem).

PCR-based DNA typing amplifies certain chromosomal regions that contain **short tandem repeats (STRs).** These consist of a core sequence of 2 to 6 base pairs that repeat a variable number of times in different people. On chromosome 2, for example, in an intron within the thyroid peroxidase gene, the sequence AATG is repeated sequentially between 5 to 14 times. In one individual, there may be 9 of these STRs in one copy of that chromosome and 7 in the other, whereas another individual may have 11 and 5 **(figure 1).** This variation, or polymorphism, makes tandem repeats a useful genetic marker for distinguishing individuals. With PCR, using primers that bind regions flanking the repeating sequences, the number of repeats can be determined. A fragment that contains 9 repeats, for example, will be longer than one that contains only 7. The PCR-amplified fragments can be quickly separated using a rapid type of gel electrophoresis called capillary electrophoresis. Their size can then be determined by comparing their positions as they move out of the gel to those of known standards.

The FBI's CODIS database catalogs the amplification pattern of 13 different STR loci (chromosomal locations). Commercially available kits contain fluorescently labeled primers that allow simultaneous amplification and subsequent recognition of each of the 13 loci. A laser detects the color of each amplified fragment as it moves out of the capillary gel, and computer analysis generates a pattern of peaks that reflect the STR profile of the DNA sample.

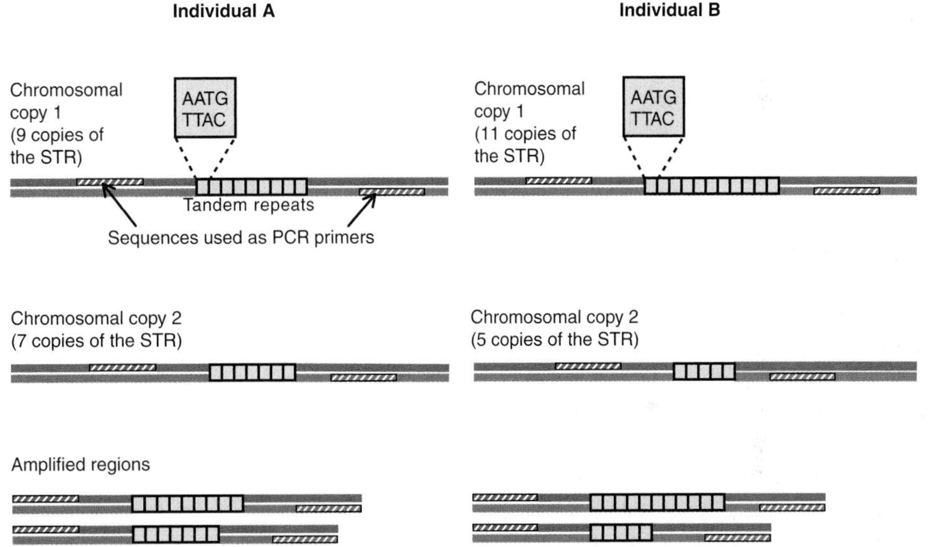

FIGURE 1 Using PCR to Type ("Fingerprint") DNA PCR is used to amplify chromosomal regions containing short tandem repeats (STRs). The number of copies of a given STR vary among people, resulting in corresponding differences in the length of the amplified fragments. Typically, at least 13 different STR locations are analyzed.

A critical factor in PCR is the heat-stable DNA polymerase of a thermophilic bacterium, *Thermus aquaticus. Taq* polymerase, unlike the DNA polymerase of *E. coli,* is not destroyed at the high temperature used to denature the DNA in the first step of each amplification cycle. If a heat-stable polymerase were not used, fresh polymerase would need to be added for every cycle of the reaction. Thus, the discovery and characterization of *T. aquaticus* through basic research was key to developing this widely used and commercially valuable method. ■ **thermophile, p. 91**

Generating a Discrete-Sized Fragment

While the preceding description explains how PCR amplifies the target DNA exponentially, it does not clarify how a discrete-length fragment, referred to as the **PCR product,** becomes the predominant product. The generation of fragments of a particular size is important, because it enables the technician to use PCR coupled with gel electrophoresis to readily detect the presence of target DNA in a sample. After PCR, the amplified target (PCR product) can be viewed as a single band on the ethidium bromide-treated gel.

To understand how discrete-sized fragments of target DNA are generated, you must consider the exact sites to which the primers anneal, and visualize at least three cycles of replication **(figure 9.16).** In the first cycle, two new fragments are generated. Note, however, that these fragments are shorter than the original full-length template molecules but longer than the target DNA. Their 5′ end is primer DNA. These mid-length

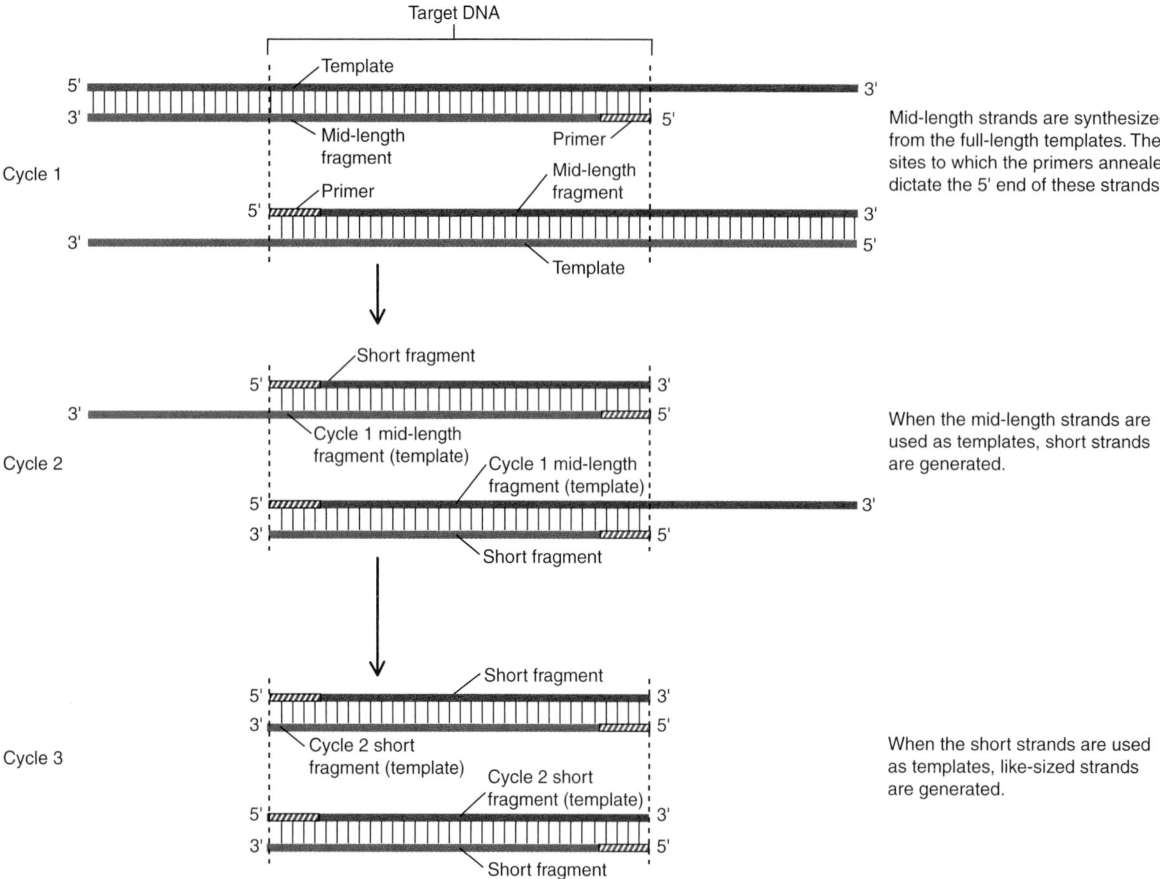

FIGURE 9.16 The PCR Product Is a Fragment of Discrete Size The positions to which the primers anneal to the template dictate the size and sequence of the fragment amplified exponentially.

products will be generated whenever the original full-length molecule is used as a template, which is one time each replication cycle.

In the next cycle, the full-length molecules will again be used as templates, repeating the process just described. More importantly, the mid-length fragments created during the first cycle will be used as templates for DNA synthesis. As before, the primers will anneal to these fragments and then nucleotides will be added to the 3′ end. Elongation, however, will stop at the 5′ end of the template molecule, because DNA synthesis requires a template. Recall that the 5′ end of the template is primer DNA. Thus, whenever a mid-length fragment is used as a template, a short fragment is generated. The 5′ and 3′ ends of this fragment are determined by the sites to which the primers initially annealed.

In the third round of replication, the full-length and the mid-length fragments again will be used as templates, repeating the processes just described. The short fragments generated in the preceding round will also be used as templates, however, generating short double-stranded molecules. Continuing to follow the events in further rounds of replication will reveal that it is this fragment that is exponentially amplified **(figure 9.17).** Ultimately, enough of these short fragments are generated to be detectable using gel electrophoresis followed by ethidium bromide staining.

Selecting Primer Pairs

The nucleotide sequences of the two primers are critical because the primers dictate which portion of the DNA is amplified. Each must be complementary to a sequence on the appropriate strand, flanking the region to be synthesized. The synthesis reaction will then add nucleotides onto the primers, elongating the DNA chain so that the DNA between those primers is copied. Thus, if a technician wants to amplify a DNA sequence that encodes a specific protein, then he or she must first determine the nucleotide sequences flanking the gene that encodes the protein and then synthesize the appropriate pair of oligonucleotides to serve as primers.

MICROCHECK 9.6

The polymerase chain reaction (PCR) is used to rapidly increase the amount of either a specific segment or total DNA in a sample.

- ✓ What happens during each of the three temperature steps of PCR (95°C, 50°C, and 70°C)?
- ✓ Explain why it is important to use a polymerase from a thermophile in the PCR reaction.
- ✓ Sequencing reactions can be done using PCR. In this case would two primers be necessary?

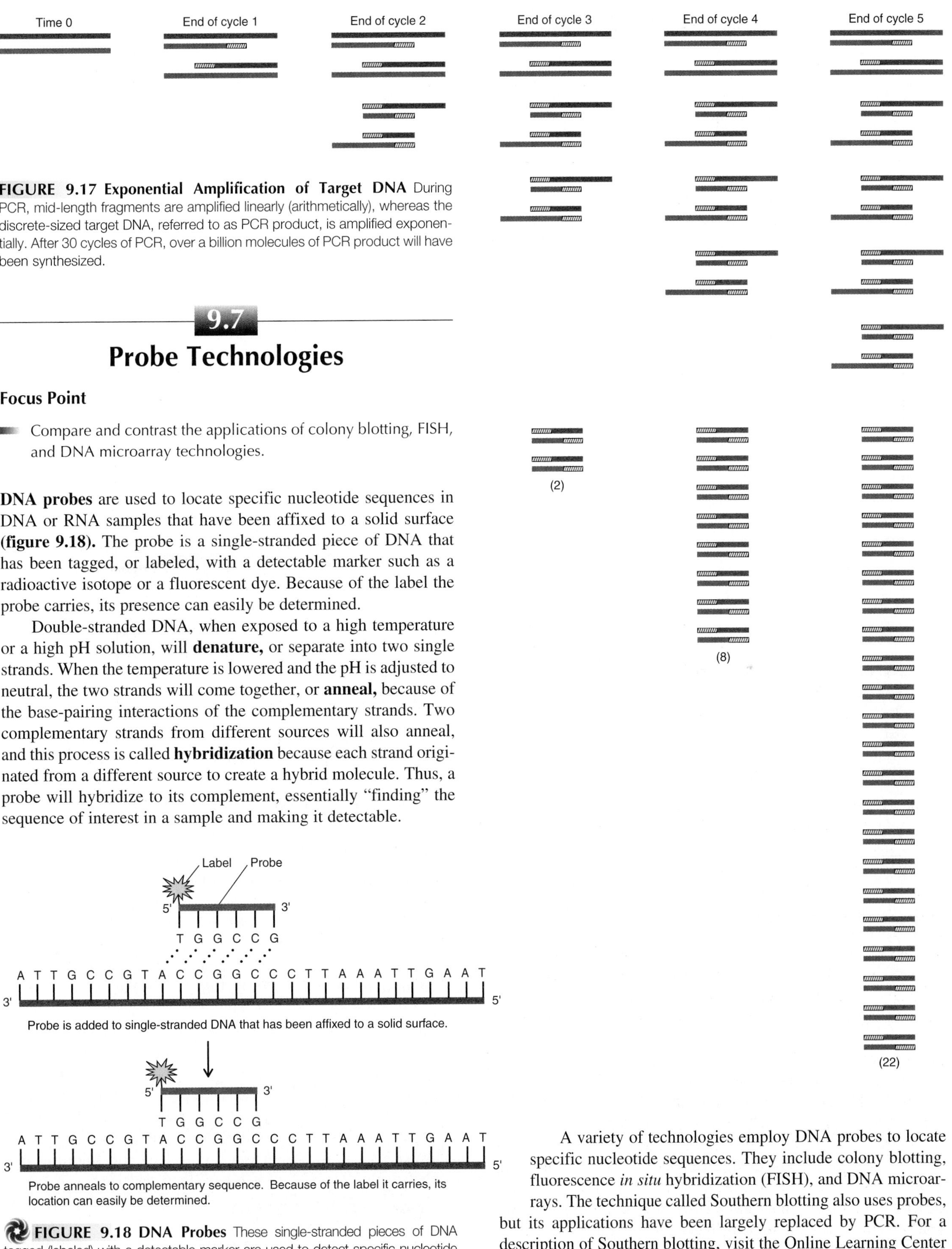

FIGURE 9.17 Exponential Amplification of Target DNA During PCR, mid-length fragments are amplified linearly (arithmetically), whereas the discrete-sized target DNA, referred to as PCR product, is amplified exponentially. After 30 cycles of PCR, over a billion molecules of PCR product will have been synthesized.

9.7 Probe Technologies

Focus Point

- Compare and contrast the applications of colony blotting, FISH, and DNA microarray technologies.

DNA probes are used to locate specific nucleotide sequences in DNA or RNA samples that have been affixed to a solid surface **(figure 9.18).** The probe is a single-stranded piece of DNA that has been tagged, or labeled, with a detectable marker such as a radioactive isotope or a fluorescent dye. Because of the label the probe carries, its presence can easily be determined.

Double-stranded DNA, when exposed to a high temperature or a high pH solution, will **denature,** or separate into two single strands. When the temperature is lowered and the pH is adjusted to neutral, the two strands will come together, or **anneal,** because of the base-pairing interactions of the complementary strands. Two complementary strands from different sources will also anneal, and this process is called **hybridization** because each strand originated from a different source to create a hybrid molecule. Thus, a probe will hybridize to its complement, essentially "finding" the sequence of interest in a sample and making it detectable.

FIGURE 9.18 DNA Probes These single-stranded pieces of DNA tagged (labeled) with a detectable marker are used to detect specific nucleotide sequences in DNA or RNA samples that have been affixed to a solid surface.

A variety of technologies employ DNA probes to locate specific nucleotide sequences. They include colony blotting, fluorescence *in situ* hybridization (FISH), and DNA microarrays. The technique called Southern blotting also uses probes, but its applications have been largely replaced by PCR. For a description of Southern blotting, visit the Online Learning Center (www.mhhe.com/nester6).

Colony Blotting

Colony blotting uses probes to detect specific DNA sequences in colonies grown on agar plates **(figure 9.19).** This method is commonly used to determine which of a collection of clones contain the DNA of interest.

The term "blot" in the name reflects the fact that the colonies are transferred in place ("blotted") onto a nylon membrane, creating a pattern of colonies identical to that of the original plate. The membrane serves as a durable, permanent support for the cells of the colonies and their DNA. After the transfer, the membrane is soaked in an alkaline solution to simultaneously lyse the cells and denature their DNA, generating single-stranded DNA molecules. A solution containing the probe is then added to the membrane and incubated under conditions that allow the probe to hybridize to complementary sequences on the filter. Any probe that has not bound is then washed off. If the probe was labeled with a radioactive isotope, a process called *autoradiography* is used to locate the position of the hybridized probe. In this procedure, X-ray film is placed on the filter containing the probe-bound DNA and incubated. After the film is developed, visible black grains can be seen at the sites where probe was located. Different methods are used to detect probes that have been labeled with other markers.

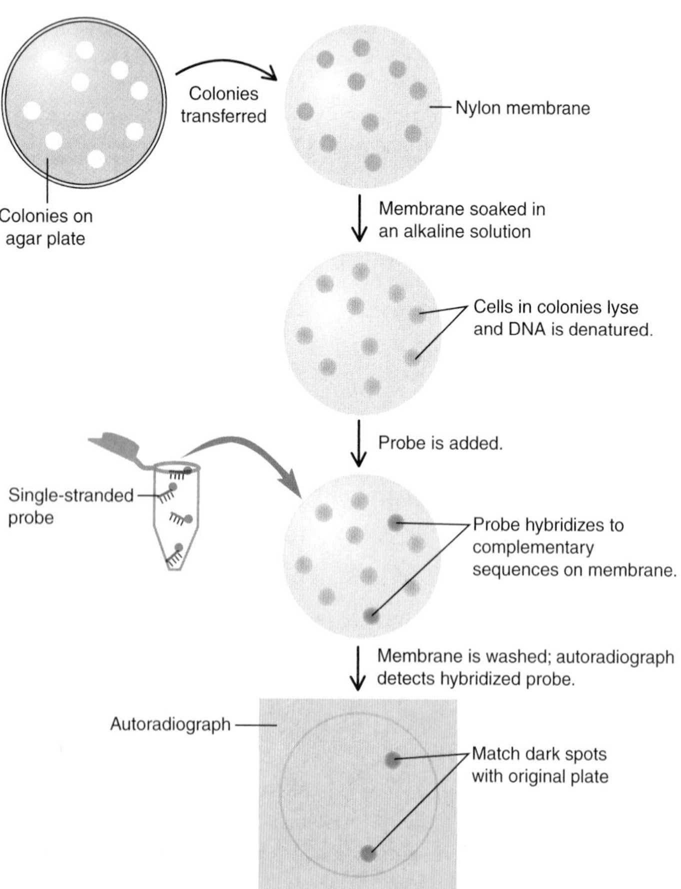

FIGURE 9.19 Colony Blotting This technique is used to determine which colonies on an agar plate contain a given DNA sequence. In this example, the probe is labeled with a radioactive isotope, which can be detected using autoradiography.

Fluorescence *in situ* Hybridization (FISH)

Fluorescence *in situ* hybridization (FISH) uses a fluorescently labeled probe to detect specific nucleotide sequences within intact cells affixed to a microscope slide. Cells containing the hybridized probe can then be viewed using a fluorescence microscope. To study prokaryotes, a probe that binds to ribosomal RNA (rRNA) is generally used. Because actively growing cells can have thousands of copies of rRNA, this increases the sensitivity of the technique. Other characteristics of rRNA that make it useful for identifying prokaryotes are discussed in chapter 10. ■ fluorescence microscope, p. 43 ■ using genotypic characteristics to identify prokaryotes, p. 241

FISH is revolutionizing the study of microbial ecology and holds great promise in clinical laboratories. It provides a means to rapidly identify microorganisms directly in a specimen, bypassing the need to grow them in culture. FISH can be used to observe either a specific species or a group of related organisms, depending on the nucleotide sequence of the probe employed. For example, FISH can determine the relative proportion of two different groups of prokaryotes in the same specimen by using two separate probes, one specific for each group, labeled with different colored fluorescent markers **(figure 9.20).** Another use is to identify and enumerate cells of *Mycobacterium tuberculosis,* the cause of tuberculosis, in a sputum specimen. ■ microbial ecology, p. 722 ■ tuberculosis, p. 514

Sample preparation is a critical aspect of fluorescence *in situ* hybridization. The sample must first be treated with chemicals to preserve the shape of the cells, inactivate enzymes that might otherwise degrade the nucleic acid, and make the cells more permeable so that the labeled probe molecules can readily enter. Once the specimen has been prepared, it is put on a glass slide, bathed with a solution containing the fluorescently labeled probe, and incubated under conditions that allow hybridization to occur. Unbound probe is then washed off. Finally, the specimen is viewed using a fluorescence microscope.

DNA Microarrays

DNA microarrays are used to study gene expression in organisms whose genome has been sequenced. They can also be used to detect a specific sequence in a DNA sample of interest.

A DNA microarray consists of a glass slide or other small, solid support that carries an arrangement of tens or hundreds of thousands of short DNA fragments. Each DNA fragment functions in a manner analogous to a probe, enabling a researcher to screen a single sample for a vast range of different sequences simultaneously. Unlike typical probes, however, the arrays do not carry a detectable label. Instead, the label must be attached to the nucleic acid of interest. For example, to study gene expression, mRNA can be isolated from an organism, and converted to labeled, single-stranded cDNA using reverse transcriptase (see figure 9.7) and fluorescently labeled nucleotides. That cDNA is then added to a microarray constructed to include a sequence specific for each gene of a particular organism. The locations of the labeled cDNA molecules are detected using a computerized scanner, allowing the researcher to discover which genes were expressed by

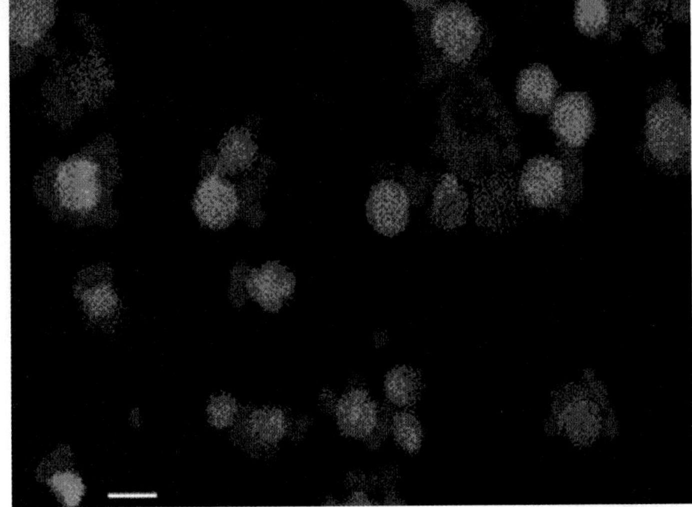

FIGURE 9.20 Fluorescence *in situ* Hybridization (FISH) Two different probes have been used to stain the cells. The probe that hybridizes to *Ignicoccus* rRNA fluoresces green; the one that hybridizes to *Nanoarchaeum* rRNA fluoresces red. Size bar = 1 μm.

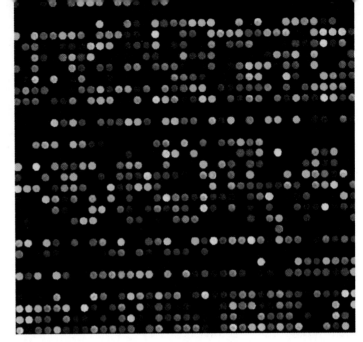

the organism. By doing the experiment using cultures grown under different sets of conditions, and labeling their cDNAs with different fluorescent markers, variations in gene expression can be revealed **(figure 9.21).** This is how researchers determined that pathogens express some genes only when they are in certain locations within the human body.

To use microarrays to detect specific sequences, the DNA is digested into small fragments, labeled with a fluorescent marker, denatured, and then added to an appropriate microarray. Because the sequence of each of the fragments that make up the array is known, the location of the label can be used to determine the presence of specific sequences in the DNA of interest.

The applications and principles of probe technologies, as well as other DNA-based technologies described in this chapter, are summarized in **table 9.3.**

FIGURE 9.21 A DNA Microarray Two different nucleic acid samples (one labeled with a red fluorescent marker and the other with a green fluorescent marker) were simultaneously hybridized to the microarray. Red dots indicate the positions to which one sample hybridized, and green dots indicate the positions of the other. Yellow dots indicate that both samples hybridized. The arrays, read with the aid of computerized scanners, are often used to study gene expression.

MICROCHECK 9.7

Colony blotting uses probes to identify colonies that contain a given sequence of DNA. Fluorescence *in situ* hybridization is used to observe individual cells that contain a given sequence. DNA microarrays enable researchers to study gene expression and to screen a sample for a vast range of different sequences simultaneously.

✓ What role does colony blotting play in cloning?

✓ What is a probe that binds to rRNA used in FISH?

✓ Why would a pathogen express different genes when inside the body?

TABLE 9.3 Summary of DNA-Based Biotechnologies

Technology	Applications	Principles of the Technique
Genetic engineering	Genetically engineered microorganisms are used to produce medically and commercially valuable proteins, to produce specific DNA sequences, and as a tool for researching gene function and regulation.	Restriction enzymes and DNA ligase are used to insert specific segments of DNA into vector molecules, which are then introduced into a new host.
DNA sequencing	Once a DNA sequence has been determined, it can be used to decipher the amino acid sequence of the encoded proteins, compare properties of the organism's genome to others that have been sequenced, and establish genetic relatedness of the organism to different isolates.	In the dideoxy chain termination method, DNA synthesis reactions generate a set of DNA fragments, each terminated when a specific dideoxynucleotide is incorporated. By separating these fragments using gel electrophoresis, the specific dideoxynucleotide that terminated each fragment can be determined, thereby revealing the position of the nucleotides in the DNA sequence.
Polymerase chain reaction (PCR)	The presence of a specific segment of DNA can be detected, and the size determined, in only a matter of hours. By amplifying DNA specific to a pathogen, PCR can be used in diagnosis. It is also used to "fingerprint" DNA for forensic evidence.	Multiple cycles of a DNA synthesis reaction using specific primers amplify a given stretch of DNA. The locations to which the primers anneal dictate which portion is amplified.
Probe technologies	Colony blots are used to detect colonies that contain a specific sequence; fluorescence *in situ* hybridization (FISH) is used to identify cells directly in a specimen; DNA microarrays are used to study gene expression.	A DNA probe is used to locate a nucleotide sequence of interest in a colony, individual cells, or mRNA extracted from a sample.

SUMMARY

9.1 Fundamental Tools Used in Biotechnology

Restriction Enzymes (figure 9.1)

Restriction enzymes cut DNA into fragments. **Cohesive ends** will **anneal** with one another, making it possible to join DNA from two different organisms.

Gel Electrophoresis (figure 9.2)

Gel electrophoresis separates DNA fragments according to their size.

9.2 Applications of Genetic Engineering (table 9.2)

Genetically Engineered Bacteria (figures 9.3, 9.4)

Bacteria can be engineered to produce pharmaceutical proteins, vaccines, and other proteins more efficiently. By **cloning** a segment of DNA into *E. coli*, an easy source of that sequence is available for study and further manipulation.

Genetically Engineered Eukaryotes

Transgenic plants have been engineered to resist pests and herbicides, have improved nutritional value, and function as edible vaccines.

9.3 Techniques Used in Genetic Engineering (figure 9.6)

Obtaining DNA

To isolate DNA, cells are lysed by adding a detergent. To obtain eukaryotic DNA without introns, reverse transcriptase is used to make a copy of DNA from an mRNA template (figure 9.7).

Generating a Recombinant DNA Molecule

DNA ligase is used to join the **vector** and the **insert** (figures 9.8, 9.9).

Introducing the Recombinant DNA into a New Host

The recombinant molecule is introduced into the new host, usually *E. coli*, using transformation or electroporation. The transformed cells are cultivated on medium that both selects for cells containing vector sequences and differentiates those that carry recombinant molecules.

9.4 Concerns Regarding Genetic Engineering and Other DNA Technologies

Advances in genomics raise ethical issues and concerns about confidentiality. Genetically modified organisms hold many promises, but concerns exist about the inadvertent introduction of allergens into a food product and adverse effects on the environment.

9.5 DNA Sequencing

By determining the DNA sequence of a genome, the encoded information can be compared to that of other organisms.

Techniques Used in DNA Sequencing

A key ingredient in a sequencing reaction is a dideoxynucleotide, a nucleotide that lacks the 3′OH and therefore functions as a chain terminator (figure 9.11). The sizes of fragments in a sequencing reaction indicate the positions of the terminating nucleotide base in the synthesized DNA strands (figures 9.12, 9.13).

9.6 The Polymerase Chain Reaction (PCR)

PCR is used to rapidly increase the amount of a specific DNA segment in a sample (figure 9.14).

Techniques Used in PCR

Double-stranded DNA is denatured, primers **anneal** to their complementary sequences, and then DNA is synthesized, amplifying the target sequence (figure 9.15). The discrete-sized fragment ultimately amplified exponentially is obtained after three cycles of replication (figures 9.16, 9.17). The primers dictate which portion of the DNA is amplified.

9.7 Probe Technologies

DNA probes are used to locate specific nucleotide sequences (figure 9.18).

Colony Blotting

Colony blotting uses a **probe** to identify colonies that contain a given sequence of DNA (figure 9.19).

Fluorescence in situ Hybridization (FISH)

Fluorescence *in situ* hybridization (FISH) uses a fluorescently labeled probe to detect specific nucleotide sequences within intact cells affixed to a microscope slide (figure 9.20).

DNA Microarrays

DNA microarrays contain tens or hundreds of thousands of oligonucleotides that each function in a manner analogous to a probe (figure 9.21).

REVIEW QUESTIONS

Short Answer

1. Why are restriction enzymes useful in biotechnology?
2. Describe three general uses of genetically engineered bacteria.
3. Describe the function of a reporter gene.
4. Describe four uses of genetically engineered plants.
5. Describe the function of a probe.
6. Explain how DNA microarray technology can be used to study gene expression.
7. What is cDNA? Why is it used when cloning eukaryotic genes?
8. Explain how gel electrophoresis separates DNA fragments.
9. How many different temperatures are used in each cycle of the polymerase chain reaction?
10. Explain how PCR eventually generates a discrete-sized fragment from a much longer piece of DNA.

Multiple Choice

1. What is the function of a vector?
   a) Destroys cells that do not contain cloned DNA
   b) Allows cells to take up foreign DNA
   c) Carries cloned DNA, enabling it to replicate in cells

d) Encodes herbicide resistance

e) Encodes Bt-toxin

2. The Ti plasmid of *Agrobacterium tumefaciens* is used to genetically engineer which of the following cell types?

   a) Animals b) Bacteria c) Plants

   d) Yeast e) All of these

3. Which of the following can be used to generate a DNA library?

   a) PCR b) Sequencing c) Colony blotting

   d) Microarrays e) Cloning

4. An ideal vector has all of the following, *except*

   a) an origin of replication.

   b) a gene encoding a restriction enzyme.

   c) a gene encoding resistance to an antibiotic.

   d) a multiple-cloning site.

   e) the *lacZ′* gene.

5. Which of the following describes the function of the *lacZ′* gene in a cloning vector?

   a) Means of selecting for cells that contain vector sequences

   b) Means of distinguishing cells that have taken up recombinant molecules

   c) Site required for the vector to replicate

   d) Mechanism by which cells take up the DNA

   e) Gene for a critical nutrient required by transformed cells

6. Which is used for cloning eukaryotic genes but not prokaryotic genes?

   a) Restriction enzymes

   b) DNA ligase

   c) Reverse transcriptase

   d) Vector

   e) Selectable marker

7. Which of the following does a dideoxynucleotide lack?

   a) $5'PO_4$ b) $3'OH$ c) $5'OH$

   d) $3'PO_4$ e) C and D

8. In a sequencing reaction, the dATP was left out of the tube. What would be the result of this error?

   a) No synthesis would occur.

   b) Synthesis would never continue past the first A.

   c) Synthesis would not stop until the end of the template.

   d) Synthesis would terminate randomly, regardless of the nucleotide incorporated.

   e) The error would have no effect.

9. The polymerase chain reaction uses *Taq* polymerase rather than a DNA polymerase from *E. coli*, because *Taq* polymerase

   a) introduces fewer errors during DNA synthesis.

   b) is heat-stable.

   c) can initiate DNA synthesis at a wider variety of sequences.

   d) can denature a double-stranded DNA template.

   e) is easier to obtain.

10. The polymerase chain reaction generates a fragment of a distinct size even when an intact chromosome is used as a template. What determines the boundaries of the amplified fragment?

    a) The concentration of one particular deoxynucleotide in the reaction.

    b) The duration of the elongation step in each cycle.

    c) The position of a termination sequence, which causes the *Taq* polymerase to fall off the template.

    d) The sites to which the primers anneal.

    e) The temperature of the elongation step in each cycle.

Applications

1. Two students in a microbiology class are arguing about the origins of biotechnology. One student argued that biotechnology started with the advent of genetic engineering. The other student disagreed, saying that biotechnology was as old as ancient civilization. What was the rationale for the argument by the second student?
2. A student wants to clone Gene X. On both sides of the gene are the recognition sequences for *AluI* and *Bam*HI (look at table 9.1). Which enzyme would be easier to use for the cloning experiment and why?

Critical Thinking

1. Discuss some potential issues regarding gene therapy, the use of genetic engineering to correct genetic defects.
2. An effective DNA probe can sometimes be developed by knowing the amino acid sequence of the protein encoded by the gene. A student argued that this is too time-consuming since the complete amino acid sequence must be determined in order to create the probe. Does the student have a valid argument? Why or why not?

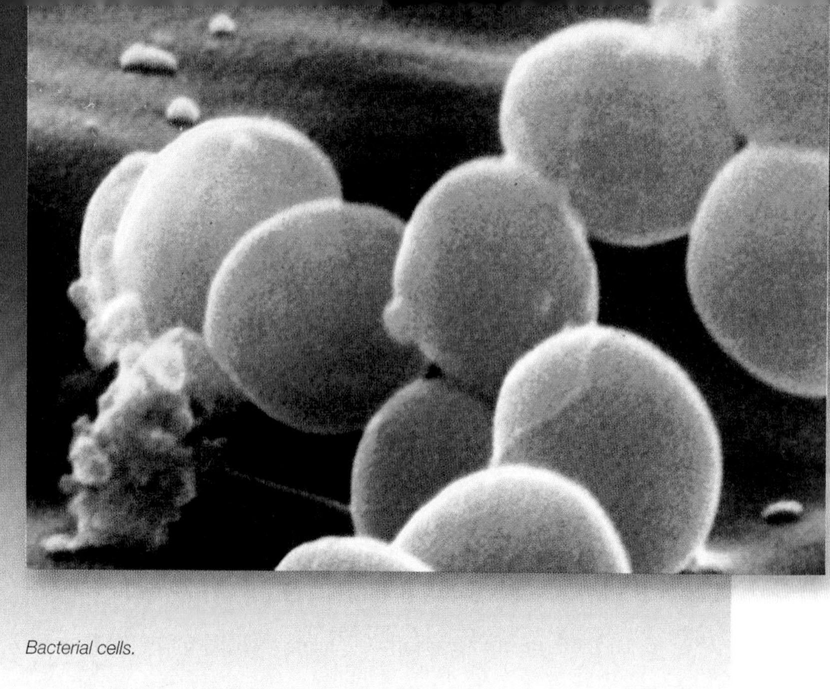

10

Bacterial cells.

Identification and Classification of Prokaryotic Organisms

A Glimpse of History

In the early 1870s, the German botanist Ferdinand Cohn published several papers on bacterial classification, grouping microorganisms according to shape: spherical, short rods, elongated rods, and spirals. However, that classification, based solely on shapes, was not adequate for categorizing all of the different bacteria. There were too many kinds and too few shapes.

The second major attempt at bacterial classification was initiated by Sigurd Orla-Jensen. His early training in Copenhagen was in chemical engineering, but he soon became interested in microbiology. In 1908, he proposed that bacteria be classified according to their physiological properties rather than morphology.

A quarter of a century later, two Dutch microbiologists, Albert Kluyver and C. B. van Niel, proposed classification systems based on presumed evolutionary relationships. They recognized a very serious problem, however: There was no way to distinguish between "resemblance" and "relatedness." The fact that two prokaryotes look alike does not mean they are genetically related.

In 1970, Roger Stanier, a microbiologist at the University of California, Berkeley, pointed out that relationships could be determined by comparing either gene products, such as proteins and cell walls, or nucleotide sequences. At that time, most microbiologists, including Stanier, assumed that all prokaryotes are basically similar. When the chemical compositions of a wide variety of prokaryotes were examined in detail, however, it was found that many had features that differed from those of *Escherichia coli,* considered a "typical" bacterium. These "unusual" features pertained to the chemical nature of the cell wall, cytoplasmic membrane, and ribosomal RNA.

In the late 1970s, Carl Woese and his colleagues at the University of Illinois determined the nucleotide sequence of ribosomal RNA in a wide variety of organisms. Based on the data, they recognized that prokaryotes could be divided into two major groups that differ from one another as much as they differ from eukaryotic cells. This led to a revolutionary system of classification that separates prokaryotes into two domains—the *Archaea* and the *Bacteria.* Each of these is on the same level as the *Eucarya,* which includes the animals, plants, and fungi (all eukaryotes).

Information that is logically organized is easier to both retrieve and understand. Newspapers, for instance, do not scatter various subjects throughout the paper; rather they are divided into sections such as local news, sports, and entertainment. A large library would be extremely difficult to use if the multitude of books were not split into sections by subject matter. Likewise, scientists have divided living organisms into different groups, the better to understand the relationships among the species.

Take a moment and think about how you would group bacteria if you were to arrange a classification system. Would you group them according to shape? Or would it make more sense to group them according to their motility? Perhaps you would group them according to their medical significance. But then, how would you classify two apparently identical bacteria that differed in their disease-causing potential?

KEY TERMS

Classification The process of arranging organisms into similar or related groups (taxa), primarily to provide easy identification and study.

Dichotomous Key Flowchart of tests used for identifying organisms.

Domain A collection of similar kingdoms; there are three domains—*Bacteria, Archaea,* and *Eucarya.*

Genus A collection of related species.

Identification The process of characterizing an isolate in order to determine the group (taxon) to which it belongs.

Lateral (Horizontal) Gene Transfer Transfer of DNA from one organism to another through conjugation, DNA-mediated transformation, or transduction.

Nomenclature The system of assigning names to organisms.

Phylogeny Evolutionary relatedness of organisms.

Signature Sequence Characteristic sequences in the ribosomal RNA genes, or their products, that can be used to classify or identify certain organisms.

Species A group of closely related isolates or strains; the basic unit of taxonomy.

Strain An isolate; subgroup within a species.

Taxonomy The science that studies organisms in order to arrange them into groups (taxa); involves three interrelated areas—identification, classification, and nomenclature.

10.1

Principles of Taxonomy

Focus Point

- Describe how prokaryotes are identified, classified, and assigned names.

Taxonomy is the science that studies organisms in order to arrange them into groups (taxa). Those organisms with similar properties are grouped together and separated from ones that are different. Taxonomy can be viewed as three separate but interrelated areas:

- **Identification**—the process of characterizing an isolate to determine the group (taxon) to which it belongs.
- **Classification**—the process of arranging organisms into similar or related groups, primarily to provide easy identification and study.
- **Nomenclature**—the system of assigning names to organisms.

Strategies Used to Identify Prokaryotes

In practical terms, identifying the genus and species of a prokaryote may be more important than understanding its genetic relationship to other microbes. For example, a food manufacturer is most interested in detecting the presence of microbial contaminants that can spoil a food product. In a clinical laboratory, it is critical to quickly identify a pathogen isolated from a patient so the best possible treatment can be given.

To characterize and identify microorganisms, a wide assortment of technologies may be used including microscopic examination, culture characteristics, biochemical tests, and nucleic acid analysis. In a clinical laboratory, the patient's disease symptoms play an important role in identifying the infectious agent. For example, pneumonia in an otherwise healthy adult is typically caused by *Streptococcus pneumoniae,* an organism that is easily differentiated from others using a few specific tests. In contrast, diagnosing the cause of a wound infection is often more difficult, because many different microorganisms could be involved. Often, however, it is only necessary to rule out the presence of organisms known to cause a particular disease, rather than to conclusively identify each and every organism in the specimen. For instance, a fecal specimen from a patient complaining of a diarrhea and fever would generally only be tested for the presence of specific organisms that cause those symptoms. ■ *Streptococcus pneumoniae,* p. 509

The various methods used to identify prokaryotes will be discussed in detail later in the chapter.

Strategies Used to Classify Prokaryotes

Understanding the evolutionary relatedness, or **phylogeny,** of prokaryotes is important in constructing a classification scheme that reflects the actual evolution and biology of these organisms. Such a scheme is more useful than one that simply groups organisms by arbitrary characteristics, because it is less prone to the bias of human perceptions. It also makes it easier to classify newly recognized organisms and allows scientists to make predictions, such as which genes are likely to be transferred between organisms.

Unfortunately, determining genetic relatedness among prokaryotes is more difficult than it is for plants and animals. Not only do prokaryotes have few differences in size and shape, they do not undergo sexual reproduction. In higher organisms such as plants and animals, the basic taxonomic unit, a species, is generally considered to be a group of morphologically similar organisms that are capable of interbreeding to produce fertile offspring. Obviously, it is not possible to apply these same criteria to prokaryotes, thus making classification problematic.

Historically, taxonomists have relied heavily on phenotypic attributes to classify prokaryotes. The development and application of molecular techniques such as nucleotide sequencing, however, is finally making it possible to determine the genetic relatedness of microorganisms. ■ phenotype, p. 186

Taxonomic Hierarchies

Taxonomic classification categories are arranged in a hierarchical order, with the species being the basic unit. The species designation gives a formal taxonomic status to a group of related isolates or **strains,** which, in turn, permits their identification. Without classification, scientists and others would not be able to communicate about organisms with any degree of accuracy. Taxonomic categories include:

- **Species**—a group of closely related isolates or strains. Note that members of a species are not all identical; individual strains may vary in minor properties. The difficulty for the taxonomist is to decide how different two isolates must be in order to be classified as separate species rather than strains of the same species.
- **Genus**—a collection of similar species.
- **Family**—a collection of similar genera. In prokaryotic nomenclature, the name of the family ends in the suffix *-aceae.*
- **Order**—a collection of similar families. In prokaryotic nomenclature, the name of the order ends in the suffix *-ales.*
- **Class**—a collection of similar orders.
- **Phylum** or **Division**—a collection of similar classes.
- **Kingdom**—a collection of similar phyla or divisions.
- **Domain**—a collection of similar kingdoms. The domain is a relatively new taxonomic category that reflects the characteristics of the cells that make up the organism.

Note, however, that microbiologists often group prokaryotes into informal categories based on one or more distinctive characteristic, rather than utilizing the higher taxonomic ranks such as order, class, and phylum. Examples of such informal groupings include the lactic acid bacteria, the anoxygenic phototrophs, the endospore-formers and the sulfate reducers. Organisms within these groupings share similar phenotypic and physiological characteristics, but may not be genetically related. ■ lactic acid bacteria, p. 255 ■ anoxygenic phototrophs, p. 256 ■ sulfate reducers, p. 254

An example of how a particular bacterial species is classified is shown in **table 10.1.** Note that the table intentionally omits the taxonomic category of kingdom. This is because the use of kingdoms within the *Bacteria* is still in a state of flux.

TABLE 10.1 Taxonomic Ranks of the Bacterium *Escherichia coli*

Formal Rank	Example
Domain	*Bacteria*
Phylum	*Proteobacteria*
Class	*Gammaproteobacteria*
Order	*Enterobacteriales*
Family	*Enterobacteriaceae*
Genus	*Escherichia*
Species	*coli*

Classification Systems

Taxonomy is still an evolving discipline, with systems of classification that change over the years as new information is discovered. There is no such thing as an "official" classification system, and, as new ones are introduced, others fall into disfavor. The classification scheme currently favored by most microbiologists is the three-domain system. This designates all organisms as belonging to one of the three domains—*Bacteria, Archaea,* and *Eucarya* **(figure 10.1).** The system is based on the work of Carl Woese and colleagues who compared the sequences of nucleotide bases in ribosomal RNA from a wide variety of organisms. They showed that prokaryotes could be divided into two major groups that differ from one another as much as they do from the eukaryotic cell. The ribosomal RNA data are consistent with other observed differences between the *Archaea* and *Bacteria,* including the chemical compositions of their cell wall and cytoplasmic membrane **(table 10.2).** ■ *Bacteria,* p. 10 ■ *Archaea,* p. 10 ■ *Eucarya,* p. 10

Before the three-domain classification system was introduced, the most widely accepted scheme was the five-kingdom system, proposed by R. H. Whittaker in 1969. The five kingdoms in this system are Plantae, Animalia, Fungi, Protista (mostly single-celled eukaryotes), and Prokaryotae. While the five-kingdom system recognizes the obvious morphological differences between plants and animals, it does not reflect the recent genetic insights of the ribosomal RNA data, which indicates that plants and animals are more closely related to each other than *Archaea* are to *Bacteria.*

Bergey's Manual of Systematic Bacteriology While there is no "official" classification of prokaryotes, microbiologists generally rely on the reference text *Bergey's Manual of Systematic Bacteriology* as a guide. All known species are described there, including those that have not yet been cultivated. If the properties of a newly isolated organism do not agree with any description in *Bergey's Manual,* then presumably a new organism has been isolated. The newest edition of this comprehensive manual is being published in five volumes and classifies prokaryotes according to the most recent information on their genetic relatedness **(table 10.3).** In some cases, this classification differs substantially from that of the previous edition, which grouped organisms according to their phenotypic characteristics.

In addition to containing descriptions of organisms, all volumes contain information on the ecology, methods of enrichment, culture, and isolation of the organisms as well as methods for their maintenance and preservation. However, the heart of the work is a description of all characterized prokaryotes and their groupings.

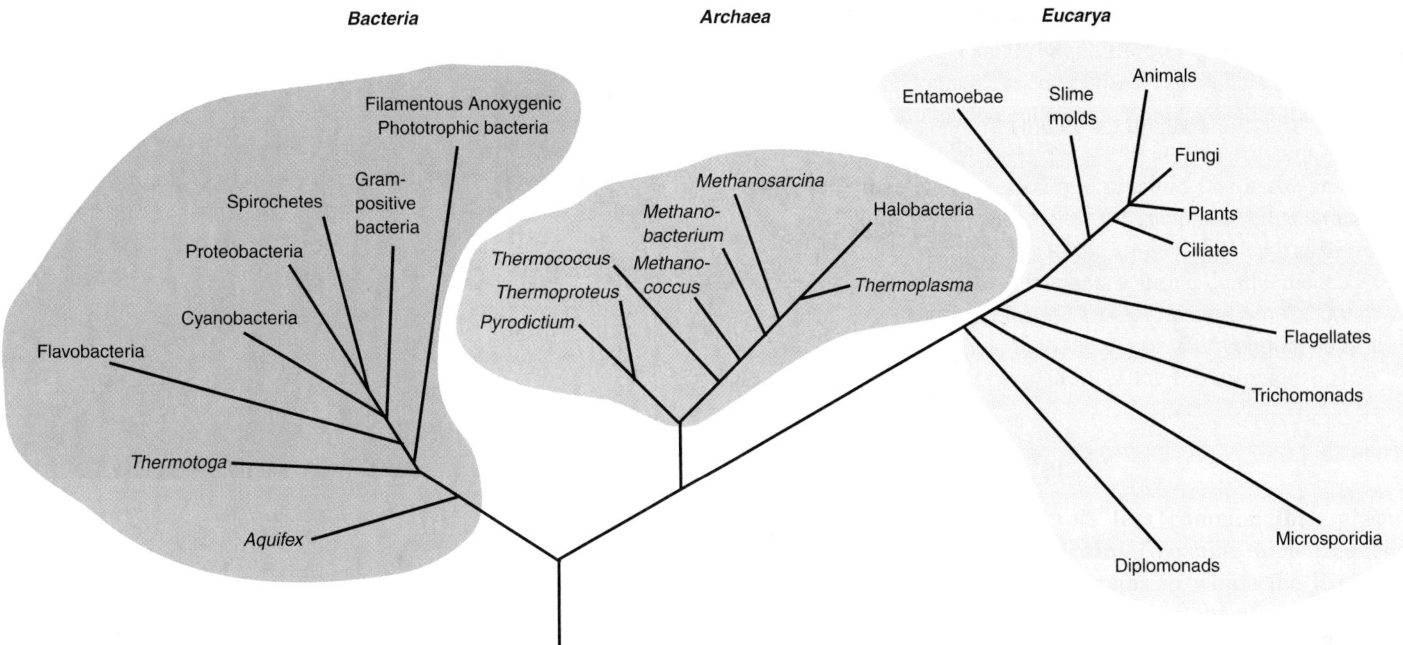

FIGURE 10.1 The Three-Domain System of Classification This classification system separates prokaryotic organisms into two domains—*Bacteria* and *Archaea*. The third domain, *Eucarya*, contains all organisms composed of eukaryotic cells. This system of classification is based on ribosomal RNA sequence data.

Nomenclature

Bacteria are given names according to an official set of internationally recognized rules, the *International Code for the Nomenclature of Bacteria.* Bacterial names may originate from any language, but they must be given a Latin suffix. In some cases the name reflects a characteristic of an organism such as its habitat, but often bacteria are named in honor of a prominent researcher.

Just as classification is always in a state of flux, so is the assignment of names. While revision of names is desirable from a scientific perspective, it is often a great source of confusion from a practical standpoint, particularly when the names of medically important bacteria are changed. To ease the transition of nomenclature changes, the former name is sometimes included in parentheses. For example, *Lactococcus lactis,* a bacterium that was included in the genus *Streptococcus,* is sometimes indicated as *Lactococcus (Streptococcus) lactis.*

MICROCHECK 10.1

Taxonomy consists of three interrelated areas: identification, classification, and nomenclature. In clinical laboratories, identifying the genus and species of an organism is more important than understanding its evolutionary relationship to other organisms.

- ✓ Why might it be easier to determine the cause of pneumonia than the cause of a wound infection?
- ✓ Why do microbiologists prefer the three-domain system of classification?
- ✓ Some biologists have been reluctant to accept the three-domain system. Why might this be?

TABLE 10.2 A Comparison of Some Properties of the Three Domains—*Archaea, Bacteria,* and *Eucarya*

Cell Feature	*Archaea*	*Bacteria*	*Eucarya*
Peptidoglycan cell wall	No	Yes	No
Cytoplasmic membrane lipids	Hydrocarbons (not fatty acids) linked to glycerol by ether linkage	Fatty acids linked to glycerol by ester linkage	Fatty acids linked to glycerol by ester linkage
Ribosomes	70S	70S	80S
Presence of introns	Sometimes	No	Yes
Membrane-bound nucleus	No	No	Yes

TABLE 10.3 Taxonomic Outline of *Bergey's Manual of Systematic Bacteriology*, 2nd edition

	Representative Genera
Volume 1: The *Archaea* and the Deeply Branching Phototrophic Bacteria	
Domain *Archaea*	
Phylum *Crenarchaeota*	*Pyrodictium, Ignicoccus*
Phylum *Euryarchaeota*	*Halobacterium, Methanococcus, Natronococcus, Picrophilus*
Domain *Bacteria*	
Phylum *Aquificae*	*Aquifex*
Phylum *Thermotogae*	*Thermotoga*
Phylum *Thermodesulfobacteria*	*Thermodesulfobacterium*
Phylum *Deinococcus–Thermus*	*Deinococcus, Thermus*
Phylum *Chrysiogenetes*	*Chrysiogenes*
Phylum *Chloroflexi*	*Chloroflexus*
Phylum *Thermomicrobia*	*Thermomicrobium*
Phylum *Nitrospira*	*Nitrospira*
Phylum *Deferribacteres*	*Deferribacter*
Phylum *Cyanobacteria*	*Anabaena, Spirulina, Synechococcus*
Phylum *Chlorobi*	*Chlorobium, Pelodictyon*
Volume 2: The *Proteobacteria*	
Phylum *Proteobacteria*	
Class *Alphaproteobacteria*	*Agrobacterium, Caulobacter, Ehrlichia, Nitrobacter, Rhodospirillum, Rickettsia, Rhizobium*
Class *Betaproteobacteria*	*Neisseria, Nitrosomonas, Thiobacillus*
Class *Gammaproteobacteria*	*Azotobacter, Chromatium, Escherichia, Legionella, Nitrosococcus, Pseudomonas, Vibrio*
Class *Deltaproteobacteria*	*Bdellovibrio, Myxococcus*
Class *Epsilonproteobacteria*	*Campylobacter, Helicobacter*
Volume 3: The Low G + C Gram-Positive Bacteria	
Phylum *Firmicutes*	
Class I. *Clostridia*	*Clostridium, Heliobacterium*
Class II. *Mollicutes*	*Mycoplasma*
Class III. *Bacilli*	*Bacillus, Streptococcus, Listeria, Staphylococcus*
Volume 4: The High G + C Gram-Positive Bacteria	
Phylum *Actinobacteria*	*Corynebacterium, Bifidobacterium, Micrococcus, Mycobacterium, Streptomyces*
Volume 5: The *Planctomycetes, Spirochaetes, Fibrobacteres, Bacteroides, Fusobacteria*	
Phylum *Planctomycetes*	*Planctomyces*
Phylum *Chlamydiae*	*Chlamydia*
Phylum *Spirochaetes*	*Borrelia, Treponema*
Phylum *Fibrobacteres*	*Fibrobacter*
Phylum *Acidobacteria*	*Acidobacterium*
Phylum *Bacteroidetes*	*Bacteroides*
Phylum *Fusobacteria*	*Fusobacterium*
Phylum *Verrucomicrobia*	*Verrucomicrobium*
Phylum *Dictyoglyomi*	*Dictyoglomus*
Phylum *Gemmatimonadetes*	*Gemmatinonas*

10.2

Using Phenotypic Characteristics to Identify Prokaryotes

Focus Point

- Describe how phenotypic characteristics including microscopic morphology, metabolic capabilities, serology, and fatty acid analysis can be used to identify prokaryotes.

Phenotypic characteristics such as cell morphology, colony morphology, biochemical traits, and the presence of specific proteins can all be used in the process of identifying microorganisms. Most of these methods do not require sophisticated equipment and can easily be done anywhere in the world. Methods used to identify prokaryotes are summarized in **table 10.4.**

Microscopic Morphology

An important initial step in identifying a microorganism is to determine its size, shape, and staining characteristics. Microscopic examination gives information very quickly and is sometimes enough to make a presumptive identification.

Size and Shape

The size and shape of a microorganism can readily be determined by microscopically examining a wet mount. Based only on the size and shape, one can readily decide whether the organism in question is a prokaryote, fungus, or protozoan. In a clinical lab, this can sometimes provide all the information needed for diagnosis of certain eukaryotic infections. For example, a wet mount of vaginal secretions is routinely used to diagnose infections caused by yeast and one of stool is examined for the eggs of parasites when certain roundworms are suspected **(figure 10.2).** ■ **wet mount, p. 50**

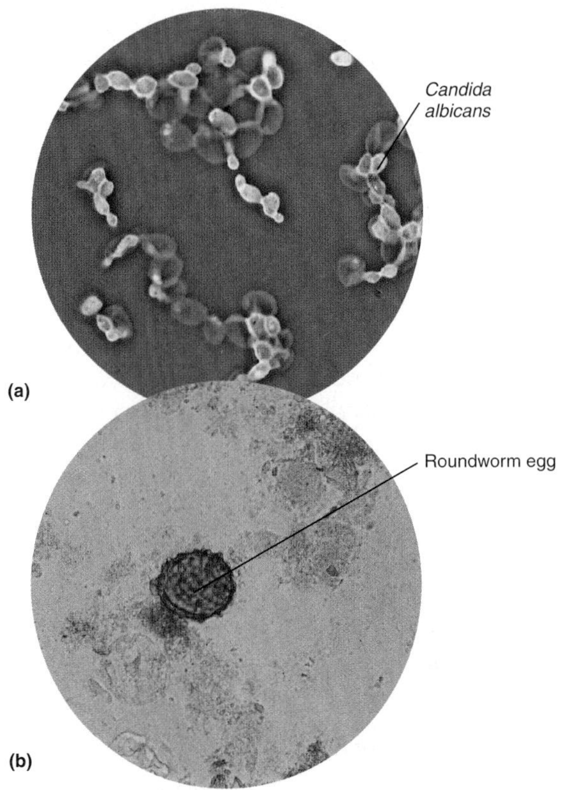

FIGURE 10.2 Wet Mounts of Clinical Specimens **(a)** Vaginal secretions containing yeast (*Candida albicans*, 410×); **(b)** roundworm (*Ascaris*) eggs in a stool (400×).

TABLE 10.4 Methods Used to Identify Prokaryotes

Method	Comments
Phenotypic Characteristics	Most of these methods do not require sophisticated equipment and can easily be done anywhere in the world.
Microscopic morphology	Size, shape, and staining characteristics such as Gram stain can give suggestive information as to the identity of the organism. Further testing, however, is needed to confirm the identification.
Metabolic capabilities	Culture characteristics can give suggestive information. A battery of biochemical tests can be used to confirm the identification.
Serology	Proteins and polysaccharides that make up a prokaryote are sometimes characteristic enough to be considered identifying markers. These can be detected using specific antibodies.
Fatty acid analysis	Cellular fatty acid composition can be used as an identifying marker and is analyzed by gas chromatography.
Genotypic Characteristics	These methods are increasingly being used to identify microorganisms.
Nucleic acid probes to detect specific nucleotide sequences	Probes can be used to identify prokaryotes grown in culture. In some cases, the method is sensitive enough to detect the organism directly in a specimen.
Amplifying specific DNA sequences using PCR	Even an organism that occurs in very low numbers in a mixed culture can be identified.
Sequencing rRNA genes	This requires amplifying and then sequencing rRNA genes, but it can be used to identify organisms that have not yet been grown in culture.

Gram Stain

The Gram stain distinguishes between Gram-positive and Gram-negative bacteria (see figure 3.14). This relatively rapid test narrows the list of possible identities of an organism and provides suggestive information that can be helpful in the identification process. ■ Gram stain, p. 49

In a clinical lab, the Gram stain of a specimen by itself is generally not sensitive or specific enough to diagnose the cause of most infections, but it is still an extremely useful tool. The clinician can see the Gram reaction, the shape and arrangement of the bacteria, and whether the organisms appear to be growing as a pure culture or with other bacteria and/or cells of the host. However, most medically important bacteria cannot be identified by Gram stain alone. For example, *Streptococcus pyogenes,* which causes strep throat, cannot be distinguished microscopically from the other streptococci that are part of the normal flora of the throat. A Gram stain of a stool specimen cannot distinguish *Salmonella* species from *E. coli.* These organisms generally must be isolated in pure culture and tested for their biochemical attributes to provide precise identification. ■ strep throat, p. 499

In certain cases, the Gram stain gives enough information to start appropriate antimicrobial therapy while awaiting more accurate identification. For example, a Gram stain of sputum showing numerous white blood cells and Gram-positive encapsulated diplococci is highly suggestive of *Streptococcus pneumoniae,* an organism that causes pneumonia **(figure 10.3a).** In certain other cases, the result of a Gram stain is enough for accurate diagnosis. For instance, the presence of Gram-negative diplococci clustered in white blood cells in a sample of a urethral secretion from a man is considered diagnostic for gonorrhea, the sexually transmitted disease caused by *Neisseria gonorrhoeae* (figure 10.3b). This diagnosis can be made because *N. gonorrhoeae* is the only Gram-negative diplococcus found inhabiting the male urethra. ■ pneumonia, p. 509 ■ gonorrhea, p. 629

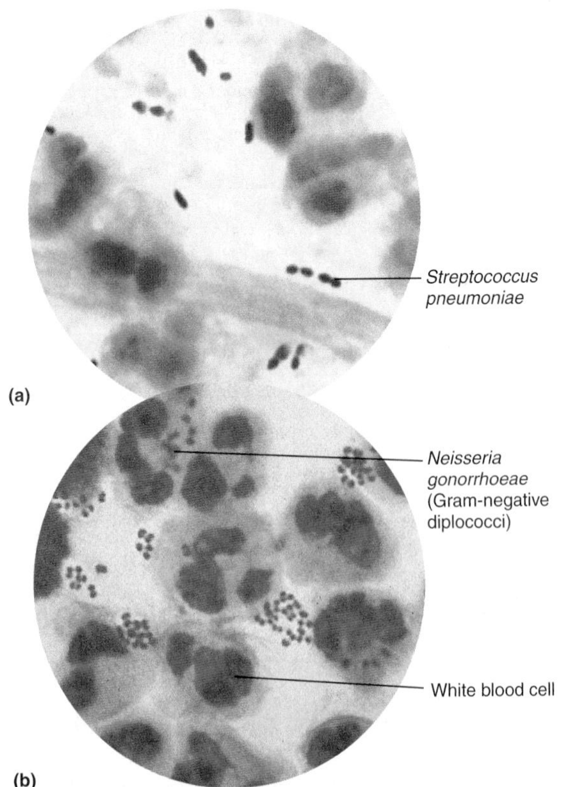

FIGURE 10.3 Gram Stains of Clinical Specimens (a) Sputum showing Gram-positive *Streptococcus pneumoniae* and **(b)** male urethra secretions showing Gram-negative *Neisseria gonorrhoeae* inside white blood cells.

Special Stains

Certain microorganisms have unique characteristics that can be detected with special staining procedures. As an example, members of the genus *Mycobacterium* are some of the few microorganisms that are acid-fast (see figure 3.15). If a patient has symptoms of tuberculosis, then an acid-fast stain will be done on a sample of his or her sputum to determine whether *Mycobacterium tuberculosis* can be detected. ■ acid-fast stain, p. 50

Metabolic Capabilities

The identification of most prokaryotes relies on analyzing their metabolic capabilities such as the types of sugars utilized or the end products produced. In some cases these characteristics are revealed by the growth and colony morphology on cultivation media, but most often they are demonstrated using biochemical tests. ■ growth of colonies, p. 86

Culture Characteristics

Microorganisms that can be grown in pure culture are the easiest to identify, because it is possible to obtain high numbers of a single type. Even the colony morphology can give initial clues to the identity of the organism. For example, colonies of streptococci are generally fairly small relative to many other bacteria such as staphylococci. Colonies of *Serratia marcescens* are often red when incubated at 22°C due to the production of a pigment. *Pseudomonas aeruginosa* often produces a soluble greenish pigment, which discolors the growth medium (see figure 11.12). In addition, cultures of *P. aeruginosa* have a distinct fruity odor.

In a clinical lab, where rapid but accurate diagnosis is essential, specimens are inoculated onto media specially designed to provide important clues as to the identity of the disease-causing organism. For instance, a specimen taken by swabbing the throat of a patient complaining of a sore throat is inoculated onto blood agar. This allows detection of the characteristic β-hemolytic colonies typical of *Streptococcus pyogenes* (see figure 4.11). Urine collected from a patient suspected of having a urinary tract infection is plated onto MacConkey agar. *E. coli,* the most common cause of urinary tract infections, forms characteristic pink colonies on MacConkey agar due to its ability to ferment lactose (see figure 4.12). ■ blood agar, p. 97 ■ MacConkey agar, p. 97

Biochemical Tests

Growth characteristics on culture media can give clues as to the identity of an organism, but biochemical tests are generally necessary for a more conclusive identification. One of the simplest is an assay for the enzyme catalase **(figure 10.4a).** Most bacteria that grow in the presence of oxygen are catalase positive. Important exceptions are the lactic acid bacteria, which include members of the genus *Streptococcus.* Thus, if a throat culture yields β-hemolytic colonies but further testing reveals they are all catalase positive, then *Streptococcus pyogenes* has been ruled out. ■ catalase, p. 92 ■ *Streptococcus pyogenes*, p. 499

(a)

(b)

(c)

FIGURE 10.4 Biochemical Tests (a) Catalase production. Bacteria that produce catalase break down hydrogen peroxide to release oxygen gas ($2\ H_2O_2 \longrightarrow 2\ H_2O + O_2$), which causes the bubbling shown on the left. A negative catalase test is shown on the right. **(b)** Sugar fermentation. The tube on the left shows acid (yellow color) and gas, indicating that the sugar was fermented and gas was produced during the process. The center tube shows no color change, indicating that the sugar was not utilized. The tube on the right is an uninoculated control. **(c)** Urease production. Breakdown of urea releases ammonia, which turns the pH indicator pink, as shown in the left tube. The tube in the center shows no color change, indicating that urease was not produced. The tube on the right is an uninoculated control.

Most biochemical tests rely on a pH indicator or chemical reaction that results in a color change when a compound is degraded. To test for the ability of an organism to ferment a given sugar, a broth medium containing that sugar and a certain pH indicator is employed. Fermentation of the sugar results in acid production, which lowers the pH, resulting in a color change; an inverted tube traps any gas produced (figure 10.4b). A medium designed to detect **urease,** an enzyme that degrades urea to produce carbon dioxide and ammonia, contains urea and a pH indicator (figure 10.4c). The characteristics of these and other important biochemical tests are summarized in **table 10.5.**

The basic strategy for identifying bacteria based on biochemical tests relies on the use of a **dichotomous key,** a flowchart of tests that give either a positive or negative result **(figure 10.5).** Because each test often requires an incubation period, however, it would be too time-consuming to proceed one step at a time. In addition, relying on a single biochemical test at each step could lead to misidentification. For example, if a strain that normally gives a positive result for a certain test lost the ability to produce a key enzyme, it would instead produce a negative result. Therefore, simultaneously inoculating multiple tests identifies the organism faster and more conclusively.

In certain cases, biochemical testing can be done without culturing the organism. *Helicobacter pylori,* the cause of most stomach ulcers, can be detected using the **breath test,** which assays for the presence of urease. The patient drinks a solution containing urea that has been labeled with an isotope of carbon. If *H. pylori* is present, its urease breaks down the urea, releasing labeled carbon dioxide, which escapes through the airway. Several hours after drinking the solution, the patient exhales into a balloon. The expired air is then tested for labeled carbon dioxide. This test is less invasive and, consequently, much cheaper and faster than the stomach biopsy that would otherwise need to be performed to culture the organism. ■ *Helicobacter pylori,* p. 589 ■ isotope, p. 26

Commercial Modifications of Traditional Biochemical Tests

Several less labor-intensive commercial modifications of traditional biochemical tests are available **(figure 10.6).** The API™ system utilizes a strip holding a series of tiny cups that contain dehydrated media. A liquid suspension of the test bacterium is inoculated into each compartment, thus rehydrating the media. Because the formulations of the media are similar in composition to those used in traditional tests, the positive results give rise to similar color changes. After a 16-hour incubation of the inoculated test strip, the results are determined by inspection. The pattern of results is converted to a numerical score, which can then be entered into a computer to identify the organism. A similar system is the Enterotube™, a tube with small compartments, each containing a different type of medium. One end of a metal rod that runs through the tube is used to touch a bacterial colony. When the rod is withdrawn, it inoculates each of the compartments. A system by Biolog uses a microtiter plate, a small tray containing 96 wells, to assay simultaneously an organism's ability to use a wide variety of carbon sources. Modifications of these plates enable researchers to characterize the metabolic capabilities of microbial communities, such as those in soil, water, or wastewater.

TABLE 10.5 Characteristics of Some Important Biochemical Tests

Biochemical Test	Principle of the Test	Positive Reaction
Catalase	Detects the activity of the enzyme catalase, which causes the breakdown of hydrogen peroxide to produce O_2 and water.	The reagent bubbles.
Citrate	Determines whether or not citrate can be used as a sole carbon source.	Growth, which is usually accompanied by the color change of a pH indicator.
Gelatinase	Detects enzymatic breakdown of gelatin to polypeptides.	The solid gelatin is converted to liquid.
Hydrogen sulfide production	Detects H_2S liberated as a result of the degradation of sulfur-containing amino acids.	A black precipitate forms due to the reaction of H_2S with iron salts in the medium.
Indole	Detects the enzymatic removal of the amino group from tryptophan.	The product, indole, reacts with a chemical reagent that is added, turning the reagent a deep red color.
Lysine decarboxylase	Detects the enzymatic removal of the carboxyl group from lysine.	The medium becomes more alkaline, causing a pH indicator to change color.
Methyl red	Detects mixed acids, the characteristic end products of a particular fermentation pathway. ■ **mixed acids, p. 148**	The medium becomes acidic (pH < 4.5); a red color develops upon the addition of a pH indicator.
Oxidase	Detects the activity of cytochrome *c* oxidase, a component of the electron transport chain of specific organisms. ■ **cytochrome *c*, p. 144**	A dark color develops upon the addition of a specific reagent.
Phenylalanine deaminase	Detects the enzymatic removal of the amino group from phenylalanine.	The product of the reaction, phenylpyruvic acid, reacts with ferric chloride to give the medium a green color.
Sugar fermentation	Detects the acidity resulting from fermentation of the sugar incorporated into the medium; also detects gas production.	The color of a pH indicator incorporated into the medium changes if acid if produced. An inverted tube traps any gas that is made.
Urease	Detects the enzymatic degradation of urea to carbon dioxide and ammonia.	The medium becomes alkaline, causing a pH indicator to change color.
Voges-Proskauer	Detects acetoin, an intermediate of the fermentation pathway that leads to the production of a 2, 3-butanediol. ■ **2, 3-butanediol, p. 148**	A red color develops upon addition of chemicals that detect acetoin.

Highly automated systems also are available. The Vitek™ system uses a miniature card that contains multiple wells with different formulations of dehydrated media. A computer then reads the growth pattern in the wells after a relatively short incubation period.

Serology

The proteins and polysaccharides that make up a bacterium are sometimes characteristic enough to be considered identifying markers. The most useful of these are the molecules that make up surface structures including the cell wall, capsule, flagella, and pili. For example, some species of *Streptococcus* contain a unique carbohydrate molecule as part of their cell wall that can be used to distinguish them from other species. These carbohydrates, as well as any distinct proteins or polysaccharides, can be detected using techniques that rely on the specificity of interaction between antibodies and antigens. Methods that exploit these interactions are called serological tests and will be discussed in more detail in later chapters. Some serological tests, such as those used to confirm the identity of *S. pyogenes,* are quite simple and rapid. ■ **antibodies, p. 371** ■ **antigens, p. 370** ■ **serology, p. 439**

Fatty Acid Analysis (FAME)

Bacteria differ in the type and relative quantity of fatty acids that make up their membranes; thus, the cellular fatty acid composition can be used as an identifying marker. In the case of Gram-negative bacteria, the fatty acids are contained in both the cytoplasmic and outer membranes. Gram-positive bacteria, however, lack an outer membrane; the cytoplasmic membrane is the source of their fatty acids. To analyze their fatty acid composition, bacterial cells are grown under standardized conditions. The cells are then chemically treated with sodium hydroxide and methanol to release the fatty acids and convert them to their more volatile methyl ester form (FAME stands for fatty acid methyl ester). The resulting fatty acid methylated esters can then be separated and analyzed using gas chromatography. By comparing the pattern of peaks, or **chromatogram,** to those of known species, an isolate can be identified.

MICROCHECK 10.2

The size, shape, and staining characteristics of a microorganism, all of which can be viewed using a microscope, yield important clues to its identity. Conclusive identification generally requires multiple biochemical tests that assay for compounds indicating the presence of specific biochemical pathways. The proteins and polysaccharides that make up a bacterium are sometimes unique enough to be considered identifying markers. Cellular fatty acid composition can be used as an identifying characteristic.

- ✓ How does MacConkey agar help to identify the cause of a urinary tract infection?
- ✓ Describe two methods to test for the enzyme urease.
- ✓ Why must a sample contain many microorganisms for any to be viewed by microscopic examination?

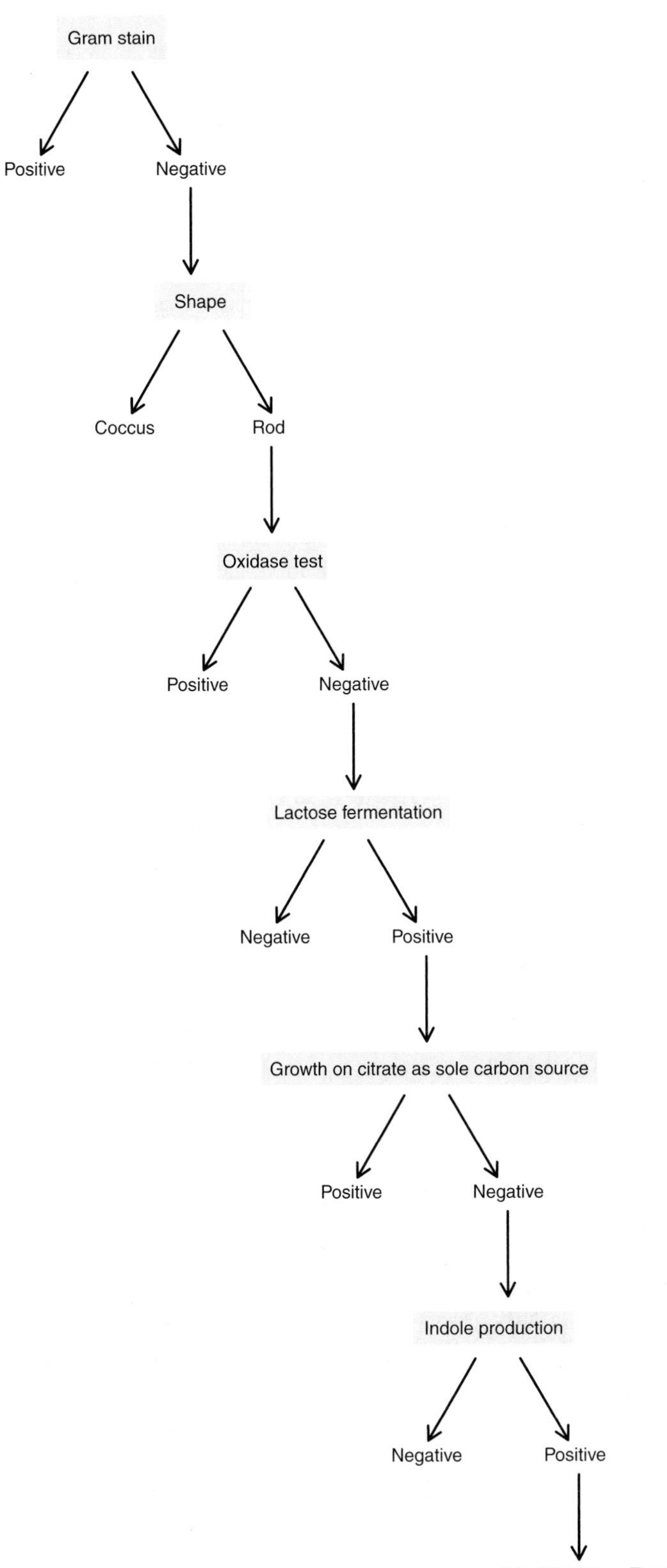

FIGURE 10.5 An Example of a Dichotomous Key Leading to the Identification of *E. coli* The biochemical tests are usually initiated simultaneously to speed identification.

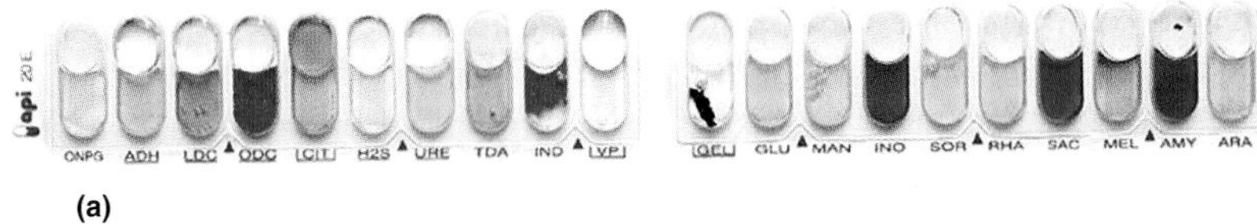

(a)

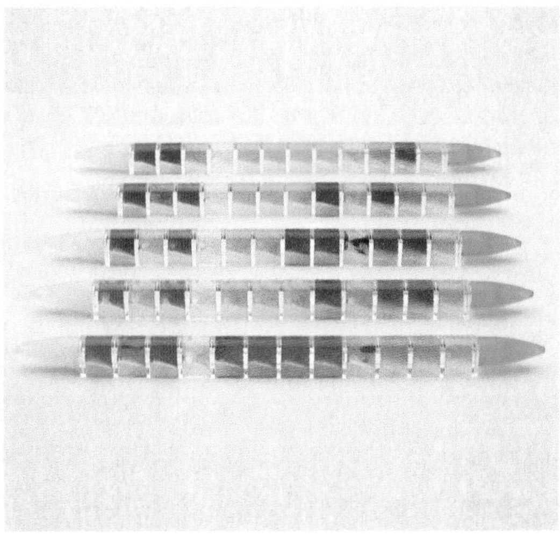

(b)

FIGURE 10.6 Commercial Modifications of Traditional Biochemical Tests These methods are less labor-intensive than traditional tests. **(a)** An API™ test strip. Each small cup contains a dehydrated medium similar in formulation to the traditional tests. A liquid suspension of the isolated test bacterium is added to each compartment; after incubation, the results are read manually. **(b)** An Enterotube™. Each compartment contains a different type of medium. One end of a metal rod that runs through the tube is used to touch a bacterial colony. When the rod is withdrawn it inoculates each of the compartments. After incubation, the results are read manually.

10.3

Using Genotypic Characteristics to Identify Prokaryotes

Focus Point

- Describe how nucleic acid probes, PCR, and DNA sequencing of 16S rDNA can be used to identify prokaryotes.

Many of the technologies discussed in chapter 9 are being used to identify microorganisms based on their genotype. For example, both DNA probes and polymerase chain reaction (PCR) can detect nucleotide sequences unique to a given species. A significant limitation, however, is that detection of specific sequence can only determine whether a given organism or group of related organisms is present. Thus, if the isolate could be one of five different possibilities, then five different probes would be required, each either used in separate reactions or labeled with different detectable markers. A distinct advantage of these methods, though, is that they make it possible to identify organisms that cannot yet be grown in culture. ■ **DNA probe, p. 227** ■ **polymerase chain reaction, p. 223**

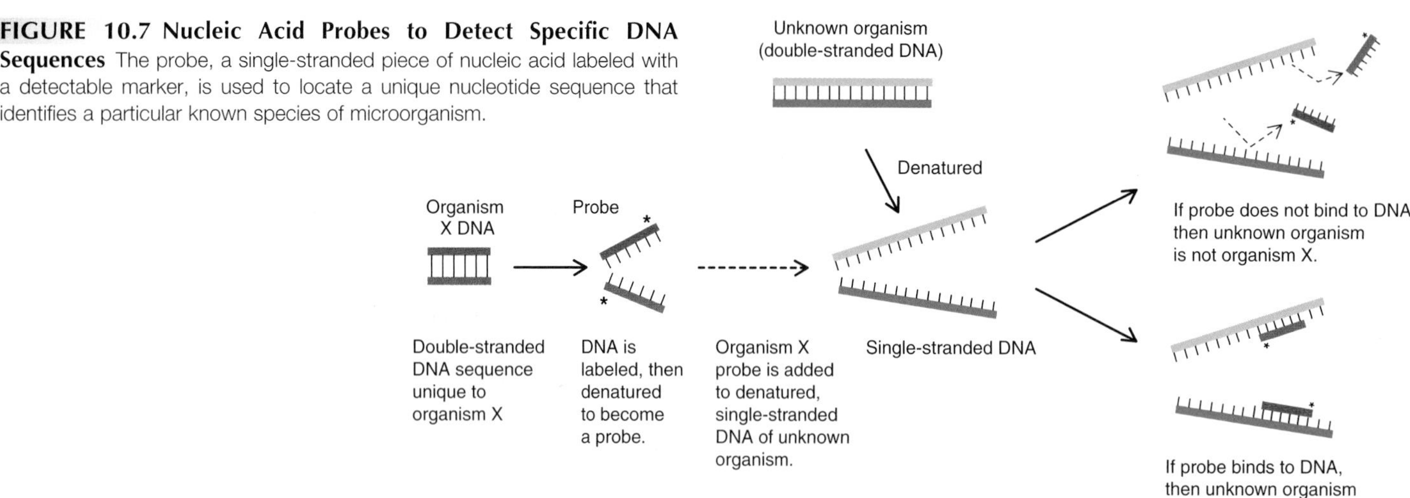

FIGURE 10.7 Nucleic Acid Probes to Detect Specific DNA Sequences The probe, a single-stranded piece of nucleic acid labeled with a detectable marker, is used to locate a unique nucleotide sequence that identifies a particular known species of microorganism.

Detecting Specific Nucleotide Sequences Using Nucleic Acid Probes

A nucleic acid probe can locate a unique nucleotide sequence that characterizes a particular species **(figure 10.7).** The probe is a single-stranded piece of nucleic acid, usually DNA, labeled with a detectable tag such as a radioisotope or a fluoresent dye. It is complementary to the sequence of interest.

Most methods that employ nucleic acid probes to detect DNA sequences rely on an amplification step. For example, in some cases the bacteria in the specimen must first be cultivated on an agar medium so that each cell can multiply to form a colony. In other cases, a preliminary *in vitro* DNA amplification step is required.

Fluorescence *in situ* hybridization (FISH) often employs probes that bind 16S ribosomal RNA (rRNA). Because of this, an amplification step is not needed, as numerous copies of rRNA are naturally present in multiplying cells. Various different probes that bind rRNA are available, each specific for a given **signature sequence** found in rRNA. A signature sequence characterizes either a certain species or a group of related organisms (see figure 9.20). Clinical and environmental applications of FISH, as well as the techniques involved, were discussed in chapter 9. ■ fluorescence *in situ* hybridization, p. 228

Amplifying Specific DNA Sequences Using the Polymerase Chain Reaction

The polymerase chain reaction (PCR) can be used to amplify target sequences of DNA, allowing researchers to detect specific sequences in samples such as body fluids, soil, food, and water (see figure 9.14). The technique can be used to detect organisms present in extremely small numbers as well as those that cannot yet be grown in culture.

To use PCR to detect a microbe of interest, a sample is first treated to release and denature the DNA. Specific primers and other ingredients are then added to the denatured DNA, forming the components of the PCR reaction (see figure 9.15). Some information about the nucleotide sequence of the organism must be known in order to select the appropriate primers. After approximately 30 cycles of PCR, the DNA region flanked by the primers will have been amplified approximately a billion-fold (see figure 9.17). In most cases, this results in a sufficient quantity for the amplified fragment to be readily visible as a discrete band on an ethidium bromide-stained agarose gel illuminated with UV light. Alternatively, a DNA probe can be used to detect the amplified DNA. ■ ethidium bromide, p. 215 ■ gel electrophoresis, p. 214

Sequencing Ribosomal RNA Genes

The nucleotide sequence of ribosomal RNA (rRNA) can be used to identify prokaryotes, particularly those difficult or currently impossible to grow in culture. The prokaryotic 70S ribosome, which plays an indispensable role in protein synthesis, is composed of protein and three different rRNAs (5S, 16S, and 23S) **(figure 10.8).** Because of their highly constrained and essential function, the nucleotide sequence changes that can occur in the rRNAs, yet still allow the ribosome to operate, are limited. This is why they have proved so useful in microbial classification and, more recently, identification. While earlier methods relied on determining the sequence of the rRNA molecule itself, newer techniques sequence the DNA that encodes rRNA, which is called **rDNA.**

Of the different rRNAs, the 16S molecule has proved most useful in taxonomy because of its moderate size (approximately 1,500 nucleotides). That molecule, as well as its eukaryotic counterpart (18S RNA), is sometimes referred to as **small subunit (SS** or **SSU) rRNA,** reflecting the fact that it is part of the small subunit of the ribosome (see figure 10.8). Once the nucleotide sequence of SSU RNA in an unknown organism has been determined, it can be compared with sequences of known organisms by searching extensive computerized databases.

Using rDNA to Identify Uncultivated Organisms

In any environment, including soil, water, and the human body, a multitude of organisms exist that cannot yet be grown in culture. Some scientists estimate that every gram of fertile soil contains

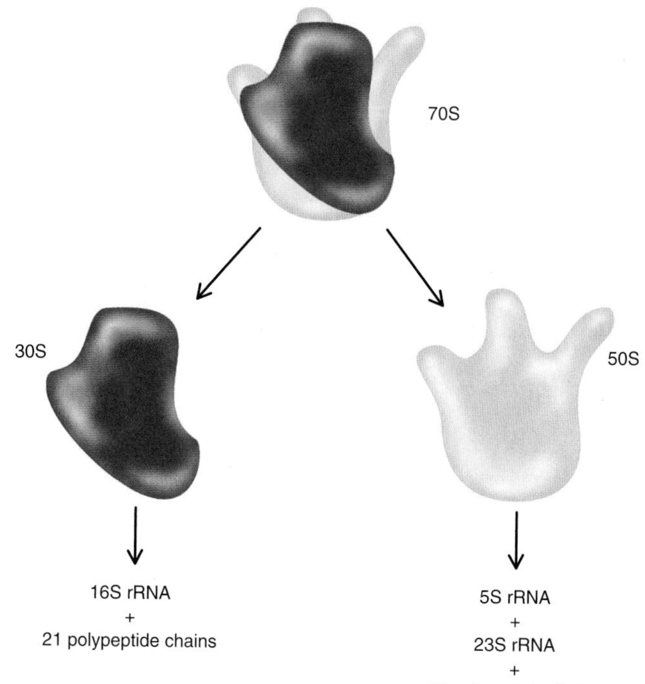

FIGURE 10.8 Ribosomal RNA The 70S ribosome of prokaryotes has three types of rRNA: 5S, 16S, and 23S.

more than 4,000 different species of prokaryotes, the vast majority of which have not been identified. With current technologies that enable amplification of specific portions of DNA, followed by the cloning and then sequencing of those fragments, it is possible not only to detect such organisms, but also to obtain information about their identity.

One of the first examples of using molecular biology to identify an organism that had not been cultivated was the characterization of the causative agent of a rare illness called Whipple's disease. This was done by using PCR to amplify bacterial 16S rDNA from intestinal tissue of patients who had symptoms of the disease. That DNA was then cloned and sequenced. The nucleotide sequence showed that the causative agent was an actinomycete unrelated to any of those previously identified. It was given the name *Tropheryma whipplei.* Even though it had never been grown in culture, a specific probe was then developed that can detect it in intestinal tissue. Since that time it has been successfully cultivated.

MICROCHECK 10.3

A microorganism can be identified by using a probe or PCR to detect a nucleotide sequence unique to that particular organism. Ribosomal RNA genes can be sequenced to identify an organism that cannot be grown in culture.

✓ When using a probe to identify an organism, why is it necessary to have some idea as to the organism's identity?

✓ How can ribosomal DNA be used to identify bacteria that cannot be grown in culture?

✓ Why are molecular methods especially useful when bacteria are difficult to cultivate?

10.4 Characterizing Strain Differences

Focus Point

- Describe five distinct methods by which different strains can be distinguished.

In some situations, distinguishing different strains of a given species is useful. In 2007, for example, reports of 52 salmonellosis cases across 17 states in the United States were found to involve a specific strain of *Salmonella enterica* serotype Enteritidis. Most patients reported consuming a puffed vegetable snack prior to their illness, leading to a widespread recall of the product. Linking 52 cases spread across the United States amid the thousands of salmonellosis cases that occur nationwide each year would not have been possible without methods to determine strain differences and then catalog those results, making it feasible to match cases from around the country.

Characterizing strain differences is not limited to investigations of foodborne illness. It also plays an instrumental role in forensic investigations of bioterrorism and other biocrimes, and in identifying cause of disease when only certain strains are pathogenic. The methods used to characterize different strains are summarized in **table 10.6.**

Biochemical Typing

Biochemical tests are used to identify various species of bacteria, but they can also be used to distinguish strains. If the biochemical variation is uncommon, it can be used for tracing the source of certain disease outbreaks. A group of strains that have a characteristic biochemical pattern is called a **biovar** or a **biotype.** A biochemical variant of *Vibrio cholerae* called Eltor caused a worldwide epidemic of cholera beginning in 1961. Because this biovar can be readily distinguished, its spread can be traced. ■ *Vibrio cholerae*, p. 596

Serological Typing

Proteins and carbohydrates that vary among strains can be used to differentiate strains. For example, *E. coli* and other Gram-negative bacteria vary in the antigenic structure of certain parts of the lipopolysaccharide portion of the cell wall, the **O antigen** (see figure 11.14). The composition of the flagella, the **H antigen,** can also vary. The "O157:H7" designation of *E. coli* O157:H7 refers to the antigenic type of its lipopolysaccharide and flagella. A group of strains that differ serologically from other strains is called a **serovar** or a **serotype.** ■ antigen, p. 370 ■ lipopolysaccharide, p. 61 ■ *E. coli* O157:H7, p. 600

Genomic Typing

Subtle differences in DNA sequences can be used to distinguish among strains that are phenotypically identical. These genomic variations are used to trace epidemics of foodborne illness. The ability to link geographically distant cases can enable public health officials to determine the source of the epidemic, leading to the recall of the implicated product and

TABLE 10.6 Summary of Methods Used to Characterize Different Strains

Method	Characteristics
Biochemical typing	Biochemical tests are most commonly used to identify various species of bacteria, but in some cases they can be used to distinguish different strains. A group of strains that have a characteristic biochemical pattern is called a biovar or a biotype.
Serological typing	Proteins and carbohydrates that vary among strains can be used to differentiate strains. A group of strains that have a characteristic serological type is called a serovar or a serotype.
Genomic typing	Molecular methods such as pulsed-field gel electrophoresis can be used to detect restriction fragment length polymorphisms (RFLPs).
Phage typing	Strains of a given species sometimes differ in their susceptibility to various types of bacteriophage.
Antibiograms	Antibiotic susceptibility patterns can be used to characterize strains.

preventing further cases of disease. One method of doing this is to compare the patterns of fragment sizes produced when the same restriction enzyme is used to digest DNA from each organism. Gel electrophoresis is used to separate the fragments so they can be observed; if the fragments are quite large, a special type of electrophoresis called pulsed-field gel electrophoresis (PFGE) is used. The different patterns of fragment sizes obtained by digesting DNA with restriction enzymes are called **restriction fragment length polymorphisms (RFLPs) (figure 10.9).** Two isolates of the same species that have different RFLPs are considered different strains. Two isolates that have identical RFLPs may or may not be the same strain. ■ restriction enzymes, p. 213

To facilitate the tracking of foodborne disease outbreaks, the Centers for Disease Control has established the **National Molecular Subtyping Network for Foodborne Disease Surveillance (PulseNet),** which catalogs the RFLPs of certain foodborne bacterial pathogens. Laboratories from around the country can submit RFLP patterns to a computer database and quickly receive information about other isolates showing the same patterns. Using this database, multistate foodborne disease outbreaks can more readily be recognized and traced. This is how the salmonellosis cases that led to the snack recall were found to be related.

Phage Typing

Strains of a given species may differ in their susceptibility to **bacteriophages.** A bacteriophage, or **phage,** is a virus that infects and multiplies within bacteria, often lysing them. Each type of phage has a limited host range. Lysis of the infected cell releases more phage particles, which in turn infect neighboring bacterial cells. The susceptibility of an organism to a particular type of phage can be readily demonstrated in the laboratory. First, a culture of the test organism is inoculated into melted, cooled nutrient agar and poured onto the surface of an agar plate, thus creating a uniform layer of cells. Then drops of different types of bacteriophage are carefully placed

PERSPECTIVE 10.1

Tracing the Source of an Outbreak of Foodborne Disease

On September 8, 2006, Wisconsin state health authorities alerted the CDC about an *E. coli* O157:H7 outbreak that appeared to involve consumption of fresh spinach. Soon thereafter, Oregon officials reported a similar outbreak. On September 12, the CDC determined that the strains from Wisconsin matched those from Oregon, as well as strains then reported from New Mexico. Within days, the CDC issued a press release advising people to not eat bagged fresh spinach, and a California company that produces several brands of bagged spinach announced a voluntary recall of all fresh spinach products. In the end, the implicated strain was found to have caused 205 cases of illness in 26 states, resulting in 3 deaths. ■ *E. coli* O157:H7, p. 600

While the sequence of events that led to the recognition of the *E. coli* O157:H7 outbreak may sound quite simple, they are actually very complex. For one thing, most strains of *E. coli* are normal inhabitants of the intestine and can be found in almost every sample of feces. How was this particular strain separated and distinguished from the hundreds of *E. coli* strains that do not cause diarrheal disease? It is also true that sporadic cases of *E. coli* O157:H7 infections occur regularly, originating from unrelated sources. How was it recognized that these cases were connected?

To identify *E. coli* O157:H7 in a stool specimen, the sample is inoculated onto a special agar medium designed to distinguish it from strains that typically inhabit the large intestine. One such medium is sorbitol-MacConkey, a modified version of MacConkey agar in which the lactose is replaced with the carbohydrate sorbitol. On this medium most *E. coli* O157:H7 isolates are colorless because they do not ferment sorbitol. In contrast, common strains of *E. coli* ferment the carbohydrate, giving rise to pink colonies. Serology is then used to determine if the colorless *E. coli* colonies are serotype O157; those that test positive are then generally tested to confirm they are serotype H7.

The next task is to determine whether or not two isolates of *E. coli* O157:H7 originated from the same source. DNA is extracted and purified from each isolate and is then digested with restriction enzymes. Pulsed-field gel electrophoresis is generally used to compare the resulting restriction fragment length polymorphism (RFLP) patterns of the isolates (see figure 10.9). Those that have identical patterns are presumed to have originated from the same source. The patients from whom those isolates originated can then be questioned to determine their likely point of contact with the disease-causing organism. Culture methods are then used to try to isolate the organism from the suspected food source. If that attempt is successful, the RFLP pattern of that isolate is then compared with those of the related cases.

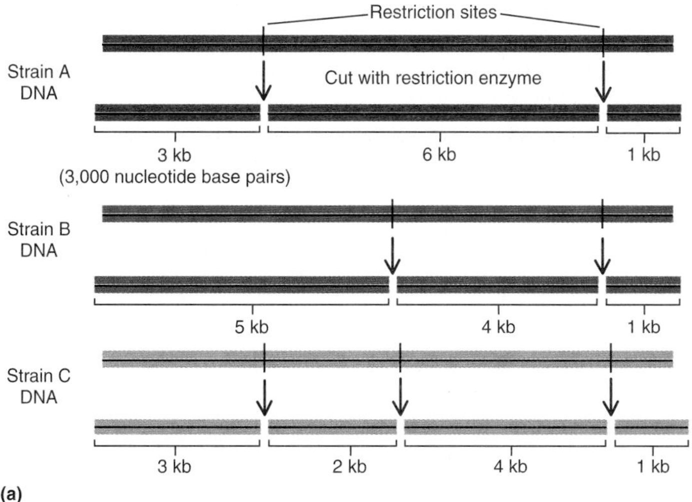

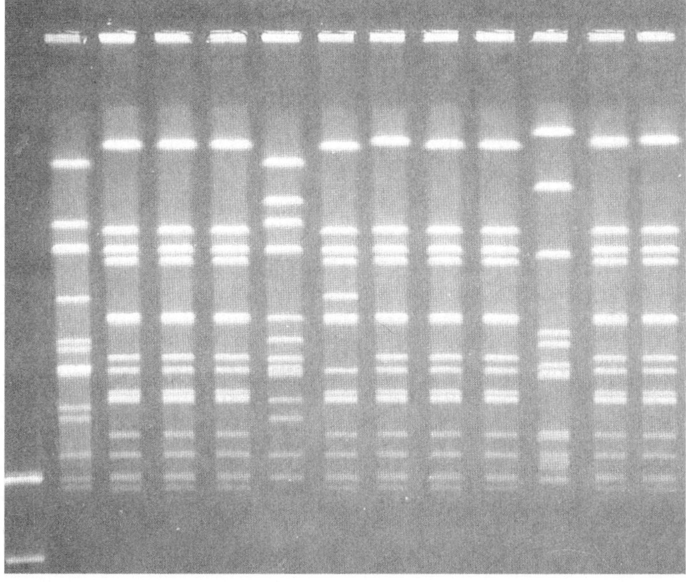

(b)

FIGURE 10.9 Restriction Fragment Length Polymorphisms (RFLPs) **(a)** Different strains of a species may have subtle variations in nucleotide sequences that give rise to a slightly different assortment of restriction fragment sizes. **(b)** In the method shown, genomic DNA is digested with a restriction enzyme that cuts infrequently; the resulting fragments are separated by pulsed-field gel electrophoresis and then stained with ethidium bromide.

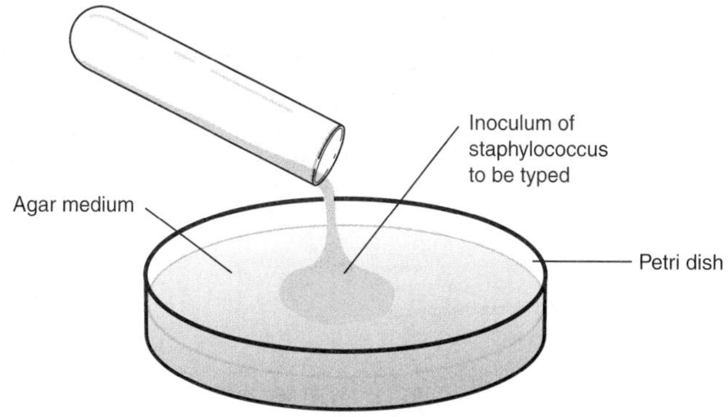

(a) An inoculum of *S. aureus* is spread over the surface of agar medium.

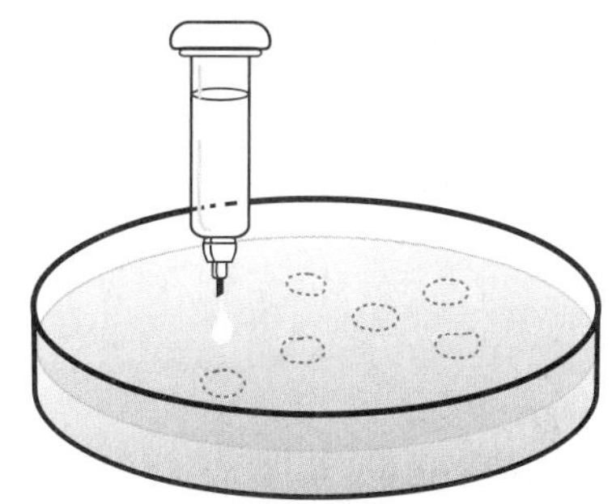

(b) 31 different bacteriophage suspensions are deposited in a fixed pattern.

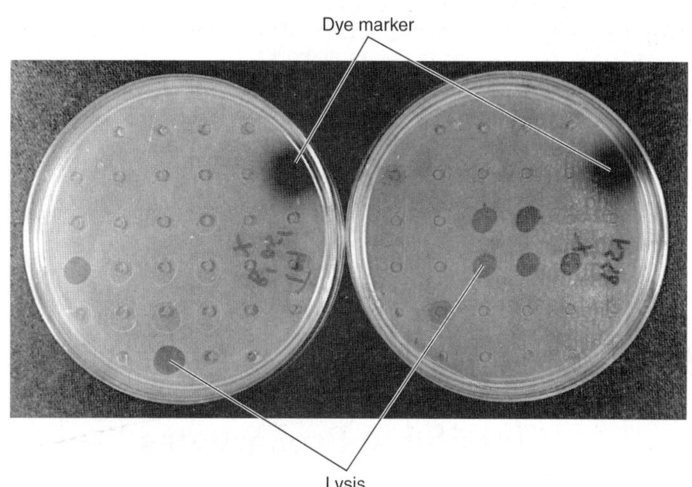

(c) After incubation, different patterns of lysis are seen with different strains of *S. aureus*.

FIGURE 10.10 Phage Typing

on the surface of the agar. During incubation, the bacteria multiply, forming a visible haze of cells. If the bacterial strain is susceptible to a specific type of phage, a clear area will form at the spot where bacteriophage was added. The patterns of clearing indicate the susceptibility of the test organism to different phages and it is these patterns that are compared to determine strain differences **(figure 10.10).** Bacteriophage typing has now largely been replaced by molecular methods that detect genomic differences, but it is still a useful tool for laboratories that lack equipment to do genomic testing. ■ **bacteriophage, p. 303** ■ **host range, p. 314**

Antibiograms

Antibiotic susceptibility patterns, or **antibiograms,** can distinguish different strains. As with phage typing, this method has now largely been replaced by molecular techniques. To determine the antibiogram, a culture is uniformly inoculated onto the surface of a nutritional agar medium. Paper discs, each of which has been impregnated with a given antibiotic, are then placed on the agar. During incubation, the organism will multiply to form a visible film of cells. A clear area, indicating lack of growth, will form around each antibiotic disc that inhibits the organism. Different

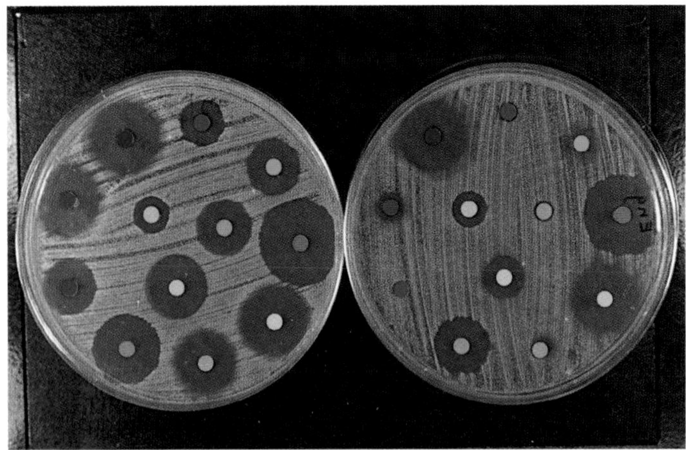

FIGURE 10.11 An Antibiogram In this example, 12 different antimicrobial drugs incorporated in paper discs have been placed on two plates containing different cultures of *Staphylococcus aureus*. Clear areas represent zones of inhibited growth. The different patterns of clearing indicate that these are two different strains of *S. aureus*.

strains will have different patterns of clearing **(figure 10.11).**

■ antibiotic, p. 470

MICROCHECK 10.4

Strains of a given species may differ in phenotypic attributes such as biochemical capabilities, protein and polysaccharide components, susceptibility to bacteriophages, and sensitivity to antimicrobial drugs. Molecular techniques can be used to detect genomic differences between strains that are phenotypically identical.

✓ Explain the difference between a biotype and a serotype.

✓ Describe the significance of RFLPs.

10.5 Classifying Prokaryotes

Focus Point

- Describe how 16S rRNA sequences, DNA hybridization, and DNA base ratios are used to classify prokaryotes.

As mentioned earlier in the chapter, the goal of phylogenetic classification is to categorize organisms according to their evolutionary relatedness. Unfortunately, this is difficult when trying to place the diverse types of prokaryotes in their proper positions with respect to the evolution of living beings. Fossilized **stromatolites,** coral-like mats of filamentous microorganisms, are available for study, but because of the relatively few sizes and shapes of prokaryotes, these remains do little to help identify or understand these ancient organisms.

Prokaryotic classification has historically relied on phenotypic attributes such as size and shape, staining characteristics, and metabolic capabilities to group organisms. Using this system, a species can loosely be defined as a group of organisms that share many properties and differ significantly from other groups. While this is a convenient approach to prokaryotic taxonomy, there are several drawbacks. For instance, observable differences may be due to only a few gene products, and a single mutation resulting in a non-functional enzyme can dramatically alter that phenotypic property. In addition, organisms that are phenotypically similar may, in fact, be distantly related. Conversely, those that appear dissimilar may be closely related.

Newer molecular techniques such as DNA sequencing circumvent some of the problems associated with phenotypic classification while also giving greater insights into the evolutionary relatedness of microorganisms. DNA sequences are viewed as **evolutionary chronometers,** meaning that sequence differences provide a relative measure of the time elapsed since the organisms diverged from a common ancestor. This is because random mutations cause sequences to change over time. Thus, the more time that has elapsed since two organisms diverged, the greater the differences in the sequences of their DNA.

DNA sequencing makes it possible to more accurately construct a **phylogenetic tree.** These trees are somewhat like a family tree, tracing the evolutionary heritage of organisms. Each line, or **branch,** represents the evolutionary distance between two species **(figure 10.12).** Individual species are represented as **nodes.** Ancient prokaryotes, those that branch at an early point in evolution, are sometimes called **deeply branching** to reflect their position in the phylogenetic tree.

While sequencing data has solved some difficulties in prokaryotic classification, it has also highlighted an important obstacle. Prokaryotic cells transfer DNA to other species, a process called **lateral (horizontal) gene transfer,** complicating insights provided by some types of DNA sequence comparison. The bacterium *Thermotoga maritima,* for example, appears to have acquired one-fourth of its genes from a hyperthermophilic archaeal species. Observations such as this have prompted some scientists to suggest that the tree of life (see figure 10.1) is more appropriately depicted as a shrub with interwoven branches **(figure 10.13).**

Table 10.7 summarizes some of the methods used to classify prokaryotes.

16S rDNA Sequence Analysis

Analyzing and comparing the nucleotide sequences of 16S ribosomal RNA (rRNA) and, more recently, the genes that encode rRNA (16S rDNA) has revolutionized the classification of organisms (see **Glimpse of History**). This is because rRNA is present in all organisms and performs a critical and functionally constant task. The number of mutations that can happen in certain regions without affecting the viability of an organism is limited. Long after organisms have diverged, the nucleotide sequences of portions of their 16S rDNA are still similar. Changes in these highly conserved regions occur very slowly over time and are thus useful for determining even distant relationships of diverse organisms. At the same time, certain regions are relatively variable. These sequences can be used to determine more recent divergence. In addition, horizontal gene transfer is unlikely to complicate the analysis because rDNA transfer appears to be rare.

Even the phylogeny of the multitude of prokaryotes that have not yet been grown in culture can be tentatively determined by 16S

Aquifex pyrophilus — Aquificae
Thermotoga maritima — Thermotogae
Deinococcus radiodurans, Thermus aquaticus — "Deinococcus-Thermus"
Chloroflexus aurantiacus — Chloroflexi
Corynebacterium glutamicum, Mycobacterium tuberculosis, Micrococcus luteus, Streptomyces griseus, Frankia sp. — Actinobacteria (High G + C Gram-positives)
Fusobacterium ulcerans — Fusobacteria
Staphylococcus aureus, Bacillus cereus, Enterococcus faecalis, Streptococcus pyogenes, Mycoplasma pneumoniae, Clostridium perfringens — Firmicutes (Low G + C Gram-positives)
Anabaena "cylindrica", Synechococcus lividus, Oscillatoria sp. — Cyanobacteria
Chlamydia trachomatis — Chlamydiae
Planctomyces maris — Planctomycetes
Chlorobium limicola — Chlorobi
Flexibacter litoralis, Cytophaga aurantiaca, Flavobacterium hydatis, Bacteroides fragilis — Bacteroidetes
Fibrobacter succinogenes — Fibrobacteres
Treponema pallidum, Borrelia burgdorferi — Spirochaetes
Campylobacter jejuni, Helicobacter pylori — ε-Proteobacteria
Desulfovibrio desulfuricans, Bdellovibrio bacteriovorus, Myxococcus xanthus — δ-Proteobacteria
Rickettsia rickettsii, Caulobacter crescentus, Rhodospirillum rubrum — α-Proteobacteria
Vibrio cholerae, Escherichia coli, Pseudomonas aeruginosa — γ-Proteobacteria
Neisseria gonorrhoeae, Alcaligenes denitrificans, Nitrosococcus mobilis — β-Proteobacteria
Proteobacteria

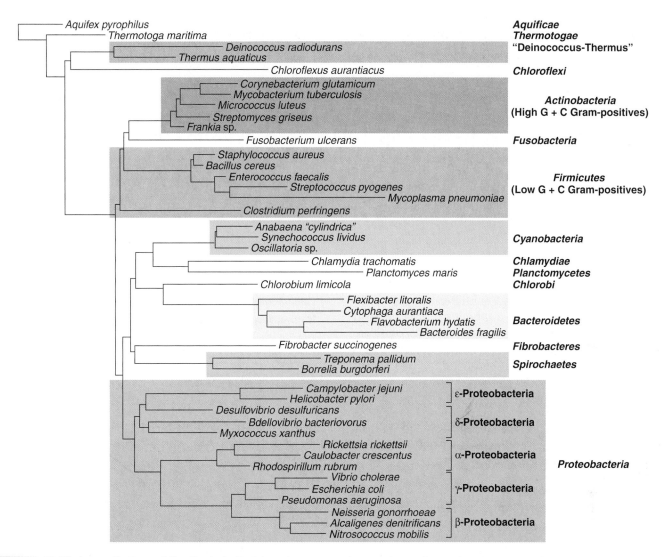

FIGURE 10.12 Phylogenetic Tree of the *Bacteria* Each branch represents the evolutionary distance between two species.
Source: The Ribosomal Database Project

rDNA sequence analysis. DNA can be extracted from environmental samples such as soil and water, and the 16S rDNA then amplified, cloned, and sequenced using techniques similar to those used to identify uncultivated organisms. The resulting sequences can be compared to databases containing 16S rDNA sequences of known organisms to assess the relatedness. The results obtained from such procedures have uncovered a variety of prokaryotes quite distinct from those already described. In fact, while most characterized archaeal genera are extremophiles, 16S rDNA sequences from soil and water samples indicate that members of this domain are common in non-extreme environments as well. Clearly, the prokaryotic world is much more diverse than previously recognized. ■ **Using rDNA to identify uncultivated organisms, p. 242**

While 16S rDNA sequence analysis has been instrumental in determining the phylogeny of distantly related organisms, it sometimes does not resolve differences at the species level. This is because closely related prokaryotes can have identical 16S rDNA sequences, even though the organisms are phenotypically distinct. In these cases, DNA hybridization (discussed next) is a better tool to assess relatedness.

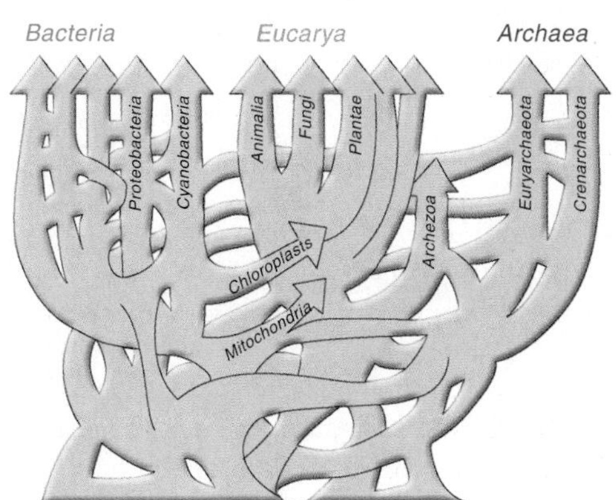

FIGURE 10.13 *"Shrub"* of Life

DNA Hybridization

The extent of nucleotide sequence similarity between two organisms can be determined by measuring how completely single strands of their DNA hybridize to one another. Just as the

TABLE 10.7 Methods Used to Determine the Relatedness of Different Prokaryotes for Purposes of Classification

Methods	Comments
Genotypic Characteristics	Differences in DNA sequences can be used to determine the point in time at which two organisms diverged from a common ancestor.
Comparing the sequences of 16S rDNA	This technique has revolutionized classification. Certain regions of the 16S rDNA can be used to determine distant relatedness of diverse organisms; other regions can be used to determine more recent divergence.
DNA hybridization	The extent of nucleotide sequence similarity between two isolates can be determined by measuring how completely single strands of their DNA hybridize to one another.
DNA base composition	Determining the G + C content offers a crude comparison of genomes. Organisms with identical G + C contents can be entirely unrelated, however.
Phenotypic Characteristics	Traditionally, relatedness of different bacteria has been decided by comparing properties such as ability to degrade lactose and the presence of flagella. These characteristics, however, do not necessarily reflect the evolutionary relatedness of organisms.

complementary strands of DNA from one organism will anneal (base-pair), so will homologous DNA of a different organism. The extent of hybridization reflects the degree of sequence similarity. Two strains that show at least 70% similarity are often considered to be members of the same species. Surprisingly, DNA hybridization studies have shown that members of the genera *Shigella* and *Escherichia,* which are quite different based on biochemical tests, should actually be grouped in the same species. Note that human and chimpanzee DNA have approximately 99% similarity by DNA hybridization studies. Therefore, by the criteria used to classify prokaryotes, humans and chimpanzees would be members of the same species! ■ **DNA hybridization, p. 227**

DNA Base Ratio (G + C Content)

One way to roughly compare the genomes of different bacteria is to determine their DNA base ratio, which is the relative portion of adenine (A), thymine (T), guanine (G), and cytosine (C). Because of base-pairing rules, the number of molecules of G in double-stranded DNA always equals the number of molecules of C. Likewise, the number of molecules of T equals that of A. The base ratio of an organism is usually expressed as the percent of guanine plus cytosine, termed the **G + C content,** or more commonly, the GC content. If the GC content of two organisms differs by more than a small percent, they cannot be closely related. A similarity of base compositions, however, does not necessarily mean that the organisms are related, however, since the nucleotide sequences and genome sizes could differ greatly.

The GC content is often measured by determining the temperature at which the double-stranded DNA denatures, or **melts.** DNA that has a high GC content melts at a higher temperature because G-C base pairs are held together by three hydrogen bonds, whereas A-T pairs are held together by only two hydrogen bonds. The temperature at which double-stranded DNA melts can readily be determined by monitoring the absorbance of UV light by a solution of DNA as it is heated. The absorbance rapidly increases as the DNA denatures **(figure 10.14).**

Phenotypic Methods

While 16S ribosomal nucleic acid sequences are playing an increasingly prominent role in prokaryotic classification, phenotypic methods are still important, particularly because they provide a foundation for prokaryotic identification. In addition, some taxonomists believe classification should be based on more than just genotypic traits.

Numerical taxonomy uses a quantitative approach to phenotypic classification, comparing a battery of characteristics. Information about this method can be obtained by visiting the Online Learning Center (www.mhhe.com/nester6).

MICROCHECK 10.5

Analyzing and comparing the sequences of ribosomal RNA genes has revolutionized the classification of organisms. Other characteristics used to classify prokaryotes include DNA hybridization, GC ratio, and phenotypic properties.

- ✓ Explain why 16S rDNA has proved so useful in determining evolutionary relatedness.
- ✓ Explain the role of DNA hybridization in classification.
- ✓ Why would it be easier to sequence rDNA than rRNA?

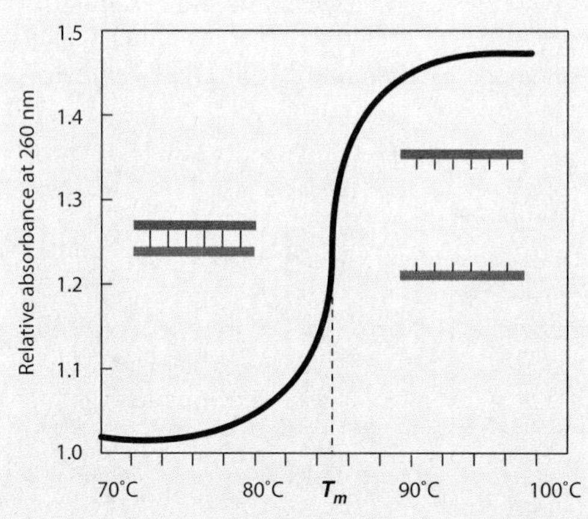

FIGURE 10.14 A DNA Melting Curve The absorbance (relative absorbance at 260 nm) rapidly increases as double-stranded DNA denatures, or melts. The T_m is the temperature at which 50% of the DNA has melted; it reflects the GC content.

FUTURE CHALLENGES

Tangled Branches in the Phylogenetic Tree

While 16S rDNA sequencing has given remarkable new insight into the evolutionary relatedness of organisms, total genome sequencing promises to reveal myriad additional information. In some cases this will serve to clarify the evolutionary picture, providing solid evidence that two organisms are, in fact, closely related. Genomic sequencing, however, can also give information that conflicts with that of rDNA data, creating an uncomfortable confusion. Already, examples exist that challenge the current rDNA phylogenetic tree. For example, sequences of some genes for highly conserved proteins show a closer relationship between some members of the *Archaea* and *Bacteria* than their rDNA sequences suggest. This conflicting data might be accounted for by evidence suggesting that horizontal DNA transfer has occurred even between distantly related organisms. Due to gene transfer, an organism from one domain may contain genes acquired from an organism in a different domain. As a result, some suggest that the phylogenetic tree may be closer to a shrub, with lateral gene transfer giving rise to intertwining branches (see figure 10.13).

The vast amount of genetic knowledge being obtained is affecting bacterial nomenclature. As more genomes are sequenced, the evolutionary relatedness among organisms will become apparent. Nomenclature is bound to change as a result of the new insights. Already, some species of what were formerly included in the genus *Streptococcus* have been removed to create two relatively new genera, *Enterococcus* and *Lactococcus*. Meanwhile, other bacteria will be moved from one existing genus to another. All of these changes will create a potential source of confusion for the scientist and layperson alike.

SUMMARY

10.1 Principles of Taxonomy

Taxonomy consists of three interrelated areas: **identification, classification,** and **nomenclature.**

Strategies Used to Identify Prokaryotes

To characterize and identify microorganisms, a wide assortment of technologies is used, including microscopic examination, cultural characteristics, biochemical tests, and nucleic acid analysis.

Strategies Used to Classify Prokaryotes

Taxonomic classification categories are arranged in a hierarchical order, with the **species** being the basic unit. Taxonomic categories include **species, genus, order, class, phylum** (or division), **kingdom,** and **domain.** Individual strains within a species vary in minor properties (table 10.1).

Nomenclature

Prokaryotes are assigned names governed by official rules.

10.2 Using Phenotypic Characteristics to Identify Prokaryotes (table 10.4)

Microscopic Morphology

The size, shape, and staining characteristics of a microorganism yield important clues as to its identity (figures 10.2, 10.3).

Metabolic Capabilities

The use of selective and differential media in the isolation process can provide information that helps identify an organism. Most biochemical tests rely on a pH indicator or chemical reaction that shows a color change when a compound is degraded. The basic strategy for identification using biochemical tests relies on the use of a **dichotomous key** (figures 10.4, 10.5, table 10.5).

Serology

Proteins and polysaccharides that make up a prokaryote's surface are sometimes characteristic enough to be identifying markers.

Fatty Acid Analysis (FAME)

Cellular fatty acid composition can be used as an identifying marker.

10.3 Using Genotypic Characteristics to Identify Prokaryotes

Detecting Specific Nucleotide Sequences using Nucleic Acid Probes

By selecting a probe complementary to a sequence unique to a given microbe, researchers can use nucleic acid hybridization to detect specific organisms (figure 10.7).

Amplifying Specific DNA Sequences Using the Polymerase Chain Reaction

By selecting primers that amplify a nucleotide sequence unique to a microbe of interest, researchers can apply PCR to determine if a particular agent is present.

Sequencing Ribosomal RNA Genes

The nucleotide sequence of ribosomal RNA (rRNA) can be used to identify prokaryotes. Newer techniques simply sequence rDNA, the DNA that encodes rRNA. Organisms that cannot yet be cultivated can be identified by amplifying, cloning, and then sequencing specific regions of rDNA.

10.4 Characterizing Strain Differences (table 10.6)

Biochemical Typing

A group of strains that has a characteristic biochemical variation is called a **biovar** or a **biotype.**

Serological Typing

A group of strains that differs serologically from other strains is called a **serovar** or a **serotype.**

Genomic Typing

Two isolates that have different **restriction fragment length polymorphisms (RFLPs)** are considered different strains (figure 10.9).

Phage Typing

The susceptibility to various types of bacteriophages can be used to demonstrate strain differences (figure 10.10).

Antibiograms

Antibiotic susceptibility patterns can be used to distinguish strains (figure 10.11).

10.5 Classifying Prokaryotes (table 10.7)

DNA sequencing enables one to more accurately construct a **phylogenetic tree** (figure 10.12). Horizontal (lateral) gene transfer can complicate insights provided by some types of DNA sequence comparison (figure 10.13).

16S rDNA Sequence Analysis

Analyzing and comparing the sequences of rRNA and more recently, rDNA, has revolutionized the classification of organisms.

REVIEW QUESTIONS

Short Answer

1. Name and describe each of the areas of taxonomy.
2. Compare and contrast the five-kingdom and three-domain systems of classification.
3. Describe how a dichotomous key is used in the identification of bacteria.
4. Describe the difference between using a probe and using PCR to detect a specific sequence.
5. Explain how signature sequences are used in bacterial identification.
6. Describe the function of PulseNet.
7. Describe how the GC content of DNA can be measured.
8. Explain why DNA sequences are viewed as evolutionary chronometers.
9. What is a phylogenetic tree?
10. Why is it preferable for a classification scheme to reflect the phylogeny of organisms?

Multiple Choice

1. Which of the following is the newest taxonomic unit?
   a) Strain b) Family c) Order
   d) Species e) Domain
2. An acid-fast stain can be used to detect which of the following organisms?
   a) *Cryptococcus neoformans* b) *Mycobacterium tuberculosis*
   c) *Neisseria gonorrhoeae* d) *Streptococcus pneumoniae*
   e) *Streptococcus pyogenes*
3. The "breath test" for *Helicobacter pylori* infection assays for the presence of which of the following?
   a) Antigens b) Catalase c) Hemolysis
   d) Lactose fermentation e) Urease
4. The "O" of *E. coli* O157:H7 refers to the
   a) biotype. b) serotype. c) phage type.
   d) ribotype. e) antibiogram.
5. PulseNet catalogs which of the following?
   a) Biotype b) Serotype c) Phage type
   d) RFLP e) Antibiogram
6. Which of the following is an example of an evolutionary chronometer?
   a) Ability to form endospores b) 16S ribosomal RNA sequence
   c) Sugar degradation d) Motility
7. If the GC content of two organisms is 70%, which of the following is true?
   a) The organisms are definitely related.
   b) The organisms are definitely not related.
   c) The AT content is 30%.
   d) The organisms likely have extensive DNA homology.
   e) The organisms likely have many characteristics in common.
8. Which of the molecular methods of assessing similarity gives the crudest approximation of relatedness?
   a) DNA hybridization b) PCR
   c) 16S rDNA sequencing d) DNA base composition
9. The sequence of which ribosomal genes are most commonly used for establishing phylogenetic relatedness?
   a) 5S b) 16S
   c) 23S d) All of these are commonly used.
10. Which of the following statements is false?
    a) *Tropheryma whipplei* could be identified before it had been grown in culture.
    b) The GC content of DNA can be measured by determining the temperature at which double-stranded DNA melts.
    c) Sequence differences between organisms can be used to assess their relatedness.
    d) Based on DNA homology studies, members of the genus *Shigella* should be in the same species as *Escherichia coli*.
    e) Gel electrophoresis is used to determine the serotype of an organism.

Applications

1. Microbiologists debate the use of biochemical similarities and cell features as a way of determining the taxonomic relationships among prokaryotes. Explain why some microbiologists believe these similarities and differences are a powerful taxonomic indicator while others think they are not very useful for that purpose.
2. A researcher interested in investigating the genetic relationship of mitochondria to bacteria must decide on the best method to study this. What advice would you give the researcher?

Critical Thinking

1. In figure 10.14, how would the curve appear if the GC content of the DNA sample were increased? How would the curve appear if the AT content were increased?
2. When DNA probes are used to detect specific sequence similarities in bacterial DNA, the probe is heated and the two strands of DNA are separated. Why must the probe DNA be heated?

Bacterial cells

The Diversity of Prokaryotic Organisms

A Glimpse of History

Cornelis B. van Niel (1897–1985) earned his Ph.D. from the Technological University in Delft, Holland, the home of an approach to the study of microbiology now commonly referred to as "the Delft School." The outstanding program there was chaired in succession by two prominent microbiologists—Martinus Beijerinck and Albert Kluyver.

As Kluyver's student, van Niel was heavily influenced by his mentor's belief that biochemical processes were fundamentally the same in all cells and that microorganisms could be important research tools, serving as a model to study biochemical process. Thirty years later, Kluyver and van Niel presented lectures that would be published in a book entitled *The Microbe's Contribution to Biology*.

Shortly after completing his dissertation in 1928, van Niel accepted a position at the Hopkins Marine Station in California. There, he continued work he started under Kluyver's direction on the photosynthetic activities of the vividly colored purple bacteria. He conclusively demonstrated that these organisms require light for growth, yet, unlike plants and algae, do not evolve O_2. He also showed that in order to fix CO_2, the purple bacteria oxidize hydrogen sulfide. Furthermore, van Niel noted that the stoichiometry of the photosynthetic reactions in all photosynthetic organisms was remarkably similar, except that the purple bacteria use hydrogen sulfide in place of water, and produce oxidized sulfur compounds instead of O_2. This finding raised the possibility that O_2 generated by plants and algae did not come from carbon dioxide, as was believed at the time, but rather from water.

In addition to his scientific contributions, van Niel was an outstanding teacher. During the summers at Hopkins Marine Station, he taught a bacteriology course, inspiring many microbiologists with his enthusiasm for the diversity of microorganisms and their importance in nature. His keen memory and knowledge of the literature, along with his appreciation for the remarkable abilities of microorganisms, enabled him to successfully impart the awe and wonder of the microbial world to his students.

Scientists are only beginning to understand the vast diversity of microbial life. Although a million species of prokaryotes are thought to exist, only approximately 6,000 of these, grouped into over 950 genera, have been actually described and classified. Traditional culture and isolation techniques have not supported the growth, and subsequent study, of the vast majority. This situation is changing as new molecular techniques aid in the discovery and characterization of previously unrecognized species. The sheer volume of the rapidly accruing information made possible by this modern technology, however, can be daunting for scientists and students alike.

This chapter covers a wide spectrum of prokaryotes, focusing primarily on their extraordinary diversity rather than concentrating on the phylogenetic relationships discussed in chapter 10. Note, however, that no single chapter could describe all known prokaryotes and, consequently, only a relatively small selection is presented.

KEY TERMS

Anoxygenic Phototrophs Photosynthetic organisms that do not produce O_2.

Chemolithotroph An organism that harvests energy by oxidizing inorganic chemicals.

Chemoorganotroph An organism that harvests energy by oxidizing organic chemicals.

Chemotroph An organism that harvests energy by oxidizing chemicals.

Cyanobacteria Gram-negative oxygenic phototrophs; genetically related to chloroplasts.

Lactic Acid Bacteria Gram-positive bacteria that generate lactic acid as a major end product of their fermentative metabolism.

Methanogens Archaea that obtain energy by oxidizing hydrogen gas, using CO_2 as a terminal electron acceptor, thereby generating methane.

Myxobacteria Gram-negative bacteria that congregate to form complex structures called fruiting bodies.

Nitrifiers Gram-negative bacteria that obtain energy by oxidizing inorganic nitrogen compounds such as ammonia or nitrate.

Oxygenic Phototrophs Photosynthetic organisms that produce O_2.

Prosthecate Bacteria Gram-negative bacteria that have extensions projecting from the cells, thereby increasing their surface area.

Spirochetes Long helical bacteria that have flexible cell walls and axial filaments.

Sulfur-Oxidizing Bacteria Gram-negative bacteria that obtain energy by oxidizing elemental sulfur and reduced sulfur compounds, thereby generating sulfuric acid.

METABOLIC DIVERSITY

As a group, prokaryotes use an impressive array of mechanisms to harvest energy in order to produce ATP. This section will highlight this metabolic diversity by describing select prokaryotes. **Table 11.1** summarizes characteristics of the archaea and bacteria covered in this section.

TABLE 11.1 Metabolic Diversity of Prokaryotes

Group/Genera	Characteristics	Phylum
Anaerobic Chemolithotrophs		
Methanogens – *Methanospirillum, Methanosarcina*	*Archaea* that oxidize hydrogen gas, using CO_2 as a terminal electron acceptor to generate methane.	*Euryarchaeota*
Anaerobic Chemoorganotrophs—Anaerobic Respiration		
Sulfur- and sulfate-reducing bacteria – *Desulfovibrio*	Use sulfate as a terminal electron acceptor, generating hydrogen sulfide. Found in anaerobic muds rich in organic material. Gram-negative.	*Proteobacteria*
Anaerobic Chemoorganotrophs—Fermentation		
Clostridium	Endospore-forming obligate anaerobes. Inhabitants of soil. Gram-positive.	*Firmicutes*
Lactic acid bacteria – *Streptococcus, Enterococcus, Lactococcus, Lactobacillus, Leuconostoc*	Produce lactic acid as the major end product of their fermentative metabolism. Aerotolerant anaerobes. Several genera are exploited by the food industry. Gram-positive.	*Firmicutes*
Propionibacterium	Obligate anaerobes that produce propionic acid as their primary fermentation end product. Used in the production of Swiss cheese. Gram-positive.	*Actinobacteria*
Anoxygenic Phototrophs		
Purple sulfur bacteria – *Chromatium, Thiospirillum, Thiodictyon*	Grow in colored masses in sulfur-rich aquatic habitats, using sulfur compounds as a source of electrons when making reducing power. Gram-negative.	*Proteobacteria*
Purple non-sulfur bacteria – *Rhodobacter, Rhodopseudomonas*	Grow in aquatic habitats, preferentially using organic compounds as a source of electrons for reducing power. Many are metabolically versatile. Gram-negative.	*Proteobacteria*
Green sulfur bacteria – *Chlorobium, Pelodictyon*	Found in habitats similar to those preferred by the purple sulfur bacteria. Gram-negative.	*Chlorobi*
Filamentous anoxygenic phototrophic bacteria – *Chloroflexus*	Characterized by their filamentous growth. Gram-negative.	*Chloroflexi*
Others – *Heliobacterium*	Have not been studied extensively.	*Firmicutes*
Oxygenic Phototrophs—Cyanobacteria		
Anabaena, Synechococcus	Important primary producers. Some fix N_2. Gram-negative.	*Cyanobacteria*

TABLE 11.1 Metabolic Diversity of Prokaryotes (*Continued*)

Group/Genera	Characteristics	Phylum
Aerobic Chemolithotrophs		
Filamentous sulfur oxidizers – *Beggiatoa, Thiothrix*	Oxidize sulfur compounds as an energy source. Found in sulfur springs, and sewage-polluted waters, Gram-negative.	*Proteobacteria*
Unicellular sulfur oxidizers – *Thiobacillus, Acidithiobacillus*	Oxidize sulfur compounds as an energy source. Some species produce enough acid to lower the pH to 1.0. Gram-negative.	*Proteobacteria*
Nitrifiers – *Nitrosomonas, Nitrosococcus, Nitrobacter, Nitrococcus*	Oxidize ammonia or nitrate as an energy source. In so doing, they convert certain fertilizers to a form readily leached from soils, and deplete O_2 in waters polluted with ammonia containing wastes. Genera that oxidize nitrite prevent the toxic buildup of this compound in soils. Gram-negative.	*Proteobacteria*
Hydrogen-oxidizing bacteria – *Aquifex, Hydrogenobacter*	Thermophilic bacteria that oxidize hydrogen gas as an energy source. One of the earliest bacterial forms to exist on earth.	*Aquifacae*
Aerobic Chemoorganotrophs—Obligate Aerobes		
Micrococcus	Widely distributed; common laboratory contaminants. Gram-positive.	*Actinobacteria*
Mycobacterium	Waxy cell wall resists staining; acid-fast.	*Actinobacteria*
Pseudomonas	Common environmental bacteria that, as a group, can degrade a wide variety of compounds. Gram-negative.	*Proteobacteria*
Thermus	*Thermus aquaticus* is the source of *Taq* polymerase, the heat-resistant polymerase used in PCR. Stains Gram-negative.	*Deinococcus-Thermus*
Deinococcus	Resistant to the damaging effects of gamma radiation. Stains Gram-positive.	*Deinococcus-Thermus*
Aerobic Chemoorganotrophs—Facultative Anaerobes		
Corynebacterium	Widespread in nature. Gram-positive.	*Actinobacteria*
The *Enterobacteriaceae* – *Escherichia, Enterobacter, Klebsiella, Proteus, Salmonella, Shigella, Yersinia*	Most reside in the intestinal tract. Those that ferment lactose are coliforms; their presence in water serves as an indicator of fecal pollution. Gram-negative.	*Proteobacteria*

Anaerobic Chemotrophs

Focus Point

- Compare and contrast the metabolism and habitats of methanogens, sulfur- and sulfate-reducing bacteria, *Clostridium* species, lactic acid bacteria, and *Propionibacterium* species.

For approximately the first 1.5 billion years that prokaryotes inhabited earth, the atmosphere was **anoxic,** or devoid of O_2. In that anaerobic environment, some early **chemotrophs** (organisms that harvest energy by oxidizing chemicals) probably used a pathway of anaerobic respiration, employing terminal electron acceptors such as carbon dioxide or elemental sulfur, which were plentiful in the environment. Others may have used fermentation, passing the electrons to an organic molecule such as pyruvate. ■ chemotrophs, p. 94 ■ anaerobic respiration, pp. 133, 145 ■ terminal electron acceptor, p. 131 ■ fermentation, pp. 133, 147

Today, anaerobic habitats still abound. Mud and tightly packed soil limit the diffusion of gases, and any O_2 that does penetrate is rapidly consumed by aerobically respiring chemotrophs. This creates anaerobic conditions just below the surface. Aquatic environments may also become anaerobic if they contain nutrients that foster the rapid growth of O_2-consuming microbes. This is evident in polluted lakes, where fish may die because of a lack of dissolved O_2. The bodies of humans and other animals also provide numerous anaerobic environments. It is estimated that 99% of the bacteria that inhabit the intestinal tract are obligate anaerobes. Even the skin and the oral cavity, which are routinely exposed to O_2, have anaerobic microenvironments. These are created via the localized depletion of O_2 by aerobes. ■ obligate anaerobes, p. 92

Anaerobic Chemolithotrophs

Chemolithotrophs oxidize reduced inorganic chemicals such as hydrogen gas (H_2) to obtain energy. Those that grow anaerobically obviously cannot use O_2 as a terminal electron acceptor and instead must employ an alternative such as carbon dioxide or sulfur. Relatively few anaerobic chemolithotrophs have been discovered, and most are members of the Domain *Archaea*. Some bacterial examples that inhabit aquatic environments will be discussed later. ■ chemolithotroph, p. 150

The Methanogens

The **methanogens** are a group of archaea that generate ATP by oxidizing hydrogen gas, using CO_2 as a terminal electron acceptor. This process generates methane (CH_4), a colorless, odorless, flammable gas:

$4\ H_2$	+	CO_2	⟶	$CH_4 + 2\ H_2O$
(energy source)		(terminal electron acceptor)		

Many methanogens can also use alternative energy sources such as formate; some can use methanol or acetate as well. Representative genera of methanogens include *Methanospirillum* and *Methanosarcina* **(figure 11.1).**

Methanogens are found in anaerobic environments where hydrogen gas and carbon dioxide are available. Because these gases are generated by bacteria that ferment organic material, methanogens often grow in association with these microbes. Methanogens, however, are generally not found in environments containing high levels of sulfate, nitrate, or other inorganic electron acceptors. This is because microorganisms that use these electron acceptors to oxidize H_2 have a competitive advantage; the use of CO_2 as an electron acceptor releases comparatively little energy (see figure 6.23). Environments from which methanogens are commonly isolated include sewage, swamps, marine sediments, rice paddies, and the digestive tracts of humans and other animals. The methane produced can present itself as bubbles rising in swamp waters or the 10 cubic feet of gas discharged from a cow's digestive system each day. As a by-product of sewage treatment plants, methane gas can be collected and used for heating, cooking, and even the generation of electricity (see Perspective 31.1).

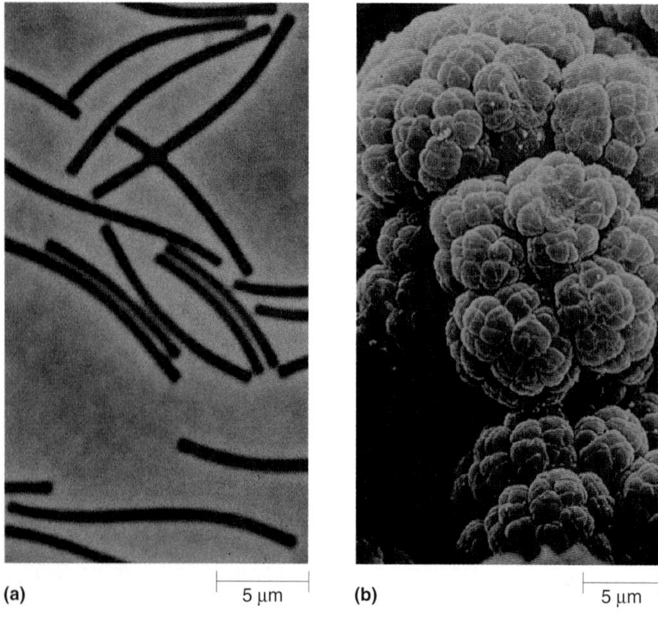

FIGURE 11.1 Methanogens (a) Phase-contrast micrograph of a *Methanospirillum* species. **(b)** Scanning electron micrograph of a *Methanosarcina* species.

The study of methanogens is quite challenging because they are exquisitely sensitive to O_2, as are many of their enzymes. Special techniques, including anaerobe jars and anaerobic chambers, are used for their cultivation. ■ culturing anaerobes, p. 98

Anaerobic Chemoorganotrophs—Anaerobic Respiration

Chemoorganotrophs oxidize organic compounds such as glucose to obtain energy. Those that grow anaerobically often employ sulfur or sulfate as a terminal electron acceptor. ■ chemoorganotrophs, p. 95

Sulfur- and Sulfate-Reducing Bacteria

When sulfur compounds are used as terminal electron acceptors, they become reduced to form hydrogen sulfide, the compound responsible for the rotten-egg smell of many anaerobic environments. An example of this reaction is:

organic compounds	+	S	⟶	$CO_2 + H_2S$
(energy source)		(terminal electron acceptor)		

In addition to the unpleasant odor, the H_2S is a problem to industry because it reacts with metals, corroding pipes and other structures. Ecologically, however, prokaryotes that reduce sulfur compounds are an indispensable component of the sulfur cycle. ■ sulfur cycle, p. 731

Sulfate- and sulfur-reducing bacteria generally live in mud rich in organic material and oxidized sulfur compounds. The H_2S they produce causes mud and water to turn black when it reacts with iron molecules. At least a dozen genera are recognized in this group, the most extensively studied of which are species of *Desulfovibrio*. These are Gram-negative curved rods.

Some representatives of the *Archaea* also use sulfur compounds as terminal electron acceptors, but the characterized examples generally do not inhabit the same environments as their bacterial counterparts. While most of the sulfur-reducing bacteria are mesophiles or thermophiles, the known sulfur-reducing archaea are hyperthermophiles, inhabiting extreme environments such as hydrothermal vents. They will be discussed later in the chapter. ■ thermophiles, p. 91 ■ hyperthermophiles, p. 91

Anaerobic Chemoorganotrophs—Fermentation

Numerous types of anaerobic bacteria obtain energy using the process of fermentation, producing ATP only by substrate-level phosphorylation. There are many variations of fermentation, using different organic energy sources and producing characteristic end products, but one example is:

glucose	⟶	pyruvate	⟶	lactic acid
(energy source)		(terminal electron acceptor)		

The Genus *Clostridium*

Members of the genus *Clostridium* are Gram-positive rods that can form endospores (see figure 3.45). They are common inhabitants of soil, where the vegetative cells live in the anaerobic microenvironments created when aerobic organisms consume available O_2. Their endospores are indifferent to O_2 and can survive for long

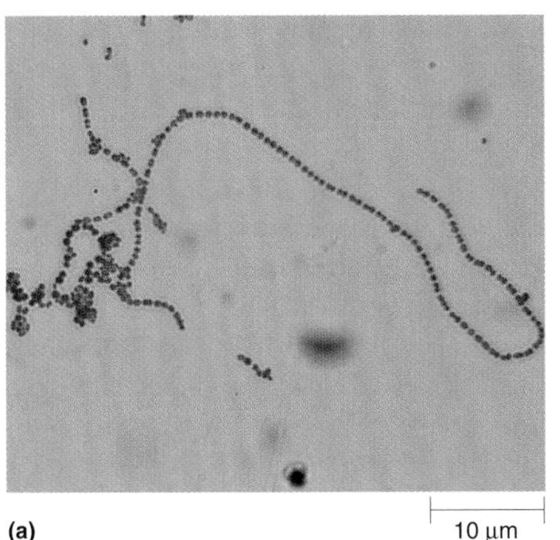

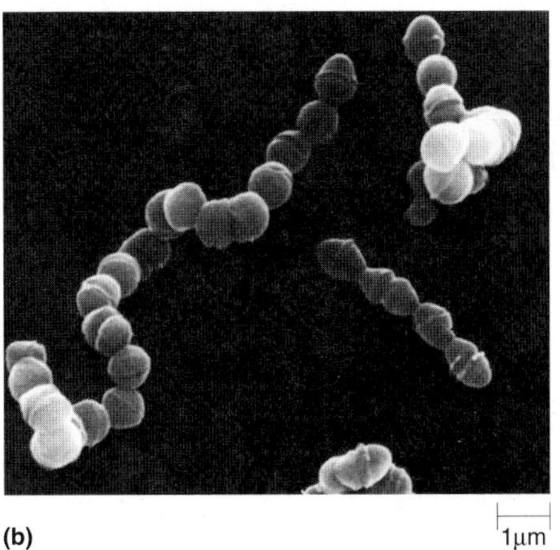

FIGURE 11.2 Streptococcus Species
(a) Gram stain. **(b)** Scanning electron micrograph.

periods by withstanding measures of heat, desiccation, chemicals, and irradiation that would kill vegetative bacteria. When the appropriate conditions are renewed, these endospores germinate, and the resulting vegetative bacteria multiply. Vegetative cells that arise from soil-borne endospores are responsible for a variety of diseases, including tetanus (caused by *C. tetani*), gas gangrene (caused by *C. perfringens*), and botulism (caused by *C. botulinum*). Some species of *Clostridium* are normal inhabitants of the intestinal tract of humans and other animals. ■ **endospores, p. 69** ■ **tetanus, p. 566** ■ **gas gangrene, p. 568** ■ **botulism, p. 657**

As a group, *Clostridium* species ferment a wide variety of compounds, including sugars and cellulose. Some of the end products are commercially valuable; for example, *C. acetobutylicum* produces acetone and butanol. Some species can ferment amino acids by an unusual process that oxidizes one amino acid, using another as a terminal electron acceptor. This generates a variety of foul-smelling end products associated with rotting flesh.

The Lactic Acid Bacteria

Gram-positive bacteria that produce lactic acid as a major end product of their fermentative metabolism make up a group called the **lactic acid bacteria.** They include members of the genera *Streptococcus, Enterococcus, Lactococcus, Lactobacillus,* and *Leuconostoc.* Most can grow in aerobic environments, but all are obligate fermenters and thus derive no benefit from O_2. They can be readily distinguished from other bacteria that grow in the presence of O_2 because they lack the enzyme catalase (see figure 10.4). ■ **obligate fermenters, pp. 92, 147** ■ **catalase, p. 92**

Streptococcus species are cocci that typically grow in chains of varying lengths **(figure 11.2).** They inhabit the oral cavity, generally as part of the normal microbiota. Some, however, are notable for their adverse effect on human health. One of the most important is *S. pyogenes* (Group A strep), which causes pharyngitis (strep throat) and other diseases. Unlike the streptococci that typically inhabit the throat, *S. pyogenes* is β-hemolytic, an important characteristic used in its identification (see figure 4.11). ■ ***Streptococcus pyogenes,* p. 499** ■ **hemolysis, p. 97**

Species of *Lactococcus* and *Enterococcus* were at one time included in the genus *Streptococcus.* The genus *Lactococcus* now includes species used by the dairy industry to produce fermented milk products such as cheese and yogurt. *Enterococcus* species typically inhabit the intestinal tract of humans and other animals. ■ **fermentation of dairy products, p. 756**

Members of the genus *Lactobacillus* are rod-shaped bacteria that grow as single cells or loosely associated chains. They are common members of the microbiota in the mouth and the healthy human vagina during child-bearing years. In the vagina, they metabolize glycogen, which has been deposited in the vaginal lining in response to the female sex hormone, estrogen. The resulting low pH helps the vagina resist infection. Lactobacilli are also often present in decomposing plant material, milk, and other dairy products. Like the lactococci, they are important in the production of fermented foods **(figure 11.3).**

The Genus Propionibacterium

Propionibacterium species are Gram-positive pleomorphic (irregular-shaped) rods that produce propionic acid as their primary fermentation end product. Significantly, they can also ferment lactic acid. Thus, *Propionibacterium* species can extract residual energy from a waste product of another group of bacteria.

Propionibacterium species are important to the dairy industry because their fermentation end products play an indispensable role

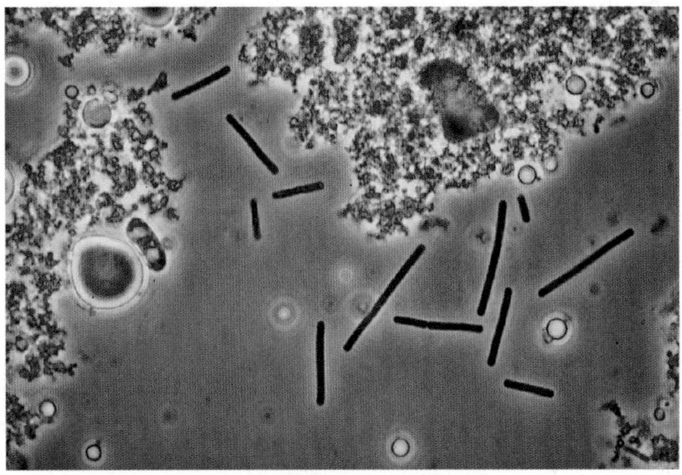

FIGURE 11.3 ***Lactobacillus* Species from Yogurt**

in the production of Swiss cheese. The propionic acid is responsible for the typical nutty flavor of the cheese. The CO_2, also a product of the fermentation, creates the signature holes in the cheese. *Propionibacterium* species are also found growing in the intestinal tract and in anaerobic microenvironments on the skin. ■ Swiss cheese production, p. 757

MICROCHECK 11.1

Methanogens are archaea that oxidize hydrogen gas, using CO_2 as a terminal electron acceptor to generate methane. The sulfur and sulfate-reducing bacteria oxidize organic compounds, with sulfur or sulfate serving as a terminal electron acceptor to generate hydrogen sulfide. *Clostridium* species, the lactic acid bacteria, and *Propionibacterium* species oxidize organic compounds, with an organic compound serving as a terminal electron acceptor.

- ✓ What metabolic process creates the rotten-egg smell characteristic of many anaerobic environments?
- ✓ Describe two beneficial contributions of the lactic acid bacteria.
- ✓ Relatively little is known about many obligate anaerobes. Why would this be so?

11.2 Anoxygenic Phototrophs

Focus Point

- Compare and contrast the metabolism and habitats of the purple bacteria and the green bacteria.

The earliest photosynthesizing organisms were likely **anoxygenic phototrophs.** These use hydrogen sulfide or organic compounds rather than water as a source of electrons when making reducing power in the form of NADPH, and therefore do not generate O_2. For example:

$$\underset{\text{(carbon source)}}{6\,CO_2} + \underset{\text{(electron source)}}{12\,H_2S} \longrightarrow C_6H_{12}O_6 + 12\,S + 6\,H_2O$$

Modern-day anoxygenic phototrophs are a phylogenetically diverse group of bacteria that inhabit a restricted ecological niche that provides adequate light penetration yet little or no O_2; most often, they are found in aquatic habitats such as bogs, lakes, and the upper layer of muds. ■ reducing power, p. 131

As discussed in chapter 6, the photosynthetic systems of the anoxygenic phototrophs are fundamentally different from those of plants, algae, and cyanobacteria. They have a unique type of chlorophyll called **bacteriochlorophyll.** This and their other light-harvesting pigments absorb wavelengths of light that penetrate to greater depths and are not used by other photosynthetic organisms. ■ photosynthetic pigments, p. 152

The Purple Bacteria

The **purple bacteria** are Gram-negative organisms that appear red, orange, or purple due to their light-harvesting pigments. Unlike other anoxygenic phototrophs, the components of their photosynthetic apparatus are all contained within the cytoplasmic membrane. Invaginations in this membrane effectively increase the surface area available for the photosynthetic processes.

Purple Sulfur Bacteria

Purple sulfur bacteria can sometimes be seen growing as colored masses in sulfur-rich aquatic habitats such as sulfur springs **(figure 11.4a).** The cells are relatively large, sometimes in excess of 5 μm in diameter, and some are motile by flagella. They may also have gas vesicles, enabling them to move up or down to their preferred level in the water column. Most accumulate sulfur in granules that are readily visible microscopically and appear to be contained within the cell (figure 11.4b). ■ gas vesicles, p. 68

The purple sulfur bacteria preferentially use hydrogen sulfide to generate reducing power, although some species can use other inorganic molecules, (such as H_2) or organic compounds, (such as pyruvate). Many are strict anaerobes and phototrophs, but some can grow in absence of light aerobically, oxidizing reduced inorganic or organic compounds as a source of energy. Representative genera of purple sulfur bacteria include *Chromatium, Thiospirillum,* and *Thiodictyon.*

(a)

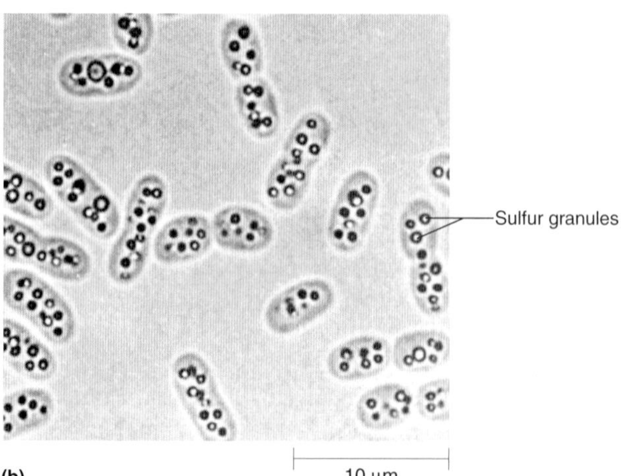

(b)

FIGURE 11.4 Purple Sulfur Bacteria (a) Photograph of bacteria growing in a bog. **(b)** Photomicrograph showing intracellular sulfur granules.

Purple Non-Sulfur Bacteria

The purple non-sulfur bacteria are found in a wide variety of aquatic habitats, including moist soils, bogs, and paddy fields. One important characteristic that distinguishes them from the purple sulfur bacteria is that they preferentially use a variety of organic molecules rather than hydrogen sulfide as a source of electrons for reducing power. In addition, they lack gas vesicles. If sulfur does accumulate, the granules form outside of the cell.

Purple non-sulfur bacteria are remarkably versatile metabolically. Not only do they grow as phototrophs using organic molecules to generate reducing power, but many can use a metabolism similar to the purple sulfur bacteria, employing hydrogen gas or hydrogen sulfide as an electron source. In addition, most can grow aerobically in the absence of light using chemotrophic metabolism. Representative genera of purple sulfur bacteria include *Rhodobacter* and *Rhodopseudomonas.*

The Green Bacteria

The **green bacteria** are Gram-negative organisms that are typically green or brownish in color.

Green Sulfur Bacteria

Green sulfur bacteria are found in habitats similar to those preferred by the purple sulfur bacteria. Like the purple sulfur bacteria, they use hydrogen sulfide as a source of electrons for reducing power and form sulfur granules. The granules, however, form outside of the cell **(figure 11.5).** The accessory pigments of the green sulfur bacteria are located in structures called **chlorosomes.** The bacteria lack flagella, but many have gas vesicles. All are strict anaerobes, and none can use a chemotrophic metabolism. Representative genera include *Chlorobium* and *Pelodictyon.* ■ **accessory pigments, p. 152**

Filamentous Anoxygenic Phototrophic Bacteria

Filamentous anoxygenic phototrophic bacteria from multicellular arrangements and exhibit gliding motility. The most thoroughly studied of this group are members of the genus *Chloroflexus*, particularly the thermophilic strains that grow in hot springs. Many of the filamentous anoxygenic phototrophs have chlorosomes, which initially led scientists to believe they were related to the green sulfur bacteria. Their 16S rDNA sequences indicate otherwise. As a group, filamentous anoxygenic phototrophs are diverse metabolically. Some preferentially use organic compounds to generate reducing power and can also grow in the dark aerobically using chemotrophic metabolism.

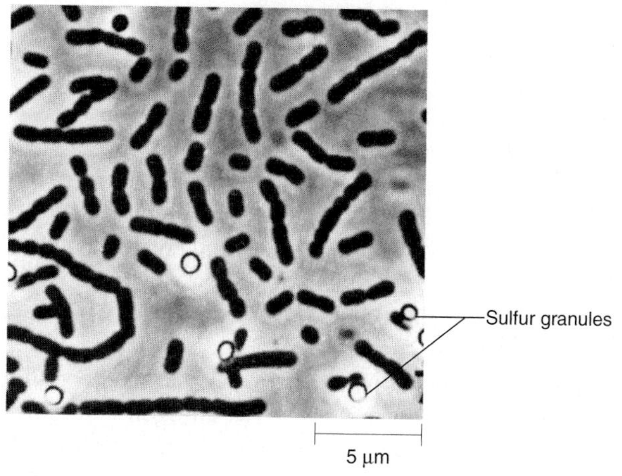

FIGURE 11.5 Green Sulfur Bacteria Note that the sulfur granules are extracellular.

Other Anoxygenic Phototrophs

While the green and purple bacteria have been studied most extensively, other types of anoxygenic phototrophs exist. Among these are members of the genus *Heliobacterium,* Gram-positive endospore-forming rods related to members of the genus *Clostridium.*

MICROCHECK 11.2

Anoxygenic phototrophs harvest the energy of sunlight, but do not generate O_2. The purple sulfur bacteria and green sulfur bacteria use hydrogen sulfide as a source of electrons to generate reducing power; the purple non-sulfur bacteria and many of the filamentous anoxygenic phototrophs preferentially use organic compounds.

- ✓ Describe a structural characteristic that distinguishes the purple sulfur bacteria from the green sulfur bacteria.
- ✓ What is the function of gas vesicles?
- ✓ Why is it beneficial to the anoxygenic phototrophs to have light-harvesting pigments that absorb wavelengths of light that penetrate to greater depths?

11.3

Oxygenic Phototrophs

Focus Point

- Describe the cyanobacteria, including how nitrogen-fixing species protect their nitrogenase enzyme from O_2.

Nearly 3 billion years ago, the earth's atmosphere began changing as O_2 was gradually introduced to the previously anoxic environment. This change was probably due to the evolution of the cyanobacteria, thought to be the earliest **oxygenic phototrophs.** These photosynthetic organisms use water as a source of electrons for reducing power, liberating O_2:

$$\underset{\text{(carbon source)}}{6\,CO_2} + \underset{\text{(electron source)}}{6\,H_2O} \longrightarrow C_6H_{12}O_6 + 6\,O_2$$

Today, cyanobacteria still play an essential role in the biosphere. As **primary producers,** they harvest the energy of sunlight, using it to convert CO_2 into organic compounds. They were initially thought to be algae and were called blue-green algae until electron microscopy revealed their prokaryotic structure. ■ **primary producers, p. 722**

The Cyanobacteria

The **cyanobacteria** are a diverse group of more than 60 genera of Gram-negative bacteria. They inhabit a wide range of environments, including freshwater and marine habitats, soils, and the surfaces of rocks. In addition to being photosynthetic, many are able to convert nitrogen gas (N_2) to ammonia, which can then be incorporated into cell material. This process, called **nitrogen fixation,** is an exclusive ability of prokaryotes. ■ nitrogen fixation, p. 730

General Characteristics of Cyanobacteria

Cyanobacteria are morphologically diverse. Some are unicellular, with typical prokaryotic shapes such as cocci, rods, and spirals. Others form filamentous multicellular associations called **trichomes** that may or may not be enclosed within a sheath, a tube that holds and surrounds a chain of cells **(figure 11.6).** Motile trichomes glide as a unit. Cyanobacteria that inhabit aquatic environments often have gas vesicles, enabling them to move vertically within the water column. When large numbers of cyanobacteria accumulate in stagnant lakes or other freshwater habitats, it is called a **bloom (figure 11.7).** In the bright, hot conditions of summer, the bouyant cells lyse and decay, creating an odoriferous scum. The ecological effects of these blooms on aquatic habitats are discussed in chapter 30. ■ aquatic habitats, p. 725

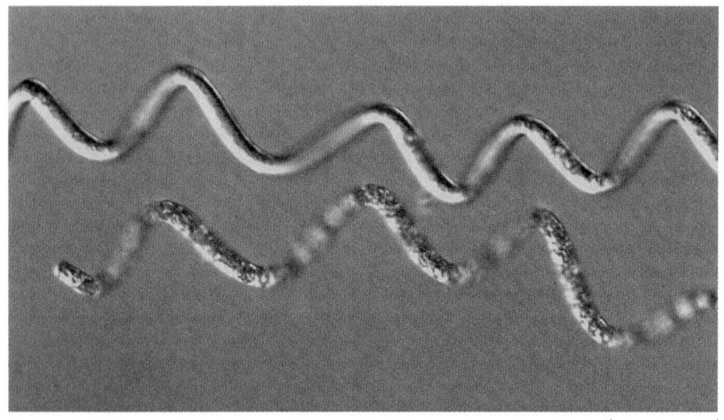

FIGURE 11.6 Cyanobacteria (a) The spiral trichome of *Spirulina* species. **(b)** Differential interference contrast photomicrograph of a species of *Oscillatoria*. Note the arrangement of the individual cells in the trichome.

FIGURE 11.7 Cyanobacterial Bloom Excessive growth of cyanobacteria is evidenced by buoyant masses of cells that have risen to the surface.

The photosynthetic systems of the cyanobacteria are like those contained within the chloroplasts of algae and plants. This is not surprising in light of the genetic evidence indicating chloroplasts evolved from a species of cyanobacteria that once resided as an endosymbiont within eukaryotic cells (see Perspective 3.1). In addition to light-harvesting chlorophyll pigments, cyanobacteria have **phycobiliproteins.** These pigments absorb energy from wavelengths of light not well absorbed by chlorophyll. They contribute to the blue-green, or sometimes reddish, color of the cyanobacteria.

Nitrogen-Fixing Cyanobacteria

Nitrogen-fixing cyanobacteria are critically important ecologically. Because they can incorporate both N_2 and CO_2 into organic material, they generate a form of these nutrients that can then be used by other organisms. Thus, their activities can ultimately support the growth of a wide range of organisms in environments that would otherwise be devoid of usable nitrogen and carbon. As an example, nitrogen-fixing cyanobacteria inhabiting the oceans are essential primary producers that support other sea life. Also, like all cyanobacteria, they help limit atmospheric carbon dioxide buildup by utilizing the gas as a carbon source.

Nitrogenase, the enzyme complex that mediates the process of nitrogen fixation, is destroyed by O_2; therefore, nitrogen-fixing cyanobacteria must protect the enzyme from the O_2 they generate. Species of *Anabaena,* which are filamentous, isolate nitrogenase by confining the process of nitrogen fixation to a specialized thick-walled cell called a **heterocyst (figure 11.8).** Heterocysts lack photosystem II and consequently, do not generate O_2. The heterocysts of some species form at very regular intervals within the filament, reflecting the ability of cells within a trichome to communicate. One species of *Anabaena, A. azollae,* forms an intimate relationship with the water fern *Azolla.* The bacterium grows and fixes nitrogen within the protected environment of a special sac in the fern, providing *Azolla* with a source of available nitrogen. *Synechococcus* species fix nitrogen only in the dark. Consequently, nitrogen fixation and photosynthesis are temporally separated. ■ photosystem II, p. 153

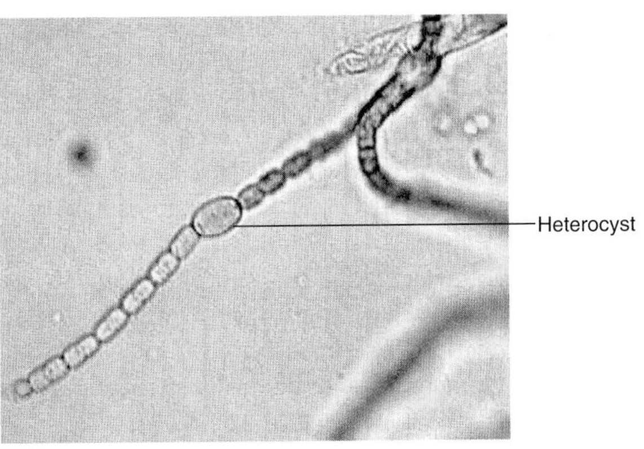

FIGURE 11.8 Heterocyst of an *Anabaena* Species Nitrogen fixation occurs within these specialized cells.

Other Notable Characteristics of Cyanobacteria

Cyanobacteria have various other notable characteristics—some beneficial, others damaging. Filamentous cyanobacteria appear to be responsible for maintaining the structure and productivity of soils in cold desert areas such as the Colorado Plateau. Their sheaths persist in soil, creating a sticky fibrous network that prevents erosion. In addition, these bacteria provide an important source of nitrogen and organic carbon in otherwise nutrient-poor soils. On the negative side, cyanobacteria growing in freshwater lakes and reservoirs used as a source of drinking water can impart an undesirable taste to the water. This is due to their production of a compound called **geosmin,** which has a distinctive "earthy" odor. Some aquatic species such as *Microcystis aeruginosa* can produce toxins. These can be deadly to animals that consume them.

MICROCHECK 11.3

The photosynthetic systems of cyanobacteria generate O_2 and are similar to those of algae and plants. Many cyanobacteria species can fix nitrogen.

✓ What is the function of a heterocyst?

✓ How do cyanobacteria prevent erosion in cold desert regions?

✓ How could heavily fertilized lawns foster the development of cyanobacterial blooms?

11.4 Aerobic Chemolithotrophs

Focus Point

- Compare and contrast the metabolism and habitats of sulfur-oxidizing bacteria, nitrifiers, and hydrogen-oxidizing bacteria.

Aerobic chemolithotrophs obtain energy by oxidizing reduced inorganic chemicals, using O_2 as a terminal electron acceptor.

The Sulfur-Oxidizing Bacteria

The **sulfur-oxidizing bacteria** are Gram-negative rods or spirals, which sometimes grow in filaments. They obtain energy by oxidizing elemental sulfur and reduced sulfur compounds, including hydrogen sulfide and thiosulfate. Molecular oxygen serves as a terminal electron acceptor, generating sulfuric acid.

An example of this reaction is:

$$\underset{\text{(energy source)}}{S} + \underset{\text{(terminal electron acceptor)}}{1\tfrac{1}{2}\,O_2} + H_2O \longrightarrow H_2SO_4$$

These bacteria play an important role in the sulfur cycle. ■ sulfur cycle, p. 731

Filamentous Sulfur Oxidizers

Species of the filamentous sulfur oxidizers *Beggiatoa* and *Thiothrix* live in sulfur springs, in sewage-polluted waters, and on the surface of marine and freshwater sediments. They accumulate sulfur, depositing it as intracellular granules. Members of the genera *Beggiatoa* and *Thiothrix* differ in the nature of their filamentous growth **(figure 11.9).** The filaments of *Beggiatoa* species move

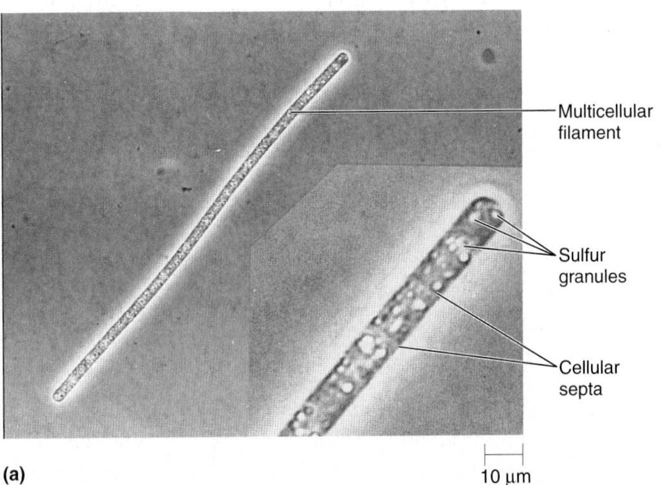

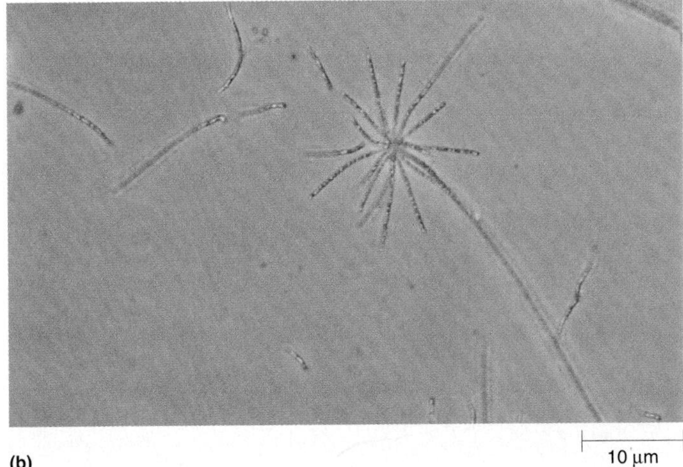

FIGURE 11.9 Filamentous Sulfur Bacteria Phase-contrast photomicrographs. **(a)** Multicellular filament of a *Beggiatoa* species; the septa separate the cells **(b)** Multicellular filaments of a *Thiothrix* species, forming a rosette arrangement.

by gliding motility, a mechanism that does not require flagella. The filaments may flex or twist to form a tuft. In contrast, the filaments of *Thiothrix* species are immobile; they fasten at one end to rocks or other solid surfaces. Often they attach to other cells, causing the filaments to form characteristic rosette arrangements. Progeny cells detach from the ends of these filaments and use gliding motility to disperse to new locations, where they form new filaments. Overgrowth of these filamentous organisms in sewage at treatment facilities causes a problem called **bulking.** Because the masses of filamentous organisms do not settle easily, bulking interferes with the separation of the solid sludge from the liquid effluent. ■ sewage treatment, p. 739

Unicellular Sulfur Oxidizers

Acidithiobacillus species (formerly included in the genus *Thiobacillus*) are found in both terrestrial and aquatic habitats, where their ability to oxidize metal sulfides is responsible for a process called **bioleaching.** In this process, insoluble metal sulfides are oxidized, producing sulfuric acid while converting the metal to a soluble form; some species produce enough acid to lower the pH to 1.0. Bioleaching can cause severe environmental problems. For example, the strip mining of coal exposes metal sulfides, which can then be oxidized by *Acidithiobacillus* species to produce sulfuric acid. The resulting runoff can acidify nearby streams, killing trees, fish, and other wildlife **(figure 11.10).** The runoff may also contain toxic metals made soluble by the bacteria. Under controlled conditions, however, bioleaching can enhance the recovery of metals (see Perspective 6.1). For example, gold can be extracted from deposits of gold sulfide. The metabolic activities of *Acidithiobacillus* species can also be used to prevent acid rain, which results from the burning of sulfur-containing coals and oils. The sulfur can be removed from the fuels by allowing the bacteria to oxidize it to sulfate, which can then be extracted.

The Nitrifiers

Nitrifiers are a diverse group of Gram-negative bacteria that obtain energy by oxidizing inorganic nitrogen compounds such as ammonia or nitrite. These bacteria are of particular interest to farmers who fertilize their crops with ammonium nitrogen, a form of nitrogen retained by soils because its positive charge enables it to adhere to negatively charged soil particles. The potency and longevity of the fertilizer are affected by nitrifying bacteria converting the ammonia to nitrate. While plants use the latter form of nitrogen more readily, it is rapidly leached from soils. Nitrifying bacteria are also an important consideration in disposal of sewage or other wastes with a high ammonia concentration. As nitrifying bacteria oxidize nitrogen compounds, they consume O_2. Because of this, waters polluted with nitrogen-containing wastes can quickly become hypoxic (low in dissolved O_2).

FIGURE 11.10 Acid Drainage from a Mine Sulfur-oxidizing bacteria oxidize exposed metal sulfides, generating sulfuric acid. The yellow-red color is due to insoluble iron oxides.

The nitrifiers encompass two metabolically distinct groups of bacteria that typically grow in close association. Together, they can oxidize ammonia to form nitrate. The **ammonia oxidizers,** which include the genera *Nitrosomonas* and *Nitrosococcus,* convert ammonia to nitrite in the following reaction:

$$\underset{\text{(energy source)}}{NH_4^+} + \underset{\text{(terminal electron acceptor)}}{1\tfrac{1}{2}O_2} \longrightarrow NO_2^- + H_2O + 2\,H^+$$

The **nitrite oxidizers,** which include the genera *Nitrobacter* and *Nitrococcus,* then convert nitrite to nitrate as follows:

$$\underset{\text{(energy source)}}{NO_2^-} + \underset{\text{(terminal electron acceptor)}}{\tfrac{1}{2}O_2} \longrightarrow NO_3^-$$

The latter group is particularly important in preventing the buildup of nitrite in soils, which is toxic and can leach into groundwater. The oxidation of ammonia to nitrate is called **nitrification** and is an important part of the nitrogen cycle. ■ nitrogen cycle, p. 730

The Hydrogen-Oxidizing Bacteria

Members of the Gram-negative genera *Aquifex* and *Hydrogenobacter* are among the few hydrogen-oxidizing bacteria that are obligate chemolithotrophs. An example of the reaction in their metabolism is:

$$\underset{\text{(energy source)}}{H_2} + \underset{\text{(terminal electron acceptor)}}{\tfrac{1}{2}O_2} \longrightarrow H_2O$$

These related organisms are thermophilic and typically inhabit hot springs. Some *Aquifex* species have a maximum growth temperature of 95°C, the highest of any bacteria. The hydrogen-oxidizing bacteria are deeply branching in the phylogenetic tree, meaning that according to 16S rRNA studies, they were one of the earliest bacterial forms to exist on earth. The fact that they require O_2 seems contradictory to their evolutionary position, but in fact, the low amount they require might have been available early on in certain niches due to photochemical processes that split water.

A wide range of aerobic chemoorganotrophs can also oxidize hydrogen gas. These organisms switch between energy sources as conditions dictate.

MICROCHECK 11.4

Sulfur oxidizers use sulfur compounds as an energy source, generating sulfuric acid. The nitrifiers oxidize nitrogen compounds such as ammonium or nitrite. Hydrogen-oxidizing bacteria oxidize H_2.

- ✓ How does the growth of *Acidithiobacillus* species cause bioleaching?
- ✓ Why would farmers be concerned about nitrifying bacteria?
- ✓ Why would sulfur-oxidizing bacteria accumulate sulfur?

11.5

Aerobic Chemoorganotrophs

Focus Points

- Compare and contrast the metabolism and habitats of two obligate aerobes and one facultative anaerobe.
- Describe members of the family *Enterobacteriaceae,* and what distinguishes coliforms from other members of this family.

Aerobic chemoorganotrophs oxidize organic compounds to obtain energy, using O_2 as a terminal electron acceptor:

$$\underset{\text{(energy source)}}{\text{organic compounds}} + \underset{\text{(terminal electron acceptor)}}{O_2} \longrightarrow CO_2 + H_2O$$

They include a tremendous variety of bacteria, ranging from some that inhabit very specific environments to others that are ubiquitous. This section will profile only representative genera found in variety of different environments. Later sections will describe examples that thrive in specific habitats.

Obligate Aerobes

Obligate aerobes obtain energy using respiration exclusively; none can use fermentation as an alternative.

The Genus *Micrococcus*

Members of the genus *Micrococcus* are Gram-positive cocci found in soil and on dust particles, inanimate objects, and skin. Because they are often airborne, they easily contaminate bacteriological media. There, they typically form pigmented colonies, a characteristic that aids in their identification. The colonies of *M. luteus,* for example, are generally yellow **(figure 11.11).** Like members of the genus *Staphylococcus,* which will be discussed later, they tolerate arid conditions and can grow in salty environments such as 7.5% NaCl.

The Genus *Mycobacterium*

Mycobacterium species are widespread in nature and include harmless **saprophytes,** which live on dead and decaying matter, as well as species that produce disease in humans and domestic animals. They stain poorly because of a waxy lipid in their unusual cell wall, but special procedures can be used to enhance the penetration of certain dyes. Once stained, they resist destaining, even when acidic decolorizing solutions are used. Because of this, *Mycobacterium* species are called **acid-fast,** and the acid-fast staining procedure is an important step in their identification (see figure 3.15). *Nocardia* species, a related group of bacteria that commonly reside in the soil, are also acid-fast. ■ acid-fast, p. 50

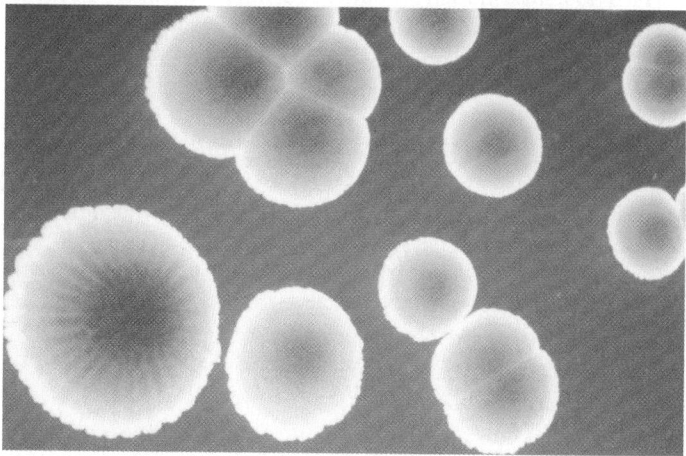

FIGURE 11.11 *Micrococcus luteus* Colonies

Mycobacterium species are generally pleomorphic rods; they often occur in chains that sometimes branch, or bunch together to form cordlike groups. Several species are notable for their effect on human health, including *M. tuberculosis,* which causes tuberculosis, and *M. leprae,* which causes Hansen's disease (leprosy). *Mycobacterium* species are more resistant to disinfectants than most other vegetative bacteria. In addition, they differ from other bacteria in their susceptibility to antimicrobial drugs.

The Genus *Pseudomonas*

Pseudomonas species are Gram-negative rods that are motile by polar flagella and often produce pigments **(figure 11.12).** Although most are strict aerobes, some can grow anaerobically if nitrate is available as a terminal electron acceptor. The fact that they do not ferment and are oxidase positive serve, in the laboratory, as important characteristics that distinguish them from *E. coli* and other members of the family *Enterobacteriaceae,* which will be discussed in a later section. ■ oxidase test, p. 144

As a group, *Pseudomonas* species have extremely diverse biochemical capabilities. Some can metabolize more than 80 different substrates, including unusual sugars, amino acids, and compounds containing aromatic rings. Because of this, *Pseudomonas* species play an important role in the degradation of many synthetic and natural compounds that resist breakdown by most other microorganisms. The ability to carry out some of these degradations is encoded by plasmids.

Pseudomonas species are widespread, typically inhabiting soil and water. While most are harmless, some cause disease in plants and animals. Medically, the most significant species is *P. aeruginosa.* It is a common **opportunistic pathogen,** meaning that it primarily infects people who have underlying medical conditions. Unfortunately, it can grow in nutrient-poor environments,

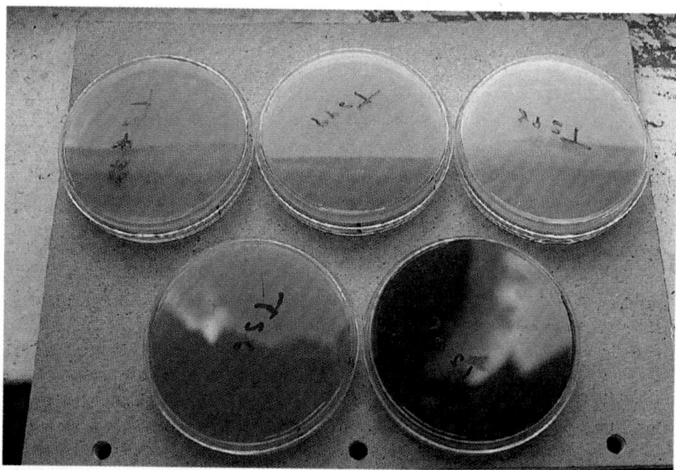

FIGURE 11.12 Pigments of *Pseudomonas* Species Cultures of different strains of *Pseudomonas aeruginosa.* Note the different colors of the water-soluble pigments.

such as water used in respirators, and is resistant to many disinfectants and antimicrobial medications. Because of this, hospitals must be diligent to prevent it from infecting patients. ■ *Pseudomonas aeruginosa*, p. 564 ■ opportunistic pathogen, p. 394

The Genera *Thermus* and *Deinococcus*

Thermus and *Deinococcus* are related genera that have scientifically and commercially noteworthy characteristics. *Thermus* species are thermophilic, as their name implies. This trait has proven to be extremely valuable because of their heat-stable enzymes; an integral part of the polymerase chain reaction (PCR) is ***Taq* polymerase,** the DNA polymerase of *T. aquaticus. Thermus species* have an unusual cell wall, and stain Gram-negative. *Deinococcus* species are unique in their extraordinary resistance to the damaging effects of gamma radiation. For example, *D. radiodurans* can survive exposure to a dose several thousand times that lethal to a human being. The dose literally shatters the organism's genome into many fragments, yet enzymes are able to repair the extensive damage. Scientists anticipate that through genetic engineering, *Deinococcus* species may eventually help clean up the soil and water contaminated by the 10 million cubic yards of radioactive waste that have accumulated in the United States. Their unusual cell wall has multiple layers, and they stain Gram-positive. ■ PCR, p. 223

Facultative Anaerobes

Facultative anaerobes preferentially use aerobic respiration if O_2 is available. As an alternative, however, they can use fermentative metabolism.

The Genus *Corynebacterium*

Members of the genus *Corynebacterium* commonly inhabit soil, water, and the surface of plants. They are Gram-positive pleomorphic rods, often club-shaped and arranged to form V shapes or palisades (*koryne* is Greek for "club") **(figure 11.13).** Bacteria that exhibit this characteristic microscopic morphology are sometimes referred to as **coryneforms** or **diphtheroids.** *Corynebacterium* species are generally facultative anaerobes, although some are strict aerobes. Many species of *Corynebacterium* reside harmlessly in the throat, but toxin-producing strains of *C. diphtheriae* can cause the disease diphtheria. ■ diphtheria, p. 502

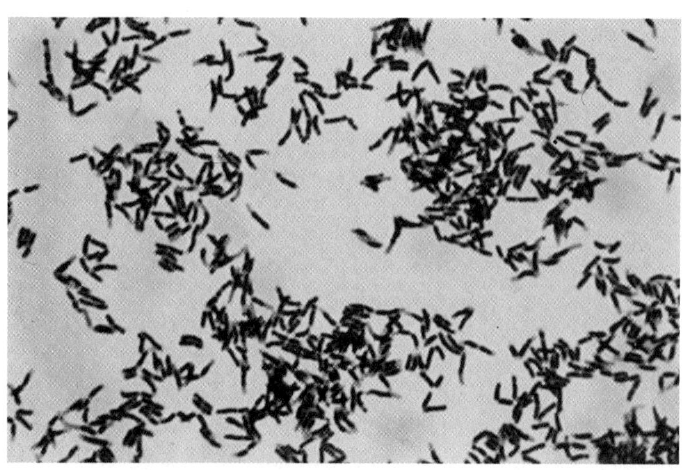

FIGURE 11.13 *Corynebacterium* The Gram-positive pleomorphic rods are often arranged to form V shapes or palisades.

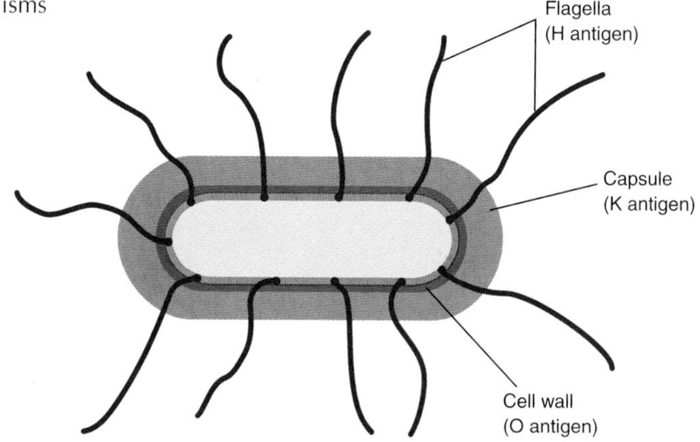

FIGURE 11.14 Schematic Drawing of a Member of the Family *Enterobacteriaceae* The cell structures used to distinguish different strains are shown.

The Family *Enterobacteriaceae*

Members of the family *Enterobacteriaceae,* frequently referred to as **enterics** or **enterobacteria,** are Gram-negative rods. Their name reflects the fact that most reside in the intestinal tract of humans and other animals (Greek *enteron* means "intestine"), although some thrive in rich soil. Enterics that are part of the normal microbiota of the intestine include *Enterobacter, Klebsiella,* and *Proteus* species as well as most strains of *E. coli.* Those that cause diarrheal disease include *Shigella* species, strains of *Salmonella enterica,* and some strains of *E. coli.* Life-threatening systemic diseases include typhoid fever, caused by *Salmonella* Typhi, and both the bubonic and pneumonic forms of plague, caused by *Yersinia pestis.* ■ diarrheal disease, p. 595 ■ typhoid fever, p. 601 ■ plague, p. 682

Members of the family *Enterobacteriaceae* are facultative anaerobes that ferment glucose and, if motile, generally have peritrichous flagella. The family includes over 40 recognized genera that can be distinguished using biochemical tests. Within a given species, many different strains have been described. These are often distinguished using serological tests that detect differences in cell walls, flagella, and capsules **(figure 11.14).** ■ peritrichous flagella, p. 66

Enteric bacteria that characteristically ferment lactose are included in a group called **coliforms.** This is an informal grouping of certain common intestinal inhabitants such as *E. coli* that are easy to detect in food and water; for years regulatory agencies have considered them to be an indicator of fecal pollution. Their presence indicates a possible health risk because fecal-borne pathogens might also be present. ■ coliform, p. 746

MICROCHECK 11.5

Members of the genera *Micrococcus, Mycobacterium, Pseudomonas, Thermus,* and *Deinococcus* are obligate aerobes that generate ATP by degrading organic compounds, using O_2 as a terminal electron acceptor. Most *Corynebacterium* species and all members of the family *Enterobacteriaceae* are facultative anaerobes.

- ✓ What is the significance of finding coliforms in drinking water?
- ✓ What unique characteristic makes members of the genus *Deinococcus* noteworthy?
- ✓ Why would it be an advantage for a *Pseudomonas* species to encode enzymes for degrading certain compounds on a plasmid rather than the chromosome?

ECOPHYSIOLOGICAL DIVERSITY

As a group, prokaryotes show remarkable diversity in their physiological adaptations to a wide range of habitats. From the hydrothermal vents of deep oceans to the frozen expanses of Antarctica, prokaryotes have evolved to thrive in virtually all environments, including many that most plants and animals would find inhospitable.

This section will highlight the physiological mechanisms prokaryotes use to thrive in terrestrial and aquatic environments; the study of these adaptations is called **ecophysiology.** It will also describe some examples of bacteria that use animals as habitats. **Table 11.2** summarizes characteristics of the bacteria covered in this section.

11.6 Thriving in Terrestrial Environments

Focus Point

- Describe two mechanisms that terrestrial bacteria use to thrive in the ever-changing environment of soil, and list two genera that use each mechanism.

TABLE 11.2 Ecophysiological Diversity

Group/Genera	Characteristics	Phylum
Thriving in Terrestrial Environments		
Endospore-formers – *Bacillus, Clostridium*	*Bacillus* species include both obligate aerobes and facultative anaerobes; *Clostridium* species are obligate anaerobes. Gram-positive.	*Firmicutes*
Azotobacter	Form a resting stage called a cyst. Notable for their ability to fix nitrogen in aerobic conditions. Gram-negative.	*Proteobacteria*
Myxobacteria – *Chondromyces, Myxococcus, Stigmatella*	Congregate to form fruiting bodies; cells within these differentiate to form dormant microcysts. Gram-negative.	*Proteobacteria*
Streptomyces	Naturally produce a wide array of medically useful antibiotics. Gram-positive.	*Actinobacteria*
Agrobacterium	Cause plant tumors. Scientists use their Ti plasmid to introduce desired genes into plant cells. Gram-negative.	*Proteobacteria*
Rhizobia – *Rhizobium, Sinorhizobium, Bradyrhizobium, Mesorhizobium, Azorhizobium*	Fix nitrogen; form a symbiotic relationship with legumes. Gram-negative.	*Proteobacteria*
Thriving in Aquatic Environments		
Sheathed bacteria – *Sphaerotilus, Leptothrix*	Form chains of cells enclosed within a protective sheath. Swarmer cells move to new locations. Gram-negative.	*Proteobacteria*
Prosthecate bacteria – *Caulobacter, Hyphomicrobium*	Appendages increase their surface area. Gram-negative.	*Proteobacteria*
Bdellovibrio	Predator of other bacteria. Gram-negative.	*Proteobacteria*
Bioluminescent bacteria – *Photobacterium, Vibrio fischeri*	Some form a symbolic relationship with specific types of squid and fish. Gram-negative.	*Proteobacteria*
Legionella	Often reside within protozoa. Gram-negative.	*Proteobacteria*
Epulopiscium	Exceptionally large cigar-shaped bacteria that multiply by increasing in size and then releasing several daughter cells; each cell has thousands of copies of the genome. Gram-positive.	*Firmicates*
Free-living spirochetes – *Spirochaeta, Leptospira* (some species)	Long spiral-shaped bacteria that move by means of an axial filament. Gram-negative.	*Spirochaetes*
Magnetospirillum	Contain a string of magnetic crystals that enable them to move up or down in water and sediments. Gram-negative.	*Proteobacteria*
Spirillum	Spiral-shaped, microaerophilic bacteria. Gram-negative.	*Proteobacteria*
Sulfur-oxidizing, nitrate-reducing marine bacteria – *Thioploca, Thiomargarita*	Use novel mechanisms to compensate for the fact that their energy source (reduced sulfur compounds) and terminal electron acceptor (nitrate) do not coexist.	*Proteobacteria*
Animals as Habitats – See table 11.3		

Microorganisms that inhabit soil must endure a variety of conditions. Daily and seasonally, soil is an environment that can routinely alternate between wet and dry as well as warm and cold. The availability of nutrients can also cycle from abundant to sparse. To thrive in this ever-changing environment, microbes have evolved mechanisms to cope with adverse conditions and to exploit unique sources of nutrients.

Bacteria That Form a Resting Stage

Several genera that inhabit the soil can form a resting stage that enables them to survive the dry periods typical in many soils. Of these various types of dormant cells, endospores are by far the most resistant to environmental extremes.

Endospore-Formers

Bacillus and *Clostridium* species are the most common Gram-positive rod-shaped bacteria that form endospores; the position of the spore in the cell can be used as an aid in identification **(figure 11.15).** *Clostridium* species, which are obligate anaerobes, were discussed earlier. *Bacillus* species include both obligate aerobes and facultative anaerobes, and some are medically important. *Bacillus anthracis* causes the disease anthrax, which can be acquired from contacting its endospores in soil or in animal hides or wool. Unfortunately, the spores have also been used as an agent of domestic bioterrorism. ■ endospores, p. 69 ■ anthrax, p. 512

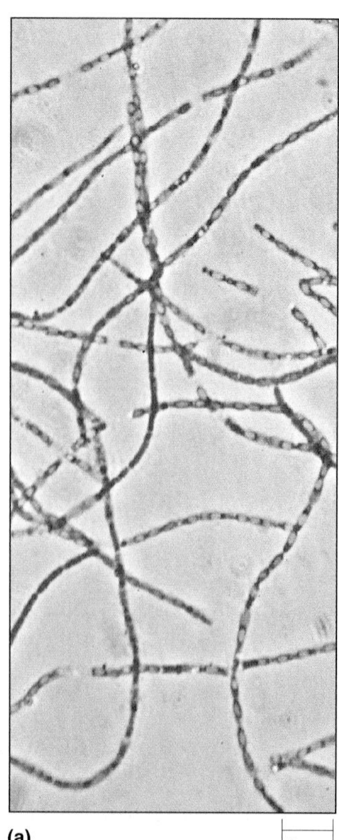

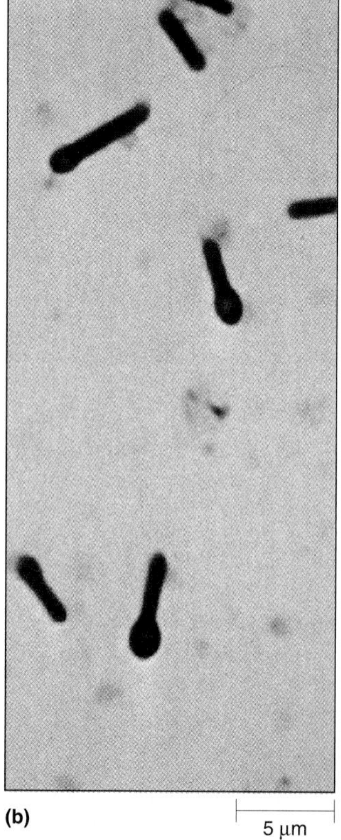

FIGURE 11.15 Endospore-Formers (a) Endospores forming in the mid-portion of the cells of *Bacillus anthracis.* **(b)** Endospores forming at the ends of the cells in *Clostridium tetani.* Both of these species can cause fatal disease, but many other species of endospore-formers are harmless.

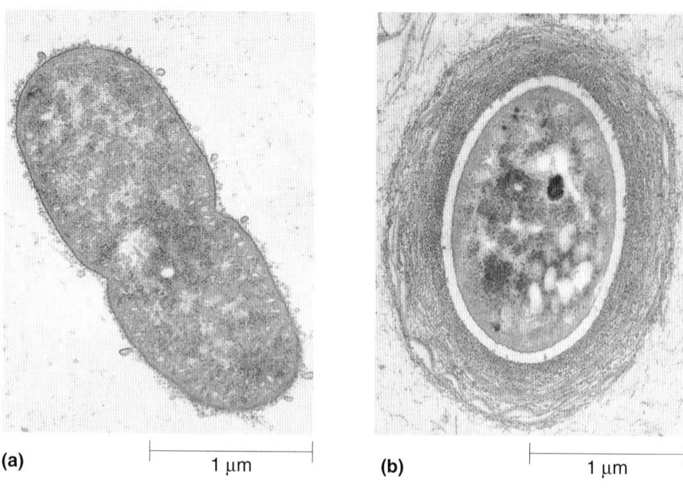

FIGURE 11.16 *Azotobacter* (a) Vegetative cells; **(b)** cyst.

The Genus Azotobacter

Azotobacter species are Gram-negative pleomorphic, rod-shaped bacteria that live in soil. They can form a type of resting cell called a **cyst (figure 11.16).** These have negligible metabolic activity and can withstand drying and ultraviolet radiation but are not highly resistant to heat.

Azotobacter species are also notable for their ability to fix nitrogen in aerobic conditions; recall that the enzyme nitrogenase is inactivated by O_2. Apparently, the exceedingly high respiratory rate of *Azotobacter* species consumes O_2 so rapidly that a low O_2 environment is maintained inside the cell. In addition, a protein in the cell binds nitrogenase, thereby protecting it from oxygen damage.

Myxobacteria

The **myxobacteria** are a group of aerobic Gram-negative rods that have a unique developmental cycle as well as a resting stage. When conditions are favorable, cells secrete a slime layer that other cells then follow, creating a **swarm** of cells. But then, when nutrients are exhausted, the behavior of the group changes. The cells begin to congregate, and then pile up to form a complex structure called a **fruiting body,** which is often brightly colored **(figure 11.17).**

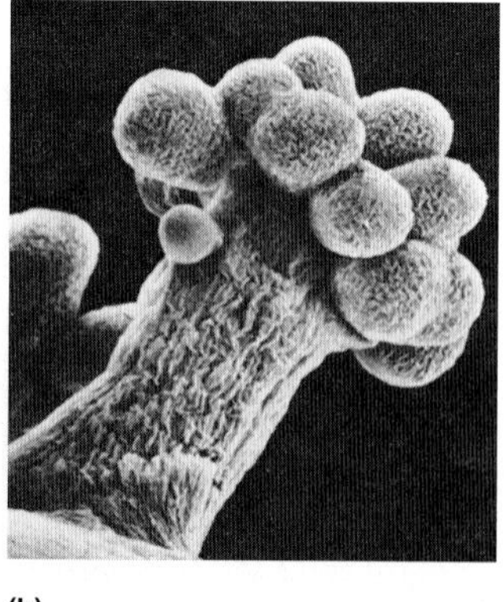

FIGURE 11.17 Fruiting Bodies of Myxobacteria These are the elaborate fruiting bodies of a species of *Chondromyces:* **(a)** photograph; **(b)** scanning electron micrograph.

In some species the fruiting body is quite elaborate, consisting of a mass of cells elevated and supported by a stalk made of a hardened slime. The cells within the fruiting body differentiate to become spherical, dormant forms called **microcysts.** These are considerably more resistant to heat, drying, and radiation than are the vegetative cells of myxobacteria, but are much less resistant than bacterial endospores.

Myxobacteria are important in nature as degraders of complex organic substances; they can digest bacteria and certain algae and fungi. Scientifically, these bacteria serve as an important model for studying developmental biology. Included in the myxobacteria are the genera *Chondromyces, Myxococcus,* and *Stigmatella.*

The Genus *Streptomyces*

The genus *Streptomyces* encompasses more than 500 species of aerobic Gram-positive bacteria that resemble fungi in their pattern of growth. Like the fungi, they form a **mycelium,** which is a visible mass of branching filaments. The filaments are called **hyphae.** At the tips of the hyphae, chains form of characteristic spores called **conidia (figure 11.18).** These dormant spores are resistant to drying and are readily dispersed in air currents to potentially more favorable locations. Note that while this pattern of growth resembles fungi, which are eukaryotes, *Streptomyces* species are much smaller and are prokaryotes.

Streptomyces species produce a variety of extracellular enzymes that enable them to degrade various organic compounds. They are also responsible for the characteristic "earthy" odor of soil; like the cyanobacteria, they produce geosmin. One species of *Streptomyces, S. somaliensis,* can cause an infection of subcutaneous tissue called an actinomycetoma.

Streptomyces species naturally produce a wide array of medically useful antibiotics, including streptomycin, tetracycline, and erythromycin. The role that these antimicrobial compounds play in the life cycle of *Streptomyces* has not been proven, but it is quite possible that they provide the organism a competitive advantage.

Bacteria That Associate with Plants

Members of two related genera use very different means to obtain the nutrients needed for growth from plants. *Agrobacterium* species are plant pathogens that cause tumorlike growths, whereas *Rhizobium* species form a mutually beneficial relationship with certain types of plants.

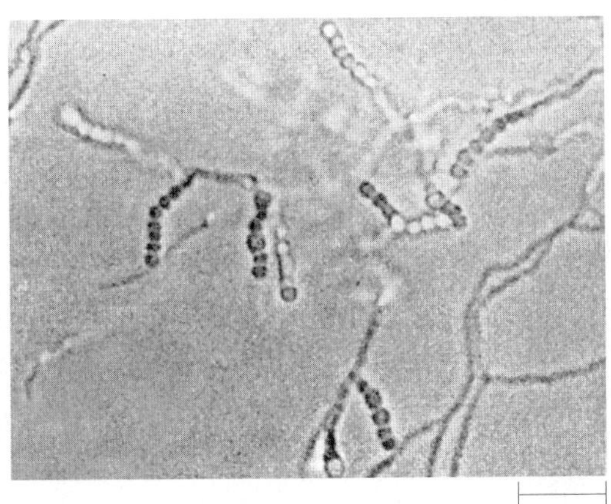

FIGURE 11.18 ***Streptomyces*** A photomicrograph showing the spherical conidia at the ends of the filamentous hyphae.

FIGURE 11.19 Plant Tumor Caused by *Agrobacterium tumefaciens*

The Genus *Agrobacterium*

Agrobacterium species have an unusual mechanism of gaining a competitive advantage in soil. They cause plant tumors, a manifestation of their ability to genetically alter plants for their own benefit **(figure 11.19).** These Gram-negative, rod-shaped bacteria enter a plant via a wound, and then transfer to the plant a portion of a plasmid; in *Agrobacterium tumefaciens* that plasmid is called the **Ti plasmid** (for "tumor-inducing"). The transferred DNA encodes the ability to synthesize a specific plant growth hormone, causing uncontrolled growth of the plant tissue and resulting in a tumor. The transferred DNA also encodes enzymes that direct the synthesis of an **opine,** an unusual amino acid derivative; *Agrobacterium* can then use this compound as a nutrient source (see Perspective 8.2).

Scientists have now modified the Ti plasmid, turning it into a commercially valuable tool. By removing those genes that cause tumor formation, the plasmid can be used as a vector to introduce DNA into plant cells. ■ vector, pp. 216, 219

Rhizobia

Rhizobia are a group of Gram-negative rod-shaped bacteria that often fix nitrogen and form intimate relationships with **legumes,** plants that bear seeds in pods. This group of bacteria includes members of the genera *Rhizobium, Sinorhizobium, Bradyrhizobium, Mesorhizobium,* and *Azorhizobium.* The bacteria live within cells in nodules formed on the root of the plant **(figure 11.20).** The plant synthesizes the protein **leghemoglobin,** which binds and controls the levels of O_2 (see figure 30.14). Within the resulting microaerobic confines of the nodule, the bacteria are able to fix nitrogen. Rhizobia residing within plant cells are examples of **endosymbionts,** organisms that provide a benefit to the cells in which they reside. ■ symbiotic nitrogen fixers, p. 734

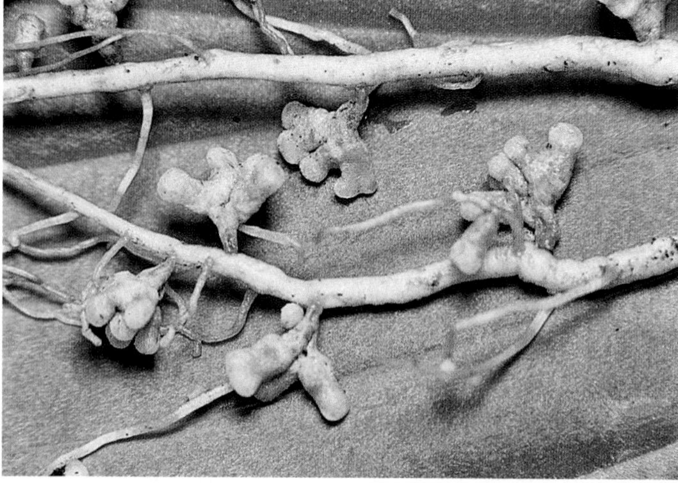

(a)

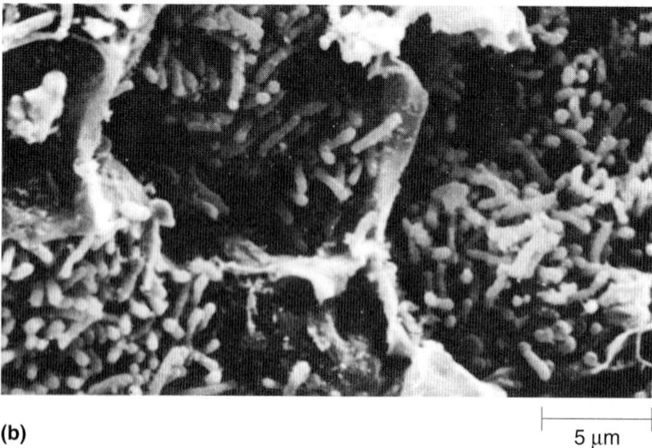

(b)

FIGURE 11.20 Symbiotic Relationship Between *Rhizobia* and Certain Plants **(a)** Root nodules. **(b)** Scanning electron micrograph of bacterial cells within a nodule.

MICROCHECK 11.6

Bacillus and *Clostridium* species produce endospores, the most resistant type of dormant cell known. *Azotobacter* species, myxobacteria, and *Streptomyces* species all produce dormant forms that tolerate some adverse conditions but are less resistant than endospores. *Agrobacterium* species and rhizobia derive nutrients from plants, although the former are plant pathogens and the latter benefit the plant.

- ✓ Why are myxobacteria important in nature?
- ✓ How does *Agrobacterium* derive a benefit from inducing a plant tumor?
- ✓ If you wanted to determine the number of endospores in a sample of soil, what could you do before plating it?

11.7

Thriving in Aquatic Environments

Focus Point

- Describe four mechanisms by which aquatic bacteria maximize nutrient acquisition and retention, and list one genus that uses each mechanism.

Most aquatic environments lack a steady supply of nutrients. To thrive in these habitats, bacteria have evolved various mechanisms to maximize nutrient acquisition and retention.

Sheathed Bacteria

Sheathed bacteria form chains of cells encased within a tube, or **sheath (figure 11.21).** This provides a protective function, helping the bacteria attach to solid objects located in favorable habitats while sheltering them from attack by predators. Masses of these filamentous sheaths can often be seen streaming from rocks or wood in flowing water polluted by nutrient-rich effluents. They often interfere with sewage treatment and other industrial processes by clogging pipes. Sheathed bacteria include species of *Sphaerotilus* and *Leptothrix,* which are Gram-negative rods.

Sheathed bacteria disperse themselves by forming **swarmer cells,** which have polar flagella and exit through the unattached end of the sheath. These motile cells move to a new solid surface, where they attach. If enough nutrients are present, they can multiply and form a new sheath, which elongates as the chain of cells grows.

Prosthecate Bacteria

The **prosthecate bacteria** are a diverse group of Gram-negative bacteria that have projections called **prosthecae,** which are extensions of the cytoplasm and cell wall. These extensions provide increased surface area to facilitate absorption of nutrients. Some prosthecae enable the organisms to attach to solid surfaces.

The Genus Caulobacter

Because of their remarkable life cycle, *Caulobacter* species have served as a model for the research of cellular differentiation. Entirely different events occur in an orderly fashion at opposite ends of the cell. *Caulobacter* cells have a single polar prostheca, commonly called a **stalk (figure 11.22).** At the tip of the stalk is an adhesive **holdfast,** which provides a mechanism for attachment. To multiply, the cell elongates and divides by binary fission, producing a motile swarmer cell at the end opposite the stalk. This swarmer cell has a single polar flagellum, located at the pole opposite the site of division. The swarmer cell detaches and moves to a new location, where it adheres via a holdfast near the base of its flagellum. It then loses its flagellum, replacing it with a stalk. Only then can the daughter cell replicate its DNA and repeat the process. In favorable

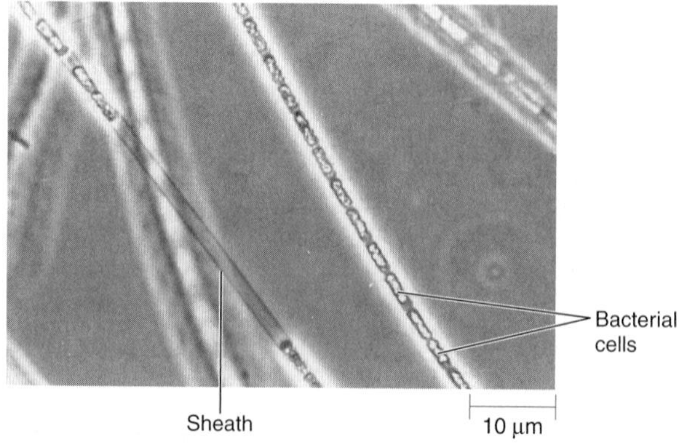

FIGURE 11.21 Sheathed Bacteria Phase-contrast photomicrograph of a *Sphaerotilus* species.

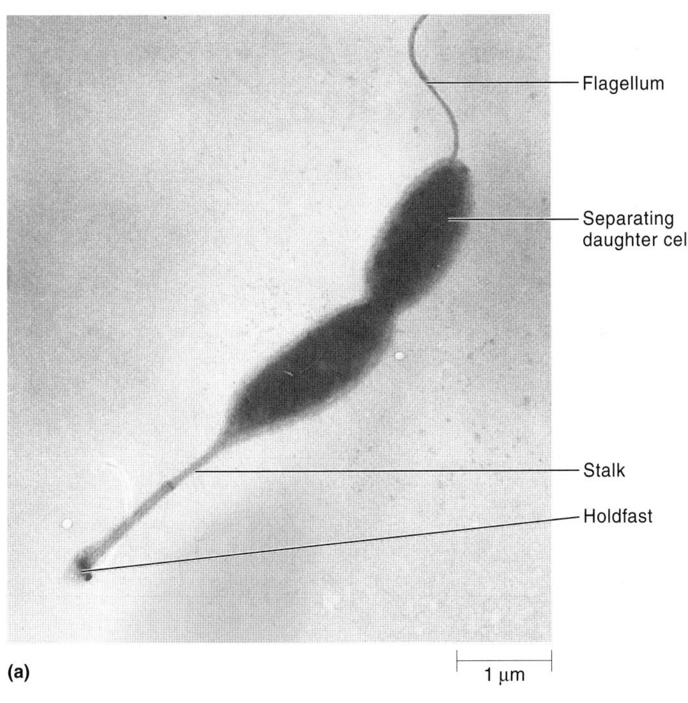

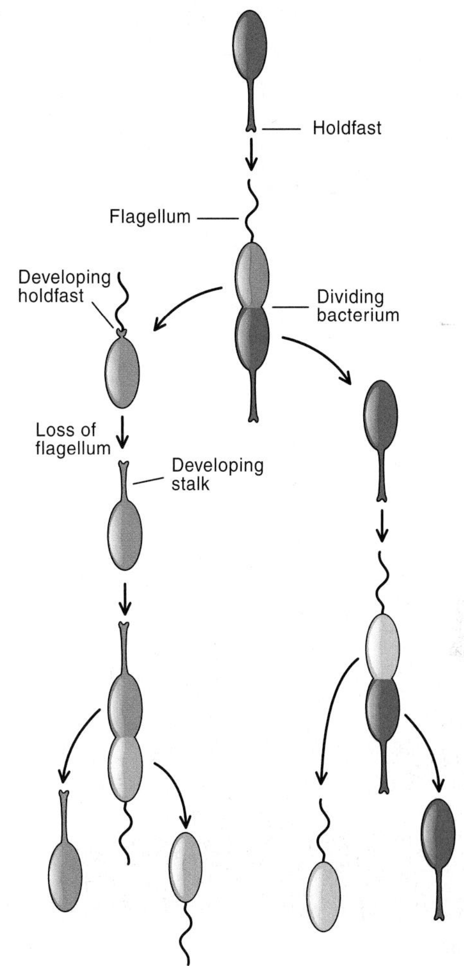

FIGURE 11.22 ***Caulobacter*** **(a)** Photomicrograph; **(b)** life cycle.

conditions, a single cell divides and produces daughter cells many times. With each division, a ring remains at the site of division, enabling a researcher to count the number of progeny.

The Genus Hyphomicrobium

Hyphomicrobium species are in many ways similar to *Caulobacter* species, except they have a distinct method of reproduction. The single polar prostheca of the parent cell enlarges at the tip to form a bud **(figure 11.23).** This continues enlarging and develops a flagellum, eventually giving rise to a motile daughter cell. The daughter cell then detaches and moves to a new location, eventually losing its flagellum and forming a polar prostheca at the opposite end to repeat the cycle. As with *Caulobacter* species, a single cell can sequentially produce multiple daughter cells.

Bacteria That Derive Nutrients from Other Organisms

Some bacteria obtain nutrients directly from other organisms. Examples include *Bdellovibrio* species, bioluminescent bacteria, *Epulopiscium* species, and *Legionella* species.

The Genus Bdellovibrio

Bdellovibrio species (*bdello,* from the Greek word for "leech") are highly motile Gram-negative bacteria that prey on *E. coli* and other Gram-negative bacteria **(figure 11.24).** They are small, curved rods, approximately 0.25 μm wide and 1 μm long, with a polar flagellum. When a *Bdellovibrio* cell attacks, it strikes its prey with such a high velocity that it actually propels the prey a short distance. The parasite then attaches to its host and rotates with a spinning motion. At the same time, it synthesizes digestive enzymes that break down lipids and peptidoglycan, eventually forming a hole in the cell wall of the prey. This allows the parasitic bacterium to penetrate, lodging in the periplasm between the cytoplasmic membrane and the peptidoglycan layer. There, over a period of several hours, *Bdellovibrio* degrades and utilizes the prey's cellular contents. It derives energy by aerobically oxidizing amino acids and acetate. The parasite increases in length as it resides in the periplasm, ultimately dividing to form several motile daughter cells. When the host cell lyses, the *Bdellovibrio* progeny are released to find new hosts, repeating the growth and reproduction cycle. ■ **periplasm, p. 61**

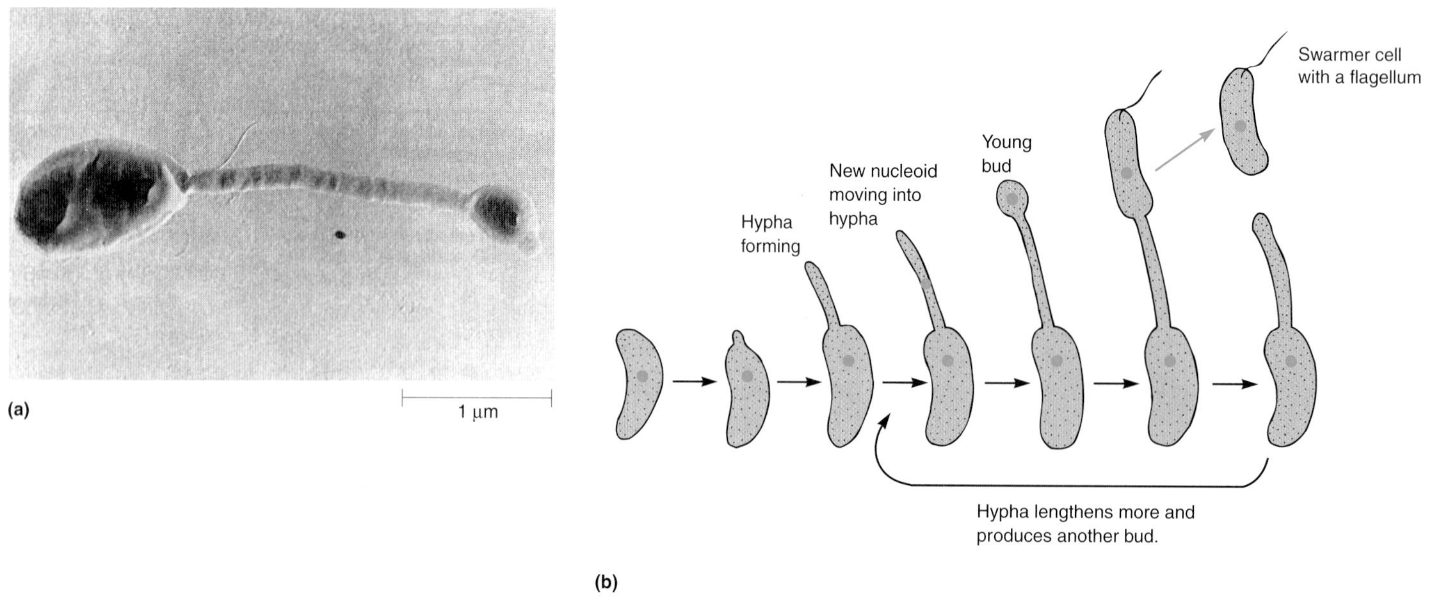

FIGURE 11.23 ***Hyphomicrobium*** **(a)** Photomicrograph. Note the bud forming at the tip of the polar prostheca. **(b)** Life cycle.

Bioluminescent Bacteria

Some species of *Photobacterium* and *Vibrio* can emit light **(figure 11.25).** This phenomenon, called **bioluminescence,** plays an important role in the symbiotic relationship between some of these bacteria and specific types of fish and squid. For example, certain types of squid have a specialized organ within their ink sac that is colonized by *Vibrio fischeri.* The light produced in the organ is thought to serve as a type of camouflage, obscuring the squid's contrast against the light from above and any shadow it might otherwise cast. The squid provides nutrients to the symbiotic bacteria, facilitating their growth. Another example is the flashlight fish, which has a light organ in a specialized pouch below its eye that harbors bioluminescent bacteria. By opening and closing a lid that covers the pouch, the fish can control the amount of light released, which is believed to confuse predators and prey.

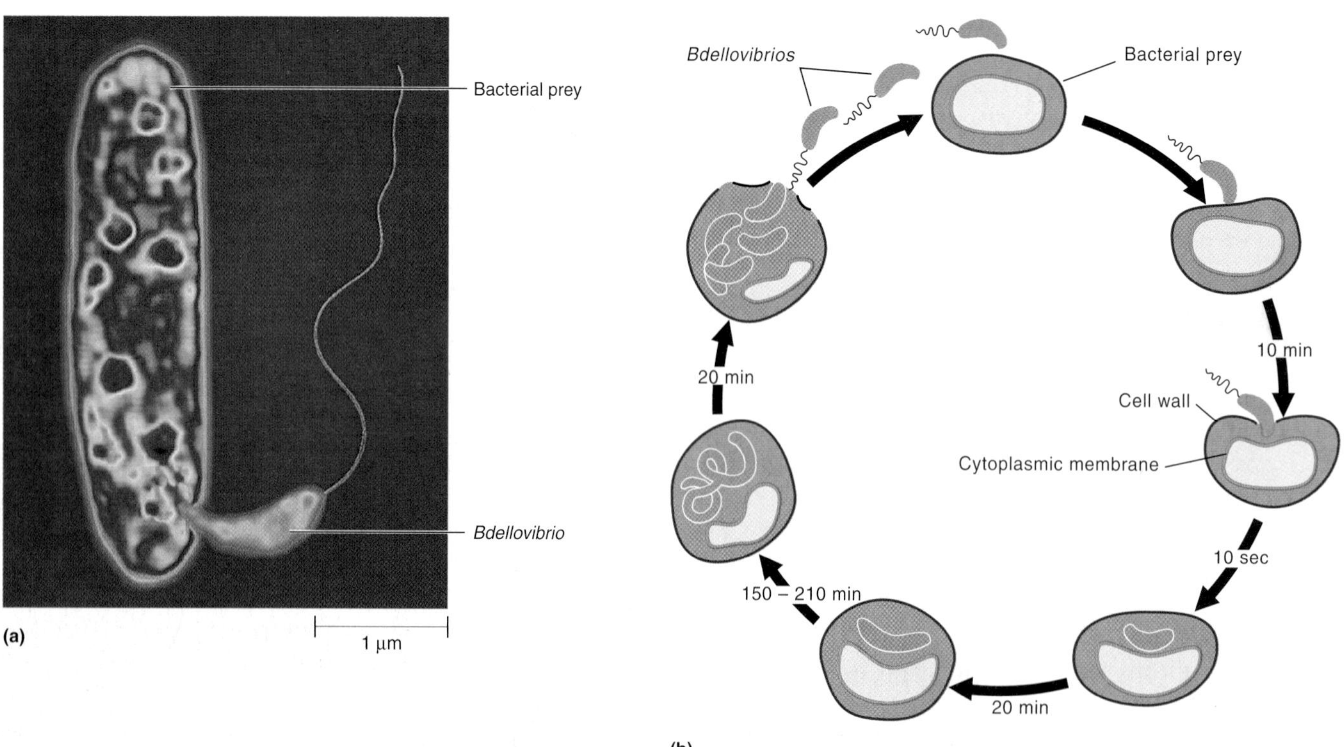

FIGURE 11.24 ***Bdellovibrio*** **(a)** Color-enhanced transmission micrograph of a *Bdellovibrio* cell attacking its prey. **(b)** Life cycle of *Bdellovibrio.* Note that the diagram exaggerates the size of the space in which *Bdellovibrio* multiplies.

(a)

(b)

FIGURE 11.25 Luminescent Bacteria (a) Plate culture of bioluminescent bacteria. **(b)** Photograph of a flashlight fish; under the eye is a light organ colonized with bioluminescent bacteria.

Luminescence is catalyzed by the enzyme **luciferase.** Studies revealed that the genes encoding it are only expressed when the density of the bacterial population reaches a critical point. This phenomenon of **quorum sensing** is now recognized as an important mechanism by which a variety of different bacteria regulate the expression of certain genes. ■ quorum sensing, p. 180

Members of the genera *Photobacterium* and *Vibrio* are Gram-negative rods (the rods of *Vibrio* species are curved) with polar flagella. They are facultative anaerobes and typically inhabit aqueous environments; species that require sodium for growth are usually found in marine environments. Not all are luminescent, and some species of *Vibrio* cause human disease. Medically important species include *V. cholerae,* which causes cholera, and *V. parahaemolyticus,* which also causes diarrheal disease; neither of these is bioluminescent. ■ *Vibrio cholerae,* p. 596

The Genus *Epulopiscium*

Epulopiscium species are Gram-positive cigar-shaped bacteria that reside in the intestinal tract of surgeon fish. They are considerably larger than most prokaryotes (600 μm × 80 μm), and each cell has thousands of copies of the genome dispersed throughout the cell. The multiple genome copies might help the organism overcome the problem of ensuring that necessary proteins are synthesized even in the far reaches of the large cell.

In addition to being noteworthy due to their large size, *Epulopiscium* species have a unique life cycle. Rather than undergoing typical binary fission, they enlarge considerably, finally lysing to release up to seven daughter cells. They have not yet been grown in culture.

The Genus *Legionella*

Legionella species are commonly found in aquatic environments, where they often reside within protozoa. They have even been isolated from water in air conditioners and produce misters. They are Gram-negative obligate aerobes that utilize amino acids but not carbohydrates as a source of carbon and energy. *Legionella pneumophila* can cause respiratory disease when inhaled in aerosolized droplets. ■ Legionnaires disease, p. 517

Bacteria That Move by Unusual Mechanisms

Some bacteria have unique mechanisms of motility that enable them to easily move to desirable locations. These organisms include the spirochetes and the magnetotactic bacteria.

Spirochetes

The **spirochetes** (Greek *spira* for "coil" and *chaete* for "hair") are a group of Gram-negative bacteria with a unique motility mechanism that enables them to move through thick, viscous environments such as mud. Distinguishing characteristics include their spiral shape, flexible cell wall, and motility by means of an **axial filament.** The axial filament is composed of sets of flagella that originate from both poles of the cell; unlike typical flagella, these are contained within the periplasm. The opposing sets of flagella extend toward each other, overlapping in the mid-region of the cell. Rotation of the flagella within the confines of the periplasm causes the cell to move like a corkscrew, sometimes deviating into flexing motions. Many spirochetes are very slender and can only be seen using special methods such as dark-field microscopy **(figure 11.26).** Many are also difficult or impossible to

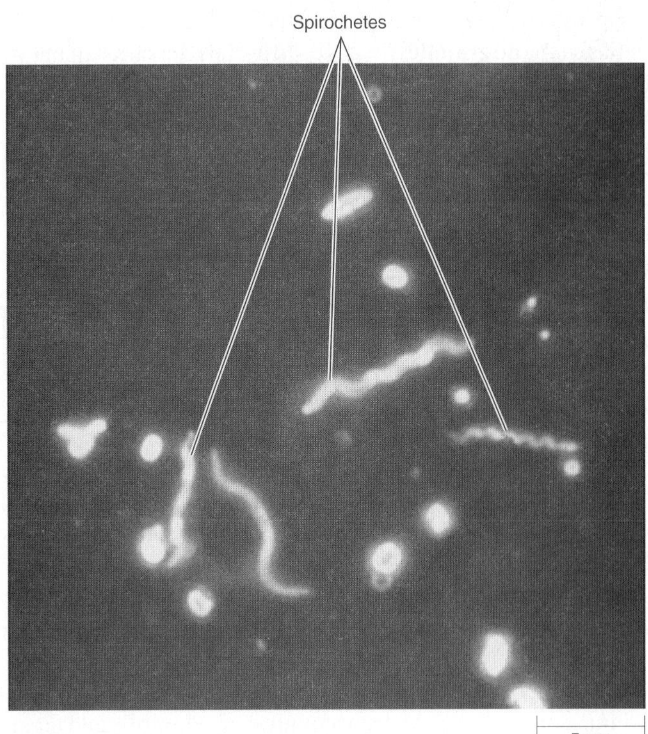

FIGURE 11.26 Spirochetes Dark-field photomicrograph of spirochetes.

cultivate, and their classification is based largely on morphology and ability to cause disease. ■ periplasm, p. 61

Spirochetes include free-living bacteria that inhabit aquatic environments, as well as species that reside on or in animals. Members of the genus *Spirochaeta* are anaerobes or facultative anaerobes that thrive in muds and anaerobic waters. *Leptospira* species are aerobic; some are free-living in aquatic environments, whereas others thrive within animals. *Leptospira interrogans* causes the disease leptospirosis, which can be transmitted in the urine of infected animals. Spirochetes adapted to reside in body fluids of humans and other animals will be discussed later. ■ leptospirosis, p. 622

Magnetotactic Bacteria

Magnetotactic bacteria such as *Magnetospirillum (Aquaspirillum) magnetotacticum* contain a string of magnetic crystals that align them with the earth's magnetism (see figure 3.40). This allows them to efficiently move up or down in the water or sediments. It is thought that this unique type of movement enables them to locate the microaerophilic habitats they require. *Magnetospirillum* species are Gram-negative, spiral-shaped organisms. ■ magnetotaxis, p. 66

Bacteria That Form Storage Granules

A number of aquatic bacteria form granules that serve to store nutrients. Recall that anoxygenic phototrophs often store sulfur granules, which can later be used as a source of electrons for reducing power. Some bacteria store phosphate, and others store compounds that can be used to generate ATP.

The Genus Spirillum

Members of the genus *Spirillum* are Gram-negative spiral-shaped, microaerophilic bacteria. *Spirillum volutans* forms volutin granules, which are storage forms of phosphate. These are sometimes called **metachromatic granules** to reflect their characteristic staining with the dye methylene blue. The cells of *S. volutans* are typically large, over 20 μm in length. In wet mounts, *Spirillum* species may be seen moving to a narrow zone near the edge of the coverslip, where O_2 is available in the optimum amount. ■ volutin, p. 68

Sulfur-Oxidizing, Nitrate-Reducing Marine Bacteria

Some marine bacteria store both sulfur, which can be oxidized as an energy source, and nitrate, which can serve as an electron acceptor. This provides an advantage to the bacteria, because anaerobic marine sediments are often abundant in reduced sulfur compounds but deficient in suitable terminal electron acceptors. In contrast, waters above those sediments lack reduced sulfur compounds but provide a source of nitrate. In other words, the energy source and terminal electron acceptor do not coexist. *Thioploca* species respond by forming long sheaths within which filamentous cells shuttle between the sulfur-rich sediments and nitrate-rich waters, storing reserves of sulfur and nitrate.

A huge bacterium was discovered in the ocean sediments off the coast of the African country of Namibia and named *Thiomargarita namibiensis,* "sulfur pearl of Namibia" (see Perspective 1.1). The cells are a pearly white color due to globules of sulfur in their cytoplasm. Each cell contains a large nitrate storage vacuole that takes up about 98% of the cell volume. It is thought that these organisms, which may reach a diameter of ¾ mm, can store a 3-month supply of sulfur and nitrate. They are not motile and instead appear to rely on occasional disturbances such as storms to bring them into contact with the nitrate-rich water.

MICROCHECK 11.7

Sheathed bacteria cluster within a tube attached to solid objects in favorable locations. Prosthecate bacteria produce extensions that maximize the absorptive surface area. *Bdellovibrio* species, bioluminescent bacteria, and *Legionella* species exploit nutrients provided by other organisms. Spirochetes and magnetotactic bacteria move by unusual mechanisms to more favorable locations. Some organisms form storage granules.

✓ What characteristic of *Caulobacter* species makes them an important model for research?

✓ How do squid benefit from having a light organ colonized by luminescent bacteria?

✓ The genomes of free-living spirochetes are larger than those of ones that live within an animal host. Why would this be so?

11.8

Animals As Habitats

Focus Point

- Describe one genus that inhabits the skin, three that inhabit mucous membranes, and two obligate intracellular parasites.

The bodies of animals, including humans, provide a wide variety of ecological habitats in which prokaryotes reside—from arid, O_2-rich surfaces to moist, anaerobic recesses. **Table 11.3** lists the medically important bacteria covered in this and other sections of the chapter.

Bacteria That Inhabit the Skin

The skin is typically dry and salty, providing an environment inhospitable to many microorganisms. Members of the genus *Staphylococcus,* however, thrive under these conditions. The propionic acid bacteria, which were discussed earlier, inhabit anaerobic microenvironments of the skin. ■ normal microbiota of the skin, p. 533

The Genus Staphylococcus

Staphylococcus species are Gram-positive cocci that are facultative anaerobes. Most, such as *S. epidermidis,* reside harmlessly as a component of the normal microbiota of the skin. Like other bacteria that aerobically respire, *Staphylococcus* species are catalase positive. This distinguishes them from *Streptococcus, Enterococcus,* and *Lactococcus* species, which are also Gram-positive cocci but lack the enzyme catalase. Several species

TABLE 11.3 Medically Important Bacteria

Organism	Medical Significance	Phylum
Gram-Negative Rods		
Bacteroides species	Obligate anaerobes that commonly inhabit the mouth, intestinal tract, and genital tract. Cause abscesses and bloodstream infections.	*Bacteroidetes*
Enterobacteriaceae		*Proteobacteria*
Enterobacter species	Normal microbiota of the intestinal tract.	
Escherichia coli	Normal microbiota of the intestinal tract. Some strains cause urinary tract infections; some strains cause specific types of intestinal disease. Causes meningitis in newborns.	
Klebsiella pneumoniae	Normal microbiota of the intestinal tract. Causes pneumonia.	
Proteus species	Normal microbiota of the intestinal tract. Cause urinary tract infections.	
Salmonella Enteritidis	Causes gastroenteritis. Grows in the intestinal tract of infected animals; acquired by consuming contaminated food.	
Salmonella Typhi	Causes typhoid fever. Grows in the intestinal tract of infected humans; transmitted in feces.	
Shigella species	Cause dysentery. Grow in the intestinal tract of infected humans; transmitted in feces.	
Yersinia pestis	Causes bubonic plague, which is transmitted by fleas, and pneumonic plague, which is transmitted in respiratory droplets of infected individuals.	
Haemophilus influenzae	Causes ear infections, respiratory infections, and meningitis in children.	*Proteobacteria*
Haemophilus ducreyi	Causes chancroid, a sexually transmitted disease.	*Proteobacteria*
Legionella pneumophila	Causes Legionnaires' disease, a lung infection. Grows within protozoa; acquired by inhaling contaminated water droplets.	*Proteobacteria*
Pseudomonas aeruginosa	Causes burn, urinary tract, and bloodstream infections. Ubiquitous in the environment. Grows in nutrient-poor aqueous solutions and is resistant to many disinfectants and antimicrobial medications.	*Proteobacteria*
Gram-Negative Rods—Obligate Intracellular Parasites		
Chlamydophila (Chlamydia) pneumoniae	Causes atypical pneumonia, or "walking pneumonia." Acquired from an infected person.	*Chlamydiae*
Chlamydophila (Chlamydia) psittaci	Causes psittacosis, a form of pneumonia. Transmitted by birds.	*Chlamydiae*
Chlamydia trachomatis	Causes a sexually transmitted disease that mimics the symptoms of gonorrhea. Also causes trachoma, a serious eye infection, and conjunctivitis in newborns.	*Chlamydiae*
Coxiella burnetii	Causes Q fever. Acquired by inhaling organisms shed by infected animals.	*Proteobacteria*
Ehrlichia chaffeenis	Causes human ehrlichiosis. Transmitted by ticks.	*Proteobacteria*
Orientia tsutsugamushi	Causes scrub typhus. Transmitted by mites.	*Proteobacteria*
Rickettsia prowazekii	Causes epidemic typhus. Transmitted by lice.	*Proteobacteria*
Rickettsia rickettsii	Causes Rocky Mountain spotted fever. Transmitted by ticks.	*Proteobacteria*
Wolbachia pipientis	Resides within the filarial worms that cause river blindness and elephantiasis in an obligate relationship.	
Gram-Negative Curved Rods		
Campylobacter jejuni	Causes gastroenteritis. Grows in the intestinal tract of infected animals; acquired by consuming contaminated food.	*Proteobacteria*
Helicobacter pylori	Causes stomach and duodenal ulcers. Neutralizes stomach acid by producing urease, resulting in the breakdown of urea to form ammonia.	*Proteobacteria*
Vibrio cholerae	Causes cholera, a severe diarrheal disease. Grows in the intestinal tract of infected humans; acquired by drinking contaminated water.	*Proteobacteria*
Vibrio parahaemolyticus	Causes gastroenteritis. Acquired by consuming contaminated seafood.	*Proteobacteria*
Gram-Negative Cocci		
Neisseria meningitidis	Causes meningitis.	*Proteobacteria*
Neisseria gonorrhoeae	Causes gonorrhea, a sexually transmitted disease.	*Proteobacteria*

(continued)

TABLE 11.3 Medically Important Bacteria (*Continued*)

Organism	Medical Significance	Phylum
Gram-Positive Rods		
Bacillus anthracis	Causes anthrax. Acquired by inhaling endospores in soil, animal hides, and wool. Bioterrorism agent.	*Firmicutes*
Bifidobacterium species	Predominant member of the intestinal tract in breast-fed infants. Thought to play a protective role in the intestinal tract of infants by excluding pathogens.	*Actinobacteria*
Clostridium botulinum	Causes botulism. Disease results from ingesting toxin-contaminated foods, typically canned foods that have been improperly processed.	*Firmicutes*
Clostridium perfringens	Causes gas gangrene. Acquired when soil-borne endospores contaminate a wound.	*Firmicutes*
Clostridium tetani	Causes tetanus. Acquired when soil-borne endospores are inoculated into deep tissue.	*Firmicutes*
Corynebacterium diphtheriae	Toxin-producing strains cause diphtheria, a frequently fatal throat infection.	*Actinobacteria*
Gram-Positive Cocci		
Enterococcus species	Normal microbiota of the intestinal tract. Cause urinary tract infections.	*Firmicutes*
Micrococcus species	Found on skin as well as in a variety of other environments; often contaminate bacteriological media.	*Actinobacteria*
Staphylococcus aureus	Leading cause of wound infections. Causes food poisoning and toxic shock syndrome.	*Firmicutes*
Staphylococcus epidermidis	Normal microbiota of the skin.	*Firmicutes*
Staphylococcus saprophyticus	Causes urinary tract infections.	*Firmicutes*
Streptococcus pneumoniae	Causes pneumonia and meningitis.	*Firmicutes*
Streptococcus pyogenes	Causes pharyngitis (strep throat), rheumatic fever, wound infections, glomerulonephritis, and streptococcal toxic shock.	*Firmicutes*
Acid-Fast Rods		
Mycobacterium tuberculosis	Causes tuberculosis.	*Actinobacteria*
Mycobacterium leprae	Causes Hansen's disease (leprosy); peripheral nerve invasion is characteristic.	*Actinobacteria*
Spirochetes		
Treponema pallidum	Causes syphilis, a sexually transmitted disease. The organism has never been grown in culture.	*Spirochaetes*
Borrelia burgdorferi	Causes Lyme disease, a tick-borne disease.	*Spirochaetes*
Borrelia recurrentis and *B. hermsii*	Causes relapsing fever. Transmitted by arthropods.	*Spirochaetes*
Leptospira interrogans	Causes leptospirosis, a waterborne disease. Excreted in urine of infected animals.	*Spirochaetes*
Cell Wall-less		
Mycoplasma pneumoniae	Causes atypical pneumonia ("walking pneumonia"). Not susceptible to penicillin because it lacks a cell wall.	*Firmicutes*

of *Staphylococcus* are notable for their medical significance. *Staphylococcus aureus* causes a variety of diseases, including skin and wound infections, as well as food poisoning. *Staphylococcus saprophyticus* causes urinary tract infections.

Bacteria That Inhabit Mucous Membranes

Mucous membranes of the respiratory, genitourinary, and intestinal tracts provide a habitat for numerous kinds of bacteria, many of which have already been discussed. For example, *Streptococcus* and *Corynebacterium* species reside in the respiratory tract, *Lactobacillus* species inhabit the vagina, and *Clostridium* species and members of the family *Enterobacteriaceae* thrive in the intestinal tract. Some of the other genera are discussed next. ■ **normal microbiota of the respiratory tract, p. 498** ■ **normal microbiota of the genitourinary tract, p. 620** ■ **normal microbiota of the intestinal tract, p. 585**

The Genus Bacteroides

Members of the genus *Bacteroides* are small, strictly anaerobic, Gram-negative rods and coccobacilli. They inhabit the mouth, intestinal tract, and genital tract of humans and other animals. *Bacteroides fragilis* and related species constitute about 30% of the bacteria in human feces and are often responsible for abscesses and bloodstream infections that follow appendicitis and abdominal surgery. Since many are killed even by brief exposure to O_2, they are difficult to study.

The Genus Bifidobacterium

Bifidobacterium species are Gram-positive, irregular, rod-shaped anaerobes that reside primarily in the intestinal tract of humans and other animals. They are the predominant members of the intestinal microbiota of breast-fed infants and are thought to provide a protective function by excluding disease-causing bacteria. Formula-fed infants are also colonized with members of this genus, but generally the concentrations are lower.

The Genera Campylobacter and Helicobacter

Members of the genera *Campylobacter* and *Helicobacter* are curved Gram-negative rods. As microaerophiles, they require specific atmospheric conditions to be successfully cultivated. *Campylobacter jejuni* causes diarrheal disease in humans. It typically resides in the intestinal tract of domestic animals, particularly poultry. *Helicobacter pylori* inhabits the stomach, where it can cause stomach and duodenal ulcers. It has also been implicated in the development of stomach cancer. An important factor in its ability to survive in the stomach is its production of the enzyme **urease.** This breaks down urea to produce ammonia, which neutralizes the acid in the immediate surroundings. ■ stomach ulcers, p. 590

The Genus Haemophilus

Members of the genus *Haemophilus* are tiny Gram-negative coccobacilli that, as their name reflects, are "blood loving." They require one or more compounds found in blood, such as hematin and NAD. Many species are common microbiota of the respiratory tract, but *H. influenzae* can also cause ear infections, respiratory infections, and meningitis, primarily in children. *Haemophilus ducreyi* causes the sexually transmitted disease chancroid. ■ ear infections, p. 504 ■ meningitis, p. 648 ■ chancroid, p. 636

The Genus Neisseria

Neisseria species are Gram-negative bacteria, typically kidney-bean-shaped cocci in pairs. They are common inhabitants of animals including humans, growing on mucous membranes. *Neisseria* species are typically aerobes, but some can grow anaerobically if a suitable terminal electron acceptor such as nitrite is present. Those noted for their medical significance include *N. gonorrhoeae,* which causes the sexually transmitted diease gonorrhea, and *N. meningitidis,* which causes meningitis; both of these are nutritionally fastidious. ■ fastidious, p. 94 ■ gonorrhea, p. 629 ■ meningitis, p. 648

The Genus Mycoplasma

Members of the genus *Mycoplasma* lack a cell wall, making them pliable and able to pass through the pores of filters that retain other bacteria. Most have sterols in their membrane, providing added strength and rigidity, thereby protecting the cells from osmotic lysis. They are among the smallest forms of life, and their genomes are thought to be the minimum size for encoding the essential functions for a free-living organism; the genome of *Mycoplasma genitalium* is only 5.8×10^5 base pairs, which is approximately one-eighth the size of the *E. coli* genome.

Medically, the most significant member of this group is *M. pneumoniae,* which, as its name implies, causes a form of pneumonia. This type cannot be treated with penicillin or other antibiotics that interfere with peptidoglycan synthesis, because these organisms lack a cell wall. Colonies of *Mycoplasma* species growing on solid media produce a characteristic "fried egg" appearance **(figure 11.27).**

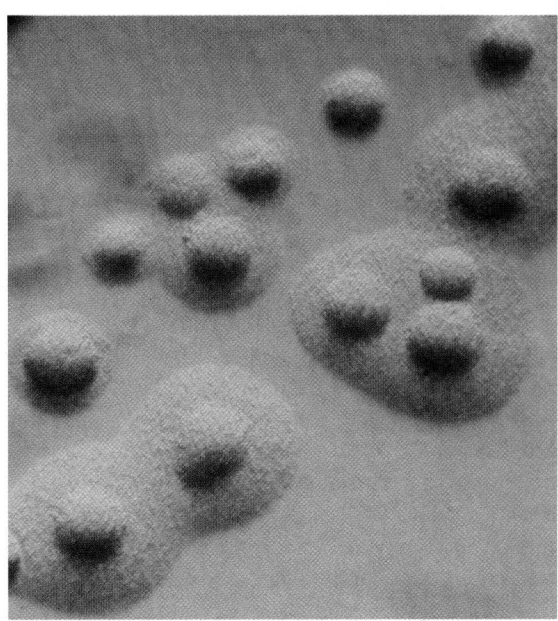

FIGURE 11.27 ***Mycoplasma pneumoniae*** **Colonies** Note the dense central portion of the colony, giving it the typical "fried egg" appearance.

The Genera Treponema and Borrelia

Members of the genera *Treponema* and *Borrelia* are spirochetes that typically inhabit body fluids and mucous membranes of humans and other animals. Recall that spirochetes are characterized by their corkscrew shape and axial filaments. Although they have a Gram-negative cell wall, they are often too thin to be viewed by conventional microscopy.

Treponema species are obligate anaerobes or microaerophiles that often inhabit the mouth and genital tract. Study of the species that causes syphilis, *T. pallidum,* is difficult because it has never been grown in culture. Its genome has been sequenced, however, providing evidence that it is a microaerophile with a metabolism highly dependent on its host. It lacks critical enzymes of the TCA cycle and a variety of other pathways.

Three species of *Borrelia* are medically significant microaerophiles transmitted by arthropods such as ticks and lice. *B. recurrentis* and *B. hermsii* both cause relapsing fever; *Borrelia burgdorferi* causes Lyme disease. A striking feature of *Borrelia* species is their genome, which is composed of a linear chromosome and many linear and circular plasmids.

Obligate Intracellular Parasites

Obligate intracellular parasites are unable to reproduce outside a host cell. By living within host cells, the parasites are supplied with a readily available source of compounds they would otherwise need to synthesize for themselves. As a result, most intracellular parasites have lost the ability to synthesize substances needed for extracellular growth. Bacterial examples include members of the genera *Rickettsia, Orientia, Ehrlichia, Coxiella, Chlamydia,* and *Wolbachia,* which are all tiny Gram-negative rods or coccobacilli.

The Genera *Rickettsia*, *Orientia*, and *Ehrlichia*

Species of *Rickettsia, Orientia,* and *Ehrlichia* are responsible for several serious human diseases spread by blood-sucking arthropods such as ticks and lice. *Rickettsia rickettsii* causes Rocky Mountain spotted fever, *R. prowazekii* causes epidemic typhus, *O. tsutsugamushi* causes scrub typhus, and *E. chaffeenis* causes human ehrlichiosis.

The Genus *Coxiella*

The only characterized species of *Coxiella, C. burnetii,* is an obligate intracellular bacterium that survives well outside the host cell and is transmissible from animal to animal without necessarily involving a blood-sucking parasite. During its intracellular growth, *C. burnetii* forms sporelike structures called small-cell variants (SCVs) that later allow it to survive in the environment. The structures, however, lack the extreme resistance to heat and disinfectants characteristic of endospores **(figure 11.28).** *Coxiella burnetii* causes Q fever of humans, a disease most often acquired by inhaling bacteria shed from infected animals.

The Genera *Chlamydia* and *Chlamydophila*

Chlamydia and *Chlamydophila* species are quite different from the other obligate intracellular parasites. They are transmitted directly from person to person rather than through the bite of a blood-sucking arthropod, and they have a unique growth cycle **(figure 11.29).** Inside the host cell, they initially exist as fragile non-infectious forms called **reticulate bodies** that reproduce by binary fission. Later in the infection, the bacteria differentiate into smaller, dense-appearing infectious forms called **elementary bodies** that are released upon death and rupture of the host cell. The cell wall of *Chlamydia* and *Chlamydophila* species is highly unusual among the *Bacteria* in that it lacks peptidoglycan, although it has the general appearance of a Gram-negative type of cell wall. *Chlamydia trachomatis* causes eye infections and a sexually transmitted disease that mimics gonorrhea, *Chlamydophila pneumoniae* causes atypical pneumonia, and *Chlamydophila psittaci* causes psittacosis, a form of pneumonia. ■ *Chlamydia trachomatis*, p. 631

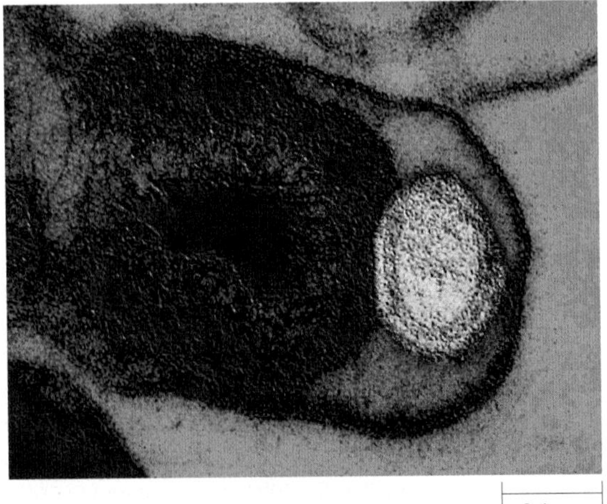

FIGURE 11.28 ***Coxiella*** Color-enhanced transmission electromicrograph of *C. burnetii*. The oval copper-colored object is the sporelike structure.

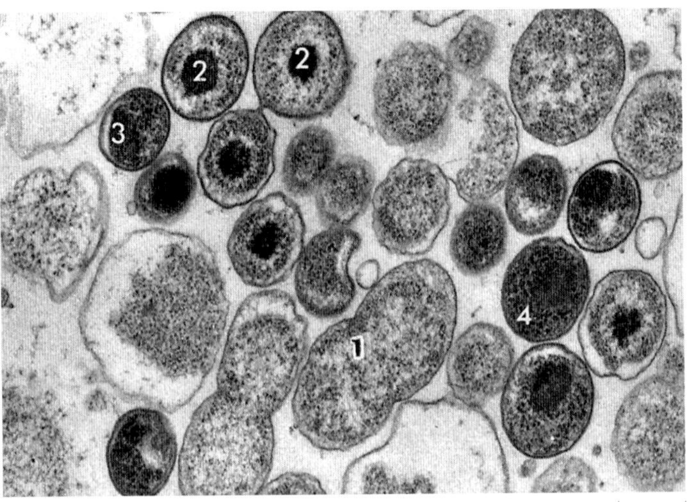

FIGURE 11.29 ***Chlamydia*** **Growing in Tissue Cell Culture** The numbers indicate the development from the dividing reticulate body to an infectious elementary body.

The Genus *Wolbachia*

The only known species of *Wolbachia, W. pipientis,* infects arthropods (including insects, spiders, and mites) and parasitic worms. It is primarily transmitted maternally, via the eggs of infected females to their offspring. In arthropods, the bacterium uses unique strategies to increase the overall population of infected females, including killing male embryos, allowing infected females to reproduce asexually, and causing infected males to gain female traits. In addition, the embryos resulting from the mating of an infected male with either an uninfected female or a female infected with a different strain of *Wolbachia* are destroyed. *Wolbachia* does not infect mammals, but its medical importance was recognized with the discovery that it resides within filarial worms that cause the diseases river blindness and elephantiasis. The chronic and debilitating inflammation associated with these diseases appears to be a result of the immune response directed against the bacterial cells carried by the invading worms. This discovery paves the way for new treatments, because eliminating the bacteria the not only lessens the symptoms but also kills the filarial worms.

MICROCHECK 11.8

Staphylococcus species are able to thrive in the dry, salty conditions of the skin. *Bacteroides* and *Bifidobacterium* species reside in the gastrointestinal tract; *Campylobacter* and *Helicobacter* species can cause disease when they reside there. *Neisseria* species, mycoplasma, and spirochetes inhabit other mucous membranes. Obligate intracellular parasites, including *Rickettsia, Orientia, Ehrlichia, Coxiella, Chlamydia, Chlamydophila,* and *Wolbachia* species, are unable to reproduce outside of a host cell.

- ✓ What characteristic of *Mycoplasma* species separates them from other bacteria?
- ✓ How is *Helicobacter pylori* able to withstand the acidity of the stomach?
- ✓ Why would breast feeding affect the composition of a baby's intestinal flora?

11.9

Archaea That Thrive in Extreme Conditions

Focus Point

- Describe the habitats of the extreme halophiles and extreme thermophiles.

Members of the *Archaea* that have been characterized typically thrive in extreme environments otherwise devoid of life **(table 11.4).** These include conditions of high heat, acidity, alkalinity, and salinity. An exception to this attribute is the methanogens, discussed earlier in the chapter; they inhabit anaerobic niches shared with members of the *Bacteria.* In addition to the characterized archaea, many others have been detected in a variety of non-extreme environments by using DNA probes that bind to ribosomal RNA (rRNA) genes. ■ DNA probes, p. 227

FIGURE 11.30 Typical Habitat of Extreme Halophiles The red color in these solar evaporation ponds is due to the pigments of extreme halophiles such as *Halobacterium* species.

Extreme Halophiles

The extreme halophiles are found in very high numbers in high-salt environments such as salt lakes, soda lakes, and brines used for curing fish. Most can grow well in a saturated salt solution (32% NaCl), and they require a minimum of about 9% NaCl. Because they produce pigments, their growth can be seen as red patches on salted fish and pink blooms in concentrated salt water ponds **(figure 11.30).**

Extreme halophiles are aerobic or facultatively anaerobic chemoheterotrophs, but some also can obtain additional energy from light. These organisms have the light-sensitive pigment **bacteriorhodopsin,** which absorbs energy from sunlight and uses it to expel protons from the cell. This creates a proton gradient that can be used to drive flagella or synthesize ATP.

Extreme halophiles come in a variety of shapes, including rods, cocci, discs, and triangles. They include genera such as *Halobacterium, Halorubrum, Natronobacterium,* and *Natronococcus;* members of these latter two genera are extremely alkaliphilic as well as halophilic.

Extreme Thermophiles

The extreme thermophiles (hyperthermophiles) are found in regions of volcanic vents and fissures that exude sulfurous gases and other hot vapors. Because these regions are thought to closely mimic the environment of early earth, scientists are particularly interested in studying the prokaryotes that thrive there. Others are found in hydrothermal vents in the deep sea and hot springs. ■ hyperthermophiles, p. 91

Methane-Generating Hyperthermophiles

In contrast to the methanogens discussed earlier, some are extreme thermophiles. Like the mesophilic methanogens, these oxidize H_2, using CO_2 as a terminal electron acceptor to yield methane. For example, *Methanothermus* species, which can grow in temperatures as high as 97°C, grow optimally at approximately 84°C.

Sulfur-Reducing Hyperthermophiles

The sulfur-reducing hyperthermophiles are obligate anaerobes that use sulfur as a terminal electron acceptor, generating H_2S. They harvest energy by oxidizing organic compounds and/or H_2. These archaea can be isolated from hot sulfur-containing environments such as sulfur hot springs and hydrothermal vents. They include some of the most thermophilic organisms known, a few even growing above 100°C. One notable example, *Pyrolobus fumarii,* was isolated from a "black smoker" 3,650 m (about 12,000 feet) deep in the Atlantic Ocean and grows between 90°C and 113°C. Another hydrothermal vent isolate, *Pyrodictium occultum,* has an

TABLE 11.4 Archaea

Group/Genera	Characteristics	Phylum
Methanogens – *Methanospirillum, Methanosarcina*	Generate methane when they oxidize hydrogen gas as an energy source, using CO_2 as a terminal electron acceptor.	*Euryarchaeota*
Extreme halophiles – *Halobacterium, Halorubrum, Natronobacterium, Natronococcus*	Found in salt lakes, soda lakes, and brines. Most grow well in saturated salt solutions.	*Euryarchaeota*
Extreme Thermophiles – *Methanothermus, Pyrodictium, Pyrolobus, Sulfolobus, Thermophilus, Picrophilus, Nanoarchaeum*	Found near hydrothermal vents and in hot springs; some grow at temperatures above 100°C. Includes examples of methane-generating, sulfur-reducing, and sulfur-oxidizing archaea, as well as extreme acidophiles.	*Crenarchaeota, Euryarchaeota,* and *Nanoarchaeota*

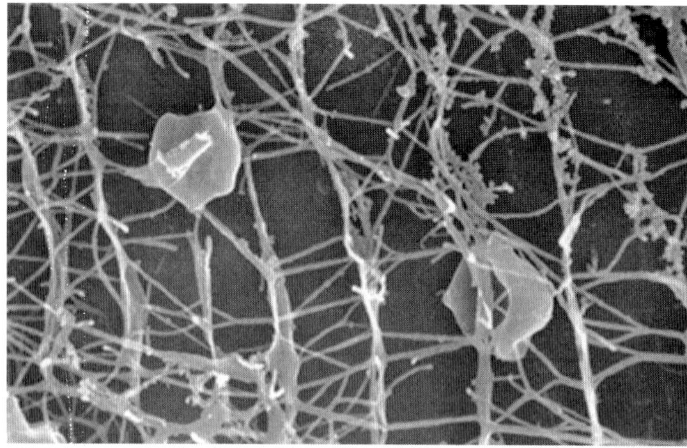

FIGURE 11.31 ***Pyrodictium*** The disc-shaped cells are connected by hollow tubes. Scanning electron micrograph.

optimum temperature of approximately 105°C, and cannot grow below 82°C. Its disc-shaped cells are connected by hollow tubes, forming a weblike network **(figure 11.31).** The record-holder for the highest maximum growth temperature, dubbed "strain 121" to reflect that it grows at 121°C, appears to be most closely related to *Pyrodictium* species, but it uses iron as an electron acceptor.

Nanoarchaea

The discovery of an archaeum so unique that it represents an entirely new phylum, *Nanoarcheota* ("dwarf archaea"), was made possible by the earlier discovery of a new genus of sulfur-reducing hyperthermophiles, *Ignicoccus* ("the fire sphere"). *Nanoarchaeum equitans* ("rides the fire sphere") grows as 400 nm spheres attached to the surface of—presumably parasitizing—*Ignicoccus* species (see figure 9.20).

Sulfur Oxidizers

Sulfolobus species are found at the surface of acidic sulfur-containing hot springs such as many of those found in Yellowstone National Park **(figure 11.32).** They are obligate aerobes that oxidize sulfur compounds, using O_2 as a terminal electron acceptor to generate sulfuric acid. In addition, they are thermoacidophilic, only growing above 50°C and at a pH between 1 and 6.

FIGURE 11.32 Typical Habitat of *Sulfolobus* Sulfur hot spring in Yellowstone National Park.

Thermophilic Extreme Acidophiles

Members of two genera, *Thermoplasma* and *Picrophilus,* are notable for their preference of growing in extremely acidic, hot environments. *Thermoplasma* species grow optimally at pH 2; in fact, *T. acidophilum* lyses at neutral pH. It was originally isolated from coal refuse piles. *Picrophilus* species tolerate conditions even more acidic, growing optimally at a pH below 1. Two species have been isolated in Japan from acidic areas in regions that exude sulfurous gases.

MICROCHECK 11.9

Many *Archaea* that have been characterized typically inhabit extreme environments that are otherwise devoid of life. These include conditions of high salinity, heat, acidity, and alkalinity.

- ✓ Why do seawater ponds sometimes turn pink as the water evaporates?
- ✓ At which relative depth in a sulfur hot spring would a sulfur reducer likely be found? How about a sulfur oxidizer?
- ✓ What characteristic of the methanogens makes it logical to discuss them with the *Bacteria* rather than the other *Archaea?*

FUTURE CHALLENGES

Astrobiology: The Search for Life on Other Planets

If life as we know it exists on other planets, it will likely be microbial. The task, then, is to figure out how to find and detect such extraterrestrial microorganisms.

Considering that we still know relatively little about the microbial life on our own planet, coupled with the extreme difficulty of obtaining or testing extraterrestrial samples, this is a daunting challenge with many as yet unanswered questions. For example, what is the most likely source of life on other planets? What is the best way to preserve specimens for study on earth? What will be the culture requirements to grow such organisms? **Astrobiology,** the study of life in the universe, is a new field that brings together scientists from a wide range of disciplines, including microbiology, geology, astronomy, biology, and chemistry, to begin answering some of these questions. The goal is to determine the origin, evolution, distribution, and destiny of life in the universe. Astrobiologists are also given the task of developing lightweight, dependable, and meaningful testing devices to be used in future space missions.

Astrobiologists believe that within our solar system, life would most likely be found either on Europa, a moon of Jupiter, or on Mars. This is because Europa and Mars appear to have, or have had, water, which is crucial for all known forms of life. Europa has an icy crust, beneath which may be liquid water or even a liquid ocean. Mars is the planet closest to earth, and it has the most similar environment. Images and data from the recent Mars missions indicate that flowing water once existed there.

To prepare for researching life on other planets, microbiologists have turned to some of the most extreme environments here on earth. These include glaciers and ice shelves, hot springs, deserts, volcanoes, deep ocean hydrothermal vents, and subterranean features such as caves. Because select microorganisms can survive in these environs, which are analogous to conditions expected on other planets, they are good testing grounds for the technology to be used on future missions.

SUMMARY

Metabolic Diversity (table 11.1)

11.1 Anaerobic Chemotrophs

Anaerobic Chemolithotrophs

The **methanogens** are a group of archaea that generate energy by oxidizing hydrogen gas (H_2), using CO_2 as a terminal electron acceptor (figure 11.1).

Anaerobic Chemoorganotrophs—Anaerobic Respiration

Desulfovibrio species reduce sulfur compounds to form hydrogen sulfide.

Anaerobic Chemoorganotrophs—Fermentation

Clostridium species form endospores. The **lactic acid bacteria** produce lactic acid as their primary fermentation end product (figures 11.2, 11.3). *Propionibacterium* species produce propionic acid as their primary fermentation end product.

11.2 Anoxygenic Phototrophs

The Purple Bacteria

The **purple bacteria** appear red, orange, or purple; the components of their photosynthetic apparatus are all contained within the cytoplasmic membrane, which has extensive invaginations.

The Green Bacteria

The **green bacteria** are typically green or brownish in color. Their accessory pigments are often located in **chlorosomes.**

Other Anoxygenic Phototrophs

Other anoxygenic phototrophs have been discovered, including some that form endospores.

11.3 Oxygenic Phototrophs

The Cyanobacteria (figures 11.6, 11.7)

Genetic evidence indicates that chloroplasts of plants and algae evolved from a species of cyanobacteria. Nitrogen-fixing cyanobacteria are critically important ecologically, because they provide an available source of both carbon and nitrogen. Filamentous species may maintain the structure and productivity of some soils. Some species of cyanobacteria produce toxins that can be deadly to animals that ingest contaminated water.

11.4 Aerobic Chemolithotrophs

The Sulfur-Oxidizing Bacteria (figure 11.10)

The filamentous sulfur oxidizers *Beggiatoa* and *Thiothrix* live in sulfur springs, in sewage-polluted waters, and on the surface of marine and freshwater sediments (figure 11.9). *Acidithiobacillus* species are found in both terrestrial and aquatic habitats.

The Nitrifiers

Ammonia oxidizers convert ammonia to nitrite and include *Nitrosomonas* and *Nitrosococcus;* nitrite oxidizers convert nitrite to nitrate and include *Nitrobacter* and *Nitrococcus.*

The Hydrogen-Oxidizing Bacteria

Aquifex and *Hydrogenobacter* species are thermophilic bacteria and thought to be among the earliest bacterial forms to exist on earth.

11.5 Aerobic Chemoorganotrophs

Obligate Aerobes

Micrococcus species are found in soil and on dust particles, inanimate objects, and skin (figure 11.11). *Mycobacterium* species are widespread in nature. They are **acid-fast.** *Pseudomonas* species are widespread in nature and have extremely diverse metabolic capabilities (figure 11.12). *Thermus aquaticus* is the source of *Taq* polymerase, an essential component in the polymerase chain reaction. *Deinococcus radiodurans* can survive exposure to a dose of gamma radiation several thousand times that lethal to a human being.

Facultative Anaerobes

Corynebacterium species commonly inhabit soil, water, and the surface of plants (figure 11.13). Members of the family *Enterobacteriaceae* typically inhabit the intestinal tract of animals, although some reside in rich soil. Enterics that ferment lactose are included in the group **coliforms** and are used as indicators of fecal pollution (figure 11.14).

Ecophysiological Diversity (table 11.2)

11.6 Thriving in Terrestrial Environments

Bacteria That Form a Resting Stage

Of the various types of dormant cells produced by soil organisms, endospores are by far the most resistant to environmental extremes. Endospore-forming genera include *Bacillus* and *Clostridium* (figure 11.15). *Azotobacter* species form a resting cell called a **cyst** and are notable for their ability to fix nitrogen in aerobic conditions (figure 11.16). **Myxobacteria** aggregate to form a **fruiting body** when nutrients are exhausted; within the fruiting body, cells differentiate to form dormant microcysts (figure 11.17). *Streptomyces* species resemble fungi in their pattern of growth; they form chains of **conidia** at the end of **hyphae.** Many species naturally produce antibiotics (figure 11.18).

Bacteria That Associate with Plants

Agrobacterium species cause plant tumors. They transfer a portion of the Ti plasmid to plant cells, genetically engineering the plant cells to produce opines and plant growth hormones. **Rhizobia** reside as **endosymbionts** in nodules formed on the roots of **legumes.** In these protected confines, they fix nitrogen (figure 11.20).

11.7 Thriving in Aquatic Environments

Sheathed Bacteria

Sheathed bacteria form chains of cells encased in a **sheath,** which enables cells to attach to solid objects in favorable habitats while sheltering them from attack by predators (figure 11.21).

Prosthecate Bacteria

Caulobacter species have a single polar prostheca called a **stalk;** at the tip of the stalk is a **holdfast.** The cells divide by binary fission (figure 11.22). *Hyphomicrobium* species divide by forming a bud at the tip of their single polar prostheca (figure 11.23).

Bacteria That Derive Nutrients from Other Organisms

Bdellovibrio species prey on other bacteria (figure 11.24). Certain species of bioluminescent bacteria establish symbiotic relationships with specific types of squid and fish (figure 11.25). *Epulopiscium* species reside within the intestinal tract of surgeon fish. *Legionella* species often reside within protozoa and can cause respiratory disease when inhaled in aerosolized droplets.

Bacteria That Move by Unusual Mechanisms

Spirochetes move by means of an axial filament (figure 11.26). Magnetotactic bacteria contain a string of magnetic crystals that enable them to move up or down in water or sediments to the microaerophilic niches they require.

Bacteria That Form Storage Granules

Spirillum volutans forms polyphosphate granules. *Thiomargarita namibiensis* is the largest bacterium known; it stores granules of sulfur and has a nitrate-containing vacuole.

11.8 Animals As Habitats (table 11.3)

Bacteria That Inhabit the Skin

Staphylococcus species are facultative anaerobes.

Bacteria That Inhabit Mucous Membranes

Bacteroides species inhabit the mouth, intestinal tract, and genital tract of humans and other animals. *Bifidobacterium* species reside primarily in the intestinal tract of animals, including humans, particularly breast-fed infants. *Campylobacter* and *Helicobacter* species are microaerophilic. *Haemophilus* species require compounds found in blood for growth. *Neisseria* species are nutritionally fastidious, obligate aerobes that grow in the oral cavity and genital tract. *Mycoplasma* species lack a cell wall; they often have sterols in their membrane that provide strength and rigidity (figure 11.27). *Treponema* and *Borrelia* species are spirochetes that typically inhabit mucous membranes and body fluids of humans and other animals.

Obligate Intracellular Parasites

Species of *Rickettsia, Orientia,* and *Ehrlichia* are spread when a blood-sucking arthropod transfers bacteria during a blood meal. *Coxiella burnetii* survives well outside the host due to the production of sporelike structures (figure 11.28). *Chlamydia* and *Chlamydophila* species are transmitted directly from person to person (figure 11.29). *Wolbachia pipientis* alters the reproductive biology of infected arthropods; although it does not infect mammals, it resides within the filarial worms that cause river blindness and elephantiasis.

11.9 Archaea That Thrive in Extreme Conditions

Extreme Halophiles

Extreme halophiles are found in salt lakes, soda lakes, and brines used for curing fish (figure 11.30).

Extreme Thermophiles

Methanothermus species are hyperthermophiles that generate methane. Sulfur-reducing hyperthermophiles are obligate anaerobes that use sulfur as a terminal electron acceptor. Nanoarchaea grow as spheres attached to *Ignicoccus* species. Sulfur-oxidizing hyperthermophiles oxidize sulfur compounds, using O_2 as a terminal electron acceptor, to generate sulfuric acid. Thermophilic extreme acidophiles have an optimum pH of 2 or below.

REVIEW QUESTIONS

Short Answer

1. What kind of bacteria might compose the subsurface scum of polluted ponds?
2. What kind of bacterium might be responsible for plugging the pipes in a sewage treatment facility?
3. Give three examples of energy sources used by chemolithotrophs.
4. Name two genera of endospore-forming bacteria. How do they differ?
5. Serological tests involving which structures help distinguish the different enterobacteria from each other?
6. What unique motility structure characterizes the spirochetes?
7. In what way does the metabolism of *Streptococcus* species differ from that of *Staphylococcus* species?
8. How have species of *Streptomyces* contributed to the treatment of infectious diseases?
9. What characteristics of *Azotobacter* species protects their nitrogenase enzyme from inactivation by O_2?
10. Compare and contrast the relationships of *Agrobacterium* and *Rhizobium* species with plants.

Multiple Choice

1. A catalase-negative colony growing on a plate that was incubated aerobically could be which of these genera?
   a) *Bacillus* b) *Escherichia* c) *Micrococcus*
   d) *Staphylococcus* e) *Streptococcus*
2. All of the following genera are spirochetes, *except*
   a) *Borrelia* b) *Caulobacter* c) *Leptospira*
   d) *Spirochaeta* e) *Treponema*
3. Which of the following genera would you most likely find growing in acidic runoff from a coal mine?
   a) *Clostridium* b) *Escherichia* c) Lactic acid bacteria
   d) *Thermus* e) *Acidithiobacillus*
4. The dormant forms of which of the following genera are the most resistant to environmental extremes?
   1. *Azotobacter* 2. *Bacillus* 3. *Clostridium*
   4. *Myxobacteria* 5. *Streptomyces*
   a) 1, 2 b) 2, 3 c) 3, 4
   d) 4, 5 e) 1, 5
5. Members of which of the following genera are coliforms?
   a) *Bacteroides* b) *Bifidobacterium* c) *Clostridium*
   d) *Escherichia* e) *Streptococcus*
6. Which of the following genera preys on other bacteria?
   a) *Bdellovibrio* b) *Caulobacter* c) *Hyphomicrobium*
   d) *Photobacterium* e) *Sphaerotilus*
7. All of the following genera are obligate intracellular parasites, *except*
   a) *Chlamydia* b) *Coxiella* c) *Ehrlichia*
   d) *Mycoplasma* e) *Rickettsia*
8. Which of the following genera fix nitrogen?
   1. *Anabaena* 2. *Azotobacter* 3. *Deinococcus*
   4. *Mycoplasma* 5. *Rhizobium*
   a) 1, 3, 4 b) 1, 2, 5 c) 2, 3, 5
   d) 2, 4, 5 e) 3, 4, 5
9. Which of the following archaea would most likely be found coexisting with bacteria?
   a) *Nanoarchaeum* b) *Halobacterium* c) *Methanococcus*
   d) *Picrophilus* e) *Sulfolobus*

10. *Thermoplasma* and *Picrophilus* grow best in which of the following extreme conditions?
    a) Low pH b) High salt c) High temperature
    d) a and c e) b and c

Applications

1. A student argues that it makes no sense to be concerned about coliforms in drinking water because coliforms are harmless members of our normal intestinal microbiota. Explain why regulatory agencies are concerned about coliforms.
2. A friend who has lakefront property and cherishes her lush green lawn complains of the green odoriferous scum on the lake each summer. Explain how her lawn might be contributing to the problem.

Critical Thinking

1. Soil often goes through periods of extreme dryness and extreme wetness. What characteristics of *Clostridium* species make them well suited for these conditions?
2. Some organisms use sulfur as an electron donor (a source of energy), whereas others use sulfur as an electron acceptor. How can this be if there must be a difference between the electron affinity of electron donors and acceptors for an organism to obtain energy?

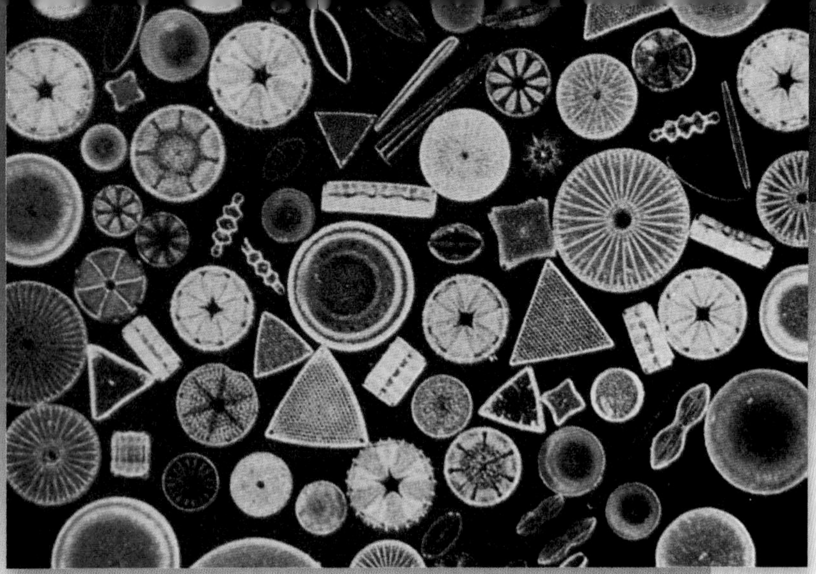

12

Diatoms, a type of algae, have complex cell walls.

The Eukaryotic Members of the Microbial World

A Glimpse of History

Sometimes, a single event or series of events can result in a change in the course of history. A fungus-like organism, *Phytophthora infestans*, which causes late blight in potatoes, triggered just such an event.

In 1845, almost half of the potato crop in Ireland was decimated by late blight and, by 1847, nearly the entire population of Ireland was facing starvation. As a result 1.5 million Irish people died and more than 1 million emigrated to other countries—primarily the United States and Canada. Eventually 5 million would emigrate from Ireland to the United States.

Potatoes and the associated late blight disease are both native to South America, where many varieties of potatoes grow in the continent's diverse climates. In 1537, the Spaniards discovered potatoes and brought them to Europe. At that time various grains such as barley, wheat, and rye were the main staple of the diet in the various countries of Europe. Most of the people did not depend on potatoes as their major source.

However, when the potato arrived in Ireland, it became a major source of food for several reasons. On the one hand, the climate made growing potatoes quite easy. In addition, with the advent of the Industrial Revolution and the move of people from the farms to the cities to work in the factories, potatoes became a convenient and good food source. One acre of potatoes can easily feed a family and some animals for a year. Potatoes have all the vitamins and enough protein to support human life. Potatoes are easy to grow because they are grown as clones. Once a farmer has sown his field with potatoes he only needs to save a few potatoes to be able to grow the next crop. Potatoes also store very well in a cool place for months.

The Irish peoples' dependence on this single food source left them open to a major disaster when that food source was destroyed. This disaster occurred in 1845 and in subsequent years.

Phytophthora infestans, the cause of late blight, affects every part of the plant including the leaves, the stems, and the tubers. The tuber turns to a black, mushy mess. Once a field is infected, it is very difficult to rid it of the infection. Because the potato tuber is under ground, the infection is often not visible until it has destroyed the entire plant. The infection is spread both by the asexual and sexual spores, and the spores, which can live for years in the soil, are spread by wind as well as by those who handle the potatoes.

Plant breeders in the late nineteenth and early twentieth centuries developed resistant strains of potatoes and fungicides to control late blight. However, in 1980s and 1990s, resistant infections of *Phytophthora infectans* again become a problem and affected many acres of potatoes. Once again, the study of this organism has become important. ■

As noted at the beginning of this book, when the ribosomal RNA (rRNA) sequences of organisms are compared, the living world can be divided into three divisions: *Bacteria, Archaea*, and *Eucarya*. In this chapter we will consider the *Eucarya*. These organisms, which include the algae, fungi, protozoa, multicellular parasites, and insect vectors, have one feature in common: they are all eukaryotic organisms. Recall from chapter 3 that the basic cell structure of the *Eucarya* is distinctly different from that of the *Bacteria* and *Archaea*. They are included in a textbook of microbiology because many members of these groups are microscopic and are studied with techniques that are similar to those used to study bacteria and archaea. In addition, many of these organisms cause disease in humans as well as plants and animals. ■ ***Eucarya*, p. 10**

KEY TERMS

Alga A unicellular or simple multicellular photosynthetic eukaryotic organism.

Arthropod Taxonomic grouping of invertebrate animals that includes insects, ticks, lice, and mites.

Convergent Evolution Process of evolution when two genetically different organisms develop similar environmental adaptations.

Eucarya Name of the domain comprising eukaryotic organisms.

Fungus A non-photosynthetic eukaryotic organism.

Helminth A parasitic worm.

Nematode Roundworm.

Neurotoxin Toxin that damages the nervous system.

Phytoplankton Floating and swimming algae and photosynthetic prokaryotic organisms of lakes and oceans.

Polymorphic Having different distinct forms.

Protozoa Group of single-celled eukaryotic organisms.

Toxin Poisonous chemical substance.

Trematodes Flatworms known also as flukes.

Yeasts Unicellular fungi.

Classification using gross anatomical characteristics of algae, fungi, protozoa, and even some multicellular parasites has always been problematic. Now, however, with modern techniques that examine these organisms at the molecular and ultrastructural levels, we know that some organisms that were traditionally grouped together are more dissimilar than similar. Instead, they arose at various times along a continuum of evolution. Therefore, in classification schemes that describe an accurate evolutionary history of organisms and are based on molecular and ultrastructure examination, the words *algae, fungi,* and *protozoa* are no longer really accurate. For the purposes of this book, however, we will use the term *algae* to describe the photosynthetic members and *fungi* and *protozoa* to describe the non-photosynthetic members that are discussed in this chapter. In addition, a discussion of arthropods and helminths is included because these eukaryotes are also implicated in human disease. **Figure 12.1** shows a phylogeny based on the ribosomal RNA sequences of the eukaryotic organisms. We will refer to this phylogeny throughout this chapter, indicating where the organisms fit on this evolutionary scale.

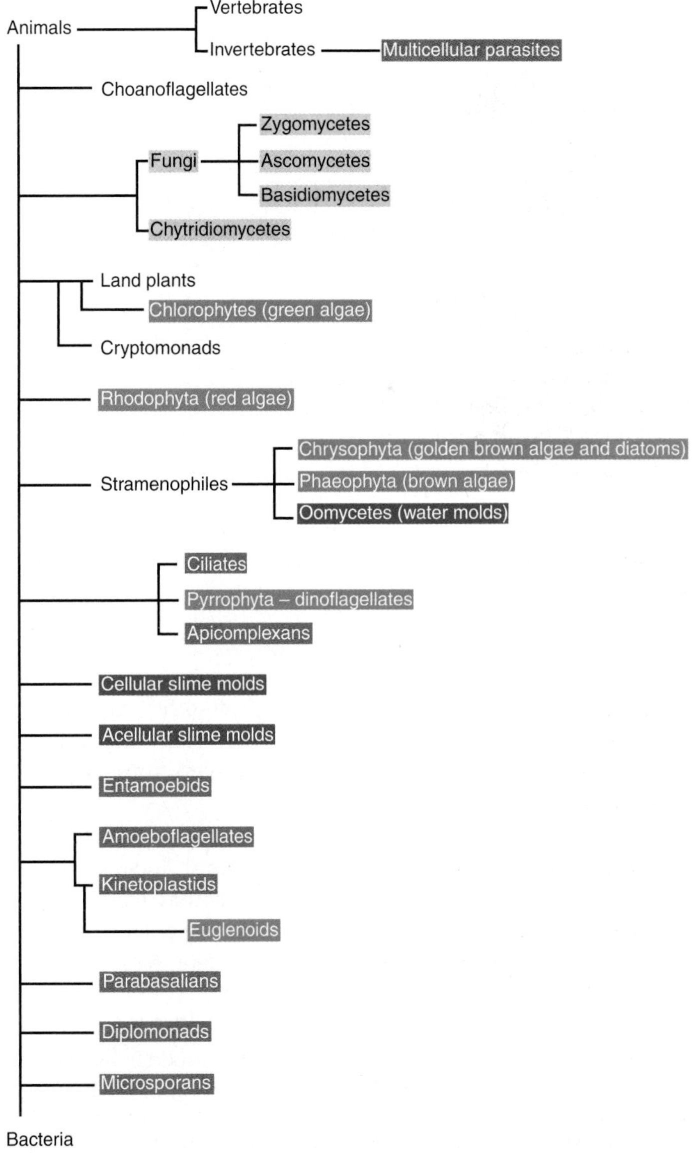

FIGURE 12.1 A Phylogeny of the Eukaryote Based on Ribosomal RNA Sequence Comparison Algae are green; fungi are yellow; protozoa are red; slime molds are blue; multicellular parasites are brown.

12.1 Algae

Focus Points

- Explain how algae differ from the other members of eukaryotic microbial world.
- Describe how algae can affect human health.

The **algae** are a diverse group of eukaryotic organisms that share some fundamental characteristics but are not necessarily related on the phylogenetic tree. These organisms are studied by **algologists** in a field known as **algology.** Algae are organisms that use light energy to convert CO_2 and H_2O to carbohydrates and other cellular products with the release of oxygen. Algae contain chlorophyll *a,* which is necessary for photosynthesis. In addition, many algae contain other pigments that extend the range of light waves that can be used by these organisms for photosynthesis. Algae include both microscopic unicellular members and macroscopic multicellular organisms

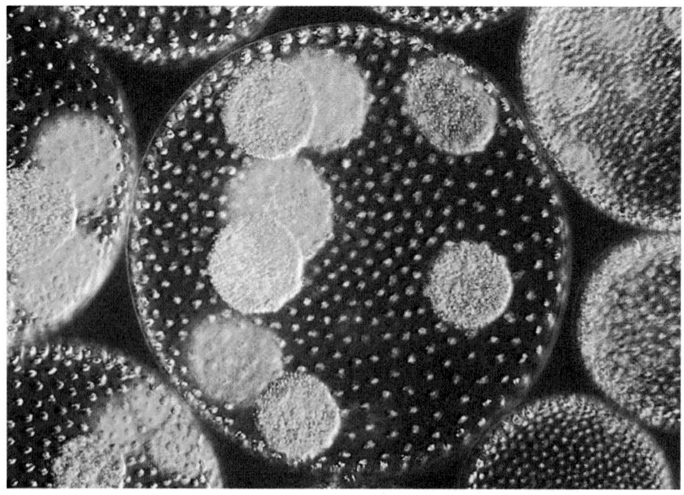

(a)

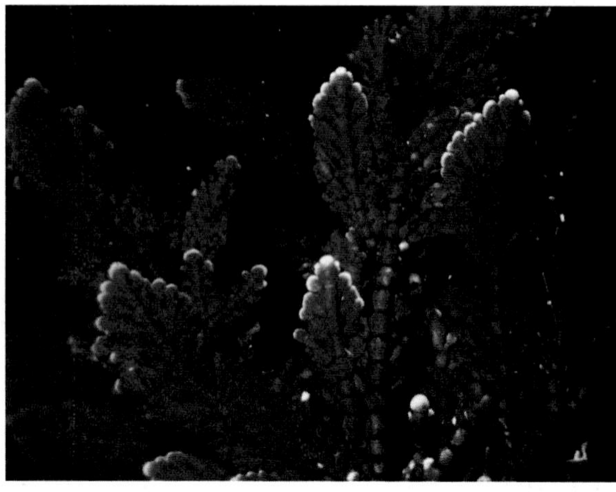

(b)

FIGURE 12.2 Algae (a) *Volvox* sp. a colony of cells formed into a hollow sphere (125x). The yellow-green circles are reproductive cells that will eventually become new colonies. **(b)** *Corallina gracilis,* red coral algae.

(figure 12.2). What distinguishes algae from other eukaryotic photosynthetic organisms such as land plants is their lack of an organized vascular system and their relatively simple reproductive structures. ■ photosynthesis, p. 151

Algae do not directly infect humans, but some produce toxins that cause paralytic shellfish poisoning. Some of these toxins do not cause illness in the shellfish that feed on the algae but accumulate in their tissues. When the shellfish are eaten by humans the toxins cause nerve damage. ■ paralytic shellfish poisoning, p. 284

As one of the primary producers of carbohydrates and other cellular products, the algae are essential in the food chains of the world. In addition, they produce a large proportion of the oxygen in the atmosphere.

Classification of Algae

As just noted, *algae* is not a strict classification term; nevertheless, organisms considered under this general heading are grouped for identification by a number of properties. These include the principal photosynthetic pigments of each group, cell wall structure, type of storage products, mechanisms of motility, and mode of reproduction. The names of the different algal groups are derived from the major color displayed by most of the algae in that group. Note that the different algal groups lie in different places along the evolutionary tree (see figure 12.1—green).

Some of the general characteristics of the algal groups are summarized in **table 12.1.**

Algal Habitats

Algae are found in both fresh and salt water, as well as in soil. Since the oceans cover more than 70% of the earth's surface, aquatic algae are major producers of molecular oxygen as well as important users of carbon dioxide. Unicellular algae make up a significant part of the **phytoplankton** (*phyto* means "plant" and *plankton* means "drifting"), the free-floating, photosynthetic organisms that are found in aquatic environments. More oxygen is produced by the phytoplankton than by all forests combined.

Phytoplankton is a major food source for many animals, both large and small. Microscopic animals in the **zooplankton** (*zoo* means "animal") graze on this phytoplankton, and then both the zooplankton and phytoplankton, for example, become food for the benthic whales, some of the world's largest mammals, as well as other animals in the sea. The unicellular algae of the phytoplankton are well adapted to this aquatic environment. As single cells, they have large, adsorptive surfaces relative to their volume and can move freely about, thus effectively using the dilute nutrients available.

Because one or more algal species can grow in almost any environment, algae often grow where other forms of life cannot thrive. Frequently, algae are among the first organisms to become established in barren environments, where they synthesize the organic materials necessary for the subsequent invasion and survival of other organisms. They are often found on rocks, preparing the surface for the growth of more complex members of the biological community.

Structure of Algae

Algae can be both microscopic and macroscopic. Microscopic algae can be single-celled organisms floating free or propelled by flagella, or they can grow in long chains or filaments. Some microscopic algae such as *Volvox* form colonies of 500 to 60,000 biflagellated cells, which can be visible to the naked eye (see figure 12.2a).

Macroscopic algae are multicellular organisms with a variety of specialized structures that serve specific functions (**figure 12.3**). Some possess a structure called a **holdfast**, which looks like a root system but primarily serves to anchor the organism to a rock or some other firm substrate. Unlike a root, it is not used to obtain water and nutrients for the organism. Nutrients and water surround the organism and do not need to be drawn up from the soil. The stalk of a multicellular alga, known as the **stipe**, usually has leaflike structures or **blades** attached to it. The blades are the principal photosynthetic portion of an alga, and some also bear

TABLE 12.1 Characteristics of Major Groups of Algae

Group and Representative Member(s)	Usual Habitat	Principal Pigments (in addition to chlorophyll a)	Storage Products	Cell Walls	Mode of Motility (if present)	Mode of Reproduction
Chlorophyta						
Green algae	Fresh water; salt water; soil; tree bark; lichens	Chlorophyll *b*; carotenes; xanthophylls	Starch	Cellulose and pectin	Mostly non-motile except one order, but some reproductive elements are flagellated	Asexual by multiple fission; spores or sexual
Phaeophyta						
Brown algae	Salt water	Xanthophylls, especially fucoxanthin	Starchlike carbohydrates; mannitol; fats	Cellulose and pectin; alginic acid	Two unequal, lateral flagella	Asexual, motile zoospores; sexual, motile gametes
Rhodophyta						
Red algae, corallines	Mostly salt water, several genera in fresh water	Phycobilins including phycoerythrin and phycocyanin; carotenes; xanthophylls	Starchlike carbohydrates	Cellulose and pectin; agar; carrageenan	Non-motile	Asexual spores; sexual gametes
Chrysophyta						
Diatoms, golden brown algae	Fresh water; salt water; soil; higher plants	Carotenes	Starchlike carbohydrates	Pectin, often impregnated with silica or calcium	Unique diatom motility; one, two, or more unequal flagella	Asexual or sexual
Pyrrophyta						
Dinoflagellates	Mostly salt water but common in fresh water	Carotenes; xanthophylls	Starch; oils	Cellulose and pectin	Two unequal, lateral flagella in different planes	Asexual; rarely sexual
Euglenophyta						
Euglena	Fresh water	Chlorophyll *b*; carotenes; xanthophylls	Fats; starchlike carbohydrates	Lacking, but elastic pellicle present	One to three anterior flagella	Asexual only by binary fission

the reproductive structures. Many large algae have gas-containing **bladders** or floats that help them maintain their blades in a position suitable for obtaining maximum sunlight.

Cell Walls

Algal cell walls are rigid and for the most part composed of cellulose, often associated with pectin. Some multicellular species such as some red algae contain large amounts of other compounds in their cell walls. The red algae cell wall components such as carrageenan and agar are harvested commercially and commonly used in foods as stabilizing compounds. Agar is also used to solidify growth media in the laboratory. ■ agar, p. 87

Diatoms are algae that have silicon dioxide incorporated into their cell walls. When these organisms die, their shells sink to the bottom of the ocean, and the silicon-containing material does not decompose. Deposits of diatoms that formed millions of years ago are mined for a substance known as **diatomaceous earth,** used for filtering systems, abrasives in polishes, insulation, and many other purposes.

Eukaryotic Cell Structures

As is true in all eukaryotes, algae have a membrane-bound **nucleus.** The genetic information is contained in a number of chromosomes that are tight packages of DNA with their associated basic protein. ■ nucleus, p. 75

In addition, algae have other organelles in their cytoplasm such as **chloroplasts** and **mitochondria**. Chloroplasts where photosynthesis occurs, contain chlorophyll as well as other light-trapping pigments such as carotenoids and phycocyanin. Respiration and oxidative phosphorylation occur in the mitochondria. ■ chloroplasts, p. 77 ■ mitochondria, p. 76

Algal Reproduction

Most single-celled algae reproduce asexually by binary fission, as do most bacteria **(figure 12.4).** The major difference between prokaryotic and eukaryotic fission involves events that take place with the genetic material within the cell. Recall from chapter 3 that in prokaryotic fission, the circular DNA replicates and each

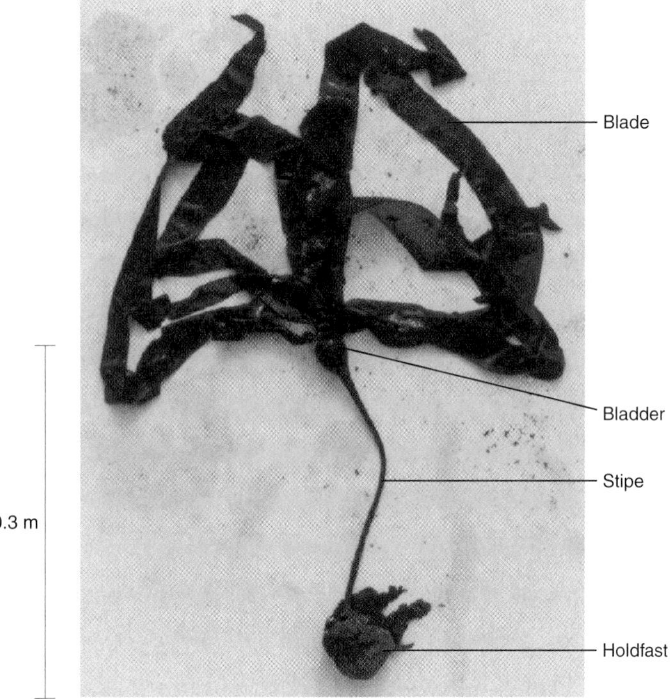

FIGURE 12.3 A Young *Nereocystis luetkeana,* the Bladder Kelp The alga has a large bladder filled with gas. This bladder keeps the blades floating on the surface of the water to maximize exposure to sunlight. The blades are the most active sites for photosynthesis. The holdfast anchors the kelp to rocks or other surfaces. In a single season, kelp can grow to lengths of 5 to 15 m.

daughter cell receives half the original double strand of DNA plus a newly replicated strand. In eukaryotic organisms with multiple chromosomes, after the DNA is replicated, the chromosomes go through a nuclear division process called **mitosis.** This process ensures that the daughter cells receive the same number of chromosomes as the original parent.

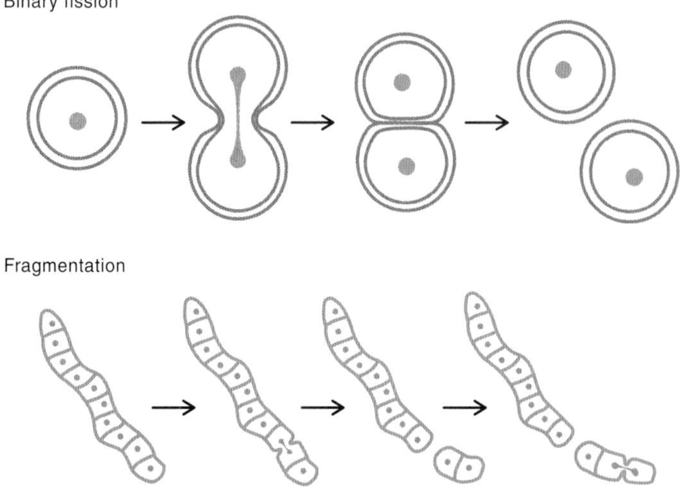

FIGURE 12.4 Binary Fission Is an Asexual Reproduction Process in Which a Single Cell Divides into Two Independent Daughter Cells Fragmentation is a form of asexual reproduction in which a filament composed of a string of cells breaks into pieces to form new organisms.

Some algae, especially multicellular filamentous species, reproduce asexually by fragmentation. In this type of reproduction, portions of the parent organisms break off to form new organisms (see figure 12.4), and the parent organism survives.

Sexual reproduction also regularly occurs in most algae. During the process known as **meiosis, haploid** cells with half the chromosome content are formed. These cells are called **gametes** and when they fuse together they form a **diploid** cell with a full complement of chromosomes known as a **zygote.** Gametes are often flagellated and highly motile. Many algae alternate between a haploid generation and a diploid generation. Sometimes, as is the case with *Ulva* (sea lettuce), the generations look physically similar and can only be distinguished by microscopic examination. In other cases, the two forms look quite different.

Paralytic Shellfish Poisoning

Although algae do not directly cause disease in humans, some do so indirectly. A number of algae produce toxins that are poisonous to humans and other animals. Several dinoflagellates of the group Pyrrophyta cause *red tides*, or algal blooms, in the ocean. Red tides were reported in the Bible along the Nile and today seem to be spreading worldwide. In the warm waters of Florida and Mexico, red tides result from an abundance of *Gymnodinium breve* **(figure 12.5).** This dinoflagellate discolors the water about 17 to 75 km from shore and produces **brevetoxin,** which kills the fish that feed on the phytoplankton. It is unclear why algae suddenly grow in such large numbers, but it is thought that sudden changes in conditions of the water are responsible. The runoff of fertilizers along the waterways and coastlines may also cause the algae to grow faster and be more prolific. In addition, an upwelling of the water often brings more nutrients as well as the cysts (the resting, resistant stages of *G. breve*) from the ocean bottom to the surface. When these cysts encounter warmer waters and additional nutrients, they are released from their resting state and begin to multiply rapidly. Persons eating fish that have ingested *G. breve* and thus contain brevetoxin may suffer a tingling sensation in their mouths and fingers, a reversal of hot and cold perceptions, reduced pulse rate, and diarrhea. The

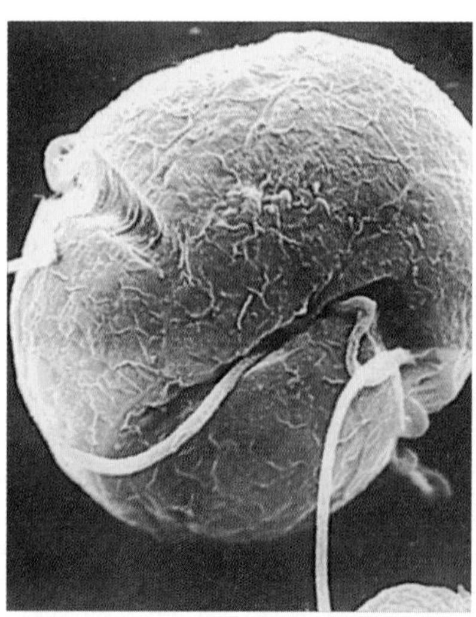

FIGURE 12.5 The Dinoflagellate *Gymnodinium* Scanning Electron Micrograph (4000×)

PERSPECTIVE 12.1

How Marine Phytoplankton Help Combat Global Warming

It has only been in recent times that the scientific community has come to realize that the small single-celled organisms that constitute the phytoplankton of the oceans and other waterways have such an influence on the climate of the world. These single-celled organisms are able to draw significant amounts of carbon dioxide (CO_2) out of the atmosphere and store it deep in the sea. In the last few years, researchers have been able to use satellite images to make these observations.

Previously it was thought that the land plants were the primary users of the CO_2 in the atmosphere. That perception changed in 1997 when NASA launched the Sea Wide Field Sensor, the first satellite capable of observing the entire planet's phytoplankton population in a single week. From their observations the scientists concluded that every year phytoplankton incorporate 45 to 50 billion metric tons of inorganic carbon into their cells, nearly double the amount previously estimated. What also surprised scientists was that the new satellite analysis showed that land plants assimilated only about 52 billion tons of carbon, half the amount earlier estimated. Thus the phytoplankton can draw out just as much CO_2 from the atmosphere as all the land plants put together.

Some scientists have suggested that it might be possible to add nutrients such as fertilizer and iron to seawater to increase the growth of the phytoplankton and thus increase the amount of CO_2 used up. This might allow for additional burning of fossil fuels and other human activities that have added CO_2 to the atmosphere, one of the causes of global warming. However, other scientists have cautioned that this might actually add more CO_2 to the atmosphere as these organisms die. Much more study of the impact of this increase in phytoplankton, as it affects the ecosystems of the ocean, is required before such projects are carried out on any large scale.

symptoms may be unpleasant but are rarely deadly, and people recover in 2 to 3 days.

Red tides caused by the dinoflagellate of the genus *Gonyaulax* are much more serious. *Gonyaulax* species produce **neurotoxins** such as **saxitoxin** and **gonyautoxins**, some of the most potent non-protein poisons known. Shellfish such as clams, mussels, scallops, and oysters feed on these dinoflagellates without apparent harm and, in the process, accumulate the neurotoxin in their tissues. Then when humans eat the shellfish, they suffer symptoms of paralytic shellfish poisoning including general numbness, dizziness, general muscle weakness, and impaired respiration. Death can result from respiratory failure. *Gonyaulax* species are found in both the North Atlantic and the North Pacific. They have seriously affected the shellfish industry on both coasts over the years. At least 200 manatees died along the coast of Florida in the spring of 1996 as a result of red tide poisoning.

Another dinoflagellate, *Pfiesteria piscida*, usually is found as a non-toxic cyst or ameba in marine sediments. This organism changes when a school of fish approaches. The fish secrete a chemical cue that alerts the *Pfiesteria* to transform into a flagellated zoospore. It then releases two toxins. One stuns the fish and the other causes its skin to slough away. After this, *Pfiesteria* enjoys a meal on the fish's red blood cells and proceeds to sexually reproduce. The toxins are so potent that researchers working with them in a laboratory have been seriously affected. This organism now must be studied using the same precautions that are used to study the virus that causes AIDS.

Another algal toxin found in some species of diatoms can cause paralytic shellfish poisoning. This poison is domoic acid and is most often associated with shellfish and crabs that accumulate the poison. Persons eating the shellfish suffer nausea, vomiting, diarrhea, and abdominal cramps as well as some neurological symptoms such as loss of memory.

State agencies constantly monitor for algal toxins, and it is wise to check with the local health department before harvesting shellfish for human consumption. Algal toxins may be present even when the water is not obviously discolored. Cooking the shellfish does not destroy these toxins.

MICROCHECK 12.1

Algae are a diverse group of terrestrial and aquatic eukaryotic organisms that all contain chlorophyll a and carry out photosynthesis. Algae are found in both fresh and salt water and are a significant part of the phytoplankton. Algae do not cause disease directly, but they can produce toxins that are harmful when ingested by humans.

- ✓ What are the primary characteristics used to distinguish algae from other organisms?
- ✓ What harmful effect can algae have on humans?
- ✓ What advantage is it that single cells have a large absorptive surface relative to their surface area?
- ✓ Would organisms that reproduce by binary fission necessarily be genetically identical? Why or why not?

12.2 Protozoa

Focus Points

- Explain how protozoa are different from other members of eukaryotic microbial world.
- Describe some diseases of humans that are caused by protozoa.

Along with the algae, the **protozoa** constitute another group of eukaryotic organisms that traditionally have been considered part of the microbial world. Protozoa are microscopic, unicellular organisms that lack photosynthetic capability, usually are motile at least at some stage in their life cycle, and generally reproduce by asexual fission.

Classification of Protozoa

As with algae, classification of protozoa according to rRNA and ultrastructure shows that they are not a unified group, but appear along the evolutionary continuum (see figure 12.1—red). The primary reason they are lumped together in a field known

TABLE 12.2 Protozoa of Medical Importance

Traditional Classification	18s rRNA Classification	Genus of Disease-Causing Protozoa	Disease Caused by Protozoa	Mode of Motility	Mode of Asexual Reproduction	Page for Additional Information
Phylum: Sarcomastigophora						
Subphylum:						
Mastigophora	Kinetoplastid	*Trypanosoma*	African sleeping sickness	Flagella	Longitudinal fission	p. 667
	Diplomonad	*Giardia*	Giardiasis			p. 609
	Parabasalian	*Trichomonas*	Trichomoniasis			p. 642
	Kinetoplastic	*Leishmania*	Leishmaniasis			
Sarcodina	Entamoebids	*Entamoeba*	Amebiasis (diarrhea)	Pseudopodia	Binary fission	p. 612–613
Phylum: Ciliophora	Ciliates	*Balantidium*	Dysentery	Cilia	Transverse fission	
Phylum: Apicomplexa	Apicomplexans	*Plasmodium*	Malaria	Flagella	Multiple fission	p. 691
		Toxoplasma	Toxoplasmosis			p. 713
		Cryptosporidium	Cryptosporidiosis			p. 610
Phylum: Microspora	Microsporans	*Microsporidium*	Diarrhea	Polar filament	?	

as **protozoology** is because they are all single-celled eukaryotic organisms that lack chlorophyll. We will concentrate on members of this group that cause human disease.

Protozoa have traditionally been divided into groups primarily based on their mode of locomotion. **Table 12.2** and figure 12.1 show where each group fits in these schemes.

The phylum **Sarcomastigophora** includes two subphyla in which most of the human disease-causing protozoa are found. The subphylum **Mastigophora** includes the flagellated protozoa. They are mainly unicellular and have one or more flagella at some time in their life cycle. These flagella are used for locomotion and food gathering as well as sensory receptors. The most important disease-causing Mastigophora are *Giardia lamblia*, *Leishmania* species, *Trichomonas vaginalis*, *Trypanosoma brucei rhodesiense*, and *Trypanosoma brucei gambiense* (see table 12.2). Diseases caused by these species will be discussed later in this book. ■ flagella, p. 65

Members of the subphylum **Sarcodina** move by means of pseudopodia. The Sarcodina change shape as they move. *Entamoeba histolytica* infects humans, causing diarrhea ranging from mild asymptomatic disease to severe dysentery. ■ *E. histolytica*, p. 612

The phylum **Ciliophora,** or the **ciliates,** includes organisms that have cilia. The cilia are similar in construction to the flagella and usually completely cover the surface of an organism. Most often, they are arranged in distinct rows and are connected to one another by fibrils known as kinetodesma. Cilia beat in a coordinated fashion in waves across the body of the protozoan. A beat of one cilium affects the cilia immediately around it, but there is no evidence that the connecting fibrils aid in this coordination. The cilia found near the oral cavity propel food into the opening. Paramecia are members of the Ciliophora. *Balantidium coli* is the only known ciliate to cause human disease. It produces ulcers in the large intestines, and pigs are its major reservoir.

Organisms in the phylum **Apicomplexa,** also referred to as **sporozoa,** cause some of the most serious protozoan diseases of humans. Malaria is caused by any of four *Plasmodium* species. It is transmitted by the female *Anopheles* mosquito. Cats are the primary host for *Toxoplasma gondii*, with humans serving as secondary hosts. Another of the Apicomplexa is *Cryptosporidium parvum*, which causes the diarrheal disease cryptosporidiosis. ■ malaria, p. 691 ■ toxoplasmosis, p. 713 ■ cryptosporidiosis, p. 610

The phylum **Microspora** includes the intracellular protozoa that infect immunocompromised humans, especially persons with AIDS. There are other protozoan phyla such as Labyrinthomorpha, Ascetospora, and Myxozoa, but they are not implicated in human disease and so we will not consider them here. They are most often found in marine habitats and are parasitic on fish and other sea life.

Protozoan Habitats

A majority of protozoa are free-living and found in marine, freshwater, or terrestrial environments. They are essential as decomposers in many ecosystems. Some species, however, are parasitic, living on or in other host organisms. The hosts for protozoan parasites range from simple organisms, such as algae, to complex vertebrates, including humans. All protozoa require large amounts of moisture, no matter what their habitat.

In marine environments, protozoa make up part of the zooplankton, where they feed on the algae of the phytoplankton and are an important part of the aquatic food chains. On land, protozoa are abundant in soil as well as in or on plants and animals. Specialized protozoan habitats include the guts of termites, roaches, and ruminants such as cattle.

Protozoa are an important part of the food chain. They eat bacteria and algae and, in turn, serve as food for larger species. The protozoa help maintain an ecological balance in the soil by devouring vast numbers of bacteria and algae. For example, a single paramecium can ingest as many as 5 million bacteria in one day. Protozoa are important in sewage disposal because most of the nutrients they consume are metabolized to carbon dioxide and water, resulting in a large decrease in total sewage solids. ■ sewage treatment, p. 739

Structure of Protozoa

Cell Wall

Protozoa lack the rigid cellulose cell wall found in algae. Most protozoa do, however, have a specific shape determined by the rigidity or flexibility of the material lying just beneath the plasma membrane. **Foraminifera** have distinct hard shells composed of silicon or calcium compounds **(figure 12.6).** The foraminiferans, which secrete a calcium shell, have through the course of millions of years formed limestone deposits such as the white cliffs of Dover on England's southern coast.

Eukaryotic Cell Structures

Protozoa are eukaryotic organisms and as such have a membrane-bound nucleus as well as other membrane-bound organelles such as mitochondria. Protozoa are not photosynthetic and thus lack chloroplasts. ■ eukaryotic organelles, p. 75

Protozoa have specialized structures for movement such as **cilia, flagella,** or **pseudopodia.** As described in chapter 3, eukaryotic flagella and cilia are distinctly different in construction from prokaryotic flagella (see figures 3.40 and 3.54). Protozoa are grouped by their mode of locomotion. For example, the Mastigophora have flagella, and Ciliophora have cilia during at least some part of their life cycle, and Sarcodina use pseudopodia for movement (see table 12.2). ■ flagella, p. 74

Feeding in Protozoa

Since protozoa live in an aquatic environment, water, oxygen, and other small molecules readily diffuse through the cell membrane. In addition, as described in chapter 3, protozoa use either pinocytosis or phagocytosis to obtain food and water (see figure 3.49).

FIGURE 12.6 A Group of Protozoa with Hard Silicon Shells

Protozoan Reproduction

The life cycles of protozoa are sometimes complex, involving more than one habitat or host. Morphologically distinct forms of a single protozoan species can be found at different stages of the life cycle. Such organisms are said to be **polymorphic (figure 12.7).** This polymorphism is comparable in some respects to the differentiation of various cell types that form plant and animal tissues.

The ability to exist in either a **trophozoite** (vegetative or feeding form) or **cyst** (resting form) is characteristic of many protozoa. Certain environmental conditions, such as the lack of nutrients, moisture, oxygen, low temperature, or the presence of toxic chemicals may trigger the development of a protective cyst wall within which the cytoplasm becomes dormant. Cysts provide a means for the dispersal and survival of protozoa under adverse conditions and can be compared to the bacterial endospore. Protozoan cysts, however, are not as resistant to heat and other adverse conditions as are bacterial endospores. When the cyst encounters a favorable environment, the trophozoite emerges. Thus, a number of parasitic protozoa are disseminated to new hosts during their cyst stage. ■ endospores, p. 69

Both asexual and sexual reproduction are common in protozoa and may alternate during the complicated life cycle of some

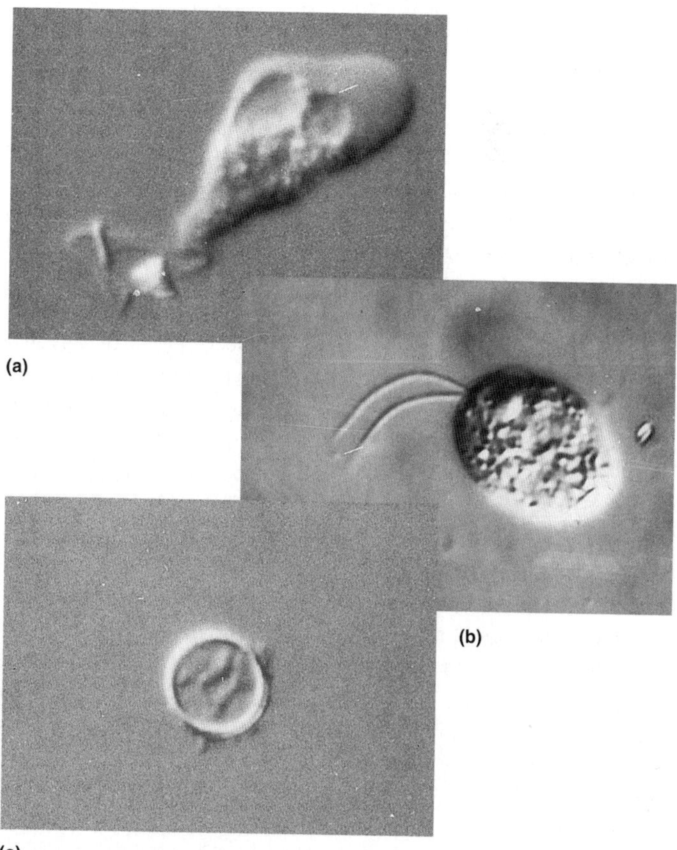

FIGURE 12.7 Polymorphism in a Protozoan The species of *Naegleria* may infect humans. **(a)** In human tissues, the organism exists in the form of an ameba (10–11 μm at its widest diameter). **(b)** After a few minutes in water, the flagellate form appears. **(c)** Under adverse conditions, a cyst is formed.

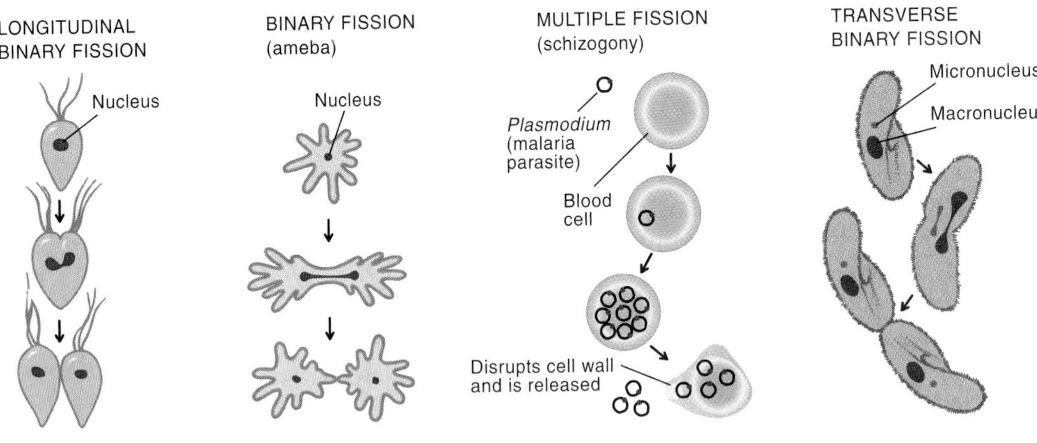

FIGURE 12.8 Various Forms of Asexual Reproduction in Protozoa

organisms. Binary fission takes place in many groups of protozoa (**figure 12.8**). In the flagellates, it usually occurs longitudinally, and in the ciliates, it occurs transversely. Since some protozoa possess both cilia and flagella, their method of asexual reproduction determines in which group they are classified.

Some protozoa divide by multiple fissions, or **schizogony,** in which the nucleus divides a number of times and then the cell produces many small single-celled organisms, each one capable of infection. Multiple fission of the asexual forms in the human host results in large numbers of parasites released into the host's circulation at regular intervals, producing the characteristic cyclic symptoms of malaria. A more detailed account of this process is described in chapter 28.

Protozoa and Human Disease

The major threat posed by protozoa results from their ability to parasitize and often kill a wide variety of animal hosts. Human infections with the protozoans *Toxoplasma gondii*, which causes toxoplasmosis; *Plasmodium* species, which are responsible for malaria; *Trypanosoma* species, which cause sleeping sickness; and *Trichomonas vaginalis*, which causes vaginitis, are common in many parts of the world. Malaria has been one of the greatest killers of humans through the ages. At least 300 million people in the world contract malaria each year, and 1 million die of it. Protozoan infections of animals are so common that it is difficult to estimate their extent or overestimate their economic importance. Some parts of tropical Africa are uninhabitable, due in large part to the presence of the tsetse fly, the carrier of the trypanosomes that cause African sleeping sickness. Humans have no natural defense mechanisms against this infection. Some animals, including cattle and horses, are reservoirs for these organisms.

Some diseases caused by protozoa are given in table 12.2.

MICROCHECK 12.2

Protozoa are microscopic, single-celled, non-photosynthetic, motile organisms. Most are free-living, but some can cause serious human disease. Protozoa are a very important part of the food chain.

- ✓ What are the primary characteristics that distinguish protozoa from other eukaryotic organisms?
- ✓ What are some important diseases caused by protozoa?
- ✓ Why would all protozoa be expected to require large amounts of water?

12.3 Fungi

Focus Points

- Explain how fungi are different from other members of eukaryotic microbial world.
- Describe the economic importance of fungi.
- Describe some of the diseases of humans caused by fungi.

The term **fungi** describes a taxonomic classification of organisms but no longer includes organisms such as slime molds and water molds that had traditionally been considered to be fungi (see figure 12.1—yellow). The slime molds and water molds once thought to be related to the fungi now appear to have evolved much earlier and will be considered separately in this chapter.

Fungi require organic compounds for energy and as a carbon source, often from dead organisms. Most fungi are aerobic or facultatively anaerobic. Only a few fungi are anaerobic. ■ aerobic, p. 92 ■ facultatively anaerobic, p. 92 ■ anaerobic, p. 91

A large number of fungi cause disease in plants. Fortunately, only a few species cause disease in animals and humans. As modern medicine has advanced to treat once-fatal diseases, however, it has left many individuals with impaired immune systems. It is these immunocompromised individuals who are most vulnerable to the fungal diseases.

The study of fungi is known as **mycology,** and a person who studies fungi is known as a **mycologist.** Along with bacteria, fungi are the principal decomposers of carbon compounds on earth. This decomposition releases carbon dioxide into the atmosphere and nitrogen compounds into the soil, which are then taken up by plants and converted into organic compounds. Without this breakdown of organic material, the world would quickly be overrun with organic waste.

Classification of Fungi

Many fungi are microscopic and can be examined using basic microbiological techniques; others are macroscopic (**figure 12.9**). All fungi have chitin in their cell walls and no flagellated cells at any time during their life cycle. There are four groups of fungi: the **Zygomycetes, Basidiomycetes, Ascomycetes,** and

FIGURE 12.9 **Fungi Range in Size from Microscopic to Macroscopic Forms** **(a)** Microscopic *Candida albicans,* showing the chlamydospores (large, round circles) that are an asexual reproductive spore. **(b)** *Polyporus sulphureus* (chicken of the woods), a shelflike fungus growing on a tree. **(c)** *Amanita muscaria,* a highly poisonous mushroom, growing in a cranberry bog on the Oregon coast.

Deuteromycetes. The classification of the first three groups is based on their method of sexual reproduction. For the fourth group, the Deuteromycetes, sexual reproduction has not been observed, and so these fungi have traditionally been lumped together. With additional rRNA analysis, however, most Deuteromycetes can now be placed in one of the other three fungal groups. Most are either Ascomycetes or Basidiomycetes who have lost the sexual part of their life cycle or it has perhaps gone unobserved by scientists **(table 12.3).**

The Zygomycetes include the common bread mold (*Rhizopus*) and other food spoilage organisms. The Ascomycetes include fungi that cause Dutch elm disease and rye smut. Smuts got this common name because the black spores that are produced give the appearance of soot. The Basidiomycetes include the common mushrooms and puffballs. In addition to these four groups of fungi, the **Chytridiomycetes** are a close relative on the evolutionary scale. They have flagellated sexual spores and more variable life cycles, however, than the fungi. Most of these organisms live in water or soil, and a few are parasitic. Black wart disease of potatoes is caused by a chytridiomycete.

Common Groupings of Fungal Forms

When people talk about fungi, they frequently use terms such as yeast, mold, and mushroom. These terms have nothing to do with the classification of fungi but instead indicate their morphological forms.

Yeasts are single-celled fungi **(figure 12.10).** Yeasts can be spherical, oval, or cylindrical and are usually 3 to 5 μm in diameter. Some yeasts reproduce by binary fission, whereas others reproduce by budding, in which a small outgrowth on the cell produces a new cell **(figure 12.11).**

TABLE 12.3 Characteristic of Major Groups of Fungi

Group and Representative Member	Usual Habitat	Some Distinguishing Characteristics	Asexual Reproduction	Sexual Reproduction
Zygomycetes *Rhizopus stolonifer* (black bread mold)	Terrestrial	Multicellular, coenocytic mycelia (with many haploid nuclei)	Asexual spores develop in sporangia on the tips of aerial hyphae	Sexual spores known as zygospores can remain dormant in adverse environment
Basidiomycetes *Agaricus campestris* (meadow mushroom) *Cryptococcus neoformans*	Terrestrial	Multicellular, uninucleated mycella. Group includes mushrooms, smuts, rusts that affect the food supply	Commonly absent	Produce basidiophores that are borne on club-shaped structures at the tips of the hyphae
Ascomycetes *Neurospora, Saccharomyces cerevisiae* (baker's yeast) *Penicillium, Aspergillus*	Terrestrial, on fruit and other organic materials	Unicellular and multicellular with septated mycelia	Is common by budding; conidiospores	Involves the formation of an ascus (sac) on specialized hyphae
Deuteromycetes (Fungi Imperfecti)	Terrestrial	A number of these are human pathogens	Budding	Absent or unknown

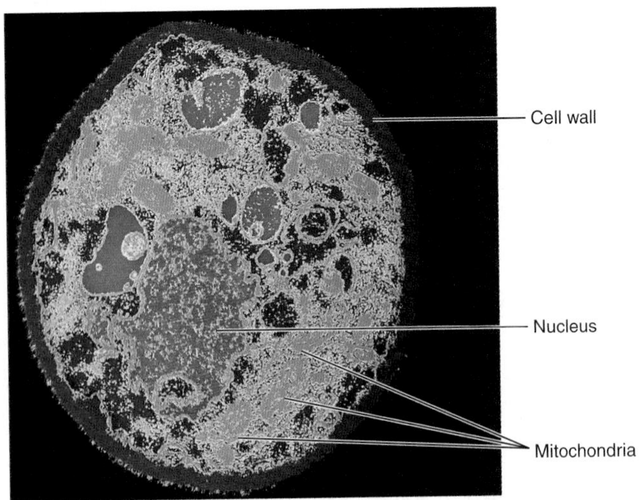

FIGURE 12.10 Morphology of a Yeast Cell As Seen with an Electron Microscope

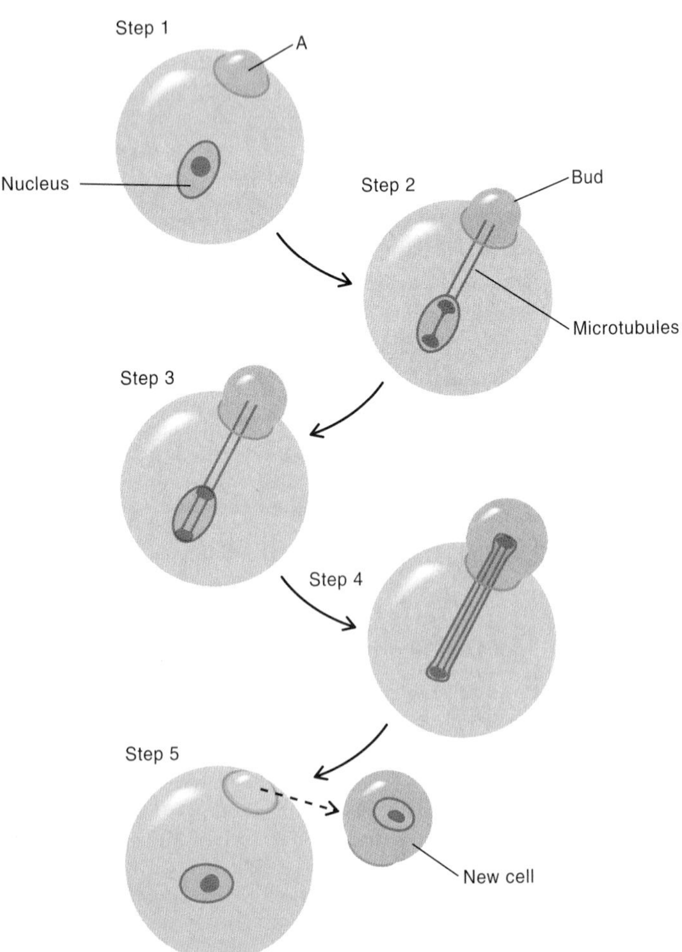

FIGURE 12.11 Budding in Yeast (1) Cell wall softens at point A, allowing the cytoplasm to bulge out. **(2, 3)** Nucleus divides by mitosis, and **(4)** one of the nuclei migrates into the bud. **(5)** Cell wall grows together and the bud breaks off, forming a new cell.

Molds are filamentous fungi. A single filament is known as a **hypha** (plural, **hyphae**), and a collection of hyphae growing in one place is known as a **mycelium.** Hyphae develop from fungal spores. A fungal reproductive spore is typically a single cell about 3 to 30 μm in diameter, depending on the species. When a fungal spore lands on a suitable substrate, it germinates and sends out a projection called a **germ tube (figure 12.12).** This tube grows at the tip and develops into a hypha. The cells divide and form new cells. In some fungi, the cell wall does not completely close off one cell from another. In that case the cells become multinucleated. The white mass seen inside the potato or on moldy bread **(figure 12.13)** is an example of a mycelium. Only a small portion of the mycelium is actually visible on the surface of the bread; the rest is buried deep within. Some mycelia appear above the surface of the substrate as a mushroom, or puffball **(figure 12.14).** These macroscopic structures produce reproductive spores. Some large mushrooms are edible.

Hyphae are well adapted to absorb nutrients. They are narrow and threadlike. With their high surface-to-volume ratio, they can absorb large amounts of nutrients. Hyphae release enzymes that break down the material into readily absorbed smaller organic compounds. In addition, these enzymes act to repel the growth of other hyphae near each other. As a result, hyphae spread throughout the food source, ensuring that each hypha will have access to adequate nutrients.

Parasitic fungi have specialized hyphae called **haustoria,** which can penetrate animal or plant cell walls to gain nutrients. Saprophytic fungi sometimes have specialized hyphae called **rhizoids,** which anchor them to the substrate.

Dimorphic Fungi

Dimorphic fungi are capable of growing either as yeastlike cells or as mycelia, depending on the environmental conditions. Some of the fungi that cause disease in humans are dimorphic. Certain fungi such as *Coccidioides immitis* grow in the soil as molds. When their spores, which are readily carried in the air, are inhaled into the warm, moist environment of the lungs, they develop into the yeast form of the organism and cause disease.

Fungal Habitats

Fungi are found in virtually every habitat on the earth where organic materials exist. Whereas algae and protozoa grow primarily in aquatic environments, the fungi are mainly terrestrial organisms. Some species occur only on a particular strain of one genus of plants, whereas others are extremely versatile in what they can degrade and use as a source of carbon and energy. Materials such as leather, cork, hair, wax, ink, jet fuel, and even some synthetic plastics like the polyvinyls can be attacked by fungi. Some species can grow in concentrations of salts, sugars, or acids strong enough to kill most bacteria. Thus, fungi are often responsible for spoiling pickles, fruit preserves, and other foods. Some fungi are resistant to pasteurization and others can grow at temperatures below the freezing point of water, rotting bulbs and destroying grass in frozen ground. Fungi are found in the thermal pools at Yellowstone National Park, in volcanic craters, and in lakes with very high salt content, such as the Great Salt Lake and the Dead Sea. ■ **food spoilage, p. 762**

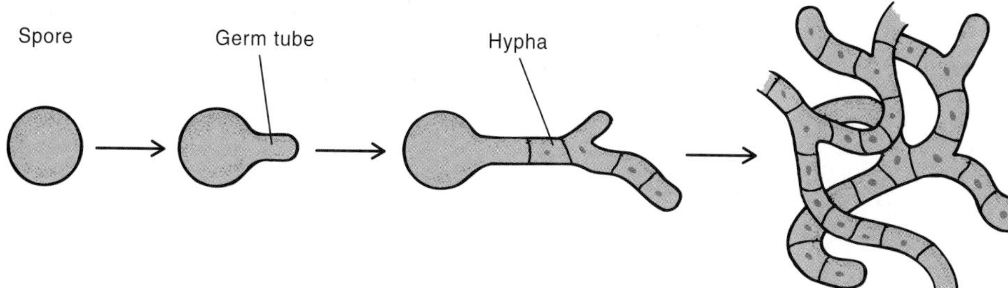

FIGURE 12.12 Formation of Hyphae and Mycelium Spores of fungi germinate to form a projection from the side of the cell called a germ tube, which elongates to form hyphae. As the hyphae continue to grow, they form a tangled mass called a mycelium.

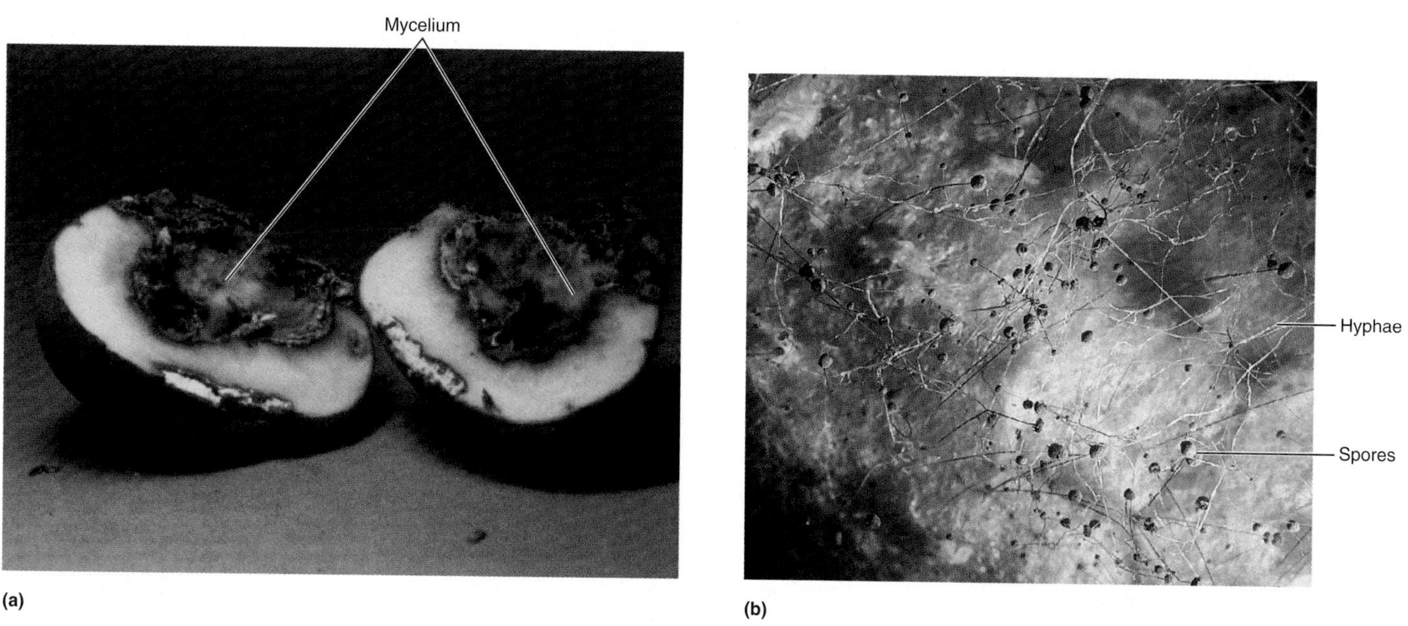

FIGURE 12.13 Examples of a Mycelium on Various Foods **(a)** The cottony white mass inside the potato is an example of a mycelium. **(b)** Magnified hyphae of *Rhizopus stolonifer,* black bread mold, showing the hyphae and the spores.

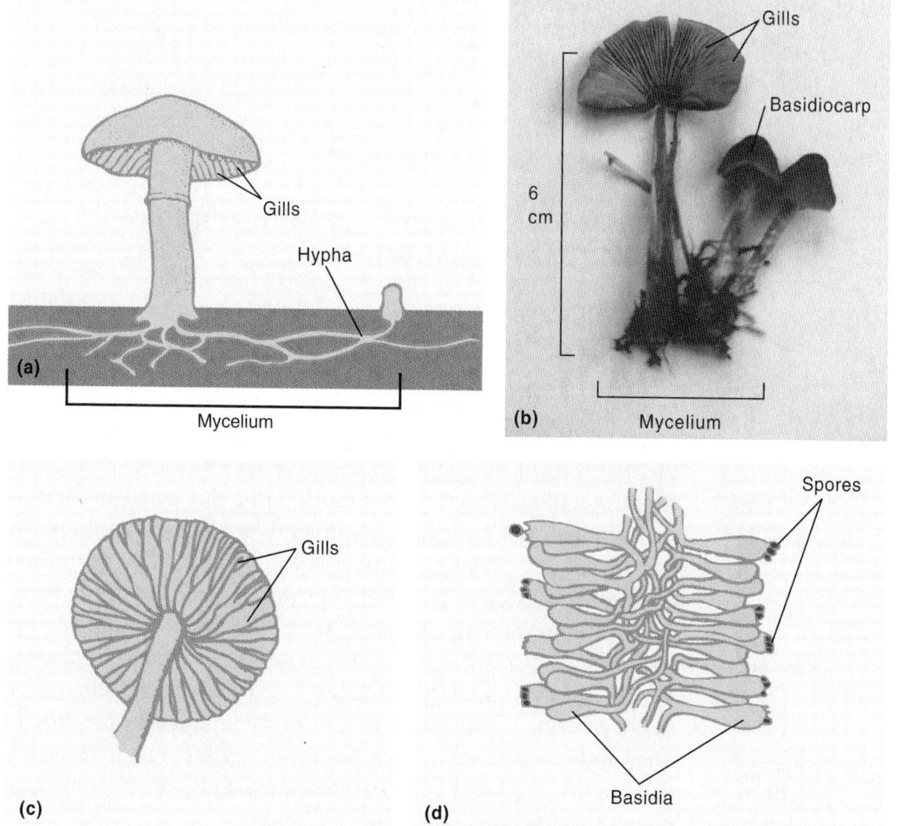

FIGURE 12.14 A Meadow Mushroom, *Agaricus campestri* **(a)** A drawing shows the extensive underground mycelium with a fruiting body emerging as a small button. **(b)** A photograph of *Lepiota rachodes,* a similar mushroom, showing the fruiting bodies and gills. **(c)** The underside of the cap is composed of radiating gills. **(d)** Magnified view of the surface of the gill showing a mass of basidia, bearing spores.

The pH at which different fungi can grow varies widely, ranging from as low as 2.2 to as high as 9.6, but fungi usually grow well at an acid pH of 5.0 or lower. This explains why fungi grow well on fruits and many vegetables that tend to be acidic.

Fungal reproductive cells, or spores, are found throughout the earth. They also occur in tremendous numbers in the air near the earth's surface as well as at altitudes of more than 7 miles. Although not as resistant as bacterial endospores, fungal spores are generally resistant to the ultraviolet rays of sunlight. Sunlight will, however, sometimes kill fungal vegetative cells. Fungal spores are a major cause of asthma. ■ asthma, p. 417

Most fungi prefer a slightly moist environment with a relative humidity of 70% or more, and various species can grow at temperatures ranging from –6°C to 50°C. The optimal temperature for the majority of fungi is in the range of 20°C to 35°C.

As heterotrophs, fungi secrete a wide variety of enzymes that degrade organic materials, especially complex carbohydrates, into small molecules that can be readily absorbed. Most fungi are aerobic, but some of the yeasts are facultative anaerobes and carry out alcoholic fermentation. Facultatively anaerobic fungi live in the intestines of certain species of fish and help degrade algae. Some fungi living in the rumen of cows and sheep are known to be obligately anaerobic. They are important in the digestion of the plant material that these animals ingest.

Fungal Disease in Humans

Fungi cause disease in humans in one of three ways. First, the fungi may actually grow on or in the human body, causing diseases, or **mycoses.** Second, a person might react to the **toxins** produced by fungi. Third, a person can develop an allergic reaction to fungal spores or vegetative cells.

Mycoses

Fungal diseases are called mycoses. The names of the individual diseases often begin with the names of the causative fungi. Thus, **histoplasmosis,** a human disease seen worldwide, is a mycosis caused by the fungus *Histoplasma capsulatum.* Similarly, **coccidioidomycosis** is a mycosis caused by *Coccidioides immitis*, a fungus unique to certain arid regions of the Western Hemisphere. Diseases caused by the yeast *Candida albicans* are called **candidiasis** and are among the most common mycoses. ■ histoplasmosis, p. 526 ■ coccidioidomycosis, p. 525 ■ candidiasis, p. 625

Infections can also be referred to by the parts of the body that they affect. **Superficial mycoses** affect only the hair, skin, or nails. **Intermediate mycoses** are limited to the respiratory tract or the skin and subcutaneous tissues. **Systemic mycoses** affect tissues deep within the body. Some diseases caused by fungi are given in **table 12.4** and are discussed in the chapters dealing with the organ systems that are affected.

Effect of Fungal Toxins

For their hallucinogenic properties, certain mushrooms have long been used as part of religious ceremonies in some cultures. The lethal effects of many mushrooms have also been known for centuries. The poisonous effects of a rye smut called **ergot** were known during the Middle Ages. The active chemical has been purified from this fungus to yield the drug ergotamine, which is now used to control uterine bleeding, relieve migraine headaches, and assist in childbirth.

TABLE 12.4 Some Medically Important Fungal Diseases

Disease	Causative Agent	Page for More Information
Candidial skin infection	*Candida albicans*	p. 554
Coccidioidomycosis	*Coccidioides immitis*	p. 525
Cryptococcal meningoencephalitis	*Cryptococcus neoformans*	p. 666
Histoplasmosis	*Histoplasma capsulatum*	p. 526
Pneumocytosis	*Pneumocystis jirovecii*	p. 712
Sporotrichosis	*Sporothrix schenckii*	p. 576
Vulvovaginal candidiasis	*Candida albicans*	p. 625

Some fungi produce toxins that are carcinogenic. The most thoroughly studied of these carcinogenic toxins, produced by species of *Aspergillus*, are called **aflatoxins.** Ingestion of aflatoxins in moldy foods, such as grains and peanuts, has been implicated in the development of liver cancer (hepatoma) and thyroid cancer in hatchery fish. Governmental agencies monitor levels of aflatoxins in foods such as peanuts, and if a certain level is exceeded, the food cannot be sold. ■ aflatoxin, p. 762

Allergic Reactions in Humans

Medical mycologists study fungi that affect humans, including fungi that cause allergic reactions. Allergic diseases such as hay fever and asthma can result from inhaling fungi or their spores if exposed humans have become sensitized. Sometimes, severe, long-term allergic lung disease results from these allergic reactions. ■ hay fever, p. 417 ■ asthma, p. 417

Symbiotic Relationships Between Fungi and Other Organisms

Fungi form several types of symbiotic relationships with other organisms. For example, lichens result from the association of a fungus with a photosynthetic organism such as an alga or a cyanobacterium **(figure 12.15).** These associations are extremely close and, in some cases, the fungal hyphae actually penetrate the cell wall of the photosynthetic partner. The fungus provides the protection and growing platform for the pair. In addition, the fungus absorbs water and minerals for the association. The photosynthetic member supplies the fungus with organic nutrients. Sometimes it is possible to grow each partner of the lichen association separately. Usually the algal partner is able to grow well when separated, but the fungal partner does not. Because of this association, lichens can grow in extreme ecosystems where neither partner could survive on its own. Lichens are often a good indicator of air quality because they are very sensitive to sulfur dioxide, ozone, and toxic metals. You will not find very many lichens in cities with air pollution.

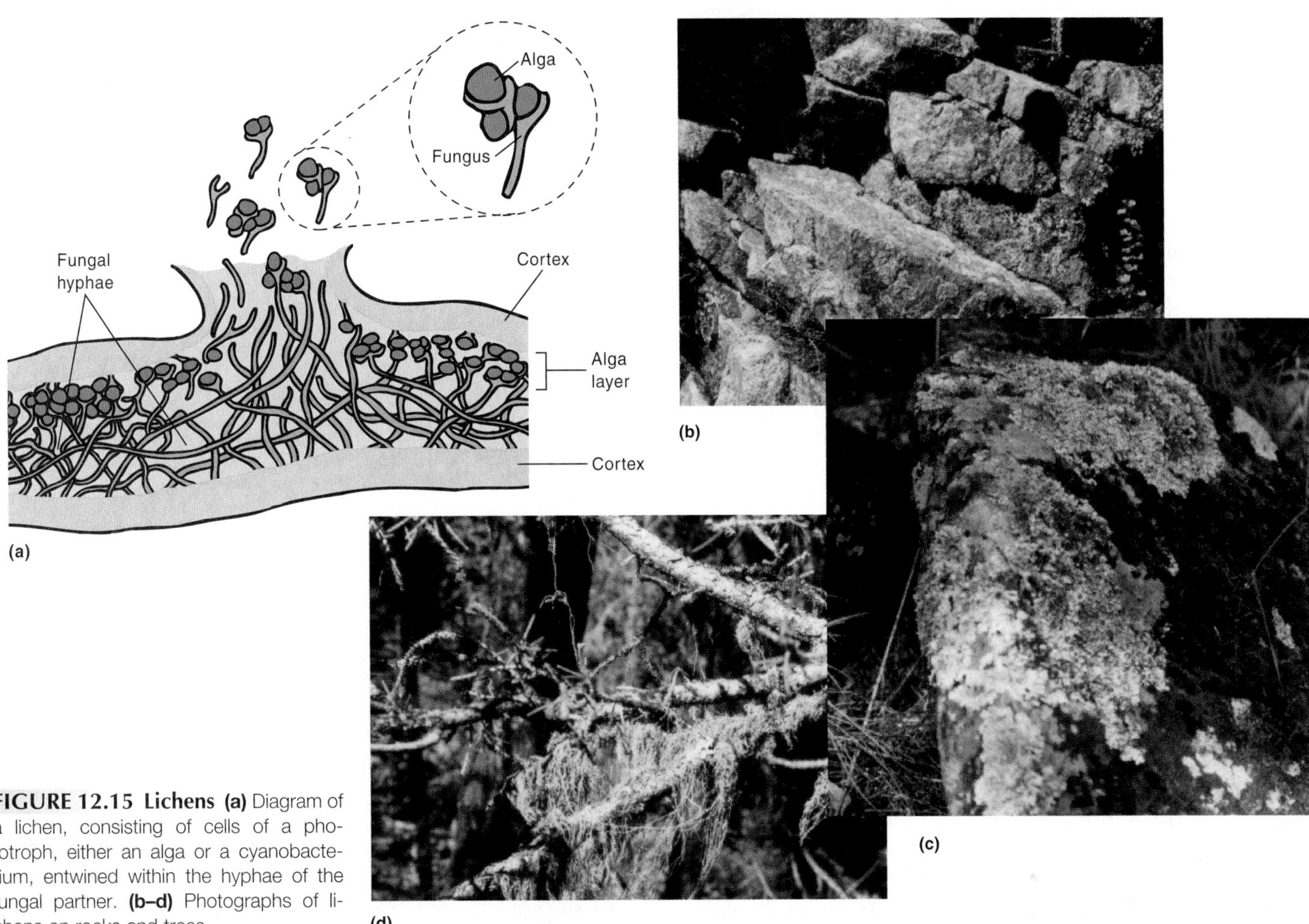

FIGURE 12.15 Lichens **(a)** Diagram of a lichen, consisting of cells of a phototroph, either an alga or a cyanobacterium, entwined within the hyphae of the fungal partner. **(b–d)** Photographs of lichens on rocks and trees.

Mycorrhizae, fungal symbioses of particular importance, are formed by the intimate association between fungi and the roots of certain plants such as the Douglas fir. It is estimated that 80% of the vascular plants have some type of mycorrhizal association with their roots. There are more than 5,000 species of fungi that form mycorrhizal associations. By increasing the absorptive power of the root, they often allow their plant partners to grow in soils where these plants could not otherwise survive. With the world facing major shortages of food and forest products, a better understanding of these symbiotic relationships is urgently needed. Many trees, including the conifers, are able to live in sandy soil because of their extensive mycelial mycorrhizas. In Puerto Rico, for example, the pine tree industry almost perished before the proper fungi were introduced to form mycorrhizal relationships with the trees. Now the industry is flourishing. Similarly, orchids cannot grow without mycorrhizal association of a fungus that helps provide nutrients to the young plant. ■ mycorrhizae, p. 727

Certain insects also depend on symbiotic relationships with fungi. For example, leaf cutter ants are estimated to bring about 15% of the tropical vegetation into their nests to use for food. The leaves are used as food by a fungus that the ants cultivate in their nests. The fungus removes the poison from the leaves of the plant, and then the ants use the fungus as their food source.

Economic Importance of Fungi

Many fungi are important commercially. The yeast *Saccharomyces* has long been used in the production of wine, beer, and bread. Other fungal species are useful in making the large variety of cheeses that are found throughout the world. Penicillin, griseofulvin, and other antimicrobial medicines are synthesized by fungi. ■ *saccharomyces*, p. 758 ■ pencillin, p. 475

Ironically, fungi are also among the greatest spoilers of food products, and large amounts of food are thrown away each year because they have been made inedible by species of *Penicillium*, *Rhizopus*, and others.

Fungi also cause many diseases of plants. Dutch elm disease caused by *Ceratocystis ulmi* is transmitted by beetles. It has destroyed the American elm trees that once shaded the streets in many cities. The wheat rust (*Puccinia graminis*) destroys tons of wheat yearly. Rust-resistant varieties have been bred to reduce the losses, but mutations in the rusts have made these advantages short-lived.

Fungi have been very useful tools for genetic and biochemical studies. *Neurospora crassa,* a common mold, has been widely used for investigating biochemical reactions. *Aspergillus nidulans* is a model for genetic studies. Yeasts have been genetically engineered to produce human insulin and the human growth hormone somatostatin, as well as a vaccine against hepatitis B. ■ biotechnology, p. 212

MICROCHECK 12.3

Fungi are organisms that have chitin in their cell walls and reproduce by both asexual and sexual methods. They are saprophytes and do not have motile cells during any stage in their life cycle. They are a source of food as well as a source of disease in plants and animals including humans.

- ✓ What are the primary characteristics that distinguish fungi from other eukaryotic organisms?
- ✓ What are some ways that fungi cause disease in humans?
- ✓ What kind of symbiotic relationships do fungi form with other organisms? How are these relationships beneficial to each partner?
- ✓ How would the narrow, threadlike structure of hyphae indicate that they have a high surface-to-volume ratio?

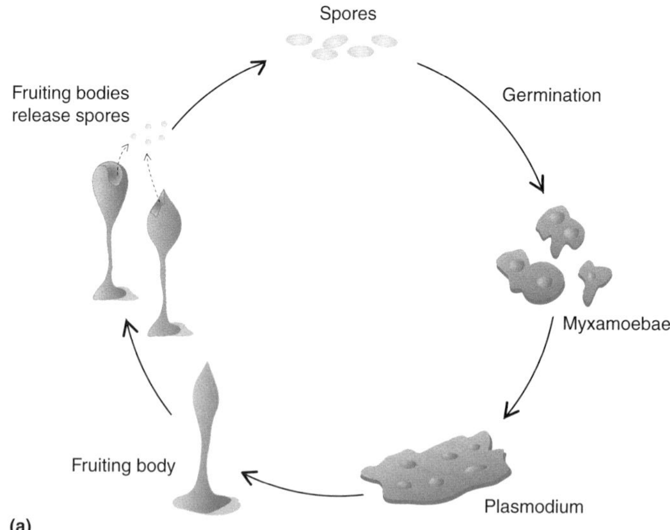

FIGURE 12.16 Slime Molds (a) Life cycle of an acellular slime mold. **(b)** Fruiting body of a cellular slime mold.

12.4

Slime Molds and Water Molds

Focus Point

- Compare and contrast slime molds and water molds.

The slime molds and water molds used to be considered types of fungi. They are, however, completely unrelated to the fungi and are good examples of **convergent evolution.** Convergent evolution occurs when two organisms develop similar characteristics because of adaptations to similar environments and yet are not related on a molecular level (see figure 12.1—blue).

Plasmodial and Cellular Slime Molds

The slime molds are terrestrial organisms that live on soil, leaf litter, or the surfaces of decaying leaves or wood. There are two groups of slime molds—the **plasmodial slime molds** and the **cellular slime molds.** They are non-motile, and reproduction depends on the formation of spores that can be dispersed. Plasmodial slime molds are widespread and readily visible in their natural environment. Following sporulation and germination of the spores, ameboid cells fuse to form a **myxameba.** The nucleus of this ameba divides repeatedly, forming a multinucleated stage called a **plasmodium.** The plasmodium oozes like slime over the surface of decaying wood and leaves. As it moves, it ingests microorganisms, spores of other fungi, and any other organic material with which it comes in contact. Eventually the plasmodium is stimulated to form a spore-bearing fruiting body, and the process begins again **(figure 12.16a).**

The cellular slime mold has a vegetative form composed of single, ameba-like cells. When cellular slime molds run out of food, the single cells congregate into a mass of cells called a slug that then forms a fruiting body and spores. These fruiting bodies and spores look very much like fungal fruiting bodies and spores (figure 12.16b).

Slime molds are important links in the food chain in the soil. They ingest bacteria, algae, and other organisms and, in turn, serve as food for larger predators. Slime molds have been valuable as unique models for studying cellular differentiation during their aggregation and formation of their fruiting bodies.

Oomycetes (Water Molds)

The **oomycetes,** or **water molds,** are members of a group of organisms known as **stramenophiles.** The oomycetes do not have chlorophyll, whereas other **stramenophiles,** which include the grass green algae, diatoms, and brown algae, have chlorophyll and other photosynthetic pigments. Oomycetes were once considered fungi because they look like fungi. They form masses of white threads on decaying material. They have flagellated reproductive cells known as **zoospores.** Oomycetes cause some serious diseases of food crops. The late blight of potato (*Phytophthora infestans*) and downy mildew of grapes are included in this group. The late blight of potato was the cause of the potato famine in Ireland in the 1840s that sent waves of immigrants to the United States (see **A Glimpse of History**).

MICROCHECK 12.4

There are three types of fungilike organisms—the plasmodial slime molds, the cellular slime molds, and the water molds. They are examples of convergent evolution.

✓ Why were the slime molds and water molds once considered to be fungi?

✓ When cellular slime molds run out of food, they form a fruiting body. What would be the advantage of this reproductive strategy?

12.5

Multicellular Parasites: Arthropods and Helminths

Focus Points

- Explain the differences between arthropods and helminths.
- Explain how each group causes disease in humans.

A number of disease-causing multicellular organisms are also studied using the same microscopic and immunological techniques that are used to study microorganisms and viruses. For that reason, they are included here. Most of the medically important multicellular parasites fall into one of two groups of invertebrates (see figure 12.1—brown): **arthropods** and **helminths.** The arthropods are more highly advanced on the evolutionary scale and include the insects, ticks, lice, and mites. Their main medical importance is that they serve as **vectors** that may transmit microorganisms and viruses to humans. The helminths, which include the **nematodes** (roundworms), **cestodes** (tapeworms), and the **trematodes** (flukes), are more primitive animals. In only a few instances do they transmit microbial infections to their host animal. Instead, they cause disease by invading the host's tissues or robbing it of nutrients.

Most multicellular parasites have been well controlled in the industrialized nations, but they still cause death and misery to many millions in the economically underdeveloped areas of the world. Our need to know about these problems has become more critical due to the global economy, more people traveling farther, more people are moving from one place to another, and more goods are being exchanged worldwide. A clear example of this occurred in New York City in the summer of 1999 when a number of people contracted West Nile fever. At least 61 persons suffered serious disease and seven people died. A significant number of crows died at the same time and were found to be carrying the disease. In addition to birds and people, horses, cats, and dogs were also found to carry the virus. It is not clear how the virus arrived in New York City, but it perhaps could have been carried by a traveler from Africa, West Asia, or the Middle East, where it is commonly found. It could possibly have been brought by an imported bird from the same areas. Worldwide travel makes us more vulnerable to diseases from other parts of the world. ■ **West Nile fever, p. 660**

In addition, worldwide climatic conditions are changing and bringing increases in certain insect populations to areas that were previously free of them. As a result, physicians are seeing more cases of multi-cellular parasitic infections in the United States.

Arthropods

The arthropods include insects, ticks, fleas, and mites. Arthropods act as vectors for transmitting diseases. In some instances, an arthropod such as a fly simply picks up a pathogen on its feet from some contaminated material such as feces and then lands on food that is then eaten by humans, thus transmitting the pathogen. In this case, the fly acts as a **mechanical vector.** In other cases, such as with *Plasmodium* sp., the cause of malaria, the vector, a mosquito, is a host for the organism before it transfers that organism to a human through a bite on the skin. In this case, the vector acts as an essential part of the life cycle of the organism and is known as a **biological vector.** The pathogen actually multiplies in number within the vector. ■ **mechanical vector, p. 455** ■ **biological vector, p. 455**

Examples of some important arthropods, the agents they transmit, and the resulting diseases are shown in **table 12.5.**

TABLE 12.5 Some Arthropods That Transmit Infectious Agents

Arthropod	Infectious Agent	Disease and Characteristic Features	Page for More Information
Insects			
Mosquito (*Anopheles* species)	*Plasmodium* species	Malaria—chills, bouts of recurring fever	p. 691
Mosquito (*Culex* species)	Togavirus	Equine encephalitis—fever, nausea, convulsions, coma	p. 660
Mosquito (*Aedes aegypti*)	Flavivirus	Yellow fever—fever, vomiting, jaundice, bleeding	p. 688
Flea (*Xenopsylla cheopis*)	*Yersinia pestis*	Plague—fever, headache, confusion, enlarged lymph nodes, skin hemorrhage	p. 682
Louse (*Pediculus humanus*)	*Rickettsia prowazekii*	Typhus—fever, hemorrhage, rash, confusion	p. 531
Arachnids			
Tick (*Dermacentor* species)	*Rickettsia rickettsii*	Rocky Mountain spotted fever—fever, hemorrhagic rash, confusion	p. 540
Tick (*Ixodes* species)	*Borrelia burgdorferi*	Lyme disease—fever, rash, joint pain, nervous system impairment	p. 542

Mosquitoes

The female mosquito needs the blood of a warm-blooded animal for the proper development of her eggs. To get this, she needs to bite such an animal. The mosquito can take in as much as twice its body weight in blood, thus giving it a relatively good chance of picking up infectious agents such as malarial parasites circulating within the host's capillaries. The anatomy of a mosquito is particularly adapted to transmit disease **(figure 12.17).** The mouthparts of the female mosquito consist of sharp stylets that are forced through the host's skin to the subcutaneous capillaries. One of these needle-like stylets is hollow, and the mosquito's saliva is pumped through it. The saliva increases blood flow and prevents clotting as the victim's blood is sucked into a tube formed by the other mouthparts of the insect. The saliva can also cause allergic reactions (the itch of a mosquito bite). After the mosquito has taken more than one blood meal, she can transmit disease from one animal to the next. Viruses and parasite found in the blood of the first animal are then transmitted to the next, and so on.

Mosquitoes in an area of arthropod-borne disease can be trapped and identified microscopically, and the blood they have ingested can be tested to see on which kinds of animals the different species are feeding. Precise identification of species and subspecies of these genera is important because different species of mosquitoes differ greatly in their breeding areas, time of feeding, and choice of host. Identification depends largely on microscopic examination of antennae, wings, claws, mating apparatus, and other features. Correct identification is often essential in designing specific control measures. ■ epidemiology, p. 450

Fleas

Fleas are wingless insects that depend on powerful hind legs to jump from place to place. Points of importance in identifying fleas include the spines (combs) about the head and thorax, the muscular pharynx, the long esophagus, and the spiny valve composed of rows of teethlike cells. Fleas are generally more of a nuisance than a health hazard, but they can transmit the bacterium *Yersinia pestis*, which causes plague, and a rickettsial disease, murine typhus, to humans. Larval fleas have a chewing type of mouth for feeding on organic matter. They ingest eggs of the common dog and cat tapeworm, *Dipylidium caninum*, serving as its intermediate host. Children acquire this tapeworm when they accidentally swallow fleas. Fleas can live in vacant buildings in a dormant stage for many months. When the building becomes inhabited, the fleas quickly mature and hungrily greet the new hosts. ■ plague, p. 682

Lice

Like fleas, lice are small, wingless insects that prey on warm-blooded animals by piercing their skin and sucking blood. The legs and claws of lice, however, are adapted for holding onto body surfaces and clothing rather than for jumping. Human lice generally survive only a few days away from their hosts.

Pediculus humanus, the most notorious of the lice, is 1 to 4 mm long, with a characteristically small head and thorax, and a large abdomen **(figure 12.18).** This louse has a membrane-like lip with tiny teeth that anchor it firmly to the skin of the host. Within the floor of the mouth is a piercing apparatus somewhat similar to that of fleas and mosquitoes. *Pediculus humanus* has only one host—humans—but easily spreads from one person to another by direct contact or by contact with personal items, especially in areas of crowding and poor sanitation.

There are two subspecies of *pediculus humanus,* popularly termed head lice and body lice. Body lice can transmit trench fever, which is caused by the bacterium *Bartonella quintana*; epidemic typhus, which is caused by the bacterium *Rickettsia prowazekii;* and relapsing fever, caused by the bacterium *Borrelia recurrentis*. Trench fever occurs episodically among severe alcoholics and the homeless of large American and European cities.

The crab louse, *Phthirus pubis*, is commonly transmitted among young adults during sexual intercourse. It is not a vector of infectious disease, but it can cause an unpleasant itch.

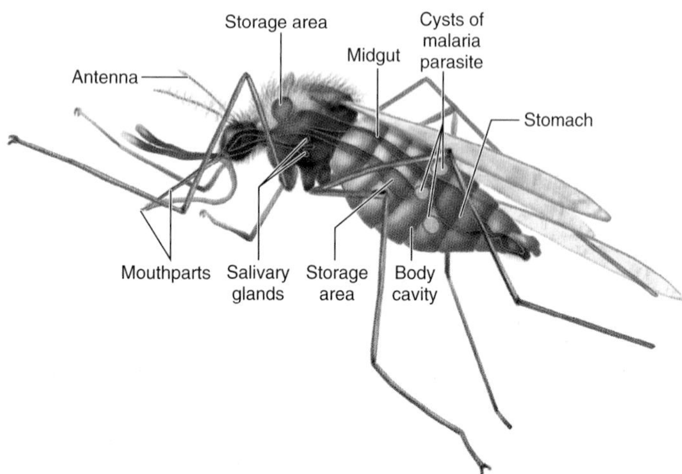

FIGURE 12.17 Internal Anatomy of a Mosquito Note the storage areas that allow ingestion of large amounts of blood and the salivary glands that discharge pathogens into the host.

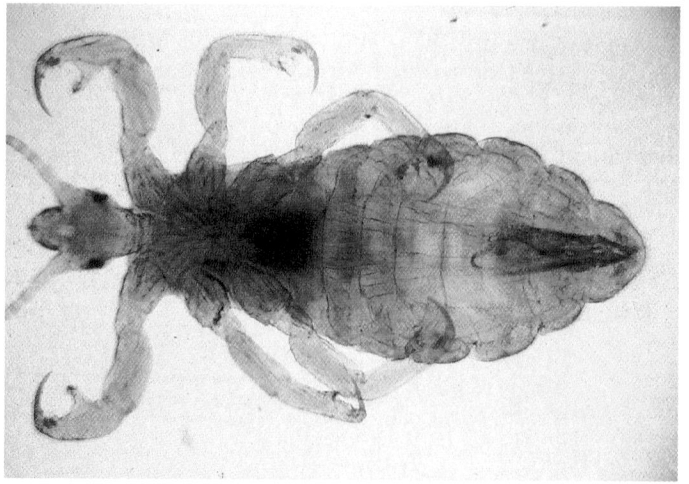

FIGURE 12.18 *Pediculus humanus* A body louse, which is the vector for *Rickettsia prowazekii,* the cause of typhus.

Ticks

Ticks are arachnids. Arachnids differ from insects in their lack of wings and antennae, and their thorax and abdomen are fused together. Although like insects the immature ticks have three pairs of legs, the adults have four pairs. *Dermacentor andersoni*, the wood tick, is the vector for Rocky Mountain spotted fever caused by the bacterium *Rickettsia rickettsii*. Another tick, *Ixodes scapularis*, transmits with its saliva *Borrelia burgdorferi,* the spirochete that causes Lyme disease. In addition, the saliva of several genera of ticks can produce a profound paralysis, especially in children on whom the tick feeds for several days. Paralyzed humans and animals usually recover rapidly following removal of the tick. ■ Rocky Mountain spotted fever, p. 540 ■ Lyme disease, p. 542

Mites

Mites, like ticks, are arachnids. They are generally tiny, fast moving, and live on the outer surfaces of animals and plants. *Demodex folliculorum* and *D. brevis* are elongated microscopic mites that live in the hair follicles or oil-producing glands usually of the face, typically without producing symptoms. Other species of mites cause human disease.

The disease scabies, caused by a mite, *Sarcoptes scabiei*, is characterized by an itchy rash most prominent between the fingers, under the breasts, and in the genital area. Scabies is easily transmitted by personal contact, and the disease is commonly acquired during sexual intercourse. The female mites burrow into the outer layers of epidermis (**figure 12.19**) feeding and laying eggs over a lifetime of about 1 month. Allergy to the mites is largely responsible for the itchy rash. The diagnosis can only be made by demonstrating the mites, since scabies mimics other skin diseases. Treatment of scabies is easily accomplished with medication applied to the skin. *Sarcoptes scabiei* is not known to transmit infectious agents.

Mites of domestic animals and birds can cause an itchy rash in humans, as can mites sometimes present in hay, grain, cheese, or dried fruits. The dust mites that often live in large numbers in bedrooms can sometimes cause asthma when the mites and their excreta are inhaled.

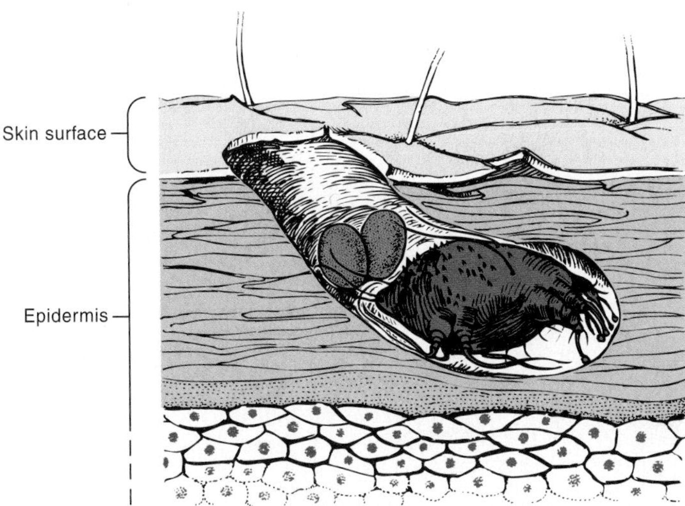

FIGURE 12.19 ***Sarcoptes scabiei*** **(Scabies Mite)** The female burrows into outer skin layers to lay her eggs, causing an intensely itchy rash.

The mites of rodents can transmit rickettsial diseases to humans. Rickettsial pox, caused by *Rickettsia akari* transmitted by mouse mites, is a mild disease characterized by fever and rash. Epidemics occur periodically in cities of the eastern United States. Serious rickettsial diseases of other parts of the world such as scrub typhus are transmitted by rodent mites.

Helminths

In addition to the arthropods that can lead to disease in humans, the other group of multicellular animals that causes human disease are the helminths. In humans, the helminths that cause disease generally belong to one of three classes: the **nematodes,** or roundworms; the **cestodes,** or the tapeworms; and the **trematodes,** or the flukes.

These multicellular parasites have been controlled in the developed nations, but they continue to kill many millions in underdeveloped parts of the world. Helminths enter the body in a number of ways. They may be eaten in contaminated food, be passed through insect bites, or directly penetrate the skin. They cause disease by invading the host tissues or robbing the host of nutrients. Some helminths have complex life cycles, involving one or more intermediate hosts where early stages of development occur, and a definitive host where the sexually mature forms occur.

Nematodes or Roundworms

The nematodes, or roundworms, have a cylindrical, tapered body with a tubular digestive tract that extends from the mouth to the anus. There are both male and female nematodes. Nematodes include a large number of species. Many nematodes are free-living in soil and water. Others are parasites of human and other animals and plants and produce serious disease.

The nematodes that cause disease can be divided into two groups—the ones that inhabit the gastrointestinal tract of the host, and the ones that are found in the blood and other tissues of the host. Generally, diagnosis of worm infestation depends on microscopic identification of the worms or their ova (eggs), or on blood tests for antibody to the worms. **Table 12.6** summarizes the major diseases caused by these parasites.

Cestodes or Tapeworms

Cestodes, or tapeworms, have flat, ribbon-shaped bodies that are segmented. The head (scolex) of the tapeworm has suckers for attachment and sometimes has hooks. Directly behind the head is a region that produces the reproductive segments (proglottids). Each segment has both male and female sex organs. The tapeworm does not have a digestive system but rather absorbs nutrients directly. Tapeworms are often associated with beef, lamb, pork, and fish. Transmission of these organisms to humans often occurs when the flesh of these animals is eaten either uncooked or undercooked **(figure 12.20).** Some tapeworms are transmitted to humans when they ingest fleas infected with dog or cat tapeworms. Table 12.6 lists specific tapeworm diseases.

TABLE 12.6 Nematodes, Cestodes, and Trematodes

Infectious Agents	Disease	Disease Characteristics
Nematodes (roundworms)		
Pinworms (*Enterobius vermicularis*)	Enterobiasis	Anal itching, restlessness, irritability, nervousness, poor sleep
Whipworm (*Trichuris trichiura*)	Trichuriasis	Abdominal pain, bloody stools, weight loss
Hookworm (*Necator americanus*) and (*Ancylostoma duodenale*)	Hookworm disease	Anemia, weakness, fatigue, physical and mental retardation in children
Threadworm (*Strongyloides stercoralis*)	Strongyloidiasis	Skin rash at site of penetration, cough, abdominal pains, weight loss
Ascaria (*Ascaris lumbricoides*)	Ascariasis	Abdominal pain, live worms vomited or passed in stools
Trichinella (*Trichinella spiralis*)	Trichinosis	Fever, swelling of upper eyelids, muscle soreness
Filaria (*Wuchereria bancrofti*) (*Brugia malayi*)	Filariasis	Fever, swelling of lymph glands, genitals, and extremities
Cestodes (tapeworms)		
Fish tapeworm (*Diphyllobothrium latum*)	Tapeworm disease	Few or no symptoms, sometimes anemia
Beef tapeworm (*Taenia saginata*)	Tapeworm disease	Few or no symptoms, sometimes anemia
Pork tapeworm (*Taenia solium*)	Cysticercosis	Variable symptoms depending on location and number of cysticerci in the body
Trematodes (flukes)		
Cercaria (*Schistosoma mansoni*)	Schistosomiasis	Liver damage, malnutrition, weakness, and accumulation of fluid in the abdominal cavity
Cercaria of birds and other animals	Swimmer's itch	Inflammation of the skin, itching

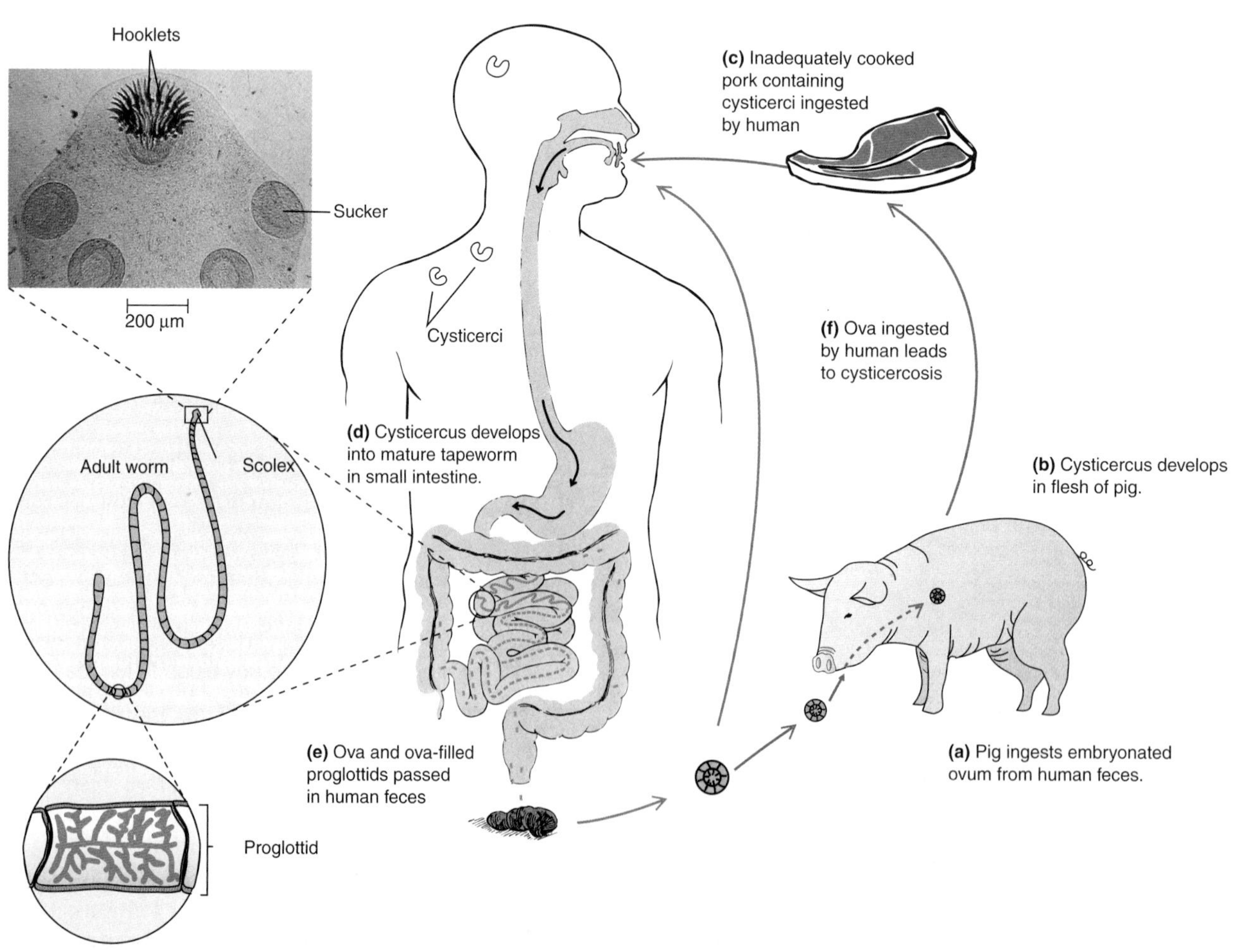

FIGURE 12.20 (a–e) Life Cycle of the Pork Tapeworm, *Taenia solium,* Acquired by Eating Inadequately Cooked Pork If worm ova hatch in the intestine, larval forms called cysticerci can develop throughout the person's tissues. A cysticercus in the brain can cause epilepsy.

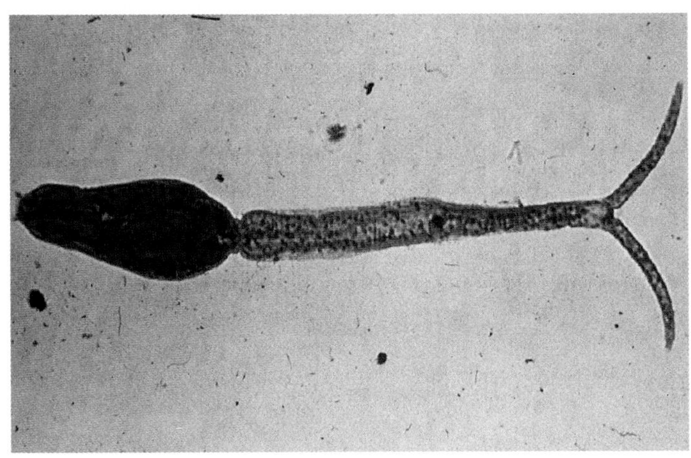

FIGURE 12.21 ***Schistosoma mansoni* Cercaria** The form that penetrates the skin and initiates the infection. Water bird schistosomes may penetrate the skin of swimmers and then die, causing swimmers itch.

Trematodes or Flukes

Trematodes, or flukes, are bilaterally symmetrical, flat, and leaf-shaped. They have suckers that hold the organism in place as well as suck fluids from the host. Most species are hermaphroditic (have both sex organs in the same worm). Most trematodes have a complicated life cycle, which may include one or more intermediary hosts. Usually, the worms begin with a larval form developing within the egg. These larvae escape into the environment, where they are taken up by one or more intermediate hosts such as a snail. Eventually, the last stage is a tail-bearing larva known as a **cercaria,** which is released from the snail and is ready to attach to the susceptible host **(figure 12.21).** For example, if a human is wading in water and the cercaria of *Schistosoma mansoni* has been discharged into that water, the cercaria can penetrate the skin and work its way through the circulation to the liver and intestine, where it matures and lays eggs.

More information about the specific diseases that the above organisms cause can be found on the website.

MICROCHECK 12.5

Arthropods such as mosquitoes, fleas, lice, and ticks act primarily as vectors for the spread of disease. On the other hand, the helminths, which include the roundworms, tapeworms, and flukes, cause serious disease in humans.

- ✓ How are arthropods able to spread disease in humans?
- ✓ What are the major differences among nematodes, cestodes, and trematodes?
- ✓ How would increased travel lead to increased spread of multicellular parasites?

FUTURE CHALLENGES

The Continued Fight to Eradicate Malaria

Malaria has affected humans for millennia. The symptoms were described in ancient Chinese writings from more that 4,500 years ago. The world map here shows how widespread malaria still is throughout certain parts of the world. It is estimated that each year a million people die from malaria; about 90% are children under the age of 5 years. The disease is still found in about 90 countries worldwide and drug-resistant strains are spreading rapidly. More people are dying today from malaria than did 30 years ago.

The economic costs to the people living in malaria-infested parts of the world are huge. Drug companies and foundations are continuing their effort to develop new vaccines and drugs to treat malaria. The Bill and Melinda Gates Foundation, among others, has pledged more than 3.6 billion dollars to aid in the effort. A multifaceted approach includes development of a vaccine, distribution of insecticide-treated bed nets, spraying interior buildings with DDT, eliminating breeding grounds for mosquitoes, and providing effective and affordable drugs.

Distribution of malaria

Map from Centers for Disease Control (CDC) 2004.

In October 2007, malaria was once again in the news headlines when more than 300 scientists, physicians, public health leaders, and government officials from all over the globe met in Seattle to discuss the best ways to combine forces and conquer the scourge, once and for all. Prior to this meeting, the *Seattle Times* ran a three-day series of articles about malaria and the efforts being made to control it. With a large infusion of money and the attention that it brings, perhaps the scourge of malaria can finally be eliminated.

SUMMARY

The Eukaryotic Members of the Microbial World

Cell structure in the *Eucarya* is different from that seen in the *Bacteria* or the *Archaea. Algae, fungi,* and *protozoa* are not accurate classification terms when the rRNA sequences of these organisms are considered (figure 12.1).

12.1 Algae

Algae are a diverse group of photosynthetic organisms that contain chlorophyll *a.*

Classification of Algae (table 12.1, figure 12.1)

Classification of algae is based on their major photosynthetic pigments. Organisms are placed on the phylogenetic tree according to rRNA sequences.

Algal Habitats

Algae are found in fresh and salt water as well as in soil. Unicellular algae make up a significant part of the **phytoplankton.**

Structure of Algae

Algae may be microscopic or macroscopic. Their cell walls are made of cellulose and other commercially important materials such as agar and carrageenan. They have membrane-bound organelles including a **nucleus, chloroplasts,** and **mitochondria.**

Algal Reproduction

Algae reproduce asexually as well as sexually (figure 12.4).

Paralytic Shellfish Poisoning

Algae do not directly cause disease, but produce toxins that are ingested by fish and shellfish (figure 12.5). When these fish and shellfish are eaten by humans, dizziness, muscle weakness, and even death may result; cooking does not destroy the toxins.

12.2 Protozoa

Protozoa are microscopic, unicellular organisms that lack chlorophyll, are motile during at least one stage in their development, and reproduce most often by binary fission.

Classification of Protozoa (figure 12.1)

In classification schemes based on rRNA, protozoa are not a single group of organisms. Protozoa have traditionally been put into groups based on their mode of locomotion (table 12.2). **Sarcomastigophora** include the **Mastigophora,** the flagellated protozoa, and **Sarcodina,** which move by means of pseudopodia. **Ciliophora** move by means of cilia. **Apicomplexa,** also referred to as the sporozoa, include *Plasmodium* sp., the cause of malaria. **Microsporidia,** an intracellular protozoa, causes disease in immunocompromised individuals.

Protozoan Habitats

Most protozoa are free-living and are found in marine and fresh water as well as terrestrial environments. They are important decomposers in many ecosystems and are a key part of the food chain.

Structure of Protozoa

Protozoa lack a cell wall, but most maintain a definite shape using the material lying just beneath the plasma membrane. Life cycles are often complex and include more than one habitat (figure 12.7). Protozoa feed by either phagocytosis or pinocytosis.

Protozoan Reproduction (figure 12.8)

Reproduction is often by binary fission; some reproduce by multiple fissions or **schizogony.**

Protozoa and Human Disease (table 12.2)

Protozoa cause diseases such as malaria, African sleeping sickness, toxoplasmosis, and vaginitis.

12.3 Fungi

Fungi can cause serious disease, primarily in plants. They produce useful food products.

Classification of Fungi (table 12.3, figure 12.1)

Zygomycetes, Ascomycetes, Basidiomycetes, and **Deuteromycetes** or **Fungi Imperfecti** are the four groups of true fungi. **Chytridiomycetes** are a close relative. **Yeast, mold,** and **mushroom** are common terms that indicate morphological forms of fungi (figures 12.9, 12.10, 12.14). Fungal filaments are called **hyphae** and a group of hyphae is called a **mycelium** (figures 12.12, 12.13). **Dimorphic fungi** can grow either as a single cell (yeast) or as mycelia.

Fungal Habitats

Fungi inhabit just about every ecological habitat and can spoil a large variety of food materials because they can grow in high concentrations of sugar, salt, and acid. Fungi can be found in moist environments at temperatures from –6°C to 50°C and pH from 2.2 to 9.6. Fungi are heterotrophs with enzymes that can degrade most organic materials.

Fungal Disease in Humans (table 12.4)

Fungi may produce an allergic reaction. Fungi can produce toxins that make humans ill. These include **ergot,** those in poisonous mushrooms, and **aflatoxin.** Fungi cause **mycoses** such as **histoplasmosis, coccidioidomycosis,** and **candidiasis.**

Symbiotic Relationships Between Fungi and Other Organisms

Lichens result from an association of a fungus with a photosynthetic organism such as an alga or a cyanobacterium (figure 12.15). **Mycorrhizas** are the result of an intimate association of a fungus and the roots of a plant.

Economic Importance of Fungi

The yeast *Saccharomyces* is used in the production of beer, wine, and bread. *Penicillium* and other fungi synthesize antibiotics. Fungi spoil many food products and cause diseases of plants such as Dutch elm disease and wheat rust. Fungi have been useful tools in genetic and biochemical studies.

12.4 Slime Molds and Water Molds (figure 12.1)

Plasmodial and **cellular slime molds** are important links in the terrestrial food chain (figure 12.16). **Oomycetes,** also known as **water molds,** cause some serious diseases of plants.

12.5 Multicellular Parasites: Arthropods and Helminths

Arthropods

Arthropods act as vectors for disease (table 12.5). Mosquitoes spread disease by picking up disease-causing organisms when the mosquito bites, and later injecting these organisms into subsequent animals that it bites (figure 12.17). Fleas transmit disease such as plague; lice can transmit trench fever, epidemic typhus, and relapsing fever (figure 12.18). Ticks are implicated in Rocky Mountain spotted fever and Lyme disease. Mites cause scabies, and dust mites are responsible for allergies and asthma (figure 12.19).

Helminths (table 12.6)

Most **nematodes** or roundworms are free-living, but they may cause serious disease such as pinworm disease, whipworm disease, hookworm disease, and ascariasis. **Cestodes** are tapeworms with segmented bodies and hooks to attach to the wall of the intestine. Most tapeworm infections occur in persons who eat uncooked or undercooked meats; some tapeworms are acquired by ingesting fleas infected with dog or cat tapeworms (figure 12.20). **Trematodes,** or flukes, often have complicated life cycles that necessarily involve more than one host. *Schistosoma mansoni* **cercaria** can penetrate the skin of persons wading in infected waters and cause serious disease (figure 12.21).

REVIEW QUESTIONS

Short Answer

1. What are the major differences among the *Eucarya,* the *Bacteria,* and the *Archaea*?
2. What distinguishes algae from all the other eukaryotic microorganisms?
3. Why are algae economically important?
4. Contrast the various modes of locomotion in protozoa.
5. Why are protozoa economically important?
6. What is the difference among a yeast, a mold, and a mushroom?
7. What are fungal diseases called?
8. Why are fungi economically important?
9. Discuss the differences and similarities in the ways algae, protozoa, fungi, helminths, and arthropods cause disease in humans.
10. What is a vector? Give two examples.

Multiple Choice

Choose one or more of these organisms that best answers the question.

1. Members of this group have chitinous cell walls.
   a) Algae b) Protozoa c) Fungi
   d) Helminths e) Arthropods
2. Members of this group are photosynthetic.
   a) Algae b) Protozoa c) Fungi
   d) Helminths e) Arthropods
3. All members of this group are single-celled.
   a) Algae b) Protozoa c) Fungi
   d) Helminths e) Arthropods
4. This group can have both single-celled and multicellular members.
   a) Algae b) Protozoa c) Fungi
   d) Helminths e) Arthropods
5. This group can only cause disease in humans through toxins.
   a) Algae b) Protozoa c) Fungi
   d) Helminths e) Arthropods
6. This group lacks a cell wall.
   a) Algae b) Protozoa c) Fungi
   d) Helminths e) Arthropods
7. These groups are found in plankton.
   a) Algae b) Protozoa c) Fungi
   d) Helminths e) Arthropods
8. Red tides are caused by this group.
   a) Algae b) Protozoa c) Fungi
   d) Helminths e) Arthropods
9. This group helps produce many of the foods that we eat.
   a) Algae b) Protozoa c) Fungi
   d) Helminths e) Arthropods
10. Without these groups the food chains of the world would not exist.
    a) Algae b) Protozoa c) Fungi
    d) Helminths e) Arthropods

Applications

1. A molecular biologist working for a government-run fishery in Vietnam is interested in controlling *Pfisteria* in fish farms. He needs to come up with a treatment that kills *Pfisteria* without harming the fish and other protista and beneficial algae that serve as food for the young fish. What strategy should the biologist consider for developing a selective treatment?
2. Paper recycling companies refuse to collect paper products that are contaminated with food or have been sitting wet for a day. A college sorority member who is running a recycling program on campus wishes to know the reason for this. What reason did the chemist who works for the recycling company probably give her for this policy?

Critical Thinking

1. Explain why it may be more difficult to treat diseases in humans caused by members of the *Eucarya* than diseases caused by the *Bacteria.*
2. Fungi are known for growing and reproducing in a wide range of environmental extremes in temperature, pH, and osmotic pressure. What does this tolerance for extremes indicate about fungal enzymes?

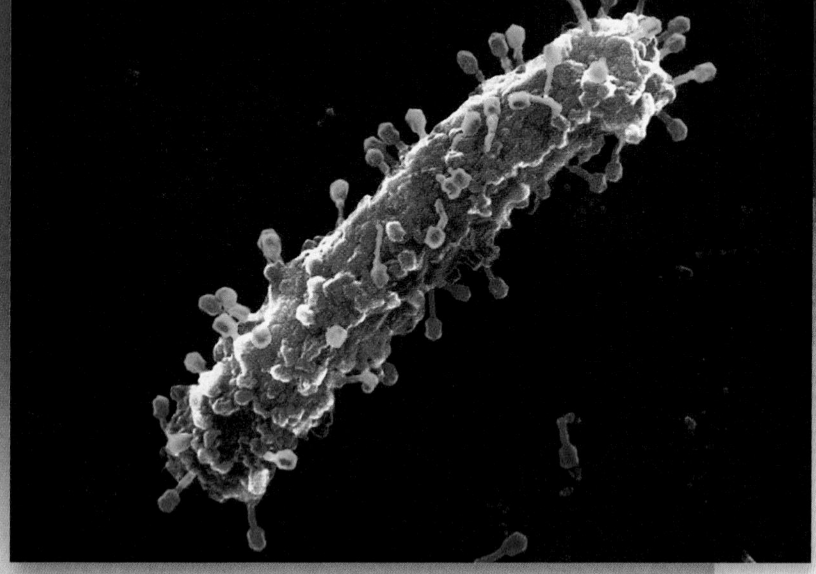

Bacterial viruses attached to a bacterium.

Viruses of Bacteria

A Glimpse of History

During the late nineteenth century, many bacteria, fungi, and protozoa were identified as infectious organisms. Most of these organisms could be readily seen with a microscope, and they generally could be grown in the laboratory. In the 1890s, D. M. Iwanowsky and Martinus Beijerinck found that a disease of tobacco plants, called mosaic disease, was caused by an agent different from anything that was known. About 10 years later, F. W. Twort in England and F. d'Herelle in France showed that infectious agents existed that had the same unusual properties and could destroy bacteria. Both agents were so small that they could not be seen with the light microscope; they passed through filters that retained almost all known bacteria, and they could be grown in media only if it contained living cells. Beijerinck called the agents filterable viruses. *Virus* means "poison," a term that once was applied to all infectious agents. With time, the adjective *filterable* was dropped and only the word *virus* was retained.

Viruses have many features that are more characteristic of complex chemicals than of cells. For example, tobacco mosaic virus (TMV) can be precipitated from a suspension with ethyl alcohol and still remain infective. A similar treatment destroys the infectivity of bacteria. Further, in 1935, Wendell Stanley of the University of California, Berkeley, crystallized tobacco mosaic virus. Its physical and chemical properties obviously differed from those of cells, which cannot be crystallized. Surprisingly, the crystallized tobacco mosaic virus could still cause the disease described by Iwanowsky and Beijerinck 40 years before.

Viruses posed a mystery to scientists as recently as 50 years ago. They were clearly smaller than known bacteria and possessed properties different from those of cells. The nature of these curious agents, some of which infected animals and others, plants, became clearer through the study of viruses that infect bacteria. Indeed, bacterial viruses are a model system for the study of viruses that infect far more complex eukaryotic cells. Because bacteria can be grown easily and multiply rapidly, viruses that infect them can be studied much more readily and easily than can viruses that infect other organisms. The information gained through the study of bacterial viruses has contributed greatly to our understanding of how animal viruses interact with their hosts.

In addition, bacterial viruses have contributed enormously to our knowledge of the molecular biology of cells. Specifically, ribosomes and messenger RNA were first discovered in virus-infected bacteria. Further, it was in virus-infected cells that the genetic code was shown to be a triplet code that was degenerate.

Bacterial viruses also play a big role in human disease. Many human pathogens synthesize deadly toxins. The genetic information for toxin synthesis is coded by bacterial virus DNA integrated in the DNA of the pathogen.

In this chapter, we will first discuss features that all viruses share and then focus on viruses that infect only bacteria. In the next chapter, we will discuss viruses that infect vertebrates and other agents of disease.

KEY TERMS

Bacteriophage A virus that infects bacteria; often shortened to *phage*.

Burst Size Number of newly formed virus particles released from a single cell following virus replication.

Carrier Cells Cells that are capable of releasing virus particles without being killed by the virus.

Host The organism infected by the virus.

Latent State The state of a phage when its DNA is integrated into the genome of the host.

Lysogen A bacterium that carries phage DNA (the prophage) integrated into its genome.

Lysogenic Conversion The change in properties of a bacterium as a result of carrying a prophage. The phage DNA codes for the new properties.

Maturation The stage in viral replication in which the various components of the virion assemble to form a whole virion; also termed *assembly*.

Productive Infection Virus infection in which more virus particles are produced as a result of infection.

Prophage Phage DNA that is integrated into the genome of a host.

Temperate Phage A phage that has the ability to integrate its DNA into the chromosome of the host.

Virion A complete virus in its inert non-replicating form.

13.1 General Characteristics of Viruses

Focus Points

- Describe the features of viruses that distinguish them from living cells.
- Draw the most common shape of a phage particle and label its parts.

Viruses are non-living entities and therefore are not organisms. Indeed, they are merely bits of nucleic acid surrounded by a protein coat. They contain no ribosomes or any structures required for the harvesting of energy. Therefore, they can only replicate inside living cells whose capabilities they use for their replication. Further, they do not undergo cell division but replicate by assembling their structural component. These agents can infect all forms of life, including all members of the *Bacteria, Archaea,* and *Eucarya* and represent another class of **mobile genetic elements.** They commonly are referred to by the organisms they infect. Those that infect bacteria are **bacteriophage,** or **phage** (phage means "to eat"). The word *phage* is both singular and plural when referring to one type of virus. The word *phages* is used when different types of phages are being referenced. Animal viruses infect animals and plant viruses infect plants. These organisms are the **hosts** for the viruses. Since phages have been isolated from extremophiles of the *Bacteria* and *Archaea,* these phages must be capable of multiplying and existing in cells exposed to extreme conditions of high temperature, high salt, and low pH. ■ mobile genetic element, p. 205

Virus Architecture

A virus particle, consists of nucleic acid (DNA or RNA) surrounded by a protective protein coat, termed a **capsid.** Different viruses have different shapes **(figure 13.1).** Some are **isometric,** and are composed of flat surfaces, forming equilateral triangles. The simplest and most common shape of viruses show **icosahedral symmetry.** This shape is a structure with 20 equilateral faces. This shell is the most efficient design for any biological container and requires the least energy to assemble. They appear spherical when viewed with the electron microscope. Others are **helical,** which gives the virion a filamentous or rodlike appearance. Most phages are more **complex** in shape, having an isometric head with a long, helical protein component, the **sheath** or **tail.**

The shape of a virus is determined by the shape of the protein capsid, either helical or spherical. The viral capsid together with the nucleic acid that is tightly packed within the protein coat is called the **nucleocapsid (figure 13.2).** The nucleic acid in some viruses is so tightly packed that the internal pressure is 10 times higher than in a champagne bottle. Each capsid is composed of many identical protein subunits, called **capsomers.** All bacterial and animal viruses, but not plant viruses, must be able to attach (adsorb) to specific receptor sites on host cells. In tailless isometric viruses, **attachment proteins,** or **spikes,** project from the capsid and are involved in attaching the virus to the host cell (see figure 13.2). In isometric viruses with tails (sheaths), fibers on the tail serve to attach the virus intially to the host cell (see figure 13.1).

There are two basic types of virions. The outer coat of most phages consists only of the protein capsid. This type of virion is called **naked.** Many virions that infect humans and other animals, however, have an additional covering over the capsid protein. This lipid is a bilayer similar in structure to the cell membrane of the eukaryotic cell. These are termed **enveloped viruses.** Just inside the lipid envelope is often a

FIGURE 13.1 Common Shapes of Viruses (a) Isometric with icosahedral symmetry (adenovirus). **(b)** Helical (tobacco mosaic virus). **(c)** Complex (T4 bacteriophage).

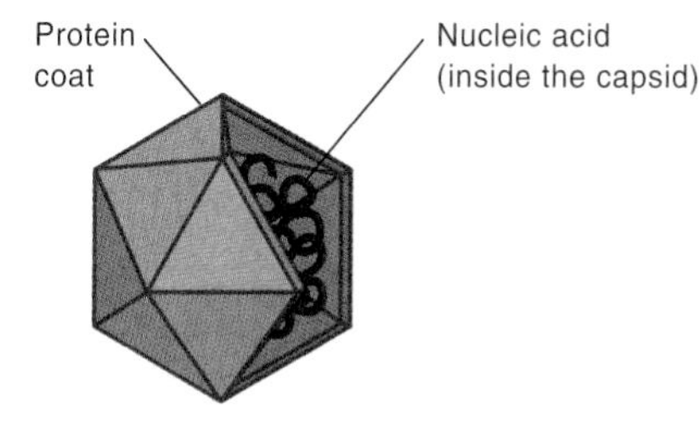

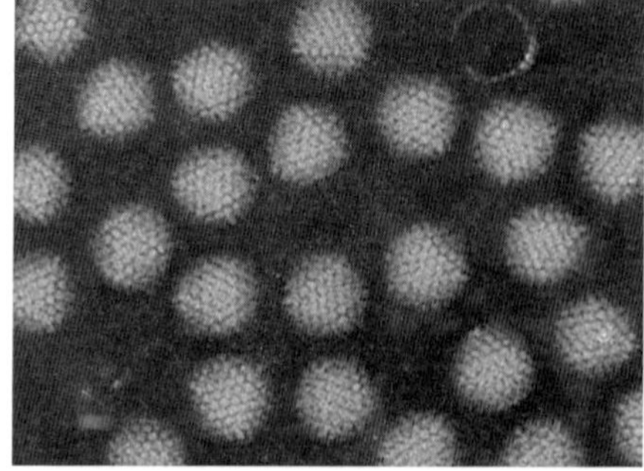

(a) Isometric (adenovirus)

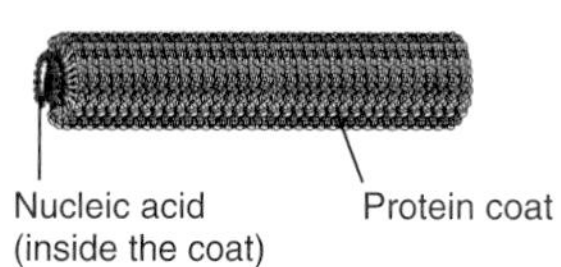

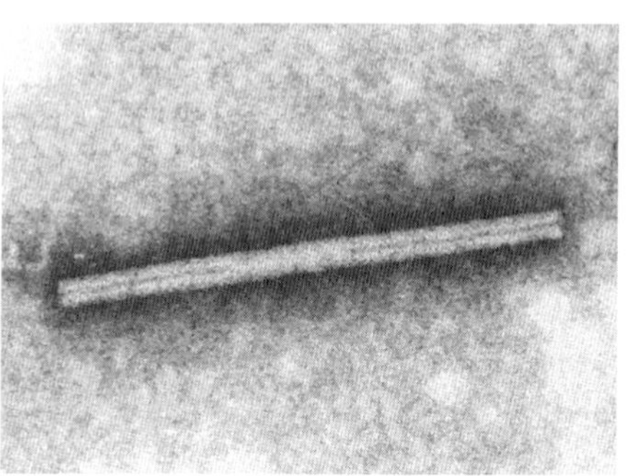

(b) Helical (tobacco mosaic virus)

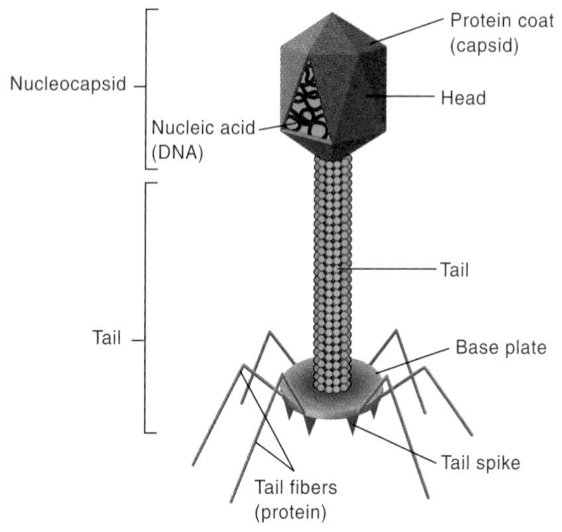

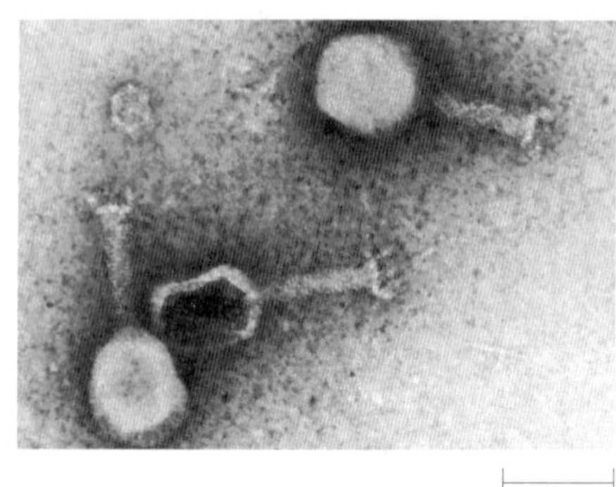

(c) Complex (T4 bacteriophage)

protein, the **matrix protein,** which is found only in enveloped viruses. The attachment spikes project from the envelope (see figure 13.2).

Most viruses are notable for their small size **(figure 13.3).** They are approximately 100- to 1,000-fold smaller than the cells they infect. The smallest viruses are about 10 nm in diameter; the largest animal viruses are about 500 nm, the size of the smallest bacterial cells. The smallest viruses contain very little nucleic acid, perhaps as few as four genes. The largest contain almost a 1,000 genes (see Perspective 13.1).

The Viral Genome

The structure of the viral genome is unusual. Viruses contain only a single type of nucleic acid—either RNA or DNA—but never both. The nucleic acids can occur in one of several different forms characteristic of the virus. DNA may be linear or circular, either double-stranded or single-stranded. RNA is usually single-stranded but a few viruses contain double-stranded RNA. Depending on the type of nucleic acid they contain, viruses are frequently referred to as RNA or DNA viruses.

Replication Cycle—Overall Features

Viruses can only multiply within living cells that are actively metabolizing because they lack the cellular components necessary to harvest energy and synthesize protein. Viruses must use structures and enzymes of cells they infect to support their own reproduction. They are **obligate intracellular parasites.** Another important characteristic is that in all viruses, the nucleic acid separates from its coat before replication begins. ■ **mitochondria, p. 76** ■ **ribosomes, p. 68**

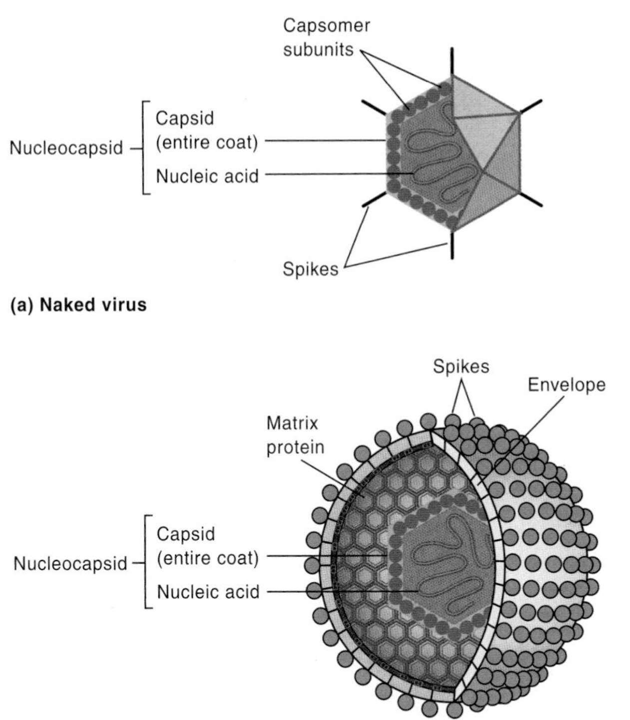

FIGURE 13.2 Two Different Types of Virions **(a)** Naked, not containing an envelope around the capsid and **(b)** enveloped, containing an envelope around the capsid and the matrix protein inside the envelope.

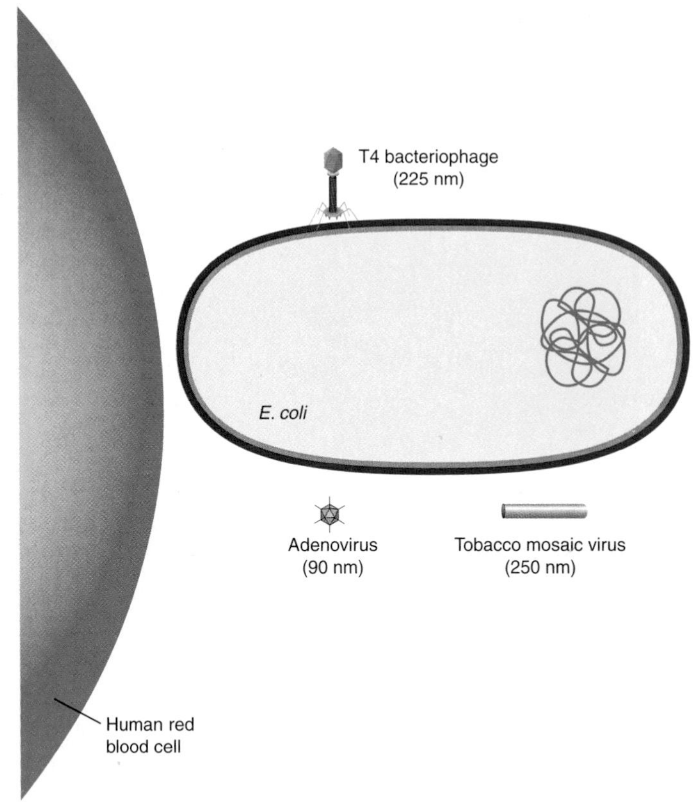

FIGURE 13.3 Virion Size Relative sizes of an animal virus (adenovirus), plant virus (tobacco mosaic virus), a bacterial virus, phage T4 and *Escherichia coli,* the host of T4. The bacterial cell is 2,000 nm in length; the red blood cell is 5,000 nm in diameter.

Because viruses contain so little nucleic acid, their few genes can code for only a very limited number of proteins, including enzymes. Every virus, however, must contain the genetic information to encode proteins required to: (1) make the viral protein coat, (2) replicate viral nucleic acid, and (3) move the virus into and out of the host cell. Some viruses require enzymes for their replication that are not present in uninfected host cells. These enzymes are coded by the nucleic acid of the virus.

Viruses exist in two distinct phases. Outside of living cells, they are metabolically inert. Essentially, they are only macromolecules. In this state, a virus is referred to as a **virion** or a **virus particle.** Inside susceptible cells, viruses are in a replication form using the metabolic machinery and pathways of their host cells along with their own genetic information to produce new virions. Viruses and cells are compared in **table 13.1.**

PERSPECTIVE 13.1

Microbe Mimicker

Like bacteria, viruses vary tremendously in size and complexity. This is becoming increasingly apparent as viruses from a variety of animal and bacterial hosts are studied in greater detail. Perhaps the most unusual virus recently characterized came from the ameba, *Acanthamoeba,* which may cause certain forms of pneumonia. The ameba is free-living and is found in air, soil, and aquatic environments. When first seen inside its *Acanthamoeba* host, the virus was thought to be a Gram-positive bacterium because of its staining characteristics and its hairy appearance, with many fibrils sticking out from its capsid. Because it was initially taken to be a bacterium, it was given the name of "mimivirus" because it mimicked the a microbe—thus "mi-mi." The mimivirus has several very unusual features. It is the largest DNA virus ever characterized—the diameter of its capsid is 500 nm. This is twice the size of a small bacterium such as *Mycoplasma.* Its genome size is enormous for a virus—1.2 million base pairs, which is large enough to encode almost a thousand proteins. These genes encode proteins never before seen in other large viruses and include enzymes of nucleic acid synthesis, DNA repair, protein translation, and polysaccharide biosynthesis. However, like all viruses, the mimi virion does not undergo cell division nor does it contain ribosomes. This virus has been placed in a group of other large DNA viruses, members of which infect disparate organisms, including vertebrates, and algae.

The mimivirus is unlikely to be the last unusual virus discovered. Many host organisms exist in nature that have not yet been cultured, and thus many likely virus families remain undiscovered. Their discovery and characterization will not only expand our knowledge of viruses but also will challenge our definition of what is living and non-living and perhaps provide an answer to the question, "Where did viruses originate?"

TABLE 13.1 Comparison Between Viruses and Cells

	Viruses	Cells
Size	10–500 nm	Bacteria: 1,000 nm Animal: 200,000 nm Plant: 400,000 nm
Multiplication	Multiply only within cells; outside the cell they are metabolically inert; protein coat and nucleic acid separate prior to multiplication	Most are free-living and multiply in the absence of other cells; entire cell remains intact during multiplication
Nucleic acid content	Contain either DNA or RNA, but never both	Always contain both DNA and RNA
Enzyme content	Contain very few, if any, enzymes	Contain many enzymes
Internal components	Lack ribosomes and enzymes for harvesting energy	Contain ribosomes and enzymes for harvesting energy

MICROCHECK 13.1

Viruses are non-living agents that consist of nucleic acid surrounded by a protein coat. Viruses only multiply within living cells, because they need certain functions of the host to multiply.

- ✓ List four features of viruses that distinguish them from free-living cells.
- ✓ List three functions that all viruses must perform.
- ✓ An antibiotic is added to a growing culture of *E. coli,* resulting in death of the cells. Bacteriophage are then added. Would the phage replicate in the *E. coli* cells? Explain your answer.

13.2

Phage Interactions with Host Cells

Focus Points

- List the steps in the replication process of a lytic phage and the major feature of each step.
- List the steps in the replication of phage in a latent stage and describe how latency is maintained.

Although viruses as a group infect all kinds of cells, the relationships between even the well-studied animal viruses and the host cells they invade are poorly understood in comparison with phage-bacteria systems. This is largely because the eukaryotic host cells of animal viruses are far more complex and grow much more slowly than the prokaryotic host cells of the phage. We will first focus on bacteriophages, which serve as excellent models for all other viruses. What you learn about them will help immensely in understanding similar relationships between viruses and the animal cells they infect, described in chapter 14.

How phages affect the cells they infect depends primarily on the type of phage. Some types multiply inside the cells they invade and escape by lysing the host cell. Because more virus is produced, this interaction is termed a **productive infection** and the phages that lyse the cell are termed **lytic** (*lysis* means dissolution). These viruses take over the metabolism of the cell and direct the cell to produce only phage. Another type of productive infection is carried out by phage that multiply, and then leak out or **extrude** without killing the host cells. These phages take over only some of the metabolic processes of the cell. An example is the filamentous phage, M13. Still other phages, termed **temperate** (*temperatus* means controlled), integrate their DNA into the genome of the bacteria they infect or the DNA replicates as a plasmid. In either case, the phage DNA replicates as the bacterial DNA replicates. The infection is termed **latent** because there may be no sign that the cells are infected. The bacterium carrying the phage DNA is a **lysogen** and the cell is in the **lysogenic state.** In the latent state, the phage DNA often codes for proteins that modify important properties of the host—the phenomenon of **lysogenic conversion,** which is discussed on page 310. An example of a temperate phage that infects *E. coli* is lambda (λ). These different relationships are shown in **figure 13.4.** ■ plasmid, p. 68

Some of the phages that undergo these three kinds of relationship with their host bacteria are listed in **table 13.2.** We will now describe in more detail phages that exhibit each of these relationships.

Lytic Phage Replication by Double-Stranded DNA Phages

In all phage infections, the phage nucleic acid enters the bacterium while its protein coat remains on the outside. The nucleic acid code

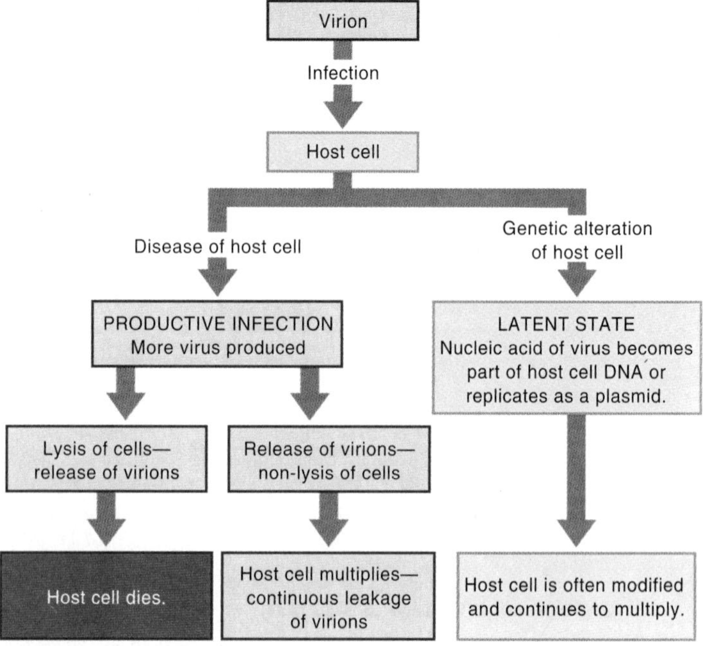

FIGURE 13.4 Major Types of Relationships Between Viruses and the Host Cells They Infect

TABLE 13.2 Important Bacteriophages

Bacteriophage	Host	Shape	Genome Structure*	Relationship to Host Cell
T4, T1–T7	*Escherichia coli*	Complex	ds DNA	Lytic
M13, fd	*Escherichia coli*	Filamentous	ss DNA	Exit by extrusion
ϕX174	*Escherichia coli*	Isometric	ss DNA	Lytic
Lambda (λ)	*Escherichia coli*	Complex	ds DNA	Latent or lytic
MS2, Qβ	*Escherichia coli*	Isometric	ss RNA	Lytic
Beta (β)	*Corynebacterium diphtheriae*	Isometric	ds DNA	Latent; codes for diphtheria toxin

*ds, double-stranded; ss, single-stranded

for the protein in the phage coat and replicates as the phage proteins are synthesized, resulting in the formation of many virions. At the end of the replication cycle, phage exit by lysing the bacterium. Phages that go through this productive life cycle are termed **virulent.** An intensively studied virulent phage is the double-stranded DNA phage, T4 (see figure 13.1). Its replication cycle illustrates how a phage can take over the life of a cell and reprogram the cells' activities solely to the synthesis of phage **(figure 13.5).**

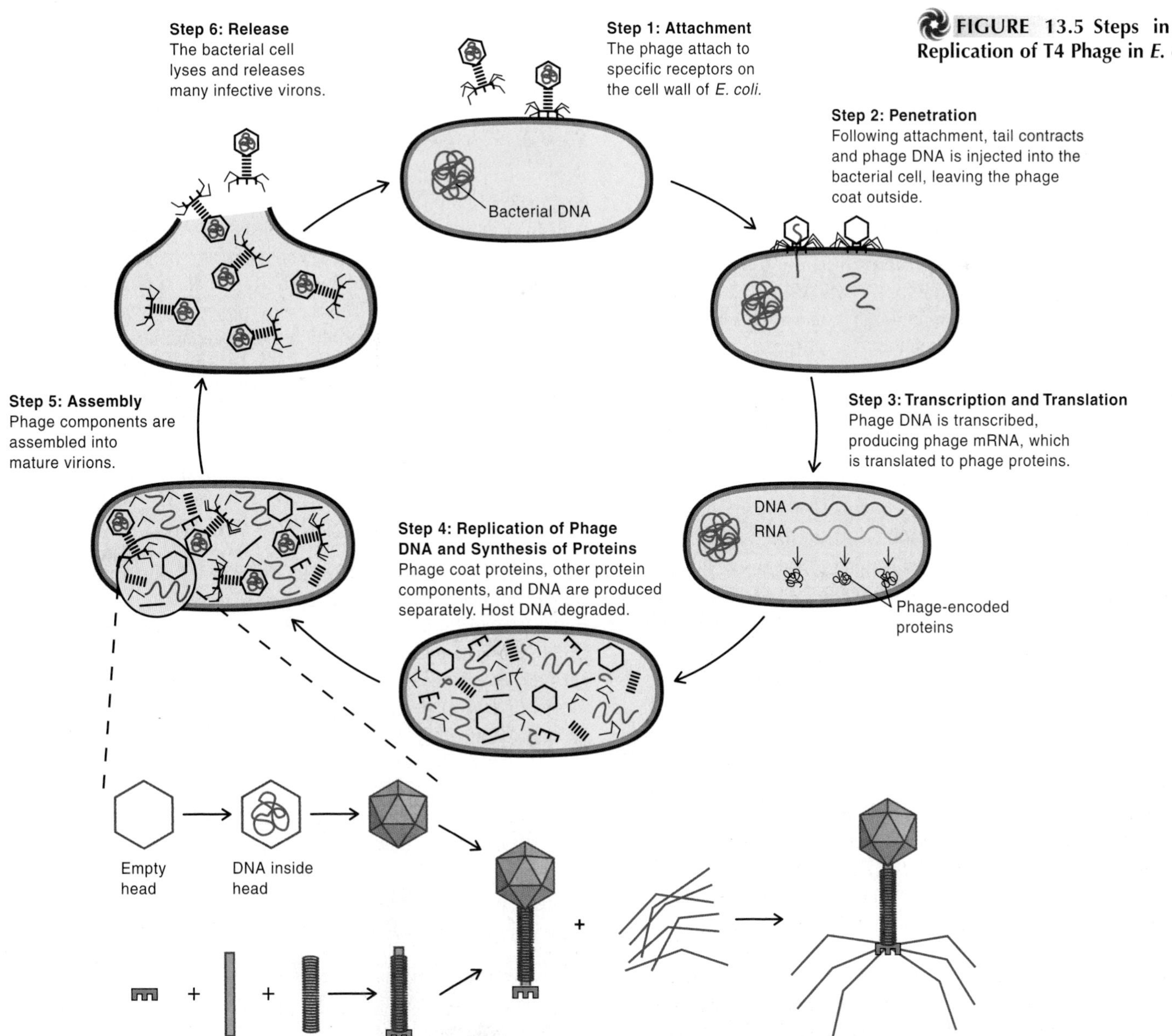
Step 1: Attachment
The phage attach to specific receptors on the cell wall of *E. coli.*
Bacterial DNA
Step 2: Penetration
Following attachment, tail contracts and phage DNA is injected into the bacterial cell, leaving the phage coat outside.
Step 3: Transcription and Translation
Phage DNA is transcribed, producing phage mRNA, which is translated to phage proteins.
DNA
RNA
Phage-encoded proteins
Step 4: Replication of Phage DNA and Synthesis of Proteins
Phage coat proteins, other protein components, and DNA are produced separately. Host DNA degraded.
Step 5: Assembly
Phage components are assembled into mature virions.
Step 6: Release
The bacterial cell lyses and releases many infective virons.
Empty head
DNA inside head

FIGURE 13.5 Steps in the Replication of T4 Phage in *E. coli*

Step 1: Attachment

When a suspension of T4 phage is mixed with a susceptible strain of *E. coli,* the phage collide by chance with the bacteria. Protein fibers at the end of the phage tail attach to specific receptors on the bacterial cell wall and the base plate with its tail spikes (see figure 13.1c) settles on the surface of the bacterium. (Step 1 in figure 13.5) The receptors of the bacteria generally perform other functions for the cell, such as transport functions. The phage use these structures for their own purposes.

Step 2: Penetration

A few minutes after attachment, an enzyme (lysozyme) located in the tip of the phage tail degrades a small portion of the bacterial cell wall. The tail contracts, and the tip of the tail opens. The double-stranded linear viral DNA in the head passes through the open channel of the phage tail (figure 13.5, Step 2). The DNA is literally injected through the cell wall and, by some unknown mechanism, passes through the cytoplasmic membrane and enters the interior of the cell. The protein coat of the phage remains on the outside of the cell. Thus, the protein and nucleic acid separate before the virus begins to replicate.

Step 3: Transcription and Translation

Within minutes of the entry of phage DNA into a host cell, a portion is transcribed into mRNA, which is then translated into proteins that are important for phage production (figure 13.5, Step 3). The first proteins produced, termed **early proteins,** are enzymes that are not normally present in the uninfected host cell. They are also known as **phage-encoded proteins,** because they are coded by phage DNA. Some are essential for the synthesis of phage DNA, which contains an unusual pyrimidine, and others are involved in the synthesis of phage coat protein. One phage-encoded early protein is a nuclease that degrades the DNA of the host cell. As a result, soon after infection, no host cell DNA is transcribed and all mRNA synthesized is transcribed from the phage DNA. In this way, the phage takes over the metabolism of the bacterial cell for its own purpose, namely the synthesis of more virions. Preexisting bacterial enzymes, however, continue to function. Some supply the energy necessary for phage replication through the breakdown of energy sources outside the cell. The host enzymes also synthesize amino acids and nucleotides for the production of phage proteins and nucleic acids, the former being synthesized on bacterial ribosomes.

Not all phage-encoded proteins are synthesized simultaneously. Rather, they are made in a sequential manner during the course of infection. Those required early are synthesized near the beginning of infection; others that are required later in infection are made later on. Examples of early phage-encoded enzymes are the nuclease that degrades the host chromosome and the enzymes of phage DNA synthesis. Late phage-encoded enzymes include those concerned with the assembly of capsids and the phage lysozyme that lyses the bacteria to release the newly formed phage.

Step 4: Replication of Phage DNA and Synthesis of Proteins

Phage protein and nucleic acid are synthesized independently of one another. The DNA of the entering phage serves two distinct functions: the template for replication of phage DNA and the template for the synthesis of mRNA, which is then translated into phage-encoded enzymes and the proteins that form the complete virion. The DNA and proteins are synthesized by the mechanisms described for bacterial DNA and protein synthesis (figure 13.5, Step 4). ■ DNA synthesis, p. 164 ■ protein synthesis, p. 168

Step 5: Assembly

The **assembly,** or **maturation,** process involves the assembly of phage protein with phage DNA to form intact or mature virions (figure 13.5, Step 5). This is a complex, multistep process in the case of T4 phage. The protein structures of the phage, such as the heads, tails (sheaths), tail spikes, and tail fibers, are synthesized independently of one another. Once the phage head is formed, it is packed with DNA; the tail is then attached, followed by the addition of the tail spikes. The assembly of some of these various components involves a **self-assembly** process, in which the protein components come together spontaneously without any enzyme catalyst to form a specific structure. In other steps, certain phage-encoded proteins serve as scaffolds on which various phage protein components associate. The scaffolds themselves do not become a part of the final structure, much as scaffolding required to build a house does not become part of the house.

Step 6: Release

During the latter stages of the infection period, the phage-encoded enzyme **lysozyme** is synthesized. This enzyme digests the host cell wall from within, resulting in cell lysis and the release of phage (figure 13.5, Step 6). If this enzyme were synthesized early in the infection process, it would lyse the host cell before any mature phage could be formed. This is why phage genes are expressed only when the enzymes for which they code are needed.

In the case of the T4 phage, the **burst size,** the number of virions released per cell, is about 200. The time required for the entire cycle from adsorption to release is about 30 minutes. These phage then infect any susceptible cells in the environment, and the process of phage replication is repeated.

Lytic Single-Stranded RNA Phages

Another group of lytic phages that undergo a productive infection are the single-stranded RNA phages. The ones most intensively studied are those that infect *E. coli.* These include MS2 and Qβ, which show many similarities to one another and share several unusual properties (see table 13.2). First, they infect only F^+ strains of *E. coli* because they attach to the sides of the F pilus. Second, they replicate rapidly and have a burst size of about 10,000. Third, the replication of this phage requires an unusual enzyme, an RNA-dependent RNA polymerase which uses RNA as a template. This enzyme is not present in *E. coli,* so the entering RNA must code for it. ■ F pilus, pp. 66, 203

Phage Replication in a Latent State—Phage Lambda

Some phages can live in harmony with their host bacteria. These latter are known as temperate phages although under certain conditions they can undergo a lytic infection like T4. Ninety percent

PERSPECTIVE 13.2

Viral Soup

For many years it has been known that viruses cause many very deadly diseases of humans, animals, and plants. Within the past few years, scientists have discovered that they may play another important role in nature—that of being a major component of **bacterioplankton,** the bacteria found in the ocean. In the 1970s and 1980s, bacterial cells were found to be much more abundant than had been believed previously. Earlier estimates of the number of bacteria in aquatic environments were based on the numbers that could be cultured in the laboratory. When direct microscopic counts were made, however, the number of heterotrophic bacteria was estimated to be 10^5 to 10^7 bacteria per ml, much higher than previously estimated. Most carbon fixation in the ocean is carried out by bacteria. The number of cyanobacteria and single-celled eukaryotic algae is also much greater than previously believed. What was most surprising, though, was the number of bacteriophages in aquatic environments. To get an accurate estimate of their numbers, scientists centrifuged large volumes of natural unpolluted waters from various locations. They then counted the number of particles in the pellet. Much to their surprise, up to 2.5×10^8 phages per ml were counted—about 1,000 to 10 million times as high as previous estimates. Why are these new numbers important? First, they may answer the question of why bacteria have not saturated the ocean. The bacteria are most likely held in check by the phages. From the estimates of the numbers of bacteria and phages, it is calculated that one-third of the bacterial population may experience a phage attack each day. Second, the interaction of phages with bacteria in natural water implies that the phages may be actively transferring DNA from one bacterium to another, by the process of transduction.

Thus, the smallest agents in natural waters may have important consequences for the ecology of the aquatic environment. Perhaps viruses have a similar function for eukaryotic organisms. Indeed, it is now known that phages can stop algal blooms.

of all phages are temperate. In the harmonious relationship, the entering phage DNA most commonly becomes integrated into the chromosome of the host cell, where it replicates as the host chromosome replicates. When integrated, the phage DNA is called a **prophage.** Some temperate phages multiply as plasmids inside the host cell, another kind of latency. ■ plasmid, p. 68

The most thoroughly studied temperate phage is lambda (λ), which lysogenizes *E. coli.* Its size and shape are similar to T4 phage. Once the phage DNA enters the host cell, one of two events occur. Some of the bacteria will be lysogenized, while others will be lysed **(figure 13.6).** Whether the lambda phage lyses the majority of the cells or lysogenizes them depends largely on chance events that can be tipped in favor of one or the other by modifying the external environment. For example, if the bacteria are growing slowly, because of nutrient deprivation, then the likelihood of the phage DNA becoming integrated

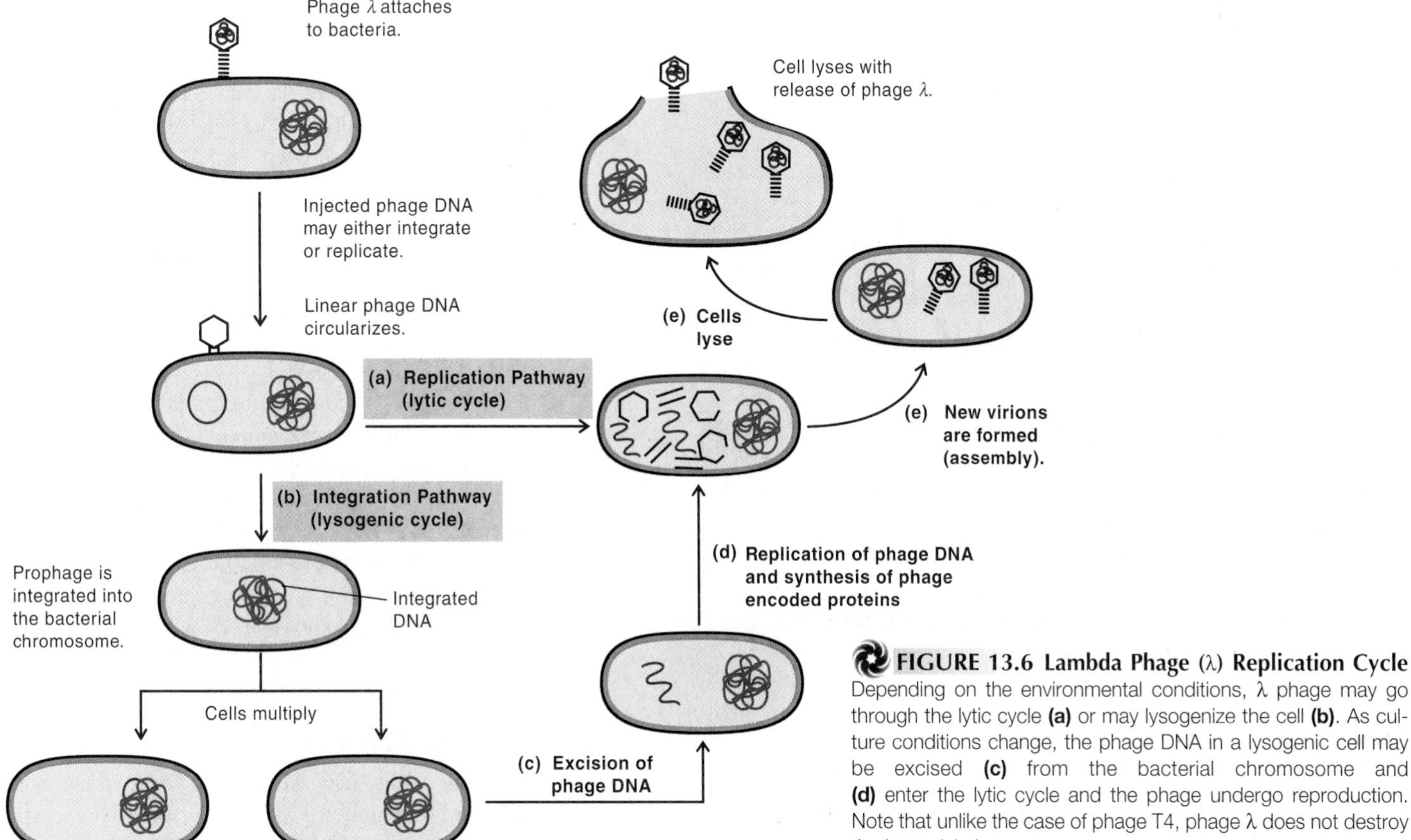

FIGURE 13.6 Lambda Phage (λ) Replication Cycle Depending on the environmental conditions, λ phage may go through the lytic cycle **(a)** or may lysogenize the cell **(b)**. As culture conditions change, the phage DNA in a lysogenic cell may be excised **(c)** from the bacterial chromosome and **(d)** enter the lytic cycle and the phage undergo reproduction. Note that unlike the case of phage T4, phage λ does not destroy the bacterial chromosome.

increases because phage replicate in a lytic cycle only in actively metabolizing cells.

The Integration of Phage DNA into the Bacterial Chromosome

Phage λ DNA integrates into the *E. coli* genome at a specific site through a process called **site-specific recombination.** Short nucleotide sequences in the DNA of phage λ and the *E. coli* host are identical (homologous), allowing the phage and the bacterial DNA to **synapse** (pair) **(figure 13.7).** This region of homology is located between the genes coding for galactose metabolism (gal) and biotin synthesis (bio). Following synapsis, the phage DNA becomes integrated into the bacterial chromosome between these genes. Note that phage DNA is added without replacing any bacterial chromosomal genes. This differs from the situation following gene transfer by transformation, transduction, or conjugation. In these cases, chromosomal genes in the recipient cells are replaced by donor DNA. ■ DNA transformation, p. 199 ■ transduction, p. 201 ■ conjugation, p. 202

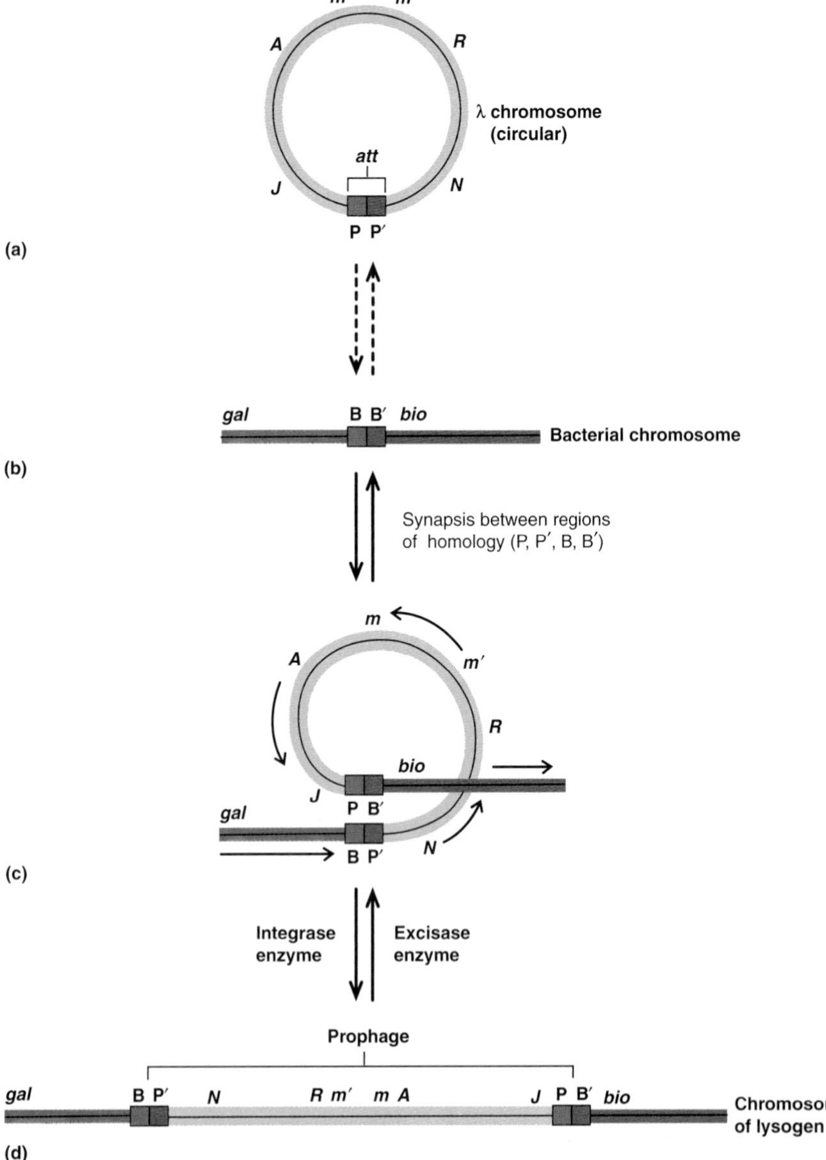

FIGURE 13.7 Reversible Insertion and Excision of Lambda (λ) Phage The λ DNA circularizes **(a)**, the phage *att* site, P,P′ synapses with an identical bacterial sequence B,B′ located between *gal* and *bio* **(b)** and is integrated between the *gal* and *bio* operons **(c)** to form the prophage, **(d)**. If the process is reversed, the lambda chromosome is excised from the bacterial chromosome and can then replicate and code for the synthesis of phage proteins.

Maintenance of the Prophage in an Integrated State

The phage DNA can remain integrated, replicate along with the bacterial DNA, and be passed on to all daughter cells indefinitely. For this to occur, however, the expression of certain genes on the integrated phage DNA must be repressed. These genes code for enzymes that **excise** (remove) the integrated DNA from the host chromosome. If excision occurs, the replication cycle of a productive phage infection ensues. How are these genes repressed to maintain the lysogenic state? One gene in the integrated viral DNA codes for a **repressor** that binds to a viral operator, which controls transcription of the integrated genes required for excision. As long as this repressor is produced, the integrated phage DNA is not excised. If the repressor is no longer synthesized or is inactivated, however, viral genes are transcribed and an enzyme, an **excisase** which removes the viral DNA from the bacterial chromosome, is synthesized. A productive infection with lysis ensues (see figure 13.6d). ■ repressor, p. 177 ■ operator, p. 177

Under ordinary conditions of growth, the phage DNA is excised from the chromosome only about once in 10,000 divisions of the lysogen. If, however, a lysogenic culture is treated with an agent that damages the bacterial DNA (such as ultraviolet light), the repair system, called SOS, comes into play. This system activates a protease, an enzyme that destroys the repressor. As a result, the prophage enter into a productive infection. In this way the phage escape from a host that is in deep trouble because its DNA has been damaged. This process, termed **phage induction,** results in complete lysis of the culture. The term *induction* as it is used here should not be confused with induced enzyme synthesis as discussed in chapter 7. ■ SOS system, p. 194 ■ ultraviolet light, p. 116

Immunity of Lysogens

In addition to maintaining the prophage in the integrated state, the repressor protein prevents a lysogenic cell from being infected by the same type of phage the cell already carries. Infections are blocked because the repressor binds to the operator in the phage DNA as it enters the cell and inhibits its replication. Consequently, the cell is immune to infection by the same phage but not to infection by other phages, to whose DNA the repressor cannot bind. In this way the phage protects its turf from closely related phages.

Lysogenic Conversion

Lysogenic cells may also differ from their nonlysogenic counterparts in other important ways.

TABLE 13.3 Some Properties Conferred by Prophage

Microorganism	Medical Importance	Property Coded by Phage
Corynebacterium diphtheriae	Causes diphtheria	Synthesis of diphtheria toxin
Clostridium botulinum	Causes botulism	Synthesis of botulinum toxin
Streptococcus pyogenes	Causes scarlet fever	Synthesis of toxin responsible for scarlet fever
Salmonella	Causes food poisoning	Modification of lipopolysaccharide of cell wall
Vibrio cholerae	Causes cholera	Synthesis of cholera toxin

The prophage can confer new properties on the cell, because of the expression of genes on the integrated phage DNA. This is the phenomenon of lysogenic conversion. For example, strains of *Corynebacterium diphtheriae* that are lysogenic for a certain phage (β phage) synthesize the toxin that causes diphtheria. Similarly, lysogenic strains of *Streptococcus pyogenes* and *Clostridium botulinum* manufacture toxins that are responsible for scarlet fever and botulism, respectively. In all of these cases, nonlysogens cannot synthesize toxin. The genes that code for these toxins are phage genes, which are expressed only when the phage DNA is integrated into the bacterial chromosome. Some examples of lysogenic conversion are given in **table 13.3.**

Extrusion Following Phage Replication—Filamentous Phages

In addition to the phages that develop productive (lytic) infections and latency, some phage can develop other relationships with their host cells. A few closely related bacterial viruses that have the appearance of long thin fibers are known as **filamentous phages.** These include the single-stranded DNA phages M13 and fd **(figure 13.8).** They do not lyse the cell; rather, they replicate their DNA and synthesize phage protein but the bacteria continue to multiply. Indeed, these few phages are the only ones that do not lyse their hosts following phage maturation. However, because the replicating phage rob the cells of molecules that normally would go into bacterial replication, the cells grow slowly.

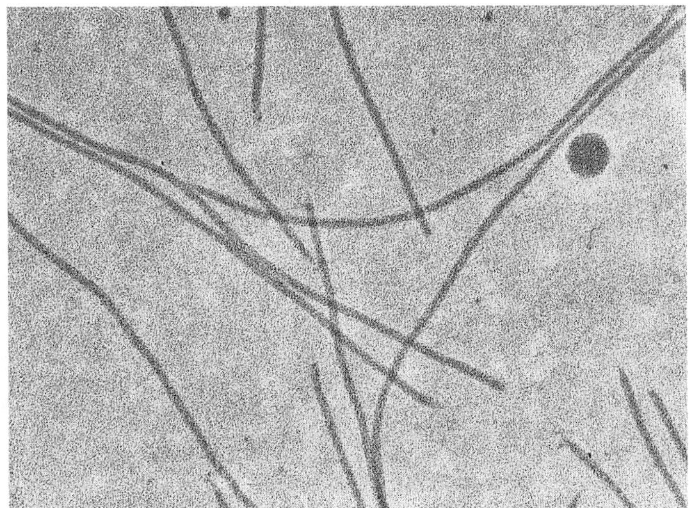

FIGURE 13.8 Electronphotomicrograph of fd Phage This phage only infects cells that have an F pilus. Each phage is about 900 nm long.

Replication of Filamentous Phages

Unlike phage T4, filamentous phages adsorb to the tip of the F^+ pilus of *E. coli* and therefore infect only F^+ cells **(figure 13.9).** The single-stranded DNA that enters the cell does not completely take over their host's metabolism exclusively for phage production because the bacteria continue to multiply. The phage DNA replicates and also codes for the synthesis of the phage coats. Interestingly, neither mature filamentous phage nor their coats can be detected in the cytoplasm of the host cells. It seems likely that after being synthesized, the phage capsomers are stored in the cytoplasmic membrane of the bacterium. The phage are assembled as they are extruded from the cell. The extrusion occurs continuously, and the phage are not released in a burst. Infected cells, termed **carrier cells,** can be subcultured and stored in the same way as non-infected cells.

Replication of Single-Stranded DNA of Filamentous Phage

The replication of single-stranded DNA of a filamentous phage has some features that are similar but other features are different from double-stranded DNA replication. The DNA that enters the cell is a positive (+) stranded molecule that is converted to a double-stranded form by enzymes of the host cell **(figure 13.10).** This double-stranded, **replicative form** consists of a positive (+) and a negative (–) strand and replicates in much the same way as double-stranded DNA of bacteria. This replicative form gives rise to the single-stranded positive (+) strand and a negative (–) strand. The single-stranded DNA that is incorporated into the coat protein is a positive (+) strand, and is derived from the double-stranded replicative form (see figure 13.10).

■ **DNA replication, p. 164** ■ **(+) and (–) strand DNA, p. 168**

Lytic Infection by Single-Stranded DNA Phages

Not all phages with single-stranded DNA are extruded. The isometric phage, ϕX174 also contains a positive (+) strand but differs in shape from the filamentous phages (see Table 13.2) and infection is not limited to F^+ cells. This phage goes through a lytic cycle.

Table 13.4 summarizes the salient features of the various interactions of the phages with their host cells.

Filamentous phage
Phage DNA
F pilus
Cytoplasmic membrane
Cell wall
Capsomer subunits of coat protein
Phage DNA
Phage adsorbs to the tip of the F pilus of the male cell.
Phage DNA enters host cell.
Capsomer subunits of coat protein
Cell wall
Phage DNA
Outside environment
Cytoplasmic membrane
Phage DNA replicates and phage coat capsomers are synthesized and embedded in the cytoplasmic membrane. The phage are released after the nucleic acid gains its protein coat as it passes through the cytoplasmic membrane. The bacteria do not lyse.
Carrier cell
Carrier cell
Carrier cell
Carrier cell
Carrier cells
Carrier cells
Carrier cells

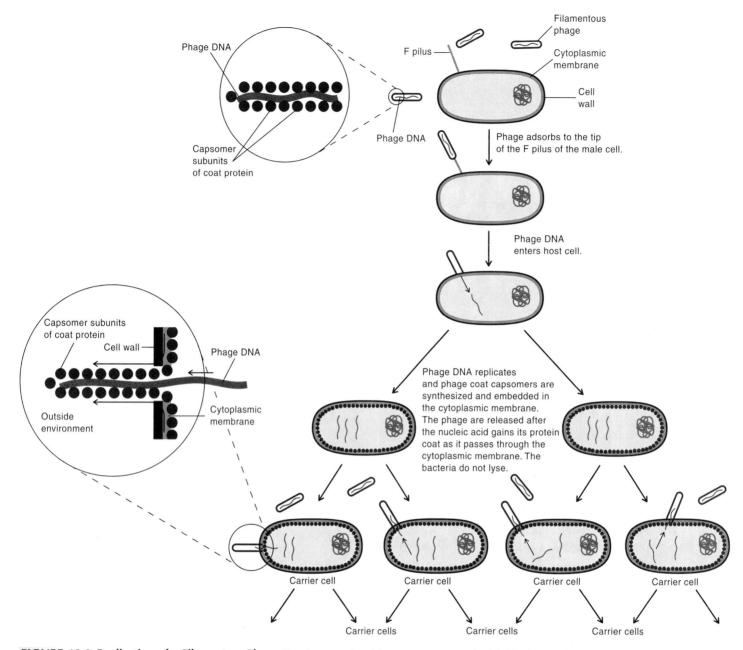

FIGURE 13.9 Replication of a Filamentous Phage The phage coat protein capsomers are embedded in the cytoplasmic membrane of the bacterium. Maturation of the phage occurs as the DNA is extruded through the cytoplasmic membrane. Carrier cells continue to extrude phage as they multiply.

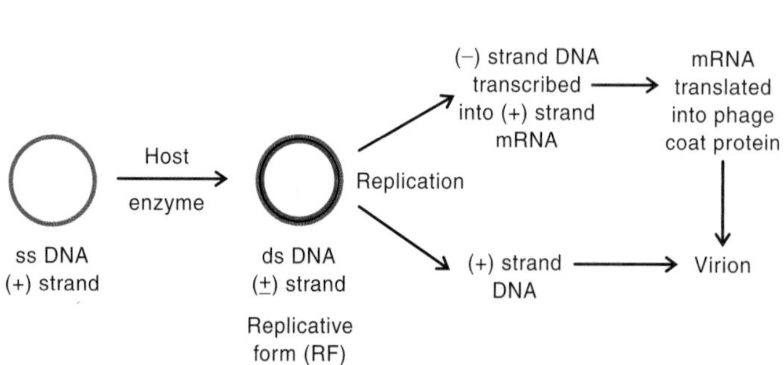

FIGURE 13.10 Macromolecule Synthesis in Filamentous Phage Replication

MICROCHECK 13.2

Some phages lyse their host cells but the DNA of most phages becomes integrated into the genome of the host cell. The viral DNA often codes for gene products that confer new properties on the host, which is termed lysogenic conversion. Some phages are extruded without killing the host.

✓ How does T4 phage prevent the infected bacterium from synthesizing bacterial proteins?

✓ What are two ways that phage can replicate in harmony with their host?

✓ Have phage that are extruded from the bacterial cell surface undergone a productive infection? Are they virulent? Explain your answers.

✓ In what way is phage induction an advantage for the phage?

TABLE 13.4 Features of Interactions of Phages with Host Cells

Phage	Type of Interaction	Type of Nucleic Acid*	Result of Interaction
T4	Productive	ds DNA	Cell lyses
MS2, Qβ	Productive	ss RNA (+) strand	Cell lyses
Lambda (λ)	Latent or Productive	ds DNA	Phage and host often live in harmony
M13, fd	Productive	ss DNA (+) strand	Phage extrudes from cell, which continues to multiply
ϕX174	Productive	ss DNA (+) strand	Cell lyses

*ds, double-stranded; ss, single-stranded

13.3 Transduction

Focus Point

- Describe the differences between generalized and specialized transduction.

Phages play an important role in the transfer of bacterial genes from one bacterium to another. As briefly discussed in chapter 8, DNA can be transferred from one bacterial cell, the donor, to another, the recipient, by phage in the process called **transduction.** There are two types of transduction. In one type, any bacterial gene can be transferred, a process called **generalized transduction.** The phages that carry out this process are termed **generalized transducing phages.** The second type is termed **specialized transduction** because only a few specific genes can be transferred by the phages, which are **specialized transducing phages.** ■ transduction, p. 201

Generalized Transduction

Virulent as well as temperate phages can serve as generalized transducing phages. Recall that some phages in their replication life cycle degrade the bacterial chromosome into many fragments at the beginning of a productive infection. These short DNA fragments can be incorporated inadvertently into the phage head in place of phage DNA during phage maturation (see figure 8.21). Following lysis and release of the phage, the phage binds to another bacterial cell and injects the bacterial DNA. Once inside the new host, the DNA from the donor cell can integrate into the recipient cell DNA by homologous recombination. These recipient cells do not lyse. Since the genetic information transferred can be any gene of the donor cell, this gene transfer mechanism is called generalized transduction. ■ generalized transduction, p. 201 ■ homologous recombination, p. 200

Why do the transducing virulent phage not lyse the cells they invade? Because the bacterial DNA replaces the phage DNA inside the phage's head, the genetic information necessary for the synthesis of phage-encoded proteins is lacking. The phage is termed **defective** because it lacks the DNA necessary to form complete phage and lyse the recipient cell.

Specialized Transduction

Specialized transduction involves the transfer of only a few specific genes and is carried out only by temperate phages. The most actively studied specialized transducing phage is λ, which integrates only at specific sites in the chromosome of *E. coli.* Lambda transduces specific genes by the following means **(figure 13.11).**

Following induction, the λ phage DNA is usually precisely excised from the bacterial chromosome. On rare occasions, however, a piece of bacterial DNA remains attached to the piece of phage DNA that is excised and a piece of phage DNA is left behind in the bacterial chromosome. This loss of phage DNA creates a defective phage. The bacterial genes attached to the phage DNA replicate as the phage DNA replicates in the bacteria. These DNA molecules, consisting of phage and bacterial DNA, then become incorporated into mature phage in the maturation process. The defective phage are released from the lysed cells. When the defective phage infect another bacterial cell, both phage and bacterial DNA enter the new host and the bacterial genes become integrated into the chromosome. Thus, the resulting lysogens contain bacterial genes from the previously lysogenized cells. However, in contrast to generalized transduction, only bacterial genes located near the site of integration of the phage DNA can be transduced.

MICROCHECK 13.3

Bacterial genes from a donor can be transferred to recipient bacteria following the incorporation of bacterial genes in place of phage genes in the phage head. After the phage lyse the donor bacteria and infect the recipient cells, the bacterial genes are integrated into the chromosome of the recipient cell, the process of transduction. There are two types of transduction: generalized, in which any gene of the host can be transferred; and specialized, in which only the genes near the site at which the phage DNA integrates into the host chromosome can be transferred. The latter process is carried out only by temperate phages.

- ✓ What is a defective phage?
- ✓ Most temperate phages integrate into the host chromosome, whereas some replicate as plasmids. Which kind of relationship do you think would be more likely to maintain the phage in the host cell? Why?

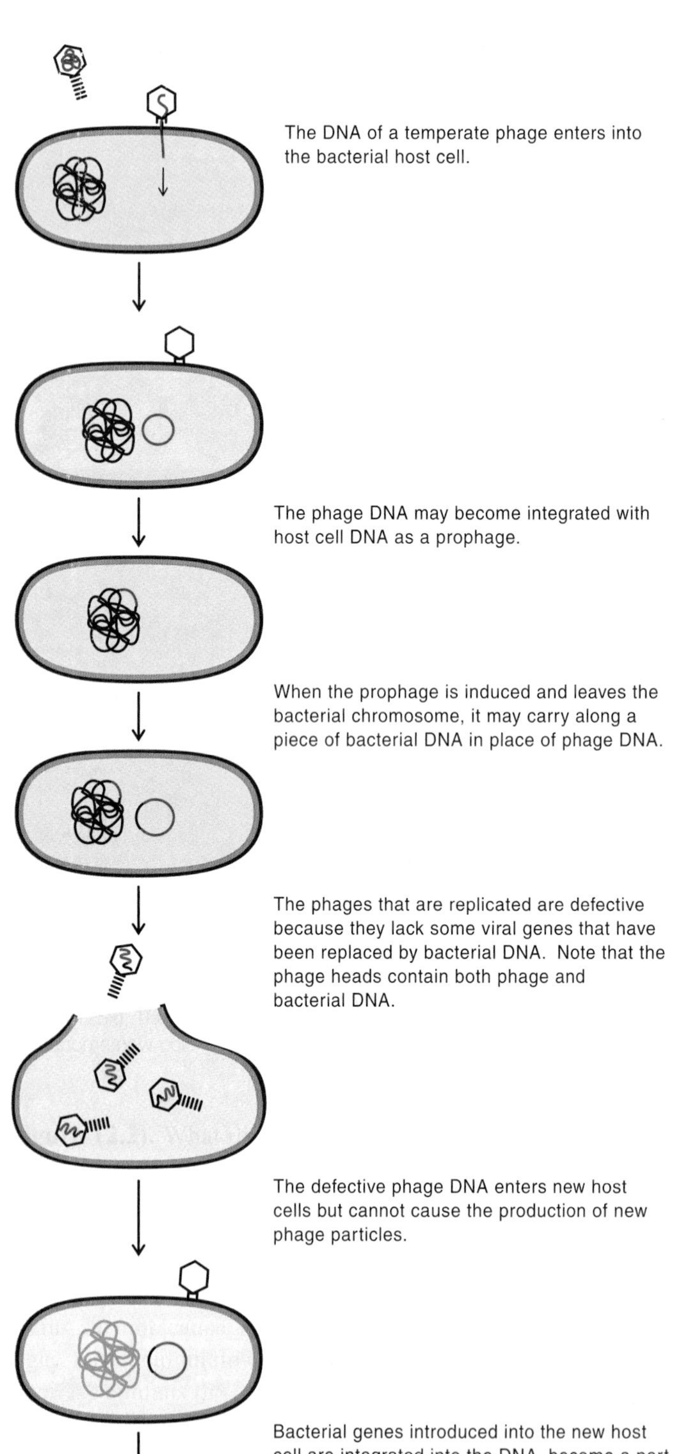

The DNA of a temperate phage enters into the bacterial host cell.
The phage DNA may become integrated with host cell DNA as a prophage.
When the prophage is induced and leaves the bacterial chromosome, it may carry along a piece of bacterial DNA in place of phage DNA.
The phages that are replicated are defective because they lack some viral genes that have been replaced by bacterial DNA. Note that the phage heads contain both phage and bacterial DNA.
The defective phage DNA enters new host cells but cannot cause the production of new phage particles.
Bacterial genes introduced into the new host cell are integrated into the DNA, become a part of the bacterial chromosome, and are replicated along with the rest of the bacterial DNA.
Integrated bacterial DNA

FIGURE 13.11 Specialized Transduction by Temperate Phage Only bacterial genes near the site where the prophage has integrated can be transduced.

13.4

Host Range of Phages

Focus Points

- List three mechanisms that reduce infection by phage.
- Describe the activities of the two enzymes in the restriction-modification system.

The number of different bacterial strains that a particular phage can infect defines its **host range.** Several thousand different phages have been isolated, and the host range of any particular phage is usually limited to a single bacterial species and often to only a few strains of that species. This fact is put to practical use in distinguishing between different bacterial strains—the technique of **phage typing.** Several factors limit the host range of phage. The two most important are (1) the requirement that the phage must attach to specific receptors on the host cell surface to start infection and (2) a restriction-modification system of the host cell must be overcome. The limited host range means that phages seldom transfer DNA between unrelated bacteria. ■ phage typing, p. 244

Receptors on the Bacterial Surface

Receptor sites vary in chemical structure and location. Receptors are usually on the bacterial cell wall, although a few phages attach to pili and a few others attach to sites on flagella **(figure 13.12).** Receptor sites can be altered by two distinct mechanisms, thereby creating a resistant cell. First, the receptor sites can be modified by mutation. In any large population of susceptible cells, some will have a modified (mutant) receptor site that confers resistance to any given phage.

Second, some, but not all, temperate phages that have lysogenized a bacterial cell can alter the cell surface, an example of lysogenic conversion. As a result the original receptor is no longer available. Thus, the prophage protects its host and, in turn, is able to continue replicating inside it.

Restriction-Modification System

Another mechanism of resistance is limited to prokaryotes and led to the biotechnology revolution that started when scientists learned how to clone genes. Virtually all prokaryotes have two genes located next to each other in the genome that are involved in the **restriction-modification system (figure 13.13).** One gene codes for a **restriction enzyme,** an endonuclease that recognizes short base sequences in double-stranded DNA and cleaves the DNA at these sequences. The second gene codes for a **modification enzyme,** which attaches methyl groups to the purine and pyrimidine bases of the nucleotide sequence recognized by the restriction enzyme. These methylated bases are not recognized by the restriction enzyme, and so it does not cleave the DNA. This is the mechanism that the cell uses to protect its own DNA from being degraded. Its own DNA is methylated and therefore not degraded, but foreign DNA entering the cell will be degraded unless it is methylated first. ■ restriction enzyme, p. 213

The restriction-modification system explains why, in general, a recipient bacterium will maintain DNA by transformation,

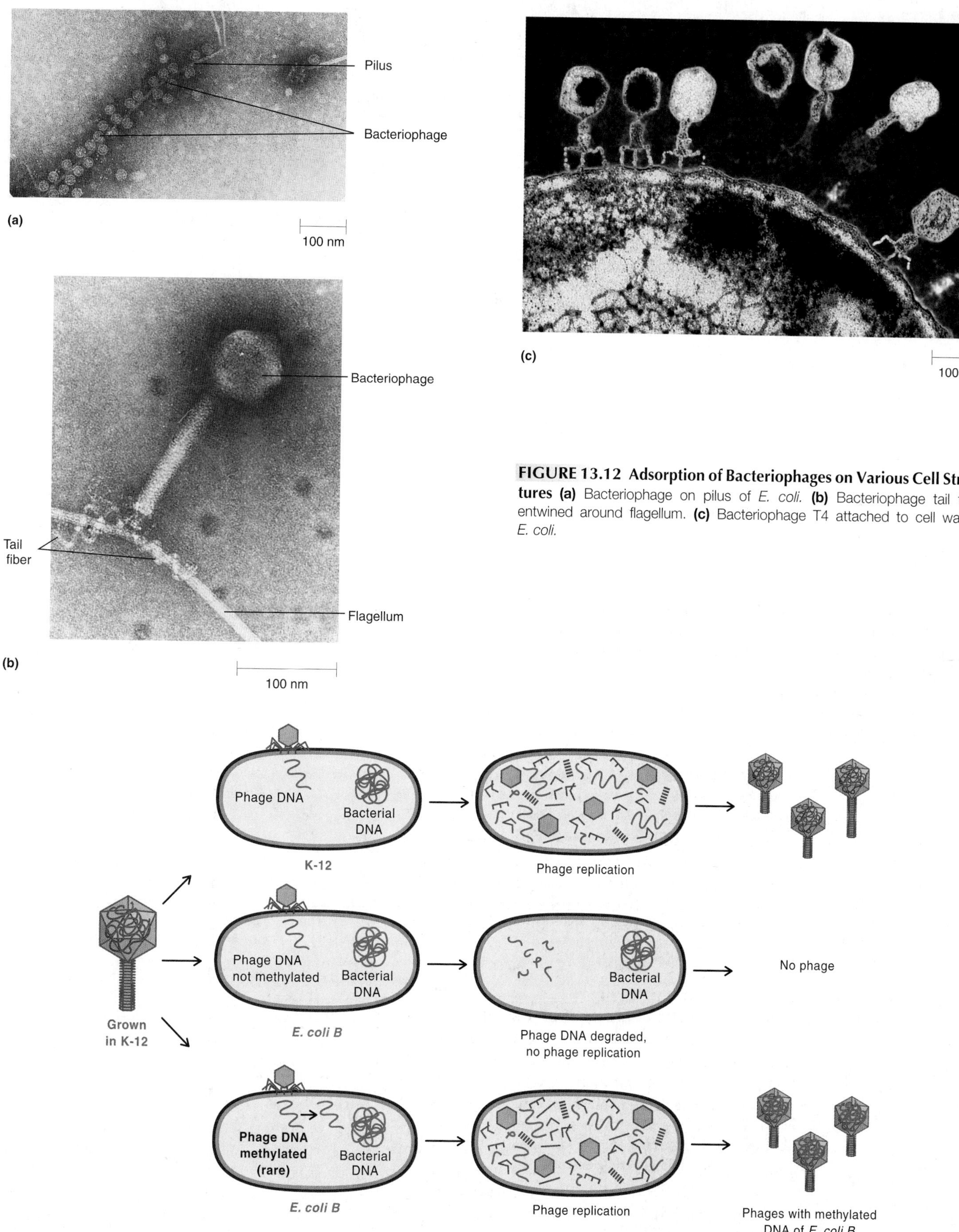

FIGURE 13.12 Adsorption of Bacteriophages on Various Cell Structures **(a)** Bacteriophage on pilus of *E. coli.* **(b)** Bacteriophage tail fiber entwined around flagellum. **(c)** Bacteriophage T4 attached to cell wall of *E. coli.*

FIGURE 13.13 Restriction-Modification System The DNA of *E. coli* B and K-12 differ in their patterns of methylation. Phage DNA entering a cell will usually be degraded by a restriction enzyme unless it has the same pattern of methylation as its own chromosomal DNA (Line 2). However, on rare occasions, a modification enzyme (a methylase) can methylate the entering to the same pattern of methylation as its own chromosome and it escapes degradation by the restriction enzyme in the cell (Line 3). The colors, which are the same, represent DNA with the same restriction pattern.

conjugation, or transduction only if the donor DNA and recipient cells are of the same species or even subspecies. Recipient cells restrict any DNA if the methylation pattern of the entering donor DNA differs from its own. Foreign DNA from different species, however, isn't always destroyed when it enters a recipient cell. Why would this be the case in light of the cells having restriction enzymes? When foreign DNA enters a cell, a race ensues between the enzyme intent on degrading the DNA (restriction enzyme) and the enzyme that is attempting to methylate the purines and pyrimidines of the entering DNA before it is degraded (modification enzymes). In most cases, the DNA is degraded. However, in rare cases, the entering DNA is methylated (modified) so that it is no longer recognized as being foreign. Now the DNA can replicate if it is a replicon or becomes integrated into the genome of the recipient cell in the case of DNA transformation and conjugation.

■ replicon, p. 198 ■ conjugation, p. 202 ■ DNA transformation, p. 199

MICROCHECK 13.4

The host range of bacteriophages is determined by the attachment proteins on the virus recognizing receptors on the bacterial host and by the ability of the entering phage DNA to escape degradation by host restriction enzymes. Most receptor sites for phage are on the cell wall of the bacterium, although some are on pili and flagella.

✓ What kind of enzymatic activity does the modification system have?

✓ You add an unknown phage to a mixture of F^+ and F^- cells of *E. coli* and plate out the bacteria. The bacterial colonies that grow are all F^-. How can you explain this phenomenon?

✓ A mutation in *E. coli* results in the loss of both restriction and modification enzymes. Would you expect any difference in the frequency of gene transfer by transduction from *Salmonella*? Explain.

FUTURE CHALLENGES

"Take Two Phage and Call Me in the Morning"

More and more bacteria are becoming resistant to an ever-greater number of antibiotics, greatly reducing the effectiveness of these medications. One interesting novel approach to treating infectious diseases is to use phage to kill the disease-causing bacteria. This is not a new idea. Felix d'Herelle, the co-discoverer of phage, believed that they were mainly responsible for natural immunity. He also claimed that phages would be a universal prophylaxis and therapy. So convinced was he that while he was a Professor of Bacteriology at Yale, he also owned a commercial laboratory in Paris that produced preparations of phage that he sold to the pharmaceutical industry. After leaving Yale, he went to the Bacteriophage Research Institute in the USSR state of Georgia. Today, much of the work on the possible use of phage as therapeutic agents is being carried out in this same Institute.

Although the use of phage sounds attractive, many problems are associated with this technology. First, a specific phage will only infect a single species or only a specific strain of a given species. Since it would take too long to isolate the causative infectious bacterium and isolate a phage that could attack it, "cocktails" containing many different phages must be used. Thus, to treat intestinal diseases, a "cocktail" of 17 different phages that attack intestinal pathogens has been developed. To treat burns that may be subject to infection by *Pseudomonas,* a kind of bandage saturated with a cocktail of five to nine different phages has been employed. Whether phages are present that will attack all of the bacteria responsible for the disease, however, is questionable.

Further great care must be taken in purifying the phage before the preparation is administered. When phages lyse bacteria, a great deal of debris from the lysed cells is mixed in with the phage. This material could be very dangerous if it got into the bloodstream.

A third problem that must be overcome is that phages are recognized as foreign by the body's immune system, and therefore they are quickly eliminated. Some success has been achieved in increasing the time the phages stick around by altering the phage coat.

The fourth hurdle is the need for in-depth animal testing and human trials. No clinical trials involving human patients have been carried out but studies from England have reported that mice, calves, and chickens could be protected from disease-causing strains of *E. coli.* Enough statements of success have been made in the use of phage in Russia and countries in the former Soviet Union, that phages are now being seriously considered as therapeutic agents in Western countries. It is the only therapy that increases in amount as it successfully carries out its job.

Although phage have not been approved for use in treating human diseases, they have been given approval for preventing disease. In 2007, the Food and Drug Administration and the U.S. Department of Agriculture approved a bacteriophage product specifically directed against a common and dangerous food pathogen, *Listeria monocytogenes.* Presumably, the phage will be added at various stages in the processing of cheeses, meats, fish and any other food product susceptible to *Listeria.*

SUMMARY

13.1 General Characteristics of Viruses (table 13.1)

Viruses are non-living agents associated with all forms of life. Each virus particle consists of nucleic acid surrounded by a protein coat. They are approximately 100- to 1,000-fold smaller than the cells they infect (figures 13.2, 13.3). They represent another class of **mobile genetic elements.**

Virus Architecture

Different viruses have different shapes. Some are **isometric,** others are **helical** and still others are complex (figure 13.1). The shape is determined by the protein coat (or capsid) that surrounds the nucleic acid. These make up the **nucleocapsid.** Each capsid is composed of **capsomers; attachment proteins** project from the capsid. (figures 13.1, 13.2). Some animal viruses have a lipid bilayer surrounding the coat. These viruses are **enveloped.** Viruses without this envelope are **naked.** Virtually all viruses that infect bacteria—**bacteriophages**—are complex and naked (figure 13.2).

The Viral Genome

Viruses contain either RNA or DNA, but never both. The nucleic acid may be single-stranded or double-stranded.

Replication Cycle—Overall Features

Viruses only multiply within living cells and use the machinery of the cells to support their own multiplication. Some viruses take over the metabolism of the host cell completely and kill it; others live in harmony with their hosts. Viruses exist in two states. Outside the cell they cannot multiply. They are called **virions.** Inside infected cells, they replicate.

13.2 Phage Interactions with Host Cells (table 13.4, figure 13.4)

Bacteriophages (phages) have the same relationships to their host as do animal viruses. Phages are much easier to study and serve as excellent model systems. Some phages multiply inside bacteria and lyse the cells; this is a **productive infection;** the phages are virulent and lytic. Other phages multiply but are extruded from the cell and do not kill it. **Temperate** phages transfer their DNA into the host cell, where it multiplies either as a plasmid, becomes integrated into the chromosome of the host, or undergoes a lytic infection. A **latent** infection may show no sign that cells are infected.

Lytic Phage Replication by Double-Stranded DNA Phages (figure 13.5)

This type of productive infection is one in which the host cell metabolism is taken over by a **virulent** phage and the cell lyses. The infection proceeds through a number of defined steps:

- **Attachment—protein tail fibers** on the tail of the phage adsorb to specific receptors on the cell wall.
- **Penetration**—the DNA passes through the open channel of the **tail** and is injected into the cell. The phage coat remains on the outside.
- **Transcription and Translation**—the phage DNA is **transcribed.** Some is transcribed early in infection and other regions are transcribed later. The mRNAs are then **translated** into **phage proteins.**
- **Replication** of phage DNA and proteins—the phage DNA and proteins replicate independently of one another. A phage enzyme degrades the bacterial chromosome so that only phage proteins are synthesized.
- **Assembly (maturation)**—this is a highly complex and ordered series of processes; some are catalyzed by enzymes, others not. The net result is the assembly of the phage components into a complete virus particle.
- **Release**—a **phage-encoded** lysozyme lyses the cells resulting in the release of many virus particles per infected cell. The number is the **burst size.**

Lytic Single-Stranded RNA Phages

These phages attach to the **F pilus.** They replicate rapidly and reach burst sizes of 10,000. The entering RNA codes for an unusual RNA polymerase that uses RNA as a substrate.

Phage Replication in a Latent State—Phage Lambda (figure 13.6)

The temperate phage λ can either go through a lytic cycle similar to T4 or integrate its DNA into a specific site in the bacterial chromosome. **Integration** of phage DNA into the bacterial chromosome as a prophage occurs by means of site-specific recombination (figure 13.7). The **prophage** is maintained in an integrated state because a repressor prevents expression of genes coding for an enzyme that **excises** the prophage from the chromosome. Destruction of the repressor by UV light results in excision of the phage DNA and its replication, the process of **induction.** The cells undergo a lytic infection. **Lysogens,** the bacteria carrying prophage, are immune to the same phage whose DNA they carry. Prophage often code for proteins that confer unique properties on the bacteria, a process called **lysogenic conversion** (table 13.3).

Extrusion Following Phage Replication—Filamentous Phages (figures 13.8, 13.9)

The **filamentous phage** attach to the F pilus of *E. coli,* and the single-stranded DNA enters the cell. The entering positive (+) single strand of DNA is converted to a double-stranded **replicative form.** This DNA gives rise to two single-stranded forms, one positive (+), the other **negative** (–). The negative (–) strand gives rise to its complementary strand, which serves as mRNA for the synthesis of phage proteins. Filamentous phages do not take over the metabolism of the host cell completely, but multiply productively as the host multiplies. Phage are released by **extrusion** through the cell wall, a process that does not kill the bacteria. As the positive strand of DNA is extruded, the DNA becomes surrounded by the capsomers in the cytoplasmic membrane.

Lytic Infection by Single-Stranded DNA Phages

Single-stranded DNA phages exist that lyse cells.

13.3 Transduction

There are two types of transduction: **generalized** and **specialized.**

Generalized Transduction (figure 8.21)

Generalized transduction involves the transfer of any piece of the bacterial chromosome from one cell to another cell of the same species. Phage genes, together with bacterial DNA, replicate and the phage DNA codes for the phage coat protein. The coat surrounds the bacterial and phage DNA, and, following lysis, the bacterial DNA is transferred to other bacteria in the environment. Generalized transduction can be carried out by virulent and temperate phages.

Specialized Transduction (figure 13.11)

Specialized transduction involves the transfer of specific genes and is carried out only by temperate phages. Lambda is a well-studied **specialized transducing phage.** Only genes located near the site at which the temperate phage integrates its DNA are transduced. Bacterial genes may remain attached to the phage DNA when the phage DNA excises from the bacterial chromosome.

13.4 Host Range of Phages

Several factors determine the **host range** of phage. These include the requirement that phage attachment proteins must bind to specific receptors on the bacteria, and the phage must circumvent the **restriction-modification** system found in all prokaryotes that degrades foreign DNA.

Receptors on the Bacterial Surface (figure 13.12)

Most receptors are found primarily on the bacterial cell wall, but some phage attach to pili and flagella.

Restriction-Modification System (figure 13.13)

All prokaryotes have **restriction enzymes** that recognize short sequences of bases in DNA and cleave, and thereby degrade the DNA at those sites. All prokaryotes have **modification enzymes** that add methyl groups to the bases in the short sequence so that they are not recognized by the endonuclease and are not cleaved. Cells use this mechanism to protect their own DNA from degradation. When DNA from one strain of *E. coli* enters another strain, a race ensues between the restriction enzyme and the modification enzyme. If the modification enzyme can methylate the sequences before the restriction endonuclease recognizes and cleaves the DNA, then the DNA will be spared from degradation and will replicate. Its pattern of methylation will be identical to the DNA of the cell in which it is replicating, but different from the cell that it came from.

REVIEW QUESTIONS

Short Answer

1. What is the name given to the phage whose DNA can be integrated into the host chromosome? What is the name of the host cell containing the phage DNA?
2. Name the process by which prophages confer new properties on their host cells.
3. Name the two types of transduction. Explain how they differ from each other in the DNA that is transferred.
4. Do all productive infections result in the complete takeover of host metabolism by the phage? Explain your answer.
5. Name three structures of bacteria that contain receptors for phage.
6. List the two mechanisms that allow bacteria to resist phage infection.
7. What is the most common shape of phages?
8. Compare double-stranded DNA phages with single-stranded filamentous DNA phages in terms of cell lysis and number of phages produced.
9. How does UV light induce the synthesis of virions from lysogenic bacteria?
10. In virulent phage infection, why is it important that not all phage-encoded enzymes be synthesized simultaneously? What general classes of enzymes are synthesized first? What class is synthesized last?

Multiple Choice

1. Capsids are composed of
   a) DNA. b) RNA. c) protein.
   d) lipids. e) polysaccharides.
2. Temperate phages often
   1. lyse their host cells.
   2. change properties of their hosts.
   3. integrate their DNA into the host DNA.
   4. kill their host cells on contact.
   5. are rare in nature.

   a) 1, 2 b) 2, 3 c) 3, 4 d) 4, 5 e) 1, 5
3. All phages must have the ability to
   1. have their nucleic acid enter the host cell.
   2. kill the host cell.
   3. multiply in the absence of living bacteria.
   4. lyse the host cell.
   5. have their nucleic acid replicate in the host cell.

   a) 1, 2 b) 2, 3 c) 3, 4 d) 4, 5 e) 1, 5
4. The tail fibers on phages are associated with
   a) attachment.
   b) penetration.
   c) transcription of phage DNA.
   d) assembly of virus.
   e) lysis of host.
5. The phages Qβ and MS2
   1. contain double-stranded RNA.
   2. infect all strains of *E. coli*.
   3. contain single-stranded RNA.
   4. have very large burst sizes.
   5. contain single-stranded DNA.

   a) 1, 2 b) 2, 3 c) 3, 4 d) 4, 5 e) 1, 5
6. T4 is
   1. a phage that contains double-stranded DNA.
   2. a phage that contains single-stranded DNA.
   3. a phage that can lysogenize cells.
   4. a phage that can carry out specialized transduction.
   5. a virulent phage.

   a) 1, 2 b) 2, 3 c) 3, 4 d) 4, 5 e) 1, 5
7. The induction of a temperate phage by ultraviolet light results from
   a) damage to the phage.
   b) formation of thymine dimers.
   c) destruction of excision enzymes.
   d) destruction of a repressor.
   e) killing of the host cell.
8. Filamentous phages
   a) attach to bacterial receptors in the cell wall.
   b) take over metabolism of the host cell.
   c) are extruded from the host cell.
   d) undergo assembly in the cytoplasm of the host cell.
   e) degrade the host cells' DNA.
9. Phages have
   1. only one kind of nucleic acid.
   2. a protective protein coat.
   3. many enzymes.
   4. a single shape.
   5. a wide host range.

   a) 1, 2 b) 2, 3 c) 3, 4 d) 4, 5 e) 1, 5
10. *E. coli* most likely becomes resistant to T4 phage through mutations in
    a) the cell wall.
    b) a restriction enzyme.
    c) a modification enzyme.
    d) the structure of pili.
    e) the cytoplasmic membrane.

Applications

1. A researcher discovered a mutation in *E. coli* that prevented phage T4 but not lambda from lysing the cells. What would you surmise is the nature of this mutation?
2. A public health physician isolated large numbers of phage from rivers used as a source of drinking water in western Africa. The physician is very concerned about humans becoming ill from drinking this water although she knows that phages specifically attack bacteria. Why is she concerned?

Critical Thinking

1. Would transduction or conjugation be the most likely mechanism of gene transfer from a Gram-negative to a Gram-positive organism? Explain the reason for your answer.
2. A filter capable of preventing passage of bacteria is placed at the bottom of a U tube to separate the two sides. Streptomycin-resistant cells of a bacterium are placed on one side of the filter and streptomycin-sensitive cells are placed on the other side. After incubation for 24 hours, the side of the tube that originally contained only strep-

tomycin-sensitive cells now contains some streptomycin-resistant cells. Give three possible reasons for this observation. What further experiments would you do to determine the correct explanation?

3. A suspension of phage when added to a culture of bacteria lyses the cells. When a suspension of the same phage is added to bacteria that have been previously agitated in a blender, no lysis occurs. Explain.

4. Is it surprising that most phages are temperate and not virulent? Explain.

5. Explain how the study of virus host range led to the biotechnology revolution. (Hint: It involves gene cloning technology.)

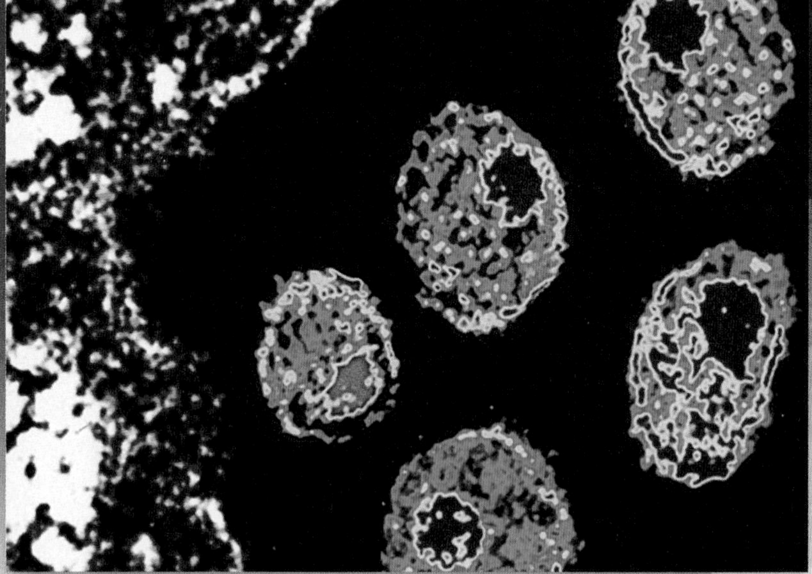

Transmission electron micrograph (TEM) of rotavirus (575,000×).

Viruses, Prions, and Viroids: Infectious Agents of Animals and Plants

A Glimpse of History

In December 2003, a cow that had crossed the border from Canada to Washington State was discovered to be suffering from an always fatal neurodegenerative condition called mad cow disease. In the fall of 2004 and January 2005, several more cows were found in Western Canada suffering from the same malady. Finding these few sick cows halted the importation of all beef from Canada into the United States. Such drastic actions were taken because of the tragedies that occurred in the United Kingdom (UK). In the mid-1980s and early 1990s, more than 180,000 cases of mad cow disease, scientific name **bovine spongiform encephalopathy,** were diagnosed in animals in the UK. The disease is so named because the brain of infected animals looks like a sponge with holes. The infected cows were apparently fed meat and bone meal from sheep and cattle that had also been suffering from this disease. The real tragedy unfolded when 11 people died of the disease after eating beef from infected animals. By early 2004, an additional 146 people were diagnosed in the UK as suffering from mad cow disease. Although many people were skeptical at the time of the first diagnosis, there is no question now that the disease was transmitted from cows to humans.

Mad cow disease is only one of a closely related group of transmissible neurodegenerative diseases that affect both humans and animals. The first example, known for 200 years in Europe, was scrapie, a disease of sheep and goats, so named because affected animals had difficulty standing up and "scraped" along fences for support. In 1936, scientists demonstrated that scrapie could be transmitted by inoculating tissue from diseased animals into healthy sheep and goats—however, it was a year before symptoms appeared. It was assumed that a virus must be the causative agent and the name slow virus disease was given to scrapie in 1954.

In the 1950s, considerable interest developed in a human neurodegenerative disease called kuru that was occurring in epidemic proportions among a native tribe in the highlands of Papua, New Guinea. The disease was characterized by symptoms similar to those of scrapie, and apparently was transmitted through cannibalism. In 1959, Dr. William Hadlow made the connection between kuru and scrapie and suggested that kuru was also a slow virus disease. Dr. Carlton Gajdusek then demonstrated that kuru was infectious since the symptoms of kuru appeared in chimpanzees 18 to 21 months after he inoculated them with brain tissue from patients with kuru.

From early on, it was obvious that these diseases were caused by a transmissible agent, but the nature of this agent was in doubt. Although it was assumed that a virus was the causative agent, no virus could be detected. Further, whatever caused these diseases was resistant to many treatments, such as ultraviolet radiation, nucleases, and treatment with formalin and heat, which inactivated most viruses. Also, patients never developed an immune response. All of these properties led some scientists to suggest that perhaps another type of agent was involved. To identify the active agent, brain homogenates were fractionated to identify the portion that was most active in causing disease. The most infectious fractions contained a protease-resistant glycoprotein, which accumulated in affected brains and often formed protein deposits. In 1982, Dr. Stanley Prusiner proposed the term **prion,** from the letters of proteinaceous infectious particle, to distinguish this infectious agent from viruses and viroids. This protein was abbreviated PrP. Initially it was assumed that this protein was encoded by a gene of the slow virus that was still believed to be the causative agent. However, sequencing the PrP

KEY TERMS

Acute Infection Infection in which the symptoms appear soon after the pathogen is introduced. Symptoms are short lived.

Antigenic Drift Minor changes that occur in the influenza virus antigen as a result of mutations.

Antigenic Shift Major changes that occur in influenza virus antigens resulting from reassortment of viral segments following infection of the same cell by different influenza virions.

Budding A mechanism of release of virions through the plasma membrane without cell death.

Chronic Infection A persistent infection in which the virion can be demonstrated at all times, with or without symptoms.

Endocytosis Process by which animal cells take up particles by enclosing them in a vesicle pinched off from the cell membrane.

Latent Infection A persistent infection in which the infectious agent is present but not active. The agent can reactivate and then cause symptoms.

Persistent Infection An infection in which the virion or its genome is continually present with or without disease.

Prion An infectious agent that causes neurodegenerative disease; consists of protein similar in amino acid sequence to a normal protein in the body.

Segmented Virus A virion with a genome consisting of multiple different fragments.

Slow Infection An infection in which the pathogen increases in number over a very long period of time before symptoms appear.

Zoonoses Natural diseases of animals that can be transmitted to humans as accidental hosts.

led to the astounding discovery that the PrP was very similar in the sequence of amino acids in a protein encoded by a normal chromosomal gene in humans, which is now designated PrP^{C} (for cellular). Significantly, PrP^{C} is protease sensitive whereas the prion protein isolated from infected animals, labeled PrP^{Sc} (for scrapie), is protease resistant. However, the amino acid sequences of PrP^{C} and PrP^{Sc} are very similar. Therefore, there must be a difference in how these proteins fold to assume their tertiary structure. This, in fact, is the case. In 1997, Dr. Prusiner was awarded the Nobel Prize for his discovery of prions, and for elucidating the principles that underlie their mode of action.

How can proteins that cannot replicate be infectious? From where did prions arise? Do the more common neurodegenerative diseases that are not infectious also have their basis in modifications in protein folding? The answers to these and other questions about these frightening diseases are beginning to be uncovered and are discussed in this chapter. ■

Much of the basic biology of bacteriophages also applies to animal and plant viruses; however, each viral group has certain unique properties. This chapter presents a general approach to the classification of animal viruses, followed by a discussion of their modes of replication and effects on host cells, including their role in causing certain tumors. A discussion of plant viruses as well as viroids and prions is also included.

14.1 Structure and Classification of Animal Viruses

Focus Points

- Distinguish between naked and enveloped viruses.
- Name three different ways by which viruses can be classified.

The structure of viruses was covered in chapter 13 (see figure 13.2); we give a brief review here. The structures of phage, and animal and plant viruses are similar, namely nucleic acid, either DNA or RNA, surrounded by a protein coat, the **capsid.** The capsid and nucleic acid together are called the **nucleocapsid.** The capsid, composed of a defined number of units called **capsomers,** is held together by non-covalent bonds. If there is no additional covering, the virus is termed **naked.** However, many viruses that infect humans and other animals have a lipid membrane or **envelope** that surrounds the protein coat. Such a virus, called an **enveloped** virus, is rarely found among phage or plant viruses. The envelope is usually acquired from the cytoplasmic membrane of the infected cell during viral release from the cell. Thus, the structure of the viral envelope is similar to the membrane of the cell, a lipid bilayer containing various proteins. In certain virus families, a **matrix protein** is found just inside the lipid envelope. The **attachment proteins,** or **spikes,** that bind the virus to the cell project from the envelope or the capsid. Plant viruses, such as tobacco mosaic virus, do not bind to specific sites on the plant cell wall; rather, they enter through wounds and have no protruding attachment proteins.

Another distinguishing feature of some RNA animal and plant viruses is that their genome is divided into more than one RNA molecule and each molecule carries different genetic information. For example, the influenza virus has eight RNA molecules, each carrying different genetic information. Such viruses are termed **segmented viruses.**

The virion can have a number of shapes **(figure 14.1).** One is **isometric** in which the protein subunits are arranged in groups of equilateral triangles, the most common arrangement being **icosahedral symmetry** in which 20 equilateral triangular faces enclose the nucleic acid. These viruses appear spherical when viewed with the electron microscope. A less common shape in animal viruses is the helical- or rod-shaped structure. Other virions are **pleomorphic**—they have an irregular shape. The most common type of phage, the complex tailed form, does not occur in animal and plant viruses. ■ complex phages, p. 303 ■ icosahedral symmetry, p. 303

Classification of Animal Viruses

The taxonomy of animal viruses changes as more is learned about their properties. The taxonomy likely will continue to evolve, and

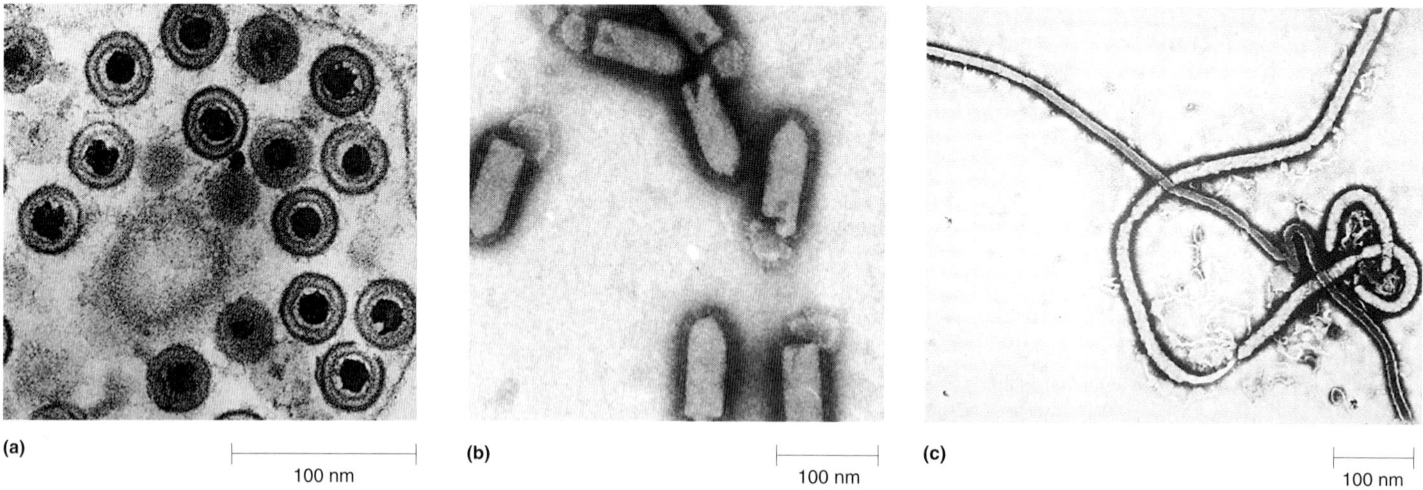

FIGURE 14.1 Shapes of Viruses (a) Electron microscopy of human papillomavirus, an isometric virus whose capsomers can be clearly seen. This virus is a cause of cancer in humans. **(b)** Electron microscopy of rhabdovirus particles, with their characteristic bullet shape. This virus causes rabies. **(c)** Ebola virus, a filamentous virus that occurs in a number of shapes. This virus causes Ebola.

so only general principles are considered here. The most widely employed taxonomic criteria for animal viruses are based on a number of characteristics:

1. Genome structure—DNA or RNA, single-stranded or double-stranded, a single molecule or segmented
2. Virus particle structure—isometric (icosahedral), helical (rod-shaped), or pleomorphic (irregular in shape)
3. Presence or absence of a viral envelope

Based on these major criteria, animal viruses are divided into a number of families, whose names end in *-viridae.* Fourteen families of RNA-containing viruses and eight families of DNA-containing viruses infect vertebrates (**tables 14.1** and **14.2,** respectively). The members of each family are derived from a common ancestor, as shown by nucleic acid hybridization. Evolutionary relationships between families however, cannot be inferred from this taxonomic scheme. The names of the families come from a variety of sources (see tables 14.1 and 14.2). In some cases, the name indicates the appearance of the virion—for example, Coronaviridae coming from *corona,* which means crown. In other cases, the virus is named from the geographic area where the virion was first isolated. Bunyaviridae is derived from *Bunyamwera,* a locality in Uganda, Africa. Each family contains numerous genera whose names end in *-virus,* making it a single word—for example, Enterovirus. The species name is the name of the disease the virus causes (for example, polio or poliovirus). The species name is one or two words. In contrast to bacterial nomenclature in which an organism is referred to by its genus and species names, viruses are commonly referred to only by their species name. The name of the disease they cause and the names are not italicized. ■

nucleic acid hybridization, p. 227

The classification of the family Picornaviridae, which infects humans, is shown in **table 14.3.** This family contains three genera. The genus Enterovirus contains four species. Each species contains numerous "types." Whether these "types" should really be different species is in dispute. Rhinovirus contains a single species, rhinovirus, which in turn contains 100 "types."

In general, viruses with a similar genome structure replicate in a similar way. For example, animal viruses generally follow the same replication strategies as phages with similar genomes.

Groupings Based on Routes of Transmission

Viruses that cause disease are often grouped according to their routes of transmission from one individual to another. These are not taxonomic groupings, and members of more than one family can be included in the same group. These groupings, summarized in **table 14.4,** provide examples of such a scheme.

The **enteric viruses** are usually ingested on material contaminated by feces, the **fecal-oral route.** They replicate primarily in the intestinal tract, where they usually remain localized. They often cause **gastroenteritis,** an inflammation of the stomach and intestine. Some, however, such as the poliovirus, replicate first in the intestines but do not cause gastroenteritis. Rather, they cause a systemic disease.

Respiratory viruses usually enter the body in inhaled droplets and replicate in the respiratory tract. The respiratory viruses include only those viruses that remain localized in the respiratory tract. Viruses that infect via the respiratory tract but then cause systemic diseases are not considered respiratory viruses. These latter include the viruses that cause mumps and measles.

Viral **zoonoses,** caused by **zoonotic viruses,** are diseases that are transmitted from an animal to a human or to another animal. Humans are accidental hosts and rarely is the disease spread from human to human. Many viruses, such as rabies, are transmitted directly from animals to humans, but humans do not generally transmit it to other humans. Others, such as canine distemper, can be transmitted from dogs to African lions. One group of viruses, the arboviruses, are so named because they infect arthropods such as mosquitoes,

TABLE 14.1 Classification of RNA Viruses Infecting Vertebrates

Family	Drawing of Virion	Virion Structure	Genome Structure*	Representative Pathogenic Members and Some Diseases They Cause
Picornaviridae (*pico,* micro; *rna,* ribonucleic acid)		Naked; isometric	1 molecule, ss RNA	Poliovirus; rhinovirus causes colds Hepatitis A virus
Caliciviridae (*calix,* cup)		Naked; isometric	1 molecule, ss RNA	Norovirus; many members cause gastroenteritis
Togaviridae (*toga,* cloak)		Lipid-containing envelope	1 molecule, ss RNA	Many multiply in arthropods and vertebrates; encephalitis in humans
Flaviviridae (flavus, yellow)		Lipid-containing envelope	1 molecule, ss RNA	Yellow fever virus; dengue virus Hepatitis C virus
Coronaviridae (*corona,* crown)		Lipid-containing envelope	1 molecule, ss RNA	Colds and respiratory tract infections, including severe acute respiratory syndrome (SARS)
Rhabdoviridae (*rhabdos,* rod)		Bullet-shaped; lipid-containing envelope	1 molecule, ss RNA	Rabies virus
Filoviridae, (*filo,* threadlike)		Long filamentous; sometimes circular; lipid-containing envelope	1 molecule, ss RNA	Marburg virus; ebola virus
Paramyxoviridae, (*para,* by the side of; *myxa,* mucus)		Pleomorphic; lipid-containing envelope	1 molecule, ss RNA	Mumps virus; parainfluenza virus measles virus
Orthomyxoviridae (*orthos,* straight; *myxa,* mucus)		Pleomorphic; lipid-containing envelope	7–8 segments of linear ss RNA	Influenza virus
Bunyaviridae (Bunyamwera, a locality in Uganda)		Lipid-containing envelope	3 molecules of ss RNA	Hantaan virus
Arenaviridae (*arena,* sand)		Pleomorphic; lipid-containing envelope	2 molecules of ss RNA with hydrogen-bonded ends	Lassa virus
Reoviridae (respiratory enteric orphan virus)		Naked; isometric	Linear ds RNA divided into 10, 11, or 12 segments	Diarrhea in animals
Birnaviridae (*bi,* two)		Naked; isometric	2 segments of linear ds RNA	No human pathogens; diseases in chickens and fish
Retroviridae (*retro,* backward)		isometric; lipid-containing envelope	2 identical molecules ss RNA	HIV

*ss, single-stranded; ds, double-stranded.

ticks, and sandflies, where they replicate. Arthropods then bite vertebrates and transmit the virus. Thus, the viruses are arthropod-borne. In many cases, viruses can invade and replicate in widely different species. The same arthropod might bite birds, reptiles, and mammals and transfer viruses among these widely different groups. More than 500 arboviruses are known, and about 80 are known to infect humans. Twenty cause significant diseases such as West Nile fever, yellow fever, Western equine encephalitis, and dengue fever.

Sexually transmitted viruses cause lesions in the genital tract. These include herpesviruses and papillomaviruses. Other viruses that cause systemic infections are often transmitted during sexual activity. They include human immunodeficiency virus (HIV) and hepatitis viruses.

TABLE 14.2 Classification of DNA Viruses Infecting Vertebrates

Family	Drawing of Virion	Virion Structure	Genome Structure*	Representative Pathogenic Members and Some Diseases They Cause
Hepadnaviridae (*hepa*, liver; *dna*, deoxyribonucleic acid)		Lipid-containing envelope	1 molecule, mainly ds DNA but with a single-stranded gap	Hepatitis B virus
Parvoviridae (*parvus*, small)		Naked; isometric	1 molecule, ss DNA	Outbreaks of gastroenteritis following eating of shellfish
Papillomaviridae (*oma*, tumor)		Naked; isometric	1 molecule, circular ds DNA	Human papillomaviruses associated with genital and oral carcinomas
Polyomaviridae (*poly*, many; *oma*, tumor)		Naked; isometric	1 molecule, circular ds DNA	SV 40 - causes tumors in monkeys
Adenoviridae (*adenos*, gland)		Naked; isometric	1 molecule, ds DNA	Some cause tumors in animals
Herpesviridae (*herpes*, creeping)		Enveloped with surface projections	1 molecule, ds DNA	Herpes simplex virus; cytomegalovirus Chickenpox virus; mononucleosis
Poxviridae (*poc*, pustule)		Enveloped; large brick-shaped	ds DNA; covalently closed ends	Smallpox virus; vaccinia virus
Iridoviridae (*irid*, rainbow)		Naked; isometric	1 molecule, ds DNA	No known human pathogens; only animal pathogens

*ss, single-stranded; ds, double-stranded.

TABLE 14.3 Classification of Human Picornaviruses

Family	Genus	Species
Picornaviridae	Enterovirus	Polioviruses 1–3 Coxsackieviruses A1–A24 (no A23), B1–B6 Echoviruses 1–34 (no 10 or 28) enteroviruses 68–71
	Rhinovirus	Rhinoviruses 1–100
	Hepatovirus	Hepatitis A virus

MICROCHECK 14.1

The genomic structure and shapes of animal viruses are similar to bacteriophages, except the complex shape is not represented. Many animal viruses have an envelope derived from the host cell membrane surrounding their capsid. Animal viruses are classified based on whether they contain DNA or RNA, whether they are enveloped, and their shape. Viruses that cause disease are often grouped by their routes of transmission.

✓ Give two differences in the structure of some animal viruses and bacteriophages.

✓ List four ways in which viruses can be transmitted from one organism to another.

✓ Why do animal viruses have envelopes and phages rarely do?

TABLE 14.4 Grouping of Human Viruses Based on Route of Transmission

Virus Group	Mechanism of Transmission	Common Viruses Transmitted
Enteric	Fecal-oral route	Enteroviruses (polio, coxsackie B); rotaviruses (diarrhea)
Respiratory	Respiratory or salivary route	Influenza; measles; rhinoviruses (colds)
Zoonotic	Vector (such as arthropods)	Sandfly fever; dengue
	Animal to human directly	Rabies; cowpox; West Nile fever
Sexually transmitted	Sexual contact	Herpes simplex virus-2 (genital herpes); HIV

14.2

Interactions of Animal Viruses with Their Hosts

Focus Points

- Name the major classes and subclasses of viral infections and describe how they differ from each other.
- Compare the steps in the infection process in an acute infection of an animal with infection of a bacterium by phage T4.

For bacterial viruses, the host organism is a single cell, and so other kinds of cells do not affect the course of the infection. In the case of animals, however, the outcome of viral infection depends on many factors that are independent of the infected cell. Of special importance are the defense mechanisms of the host, such as the presence of protective antibodies that can confer immunity against a virus ordinarily lethal to an individual without such immunity. Devastating epidemics of measles and smallpox, which decimated the indigenous native population following the arrival of Europeans to the Americas, are good examples of the consequences of the lack of immunity to particular viruses.

Sudden epidemics causing widespread deaths are the most dramatic events of virus interactions with humans. The death of the host, however, also means that the virus can no longer multiply. Obviously, this is not in the best interests of the virus. Just as with the vast majority of bacterial viruses, which are temperate, animal viruses may develop a relationship with their normal hosts in which they cause no obvious harm or disease. Thus, the virus infects and persists within the host, in a state of **balanced pathogenicity,** in which neither the virus nor the host is in serious danger. Indeed, most healthy animals, including humans, carry a number of viruses as well as antibodies against these viruses without suffering any ill effects. If, however, a virus is transmitted to an animal that has no immunity against it, disease may result. ■ **temperate phage, p. 306**

Many viruses that are carried by one group of organisms without causing disease may cause serious disease when transferred to another group. For example, Lassa fever virus does not cause disease in rodents, in which it is normally found, but when transferred to humans, it kills a large percentage of the infected population.

The relationship between disease-causing viruses and their hosts can be divided into two major categories based on the disease and the state of the virion in the host. These are **acute** and **persistent.** Acute infections are usually self-limited diseases in which the virus often remains localized and disappears when the disease ends. In persistent infections the virus establishes infections and remains for years or even life, often without causing any disease symptoms.

Acute Infections (figure 14.2)

Acute infections are usually of relatively short duration, and the host organism may develop long-lasting immunity. Viruses that cause acute infections result in productive infections. The infected cells die and may or may not lyse with the release of virions. Viruses that cause lysis of host cells are usually naked, whereas those that do not cause lysis are frequently enveloped. Although infected cells die, this does not mean that the host dies. Disease symptoms result from localized or wide-spread tissue damage following lysis of cells and spreading and infection of new cells. With recovery, the defense mechanisms of the host gradually eliminate the virus over a period of days to months. Examples of acute infections are mumps, measles, influenza, and poliomyelitis. ■ **mumps, p. 593** ■ **measles, p. 547** ■ **influenza, p. 519** ■ **polio, p. 661**

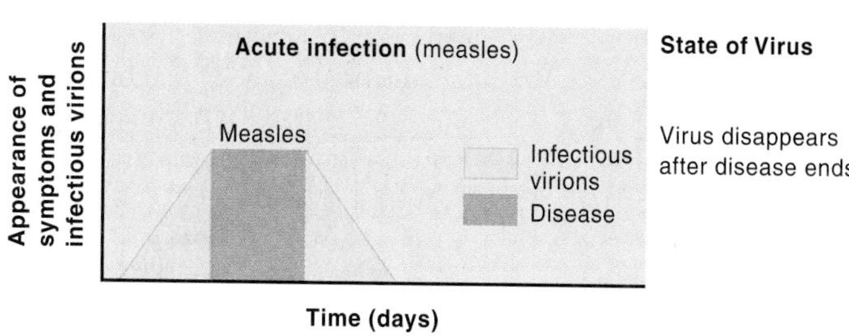

FIGURE 14.2 Acute Infection-Measles A Time Course of Appearance of Symptoms of Measles and the Measles Virions.

The reproductive cycle of an animal virus that results in an acute infection with cell lysis can be compared with the productive infection of a bacterium with a virulent phage. Basically the steps are the same except that infection of more complex eukaryotic cells in a multicellular host requires several additional steps. The essential steps include:

- Attachment
- Entry into susceptible cells following attachment
- Targeting of the virion to the site where it will reproduce
- Uncoating of the virion—separation of protein coat from nucleic acid
- Synthesis of protein and replication of nucleic acid
- Maturation of the viral particles
- Cell lysis
- Spreading of the virus within the host
- Shedding of the infectious virions outside the host
- Transmission to the next host, thereby repeating the infection cycle ■ **lytic phage replication, p. 306**

Step 1: Attachment

The process of attachment (adsorption) is basically the same in all virus-cell interactions, except that the process is more complex in animal viruses than in phages. Animal viruses usually do not contain a single specific attachment appendage, a tail with fibers, as does phage T4, for instance. Rather, surface projections containing attachment proteins or spikes protrude all over the surface of a virion (see figure 13.2). Frequently, several different attachment proteins exist. The receptors to which the viral attachment proteins bind are usually glycoproteins located on the plasma membrane, and often more than one receptor is required for effective attachment. For example, HIV must bind to two key molecules on the cell surface before it can enter the cell. The receptors number in the tens to hundreds of thousands per host cell. As with bacterial phage receptors, the normal function of these glycoproteins is completely unrelated to their role in virus attachment. For example, many receptors are immunoglobulins; other viruses use hormone receptors and permeases. ■ immunoglobulins, p. 371 ■ glycoproteins, p. 30

Different viruses may use the same receptor, and related viruses may use different receptors. Certain viruses can bind to more than one type of receptor and thus be able to invade different kinds of cells. The binding of the attachment proteins to their receptors often changes the shape of viral proteins concerned with entry of the virion and facilitates their entry. Because a virion must bind to specific receptors, frequently a particular virus can infect only a single or a limited number of cell types within a host species, and most viruses can infect only a single species. This may account for the resistance that some animals have to certain diseases. For example, dogs do not contract measles from humans, and humans do not contract distemper from cats. Some viruses, however—for example, those that cause zoonoses—can infect unrelated animals such as horses and humans with serious consequences in both.

Step 2: Entry

The mechanism of entry of animal viruses into host cells depends on whether the virion is enveloped or naked. In the case of enveloped viruses, two mechanisms exist. In one mechanism, the envelope of the virion fuses with the plasma membrane of the host after attachment to a host cell receptor **(figure 14.3a)**. This fusion is promoted by a specific **fusion protein** on the surface of the virion. In some viruses, such as measles, mumps, influenza, and HIV, the protein, which recognizes a target protein on the cell, changes its shape when it contacts the host cell. Following fusion, the nucleocapsid is released directly into the cytoplasm, where the nucleic acid separates from the protein coat.

In another mechanism, enveloped viruses adsorb to the host cell with their protein spikes, and the virions are taken into the cell in a process termed **endocytosis** (figure 14.3b). In this process, the host cell plasma membrane surrounds the whole virion and forms a vesicle. Then, the envelope of the virion fuses with the plasma membrane of the vesicle. The nucleocapsid is then released into the host's cytoplasm. ■ endocytosis, p. 74

In the case of naked virions, the virion also enters by endocytosis. Since the virus has no envelope, however, it cannot fuse with the plasma membrane. Rather, after being engulfed, the virus dissolves the vesicle, resulting in release of the nucleocapsid into the cytoplasm.

Entry by animal viruses differs from phage penetration in two ways. First, the envelope of the virion and the plasma membrane of the host may fuse. Such fusion is not possible when the outside covering of the host has a rigid cell wall. Second, the entire virion is taken into the cell, whereas in the case of phages, the protein coat remains on the outside of the bacterium.

Step 3: Targeting to the Site of Viral Replication

Following penetration, the virion must be targeted to the site where it will multiply. Most DNA viruses multiply in the nucleus, but how the virion gets to the nucleus is not known.

Step 4: Uncoating

In all viruses, the nucleic acid separates from its protein coat prior to the start of replication, the process termed **uncoating.**

Step 5: Nucleic Acid Replication and Protein Synthesis

The first step in replication is transcription of the nucleic acid of the virion. Diverse strategies are followed by viruses of different families for the synthesis of mRNA. In large part, the transcription strategy depends on whether the virus is RNA or DNA and whether the nucleic acid is single- or double-stranded **(figure 14.4)**. For example, in the case of positive single-stranded RNA viruses, the RNA itself functions as a messenger, whereas in the case of negative single-stranded RNA viruses, the RNA must be transcribed into a positive strand before transcription starts.

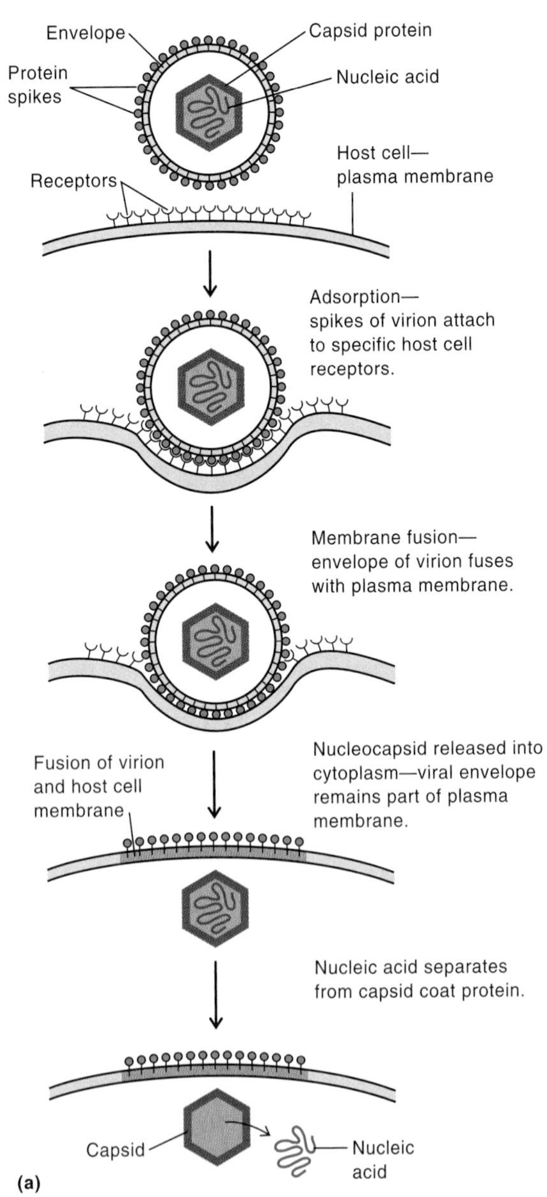

FIGURE 14.3 Entry of Enveloped Animal Viruses into Host Cells (a) Entry following membrane fusion.

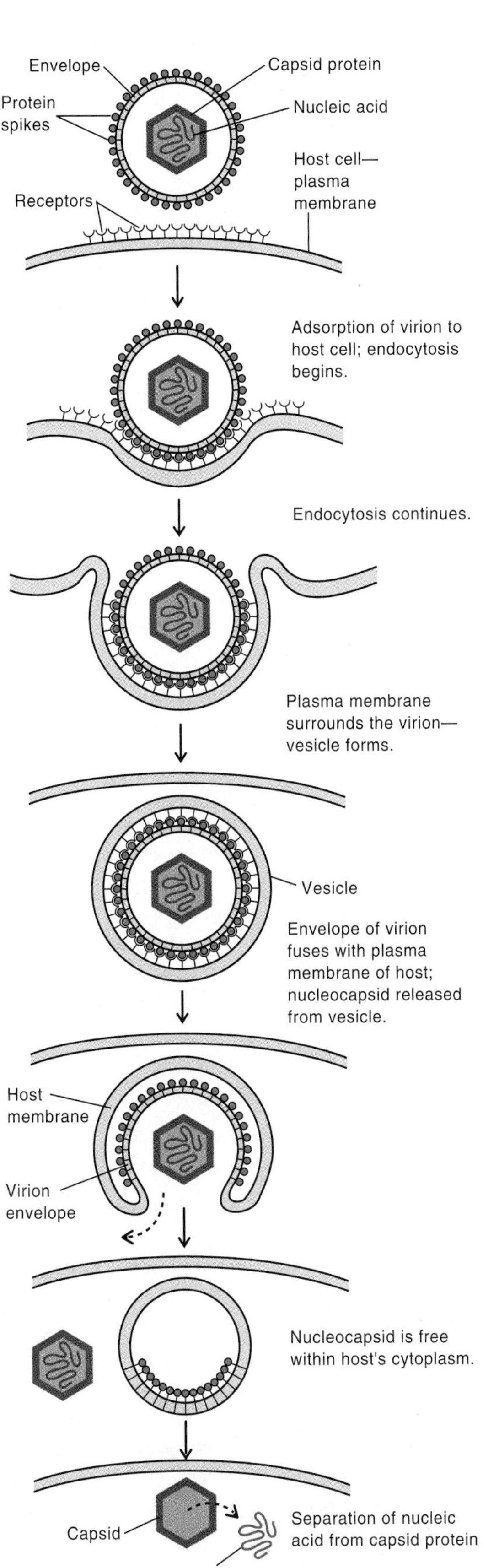

(b) Entry by endocytosis.

Double-stranded (±) DNA
(+) mRNA → (−) mRNA
Synthesis of complementary DNA strand
(±) DNA ← (+) DNA [or (−) DNA]
Transcription of (−) DNA strand
(+) mRNA → (−) DNA → (±) DNA → (+) mRNA ← (−) mRNA
Reverse transcription
Transcription of (−) DNA strand
Synthesis of complementary RNA strand
Double-stranded (±) mRNA (Rare)

Key:
(±) = double-stranded
(+) = positive single strand
(−) = negative single strand

FIGURE 14.4 Strategies of Transcription Employed by Different Viruses Viruses that have the same genome structure follow the same strategy for the synthesis of (+) mRNA, the form of mRNA which is translated into protein. Only (+) mRNA can be translated into protein.

In this case, the RNA-dependent RNA polymerase required for transcription enters the cell as part of the virion since the uninfected cell does not have such an enzyme. Replication of RNA molecules is unique to viruses and generally occurs in the cytoplasm. Replication of viral DNA generally occurs in the nucleus.
■ nomenclature for nucleic acid strandedness, p. 168

In all cases, the patterns of transcription of the viral genome follow the same patterns as phages having the same type of genome. If a viral enzyme is required for early transcription, it may be carried into the cell in the virion or the viral genome may encode the enzyme early in replication. In the case of phages, enzymes do not enter the cell; thus, if host cell enzymes are not available, the enzymes are encoded by the entering viral genome.

In all virus systems, whether phage, animal, or plant, the nucleic acids replicate and proteins are synthesized independently of one another. The replication of viral nucleic acid often depends on enzymes present in the host cell prior to infection. Whether enzymes of the host or viral-encoded enzymes are used varies with the virus. In some cases, viral enzymes concerned with virus replication which are not present in the host enter the host cells along with the virion. As a general rule, the larger the viral genome, the fewer host cell enzymes are involved in replication. This is not surprising since enough DNA is present in large viral genomes to encode most enzymes of nucleic acid synthesis. For example, the largest of the DNA animal viruses, the poxviruses, like T4 phage, are totally independent of host cell enzymes for the replication of their nucleic acid. On the other hand, the very small parvoviruses depend so completely on the biosynthetic machinery of the host cell that they require that the host cell actually be synthesizing its own DNA at the time of infection so that viral DNA can also be synthesized. Most animal viruses are between these two extremes. The replication of the genome of many RNA viruses requires enzymes that are not required in the uninfected cell. Obviously, these must be encoded by the virus and they often enter the cell with the virion. However, all viruses require that host cells supply the machinery for the generation of energy and the biosynthesis of macromolecules.

In some viruses, a polycistronic message is translated into a **polyprotein,** which consists of many proteins strung together. A virus-encoded protease then cleaves the polyprotein to yield individual proteins. ■ protease, p. 37

Step 6: Maturation

The final assembly of the nucleic acid with its coat protein, the process of maturation, is preceded by formation of the protein capsid structure that surrounds the viral genome. The maturation process and multistep formation of the viral coat involve the same general principles in all kinds of viruses. In animal viruses, maturation takes place in a variety of organelles such as the nucleus and microtubules, depending on the virus. This process has already been discussed for bacteriophage T4. ■ assembly, p. 308 ■ microtubules, p. 74

The maturation of a tobacco mosaic virus (TMV), a cylindrically shaped plant virus, has been studied extensively and serves as a model for both animal and plant viruses **(figure 14.5).** For TMV, many identical protein structural subunits, the capsomers, are first formed and then are added one by one to the growing coat structure that surrounds the viral RNA. The coat elongates in both directions, starting from a specific site on the single-stranded viral RNA. The RNA interacts with each protein disc as it is added, and when the end of the long RNA molecule is reached, the discs are no longer added. Enzymes are not required for the process, since the coat self assembles. Recall that the maturation of bacteriophage T4 is also, in part, a self-assembly process. It is a far more complicated process in T4 because this virion has many more different parts than the coat of TMV.

Assembly of many animal viruses takes place at specific regions of the host cell membrane that have embedded specific protein components. These regions are termed **lipid rafts.**

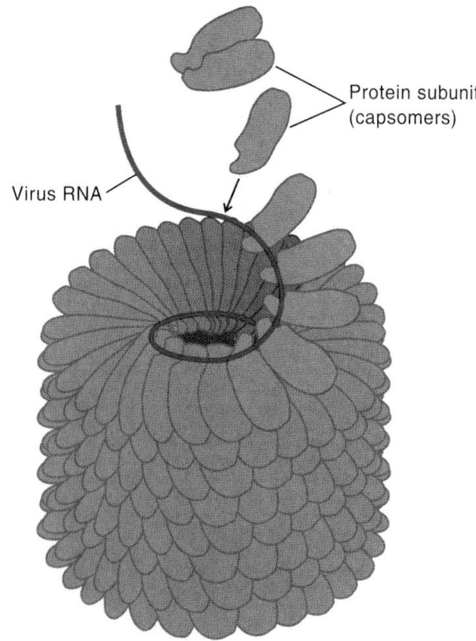

FIGURE 14.5 Tobacco Mosaic Virus Assembly Starting at a specific site on the RNA, the capsomers are added, one by one, to the coat structure to enclose the viral nucleic acid (RNA).

Step 7: Release From Cells

Depending on where the virion is assembled, most non-enveloped viruses accumulate within the cytoplasm or nucleus following their assembly. Unlike virulent phages, animal viral nucleic acid does not always code for enzymes that lyse the host cells. Infected cells often die with the release of virions because viral DNA and proteins rather than host cell material are synthesized. Thus, functions required for cell survival are not carried out, and cells die. Cell degradation and lysis may also result from the release of degradative enzymes contained in cellular lysosomes. This degradation and release of the virions give rise to **cytopathic effects,** changes in the appearance of cells (see figure 14.13). Dead cells lyse, releasing virions, which may then invade any healthy cells in the vicinity. ■ lysosomes, p. 79

Another mechanism for release is by **budding** from the plasma membrane **(figure 14.6).** This process is frequently associated with persistent infections, but the process may kill infected cells. An example of the latter is the killing by HIV as it buds from cells. This process involves a number of steps. First, the region of the host cell plasma membrane where budding is going to take place acquires the protein spikes coded by the viruses, which eventually are attached to the outside of the virion. These are at the lipid raft regions. Then, the inside of the plasma membrane becomes coated with the matrix protein of the virus. In the next step, the nucleocapsid becomes completely enclosed by the lipid raft region of the plasma membrane into which the spikes and matrix protein are embedded. Most enveloped viruses obtain their envelopes as they exit the cell through the plasma membrane. Some viruses, however, bud through the Golgi apparatus or rough endoplasmic reticulum. Vesicles containing the virus then migrate to the plasma membrane, with which they fuse. The virions are released by **exocytosis.** Thousands of virions can be released over hours or days, often without significant cell damage. For all enveloped viruses, budding is part of the maturation process. The process of budding may not lead to cell death, because the plasma membrane can be repaired following budding. As discussed in chapter 13, filamentous phages also are released from bacterial cells by budding or extrusion, without killing the bacterial cells. ■ filamentous phage, p. 311 ■ Golgi apparatus, p. 78 ■ rough endoplasmic reticulum, pp. 77, 78

Step 8: Shedding From Host

To be maintained in nature, infectious virions must exit or be shed from the host. Shedding usually occurs from the same openings or surfaces that viruses use to gain entry. These include mucus or saliva from the respiratory tract during coughing or sneezing, and from feces, urine, skin, genital secretions, or blood.

Step 9: Transmission To Other Hosts

Once an infectious virion has been shed from a host, it must be transmitted to another host, whether the same or another species. It enters into the new host and begins the infection cycle again. As previously discussed, human viruses can be classified based on their route of transmission (see table 14.4).

The various steps in the replication cycle of virulent animal viruses and phages are compared in **table 14.5.**

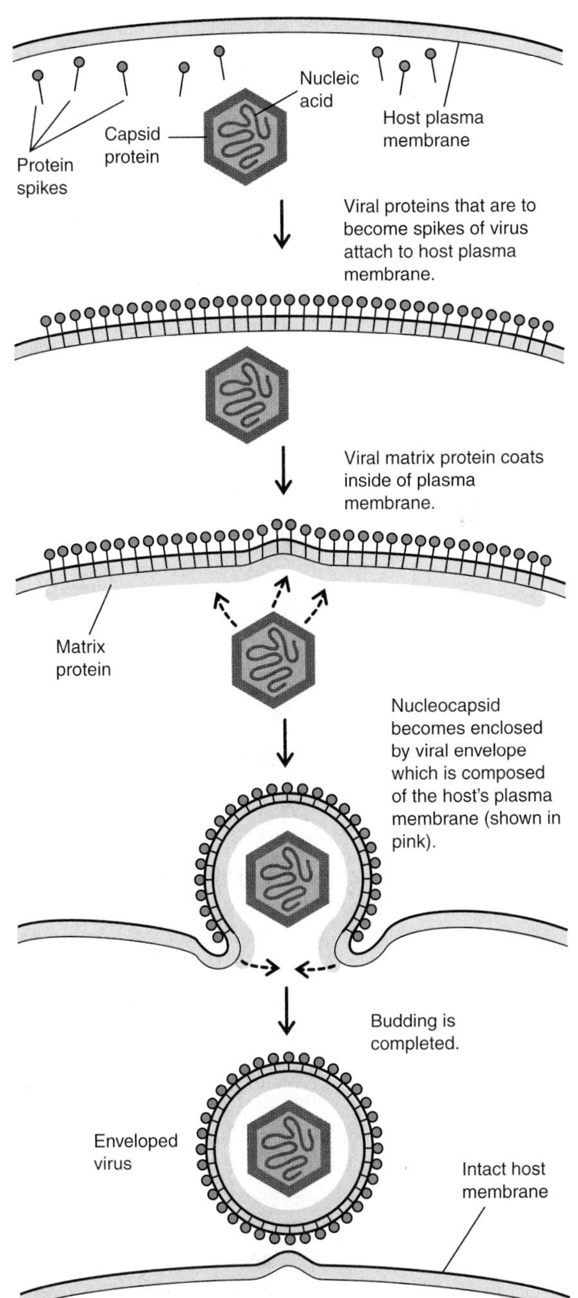

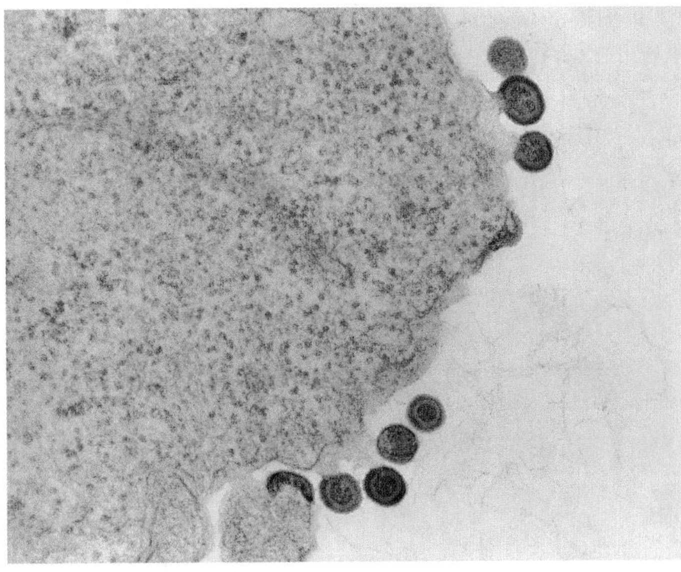

(b)

(b) Electron micrograph of virus particles budding from the surface of a human cell. The virion on the left has completed the process. The other three are in various degrees of completion. It is clear from the micrograph how the virions gain the plasma membrane of the host cell. Note that the membrane of the host remains intact after budding has been completed.

FIGURE 14.6 A Mechanism for Releasing Enveloped Virions
(a) Process of budding.

Persistent Infections (figure 14.7)

In persistent infections, the viruses or their genomes are continually present in the body and virions are released from infected cells by budding. Persistent infections can be divided conveniently into three major categories. These are (a) **latent infections,** (b) **chronic infections,** and (c) **slow infections.** The categories are distinguished from one another largely by whether a virus can be detected in the body during the long period of persistence. Some persistent infections have features of more than one of these categories. For example, infection by HIV has features of latent, chronic, and slow infections.

A persistent infection may or may not cause disease, but since the infected person carries the virus, he or she is a potential source

TABLE 14.5 Comparison of Replication Cycle of Bacteriophages and Animal Viruses in Virulent Infections

Stage	Bacteriophages	Animal Viruses
Attachment	Fusion of capsid with host membrane does not occur.	Fusion of viral envelope and host membrane is common.
Entry	Only nucleic acid enters cell—no enzymes.	Entire virion enters cell, including enzymes of replication.
Targeting of virion	Targeting is unnecessary.	Targeting to site of viral replication.
Uncoating	Takes place at surface of cell.	Takes place inside the cell.
Replication cycle	Depends on whether nucleic acid is DNA or RNA, double- or single-stranded.	Same pattern of replication as phage with the same genome.
Exit	In lytic infection, phage codes for lytic enzyme, which lyses the cell. Budding is rare—cells are not killed.	Some viruses encode a lytic enzyme or the cell dies and lyses with release of virus. Budding is common—cells may or may not be killed.

FIGURE 14.7 Persistent Infections A Time Course of Appearance of Disease Symptoms and Infectious Virions in Various Kinds of persistent Viral Infections.

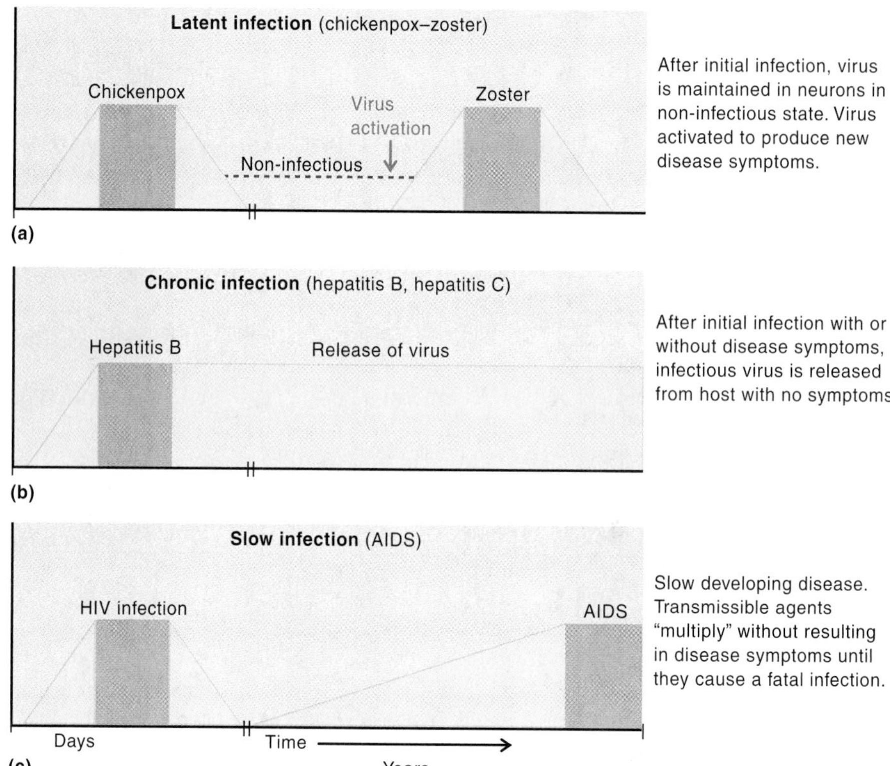
Latent infection (chickenpox–zoster)
Chickenpox
Virus activation
Zoster
Non-infectious
After initial infection, virus is maintained in neurons in non-infectious state. Virus activated to produce new disease symptoms.
(a)
Chronic infection (hepatitis B, hepatitis C)
Hepatitis B
Release of virus
After initial infection with or without disease symptoms, infectious virus is released from host with no symptoms.
(b)
Slow infection (AIDS)
HIV infection
AIDS
Slow developing disease. Transmissible agents "multiply" without resulting in disease symptoms until they cause a fatal infection.
Days
Time
Years
(c)

of infection to others. A person who sheds the virus is a **carrier** and is able to spread disease.

Latent Infections

Latent infections are persistent infections in which a symptomless period is followed by reactivation of the virus with accompanying symptoms. Infectious virus particles cannot be detected until the disease is reactivated (see figure 14.7a). The symptoms of the initial and reactivated forms of the disease may differ. The viruses causing latent infections can be either DNA or RNA viruses. The best known examples are caused by members of the herpesvirus family (Herpesviridae), which is divided into two herpes simplex types: **HSV-1** and **HSV-2.** The latter, frequently called **genital herpes,** is an important and common sexually transmitted disease. ■ HSV-1, p. 592 ■ HSV-2, p. 638

Initial infection of young children with herpes simplex type 1 (HSV-1) may not lead to any symptoms, but cold sores and fever blisters often result. After this initial acute infection, the HSV-1 infects the sensory nerve cells, where it remains in a non-infectious form without causing symptoms of disease **(figure 14.8).** Replication of this virus in the nerve cells is repressed by a mechanism that may involve silencing of gene expression by a micro RNA encoded by the virus. This RNA is similar in its action to siRNA. Virus replication can be activated by such stress conditions as menstruation, fever, or sunburn. After the start of replication, mature infectious virions are produced and are carried to the skin or mucous membranes by the nerve cells, once again resulting in cold sores. After these sores have healed, the virus and host cells once again exist in harmony and, as with other latent infections, no virions are synthesized until the disease recurs. ■ herpes simplex type I, p. 592 ■ SiRNA, p. 179

Another example of a latent infection is provided by another member of the herpesvirus family, varicella-zoster virus, the cause of **chickenpox (varicella).** Initial infection of normal children results in a rash termed chickenpox. This virus then can remain latent for years without producing any disease symptoms. It can then be reactivated and produce the disease called **shingles,** or **herpes zoster.** Thus, chickenpox and shingles are different diseases caused by the same virus. Most infections by this virus, however, never reactivate. ■ chickenpox, p. 545 ■ shingles, p. 546

Most herpesviruses, including HSV-2, tend to become latent under various conditions. It appears that part or all of the viral DNA becomes integrated into the genome of the host, or copies of the nucleic acid of some herpesviruses may replicate as plasmids in the host cell. **Table 14.6** gives some examples of latent infections. ■ temperate phages, p. 306 ■ plasmids, pp. 68, 205

Chronic Infections

In chronic infections, the infectious virus can be demonstrated at all times (see figure 14.7b). Disease symptoms may be present or absent during an extended period of time or may develop late. The best known chronic human infection is caused by the **hepatitis B virus,** formerly called **serum hepatitis virus.** This disease is transmitted sexually or from the blood of a chronic carrier who shows no symptoms. However, some people who contract the virus develop an acute illness marked by nausea, fever, and jaundice. About 300 million people worldwide are carriers of the

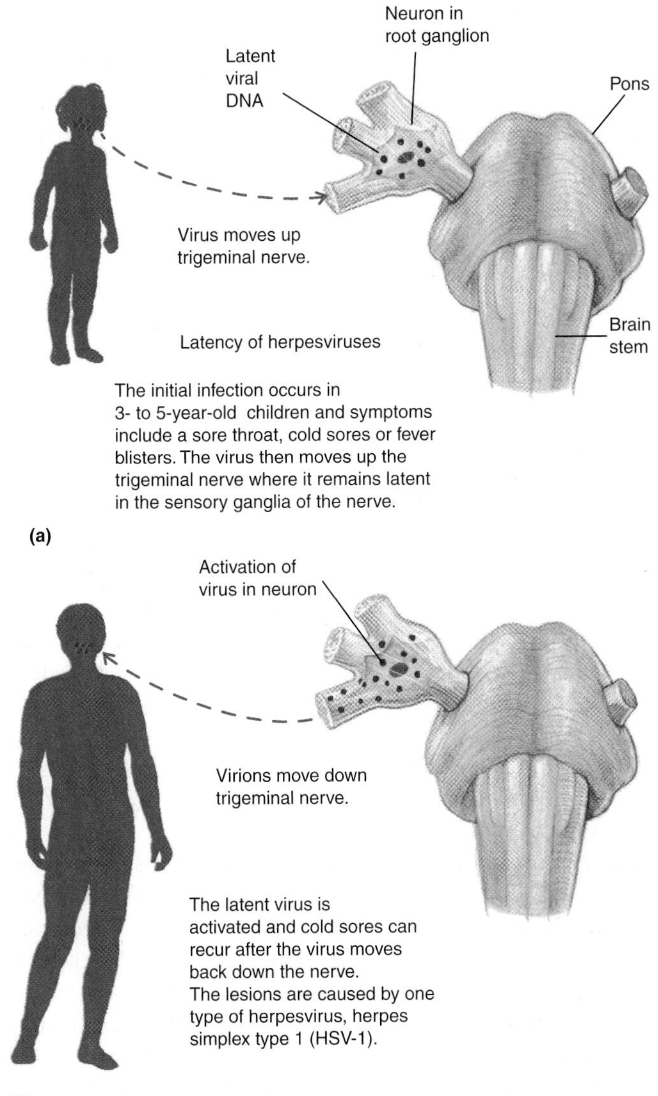

Neuron in root ganglion
Latent viral DNA
Pons
Virus moves up trigeminal nerve.
Latency of herpesviruses
Brain stem
The initial infection occurs in 3- to 5-year-old children and symptoms include a sore throat, cold sores or fever blisters. The virus then moves up the trigeminal nerve where it remains latent in the sensory ganglia of the nerve.
(a)
Activation of virus in neuron
Virions move down trigeminal nerve.
The latent virus is activated and cold sores can recur after the virus moves back down the nerve.
The lesions are caused by one type of herpesvirus, herpes simplex type 1 (HSV-1).
(b)

FIGURE 14.8 Infection Cycle of Herpes Simplex Virus, HSV-1

virus, and a significant number develop cirrhosis or cancer of the liver; more than 1 million people die each year from hepatitis B. ■ hepatitis B, p. 606

In the **carrier state,** infectious virions of hepatitis B are continually produced and can be detected in the bloodstream, saliva, and semen. The viral DNA genome may also occur as a plasmid in liver cells (hepatocytes), where it replicates, encodes viral proteins and assembles into many infectious virions. The plasmid can also integrate into the cells of the liver. In this state, only some of the protein components of the virion are synthesized and infectious virions are not produced. Some examples of viruses that cause chronic infections are summarized in **table 14.7.**

Hepatitis B virus replicates its DNA by a very unusual mechanism. Unlike all other double-stranded DNA viruses, which replicate their DNA using viral DNA as a template, the hepatitis B virus first synthesizes RNA from the DNA. The RNA is then used as a template for the synthesis of DNA, using a viral-encoded DNA polymerase. This enzyme acts as a **reverse transcriptase,** an unusual enzyme associated with the retroviruses, a group of viruses that will be discussed shortly. Replication can be simply diagrammed as shown here:

$$\text{DNA} \longrightarrow \text{RNA} \xrightarrow{\text{Reverse transcriptase}} \text{DNA}$$

A more detailed discussion of the replication of hepatitis B virus is presented in chapter 25. ■ hepatitis B replication, p. 607

Slow Infections

In slow infections, following the initial infection, the infectious agent gradually increases in amount over a very long time during which no significant symptoms are apparent. Eventually, a slowly progressive lethal disease ensues (see figure 14.7c). HIV disease has features of slow virus infections. ■ AIDS, p. 698

Two groups of unusual agents that cause slow infections have been identified. One genus is the Lentivirus (*lenti* means "slow"), which is in the family Retroviridae (retroviruses; *retro* means backward). Other members of this family cause tumors in animals. The second is the protein infectious particles called **prions.** Both groups cause diseases that have long preclinical phases and result

TABLE 14.6 Examples of Latent Infections

Virus	Primary Disease	Recurrent Disease	Cells Involved in Latent State
Herpes simplex virus			
HSV-1	Primary oral herpes	Recurrent herpes simplex	Neurons of sensory ganglia
HSV-2	Genital herpes	Recurrent herpes genitalis	Neurons of sensory ganglia
Varicella-zoster virus (herpesvirus family)	Chickenpox	Herpes zoster (shingles)	Satellite cells of sensory ganglia
Cytomegalovirus (CMV; herpesvirus family)	Usually subclinical except in fetus or immunocompromised host	CMV pneumonia, eye infections, mononucleosis-like symptoms	Salivary glands, kidney epithelium, leukocytes
Epstein-Barr virus (herpesvirus family)	Mononucleosis	Burkitt's lymphoma	B cells, which are involved in antibody production

TABLE 14.7 Examples of Chronic Infections

Virus	Site of Infection	Location of Infectious Virions in Carrier State	Disease
Hepatitis B	Liver	Plasma, saliva, genital secretions	Hepatitis, cirrhosis, carcinoma
Hepatitis C	Liver	Plasma, saliva, genital secretions	Hepatitis, cirrhosis, carcinoma
Rubella virus	Many organs	Urine, saliva	Congenital rubella syndrome

in progressive, invariably fatal diseases. In both groups, the infectious agent can be recovered from infected animals during both the preclinical years when no symptoms are evident and the time that clinical symptoms are present. ■ prions, p. 12

The most common slow virus infection is caused by the human immunodeficiency virus (HIV), which causes AIDS. AIDS results from the invasion and destruction of T lymphocytes and macrophages, important components in the immune system of the body. Without a healthy level of T lymphocytes and macrophages, the body becomes susceptible to a wide variety of infectious diseases. AIDS and HIV infection will be covered in detail in chapter 29. ■ macrophages, p. 350 ■ T lymphocytes, p. 368

The retroviruses are single-stranded, enveloped RNA viruses, many of which infect humans. Many members cause tumors in animals; one causes a leukemia in humans, but the most prominent retrovirus is HIV. The HIV genome is unusual in that it consists of two duplicate copies of RNA.

The replication of retroviruses is unusual and has no counterpart in phages. The major feature is that the genetic information in its RNA is converted into DNA, which is then integrated into the genome of the host cell. Its replication is shown in **figure 14.9.** A more complete diagram of all of the steps of the replication cycle is presented in chapter 29. ■ HIV replication cycle, p. 703

Like all retroviruses, the replication of HIV requires that its single-stranded RNA genome be converted into a double-stranded DNA copy. Two Americans, Howard Temin and David Baltimore, independently demonstrated in 1970 that retroviruses contain, in their capsid, an unusual enzyme, **reverse transcriptase,** that enters the host cell as part of the entering nucleocapsid at the time of infection. Reverse transcriptase is not found in uninfected cells and therefore must be encoded by the viral genome. This enzyme copies the single-stranded viral RNA into a complementary strand of DNA. A second strand of DNA complementary to the first DNA strand is then synthesized. The double-stranded DNA is integrated permanently into a chromosome of the host cell as a **provirus.** This provirus is superficially analogous to the phage lambda (λ) when it is present as a prophage in *E. coli.* Recall, however, that lambda integrates at specific sites in the chromosome of *E. coli* whereas HIV DNA integrates randomly into host cell chromosomes. Further, once integrated the HIV DNA cannot be excized from the host chromosome, which differs from the prophage which can be excized. ■ prophage λ, p. 309

The details of how the activity of the provirus is regulated are not nearly so clear as they are in the case of lambda. It is known that some of the infected cells continuously synthesize new virions that bud from the cell. In this situation, the RNA is transcribed by the host cell RNA polymerase to produce one long mRNA molecule that contains all of the viral information, a **polygenic** or **polycistronic** mRNA. This mRNA molecule is translated into a long **polyprotein,** which is then cleaved by a viral-encoded protease to yield the individual proteins. Following cleavage, the individual proteins fold to their proper conformation resulting in the proteins that make up the virion. If the action of this protease is inhibited, then the virus cannot be assembled. Consequently, inhibitors of this viral protease, termed **protease inhibitors,** created in the laboratory are a major weapon against HIV. ■ polycistronic mRNA, p. 168 ■ protease inhibitor, pp. 487, 488

Many HIV infected cells can carry the provirus in the latent state, and no virions are produced. Various agents can activate the provirus, however, so that it results in a productive infection in which the virions are released. What these agents are in nature is not known. Small amounts of the virus are present continuously or intermittently in the blood and genital secretions, and carriers can transmit the infection through sexual contact.

DNA polymerase makes very few mistakes in the replication of DNA because of its proofreading ability. Reverse transcriptase, however, has no proofreading activity and makes many mistakes when copying RNA into DNA. As a result, the DNA encodes a variety of capsid proteins. Many of these proteins are no longer recognized by antibodies that recognized the protein capsid of the original virus. These errors in copying help explain why the virus becomes resistant very quickly to antiviral drugs such as AZT and probably protease inhibitors. ■ proofreading, DNA polymerase, p. 166 ■ AZT, p. 487

MICROCHECK 14.2

Most animal viruses live in harmony with their natural hosts and do not cause serious illness. If the virus infects an unnatural host, a serious disease may result. The various kinds of relationships of animal viruses with their hosts are, in general, similar to those seen with bacterial viruses and bacteria. Replication of viral nucleic acid depends to varying degrees on the enzymes of nucleic acid replication of the host cells.

A retrovirus, HIV, is responsible for the disease AIDS, which has features of a latent, chronic, and slow infection. The replication of HIV involves an enzyme, reverse transcriptase, which copies the single-stranded RNA of the virion into DNA, which then becomes integrated into certain cells of the immune system.

✓ Name the two *major* kinds of infections that viruses cause.

✓ What is a major difference in the entry of animal viruses and bacterial viruses into their host cells?

✓ Name two ways by which animal viruses cause lysis of host cells.

✓ Explain why HIV becomes resistant to drugs so quickly.

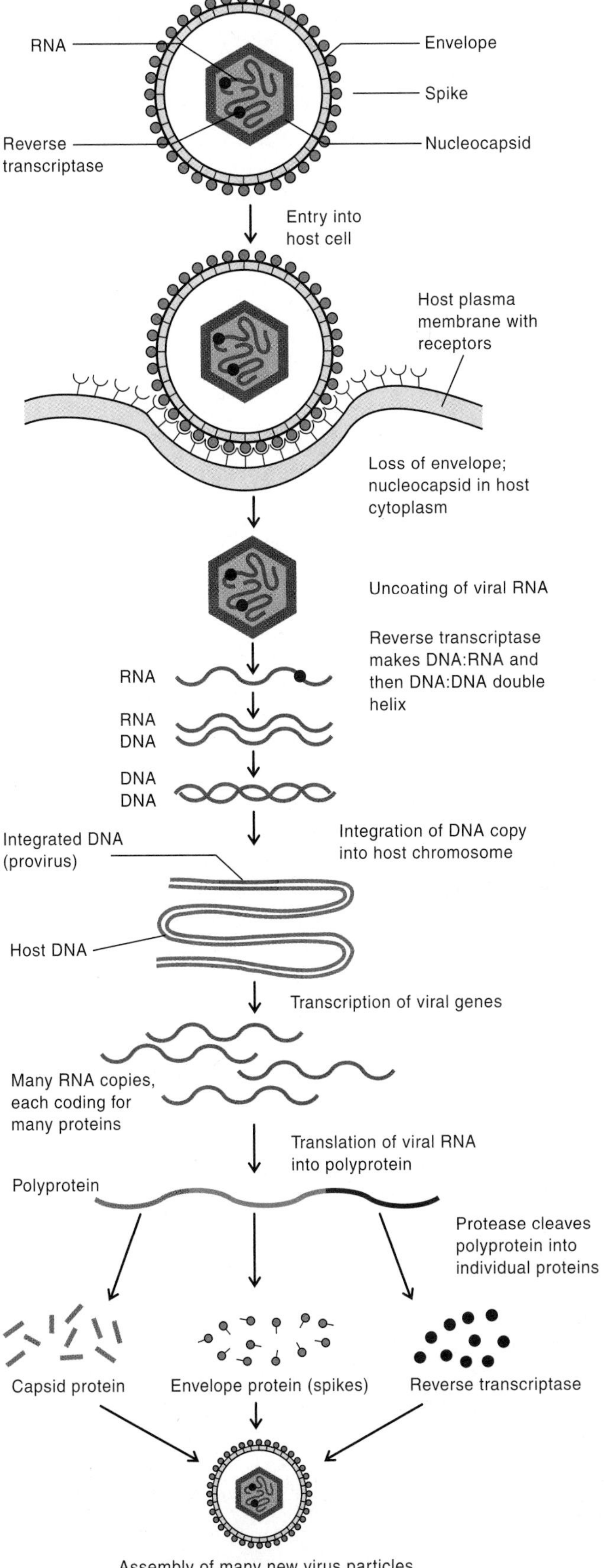
RNA
Envelope
Spike
Reverse transcriptase
Nucleocapsid
Entry into host cell
Host plasma membrane with receptors
Loss of envelope; nucleocapsid in host cytoplasm
Uncoating of viral RNA
Reverse transcriptase makes DNA:RNA and then DNA:DNA double helix
RNA
RNA
DNA
DNA
DNA
Integrated DNA (provirus)
Integration of DNA copy into host chromosome
Host DNA
Transcription of viral genes
Many RNA copies, each coding for many proteins
Translation of viral RNA into polyprotein
Polyprotein
Protease cleaves polyprotein into individual proteins
Capsid protein
Envelope protein (spikes)
Reverse transcriptase
Assembly of many new virus particles each containing reverse transcriptase

FIGURE 14.9 Replication Cycle of a Retrovirus A retrovirus is the cause of AIDS and many tumors in animals.

14.3

Viruses and Human Tumors

Focus Points

- Explain how an oncogene can cause cancer.
- Name the two states in which viral DNA can exist inside a cell and cause cancer.

Most human tumors are not caused by viruses, despite intensive efforts to prove otherwise. The numbers are increasing, however, because of the common occurrence of a viral-induced tumor, Kaposi's sarcoma, in AIDS patients. It is now estimated that viruses play a role in the development of 20% of all human tumors.

Considering all viruses, double-stranded DNA viruses are the main cause of virus-induced tumors in humans **(table 14.8).** DNA tumor viruses interact with their host cells in one of two ways. They can go through a productive infection in which they lyse the cells. Other viruses can transform the cells and modify their properties and a few chronic lytic viruses, such as hepatitis B and C, can cause cancers without killing the cells. In this case, the cancers caused by the DNA viruses result from the integration of all or part of the virus genome into the host chromosome. Following integration, the transforming genes, or **oncogenes** (from the word *onkos,* which means "mass" or "lump"), are expressed, resulting in uncontrolled growth of the host cells. Oncogenes are often altered forms of the normal cells' genes coding for proteins involved in regulating cell growth. Thus, these cases of abnormal growth are analogous to lysogenic conversion observed in certain temperate phage infections of bacteria. In both cases, the expression of viral genes integrated into the host's chromosome confers new properties on the host cells. ■ lysogenic conversion, p. 310

In the case of some DNA viruses, such as papillomaviruses and herpesviruses, the viral DNA is not integrated but apparently replicates as a plasmid. In the case of the papillomaviruses, on rare occasions, the plasmid may integrate into the host chromosome and this may cause tumors. Certain types of human papillomavirus are linked to most cases of cervical cancer, as well as with vulval, penile, and anal cancers. ■ papillomavirus, p. 639 ■ herpesvirus, p. 638 ■ plasmid, pp. 68, 205

Kaposi's sarcoma, a cancer of the skin and internal organs common in AIDS patients, is caused by a herpesvirus. How this particular virus causes normal cells to become tumorous is not known. Note that in all cases of virus-induced tumors, the virus does not kill the host cell but instead changes its properties. ■ Kaposi's sarcoma, p. 710

The various interactions that viruses display with their hosts are illustrated in **figure 14.10.**

Retroviruses and Human Tumors

Although retroviruses are the main class of viruses causing tumors in other animals, it was not until 1980 that a

TABLE 14.8 Viruses Associated with Cancers in Humans

Virus	Type of Nucleic Acid	Kind of Tumor
Human papillomaviruses (HPV)	DNA	Different kinds of tumors, including squamous cell and genital carcinomas, caused by different HPV types
Hepatitis B	DNA	Hepatocellular carcinoma
Epstein-Barr	DNA	Burkitt's lymphoma; nasopharyngeal carcinoma; B-cell lymphoma
Hepatitis C	RNA	Hepatocellular carcinoma
Human herpes, virus 8	DNA	Kaposi's sarcoma
HTLV-1	RNA (retrovirus)	Adult T-cell leukemia (rare)

rare human leukemia was also shown to be caused by a retrovirus. It was named human T-cell lymphotrophic virus type 1 (HTLV-1), and the leukemia this virus causes is restricted to certain geographic areas. The virus causes tumors by a somewhat different mechanism than discussed thus far. The virus carries an oncogene that, when integrated into the host cell genome, codes for an activator protein that activates regulatory genes of the cell.

MICROCHECK 14.3

Most human tumors are not caused by viruses but by mutations in certain host genes. The most common viral cause of tumors in humans is DNA tumor viruses. One retrovirus is known to cause a rare human leukemia.

✓ Name three viruses that cause tumors in humans.

✓ Name a tumor common in AIDS patients and the virus that causes it.

✓ Why is it not surprising that AIDS patients frequently suffer a viral-induced tumor?

FIGURE 14.10 Various Effects of Animal Viruses on the Cells They Infect

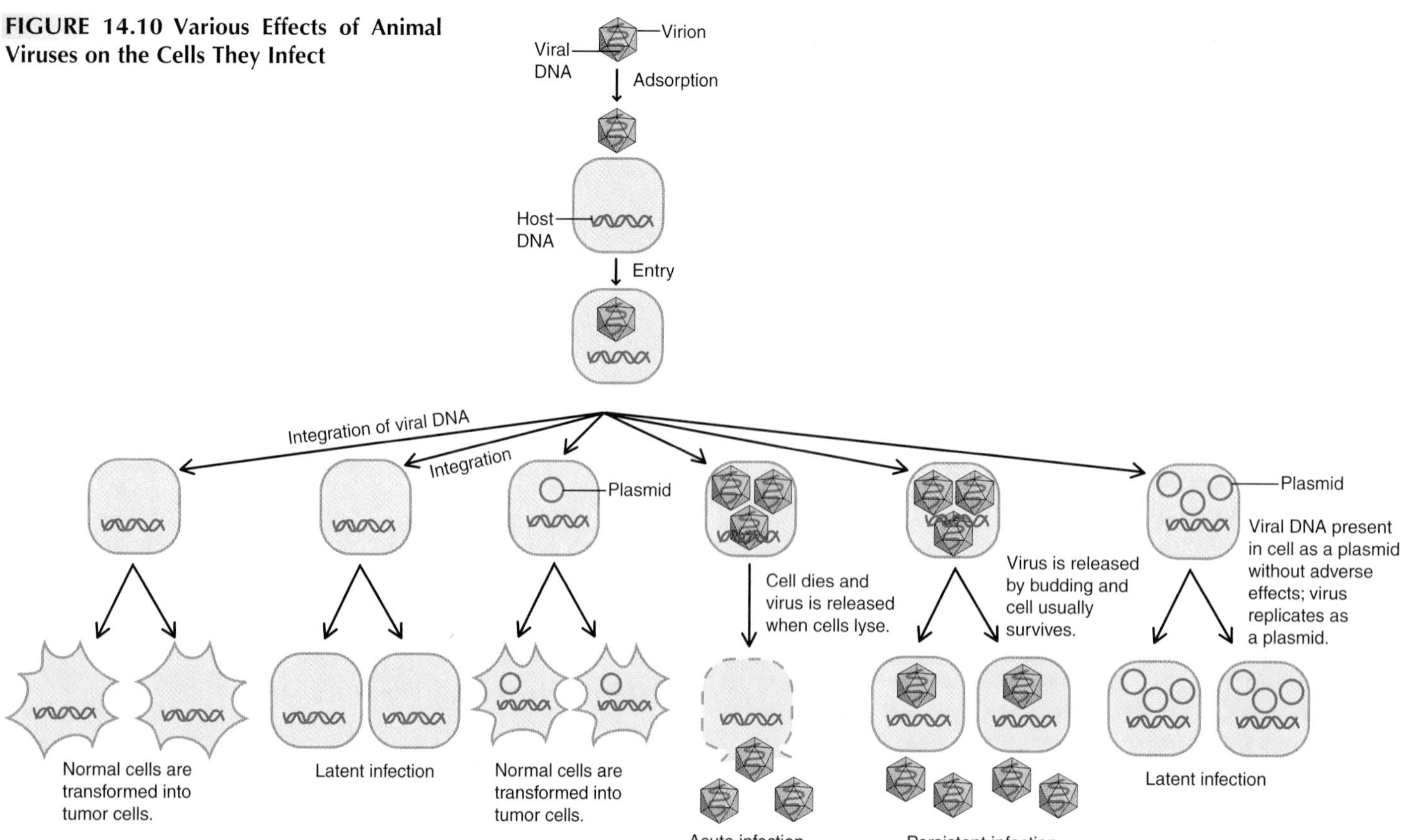

14.4

Viral Genetic Alterations

Focus Points

- Describe the process of genetic reassortment.
- What protein of influenza is critical for attachment to host cells?

Most viruses can infect a single species or even only certain cells within an organism. The major limiting factor in host range is the need for the attachment proteins of the virus to bind to specific receptors on the surface of the host. Another less important limitation is that each virion needs different cellular factors and machinery in order to replicate. Some viruses, however, especially those that cause zoonoses, can multiply in widely divergent species. For example, the West Nile virus that has now infected thousands of people across the United States is primarily a disease of birds that have been bitten by infected mosquitoes. The virus can also infect a wide variety of animals.

Genetic alterations in strains as a result of mutation can result in increased virulence for their hosts as well as increased transmissibility between hosts.

Genome Exchange in Segmented Viruses

In addition to mutations, segmented viruses can alter their properties by a process called **genetic reassortment.** This process results from the infection of the same cell by two different segmented viruses. During virus maturation, one segment of one of the infecting virions becomes incorporated into the coat of the other virion with the other segments coming from the latter virion **(figure 14.11).** This is well illustrated in the case of the influenza virus.

A number of different strains of the influenza virus exist. They can be distinguished by the fact that they are specific for infecting different species of animals and birds. The genome of the influenza virus is divided into eight segments of RNA, each one containing different genetic information. One of the segments codes for **hemagglutinin,** a key protein involved in attachment to host cells. A human with antibodies against hemagglutinin from the human strain is protected against the disease. If, however, the structure of the hemagglutinin gene changes, then the antibodies will not recognize the protein and will not protect against the virus. Experimental data suggest that avian and human influenza virions can simultaneously infect the same cells in a permissive host such as a pig or duck. The cells then serve as a mixing vessel for the 16 RNA segments of the two virions. On occasion, the RNA segment encoding the avian hemagglutinin is incorporated into the same protein coat along with the seven RNA segments from the human strain. Such a strain is still able to infect humans, but its avian hemagglutinin makes it a new strain now able to evade the host's antibody defense. This is called an **antigenic shift.** This phenomenon likely explains how the virulence of the human influenza virion changes so dramatically every 10 to 30 years (see **Perspective 14.1**). These sudden changes result in deadly worldwide epidemics, called **pandemics,** because the global population does not have protective antibodies. Four influenza pandemics occurred. In the twentieth century the most devastating occurring from 1918 to 1920.

In addition to hemagglutinin experiencing this major genetic change, the hemagglutinin gene, along with other viral genes, can undergo point mutations that result in relatively small changes in the protein. Since RNA replication by RNA dependent RNA polymerase does not involve proofreading of the RNA product, the RNA synthesized suffers numerous mistakes. These changes are termed **antigenic drift.** Both of these processes will be discussed later in terms of the epidemiology of influenza. ■ **antigenic drift and shift, p. 520** ■ **antibody, p. 367** ■ **antigen, p. 370** ■ **point mutation, p. 188**

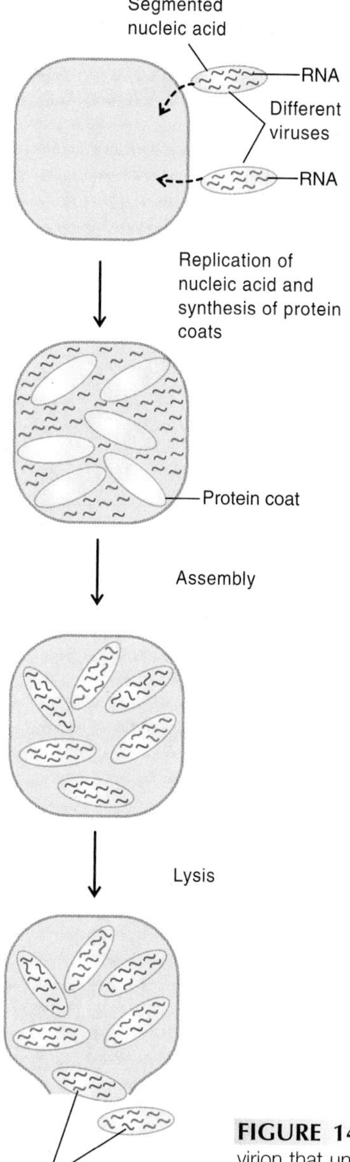

FIGURE 14.11 Genetic Reassortment The virion that undergoes genetic reassortment when it invades another cell gives rise to progeny with the altered characteristics.

MICROCHECK 14.4

The properties of animal viruses can be altered if two viruses with different host ranges infect the same cell. Genes of one virus can be incorporated into the protein coat of the other virus in the process of viral assembly, the phenomena of antigenic shift.

✓ Differentiate between antigenic shift and antigenic drift in the influenza virus.

✓ Is antigenic shift alone likely to lead to influenza pandemics? Explain.

PERSPECTIVE 14.1

A Whodunit in Molecular Virology

Influenza is a disease that results in symptoms of headache, fever, muscle pain, and coughing. Within a week, these symptoms go away and usually only the elderly or others with a weak immune system die. However, the consequences were much more devastating in two influenza outbreaks in the twentieth century. The "Spanish flu" pandemic of 1918 resulted in more than 20 million deaths around the world, and many of the victims were young adults. In 1997, another deadly influenza virus appeared in Hong Kong. Of 18 cases that were diagnosed, 6 were fatal. As in the case of the 1918 "Spanish flu" pandemic, many of the victims were young adults. The Hong Kong virus was transmitted from chickens to humans, but rarely from humans to humans. Therefore, the epidemic was stopped in its tracks by killing all of the chickens in Hong Kong. Clearly, the influenza virions that caused each of these epidemics must differ from the influenza virions that result in the usual unpleasant but short-term symptoms of influenza.

What made the virions that caused the "Spanish flu" and Hong Kong outbreaks so deadly? By sequencing the RNA genome of the various strains of the influenza virion, some answers are now beginning to emerge. To study the 1918 virions, tissues were obtained from preserved bodies of the 1918 victims who had died from influenza. These included several soldiers and an Eskimo woman whose body was buried in the Alaskan permafrost. The sequencing of the RNA genome focused on the hemagglutinin gene since this gene is very important in determining the virulence of the virion. What investigators found was that the hemagglutinin gene of the "Spanish flu" virion originated by recombination between a swine influenza virion and a human-lineage virion. The end of the protein that binds to receptors on the host cell was encoded by the swine-lineage influenza, whereas the rest of the molecule was encoded by a gene that came from a human-lineage influenza. Apparently the recombination occurred shortly before the pandemic began in 1918 and may have triggered the pandemic. However, this story is still not complete and investigations are continuing to unravel the mystery.

The highly virulent virion that caused the Hong Kong epidemic in 1997 appears to be a different, more complicated story. By mixing various combinations of the eight segments of the influenza genome, sequencing the segments and determining the virulence of the various strains, investigators determined that several other genes in addition to the hemagglutinin gene were responsible for the virulence of the Hong Kong strain. These genes included the gene encoding the enzyme neuraminidase, which is required for the spread of the virus within the body, and the gene encoding a protein that blocks the synthesis of interferon, a known viral antagonist.

How these changes in the 1918 "Spanish flu" and the Hong Kong 1997 virions created such deadly strains is not totally understood. However, it seems likely that the new strains were able to circumvent the immune response of the host by expressing new proteins, which humans had not encountered before. Thus the immune cells of the body did not recognize the virions and the body was defenseless. We will likely face new killer strains in the future.

14.5 Methods Used to Study Viruses

Focus Point

- Describe methods for quantifying numbers of animal virions.

A variety of techniques are available to recognize the presence of viruses, identify them, and grow them in large quantities. We focus here on methods for studying animal viruses, which are far more expensive and time-consuming than the methods used in studying phage.

Cultivation of Host Cells

Since viruses can multiply only inside living cells, such cells are needed to study virus growth. The study of bacterial viruses has advanced much more rapidly than investigations on animal and plant viruses, in large part because bacteria are much easier to grow in large quantities in short time periods. The primary difficulty in studying animal viruses is not so much in purifying the virions as it is in obtaining enough cells to infect. Some viruses can only be cultivated in living animals. Others may be grown in **embryonated chicken eggs**—those that contain developing chicks. Some animal viruses can be grown in cells taken from a human or another animal.

When animal viruses can be grown in isolated animal cells, the host cells are cultivated in the laboratory by a technique called **cell culture,** or **tissue culture.** To prepare cells for growth outside the body of the animal (*in vitro*), a tissue is removed from an animal and minced into small pieces **(figure 14.12).** The cells are separated from one another by treating them with a protease enzyme, such as trypsin, that breaks down proteins, holding the cells together. The suspension of cells is then placed in a screw-capped flask in a medium containing a mixture of amino acids, minerals, vitamins, and sugars as a source of energy. The growth of animal cells also requires a number of additional growth factors that have yet to be identified but are present in blood serum. Tissue cultures prepared directly from the tissues of an animal are termed **primary cultures.** ■ **protease, pp. 37, 698**

The cells bathed in the proper nutrients attach to the bottom of the flask and divide every several days, eventually covering the surface of the dish with a single layer of cells, a **monolayer.** When cells become crowded, they stop dividing and enter a resting state. One can continue to propagate the cells by treating them with trypsin, removing them from the primary culture, diluting them, and putting the diluted suspension into another flask containing the required nutrients. This can result in an **established** cell line.

Most cells taken from normal vertebrate tissue die after a certain number of divisions in culture. For example, human skin cells divide 50 to 100 times and then die, even when they are diluted into fresh media. Accordingly, cells must once again be taken from the animal and a new primary culture started.

Cells taken from a tumor, however, can be cultivated *in vitro* indefinitely. Accordingly, they are much easier to use for growing viruses than normal tissue, and several tissue lines have been established from tumors.

Tissue culture is important for growing viruses in the laboratory. The virus is mixed with susceptible cells, and the mixture is incubated until the infected cells lyse. Following lysis, the unlysed cells and the cell debris are removed by centrifugation. The cells and debris go to the bottom of the centrifuge tube while the light, small virions remain in the liquid, the **supernatant.** This liquid containing the virions is also termed a **lysate.**

Tissue culture cells can also be used in virus detection. When a virus is propagated in tissue culture cells, it often changes the cells' appearance. Often, these changes are characteristic for a particular

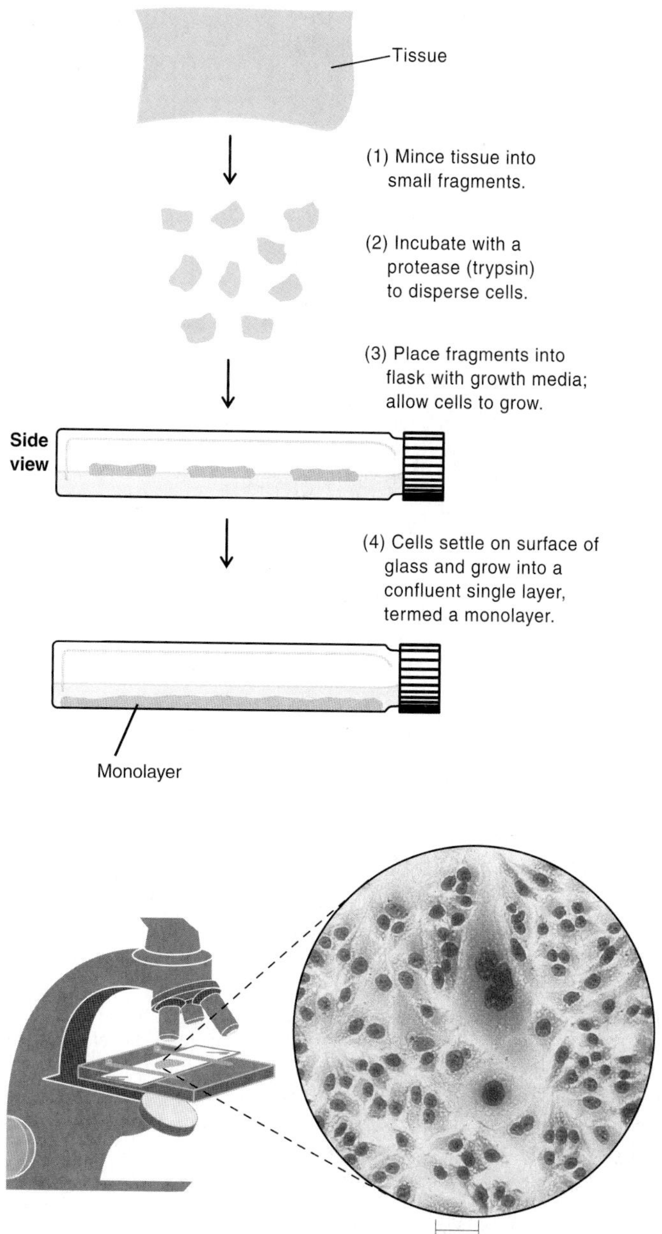

FIGURE 14.12 Preparation of Primary Cell Culture

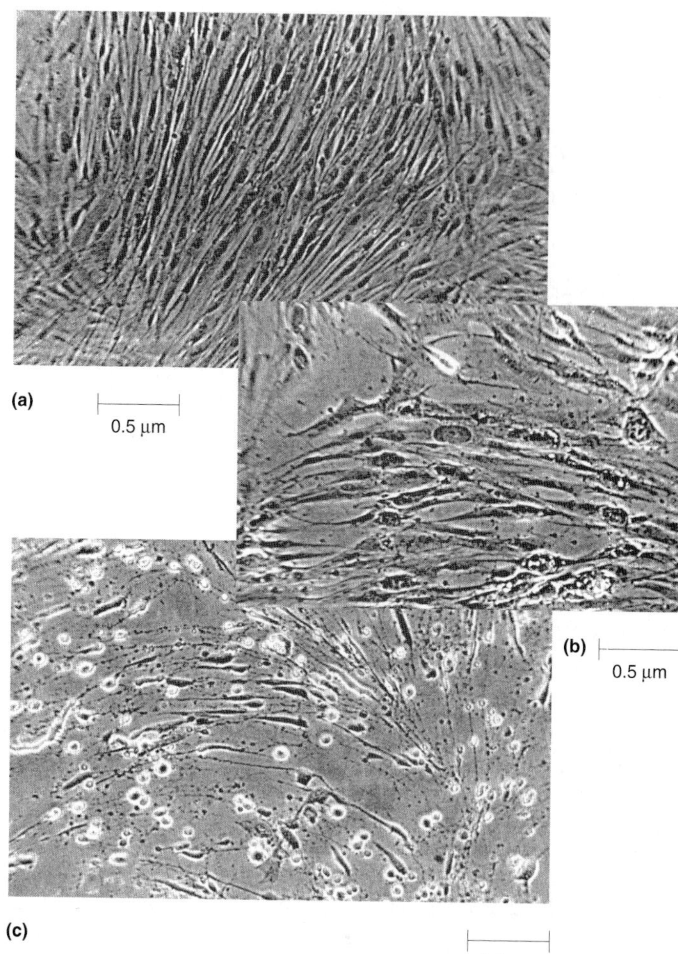

FIGURE 14.13 Cytopathic Effects of Virus Infection on Tissue Culture (a) Fetal tonsil diploid fibroblasts growing as a monolayer, uninfected. **(b)** Same cells infected with adenovirus. **(c)** Same cells infected with herpes simplex virus. Note that the monolayer is totally destroyed.

virus and are referred to as the **cytopathic effect** of the virus **(figure 14.13).** Sometimes the cytopathic effect is localized to particular sites within infected cells. The most common cytopathic effect of this type is termed an **inclusion body.** It is the site at which the virus is being assembled, viral components are being actively synthesized, or cellular damage has occurred. The sites in the cell where inclusion bodies localize vary depending on the virus. Thus, the presence of a virus in an unknown sample and some idea of its identity can often be gained by culturing the specimen on cells in tissue culture.

Quantitation

Tissue culture is also used in virology to study the number of virions in a sample. The most commonly used method for detecting and quantifying the amount of virus present in any sample is the **plaque assay.** A number of other methods can be used for quantitating the number of virions. These include the counting of virions using an electron microscope, **quantal assays,** and in the case of some animal viruses, **hemagglutination.**

Plaque Assay

The plaque assay involves determining the number of viruses in a suspension by adding a known volume of various dilutions to actively metabolizing cell cultures in a Petri dish. The infection, lysis, and subsequent infection of surrounding cells lead to a clear zone or plaque surrounded by the uninfected cells **(figure 14.14).** Each plaque represents one virion, initially infecting one cell, and so the number of virions in the original solution can be readily determined by counting the number of plaques and knowing the dilution of the viral suspension that was plated. Plaques are only formed by infective viruses and can be used with any viruses that lyse their host cells.

It is possible to prepare large numbers of viruses for future studies by adding some liquid to the plate, scraping the surface of the plate with a glass rod, and harvesting the virions.

Counting of Virions with the Electron Microscope

If reasonably pure preparations of virions are obtainable, their concentration can be readily determined by counting the number

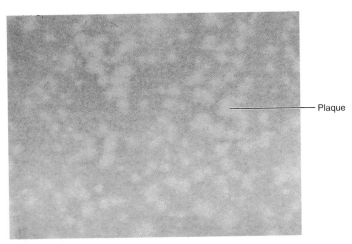

FIGURE 14.14 Viral Plaques Plaques formed by the poliovirus infecting a monolayer of cells that have been stained.

of virions in a specimen prepared for the electron microscope **(figure 14.15).** This method often distinguishes between infective and non-infective virions (see figure 14.15). From their shape and size, it also provides clues as to the identity of the virus.

Quantal Assays

Quantal assays can often provide an approximate virus concentration. In this assay, several dilutions of the virus preparation are administered to a number of animals, cells, or chick embryos, depending on the host specificity of the virus. The **titer** of the virus, or the **endpoint,** is the dilution at which 50% of the inoculated hosts are infected or killed. This titer can be reported as either the **ID_{50}, infective dose,** or the **LD_{50}, lethal dose.**

Hemagglutination

Some animal viruses clump, or **agglutinate,** red blood cells because they interact with the surfaces of the cells. This phenomenon is called **hemagglutination.** In this process, a virion attaches to two red blood cells simultaneously and causes clumping **(figure 14.16a).** Sufficiently high concentrations of virus cause aggregation of red blood cells, which is readily visible (figure 14.16b). Hemagglutination can be measured by mixing serial dilutions of the viral suspension with a standard amount of red blood cells. The highest dilution showing maximum agglutination is the titer of the virus. One group of animal viruses that can agglutinate red blood cells is the myxoviruses, of which the influenza virus is a member. ■ **hemagglutination, p. 443**

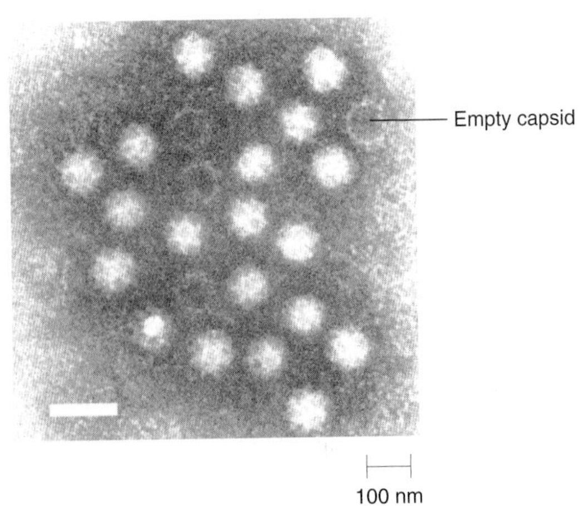

FIGURE 14.15 Electron Micrograph of Calicivirus The number of virions can be counted, and the number of empty capsids also can be readily determined. The latter are not infectious.

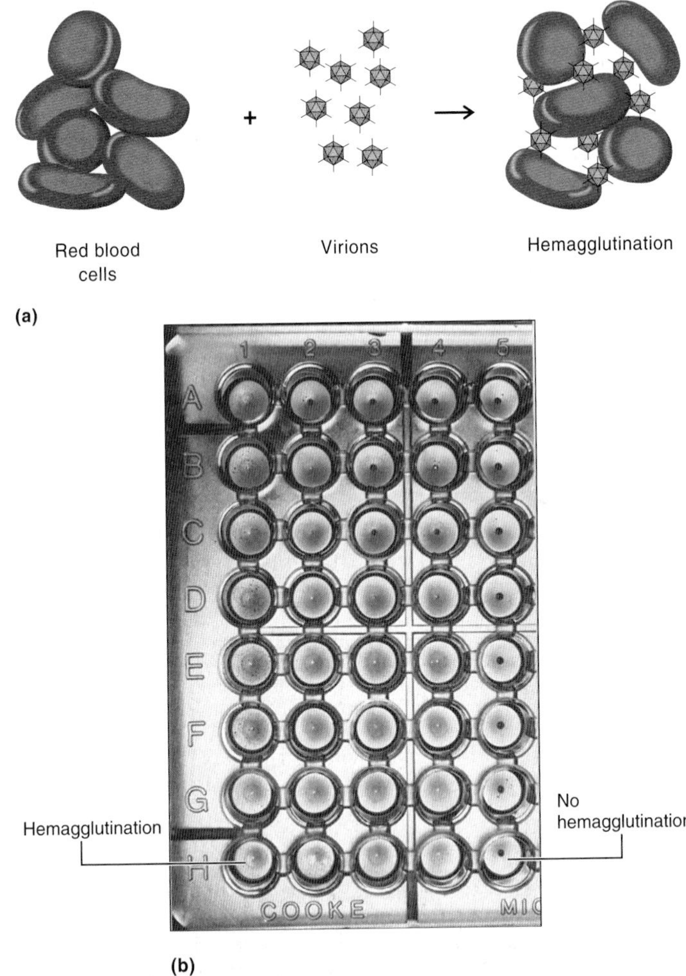

FIGURE 14.16 Hemagglutination (a) Diagram showing virions combining with red blood cells resulting in hemagglutination. **(b)** Assay of viral titer by hemagglutination. Each vertical series of cups in vertical lanes (A–H) represents serial twofold dilutions of different preparations of influenza virus mixed with a suspension of red blood cells. Hemmaglutination is positive when the clumping of the red blood is diffuse at the bottom of the cup and negative when the red blood cells clump into a small, compact button.

MICROCHECK 14.5

Various hosts are required to grow different viruses. These include live animals, embryonated chicken eggs, and cells taken from vertebrates. In contrast to normal cells, which divide 50 to 100 times, cells taken from tumors divide indefinitely. The animal viruses that lyse their host cells can be assayed by counting plaques. Other assays include the counting of virions, quantal assays and hemagglutination.

✓ Which of the methods used in quantifying viruses requires an electron microscope?

✓ Why is it necessary to continually make new primary cultures of normal cells?

✓ Would you expect the number of virions to be the same if you measured them by the plaque assay or by counting using the electron microscope? Explain your answer.

14.6

Plant Viruses

Focus Points

- Compare and contrast the mechanisms by which plant and animal viruses enter host cells.
- Compare and contrast how plant and animal viruses spread from cell to cell.

A great number of plant diseases are caused by viruses. These can be of major economic importance, particularly when they occur in crop plants such as corn, wheat, and rice. Virus infections are especially prevalent among perennial plants (those that live for many seasons), such as tulips, and those propagated vegetatively (not by seeds), such as potatoes. Other crops in which viruses cause considerable damage are soybeans and sugar beets. A serious virus infection can reduce yields of these crops in a field by more than 50%.

Infection of plants by viruses can be recognized through various outward signs **(figure 14.17).** Localized abnormalities may result in a loss of green pigment, and entire leaves may turn color. In many cases, rings or irregular lines appear on the leaves and fruits of the infected plant. Individual cells or specialized organs of the plant may die, and tumors may appear. Usually, infected plants become stunted in their growth, although in a few cases growth is stimulated, leading to deformed structures. In the vast majority of cases, plants do not recover from viral infections, for unlike animals, plants are not capable of developing specific immunity to rid themselves of invading viruses. On occasion, however, infected plants produce new growth in which visible signs of infection are absent, even though the infecting virus is still present.

In severely infected plants, virions may accumulate in enormous quantities. As much as 10% of the dry weight of a tobacco mosaic virus (TMV)-infected tobacco plant might consist of virus.

In a few instances, plants have been purposely maintained in a virus-infected state. The best known example of this involves tulips, in which a virus transmitted through the bulbs can cause a desirable color variegation of the flowers **(figure 14.18).** The infecting virus was transmitted through bulbs for a long time before the cause of the variegation was even suspected. The multiplication of viruses within a plant cell is analogous to that of bacterial and animal viruses in most respects.

(a)

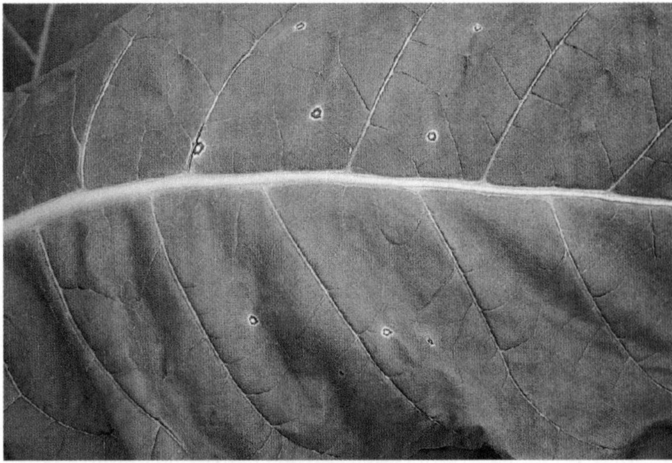

(b)

(c)

FIGURE 14.17 Signs of Viral Diseases of Plants (a) A healthy wheat leaf can be seen in the center. The yellowed leaves on either side are infected with wheat mosaic virus. **(b)** Typical ring lesions on a tobacco plant leaf resulting from infection by tobacco mosaic virus. **(c)** Stunted growth (right) in a wheat plant caused by wheat mosaic virus.

Spread of Plant Viruses

In contrast to phages and animal viruses, plant viruses do not attach to specific receptors on host plants. Instead, they enter through wound sites in the cell wall, which is very tough and rigid. Once started, infection in the plant can spread from cell to cell through openings, the **plasmodesmata,** that interconnect cells.

Many plant viruses are extraordinarily resistant to inactivation. Tobacco mosaic virus apparently retains its infectivity for up to 50 years, which explains why it is usually difficult to eradicate the virions from a contaminated area. Smokers who garden can transmit TMV to susceptible plants from their cigarette tobacco. The virions are very stable, which is important in

FIGURE 14.18 Tulips The variegated colors of tulips result from viral infection. The virus is transmitted directly from plant to plant usually by invertebrate animals such as insects and nematodes.

maintaining the virus because the usual processes of infection are generally inefficient.

Some viruses are transmitted through soil contaminated by prior growth of infected plants. Some 10% of the known plant viruses transmit disease through contaminated seeds or tubers, or by pollination of flowers on healthy plants with contaminated pollen from diseased plants. Virus infections may also spread through grafting of healthy plant tissue onto diseased plants. Another more exotic transmission mechanism is employed via the parasitic vine, **dodder (figure 14.19).** This vine can establish simultaneous connections with the vascular tissues of two host plants, which serve as conduits for transfer of viruses from one host plant to the other.

FIGURE 14.19 Dodder Orange-brown twining stems of the parasitic vine dodder wrap around two different hosts so that virions can pass from one host to the other through dodder.

Other important infection mechanisms involve vectors of various types. These include insects, worms, fungi, and humans. For example, TMV has no known insect vector. Humans are the major vectors of this disease. Viruses are transmitted to healthy seedlings on the hands of workers who have been in contact with the virus from infected plants or by people who smoke. The most important plant virus vectors are probably insects; thus, insect control is a potent tool for controlling the spread of plant viruses.

■ vector, p. 295

Insect Transmission of Plant Viruses

Plant viruses can be transmitted by insects in several ways. First, in **external,** or **temporary, transmission,** a virus is associated with the external mouthparts of the vector. In this case, the ability to transmit the virus lasts only a few days. Second, in **circulative transmission,** the virus circulates but does not multiply in the body of the insect; the virus can be transmitted during the lifetime of the insect. Third, the transmission may involve actual multiplication of the virus within the insect. In this case, the virus is infectious for both the insect and plant cells.

In many instances of insect infection with plant viruses, the viruses are passed from generation to generation of the insect and may be transmitted to plants at any time. The existence of insect-transmitted plant viruses raises several interesting questions about virus evolution. In particular, plant and animal viruses may not be as different as they first appear.

MICROCHECK 14.6

Plant viruses cause many plant diseases and are of major economic importance. They invade through wound sites in the cell wall, and humans are major vectors for their transmission. Some viruses can multiply in both plants and insects.

- ✓ Why is it especially important for plant viruses to be stable outside the plant?
- ✓ How does a plant virus penetrate the tough outer coat of the plant cell?
- ✓ Why is it surprising that some viruses have the ability to replicate in both plants and insects?

14.7 Other Infectious Agents

Focus Points

- What is the chemical structure of prions and viroids?
- Describe the process by which prions replicate.
- What is the source of infectious prions?

Although viruses are composed of only one type of nucleic acid surrounded by a protective protein coat, other agents that cause serious diseases are even simpler in structure. These are the prions and the viroids.

Prions

In addition to the lentiviruses, which include HIV, another group of agents that causes slow diseases are prions—proteinaceous infectious particles that apparently contain only protein and no nucleic acid (see figure 1.14). They are very similar in amino acid sequence to a protein, the normal prion protein, that is found in the brains of all vertebrates. The prion protein a membrane-bound glycoprotein whose function is not known, is either normal when it is in the vertebrate or abnormal when it causes disease. These agents have been linked to a number of fatal human diseases as well as to diseases of animals. In all of these afflictions, brain function degenerates as neurons die, and brain tissue develops spongelike holes **(figure 14.20).** Thus, the general term **transmissible spongiform encephalopathies** has been given to all of these diseases. The time after infection before symptoms appear is many years. Some of the slow but always fatal infections that have been attributed to prions are listed in **table 14.9.** Before the true nature of prions was understood, the name **slow virus disease** was given to these diseases. ■ **prions, p. 12** ■ **glycoprotein, p. 30**

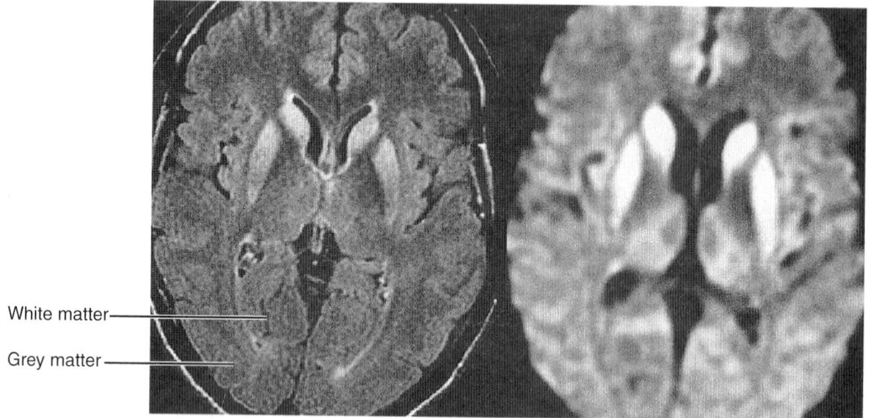

FIGURE 14.20 Appearance of Brain with Spongiform Encephalopathy Left—Normal brain. Right—Brain infected with a prion. Note its spongelike appearance.

Prions differ from viruses in several ways. First, unlike viruses, prions contain no nucleic acid. Second, prions apparently arose following mutations in the gene encoding the normal prion protein in animals and humans. The normal prion protein is abbreviated **PrP^C.** The mutation caused the protein to have different folding properties from the PrP^C. It is designated **PrP^{Sc},** in which the Sc stands for scrapie, the first recognized prion disease. The PrP^{Sc} is usually resistant to protease whereas PrP^C is sensitive. Both forms are resistant to UV light and nucleases because, they do not contain nucleic acid. However, they are inactivated by chemicals that denature proteins.

One of the most intriguing questions regarding prions is how can they replicate if they do not contain any nucleic acid. Recall that only nucleic acids can replicate and code for proteins. The answer to this question is now clear. The prion protein replicates by converting the normal host protein into prion protein thereby creating more prion protein molecules **(figure 14.21).** Thus, the prion protein is "infectious" because it catalyzes in some unknown way the conversion of normal protein into prion protein by changing its folding

TABLE 14.9 Slow Infections Caused by Prions

Agent	Host	Site of Infection	Disease
Scrapie agent	Sheep	Central nervous system	Scrapie spongiform encephalopathy
Kuru agent	Humans	Central nervous system	Kuru spongiform encephalopathy
Creutzfeldt-Jakob agent	Humans	Central nervous system	Creutzfeldt-Jakob disease
Gerstmann-Sträussler agent	Humans	Central nervous system	Gerstmann-Sträussler syndrome
Mad cow agent	Cows and humans	Central nervous system	Mad cow spongiform encephalopathy
Chronic wasting disease agent	Deer and elk	Central nervous system	Chronic wasting disease
Transmissible mink agent	Ranched mink	Central nervous system	Transmissible mink encephalopathy

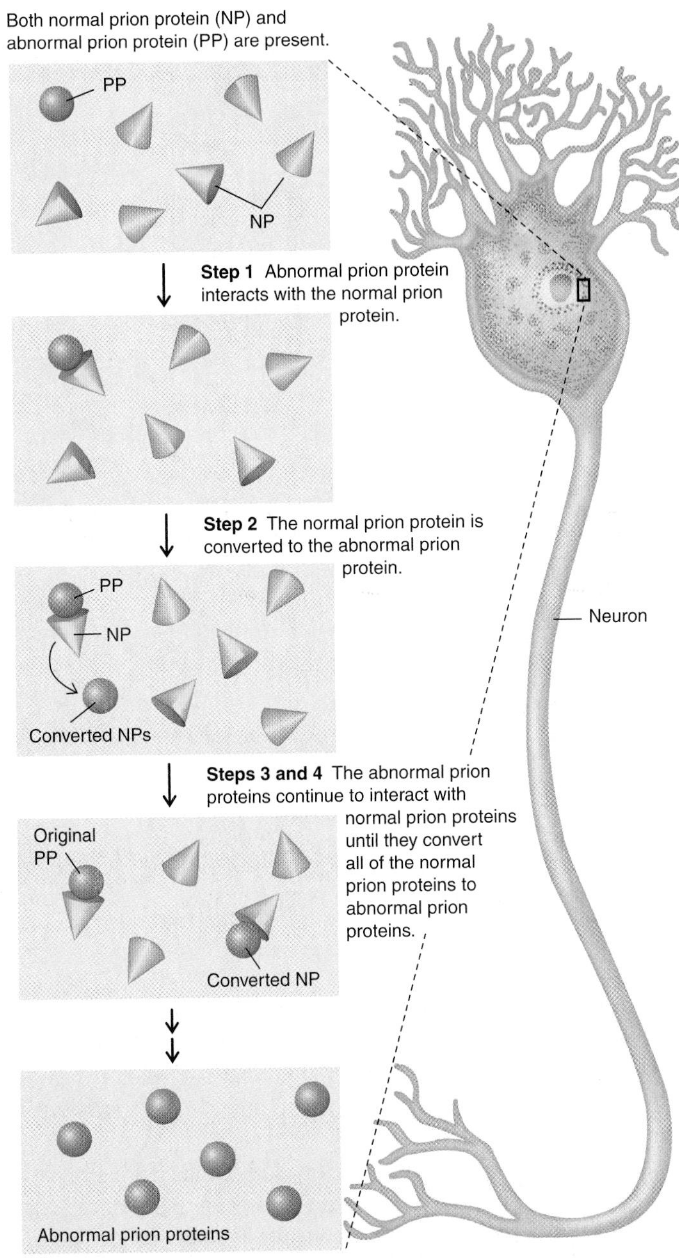
Both normal prion protein (NP) and abnormal prion protein (PP) are present.
PP
NP
Step 1 Abnormal prion protein interacts with the normal prion protein.
Step 2 The normal prion protein is converted to the abnormal prion protein.
PP
NP
Converted NPs
Neuron
Steps 3 and 4 The abnormal prion proteins continue to interact with normal prion proteins until they convert all of the normal prion proteins to abnormal prion proteins.
Original PP
Converted NP
Abnormal prion proteins

FIGURE 14.21 Proposed Mechanism by Which Prions Replicate The normal and abnormal prion proteins differ in their tertiary structures.

properties. Therefore, only if the cell is synthesizing normal prion protein is the prion protein able to replicate. Further, because the amino acid sequence of the abnormal prion protein and the normal prion protein found in all vertebrates are very similar, the notion that only the amino acid sequence of a protein determines its conformation cannot be always true. ■ **tertiary structure, p. 28**

In most cases, the prion disease is only transmitted to members of the same species, because the amino acid sequence of different prion proteins in different species differs from one another. However, the barrier to prion transmission between species also depends on the strain of prion. It is now clear that the prion that caused mad cow disease in England has killed more than 100 people by causing a disease very similar to Creutzfeldt-Jakob disease. Presumably these people ate beef of infected animals. Thus far, no human deaths have been attributed to eating sheep infected with the scrapie agent or deer and elk infected with the prion causing chronic wasting disease. However, because the incubation period extends over many years, the possibility that cross infection can occur in these situations also has not been ruled out entirely. Additional information about the role of prions in human disease is covered in chapter 27. A brief history was covered in the **"A Glimpse of History"** in this chapter.

Viroids

The term **viroid** defines a group of pathogens that are also much smaller and distinctly different from viruses (see figure 1.13). Viroids consist solely of a small, single-stranded RNA molecule that varies in size from 246 to 375 nucleotides. This is about one-tenth the size of the smallest infectious viral RNA known. They have no protein coat and therefore are resistant to proteases.

Their other properties include:

- Viroids replicate autonomously within susceptible cells. No other virions or viroids are required for their replication.
- A single viroid RNA molecule is capable of infecting a cell.
- The viroid RNA is circular and is resistant to digestion by nucleases.

All viroids that have been identified infect only plants, where they cause serious diseases. These diseases include potato spindle tuber, chrysanthemum stunt, citrus exocortis, cucumber pale fruit, hopstunt, and cadang-cadang.

A great deal is known about the structure of viroid RNA, but many questions remain. How do viroids replicate? How do they cause disease? How did they originate? Do they have counterparts in animals, or are they restricted to plants? The answers to these questions will provide insights into new and fascinating features of another unusual member of the microbial world.

MICROCHECK 14.7

Two infectious agents that are structurally simpler than viruses are prions and viroids. Prions contain protein and no nucleic acid; viroids contain only single-stranded RNA and no protein.

✓ Distinguish between a prion and a viroid.
✓ What are the hosts of prions? Of viroids?
✓ Why are viroids resistant to nucleases?
✓ Must all prion diseases result from eating infected food? Explain.

FUTURE CHALLENGES

Great Promise, Great Challenges

Gene therapy, the treatment of disease by introducing new genetic information into the body, is a procedure with tremendous promise, but with limited success thus far. In large part, this lack of total success and acceptance results from a number of basic biological and technological problems that have not yet been solved. However, gene therapy has restored the health of about 20 children suffering from severe immunodeficiency disease.

Gene therapy is rooted in the advances that have been made in microbial genetics, molecular biology, and virology. The idea that it might be possible to introduce genes into mammalian cells and correct life-threatening conditions was first suggested by a number of scientists who worked with microbes and studied gene transfer in bacteria. These included Joshua Lederberg and Edward Tatum, co-discoverers of conjugation in *E. coli.* Lederberg also discovered transduction by a bacterial virus. It was a small step for these investigators to suggest the possibility of gene transfer into mammalian cells by an animal virus.

Viruses are the most popular vectors for introducing genes into cells. They have the ability to be taken up by specific tissues and then induce the cells' machinery to synthesize protein from the introduced genes. The desired genes are merely cloned into the viral genome and these will be introduced into all cells the virus infects.

The major challenge of gene therapy is to design a viral vector that can deliver and express genes in mammalian cells with great efficiency and absolute safety. A long-term goal is to deliver useful genes to the right spot and have them turn on and off normally. Considerable progress is being made to achieve these goals.

The two most common viral vectors currently being used are retroviruses, which include the HIV virus, and adenoviruses, a common cause of colds. The former results in integration of nucleic acid into the host cells, whereas the latter leads only to expression of the viral genes and protein synthesis for a short period of time while the virus replicates, referred to as **transient expression.** However, a stable protein can remain active for a long period of time.

The exciting promise of gene therapy has been on a roller coaster. First there was great hope of curing many diseases caused by defective genes, such as cystic fibrosis and hemophilia. However, the issues of safety, resulting from the deaths of young patients, and the development of cancer in several others, temporarily on several occasions stopped all clinical trials. Another hopeful approach is the use of **RNA interference,** to shut off genes synthesizing harmful products. Using this technique, scientists hope to treat such conditions as macular degeneration, high cholesterol levels, and HIV. Has a new era in this promising field begun? ■ **adenoviral pharyngitis, p. 508** ■ **RNA Interference, p. 179**

SUMMARY

14.1 Structure and Classification of Animal Viruses

Phages and animal and plant viruses are nucleic acid surrounded by a protein coat, the **capsid.** Many animal viruses have an additional covering, an **envelope,** a lipid bilayer similar to the plasma membrane of the host (figure 13.2). Animal and plant viruses containing more than one RNA molecule are **segmented viruses.**

Classification of Animal Viruses

The criteria for classifying animal viruses are **genome structure, particle structure,** and the presence or absence of a **viral envelope** (tables 14.1, 14.2).

Groupings Based on Routes of Transmission

Viruses that cause disease can be grouped according to their **routes of transmission**—the **enteric** viruses, **respiratory** viruses, **zoonotic** viruses (transferred from animals to other animals), and **sexually transmitted viruses** (table 14.4).

14.2 Interactions of Animal Viruses with Their Hosts

Most viruses infect and persist within their hosts in a state of **balanced pathogenicity** in which the host is not killed. Viruses cause diseases that can be classified as **acute** or **persistent.**

Acute Infections

Acute infections are self-limited; the virus often remains localized, and diseases are of short duration and lead to lasting immunity (figure 14.2). The replication cycle of virulent animal virus that causes an acute infection is similar to the lytic cycle shown by phage T4 (table 14.5, figure 13.5). The steps in the infection process include **attachment** to specific receptors; **entry of the virion; targeting** to the site of viral replication; **uncoating** of the virion (figure 14.3); **replication** of virus nucleic acid and synthesis of protein (figure 14.4); **maturation** of the virion (figure 14.5); **release** of the virion from cells and infection of other cells (figure 14.6); and **shedding** of the virions from one host and **transmission** to other hosts (table 14.5).

Persistent Infections

In persistent infections, the virions or their nucleic acids are continually present in the body and virions are released from cells by budding (figure 14.6). Persistent infections can be (1) **latent infections** (table 14.6), (2) **chronic infections** (table 14.7), and (3) **slow infections.** These categories are distinguished from one another largely by whether a virus can be detected in the body during the period of persistence (figure 14.7). An example of a **latent** infection is caused by herpes simplex, HSV-1, which causes cold sores in childhood and becomes latent in nerve cells (figure 14.8). The most common **slow** infection is AIDS, caused by HIV. In HIV infection, the viral RNA is converted to double-stranded DNA by the viral enzyme **reverse transcriptase.** The DNA is then integrated into chromosomes of host cells concerned with the immune response, where it can remain latent or give rise to intact virions (figure 14.9). Unlike DNA polymerase, the enzyme reverse transcriptase has no proofreading ability, and so virions undergo numerous uncorrected mutations.

14.3 Viruses and Human Tumors

Twenty percent of human tumors are caused by viruses, primarily double-stranded DNA tumor viruses (table 14.8).

Retroviruses and Human Tumors

A rare tumor, a leukemia, is caused by a retrovirus.

14.4 Viral Genetic Alterations

Most viruses can infect only a single species and only certain cells within an organism. Viruses causing **zoonoses** can multiply in widely divergent species such as birds, mosquitoes, and humans. Viruses can modify their properties if two viruses with different host ranges infect the same cell.

Genome Exchange in Segmented Viruses (figure 14.11)

Segmented viruses like influenza can alter their properties through an exchange of genomes following infection of the same cell by two viruses with different host ranges—the process of **genetic reassortment.** This results in changes of many properties of the virion, as a result of **genetic shift.** This type of exchange likely accounts for the pandemics caused by the influenza virus every 10 to 30 years. The genome also undergoes small changes as a result of mutations, the process of **genetic drift.**

14.5 Methods Used to Study Viruses

Cultivation of Host Cells

Some viruses can only be cultivated in living animals; others can be grown in **tissue culture** (figures 14.12, 14.13).

Quantitation

The **plaque assay** is commonly used to determine the number of infective virions in a sample (figure 14.14). Virions can be counted with an **electron microscope** (figure 14.15). **Quantal assays** determine the **infective** or **lethal** dose in a viral preparation. Some viruses can clump red blood cells and their concentration can be measured by determining the dilution of virus able to clump the cells, **hemagglutination** (figure 14.16).

14.6 Plant Viruses

Many plant diseases are caused by viruses (figure 14.17). Virions do not bind to receptor sites on plant cells, but enter through wound sites and move to other cells via **plasmodesmata.**

Spread of Plant Viruses

Many plant viruses are very resistant to inactivation. Viruses are spread in large part by humans, by planting seeds in contaminated soils, through transfer from infected plants by grafting, and through the parasitic plant **dodder,** which can establish connections between an infected and uninfected plant (figure 14.19). Various vectors, like insects and worms, can also spread virions.

Insect Transmission of Plant Viruses

Viruses may be associated with the external mouthparts of an insect or they may multiply within the insect.

14.7 Other Infectious Agents

Prions (figure 1.11, table 14.9)

Prions consist of **protein** and no **nucleic acid,** and they cause a number of fatal neurodegenerative diseases called **transmissible spongiform encephalopathies.** Uninfected cells synthesize a protein almost identical in amino acid sequence to the abnormal prion protein. Prions replicate by converting the normal cellular prion protein to the abnormal prion protein, which has a different conformation (figure 14.21).

Viroids (figure 1.10)

Viroids are plant pathogens that consist of circular, single-stranded RNA molecules; they are about one-tenth the size of the smallest infectious viral RNA known. Many unanswered questions remain regarding their origin, how they multiply, and how they cause disease.

REVIEW QUESTIONS

Short Answer

1. What are the criteria on which animal viruses are classified?
2. What are the substrate and product of reverse transcriptase?
3. Name one similarity between prions and HIV in terms of time of onset of symptoms appearing.
4. Distinguish between genetic shift and genetic drift.
5. Compare the mechanisms by which phages and animal and plant viruses enter their host cells.
6. What is the major difference between a latent and chronic infection in terms of the virus in the host cell?
7. Name two DNA viruses that cause tumors in humans. Name one retrovirus that causes tumors in humans.
8. Distinguish between a prion and a viroid. What hosts does each infect?
9. What family of animal viruses is most closely related to temperate phages in their interaction with host cells?
10. Define genetic reassortment. Explain how it can alter the properties of a virus.

Multiple Choice

Questions 1 to 7 concern the differences and similarities between animal, plant, and bacterial viruses. Answer each question based on the following possibilities by circling the correct letter:

a) Bacteriophages only
b) Animal viruses only
c) Plant viruses only
d) Bacteriophages and animal viruses
e) Animal and plant viruses

1. Bind to specific receptors on the host cell to initiate infection.
   a) b) c) d) e)
2. Only the nucleic acid enters the host cell.
   a) b) c) d) e)
3. Lipid envelopes are common.
   a) b) c) d) e)
4. Reverse transcriptase is involved in the replication cycle.
   a) b) c) d) e)
5. Integration of viral nucleic acid leads to changes in properties of the host.
   a) b) c) d) e)
6. Tumors are caused by these viruses.
   a) b) c) d) e)
7. Restriction enzymes play a role in the host range of these viruses.
   a) b) c) d) e)
8. Prions
   a) contain only nucleic acid without a protein coat.
   b) replicate like HIV.
   c) integrate their nucleic acid into the host genome.
   d) cause diseases of humans.
   e) cause diseases of plants.
9. Viroids
   1. contain only single-stranded RNA and no protein coat.
   2. use reverse transcriptase in their replication.
   3. are similar in structure to bacteriophages.
   4. cause diseases in animals.
   5. cause diseases in plants.

   a) 1, 2 b) 2, 3 c) 3, 4 d) 4, 5 e) 1, 5

10. Acute infections in animals
    1. are a result of productive infection.
    2. generally lead to long-lasting immunity.
    3. result from integration of viral nucleic acid into the host.
    4. are usually followed by chronic infections.
    5. often lead to tumor formation.

    a) 1, 2 b) 2, 3 c) 3, 4 d) 4, 5 e) 1, 5

Applications

1. You are a scientist at a pharmaceutical company in charge of developing drugs against HIV. Discuss four possible targets for drugs that might be effective against this virus.
2. Researchers debate the evolutionary value to the virus of its ability to cause disease. Many argue that viruses accidentally cause disease and only in animals that are not the natural host. They state that this strategy may eventually prove fatal to the virus's future in that host. It is reasoned that the animals will eventually develop immune mechanisms to combat the virus and prevent its spread. Another group of researchers supports the view that disease is a way to enhance the survival of the virus. What rationale would this group use to support its view?

Critical Thinking

1. Would ID_{50} and LD_{50} necessarily be the same for a given virus? Why or why not?
2. The observation that viruses can agglutinate red blood cells suggests to some people that both viruses and red blood cells must have multiple binding sites. Is this a good argument? Why or why not?
3. An agricultural scientist is investigating ways to prevent viral infection of plants. Is preventing the specific attachment of the virus to its host cells a possible way to prevent infection? Why or why not?
4. Why is it virtually impossible to stamp out a disease caused by a zoonotic virus?

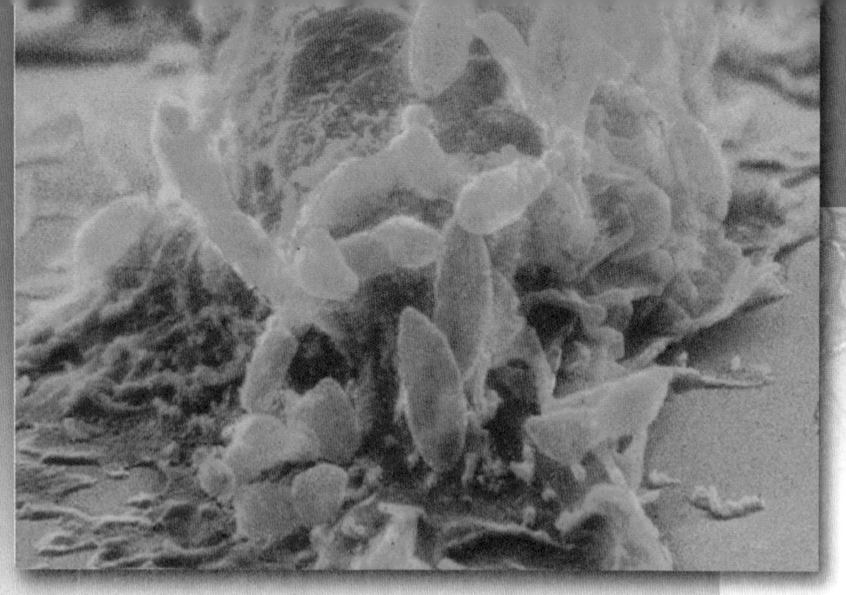

15

Phagocytic cells engulfing bacteria.

The Innate Immune Response

A Glimpse of History

Once microorganisms were shown to cause disease, scientists worked to explain how the body defended itself against their invasion. Elie Metchnikoff, a Russian-born scientist, hypothesized that specialized cells within the body destroyed invading organisms. His ideas arose from observations he made while studying the transparent immature larval form of starfish in Sicily in 1882. As he looked at the larvae under the microscope, he could see ameba-like cells within the bodies. He described his observations:

> . . . I was observing the activity of the motile cells of a transparent larva, when a new thought suddenly dawned on me. It occurred to me that similar cells must function to protect the organism against harmful intruders. . . . I thought that if my guess was correct a splinter introduced into the larva of a starfish should soon be surrounded by motile cells much as can be observed in a man with a splinter in his finger. No sooner said than done. In the small garden of our home . . . I took several rose thorns that I immediately introduced under the skin of some beautiful starfish larvae which were as transparent as water. Very nervous, I did not sleep during the night, as I was waiting for the results of my experiment. The next morning, very early, I found with joy that it had been successful.

Metchnikoff reasoned that certain cells in animals were responsible for ingesting and destroying foreign material. He called these cells phagocytes, meaning "cells that eat," and he proposed that they were primarily responsible for the body's ability to destroy invading microorganisms.

When Metchnikoff returned to Russia, he looked for a way to study the ingestion of materials by phagocytes. A water flea that could be infected with a yeast provided a vehicle for such studies. He observed phagocytes ingesting and destroying invading yeast cells within the experimentally infected, transparent water fleas. In 1884, Metchnikoff published a paper that strongly supported his contention that phagocytic cells were primarily responsible for destroying disease-causing organisms. He spent the rest of his life studying this process and other biological phenomena; in 1908, he was awarded the Nobel Prize for these studies of immunity.

From a microorganism's standpoint, the tissues and fluids of the human body are much like a warm culture flask filled with a nutrient-rich solution. Considering this, it may be surprising that the interior of the body—including blood, muscles, bones, and organs—is generally sterile. If this were not the case, microbes would simply degrade our tissues, just as they readily decompose the carcasses of dead animals.

How does the interior of the body remain sterile in this world full of microbes? Like other multicellular organisms, humans have evolved several mechanisms of defense. First, we are covered with skin and mucous membranes that prevent entry of most foreign material, including microbes, into the body. Ready in case the barriers are breached are sensor systems that detect molecules associated with microbes. The sensors can direct and assist other host defenses, facilitating the destruction of the invaders. Also lying in wait are defense cells that specialize in ingesting and digesting microbes and other foreign material; if needed, additional reinforcements can be recruited to the site of breach. The protection provided by these systems is termed **innate immunity.**

Innate immunity had been called a non-specific defense, but recent discoveries have shown that most of the components rely on the detection of certain molecules associated with invading microbes, a feature called **pattern recognition.** The molecules recognized include various compounds in bacterial cell walls (such as lipopolysaccharide, lipoteichoic acid, and peptidoglycan) and other features unique to microbes. ■ **bacterial cell walls, p. 59**

In addition to the innate response, vertebrates have evolved a more specialized defense system, providing protection called

KEY TERMS

Apoptosis Programmed cell death of "self" cells that does not elicit inflammation.

Complement System Series of proteins in blood and tissue fluids that can be activated to facilitate the removal and destruction of invading microbes.

Cytokines Proteins that function as chemical messengers, allowing cells to communicate.

Inflammation Coordinated innate response characterized by swelling, heat, redness, and pain in the infected area, aimed at containing a site of damage, localizing the response, and restoring tissue function.

Innate Immunity Host defenses involving anatomical barriers, sensor systems that recognize patterns associated with microbes, and phagocytic cells.

Macrophage Type of phagocytic cell present in tissues; can be activated to gain more killing power.

Membrane Attack Complex (MAC) Complement system components that assemble to form pores in membranes of invading cells.

Neutrophil Type of phagocytic cell that quickly moves to an infected area to attack offending microbes.

Opsonization Coating of an object with particles for which phagocytes have receptors, making it easier for phagocytosis to occur.

Phagocyte Cells that specialize in engulfing and digesting microbes and cell debris.

Phagocytosis The process by which a phagocyte engulfs an invader.

Toll-Like Receptors (TLRs) Receptors on cell surfaces and within endosomes that recognize specific compounds unique to microbes, enabling cells to sense the presence of invading microbes and then alert other components of the defense systems.

adaptive immunity. This develops throughout life and substantially increases the ability of the host to defend itself. Each time the body is exposed to a microbe or certain other types of foreign material, the adaptive defense system first "learns" and then "remembers" the most effective response to that specific material; it then reacts accordingly if the material is encountered again. The material that evokes an immune response is called an **antigen.** An important action of the adaptive immune response is the production of **antibodies.** These protein molecules bind specifically to the antigen, directing other host defenses to remove or destroy it. The adaptive immune response can also destroy a body cell, referred to as a **host cell** or a **"self" cell,** if it harbors a virus or other invader. The initial adaptive response to an antigen develops relatively slowly, during which time the invader may cause illness or death if the innate defenses are unable to contain it. Successive exposures, however, lead to a swift and greater repeat response, generally eliminating the microbe before it causes obvious harm.

To simplify the description of a network as complex and intricate as the immune system, it is helpful to consider it as a series of individual parts. This chapter, for example, will focus almost exclusively on innate immunity. Bear in mind, however, that although the various parts are discussed separately, in the body their actions are intimately connected and coordinated. In fact, as you will see in chapter 16, certain components of the innate defenses are instrumental in educating the adaptive defenses, helping them to recognize that a particular antigen represents a microbial invader.

15.1 Overview of the Innate Defenses

Focus Point

- Outline the essential components of the innate defenses.

First-line defenses are the barriers that separate and shield the interior of the body from the surrounding environment; they are the initial obstacles microorganisms must overcome to invade the tissues. The anatomical barriers, which include the skin and mucous membranes, not only provide physical separation, but are often bathed in secretions that have antimicrobial properties.

Sensor systems within the body recognize when the first-line barriers have been breached and then relay that information to other components of the host defenses. Two important groups of sensors are the **toll-like receptors** and **NOD (nucleotide-binding oligomerization domain) proteins,** which are found on or within a variety of different cell types. These receptors recognize families of compounds unique to microbes, enabling the cell to sense invaders and then send chemical signals to alert other components of the host's defense. Another type of sensor is a series of proteins always present in blood and tissue fluids; these proteins are collectively called the **complement system** because they can "complement," or act in conjunction with, the adaptive immune defenses. In response to certain stimuli, the complement proteins become activated, setting off a chain of events that results in removal and destruction of invading microbes.

Phagocytes, cells that specialize in engulfing and digesting microbes and cell debris, act as sentries, alert for signs of invasion of the body. More can be recruited from the bloodstream, serving as reinforcements at the sites in tissues where first-line defenses have been breached.

Cells of the immune system communicate with one another by producing proteins that function as chemical messengers, called **cytokines.** A cytokine produced by one cell diffuses to another and binds to the appropriate cytokine receptor of that cell. When a cytokine binds a receptor, a signal is transmitted to the interior of the cell, inducing certain changes in the activities of the cell. Some types of cytokines endow cells with enhanced powers; others prompt cells to migrate to specific locations within the body.

When invading microorganisms or tissue damage is detected, **inflammation** ensues; this is a coordinated response involving many aspects of the innate defenses. During inflammation, the cells that line local blood vessels near the area of invasion or damage undergo changes that allow proteins such as those of the complement system to leak into tissues. Other changes allow phagocytic cells in the blood to adhere to the vessels and then squeeze between cells, exiting the bloodstream. Phagocytic cells then migrate to the area of infection or damage where they ingest and destroy foreign material. Some types of phagocytes play a dual role, destroying invaders while also communicating with cells of the adaptive immune system, enlisting their far more powerful effects.

Fever is another of the body's innate defense mechanisms. This increase in internal body temperature acts in several ways to discourage infection.

MICROCHECK 15.1

First-line defenses are the initial obstacles that microbes must overcome to invade the tissues. Within the body are sensor systems such as toll-like receptors, NOD proteins, and the complement system that recognize when the barriers have been breached. Phagocytic cells engulf foreign material; they communicate with other cells via cytokines. Inflammation is a coordinated response to invasion or tissue damage.

✓ How do cytokines function?
✓ Describe the dual roles played by some types of phagocytes.
✓ What molecules unique to microbes might toll-like receptors recognize?

15.2 First-Line Defenses

Focus Point

- Describe the first-line defenses, including the physical barriers, antimicrobial substances, and normal microbiota.

The body's first line of defense against invading microbes includes physical barriers such as the skin and mucous membranes, and the antimicrobial substances in secretions that bathe those barriers **(figure 15.1)**. Some of these barriers are often considered to be "inside" the body, but actually they are in direct contact with external environment. For example, the digestive tract, which begins at the mouth and ends at the anus, is simply a hollow tube that runs through the body, allowing intestinal cells to absorb nutrients from food that passes through (see figure 25.1); the respiratory tract is a cavity that allows O_2 and CO_2 to be exchanged (see figure 22.1). In addition to the protection provided by the anatomical barriers, the **normal microbiota**—the population of microbes that routinely inhabit the body surfaces—helps prevent harmful microbes from colonizing the body surfaces. ■ normal microbiota, p. 9

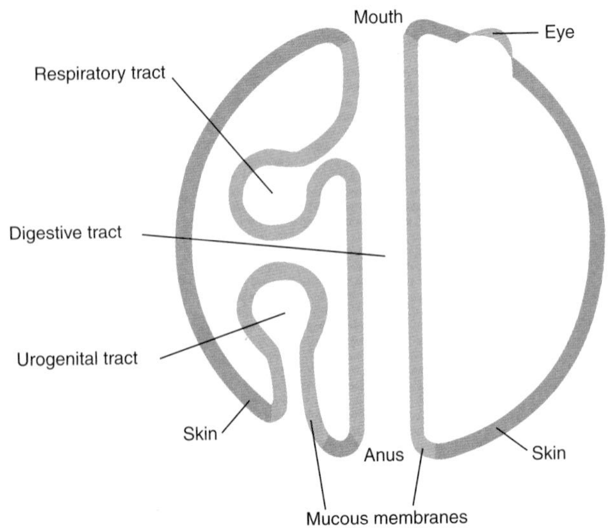

FIGURE 15.1 Physical Barriers These barriers separate the interior of the body from the surrounding environment; they are the initial obstacles microorganisms must overcome to invade tissues. The skin is shown in purple, and mucous membranes in pink.

In this section we will describe the general physical and chemical aspects of the anatomical barriers, as well as the protective contributions of the normal microbiota. These are described in more detail in the chapters dealing with each body system.

Physical Barriers

All exposed surfaces of the body are lined with **epithelial cells (figure 15.2).** These cells are tightly packed together and rest on a thin layer of fibrous material, the **basement membrane.**

Skin

The skin is the most obvious visible barrier and is the most difficult for microbes to penetrate. It is composed of two main layers—the dermis and the epidermis (see figure 23.1). The **dermis** contains tightly woven fibrous connective tissue, making it extremely tough and durable (the dermis of cattle is used to make leather). The **epidermis** is composed of many layers of epithelial cells that become progressively flattened toward the exterior. The outermost sheets are made up of dead cells embedded with a water-repelling protein called **keratin,** resulting in the skin being an arid environment. The cells continually slough off, taking with them any microbes that might be adhering. ■ anatomy and physiology of the skin, p. 531

Mucous Membranes

Mucous membranes line the digestive tract, respiratory tract, and genitourinary tract. They are constantly bathed with mucus or other secretions that help wash microbes from the surface. Some mucous membranes have mechanisms that propel microbes, directing them toward areas where they can be eliminated more easily. For example, flow of urine regularly flushes organisms from the urinary tract. **Peristalsis,** the rhythmic contractions of the intestinal tract, propels food and liquid and also helps to expel microbes. The respiratory tract is lined with ciliated cells; the hairlike cilia constantly beat in an upward motion, propelling materials away from the lungs to the throat where they can then be swallowed. This "free ride" out of the respiratory tract is referred to as the **mucociliary escalator.** ■ cilia, p. 74

Antimicrobial Substances

Both the skin and mucous membranes are protected by a variety of antimicrobial substances that inhibit or kill microorganisms **(figure 15.3).** Sweat, for example, is high in salt; as it evaporates

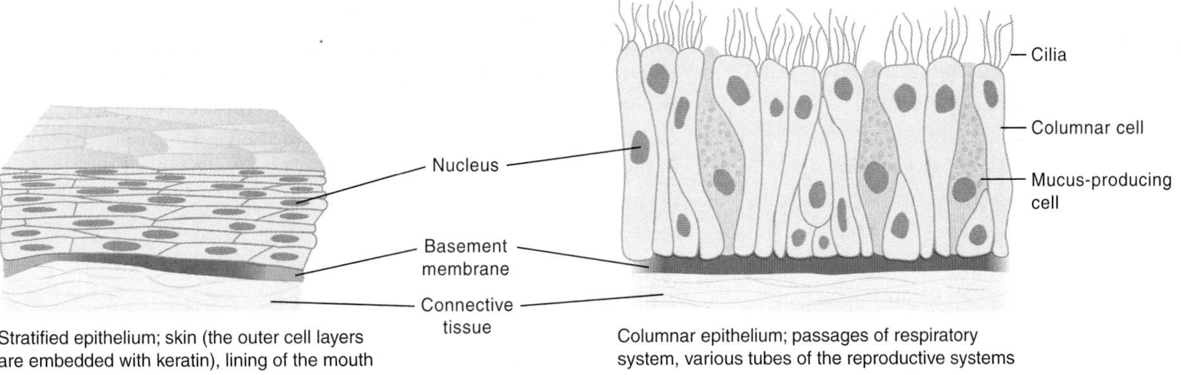

FIGURE 15.2 Epithelial Barriers Cells of these barriers are tightly packed together and rest on a layer of thin fibrous material, the basement membrane. Note that some epithelial cells have cilia which propel material to an area where it can be eliminated.

it leaves a salty residue, inhibiting many organisms that might otherwise proliferate on the skin.

Lysozyme, the enzyme that degrades peptidoglycan, is found in tears, saliva, and in mucus that bathes mucous membranes. It is also found within the body, in phagocytic cells, blood, and the fluid that bathes tissues. ■ lysozyme, p. 62

Peroxidase enzymes are found in saliva and milk; they are also found within body tissues and inside phagocytes. These enzymes break down hydrogen peroxide and, in the process, produce potent oxidizing compounds. Bacteria that produce the enzyme catalase, however, may avoid the damaging products associated with peroxidase activity; catalase breaks down hydrogen peroxide, potentially destroying the compound before it can interact with peroxidase. Catalase-negative organisms are more sensitive to peroxidase killing. ■ catalase, p. 92

Lactoferrin is an iron-binding protein found in saliva, mucus, and milk; it is also found in some types of phagocytic cells. A similar compound, **transferrin** is found in blood and tissue fluids. Iron, an important part of some enzymes, is one of the major elements required for growth (see table 4.4). By sequestering iron, the lactoferrin and transferrin effectively withhold the essential element from most microbes. Some bacteria, however, make compounds that capture iron from the host, thus circumventing this defense.

Defensins are short antimicrobial peptides produced by neutrophils and epithelial cells. They function by inserting into bacterial membranes, forming pores that disrupt the integrity of the cell.

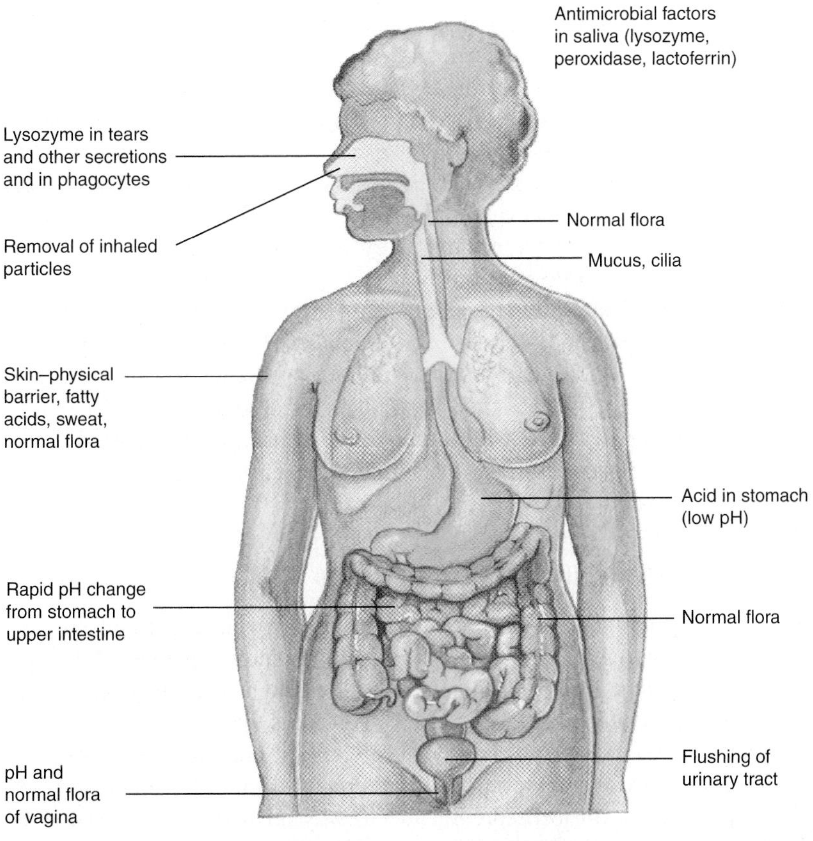

Antimicrobial factors in saliva (lysozyme, peroxidase, lactoferrin)
Lysozyme in tears and other secretions and in phagocytes
Removal of inhaled particles
Normal flora
Mucus, cilia
Skin–physical barrier, fatty acids, sweat, normal flora
Acid in stomach (low pH)
Rapid pH change from stomach to upper intestine
Normal flora
Flushing of urinary tract
pH and normal flora of vagina

FIGURE 15.3 Antimicrobial Substances and Other First-Line Defenses Physical barriers, such as skin and mucous membranes, antimicrobial secretions, and normal flora work together to prevent entry of microorganisms into the host's tissues.

Normal Microbiota (Flora)

The population of microorganisms routinely found growing on the body surfaces of healthy individuals is called the **normal microbiota (flora)** (see figure 17.1). Although these organisms are not technically part of the immune system, the protection they provide is considerable.

One protective effect of the normal microbiota is competitive exclusion of pathogens. For example, the normal microbiota prevents invading organisms from adhering to host cells by covering binding sites that might otherwise be used for attachment. The population also consumes available nutrients that could otherwise be used by less desirable organisms. Members of the normal microbiota also produce compounds toxic to other bacteria. In the hair follicles of the skin, for example, *Propionibacterium* species degrade the lipids found in skin glands, releasing fatty acids that inhibit the growth of many pathogens. In the gastrointestinal tract, some strains of *E. coli* synthesize colicins, proteins toxic to other strains of bacteria. *Lactobacillus* species growing in the vagina produce lactic acid as a fermentation end product, resulting in an acidic pH that inhibits the growth of some potential disease-causing organisms. Disruption of the normal microbiota, which occurs when antibiotics are used, can predispose a person to various infections. Examples include antibiotic-associated diarrhea, caused by the growth of toxin-producing strains of *Clostridium difficile* in the intestine, and vulvovaginitis, caused by excessive growth of *Candida albicans* in the vagina. ■ antibiotic-associated diarrhea, p. 586 ■ vulvovaginitis, p. 625

The normal microbiota is also essential to the development of the immune system in infants. Other aspects of the normal microbiota will be discussed in chapter 17.

MICROCHECK 15.2

Physical barriers that prevent entry of microorganisms into the body include the skin and mucous membranes. Various antimicrobial substances, including lysozyme, peroxidase enzymes, lactoferrin, and defensins, are found on the body surfaces. The normal microbiota plays a protective role by excluding certain other microbes and enhancing development of the immune system.

- ✓ What is peristalsis?
- ✓ What is the role of lactoferrin?
- ✓ How would damage to the ciliated cells of the respiratory tract predispose a person to infection?

15.3

The Cells of the Immune System

Focus Point

- Describe the role of granulocytes, mononuclear phagocytes, dendritic cells, and lymphocytes in immunity.

The cells of the immune system can move from one part of the body to another, traveling through the body's circulatory systems like vehicles on an extensive interstate highway system. They are always found in normal blood, but their numbers usually increase during infections, recruited from reserves of immature cells that develop in the bone marrow. Some cells play dual functions, having crucial roles in both innate and adaptive immunity.

The formation and development of blood cells is called **hematopoiesis** (Greek for "blood" and "to make"). All blood cells, including those important in the body's defenses, originate from the same type of cell, the **hematopoietic stem cell,** found in the bone marrow **(figure 15.4).** These stem cells are induced to develop into the various types of blood cells by a group of cytokines called **colony-stimulating factors.** Some of the cells of the immune system are already mature as they circulate in the bloodstream, but others **differentiate,** developing functional properties, after they leave the blood and enter the tissues.

The general categories of blood cells and their derivatives include red blood cells, platelets, and white blood cells. Red blood cells, or **erythrocytes,** carry oxygen in the blood. **Platelets,** which are actually fragments arising from large cells called **megakaryocytes,** are important for blood clotting. White blood cells, or **leukocytes,** are important in all host defenses. Leukocytes can be divided into four broad groups—granulocytes, mononuclear phagocytes, dendritic cells, and lymphocytes **(table 15.1).**

Granulocytes

All **granulocytes** contain prominent cytoplasmic granules filled with biologically active chemicals. There are three types of granulocytes—neutrophils, basophils, and eosinophils; their names reflect the staining properties of their cytoplasmic granules.

Neutrophils are highly efficient at phagocytizing and destroying foreign material, particularly bacteria, and damaged cells. The contents of their granules, which stain poorly, include many antimicrobial substances and degradative enzymes essential for destruction of materials that the cells engulf. They are the most abundant and important granulocytes of the innate responses. Neutrophils are sometimes called **polymorphonuclear neutrophilic leukocytes, polys,** or **PMNs,** names that reflect the appearance of multiple lobes of their single nucleus. They normally account for over 50% of circulating leukocytes, and their numbers increase during most acute bacterial infections. There are generally few in tissues except during inflammation. Because of the importance of neutrophils in innate immunity, they will be described in more detail later in the chapter. ■ specialized attributes of neutrophils, p. 359

Basophils are blood cells involved in allergic reactions and inflammation. Their granules, which stain dark purplish-blue with the basic dye methylene blue, contain histamine and other chemicals that increase capillary permeability during inflammation. **Mast cells** are similar in appearance and function to basophils but are found in virtually all tissues, rather than in blood. They do not come from the same precursor cells as basophils. Mast cells are important in the inflammatory response and are responsible for many allergic reactions.

Eosinophils are thought to be primarily important in expelling parasitic worms from the body. They seem to be involved in allergic reactions, causing some of the symptoms associated with allergies, but reducing others. The granules of eosinophils, which stain red with the acidic dye eosin, contain antimicrobial substances and also histaminase, an enzyme that breaks down histamine.

Mononuclear Phagocytes

Mononuclear phagocytes constitute a widespread collection of important phagocytic cells called the **mononuclear phagocyte system (MPS) (figure 15.5).** They include **monocytes,** which circulate in the blood, and the cell types that develop from monocytes as they leave the bloodstream and migrate into tissues.

Macrophages, a differentiated form of monocytes, are phagocytic cells present in virtually all tissues to at least some extent. They are particularly abundant in liver, spleen, lymph

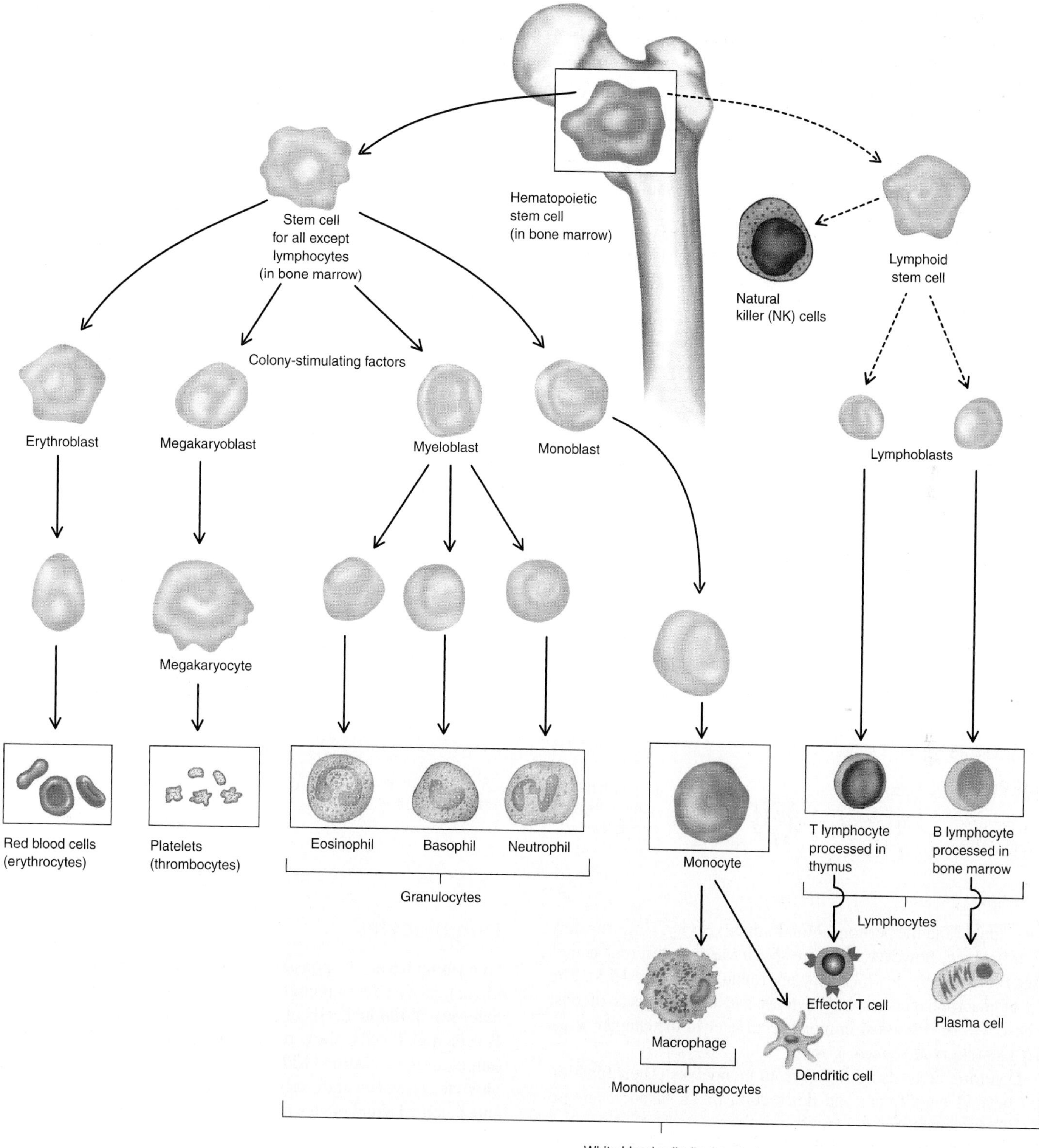

FIGURE 15.4 Blood Cells and Their Derivatives All these descend from hematopoietic stem cells found in the bone marrow. Some of the steps not yet clearly defined are indicated by dotted arrows. Multiple steps occur between the stem cell and the final cells produced. The role of these cells in the immune response will be explained in this chapter and chapter 16.

TABLE 15.1 Leukocytes

Cell Type (% of Blood Leukocytes)	Location in Body	Functions
Granulocytes		
Neutrophils (polymorphonuclear neutrophilic leukocytes or PMNs, often called polys; 55–65%)	Account for most of the circulating leukocytes; few in tissues except during inflammation	Phagocytize and digest engulfed materials
Eosinophils (2–4%)	Few in tissues except in certain types of inflammation and allergies	Participate in inflammatory reaction and immunity to some parasites
Basophils (0–1%), mast cells	Basophils in circulation; mast cells present in most tissues	Release histamine and other inflammation-inducing chemicals from the granules
Mononuclear Phagocytes		
Monocytes (3–8%)	In circulation; they differentiate into either macrophages or dendritic cells when they migrate into tissue	Phagocytize and digest engulfed materials
Macrophages	Present in virtually all tissues; given various names based on the tissue in which they are found	Phagocytize and digest engulfed materials
Dendritic cells	Initially in tissues, but they migrate to secondary lymphoid organs (such as lymph nodes, spleen, thymus, appendix, tonsils)	Gather antigen from the tissues and then bring it to lymphocytes that congregate in the secondary lymphoid organs
Lymphocytes		
Several types (25–35%)	In lymphoid organs (such as lymph nodes, spleen, thymus, appendix, tonsils); also in circulation	Participate in adaptive immune responses

nodes, lungs, and the peritoneal (abdominal) cavity. Unfortunately for the novice, however, they are given various different names based on the tissue in which they are found (see figure 15.5). The role of macrophages in phagocytosis and other aspects of host defense will be discussed in more detail later in the chapter. ■ specialized attributes of macrophages, p. 358

Dendritic cells also develop from monocytes. Their function goes beyond engulfment and destruction of an invader, and so they will be considered separately.

Dendritic Cells

Dendritic cells are found in various tissues throughout the body, where they function as "scouts." They routinely engulf material in the tissues, and then bring it to the cells of the adaptive immune system for "inspection." Dendritic cells develop from monocytes, but some also appear to descend from other cell types. Details regarding the interactions of dendritic cells with the cells of adaptive immunity will be discussed in chapter 16.

Lymphocytes

Lymphocytes are responsible for adaptive immunity. In contrast to the generic pattern recognition of antigens by cells of the innate defenses, individual cells of the two major groups of lymphocytes, **B cells** and **T cells,** show remarkable specificity in their recognition of antigen. **Natural killer (NK) cells** are another type of lymphocyte; however, they lack the specificity exhibited by B cells and T cells. Lymphocytes are the center of focus of chapter 16.

MICROCHECK 15.3

Granulocytes include neutrophils, basophils, and eosinophils. Mononuclear phagocytes include monocytes and macrophages. Dendritic cells function as scouts for the adaptive immune system. Lymphocytes are responsible for adaptive immunity.

- ✓ Which type of granulocyte is the most abundant?
- ✓ How are macrophages related to monocytes?
- ✓ Why can stem cell transplants be used to replace defective lymphocytes?

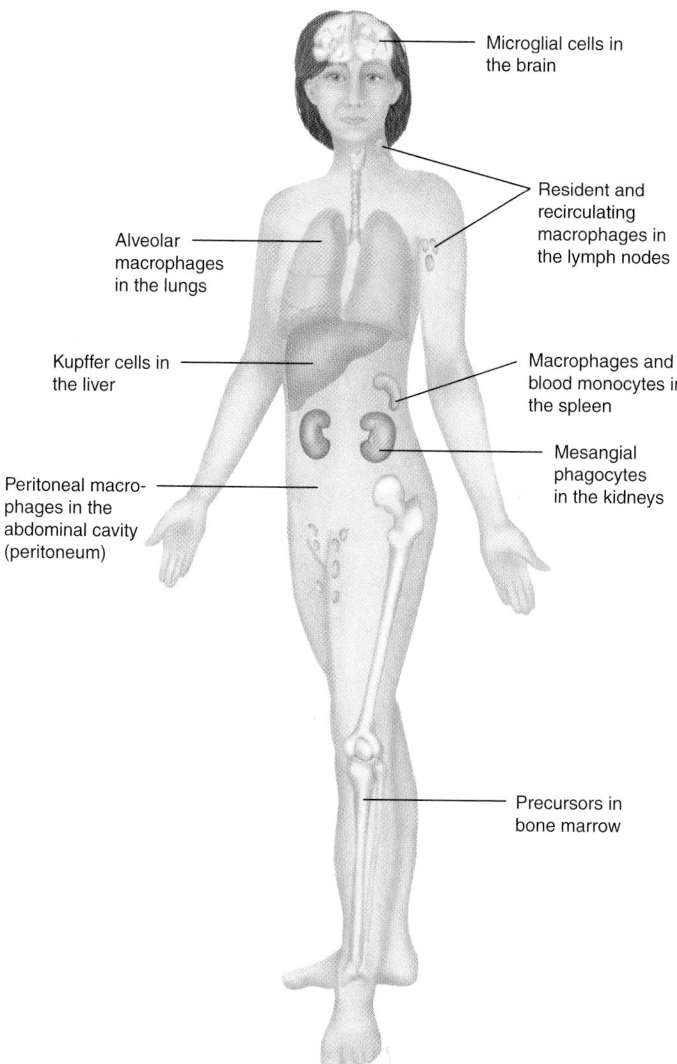

FIGURE 15.5 Mononuclear Phagocyte System Cells in this system of monocytes and macrophages sometimes have special names to denote their location—for example, Kupffer cells (in the liver) and alveolar macrophages (in the lung).

15.4

Cell Communication

Focus Point

- Describe the role of surface receptors, cytokines, and adhesion molecules in immunity.

In order for the various cells of the immune system to respond to trauma or invasion in a cooperative fashion, cells must communicate both with their immediate environment and with each other. They do this through surface receptors, cytokines, and adhesion molecules.

Surface Receptors

Surface receptors can be viewed as the "eyes" and "ears" of a cell. They are proteins that generally span the cell membrane, connecting the outside of the cell with the inside, enabling the inner workings of the cell to sense and respond to signals outside of the cell. Each surface receptor is specific with respect to the compound or compounds it will bind; a molecule that can bind to a given receptor is called a **ligand** for that receptor. When a ligand binds to its surface receptor, the internal portion of the receptor becomes modified in some manner. This modification then elicits some type of response by the cell, such as chemotaxis. Cells can alter the types and numbers of surface molecules they make, enabling them to respond to signals relevant when the cell is in a certain location or developmental stage. ■ chemotaxis, p. 66

Cytokines

Cytokines can be viewed as the "voices" of a cell. These proteins are made by certain cells as a mechanism to communicate with other cells. They bind to certain surface receptors called **cytokine receptors,** which are found on the cells that cytokines regulate. Binding of a cytokine to its receptor induces a change in the cell such as growth, differentiation, movement, or cell death. Although cytokines are short-lived, they are very powerful, acting at extremely low concentrations. They can act locally, regionally, or systemically. Often, they act together or in sequence, in a complex fashion. The source and effects of representative cytokines are summarized in **table 15.2.**

Chemokines are cytokines important in chemotaxis of immune cells. Certain types of defense cells have receptors for chemokines, thereby enhancing their ability to migrate to the appropriate region of the body, such as an area of inflammation. Microbial invaders can subvert the function of receptors. For example, two chemokine receptors, CCR5 and CXCR4, play a critical role in HIV infection; they serve as co-receptors for the virus, influencing which cell types are most likely to become infected. ■ chemotaxis, p. 66 ■ HIV co-receptors, p. 702

Colony-stimulating factors (CSFs) are important in the multiplication and differentiation of leukocytes (see figure 15.4). During the immune response when more leukocytes are needed, a variety of different colony-stimulating factors direct immature cells into the appropriate maturation pathways.

Interferons (IFNs) are important in the control of viral infections. In addition to being antiviral, IFN-gamma helps regulate the function of cells involved in the inflammatory response, particularly mononuclear phagocytes, and modulates certain responses of adaptive immunity. The role of interferons in the containment of viral infections will be described in more detail later in the chapter.

Interleukins (ILs), produced by leukocytes, have diverse functions. As a group, they are important in both innate immunity, including the inflammatory response, and adaptive immunity. Their activities often overlap.

Tumor necrosis factors (TNFs) were discovered because of their activities in killing tumor cells, which is how they acquired their name, but they actually have multiple roles. TNF-alpha,

TABLE 15.2 Some Important Cytokines

Cytokine	Source	Effects
Chemokines	Various cells	Chemotaxis
Colony-Stimulating Factors (CSFs)	Fibroblasts, endothelium, other cells	Stimulation of growth and differentiation of different kinds of leukocytes
Interferons		
Interferon alpha	Leukocytes	Antiviral; induces fever, contributes to inflammation
Interferon beta	Fibroblasts	Antiviral
Interferon gamma	T lymphocytes	Antiviral; macrophage activation; development and regulation of adaptive immune responses
Interleukins (ILs)		
IL-1	Macrophages, epithelial cells	Proliferation of lymphocytes; macrophage production of cytokines, induce adhesion molecules for PMNs on blood vessel cells; induce fever
IL-2 (T-cell growth factor)	T lymphocytes	Changes in growth of lymphocytes; activation of natural killer cells; promote adaptive cell-mediated immune responses
IL-3	T lymphocytes, mast cells	Changes in growth of precursors of blood cells and also of mast cells
IL-4, IL-5, IL-10, IL-14	T lymphocytes, mast cells, other cells	Promote antibody responses
IL-6	T lymphocytes, macrophages	T- and B-cell growth; inflammatory response; fever
Tumor Necrosis Factors (TNFs)		
Alpha	Macrophages, T lymphocytes, other cell types, mast cell granules	Initiation of inflammatory response; cytotoxicity for some tumor cells; regulation of certain immune functions; induce fever; chemotactic for granulocytes
Beta	T lymphocytes	Killing of target cells by cytotoxic T cells and natural killer (NK) cells

which is produced by macrophages and other cell types, plays an instrumental role in initiating the inflammatory response. Tumor necrosis factors can also initiate the process of programmed cell death, or apoptosis. ■ apoptosis, p. 362

Groups of cytokines often act together to facilitate a particular response by the host defenses. For example, certain cytokines referred to as **pro-inflammatory cytokines** (TNF-alpha, IL-1, IL-6, and others) contribute to inflammation. Others are especially involved in promoting antibody responses (IL-4, IL-5, IL-10, and IL-14). A different group promotes responses that involve certain types of T cells (IL-2, and IFN-gamma, and others).

Adhesion Molecules

Adhesion molecules on the surface of cells allow those cells to adhere to other cells. Some cells use adhesion molecules to "grab" certain cells as they pass by. For example, when phagocytic cells in the blood are needed in tissues, the **endothelial cells** that line the blood vessels synthesize adhesion molecules, to snare passing phagocytic cells. This slows the rapidly moving phagocytic cells, providing them with the opportunity to exit the bloodstream. Other types of adhesion molecules allow cells to make direct contact with one another, thereby enabling cells to target the delivery of cytokines or other compounds to a particular cell.

MICROCHECK 15.4

Surface receptors allow a cell to detect molecules present outside of that cell. Cytokines provide cells with a mechanism of communication. Adhesion molecules allow a cell to adhere to other cells.

- ✓ What is a ligand?
- ✓ What is the function of colony-stimulating factors?
- ✓ How could colony-stimulating factors be used as a therapy?

15.5 Sensor Systems

Focus Point

- Describe the outcomes of toll-like receptor engagement, complement system activation, and detection of long dsRNA.

Sensor systems within the blood and tissues lie ready to detect signs of microbial invasion. They include toll-like receptors, NOD proteins, the complement system, and sensors that detect long, double-stranded RNA.

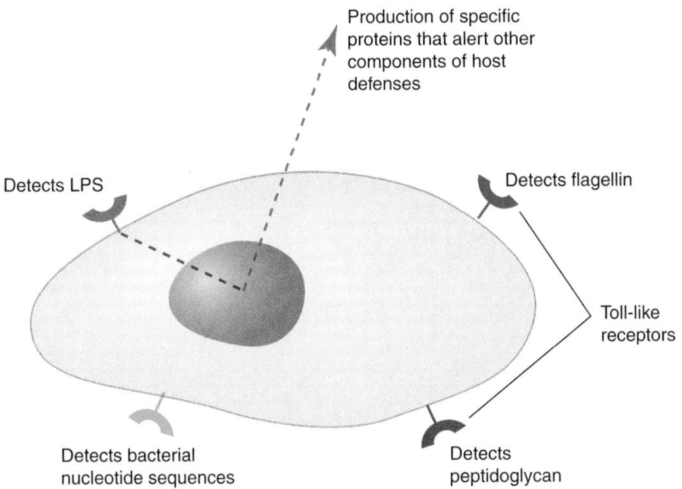

FIGURE 15.6 Toll-Like Receptors These surface receptors are used by the body's cells to detect the presence of microbial products. Engagement of a toll-like receptor transmits a signal to the cell's nucleus, inducing the cell to begin producing certain proteins such as cytokines, alerting other components of host defenses.

Toll-Like Receptors and NOD Proteins

Toll-like receptors (TLRs) are surface receptors that allow cells to "see" molecules that signify the presence of microbes outside of the cell **(figure 15.6); NOD proteins** do the same for the inside of a cell (the cell's cytoplasm). TLRs and NOD proteins are pattern recognition receptors that have only recently been discovered, and much is still being learned about them, but they have caused a tremendous resurgence of interest in innate immunity, which many scientists had thought was well understood.

TLRs are part of a family of receptors called Toll receptors, first identified in *Drosophila* species (fruit flies). The name of this family was coined when one of the researchers involved in their discovery exclaimed "toll!" (a German word meaning the equivalent of "awesome").

TLRs are found on a variety of cell types including macrophages and cells that line normally sterile body sites. Some are also within endosomes. A number of TLRs have been described, and each recognizes a distinct compound or group of compounds associated with microbes. For example, one recognizes peptidoglycan and another is triggered by lipopolysaccharide. Other bacteria-specific compounds that activate the receptors include flagellin and certain nucleotide sequences that typify bacterial DNA. ■ endosome, p. 74

When a compound engages a TLR, a signal is transmitted to the nucleus of the host cell, inducing that cell to alter the expression of certain genes. For example, lipopolysaccharide triggers a toll-like receptor on monocytes and macrophages, causing the cells to begin producing chemokines that attract additional phagocytes to the area. Engagement of TLRs on endothelial cells that line blood vessels causes those cells to produce pro-inflammatory cytokines.

The NOD proteins are intracellular receptors that recognize bacterial cell wall components within the cytoplasm. Although details regarding the NOD proteins are still being elucidated, a defect in one of the proteins appears to be a predisposing factor in the development of Crohn's disease, an inflammatory bowel disease.

The Complement System

The **complement system** is a series of proteins that constantly circulate in the blood and the fluid that bathes the tissues. Early studies showed that these proteins augment the activities of the adaptive immune response; in fact, their name is derived from observations that they "complement" the activities of antibodies. They routinely circulate in an inactive form, but in response to certain stimuli indicating the presence of foreign material, a cascade of reactions occurs. This results in the rapid activation of critical complement system components. These activated forms have specialized functions that cooperate with other host defenses to quickly remove and destroy the offending material.

Three pathways lead to the activation of the complement system **(figure 15.7):**

- **Alternative pathway.** The name of this pathway may seem to imply that it is "second choice," but it actually reflects the fact that it was not discovered first. In fact, the pathway is quickly and easily initiated, providing vital early warning that an invader is present. The alternative pathway relies on the binding of the complement protein C3b to cell surfaces, allowing other complement proteins to subsequently attach and form a complement activating complex. C3b is always present in blood and tissues to at least some extent, so nearly any cell surface can trigger the pathway. The body's own cells, prevent their surfaces from activating the complement system by binding regulatory proteins that inactivate bound C3b before the other complement proteins can attach. As we will discuss in chapter 17, some pathogens have developed mechanisms to attract the regulatory proteins to their own surfaces, thwarting complement activation by this pathway.
- **Lectin pathway.** Activation of the lectin pathway requires **mannan-binding lectins (MBLs);** these are pattern-recognition molecules the body uses to detect mannan, a polymer of mannose often found on microbial but rarely on mammalian cells. When MBL attaches to a surface, it can then interact with one of the complement components, activating it. This, in turn, leads to activation of other complement proteins. ■ mannose, p. 31
- **Classical pathway.** Activation by the classical pathway requires antibodies, a component of adaptive immunity. When antibodies bind to antigen, they can then interact with the same complement component involved in activating the lectin pathway. This activates that protein, leading to activation of other complement proteins.

The nature of the complement system allows an exceedingly rapid and powerful response. Its activation occurs by a cascade of reactions; once a specific complement protein becomes activated, it functions as an enzyme, cleaving and therefore activating millions of molecules of the next complement protein in the cascade. In turn, each molecule activates multiple molecules of the next protein in the cascade, and so on. Generally, activation involves splitting the protein into two parts, each of which then carries out a specific function. Stringent mechanisms control complement system activation at various points.

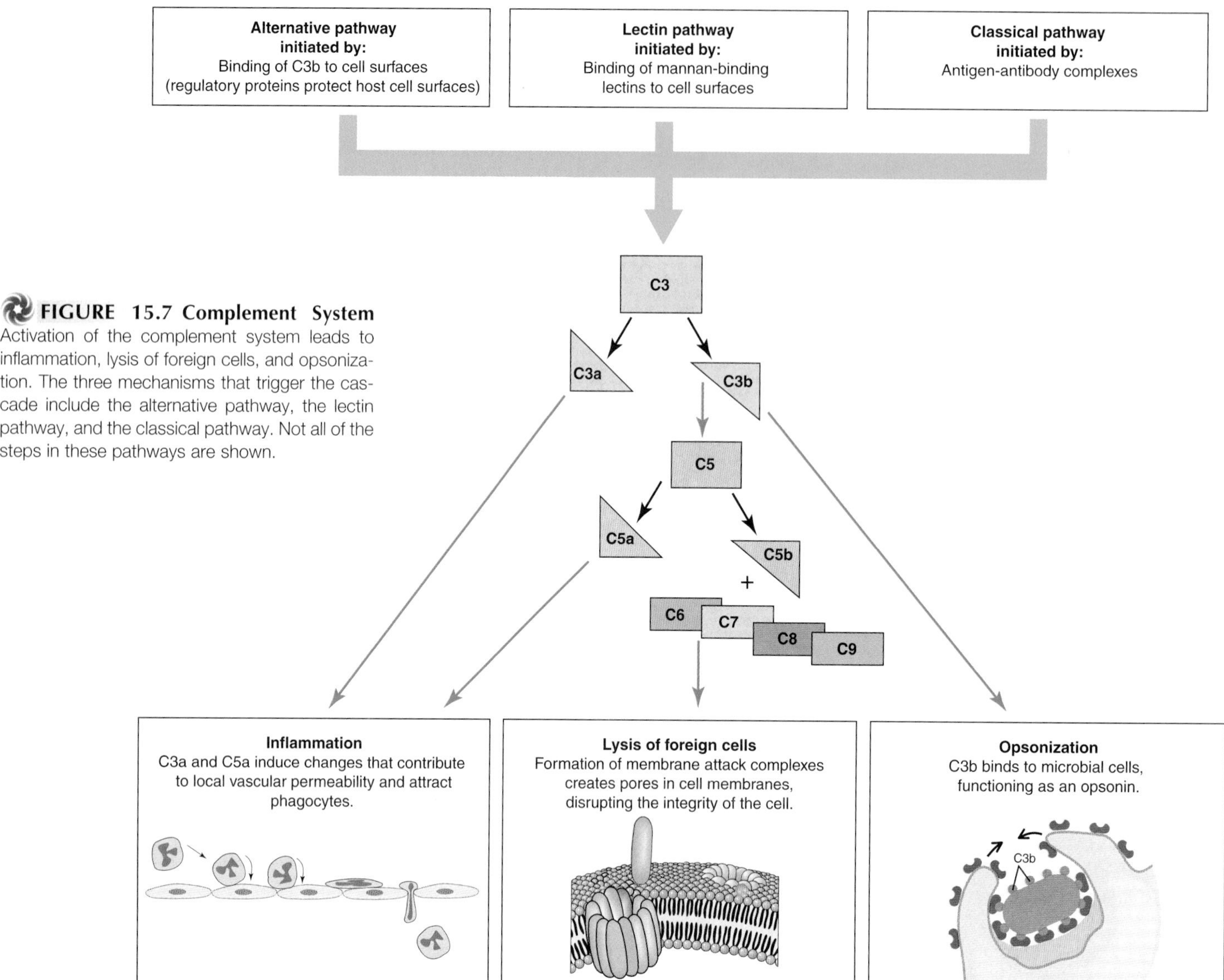

FIGURE 15.7 Complement System Activation of the complement system leads to inflammation, lysis of foreign cells, and opsonization. The three mechanisms that trigger the cascade include the alternative pathway, the lectin pathway, and the classical pathway. Not all of the steps in these pathways are shown.

Each of the major complement proteins has been given a number along with the letter C, for complement. The nine major proteins, C1 through C9, were numbered in the order in which they were discovered and not the order in which they react. When one of these is split into two fragments, a lowercase letter is added to the name. For example, the activation of C3 splits it into C3a and C3b. Note that C3 spontaneously splits into C3a and C3b even when the complement system has not been activated, but does so at a very low rate; this spontaneous hydrolysis allows enough C3b to be present to potentially trigger the alternative pathway of complement activation. ■ hydrolysis, p. 25

Activation of the complement system eventually leads to three major protective outcomes:

- **Inflammation.** The complement components C3a and C5a induce changes in endothelial cells that line the blood vessels, and in mast cells. These effects contribute to the vascular permeability associated with inflammation. C5a is also a potent chemoattractant, drawing phagocytes into the area where complement was activated.
- **Lysis of foreign cells.** Complexes of C5b, C6, C7, C8, and multiple C9 molecules spontaneously assemble in the membranes of cells, forming doughnut-shaped structures each called a **membrane attack complex (MAC) (figure 15.8).** This creates pores in the membrane, disrupting the integrity of the cell. Note that the membrane attack complex has little effect on Gram-positive bacteria because their peptidoglycan layer prevents the complement components from reaching their cytoplasmic membrane. The outer membrane of Gram-negative bacteria, however, renders them susceptible.
- **Opsonization.** The complement protein C3b binds to foreign material. Phagocytes more easily "grab" particles coated with C3b because phagocytic cells have receptors for the molecule on their surface. The material that C3b has coated is said to be **opsonized** (which means "prepared for eating"); compounds such as C3b that can opsonize material are **opsonins.** Opsonized material may be viewed as carrying a giant "eat me" sign that can be read by phagocytes. Our own cells are protected from the effects of C3b because our membranes bind regulatory molecules, leading to the inactivation of C3b when it binds.

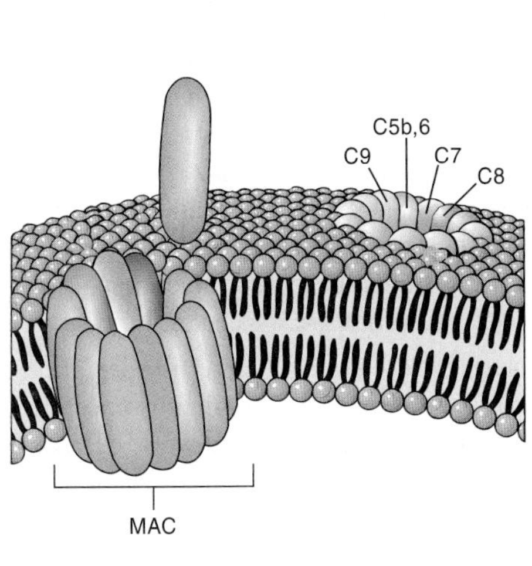
C5b,6
C9
C7
C8
MAC
(a)

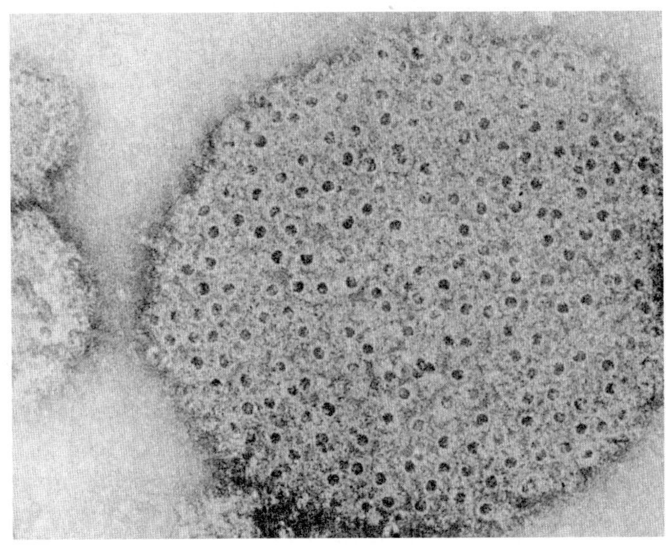
(b)

FIGURE 15.8 Membrane Attack Complex (MAC) of the Complement System (a) The MAC is formed after C5b, C6, and C7 combine into a complex on the cell surface of bacteria or other foreign cells. This complex, together with C8, causes changes in C9, allowing it to polymerize with the complex and form a MAC. The MAC forms a pore in the membrane, resulting in lysis of the cell. **(b)** An electron micrograph of MACs; each dark dot is a MAC.

Sensors That Detect Long Double-Stranded RNA (dsRNA)

Most cells typically do not contain long dsRNA (>30 bp) because only one DNA strand in a gene is used as a template for mRNA synthesis. In contrast, cells infected with RNA viruses other than retroviruses routinely have long dsRNA as a result of viral replication. Even cells infected with DNA viruses often have long dsRNA as a consequence of the viruses' efficient use of their relatively small genomes; in some regions of a viral genome, both strands are transcribed into mRNA, generating complementary RNA molecules. Thus, long dsRNA serves as a signal to a cell that it is infected with a virus. Various sensors within cells trigger an antiviral response when they detect the molecule.

Long dsRNA in an animal cell induces synthesis and subsequent secretion of alpha and beta interferons **(figure 15.9).** These molecules then attach to specific receptors on both the infected cell and neighboring cells, causing the cells to express several proteins (protein kinase R, RNAse L, and others) that can be viewed as inactive "suicide enzymes." For convenience, we will refer to these collectively as inactive antiviral proteins (iAVPs). The activated forms of the antiviral proteins (AVPs) degrade mRNA and stop protein synthesis, leading to a programmed cell death ("cell suicide") called **apoptosis.** A key feature of this system is that the iAVPs are activated by

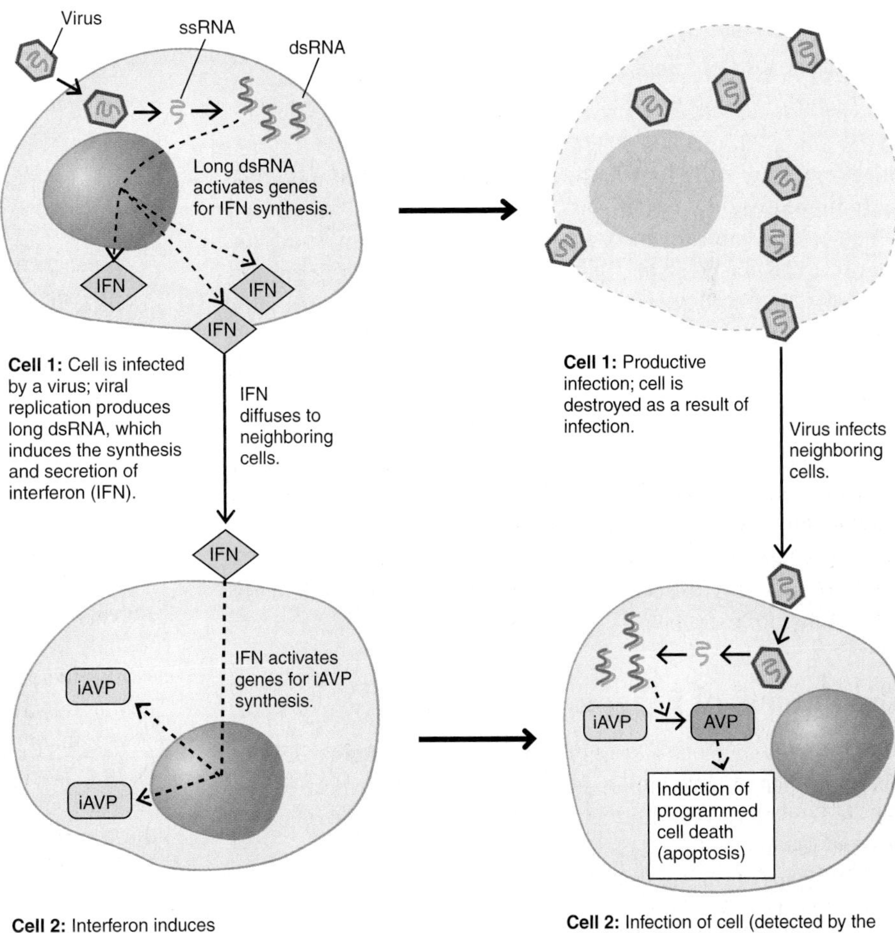
Virus
ssRNA
dsRNA
Long dsRNA activates genes for IFN synthesis.
IFN
IFN
IFN
Cell 1: Cell is infected by a virus; viral replication produces long dsRNA, which induces the synthesis and secretion of interferon (IFN).
IFN diffuses to neighboring cells.
IFN
IFN activates genes for iAVP synthesis.
iAVP
iAVP
Cell 2: Interferon induces synthesis of a group of inactive antiviral proteins (iAVP). These have no effect on the cell unless they are activated.
Cell 1: Productive infection; cell is destroyed as a result of infection.
Virus infects neighboring cells.
iAVP
AVP
Induction of programmed cell death (apoptosis)
Cell 2: Infection of cell (detected by the presence of long dsRNA) results in activation of the antiviral proteins, triggering the destruction of mRNA and inhibition of translation, leading to apoptosis. Although the cell dies, the virus does not have the opportunity to replicate, thus preventing viral spread.

FIGURE 15.9 Antiviral Effects of Interferon

dsRNA. Thus, when cells bind interferon, only the infected ones are sacrificed. Their uninfected counterparts remain functional, but are poised to cease vital functions should they too become infected. ■ apoptosis, p. 362

MICROCHECK 15.5

Toll-like receptors and NOD proteins enable cells to detect molecules that signify the presence of a microbe. Complement proteins can be activated by three mechanisms, leading to opsonization, lysis of foreign cells, and inflammation. Long dsRNA functions as a signal to a cell that it is infected with a virus, triggering an interferon response.

✓ What is the role of C3b in opsonization?

✓ What is the role of C3b in complement activation?

✓ Why would the discovery of toll-like receptors alter the view that innate immunity is non-specific?

15.6 Phagocytosis

Focus Points

- Outline the steps of phagocytosis.
- Compare and contrast the roles of macrophages and neutrophils.

Phagocytes are cells that routinely engulf and digest material, including invading organisms. How do the cells determine which of the multitude of different particles in the body to engulf? The answer lies in the various pattern recognition receptors that stud the phagocyte surface. Binding of a substance to certain phagocyte receptors induces the cell to engulf that material. A receptor called the **scavenger receptor,** for example, facilitates the engulfment of various materials that have charged molecules on their surface.

In routine situations, such as when organisms are introduced through a minor skin wound, macrophages residing in the tissues readily destroy the relatively few bacteria that enter. If the invading microbes are not rapidly cleared, however, macrophages produce cytokines to recruit additional phagocytes, particularly neutrophils, for extra help.

The Process of Phagocytosis

Phagocytosis involves a series of complex steps **(figure 15.10).** These are particularly important medically, because most pathogens have evolved the ability to evade one of them. This topic will be explored in chapter 17.

The steps of phagocytosis include:

- **Chemotaxis.** The phagocytic cells are recruited to the site of infection or tissue damage by certain chemical stimuli that act as chemoattractants. These include products of microorganisms, phospholipids released by injured mammalian cells, and the complement component C5a.
- **Recognition and attachment.** Phagocytic cells use various receptors to bind invading microbes either directly or indirectly. Direct binding occurs through receptors that recognize patterns associated with compounds found on microbes. For example, one type of receptor on phagocytic cells binds mannose, a sugar found on the surface of some bacteria and yeasts. Indirect binding occurs when a particle has first been opsonized, dramatically enhancing the phagocytes' ability to attach and subsequently engulf the material. Opsonins include the complement component C3b and certain classes of antibody molecules; phagocytes have receptors for specific parts of these molecules.
- **Engulfment.** The phagocytic cell engulfs the invader, forming a membrane-bound vacuole called a **phagosome.** This process involves rearrangement of the phagocyte's cytoskeleton, forming armlike extensions called **pseudopods** that surround the material being engulfed. Engulfment itself does not destroy the microbe. ■ cytoskeleton, p. 74 ■ pseudopod, p. 74
- **Fusion of the phagosome with the lysosome.** Within the phagocyte, the phagosome is transported along the cytoskeleton to a point where it can fuse with **lysosomes,** membrane-bound bodies filled with various digestive enzymes, including lysozyme and proteases. The fusion results in the formation of a **phagolysosome.**
- **Destruction and digestion.** Within the phagolysosome, oxygen consumption increases enormously as sugars are metabolized via aerobic respiration, with the production of highly toxic oxygen products such as superoxide, hydrogen peroxide, singlet oxygen, and hydroxyl radicals. As the available oxygen in the phagolysosome is consumed, the metabolic pathway switches to fermentation with the production of lactic acid, lowering the pH. Various enzymes degrade peptidoglycan and other components of the bacterial cell. ■ superoxide, p. 92 ■ hydrogen peroxide, p. 92
- **Exocytosis.** Following digestion of the microorganisms, the membrane-bound vesicle fuses with the plasma membrane, expelling the digested material to the external environment. ■ exocytosis, p. 73

Specialized Attributes of Macrophages

Macrophages can be viewed as the scavengers and sentries—routinely phagocytizing dead cells and debris, but always ready to destroy invaders and able to call in reinforcements when needed. They are always present in tissues, where they either slowly wander or remain stationary. These phagocytic cells play an essential role in every major tissue in the body.

Macrophages live for weeks to months, and maintain their killing power by continually regenerating their lysosomes. As macrophages die they are continually replaced by circulating monocytes that leave the blood and migrate to the tissues; recall that monocytes can differentiate into macrophages. Migration of monocytes is enhanced by certain stimuli associated with invasion and tissue damage.

Macrophages have several important characteristics that allow them to accomplish their diverse tasks. Various toll-

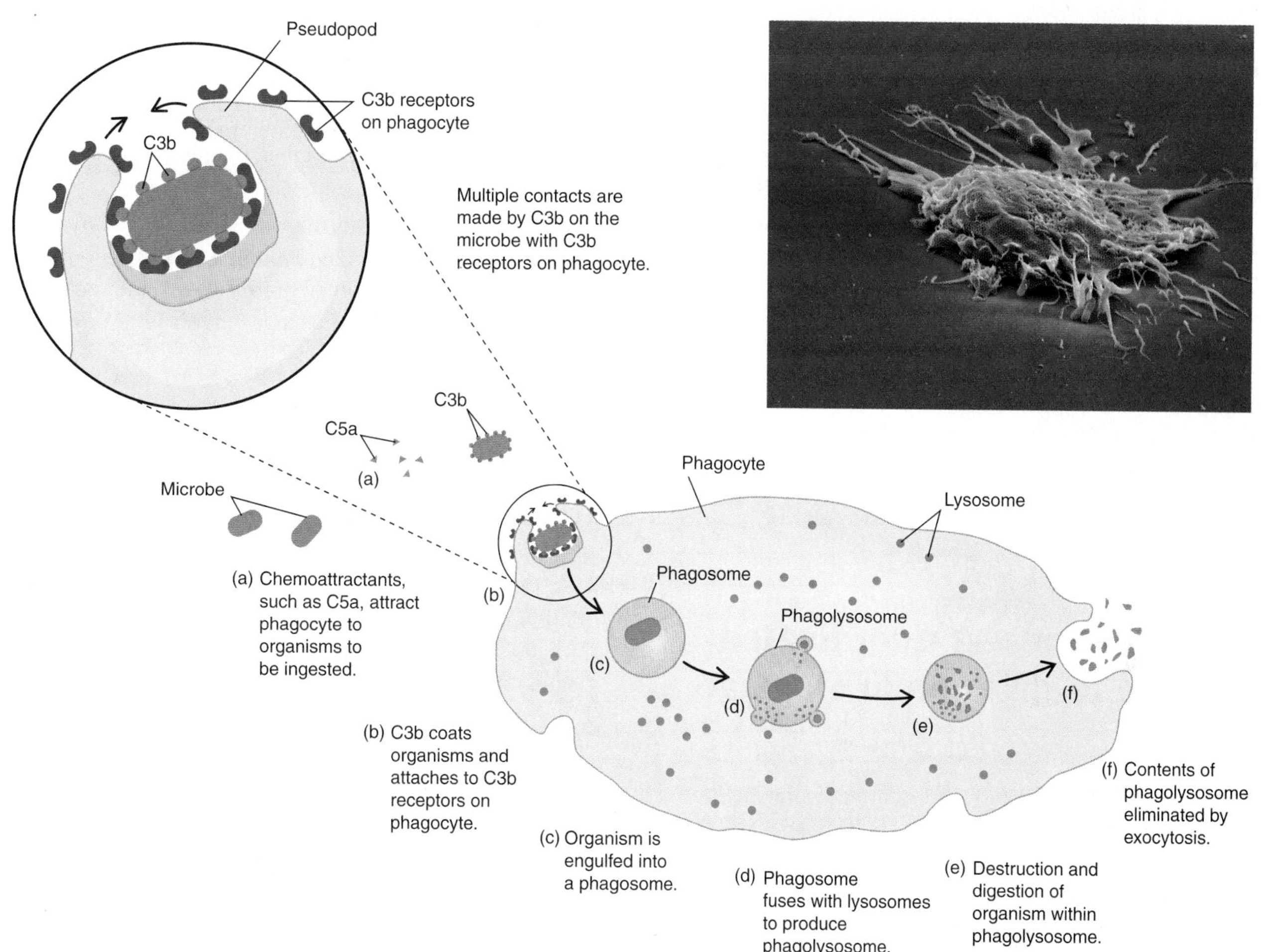

FIGURE 15.10 Phagocytosis and Intracellular Destruction of Phagocytized Material This diagram shows a microbe that has been opsonized by the complement protein C3b; certain classes of antibodies can also function as opsonins.

like receptors enable them to sense microbial invaders. When these receptors are triggered, the macrophage produces pro-inflammatory cytokines to alert and stimulate various other cells of the immune system. Macrophages can increase their otherwise limited killing power with the assistance of certain T cells to become **activated macrophages.** This is an example of the cooperation between the innate and adaptive host defenses. Activation of macrophages induces them to produce nitric oxide (NO) and oxygen radicals, which more effectively destroy microbes. These products also damage tissues when they are released, a reason why it would be detrimental for macrophages to continually maintain an activated state. Details of the activation process, including the roles of T cells, will be discussed in chapter 16. ■ macrophage activation, p. 383

If activated macrophages fail to destroy microbes and chronic infection ensues, large numbers of them can fuse together to form **giant cells.** Macrophages, giant cells, and T cells form concentrated groups called **granulomas** that wall off and retain organisms or other material that cannot be destroyed by the cells; again, this is an example of the cooperation between defense systems. Granulomas, which are part of the disease process in tuberculosis, histoplasmosis, and other illnesses, prevent the microbes from escaping to infect other cells (see figure 22.18). ■ tuberculosis, p. 514 ■ histoplasmosis, p. 526

Specialized Attributes of Neutrophils

Neutrophils can be viewed as the rapid response team—quick to move into an area of trouble and ready to eliminate the offending invaders. They play a critical role during the early stages of inflammation, being the first cell type recruited from the bloodstream to the site of damage. They inherently have more killing power than macrophages, including those that have been activated. The cost for their effectiveness, however, is a relatively short life span of only 1 to 2 days in the tissues; once they have expended their granules, they die. Many more are in reserve, however, for it is estimated that for every neutrophil in the circulatory system, 100 more are waiting in the bone marrow, ready to be mobilized when needed.

Neutrophils not only kill microbes through phagocytosis, they can also release the contents of their granules along with DNA to form **neutrophil extracellular traps (NETs).** The DNA backbone of the NET ensnares microbes, allowing the granule contents (enzymes and peptides) that accumulate within the NET to destroy them.

MICROCHECK 15.6

The process of phagocytosis includes chemotaxis, recognition and attachment, engulfment, fusion of the phagosome with the lysosome, destruction and digestion of the ingested material, and exocytosis. Macrophages are long-lived phagocytic cells that are always present in tissues; they can be activated to enhance their killing power. Neutrophils are highly active, short-lived phagocytic cells that must be recruited to the site of damage.

✓ How does a phagolysosome differ from a phagosome?
✓ What is a granuloma?
✓ What could a microorganism do to avoid engulfment?

15.7

Inflammation—A Coordinated Response to Invasion or Damage

Focus Point

- Describe the inflammatory process, focusing on the factors that initiate the response and the outcomes of inflammation.

When tissues have been damaged, such as when an object penetrates the skin or when microbes are introduced, a coordinated response called the **inflammatory response,** or **inflammation,** occurs. Everyone has experienced the signs of inflammation; in fact, the Roman physician Celsus described the four cardinal signs in the first century A.D.—swelling, redness, heat, and pain. A fifth sign, loss of function, is sometimes present.

The vital role of inflammation is to contain a site of damage, localize the response, and restore tissue function. The process recruits neutrophils, followed by monocytes and other cells, to assist the local macrophages and eosinophils at the site of damage.

Factors That Initiate the Inflammatory Response

Inflammation is initiated in response to invading microbes or tissue damage. In the case of a surface wound, the action causing the tissue damage is likely to also introduce microbes either residing on the offending instrument or on the skin's surface. Therefore, both factors are often involved in eliciting the response. Events that initiate inflammation include, either singly or in combination:

- Microbial products such as LPS, flagellin, and bacterial DNA trigger the toll-like receptors of macrophages, causing the cells to produce pro-inflammatory cytokines. One of these, tumor necrosis factor alpha, induces the liver to synthesize **acute-phase proteins,** a group of proteins, that facilitate phagocytosis and complement activation.
- Microbial cell surfaces can trigger the complement cascade, leading to the production of C3a and C5a, both of which stimulate changes associated with inflammation. The complement components also induce mast cells to release various pro-inflammatory cytokines (including tumor necrosis factor α), histamine, and other substances.
- Tissue damage results in the activation of two enzymatic cascades. One is the coagulation cascade, which results in blood clotting, and the other produces several molecules such as bradykinin that elicit changes involved in inflammation.

The Inflammatory Process

Initiation of the inflammatory process leads to a cascade of events that result in dilation of small blood vessels, leakage of fluids from those vessels, and the migration of leukocytes out of the bloodstream and into the tissues (**figure 15.11**).

The diameter of local blood vessels increases during inflammation due to the action of certain pro-inflammatory chemicals. This results in an increase in blood flow to the area, causing the heat and redness associated with inflammation, accompanied by a decrease in the velocity of blood flow in the capillaries. Because of the dilation, normally tight junctions between endothelial cells are disrupted, allowing fluid to leak from the vessels and into the tissue. This fluid contains various substances such as transferrin, complement system proteins, and antibodies, and thus helps to counteract invading microbes. The increase of fluids in the tissues causes the swelling and pain associated with inflammation. The direct effects of chemicals on sensory nerve endings also cause pain.

Some of the pro-inflammatory cytokines cause endothelial cells in the local area to produce adhesion molecules that loosely adhere to phagocytes. The phagocytes normally flow rapidly through the vessels, but slowly tumble to a halt as they attach to the adhesion molecules. The phagocytic cells themselves then begin producing a different type of adhesion molecule that strengthens the attachment. Then, in response to other cytokines and complement components that function as chemoattractants, phagocytes migrate from the blood vessels into the area. They do this by squeezing between the cells of the dilated permeable vessel, the process of **diapedesis.** Neutrophils (PMNs) are the first to be lured from the circulation, and soon they predominate in the area. After the influx of neutrophils, monocytes and lymphocytes then accumulate. The monocytes, which mature into macrophages at the site of infection, and neutrophils actively phagocytize foreign material. Clotting factors in the fluid that leaks into the tissues initiate clotting reactions. This helps prevent bleeding and halts spread of invading microbes. As the inflammatory process continues, large quantities of dead neutrophils accumulate. Along with tissue debris, these dead cells make up **pus.** A large amount of pus constitutes an **abscess** (see figure 24.2). ■ abscess, p. 560

The extent of inflammation varies, depending on the nature of the injury, but the response is localized, begins immediately upon injury, and increases over a short period of time. This short-term inflammatory response is called **acute inflammation** and

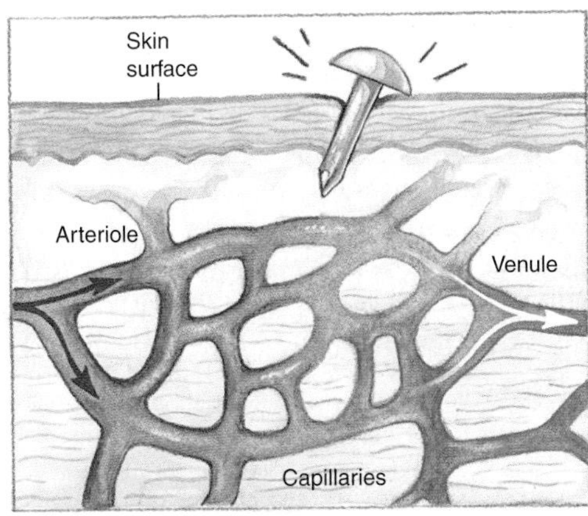

(a) Normal blood flow in the tissues as injury occurs

- Microbial products
- Microbes
- Tissue damage

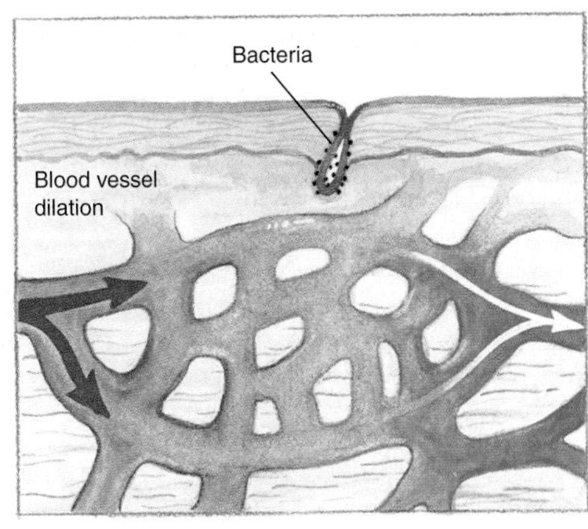

(b) Substances released cause dilation of small blood vessels and increased blood flow in the immediate area.

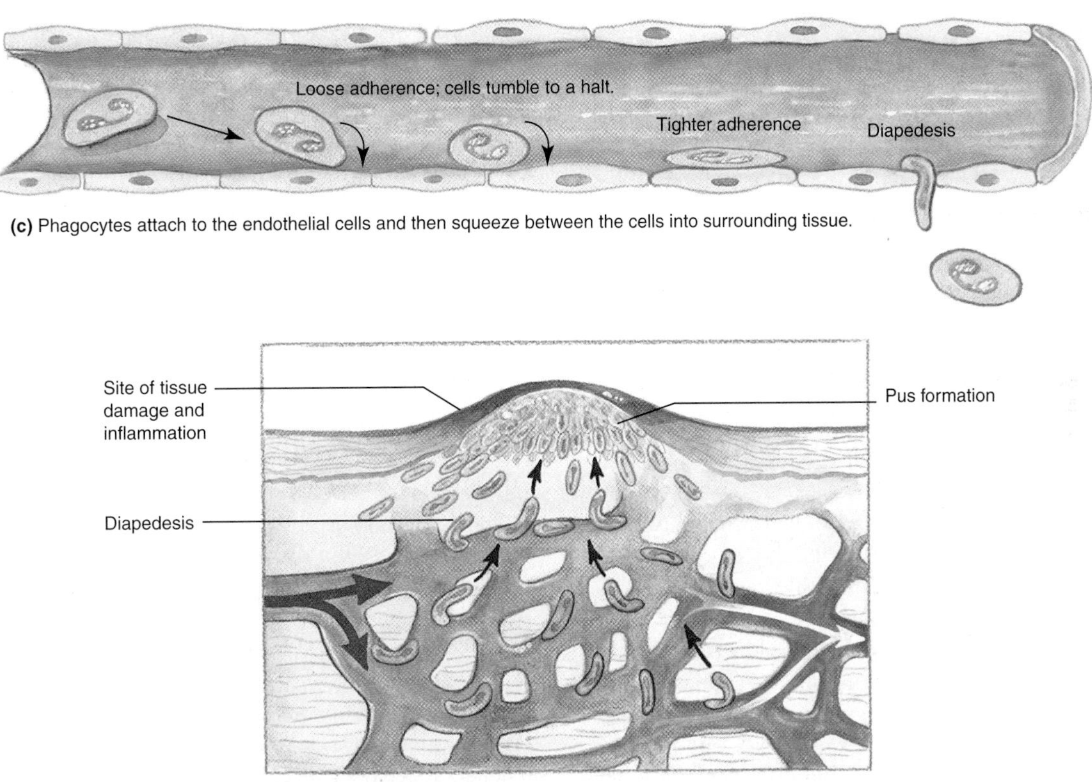

(c) Phagocytes attach to the endothelial cells and then squeeze between the cells into surrounding tissue.

(d) The attraction of phagocytes causes them to move to the site of damage and inflammation. Collections of dead phagocytes and tissue debris make up the pus often found at sites of an active inflammatory response.

FIGURE 15.11 The Inflammatory Process This coordinated response to microbial invasion or tissue damage brings phagocytes and other leukocytes to the site. The role of inflammation is to contain a site of damage, localize the response, and restore tissue function.

is marked by a prevalence of neutrophils. Then, as inflammation subsides, healing occurs. During healing, new capillaries grow into the area and destroyed tissues are replaced; eventually, scar tissue is formed. If acute inflammation cannot limit the infection, **chronic inflammation** occurs. This is a long-term inflammatory process that can last for years. Chronic inflammation is characterized by the prevalence of macrophages, giant cells, and granulomas. ■ giant cells, p. 359 ■ granulomas, p. 359

Outcomes of Inflammation

The inflammatory process can be likened to a sprinkler system that prevents fire from spreading in a building. While the process usually limits damage and restores function, the response itself can cause significant harm. One undesirable consequence is that some enzymes and toxic products contained within phagocytic cells are inevitably released, damaging tissues.

PERSPECTIVE 15.1

For *Schistosoma*, the Inflammatory Response Delivers

The parasitic flatworms that cause schistosomiasis do not shy from the immune response when it comes to procreation; instead they appear to use it to deliver their ova to an environment where they might hatch. Adult females of *Schistosoma* species, which live in the bloodstream of infected hosts, lay their ova in veins near the intestine or bladder; they seem to rely on a robust inflammatory response to expel the ova, completing one portion of a complex life cycle. The ova released in feces or urine can hatch, liberating a larval stage called a *miracidium,* if untreated sewage reaches water. The miracidium then infects a specific freshwater snail host and undergoes asexual multiplication. The infected snail then releases large numbers of another larval form, *cercariae,* which swim about in search of a human host.

The parasite is acquired when a person wades or swims in contaminated water. The cercariae penetrate the skin by burrowing through it with the aid of digestive enzymes; schistosomes are rare among pathogens because they can actually penetrate intact skin. The larvae then enter the circulatory system where they can live for over a quarter of a century. *Schisotosoma* species have separate sexes and, remarkably, the male and female worms locate one another in the bloodstream. The male's body has a deep longitudinal groove in which he clasps his female partner to live in copulatory embrace (*shisto-soma* means "split-body," referring to the long slit). The adult worms effectively mask themselves from the immune system by adsorbing various blood proteins; this provides them with a primitive stealth "cloaking device."

Depending on the species, the female worm migrates to the veins of either the intestine or bladder to lay hundreds of ova per day. The body responds vigorously to the highly antigenic eggs, ejecting them in manner similar to what is experienced as a sliver in the skin works its way to the surface. Over half of the ova are not expelled, however, and many of these are instead swept away by the bloodstream to the liver. The inflammatory process and granuloma formation there gradually destroy liver cells, replacing the cells with scar tissue. Malfunction of the liver results in malnutrition and a buildup of pressure in intestinal and esophageal veins. Fluid accumulates in the abdominal cavity and hemorrhage occurs if the engorged esophageal veins rupture.

Despite their complex life cycle, *Schistosoma* species are highly successful. Not only are they adept at avoiding certain immune responses that would otherwise lead to their destruction, they have learned to exploit inflammation for their own dissemination. Over 200 million people worldwide are infected with these parasites, resulting in the deaths of over 500,000 people each year.

If inflammation is limited, such as in a response to a cut finger, the damage caused by the process is normally minimal. If the process occurs in a delicate system, however, such as the membranes that surrounds the brain and spinal cord, the consequences can be much more severe, even life threatening.

Apoptosis—Controlled Cell Death That Circumvents the Inflammatory Process

The inflammatory response represents a potential problem for the host; that is, how to distinguish cell death caused by abnormal events, such as injury, from that caused by normal events such as tissue remodeling that render certain cells unnecessary or potentially harmful. The former merits an inflammatory response whereas the latter does not and, in fact, would be unnecessarily destructive to normal tissue. **Apoptosis** (Greek, *apo* for "falling"; *ptosis* for "off"), or programmed cell death, is a process that destroys "self" cells without eliciting inflammation. During apoptosis, the dying cells undergo certain changes. For example, the shape of the cell changes, enzymes cut the DNA, and portions of the cell bud off, effectively shrinking the cell. Some changes appear to signal to macrophages that the remains of the cell are to be engulfed without the commotion associated with inflammation.

MICROCHECK 15.7

Inflammation is a cascade of events initiated in response to invading microbes or tissue damage. The outcome is dilation of small blood vessels, leakage of fluids from those vessels, and migration of leukocytes out of the bloodstream and into the tissue. Inflammation can help contain an infection, but the response itself can cause damage. Apoptosis provides a mechanism for the destruction of "self" cells without initiating inflammation.

- ✓ Describe three general events that can initiate inflammation.
- ✓ Describe the changes that characterize cells undergoing apoptosis.
- ✓ How could infection of the fallopian tubes lead to sterility and ectopic pregnancy?

15.8 Fever

Focus Point

- Describe the induction and outcomes of fever.

Fever is one of the strongest indications of infectious disease, especially those of bacterial origin, although significant infections can occur without it. There is abundant evidence that fever is an important host defense mechanism in a number of vertebrates, including humans. Within the human body, the temperature is normally kept within a narrow range, around 37°C, by a temperature-regulation center in the brain. During an infection, the regulating center continues to function but the body's thermostat is "set" at higher levels. An oral temperature above 37.8°C is regarded as fever.

A higher temperature setting occurs as a result of certain pro-inflammatory cytokines released by macrophages when their toll-like receptors detect microbial products. The cytokines are carried in the bloodstream to the brain, where they act as messages that microbes have invaded the body. These cytokines and other fever-inducing substances are **pyrogens.** Fever-inducing cytokines are called **endogeneous pyrogens,** indicating the body makes them, whereas microbial products, are called **exogenous pyrogens,** indicating they are introduced from external sources. The temperature-regulating center responds to pyrogens by raising body temperature. The resulting fever inhibits the growth of many pathogens by at least two mechanisms: (1) elevating the temperature above the optimum growth temperature of the pathogen, and (2) activating and speeding up a number of other body defenses.

The adverse effects of fever on pathogens correlates in part with their ideal growth temperature. Bacteria that grow best at 37°C are less likely to cause disease in people with fever. The growth rate of bacteria often declines sharply as the temperature rises above their optimum temperature. A slower growth rate allows more time for other defenses to destroy the invaders. ■ temperature requirements, p. 90

A moderate rise in temperature increases the rate of enzymatic reactions. It is thus not surprising that fever has been shown to enhance the inflammatory response, phagocytic killing by leukocytes, multiplication of lymphocytes, release of substances that attract neutrophils, and production of interferons and antibodies. Release of leukocytes into the blood from the bone marrow is also enhanced. For all these reasons, it is wise to consult a physician before taking drugs to reduce the fever of infectious disease.

MICROCHECK 15.8

Fever results when macrophages release pro-inflammatory cytokines; this occurs when the toll-like receptors on the macrophages are engaged by microbial products.

✓ What is an endogenous pyrogen? What is an exogenous pyrogen?

✓ How does fever inhibit the growth of pathogens?

✓ Syphilis was once treated by intentionally infecting the patient with the parasite that causes malaria, a disease characterized by repeated bouts of fever, shaking, and chills. Why would this treatment cure syphilis?

SUMMARY

15.1 Overview of the Innate Defenses

The innate defense system is composed of **first-line defenses,** sensor systems, and **phagocytes.**

15.2 First-Line Defenses (figures 15.1, 15.2, 15.3)

Physical Barriers

The skin is composed of two main layers—the **dermis** and the **epidermis. Mucous membranes** are constantly bathed with mucus and other secretions that help wash microbes from the surfaces.

Antimicrobial Substances

Lysozyme, peroxidase enzymes, lactoferrin, and **defensins** inhibit or kill microorganisms.

Normal Microbiota (Flora)

Members of the **normal microbiota** competitively exclude pathogens and stimulate the host defenses.

15.3 The Cells of the Immune System (figure 15.4, table 15.1)

Granulocytes

There are three types of **granulocytes—neutrophils, basophils,** and **eosinophils.**

Mononuclear Phagocytes

Monocytes circulate in blood; **macrophages** are found in tissues (figure 15.5).

Dendritic Cells

Dendritic cells develop from monocytes; some appear to have other origins.

Lymphocytes

Lymphocytes, which include **B cells, T cells,** and **natural killer (NK) cells,** are involved in adaptive immunity.

15.4 Cell Communication

Surface Receptors

Surface receptors bind ligands that are on the outside of the cell, enabling the cell to detect that the ligand is present.

Cytokines (table 15.2)

Cytokines include **interleukins (ILs), colony-stimulating factors (CSFs), tumor necrosis factors (TNFs), chemokines,** and **interferons.**

Adhesion Molecules

Adhesion molecules allow cells to adhere to other cells.

15.5 Sensor Systems

Toll-Like Receptors and NOD Proteins (figure 15.6)

Toll-like receptors and **NOD proteins** enable cells to detect molecules that signify the presence of invading microbes.

The Complement System (figure 15.7)

Complement proteins circulate in the blood and the fluid that bathes tissues; in response to certain stimuli that indicate the presence of foreign material, they become activated. The major protective outcomes of complement activation include **opsonization,** lysis of foreign cells, and initiation of **inflammation.**

Sensors That Detect Long Double-Stranded RNA (dsRNA) (figure 15.9)

Long dsRNA signifies to a cell that it has been infected with a virus. This induces synthesis of interferons, causing cells in the vicinity to prepare to cease vital cell functions in the event they become infected with a virus.

15.6 Phagocytosis

The Process of Phagocytosis (figure 15.10)

The steps of phagocytosis include chemotaxis, recognition and attachment, engulfment, destruction and digestion, and exocytosis.

Specialized Attributes of Macrophages

Macrophages are always present in tissues to some extent, but are able to call in reinforcements when needed. A macrophage can increase its killing power, becoming an **activated macrophage.** Macrophages, **giant cells,** and T cells form **granulomas** that wall off and retain organisms or other material that cannot be destroyed by macrophages.

Specialized Attributes of Neutrophils

Neutrophils play a critical role during the early stages of inflammation, being the first cell type recruited from the bloodstream to the site of damage.

15.7 Inflammation—A Coordinated Response to Invasion or Damage (figure 15.11)

Swelling, redness, heat, and pain are the signs of inflammation, the attempt by the body to contain a site of damage, localize the response, and restore tissue function.

Factors That Initiate the Inflammatory Response

Inflammation is initiated when pro-inflammatory cytokines or other inflammatory mediators are released as a result of the engagement of toll-like receptors or activation of the complement system by invading microbes, or when tissue damage occurs.

The Inflammatory Process

The inflammatory process results in dilation of small blood vessels, leakage of fluids from those vessels, and the migration of leukocytes out of the bloodstream and into the tissues. **Acute inflammation** is marked by a preponderance of neutrophils; **chronic inflammation** is characterized by the prevalence of macrophages, giant cells, and granulomas.

Outcomes of Inflammation

Inflammation can contain an infection, but the process itself can cause damage; a systemic response can be life threatening.

Apoptosis—Controlled Cell Death That Circumvents the Inflammatory Process

Apoptosis is a mechanism of eliminating "self" cells without evoking an inflammatory response.

15.8 Fever

Fever occurs as a result of certain pro-inflammatory cytokines released by macrophages. It inhibits the growth of many pathogens and increases the rate of various body defenses.

REVIEW QUESTIONS

Short Answer

1. Why is iron metabolism important in body defenses?
2. How do phagocytes get into tissues during an inflammatory response?
3. Describe how the skin protects against infection.
4. What are the benefits of saliva in protection against infection? What factors found in saliva aid in protection?
5. Name two categories of cytokines and give their effects.
6. Contrast the classical and alternative pathways of complement activation.
7. How does the activation of a few molecules in early stages of the complement cascade result in the cleavage of millions of molecules of later ones?
8. How do complement proteins cause foreign cell lysis?
9. Describe three mechanisms of triggering inflammation.
10. Describe the function of apoptosis.

Multiple Choice

1. Lysozyme does which of the following?
   a) Disrupts cell membranes
   b) Hydrolyzes peptidoglycan
   c) Waterproofs skin
   d) Propels gastrointestinal contents
   e) Propels the cilia of the respiratory tract
2. The hematopoietic stem cells in the bone marrow can become which of the following cell types?
   1. red blood cell 2. T cell 3. B cell
   4. monocyte 5. macrophage
   a) 2, 3 b) 2, 4 c) 2, 3, 4, 5
   d) 1, 4, 5 e) 1, 2, 3, 4, 5
3. All of the following refer to the same type of cell *except*
   a) macrophage. b) neutrophil.
   c) poly. d) PMN.
4. Toll-like receptors are triggered by all of the following compounds *except*
   a) peptidoglycan.
   b) glycolysis enzymes.
   c) lipopolysaccharide.
   d) flagellin.
   e) certain nucleotide sequences.
5. A pathogen that can avoid the complement component C3b would directly protect itself from
   a) opsonization. b) triggering inflammation. c) lysis.
   d) inducing interferon. e) antibodies.
6. Which of the following statements about phagocytosis is *false?*
   a) Phagocytes move toward an area of infection by a process called chemotaxis.
   b) The vacuole in which bacteria are exposed to degradative enzymes is called a phagolysosome.
   c) Phagocytes have receptors that recognize complement proteins bound to bacteria.
   d) Phagocytes have receptors that recognize antibodies bound to bacteria.
   e) Macrophages die after phagocytizing bacteria but neutrophils regenerate their lysosomes and survive.
7. All of the following cell types are found in a granuloma *except*
   a) neutrophils. b) macrophages.
   c) giant cells. d) T cells.
8. All of the following trigger inflammation *except*
   a) engagement of toll-like receptors.
   b) activation of complement.
   c) interferon induction of antiviral protein synthesis.
   d) tissue damage.
9. Which of the following statements about inflammation is *false?*
   a) Vasodilation results in leakage of blood components.
   b) The process can cause damage to host tissue.
   c) Neutrophils predominate at the site during the early stages of acute inflammation.
   d) Apoptosis induces inflammation.
   e) The cardinal signs of inflammation are redness, swelling, heat, and pain.
10. The direct/immediate action of interferon on a cell is to
   a) interfere with the replication of the virus.
   b) prevent the virus from entering the cell.
   c) stimulate synthesis of inactive "suicide enzymes."
   d) stimulate the immune response.
   e) stop the cell from dividing.

Applications

1. Physicians regularly have to treat recurrent urinary tract infections in paralyzed paraplegic patients. What explanation would the physician provide to a patient who asked why the condition keeps coming back in spite of repeated treatment?
2. A cattle farmer sees a sore on the leg of one of his cows. The farmer feels the sore and notices that the area just around the sore is warm to the touch. A veterinarian examines the wound and explains that the warmth may be due to inflammation. The farmer wants an explanation of the difference between the localized warmth and fever. What would be the vet's explanation to the farmer?

Critical Thinking

1. A student argues that phagocytosis is a wasteful process because after engulfed organisms are digested and destroyed, the remaining material is excreted from the cell (see figure 15.10). A more efficient process would be to release the digested material *inside* the cell. This way, the material and enzymes could be reused by the cell. Does the student have a valid argument? Why or why not?
2. According to figure 15.9, *any* cell infected by viruses may die due to the action of interferons. This strategy, however, seems counterproductive. The same result would occur without interferon—any cell infected by a virus might die directly from the virus. Is there any apparent benefit from the interferon action?

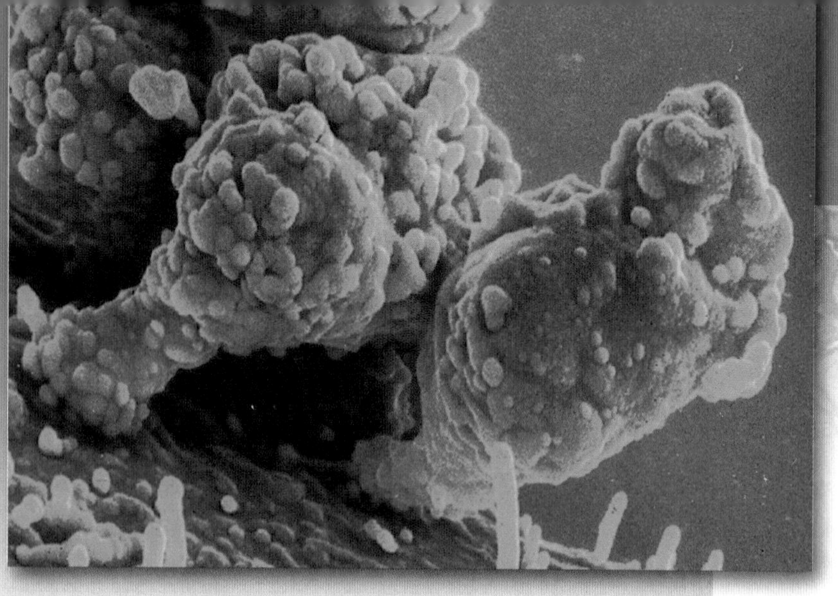

16

Immune cells.

The Adaptive Immune Response

A Glimpse of History

Near the end of the nineteenth century, diphtheria was a terrifying disease that killed many infants and small children. The first symptom was a sore throat, often followed by the development of a gray membrane in the throat that could come loose and obstruct the airway. Death sometimes occurred rapidly, even in the absence of a membrane. Frederick Loeffler, working in Robert Koch's laboratory in Berlin, found club-shaped bacteria growing in the throats of people with the disease but not elsewhere in their bodies. He guessed that the organisms were making a poison that spread through the bloodstream. In Paris, at the Pasteur Institute, Emile Roux and Alexandre Yersin followed up by growing the bacteria in quantity and extracting the poison, or toxin, from culture fluids. When injected into guinea pigs, the toxin generally killed the animals. ■ Alexandre Yersin, p. 674

Back in Berlin, Emil von Behring injected the diphtheria toxin into guinea pigs that had been previously inoculated with the agent and had recovered from diphtheria. These guinea pigs did not become ill from the toxin, suggesting to von Behring that something in their blood, which he called *antitoxin,* protected against the toxin. To test this idea, he mixed toxin with serum from a guinea pig that had recovered from diphtheria and injected this mixture into an animal that had not had the disease. The guinea pig remained well. In further experiments, he cured animals with diphtheria by giving them antitoxin.

The results of these experiments in animals were put to the test in people in late 1891, when an epidemic of diphtheria occurred in Berlin. On Christmas night of that year, antitoxin was first given to an infected child, who then recovered from the dreaded disease. The substances in blood with antitoxin properties soon were given the more general name of *antibodies,* and materials that generated antibody production were called *antigens.*

Emil von Behring received the first Nobel Prize in Medicine in 1901 for this work on antibody therapy. It took many more decades of investigation before the biochemical nature of antibodies was elucidated. In 1972, Rodney Porter and Gerald Edelman were awarded the Nobel Prize for their part in determining the chemical structure of antibodies.

In contrast to the innate immune response, which is always ready to respond to patterns that signify damage or invasion, the adaptive immune response matures throughout life, developing from the immune system arsenal the most effective response against specific invaders as each is encountered. The protection provided by the response is called **adaptive immunity.** An important hallmark of the adaptive immune response is that it has memory, a greatly enhanced response to re-exposure. Individuals who survived diseases such as measles, mumps, or diphtheria generally never developed the acute disease again. Vaccination now prevents these diseases by exposing a person's immune system to harmless forms of the causative microbe or its products. While it is true that some diseases can be contracted repeatedly, that phenomenon is generally due to the causative agent's ability to evade the host defenses, a topic we will discuss in chapter 17. The adaptive immune response also has molecular specificity. The response that protects an individual from developing symptoms of measles does not prevent the person from contracting a different disease; for example, chickenpox. The immune system can also discriminate healthy "self," your own normal cells, from "dangerous," such as invading bacteria. If this were not the case, the immune system would routinely turn against the body's own cells, attacking them just as it does an invading microbe. This is not a fail-safe system, however, which is why autoimmune diseases can occur. ■ acute disease, p. 395 ■ vaccination, p. 433 ■ autoimmune disease, p. 424

The adaptive immune system is extraordinarily complex, involving an intricate network of cells, cytokines, and other compounds. In fact, immunologists are still working out many of its secrets.

In this chapter, we will first cover the general strategies the adaptive immune response uses to eliminate invading microbes. This will then lead to a more detailed description of the various cells and molecules involved. At the end of the chapter, we will focus on the development of the immune system, concentrating on how the cells involved in adaptive immunity gain the specificity

KEY TERMS

Adaptive Immunity Protection provided by immune responses that mature throughout life; involves B cells and T cells.

Antibody Y-shaped protein that binds antigen.

Antigen Molecule that reacts specifically with either an antibody or an antigen receptor on a lymphocyte.

Antigen-Presenting Cells (APCs) Cells such as B cells, macrophages, and dendritic cells that can present exogenous antigens to naive or memory T cells, activating them.

B Cell Type of lymphocyte programmed to make antibodies.

Clonal Selection Process in which a specific antigen receptor on a lymphocyte binds to a given antigen, allowing the lymphocyte to proliferate.

Cytotoxic T Cell Type of lymphocyte programmed to destroy corrupt "self" cells.

Dendritic Cell Cell type responsible for activating naive T cells.

Helper T Cell Type of lymphocyte programmed to activate B cells and macrophages, and assist other aspects of the adaptive immune response.

Lymphocytes White blood cells (leukocytes) that have antigen-specific receptors on their surface; B cells are lymphocytes that mediate humoral immunity, and T cells are lymphocytes that mediate cellular immunity.

Major Histocompatibility Complex (MHC) Molecules Cell surface molecules that present antigen to T cells. MHC class I molecules present endogenous antigen to cytotoxic T cells; MHC class II molecules present exogenous antigen to helper T cells.

Memory Lymphocytes Long-lived descendants of activated lymphocytes that can quickly respond when specific antigen is encountered again.

Plasma Cell Effector form of a B cell; it functions as an antibody-secreting factory.

T_C Cell Effector form of a cytotoxic T cell; it induces apoptosis in infected or cancerous "self" cells.

T_H Cell Effector form of a helper T cell; it activates B cells and macrophages, and releases cytokines that stimulate other aspects of the immune system.

required to respond to an incredibly diverse and ever-changing assortment of microbes. Throughout the chapter, we will describe some of the mechanisms used by the adaptive immune system to build **tolerance.** This is the ability to ignore any given molecule, particularly those that make up the tissues.

16.1 Strategy of the Adaptive Immune Response

Focus Point

- Compare and contrast the general aspects of humoral immunity and cellular immunity.

On first exposure to a given microbe or any other antigen, evidence of the adaptive immune response takes a week or more to develop. During this delay the host depends on the protection provided by innate immunity, which may not be sufficient to prevent disease. This first response to a particular antigen is called the **primary response.** As a result of that initial encounter, the adaptive immune system is able to "remember" the mechanism that proved effective against that specific antigen. As a result, when the same antigen is encountered later in life, there is an enhanced antigen-specific immune response called the **secondary,** or **anamnestic, response.** The efficiency of the secondary response reflects the memory of the immune system. ■ antigen, p. 347

The adaptive immune response uses two basic strategies for countering foreign material. One response, **humoral immunity,** works to eliminate antigens that are extracellular (not within a host cell); for example, bacteria, toxins, or viruses in the bloodstream or in tissue fluids **(figure 16.1).** The other, called **cellular immunity,** or **cell-mediated immunity,** deals with antigens residing within a host cell, such as a virus that has infected a cell. Humoral and cellular immunity are both powerful and, if misdirected, can cause a great deal of damage to the body's own tissues. Because of this, the adaptive immune response is tightly regulated; each **lymphocyte,** the primary participants in the adaptive response, generally requires a "second opinion" from a different type of cell before it can unleash its power. ■ lymphocytes, p. 352

Overview of Humoral Immunity

Humoral immunity is mediated by **B lymphocytes,** or **B cells.** Their name reflects the fact that they develop in an organ in birds called the bursa. In humans, however, B cells develop in the bone marrow. In response to extracellular antigens, B cells may be triggered to proliferate and then differentiate into **plasma cells,** which function as factories that produce Y-shaped proteins called **antibodies.** These molecules bind to antigens, providing protection to the host by mechanisms that will be described shortly. A high degree of specificity is involved in the binding, so a multitude of different antibody molecules are needed to bind to the wide array of antigens encountered throughout life. Some of the B cells form **memory B cells,** long-lived B cells that respond more quickly if the antigen is encountered again.

Antibody molecules have two functional regions—the two identical arms and the stem of the molecule. It is the arms of the Y that bind to a specific antigen; the amino acid sequence of the end of the arms varies from antibody to antibody, providing the basis for their specificity. The stem of the Y functions as a "red flag," tagging antigen bound by antibody and enlisting other components of the immune system to eliminate the bound molecule.

Antibodies that bind to an antigen protect the host by both direct and indirect mechanisms. Simply by coating an antigen, the antibodies prevent that molecule from attaching to critical sites on host cells. For example, a viral particle that has been coated with antibody molecules cannot bind to its intended

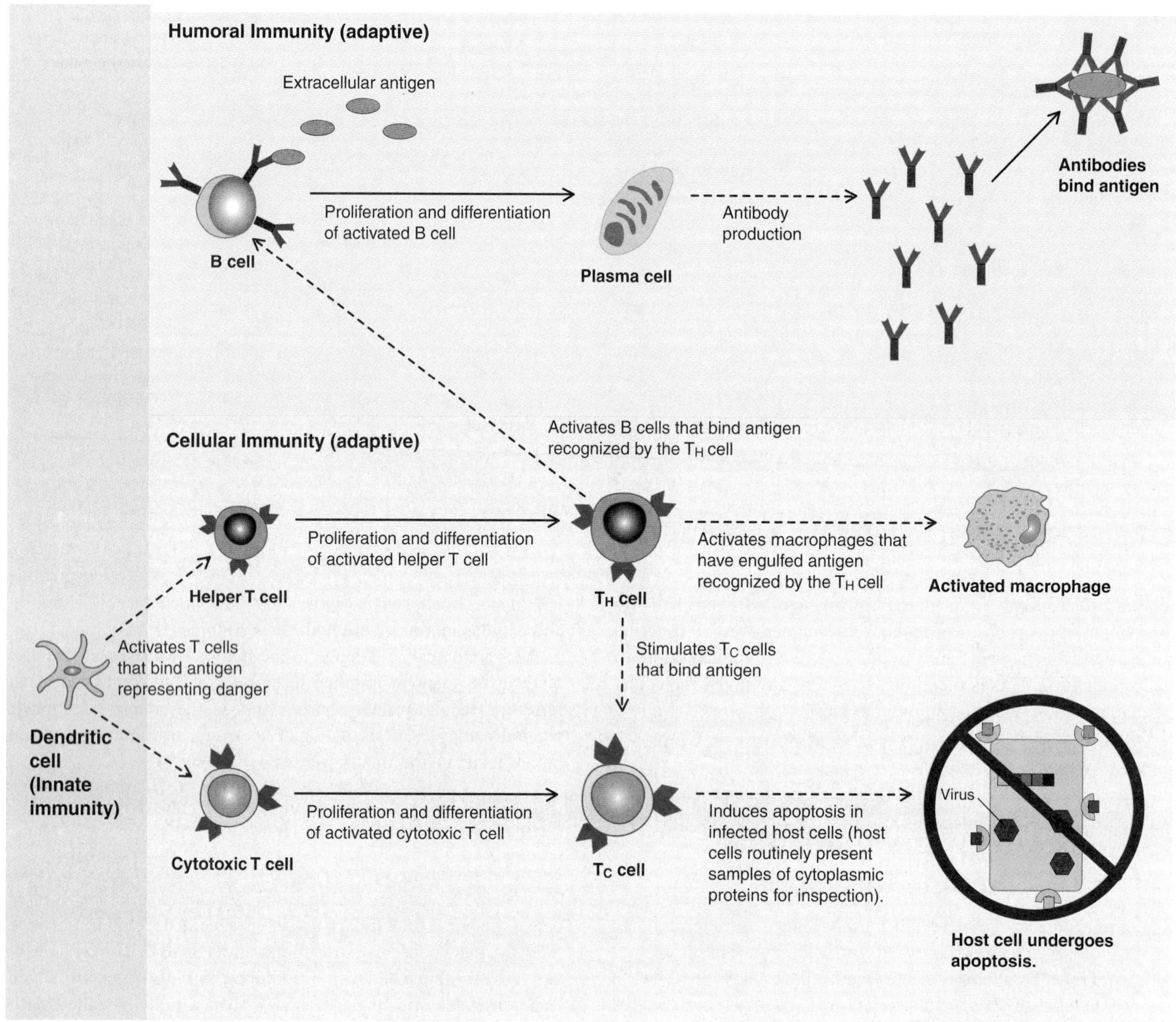

FIGURE 16.1 Overview of Humoral and Cellular Immunity Memory cells are not shown in this diagram. Cellular immunity is also called cell-mediated immunity (CMI).

receptor on a host cell. Because the virus cannot attach, it is unable to enter the cell. The indirect protective effect is due to the "red flag" region that facilitates elimination of the antigen by the innate defenses. Phagocytes, for example, have receptors for that region of an antibody molecule, enabling them to more easily engulf an antigen coated with bound antibodies; this is the process of opsonization described in chapter 15. ■ **attachment of viruses, p. 325** ■ **opsonization, p. 356**

How does a B cell know when to replicate in order to eventually produce antibodies? Each B cell carries on its surface multiple copies of a membrane-bound derivative of the specific antibody it is programmed to make; each of these molecules is called a **B-cell receptor.** If the B cell encounters an antigen that its B-cell receptors bind, then the cell may gain the capacity to multiply. **Clones,** or copies, of the cell are produced that can eventually differentiate to become plasma cells that make and secrete copious amounts of antibody. Generally, however, before the B cell can multiply, it needs confirmation by another lymphocyte, a T_H cell, that the antigen is indeed dangerous.

Overview of Cellular Immunity

Cellular immunity is mediated by **T lymphocytes,** or **T cells;** their name reflects the fact that they mature in the thymus. T cells involved in eliminating antigen include two subsets—**cytotoxic T cells** and **helper T cells.** Both of these have multiple copies of a surface molecule called a **T-cell receptor,** which is functionally analogous to the B-cell receptor; it enables the cell to recognize a specific antigen. Unlike the B-cell receptor, however, the T-cell receptor does not recognize free antigen. Instead, the antigen must be presented by one of the body's own cells. A third T-cell subset, **regulatory T cells** (formerly T suppressor cells), has recently been described and is currently the focus of a great deal of research. Regulatory T cells are similar to the other T cells in that they have a T-cell receptor, but their role is entirely different. Instead of fostering a response, they appear to be critical in preventing the immune system from mounting a response against "self" molecules, thereby preventing autoimmune diseases. Because of their involvement in preventing autoimmunity, they will be discussed further in chapter 18. The description of T cells here will focus exclusively on the cytotoxic and helper subsets.

Like B cells, a T cell must receive confirmation from another cell that the antigen it recognizes signifies danger before it can be triggered to multiply. The cell type responsible for providing the "second opinion" to T cells is the dendritic cell, a component of innate immunity. Once activated, the proliferating T cells differentiate to form effector T cells, which are armed to perform distinct protective roles. For simplicity and clarity, we will refer to effector helper T cells as **T_H cells,** and effector cytotoxic cells as **T_C cells.** Like B cells, both types of T cells are able to form memory cells following activation. These quickly respond if the same antigen is encountered later in life.

In response to intracellular agents such as viruses, T_C cells induce the cells harboring the intruder to undergo apoptosis. While this response obviously harms one's own cells, the sacrifice of infected "self" cells ultimately protects the body. For example, destroying virally infected cells prevents those cells from being used by the virus to produce and release more viral particles. Sacrifice of the cells also releases unassembled viral components. This can strengthen the overall immune response by stimulating production of more antibodies that can then block further cellular infection. ■ apoptosis, p. 362

The critical task for the immune system is to distinguish and destroy only those "self" cells that are infected or otherwise tainted; failure to do this can result in an autoimmune disease. As a means to facilitate detection of intracellular "corruption," all nucleated cells of the body regularly display short fragments of proteins present within their cytoplasm in specialized molecules on their cell surface. T_C cells inspect the peptides being presented. If a "self" cell presents an abnormal protein that signifies danger, such as a viral protein, a T_C cell will induce the presenting cell to sacrifice itself.

T_H cells help orchestrate the various responses of humoral and cellular immunity. They provide direction and support to B cells and T cells, and they direct the activation of macrophages.

MICROCHECK 16.1

B cells are programmed to eliminate extracellular antigens. Cytotoxic T cells are programmed to eliminate intracellular antigens. Helper T cells are programmed to orchestrate the adaptive immune response.

- ✓ What are plasma cells?
- ✓ How is the T-cell receptor functionally analogous to the B-cell receptor? How is it different?
- ✓ How would you expect a T_C cell to respond if it encountered a T_H cell that was infected with a virus?

16.2 Anatomy of the Lymphoid System

Focus Point

- Compare and contrast the roles of lymphatic vessels, secondary lymphoid organs, and primary lymphoid organs.

The **lymphoid system** is a collection of tissues and organs strategically designed and located to bring the population of B cells and T cells into contact with any and all antigens that enter the body **(figure 16.2).** This is important because lymphocytes are highly specific, recognizing only one or a few different antigens. In order for the body to mount an effective response, the appropriate lymphocyte must actually encounter the given antigen. ■ tissues, p. 71 ■ organs, p. 71

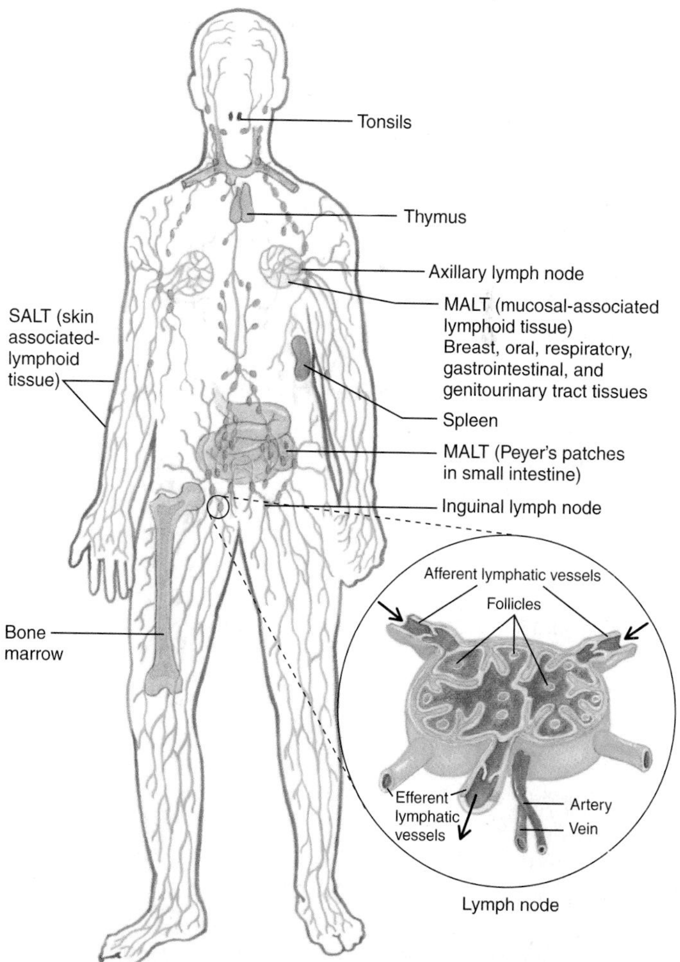

FIGURE 16.2 Anatomy of the Lymphoid System Lymph is distributed through a system of lymphatic vessels, passing through many lymph nodes and lymphoid tissues. For example, lymph enters a lymph node (inset) through the afferent lymphatic vessels, percolates through and around the follicles in the node, and leaves through the efferent lymphatic vessels. The lymphoid follicles are the site of cellular interactions and extensive immunologic activity.

Lymphatic Vessels

Flow within the lymphoid system occurs via the **lymphatic vessels,** or **lymphatics.** These vessels carry a fluid called **lymph.** This fluid is formed as a result of the body's circulatory system (see figure 28.1). As oxygenated blood travels from the heart and lungs through the capillaries, much of the fluid portion filters into the surrounding tissues, supplying them with the oxygen and nutrients carried by the blood **(figure 16.3).** The majority of the tissue fluid then reenters the capillaries as they return to the heart and lungs, but some enters the lymphatic vessels instead. The lymph, which also contains white blood cells and antigens that have entered the tissues, travels via the lymphatics to the lymph nodes, where materials including protein and cells are removed. The lymph then empties back into the blood circulatory system at a large vein behind the left collarbone. Note that the inflammatory response results in greater accumulation of fluid in the tissues at the site of inflammation; this causes a corresponding increase in the antigen-containing fluids that enter lymphatic vessels.

Secondary Lymphoid Organs

Secondary lymphoid organs are the sites where lymphocytes gather to contact the various antigens that have entered the body. Examples include the lymph nodes, spleen, tonsils, adenoids, and appendix. They are situated at strategic positions in the body so that immune responses can be initiated at almost any location. For example, lymph nodes capture materials from the lymphatics, and the spleen collects materials from blood.

The secondary lymphoid organs are like busy, highly organized lymphoid coffee shops where many cellular meetings take place and information is exchanged. The anatomy of the organs provides a structured center to facilitate the interactions and transfer of cytokines between various cells of the immune system. No other places in the body have the structure necessary to facilitate the complex interactions required, which is why these organs are the only sites where productive adaptive immune responses can be initiated.

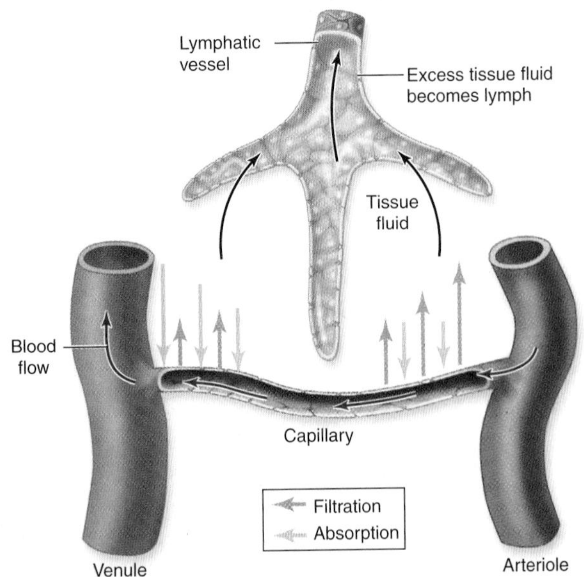

FIGURE 16.3 Formation of Lymph

When lymphocytes make contact with a given antigen and receive the required "second opinion," they respond by proliferating to form clones of cells specific for that antigen. The metabolically active and dividing lymphocytes have larger nuclei and more abundant cytoplasm than their resting counterparts.

Some secondary lymphoid organs are less organized in structure than the lymph nodes and spleen, but their purpose remains the same—to capture antigens, bringing them into contact with lymphocytes. Among the most important of these organs are the **Peyer's patches,** which sample antigens collected from the small intestine, allowing presentation of the antigen to lymphocytes below the mucosal surface (see figure 17.6). **M cells,** specialized epithelial cells lying over the Peyer's patches, collect material in the intestine and transfer it to the lymphoid tissues beneath. Peyer's patches are part of a network of lymphoid tissues called **mucosal-associated lymphoid tissue (MALT).** These play a critical role in **mucosal immunity,** the element of adaptive immunity that prevents microbes from invading the body via the mucous membranes. Lymphoid tissues under the skin are called **skin-associated lymphoid tissue (SALT).**

Primary Lymphoid Organs

The bone marrow and thymus are the **primary lymphoid organs.** This is where the hematopoietic stem cells destined to become B cells and T cells mature (see figure 15.4). Both B cells and T cells originate in the bone marrow but only B cells mature there; immature T cells migrate to the thymus. Once mature, the lymphocytes gather in the secondary lymphoid organs just described, waiting to encounter antigen. ■ hematopoietic stem cells, p. 350

MICROCHECK 16.2

The lymphatic vessels carry the fluid collected from tissues to the lymph nodes; these and other secondary lymphoid organs are where lymphocytes gather to encounter antigens that have entered the body. The primary lymphoid organs, the bone marrow and thymus, are where hematopoietic stem cells destined to become B cells and T cells mature.

✓ How is lymph formed?
✓ What are Peyer's patches?

16.3 The Nature of Antigens

Focus Point

- Distinguish between an antigen and an epitope.

The term **antigen** was initially coined in reference to compounds that elicit the production of antibodies; it is derived from the descriptive expression **anti**body **gen**erator. The compounds observed to induce the antibody response are recognized as being foreign to the host by the adaptive immune system. They include an enormous variety of materials, from invading microbes and their various products to plant pollens. Today, the term *antigen* is used more broadly to describe any molecule that reacts specifically with an antibody or an antigen receptor on a lymphocyte; it does not necessarily imply that the molecule can induce an immune response. When referring specifically to an antigen that elicits an immune response in a given situation, the more restrictive term **immunogen** may be used. The distinction between the terms *antigen* and *immunogen* helps clarify discussions in which a normal protein from host "A" elicits an immune response when transplanted into host "B"; the protein is an antigen because it can react with an antibody or lymphocyte, but it is an immunogen only for host "B," not for host "A."

Various antigens differ in their effectiveness in stimulating an immune response. Proteins and polysaccharides, for example, generally induce a strong response, whereas lipids and nucleic acids often do not. The terms **antigenic** and **immunogenic** are used interchangeably to describe the relative ability of an antigen to elicit an immune response. Substances with a molecular weight of less than 10,000 daltons are generally not immunogenic.

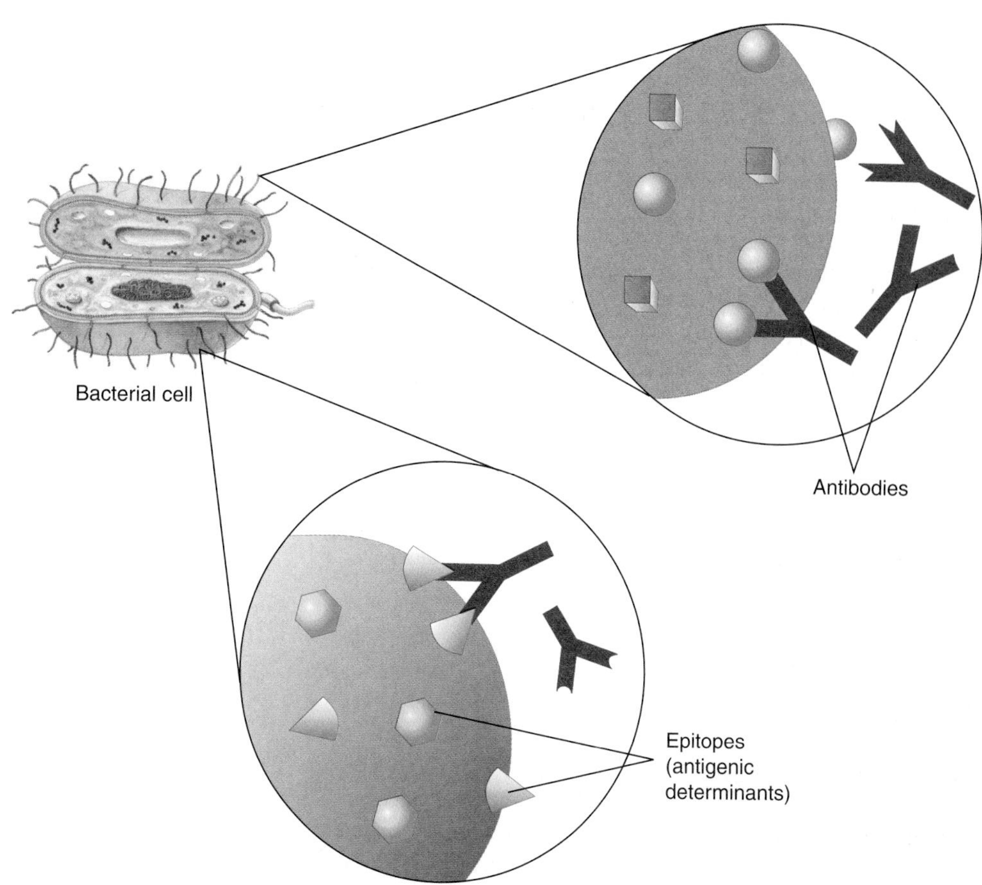

FIGURE 16.4 Antibodies Binding to Epitopes on a Bacterial Cell

Although antigens are generally large molecules, the adaptive immune response directs its recognition to discrete regions of the molecule known as **antigenic determinants** or **epitopes (figure 16.4).** Some epitopes are stretches of 10 or so amino acids, whereas others are three-dimensional shapes such as a protrusion in a globular molecule. A bacterial cell usually has many macromolecules on its surface, each with a number of different epitopes, so that the entire cell has a multitude of various epitopes.

MICROCHECK 16.3

The immune response is directed to epitopes on antigens.

✓ Which macromolecules are most antigenic?

✓ Would a denatured antigen be expected to have the same epitopes as its native (undenatured) counterpart?

16.4 The Nature of Antibodies

Focus Points

- Diagram an antibody, labeling the Fab regions, Fc region, heavy chain, light chain, constant region, variable region, and antigen-binding site.
- Describe six protective outcomes of antibody-antigen binding.
- Compare and contrast the five classes of immunoglobulins.

Antibodies, also called **immunoglobulins,** are Y-shaped proteins that have two functional parts—the arms and the stem **(figure 16.5a).** The two identical arms, called the **Fab regions,** bind antigen. The stem is the **Fc region,** functioning as a "red flag" that enlists other components of the immune system. These names were assigned following early studies that showed that enzymatic digestion of antibodies yielded two types of fragments—fragments that were antigen-binding (Fab) and fragments that could be crystallized (Fc).

Structure and Properties of Antibodies

All antibodies have the same basic Y-shaped structure, called an antibody monomer. It consists of two high-molecular-weight polypeptide chains, called the **heavy chains,** and two lower-molecular-weight polypeptide chains, called the **light chains** (see figure 16.5b). Disulfide bonds (S—S) join the two chains, creating a Y-shaped molecule with identical halves. Disulfide bonds also join the two halves of the molecule. At the fork of the Y is a flexible, or "hinge," region.

There are five major classes of human immunglobulin (Ig) molecules—IgM, IgG, IgA, IgD, and IgE. Each class shares the same basic monomeric structure, but is distinguished by a characteristic amino acid sequence in the constant portion of the heavy chain. Since this is the part of the molecule that interacts with other "players" of the immune system, the various classes differ in their functional properties. The specialized attributes of each class will be described later, after we consider some of the general characteristics of antibodies.

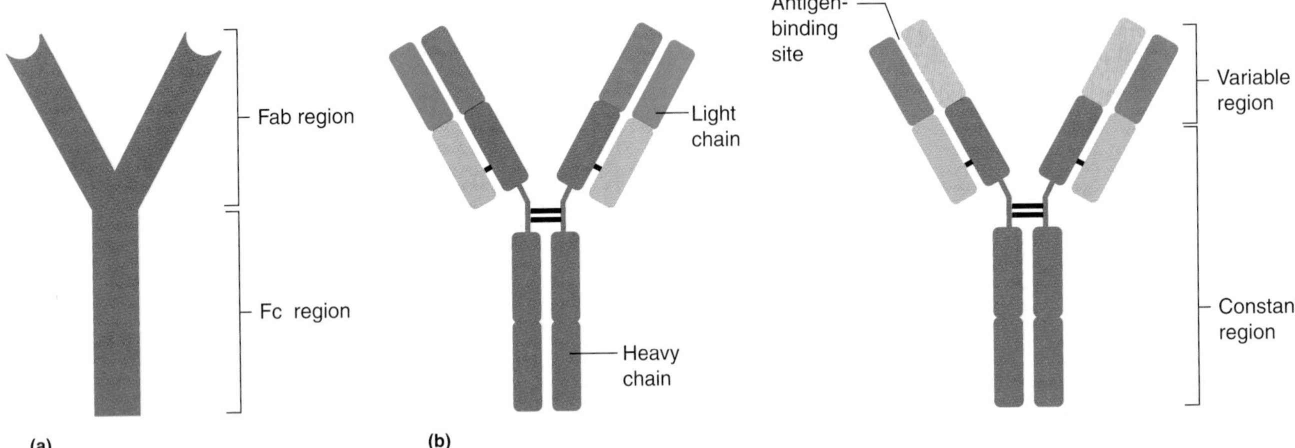

FIGURE 16.5 Basic Structure of an Antibody Molecule **(a)** The Y-shaped molecule; the arms of the Y make up the Fab regions, and the stem is the Fc region. **(b)** The molecule is made up of two identical heavy chains and two identical light chains. Disulfide bonds join the two chains as well as the two halves of the molecule. The constant region is made up of the regions depicted in shades of red. The variable regions differ among antibody molecules, and account for the antigen-binding specificity of antibody molecules.

Variable Region

When the amino acid sequence of antibody molecules that bind to different epitopes are compared, tremendous variation is seen in the parts that form the ends of the Fab regions. These parts make up the **variable regions** of the antibody molecule; they contain the antigen-binding sites (see figure 16.5b). The differences in the variable regions account for the specificity of antibody molecules.

Each antibody binds via the antigen-binding site to the antigen that induced its production. The interaction depends on close complementarities between the antigen-binding site and the specific antigenic epitope. The fit must be precise, because the bonds that hold the antibody and antigen together are non-covalent and are therefore weak. However, many such bonds are formed, usually keeping the two together very effectively. Nevertheless, the antigen-antibody interaction is reversible. Upon reversal, both antigen and antibody are unchanged.

Constant Region

The **constant region** encompasses the entire Fc region, as well as part of both the heavy and light chains in the two Fab regions (see figure 16.5b). The amino acid sequence of this region is the same for all antibody molecules of a given class and it imparts the distinct functional properties of the class. The consistent nature of this amino acid sequence allows other components of the immune system to recognize the otherwise diverse antibody molecules.

Protective Outcomes of Antibody-Antigen Binding

The protective outcomes of antibody-antigen binding depend partly on the class of the antibody, and may include these mechanisms **(figure 16.6)**:

- **Neutralization.** In order to damage a host cell, toxins and viruses typically must bind specific molecules on the cell surface. A toxin or virus coated with antibodies is prevented from interacting with a cell, and therefore can no longer cause damage.
- **Immobilization and prevention of adherence.** Binding of antibodies to surface structures such as flagella and pili on a

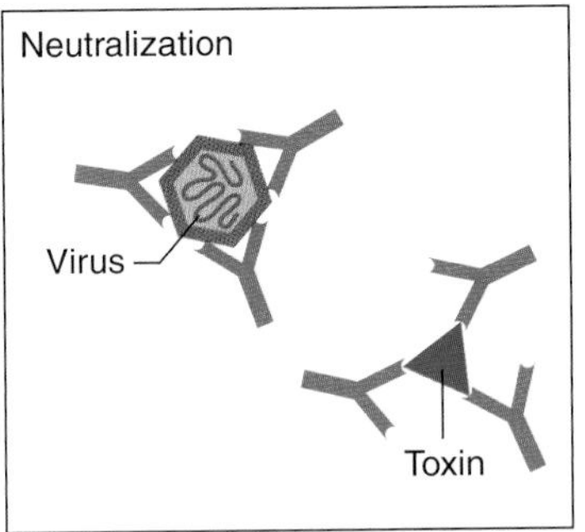

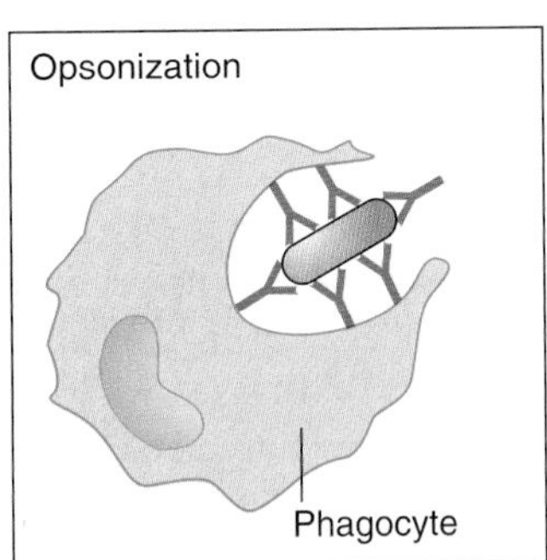

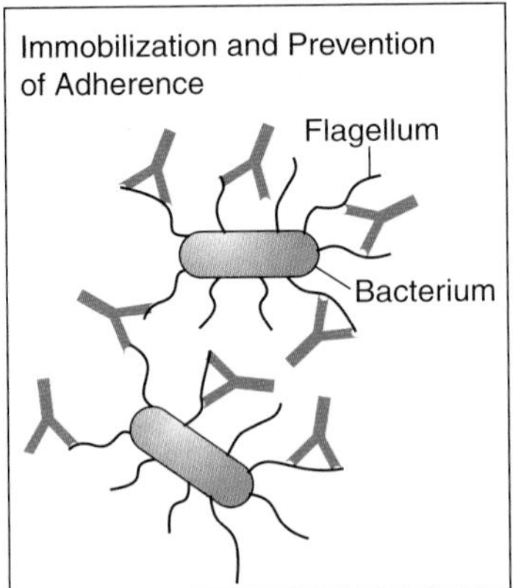

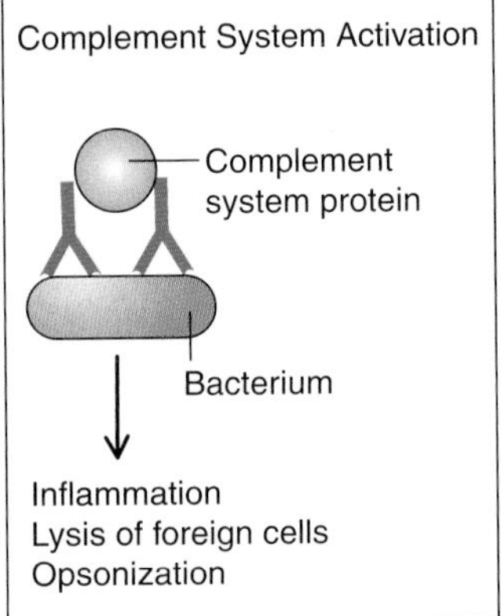

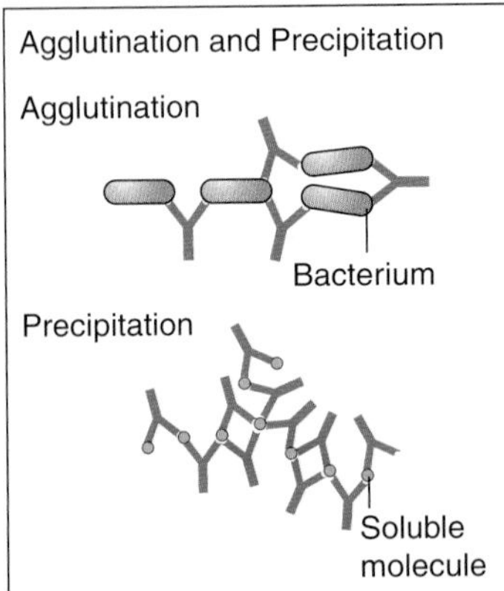

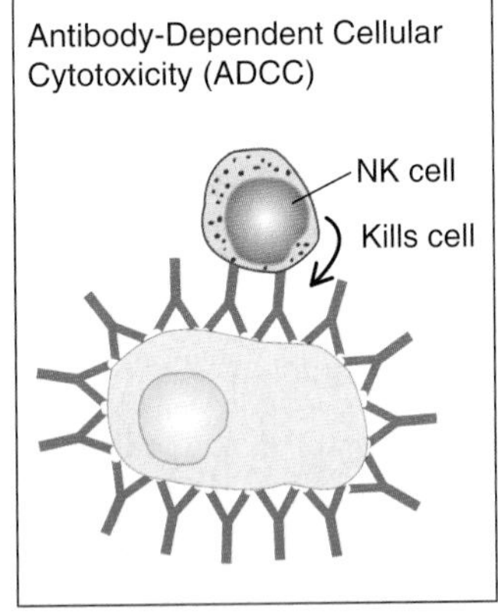

FIGURE 16.6 Protective Outcomes of Antibody-Antigen Binding

bacterium can interfere with such functions as motility and attachment. If these abilities are necessary for the microbe's interaction with the host, binding of antibodies to these structures can protect the host.

- **Agglutination and precipitation.** Binding of antibodies to multiple molecules of antigen causes large antibody-antigen complexes to form, effectively rounding up dispersed antigens to create one large "mouthful" for a phagocytic cell. The complexes form because single antibody molecules can bind adjacent antigens, interconnecting individual antigens to form a network. In the laboratory, the aggregation of antigen and antibody molecules can be seen as either agglutination of particulate antigens or precipitation of soluble antigens. ■ agglutination reactions, p. 443 ■ precipitation reactions, p. 441
- **Opsonization.** Recall from chapter 15 that antigens bound by the complement protein C3b are more easily engulfed by phagocytic cells; the C3b-coated antigens are said to be opsonized. IgG molecules have an analogous effect when they bind to antigen. Macrophages and neutrophils both have receptors for the Fc region of these antibodies on their surface, facilitating the attachment of the phagocytic cell to the antibody-coated antigen as a prelude to engulfment. ■ opsonization by C3b, p. 356
- **Complement system activation.** The binding of antibody to antigen can trigger the classical pathway of the complement cascade. When multiple antibodies of certain antibody classes are bound to a cell surface, a specific complement protein attaches to their Fc regions, initiating the cascade. Recall that activation of the complement system results in formation of membrane attack complexes, stimulation of the inflammatory response, and production of the opsonin C3b. ■ complement, p. 355
- **Antibody-dependent cellular cytotoxicity (ADCC).** When multiple IgG molecules bind to a cell, that cell becomes a target for destruction by certain cells. For example, **natural killer (NK) cells** can attach to the Fc regions of those antibodies and once attached, release compounds directly to the target cell, killing that cell. The mechanism by which NK cells destroy target cells will be described later. ■ natural killer cells, pp. 352, 384

Immunoglobulin Classes

All five major classes of immunoglobulin molecules have the same basic structure: two identical light chains connected by disulfide bonds to two identical heavy chains. Each class, however, has a different constant portion of the heavy chain, characterized by distinct amino acid sequences. Some of the immunoglobulins form multimers of the basic monomeric structure. Characteristics of the various classes of human immunoglobulins are summarized in **table 16.1.**

TABLE 16.1 Characteristics of the Various Classes of Human Immunoglobins

Class and Molecular Weight (daltons)	Structure	Percent of Total Serum Immunoglobulin (Half-life in serum)	Properties and Functions
IgM 970,000		5–13% (10 days)	First antibody class produced during the primary immune response. Principle class produced in response to T-independent antigens. Provides direct protection by neutralizing viruses and toxins, immobilizing motile organisms, preventing the adherence of microbes to cell surfaces, and agglutinating/precipitating antigens. Binding of IgM to antigen leads to activation of the complement system (classical pathway).
IgG 146,000		80–85% (21 days)	Most abundant class in the blood and tissue fluids. Provides longest term protection because of its long half-life. Transported across the placenta, providing protection to a developing fetus; long half-life extends the protection through the first several months after birth. Provides direct protection by neutralizing viruses and toxins, immobilizing motile organisms, preventing the adherence of microbes to cell surfaces, and agglutinating/precipitating antigens. Binding of IgG to antigen facilitates phagocytosis, leads to activation of the complement system (classical pathway) and elicits antibody-dependent cellular cytotoxicity.
IgA monomer 160,000; secretory IgA 390,000		10–13% (6 days)	Most abundant class produced, but most of it is secreted into mucus, tears, and saliva, providing mucosal immunity. Also found in breast milk, protecting the intestinal tract of breast-fed infants. Protects mucous membranes by neutralizing viruses and toxins, immobilizing motile organisms and preventing attachment of microbes to cell surfaces.
IgD 184,000		< 1% (3 days)	Involved in the development and maturation of the antibody response. Its functions in blood have not been clearly described.
IgE 188,000		< 0.01% (2 days)	Binds via the Fc region to mast cells and basophils. This bound IgE allows those cells to detect parasites and other antigens and respond by releasing their granule contents. Involved in many allergic reactions.

Immunoglobulin M (IgM)

IgM accounts for 5% to 13% of the circulating antibodies and is the first class produced during the primary response to an antigen. It is the principle class produced in response to T-independent antigens, a group of antigens that will be discussed later.

IgM is a pentamer. Its large size normally prevents it from crossing from the bloodstream into tissues, so its role is primarily to control bloodstream infections. The five monomeric subunits give IgM a total of 10 antigen-binding sites, making it very effective in agglutination and precipitation. It is the most efficient class in initiating the classical pathway of the complement cascade.

As a fetus is normally sterile until the birth membrane is ruptured, IgM generally begins being made about the time of birth. However, a fetus that is infected *in utero* is capable of making IgM antibodies.

Immunoglobulin G (IgG)

IgG accounts for about 80% to 85% of the total serum immunoglobulin. It circulates in the blood, but readily exits the vessels into tissues with the assistance of receptors on endothelial cells that recognize its Fc region. IgG provides the longest-term protection of any antibody class; its half-life is 21 days, meaning that a given number of IgG molecules will be reduced by approximately 50% after 21 days. In addition, IgG is generally the first and most abundant circulating class produced during the secondary response. The basis for this phenomenon will be discussed later in the chapter. IgG antibodies provide protection by neutralization, agglutination and precipitation, opsonization, complement activation, and antibody-dependent cellular cytotoxicity. ■ serum, p. 439

An important distinguishing characteristic of IgG is that, unlike other immunoglobulin classes, it can cross the placenta, thereby protecting a developing fetus. Its Fc region is recognized by receptors in the placenta, permitting transport across to the fetus. Since IgG production is not optimal until the secondary response, women who lack immunity to certain disease-causing agents that can infect and damage the fetus are warned to take extra precautions during pregnancy. For example, pregnant women are advised not to eat raw meat or become first-time cat owners; this is to avoid a primary infection by *Toxoplasma gondii,* a parasite that can be transmitted in raw meat and the feces of infected cats.

Maternal IgG not only protects the developing fetus against infections, but also the newborn because of the relatively long half-life of this class **(figure 16.7).** The protection provided by these maternal antibodies wanes after about 3 to 6 months, but by this time the infant has begun to produce its own protective antibodies.

IgG is also present in **colostrum,** the first breast milk produced after birth. The intestinal tract of newborns is able to absorb this antibody.

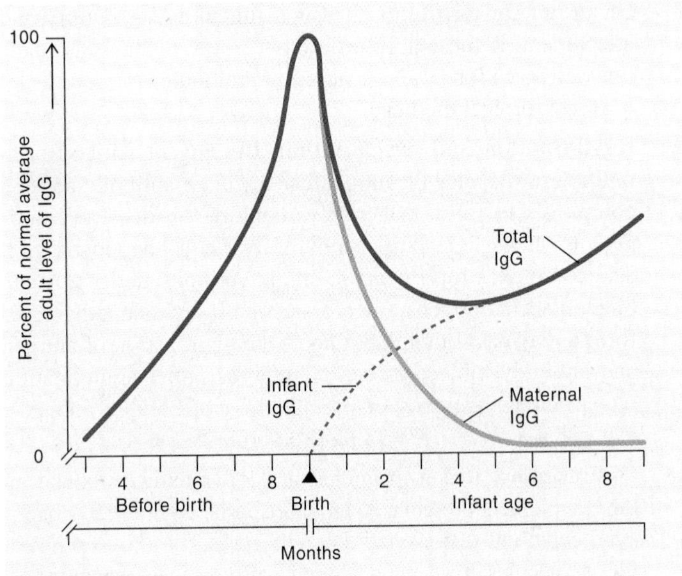

FIGURE 16.7 Immunoglobulin G Levels in the Fetus and Infant During gestation, maternal IgG is transported across the placenta to the fetus. Colostrum also contains IgG. Normally, the fetus does not make appreciable amounts of immunoglobulin; it relies on the maternal antibodies that are transferred passively. After birth, the infant begins to produce immunoglobulins. The maternal antibodies gradually disappear over a period of about 6 months. Usually by 3 to 6 months, most of the antibodies present are those produced by the infant.

IgA

The monomeric form of IgA accounts for about 10% to 13% of antibodies in the serum. Most IgA, however, is the secreted form, a dimer called **secretory IgA** (sIgA). In fact, IgA is the most abundant immunoglobulin class produced, even though it makes up only a small fraction of the antibodies in blood. The secreted form is important in mucosal immunity and is found on the mucous membranes that line the gastrointestinal, genitourinary, and respiratory tracts and in secretions such as saliva, tears, and breast milk. Secretory IgA in breast milk protects breast-fed infants against intestinal pathogens. ■ mucosal immunity, p. 370

Protection by secretory IgA is primarily due to the direct effect that binding of antibody has on antigens. These include neutralization of toxins and viruses and interference with the attachment of microorganisms to host cells.

IgA is produced by the plasma cells that reside in the mucosal-associated lymphoid tissues (MALT). Recall that plasma cells are the antibody-secreting form of B cells. As IgA is transported to the mucous membrane, or **mucosa,** a polypeptide called the secretory component is added; this component may help protect the antibody from being destroyed by most proteolytic enzymes that might be encountered in the mucosa. ■ MALT, p. 370

IgD

IgD accounts for less than 1% of all serum immunoglobulins. It is involved with the development and maturation of the antibody response, but its functions in blood have not yet been clearly defined.

IgE

IgE is barely detectable in normal blood, because most is tightly bound via the Fc region to basophils and mast cells, rather than being free in the circulation. The bound IgE molecules allow these cells to detect and respond to antigens. For example, when antigen binds to two adjacent IgE molecules carried by a mast cell, the cell releases a mixture of potent chemicals including histamine, cytokines, and various compounds that contribute to the inflammatory response. Evidence suggests that these responses are

important in the elimination of parasites, particularly helminths.
■ helminths, p. 295

Unfortunately for allergy sufferers, basophils and mast cells also release their chemicals when IgE binds to normally harmless materials such as foods, dusts, and pollens, leading to immediate reactions such as coughing, sneezing, and swelling. In some cases these allergic, or hypersensitivity, reactions can be life-threatening.
■ hypersensitivity reactions, p. 414

MICROCHECK 16.4

The results of antibody-antigen binding include neutralization, immobilization and prevention of adherence, agglutination and precipitation, opsonization, complement activation, and antibody-dependent cytotoxicity. Immunoglobulin classes include IgM, IgG, IgA, IgD, and IgE.

- ✓ Why is IgM particularly effective at agglutinating antigens?
- ✓ Which two maternal antibody classes protect a newborn that is breast-fed?
- ✓ In opsonization with IgG, why would it be important that IgG react with the antigen *before* a phagocytic cell recognizes the antibody molecule?

16.5 Clonal Selection and Expansion of Lymphocytes

Focus Point

- Outline the process of clonal selection and expansion.

Early on, immunologists recognized that the immune system is capable of making a seemingly infinite array of antibody specificities. A model for how this occurs, proposed in the 1950s, states that each cell in a large population of antibody-producing cells makes only a single specificity of antibody molecule. Then, when antigen is introduced, only the cells capable of making the appropriate antibody can bind to the antigen; this process is called **clonal selection (figure 16.8).** The cells that bind antigen then begin multiplying, thereby generating a population of copies, or **clones,** of the initial cell; this process is called **clonal expansion.**

The model of clonal selection and expansion, now called the **clonal selection theory,** has been shown to be a critical theme in the adaptive immune response, pertaining to both B cells and T cells. As lymphocytes mature in the primary lymphoid organs, a population of cells able to recognize a functionally limitless variety of antigens is generated; each individual cell, however, is able to recognize and respond to only one epitope. Thus, if a person's immune system can make antibodies to billions of different epitopes, that person must have billions of different B cells, each interacting with a single epitope. In fact, the body is estimated to have approximately 10^9 (1 billion) B cells, and only one or a few will recognize a given epitope. Since a pathogen has multiple different epitopes, a number of distinct B cells will recognize it. The process of generating the diversity in antigen recognition is random and does not require previous exposure to antigen; the mechanisms will be described later.

Each lymphocyte residing in the secondary lymphoid organs is waiting for the "antigen of its dreams," an antigen that has an epitope to which that particular lymphocyte is programmed to respond. When an antigen enters a lymphoid organ, only those rare lymphocytes that specifically recognize it may respond; the specificity of the antigen receptor they carry on their surface (B-cell receptor or T-cell receptor) governs this recognition. Lymphocytes that do not recognize the antigen remain inactive. Recall that in most cases, lymphocytes that recognize antigen require accessory signals, a "second opinion" by another cell type, in order to multiply. This provides a mechanism to help the immune system avoid mounting a response against "self" molecules.

Some progeny of the lymphocytes that encountered their "dream antigen" leave the secondary lymphoid organs and migrate to the tissues where they continue responding for as long as the antigen is present. Without sustained stimulation by antigen, these cells will undergo apoptosis, curtailing the immune response.
■ apoptosis, p. 362

The activities of individual lymphocytes change over the lifetime of the cell, particularly as the cell encounters specific antigen. As a means of clarifying discussions of lymphocyte characteristics, descriptive terms are sometimes used:

- **Immature lymphocytes** have not fully developed their antigen specific receptors.
- **Naive lymphocytes** have antigen receptors, but have not yet encountered the antigen to which they are programmed to respond.
- **Activated lymphocytes** are able to proliferate; they have bound antigen by means of their antigen receptor and have received any required accessory signals from another cell, confirming the danger of the antigen.
- **Effector lymphocytes** are descendants of activated lymphocytes that have become armed with the ability to produce specific cytokines or other substances. This endows the cell with specific protective attributes, or **effector functions.** Plasma cells are effector B cells, T_C cells are effector cytotoxic T cells, and T_H cells are effector helper T cells.
- **Memory lymphocytes** are long-lived descendants of activated lymphocytes; they can quickly change back to the activated form when antigen is encountered again. Memory lymphocytes are responsible for the speed and effectiveness of the secondary response.

MICROCHECK 16.5

In response to antigen, only those lymphocytes that recognize the antigen proliferate. This process gives rise to a population of clones of the original cell. Depending on their developmental stage, lymphocytes may be referred to as immature, naive, activated, effector, or memory cells.

- ✓ Describe the clonal selection theory.
- ✓ How does a naive lymphocyte differ from an activated one?
- ✓ If the heavy chain of an antibody is approximately 450 amino acids long, how much DNA would be required to encode 10^9 separate heavy chain genes?

FIGURE 16.8 Clonal Selection and Expansion During the Antibody Response

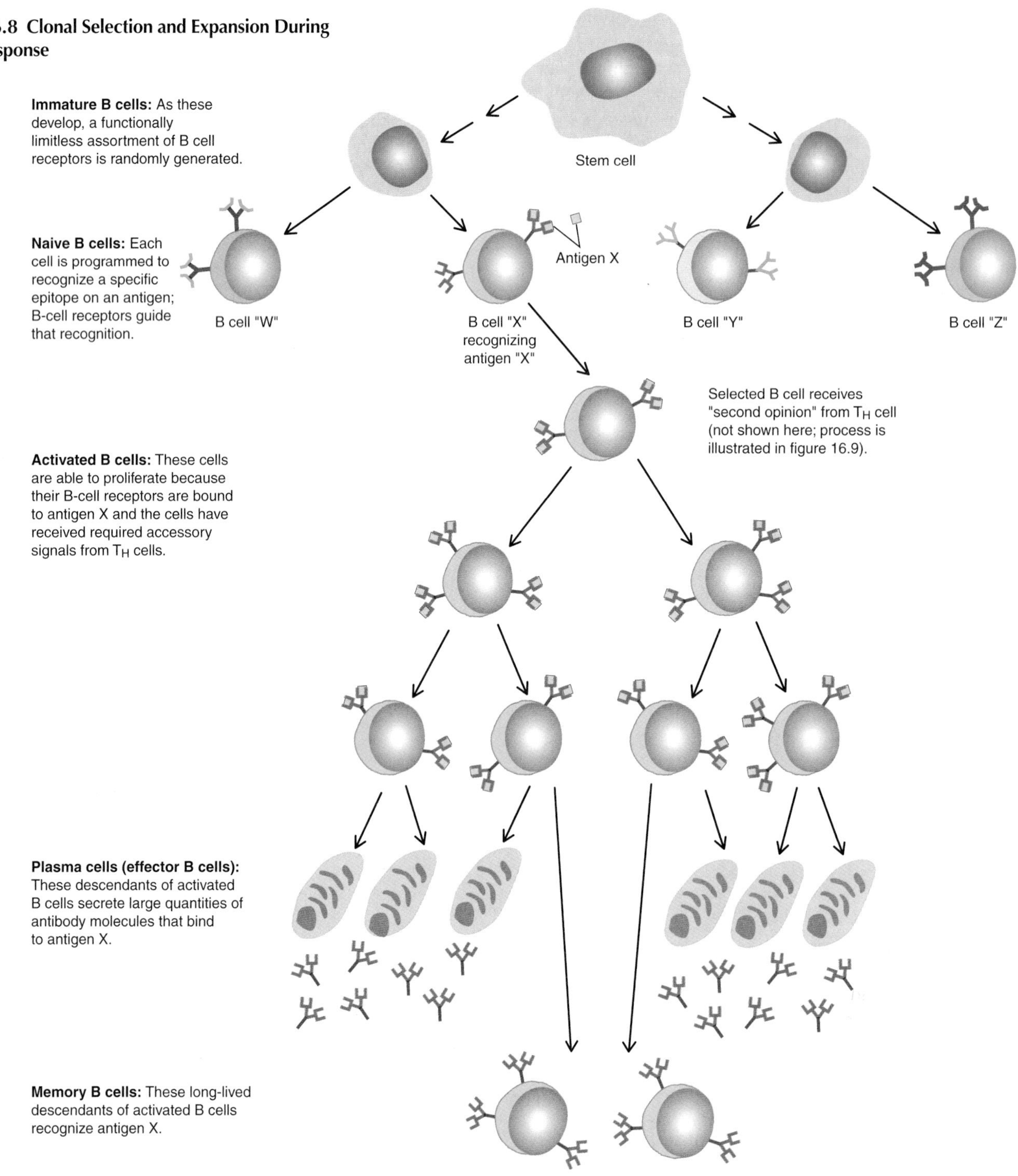

Immature B cells: As these develop, a functionally limitless assortment of B cell receptors is randomly generated.
Stem cell
Naive B cells: Each cell is programmed to recognize a specific epitope on an antigen; B-cell receptors guide that recognition.
Antigen X
B cell "W"
B cell "X" recognizing antigen "X"
B cell "Y"
B cell "Z"
Selected B cell receives "second opinion" from T_H cell (not shown here; process is illustrated in figure 16.9).
Activated B cells: These cells are able to proliferate because their B-cell receptors are bound to antigen X and the cells have received required accessory signals from T_H cells.
Plasma cells (effector B cells): These descendants of activated B cells secrete large quantities of antibody molecules that bind to antigen X.
Memory B cells: These long-lived descendants of activated B cells recognize antigen X.

16.6 B Lymphocytes and the Antibody Response

Focus Points

- Describe the role of T_H cells in B-cell activation.
- Compare and contrast the primary and the secondary responses.
- Compare and contrast the response to T-dependent antigens and T-independent antigens.

When antigen binds to a B-cell receptor, that B cell becomes poised to respond. In most cases, however, the B cell requires confirmation by a T_H cell that a response is truly warranted. Only when this occurs can the B cell become activated to begin dividing, differentiating, and, finally, producing antibodies. Compounds that evoke a response by B cells

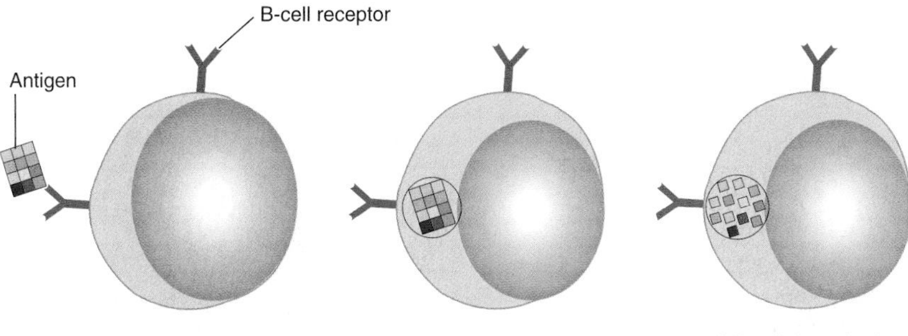
B-cell receptor
Antigen
1) B-cell receptor binds to antigen.
2) B cell internalizes antigen.
3) B cell degrades internalized antigen into peptide fragments.

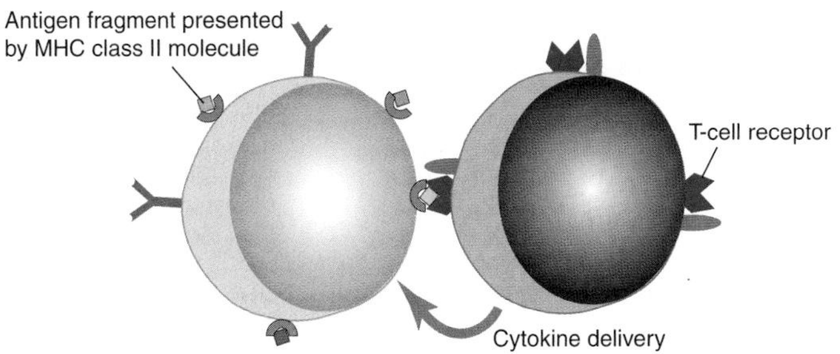
Antigen fragment presented by MHC class II molecule
T-cell receptor
Cytokine delivery
4) B cell presents peptide fragments in the groove of MHC class II molecules on the B-cell surface.
5) If the T-cell receptor of a T_H cell binds to one of the fragments, then cytokines are delivered to the B cell, initiating the process of clonal expansion.

FIGURE 16.9 Antigen Presentation by a B Cell This process enlists the assistance of a T_H cell, which can activate the B cell, allowing it to undergo clonal expansion. The T_H cell also directs affinity maturation, class switching, and the formation of memory cells.

only with the assistance of T_H cells are called **T-dependent antigens;** these antigens are generally proteins and are the primary focus of this section. We will begin by describing the role of T_H cells in B-cell activation. Later in the chapter, we will explain how naive helper T cells become activated to attain their effector functions. Some carbohydrates and lipids can activate B cells without the aid of T_H cells and are called **T-independent antigens;** they will be covered at the end of this section.

B-Cell Activation

When a T-dependent antigen binds to a B-cell receptor, the B cell internalizes the antigen, enclosing it within a membrane-bound vacuole inside the B cell. Within that vacuole the antigen is degraded into peptide fragments that are delivered to proteins called **MHC class II** molecules that then move to the B-cell surface **(figure 16.9).** This process, called **antigen presentation,** "presents" pieces of the antigen for inspection by T_H cells. Recall that T cells have on their surface multiple copies of an antigen-specific receptor called a T-cell receptor, which is functionally analogous to a B-cell receptor. If the receptor of a T_H cell binds to one of the peptide fragments being presented by the B cell, then that T cell activates the B cell. It does this by delivering cytokines to the B cell, initiating the process of clonal expansion of that particular B cell. If the population of T_H cells fails to recognize any of the fragments being presented by the B cell, then that B cell may become unresponsive to future exposure to the antigen. This results in tolerance to that antigen, endowing the adaptive immune system with a mechanism to avoid erroneous responses against "self" and other antigens. ■ tolerance, p. 367

Characteristics of the Primary Response

A lag period of approximately 10 days to 2 weeks occurs before a substantial amount of antibody can be detected in the blood following the first (primary) exposure to an antigen **(figure 16.10).** During this delay, the individual could very well experience symptoms of an infection, which could be life-threatening. However, the immune system is actively responding; naive B cells present antigen to T_H cells, resulting in B-cell activation. The activated B cells multiply, generating a population of cells that

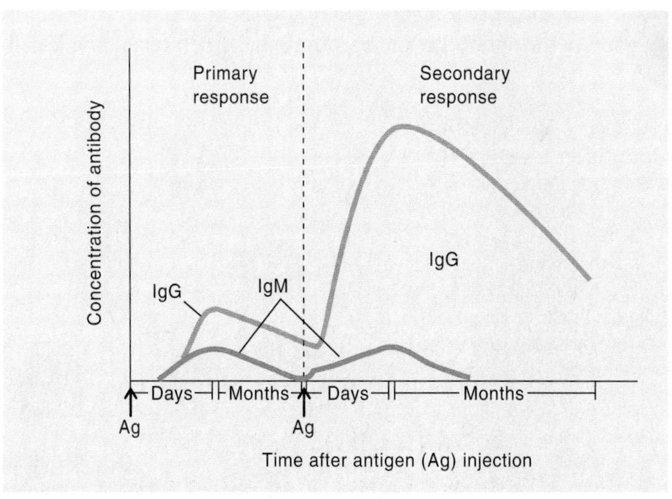
Primary response
Secondary response
Concentration of antibody
IgG
IgM
IgG
Days
Months
Days
Months
Ag
Ag
Time after antigen (Ag) injection

FIGURE 16.10 The Primary and Secondary Responses to Antigen The first exposure to antigen elicits relatively low amounts of first IgM, followed by IgG in the blood. The second exposure, which characterizes the memory of the adaptive immune system, elicits rapid production of relatively large quantities of IgG.

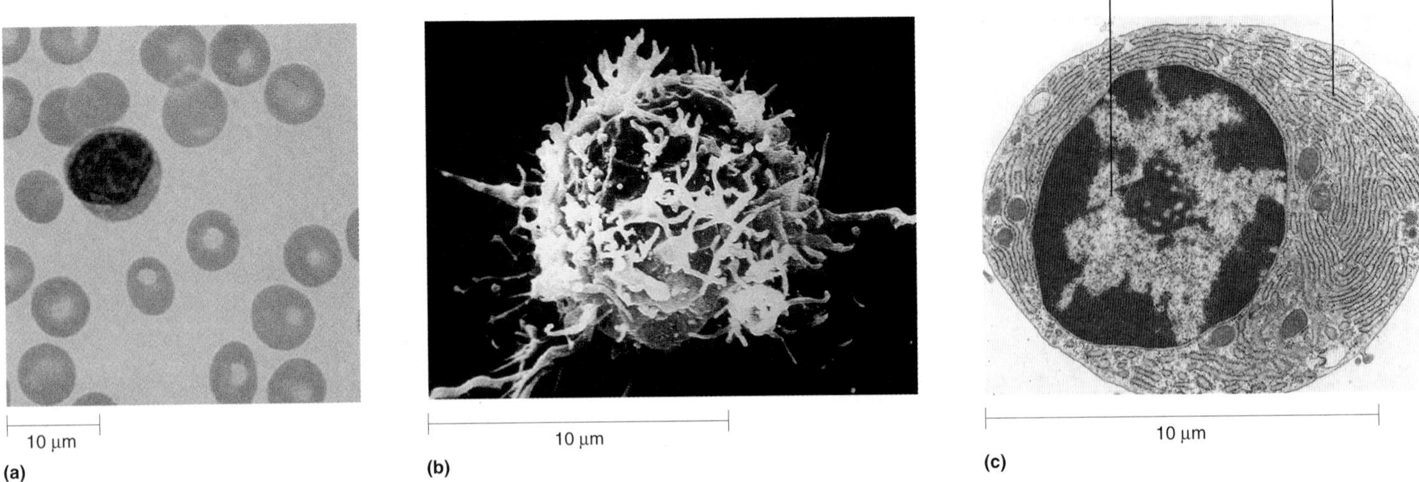

FIGURE 16.11 Lymphocytes and Plasma Cells **(a)** Light micrograph of a T lymphocyte. The morphology is the same as that of a B lymphocyte. **(b)** Scanning electron micrograph of a T lymphocyte. **(c)** Plasma cell, an effector B cell, which produces large amounts of antibody. Note the extensive rough endoplasmic reticulum, the site of protein synthesis. All of the antibody molecules produced by a single plasma cell have the same specificity.

recognize the antigen. As some of the activated B cells continue dividing, others differentiate to form plasma cells, which secrete thousands of antibody molecules per second **(figure 16.11).** Each plasma cell generally undergoes apoptosis after several days, but activated B cells continue proliferating and differentiating, generating increasing numbers of plasma cells as long as antigen is present. The net result is the slow but steady increase in the **titer,** or concentration, of antibody molecules. Over time, some of the proliferating B cells undergo changes, enhancing the immune response. These include:

- **Affinity maturation.** This is a form of natural selection that occurs among proliferating B cells, effectively fine-tuning the quality of the response with respect to antigen binding **(figure 16.12).** An inordinately large number of mutations naturally occur in certain regions of the antibody genes as the activated B cells replicate their DNA in preparation for division. Some of the mutations result in alterations in the antigen-binding site of the antibody (and therefore the B-cell receptor). B cells that bind antigen for the longest duration are most likely to proliferate; others undergo apoptosis.
- **Class switching.** All B cells are initially programmed to differentiate into plasma cells that secrete IgM. Under the direction of cytokines produced by T_H cells, however, some activated B cells switch that genetic program, allowing them to differentiate into plasma cells that secrete another class of antibody. This allows the rare naive B cell that recognized antigen to give rise to an antibody response of the class most effective for a given situation. Circulating B cells most commonly switch to IgG production **(figure 16.13),** whereas B cells that reside in the mucosal-associated lymphoid tissues generally switch to IgA production, providing mucosal immunity.
- **Formation of memory cells.** Some of the B cells that have undergone class switching form memory cells. Memory B cells persist in the body for years and are present in numbers

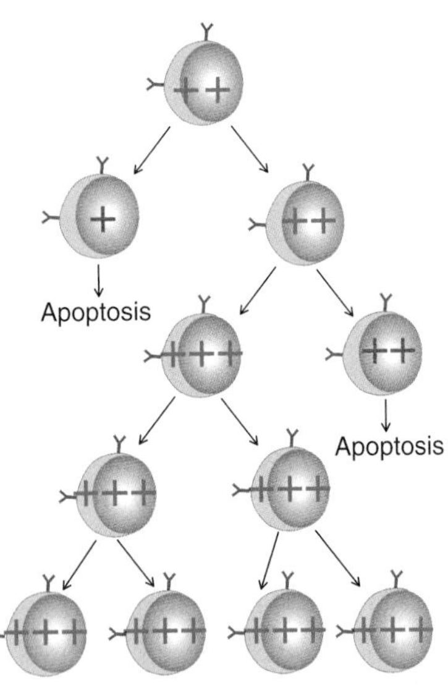

FIGURE 16.12 Affinity Maturation B cells that bind antigen for the longest duration are the most likely to proliferate. The plus signs indicate the relative quality of binding of the antibody to the antigen; those in green indicate the most "fit" to continue proliferating.

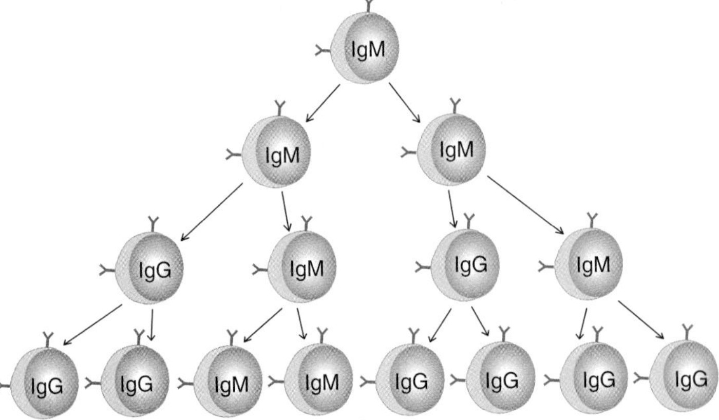

FIGURE 16.13 Class Switching B cells are initially programmed to produce IgM antibodies. With the direction of T_H cells, the activated B cells can switch to express a different class. The plasma cells descended from circulating B cells that have undergone class switching most commonly produce IgG. Plasma cells that descend from B cells residing in the mucosal-associated lymphoid tissues most commonly produce IgA. Note that class switching does not alter the antigen specificity.

sufficient to give a prompt and effective secondary response when the same antigen is encountered again at a later time.

The antibody response begins to wane as the accumulating antibodies clear the antigen. Progressively fewer molecules of antigen remain to stimulate the lymphocytes, and, as a result, the activated lymphocytes undergo apoptosis. Memory B cells, however, are long-lived even in the absence of antigen.

Characteristics of the Secondary Response

Memory B cells are responsible for the swift and effective reaction of the secondary response, eliminating identical repeat invaders before they cause noticeable harm. Thus, once a person has recovered from a particular disease, he or she generally has long-lasting immunity to that disease. Vaccination exploits this naturally occurring phenomenon. ■ vaccination, p. 433

Memory B cells that bind antigen can promptly become activated. Compared to the few naive B cells that initiated the primary response, they are markedly faster and more effective. For one thing, there are more cells able to respond to a specific antigen. In addition, the memory cells are able to scavenge even low concentrations of antigen because their receptors have been fine-tuned through affinity maturation to bind antigen more tightly. Likewise, the antibodies coded for by these cells more effectively bind antigen.

Some of the memory B cells that become activated will quickly differentiate to form plasma cells, resulting in the rapid production of antibodies. Because of class switching, most of the circulating antibodies produced are IgG (see figure 16.10). Other activated memory cells begin proliferating, once again undergoing affinity maturation to further enhance the effectiveness of the antibodies they encode. Subsequent exposures to antigen lead to an even stronger response.

The Response to T-Independent Antigens

T-independent antigens can stimulate an antibody response by activating B cells without the aid of T_H cells. Relatively few antigens are T-independent, but they can be very important medically.

Molecules such as polysaccharides that have numerous identical evenly spaced epitopes characterize one type of T-independent antigen. Because of the arrangement of epitopes on the antigen, clusters of B-cell receptors bind the antigen simultaneously, which leads to B-cell activation without the involvement of helper T cells **(figure 16.14)**. These antigens are particularly significant because the immune systems of young children respond poorly to them. This is why children less than 2 years of age are more susceptible to diseases caused by organisms such as *Streptococcus pneumoniae* and *Haemophilus influenzae,* which cloak themselves in polysaccharide capsules. Antibodies against the capsules would be protective, but children do not effectively make antibodies against them. Vaccines made from purified capsules are available, but, likewise, they do not elicit a protective response in young children. Fortunately, newer vaccines designed to evoke a T-dependent response have been developed. They will be discussed later, when the role of T_H cells in the antibody response is described in more detail.

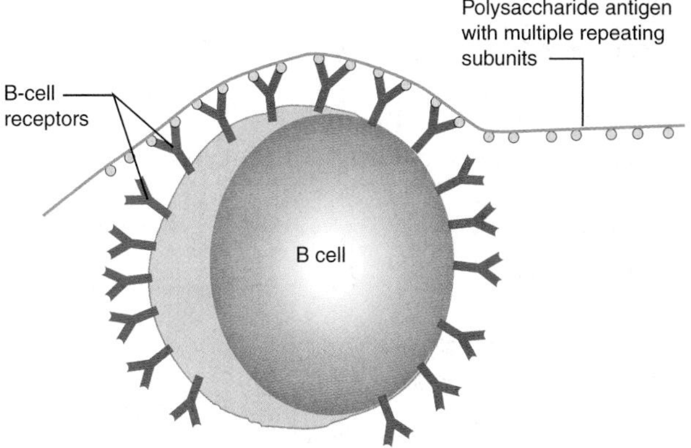

FIGURE 16.14 T-Independent Antigens Antigens such as some polysaccharides have multiple repeating epitopes. Because of the arrangement of epitopes, clusters of B-cell receptors bind to the antigen simultaneously, leading to B-cell activation without the involvement of T_H cells.

Another type of T-independent antigen is lipopolysaccharide (LPS), a component of the outer membrane of Gram-negative bacteria. The constant presence of antibodies against LPS, evoked without the need for T-cell help, is thought to provide an early defense against Gram-negative bacteria that breach the body's barriers.

MICROCHECK 16.6

In most cases, B cells that bind antigen require accessory signals from T_H cells to become activated. Activated B cells proliferate, ultimately producing plasma cells that secrete antibody molecules, and long-lived memory cells. Affinity maturation and class switching occur in the primary response; these enable a swift and more effective secondary response. T-independent antigens can stimulate an antibody response by activating B cells without the aid of T_H cells.

✓ Describe the significance of class switching.

✓ How do B cells increase their ability to bind to antigen?

✓ Why should B cells residing in the mucosal-associated lymphoid tissues produce IgA?

16.7

T Lymphocytes: Antigen Recognition and Response

Focus Points

- Describe the importance of T-cell receptors and CD markers.
- Describe the role of dendritic cells in T-cell activation.
- Compare and contrast T_H and T_C cells with respect to antigen recognition and the response to antigen.

A discussion of T cells encompasses not only their traits, but also the processes that lead to T-cell activation, and the functions of the resulting effector cells. We will begin by describing the general

TABLE 16.2 Characteristics of T Cells

T Cell Type/CD Marker	Effector Form	Effector Function	Potential Target Cell	Antigen Recognition	Source of Antigen
Cytotoxic T cell/CD8	T_C cell	Induces target cell to undergo apoptosis	All nucleated cells	Peptides presented by MHC class I molecules	Endogenous (produced within the target cell)
Helper T cell/CD4	T_H cell	Activates target cell	B cells, macrophages	Peptides presented by MHC class II molecules	Exogenous (produced outside of the target cell)

characteristics of T cells **(table 16.2).** Recognize, however, that the importance of some of these will be more evident as the specific roles of the cells are explained later.

General Characteristics of T Cells

T cells share several important characteristics with B cells. Like B cells, T cells have multiple copies of a receptor on their surface that recognizes a specific antigen. The **T-cell receptor (TCR)** consists of two polypeptide chains (a set of either alpha and beta or gamma and delta), each with a variable and constant region **(figure 16.15).** As in the B-cell receptor, the variable regions make up the antigen-binding sites. The specificity of the T-cell receptor is like that of the B-cell receptor; of the approximately 10^{10} T cells in the body, only a few will recognize a given epitope that appears in the body for the first time.

Despite the similarities in certain characteristics, the role of T cells is very different from that of B cells. For one thing, T cells never produce antibodies. Instead, the naive cells become armed as effectors that directly interact with other cells, **target cells,** to cause distinct changes in those cells. Another important difference is that the T-cell receptor does not interact with free antigen. Instead, the antigen must be "presented" by another host cell. The host cell does this by partly degrading, or **processing,** the antigen and then displaying, or **presenting,** individual peptides of the proteins that make up the antigen. This process is called **antigen presentation.** Dendritic cells present antigen to naive T cells as part of the T-cell activation process. Other cells present antigen to effector T cells, becoming targets if the antigen is recognized by any of those effector cells. ■ dendritic cell, p. 352

During antigen presentation, the peptides from the antigen are cradled in the groove of proteins called **major histocompatibility complex molecules,** or **MHC molecules,** which are on the surface of the presenting cell. There are two types of MHC molecules involved in antigen presentation—MHC class I and MHC class II **(figure 16.16).** Each is shaped somewhat like an elongated bun; it holds the peptide lengthwise, like a bun holds a hot dog. T cells will recognize an antigen only when it is presented within the groove of an MHC molecule; the T cell is actually recognizing both the peptide and MHC molecule simultaneously. In other words, the T-cell receptor recognizes the "whole sandwich"—the peptide:MHC complex. **Endogenous antigens,** those that have been made within the cell, are presented by MHC class I molecules. **Exogenous antigens,** those that have been taken up by a cell, are presented by MHC class II molecules. All nucleated cells produce MHC class I molecules, but only specialized cell types (dendritic cells, B cells, and macrophages), collectively referred to as **antigen-presenting cells,** produce MHC class II molecules.

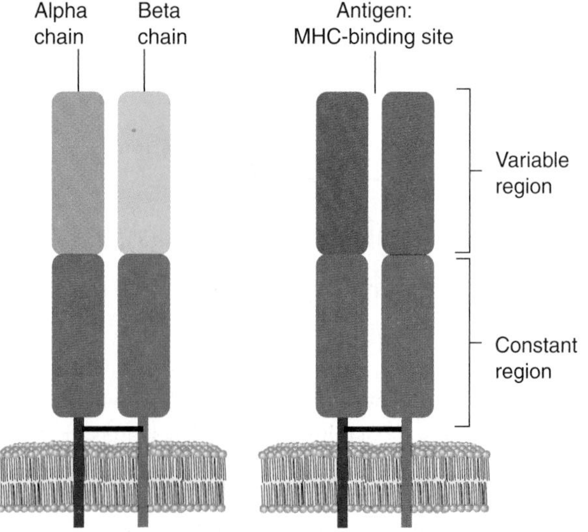

FIGURE 16.15 T-Cell Receptors Each chain has one variable and one constant region. The two chains are connected by a disulfide bond. Unlike antibodies, which have two binding sites for antigen, T-cell receptors have only one.

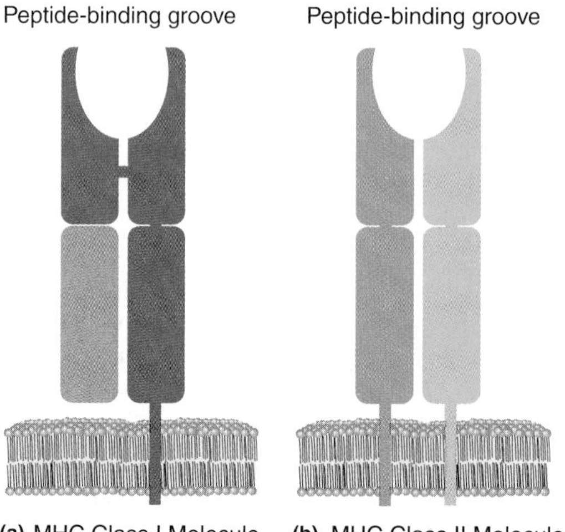

FIGURE 16.16 MHC Molecules (a) MHC class I molecule; cytoplasmic proteins (endogenous antigens) are presented in the groove of these molecules. **(b)** MHC class II molecule; proteins taken in by the cell (exogenous antigens) are presented in the groove of these molecules.

Two distinct major functional populations of T cells involved in eliminating antigen have been characterized: cytotoxic T cells and helper T cells. Upon activation, naive cytotoxic T cells proliferate and differentiate to become T_C cells, which destroy infected or cancerous "self" cells. In contrast, activated helper T cells multiply and develop into T_H cells, which then activate B cells and macrophages, stimulate other T cells, and orchestrate other immune responses. Cytotoxic T cells (including T_C cells) recognize antigen presented by MHC class I molecules; recall that endogenous antigens are presented by these molecules **(figure 16.17).** In contrast, helper T cells (including T_H cells) recognize antigen displayed by MHC class II molecules; these present exogenous antigens.

The most practical way for scientists to distinguish functionally different T cells, which are identical microscopically, is to examine surface proteins called **cluster of differentiation (CD) markers** (or molecules). Most cytotoxic T cells have the CD8 marker and are frequently referred to as CD8 T cells; most helper T cells carry the CD4 marker and are often called CD4 T cells. Note that CD4 is also a receptor for HIV, which explains why the virus infects helper T cells.

There are various subsets of the major cell types we will describe in this section (for example, T_H cells include three subsets—T_H1, T_H2, and T_H17). Initially, we will focus on only the general characteristics of each cell type; later, we will describe the different roles of the various subsets.

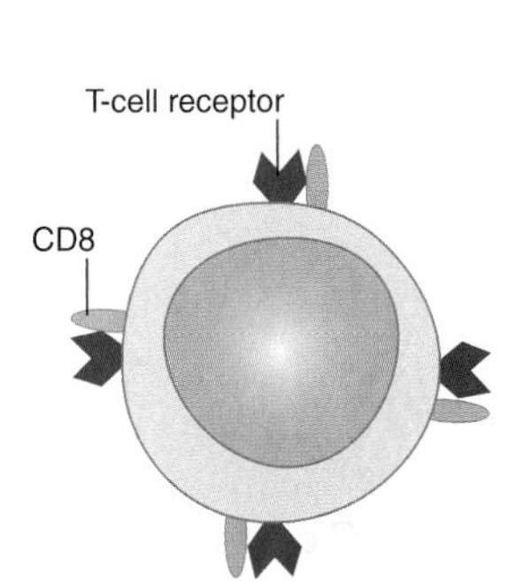

Cytotoxic T cell; recognizes antigen presented by MHC class I molecules

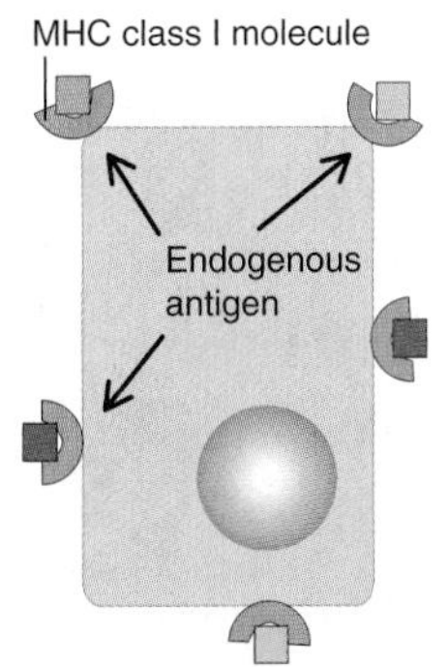

All nucleated cells present endogenous antigen (originated inside of the cell) in the groove of MHC class I molecules.

(a)

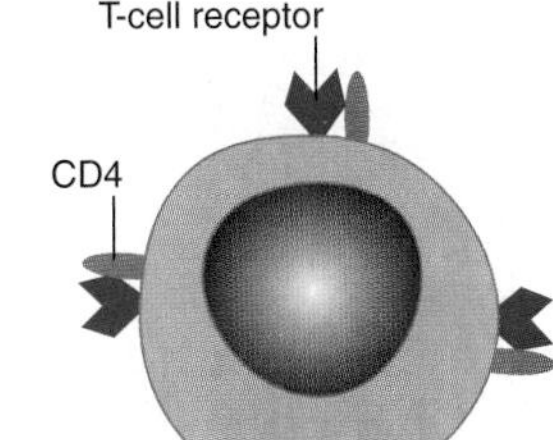

Helper T cell; recognizes antigen presented by MHC class II molecules

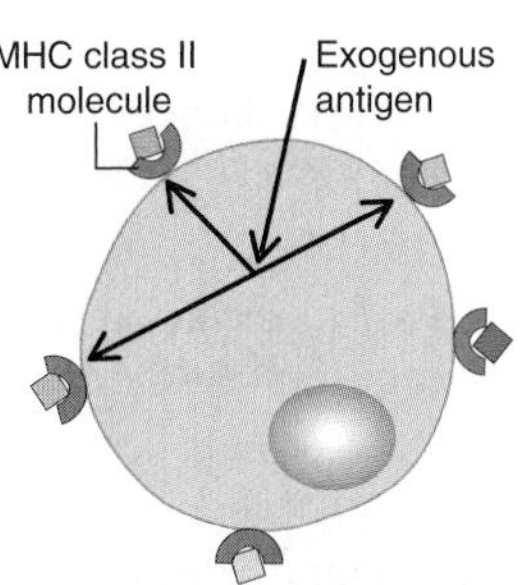

Antigen-presenting cells (dendritic cells, B cells, and macrophages) present exogenous antigen (originated outside of the cell) in the groove of MHC class II molecules.

(b)

FIGURE 16.17 Antigen Recognition by T Cells (a) Cytotoxic T cell. **(b)** Helper T cell.

Activation of T Cells

Dendritic cells, the scouts of innate immunity, play a crucial role in T-cell activation **(figure 16.18).** Immature dendritic cells reside in peripheral tissues, such as beneath the skin, gathering various materials from those areas. The cells use both phagocytosis and pinocytosis to take up particulate and soluble material that could contain foreign protein. Dendritic cells located just below the mucosal barriers are even able to send tentacle-like extensions between the epithelial cells of the barriers. Using this action, the dendritic cells are able to sample material in the respiratory tract and the lumen of the intestine. After collecting substances from the periphery and mucous membranes, the dendritic cells travel to the secondary lymphoid organs where they encounter naive T cells; the inflammatory process can trigger the migration.

En route to the secondary lymphoid organs, the dendritic cells mature into a form able to present antigen to naive T cells as a

FIGURE 16.18 Activation of T Cells

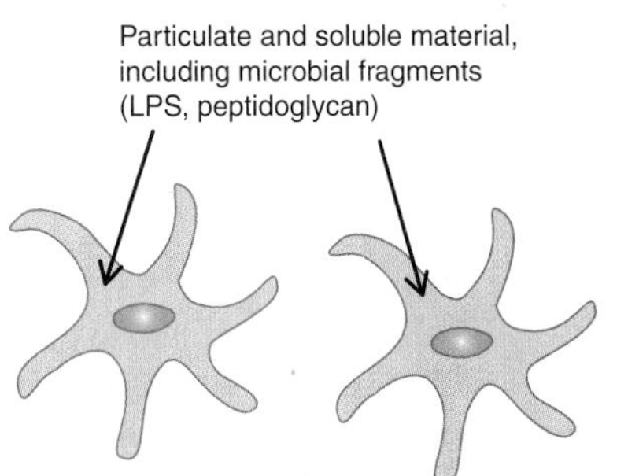

Dendritic cells in the tissues collect particulate and soluble antigen. When a toll-like receptor is engaged, the cell produces co-stimulatory molecules.

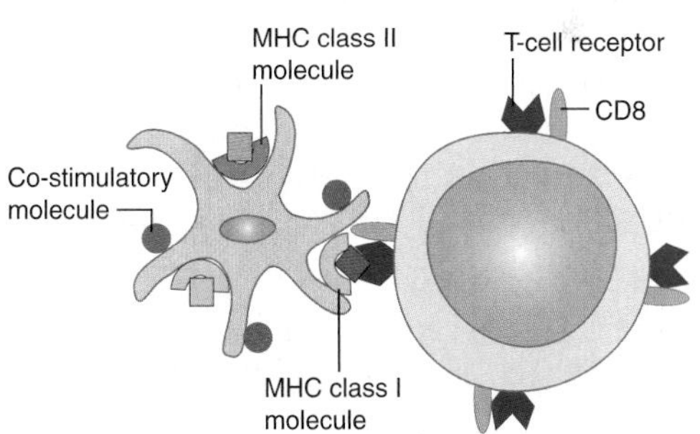

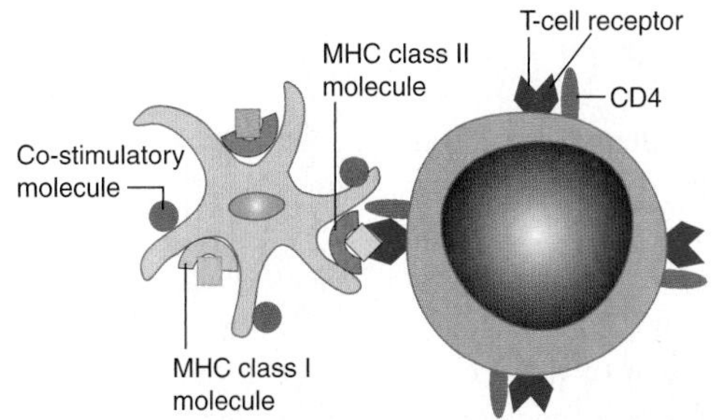

En route to the secondary lymphoid organs, the dendritic cells mature to become antigen-presenting cells. Peptides from the collected material can be presented by both MHC class I and MHC class II molecules. Dendritic cells that have engulfed microbial fragments produce co-stimulatory molecules. Naive T cells that recognize antigen presented by a dendritic cell that expresses co-stimulatory molecules may become activated, allowing them to proliferate and develop their effector functions.

prelude to T-cell activation. A dilemma lies in the fact that the dendritic cell must only activate a given T cell if the material recognized by it represents danger. Antigen presentation alone does not convey the significance of the material being displayed; the fragments could be parts of an invading microbe, which would merit an adaptive immune response, or routine cellular debris, which would not. In fact, a response to cellular debris would harm the host. The sensors of the innate immune response, such as the toll-like receptors, help solve this problem. They enable the dendritic cells to sense the presence of molecules that signify an invading microbe, which, in turn, allows dendritic cells to relay that fact to T cells. When a toll-like receptor on a dendritic cell is engaged, the cell produces surface proteins called **co-stimulatory molecules.** In essence, the co-stimulatory molecules function as "flashing red lights" that interact with the T cell, communicating that the material being presented by the dendritic cell indicates danger. Dendritic cells displaying co-stimulatory molecules while presenting antigen are able to activate T cells. In contrast, T cells that recognize antigen presented by a dendritic cell not displaying co-stimulatory molecules generally undergo apoptosis, but may simply become unresponsive to future encounters with the antigen. Recall that inducing unresponsiveness is one mechanism by which the adaptive immune response eliminates those lymphocytes that recognize "self" proteins. ■ toll-like receptor, p. 355

Dendritic cells present the processed antigen on both types of MHC molecules—class I and class II. This enables them to present antigen to, and therefore activate, cytotoxic T cells as well as helper T cells. T cells that recognize the antigen presented by dendritic cells undergo clonal selection and expansion as described previously, eventually forming effector cells and memory cells that can leave the lymph nodes and react against antigen. Like dendritic cells, macrophages and B cells also present antigen on MHC class II molecules (in other words, they are antigen-presenting cells), and they can produce some co-stimulatory molecules. They do not appear able to contact naive T cells, however; thus they are only able to reactivate memory T cells, which they can encounter in the bloodstream and tissues.

Upon activation, T cells produce the cytokine that stimulates T-cell growth (IL-2) and the receptor for that cytokine. They also begin producing additional cytokines and adhesion molecules that allow them to acquire their effector functions. The effector T cells can leave the secondary lymphoid organs and circulate in the bloodstream. They can also enter tissues, particularly at sites of infection.

Functions of T_C (CD8) Cells

T_C cells induce apoptosis in "self" cells infected with a virus or other intracellular microbe; they also can destroy cancerous "self" cells. How do T_C cells distinguish dysfunctional cells from their normal counterparts? The answer lies in the significance of antigen presentation on MHC class I molecules **(figure 16.19).** All nucleated cells routinely degrade a portion of the proteins they have produced (endogenous proteins) and load peptides from those proteins into the groove of MHC class I molecules to be delivered to the surface of the cell (see figure 16.17a). If a host cell is infected with a virus or a bacterium that resides in the cytoplasm, or if the cell is producing certain abnormal proteins such as those that characterize cancerous cells, then some of the peptides presented on MHC class I molecules will be recognized by circulating T_C cells. This makes the presenting cell a target for the lethal effector functions of T_C cells. In contrast, the peptides presented

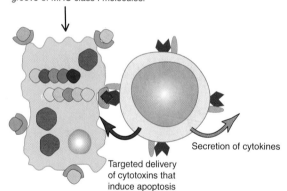

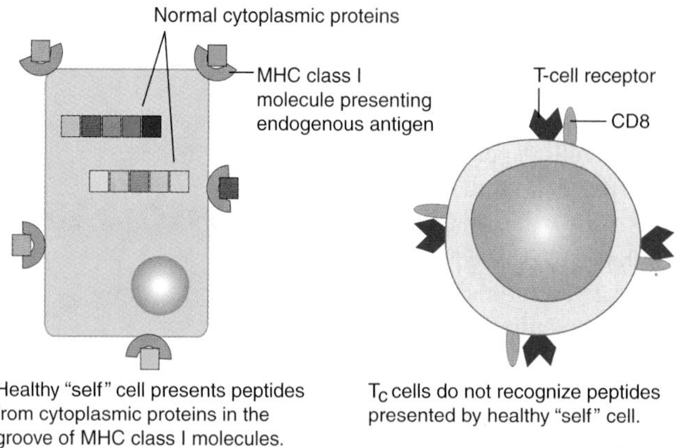
Normal cytoplasmic proteins
MHC class I molecule presenting endogenous antigen
T-cell receptor
CD8
Healthy "self" cell presents peptides from cytoplasmic proteins in the groove of MHC class I molecules.
T_C cells do not recognize peptides presented by healthy "self" cell.
(a)

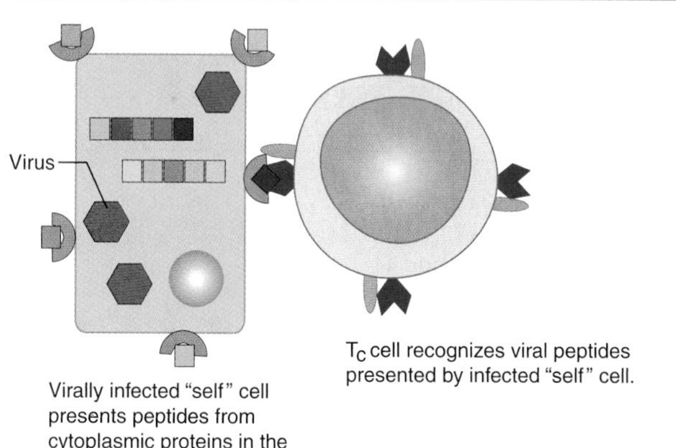
Virus
T_C cell recognizes viral peptides presented by infected "self" cell.
Virally infected "self" cell presents peptides from cytoplasmic proteins in the groove of MHC class I molecules.
Secretion of cytokines
Targeted delivery of cytotoxins that induce apoptosis
Virally infected "self" cell undergoes apoptosis.
T_C cell delivers preformed cytotoxins to the infected "self" cell and produces cytokines that allows neighboring cells to become more vigilant against intracellular pathogens.
(b)

FIGURE 16.19 Functions of T_C Cells (a) T_C cells ignore healthy "self" cells. **(b)** T_C cells induce apoptosis in virally infected "self" cells.

on MHC class I molecules of normal cells will be parts of standard proteins typically found in the cell. There should be no T_C cells that recognize these because of the constraints of the T-cell activation process. In addition, as we will describe later, most "self"-recognizing T cells are eliminated during T-cell development in the thymus.

When a T_C cell encounters a cell displaying a peptide:MHC class I complex it recognizes, it establishes intimate contact with that cell. The T_C cell then releases several pre-formed **cytotoxins**—molecules lethal to cells—directly to the target cell. The cytotoxins include perforin, a molecule that forms pores in cell membranes, and a group of proteases. Evidence indicates that at the concentrations released *in vivo,* perforin simply allows the proteases to enter the target cell.

Once inside that cell, the proteases facilitate reactions that induce the target cell to undergo apoptosis. In addition, a specific molecule on the T_C cell can engage a "death receptor" on the target cell, also initiating apoptosis. The remains of the apoptotic cell are then quickly removed by macrophages; the T_C cell survives and can go on to kill other targets. Killing the target cell by inducing apoptosis rather than lysis minimizes the number of intracellular microbes that might spill into the surrounding area and infect other cells. Most microbes remain in cell remnants until they are ingested by macrophages.

In response to antigen recognition, T_C cells also produce various cytokines that allow neighboring cells to become more vigilant against intracellular invaders. One of the cytokines, for example, increases antigen processing and presentation in nearby cells, facilitating detection of other infected cells. Another cytokine selectively activates local macrophages whose toll-like receptors have been triggered. Note that a more efficient mechanism of macrophage activation involves T_H cells and will be discussed in more detail shortly.

Functions of T_H (CD4) Cells

T_H cells orchestrate the immune response, directing the activities of B cells, macrophages, and T cells. They recognize antigen presented by MHC class II molecules, which are found only on antigen-presenting cells (APCs). These cells, which include B cells and macrophages, gather, process, and present exogenous antigens (see figure 16.17b). When a T_H cell recognizes antigen presented by a B cell or macrophage, it delivers cytokines that activate the presenting cell. Various cytokines are also released; their array depends on the subset of T_H cell.

The Role of T_H Cells in B Cell Activation

When a naive B cell binds antigen via its B-cell receptors, the cell takes the antigen in by endocytosis. Proteins within the endosome are degraded to produce short peptides that can then be loaded into the groove of MHC class II molecules (see figure 16.9). If a T_H cell encounters a B cell bearing the peptide:MHC class II complex it recognizes, it responds by synthesizing cytokines and delivering them to that cell. The cytokines activate the B cell, enabling it to proliferate and undergo class switching. The cytokines also drive the formation of memory B cells. Note that the T-cell receptor could be recognizing any of the various peptides generated from the antigen during antigen processing and presentation. Thus, the epitope to which the T_H cell responds is most likely different from the one that the B-cell receptor recognized. In fact, a B cell that binds to a bacterium is probably recognizing an epitope on the surface of that cell, whereas the T_H cell could very well be responding to a peptide from one of the bacterium's cytoplasmic proteins being presented by the B cell. ■ endocytosis, pp. 73, 74

Understanding the mechanisms used in antigen processing and presentation is what led to an effective vaccine for children against what was the most common cause of meningitis in children, *Haemophilus influenzae.* Recall that young children are particularly susceptible to meningitis caused by this organism because it produces a polysaccharide capsule, an example of a T-independent antigen to which this age group responds poorly. Polysaccharide antigens can be converted to T-dependent antigens by covalently attaching, or conjugating, them to large protein molecules; this is done to make what is called a conjugate vaccine. The polysaccharide component of the vaccine binds to the B-cell receptor and the entire molecule is taken in. The protein component will then be processed and presented to a T_H cell. Although the B cell recognizes the polysaccharide component of the vaccine, the T cell recognizes peptides from the protein component. The T_H cell then activates the B cell, leading to production of antibodies that bind the capsule. ■ conjugate vaccines, p. 435

The requirement for antigen processing and presentation also explains how some people develop allergies to penicillin. This medication is a **hapten,** a molecule that binds a B-cell receptor yet does not elicit the production of antibodies unless it is attached to a protein carrier. In the body, penicillin can react with proteins, forming a penicillin-protein conjugate. This functions in a manner analogous to the *H. influenzae* conjugate vaccine, resulting in antibodies that bind penicillin. The reaction of IgE antibodies with penicillin can result in allergic reactions, precluding the further use of the antimicrobial medication in these reactive individuals. ■ allergy, p. 414, ■ penicillin, pp. 62, 475

The Role of T_H Cells in Macrophage Activation

As discussed in chapter 15, macrophages routinely engulf and degrade invading microbes, rapidly clearing most organisms even before an adaptive response is mounted. Some microbes, however, can evade this method of destruction, enabling them to survive and actually multiply within the phagocytic cell. T_H cells recognize macrophages harboring engulfed microbes resistant to such killing, and then activate those macrophages by delivering cytokines that induce more potent destructive mechanisms.

The steps that lead to macrophage activation are very similar to those described for B-cell activation. When macrophages engulf material, they bring the substance into the cell enclosed within a membrane-bound phagosome **(figure 16.20).** The fate of the proteins within the phagosome is identical to that of proteins within the B cell's endosome, resulting in peptides being loaded into the groove of an MHC class II molecule. If a T_H cell recognizes a peptide presented by a class II molecule on a macrophage, it delivers cytokines directly to that macrophage. This activates the macrophage, leading to several morphological and physiological changes. The macrophage enlarges, the plasma membrane becomes ruffled and irregular, and the cell increases its metabolism so that the lysosomes, each containing antimicrobial substances, increase in number. The activated macrophage also begins producing nitric oxide, a potent antimicrobial chemical, along with various compounds that can be released to destroy extracellular microorganisms. ■ phagosome, p. 358

If the response is still not sufficient to control the infection, activated macrophages fuse together, forming **giant cells.** These, along with other macrophages and T cells, can form granulomas that wall off the offending agent, preventing infectious microbes from escaping to infect other cells. Activated macrophages are an important aspect of the immune response against diseases such as tuberculosis that are caused by organisms capable of surviving within macrophages. ■ giant cell, p. 359 ■ granuloma, p. 359

Subsets of Dendritic Cells and T Cells

Various subsets of dendritic cells (DC1 and DC2) and effector helper T cells (T_H1, T_H2, T_H17) appear to steer the immune system toward an

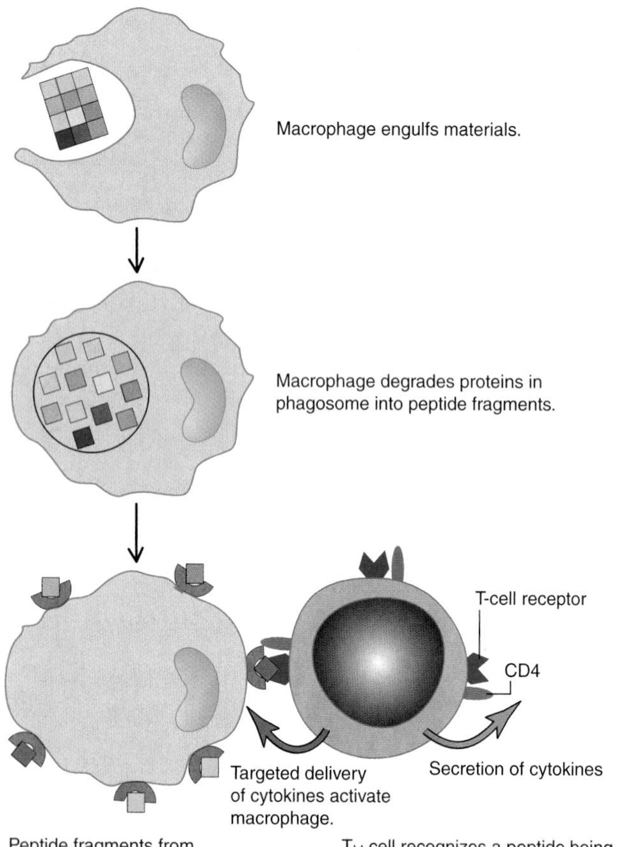

Macrophage engulfs materials.
Macrophage degrades proteins in phagosome into peptide fragments.
T-cell receptor
CD4
Targeted delivery of cytokines activate macrophage.
Secretion of cytokines
Peptide fragments from engulfed material are presented by MHC class II molecules.
T_H cell recognizes a peptide being presented by the macrophage and responds by activating the macrophage.

FIGURE 16.20 The Role of T_H Cells in Macrophage Activation

appropriate response. The roles of these subsets are still being clarified, but the different types of dendritic cells produce specific cytokines that cause activated helper T cells to differentiate into specialized effector subsets. For example, T_H1 cells activate macrophages, thereby promoting a response against intracellular pathogens; T_H2 cells direct a response against multicellular pathogens by recruiting eosinophils and basophils; and the recently discovered T_H17 cells recruit neutrophils, thereby directing a response against extracellular pathogens. The outcome of some conditions, such as Hansen's disease (leprosy), appears to correlate with the type of helper T cell response.

MICROCHECK 16.7

T cells are activated when they recognize antigen presented by dendritic cells expressing co-stimulatory molecules. T_C (CD8) cells recognize antigen presented by MHC class I molecules and respond by inducing apoptosis in the target cell and secreting cytokines that stimulate surrounding cells to be more vigilant against intracellular invaders. T_H (CD4) cells recognize antigen presented by MHC class II molecules (found on B cells and macrophages) and respond by activating the target cell and secreting various cytokines that orchestrate the immune response.

- ✓ Name three types of antigen-presenting cells.
- ✓ If an effector CD8 cell recognizes antigen presented by an MHC class I molecule, how should it respond?
- ✓ Why would a person who has AIDS be more susceptible to the bacterium that causes tuberculosis?

16.8 Natural Killer (NK) Cells

Focus Point

- Describe two distinct protective roles of NK cells.

Natural killer (NK) cells descend from lymphoid stem cells, but they lack the antigen-specific receptors that characterize B cells and T cells. Their activities, however, augment the adaptive immune responses. ■ lymphoid stem cells, p. 351

NK cells are important in the process of antibody-dependent cellular cytotoxicity (ADCC), which is a means of killing cells that have been bound by antibody (see figure 16.6). This enables the killing of host cells that have foreign proteins inserted into their membrane, such as those that have been infected by certain types of viruses. NK cells recognize their target by means of Fc receptors for IgG antibodies on their surface; recall that Fc receptors bind the "red flag" portion of antibody molecules. These Fc receptors enable the NK cell to detect and attach to an antibody-coated cell. When multiple receptors on an NK cell bind the Fc regions, the NK cell delivers granules that contain perforin and proteases directly to the target cell, inducing apoptosis in that cell.

NK cells also recognize and destroy host cells that do not have MHC class I molecules on their surface and are under stress. This is important because some viruses have evolved mechanisms to circumvent the action of cytotoxic T cells by interfering with the process of antigen presentation; cells infected with such a virus will be essentially bare of MHC class I molecules and thus cannot be a target of cytotoxic T cells. The NK cells can recognize the absence of MHC class I molecules on those cells, along with certain molecules that indicate the cells are under stress, and induce the infected cells to undergo apoptosis. This can occur because NK cells are actually programmed to destroy "self" cells under stress, but recognition of the MHC class I molecules suppresses that killing action. In the absence of MHC class I molecules, the action can proceed.

Recent evidence indicates that the role of NK cells goes far beyond that of simple killing machines. For example, they produce cytokines that help regulate and direct certain immune responses. Unfortunately, studying these actions of NK cells is difficult because there are different subsets, and the activities of the various subsets are influenced by cues in their local environment.

MICROCHECK 16.8

Natural killer (NK) cells can mediate antibody-dependent cellular cytotoxicity (ADCC). NK cells also kill cells not bearing MHC class I molecules on their surface.

- ✓ What mechanism do NK cells use to kill "self" cells?
- ✓ What can cause a "self" cell to not bear MHC class I molecules on its surface?
- ✓ What selective advantage would a virus have if it interferes with the process of antigen presentation?

16.9

Lymphocyte Development

Focus Points

- Describe the roles of gene rearrangement, imprecise joining, and combinatorial associations in the generation of diversity of antibody molecules.
- Describe positive and negative selection of self-reactive lymphocytes.

During lymphocyte development, as the cells differentiate from hematopoietic stem cells into either B cells or T cells, they acquire their ability to recognize distinct epitopes. Then, once they have committed to that specificity, they pass through rigorous checkpoints intended to ensure that their antigen-specific receptors are functional yet will not evoke a response against "self" molecules. Most developing lymphocytes fail these tests and, as a consequence, are induced to undergo apoptosis. ■ **hematopoietic stem cell, p. 350**

B cells undergo the developmental stages in the bone marrow; T cells go through the maturation processes described in this section in the thymus. The events involved in the adaptive immune response, from the maturation of lymphocytes to the development of their effector functions, are summarized in **figure 16.21.**

PERSPECTIVE 16.1

What Flavor Are Your Major Histocompatibility Complex Molecules?

The major histocompatibility molecules were discovered over half a century ago, long before their critical role in adaptive immunity was recognized. During World War II, bombing raids caused serious burns in many people, stimulating research into skin transplants to replace burned tissue. That research quickly expanded to include transplants of a variety of other tissues and organs. Unfortunately, the recipient's immune system generally rejected the transplanted tissue, mounting a vigorous response against it. The tissue was perceived as an "invader" because certain molecules on the cells of the donor tissue differed from those on the recipient's cells. To overcome this problem, tissue-typing tests were developed to allow researchers to more closely match tissue of donors with those of recipients. The typing tests exploit surface molecules on leukocytes that serve as markers for tissue compatibility; the molecules were called **human leukocyte antigens,** or **HLAs.** Later, researchers determined that HLAs were encoded by a cluster of genes, now called the major histocompatibility complex. Unfortunately, the terminology can be confusing because the molecules that transplant biologists refer to as HLAs are called MHC molecules by immunologists.

It is highly unlikely that two random individuals will have identical MHC molecules. This is because the genes encoding them are **polygenic,** meaning they are encoded by more than one **locus,** or position on the chromosome, and each locus is highly **polymorphic,** meaning there are multiple variations **(figure 1).** There are three loci of MHC class I genes, designated HLA-A, HLA-B, and HLA-C. As an analogy, if MHC molecules were candy, each cell would be covered with pieces of chocolate (HLA-A), taffy (HLA-B), and lollipop (HLA-C). There are more than 100 different **alleles,** or forms, of each of the three genes. Continuing with the candy analogy, the flavor at the chocolate locus could be dark, white, or milk chocolate; the taffy could be peppermint or cinnamon, and so on. The list of known alleles continues to increase, but currently there are at least 229 alleles for HLA-A, 464 for HLA-B, and 111 for HLA-C. In addition, all of the loci are co-dominantly expressed. In other words, you inherited one set of the three genes from your mother and one set from your father; both sets are expressed. Putting this all together, and assuming that you inherited two completely different sets of alleles from your parents, your cells express six different MHC class I molecules—two of the over 229 known HLA-A possibilities, two of the over 464 HLA-B possibilities, and two of the over 111 HLA-C possibilities. As you can imagine, the likelihood that any person you encounter in a day will have those same MHC class I molecules is extremely unlikely, unless you have an identical twin.

Set of MHC genes inherited from your mother:

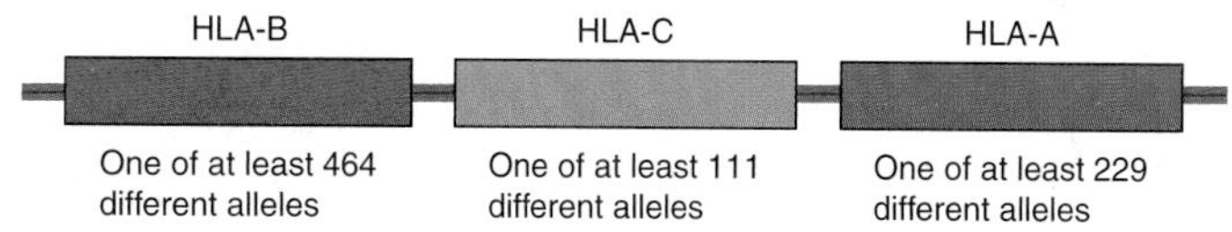

Set of MHC genes inherited from your father:

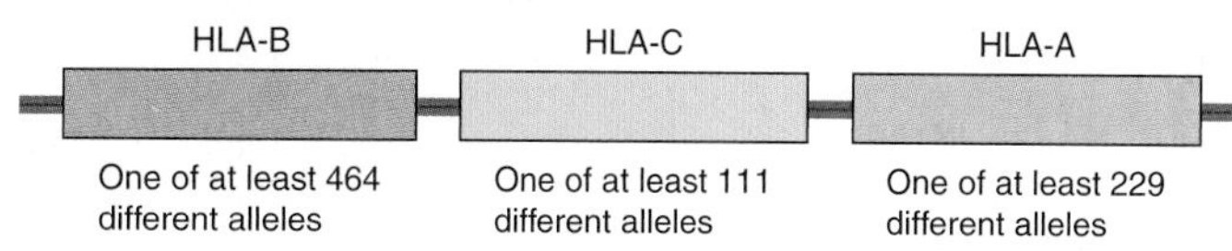

FIGURE 1 MHC Polymorphisms The order of the MHC class I genes on the chromosome is B, C, and A.

Why is there so much diversity in MHC molecules? The answer lies in the complex demands of antigen presentation. MHC class I molecules bind peptides that are only 8 to 10 amino acids in length; MHC class II molecules bind peptides that are only 13 to 25 amino acids in length. Somehow, within that constraint, the MHC molecules must bind as many different peptides as possible in order to ensure that a representative selection from the proteins within a cell can be presented to T cells. The ability to bind a wide variety of peptides is particularly important considering how readily viruses and bacteria can evolve in response to selective pressure. For example, if a single alteration in a viral protein prevented all MHC molecules from presenting peptides from that protein, then a virus with that mutation could routinely overwhelm the body. The ability to bind different peptides is not enough, however, because, ideally, a given peptide should be presented in several slightly different orientations so that distinct aspects of the three-dimensional structures can be inspected by T-cell receptors. No single variety of MHC molecules can accomplish all of these aims, which probably accounts for the diversity in MHC molecules.

The variety of MHC molecules that a person has on his or her cells impacts that individual's adaptive response to certain antigens. This is not surprising since MHC molecules differ in the array of peptides they can bind and manner in which those peptides are held in the molecule. Thus, they impact what the T cells actually "see." In fact, the severity of certain diseases has been shown to correlate with the MHC type of the infected individual. For example, rheumatic fever, which can occur as a consequence of *Streptococcus pyogenes* infection, develops more frequently in individuals with certain MHC types. The most serious manifestations of schistosomiasis have also been shown to correlate with certain MHC types. Epidemics of life-threatening diseases such as plague and smallpox have dramatically altered the relative proportion of MHC types in certain populations, killing those whose MHC types ineffectively present peptides from the causative agent. ■ **rheumatic fever, p. 501** ■ **schistosomiasis, p. 362**

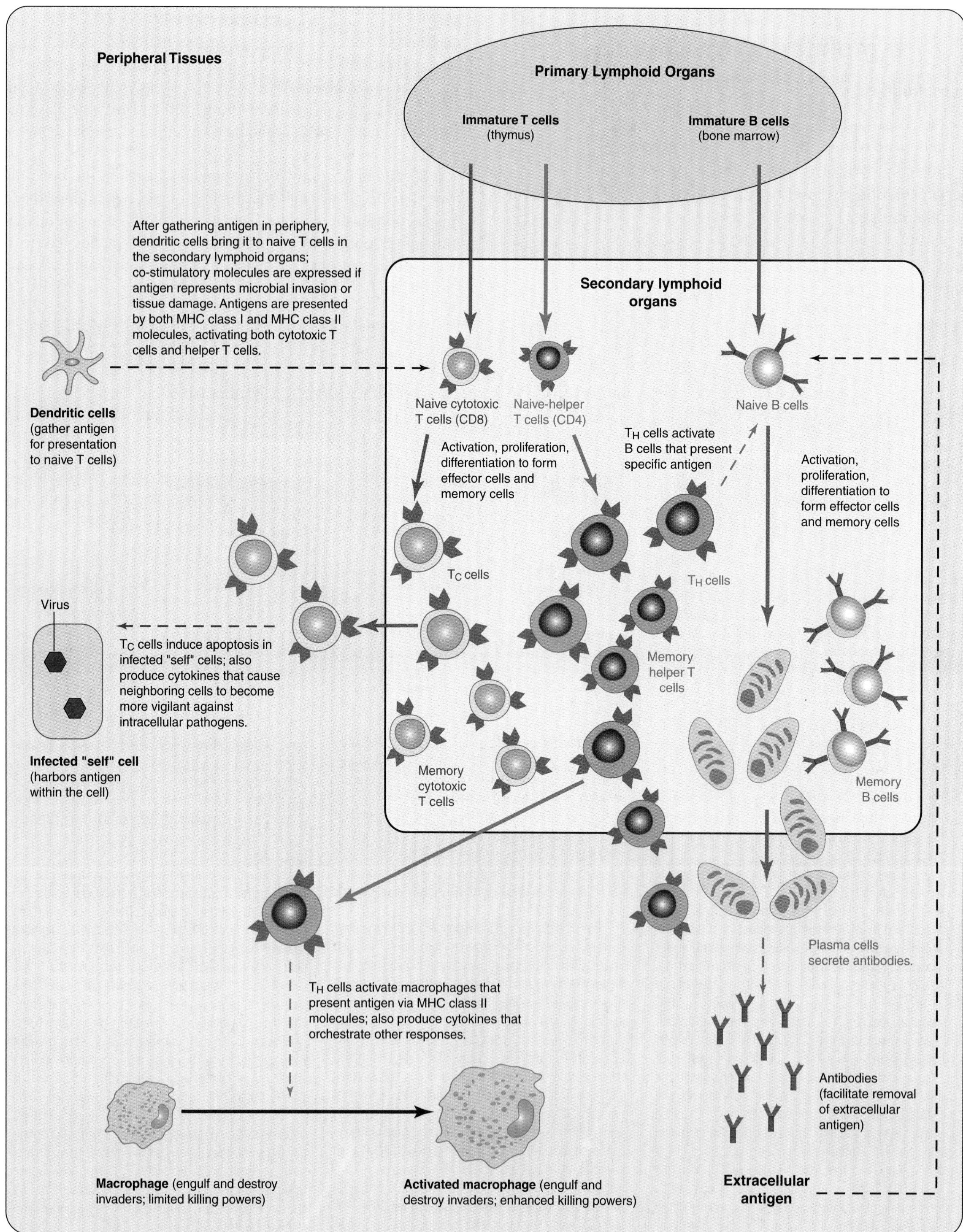

FIGURE 16.21 Summary of the Adaptive Immune Response

Generation of Diversity

The mechanisms lymphocytes use to produce a seemingly limitless assortment of antibodies and antigen-specific receptors were first revealed in studies using B cells. Because the processes are markedly similar to those employed by T cells, we will use them as a general model to describe the generation of diversity with respect to specificity for antigen.

Each B cell responds to only one epitope, yet it is estimated that the population of B cells within the body can respond to more than 100 million different epitopes. Based on the information presented in chapter 7, it might seem logical to assume that humans have over 100 million different antibody genes, each encoding specificity for a single epitope. This is impossible, however, because the human genome has only 3 billion nucleotides and codes for only about 25,000 genes.

The question of how such tremendous diversity in antibodies could be generated perplexed immunologists until Dr. Susumu Tonegawa solved the mystery. For this work, he was awarded a Nobel Prize in 1987. Diversity in antibodies involves a combination of gene rearrangement, imprecise joining of gene fragments, and the association of light chains and heavy chains coded for by the rearranged genes.

Gene Rearrangement

A primary mechanism for generating a wide variety of different antibodies using a limited-size region of DNA employs a strategy similar to that of a savvy and well-dressed traveler living out of a small suitcase. By mixing and matching different shirts, pants, and shoes, the traveler can create a wide variety of unique outfits from a limited number of components. Likewise, the maturing B cell selects three gene segments, one each from DNA regions called V (variable), D (diversity), and J (joining), to form an ensemble that encodes a nearly unique variable region of the heavy chain of an antibody **(figure 16.22).**

A human lymphoid stem cell has about 65 different V segments, 27 different D segments, and 6 different J segments in the DNA that encodes the variable region of the heavy chain. As a B cell develops, however, two large regions of DNA are permanently removed, effectively joining discrete V, D, and J regions. The joined segments encode the heavy chain of the antibody that the mature B cell is programmed to make. Thus, one B cell could express the combination V5, D3, and J6 to produce its heavy chain, whereas another B cell might use V19, D27, and J6; each combination would result in a unique antibody specificity. Similar rearrangements occur in the genes that encode the light chain of the antibody molecule.

Imprecise Joining

As the various segments are joined during gene rearrangement, nucleotides are often deleted or added between the sections. This imprecise joining changes the reading frame of the encoded protein so that two B cells that have the same V, D, and J segments for their heavy chain could potentially give rise to antibodies with very different specificities.

Combinatorial Associations

Combinatorial association refers to the specific groupings of light chains and heavy chains that make up the antibody molecule. Both types of chains acquire diversity through gene rearrangement and imprecise joining. Additional diversity is then introduced when these two molecules join; it is the combination of the two chains that creates the antigen-binding site (see figure 16.5b).

Negative Selection of Self-Reactive B Cells

Negative selection is the process of eliminating lymphocytes, including B cells, that recognize "self" molecules. The result is called **clonal deletion.** Failure to eliminate such B cells results in the production of **autoantibodies,** which are antibodies that bind to host components, causing the immune system to attack "self" substances.

After a developing B cell begins producing a functional B-cell receptor, it is then exposed to various other cells and material in the bone marrow. Because the bone marrow is normally free of foreign substances, any B cell that binds material there must be recognizing "self" and therefore ought to be eliminated. This occurs by inducing the cell to undergo apoptosis.

Positive and Negative Selection of Self-Reactive T Cells

Developing T cells have two phases of trials—positive and negative selection—that seal their fate. **Positive selection** is a process that permits only those T cells that recognize MHC to

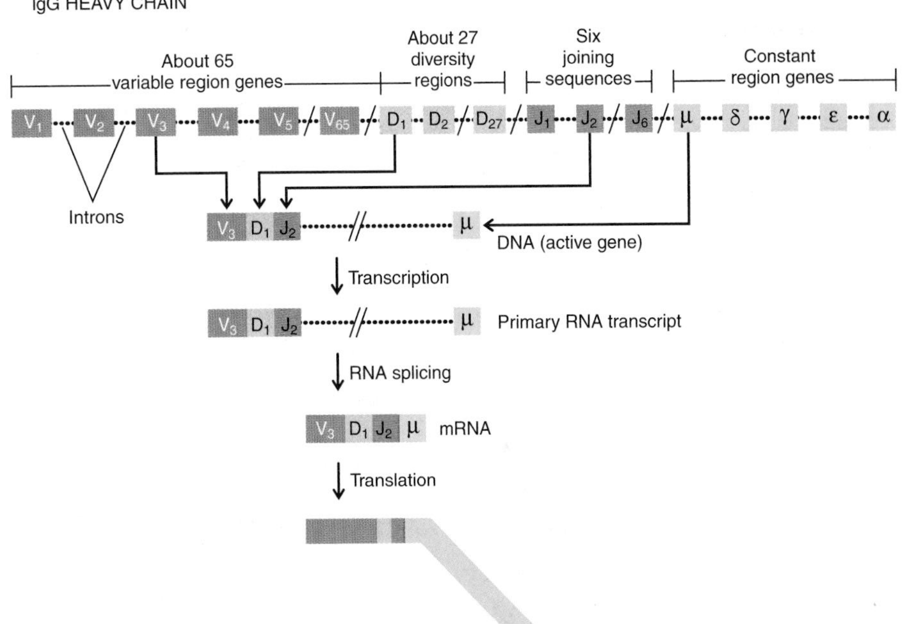

FIGURE 16.22 Antibody Diversity Immunoglobulin gene arrangement in an immature lymphocyte and the mechanism of active gene formation. Only the heavy chain is shown.

some extent to develop further. Recall that the T-cell receptor, unlike the B-cell receptor, recognizes a peptide:MHC complex. T cells, therefore, must show at least some recognition of the MHC molecules regardless of the peptide they are carrying. T cells that show insufficient recognition fail positive selection and, as a consequence, are eliminated. Each T cell that passes positive selection is also subjected to negative selection, analogous to that which occurs during B-cell development. T cells that recognize "self" peptides presented by MHC molecules are eliminated. Positive and negative selection processes are so stringent that over 95% of developing T cells undergo apoptosis in the thymus.

MICROCHECK 16.9

Mechanisms used to generate the diversity of antigen specificity in lymphocytes include rearrangement of gene segments, imprecise joining of those segments, and combinatorial associations of heavy chains and light chains. Negative selection eliminates B cells and T cells that recognize normal "self" molecules. Positive selection permits only those T cells that show moderate recognition of the MHC molecules to develop further.

- ✓ What three gene segments encode the variable region of the heavy chain of an antibody molecule?
- ✓ What are autoantibodies?
- ✓ How could imprecise joining be considered a type of frameshift mutation?

SUMMARY

16.1 Strategy of the Adaptive Immune Response (figure 16.1)

Overview of Humoral Immunity

Humoral immunity is mediated by **B cells;** in response to extracellular antigens, these proliferate and then differentiate into **plasma cells** that function as antibody-producing factories. **Memory B cells** are also formed.

Overview of Cellular Immunity

Cellular immunity is mediated by **T cells;** in response to intracellular antigens, **cytotoxic T cells** proliferate and then differentiate into **T_C cells** that induce apoptosis in "self" cells harboring the intruder. **Memory cytotoxic T cells** are also formed. **Helper T cells** proliferate and then differentiate to form **T_H cells** that help orchestrate the various responses of humoral and cellular immunity. **Memory helper T cells** are also formed.

16.2 Anatomy of the Lymphoid System (figure 16.2)

Lymphatic Vessels

Lymph, which contains antigens that have entered tissues, flows in the lymphatic vessels to the lymph nodes (figure 16.3).

Secondary Lymphoid Organs

Secondary lymphoid organs are the sites at which lymphocytes gather to contact antigens.

Primary Lymphoid Organs

Primary lymphoid organs are the sites where B cells and T cells mature.

16.3 The Nature of Antigens

Antigens are molecules that react specifically with an antibody or lymphocyte; **immunogen** refers specifically to an antigen that elicits an immune response. The immune response is directed to **antigenic determinants,** or **epitopes,** on the antigen (figure 16.4).

16.4 The Nature of Antibodies

Structure and Properties of Antibodies (figure 16.5)

Antibody monomers have a Y shape with an antigen-binding site at the end of each arm. The tail of the Y is the **Fc region.** The antibody monomer is composed of two identical **heavy chains** and two identical **light chains.** The **variable region** contains the antigen-binding site; the **constant region** encompasses the entire Fc region as well as part of the Fab regions.

Protective Outcomes of Antibody-Antigen Binding (figure 16.6)

Antibody-antigen binding results in **neutralization, immobilization** and **prevention of adherence, agglutination** and **precipitation, opsonization, complement activation,** and **antibody-dependent cytotoxicity.**

Immunoglobulin Classes (table 16.1)

There are five major antibody classes: **IgM, IgG, IgA, IgD,** and **IgE,** and each has distinct functions.

16.5 Clonal Selection and Expansion of Lymphocytes (figure 16.8)

When antigen enters a secondary lymphoid organ, only the lymphocytes that specifically recognize the antigen will respond; the antigen receptor they carry on their surface governs this recognition. Lymphocytes may be **immature, naive, activated, effector,** or **memory cells.**

16.6 B Lymphocytes and the Antibody Response

B-Cell Activation

B cells present antigen to T_H cells for inspection. If a T_H cell recognizes the antigen, it will deliver cytokines to the B cell, initiating the process of clonal expansion, which ultimately gives rise to plasma cells that produce antibody (figure 16.9).

Characteristics of the Primary Response

Under the direction of T_H cells, the expanding B-cell population will undergo affinity maturation and class switching, and form memory cells (figures 16.12, 16.13).

Characteristics of the Secondary Response

Memory cells are responsible for the swift and effective **secondary response,** eliminating invaders before they cause noticeable harm (figure 16.10).

The Response to T-Independent Antigens

T-independent antigens include polysaccharides that have multiple identical evenly spaced epitopes, and LPS (figure 16.14).

16.7 T Lymphocytes: Antigen Recognition and Response

General Characteristics of T Cells (table 16.2, figure 16.17)

Cytotoxic T cells (CD8) recognize antigen presented by **major histocompatibility complex (MHC) class I molecules.** Helper T cells (CD4) recognize antigen presented by **major histocompatibility complex (MHC) class II molecules.**

Activation of T Cells (figure 16.18)

Dendritic cells sample material in tissues and then travel to secondary lymphoid organs to present antigens to naive T cells. The dendritic cells that detect molecules associated with danger produce **co-stimulatory molecules** and are able to activate both subsets of T cells.

Functions of T_C (CD8) Cells

T_C cells induce apoptosis in cells that present peptides they recognize in MHC class I; they also produce cytokines that allow neighboring cells to become more vigilant against intracellular invaders (figure 16.19). All nucleated cells present peptides from endogenous proteins in the groove of MHC class I molecules.

Functions of T_H (CD4) Cells (figures 16.9, 16.20)

T_H cells activate cells that present peptides they recognize in MHC class II; various cytokines are released, depending on subset of the responding T_H cell. Macrophages and B cells present peptides from exogenous proteins in the groove of MHC class II molecules.

Subsets of Dendritic Cells and T Cells

Subsets of dendritic cells and T_H cells direct the immune system to an appropriate response.

16.8 Natural Killer (NK) Cells

NK cells mediate **antibody-dependent cellular cytotoxicity (ADCC).** NK cells induce apoptosis in host cells that are not bearing MHC class I molecules on their surface.

16.9 Lymphocyte Development

Generation of Diversity

Mechanisms used to generate the diversity of antigen specificity in lymphocytes include rearrangement of gene segments, imprecise joining of those segments, and combinatorial associations of heavy and light chains (figure 16.22).

Negative Selection of Self-Reactive B Cells

Negative selection occurs as B cells develop in the bone marrow; cells to which material binds to their B-cell receptor are induced to undergo apoptosis.

Positive and Negative Selection of Self-Reactive T Cells

Positive selection permits only those T cells that show moderate recognition of the MHC molecules to develop further. Negative selection also occurs.

REVIEW QUESTIONS

Short Answer

1. Which antibody class neutralizes viruses in the intestinal tract?
2. Which antibody class is the first produced during the primary response?
3. Diagram an IgG molecule and label (a) the Fc area and (b) the areas that combine with antigen.
4. How do natural killer cells differ from cytotoxic T cells?
5. How do T-independent antigens differ from T-dependent antigens?
6. What is a secondary lymphoid organ?
7. Describe clonal selection and expansion in the immune response.
8. Describe the role of dendritic cells in T-cell activation.
9. What are the protective outcomes of antibody-antigen binding?
10. What are antigen-presenting cells (APCs)?

Multiple Choice

1. The variable regions of antibodies are located in the
   1. Fc region. 2. Fab region. 3. light chain.
   4. heavy chain. 5. light chain *and* heavy chain.
   a) 1, 3 b) 1, 5 c) 2, 3 d) 2, 4 e) 2, 5
2. Which of the following statements about antibodies is *false?*
   a) If you removed the Fc portion, antibodies would no longer be capable of opsonization.
   b) If you removed the Fc portion, antibodies would no longer be capable of activating the complement system.
   c) If you removed the Fab portion, an antibody would no longer be capable of agglutination.
   d) If IgG were a pentamer, it would be more effective at agglutinating antigens.
   e) If IgE had longer half-life, it would protect newborn infants.
3. Which class of antibody can cross the placenta?
   a) IgA b) IgD c) IgE d) IgG e) IgM
4. A person who has been vaccinated against a disease should have primarily which of these types of antibodies against that agent 2 years later?
   a) IgA b) IgD c) IgE d) IgG e) IgM
5. Which of the following statements about B cells/antibody production is *false?*
   a) B cells of a given specificity initially have the potential to make more than one class of antibody.
   b) In response to antigen, all B cells located close to the antigen begin dividing.
   c) Each B cell is programmed to make a single specificity of antibody.
   d) The B-cell receptor enables B cells to "sense" that antigen is present.
   e) The cell type that makes and secretes antibody is called a plasma cell.
6. Which term describes the loss of specific heavy chain genes?
   a) Affinity maturation
   b) Apoptosis
   c) Clonal selection
   d) Class switching

7. Which of the following cell types cannot *replicate* in response to a specific antigen?

   a) B cells b) Cytotoxic T cells c) Helper T cells

   d) Plasma cells

8. Which markers are found on all nucleated cells?

   a) MHC class I molecules

   b) MHC class II molecules

   c) CD4

   d) CD8

9. Which of the following are examples of an antigen-presenting cell (APC)?

   1. Macrophage 2. Neutrophil 3. B cell

   4. T cell 5. Plasma cell

   a) 1, 2 b) 1, 3 c) 2, 4 d) 3, 5 e) 1, 2, 3

10. What is the appropriate response when antigen is presented by MHC class II molecules?

    a) An effector CD8 cell should kill the presenting cell.

    b) An effector CD4 cell should kill the presenting cell.

    c) An effector CD8 cell should activate the presenting cell.

    d) An effector CD4 cell should activate the presenting cell.

Applications

1. Many dairy operations keep cow's milk for sale and use formula and feed to raise any calves. One farmer noticed that calves raised on the formula and feed needed to be treated for diarrhea more frequently than calves left with their mothers to nurse. He had some tests run on the diets and discovered no differences in the calories or nutritional content. The farmer called a veterinarian and asked him to explain the observations. What was the vet's response?
2. What kinds of diseases would be expected to occur as a result of lack of T or B lymphocytes?

Critical Thinking

1. The development of primary and secondary immune responses to an antigen differ significantly. The primary response may take a week or more to develop fully and establish memory. The secondary response is rapid and relies on the activation of clones of memory cells. Wouldn't it be better if clones of reactive cells were maintained regardless of prior exposure? In this way, the body could always respond rapidly to *any* antigen exposure. Would there be any disadvantages to this approach? Why?
2. Early investigators proposed two hypotheses to explain the specificity of antibodies. The clonal selection hypothesis states that each lymphocyte can produce only one specificity of antibody. When an antigen appears that binds to that antibody, the lymphocyte is selected to give rise to a clone of plasma cells producing the antibody. The template hypothesis states that any antigen can interact with any lymphocyte and act as a template, causing newly forming antibody to be specific for that antigen. In one experiment to test these hypotheses, an animal was immunized with two different antigens. After several days, lymphocytes were removed from the animal and individual cells placed in separate small containers. Then, the original two antigens were placed in the containers with each cell. What result would support the clonal selection hypothesis? The template hypothesis?

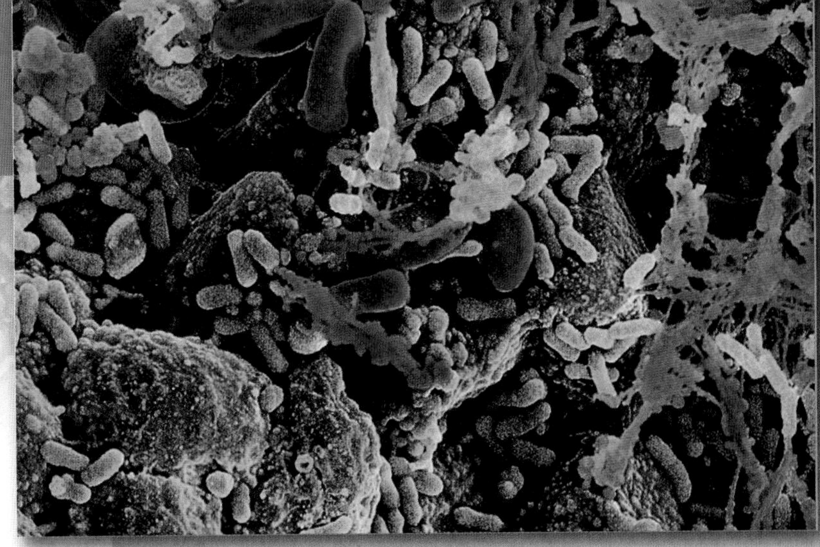

Bacterial cells adhering to a body surface.

Host-Microbe Interactions

A Glimpse of History

The ancients thought epidemics and diseases were divine punishment of the people for their sins. By the time of Moses, however, the Egyptians and Hebrews had come to believe that leprosy could be transmitted by contact with lepers. In Europe, around 430 B.C. Thucydides had concluded that some plagues were contagious. By the Middle Ages, many accepted this, and fled cities to escape the diseases. Fracastorius, in 1546, first proposed that communicable diseases were caused by living agents passed from one person or animal to another. He had no way to test this idea, however.

With Leeuwenhoek's discovery of microorganisms in the late seventeenth century, people began to suspect that microorganisms might cause disease, but the techniques of the times could not prove this. It was not until 1876 that Robert Koch offered convincing proof of the "germ theory" of disease. He showed that *Bacillus anthracis* is the cause of anthrax, an often fatal disease of humans, sheep, and other animals. With his microscope, he observed *B. anthracis* cells in the blood and spleen of dead sheep. He then inoculated mice with the infected sheep blood and was able to recover *B. anthracis* from the blood of the mice. In addition, he grew the bacteria in pure culture and showed that they caused anthrax when injected into healthy mice. From these experiments and later work with *Mycobacterium tuberculosis,* Koch formalized a group of criteria for establishing the cause of an infectious disease, known as Koch's Postulates.

Every day we contact an enormous number and variety of microorganisms. Some enter our respiratory system as we breathe; other we ingest with each bite of food or sip of drink; and still more adhere to our skin whenever we touch an object or surface. It is important to recognize, however, that the vast majority of these microbes generate no ill effects whatsoever. Some may colonize the body surfaces, taking up residence with the variety of other harmless microbes that live there; others are sloughed off with dead epithelial cells. Most of those swallowed are either killed in the stomach or eliminated in feces.

Relatively few microbes are able to inflict any noticeable damage, such as invading tissues or producing toxic substances. Those that can, and do, are called **pathogens.** They have distinct characteristics that allow them to elude at least some of the body's defenses that would otherwise keep the invaders at bay. Research into the how these pathogens evade our innate and adaptive defenses is unraveling an impressive array of ploys that microbes have developed for subverting and circumventing our sophisticated systems. The knowledge we are gaining in areas such as genomics and immunology have given new insights into the pathogenic strategies, fueling hope that therapies targeted to specific disease-causing microbes can be developed.

This chapter will explore some of the ways in which microbes colonize the human host, living either as members of the normal microbiota in harmony with the host or subverting the host defenses and causing disease. We will also discuss how pathogens are able to evade or overcome the host responses and damage the host.

KEY TERMS

Acute Infection An infection characterized by symptoms that have a rapid onset but last a relatively short time.

Chronic Infection An infection that generally develops slowly and lasts for months or years.

Colonization Establishment and growth of a microorganism on a body surface.

Disease Condition that results in noticeable impairment of body function.

Endotoxin The lipopolysaccharide (LPS) component of the outer membrane of Gram-negative bacteria; lipid A is responsible for the toxic properties of LPS.

Exotoxin A toxic protein produced by a microorganism; often simply referred to as a *toxin.*

Immunocompromised A host with weaknesses or defects in the innate or adaptive defenses.

Infection Colonization by a pathogen on or within the body.

Latent Infection Infection in which the infectious agent is present but not causing symptoms.

Normal Microbiota The population of microorganisms routinely found growing on the body surfaces of healthy individuals.

Opportunistic Pathogen A microorganism or virus that causes disease only when introduced into an unusual location or into an immunocompromised host.

Primary Pathogen A microorganism or virus that is able to cause disease in an otherwise healthy individual.

Virulence Factors (Determinants) Attributes of a microorganism or virus that promote pathogenicity.

MICROBES, HEALTH, AND DISEASE

Many people think of microorganisms as "germs" that should routinely be killed or avoided. Most microbes are harmless, however, and many even provide beneficial aspects. The organisms that routinely reside on the body's surfaces are called the **normal microbiota,** or **normal flora.** This relationship is a delicate balancing act, though, because some members of the normal microbiota, as well as microbes that make incidental contact with humans, are quite able to exploit body fluids and tissues as a source of nutrients should the opportunity arise. Weaknesses or defects in the innate or adaptive defenses can leave people vulnerable to invasion; these individuals are said to be **immunologically compromised,** or **immunocompromised.** Factors that can lead to an individual becoming immunocompromised include malnutrition, cancer, AIDS or other diseases, surgery, wounds, genetic defects, alcohol or drug abuse, and immunosuppressive therapy that accompanies procedures such as organ transplants.

17.1 The Anatomical Barriers As Ecosystems

Focus Point

- Compare and contrast mutualism, commensalism, and parasitism.

The skin and mucous membranes provide anatomical barriers against invading microorganisms, but they also supply the foundation for a complex **ecosystem,** an interacting biological community. The intimate relationships between the microorganisms and the human body are an example of **symbiosis,** meaning "living together;" symbionts are different organisms that live close together on more or less a permanent basis.

Symbiotic Relationships Between Microorganisms and Hosts

Microorganisms that inhabit the body can have a variety of symbiotic relationships with each other and with the human host. These relationships may take on different characteristics depending on the closeness of the association and the relative advantages to each partner. Symbiotic associations can be one of several forms, and these may change, depending on the state of the host and the attributes of the microbes:

- **Mutualism** is an association in which both partners benefit. In the large intestine, for example, some bacteria synthesize vitamin K and certain B vitamins. These nutrients are then available for the host to absorb, providing an important source of essential vitamins, particularly for hosts lacking a well-balanced diet. The bacteria residing in the intestine benefit as well, as they are supplied with warmth and a variety of different energy sources.
- **Commensalism** is an association in which one partner benefits but the other remains unharmed. Many bacteria living on the skin are neither harmful nor advantageous to the human host, but the bacteria gain by obtaining food and other necessities from the host.
- **Parasitism** is an association in which one organism, the **parasite,** derives benefit at the expense of the other organism, the **host.** All pathogens are parasites, but medical microbiologists often reserve the word *parasite* for eukaryotic organisms such as protozoa and helminths that cause disease.

MICROCHECK 17.1

Depending on the relative benefit to each partner in a relationship, such as a human host and a microbial species, the relationship can be described as mutualism, commensalism, or parasitism.

✓ How is mutualism different from commensalism?

17.2

The Normal Microbiota

Focus Points

- List two ways that our normal microbiota plays a protective role in our overall health.
- Describe how the composition of normal microbiota can change over time.

The normal microbiota is the population of microorganisms routinely found growing on the body of healthy individuals **(figure 17.1).** Microbes that typically inhabit body sites for extended periods are **resident microbiota,** whereas those that are only temporary are **transient microbiota.** Many different species make up the normal microbiota, and they occur in large numbers. In fact, there are more bacteria in just one person's mouth than there are people in the world!

The Protective Role of the Normal Microbiota

The most significant contributions of the normal microbiota to health include protection against potentially harmful microorganisms and stimulation of the immune system. When members of the normal microbiota are killed or their growth suppressed, as can happen during antibiotic treatment, harmful organisms may colonize and cause disease. For instance, the *Lactobacillus* species that predominate in the vagina of mature females normally suppress growth of the yeast *Candida albicans*. During intensive, unrelated treatment with certain antibiotics, however, the normal bacterial population may be inhibited, allowing the fungi to overgrow and cause disease. As another example, oral administration of antibiotics can suppress members of the normal intestinal microbiota, allowing the overgrowth of toxin-producing strains of *Clostridium difficile* that cause antibiotic-associated diarrhea. ■ antibiotic-associated diarrhea, p.586

As we discussed in chapter 15, the presence and multiplication of normal microbiota competitively excludes pathogens via several different mechanisms. These include covering binding sites that might otherwise be used for attachment, consuming available nutrients, and producing compounds that are toxic to other bacteria. ■ normal microbiota, p. 349

The normal microbiota appears to play an instrumental role in the development of oral tolerance by the immune system. In a complex series of events that is not entirely understood, our defenses learn to dampen the immune response to the multitude of microbes that routinely inhabit the gut, as well as the foods that pass through. Recent studies into the actions of regulatory T cells suggest that early and consistent exposure to certain microbes in the gut stimulates those modulating T cells, thereby preventing the immune system from overreacting to harmless microbes and substances. This idea is the basis of the **hygiene hypothesis,** which proposes that lack of exposure to microbes can promote development of allergies. It is a fine balance, however, because contact with certain pathogens can be deadly. ■ tolerance, p. 367

The normal microbiota also primes the adaptive immune system. The response mounted against those members that routinely breach the body's anatomical barriers in small numbers may cross-react with pathogens that could be encountered later. The importance of the normal microbiota in the development of immune responses is shown in mice reared in a microbe-free environment. These animals lack a normal microbiota and have greatly underdeveloped mucosal-associated lymphoid tissues (MALT). ■ MALT, p. 370

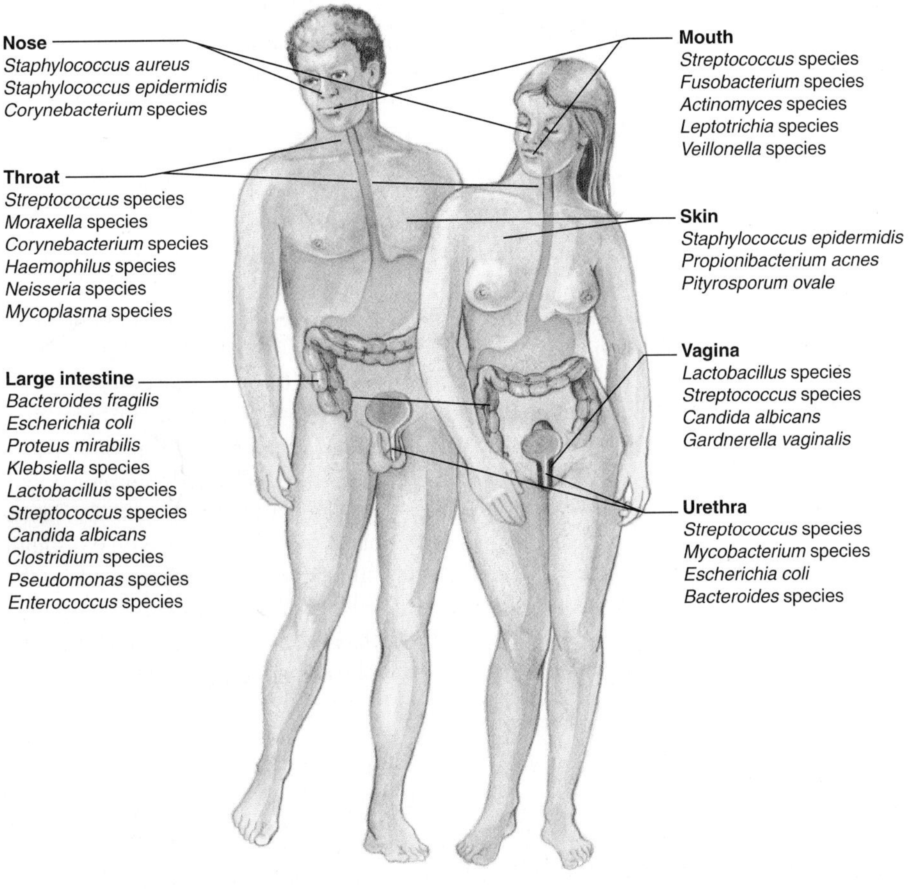

FIGURE 17.1 Normal Microbiota Many different organisms are part of the normal microbiota of the male and female human body.

The Dynamic Nature of the Normal Microbiota

A healthy human fetus is sterile until the protective membrane that surrounds it ruptures as a prelude to birth. During the passage through the birth canal, the baby is exposed to a variety of microbes that take up residence on its skin and in its digestive tract. Various microorganisms in food, on other humans, and in the environment soon also become established as residents on the newborn.

Once established, the composition of the normal microbiota is dynamic. At any one time the makeup of this complex ecosystem represents a balance of many forces that can dramatically or discreetly alter the bacterial population's quantity and composition. Changes occur in response to physiological variations within the host, such as hormonal changes, and as a direct result of the activities of the human host, such as the type and amount of food consumed. An intriguing example was the recent discovery that the intestinal microbiota of obese and lean people differs. Obese people have more members of the *Firmicutes,* a phylum that includes *Clostridium* and *Bacillus* species, whereas thin individuals typically have more members of the *Bacteriodetes,* a phylum that includes *Bacteriodes* species. As obese people lost weight, their intestinal microbiota changed to resemble that of typically lean people. As part of the Human Microbiome Project, further studies are underway to track changes in the composition of the normal microbiota in both health and disease. ■ Human Microbiome Project, p. 221

MICROCHECK 17.2

The normal microbiota provides protection against potentially harmful organisms and stimulates the immune system.

- ✓ What factor favors the growth of *Clostridium difficile* in the intestine?
- ✓ Why would the immune response to members of the normal microbiota cross-react with pathogens?

17.3

Principles of Infectious Disease

Focus Points

- Define the terms *primary pathogen, opportunist,* and *virulence.*
- Compare and contrast acute, chronic, and latent infections.

Colonization of a host by a microorganism implies that the microbe has become established and is multiplying on a body surface. If the microbe has a parasitic relationship with the host, then the term **infection** can be used. That is, a member of the normal microbiota is said to have colonized the host, but an organism capable of causing illness is described as having either colonized or infected the host. Infection does not always lead to noticeable adverse effects. It can be **subclinical** or **inapparent,** meaning that symptoms either do not appear or are mild enough to go unnoticed.

An infection that results in **disease,** a noticeable impairment of body function, is called an **infectious disease.** Diseases are characterized by symptoms and signs; **symptoms** are the subjective effects of the disease experienced by the patient, such as pain and nausea, whereas **signs** are the objective evidence, such as rash, pus formation, and swelling.

Effects of one disease may leave a person predisposed to developing another. For example, various respiratory illnesses that damage the mucociliary escalator make a person more likely to develop pneumonia. The initial infection is a **primary infection;** an additional infection that occurs as a result of the primary infection is a **secondary infection.** ■ mucociliary escalator, p. 348

Pathogenicity

Some microorganisms and viruses are able to cause disease in otherwise healthy individuals; a microbe able to do this is referred to as a **primary pathogen** or, more simply, a **pathogen.** Diseases such as plague, malaria, measles, influenza, diphtheria, tetanus, and tuberculosis are caused by primary pathogens. In contrast, a microbe able to cause disease only when the body's innate or adaptive defenses are compromised, or when introduced into an unusual location, is called an **opportunistic pathogen,** or **opportunist.** Opportunists can be members of the normal microbiota or they can be common in the environment. For instance, *Pseudomonas* species are environmental microbes that routinely come into contact with healthy individuals without harmful effect, yet they can cause fatal infections in individuals who have the genetic disease cystic fibrosis and also in burn patients (see figure 24.6). Ironically, as our health care systems improve, extending the life span of many patients through surgery and immunosuppressive drugs, diseases caused by opportunists are becoming more common. Also, many organisms not previously recognized as able to cause disease have now been shown to do so in some severely immunocompromised patients.

The term **virulence** refers to the degree of pathogenicity of an organism. An organism described as highly virulent is more likely to cause disease, particularly severe disease, than might otherwise be expected. *Streptococcus pyogenes* causes strep throat, for example, but certain strains are particularly virulent, causing diseases such as necrotizing fasciitis ("flesh-eating disease"). **Virulence factors,** or **virulence determinants,** are the attributes of a microorganism that promote pathogenicity. Unfortunately, the DNA encoding these virulence factors can sometimes be transferred to other bacteria. ■ mobile genetic elements, p. 205 ■ necrotizing fasciitis, p. 563

Characteristics of Infectious Disease

Infectious diseases that spread from one host to another are called **communicable,** or **contagious, diseases.** Some contagious diseases, such as colds and measles, are easily transmitted. The ease with which a contagious disease spreads partly reflects the **infectious dose**—or the number of microbes necessary to establish an infection. For example, the intestinal disease shigellosis is quite contagious in humans because only 10 to 100 cells of a *Shigella* species need be ingested to establish infection; in contrast, salmonellosis, which is not as contagious, requires ingestion of as many as 10^6 cells of *Salmonella* Enteritidis. The difference in these infectious doses reflects, in part, the ability of *Shigella* species to survive the acidic conditions encountered during passage through the stomach. Generally, the infectious dose is expressed as the **ID_{50},** an experimentally derived figure that indicates the number of microbes administered that resulted in disease in 50% of a population. ■ *Shigella,* p. 598 ■ *Salmonella* Enteritidis, p. 601

Course of Infectious Disease

The course of an infectious disease includes several stages **(figure 17.2).** The interval between introduction of an organism to a susceptible host and the onset of illness is the **incubation period.** The incubation period may vary considerably, from only a few days for the common cold, to several weeks for hepatitis A, to many months for rabies, and even years for Hansen's disease (leprosy). The

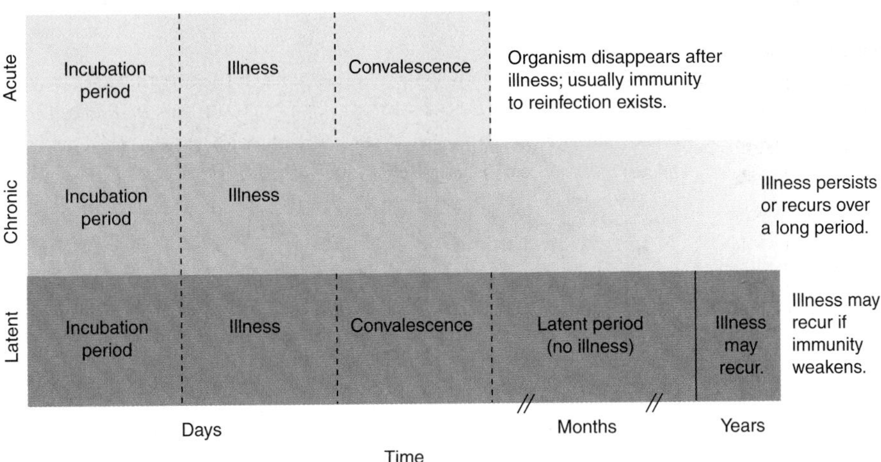

FIGURE 17.2 The Course of Infectious Diseases Diseases may be acute, chronic, or latent.

length of the incubation period depends on the number of organisms encountered, the condition of the host, and many other factors.

A phase of **illness** follows the incubation period. During this period a person will experience the signs and symptoms of the disease. In some cases, onset of illness is heralded by a **prodromal** phase—the early, vague symptoms of disease such as malaise and headache. After the illness subsides, there is a period of **convalescence,** the stage of recuperation and recovery from the disease. Even though there is no indication of infection during the incubation and convalescent periods, many infectious agents can still be spread during these stages. Some individuals, called **carriers,** may harbor infectious agents for months or years and continue to spread pathogens, even though they themselves show no signs or symptoms of the disease. The impact of carriers on spread of disease will be discussed in chapter 20. ■ carriers, p. 453

Following recovery from infection, or after immunization, the host normally has accumulated specific antibodies and memory lymphocytes that prevent reinfection with the same organism or virus. In most cases, the host is no longer susceptible to infection with that particular infectious agent.

Duration of Symptoms

Infections and the associated diseases are often described according to the timing and duration of the symptoms (see figure 17.2):

- **Acute infections** are characterized by symptoms that have a rapid onset but last only a short time; an example is strep throat.
- **Chronic infections** develop slowly and last for months or years; an example is tuberculosis.
- **Latent infections** are never completely eliminated; the microbe continues to exist in host tissues, often within host cells, for years without causing any symptoms. If there is a decrease in immune response, the latent infection may become reactivated and symptomatic. Note that the symptomatic phase of the disease may be either acute or chronic. For example, the infection caused by the varicella-zoster virus results in the characteristic symptoms of chickenpox, an acute illness. The illness is halted by an effective immune response, leaving the host immune against reinfection. The virus, however, is not completely eliminated. It takes refuge in sensory nerves, held in check by the immune system. Later in life, if there is a decrease in immunity, infectious viral particles are produced, causing the skin disease called shingles (herpes zoster). In tuberculosis, the mycobacteria are often initially confined within a small area by host defense mechanisms, causing no symptoms; much later, if the host becomes immunocompromised, the bacteria may begin multiplying and destroying tissue, resulting in a chronic illness. Other diseases in which the causative agent becomes latent include cold sores, genital herpes, and typhus. ■ varicella virus, p. 545 ■ tuberculosis, p. 514

Distribution of the Pathogen

Infections are often described according to the distribution of the causative agent in the body. In a **localized** infection, the microbe is limited to a small area; an example is a boil caused by *Staphylococcus aureus.* In a **systemic,** or **generalized,** infection, the infectious agent is spread, or **disseminated,** throughout the body; an example is measles.

The suffix -emia means "in the blood." Thus, **bacteremia** indicates that bacteria are circulating in the bloodstream. Note that this term does not necessarily imply a disease state. A person can become transiently bacteremic after vigorous tooth brushing. **Toxemia** indicates that toxins are circulating in the bloodstream. The organism that causes tetanus, for instance, produces a localized infection yet its toxins circulate in the bloodstream. The term **viremia** indicates that viral particles are circulating in the bloodstream. **Septicemia** is an acute, life-threatening illness caused by infectious agents or their products circulating in the bloodstream.

MICROCHECK 17.3

A primary pathogen can cause disease in an otherwise healthy individual; an opportunist causes disease in an immunocompromised host. The course of infectious disease includes an incubation period, illness, and a period of convalescence. Infections can be acute or chronic, latent, localized, or systemic.

✓ Why are diseases caused by opportunists becoming more frequent?
✓ Give an example of a latent disease.
✓ What factors might contribute to a long incubation period?

17.4 Establishing the Cause of Infectious Disease

Focus Point

- List Koch's postulates, and compare them to the Molecular postulates.

Criteria are needed to guide scientists as they try to determine the causative agents of disease and study the disease process. **Koch's postulates,** the criteria that Robert Koch used to establish that

Bacillus anthracis causes anthrax (see *A Glimpse of History*), provide a foundation for establishing that a given microbe causes a specific disease. With the advent of precise genetic techniques, **Molecular Koch's postulates** were proposed as a means to identify the critical virulence factors of microorganisms and viruses.

Koch's Postulates

Robert Koch proposed that in order to conclude that a microbe causes a particular disease, these postulates must be fulfilled **(figure 17.3):**

1. The microorganism must be present in every case of the disease.
2. The organism must be grown in pure culture from diseased hosts.
3. The same disease must be produced when a pure culture of the organism is introduced into susceptible hosts.
4. The organism must be recovered from the experimentally infected hosts.

When Koch studied anthrax, he grew *B. anthracis* from all cases examined; he introduced pure cultures of the organisms grown in the laboratory into healthy susceptible mice, again causing the disease anthrax. Finally, he recovered the organism from the experimentally infected mice.

It is important to note, however, that there are many situations in which the postulates cannot be carried out. For example, the second postulate cannot be fulfilled for organisms such as the causative agent of syphilis, *Treponema pallidum,* which has never been grown in laboratory medium. In addition, the third postulate does not always hold true. There are many examples, including cholera and polio, in which some infected people do not have symptoms of disease. Also, suitable experimental animal hosts are not available for some disease and it would not be ethical to test the postulates on humans because of safety concerns. Nevertheless, despite the limitations of the postulates, they have provided scientists with a logical framework for determining the causes of a disease.

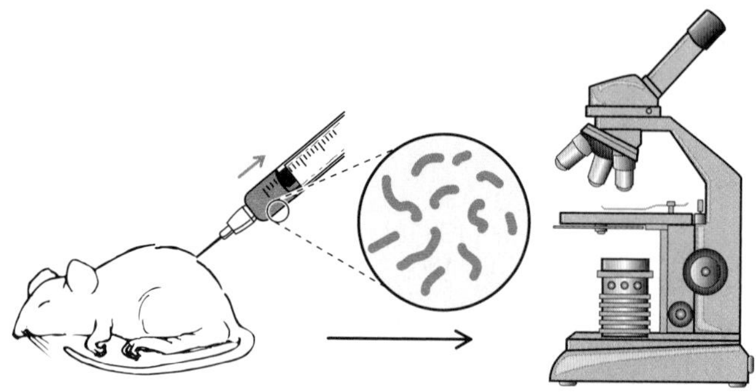

(1) The microorganism must be present in every case of the disease, but not in healthy people.

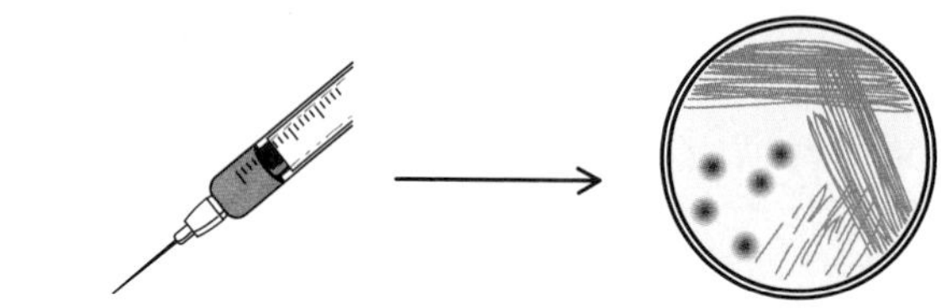

(2) The organism must be grown in pure culture from diseased hosts.

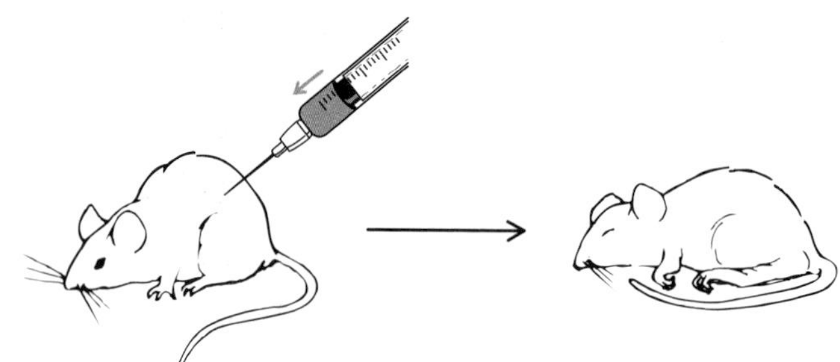

(3) The same disease must be produced when a pure culture of the organism is introduced into susceptible hosts.

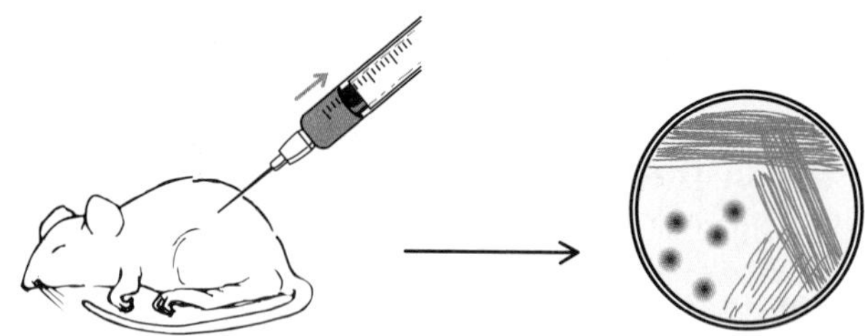

(4) The organism must be recovered from the experimentally infected hosts.

FIGURE 17.3 Koch's Postulates These criteria provide a foundation for establishing that a given microbe causes a specific disease.

Molecular Koch's Postulates

Molecular Koch's postulates are similar in principle to Koch's postulates, but they rely on molecular techniques to study a microbe's virulence factors. They are particularly relevant in the study of bacteria such as *E. coli* and *Streptococcus pyogenes,* which can cause a number of different diseases depending on the virulence factors of a given strain. Molecular Koch's postulates are:

1. The virulence factor gene or its product should be found in pathogenic strains of the organism.
2. Mutating the virulence gene to disrupt its function should reduce the virulence of the pathogen.
3. Reversion of the mutated virulence gene or replacement with a wild-type version should restore virulence to the strain.

As with the traditional Koch's postulates, it is not always possible to apply all of these criteria, but they provide an approach to studying how infectious agents cause disease.

MICROCHECK 17.4

Koch's postulates can be used to prove the cause of an infectious disease. Molecular Koch's postulates are used to identify the virulence factors responsible for disease.

- ✓ How were Koch's postulates used to prove the cause of anthrax?
- ✓ In two of Koch's postulates (#2 and #3), a pure culture of the organism is required. What would be the consequence of using a culture that was not pure?

MECHANISMS OF PATHOGENESIS

From a microbe's perspective, the interior of the human body is a lucrative source of nutrients, provided that some obstacles of the innate and adaptive defenses can be overcome. The ability to subvert these defenses and cause damage is what separates pathogenic microbes from the multitudes of other microorganisms that inhabit this planet. The methods that disease-causing microbes use to evade the host defenses and then cause damage are called **mechanisms of pathogenicity.**

A successful pathogen only needs to overcome the innate and adaptive defenses long enough for the pathogen to multiply and then successfully exit the host. In fact, a pathogen that is too adept at overcoming the host defenses and causing damage is actually at a disadvantage because its opportunity to be transmitted may be limited and it loses an exclusive source of nutrients if the host dies. Pathogens and their hosts generally evolve over time to a state of **balanced pathogenicity.** The pathogen becomes less virulent while the host simultaneously becomes less susceptible. This was demonstrated when the myxoma virus was intentionally introduced into Australia in the early 1950s to kill the burgeoning rabbit population. Shortly after introduction of the virus, the rabbit population plummeted. Eventually, however, the numbers of rabbits again began increasing. Viruses recovered from surviving rabbits were shown to be less virulent than the original strain, while at the same time the rabbits themselves were more resistant to the original virus strain.

The mechanism by which a microorganism causes disease generally follows one of several patterns:

- **Production of toxins that are then ingested.** The microorganism does not grow on or in the host, so this is not an infection but rather a **foodborne intoxication,** a form of food poisoning. The only virulence determinant these microbes require is toxin production. Relatively few organisms cause foodborne intoxication; these include *Clostridium botulinum,* which causes botulism; and toxin-producing strains of *Staphylococcus aureus,* which cause staphylococcal food poisoning. ■ botulism, p. 763 ■ *Staphylococcus aureus* foodborne intoxication, p. 763
- **Colonization of mucous membranes of the host, followed by toxin production.** The microorganism multiplies to high numbers on a mucous membrane, such as in the intestinal or upper respiratory tract. There, the microbes produce a toxin that interferes with cell function, sometimes structurally damaging the cell. Examples of bacteria that use this strategy include *Vibrio cholerae,* which causes cholera, *E. coli* O157:H7, which causes bloody diarrhea, and *Corynebacterium diphtheriae,* which causes diphtheria. ■ cholera, p. 596 ■ *E. coli* O157:H7 diarrhea, p. 600 ■ diphtheria, p. 502
- **Invasion of host tissues.** The microbe penetrates the first-line defenses, breaching the body's barriers, and then multiplies within the tissues. Organisms that use this strategy generally have mechanisms to avoid destruction by macrophages; additionally, some have mechanisms to avoid antibodies. There are numerous examples of bacteria that use this strategy, including *Mycobacterium tuberculosis, Yersinia pestis, Salmonella* species, and *Streptococcus pyogenes.* ■ *Mycobacterium tuberculosis,* p. 515 ■ *Yersinia pestis,* p. 682 ■ *Salmonella,* p. 601 ■ *Streptococcus pyogenes,* p. 499
- **Invasion of host tissues, followed by toxin production.** These organisms are similar to those in the previous category, but in addition to their other attributes, they also make toxins. Examples include *Shigella dysenteriae* and certain strains of *Streptococcus pyogenes.* ■ *Shigella,* p. 598

Awareness of the pathogenic strategies of microorganisms and viruses is important because they illustrate why only certain microbes are able to cause disease in a healthy host. They also help explain the epidemiology and symptoms associated with the various diseases described later in the textbook.

In the next sections we will describe mechanisms that pathogens use to adhere to and colonize host tissue, avoid innate defenses, avoid adaptive defenses, and, finally, cause the damage associated with disease. We will focus on mechanisms used by bacterial pathogens because these are by far the most thoroughly characterized; later in the chapter we will discuss mechanisms of pathogenicity of viruses and eukaryotic organisms. As we describe various virulence determinants, recognize that their attributes are not mutually exclusive—a single structure on a cell can serve more than one purpose. Also note that a single microbial species can have more than one virulence determinant.

17.5 Establishment of Infection

Focus Points

- Compare and contrast adherence and colonization.
- Explain the role of type III secretion systems in infection.

In order to cause disease, most pathogens must first adhere to a body surface; once they have attached, they then multiply to high enough numbers to produce an appreciable amount of toxin, to invade, or both. In some cases, they deliver molecules to epithelial cells, inducing a specific change in those cells.

Adherence

Because the first-line defenses are very effective in removing organisms, a pathogen must adhere to host cells as a necessary first step in the establishment of infection. Microorganisms that attach to cells, however, do not necessarily cause disease. Many members of the normal microbiota adhere to epithelial cells with no ill effect whatsoever. Other factors such as toxin production and invasive ability of the microbe generally must come into play before disease results. ■ first-line defenses, p. 348

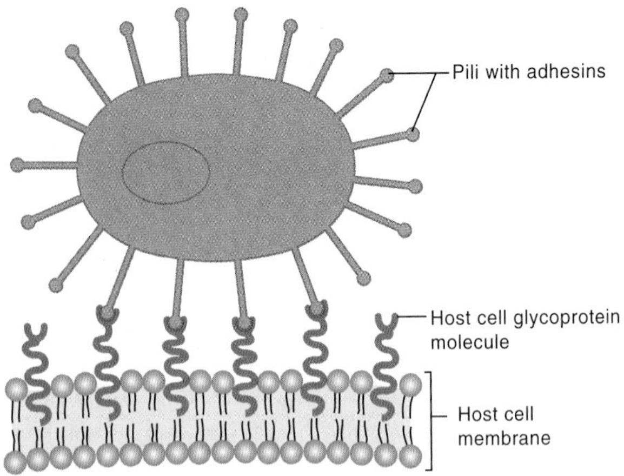

FIGURE 17.4 Pili Attachment to Host Cell Protein adhesins of microorganisms attach to glycoprotein or glycolipid molecules on host cells (the molecules serve as receptors).

To bind to host cells, bacteria use **adhesins.** These are often located at the tips of pili, the filamentous protein structures found on the surface of the cell (**figure 17.4;** see figure 3.41). Pili used for attachment are often called **fimbriae.** Adhesins can also be a component of other surface structures such as capsules or various cell wall proteins (see figure 3.36). The streptococci that cause tooth decay attach to the tooth surface and then convert dietary sugar to an insoluble glue. Other bacteria then adhere to this sticky surface, forming a biofilm. ■ pili, p. 66 ■ glycocalyx, p. 64 ■ biofilm, p. 85

The surface receptors on animal cells to which the bacterial adhesins attach are typically glycoproteins (protein molecules that have various sugars attached) or glycolipids; the adhesin binds to the sugar component of the receptor. Note that the surface receptor serves a distinct role for the cell that produces it; the microbe merely exploits the molecule for its own use. *Neisseria gonorrhoeae* adheres to mucosal epithelial cells by means of a molecule called CD46; the normal function of this molecule is to bind certain complement system components, facilitating their degradation before they trigger responses that can damage the host cell. ■ complement system, p. 355

The binding of an adhesin to a surface receptor is highly specific, dictating the type of cells to which the bacterium can attach. The adhesin of common strains of *E. coli* allows them to adhere to cells that line the large intestine. Pathogenic strains have other additional adhesins, broadening the range of tissues to which they can attach. Strains of *E. coli* that cause urinary tract infections generally produce a type of pili called P pili; these pili mediate attachment to cells that line the bladder. *E. coli* strains that cause watery diarrhea produce a type of pili that allows the bacteria to attach specifically to cells of the small intestine. These strains, enterotoxigenic *E. coli* (ETEC), are often species-specific; some infect humans, whereas others infect only certain domestic animals. ■ enterotoxigenic *E. coli*, p. 600

Colonization

A microorganism must multiply in order to colonize the host, a prerequisite for infectious disease. To colonize a site already populated by normal microbiota, the new arrival must compete successfully with established organisms for space and nutrients, and overcome their toxic products such as fatty acids. They also must counter the body's defenses aimed at protecting the surfaces.

Secretory IgA antibodies that bind to adhesin molecules on pathogens prevent attachment of the organisms to mucosal surfaces, thwarting their attempts at colonization. Microbes, however, have developed counterstrategies. These include rapid turnover of pili, which effectively sheds bound antibody; antigenic variation, which can alter the type of pili produced; and **IgA proteases,** which cleave the antibody molecule. ■ IgA, p. 374 ■ antigenic variation, p. 181

One of the limiting nutrients for bacteria is often iron; recall that the body uses lactoferrin and transferrin to sequester iron. Some pathogens produce their own iron-binding molecules, **siderophores,** which compete with the host proteins; others are able to use the iron bound to the host proteins. ■ lactoferrin and transferrin, p. 349

Delivery of Effector Molecules to Host Cells

Once they have colonized a surface, some bacteria are able to deliver certain molecules directly to host cells, inducing changes in those cells. This characteristic appears to be particularly common among gastrointestinal pathogens. In some cases, the compounds delivered induce changes that damage the recipient cell, such as loss of microvilli; in others, they direct the uptake of the bacterial cell, a process that will be discussed in the next section. ■ microvilli, p. 584

Gram-negative bacteria use what is called a **type III secretion system** or **injectisome,** to deliver proteins to eukaryotic cells. These structures resemble short flagella and function as microscopic hypodermic needles **(figure 17.5).** The secretion systems and transferred proteins are often encoded by pathogenicity islands. ■ secretion system, p. 61 ■ pathogenicity island, p. 208

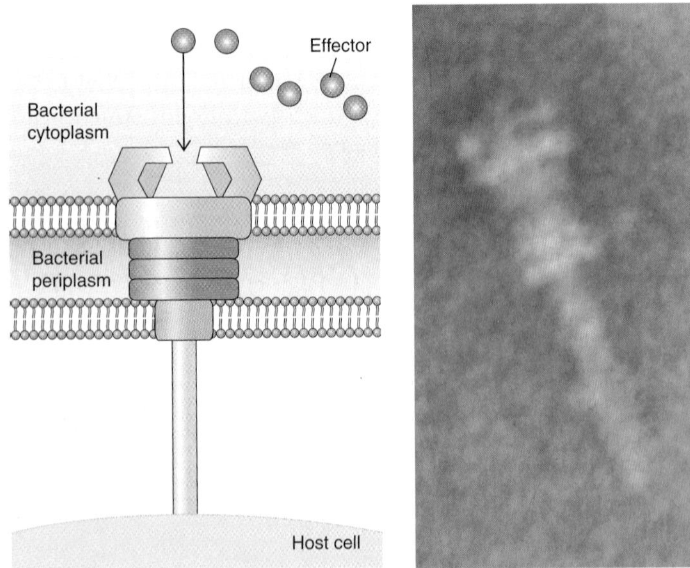

FIGURE 17.5 Type III Secretion Systems Gram-negative bacteria use type III secretion systems, or injectisomes, to deliver certain molecules directly to host cells, inducing changes in those cells.

MICROCHECK 17.5

Pathogens use adhesins, often on pili, to bind to a body surface. To colonize a surface, the pathogen must often compete with the normal microbiota, prevent binding of secretory IgA, and obtain iron. Some bacteria deliver certain molecules to epithelial cells, inducing a specific change in those cells.

✓ What are siderophores?

✓ What is a type III secretion system?

✓ Why is it a good strategy for a microbe to adhere to a receptor that plays a critical function for a host cell?

17.6

Invasion—Breaching the Anatomical Barriers

Focus Point

- Describe the mechanisms pathogens use to penetrate the skin and mucous membranes.

Some bacterial pathogens cause disease while remaining on the mucosal surfaces, but many others breach the anatomical barriers. By traversing the epithelial cell barrier and accessing the nutrient-rich tissue, these invading microbes obtain an exclusive source of nutrients, multiplying without competition.

Penetration of Skin

Skin is the most difficult anatomical barrier for a microbe to penetrate. Bacterial pathogens that invade via this route rely on trauma of some sort that destroys the integrity of the skin. *Staphylococcus aureus,* a common cause of wound infections, accesses tissues via the lesion. *Yersinia pestis,* the causative agent of plague, exploits fleas, relying on them to inject the microbe through the skin when they bite. ■ plague, p. 682

Penetration of Mucous Membranes

Invasion of mucous membranes is the most common route of entry for most pathogens, but the invasive processes are complex and inherently difficult to study. It appears, however, that there are at least two general mechanisms used for invasion—directed uptake by cells and exploitation of antigen-sampling processes.

Directed Uptake by Cells

Some pathogens induce non-phagocytic cells to take up the bacterial cells by endocytosis. The pathogen first attaches to a cell, then triggers that cell to engulf the bacterium. In many cases, specialized proteins that cause changes in the host cell's cytoskeleton are delivered by the type III secretion systems to induce the uptake. The directed disruption of the cytoskeleton causes obvious morphological disturbances in the cell membrane called **ruffling** **(figure 17.6).** The bacteria appear to sink into the ruffles, leading to their eventual engulfment. *In vitro* studies show that *Salmonella* species can induce uptake by intestinal epithelial cells. *Shigella*

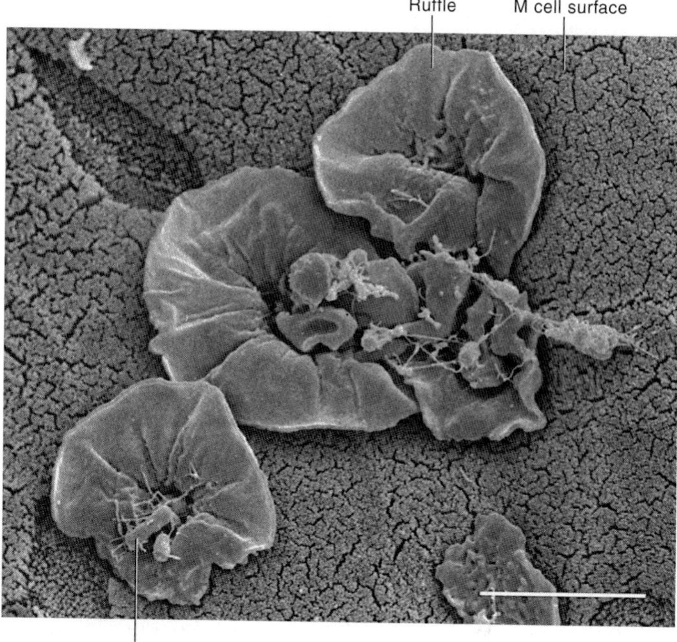

FIGURE 17.6 Ruffling *Salmonella* Typhimurium inducing ruffles on an M cell (a specialized epithelial cell) leading to uptake of the bacterial cells (scale bar = 10 μm).

species do the same, but they enter the mucosal epithelial cells at their base; they must first pass through an M cell to access the basement membrane side of epithelial cells. The mechanism they use to do this will be described in the next section. ■ endocytosis, p. 73 ■ cytoskeleton, p. 74 ■ basement membrane, p. 348 ■ M cell, p. 370

Exploitation of Antigen-Sampling Processes

Recall that secondary lymphoid tissues are located at strategic sites so that antigens that enter the body or pass through the intestine can be sampled. Some bacteria exploit the sampling process in order to access deeper tissues. ■ secondary lymphoid tissues, p. 370

Several intestinal pathogens penetrate the mucous membrane by way of the M cells that overlie the Peyer's patches in the intestine. Recall that M cells are specialized cells that function as a conduit between the intestinal lumen and the lymphoid tissues of the Peyer's patches. They routinely sample intestinal contents and deliver them to macrophages in the lymphoid tissues beneath; this process of transferring material from one side of the cell to the other is called **transcytosis.** Most microbes transcytosed by M cells are destroyed by the macrophages that receive them, but pathogens have evolved mechanisms to avoid this demise. *Shigella* species use M cells to traverse the epithelial barrier; once on the other side, the macrophages ingest them, but the bacteria survive the phagocytic process, eventually escaping by inducing apoptosis in the phagocytic cells **(figure 17.7).** *Shigella* cells then adhere to specific receptors at the base of the epithelial cells and then induce these non-phagocytic cells to engulf them. ■ survival within the phagocyte, p. 403 ■ Peyer's patch, p. 370

Some pathogens invade by means of alveolar macrophages, which engulf material that enters the lungs. *Mycobacterium tuberculosis* produces surface proteins that facilitate their uptake by the

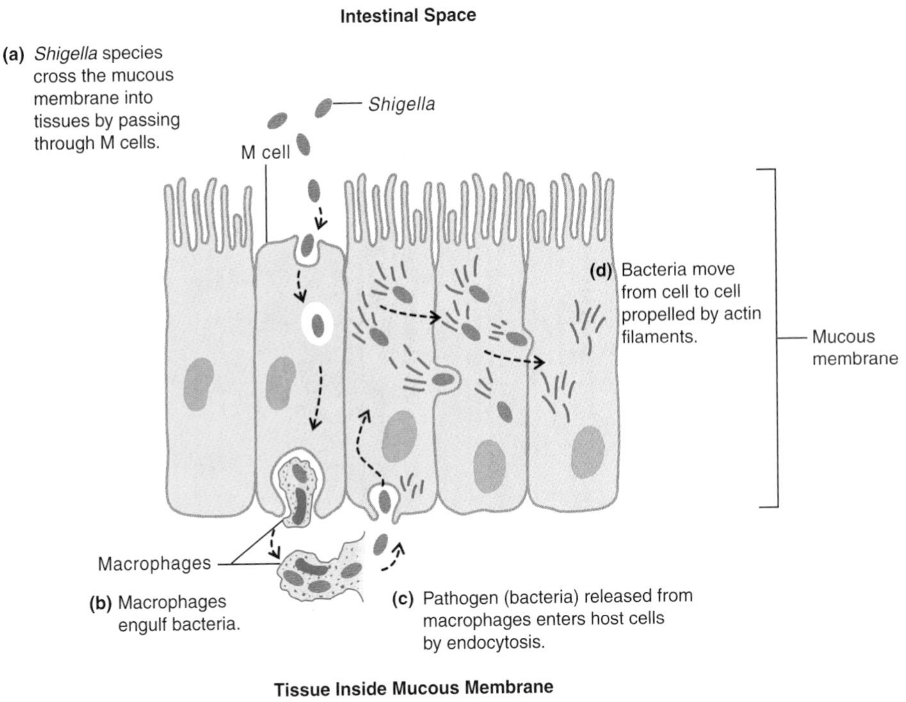

Intestinal Space

(a) *Shigella* species cross the mucous membrane into tissues by passing through M cells.

Shigella

M cell

(d) Bacteria move from cell to cell propelled by actin filaments.

Mucous membrane

Macrophages

(b) Macrophages engulf bacteria.

(c) Pathogen (bacteria) released from macrophages enters host cells by endocytosis.

Tissue Inside Mucous Membrane

FIGURE 17.7 Antigen-Sampling Processes Provide a Mechanism for Invasion *Shigella* species use M cells to traverse the epithelial barrier. Once on the other side, macrophages ingest them, but the bacteria are able to escape and then infect other cells.

macrophages. While this might seem disadvantageous, it actually allows the organism to avoid a process that could otherwise lead to macrophage activation. *Mycobacterium* cells can survive within macrophages that have not been activated.

MICROCHECK 17.6

Skin is the most difficult barrier for a microbe to penetrate. Some pathogens induce mucosal epithelial cells to engulf the bacterial cells. Some exploit antigen-sampling processes, including transcytosis by M cells and phagocytosis.

✓ How do *Shigella* species enter intestinal epithelial cells?

✓ Why does *Mycobacterium tuberculosis* facilitate its own uptake by macrophages?

✓ Why would some pathogens behave differently in a non-human host?

17.7 Avoiding the Host Defenses

Focus Point

- Describe mechanisms that bacteria use to avoid complement system proteins, antibodies, and destruction by phagocytes.

Inside the body, invading microorganisms soon encounter the innate and adaptive immune defenses. Pathogens as a group have evolved a variety of mechanisms to circumvent the otherwise lethal effects of these defenses.

Hiding Within a Host Cell

Some bacteria evade critical components of the host defenses by remaining inside host cells, out of the reach of the phagocytes, complement proteins, and antibodies. An example is *Shigella* species; recall that these organisms induce their own uptake by intestinal epithelial cells. Once inside a host cell, the bacteria can orchestrate their transfer to adjacent cells. They do this by causing the rapid polymerization of host cell actin at one end of the bacterial cell, effectively forming an "actin tail." This action propels the bacterium within the cell. The force of the propulsion is so great that the bacterial cells are often driven into neighboring cells. *Listeria monocytogenes* also polymerizes host cell actin to facilitate its own movement **(figure 17.8).** ■ actin, p. 74

Avoiding Killing by Complement System Proteins

As we discussed in chapter 15, activation of the complement system leads to three primary outcomes—lysis of foreign cells by the membrane attack complex (MAC), opsonization, and inflammation (see figure 15.7). Because the latter two outcomes are associated with phagocytosis, mechanisms that bacteria use to subvert them will be discussed in the next section. Here, we will focus on mechanisms bacteria use to avoid the lethal effects of the MAC. Recall that Gram-negative bacteria are susceptible to the MAC because their outer membrane serves as a target; the MAC has little effect on Gram-positive organisms. ■ complement system, p. 355

Gram-negative bacteria that circumvent killing by the complement proteins are said to be **serum resistant.** Strains

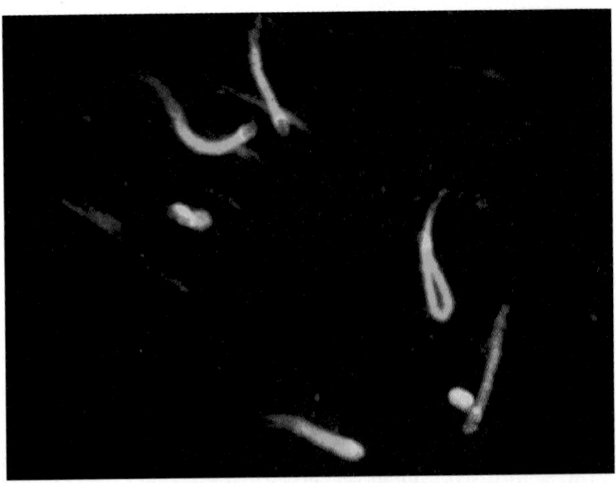

FIGURE 17.8 Actin Tail of Intracellular *Listeria monocytogenes* Rapid polymerization of host cell actin (green) at one end of the bacterial cell (orange) propels the bacterium within the cell.

of *Neisseria gonorrhoeae* that cause disseminated gonococcal infection, a systemic disease characterized by symptoms including fever, rash, and arthritis, are serum resistant. These strains are able to capture the mechanism that host cells use to prevent their own surfaces from activating the complement system, thereby preventing MAC formation. Recall that the alternative pathway of complement system activation can be initiated when the complement component C3b binds to a cell surface (see figure 15.7). Host cells are protected because they bind complement regulatory proteins that quickly inactivate any C3b that binds to the cell, preventing the alternative pathway from being triggered. Invasive *N. gonorrhoeae* strains effectively hijack the protective mechanism normally used by host cells, thwarting the events that would otherwise lead to formation of MAC (**figure 17.9**). ■ disseminated gonococcal infection, p. 631

Avoiding Destruction by Phagocytes

Phagocytosis involves multiple steps—including chemotaxis, recognition and attachment, engulfment, and fusion of the phagosome with the lysosome—that lead to the destruction and digestion of an invading microbe (see figure 15.10). Pathogenic bacteria have evolved a variety of ingenious mechanisms to avoid the destructive effects of phagocytosis (**figure 17.10**).

Preventing Encounters with Phagocytes

Some pathogens prevent phagocytosis by avoiding macrophages and neutrophils altogether. One way they do this is by destroying the complement component that attracts phagocytes; the other is to kill phagocytic cells as they arrive. The mechanisms involved include:

- **C5a peptidase.** This enzyme degrades the complement component C5a, a chemoattractant that recruits phagocytic cells to an area where the complement system has been activated. *Streptococcus pyogenes,* which causes strep throat, is an example of a bacterium that makes C5a peptidase. ■ C5a, p. 356 ■ *Streptococcus pyogenes*, p. 499
- **Membrane-damaging toxins.** These kill phagocytes and other cells, often by forming pores in their membranes. *S. pyogenes* makes a membrane-damaging toxin called streptolysin O. ■ membrane-damaging toxins, p. 405

Avoiding Recognition and Attachment

Recall that recognition and attachment of phagocytes to foreign material is most efficient if either the complement system component C3b or certain classes of antibody molecules have first opsonized the material. Mechanisms that bacteria use to avoid recognition and attachment include:

- **Capsules.** These have long been recognized for their ability to prevent phagocytosis. This trait is often due to their capacity to interfere with the alternative pathway of complement activation. They accomplish this by binding to host cell complement regulatory proteins that inactivate C3b that has bound to a surface; this mechanism is identical to that described earlier for serum-resistant bacteria (see figure 17.9). Rapid inactivation of

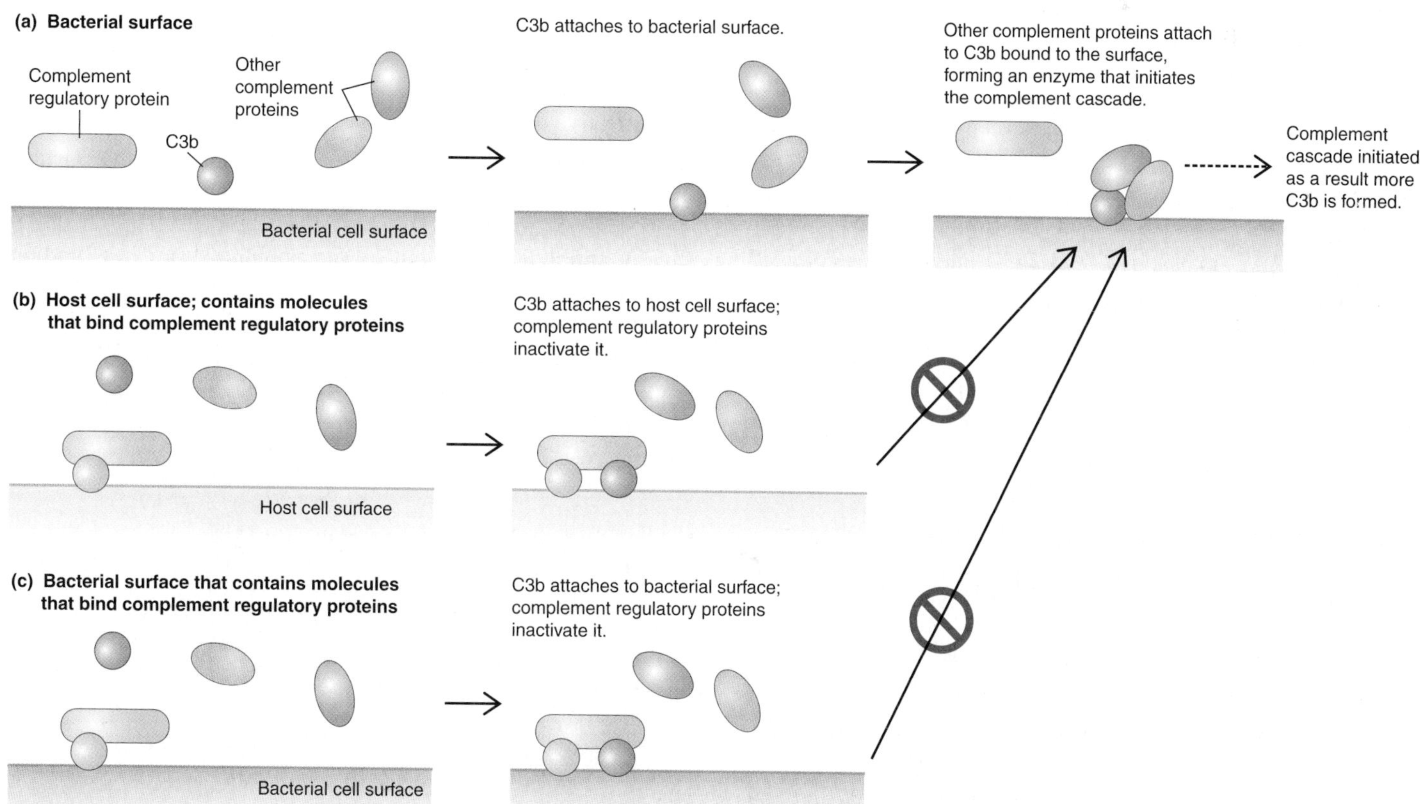

(a) Bacterial surface
Complement regulatory protein
Other complement proteins
C3b
Bacterial cell surface
C3b attaches to bacterial surface.
Other complement proteins attach to C3b bound to the surface, forming an enzyme that initiates the complement cascade.
Complement cascade initiated; as a result more C3b is formed.
(b) Host cell surface; contains molecules that bind complement regulatory proteins
Host cell surface
C3b attaches to host cell surface; complement regulatory proteins inactivate it.
(c) Bacterial surface that contains molecules that bind complement regulatory proteins
Bacterial cell surface
C3b attaches to bacterial surface; complement regulatory proteins inactivate it.

FIGURE 17.9 Avoiding the Alternative Pathway of Complement Activation Some bacterial pathogens foil the activation of the complement system by carrying surface components that attach to host cell regulatory proteins. The regulatory proteins inactivate C3b that has bound to a surface.

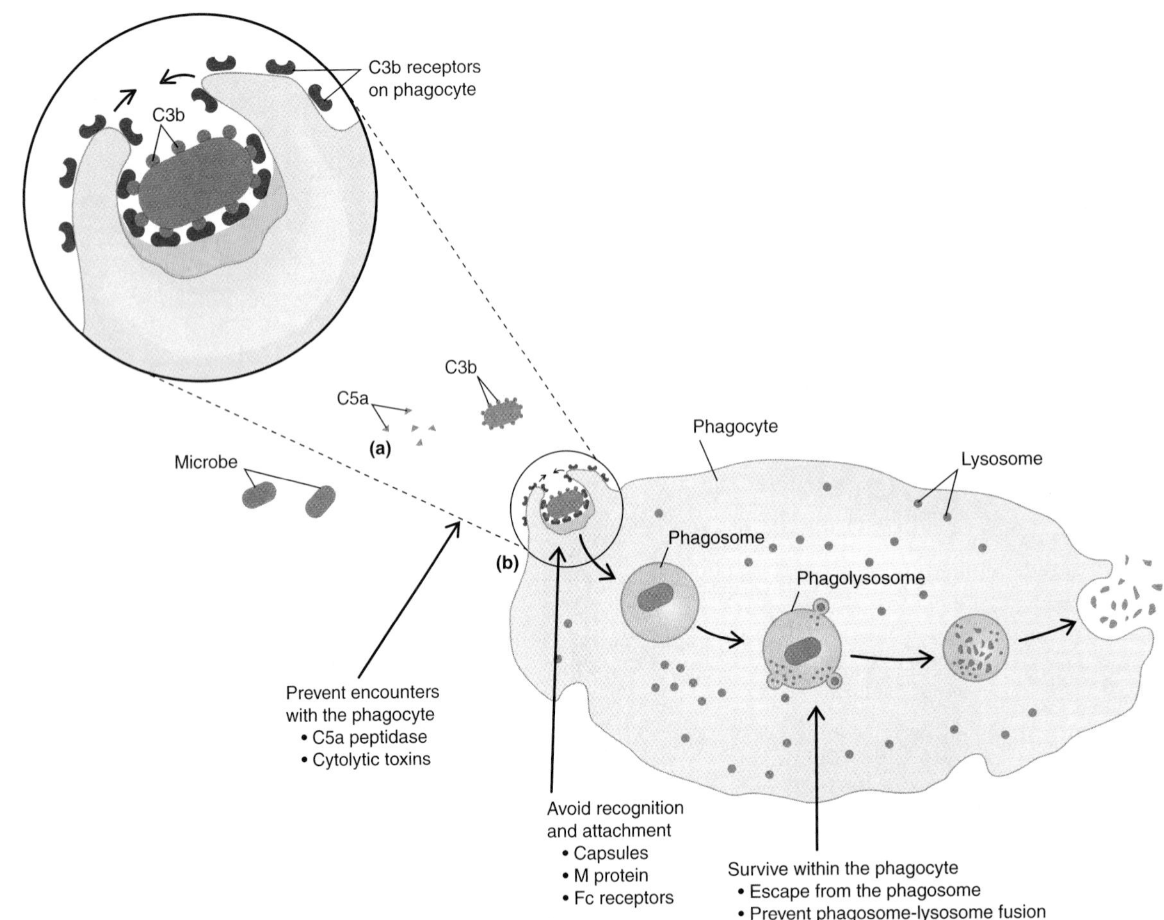

FIGURE 17.10 Avoiding Destruction by Phagocytes

C3b also prevents the molecule from being an effective opsonin. The pneumonia-causing bacterium *Streptococcus pneumoniae* produces a capsule that prevents phagocytosis.

- **M protein.** This component of the cell wall of *Streptococcus pyogenes* functions in a manner similar to that of the *Streptococcus pneumoniae* capsule; it binds to a complement regulatory protein that inactivates C3b, interfering with complement activation and the subsequent formation of more C3b. Rapid inactivation of C3b also prevents it from being an effective opsonin.
- **Fc receptors.** These proteins are found on the surface of *Staphylococcus aureus* (protein A) and *Streptococcus pyogenes* (protein G). The Fc receptors foil opsonization by antibodies; they outfox the immune system by binding the Fc region of antibody molecules, reversing the intended orientation of the molecule on the cell **(figure 17.11).** Recall

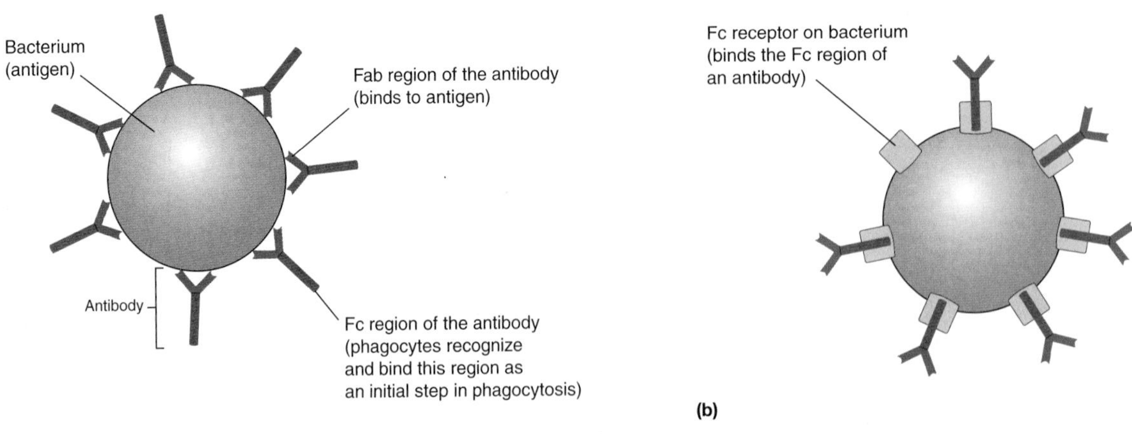

FIGURE 17.11 Fc Receptors Foil Opsonization by Antibodies **(a)** The defense system's intended orientation of antibody molecules on the surface of a bacterium; note that the Fc region of the antibody projects from the bacterial cell, making it available for a phagocyte to recognize and bind. **(b)** The effect of Fc receptors on a bacterial cell's surface; the receptors reverse the defense system's intended orientation of the antibody molecules on the cell, preventing the Fc region from interacting with the phagocyte cell.

that antibodies have two parts—the Fab region, which binds specifically to antigens, and the Fc region, which functions as a "red flag," directing the fate of the antigen. Bacterial cells that have Fc receptors are able to coat themselves with antibody molecules in a manner that obscures the cell from phagocytes. The phagocytic cell has no mechanism for recognizing the Fab region, which will be projecting outward on these cells.

Surviving Within the Phagocyte

Some bacteria make no attempt to avoid engulfment, instead using phagocytosis as an opportunity. It allows them to hide from antibodies, control some aspects of the immune response, and obtain a "free ride" to other locations in the body. Mechanisms used to survive within phagocytes include:

- **Escape from the phagosome.** Some bacteria are able to escape from the phagosome before it fuses with the lysosome; these organisms then multiply within the cytoplasm of the phagocyte, protected from other host defenses. *Listeria monocytogenes* produces a molecule that forms pores in the phagosomal membrane, allowing the bacteria to escape to the cytoplasm. *Shigella* species are also able to lyse the phagosome before it fuses with the lysosome.
- **Preventing phagosome-lysosome fusion.** Bacteria that prevent phagosome-lysosome fusion avoid the otherwise inevitable exposure to the degradative enzymes and other toxic components of the lysosomes. *Salmonella* species are able to sense they have been ingested by a macrophage, and respond by producing a protein that blocks the fusion process.
- **Surviving within the phagolysosome.** Relatively few organisms are able to survive the destructive environment within the phagolysosome. *Coxiella burnetii,* however, an obligate intracellular parasite that causes Q fever, is able to withstand the conditions. It appears that once the organism has been ingested by a macrophage, it can delay fusion of the phagosome with the lysosome, allowing the microbe additional time to equip itself for growth within the phagolysosome.

Avoiding Antibodies

Pathogens that survive the initial assault of the innate defenses soon encounter an additional obstacle, the adaptive defenses. For most bacteria, the most formidable of these are antibodies. Mechanisms for avoiding antibodies include:

- **IgA protease.** This enzyme cleaves, IgA, the class of antibody found in mucus and other secretions. *Neisseria gonorrhoeae* and a variety of other pathogens produce IgA protease. This enzyme may also have other roles.
- **Antigenic variation.** Some pathogens routinely alter the structure of their surface antigens. This allows them to stay ahead of antibody production by altering the very molecules antibodies would otherwise recognize. *Neisseria gonorrhoeae* is able to vary the antigenic structure of its pili; antibodies produced by the infected host in response to one variation of the pili cannot bind effectively to another. ■ antigenic variation, p. 181
- **Mimicking host molecules.** An essential aspect of the immune system is the development of tolerance for molecules that the body perceives as healthy "self." Pathogens can exploit this by covering themselves with molecules that resemble normal "self" molecules. Certain strains of *Streptococcus pyogenes* have a capsule composed of hyaluronic acid, a polysaccharide found in tissues.

MICROCHECK 17.7

Serum-resistant bacteria avoid the killing effects of complement system proteins. Mechanisms bacteria use to avoid destruction by phagocytes include preventing encounters with phagocytes, avoiding recognition and attachment, and surviving within the phagocyte. Mechanisms for avoiding antibodies include IgA protease, antigenic variation and mimicking host molecules.

✓ Describe the effects of the Fc receptor on the surface of a bacterial cell.

✓ Describe three mechanisms that bacteria can use to survive within a phagocytic cell.

✓ Encapsulated organisms can be phagocytized once antibodies against the capsule have been produced. Why would this be so?

17.8

Damage to the Host

Focus Points

- Describe the difference between exotoxins and endotoxins.
- Compare and contrast neurotoxins, enterotoxins, and cytotoxins, giving two examples of each.
- Explain how inflammation and antibody production can be damaging.

In order to cause disease, a pathogen must evoke some type of damage to the host. In many cases, the damage facilitates dispersal of the organism, enabling it to infect other hosts. *Vibrio cholerae,* which causes cholera, induces watery diarrhea; up to 20 liters of microbe-containing fluid can be excreted in one day. In areas of the world without adequate sewage treatment, this can lead to contaminated water supplies and widespread outbreaks. *Bordetella pertussis,* which causes whooping cough, facilitates its airborne dispersal by causing severe bursts of coughing. Damage due to infection can be the result of direct effects of the pathogen, such as toxins produced, or indirect effects, such as the immune response.

Exotoxins

A number of Gram-positive and Gram-negative bacterial pathogens produce **exotoxins**—proteins that have very specific damaging effects **(table 17.1).** These proteins are among

TABLE 17.1 Exotoxins Produced by Various Primary Pathogens

Toxins	Name of Disease; Name of Toxin	Characteristics of the Disease	Mechanism	Page Reference
A-B TOXINS—Composed of two subunits, A and B. The A subunit is the toxic, or active, part; the B subunit binds to the target cell.				
Neurotoxins				
Clostridium botulinum	Botulism; botulinum toxin	Flaccid paralysis	Blocks transmission of nerve signals to the muscles by preventing the release of acetylcholine.	p. 657
Clostridium tetani	Tetanus; tetanospasmin	Spastic paralysis	Blocks the action of inhibitory neurons by preventing the release of neurotransmitters.	p. 566
Enterotoxins				
Enterotoxigenic *E. coli*	Traveler's diarrhea; heat-labile enterotoxin (cholera-like toxin)	Severe watery diarrhea	Modifies a regulatory protein in intestinal cells, causing those cells to continuously secrete electrolytes and water.	p. 600
Vibrio cholerae	Cholera; cholera toxin	Severe watery diarrhea	Modifies a regulatory protein in intestinal cells, causing those cells to continuously secrete electrolytes and water.	p. 596
Cytotoxins				
Bacillus anthracis	Anthrax; edema factor, lethal factor	Inhaled form—septic shock; cutaneous form—skin lesions	Edema factor modifies a regulatory protein in cells, causing accumulation of fluids. Lethal factor inactivates proteins involved in cell signaling functions.	p. 512
Bordetella pertussis	Pertussis (whooping cough); pertussis toxin	Sudden bouts of violent coughing	Modifies a regulatory protein in respiratory cells, causing accumulation of respiratory secretions and mucus. Other factors also contribute to the symptoms.	p. 513
Corynebacterium diphtheriae	Diphtheria; diphtheria toxin	Pseudomembrane in the throat; heart, nervous system, kidney damage	Inhibits protein synthesis by inactivating an elongation factor of eukaryotic cells. Kills local cells (in the throat) and is carried in the bloodstream to various organs.	p. 503
E. coli O157:H7	Bloody diarrhea, hemolytic uremic syndrome; shiga toxin	Diarrhea that may be bloody; kidney damage	Inactivates the 60S subunit of eukaryotic ribosomes, halting protein synthesis.	p. 600
Shigella dysenteriae	Dysentery, hemolytic uremic syndrome; shiga toxin	Diarrhea that contains blood, pus, and mucus; kidney damage	Inactivates the 60S subunit of eukaryotic ribosomes, halting protein synthesis.	p. 598
MEMBRANE-DAMAGING TOXINS (cytotoxins)—Disrupts plasma membranes.				
Clostridium perfringens	Gas gangrene; α-toxin	Extensive tissue damage	Removes the polar head group on the phospholipids in the membrane, destablizing membrane integrity of cells.	p. 568
Staphylococcus aureus	Wound and other infections; leukocidin	Accumulation of pus	Inserts into membranes, forming pores that allow fluids to enter the cells.	p. 562
Streptococcus pyogenes	Pharyngitis and other infections; streptolysin O	Accumulation of pus	Inserts into membranes, forming pores that allow fluids to enter the cells.	p. 499
SUPERANTIGENS—Overrides the specificity of the T-cell response.				
Staphylococcus aureus (certain strains)	Foodborne intoxication; staphylococcal enterotoxins	Nausea and vomiting	How the ingested toxins lead to the characteristic symptoms of foodborne intoxication is not understood.	p. 763
Staphylococcus aureus (certain strains)	Staphylococcal toxic shock; toxic shock syndrome toxin (TSST)	Fever, vomiting, diarrhea, muscle aches, rash, low blood pressure	Systemic toxic effects due to the resulting massive release of cytokines.	p. 626
Streptococcus pyogenes (certain strains)	Streptococcal toxic shock; streptococcal pyrogenic exotoxins (SPE)	Fever, vomiting, diarrhea, muscle aches, rash, low blood pressure	Systemic toxic effects due to the resulting massive release of cytokines.	p. 501

TABLE 17.1 Exotoxins Produced by Various Primary Pathogens (*continued*)

Toxins	Name of Disease; Name of Toxin	Characteristics of the Disease	Mechanism	Page Reference
OTHER TOXIC PROTEINS				
Staphylococcus aureus	Scalded-skin syndrome; exfoliatin	Separation of the outer layer of skin	Thought to break ester bonds that hold the layers of skin together.	p. 537
Various organisms	Various diseases; proteases, lipases, and other hydrolases	Tissue damage	Degrades proteins, lipids, and other compounds that make up tissues.	

the most potent toxins known and are often a major cause of damage to an infected host.

Exotoxins are either secreted by the bacterium or leak into the surrounding fluid following lysis of the bacterial cell. In most cases, the pathogen must colonize a body surface or tissue to produce enough toxin to cause damage. With foodborne intoxication, however, the bacterial cells multiply in a food product where they produce toxin that is then consumed. In the case of botulism, ingestion of minute amounts of botulinum toxin is sufficient to cause paralysis. Like most other exotoxins, this toxin can be destroyed by heating. ■ botulism, p. 763

Exotoxins can act locally, or they may be carried in the bloodstream throughout the body, causing systemic effects. *Corynebacterium diphtheriae,* the organism that causes diphtheria, grows and releases its exotoxin in the throat. There, the toxin destroys local cells, leading to the formation of a pseudomembrane composed of dead host cells, pus, and blood. This membrane can sometimes dislodge and obstruct the airway. The toxin is absorbed and carried to the heart, nervous system, and other organs, causing additional damage. ■ diphtheria, p. 502

Because exotoxins are proteins, the immune system can generally produce protective antibodies. Unfortunately, many exotoxins are so powerful that fatal damage can occur before an adequate immune response is mounted. This is why vaccination, using a toxoid (an inactivated toxin), is important in preventing otherwise common diseases such as tetanus and diphtheria (see table 19.1). Passive immunity to a given toxin can be provided by administering an antitoxin; for example, a person who develops symptoms of botulism is treated with botulinum antitoxin. Although vaccines against botulinum toxin exist, they are not routinely used because the risks of developing the disease are negligible if prudent food preparation procedures are followed. ■ toxoid, p. 435 ■ antitoxin, p. 433

Many exotoxins can be grouped into functional categories according to the tissues they adversely impact (see table 17.1). **Neurotoxins** damage the nervous system, causing symptoms such as paralysis. **Enterotoxins** cause symptoms associated with intestinal disturbance, such as diarrhea and vomiting. **Cytotoxins** damage a variety of different cell types, either by interfering with essential cellular mechanisms or by lysing the cell. Some exotoxins do not fall into any of these groups, instead causing symptoms associated with excessive stimulation of the immune response.

Most exotoxins fall into three general categories that reflect their structure and general mechanism of action—A-B toxins, membrane-damaging toxins, and superantigens.

A-B Toxins

A-B toxins consist of two parts—one constitutes the toxic, or active, part (the A subunit) and another (the B subunit) binds to specific receptors on cells **(figure 17.12).** In other words, the A subunit, usually an enzyme, is responsible for the effects the toxin has on a cell, whereas the B subunit dictates the type of cell to which the toxin is delivered.

The structure of A-B toxins offers novel approaches for the development of vaccines and therapies. For example, a fusion protein that contains diphtheria toxin is now being used to treat cutaneous T-cell lymphoma. By fusing the toxin to the cytokine IL-12, the toxin is delivered to T cells, including cancerous ones. In some countries, the B subunit of cholera toxin is used as an orally administered vaccine against cholera. Antibodies that bind the B subunit prevent cholera toxin from attaching to intestinal cells, thus protecting the vaccine recipient. Researchers are now experimenting with joining medically useful compounds to B subunits, allowing those compounds to be delivered specifically to the cell type targeted by the B subunit.

Membrane-Damaging Toxins

Membrane-damaging toxins are cytotoxins that disrupt plasma membranes, causing the cell to lyse. Many also lyse red blood cells, causing hemolysis that can be observed when the organisms are grown on blood agar; thus, many of these toxins are also referred to as **hemolysins.** They are also called **cytolysins.** ■ hemolysis, p. 97 ■ blood agar, p. 96

Some membrane-damaging toxins insert themselves into membranes, forming pores that allow fluids to enter the cell. One pore-forming toxin is streptolysin O, the compound responsible for the characteristic β-hemolysis of *Streptococcus pyogenes* grown anaerobically on blood agar (see figure 4.11). Recall that this membrane-damaging toxin enables *S. pyogenes* to avoid phagocytosis.

Phospholipases are a group of membrane-damaging toxins that enzymatically remove the polar head group of the phospholipids in the plasma membrane, destabilizing its integrity. The α-toxin of *Clostridium perfringens,* which causes gas gangrene, is a phospholipase. ■ phospholipid, p. 36 ■ gas gangrene, p. 568

Superantigens

Superantigens override the specificity of the T-cell response, causing toxic effects due to the massive release of cytokines by an inordinate number of T_H cells (effector helper T cells). They include toxic shock syndrome toxin (TSST) as well as several other toxins produced by *Staphylococcus aureus* and *Streptococcus pyogenes*. ■ helper T cells, p. 368

Superantigens effectively short-circuit the normal control mechanisms inherent in antigen processing and presentation. They do this by binding simultaneously to the outer portion of the major histocompatibility (MHC) class II molecule on antigen-presenting cells and the T-cell receptor **(figure 17.13).** Whereas most antigens stimulate about one in 10,000 T cells, superantigens stimulate as many as one in five T cells. The resulting release of cytokines by an inordinate number of T cells leads to symptoms including fever, nausea, vomiting, and diarrhea. Shock may occur, with failure of many organ systems, circulatory collapse, and even death. In addition, the immune response is suppressed because many T cells undergo apoptosis following the excessive stimulation. Superantigens are also suspected of contributing to autoimmune diseases; by overriding the control mechanisms that normally characterize adaptive immunity, they may induce the proliferation of those few T cells that respond to healthy "self." ■ MHC class II molecules, p. 380 ■ apoptosis, p. 362

The family of related exotoxins produced by *Staphylococcus aureus* strains that cause foodborne intoxication are superantigens. Although they cause nausea and vomiting and are therefore referred to as enterotoxins, the manner in which they function and

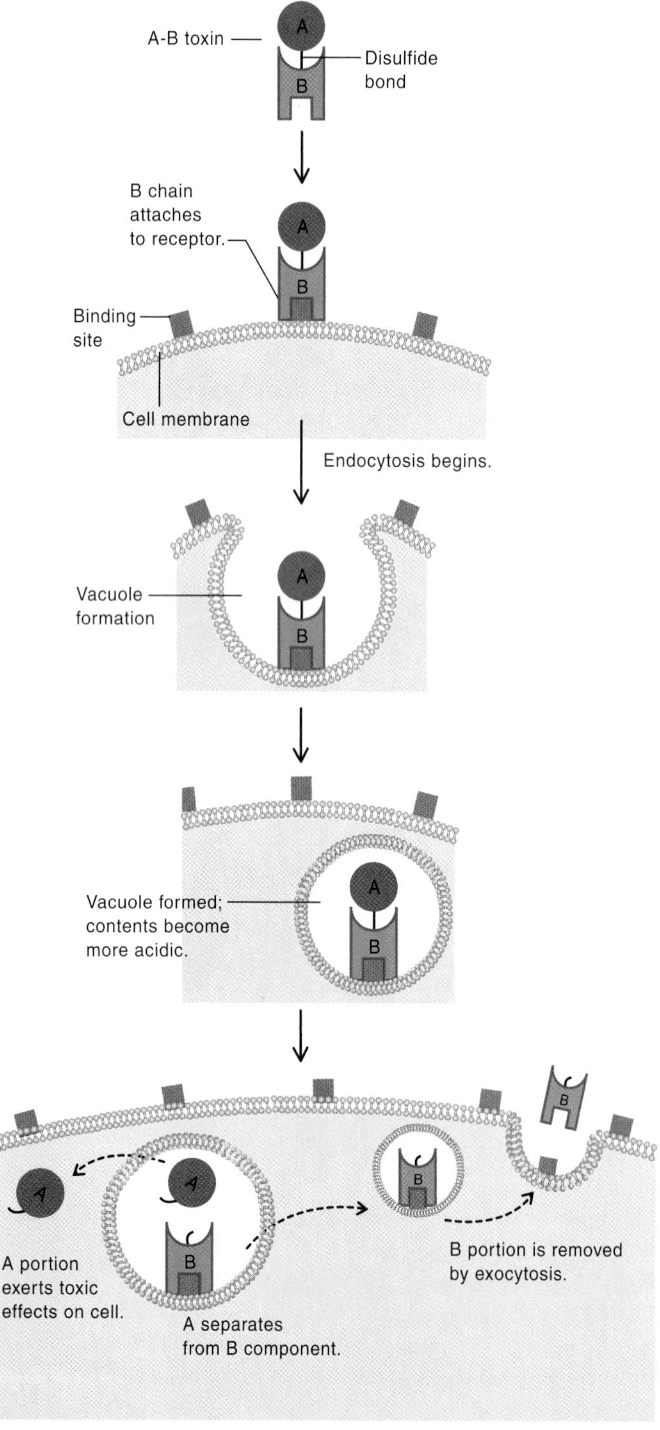

FIGURE 17.12 The Action of A-B Toxins The A portion is toxic and the B portion binds to specific receptors on cells. After binding, the A-B molecule is taken up by endocytosis. After endocytosis, the contents of the vacuole become more acidic, causing the A and B portions to separate. The B portion remains in the vacuole and is removed by exocytosis, while the A portion enters the cytoplasm of the cell and exerts its toxic effect.

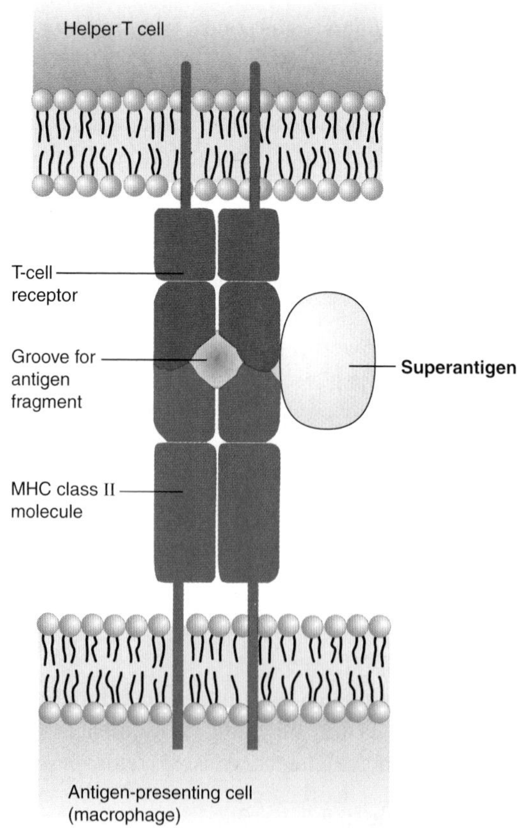

FIGURE 17.13 Superantigens These exotoxins override the specificity of the T-cell response, causing systemic toxic effects due to the massive release of cytokines by an inordinate number of effector T cells. The superantigens effectively short-circuit the normal control mechanisms inherent in antigen processing and presentation by binding simultaneously to the outer portion of the MHC class II molecule and the T-cell receptor.

the symptoms they cause are distinct from those of the enterotoxins of *Vibrio cholerae* and certain *E. coli* strains. The spectrum of symptoms associated with the ingestion of a staphylococcal enterotoxin is also different from that of other superantigens. The precise mechanism by which they induce vomiting is poorly understood, but they seem to affect inflammatory functions of subepithelial macrophages and cause a change in local vascular permeability. Unlike most other exotoxins, the enterotoxins produced by *Staphylococcus aureus* are heat-stable. Even thorough cooking of foods that have been contaminated with these toxins will not prevent illness. ■ *Staphylococcus aureus* food poisoning, p. 763

Other Toxic Proteins

Various proteins that are not A-B toxins, superantigens, or membrane-damaging toxins can have detrimental effects. An important example is the toxin produced by strains of *Staphylococcus aureus* that cause scalded skin syndrome. This toxin, **exfoliatin,** destroys material that binds together the layers of skin, causing the outer layer to separate (see figure 23.4). The organism might be growing in a small, localized lesion but the toxin spreads systemically.

Various hydrolytic enzymes including proteases, lipases, and collagenases break down tissue components. Along with destroying tissues, some of these enzymes facilitate the spread of the organism.

Endotoxin and Other Bacterial Cell Wall Components

The host defenses are primed to respond to various bacterial cell wall components, including lipopolysaccharide and peptidoglycan, in order to contain an infection. A systemic response to these compounds, however, can overwhelm the system and cause toxic effects.

Endotoxin

Endotoxin is lipopolysaccharide (LPS), the molecule that makes up the outer leaflet of the outer membrane of the Gram-negative cell wall (see figure 3.34). Thus, endotoxin is a fundamental part of Gram-negative bacteria. The nomenclature is somewhat unfortunate, because it implies that endotoxin is "inside the cell" and exotoxins are "outside the cell"; in fact, endotoxin is an integral part of the outer membrane, whereas exotoxins are proteins produced by a bacterium that may or may not be secreted. Unlike most exotoxins, endotoxin cannot be converted to an effective toxoid for immunization. **Table 17.2** summarizes some of the other differences between exotoxins and endotoxin. ■ lipopolysaccharide molecule, p. 61

Recall from chapter 3 that the lipopolysaccharide molecule is composed of two medically important parts—lipid A and the O-specific polysaccharide side chain. The lipid A component, which is responsible for the adverse effects of LPS, triggers a vigorous innate immune response. When lipid A is present in a localized region, the magnitude of the response helps clear an infection. It is a different situation entirely, however, when the infection is systemic, as in septicemia. Imagine a state in which inflammation occurs throughout the body—extensive leakage of fluids from permeable blood vessels and widespread activation of the coagulation cascade. The overwhelming response in such bloodstream infections has profound effects, causing fever, a dramatic drop in blood pressure and disseminated intravascular coagulation (see figure 28.3). This array of symptoms associated with systemic bacterial infection is called **septic shock;** when it is caused by endotoxin, it may also be called **endotoxic shock.** ■ septic shock, pp. 675, 679

Lipid A is embedded in the outer membrane and does not evoke a response unless it is released. This occurs primarily when the bacterium lyses, which can happen as a result of phagocytosis, formation of membrane attack complexes by complement components, and treatment with certain types of antibiotics.

Once released from a cell, the LPS molecules can activate the innate and adaptive defenses by a variety of mechanisms. Monocytes, macrophages, and other cells have toll-like receptors that detect liberated LPS, inducing the cells to produce proinflammatory cytokines. LPS also functions as a T-independent antigen; at high concentrations it activates a variety of different B cells, regardless of the specificity of their B-cell receptor. ■ toll-like receptors, p. 355 ■ T-independent antigens, p. 379

Endotoxin is heat-stable; it is not destroyed by autoclaving. Consequently, solutions intended for intravenous administration must not only be sterile, but free of endotoxin as well. Disastrous

TABLE 17.2 Comparison of Exotoxins and Endotoxin

Property	Exotoxins	Endotoxin
Bacterial source	Gram-positive and Gram-negative species	Gram-negative species only
Location in the bacterium	Synthesized in the cytoplasm; may or may not be secreted	Component of the outer membrane of the Gram-negative cell wall
Chemical nature	Protein	Lipopolysaccharide (the lipid A component)
Ability to form a toxoid	Generally	No
Heat stability	Generally inactivated by heat	Heat-stable
Mechanism	A distinct toxic mechanism for each	Innate immune response; a systemic response leads to fever, a dramatic drop in blood pressure, and disseminated intravascular coagulation
Toxicity	Generally very potent; some are among the most potent toxins known.	Small amounts in a localized area lead to an appropriate response that helps clear an infection, but systemic distribution can be deadly

results including death have resulted from administering intravenous fluids contaminated with endotoxin. To verify that fluids are not contaminated, a very sensitive test known as the *Limulus* amoebocyte lysate (LAL) assay is used. This employs proteins extracted from blood of the horseshoe crab *Limulus polyphemus* that form a gel-like clot in the presence of endotoxin; as little as 10 to 20 picograms (1 picogram = 10^{-12} grams) of endotoxin per milliliter can be detected using the LAL. Horseshoe crabs are one of this planet's more unique and ancient life forms. The critical role they play in this test has led to an increased awareness of the importance of their habitat. A non-lethal system of capture, blood sampling, and release has been developed.

Other Bacterial Cell Wall Components

Peptidoglycan and other bacterial cell wall components can elicit symptoms similar to those that characterize the response to endotoxin. The systemic response leads to septic shock.

Damaging Effects of the Immune Response

Although the intent of the immune response is to eliminate the invading microbe, host tissues can inadvertently be damaged as well. Note that the reactions to endotoxin and other cell wall components can be viewed as damaging effects of the immune response, but are typically considered a toxic effect of the bacterium because the reactions can be immediate and overwhelming. The damaging responses discussed next generally manifest themselves more slowly.

Damage Associated with Inflammation

The inflammatory response itself can destroy tissue because phagocytic cells recruited to the area inevitably release some of the enzymes and toxic products they contain. The life-threatening aspects of bacterial meningitis, for example, are due to the inflammatory response itself. Complications of certain sexually transmitted diseases are also due to the damage associated with inflammation. If *Neisseria gonorrhoeae* or *Chlamydia trachomatis* infections ascend from the cervix to involve the fallopian tubes, the inflammatory response can lead to scarring that obstructs the tubes, either predisposing a woman to an ectopic pregnancy or preventing fertilization altogether.

Damage Associated with Antibodies

Antibodies generated during an immune response can also lead to damaging effects by these mechanisms:

- **Antigen-antibody complexes.** These complexes can form during the immune response and settle in the kidneys and joints. There they may activate the complement system, causing destructive inflammation. Acute glomerulonephritis, a complication that can follow strep throat and impetigo caused by *Streptococcus pyogenes,* is due to antigen-antibody complexes that settle in the kidney, eliciting a response that damages kidney structures called glomeruli. ■ acute glomerulonephritis, p. 501
- **Cross-reactive antibodies.** Certain antibodies produced in response to an infection bind to the body's own tissues, promoting an autoimmune response. Some evidence indicates that acute rheumatic fever, a complication that can follow strep throat, may be due to the binding of antibodies against *Streptococcus pyogenes* to a normal tissue protein. This occurs most frequently in people with certain MHC types (see Perspective 16.1). ■ acute rheumatic fever, p. 501

MICROCHECK 17.8

Damage can be due to exotoxins, including A-B toxins, superantigens, membrane-damaging toxins, and other toxic proteins. Endotoxin and other cell wall components can elicit a host response so strong that septic shock may result. The inflammatory response can cause tissue damage that leads to scarring. Antigen-antibody complexes can trigger damaging inflammation in the kidneys and joints; cross-reactive antibodies can lead to an autoimmune response.

✓ What is an A-B toxin?

✓ How is an enterotoxin different from endotoxin?

✓ Home-canned foods should be boiled before consumption to prevent botulism. Considering that this treatment does not destroy endospores, why would it prevent the disease?

17.9

Mechanisms of Viral Pathogenesis

Focus Point

- Describe three general mechanisms of viral pathogenicity.

The mechanisms of viral pathogenesis are somewhat different from those of bacteria. First, viruses "live" only within the context of a cell. All viruses are parasites, but some have long-term relationships with their hosts—perhaps even across generations.

To infect a host, a virus must recognize and enter an appropriate cell, exploit the cell's replication processes, keep the host from recognizing and destroying the infected cell during viral replication, and then move to new cells or hosts. In carrying out these functions, many viruses damage host cells and induce inflammatory responses, causing the signs and symptoms of disease.

Binding to Host Cells and Invasion

As discussed in chapter 14, viruses attach to target cells via specific receptors. The receptor used by a particular virus influences the host range and tissue specificity of that virus; only cells that bear the specific receptor can be infected. HIV uses CD4 as a receptor; recall that this is found on helper T cells. Picorna viruses and reoviruses bind to receptors on M cells in the Peyer's patches of the intestinal tract. ■ CD4, p. 381

To enter a cell, some viruses exploit the normal cell process of receptor-mediated endocytosis. When these viruses bind to the appropriate receptor, the cell internalizes the receptors, along with the attached viruses. Other viruses have the specific proteins on their surface that lead to fusion of the viral envelope with the cell membrane, releasing the viral capsid into the cell (see figure 14.3a). ■ receptor-mediated endocytosis, p. 74

The viruses released from the infected cell can infect neighboring cells or they can disseminate in the bloodstream or lymphatic system to other tissues. Polioviruses, for example, initially infect cells in the throat and intestinal tract. Upon release from these cells, the virus may spread via the bloodstream to infect motor nerve cells of the brain and spinal cord causing paralysis. ■ polio, p. 661

Avoiding Immune Responses

As obligate intracellular parasites, viruses must subvert or avoid mechanisms the body uses to detect and eliminate invaders. These include the production of interferon, induction of apoptosis (programmed cell death), and antibody production. ■ apoptosis, p. 362

Avoiding the Antiviral Effects of Interferons

Early in viral infection, interferons play an important role in limiting viral spread (see figure 15.9). They induce cells to produce enzymes that, when activated, prevent viral replication. To avoid this, some viruses encode proteins that shut down expression of host genes. Others circumvent interferon's effects by interfering with activation of the enzymes. ■ interferon, p. 357

Regulation of Host Cell Death by Viruses

Some types of viruses kill the host cell after it produces a large number of viral particles; others control certain aspects of immune surveillance to prevent the host cell from dying prematurely. Recall that the interferon response induces apoptosis in virally infected cells. Some viruses prevent this from happening, giving the virus more time to replicate. They do this by controlling a protein called p53 that regulates apoptosis activation in the cell. The inhibition of the p53 protein in the host cell is often associated with tumor development. Papillomaviruses, a group of viruses that cause various types of warts, interfere with the normal function of p53. ■ papillomavirus, p. 639

To avoid being detected by the cellular immune response, many types of viruses interfere with antigen presentation by MHC class I molecules; recall that this mechanism allows T_C cells (effector cytotoxic T cells) to recognize infected "self" cells (see figure 16.19). Herpesviruses, for example, block the processing and movement of MHC class I molecules to the surface of the infected cell, thereby shielding that cell from T_C cells surveillance. Our immune defenses, in turn, have evolved methods to thwart such evasion. Natural killer (NK) cells recognize cells that have "too few" MHC class I molecules and destroy them. Perhaps not surprisingly, some viruses have countered by developing methods to trick NK cells. Cytomegalovirus (CMV), a herpesvirus that causes severe problems in immunocompromised people, encodes production of "counterfeit" MHC class I molecules that are displayed on the surface of its host cells to trick the body into believing "all is well in this cell" **(figure 17.14).** These counterfeit molecules do not identify the infection to the immune system and buy time for the virus to replicate. ■ antigen presentation by MHC class I, p. 380 ■ herpesviruses, p. 330 ■ natural killer cells, p. 384 ■ cytomegaloviruses, p. 715

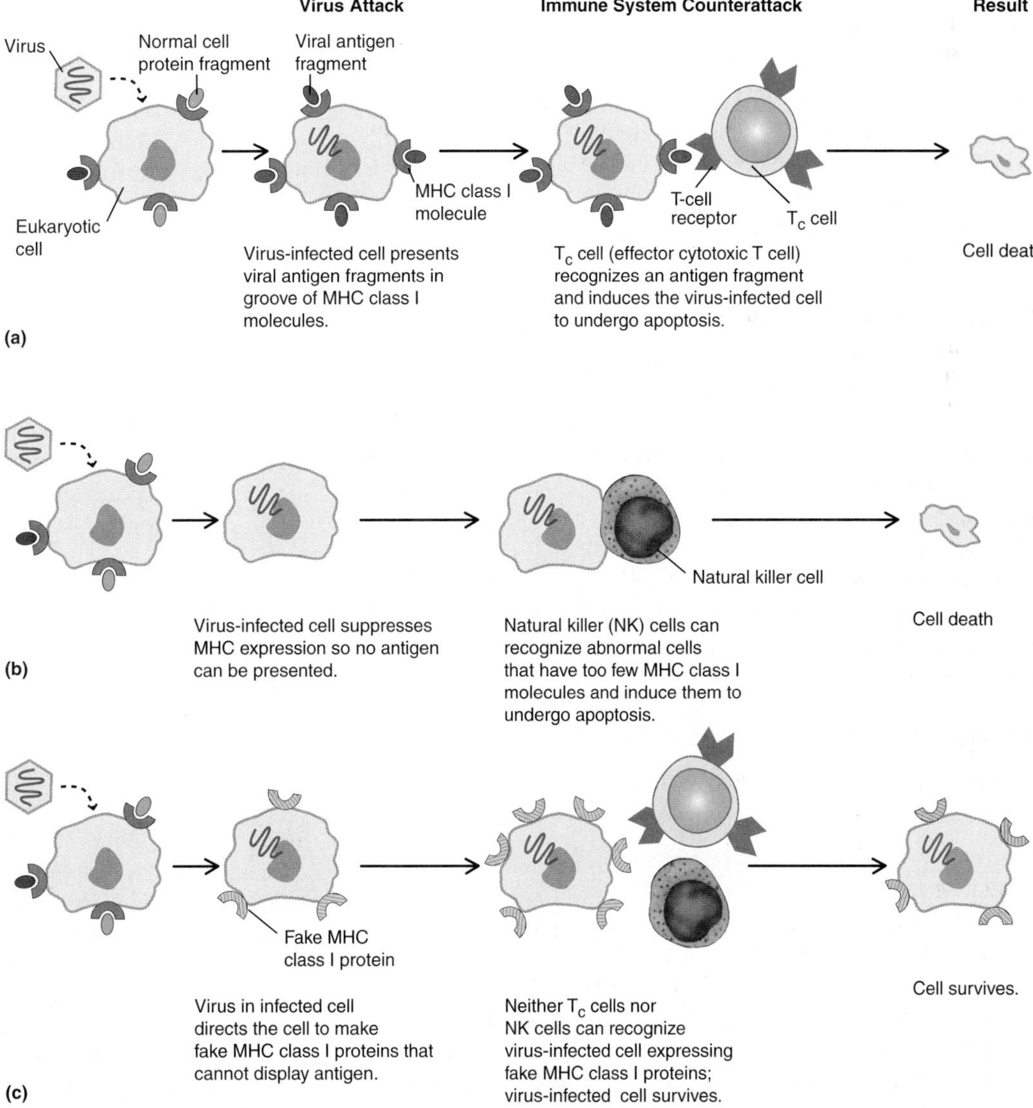

FIGURE 17.14 Cytomegalovirus-Immune Cell Interactions **(a)** Normal reaction to virus-infected cell by a T_C cell. **(b)** Some viruses can evade destruction of their host cell by suppressing MHC expression, but NK cells can induce apoptosis in cells lacking MHC molecules. **(c)** Virus causes cell to produce fake MHC that neither T_C cells nor NK cells can recognize, and so the virus-infected cell survives.

Antibodies and Viruses

Antibodies generally control the spread of virus from cell to cell by neutralizing extracellular viral particles (see figure 16.6). To avoid antibodies, some viruses transfer directly from one cell to its immediate neighbors. Other viruses remain intracellular by forcing all the cellular neighbors to join a "commune," fusing to from what is called a **syncytium.** HIV is an example of a virus that avoids the neutralizing activity of antibodies by inducing syncytia formation.

Another way viruses deal with the antibodies is to routinely modify the viral surface antigens, outpacing the body's capacity to produce an effective neutralizing antibody. RNA viruses contain, on average, one mutation per new viral particle. Their polymerases copy RNA without any proofreading, so the mutation frequency is about one base in 100,000. There are more than 100,000 bases in the average RNA virus, so there is about one error per new virus. The accumulation of mutations by these viruses generates pools of genetically altered viruses. These mutants come into dominance as the "parent" viruses are rapidly neutralized by antibody. Thus, the body keeps selecting for new viral "strains" over the course of infection. Many of these viruses are eventually cleared as a result of an immune response against essential proteins that cannot tolerate extensive mutations. Other viruses, like HIV, exploit this method to infect the host for a very long time. An HIV-infected person produces many antibodies against HIV, but none are effective in clearing the virus.

Some viruses actually exploit antibodies, using them to facilitate infection of macrophages. These viruses use the "free ride" into macrophages granted when opsonized viral particles are engulfed by the phagocyte. Development of dengue hemorrhagic fever, a severe illness caused by dengue viruses, is associated with antibody-dependent enhancement of infection.

MICROCHECK 17.9

Proteins on the surface of virus particles function as adherence factors. Viruses, as a group, have evolved mechanisms to avoid the antiviral effects of interferon and to control apoptosis of the host cell. Damage is often due to the host response.

- ✓ From a virus's perspective, why would it be beneficial to prevent apoptosis?
- ✓ How do cytomegaloviruses avoid the cellular immune response?
- ✓ Why would various unrelated viral infections often include a similar set of symptoms (fever, headache, fatigue, and runny nose)?

17.10 Mechanisms of Eukaryotic Pathogenesis

Focus Point

- Describe some of the virulence determinants of fungi and eukaryotic parasites.

Pathogenesis of eukaryotic cells including fungi and protozoa involve the same basic scheme as that of bacterial pathogens—colonization, evasion of host defenses, and damage to the host. The mechanisms, however, are generally not as well understood.

Fungi

Most fungi, such as yeasts and molds, are **saprophytes,** meaning they acquire nutrients from dead and dying material; those that can cause disease are generally opportunists, although notable exceptions exist.

Members of a group of fungi referred to as **dermatophytes** cause superficial infections of hair, skin, and nails, but do not invade deeper tissues. These fungi have keratinase enzymes that break down the keratin in superficial tissues, resulting in diseases such as ringworm and athlete's foot.

Fungi in the normal microbiota, especially the yeast *Candida albicans,* can cause disease in immunocompromised hosts. Generally this leads to infection of the mucous membranes, causing thrush, (an infection of the throat and mouth), or vaginitis. Factors that may lead to excessive growth of *C. albicans* include disruption of normal microbiota due to hormonal influences or antibiotic treatment, AIDS, uncontrolled diabetes, and severe burns.

The most serious fungal infections are caused by a group of fungi that are **dimorphic;** they occur as molds in the environment, but assume other forms, usually yeasts, when they invade tissues. Infection occurs when the small airborne spores of the molds are inhaled, lodging deep within the lungs. These infections are generally controlled by the immune system and do not cause serious disease unless there is an overwhelming infection or if the person is immunocompromised.

Some fungi produce toxins, collectively referred to as mycotoxins, which can cause disease. *Aspergillus flavus,* for example, a fungus that grows on certain grains and nuts, produces aflatoxin. Ingestion of this toxin can damage the liver, perhaps leading to cancer. Fungal spores and other products can cause hypersensitivities in some people.

Protozoa and Helminths

Most protozoa and helminths either live within the intestinal tract or enter the body via the bite of an arthropod. *Schistosoma* species, however, can enter the skin directly (see Perspective 15.1).

Like bacteria and viruses, eukaryotic parasites attach to host cells via specific receptors. *Plasmodium vivax,* one of the two most common causes of malaria, attaches to the Duffy blood group antigen on red blood cells. Most people of West African ancestry lack this antigen and are therefore resistant to infection by this species. *Giardia lamblia* uses a disc that functions as a suction cup to attach to the intestine; it appears that it also has an adhesin associated with the disc that facilitates the initial attachment.

Protozoa and helminths use a variety of mechanisms to avoid antibodies. Some hide within cells, thus avoiding exposure to antibodies as well as certain other defenses. Malarial parasites produce enzymes that allow them to penetrate red blood cells. These host cells do not present antigen to T_C cells, enabling the parasite to escape the cellular immune defenses as well. *Leishmania* species are able to survive and multiply within macrophages when phagocytized. Parasites such as the African trypanosomes, the cause of sleeping sickness, escape from the effects of antibody by routinely varying their surface antigens, repeatedly activating different genes. *Schistosoma* species coat themselves with host proteins, effectively disguising themselves. Some parasitic worms appear to suppress immune responses in general.

The extent and type of damage caused by parasites varies tremendously. In some cases, the parasites compete for nutrients in the intestinal tract, contributing to malnutrition of the host. Helminths may accumulate in high enough numbers or grow long enough to cause blockage of the intestines or other organs. Some parasites produce enzymes that digest host tissue, causing direct damage. In other cases, damage is due to the immune response; high fevers that characterize malaria, and the granulomatous response to schistosoma eggs are examples.

MICROCHECK 17.10

Pathogenic mechanisms of fungi, protozoa, and helminths are not as well understood as those of bacteria; however, they involve the same basic scheme—colonization, evasion of host defenses, and damage to the host.

✓ What is the importance of keratinase?
✓ How do the African trypanosomes avoid the effects of antibodies?
✓ Why would relatively few people of West African ancestry have the Duffy blood group antigen?

FUTURE CHALLENGES

The Potential of Probiotics

Considering the important protective roles the normal microbiota play in human health, it is not surprising that administering live beneficial microbes, referred to as **probiotics,** has been suggested as therapy for diarrheal and other diseases. *In vitro* studies certainly support their use, indicating that some strains of bacteria interfere with the growth or toxin production of certain pathogens. Animal studies using probiotics have also shown promising results. For example, chicks fed a mixture of *Lactobacillus* species that normally inhabit poultry intestines were less likely to be subsequently colonized by the pathogen *Salmonella enterica.* But which probiotics will be successful in treating humans?

Marketing claims touting probiotics abound, but there are still many unanswered questions. For example, will a microbe that has beneficial effects *in vitro* be stable in a food or supplement so that an adequate dose can be consumed? And even if it is stable in the package, will it survive passage through the stomach so that viable cells enter the intestinal tract? If it does access the intestines, can it colonize there, or will it simply pass through as part of feces?

Beneficial effects of probiotics observed in humans thus far are species and strain specific. One species may be protective, whereas another closely related organism is not. For example, several small studies indicate that consumption of *Lactobacillus rhamnosus* GG can shorten the duration of certain diarrheal illnesses. In contrast, studies that used a mixture containing *L. bulgaricus* and *L. acidophilus* generally showed no helpful effect. Because of this, products that list ingredients such as "live active culture" or "lactic acid bacteria" do not necessarily contain a beneficial strain.

Complicating matters is that probiotics are considered "dietary supplements," so there is little regulatory control over health claims. Early studies such as those using *L. rhamnosus* GG certainly highlight their potential, but larger and well-controlled scientific studies of probiotics are needed to provide more data as to their efficacy in curing and preventing various medical conditions.

SUMMARY

Microbes, Health, and Disease

17.1 The Anatomical Barriers As Ecosystems

Symbiotic Relationships Between Microorganisms and Hosts

In **mutualism,** both partners benefit, in **commensalism** one partner benefits while the other is unaffected, and in **parasitism,** the **parasite** benefits at the expense of the host.

17.2 The Normal Microbiota (figure 17.1)

The Protective Role of Normal Microbiota

The **normal microbiota** excludes pathogens by covering binding sites that might otherwise be used for attachment, consuming available nutrients, and producing compounds toxic to other bacteria. The normal flora primes the adaptive immune system.

The Dynamic Nature of the Normal Microbiota

The composition of the normal microbiota continually changes in response to host factors including hormonal changes, and type and quantity of food consumed.

17.3 Principles of Infectious Disease

Pathogenicity

A **primary pathogen** causes disease in otherwise healthy individuals; an **opportunist** causes disease only when the body's innate or adaptive defenses are compromised. **Virulence** refers to the degree of pathogenicity of an organism.

Characteristics of Infectious Disease

Communicable or **contagious diseases** spread from one host to another; ease of spread partly reflects the **infectious dose.** Stages of infectious disease include the **incubation period, illness,** and **convalescence;** during the illness a person experiences **signs** and **symptoms** of the disease (figure 17.2). Infections can be described as **acute, chronic,** or **latent,** depending on the timing and duration of symptoms. Infections can be **localized** or **systemic.**

17.4 Establishing the Cause of Infectious Disease

Koch's Postulates

Koch's postulates establish the cause of infectious disease (figure 17.3).

Molecular Koch's Postulates

Molecular Koch's postulates identify virulence factors that contribute to disease.

Mechanisms of Pathogenesis

17.5 Establishment of Infection

Adherence

Bacteria use **adhesins** to bind to host cells (figure 17.4).

Colonization

Rapid turnover of pili, antigenic variation, and **IgA proteases** enable bacteria to avoid the effects of secretory IgA. **Siderophores** enable microbes to scavenge iron.

Delivery of Effector Molecules to Host Cells

Type III secretion systems of Gram-negative bacteria allow them to deliver compounds directly to host cells (figure 17.5).

17.6 Invasion—Breaching the Anatomical Barriers

Penetration of Skin

Bacteria take advantage of trauma that destroys the integrity of the skin or rely on arthropods to inject them.

Penetration of Mucous Membranes

Some pathogens induce mucosal epithelial cells to engulf bacterial cells; others exploit antigen-sampling processes (figures 17.6, 17.7).

17.7 Avoiding the Host Defenses

Hiding Within a Host Cell

Some bacteria can evade the innate defenses, as well as some aspects of the adaptive defenses, by remaining inside host cells.

Avoiding Killing by Complement System Proteins

Some **serum-resistant** bacteria postpone the formation of the membrane attack complex by interfering with activation of the complement system via the alternative pathway (figure 17.9).

Avoiding Destruction by Phagocytes (figure 17.10)

Mechanisms to prevent encounters with phagocytes include **C5a peptidase** and **membrane-damaging toxins.** Mechanisms to avoid recognition and attachment by phagocytes include **capsules, M protein,** and **Fc receptors** (figure 17.11). Mechanisms to survive within the phagocyte include escape from the phagosome, preventing phagosome-lysosome fusion, and surviving within the phagolysosome.

Avoiding Antibodies

Mechanisms to avoid antibodies include **IgA protease, antigenic variation,** and mimicking "self."

17.8 Damage to the Host

Exotoxins (table 17.1)

Exotoxins are proteins that have very specific damaging effects; they may act locally or cause dramatic systemic effects. Many can be grouped into categories such as **neurotoxins, enterotoxins,** or **cytotoxins.** The toxic activity of **A-B toxins** is mediated by the A subunit; binding to specific cells is mediated by the B subunit (figure 17.12). **Superantigens** override the specificity of the T-cell response, causing systemic effects due to the massive release of cytokines (figure 17.13). Membrane-damaging toxins disrupt cell membranes either by forming pores or by removing the polar head group on phospholipids in the membrane.

Endotoxin and Other Bacterial Cell Wall Components

The symptoms associated with endotoxin are due to a vigorous host response. **Lipid A** of lipopolysaccharide is responsible for its toxic properties. Peptidoglycan and certain other components induce various cells to produce pro-inflammatory cytokines.

Damaging Effects of the Immune Response

The release of enzymes and toxic products from phagocytic cells can damage tissues. Antigen-antibody complexes can cause kidney and joint damage; cross-reactive antibodies can promote an autoimmune response.

17.9 Mechanisms of Viral Pathogenesis

Binding to Host Cells and Invasion

Viruses attach to specific receptors on the target cell.

Avoiding Immune Responses

Some viruses can avoid the effects of interferon; some can regulate apoptosis of the host cell (figure 17.14). To subvert the role of antibodies, some viruses transfer directly from cell to cell; the surface antigens of some viruses change quickly, outpacing the production of antibodies.

17.10 Mechanisms of Eukaryotic Pathogenesis

Fungi

Saprophytes are generally opportunists; dermatophytes cause superficial infections of skin, hair, and nails. The most serious fungal infections are caused by dimorphic fungi.

Protozoa and Helminths

Eukaryotic parasites attach to host cells via specific receptors. They use a variety of mechanisms to avoid antibodies; the extent and type of damage they cause varies tremendously.

REVIEW QUESTIONS

Short Answer

1. Describe three types of symbiotic relationships.
2. Describe two situations that can lead to changes in the composition of the normal microbiota.
3. How are acute, chronic, and latent infections different?
4. Why are Koch's postulates not sufficient to establish the cause of all infectious diseases?
5. Describe the four general mechanisms by which microorganisms cause disease.
6. Describe two mechanisms that bacteria use to invade mucous membranes.

7. Explain how capsules enable an organism to be serum resistant and avoid phagocytosis.
8. Give an example of a neurotoxin, an enterotoxin, and a cytotoxin.
9. Describe two mechanisms a virus might use to prevent the induction of apoptosis in an infected cell.
10. How do *Schistosoma* species avoid antibodies?

Multiple Choice

1. Opportunistic pathogens are *least* likely to affect which of the following groups?

   a) AIDS patients b) Cancer patients c) College students
   d) Drug addicts e) Transplant recipients

2. Capsules and M protein are thought to interfere with which of the following?

   a) Opsonization by complement proteins
   b) Opsonization by antibodies
   c) Recognition by T cells
   d) Recognition by B cells
   e) Phagosome-lysosome fusion

3. The C5a peptidase enzyme of *Streptococcus pyogenes* breaks down C5a, resulting in

   a) lysis of the *Streptococcus* cells.
   b) lack of opsonization of *Streptococcus* cells.
   c) killing of phagocytes.
   d) decreased accumulation of phagocytes.
   e) inhibition of membrane attack complexes.

4. All of the following are known mechanisms of avoiding the effects of antibodies *except*

   a) antigenic variation.
   b) mimicking "self".
   c) synthesis of an Fc receptor.
   d) synthesis of IgG protease.
   e) remaining intracellular.

5. Which of the following statements about diphtheria toxin is *false?* It

   a) is an example of an endotoxin.
   b) is produced by a species of *Corynebacterium.*
   c) inhibits protein synthesis.
   d) can cause local damage to the throat.
   e) can cause systemic damage (that is, to organs such as the heart).

6. Which of the following statements about botulism is *true?*

   a) It is caused by *Bacillus botulinum,* an obligate aerobe.
   b) The toxin is resistant to heat, easily withstanding temperatures of 100°C.
   c) The organism that causes botulism can cause disease without avoiding the immune response.
   d) Vaccinations are routinely given to prevent botulism.
   e) Symptoms of botulism include uncontrolled contraction of muscles.

7. Superantigens

   a) are exceptionally large antigen molecules.
   b) cause a very large antibody response.
   c) elicit a response from a large number of T cells.
   d) attach non-specifically to B-cell receptors.
   e) assist in a protective immune response.

8. Which of the following statements about endotoxin is *true?* It

   a) is an example of an A-B toxin.
   b) is a component Gram-positive bacteria.
   c) can be converted to a toxoid.
   d) is heat-stable.
   e) causes T cells to release cytokines.

9. The tissue damage caused by *Neisseria gonorrhoeae* is primarily due to

   a) cross-reactive antibodies.
   b) exotoxins.
   c) hydrolytic enzymes.
   d) the inflammatory response.
   e) all of the these.

10. Which of the following statements about viruses is *false?* They may

    a) colonize the skin.
    b) enter host cells by endocytosis.
    c) enter host cells by fusion of the viral envelope with the cell membrane.
    d) induce apoptosis in infected host cells.
    e) suppress expression of MHC class I molecules on host cells.

Applications

1. A group of smokers suffering from the outcome of severe *Staphylococcus aureus* infections are suing the cigarette companies. They claim that the disease was aggravated by cigarette smoking. The group is citing studies indicating that phagocytes are inhibited in their action by compounds in cigarette smoke. A statement prepared by their lawyers states that the *S. aureus* would not have caused such a severe disease if the phagocytes and immune system overall were functioning properly. During the proceedings, a microbiologist was called in as a professional witness for the court. What were her conclusions about the validity of the claim?
2. A microbiologist put forth a grant proposal to study the molecules bacteria use to communicate. His principal rationale was that the damaging effects of many pathogenic microorganisms could be prevented by inactivating the molecules these bacteria use to communicate. Is this a reasonable proposal? Why or why not?

Critical Thinking

1. A student argued that no distinction should be made between commensalism and parasitism. Even in commensalism, the microorganisms are gaining some benefit (such as nutrients) from the host and this represents a loss to the host. In this sense the host is being damaged. Does the student have a valid argument? Why or why not?
2. A microbiologist argued that there is no such thing as "normal" microbiota in the human body, since the population is dynamic and is constantly changing depending on diet and external environment. What would be an argument against this microbiologist's view?

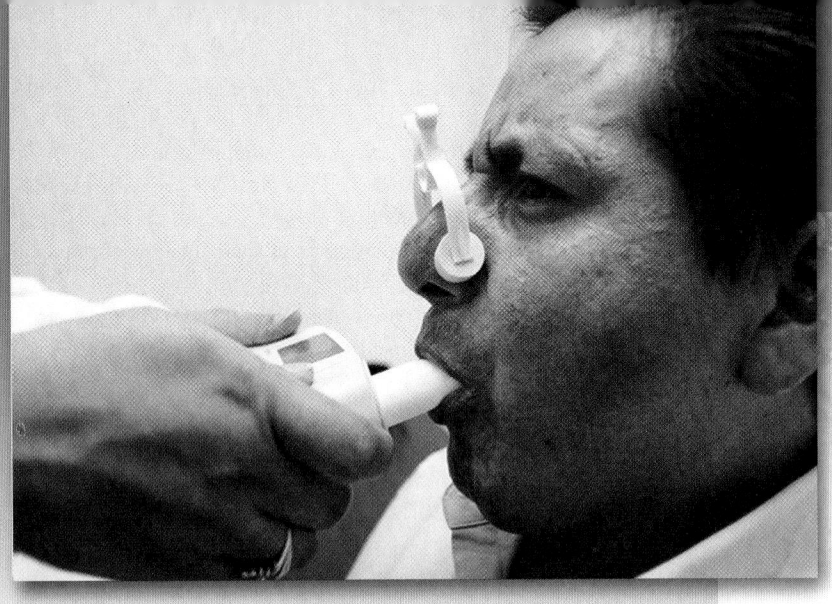

Asthmatic individual using a spirometer.

18

Immunologic Disorders

A Glimpse of History

Pasteur is widely quoted as saying, "Chance favors the prepared mind." This was certainly the case with the physiologist Charles Richet in his discovery of **hypersensitivities,** commonly called allergies, near the end of the nineteenth century. Richet performed some of the early experiments with toxins and antisera. He and his colleague, Paul Portier, while cruising in the South Seas on Prince Albert of Monaco's yacht, hypothesized that the Portuguese man-of-war jellyfish must have a toxin responsible for its ugly stings. They made an extract of the jellyfish tentacles and showed that it was very toxic to rabbits and ducks.

Upon returning to France they began studying the effects of toxins from the tentacles of sea anemones on dogs. Richet and Portier noted that some dogs in their experiments survived the potent toxin, but when these animals were tested with the toxin again, they died a few minutes after receiving a small dose. One very healthy dog survived the first challenge with the toxin. When given a second dose 22 days later, the dog became very ill within a few seconds. It could not breathe, and laid on its side, had violent diarrhea, vomited blood, and died within 25 minutes.

Because of the insights they gained from their early work with antisera, Richet and Portier had prepared minds. They recognized that the dogs' reactions were probably immunologically mediated. The reactions, however, represented the opposite effect of prevention of disease—**prophylaxis**—conferred by antisera. Therefore, they named this development of hypersensitivity upon repeat exposure to antigenic substances **anaphylaxis,** the extreme opposite of prophylaxis. For this and other outstanding contributions to medicine, Richet received the Nobel Prize in 1913.

For the most part, the immune system does a superb job protecting the body from invasion by various microorganisms and viruses; however, as Richet showed, the same mechanisms that are so effective in protecting us can be harmful under some circumstances. Our immune response can be likened to fire. Fire is essential for warmth and cooking and in many ways it is beneficial, but the same fire is destructive if it starts in the wrong place or becomes uncontrolled. Similarly, immune responses are essential for protection, but can be destructive if out of control. The protective responses confer immunity; immune responses that cause tissue injury are referred to as **hypersensitivities.** In addition to hypersensitivities, a second type of immunologic disorder, **autoimmune disease,** results from responses against self-antigens. A third type of immunologic disorder occurs when the immune system responds too little, resulting in **immunodeficiency.**

Hypersensitivity reactions to usually harmless substances are often called **allergies** or allergic reactions. Allergic reactions occur only in **sensitized** individuals—that is, those who have been immunized or sensitized by prior exposure to that specific antigen. Antigens that cause allergic reactions are **allergens.** For example, in hay fever, the responsible allergen is often a grass or tree pollen. In food allergies, the allergens are antigens in food—most commonly from nuts, eggs, cow's milk or fish—but hundreds of other substances can also be allergenic. One of the interesting questions immunologists are investigating is why some people are much more likely to develop allergies than others. For example, food allergies affect only about one out of every five children, and these allergic conditions often go away by the time the children reach school age. These differences can be explained by the failure of a child's immune system to develop tolerance to normally harmless substances. An infant's immune system is immature at birth and normal development depends on heredity, exposure to normal commensal microbiota, and probably other unknown factors. ■ **tolerance, p. 367**

Hypersensitivities are categorized according to which parts of the immune response are involved and how quickly the response occurs. Most allergic or hypersensitivity reactions fall into one of four major types:

- Type I: Immediate IgE-mediated
- Type II: Cytotoxic

KEY TERMS

Allergy Hypersensitivity to an allergen, generally mediated by IgE.

Allergen Substance that causes an allergic reaction.

Allograft Transplanted tissue from a non-identical individual of the same species.

Anaphylaxis Harmful, as opposed to protective, reactions resulting from prior exposure to a foreign substance. Usually refers to IgE-mediated reactions that potentially could cause shock.

Autograft Tissue transplanted from one site to another on a person's body, as normal skin from the thigh to a burn on the arm.

Autoimmune Refers to conditions caused by reactions of the immune system against self-antigens.

Hemolytic Disease Conditions resulting from destruction of red blood cells, as in hemolytic disease of the newborn.

Immunodeficiency A condition in which one or more components of the immune system is defective, thereby rendering an individual susceptible to infection.

Immunotherapy Techniques used to modify the immune system action for a favorable effect.

Isograft Tissue transplanted from an identical sibling.

Oral Tolerance Decreased reactivity of immune cells resulting from feeding an antigen.

Serum Sickness A condition caused by immune complexes arising from antibody formed in response to an antigen (such as snakebite antivenin) injected into the bloodstream.

Urticaria Hives, an IgE-mediated skin reaction characterized by a wheal and flare; an example of type I anaphylaxis.

- Type III: Immune complex-mediated
- Type IV: Delayed cell-mediated

The main characteristics of the four types of hypersensitivities are shown in **table 18.1.**

18.1 Type I Hypersensitivities: Immediate IgE-Mediated

Focus Points

- Describe the immunologic reactions involved in type I hypersensitivities.
- Give four examples of type I hypersensitivities.

Antibodies of the class IgE have a protective role, especially against certain parasitic worms. However, IgE antibodies also cause immediate (type I) hypersensitivity, characterized by a reaction in a sensitized individual within minutes of exposure to antigen. Allergic people develop such a reaction when exposed to substances such as dust, pollens, animal dander, and molds, which do not evoke a response in non-sensitized people. ■ IgE, p. 374

A person becomes sensitized when the allergen makes contact with some part of the body (pollen grains contact the mucous membrane of the respiratory tract, for example) and induces an antibody response. Tissues under the mucous membranes are rich in B cells committed to IgA and IgE production, and IgE-producing cells are more abundant in allergic than in non-allergic individuals. As IgE is produced in this area, which is rich in mast cells, the IgE molecules attach via their Fc portion to receptors on the mast cells and

TABLE 18.1 Some Characteristics of the Major Types of Hypersensitivities

Characteristic	Type I Hypersensitivity Immediate; IgE-Mediated	Type II Hypersensitivity Cytotoxic	Type III Hypersensitivity Immune Complex-Mediated	Type IV Hypersensitivity Delayed Cell-Mediated
Cell type responsible	B cells	B cells	B cells	T cells
Type of antigen	Soluble	Cell-bound	Soluble	Soluble or cell-bound
Type of antibody	IgE	IgG, IgM	IgG	None
Other cells involved	Basophils, mast cells	Red blood cells, white blood cells, platelets	Various host cells	Various host cells
Mediators	Histamine, serotonin, leukotrienes	Complement, ADCC	Complement, neutrophil proteases	Cytokines
Transfer of hypersensitivity	By serum	By serum	By serum	By T cells
Time of reaction after challenge with antigen	Immediate, up to 30 minutes	Hours to days	Hours to days	Peaks at 48 to 72 hours
Skin reaction	Wheal and flare	None	Arthus	Induration, necrosis
Examples	Anaphylactic shock, hay fever, hives, asthma	Transfusion reaction, hemolytic disease of newborns	Serum sickness, farmer's lung, malarial kidney damage	Tuberculin reaction, contact dermatitis, tissue transplant rejection

First Exposures to Allergen

B cell receptors
Allergen
IgE antibodies
B cell
IgE receptor
Helper T cell
Mast cell
Cytokines
Plasma cell

Cytokines from an effector helper T cell cause class switching resulting in a clone of B cells destined to produce IgE.

The B cells differentiate into IgE-producing plasma cells and memory cells.

IgE antibodies bind to mast cell receptors; the individual is sensitized.

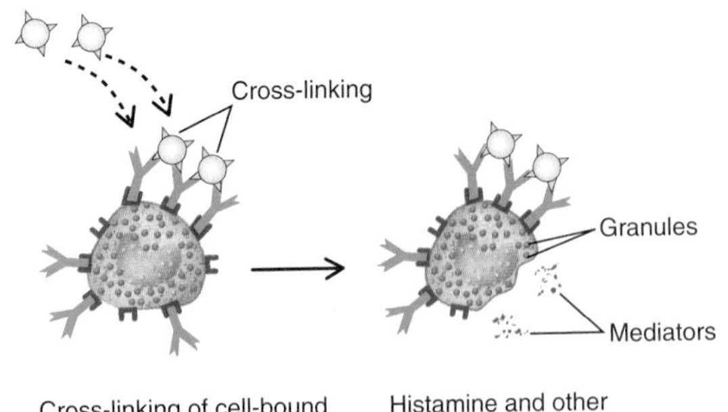
Subsequent Exposures to Allergen
Cross-linking
Granules
Mediators
Cross-linking of cell-bound IgE causes the mast cell to degranulate.
Histamine and other mediators are released.

FIGURE 18.1 Mechanisms of Type I Hypersensitivity: Immediate IgE-Mediated First exposure to an allergen induces an IgE antibody response leading to sensitization. With subsequent exposures to the allergen, cross-linking of IgE molecules on mast cells results in release of histamine and other substances that can cause dilation and increased leakage of plasma from capillaries, airway constriction, and increased mucus production. These effects can cause itching, swelling, and pain, and conditions such as asthma, hay fever, and anaphylactic shock.

also on circulating basophils **(figure 18.1).** Once attached, the IgE molecules can survive for many weeks with their reaction sites ready to interact with antigen. The individual is thereby sensitized to that antigen. ■ **antibody response, p. 376**

People with a type I hypersensitivity will have many IgE antibodies fixed to mast cells throughout their bodies. When such a person is not exposed to allergen, these antibodies are harmless. Upon exposure, however, the allergen readily combines with the cell-fixed IgE antibodies. At least two cell-bound IgE molecules must react with specific antigen, cross-linking the IgE molecules, in order for a reaction to occur (see figure 18.1). Within seconds, the IgE-antigen attachment and cross-linking of IgE in the cell membrane cause the mast cell to release histamine, leukotrienes, prostaglandins, and cytokines in a process called **degranulation.** These **chemical mediators** released from the mast cell granules are the direct causes of hives (urticaria), hay fever (allergic rhinitis), asthma, anaphylactic shock, and other allergic manifestations. ■ **mast cell, p. 350**

The tendency to have type I allergic reactions is inherited. The reactions occur in at least 20% to 30% of the population of the United States. Susceptible people often have higher than normal levels of IgE in the circulation, and increased numbers of eosinophils.

Localized Anaphylaxis

Anaphylaxis is the name given to allergic reactions caused by IgE-mediated release of chemical mediators from mast cell granules. Although anaphylaxis can be generalized or local, by far the most usual allergic reactions are examples of localized anaphylaxis. **Hives (urticaria)** is an allergic skin condition characterized by the formation of a wheal and flare; the wheal is an itchy swelling generally resembling a mosquito bite, surrounded by redness, the flare. The wheal and flare reaction is seen also in positive skin tests for allergens **(figure 18.2).** Hives may occur, for example, when a person allergic to lobster eats some of the seafood. Lobster antigen absorbed from the intestinal tract enters the bloodstream and is carried to tissues such as skin, where it reacts with mast cells that have anti-lobster IgE antibody attached to them. Reaction between antibody and antigen on the mast cell surface releases histamine, which in turn causes dilation of tiny blood vessels and the leaking of plasma into skin tissues. Life-threatening respiratory obstruction is possible during a reaction if there is extensive tissue swelling in the throat and larynx. Because histamine is a major mediator in this situation, the reaction is blocked by antihistamine medications.

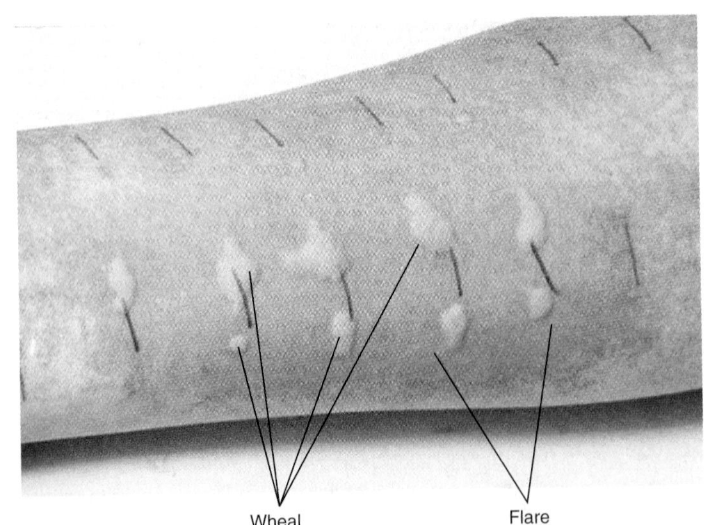
Wheal
Flare

FIGURE 18.2 The Wheal and Flare Skin Reaction of IgE-Mediated Allergies A variety of antigens are injected or placed in small cuts in the skin to test for sensitivity. Immediate wheal and flare reactions occur with antigens to which the person is sensitive.

Hay fever (allergic rhinitis), marked by itching, teary eyes, sneezing, and runny nose, occurs when allergic persons inhale an antigen such as ragweed pollen to which they are sensitive. The mechanism is similar to that of hives and is also blocked by antihistamines.

Asthma is another type of immediate respiratory allergy. An allergen reacting with IgE-sensitized mast cells causes their degranulation and release of chemical mediators. The mediators cause spasms of the bronchi, which interfere with breathing. Mediators other than histamine, mainly lipids such as leukotrienes and prostaglandins, and protein cytokines, are responsible for bronchospasm and increased mucus production. Eosinophils recruited to the area also release leukotrienes and contribute to an inflammatory response. Antihistamines are therefore not effective in treating asthma, but several other drugs are available to block the reaction, such as bronchodilating drugs that relax constricted muscles and relieve the bronchospasm. An example is albuterol, a bronchodilator usually delivered by an inhaler. Cortisone-like steroids are often used to decrease the inflammatory reaction. An anti-IgE therapy is described later.

Generalized Anaphylaxis

Generalized systemic anaphylaxis is a rare but serious form of IgE-mediated allergy. This is the form of anaphylaxis that killed Richet's dog (see A Glimpse of History). Antigen enters the bloodstream and becomes wide-spread. Instead of being localized in areas of the skin or respiratory tract, the reaction affects almost the entire body, causing a drop in blood pressure that can culminate in **anaphylactic shock.** Shock is a state in which the blood pressure is too low to supply the blood flow required to meet the oxygen needs of vital body tissues. In the case of generalized anaphylaxis, large amounts of released mediators cause extensive blood vessel dilation and loss of fluid from the blood, so that blood pressure falls dramatically and there is insufficient blood flow to vital organs such as the brain. Suffocation also occurs, the result of marked constriction of the bronchial tubes. Anaphylactic reactions may be fatal within minutes. Bee stings, peanuts, and penicillin injections probably account for most cases of generalized anaphylaxis. Penicillin molecules are changed in the body to a haptenic form that can react with body proteins, forming a penicillin-protein conjugate. In a small percentage of people, this hapten-protein complex elicits the production of IgE antibodies that react with penicillin. Fortunately, only a tiny fraction of people who make the antibodies are prone to anaphylactic shock. Peanuts and their products, such as peanut oil, are found in so many foods that it is often hard to predict where they will be encountered. The problem is widespread enough that peanuts have been barred by some airlines as a snack during flights. Generalized anaphylaxis can usually be controlled by epinephrine (Adrenalin) injected immediately. Anyone who has had a local allergic reaction to bee sting, penicillin, peanuts, or other substance is at risk of anaphylactic shock if exposed to the same substances and should wear a medical-alert bracelet and carry emergency medications. ■ **hapten, p. 383**

Treatments to Prevent Allergic Reactions

One way to treat allergies involves **immunotherapy,** a technique used to modify immune responses for a favorable effect. A procedure called **desensitization,** or **hyposensitization,** causes the immune system to mount an IgG response rather that IgE. Over a period of many months, the person is injected with the allergen in extremely diluted, but gradually increasing concentrations. The antigen must be diluted enough to avoid an anaphylactic reaction. As the concentration in the injections increases during the course of treatment, the individual becomes less and less sensitive and may even lose the hypersensitivity entirely. Concurrently, there is an increase in IgG antibody levels and a decrease in the IgE response to the allergen. The IgG antibodies probably protect the patient by binding to the offending antigen, thus preventing its attachment to cell bound IgE (**figure 18.3**). Activation of regulatory T cells may also play a role through release of cytokines that suppress the IgE response.

Another way of treating allergies is to prevent IgE antibodies from binding to mast cells and basophils. A medication called omalizumab uses a genetically engineered form of an IgG molecule that binds specifically to the Fc portion of the IgE molecules, thereby blocking the site that would otherwise attach to mast cells and basophils. Injections of omalizumab are generally well tolerated by asthmatic people and are very effective in treating their asthma, reducing their need for more toxic medications. Omalizumab is used mainly in the most serious cases of asthma because it is expensive and can paradoxically cause anaphylactic

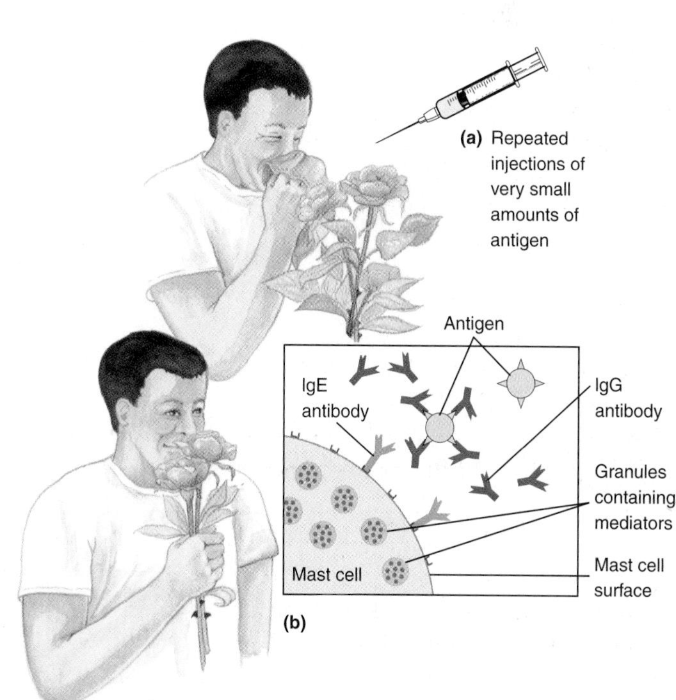

FIGURE 18.3 Immunotherapy for IgE Allergies (a) Repeated injections of very small amounts of antigen are given over several months. **(b)** This regimen leads to the formation of specific IgG antibodies. The IgG reacts with antigen before it can bind to IgE, and therefore it blocks the IgE reaction.

shock on rare occasions. It is an example of a **rhuMab** (**r**ecombinant **hu**manized **M**onoclonal **a**nti**b**ody) used in treating various other conditions. ■ rhuMab, p. 439

MICROCHECK 18.1

Hypersensitivity reactions are immunological reactions that cause tissue damage. Type I hypersensitivity reactions mediated by cell-bound IgE antibodies occur immediately after exposure to antigen. The reactions are caused by the release of mediators from mast cell granules. Localized anaphylactic reactions include hives, hay fever, and asthma; generalized reactions lead to anaphylactic shock. Immunotherapy is directed toward inducing IgG rather than IgE. Engineered anti-IgE (omalizumab) is useful in treating severe asthma.

✓ How do localized and generalized anaphylactic reactions differ?

✓ Define *allergen* and give five common examples of substances that act as allergens.

✓ Why would leaking of plasma from blood vessels during an allergic reaction cause a wheal?

18.2

Type II Hypersensitivities: Cytotoxic

Focus Point

- Outline the main features of transfusion reactions and hemolytic disease of the newborn.

In type II hypersensitivity reactions, antibodies react with molecules on the surface of a person's cells and trigger destruction of the cells. The antibodies recognize and bind some normal component of the target cells or drug or other foreign substance attached to the target cells. Cell destruction can occur through activation of the complement system or by antibody-dependent cellular cytotoxicity (ADCC), in which the antibody-coated target cells are destroyed by the mononuclear phagocyte system. Two common examples of type II sensitivity reactions are transfusion reactions and hemolytic disease of the newborn. ■ antibody dependent cellular cytotoxicity (ADCC), p. 373 ■ mononuclear phagocyte system, p. 350

Transfusion Reactions

Erythrocytes (red blood cells) have various antigenic determinants on their surface that differ from one person to another. When an individual receives a transfusion of red blood cells antigenically different from his or her own, the immune system attacks those cells. Occasionally, the complement system is activated via the classic pathway and life-threatening rapid destruction of transfused red blood cells occurs by membrane attack complexes. More often the antibody-coated cells are removed by ADCC, with symptoms that include fever, pain, nausea, and vomiting. Matching the blood of donor and recipient to ensure compatibility minimizes the possibility of a transfusion reaction.

Transfusion reactions generally involve antigens of the ABO blood group system. People are designated as having blood type A, B, or O, depending on which, if any, specific polysaccharides are present on their red blood cells **(table 18.2).** People who lack A or B antigens have antibodies against those antigens they do not have. Thus, people with blood type O have both anti-A and anti-B antibodies; those who have blood type A have anti-B antibodies, and those with blood type B have anti-A antibodies. These antibodies are called **natural antibodies** because they are present without any obvious cause. Anti-A and anti-B antibodies are not present at birth but generally appear before the age of six months. These natural antibodies are mostly of the class IgM, and therefore cannot cross the placenta. These probably arise because of multiple exposures to substances similar to blood group antigens that occur in environmental materials such as bacteria, dust, and food.

Hemolytic Disease of the Newborn

Antigens of the rhesus (Rh) system represent another important kind of erythrocyte antigen, first described in rhesus monkeys. If an Rh antigen is present on a person's erythrocytes, he or she is termed "Rh-positive"; if no Rh antigens are present, that person is "Rh-negative." Just as donor and recipient bloods are matched for the ABO system, they are also matched for Rh system antigens, so that serious transfusion reactions are uncommon.

A problem can develop when an Rh-negative woman is pregnant with an Rh-positive fetus. If the woman has anti-Rh

TABLE 18.2 Antigens and Antibodies in Human ABO Blood Groups

			Incidence of Type in United States		
Blood Type	**Antigen Present on Erythrocyte Membranes**	**Antibody in Plasma**	**Among Whites**	**Among Asians**	**Among Blacks**
A	A	Anti-B	41%	28%	27%
B	B	Anti-A	10%	27%	20%
AB	A and B	Neither anti-A nor anti-B	4%	5%	7%
O	Neither	Anti-A and anti-B	45%	40%	46%

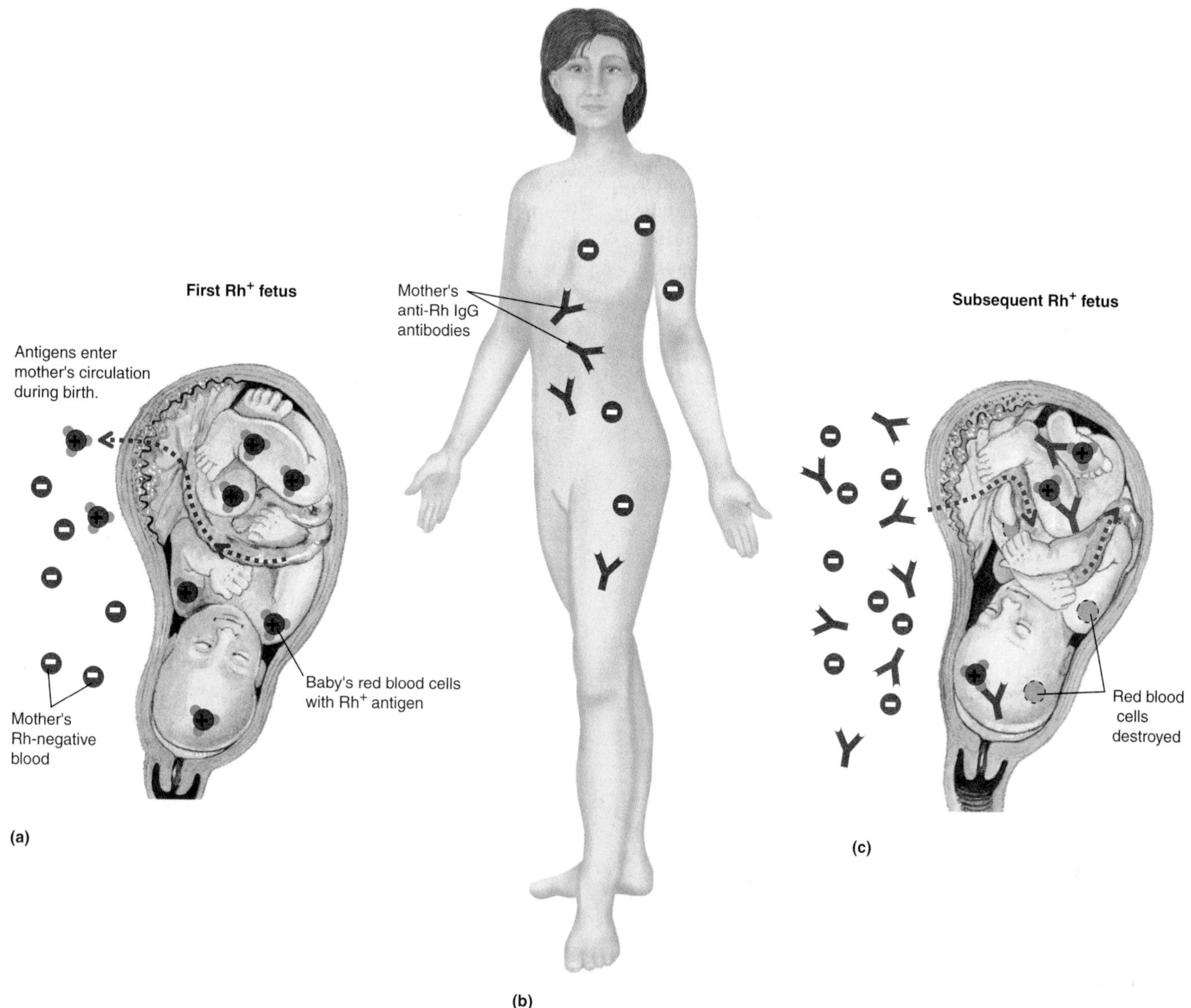

FIGURE 18.4 Hemolytic Disease of the Newborn **(a)** The fetus of an Rh-negative (Rh^-) mother may inherit paternal genes for Rh antigens and have Rh-positive (Rh^+) blood. During delivery of a first Rh^+ baby, enough Rh^+ red cells can enter the Rh^- mother to induce an anti-Rh response. **(b)** The mother makes a primary response to Rh antigens and develops memory cells for anti-Rh antibodies. **(c)** During subsequent pregnancies with an Rh^+ fetus, the very few fetal red cells that cross the placenta can cause a vigorous secondary response. The IgG anti-Rh antibodies cross the placenta and destroy the baby's red blood cells.

IgG antibodies, they can cross the placenta and damage the developing fetus, causing **hemolytic disease of the newborn (erythroblastosis fetalis) (figure 18.4).** An Rh-negative woman is likely to become sensitized to the Rh antigen during delivery of her first Rh-positive baby when enough of the baby's erythrocytes enter her circulation to cause an immune response. Sensitization can also occur during pregnancy in some cases, and after induced or spontaneous abortion. The anti-Rh antibodies formed by the mother do her no harm because her red blood cells lack the Rh antigen. However, the antibody response can damage her fetus during subsequent pregnancies. With each new pregnancy with an Rh-positive fetus, the mother produces large quantities of anti-Rh IgG antibodies that cross the placenta and enter the fetal circulation. Extensive damage to the fetus results. To save the fetus it is sometimes necessary to transfuse it repeatedly *in utero,* using Rh-negative blood. However, much more commonly, the disease only becomes life-threatening shortly after birth.

The fetus can usually survive the attack by anti-Rh antibodies while in the uterus because harmful products of red cell destruction are eliminated by maternal enzymes. However, the newborn baby has very little of these enzymes, and as soon as 36 hours after birth can become jaundiced and seriously ill as a result of the toxic products of red blood cell destruction. In this critical situation, immediate treatment is sometimes required to correct anemia and prevent permanent brain damage. It may be necessary to withdraw part of the baby's blood and replace it with Rh-negative blood. Irradiation of the baby with light of wavelengths 420 to 480 nm converts the toxic erythrocyte breakdown products to a form that is more readily excretable.

To prevent hemolytic disease of the newborn, Rh-negative women who are carrying an Rh-positive fetus are injected with anti-Rh antibodies once during pregnancy and again shortly after delivery. The anti-Rh antibodies bind to any Rh-positive erythrocytes that may have entered the mother's circulation from the baby, thereby preventing the development of a primary immune response. Injecting anti-Rh antibody is not effective if the individual is already sensitized to Rh antigen.

MICROCHECK 18.2

Type II cytotoxic hypersensitivity reactions are mediated by antibodies, either by lysis of cells via the complement system or by antibody-dependent cellular cytotoxicity. Blood transfusion reactions and hemolytic disease of the newborn are examples.

- ✓ Describe the mechanism of cell damage in a blood transfusion where ABO antigens are mismatched.
- ✓ Why do Rh-negative but not Rh-positive mothers sometimes have babies with hemolytic disease of the newborn?
- ✓ Why is the finding surprising that people lacking the A or B antigen are found to have antibodies to the corresponding antigen (anti-A or anti-B)?

18.3 Type III Hypersensitivities: Immune Complex–Mediated

Focus Points

- Describe the importance of immune complexes in type III hypersensitivity reactions.
- List four medical conditions in which type III hypersensitivities play an important role.

An **immune complex** consists of IgG or IgM antibody bound to antigen, often with some complement components. Immune complexes usually adhere to Fc receptors on cells of the mononuclear phagocyte system and are engulfed and destroyed intracellularly, especially if they are large. Under ordinary circumstances, this rapidly removes them from circulation. In conditions where a moderate excess of antigen over antibody exists, however, small complexes form. These are not quickly removed and destroyed but persist in circulation or at their sites of formation in tissue. Circulating immune complexes lodge in blood vessel walls and in various tissues. Immune complexes possess considerable biological activity. They initiate the blood clotting mechanism, and they activate components of the complement system that attract neutrophils into the area and contribute to inflammation **(figure 18.5).** Proteases released from the neutrophils then damage tissue. Circulating immune complexes are commonly deposited in skin, joints, and the kidneys. Immune complexes are responsible for the rashes, joint pains, and other symptoms seen in a number of diseases, such as farmer's lung, bacterial endocarditis, early rubella infection, and malaria. When they are deposited in the kidneys they cause glomerulonephritis. Immune complexes can also precipitate a devastating condition, **disseminated intravascular coagulation,** in which clots form in small blood vessels, leading to failure of vital organs. ■ disseminated intravascular coagulation, p. 679 ■ glomerulonephritis, pp. 539, 678 ■ Fc receptors, p. 402

Immune complex disease may arise during a variety of bacterial, viral, and protozoan infections, as well as from inhaled dusts or bacteria and injected medications such as penicillin.

Immune complex formation is also responsible for the localized injury or death of tissue, known as the **Arthus reaction,** that occurs if antigen is injected into the tissue of a previously immunized person with high levels of circulating specific antibody. This can happen, for example, when tetanus-diptheria booster vaccine is given to a person too frequently. The immune complexes form outside the blood vessels, in the tissues, and activate complement, producing complement components that attract neutrophils. The release of neutrophil enzymes contributes to a local inflammatory response that peaks in 6 to 12 hours.

Serum sickness is an immune complex disease caused by passive immunization when an antibody-containing serum from a horse or other animal is injected into humans to prevent or treat a disease such as diphtheria or tetanus. The recipient of the animal serum may make an immune response to antigens in the foreign serum, and after 7 to 10 days enough immune complexes form to cause signs of disease, which include fever, inflammation of blood vessels, arthritis, and kidney damage. The disease generally resolves as the antigens of the animal serum are cleared. Of course, horses are no longer used to produce antibodies against diphtheria and tetanus; instead, hyperimmune human globulin is used. The serum sickness form of hypersensitivity is rarely seen now, but it can occur following treatment of heart attack patients with the bacterial enzyme streptokinase to dissolve clots, after the use of serum from horses immunized with snake venom to treat snakebites in people, or in a few other rare instances. ■ passive immunity, p. 432 ■ hyperimmune globulin, p. 433

MICROCHECK 18.3

Type III immune complex–mediated hypersensitivities are caused by small complexes of antigen-antibody and complement that persist in tissues or blood and that cause inflammation.

- ✓ Why do immune complexes remain in the circulation or in the tissues in immune complex diseases?
- ✓ How can immune complexes cause tissue damage?
- ✓ Why would the kidneys be particularly prone to immune complex damage?

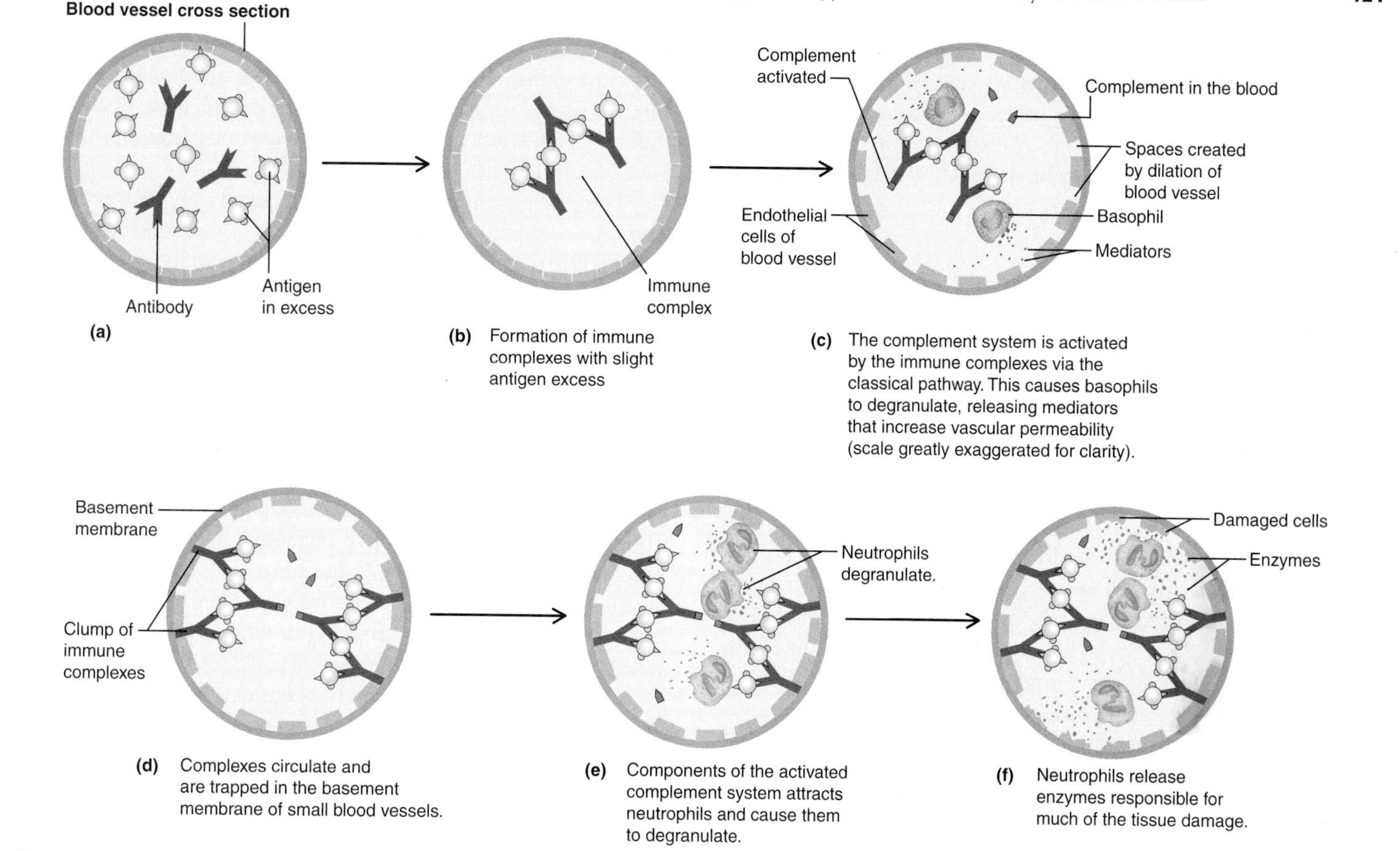

Blood vessel cross section

Antibody

Antigen in excess

(a)

Immune complex

(b) Formation of immune complexes with slight antigen excess

Complement activated

Complement in the blood

Spaces created by dilation of blood vessel

Endothelial cells of blood vessel

Basophil

Mediators

(c) The complement system is activated by the immune complexes via the classical pathway. This causes basophils to degranulate, releasing mediators that increase vascular permeability (scale greatly exaggerated for clarity).

Basement membrane

Clump of immune complexes

(d) Complexes circulate and are trapped in the basement membrane of small blood vessels.

Neutrophils degranulate.

(e) Components of the activated complement system attracts neutrophils and cause them to degranulate.

Damaged cells

Enzymes

(f) Neutrophils release enzymes responsible for much of the tissue damage.

FIGURE 18.5 Type III Hypersensitivity: Immune Complex-Mediated

18.4

Type IV Hypersensitivities: Delayed Cell-Mediated

Focus Points

- Describe the key immunologic reactions involved in type IV hypersensitivities.
- Define the importance of type IV hypersensitivity reactions in infectious diseases.

Harmful effects produced by the mechanisms of cell-mediated immunity are referred to as **delayed hypersensitivity.** The name reflects the slowly developing response to antigen; reactions peak at 2 to 3 days rather than in minutes as in immediate hypersensitivity. As would be expected with cell-mediated responses, T cells are responsible and antibodies are not involved. Delayed hypersensitivity reactions can occur almost anywhere in the body. They are wholly or partly responsible for **contact dermatitis** (such as from poison ivy and poison oak), tissue damage in a variety of infectious diseases, rejection of tissue grafts, and some autoimmune diseases directed against antigens of self.

Tuberculin Skin Test

A familiar example of a delayed hypersensitivity reaction is the positive reaction to a tuberculin skin test that occurs in most people who have been infected with *Mycobacterium tuberculosis.* This test involves the introduction of very small quantities of protein antigens of the organism into the skin. In people with delayed hypersensitivity to *M. tuberculosis,* the site of injection reddens and gradually becomes indurated (thickened) within 6 to 24 hours. The reaction reaches its peak at 2 to 3 days (**figure 18.6**). There is no wheal formation, as would be seen with IgE-

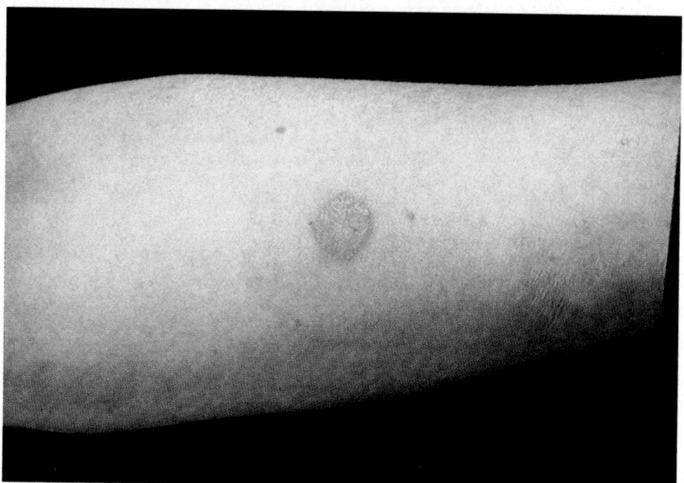

FIGURE 18.6 Positive Delayed (Type IV) Hypersensitivity Skin Test Injection of tuberculin protein into the skin of a person sensitized to *Mycobacterium tuberculosis* causes a firm red plaque to form by 48 to 72 hours. A reaction greater than 10 mm in diameter is considered positive.

mediated reactions. The redness and induration of delayed skin hypersensitivity reactions are the result mainly of the reaction of sensitized T cells with specific antigen, followed by the release of cytokines and the influx of macrophages to the injection site. The positive skin test mirrors what happens at the site of infection in the body.

Delayed Hypersensitivity in Infectious Diseases

Cell-mediated immunity plays an important role in combating intracellular infections through the cell-destroying activity of activated macrophages and T lymphocytes. Although these functions are protective, tissue damage or hypersensitivity also results. These infections may be caused by viruses, mycobacteria and certain other bacteria, protozoa, and fungi. They include leprosy, tuberculosis, leishmaniasis, and herpes simplex, among many others. In particularly slowly progressing infections, delayed hypersensitivity causes extensive host cell destruction and progressive impairment of tissue function, such as the damaged sensory nerves in leprosy. The immune response is a two-edged sword, protecting on the one side, but causing damage on the other. ■ immunity to intracellular infections, p. 367 ■ leprosy, p. 655

Contact Hypersensitivities

Contact hypersensitivity is also mediated by T cells. Here, T cells that have become sensitized to an allergen contacting the skin release cytokines when they come into contact again with the same antigen. These cytokines cause inflammatory reactions that attract macrophages to the site. The macrophages then release mediators that add to the inflammatory response, resulting in allergic dermatitis.

Familiar examples of contact hypersensitivity (contact allergy or contact dermatitis) are poison ivy and poison oak **(figure 18.7)** and allergic reactions to the nickel of metal jewelry, the chromium salts in certain leather products, or components of some cosmetics. In the case of poison oak or ivy, a hapten from an oily product of the plant is responsible. With a metal, a soluble salt of the metal acts as a hapten.

Latex products are a frequent cause of contact hypersensitivity reactions, and they also can cause IgE-mediated reactions. Latex, a product of the rubber tree, contains a plant protein that readily induces sensitization. Many products contain latex, such as fabrics, elastics, toys, and contraceptive condoms, but latex gloves probably account for most latex sensitization. Gloves are used extensively by health care and laboratory workers, food preparers, and many others. Typically, a person will notice redness, itching, and a rash on the hands after wearing gloves. To prevent the reaction, latex gloves should be replaced by vinyl or other synthetic gloves. Topical cortisone-like medications are effective treatment.

The causative substance in contact hypersensitivity is commonly detected by patch tests, in which suspect substances are applied to the skin under an adhesive bandage. Positive reactions reach their maximum in about 3 days and consist of redness, itching, and blisters of the skin. **Figure 18.8** shows a severe contact hypersensitivity skin rash.

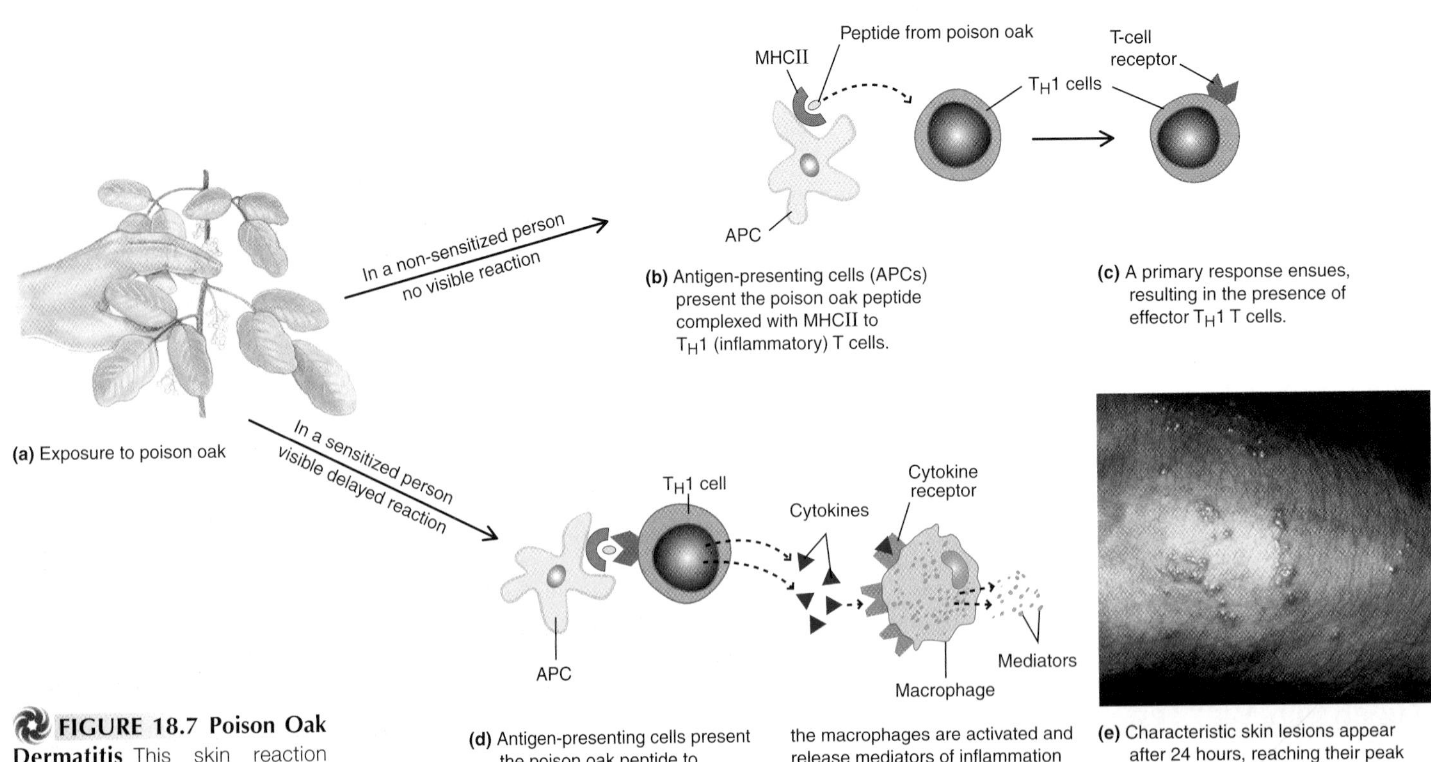

Peptide from poison oak
MHCII
T-cell receptor
T_H1 cells
APC
In a non-sensitized person
no visible reaction
(b) Antigen-presenting cells (APCs) present the poison oak peptide complexed with MHCII to T_H1 (inflammatory) T cells.
(c) A primary response ensues, resulting in the presence of effector T_H1 T cells.
(a) Exposure to poison oak
In a sensitized person
visible delayed reaction
T_H1 cell
Cytokine receptor
Cytokines
Mediators
APC
Macrophage
(d) Antigen-presenting cells present the poison oak peptide to sensitized T_H1 cells, which release cytokines and attract macrophages; the macrophages are activated and release mediators of inflammation that cause skin lesions.
(e) Characteristic skin lesions appear after 24 hours, reaching their peak at 48–72 hours after exposure to the plant.

FIGURE 18.7 Poison Oak Dermatitis This skin reaction results from a delayed (Type IV) hypersensitivity.

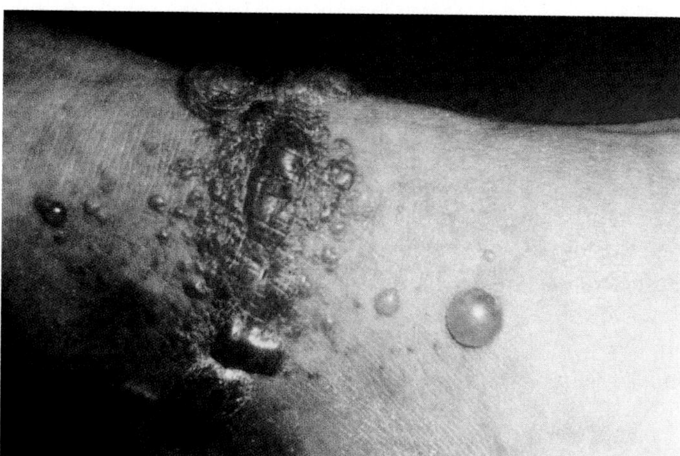

FIGURE 18.8 Severe Contact Hypersensitivity Note the skin rash showing redness, blisters and scaling.

MICROCHECK 18.4

Type IV delayed hypersensitivity depends on the action of sensitized T cells. The reaction peaks 2 to 3 days after exposure to antigen. Examples are contact dermatitis, damage in a variety of infectious diseases, rejection of tissue grafts, and some autoimmune diseases.

✓ Explain the events that occur in the skin during a positive delayed hypersensitivity skin test.
✓ Describe a patch test for contact hypersensitivity.
✓ In the tuberculin skin test, why would there be no reaction if there is no infection?

18.5 Rejection of Transplanted Tissues

Focus Points

- Explain why rejection of transplanted tissues occurs.
- Describe the mode of action of medications used to prevent rejection of transplanted tissue.

Transplantation of organs, bone marrow, and other tissues between genetically non-identical humans is a well-established clinical procedure. These grafts are called **allografts. Xenografts,** from non-human species, such as pigs, may become more widely available with the development of cloned animals having tissues more closely compatible with humans. Except for **autografts** (tissue transplanted from elsewhere on the recipient's body) and **isografts** (grafts donated by an identical sibling), the major drawback to graft transplantation is possible immunological rejection of the transplant. Body cells vary antigenically from individual to individual, and differences between tissues of the transplant donor and recipient lead to rejection of the graft. Transplantation rejection in this situation is predominantly a type IV cellular immunological reaction showing both specificity and memory. Killing of the graft cells occurs through a complex combination of mechanisms, including direct contact with effector cytotoxic T-lymphocytes and natural killer (NK) cells. The ability to overcome immunologic rejection by treatment with immunosuppressive agents allows the grafts to survive. **Perspective 18.1** discusses a very special transplantation situation, the fetus as an allograft. ■ cytotoxic T cells, p. 381 ■ NK cells, p. 384

PERSPECTIVE 18.1

The Fetus As an Allograft

Grafts between non-identical members of the same species are allografts, and they are normally rejected by immunological mechanisms. The rejection time for grafts that differ in their major histocompatibility molecules is measured in days, about 10 to 14 days in most instances. The rejection process is complex, but it depends principally on host T-cell destruction of grafted cells. For example, when skin allografts are transplanted, the grafted skin appears normal for about a week. Gradually it begins to look bruised and unhealthy, until 10 days to 2 weeks after transplantation. By that time the grafted skin becomes dried and is sloughed off. Microscopic examination of the graft shows that by the end of the first week T cells invade the tissues and within the next few days these lymphocytes have killed the grafted cells. The same events occur in allografts of kidney or other organs, unless effective immunosuppressive therapy is given. There is one kind of allograft that does not follow this sequence and that is not rejected, however—the mammalian fetus.

The fetus is an allograft, with half of its antigens of paternal origin and likely to differ from the other half contributed by the mother. In spite of these major immunological differences, the fetus lives in the uterus for 9 months and is not rejected. In fact, over a period of years a mother often has several (or even occasionally as many as 20!) children without showing any signs of immunological rejection of the fetus. The mechanisms for survival of the fetal allograft have been the subject of research for many years, but they are not yet fully understood.

It cannot be that paternal antigens from the fetus do not reach the mother's immune system to cause a response. Mothers are known to make antibodies to paternal antigens, such as the Rhesus red blood cell antigens. Also the antibodies used for typing major histocompatibility antigens have long been obtained from women who have borne several children by the same father and have made antibodies to his MHC molecules. Furthermore, various techniques show the presence of small numbers of fetal cells in the maternal circulation during pregnancy. Clearly, paternal antigens can reach the mother's immune system and cause a response. The placenta, however, does prevent most fetal cells from entering the mother and most maternal T cells from reaching the fetus.

The outer layer of the placenta, the trophoblast, forms sort of a buffer zone between the fetus and the mother. The trophoblast does not express MHC class I or II molecules and is not subject to T-cell attack; it also has a mechanism for avoiding destruction by natural killer cells. Thus, the fetus is protected by being in an immunologically privileged site. A few other areas in the body—the brain, the eyes, and the testes—are also immunologically privileged sites. Antigens leaving these sites do not drain through lymphatic vessels and reach lymphoid tissues where antigen-presenting cells are abundant. Also, antigen leaves these privileged sites accompanied by cytokines that are immunosuppressive and direct the immune response toward tolerance, rather than toward active harmful responses. It is well recognized that maternal immune responses are suppressed to some extent during pregnancy, though the reasons for this are not clear.

Thus, one major factor responsible for preventing the rejection of the fetal allograft is the location of the fetus in the uterus, protected by the placental barrier. A second factor is the ability of the pregnancy to cause an immunosuppressive response in the mother.

Special problems are involved in the many thousands of bone marrow transplants performed each year to save the lives of individuals with malignancies such as multiple myeloma and leukemia, and other serious conditions. Generally, the patient's bone marrow and immune system have been intentionally destroyed by intense radiation and chemotherapy. In most cases, the marrow transplant is an allograft from a carefully matched, healthy person. Under general anesthesia, a liter or more of marrow is obtained from the donor using syringes and a large, sturdy needle pushed into the iliac bone of the pelvis. The marrow is treated in various ways (for example, to remove erthrocytes and bone particles) and stored frozen if necessary. It is then infused into the patient through the large subclavian vein under the left collar bone. The marrow hematopoetic stem cells leave the circulation, repopulate the patient's marrow spaces, and begin producing hematopoetic cells. Close nursing supervision is usually required for months to watch for infection and bleeding until the grafted marrow produces enough immune cells, erythrocytes, leukocytes, and platelets. Recovery is complicated in most cases by a **graft-versus-host reaction** in which the immune system arising from the graft sees the patient's tissues as "foreign" and attacks them. Therefore, many of these patients require immunosuppressive treatment. ■ hematopoetic stem cells, p. 350

There are many different tissue antigens, but those of the major histocompatibility complex (MHC) system (see Perspective 16.1) are most commonly involved in transplantation graft rejections. Although these MHC molecules are found on many human cells, they are abundant on leukocytes and are called human leukocyte antigens or HLAs. MHC tissue typing is done in an effort to minimize major tissue incompatibility between a prospective tissue donor and the recipient patient. In addition to carefully matching donor and recipient tissue antigens, it is necessary to use immunosuppressant drugs indefinitely to prevent graft rejection. These drugs are needed because many minor antigens exist and except with identical siblings it is impossible to find a donor compatible in all of these tissue antigens. ■ MHC molecules, p. 380

Radiation and various cytotoxic immunosuppressive drugs interfere with the rejection process, but at the same time they make the patient susceptible to opportunistic infections and also more likely to develop cancer. Cyclosporin A (produced by a fungus) and tacrolimus (produced from a species of *Streptomyces*) are examples of effective immunosuppressants. They interfere with cellular signaling and thereby inhibit clonal expansion of activated T lymphocytes. These drugs specifically suppress T-cell proliferation, and thus they have fewer side effects than other immunosuppressants that affect many cell types.

Combinations of agents are commonly used to prevent allograft rejection. For example, cyclosporin A and a cortisone-like steroid may be given, along with a monoclonal antibody preparation that blocks IL-2 binding to T-cell receptors thereby preventing activation of the cells.

MICROCHECK 18.5

Successful organ and tissue transplantation depend on matching major histocompatibility antigens and using immunosuppressive agents such as cyclosporin and tacrolimus to minimize the immune response to other antigens of the graft. Rejection of transplants is complex, but type IV cellular immune responses are the major mechanism of rejection for allografts.

- ✓ What are the antigens primarily responsible for allograft rejection?
- ✓ How are allografts rejected?
- ✓ Why is matching of transplant donors and recipients important?
- ✓ What would happen if administration of the antirejection drugs were discontinued?

18.6 Autoimmune Diseases

Focus Points

- Outline the concept of autoimmunity.
- Give four examples of autoimmune diseases and the mechanism of tissue injury in each.

Usually, the body's immune system recognizes its self-antigens and eliminates developing lymphocytes that would respond and attack its own tissues. A growing number of diseases are suspected of being caused by an **autoimmune process,** however, meaning that the immune system responds to tissues of the body as if they were foreign. Some of these diseases are listed in **table 18.3.** Susceptibility to many of them is influenced by

TABLE 18.3 Characteristics of Some Autoimmune Diseases

Disease	Organ Specificity	Major Mechanism of Tissue Damage
Graves' disease	Thyroid	Autoantibodies bind thyroid-stimulating hormone receptor, causing overstimulation of thyroid
Myasthenia gravis	Muscle	Autoantibodies bind to acetylcholine receptor on muscle, preventing muscle contraction
Type 1 diabetes mellitus	Pancreas	T-cell destruction of pancreatic β cells
Autoimmune hemolytic anemia	Red blood cells	Antibody, complement, and phagocyte destruction of red cells
Rheumatoid arthritis	Widespread, especially joints	Lymphocyte destruction of joint tissues; immune complexes of IgG and anti-IgG
Systemic lupus erythematosus	Widespread (glomerulonephritis, vasculitis, arthritis)	Autoantibodies to DNA and other nuclear components form immune complexes in small blood vessels

the major histocompatibility makeup of the patient, and so, not surprisingly, they often occur in the same family (see Perspective 16.1). ■ negative selection, p. 387

Autoimmune diseases may result from reaction to antigens that are similar though not identical to antigens of self. Some bacterial and viral agents evade destruction by the immune system by developing amino acid sequences that are similar to self-antigens. As a result, the immune system may be unable to discriminate between the agent and self. The immune system then destroys the substances of self as well as those of the bacterium or virus. The likeness of amino acid sequences need not be exact for this self-destruction; even a 50% likeness may lead to an autoimmune response. Autoimmune responses may also occur after tissue injury in which self-antigens are released from the injured organ, as in the case of a heart attack. The autoantibodies formed react with heart tissue and evidence suggests that they can cause further damage.

The Spectrum of Autoimmune Reactions

Autoimmune reactions occur over a spectrum ranging from organ-specific to widespread responses not limited to any one tissue. Examples of organ-specific autoimmune reactions are several kinds of thyroid disease, in which only the thyroid is affected. Widespread responses include lupus erythematosus and rheumatoid arthritis. Lupus is a disease in which autoantibodies are made against nuclear constituents of all body cells. In **rheumatoid arthritis (figure 18.9),** the response is against the collagen protein of supporting connective tissues. In these widespread diseases, many organs are affected. In both organ-specific and widespread autoimmune diseases, the damage may be caused either by antibodies, immune cells, or both.

Myasthenia gravis, characterized by muscle weakness, is an example of an **autoantibody-mediated disease.** The disease is caused by the production of antibodies to the acetylcholine receptor proteins that are present on muscle membranes where the nerve contacts the muscle. Normally, transmission of the impulses from the nerve to the muscle takes place when acetylcholine is released from the end of the nerve and crosses the gap to the muscle fiber, causing muscle contraction. In myasthenia gravis IgG autoantibodies bind to the acetylcholine receptors, blocking access of acetylcholine to the receptors. These antibodies along with complement system components cause many of the receptors to be degraded, so that fewer receptors are present on the muscle membranes. Babies born to mothers with myasthenia gravis also experience muscle weakness, since IgG antibodies cross the placenta. Fortunately, the effect is not permanent, as these IgG antibodies decay within a few months and the babies are no longer affected. Treatment of myasthenia gravis includes the administration of drugs that inhibit the enzyme cholinesterase, allowing acetylcholine to accumulate so some contact with receptors can occur. Immunosuppressive medications and removal of the thymus gland are helpful in many cases. The role of the thymus in this disease is not understood.

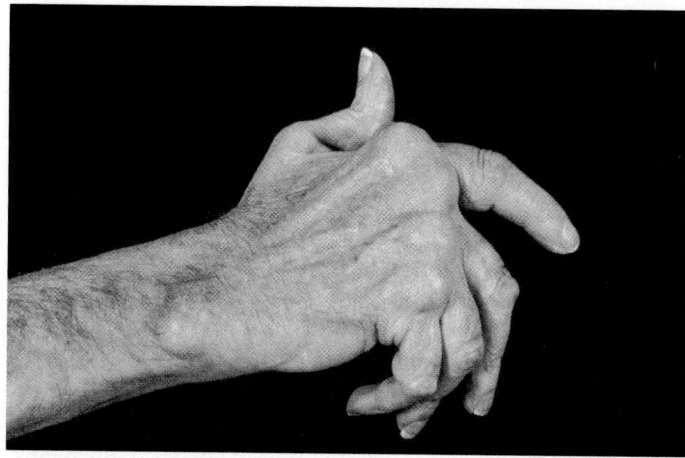

FIGURE 18.9 Rheumatoid Arthritis Autoimmune reactions cause chronic inflammation and destruction of the Joints.

Type 1 diabetes mellitus, also known as insulin-dependent or juvenile onset diabetes, is an autoimmune disease caused by cellular mechanisms. The major damage is destruction of the insulin-producing β cells of the pancreas by cytotoxic T cells. When the β cells are destroyed there is a lack of insulin, resulting in an increase of glucose in the blood. This leads to marked thirst and increased urine production. Although type 1 diabetes can occur at any age, the peak onset of this disease is about 12 years of age. Persons with type 1 diabetes must receive multiple injections of insulin daily to forestall complications such as blindness and kidney failure. In the United States there are an estimated one million victims of type 1 diabetes, representing a small fraction of all diabetes cases.

A combination of antibody and cellular mechanisms is active in some autoimmune diseases, such as rheumatoid arthritis. This crippling inflammatory condition is one of the most common autoimmune diseases, occurring in both sexes and in adults and children all over the world. About 1% of males and 3% of females in the United States are affected. Rheumatoid arthritis is most common in women ages 30 to 50. T_H 1 cells, the inflammatory CD4 T cells, infiltrate the joints. When stimulated by specific antigens, the T cells release cytokines that cause inflammation. Autoantibodies are also formed and contribute to widespread tissue damage by forming immune complexes. ■ T_H cells, p. 369

In **Graves' disease,** autoantibody is directed at an antigen of the thyroid gland. Normally, this antigen is a receptor for the pituitary hormone TSH (Thyroid Stimulating Hormone). Attachment of the autoantibody does not damage the thyroid gland, but causes it to enlarge and markedly overproduce thyroid hormone. Affected individuals suffer weight loss, heat intolerance, muscle weakness, rapid heartbeat, and other effects of too much thyroid hormone.

Treatment of Autoimmune Diseases

Autoimmune diseases are usually treated with immunosuppressants that kill dividing cells and thus control the response, or drugs that interfere with T-cell signaling, such as cyclosporin. Also, cortisone-like steroids and other anti-inflammatory medications are often used. Replacement therapy is necessary in some of the diseases, such as insulin in type 1 diabetes.

In type 1 diabetes, attempts are made to cure the disease by replacing the tissues destroyed by immune cells. Transplantation of the pancreas or insulin-producing cells of the pancreas has been successful in many cases, but dangerous immunosuppressive agents must be given to prevent rejection. Generally, only those with advanced diabetes who require a kidney transplant, and must take the immunosuppressive drugs anyway, are given pancreas transplants. Research efforts are directed toward developing better methods of transplantation, such as injecting cells harvested from cadaver pancreases into the vein that carries blood to the liver. The cells establish themselves in the liver and produce insulin, often eliminating or decreasing the need for injected insulin. However, the cells usually die off in months or a few years, and the immunosuppressant side effects are often severe. Stem cell research holds greater promise for a cure of type 1 diabetes. ■ stem cells, p. 350

Ideally, it would be better to induce tolerance to the specific antigen causing the immune response. Experimentally, diabetic mice fed with insulin were protected from their diabetes. Rheumatoid arthritis patients often have an active immune response to collagen, a protein present in the joints and surrounding tissues. An experimental treatment involves feeding solutions of animal collagen daily to the patients. The rationale of these experiments depends on a well-known phenomenon called **oral tolerance.** Antigen introduced by the oral route can cause a local intestinal immune response with release of cytokines, down-regulation of antigen receptors, and elimination of immune cells. Trials of oral administration of antigens in people have shown benefit in several autoimmune diseases, but there is much still to be learned about the immunological mechanisms, antigen preparations, doses, and duration of treatment. ■ immunological tolerance, p. 367

MICROCHECK 18.6

Autoimmune disease can result when the immune system reacts to substances of self. Autoimmune diseases cover a spectrum from organ-specific to generalized. They are treated with drugs that suppress immune or inflammatory responses. Attempts are being made to induce specific tolerance to the causative substances.

- ✓ How could viruses or bacteria be implicated in causing autoimmune diseases?
- ✓ Explain what is meant by the "spectrum of autoimmune diseases," from organ-specific to generalized.
- ✓ What advantages might stem cells have over pancreatic cell allografts in treating type I diabetes?

18.7 Immunodeficiency Disorders

Focus Points

- Contrast the two main categories of immunodeficiency disorders.
- Explain how immunodeficiency can lead to infection.

The immunologic disorders discussed up to now can be viewed as resulting from the overreaction of the body's immune systems. By contrast, in immunodeficiencies, the body is incapable of making or sustaining an adequate immune response. There are two basic types of immunodeficiency diseases: primary, or congenital; and secondary, or acquired. Primary immunodeficiency can be inborn as the result of a genetic defect or can result from developmental abnormalities. Secondary immunodeficiency can be acquired as the result of infection or other stresses on the immune system such as malnutrition. People with either type of immunodeficiency are subject to repeated infections. The types of these infections will often depend on which part of the immune system is absent or malfunctioning. Some of the more important immunodeficiency diseases are listed in **table 18.4.**

Primary Immunodeficiencies

The genetic or developmental abnormalities that cause primary immunodeficiencies affect B cells, T cells, or both. Some primary immunodeficiencies affect natural killer (NK) cells, phagocytes, or complement components. Primary immunodeficiencies are generally rare. **Agammaglobulinemia,** a disease in which few or no antibodies are produced, occurs in one in about 50,000 people, and **severe combined immunodeficiency (SCID),** where neither T nor B lymphocytes are functional, occurs in only about one of 500,000 live births.

Selective IgA deficiency, in which very little or no IgA is produced, is the most common primary immunodeficiency known. It has been reported in different studies to occur as often as one per 333 to 700 people. Although people with this disorder may appear healthy, many have repeated bacterial infections of the respiratory, gastrointestinal, and genitourinary tracts, where secretory IgA normally protects.

Primary deficiencies may occur in various components of the complement system. For example, the few individuals who lack C3 are prone to develop severe, life-threatening infections with encapsulated and pyogenic (pus-producing) bacteria. Patients with deficiencies in the early components of complement such as C1 and C2, may develop immune complex diseases, because these components normally help clear immune complexes from the circulation. Patients who lack late components of the classical pathway of C activation (C5, C6, C7, C8) have recurrent *Neisseria* infections. Immunity to these bacteria is associated with destruction of the organisms by complement-dependent bactericidal antibodies. People who lack one of the important control proteins of the sequence, C1-inhibitor, experience uncontrolled complement activation. This causes fluid accumulation and potentially fatal tissue swelling, a condition called **hereditary angioneurotic edema.** ■ complement system, p. 355

In children with DiGeorge syndrome, the thymus fails to develop in the embryo. As a result, T cells do not differentiate and are absent. Affected individuals have other developmental defects as well, such as heart and blood vessel abnormalities, and a characteristic appearance with low-set deformed ears, small mouth, and wide-set eyes. As expected from a lack of T cells, affected people are very susceptible to infections by eukaryotic pathogens, such as *Pneumocystis jirovecii* and other fungi, as well as viruses and obligate intracellular bacteria. ■ *Pneumocystis jirovecii*, p. 712

Severe combined immunodeficiency (SCID) results when neither T nor B lymphocytes are produced from bone marrow stem cells. Children with SCID die of infectious diseases at an early age unless they are successfully treated by receiving a bone marrow transplant to reconstitute the bone marrow with healthy

TABLE 18.4 Immunodeficiency Diseases

	Disease	Part of the Immune System Involved
Primary Immunodeficiencies		
	DiGeorge syndrome	T cells (deficiency)
	Congenital agammaglobulinemia	B cells (deficiency)
	Infantile X-linked agammaglobulinemia	Early B cells (deficiency)
	Selective IgA deficiency	B cells making IgA (deficiency)
	Severe combined immunodeficiency (SCID)	Bone marrow stem cells (defect)
	Chediak-Higashi disease	Phagocytes (defect)
	Chronic granulomatous disease	Phagocytes (defect)
Secondary Immunodeficiencies		
	Acquired immunodeficiency syndrome (AIDS)	T cells (destroyed by virus)
	Monoclonal gammopathy	B cells (multiply out of control)

cells. There are a variety of gene defects that can cause SCID. One defect is in an enzyme necessary for V, D, and J chain recombination to form B- and T-cell receptors for antigen. Without these receptors, there are no functioning B and T cells. SCID results from mutation in a gene for the interleukin-2 receptor on lymphocytes, such that the cells do not receive the signal to proliferate. Other individuals with SCID lack adenosine deaminase, an enzyme important in the proliferation of B and T cells. A number of these people have responded well to repeated replacement of the adenosine deaminase enzyme. It has been possible to correct this condition temporarily in a few children by collecting their own defective T cells, inserting the adenosine deaminase gene linked to a retrovirus vector, and returning the cells to them. Unfortunately, the genetically altered cells do not live long, and the treatment must be repeated. Still, these results are promising, and there is much excitement about the possibility of treating other severe disorders with gene therapy. Many of the gene defects that cause primary immunological disorders are known and work is under way to correct them. **Table 18.5** lists some of the primary immunodeficiencies for which gene defects have been identified. As the table indicates, deficiencies and defects can occur at any point in the complex steps that lead to an effective immune response. ■ **interleukin-2, p. 354** ■ **V, D, J gene segments, p. 387** ■ **antigen receptors on B and T cells, p. 368** ■ **gene therapy, p. 343**

Chronic granulomatous disease involves the phagocytes, which fail to produce hydrogen peroxide and certain other active products of oxygen metabolism, due to a defect in an oxidase system normally activated by phagocytosis. Hence, the phagocytes are unable to kill some organisms, especially the catalase-positive *Staphylococcus aureus.* In Chediak-Higashi disease, the phagocyte lysosomes are deficient in certain enzymes and thus cannot destroy phagocytized bacteria. People with this condition suffer from recurring pyogenic bacterial infections. In another primary immunodeficiency involving phagocytes, leukocyte adhesion deficiency, white blood cells fail to localize at sites of infection. Most victims of this disease have very high white blood cell counts and infections that are difficult to detect in time for effective treatment. ■ **pyogenic bacterial infections, p. 562**

TABLE 18.5 Some Primary Immunodeficiency Diseases for Which Genetic Defects Are Known

Severe combined immunodeficiency (SCID)	X-linked hyper-IgM syndrome
X-linked SCID	Wiscott-Aldrich syndrome
MHC class II deficiency	Ataxia telangiectasia
CD3 deficiency	Chronic granulomatous disease
CD8 deficiency	Leukocyte adhesion deficiency
X-linked agammaglobulinemia	Many complement deficiencies

Secondary Immunodeficiencies

Secondary, or acquired, immunodeficiency diseases result from environmental rather than genetic factors. Malignancies, advanced age, certain infections (especially viral infections), immunosuppressive drugs, or malnutrition may all lead to secondary immunodeficiencies. Often, an infection will cause a depletion of certain cells of the immune system. The measles virus, for example, replicates in lymphoid cells, killing many of them and leaving the body temporarily open to other infections. Syphilis, leprosy, and malaria affect the T-cell population and also macrophage function, causing defects in cell-mediated immunity. Malnutrition also causes decreased immune responses, especially the cell-mediated response.

Malignancies involving the lymphoid system often decrease effective antibody-mediated immunity. For example, multiple myeloma is a malignancy arising from a single plasma cell that proliferates out of control and in most cases produces large quantities of immunoglobulin. This overproduction of a single kind of molecule results in the body using its resources to produce immunoglobulin of a single specificity at the expense of those needed to fight infection. The result is an overall immunodeficiency. Other lymphoid disorders include macroglobulinemia (overproduction of IgM) and some forms of leukemia.

One of the most serious and widespread secondary immunodeficiencies is AIDS (acquired immunodeficiency syndrome), caused by human immunodeficiency virus (HIV). This RNA virus of the retrovirus group infects and destroys helper T cells, leaving the affected person highly susceptible to infections, especially with opportunistic agents. AIDS and opportunistic infections are covered in chapter 29. ■ retrovirus, p. 332

MICROCHECK 18.7

Immunodeficiencies may be primary, either genetic or developmental defects in any components of the immune response, or they may be secondary, acquired as a result of infection or environmental influences.

✓ What is the defect in severe combined immunodeficiency? What could cause it?

✓ Multiple myeloma is a plasma cell tumor in which a clone of malignant plasma cells produces large amounts of immunoglobulin. With all this excess immunoglobulin, how can a person with multiple myeloma be immunodeficient?

FUTURE CHALLENGES

New Approaches to Correcting Immunologic Disorders

In recent years many of the genes responsible for immunodeficiency diseases have been identified. It has been possible to correct some of these gene defects in cells in the laboratory and, rarely, in patients. In the near future, research will be directed toward developing the existing technology for gene transfer to make it more effective in correcting these gene defects in human patients. It is important also to continue the search for other defective genes; it is likely that with increasing knowledge of the human genome more will be found soon. A continuing challenge is finding ways to overcome graft rejection to make bone marrow and other transplants more acceptable. The challenge of treating cancer and of preventing rejection of essential transplants may be met, at least in part, by the development of gene transfer technology and by better understanding of the mechanisms of cellular immunological mechanisms.

One interesting line of research stems from the observation that parasitic worm infestations appear to protect people from allergies and autoimmune diseases. Experiments with mice support this idea and also show that feeding mice parasitic worms effectively treats experimental autoimmune disease. The effect appears to be due at least in part to down-regulation of the immune response. Research into understanding the regulation of immune responses will help in controlling autoimmune and immunodeficiency disease, and it will permit development of improved vaccines.

A promising and challenging area is the development of human stem cell research. Stem cells have an almost unlimited capacity to divide, and some of them can differentiate into most of the tissues in the body. They could be used to generate cells for transplantation and to replace defective or injured tissues such as nerve tissue. They might also be used to test the effects of drugs on human cells, without the danger of testing on human beings. A major stumbling block is the fact that the stem cells come from fetal material, either from early stage embryos obtained from fertility treatments or from non-living fetuses from terminated pregnancies. The legal and ethical guidelines for the use of these cells are matters of considerable debate.

SUMMARY

18.1 Type I Hypersensitivities: Immediate IgE-Mediated (table 18.1)

IgE attached to mast cells or basophils reacts with specific antigen, resulting in the release of powerful mediators of the allergic reaction (figure 18.1).

Localized Anaphylaxis

Localized anaphylactic (type I) reactions include **hives (urticaria), hay fever (allergic rhinitis),** and **asthma** (figure 18.2).

Generalized Anaphylaxis

Generalized or systemic anaphylaxis is a rare but serious reaction that can lead to **shock** and death.

Treatments to Prevent Allergic Reactions

Desensitization or immunotherapy is often effective in decreasing the type I hypersensitivity state (figure 18.3). A new treatment, using an engineered anti-IgE, promises to be effective in treating asthma.

18.2 Type II Hypersensitivities: Cytotoxic (table 18.1)

Type II hypersensitivity reactions, or cytotoxic reactions, are caused by antibodies that can destroy normal cells by complement lysis or by **antibody-dependent cellular cytotoxicity (ADCC).**

Transfusion Reactions (table 18.2)

The ABO blood group antigens have been the major cause of **transfusion reactions.**

Hemolytic Disease of the Newborn (figure 18.4)

The Rhesus blood group antigens are usually responsible for this potentially fatal disease.

Injected anti-Rh antibody helps prevent Rh sensitization of Rh-negative mothers.

18.3 Type III Hypersensitivities: Immune Complex-Mediated (table 18.1)

Type III hypersensitivity reactions are mediated by small antigen-antibody complexes that activate complement and other inflammatory systems, attract neutrophils, and contribute to inflammation. The small **immune complexes** are often deposited in small blood vessels in organs, where they cause inflammatory disease—for example, glomerulonephritis in the kidney or arthritis in the joints (figure 18.5).

18.4 Type IV Hypersensitivities: Delayed Cell-Mediated (table 18.1)

Delayed hypersensitivity reactions depend on the actions of sensitized T lymphocytes.

Tuberculin Skin Test (figure 18.6)

A positive reaction to protein antigens of the tubercle bacillus introduced under the skin peaks 2 to 3 days after exposure to antigen.

Delayed Hypersensitivity in Infectious Diseases

Delayed hypersensitivity is important in responses to many chronic, long-lasting infectious diseases.

Contact Hypersensitivities

Contact allergy, or **contact dermatitis,** occurs frequently in response to substances such as poison ivy, nickel in jewelry, and chromium salts in leather products (figures 18.7, 18.8).

18.5 Rejection of Transplanted Tissues

Transplantation rejection of **allografts** is caused largely by type IV cellular reactions.

18.6 Autoimmune Diseases

Responses against substances of self can lead to **autoimmune diseases** (table 18.3).

The Spectrum of Autoimmune Reactions

Autoimmune diseases can be organ-specific or widespread. Some autoimmune diseases are caused by antibodies produced to body components, and others by cell-mediated reactions or a combination of antibodies and immune cells.

Treatment of Autoimmune Diseases

Autoimmune diseases are usually treated with drugs that suppress the immune and/or inflammatory responses.

18.7 Immunodeficiency Disorders

Immunodeficiencies can be primary genetic or developmental defects in any components of the immune response, or they can be secondary and acquired (table 18.5).

Primary Immunodeficiencies (table 18.6)

B-cell immunodeficiencies result in diseases involving a lack of antibody production, such as agammaglobulinemias and selective IgA deficiency.

T-cell deficiencies result in diseases such as DiGeorge syndrome.

Lack of both T- and B-cell functions results in combined immunodeficiencies, which are generally severe.

Defective phagocytes are found in chronic granulomatous disease and Chediak-Higashi disease.

Secondary Immunodeficiencies

Acquired immunodeficiencies can result from malnutrition, immunosuppressive agents, infections (such as AIDS), and malignancies such as multiple myeloma.

REVIEW QUESTIONS

Short Answer

1. Why are antihistamines useful for treating many IgE-mediated allergic reactions but not effective in treating asthma?
2. Penicillin is a very small molecule, yet it can cause any of the types of hypersensitivity reactions, especially type I. How can this occur?
3. What are some major differences between an IgE-mediated skin reaction, such as hives, and a delayed hypersensitivity reaction, such as a positive tuberculin skin test?
4. What causes insulin-dependent diabetes mellitus?
5. Give the evidence showing that myasthenia gravis results from antibody activities.
6. Give an example of an organ-specific autoimmune disease and one that is widespread, involving a variety of tissues and organs.
7. Describe an Arthus reaction.
8. Why might malnutrition lead to immunodeficiencies?
9. What is the most common primary immunodeficiency disorder?
10. How can genetic abnormalities leading to immunodeficiency disorders be corrected? Give an example.

Multiple Choice

1. An IgE-mediated allergic reaction
   a) reaches a peak within minutes after exposure to antigen.
   b) occurs only to polysaccharide antigens.
   c) requires complement activation.
   d) requires considerable macrophage participation.
   e) is characterized by induration.
2. Which of the following statements is true of the ABO blood group system in humans?
   a) A antigen is present on type O red cells.
   b) B antigen is the most common antigen in the population of the United States.
   c) Natural anti-A and anti-B antibodies are of the class IgG.
   d) People with blood group O do not have natural antibodies against A and B antigens.
   e) In blood transfusions, incompatibilities cause complement lysis of red blood cells.
3. All of the following are true of immune complexes, *except*
   a) the most common complexes consist of antigen-IgE-complement.
   b) an immune complex consists of antigen attached to antibody.
   c) usually complement components are included in antigen-antibody complexes.
   d) immune complexes activate strong inflammatory reactions.
   e) immune complexes deposit in kidneys, joints, and skin.
4. Delayed hypersensitivity reactions in the skin
   a) are characterized by a wheal and flare reaction.
   b) peak at 4 to 6 hours after exposure to antigen.
   c) require complement activation.
   d) show induration because of the influx of sensitized T cells and macrophages.
   e) depend on activities of the Fc portion of antibodies.
5. Organ transplants, such as of kidneys
   a) are experimental at present.
   b) can be successful only if there are exact matches between donor and recipient.
   c) survive best if radiation is used for immunosuppression.
   d) survive best if B cells are suppressed.
   e) are rejected by a complex process in which cellular mechanisms predominate.

6. All of the following are true of autoimmune disease, *except*
   a) many of them show association with particular major histocompatibility types.
   b) some of them often occur in members of the same family.
   c) during heart attacks heart antigens are released, but no response occurs to them.
   d) disease may result from reaction to viral antigens that are similar to antigens of self.
   e) some are organ-specific and some are widespread in the body.
7. Autoantibody-induced autoimmune diseases
   a) can sometimes be passively transferred from mother to fetus.
   b) include diabetes mellitus.
   c) are always organ-specific.
   d) are never organ-specific.
   e) cannot be treated.
8. All of the following are approaches being used to treat autoimmune diseases, *except*
   a) cytotoxic drugs to prevent lymphoid cell proliferation.
   b) cyclosporin.
   c) antibiotics.
   d) steroid with anti-inflammatory action.
   e) replacement therapy, as with insulin in diabetes.
9. Patients with primary immunodeficiencies in the complement system
   a) who lack late-acting components (C5, C6, C7, C8) show increased susceptibility to *Neisseria* infections.
   b) who lack C3 are prone to develop tuberculosis.
   c) generally have no symptoms.
   d) only show defects in the major components, C1 through C9.
   e) usually handle infections normally.
10. One of the most serious of the secondary immunodeficiencies is
   a) acquired immunodeficiency syndrome, caused by the human immunodeficiency virus.
   b) severe combined immunodeficiency.
   c) DiGeorge syndrome.
   d) chronic granulomatous disease.
   e) Chediak-Higashi disease.

Applications

1. Patients with advanced leprosy do not give positive type IV reactions to any of a variety of antigens that normal people are exposed to and usually respond to with a positive delayed hypersensitivity skin reaction. Why might this be? What is another disease where lack of the ability to give a positive delayed hypersensitivity reaction could be a useful diagnostic criterion?
2. There has been debate about keeping smallpox virus stored, since the disease has been eradicated. What would be an argument for keeping the virus? What should be done to protect against use of the virus in biological warfare?

Critical Thinking

1. Hypersensitivity reactions, by definition, lead to tissue damage. Can they also be beneficial? Explain.
2. Why does blood need to be cross-matched before transfusion? Why not just type the blood and use compatible ABO types?
3. What hypothesis could explain why primary immunodeficiencies are generally rare?

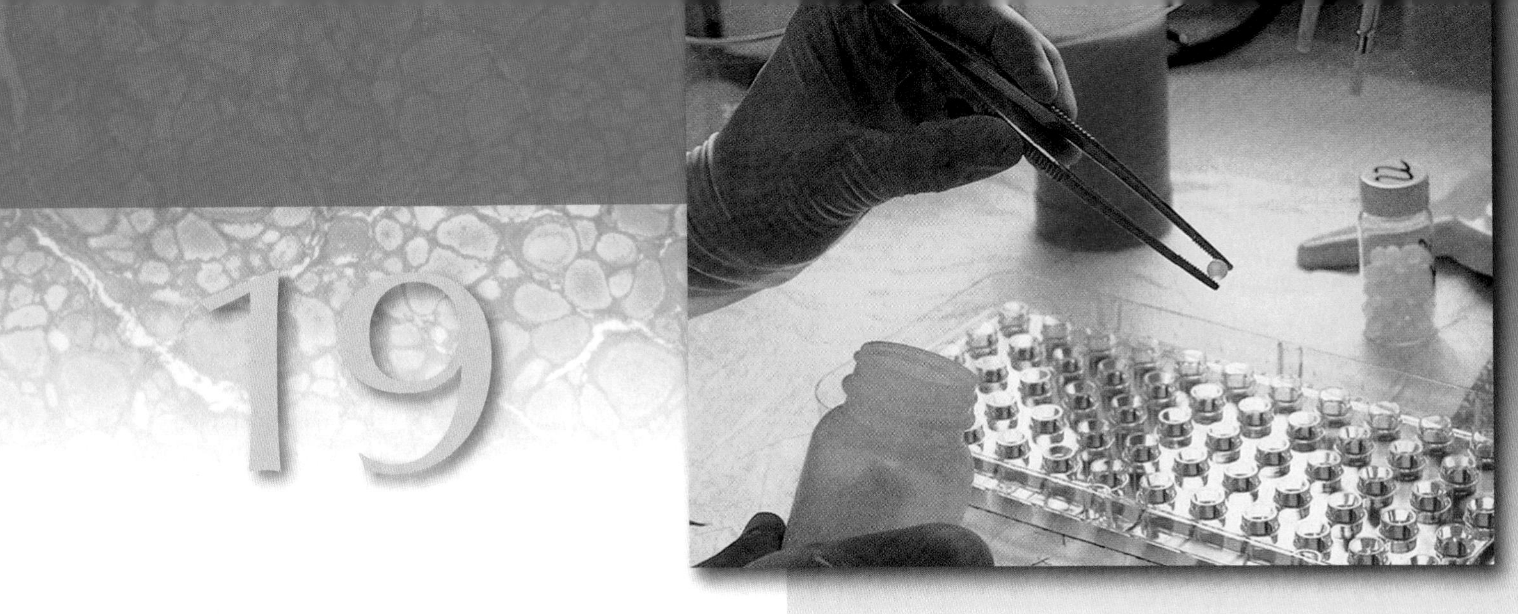

An immunoassay.

Applications of Immune Responses

A Glimpse of History

Even before people knew that microbes caused disease, it was recognized that individuals who recovered from a disease such as smallpox rarely contracted it a second time. Old Chinese writings dating from the Sung dynasty (A.D. 960–1280) describe a procedure known as variolation, in which small amounts of the powdered crusts of smallpox pustules were inhaled or placed into a scratch made in the skin. Usually the resulting disease was mild, and a permanent immunity to smallpox resulted. Occasionally, however, severe disease developed, often resulting in death.

Although variolation was practiced in China and the Mideast a thousand years ago, it was not widely used in Europe until after 1719. At that time, Lady Mary Wortley Montagu, wife of the British ambassador to Turkey, had their children immunized against smallpox in this way. Variolation subsequently became popular in Europe. Although a person exposed to smallpox through variolation would usually completely recover, he or she would become contagious. Because of the danger of contagion and because the procedure was reasonably expensive, large segments of the population in Europe remained unprotected.

As an apprentice physician, Edward Jenner noted that milkmaids who had suffered cowpox infections rarely got smallpox. Cowpox was a disease of cows that caused few or no symptoms in humans. In 1796, long before viruses had been discovered, Jenner conducted a classic experiment in which he deliberately transferred material from a cowpox lesion on the hand of a milkmaid, Sarah Nelmes, to a scratch on the arm of a young boy named James Phipps. Six weeks later, when exposed to pus from a smallpox victim, Phipps did not develop the disease. The boy had been made immune to smallpox when he was inoculated with pus from the cowpox lesion. Using the less dangerous cowpox material in place of the pustules from smallpox cases, Jenner and others worked to spread the practice of variolation. Later, Pasteur used the word vaccination (from the Latin vacca for "cow") to describe any type of protective inoculation. By the twentieth century, most of the industrialized world was generally free of smallpox as the result of routine vaccination of large populations.

In 1967, the World Health Organization (WHO) initiated a program of intensive smallpox vaccination. Since there were no animal hosts and no non-immune humans to whom it could be spread, the disease died out. The last case of naturally contracted smallpox occurred in Somalia, Africa, in 1977. Two years later in a ceremony in Nairobi, Kenya, WHO declared the world free of smallpox. Nevertheless, a few laboratories around the world still have the virus. In this age of bioterrorism concerns, some see smallpox as a major threat should the deadly virus ever be released into the largely unprotected populations of the world. Because of this, vaccine stores in the United States have been increased.

In chapters 15 and 16, we discussed both innate and adaptive defense systems, becoming acquainted with antibodies and lymphocytes. This chapter will consider how **immunization,** the process of inducing immunity, can be used to protect against disease and how immunization techniques have advanced remarkably in recent years to become safer and more effective. In fact, immunization has probably had the greatest impact on human health of any medical procedure, and even better means of immunization are likely in the near future. We will also explore some useful applications of immunological reactions in diagnostic tests.

KEY TERMS

Active Immunity Immunity that results from an immune response in an individual upon exposure to an antigen.

Adjuvant Substance that increases the immune response to antigens.

Agglutination Reaction Technique that relies on the clumping observed when antibody molecules bind to and cross-link insoluble antigen molecules as a means to detect antibodies or antigens.

Antiserum A preparation of serum that contains protective antibodies.

Attenuated Vaccine Vaccine composed of a weakened form of the pathogen that is generally unable to cause disease.

Enzyme-Linked Immunosorbent Assay (ELISA) Technique that uses enzyme-labeled antibodies to detect antigens or antibodies.

Fluorescent Antibody (FA) Test Technique that uses fluorescence-labeled antibodies to detect specific antigens in cells affixed to a microscope slide.

Inactivated Vaccine Vaccine composed of killed bacterial cells, inactivated virus, or fractions of the pathogen.

Passive Immunity Immunity that results when antibodies are transferred to an individual.

Precipitation Reaction Technique that relies on the visible insoluble complexes that form when antibody molecules bind to and cross-link soluble antigens as a means to detect specific antibodies or antigens.

Serology The study of *in vitro* antibody-antigen reactions.

Western Blotting (Immunoblotting) Procedure that uses labeled antibodies to detect specific antigens in a mixture of proteins separated according to their molecular weight.

IMMUNIZATION

19.1 Principles of Immunization

Focus Point

- Give examples of naturally acquired active immunity, artificially acquired active immunity, naturally acquired passive immunity, and artificially acquired passive immunity.

Naturally acquired immunity is the acquisition of adaptive immunity through normal events, such as exposure to an infectious agent. Immunization mimics those same events, protecting against disease by inducing what is termed **artificially acquired immunity (figure 19.1).** The protection provided by immunization can be either active or passive.

Active Immunity

Active immunity is the result of an immune response in an individual upon exposure to antigen. Specific B and T cells are activated and then proliferate, providing the individual with the lasting protection associated with immunological memory. Active immunity can develop either naturally from an actual infection or artificially from administration of a vaccine. ■ memory, p. 367

Passive Immunity

Passive immunity occurs naturally during pregnancy; the mother's IgG antibodies cross the placenta and protect the fetus. These antibodies remain active in the newborn infant during the first few months of life, when the neonate's own immune responses are still developing. Consequently, a number of infectious diseases normally do not occur until a baby is three to six months of age, when the maternal antibodies have been degraded. Passive immunity also occurs as a result of breast feeding; the secretory IgA in breast milk protects the digestive tract of the child. Note that passive immunity provides no memory; once the transferred antibodies are degraded, the protection is lost.

Artificially acquired passive immunity involves transferring antibodies produced by other people or animals. This type

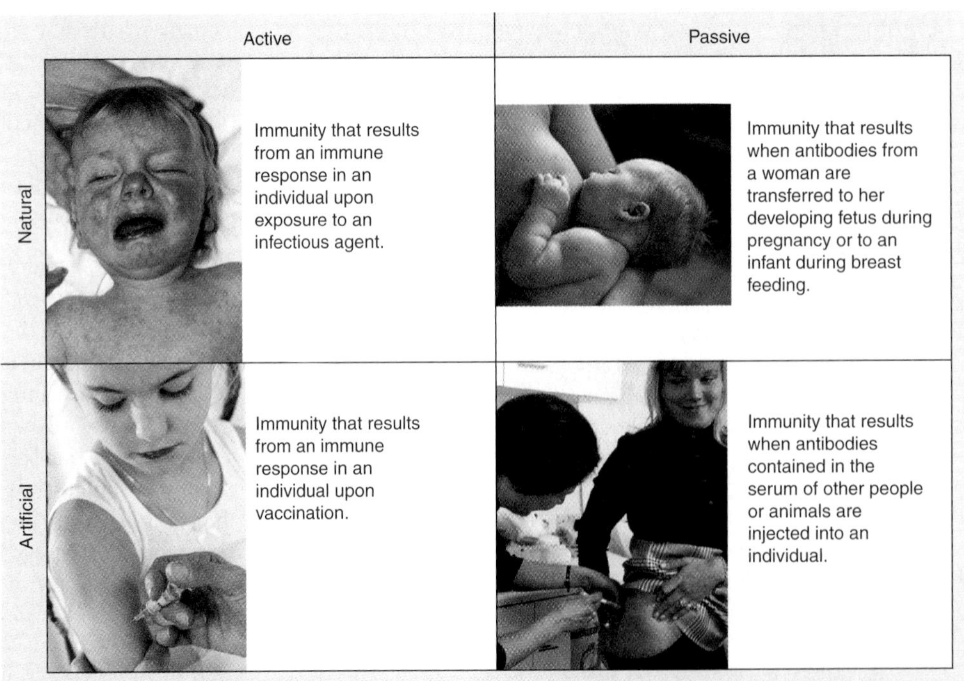
Active
Passive
Natural
Artificial
Immunity that results from an immune response in an individual upon exposure to an infectious agent.
Immunity that results when antibodies from a woman are transferred to her developing fetus during pregnancy or to an infant during breast feeding.
Immunity that results from an immune response in an individual upon vaccination.
Immunity that results when antibodies contained in the serum of other people or animals are injected into an individual.

FIGURE 19.1 Acquired Immunity Acquired immunity can be natural or artificial, active or passive.

of immunity can be used to prevent disease before or after likely exposure to an infectious agent, to limit the duration of certain diseases, and to block the action of microbial toxins. A preparation of serum (the fluid portion of blood that remains after blood clots) containing the protective antibodies is referred to as **antiserum.** One that protects against a given toxin is called an **antitoxin.** Passive immunization preparations include **immune globulin,** or gamma globulin, which is the immunoglobulin G (IgG) fraction of pooled blood plasma from many donors. This contains a variety of antibodies that the immune systems of various donors have made because of infections and vaccination. Immune globulin can be given to an unvaccinated individual who has been recently exposed to the measles virus. It is also useful for protecting immunosuppressed people who have insufficient levels of antibodies. **Hyperimmune globulin,** prepared from the sera of donors with high amounts of antibodies to certain disease agents, is used to prevent or treat specific diseases. Examples include tetanus immune globulin (TIG), rabies immune globulin (RIG), and hepatitis B immune globulin (HBIG). These preparations, given during the incubation period—after exposure, but before disease develops—can often prevent severe diseases from developing.

MICROCHECK 19.1

Immunity is natural or artificial, active or passive. Active immunity occurs naturally in response to infections or other natural exposure to antigens, and artificially in response to vaccine administration. Passive immunity occurs naturally during pregnancy and breast feeding, and artificially through administration of immune serum globulin or hyperimmune globulin.

✓ What is an antitoxin?

✓ What would be a primary advantage of passive immunity with diseases such as tetanus?

19.2 Vaccines and Immunization Procedures

Focus Points

- Compare and contrast the attributes and risks of attenuated vaccines and inactivated vaccines.
- List six diseases that routine childhood immunizations have reduced in occurrence by at least 95%.

A **vaccine** is a preparation of a pathogen or its products used to induce active immunity. Vaccines not only protect an individual against disease, they can also prevent diseases from spreading in a population. When a critical portion of a population is immune to a disease, either through natural immunity or vaccination, a phenomenon called **herd immunity** develops. This is the inability of an infectious disease to spread because of the lack of a critical concentration of susceptible hosts. Herd immunity is responsible for dramatic declines in childhood diseases, both in the United States and in developing countries. Unfortunately, we periodically see some of these diseases reappear and spread as a direct consequence of parents failing to have their children vaccinated. **Table 19.1** lists a number of human diseases for which vaccines are available. As the table indicates, some are routinely used, whereas others are employed only in special circumstances.

Effective vaccines should be safe, with few side effects, while giving lasting protection against the specific illness. They should induce protective antibodies or immune cells, or both, as appropriate. For example, polio vaccine should induce antibodies that neutralize the virus, thus preventing it from reaching and attaching to nerve cells to cause the paralysis of severe poliomyelitis. On the other hand, an effective vaccine against tuberculosis would induce cellular immunity that can limit growth of the intracellular bacterium. Of course, vaccines ideally should be low in cost, stable with a long shelf life, and easy to administer.

Vaccines fall into two general categories—attenuated and inactivated, based on whether or not the immunizing agent can replicate. Each type has characteristic advantages and disadvantages **(table 19.2).**

Attenuated Vaccines

An **attenuated vaccine** is a weakened form of the pathogen that is generally unable to cause disease. The attenuated strain replicates in the vaccine recipient, causing an infection with undetectable or mild disease that typically results in long-lasting immunity. Because infection with the attenuated strain mimics that of the wild-type strain, the type of immunity it evokes is generally appropriate for controlling the infection. For example, attenuated vaccines given orally induce mucosal immunity (an IgA response), protecting against disease-causing agents that infect via the gastrointestinal tract. Some attenuated vaccines are able to stimulate cytotoxic T cells, inducing cellular immunity.

Production of an attenuated strain often involves successively cultivating the microbe under a given set of conditions, resulting in a gradual accumulation of mutations that make it less able to cause disease. Viruses of humans can be attenuated by growing them in cells of a different animal species; mutations occur so that the virus then grows poorly in human cells. Genetic manipulation is now being used to produce strains of pathogens with low virulence. Specific genes are mutated and used to replace wild-type genes. The inserted mutant genes are engineered so they cannot revert to the wild type.

Attenuated vaccines have several advantages compared to their inactivated counterparts. For example, a single dose of an attenuated agent can be sufficient to induce long-lasting immunity. This is because the microbe multiplies in the body, causing the immune system to be exposed to the antigen for a longer period and in greater amounts than with inactivated agents. In addition, the vaccine strain has the added potential of being spread from an individual being immunized to other non-immune people, inadvertently immunizing the contacts of the vaccine recipient.

The disadvantage of attenuated agents is that they have the potential to cause disease in immunosuppressed people, and they can occasionally revert or mutate to become pathogenic again. Care must be taken to avoid giving attenuated vaccines to pregnant women, because some microbes can cross the placenta and

TABLE 19.1 Some Important Vaccines for Humans

Disease or Infectious Agent	Type of Vaccine	Persons Who Should Receive the Vaccine
Anthrax	Acellular	People in occupations that put them at risk of exposure, such as military personnel
Diphtheria	Toxoid	Children (the "D" in the DTaP vaccine given to children); adolescents and adults receive a booster every 10 years (the "d" in the Td and Tdap booster vaccines)
Haemophilus influenzae type b infections	Polysaccharide-protein conjugate	Children
Hepatitis A	Inactivated virus	Children; adolescents who live in selected areas; adults with indications that put them at increased risk (such as traveling to certain countries; men who have sex with men)
Hepatitis B	Protein subunit	Newborns and children; also adults with indications that put them at increased risk (such as health care workers who might be exposed to blood, people who have multiple sexual partners, and contacts of infected people)
Human papillomavirus infection	Protein subunits (two or four serotypes)	Girls/women ages 11–26
Influenza	Two types—inactivated virus (TIV), given by injection, and attenuated virus (LIAV), given as a nasal mist	Children, adults age 50 and over, medical personnel, and people at increased risk for complications; given yearly, as the antigens of the virus change frequently
Measles	Attenuated virus	Children (an "M" in the MMRV vaccine given to children); booster(s) for adults born after 1956 who do not have evidence of immunity and do not have a medical contraindication
Meningococcal disease	Two types active against four serotypes—meningococcal conjugate vaccine (MCV4) and meningococcal polysaccharide vaccine (MPSV4)	Adolescents; also children and adults with certain medical conditions that put them at greater risk (for example, those without a spleen or who have certain complement system defects); also adults in certain high-risk groups (such as college students living in dormitories and people traveling to sub-Saharan Africa)
Mumps	Attenuated virus	Children (an "M" in the MMRV vaccine given to children); booster(s) for adults born after 1956 who do not have evidence of immunity and do not have a medical contraindication
Pertussis (whooping cough)	Acellular	Children (the "aP" in the DTaP vaccine given to children); adolescents should receive a booster (the "ap" in the Tdap booster vaccine); adults younger than age 65 may receive a booster
Pneumococcal infection	Two forms—purified polysaccharide (PPV) and polysaccharide protein conjugate (PCV)	Children; adults age 65 and over, people with certain chronic infections, and others in high-risk groups
Polio	Two forms—inactivated virus (Salk vaccine) and attenuated virus (Sabin vaccine)	Children; attenuated virus is used for global control
Rabies	Inactivated virus	People exposed to the virus, people at high risk for exposure, such as veterinarians and other animal handlers
Rotavirus infection	Attenuated virus	Children
Rubella (German measles)	Attenuated virus	Children (the "R" in the MMRV vaccine given to childen); women who do not have evidence of immunity and do not have a medical contraindication
Shingles	Attenuated virus	Adults age 60 and over
Tetanus	Toxoid	Children (the "T" in the DTaP vaccine given to children); adults receive a booster every 10 years (the "T" in the Td and Tdap vaccines)
Tuberculosis	Attenuated bacterium (BCG strain)	Used only in special circumstances in the United States; widely used in other countries
Typhoid fever	Two forms—attenuated bacterium (Ty21a strain; taken orally) and purified polysaccharide (ViCPS)	People traveling to certain parts of the world
Varicella-zoster (chickenpox)	Attenuated virus	Children ("V" in the MMRV vaccine given to children); also adults without evidence of immunity
Yellow fever	Attenuated virus	Travelers to endemic areas

TABLE 19.2 A Comparison of Characteristics of Attenuated and Inactivated Vaccines

Characteristic	Attenuated Vaccine	Inactivated Vaccine
Antibody response (memory)	IgG; secretory IgA if administered orally or nasally	IgG
Cellular immune response	Good	Poor
Duration of protection	Long-term	Short-term
Need for adjuvant	No	Yes
Number of doses	Usually single	Multiple
Risk of mutation to virulence	Very low	Absent
Risk to immunocompromised recipient	Can be significant	Absent
Route of administration	Injection, oral, or nasal	Injection
Stability in warm temperatures	Poor	Good
Types	Attenuated viruses, attenuated bacteria	Inactivated whole agents, toxoids, subunit vaccines, polysaccharide vaccines

damage the developing fetus. Another disadvantage of attenuated vaccines, especially in developing countries where they are desperately needed, is that they usually require refrigeration to keep them active. Attenuated vaccines currently in widespread use include those against measles, mumps, rubella, and yellow fever. The Sabin vaccine against polio is also an attenuated vaccine.

Inactivated Vaccines

An **inactivated vaccine** is unable to replicate, but retains the immunogenicity of the infectious agent or toxin. The advantage of inactivated vaccines is that they cannot cause infections or revert to pathogenic forms. Because they do not replicate, however, the magnitude of the immune response is limited because there is no amplification of the dose *in vivo*. To compensate for the relatively low effective dose, it is usually necessary to give several booster doses of the vaccine to induce protective immunity. Inactivated vaccines fall into two general categories—whole agents and fractions of the agent.

Inactivated whole agent vaccines contain killed microorganisms or inactivated viruses. The vaccines are made by treating the infectious agent with a chemical such as formalin, which does not significantly change the surface epitopes. Such treatments leave the agent immunogenic even though it cannot reproduce. Inactivated whole agent vaccines include those against cholera, influenza, rabies, and the Salk vaccine against polio. ■ formalin, p. 119

Toxoids are inactivated toxins used to protect against diseases due to toxins produced by the invading bacterium. They are prepared by treating the toxins to destroy the toxic part of the molecules while retaining the antigenic epitopes. Diphtheria and tetanus vaccines are toxoids.

Protein subunit vaccines are composed of key protein antigens or antigenic fragments of an infectious agent, rather than whole cells or viruses. Obviously, they can only be developed after research has revealed which of the components of the microbe are most important in eliciting a protective immune response. Their advantage is that parts of the microbe that sometimes cause undesirable side effects are not included. For example, the whooping cough (pertussis) killed vaccine that was previously used for immunizing babies and young children often caused reactions such as pain, tenderness at the site of the injection, fever, and occasionally, convulsions. A subunit vaccine, referred to as the acellular pertussis (aP) vaccine, does not cause these side effects and has now replaced the killed whole cell vaccine. A **recombinant vaccine** is a subunit vaccine produced by a genetically engineered microorganism. An example is the vaccine against the hepatitis B virus; it is produced by yeast cells that have been engineered to produce part of the viral protein coat. A **VLP** (virus-like particle) **vaccine** is essentially an empty capsid. Laboratory organisms are genetically engineered to produce the major capsid proteins of a virus, which then self-assemble. The human papilloma virus (HPV) vaccines are VLPs.

Polysaccharide vaccines are composed of the polysaccharides that make up the capsules of certain organisms. Recall that polysaccharides are T-independent antigens; they generally elicit a poor response in young children. **Conjugate vaccines** represent an improvement over purified polysaccharide vaccines because they are effective in young children. Scientists intentionally converted polysaccharides into T-dependent antigens by chemically linking the polysaccharides to proteins. The first conjugate vaccine developed was against *Haemophilus influenzae* type b; it has nearly eliminated meningitis caused by this organism in children. The conjugate vaccine developed against certain *Streptococcus pneumoniae* strains promises to do the same for a variety of infections caused by those strains. ■ *Haemophilus influenzae* type b, p. 650 ■ *Streptococcus pneumoniae*, p. 509

Many inactivated vaccines contain an **adjuvant,** a substance that enhances the immune response to antigens (*adjuvare* means "to help"). These are necessary additives because purified antigens such as toxoids and subunit vaccines are often poorly immunogenic by themselves because they lack the "danger" signals—the patterns associated with tissue damage or invading microbes. These patterns trigger dendritic cells to produce co-stimulatory molecules, allowing them to activate helper T cells, which, in turn, activate B cells. Adjuvants are thought to function by providing the "danger" signals to dendritic cells. Some adjuvants appear to adsorb the antigen, releasing it at a slow but constant rate to the tissues and surrounding blood vessels. Unfortunately, many effective adjuvants evoke an intense inflammatory response, making them unsuitable for use in vaccines for humans. Alum (aluminum hydroxide and aluminum phosphate) is the most common adjuvant used, but others, including one that uses a derivative of lipid A, have recently been developed. ■ pattern recognition, p. 346 ■ dendritic cells, pp. 352, 381 ■ lipid A, p. 61

An Example of Vaccination Strategy— The Campaign to Eliminate Poliomyelitis

Vaccines against poliomyelitis provide an excellent illustration of the complexity of vaccination strategies. The virus that causes this disease enters the body orally, infects the throat and intestinal tract, and then invades the bloodstream. From there, it can invade nerve cells and cause the disease poliomyelitis (see figure 27.15). There are three types of poliovirus, any of which can cause poliomyelitis. The Salk vaccine, developed in the mid-1950s, consists of inactivated viruses of all three types. It was a huge success in lowering the rate of the disease, but it had the disadvantage of requiring a series of injections over a period of time for maximum protection. In 1961, the Sabin vaccine became available, with the advantage of cheaper oral administration. Even though this attenuated poliovirus vaccine replicates in the intestine, however, it still has to be given in a series of three doses rather than one because of interactions among the three types of virus included in the vaccine. Attenuated and inactivated polio vaccines both induce circulating antibodies and protect against viral invasion of the central nervous system and consequent paralytic poliomyelitis. The Sabin vaccine has a distinct advantage over the Salk vaccine in that it induces better mucosal immunity (secretory IgA response), and thus potentially provides herd immunity. ■ poliomyelitis, p. 661

Polio vaccination was so successful that by 1980 the United States was free of wild-type poliovirus (see figure 27.16). Ironically, poliomyelitis still occurred occasionally, caused by the vaccine strain; approximately one case of poliomyelitis arises for every 2.4 million doses of Sabin vaccine administered. An obvious way to avoid these vaccine-related illnesses is to abandon the Sabin vaccine in favor of the Salk vaccine. As usual, however, the situation is not as simple as it might seem. The Sabin vaccine, unlike the Salk vaccine, prevents transmission of the wild-type virus should it ever be reintroduced to the population. If only the inactivated vaccine is given, the virus can still replicate in the gastrointestinal tract and be transmitted to others, rapidly spreading in a population. Eventually the virus may infect individuals who are susceptible, potentially causing an outbreak of poliomyelitis.

A campaign to eliminate polio worldwide was so successful that by 1991, wild poliovirus had been eliminated from the Western Hemisphere. By 1997, the worldwide incidence of polio had decreased substantially, minimizing the risk that wild-type polio would be reintroduced into the United States. Because of the continued risk of vaccine-associated paralytic polio, a vaccine strategy that attempted to capture the best of both vaccines was adopted. Children first received doses of the Salk vaccine, protecting them from poliomyelitis. Following these doses, the Sabin vaccine was given, providing mucosal protection while also boosting immunity. In mid-1999 the routine use of the Sabin vaccine was discontinued altogether in the United States. Although the original goal of global eradication of polio by 2000 was not achieved, global eradication efforts continue.

The Importance of Routine Immunizations for Children

Before vaccination was available for common childhood diseases, thousands of children died or were left with permanent disability from these diseases. **Table 19.3** illustrates how dramatically vaccination has decreased the occurrence of certain infectious diseases. Unfortunately, even now, many people become ill or even die every year from diseases that are readily prevented by vaccines.

One reason some children are not protected is that parents have refused to have their children vaccinated, fearing the rare chance that immunization procedures might be harmful. Vaccines have, in these cases, become victims of their own success. They have been so effective at preventing diseases that people have been lulled into a false sense of security. Reports of adverse effects of vaccination have led some people to falsely believe that the risk of vaccination is greater than the risk of diseases. Although there is some risk associated with almost any medical procedure, there is no question that the benefits of routine immunizations greatly outweigh the very slight risks. Data show that a child with measles has a 1:2,000 chance of developing serious encephalitic involvement of the nervous system, compared with a 1:1,000,000 chance from the measles vaccine. Between 1989 and 1991 measles immunization rates dropped 10% and an outbreak of 55,000 cases occurred, with 120 deaths. Now that immunization rates have increased again, measles outbreaks are rarely seen. The suggestion that vaccines are associated with autism in young children, however, has again threatened the acceptance of immunization. Studies so far have not shown evidence of this association. Routine immunization against pertussis (whooping cough) caused a marked decrease in its incidence in the United States and saved many lives. Because of some adverse reactions to the killed whole cell vaccine being used at the time, however, many parents refused to allow their babies to get the vaccine. By 1990, this refusal of vaccination resulted in the highest incidence of pertussis cases in 20 years and the deaths of some children, mostly those under one year of age. Currently an acellular subunit pertussis vaccine is used, usually in combination with diphtheria and tetanus toxoids (DTaP). Several large-scale studies have shown the acellular pertussis

TABLE 19.3 The Effectiveness of Universal Immunization in the United States

Disease	Cases per Year Before Immunization	Decrease After Immunization
Smallpox	48,164 (1900–1904)	100%
Diphtheria	175,885 (1920–1922)	Nearly 100%
Pertussis (whooping cough)	147,271 (1922–1925)	93.4%
Tetanus	1,314 (1922–1926)	98.1%
Paralytic poliomyelitis	16,316 (1951–1954)	100%
Measles	503,282 (1958–1962)	Nearly 100%
Mumps	152,209 (1968)	99.8%
Rubella (German measles)	50,230 (1966–1969)	98%
Haemophilus influenzae type b invasive disease in children	20,000 (estimated)	99.8%

vaccine to be more effective and have fewer side effects than the whole cell vaccine.

The U.S. Centers for Disease Control regularly publishes recommended immunization schedules for children, adolescents, and adults. The schedules are updated regularly as vaccines are developed and modified. Because of the complexity of the schedules and the frequency at which they have been update recently, it is important to know how to access the most current schedule. A link can be found in the chapter 19 readings at the Online Learning Center (www.mhhe.com/nester6).

Vaccines included in the CDC recommended immunization schedule are generally covered by the National Vaccine Injury Compensation Program. This no-fault alternative for resolving vaccine injury claims was established in 1988 to stabilize the U.S. vaccine market. An excise tax on every vaccine dose purchased is used to fund the program.

Current Progress in Immunization

Recent advances in understanding the immune system are enabling researchers to make safer and more effective vaccines. Progress is occurring in several fronts—enhancement of the immune response to vaccines, development of new or improved vaccines against certain diseases, and development of new types of vaccines.

An excellent example of how a better understanding of the immune response can lead to the development of more effective vaccines is the introduction of conjugate vaccines, which enlist T-cell help. Another way in which the immune response can be bolstered is to administer certain cytokines along with vaccines, guiding the immune response. The discovery and characterization of toll-like receptors is giving insights into adjuvants that are being incorporated into vaccines to enhance their effectiveness.

■ **toll-like receptors, p. 355**

Novel types of vaccines being actively studied include peptide vaccines, edible vaccines, and DNA-based vaccines. Because none of these relies on whole cells, the procedures eliminate the possibility of infection with the immunizing agent; however, some of these vaccines are weakly immunogenic. **Peptide vaccines** are composed of key antigenic peptides from disease-causing organisms. They are stable to heat and do not contain extraneous materials to cause unwanted reactions or side effects. **Edible vaccines** are created by transferring genes encoding key antigens from infectious agents into plants. If appropriate plants can be genetically engineered to function as vaccines, they could potentially be grown throughout the world, eliminating difficulties involving transport and storage. **DNA-based vaccines** are segments of naked DNA from infectious organisms that can be introduced directly into muscle tissue. The host tissue actually expresses the DNA for a short period of time, producing the microbial antigens encoded by the DNA, which induces an immune response.

TABLE 19.4 Some Diseases for Which New or Improved Vaccines Are Sought

Disease	Estimated Impact
HIV/AIDS	40 million infected worldwide, with approximately 14,000 new infections daily
Malaria	300–500 million cases/yr and up to 3 million deaths/yr worldwide
Influenza	30–50 million cases/yr worldwide; 10,000–40,000 deaths/yr in the United States
Strep throat	20 million cases/yr in the United States
Genital herpes	45 million infected and 500,000 new infections/yr in the United States
Hepatitis C	170 million infected worldwide
Cancer	1 in 3 in the United States may get cancer, resulting in 560,000 deaths/yr

There are several serious and widespread diseases for which new or more effective vaccines are currently being sought **(table 19.4).** Many of these disease-causing agents have been shown to be particularly adept at avoiding the host defenses, complicating the development of long-lasting effective vaccines. In addition to seeking vaccines that protect against infectious diseases, other uses of vaccines are also being studied. Attempts are being made to develop vaccines to control fertility and hormone activity, and to prevent diabetes and cancer, among other conditions. Vaccines are also being used experimentally in the treatment of cancer.

MICROCHECK 19.2

Attenuated vaccines are weakened forms of the disease-causing agent. Inactivated vaccines are unable to replicate, but they retain the immunogenicity of the infectious agent; they include killed microorganisms, inactivated viruses, and fractions of the agents, including toxoids. Routine childhood immunizations have prevented millions of cases of disease and many deaths during the past decades. Many experimental vaccines are under study or in clinical trials.

- ✓ Compare and contrast attenuated and inactivated vaccines.
- ✓ Since many childhood diseases such as measles and mumps are rare now, why is it important for children to be immunized against them?
- ✓ What would be a primary advantage of using an attenuated agent rather than just an antigen from that agent?

IMMUNOLOGICAL TESTING

The specificity of immunological reactions described in chapter 16 is exploited in **immunoassays,** which use antigen-antibody interactions for diagnosis. For example, antibodies that bind specifically to *Treponema pallidum,* the bacterium that causes syphilis, can be added to fluid collected from a genital lesion that characterizes the disease **(figure 19.2).** Binding of the antibodies to a bacterium in the specimen identifies that organism as *T. pallidum,* indicating that the patient does indeed have syphilis. Likewise, specific antibodies in a patient's body fluids or tissues can be detected using a known antigen. For instance, if a specimen from a patient suspected of having syphilis is added to proteins specific for *T. pallidum,* and antibodies in that specimen bind to those proteins, then the patient's immune system must have responded to the bacterium at some point, suggesting either previous or current infection.

One of the earliest examples of immunological testing is the PPD skin test (also called the Mantoux test), which is still used for diagnosing tuberculosis. People who have been infected with *Mycobacterium tuberculosis* develop a strong cellular response to the bacterium and its products, which is the basis of the test. If a very small amount of purified protein derivative (PPD) from cultures of *M. tuberculosis* is injected into the skin of someone who has been infected with the organism, redness and a firm swelling usually develop at the site (see figure 18.6). In contrast, people who have not been infected show little, if any, response.

Just as the field of immunology has advanced markedly in the last few decades, so has the technology of immunological testing. New tests are continually being developed, augmenting or gradually replacing many of the older methods. This section will focus primarily on tests commonly used today; information about other tests such as complement fixation, radioimmunoassay and hemagglutination inhibition—which are declining in use but are historically important and illustrate key immunological principles—can be found in the chapter 19 readings at the Online Learning Center (www.mhhe.com/nester6).

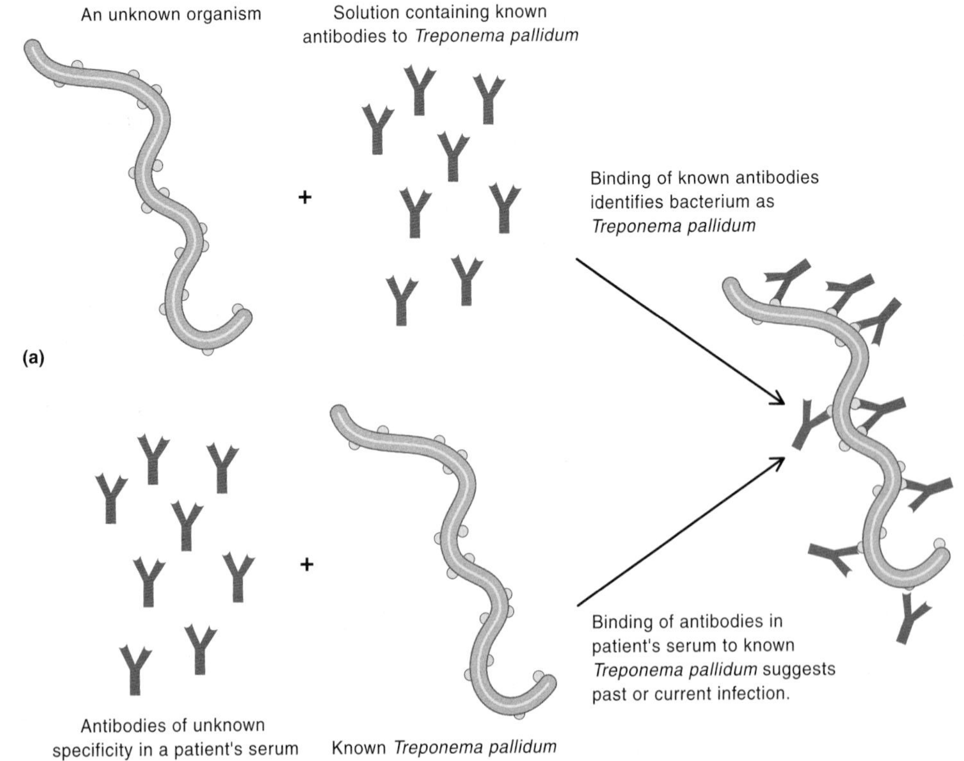

An unknown organism
Solution containing known antibodies to *Treponema pallidum*
+
Binding of known antibodies identifies bacterium as *Treponema pallidum*
(a)
+
Binding of antibodies in patient's serum to known *Treponema pallidum* suggests past or current infection.
Antibodies of unknown specificity in a patient's serum
Known *Treponema pallidum*
(b)

FIGURE 19.2 Principles of Immunoassays These assays can be used to **(a)** identify unknown bacteria (or other antigens); **(b)** detect specific antibodies.

19.3

Principles of Immunological Testing

Focus Points

- Describe the difference between polyclonal and monoclonal antibodies.
- Describe how the antibody titer is determined.

A person who has not been exposed to a given pathogen has no specific antibodies against the agent in their serum, and is referred to as **seronegative.** Once infected, that person will begin producing specific antibodies about a week to 10 days later, becoming **seropositive.** This change from seronegative to seropositive is referred to as **seroconversion.** As the infection progresses, increasing amounts of specific antibodies are produced, causing the amount of those antibodies in the blood to increase. A rise in the amount of specific antibodies, or **titer,** is characteristic of an active infection. In contrast, small but steady amounts of antibodies indicate a previous infection or vaccination.

Obtaining Antibodies

To determine if a patient has antibodies in the blood against a specific infectious agent, then either the patient's serum or plasma is tested. **Serum** is the fluid portion of blood that remains after blood clots; **plasma** is the fluid portion of blood treated with an anticoagulant to prevent clotting. Because serum is so often used as a source of antibodies, the study of *in vitro* antibody-antigen interactions is referred to as **serology.** Cerebrospinal fluid, tissues, and other clinical specimens can also be tested for antibodies.

To obtain antibodies known to bind a certain infectious agent, laboratory animals are used. The animals are immunized with either the whole agent or part of the agent, and the resulting antibodies are then collected by harvesting the animal's serum. The antibody preparation will be **polyclonal,** meaning that multiple naive B cells responded to the immunization, producing a mix of different antibodies that together recognize a variety of epitopes on the antigen. The more complex the antigen, the greater the number of different epitopes recognized by the antibody preparation. For instance, injection of whole bacteria will result in a wider array of antibody specificities than injection of purified toxin. One problem with polyclonal antibodies is that some may bind to closely related organisms, resulting in a false positive reaction. As an example, *Shigella* species have outer membrane proteins in common with *E. coli,* so an animal immunized with whole *Shigella* cells would produce some antibodies that also bind *E. coli* cells. If those antibodies were used in a diagnostic test for *Shigella,* a specimen containing *E. coli* cells but not *Shigella* would yield a false positive result.

Monoclonal antibodies recognize only a single epitope. They are obtained through a laborious process that involves isolating individual B cells from an immunized animal, and then fusing those short-lived B cells with other cells that will divide repeatedly in culture (see **Perspective 19.1**). Because each B cell is programmed to produce antibody molecules that recognize only a single epitope, antibody preparations produced by the descendants (clones) of a single B cell are all identical. Monoclonal antibodies are difficult and expensive to develop and thus are generally available only if they have commercial value due to their widespread use. In addition to the diagnostic value of monoclonal antibodies, some have been developed that can be used to treat disease. Monoclonal antibodies derived from an animal such as a mouse can be "humanized," meaning that recombinant DNA techniques were used to replace at least part of the mouse-derived molecule with the human equivalents. This gives the molecule a longer half-life in humans because the human immune system is less apt to recognize it as foreign. Humanized monoclonal antibodies are referred to as **rhu-Mabs** (recombinant **hu**man **m**onoclonal **a**nti**b**odies). Some rhu-Mabs are given as a form of passive immunity and can be used to treat certain types of cancers. Others are tagged with a drug or toxic substance and then used to deliver that tag to a specific cell type. These tagged monoclonal antibodies have been used to treat non-Hodgkin's lymphoma that has not responded to traditional treatment.

Certain serological tests we will be discussing use antibodies that bind to human IgG molecules. These are referred to as **anti-human IgG antibodies.** They are produced by animals that have been immunized with IgG from human serum. Anti-human IgG antibodies are readily available commercially as are antibodies that bind to the other immunoglobulin classes.

Quantifying Antigen-Antibody Reactions

The concentration of antibody molecules in a specimen such as serum is usually determined by making serial dilutions similar to those done to make plate counts used to enumerate bacterial cells (see figure 4.18). The specimen is prepared by making a series of two-fold or ten-fold dilutions, and then antigen is added to each dilution. The **titer** (concentration) is expressed as the reciprocal of the last dilution that gives a detectable antigen-antibody reaction. Thus, if a positive reaction is observed in the dilution 1:256 but not in 1:512, then the antibody titer is 256.

Serology tests can be done in test tubes, but this requires many tubes and large quantities of reagents. Therefore, the tests are usually done using plastic **microtiter plates,** which have 96, 384, or 768 wells **(figure 19.3).** Proteins, either antigen or antibody, can be permanently affixed to the tiny plastic wells. The volumes used in each well are a mere fraction of the volumes needed for even a small test tube, so that tests can be done on minute samples. Special equipment allows rapid dilution and mixing of reagents, as well as accurate reading of results.

MICROCHECK 19.3

Antibodies used in serological tests can be either polyclonal or monoclonal. Serial dilution of serum or plasma specimens permits quantification of antibodies in the specimen.

- ✓ What is the significance of a rise in titer of specific antibodies in serum samples taken early and later during an infectious disease?
- ✓ Would antibodies produced by a patient in response to infection be monoclonal, or polyclonal?

PERSPECTIVE 19.1

Monoclonal Antibodies

In 1975, an exciting breakthrough occurred in immunology. Georges Köhler and Cesar Milstein developed techniques that fused normal antibody-producing B lymphocytes with malignant plasma cells (myeloma tumor cells), resulting in clones of cells they termed **hybridomas.** Because these hybridomas are clones, they produce antibody molecules with a single specificity, known as **monoclonal antibodies (figure 1).**

Plasma cells produce large amounts of antibody, and when they become malignant they grow profusely and indefinitely. Special myeloma cells are used to make hybridomas; they have lost the ability to make their own specificity of antibody but have retained the ability to produce large amounts of immunoglobulin. The normal B cell in the hybridoma supplies the genes for the specific antibody to be produced; the myeloma cell supplies the cellular machinery, the rough endoplasmic reticulum, for producing the antibodies.

Usually, when an animal is injected with an immunizing agent, it responds by making a variety of antibodies directed against different epitopes on the antigen. Therefore, even though there is a single antigen, the result is a mixture of different antibodies (i.e., polyclonal). When these antisera are used in immunological tests, standardizing the results is difficult because there are differences each time the antiserum is made. Monoclonal antibodies, however, will be of the same immunoglobulin class and have the same variable regions and, thus, the same specificity and other characteristics. With such specificity, tests can be standardized much more easily and with greater reliability.

In the laboratory, monoclonal antibodies are the basis of a number of diagnostic tests. For example, monoclonal antibodies against a hormone can detect pregnancy only 10 days after conception. Specific, monoclonal antibodies are used for rapid diagnosis of hepatitis, influenza, herpes simplex, and chlamydia infections. Köhler and Milstein won the Nobel Prize in 1984 for their work.

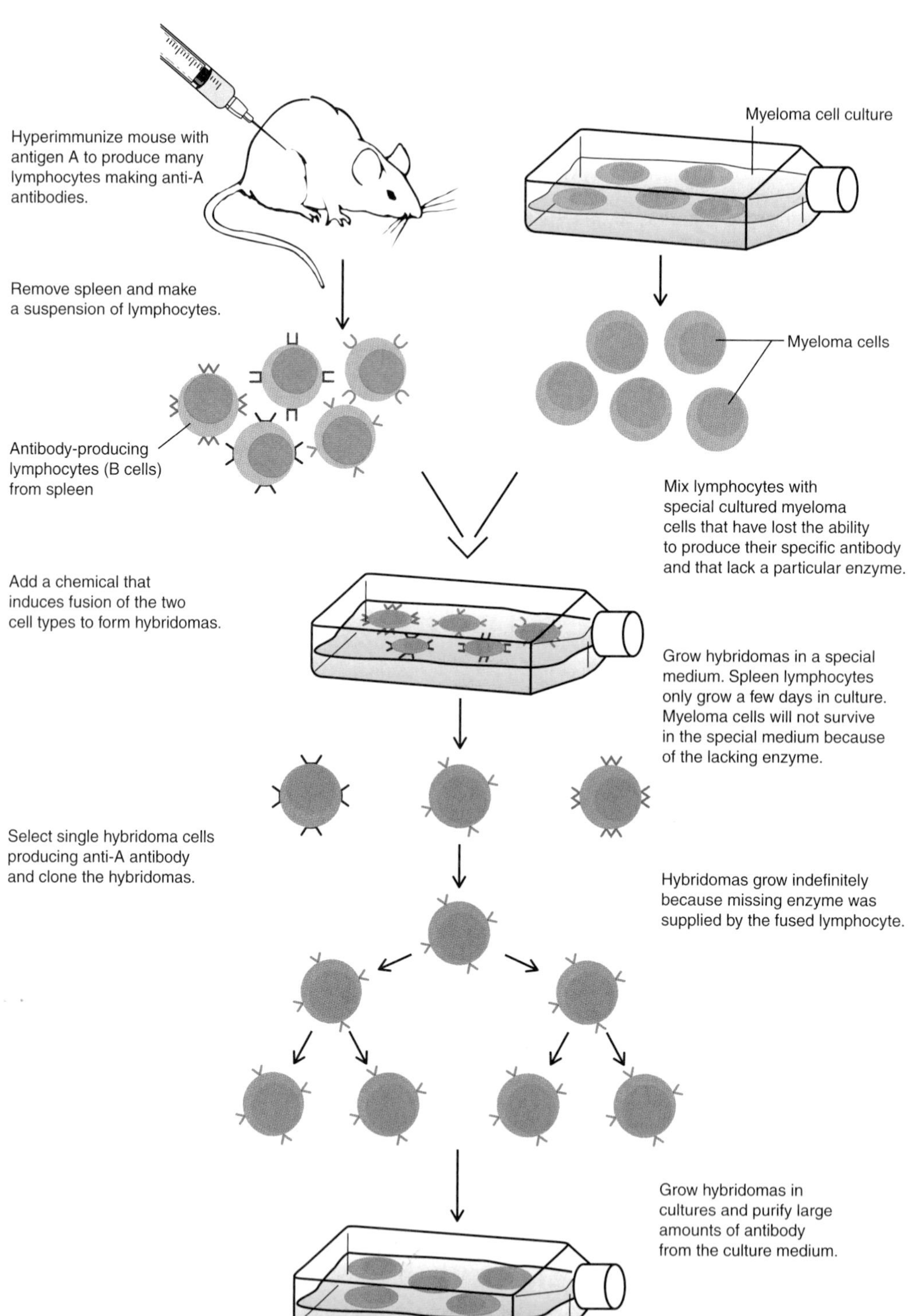

Hyperimmunize mouse with antigen A to produce many lymphocytes making anti-A antibodies.
Myeloma cell culture
Remove spleen and make a suspension of lymphocytes.
Myeloma cells
Antibody-producing lymphocytes (B cells) from spleen
Mix lymphocytes with special cultured myeloma cells that have lost the ability to produce their specific antibody and that lack a particular enzyme.
Add a chemical that induces fusion of the two cell types to form hybridomas.
Grow hybridomas in a special medium. Spleen lymphocytes only grow a few days in culture. Myeloma cells will not survive in the special medium because of the lacking enzyme.
Select single hybridoma cells producing anti-A antibody and clone the hybridomas.
Hybridomas grow indefinitely because missing enzyme was supplied by the fused lymphocyte.
Grow hybridomas in cultures and purify large amounts of antibody from the culture medium.

FIGURE 1 Production of Monoclonal Antibodies Antibody-producing cells from the spleen of a mouse immunized with the desired antigen are fused with myeloma tumor cells. The hybrid clone is grown as a line of cells *in vitro,* all producing large amounts of homogeneous antibody.

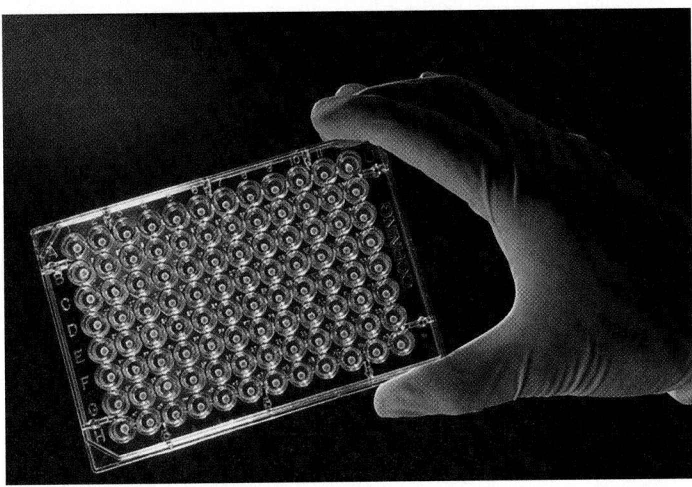

FIGURE 19.3 Microtiter Plate Serological tests can be done in the wells of these small plates, minimizing the volumes of reagents required.

19.4

Observing Antigen-Antibody Aggregations

Focus Point

- Compare and contrast precipitation reactions and agglutination reactions.

Recall from chapter 16 that antigen-antibody complexes form aggregates, creating one large "mouthful" for a phagocytic cell (see figure 16.6). This type of antigen-antibody binding can be readily observed in precipitation and agglutination reactions.

Precipitation Reactions

When antibodies bind to soluble antigens that have multiple epitopes, extensive cross-linking of the two types of molecules may occur, forming latticelike insoluble complexes that then precipitate out of solution. This is the basis of **precipitation reactions,** which are used to detect specific antibodies or antigens. Complete formation of the aggregates can take several hours, and only occurs at certain relative concentrations of antibody and antigen molecules. If there is a great excess of either, the aggregate does not form, and consequently, no precipitate will be seen **(figure 19.4).** The easiest way to achieve this concentration is to place separate antigen and antibody suspensions side by side in a gel, and let the two types of soluble molecules diffuse toward each other. A precipitate will form in a distinct region called the **zone of optimal proportions.**

Immunodiffusion tests

Immunodiffusion tests are simply precipitation reactions carried out in a gel such as agarose. One method used is the **Ouchterlony** technique, which can be done in a Petri dish **(figure 19.5).** Antigen and antibody solutions are placed into separate wells cut in the gel contained in the dish. These solutions will gradually diffuse outward in the gel, meeting between the wells. If the antibody molecules recognize the antigen, they will form antigen-antibody complexes, resulting in a line of precipitate at the zone of optimal proportions. Because there is often more than one antigen present in the sample and different specificities of antibodies in the serum, more than one line can form, each in its area of optimal proportions. The Ouchterlony test is commonly used to detect autoantibodies associated with certain connective tissue disorders.

■ autoantibody, p. 425

Unlike the Ouchterlony technique, the **radial immunodiffusion test** is quantitative. To do the test, the antibody is added to the melted, cooled agarose before it hardens, generating a uniform concentration of antibody molecules throughout the gel. Then, once the gel has solidified, antigen samples can be added into wells cut in the gel. The antigens will diffuse outward, forming a concentration gradient of antigen around the well. Because antibody molecules have been incorporated into the agar, a ring will form around the well as antibody-antigen aggregates precipitate

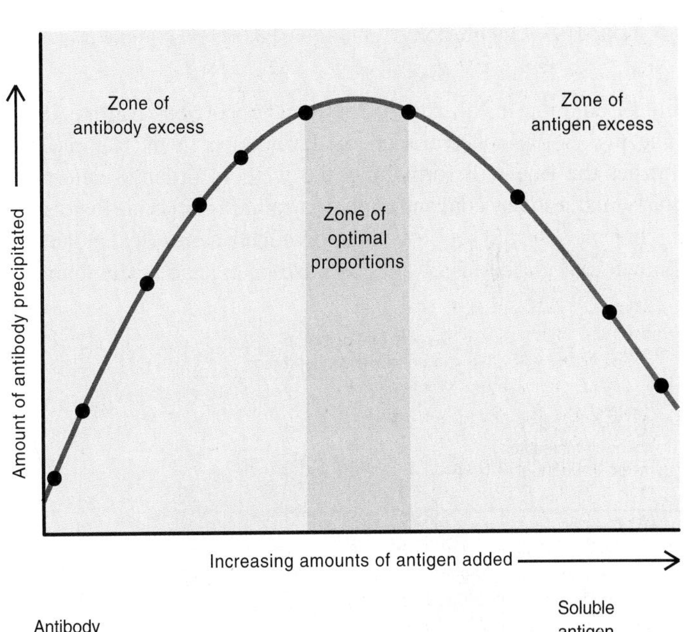

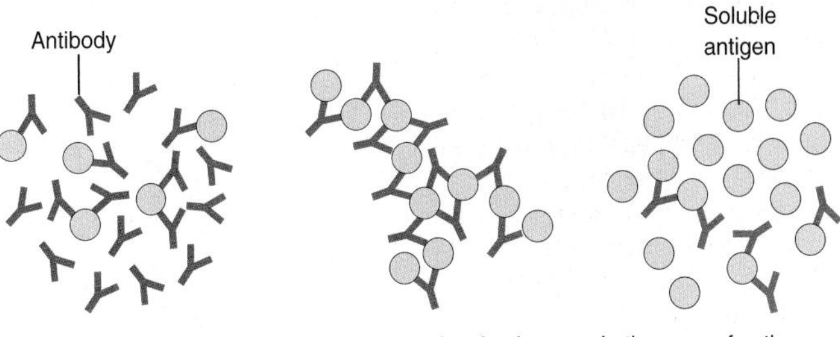

FIGURE 19.4 Antigen-Antibody Precipitation Reactions The maximum amount of precipitate forms in the zone of optimal proportion.

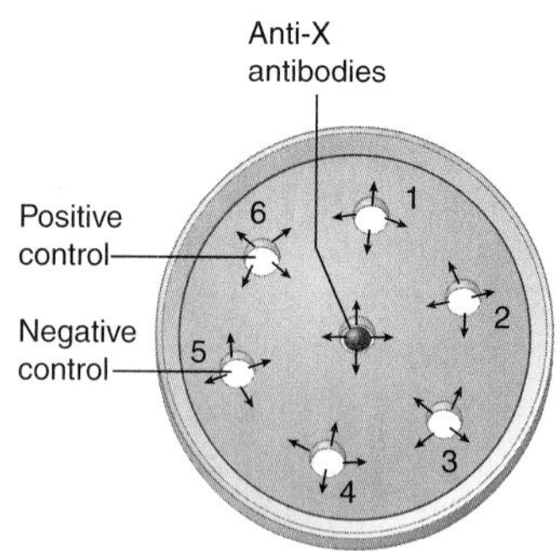

Antibody and antigen solutions are placed into separate wells cut into the gel. The antigens and antibodies diffuse toward each other. In this example, antibodies that bind antigen X (called anti-X antibodies) were added to the well in the center of the gel. Samples that contain unknown antigens are added to wells 1 through 4. A negative control (contains no antigen X) was added to well 5, and a positive control (contains known antigen X) was added to well 6.

When antibody molecules that recognize the antigen meet at the zone of optimal proportions, antigen-antibody complexes precipitate out of solution, forming a visible line. In this example, a line has formed between the center well and well 6 (the positive control). A line has also formed between the center well and the sample in well 4, indicating that the sample contains antigen X. The other samples do not contain detectable amounts of antigen X.

(a)

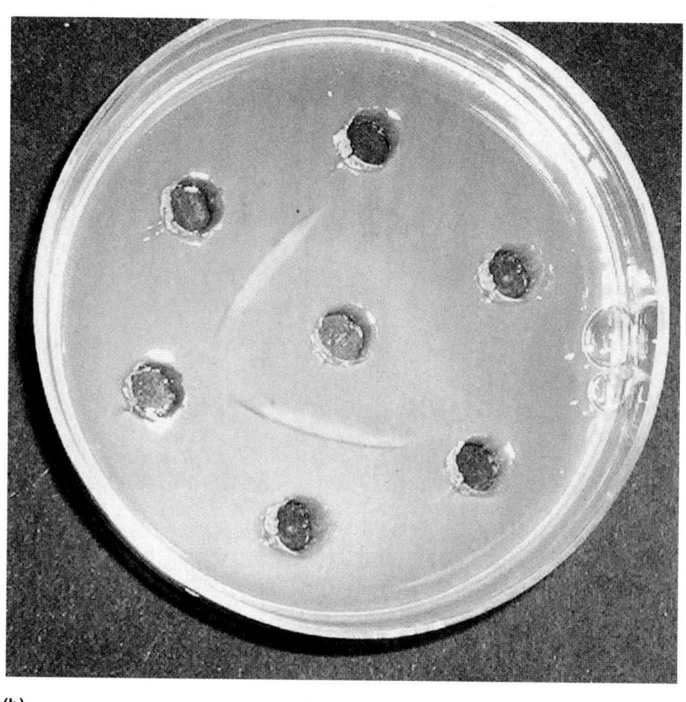

(b)

FIGURE 19.5 Ouchterlony Technique (a) Method **(b)** Photograph of results.

out of solution in the zone of optimal proportions **(figure 19.6).** The higher the concentration of the antigen in the sample, the further the ring will form from the well. In order to determine the actual antigen concentration, the radial immunodiffusion test includes a separate set of wells into which a series of standards of known concentrations of antigen have been added. The diameter of the ring of precipitation around each of the standards can be used to construct a standard curve, which can then be used to determine the concentration of the unknown. The radial immunodiffusion test is commonly used to measure the level of the various complement system proteins in order to diagnose inherited deficiencies. ■ **complement system, p. 355**

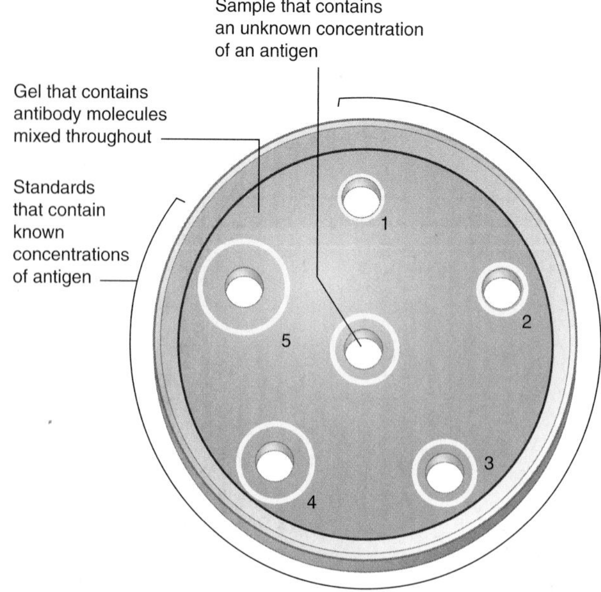

The gel contains a uniform concentration of anti-X antibodies (antibodies known to bind antigen X) mixed throughout. Standards that contain known concentrations of antigen X have been added to wells 1 through 5. A sample that contains an unknown concentration of antigen X has been added to the center well. A ring has formed around each well because antigen-antibody aggregates precipitated out of solution in the zone of optimal proportions. The higher the concentration of the antigen in the sample, the further the ring forms from the well.

(a)

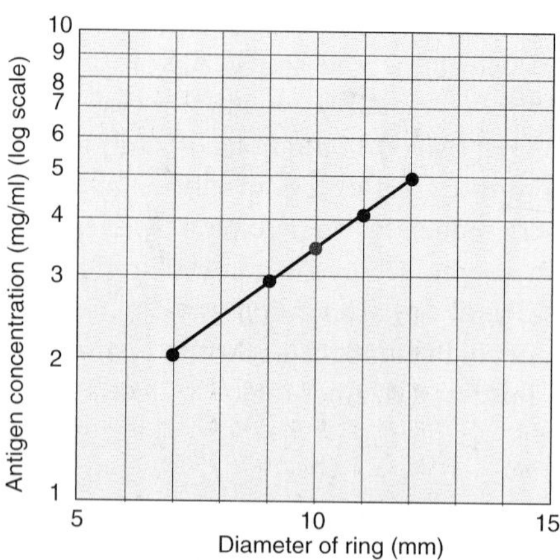

A standard curve is constructed that correlates the size of the ring with the concentration of antigen. This is done by measuring the diameter of the ring around each of the standards and then plotting the values against the concentrations of the respective standards on semi-log graph paper. The standard curve is then used to determine the concentration of antigen in the unknown.

(b)

FIGURE 19.6 Radial Immunodiffusion This quantitative test is used to measure the concentration of an antigen in a sample. **(a)** Results, **(b)** standard curve generated from the results.

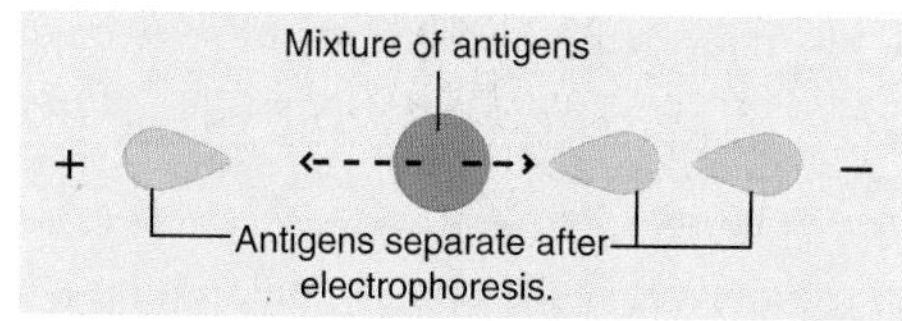
Mixture of antigens
+
−
Antigens separate after electrophoresis.
Protein antigens are separated by electrophoresis (movement through an electrical field).

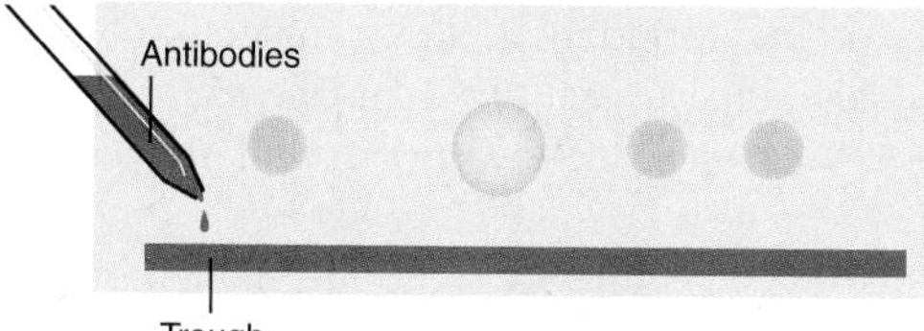
Antibodies
Trough
A trough is cut in the gel and antiserum containing a mixture of antibodies against the antigens in the gel is added.

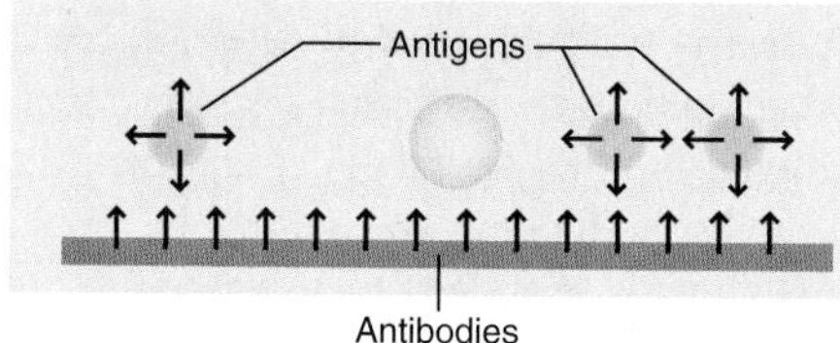
Antigens
Antibodies
Antigens and antibodies diffuse toward each other.

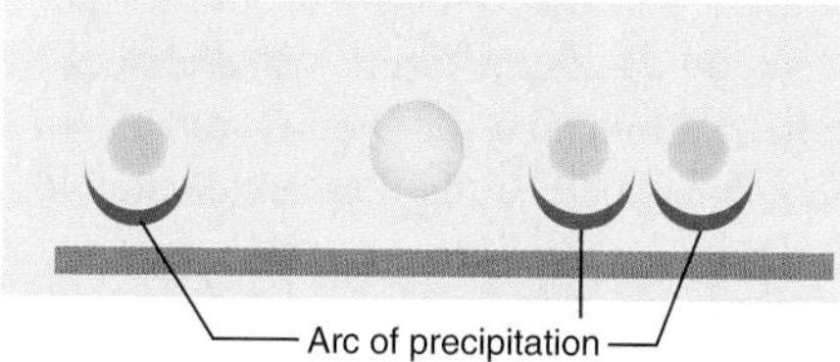
Arc of precipitation
An arc of precipitate forms where antigen and antibodies meet in optimal proportions.

FIGURE 19.7 Immunoelectrophoresis Electrophoresis is used to separate the antigens; then, immunodiffusion is used to identify them.

Immunoelectrophoresis

In **immunoelectrophoresis,** the proteins are first separated using gel electrophoresis **(figure 19.7).** Then, the antibodies are placed in a trough and allowed to diffuse toward the separated proteins. A line of precipitation forms at the location of each protein (antigen) recognized by antibodies. This test is most often used to determine if a patient is producing abnormally high or low levels of certain classes of immunglobulins. High levels of certain classes could indicate myeloma, in which a single plasma cell gives rise to a tumor.

Agglutination Reactions

Agglutination and precipitation reactions are similar in principle—both depend on cross-linking and lattice formation. In **agglutination reactions,** however, relatively large insoluble particles are involved rather than soluble molecules. Because of this, obvious aggregates are formed, which are much easier to see.

In **direct agglutination tests,** a suspension of specific antibody is mixed with the insoluble antigen, such as red blood cells, bacteria, or fungi. Readily visible clumping is a positive test. The agglutination of red blood cells by antibody binding or other means is referred to as **hemagglutination,** and is used in blood typing **(figure 19.8).**

Passive agglutination tests amplify the outcome of antibody-antigen aggregate formation by attaching either the antibody or the antigen to latex beads or other particles. Agglutination of these insoluble particles is much easier to see than a precipitate of soluble molecules. Latex beads to which specific antibodies have been

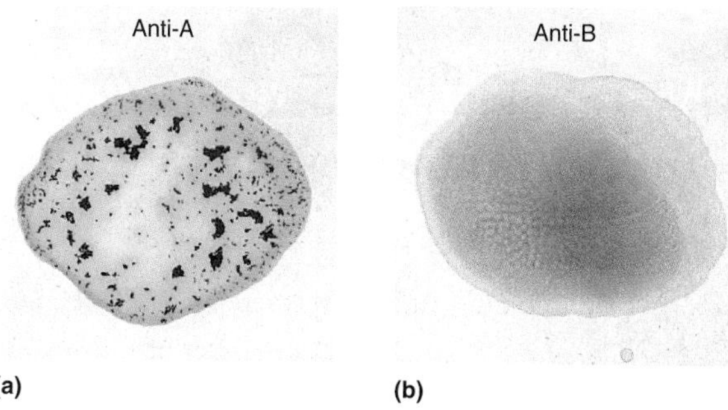
Anti-A
Anti-B
(a)
(b)

FIGURE 19.8 ABO Blood Typing This method tests for two different antigens (A and B) on red blood cells, using separate antibody suspensions (anti-A and anti-B). The red blood cells in the photographs agglutinated when mixed with anti-A antibodies, but not with anti-B antibodies, indicating that the blood group is Type A.

attached are produced commercially and used to test for various bacteria, fungi, viruses, and parasites, as well as hormones, drugs, and other substances. The beads are mixed with a drop of a body fluid or suspended microbial culture. If the specific antigen is present, easily visible clumps will form. Latex agglutination tests are commonly used to identify beta-hemolytic *Streptococcus* species **(figure 19.9).**

MICROCHECK 19.4

Agglutination and precipitation reactions both depend on cross-linking and lattice formation of antigen-antibody complexes. Large complexes of antibodies and soluble antigens will precipitate out of solution. Complexes of either antibodies and insoluble antigens or soluble antigens and antibody-coated beads will agglutinate.

- ✓ What is the advantage of the radial immunodiffusion test over the Ouchterlony test?
- ✓ How do the antigens used in precipitation reactions differ from those in direct agglutination tests?
- ✓ In precipitation reactions, why can cross-linked lattices not form when there is an excess of antibody?

FIGURE 19.9 Latex Agglutination In this example, latex beads coated with antibodies that bind specifically to cell wall antigens of *Streptococcus pyogenes* are mixed with a suspension of a culture suspected of being that species. Visible clumping (shown on the left), confirms that the organism is *S. pyogenes.* Negative test results are shown on the right.

19.5

Using Labeled Antibodies to Detect Antigen-Antibody Interactions

Focus Points

- Compare and contrast fluorescent antibody tests, ELISAs, and Western blots.
- Describe how the fluorescence-activated cell sorter is used in immunoassays.

Detectable markers such as enzymes, fluorescent dyes, and radioactive tags can be attached to specific antibodies, which are then used to detect the presence of given antigens. Examples of tests that use labeled antibodies include the fluorescent antibody test, enzyme-linked immunosorbent assay (ELISA), and Western blotting. Marking antigens with fluorescently-labeled antibodies also provides a mechanism for sorting antigens.

Fluorescent Antibody (FA) Tests

Fluorescent antibody (FA) tests employ fluorescence microscopy to locate fluorescently-labeled antibodies bound to antigens fixed to a microscope slide. The antigens are often bacterial cells, and the antibodies may be bound directly or indirectly **(figure 19.10).** ■ fluorescent dyes and tags, p. 51 ■ fluorescence microscopy, p. 43

The **direct FA test** is typically used to identify an unknown antigen fixed to the slide (see figure 19.10a). Labeled antibodies of known specificity are added, the mix is incubated, and then the slide is washed to remove unbound antibodies. The labeled antibodies bound to the antigen will remain, making the antigen visible under the fluorescence microscope. Since the specificity of the antibodies was known, binding of those molecules to the antigen identifies the antigen. Several different fluorescent dyes, including fluorescein (fluoresces yellow-green) and rhodamine (fluoresces orange-red) can be used to label the antibody. By using various fluorescent dyes, it is possible to locate different antigens in the same preparation. Labeled antibodies that bind specifically to common pathogens are readily available commercially.

The **indirect FA test** is typically used to determine the presence of a given antibody specificity in human serum (see figure 19.10b). A known antigen is fixed to the slide, and then the test serum is added. If specific antibodies are present in the serum, they will bind to the antigen. The slide is then washed to remove unbound antibodies. The antigen-antibody complexes cannot be seen however, until they are labeled or tagged with a fluorescent dye. This is done by adding fluorescently-labeled anti-human IgG antibodies, which will bind any antibodies remaining from the human serum. These labeled antibodies are available commercially. Unbound labeled antibodies are washed away before microscopic examination of the slide. ■ anti-human IgG antibodies, p. 439

Fluorescence polarization immunoassay is a highly sensitive test for antigen-antibody interactions that uses a beam of polarized light to determine the rate of spin of fluorescent anti-

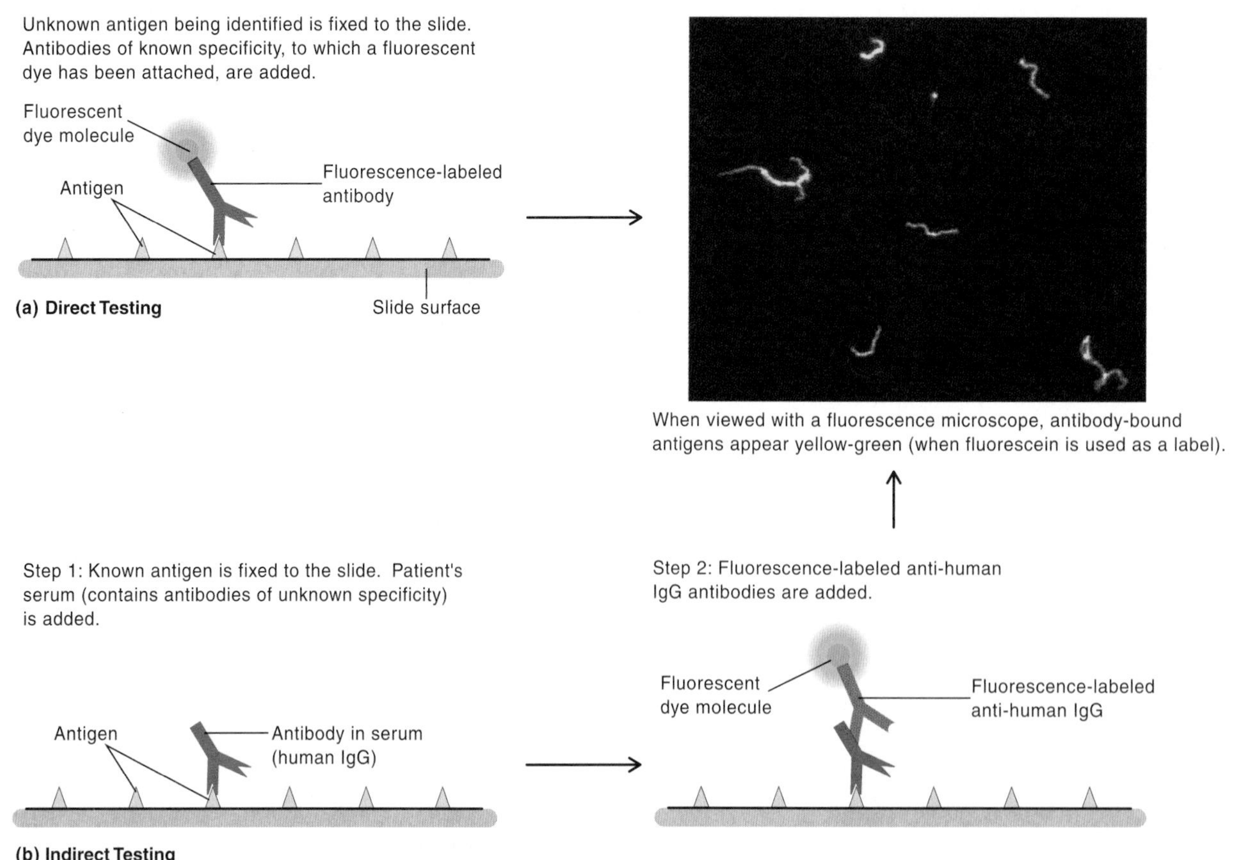
Unknown antigen being identified is fixed to the slide. Antibodies of known specificity, to which a fluorescent dye has been attached, are added.
Fluorescent dye molecule
Antigen
Fluorescence-labeled antibody
(a) Direct Testing
Slide surface
When viewed with a fluorescence microscope, antibody-bound antigens appear yellow-green (when fluorescein is used as a label).
Step 1: Known antigen is fixed to the slide. Patient's serum (contains antibodies of unknown specificity) is added.
Antigen
Antibody in serum (human IgG)
Step 2: Fluorescence-labeled anti-human IgG antibodies are added.
Fluorescent dye molecule
Fluorescence-labeled anti-human IgG
(b) Indirect Testing

FIGURE 19.10 Fluorescent Antibody (FA) Tests (a) Direct testing. **(b)** Indirect testing.

bodies. The assay exploits the fact that large molecules, such as antibody-antigen complexes, spin in a liquid solution more slowly than small molecules, such as free antibodies. As the antigen-antibody complexes form, the molecular aggregates become much larger and the complex spins more slowly. Fluorescently-labeled antigen can be used to detect antibodies in serum or plasma.

Enzyme-Linked Immunosorbent Assay (ELISA)

The **enzyme-linked immunosorbent assay (ELISA)** employs antibodies that have been labeled with a detectable enzyme such as peroxidase from the horseradish plant. When the labeled antibodies bind directly or indirectly to an antigen fixed to a surface, their location can be determined using a colorimetric assay **(figure 19.11)**. This extremely sensitive test is often done in microtiter plates, and is widely used to screen large numbers of specimens for either antigen or antibody. A variety of prepared ELISA plates are available from various companies.

Direct ELISA

The **direct ELISA** is typically used to detect a given antigen. A specimen suspected of containing that antigen is affixed to a surface, such as a well in a microtiter plate. Oftentimes, the antigen is not affixed directly, but is instead "captured" by antibodies that have been attached to the surface (see figure 19.11a). This modification is referred to as the sandwich ELISA. After attachment of the antigen, enzyme-labeled antibodies known to bind the antigen are added. The unbound antibodies are then washed away. To detect remaining labeled antibodies, a colorless substrate is added

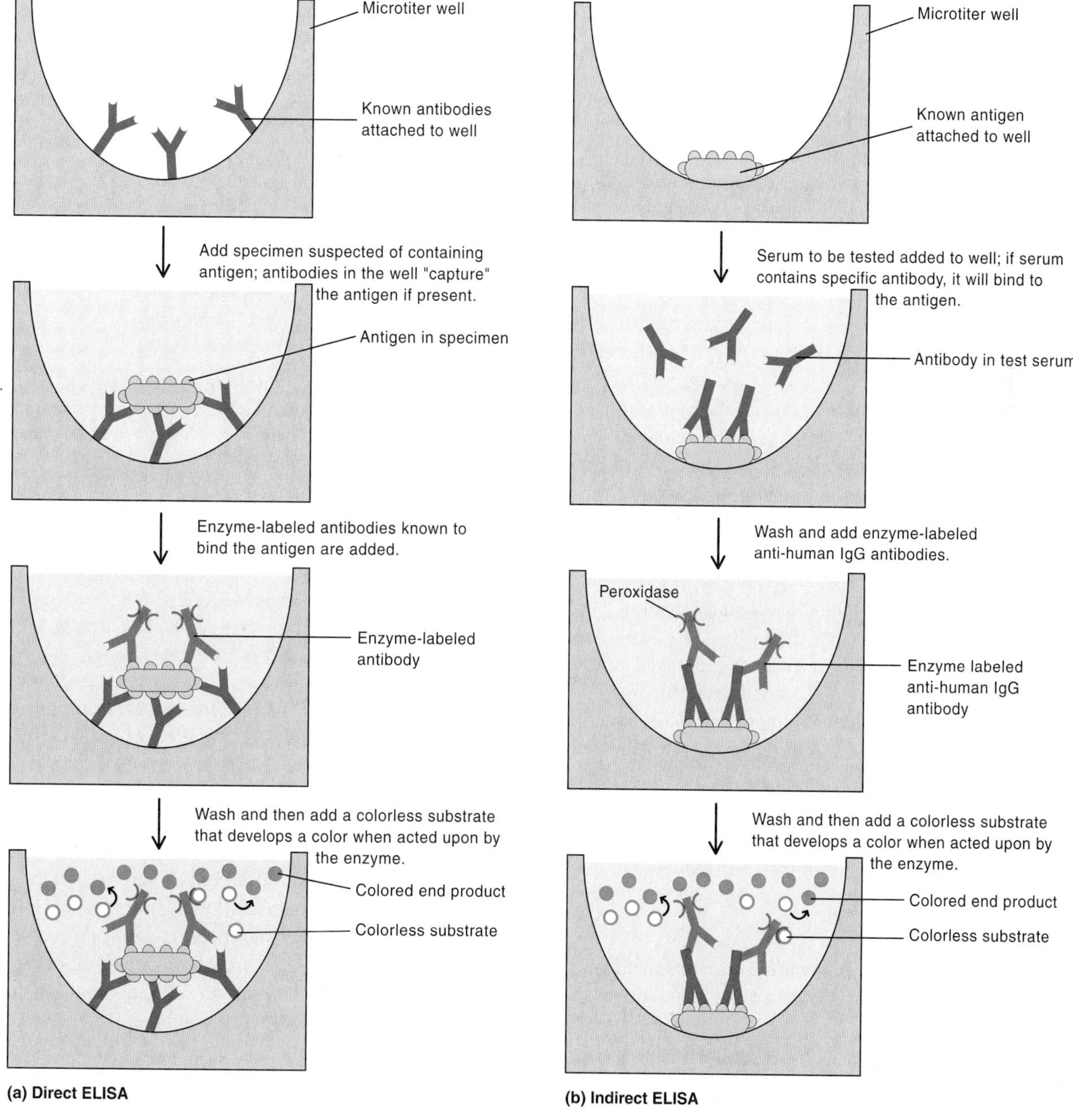

FIGURE 19.11 Enzyme-Linked Immunosorbent Assay (ELISA) **(a)** Direct ELISA. **(b)** Indirect ELISA.

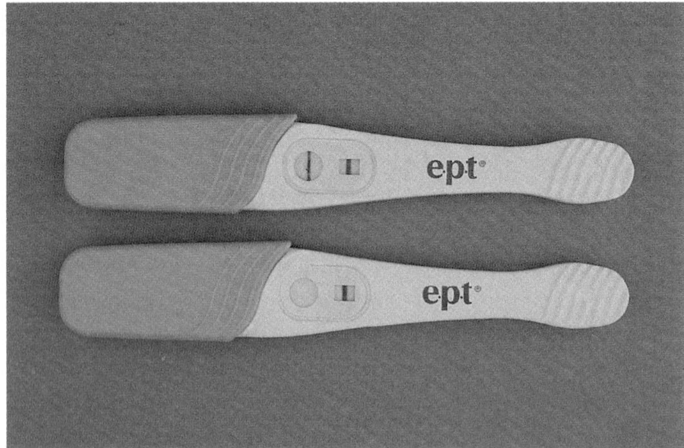

FIGURE 19.12 ELISA Test for Pregnancy The test detects human chorionic gonadotropin (HCG), an antigen present only in pregnant women. A urine sample is applied on the left. Two pink lines indicate reaction of HCG with antibodies, a positive test. A single line indicates absence of HCG in the urine, a negative test.

that develops a color if acted upon by the enzyme. Examples of commercial modifications of the ELISA test include rapid Group A strep tests done in doctors' clinics, and home pregnancy kits **(figure 19.12).**

Indirect ELISA

The **indirect ELISA** is typically used to detect the presence of a given antibody specificity. To do this, known antigen is affixed to a surface, such as a well in a microtiter plate (see figure 19.11b). The serum to be tested is added, and then a washing step removes unbound antibodies. To allow the antigen-antibody complex to be detected, enzyme-labeled anti-human IgG antibodies are added. These bind any human IgG antibodies remaining from the serum. Unbound labeled antibodies are then washed away. As with the direct ELISA, a colorless substrate of the enzyme is then added. Development of color indicates a positive reaction. The intensity of the color relative to that of a standard is usually quantitative.

■ **anti-human IgG antibodies, p. 439**

The indirect ELISA is routinely used to test donated blood for antibodies against HIV before the blood is used for transfusion. The presence of specific antibodies implies that the virus is present. The ELISA is also used to screen patient serum samples to determine HIV status. A small percentage of the results will be false positive, however. Because of this, a complicated but more reliable test, Western blotting, which will be described next, is used to confirm the positive ELISA results.

Western Blotting

In the **Western blotting** technique, the various proteins that make up an antigen are separated by size before reacting them with antibody. This makes it possible to determine exactly which proteins the antibodies are recognizing, an essential aspect of accurate HIV testing **(figure 19.13).**

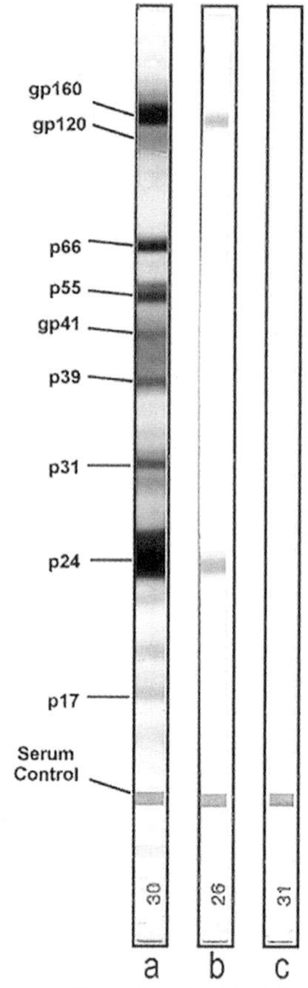

FIGURE 19.13 Western Blotting Results Each strip in this photograph contains HIV proteins that have been separated by gel electrophoresis and then transferred in-place to the strips, along with a serum control. The dark portions of the strips indicate the locations of bound antibodies. The results for the strong reactive control (lane a) indicate that antibodies have bound all HIV proteins on the strip. The results for the weak reactive control (lane b) show binding of antibodies to only certain critical HIV proteins (gp160/120 and p24). The results for the non-reactive control (lane c) show no binding to HIV proteins. When this test is done using antibodies from a patient, strict criteria must be used when interpreting the results.

To separate the proteins, a special type of gel electrophoresis called SDS-PAGE is used. This resolves proteins of different sizes into a series of bands, as smaller proteins move faster than the larger ones. The separated proteins are then transferred to a nylon membrane filter to immobilize them before a solution of antibody molecules is added. The steps after that are very similar to those of an ELISA. To test for antibodies specific for the separated proteins in a patient, a serum sample is added to the blot, and then unbound antibodies are washed off. Enzyme- (particularly luciferase from fireflies) or radioactively-labeled anti-human IgG antibodies are then added, which attach to any antibodies remaining from the serum. Unbound labeled antibodies are then washed off. Finally, the label is detected, generating a bar code-like pattern that indicates which proteins were recognized by the patient's antibodies.

Fluorescence-Activated Cell Sorter (FACS)

As described in chapter 4, flow cytometry counts cells or other particles by measuring the light scattered as they pass single-file by a laser. A specialized version called the **fluorescence-activated cell sorter (FACS)** can be used to count and separate cells labeled with fluorescent antibodies. For example, subsets of T cells ($CD4^+$ and $CD8^+$) can be counted and even separated by first mixing the cells with two preparations of monoclonal antibodies (one binds CD4 markers and the other CD8 markers), each labeled with a different fluorescent dye. ■ flow cytometry, p. 100

MICROCHECK 19.5

The fluorescence antibody test, ELISA, and Western blot use antibodies labeled with a detectable marker to locate a given antigen. Direct tests use the labeled antibody to identify an antigen. Indirect tests use labeled antibodies to detect a patient's antibodies bound to a known antigen. Flow cytometry uses a fluorescence activated cell sorter to separate and count cells labeled with fluorescent antibodies.

✓ Compare and contrast the fluorescent antibody test and the ELISA.

✓ In HIV testing, why is the Western blot test used to confirm ELISA results?

✓ Why is a false positive more significant in HIV testing of patients than in screening donated blood for transfusions?

FUTURE CHALLENGES

Global Immunization

In addition to developing new safe and effective vaccines, a major challenge for the future is delivering available vaccines to populations worldwide. When this was done in the case of smallpox, the disease was eliminated from the world; and poliomyelitis is almost eradicated now. The World Health Organization and the governments that support it deserve much of the credit for these achievements. Much remains to be done, however, and gifts totaling well over a billion dollars from Bill Gates, one of the founders of Microsoft, and his wife have stimulated further progress in this area.

To immunize universally, it is necessary to have vaccines that are easily administered, inexpensive, stable under a variety of environmental conditions, and preferably painless. Instead of expensive needles and syringes and painful injections, vaccines may be delivered in a number of easy ways. For example, naked DNA can be coated onto microscopic gold pellets, which are shot from a gunlike apparatus directly through the skin into the muscle, painlessly. Skin patches deliver antigens slowly through the skin. Vaccines against mucous membrane pathogens can be delivered by a nasal spray, as in some new influenza vaccines, or by mouth. Time-release pills introduce antigen steadily to give a sustained immune response.

In addition to sprays and pills that deliver antigens to mucous membranes in order to induce mucosal immunity, techniques are being developed to get antigens directly to M cells in the gastrointestinal mucosa. These are the cells that can endocytose antigens and deliver them across the membrane to the lymphoid tissue of the Peyer's patches, where immune responses occur. Antigens are incorporated into substances known to bind to M cells, thereby facilitating entry into the M cells and the Peyer's patches. ■ M cells, p. 370

One of the major remaining challenges is development of an effective vaccine against HIV that can be administered universally. Although anti-HIV vaccines are being studied, there is no indication that a truly effective immunization program for HIV disease is imminent. It is also probable that when HIV disease is brought under control, a new and currently unforeseen challenge will arise during the twenty-first century.

SUMMARY

Immunization

19.1 Principles of Immunization (figure 19.1)

Active Immunity

Active immunity occurs naturally in response to infections or other natural exposure to antigens, and artificially in response to vaccination.

Passive Immunity

Passive immunity occurs naturally during pregnancy and through breast feeding, and artificially by transfer of preformed antibodies, as in **immune globulin** and **hyperimmune globulin.**

19.2 Vaccines and Immunization Procedures

A **vaccine** is a preparation of a disease-causing agent or its products used to induce active immunity. It protects an individual against disease, and can also provide **herd immunity** (table 19.1).

Attenuated Vaccines

An **attenuated vaccine** is a weakened form of the pathogen that can replicate but is generally unable to cause disease.

Inactivated Vaccines

Inactivated vaccines are unable to replicate but they retain the immunogenicity of the infectious agent or toxin. They include inactivated whole agents, **toxoids,** protein subunits, purified polysaccharides, and polysaccharide-protein **conjugates. Adjuvants** increase the intensity of the immune response to the antigen in a vaccine.

An Example of Vaccination Strategy—The Campaign to Eliminate Poliomyelitis

Both the Sabin and the Salk polio vaccines protect against paralytic poliomyelitis; the Sabin vaccine induces mucosal immunity.

The Importance of Routine Immunizations for Children (table 19.3)

Routine childhood immunizations have prevented millions of cases of disease and many deaths during the past decades.

Current Progress in Immunization

Progress in vaccination includes enhancement of the immune response to vaccines, development of new or improved vaccines against certain diseases, and development of new types of vaccines (table 19.4).

Immunological Testing

19.3 Principles of Immunological Testing

A **seronegative** individual will **seroconvert,** becoming **seropositive,** a week to 10 days after initial infection with an agent.

Obtaining Antibodies

To determine if a patient has antibodies in the blood against a specific infectious agent, either the patient's **serum** or **plasma** is tested.

Polyclonal antibodies recognize multiple epitopes whereas **monoclonal antibodies** recognize only a single epitope. **Anti-human IgG antibodies** are used to detect IgG molecules in a patient specimen.

Quantifying Antigen-Antibody Reactions

The concentration of antibody molecules in a specimen is usually determined by making serial dilutions; the last dilution that gives a detectable antigen-antibody reaction reflects the **titer.**

19.4 Observing Antigen-Antibody Aggregations

Precipitation Reactions (figure 19.4)

When antibodies bind to soluble antigens that have multiple epitopes, the complexes precipitate out of solution at the **zone of optimal proportion;** examples of tests that involve **precipitation reactions** include the **Ouchterlony** procedure, the **radial immunodiffusion test,** and **immunoelectrophoresis** (figures 19.5, 19.6, 19.7).

Agglutination Reactions (figures 19.8, 19.9)

Agglutination reactions are similar in principle to precipitation reactions, but insoluble particles are involved, resulting in the formation of obvious aggregates; examples include direct agglutination tests and passive agglutination tests.

19.5 Using Labeled Antibodies to Detect Antigen-Antibody Interactions

Fluorescent Antibody (FA) Tests (figure 19.10)

Fluorescent antibody (FA) tests rely on fluorescence microscopy to locate fluorescently-labeled antibodies bound to antigens fixed to a microscope slide; examples of FA procedures include the **direct FA test, indirect FA test,** and **fluorescence polarization immunoassay.**

Enzyme-Linked Immunosorbent Assay (ELISA) (figure 19.11)

The **enzyme-linked immunosorbent assay (ELISA)** employs an antibody labeled with a detectable enzyme; the procedure can be direct or indirect.

Western Blotting (figure 19.13)

In the **Western blotting** technique, the various proteins that make up an antigen are separated by size before reacting them with antibody.

Fluorescence-Activated Cell Sorter (FACS)

The **fluorescence-activated cell sorter (FACS)** can be used to count and separate antigens labeled with fluorescent antibodies.

REVIEW QUESTIONS

Short Answer

1. Why are acellular subunit vaccines replacing some whole cell vaccines? Discuss an example.
2. What are some ways in which the number of injections needed for childhood immunization could be lessened?
3. Can DNA vaccines cause the disease they are meant to protect against?
4. Which would be expected to be more effective against common childhood diseases: active or passive immunity? Why? Answer in terms of protection and cost-effectiveness.
5. Describe how both active and passive immunization can be used to combat tetanus.
6. In a precipitation reaction, what is meant by "optimal proportions"?
7. To determine a person's blood type, antibodies against red blood cells bearing the A and/or B antigens are mixed with the person's red cells. Is this an example of a precipitation reaction or an agglutination reaction?
8. What are the advantages of the ELISA test?
9. An ELISA test is used to screen patient specimens for HIV. A positive ELISA test is confirmed by a Western blot test. Why not the other way around, with the ELISA second?
10. What is the purpose of anti-human IgG antibodies in immunological testing?

Multiple Choice

1. All of the following are attenuated vaccines, *except* that against
   a) measles. b) mumps. c) rubella.
   d) Salk polio. e) yellow fever.
2. Examples of active immunization include
   a) giving antibodies against diphtheria.
   b) immune globulin injections to prevent hepatitis.
   c) Sabin polio immunization.
   d) rabies immune globulin.
   e) tetanus immune globulin.
3. Disease may be inadvertently caused in immunosuppressed individuals by administration of
   a) inactivated whole agent vaccines.
   b) toxoids.
   c) subunit vaccines.
   d) genetically engineered vaccine against hepatitis B.
   e) attenuated vaccines.
4. Vaccines ideally should be all of the following, *except*
   a) effective in protecting against the disease.
   b) inexpensive.
   c) stable.

d) living.
e) easily administered.

5. An important subunit vaccine that is widely used is the
   a) pertussis vaccine. b) Sabin vaccine. c) Salk vaccine.
   d) measles vaccine. e) mumps vaccine.
6. Which of the following about immunological testing is *false?*
   a) Polyclonal antibodies recognize multiple epitopes.
   b) Monoclonal antibodies recognize a single epitope.
   c) Serum and plasma can both be tested for antibodies.
   d) The direct ELISA employs antihuman IgG antibodies.
   e) A rise in specific antibody titer indicates an active infection.
7. Precipitation tests include all of the following, *except*
   a) radial immunodiffusion. b) Ouchterlony.
   c) ELISA. d) immunoelectrophoresis.
8. All of the following are matching pairs *except?*
   a) ELISA—radioactive label
   b) Fluorescence-activated cell sorter—flow cytometry
   c) Radial immunodiffusion—quantitative
   d) Fluorescent antibody test—microscopy
   e) Western blot—gel electrophoresis
9. Which of the following would be most useful for screening thousands of patient specimens for antibodies that indicate a certain disease?
   a) Western blot b) Fluorescent antibody c) ELISA
   d) All of the above e) None of the above
10. In quantifying antibodies in a patient's serum
   a) total protein in the serum is measured.
   b) the antibody is usually measured in grams per ml.
   c) the serum is serially diluted.
   d) both antigen and antibody are diluted.
   e) the titer refers to the amount of antigen added.

Applications

1. A new parent asks you which vaccines the CDC recommends for a 2-month-old infant. What is your answer? The chapter 19 readings at the Online Learning Center (www.mhhe.com.nester6) provide a link to the CDC's recommended immunization schedules.
2. There has been debate about keeping smallpox virus stored, since the disease has been eradicated. What would be an argument for keeping the virus? What should be done to protect against use of the virus in biological warfare?

Critical Thinking

1. In figure 19.4, how would the curve change if the concentration of antibody in the original sample were increased? (Would the shape of the curve change? Would the curve be shifted left, right, up, or down?) Briefly explain your answer.
2. *Staphylococcus aureus* makes a protein called protein A, which binds to the Fc region of antibody molecules from a wide variety of species. How could protein A be exploited in immunoassays?

Community water supply in a developing country.

Epidemiology

A Glimpse of History

Puerperal fever, an illness we now recognize is the result of a bacterial infection of the uterus following childbirth, has occurred at least since the time of Hippocrates (460–377 B.C.). In the eighteenth century, when it became popular for women to deliver their babies in hospitals, the incidence of the disease rose to epidemic proportions. By the middle of the nineteenth century in the hospitals of Vienna, the major medical center of the world at that time, about one of every eight women died of puerperal fever following childbirth.

In 1841, Ignaz Semmelweis, a Hungarian, traveled to Vienna to study medicine. After finishing medical school, he became the first assistant to Professor Johann Klein at the Lying-In Hospital. There were two sections of the hospital. The first section was under the management of Professor Klein and the medical students. Midwives and midwifery students served the second section. Being an astute observer, Semmelweis noticed that the incidence of puerperal fever in the first section rose as high as 18%, four times that in the second section. When he investigated the conditions in the two sections, they appeared to be the same except in terms of their management. Semmelweis was dismissed from his post when he implicated Professor Klein and the students in the spread of the disease, but was reinstated a few months later when others intervened on his behalf.

Semmelweis recognized that disease symptoms that led to the death of a friend who had incurred a scalpel wound while doing an autopsy were similar to puerperal fever. He reasoned that the "poison" that killed his friend probably also contaminated the hands of the medical students who did autopsies. These students were transferring the "poison" from the cadavers to the women in childbirth. Midwives did not perform autopsies. Since this was before Pasteur and Koch established the germ theory of disease, Semmelweis had no way of knowing that the "poison" being transferred was probably *Streptococcus pyogenes,* a common cause of many infections, including puerperal fever. Those being autopsied had likely died of streptococcal disease as well. To test his hypothesis, Semmelweis instituted the practice of having physicians and students wash their hands with a solution of chloride of lime, a strong disinfectant, before attending their patients. The result was a drop in the incidence of puerperal fever to one-third its previous level. Instead of accepting these findings, Semmelweis's colleagues refused to admit responsibility for the deaths of so many patients. His work was so fiercely attacked that he was forced to leave Vienna and return to his native Hungary. There he was again able to use disinfection techniques to achieve a remarkable reduction in the number of deaths from puerperal fever.

Semmelweis became increasingly outspoken and bitter, finally becoming so deranged that he was confined to a mental institution. Ironically, he died one month later of a generalized infection similar to the kind that had killed his friend and the many women who had contracted puerperal fever following childbirth. The infection originated from a finger wound received before his confinement. Some said he deliberately infected himself from a cadaver while performing an autopsy.

Epidemiology is the study of the spread, control, and prevention of disease in populations. Many of our habits in the daily routine of life, from handwashing to waste disposal, reflect the importance of this field. **Epidemiologists,** the "health detectives," collect and compile data to describe disease outbreaks, much as a criminal detective describes the scene of a crime. Their expertise involves diverse disciplines including ecology, microbiology, sociology, statistics, and psychology.

KEY TERMS

Attack Rate The proportion of individuals developing illness in a population exposed to an infectious agent.

Communicable Disease An infectious disease that can be transmitted from one host to another.

Endemic A disease or other occurrence that is constantly present in a population.

Epidemic A disease or other occurrence that has a much higher incidence than usual.

Herd Immunity A phenomenon that occurs when a critical concentration of immune hosts prevents the spread of an infectious agent, thereby protecting the entire population.

Incidence The number of new cases of a disease in a population at risk during a specified time period.

Index Case The first identified case of a disease in an outbreak or epidemic.

Morbidity Illness; most often expressed as the rate of illness in a given population at risk.

Mortality Death; most often expressed as a rate of death in a given population at risk.

Outbreak A cluster of cases occurring during a brief time interval and affecting a specific population; an outbreak may herald the onset of an epidemic.

Pandemic A worldwide epidemic.

Prevalence The total number of cases of a disease in a given population at risk at a point in time.

Reservoir of Infection The natural habitat of a pathogen; sum of the potential sources of an infectious agent.

20.1 Principles of Epidemiology

Focus Points

- Describe why epidemiologists are most concerned with the rate of disease rather than the absolute number of cases.
- Describe how a pathogen gets from its reservoir to the next host. Focus on the types of reservoirs, portals of exit, mechanisms of transmission, and portals of entry.
- Describe three factors that influence the epidemiology of disease.

Diseases that can be transmitted from one host to another, such as measles, colds, and influenza, are **contagious,** or **communicable, diseases.** In order for a communicable disease to spread, a specific chain of events must occur. First, the pathogen must have a suitable environment in which to live. That natural habitat, the **reservoir of infection** may be on or in an animal, including humans, or in an environment such as soil or water **(figure 20.1a).** A pathogen must then leave its reservoir in order to be transmitted to a susceptible host. If the reservoir is an animal, the body orifice or surface from which a microbe is shed is called the **portal of exit.** Disease-causing organisms must then be transmitted to the next host, usually through direct contact or via contaminated food, water, or air (figure 20.1b). They enter the next host through a body surface or orifice called the **portal of entry** (figure 20.1c).

Diseases that do not spread from one host to another are called **non-communicable diseases.** Microorganisms that cause these diseases most often arise from an individual's normal microbiota or from an environmental reservoir. Prevention of non-communicable illness is often disease-specific.

Rate of Disease in a Population

Epidemiologists are most concerned with the **rate** of a disease (the proportion of a given population infected), rather than the absolute number of cases. For example, 100 people in a large city developing disease X in a given period may not be abnormal, whereas 100 people in a small rural community developing the same disease would indicate a much higher rate and thus be of greater concern to the epidemiologist. A related concept is the **attack rate**—the number of cases developing in a group of people who were exposed to the infectious agent. For example, if 100 people at a party ate chicken that was contaminated with *Salmonella,* and 10 people came down with symptoms of salmonellosis, then the attack rate was 10%. The attack rate reflects many factors, including the infectious dose of the organism and the immunity of the population.

Other rates are also often used to express the effect of a disease on a population. **Morbidity rate** is calculated as the number of cases of an illness in a given time period divided by the population at risk. Contagious diseases such as influenza often have a high morbidity rate because each infected individual may transmit the infection to several others. **Mortality rate** reflects the percentage of a population that dies from the disease. Diseases such as plague and Ebola hemorrhagic fever are feared because of their very high mortality rate. The **incidence** of a disease reflects the number of new cases in a specific time period in a given population at risk, whereas the **prevalence** reflects the number of total existing cases both old and new in a given population at risk. Usually these rates are expressed as the number of cases per 100,000 people. The prevalence is useful to assess the overall impact of the disease on society because it takes the duration of the disease into account, whereas the incidence provides a means of measuring the risk of an individual contracting the disease.

Diseases that are constantly present in a given population are called **endemic.** Both the common cold and influenza are endemic in the United States. An unusually large number of cases in a population constitutes an **epidemic.** Epidemics may be caused by diseases not normally present in a population, such as cholera being reintroduced to the Western Hemisphere, or by endemic diseases such as influenza and pneumonia **(figure 20.2).** A related term is **outbreak,** which generally implies a cluster of cases occurring during a brief time interval and affecting a specific population; an outbreak may herald the onset of an epidemic. When an epidemic spreads worldwide, as AIDS has, it is called a **pandemic.**

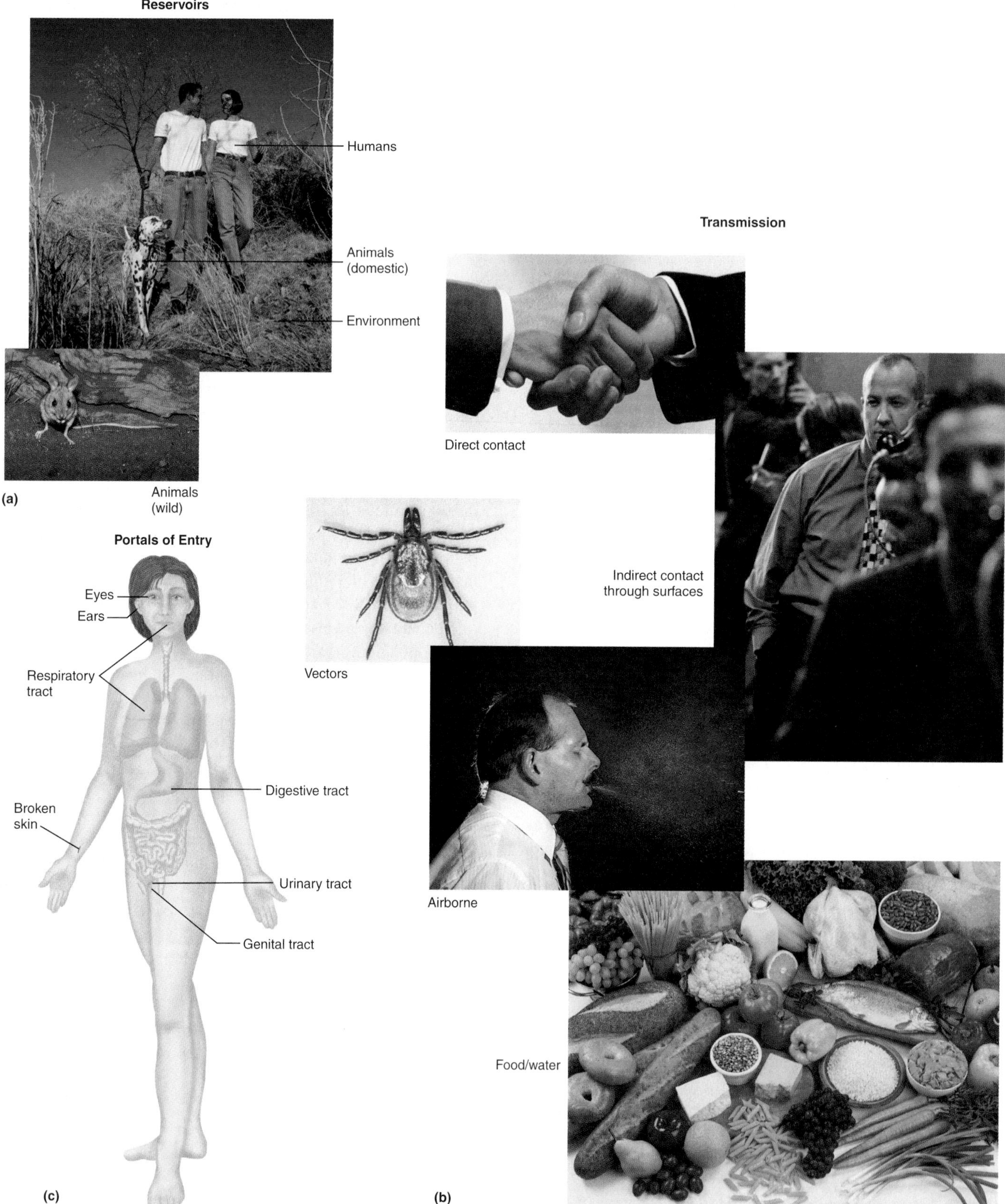

FIGURE 20.1 Spread of Pathogens **(a)** Reservoirs of infection, **(b)** transmission, and **(c)** portals of entry are necessary for the spread of communicable diseases.

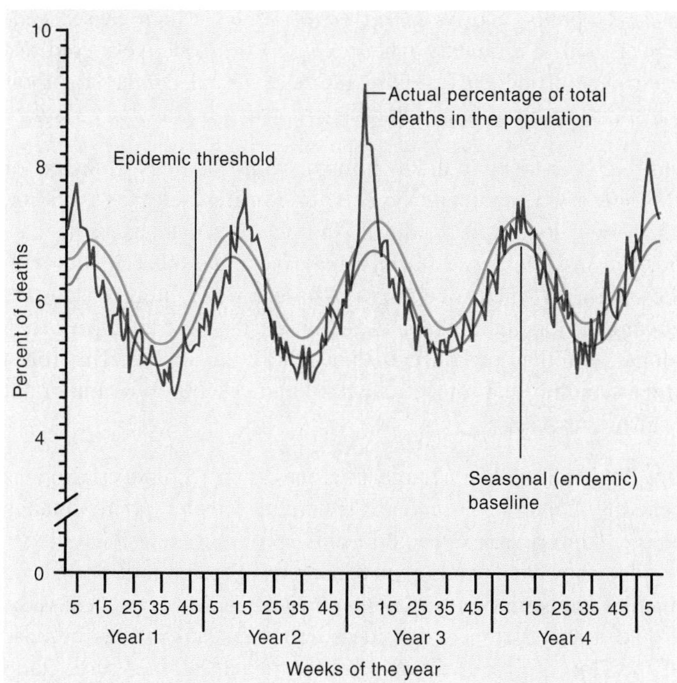

FIGURE 20.2 Endemic Disease that Can Be Epidemic Example of yearly fluctuation of pneumonia and influenza mortality (expressed as a percentage of all deaths).

Reservoirs of Infection

The reservoir of infection is important because it affects the extent and distribution of a disease. Identifying the reservoir can help protect a population from disease, because measures can then be instituted to prevent the people from coming into contact with that source. The fact that the United States does not have epidemics of plague, the disease that killed over one-quarter of the European population in the fourteenth century, is in part because we recognize the importance of controlling the rodent reservoir. Wild rats, mice, and prairie dogs are the natural reservoir of *Yersinia pestis,* the bacterium that causes plague. By avoiding a buildup of garbage in cities and homes, we help prevent rodent infestation of our living quarters. ■ *Yersinia pestis,* p. 682

Human Reservoirs

Infected humans are the most significant reservoir of the majority of communicable diseases. In some cases, humans are the only reservoir. In other instances, the pathogenic organism can exist not only in humans but in other animals and, occasionally, the environment as well. If infected humans are the only reservoir, then theoretically the disease is easier to control. This is because it is more feasible to institute prevention and control programs in humans than it is in wild animals. The eradication of smallpox is an excellent example. The combined effects of widespread vaccination programs, which resulted in fewer susceptible people, and the successful isolation of those who did become infected, eliminated the smallpox virus from nature. The virus no longer had a reservoir in which to multiply.

Symptomatic Infections People who have symptomatic illnesses are an obvious source of infectious agents, and ideally they understand the importance of taking precautions to avoid transmitting their illness to others. Staying home and resting while ill both helps the body recover and protects others from exposure to the disease-causing agent. Even conscientious people, however, can unintentionally be a source of infection to others. For example, people who are in the incubation period of mumps shed virus. ■ mumps, p. 593 ■ incubation period, p. 394

Asymptomatic Carriers A person can harbor a pathogen with no ill effects whatsoever, acting as a **carrier** of the disease agent. These people may shed the organism intermittently or constantly for months, years, or even a lifetime.

Some carriers have an **asymptomatic infection;** their immune system is actively responding to the invading microorganism, but they have no obvious clinical symptoms. Because these people often have no reason to consider themselves a reservoir, they move freely about, spreading the pathogen. People with asymptomatic infections are a significant complicating factor in the control of sexually transmitted diseases such as gonorrhea. Up to 50% of women infected with *Neisseria gonorrhoeae* have no symptoms, which means they often unknowingly transmit the disease to their sexual partners. In contrast, infected men are more often symptomatic and therefore seek medical treatment. Gonorrhea infections can be treated with antibacterial drugs, but tracking down sexual contacts of infected people is difficult and costly. ■ *Neisseria gonorrhoeae,* p. 629

Some potentially pathogenic microbes can colonize the skin or mucosal surfaces, establishing themselves as part of a person's microbiota. For instance, many people carry *Staphylococcus aureus* as a part of their nasal or skin microbiota. Carriers of *S. aureus* may never have any illness or disease as a result of the organism, but they remain a potential source of infection to themselves and others. Unfortunately, ridding a colonized carrier of the infectious organism is often difficult, even with the use of antimicrobial drugs. ■ *Staphylococcus aureus,* p. 535

Non-Human Animal Reservoirs

Non-human animal reservoirs are the source of many pathogens. Poultry are a reservoir of gastrointestinal pathogens such as species of *Campylobacter* and *Salmonella.* In the United States, raccoons, skunks, and bats are the reservoir of the rabies virus. Recall that rodents are the reservoir of *Yersinia pestis,* the causative agent of plague. Occasional transmission of plague to humans is still reported in the southwestern states where *Y. pestis* is endemic in prairie dogs and other rodents. Rodents, particularly the deer mouse, are the reservoir for hantavirus. ■ *Campylobacter,* p. 602 ■ *Salmonella,* p. 601

Diseases such as plague and rabies that can be transmitted to humans but primarily exist in other animals are called **zoonotic** diseases, or **zoonoses.** Zoonotic diseases are often more severe in humans than in the normal animal host because the infection in humans is accidental; there has been no evolution toward the balanced pathogenicity that normally exists between host and parasite. ■ balanced pathogenicity, p. 397

Environmental Reservoirs

Some pathogens have environmental reservoirs. *Clostridium botulinum,* which causes botulism, and *Clostridium tetani,* which causes tetanus, are both widespread in soils. Pathogens that have environmental reservoirs are difficult or impossible to eliminate. ■ *Clostridium botulinum*, p. 657 ■ *Clostridium tetani*, p. 566

Portals of Exit

Microorganisms must leave one host in order to be transmitted to another. Those that inhabit the intestinal tract are shed in the feces. Pathogens such as *Vibrio cholerae* that cause massive volumes of watery diarrhea may have an evolutionary advantage because the large volumes discharged can contaminate drinking water and food. Respiratory organisms are expelled in droplets of saliva and mucus when people talk, laugh, sing, sneeze, or cough (see figure 20.1b). Pathogens such as *Mycobacterium tuberculosis* and various respiratory viruses exit the body via this route. Meanwhile, organisms that inhabit the skin are constantly shed on skin cells. Even as you read this text you are shedding skin cells, some of which may have *Staphylococcus aureus* on their surface. Genital pathogens such as *Neisseria gonorrhoeae* can be carried in semen and vaginal secretions. Hantavirus is found in the saliva, urine, and droppings of deer mice. ■ *Vibrio cholerae*, p. 596 ■ *Neisseria gonorrhoeae*, p. 629 ■ *Mycobacterium tuberculosis*, p. 515 ■ *Staphylococcus aureus*, p. 535

Transmission

A successful pathogen must somehow be passed from its reservoir to the next susceptible host. Transmission of a pathogen from one person to another through the air, by physical contact, by ingestion of food or water, or via a living agent such as an insect is called **horizontal transmission.** This contrasts with **vertical transmission,** which is the transfer of a pathogen from a pregnant woman to her fetus, or from a mother to her infant during childbirth or breast feeding.

Contact

Transmission of a pathogen from one host to another often involves direct or indirect contact.

Direct Contact **Direct contact** occurs when one person physically touches another. It can be an act as simple as a handshake, or a more intimate contact such as sexual intercourse. In some cases, direct contact is the primary way in which a microbe is transmitted. This is particularly true if the transfer of even very low numbers of the microbe can initiate an infection. For example, the infectious dose of *Shigella* species, which are intestinal pathogens, is approximately 10 to 100 cells, a number easily passed when shaking hands. Once on the hands, the organisms can inadvertently be ingested. This is just one example of how **fecal-oral transmission,** the consumption of organisms that originate from the intestine, can occur. Handwashing, a fairly simple routine that physically removes microbes, is important in preventing this type of spread of disease. Even washing in plain water reduces the numbers of potential pathogens on the hands, which in turn decreases the possibility of transferring or ingesting sufficient numbers of cells to establish an infection. In fact, routine handwashing is considered to be the single most important measure for preventing the spread of infectious disease. ■ *Shigella* sp., p. 598

Pathogens that cannot survive for extended periods in the environment must generally, because of their fragile nature, be transmitted through direct contact. *Treponema pallidum,* which causes syphilis, and *Neisseria gonorrhoeae,* which causes gonorrhea, both die quickly when exposed to a relatively cold dry environment and thus require intimate sexual contact for their transmission. ■ *Treponema pallidum*, p. 634 ■ *Neisseria gonorrhoeae*, p. 629

Indirect Contact Indirect transmission involves transfer of pathogens via inanimate objects, or **fomites,** such as clothing, table-tops, doorknobs, and drinking glasses. As an example, carriers of *Staphylococcus aureus* may inoculate their fingers with the organism when touching a skin lesion or colonized nostril. Organisms on the fingers can then easily be transferred to a fomite. Another person can then readily acquire the microbes when handling that object. Again, handwashing is an important control measure.

Droplet Transmission Large microbe-laden respiratory droplets generally fall to the ground no farther than a meter (approximately 3 feet) from release. People in close proximity can inhale those droplets, however, resulting in the spread of respiratory disease via **droplet transmission.** Although physical contact is not necessary, droplet transmission is considered contact transmission because of the close range involved. Droplet transmission is particularly important as a source of contamination in densely populated buildings such as schools and military barracks. Desks or beds in such locations ideally should be spaced more than 4 feet and preferably 8 to 10 feet apart to minimize the transfer of infectious agents. Another way to minimize the spread of respiratory diseases is to educate people about the importance of covering their mouths with a tissue when they cough or sneeze.

Food and Water

Pathogens, particularly those that infect the digestive tract, can be transmitted through contaminated food or water. Foods can become contaminated in a number of different ways. Animal products such as meat and eggs can harbor pathogens that originated from the animal itself. This is the case with poultry contaminated with species of *Salmonella* or *Campylobacter* and hamburger contaminated with *E. coli* O157:H7. Pathogens can also be inadvertently added during food preparations. *Staphylococcus aureus* carriers who do not wash their hands thoroughly prior to preparing food can easily contaminate the food. **Cross-contamination** results when pathogens from one food are transferred to another. A cutting board used first to carve raw chicken and then to cut cooked potatoes can serve as a fomite, transferring *Salmonella* species from the chicken onto the potatoes. Because many foods are a rich nutrient source, microorganisms can multiply to high numbers if the contaminated food is not refrigerated. Sound food-handling methods, including sanitary preparation as well as thorough cooking and proper storage, can prevent foodborne diseases.

Waterborne disease outbreaks can involve large numbers of people because municipal water systems distribute water to large areas. The 1993 waterborne outbreak of *Cryptosporidium parvum,* an intestinal parasite, in Milwaukee, Wisconsin, was estimated to have involved approximately 400,000 people. Prevention of waterborne diseases requires disinfection and filtration of drinking water and proper disposal and treatment of sewage. ■ *Cryptosporidium parvum*, p. 611 ■ drinking water treatment, p. 744 ■ sewage treatment, p. 739

Air

Respiratory diseases can also be transmitted through the air. When particles larger than 10 μm are inhaled, they are usually trapped in the mucus lining of the nose and throat and eventually swallowed. Smaller particles, however, can enter the lungs, where any pathogens they carry can potentially cause disease.

As mentioned earlier, when people talk, laugh, or sneeze they continually discharge microorganisms in liquid droplets. Whereas large droplets quickly fall to the ground, the smaller droplets dry, leaving small numbers of organisms attached to a thin coat of the dried material, creating **droplet nuclei.** The droplet nuclei can remain suspended indefinitely in the presence of even slight air currents. Other airborne particles, including dead skin cells, household dust, and soil disturbed by the wind, may also carry respiratory pathogens. An air conditioning system can distribute air contaminated by people or with organisms growing within the system.

The number of viable organisms in air can be estimated by using a machine that pumps a measured volume of air, including any suspended dust and particles, against the surface of a nutrient-rich medium in a Petri dish. This technique has shown that the number of bacteria in the air sampled rises in proportion to the number of people in a room **(figure 20.3).**

Understandably, airborne transmission of pathogens is very difficult to control. To prevent the buildup of airborne pathogens, modern public buildings have ventilation systems that constantly change the air. Hospital microbiology laboratories can be kept under a slight vacuum so that air flows in from the corridors, preventing microorganisms and viruses from being swept out of the lab to other parts of the building. Air in some laboratories, specialized hospital rooms, and jetliners is circulated through high-efficiency particulate air (HEPA) filters to remove airborne organisms that may be present. ■ HEPA filters, p. 115

Vectors

A **vector** such as a mosquito or flea can transmit certain diseases. The term "vector" applies to any living organism that can carry a disease-causing microbe, but most commonly these are arthropods such as mosquitoes, fleas, lice, and ticks. A vector can carry a pathogen externally or internally.

(a)

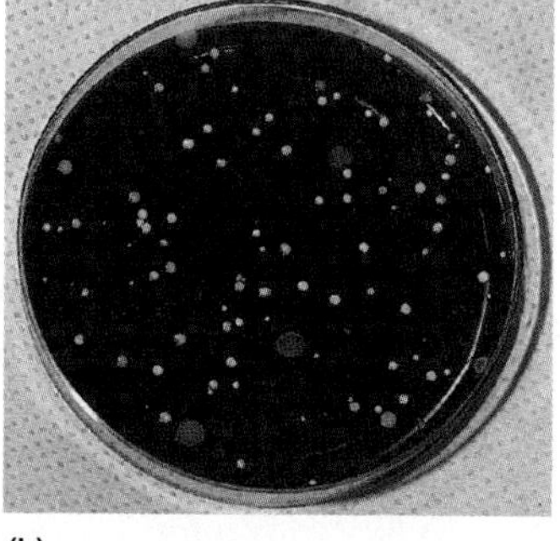
(b)

FIGURE 20.3 Air Sample Cultures (a) Air from a clean, empty hospital room. **(b)** Air from a small room containing 12 people. In both situations, 5 cubic feet of air was sampled.

Flies that land on feces can pick up intestinal pathogens such as *Escherichia coli* O157:H7 and *Shigella* species on their legs. If the fly then moves to a food, it transfers the microorganisms. In this case, the fly serves as a **mechanical vector,** carrying the microbe on its body from one place to another.

Diseases such as malaria, plague, and Lyme disease are transmitted via arthropods that harbor the pathogen internally. The vector either injects the infectious agent while taking a blood meal or defecates, depositing the pathogen onto a person's skin where it can then be inadvertently inoculated when the individual scratches the bite. For example, infected fleas inject *Yersinia pestis* while attempting to take a blood meal. In the case of malaria, caused by species of the eukaryotic pathogen *Plasmodium,* the mosquito is not only the transmitter of the parasite but also serves as an essential part of its reproductive life cycle. A vector that is required as a part of a parasite's life cycle is called a **biological vector.** An important significance of a biological vector is that the pathogen can multiply to high numbers within the vector.

Prevention of vector-borne disease relies on control of arthropods. Malaria, once endemic in the continental United States, was successfully eliminated from the nation through a combination of mosquito elimination and prompt treatment of infected patients. Unfortunately, worldwide eradication efforts that initially showed great promise ultimately failed, in part due to the decreased vigilance that accompanied the dramatic but short-lived decline of the disease.

Portals of Entry

To cause disease, not only must a pathogen be transmitted from its reservoir to a new host, it must also colonize a surface of or enter that new host. Colonization is generally a prerequisite for causing disease. Cells of a *Shigella* species can be transferred via a handshake, for example, but they will only cause disease if the person then transfers them to his or her mouth or to food that is then eaten. In this case the mouth is the portal of entry. Respiratory pathogens released into the air during a cough generally cause disease only when someone inhales them. In this case the nose is the portal of entry. Many organisms that cause disease if they enter one body site are harmless if they enter another. For instance, *Enterococcus faecalis* may cause a bladder infection if it enters the urinary tract, but it is harmless when it enters the mouth and colonizes the intestine, where it frequently resides as a member of the normal microbiota.

Factors That Influence the Epidemiology of Disease

An infectious agent transmitted to a new host can potentially cause disease in that host. The outcome of transmission, however, is affected by many different factors including the dose, the incubation period, and characteristics of the host.

The Dose

The probability of infection and disease is generally lower when an individual is exposed to small numbers of a pathogen. This is because a certain minimum number of cells of the pathogen must enter the body and produce enough damage to cause disease. If

30 *Salmonella* Typhi cells are ingested in contaminated drinking water yet 1,000,000 are required to produce typhoid fever symptoms, then it will take some time for the bacterial population to increase to that number. Because host defenses are being mobilized at the same time and are racing to eliminate the invader, small doses often result in a high percentage of asymptomatic infections; the immune system eliminates the organism before symptoms appear. On the other hand, there are few if any infections for which immunity is absolute. An unusually large exposure to a pathogen, such as can occur in a laboratory accident, may produce serious disease in a person who has immunity to ordinary doses of that microbe. Therefore, even immunized persons should take precautions to minimize exposure to infectious agents. This principle is especially important for medical workers who attend patients with infectious diseases. ■ *Salmonella* Typhi, p. 601

The Incubation Period

The extent of the spread of an infectious agent is influenced by the incubation period. Diseases with typically long incubation periods can spread extensively before the first cases appear. An excellent example of this was the spread of typhoid fever from a ski resort in Switzerland in 1963. As many as 10,000 people had been exposed to drinking water containing small numbers of *Salmonella* Typhi, the causative agent of the disease. The long incubation period of the disease, 10 to 14 days, allowed widespread dissemination of the organisms by the skiers, who flew home to various parts of the world before they became ill. As a result, there were more than 430 cases of typhoid fever in at least six countries. ■ incubation period, p. 394

Population Characteristics

Certain population groups are more likely to be affected by a given disease-causing agent. Population characteristics that influence the occurrence of disease include:

- **Immunity to the pathogen.** Previous exposure or immunization of a population to a disease agent or an antigenically related agent influences the number of people who become ill. A disease is unlikely to spread very widely in a population in which 90% of the individuals are immune to the disease agent as a result of prior exposure or immunization. When an infectious agent cannot spread in a population because it lacks a critical concentration of non-immune hosts, a phenomenon called **herd immunity** results. The non-immune individuals are essentially protected by the immune individuals. Unfortunately, some infectious agents are able to undergo antigenic variation so that they can overcome herd immunity. Currently, humans have no herd immunity to the H5N1 (avian) influenza strain. ■ antigenic variation, pp. 181, 403
- **General health.** Malnutrition, overcrowding, and fatigue increase the susceptibility of people to infectious diseases and enhance the diseases' spread. Infectious diseases have generally been more of a problem in poor areas of the world where individuals are crowded together without proper food or sanitation. Factors that promote good general health result in increased resistance to diseases such as tuberculosis. When infection does occur in a healthy individual, it is more likely to be asymptomatic or to result in mild disease.
- **Age.** The very young and the elderly are generally more susceptible to infectious agents. The immune system of young children is not fully developed and, consequently, they are predisposed to certain diseases. For example, young children are particularly susceptible to meningitis caused by *Haemophilus influenzae.* The elderly are more prone to disease because immunity wanes over time. Influenza outbreaks in nursing homes often have high mortality rates. Older adults are also less likely to be current on their immunizations, making them more susceptible to diseases such as tetanus **(figure 20.4).** To prevent tetanus, a booster vaccine is needed once every 10 years. ■ meningitis, p. 650 ■ influenza, p. 519, ■ tetanus, p. 566
- **Gender.** In some cases, gender influences disease distribution. Women are more likely to develop urinary tract infections because their urethra, the tube that connects the bladder to the external environment, is relatively short. Microbes can ascend the urethra into the bladder. Pregnant women are more susceptible to listeriosis, caused by *Listeria monocytogenes.* ■ *Listeria monocytogenes*, p. 653
- **Religious and cultural practices.** The distribution of disease is also influenced by religious and cultural practices. An infant who is breast-fed is less likely to have infectious diarrhea because of the protective effects of antibodies in the mother's milk. Groups who eat traditional dishes made from raw freshwater fish are more likely to acquire tapeworm, a parasite normally killed by cooking.
- **Genetic background.** Natural immunity can vary with genetic background, but it is usually difficult to determine

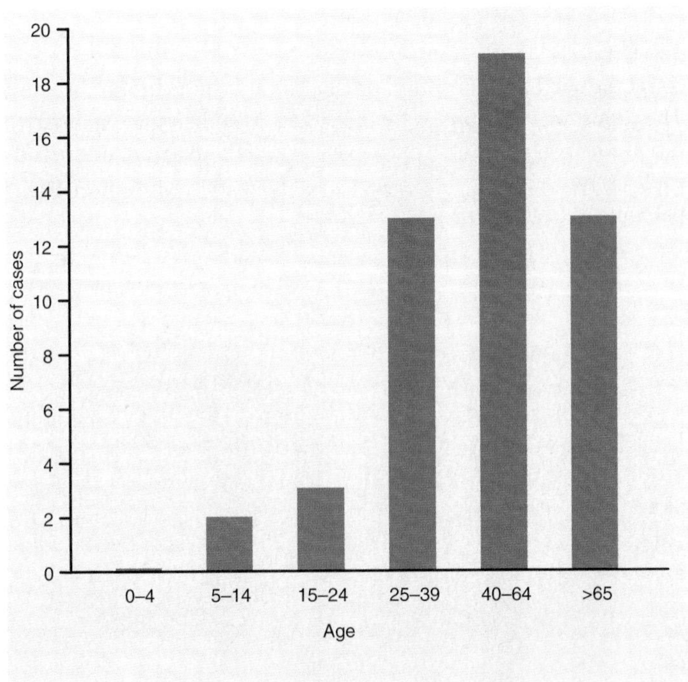

FIGURE 20.4 Typical Incidence of Tetanus by Age Group

the relative importance of genetic, cultural, and environmental factors. In a few instances, however, the genetic basis for resistance to infectious disease is known. For example, many people of black African ancestry are not susceptible to malaria caused by *Plasmodium vivax* because they lack a specific red blood cell receptor used by the organism. Some populations of Northern European ancestry are less susceptible to HIV infection because they lack a certain receptor on their white blood cells.

MICROCHECK 20.1

The reservoir of a disease agent can be infected people, other animals, or the environment. In order to spread, infectious microbes must exit a host. Handwashing and vector control can prevent many diseases; airborne transmission of pathogens is difficult to control. The portal of entry can affect the outcome of disease. The outcome of transmission is affected by the dose of the infecting agent, the incubation period, and characteristics of the host population.

- ✓ Explain why smallpox was successfully eradicated but rabies probably never will be.
- ✓ Explain how the incubation period can influence the spread of an infectious agent.
- ✓ Considering that circulating blood is not normally released from the body, describe how blood-borne microbes might exit.

20.2 Epidemiological Studies

Focus Point

- Compare and contrast descriptive studies, analytical studies, and experimental studies.

Epidemiologists investigate a disease outbreak to determine the causative agent as well as its reservoir and route of transmission so as to recommend ways to minimize the spread. The British physician John Snow illustrated the power of a well-designed epidemiological study over a century ago. Years before the relationship between microbes and disease was accepted, he documented that the cholera epidemics plaguing England from 1849 to 1854 were due to contaminated water supplies. He did this by carefully comparing the conditions of households that were affected by cholera to those that were not, eventually determining that the primary difference was their water supply. At one point, he ordered the removal of the handle of a public water pump in the neighborhood of an outbreak; this simple act helped halt an epidemic that in 10 days had killed more than 500 people. ■ cholera, p. 596

Descriptive Studies

When a disease outbreak occurs, epidemiologists conduct a **descriptive study** to define characteristics such as the person, the place, and the time. That information is used to compile a list of possible risk factors involved in the spread of disease.

The Person

Determining the profile of those who become ill is critical to defining the population at risk. Variables such as age, sex, race/ethnicity, occupation, personal habits, previous illnesses, socioeconomic class, and marital status may all yield clues about risk factors for developing the disease. For example, in the Swiss ski resort epidemic of typhoid fever mentioned earlier, cases occurred only in tourists because the local people rarely drank water, preferring wine instead.

The Place

The geographic location of disease acquisition identifies the general site of contact between the person and the infectious agent. This helps pinpoint the exact source. The location may also give clues about potential reservoirs, vectors, or geographical boundaries that might affect disease transmission. For example, malaria can only be transmitted in regions that have the appropriate mosquito vector.

The Time

The timing of the outbreak may also yield helpful clues. A rapid rise in the numbers of people who became ill suggests that they were all exposed to a single, common source of the infectious agent, such as contaminated chicken at a picnic. This is called a **common-source epidemic.** In contrast, if the numbers of ill people rise gradually, then the disease is likely contagious, with one person transmitting it to several others, who each then transmit it to several more, and so on. This is called a **propagated epidemic (figure 20.5).** The first case in such an outbreak is called the **index**

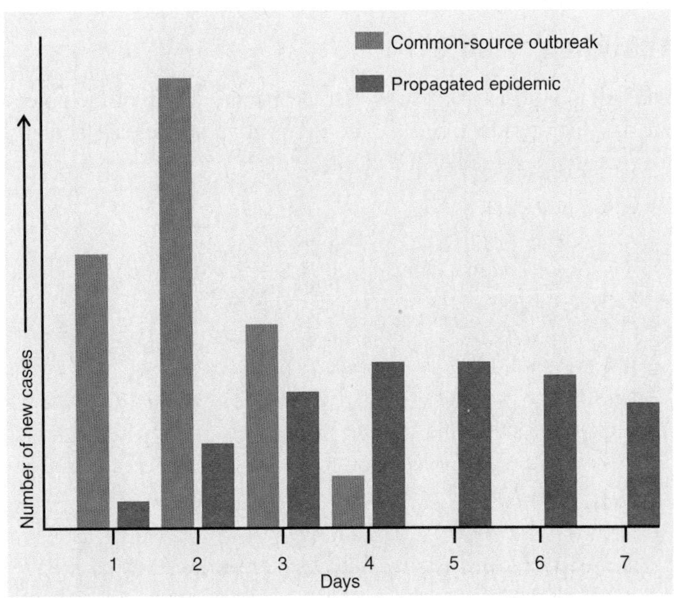

FIGURE 20.5 Comparison of Propagated Versus Common-Source Epidemics The graph depicts the number of new cases that develop over a period of days.

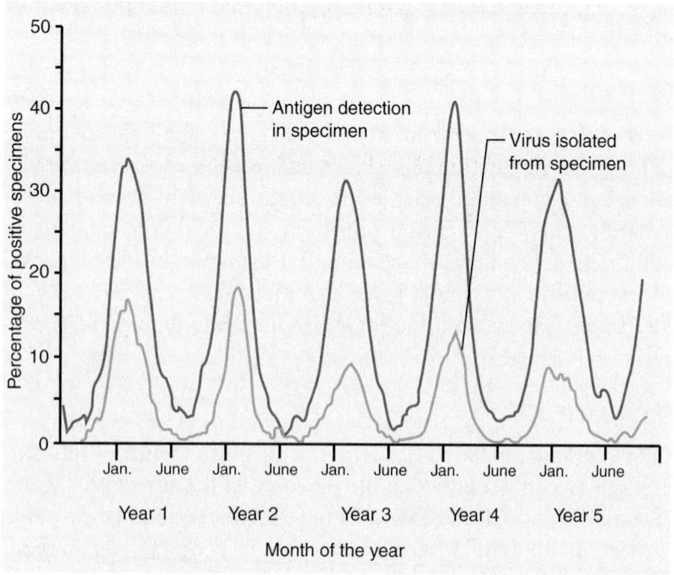

FIGURE 20.6 Typical Seasonal Occurrence of Respiratory Infections Caused by Respiratory Syncytial Virus

case. In a propagated epidemic, if a direct chain of contacts can be established, the time between the onset of symptoms in one case and the next reflects the incubation period of the disease.

The season in which the epidemic occurs may also be significant. Respiratory diseases including influenza, respiratory syncytial virus infections, and the common cold are more easily transmitted in crowded indoor conditions during the winter **(figure 20.6).** Conversely, vector and foodborne diseases are more often transmitted in warm weather when people are more likely to be exposed to mosquitoes and ticks, or eat picnic food that has not been stored properly **(figure 20.7).**

Analytical Studies

Analytical studies are designed to determine which of the potential risk factors identified by the descriptive studies are actually relevant in the spread of the disease.

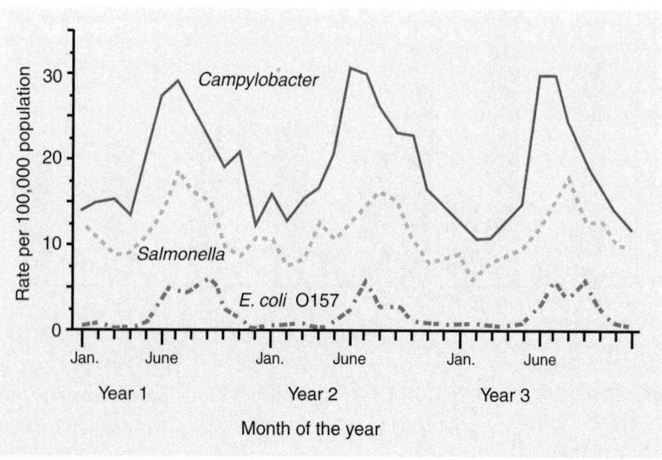

FIGURE 20.7 Seasonal Occurrence of Gastrointestinal Diseases

Cross-Sectional Studies

A **cross-sectional study** surveys a range of people to determine the prevalence of any of a number of characteristics including disease, risk factors associated with disease, or previous exposure to a disease-causing agent. This survey provides a rapid assessment of the features of a population at a given point in time and may suggest associations between risk factors and disease. The cross-sectional survey does not attempt to follow a certain group, nor does it establish cause of disease.

Retrospective Studies

A **retrospective study** is done following a disease outbreak. It compares the actions and events surrounding clinical **cases** (individuals who developed the disease) against appropriate **controls** (those who remained healthy). Thus, a **case-control study** starts by looking at the effect, which is the disease, and attempts to identify the causative chain of events. The activity or event that was common among the cases but not the controls is likely to have been a factor in the development of the disease. It is important to select controls that match the cases with respect to variables not thought to be associated with disease. Matching these variables, which might include factors such as age, sex, and socioeconomic status, ensures that all controls had equal probability of coming in contact with the disease agent.

Prospective Studies

A **prospective study** looks ahead to see if the risk factors identified by the retrospective study predict a tendency to develop the disease. **Cohort groups,** which are study groups that have a known exposure to the risk factor, are selected and then followed over time. The incidence of disease in those who were exposed to the risk factor and those who were not is then compared. By following cohort groups, the study attempts to determine if the hypothesized cause does indeed correlate with the expected effect. This type of study is less prone to the bias of inaccurate recall than a retrospective study. It is generally more time-consuming and expensive, however, particularly when examining a disease that has a long incubation period. Also, an error in the initial identification of the risk factor renders the entire study useless.

Experimental Studies

An **experimental study** is used to judge the cause-and-effect relationship of the risk factors or, more commonly, the preventative factors and the development of disease. Experimental studies are done most frequently to assess the value of a particular intervention or treatment, such as antimicrobial drug therapy. The effectiveness of the treatment is compared with one of known value or with a **placebo.** A placebo is a mock drug—it looks and tastes like the experimental drug but has no medicinal value. To assess the value of the experimental drug, a group of patients is divided into two subgroups, one of which will be given the treatment and the other an alternative or a placebo. To avoid bias, the study should ideally be **double-blind,** where neither the physicians nor the patients know who is receiving the actual treatment. Ethical issues sometimes necessitate the use of experimental animals rather than patients in experimental studies.

MICROCHECK 20.2

Descriptive epidemiological studies attempt to identify the potential risk factors that lead to disease. Analytical studies try to determine which factors are actually relevant to disease development. Experimental studies are generally used to evaluate the effectiveness of a treatment or an intervention in preventing disease.

✓ On what three factors does a descriptive study focus?

✓ What is the value of a double-blind experimental study?

✓ Why is it important to include a placebo in a scientific study to assess the effectiveness of a drug?

20.3 Infectious Disease Surveillance

Focus Point

- Compare and contrast the roles of the Centers for Disease Control and Prevention, state public health departments, and the World Health Organization.

Infectious disease surveillance, nationally and worldwide, is one of the most important aspects of disease prevention.

National Disease Surveillance Network

Infectious disease control nationwide depends heavily on a network of agencies across the country that monitors disease development. It is partly because of this network that infectious diseases do not claim more lives in the United States.

Centers for Disease Control and Prevention

The Centers for Disease Control and Prevention (CDC) is part of the U.S. Department of Health and Human Services and is located in Atlanta, Georgia. It provides support for infectious disease laboratories in the United States and abroad and collects data on diseases of public health importance. Each week, the CDC publishes the *Morbidity and Mortality Weekly Report (MMWR)*, which summarizes the status of a number of diseases. The *MMWR* is available online (http://www.cdc.gov/mmwr/), making it readily accessible to anyone in the world.

The number of new cases of over 50 **notifiable diseases** is reported to the CDC by individual states **(table 20.1).** The list of diseases considered notifiable is determined through collaborative efforts of the CDC and state health departments. Typically the diseases are of relatively high incidence or otherwise a potential danger to public health. The data collected by the CDC are published in the *MMWR* along with historical numbers to reflect any trends. Potentially significant case reports, such as the 1981 report

TABLE 20.1 Notifiable Infectious Diseases

Individual states and territories require physicians to report cases of these notifiable diseases. In turn, the number of cases is reported to the CDC, where they are collated and published in the *MMWR*.

- AIDS
- Anthrax
- Arboviral neuroinvasive and non-neuroinvasive diseases
- Botulism
- Brucellosis
- Chancroid
- *Chlamydia trachomatis,* genital infections
- Cholera
- Coccidioidomycosis
- Cryptosporidiosis
- Cyclosporiasis
- Diphtheria
- Ehrlichiosis/Anaplasmosis
- Giardiasis
- Gonorrhea
- *Haemophilus influenzae,* invasive disease
- Hansen's disease (leprosy)
- Hantavirus pulmonary syndrome
- Hemolytic uremic syndrome, post-diarrheal
- Hepatitis, viral, acute and chronic
- HIV infection
- Influenza-associated pediatric mortality
- Legionellosis (Legionnaires' disease)
- Listeriosis
- Lyme disease
- Malaria
- Measles
- Meningococcal disease
- Mumps
- Novel influenza A infections
- Pertussis
- Plague
- Poliomyelitis, paralytic
- Poliovirus infection, nonparalytic
- Psittacosis
- Q fever
- Rabies, animal and human
- Rocky Mountain spotted fever
- Rubella
- Rubella, congenital syndrome
- Salmonellosis
- Severe acute respiratory syndrome-associated coronavirus (SARS-CoV) disease
- Shiga toxin producing *Escherichia coli* (STEC)
- Shigellosis
- Smallpox
- Streptococcal disease, invasive, group A
- Streptococcal toxic shock syndrome
- *Streptococcus pneumoniae,* drug-resistant, invasive disease
- *Streptococcus pneumoniae,* invasive in children < 5 years
- Syphilis
- Syphilis, congential
- Tetanus
- Toxic shock syndrome (other than streptococcal)
- Trichinosis
- Tuberculosis
- Tularemia
- Typhoid fever
- Vancomycin-intermediate *Staphylococcus aureus* (VISA)
- Vancomycin-resistant *Staphylococcus aureus* (VRSA)
- Varicella
- Vibriosis
- Yellow fever

of a cluster of opportunistic infections in young homosexual men that heralded the AIDS epidemic, are also included in the *MMWR*. This publication is an invaluable aid to physicians, public health agencies, teachers, students, and anyone else concerned about infectious disease or public health. In fact, many of the epidemiological charts and stories in this textbook are from the *MMWR*.

The CDC also conducts research relating to infectious diseases and can dispatch teams worldwide to assist with identifying and controlling epidemics. This is important because an epidemic anywhere in the world is a potential threat. In addition, the CDC provides refresher courses that update the knowledge of laboratory and infection control personnel.

Public Health Departments

Each state has a public health laboratory responsible for infection surveillance and control as well as other health-related activities. Individual states have the authority to mandate which diseases must be reported by physicians to the state laboratory. The prompt response of health authorities in Washington State that led to the cessation of an outbreak of *Escherichia coli* O157:H7 in 1993 caused by contaminated hamburger patties was partly because Washington then was one of the few states with surveillance and reporting measures for the organism. The epidemic had actually started in other states but had gone unrecognized.

Other Components of the Public Health Network

The public health network also includes public schools, which report absentee rates, and hospital laboratories, which report on the isolation of pathogens that have epidemiological significance for the community. In conjunction with these local activities, the news media alert the general public of the presence of infectious disease.

Worldwide Disease Surveillance

The World Health Organization (WHO) is an international agency devoted to achieving the highest possible level of health for all peoples. An agency of the United Nations, the WHO has 192 member countries. It has four main functions: (1) provide worldwide guidance in the field of health; (2) set global standards for health; (3) cooperatively strengthen national health programs; and (4) develop and transfer appropriate health technology. To accomplish its goals, the WHO provides education and technical assistance to member countries.

The WHO disseminates information through a series of periodicals and books. For example, the *Weekly Epidemiological Record* reports timely information about epidemics of public health importance, particularly those of global concern.

MICROCHECK 20.3

Across the country, a network of agencies including the Centers for Disease Control and Prevention and state and local public health departments monitors disease development. The WHO is devoted to achieving the highest possible level of health for all peoples.

✓ What is the *MMWR*?

✓ Explain why we have relatively accurate data on the number of cases of measles that occur in the United States but not on the number of cases of the common cold.

20.4

Trends in Disease

Focus Points

- Name one disease that has been eradicated.
- Describe six factors that contribute to the emergence and reemergence of disease.

The rapid scientific advances made over the past several decades led some people to speculate at one point that the war against infectious diseases, particularly those caused by bacteria, had been won. Microorganisms have occupied this planet far longer than have humans, however, evolving to reside in every habitat having the potential for life, including the human body. Perhaps it should be no surprise, then, that new and previously unrecognized pathogens are emerging, and that some of those we had previously controlled are now making a comeback.

Reduction and Eradication of Disease

Humans have been enormously successful in developing the means to eliminate or reduce the occurrence of certain diseases through such efforts as improved sanitation, reservoir and vector control, vaccination, and antibiotic treatment (**figure 20.8**). One disease, smallpox, has been globally eradicated, eliminating the natural occurrence of a disease that had a 25% mortality rate and left many of those who survived permanently disfigured (**figure 20.9**). The World Health Organization hopes to eradicate polio, measles, and dracunculiasis soon. Political and social upheaval, complacency, and lack of financial support can result in a resurgence of diseases unless the pathogens are completely eliminated. ■ **polio, p. 661** ■ **measles, p. 547**

In the United States, many diseases that were once common and claimed many lives are now relatively rare, Successful vaccination programs have led to dramatic decreases in the number of

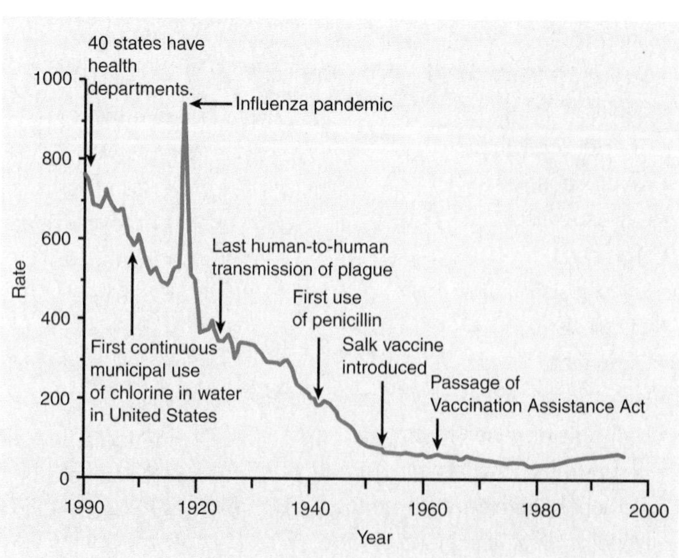

FIGURE 20.8 Crude Death Rate for Infectious Diseases, United States, Per 100,000 Population per Year

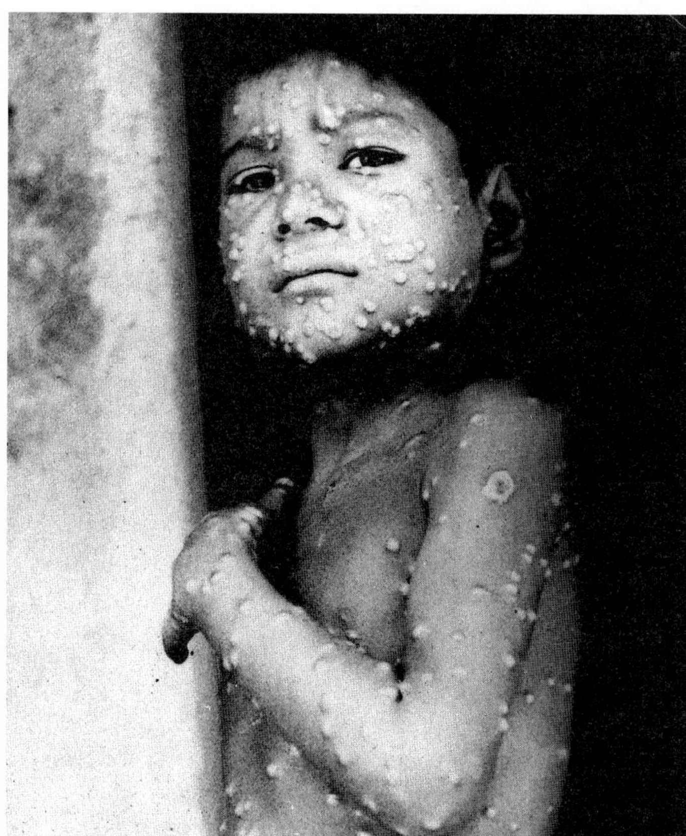

FIGURE 20.9 A Case of Smallpox, a Disease That Has Now Been Eradicated

deaths caused by *Haemophilus influenzae, Corynebacterium diphtheriae, Clostridium tetani,* and *Bordetella pertussis.* Meanwhile, recognizing and controlling the source of diseases such as malaria, plague, and cholera have been effective in limiting their spread.

Emerging Diseases

Just as humans have been successful in reducing and eliminating certain diseases, microorganisms are equally adept at taking advantage of new opportunities in which to thrive and multiply. As human lifestyles change due to advancing technologies, increasing populations, and shifting social behaviors, new diseases emerge while those that have been controlled in the past sometimes make a comeback.

Diseases that have increased in incidence in the past two decades are referred to as **emerging diseases.** These include new or newly recognized diseases such as SARS, mad cow disease, and avian flu, as well as familiar ones such as malaria, that are reemerging after years of decline (see figure 1.4). Some of the factors that contribute to the emergence and reemergence of diseases include:

- **Microbial evolution.** The emergence of some diseases is due to the natural evolution of microorganisms. For example, a new serotype of *Vibrio cholerae,* designated O139, appears to be nearly identical to the strain that most commonly causes cholera epidemics, *V. cholerae* O1, except that it has gained the ability to produce a capsule. The consequence of the new serotype is that even people who have immunity against the earlier strain are susceptible to the new one. Another example is avian influenza. Because the virus can undergo antigenic variation, scientists are concerned that it will evolve to spread from person to person. Resistance to the effects of antimicrobial drugs is contributing to the reemergence of many diseases, including malaria. ■ *V. cholerae*, p. 596 ■ malaria, p. 691 ■ avian flu, p. 520
- **Complacency and the breakdown of public health infrastructure.** As infectious diseases are controlled and therefore of lessening concern, complacency can develop, paving the way for the resurgence of a disease. The preliminary success of the plan to eliminate tuberculosis in the United States by the year 2000 resulted in less public attention being paid to the disease. News reports, education, and research money were all diverted to more common diseases. Simultaneously, the AIDS crisis developed and funding for some social welfare programs was curtailed, resulting in an increase in the number of people at risk of developing active tuberculosis due to poor health and living conditions. Consequently, tuberculosis reemerged as an increasing threat. Fortunately, increased public health measures, brought the disease back under at least temporary control (see figure 22.15). ■ tuberculosis, p. 514
- **Changes in human behavior.** Changes in society's behavior can inadvertently create opportunities for microorganisms to spread and flourish. For example, day care centers, where diapered infants mingle, oblivious to sanitation and hygiene, are a relatively new component of American society. For obvious reasons, the centers can be hotbeds of contagious diseases. Many young children have not yet acquired immunity to common communicable diseases. As a consequence, illnesses such as colds and diarrhea are readily transmitted among this susceptible population. This is particularly true for the intestinal pathogens *Giardia* and *Shigella,* which have a low infectious dose, because infants often explore through taste and touch and are thus likely to ingest fecal organisms. ■ *Giardia*, p. 609 ■ *Shigella*, p. 598
- **Advances in technology.** Technology can make life easier but can inadvertently create new habitats for microorganisms. The advent of contact lenses to correct vision gave microorganisms the opportunity to grow in a new location—the lenses and storage solutions of users who did not employ proper disinfection techniques. In turn, this resulted in new types of eye infections.
- **Population expansion.** The increase in world population and increasingly crowded living conditions create situations in which diseases can be more readily transmitted. In areas where the population has expanded outward, people are coming in contact with previously unknown reservoirs of disease such as that of the Ebola virus. Recent evidence suggests that fruit bats might be the reservoir of that virus.
- **Development.** Dams, which provide important sources of power necessary for economic development, have inadvertently extended the range of certain diseases. Transmission of the disease schistosomiasis relies on the presence of an aquatic snail that serves as a host for the *Schistosoma* parasite.

Construction of dams such as the Aswan dam on the Nile River has increased the habitat for the snail, thus extending the distribution of the disease. ■ schistosomiasis, p. 362

- **Mass production, widespread distribution, and importation of food.** Foodborne illness has always existed, but the ease with which we can now transport items worldwide can create new problems. Widespread distribution of foods contaminated with pathogens can result in a similarly broad outbreak of disease. Contaminated raspberries grown in Guatemala were linked to a 1996 multistate outbreak of diarrheal disease in the United States, affecting more than 900 people, caused by the intestinal parasite *Cyclospora cayetanensis.* ■ Cyclospora, p. 612
- **War and civil unrest.** Wars and civil unrest can disrupt the infrastructure on which disease prevention relies. Refugee camps that crowd people into substandard living quarters lacking toilet facilities and safe drinking water are hotbeds of infectious diseases such as cholera and dysentery. Unfortunately, war also disrupts disease eradication efforts. ■ cholera, p. 596 ■ dysentery, p. 595
- **Climate changes.** Changes in temperature and rainfall may affect the incidence of certain diseases. Warm temperatures favor the reproduction and survival of some arthropods, which in turn can serve as vectors for diseases such as malaria. The heavy rainfall and flooding that resulted in a surge of cholera cases in Africa may have been due to the effects of El Niño. ■ malaria, p. 691 ■ cholera, p. 596

MICROCHECK 20.4

Humans have been successful in reducing and eliminating certain diseases, but microorganisms are adept at taking advantage of new opportunities in which to thrive and multiply.

- ✓ Explain why diarrheal diseases spread so easily in day care centers.
- ✓ Describe two factors that can contribute to the reemergence of an infectious disease.
- ✓ What political and societal factors might lead to a decrease in childhood immunizations?

20.5

Healthcare-Associated Infections

Focus Points

- List four species that commonly cause healthcare-associated infections.
- Describe four reservoirs of infectious agents in healthcare settings and three mechanisms by which the agents can be transferred to patients.
- Describe the role of Infection Control Committees in preventing nosocomial infections.

Healthcare-associated infections (HAIs), infections individuals acquire while receiving treatment in a hospital or other healthcare setting, are one of the top 10 causes of death in the United States. Perhaps this should not be surprising—hospitals, in particular, can be viewed as high-population-density communities made up of unusually susceptible people where the most antimicrobial resistant and virulent pathogens can potentially circulate. In fact, hospital-acquired infections, or **nosocomial infections,** have been a problem since hospitals began (*nosocomial* is derived from the Greek word for hospital). Modern medical practices, however, including the extensive use of antimicrobial drugs and invasive therapeutic procedures, have changed the nature of the problem. In the United States alone, it is estimated that 5% to 10% of patients admitted to a hospital develop a nosocomial infection, adding over $4.5 billion to the price of health care. Many of these infections originate from the patient's own normal flora, but approximately one-third are potentially preventable. **Figure 20.10** shows the relative frequency of different types of nosocomial infections.

The severity of healthcare-associated infections can range from very mild to fatal. Sometimes, because of a long incubation period, the infection may not be discovered until after the patient has been discharged. Many factors determine which microorganisms or viruses are responsible for these infections. Examples include the length of time the person is exposed, the manner in which a patient is exposed, the virulence and number of organisms, and the state of the patient's host defenses. Characteristics of some of the microorganisms and viruses commonly implicated in healthcare-associated infections are summarized in **table 20.2.**

Reservoirs of Infectious Agents in Healthcare Settings

The organisms that cause healthcare-associated infections can originate from a number of different sources, including other patients, the healthcare environment, medical personnel, and the

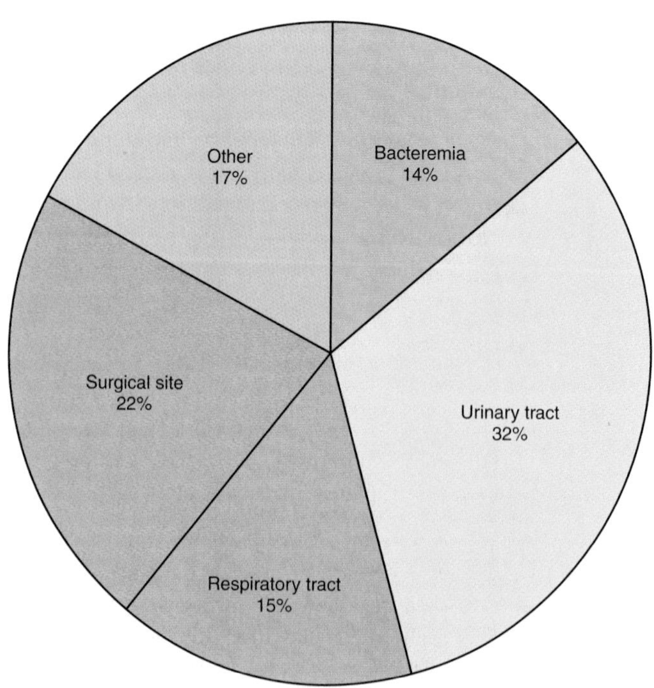

FIGURE 20.10 Relative Frequency of Different Types of Nosocomial Infections

TABLE 20.2 Common Causes of Healthcare-Associated Infections

Infectious Agent	Comments
Acinetobacter baumannii	This environmental bacterium can be found on skin of healthy people and is resistant to a number of antimicrobial medications. It causes a variety of healthcare-associated infections including bloodstream and surgical site infections and pneumonia.
Candida species	These yeasts, part of the normal microbiota, are a common cause of healthcare-associated bloodstream infections. Some are resistant to a number of antifungal medications. ■ *Candida albicans*, p. 555
Clostridium difficile	Toxin-producing strains of this bacterium can cause diarrhea and colitis in people taking antibiotics. Because the bacterium produces endospores, which are not killed by disinfectants, thorough handwashing is an important means of preventing transmission. ■ antibiotic-associated diarrhea, p. 586
Enterococcus species	These bacteria, part of the normal intestinal microbiota, are a common cause of nosocomial urinary tract infections as well as wound and bloodstream infections. Some strains are resistant to all conventional antimicrobial drugs. ■ enterococci, p. 255
Escherichia coli and other members of the *Enterobacteriaceae*	Some of these members of the normal intestinal microbiota (particularly *E. coli*) commonly cause healthcare-associated urinary tract infections. ■ *E. coli*, p. 621 ■ *Enterobacteriaceae*, p. 262
Norovirus	This virus infects the gastrointestinal tract, causing vomiting and diarrhea. It has a very low infectious dose, so scrupulous cleaning and handwashing are necessary to prevent its transmission. ■ norovirus, p. 604
Pseudomonas species	These bacteria grow readily in many moist, nutrient-poor environments such as the water in the humidifier of a mechanical ventilator. *Pseudomonas* species are resistant to many disinfectants and antimicrobial drugs. They are a common cause of healthcare-associated pneumonia, and infections of the urinary tract and burn wounds. ■ *Pseudomonas*, p. 261
Respiratory syncytial virus (RSV)	This virus is easily transmitted, and causes outbreaks of healthcare-associated lower respiratory tract infections, particularly at times when the virus is circulating in non-healthcare settings. ■ respiratory syncytial virus, p. 521
Staphylococcus aureus	Many people, including healthcare personnel, are carriers of this bacterium. Because it survives for prolonged periods in the environment, it is readily transmissible on fomites. It is a common cause of healthcare-associated pneumonia and surgical site infections. Hospital-acquired strains are often resistant to a variety of antimicrobial drugs. ■ *Staphylococcus aureus*, p. 535
Staphylococcus species other than *S. aureus*	These members of the normal skin microbiota can colonize the tips of intravenous catheters (small plastic tubes inserted into the veins). The resulting biofilms continuously seed organisms into the bloodstream and increase the likelihood of a systemic infection. ■ biofilm, p. 85

patient's own microbiota. Because of the widespread use of antimicrobial drugs in hospitals, many organisms that cause nosocomial infections are resistant to these medications.

Other Patients

Because of the very nature of healthcare settings, infectious agents are always present. Patients are often hospitalized because they have a severe infectious disease. The pathogens that these patients harbor can be discharged into the environment via skin cells, respiratory droplets, and other body secretions and excretions. Scrupulous cleaning and the use of disinfectants minimize the spread of these pathogens.

Healthcare Environment

Some Gram-negative rods, particularly the common opportunistic pathogen *Pseudomonas aeruginosa,* can thrive in healthcare environments such as sinks, respirators, and toilets. Not only is *P. aeruginosa* resistant to the effects of many disinfectants and antimicrobial drugs, it requires few nutrients, enabling it to multiply in environments containing little other than water. Many nosocomial infections have been traced to soaps, disinfectants, and other aqueous solutions that have become contaminated with the organism.

Healthcare Workers

Outbreaks of healthcare-associated infections are sometimes traced to infected personnel. Clearly, those who report to work with even a mild case of influenza can expose patients to an infectious agent that has serious or fatal consequences to those with impaired health. A more troublesome source of infection is a healthcare worker who is a carrier of a pathogen such as *Staphylococcus aureus* or *Streptococcus pyogenes.* These personnel often do not recognize they pose a risk to patients until they have been implicated in an outbreak. Carriers who are members of a surgical team pose a particular threat to patients, because inoculation of a pathogen directly into a surgical site can result in a systemic infection.

Patient's Own Microbiota

Many healthcare-associated infections originate from the patient's own microbiota. Nearly any invasive procedure can transmit these organisms to otherwise sterile body sites. When intravenous fluids are administered, for example, *Staphylococcus epidermidis,* a common member of the normal skin microbiota, can potentially gain access to the bloodstream. While the immune system can usually readily eliminate these normally benign organisms, the underlying illness of many hospitalized patients compromises their immunity and they can develop a bloodstream infection. Patients who undergo intestinal surgery are prone to surgical site infection by their normal bowel microbiota. Similarly, patients who are on certain medications or have impaired cough reflexes can inadvertently inhale their normal oral microbiota, resulting in healthcare-associated pneumonia.

Severely immunocompromised patients, such as people who have undergone cancer chemotherapy or are on immunosuppressive drugs, are prone to activation of latent infections their immune system was previously able to control. For example, latent infections of *Toxoplasma gondii,* a protozoan parasite commonly acquired during childhood, can become activated and cause a life-threatening disease. ■ latent infection, p. 395

Transmission of Infectious Agents in Healthcare Settings

Diagnostic and therapeutic procedures can potentially transmit infectious agents to patients. This is particularly true in intensive care units (ICUs), where patients generally have indwelling catheters used to deliver intravenous fluids or monitor the patient's condition **(figure 20.11).**

Medical Devices

Healthcare-associated infections most often result from medical devices that breach the first-line barriers of the normal host defense. Catheterization of the urinary tract can readily introduce microorganisms into the normally sterile bladder. Because urine is an excellent growth medium, the urinary tract often becomes infected. Urinary tract infections are the most common type of nosocomial infection (see figure 20.10).

Just as urinary catheters can introduce bacteria into the bladder, intravenous (IV) catheters can introduce microorganisms into the bloodstream. This can happen when normal skin microbiota colonize the tip of an indwelling catheter or when environmental organisms contaminate IV fluids or the lines that deliver them. Even normally benign skin microbiota can cause life-threatening bacteria when they gain access to the bloodstream.

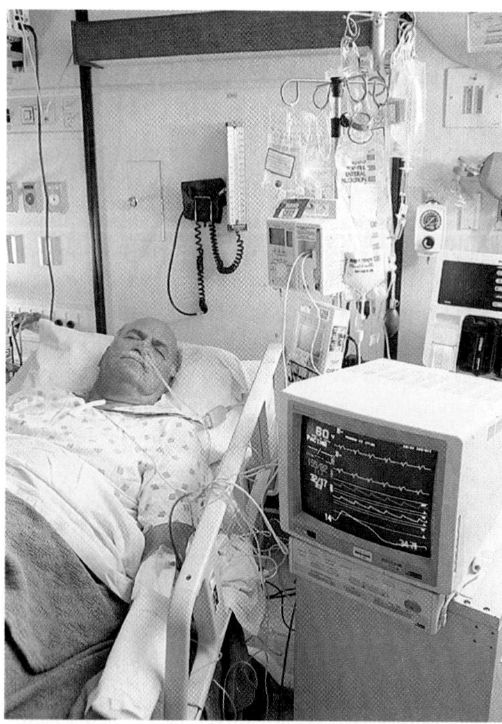

FIGURE 20.11 Patient in an Intensive Care Unit

Mechanical respirators that assist a patient's breathing by pumping air directly into the trachea can potentially deliver microorganisms to the lungs. This is particularly a problem if a nutritionally versatile organism such as *Pseudomonas* can gain access to water droplets in the machine, allowing it to multiply.

Inadequately sterilized instruments that are used in invasive procedures such as surgery or biopsy can also transmit infectious agents. Endoscopes and other heat-sensitive instruments are often treated with chemical sterilants to render them microbe-free. Improper use of these chemicals, however, can result in the survival of some organisms. ■ sterilants, p. 117

Healthcare Personnel

Healthcare personnel must be extremely vigilant to avoid transmitting infectious disease agents, particularly from patient to patient. What Ignaz Semmelweis found to be true in the 1800s is equally true today—handwashing between contact with individual patients helps prevent the spread of disease (see **A Glimpse of History**). Unfortunately, this relatively simple procedure is too often overlooked.

Healthcare personnel should routinely wash or disinfect their hands after touching one patient before going on to the next. A more thorough hand scrubbing, lasting 10 minutes and using a strong disinfectant, should be performed by nurses, physicians, and other personnel before participating in an operation, or when working in a nursery, or an intensive care or isolation unit. Healthcare workers wear gloves whenever they have contact with blood, mucous membranes, broken skin, or body fluids.

Airborne

Most hospitals are designed to minimize the airborne spread of microorganisms. Airflow to operating rooms is usually regulated so that it is supplied under slight pressure, thereby preventing contaminated air in the corridors from flowing into the room. Floors are washed with a damp mop or floor washer rather than swept in order to avoid resuspending microbes into the air. To exclude airborne microorganisms and viruses from rooms in which exquisitely susceptible patients reside, such as those who have recently undergone a bone marrow transplant, high-efficiency particulate air filters (HEPA) are employed. These filter out most airborne particles, including microorganisms. ■ HEPA filters, p. 115

Preventing Healthcare-Associated Infections

The most important steps in preventing healthcare-associated infections are to first detect their occurrence and then establish policies to prevent their development. To accomplish this, nearly every hospital has an Infection Control Committee, composed of representatives of the various professionals in the hospital, including nurses, physicians, dietitians, housekeeping staff, and microbiology laboratory personnel. On this committee, and sometimes chairing it, is often a **hospital epidemiologist,** a professional specially trained in hospital infection control. Hospitals may employ an **infection control practitioner (ICP),** whose

PERSPECTIVE 20.1

Standard Precautions—Protecting Patients and Healthcare Personnel

One of the biggest challenges for a hospital has always been the prevention of spread of disease within that confined setting. A century ago, patients with infectious diseases were segregated in separate hospitals; those with similar diseases were sometimes housed in clusters on the same floor. In 1910, a cubicle system of isolation was introduced in which patients were placed in multiple-bed wards. Aseptic nursing procedures were aimed at preventing the transmission of infectious agents to other patients and personnel. These scrupulous measures were so successful that general hospitals were able to accommodate infectious disease patients, ultimately resulting in the closure of many infectious disease hospitals beginning in the 1950s. To assist general hospitals with isolation procedures, in 1970 the CDC began publishing a manual that recommended a category system of seven isolation procedures for patients based on their diagnosis. These procedures included Strict Isolation, Respiratory Precautions, Protective Isolation, Enteric Precautions, Wound and Skin Precautions, Discharge Precautions, and Blood Precautions. Over the years, some of these categories were changed or deleted.

In the early 1980s it became increasingly apparent that healthcare workers were acquiring hepatitis B and, later, HIV from contact with the blood or other body fluids of patients, including those who were not suspected of having blood-borne disease. Existing guidelines were primarily aimed at preventing patient-to-patient transmission of disease, rather than patient-to-personnel. In response, the CDC recommended an additional set of guidelines, the **Universal (Blood and Body Fluid) Precautions,** to be followed when working with any patient, regardless of the diagnosis. These defined the situations in which gloves, gowns, masks, and eye protection were required to prevent contact with blood. Many hospitals then broadened this concept, requiring the use of gloves to isolate all moist and potentially infectious body substances. This approach, **Body Substance Isolation,** made the traditional diagnosis-dependent isolation procedures largely unnecessary.

The advent of Universal Precautions and Body Substance Isolation, in addition to the previous diagnosis-dependent isolation procedures, resulted in a such a mix of recommendations that it generated a great deal of confusion. No existing, single set of guidelines was sufficient, and it was not clear when each should be used. In response, the CDC and the Hospital Infection Control Practices Advisory Committee (HICPAC) established a new set of guidelines in 1996 that incorporated the strengths of each of the alternatives. The guidelines were updated and expanded in 2007 to include recommendations that can be applied to all healthcare facilities. The advisory committee has also been renamed, substituting the word "healthcare" for "hospital," to reflect the expanded scope of the infection control recommendations.

The current guidelines have two tiers of isolation procedures: The fundamental measures are the **Standard Precautions,** designed for the care of all patients. The **Transmission-Based Precautions** are supplementary measures to be used in addition to the Standard Precautions if a patient is, or might be, infected with a highly transmissible or epidemiologically important pathogen. These are separated into three sets—Airborne Precautions, Droplet Precautions, and Contact Precautions—that are used singly or in combination as appropriate.

The Standard Precautions can be summarized as:

Hand hygiene. During the delivery of healthcare, avoid unnecessary touching of surfaces in close proximity to the patient. When hands are visibly dirty or contaminated, wash them with soap and water; if non-antimicrobial soap is used, decontaminate hands with an alcohol-based hand rub. If hands are not visibly soiled, decontaminate hands before direct contact with patients; after contact with blood, body fluids, secretions, excretions, contaminated items; immediately after removing gloves; and between patient contacts. If contact with spores is likely to have occurred, wash with soap and water.

Personal protective equipment. Gloves are worn for touching blood, body fluids, secretions, excretions, and contaminated items; and for touching mucous membranes and non-intact skin. A gown is worn during procedures and patient-care activities when contact of clothing/exposed skin with blood/body fluids, secretions, and excretions is anticipated. Mask, goggles, or a face shield is worn during procedures and patient-care activities likely to generate splashes or sprays of blood, body fluids, or secretions.

Respiratory hygiene/cough etiquette. Instruct symptomatic persons to cover mouth/nose when sneezing/coughing; use tissues and dispose in a no-touch receptacle; observe hand hygiene after soiling of hands with respiratory secretions; wear surgical mask if tolerated or maintain spatial separation >3 feet if possible.

Patient placement. Prioritize for single-patient room if patient is at increased risk of transmission, is likely to contaminate the environment, does not maintain appropriate hygiene, or is at increased risk of acquiring infection or developing adverse outcome following infection.

Patient-care equipment and instruments/devices. If the equipment is soiled, handle in a manner that prevents transfer of microorganisms to others and to the environment; wear gloves if visibly contaminated; perform hand hygiene.

Care of the environment. Develop procedures for routine care, cleaning, and disinfection of environmental surfaces, especially frequently touched surfaces in patient-care areas.

Textiles and laundry. Handle in a manner that prevents transfer of microorganisms to others and to the environment.

Safe injection practices. Use aseptic technique to avoid contamination of sterile injection equipment. Specific precautions describe how medications and IV solutions are stored and administered.

Infection control practices for special lumbar puncture procedures. Wear a surgical mask when placing a catheter or injection material into the spinal canal or subdural space.

Worker safety. Adhere to federal and state requirements for protection of healthcare personnel from exposure to bloodborne pathogens.

From Siegel, J.D., Rhinehart, E., Jackson, M., Chiarello, L., and the Healthcare Infection Control Practices Advisory Committee, *2007 Guidelines for Isolation Precautions: Preventing Transmission of Infectious Agents in Healthcare Settings,* June 2007. http://www.cdc.gov/ncidod/dhqp/pdf/isolation2007.pdf

role is to perform active surveillance of the types and numbers of infections that arise in the hospital. The Infection Control Committee, in conjunction with the ICP, drafts and implements preventative policies following the guidelines suggested by the Standard Precautions and the Transmission-Based Precautions (see **Perspective 20.1**).

The CDC also takes an active role in preventing healthcare-associated infections, and has established the Healthcare Infection Control Practices Advisory Committee (HICPAC). The role of this national committee is to provide advice to hospitals and recommend guidelines for surveillance, prevention, and control of healthcare-associated infections.

MICROCHECK 20.5

Healthcare-associated infections may originate from other patients, the healthcare environment, medical personnel, or the patient's own normal flora. Diagnostic and therapeutic procedures can potentially transmit infectious agents. The most important steps in preventing healthcare-associated infections are to first detect their occurrence and then establish policies to prevent their development.

- ✓ Explain why an IV catheter poses a risk to a patient.
- ✓ Describe two ways in which infectious agents can be transmitted to a patient.
- ✓ The rate of nosocomial infections is often relatively high in emergency room settings. Explain why this might be so.

FUTURE CHALLENGES

Maintaining Vigilance Against Bioterrorism

Today, an unfortunate challenge in epidemiology is to maintain vigilance against **bioterrorism**—the deliberate release of infectious agents or their toxins as a means to cause harm. Even as we work to control, and seek to eradicate, some diseases, we must be aware that microbes pose a threat as agents of bioterrorism. Hopefully, future attacks will never occur, but it is crucial to be prepared for the possibility. Prompt recognition of such an event, followed by rapid and appropriate isolation and treatment procedures, can help to minimize the consequences. The CDC, in cooperation with the Association for Professionals in Infection Control and Epidemiology (APIC), has prepared a bioterrorism readiness plan to be used as a template by healthcare facilities. Many of the recommendations are based on the Standard Precautions already employed by hospitals to prevent the spread of infectious agents (see Perspective 20.1).

The CDC separates bioterrorism agents into three categories based on the ease of spread and severity of disease. **Category A agents** pose the highest risk because they are easily spread or transmitted from person to person and result in high mortality. These agents include:

- ***Bacillus anthracis.*** Endospores of this bacterium were used in the bioterrorism events of 2001. The most severe outcome, **inhalational anthrax,** results when an individual breathes in the airborne spores. It can lead to a rapidly fatal systemic illness. **Cutaneous anthrax,** which occurs when the organism enters the skin, manifests as a blister that develops into a skin ulcer with a black center. Although this usually heals without treatment, it can also progress to a fatal bloodstream infection. **Gastrointestinal anthrax** results from consuming contaminated food, leading to vomiting of blood and severe diarrhea; it is not common but has a high mortality rate. Anthrax can be prevented by vaccination, but that option is not widely available. Prophylaxis with antimicrobial medications is possible for those who might have been exposed, but this requires prompt recognition of exposure. Fortunately, person-to-person transmission of the agent is not likely.
- **Botulism.** Botulism is caused naturally by the ingestion of botulinum toxin, produced by *Clostridium botulinum.* Any mucous membrane can absorb the toxin, so aerosolized toxin could be used as a weapon. Botulism can be prevented by vaccination, but that option is not widely available. An antitoxin is also available in limited supplies. Botulism is not contagious.
- ***Yersinia pestis.*** Pneumonic plague, caused by inhalation of *Yersinia pestis,* is the most likely form of plague to result from a biological weapon. Although no effective vaccine is available, post-exposure prophylaxis with antimicrobial medications is possible. Special isolation precautions must be used for patients who have pneumonic plague because the disease is easily transmitted by respiratory droplets.
- **Smallpox.** Although a vaccine is available to prevent infection with this virus, routine immunization was stopped over 30 years ago because the natural disease has been eradicated. As is the case with nearly all infections caused by viruses, effective drug therapy is not available. Special isolation precautions must be used for smallpox patients because the virus can be acquired through droplet, airborne, or contact transmission.
- ***Francisella tularensis.*** This bacterium, naturally found in animals such as rodents and rabbits, causes the disease tularemia. Inhalation of the bacterium results in severe pneumonia, which is incapacitating but would probably have a lower mortality rate than inhalational anthrax or plague. A vaccine is not available, but post-exposure prophylaxis with antimicrobial medications is possible. Fortunately, person-to-person transmission of the agent is not likely.
- **Viruses that cause hemorrhagic fevers.** These include various viruses such as Ebola and Marburg. Symptoms vary depending on the virus, but severe cases show signs of bleeding from many sites. There are no vaccines against these viruses, and generally no treatment. Some, but not all, of these viruses can be transmitted from person to person, so patient isolation in these cases is important.

Category B agents pose moderate risk because they are relatively easy to spread and cause moderate morbidity. These agents include organisms that cause food- and waterborne illness, various biological toxins, *Brucella* species, *Burkholderia mallei* and *pseudomallei, Coxiella burnetii,* and *Chlamydophila* (*Chlamydia*) *psittaci.* **Category C agents** are emerging pathogens that could be engineered for easy dissemination. These include Nipah virus, which was first recognized in 1999, and hantavirus, first recognized in 1993.

SUMMARY

20.1 Principles of Epidemiology

Epidemiologists study the frequency and distribution of disease in order to identify its cause, source, and route of transmission.

Rate of Disease in a Population

Epidemiologists focus on the rate of disease. Diseases that are constantly present in a population are **endemic;** an unusually large number of cases in a population constitutes an **epidemic** (figure 20.2).

Reservoirs of Infection

Preventing susceptible people from coming in contact with a **reservoir of infection** can prevent infectious disease. People who have asymptomatic infections or are colonized with a pathogen are **carriers** of the infectious agent. **Zoonotic** diseases are those such as plague and rabies that can be transmitted to humans but exist primarily in other animals. Pathogens that have environmental reservoirs are probably impossible to eliminate.

Portals of Exit

Pathogens may be shed in feces, in respiratory droplets, on skin cells, in genital secretions, and in urine.

Transmission

Handwashing is a key control measure in preventing diseases that are spread through direct or indirect contact, as well as those that spread via contaminated food. **Direct contact** occurs when one person physically touches another. **Indirect contact** involves transfer of pathogens via **fomites. Droplet transmission** of respiratory pathogens is considered contact because of the close proximity involved. Foodborne pathogens can originate from the animal reservoir or from contamination during food preparation. Waterborne pathogens often originate from sewage contamination. Airborne transmission of pathogens is the most difficult to control. Prevention of vector-borne disease relies on mosquito, tick, and insect control.

Portals of Entry

The portal of entry of a pathogen can affect the outcome of disease.

Factors That Influence the Epidemiology of Disease

The probability of infection and disease is generally lower if an individual is exposed to small numbers of pathogens. Diseases with a long incubation period can spread extensively before the first cases appear. A disease is unlikely to spread very widely in a population in which 90% of the people are immune to the disease agent. Malnutrition, overcrowding, and fatigue increase the susceptibility of people to infectious diseases. The very young and the elderly are generally more susceptible to infectious agents (figure 20.4). Natural immunity can vary with genetic background, but it is difficult to determine the relative importance of genetic, cultural, and environmental factors.

20.2 Epidemiological Studies

Descriptive Studies

Descriptive studies attempt to identify potential risk factors that correlate with the development of disease. Determining the time that the illness occurred helps distinguish a **common-source epidemic** from **a propagated epidemic** (figure 20.5).

Analytical Studies

Analytical studies try to determine which risk factors are actually relevant to disease development. A **retrospective study** compares the activities of **cases** with **controls** to determine the cause of the epidemic. A **prospective study** compares **cohort groups,** to determine if the identified risk factors predict a tendency to develop disease.

Experimental Studies

Experimental studies are generally used to evaluate the effectiveness of a treatment or intervention in preventing disease.

20.3 Infectious Disease Surveillance

National Disease Surveillance Network

The Centers for Disease Control and Prevention (CDC) collects data on diseases of public health importance and summarizes their status in the ***Morbidity and Mortality Weekly Report*** *(MMWR);* other activities of the CDC include research, assistance in controlling epidemics, and support for infectious disease laboratories. State public health departments are involved in infection surveillance and control.

Worldwide Disease Surveillance

The World Health Organization (WHO) is devoted to achieving the highest possible level of health for all peoples.

20.4 Trends in Disease

Reduction and Eradication of Disease

Smallpox has been eradicated. The WHO hopes to soon eliminate polio, measles, and dracunculiasis.

Emerging Diseases

Emerging diseases include those that are new or newly recognized and familiar ones that are reemerging after years of decline. Factors that contribute to the emergence and reemergence of diseases include microbial evolution, the breakdown of public health infrastructure, changes in human behavior, advances in technology, population expansion, economic development, mass distribution and importation of food, war, and climate changes.

20.5 Healthcare-Associated Infections

Reservoirs of Infectious Agents in Healthcare Settings

The organisms that cause healthcare-associated infections may originate from other patients, the healthcare environment, healthcare personnel, or the patient's own flora.

Transmission of Infectious Agents in Healthcare Settings

Healthcare-associated infections can result from medical devices that breach the first-line barriers of the normal host defense. Health care personnel should routinely wash or disinfect their hands after touching one patient before going on to the next in order to prevent transmission of disease-causing organisms.

Preventing Healthcare-Associated Infections

The most important steps in preventing healthcare-associated infections are to first recognize their occurrence and then establish policies to prevent both their development and spread.

REVIEW QUESTIONS

Short Answer

1. Explain the difference between incidence and prevalence.
2. What is the epidemiological significance of people who have asymptomatic infections?
3. Explain why zoonotic diseases are often severe in humans.
4. Name the most important control measure for preventing person-to-person transmission of a disease.
5. Explain why *Shigella* and *Giardia* species are readily transmitted in day care centers.
6. Explain how smallpox was eradicated.
7. Describe three factors that contribute to the emergence of disease.
8. Draw a representative graph (time versus number of people ill) depicting both a propagated and common-source epidemic.
9. What are three important factors that a descriptive epidemiological study attempts to determine?
10. Describe the difference between a retrospective (case-control) study and a prospective (cohort) study.

Multiple Choice

1. Which of the following is an example of a fomite?
   a) Table b) Flea
   c) *Staphylococcus aureus* carrier
   d) Water e) Air
2. Which of the following would be the easiest to eradicate?
   a) A pathogen that is common in wild animals but sometimes infects humans
   b) A disease that occurs exclusively in humans, always resulting in obvious symptoms
   c) A mild disease of humans that often results in no obvious symptoms
   d) A pathogen found in marine sediments
   e) A pathogen that readily infects both wild animals and humans
3. Which of the following methods of disease transmission is the most difficult to control?
   a) Airborne b) Foodborne c) Waterborne
   d) Vector-borne e) Direct person to person

4. Which of the following statements is *false?*
   a) A botulism epidemic that results from improperly canned green beans is an example of a common-source outbreak.
   b) Droplet nuclei fall quickly to the gound.
   c) Congenital syphilis is an example of a disease acquired through vertical transmission.
   d) Plague is endemic in the prairie dog population in parts of the United States.
   e) The first case in an outbreak is called the index case.
5. Which of the following statements is *false?*
   a) A disease with a long incubation period might spread extensively before an epidemic is recognized.
   b) A person exposed to a low dose of a pathogen might not develop disease.
   c) The young and the aged are more likely to develop certain disease.
   d) Malnourished populations are more likely to develop certain diseases.
   e) Herd immunity occurs when a population does not engage in a given behavior, such as eating raw fish, that would otherwise increase their risk of disease.
6. The purpose of an analytical study is to
   a) identify the person, place, and time of an outbreak.
   b) identify risk factors that result in high frequencies of disease.
   c) assess the effectiveness of preventative measures.
   d) determine the effectiveness of a placebo.
   e) None of the above
7. Which of the following causes of emerging diseases is thought to be a new pathogen?
   a) *Giardia* b) *Vibrio cholerae* O139
   c) *Mycobacterium tuberculosis* d) *Shigella dysenteriae*
   e) *Schistosoma*
8. All of the following are thought to contribute to the emergence of disease, *except*
   a) advances in technology.
   b) breakdown of public health infrastructure.
   c) construction of dams.
   d) mass distribution and importation of food.
   e) widespread vaccination programs.
9. Which of the following common causes of healthcare-associated infections is an environmental organism that grows readily in nutrient-poor solutions?
   a) *Enterococcus* b) *Escherichia coli*
   c) *Pseudomonas aeruginosa* d) *Staphylococcus aureus*
10. What is the most common type of nosocomial infection?
    a) Bloodstream infection b) Gastrointestinal infection
    c) Pneumonia d) Surgical wound infection
    e) Urinary tract infection

Applications

1. A news station reported about a potentially fatal epidemic disease occurring in a small Laotian village. An epidemiologist from the CDC was interviewed to discuss the disease and was very distressed that it was not being contained. Why did the epidemiologist feel the disease was a concern for people in North America?
2. An international team was gathered to discuss how funding should be spent to eliminate human infectious disease. There is only enough funding to eliminate one disease. How would the scientists go about choosing the next disease to be eliminated from the planet?

Critical Thinking

1. *Yersinia pestis* and hantavirus are both found in wild rodents in the southwestern United States. What is the risk of trying to stop a hantavirus epidemic by destroying rodents in that region?
2. A student disagreed with the presentation of the examples in figure 20.5. She claimed that the number of cases from a common-source outbreak could remain high over a much longer period of time in some cases and not decrease to zero. Is the student's claim reasonable? Why or why not?

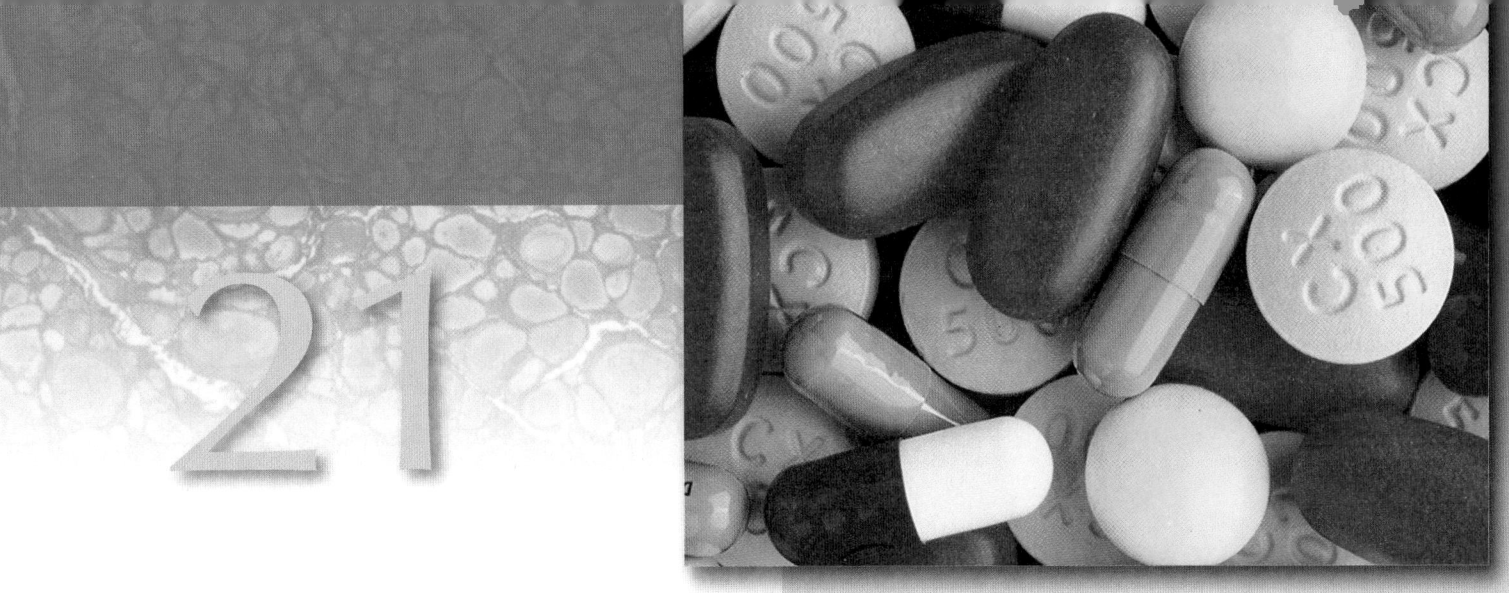

A few of the many important antimicrobial medications.

21 Antimicrobial Medications

A Glimpse of History

Paul Ehrlich (1854–1915), a German physician and bacteriologist, was born into a wealthy family. As the only son after many daughters, family and servants indulged his interests, even though his large collection of frogs and snakes occasionally entered the laundry room. As an adult, he was rarely without a good cigar and habitually scribbled notes on his shirt cuffs. After receiving a degree in medicine in 1878, he became intrigued with the way various types of body cells differ in their ability to take up dyes and other substances. When he observed that certain dyes stain bacterial cells but not animal cells, indicating that the two cell types are somehow fundamentally different, it occurred to him that it might be possible to find a chemical that selectively harms bacteria without affecting human cells.

Ehrlich began a systematic search attempting to find a "magic bullet," a term he used to describe a drug that would kill a microbial pathogen without harming the human host. He began by looking for a chemical that would cure the sexually transmitted disease syphilis, which is caused by the bacterium *Treponema pallidum.* Much of the mental illness during this time resulted from tertiary syphilis, a late stage of the disease. Ehrlich knew that an arsenic compound had shown some success in treating a protozoan disease of animals, and so he and his colleagues began tediously synthesizing hundreds of different arsenic compounds in search for a cure for syphilis. In 1910, the 606th compound tested, arsphenamine, proved to be highly effective in treating the disease in laboratory animals. Although the drug itself was potentially lethal for patients, it did cure infections that were previously considered hopeless. The drug was given the name Salvarsan, a term derived from the words salvation and arsenic. The use of Salvarsan to cure syphilis proved that chemicals could selectively kill pathogens without permanently harming the human host.

Think back to the last time you were prescribed an antimicrobial medication. Could you have recovered from the infection without the drug? The prognosis for people with common diseases such as bacterial pneumonia and severe staphylococcal infection was grim before the discovery and widespread availability of penicillin in the 1940s. Physicians were able to identify the cause of the disease, but were generally unable to recommend treatments other than bed rest. Today, however, antimicrobials are routinely prescribed, and the simple cure they provide for so many infectious diseases is often taken for granted. Unfortunately, the misuse of these life-saving medications, coupled with bacteria's amazing ability to adapt, has led to an increase in the number of drug-resistant organisms. Some people even speculate that we are in danger of seeing an end to the era of antimicrobial medications. In response, scientists are scrambling to develop new drugs.

21.1 History and Development of Antimicrobial Drugs

Focus Points

- Describe the discovery of antimicrobial drugs and antibiotics.
- Explain how new generations of antimicrobial drugs are developed.

To appreciate the unique antibiotic era in which we now live, it is important to understand the history and development of these life-saving remedies.

Discovery of Antimicrobial Drugs

The development of Salvarsan by Paul Ehrlich was the first documented example of a chemical used successfully as an antimicrobial medication. The next breakthrough in the developing science of antimicrobial chemotherapy came almost 25 years later.

KEY TERMS

Acquired Resistance Resistance that develops through mutation or acquisition of new genes.

Antibiotic A compound naturally produced by molds or bacteria that inhibits the growth of or kills other microorganisms.

Antimicrobial Drug A chemical that inhibits the growth of or kills microorganisms; the term encompasses antibiotics and chemically synthesized drugs.

Antiviral Drug A drug that interferes with the replication of viruses.

Bactericidal Drug An antimicrobial drug that kills bacteria.

Bacteriostatic Drug An antimicrobial drug that inhibits the growth of bacteria.

Broad-Spectrum Antimicrobial An antimicrobial drug that is effective against a wide range of microorganisms, often including both Gram-positive and Gram-negative bacteria.

Chemotherapeutic Agent A chemical used to treat disease.

Innate (Intrinsic) Resistance Resistance due to inherent characteristics of the organism.

Narrow-Spectrum Antimicrobial An antimicrobial drug that is effective against a limited range of microorganisms.

R Plasmid A plasmid that encodes resistance to one or more antimicrobial drugs.

In 1932, the German chemist Gerhard Domagk, using the same dogged persistence demonstrated by Ehrlich, discovered that a red dye called Prontosil was dramatically effective in treating streptococcal infections in animals. Surprisingly, Prontosil had no effect on streptococci growing in test tubes. It was later discovered that enzymes in the blood of the animal split the Prontosil molecule, producing a smaller molecule called **sulfanilamide;** this breakdown product acted against the infecting streptococci. Thus, the discovery of sulfanilamide, the first **sulfa drug,** was based on luck as well as scientific effort. If Prontosil had only been screened against bacteria in test tubes and not given to infected animals, its effectiveness might never have been discovered.

Salvarsan and Prontosil were the first documented examples of chemicals used successfully as antimicrobial medications. Any chemical that is used to treat a disease is called a **chemotherapeutic agent.** One used to treat microbial infections can also be called an **antimicrobial drug** or, more simply, an **antimicrobial.**

Discovery of Antibiotics

In 1928, Alexander Fleming, a British scientist, was working with cultures of *Staphylococcus* when he noticed that colonies growing near a contaminating mold looked as if they were dissolving. Recognizing that the mold might be secreting a substance that killed the bacteria, he proceeded to study it more carefully. He identified the mold as a species of *Penicillium* and found it was indeed producing a bacteria-killing substance; he called this **penicillin.** Even though Fleming was unable to purify penicillin, he showed it was remarkably effective in killing many different kinds of bacteria and could be injected into rabbits and mice without adverse effects. Fleming recognized the potential medical significance of his discovery, but became discouraged with his inability to purify the compound and eventually abandoned his study of it.

Approximately ten years after Fleming's discovery, two other scientists in Britain, Ernst Chain and Howard Florey, were successful in their attempts to purify penicillin. In 1941, the drug was tested for the first time on a police officer with a life-threatening *Staphylococcus aureus* infection. He improved so dramatically that within 24 hours his illness seemed under control. Unfortunately, the supply of purified penicillin ran out, and the man eventually died of the infection. Later, with greater supplies of the drug, two deathly ill patients were successfully cured. World War II spurred cooperation of British and American scientists to determine the chemical structure of penicillin and to develop the means for its large-scale production so that it could be used to treat infected soldiers and workers. Several different penicillins were found in the *Penicillium* cultures, and were designated alphabetically. **Penicillin G** (or benzyl penicillin) was found to be the most suitable for treating infections. This was the first of what we now call **antibiotics**—antimicrobial drugs naturally produced by microorganisms.

Soon after the discovery of penicillin, Selman Waksman isolated a bacterium from soil, *Streptomyces griseus,* that produced an antibiotic he called **streptomycin.** The realization that bacteria as well as molds could produce medically useful antimicrobial drugs prompted researchers to begin laboriously screening hundreds of thousands of different strains of microorganisms for antibiotic production. Even today, pharmaceutical companies examine soil samples from around the world for organisms that produce novel antibiotics.

Development of New Generations of Drugs

In the 1960s scientists discovered they could alter the chemical structure of drugs such as penicillin G, giving them new properties. For example, penicillin G, which is active mainly against Gram-positive bacteria, can be altered to produce ampicillin, a drug that kills a variety of Gram-negative species as well. Other changes to penicillin created the drug methicillin, which is less susceptible to enzymes used by some bacteria to inactivate penicillin. Thus, methicillin can be used to treat infections caused by certain penicillin-resistant organisms. Today a variety of penicillin-like medications exist, making up what is referred to as the family of penicillins **(figure 21.1).** Other unrelated antimicrobial drugs have also been altered to give them new characteristics.

MICROCHECK 21.1

Antimicrobials are chemotherapeutic agents that are effective in treating microbial infections. Antibiotics are antimicrobial chemicals naturally produced by particular microorganisms.

- ✓ How is the microbe that makes penicillin different from the one that makes streptomycin?
- ✓ Define and contrast the terms *chemotherapeutic agent, antimicrobial,* and *antibiotic.*
- ✓ How might *Streptomyces griseus* cells protect themselves from the effects of streptomycin?

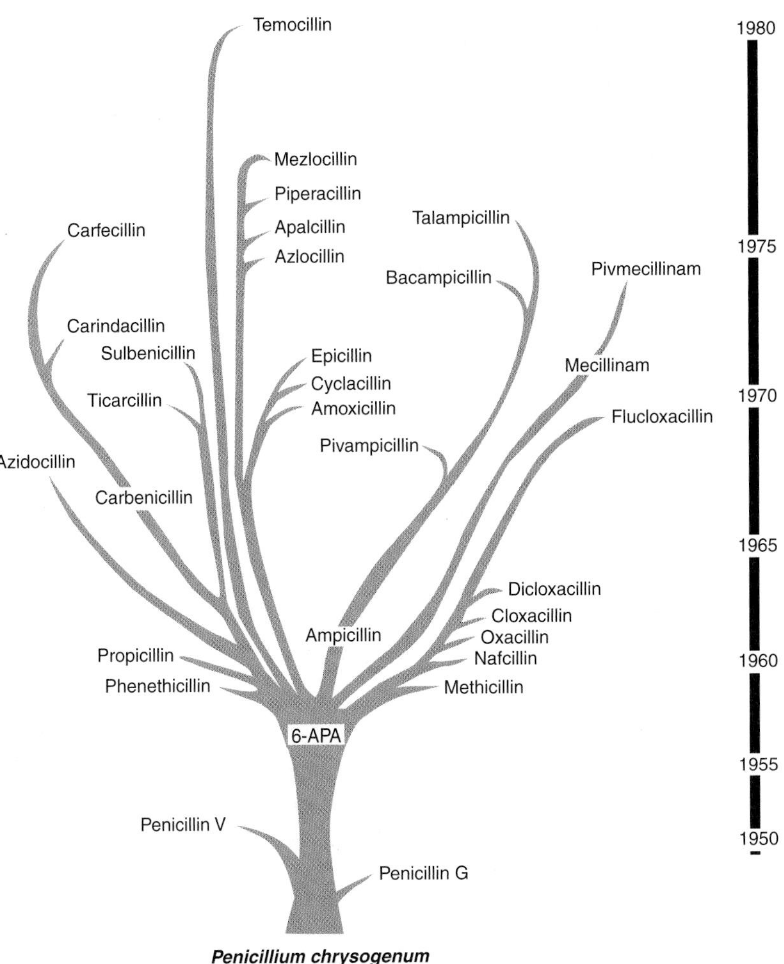

FIGURE 21.1 Family Tree of Penicillins All of the derivatives contain 6-aminopenicillanic acid (6-APA), the core portion of penicillin G.

21.2 Features of Antimicrobial Drugs

Focus Point

- Describe the important features of antimicrobial drugs that physicians must consider when prescribing an appropriate medication, including selective toxicity; antimicrobial action; spectrum of activity; tissue distribution, metabolism, and excretion; effects of combinations of antimicrobial drugs; adverse effects; and resistance to antimicrobials.

Most modern antibiotics come from microorganisms that normally reside in the soil; these include species of *Streptomyces* and *Bacillus* (bacteria), and *Penicillium* and *Cephalosporium* (fungi). To commercially produce an antibiotic, a carefully selected strain of the appropriate species is inoculated into a broth medium and incubated in a huge vat. As soon as the maximum antibiotic concentration is reached, the drug is extracted from the medium and extensively purified. In many cases, the antibiotic is chemically altered after purification to impart new characteristics such as increased stability. These chemically modified compounds are called **semisynthetic.** In some cases, the entire drug can be synthesized in the laboratory. By convention, these partially or totally synthetic chemicals are still called antibiotics because microorganisms produce them naturally. Numerous different antimicrobial drugs are now available, each with characteristics that make it more or less suitable for a given clinical situation. Hundreds of tons and many millions of dollars' worth of antibiotics are now produced each year.

Selective Toxicity

Medically useful antimicrobial drugs exhibit **selective toxicity,** causing greater harm to microorganisms than to the human host. They do this by interfering with essential biological structures or biochemical processes that are common in microorganisms but not human cells.

While the ideal antimicrobial drug is non-toxic to humans, most can be harmful at high concentrations. In other words, selective toxicity is a relative term. The toxicity of a given drug is expressed as the **therapeutic index,** which is the lowest dose toxic to the patient divided by the dose typically used for therapy. Antimicrobials that have a high therapeutic index are less toxic to the patient, often because the drug acts against a vital biochemical process of bacteria that does not exist in human cells. For example, penicillin G, which interferes with bacterial cell wall synthesis, has a very high therapeutic index. When an antimicrobial that has a low therapeutic index is administered, the concentration in the patient's blood must be carefully monitored to ensure it does not reach a toxic level. Drugs that are too toxic for systemic use can sometimes be used for topical applications, such as first-aid antibiotic skin ointments.

Antimicrobial Action

Antimicrobial drugs either kill microorganisms or inhibit their growth. Those that inhibit bacterial growth are called **bacteriostatic.** These drugs depend on the normal host defenses to kill or eliminate the pathogen after its growth has been inhibited. For example, sulfa drugs, which are frequently prescribed for urinary tract infections, inhibit the growth of bacteria in the bladder until they are eliminated by the body's defenses. Drugs that kill bacteria are **bactericidal.** These are particularly useful in situations in which the normal host defenses cannot be relied on to remove or destroy pathogens. A given drug can be bactericidal in one situation yet bacteriostatic in another, depending on the concentration of the drug and the growth stage of the microorganism.

Spectrum of Activity

Antimicrobial drugs vary with respect to the range of microorganisms they kill or inhibit. Some kill or inhibit a narrow range of microorganisms, such as only Gram-positive bacteria, whereas others affect a wide range, generally including both Gram-positive and Gram-negative organisms. Antimicrobials that affect a wide range of bacteria are called **broad-spectrum** antimicrobials. These are very important in the treatment of acute life-threatening diseases when immediate antimicrobial therapy is essential and there is no time to culture and identify the disease-causing agent. The disadvantage of broad-spectrum antimicrobials is that, by affecting a wide range of organisms, they disrupt the normal microbiota that play an important role in excluding pathogens. This in turn can leave the patient predisposed to other infections. Antimicrobials that affect a limited range of bacteria are **narrow-spectrum** antimicrobials. Their use requires identification of the pathogen and testing of its susceptibility to antimicrobials, but they cause less disruption to the normal microbiota. ■ normal microbiota, pp. 349, 393

Effects of Combinations of Antimicrobial Drugs

Combinations of antimicrobials are sometimes used to treat infections, but care must be taken when selecting the combinations because some drugs will counteract the effects of others. For example, bacteriostatic drugs typically interfere with the effects of other drugs that kill only actively dividing cells. Counteracting combinations such as this are called **antagonistic.** In contrast, combinations in which the activity of one drug enhances the activity of the other are called **synergistic.** Combinations that are neither synergistic nor antagonistic are called **additive.**

Tissue Distribution, Metabolism, and Excretion of the Drug

Antimicrobials differ not only in their action and activity, but also in how they are distributed, metabolized, and excreted by the body. For example, only some drugs are able to cross from the blood into the cerebrospinal fluid, an important factor for a physician to consider when prescribing a drug to treat meningitis. Drugs that are unstable at low pH are destroyed by stomach acid when taken orally, and so these drugs must instead be administered through intravenous or intramuscular injection. ■ meningitis, p. 650

Another important characteristic of an antimicrobial is its rate of elimination, which is expressed as the **half-life.** The half-life of a drug is the time it takes for the body to eliminate one-half of the original concentration in the serum. The half-life of a drug dictates the size and frequency of doses required to maintain an effective level in the body. Penicillin V, which has a very short half-life, needs to be taken four times a day, whereas azithromycin, with a half-life of over 24 hours, is taken only once a day or less. Patients who have kidney or liver dysfunction often excrete or metabolize drugs more slowly, and so their drug dosages must be adjusted accordingly to avoid toxic levels.

Adverse Effects

As with any medication, several concerns and dangers are associated with antimicrobial drugs. It is important to remember, however, that antimicrobials are extremely valuable drugs that save countless lives when properly prescribed and used.

Allergic Reactions

Some people develop hypersensitivities or allergies to antimicrobials. An allergic reaction to penicillin or other related drugs usually results in a fever or rash but can abruptly cause life-threatening anaphylactic shock. For this reason, people who have allergic reactions to a given antimicrobial must alert their physicians and pharmacists so that an alternative drug can be prescribed. They should also wear a bracelet or necklace that records that information in case of emergency. ■ anaphylactic shock, p. 417

Toxic Effects

Several antimicrobials are toxic at high concentrations or occasionally cause adverse reactions. Aminoglycosides such as streptomycin can damage kidneys, impair the sense of balance, and even cause irreversible deafness. Patients taking these drugs must be closely monitored because of the low therapeutic index. Some antimicrobials have such severe potential side effects that they are reserved for only life-threatening conditions. For example, in rare cases, chloramphenicol causes the potentially lethal condition **aplastic anemia,** in which the body is unable to make white and red blood cells. For this reason, chloramphenicol is usually used only when no other alternatives are available.

Suppression of the Normal Microbiota

The normal microbiota plays an important role in host defense by excluding pathogens. When the composition of the normal microbiota is altered, which happens when a person takes an antimicrobial, pathogens normally unable to compete may multiply to high numbers. Patients who take broad-spectrum antibiotics orally sometimes develop the life-threatening disease called **antibiotic-associated colitis,** caused by the growth of toxin-producing strains of *Clostridium difficile.* This organism generally is not able to establish itself in the intestine due to competition from other bacteria. When members of the normal intestinal microbiota are inhibited or killed, however, *C. difficile* can sometimes flourish and cause serious intestinal damage. ■ normal microbiota, pp. 349, 393 ■ *Clostridium difficile*, p. 586

Resistance to Antimicrobials

Just as humans are assembling a vast array of antimicrobial drugs, microorganisms have their own genetic toolbox of mechanisms to avoid their effects. In some cases, certain types of bacteria are inherently resistant to the effects of a particular drug; this is called **innate,** or **intrinsic, resistance.** Members of the genus *Mycoplasma* lack a cell wall, so, not surprisingly, they are resistant to any drug such as penicillin that exerts its action by interfering with cell wall synthesis. Many Gram-negative organisms are intrinsically resistant to certain drugs because the lipid bilayer of their outer membrane excludes entry of the drug. In other instances, previously sensitive organisms develop resistance through spontaneous mutation or the acquisition of new genetic information; this is called **acquired resistance.** The mechanisms and acquisition of resistance will be discussed later. ■ characteristics of bacteria that lack a cell wall, p. 63 ■ Gram-negative cell wall, p. 61

MICROCHECK 21.2

When choosing an antimicrobial to prescribe, a physician must consider a variety of factors including the therapeutic index, antimicrobial action, spectrum of activity, effects of combinations of antimicrobials, tissue distribution, half-life, adverse effects, and resistance of the microbe.

✓ Which would you rather take: an antimicrobial that has a low therapeutic index or one that has a high therapeutic index? Why?

✓ In what clinical situation is it most appropriate to use a broad-spectrum antimicrobial?

✓ Why would antimicrobials that have toxic side effects be used at all?

Mechanisms of Action of Antibacterial Drugs

Focus Points

- Compare and contrast the antimicrobial drugs that inhibit cell wall synthesis.
- Compare and contrast the antimicrobial drugs that inhibit protein synthesis.
- Compare and contrast the antimicrobial drugs that inhibit nucleic acid synthesis.
- Compare and contrast the antimicrobial drugs that inhibit metabolic pathways.
- Describe the antimicrobial drugs that interfere with cell membrane integrity.
- Describe the antibacterial medications used to treat infections caused by *Mycobacterium tuberculosis*.

A number of bacterial processes utilize enzymes or structures that are different or absent, or are not commonly found in eukaryotic cells. Several microbial processes, including the synthesis of bacterial cell walls, proteins, and nucleic acids, metabolic pathways, and the integrity of the cytoplasmic membrane, are the targets of most antimicrobial drugs **(figure 21.2).**

This section will discuss the bacterial processes commonly targeted by antimicrobial medications. To illustrate how these targets are affected, the mechanism of action of some of the most widely used antimicrobials will be described **(table 21.1).** A group of antibiotics called β-lactam drugs will be covered in the greatest detail, because they serve as excellent examples of some of the important features of antimicrobials.

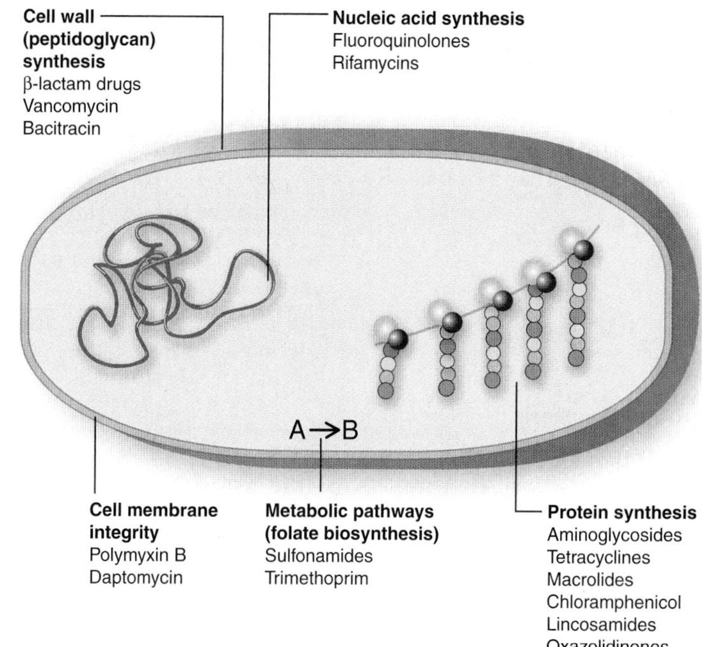

FIGURE 21.2 Targets of Antibacterial Medications

TABLE 21.1 Characteristics of Antibacterial Drugs

Target/Drug	Comments/Characteristics
Cell Wall Synthesis	
β-lactam drugs	Bactericidal against a variety of bacteria; inhibits penicillin-binding proteins. Resistance is due to synthesis of β-lactamases, decreased affinity of penicillin-binding proteins, or decreased uptake.
Penicillins	A family of antibacterial medications; different groups vary in their spectrum of activity and their susceptibility to β-lactamases.
Natural penicillins: penicillin G, penicillin V	Active against Gram-positive and a few Gram-negative bacteria. Penicillin G is destroyed by stomach acid, and so it usually must be administered by injection. Penicillin V can be taken orally.
Penicillinase-resistant: methicillin, dicloxicillin	Similar to natural penicillins, but resistant to inactivation by the penicillinase of staphylococci.
Broad-spectrum: ampicillin, amoxicillin	Similar to the natural penicillins, but more active against Gram-negative organisms.
Extended-spectrum: ticarcillin, piperacillin	Increased activity against Gram-negative rods, including *Pseudomonas* species.
Cephalosporins Cephalexin, cephradine, cefaclor, cefprozil, cefixime, cefibuten, cefepime	A family of antibacterial medications. The later generations are generally more effective against Gram-negative bacteria and less susceptible to destruction by β-lactamases.
Carbapenems Imipenem, meropenem	Resistant to inactivation by β-lactamases. Imipenem must be given in combination with a drug that inhibits certain kidney enzymes in order to avoid its inactivation.

(continued)

TABLE 21.1 Characteristics of Antibacterial Drugs (*continued*)

Target/Drug	Comments/Characteristics
Cell Wall Synthesis (continued)	
Monobactams Aztreonam	Resistant to β-lactamases; can be given to patients who are allergic to penicillin. Primarily active against members of the family *Enterobacteriaceae.*
Vancomycin	Bactericidal against Gram-positive bacteria; binds to the peptide side chain of *N*-acetylmuramic acid. Used to treat serious systemic infections and antibiotic-associated colitis. In enterococci, resistance is due to a plasmid-encoded altered target.
Bacitracin	Bactericidal against Gram-positive bacteria; interferes with the transport of peptidoglycan precursors. Common ingredient in non-prescription antibiotic ointments.
Protein Synthesis	
Aminoglycosides Streptomycin, gentamicin, tobramycin, amikacin, neomycin	Bactericidal against aerobic and facultative bacteria; binds to the 30S ribosomal subunit, blocking the initiation of translation and causing the misreading of mRNA. Toxicity limits the use. Resistance is due to a plasmid-encoded inactivating enzyme, alteration of the target molecule, or decreased uptake by a cell. Neomycin is commonly used in non-prescription topical antibiotic ointments.
Tetracyclines Tetracycline, doxycycline, glycylcyclines	Bacteriostatic against some Gram-positive and Gram-negative bacteria; binds to the 30S ribosomal subunit, blocking the attachment of tRNA. Resistance is generally due to decreased accumulation, either through decreased uptake or increased efflux.
Macrolides Erythromycin, clarithromycin, azithromycin	Bacteriostatic against many Gram-positive bacteria as well as the most common causes of atypical pneumonia; binds to the 50S ribosomal subunit, preventing the continuation of protein synthesis. Used for treating patients who are allergic to β-lactam drugs. Resistance is due to an inactivating enzyme, alteration of the target molecule, or decreased uptake by a cell.
Chloramphenicol	Bacteriostatic and broad-spectrum; binds to the 50S ribosomal subunit, preventing peptide bonds from being formed. Generally used only as a last resort for life-threatening infections. Resistance is often due to a plasmid-encoded inactivating enzyme.
Lincosamides Lincomycin, clindamycin	Bacteriostatic against a variety of Gram-positive and Gram-negative bacteria, including the anaerobe *Bacteroides fragilis.* Binds to the 50S ribosomal subunit, preventing the continuation of protein synthesis. Associated with an even greater risk of developing antibiotic-associated colitis.
Oxazolidinones Linezolid	Bacteriostatic against a variety of Gram-positive bacteria. Binds to the 50S ribosomal subunit, interfering with the initiation of protein synthesis.
Streptogramins Quinupristin, dalfopristin	A synergistic combination of two drugs that bind to two different sites on the 50S ribosomal subunit, inhibiting distinct steps of protein synthesis. Individually each drug is bacteriostatic, but together they are bacteriocidal. Effective against a variety of Gram-positive bacteria.
Nucleic Acid Synthesis	
Fluoroquinolones Ciprofloxacin, moxifloxacin	Bactericidal against a wide variety of Gram-positive and Gram-negative bacteria; inhibits topoisomerases. Resistance is most often due to structural alterations in the topoisomerase target.
Rifamycins Rifampin	Bactericidal against Gram-positive and some Gram-negative bacteria. Binds RNA polymerase, blocking the initiation of RNA synthesis. Primarily used to treat infections caused by *Mycobacterium tuberculosis* and as prophylaxis for patients who have been exposed to *Neisseria meningitidis.*
Folate Biosynthesis	
Sulfonamides	Bacteriostatic against a variety of Gram-positive and Gram-negative bacteria. Structurally similar to para-aminobenzoic acid (PABA) and therefore inhibits the enzyme for which PABA is a substrate. Resistance is most commonly due to a plasmid-encoded alternative enzyme.
Trimethoprim	Often used in combination with a sulfa drug for a synergistic effect; inhibits the enzyme that catalyzes a step following the one inhibited by the sulfonamides. Resistance is commonly due to a plasmid-encoded alternative enzyme; the genes that encode resistance to sulfa drugs are often carried on the same plasmid.
Cell Membrane Integrity	
Polymyxin B	Bactericidal against Gram-negative bacteria by damaging cell membranes. Its toxicity limits its use primarily to topical applications, but it is a common ingredient in non-prescription antibiotic ointments.
Daptomycin	Bactericidal against Gram-positive bacteria by damaging the cytoplasmic membrane.
Mycobacterium tuberculosis	
Ethambutol	Inhibits the synthesis of a component of the mycobacterial cell wall.
Isoniazid	Inhibits synthesis of mycolic acid, a major component of the mycobacterial cell wall.
Pyrazinamide	Mechanism unknown.

Antibacterial Medications That Inhibit Cell Wall Synthesis

Bacterial cell walls are unique in that they contain peptidoglycan (see figures 3.33 and 3.34). Because of this, antimicrobial medications that interfere solely with synthesis of this cell wall component do not affect eukaryotic cells, generally resulting in a very high therapeutic index **(figure 21.3).**

Penicillins, Cephalosporins and Other β-Lactam Drugs

Penicillins and cephalosporins are members of a group of antimicrobial medications collectively referred to as **β-lactam drugs.** This group, which also includes the monobactams and carbapenems, all have a shared chemical structure called a **β-lactam ring (figure 21.4).**

The β-lactam drugs competitively inhibit a group of enzymes that catalyze formation of peptide bridges between adjacent glycan strands, an essential step in the final stages of peptidoglycan synthesis (see figure 3.32). These enzymes are commonly called **penicillin-binding proteins (PBPs),** reflecting the fact that they bind penicillin and were initially discovered during experiments to study the effects of the medication. The disruption in cell wall biosynthesis leads to weakening of the wall and ultimately results in cell lysis **(figure 21.5).** Because cell walls are synthesized only in actively multiplying cells, the β-lactam drugs are effective only against growing bacteria. ■ competitive enzyme inhibition, p. 138

The different β-lactam drugs vary in their spectrum of activity. Some are more active against Gram-positive bacteria, whereas others are more active against Gram-negative organisms. One reason for this difference arises from the architecture of the cell wall. The peptidoglycan layer of Gram-positive organisms directly contacts the outside environment, making the enzymes that synthesize it readily accessible to drugs. In contrast, the outer membrane of Gram-negative bacteria excludes many antimicrobials, making many of these organisms innately resistant to many medications, including certain β-lactam drugs. Another difference is the affinity of an organism's penicillin-binding proteins for a particular β-lactam drug. The PBPs of Gram-positive bacteria differ somewhat from those of Gram-negative bacteria, and the PBPs of obligate anaerobes differ from those of aerobes. Differences in affinity can even exist among related organisms such as Gram-positive cocci.

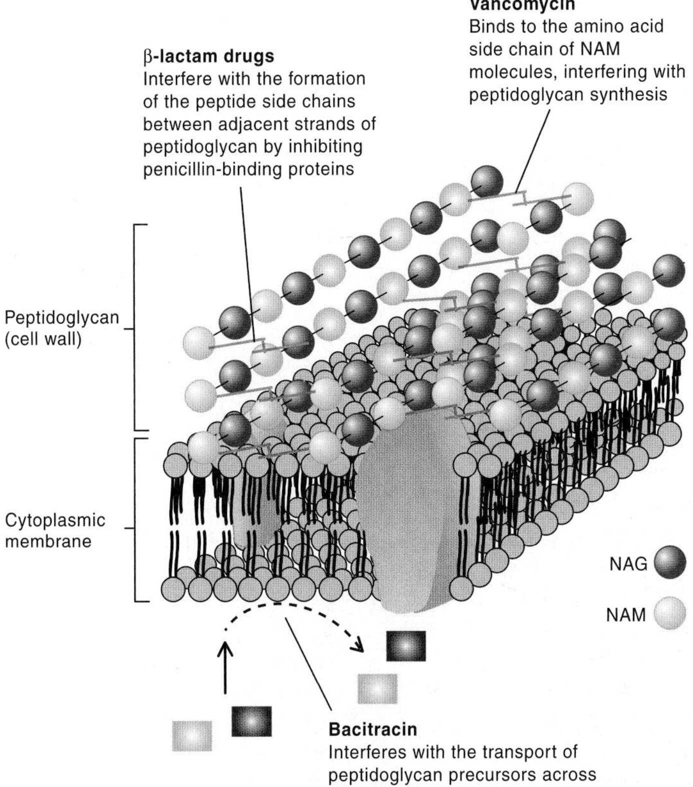

FIGURE 21.3 Antibacterial Medications that Interfere with Cell Wall Synthesis

(a) Penicillin

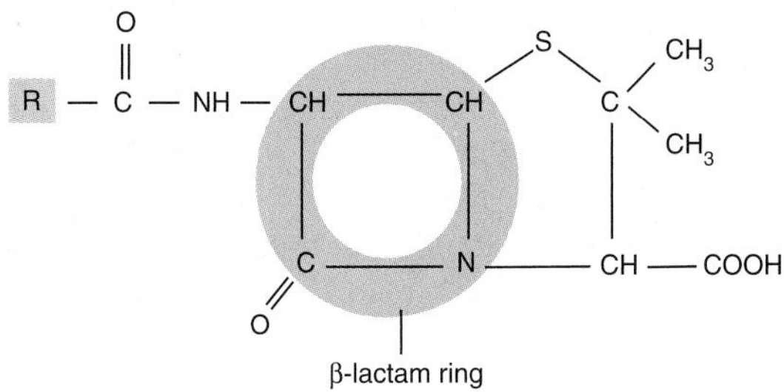

(b) Cephalosporin

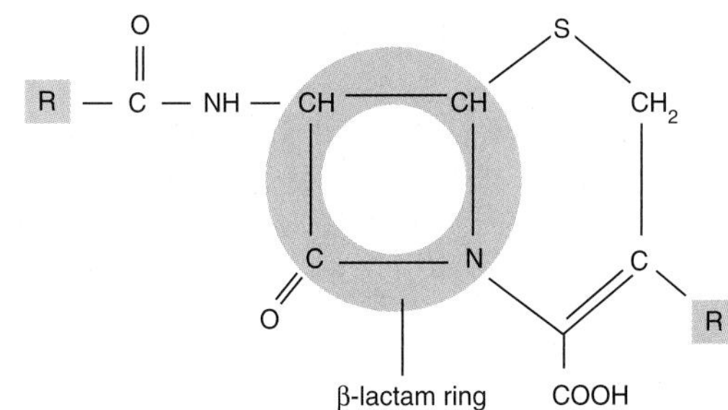

FIGURE 21.4 The β-Lactam Ring of Penicillins and Cephalosporins The core chemical structure of **(a)** a penicillin; **(b)** a cephalosporin. The β-lactam rings are marked by an orange circle. The R groups vary among different penicillins and cephalosporins.

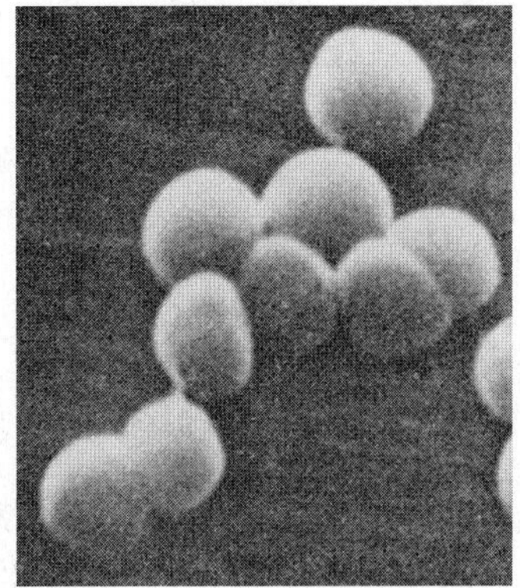
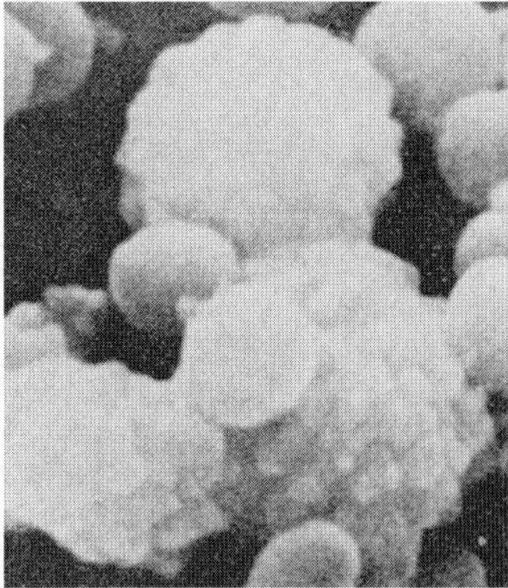

FIGURE 21.5 Effect of a β-Lactam Drug on a Cell The drug disrupts cell wall synthesis, leading to weakening of the cell wall, ultimately causing the cells to lyse.

Some bacteria can resist the effects of certain β-lactam drugs by synthesizing an enzyme, a **β-lactamase,** that breaks the critical β-lactam ring, destroying the activity of the antibiotic. Just as there are many β-lactam drugs, there are various β-lactamases; these differ in the range of drugs they destroy. A β-lactamase that was originally detected in some strains of staphylococci only inactivates members of the penicillin family. To reflect this fact, it is often called a **penicillinase.** In contrast, some of the β-lactamases produced by Gram-negative organisms inactivate a wide variety of β-lactam drugs. The **extended-spectrum β-lactamases** inactivate both the penicillins and the cephalosporins. As a whole, Gram-negative bacteria can produce a much more extensive array of β-lactamases than can Gram-positive organisms.

The Penicillins All members of the family of penicillins share a common basic structure. Only the side chain has been modified in the laboratory to create penicillin derivatives, each with unique characteristics **(figure 21.6).** Currently the family of penicillins can be loosely grouped into several categories, each of which consists of several different drugs:

- **Natural penicillins.** These are the original penicillins produced naturally by the mold *Penicillium chrysogenum.* Natural penicillins are narrow-spectrum antibiotics, effective against Gram-positive and a few Gram-negative bacteria. Strains of bacteria that produce penicillinase are resistant to the natural penicillins. Penicillin V is more stable in acid and, therefore, better absorbed than penicillin G when taken orally.
- **Pencillinase-resistant penicillins.** These drugs were developed in the laboratory as a response to the problem of penicillinase-producing staphylococci. Side chains of the drugs prevent penicillinase from inactivating them. Unfortunately, some strains of penicillinase-producing *Staphylococcus aureus* synthesize altered PBPs to which β-lactam drugs, including the penicillins, no longer bind. Penicillinase-resistant penicillins include methicillin and dicloxicillin.
- **Broad-spectrum penicillins.** The modified side chains of these drugs give them a broad spectrum of activity. They retain their activity against penicillin-sensitive, Gram-positive bacteria, yet they are also active against Gram-negative organisms. Unfortunately, they can be inactivated by many β-lactamases. Broad-spectrum penicillins include ampicillin and amoxicillin.
- **Extended-spectrum penicillins.** These have greater activity against *Pseudomonas* species—Gram-negative bacteria that are unaffected by many conventional antimicrobial drugs. The extended-spectrum penicillins, however, have less activity against Gram-positive organisms. Like the other broad-spectrum penicillins, they are destroyed by many β-lactamase-producing organisms. Extended-spectrum penicillins include ticarcillin and piperacillin.
- **Penicillins + β-lactamase inhibitor.** Rather than a new drug, this is a novel combination of therapeutic agents. β-lactamase inhibitors are chemicals that interfere with the activity of some types of β-lactamases. When a β-lactamase inhibitor is administered with one of the penicillins, the medication is protected against enzymatic destruction. An example is Augmentin®, a combination of amoxicillin and clavulanic acid.

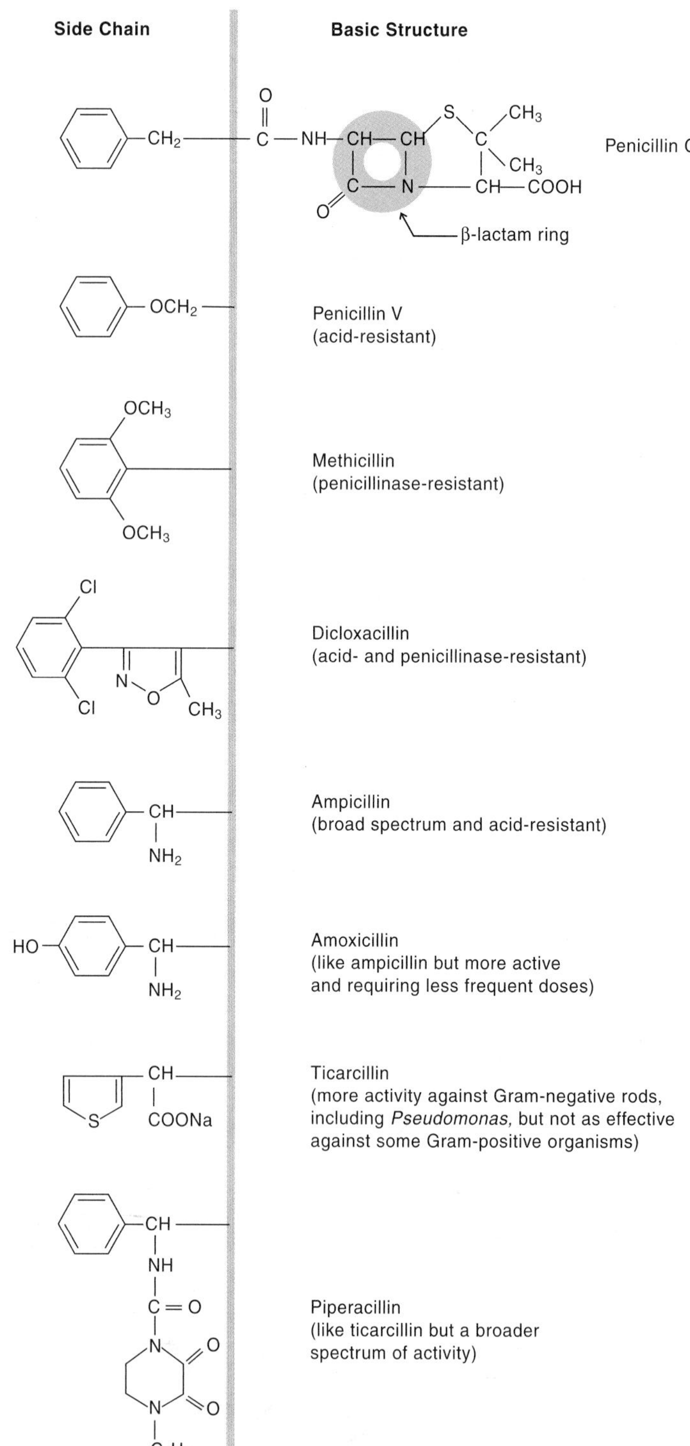

FIGURE 21.6 Chemical Structures and Properties of Representative Members of the Penicillin Family The entire structure of penicillin G and the side chains of other penicillins are shown.

The Cephalosporins The **cephalosporins** are derived from an antibiotic produced by the fungus *Acremonium cephalosporium* (formerly called *Cephalosporium acremonium*). Generally included in this family of drugs is a closely related group of antibiotics made by members of a genus of filamentous bacteria related to *Streptomyces.*

The chemical structure of the cephalosporins makes them resistant to inactivation by certain β-lactamases, but some have a low affinity for penicillin-binding proteins of Gram-positive bacteria, thus limiting their effectiveness against these organisms. Like the penicillins, the cephalosporins have been chemically modified to produce a family of various related antibiotics. They are grouped as the first-, second-, third-, and fourth-generation cephalosporins. These include cephalexin and cephradine (first generation), cefaclor and cefprozil (second generation), cefixime and cefibuten (third generation), and cefepime (fourth generation). The later generations are generally more effective against Gram-negative bacteria and are less susceptible to destruction by β-lactamases.

Other β-Lactam Antibiotics Two other groups of β-lactam drugs, carbapenems and monobactams, are very resistant to β-lactamases. The carbapenems are effective against a wide range of Gram-negative and Gram-positive bacteria. Two types are available—imipenem and meropenem. Imipenem is rapidly destroyed by a kidney enzyme and is therefore administered in combination with a drug that inhibits that enzyme. The only monobactam used therapeutically, aztreonam, is primarily effective against members of the family *Enterobacteriaceae,* which are Gram-negative rods. Structurally, it is slightly different from other β-lactam drugs; this characteristic is important because aztreonam can be given to patients who have developed an allergy to penicillin. ■ *Enterobacteriaceae*, p. 262

Vancomycin

Vancomycin binds to the terminal amino acids of the peptide side chain of NAM molecules that are being assembled to form glycan chains. By doing so, it blocks synthesis of peptidoglycan, resulting in weakening of the cell wall and, ultimately, cell lysis. Vancomycin does not cross the outer membrane of Gram-negative bacteria; consequently, these organisms are innately resistant. Although vancomycin resistance is increasing, it is still a very important medication for treating infections caused by Gram-positive bacteria that are resistant to β-lactam drugs. In addition, it is sometimes the preferred drug for treating severe cases of antibiotic-associated colitis. Because vancomycin is poorly absorbed from the intestinal tract, it must be administered intravenously except when used to treat intestinal infections. Acquired resistance to vancomycin is most often due to an alteration in the peptide side chain of the NAM molecule that prevents vancomycin from binding.

Bacitracin

Bacitracin inhibits cell wall biosynthesis by interfering with the transport of peptidoglycan precursors across the cytoplasmic membrane. Its toxicity limits its use to topical applications; however, it is a common ingredient in non-prescription first-aid ointments.

Antibacterial Medications That Inhibit Protein Synthesis

Several types of antibacterial drugs inhibit prokaryotic protein synthesis **(figure 21.7).** While all cells synthesize proteins, the structure of the prokaryotic 70S ribosome, which is composed of a 30S and a 50S subunit, is different enough from the eukaryotic 80S ribosome to make it a suitable target for selective toxicity. The mitochondria of eukaryotic cells also have 70S ribosomes, however, which may partially account for the toxicity of some of these drugs. ■ ribosome structure, p. 68 ■ protein synthesis, p. 164

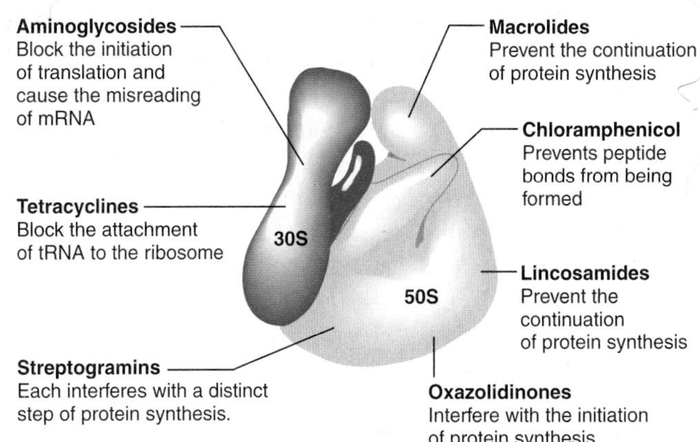
Aminoglycosides
Block the initiation of translation and cause the misreading of mRNA
Tetracyclines
Block the attachment of tRNA to the ribosome
30S
50S
Streptogramins
Each interferes with a distinct step of protein synthesis.
Macrolides
Prevent the continuation of protein synthesis
Chloramphenicol
Prevents peptide bonds from being formed
Lincosamides
Prevent the continuation of protein synthesis
Oxazolidinones
Interfere with the initiation of protein synthesis

FIGURE 21.7 Antibacterial Medications that Inhibit Prokaryotic Protein Synthesis These medications bind to the 70S ribosome.

The Aminoglycosides

The **aminoglycosides** are bactericidal drugs that irreversibly bind to the 30S ribosomal subunit, causing it to distort and malfunction. This blocks the initiation of translation and causes misreading of mRNA by ribosomes that have already passed the initiation step. Aminoglycosides are actively transported into bacterial cells by a process that requires respiratory metabolism. Consequently, they are generally not effective against anaerobes, enterococci, and streptococci. To extend their spectrum of activity, the aminoglycosides are sometimes used in a synergistic combination with a β-lactam drug. The β-lactam drug interferes with cell wall synthesis, which, in turn, allows the aminoglycoside to more easily enter cells that would otherwise be resistant. Examples of aminoglycosides include streptomycin, gentamicin, tobramycin, and amikacin. Unfortunately, these all can cause severe side effects including hearing loss and kidney damage; consequently, they are generally used only when other alternatives are not available. A form of tobramycin that can be administered through inhalation rather than injection makes treatment of lung infections in cystic fibrosis patients caused by *Pseudomonas aeruginosa* safer and more effective. Another aminoglycoside, neomycin, is too toxic for systemic use; however, it is a common ingredient in non-prescription topical ointments. ■ active transport, p. 58

The Tetracyclines

The **tetracyclines** reversibly bind to the 30S ribosomal subunit, blocking the attachment of tRNA to the ribosome and preventing the continuation of protein synthesis. These bacteriostatic drugs are actively transported into prokaryotic but not animal cells, which effectively concentrates them inside bacteria. This, in part, accounts for their selective toxicity. The tetracyclines are effective against certain Gram-positive and Gram-negative bacteria. Tetracyclines such as doxycycline have a longer half-life, allowing less-frequent doses. The glycylcyclines are a new group of semisynthetic tetracycline derivatives. They have a wide spectrum of activity and are effective against some bacteria that have acquired resistance against tetracycline. Tigecycline is the only example currently approved. Resistance to the tetracyclines is primarily due to a decrease in their accumulation by the bacterial cell, either by decreased uptake or increased excretion. Tetracyclines can cause discoloration in teeth when used by young children.

The Macrolides

The **macrolides** reversibly bind to the 50S ribosomal subunit and prevent the continuation of protein synthesis. Macrolides as a group are bacteriostatic against a variety of bacteria, including many Gram-positive organisms as well as the most common causes of atypical pneumonia ("walking pneumonia"). They often serve as the drug of choice for patients who are allergic to penicillin. Macrolides are not effective against members of the family *Enterobacteriaceae,* however, because the outer membrane of these organisms excludes the drug. Examples of macrolides include erythromycin, clarithromycin, and azithromycin. Both clarithromycin and azithromycin have a longer half-life than erythromycin, so that they can be taken less frequently. Resistance to all the macrolides can occur through modification of the ribosomal RNA target. Other mechanisms of resistance include the production of an enzyme that chemically modifies the drug and alterations that result in decreased uptake of the drug. ■ walking pneumonia, p. 511

Chloramphenicol

Chloramphenicol binds to the 50S ribosomal subunit, preventing peptide bonds from being formed and, consequently, blocking protein synthesis. Although it is bacteriostatic against a wide range of bacteria, it is generally only used as a last resort for life-threatening infections in order to avoid a rare but lethal side effect. This complication, aplastic anemia, can occur in response to even a small amount of chloramphenicol and is characterized by the inability of the body to form white and red blood cells.

The Lincosamides

The **lincosamides** bind to the 50S ribosomal subunit and prevent the continuation of protein synthesis, inhibiting a variety of Gram-negative and Gram-positive bacteria. They are particularly useful for treating infections resulting from intestinal perforation because they inhibit *Bacterioides fragilis,* a member of the normal intestinal microbiota that is frequently resistant to other antimicrobials. Unfortunately, the risk of developing antibiotic-associated colitis for people taking lincosamides is greater than for some other antimicrobials because *Clostridium difficile* is generally resistant to these drugs. The most commonly used lincosamide is clindamycin.

The Oxazolidinones

The **oxazolidinones** also bind to the 50S ribosomal subunit and interfere with the initiation of protein synthesis. They are bacteriostatic against a variety of Gram-positive bacteria and are useful in treating infections caused by bacteria that are resistant to β-lactam drugs and vancomycin. Linezolid is an example.

The Streptogramins

Two **streptogramins,** quinupristin and dalfopristin, are administered together in a medication called Synercid®. These act as a synergistic combination, binding to two different sites on the 50S ribosomal subunit and inhibiting distinct steps of protein synthesis. Individually, each drug is bacteriostatic but together they are bactericidal. Synercid® is effective against a variety of Gram-positive bacteria, including some that are resistant to β-lactam drugs and vancomycin.

Antibacterial Medications That Inhibit Nucleic Acid Synthesis

Enzymes required for nucleic acid synthesis are the targets of some groups of antimicrobial drugs. These groups include the fluoroquinolones and the rifamycins.

The Fluoroquinolones

The synthetic drugs called the **fluoroquinolones** inhibit one or more of a group of enzymes called **topoisomerases,** which maintain the supercoiling of closed circular DNA within the bacterial cell. One type of topoisomerase, called **DNA gyrase,** or **topoisomerase II,** breaks and rejoins strands to relieve the strain caused by the localized unwinding of DNA during replication and transcription. Consequently, inhibition of this enzyme prevents these essential cell processes. The fluoroquinolones are bactericidal against a wide variety of bacteria, including both Gram-positive and Gram-negative organisms. Examples of fluoroquinolones include ciprofloxacin and moxifloxacin. Acquired resistance is most commonly due to an alteration in the DNA gyrase target. ■ supercoiled DNA, p. 67

The Rifamycins

The **rifamycins** block prokaryotic RNA polymerase from initiating transcription. Rifampin, which is the most widely used rifamycin, exhibits bactericidal activity against many Gram-positive and some Gram-negative bacteria as well as members of the genus *Mycobacterium.* It is primarily used to treat tuberculosis and Hansen's disease (leprosy) and to prevent meningitis in people who have been exposed to *Neisseria meningitidis.* In some patients, a reddish-orange pigment appears in urine and tears. Resistance to rifampin develops rapidly and is due to a mutation in the gene that encodes RNA polymerase.

Antibacterial Medications That Inhibit Metabolic Pathways

Relatively few antibacterial medications interfere with metabolic pathways. Among the most useful are the folate inhibitors—sulfonamides and trimethoprim. These each inhibit different steps in the pathway that leads initially to the synthesis of folic acid and ultimately to the synthesis of a coenzyme

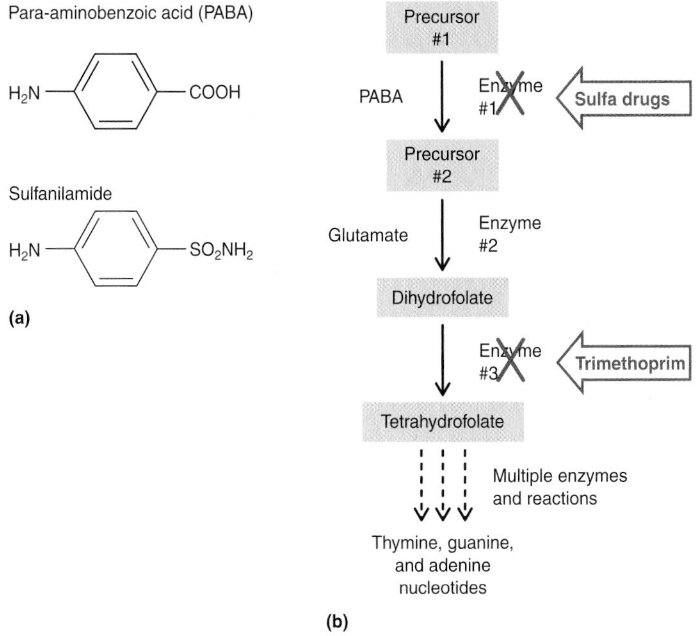

FIGURE 21.8 Inhibitors of the Folate Pathway (a) The chemical structure of PABA and a sulfa drug (sulfanilamide). **(b)** The sulfonamides and trimethoprim interfere with different steps of the pathway that leads initially to the synthesis of folic acid and ultimately to the synthesis of a coenzyme required for nucleotide biosynthesis.

required for nucleotide biosynthesis (**figure 21.8**). Animal cells lack the enzymes in the folic acid synthesis portion of the pathway, which is why folic acid is a dietary requirement. ■ coenzyme, p. 135

The Sulfonamides

Sulfonamides and related compounds, collectively referred to as **sulfa drugs,** inhibit the growth of many Gram-positive and Gram-negative bacteria. They are structurally similar to para-aminobenzoic acid (PABA), a substrate in the pathway for folic acid biosynthesis. Because of this similarity, the enzyme that normally binds PABA binds sulfa drugs instead, an example of competitive inhibition (see figure 6.13). Human cells lack this enzyme, providing the basis for the selective toxicity of the sulfonamides. Resistance of bacteria to the sulfonamides is often due to the acquisition of a plasmid-encoded enzyme that has a lower affinity for the drug. ■ competitive inhibition, p. 138

Trimethoprim

Trimethoprim inhibits the bacterial enzyme that catalyzes a metabolic step following the one inhibited by sulfonamides. Fortunately, the drug has little effect on the analogous enzyme in human cells. The combination of trimethoprim and a sulfonamide has a synergistic effect, and they are often used together to treat urinary tract infections. The most common mechanism of resistance is a plasmid-encoded alternative enzyme that has a lower affinity for the drug. Unfortunately, the genes encoding resistance to trimethoprim and sulfonamide are often carried on the same plasmid.

Antibacterial Medications That Interfere with Cell Membrane Integrity

A few antimicrobial drugs damage bacterial membranes. **Polymyxin B,** a common ingredient in first-aid skin ointments, binds to the membranes of Gram-negative cells and alters the permeability, leading to leakage of cellular contents and eventual death of the cells. Unfortunately, these drugs also bind to eukaryotic cells, though to a lesser extent, which generally limits their use to topical applications. **Daptomycin** inserts into the bacterial cytoplasmic membrane, resulting in death of the cell. Its approval increases the arsenal of antimicrobial medications available for treating infections caused by Gram-positive bacteria resistant to other drugs. It is not effective against Gram-negative bacteria, however, because it cannot penetrate the outer membrane.

Antibacterial Medications That Interfere with Processes Essential to *Mycobacterium tuberculosis*

Only a limited range of antimicrobials can be used to treat infections caused by *Mycobacterium tuberculosis* and related species. This is due to several factors, including the chronic nature of the disease caused by these organisms, their slow growth, and their waxy cell wall, which is impervious to many drugs. A group of five medications, called the **first-line** drugs, are preferred because they are the most effective as well as the least toxic. These are generally given in combination of two or more to patients who have active tuberculosis. This combination therapy prevents the development of resistant mutants; if some cells in the infecting population spontaneously develop resistance to one drug, the other drug will eliminate them. The **second-line** medications are used if the first-line drugs are not an option because of resistance; however, they are either less effective or have greater risk of toxicity. ■ *Mycobacterium tuberculosis*, p. 515

Of the first-line medications, some specifically target the unique cell wall that characterizes the mycobacteria. **Isoniazid** inhibits the synthesis of mycolic acids, a primary component of the cell wall. **Ethambutol** inhibits enzymes that are required for synthesis of other mycobacterial cell wall components. The mechanism of **pyrazinamide** is unknown. Other first-line drugs include rifampin and streptomycin, which have already been discussed.

MICROCHECK 21.3

Bacterial processes that utilize enzymes or structures that are different, absent, or not commonly found in eukaryotic cells are the targets of most medically useful antimicrobial drugs. The targets of antimicrobial drugs include biosynthetic pathways for peptidoglycan, protein, nucleic acid, and folic acid and the integrity of membranes. Drugs used to treat tuberculosis often interfere with processes unique to *Mycobacterium tuberculosis.*

- ✓ Explain the normal biological role of penicillin-binding proteins.
- ✓ What is the target of the macrolides?
- ✓ Why would co-administration of a bacteriostatic drug interfere with the effects of penicillin?

21.4

Determining the Susceptibility of a Bacterial Strain to an Antimicrobial Drug

Focus Points

- Describe how the minimum inhibitory concentration (MIC) and the minimum bactericidal concentration (MBC) of an antimicrobial drug are determined.
- Compare and contrast the Kirby-Bauer disc diffusion test with commercial modifications of antimicrobial susceptibility testing.

In many cases, susceptibility of a pathogenic organism to a specific antimicrobial drug is unpredictable. Unfortunately, it has often been the practice to try one drug after another until a favorable response is observed or, if the infection is very serious, to give several together. Both approaches are undesirable. With each unnecessary drug given, needless risks of toxic or allergic effects arise and the normal microbiota may be altered, permitting the overgrowth of pathogens resistant to the drug. A better approach is to determine the susceptibility of the specific pathogen to various antimicrobial drugs and then choose the drug that acts against the offending organism but against as few other bacteria as possible.

Determining the Minimum Inhibitory and Bactericidal Concentrations

The **minimum inhibitory concentration (MIC)** is the lowest concentration of a specific antimicrobial drug needed to prevent the growth of a given organism *in vitro*. It is determined by examining the test bacterial strain's ability to grow in broth cultures containing different concentrations of the antimicrobial. Serial dilutions generating decreasing concentrations of the drug are first prepared in tubes containing a suitable growth medium **(figure 21.9).** Then, a fixed concentration of the organism is added to each tube. The tubes are incubated for at least 16 hours and then examined for visible growth or turbidity. The lowest concentration of the drug that prevents growth of the microorganism is the minimum inhibitory concentration. The fact that a strain is inhibited by a given concentration of drug, however, does not necessarily mean that an infection caused by that

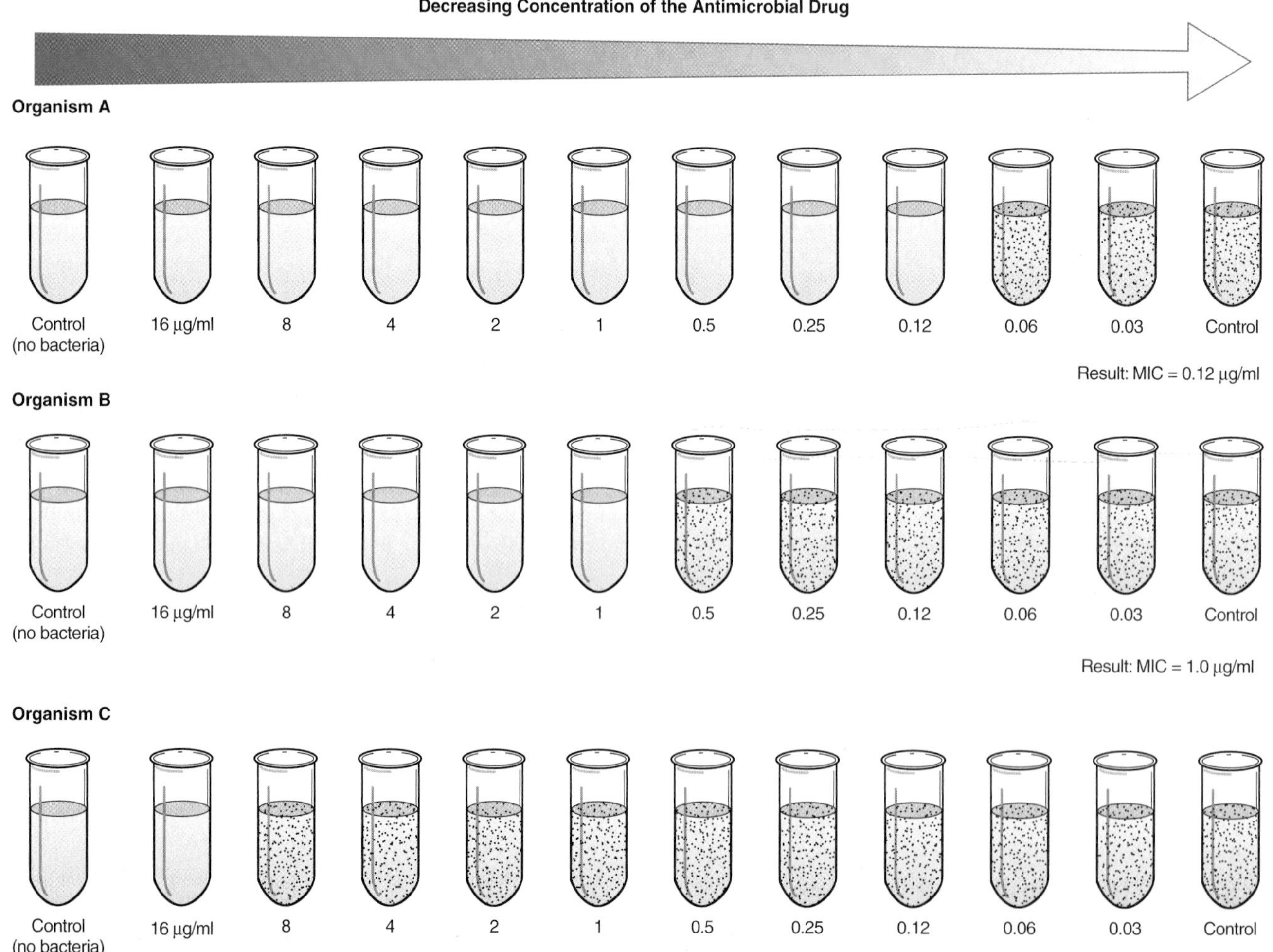

FIGURE 21.9 Determining the Minimum Inhibitory Concentration (MIC) of an Antimicrobial Drug The lowest concentration of drug that prevents growth of the culture is the MIC.

PERSPECTIVE 21.1

Measuring the Concentration of an Antimicrobial Drug in Blood or Other Body Fluids

There are many situations in which it is necessary to determine the concentration of an antimicrobial drug in a patient's blood or other body fluid. For example, patients who are being administered an aminoglycoside must often be carefully monitored to ensure that the concentration of drug in their blood does not reach an unsafe level, particularly if they have kidney or liver dysfunction that interferes with normal elimination. Likewise, new drugs must be tested to determine achievable levels in the blood, urine, or other body fluids.

A technique called the **diffusion bioassay** is used to measure the concentration of an antimicrobial drug in a fluid specimen. The test relies on the same principle as the Kirby-Bauer test, except in this case it is the concentration of drug, not the sensitivity of organism, being assayed. A culture of a stock organism highly susceptible to the drug is added to melted cooled agar, and the mixture poured into an agar plate and allowed to solidify. This results in a solid medium uniformly inoculated throughout with the sensitive organism. Cylindrical holes are then punched out of the agar, creating wells. **Standards,** or fluids containing known concentrations of the drug, are then added to some of the wells, while others are filled with the body fluid being tested. Following overnight incubation, zones of inhibition form around the agar wells, the sizes of which correspond to the concentrations of the drug **(figure 1).** The higher the concentration of the antimicrobial drug, the larger the zone of inhibition. The zone sizes around the standards are measured, and from this a **standard curve** is constructed by plotting the zone sizes against the corresponding drug concentration. A line relating zone size to concentration is obtained, from which the concentration of the antimicrobial drug in the body fluid can be read.

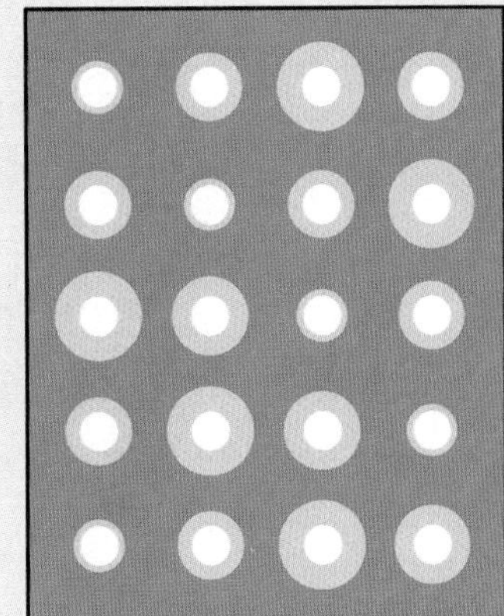

(a) Standards and patient's serum are added to agar that has been seeded with susceptible strain of bacteria.

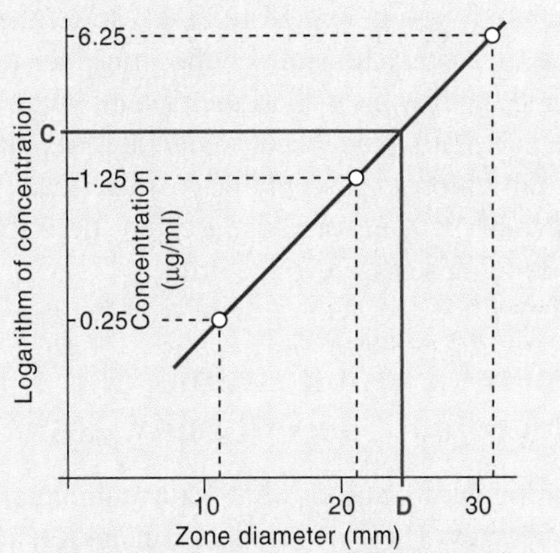

(b) A standard curve that correlates the size of the zone with the concentration of antibiotic is constructed. The concentration of the antimicrobial drug in the body fluid can be read from the line relating zone size to concentration.

FIGURE 1 Biological Assay to Determine the Concentration of an Antimicrobial Drug in a Patient's Body Fluid

strain can be successfully treated with the drug. For example, an organism with an MIC of 16 μg/ml of a drug is sensitive to that concentration *in vitro*. Nevertheless, the organism would be considered resistant for the purposes of treatment if the level that can be achieved in a person's blood were less than that. Microbes that have an MIC on the borderline between susceptible (treatable) and resistant (untreatable) are called intermediate.

The **minimum bactericidal concentration (MBC)** is the lowest concentration of a specific antimicrobial drug that kills 99.9% of cells of a given bacterial strain. The MBC is determined by assaying for live organisms in tubes from the MIC test that showed no growth. A small sample from each of those tubes is transferred to fresh, antibiotic-free medium. If growth occurs, then living organisms remained in the original tube. Conversely, if no growth occurs, then no living organisms remained, indicating that the antibiotic was bactericidal at that concentration.

Determining the MIC and MBC using these conventional methods gives precise information regarding an organism's susceptibility. The techniques, however, are labor-intensive and consequently expensive. In addition, individual sets of tubes must be inoculated to determine susceptibility to each different antimicrobial tested.

Conventional Disc Diffusion Method

The **Kirby-Bauer disc diffusion test** is routinely used to qualitatively determine the susceptibility of a given bacterial strain to a battery of antimicrobial drugs. A standard concentration of the strain is first uniformly spread on the surface of an agar plate. Then 12 or so discs, each impregnated with a specified amount of a selected antimicrobial drug, are placed on the surface of the medium **(figure 21.10).** During incubation, the various drugs diffuse outward

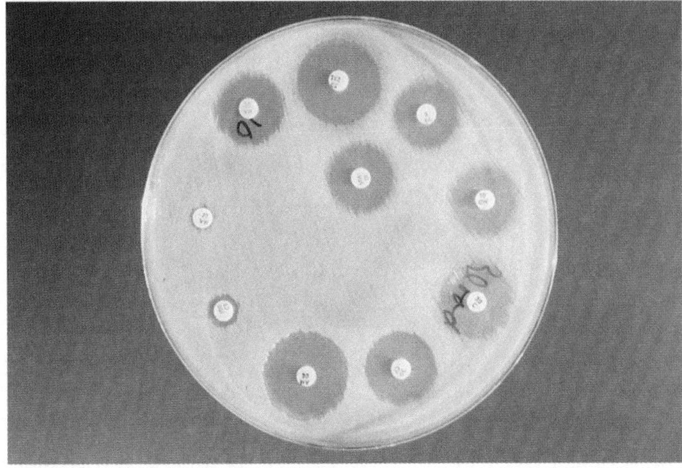

FIGURE 21.10 Kirby-Bauer Method for Determining Drug Susceptibility The size of the zone of inhibition surrounding the disc reflects, in part, the sensitivity of the bacterial strain to the drug. Because zone size is influenced by characteristics of the drug such as molecular weight, however, a chart must be consulted that correlates zone size to susceptibility.

from the discs, forming a concentration gradient around each disc. Meanwhile the bacterial cells multiply, eventually forming a film of growth on the plate, except in regions around the discs where antimicrobial drugs kill the cells or inhibit their growth. A clear **zone of inhibition** around an antimicrobial disc reflects, in part, the degree of susceptibility of the organism to the drug. The zone size is also influenced by characteristics of the drug including its molecular weight and stability, as well as the amount in the disc.

Special charts have been prepared correlating the size of the zone of inhibition to susceptibility of bacteria to the drug. Based on the size of the zone, organisms can be described as susceptible, intermediate, or resistant to the drug.

Commercial Modifications of Antimicrobial Susceptibility Testing

Commercial modifications of the conventional methods offer certain advantages. They are less labor-intensive, and the results can be obtained in as little as 4 hours. One system utilizes a small card with miniature wells containing specific antimicrobial concentrations. The highly automated system inoculates and incubates the cards, determines the growth rate by reading the turbidity, and uses mathematical formulas to interpret the results and derive the MICs in 6 to 15 hours **(figure 21.11).**

The E test, a modification of the disc diffusion test, utilizes a strip impregnated with a gradient of concentrations of an antimicrobial drug. Multiple strips, each containing a different drug, are placed on the surface of an agar medium that has been uniformly inoculated with the test organism. During incubation the test organism will grow, and a zone of inhibition will form around the strip, but because of the gradient of drug concentrations, the zone of inhibition will be shaped somewhat like a teardrop that intersects the strip at some point **(figure 21.12).** The MIC is determined by reading the printed number at the point where the bacterial growth intersects the strip.

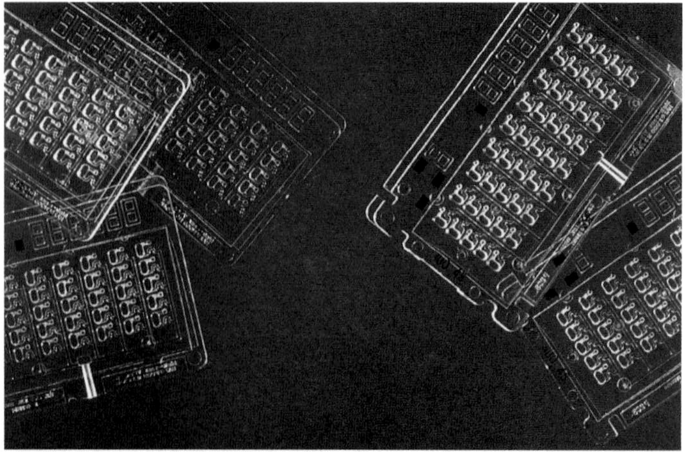

FIGURE 21.11 Automated Tests Used to Determine Antimicrobial Susceptibility The miniature wells in the card contain specific concentrations of an antimicrobial drug. An automated system inoculates and incubates the cards, determines the growth rate by reading turbidity, and uses mathematical formulas to interpret the results and derive the MICs.

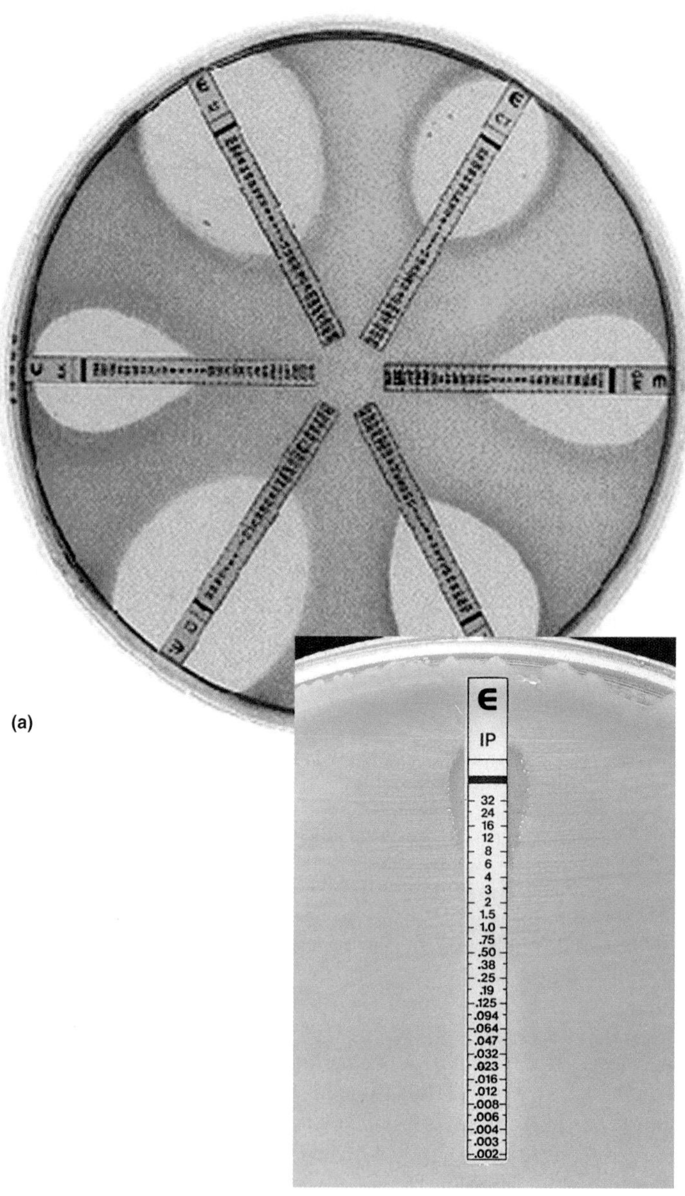

FIGURE 21.12 The E Test The strip is impregnated with a gradient of concentrations of a given antimicrobial drug. The MIC is determined by reading the number on the strip at the point at which growth intersects the strip.

MICROCHECK 21.4

The minimum inhibitory concentration and minimum bactericidal concentration are quantitative measures of a bacterial strain's susceptibility to an antimicrobial drug. Disc diffusion tests can determine whether an organism is sufficiently susceptible to a specific drug for it to successfully be used in treatment. Commercial tests for determining antimicrobial sensitivity are less labor-intensive and often more rapid.

- ✓ Explain the difference between the MIC and the MBC.
- ✓ List two factors other than an organism's sensitivity to a drug that can influence the size of the zone of inhibition around an antimicrobial disc.
- ✓ Why would it be important for the Kirby-Bauer disc diffusion test to use a standard concentration of the bacterial strain being tested?

21.5

Resistance to Antimicrobial Drugs

Focus Points

- Describe four general mechanisms of antimicrobial resistance.
- Describe how antimicrobial resistance can be acquired.
- List four examples of emerging antimicrobial resistance.
- Describe how the emergence and spread of antimicrobial resistance can be slowed.

After the introduction of sulfa drugs and penicillin, there was great hope that such drugs would soon eliminate most bacterial diseases. It is now recognized, however, that resistance to antimicrobial medications limits the usefulness of all known antimicrobials. As these drugs are increasingly used and misused, the resistant bacterial strains have a selective advantage over their sensitive counterparts **(figure 21.13).** For example, when penicillin G was first introduced, less than 3% of *Staphylococcus aureus* strains were resistant to its effects. Heavy use of the drug, measured in hundreds of tons per year, progressively eliminated sensitive strains, so that 90% or more are now resistant. This development is understandably of great concern to health professionals because of the impact on the cost, complications, and outcomes of treatment. Understanding the mechanisms and the spread of antimicrobial resistance is an important step in curtailing the problem, and it may also allow pharmaceutical companies to develop new drugs that foil common resistance mechanisms.

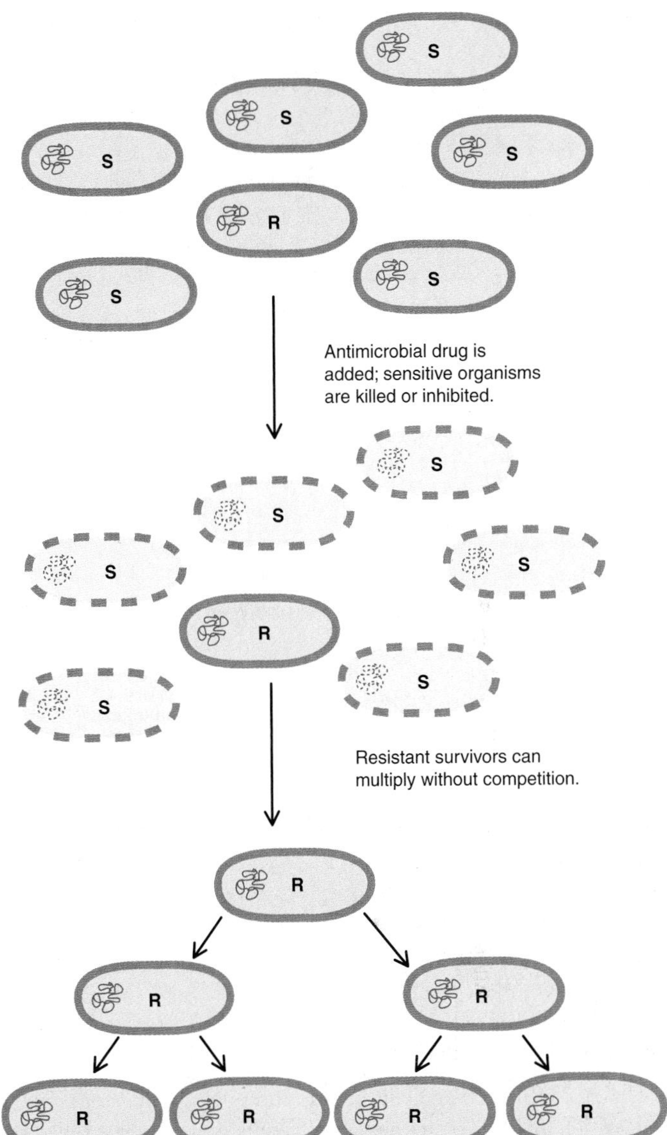

FIGURE 21.13 The Selective Advantage of Drug Resistance When antimicrobial drugs are used, bacterial strains that are resistant (R) to their effects have a selective advantage over their sensitive (S) counterparts.

Mechanisms of Acquired Resistance

Bacteria can resist the effects of antimicrobial drugs through a variety of mechanisms. In some cases this resistance is innate, but oftentimes it is acquired. **Figure 21.14** depicts the most common mechanisms of acquired resistance to antimicrobial drugs.

Drug-Inactivating Enzymes

Some organisms produce enzymes that chemically modify a specific drug in such a way as to render it ineffective. Recall that bacteria that synthesize the enzyme penicillinase are resistant to penicillin. As another example, the enzyme chloramphenicol acetyltransferase chemically alters the antibiotic chloramphenicol, making it ineffective.

Alteration in the Target Molecule

An antimicrobial drug generally acts by recognizing and binding to a specific target molecule in a bacterium, interfering with its function. Minor structural changes in the target can prevent the drug from binding. Alterations in the penicillin-binding proteins prevent β-lactam drugs from binding to them. Similarly, a change in the ribosomal RNA, the target for the macrolides, prevents those drugs from interfering with ribosome function.

Decreased Uptake of the Drug

The porin proteins in the outer membrane of Gram-negative bacteria selectively permit small hydrophobic molecules to enter a cell. Alterations in these proteins can therefore alter permeability and prevent certain drugs from entering the cell. By excluding entry of a drug, an organism avoids its effects. ■ porins, p. 61

Increased Elimination of the Drug

The systems that bacteria use to transport detrimental compounds out of a cell are called **efflux pumps.** Alterations that result in the increased expression of these pumps can increase the overall capacity of a cell to eliminate a drug, thus enabling the organism to resist higher concentrations of that drug. In addition, structural changes influence the array of drugs that can be actively pumped

NON-RESISTANT CELL

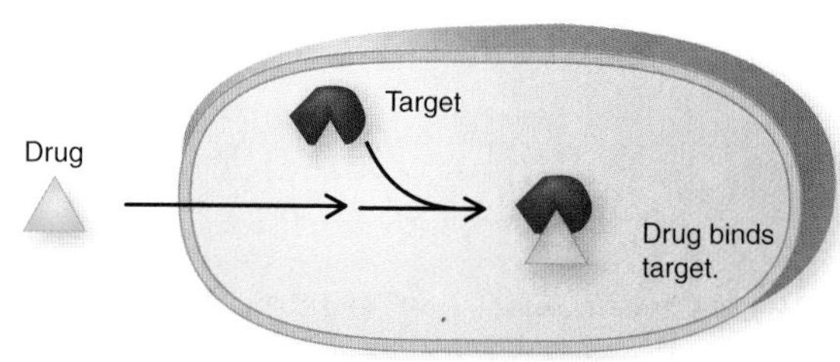
Target
Drug
Drug binds target.

RESISTANT CELL

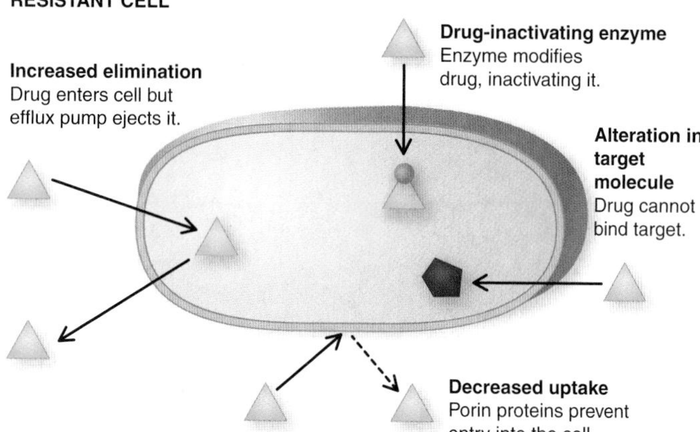
Drug-inactivating enzyme
Enzyme modifies drug, inactivating it.
Increased elimination
Drug enters cell but efflux pump ejects it.
Alteration in target molecule
Drug cannot bind target.
Decreased uptake
Porin proteins prevent entry into the cell.

FIGURE 21.14 Common Mechanisms of Acquired Antimicrobial Drug Resistance

out. Resistance that develops by this mechanism is particularly worrisome because it potentially enables an organism to become resistant to several drugs simultaneously. ■ efflux pumps, p. 58

Acquisition of Resistance

Antimicrobial resistance can be due to either spontaneous mutation, which alters existing genes, or acquisition of new genes (see figure 8.1).

Spontaneous Mutation

As cells replicate, spontaneous mutations occur at a relatively low rate. Even at a low rate, however, such mutations can ultimately have a profound effect on the resistance of a bacterial population to an antimicrobial drug. For instance, resistance to the aminoglycoside streptomycin results from a single base-pair change in the gene encoding the ribosomal protein to which streptomycin binds; that point mutation alters the target sufficiently to render the drug ineffective. When a streptomycin-sensitive bacterium is grown in streptomycin-free medium to a population of 10^9 cells, it is probable that at least one cell in the population has that particular mutation in its genome. Through spontaneous mutation, that particular cell has acquired resistance to streptomycin. If streptomycin is then added to the medium, only that cell and its progeny will be able to replicate, generating a population of streptomycin-resistant clones.

When an antimicrobial drug has several different potential targets or has multiple binding sites on a single target, it is more difficult for an organism to develop resistance through spontaneous mutation. This is because several different mutations are required to prevent binding of the drug. Unlike streptomycin, the newer aminoglycosides bind to several sites on the ribosome, making resistance due to spontaneous mutation less likely.

Drugs such as streptomycin to which single point mutations can confer resistance are sometimes used in combination with one or more other drugs to prevent survival of resistant mutants. If any organism spontaneously develops resistance to one drug, another drug will still kill it. The chance of an organism simultaneously developing mutational resistance to multiple drugs is extremely low.

Gene Transfer

Given the mobility of DNA, genes encoding resistance to an antimicrobial drug can spread to different strains, species, and even genera. The most common mechanism of transfer of resistance is through the conjugative transfer of **R plasmids.** R plasmids frequently carry several different resistance genes, each one mediating resistance to a specific antimicrobial drug. Thus, when an organism acquires an R plasmid, it acquires resistance to several different medications simultaneously. ■ R plasmid, p. 205 ■ conjugation, p. 202

In some cases, the resistance gene that was transferred from one organism to another originated through spontaneous mutation of a common bacterial gene, such as one encoding the target of the drug. In other cases, the gene may have originated from the soil microbe that naturally produces that antibiotic. For example, a gene coding for an enzyme that chemically modifies an aminoglycoside likely originated from the *Streptomyces* species that produces that aminoglycoside.

Examples of Emerging Antimicrobial Resistance

Some of the problems associated with the increasing resistance of bacteria to antimicrobial drugs are highlighted by the following examples.

Enterococci

One of the most dramatic examples of antimicrobial resistance is the enterococci, a group of bacteria that is part of the normal intestinal microbiota and a common cause of healthcare-associated infections. Enterococci are intrinsically less susceptible to many common antimicrobials. For example, their penicillin-binding proteins have low affinity for certain β-lactam antibiotics. In addition, many enterococci have plasmid-borne resistance genes. Some strains, called **vancomycin-resistant enterococci (VRE),** are even resistant to vancomycin. This drug is usually reserved as a last resort for treating life-threatening infections caused by Gram-positive organisms that are resistant to all β-lactam drugs. Because vancomycin resistance in these strains is encoded on a plasmid, the resistance is transferable to other organisms. ■ healthcare-associated infections, p. 462

Staphylococcus aureus

Staphylococcus aureus, another common cause of healthcare-associated infections, is becoming increasingly resistant to antimicrobials. Over the past 50 years, most strains have acquired resistance to penicillin due to their acquisition of a gene encoding the enzyme penicillinase. Until recently, infections caused by these strains could be treated with methicillin or other penicillinase-resistant penicillins. New strains have emerged, however, that not only produce penicillinase, but also have penicillin-binding proteins with low affinity for all β-lactam drugs. These strains, called **methicillin-resistant *Staphylococcus aureus,*** or **MRSA,** are resistant to methicillin as well as all other β-lactam drugs. Healthcare-associated infections caused by these strains are generally treated with vancomycin. A few hospitals, however, have reported isolates that are no longer susceptible to normal levels of vancomycin. So far, strict hospital guidelines designed to immediately halt the spread of these **vancomycin-intermediate *S. aureus* (VISA)** and **vancomycin-resistant *S. aureus* (VRSA)** strains have been successful. MRSA strains that cause community-acquired infections are generally susceptible to antibiotics other than β-lactam drugs. ■ *Staphylococcus aureus,* p. 535

Streptococcus pneumoniae

Until recently, *Streptococcus pneumoniae,* the leading cause of pneumonia in adults, has remained exquisitely sensitive to penicillin. Some isolates, however, are now resistant to the drug. This acquired resistance is due not to the production of a β-lactamase, but rather to modifications in the chromosomal genes coding for four different penicillin-binding proteins, decreasing their affinities for the drug. The nucleotide changes do not appear to have arisen through a series of point mutations as one might expect; instead, they are due to the acquisition of chromosomal DNA from other species of *Streptococcus.* As you may recall from earlier reading, *S. pneumoniae* can acquire DNA through DNA-mediated transformation. ■ *Streptococcus pneumoniae,* p. 509 ■ DNA-mediated transformation, p. 199

Mycobacterium tuberculosis

Treatment of *Mycobacterium tuberculosis* active infections has always been a long and complicated process, requiring a combination of two or more different drugs taken for a period of 6 months or more. Unfortunately, the first-line drugs (the preferred therapeutic agents) are often those to which spontaneous mutations readily occur. Because large numbers of bacterial cells are found in an active infection, the probability that one has developed spontaneous resistance to a single drug is likely, which is why multiple drug therapy is required. The length of treatment is due to the tediously slow growth of *M. tuberculosis.* ■ *Mycobacterium tuberculosis,* p. 514

Many tuberculosis patients do not comply with the complex treatment regime, skipping doses or stopping treatment too soon, and as a consequence, strains of *M. tuberculosis* are developing resistance to the first-line drugs. This results in even lengthier, more costly treatments that are also less effective. Strains that are resistant to two of the favored drugs for tuberculosis treatment, isoniazid and rifampin, are called **multiple-drug-resistant *M. tuberculosis,*** or **MDR-TB.** To prevent the emergence of these strains, some cities are utilizing **directly observed therapy** in which healthcare workers routinely visit patients in the community and watch them take their drugs to ensure they comply with their prescribed antimicrobial treatment. Of even greater concern, **extensively-drug-resistant *M. tuberculosis,*** or **XDR TB,** strains have been identified. These are defined as *M. tuberculosis* strains that are resistant to isoniazid and rifampin, plus three or more of the second-line drugs.

Slowing the Emergence and Spread of Antimicrobial Resistance

To reverse the alarming trend of increasing antimicrobial drug resistance, everyone must cooperate. On an individual level, physicians as well as the general public must take more responsibility for the appropriate use of these life-saving medications. On a global scale, countries around the world need to make important policy decisions about what is, and what is not, an appropriate use of these drugs.

The Responsibilities of Physicians and Other Healthcare Workers

Physicians and other healthcare workers need to increase their efforts to identify the specific causative agent of a given infection and only if appropriate, prescribe suitable antimicrobials. They must also educate their patients about the proper use of prescribed drugs in order to increase patient compliance. While these efforts may be more expensive in the short term, they will ultimately save both lives and money.

The Responsibilities of Patients

Patients need to carefully follow the instructions that accompany their prescriptions, even if those instructions seem inconvenient. It is essential to maintain the concentration of the antimicrobial in the blood at the required level for a specific time period. When a patient skips a scheduled dose of a drug, the blood level of the drug may not remain high enough to inhibit the growth of the least-sensitive members of the population. If these less-sensitive organisms then have a chance to grow, they will give rise to a population that is not as sensitive as the original. Likewise, failure to complete the prescribed course of treatment may not kill the least-sensitive organisms, allowing their subsequent multiplication. Misusing antimicrobials by skipping doses or failing to complete the prescribed duration of treatment promotes the gradual emergence of resistant organisms. In essence, the patient is selecting for the step-wise development of resistant mutants.

The Importance of an Educated Public

A greater effort must also be made to educate the public about the appropriateness and limitations of antibiotics in order to ensure they are utilized wisely. First and foremost, people need to understand that antibiotics are not effective against viruses. Taking antibiotics will not cure the common cold or any other viral illness. A few antiviral drugs are available, but they are effective against only a limited group of viruses such as HIV or herpesviruses. Unfortunately, surveys indicate that far too many people erroneously believe that antibiotics are effective against viruses and often seek prescriptions to "cure" viral infections. This misuse only selects for antibiotic-resistant bacteria in the normal microbiota. Even though these organisms are not pathogenic themselves, they can serve as a reservoir for R plasmids, eventually transferring their resistance genes to an infecting pathogen.

Global Impacts of the Use of Antimicrobial Drugs

Worldwide, there is growing concern about the overuse of antimicrobial drugs. Countries may vary in their laws and social norms, but antimicrobial resistance recognizes no political boundaries. An organism that develops resistance in one country can quickly be transported globally. In many parts of the world, particularly in developing countries, antimicrobial drugs are available on a non-prescription basis. Because of the consequences of inappropriate use, there is a growing opinion that over-the-counter availability of these drugs should be curtailed or eliminated. Another worldwide concern is the use of antimicrobial drugs in animal feeds. Low levels of these drugs in feeds result in larger, more economically productive animals, and therefore, less expensive meat, a seemingly attractive option. This use, like any other, however, selects for drug-resistant organisms, which has caused some scientists to question its ultimate wisdom. In fact, infections caused by drug-resistant *Salmonella* strains have been linked to animals whose feed was supplemented with those drugs. In response to these concerns, there is growing pressure worldwide to ban the use of antimicrobial drugs in animal feeds.

MICROCHECK 21.5

Mutations and transfer of genetic information have enabled microorganisms to develop resistance to each new antimicrobial drug developed. Drug resistance affects the cost, complications, and outcomes of medical treatment. Slowing the emergence and spread of resistant microbes involves the cooperation of health care personnel, educators, and the general public.

- ✓ Explain how using a combination of two antimicrobial drugs helps prevent the development of spontaneously resistant mutants.
- ✓ Explain the significance of a member of the normal flora that harbors an R plasmid.
- ✓ A student argued that "spontaneous mutation" meant that a drug could cause mutations. Is the student correct? Why or why not?

21.6 Mechanisms of Action of Antiviral Drugs

Focus Points

- Describe the antiviral drugs that interfere with viral uncoating.
- Describe the antiviral drugs that interfere with nucleic acid synthesis.
- Describe the antiviral drugs that interfere with the assembly and release of viral particles.

Viruses rely almost exclusively on the host cell's metabolic machinery for their replication, making them extremely difficult targets for selective toxicity. Viruses have no cell wall, ribosomes, or any other structure targeted by commonly used antibiotics. Thus, they are completely unaffected by antibiotics. Some viruses encode their own polymerases, however, and these are potential targets of antiviral drugs **(figure 21.15).** Relatively few other targets have been discovered.

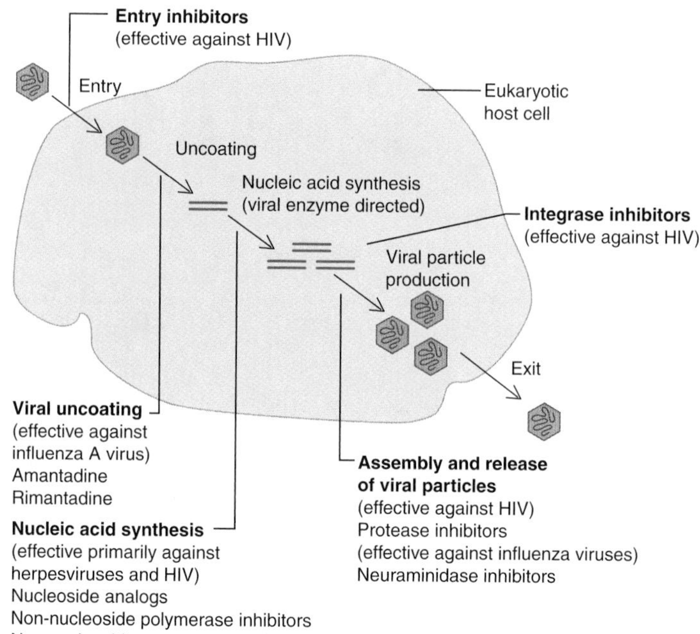

FIGURE 21.15 Targets of Antiviral Drugs

Many researchers and pharmaceutical companies are currently trying to develop more effective **antiviral drugs,** medications that interfere with viral replication. The relatively few drugs available are generally effective against only a specific type of virus; none can eliminate latent viral infections. **Table 21.2** summarizes the characteristics of the most common antiviral drugs. ■ latent viral infections, p. 330

Entry Inhibitors

A new group of drugs in the anti-HIV arsenal are the **entry inhibitors,** which prevent the virus from entering host cells. Enfuvirtide does this by binding to an HIV protein that mediates fusion of the viral envelope with the cell membrane. Maraviroc blocks the HIV coreceptor CCR5. ■ HIV attachment and entry, p. 702

Viral Uncoating

After a virus enters a host cell, the protein coat must dissociate from the nucleic acid in order for replication to occur. Drugs that interfere with the uncoating step thus prevent viral replication. The only effective drugs that target this step, however, are those that block influenza A viruses. ■ influenza A virus, p. 519

Amantadine and Rimantadine

Two drugs, **amantadine** and **rimantadine,** which are similar in both their chemical structure and mechanism of action, block the uncoating of influenza A virus after it enters a cell. Consequently, they prevent or reduce the severity and duration of the disease. The drugs are only used until active immunization can be achieved, since immunization is important for long-term protection. Unfortunately, resistance develops frequently and limits the usefulness of the drugs.

TABLE 21.2 Characteristics of Antiviral Drugs

Target/Drug Examples	Comments/Characteristics
Entry Inhibitors	
Enfuvirtide, Maraviroc	Used to treat HIV infections.
Viral Uncoating	
Amantadine and rimantadine	Reduces severity and duration of influenza A infections, but resistance limits their use.
Nucleic Acid Synthesis	
Nucleoside analogs Acyclovir, ganciclovir, ribavirin, zidovudine (AZT), didanosine (ddI), lamivudine (3TC)	Primarily used to treat infections caused by herpesviruses and HIV; they do not cure latent infections. The drugs are converted within eukaryotic cells to a nucleotide analog; virally encoded enzymes are prone to incorporate these, resulting in premature termination of synthesis or improper base-pairing of the viral nucleic acid. Acyclovir is used to treat herpes simplex virus (HSV) and varicella-zoster virus (VZV) infections. Ganciclovir is used to treat cytomegalovirus infections in immunocompromised patients. Ribavirin is used to treat respiratory syncytial virus (RSV) infections in newborns. Combinations of nucleoside analogs such as zidovudine (AZT), didanosine (ddI), and lamivudine (3TC) are used to treat HIV infections.
Non-nucleoside polymerase inhibitors Foscarnet	Primarily used to treat infections caused by herpesviruses. They inhibit the activity of viral polymerases by binding to a site other than the nucleotide-binding site. Foscarnet is used to treat ganciclovir-resistant cytomegalovirus (CMV) and acyclovir-resistant herpes simplex virus (HSV).
Non-nucleoside reverse transcriptase inhibitors Nevirapine, delavirdine, efavirenz	Used to treat HIV infections. They inhibit the activity of reverse transcriptase by binding to a site other than the nucleotide-binding site and are often used in combination with nucleoside analogs.
Integrase Inhibitors	
Raltegravir	Used to treat HIV infections.
Assembly and Release of Viral Particles	
Protease inhibitors Indinavir, ritonavir, saquinavir, nelfinavir	Used to treat HIV infections. They inhibit protease, an essential enzyme of HIV, by binding to its active site.
Neuraminidase inhibitors Zanamivir, oseltamivir	Used to treat influenza virus infections.

Nucleic Acid Synthesis

Many of the most effective antiviral drugs exploit the error-prone, virally encoded enzymes used to replicate viral nucleic acid. With few exceptions, however, the use of these drugs is generally limited to treating infections caused by the herpesviruses and HIV.

Nucleoside Analogs

A growing number of antiviral drugs are **nucleoside analogs,** compounds similar in structure to a nucleoside. These analogs can be phosphorylated *in vivo* by a virally encoded or normal cellular enzyme to form a **nucleotide analog,** a chemical structurally similar to the nucleotides of DNA and RNA. In some cases, incorporation of the nucleotide analog results in termination of the growing nucleotide chain. In other cases, incorporation of the analog results in a defective strand with altered base-pairing properties. ■ nucleotide, p. 32

The basis for the selective toxicity of most nucleoside analogs is the fact that virally encoded enzymes are much more prone to incorporating nucleotide analogs than are host polymerases. Thus, more damage is done to the rapidly replicating viral genome than to host cells. The analogs, however, are only effective against replicating viruses. Because viruses such as herpesvirus and HIV can remain latent in cells, the drugs do not cure these infections; they simply limit the duration of the active infection. Latent virus can still undergo reactivation, causing a reoccurrence of symptoms.

Most nucleoside analogs are reserved for severe infections because of their significant side effects. An important exception to this principle is **acyclovir,** a drug used to treat herpesvirus infections. This drug causes little harm to uninfected cells because the conversion of acyclovir to a nucleotide analog does not utilize a normal cellular enzyme but instead requires a virally encoded enzyme. The enzyme is only present in cells infected by herpesviruses such as herpes simplex virus (HSV) and varicella-zoster virus (VZV). Thus, acyclovir does not get converted to a nucleotide analog in uninfected cells. Other nucleoside analogs include ganciclovir, which is used to treat life-threatening or sight-threatening cytomegalovirus (CMV) infections in immunocompromised patients, and ribavirin, which is used to treat respiratory syncytial virus infections (RSV) in newborns. ■ herpes simplex virus, pp. 592, 638 ■ varicella-zoster, p. 546

A number of different nucleoside analogs are used to treat HIV infection by interfering with the activity of reverse transcriptase. Unfortunately, the virus rapidly develops mutational resistance against these drugs, which is why they are often used in combination with other anti-HIV drugs. Nucleoside analogs that interfere with reverse transcription include zidovudine (AZT), didanosine (ddI), and lamivudine (3TC). Two of these are often used in combination for HIV therapy.

Non-Nucleoside Polymerase Inhibitors

Non-nucleoside polymerase inhibitors are compounds that inhibit the activity of viral polymerases by binding to a site other than the nucleotide-binding site. One example, foscarnet, is used to treat infections caused by ganciclovir-resistant CMV and acyclovir-resistant HSV.

Non-Nucleoside Reverse Transcriptase Inhibitors

Non-nucleoside reverse transcriptase inhibitors inhibit the activity of reverse transcriptase by binding to a site other than the nucleotide-binding site. They are often used in combination with nucleoside analogs to treat HIV infections. The medications include nevirapine, delavirdine, and efavirenz.

Integrase Inhibitors

Integrase inhibitors offer a new option for treating HIV infections. They inhibit the HIV-encoded enzyme integrase, thereby preventing the virus from inserting the DNA copy of its genome into that of the host cell. Raltegravir is the first approved drug of this class. ■ HIV integrase, p. 701

Assembly and Release of Viral Particles

Virally encoded enzymes required for the production and release of viral particles are the targets of medications used to treat certain viral infections.

Protease Inhibitors

Protease inhibitors are used to treat HIV infections. These medications inhibit the HIV-encoded enzyme protease, which plays an essential role in the production of viral particles. When HIV replicates, several of its proteins are translated as a single amino acid chain, or polyprotein; protease then cleaves the polyprotein into individual proteins. The various protease inhibitors, including indinavir, ritonavir, saquinavir, and nelfinavir, differ in several aspects including dosage and side effects. ■ HIV protease, p. 701

Neuraminidase Inhibitors

Neuraminidase inhibitors inhibit **neuraminidase,** an enzyme encoded by influenza viruses that is essential for the release of infectious viral particles from infected cells. Two neuraminidase inhibitors are currently available—zanamivir, which is administered by inhalation, and oseltamivir, which is administered orally. Both limit the duration of influenza infections when taken within two days of the onset of symptoms.

MICROCHECK 21.6

Viral replication generally uses host cell machinery; because of this there are few targets for selectively toxic antiviral drugs. Available antiviral drugs are virus-specific; targets include viral entry, viral uncoating, nucleic acid synthesis, integrase, and the assembly and release of viral particles.

- ✓ Explain why acyclovir has fewer side effects than do other nucleoside analogs.
- ✓ Why are nucleoside analogs active only against replicating viruses?

21.7 Mechanisms of Action of Antifungal Drugs

Focus Points

- Describe the antifungal drugs that interfere with plasma membrane synthesis and function.
- Describe the antifungal mechanisms of echinocandins, griseofulvin, and flucytosine.

Eukaryotic pathogens such as fungi more closely resemble human cells than do bacteria. It is not surprising, therefore, that relatively few drugs are available for systemic use against fungal pathogens. The targets of antifungal drugs are illustrated in **figure 21.16.** **Table 21.3** summarizes the characteristics of the most common antifungal drugs.

Plasma Membrane Synthesis and Function

The target of most antifungal drugs is **ergosterol,** a sterol found in the plasma membrane of fungal but not human cells. ■ sterol, p. 36

Polyenes

The **polyenes,** a group of antibiotics produced by certain species of *Streptomyces,* bind to ergosterol. This disrupts the fungal membrane, causing leakage of the cytoplasmic contents and leading to cell death. Unfortunately, the polyenes are quite toxic to humans, which limits their systemic use to life-threatening infections. Amphotericin B causes severe side effects, but it is the most effective drug for treating many systemic infections. Newer lipid-based emulsions are less toxic but are more expensive. Nystatin is too toxic to be given systemically, but is used topically.

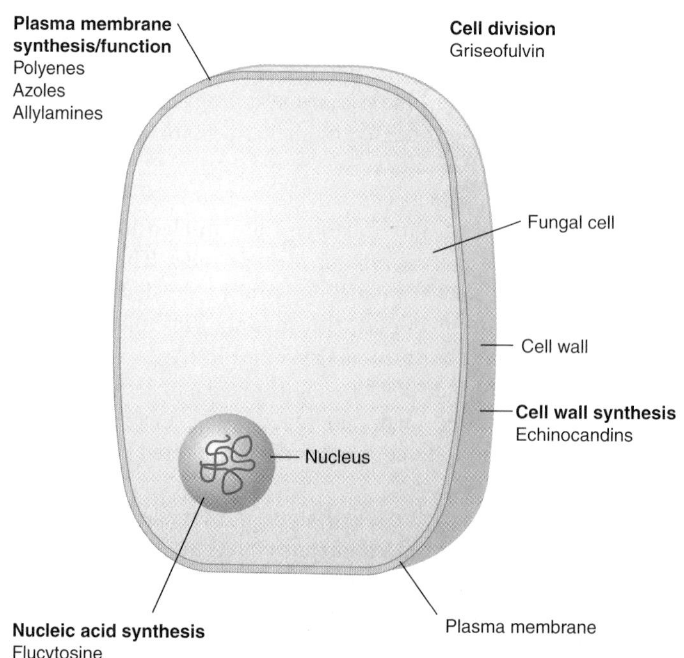

FIGURE 21.16 Targets of Antifungal Drugs

TABLE 21.3 Characteristics of Antifungal Drugs

Drug Target/Drug	Comments
Plasma Membrane	
Polyenes Amphotericin B, nystatin	Bind to ergosterol, disrupting the plasma membrane and causing leakage of the cytoplasm. Amphotericin B is very toxic but the most effective drug for treating life-threatening infections; newer lipid-based emulsions are less toxic but very expensive. Nystatin is too toxic for systemic use, but can be used topically.
Azoles *Imidazoles:* ketoconazole, miconazole, clotrimazole *Triazoles:* fluconazole, itraconazole	Interfere with ergosterol synthesis, leading to defective cell membranes; active against a wide variety of fungi. Used to treat a variety of systemic and localized fungal infections. Triazoles are less toxic than imidazoles.
Allylamines Naftifine, terbinafine	Inhibit an enzyme in the pathway of ergosterol synthesis. Administered topically to treat dermatophyte infections. Terbinafine can be taken orally.
Cell Wall Synthesis	
Echinocandins Capsofungin	Interfere with β-1,3 glucan synthesis. Used to treat *Candida* infections as well as invasive aspergillosis that resists other treatments.
Cell Division	
Griseofulvin	Used to treat skin and nail infections. Taken orally for months; concentrates in the dead keratinized layers of the skin; taken up by fungi invading those cells and inhibits their division. Active only against fungi that invade keratinized cells.
Nucleic Acid Synthesis	
Flucytosine	Used to treat systemic yeast infections; enzymes within yeast cells convert the drug to 5-fluorouracil, which inhibits an enzyme required for nucleic acid synthesis; not effective against most molds; resistant mutants are common.

Azoles

The **azoles** are a large family of chemically synthesized drugs, some of which have antifungal activity. They include two classes—the imidazoles and the newer triazoles; the latter are generally less toxic. Both classes inhibit the synthesis of ergosterol, resulting in defective fungal membranes that leak cytoplasmic contents. Fluconazole and itraconazole, which are triazoles, are increasingly being used to treat systemic fungal infections. Ketoconazole, an imidazole, is also used systemically, but it is associated with more severe side effects. Other imidazoles, including miconazole and clotrimazole, are commonly used in nonprescription creams, ointments, and suppositories to treat vaginal yeast infections. They are also used topically to treat dermatophyte infections. ■ dermatophyte, p. 554

Allylamines

The **allylamines** inhibit an enzyme in the pathway of ergosterol synthesis. Naftifine and terbinafine can be administered topically to treat dermatophyte infections. Terbinafine can also be taken orally.

Cell Wall Synthesis

Fungal cell walls contain some components not produced by animal cells. Synthesis of these is a target for some antifungal medications.

Echinocandins

Echinocandins are a family of antifungal agents that interfere with synthesis of the β-1, 3 glucan component of fungal cell walls. Caspofungin, the first member to be approved, is used to treat *Candida* infections as well as invasive aspergillosis that resists other treatments.

Cell Division

The target of one antifungal drug, griseofulvin, is cell division.

Griseofulvin

Griseofulvin interferes with the action of tubulin, a necessary factor in nuclear division. Because tubulin is a part of all eukaryotic cells, the selective toxicity of this drug may be due to its greater uptake by fungal cells. When the drug is taken orally for months, it is absorbed and eventually concentrated in the dead keratinized layers of the skin. The fungi that then invade keratin containing structures such as skin and nails take up the drug, which prevents their multiplication. It is only active against fungi that invade keratinized cells and is used to treat skin and nail infections. ■ tubulin, p. 74 ■ keratin, p. 532

Nucleic Acid Synthesis

Nucleic acid synthesis is a common feature of all eukaryotic cells, which generally makes it a poor target for antifungal drugs. The drug flucytosine, however, is taken up by yeast cells and then converted by yeast enzymes to an active, inhibitory form.

Flucytosine

Flucytosine is a synthetic derivative of cytosine, one of the pyrimidines found in nucleic acids. Enzymes within infecting yeast cells convert flucytosine to 5-fluorouracil, which inhibits an enzyme required for nucleic acid synthesis. Unfortunately, resistant mutants are common, and therefore, flucytosine, is used mostly in combination with amphotericin B or as an alternative drug for patients with systemic yeast infections who are unable to tolerate amphotericin B. Flucytosine is not effective against molds.

MICROCHECK 21.7

Because fungi are eukaryotic cells, there are relatively few targets for selectively toxic antifungal drugs. Most antifungal drugs interfere with the function or synthesis of ergosterol, which is found in the membrane of fungal but not human cells. Other targets of antifungal drugs include cell wall synthesis, cell division, and nucleic acid synthesis.

- ✓ Why is amphotericin B, a polyene, used only for treating life-threatening infections?
- ✓ Why is flucytosine generally used only in combination with other drugs?
- ✓ If griseofulvin were not concentrated in keratin containing structures, would it still be toxic to fungi that invade these structures? Why or why not?

21.8 Mechanisms of Action of Antiprotozoan and Antihelminthic Drugs

Focus Point

- Describe five antiprotozoan drugs and four antihelminthic drugs.

Most antiparasitic drugs probably interfere with biosynthetic pathways of protozoan parasites or the neuromuscular function of worms. Unfortunately, compared with antibacterial, antifungal, and antiviral drugs, little research and development goes into these drugs, because most parasitic diseases are concentrated in the poorer areas of the world where people simply cannot afford to spend money on expensive medications.

Some of the most important antiparasitic drugs and their characteristics are summarized in **table 21.4.**

TABLE 21.4 Characteristics of Some Antiprotozoan and Antihelminthic Drugs

Causative Agent/Drug	Comments
Intestinal protozoa	
Iodoquinol	Mechanism unknown; poorly absorbed but taken orally to eliminate amebic cysts in the intestine.
Nitroimidazoles Metronidazole	Activated by the metabolism of anaerobic organisms. Interferes with electron transfer and alters DNA. Does not reliably eliminate the cyst stage. Metronidazole is also used to treat infections caused by anaerobic bacteria.
Quinacrine	Mechanism of action is unknown, but may be due to interference with nucleic acid synthesis.
Plasmodium (Malaria) and Toxoplasma	
Folate antagonists Pyrimethamine, sulfonamide	Interferes with folate metabolism; used to treat toxoplasmosis and malaria.
Malarone®	A synergistic combination of atovaquone and proguanil hydrochloride used to treat malaria. Atovaquone interferes with mitochondrial electron transport while proguanil disrupts folate synthesis. The combination is active against both the blood stage and early liver stage of *Plasmodium* species.
Quinolones Chloroquine, mefloquine, primaquine, tafenoquine	The mechanism of action is not completely clear. Chloroquine is concentrated in infected red blood cells and is the drug of choice for preventing or treating the red blood cell stage of the malarial parasite. Its effects may be due to inhibition of an enzyme that protects the parasite from the toxic by-products of hemoglobin degradation. Primaquine and tafenoquine destroy the liver stage of the parasite and are used to treat relapsing forms of malaria. Mefloquine is used to treat infection caused by chloroquine-resistant strains of the malarial parasite.
Trypanosomes and Leishmania	
Eflornithine	Used to treat infections caused by some types of *Trypanosoma;* inhibits the enzyme ornithine decarboxylase.
Heavy metals Melarsoprol, sodium stibogluconate, meglumine antimonate	These inactivate sulfhydryl groups of parasitic enzymes, but are very toxic to host cells as well. Melarsoprol is used to treat trypanosomiasis, but the treatment can be lethal. Sodium stibogluconate and meglumine antimonate are used to treat leishmaniasis.
Nitrofurtimox	Widely used to treat acute Chagas' disease; forms reactive oxygen radicals that are toxic to the parasite as well as the host.

TABLE 21.4 Characteristics of Some Antiprotozoan and Antihelminthic Drugs *(continued)*

Causative Agent/Drug	Comments
Intestinal and Tissue Helminths	
Avermectins	Ivermectin causes neuromuscular paralysis in parasites; used to treat infections caused by *Strongyloides* and tissue nematodes.
Ivermectin	
Benzimidazoles	Mebendazole binds to tubulin of helminths, blocking microtubule assembly and inhibiting glucose uptake. It is poorly absorbed in the intestine, making it effective for treating intestinal, but not tissue, helminths. Thiabendazole may have a similar mechanism, but it is well absorbed and has many toxic side effects. Albendazole is used to treat tissue infections caused by *Echinococcus* and *Taenia solium*.
Mebendazole, thiabendazole, albendazole	
Phenols	Absorbed by cestodes in the intestinal tract, but not by the human host.
Niclosamide	
Piperazines	Piperazine causes a flaccid paralysis in worms and can be used to treat *Ascaris* infections. Diethylcarbamazine immobilizes filarial worms and alters their surface, which enhances killing by the immune system. The resulting inflammatory response, however, causes tissue damage.
Piperazine, diethylcarbamazine	
Pyrazinoisoquinolines	A single dose of praziquantel is effective in eliminating a wide variety of trematodes and cestodes. It is taken up but not metabolized by the worm, ultimately causing sustained contractions of the worm.
Praziquantel	
Tetrahydropyrimidines	Pyrantel pamoate interferes with neuromuscular activity of worms, causing a type of paralysis. It is not readily absorbed from the gastrointestinal tract and is active against intestinal worms including pinworm, hookworm, and *Ascaris*. Oxantel can be used to treat *Trichuris* infections.
Pyrantel pamoate, oxantel	

FUTURE CHALLENGES

War with the Superbugs

With respect to antimicrobial drugs, the future challenge is already upon us. The challenge is to maintain the effectiveness of antimicrobials by (1) preventing the continued spread of resistance, and (2) developing new drugs that have even more desirable properties. New ways must be developed to fight infections caused by the ever greater numbers of bacterial strains that are resistant to the effects of conventional antimicrobial drugs. Several strategies are being used to develop potential weapons against these "superbugs."

One way to combat resistance is to continue developing modifications of existing antimicrobial drugs. Researchers constantly work to modify drugs chemically, trying to keep at least one step ahead of bacterial resistance. Another method to foil drug resistance is to interfere with the resistance mechanisms, as is done currently in using β-lactamase inhibitors in combination with β-lactam drugs to protect the antimicrobial drug from enzymatic destruction. Likewise, it may be possible to thwart resistance using other mechanisms—for example, developing chemicals that can be used to inactivate or interfere with bacterial efflux systems.

Other researchers are focusing on developing medications entirely unrelated to conventional antimicrobials. One example is a class of compounds called **defensins,** which are short peptides, approximately 29 to 35 amino acids in length, produced naturally by a variety of eukaryotic cells to fight infections. Various defensins and related compounds are being intensively studied as promising antimicrobials.

Accumulating knowledge of the genetic sequences of pathogens and identification of genes associated with pathogenicity may allow development of antimicrobials that interfere directly with those processes. The "master switch" that controls expression of virulence determinants is one such potential target. Nucleotide sequence information and determination of the three-dimensional conformation of proteins may uncover new targets for antimicrobial drug therapy.

SUMMARY

21.1 History and Development of Antimicrobial Drugs

Discovery of Antimicrobial Drugs

Salvarsan, developed by Paul Ehrlich, was the first documented example of an antimicrobial medication.

Discovery of Antibiotics

Alexander Fleming discovered that a species of the fungus *Penicillium* produces **penicillin,** which kills some bacteria.

Development of New Generations of Drugs

Antimicrobial drugs can be chemically modified to give them new properties.

21.2 Features of Antimicrobial Drugs

Most modern antibiotics come from species of *Streptomyces* and *Bacillus* (bacteria) and *Penicillium* and *Cephalosporium* (fungi).

Selective Toxicity

Medically useful antimicrobials are **selectively toxic;** the relative toxicity of a drug is expressed as the **therapeutic index.**

Antimicrobial Action

Bacteriostatic drugs inhibit the growth of bacteria; drugs that kill bacteria are **bactericidal.**

Spectrum of Activity

Broad-spectrum antimicrobials affect a wide range of bacteria; those that affect a narrow range are called **narrow-spectrum.**

Effects of Combinations of Antimicrobial Drugs

Combinations of antimicrobial drugs can be **synergistic, antagonistic,** or **additive.**

Tissue Distribution, Metabolism, and Excretion of the Drug

Some antimicrobials cross the blood-brain barrier into the cerebrospinal fluid; these can be used to treat meningitis. Drugs that are unstable in acid must be administered through injection. Drugs that have a long **half-life** need to be administered less frequently.

Adverse Effects

Some people develop allergies to certain antimicrobials. Some antimicrobials can have potentially damaging side effects such as kidney damage. When the composition of the normal microbiota is altered, which happens when a person takes antimicrobials, pathogens normally unable to compete may grow to high numbers.

Resistance to Antimicrobials

The resistance of certain types of bacteria to a particular drug is **intrinsic** or **innate.** Previously sensitive microorganisms can develop resistance through spontaneous mutation or the acquisition of new genetic information.

21.3 Mechanisms of Action of Antibacterial Drugs (table 21.1, figure 21.2)

Antibacterial Medications That Inhibit Cell Wall Synthesis (figure 21.3)

The β-lactam drugs, which include the **penicillins, cephalosporins, carbapenems,** and **monobactams,** irreversibly inhibit **penicillin-binding proteins (PBPs),** ultimately leading to cell lysis (figure 21.4). **Vancomycin** binds to the terminal amino acids of the peptide side chain of NAM, blocking peptidoglycan synthesis. **Bacitracin** interferes with the transport of peptidoglycan precursors across the cytoplasmic membrane; it is a common ingredient in non-prescription ointments.

Antibacterial Medications That Inhibit Protein Synthesis (figure 21.7)

Antibiotics that inhibit protein synthesis by binding to the 70S ribosome include the **aminoglycosides,** the **tetracyclines,** the **macrolides, chloramphenicol,** the **lincosamides,** the **oxazolidinones,** and the **streptogramins.**

Antibacterial Medications That Inhibit Nucleic Acid Synthesis

The **fluoroquinolones** interfere with DNA replication and transcription by inhibiting one or more topoisomerases. The **rifamycins** block prokaryotic RNA polymerase from initiating transcription.

Antibacterial Medications That Inhibit Metabolic Pathways (figure 21.8)

Sulfa drugs competitively inhibit an enzyme in the metabolic pathway that leads to folic acid synthesis. **Trimethoprim** inhibits the enzyme that catalyzes a metabolic step following the one inhibited by sulfonamides.

Antibacterial Medications That Interfere with Cell Membrane Integrity

Polymyxin B damages membranes of Gram-negative bacteria. **Daptomycin** damages the cytoplasmic membrane.

Antibacterial Medications That Interfere with Processes Essential to Mycobacterium tuberculosis

First-line medications that specifically target *Mycobacterium* species include **isoniazid, ethambutol,** and **pyrazinamide.**

21.4 Determining the Susceptibility of a Bacterial Strain to an Antimicrobial Drug

Determining the Minimum Inhibitory and Bactericidal Concentrations (figure 21.9)

The **minimum inhibitory concentration (MIC)** is the lowest concentration of a specific antimicrobial drug needed to prevent the growth of a bacterial strain *in vitro.* The **minimum bactericidal concentration (MBC)** is the lowest concentration of a specific antimicrobial drug that kills 99.9% of cells of a given strain of bacterial *in vitro.*

Conventional Disc Diffusion Method (figure 21.10)

The **Kirby-Bauer disc diffusion test** qualitatively determines the susceptibility of a bacterial strain to a battery of antimicrobial drugs.

Commercial Modifications of Antimicrobial Susceptibility Testing (figure 21.11)

Automated methods can determine antimicrobial susceptibility in as little as 4 hours.

21.5 Resistance to Antimicrobial Drugs (figure 21.13)

Mechanisms of Acquired Resistance (figure 21.14)

Enzymes that chemically modify a drug render it ineffective. Structural changes in the target can prevent the drug from binding. Altered porin proteins prevent drugs from entering cells. **Efflux pumps** actively pump drugs out of cells.

Acquisition of Resistance

Resistance can be acquired through spontaneous mutation or horizontal gene transfer. The most common mechanism of transfer of antibiotic resistance genes is through the conjugative transfer of **R plasmids.**

Examples of Emerging Antimicrobial Resistance

Vancomycin-resistant enterococci (VRE), methicillin-resistant *Staphylococcus aureus* (MRSA), vancomycin-intermediate *S. aureus* (VISA), vancomycin-resistant *S. aureus* (VRSA), penicillin-resistant *Streptococcus pneumoniae,* multiple-drug-resistant *Mycobacterium tuberculosis* (MDR-TB), and extensively drug resistant *Mycobacterium tuberculosis* (XDR-TB) are all examples of emerging antimicrobial resistance.

Slowing the Emergence and Spread of Antimicrobial Resistance

Physicians should prescribe antimicrobial medications only when appropriate. The public must be educated about the appropriateness and limitations of antimicrobial therapy. Patients need to carefully follow prescribed instructions when taking antimicrobials.

21.6 Mechanisms of Action of Antiviral Drugs (figure 21.15, table 21.2)

Entry Inhibitors

Enfuvirtide and maraviroc prevents HIV from entering cells.

Viral Uncoating

Amantadine and **rimantadine** block the uncoating of influenza A virus after it enters a cell.

Nucleic Acid Synthesis

Most antiviral drugs take advantage of the error-prone virally encoded enzymes used to replicate viral nucleic acid. **Nucleoside analogs** are phosphorylated *in vivo* to form **nucleotide analogs;** when these are incorporated into viral DNA they interfere with replication.

Integrase Inhibitors

Raltegravir inhibits HIV integrase.

Assembly and Release of Viral Particles

Protease inhibitors bind to and inhibit protease, the enzyme required for the production of infectious HIV particles. **Neuraminidase inhibitors** interfere with the release of influenza virus particles from a cell.

21.7 Mechanisms of Action of Antifungal Drugs (figure 21.16, table 21.3)

Plasma Membrane Synthesis and Function

The **polyenes** disrupt fungal cell membranes by binding to ergosterol. The **azoles** inhibit the synthesis of ergosterol. The **allylamines** inhibit an enzyme in the pathway of ergosterol synthesis.

Cell Wall Synthesis

Echinocandins interfere with the synthesis of β-1, 3 glucan.

Cell Division

Griseofulvin is concentrated in keratinized skin cells, where it inhibits fungal cell division.

Nucleic Acid Synthesis

Flucytosine is taken up by yeast cells and converted by yeast enzymes to an active form.

21.8 Mechanisms of Action of Antiprotozoan and Antihelminthic Drugs (table 21.4)

Most antiparasitic drugs are thought to interfere with biosynthetic pathways of protozoan parasites or the neuromuscular function of worms.

REVIEW QUESTIONS

Short Answer

1. Describe the difference between the terms *antibiotic* and *antimicrobial.*
2. Define *therapeutic index* and explain its importance.
3. Explain the role of penicillin-binding proteins in drug susceptibility.
4. Name three of the first-line drugs used to treat tuberculosis.
5. Name three antimicrobial medications that target ribosomes.
6. Compare and contrast the method for determining the minimum inhibitory concentration of an antimicrobial drug with the Kirby-Bauer disc diffusion test.
7. Name three targets that can be altered sufficiently via spontaneous mutation to result in resistance to an antimicrobial drug.
8. What is MRSA? Why is it significant?
9. Why is it difficult to develop antiviral drugs?
10. Explain the difference between the mechanism of action of an azole and that of a polyene.

Multiple Choice

1. Which of the following targets would you expect to be the most selective with respect to toxicity?
   a) Cytoplasmic membrane function
   b) DNA synthesis
   c) Glycolysis
   d) Peptidoglycan synthesis
   e) 70S ribosome
2. Penicillin has been modified to make derivatives that differ in all of the following, *except*
   a) spectrum of activity.
   b) resistance to β-lactamases.
   c) potential for allergic reactions.
   d) A and C.
3. Which of the following is the target of β-lactam antibiotics?
   a) Peptidoglycan synthesis b) DNA synthesis
   c) RNA synthesis d) Protein synthesis
   e) Folic acid synthesis
4. Which of the following statements is *false?*
   a) A bacteriostatic drug stops the growth of a microorganism.
   b) The lower the therapeutic index, the less toxic the drug.
   c) Broad-spectrum antibiotics are associated with the development of antibiotic-associated colitis.
   d) Azithromycin has a longer half-life than does penicillin V.
   e) Chloramphenicol can cause a life-threatening type of anemia.
5. All of the following interfere with the function of the ribosome, *except*
   a) fluoroquinolones. b) lincosamides. c) macrolides.
   d) streptogramins. e) tetracyclines.
6. The target of the sulfonamides is
   a) cytoplasmic membrane proteins.
   b) folic acid synthesis.
   c) gyrase.
   d) peptidoglycan biosynthesis.
   e) RNA polymerase.
7. Routine antimicrobial therapy to treat tuberculosis involves taking
   a) one drug for 10 days.
   b) two or more drugs for 10 days.
   c) one drug for at least 6 months.
   d) two or more drugs for at least 6 months.
   e) five drugs for 2 years.

8. Strains of *Staphylococcus aureus* referred to as MRSA are sensitive to
   a) methicillin. b) penicillin. c) cephalosporin.
   d) vancomycin. e) none of the above.
9. Acyclovir is
   a) a nucleoside analog.
   b) a non-nucleoside polymerase inhibitor.
   c) a protease inhibitor.
   d) none of the above.
10. The antifungal drug griseofulvin is used to treat
    a) vaginal infections. b) systemic infections.
    c) nail infections. d) hair infections.

Applications

1. A physician was treating one young woman and one elderly patient for urinary tract infections caused by the same type of bacterium. Although the patients had similar body dimensions and weight, the physician gave a smaller dose of drug to the older patient. What was the physician's rationale for this decision?
2. An advocacy group in Washington, D.C., is petitioning the United States Department of Agriculture (USDA) to stop the use of low-dosage antimicrobial agents used to enhance the growth of cattle and chickens. Why is the group against this practice? Why does the USDA permit it?

Critical Thinking

1. Figure 21.12 shows the E-test procedure for determining an MIC value. How would the zone of inhibition appear if the drug concentrations in the strip were decreased slightly?
2. Why is acyclovir converted to a nucleotide analog only in cells infected with herpes simplex virus?

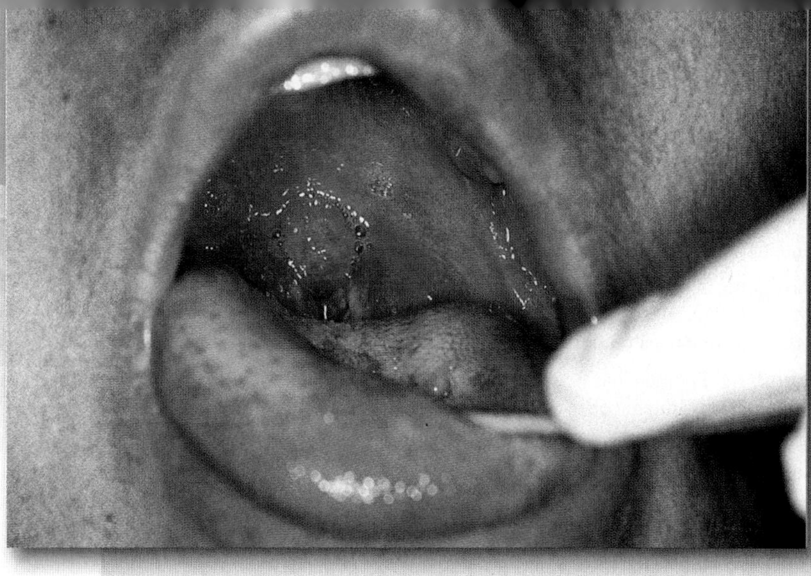

Individual with streptococcal pharyngitis.

Respiratory System Infections

A Glimpse of History

Rebecca Lancefield's mother was a direct descendent of Lady Mary Wortley Montagu, who promoted smallpox variolation in England more than 75 years before Jenner. Lancefield was educated in various schools as her parents moved from one army base to another. In 1912, she entered Wellesley College and became interested in biology. She was awarded a scholarship to Columbia University and studied under the famous microbiologist Hans Zinsser, a pioneer in the science of immunology. She received a Master's degree, but her studies were interrupted when her husband was drafted into the armed services in World War I. Fortunately he was assigned to the Rockefeller Institute, where Rebecca got a job as a laboratory technician for the distinguished microbiologists O. T. Avery and A. R. Dochez.

Well known for their studies of pneumococci, Avery and Dochez had been commissioned to study streptococcal cultures from personnel at army camps. At that time, classification of the numerous kinds of streptococci was based largely on whether their colonies on blood agar produced β-hemolysis or a green discoloration. By using different streptococcal strains to immunize mice and produce antisera, Avery and Dochez showed that their cultures could be divided into groups based on their antigens.

Later, the Lancefields returned to Columbia, where Lancefield received her Ph.D. in 1925 for her studies of α-hemolytic streptococci. At that time, α-hemolytic streptococci were considered a possible cause of rheumatic fever, a common cause of serious heart disease, but she showed that these streptococci occurred just as often in healthy people as in those with rheumatic fever. Thereafter, Lancefield returned to the Rockefeller Institute where she spent the remainder of her scientific career studying streptococci. Building on the discoveries of others, she showed that almost all the strains of β-hemolytic streptococci from human infections had the same cell wall carbohydrate "A." Streptococci from other sources had different carbohydrates: "B" from cattle infections; "C" from cattle, horses, and guinea pigs; "D" from cheese and human normal flora; and so forth. The grouping of streptococci by their cell wall carbohydrates, now referred to as "Lancefield grouping," proved to be a much better predictor of pathogenic potential than hemolysis on blood agar. Lancefield, in 1960, became the first woman president of the American Association of Immunologists. In 1970, Lancefield was elected to the prestigious National Academy of Sciences. She died in 1981 at the age of 86. ■ **Mary Wortley Montagu, p. 431**

Respiratory infections encompass an enormous variety of illnesses ranging from the trivial to the fatal. For convenience, they can be divided into infections of the upper part of the respiratory system, primarily the head and neck, and infections of the lower respiratory system, in the chest. Most upper respiratory infections are uncomfortable, but not life threatening and go away without treatment in about a week. However, they are so common that they far outweigh other infections in terms of the cumulative misery they cause. Some illnesses, such as the childhood rashes discussed in chapter 23, have a minor upper respiratory component but injure the skin, lung, nervous system, or other parts of the body. Still others cause major symptoms involving the eye, nose, throat, middle ear, sinuses, and other body systems. The lower respiratory system is usually sterile, well protected from colonization by microorganisms. However, pathogens sometimes evade the body's defenses and cause serious diseases, such as pneumonia, tuberculosis, and whooping cough.

22.1 Anatomy and Physiology

Focus Points

- Outline the functions of the upper respiratory tract.
- Describe the mucociliary escalator.

KEY TERMS

Antigenic Drift Slight changes that occur in the antigens of a virus that reduce its susceptibility to antibodies existing before the antigenic changes.

Antigenic Shift Major changes in the antigens of a virus that make antibodies that existed before the changes ineffective against the virus.

Autolysis Spontaneous disintegration as a result of enzymes within a cell.

Conjunctivitis Inflammation of the outer surface of the eye.

Croup Acute obstruction of the larynx occurring mainly in infants and young children, often resulting from respiratory syncytial or other viral infection.

Directly Observed Therapy (DOT) Method used to ensure that patients comply with their treatment; the healthcare worker watches while the patient takes each dose.

Granuloma Found in a chronic inflammatory response, collections of lymphocytes and macrophages; an attempt by the body to wall off and contain persistent organisms and antigens.

Mucociliary Escalator Moving layer of mucus and cilia lining the respiratory tract that traps bacteria and other particles and moves them into the throat.

Otitis Media Infection of the middle ear.

Pharyngitis Inflammation of the throat.

Pneumonia Inflammation of the lungs accompanied by filling of the air sacs with fluids such as pus and blood.

Pyrogenic Fever-causing.

Sputum Material coughed up from the lungs.

Tubercle Granuloma formed in tuberculosis.

Virulence Relative ability of a pathogen to overcome body defenses and cause disease.

Microorganisms that colonize the respiratory system enter hot, humid, dark caverns and tortuous passages of the respiratory system. The atmosphere contains less oxygen and more carbon dioxide than occurs in air. The eyes and the nose are the two main "portals of entry" to the respiratory system. Respiratory infections often first establish themselves there and then spread to other parts of the system. ■ portals of entry, p. 455

Structures commonly involved in infections of the upper respiratory system **(figure 22.1)** are:

- the conjunctiva, composing the moist surfaces of the eyes and eyelids. Infection of the conjunctiva is called **conjunctivitis;**
- the nasolacrimal, or tear ducts, from the eyes to the nasal chamber. Infection of the tear ducts is called **dacryocystitis;**
- the middle ear. Infection of the middle ear is called **otitis media.** The external ear canal is not part of the respiratory system. It is a blind tunnel lined with skin and is part of the skin ecosystem; infection there is called **external otitis;**
- the air-filled chambers of the skull, the sinuses and mastoid air cells. Infections of these chambers are called **sinusitis** and **mastoiditis,** respectively;
- the nose. Infection of the nose is called **rhinitis;**
- the throat, or pharynx. Infections of the throat are called **pharyngitis;**
- the epiglottis is a little muscular flap that covers the opening to the lower respiratory system during swallowing, thereby preventing material from entering. Infection of the epiglottis is called **epiglottitis.**

The tonsils are secondary lymphoid organs, strategically located to come into contact with incoming microorganisms. The tonsils are important in the immune response but paradoxically they can also be the sites of infection, resulting in **tonsillitis.** Enlargement of the pharyngeal tonsil, also called **adenoids,** can contribute to ear infections by interfering with normal drainage from the eustachian tubes. These tubes, which extend from the middle ear to the nasopharynx, equalize the pressure in the middle ear and drain normal mucous secretions. ■ lymphoid system, p. 369

A person normally breathes about 16 times per minute, inhaling about 0.5 liters of air with each breath, or more than 11,500 liters of air per day, with any accompanying microorganisms or other pollutants. The air enters the respiratory system at the nostrils, flows into the nasal cavity, and is deflected downward through the throat.

One of the main functions of the upper respiratory tract is to regulate the temperature and water content of inspired air. When cold air enters the upper respiratory tract, nervous reflexes immediately increase the blood flow to the spongy tissues in the nose, thereby transferring heat to the air. This mechanism usually adjusts the temperature to within 2 to 3 degrees of body temperature by the time it reaches the lungs. Inspired air also becomes saturated with water vapor, the nasal tissues giving up as much as a quart of water per day to keep the air humidified. The warmth and moisture provide optimum conditions for the body's defenses against infection.

The air then enters the lower respiratory tract below the epiglottis. Structures commonly involved in lower respiratory system infections include:

- the voice box, or larynx (see figure 22.1). Inflammation of the larynx is called **laryngitis,** and is manifest as hoarseness.
- The larynx is continuous with the windpipe, or trachea, which branches into two bronchi. Inflammation of the bronchi is called **bronchitis,** commonly the result of viral infection or smoking tobacco.
- The bronchi branch repeatedly, becoming bronchioles, site of an important viral infection called **bronchiolitis.**
- The smallest branches of the bronchioles end in the tiny, thin-walled air sacs called alveoli that comprise the bulk of lung tissue. Inflammation of the lungs is called **pneumonitis,** often the result of viral infections. Pneumonitis that results in the filling of alveoli with pus and fluid is called **pneumonia.** Macrophages are numerous in the lung tissues and readily move into the alveoli and airways to engulf infectious agents, thus helping to prevent pneumonia from developing.

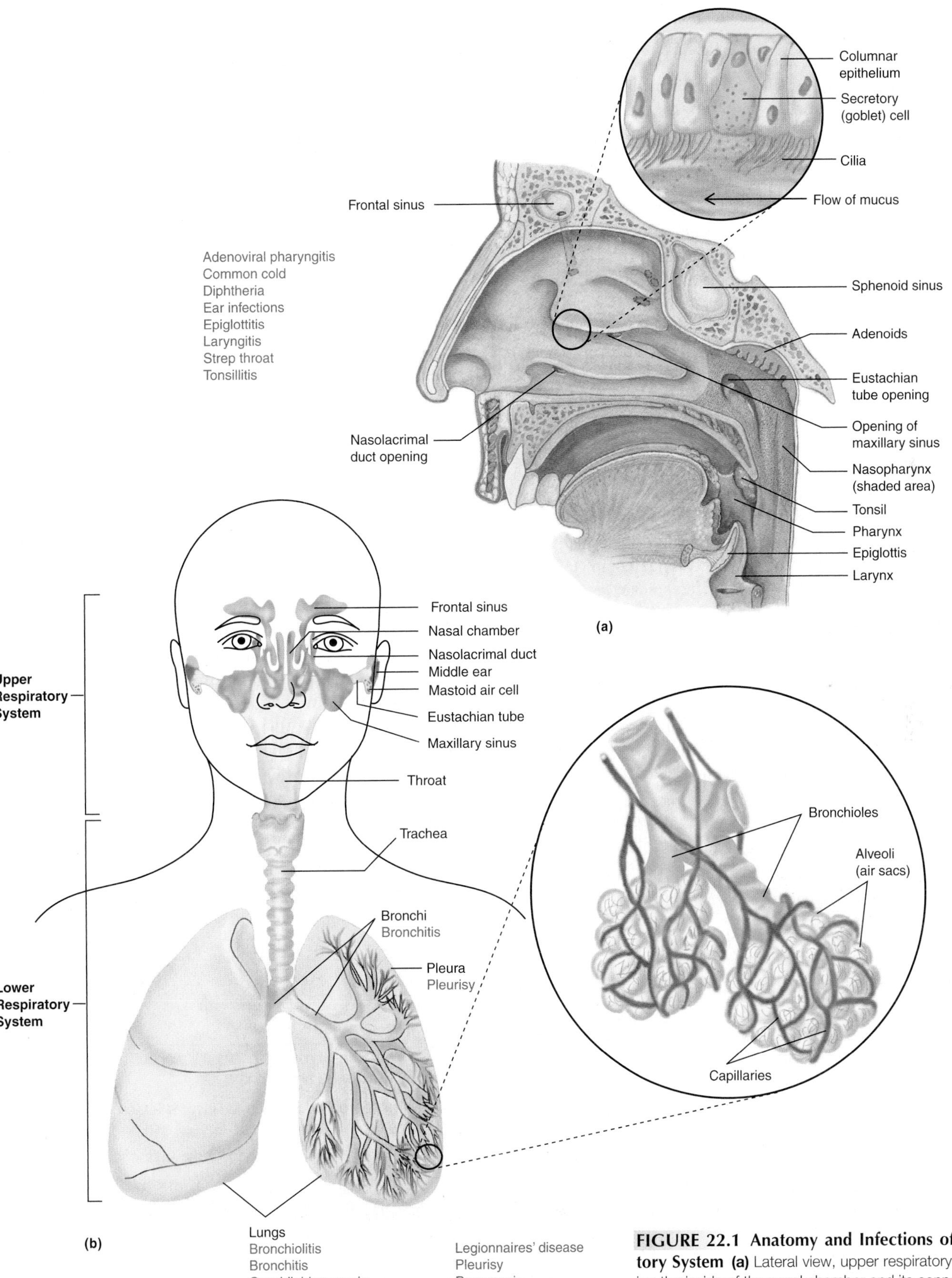

FIGURE 22.1 Anatomy and Infections of the Respiratory System **(a)** Lateral view, upper respiratory system, showing the inside of the nasal chamber and its connections with the eyes, middle ears, and sinuses, and details of the respiratory epithelium. **(b)** Frontal view of the upper and lower respiratory system, including details of the alveoli.

- The lungs are surrounded by two membranes: One adheres to the lung and the other to the chest wall and diaphragm. These membranes, termed **pleura,** normally slide against each other as the lung expands and contracts. Inflammation of the pleura is called **pleurisy,** characterized by severe chest pain aggravated by breathing or coughing.

The Mucociliary Escalator

The moist membranes of the respiratory system are coated with a slimy glycoprotein material called **mucus,** which is produced by specialized unicellular glands called goblet cells. **Goblet cells,** so called because of their shape, narrow at the base and wide at surface, are sprinkled among the other cells composing the membrane. Most of the mucous membranes of the respiratory system are composed of ciliated epithelium. Ciliated cells have tiny, hair-like projections called **cilia** along their exposed free border. The cilia beat synchronously at a rate of about 1,000 times a minute, continually propelling the mucous film out of the mastoids, middle ear, nasolacrimal duct, sinuses, and the lungs. The mucus is then swallowed, and any entrapped microorganisms and viruses are exposed to the killing action of stomach acid and enzymes. This mechanism, called the **mucociliary escalator,** normally keeps the middle ears, sinuses, mastoids, and lungs completely free of microorganisms. Viral infection, tobacco smoke, alcohol, and narcotics, however, impair ciliary movement and increase the chance of infection. ■ cilia, p. 74

MICROCHECK 22.1

The respiratory system provides a habitat for microorganisms that is warm and moist, with an atmosphere containing less oxygen and more carbon dioxide than the external environment. It is guarded by tonsils and adenoids representing the immune system, and by the mucociliary escalator.

✓ What is the normal function of the tonsils and adenoids?

✓ Why would the respiratory passages contain less oxygen than air?

22.2 Normal Microbiota

Focus Points

- List the parts of the respiratory system that are normally free of bacteria.
- Describe the important opportunistic pathogen that commonly inhabits the nose.

The mastoid air cells, middle ear, sinuses, trachea, bronchi, bronchioles, and alveoli are normally sterile. The nasal cavity, nasopharynx, and pharynx are colonized by numerous bacterial species. Aerobes, facultative anaerobes, aerotolerant organisms, and anaerobes are all represented. Although generally harmless, members of the normal bacterial microbiota are opportunists and can cause disease when host defenses are impaired. ■ oxygen requirements, p. 91 ■ opportunists, p. 394

Surprisingly, even though the eyes are constantly exposed to a multitude of microorganisms, normal healthy people commonly have no bacteria on their conjunctivae, the moist membrane covering the eye and inner surfaces of the eyelids. Presumably, this results from the frequent automatic washing of the eye with lysozyme-rich tears and from the eyelid's blinking reflex, which cleans the eye surface like a windshield wiper sweeps a windshield. Unless they are able to attach to the epithelium, the viruses and microorganisms that impinge on the conjunctiva are swept into the nasolacrimal duct and into the nasopharynx. Organisms recovered from the normal conjunctiva are usually few in number and originate from the skin flora. They are adapted to live in a different environment and are generally unable to colonize the respiratory system. ■ lysozyme, p. 62

Some of the bacterial genera that inhabit the upper respiratory system are listed in **table 22.1.** The secretions of the nasal entrance usually contain diphtheroids and staphylococci. About 20% of healthy people constantly carry *Staphylococcus aureus* in their noses, and an even higher percentage of hospital personnel are likely to be carriers of these important opportunistic patho-

TABLE 22.1 Normal Microbiota of the Respiratory System

Genus	Characteristics	Comments
Staphylococcus	Gram-positive cocci in clusters	Commonly includes the potential pathogen *Staphylococcus aureus*, inhabiting the nostrils. Facultative anaerobes.
Corynebacterium	Pleomorphic, Gram-positive rods; non-motile; non-spore-forming	Aerobic or facultatively anaerobic. Diphtheroids include anaerobic and aerotolerant organisms.
Moraxella	Gram-negative diplococci and diplobacilli	Aerobic. Some microscopically resemble pathogenic *Neisseria* species such as *N. meningitidis*.
Haemophilus	Small, Gram-negative rods	Facultative anaerobes. Commonly include the potential pathogen *H. influenzae*.
Bacteroides	Small, pleomorphic, Gram-negative rods	Obligate anaerobes.
Streptococcus	Gram-positive cocci in chains	α (especially viridans, meaning green hemolysis), β (clear hemolysis), and γ (non-hemolytic) types; the potential pathogen, *S. pneumoniae* is often present. Aerotolerant (obligate fermenters).

gens. Farther inside the nasal passages, the microbial population increasingly resembles that of the **nasopharynx** (the part of pharynx behind the nose). The nasopharynx contains mostly α-hemolytic viridans streptococci, non-hemolytic streptococci, *Moraxella catarrhalis,* and diphtheroids. Anaerobic Gram-negative bacteria, including species of *Bacteroides,* are also present in large numbers in the nasopharynx. In addition, commonly pathogenic bacteria such as *Streptococcus pneumoniae, Haemophilus influenzae,* and *Neisseria meningitidis* are often found, especially during the cooler seasons of the year. ■ viridans streptococci, p. 97

MICROCHECK 22.2

Except in parts of the upper respiratory tract, the respiratory system is free of a normal microbiota. The upper respiratory tract microbiota is highly diverse, including aerobes, anaerobes, facultative anaerobes, and aerotolerant bacteria. Although most of them are of low virulence, these organisms can sometimes cause disease.

✓ What are some possible advantages to the body of providing a niche for normal flora in the upper respiratory tract?

✓ How can strict anaerobes exist in the upper respiratory tract?

INFECTIONS OF THE UPPER RESPIRATORY SYSTEM

22.3 Bacterial Infections of the Upper Respiratory System

Focus Points

- Compare the distinctive characteristics of strep throat and diphtheria.
- List the parts of upper respiratory system commonly infected by *Streptococcus pneumoniae* and *Haemophilus influenzae.*

A number of different species of bacteria can infect the upper respiratory system. Some, such as *Haemophilus influenzae* and β-hemolytic streptococci of Lancefield group C, can cause sore throats but generally do not require treatment because the bacteria are quickly eliminated by the immune system. Other infections require treatment because they are not so easily eliminated and can cause serious complications.

Strep Throat (Streptococcal Pharyngitis)

Sore throat is one of the most common reasons that people in the United States seek medical care, resulting in about 27 million doctor visits per year. Many of these visits are due to a justifiable fear of streptococcal **pharyngitis,** commonly known as strep throat.

Symptoms

Streptococcal pharyngitis typically is characterized by pain, difficulty swallowing, and fever. The throat is red, with patches of adhering pus and scattered tiny hemorrhages. The lymph nodes in the neck are enlarged and tender. Abdominal pain or headache may be prominent in older children and young adults. Not usually present are red, weepy eyes, cough, or runny nose, common symptoms with viral pharyngitis. Most patients with streptococcal sore throat recover spontaneously after about a week. In fact, many infected people have only mild symptoms or no symptoms at all.

Causative Agent

Streptococcus pyogenes, the cause of strep throat, is a Gram-positive coccus that grows in chains of varying lengths **(figure 22.2).** It can be differentiated from other streptococci that normally inhabit the throat by its characteristic colonial morphology when grown on blood agar. *Streptococcus pyogenes* produces hemolysins, enzymes that lyse red blood cells, which result in the colonies being surrounded by a zone of β-hemolysis **(figure 22.3).** Because of their characteristic hemolysis, *S. pyogenes* and other streptococci that show a similar phenotype are called β-hemolytic streptococci. In contrast, species of *Streptococcus* that are typically part of the normal throat microbiota are either non-hemolytic, or they produce α-hemolysis, characterized by a zone of incomplete, often greenish clearing around colonies grown on blood agar. ■ streptococcal hemolysis, p. 97

Streptococcus pyogenes is commonly referred to as the group A streptococcus. The group A carbohydrate in the cell wall of *S. pyogenes* differs antigenically from that of most other streptococci and serves as a convenient basis for identification (see figure 19.9). Lancefield grouping uses antibodies to differentiate the various

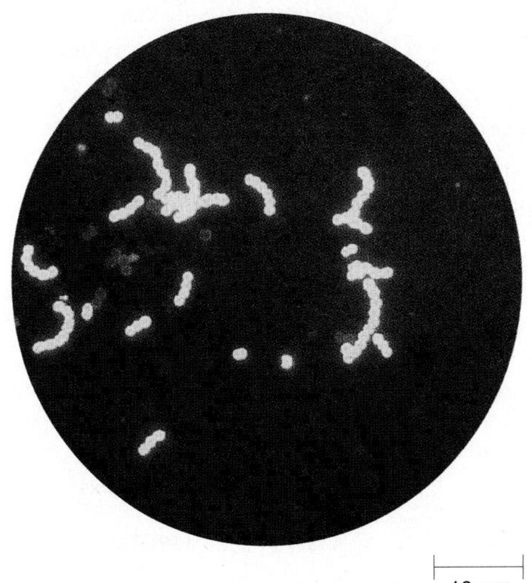

FIGURE 22.2 ***Streptococcus pyogenes*** Chain formation in fluid culture as revealed by fluorescence microscopy.

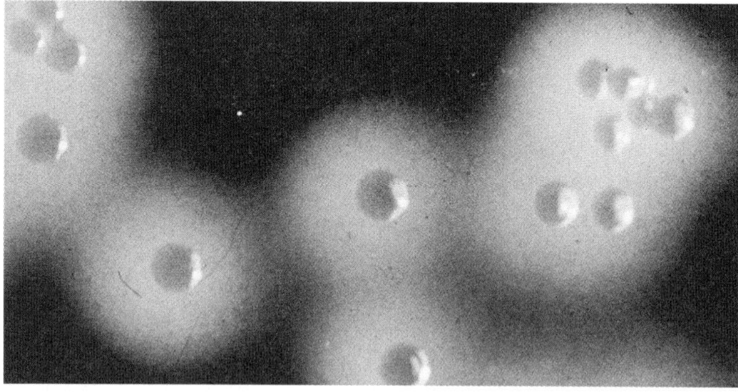

FIGURE 22.3 ***Streptococcus pyogenes*** **Growing on Blood Agar** The colonies are small and surrounded by a wide zone of β-hemolysis.

species of streptococci based on their cell wall carbohydrate. These antibodies are also the basis for many of the rapid diagnostic tests done on throat specimens in a physician's office. Other rapid tests utilize a DNA probe. ■ DNA probe, p. 227

Pathogenesis

Streptococcus pyogenes causes a wide variety of illnesses, which is not surprising considering its vast arsenal of virulence factors. Some of these disease-causing mechanisms are structural components of the outer cell wall that enable the bacterium to foil host defenses **(figure 22.4),** while others are destructive enzymes and toxins released by the bacterial cell. Some of the virulence factors of *S. pyogenes* are listed here and summarized in **table 22.2:**

- C5a peptidase is an enzyme released by *S. pyogenes.* This enzyme destroys the complement system component C5a, a substance responsible for attracting phagocytes to the site of a bacterial infection. ■ complement system, p. 355 ■ C5a Peptidase, p. 401
- A capsule composed of hyaluronic acid, present on only some strains of *S. pyogenes,* inhibits phagocytosis and aids penetration of tissues by causing epithelial cells to separate. Hyaluronic acid is a normal component of human tissue, and its presence as a *S. pyogenes* capsule may impair recognition of the bacterium by the body's immune system.

TABLE 22.2 Virulence Factors of *Streptococcus pyogenes*

Product	Effect
C5a peptidase	Inhibits attraction of phagocytes by destroying C5a
Hyaluronic acid capsule	Inhibits phagocytosis; aids penetration of epithelium
M protein	Interferes with phagocytosis by causing breakdown of C3b opsonin
Protein F	Responsible for attachment to host cells
Protein G	Interferes with phagocytosis by binding Fc segment of IgG
Streptococcal pyrogenic exotoxins (SPEs)	Superantigens responsible for scarlet fever, toxic shock, "flesh-eating" fasciitis
Streptolysins O and S	Lyse leukocytes and erythrocytes
Tissue degrading enzymes	Enhance spread of bacteria by breaking down DNA, proteins, blood clots, tissue hyaluronic acid

- M protein facilitates the degradation of complement system component C3b, an opsonin that would otherwise promote phagocytosis of the bacteria. M protein is essential for the virulence of *S. pyogenes,* because antibody to it prevents infection from occurring. Unfortunately, more than 80 different kinds of M protein exist among the many strains of *S. pyogenes,* and antibody to one type of M protein does not prevent infection by a *S. pyogenes* strain that has another kind of M protein. ■ opsonins, p. 373 ■ M protein, p. 402
- Protein F of the cell wall mediates attachment of *S. pyogenes* to the throat by adhering to a protein found on the surface of epithelial cells.
- Protein G of the *S. pyogenes* cell wall is an Fc receptor (see figure 17.11). It has a function like that of the protein A of *Staphylococcus aureus* in binding the Fc segment of immunoglobulin G. This prevents phagocytosis mediated by specific antibody against the bacterium. ■ staphylococcal protein A, p. 402

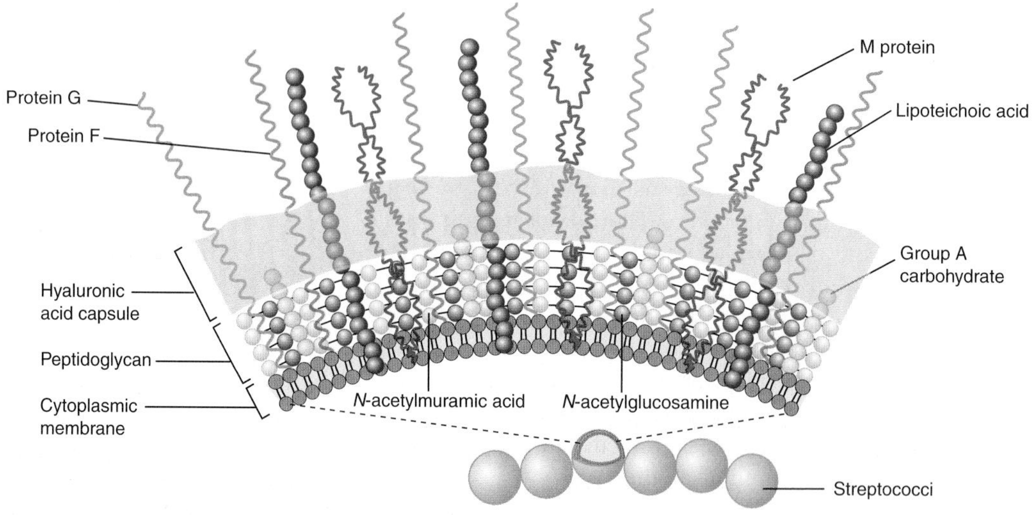

FIGURE 22.4 Components of the Cell Envelope of *Streptococcus pyogenes* The M protein is essential for virulence, while the group carbohydrate helps identify the species. Protein F binds the bacterium to epithelial cells, while protein G helps interfere with phagocytosis by binding to the Fc portion of antibodies. Hyaluronic acid interferes with phagocytosis and may aid penetration of epithelium.

- **Streptococcal pyrogenic exotoxins (SPEs)** are a family of genetically similar exotoxins produced by strains of *S. pyogenes* that cause severe streptococcal diseases frequently unrelated to strep throat. These diseases include scarlet fever, streptococcal toxic shock syndrome, and "flesh-eating" necrotizing fasciitis. Only scarlet fever is usually preceded by strep throat symptoms. These exotoxins are superantigens, causing massive activation of T cells. The resulting uncontrolled release of cytokines is probably responsible for the seriousness of these infections. ■ superantigens, p. 406 ■ cytokines, p. 353 ■ necrotizing fasciitis, p. 563
- Streptolysins O and S are the proteins responsible for β-hemolysis. These hemolysins lyse erythrocytes and leukocytes by making holes in their cell membranes and probably help to liberate iron from hemoglobin. ■ membrane-damaging toxins, p. 405
- A number of other enzymes produced by *S. pyogenes,* including DNase, protease, and hyaluronidase, degrade tissue and probably enhance the spread of the bacterium.
- Streptokinase, responsible for dissolving clots, aids spread into tissue and bloodstream. ■ streptokinase, p. 420

Complications of strep throat are uncommon in Western industrialized nations because of prompt identification of *Streptococcus pyogenes* and treatment with penicillin. However, complications can be serious if overlooked. Globally, they occur much more frequently. Some of the complications that can occur during the acute illness include:

- **Scarlet fever** is characterized by a roughening of the skin and red rash that spares the area around the mouth and causes the tongue to look like the surface of a ripe strawberry. The rash is caused by **erythrogenic toxin,** an SPE released by certain strains of *S. pyogenes* at the site of infection. The toxin enters the bloodstream and is carried throughout the body.
- **Peritonsilar abscess** causes a painful swelling in the area of one of the tonsils, often with fever and difficulty swallowing. Without prompt treatment the abscess can erode into the structures of the neck including blood vessels, potentially causing massive blood loss.
- Other pyogenic complications include **sinusitis, otitis media, mastoiditis, pneumonia,** and **meningitis.**

Post-streptococcal sequelae are complications that occur a week to several months after recovery from the symptoms of acute pharyngitis. They are all considered to be the result of immune responses to *Streptococcus pyogenes.* The most important sequelae are:

- **Acute glomerulonephritis** usually begins abruptly about 10 days after strep throat, with fever, fluid retention, high blood pressure, and blood and protein in the urine. The symptoms arise from inflammation of structures within the kidneys, glomeruli (singular, glomerulus), small tufts of tiny blood vessels, and nephrons, responsible for the formation and composition of urine. Only a few of the many strains of *S. pyogenes* cause the condition, so it is rare to get glomerulonephritis twice. The streptococci are absent from the urine and the diseased kidney tissues. Indeed, *S. pyogenes* has generally been eliminated from the throat by the body's immune response by the time the symptoms of glomerulonephritis appear. The damage to the kidneys is due to immune complexes that are deposited in the glomeruli and provoke an inflammatory reaction.
- **Acute rheumatic fever** usually begins about 3 weeks after recovery from strep throat, with fever, joint pains, chest pains, rash, and nodules under the skin. Acute behavioral changes and uncontrollable movement of various parts of the body can occur and, when present, are major criteria for diagnosing acute rheumatic fever.

These symptoms result from an inflammatory process involving various tissues, especially the joints, heart, skin, and brain. Heart failure and death can occur during acute rheumatic fever, but usually symptoms subside with rest and anti-inflammatory medicines such as aspirin. Rheumatic fever often results in damage to the heart valves, causing one or more to leak and lead to heart failure later in life. Indeed, acute rheumatic fever is the major cause of acquired heart valve disease worldwide. The damaged valves are also subject to infection, usually by bacteria from the normal skin or mouth flora, causing **subacute bacterial endocarditis.** ■ inflammation, p. 360 ■ subacute bacterial endocarditis, p. 677

Acute rheumatic fever is generally a complication of throat, not skin, *S. pyogenes* infections. Despite many years of study, its pathogenesis is not understood. Symptoms usually begin about 3 weeks after the onset of untreated streptococcal pharyngitis. Cultures of the blood, heart, and joint tissues are negative for *S. pyogenes,* and antibodies against the bacterium are present in the blood of the patient. A popular idea is that these antibodies attack tissue antigens that are similar to *S. pyogenes* antigens, starting a self-limited autoimmune process. Another hypothesis is that some streptococcal product, present in only a small percentage of strains, damages the tissue directly or makes it susceptible to attack by the immune system. Genetic factors probably play a role in the development of rheumatic fever, because it occurs more commonly in individuals with certain MHC types. Overall, the risk of developing acute rheumatic fever after severe, untreated streptococcal pharyngitis is 3% or less. Those with untreated mild pharyngitis have a lower risk. Mild infections result in many cases of rheumatic fever, however, because they are common and much more likely to go untreated. Although outbreaks still occur in the United States, rheumatic fever has generally declined in incidence over many years, probably as a result of timely treatment of streptococcal pharyngitis with antibiotics, and a decline in the streptococcal strains associated with the disease. Globally, 10 to 20 million new cases occur each year, mostly in economically disadvantaged countries. ■ major histocompatibility complex (MHC), p. 380

Epidemiology

Streptococcus pyogenes infects only humans under natural conditions, probably because streptokinase reacts specifically with a human clotting factor. Streptococcal infections spread readily by respiratory droplets generated by yelling, coughing, and sneezing, especially in the range of about 2 to 5 feet from an infected individual. If strep throat is untreated, the person may be an asymptomatic carrier for weeks. People who carry the organism in their nose spread the streptococci more effectively than do pharyngeal carriers. Anal carriers are not common but can be a dangerous source of nosocomial infections. Epidemics of strep throat can originate from food contaminated by *S. pyogenes* carriers. Some people become long-term carriers of *S. pyogenes.* In these cases, the infecting strain usually becomes deficient in M protein and is not a threat to the carrier

TABLE 22.3 Strep Throat (Streptococcal Pharyngitis)

① *Streptococcus pyogenes* enters by inhalation (nose), or by ingestion (mouth).

② Pharyngitis, fever, enlarged lymph nodes; sometimes tonsillitis, abcess; scarlet fever with strains that produce erythrogenic toxin.

Symptoms go away.

③ *S. pyogenes* exits by nose and mouth.

Late complications appear:

④ glomerulonephritis

⑤ rheumatic fever

⑥ neurological abnormalities

Complications subside.

⑦ Damaged heart valves leak, heart failure develops.

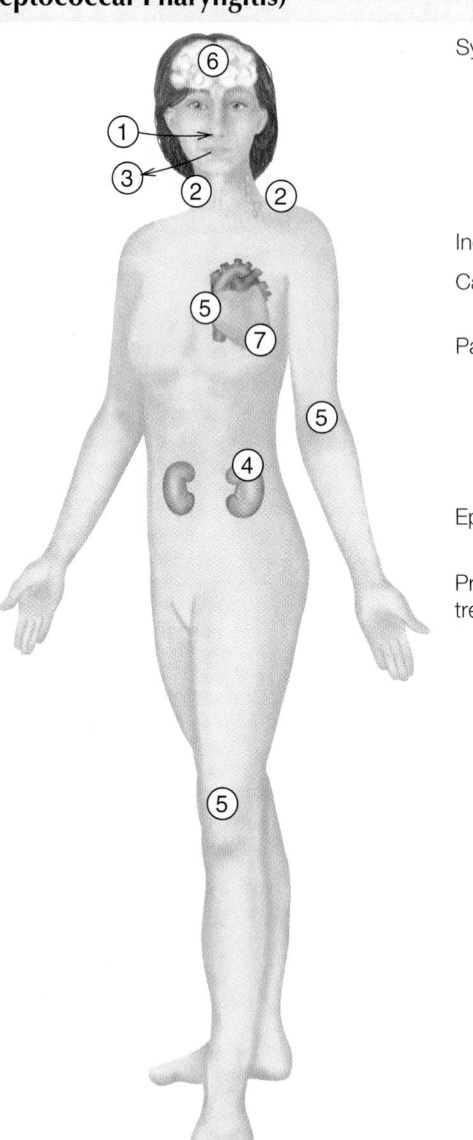

Symptoms	Sore, red throat, with pus and tiny hemorrhages, enlargement and tenderness of lymph nodes in the neck; less frequently, abscess formation involving tonsils; occasionally, rheumatic fever and glomerulonephritis as sequels
Incubation period	2 to 5 days
Causative agent	*Streptococcus pyogenes,* Lancefield group A β-hemolytic streptococci
Pathogenesis	Virulence associated with hyaluronic acid capsule and M protein, both of which inhibit phagocytosis; protein G binds Fc segment of IgG; protein F for mucosal attachment; multiple enzymes.
Epidemiology	Direct contact and droplet infection; ingestion of contaminated food.
Prevention and treatment	Avoidance of crowding; adequate ventilation; daily penicillin to prevent recurrent infection in those with a history of rheumatic heart disease. Treatment: 10 days of penicillin or erythromycin.

or to others. The peak incidence of strep throat occurs in winter or spring and is highest in grade school children. Among students visiting a clinic because of sore throat at a large West Coast university, less than 5% had strep throat. With some groups of military recruits, however, the incidence has been above 20%.

Prevention and Treatment

Adequate ventilation and avoidance of crowding help to control the spread of streptococcal infections. No vaccine is available despite many years of studies. However, genome sequencing has revealed some new possibilities. Persons with fever and sore throat should have a throat culture for *S. pyogenes* so that antibiotic treatment can be given promptly and complications prevented. Individuals with a history of having had acute rheumatic fever with cardiac damage take penicillin daily for years to prevent recurrence of the disease. Confirmed streptococcal pharyngitis is treated with a full 10 days of therapy with penicillin or erythromycin, which eliminates the organism in about 90% of the cases. Treatment given as late as 9 days after the onset prevents rheumatic fever from developing.

Table 22.3 summarizes some important facts about streptococcal pharyngitis.

Diphtheria

Diphtheria, a deadly toxin-mediated disease, is now rare in the United States because of childhood immunization. Events in other parts of the world, however, are a reminder of what can happen when public health is neglected. In 1990, a diphtheria epidemic began in the Russian Federation. Over the next 5 years, it spread to all the newly independent states of the former Soviet Union. By the end of 1995, 125,000 cases and 4,000 deaths had been reported. The social and economic disruption following the breakup of the Soviet Union had allowed diphtheria to reemerge after being well controlled over the previous quarter of a century.

Symptoms

Diphtheria usually begins with a mild sore throat and slight fever, accompanied by a great deal of fatigue and malaise. Swelling of the neck is often dramatic. A whitish gray membrane forms on the tonsils and the throat or in the nasal cavity. Heart and kidney failure and paralysis may follow these symptoms.

Causative Agent

Diphtheria is caused by *Corynebacterium diphtheriae,* a variably shaped, non-motile, non-spore-forming, Gram-positive rod that often stains irregularly. Slide preparations of *C. diphtheriae* commonly show the organisms arranged in "Chinese letter" patterns **(figure 22.5)** or side by side in "palisades." Different colony types can be identified—a property that is useful in tracing epidemic spread of the organisms. These bacteria require special media and microbiological techniques to aid their recovery from the mixture of bacteria present in throat material. It is unlikely that they would be recovered by the usual methods for culturing *S. pyogenes* from sore throats.

Most but not all strains of *C. diphtheriae* release diphtheria toxin, a powerful exotoxin responsible for the seriousness of diphtheria. Production of this toxin requires that the bacterium be lysogenized by a specific bacteriophage, an example of lysogenic conversion. Strains that are non-toxigenic become toxin producers when lysogenized with the bacteriophage. Toxin production can be demonstrated *in vitro,* using immunoassays. The toxin is produced only when the medium on which *C. diphtheriae* is growing has too little iron for optimal growth. The gene for toxin production is under the control of a repressor that is only active when bound to iron. In a medium containing adequate iron, the repressor shuts down toxin production; when the concentration of iron is very low, the iron molecule leaves the repressor and the toxin gene is expressed. ■ lysogenic conversion, p. 310 ■ repressors, p. 177 ■ immunoassays, p. 438

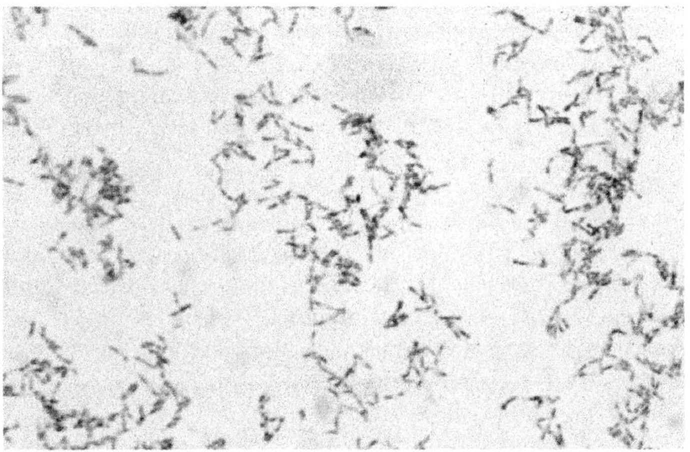

FIGURE 22.5 ***Corynebacterium diphtheriae*** Note the "Chinese letter" pattern.

Pathogenesis

Corynebacterium diphtheriae has little invasive ability, rarely entering the blood or tissues, but the powerful exotoxin it releases is absorbed by the bloodstream. The gray-white pseudomembrane that forms in the throat of people with diphtheria is made up of clotted blood along with dead epithelial cells of the host mucous membrane and the leukocytes that congregate during inflammation. The cells are killed by exotoxin released from *C. diphtheriae.* This membrane may come loose and obstruct the airways, causing the patient to suffocate. Entry of the toxin into the bloodstream results in damage to the heart, nerves, and kidneys.

The diphtheria toxin is a large protein released from the bacterium in an inactive form. It is cleaved extracellularly into two chains, A and B, which remain joined by a disulfide bond. The B chain attaches to specific receptors on a host cell membrane, and the entire molecule is taken into the cell by endocytosis **(figure 22.6).** Cells that lack the appropriate receptors do not take up the toxin and are unaffected by it. This receptor specificity explains why some tissues of the body are not affected in diphtheria, while others such as the heart, kidneys, and nerves are severely damaged. ■ endocytosis, p. 73

Once the toxin molecule is inside the cell, the A chain separates from the B chain and becomes an active enzyme. This enzyme catalyses a chemical reaction that inactivates elongation factor 2 (EF-2), a substance required for movement of the eukaryotic ribosome on mRNA (see figure 22.6). This halts protein synthesis, and the cell dies. Since the toxin A chain is an enzyme, it is not consumed in the process, so one or two molecules of toxin can inactivate essentially all the cell's elongation factor 2. This explains the extreme potency of diphtheria toxin. ■ A-B toxins, p. 405 ■ elongation factor, p. 173

Epidemiology

Humans are the primary reservoir for *C. diphtheriae.* Sources of infection include carriers who have recovered from an infection, new cases not yet exhibiting symptoms, people with active disease, and contaminated articles. The bacterium is carried in chronic skin ulcers—**cutaneous diphtheria**—in some indigent populations. Typically, the organisms are spread by air and acquired by inhalation.

Prevention and Treatment

Because diphtheria results primarily from toxin absorption rather than microbial invasion, its control can be accomplished most effectively by immunization with toxoid. Toxoid, prepared by formalin treatment of diphtheria toxin, causes the body to produce antibodies that specifically neutralize the diphtheria toxin. The well-known childhood vaccination DPT consists of diphtheria and tetanus toxoids and pertussis vaccine, all three generally given together. Unfortunately, these immunizations have often been neglected, particularly among socioeconomically disadvantaged groups, and serious epidemics of the diseases have occurred periodically. Since the 1980s, there has been an active campaign in most of the United States to ensure that children who are entering school are immunized against diphtheria. As a result, today only a few cases of diphtheria are reported annually in the United States, as compared with 30,000 cases reported in 1936. Of concern is the finding that widespread childhood immunization against

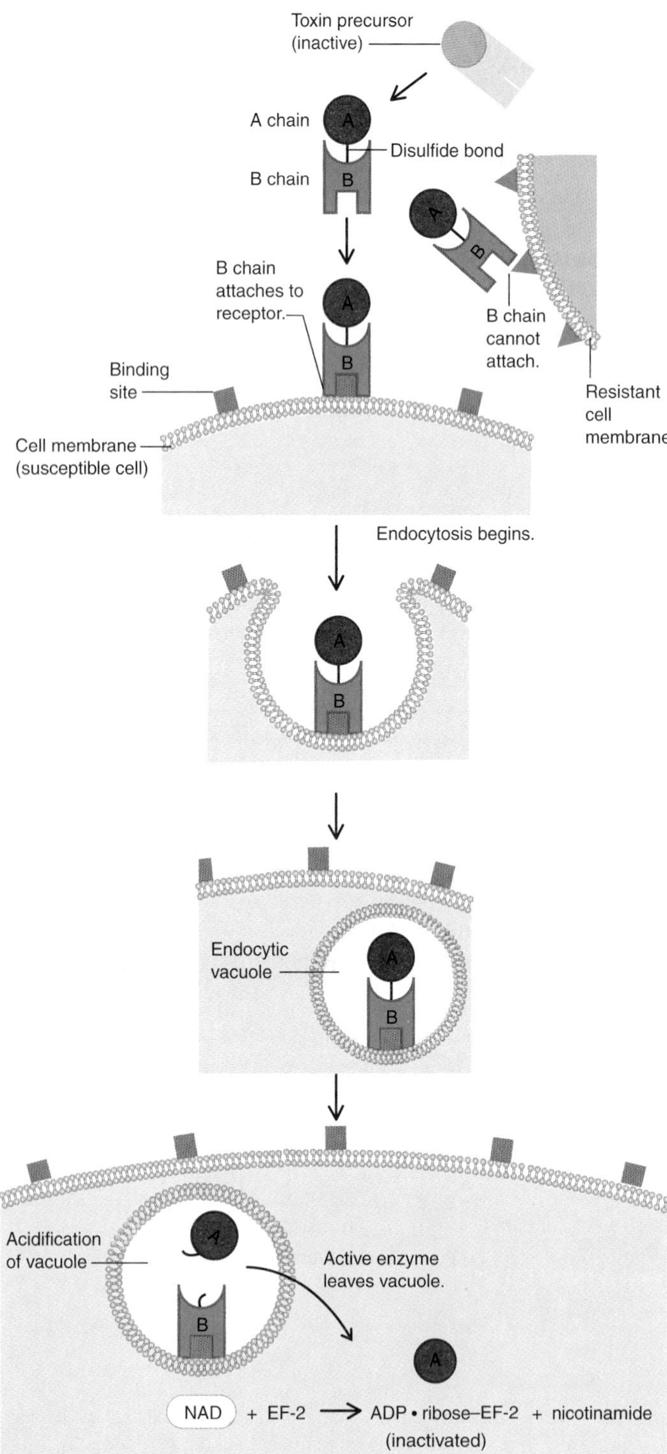

FIGURE 22.6 Mode of Action of Diphtheria Toxin The toxin precursor released from the bacterium is cleaved extracellularly into A and B chains joined only by a disulfide bond; the B chain attaches to a specific receptor on the cell membrane of a susceptible cell. The toxin enters the cell by endocytosis. With acidification of the endocytic vacuole, the A chain separates from the B chain and then enters the cytoplasm as an active enzyme that inactivates EF-2 (elongation factor 2) by ADP ribosylation. Since EF-2 is required for moving the ribosome on mRNA, protein synthesis ceases and the cell dies.

diphtheria has shifted susceptibility to the disease to adolescents and adults. Immunity wanes after childhood because of lack of exposure to now scarce *C. diphtheriae* to boost immunity. To prevent buildup of a large population of susceptible older individuals, booster injections should be given every 10 years following childhood immunization.

Effective treatment of diphtheria depends on injecting antiserum against diphtheria toxin into the patient as soon as possible. If the disease is suspected, antiserum must be administered without waiting several days for culture results, because the delay could be fatal. The bacteria are sensitive to antibiotics such as erythromycin and penicillin, but such treatment only stops transmission of the disease; it has no effect on toxin that has already been absorbed. Even with treatment, about 1 of 10 diphtheria patients dies.

Table 22.4 summarizes some important facts about diphtheria.

Pinkeye, Earache, and Sinus Infections

Bacterial infection of the eye's surface, generally known as conjunctivitis or "pinkeye," and infections of the middle ear (otitis media) and sinuses are very common, often occur together, and frequently have the same causative agent. This is not surprising in view of their intimate association with the nasal chamber and nasopharynx (see figure 22.1). Otitis media is a formidable problem, especially in children, and is responsible for 30 million doctor visits per year in the United States, at an estimated cost of $1 billion. Pinkeye, because of its high communicability, is a frequent source of panic when it appears in a day care facility or classroom. Sinusitis, inflammation of the sinuses, is common in both adults and children.

Symptoms

The symptoms of acute bacterial conjunctivitis include increased tears, redness of the conjunctiva, swelling of the eyelids, sensitivity to bright light, and large amounts of pus **(figure 22.7).** Acute bacterial conjunctivitis must be distinguished from conjunctivitis caused by a large number of viruses. In the latter cases, eyelid swelling and pus are usually minimal. In a typical case of otitis media, a young child wakes from sleep screaming with earache pain. Fever is generally mild or absent. Vomiting often occurs at the height of the pain. Sometimes the pain ends abruptly because the eardrum has ruptured, and drainage of fluid appears in the external ear canal. In sinusitis, pain and a pressure sensation characteristically occur in the region of the involved sinus. Tenderness is also present over the sinus in many instances. Sinus infections are commonly associated with a headache and severe malaise.

Causative Agents

In all three conditions the most common bacterial pathogens are *Haemophilus influenzae,* a tiny Gram-negative rod, and *Streptococcus pneumoniae,* the Gram-positive encapsulated diplococcus known as the pneumococcus. Strains that infect the conjunctiva have adhesins that allow firm attachment to the epithelium. ■ **adhesins, p. 398** ■ ***Streptococcus pneumoniae*, p. 509** ■ ***Haemophilus influenzae*, pp. 373, 650**

Less commonly, *Moraxella lacunata,* enterobacteria, *Neisseria gonorrhoeae,* and many others infect the conjunctiva. Among these are agents that can contaminate eye medications

TABLE 22.4 Diphtheria

① *Corynebacterium diphtheriae* enters by inhalation.

② Infection established in nasal cavity and/or throat.

③ Toxin released, pseudomembrane forms.

④ Toxin causes paralysis, damages heart muscle, kidneys, nerves.

⑤ Membrane may come loose and obstruct breathing.

⑥ Exit from body by respiratory secretions.

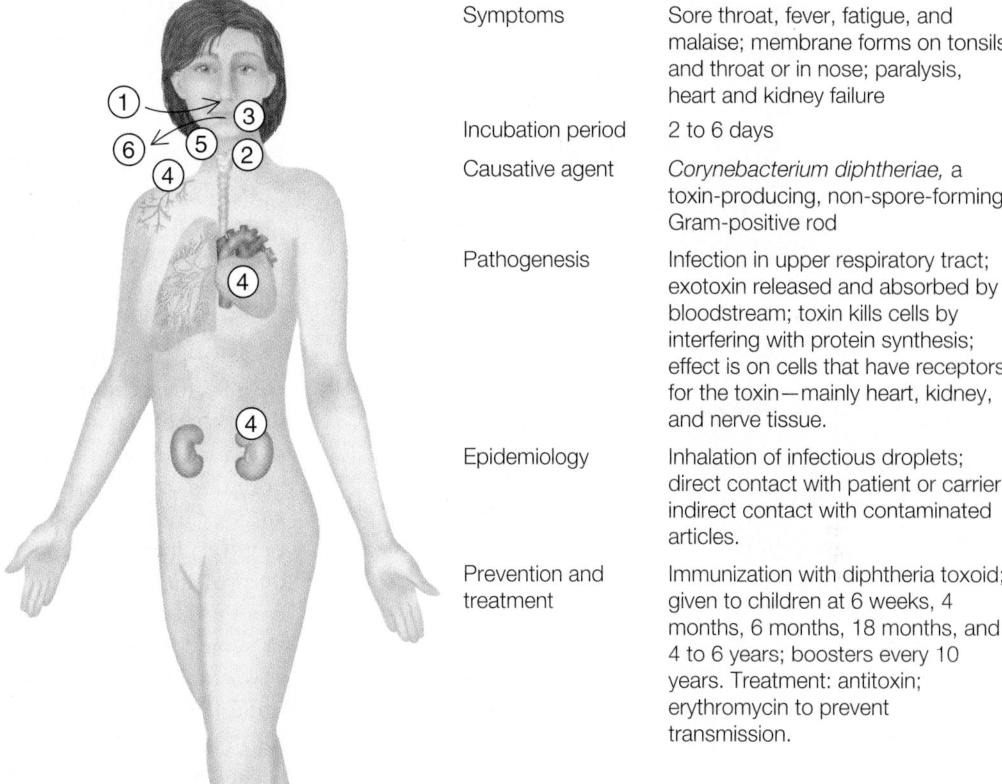

Symptoms	Sore throat, fever, fatigue, and malaise; membrane forms on tonsils and throat or in nose; paralysis, heart and kidney failure
Incubation period	2 to 6 days
Causative agent	*Corynebacterium diphtheriae,* a toxin-producing, non-spore-forming Gram-positive rod
Pathogenesis	Infection in upper respiratory tract; exotoxin released and absorbed by bloodstream; toxin kills cells by interfering with protein synthesis; effect is on cells that have receptors for the toxin—mainly heart, kidney, and nerve tissue.
Epidemiology	Inhalation of infectious droplets; direct contact with patient or carrier; indirect contact with contaminated articles.
Prevention and treatment	Immunization with diphtheria toxoid; given to children at 6 weeks, 4 months, 6 months, 18 months, and 4 to 6 years; boosters every 10 years. Treatment: antitoxin; erythromycin to prevent transmission.

and contact lens solutions such as *Bacillus* sp. (free-living aerobic Gram-positive spore-forming rods), *Pseudomonas* sp. (free-living Gram-negative aerobic rods), and *Acanthamoeba* sp. (free-living protozoa that live in soil, and fresh and salt water that have a cyst form carried by dust). Infections caused by these organisms are very uncommon, but when they occur they are usually serious and can lead to destruction of the eye. Contact lens wearers need to be very careful to follow exactly the instructions on use of cleansing solutions, and cloudy or outdated solutions and eye medications should be discarded.

FIGURE 22.7 Bacterial Conjunctivitis Redness, swollen eyelids and pus, characteristic of bacterial conjunctivitis.

Besides *Streptococcus pneumoniae* and *Haemophilus influenzae, Mycoplasma pneumoniae, Streptococcus pyogenes, Moraxella catarrhalis,* and *Staphylococcus aureus* can cause otitis media. About one-third of the cases are caused by respiratory viruses. This helps explain why some infections fail to respond to antimicrobial medications, which have no effect on viruses. The same infecting agents are often involved in sinusitis.

Pathogenesis

Few details are known about the pathogenesis of acute bacterial conjunctivitis. Probably, the organisms are inoculated directly onto the conjunctiva from airborne respiratory droplets or rubbed in from contaminated hands. Like most bacterial pathogens, they resist destruction by lysozyme. Otitis media is sometimes developing at the time conjunctivitis is diagnosed. Generally, otitis media and sinusitis are preceded by infection of the nasal chamber and nasopharynx. Probably, infection spreads upward through the eustachian tube to the middle ear **(figure 22.8).** The infection damages the ciliated cells, resulting in inflammation and swelling. Because the damaged eustachian tube cannot move secretions from the middle ear, pressure builds up from fluid and pus collecting behind the eardrum. The throbbing ache of a middle ear infection is produced by pressure on nerves supplying the

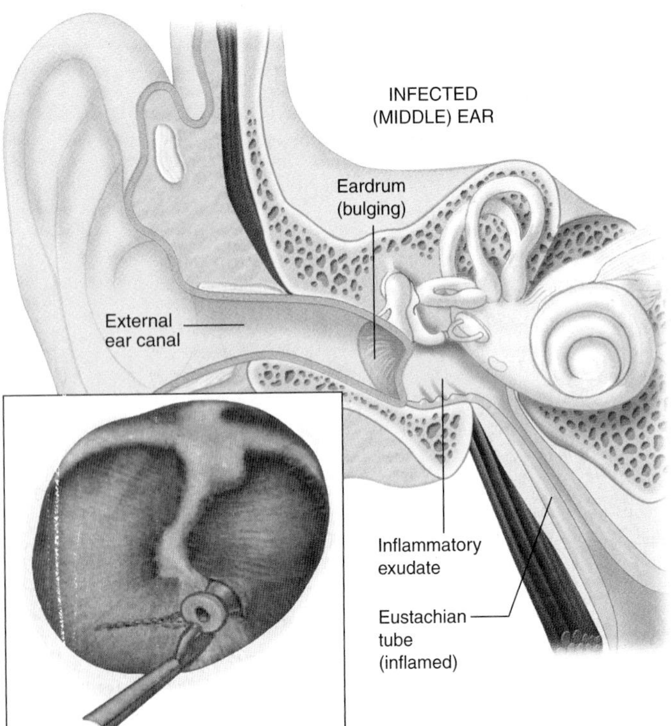

FIGURE 22.8 Otitis Media A stylized view of the infected middle ear showing accumulation of fluid, swelling of the eustachian tube, and outward bulging of the eardrum. The inset shows a ventilation tube placed in the eardrum to equalize middle ear pressure in individuals with chronically malfunctioning eustachian tubes.

middle ear. Bacterial infections of the middle ear can cause the eardrum to perforate, giving discharge of blood or pus from the ear and immediate relief from pain. With treatment, the eardrum perforations usually heal promptly. The pressure in the middle ear can force infected material into the mastoid air cells, resulting in mastoiditis. The fluid behind the eardrum impairs movement of the drum, thereby decreasing ability to hear. Because of the fluid, some young children cannot hear clearly and show a delay in normal speech development. Spread of infection from the nasopharynx to the sinuses probably occurs by a mechanism similar to that in otitis media. Both middle ear and sinus infections sometimes spread to the coverings of the brain, causing meningitis. ■ lysozyme, p. 62

Epidemiology

The carrier rates for *H. influenzae* and *S. pneumoniae* can sometimes reach 80% in the absence of disease, and the ecological factors involved in the appearance and spread of the diseases caused by these organisms are largely unknown. Epidemics of bacterial conjunctivitis commonly occur among schoolchildren. Probably the virulence of the bacterium, crowding of susceptible children, and the presence of respiratory viruses all play important roles. A preceding or concomitant viral illness is common in otitis media and sinusitis; probably the virus damages the mucociliary mechanism that would normally protect against bacterial infection. Although otitis media is rare in the first month of life, it becomes very common in early childhood. Older children develop immunity to *H. influenzae,* and it becomes an increasingly uncommon cause of otitis media after about five years of age. Sinusitis tends to involve adults and older children in whom the sinuses are more fully developed.

Prevention and Treatment

Individuals suspected of having bacterial conjunctivitis are removed from school or day care settings for diagnosis and start of treatment. General preventive measures include handwashing, and avoidance of rubbing or touching the eyes and sharing towels.

Use of pacifiers beyond the age of 2 years is associated with a substantial increase in otitis media. Viral infections and other conditions that cause inflammation of the nasal mucosa play a role in some cases. Nasal allergies and exposure to air pollution and cigarette smoke are examples of nasal irritants. Administration of influenza vaccine to infants in day care facilities substantially decreases the incidence of otitis media during the "flu" season. Ampicillin or sulfasoxazole given continuously over the winter and spring are useful preventives in people who have three or more bouts of otitis media within a six-month period. Surgical removal of enlarged adenoids improves drainage from the eustachian tubes and can be helpful in preventing recurrences in certain patients. In those with chronically malfunctioning eustachian tubes and hearing loss, plastic ventilation tubes are often installed in the eardrums so that pressure can equalize on both sides of the drum (see figure 22.8). Widespread use of pneumococcal and *Haemophilus* conjugate vaccines has resulted in only a modest overall decrease in otitis media cases. There are no proven preventive measures for sinusitis. ■ conjugate vaccines, p. 435

Bacterial conjunctivitis is effectively treated with eyedrops or ointments containing an antibacterial medicine to which the infecting strain is sensitive. Antibacterial therapy with amoxacillin is generally effective against otitis media; alternative medications are available for communities where antibiotic-resistant strains of *H. influenzae* and *S. pneumoniae* are common. In general, antibiotics have been used indiscriminately in treating otitis media. However, properly used, they decrease the risk of serious complications such as mastoiditis and meningitis. Decongestants and antihistamines generally are ineffective and can be harmful.

MICROCHECK 22.3

Untreated *Streptococcus pyogenes* infections can sometimes cause injury to other parts of the body that appear long after the infections have healed. In diphtheria, toxin is absorbed into the bloodstream and circulates throughout the body, selectively damaging certain tissues that have appropriate receptors such as heart, kidneys, and nerves. Viral infections often pave the way for bacterial infections.

- ✓ Name two post-streptococcal sequlae.
- ✓ How does diphtheria toxin kill cells?
- ✓ What purpose would adequate ventilation serve in preventing the spread of streptococcal infections?

22.4

Viral Infections of the Upper Respiratory System

Focus Points

- List the strategies helpful in avoiding common colds.
- Give the distinctive characteristics of adenoviral pharyngitis.

The average person in the United States suffers two to five episodes of viral upper respiratory infections each year. They generally subside without any treatment and rarely cause permanent damage. Most of the causative agents are highly successful parasites, using us to replicate themselves to astronomical numbers, then moving on without killing us or even leaving long-standing immunity. Although hundreds of kinds of viruses are involved, the range of symptoms they produce is similar. Their major importance to health is that they damage respiratory tract defenses and thus pave the way for more serious bacterial diseases.

The Common Cold

Medical practitioners have been taught for generations that the symptoms of the common cold, "if treated vigorously, will go away in seven days, whereas if left alone, they will disappear over the course of a week." This pessimistic outlook may now be changing with development of new antiviral medications. On average, a person in the United States has two to three colds per year. Colds are the leading cause of absences from school, and result in loss of about 150 million work days per year. Immunity generally lasts less than a few years.

Symptoms

Colds generally begin with malaise, followed by a scratchy or mildly sore throat, runny nose, cough, and hoarseness. The nasal secretions, initially profuse and watery, thicken in a day or two, then become cloudy and greenish. Unless secondary bacterial infection occurs, there is no fever; symptoms are mostly gone within a week, but a mild cough sometimes continues for longer periods of time.

Causative Agents

Between 30% and 50% of colds are caused by the 100 or more types of rhinoviruses (*rhino* means "nose," as in rhinoceros, or "horny nose"), members of the picornavirus family (*pico* is Italian for "small" and *rna* is ribonucleic acid [RNA]; thus, "small RNA viruses") **(figure 22.9).** The viruses are non-enveloped and their RNA is single-stranded. The rhinoviruses can usually be cultivated in cell cultures incubated at 33°C instead of at body temperature (37°C) and at a slightly acid pH instead of at the alkaline pH of body tissues. These conditions of lower temperature and pH normally exist in the upper respiratory tract. Rhinoviruses, however, are killed if the pH drops below 5.3, and therefore, they are usually destroyed in the human stomach. Many other viruses and some bacterial species can also produce the symptoms of a cold.

Pathogenesis

Rhinoviruses attach to specific receptors on respiratory epithelial cells and infect them. After replicating, large numbers of virions are released and infect other cells. Infected cells cease ciliary motion and may slough off. The injury causes the release of inflammatory mediators and stimulates nervous reflexes. The result is an increase in nasal secretions, sneezing, and swelling of the tissue. This swelling partially or completely obstructs the airways. Later, in the inflammatory response, dilation of blood vessels, oozing of plasma, and congregation of leukocytes in the infected area occur. Secretions from the area may then contain pus and blood. The infection is eventually halted by the inflammatory response, interferon release, and cellular and humoral immunity, but it can extend into the ears, sinuses, or even the lower respiratory tract before it is stopped. Rhinoviruses can even cause life-threatening pneumonia in individuals with AIDS. ■ interferon, p. 357

Epidemiology

Humans are the only source of cold viruses, and close contact with an infected person is generally necessary for viruses to be transmitted. A person with severe symptoms early in the course of a cold is much more likely to transmit the viruses than is someone whose symptoms are mild or who is in the late stage of a cold. Very high concentrations of virus are found in the nasal secretions and often on the hands of infected people during the first 2 or 3 days of a cold. By the fourth or fifth day, virus levels are often undetectable, but low levels can be present for 2 weeks. A few virions are sufficient to infect the nasal mucosa, but the mouth is quite resistant to infection. In adults the disease is usually contracted when airborne droplets containing virus particles are inhaled. Transmission can also occur when virus-containing secretions are unwittingly rubbed into the eyes or nose by contaminated hands. Virus introduced into the eye is promptly transmitted to the nasal passage via the nasolacrimal duct. With reasonable caution, however, colds are not highly contagious. In a study of non-immune adults exposed in a family or dormitory setting, less than half contracted colds. Young children, however, transmit cold and other respiratory viruses very effectively because they are often careless with their respiratory secretions. Experimental and epidemiological studies show that there is no relationship between exposure to cold temperature and development of colds, contrary to popular belief. Emotional stress, however, can almost double the risk of catching a cold.

Prevention and Treatment

Because such a large number of immunologically different viruses cause colds, vaccines are impractical. Rhinoviruses, like all other

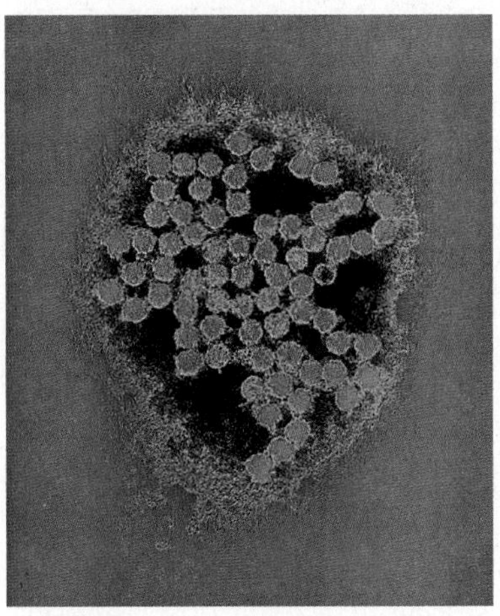

FIGURE 22.9 Rhinovirus Transmission electron micrograph. The red spheres are the rhinovirus virions, and the blue object is the cell in which they are replicating.

viruses, are not affected by antibiotics and other antibacterial medications. Prevention of the spread of rhinoviruses includes handwashing, even in plain water, which readily removes rhinoviruses, and keeping hands away from the face. In addition, one should avoid crowds and crowded places such as commercial airplanes when respiratory diseases are prevalent, and especially avoid people with colds during the first couple of days of their symptoms. In experimental rhinovirus infections, both aspirin and acetaminophen somewhat prolonged symptoms and duration of virus excretion and delayed antibody production.

Table 22.5 summarizes some facts about the common cold.

TABLE 22.5 The Common Cold

Symptoms	Scratchy throat, nasal discharge, malaise, headache, cough
Incubation period	1 to 2 days
Causative agent	Mainly rhinoviruses—more than 100 types; many other viruses, some bacteria
Pathogenesis	Viruses attach to respiratory epithelium, starting infection that spreads to adjacent cells; ciliary action ceases and cells slough; mucus secretion increases, and inflammatory reaction occurs; infection stopped by interferon release, cellular and humoral immunity.
Epidemiology	Inhalation of infected droplets; transfer of infectious mucus to nose or eye by contaminated fingers; children initiate many outbreaks in families because of lack of care with nasal secretions.
Prevention and treatment	Handwashing; avoiding people with colds and touching face. No generally accepted treatment except for control of symptoms.

Adenoviral Pharyngitis

Adenoviruses are widespread and can cause epidemics of febrile upper respiratory illness throughout the year. Although they typically produce sore throat and enlarged lymph nodes with or without conjunctivitis, they can sometimes also cause pneumonia or diarrhea. They are representative of the many viruses that cause upper respiratory symptoms with fever.

Symptoms

Adenoviral infections often cause a runny nose, but unlike the common cold, fever is commonly present. Typically, the throat is sore, and gray-white pus present on the pharynx and tonsils can cause confusion with strep throat. The lymph nodes of the neck enlarge and become tender. In some epidemics, conjunctivitis is a prominent symptom, sometimes showing multiple conjunctival nodules composed of infiltrating lymphocytes, and long-lasting opacities in the cornea, the clear central portion of the eye. Other strains of the virus cause epidemics of hemorrhagic conjunctivitis. A mild cough is common; sometimes a severe cough develops, indicating possible pneumonia or whooping cough, or chest pain indicating pleurisy. With or without treatment, recovery usually occurs in 1 to 3 weeks.

Causative Agent

More than 45 antigenic types of adenoviruses infect humans. The viruses are non-enveloped, 70 to 90 nm in diameter, with double-stranded DNA. They can remain infectious in the environment for long periods of time and can be transferred readily from one patient to another on medical instruments. They are easily inactivated, however, by heat at 56°C, chlorine, and various other disinfectants.

Pathogenesis

Adenoviruses infect epithelial cells by attaching to receptors near the basement membrane. Inside the cells of the host, the virus multiplies in the nuclei. The virus escapes to the epithelial surface by overproducing its attachment protein, causing the epithelial cells to separate from each other. In severe infections, extensive cell destruction and inflammation occur. Different types of adenoviruses vary with respect to the tissues they affect. For example, some cause illness characterized by sore throat and enlarged lymph nodes, whereas others cause eye infection. ■ **basement membrane, p. 348**

Epidemiology

Humans are the only source of infection. Adenoviral illness is prevalent among schoolchildren, usually in sporadic cases, but occasionally occurring as outbreaks of respiratory sickness in winter and spring. Summertime epidemics occur as a result of transmission of the viruses in inadequately chlorinated swimming pools. These cases typically have fever, conjunctivitis, pharyngitis, enlargement of the lymph nodes of the neck, a rash, and diarrhea. Adenoviral disease is spread by respiratory droplets and is likely to occur in groups of young people living together in crowded conditions. It has been a big problem in military recruits. Epidemic spread is fostered by a high percentage of asymptomatic infections. The viruses are shed from the upper respiratory tract during the acute illness and continue to be eliminated in the feces for months thereafter.

Prevention and Treatment

Formerly, an attenuated vaccine administered orally was helpful for preventing acute respiratory disease in military recruits, but the vaccine program was abandoned in 1996 because of cost. Now, with more than 2,500 adenoviral illnesses monthly with some deaths, reinstitution of adenoviral vaccination of recruits is planned. As with colds, there is no treatment, and most immunocompetent patients get well on their own. Secondary bacterial infections, however, may occur and require antibacterial medication.

Table 22.6 summarizes some important facts about adenoviral pharyngitis.

TABLE 22.6 Adenoviral Pharyngitis

Symptoms	Fever, very sore throat, severe cough, swollen lymph nodes of neck, pus on tonsils and throat, sometimes conjunctivitis; less frequently, pneumonia
Incubation period	5 to 10 days
Causative agent	Adenoviruses—more than 45 types
Pathogenesis	Virus multiplies in host cells; cell destruction and inflammation occur; different types produce different symptoms.
Epidemiology	Inhalation of infected droplets; possible spread from gastrointestinal tract.
Prevention and treatment	Live virus vaccine formerly used by the military is no longer produced. No treatment except for relief of symptoms.

MICROCHECK 22.4

Many different kinds of infectious agents can produce the same symptoms and signs of respiratory disease. Emotional stress significantly increases the risk of contracting a cold, but exposure to cold temperatures probably does not. Persons suffering a cold are most likely to transmit it if symptoms are severe, and during the first few days of illness. Adenovirus infections can mimic colds, pinkeye, strep throat, and whooping cough.

✓ How is an adenovirus infection treated?

✓ People who staff polar ice stations where they are isolated from other human contact for long periods often do not develop colds. Is this an expected observation? Why or why not?

INFECTIONS OF THE LOWER RESPIRATORY SYSTEM

22.5 Bacterial Infections of the Lower Respiratory System

Focus Points

- Compare the distinctive features of pneumococcal, *Klebsiella,* and mycoplasmal pneumonia.
- Outline the pathogenesis of pertussis.
- List the main steps in the pathogenesis of tuberculosis.
- Describe the epidemiology of Legionnaires' disease.

Bacterial infections of the lower respiratory system are less common than those of the upper system, largely because they are stopped by body defenses at the portal of entry. Lower tract infections, however, are generally much more serious. An earache or sore throat is unlikely to be life threatening, but the causative organisms of both can endanger life when they infect the lung. Distinctive patterns of signs and symptoms are produced by the different kinds of organisms that infect the lower respiratory system. The pneumonias are inflammatory diseases of the lung in which fluid fills the alveoli. They top the list of infectious killers in the general population of the United States, and they are important as nosocomial infections. Whooping cough, tuberculosis, and Legionnaires' disease are other distinctive types of infection. ■ nosocomial infection, p. 462

Pneumococcal Pneumonia

Pneumococci are an important cause of pneumonia acquired in the community, accounting for about 60% of the adult pneumonia victims requiring hospitalization.

Symptoms

The typical symptoms of pneumococcal pneumonia are cough, fever, chest pain, and sputum production. These symptoms are usually preceded by a day or two of runny nose and upper respiratory congestion, ending with an abrupt rise in temperature and a single body-shaking chill. The sputum quickly becomes pinkish or rust colored, and severe chest pain develops, aggravated by each breath or cough. As a result of the pain, breathing becomes shallow and rapid. The patient becomes short of breath and develops a dusky color because of poor oxygenation. Without treatment, after 7 to 10 days people who survive show profuse sweating and a rapid fall in temperature to normal.

Causative Agent

Streptococcus pneumoniae, the pneumococcus, is a Gram-positive diplococcus **(figure 22.10).** The most striking characteristic of *S. pneumoniae* is its thick polysaccharide capsule, which is responsible for the organism's virulence. There are 90 different types of *S. pneumoniae,* as determined by differing capsular antigens. ■ capsules, p. 64

Pathogenesis

Pneumococcal pneumonia develops when encapsulated pneumococci, the virulent form, are inhaled into the alveoli of a susceptible host, multiply rapidly, and cause an inflammatory response. The bacteria are resistant to phagocytosis because their capsules interfere with the action of C3b, the fraction of complement system component responsible for opsonization. A surface protein, PspA, also interferes with complement-mediated opsonization. A hemolysin called **pneumolysin,** antigenically related to streptolysin O of *S. pyogenes,* is released by **autolysis,** spontaneous enzymatic disintegration, and is

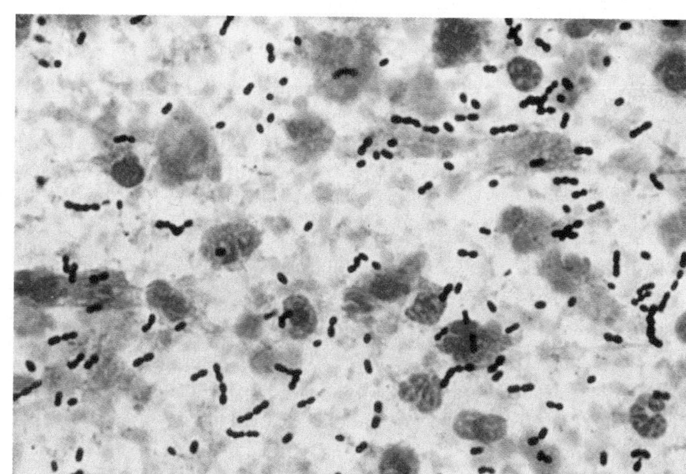

FIGURE 22.10 ***Streptococcus pneumoniae*** Gram stain of sputum from a person with pneumococcal pneumonia showing Gram-positive diplococci and polymorphonuclear neutrophils (PMNs).

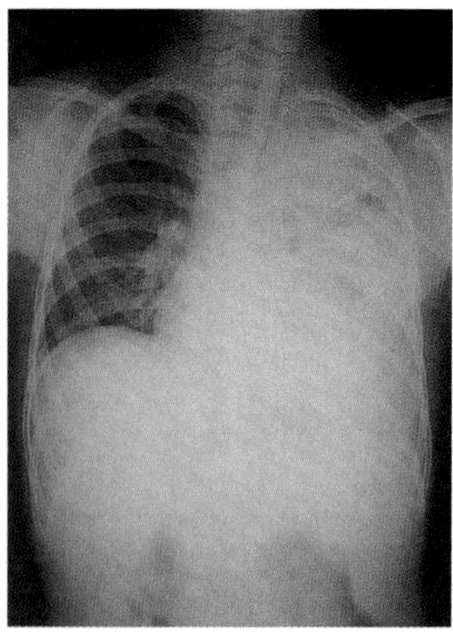

(a)

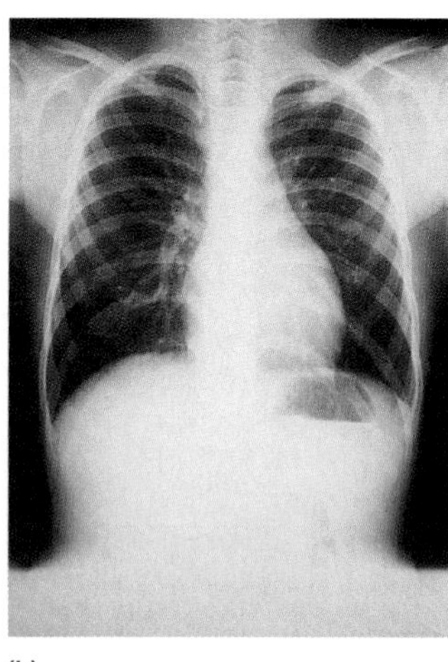

(b)

FIGURE 22.11 Chest X-Ray Appearance in Pneumococcal Pneumonia **(a)** Pneumonia. The left lung (right side of figure) appears white because fluid-filled alveoli stop the X-ray beam from reaching the X-ray film and turning it black. **(b)** Normal X-ray film after recovery. The X-ray beam passes through air-filled alveoli and turns the film black.

toxic to ciliated epithelium. Serum and phagocytic cells pour into the air sacs of the lung, causing difficulty breathing. This increase in fluid produces abnormal shadows on chest X-ray films of patients with pneumonia **(figure 22.11).** Material coughed from the lungs, called **sputum,** increases in amount and contains pus, blood, and many pneumococci. ■ opsonization by C3b, p. 356

The inflammatory response to the infection often involves nerve endings, causing pain; the condition in which pain arises from an inflamed pleura is called **pleurisy.** Pneumococci that enter the bloodstream from the inflamed lung are responsible for three often fatal complications: **septicemia,** a symptomatic infection of the bloodstream; **endocarditis,** an infection of the heart valves; and **meningitis,** an infection of the membranes covering the brain and spinal cord. Individuals who do not develop such complications usually develop sufficient specific anticapsular antibodies within about a week to permit phagocytosis and destruction of the infecting organisms. Complete recovery usually results. Most pneumococcal strains do not destroy lung tissue.

Epidemiology

Up to 30% of healthy people carry encapsulated pneumococci in their throat. Because of the effectiveness of the mucociliary escalator, these bacteria rarely reach the lung. The risk of pneumococcal pneumonia rises dramatically when this defense mechanism is impaired, however, as it is with alcohol and narcotic use, and with respiratory viral infections such as influenza. There is also an increased risk of the disease with underlying heart or lung disease, diabetes, and cancer, and with age over 50.

Prevention and Treatment

A vaccine is available that stimulates production of anticapsular antibodies and gives immunity to 23 strains that account for over 90% of serious pneumococcal disease. A conjugate vaccine against seven types is available for infants and children up to the age of 5 years. Most pneumococcal infections can be cured if penicillin or erythromycin is given early in the illness. Strains of pneumococci resistant to one or more antibiotics, however, are increasingly being encountered. ■ conjugate vaccine, p. 383

See **table 22.7** for a description of some features of pneumococcal pneumonia.

TABLE 22.7 Pneumococcal, *Klebsiella,* and Mycoplasmal Pneumonias Compared

	Pneumococcal Pneumonia	*Klebsiella* Pneumonia	Mycoplasmal Pneumonia
Symptoms	Cough, fever, single shaking chill, rust-colored sputum from degraded blood, shortness of breath, chest pain	Chills, fever, cough, chest pain, and grossly bloody, mucoid sputum	Gradual onset of cough, fever, sputum production, headache, fatigue, and muscle aches
Incubation period	1 to 3 days	1 to 3 days	2 to 3 weeks
Causative agent	The pneumococcus, *Streptococcus pneumoniae,* encapsulated strains	*Klebsiella pneumoniae,* an enterobacterium	*Mycoplasma pneumoniae;* lacks cell wall
Pathogenesis	Inhalation of encapsulated pneumococci; colonization of the alveoli incites inflammatory response; plasma, blood, and inflammatory cells fill the alveoli; pain results from involvement of nerve endings.	Aspiration of colonized mucus droplets from the throat. Destruction of lung tissue and abscess formation common; infection spreads via blood to other body tissues.	Cells attach to specific receptors on the respiratory epithelium; inhibition of ciliary motion and destruction of cells follow.
Epidemiology	High carrier rates for *S. pneumoniae.* Risk of pneumonia increased with conditions such as alcoholism, narcotic use, chronic lung disease, and viral infections that impair the mucociliary escalator. Other predisposing factors are chronic heart disease, diabetes, and cancer.	Often resistant to antibiotics, and colonize individuals who are taking them. *Klebsiella* sp. and other Gram-negative rods are common causes of fatal nosocomial pneumonias.	Inhalation of infected droplets; mild infections common and foster spread of the disease.
Prevention and treatment	Capsular vaccine available contains 23 capsular antigens; conjugate vaccine for infants. Treatment: penicillin, erythromycin, and others.	No vaccine available. A cephalosporin with an aminoglycoside.	No vaccine available; avoidance of crowding in schools and military facilities advisable; tetracycline or erythromycin for treatment.

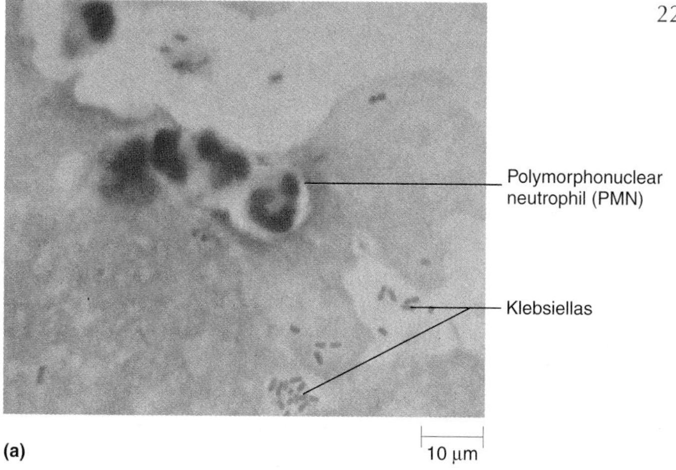

(a)

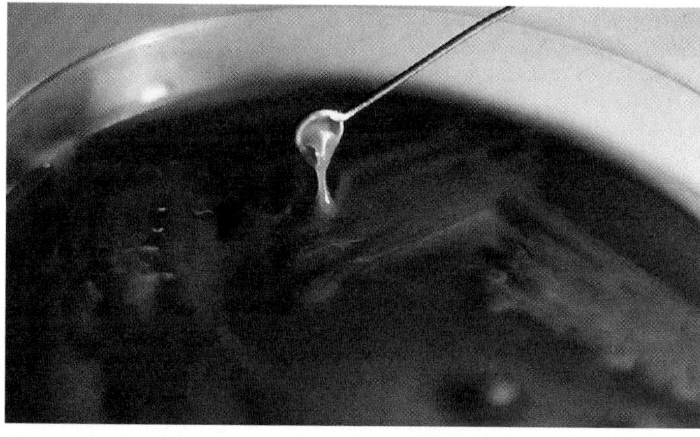

(b)

FIGURE 22.12 ***Klebsiella pneumoniae*** **(a)** In sputum from a pneumonia victim. **(b)** Colonies. When the colony is touched with a microbiological loop, it "strings out."

Klebsiella Pneumonia

Enterobacteria such as *Klebsiella* sp., and other Gram-negative rods, can cause pneumonia, especially if host defenses are impaired. These pneumonias attack the very old, the very young, alcoholics, nursing home patients, those debilitated by other diseases, and immunocompromised persons. Pneumonias caused by Gram-negative rods such as *Klebsiella* sp., cause most of the deaths from nosocomial infections. ■ **enterobacteria, p. 262**

Symptoms

In general, the symptoms of *Klebsiella* pneumonia—cough, fever, and chest pain—are indistinguishable from those of pneumococcal pneumonia. *Klebsiella* pneumonia patients typically have repeated chills, however, and their sputum is red and gelatinous, resembling currant jelly. Their mortality rate without treatment is 50% to 80%, and they tend to die sooner than other pneumonia patients.

Causative Agent

Several species of *Klebsiella* cause pneumonia, but *Klebsiella pneumoniae* is the best known. It is a Gram-negative rod with a large capsule and big, strikingly mucoid colonies **(figure 22.12).**

Pathogenesis

Inhaled organisms first colonize the throat and then are carried to the lung by inspired air or aspirated mucus. Specific adhesins aid colonization, but their exact nature is not yet known. The capsule is an essential **virulence** factor, probably functioning like the pneumococcal capsule in interfering with the action of C3b, a critically important opsonin before the appearance of antibody. Unlike *Streptococcus pneumoniae, K. pneumoniae* causes death of tissue and rapid formation of lung abscesses. The infection often enters the bloodstream, causing abscesses in other tissues, and endotoxic shock. Therefore, even with effective antibacterial medication, the lung can be permanently damaged and death may occur. ■ **endotoxic shock, p. 407** ■ **abscess, p. 560**

Epidemiology

Klebsiellas are part of the normal flora of the intestine in a small percentage of individuals. Colonization of the mouth and throat is common in debilitated individuals, especially in an institutional setting. In hospitals and nursing homes, the organisms are often resistant to antimicrobial medications, and they readily colonize patients taking the medications. *Klebsiella* resistance to antimicrobial medications can be chromosomal or plasmid-mediated. The organisms easily acquire and are a source of R factors, all of which contain transposons. This allows resistance genes to spread readily to other species and genera when antibacterial medications are used. ■ **R plasmids and transposons, pp. 189, 205**

Prevention and Treatment

There are no specific preventive measures. Disinfection of the environment, use of sterile respiratory equipment, and use of antimicrobial medications only when necessary help control the organisms in hospitals.

See table 22.7 for a description of the main features of this disease.

Mycoplasmal Pneumonia

Mycoplasmal pneumonia is the leading kind of pneumonia in college students and is also common among military recruits. The disease is generally mild, is often referred to by the popular name, "walking pneumonia," and usually does not require hospitalization.

Symptoms

The onset of mycoplasmal pneumonia is typically gradual. The first symptoms are fever, headache, muscle pain, and fatigue. After several days, a dry cough begins, but later mucoid sputum may be produced. About 15% of cases also have middle ear infections.

Causative Agent

The causative agent of mycoplasmal pneumonia is *Mycoplasma pneumoniae,* a small (0.2 μm diameter), easily deformed bacterium that has no cell wall (see figure 3.35). Distinguishing characteristics are slow growth and aerobic metabolism. As with other mycoplasmas, the central portion of *M. pneumoniae* colonies grows down into the medium, producing a fried egg appearance (see figure 11.27). ■ **mycoplasmas, p. 273**

Pathogenesis

Only a few inhaled *M. pneumoniae* cells are necessary to start an infection. The organisms attach to specific receptors on the respiratory

PERSPECTIVE 22.1

Terror by Mail: Inhalation Anthrax

During October and November 2001, 22 human cases of anthrax were reported in the United States, half due to inhalation of *Bacillus anthracis* spores (inhalation anthrax) and half due to skin infections (cutaneous anthrax). Five deaths occurred, all among the inhalation cases, and more than 30,000 individuals potentially exposed to the spores were given antibiotic treatment as a preventive. Most of the anthrax cases could reasonably be linked to four envelopes containing purified spores of *B. anthracis,* which contaminated people, air, and surfaces during their journey through the postal system to their destinations. Only 18 cases of inhalation anthrax were identified in the United States during the entire twentieth century.

Sometimes called "anthrax pneumonia," inhalation anthrax victims generally fail to show the hallmarks of pneumonia. Inhaled *B. anthracis* spores are quickly taken up by lung phagocytes and carried to the regional lymph nodes, where they germinate, kill the phagocytes, and invade the bloodstream. Although the organisms are susceptible to various antibacterial medications, treatment must be given as soon as possible after infection, because the bacteria produce a powerful toxin that usually kills the victim within a day or two after bloodstream invasion occurs. Cutaneous anthrax and anthrax acquired by ingesting contaminated meat are much less likely to be fatal.

Bacillus anthracis endospores have long been considered for use in biological warfare. The organism is readily available; anthrax is endemic in livestock in most areas of the world including the United States and Canada. The spores are easy to produce and remain viable for years. When anthrax is acquired by inhalation, the fatality rate approaches 90%, yet the disease is easily confined to the attack area because it does not spread person-to-person. *Bacillus anthracis* spores were employed as a weapon as early as World War I, although ineffectively, in an attack on livestock used for food. Just prior to World War II, Japan, the United States, USSR, Germany, and Great Britain secretly began to develop and perfect anthrax weapons, but they were not used during the war. During the "cold war" that followed, both the United States and the USSR developed massive biological warfare programs that employed many thousands of people, perfecting techniques for preparing the spores and delivering them to enemy targets. An executive order by President Nixon ended the U.S. program in 1969, and the stockpiled weapons were ordered destroyed, but other countries continued weapon development. In 1979, an accidental discharge of *B. anthracis* spores, a weight equaling about two paper clips, from a biological warfare plant at Sverdlovsk, USSR, caused more than 90 deaths downwind of the facility. The 1990 war with Iraq exposed their sizeable anthrax weapon program, and several times during the 1990s a Japanese terrorist group tried ineffectively to attack Tokyo institutions with anthrax spores.

During this long history of anthrax weapon development, remarkably little was accomplished that would help defend against an attack, presumably because the best defense was considered an opponent's fear of counterattack. As a result of the 2001 anthrax-by-mail incident a number of questions came into focus. How do you diagnose anthrax quickly, by history, physical examination, and X-ray, and what is the best way to teach and organize medical practitioners to meet an anthrax attack? What tests are available to identity *B. anthracis* rapidly and reliably, and what is the best way to sample people and the environment? How can decontamination of buildings and other objects be accomplished? Is there sufficient vaccine available, can better vaccines be developed, and what is their role before or after exposure to *B. anthracis?* What is the best antimicrobial treatment, is there enough of it, is it readily accessible, and how long should it be given to exposed individuals? Are other potential treatments such as antitoxins and designer medications to block the lethal effect of *B. anthracis* toxin being developed?

The mail attack has stimulated remarkable progress toward answering these questions, especially in the area of using the latest technology for early detection, and teaching medical personnel about symptoms, signs, and X-ray findings. Antibacterial treatment proves highly effective if given soon after exposure. Approaches to combatting the toxin are under development, using designer drugs and antibodies.

More information about anthrax is available at the Online Learning Center at www.mhhe.com/nester6.

epithelium **(figure 22.13),** interfere with ciliary action, and cause the ciliated cells to slough off. An inflammatory response characterized by infiltration of lymphocytes and macrophages causes the walls of the bronchial tubes and alveoli to thicken.

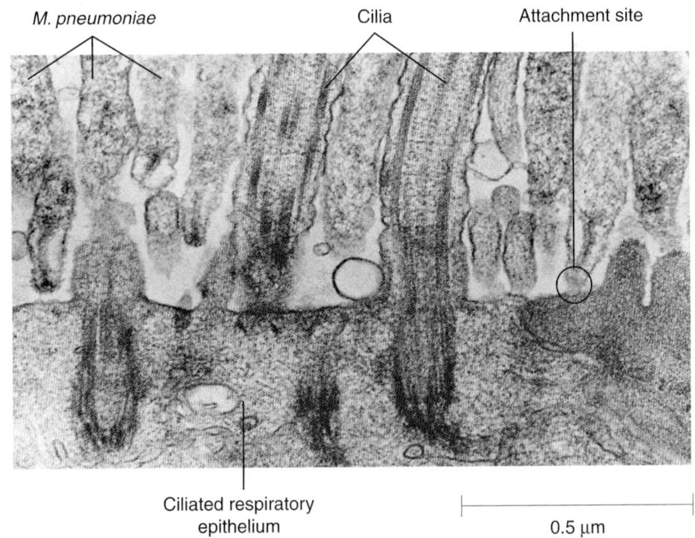

FIGURE 22.13 ***Mycoplasma pneumoniae*** **Infecting Respiratory Epithelium** Transmission electron micrograph. Notice the distinctive appearance of the tips of the mycoplasmas adjacent to the host epithelium. The tips probably represent a site on the microorganism that is specialized for attachment.

Epidemiology

The mycoplasmas are spread by aerosolized droplets of respiratory secretions. Transmission from person to person is aided by the long time period in which *M. pneumoniae* is present in respiratory secretions, ranging from about 1 week before symptoms begin to many weeks afterward. Mycoplasmal pneumonia accounts for about one-fifth of bacterial pneumonias, and it has a peak incidence in young people. Immunity after recovery is not permanent, and repeat attacks have occurred within 5 years.

Prevention and Treatment

No practical preventive measures exist for mycoplasmal pneumonia, except avoiding crowding in schools and military facilities. Antibiotics that act against bacterial cell walls, such as the penicillins and cephalosporins, are of course not effective. Tetracycline and erythromycin shorten the illness if given early, but they are bacteriostatic, meaning they only inhibit the growth of the bacteria without killing them.

Some features of pneumococcal, *Klebsiella,* and mycoplasmal pneumonias are compared in table 22.7.

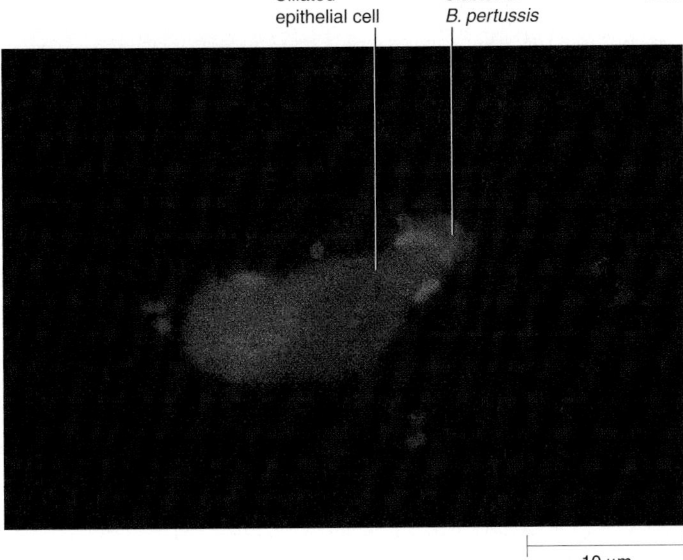

FIGURE 22.14 ***Bordetella pertussis*** Fluorescent antibody stain of respiratory secretions from an individual with whooping cough. The bacteria stain a greenish-yellow color. The large orange object is a ciliated epithelial cell. Notice the cluster of *B. pertussis* at one end of the cell, the location of the cell's cilia.

Whooping Cough (Pertussis)

Whooping cough is the common name for **pertussis,** a disease that is now uncommon in the United States because of childhood immunization. The causative bacterium is widespread, however, and remains a threat to those people who lack immunity. Worldwide, the disease causes 300,000 to 500,000 deaths yearly.

Symptoms

Pertussis typically begins with a number of days of runny nose followed by sudden bouts of violent, uncontrollable coughing. This symptom, termed **paroxysmal coughing,** is severe enough to rupture small blood vessels in the eyes. The coughing spasm is followed by forceful attempts to inhale—the "whoop." Vomiting and seizures can occur during this phase of the illness.

Causative Agent

The disease is caused by *Bordetella pertussis* **(figure 22.14),** a tiny, encapsulated, strictly aerobic, Gram-negative rod. These organisms do not tolerate drying or sunlight and die quickly outside the host.

Pathogenesis

Bordetella pertussis enters the respiratory tract with inspired air and attaches specifically to ciliated cells of the respiratory epithelium. Among the proteins present on the bacterial surface, pertussis toxin (Ptx) and filamentous hemagglutinin (Fha) are probably the main ones involved in attachment.

The areas colonized by *B. pertussis* include the nasopharynx, trachea, bronchi, and bronchioles. The violent symptoms arise because of involvement of the passageways of the lower respiratory tract. The organisms grow in dense masses on the epithelial surface, but they do not invade. Mucus secretion increases markedly, while ciliary action declines precipitously, and patches of ciliated cells slough off. Only the cough reflex remains for clearing the secretions. Some of the bronchioles become completely obstructed, resulting in small areas of collapsed lung. Others, because of spasm or mucus plugging, let air enter but not escape, causing hyperinflation. Paroxysmal coughing causes hemorrhages in the brain, and seizures can occur. Pneumonia due to *B. pertussis* or, more commonly, secondary bacterial infection is the chief cause of death.

A number of toxic products of *B. pertussis* probably play a role in its pathogenesis. Pertussis toxin (Ptx), mentioned earlier for its role in colonization, is an A-B toxin. **Figure 22.15** shows

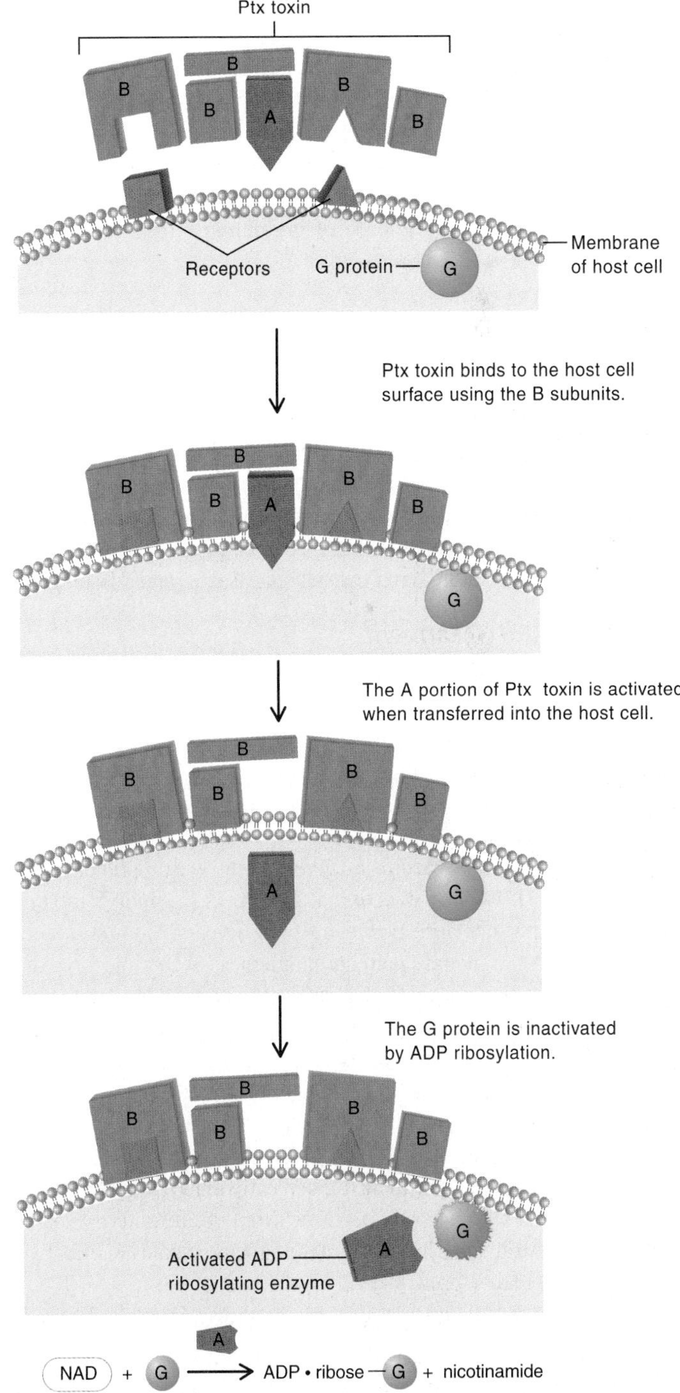

FIGURE 22.15 Mode of Action of Pertussis Toxin The B portion of this A-B toxin attaches to receptors on the cell membrane. The B portion stays behind as the A portion enters the cell, becoming an enzyme activated by intracellular conditions. Activated A enzyme inactivates a G protein by ADP ribosylation, thereby maximizing cAMP production.

the steps involved in the toxin's action. First, the B portion attaches specifically to receptors on the host cell surface. Second, the A portion is transferred through the cytoplasmic membrane of the host cell, becoming an active ribosylating enzyme in the process. The activated A portion of the Ptx toxin inactivates the G protein normally responsible for initiating a decrease in cyclic adenosine monophosphate (cAMP) synthesis, thereby allowing maximum production of cAMP. Unregulated cAMP production causes marked increase in mucus output, decreased killing ability of phagocytes, massive release of lymphocytes into the bloodstream, ineffectiveness of natural killer cells, and low blood sugar. Invasive adenylate cyclase, another *B. pertussis* toxin, also causes an increase in cAMP production. ■ A-B toxins, p. 405 ■ cAMP, p. 179

Tracheal cytotoxin produced by *B. pertussis* acts with endotoxin to cause release of nitric oxide (NO) from goblet cells. The NO causes ciliated epithelial cells to die and slough off. Tracheal cytotoxin also causes the release of interleukin-1 (IL-1), a fever-causing cytokine. ■ interleukin-1, p. 353

Epidemiology

Pertussis spreads via respiratory secretions suspended in air. Patients are most infectious during the runny nose period, the numbers of expelled organisms decreasing substantially with the onset of violent coughing. Pertussis is classically a disease of infants. It can occur in a milder form in older children and adults, however, overlooked as a persistent cold, thus fostering transmission. During 2003 and 2004, small outbreaks in hospitals involving doctors, nurses, and other staff occurred in Kentucky, Oregon, Pennsylvania, Texas, and Washington.

Prevention and Treatment

Intensive vaccination of infants with killed *B. pertussis* cells can prevent the disease in about 70% of individuals. Its widespread administration to young children is responsible for the drop from 235,239 reported cases of pertussis in the United States in 1934 to only 1,248 in 1981. Since then, reported cases have increased in number, reaching a peak of 25,827 cases in 2004.

Acellular pertussis vaccine is given along with diphtheria and tetanus toxoids (DTaP) by injection at 2, 4, 6, 12–15 and 18 months of age, with a booster dose between 4 and 6 years. Common reactions to the vaccine include pain and tenderness at the injection site and fever. Acellular vaccines that use only part of the bacterium instead of the whole cells, first approved for use in the United States in 1991, are now advised for all five doses and give good immunity with fewer side effects. A pertussis vaccine with a decreased dose of pertussis antigen (Tdap) is advised at age 11–12 years and for older individuals needing a booster. Erythromycin given for treatment reduces the duration of symptoms if given early in the disease, and the antibiotic usually eliminates *B. pertussis* from the respiratory secretions. Azithromycin and trimethoprim-sulfamethoxazole are effective alternatives. ■ acellular vaccines, p. 435

The main features of this disease are shown in **table 22.8.**

TABLE 22.8 Pertussis

Symptoms	Runny nose followed after a number of days by spasms of violent coughing; vomiting and possible convulsions
Incubation period	7 to 21 days
Causative agent	*Bordetella pertussis,* a tiny Gram-negative rod
Pathogenesis	Colonization of the surfaces of the upper respiratory tract and tracheobronchial system; ciliary action slowed; toxins released by *B. pertussis* cause death of epithelial cells and increased cAMP; fever, excessive mucus output, and a rise in the number of lymphocytes in the bloodstream result.
Epidemiology	Inhalation of infected droplets; older children and adults have mild symptoms.
Prevention and treatment	Acellular vaccines, for immunization of infants and children; erythromycin, somewhat effective if given before coughing spasms start, eliminates *B. pertussis.*

Tuberculosis

Over the last hundred or more years, tuberculosis gradually declined in industrialized countries in association with improved living standards. In 1985, however, the rate of decline slowed, and then the incidence of tuberculosis began to rise **(figure 22.16).** The rise continued for the next 6 years in association with the expanding AIDS epidemic and increases in treatment-resistant cases. Fortunately, starting in 1993, with better funding of control measures and more aggressive treatment, tuberculosis resumed its slow retreat. In 1989, a Strategic Plan for the Elimination of Tuberculosis in the United States was published by the Centers for Disease Control and Prevention. The plan depends largely on increased efforts in identifying and treating people with the disease among the high-risk groups, particularly poor people, people with AIDS, prisoners, and immigrants from countries with high rates of tuberculosis. Although the decline in the incidence rate of tuberculosis has continued, it is much slower than anticipated by the strategic plan, calling into question the goal of eliminating the disease from the United States by the year 2010. Wide variation exists among different segments of the population—the rate of disease among the foreign born approaches 10 times the

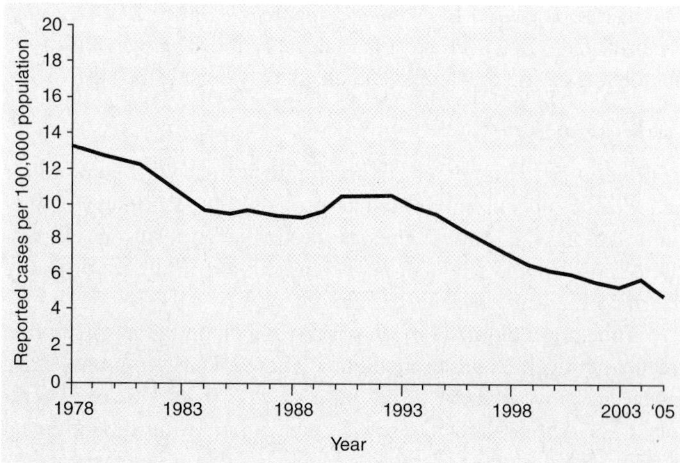

FIGURE 22.16 Incidence of Tuberculosis, United States, 1978–2005 Resurgence 1988 to 1994 responded to better funding of control measures.

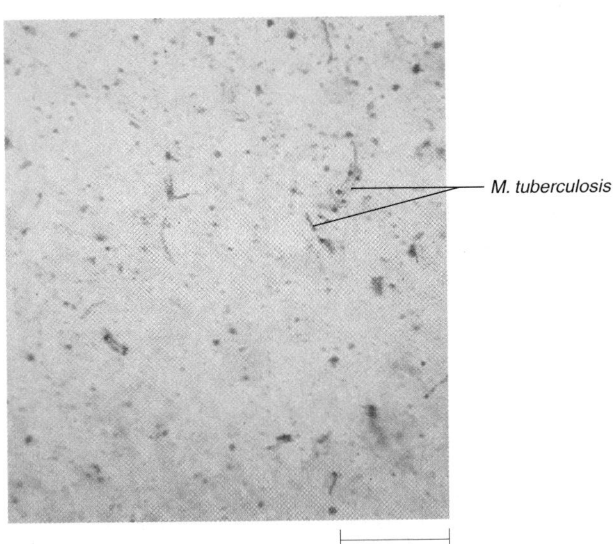

FIGURE 22.17 ***Myobacterium tuberculosis*** **in Sputum from an Individual with Tuberculosis** The sputum has been "digested," meaning that it has been treated with strong alkali to kill the other bacteria that are invariably present.

rate among the native born, and the rates in some cities such as Washington, D.C., are even higher. Worldwide, it is estimated that one-third of the population is infected with *Mycobacterium tuberculosis* and two million die of the disease annually.

Symptoms

Tuberculosis is a chronic illness characterized by slight fever, progressive weight loss, sweating at night, and chronic cough, often producing blood-streaked sputum.

Causative Agent

Today, most cases of tuberculosis are caused by *Mycobacterium tuberculosis,* although in the years before widespread milk pasteurization, the cattle-infecting species *Mycobacterium bovis* was a common cause. *Mycobacterium tuberculosis,* commonly called the tubercle bacillus, is a slender, acid-fast, rod-shaped bacterium **(figure 22.17).** The organism is a strict aerobe that grows very slowly, with a generation time of 12 hours or more. This slow growth makes it difficult to diagnose tuberculosis quickly. Although easily killed by pasteurization, the tubercle bacillus is unusually resistant to drying, disinfectants, and strong acids and alkali. The bacterium has a Gram-positive cell wall, but bound covalently to the cell wall peptidoglycan is a thick layer composed of complex glycolipids. Up to 60% of the dry weight of the *M. tuberculosis* cell wall consists of lipids, a much higher percentage than that in most other bacteria. The lipid-containing cell wall is largely responsible for its acid-fast staining, resistance to drying and disinfectants, and its pathogenicity. ■ **pasteurization, p. 112** ■ **lipids, p. 35**

Pathogenesis

Individuals usually contract tuberculosis by inhaling airborne organisms from a person who has tuberculosis. In the lungs, the bacteria are taken up by pulmonary macrophages, where they resist destruction by releasing an enzyme that prevents fusion of the phagosome with lysosomes. They also possess a mechanism that protects them from nitric oxide. Evidence indicates that they leave the phagosome and multiply within the cytoplasm of the macrophages and are carried to nearby lymph nodes. At this stage, there is little inflammatory response to the infection. The organisms continue to multiply, lyse the macrophages, and spread throughout the body. After about 2 weeks, delayed hypersensitivity to the tubercle bacilli develops. An intense reaction then occurs at sites where the bacilli have lodged. Macrophages, now activated, collect around the bacteria, and some macrophages fuse together to form large multinucleated giant cells. Lymphocytes and macrophages then collect around these multinucleated cells and wall off the infected area from the surrounding tissue. The localized collection of inflammatory cells is called a **granuloma (figure 22.18),** which is the characteristic response of the body to microorganisms and other foreign substances that resist digestion and removal. The granulomas of tuberculosis are called tubercles. Some of the mycobacteria in the tubercles remain

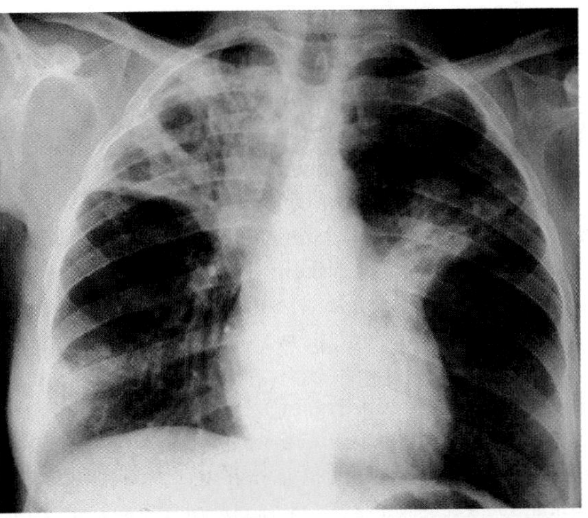

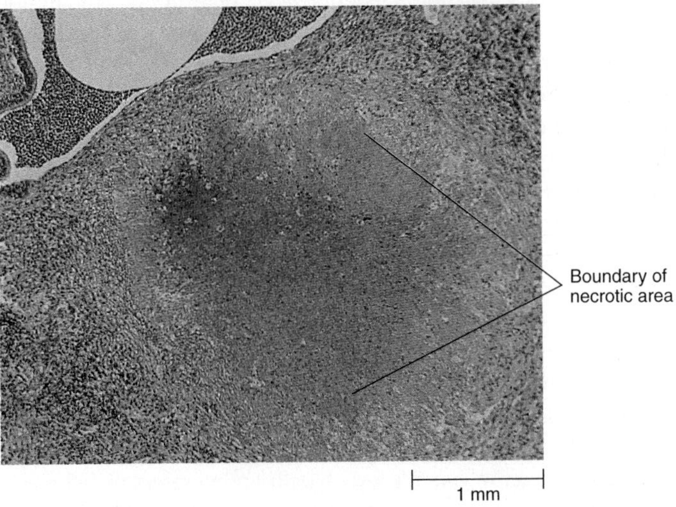

FIGURE 22.18 Stained Lung Tissue Showing a Tubercle A tubercle is a kind of granuloma caused by the body's reaction to *Myobacterium tuberculosis*. The innumerable dark dots around the outer portion of the picture are nuclei of lung tissue and inflammatory cells. Centrally and extending to the right of the photograph, most of the nuclei have disappeared because the cells are dead and the tissue has begun to liquefy. The photograph depicts a chest X-ray film of an individual with tuberculosis.

alive, but they are kept from multiplying by conditions in the tubercle including low pH and low available oxygen. They remain alive in this state for many years, causing a **latent infection.** Individuals with latent tuberculosis infection are asymptomatic.

■ delayed hypersensitivity, p. 422 ■ lysosomes, p. 79 ■ latent infection, p. 330

Sometimes the mycobacteria are not contained by the inflammatory response and cause symptomatic **active tuberculosis.** Lysis of activated macrophages attacking *M. tuberculosis* releases their enzymes into the infected tissue. The result is death of tissue, with the formation of a cheesy material by a process called **caseous necrosis.** If this process involves a bronchus, the dead material may discharge into the airways, causing a large lung defect called a **cavity** and spread of the bacteria to other parts of the lung. Lung cavities characteristically persist, slowly enlarging for months or years and shedding tubercle bacilli into the bronchi. Coughing and spitting transmit the organisms to other people.

Wherever living organisms persist, they can resume growing if the person's immunity becomes impaired by stress, advanced age, or AIDS. Disease resulting from renewed growth of the organisms is called **reactivation tuberculosis.**

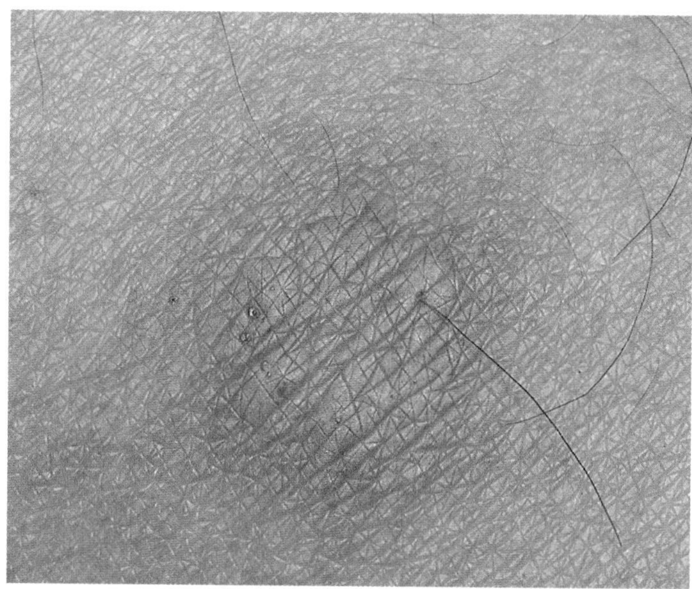

FIGURE 22.19 Tuberculin Test A positive test is the result of delayed hypersensitivity to *Myobacterium tuberculosis* antigens injected into the skin, and it gives evidence of past or current infection with the bacterium.

Epidemiology

Most of the estimated 10 million Americans who are infected by *M. tuberculosis* have latent tuberculosis infection and are asymptomatic. However, about 10% of them will develop reactivation tuberculosis at some time during their life. Infection rates are highest among non-whites and elderly poor people. Foreign-born U.S. residents have rates of tuberculosis much higher than those born in the United States. Transmission of tuberculosis occurs almost entirely by the respiratory route; 10 or even fewer inhaled organisms are enough to cause infection. Factors important in transmission include the frequency of coughing, the adequacy of ventilation (transmission is unlikely to occur outdoors), and the degree of crowding. Immunodeficiency can result in activation of latent tuberculosis, and over 5% of AIDS patients have developed active tuberculosis.

The tuberculin skin test (TST), also known as the Mantoux (pronounced man-too) test, is an extremely important tool for studying the epidemiology of the disease and in detecting those who are infected with *M. tuberculosis.* The test is carried out by injecting into the skin a very small amount of a sterile fluid called purified protein derivative, or PPD, derived from cultures of *M. tuberculosis.* People who are infected with the bacterium develop redness and a firm swelling at the injection site, reaching a peak intensity after 48 to 72 hours **(figure 22.19).** This reaction is due to the congregation of macrophages and T lymphocytes at the injection site, a manifestation of delayed hypersensitivity to the tubercle bacillus. A positive reaction to the test generally indicates that living *M. tuberculosis* bacilli are present somewhere in the body of the person tested. A positive TST does not necessarily mean that the person has tuberculosis—only that he or she has been infected by *M. tuberculosis* at some time in the past.

Beside the TST, a blood test (QFT-G) is available for diagnosing *M. tuberculosis* infection. QFT-G generally gives results similar to TST, but sooner and with greater specificity. It is based on the fact that lymphocytes in the blood of *M. tuberculosis*-infected individuals release INF-gamma when they contact specific *M. tuberculosis* antigens added to the blood. ■ IFN-gamma, p. 353

Prevention and Treatment

Vaccination against tuberculosis has been widely used in many parts of the world with varying success. The vaccinating agent, a living attenuated mycobacterium known as Bacille Calmette-Guérin, or BCG, is derived from *M. bovis.* Repeated subculture in the laboratory over many years resulted in selection of this strain of *M. bovis,* which has little virulence in humans but produces some immunity to tuberculosis. Use of the vaccine is discouraged in the United States, because people who receive the vaccine usually develop a positive tuberculin test. By causing a positive test, BCG vaccination eliminates an important way of diagnosing tuberculosis early in the disease when it can most easily be treated. BCG is not safe to use in severely immunocompromised patients because the vaccine bacillus can spread throughout the body and cause disease. Genetically engineered vaccines are under development. Control of tuberculosis is aided by identifying unsuspected cases using skin tests and lung Xrays. Individuals with active disease are then treated, thus interrupting the spread of *M. tuberculosis.* People whose TST has changed from negative to positive are also treated, as are all high-risk individuals, even when no evidence of active disease exists. High-risk individuals include babies, young children, and the elderly; those whose TST converted to positive within 2 years; those recently exposed to active tuberculosis; and those with underlying conditions such as HIV infection, drug abuse, and diabetes. Treatment reduces the risk that these people will develop active disease later in life.

Mutants resistant to antibacterial medications frequently occur among sensitive *M. tuberculosis* strains. Because mutants simultaneously resistant to more than one antimicrobial medication occur with a very low frequency, two or more of the medications are always given together in treating tuberculosis. The combination of rifampin and isoniazid (INH) is favored because both drugs are bactericidal against actively growing organisms in cavities as well as metabolically inactive intracellular organisms. Because of the long generation times of *M. tuberculosis* and its resistance to destruction by body defenses, drug treatment of tuberculosis must generally be continued for a minimum of 6 months to cure the disease. During the prolonged treatment, symptoms usually disappear and many individuals become careless about continuing to take their medications. Such negligence allows the rare, resistant mutants to multiply and can lead to high rates of relapse. Up to two-thirds of the *M. tuberculosis* strains obtained from inadequately treated patients are resistant to one or more antitubercular medications. These resistant strains can infect others in the community. DOTS (directly observed therapy short-course) programs employ three or four antitubercular medications given daily or thrice weekly for 6 months. They have been effective in assuring that medications are taken properly. Nevertheless, the problem of drug-resistant *M. tuberculosis* strains has reached alarming proportions in some areas. During the 1990s, multiple drug-resistant tuberculosis (MDR-TB) became an increasing problem. These cases resisted treatment with rifampin and isoniazid, the two most effective first-line drugs available, and were often treated with less effective, more toxic, and more expensive second-line medications. By the end of the decade, MDR-TB cases were appearing that also resisted treatment with many of the second-line drugs. These extensively drug-resistant tuberculosis (XDR-TB) cases now threaten tuberculosis-control efforts around the world. Fortunately, promising new antituberculosis medications representing at least five different chemical families are being developed. One entirely new and highly active antitubercular medication, a diarylquinoline (DARQ), is among those under evaluation. ■ antibiotic resistance, p. 483

The main features of tuberculosis are shown in **table 22.9.**

Legionnaires' Disease

Legionnaires' disease was unknown until 1976, when a number of people attending an American Legion Convention in Philadelphia developed a mysterious pneumonia that was fatal in many cases. Months of scientific investigation eventually paid off when the cause was discovered to be a previously unknown bacterium commonly present in the natural environment.

Symptoms

Legionnaires' disease typically begins with headache, muscle aches, rapid rise in temperature, confusion, and shaking chills. A dry cough develops that later produces small amounts of sputum, sometimes streaked with blood. Pleurisy, manifest as chest pain brought on by coughing or deep breathing, can also occur. About one-fourth of the cases also have some alimentary tract symptoms such as diarrhea, abdominal pain, and vomiting. Shortness of breath is common, and oxygen therapy is often needed. Recovery is slow, and weakness and fatigue last for weeks.

TABLE 22.9 Tuberculosis

① Airborne *Mycobacterium tuberculosis* bacteria are inhaled and lodge in the lungs.

② The bacteria are phagocytized by lung macrophages and multiply within them, protected by lipid-containing cell walls and other mechanisms.

③ Infected macrophages are carried to various parts of the body such as the kidneys, brain, lungs, and lymph nodes; release of *M. tuberculosis* occurs.

④ Delayed hypersensitivity develops; wherever infected *M. tuberculosis* has lodged, an intense inflammatory reaction develops.

⑤ The bacteria are surrounded by macrophages and lymphocytes; growth of the bacteria ceases.

⑥ Intense inflammatory reaction and release of enzymes can cause caseation necrosis and cavity formation.

⑦ With uncontrolled or reactive infection, *M. tuberculosis* exits the body through the mouth with coughing or singing.

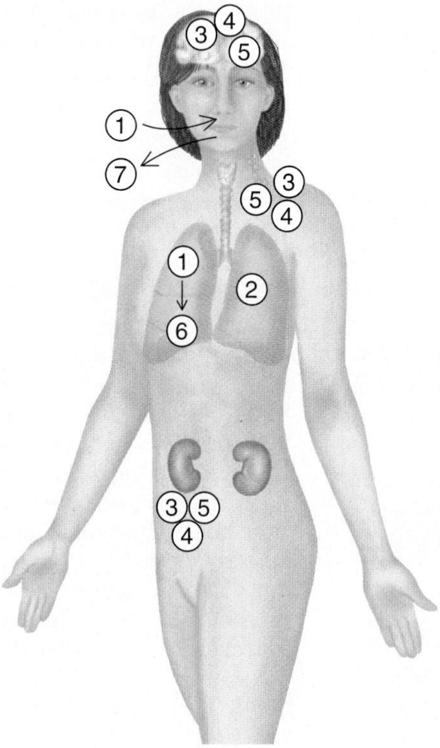

Symptoms	Chronic fever, weight loss, cough, sputum production
Incubation period	2 to 10 weeks
Causative agent	*Mycobacterium tuberculosis;* unusual cell wall with high lipid content
Pathogenesis	Colonization of the alveoli incites inflammatory response; ingestion by macrophages follows; organisms survive ingestion and are carried to lymph nodes, lungs, and other body tissues; tubercle bacilli multiply; granulomas form.
Epidemiology	Inhalation of airborne organisms; latent infections can reactivate.
Prevention and treatment	BCG vaccination, not used in the United States; tuberculin (Mantoux) test for detection of infection, allows early therapy of cases; treatment of all high-risk cases including young people with positive tests and individuals whose skin test converts from negative to positive. Treatment: two or more antitubercular medications given simultaneously long term, such as isoniazid (INH) and rifampin; DOTS.

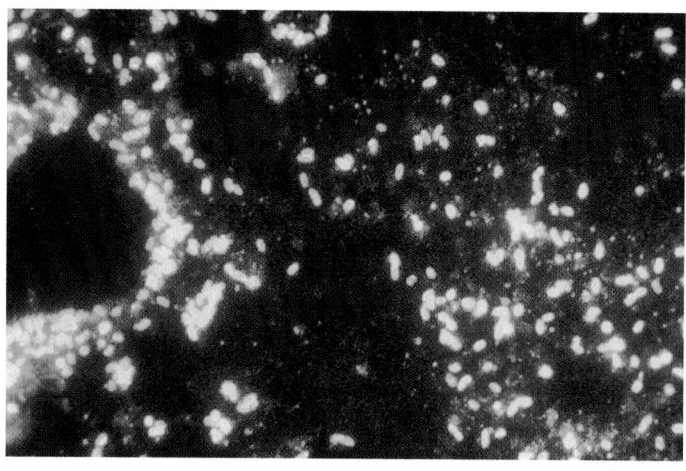

FIGURE 22.20 ***Legionella pneumophila*** **in Lung Tissue Stained with Fluorescent Antibody** When present in tissue or sputum, the bacterium fails to stain with most of the usual microbiological stains.

Causative Agent

The causative agent of Legionnaires' disease is *Legionella pneumophila* **(figure 22.20).** The organism is rod-shaped, Gram-negative, and requires a special medium for laboratory culture, which partly explains why it escaped detection for so long. Also, in tissue, *L. pneumophila* stains poorly with many of the usual microbiological stains. Genomic studies have revealed genes that aid its survival in protozoa, macrophages, and in adverse environments. The studies also give evidence of horizontal gene transfer. ■ horizontal gene transfer, p. 186

Pathogenesis

Legionella pneumophila infection is acquired by breathing aerosolized water contaminated with the organism. Healthy people are quite resistant to infection, but smokers and those with impaired host defenses from chronic diseases such as cancer and heart or kidney disease are susceptible. The organisms lodge in and near the alveoli of the lung, and their porin proteins bind complement component C3b, which aids phagocytosis of *L. pneumophila* by macrophages. A surface protein of *L. pneumophila,* macrophage invasion potentiator (Mip), also aids entry into the macrophages. The bacteria survive in the phagocytes by preventing phagosome-lysosome fusion, multiply and are released upon death of the macrophages to infect other tissues. Necrosis (tissue death) of alveolar cells and an inflammatory response result, causing multiple small abscesses, pneumonia, and pleurisy. Bacteremia is often present. Fatal respiratory failure, meaning that the lungs can no longer adequately oxygenate the blood or expel carbon dioxide, occurs in about 15% of hospitalized cases. Curiously, *L. pneumophila* infections remain confined to the lung in most cases. ■ porin proteins, p. 483

Epidemiology

The organism is widespread in warm natural waters where other microorganisms are present, and where it is taken up and multiplies within amebas, protozoa mobile by pseudopods, in the same way it grows in human macrophages. *Legionella pneumophila* survives well in the water systems of buildings, particularly in hot water systems, where chlorine levels are generally low. However, even increased levels of chlorine fail to decontaminate water systems consistently, probably because the bacteria are protected by their growth inside amebas.Legionnaires' disease cases have originated from contaminated aerosols from air conditioner cooling towers, nebulizers, sprays to freshen produce, and even from showers and water faucets. Direct person-to-person spread, however, does not occur. Monoclonal antibodies against different *L. pneumophila* strains help trace the source of epidemics. ■ monoclonal antibodies, p. 439 ■ protozoa, p. 285

Prevention and Treatment

Most efforts at control have focused on designing equipment to minimize the risk of infectious aerosols and on disinfecting procedures. Environmental surveillance is not practical because of the lack of a simple method for detecting virulent strains. Legionnaires' disease is treated with high doses of erythromycin, sometimes concurrently with rifampin. Like many other pathogens, *L. pneumophila* produces β-lactamase, which makes it resistant to many penicillins and cephalosporins. ■ β-lactamase, p. 476

The main features of Legionnaires' disease are given in **table 22.10.**

TABLE 22.10	**Legionnaires' Disease**
Symptoms	Muscle aches, headache, fever, cough, shortness of breath, chest and abdominal pain, diarrhea
Incubation period	2 to 10 days
Causative agent	*Legionella pneumophila,* a Gram-negative bacterium that stains poorly in clinical specimens.
Pathogenesis	Organism multiplies within phagocytes; released with death of the cell; necrosis of cells lining the alveoli; inflammation, and formation of microabscesses.
Epidemiology	Originates mainly from warm water contaminated with other microorganisms, such as found in air conditioning systems.
Prevention and treatment	Avoidance of contaminated water aerosols; regular cleaning and disinfection of humidifying devices. Treatment: erythromycin and rifampin.

MICROCHECK 22.5

Pneumococcal pneumonia is typically acquired in the community and leads the list of pneumonias in adults requiring hospitalization. Pneumonia due to *Klebsiella* sp. and other Gram-negative rods is mainly nosocomial and leads the causes of death from nosocomial infections. Mycoplasmal pneumonia usually does not require hospitalization. Whooping cough (pertussis) is mainly a threat to infants; childhood immunization against the disease protects them, but immunity often does not persist to adulthood. Tuberculosis is a chronic disease spread from one person to another by coughing and singing. Most *Mycobacterium tuberculosis* infections become latent, posing the risk of reactivation throughout life. *Legionella pneumophila,* the bacterium that causes Legionnaires' disease, originates from waters containing other microorganisms, where it can grow within protozoa. In the human lung, the bacterium readily multiplies within macrophages.

- ✓ What feature of the *S. pneumoniae* cell is responsible for its virulence?
- ✓ Why isn't pertussis toxin eliminated by the mucociliary escalator?

22.6

Viral Infections of the Lower Respiratory System

Focus Points

- Describe the principal factors driving the epidemiology of influenza.
- Give the symptoms of respiratory syncytial virus infection of infants.
- List measures for the prevention of hantavirus pulmonary syndrome.

DNA viruses such as varicella-zoster virus and the adenoviruses sometimes cause serious pneumonias, and some adenovirus infections can mimic pertussis. RNA viruses are of greater overall importance, however, because of the large number of people they infect and their potential for serious outcomes. The following examples come from the orthomyxovirus, paramyxovirus, and bunyavirus families of RNA viruses. ■ RNA viruses, p. 322

Influenza

Influenza is a good example of the constantly changing interaction between people and the agents that infect them. Antigenic changes in the influenza viruses are responsible for serious epidemics of the disease that recur in human populations. Epidemics of influenza also occur in animals such as birds, seals, and pigs, and antigenic variability of influenza viruses is partly due to movement of viral genes from one animal species to another. There are three major influenza types, designated A, B, and C, based on differences in their nucleoprotein antigens. Type A, considered here, causes the most severe disease and its epidemics are the most widespread. Outbreaks due to type B strains occur each year, but they are less extensive and the severity of the disease is less. Type C strains are of relatively little importance. Influenza viruses do not cause "stomach flu."

Symptoms

After a short incubation period, averaging 2 days, influenza typically begins with headache, fever, and muscle pain, which reach a peak in 6 to 12 hours. A dry cough worsens over a few days. Usually these acute symptoms go away within a week, leaving the patient with a lingering hacking cough, fatigue, and generalized weakness for additional days or weeks. Infection by influenza viruses is associated with Reye's syndrome on rare occasions. ■ Reye's syndrome, p. 546

Causative Agents

Influenza A virus belongs to the orthomyxovirus family; its genome consists of eight segments of single-stranded RNA. The virion is surrounded with a lipid-containing envelope derived from the host cell membrane **(figure 22.21).** Projecting from the envelope are two kinds of glycoprotein spikes—hemagglutinin (H) and neuraminidase (N). Hemagglutinin attaches the virus to specific receptors on ciliated epithelial cells of the host and thus initiates infection. Neuraminadase, on the other hand, is an enzyme that destroys the cell receptor to which the hemagglutinin attaches. This aids in the release of newly formed virions from the infected host cell and fosters the spread of the virus to uninfected host cells.

Pathogenesis

Individuals acquire influenza by inhaling aerosolized respiratory secretions from a person who has the disease. The virions attach by their hemagglutinin to specific receptors on ciliated epithelial cells, their envelope fuses with the cell membrane, and the virus enters the cell by endocytosis. Viral RNA is released into the cytoplasm following fusion with the endosomal membrane. Host cell protein and nucleic acid synthesis cease, and rapid synthesis of viral nucleoproteins begins. Regions of the cell membrane become embedded with viral glycoproteins, specifically hemagglutinin and neuraminidase. Within 6 hours, mature virions bud from the host cell, receiving an envelope of cell membrane containing viral hemagglutinin and neuraminidase as they are released. The virus spreads rapidly to nearby cells, including mucus-secreting cells and cells of the alveoli. Infected cells ultimately die and slough off, thus destroying the mucociliary escalator and severely impairing one of the body's major defenses against infection. The immune response quickly controls the infection in the vast majority of cases, although complete recovery of the respiratory epithelium may take 2 months or more. Usually, only a small percentage of people with influenza die, but even so, epidemics are so widespread that the total number of deaths is high. Influenza virus infection alone can kill apparently normal, healthy people. More often, however, death occurs because of bacterial secondary infections, usually by *Staphylococcus aureus, Streptococcus pyogenes,* or *Haemophilus influenzae.* Influenza takes a heavy toll on people whose hearts or lungs are weak.

Epidemiology

Outbreaks of influenza occur every year in the United States and are associated with an estimated 10,000 to 40,000 deaths. Figure 20.2 shows the variations in the percentage of total deaths attributed to pneumonia and influenza. Pandemics occur periodically over the years, marked by rapid spread of influenza viruses around the globe and higher than normal morbidity. ■ pandemics, p. 451

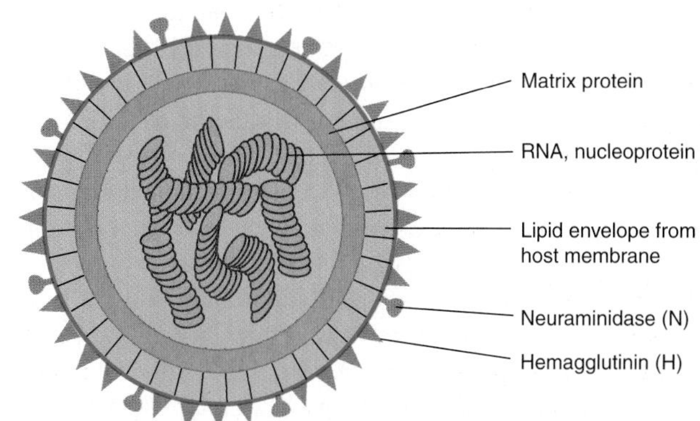

FIGURE 22.21 Diagrammatic Representation of Influenza Virus Note the surface structures, hemagglutinin and neuraminidase, the eight-segment genome, and the nucleoprotein, which distinguishes the three viral types, A, B, and C.

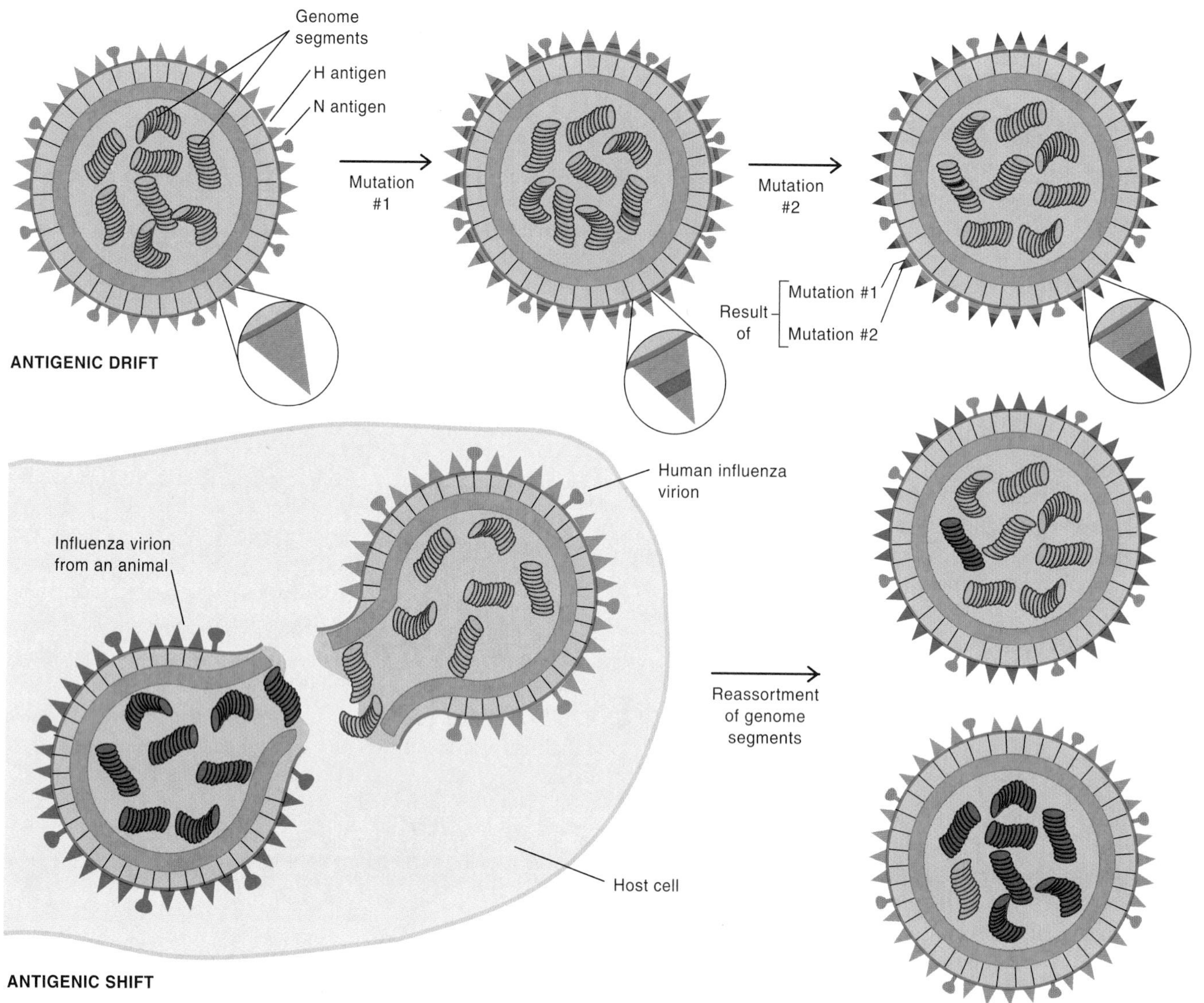

FIGURE 22.22 Influenza Virus: Antigenic Drift and Antigenic Shift With drift, repeated mutations cause a gradual change in the antigens composing the hemagglutinin, so that antibody against the original virus becomes progressively less effective. With shift, there is an abrupt, major change in the hemagglutinin antigens because the virus acquires a new genome segment, which in this case codes for hemagglutinin. Changes in neuraminidase could occur by the same mechanism.

Although other factors are involved in the spread of influenza viruses, major attention has focused on their antigenic changeability. Two types of variation are seen: antigenic drift and antigenic shift **(figure 22.22). Antigenic drift** is seen in interepidemic years and consists of minor mutations in the hemagglutinin (H) antigen that make immunity developed during prior years less effective and ensure that enough susceptible people are available for the virus to survive. This type of change is exemplified by A/Texas/77 (H3N2) and A/Bangkok/79 (H3N2), wherein there have been mutations that have altered the H antigen slightly. (The geographic names are the places where the virus was first isolated, the next two numbers represent the year.) The antibody produced by people who have recovered from A/Texas/77 (H3N2) is only partially effective against the mutant H3 antigen of A/Bangkok/79 (H3N2). Thus, the newer Bangkok strain might be able to spread and cause a minor epidemic in a population previously exposed to the Texas strain. ■ antigenic shift and drift, p. 335

Antigenic shifts are represented by more dramatic changes; virus strains appear that are markedly different antigenically from the strains previously seen. Most likely they arise as a result of genetic reassortment when two different viruses infect a cell at the same time. Animal strains of influenza virus can occasionally infect humans and cause dual infections with human strains. When genetic mixing results in a new virus that is infectious, virulent, and possesses a hemagglutinin for which a population has no immunity, it can cause widespread disease. Ecological studies show that all the known influenza A virus types exist in aquatic birds, generally causing chronic intestinal infections. These bird influenza (avian flu) viruses readily infect domestic fowl, and from them can infect other domestic animals and humans. A well-studied episode in Hong Kong in 1997 involved an H5N1 virus from chickens that caused fatal infections in humans, but fortunately it did not spread easily from person to person. Subsequent outbreaks show that H5N1 viruses are now endemic in a number of Asian communities. ■ genetic reassortment, p. 335

Prevention and Treatment

Inactivated vaccines can be 80% to 90% effective in preventing disease when the vaccine is produced from the epidemic strain. Subunit vaccines consisting mostly of hemagglutinin are now widely used instead of whole virus vaccines. Attenuated nasally administered vaccines appear to be safe and effective. Because of antigenic drift, vaccines produced from earlier strains having the same H type may be decreasingly effective, so that new vaccines must be produced each year. It takes 6 to 9 months from the appearance of a new influenza strain before adequate amounts of vaccine can be manufactured.

A serious reaction was observed during a nationwide immunization program with the swine influenza vaccine produced against the 1976 Fort Dix H1N1 influenza A strain. This reaction, called the **Guillain-Barré syndrome,** occurred in about 1 of every 100,000 persons vaccinated. It is characterized by severe paralysis, and although most people recover completely, about 5% die of the paralysis. The possibility that this or other side effects will generate lawsuits is one reason vaccine production is a risky business.

Medications such as amantadine (Symmetrel) and rimantadine (Flumadine) have been employed for short-term prevention of influenza A disease when vaccine is not available. However, they are not currently recommended because many circulating viruses are resistant. Zanamivir (Relenza) and oseltamivir (Tamiflu), first approved for use in 1999, represent neuraminidase inhibitors, active against both A and B viruses. Some experts have advised stockpiling anti-influenza medications to protect populations faced with a new virulent influenza virus. These medicines are generally used in conjunction with vaccination to protect exposed individuals until they can develop immunity. ■ **neuraminidase inhibitors, p. 488**

The main features of influenza are summarized in **table 22.11.**

Respiratory Syncytial Virus Infections

Respiratory syncytial virus (RSV) infection is first among serious lower respiratory tract infections of infants and young children, causing an estimated 90,000 hospitalizations and 4,500 deaths in the United States each year. It is also responsible for serious disease in elderly people, and for nosocomial epidemics.

Symptoms

Symptoms begin after an incubation period of 1 to 4 days with runny nose followed by cough, wheezing, and difficulty breathing. Fever may or may not be present. Victims often develop a dusky color, indicating that they are not getting enough oxygen. Death occurs in 0.5% to 1% of hospitalized infants but is much higher in those with underlying diseases such as heart and lung disease, cancer, and immunodeficiency. RSV is one of the causes of **croup,** seen commonly in older infants and manifest as a loud high-pitched cough and noisy inspiration due to airway obstruction at the larynx. Infections of healthy older children and adults generally cause symptoms of a bad cold.

Causative Agent

Respiratory syncytial virus is a member of the paramyxovirus family and contains a single-stranded RNA genome. Like all

TABLE 22.11 Influenza

① Influenza virus is inhaled and carried to the lungs.

② Viral hemagglutinin attaches to specific receptors on ciliated epithelial cells, the viral envelope fuses with the epithelial cell, and the virus enters the cell by endocytosis.

③ Host cell synthesis is diverted to synthesizing new virus.

④ Newly formed virions bud from infected cells, they are released by viral neuraminidase and infect ciliated epithelium, mucus-secreting, and alveolar cells.

⑤ Infected cells ultimately die and slough off; recovery of the mucociliary escalator may take weeks.

⑥ Secondary bacterial infection of the lungs, ears, and sinuses is common.

⑦ The virus exits with coughing.

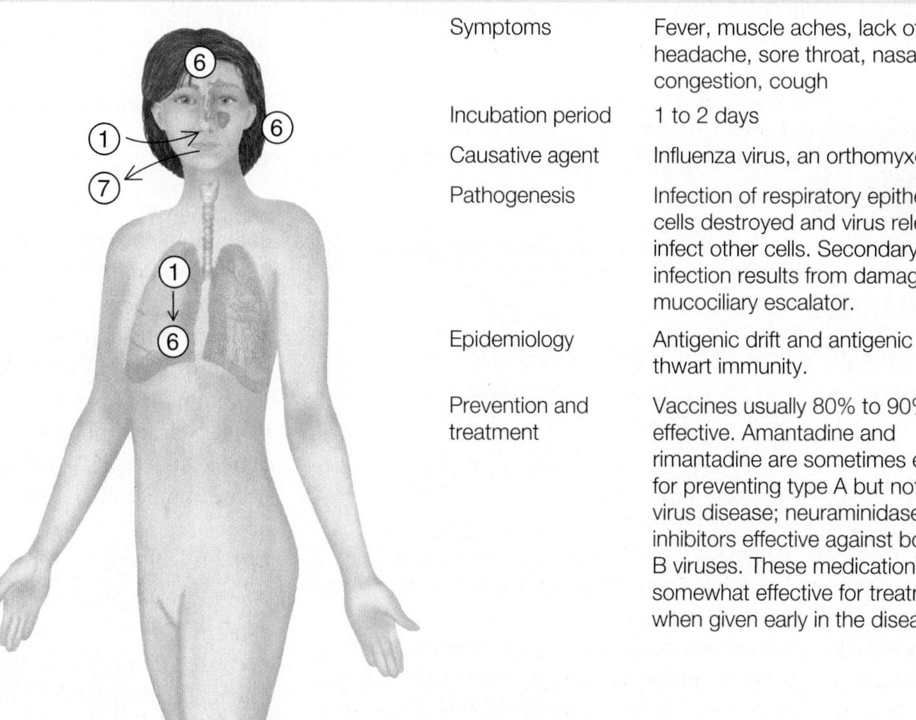

Symptoms	Fever, muscle aches, lack of energy, headache, sore throat, nasal congestion, cough
Incubation period	1 to 2 days
Causative agent	Influenza virus, an orthomyxovirus
Pathogenesis	Infection of respiratory epithelium; cells destroyed and virus released to infect other cells. Secondary bacterial infection results from damaged mucociliary escalator.
Epidemiology	Antigenic drift and antigenic shift thwart immunity.
Prevention and treatment	Vaccines usually 80% to 90% effective. Amantadine and rimantadine are sometimes effective for preventing type A but not type B virus disease; neuraminidase inhibitors effective against both A and B viruses. These medications somewhat effective for treatment when given early in the disease.

paramyxoviruses, its virion is enveloped. It causes cells in cell cultures to fuse together. The clumps of fused cells are known as **syncytia,** thus the name of the virus. ■ paramyxovirus family, p. 323

Pathogenesis

The virus enters the body by inhalation and infects the respiratory tract epithelium, causing death and sloughing of the cells. Bronchiolitis is a common feature of the disease; the inflamed bronchioles become partially plugged by sloughed cells, mucus, and clotted plasma that has oozed from the walls of the bronchi. The initial obstruction causes wheezing when air rushes through the narrowed passageways, sometimes causing the condition to be confused with asthma. The obstruction often acts like a one-way valve, allowing air to enter the lungs, but not leave them. In many cases the inflammatory process extends into the alveoli, causing pneumonia. There is a high risk of secondary infection because of the damaged mucociliary escalator. ■ mucociliary escalator, p. 498

Epidemiology

RSV outbreaks are common from late fall to late spring, peaking in mid-winter. Recovery from infection produces only weak and short-lived immunity, so that infections can recur throughout life. Healthy children and adults usually have mild illness and readily spread the virus to others.

Prevention and Treatment

No vaccines are available although there are at least six programs underway to develop a vaccine. Preventing nosocomial RSV illness requires strict isolation technique. Subjects with underlying illnesses can be protected from the disease by monthly injections of immune serum globulin or a monoclonal antibody called palivizumab. There are no effective antiviral medications. ■ monoclonal antibody, p. 439

The main features of RSV infections are summarized in **table 22.12.**

TABLE 22.12 Respiratory Syncytial Virus (RSV) Infections

Symptoms	Runny nose, cough, fever, wheezing, difficulty breathing, dusky color
Incubation period	1 to 4 days
Causative agent	RSV, a paramyxovirus that produces syncytia
Pathogenesis	Sloughing of respiratory epithelium and inflammatory response plug bronchioles, cause bronchiolitis; pneumonia results from bronchiolar and alveolar inflammation, or secondary infection.
Epidemiology	Yearly epidemics during the cool months; readily spread by healthy older children and adults who often have mild symptoms; no lasting immunity.
Prevention and treatment	No vaccine. Preventable by injections of immune serum globulin or a monoclonal antibody; no satisfactory antiviral treatment.

Hantavirus Pulmonary Syndrome

In the spring of 1993 a newly emerging disease made a dramatic appearance in the American Southwest near the place where the corners of four states, Arizona, Colorado, New Mexico, and Utah, come together. The initial outbreak of the disease was alarming. It involved less than a dozen people, mostly Navajo, and most were vigorous young adults who started with influenza-like symptoms, but were dead in a few days. Scientists from the Centers for Disease Control and Prevention rushed to join local epidemiologists and health officials investigating the outbreak. Their studies quickly established that the disease was associated with exposure to mice. By allowing the serum from victims to react with many different known viruses, they learned within a month that antibodies in the blood of the victims reacted with a hantavirus that had plagued American troops during the 1950s Korean War. Using the polymerase chain reaction (PCR), the scientists recovered a viral genome from the lungs of patients who had died, and they showed it was identical to one from mice captured in the area. This virus proved to be a close relative of the Korean virus. ■ polymerase chain reaction, p. 223

Symptoms

Hantavirus pulmonary syndrome usually begins with fever, muscle aches (especially in the lower back), nausea, vomiting, and diarrhea. Unproductive cough and increasingly severe shortness of breath appear within a few days, followed by shock and death.

Causative Agents

The causative agents are the Sin Nombre, Spanish for "no name," virus and various related hantaviruses that exist at different locations in the Western Hemisphere. The hantaviruses are enveloped viruses of the bunyavirus family. Their genome consists of three segments of single-stranded RNA. In nature these viruses cause lifetime infections, primarily of rodents, without any apparent harm to the animals. ■ bunyavirus family, p. 323

Pathogenesis

The virus enters the body by inhalation of air containing dust contaminated with the urine, feces, or saliva of infected rodents. The virus enters the circulation and is carried throughout the body, infecting the cells that line tissue capillaries. Massive amounts of the viral antigen appear in lung capillaries, but the antigen is also demonstrable in capillaries of the heart and other organs. The inflammatory response to the viral antigen causes the capillaries to leak large amounts of plasma into the lungs, suffocating the patient and causing the blood pressure to fall. Shock and death occur in more than 40% of the cases.

Epidemiology

Hantavirus pulmonary syndrome **(figure 22.23)** is another example of a zoonosis. It is called an "emerging disease" because of its recent discovery and apparent increase in frequency. Nevertheless, this zoonosis has undoubtedly existed for centuries, occasionally claiming human victims when mice

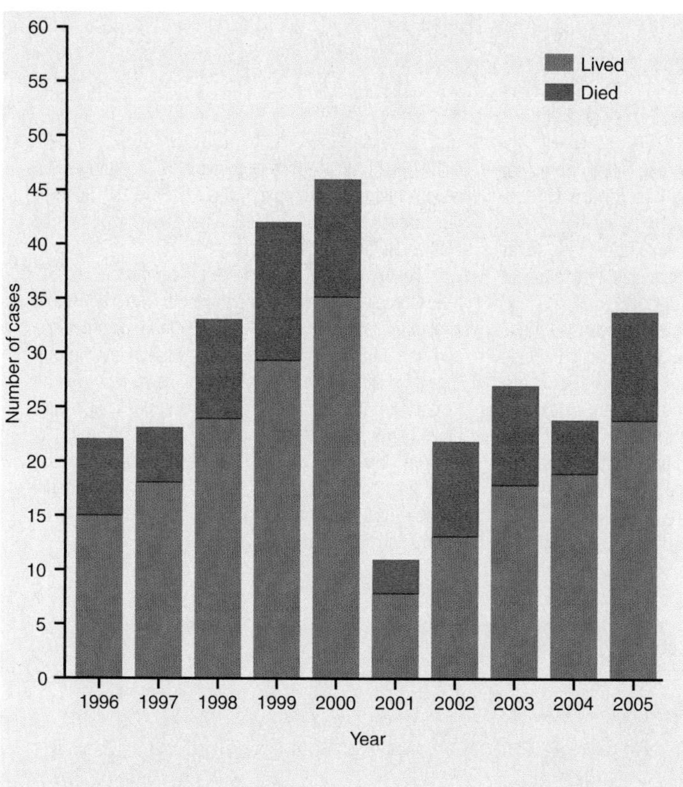

FIGURE 22.23 Total Hantavirus Pulmonary Syndrome Cases, United States, 1996–2005
(Source: MMWR, CDC 2007 54(53): 2–92).

and humans get too close together. Since the description of the syndrome, cases have been identified from Canada to Argentina, including several hundred cases from the United States. Most of the cases have occurred west of the Mississippi River and were due to Sin Nombre virus carried by deer mice. Others have been caused by related viruses carried by other rodents. Outbreaks of the disease correlate with marked increases in mouse populations adjacent to impoverished communities with substandard housing. The virus spreads more easily when the mouse population density is high. Thirty percent or more of the mice can become carriers of the disease. These mice eagerly and easily invade the houses of the poor. Complex ecological factors control the size of mouse populations. The numbers of foxes, owls, snakes, and other predators play a role. Moreover, two outbreaks of hantavirus pulmonary syndrome occurred in association with El Niño weather patterns, which brought increased rainfall to the area, yielding increased plant growth and seed production for mice to feed on. The emergence of hantavirus pulmonary syndrome is a convincing example of how environmental change can result in infectious human disease. Luckily, despite the large amount of viral antigen in the lung capillaries, few mature infectious virions enter the air passages of the lung; thus, person-to-person transmission occurs rarely, if ever. ■ **emerging diseases, p. 461**
■ **zoonoses, p. 453**

TABLE 22.13	Hantavirus Pulmonary Syndrome
Symptoms	Fever, muscle aches, vomiting, diarrhea, cough, shortness of breath, shock
Incubation period	3 days to 6 weeks
Causative agent	Sin Nombre and related hantaviruses of the bunyavirus family
Pathogenesis	Viral antigen localizes in capillary walls in the lungs; inflammation.
Epidemiology	Zoonosis likely to involve humans in proximity to booming mouse populations; generally no person-to-person spread.
Prevention and treatment	Avoid contact with rodents; seal access to houses, food supplies; good ventilation, avoid dust, use disinfectants in cleaning rodent-contaminated areas. No proven antiviral treatment.

Prevention and Treatment

Prevention of hantavirus pulmonary syndrome is based on minimizing exposure to rodents and dusts contaminated by them. Food for humans and pets should be kept in containers. Buildings should be made mouse-proof if possible, and maximal ventilation is advisable when cleaning a rodent-infested area. Mopping with a disinfectant solution is preferable to use of brooms and vacuum cleaners, because the latter two stir up dust. Lethal traps and poisons may be necessary to decrease the rodent population in the area. Rubber gloves and a mask are advisable for those doing field studies on wild rodents. When camping, a tent with a floor should be used. There is no proven antiviral treatment for this highly fatal disease.

The main features of hantavirus pulmonary syndrome are summarized in **table 22.13.**

MICROCHECK 22.6

Because of the speed with which influenza travels around the world, and the potential for development of virulent influenza virus strains, the disease poses an extremely serious threat to humankind. Most deaths from influenza are caused by secondary bacterial infections. Respiratory syncytial virus is the leading cause of serious respiratory disease in infants and young children. Hantavirus pulmonary syndrome, first recognized in 1993, is often fatal. It is contracted from inhalation of dust contaminated by mice infected with certain hantaviruses.

- ✓ Why are there so many deaths from influenza when it is generally a mild disease?
- ✓ What is the source of the virus that causes hantavirus pulmonary syndrome?
- ✓ Why might you expect an influenza epidemic to be more severe following an antigenic shift in the virus than after antigenic drift?

PERSPECTIVE 22.2

What to Do About Bird Flu

The known antigenic types of influenza A are 16 different hemaglutinins, designated H 1–16, and nine neuraminidases, N types 1–9. All of these are represented among the influenza A viruses infecting wild waterfowl. Generally, only H types 1–3, and N types 1 and 2 viruses have been important in human disease. Occasionally, however, avian viruses infect humans and cause illness. For example, there were 123 known human cases caused by avian viruses between 1996 and 2004, caused by five avian strains—H7N7, H5N1, H9N2, H7N2, and H7N3. There was a high mortality among the 29 cases caused by the H5N1 strains (seven died), whereas only one death occurred among the remaining 103 cases due to the other four avian viruses. The high virulence of the H5N1 strain has been apparent since the first H5N1 outbreak occurred in 1997. Moreover, the virus has spread rapidly from Asia to the Middle East, Europe, and Africa **(figure 1)**, devastating flocks of domestic fowl and causing several hundreds of deaths among humans. Fortunately, so far there have been only a few instances that suggest person-to-person spread of the disease.

The H5N1 avian virus concerns us because it has two of the three characteristics required of a pandemic influenza virus: (1) It is infectious for humans who have no herd immunity; and (2) it causes severe disease in humans. The virus does not yet meet the third requirement—easy spread from person to person, but it might acquire this ability through mutation or antigenic shift.

The H5N1 virus may never become a pandemic strain, but even if it doesn't, it is highly likely that other avian strains will do so sometime in the future. We have a historical example in the 1918-1919 "Spanish flu" that killed 40 to 100 million people. This pandemic was caused by an avian virus.

Today, we are using some important tools to defend ourselves that weren't available in 1918. Rapid viral diagnosis is now possible, and global communication can inform us promptly of outbreaks anywhere in the world so we can take defensive measures. We can also make and stockpile influenza vaccines, and vaccines against bacterial secondary invaders such as *Streptococcus pneumoniae* and *Haemophilus influenzae.* We can also place antiviral and antibacterial medications at strategic locations. A lot depends on international cooperation.

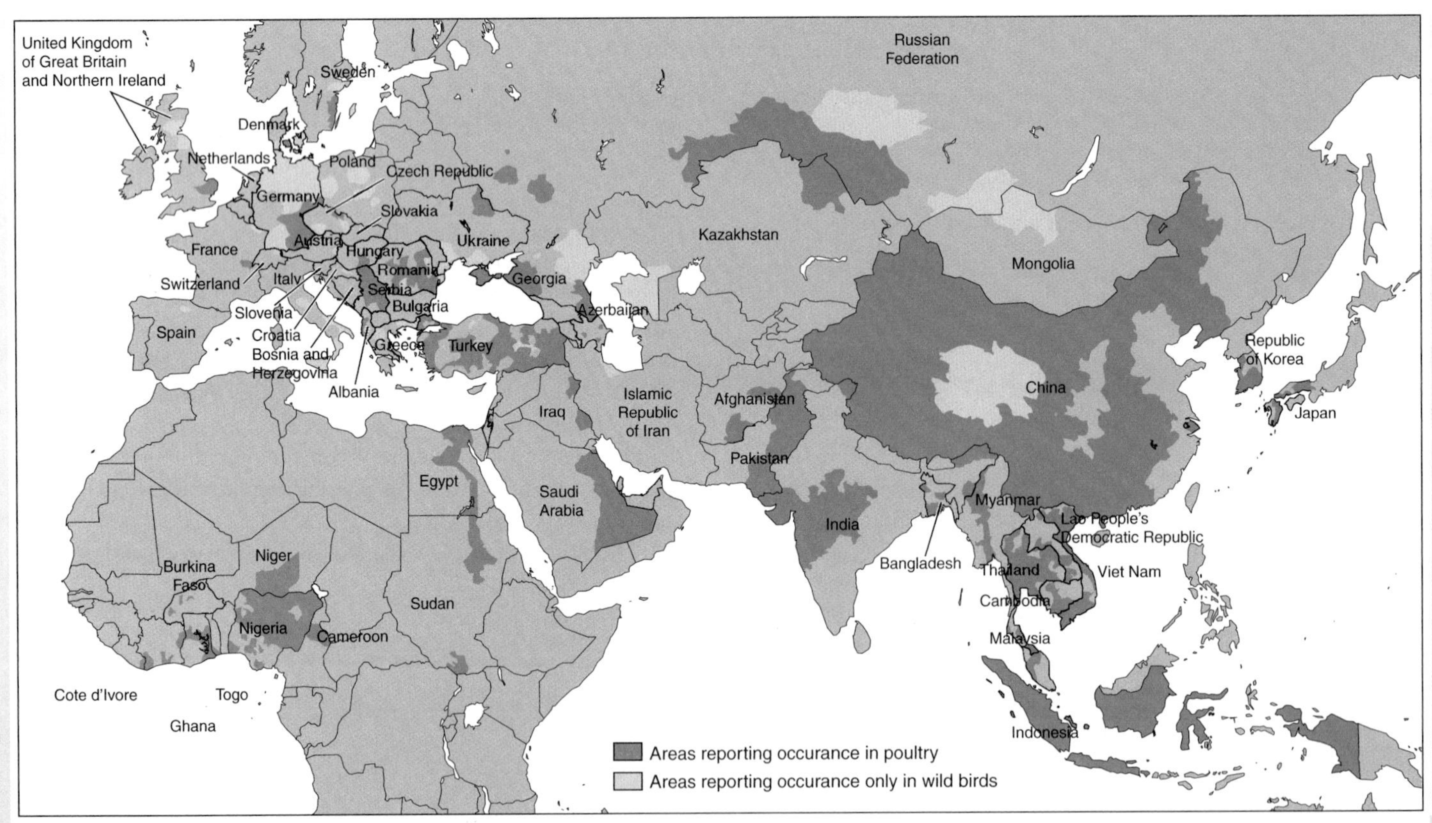

FIGURE 1 Areas Reporting H5N1 Avian Influenza in Poultry and Wild Birds. Confirmed reports since 2003, last updated September 2007. Source: World Organization for Animal Health, World Health Organization

22.7

Fungal Infections of the Lung

Focus Points

- Describe in general terms the distribution of histoplasmosis in the United States.
- Outline the epidemiology of coccidioidomycosis

Serious lung diseases caused by fungi are quite unusual in healthy, immunocompetent individuals. Symptomatic and asymptomatic infections that subside without treatment, however, are common. One fungus infects most of us in childhood without causing symptoms. This organism, *Pneumocystis jirovecii,* will be discussed in chapter 29 on AIDS. Coccidioidomycosis and histoplasmosis are two other examples of widespread mycoses.

Valley Fever (Coccidioidomycosis)

In the United States, coccidioidomycosis occurs mainly in California, Arizona, Nevada, New Mexico, Utah, and West Texas. People who are exposed to dust and soil, such as farm workers, are most likely to become infected, but only 40% develop symptoms.

Symptoms

"Flu"-like symptoms such as fever, cough, chest pain, and loss of appetite and weight are common symptoms of coccidioidomycosis. About 10% of victims experience symptoms caused by hypersensitivity to the fungal antigens, manifested by tender nodules often localized to the shins and pain in the joints. The majority of people afflicted with coccidioidomycosis recover spontaneously within a month. A small percentage of victims develop chronic disease.

Causative Agent

The causative agent of coccidioidomycosis, *Coccidioides immitis,* is a dimorphic fungus. The mold form of the organism grows in soil. The mold's hyphae develop numerous barrel-shaped, highly infectious arthrospores **(figure 22.24a),** which separate easily and become airborne. In infected tissues, the arthrospores develop into thick-walled, spherical spores that may contain several hundred small cells called endospores (figure 22.24b), not to be confused with bacterial endospores. ■ **dimorphic fungi, p. 290**

Pathogenesis

Arthrospores enter the lung with inhaled air and develop into thick-walled spheres that mature and rupture, spilling numerous endospores. The endospores develop into more large, thick-walled spores containing endospores and repeat the process, each time provoking an inflammatory response and subsequent immune response. Symptoms and tissue injury are caused mainly by the host's immune response to coccidioidal antigens. Usually, the organisms are eliminated by body defenses, but in a small percentage of individuals caseous necrosis occurs and results in a lung cavity mimicking tuberculosis. Rarely, organisms are carried throughout the body by the bloodstream and infect the skin, mucous membranes, brain, and other organs. This disseminated form of the disease occurs more often in people with AIDS or other immunodeficiencies and is fatal without treatment. ■ **caseous necrosis, p. 516**

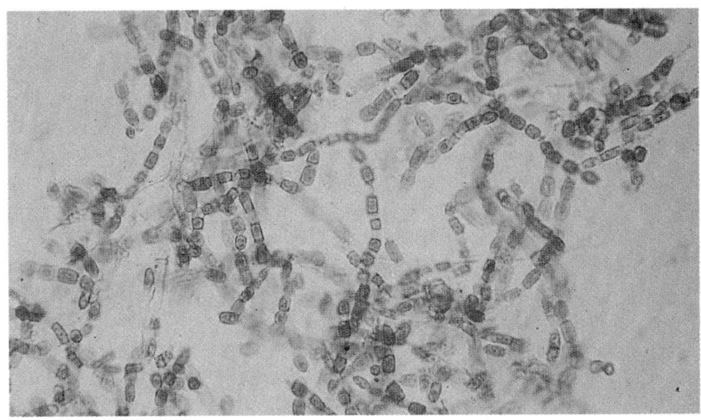

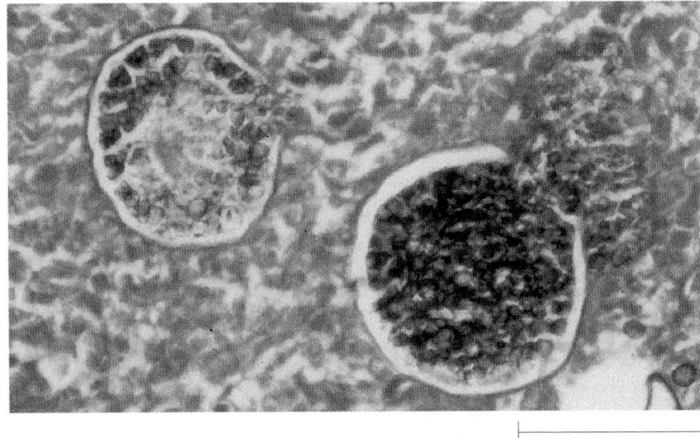

FIGURE 22.24 ***Coccidioides immitis*** **(a)** Mold-phase hyphae fragmenting into barrel-shaped arthrospores, the infectious form of the fungus. **(b)** Spherical spores containing endospores, the form of the fungus found in tissues.

Epidemiology

Coccidioides immitis grows only in semi-arid desert areas of the Western Hemisphere **(figure 22.25).** In these areas, infections occur only during the hot, dry, dusty seasons when airborne arthrospores are easily dispersed from the soil. Dust stirred up by earthquakes can result in epidemics. Rainfall promotes growth of the fungus, which then produces increased numbers of spores when dry conditions return. People can contract coccidioidomycosis by simply traveling through the endemic area. Infectious spores have unknowingly been transported to other areas, but the organism apparently is unable to establish itself in moist climates.

Prevention and Treatment

Preventive measures include avoiding dust in the endemic areas. Watering and planting vegetation aid in dust control. Medications approved for treatment of the serious cases include amphotericin

FIGURE 22.25 Area of Distribution of *Coccidioides immitis*

B and fluconazole or itraconazole. They markedly improve the prognosis but must be given for long periods of time, and cause troublesome side effects. Even with treatment, disseminated disease can reactivate months or years later.

Table 22.14 describes the main features of coccidioidomycosis.

TABLE 22.14 Coccidioidomycosis (Valley Fever)

Symptoms	Fever, cough, chest pain, loss of appetite and weight; less frequently, painful nodules on extremities, pain in joints; skin, mucous membranes, brain, and internal organs sometimes involved
Incubation period	2 days to 3 weeks
Causative agent	*Coccidioides immitis,* a dimorphic fungus
Pathogenesis	After lodging in lung, arthrospores develop into spheres that mature and discharge endospores, each of which then develops into another sphere; inflammatory response damages tissue; hypersensitivity to fungal antigens causes painful nodules and joint pain.
Epidemiology	Inhalation of airborne *C. immitis* spores with dust from soil growing the organism. Occurs only in certain semi-arid regions of the Western Hemisphere.
Prevention and treatment	Dust control methods such as grass planting and watering. Treatment: amphotericin B and fluconazole or itraconazole.

Spelunkers' Disease (Histoplasmosis)

Histoplasmosis, like coccidioidomycosis, is usually benign but occasionally mimics tuberculosis. Rare, serious forms of the disease suggest that AIDS or another immunodeficiency may also be present. The distribution is more widespread than that of coccidioidomycosis and is associated with different soil and climate.

Symptoms

Most infections are asymptomatic. Fever, cough, and chest pain are the most common symptoms, sometimes with shortness of breath. Mouth sores may develop, especially in children.

Causative Agent

Histoplasmosis is caused by the dimorphic fungus *Histoplasma capsulatum.* This organism prefers to grow in soils contaminated by bat or bird droppings, but it is not pathogenic for these animals. In pus or tissue from people with active disease, *H. capsulatum* is a tiny oval yeast that grows within host macrophages

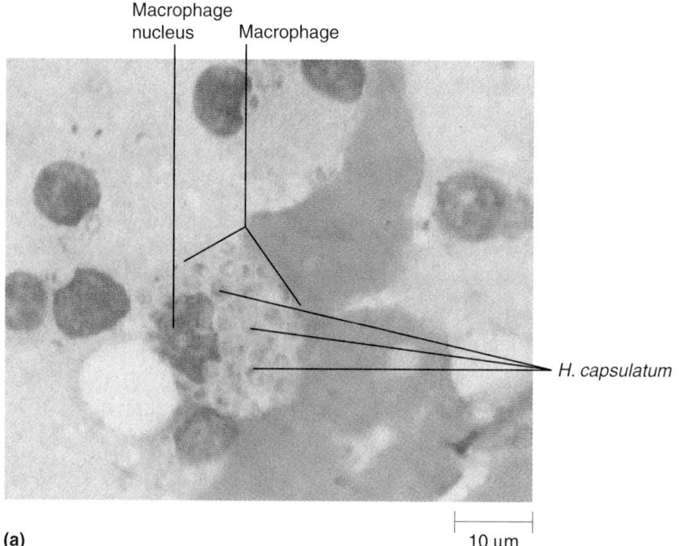

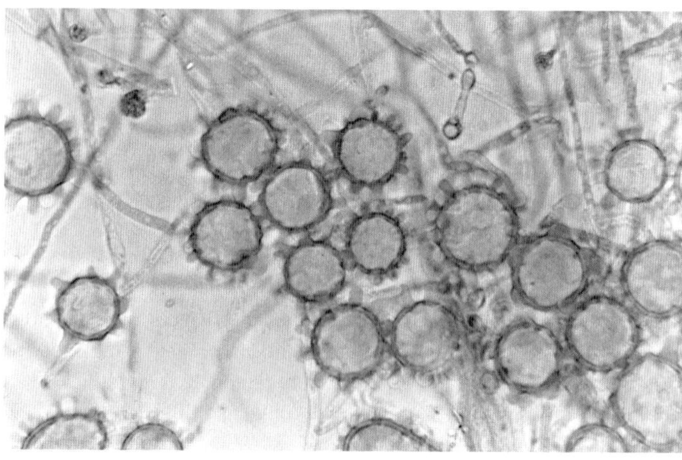

FIGURE 22.26 ***Histoplasma capsulatum*** **(a)** Yeast-phase organisms packing the cytoplasm of a macrophage. **(b)** Mold phase, showing large conidia with projecting knobs.

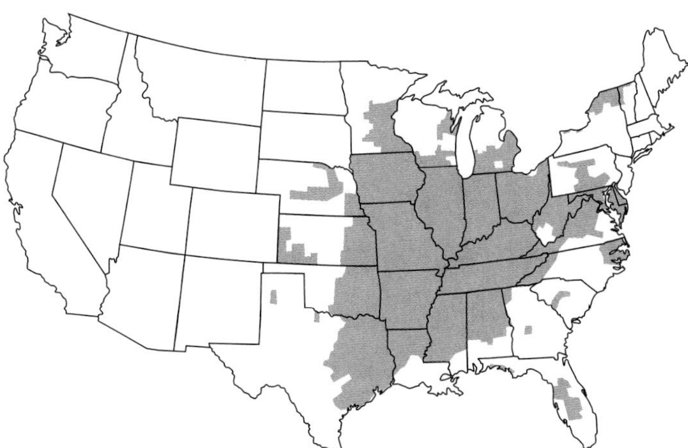

FIGURE 22.27 Geographical Distribution of *Histoplasma capsulatum* in the United States as Revealed by Positive Skin Tests Human histoplasmosis has been reported in more than 40 countries besides the United States, including Argentina, Italy, South Africa, and Thailand.

TABLE 22.15 Histoplasmosis (Spelunkers' Disease)

Symptoms	Mild respiratory symptoms; less frequently, fever, chest pain, cough, chronic sores
Incubation period	5 to 8 days
Causative agent	*Histoplasma capsulatum,* a dimorphic fungus
Pathogenesis	Spores inhaled, change to yeast phase, multiply in macrophages; granulomas form; disease spreads in individuals with AIDS or other immunodeficiencies.
Epidemiology	The fungus prefers to grow in soil contaminated by bird or bat droppings, especially in Ohio and Mississippi River valleys, and in the U.S. Southeast. Spotty distribution in many other countries around the world. Spelunkers are at risk of infection.
Prevention and treatment	Avoidance of soils contaminated with chicken, bird, or bat droppings. Treatment: amphotericin B and itraconazole for serious infections.

(figure 22.26a). Contrary to its name, the fungus is not encapsulated. The mold form of the organism characteristically produces two kinds of spores: macroconidia, which often have numerous projecting knobs (figure 22.26b) and tiny pear-shaped or spherical microconidia.

Pathogenesis

Individuals inhale infectious conidia of *H. capsulatum* with dust from contaminated soils and the conidia develop into the yeast form. The organisms promptly enter macrophages and grow intracellularly. Granulomas develop in infected areas, closely resembling those seen in tuberculosis, sometimes even showing caseation necrosis. Eventually, the lesions are replaced with scar tissue, and many calcify, meaning that the body deposits calcium compounds in them that show up on X-rays. In rare cases, the disease is not controlled and spreads throughout the body. This suggests AIDS or another immunodeficiency.

Epidemiology

The distribution of histoplasmosis is quite different from that of coccidioidomycosis. **Figure 22.27** shows the distribution of histoplasmosis in the United States, but the disease also occurs in tropical and temperate zones scattered around the world. Cave explorers, spelunkers, are at risk for contracting the disease because many caves contain soil enriched with bat droppings. Most cases of histoplasmosis in the United States have occurred in the Mississippi and Ohio River drainage area and in South Atlantic states. Skin tests reveal that millions of people living in these areas have been infected.

Prevention and Treatment

No proven preventive measures are known other than to avoid areas where soil is heavily enriched with bat, chicken, and bird droppings, especially if they have been left undisturbed for a long period. Some researchers have recommended that several inches of clay soil be placed over soils containing large quantities of old droppings. Treatment of histoplasmosis is similar to that of coccidioidomycosis. Amphotericin B and itraconazole are used for treating severe disease, but both medications have potentially serious side effects.

The main features of histoplasmosis are described in **table 22.15.**

MICROCHECK 22.7

Coccidioidomycosis and histoplasmosis are two diseases caused by fungi that live in the soil. The body responds to these infections by forming granulomas, mimicking tuberculosis. Each fungus has its own ecological niche, *Coccidioides immitis* in semi-arid regions of the Western Hemisphere, and *Histoplasma capsulatum* in moist soils enriched with bird or bat droppings around the world.

- ✓ Why should an immunodeficient person avoid traveling through hot, dry, dusty areas of the Southwest?
- ✓ Why might cave exploration increase the risk of histoplasmosis?
- ✓ In March 2001, 221 students from 37 colleges and universities in 18 states developed one or more of the following symptoms: fever, cough, shortness of breath, and chest pain. Serological tests indicated acute histoplasmosis. Most of the students had just returned from spring break in Mexico where they stayed in a hotel next door to a bulldozing operation. How might the bulldozing explain the epidemic?

FUTURE CHALLENGES

Global Preparedness vs. Emerging Respiratory Viruses

In November 2002, a frightening new respiratory disease emerged in Guangdong province, China. Symptoms included cough and fever, and Xrays showed characteristics of viral pneumonia. The disease, labeled "SARS" (severe acute respiratory syndrome), quickly spread around the world, involving 25 countries and causing more than 700 deaths. Many of its victims were medical personnel. It had two characteristics that quickly led to its control—a long incubation of 6 days, and dramatic symptoms that were easily recognized. This allowed healthcare personnel time to identify contacts and institute quarantines to stop the spread of the disease. The causative agent proved to be a previously unknown coronavirus, and its source was never identified, although it was probably a wild animal sold for meat in the markets. On three occasions, SARS escaped from laboratories where the virus was being studied, and the disease may arise again from its natural source. However, its main importance is that it caused many countries to learn to work together to help forge a more effective global response to emerging pandemic diseases. More details on SARS can be found at www.sarsreference.com.

The problem of avian influenza is more difficult to solve than SARS because the avian viruses are carried in the intestines of wild waterfowl that readily infect domestic flocks, which then infect humans. Epidemic influenza in humans generally has a short incubation period, and many victims have mild, coldlike symptoms not easily recognized as part of an influenza epidemic. Therefore, quarantines are not helpful in disease control. Many experts think we have been lucky that no virus capable of causing a pandemic has appeared. However, thanks to the battles against the H5N1 avian virus and the SARS outbreaks, the world is much closer to being prepared for the next pandemic. The challenge is to sustain and improve advances in cooperative surveillance, and action planning to avoid a global catastrophe.

SUMMARY

22.1 Anatomy and Physiology (figure 22.1)

The moist lining of the eyes (conjunctiva), nasolacrimal duct, middle ears, sinuses, mastoid air cells, nose, and throat make up the main structures of the upper respiratory system. The functions of the upper respiratory tract include temperature and humidity regulation of inspired air and removal of microorganisms. The lower respiratory system includes the trachea, bronchi, bronchioles, and alveoli. Pleural membranes cover the lungs and line the chest cavity.

The Mucociliary Escalator

Ciliated cells line much of the respiratory tract and remove microorganisms by constantly propelling mucus out of the respiratory system (figure 22.1a).

22.2 Normal Microbiota (table 22.1)

A wide variety of microorganisms colonize parts of the upper respiratory system, often including the important opportunistic pathogen, *Staphylococcus aureus.* Viruses and microorganisms are normally absent from the lower respiratory system.

Infections of the Upper Respiratory System

22.3 Bacterial Infections of the Upper Respiratory System

Strep Throat (Streptococcal Pharyngitis) (tables 22.2, 22.3)

Streptococcus pyogenes causes strep throat, a significant bacterial infection that may lead to **scarlet fever,** toxic shock, or sequelae due to the immune response, glomerulonephritis, or rheumatic fever (figures 22.2, 22.3).

Diphtheria (table 22.4)

Diphtheria, caused by *Corynebacterium diphtheriae,* is a toxin-mediated disease that can be prevented by immunization (figures 22.5, 22.6).

Pinkeye, Earache, and Sinus Infections

Conjunctivitis (pinkeye) is usually caused by *Haemophilus influenzae* or *Streptococcus pneumoniae,* (pneumococcus) (figure 22.6). Viral causes, including adenoviruses and rhinoviruses, usually result in a milder illness. Otitis media and sinusitis develop when infection extends from the nasopharynx (figure 22.7).

22.4 Viral Infections of the Upper Respiratory System

The Common Cold (table 22.5)

The common cold can be caused by many different viruses, rhinoviruses being the most common (figure 22.8).

Adenoviral Pharyngitis (table 22.6)

Adenoviruses cause illnesses varying from mild to severe, which can resemble a common cold or strep throat.

Infections of the Lower Respiratory System

22.5 Bacterial Infections of the Lower Respiratory System

Pneumococcal Pneumonia (table 22.7, figures 22.9, 22.10)

Streptococcus pneumoniae, one of the most common causes of pneumonia, is virulent because of its capsule.

Klebsiella Pneumonia (table 22.7)

Klebsiella pneumonia, is representative of many nosocomial pneumonias that cause permanent damage to the lung (figure 22.11). Serious complications such as lung abscesses and bloodstream infection are more common than with many other bacterial pneumonias. Treatment is more difficult, partly because klebsiellas often contain R factor plasmids.

Mycoplasmal Pneumonia (table 22.7, figure 22.12)

Mycoplasmal pneumonia is often called walking pneumonia; serious complications are rare. Penicillins and cephalosporins are not useful in treatment because the cause, *M. pneumoniae,* lacks a cell wall.

Whooping Cough (Pertussis) (table 22.8, figures 22.13, 22.14)

Whooping cough is characterized by violent spasms of coughing and gasping. Childhood immunization against the Gram-negative rod, *Bordetella pertussis,* prevents the disease.

Tuberculosis (table 22.9, figures 22.15, 22.16, 22.17, 22.18)

Tuberculosis, caused by the acid-fast rod *Mycobacterium tuberculosis,* is generally slowly progressive or heals and remains latent, presenting the risk of later reactivation.

Legionnaires' Disease (table 22.10, figure 22.19)

Legionnaires' disease occurs when there is a high infecting dose of the causative microorganisms or an underlying lung disease. The cause, *Legionella pneumophila,* is a rod-shaped bacterium common in the environment.

22.6 Viral Infections of the Lower Respiratory System

Influenza (table 22.11, figure 22.20)

Widespread epidemics are characteristic of influenza A viruses. Antigenic shifts and drifts are responsible. Deaths are usually but not always caused by secondary infection. Reye's syndrome may rarely occur during recovery from influenza and other viral infections but is probably not caused by the virus itself.

Respiratory Syncytial Virus Infections (table 22.12)

RSV is the leading cause of serious respiratory disease in infants and young children.

Hantavirus Pulmonary Syndrome (table 22.13, figure 22.22)

Hantavirus pulmonary syndrome is contracted from inhalation of dust contaminated by mice infected with the hantavirus and is often fatal.

22.7 Fungal Infections of the Lung

Valley Fever (Coccidioidomycosis) (table 22.14)

Coccidioidomycosis occurs in hot, dry areas of the Western Hemisphere and is initiated by airborne spores of the dimorphic soil fungus *Coccidioides immitis* (figures 22.23 and 22.24).

Spelunker's Disease (Histoplasmosis) (table 22.15)

Histoplamosis is similar to coccidioidomycosis but occurs in tropical and temperate zones around the world (figure 22.25).

The causative fungus, *Histoplasma capsulatum,* is dimorphic and found in soils contaminated by bat or bird droppings (figure 22.26).

REVIEW QUESTIONS

Short Answer

1. How does contamination of the eye lead to upper respiratory infection?
2. After you recover from strep throat, can you get it again? Explain why or why not.
3. Where is the gene for diphtheria toxin production located?
4. Describe two ways to decrease the chance of contracting a cold.
5. What kinds of diseases are caused by adenoviruses?
6. How do alcoholism and cigarette smoking predispose a person to pneumonia?
7. Give a mechanism by which *Klebsiella* sp. become antibiotic-resistant.
8. Why does the incidence of whooping cough rise promptly when pertussis immunizations are stopped?
9. Why are two or more antitubercular medications used together to treat tuberculosis?
10. Why did it take so long to discover the cause of Legionnaires' disease?

Multiple Choice

1. The following are all complications of streptococcal pharyngitis, *except*
   a) throat abscess.
   b) scarlet fever.
   c) otitis media.
   d) acute rheumatic fever.
   e) Reye's syndrome.
2. All of the following are true of diphtheria, *except*
   a) a membrane that forms in the throat can cause suffocation.
   b) a toxin is produced that acts by ADP ribosylation.
   c) the causative organism typically invades the bloodstream.
   d) immunization with a toxoid prevents the disease.
   e) nerve injury with paralysis is common.
3. Adenoviral infections generally differ from the common cold in all the following ways, except adenoviral infections are
   a) not caused by picornaviruses.
   b) often associated with fever.
   c) likely to extensively involve the cornea and conjunctiva.
   d) much more likely to cause pneumonia.
   e) associated with negative cultures for *Streptococcus pyogenes.*
4. All are true of mycoplasmal pneumonia, *except*
   a) it is a mycosis.
   b) it usually does not require hospitalization.
   c) penicillin is ineffective for treatment.
   d) it is the leading cause of bacterial pneumonia in college students.
   e) the infectious dose of the causative organism is low.
5. All of the following are true of Legionnaires' disease, *except*
   a) the causative organism can grow inside amebas.
   b) it spreads readily from person to person.
   c) it is more likely to occur in long term cigarette smokers than in nonsmokers.
   d) it is often associated with diarrhea or other intestinal symptoms.
   e) it can be contracted from household water supplies.
6. Which of the following infectious agents is most likely to cause a pandemic?
   a) Influenza A virus
   b) *Streptoccus pyogenes*
   c) *Histoplasma capsulatum*
   d) Sin Nombre virus
   e) *Coccidioides immitis*
7. Respiratory syncytial virus
   a) is a leading cause of bronchiolitis in infants.
   b) is an enveloped DNA virus of the adenovirus family.
   c) attaches to host cell membranes by means of neuraminidase.

d) poses no threat to elderly people.
e) mainly causes disease in the summer months.

8. In the United States, hantaviruses
   a) are limited to southwestern states.
   b) are carried only by deer mice.
   c) infect human beings with a fatality rate above 40%.
   d) were first identified in the early 1970s.
   e) are contracted mainly in bat caves.
9. All of the following are true of coccidioidomycosis, *except*
   a) it is contracted by inhaling arthrospores.
   b) it is caused by a dimorphic fungus.
   c) endospores are produced within a spore.
   d) it is more common in Maryland than in California.
   e) it is often associated with painful nodules on the legs.
10. The disease histoplasmosis
    a) is caused by an encapsulated bacterium.
    b) is contracted by inhaling arthrospores.
    c) occurs mostly in hot, dry, and dusty areas of the American Southwest.
    d) is a threat to AIDS patients living in areas bordering the Mississippi River.
    e) is commonly fatal for pigeons and bats.

Applications

1. A physician is advising the family on the condition of a diphtheria patient. How would the physician explain why the disease affects some tissues and not others?
2. How should a physician respond to a mother who asks if her daughter can get pneumococcal pneumonia again?

Critical Thinking

1. If all transmission of *Mycobacterium tuberculosis* from one person to another were stopped, how long would it take for the world to be rid of the disease?
2. Medications that prevent and treat influenza by binding to neuraminidase on the viral surface, act against all the kinds of influenza viruses that infect humans. What does this imply about the nature of the interaction between the medications and the neuraminidase molecules?

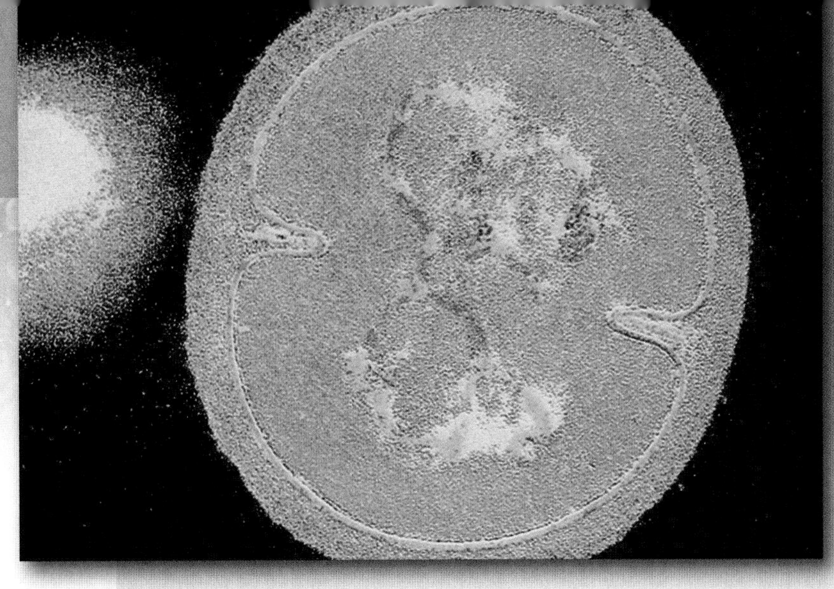

A dividing *Staphylococcus epidermidis* cell.

Skin Infections

A Glimpse of History

Howard T. Ricketts was born in Ohio in 1871. He studied medicine in Chicago, and then specialized in pathology—the study of the nature of disease and its causes. In 1902, he was appointed to the faculty of the University of Chicago, where his research interests turned to Rocky Mountain spotted fever, an often fatal and little understood disease characterized by a dramatic rash. The disease could be transmitted to laboratory animals by injecting them with blood from an infected person, and Ricketts noticed that people and laboratory animals with the disease had tiny bacilli in their blood. Ricketts was sure that these tiny bacteria were the cause of the disease, but he was never able to cultivate them on laboratory media. Based on observations of victims with the disease, Ricketts and others suspected that Rocky Mountain spotted fever was contracted from tick bites, and Ricketts went on to prove that certain species of ticks could transmit the disease from one animal to another. The infected ticks remained healthy but capable of transmitting the disease for long periods of time, and oftentimes the offspring of infected ticks were also infected. Ricketts was able to explain this by showing that the eggs of infected ticks often contained large numbers of the tiny bacilli—an example of *transovarial,* meaning via the eggs, passage of an infectious agent. Frustrated by his inability to cultivate the bacilli for further studies, Ricketts declined to give them a scientific name and went off to Mexico to study a very similar disease—louse-borne typhus (Rocky Mountain spotted fever is also known as tick-borne typhus). Unfortunately, Ricketts contracted the disease and died at the age of 39. Five years later, a European scientist, Stanislaus Prowazek, studying the same disease in Serbia and Turkey, met the same fate at almost the same age. The martyrdom of the two young scientists struggling to understand infectious diseases is memorialized in the name of the louse-borne agent, *Rickettsia prowazekii.* Both the genus and species names of the Rocky Mountain spotted fever agent, *Rickettsia rickettsii,* recognize Howard Ricketts. We now know that these bacteria are obligate intracellular parasites, which explains why they could not be cultivated on ordinary laboratory media. Antibiotics, which could have saved these men, had not yet been discovered.

Much of the body's contact with the outside world occurs at the surface of the skin. As long as skin is intact, this tough, flexible outer covering is remarkably resistant to infection. Because of its exposed state, however, it is frequently subject to cuts, punctures, burns, chemical injury, reactions, and insect or tick bites. These injure the skin and provide a way for pathogens to enter and infect the skin and underlying tissues. For example, *Staphylococcus aureus* can enter a surgical wound and then invade the bloodstream; sandfly bites can introduce *Leishmania* species, the cause of leishmaniasis cases seen in Americans returning from Afghanistan, Iraq, and Kuwait (more can be learned about this disease at the Online Learning Center, www.mhhe.com/nester6). Skin infections also occur when microorganisms or viruses are carried to the skin by the bloodstream after entering the body from another site, such as the respiratory or gastrointestinal systems. ■ **pediculosis, p. 296** ■ **scabies, p. 297**

23.1 Anatomy and Physiology

Focus Points

- Explain why the skin is more than a wrapping for the body.
- Describe the importance of skin glands in health and disease.

The skin is far more than an inert wrapping for the body. Control of body temperature and prevention of loss of fluid from body tissues are among its vitally important functions. It also plays an important role in the synthesis of vitamin D, which is needed for development of normal teeth and bones. Numerous sensory receptors of various types occur in the skin, providing the central nervous system with information about the environment. The skin also produces cytokines that aid the development and function of

KEY TERMS

Abscess A localized collection of pus within a tissue.

Carbuncle Painful infection of the skin and subcutaneous tissues; manifests as a cluster of boils.

Epidermis The outermost layer of skin.

Exanthem A skin rash, such as measles.

Exfoliatin A bacterial toxin that causes sloughing of the outer epidermis.

Folliculitis Inflammation of a hair follicle.

Furuncle A boil; a localized skin infection that penetrates into subcutaneous tissue.

Impetigo A superficial skin infection characterized by thin-walled vesicles, weeping of plasma, and yellow crusts; caused by *Staphylococcus aureus* and *Streptococcus pyogenes*.

Pyoderma Any skin disease characterized by production of pus.

Sebaceous Gland Skin glands that produce an oily substance called sebum.

immunity. In addition, collections of lymphocytes are found in skin-associated lymphoid tissues. Because of its exposed location, the temperature of the skin is generally lower than that of the rest of the body. ■ cytokines, p. 353 ■ skin-associated lymphoid tissue, p. 370

The skin is composed of two main layers—the epidermis and the dermis **(figure 23.1).** The **epidermis,** the surface layer, is stratified squamous epithelium and ranges from 0.007 to 0.12 mm thick (see figure 15.1). The outer portion is composed of scaly material made up of flat cells containing **keratin,** a durable protein also found in hair and nails. The cells on the skin surface are dead and, along with any resident organisms, continually peel off, replaced by cells from deeper in the epidermis. These cells, in

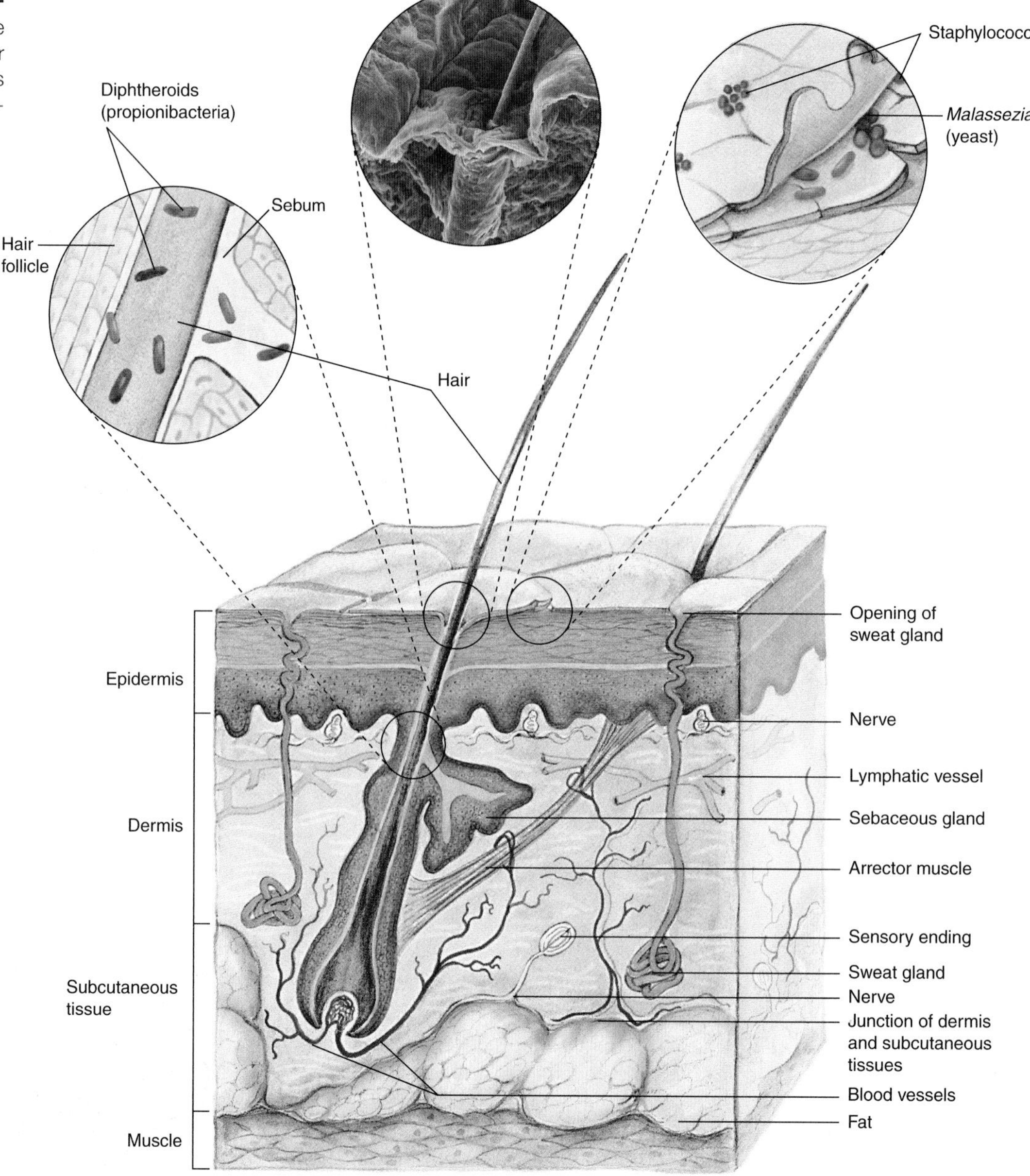

FIGURE 23.1 Microscopic Anatomy of the Skin Notice that the sebaceous unit, composed of the hair follicle and the attached sebaceous gland, almost reaches the subcutaneous tissue.

turn, become flattened and die as keratin is formed within them. This process results in a complete regeneration of the skin about once a month. Dandruff represents excessive shedding of skin cells, but mostly the shedding process is unnoticed and represents one of the skin's defenses against infection.

The epidermis is supported by the **dermis,** a deeper layer of skin cells through which many tiny nerves, blood vessels, and lymphatic vessels penetrate. The dermis adheres in a very irregular fashion to the fat and other cells that make up the subcutaneous tissue (see figure 23.1).

Fine tubules of sweat glands and hair follicles traverse the dermis and epidermis (see figure 23.1). Since sweat is a salty solution, high concentrations of salt occur as it evaporates. Evaporation of sweat and regulation of the amount of blood flow through the skin's blood vessels are critically important in controlling body temperature. **Sebaceous glands** produce an oily secretion called **sebum** that feeds into the hair follicles. This secretion flows up through the follicles and spreads out over the skin surface, keeping the hair and skin soft, pliable, and water-repellent. The hair follicles provide passage for certain salt-tolerant bacteria to penetrate the skin and reach deeper tissues.

The secretions of the sweat and sebaceous glands are essential to the normal microbial population of the skin because they supply water, amino acids, and lipids, which serve as nutrients for microbial growth. Breakdown of the lipids by the microbial residents of normal skin results in fatty-acid by-products that inhibit the growth of many potential disease-producers. The normal acidity of the skin (pH 4.0–6.8) also inhibits some microorganisms. In fact, the normal skin surface is an unfriendly habitat for most potential pathogens, being too dry, salty, unstable because of shedding, acidic, and toxic for their survival.

MICROCHECK 23.1

The skin is a large, complex organ that covers the external surface of the body. Properties of the skin cause it to resist colonization by most microbial pathogens, pose a physical barrier to infection, provide sensory input from the environment, and assist the body's regulation of temperature and fluid balance.

- ✓ Give three routes by which microorganisms invade the skin.
- ✓ Give four characteristics of skin that help it resist infection.
- ✓ Would a person living in the tropics or in the desert have larger numbers of bacteria living on the surface of their skin?

23.2 Normal Microbiota of the Skin

Focus Points

- Describe the role of normal skin microbiota in health and disease.
- Outline the pathogenesis of acne.

The skin represents a distinct ecological habitat, analogous to a cool desert, compared to the warm, moist tropical conditions that exist in other body systems. Depending on the body location and amount of skin moisture, the number of bacteria on the skin surface may range from only about 1,000 organisms per square centimeter on the back to more than 10 million in the groin and armpit, where moisture is more plentiful. The numbers actually increase after a hot shower because of increased flow from the skin glands where many bacteria reside. Most of the microbial skin inhabitants can be categorized in three groups: diphtheroids, staphylococci, and yeasts (**table 23.1;** see figure 23.1). Although generally harmless, skin organisms are opportunistic pathogens that can only cause disease in people with impaired body defenses. AIDS patients and others with impaired immunity are especially vulnerable. ■ normal microbiota, p. 393 ■ oportunistic pathogens, p. 394

TABLE 23.1 Principal Members of the Normal Skin Microbiota

Name	Characteristics
Diphtheroids	Variably shaped, non-motile, Gram-positive rods of the *Corynebacterium* and *Propionibacterium* genera
Staphylococci	Gram-positive cocci arranged in packets or clusters; coagulase negative; facultatively anaerobic
Fungi	Small yeasts of the genus *Malassezia* that require oily substances for growth

Diphtheroids

Diphtheroids are a group of Gram-positive, pleomorphic bacteria named for the fact that their microscopic appearance resembles

By permission of John Deering and Creators Syndicate, Inc.

the bacterium that causes diphtheria, *Corynebacterium diphtheriae.* Diphtheroids are responsible for body odor, caused by their breakdown of substances in sweat, which is odorless when it is first secreted. A diphtheroid found on the skin in large numbers is *Propionibacterium acnes,* which is present on virtually all humans. This bacterium grows primarily within the hair follicles, where conditions are anaerobic. Growth of *P. acnes* is enhanced by the oily secretion of the sebaceous glands, and the organisms are usually present in large numbers only in areas of the skin where these glands are especially large and numerous—on the face, upper chest, and back. These are also the areas of the skin where acne most commonly develops, and the association of *P. acnes* with acne inspired its name, even though most people who carry the organisms do not have acne. ■ aerotolerance, p. 92 ■ *Corynebacterium,* p. 262

Acne in its most common form begins at puberty in association with a rise in sex hormones, enlargement of the sebaceous glands, and enhanced secretion of sebum. The hair follicle epithelium thickens and sloughs off in clumps, causing increasing obstruction to the flow of sebum to the skin surface. Continued sebum production by the gland can force a plug of material to the surface, where it is visible as a blackhead. With complete obstruction the follicle becomes distended with sebum, which causes the epidermis to bulge outward, producing a whitish lesion called a whitehead. The *P. acnes* that normally reside in the gland multiply to enormous numbers in the trapped sebum. Lipases of the bacteria degrade the sebum, releasing fatty acids and glycerol, a growth requirement of the organisms. The metabolic products of the bacteria cause an inflammatory response, attracting leukocytes (white blood cells) whose enzymes damage the wall of the distended follicle. The inflammatory process can cause the follicle to rupture, releasing the follicle contents into the surrounding tissue. The result is an **abscess**—a collection of white blood cells, bacteria, and cellular debris, that eventually heals and leaves a scar. Squeezing acne lesions is ill-advised, because it promotes rupture of the inflamed follicles and therefore more acne scars. Usually acne can be controlled until it goes away by itself, by using medications such as antibiotics and benzoyl peroxide that inhibit the growth of *P. acnes,* or by those such as azelaic acid (Azelex) and isotretinoin (Accutane) that act primarily to reverse the hair follicle abnormalities. The latter medication is reserved for the most serious cases of acne because it has potentially serious side effects. ■ inflammation, p. 360 ■ leukocytes, p. 350

Staphylococci

The second group of microorganisms universally present on the normal skin is composed of members of the genus *Staphylococcus.* They are salt-tolerant organisms that grow well on the salty skin surface. As with the diphtheroids, most of these bacteria have little virulence, although they can cause serious disease if host defenses are breached. Generally, staphylococci are the most common of the skin bacteria able to grow aerobically. The principal species is *Staphylococcus epidermidis.* ■ *Staphylococcus,* p. 270

Important functions of the skin's staphylococci are to prevent colonization by pathogens and to maintain a balance among the microbial inhabitants of the skin ecosystem. These Gram-positive cocci compete for nutrients with other potential skin colonizers, and they also produce antimicrobial substances highly active against *P. acnes* and other Gram-positive bacteria.

Fungi

Tiny lipophilic, meaning oil-requiring, yeasts almost universally inhabit the normal human skin from late childhood onward. Their shape varies with different strains, being round, oval, or sometimes short rods. These yeasts can be cultivated on laboratory media containing fatty substances such as olive oil. They belong to the genus *Malassezia* and are generally harmless. In some people, however, they cause skin conditions such as a scaly face rash, dandruff, or **tinea versicolor (figure 23.2).** The latter is a common skin disease characterized by patchy scaliness and increased pigment in light-skin persons, or a decrease in pigment in dark-skin people. Scrapings of the affected skin show large numbers of *Malassezia furfur* both in its yeast form and as

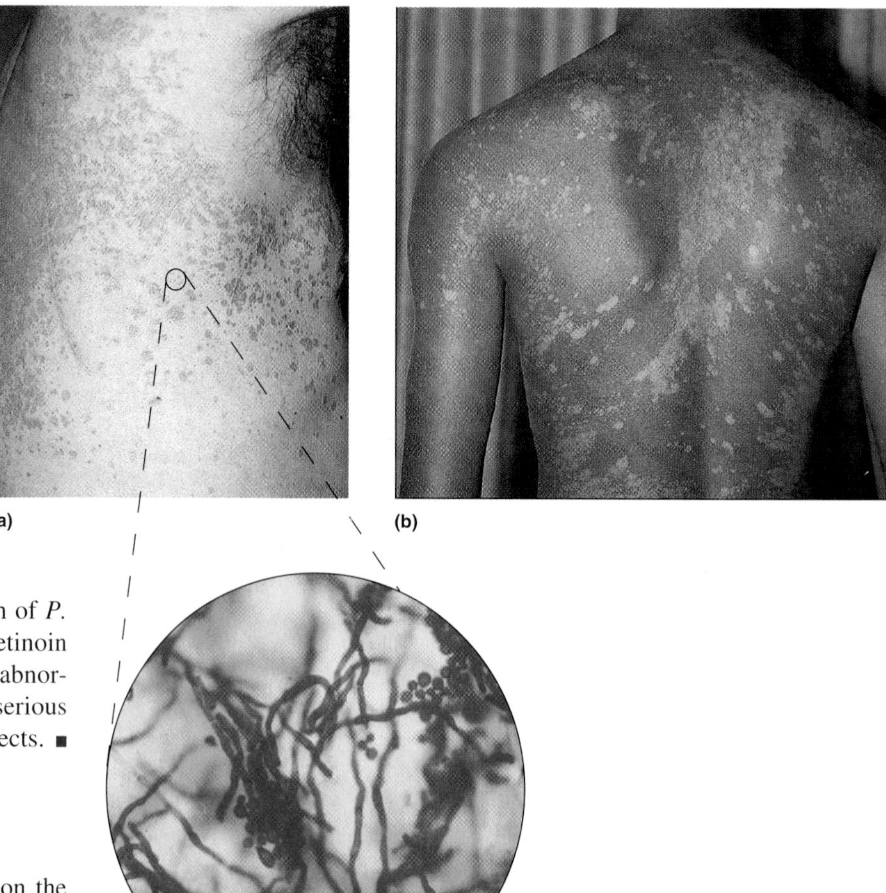

FIGURE 23.2 Tinea Versicolor Appearance in **(a)** a fair-skin individual and **(b)** a dark-skin individual. **(c)** Microscopic appearance of skin scraping showing *Malassezia furfur* yeast and filamentous forms.

short filaments called pseudohyphae. Unknown factors, probably relating to the host, are important in these diseases because most people carry the organism on their skin without any disease. AIDS patients often have a severe rash with pus-filled pimples caused by *Malassezia* yeasts, and the organisms may even infect internal organs in patients receiving lipid-containing intravenous feedings.

■ yeasts, p. 289

MICROCHECK 23.2

The normal skin microbiota are important because they help protect against colonization by pathogens. Occasionally, they cause disease when body defenses are impaired. They are responsible for body odor, and probably contribute to acne.

✓ Name and describe the three groups of organisms generally present on normal skin.

✓ Under what circumstances is *Malassezia furfur* most likely to be pathogenic?

✓ Would frequent showering tend to increase or decrease the numbers of *Staphylococcus* on the surface of the skin? Why?

23.3 Bacterial Skin Diseases

Focus Points

- Name and give a distinctive characteristic of five bacterial skin diseases.
- Give three mechanisms of pathogenesis seen in bacterial skin diseases.

Only a few species of bacteria commonly invade the intact skin directly, which is not surprising in view of the anatomical and physiological features discussed earlier. Hair follicle infections exemplify direct invasion.

Hair Follicle Infections

Infections originating in hair follicles commonly clear up without treatment. In some instances, however, they progress into severe or even life-threatening disease.

Symptoms

Folliculitis, furuncles, and carbuncles represent different outcomes of hair follicle infections. In **folliculitis,** a small red bump, or pimple, develops at the site of the involved hair follicle. Often, the hair can be pulled from its follicle, accompanied by a small amount of pus, and then the infection goes away without further treatment. If, however, the infection extends from the follicle to adjacent tissues, causing localized redness, swelling, severe tenderness, and pain, the lesion is called a **furuncle** or boil. Pus may drain from the boil along with a plug of inflammatory cells and dead tissue. A **carbuncle** is a large area of redness, swelling, and pain punctuated by several sites of draining pus. Carbuncles usually develop in areas of the body where the skin is thick, such as the back of the neck. Fever is often present, along with other signs of a serious infection.

Causative Agent

Most furuncles and carbuncles, as well as many cases of folliculitis, are caused by *Staphylococcus aureus.* It is much more virulent than the staphylococci normally found on the skin. The name derives from *staphyle,* "a bunch of grapes," referring to the arrangement of the bacteria as seen on stained smears, and *aureus,* "golden," referring to the typical color of the *S. aureus* colonies. This bacterium is an extremely important pathogen and is mentioned frequently throughout this text as the cause of a number of medical conditions **(table 23.2).**

One of the most useful identifying characteristics of *S. aureus* is that it produces **coagulase.** Despite its *ase* ending, coagulase is not an enzyme. This protein product of *S. aureus* is largely extracellular, meaning it is released from the bacterium. It reacts with a substance in blood called prothrombin. The resulting complex, called staphylothrombin, causes blood to clot by converting fibrinogen to fibrin. Some coagulase is tightly bound to the surface of the bacteria and coats their surface with fibrin upon contact with blood. Fibrin-coated staphylococci resist phagocytosis.

Generally, *S. aureus* also possesses **clumping factor,** often called "slide coagulase," because it causes a suspension of the bacteria to clump together when mixed with a drop of blood plasma on a microscope slide. Like coagulase, it is a useful identifying characteristic of *S. aureus*. The clumping is caused by a cell-fixed protein that attaches specifically to fibrinogen in the plasma. Clumping factor protein is a virulence factor for *S. aureus* because it attaches to fibrinogen and fibrin present in wounds, thus aiding colonization of wound surfaces. Plastic devices, such as intravenous catheters and heart valves, become coated with fibrinogen shortly after insertion, thus making them, too, a target for colonization. The gene for clumping factor is distinct from the one controlling coagulase.

Other virulence factors possessed by *S. aureus* that aid colonization of wounds include binding proteins for fibronectin, fibrin, fibrinogen, and collagen. Most strains of *S. aureus* also produce **α-toxin,** a membrane-damaging toxin that kills cells by attaching to specific receptors

TABLE 23.2 Some Diseases Often Caused by *Staphylococcus aureus*

Disease	Page for More Information
Carbuncles	p. 535
Endocarditis	p. 677
Folliculitis	p. 535
Food poisoning	p. 762
Furuncles	p. 535
Impetigo	p. 538
Osteomyelitis (bone infection)	p. 682
Scalded skin syndrome	p. 537
Toxic shock syndrome	p. 626
Wound infections	p. 562

on host cell membranes and making holes in them. A relatively small percentage of *S. aureus* strains produce one or more additional toxins. ■ IgG, p. 374 ■ Fc region, p. 371 ■ membrane-damaging toxin, p. 405

Pathogenesis

Infection begins when *Staphylococcus aureus* attaches to the cells of a hair follicle, multiplies, and spreads inward to involve the follicle and sebaceous glands. The infection induces an inflammatory response with swelling and redness, followed by attraction and accumulation of polymorphonuclear leukocytes. If the infection continues, the follicle becomes a plug of inflammatory cells and necrotic tissue overlying a small abscess **(figure 23.3).** The infectious process spreads deeper, reaching the subcutaneous tissue where a large abscess forms. This subcutaneous abscess is responsible for the painful localized swelling that constitutes the boil. Without effective treatment, pressure within the abscess increases, causing it to expand to other hair follicles, causing a carbuncle. If organisms enter the bloodstream, the infection can spread to other parts of the body, such as the heart, bones, or brain. ■ abscess, p. 560

The properties of *S. aureus* that probably contribute to its virulence are shown in **table 23.3.** Virtually all strains possess an unusual cell wall component called **protein A.** This protein, is an Fc receptor that prevents antibody from attaching to Fc receptors on phagocytes (see figure 17.11). Thus, a major effect of protein A is to interfere with phagocytosis. Many strains of *S. aureus* growing in body tissues synthesize a polysaccharide capsule that also inhibits phagocytosis. The *S. aureus* genes responsible for capsule formation are activated following invasion of tissue. *S. aureus* also produces numerous extracellular products that might contribute to virulence. These products include **leukocidins,** which kill white blood cells; **hyaluronidase,** which degrades hyaluronic acid, a component of host tissue that helps hold the cells together; **proteases,** which degrade various host proteins including collagen, the white fibrous protein found in skin, tendons, and connective tissue; and **lipases,** which degrade lipids. Lipases may assist colonization of the oily hair follicles by strains of *S. aureus* that cause follicle infections. ■ protein A, p. 402 ■ Fc receptors, p. 373

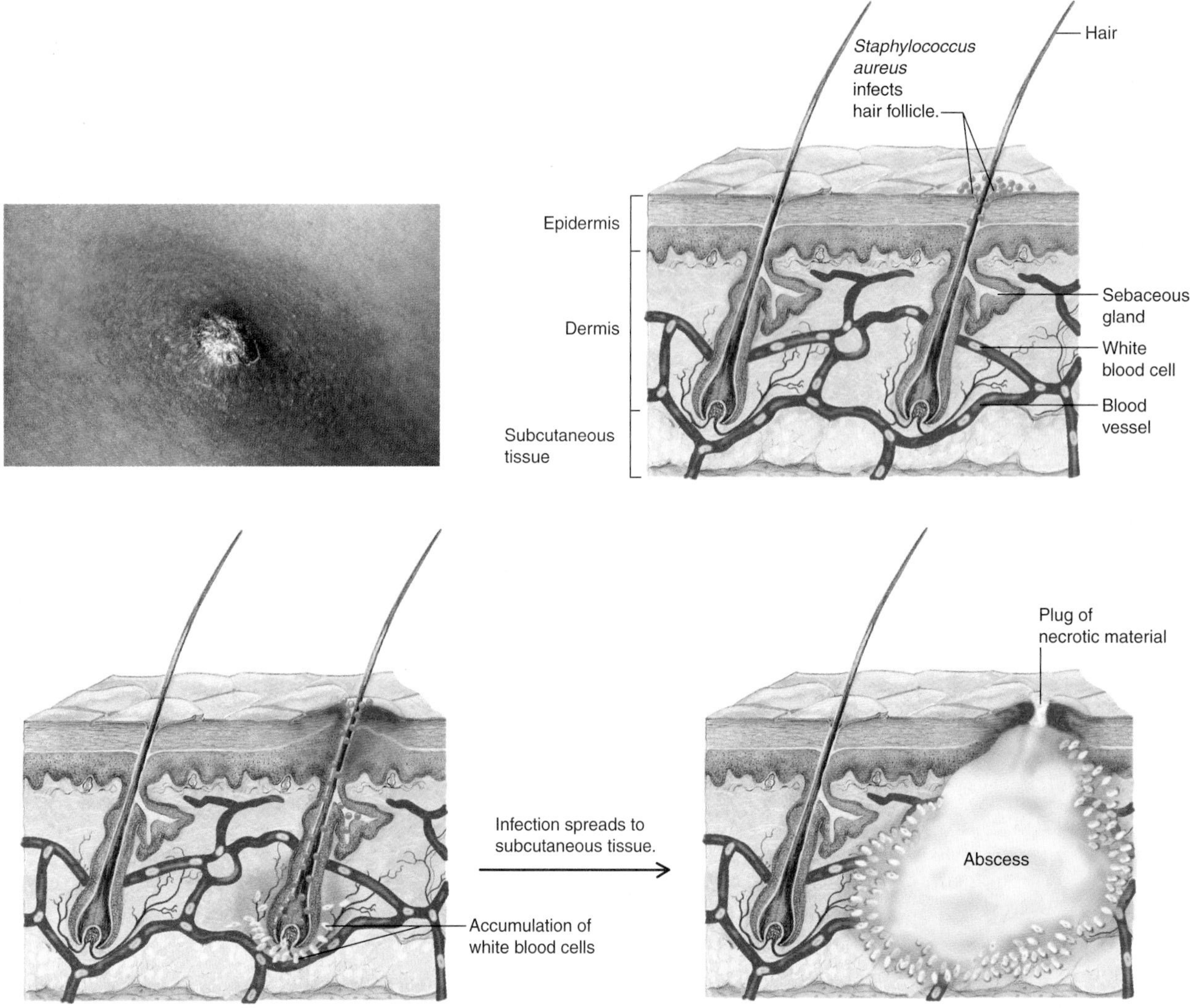

FIGURE 23.3 Pathogenesis of a Boil (Furuncle) *Staphylococcus aureus* infects a hair follicle through its opening on the skin surface. The infection produces a plug of necrotic material, a small abscess in the dermis, and, finally, a larger abscess in the subcutaneous tissue.

TABLE 23.3 Properties of *Staphylococcus aureus* Implicated in Its Virulence

Product	Effect
Capsule	Inhibits phagocytosis
Clumping factor	Attaches the bacterium to fibrin, fibrinogen, plastic devices
Coagulase	May impede progress of leukocytes into infected area by producing clots in the surrounding capillaries
Enterotoxins	Superantigens cause food poisoning if ingested, cause toxic shock if systemic
Exfoliatin	Separates layers of epidermis, causing scalded skin syndrome
Fibronectin-binding protein	Attaches bacterium to acellular tissue substances endothelium, epithelium, clots, indwelling plastic devices
Hyaluronidase	Breaks down hyaluronic acid component of tissue, thereby promoting extension of infection
Leukocidin	Kills neutrophils or causes them to release their enzymes
Lipase	Breaks down fats by hydrolyzing the bond between glycerol and fatty acids
Proteases	Degrade collagen and other tissue proteins
Protein A	Binds to Fc portion of antibody; coats bacteria with host's immunoglobulin thereby inhibiting phagocytosis
Toxic shock syndrome toxin	Causes rash, diarrhea, and shock
α-toxin	Makes holes in host cell membranes.

Epidemiology

Staphylococcus aureus inhabits the nostrils of virtually everyone at one time or another, each nostril containing as many as 10^8 bacteria. About 20% of healthy adults have continually positive nasal cultures for a year or more, while over 60% will be colonized at some time during a given year. The organisms are mainly disseminated to other parts of the body and to the environment by the hands. Although the nostrils seem to be the preferred habitat of *S. aureus,* moist areas of skin are also frequently colonized. People with boils and other staphylococcal infections shed large numbers of *S. aureus* and should not work with food, or near patients with surgical wounds or chronic illnesses. Staphylococci survive well in the environment, which favors their transmission from one host to another. Since *S. aureus* is so commonplace and there are many different strains in the population, epidemics of staphylococcal disease can generally be traced to their sources only by precise identification of the epidemic strain. Techniques for characterizing strains of *S. aureus* include determining the antibiogram, phage typing, and genome typing (see section 10.4). ■ **characterizing strain differences, p. 243**

Prevention and Treatment

Prevention of staphylococcal skin disease is very difficult. Attempts are made to eliminate the carrier state by applying an antistaphylococcal cream to the nostrils, and using soaps containing an antistaphylococcal agent such as hexachlorophene to bathe the skin. Effective treatment of boils and carbuncles often requires that the pus be surgically drained from the lesion and an antistaphylococcal medicine be given. Treatment of staphylococcal infections is complicated because so many strains of *S. aureus* are resistant to antibacterial medications. When penicillin was first introduced, more than 95% of *S. aureus* strains were susceptible to it. However, these strains soon largely disappeared with the widespread use of the antibiotic. Now, about 90% of *S. aureus* strains are resistant to penicillin because they produce a plasmid-encoded β-lactamase. Treatment became much easier with the development of penicillins and cephalosporins resistant to β-lactamase. Soon thereafter, however, strains of *S. aureus,* referred to as MRSAs, methicillin-resistant *Staphylococcus aureus,* appeared that were resistant because of modified penicillin-binding proteins. These strains were reliably treated with vancomycin until 1997, when the first vancomycin-resistant strain was identified. Because many MRSAs had R plasmids, making them resistant to most other antistaphylococcal medications, the appearance of vancomycin resistance was an alarming development. Beginning in 1999, several medications active against vancomycin-resistent bacteria were marketed. Quinupristin/dalfopristin (trade name Synercid) is a combination of two substances that act synergistically to block bacterial protein synthesis, and linezolid (trade name Zyvox), which represented a new class of antibacterials, the oxazolidinons, were followed by daptomycin (trade name Cubicin), a lipopeptide. Hopefully, development of new medications will keep pace with the growth of bacterial resistance, but this will depend on humans avoiding overuse of these valuable substances. ■ **penicillin-binding proteins, p. 475**

Most multiply-resistant *S. aureus* strains can be traced to hospitals and clinics, but more recently completely different strains have become widespread among healthy carriers in the community. Generally, these community MRSAs are resistant only to methicillin and related β-lactam antibiotics, and the macrolides, but they are highly virulent because they possess a cluster of genes that codes for a leukocyte-destroying leukocidin.

In an effort to control the spread of MRSAs, many hospitals are screening patients at admission. If they carry MRSAs, they are isolated and given appropriate antibacterial treatment to limit spread of the strain to other patients or staff. Some hospitals also screen patients for MRSAs, at discharge from the hospital to make sure they don't take a strain home with them. ■ **penicillins and other β-lactam antibiotics, p. 475**

Scalded Skin Syndrome

Staphylococcal scalded skin syndrome (SSSS), is a potentially fatal toxin-mediated disease that occurs mainly in infants.

FIGURE 23.4 Staphylococcal Scalded Skin Syndrome (SSSS) A toxin called exfoliatin, produced by certain strains of *Staphylococcus aureus,* causes the outer layer of skin to separate.

Symptoms

As the name suggests, the skin appears to be scalded **(figure 23.4).** SSSS begins as a generalized redness of the skin affecting 20% to 100% of the body. Other symptoms, such as **malaise**—a vague feeling of discomfort and uneasiness—irritability, and fever are also present. The nose, mouth, and genitalia may be painful for one or more days before the typical features of the disease become apparent. Within 48 hours after the redness appears, the skin becomes wrinkled, and large blisters filled with clear fluid develop. The skin is tender to the touch and looks like sandpaper.

Causative Agent

Scalded skin syndrome is caused by *Staphylococcus aureus* strains that produce a toxin called **exfoliatin.** Only about 5% of *S. aureus* strains produce this toxin, which destroys material that binds together the outer layers of epidermis. At least two kinds of exfoliatins exist: one is coded by a plasmid gene, and the other is chromosomal. ■ plasmids, p. 68

Pathogenesis

If an exfoliatin-producing strain of *S. aureus* is growing in a lesion, even one too small to be readily apparent, the toxin can be carried by the bloodstream to large areas of the skin. In the skin, it causes a split in the cellular layer of the epidermis just below the dead keratinized outer layer. *Staphylococcus aureus* is usually not present in the blister fluid. Because the outer layers of skin are lost as in a severe burn, there is marked loss of body fluid and danger of secondary infection with Gram-negative bacteria such as *Pseudomonas* sp., or with fungi such as *Candida albicans.* **Secondary infection** means invasion by a new organism of tissues damaged by an earlier infection. Mortality can range up to 40%, depending on how promptly the disease is diagnosed and treated, and the patient's age and general health. ■ *Candida albicans,* p. 626

Epidemiology

The disease can appear in any age group but occurs most frequently in newborn infants. The elderly and immunocompromised individuals are also at increased risk. Transmission is generally from person-to-person. Staphylococcal scalded skin syndrome usually appears in isolated cases, although small epidemics in nurseries sometimes occur.

Prevention and Treatment

There are no preventive measures except to place patients suspected of having SSSS in protective isolation. These measures help to limit spread of the pathogen to others and help prevent secondary infection of the isolated patient. Initial therapy includes a bactericidal antibiotic such as methicillin. Dead skin is removed to help prevent secondary infection. Although scalded skin syndrome can be fatal, prompt therapy usually leads to full recovery.

Table 23.4 describes the main features of this disease.

Streptococcal Impetigo

Impetigo is the most common type of **pyoderma,** a skin infection characterized by pus production **(figure 23.5).** Pyodermas can result from infection of an insect bite, burn, scrape, or other wound. Sometimes, the injury is so slight that it is not apparent.

Symptoms

Impetigo is a superficial skin infection, involving patches of epidermis just beneath the dead, scaly outer layer. Thin-walled blisters first develop, then break, and are replaced by yellowish crusts that form from the drying of plasma that weeps through the skin. Usually, little fever or pain develop, but lymph nodes near the involved areas often enlarge, indicating that bacterial products

TABLE 23.4 Staphylococcal Scalded Skin Syndrome

Symptoms	Tender red rash with sandpaper texture, malaise, irritability, fever, large blisters, peeling of skin
Incubation period	Variable, usually days
Causative agent	Strains of *Staphylococcus aureus* that produce exfoliatin toxin
Pathogenesis	Exfoliatin toxin is produced by staphylococci at an infection site, usually of the skin, and carried by the bloodstream to the epidermis, where it causes a split in a cellular layer; loss of body fluid and secondary infections contribute to mortality.
Epidemiology	Person-to-person transmission; seen mainly in infants, but can occur at any age.
Prevention and treatment	Isolation of the victim to protect from environmental potential pathogens; penicillinase–resistant penicillins; removal of dead tissue.

condition. Streptococci are absent from the urine and diseased kidney tissues. Indeed, the bacteria have generally been eliminated from the infection site in the skin by the immune response by the time symptoms of glomerulonephritis appear. Damage to the kidney is caused by immune complexes that settle in the glomeruli and provoke an inflammatory reaction. Both streptococcal skin and throat infections can sometimes cause acute glomerulonephritis. Rheumatic fever, a serious complication of strep throat, is not generally a complication of streptococcal pyoderma. ■ immune complexes, p. 420 ■ rheumatic fever, p. 501

Epidemiology

Impetigo is most prevalent among poor children of the tropics or elsewhere during the hot, humid season. Children two to six years are mainly afflicted. Person-to-person contact spreads the disease, as do flies and other insects, and fomites—inanimate objects such as toys or towels. Impetigo patients often become throat and nasal carriers of *S. pyogenes*.

Prevention and Treatment

General cleanliness and avoiding people with impetigo help prevent the disease. Prompt cleansing of wounds and application of antiseptic probably also decrease the chance of infection. So far, *S. pyogenes* strains remain susceptible to penicillin. In patients allergic to penicillin, erythromycin can be substituted.

Table 23.6 summarizes the main features of impetigo.

TABLE 23.6 Impetigo

Symptoms	Blisters that break and "weep" plasma and pus; formation of golden-colored crusts; lymph node enlargement
Incubation period	2 to 5 days
Causative organisms	*Streptococcus pyogenes, Staphylococcus aureus*
Pathogenesis	Initiated by organisms entering the skin through minor breaks; certain strains of *S. pyogenes* that are prone to cause impetigo can cause glomerulonephritis.
Epidemiology	Spread by direct contact with carriers or patients with impetigo, insects, and fomites.
Prevention and treatment	Cleanliness; care of skin injuries. An oral penicillin if cause is known to be *S. pyogenes;* otherwise, an anti-staphylococcal antibiotic orally or topically.

Rocky Mountain Spotted Fever

Rocky Mountain spotted fever was first recognized in the Rocky Mountain area of the United States—thus its name. The disease is representative of a group of serious rickettsial diseases that occur worldwide and are transmitted by certain species of ticks, mites, or lice.

Symptoms

Rocky Mountain spotted fever generally begins suddenly with a headache, pains in the muscles and joints, and fever. Within a few days, a rash consisting of faint pink spots appears on the palms, wrists, ankles, and soles. This rash spreads up the arms and legs to the rest of the body and becomes raised and hemorrhagic **(figure 23.7),** meaning that it is due to blood leaking from damaged blood vessels. Bleeding may occur at various other sites, such as the mouth and nose. Involvement of the heart, kidneys, and other body tissues can result in drop in blood pressure with shock and death unless treatment is given promptly.

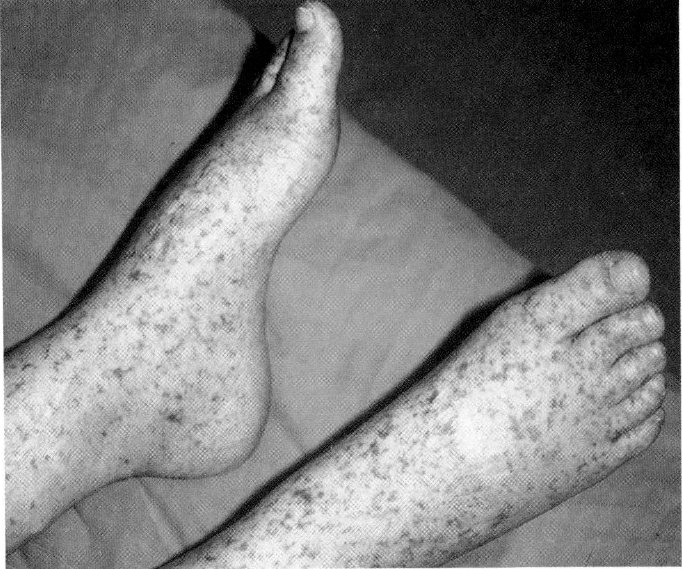

FIGURE 23.7 Rash Caused by Rocky Mountain Spotted Fever Characteristically, the rash begins on the arms and legs, spreads centrally, and as shown in this photo, becomes hemorrhagic.

Causative Agent

Rocky Mountain spotted fever is caused by *Rickettsia rickettsii* **(figure 23.8),** an obligate intracellular bacterium. The organisms are tiny, Gram-negative, non-motile coccobacilli. *Rickettsia rickettsii* can sometimes be identified early in an infection by demonstrating the organisms in **biopsies**—bits of tissue removed

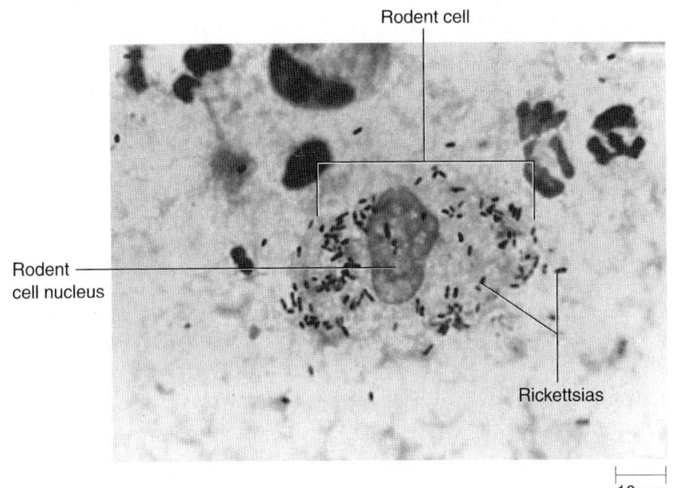

FIGURE 23.8 The Obligate Intracellular Bacterium *Rickettsia rickettsii* Growing Within a Rodent Cell

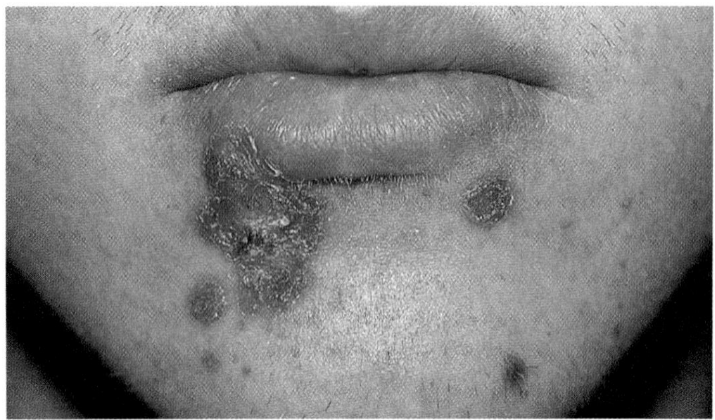

FIGURE 23.5 Impetigo This type of pyoderma is often caused by *Streptococcus pyogenes* and may result in glomerulonephritis.

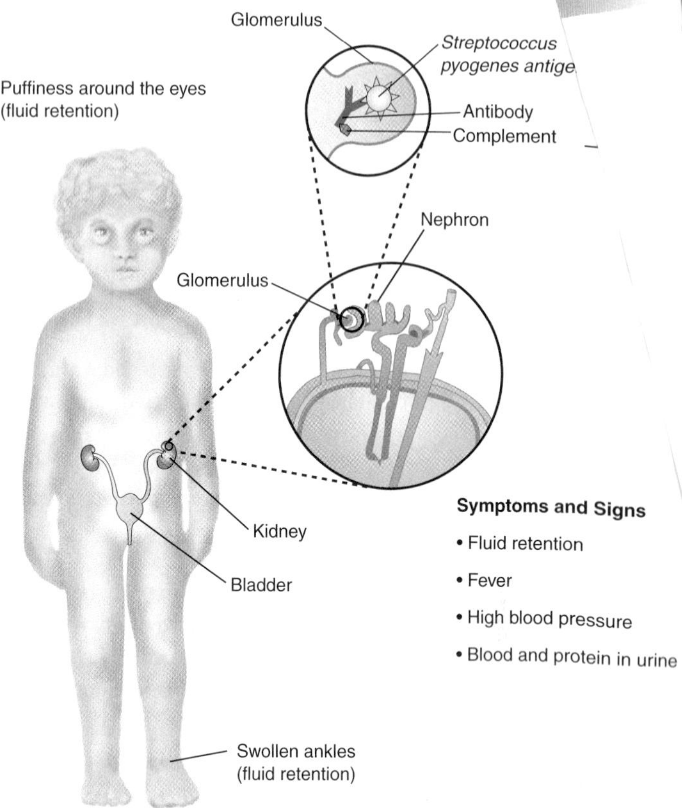

FIGURE 23.6 Pathogenesis of Post-Streptococcal Acute Glomerulonephritis Immune complexes are deposited in the kidney glomeruli, inciting an inflammatory response.

have entered the lymphatic system and an immune response is occurring. ■ plasma, p. 439

Causative Agent

Although *Staphylococcus aureus* often causes impetigo, many cases, even epidemics, are due to *Streptococcus pyogenes.* These Gram-positive, chain-forming cocci are β-hemolytic (see figure 22.3) and are frequently referred to as group A streptococci because their cell walls contain a polysaccharide called group A carbohydrate. Like *Staphylococcus aureus, Streptococcus pyogenes* causes a variety of different diseases and is mentioned in numerous places throughout this text. The most detailed description is in chapter 22. ■ hemolysis, p. 97 ■ *Streptococcus pyogenes,* p. 499

Table 23.5 compares *Staphylococcus aureus* and *Streptococcus pyogenes.*

Pathogenesis

Many different strains of *Streptococcus pyogenes* exist, some of which can colonize the skin. Infection is probably established by scratches or other minor injuries that introduce the bacteria into the deeper layer of epidermis. In impetigo, even though the infection is limited to the epidermis, streptococcal products are absorbed into the circulation.

As with *Staphylococcus aureus,* a number of extracellular products may contribute to the virulence of *Streptococcus pyogenes.* These products include enzymes such as proteases, nucleases, and hyaluronidase. The enzymes probably contribute to streptococcal pathogenicity. None of them appear to be essential, however, because antibody against them fails to protect experimental animals. On the other hand, the surface components of *S. pyogenes,* notably a hyaluronic acid capsule and a cell wall component known as the M protein, are very important in enabling this organism to cause disease because they interfere with phagocytosis. ■ M protein, p. 402

Acute glomerulonephritis is a serious complication of *S. pyogenes* pyoderma due to the immune response to the organism. This condition may appear abruptly during convalescence from untreated *S. pyogenes* infections, with fever, fluid retention, high blood pressure, and blood and protein in the urine. Acute glomerulonephritis is caused by inflammation of structures within the kidneys, the glomeruli (singular: glomerulus), small tufts of tiny blood vessels, and the nephrons, responsible for the formation and composition of urine **(figure 23.6).** Only a few of the many *S. pyogenes* strains cause the

TABLE 23.5 ***Streptococcus pyogenes* vs. *Staphylococcus aureus***

	Streptococcus pyogenes	Staphylococcus aureus
Characteristics	Gram-positive cocci in chains; β-hemolytic colonies; cell wall contains group A polysaccharide, an Fc receptor (protein G), and M protein	Gram-positive cocci in clusters; golden hemolytic colonies; cell wall contains an Fc receptor (protein A)
Extracellular Products	Hemolysins: streptolysins O and S; streptokinase, DNase, hyaluronidase, and others	Hemolysins, leukocidin, hyaluronidase, nuclease, protease, penicillinase, and others
Disease Potential	Causes impetigo, strep throat, wound infections, scarlet fever, puerperal fever, toxic shock, and flesh-eating fasciitis. Complications: glomerulonephritis, rheumatic fever, and neurological involvement	Causes boils, staphylococcal scalded skin syndrome, wound infections, abscesses, bone infections, impetigo, food poisoning, and staphylococcal toxic shock syndrome

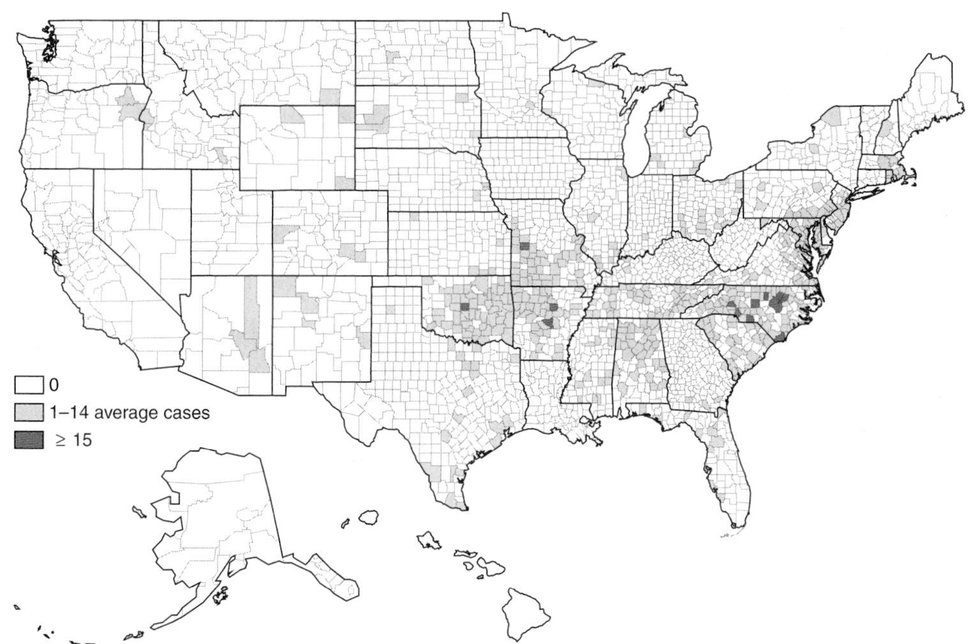

FIGURE 23.9 Rocky Mountain Spotted Fever Number of reported cases, by county—United States, 2005.
(Source: Centers for Disease Control and Prevention. *MMWR* 54(53):2–92, March 30, 2007.)

surgically—of skin lesions. Also, their DNA can be amplified by the polymerase chain reaction (PCR) and identified with a probe. ■ the genus *Rickettsia*, p. 274 ■ PCR, p. 223 ■ probe, p. 227

Pathogenesis

Rocky Mountain spotted fever is acquired from the bite of a tick infected with *R. rickettsii.* The bite is usually painless and unnoticed; the tick remains attached for hours while it feeds on capillary blood. Rickettsias are not immediately released into tick saliva from the tick's salivary glands. Therefore, the infection is not usually transmitted until the tick has fed for 4 to 10 hours. When the bacteria are released into capillary blood with the tick saliva, they are taken up preferentially by the cells lining the small blood vessels. Following attachment to host cells, *R. rickettsii* is taken into the cells by endocytosis. Inside the cell, the bacteria leave their phagosome and multiply in both the cytoplasm and nucleus without being enclosed in vacuoles. Early in the infection the bacteria enter and then lyse fingerlike host cell cytoplasmic projections. Eventually, the cell membrane is so damaged by this process, the cell takes in water, lyses, and releases the remaining rickettsias. These seed the blood-stream, infecting even more cells. Infection can also extend into the walls of the small blood vessels, causing an inflammatory reaction, clotting of the blood vessels, and small areas of necrosis, or death of tissue. This process is readily apparent in the skin as a hemorrhagic rash but, more ominously, occurs throughout the body, resulting in damage to vital organs such as the kidneys and heart. Potentially even more serious is the release of endotoxin into the bloodstream from the rickettsial cell walls, causing shock and generalized bleeding because of **disseminated intravascular coagulation.** ■ endotoxin, pp. 61, 407 ■ disseminated intravascular coagulation, p. 679

Epidemiology

Rocky Mountain spotted fever occurs in a spotty distribution across the contiguous United States and extends into Canada, Mexico, and a few countries of South America. The involved areas change over time, but despite the name of the disease, in the United States the highest incidence has generally been in the south Atlantic and south-central states **(figure 23.9).** Rocky Mountain spotted fever is a zoonotic disease maintained in nature in various species of ticks and mammals. Generally, little or no illness develops in these natural hosts, but humans, being an accidental host, often develop severe disease. Several species of ticks transmit the disease to humans. The main vector in the western United States is the wood tick, *Dermacentor andersoni* **(figure 23.10),** while in the East it is the dog tick, *Dermacentor variabilis.* Once infected, ticks remain infected for life, transmitting *R. rickettsii* from one generation to the next through their eggs. Ticks are most active from April to September, and it is during this time period that most cases of Rocky Mountain spotted fever occur. ■ zoonotic disease, p. 453

FIGURE 23.10 *Dermacentor andersoni,* the Wood Tick The wood tick is the principal vector of Rocky Mountain spotted fever in the western United States.

Prevention and Treatment

No vaccine against Rocky Mountain spotted fever is currently available to the public. The disease can be prevented if people take the following precautions: (1) avoid tick-infested areas when possible; (2) use protective clothing; (3) use tick repellents such as dimethyltoluamide; (4) carefully inspect their bodies, especially the scalp, armpits, and groin, for ticks several times daily; and (5) remove attached ticks carefully to avoid crushing them and thereby contaminating the bite wound with their infected tissue fluids. Gentle traction with blunt tweezers applied at the mouthparts is the safest method of removal. Touching the tick with a hot object, gasoline, or whiskey is ineffectual. After removing the tick, the site of the bite should be treated with an antiseptic.

The antibiotics doxycycline and chloramphenicol are highly effective in treating Rocky Mountain spotted fever if given early in the disease, before irreversible damage to vital organs has occurred. Without treatment, the overall mortality from the disease is about 20%, but it can be considerably higher in elderly patients. With early diagnosis and treatment, the mortality rate is less than 5%.

The main features of Rocky Mountain spotted fever are summarized in **table 23.7.**

Lyme Disease

In the mid-1970s, studies of a group of cases in Lyme, Connecticut, led to the recognition of Lyme disease as a distinct entity. It was not until 1982 that the causative agent was first identified in ticks from New York State by Dr. Willy Burgdorfer at the Rocky Mountain Laboratories in Hamilton, Montana. We now know that Lyme disease was present in many areas of the world long before its recognition at Lyme. The ecology of the disease is complex and still incompletely understood, but we are beginning to get some answers as to why the disease has increased and extended its range. About 20,000 new cases occur each year, making it the most common vector-borne disease in the United States.

Symptoms

Symptoms of Lyme disease can be divided roughly into three stages, although individual patients may lack symptoms in one or more of the three.

- The first stage typically begins a few days to several weeks after a bite by an infected tick. It is characterized by a skin rash called **erythema migrans (figure 23.11)** and enlargement of nearby lymph nodes. The rash begins as a red spot or bump at the site of the tick bite and slowly enlarges to a median diameter of 15 cm (about 6 inches). The advancing edge is bright red, leaving behind an area of fading redness as the lesion enlarges. About half of these cases develop smaller satellite lesions that behave similarly. The characteristic rash is the hallmark of Lyme disease but is present in only two-thirds of the cases. Most of the other symptoms that occur during this stage are influenza-like—malaise, chills, fever, headache, stiff neck, joint and muscle pains, and backache.
- Symptoms of the second stage generally begin 2 to 8 weeks after the appearance of erythema migrans and involve the heart and the nervous system. Electrical conduction within the heart is impaired, leading to dizzy spells or fainting, and a temporary pacemaker is sometimes required to maintain a normal heartbeat. Involvement of the nervous system can cause one or more of the following symptoms: paralysis of the face, severe headache, pain when moving the eyes, difficulty concentrating, emotional instability, fatigue, and impairment of the nerves of the legs or arms.
- The symptoms of the third stage are characterized by arthritis, manifest as joint pain, swelling, and tenderness, usually of a large joint such as the knee. These symptoms develop in 60% of untreated cases, beginning on the average 6 months after

TABLE 23.7 Rocky Mountain Spotted Fever

Symptoms	Headache, pains in muscles and joints, and fever, followed by a hemorrhagic rash that begins on the extremities
Incubation period	4 to 8 days
Causative organism	*Rickettsia rickettsii,* an obligate intracellular bacterium
Pathogenesis	Organisms multiply at site of tick bite; the bloodstream is invaded and endothelial cells of blood vessels are infected; vascular lesions and endotoxin account for pathologic changes.
Epidemiology	A zoonosis transmitted by bite of infected tick, usually *Dermacentor* sp.
Prevention and treatment	Avoidance of tick-infested areas, use of tick repellent, removal of ticks within 4 hours of exposure. Treatment: doxycycline or chloramphenicol.

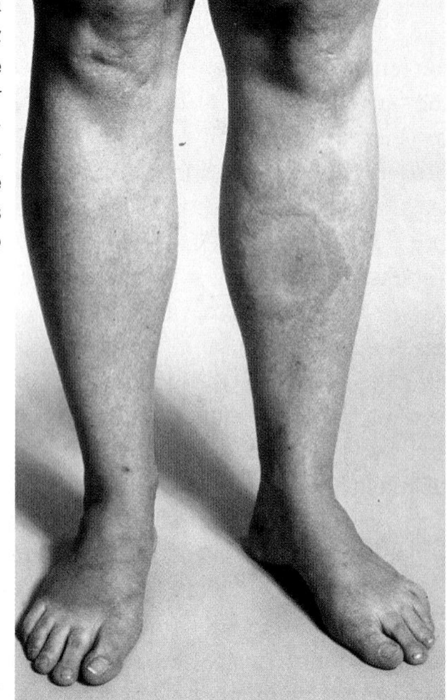

FIGURE 23.11 Erythema Migrans, the Characteristic Rash of Lyme Disease The rash usually has a targetlike or bull's-eye appearance. It generally causes little or no discomfort. While highly suggestive of Lyme disease, many victims of the disease fail to develop the rash.

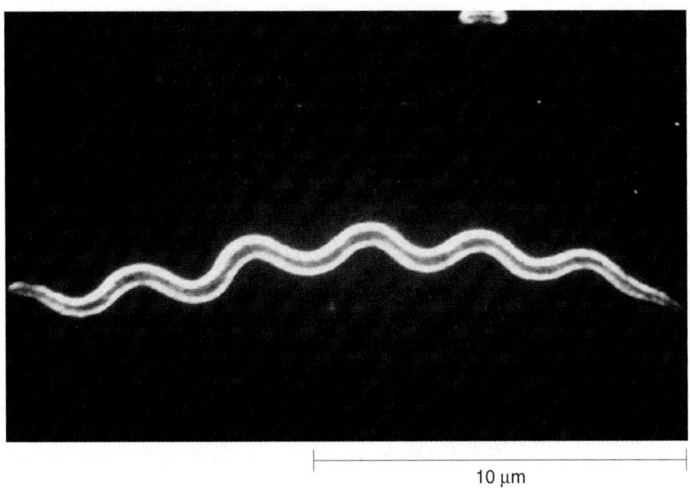

FIGURE 23.12 Scanning Electron Micrograph of *Borrelia burgdorferi,* the Cause of Lyme Disease

the skin rash, and slowly disappear over subsequent years. Chronic nervous system impairments such as localized pain, paralysis, and depression can occur.

Causative Agent

Lyme disease is caused by *Borrelia burgdorferi,* a large, microaerophilic spirochete **(figure 23.12),** 11 to 25 μm in length, with a number of axial filaments wrapped around its body and enclosed in the outer sheath of the cell wall. Surprisingly, the *Borrelia* chromosome is linear and present in multiple copies, completely unlike *E. coli* and most other prokaryotes, which have a single copy of a circular chromosome. The organism also contains numerous different plasmids both circular and linear, peculiar in that they contain genes usually found on bacterial chromosomes. These findings may lead to an understanding of how *B. burgdorferi* can infect such widely differing species as mice, lizards, and ticks. ■ spirochetes, p. 269 ■ microaerophilic conditions, p. 98 ■ axial filaments, p. 269

Pathogenesis

The spirochetes are introduced into the skin by an infected tick, multiply, and migrate outward in a radial fashion. Their Gram-negative cell walls cause an inflammatory reaction in the skin, which produces the expanding rash. The host's immune response is initially suppressed, allowing continued multiplication of the spirochete. The organisms then enter the bloodstream and become disseminated to all parts of the body but generally do not cross the placenta of pregnant women. Wide dissemination of the organisms accounts for the influenza-like symptoms of the first stage. After the first few weeks, an intense immune response occurs, and thereafter, it becomes very difficult to recover *B. burgdorferi* from blood or body tissues. The immune response against the bacterial antigens is probably responsible for the symptoms of the second stage. The third stage of Lyme disease is characterized by arthritis, and the affected joints have high concentrations of reactive immune cells and immune complexes. The joint and chronic nervous system symptoms of the third stage probably result from immune responses against persisting bacterial antigens, but evidence suggests a role for autoimmunity in some cases.

Epidemiology

Like Rocky Mountain spotted fever, Lyme disease is a zoonosis, and humans are accidental hosts. The disease is widespread in the United States **(figure 23.13),** and its incidence depends on complex ecological factors. Several species of ticks have been implicated as vectors, but the most important in the eastern United States is the black-legged (deer) tick, *Ixodes scapularis* **(figure 23.14).** In some areas of the East Coast, 80% of these ticks are infected with *Borrelia burgdorferi.* Because of their small size (1 to 2 mm before feeding; 3 to 5 mm when fully engorged with blood), these ticks

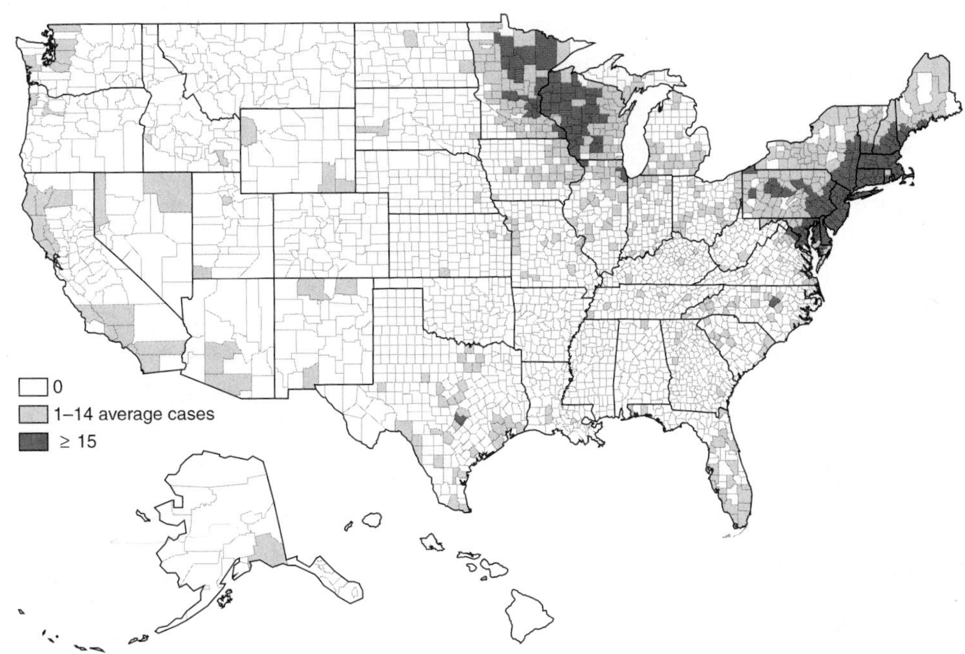

FIGURE 23.13 Number of Reported Lyme Disease Cases–2005
(Source: Centers for Disease Control and Prevention. *MMWR* 54(53):2–92, March 30, 2007.)

IN GOD WE TRUST
LIBERTY
1991
2 mm

FIGURE 23.14 The Black-Legged (Deer) Tick, *Ixodes scapularis,* Adult and Nymph This tick is the most important vector of Lyme disease in the eastern and north-central United States.

often feed and drop off their host without being detected, so that two-thirds of Lyme disease patients are unable to recall a tick bite. The ticks mature during a 2-year cycle **(figure 23.15).** A six-legged larval form emerges from the egg. After growing, it molts, shedding its outer covering to become an eight-legged form called a nymph. After another molt as the tick grows in size, the nymph becomes the sexually mature adult form. The nymph avidly seeks blood meals and is therefore mainly responsible for transmitting Lyme disease. The preferred host of *I. scapularis* is the white-footed mouse, which acquires *Borrelia burgdorferi* from an infected tick and develops a sustained bacteremia. The mouse thus becomes a source of infection for other ticks. Passage of the spirochete from adult tick to its offspring via its eggs rarely occurs. Infected ticks and mice constitute the main reservoir of *B. burgdorferi,* but deer, while not a significant reservoir, are important because they are the preferred host of the adult ticks and the site where mating occurs. Moreover, deer can quickly spread the disease over a wide area. Tick nymphs are the most active from May to September, corresponding to the peak occurrence of Lyme disease cases. Adult ticks sometimes bite humans late in the season and transmit the disease. Infectious ticks can be present in well-mowed lawns as well as in wooded areas. Expanding human populations continually intrude into the zoonotic life cycle. ■ **bacteremia, p. 395** ■ **reservoirs, p. 453**

Prevention and Treatment

General preventive measures for Lyme disease are the same as those for Rocky Mountain spotted fever. There is no available vaccine. Persons are advised to use every other means to avoid infection. Several antibiotics are effective in patients with early disease. In late disease, the response to treatment is less satisfactory, presumably because the spirochetes are not actively multiplying and antibacterial medications are usually ineffective against non-growing bacteria. Nevertheless, prolonged treatment with intravenous ampicillin or ceftriaxone has been curative in many cases.

Table 23.8 summarizes some features of Lyme disease.

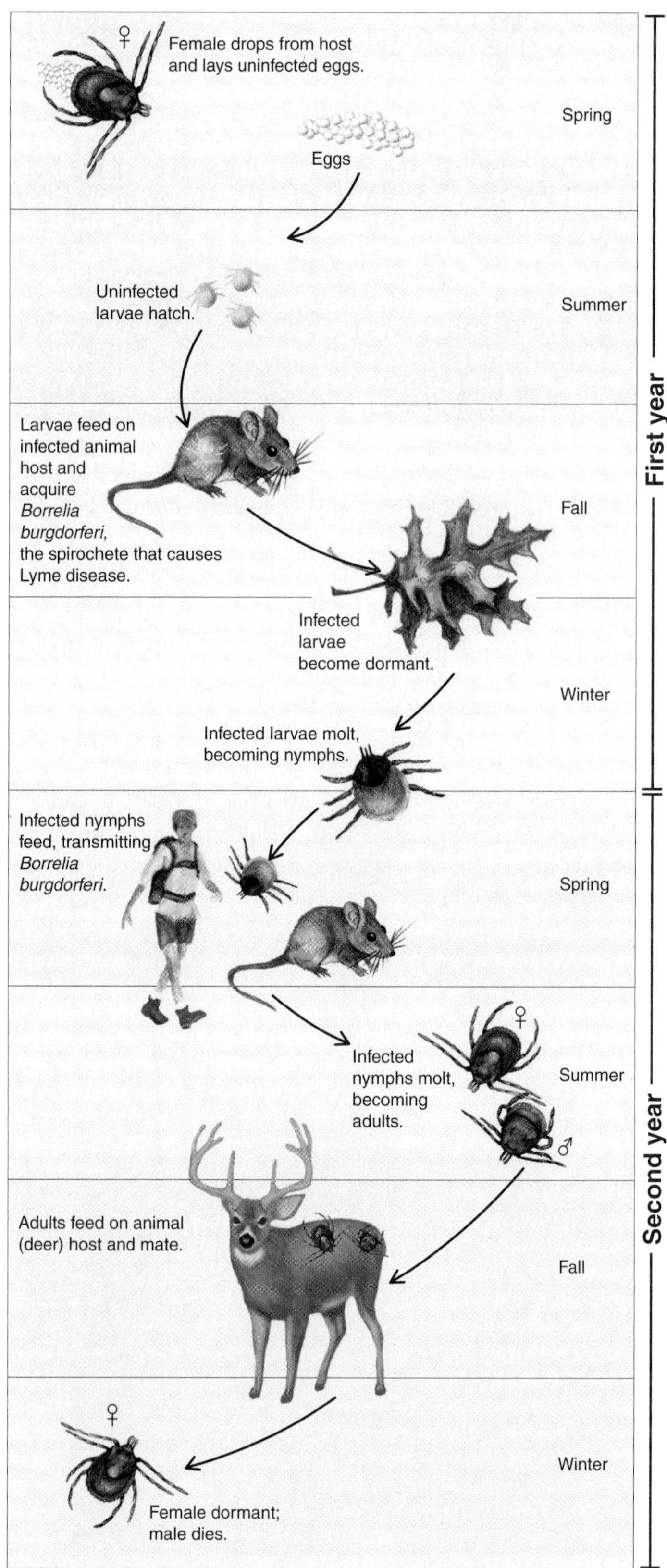
♀
Female drops from host and lays uninfected eggs.
Spring
Eggs
Uninfected larvae hatch.
Summer
Larvae feed on infected animal host and acquire *Borrelia burgdorferi,* the spirochete that causes Lyme disease.
Fall
First year
Infected larvae become dormant.
Winter
Infected larvae molt, becoming nymphs.
Infected nymphs feed, transmitting *Borrelia burgdorferi.*
Spring
Infected nymphs molt, becoming adults.
♀
♂
Summer
Adults feed on animal (deer) host and mate.
Fall
Second year
♀
Female dormant; male dies.
Winter

FIGURE 23.15 Life Cycle of the Black-Legged (Deer) Tick, *Ixodes scapularis,* the Principal Vector of *Borrelia burgdorferi,* Cause of Lyme Disease Note that the life cycle covers 2 years, during which the tick obtains three blood meals. The males die soon after mating, the females after depositing their eggs in the following spring. Variations in the life cycle occur, probably dependent on climate and food availability for the natural hosts.

TABLE 23.8 Lyme Disease

① Bite of tick infected with *Borrelia burgdorferi* introduces the bacteria into the skin.

② *B. burgdorferi* reproduce and spread radially in the skin, causing an expanding red rash which tends to clear centrally.

③ The bacteria enter the bloodstream, cause fever, acute injury to the heart and nervous system.

④ Chronic symptoms develop, such as arthritis and paralysis due to persisting bacteria and the immune response to them.

⑤ No person-to-person transmission.

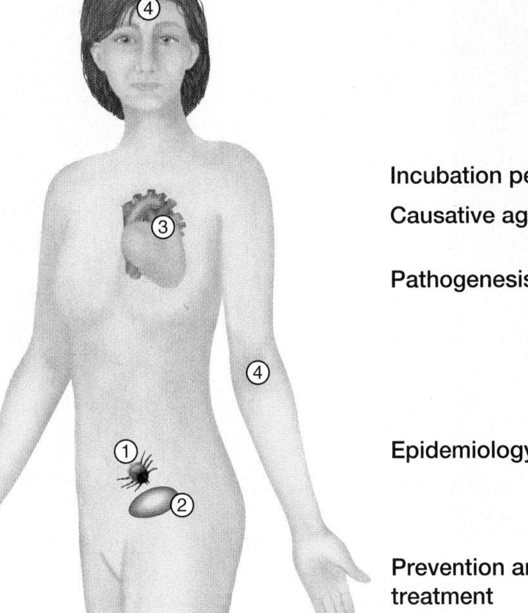

Symptoms	*Stage 1:* Enlarging, red rash at the site of the bite; fever, malaise, headache, general achiness, enlargement of lymph nodes near bite, joint pains. *Stage 2:* Acute involvement of heart and nervous system. *Stage 3:* Chronic arthritis and impairment of the nervous system.
Incubation period	Approximately 1 week
Causative agent	*Borrelia burgdorferi,* a spirochete
Pathogenesis	Spirochetes injected into the skin by an infected tick multiply and spread radially; the spirochetes enter the bloodstream and are carried throughout the body; the immune reaction to bacterial antigen causes tissue damage.
Epidemiology	Spread by the bite of ticks, *Ixodes* sp., usually found in association with animals such as white-footed mice and white-tailed deer living in wooded areas.
Prevention and treatment	Protective clothing; tick repellents. Early treatment with doxycycline and others; prolonged antibiotic therapy in chronic cases.

MICROCHECK 23.3

Folliculitis, furuncles, and carbuncles are usually caused by *Staphylococcus aureus.* Some strains of *S. aureus* produce exfoliatin and can cause scalded skin syndrome. Impetigo is a kind of pyoderma that is often caused by *Streptococcus pyogenes.* Acute glomerulonephritis is a possible sequel to streptococcal pyoderma. Rocky Mountain spotted fever is caused by *Rickettsia rickettsii* and is transmitted by ticks. Lyme disease is caused by *Borrelia burgdorferi* and is also transmitted by ticks. It is the most common vector-borne disease in the United States.

- ✓ List four extracellular products of *Staphylococcus aureus* that contribute to its virulence.
- ✓ Describe the characteristic rash of Lyme disease.
- ✓ The existence of extensive scalded skin syndrome does not indicate that *Staphylococcus* is growing in all the affected areas. Why?

23.4 Skin Diseases Caused by Viruses

Focus Points

- Name and give a distinctive characteristic of six viral skin diseases.
- Give three reasons why skin diseases controllable by vaccines have not yet been eradicated from the world.

Several childhood diseases are characterized by distinctive skin rashes called **exanthems.** The viruses that cause these rashes initially infect the upper respiratory tract but are then carried to the skin by the blood. These diseases are usually diagnosed by inspection of the rash and other clinical findings. When the disease is not typical, however, tests can be performed to identify specific antibody against the virus, and the virus can often be cultivated from skin lesions, upper respiratory secretions, or other material.

Chickenpox (Varicella)

Chickenpox is the popular name for **varicella,** a rash that was common in childhood before the introduction of varicella vaccine in 1995. The causative virus is a member of the herpesvirus family and, like others in that group, produces a latent infection that can reactivate long after recovery from the initial illness. ■ **latent infections, p. 325**

Symptoms

Most cases of chickenpox are mild, sometimes unnoticed, and recovery is usually uncomplicated. The typical case has a rash that is diagnostic. It begins as small, red spots called **macules,** little bumps called **papules,** and small blisters called **vesicles,** surrounded by a narrow zone of redness. The lesions can erupt anywhere on the body, although usually they first appear on the back of the head, then the face, mouth, main body, and arms and legs, ranging from only a few lesions to many

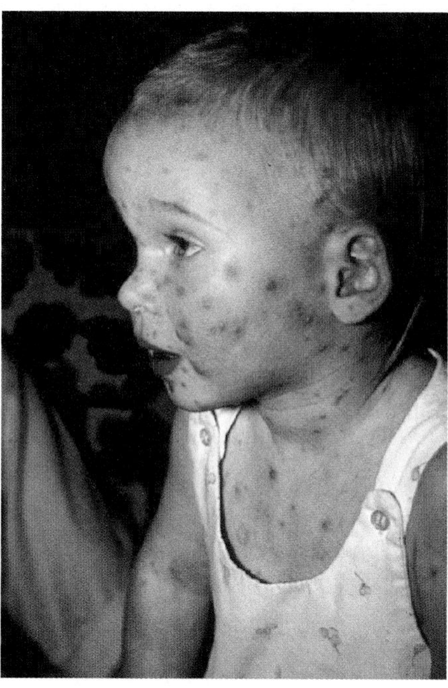

FIGURE 23.16 A Child with Chickenpox (Varicella) Characteristically, lesions in various stages of evolution—macules, papules, vesicles, and pustules—are present.

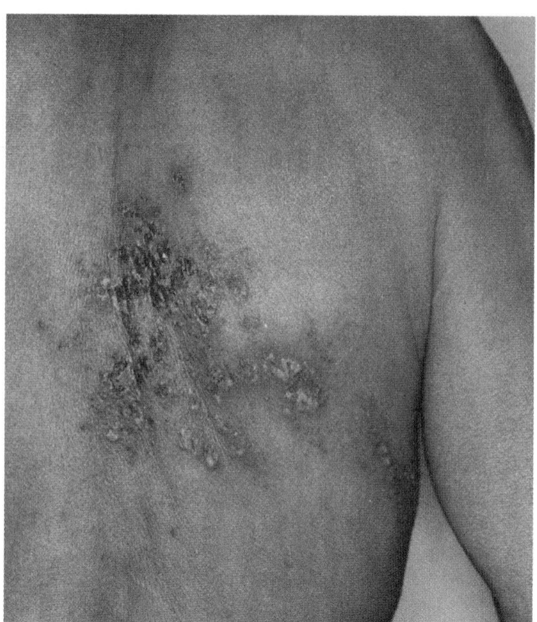

FIGURE 23.17 Shingles (Herpes Zoster) The rash mimics that of chickenpox, except that it is limited to a sensory nerve distribution on one side of the body.

hundreds. The lesions appear at different times, and within a day or so they go through a characteristic evolution from macule to papule to vesicle to **pustule,** a pus-filled blister. After the pustules break, leaking virus-laden fluid, a crust forms, and then healing takes place. At any time during the rash, lesions are at various stages of evolution **(figure 23.16).** The lesions are pruritic, meaning itchy, and scratching may lead to serious, even fatal, secondary infection by *Streptococcus pyogenes* or *Staphylococcus aureus.*

Symptoms of varicella tend to be more severe in older children and adults. In about 20% of adults, pneumonia develops, causing rapid breathing, cough, shortness of breath, and a dusky skin color. The pneumonia subsides with the rash, but respiratory symptoms often persist for weeks. Varicella is also a major threat to newborn babies if the mother develops the disease within 5 days before delivery to 2 days afterward. Mortality in these babies has been as high as 30%. Also, **congenital varicella syndrome** develops in a fraction of a percent of babies whose mothers contract varicella earlier in pregnancy. These babies are born with such defects as underdeveloped head and limbs, and cataracts. In addition, the disease is a threat to immunocompromised patients of any age. The virus can damage the lungs, heart, liver, kidneys, and brain, resulting in death in about 20% of the cases.

Reactivation of chickenpox results in a disease called **shingles, (herpes zoster).** It can occur at any age but becomes increasingly common with advancing age. It begins with pain in the area supplied by a nerve of sensation, often on the chest or abdomen but sometimes on the face or an arm or leg. After a few days to 2 weeks, a rash characteristic of chickenpox appears, but unlike chickenpox the rash is usually restricted to an area supplied by the branches of the involved sensory nerve **(figure 23.17).** The rash generally subsides within a week, but pain may persist for weeks, months, or longer. In people with AIDS or other serious immunodeficiency, instead of being confined to one area the rash often spreads to involve the entire body, as in a severe case of chickenpox.

A curious, rare affliction known as **Reye's syndrome** occasionally occurs in association with chickenpox and a number of other viral infections, usually within 2 to 12 days of the onset of the infection. The patients begin vomiting and slip into a coma. The syndrome occurs predominantly in children between 5 and 15 years old and is characterized by liver and brain damage and a death rate as high as 30%. Epidemiologic evidence suggesting that aspirin therapy increases the risk of Reye's syndrome has led physicians to use this drug sparingly in children with fever.

Causative Agent

Chickenpox is caused by the **varicella-zoster virus,** a member of the herpesvirus family. It is an enveloped, medium-sized (150 to 200 nm), double-stranded DNA virus, indistinguishable from other herpesviruses in appearance **(figure 23.18).**

Pathogenesis

The virus enters the body by the respiratory route, establishes an infection, replicates, and disseminates to the skin via the blood-stream. After the living layers of skin cells are infected, the virus spreads directly to adjacent cells, and the characteristic skin lesions appear.

Stained preparations of infected cells show **intranuclear inclusion bodies** where the virus reproduces, visible as pink

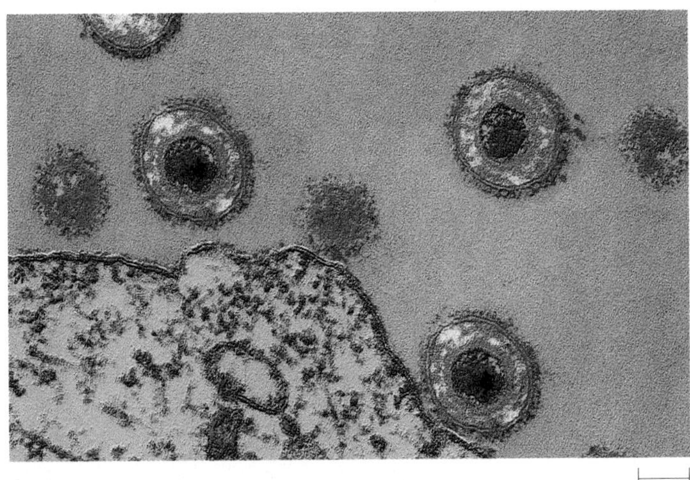

FIGURE 23.18 Electron Micrograph of Varicella-Zoster Virus, Cause of Chickenpox and Shingles

staining bodies in the nucleus. Some infected cells fuse together, forming multinucleated giant cells. The infected cells swell and ultimately lyse. The virus enters the sensory nerves, presumably when an area of skin infection advances to involve a sensory nerve ending. Conditions inside the nerve cell do not permit full expression of the viral genome; however, viral DNA is present in the ganglia (singular: ganglion) of the nerves and is fully capable of coding for mature infectious virus. Ganglia are small bulges in sensory nerves located near the spine; they contain the nuclei and cell bodies of the nerves. The mechanism of suppression of viral replication within the nerve cell is not known but is probably under the control of immune cells.

Shingles is most likely to occur when cellular immunity declines. With the decline, infectious varicella-zoster virus is presumably produced in the nucleus of the nerve cells and is carried to the skin by the normal circulation of cytoplasm within the nerve cell. With the appearance of the skin lesions, a prompt, intense secondary (memory) response of both cellular and humoral immunity ensues. A marked inflammatory reaction occurs in the ganglion with an accumulation of immune cells, and shingles quickly disappears, although sometimes leaving scars and chronic pain. ■ secondary response, p. 379

Epidemiology

The annual incidence of chickenpox in the United States, once estimated at several million, is probably less than one-tenth that number now that immunization is widespread. Reporting the disease is not required, so most cases go unreported and many are so mild that they go unnoticed.

Respiratory secretions and skin lesions are both infectious; as with many diseases transmitted by the respiratory route, most cases occur in the winter and spring months. Humans are the only reservoir, and the disease is highly contagious. The incubation period averages about 2 weeks, with a range of 10 to 21 days. Cases are infective from 1 to 2 days before the rash appears until all the lesions have crusted (usually 4 days after the onset). ■ reservoir, p. 453

The mechanism by which the varicella-zoster virus persists in the body allows it to survive indefinitely in small, isolated populations. By contrast, when a virus such as measles is introduced into an isolated community, it spreads quickly and infects most of the susceptible individuals, who either become immune or die. If susceptible victims are unavailable, the measles virus will disappear from the community. On the other hand, varicella-zoster virus will reappear from cases of shingles whenever sufficient numbers of susceptible children have been born. Shingles occurs in about 1% of elderly people.

Prevention and Treatment

A safe attenuated chickenpox vaccine has been used in the United States since 1995. The vaccine is given routinely to all healthy children in a two-dose regimen, the first dose at 12 to 15 months of age and the second at 4 to 6 years. All healthy children and adults without a history of chickenpox are also advised to receive two doses of the vaccine. Likewise, asymptomatic, HIV-infected children and adults should receive the vaccine if their immune system is still intact. In general, the vaccine should not be administered to people with immunodeficiencies. Healthy, non-immune contacts of such people, however, should be vaccinated. ■ attenuated vaccines, p. 433

By preventing chickenpox, the vaccine markedly decreases the chance of developing shingles. Another vaccine is available to prevent shingles in individuals 60 years old or older. This vaccine is composed of the same attenuated virus used in the chickenpox vaccine, but in a much higher dose. It is given in a single dose and halves the risk of developing shingles.

Increasing numbers of individuals with impaired immunity are at risk of severe disseminated varicella-zoster virus infections. These include persons with cancer, AIDS, and organ transplants and newborn babies whose mothers contracted chickenpox near the time of delivery. They can be partially protected from severe disease if they are passively immunized by injecting them with varicella zoster immune globulin (VZIG) derived from the blood of healthy individuals with a high titer of antibody to varicella-zoster virus. The antiviral medications acyclovir and famciclovir, among others, are helpful in preventing and treating varicella-zoster infections. ■ passive immunity, p. 432

The main features of chickenpox are summarized in **table 23.9.**

Measles (Rubeola)

Measles, "hard measles," and "red measles" are common names for **rubeola.** One of the great success stories of the last half of the twentieth century was the dramatic reduction in measles cases as a result of immunizing children with an attenuated vaccine against the disease. Now there is reason to hope that the disease can be entirely eliminated from the world by 2015.

Symptoms

Measles begins with fever, runny nose, cough, and swollen, red, weepy eyes. Within a few days, a fine red rash appears on the

TABLE 23.9 Chickenpox (Varicella)

① Varicella-zoster virus is inhaled; infects nose and throat.

② The virus infects nearby lymph nodes, reproduces, and seeds the bloodstream.

③ Infection of other body cells occurs, resulting in showers of virions into the bloodstream.

④ These virions cause successive crops of skin lesions, which evolve into blisters and crusts.

⑤ Immune system eliminates the infection except for some virions inside the nerve cells.

⑥ If immunity wanes with age or other reason, the virus persisting in the nerve ganglia can infect the skin, causing herpes zoster.

⑦ Transmission to others occurs from respiratory secretions and skin.

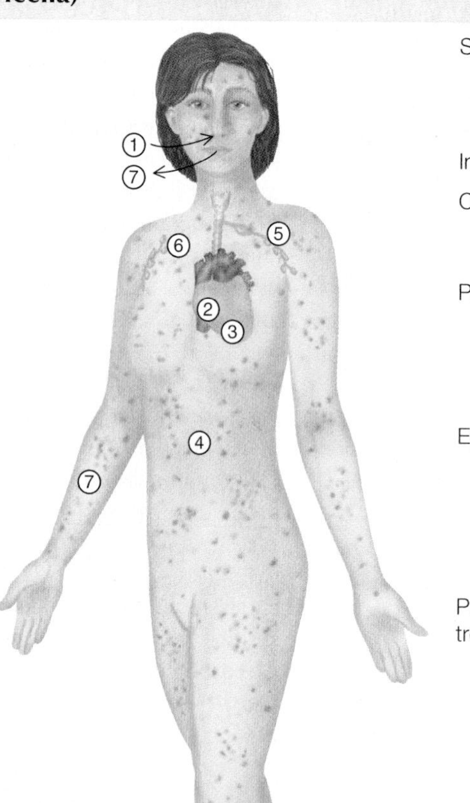

Symptoms	Itchy bumps and blisters in various stages of development, fever; latent infections can become manifest as shingles (herpes zoster) years later.
Incubation period	10 to 21 days
Causative agent	Varicella-zoster virus; enveloped double-stranded DNA virus of the herpesvirus family.
Pathogenesis	Upper respiratory virus multiplication followed by dissemination via bloodstream to the skin; cytopathic effect of virus includes the formation of giant cells.
Epidemiology	Highly infectious. Acquired by the respiratory route; humans, both individuals with chickenpox and those with shingles, the only source; dissemination is from skin lesions and respiratory secretions.
Prevention and treatment	Attenuated vaccine. Passive immunization with zoster immune globulin (ZIG) for immunocompromised individuals; acyclovir or similar antiviral medication for prevention and treatment.

forehead and spreads outward over the rest of the body **(figure 23.19).** Unless complications occur, symptoms generally disappear in about 1 week.

Unfortunately, many cases are complicated by secondary infections caused mainly by *Staphylococcus aureus, Streptococcus pneumoniae, Streptococcus pyogenes,* and *Haemophilus influenzae.* These opportunistic pathogens readily invade the body because measles damages the normal body defenses. Secondary infections most commonly cause earaches and pneumonia.

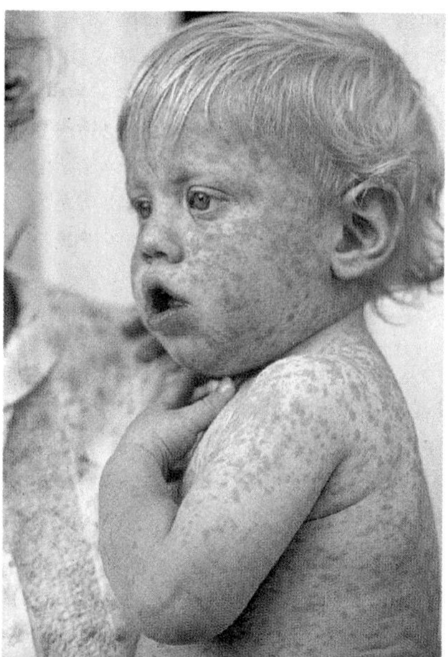

FIGURE 23.19 A Child with Measles (Rubeola) The rash is usually accompanied by fever, runny nose, and a bad cough.

In about 5% of cases, the rubeola virus itself causes pneumonia, with rapid breathing, shortness of breath, and dusky skin color from lack of adequate oxygen exchange in the lungs. Encephalitis, inflammatory disease of the brain, is another serious complication, marked by fever, headache, confusion, and seizures. This complication occurs in about one out of every 1,000 cases of measles. Permanent brain damage, with mental retardation, deafness, and epilepsy, commonly results from measles encephalitis.

Very rarely, rubeola is followed 2 to 10 years later by a disease called **subacute sclerosing panencephalitis (SSPE),** which is marked by slowly progressive degeneration of the brain, generally resulting in death within 2 years. A defective measles virus can be detected in the brains of these patients, and high levels of measles antibody are present in their blood. This is an example of a "slow virus" disease. It has all but disappeared from the United States with widespread vaccination against measles. ■ slow virus infections, p. 331

Measles that occurs during pregnancy results in an increased risk of miscarriage, premature labor, and low birth weight. Birth defects, however, are generally not seen.

Causative Agent

Measles is caused by rubeola virus, a pleomorphic, medium-sized (120 to 200 nm diameter), enveloped, single-stranded RNA virus of the paramyxovirus family. The viral envelope has two biologically active projections. One, H, is responsible for viral attachment to host cells, and the other, M, is responsible for fusion of the viral outer membrane with the host cell. The M antigen also causes adjacent infected host cells to fuse together, producing multinucleated giant cells.

Pathogenesis

Rubeola virus is acquired by the respiratory route. It replicates in the upper respiratory epithelium, spreads to lymphoid tissue, and following further replication, eventually spreads to all parts of the body. Mucous membrane involvement is responsible for an important diagnostic sign, **Koplik spots (figure 23.20),** which are usually best seen opposite the molars, located in the back part of the mouth. Koplik spots look like grains of salt lying on an oral mucosa that is red and rough, resembling red sandpaper. Damage to the respiratory mucous membranes partly explains the markedly increased susceptibility of measles patients to secondary bacterial infections, especially infection of the middle ear and lung. Involvement of the intestinal epithelium may explain the diarrhea that sometimes occurs in measles and contributes to high measles death rates in impoverished countries. In the United States, deaths from measles occur in about one to two of every 1,000 cases, mainly from pneumonia and encephalitis.

The skin rash of measles results from the effect of rubeola virus replication in skin cells and the cellular immune response against the viral antigen in the skin. The measles virus temporarily suppresses cellular immunity, causing cold sores to appear and latent tuberculosis to activate. ■ **cold sores, p. 592** ■ **tuberculosis, p. 514**

Epidemiology

Humans are the only natural host of rubeola virus. Spread is by the respiratory route. Before vaccination became widespread in the 1960s, probably less than 1% of the population escaped infection with this highly contagious virus. Continued use of measles vaccine resulted in a progressive decline in cases, so that endemic measles no longer occurs in the Western Hemisphere. Small outbreaks of the disease, however, continue to be seen as a result of introductions from other countries.

These outbreaks occur due to the presence of non-immune populations including (1) children too young to be vaccinated; (2) preschool children never vaccinated; (3) children and adults inadequately vaccinated; and (4) persons not vaccinated for religious or medical reasons. Worldwide, about 750,000 children still die from measles, which ranks among the leading causes of death and disability among the impoverished, where the mortality rate may reach 15% and secondary infections may reach 85%. Measles vaccination was a high priority following the tsunami disaster of 2004.

Prevention and Treatment

Measles can be prevented by injecting an attenuated rubeola virus vaccine. At present, less than 100 cases per year are generally reported. In 1980 the worldwide incidence of rubeola was estimated to be 100 million with 5.8 million deaths. Globally, vaccination programs have lowered the number of cases dramatically since then.

The measles vaccine is usually given together with mumps, rubella and varicella vaccines (MMRV). The first injection of vaccine is given near an infant's first birthday. Since 1989, a second injection of vaccine is given at entry into elementary school. The two-dose regimen has resulted in at least 99% of the recipients becoming immune. In an epidemic, vaccine is given to babies as young as 6 months, who are then reimmunized before their second birthday. Students entering high school or college are advised to get a second dose of vaccine if they have not received one earlier. Those at special risk of acquiring rubeola, such as medical personnel, should be immunized regardless of age, unless they definitely have had measles or have laboratory proof of immunity.

No antiviral treatment exists for rubeola at present. Some features of rubeola are summarized in **table 23.10.**

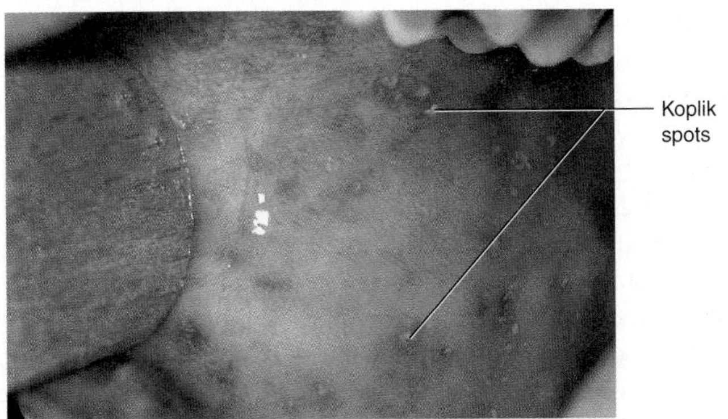

FIGURE 23.20 Koplik Spots, Characteristic of Measles (Rubeola), Are Usually Transitory They resemble grains of salt on a red base.

German Measles (Rubella)

German measles and three-day measles are common names for **rubella.** The term *German measles* arose because the disease was first described in Germany. In contrast to varicella and rubeola, rubella is typically a mild, often unrecognized disease that is difficult to diagnose. Nevertheless, infection of pregnant women can have tragic consequences.

Symptoms

Characteristic symptoms of German measles are slight fever, mild cold symptoms, and enlarged lymph nodes behind the ears and on the back of the neck. After about a day, a faint rash consisting of innumerable pink spots appears over the face, chest, and abdomen

TABLE 23.10 Measles (Rubeola)

① Airborne rubeola virus infects eyes and upper respiratory tract, then the lymph nodes in the region.

② Virus enters the bloodstream and is carried to all parts of the body including the brain, lungs, and skin.

③ Skin cells infected with the rubeola virus are attacked by immune T cells, causing a generalized rash.

④ Virus replicating in the lungs can cause pneumonia; the brain can also be infected.

⑤ In rare cases, virus persisting in the brain causes subacute sclerosing panencephalitis, months or years after the acute infection.

⑥ Secondary infection of the ears and lungs is common.

⑦ Transmission is by respiratory secretions.

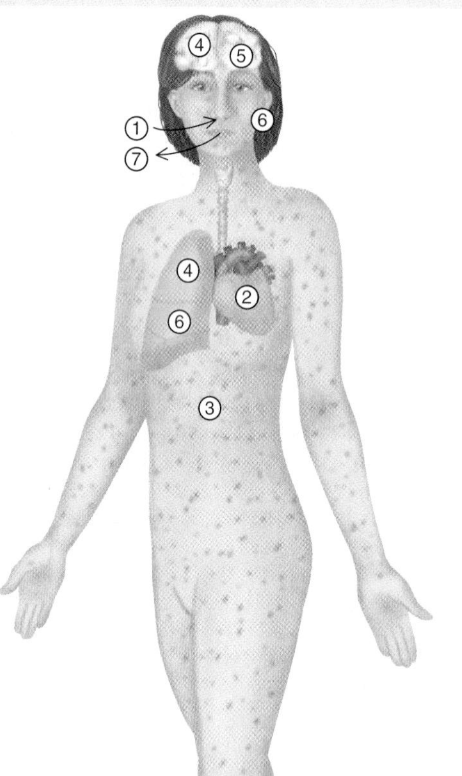

Symptoms	Rash, fever, weepy eyes, cough, and nasal discharge
Incubation period	10 to 12 days
Causative agent	Rubeola virus, a single-stranded RNA virus of the paramyxovirus family
Pathogenesis	Virus multiplies in respiratory tract; spreads to lymphoid tissue, then to all parts of body, notably skin, lungs, and brain; damage to respiratory tract epithelium leads to secondary infection of ears and lungs.
Epidemiology	Acquired by respiratory route; highly contagious; humans only source.
Prevention and treatment	Attenuated virus vaccine after age 12 months; second dose upon entering elementary school or at adolescence. No antiviral treatment available at present.

(figure 23.21). Unlike rubeola, there are no diagnostic mouth lesions. Adults commonly develop painful joints, with pain generally lasting 3 weeks or less. Other symptoms generally last only a few days. The significance of rubella, however, lies not with these symptoms but rather with rubella's threat to the fetuses of pregnant women.

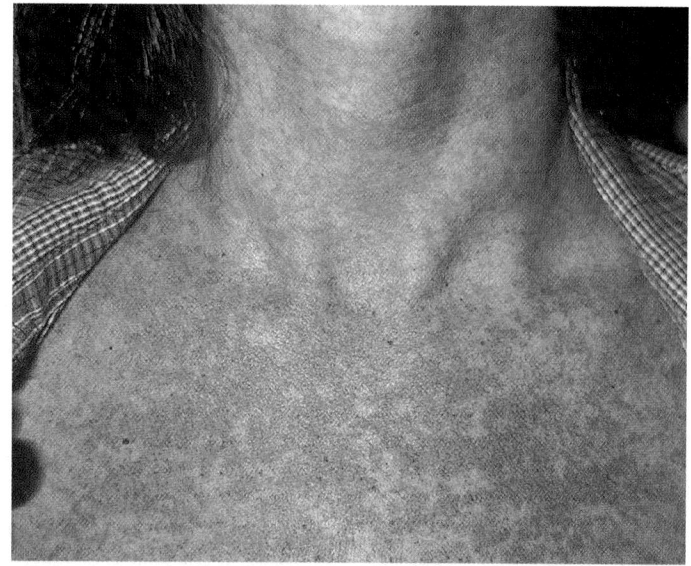

FIGURE 23.21 Adult with German Measles (Rubella) Symptoms are often very mild, but the effects on a fetus can be devastating.

Causative Agent

German measles is caused by the rubella virus—a member of the togavirus family. It is a small (about 60 nm in diameter), enveloped, single-stranded RNA virus that can readily be cultivated in cell cultures. Surface glycoproteins give the virus *in vitro* hemagglutinating ability, which is inhibited by specific antibody, allowing serological identification of the virus.

Pathogenesis

The rubella virus enters the body via the respiratory route. It multiplies in the nasopharynx and enters the bloodstream, causing a sustained viremia (meaning viruses circulating in the bloodstream). The blood transports the virus to various body tissues, including the skin and joints. Humoral and cellular immunity develop against the virus, and the resulting antibody-antigen complexes probably account for the rash and joint symptoms. ■ **viremia, p. 395**

In pregnant women, the placenta becomes infected during the period of viremia. Early in pregnancy, the virus readily crosses the placenta and infects the fetus, but as the pregnancy progresses, fetal infection becomes increasingly less likely. Virtually all types of fetal cells are susceptible to infection; some cells are killed, whereas others develop a persistent infection in which cell division is impaired and chromosomes are damaged. The result is a characteristic pattern of fetal abnormalities that is referred to as the **congenital rubella syndrome.** The abnormalities include cataracts and other abnormalities of the eyes, brain damage, deafness, heart defects, and low birth

CASE PRESENTATION

The patient was a 20-year-old asymptomatic man who was immunized against measles as a requirement for starting college. He had received his first dose of measles vaccine at approximately 1 year of age. Past medical history revealed that he was a hemophiliac and had contracted the human immunodeficiency virus from clotting factor (a blood product given to control bleeding) contaminated with the virus. Laboratory tests showed that he had a very low $CD4^+$ lymphocyte count, indicating a severely damaged immune system.

About a month after his precollege immunization, he developed pneumocystosis, a lung infection characteristic of AIDS, was hospitalized, had a good response to treatment, and was discharged. Ten months later, he was again hospitalized for symptoms of a severe lung infection. He had no rash. Multiple laboratory tests to determine the cause of his infection were negative. Finally, a lung biopsy was performed and revealed "giant cells"—very large cells with multiple nuclei. Cytoplasmic and intranuclear inclusion bodies were also present. This picture was highly suggestive of measles pneumonia, and measles virus subsequently was recovered from cell cultures of the biopsy material. Other studies showed it to be the vaccine strain of measles virus. The patient received intravenous gamma globulin and an antiviral medication—ribavirin—and improved. Subsequently, however, his condition deteriorated, and he died of presumed complications of AIDS.

1. Is measles immunization a good idea for people with immunodeficiency?
2. Is it surprising that the vaccine virus was still present in this patient 11 months after vaccination? Explain.
3. Despite the severe infection, there was no rash. Why?

Discussion

1. Measles is often disastrous for persons with AIDS or other immunodeficiencies. They should be immunized as soon as possible in their illness, before the immune system becomes so weakened it cannot respond effectively to the vaccine. Also, as this and other cases have shown, the vaccine virus can itself be pathogenic when immunodeficiency is severe. With the worldwide effort to eliminate measles, the risk of exposure to the wild-type measles virus, as opposed to the laboratory-derived vaccine virus, is declining, but outbreaks in colleges and other institutions still occur. A severely immunodeficient individual can be passively immunized against measles with immune globulin if exposure to the wild-type virus occurs.
2. Measles is often given as an example of a persistent viral infection, meaning that following infection the virus can persist in the body for months or years in a slowly replicating form. It has been suggested but not proven that this explains the lifelong immunity conferred by measles infection in normal people. Rarely in presumably normal individuals and, more commonly, in malnourished or immunodeficient individuals, persistent infection leads to damage to the brain, lung, liver, and possibly, the intestine. Subacute sclerosing panencephalitis can follow measles vaccination, but at a much lower rate than after wild virus infection.
3. Following acute infection, the measles virus floods the bloodstream and is carried to various tissues of the body, including the skin. The rash that characterizes measles is caused by T lymphocytes attacking measles virus antigen lodged in the skin capillaries. In the absence of functional T lymphocytes, the rash does not occur.

weight despite normal gestation. Babies may be stillborn. Those that live continue to excrete rubella virus in throat secretions and urine for many months. The likelihood of the syndrome varies according to the age of the fetus when infection occurs. Infections occurring during the first 6 weeks of pregnancy result in almost 100% of the fetuses having a detectable injury, most commonly minor deafness. Even infants who are apparently normal, however, excrete rubella virus for extended periods and thus can infect others.

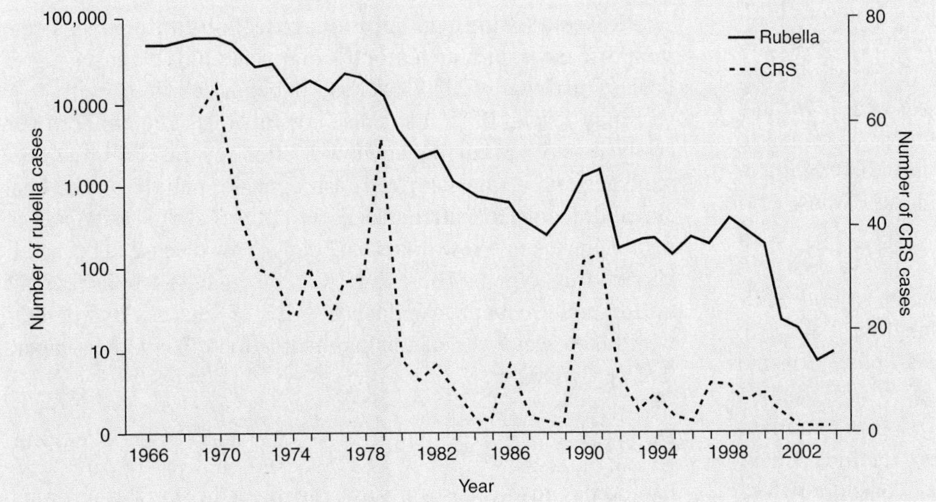

FIGURE 23.22 Reported Cases of German Measles (Rubella) and Congenital Rubella Syndrome (CRS), United States, 1966–2004
(Source: Centers for Disease Control and Prevention. *MMWR* 2005: 54(11): 27.)

Epidemiology

Humans are the only natural host for rubella virus. The disease is highly contagious although less so than rubeola; it is estimated that in the prevaccine era, 10% to 15% of people reached adulthood without being infected. Complicating the epidemiology of rubella is the fact that over 40% of infected individuals fail to develop symptoms, but can spread the virus. People who develop typical rubella can be infectious for as long as 7 days before the rash appears until 7 days afterward. Before widespread use of the vaccine began in 1969, periodic major epidemics arose. One epidemic in 1964 resulted in about 30,000 cases of congenital rubella syndrome.

Prevention and Treatment

Prevention of German measles depends on subcutaneous injection of an attenuated rubella virus administered to babies at 12 to 16 months of age with a second dose at age 4 to 6 years. The vaccine produces long-lasting immunity in about 95% of recipients. The vaccine is not given to pregnant women for fear it might result in congenital defects. As an added precaution, women are advised not to become pregnant for 28 days after receiving the vaccine.

Use of the vaccine has markedly reduced the incidence of rubella in the United States to generally less than 50 cases per year (**figure 23.22**). There were less than 1,000 confirmed cases in the entire Western Hemisphere in 2004.

TABLE 23.11 German Measles (Rubella)

① Airborne rubella virus infects nose and throat.

② Virus taken up by lymph nodes in the region.

③ Rubella virus multiples and enters the bloodstream.

④ Circulating virus reacts with antibodies, resulting in antibody-antigen complexes.

⑤ Antibody-antigen complexes lodge in the skin, causing a rash, and in the joints, causing pain.

⑥ In women during pregnancy, rubella virus crosses the placenta, infecting the fetus, resulting in congenital rubella syndrome.

⑦ Transmission to others is by respiratory secretions.

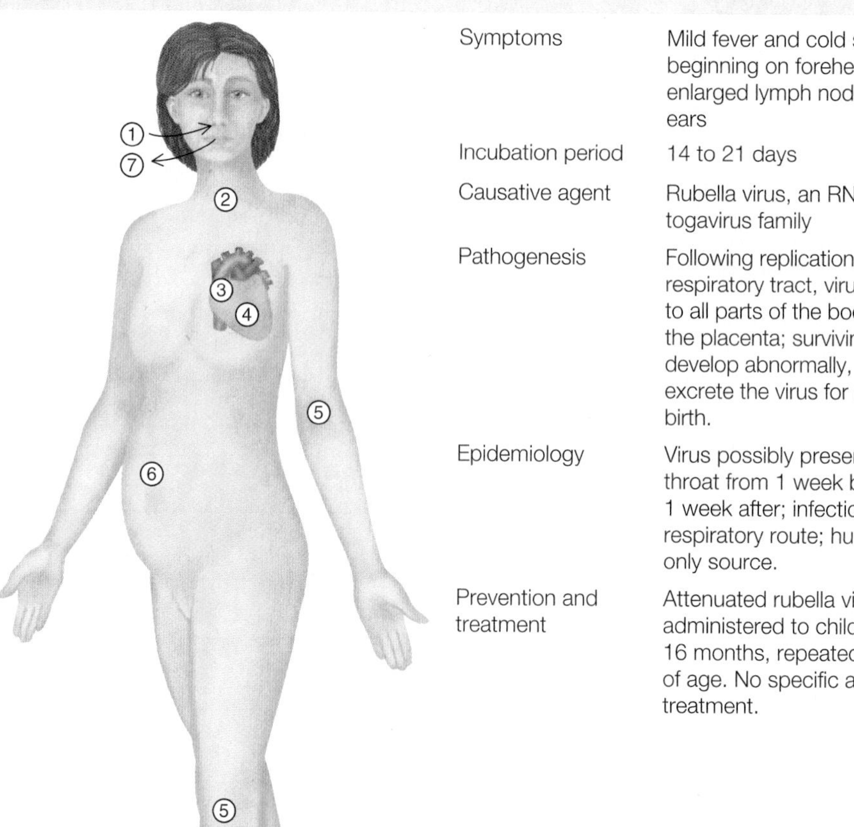

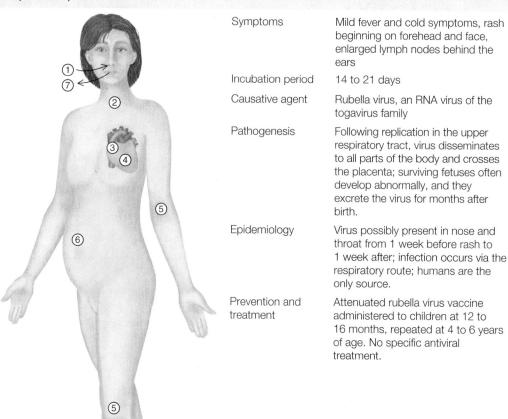

Symptoms	Mild fever and cold symptoms, rash beginning on forehead and face, enlarged lymph nodes behind the ears
Incubation period	14 to 21 days
Causative agent	Rubella virus, an RNA virus of the togavirus family
Pathogenesis	Following replication in the upper respiratory tract, virus disseminates to all parts of the body and crosses the placenta; surviving fetuses often develop abnormally, and they excrete the virus for months after birth.
Epidemiology	Virus possibly present in nose and throat from 1 week before rash to 1 week after; infection occurs via the respiratory route; humans are the only source.
Prevention and treatment	Attenuated rubella virus vaccine administered to children at 12 to 16 months, repeated at 4 to 6 years of age. No specific antiviral treatment.

No specific antiviral therapy is available. Some features of German measles are summarized in **table 23.11.**

Other Viral Rashes of Childhood

The kinds of viruses that can cause childhood rashes probably number in the hundreds. One group alone, the enteroviruses, has about 50 members that have been associated with skin lesions. In the early 1900s the causes of the common childhood rashes were largely unknown, and it was the practice to number them 1 to 6 as follows: (1) rubeola, (2) scarlet fever, (3) rubella, (4) Duke's disease—a mild disease with fever and bright red generalized rash, now thought to have been due to an enterovirus, (5) erythema infectiosum, and (6) exanthem subitum. The causes of two common childhood rashes, erythema infectiosum and exanthem subitum, have only been established in recent years.

Fifth disease (erythema infectiosum) occurs in both children and young adults. The illness begins with fever, malaise, and head and muscle aches. A diffuse redness appears on the cheeks, giving the appearance of the face as if it were slapped **(figure 23.23a).** The rash commonly spreads in a lacy pattern to involve other parts of the body, especially the extremities (figure 23.23b). The rash may come and go for 2 weeks or more before recovery. Joint pains are a prominent feature of some adult infections. The disease is caused by parvovirus B-19, a small (18 to 28 nm), non-enveloped, single-stranded DNA virus of the parvovirus family. The virus preferentially infects certain bone marrow cells and is a major threat to persons with sickle cell and other anemias because the infected marrow sometimes stops producing blood cells, a condition known as **aplastic crisis.** Also, about 10% of women infected with the virus during pregnancy suffer spontaneous abortion.

Roseola (exanthem subitum, roseola infantum), is a common disease in infants 6 months to 3 years old. It causes a great deal of parental anxiety because it begins abruptly with fever that may reach 105°F and cause convulsions. The children generally do not appear ill, however. After several days, the fever vanishes and a transitory red rash appears, mainly on the chest and abdomen. The patient has no symptoms at this point, and the rash vanishes in a few hours to 2 days. This disease is caused by herpesvirus, type 6. There is no vaccine against the disease, and no treatment except to reduce the risk of seizures by sponging with lukewarm water and using medication to keep the temperature below 102°F.

Warts

Papillomaviruses, cause of warts, can infect the skin through minor abrasions. **Warts** are small tumors called papillomas that consist

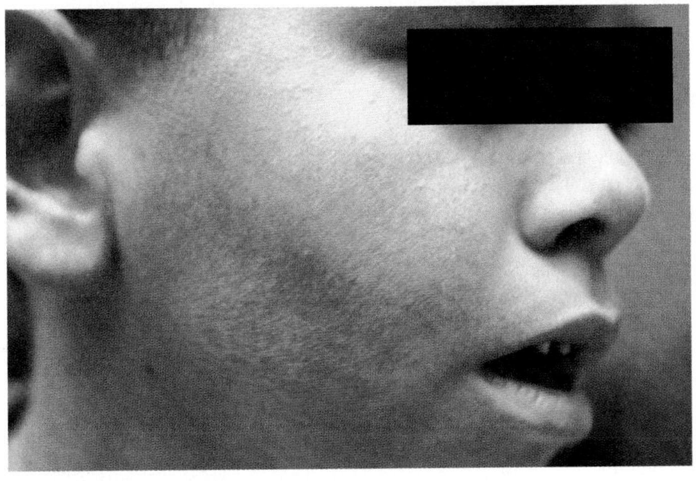

(a)

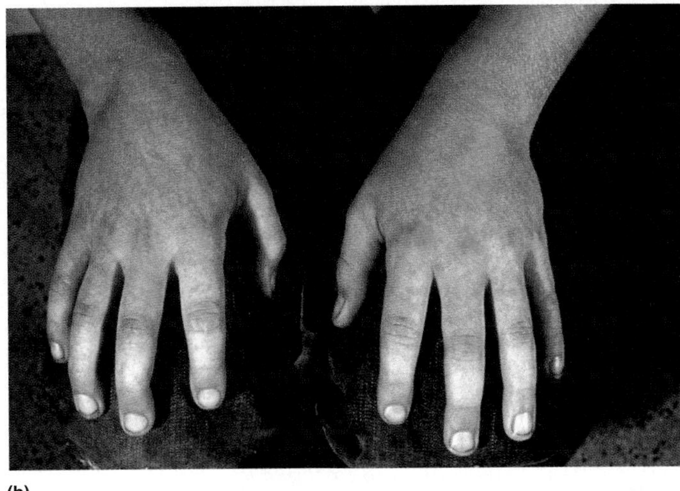

(b)

FIGURE 23.23 Erythema Infectiosum (Fifth Disease) **(a)** "Slapped cheek" appearance of the rash on the face. **(b)** Appearance of the rash on the extremities.
(Source: CDC Public Health Image Library.)

of multiple nipplelike protrusions of tissue covered by skin or mucous membrane. Warts rarely become cancers, although some sexually transmitted papillomaviruses are strongly associated with cervical cancer. About 50% of the time, warts on the skin disappear within 2 years without any treatment. Papillomaviruses **(figure 23.24)** belong to the papovavirus family. They are small (about 50 nm diameter), non-enveloped, double-stranded DNA viruses. More than 50 different papillomaviruses are known to infect humans. Warts of other animals are generally not infectious for humans.

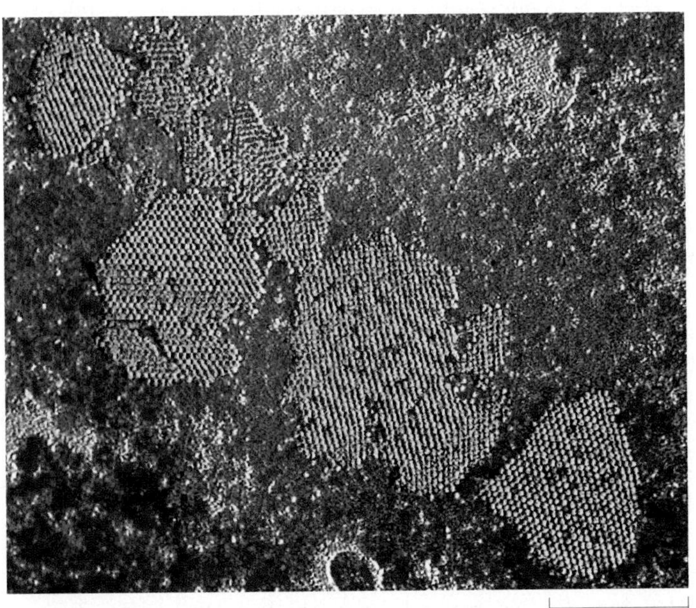

1 µm

FIGURE 23.24 Wart Virus The virions appear yellow in this color-enhanced transmission electron micrograph.

Papillomaviruses have been very difficult to study because they grow poorly in cell cultures or experimental animals. The viruses that cause warts can survive on inanimate objects such as wrestling mats, towels, and shower floors, and infection can be acquired from such contaminated objects. The virus infects the deeper cells of the epidermis and reproduces in the nuclei. Some of the infected cells grow abnormally, forming the wart. The incubation period ranges from 2 to 18 months. Infectious virus is present in the wart and can contaminate fingers or objects that pick or rub the lesions. Like other tumors, warts can only be treated effectively by killing or removing all of the abnormal cells. This can usually be accomplished by freezing the wart with liquid nitrogen, by cauterization, meaning burning the tissue usually with an electrically heated needle, or by surgical removal. Virus generally remains in the adjacent normal-appearing skin, however, and may cause additional warts. Warts that grow on the soles of the feet are called **plantar warts** (often mistakenly called planter's warts; *plantar* is a word meaning "referring to the sole of the foot"). These warts are very difficult to treat because the pressure of standing on them causes them to grow wide and deep. ■ **cervical cancer, p. 639**

MICROCHECK 23.4

Chickenpox, measles, and rubella can be controlled by vaccines. Viral diseases that may only inconvenience a pregnant woman can be disastrous to her fetus. A viral infection acquired in childhood can remain latent for years only to reactivate in a different form. One group of viruses causes benign skin tumors.

- ✓ What important diagnostic sign is often present in the mouth of measles (rubeola) victims?
- ✓ What is the epidemiological significance of shingles?
- ✓ Why is it a good idea to immunize little boys against rubella?

PERSPECTIVE 23.1

The Ghost of Smallpox, An Evil Shade

Historically, smallpox epidemics have been devastating to the Americas. In the 1500s smallpox virus introduced into Central and South America caused horrendous loss of life and may have contributed to the downfall of the Inca and Aztec nations. An epidemic that swept the Massachusetts coast in the 1600s killed so many Native Americans that in some communities there were not enough survivors to bury the dead. Even in the latter 1700s, during the Revolutionary War, a smallpox epidemic raged through the American colonies. General George Washington suspected that the virus had been deliberately introduced by the British. So many of his men were ill after his defeat at Quebec in 1777 that he ordered the mass variolation of remaining troops. ■ **smallpox, pp. 431, 460** ■ **variolation, p. 431**

Why does the ghost of smallpox concern us now, when the last case, acquired through a laboratory accident, occurred in 1978? The answer is that the smallpox virus still exists, locked in high-security laboratories in the United States and the Russian Federation, and perhaps held in secret locations by countries or individuals that could use it to harm others. A number of factors need to be considered in choosing the smallpox virus as an agent of bioterror.

Factors That Might Encourage Its Use:

- It spreads easily from person to person, mainly through close contact with respiratory secretions, but also by airborne virus from the respiratory tract, skin lesions, and contaminated bedding or other objects.
- It can be highly lethal, with mortality rates generally above 25%. After the virus establishes infection of the respiratory system, it enters the lymphatics and bloodstream, finally causing lesions of the skin and throughout the body.
- The virus is relatively stable, probably remaining infective for hours in the air of a building; viable smallpox has been demonstrated in dried crusts from skin lesions after storage for 10 years at room temperature in ordinary envelopes.
- Large numbers of people are highly susceptible to the virus. Routine vaccination against smallpox was discontinued in the United States several decades ago.
- The relatively large genome of the smallpox virus probably permits genetic modifications that could enhance its virulence.

Factors That Discourage Its Use:

- Propagating the smallpox virus is dangerous and requires advanced knowledge and laboratory facilities.
- A proven highly effective vaccine (vaccinia virus) is available. A protective antibody response occurs rapidly and prevents fatalities for 10 years or more. Smallpox is prevented even when the vaccine is administered up to four days after exposure to the virus.
- Infected persons do not spread the disease during the long incubation period, generally 12 to 14 days; they only become infectious with the onset of fever.
- The disease can usually be diagnosed rapidly, by the characteristic appearance of skin lesions that predominate on the face and hands, and by laboratory examination of material from skin lesions.
- There is already widespread experience on how to watch for and contain the disease.

In a simulated attack on an American city in the summer of 2001, only 24 primary cases of smallpox increased in 2 months to 3 million, with 1 million deaths. This study probably greatly exaggerated the risk. Nevertheless, even though unlikely, the potential danger from a smallpox introduction has caused the United States to begin preparations for this possibility, including markedly expanding its stockpile of smallpox vaccine. More information on smallpox is available at the Online Learning Center, www.mhhe.com/nester6.

23.5 Skin Diseases Caused by Fungi

Focus Points

- Explain why dermatophytes usually invade only nails, hair, and outermost skin cells.
- Describe the role of skin moisture in the pathogenesis of superficial cutaneous mycoses.

Diseases caused by fungi are called mycoses. Earlier in this chapter, we mentioned the role of normal microbiota yeast of the genus *Malassezia* in causing mild skin diseases. Other fungi are responsible for more serious infections of the skin, although even in these cases the condition of the host's defenses against infection is often crucial. The yeast *Candida albicans* **(figure 23.25)** may live harmlessly among the normal flora of the skin, but in some people it invades the deep layers of the skin and subcutaneous tissues. In many people with candidal skin infections, no precise cause for the invasion can be determined. Certain molds also cause cutaneous mycoses, but they are not as likely as *C. albicans* to invade the deep skin layers.

Superficial Cutaneous Mycoses

Certain species of molds can invade hair, nails, and the keratinized portion of the skin. The resulting mycoses have colorful names such as jock itch, athlete's foot, and ringworm, and more traditional latinized names that describe their location: tinea capitis—scalp, tinea barbae—beard, tinea axillaris—armpit, tinea corporis—body, tinea cruris—groin, and tinea pedis—feet, to list a few. Tinea just means "worm," which probably reflects early erroneous ideas about the cause.

Symptoms

Most people colonized by these molds have no symptoms at all. Others complain of itching, a bad odor, or a rash. In ringworm, a rash occurs at the site of the infection and consists of a scaly area surrounded by redness at the outer margin, producing irregular rings or a lacy pattern on the skin. On the scalp, patchy areas of hair loss can occur, with a fine stubble of short hair left behind. Involved nails become thickened and brittle and may separate from the nailbed. Sometimes, a rash consisting of fine papules and vesicles develops distant from the infected area. This rash is referred to as a dermatophytid, or "id" reaction, a reflection of allergy to products of the infecting fungus.

Causative Agents

The skin-invading molds belong mainly to the genera *Epidermophyton, Microsporum,* and *Trichophyton* and are collectively termed **dermatophytes (figure 23.26).** They can be cultivated on media especially designed for molds and are identified by their colonial and microscopic appearance, their nutritional requirements, and biochemical tests.

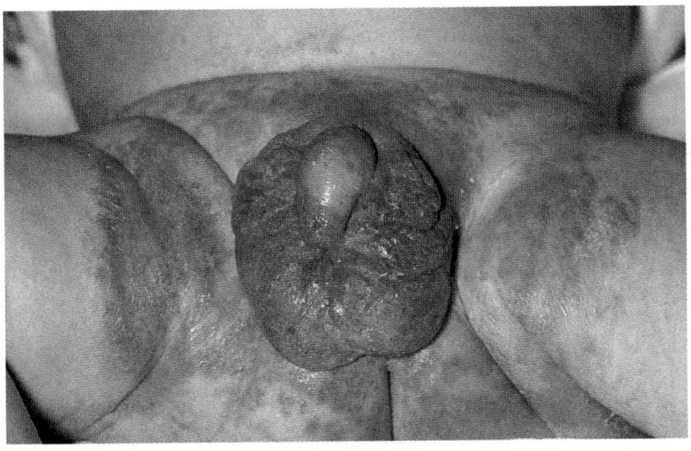
(a)

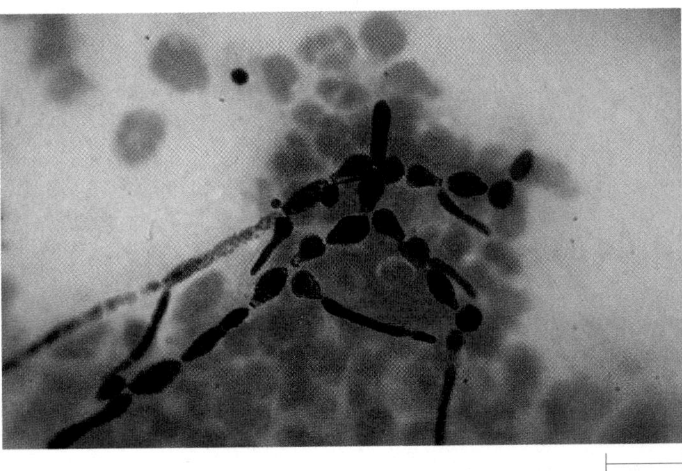
(b) 10 µm

FIGURE 23.25 ***Candida albicans*** **(a)** Causing a diaper rash; **(b)** Gram stain of pus showing *C. albicans* yeast forms and filamentous forms called pseudohyphae.

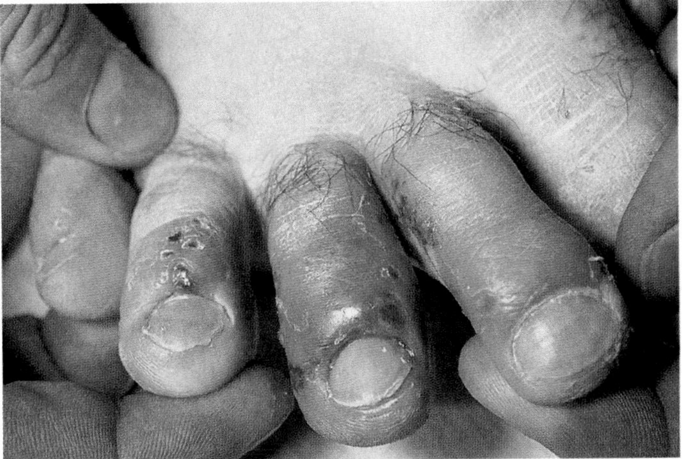
(a)

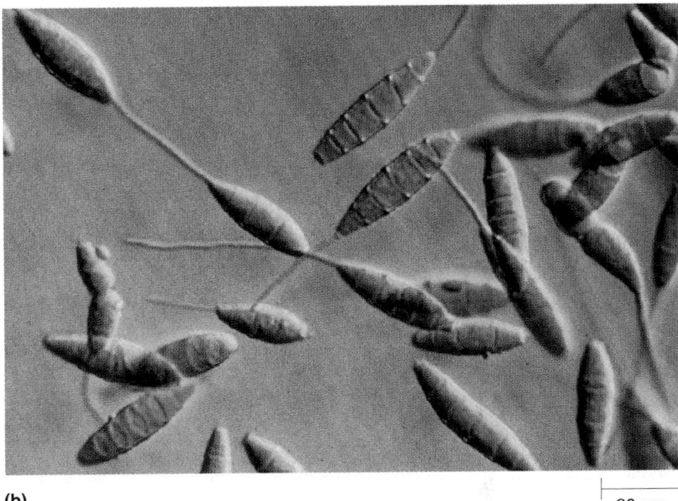
(b) 20 µm

FIGURE 23.26 Dermatophytosis (a) Tinea pedis, usually caused by species of *Trichophyton*. **(b)** Large boat-shaped spores of *Microsporum gypseum,* a cause of scalp ringworm in children.

Pathogenesis

The normal skin is generally resistant to invasion by dermatophytes. Some species, however, are relatively virulent and can even cause epidemic disease, especially in children. In conditions of excessive moisture, dermatophytes can invade keratinized structures, including the epidermis down to the level of the keratin-producing cells. A keratinase enables them to dissolve keratin and use it as a nutrient. Hair is invaded at the level of the hair follicle because the follicle is relatively moist. Fungal products diffuse into the dermis and provoke an immune reaction, which probably explains why adults tend to be more resistant to infection than children. It also explains why some people develop the allergic "id" reactions.

Epidemiology

As mentioned, age, virulence of the infecting strain of mold, and excessive moisture are important factors in causing infections. Common causes of excessive moisture include obesity causing folds of skin to lie together, tight clothing, and plastic or rubber footware. Potentially pathogenic molds may be present in soil and on pets such as young cats and dogs.

Prevention and Treatment

Attention to cleanliness and maintenance of normal dryness of the skin and nails effectively prevent most dermatophyte infections. Powders, open shoes, changing of socks, and application of alcohol after bathing may help prevent toenail infections. Numerous prescription and over-the-counter medications are promoted for treating dermatophytoses, and most are effective for treating superficial skin infections. Nail infections are often very difficult to cure, requiring taking medication by mouth for months, and sometimes surgical removal of the nail.

MICROCHECK 23.5

The causes of skin mycoses commonly colonize skin without causing symptoms. The best protection against fungal skin infections is to maintain normal skin dryness.

- ✓ What is a mycosis?
- ✓ What kinds of structures are invaded by dermatophytes?
- ✓ Would you be surprised if a child contracted ringworm from a pet that showed no signs of the disease? Why, or why not?

FUTURE CHALLENGES

The Ecology of Lyme Disease

Lyme disease is often referred to as one of the emerging diseases. Unrecognized in the United States before 1975, it is now the most commonly reported vector-borne disease. Because of the seeming explosion in the numbers of Lyme disease cases, and its apparent extension to new geographical areas, the ecology of Lyme disease is under intense study. In the northeastern United States, large increases in white-footed mouse populations occur in oak forests during years in which there is a heavy acorn crop, with a corresponding increase in *Ixodes scapularis* ticks. Both deer and mice feed on the acorns and subsequently spread the disease to adjacent areas. Variations in weather conditions, and their effect on food supply for these animals, might therefore be an important ecological factor, although it is not clear that weather cycles completely explain the emerging nature of the disease. The presence of animals other than white-footed mice for the ticks to feed on is another factor. Alternative tick hosts usually do not have a sustained *Borrelia burgdorferi* bacteremia following infection from a tick, and the blood of a common lizard host along the West Coast even kills the spirochetes. The role of snakes, foxes, and birds of prey that control mouse populations and that of birds, spiders, and wasps that feed on ticks are also under study. The challenge is to define more completely the ecology of Lyme and other tick-borne diseases in order to predict their emergence and find new ways for their prevention.

SUMMARY

23.1 Anatomy and Physiology (figure 23.1)

The skin is a large complex organ with many functions, including temperature regulation and vitamin D synthesis, and aiding cellular immunity. The skin repels potential pathogens by shedding and being dry, acidic, and toxic.

23.2 Normal Microbiota of the Skin

Skin is inhabited by large numbers of bacteria of little virulence that help prevent colonization by more dangerous species (table 23.1).

Diphtheroids

Diphtheroids are Gram-positive, pleomorphic, rod-shaped bacteria that play a role in acne and body odor. Fatty acids, produced from their metabolism of the oily secretion of sebaceous glands, keep the skin acidic.

Staphylococci

Staphylococci are Gram-positive cocci arranged in clusters. Universally present, they help prevent colonization by potential pathogens and maintain the balance among flora of the skin. The principal species, *Staphylococcus epidermidis,* can sometimes cause disease.

Fungi

Malassezia sp. are yeasts found on the skin. Usually harmless, they can cause tinea versicolor, probably some cases of dandruff, and serious skin disease in AIDS patients (figure 23.2).

23.3 Bacterial Skin Diseases

Hair Follicle Infections (figure 23.3)

Folliculitis, boils and **carbuncles** are caused by *Staphylococcus aureus* (table 23.3), which is coagulase-positive and often resists penicillin and other antibiotics. A carbuncle is more serious because the infection is more likely to be carried to the heart, brain, or bones.

Scalded Skin Syndrome (figure 23.4, table 23.4)

Staphylococcal scalded skin syndrome results from exfoliatin produced by certain strains of *Staphylococcus aureus.*

Streptococcal Impetigo (figure 23.5, table 23.6)

Impetigo is a superficial skin infection caused by *Streptococcus pyogenes* and *Staphylococcus aureus.* **Acute glomerulonephritis,** a kidney disease, caused by an antibody-antigen reaction, is an uncommon complication of *S. pyogenes* infections (figure 23.7).

Rocky Mountain Spotted Fever (figures 23.8, 23.10, table 23.7)

Rocky Mountain spotted fever, caused by the obligate intracellular bacterium *Rickettsia rickettsii,* is an often fatal disease transmitted to humans by the bite of an infected tick (figures 23.9, 23.11).

Lyme Disease (figures 23.12, 23.14, table 23.8)

Lyme disease can imitate many other diseases. It is caused by a spirochete, *Borrelia burgdorferi,* transmitted to humans by certain ticks. A target-shaped rash, present in most victims, is the hallmark of the disease (figures 23.12, 23.13, 23.15, 23.16).

23.4 Skin Diseases Caused by Viruses

Chickenpox (Varicella) (figure 23.17, table 23.9)

Chickenpox, once a common disease of childhood is caused by the varicella-zoster virus (figure 23.19). **Shingles,** or **herpes zoster** (figure 23.18), can occur months or years after chickenpox. It is a reactivation of the varicella-zoster virus infection in the distribution of a sensory nerve. Shingles cases can be sources of chickenpox epidemics.

Measles (Rubeola) (figures 23.20, 23.21, table 23.10)

Measles (rubeola) is a potentially dangerous viral disease that can lead to serious secondary bacterial infections, and fatal lung or brain damage. Measles can be controlled by vaccinating with an attenuated vaccine.

German Measles (Rubella) (figure 23.22, table 23.11)

German measles (rubella), if contracted by a woman in the first 6 weeks of pregnancy, often results in birth defects making up the **congenital rubella syndrome.** Immunization with an attenuated virus protects against this disease (figure 23.23).

Other Viral Rashes of Childhood

Numerous other viruses can cause rashes. **Fifth disease (erythema infectiosum)** is characterized by a "slapped check" rash. It is caused by parvovirus B-19. It can be fatal to people with certain anemias. **Roseola (exanthem subitum)** is marked by several days of high fever and a transitory rash that appears as the temperature returns to normal. It occurs mainly in infants six months to three years old. The disease is caused by human herpesvirus, type 6.

Warts

Warts are skin tumors caused by a number of papillomaviruses (figure 23.24). They are generally benign, but some sexually transmitted papillomaviruses are associated with cancer of the uterine cervix.

23.5 Skin Diseases Caused by Fungi

Superficial Cutaneous Mycoses

Invasive skin infections, such as diaper rashes, are sometimes caused by the yeast *Candida albicans* (figure 23.25). Certain species of molds that feed on keratin cause athlete's foot, ringworm, and invasions of the hair and nails (figure 23.26).

REVIEW QUESTIONS

Short Answer

1. What is the difference between a furuncle and carbuncle?
2. Why is the blister fluid of staphylococcal scalded skin syndrome free of staphylococci?
3. Give three ways in which *Streptococcus pyogenes* resembles *Staphylococcus aureus,* and three ways in which it differs.
4. What is the epidemiological importance of passage through tick eggs by *Rickettsia rickettsii?*
5. Describe the causative agent of Lyme disease.
6. What is characteristic about the rash of varicella?
7. What is the relationship between chickenpox (varicella) and shingles (herpes zoster)?
8. Why do so many people suffer permanent damage or die from measles?
9. What viral disease might be associated with an aplastic crisis? Describe its characteristic rash.
10. What is the significance of rubella viremia during pregnancy?
11. How does a person contract warts?
12. What is the allergic rash called that appears in response to ringworm and distant to it?

Multiple Choice

1. Which of the following conditions is important in the ecology of the skin?
   a) Temperature b) Salt concentration c) Lipids
   d) pH e) All of the above
2. *Staphylococcus aureus* can be responsible for which of these following conditions?
   a) Impetigo b) Food poisoning
   c) Toxic shock syndrome d) Scalded skin syndrome
   e) All of the above
3. The main effect of staphylococcal protein A is to
   a) interfere with phagocytosis.
   b) enhance the attachment of the Fc portion of antibody to phagocytes.
   c) coagulate plasma.
   d) kill white blood cells.
   e) degrade collagen.
4. Which of the following is essential for the virulence of *Streptococcus pyogenes?*
   a) Protease b) Hyaluronidase c) DNase
   d) All of the above e) None of the above
5. Which of the following statements is true of streptococcal acute glomerulonephritis?
   a) It is a streptococcal infection of the kidneys.
   b) It is caused by immune complexes containing streptococcal antigen.
   c) It is caused by most strains of β-hemolytic group A streptococci.
   d) It is the result of a streptococcal toxin directed against the kidneys.
   e) All of the above.
6. All of the following are true of Rocky Mountain spotted fever, *except*
   a) the disease is most prevalent in the western United States.
   b) it is caused by an obligate intracellular bacterium.
   c) it is a zoonosis transmitted to human beings by ticks.
   d) those with the disease characteristically develop a hemorrhagic rash.
   e) antibiotic therapy is usually curative if given early in the disease.
7. All of the following are true of Lyme disease, *except*
   a) it is caused by a spirochete.
   b) it is transmitted by certain species of ticks.
   c) it occurs only in the region around Lyme, Connecticut.
   d) most cases get a rash that looks like a target.
   e) it can cause heart and nervous system damage.
8. Which of the following statements is more likely to be true of measles (rubeola) than German measles (rubella)?
   a) Koplik spots are present.
   b) It causes birth defects.
   c) It causes only a mild illness.
   d) Human beings are the only natural host.
   e) Attenuated virus vaccine is available for prevention.
9. All of the following must be cultivated in cell cultures instead of cell-free media, *except*
   a) *Rickettsia rickettsii.* b) rubella virus.
   c) varicella-zoster virus. d) *Borrelia burgdorferi.*
   e) rubeola virus.
10. All of the following might contribute to development of ringworm or other superficial cutaneous mycoses, *except*
    a) obesity. b) playing with kittens.
    c) rubber boots. d) using skin powder.
    e) dermatophyte virulence.

Applications

1. A school administrator in a small Iowa community prohibited a child with chickenpox from attending school. He claimed that this was the first case of chickenpox seen in the school in 6 years and that he did not want to have an outbreak at the school. Several parents argued to the school board that an outbreak would benefit the school in the long term. Discuss the pros and cons of allowing this child to attend school.
2. A public health official was asked to speak about immunization during a civic group luncheon. One parent asked if rubella was still a problem. In answering the question, the official cautioned women planning to have another child to have their present children immunized against rubella. Why did the official make this statement to the group?

Critical Thinking

1. A microbiology instructor stated that the presence of large numbers of *Propionibacterium acnes* in the same areas where acne develops illustrates that occurrence of a bacterium and a disease together does not necessarily imply cause and effect. Why would the instructor make this statement?
2. When Lyme disease was first being investigated, the observation that frequently only one person in a household was infected was a clue leading to the discovery that the disease was spread by arthropod bites. Why was this so?
3. Why might it be more difficult to eliminate a disease like Lyme disease or Rocky Mountain spotted fever from the earth than rubeola or rubella?

Shotgun wound of the torso.

Wound Infections

A Glimpse of History

Spores of *Clostridium tetani,* the cause of lockjaw (tetanus), are found in soil and dust—thus, virtually everywhere. Before its cause and pathogenesis were understood, the disease was widespread, often ending in agonizingly painful death. Dr. Shibasaburo Kitasato (1856–1931), working in Robert Koch's laboratory in Germany, was the first to discover how to cultivate *C. tetani,* and subsequently he made a startling discovery that paved the way for control of the disease. Kitasato was born in a mountainous village in southern Japan in 1856. He was sent by his family to medical schools in Kumamoto and Tokyo. Following his graduation in 1883, Kitasato went to work for the Central Hygienic Bureau in Japan. The Japanese government needed to control epidemics of typhoid, cholera, and black-leg, a clostridial disease of calves, and so in 1885, Kitasato was sent to Germany to study with the famous Dr. Koch.

In Koch's laboratory, Kitasato was given the tetanus problem to study. In the course of his experiments, he discovered that *C. tetani* would only grow under strictly anaerobic conditions. Once he had isolated the organism in pure culture, he was able to show that it produced tetanus in laboratory animals. He was puzzled, however, by a surprise finding: although the animals died of generalized disease, there was no *C. tetani* anyplace other than the site where the organism had been injected. By doing experiments in which he injected the tails of mice and then removed the injected tissue at hourly intervals after injection, he showed that the animals only developed tetanus if the organisms were allowed to remain for more than an hour. He further showed that the organisms remained at the site of inoculation; at no time were they found in the rest of the body. Kitasato reasoned that something other than bacterial invasion was causing the disease.

About this time, another investigator, Emil von Behring, was busy investigating how *Corynebacterium diphtheriae* caused the disease diphtheria. Together, Kitasato and von Behring were able to show that both diseases were caused by poisonous substances—toxins—that were produced by the bacteria. The concept that a bacterial toxin could cause disease in the absence of bacterial invasion was an extremely important advance in the understanding and control of infectious diseases.

Kitasato published his studies in 1890. In 1892, even though urged to stay in Germany, he returned to Japan. The Japanese government was not prepared to support basic research at that time, and so Kitasato established his own institute for infectious diseases where he worked and trained Japanese scientists for the rest of his life. In 1908, Koch paid a visit to Kitasato in Japan and a Shinto shrine was built in Koch's honor. During his later years, Kitasato was instrumental in establishing laws regulating health practices in Japan. He died in 1931 at age 75. To honor him, a shrine was erected next to Koch's.

Most people occasionally sustain wounds that produce breaks in the skin or mucous membranes. Almost always, microorganisms contaminate these wounds from the air, fingers, normal microbiota, or the object causing the wound. Whether these microorganisms cause disease depends on (1) how virulent they are, (2) how many there are, (3) the status of host defenses, and (4) the nature of the wound, especially whether it contains crushed tissue or foreign material.

Wounds that contain materials such as dirt, leaves, bits of rubber, and cloth usually become infected and do not heal until the foreign material is removed. Often such a wound provides places for microorganisms to multiply and produce injurious substances, out of the reach of phagocytes and other body defenses. Foreign materials may also provide surfaces for biofilms, or the materials reduce available oxygen, thereby impairing phagocytic function and allowing growth of anaerobic pathogens. Clean wounds often heal without treatment despite microbial colonization, but sometimes even a trivial wound can result in a severe, or even fatal, infection by providing an entryway for infection or microbial toxins to spread throughout the body.

KEY TERMS

Abscess A localized collection of pus within a tissue.

Fasciitis Inflammation of the fascia, bands of fibrous tissue that underlie the skin and surround muscle and body organs. When it leads to death of tissue, it is called **necrotizing fasciitis.**

Granulation Tissue New tissue formed during healing of an injury, consisting of small, red, translucent nodules containing abundant blood vessels.

MRSA Methicillin-resistant *Staphylococcus aureus;* many strains have acquired R plasmids making them resistant to multiple anti-staphylococcal medications.

Neonatal Tetanus Tetanus that develops in newborn babies, often from cutting the umbilical cord with an unsterile instrument; preventable by immunizing the mother before delivery.

Pyogenic Pus-producing

Superantigen Molecules that bind to and stimulate T lymphocytes, resulting in activation of many T cells, overproduction of cytokines, severe reactions, and sometimes fatal shock.

Synergistic Infection An infection in which two or more species of pathogens act together to produce an effect greater than the sum of effects if each pathogen were acting alone.

Toxic Shock Life-threatening drop in blood pressure resulting from certain types of circulating bacterial exotoxins.

Wounds can be classified as:

- *burns;*
- *incised*, as when produced by a knife or other sharp object, as in surgery;
- *punctured,* from penetration of a small sharp object, as when an individual steps on a nail;
- *lacerated,* when the tissue is torn;
- *contused,* as when caused by a blow that crushes tissue.

Burns represent an important category of accidental wounds. In the United States each year, more than 2.5 million people sustain thermal burns severe enough for them to seek medical attention. Of these, more than 100,000 require hospitalization, and 12,000 die from their burns. Burns present special problems. Although initially sterile, thermal burns are often extensive, with a large area of weeping devitalized tissue, representing an enormous and nutritious feast for microorganisms.

Intentional wounds inflicted by surgery represent about 23 million cases annually in the United States. From 1% to 9% of them become infected; the percentages vary according to the type of surgery and the status of the patients' host defenses. Costs related to postoperative wound infections amount to approximately $1.5 billion each year, more than half of the total expense for nosocomial infections. They result in about 13,000 deaths each year. ■ nosocomial infections, p. 462

24.1 Anatomy and Physiology

Focus Points

- Name three tissue components exposed by wounds to which pathogens specifically attach.
- Describe the potentially beneficial and harmful aspects of abscess formation.

Wounds expose components of tissue normally protected from the outside world by skin or mucous membranes. These exposed tissue components, including collagen, fibronectin, fibrin, and fibrinogen, provide receptors to which potential pathogens specifically attach. **Collagen** is a fibrous material—the main supportive protein of skin, tendons, scars, and other body structures. **Fibronectin** is a glycoprotein that occurs both as a circulating form and as a component of tissue, where it ties cells and other tissue substances together. Shortly after a wound occurs, the soluble blood protein **fibrinogen** is converted to the fibrous material **fibrin,** thereby forming clots in the damaged vessels. This stops the flow of blood as the first step in the repair process.

Wound healing begins with the outgrowth of connective tissue cells, called fibroblasts, and capillaries from the surfaces of the wound, producing a nodular, red, translucent material called **granulation tissue.** In the absence of dirt or infection, granulation tissue fills the void created by the wound, contracts, and is converted to collagen that composes the scar tissue eventually covered by overlying skin or mucous membrane **(figure 24.1).** Sometimes in the presence of dirt or infection, the granulation tissue overgrows, bulging from the wound to form a **pyogenic granuloma.**

Wound Abscesses

An **abscess (figure 24.2)** is a localized collection of pus surrounded by body tissue. It is composed of living and dead leukocytes, components of tissue breakdown, and infecting organisms. No blood vessels are in abscesses because they have been destroyed or pushed aside. An area of inflammation and clots in adjacent blood vessels separate the abscess from normal tissue. Consequently, abscess formation helps to localize an infection and prevent its spread. Microorganisms in abscesses often are not killed by antimicrobial medications because the microorganisms cease multiplying, and active multiplication is generally required for microbial killing by the medications. In addition, the chemical nature of pus interferes with the action of some antibiotics, and many antimicrobials diffuse poorly into abscesses because of the absence of blood vessels. Microorganisms in abscesses are a potential source of infection of other parts of the body if they escape the surrounding area of inflammation and enter the blood or lymph vessels. Generally, abscesses must burst to a body surface or be drained surgically in order to effect a cure. ■ antimicrobial medications, p. 469

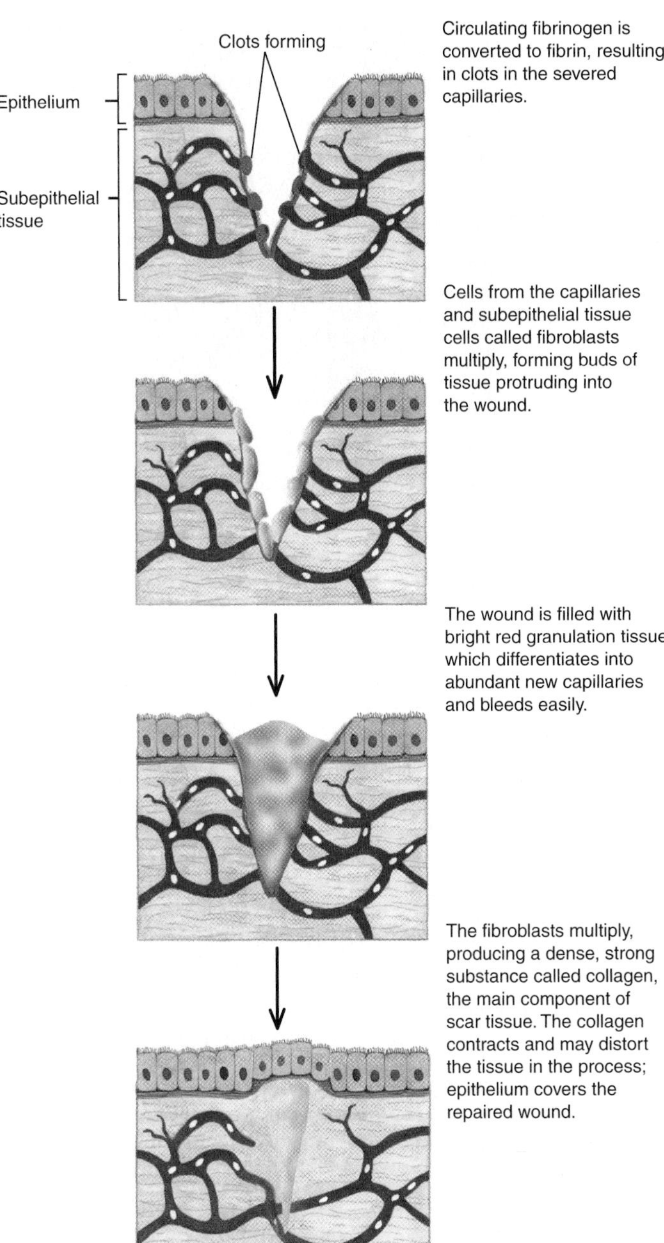

Clots forming
Epithelium
Subepithelial tissue
Circulating fibrinogen is converted to fibrin, resulting in clots in the severed capillaries.
Cells from the capillaries and subepithelial tissue cells called fibroblasts multiply, forming buds of tissue protruding into the wound.
The wound is filled with bright red granulation tissue, which differentiates into abundant new capillaries and bleeds easily.
The fibroblasts multiply, producing a dense, strong substance called collagen, the main component of scar tissue. The collagen contracts and may distort the tissue in the process; epithelium covers the repaired wound.

FIGURE 24.1 The Process of Wound Repair

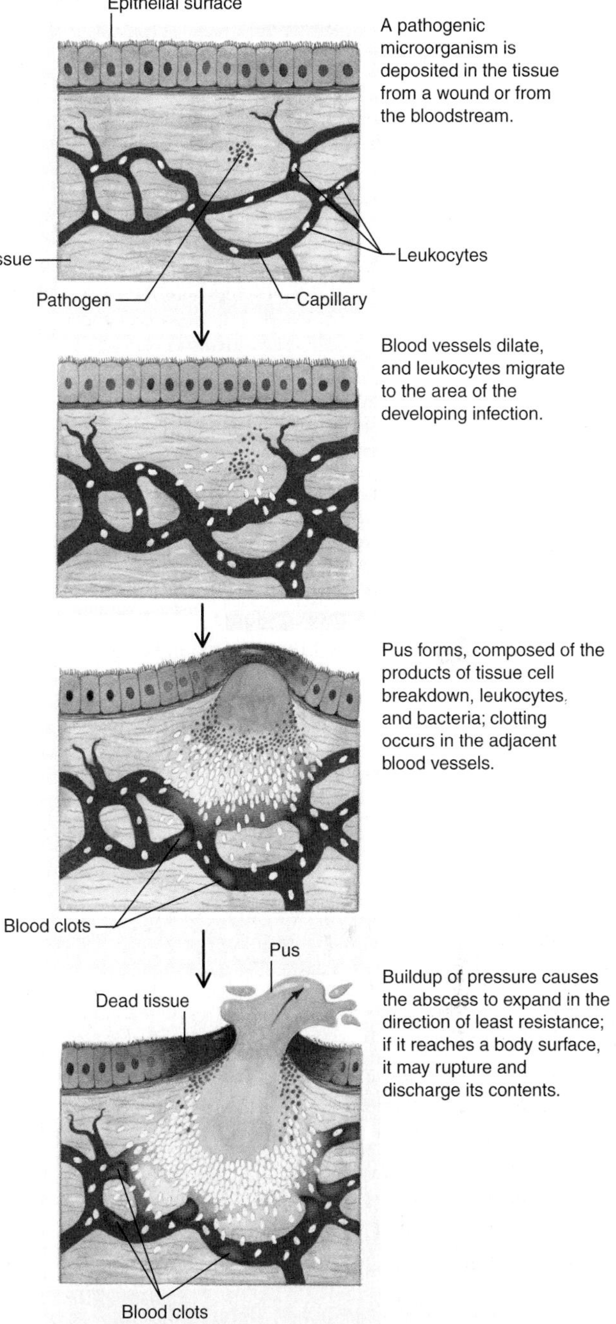

Epithelial surface
A pathogenic microorganism is deposited in the tissue from a wound or from the bloodstream.
Tissue
Pathogen
Capillary
Leukocytes
Blood vessels dilate, and leukocytes migrate to the area of the developing infection.
Pus forms, composed of the products of tissue cell breakdown, leukocytes, and bacteria; clotting occurs in the adjacent blood vessels.
Blood clots
Pus
Dead tissue
Buildup of pressure causes the abscess to expand in the direction of least resistance; if it reaches a body surface, it may rupture and discharge its contents.
Blood clots

FIGURE 24.2 Abscess Formation

Anaerobic Wounds

An important feature of many wounds is that they are relatively anaerobic, thus allowing colonization by dangerous anaerobic pathogens such as *Clostridium tetani.* Anaerobic conditions are especially likely in dirty wounds, wounds with crushed tissue, and puncture wounds. Puncture wounds caused by nails, thorns, splinters, and other sharp objects can introduce foreign material and microorganisms deep into the body. Bullets and other projectiles can carry fragments of skin or cloth contaminated with microorganisms into the tissues. Projectiles, although causing relatively small breaks in the skin, often produce extensive tissue damage because of the force with which they enter.

MICROCHECK 24.1

Wounds expose components of tissue to which pathogens specifically attach. Wounds can be classified as incised, punctured, lacerated, contused, or burns. Healing involves the outgrowth of fibroblasts and capillaries from the sides of the wound to produce granulation tissue that fills the defect. Abscess formation provides a way of isolating infections enclosed by tissue. Anaerobic conditions in wounds are created by the presence of dead tissue and foreign material.

- ✓ Name two substances in wounds to which pathogens specifically attach.
- ✓ Give two reasons why an abscessed wound might not respond to antibiotic treatment.
- ✓ What is the advantage of the contraction of granulation tissue?

24.2

Common Bacterial Wound Infections

Focus Points

- Give distinctive characteristics of three common wound infections caused by bacteria that grow in the presence of air.
- Outline the evolution and current significance of MRSA.

The possible consequences of wound infections include (1) delayed healing, (2) formation of abscesses, and (3) spread of the bacteria or their products into adjacent tissues or the bloodstream. Infected surgical wounds often split open as swelling causes the stitches to pull through tissues softened by the infection. Also, the infection can extend to involve devices such as an artificial hip, often requiring that the device be removed pending control of the infection.

The following sections present some aspects of wound infections caused by staphylococci, streptococci, and *Pseudomonas* sp.

Staphylococcal Wound Infections

Staphylococci lead the causes of wound infections, both surgical and accidental **(figure 24.3).** Staphylococci are commonly present in the nostrils or on the skin. Of the 30 or more recognized species of staphylococci, only two account for most human wound infections. ■ the genus *Staphylococcus*, p. 270

FIGURE 24.3 Surgical Wound Infection Due to *Staphylococcus aureus* The stitches and staples pull through the infected tissue, causing the wound to open.

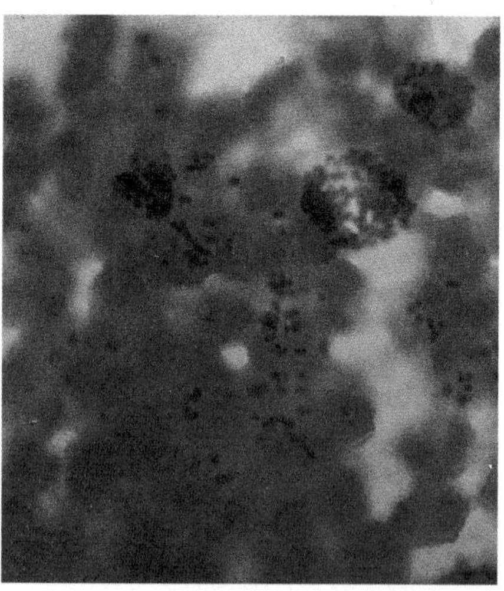

FIGURE 24.4 *Staphylococcus aureus* in Pus The dark-colored dots are staphylococci; the red objects, leukocytes.

Symptoms

Staphylococci are **pyogenic,** meaning that they characteristically cause the production of a purulent discharge, otherwise known as pus. They usually cause an inflammatory reaction, with swelling, redness, and pain. If the infected area is extensive or if the infection has spread to the general circulation, fever is a prominent symptom. Wound infections by toxin-producing strains can result in **toxic shock** syndrome, characterized by high fever, muscle aches, and life-threatening shock, sometimes accompanied by a rash and diarrhea. ■ staphylococcal toxic shock syndrome, p. 626

Causative Agents

Staphylococci are Gram-positive cocci that grow as clusters **(figure 24.4).** They grow readily aerobically or anaerobically and are salt tolerant, probably because they evolved on the skin where evaporation concentrates the salt in sweat. The most important species are *S. aureus* and *S. epidermidis.* Both species survive well in the environment, making it easy for them to transfer from one person to another.

Staphylococcus aureus Characteristics of this common pathogen were discussed in chapter 23.

Staphylococcus epidermidis Most strains of *S. epidermidis* have little or no invasive ability for healthy people but they commonly cause small abscesses of little consequence around the stitches used in surgery. Most strains bind fibronectin, however, and can therefore colonize the plastic intravenous catheters, heart valves, and other devices employed in modern medicine. Following adherence, the bacteria may produce a kind of slime or glycocalyx that cements the growing colony to the plastic in a biofilm, protecting it from attack by phagocytes and other host defense mechanisms, and antibacterial medications. ■ glycocalyx, p. 64 ■ biofilm, p. 85

Pathogenesis

Staphylococcus aureus Many years ago, it was shown that it takes more than 100,000 staphylococci injected into the skin to produce a small abscess. When injected along with a suture, however, only about 100 *S. aureus* are required to produce the same lesion. These studies, which used students as human guinea pigs, dramatized the effect of foreign material in the pathogenesis of staphylococcal infections.

Studies using genetically engineered strains have indicated that multiple virulence factors act together to produce the usual wound infection. Clumping factor and the other binding proteins attach the organisms to clots and tissue components, fostering colonization. Clumping factor, coagulase, and protein A serve to coat the organisms with host proteins, providing a disguise that hides them from attack by phagocytes and the immune system. This may explain why immunity to staphylococcal infection is generally weak or non-existent. Some protein A is released from the bacterial surface and reacts with circulating immunoglobulin. The resulting complexes activate the complement system and probably contribute to the intense inflammatory response and accumulation of pus.

Colonization of plastics and other foreign materials occurs because they quickly become coated with fibrinogen and fibronectin to which the staphylococci attach. Systemic spread of wound infections can lead to abscesses in other tissues, such as the heart and joints. Staphylococcal toxins that enter the circulation are **superantigens,** causing the widespread release of cytokines, thereby producing toxic shock. ■ superantigens, p. 406 ■ cytokines, p. 353 ■ complement system, p. 355

Staphylococcus epidermidis The bacteria growing as biofilms on indwelling plastic catheters can come loose from and be carried by the bloodstream to the heart and other tissues. This can result in subacute bacterial endocarditis or multiple tissue abscesses in people with impaired host defenses as from cancer, diabetes mellitus, or other causes. Wound infections by *S. epidermidis* in healthy people are frequently cleared by host defenses alone. ■ bacterial endocarditis, p. 677

Epidemiology

The epidemiology of *S. aureus* is discussed in chapter 23. Various studies have shown that in the case of surgical wound infections, nasal carriers have a two to seven times greater risk of infection than do those who are not nasal carriers. From 30% to 100% of the infections in different studies were due to a patient's own staphylococcus strain. Advanced age, poor general health, immunosuppression, prolonged preoperative hospital stay, and infection at a site other than the site of surgery increase the risk of infection. ■ *S. aureus* epidemiology, p. 537

Prevention and Treatment

Cleansing and removal of dirt and devitalized tissue from accidental wounds minimizes the chance of infection, as does prompt closure of clean wounds by sutures. Trying to eliminate the nasal carrier state with anti-staphylococcal medications is occasionally successful. Infections in surgical wounds can be reduced by half by administering an effective anti-staphylococcal medication immediately before surgery. For unknown reasons, the infection rate is actually increased if the medication is given more than 3 hours before or 2 hours after the surgical incision.

As described in the chapter 23, treating staphylococcal infections can often be problematic due to wide spread antibiotic resistance. ■ MRSAs, pp. 485, 537 ■ antibacterial resistance, p. 483

Group A Streptococcal "Flesh Eaters"

Streptococcus pyogenes was introduced under "Strep Throat" in chapter 22 and under "Streptococcal Impetigo," in chapter 23. It is also a common cause of wound infections, which have generally been easy to treat since the bacteria are consistently susceptible to penicillin. Occasionally, however, *S. pyogenes* infections can progress rapidly, even leading to death despite antimicrobial treatment. These more severe infections are called invasive because they extend beyond the skin into underlying tissues and organs, and include pneumonia, meningitis, puerperal or childbirth fever, necrotizing fasciitis or "flesh-eating" disease, and worst of all, streptococcal toxic shock, which is similar to staphylococcal toxic shock. Deaths from invasive strep infections have caused widespread popular concern as *S. pyogenes* became the "flesh-eating" bacterium of the tabloid press. This section will focus on **necrotizing fasciitis (figure 24.5),** a rare but dramatic complication of *S. pyogenes* infection. ■ *Streptococcus pyogenes,* p. 499 ■ staphylococcal toxic shock, p. 626

Symptoms

At the site of a surgical wound or accidental trauma, sometimes even without an obvious break in the skin, severe pain develops acutely. Within a short time, swelling becomes apparent, and the injured person develops fever and confusion. The overlying skin becomes stretched and discolored because of the swelling. Unless treatment is initiated promptly, shock and death usually follow in a short time.

Causative Agent

Streptococcus pyogenes is a β-hemolytic, aerotolerant, Gram-positive, chain-forming coccus with Lancefield group A cell wall polysaccharide. Strains of *S. pyogenes* that cause invasive disease are more virulent than other strains by virtue of at least two extracellular products: pyrogenic exotoxin A, is a superantigen that causes streptococcal toxic shock, and exotoxin B, a protease that destroys tissue by breaking down protein. ■ aerotolerance, p. 92 ■ β-hemolysis, p. 97 ■ pyrogens, p. 362

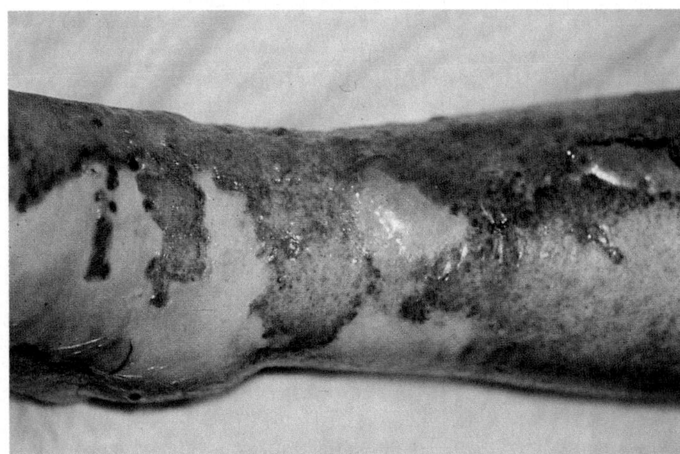

FIGURE 24.5 Individual with *Streptococcus pyogenes* "Flesh-Eating" Disease (Necrotizing Fasciitis)

Pathogenesis

Like *Staphylococcus aureus, S. pyogenes* has a fibronectin-binding protein that aids colonization of wounds. In necrotizing fasciitis, the subcutaneous fascia and fatty tissue are destroyed. **Fascia** are bands of fibrous tissue that underlie the skin and surround muscle and body organs. In some cases, the fascia surrounding muscle is penetrated, and muscle tissue is also destroyed. Intense swelling occurs as fluid is drawn into the area because of increased osmotic pressure from the breakdown of tissue into small molecules. The organisms continue to multiply and produce toxic products in the mass of dead tissue, using the breakdown products as nutrients. In most cases, toxic products and organisms enter the bloodstream. Superantigens and other streptococcal products cause shock by inducing T cells to release cytokines and other mechanisms.

Epidemiology

"Flesh-eating" infections by *S. pyogenes* have probably occurred at least since the fifth century B.C. based on descriptions of necrotizing fasciitis by Hippocrates. More than 2,000 cases of this condition were reported among soldiers during the Civil War. Cases in this country are generally sporadic, although small epidemics have occurred such as a 1996 outbreak in San Francisco among injected-drug abusers using contaminated "black tar" heroin. The number of cases with invasive *S. pyogenes* in the United States was estimated at about 9,000 for 2002, resulting in 1,080 deaths, 135 of which were due to necrotizing fasciitis. There is no firm evidence of a trend toward higher incidence. Underlying conditions that increase the risk of necrotizing fasciitis and other invasive *S. pyogenes* infections include diabetes, cancer, alcoholism, AIDS, recent surgery, abortion, childbirth, chickenpox, and injected-drug abuse; invasive infections rarely occur in healthy individuals with minor injuries.

Prevention and Treatment

There are no proven preventive measures. Because of the rapidity with which the toxins spread, urgent surgery is mandatory to relieve the pressure of the swollen tissue and to remove dead tissue. Amputation is sometimes necessary, to promptly rid the patient of the source of toxins. Penicillin is the drug of choice for early infection, but it has little or no effect on streptococci in necrotic tissue and no effect on toxins, and therefore it cannot substitute for surgery.

Pseudomonas aeruginosa Infections

Pseudomonas aeruginosa is an opportunistic pathogen that is a major cause of nosocomial infections. In addition, it is an occasional cause of community-acquired infections, meaning infections acquired outside hospital or other medical facilities. Community-acquired infections include skin rashes from contaminated swimming pools and hot tubs, serious infections of the foot bones from stepping on nails, serious eye infections from contaminated contact lens solutions, heart valve infections in injected-drug abusers, external ear canal infections in swimmers and others who fail to dry their ears properly, disfiguring infections from ear piercing, especially of the ear cartilage, and biofilms in the lungs of individuals with the inherited disease **cystic fibrosis.** ■ *Pseudomonas*, p. 261

In hospitals, the bacterium is an important cause of lung infections and a common cause of wound infections, especially of thermal burns. Burns have large exposed areas of dead tissue free of any body defenses and, therefore, are ideal sites for infection by bacteria from the environment or normal microbiota. Almost any opportunistic pathogen can infect burns, but *Pseudomonas aeruginosa* is among the most common and hardest to treat.

Symptoms

In burns and other wounds, *P. aeruginosa* can often color the tissues green from pigments released from the bacterial cells **(figure 24.6a,b).** This bacterium is especially dangerous because

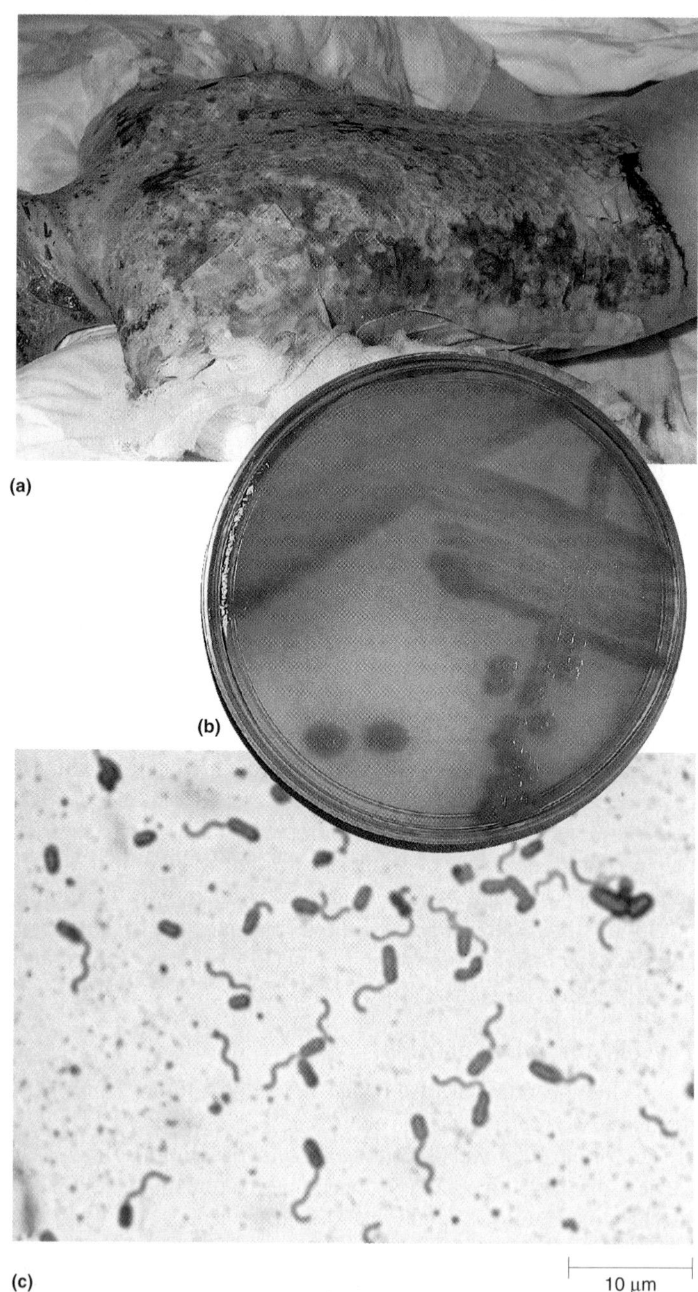

FIGURE 24.6 *Pseudomonas aeruginosa* (a) Extensive burn infected with *P. aeruginosa.* Notice the green discoloration. **(b)** Culture. The green discoloration of the medium results from water-soluble pigments diffusing from the *P. aeruginosa* colonies. **(c)** *P. aeruginosa* has a single polar flagellum.

it often invades the bloodstream and causes chills, fever, skin lesions, and shock.

Causative Agent

Pseudomonas aeruginosa is found in a wide variety of environments such as soil and water. It is motile by means of a single polar flagellum (figure 24.6c). The organisms grow readily and rapidly; some strains can grow in a variety of aqueous solutions, even in distilled water. The bacterium generally utilizes O_2 as the terminal electron acceptor in the breakdown of nutrients. Under certain circumstances, however, it can also grow anaerobically, as when nitrate is available. Nitrate substitutes for O_2 as a terminal electron acceptor, an example of anaerobic respiration. Strains of *P. aeruginosa* often produce one or more water-soluble pigments, which can be red, yellow, blue, or dark brown. Commonly they produce a fluorescent yellowish pigment called pyoverdin, which combines with a blue pigment, pyocyanin, to produce the striking green color characteristically seen in infected wounds and in growth media. ■ **anaerobic respiration, pp. 133, 354 ■ the genus *Pseudomonas,* p. 261**

Pathogenesis

The overall effect of *P. aeruginosa* infection of burns and other wounds is to produce tissue damage, prevent healing, and increase the risk of septic shock. Virulence depends mainly on production of two extracellular enzymes—exoenzyme S and toxin A. Exoenzyme S modifies various critical processes in the host cells, leading to their death. It functions in a way similar to the toxins of *Vibrio cholerae* and *Bordetella pertussis.* Most strains of *P. aeruginosa* also produce toxin A, an exoenzyme that has a mode of action identical to that of the toxin of *Corynebacterium diphtheriae,* a Gram-positive rod. The two toxins, however, are antigenically distinct and target different cells. These toxins halt protein synthesis by the host cell and are enhanced by lowering the concentration of iron.

Most strains of *P. aeruginosa* also produce proteases that cause localized hemorrhages and tissue necrosis. Also, many strains produce a heat-labile hemolysin, a phospholipase C, identical in mode of action to the principal toxin of the gas gangrene bacillus *Clostridium perfringens* to be discussed later in this chapter. Phospholipase C hydrolyses **lecithin,** an important lipid component of cell membranes. The pigments released from *P. aeruginosa* may also contribute to pathogenicity by inhibiting competing bacteria and helping it acquire iron. ■ ***Bordetella pertussis,* p. 513 ■ *Corynebacterium diphtheriae,* p. 503 ■ elongation factor, p. 173 ■ required elements p. 93 ■ *Vibrio cholerae,* p. 596**

Epidemiology

Pseudomonas aeruginosa is widespread in nature. It is introduced into hospitals on shoes, ornamental plants and flowers, and produce. It can persist in most places where there is dampness or water, and it can contaminate soaps, ointments, eye-drops, contact lens solutions, cosmetics, disinfectants, many kinds of hospital equipment, swimming pools, hot tubs, the inner soles of shoes, and illegal injectable drugs, all of which have been sources of infections.

Prevention and Treatment

Prevention of *P. aeruginosa* infections involves elimination of potential sources of the bacterium and prompt care of wounds. Careful removal of dead tissue from burn wounds, followed by application of an antibacterial cream such as silver sulfadiazine, is often effective in preventing infection with *P. aeruginosa.* Established infections are notoriously difficult to treat because *P. aeruginosa* is usually resistant to multiple antibacterial medications. Antibiotic susceptibility tests are done to guide the selection of an effective regimen. For systemic infection, antibacterial medications must usually be administered intravenously in high doses.

Table 24.1 compares the leading causes of wound infections.

MICROCHECK 24.2

Staphylococcus aureus is the most important cause of wound infections because it is so commonly carried by humans, transfers easily from one person to another, and possesses multiple virulence factors. *Staphylococcus epidermidis* forms biofilms on foreign materials, protecting the bacteria from body defenses and antibacterial medications. "Flesh-eating" strains of *Streptococcus pyogenes* are uncommon, but are often life threatening. *Pseudomonas aeruginosa,* a Gram-negative, pigment-producing rod, is widespread in the environment and is a major cause of nosocomial infections.

✓ Why are antibacterial medications not effective for treating "flesh-eating" disease (necrotizing fasciitis)?

✓ Why do wound infections due to *Pseudomonas aeruginosa* sometimes produce green pus?

✓ Why is it not surprising that staphylococci are the leading cause of wound infections?

TABLE 24.1 Leading Causes of Wound Infections

Causative Organism	Characteristics	Consequences
Staphylococcus aureus	Gram-positive cocci in clusters, coagulase-positive	Delayed healing; abscess formation; extension into tissues, artificial devices, or bloodstream; occasional strains can cause toxic shock syndrome
Streptococcus pyogenes	Gram-positive cocci in chains; Lancefield group A	Same, except occasional strains can cause "flesh-eating" necrotizing fasciitis
Pseudomonas aeruginosa	Aerobic Gram-negative rod, green pigment	Delayed healing; abscess formation; extension into tissues, artificial devices, or bloodstream; septic shock

24.3

Diseases Due to Anaerobic Bacterial Wound Infections

Focus Points

- Describe the conditions that favor the development of anaerobic wound infections.
- Outline the special difficulties in treating wounds infected with toxin-producing bacteria.

Wounds that provide anaerobic conditions allow colonization by certain strictly anaerobic species of bacteria—organisms with their own unique pathogenic abilities. This section describes three distinctive diseases that result from anaerobic bacterial wound infections: lockjaw, gas gangrene, and lumpy jaw.

"Lockjaw" (Tetanus)

Tetanus is frequently fatal; fortunately, it is rare in economically advanced countries. There is no reasonable way to avoid exposure to the causative organism because its endospores are widespread in dust and dirt, frequently contaminating clothing, skin, and wounds. Even a trivial wound in a non-immunized person can result in tetanus if the wound provides conditions sufficiently anaerobic to allow germination of the spores. ■ endospores, p. 69

Symptoms

Tetanus is characterized by sustained, painful, and uncontrollable cramplike muscle spasms, which are usually generalized **(figure 24.7)** but can be limited to one area of the body. The spasms often begin with the jaw muscles, giving the disease the popular name "lockjaw." The early symptoms include restlessness, irritability, difficulty swallowing, contraction of the muscles of the jaw, and sometimes convulsions, particularly in children. As more muscles tense, the pain grows more severe and is similar to that of a severe leg cramp. Breathing becomes labored, and after a period of almost unbearable pain, the infected person often dies of pneumonia or from stomach contents regurgitated into the lung.

FIGURE 24.7 Infant with Neonatal Tetanus

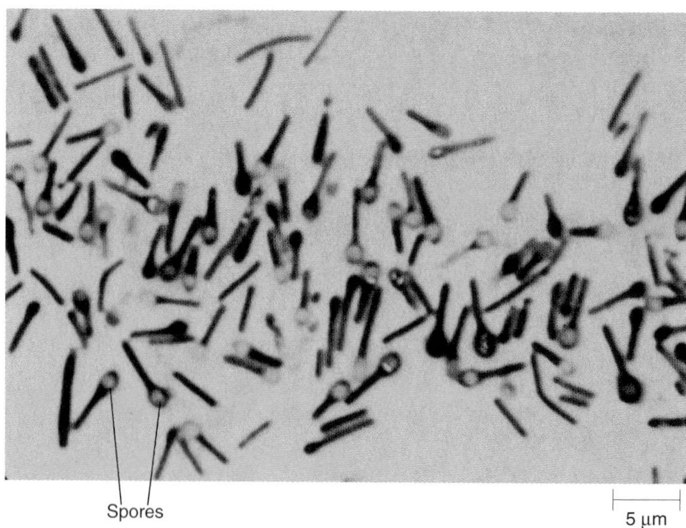

FIGURE 24.8 *Clostridium tetani* Terminal endospores are characteristic of this species.

Causative Agent

The causative agent, *Clostridium tetani,* an anaerobic, sporeforming, Gram-positive, rod-shaped bacterium **(figure 24.8),** shows two striking features: (1) a spherical endospore that forms at the end of the bacillus, in contrast to the oval endospore that develops near the center of the cell in other pathogenic species of *Clostridium,* and (2) swarming growth that quickly spreads over the surface of solid media, making it easy to obtain pure cultures. Final identification of *C. tetani* depends on identifying its toxin, which is coded by a plasmid. ■ bacterial toxins, p. 403 ■ *Clostridium*, p. 254 ■ plasmids, p. 68 ■ *C. botulinum* spores, figure 27.10, p. 657

Isolation of *C. tetani* from wounds does not prove that a person has tetanus. This is because tetanus endospores may contaminate wounds that are not sufficiently anaerobic to allow germination and toxin production, or the person may simply be immune to the toxin by prior vaccination. Only vegetative cells, not endospores, synthesize toxin. Conversely, failure to find the organism in a person's wound cultures does not eliminate the possibility of tetanus.

Pathogenesis

Clostridium tetani is not invasive, and colonization is generally localized to a wound. Its pathologic effects are entirely the result of an exotoxin called **tetanospasmin,** released from the organism during the stationary phase of growth. It is an A-B toxin. The B portion attaches specifically to receptors on motor neurons, which then take up the A portion by endocytosis. The toxin is carried by the neuron to its cell body in the spinal cord. There, the motor neuron is in contact with other neurons that normally control its action. Some of these other neurons function to stimulate the motor nerve cell, thereby producing a muscle contraction, and others make the motor neuron resistant to stimulation, thereby inhibiting muscle contraction. Tetanospasmin blocks the action of the inhibitory neurons, so that the muscles contract without control **(figure 24.9).**

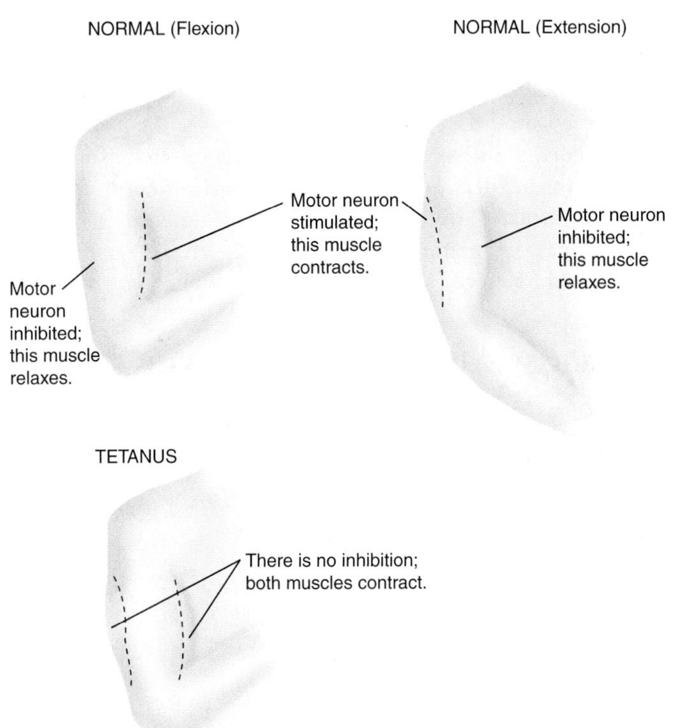

NORMAL (Flexion)
NORMAL (Extension)
Motor neuron stimulated; this muscle contracts.
Motor neuron inhibited; this muscle relaxes.
Motor neuron inhibited; this muscle relaxes.
TETANUS
There is no inhibition; both muscles contract.

FIGURE 24.9 Tetanus and Inhibitory Neuron Function The exotoxin tetanospasmin blocks the action of inhibitory neurons, allowing all muscles to contract at the same time.

The toxin generally spreads across the spinal cord to the side opposite the wound and then downward. Thus, typically, spastic muscles first appear on the side of the wound, then on the opposite side, and then downward, depending on the amount of toxin. Tetanospasmin released from the infected wound often enters the bloodstream, which carries it to the central nervous system. In these cases, inhibitory neurons of the brain are first affected, and the muscles of the jaw are among the first to become spastic. ■ A-B toxins, p. 405

Neurons exert their effect on other nerve cells by releasing chemicals called neurotransmitters. Tetanospasmin prevents the release of neurotransmitters from inhibitory neurons.

Epidemiology

Clostridium tetani occurs not only in dirt and dust, but in the gastrointestinal tract of humans and other animals that have eaten foods contaminated with its spores. Therefore, fecal contamination is a potential source of infection. About half of the cases result from puncture wounds, including stepping on a nail, body piercing, tattooing, animal bites, splinters, injected-drug abuse, and insect stings. Cases also occur following medical procedures such as hemorrhoid removal or knee surgery. Surface abrasions and burns can result in tetanus if the wound is anaerobic by virtue of dirt, dead tissue, or exclusion of air. Ordinarily, about 30 cases are reported annually in the United States, with a mortality rate around 25%. Age is an important factor, probably because many older people fail to get recommended booster injections of vaccine. Due to lack of immunization and proper care of wounds, tetanus occurs much more frequently on a global basis than it does in the economically advanced countries. In some parts of the world, babies commonly die of **neonatal tetanus** as a result of their umbilical cord being cut with unsterile instruments contaminated with *C. tetani.*

Prevention and Treatment

Active immunization with tetanus toxoid, which is inactivated tetanospasmin, is by far the best preventive weapon against tetanus. Immunization is usually begun during the first year of life. For infants and young children, the tetanus toxoid is given at two, four, six and 18 months of age in combination with diphtheria toxoid and acellular pertussis vaccine. The three together are commonly known as DTaP. A booster dose is given when children enter school. Once immunity has been established by this regimen, additional booster doses of tetanus toxoid are given at 10-year intervals to maintain an adequate level of protection. Any injury, burn, or contemplated surgery is an occasion to make sure immunization is up to date. Updating of booster doses of toxoid more frequently than every 10 years is not recommended because there is danger of an allergic reaction following intensive immunization. This type of reaction can occur when the toxoid is injected into an individual who already has large quantities of antitoxin, meaning antibodies against the toxoid. Individuals who have recovered from tetanus are not immune to the disease and must be immunized. ■ toxoid, p. 435

Even though tetanus is easily prevented by immunization with tetanus toxoid, about 97% of the people who developed tetanus in recent years were never immunized. To help prevent tetanus after a wound has been sustained, active immunization with toxoid, and passive immunization with tetanus immune globulin (TIG), may be indicated **(table 24.2).**

Neonatal tetanus can be prevented by immunizing expectant mothers. Their infants are then protected by antitoxin crossing the

TABLE 24.2 Tetanus Prevention in the Management of Wounds

	Clean Minor Wounds		All Other Wounds	
Immunization History	**Toxoid**	**TIG**	**Toxoid**	**TIG**
Unknown, or fewer than three injections	Yes	No	Yes	Yes
Fully immunized (three or more injections of toxoid)				
• 5 years or less since last dose	No	No	No	No
• 5 to 10 years since last dose	No	No	Yes	No
• more than 10 years since last dose	Yes	No	Yes	No

placenta. Since 1989, when almost 30,000 cases were reported, the World Health Organization has been attempting to eliminate neonatal tetanus. By 2006, the number of reported cases had fallen to 8,376. The decrease in cases largely resulted from immunizing millions of expectant mothers in developing countries against tetanus and educating them in the care of the severed umbilical cord.

Tetanus is treated by administering tetanus antitoxin, which is antibody against tetanospasmin, to neutralize any of the toxin not yet attached to motor nerve cells. The antitoxin of choice is **tetanus immune globulin (TIG),** which is prepared from blood of humans immunized with tetanus toxoid. This antitoxin generally does not cause hypersensitivity reactions in humans and is much more effective than the antitoxin formerly used which was derived from the blood of horses immunized with toxoid. TIG, however, cannot neutralize tetanospasmin that is already bound to nerve tissue. This explains why tetanus antitoxin is often ineffective in treating the disease. In addition to antitoxin treatment, the wound is thoroughly cleaned of all dead tissue and foreign material that could provide anaerobic conditions. An antibacterial medication such as metronidazole is given to kill any actively multiplying clostridia and thereby prevent the formation of more tetanospasmin. Antibacterials do not kill endospores or nongrowing bacteria, however.

Table 24.3 describes the main features of tetanus.

Gas Gangrene (Clostridial Myonecrosis)

Almost every sample from soil and dusty surfaces has endospores of *Clostridium perfringens,* the bacterium that causes gas gan-

TABLE 24.3 "Lockjaw" (Tetanus)

① *Clostridium tetani* spores from dust or dirt enter a wound.
② In wounds sufficiently anaerobic, the spores germinate, vegetative bacteria release an exotoxin called tetanospasmin.
③ Tetanospasmin is carried to the central nervous system by motor nerve axons or by the bloodstream.
④ The toxin prevents any inhibitory neurons it reaches from functioning.
⑤ The corresponding neurons, which cause muscles to contract, act unopposed by inhibitory neurons.
⑥ The result is a sustained, painful cramplike muscle spasm.

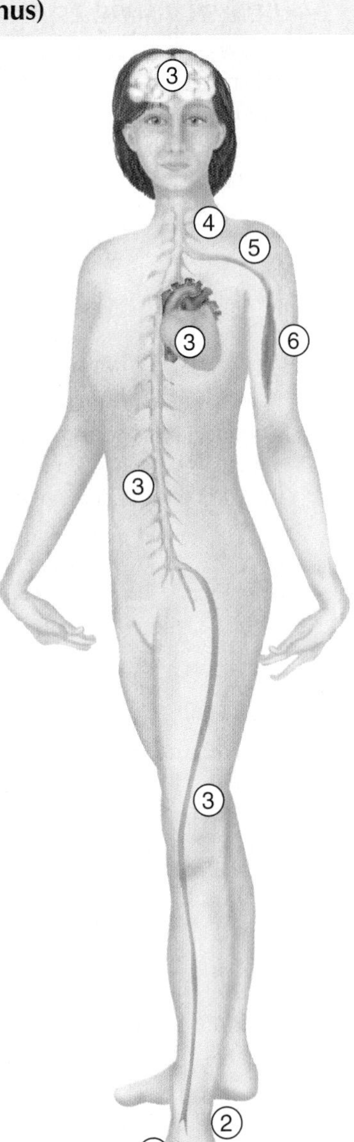

Symptoms	Restlessness, irritability, difficulty swallowing; muscle pain and spasm in jaw, abdomen, back, or entire body
Incubation period	3 days to 3 weeks; average 8 days
Causative agent	*Clostridium tetani,* an anaerobic, spore-forming, Gram-positive rod
Pathogenesis	Tetanus results from tetanospasmin, an exotoxin produced by the bacterium. The toxin is carried to brain and spinal cord by motor nerve axons or circulating blood; toxin acts against nerve cells that normally inhibit muscle contraction. Other nerves that normally cause muscle contraction then act unopposed, causing muscle spasms.
Epidemiology	Organisms common in soil; spores contaminate wounds, germinate in those having anaerobic conditions, particularly dirty or puncture wounds.
Prevention and treatment	Immunization of children at ages 2 months, 4 months, 6 months, 18 months; booster dose at time of entering school and at 10-year intervals after that; tetanus immune globulin (TIG), cleaning wound. Treatment: metronidazole, tetanus antitoxin.

grene bacillus, and the organism can frequently be recovered from cultures of wounds. Only rarely does contamination of a wound result in gas gangrene, however, because anaerobic conditions are necessary for the disease to develop. Primarily a disease of wartime, gas gangrene occurs mainly in neglected wounds with fragments of bone, foreign material, and extensive tissue damage, as might occur from schrapnel. Gas gangrene is highly unusual in peace-time gunshot wounds, but it occurs occasionally in abdominal and other surgeries, especially in persons with underlying diseases.

Symptoms

Gas gangrene **(figure 24.10)** begins abruptly, with pain rapidly increasing in the infected wound. Increased swelling occurs in the area, and a thin, bloody or brownish fluid leaks from the wound. The fluid may have a frothy appearance due to gas formation by the organism. The overlying skin becomes stretched tight and mottled with black. The victim appears very ill and apprehensive, but remains quite alert until late in the illness when, near death, he or she becomes delirious and lapses into a coma.

Causative Agent

Several species of *Clostridium* can produce life-threatening gas gangrene when they invade injured muscle, but by far the most common offender is *C. perfringens.* These encapsulated Gram-positive rods are shorter and fatter than *C. tetani* and usually do not exhibit spores in material from wounds or cultures.

Pathogenesis

Two main factors foster the development of gas gangrene: (1) the presence of dirt and dead tissue in the wound and (2) long delays before the wound gets medical attention. The toxin implicated in pathogenicity is α-toxin, an enzyme that attacks a vital component of host cell membranes called lecithin.

Clostridium perfringens is unable to infect healthy tissue but grows readily in dead and poorly oxygenated tissue, releasing α-toxin. Growth is fostered by anaerobic conditions, as well as by the presence of growth factors and amino acids in dead tissue. Curiously, these infections are generally not ominous until they invade muscle. The toxin diffuses from the area of infection, killing leukocytes and tissue cells. Several enzymes produced by the pathogen, including collagenase and hyaluronidase, break down macromolecules in the dead tissues to smaller ones, thereby promoting swelling. The organisms grow readily in the fluids of the dead tissue, producing hydrogen and carbon dioxide from fermentation of amino acids and muscle glycogen. These gases accumulate in the tissue and contribute to the rise in pressure, thereby fostering spread of the infection. Without prompt surgical treatment, massive amounts of α-toxin diffuse into the bloodstream and destroy red blood cells, tissue capillaries, and other structures throughout the body and cause death. The reason for the rapid onset of severe toxicity when muscle becomes involved is unclear. It is not reversed by administering antibody to α-toxin. ■ growth factors, p. 94 ■ membrane-damaging toxin, p. 405

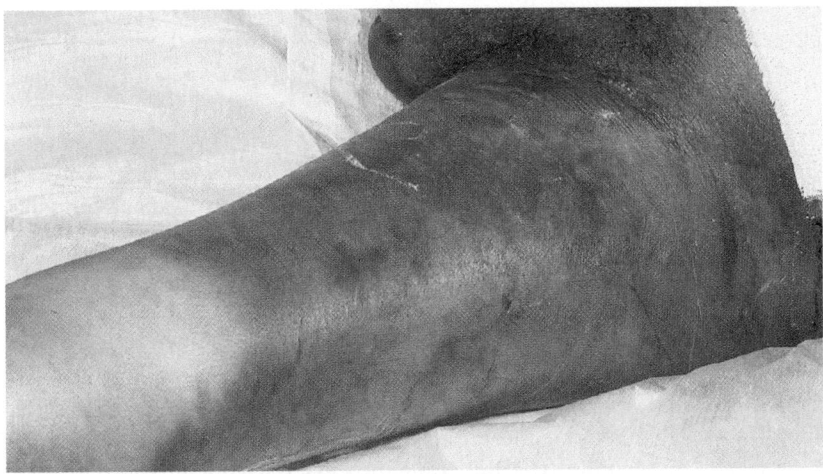

FIGURE 24.10 Individual with Gas Gangrene (Clostridial Myonecrosis) Fluid seeping from the involved area typically shows bits of muscle digested by *Clostridium perfringens.* Leukocytes are absent because the clostridial toxin kills them.

Epidemiology

Besides being widespread in soil, *C. perfringens* is present in the feces of many animals and humans, and sometimes in the vagina of healthy women. Besides neglected battlefield wounds and occasional surgical wounds, gas gangrene of the uterus is fairly common after self-induced abortions, and rarely, it can occur after miscarriages and childbirth. In other cases, impaired oxygenation of tissue from poor blood flow because of arteriosclerosis, otherwise known as hardening of the arteries, and diabetes are predisposing factors. Cancer patients also have increased susceptibility to gas gangrene.

Prevention and Treatment

Neither toxoid nor vaccine is available for immunization against gas gangrene. Prompt cleaning and removal of dead tissue from wounds, is highly effective in preventing the disease. Treatment of gas gangrene depends primarily on the prompt surgical removal of all dead and infected tissues, and it may require amputations. Some authorities recommend hyperbaric (under increased pressure) oxygen treatment because it inhibits growth of the clostridia, thereby stopping release of toxin, and it also improves oxygenation of injured tissues. In this treatment the patient is placed in a special chamber and breathes pure oxygen under three times normal atmospheric pressure. Some people appear to improve dramatically with hyperbaric oxygen, but others show no benefit and develop complications from the treatment. Antibiotics such as penicillin are given to help stop bacterial growth and toxin production, but they are of minor if any value in treating the disease because they do not diffuse well into large areas of dead tissue and they do not inactivate toxin.

Table 24.4 describes the main features of this disease.

TABLE 24.4 Gas Gangrene (Clostridial Myonecrosis)

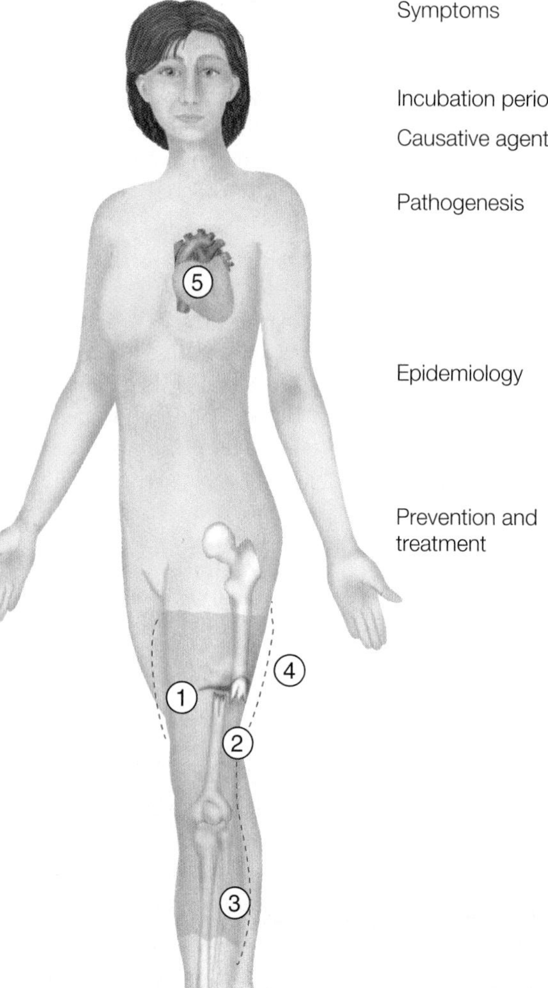
① *Clostridium perfringens* spores enter a wound having two essential characteristics: dead tissue and anaerobic conditions.
② The spores germinate, and the vegetative bacteria multiplying in dead tissue produce α-toxin.
③ α-toxin diffuses into normal tissue and kills it. The infection expands into the newly killed tissue, the bacteria utilizing amino acids and growth factors released from the tissue by bacterial enzymes.
④ Swelling and gas produced by fermenting amino acids and muscle glycogen aid rapid progress of the infection.
⑤ Massive amounts of α-toxin are produced and diffuse into the bloodstream, destroying blood and other cells throughout the body.
5
1
4
2
3

Symptoms	Severe pain, gas and fluid seep from wound, blackening of overlying skin; shock and death commonly follow
Incubation period	Usually 1 to 5 days
Causative agent	Usually *Clostridium perfringens;* other clostridia less frequently
Pathogenesis	Organism grows in dead and poorly oxygenated tissue and releases α-toxin; toxin kills leukocytes and normal tissue cells by degrading the lecithin component of their cell membranes; involvement of muscle causes shock by unknown mechanism.
Epidemiology	Wounds of war; dirt contamination of wounds, tissue death, impaired circulation to tissue as in persons with poor circulation from diabetes and arteriosclerosis; self-induced abortions.
Prevention and treatment	Prompt cleaning and debridement of wounds is preventive; no vaccine available. Treatment: surgical removal of dirt and dead tissues of primary importance; hyperbaric oxygen of possible value; antibiotics to kill vegetative *C. perfringens* of marginal value.

"Lumpy Jaw" (Actinomycosis)

Many wound infections are caused by anaerobes other than clostridia. The main offenders are members of the normal microbiota of the mouth and intestine. Actinomycosis is not a mycosis because it is not caused by a fungus, but its name persists from when it was thought to be a fungal disease.

Symptoms

Actinomycosis is characterized by slowly progressive, sometimes painful swellings under the skin that eventually open and intermittantly drain pus. The openings usually heal, only to reappear at the same or nearby areas days or weeks later. Most cases involve the area of the jaw and neck; the resulting scars and swellings gave rise to the popular name "lumpy jaw" **(figure 24.11).** In other cases the recurrent swellings and drainage develops on the chest or abdominal wall, or in the genital tract of women.

Causative Agent

Most cases of actinomycosis are caused by *Actinomyces israelii,* a Gram-positive, filamentous, branching, anaerobic bacterium that grows slowly on laboratory media. A number of similar species can cause the disease in animals and humans.

Pathogenesis

Actinomyces israelii cannot penetrate the normal mucosal surface, but it can establish an infection in association with other organisms of the normal microbiota if introduced into tissue by wounds. The infectious process is characterized by cycles of abscess formation, scarring, and formation of passageways that generally ignore tissue boundaries. The disease usually progresses to the skin where pus is discharged, but occasionally it penetrates into bone, or into the central nervous system. In the tissue, *A. israelii* grows as dense yellowish colonies

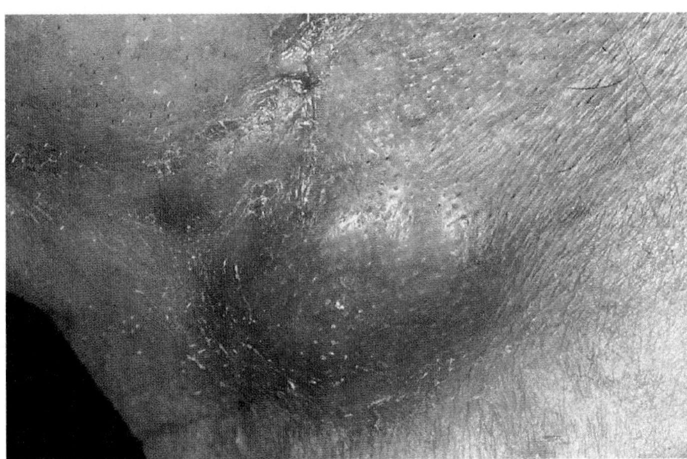

FIGURE 24.11 "Lumpy Jaw" (Actinomycosis) of the Face Abscesses, such as the one shown, form, drain, and heal, and then they recur at the same or different location. Sulfur granules, colonies of *Actinomyces israelii*, can be found in the drainage.

called **sulfur granules (figure 24.12)** because they are the color and size of particles of elemental sulfur. Because other more rapidly growing bacteria are invariably present, finding sulfur granules in the sinus drainage is a great aid in establishing the diagnosis. Almost half of the cases originate from wounds in the mouth, the remainder from the lung, intestine, or vagina. The detailed mechanisms by which the organism produces the disease have not yet been established.

Epidemiology

Actinomyces israelii can be part of the normal microbiota of the mucosal surfaces of the mouth, upper respiratory tract, intestine, and sometimes the vagina. It is common in the gingival crevice, particularly with poor dental care. Pelvic actinomycosis can complicate use of intrauterine contraceptive devices (IUDs). Other species of *Actinomyces* responsible for actinomycosis of dogs, cattle, sheep, pigs, and other animals generally do not cause human disease. The disease is sporadic and is not transmitted person to person. ■ gingival crevice, p. 582

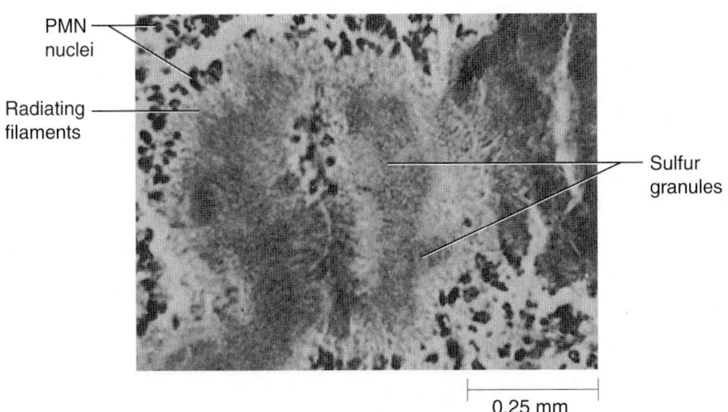

FIGURE 24.12 *Actinomyces israelii* Sulfur Granule Microscopic view of a stained smear of pus from an infected person. Filaments of the bacteria can be seen radiating from the edge of the colony.

TABLE 24.5 "Lumpy Jaw" Actinomycosis

Symptoms	Chronic disease; recurrent, sometimes painful swellings open and drain pus, heal with scarring; usually involves face and neck; chest, abdomen, and pelvis other common sites
Causative agent	*Actinomyces israelii,* a filamentous, branching, Gram-positive, slow-growing, anaerobic bacterium
Incubation period	Months (usually indeterminate)
Pathogenesis	Usually begins in a mouth wound, extends without regard to tissue boundaries to the face, neck, or upper chest; sometimes begins in the lung, intestine, or female pelvis. *A. israelii* always accompanied by normal microbiota. In tissue, grows as dense yellowish colonies called sulfur granules.
Epidemiology	No person-to-person spread. *A. israelii* commonly part of normal mouth, upper respiratory, intestine, and vagina microbiota. Dental procedures, intestinal surgery, insertion of IUDs can initiate infections.
Prevention and treatment	No proven preventive measures. Because of its slow growth. *A. israelii* infections require prolonged treatment; a number of antibacterial medications are effective.

Prevention and Treatment

There are no proven preventive measures. Actinomycosis responds to treatment with a number of antibacterial medications, including penicillin and tetracycline. To be successful, however, treatment must be given over weeks or months. This need for prolonged therapy probably is due to the slow growth of the organisms and their tendency to grow in dense colonies.

The main features of actinomycosis are presented in **table 24.5.**

MICROCHECK 24.3

Lockjaw (tetanus) is caused by an anaerobic, spore-forming bacterium with essentially no invasive ability, yet it causes death by releasing an exotoxin that is carried to the nervous system. Gas gangrene (clostridial myonecrosis), caused by another anaerobic spore-former, usually arises from neglected wounds containing dead tissue and foreign material and advances into normal muscle tissue. Lumpy jaw (actinomycosis) is caused by a branching, filamentous, slow-growing, anaerobic bacterium, a member of the normal flora of the mouth, upper respiratory tract, intestine, and vagina. The disease is characterized by recurrent abscesses that drain sulfur granules.

- ✓ Why do many tetanus victims fail to respond to tetanus antitoxin?
- ✓ What factors favor the development of gas gangrene (clostridial myonecrosis)?
- ✓ Would babies need to be immunized against lockjaw (tetanus) if their mother had been immunized against the disease? Explain.

24.4

Bacterial Bite Wound Infections

Focus Points

- Describe two zoonotic wound infections.
- Explain why human mouth microbiota can cause serious bite wound infections.

Each year, more than 3 million animal bites occur in the United States, the most feared result being the viral disease, rabies. The infection risk following an animal bite depends partly on the type of injury (crushing, lacerated, or puncture), and partly on the kinds of infectious agents in the animal's mouth. ■ rabies, p. 663

Pasteurella multocida Bite Wound Infections

Infections caused by bacteria that live in the mouth of the biting animal are very common. Surprisingly, a single species, *Pasteurella multocida,* is responsible for bite infections from a number of kinds of animals, including dogs, cats, monkeys, and humans, among others.

Symptoms

There are no reliable symptoms or signs that distinguish among most bacterial bite wound infections. Spreading redness and tenderness and swelling of tissues adjacent to the wound, followed by the discharge of pus, are early indications of infection.

Causative Agent

Pasteurella multocida is a Gram-negative, facultatively anaerobic coccobacillus that is easily cultivated in the laboratory. Most isolates are encapsulated. There are a number of different antigenic types.

Pathogenesis

The details of pathogenesis are not yet known. There is evidence for one or more adhesins. Some strains produce a toxin that kills cells. The capsules of *P. multocida* are antiphagocytic. When opsonized by specific antibody, the organisms are ingested and killed by phagocytes. Abscesses commonly form. Without prompt treatment, infection can lead to bloodstream invasion or permanent loss of function.

Epidemiology

Pasteurella multocida is best known as the cause of a devastating disease of chickens called **fowl cholera.** This disease is historically significant because while studying it, Pasteur first discovered that the virulence of a pathogen could be attenuated by repeated laboratory culture, and the attenuated organism used as a vaccine. *Pasteurella multocida* also causes diseases in a number of other animal species. Epidemics of fatal pneumonia and bloodstream infection occur in rabbits, cattle, sheep, and mice.

Many healthy animals, however, carry the bacterium among their normal oral and upper respiratory microbiota. Both diseased animals and healthy carriers constitute a reservoir for human infections. Cats are more likely to carry. *P. multocida* than dogs, and so cat bites are more likely than dog bites to cause the infection.

Prevention and Treatment

No vaccines are available for use in humans. Immediate cleansing of bite wounds and prompt medical attention usually prevent the development of serious infection and possible permanent impairment of function. Unlike many Gram-negative pathogens, *P. multocida* is susceptible to penicillin. Usually, before the cultural diagnosis is known, amoxicillin plus a β-lactamase inhibitor are administered, available in a single tablet under the trade name Augmentin. This combination is used because amoxicillin is active against *P. multocida,* and with the addition of the β-lactamase inhibitor, the antibiotic is also active against many strains of β-lactamase-producing *Staphylococcus aureus,* another common cause of bite wound infections. This and other antibacterial medications are effective if given early in the infection.

Table 24.6 gives the main features of *Pasteurella multocida* bite wound infections.

Cat Scratch Disease

In the United States, cat scratch disease is the most common cause of chronic lymph node enlargement at one body site in young children **(figure 24.13).** Despite its name, this disease can be transmitted by bites, and probably other means. Typically the infection is mild and localized, lasting from several weeks to a few months. Of the estimated 22,000 annual cases, less than 2,000 require hospitalization.

Causative Agent

Cat scratch disease is caused by *Bartonella* (formerly *Rochalimaea) henselae,* a tiny, slightly curved, Gram-negative rod cultivable on laboratory media.

Symptoms

In about half the cases, the disease begins within a week of a scratch or bite with the appearance of a pus-filled pimple at the

TABLE 24.6	*Pasteurella multocida* Bite Wound Infections
Symptoms	Spreading redness, tenderness, swelling, discharge of pus
Incubation period	24 hours or less
Causative agent	*Pasteurella multocida,* a Gram-negative, facultatively anaerobic, encapsulated coccobacillus
Pathogenesis	Introduced by bite, *P. multocida* attaches to tissue, resists phagocytes because of its capsule; probable cell-destroying toxin. Extensive swelling, abscess formation. Opsonins develop, allow phagocytic killing, limit spread.
Epidemiology	Carried by many animals in their mouth or upper respiratory tract.
Prevention and treatment	No vaccines to use in humans. Prompt wound care is preventive. Treatment with penicillin, other antibacterials, effective if given promptly.

CASE PRESENTATION

A 63-year-old woman, healthy except for mild diabetes, underwent surgery for a diseased gallbladder. The surgery went well, but within 72 hours the repaired surgical incision became swollen and pale. Within hours the swollen area widened and developed a bluish discoloration. The woman's surgeon suspected gangrene. Antibiotic therapy was started and she was rushed back to the operating room where the entire swollen area, including the repaired operative incision, was surgically removed. After that, the wound healed normally, although she required a skin graft to close the large skin deficit. Large numbers of *Clostridium perfringens* grew from the wound culture.

Six days later, a 58-year-old woman underwent surgery in the same operating room for a malignant tumor of the colon. The surgery was performed without difficulty, but 48 hours later she developed rapidly advancing swelling and bluish discoloration of her surgical wound. As with the first case, gangrene was suspected and she was treated with antibiotics and surgical removal of the affected tissue. She also required skin grafting. Her wound culture also showed a heavy growth of *Clostridium perfringens.*

Because the surgery department had never had any of its patients develop surgical wound infections with *Clostridium perfringens,* much less two cases so close together, the hospital epidemiologist was asked to do an investigation. Among the findings of the investigation:

1. Cultures of horizontal surfaces in the operating room grew large numbers of *Clostridium perfringens:*
2. Unknown to the medical staff, a workman had recently serviced a fan in the ventilation system of the operating room, and for a time air was allowed to flow into the operating room, rather than out of it.
3. Heavy machinery was doing grading outside the hospital, creating clouds of dust.

As a result of these findings, the operating room and its ventilating system were cleaned and upgraded. No further cases of surgical wound gangrene developed.

1. Was the surgeon's diagnosis correct?
2. Many other patients had surgery in the same operation room. Why did only these two patients develop wound gangrene?
3. What could be done to help identify the source of the patients' infections?

Discussion

1. *Clostridium perfringens* is commonly cultivated from wounds without any evidence of infection. However, in these cases, there was not only a heavy growth of the organism but a clinical picture compatible with gangrene. The surgeon's diagnosis was undoubtedly correct.
2. In both cases there was an underlying condition that increased the two patients' susceptibility to infection—cancer in one, and diabetes in the other. Moreover, both had a recognized source for the organism. Cultures of as many as 20% of diseased gallbladders are positive for *Clostridium perfringens,* while the organism is commonly found in large numbers in the human intestine—a potential source in the case involving removal of the bowel malignancy.
3. The surgeon favored the idea that the infecting organism came from the patients themselves because such strains tend to be much more virulent than strains that live and sporulate in the soil. One the other hand, the gross contamination of the operating room as revealed by the cultures of its surfaces could indicate a very large infecting dose at the operative site, possibly compensating for lesser virulence. Moreover, no further cases occurred after cleaning the operating room and fixing the ventilation system. Unfortunately, in this case, no cultures of the excised gallbladder or bowel tumor were done, nor were the strains isolated from the wounds and the environment compared. Comparing the antibiotic susceptibility, toxin production, and other characteristics of the different isolates could have helped identify the source of the infections.

site of injury. Painful enlargement of the lymph nodes of the region develops in 1 to 7 weeks. About one-third of the patients develop fever, and in about half that number the lymph nodes become pus-filled and soften. The disease generally disappears without treatment in 2 to 4 months. About 10% of cases develop irritation of an eye with local lymph node enlargement, epileptic seizures and coma due to encephalitis, or acute or chronic fever associated with bloodstream or heart valve infection.

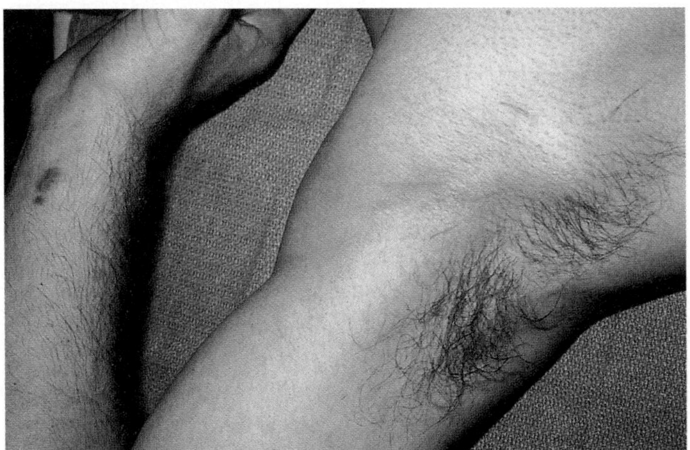

FIGURE 24.13 Individual with Cat Scratch Disease, Showing Enlarged Lymph Node and Wrist Lesion

Pathogenesis

The virulence factors of *B. henselae* and the process by which it causes disease are not yet known. The organisms can enter the body via a cat bite or scratch, or when cat saliva contaminates a mucous membrane. They are carried to the lymph nodes, and the disease is arrested by the immune system in most cases. Spread by the bloodstream, however, occurs in some individuals. **Peliosis hepatis** and **bacillary angiomatosis** are two complicating conditions seen mostly in people with AIDS. In peliosis hepatis, blood-filled cysts form in the liver. In bacillary angiomatosis, nodules composed of proliferating blood vessels develop in the skin and other parts of the body. *Bartonella henselae* identifiable with fluorescent antibody is present in these lesions and those of typical cat scratch disease.

Epidemiology

Cat scratch disease mainly occurs in people under 18 years old. The disease is a zoonosis of cats, transmitted to humans mainly by kittens. Person-to-person spread does not occur. Bites and scratches represent the usual mode of transmission of *B. henselae* from cats to people. Asymptomatic bacteremia is common in cats, however, and transmission from cat to cat occurs by cat fleas. The fleas are biological vectors, and they discharge *B. henselae* in their feces. Cat fleas probably are responsible for transmission in cases where people develop the disease after they handle cats but are not scratched or bitten. The disease occurs worldwide. ■ **biological vectors, p. 455**

Prevention and Treatment

There are no proven preventive measures for cat scratch disease. It is prudent to avoid handling stray cats, however, especially young ones and those with fleas. Any cat-inflicted wound should be promptly cleaned with soap and water and then treated with an antiseptic. If signs of infection develop, or if the cat's immunization status against rabies is uncertain, prompt medical evaluation is indicated. Severely immunodeficient persons should avoid cats if possible and, if not, control of fleas and abstinence from rough play are advisable. Severe *B. henselae* infections can usually be treated with antibacterial medications such as ampicillin, but some strains are resistant.

The main features of cat scratch disease are presented in **table 24.7.**

Streptobacillary Rat Bite Fever

Rat bites are fairly common among the poor people of large cities and among workers who handle laboratory rats. In the past, as many as one out of 10 bites resulted in rat bite fever. Currently, the disease is not reportable in any state and the incidence of the disease in the United States is unknown.

Symptoms

Usually, the bite wound heals promptly without any problem noted. Two to 10 days later, however, chills and fever, head and muscle aches, and vomiting develop. The fever characteristically comes and goes. A rash usually appears after a few days, followed by pain on motion of one or more of the large joints.

TABLE 24.7 Cat Scratch Disease

Symptoms	Pimple appears at the bite or scratch site, followed by local lymph node enlargement, fever; nodes may soften and drain pus; prolonged fever, convulsions, indicate spread to other body parts
Incubation period	Usually less than 1 week
Causative agent	*Bartonella henselae,* a tiny Gram-negative rod
Pathogenesis	The bacteria enter with cat bite or scratch and reach lymph nodes, where the disease is usually arrested. May spread by bloodstream, cause infections of the heart, brain, or other organs. Peliosis hepatis, bacillary angiomatosis mostly in those with AIDS.
Epidemiology	A zoonosis of cats, spread from one to another by cat fleas, which are biological vectors and have *B. henselae* in their feces. Infected cats usually asymptomatic, but often bacteremic. Humans are accidental hosts. No person-to-person spread. Mainly a disease of those under 18 years old.
Prevention and treatment	Avoiding rough play. Promptly wash skin breaks, apply antiseptic. Flea control in cats. Most *B. henselae* susceptible to antibacterial treatment.

Causative Agent

The cause of streptobacillary rat bite fever is *Streptobacillus moniliformis,* a facultatively anaerobic, Gram-negative rod. Stained smears show a multiplicity of forms ranging from small coccobacilli to unbranched filaments more that 100 μm long. The organism is unique in that it spontaneously develops **L-forms.** L-forms are cell wall-deficient variants, first identified at the famous Lister Institute (hence, the *L* in *L-form*). As might be expected, L-form colonies resemble those of mycoplasmas—bacteria that lack a cell wall. ■ mycoplasmas, p. 273

Pathogenesis

The organisms enter the body through a bite or scratch, and sometimes by ingestion. There is typically little enlargement of the local lymph nodes, and *S. moniliformis* quickly enters the bloodstream and spreads throughout the body. The majority of cases recover without treatment in about 2 weeks, but some are rapidly fatal, and others develop serious complications such as brain abscesses or infection of the heart valves. About 7% to 10% of untreated cases are fatal.

Epidemiology

From 50% to 100% of wild, and 10% to 100% of healthy laboratory rats, as well as mice and other rodents, carry the organism in their nose and throat. Laboratory and pet store workers are at increased risk of contracting the disease. Rat bites and scratches are the usual source of human infections and can occur while the victim is sleeping. However, 30% of rat bite fever victims do not report bites or scratches. Epidemics of the disease have arisen from ingesting milk, water, or food contaminated with *S. moniliformis* from rodent droppings. The foodborne disease is called **Haverhill fever** from a 1926 epidemic in Haverhill, Massachusetts. Cases of rat bite fever have also been associated with exposure to animals that prey on rodents, including cats and dogs.

Prevention and Treatment

Wild rat control and care in handling laboratory rats and their droppings are reasonable preventive measures. Penicillin, given intravenously, is the treatment of choice for cases of rat bite fever. This shows that either *S. moniliformis* L-forms do not occur *in vivo,* or if they do, they are avirulent.

The main features of streptobacillary rat bite fever are presented in **table 24.8.**

Human Bites

Wounds caused from human bites, striking the teeth of another person, or resulting from objects that have been in a person's mouth are common and can result in very serious infections. Rarely, diseases such as syphilis, tuberculosis, and hepatitis B are transmitted this way. Much more commonly, it is the normal mouth microbiota that cause trouble.

Symptoms

The wound may appear insignificant at first but then becomes painful and swells massively. Discharged pus often has a foul smell. Most of the wounds are on the extensor surface on the hand, and here the swelling may soon involve the palm also, and movement of some or all of the fingers becomes difficult or impossible.

PERSPECTIVE 24.1

Infection Caused by a Human "Bite"

A 26-year-old man injured the knuckle of the long finger of his right hand during a tavern brawl when he punched an assailant in the mouth. Due to his inebriated condition, a night spent in jail, and the insignificant early appearance of his knuckle, the man did not seek medical help until more than 36 hours later. At that time, his entire hand was massively swollen, red, and tender. Furthermore, the swelling was spreading to his arm. The surgeon cut open the infected tissues, allowing the discharge of pus. He removed the damaged tissue and washed the wound with sterile fluid. Smears and cultures of the infected material showed aerobic and anaerobic bacteria characteristic of mouth microbiota, including species of *Bacteroides* and *Streptococcus.* The patient was given antibiotics to combat infection, but the wound did not heal well and continued to drain pus. Several weeks later, X-rays revealed that infection had spread to the bone at the base of the finger. To cure the infection, the finger had to be amputated.

Causative Agents

Stains and cultures usually show members of the normal mouth microbiota, including anaerobic streptococci, fusiforms, spirochetes, and *Bacteroides* sp., often in association with *Staphylococcus aureus.*

Pathogenesis

The crushing nature of bite wounds provides suitable conditions for anaerobic bacteria to establish infection. Although most members of the mouth microbiota are harmless alone, together they produce an impressive number of toxins and destructive enzymes. These include leukocidin, collagenase, hyaluronidase, ribonuclease, various proteinases, neuraminidase, and enzymes that destroy proteins of the complement system and antibody. Capsules of some species inhibit phagocytosis. Facultatively anaerobic organisms reduce available oxygen and thus encourage the growth of anaerobes. The result of all these factors is a **synergistic infection,** meaning that the sum effect of all the organisms acting together is greater than the sum of their individual effects. Irreversible destruction of tissues such as tendons and permanent loss of function can be the result. ■ facultative anaerobes, p. 92

Epidemiology

Most of the serious human bite infections occur in association with violent confrontations related to alcohol ingestion, or during forcible restraint, as in law enforcement and in mental institutions. The risk is greatly increased when the biting individual has poor mouth care and extensive dental disease. Bites by little children are usually inconsequential.

Prevention and Treatment

Prevention involves avoiding situations that lead to uncivilized behavior such as biting and hitting. Prompt cleansing of wounds followed by application of an antiseptic are advised, and most important is immediate medical attention if there is any suspicion of developing infection. Treatment of infected wounds consists of opening the infected area widely with a scalpel, washing the wound thoroughly with sterile fluid, and removing dirt and dead tissue. The choice of antibacterial medication includes one effective against anaerobes.

The main features of human bite wound infections are presented in **table 24.9.**

TABLE 24.8 Streptobacillary Rat Bite Fever

Symptoms	Chills, fever, muscle aches, headache, and vomiting; later, rash and pain in one or more of the large joints
Incubation period	Usually 2 to 10 days (range, 1 to 22) after a rat bite
Causative agent	*Streptobacillus moniliformis,* a highly pleomorphic, Gram-negative bacterium that spontaneously produces L-forms
Pathogenesis	Bite wound heals without treatment; *S. moniliformis* quickly invades bloodstream. Fevers come and go irregularly. Most victims recover without treatment; in others, infection established in various body organs, results in death if treatment is not given.
Epidemiology	Wild and laboratory rats can carry *S. moniliformis.* Bites of other rodents and animals that prey on them can transmit the disease to humans. Food or drink contaminated with rodent excreta can also transmit the infection; foodborne disease is called Haverhill fever.
Prevention and treatment	Control wild rats and mice, care in handling laboratory animals. Effectively treated with penicillin, other antibacterial medications.

TABLE 24.9 Human Bite Wound Infections

Symptoms	Rapid onset, pain, massive swelling, drainage of foul-smelling pus
Incubation period	Usually 6 to 24 hours
Causative agent	Mixed mouth flora: anaerobic streptococci, fusiforms, spirochetes, anaerobic Gram-negative rods; sometimes *Staphylococcus aureus*
Pathogenesis	Various mouth bacteria act synergistically to destroy tissue.
Epidemiology	Alcohol-related violence; forcible restraint; poor mouth care and extensive dental disease.
Prevention and treatment	No proven preventive measures except to avoid altercations. Prompt cleansing of wound and application of antiseptic is advised. Treatment is usually surgical.

MICROCHECK 24.4

Pasteurella multocida can infect bite wounds caused by a number of different animals, notably cats. Cat bites and scratches can also transmit *Bartonella henselae,* cause of cat scratch disease, characterized typically by local lymph node enlargement, but the disease may involve other parts of the body. Streptobacillary rat bite fever, caused by *Streptobacillus moniliformis,* is acquired from bites of rats and mice, and animals that prey on them. It is marked by fever that comes and goes and a rash. The causative bacterium, *S. moniliformis,* spontaneously develops L-forms. Human bite infections can be dangerous because certain members of the mouth microbiota with little invasive ability when growing alone can invade and destroy tissue when growing together.

✓ What Gram-negative organism commonly infects wounds caused by animal bites?

✓ What is the most common cause of chronic localized lymph node enlargement in young children?

✓ Why would normal mouth flora be a more common cause of serious human bite infections than the causative agents of syphilis, tuberculosis, and hepatitis B, which can also be transmitted by human bites?

24.5

Fungal Wound Infections

Focus Points

- Give distinctive features of two wound infections caused by fungi originating from soil.
- Give evidence of the role of the immune system in controlling wound infections caused by fungi.

Fungal infections of wounds are unusual in economically developed countries, except that the yeast *Candida albicans* can be troublesome in severe burns and in those with wounds and underlying diseases such as diabetes and cancer. This yeast, commonly present among the normal microbiota and kept in check by it, becomes pathogenic when the competing microorganisms are eliminated, as in individuals receiving antibacterial therapy. Other fungal wound infections are much more common in impoverished people around the world. For example, Madura foot, a condition caused by various species of soil fungi, occurs in areas of the world where foot injuries are common, resulting from lack of shoes. Named after the city in India where it was first described, Madura foot is characterized by swellings and draining passageways that discharge yellow or black granules of fungal material. Only a minority of those with foot injuries contract the disease despite exposure to the same fungi, suggesting that other factors such as malnutrition may play a role. Sporotrichosis, another kind of fungal wound infection, occurs worldwide and is not poverty-related.

"Rose Gardener's Disease" (Sporotrichosis)

Sporotrichosis, also known as "rose gardener's disease," is widely distributed around the world and is associated with activities that lead to puncture wounds from vegetation. Although many cases are sporadic, the disease can occur in groups of people engaged in the same occupation. Thousands of workers in the warm, humid mines of South Africa have contracted the disease from splinters on mine timbers. Epidemics have occurred in the United States among handlers of sphagnum moss from Wisconsin.

Symptoms

In most cases, a hand or arm is involved, but the trunk, legs, and face can also be sites of infection. Typically, a chronic ulcer forms at the wound site, followed by a slowly progressing series of ulcerating nodules that develop sequentially toward the center of the body **(figure 24.14).** Lymph nodes in the region of the wound enlarge, but patients generally do not become ill. If they have AIDS or other immunodeficiency, however, the disease can spread throughout the body, threatening life.

Causative Agent

Sporotrichosis is caused by the dimorphic fungus *Sporothrix schenckii* **(figure 24.15),** which lives in soil and on vegetation.

■ dimorphic fungus, p. 290

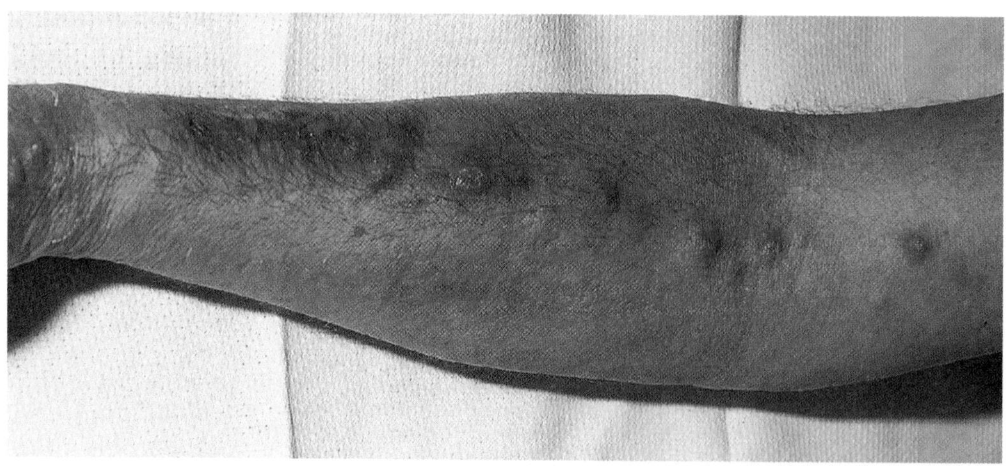

FIGURE 24.14 Individual with "Rose Gardener's Disease" (Sporotrichosis) Notice the multiple abscesses along the course of the lymphatic drainage from the hand.

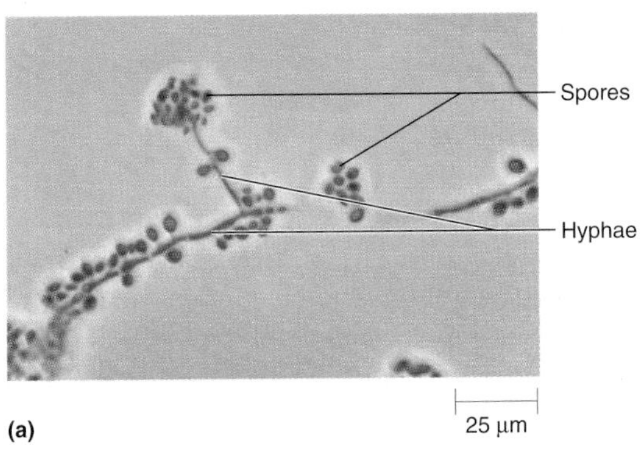

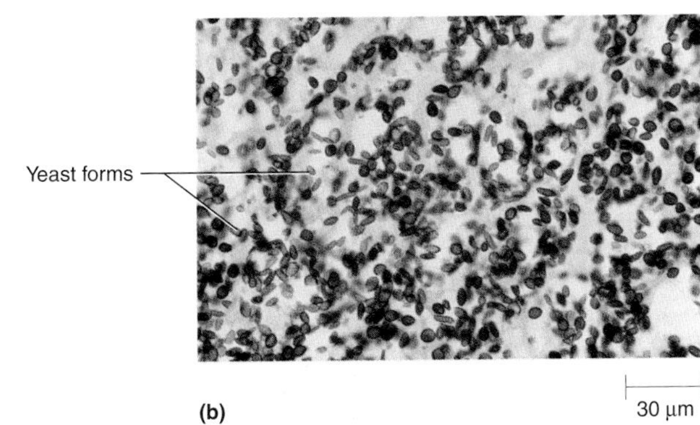

FIGURE 24.15 ***Sporothrix schenckii,*** **Showing the Two Forms of This Dimorphic Fungus** **(a)** Mold form, **(b)** Yeast form, as seen in infected tissue.

Pathogenesis

Sporothrix schenckii spores are usually introduced with an injury caused by plant material. After an incubation period that usually ranges from 1 to 3 weeks but can be much longer, the multiplying fungi cause a small nodule or pimple to form at the site of the injury. This lesion slowly enlarges and ulcerates, producing a red, easily bleeding skin defect. Unless the ulcer becomes secondarily infected with bacteria, there is little or no pus, and the lesion is pain-free. After a week or longer, the process repeats itself—progression of the disease usually follows the flow of a lymphatic vessel. In healthy individuals, the process does not proceed beyond the lymph node. Sometimes, however, satellite lesions appear irrespective of the lymphatic vessels. Without treatment, the disease can go on for years. ■ **lymphatic vessel, p. 370**

Epidemiology

Sporotrichosis is distributed worldwide, mostly in the warmer regions but extending into temperate climates. In the United States, most cases occur in the Mississippi and Missouri river valleys. It is an occupational disease of farmers, carpenters, gardeners, greenhouse workers, and others who deal with plant materials. Public health authorities do not require reporting of cases of sporotrichosis and its incidence is unknown. Risk factors for the disease besides occupation include diabetes, immunosuppression, and alcoholism. Children can contract the disease from playing in baled hay. Individuals with chronic lung disease can contract

TABLE 24.10	"Rose Gardener's Disease" (Sporotrichosis)
Symptoms	Painless, ulcerating nodules appearing in sequence in a linear pattern
Incubation period	Usually 1 to 3 weeks
Causative agent	*Sporothrix schenckii,* a dimorphic fungus
Pathogenesis	Spores multiply at site of introduction by thorn, splinter, or other plant material, causing a small nodule that ulcerates. Spores carried by lymph flow repeat the process along the course of the lymphatic vessel. May spread beneath the skin irrespective of lymphatic vessels.
Epidemiology	Distributed worldwide in tropical and temperate climates. Occupations requiring contact with sharp plant materials at particular risk.
Prevention and treatment	Protective gloves and clothing. Most cases effectively treated with potassium iodide. Generalized infections require amphotericin B or itraconazole.

S. schenckii lung infections from inhaling dust from hay or cattle feed. Deaths from sporotrichosis are rare.

Prevention and Treatment

Protective gloves and a long-sleeved shirt can help prevent sporotrichosis, especially when individuals are handling evergreen seedlings and sphagnum moss. This chronic disease is often misdiagnosed, leading to delayed and inappropriate treatment. Surprisingly, unlike other infections, sporotrichosis can usually be cured by oral treatment with the simple chemical compound potassium iodide (KI). KI is not active against *S. schenckii in vitro,* but somehow enhances the body's ability to reject the fungus. Itraconazole or the antibiotic amphotericin B is used in rare cases when the disease spreads throughout the body.

Table 24.10 presents some of the main features of sporotrichosis.

MICROCHECK 24.5

Candida albicans, often a member of the normal microbiota, can infect burns and other wounds, especially in individuals with underlying conditions. Madura foot, characterized by draining passageways that discharge colored granules, occurs mainly in areas of the world where shoes are scarce and foot injuries common. Rose gardener's disease (sporotrichosis), caused by the dimorphic fungus *Sporothrix schenckii,* is widely distributed around the world, affecting mainly those associated with occupations that expose them to splinters and sharp vegetation. Unlike most invasive fungal infections, sporotrichosis is usually easy to treat.

- ✓ What fungal disease victimizes people who can't afford shoes?
- ✓ What fungus infection is likely to result from thorn or splinter injuries? What is unique about the treatment of the disease?
- ✓ Why might *Candida albicans* become pathogenic in an individual receiving antibacterial medications?

FUTURE CHALLENGES

Staying Ahead in the Race with *Staphylococcus aureus*

Over the last half century or so, scientists have generally, but not always, kept us at least one step ahead in the race between *Staphylococcus aureus* and human health. Today, "staph" seems again poised to challenge us for the lead, but new knowledge from molecular biology promises to help us leave this microscopic menace in the dust.

In the past, we have relied heavily on developing antibacterial medications to kill or inhibit the growth of *S. aureus* but there is continuing investigation of alternative approaches. This search is aided by improvement in understanding staphylococcal pathogenesis. We now know in more detail the mechanism by which *S. aureus* obtains iron from hemoglobin. In another example, the bacterium's protein A has been shown to attach to TNF receptors, and to recruit enormous numbers of pus cells. There have also been advances in vaccine development. An experimental vaccine composed of staphylococcal capsular material conjugated with a nontoxic form of *Pseudomonas* exotoxin A has been shown to reduce substantially staphylococcal infections in kidney dialysis patients.

Other options now seem to be within reach. For example, it should be possible to determine the structure of the active portion of toxin molecules and design medications that would bind to that site and inactivate it. Also, working with the staphylococcal genome, it is possible to determine which staphylococcal genes are responsible for virulence. The products of these genes can prove to be unexpected vaccine candidates.

The race continues!

SUMMARY

24.1 Anatomy and Physiology

Wounds expose components of tissues to which pathogens specifically attach. Wounds heal by forming **granulation tissue,** which in the absence of dirt and infection, fills the defect, and subsequently contracts to minimize scar tissue (figure 24.1). Thermal burns often present large areas of dead tissue devoid of competing organisms and body defenses, ideal conditions for microbial growth.

Wound Abscesses

An **abscess** is composed of a collection of pus, which is composed of leukocytes, components of tissue breakdown, and infecting organisms. Abscess formation localizes an infection within tissue, preventing its spread; inflammatory cells and clotted blood vessels separate the abscesses from normal tissue (figure 24.2).

Anaerobic Wounds

Anaerobic conditions are likely to occur in wounds containing dead tissue or foreign material, and those with a narrow opening to the air. Anaerobic conditions permit infection by particularly dangerous pathogens.

24.2 Common Bacterial Wound Infections (table 24.1)

Possible consequences of wound infections include delayed healing, abscess formation, and extension of infection or toxins into adjacent tissue or the bloodstream. Infections can cause surgical wounds to split open, and they can spread to create biofilms on artificial devices.

Staphylococcal Wound Infections (figures 24.3, 24.4)

Staphylococci are the leading cause of wound infections, both surgical and accidental; *Staphylococcus aureus* and *S. epidermidis* are the most common wound-infecting species. *Staphylococcus aureus* possesses many virulence factors; occasional strains release a toxin that causes toxic shock syndrome (table 24.1). *Staphylococcus epidermidis* is less virulent but can form biofilms on blood vessel catheters and other devices.

Group A Streptococcal "Flesh Eaters"

Streptococcus pyogenes (Group A, β-hemolytic streptococcus) causes strep throat, scarlet fever, wound infections, and other conditions. Necrotizing fasciitis-causing strains of *S. pyogenes* produce exotoxin B, a protease thought to be responsible for the tissue destruction (figure 24.5).

Pseudomonas aeruginosa Infections

Pseudomonas aeruginosa, an aerobic, Gram-negative rod with a single polar flagellum, is an opportunistic pathogen widespread in the environment, and a cause of both nosocomial infections and those acquired outside the hospital. Production of two pigments by the bacterium often colors infected wounds green (figure 24.6). Toxin A of *P. aeruginosa* has a mode of action identical to that of diphtheria toxin, but it is antigenically distinct and attaches to different cells.

24.3 Diseases Due to Anaerobic Bacterial Wound Infections

"Lockjaw" (Tetanus) (table 24.3)

The characteristic symptom of tetanus is sustained, painful, cramplike spasms of one or more muscles (figure 24.7). The disease is often fatal. Tetanus is caused by an exotoxin, **tetanospasmin,** produced by *Clostridium tetani,* a non-invasive, anaerobic, Gram-positive rod (figure 24.8). The toxin renders the nerve cells that normally inhibit muscle contraction inactive by blocking release of their neurotransmitter (figure 24.9). The spores of *C. tetani* are widespread in dust and dirt, and most wounds are probably contaminated with them. Tetanus can be prevented by vaccination with toxoid (inactivated tetanospasmin), and maintaining immunity throughout life. Any wound, including surgeries, no matter how trivial, is an occasion to make sure immunizations are current (table 24.2).

Gas Gangrene (Clostridial Myonecrosis) (table 24.4)

Usually caused by the anaerobe *Clostridium perfringens,* symptoms begin abruptly with pain, and swelling; a thin, brown, bubbly discharge; and dark blue mottling of the tightly stretched overlying skin (figure 24.10). The toxin causes tissue necrosis; hydrogen and carbon dioxide gases are produced from fermentation of amino acids and glycogen in the dead tissue. There is no vaccine or toxoid. Prevention depends on prompt medical care of dirty wounds containing dead tissue. Treatment depends on urgent surgical removal of dead and infected tissue, and it may require amputation.

"Lumpy law" (Actinomycosis) (table 24.5)

Actinomycosis is a chronic, slowly progressive disease characterized by repeated swellings, discharge of pus, and scarring, usually of the face and neck (figure 24.11).

The causative agent is *Actinomyces israelii,* a member of the normal mouth, intestinal, and vaginal flora that enters tissues with wounds such as those with dental and intestinal surgery (figure 24.12).

The organism is slow growing; treatment must be continued for weeks or months.

24.4 Bacterial Bite Wound Infections

The kind of bite wound infection depends on the kinds of infectious agents in the mouth of the biting animal, and the nature of the wound—whether punctured, crushed, or torn.

Pasteurella multocida Bite Wound Infections (table 24.6)

Pasteurella multocida, a small, Gram-negative rod, can infect bite wounds inflicted by a number of animal species. *P. multocida* causes **fowl cholera** and diseases in other animals, but many animals are asymptomatic carriers.

Cat Scratch Disease (table 24.7, figure 24.13)

Cat scratch disease is the most common cause of chronic, localized lymph node enlargement in children. Caused by *Bartonella henselae,* it begins with a pimple at the site of a bite or scratch, followed by enlargement of local lymph nodes, which often become pus-filled. Most individuals with cat scratch disease recover without treatment, but persons with AIDS are susceptible to two serious conditions caused by *B. henselae*—**peliosis hepatis** and **bacillary angiomatosis.**

Streptobacillary Rat Bite Fever (table 24.8)

Streptobacillary rat bite fever is characterized by relapsing fevers, head and muscle aches, and vomiting, following a rat bite. A rash and joint pains often develop. It is usually caused by *Streptobacillus moniliformis,* a highly pleomorphic, Gram-negative rod that characteristically produces cell wall deficient variants called L-forms. People who live in rat-infested dwellings and those who handle laboratory rats are at greatest risk of contracting the disease.

Human Bites (table 24.9)

Human bites, injuries from objects that have been in a person's mouth, and those from striking a person in the teeth often result in severe infections, with pain, massive swelling, and foul-smelling pus. The infections are usually caused by members of the normal mouth flora acting synergistically, including anaerobic streptococci, fusiforms, spirochetes, and *Bacteroides* sp., often with *Staphylococcus aureus.* The crushing nature of bite wounds causes death of tissue and conditions suitable for growth of anaerobes; prompt cleansing and application of an antiseptic are advised.

24.5 Fungal Wound Infections

Fungal wound infections are unusual in economically developed countries, except for *Candida albicans* infections of burns and other wounds in individuals receiving antibacterial therapy. Madura foot occurs in many impoverished areas of the world where people do not wear shoes.

"Rose Gardener's Disease" (Sporotrichosis) (table 24.10)

Rose gardener's disease, also called sporotrichosis, is a chronic fungal disease mainly of people who work with wood or vegetation. The usual case is characterized by painless, ulcerating nodules that develop one after the other along the course of a lymphatic vessel (figure 24.14). The causative organism is the dimorphic fungus, *Sporothrix schenckii* (figure 24.15), usually introduced into wounds caused by thorns or splinters. The fungus is distributed worldwide in tropical and temperate climates; people at risk include farmers, carpenters, gardeners, and greenhouse workers.

REVIEW QUESTIONS

Short Answer

1. Give three undesirable consequences of wound infection.
2. What property of *Staphylococcus epidermidis* aids its ability to colonize plastic materials used in medical practice?
3. What is the relationship between the superantigens of *S. aureus* and the organism's production of toxic shock?
4. Name two underlying conditions that predispose to *Streptococcus pyogenes* "flesh-eating" disease.
5. Give two sources of *Pseudomonas aeruginosa.*
6. Outline the pathogenesis of lockjaw.
7. What characteristic of *Clostridium tetani* aids its isolation when other bacteria are present?
8. Explain why *C. tetani* can be cultivated from wounds in the absence of tetanus.
9. What characteristics of bite wounds favor anaerobic infections?
10. What is the causative agent of cat scratch disease? Why is it a threat to patients with AIDS?
11. What is a synergistic infection? How might one be acquired?
12. Why is sporotrichosis sometimes called rose gardener's disease?

Multiple Choice

1. All of the following are true of *Staphylococcus aureus, except*
   a) it is generally coagulase-positive.
   b) its infectious dose is increased in the presence of a foreign body.
   c) some strains infecting wounds can cause toxic shock.
   d) nasal carriage increases the risk of surgical wound infection.
   e) it is pyogenic.
2. Which of these statements about *Streptococcus pyogenes* is *false?*
   a) It is a Gram-positive coccus occurring in chains.
   b) Some strains that infect wounds can cause toxic shock.
   c) Some strains that infect wounds can cause necrotizing fasciitis.
   d) It can cause puerperal sepsis.
   e) A vaccine is available for preventing *S. pyogenes* infections.
3. Choose the one *false* statement about *Pseudomonas aeruginosa.*
   a) It is widespread in nature.
   b) Some strains can grow in distilled water.
   c) It is a Gram-positive rod.
   d) It produces a hemolytic toxin that has the same mode of action as α-toxin of *Clostridium perfringens.*
   e) Under certain circumstances it can grow anaerobically.

4. Which of these statements concerning tetanus is *true?*
   a) It can originate from a bee sting.
   b) Immunization is carried out using tiny doses of tetanospasmin.
   c) Those who recover from the disease are immune for life.
   d) Tetanus immune globulin is derived from the blood of immunized sheep.
   e) It is easy to avoid exposure to spores of the causative organism.
5. Choose the one true statement about gas gangrene.
   a) There are few or no leukocytes in the wound drainage.
   b) It is best to rely on antibacterial medications and avoid disfiguring surgery.
   c) A toxoid is generally used to protect against the disease.
   d) Only one antitoxin is used for treating all cases of the disease.
   e) It is easy to avoid spores of the causative agent.
6. All the following statements about actinomycosis are true, *except*
   a) it can occur in cattle.
   b) it is caused by a branching filamentous bacterium.
   c) it always appears on the jaw.
   d) it can arise from intestinal surgery.
   e) its abscesses can burrow into bone.
7. Which of the following statements about *Pasteurella multocida* is *false?*
   a) Infections generally respond to a penicillin.
   b) It can cause epidemics of fatal disease in domestic animals.
   c) It is commonly found in the mouths of biting animals, including human beings.
   d) A vaccine is, in general, used to prevent *P. multocida* disease in people.
   e) Cat bites are more likely to result in *P. multocida* infections than dog bites.
8. Which of these statements about cat scratch disease is *false?*
   a) It is a common cause of chronic lymph node enlargement in children.
   b) It is a serious threat to individuals with AIDS.
   c) Cat scratches are the only mode of transmission to humans.
   d) It is a zoonosis of cats transmitted by fleas.
   e) It can affect the brain or heart valves in a small percentage of cases.
9. The following statements about *Streptobacillus moniliformis* are all true, *except*
   a) it can be transmitted by food.
   b) its colonies can resemble those of mycoplasmas.
   c) it can be transmitted by the bites of animals other than rats.
   d) human infection is characterized by irregular fevers, rash, and joint pain.
   e) it is a Gram-positive spore-forming rod.
10. Pick out the statement concerning sporotrichosis that is *wrong.*
    a) It is characterized by ulcerating lesions along the course of a lymphatic vessel.
    b) Person-to-person transmission is common.
    c) It can occur in epidemics.
    d) It can persist for years if not treated.
    e) The causative organism is a dimorphic fungus.

Applications

1. Clinicians become concerned when the laboratory reports that organisms capable of digesting collagen and fibronectin are present in a wound culture. What is the basis of their concern?
2. ABC pharmaceutical company produces a drug that inhibits new blood vessel growth into cancers, thus depriving the cancer cells of the nutrients and hormones needed for growth of the tumor. A researcher in the company laboratory has now isolated a compound having the opposite effect. Do you see any possible role for the new compound in the treatment of wounds?
3. An army field nurse performing triage at a mobile surgical hospital asks this question of all the ambulance drivers: "Was the soldier wounded while in a pasture?" Why did he ask this question?

Critical Thinking

1. In what way would the incidence of tetanus at various ages in an economically disadvantaged country differ from age incidence in western industrialized countries?
2. Could colonization of a wound by a noninvasive bacterium cause disease? Explain your answer.
3. Surgical wound infections are frequently due to *Staphylococcus aureus,* a bacterium often carried in the nostrils and sometimes on the skin. How would you study the relationship, if any, between carriage of *S. aureus* and the chance of getting a wound infection?

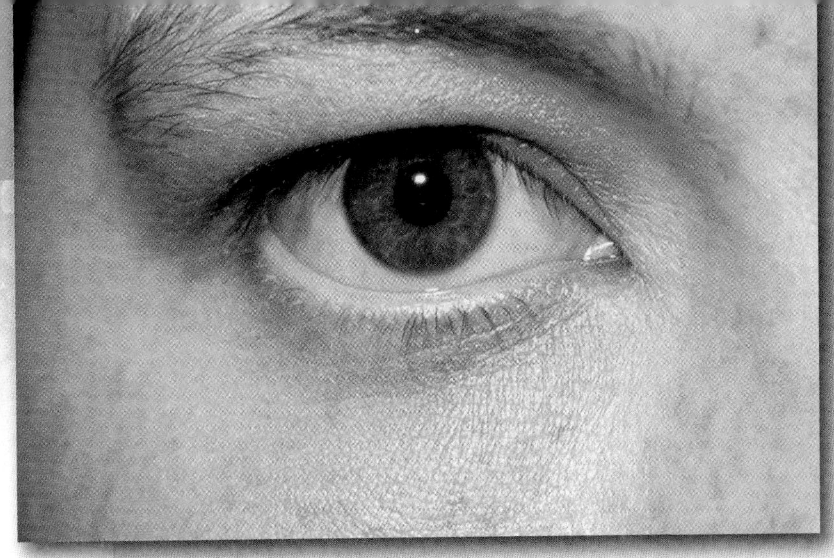

Yellow color of eye and skin as seen in jaundice.

Digestive System Infections

A Glimpse of History

"The face was sunken as if wasted by lingering consumption, perfectly angular, and rendered peculiarly ghastly by the complete removal of all the soft solids, in their places supplied by dark lead-colored lines. The hands and feet were bluish white, wrinkled as when long macerated in cold water; the eyes had fallen to the bottom of their orbes, and envinced a glaring vitality, but without mobility, and the surface of the body was cold."

This vivid description of cholera was written by Army surgeon S. B. Smith in 1832 when the disease first appeared in the United States. "Lingering consumption" refers to the marked wasting of the body seen in people with chronic tuberculosis. Cholera is a very old disease and is thought to have originated in the Far East thousands of years ago. Sanskrit writings indicate that it existed endemically in India many centuries before Christianity. With the increased shipping of goods, and the mobility of people during the nineteenth century, cholera spread from Asia to Europe and then to North America. Cholera was a major epidemic disease of the nineteenth century and appeared in almost every part of the world.

In 1854, John Snow, a London physician, demonstrated that cholera was transmitted by contaminated water. He observed that almost all people who contracted cholera got their water from a well on Broad Street. When the handle of the Broad Street pump was removed, people were forced to obtain their water elsewhere and the cholera epidemic in that area subsided. Snow's explanation was not generally accepted by other doctors, mostly because disease-causing "germs" had yet to be discovered. It was not until 1883 that Robert Koch isolated *Vibrio cholerae,* the bacterium that causes cholera.

In the United States, cholera epidemics occurred in 1832, 1849, and 1866. The disease seemed to affect poorer people—those unfortunate enough to be crowded together in cities where cleanliness was impossible to maintain. Many doctors were of the opinion that not only were the personal habits of these people "rash and excessive," but also that they insisted on taking the "wrong" medicines. By 1866, however, it was evident that where cholera appeared, the lack of sanitation was at fault. Public health agencies then played a major role in preventing epidemic cholera.

In the spring of 1997, an epidemic of severe diarrhea broke out among 90,000 sick and malnourished refugees in the Democratic Republic of Congo. More than 1,500 deaths occurred over a 3-week period. Volunteers from the medical relief organization Medecins Sans Frontieres rushed to set up medical facilities, while people from the World Health Organization, Red Cross, and local health agencies worked to provide a filtered and chlorinated water source, constructed latrines, and educated the people about sanitary measures. Treatments were started using oral and intravenous rehydration fluids. Unfortunately, at that point, the refugees were all scattered and driven away by unidentified soldiers who were afraid of the disease.

Fear generated by a diarrhea epidemic, compounded by ignorance, can defeat efforts based on reason and understanding of disease.

KEY TERMS

Bile Yellow-colored fluid produced by the liver that aids in the absorption of nutrients from the intestine.

Cariogenic Causing dental caries, tooth decay.

Cholera Severe, watery diarrhea caused by *Vibrio cholerae.*

Cirrhosis A condition marked by extensive scarring of the liver.

Dysentery Condition characterized by crampy abdominal pain and bloody diarrhea.

Gastroenteritis Acute inflammation of the stomach and intestines; the syndrome of nausea, vomiting, diarrhea, and abdominal pain.

Gingivitis Inflammation of the gums.

Glucans Polysaccharides composed of repeating subunits of glucose; involved in formation of dental plaque.

Hemolytic Uremic Syndrome (HUS) Serious condition characterized by red cell breakdown and kidney failure; a sequel to infection by certain Shiga toxin-producing strains of *Shigella dysenteriae* and *Escherichia coli.*

Hepatitis Inflammation of the liver.

Microvilli Tiny tubular processes from the luminal surfaces of cells such as those lining the intestine; increases surface area of the cell.

Pseudomembranous Colitis Disease of the colon caused by *Clostridium difficile* in which patches called pseudomembranes, composed of dead epithelium, inflammatory cells, and clotted blood, form on the intestinal lining.

Reverse Transcriptase Enzyme that synthesizes DNA complementary to an RNA template.

Secretion System A mechanism by which bacterial pathogens transfer gene products directly into host cells.

25.1 Anatomy and Physiology

Focus Points

- List the functions of the main components of the upper digestive tract.
- Describe the role of the liver in health and disease.

The digestive tract, sometimes referred to as the alimentary or gastrointestinal (GI) tract, is the passageway that extends from the mouth to the anus. Like the skin, it is one of the body's boundaries with the environment, and it is one of the major routes into the body for invading microbial pathogens. The digestive tract and its accessory organs—the saliva-producing glands, liver, and pancreas—together compose the digestive system, the main purpose of which is to provide nourishment for the body. The upper part consists of the mouth, salivary glands, esophagus, and stomach, and the lower part includes the intestines, pancreas, and liver. **Figure 25.1** illustrates the relationships among the various parts of the digestive system. The main functions of the system are:

- to grind food in the mouth into small particles that can readily react with digestive juices that break down large, complex molecules into absorbable components;
- to move the food through the esophagus to the stomach, where it undergoes preliminary treatment with acid and enzymes;
- to discharge stomach contents at a controlled rate into the small intestine, where they are digested under alkaline conditions, and nutrients are absorbed into the bloodstream;
- to move undigested material into the large intestine, where water is absorbed from the intestinal contents along with any remaining vitamins, minerals, or nutrients; and
- to discharge the waste as feces.

The Mouth

The mouth is mainly a grinding apparatus. The tongue and cheeks move bites of food to be ground up by the teeth, which are composed of a calcium compound harder than bone **(figure 25.2).** The outer portion of the teeth, the **enamel,** is a hard, protective layer. Pits and crevices normally present on the enamel surfaces of the teeth collect food particles and offer protected sites for microbial colonization. Beneath the enamel lies the **dentin,** a softer, more easily penetrated layer. The **gingival crevice,** the space between the tooth and gum, is important because inflammation at this site can lead to loss of the tooth.

Salivary Glands

Salivary glands are located under the tongue and laterally in the floor of the mouth. There are two additional saliva-producing glands, called the **parotids,** one on either side of the face below the ears. About 1,500 ml of saliva is secreted from the salivary glands each day. It moistens and lubricates the mouth, and

FRANK & ERNEST: © Thaves/Dist. By Newspaper Enterprise Association, Inc.

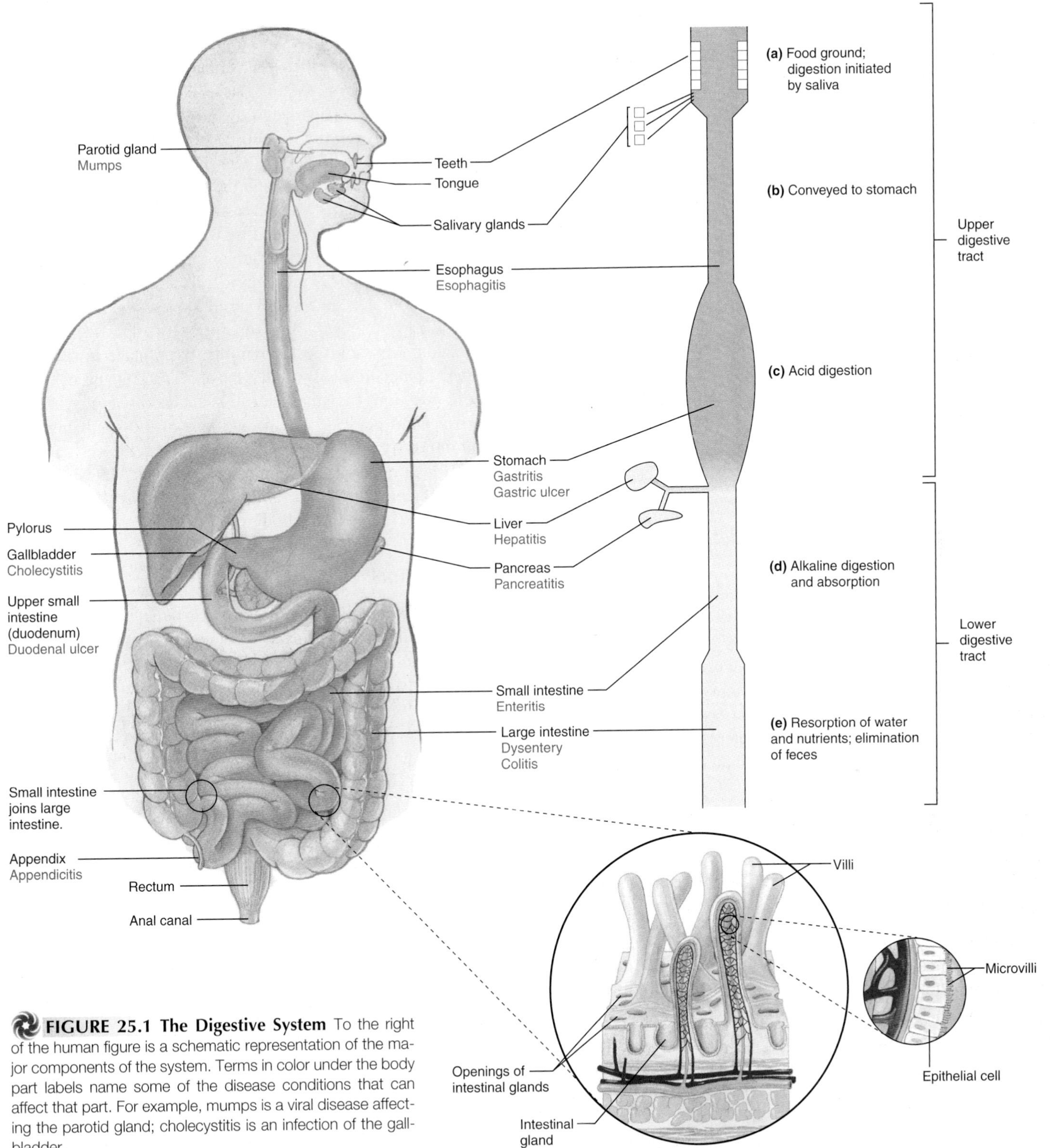

FIGURE 25.1 The Digestive System To the right of the human figure is a schematic representation of the major components of the system. Terms in color under the body part labels name some of the disease conditions that can affect that part. For example, mumps is a viral disease affecting the parotid gland; cholecystitis is an infection of the gallbladder.

contains the enzyme **amylase,** which begins the breakdown of starches present in food. Saliva is extremely important in protecting the teeth from decay because it is saturated with calcium and contains buffers that help neutralize acids. People with poor saliva production are subject to severe tooth decay. Saliva also contains antibacterial substances including **lysozyme,** the enzyme that attacks the peptidoglycan of bacterial cell walls; **lactoferrin,** a substance that binds iron ions critically required by bacteria; and secretory IgA antibodies that inhibit bacterial attachment. ■ lactoferrin, p. 349 ■ IgA, p. 374 ■ lysozyme, p. 62

The Esophagus

The tongue and throat muscles work together in swallowing to propel food from the mouth and throat into the esophagus, a collapsible tube about 10 inches long, located behind the

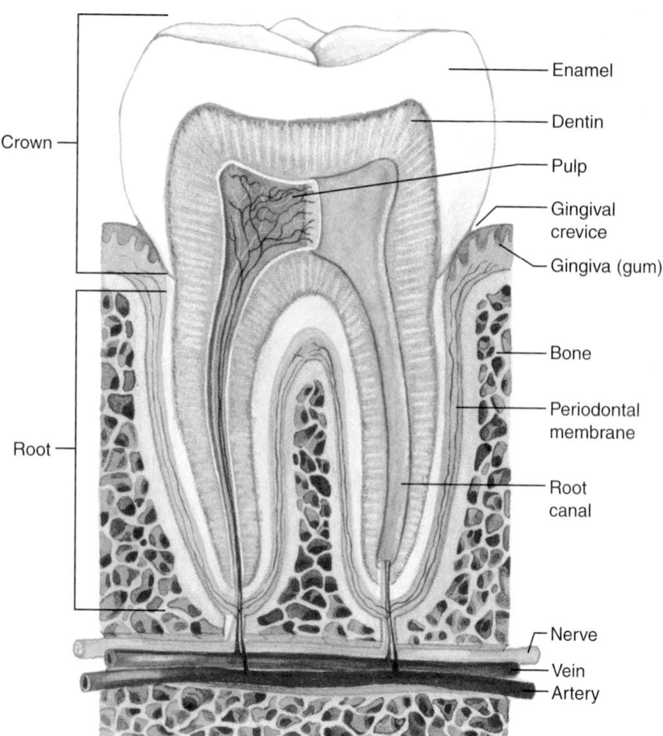

FIGURE 25.2 Structure of a Tooth and Its Surrounding Tissues

windpipe. The esophagus has a muscular wall that contracts rhythmically to move food and liquid in a process called **peristalsis.** The lower portion of the esophagus normally closes to prevent regurgitation of stomach contents. The esophagus rarely becomes infected except in individuals with AIDS or other immunodeficiencies.

The Stomach

The lower portion of the esophagus leads into the stomach, an elastic, saclike structure with a muscular wall. Some of the cells that line it produce hydrochloric acid, whereas others produce pepsinogen, which becomes the protein-splitting enzyme **pepsin** upon contact with acid. The stomach itself is protected from the acid and enzymes by a thick layer of mucus secreted by the stomach lining. Stomach emptying is controlled by a complex system of nerve impulses and hormones acting on a muscular valve, the **pylorus,** which determines the rate at which the stomach contents enter the intestine.

The Small Intestine

Although it is only 8 or 9 feet long, the small intestine has an enormous surface area, about 30 square feet, which facilitates nutrient absorption. The inside surface of the intestine is covered with many small, fingerlike projections called **villi,** each 0.5 to 1 mm long. Each of these villi is, in turn, covered with cells that have cytoplasmic projections called **microvilli** (see figure 25.1). Many intestinal digestive enzymes are present as integral parts of the plasma membranes of these microvilli. This epithelial surface has both secretory and absorptive functions. Tiny intestinal glands continuously secrete large amounts of fluid into the **lumen,** the space enclosed by the walls of the intestine. This fluid is then reabsorbed along with the products of food digestion. Small intestine function depends on active transport of nutrients and electrolytes across the plasma membranes of the epithelial cells. For example, sodium ions (Na^+) are taken up by the cell when glucose or amino acids are absorbed; hydrogen ions (H^+) and bicarbonate ions (HCO_3^-) are secreted to adjust the pH of the intestinal contents. Movement of these substances is accompanied by water molecules. Digestion takes place as the cells secrete juices rich in enzymes, such as those that break peptides into amino acids, and disaccharidases that convert complex sugars into simpler ones. The major part of nutrient absorption, including absorption of monosaccharides, amino acids, fatty acids, and vitamins, takes place in the small intestine. Some minerals, such as iron, can only be absorbed there. In addition, the small intestine reabsorbs approximately 9 liters of fluid per day. The lining cells of the small intestine are continuously shed and replaced by new cells, so that cells poisoned by a microbial toxin are soon replaced. Intestinal epithelial cells are completely replaced every 9 days—one of the fastest turnover rates in the body.

The Pancreas

The pancreas is a large, glandular organ located behind the stomach. Some cells of the pancreas produce hormones, and others produce digestive enzymes. The hormones are released directly into the bloodstream and help regulate various body functions—for example, **insulin** helps regulate the amount of glucose in the blood. On the other hand, about 2 liters of pancreatic digestive juices discharge directly into the upper portion of the small intestine each day. These fluids, as well as digestive juices of the small intestine itself, are alkaline and neutralize the stomach acid as it passes into the intestine.

The Liver

The liver is a large, dark-red, glandular organ located in the upper abdomen on the right side. The liver aids digestion, neutralizes poisons, degrades medications, and removes a normal breakdown product of hemoglobin from the bloodstream. This yellowish product, called **bile pigment,** is discharged with the **bile,** the fluid produced by the liver. Each day, about 500 ml of bile flows through a system of tubes into the upper small intestine. Severe liver disease and obstruction of the bile ducts can produce jaundice, a yellow color of the skin and eyes caused by buildup of bile pigment in the blood. The gallbladder is a saclike structure in which bile is concentrated and stored. Substances in the bile called bile salts help the intestine absorb oils and fats and the fat-soluble vitamins A, D, E, and K. The normal brown color of feces results from the action of intestinal bacteria on bile pigment.

The liver also inactivates poisonous substances that enter the bloodstream. For example, ammonia produced by intestinal bacteria and absorbed into the bloodstream could poison the body if it were not detoxified by the liver. The liver also chemically alters and excretes many medications; if the liver is damaged, as it might be by a viral infection, lower medication doses might be needed to avoid the buildup of toxic levels of the medication.

The Large Intestine

The main function of the large intestine is to recycle water and absorb nutrients. Because the small intestine absorbs so much water, only 300 to 1,000 ml of fluid normally reach the large intestine per day. From this volume of fluid, the large intestine absorbs water, electrolytes, vitamins, and amino acids. The semisolid feces, composed of indigestible material and bacteria, remain. Infection of the large intestine can interfere with absorption and stimulate the painful peristaltic contractions known as "stomach cramps."

MICROCHECK 25.1

The digestive tract is a major route for pathogenic microorganisms to enter the body. People with poor saliva production risk severe tooth decay. Infections of the esophagus are so unusual in normal people that their occurrence suggests immunodeficiency. The stomach is responsible for acid digestion of food; the small intestine carries out alkaline digestion and absorption of nutrients. Undigested material becomes feces in the large intestine.

- ✓ What makes the surface area of the small intestine so large, when it is only 8 to 9 feet long?
- ✓ What is the main function of the large intestine?
- ✓ What might occur to the blood levels of a medication if another medication interferes with its degradation by the liver?

25.2 Normal Microbiota

Focus Points

- Describe how suppression of the normal microbiota can result in pseudomembranous colitis.
- List activities of the normal microbiota potentially of value to the host.

This section focuses mainly on microbiota of the oral cavity and intestine. The esophagus has a relatively sparse population, consisting mostly of bacteria from the mouth and upper respiratory tract. When empty of food, the normal stomach has few microorganisms because most are killed by the action of acid and pepsin. ■ **normal microbiota, p. 462**

The Mouth

Of all species of bacteria introduced into the mouth from the time of birth onward, relatively few can colonize the oral cavity. Streptococcal species—Gram-positive, chain-forming cocci that produce lactic acid as a by-product of carbohydrate metabolism—are the most numerous. One species of *Streptococcus* preferentially colonizes the upper part of the tongue, another colonizes the teeth, and still another colonizes the mucosa of the cheek. Streptococci and other bacteria attach specifically to receptors on host tissues, allowing the microorganisms to resist the scrubbing action of food and the tongue and the flushing action of salivary flow. The fact that the cells lining various parts of the mouth have different receptors accounts for the distribution of the various species of streptococci. The host limits the numbers of bacteria on its mucous membranes by constantly shedding the superficial layers of cells and replacing them with new ones; the rate of this shedding correlates with the number of microorganisms present on the surface.

Because teeth are a non-shedding surface, large collections of bacteria can build up on them in a **biofilm.** These masses of bacteria, called **dental plaque (figure 25.3),** form because the bacteria attach to specific receptors on each other or on the tooth, and they can be bound together by extracellular polysaccharides. There can be up to 100 billion bacteria per gram of plaque. Metabolic by-products of one species are utilized by another. Plaque organisms consume O_2, thereby creating conditions that permit the growth of strict anaerobes. In fact, colonization of the mouth by strictly anaerobic microbes requires the presence of teeth, because only teeth provide sufficiently anaerobic habitats for growth of these organisms. Dental plaque, gingival crevices, and fissures in the teeth are such habitats. ■ **lactic acid production, p. 147** ■ **biofilms, p. 85**

The Intestines

Only small numbers of bacteria live in the upper small intestine because they are continually flushed away by the rapid passage of digestive juices. The predominant organisms are usually aerobic and facultatively anaerobic, Gram-negative rods and some streptococci. Lactobacilli and yeasts such as *Candida albicans* are found in small numbers. The bacterial population increases as the intestinal contents move toward the large intestine.

In contrast to the relatively scanty numbers of organisms in the small intestine, the large intestine contains high numbers of microorganisms—approximately 10^{11} bacteria per gram of feces. These large numbers occur because of the abundance of nutrients

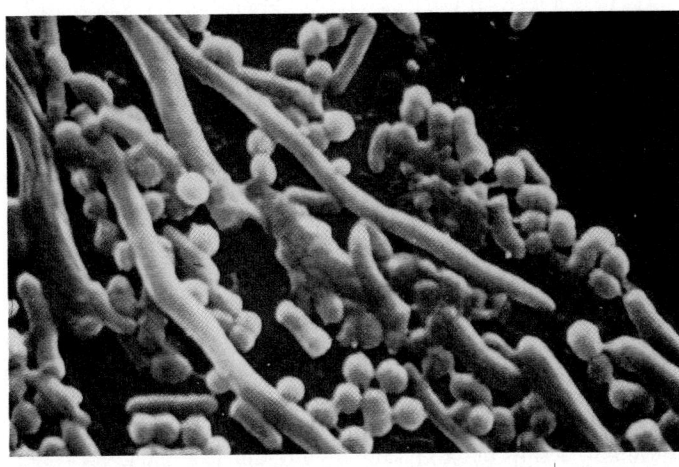

FIGURE 25.3 Scanning Electron Micrograph of Dental Plaque The many kinds of bacteria composing the plaque exhibit specific attachments to the tooth and to each other, and may be bound together by extracellular polysaccharides.

in undigested and indigestible food material. Bacteria make up about one-third of the fecal weight. The numbers of anaerobic bacteria, notably including members of the genus *Bacteroides*, generally exceed the numbers of other organisms by about 100-fold. Of the fecal microorganisms able to grow in the presence of air, facultatively anaerobic, Gram-negative rods, particularly *Escherichia coli* and other enterobacteria, predominate. Fecal organisms are an important source of opportunistic pathogens, especially for the urinary tract. The enormous population of bacteria comprises species that can work together enzymatically to change numerous materials. The currently recommended high-fiber diets contain substances that are indigestible by the gastric and intestinal juices but are readily degraded by intestinal organisms, which often produce large amounts of carbon dioxide, hydrogen, and methane gas. Abdominal discomfort and discharge of intestinal gas (flatus) from the anus is an unfortunate result. Bacterial enzymes can also convert various substances in food to carcinogens and therefore may be involved, along with diet, in the production of intestinal cancer. ■ opportunistic pathogens, p. 394 ■ enterobacteria, p. 262

Intestinal bacteria synthesize a number of useful vitamins, including niacin, thiamine, riboflavin, pyridoxine, vitamin B_{12}, folic acid, pantothenic acid, biotin, and vitamin K. These vitamins are important when a person's diet is inadequate.

Antibiotic-Associated Diarrhea

The normal microbiota helps prevent colonization of the large intestine by pathogens. Antibiotic treatment, especially with broad-spectrum medications, disrupts the normal microbiota and results in mild to severe diarrhea. A sometimes life-threatening disease called **antibiotic-associated,** or **pseudomembranous, colitis** can follow antibiotic therapy. The disease is caused by the toxin-producing, anaerobic bacterium *Clostridium difficile*, which readily colonizes the intestine of people whose normal intestinal microbiota has been reduced by antimicrobial chemotherapy. The toxins of *C. difficile* are lethal to intestinal epithelium and cause small patches called pseudomembranes, composed of dead epithelium, inflammatory cells, and clotted blood, to form on the intestine. Fever, abdominal pain, and profuse diarrhea result. Suppression of intestinal microbiota with antibacterial medications can also increase susceptibility to other pathogens such as *Salmonella enterica*, to be discussed later.

MICROCHECK 25.2

The distribution of bacterial microbiota in the mouth is governed by different receptors on the epithelium. The quantity of microbiota on mouth epithelium is limited by shedding. Because the teeth are a non-shedding surface, enormous populations of bacteria, manifest as dental plaque, form a biofilm on teeth. A properly functioning stomach destroys most microorganisms before they reach the intestine. The normal intestinal microbiota helps protect the body from infection; disruption of the microbiota by antibiotics can lead to infection. Metabolic activity of intestinal microbiota produces vitamins, but it causes gas and can contribute to cancer.

- ✓ Why must teeth be present in order for anaerobic bacteria to colonize the mouth?
- ✓ What is the advantage to the body of having pepsin becoming active only on exposure to acid?
- ✓ Why is it that the tongue and cheek epithelium doesn't provide sufficiently anaerobic conditions for plaque anaerobes to grow?

UPPER DIGESTIVE SYSTEM INFECTIONS

25.3 Bacterial Diseases of the Upper Digestive System

Focus Points

- Compare and contrast dental caries, periodontal disease, and trench mouth.
- Give the evidence that *Helicobacter pylori* infection predisposes people to stomach cancer.

It may be surprising that the most common bacterial disease of human beings occurs in the mouth, and that bacterial infections of the stomach are not only common but can lead to ulcers and cancer. These infections often go unnoticed for years but have consequences both locally and for the rest of the body. Oral microbiota, for example, can enter the bloodstream during dental procedures and cause subacute bacterial endocarditis, and some studies suggest that chronic gum infections play a role in hardening of the arteries and arthritis. The most important bacterial diseases of the upper alimentary system concern the teeth, the gums, and the stomach. ■ subacute bacterial endocarditis, p. 677

Tooth Decay (Dental Caries)

Dental caries, generally known as tooth decay, is the most common infectious disease of human beings. The cost to restore and replace teeth damaged by this disease is estimated to be about $20 billion per year in the United States. Dental caries is the main reason for tooth loss. Across the United States, from 14% to 47% of people aged 65 or older have lost all their teeth.

Symptoms

Dental caries is usually far advanced before any symptoms develop. The severe, throbbing pain of a toothache is often the first symptom. Sometimes there is noticeable discoloration, roughness, or defect, and a tooth can break during chewing.

Causative Agent

Dental caries is caused principally by *Streptococcus mutans* and closely related species. These bacteria live only on the teeth and cannot colonize the mouth in the absence of teeth. They produce lactic acid as a by-product of their metabolism of sugars and, unlike many other bacteria, they thrive under acidic conditions below pH 5. Another important feature is that they produce insoluble extracellular **glucans** from sucrose, but not from other sugars. Glucans—polysaccharides composed of repeating subunits of glucose—are essential for the production of dental caries on smooth tooth surfaces. ■ polysaccharides, p. 37

Pathogenesis

The first step in the formation of a **cariogenic** plaque, meaning a plaque that causes tooth decay, is the adherence of oral streptococci to specific receptors on the tooth pellicle. The pellicle is a thin film of proteinaceous material adsorbed on the tooth from the saliva. Other species of bacteria attach specifically to earlier arrivals. Two species that might not attach to each other can attach to a third. If dietary sucrose is present, *S. mutans* attaches to the bacterial mass and produces glucans from sucrose through the action of extracellular enzymes. Sucrose is split by the enzymes to the monosaccharides glucose and fructose. The glucose is polymerized, yielding glucan, and the fructose is metabolized, producing lactic acid. The glucans bind the organisms together and to the tooth to form a biofilm, making the plaque impenetrable to saliva.

When sugar enters the mouth, the pH of cariogenic plaques drops from its normal value of about 7 to below 5 within minutes **(figure 25.4).** This 100-fold increase in acidity begins dissolving the calcium phosphate of the teeth. The duration of this acidic state depends on how long the teeth are exposed to sugars and on the concentration of the sugars. After food leaves the mouth, the pH of the plaque rises slowly to neutrality. The delay in return of the pH to neutrality is due to the ability of *S. mutans* to store a portion of its food as an intracellular, starchlike polysaccharide that is later metabolized with the production of acid. Cariogenic plaque thus acts as a tiny, acid-soaked sponge closely applied to the tooth. Both *S. mutans* and a suitable sucrose-rich diet are required to produce dental caries on smooth surfaces of the teeth. In deep fissures or pits in the teeth, plaque can accumulate in the absence of *S. mutans* but can still be cariogenic as long as lactic acid-producing bacteria and fermentable substances are present.

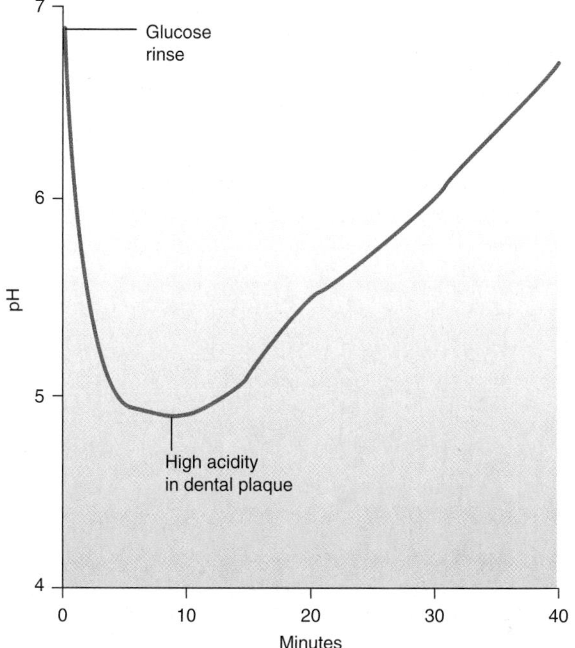

FIGURE 25.4 Increase in Acidity in Cariogenic Dental Plaque After Rinsing the Mouth with a Glucose Solution Tooth enamel begins to dissolve at about pH 5.5, continuing as pH decreases.

Epidemiology

Dental caries is worldwide in distribution, but the incidence varies markedly depending mainly on intake of dietary sucrose and access to preventive dental care. Heredity also plays an important role—some individuals inherit resistance to the disease. Young people are generally much more susceptible than older people, probably because the pits and fissures that are sites for dental caries wear down with time.

Prevention and Treatment

The most important method for controlling dental caries is restricting sucrose and other refined dietary carbohydrates, thereby reducing *Streptococcus mutans* colonization of teeth and acid production by cariogenic plaques. Dental caries can be reduced by 90% if sucrose-containing sweets are eliminated from the diet. It is not, however, simply the quantity of sugar in the diet that is important. The frequency of eating and the length of time food stays on the teeth are more critical than the actual quantity of sucrose ingested. Interestingly, chewing paraffin or sorbitol-sweetened gum reduces dental caries, probably because it increases the flow of saliva.

Trace amounts of fluoride are required for teeth to resist the acid of cariogenic plaques. Fluoride makes tooth enamel harder and more resistant to dissolving in acid. In the United States, more than 100 million people are currently supplied by fluoridated public drinking water, which has resulted in a 60% reduction in dental caries. In areas where fluoridated drinking water is not available, fluoride tablets or solutions can be used. To have optimum effect, children should begin receiving fluoride before their permanent teeth erupt. Fluoride applied to tooth surfaces in the form of mouthwashes, gels, or toothpaste is generally less effective.

Mechanical removal of plaque by toothbrushing and use of dental floss is another important preventive measure, reducing the incidence of dental caries by about 50%. Toothbrush bristles cannot remove plaque from the pits and fissures normally present in children's teeth, however, because they are too deep and narrow. Caries commonly develops in these pits and fissures and can be prevented by using a sealant—a kind of epoxy glue that seals the fissures, kills the bacteria in plaque, and prevents bacterial recolonization. Older people have less fissure-related caries; however, receding gums can expose root surfaces that become sites for dental caries. Treatment of dental caries requires drilling out the cavity, filling the defect with **amalgam** (an alloy of mercury and some other metal) or other material, and restoring the contour of the tooth.

Periodontal Disease

Periodontal disease is a chronic inflammatory process involving the gums and tissues around the roots of the teeth. It usually develops slowly over many years, and is an important cause of tooth loss from middle age onward.

Symptoms

The majority of individuals with periodontal disease are asymptomatic. Common symptoms are bleeding and sensitive gums, bad breath, and loosening of the teeth. Discoloration, ranging from yellowish to black, occurs at the base of the teeth. The gums generally recede and expose the roots of the teeth to dental caries.

Causative Agent

Periodontal disease is caused by dental plaque that forms at the point where the gum joins the tooth. The plaque may or may not be cariogenic. Hundreds of kinds of bacteria have been identified in plaques associated with periodontal disease.

Pathogenesis

Plaque forms on teeth at the gum margin, especially in hard-to-clean areas between the teeth. Calcium salts deposited in the plaque result in hard-to-remove **dental calculus,** often referred to as **tartar.** Plaque gradually extends into the gingival crevice and bacterial products incite an inflammatory and immune response manifested by swelling and redness of the gingiva **(figure 25.5).** This gum inflammation is known as **gingivitis.** If the plaque remains small, neutrophils along with cellular and humoral immunity limit the process at this stage. Large populations of microorganisms, however, release the enzymes collagenase and hyaluronidase, which weaken the gingival tissue and cause the gingival crevice to widen and deepen. As the plaque enlarges, the proportion of anaerobic, Gram-negative bacteria such as *Porphyromonas gingivalis* increases. These organisms release endotoxin and a variety of exotoxins that attack leukocytes and host tissue, but they generally do not invade the tissues of the host. The membrane that attaches the root of the tooth to the bone weakens, and the bone surrounding the tooth gradually softens. The tooth becomes loose and may be lost. ■ neutrophils, p. 350

Epidemiology

Periodontal disease is mainly a disease of those over 35 years of age. After age 65, almost 90% of individuals have some degree of periodontal disease. Persons with AIDS and other immunodeficiencies, and those with defective neutrophils, often have severe periodontal disease that leads to loss of their teeth.

Prevention and Treatment

Careful flossing and toothbrushing can prevent periodontal disease, especially if combined with twice-yearly polishing and removal of calculus at a dental office. Periodontal disease can be treated in its earlier stages by cleaning out the inflamed gingival crevice and removing plaque and calculus. In advanced cases, surgery is usually required to expose and clean the roots of the teeth.

Trench Mouth

Trench mouth **(figure 25.6),** also known as **Vincent's disease** or **acute necrotizing ulcerative gingivitis (ANUG),** is a severe, acute

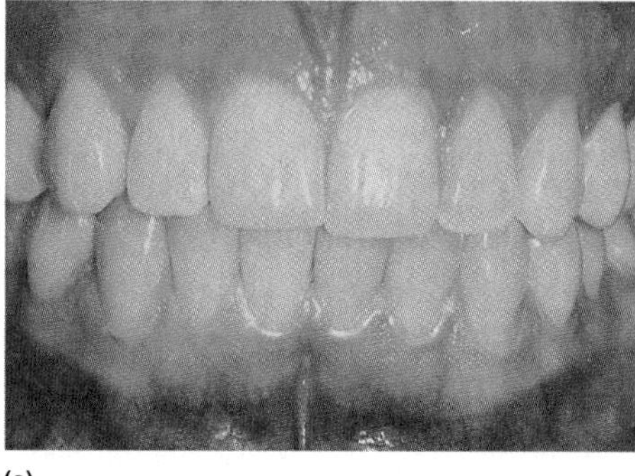

(a)

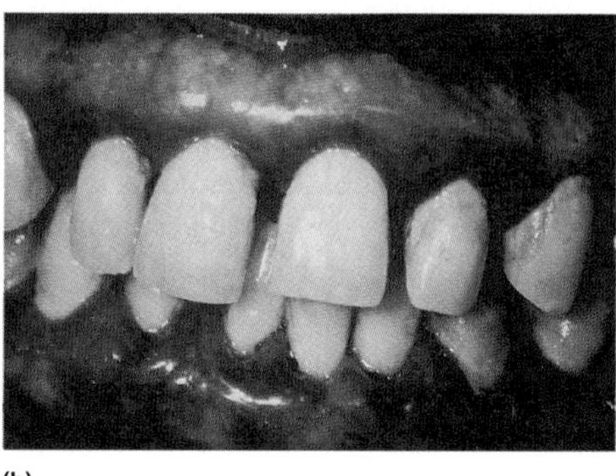

(b)

FIGURE 25.5 Periodontal Disease (a) Normal gingival. **(b)** Periodontal disease, with plaque, inflammatory changes, bleeding, and shortening of the gingival between the teeth.

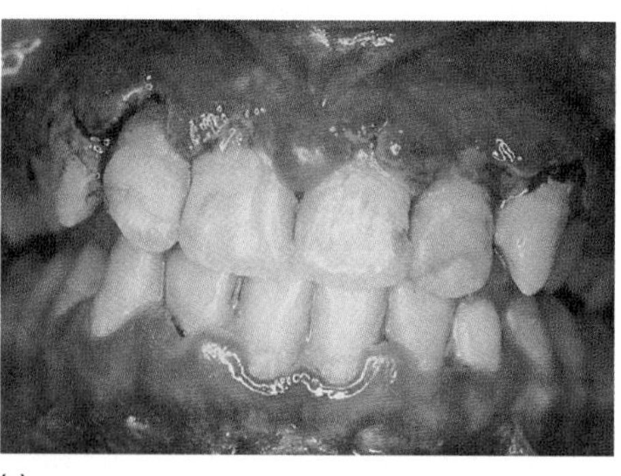

(a)

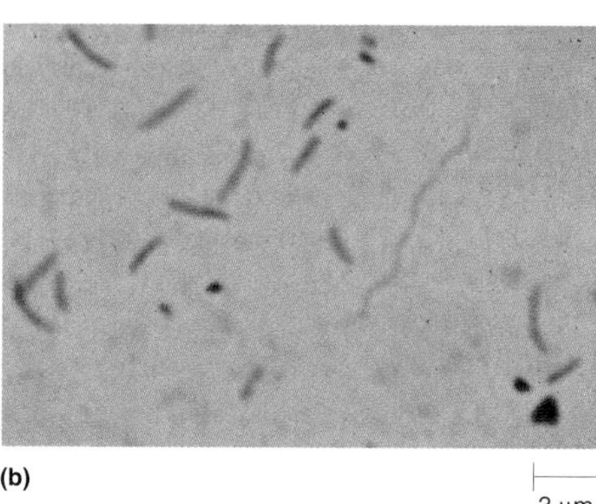

(b)

FIGURE 25.6 Trench Mouth (Acute Necrotizing Ulcerative Gingivitis, or ANUG) (a) Red, swollen gingiva with loss of tissue, especially between the teeth. **(b)** Gram stain of exudate showing a spirochete and rod-shaped bacteria.

condition distinct from other forms of periodontitis. The disease was rampant among soldiers living in trenches during World War I because they were unable to attend to mouth care—thus, its name.

Symptoms

Trench mouth is characterized by an abrupt onset, fever, bleeding and painful gums, and a foul odor.

Causative Agent

The suspected causative agent is an oral, unculturable spirochete of the *Treponema* genus, antigenically related to *T. pallidum,* the cause of syphilis. These trench mouth spirochetes probably act synergistically with other anaerobic species, with which they are always associated. ■ the genus *Treponema,* p. 273 ■ syphilis, p. 633

Pathogenesis

The spirochetes and the other anaerobes are presumed to act together to destroy tissue, but the precise mechanisms are unknown. Plaque is always present, but its bacterial composition shows much larger numbers of spirochetes and other anaerobes than present in chronic periodontal disease. The spirochetes invade the tissue, causing necrosis and ulceration, mainly of the gums between the teeth.

Epidemiology

The disease can occur at any age in association with poor mouth care, especially with stress, malnutrition, or immunodeficiency. It is not contagious.

Prevention and Treatment

Prevention begins with daily brushing and flossing, and twice-yearly professional cleaning. Antibacterial treatment directed against the spirochetes and anaerobic rods rapidly relieves the acute symptoms, but this must be followed by extensive removal of plaque and calculus.

The main features of teeth and gum infections are presented in **table 25.1.**

Helicobacter pylori Gastritis

Gastritis, meaning inflammation of the stomach, is commonly present in otherwise healthy, asymptomatic people. It was not until the early 1980s that a bacterial cause of this condition was identified and its association with ulcers and gastric malignancy became apparent. These ulcers occur in the stomach and uppermost part of the duodenum. They are localized, roughly circular erosions of the epithelium that can extend deeply into the underlying tissue.

Symptoms

The initial infection can cause symptoms ranging from belching to vomiting. Most people, however, have no symptoms except when infection is complicated by ulcers or cancer. Localized abdominal pain, tenderness, and bleeding are manifestations of these complications.

TABLE 25.1 Important Infections of the Teeth and Gums

	Dental Caries	Periodontal Disease	Trench Mouth
Symptoms	None until advanced disease. Late: discoloration, roughness, broken tooth, throbbing pain	Most cases asymptomatic until advanced disease. Bleeding, sensitive gums, bad breath, loosening of the teeth. Receding gums with exposed discolored tooth roots.	Abrupt onset of fever, painful bleeding gums, and a foul mouth odor
Incubation period	1 to 24 months before cavity is detectable	Months or years	Undetermined
Causative agent	Dental plaque populated with *Streptococcus mutans*	Dental plaque, cariogenic or not	Probably a spirochete of the genus *Treponema* acting with *Fusobacterium, Prevotella,* or other anaerobes
Pathogenesis	Bacteria in plaque produce acid from dietary sugars; slowly dissolves the calcium phosphate crystals composing the tooth; sucrose critical for cariogenic plaque formation.	Plaque forms at the gum margins and gradually extends into the gingival crevices. Bacterial products incite an inflammatory response. The crevices widen and deepen, and the proportion of anaerobes increases. Toxins and enzymes weaken the tissues holding the teeth and cause them to become loose.	The spirochetes and certain other anaerobes act synergistically to cause death of tissue, ulceration, and tissue invasion by spirochetes.
Epidemiology	Worldwide distribution, incidence depending on dietary sucrose, natural or supplemental fluoride. The young are more susceptible than the old.	Primarily a disease of those older than 35 years. Immunodeficient individuals are at increased risk of severe disease.	All ages are susceptible in association with poor mouth care, malnutrition, or immunodeficiency. It is not contagious.
Prevention and treatment	Restriction of dietary sucrose, supplemental fluoride, mechanical removal of plaque, sealing pits and fissures in childhood teeth.	Avoid buildup of plaque. Surgical treatment in severe cases to expose tooth roots and remove plaque and calculus.	Avoid buildup of plaque. Antibiotic treatment acutely, followed by removal of plaque and calculus.

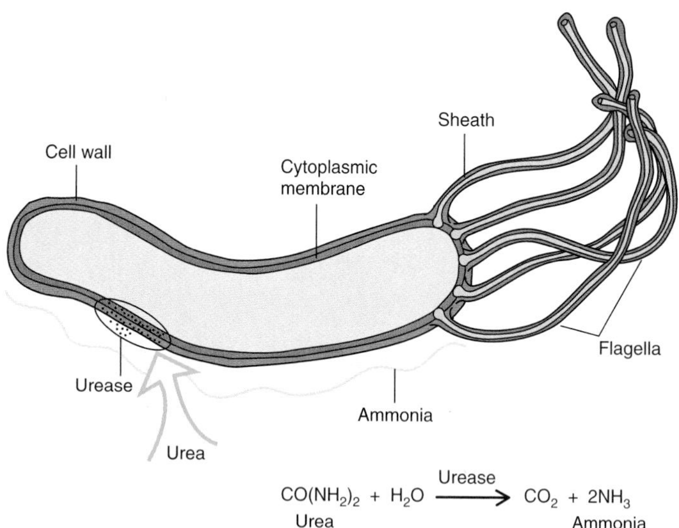

FIGURE 25.7 ***Helicobacter pylori*** This bacterium, which has unusual sheathed flagella with terminal knobs, produces a powerful urease that allows it to survive stomach acid by converting urea to ammonia, which neutralizes acidity.

Causative Agent

Helicobacter pylori is a short, spiral, Gram-negative, microaerophilic bacterium with multiple unusual-appearing, sheathed polar flagella **(figure 25.7).** ■ the genus *Helicobacter*, p. 273

Pathogenesis

These remarkable organisms survive the extreme acidity of the stomach because of their powerful urease. This enzyme creates an alkaline microenvironment by hydrolyzing **urea** to ammonia. Urea is a waste product of protein catabolism by the body's cells and is normally present in the gastric juices.

Once the bacteria reach the mucus that coats the stomach or duodenal lining, they use their flagella to corkscrew through the mucus to the epithelial cells. In this location, the pH of the mucus is nearly neutral, and the bacteria attach to the mucus-secreting epithelium or multiply adjacent to it. Bacterial products incite an inflammatory response in the wall of the stomach, and mucus production decreases. Once infection occurs it persists for years, often for life. From 10% to 20% of infected persons develop ulcers; 65% to 80% of patients with gastric ulcers and 95% of those with duodenal ulcers are infected with *H. pylori*. The thinning of the protective mucus layer at the site of infection **(figure 25.8)** probably accounts for the development of peptic (meaning "caused by digestive juices") ulcers of the stomach and duodenum. A very small percentage of individuals infected with *H. pylori* develop cancer of the stomach, but more than 90% of those with stomach cancer are infected by the bacterium. Virulent *H. pylori* strains produce a protein, CagA, which they transfer into host cells, resulting in changes in shape and surface characteristics of the cells. Evidence suggests these changes represent a prelude to malignancy. Another bacterial product, VacA, acts on mucosal cells to promote flow of urea into the stomach.

Epidemiology

Infections tend to cluster in families. Transmission of *H. pylori* probably occurs by the fecal-oral route, and the bacteria have been found in well water. Also, flies are capable of transmitting the organisms. Overall, about 20% of the adult U.S. population is infected with the bacterium, but the incidence progressively

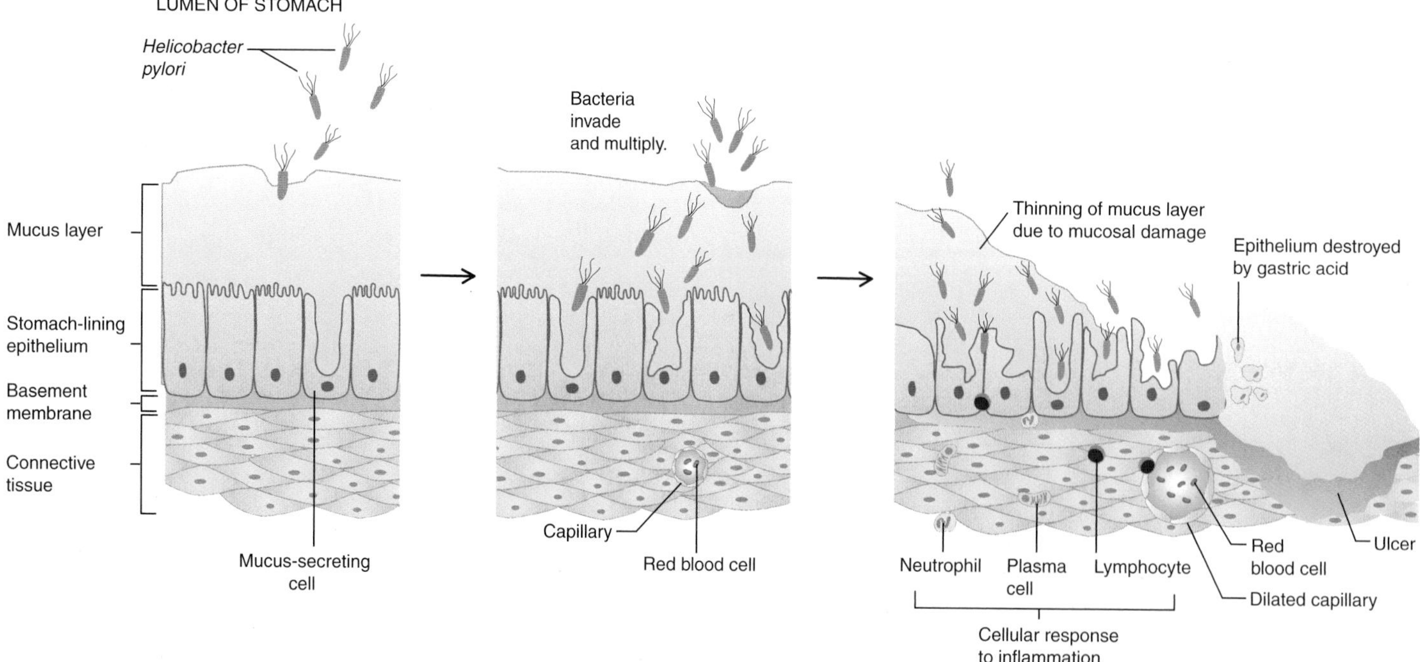

FIGURE 25.8 Gastric Ulcer Formation Associated with *Helicobacter pylori* Infection Thinning of the protective layer of mucus probably results from an inflammatory and immune response to the bacteria growing on or near the epithelium. The acid and pepsin of the stomach juices can then attack the epithelium. Neutrophils are phagocytic early responders to inflammation: lymphocytes and plasma cells represent the immune response.

increases with age, reaching almost 80% for those over age 75. Infection rates are highest in low-socioeconomic groups.

Prevention and Treatment

There are no proven preventive measures. *Helicobacter pylori* infections can usually be eradicated by combined treatment with two antibiotics and a medication that inhibits stomach acid production, with complete clearing of the gastritis and healing of any ulcers.

The main features of *H. pylori* gastritis are shown in **table 25.2.**

MICROCHECK 25.3

The pathogenesis of tooth decay depends on both dietary sucrose and acid-forming bacteria that are adapted to colonize hard, smooth surfaces within the mouth. Periodontal disease is a chronic condition caused by plaque at the gum margin that can lead to loosening of teeth. Trench mouth is an acute illness with destruction of gingival tissue. Chronic infection by *Helicobacter pylori* is a key factor in the development of gastric and duodenal ulcers and gastric cancer.

- ✓ Why is loss of teeth so common in people over age 65?
- ✓ What enzyme produced by *Heliobacter pylori* assists its survival in stomach acid?
- ✓ Discuss the pros and cons of the statement "*Helicobacter pylori* causes stomach cancer."

TABLE 25.2 *Heliobacter pylori* Gastritis

Symptoms	Initial infection: range from belching to vomiting. Localized abdominal pain and tenderness, bleeding when complicated by ulcer or cancer
Incubation period	Usually undetermined
Causative agent	*Helicobacter pylori,* a spiral, Gram-negative, microaerophilic bacterium, with sheathed flagella
Pathogenesis	Organisms survive the acidity of stomach juices by producing a powerful urease. Upon reaching the layer of mucus, they penetrate to the epithelial surface, where bacterial products incite an inflammatory response. Thinning of the mucus layer occurs, and 10% to 20% of infected individuals develop ulcerations. Only a small percentage develop cancer, but more than 90% of individuals with stomach cancers are infected with *H. pylori.*
Epidemiology	Probably fecal-oral transmission. Progressive increase with age, reaching almost 80% of those over 75.
Prevention and treatment	No proven preventive. Most infections are cured using two antibiotics together, plus a medication to suppress stomach acid.

CASE PRESENTATION

The patient was a 35-year-old man who consulted his physician because of upper abdominal pain. The pain was described as a steady burning or gnawing sensation, like a severe hunger pain. Usually it came on 1½ to 3 hours after eating, and sometimes it woke him from sleep. Generally, it was relieved in a few minutes by food or antacid medicines.

On examination, the patient appeared well, without evidence of weight loss. The only positive finding was tenderness slightly to the right of the midline in the upper part of the abdomen. A test of the patient's feces was positive for blood. The remaining laboratory tests were normal.

Endoscopy, a procedure that employs a long, flexible fiber-optic device passed through the mouth, showed a patchy redness of parts of the stomach lining. A biopsy was taken. The endoscopy tube was passed through the pylorus and into the duodenum. About 2 cm into the duodenum, there was a lesion 8 mm in diameter that lacked a mucous membrane and appeared to be "punched out." The base of the lesion was red and showed adherent blood clot. After the endoscopy, a portion of material obtained by biopsy was placed on urea-containing medium. Within a few minutes, the medium began to turn color, indicating a developing alkaline pH.

1. What is the patient's diagnosis?
2. What would you expect microscopic examination and culture of the gastric mucosa biopsy to show?
3. Outline the pathogenesis of this patient's disease.
4. It took a long time for doctors to accept that this condition had an infectious etiology. Why?

Discussion

1. This patient had a duodenal ulcer. The ulcer had penetrated deeply beyond the mucosa, involving small blood vessels and causing bleeding. This was apparent from the clot that was visualized at endoscopy and the positive test for blood in the stool.
2. Microscopic examination of the biopsy showed curved bacteria, confirmed by culture to be *Helicobacter pylori.*
3. *Helicobacter pylori* enter the gastrointestinal tract by the fecal-oral route. In the stomach, they escape the lethal effect of gastric acid because they produce urease. Highly motile, they enter the gastric mucus and follow a gradient of acidity ranging from pH 2 in the gastric juices to pH 7.4 at the epithelial surface. Mutant strains that lack the ability to produce urease are only infectious if they are introduced directly into the mucus layer. Multiplication occurs just above the epithelial surface, but some of the bacteria attach to the epithelial cells and cause a loss of microvilli and thickening at the site of attachment. An inflammatory reaction develops beneath the affected mucosa. Two genes, *vacA* and *cagA*, correlate with virulence. The gene product VacA is a toxin similar to the adenylate cyclase of *Bordetella pertussis.* The bacteria inject CagA into host cells, causing the cells to elongate and spread out. CagA also provokes a strong immune response. Once established, *H. pylori* infections persist for years and often for a lifetime. It is not known why some people develop gastric or duodenal ulcers and others do not. Both host and bacterial factors are almost certainly involved. For example, strains of *H. pylori* isolated from peptic ulcer patients tend to be more virulent than those from patients who just have gastritis; patients with blood group O have more receptors for the bacterium and a higher incidence of peptic ulcers than do other people. Stomach acid and peptic enzymes probably play a role in ulcer formation by acting on damaged epithelium unprotected by normal mucus.
4. Claude Bernard, a scientist of Pasteur's time, put it this way: "It is that which we do know which is the greatest hindrance to our learning that which we do not know." In 1983, when Dr. Barry J. Marshall proclaimed before an international gathering of infectious disease experts that a bacterium caused stomach and duodenal ulcers, everyone "knew" it could not be true because no organism was thought to exist that could survive stomach acidity and enzymes. Indeed, almost everyone already "knew" the cause of ulcers to be psychosomatic. There is much still to be learned about the cause of ulcers, however, and Bernard's statement remains relevant.

25.4

Viral Diseases of the Upper Digestive System

Focus Points

- Explain why HSV-1 can spread easily from person to person.
- Give the characteristics of the mumps virus that make it a good candidate for eradication.

Some viral diseases involve the upper alimentary system but produce more dramatic symptoms elsewhere in the body. For example, measles produces Koplik spots in the mouth, respiratory symptoms, and a dramatic skin rash; chickenpox causes oral blisters, ulcers, and a striking skin rash; infectious mononucleosis can cause multiple oral ulcers, bleeding gums, and impressively enlarged lymph nodes and spleen. In this section, we focus on two viruses that produce dramatic symptoms in the upper digestive system, herpes simplex, with its characteristically painful oral ulcers, and mumps, with its enlarged, painful parotid glands. ■ chickenpox, p. 545 ■ infectious mononucleosis, p. 686 ■ measles, p. 547

Herpes Simplex (Cold Sores or Fever Blisters)

Herpes simplex is an extremely widespread disease with many manifestations. In its most common form, it begins in the mouth and throat. Involvement of the esophagus is suggestive of AIDS or other immunodeficiency. The infection persists for life; its causative virus is transmissible with saliva. Although the disease is usually insignificant, it can have tragic consequences if it results in infection of newborn infants or people with immunodeficiency.

Symptoms

Herpes simplex typically begins during childhood with fever, and blisters and ulcers in the mouth and throat so painful that it is difficult for those infected to eat or drink. The first lesions are small blisters that break within a day or two, leaving superficial, painful ulcers that heal without treatment within about 10 days. Thereafter, the infection becomes latent and the affected person may suffer the recurrent disease, **herpes simplex labialis** (the word *labialis* indicates location of the lesions on the lips), otherwise known as "cold sores" or "fever blisters" **(figure 25.9).** The symptoms of recurrences usually begin on the lips and include a tingling, itching, burning, or painful sensation. Blisters then appear, followed by painful ulcerations. Healing occurs within 7 to 10 days.

Causative Agent

Herpes simplex is caused by the herpes simplex virus (HSV), a medium-sized, enveloped virus containing double-stranded, linear DNA. There are two types of the virus—HSV-1 and HSV-2. Most oral infections are due to HSV-1; HSV-2 usually causes genital infections. ■ genital herpes simplex, p. 638

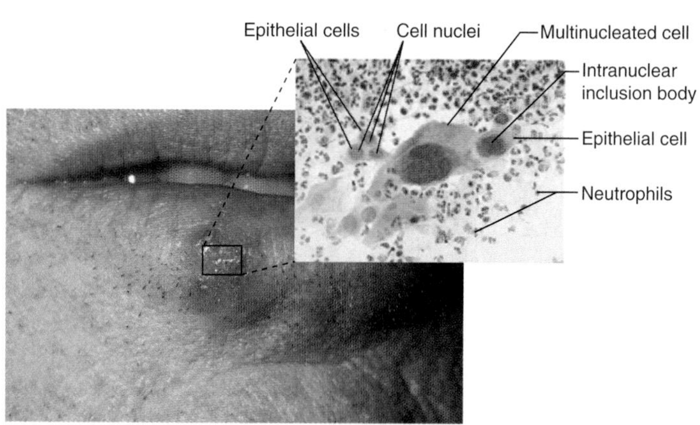

FIGURE 25.9 Herpes Simplex Labialis, (Cold Sores or Fever Blisters) The photomicrograph is of stained material from a herpes simplex lesion. It shows a multinucleated giant cell and intranuclear inclusion bodies. The pink areas within the epithelial cell nuclei are inclusion bodies, the sites of viral replication.

Pathogenesis

The virus multiplies in the epithelium of the mouth or throat, and destroys the cells. Some epithelial cells fuse together, producing large, multinucleated giant cells. Nuclei of infected cells characteristically contain a deeply staining area called an intranuclear inclusion body (see figure 25.9), which is the site of earlier viral replication. The blisters that form contain large numbers of infectious virions. Although an immune response develops and quickly limits the infection, some of the virions enter the sensory nerves in the area. Viral DNA persists in these nerve cells in a non-infectious, non-replicating form. This **latent virus** can, from time to time, become infectious. It is then carried by the nerves to skin or mucous membranes and produces recurrent disease. Stresses that can precipitate recurrences include menstruation, sunburn, and any illness associated with fever.

Epidemiology

HSV is extremely widespread and infects up to 90% of some U.S. inner-city populations, usually resulting in mild, if any, symptoms. An estimated 20% to 40% of Americans suffer recurrent herpes simplex. The virus is transmitted primarily by close physical contact, although it can survive for several hours on plastic and cloth. The greatest risk of infection is from contact with lesions or saliva from patients within a few days of disease onset, because at this time large numbers of virions are present. The saliva of people with no symptoms can be infectious, however, posing a risk to healthcare workers or therapists who have contact with saliva, such as nurses and dental workers. HSV can infect almost any body tissue—for example, **herpetic whitlow,** a painful finger infection, is not uncommon among nurses; and wrestlers can develop infections at almost any skin site because saliva containing HSV can contaminate wrestling mats and get rubbed into abrasions. Blindness can result if the virus is rubbed into the eye. Although uncommon, HSV is the most frequently identified cause of sporadic, viral encephalitis, a serious brain disease.

Prevention and Treatment

Acyclovir, penciclovir, and similar medications target HSV DNA polymerase and halt the progress of herpes simplex, but they do not affect the latent virus and thus cannot rid the body of HSV infection. These medications are useful for treating severe cases and for preventing disabling recurrences. Because the ultraviolet portion of sunlight can trigger recurrent disease, sunscreens are sometimes a helpful preventive. ■ DNA polymerase, p. 166 ■ acyclovir, p. 487

Some of the main features of herpes simplex are shown in **table 25.3.**

Mumps

Mumps is an acute viral illness that preferentially attacks glands such as the parotids. Formerly common in the United States, the disease is now relatively rare because of routine childhood immunization. However, outbreaks occur due to waning immunity, mostly in college students and other young adults. A large outbreak in 2006 resulted in almost 6,000 reported cases.

Symptoms

The onset of mumps is marked by fever, loss of appetite, and headache. Typically these symptoms are followed by painful swelling of one or both parotid glands **(figure 25.10).** Spasm of the underlying muscle makes it difficult to chew or talk. The word *mumps* probably derives from the verb *mump,* meaning "to mumble" or "whisper." Symptoms of mumps usually disappear in about a week.

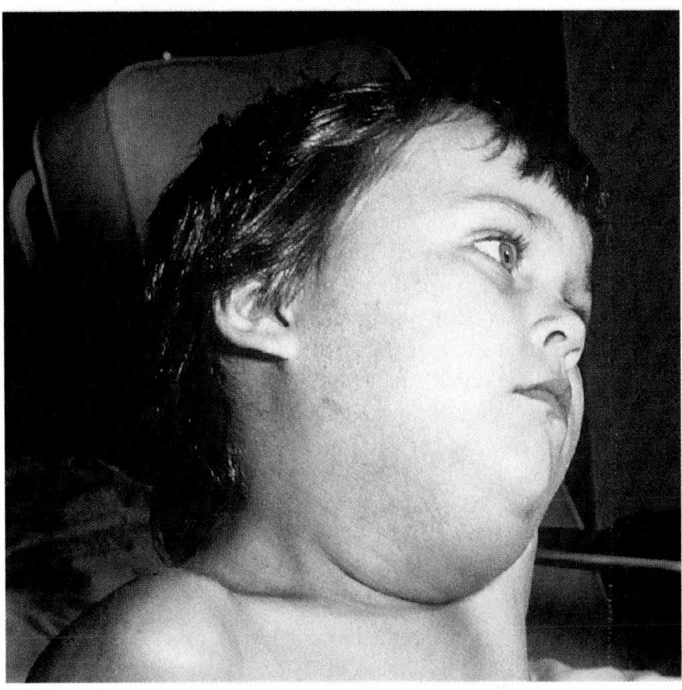

FIGURE 25.10 A Child with Mumps The swelling directly below the earlobe is due to enlargement of the parotid gland.

TABLE 25.3	Herpes Simplex
Symptoms	Initial infection: fever, severe throat pain, ulcerations of the mouth and throat. Recurrences: itching, tingling or pain usually localized to the lip, followed by blisters that break leaving a painful sore, which usually heals in 7 to 10 days
Incubation period	2 to 20 days
Causative agent	Herpes simplex virus (HSV), usually type 1
Pathogenesis	The virus multiplies in the epithelium, producing cell destruction and blisters containing large numbers of infectious virions. An immune response quickly limits the infection, but non-infectious HSV DNA persists in sensory nerves. This DNA becomes the source of infectious virions that are carried to the skin or mucous membranes, usually of the lip, causing recurrent sores.
Epidemiology	Widespread virus, transmitted by close physical contact. The saliva of asymptomatic individuals is commonly infectious.
Prevention and treatment	Acyclovir, penciclovir, and similar medications that target HSV DNA polymerase can shorten the duration of the illness or prevent recurrences. Sunscreens are helpful in preventing recurrences due to ultraviolet exposure.

Although painful parotid swelling is characteristic of mumps, up to half of the cases of mumps virus infection show no obvious parotid involvement. Symptoms can arise elsewhere in the body with or without parotid swelling. For example, headache and stiff neck indicate that the virus is causing **meningitis**—infection of the coverings of the brain, a common manifestation of mumps virus infection. Generally, mumps symptoms are much more severe in individuals past the onset of puberty. For example, about one-quarter of cases of mumps in post-pubertal boys and men are complicated by **orchitis**—a rapid, intensely painful swelling of one or both testicles to three to four times their normal size. Atrophy, or shrinkage, of the involved testicles commonly develops after recovery from the illness and, in rare cases, sterility results. In women and post-pubertal girls, ovarian involvement occurs in about one of 20 cases, and is manifested by pelvic pain. Pregnant women with mumps commonly miscarry, but birth defects do not result from mumps as they do from rubella. Serious consequences of mumps are rare and are most likely to occur in older people. These consequences include deafness and death from **encephalitis** (brain infection). ■ rubella, p. 549

Causative Agent

Mumps virus has a lipid-containing envelope and is classed in the paramyxovirus family, a group of single-stranded RNA viruses that includes the rubeola and respiratory syncytial viruses. Only one antigenic type of the mumps virus is known.

Pathogenesis

Infection occurs when a person who lacks immunity to mumps inhales virus-laden droplets of saliva. The incubation period is long, generally 15 to 21 days, because the virus reproduces first in the upper respiratory tract, then is spread throughout the body by the bloodstream, and produces symptoms only after infecting tissues such as the parotid glands, meninges, pancreas, ovaries, or testicles. In the salivary glands, the virus multiplies in the epithelium of ducts that convey saliva to the mouth. This destroys the epithelium, which releases enormous quantities of virus into the saliva. The body's inflammatory response to the infection is responsible for the severe swelling and pain. A similar sequence of events occurs in the testicles, where the virus infects the system of tubules that convey the sperm. The marked swelling and pressure often impair the blood supply, leading to hemorrhages and death of testicular tissue. Kidney tubules are also infected, and the virus can be cultivated from the urine for 10 or more days following the onset of illness.

Epidemiology

Humans are the only natural host of mumps virus, and natural infection confers lifelong immunity. Individuals sometimes claim to have had mumps more than once, probably because other infectious and non-infectious diseases can cause parotid swelling. The virus is spread by the high percentage, about 30%, of individuals who have asymptomatic infections but secrete mumps virus in their saliva and continue to mingle with other people. In symptomatic patients, the virus can be present in saliva from almost a week before symptoms appear to 2 weeks afterward. Peak infectivity however, is from 1 to 2 days before parotid swelling until the gland begins to return to normal size.

Prevention and Treatment

An effective attenuated mumps vaccine has been available in the United States since 1967, usually given in the same injection as measles, mumps, and rubella varicella vaccines (MMRV). **Figure 25.11** shows how the incidence of mumps has generally declined, although it increased in the 1980s because of a decline in funding for vaccinations. Because its host range is limited to humans, there is only a single viral serotype, and latent recurrent infections do not occur, mumps is a good candidate for eradication. There is no effective antiviral medication.

The main features of mumps are presented in **table 25.4.**

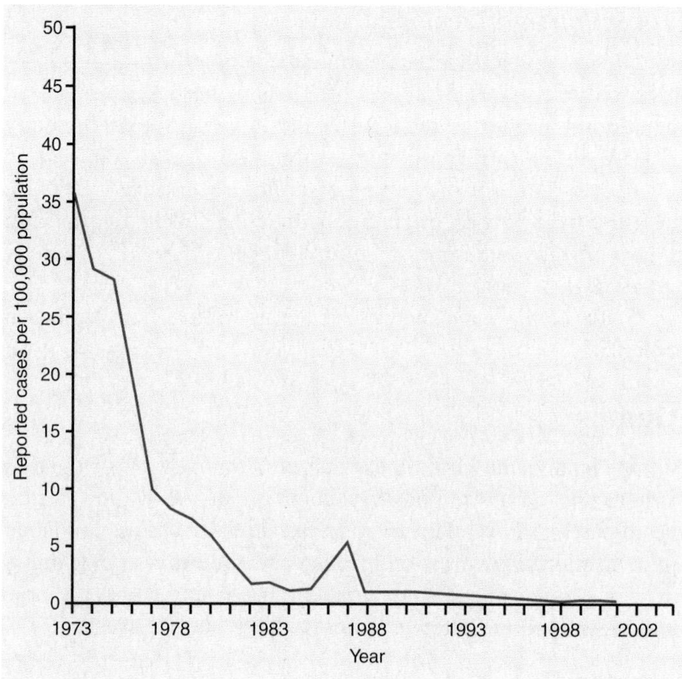

FIGURE 25.11 Reported Cases of Mumps per 100,000 Population, United States, 1973–2001 Mumps vaccine was licensed in 1967. There were only 236 cases reported for the year 2004 but more than six thousand in 2006 because of an epidemic among college students.

MICROCHECK 25.4

Herpes simplex is characterized by acute infection followed by lifelong latency and the possibility of recurrent disease. Infectious virus is often present in saliva in the absence of symptoms. Mumps virus infections characteristically cause enlargement of the parotid glands, but they can involve the brain, testicles, and ovaries, and cause miscarriages. Mumps is a good candidate for eradication.

- ✓ What infections are caused by the two types of herpes simplex viruses?
- ✓ Why is mumps a good candidate for eradication from the world?
- ✓ Why would you expect acyclovir to be ineffective against latent HSV infections?

TABLE 25.4	Mumps
Symptoms	Fever, headache, loss of appetite, typically followed by painful swelling of one or both parotid glands. Painful enlargement of the testicles, pelvic pain in women, and symptoms arising from brain involvement are likely to occur in individuals past the age of puberty.
Incubation period	Generally 15 to 21 days
Causative agent	Mumps virus, a single-stranded RNA virus of the paramyxovirus family
Pathogenesis	The virus initially replicates in the upper respiratory tract, then disseminates to the parotids and other organs of the body via the bloodstream. The Inflammatory response to cell destruction is responsible for the marked swelling and pain.
Epidemiology	Humans are the only source of the virus. Spread is likely because of a high percentage of asymptomatic infections.
Prevention and treatment	An effective attenuated vaccine has been used since 1967. No antiviral therapy is available.

LOWER DIGESTIVE SYSTEM INFECTIONS

Each year, in developing countries, one out of five children dies of diarrhea before the age of 5 years; most of the 5 million individuals who die of diarrhea each year are infants, but no age group is spared. Fatal cases are less common in the United States than in the developing world, but there are millions of diarrhea cases annually.

The all-too-familiar symptoms of lower alimentary system diseases include diarrhea, loss of appetite, nausea and vomiting, and sometimes fever—or any combination of these symptoms can occur. Members of the medical profession often loosely ascribe these symptoms to **gastroenteritis** (*gastro-*, "stomach," *entero-*, "intestine," *-itis*, "inflammation"); others prefer the term "stomach flu." (Note that "stomach flu" has no relationship to "the flu," influenza). Diarrhea can be copious and watery with infections of the **small intestine,** or in smaller amounts containing mucus, pus, and sometimes blood when invasion of the large intestine is involved. The name **dysentery** is given to diarrheal illnesses when pus and blood are present in the feces.

Various causative agents can be responsible for these symptoms, including poisonings by microbial toxins in food (foodborne intoxication) and infections. Chapter 32 gives examples of microbial poisonings; this chapter focuses on infectious causes. Bacterial causes will be described first, and viral and protozoal causes will be discussed later. Pork tapeworm, one of a large number of intestinal helminth infestations, is discussed in chapter 12. Others can be found at the Online Learning Center (www.mhhe.com/nester6). ■ foodborne intoxication, p. 763

25.5 Bacterial Diseases of the Lower Digestive System

Focus Points

- List the mechanisms by which bacterial pathogens cause diarrheal disease.
- Give the most common sources of bacteria pathogenic for the intestine.

Diarrheal illness is a common result of bacterial infection of the intestine, but bacteria can also use the intestine as an entryway to the rest of the body, thereby causing other types of illness. A minority of bacterial pathogens that invade the body from intestinal infections can cause severe headache, high temperature, abscesses throughout the body, intestinal rupture, shock, and death. This kind of illness, called **enteric fever,** is generally seen after infection by certain strains of *Salmonella.* **Typhoid fever** is an example.

Because of the wide variety of bacterial intestinal diseases, we will first consider some generalities.

The pathogenesis of bacterial intestinal disease involves a number of mechanisms. Different strains within the same species can differ in the way they cause disease, and the same strain can employ more than one mechanism. Moreover, the responsible genes can be transferred from one species to another by conjugation or by bacteriophages. Microbiologists are discovering exciting details at the molecular level of how bacteria cause disease, and finding unexpected complexity in these seemingly simple creatures. Many of these pathogens, upon contact with a host cell, activate a type III **secretion system** that employs a structure resembling a short flagellum to join the bacterial cytoplasm with that of the host cell. Gene products from pathogenicity islands then pass directly into the host cell and activate endocytosis or other processes to the benefit of the bacterium. ■ secretion system, p. 378 ■ pathogenicity islands, p. 208

The main pathogenic mechanisms can be summarized as:

Attachment. Infecting bacteria must have a means of attaching to intestinal cells to avoid being swept away by the flow of intestinal contents. Initial attachment is often by means of pili (see figure 17.4); but other adhesins are usually required for intimate attachment. Secretion systems sometimes transfer substances into the host cell that become receptors for bacterial attachment. ■ adhesins, p. 398

Toxin production. Toxins involved in intestinal infections fall into two groups: (1) toxins that increase secretion of water and electrolytes, as seen with *Vibrio cholerae*; and (2) toxins that cause cell death by halting protein synthesis, as seen with *Shigella dysenteriae*.

Alterations in the host cells. Some infecting bacteria use type III secretion systems to deliver molecules into intestinal epithelial cells (see figure 17.5), causing the cells to change. For example, certain pathogenic strains of *E. coli* deliver substances to intestinal epithelial cells that cause rearrangement of the actin filaments in those cells, resulting in loss of microvilli and creation of a platform or pedestal under the bacterium. The secretion system can also transfer substances that become receptors for intimate bacterial attachment.

Cell invasion. Some infecting bacteria use a type III secretion system to deliver molecules to intestinal epithelial cells, causing the cells to engulf the bacteria (see figure 17.6). In the example of invasion by *Shigella* species, the bacteria then multiply in the cell's cytoplasm, leading to destruction of the cell.

The epidemiology of intestinal diseases involves transmission by the fecal-oral route, usually when a person ingests food or drinking water contaminated with animal or human feces. Generally, with certain important exceptions, bacterial intestinal pathogens from animal sources have a high infecting dose and are not transmitted from person to person. Pathogens from a human source usually have a low infecting dose and are transmitted from person to person.

Regarding prevention and treatment, only a few vaccines are currently available, and they are of limited effectiveness because they fail to elicit a sufficient secretory IgA response on the intestinal mucosa. Sanitary and hygienic measures, and chlorination of drinking water, are important control measures. Antibacterial medications are not helpful for most cases originating from animal sources, and they often prolong the illness because they depress the normal microbiota. On the other hand, antibacterial medications can be life-saving in cases where the bacterium invades beyond the intestine. ■ IgA antibodies, p. 374

Cases of severe, watery diarrhea can cause rapid loss of water and electrolytes, so that the blood becomes reduced in volume and there is insufficient blood flow to keep vital organs, such as the kidneys, working properly. Severe, watery diarrhea can be rapidly fatal unless the lost fluid can be replaced promptly. Fortunately, despite the inability of the small intestine to control fluid secretion, if glucose is supplied it is still able to absorb fluid and electrolytes in most diarrheal diseases. An **oral rehydration solution (ORS)** that consists of water, glucose, and electrolytes is commercially available, and recipes for homemade ORS have been approved by the World Health Organization. ORS is a highly effective lifesaver in severe diarrheas regardless of cause.

Only three kinds of bacteria account for almost all bacterial intestinal infections: *Vibrio* species, *Campylobacter jejuni,* and enterobacteria. ■ enterobacteria, p. 262

Cholera

There have been seven **cholera** pandemics since the early 1800s. Between 1832 and 1836, more than 200,000 Americans died when the second and fourth pandemics of cholera swept across North America. The seventh pandemic began in 1961 in Indonesia and spread to South Asia, the Middle East, and parts of Europe and Africa. South America had remained cholera-free for 100 years until January 1991, when the disease abruptly appeared in Peru. Introduction probably occurred when a freighter discharged bilgewater into a Lima harbor. The municipal water supply was not chlorinated and quickly became contaminated. The disease then spread rapidly, so that in 2 years more than 700,000 cases and 6,323 deaths had been reported from South and Central America. Cases also appeared in 14 states in the United States. ■ pandemic disease, p. 451

Symptoms

Cholera causes the classic example of severe watery diarrhea. The diarrheal fluid can amount to 20 liters a day and because of its appearance, has been described as "rice water stool." Vomiting also occurs in most people at the onset of the disease, and many people suffer muscle cramps caused by loss of fluid and electrolytes.

Causative Agent

The causative agent is *Vibrio cholerae,* a curved, Gram-negative rod able to tolerate strong alkaline conditions and high salt concentrations.

Pathogenesis

Vibrio cholerae is killed by acid, and so large numbers of the organism must be ingested before enough survive stomach passage to establish infection. Those that make it to the small intestine adhere to the small intestinal epithelium by means of pili and other surface proteins. The organisms multiply on the epithelial cells **(figure 25.12)** but do no visible damage to them. The bacteria, however, produce **cholera toxin,** a potent exotoxin. This enterotoxin causes the epithelial cells to continuously secrete chloride ions into the intestinal lumen. Sodium and other ions and water follow the chloride, resulting in an outpouring of fluid and electrolytes from the cells. Although the large intestine is not affected by the toxin, it cannot absorb the large volume of fluid that rushes through it, and diarrhea results.

Cholera toxin is an A-B toxin **(figure 25.13).** The B portion binds irreversibly to specific receptors on the microvilli of the epithelial cells whereas the A portion enters the intestinal cell and causes the symptoms of cholera. It does this by modifying a protein that normally controls the level of cyclic adenosine monophosphate (cAMP) in the cells, effectively making a key enzyme, adenylate cyclase, continuously active. Adenylate cyclase converts ATP to cAMP and its continuous activity results in maximum levels of cAMP in the intestinal cells, causing the cells to constantly pump water and electrolytes into the intestinal lumen. The normal shedding of intestinal cells eventually gets rid of the toxin.

Synthesis of cholera toxin is an example of lysogenic conversion. It is coded by a filamentous bacteriophage that infects *Vibrio cholerae* via the pili by which the bacteria attach to the intestinal cells. Synthesis of both cholera toxin and the pili is regulated by the same bacterial gene, so that toxin and pili, the two factors required for disease production, are synthesized at the same time. ■ lysogenic conversion, p. 310

Epidemiology

Fecally contaminated water is the most common source of cholera infection, although foods such as crab and vegetables fertil-

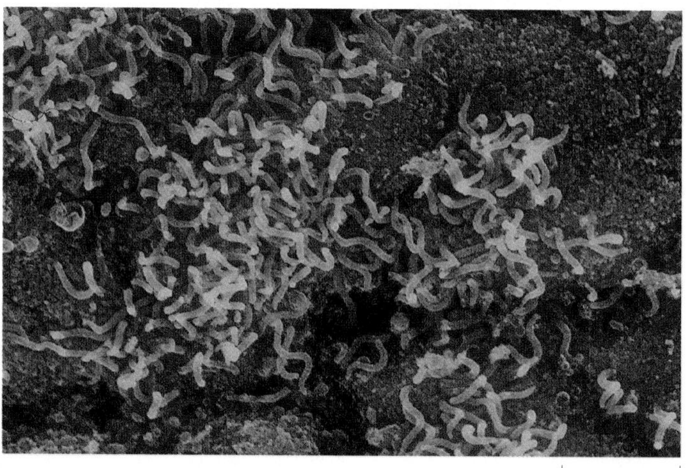

FIGURE 25.12 Scanning Electron Micrograph of *Vibrio cholerae* Attached to Small Intestinal Mucosa Attachment occurs by means of pili.

PERSPECTIVE 25.1

Ecology of Cholera

The ecology of cholera is fascinating but still incompletely understood. The 0139 *Vibrio cholerae* strain arose in India in the fall of 1992 from a serotype 01 strain, the same serotype as the pandemic and American Gulf Coast strains. Interestingly, the change in O antigen was accompanied by acquisition of a capsule—these changes probably resulting from genes acquired by transduction or conjugation. Various strains of *V. cholerae*, most of which are not pathogenic for humans, live in the coastal seas around the world, largely in association with zooplankton. These zooplankton can sometimes also harbor the *V. cholerae* strains responsible for pandemic cholera. Moreover, the zooplankton feed on phytoplankton and therefore increase markedly in number when warm seas and abundant nutrients cause explosive growth of phytoplankton. These findings could explain (1) how new pandemic strains could arise through genetic interchange; (2) an association between cholera and climatic changes such as El Niño; and (3) the onset and rapid spread of epidemic cholera by ocean currents along coastal areas. ■ **zooplankton, p. 286**

ized with human feces have also been implicated in outbreaks. A person with cholera can discharge a million or more *V. cholerae* organisms in each milliliter of feces. Although cholera is relatively common worldwide, few cases due to the pandemic strain have been acquired in the United States since the early 1900s. Since 1973, however, sporadic cases have occurred along the Gulf of Mexico. Most of these infections have been traced to consumption of coastal marsh crabs. These cases were due to a strain of *V. cholerae* different from the pandemic OI strain. Ominously, in September 1992, a new strain of *V. cholerae,* belonging to a different O group (O139), appeared in India and spread rapidly across South Asia, attacking even those people with immunity to the seventh pandemic strain. This new strain and its rapid spread suggested it could initiate another pandemic, but after about a year it declined in incidence and remained endemic, causing resurgent localized epidemics. ■ **O antigen, pp. 243, 262**

Prevention and Treatment

Control of cholera depends largely on adequate sanitation and the availability of safe, clean water supplies. Travelers to areas where cholera is occurring are advised to cook food immediately before eating it. Crabs should be cooked for no less than 10 minutes. No fruit should be eaten unless peeled personally by the traveler, and ice should be avoided unless it is known to be made from boiled

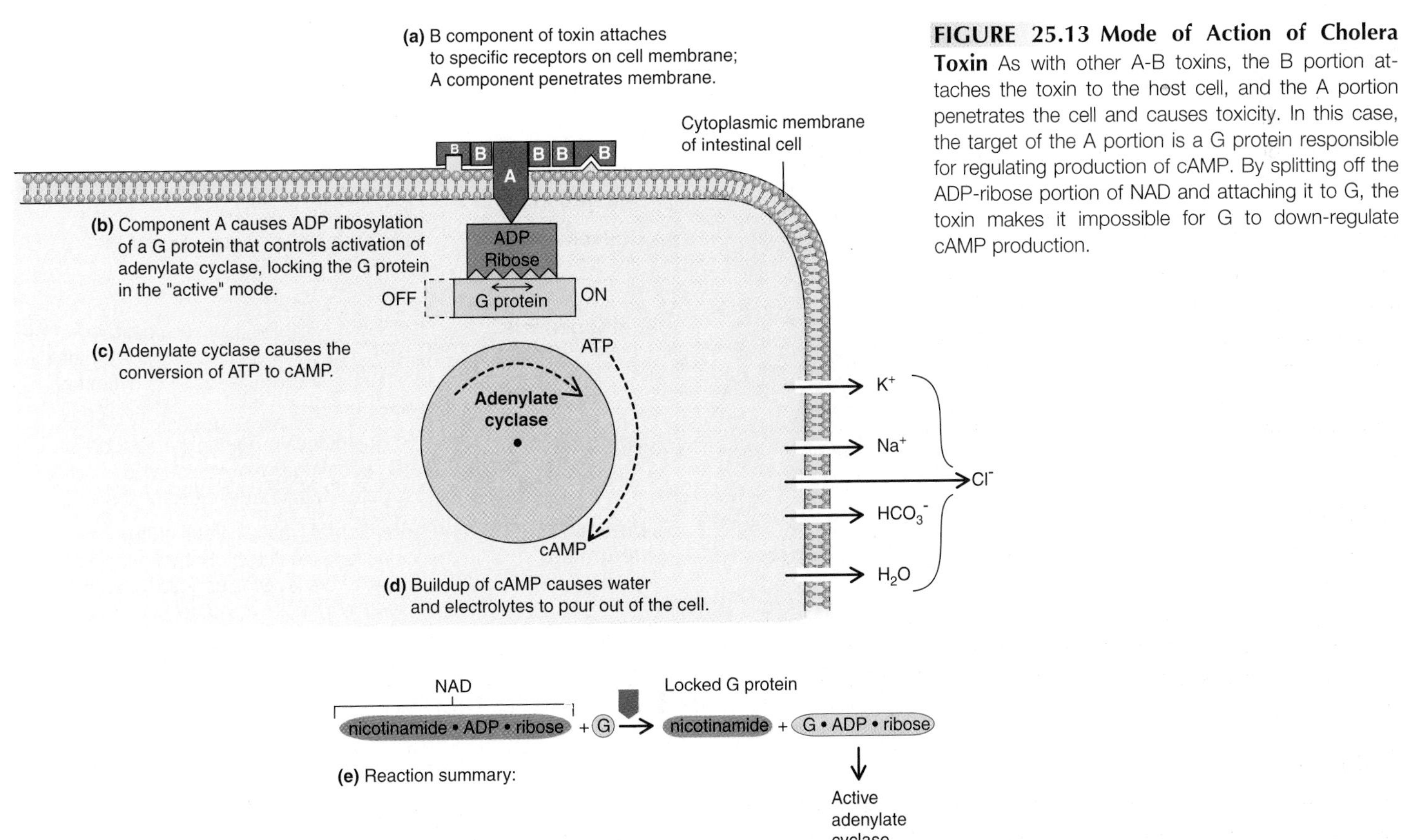
(a) B component of toxin attaches to specific receptors on cell membrane; A component penetrates membrane.
Cytoplasmic membrane of intestinal cell
B
B
A
B
B
B
(b) Component A causes ADP ribosylation of a G protein that controls activation of adenylate cyclase, locking the G protein in the "active" mode.
ADP Ribose
OFF
G protein
ON
(c) Adenylate cyclase causes the conversion of ATP to cAMP.
ATP
Adenylate cyclase
cAMP
K+
Na+
Cl-
HCO3-
H2O
(d) Buildup of cAMP causes water and electrolytes to pour out of the cell.
NAD
Locked G protein
nicotinamide • ADP • ribose + G → nicotinamide + G • ADP • ribose
(e) Reaction summary:
Active adenylate cyclase
ATP → cAMP

FIGURE 25.13 Mode of Action of Cholera Toxin As with other A-B toxins, the B portion attaches the toxin to the host cell, and the A portion penetrates the cell and causes toxicity. In this case, the target of the A portion is a G protein responsible for regulating production of cAMP. By splitting off the ADP-ribose portion of NAD and attaching it to G, the toxin makes it impossible for G to down-regulate cAMP production.

water. Orally administered vaccines are commercially available in a number of countries outside the United States. One consists of a living, genetically altered strain of *V. cholerae,* and another, killed *V. cholerae* in combination with the purified recombinant B subunit of cholera toxoid. Treatment of cholera depends on the rapid replacement of electrolytes and water before irreversible damage to vital organs can occur. The prompt administration of intravenous or oral rehydration fluid decreases the mortality of cholera to less than 1%.

The main features of cholera are summarized in **table 25.5.**

Shigellosis

Shigellosis is distributed worldwide wherever sanitary practices are lacking. Reported cases in the United States have averaged about 21,500 per year, but the true prevalence is much higher.

Symptoms

The classic symptom of shigellosis is dysentery. Other symptoms are headache, vomiting, fever, stiff neck, convulsions, and joint pain. Shigellosis is commonly fatal for infants in developing countries.

Causative Agents

There are four species of *Shigella—S. flexneri, S. boydii, S. sonnei,* and *S. dysenteriae.* Shigellas are Gram-negative enterobacteria. Virulent strains contain a large plasmid that is essential for the attachment and entry of *Shigella* into host cells.

Pathogenesis

Shigella species invasion of intestinal epithelial cells causes a strong inflammatory response. Invasion is accomplished by exploiting the antigen sampling function of M cells **(figure 25.14).** Recall that M cells normally take in microbes and transfer them to macrophages in Peyer's patches. *Shigella* species selectively attach to M cells and are thereby transported across the epithelial barrier. Upon engulfment by macrophages, the bacteria escape from the phagosome and multiply in the cytoplasm of the macrophages. The bacteria are released when the infected macrophages undergo apoptosis. The released bacteria then attach to specific receptors near the bases of the epithelial cells and induce those cells to take them in. Once inside, the *Shigella* bacteria escape from the endosome and multiply in the cytoplasm. These nonmotile bacteria can nevertheless propel themselves into neighboring intestinal cells by producing a protein that causes the host cell actin to rapidly polymerize at one end of the bacterial cell, forming an "actin tail." The overall result of infection is death and sloughing of patches of epithelium. The denuded areas become intensely inflamed, covered with pus and blood, accounting for the pus and blood in the diarrhea of dysentery. ■ actin, p. 74

Some strains of *Shigella dysenteriae,* a species rarely encountered in the United States, produce a potent toxin known as the **Shiga toxin,** a cytotoxin. It is a chromosomally coded A-B toxin. The B portion binds to endothelial cells of small blood vessels, and the A subunit enters the endothelial cell and reacts with its ribosome, thereby halting protein synthesis. The importance of this toxin is that it is responsible for the **hemolytic uremic syndrome (HUS),** an often fatal condition that can follow *Shigella*

TABLE 25.5 Cholera

① *Vibrio cholerae,* the causative bacterium, enters the mouth with fecally contaminated food or drink.

② The bacteria attach to epithelial cells of the small intestine.

③ *V. cholerae* toxin enters the cells and prevents them from down-regulating secretion of water and electrolytes.

④ The epithelial cells pump water and electrolytes from the blood into the intestinal lumen, causing watery diarrhea.

⑤ Shock and death occur because of fluid loss from the circulatory system, unless the fluid can be replaced.

⑥ The bacteria exit the body with feces.

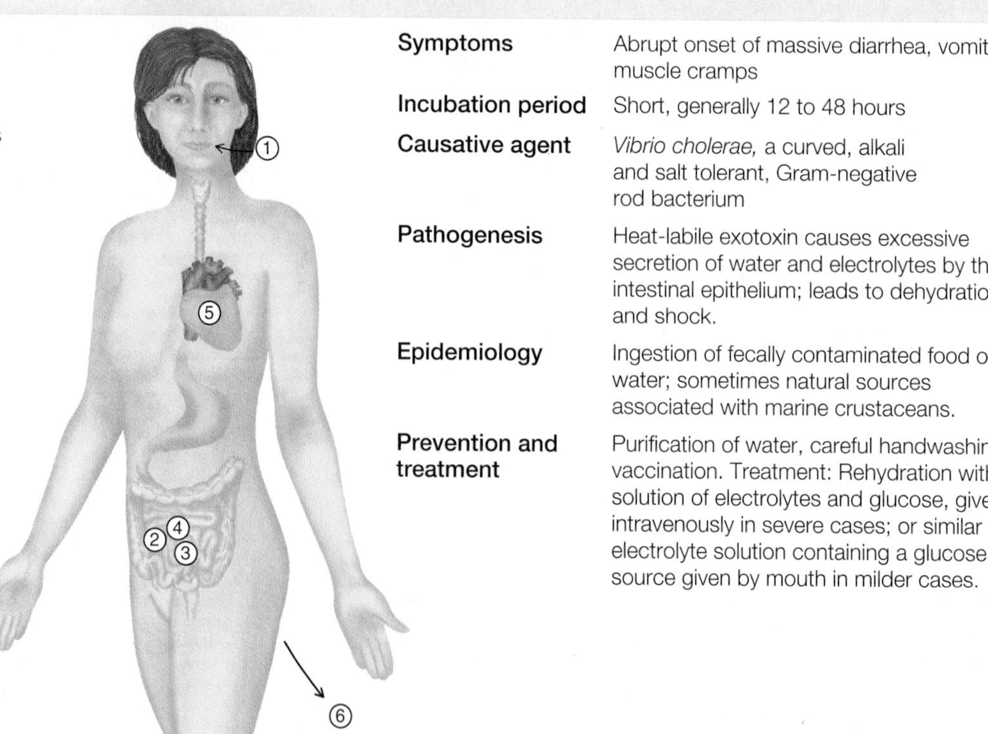

Symptoms	Abrupt onset of massive diarrhea, vomiting, muscle cramps
Incubation period	Short, generally 12 to 48 hours
Causative agent	*Vibrio cholerae,* a curved, alkali and salt tolerant, Gram-negative rod bacterium
Pathogenesis	Heat-labile exotoxin causes excessive secretion of water and electrolytes by the intestinal epithelium; leads to dehydration and shock.
Epidemiology	Ingestion of fecally contaminated food or water; sometimes natural sources associated with marine crustaceans.
Prevention and treatment	Purification of water, careful handwashing; vaccination. Treatment: Rehydration with a solution of electrolytes and glucose, given intravenously in severe cases; or similar electrolyte solution containing a glucose source given by mouth in milder cases.

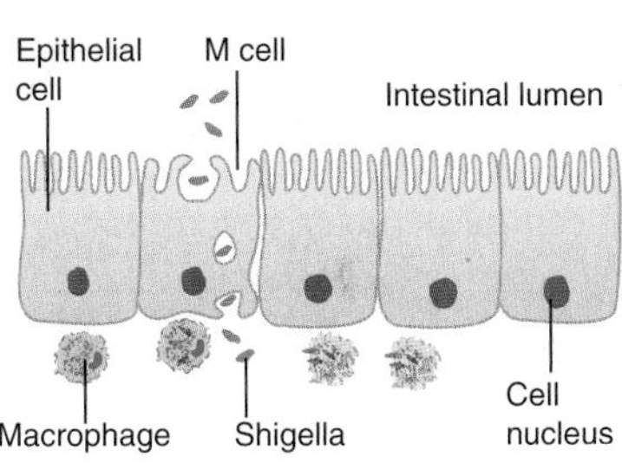

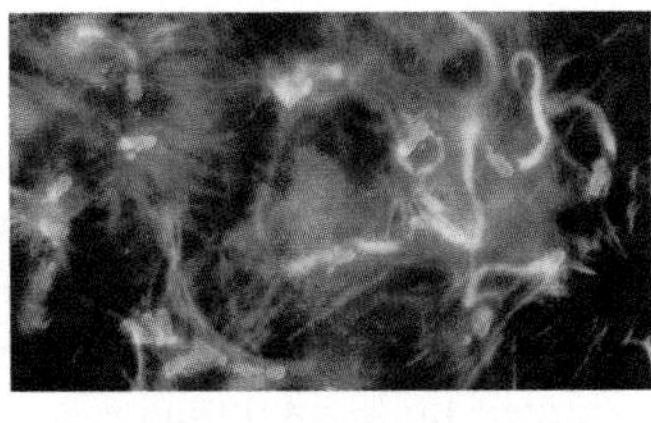

(1) Shigellas are taken up by M cells and transported beneath the epithelium. Macrophages take up shigellas, die and release the bacteria.

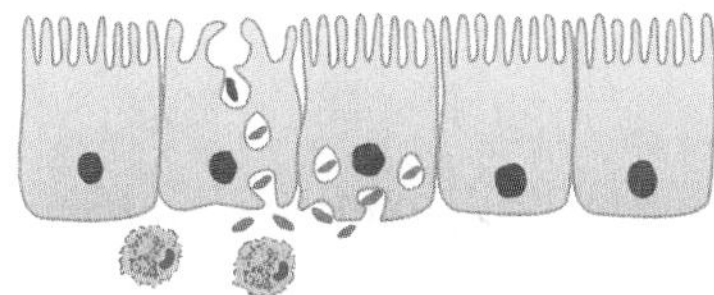

(2) The bacteria enter the inferior and lateral aspects of the epithelial cells by inducing endocytosis. The endosomes are quickly lysed, leaving the shigellas free in the cytoplasm.

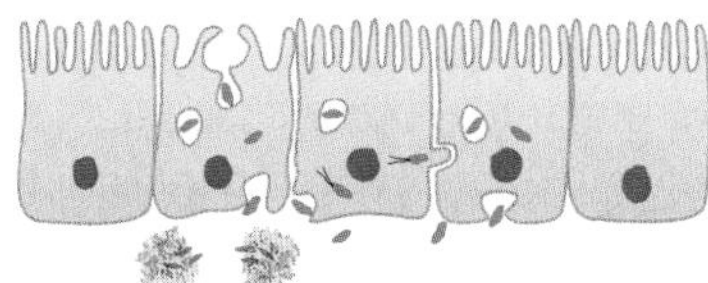

(3) Actin filaments quickly form a tail, pushing the shigellas into the next cell.

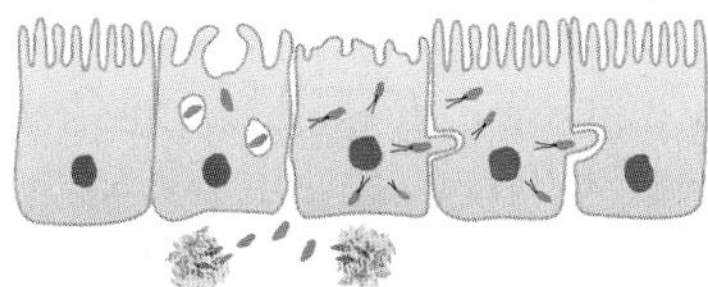

(4) Shigellas multiply in the cytoplasm, and the infection extends to the next cell.

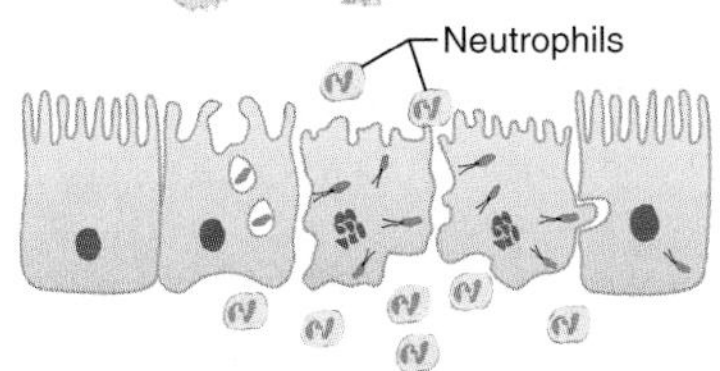

(5) Infected cells die and slough off. Intense response of acute inflammatory cells (neutrophils), bleeding and abscess formation.

FIGURE 25.14 Pathogenesis of Shigellosis The bacteria do not invade the intestinal epithelium directly, but are transported to the area underneath the epithelial cells by M cells. They then invade the epithelial cells from their inferior and lateral borders. The photomicrograph shows the actin tails (green) that form on intracellular shigellas (orange) and rapidly push these non-motile bacteria from cell to cell.

dysenteriae dysentery. In this syndrome, red blood cells break up in the tiny blood vessels of the body, resulting in anemia and kidney failure, sometimes accompanied by paralysis or other signs of nervous system injury. Shiga toxin is also produced by strains of *Escherichia coli* that cause the hemolytic uremic syndrome.

Epidemiology

Shigellosis generally has a human source and is transmitted by the fecal-oral route. The bacteria are not easily killed by stomach acid, so that the infectious dose is small and as few as 10 of the bacterial cells can initiate infection. Transmission occurs most readily in overcrowded populations with poor sanitation. It is a common problem in day care centers and among homosexual men. Fecally contaminated food and water have also caused outbreaks.

Prevention and Treatment

The spread of *Shigella* sp. is controlled by sanitary measures and surveillance of food handlers and water supplies. There is no vaccine. Antimicrobial medications such as ampicillin and cotrimoxazole are useful against susceptible strains because they shorten the duration of symptoms and the time during which shigellas are discharged in the feces. Almost 20% of *Shigella* strains, however, are resistant to these two commonly employed medications. R plasmids that convey resistance to several antibacterial medications are often present. Two specimens of feces, collected at least 48 hours after stopping antimicrobial medicines, must be negative for *Shigella* sp. before a person is allowed to return to a day care center or food-handling job. ■ R plasmids, p. 205

Table 25.6 describes the main features of shigellosis.

Escherichia coli Gastroenteritis

Escherichia coli generally ferments lactose, in contrast to most shigellas and salmonellas. It is an almost universal member of the normal intestinal flora of humans and a number of other animals. Long ignored as a possible cause of gastrointestinal disease, certain strains were shown in 1945 to cause life-threatening epidemic gastroenteritis in hospitalized infants. Later, *E. coli* strains were shown to cause gastroenteritis in adults, notably as agents responsible for traveler's diarrhea (also called "Delhi belly," "Montezuma's revenge," "Turkey trots," etc.). Still later, *E. coli* strains were identified as causative agents in dysentery, cholera-like illnesses, and diarrheas associated with the hemolytic uremic syndrome.

Symptoms

Symptoms depend largely on the virulence of the infecting *E. coli* strain. They range from vomiting and a few loose bowel movements, to profuse watery diarrhea, to severe cramps and bloody diarrhea. Fever is not usually prominent and recovery usually occurs within 10 days. Hemolytic uremic syndrome, however, develops after some bloody diarrhea cases.

TABLE 25.6 Shigellosis

Symptoms	Fever, dysentery, vomiting, headache, stiff neck, convulsions, and painful joints
Incubation period	3 to 4 days
Causative agent	Species of *Shigella,* Gram-negative, non-motile enterobacteria
Pathogenesis	Invasion of and multiplication within intestinal epithelial cells; death of cells, intense inflammation and ulcerations of intestinal lining. Some species produce a cytotoxin.
Epidemiology	Transmission via fecal-oral route; sometimes by fecally contaminated food or water; humans generally the only source.
Prevention and treatment	Sanitary measures; surveillance of food handlers and water supplies. Treatment: Antibacterial medications such as ampicillin and cotrimoxazole (trimethoprim plus sulfamethoxazole) shorten duration of symptoms and time shigellas are excreted; multiple R factor resistances may be present.

Causative Agent

Escherichia coli are Gram-negative, generally lactose-fermenting rods of the enterobacteria group. Most of the diarrhea-causing *E. coli* fall into five groups: enterotoxigenic (ETEC), enteroinvasive (EIEC), enteropathogenic (EPEC), enteroaggregative (EAEC), and Shiga toxin-producing *E. coli* (STEC), also known as enterohemorrhagic *E. coli*. In contrast to other strains of *E. coli,* strains in these groups possess virulence factors that cause them to be intestinal pathogens. Other strains of *E. coli* cause urinary infections, septicemia, and meningitis.

Enterotoxigenic *E. coli* (ETEC) is a common cause of traveler's diarrhea and diarrhea in infants. Some members of this group are responsible for significant mortality due to diarrhea in young livestock. ETEC strains usually possess adhesins that allow them to attach to and colonize the intestinal epithelium, where they secrete one or more toxins. One such toxin is nearly identical to cholera toxin in action and antigenicity.

Enteroinvasive *E. coli* (EIEC) invade the intestinal epithelium, resulting in a disease closely resembling that caused by *Shigella* sp.

Enteropathogenic *E. coli* (EPEC) strains cause diarrheal outbreaks in hospital nurseries and in bottle-fed infants in developing countries and can also cause chronic diarrhea in infants. They possess plasmid-encoded adhesins; the bacteria cause loss of microvilli and a thickening of the cell surface at the site where the organisms attach. Enteropathogenic *E. coli* strains can be distinguished from other *E. coli* strains by the fluorescent antibodies that identify their somatic and capsular antigens.

Enteroaggregative *E. coli* (EAEC) causes epidemic diarrhea in children, diarrhea in patients with AIDS, and travelers diarrhea. Originally, these strains were identified by their distinctive aggregation on tissue culture cells, but now can best be identified by a DNA probe.

Shiga toxin-producing *E. coli* (STEC), the fifth group of diarrhea-producing strains, was discovered in 1982. Most of these organisms belong to a single serological type, O157:H7. They often produce severe illness, including bloody diarrhea. These strains produce potent cytotoxins that cause the death of intestinal epithelium by interfering with protein synthesis. Some of the toxins are identical to the Shiga toxins of *Shigella dysenteriae,* mentioned earlier. Unlike shigellas, however, enterohemorrhagic *E. coli* generally do not penetrate intestinal epithelial cells. Their toxin production depends on lysogenic conversion by a distinct bacteriophage. About 4% of infected individuals develop hemolytic uremic syndrome, marked by anemia due to lysis of red blood cells, kidney failure, and central nervous system damage. STEC epidemics have been traced to ground beef, sausage, unpasteurized milk, apple juice, spinach, and unchlorinated city water. In most instances, cow manure was the source of the offending strain. ■ enterohemorrhagic *E. coli* epidemics, p. 244

Characteristics of diarrhea-causing *E. coli* are summarized in **table 25.7.**

Pathogenesis

Of hundreds of different strains of *E. coli,* only those possessing certain virulence factors cause gastrointestinal disease. Some of these factors were discussed in the previous section; there is strong evidence that not all factors have been identified. Two important virulence factors, enterotoxin production and the ability to adhere to the small intestine, are coded by plasmids. These plasmids can be transferred to other *E. coli* organisms by conjugation, thereby conferring virulence on the recipient strain. Gastroenteritis-producing strains of *E. coli* often have more than one type of virulence plasmid. It is interesting that the heat-labile toxin of *E. coli* is antigenically closely related to the cholera toxin of *Vibrio cholerae.* This similarity suggests that the genes responsible for the two toxins have a common ancestry. ■ conjugation, p. 202 ■ plasmids, p. 68

Epidemiology

Epidemics occur from person-to-person spread, contamination of foods, unpasteurized milk and juices, and water sources contaminated with feces. Human beings, and domestic and wild animals can all be sources of pathogenic strains. The infecting dose is low—only 200 organisms or less.

Prevention and Treatment

Preventive measures include handwashing, pasteurization of drinks, and thorough cooking of food. Treatment of *E. coli* gastroenteritis includes replacing the fluid lost from vomiting and diarrhea. In addition, infants may require antibiotic treatment for a few days. Traveler's diarrhea can usually be prevented with bismuth preparations (such

TABLE 25.7 Characteristics of Diarrhea-Causing *Escherichia coli*

Designation	Characteristic Features	Clinical Picture
Enterotoxigenic *E. coli* (ETEC)	Can release two kinds at toxin, one similar to cholera toxin; small intestinal location	Nausea, vomiting, abdominal cramps, massive watery diarrhea leading to dehydration
Enteroinvasive *E. coli* (EIEC)	Entry into and growth in intestinal epithelium, with cell destruction; large intestinal location	Fever, cramps, blood and pus in the feces
Enteropathogenic *E. coli* (EPEC)	Attachment of the bacterium is followed by loss of microvilli and formation of a platform or pedestal of actin fibrils under the bacterium; small intestinal location	Fever, vomiting, watery diarrhea containing mucus; associated with a limited number of serotypes
Shiga toxin-producing *E. coli* (STEC)	Same as above, except large intestine location and release of Shiga toxin	Fever, abdominal cramps, bloody diarrhea without pus; 2% to 7% develop hemolytic uremic syndrome; most cases due to serotype O157:H7

as Pepto-Bismol). Antibacterial medications such as ciprofloxacin (Cipro) and the non-absorbable rifaximin (Xifaxan) are no longer routinely recommended. The widespread use of an antibiotic to prevent diarrhea has promoted development of resistant strains.

Some features of *E. coli* gastroenteritis are summarized in **table 25.8.**

Salmonellosis

Salmonellosis, a disease caused by bacteria of the *Salmonella* genus, can be contracted from many animal sources. There are an estimated 1.4 million cases annually in the United States, resulting in 400 deaths. Large outbreaks are usually due to commercially distributed produce contaminated by animal feces.

A few *Salmonella* strains, fortunately now rare in the United States, originate from humans. These human strains generally cause systemic disease, as in **typhoid fever.**

Symptoms

Gastroenteritis. Most cases of salmonellosis have gastroenteritis symptoms, with diarrhea (sometimes with blood in the stool), abdominal cramps, nausea, vomiting, headache, and fever. The symptoms vary depending on the virulence of the strain of *Salmonella* and the number of infecting organisms. The symptoms are usually short-lived and mild.

Enteric fever. A few *Salmonella* strains cause symptoms of enteric fever, with progressively increasing fever over a number of days, severe headache, constipation, and abdominal pain, followed in some cases by intestinal rupture, internal bleeding, shock, and death. *Salmonella* Typhi, the cause of typhoid fever, is an example.

Causative Agents

Salmonellas are motile, Gram-negative enterobacteria. It is generally recognized that there are only two species of *Salmonella*—*S. enterica* and *S. bongori,* the latter species rarely isolated from humans. The *Salmonella* are subdivided into more than 2,400 serotypes based on differences in their somatic (O), flagellar (H), and capsular (K) antigens. They are referred to as serotypes because the *Salmonella* strains are identified by sera that contain antibody against known strains. Formerly, many of these serotypes were assigned species names; for example, *Salmonella dublin* and *Salmonella heidelberg.* (In this discussion we will not italicize the serotype name, to distinguish it from a species name, as in *Salmonella* Dublin.) *Salmonella* Typhimurium and *Salmonella* Enteritidis are the serotypes most commonly isolated in the United States. Typhoid fever, an example of enteric fever, is caused by *Salmonella* Typhi.

Pathogenesis

Most salmonellas are killed by acid, and so a large number must generally be ingested to survive passage through the stomach. Upon reaching the lower small and large intestines, an adhesin on the bacterial surface attaches the bacteria to specific receptors on the surface of the epithelial cells. Contact with intestinal epithelium activates a type III secretion system (see figure 17.6). In only

TABLE 25.8 *Escherichia coli* **Gastroenteritis**

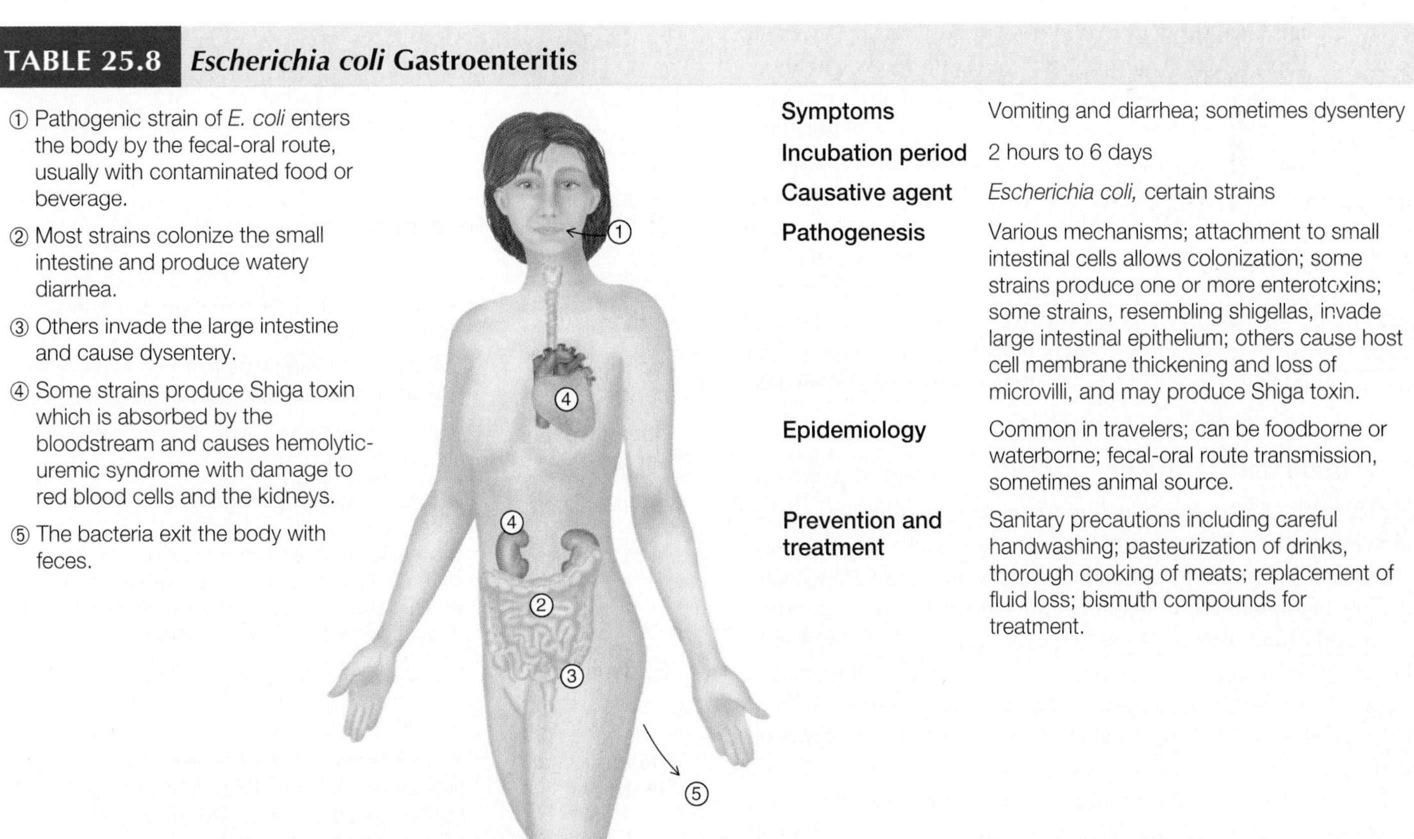
① Pathogenic strain of E. coli enters the body by the fecal-oral route, usually with contaminated food or beverage.
② Most strains colonize the small intestine and produce watery diarrhea.
③ Others invade the large intestine and cause dysentery.
④ Some strains produce Shiga toxin which is absorbed by the bloodstream and causes hemolytic-uremic syndrome with damage to red blood cells and the kidneys.
⑤ The bacteria exit the body with feces.
①
②
③
④
④
⑤

Symptoms	Vomiting and diarrhea; sometimes dysentery
Incubation period	2 hours to 6 days
Causative agent	*Escherichia coli,* certain strains
Pathogenesis	Various mechanisms; attachment to small intestinal cells allows colonization; some strains produce one or more enterotoxins; some strains, resembling shigellas, invade large intestinal epithelium; others cause host cell membrane thickening and loss of microvilli, and may produce Shiga toxin.
Epidemiology	Common in travelers; can be foodborne or waterborne; fecal-oral route transmission, sometimes animal source.
Prevention and treatment	Sanitary precautions including careful handwashing; pasteurization of drinks, thorough cooking of meats; replacement of fluid loss; bismuth compounds for treatment.

a matter of minutes, transfer of bacterial proteins into the epithelial cell tricks it into taking in the bacterium by endocytosis. The bacteria multiply within a phagosome and are discharged from the base of the cell by exocytosis. In most instances, the bacteria are quickly taken up and killed by macrophages, but the inflammatory response to the infection increases fluid secretion, causing diarrhea. ■ endocytosis, exocytosis, p. 73 ■ type III secretion system, p. 398

Some strains, such as *Salmonella* Typhi, are not so easily eliminated by host defenses. They cross the mucous membrane via M cells, resist killing by macrophages, multiply within them, and are carried by the bloodstream throughout the body. Released by death of the macrophages, the bacteria can invade tissues, cause prolonged fever, abscesses, septicemia, and shock, often with little or no diarrhea. *Salmonella* Typhi can also cause destruction of Peyer's patches, leading to rupture of the intestine and hemorrhage. ■ septicemia, p. 678 ■ Peyer's patches, p. 370

Epidemiology

Salmonellas survive for months in soil, water, and other environments. Children are commonly infected by seemingly healthy pets, such as turtles, iguanas, baby chickens, and ducks, that discharge salmonellas in their feces. Eggs and poultry are often contaminated with *Salmonella* strains. Other outbreaks have resulted from contaminated tomatoes, brewer's yeast, alfalfa sprouts, protein supplements, dry milk, and even a red dye used to diagnose intestinal disease. Most cases of *Salmonella* gastroenteritis have an animal source rather than a human source. Enteric fever strains such as *Salmonella* Typhi generally can be traced to long-term human carriers.

Although fecal discharge of gastroenteritis-causing strains is usually of short duration, carriers of *Salmonella* Typhi can eliminate up to 10 billion of the bacteria per gram of their feces for years. The source of these organisms is usually the gallbladder, which *Salmonella* Typhi colonizes free of competition because much of the normal flora is killed or inhibited by concentrated bile. Mary Mallone, "Typhoid Mary," a young Irish cook living in New York State in the early 1900s, was a notorious carrier. She was responsible for at least 53 cases of typhoid fever over a 15-year period. In her day, about 350,000 cases of typhoid occurred in the United States each year. The low incidence of typhoid fever in the United States today can be attributed to improved sanitation and public health surveillance measures.

Prevention and Treatment

Control of *Salmonella* infections depends on reporting cases of salmonellosis, tracing sources, sanitary handling of animal carcasses, pasteurizing animal products or irradiating them, and testing them for contamination. Adequate cooking effectively kills salmonellas. Note, however, that heat penetration to the center of a carcass may be inadequate to kill salmonellas, especially in frozen fowl, even when the outside appears "well done." Unfortunately, the number of reported cases of salmonellosis has tended to rise, due in part to mass distributions of food and water that become contaminated from time to time. Most people with salmonellosis recover without antimicrobial treatment; this is fortunate because *S. enterica* has shown increasing plasmid-mediated resistance to antibacterial treatment. About 35% of *Salmonella* Typhimurium isolates are resistant to five or more antibacterial medications. This resistance is partly due to selection of resistant strains of *Salmonella* by the widespread, ill-advised addition of antibiotics to animal feeds. Antibacterial medications are not advised for treating humans except when there is tissue or bloodstream invasion.

An attenuated oral vaccine for preventing typhoid fever is about 50% to 75% effective. An equally effective injectable vaccine composed of *Salmonella* Typhi capsular polysaccharide is also available. Surgical removal of the gallbladder is often necessary to rid *Salmonella* Typhi carriers of their infection.

Table 25.9 summarizes some of the features of salmonellosis.

Campylobacteriosis

It was not until 1972 that *Campylobacter jejuni* was isolated from a diarrheal stool. It took another five years for a suitable culture medium to be developed and widely used, which led to recognition of campylobacteriosis as a leading bacterial diarrheal illness in the United States, with an estimated 2.1 to 2.4 million cases per year. There are generally less than 1,000 fatalities per year, mostly in the elderly and those with AIDS or other immunodeficiencies. ■ *Campylobacter*, p. 273

Symptoms

Fever, vomiting, diarrhea, and abdominal cramps are typical. Dysentery occurs in about half the cases.

Causative Agent

Campylobacter jejuni is a motile, curved, Gram-negative rod **(figure 25.15)** that can be cultivated from feces under microaero-

TABLE 25.9	Salmonellosis
Symptoms	Diarrhea and vomiting; rarely, prolonged fever, headache, abdominal pain, abscesses, and shock
Incubation period	Usually 6 to 72 hours; can be 1 to 3 weeks in typhoid fever
Causative agent	*Salmonella enterica,* motile, Gram-negative, enterobacteria
Pathogenesis	Invasion of the lining cells of lower small and large intestine, with penetration to underlying tissues; body's inflammatory response causes increase in fluid secretion. Sometimes survival within macrophages and spread throughout the body, destruction of Peyer's patches.
Epidemiology	Ingestion of food contaminated by animal feces, especially poultry. Human fecal source in typhoid fever-like illnesses.
Prevention and treatment	Adequate cooking and handling of food; attenuated vaccine against typhoid fever. Usually no antimicrobial advised unless invasion of tissues or blood occurs.

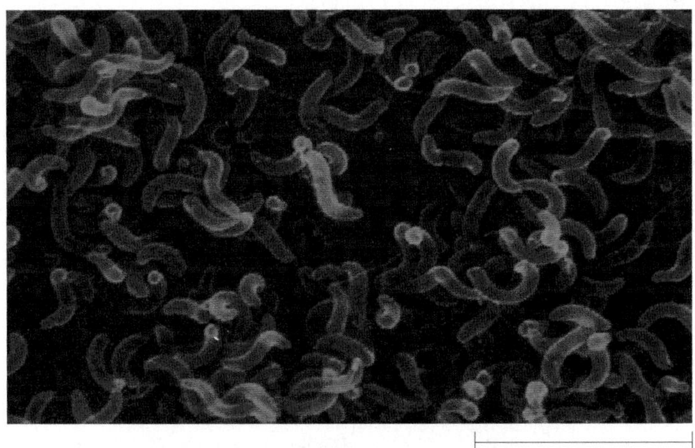

FIGURE 25.15 Scanning Electron Micrograph of *Campylobacter jejuni* *C. jejuni* is a common bacterial cause of diarrhea in the United States.

philic conditions using a selective medium to suppress the other intestinal organisms. ■ selective media, p. 96

Pathogenesis

The number of organisms required to produce symptomatic infection is 500 or less. The bacteria penetrate the intestinal epithelial cells of the small and large intestines, multiply within and beneath the cells, and cause an inflammatory reaction. As with most strains of *Shigella* and *Salmonella,* penetration into the bloodstream is uncommon. A mysterious sequel to *C. jejuni* infections, **Guillain-Barré syndrome,** occurs in about 0.1% of cases. Although *C. jejuni* infections are probably not a direct cause of this syndrome, almost 40% of all cases are preceded by campylobacteriosis, and most of the rest by viral infections. Evidence suggests that autoimmunity could be responsible for *C. jejuni*-associated cases. The syndrome begins abruptly within about 10 days of the onset of diarrhea, with tingling of the feet followed by progressive paralysis of the legs, arms, and rest of the body. Most victims require hospitalization, but over 80% recover completely. About 5% of patients die of the syndrome despite treatment. ■ Guillain-Barré syndrome, p. 521

Epidemiology

Numerous foodborne and waterborne outbreaks of *C. jejuni* have been reported, involving as many as 3,000 people. Most cases, however, are sporadic. Like the salmonellas, *C. jejuni* lives in the intestines of a variety of domestic animals. Poultry is a common source of infection, and as many as 89% of raw poultry products have harbored the organism. One drop of juice from raw chicken meat can easily contain an infectious dose. Cats have also been implicated as a source. Epidemics have resulted from ingesting unpasteurized cow and goat milk and from drinking non-chlorinated surface water. Despite its low infectious dose, person-to-person spread of *C. jejuni* rarely occurs.

Prevention and Treatment

Prevention of outbreaks depends mostly on pasteurization of milk and on chlorination of water. Proper cooking and handling of raw poultry to avoid contamination of hands and kitchen surfaces can prevent most of the other cases. Chicken should be cooked until it is no longer pink. Most cases of *C. jejuni* gastroenteritis subside without antimicrobial treatment in 10 days or less and leave the infected individual immune to further infection. The antibiotic erythromycin is recommended for severe cases.

The main features of campylobacteriosis are shown in **table 25.10.**

MICROCHECK 25.5

Vibrios, *Campylobacter jejuni,* and enterobacteria account for most intestinal bacterial infections. Pathogenic mechanisms include attachment, toxin production, cell invasion, and destruction of microvilli. Some microbial toxins alter the secretory function of cells in the small intestine without killing or visibly damaging them. Unsuspected human carriers can remain sources of enteric infections for many years. Most vaccines are of limited value because they induce little secretory IgA production.

- ✓ Explain how *Vibrio cholerae* causes cholera without apparent damage to the intestinal epithelium.
- ✓ How does the epidemiology of *Salmonella* Typhi differ from that of most other *S. enterica*?
- ✓ What makes *Campylobacter jejuni* a leading cause of bacterial diarrhea in the United States?
- ✓ How would you devise a selective medium for *Vibrio cholerae?*

TABLE 25.10	Campylobacteriosis
Symptoms	Diarrhea, fever, abdominal cramps, nausea, vomiting, bloody stools
Incubation period	Usually 3 days (range, 1 to 5 days)
Causative agent	*Campylobacter jejuni,* a curved Gram-negative, microaerophilic rod
Pathogenesis	Low infecting dose. The bacteria multiply within and beneath the epithelial cells and incite an inflammatory response. Bloodstream invasion is uncommon. Complicated by a generalized paralysis, Guillain-Barré syndrome, on rare occasions.
Epidemiology	Large foodborne and waterborne outbreaks originating from chickens, cows and other animals. Person-to-person spread rarely occurs.
Prevention and treatment	Avoiding contamination of hands and food preparation areas with uncooked poultry, and cooking of poultry until it is no longer pink are also effective. Water chlorination and pasteurization of beverages are effective control measures. Most victims recover within 10 days without antibacterial medications.

25.6

Viral Diseases of the Lower Digestive System

Focus Points

- Describe measures for the prevention of hepatitis A, B, and C.
- Outline the significance of hepatitis B viral antigen or antibody in the blood of a hepatitis victim.

Viral infections of the lower digestive system are common in all age groups. The intestines and accessory organs such as the liver can be affected. At least five different groups of viruses can cause epidemic gastroenteritis, resulting in millions of cases each year in the United States. More than six different viruses can cause **hepatitis,** meaning inflammation of the liver. Hepatitis A virus (HAV), hepatitis B virus (HBV), and hepatitis C virus (HCV) account for most cases of viral hepatitis.

Rotaviral Gastroenteritis

Rotaviruses cause most cases of viral gastroenteritis in infants and children. In the United States, almost every child is infected at some time before they reach the age of 5 years. Each year in the United States, rotavirus infections result in 500,000 emergency department or clinic visits, and 49,000 hospital admissions, but few deaths. Worldwide, more than 600,000 deaths are attributed to rotaviruses each year.

Symptoms

Illness begins abruptly with vomiting and slight fever, followed in a short time by profuse, watery diarrhea. Symptoms generally are gone in about a week. Unless adequate replacement fluid is given, death can occur from dehydration.

Causative Agents

Rotaviruses **(figure 25.16)** are about 70 nm in diameter, have a double-walled capsid, and a double-stranded, 11-segment RNA genome. The viruses represent a major subgroup of the reovirus family. Their name comes from *rota,* Latin for "wheel." ■ reovirus family, p. 323

Pathogenesis

The viruses infect mainly the epithelial cells that line the upper part of the small intestine, causing cell death and decreased production of digestive enzymes. These changes gradually return to normal over a number of weeks.

Epidemiology

Rotaviruses spread by the fecal-oral route. Childhood epidemics generally occur in winter in temperate climates, perhaps because children are more apt to be confined indoors in groups where the viruses can spread easily. Generally, by the age of 4 years, children have acquired some immunity to these viruses. Rotaviruses also cause about 25% of traveler's diarrhea cases. Rotaviruses that infect a wide variety of young wild and domestic animals do not cause human disease. Experimentally, however, reassortment of genetic segments can occur with dual infections. It is not known whether new human pathogenic rotaviruses arise by this mechanism.

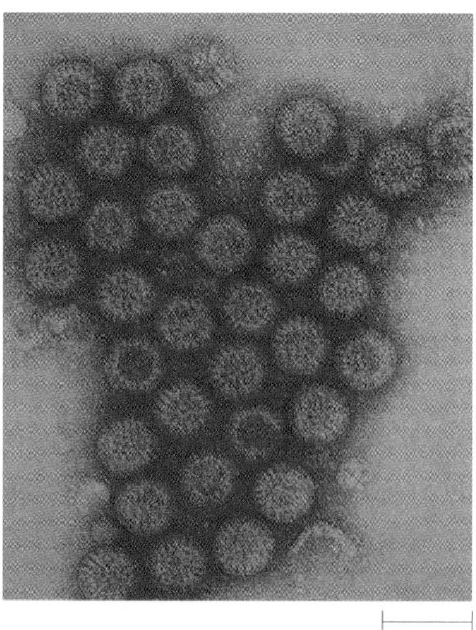

FIGURE 25.16 Rotavirus in Human Feces Transmission electron micrograph.

Prevention and Treatment

Handwashing, disinfectant use, and other sanitary measures help limit the spread of rotaviruses. An effective attenuated vaccine was approved in 1998, but was withdrawn the next year because it increased the risk of a rare, potentially fatal kind of bowel obstruction. A new, safer attenuated vaccine was approved in 2006 and is administered to infants at ages 2, 4, and 6 months.

Norovirus Gastroenteritis

Noroviruses are responsible for an estimated 23 million cases of viral gastroenteritis annually in the United States. Originally they were called "Norwalk viruses," from Norwalk, Ohio, the place where they were first implicated in an epidemic of gastroenteritis. Because noroviruses tend to cause large, demoralizing outbreaks that affect people of all ages, they have been designated a category B bioterrorism agent. ■ bioterrorism agents, p. 466

Symptoms

The symptoms of nausea, vomiting, and watery diarrhea of abrupt onset are identical to those of rotaviruses. Symptoms with norovirus infections, however, generally last only 12 to 60 hours.

Causative Agent

Noroviruses are small (about 30 nm diameter), non-enveloped, single-stranded RNA viruses. Their spherical surface is covered with small depressions **(figure 25.17).** These viruses represent a group of gastroenteritis-producing viruses within the calcivirus family, none of which has been cultivated in the laboratory. ■ calcivirus family, p. 323

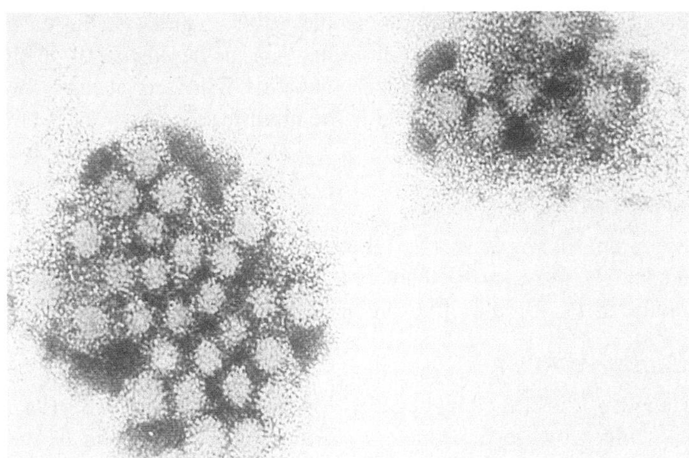

FIGURE 25.17 Electron Micrograph of a Norovirus

Pathogenesis

Noroviruses infect the upper small intestinal epithelium and produce changes similar to those of rotaviruses. The epithelium generally recovers fully within about 2 weeks.

Epidemiology

Transmission is by the fecal-oral route; the viruses are highly contagious and survive well in the environment. The incubation period is 12 to 48 hours. Although some viral strains cause gastroenteritis in infants, noroviruses typically infect children and adults. Infected individuals generally eliminate virus in their feces for only a few days. Individuals sometimes contract the illness by eating shellfish that have no known exposure to human feces. Norovirus epidemics are common on cruise ships and college dormatories. Calciviruses are widespread among marine and other animals, raising the question whether they could be involved in the ecology of noroviruses.

Prevention and Treatment

Handwashing, disinfectants, and other sanitary measures can minimize transmission of noroviruses. Infected foodworkers should be restricted from working for 72 hours after their symptoms subside. There is no vaccine and no proven antiviral medication. **Table 25.11** compares noroviruses and rotaviruses and the illnesses they produce.

TABLE 25.11 Rotavirus and Norovirus Compared

Causative Agent	Rotavirus	Norovirus
Characteristics	70 nm diameter double-walled capsid; double-stranded RNA	30 nm diameter, single-stranded RNA
Incubation period	24 to 48 hrs	12 to 48 hrs
Typical symptoms	Vomiting, abdominal cramps, diarrhea, lasting 5 to 8 days	Vomiting, abdominal cramps, diarrhea, lasting 12 to 60 hours
Prevention	Attenuated vaccine	No vaccine

Hepatitis A

Hepatitis A, formerly called infectious hepatitis, is endemic around the world. The distribution of reported cases in the United States is shown in **figure 25.18,** but its true incidence is undoubtedly much higher. With its lack of an animal reservoir for the causative virus, its antigenic simplicity, and the availability of an effective vaccine, the disease could be considered for eradication. Hepatitis A is only one example of viral hepatitis; at least five other kinds of viral hepatitis—B, C, D, E, and G—are known.

Symptoms

Typical symptoms of hepatitis A are fatigue, fever, loss of appetite, nausea, right-sided abdominal pain, dark-colored urine, clay-colored feces, and jaundice. In children less than 6 years old, 70% have no symptoms and those with symptoms rarely have jaundice. Older children and adults usually develop symptomatic infections and most of them develop jaundice. About 20% of adults require hospitalization. Usually patients recover within 2 months, but 10% to 15% require up to 6 months for full recovery.

Causative Agent

Hepatitis A is caused by the hepatitis A virus (HAV), a small (27 nm), single-stranded RNA virus of the picornavirus family. HAV differs from other members of the family and has been given the name *hepatovirus*. There is only one serotype. ■ picornaviruses, p. 324

Pathogenesis

Following ingestion, the virus reaches the liver by an unknown route. The liver is the main site of replication and the only tissue known to be damaged by the infection. The virus is released into the bile and eliminated with the feces.

Epidemiology

Hepatitis A virus spreads by the fecal-oral route, principally through fecal contamination of hands, food, or water. Outbreaks

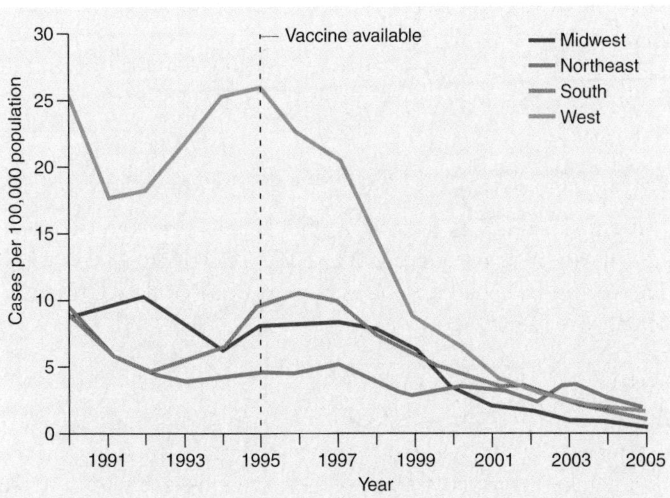

FIGURE 25.18 Acute Hepatitis A Reported Cases, different regions of the United States, 1991–2005.
Source: Centers for Disease Control and Prevention 2007. *Morbidity and Mortality Weekly Report Surveillance Summary* 56 (SS 3).

of the disease have originated from restaurants because food-handlers who carried the virus failed to wash their hands. A large 2003 epidemic caused by contaminated green onions imported from Mexico involved 600 people in Pennsylvania and smaller outbreaks in several other states. Consumption of raw shellfish is a frequent source of infection because these animals concentrate the hepatitis A virus from fecally polluted seawater. Other important risk factors are international travel and sexual contact with someone with the disease. Other groups at high risk of hepatitis A include children in day care centers and residents in nursing homes, and homosexual men. Infants and children with hepatitis A can shed the virus in their feces for several months after symptoms begin, but the amount of virus in feces usually drops markedly with the appearance of jaundice.

Prevention and Treatment

An effective, inactivated virus vaccine against hepatitis A has been available since 1995. It is indicated for travelers to economically deprived regions, homosexual men, those who work on sewers, healthcare workers, individuals with chronic liver disease, food-handlers, and all persons 12 months to 40 years of age. Since the introduction of vaccination, the number of reported cases of hepatitis A has dropped to historic lows (1.5 acute cases per 100,000). Immune globulin containing antibody to HAV can be given by injection for passive immunization of travelers under the age of 12 months or more than 40 years old, and those at high risk for other reasons. It gives short-term protection against the disease if given within 2 weeks of exposure. No antiviral treatment is available. Victims of the disease are advised to avoid alcohol and other hepatic toxins.

Hepatitis B

Hepatitis B, formerly known as **serum hepatitis,** represents about 40% of the cases of viral hepatitis in the United States. Approximately 50,000 new cases occur each year, and as many as 1% to 6% of the infected adults will become carriers of the disease. About 1.25 million people in the United States have chronic hepatitis B and therefore have an increased risk of death from cirrhosis and liver cancer. Hepatitis B infects at least 5% of the world's population and is the ninth-leading cause of death worldwide.

Symptoms

Symptoms of hepatitis A and B are similar, except that hepatitis B tends to be more severe than hepatitis A, causing death from liver failure in 1% to 10% of hospitalized cases.

Causative Agent

Hepatitis B is caused by hepatitis B virus (HBV) **(figure 25.19a),** a member of the hepadnavirus family (*hepa-,* referring to the liver, and *-dna-* "DNA"). The virus contains double-stranded DNA and has a lipid-containing outer envelope. Three important HBV antigens are (1) surface antigen (HBsAg); (2) core antigen (HBcAg), a protein of the nucleocapsid; and (3) e antigen (HBeAg), a soluble component of the viral core. HBsAg is produced during viral replication in amounts far in excess of that needed for virus production. It occurs in the bloodstream on small spheres and filaments of viral envelope material empty of DNA (figure 25.19b) in quantities often 1,000 or more times as great as complete virions. HBsAg is responsible for the ability of the virus to attach to and infect host cells; antibody to surface antigen (anti-HBsAg) confers immunity.

Pathogenesis

Following entry into the body, HBV is carried to the liver by the bloodstream. The mechanism by which HBV causes liver injury is not known but most likely results from the body's immune system attacking the infected liver cells. After entering the liver cell, the virus replicates by a mechanism involving **reverse transcriptase (figure 25.20),** unusual for a DNA virus. The double-stranded viral DNA genome is transported to the host cell nucleus, where messenger RNA and a RNA copy of the genome are synthesized. This RNA copy of the genome is

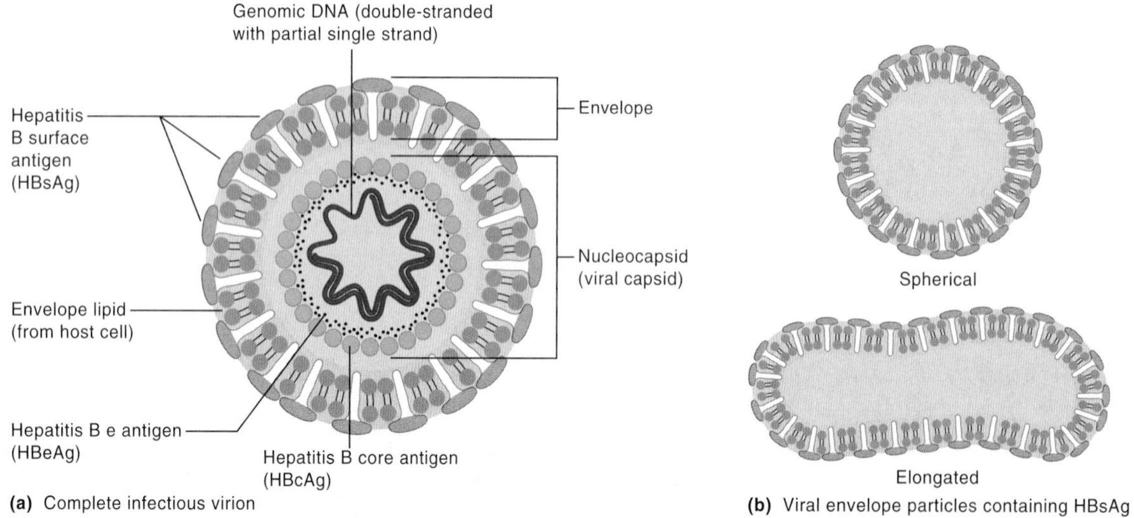

FIGURE 25.19 Hepatitis B Virus Components Found in the Blood of Infected Individuals **(a)** Complete infectious virion. **(b)** Smaller spherical and elongated envelope particles lacking DNA.

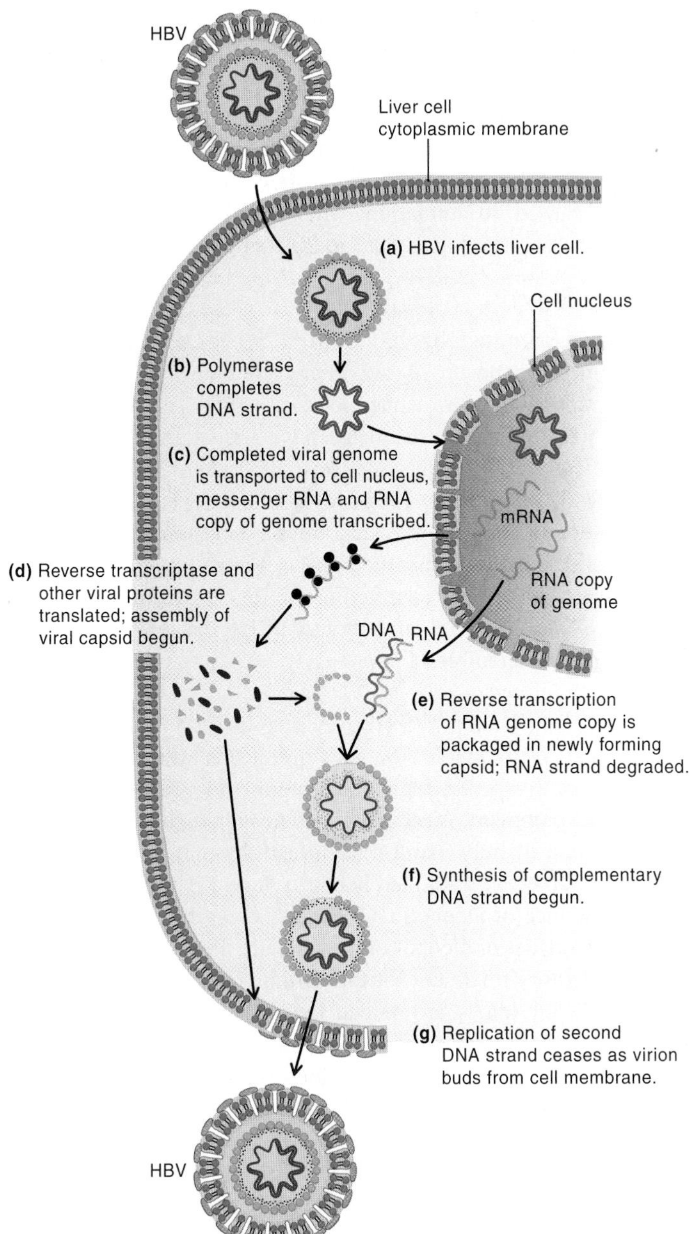

FIGURE 25.20 Replication of Hepatitis B Virus Notice that this DNA virus employs reverse transcriptase in its replication cycle.

subsequently transcribed into DNA by reverse transcriptase, and when a double strand is formed by DNA polymerase, this DNA becomes the genome for the newly forming virions. Double-strand formation ceases when the virions bud from the host cell, however, generally leaving part of the genome single-stranded. ■ **reverse transcriptase, p. 332**

HBsAg appears in the bloodstream days or weeks after infection, often long before signs of liver damage are evident. In 1% to 6% of cases, the infection smolders in the liver, unsuspected for years. About 40% of chronically infected people eventually die from **cirrhosis,** meaning scarring, of the liver, or from liver cell cancer. Evidence indicates that these cancers result from HBV transformation of liver cells. ■ **transformed cells, p. 333**

Epidemiology

From 1965 to 1985, a progressive rise in reported hepatitis B cases occurred. Since then, the incidence of the disease has appeared to plateau and decline **(figure 25.21).** HBV is spread mainly by blood, blood products, and semen. Persisting viremia, meaning virus circulating in the bloodstream, can follow both symptomatic and asymptomatic cases, and the virus may continue to circulate in the blood for many years. Carriers are of major importance in the spread of hepatitis B because they are often unaware of their infection. If only a minute amount of blood from an infected person is injected into another person's bloodstream or rubbed into minor wounds, infection can result. Blood and other body fluids can be infectious by mouth, the virus probably infecting the recipient through small scratches or abrasions. Many hepatitis B virus infections result from sharing of needles by drug abusers. Unsterile tattooing and ear-piercing instruments and shared toothbrushes, razors, or towels can also transmit HBV infections.

Sexual intercourse is responsible for transmission in nearly half of hepatitis B cases in the United States. HBV antigen is often present in saliva and breast milk, but the quantity of infectious virus and the risk of transmission are low. Five percent or more of pregnant women who are HBV carriers transmit the disease to their babies at delivery, and more than two-thirds of women who develop hepatitis late in pregnancy or soon after delivery do so. Most of these babies have asymptomatic infections and become long-term carriers, but some die of liver failure.

Prevention and Treatment

The first vaccine against hepatitis B, approved in the early 1980s, consisted of HBsAg obtained from the blood of chronic carriers; 1 ml of their blood often had enough antigen to immunize eight people! Since 1986, however, a subunit vaccine produced in

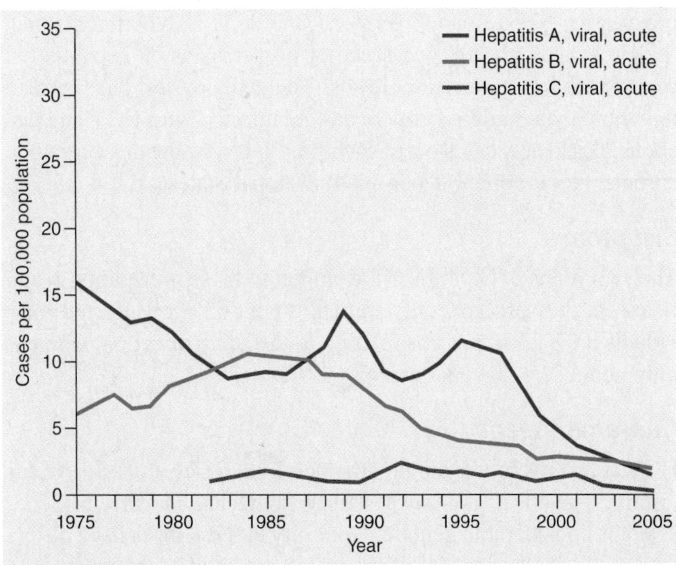

FIGURE 25.21 Incidence of Viral Hepatitis in the United States, 1975–2005 The first hepatitis B vaccine was licensed in 1982, the first hepatitis A vaccine in 1995. A test for hepatitis C antibody became available in 1990.

Source: Centers for Disease Control and Prevention, 2007. Summary of Motifiable Diseases-United States–2005, 54(53).

genetically engineered yeast has been available against hepatitis B and a combined vaccine against both hepatitis A and B has also been approved. It is now recommended that all healthy newborns receive hepatitis B vaccine before they leave the hospital. Infants born to an infected mother are injected at birth with hepatitis B immune globulin (HBIG) containing a high titer of antibody to HBV. Active immunization is also started, with injections of the vaccine at 1 week, 1 month, and 6 months of age. Vaccination against hepatitis B prevents the disease, and it can also help prevent many of the 500,000 to 1 million new liver cell cancers that occur worldwide each year. Education of groups at high risk of contracting hepatitis B can help prevent the disease. Those likely to be exposed to blood are taught to consider all blood to be infectious, to wash their hands contaminated with blood, to wear gloves and protective clothing, and to handle contaminated, sharp objects such as needles and scalpel blades carefully (see Perspective 20.1). Teaching the importance and proper use of condoms also helps limit spread of the infection. Victims are advised to avoid alcohol and other hepatic toxins. There is no curative antiviral treatment yet for hepatitis B. Remarkable improvement can be achieved in many cases, however, by giving injections of genetically engineered interferon along with an antiviral medication such as lamivudine, a reverse transcriptase inhibitor, or adefovir dipivoxil. ■ interferon, p. 357

Hepatitis C

Even when blood that tested positive for HBV was excluded from transfusions, post-transfusion hepatitis continued to be common, indicating the presence of another bloodborne hepatitis virus. After a number of years of trying to isolate a non-A, non-B hepatitis virus, scientists in 1989 were able to clone parts of the genome of a transfusion-associated virus, now known as hepatitis C virus (HCV). One of the gene products of these clones proved to be an HCV antigen satisfactory for detecting antibody to HCV in the blood of prospective blood donors. By not using donated blood containing the HCV antibody, the incidence of post-transfusion hepatitis fell from about 8% to very low levels. The antibody test has revealed that more than 3 million Americans are infected with HCV and that about 20,000 new cases occur each year. It is now the most common chronic, blood-borne infection in the United States.

Symptoms

The symptoms of hepatitis C are similar to those of hepatitis A and B except they are generally milder. About 65% of infected individuals have no symptoms relating to the acute infection, whereas only about 25% have jaundice.

Causative Agent

HCV is an enveloped, single-stranded RNA virus of the flavivirus family. The virus was successfully cultivated *in vitro* in 2005. There is considerable genetic variability in the viruses from different people, leading to classification into types and subtypes having differing pathogenicity. ■ flavivirus family, p. 323

Pathogenesis

Few details are known about the pathogenesis of hepatitis C. Infection generally occurs from exposure to contaminated blood. The incubation period averages about 6 weeks (range, 2 weeks to 6 months). Despite the lack of symptoms in most people with acute infection, more than 80% develop chronic infections. The virus infects the liver and incites inflammatory and immune responses. The disease process in the liver waxes and wanes, at times seeming to return to normal, then weeks or months later showing marked inflammation. After months or years, cirrhosis and liver cancer develop in 10% to 20% of patients.

Epidemiology

Although transmitted by blood from an infected person, the mechanism of exposure is not always obvious. Sharing toothbrushes, razors, and towels can be responsible. Tattoos and body piercing with unclean instruments have transmitted the disease. Approximately 50% of transmissions in the United States are due to sharing of syringes by illegal drug abusers. Transmission by sexual intercourse is probably uncommon, but those with multiple partners and sexually transmitted diseases are at increased risk for the disease. The risk of contracting the disease from transfusion of a unit of blood is now only about 0.001% because of effective screening of donated blood.

Prevention and Treatment

No vaccine is available for preventing hepatitis C although new methods for producing viral proteins have raised hopes for vaccine development. Vaccination against hepatitis A and B is recommended to help avoid dual infections that might severely damage the liver. A combined hepatitis A and B vaccine is available. Avoidance of alcoholic beverages is recommended because of the toxic effect of alcohol on the already damaged liver. There is no satisfactory treatment for the chronic disease, although 30% to 50% of individuals are helped by interferon injections combined with ribavirin, an antiviral nucleoside derivative. Prolonged interferon treatment of the acute illness may prevent the chronic disease. ■ ribavirin, p. 487

Besides HAV, HBV, and HCV, hepatitis can be caused by several other hepatitis viruses, and can also occur as part of viral diseases such as yellow fever and infectious mononucleosis.

Table 25.12 compares the known hepatitis viruses—A, B, C, D, E, and G.

MICROCHECK 25.6

Rotaviruses are the leading cause of viral gastroenteritis in infants and children, and they also are a common cause of traveler's diarrhea. Noroviruses cause almost half the viral gastroenteritis in children and adults. Hepatitis A is transmitted by the fecal-oral route and is preventable by gamma globulin and an inactivated vaccine. Hepatitis B, transmitted by exposure to blood and by sexual intercourse, is preventable by a vaccine produced in yeast. A combination hepatitis A and hepatitis B vaccine is available. Hepatitis C is transmitted by blood and perhaps occasionally by sexual intercourse. Chronic hepatitis often results in cirrhosis and cancer.

- ✓ What two serious complication can occur late in the course of both chronic hepatitis B and C?
- ✓ At what stage in the replication of hepatitis B would a reverse transcriptase inhibitor act?
- ✓ Why might it be more difficult to prepare a vaccine against noroviruses than against rotaviruses?

TABLE 25.12 Viral Hepatitis

	Hepatitis A	Hepatitis B	Hepatitis C
Causative agent	Non-enveloped, single-stranded RNA picornavirus, HAV	Enveloped, double-stranded DNA hepadnavirus, HBV	Enveloped, single-stranded RNA flavivirus, HCV
Mode of spread	Fecal-oral	Blood, semen	Blood, possibly semen
Incubation period	3 to 5 weeks (range, 2 to 7 weeks)	10 to 15 weeks (range, 6 to 23 weeks)	6 to 7 weeks (range, 2 to 24 weeks)
Prevention	Inactivated vaccine; immune globulin	Subunit-vaccine; immune globulin	No vaccine
Comments	Usually mild symptoms, but often prolonged; full recovery; no long-term carriers; combined hepatitis A and hepatitis B vaccine available	Symptoms often more severe than hepatitis A; progressive liver damage in 1% to 6% can lead to cirrhosis and cancer, chronic carriers; can cross the placenta; combined hepatitis A and hepatitis B vaccine available	Usually few or no symptoms; progressive liver damage or cancer in 10% to 20% of cases; chronic carriers
	Hepatitis D	**Hepatitis E**	**Hepatitis G**
Causative agent	Defective single-stranded RNA virus, HDV	Non-enveloped, single-stranded RNA calcivirus, HEV	Single-stranded RNA flavivirus
Mode of spread	Blood, semen	Fecal-oral	Blood, possibly semen
Incubation period	2 to 12 weeks	2 to 6 weeks	Weeks
Prevention	No vaccine	No vaccine	No vaccine
Comments	Prior or concurrent HBV infection necessary; can cause worsening of hepatitis B; can cross the placenta	Similar to hepatitis A, except severe disease in pregnant women, same or related virus in rats	Usually mild symptoms; persistent viremia for months or years

25.7 Protozoan Diseases of the Lower Digestive System

Focus Points

- Describe the causative agents of giardiasis, cryptosporidiosis, cyclosporiasis, and amebiasis.
- Compare and contrast the epidemiology of these protozoan intestinal diseases.

Protozoa are important causes of human intestinal disease. All the major protozoan motility groups are represented: Mastigophora, those with flagella; Ciliophora, those with cilia; Apicomplexa, those that are usually immobile; and Sarcodina, those that move by pseudopods. All are transmitted by the fecal-oral route. ■ protozoa, p. 285

Giardiasis

Disease resulting from *Giardia lamblia* infection, called giardiasis, is the most commonly identified waterborne illness in the United States. It can be contracted from clear mountain streams or from chlorinated city water and can be transmitted by person-to-person contact. Giardiasis occurs worldwide and is responsible for many cases of traveler's diarrhea. In Americans, *G. lamblia* is the most commonly identified intestinal parasite, and in economically underdeveloped areas of the world, its incidence often exceeds 10%.

Symptoms

In the usual epidemic of giardiasis, about two-thirds of exposed individuals develop symptoms. The incubation period is generally 6 to 20 days. Symptoms can range from mild (indigestion, "gas," and nausea) to severe (vomiting, explosive diarrhea, abdominal cramps, fatigue, and weight loss). The symptoms usually disappear without treatment in 1 to 4 weeks, but some cases become chronic. Both symptomatic and asymptomatic persons can become long-term carriers, unknowingly excreting infectious cysts with their feces.

Causative Agent

Giardia lamblia is a flagellated protozoan shaped like a pear cut lengthwise, with two side-by-side nuclei that resemble eyes. These features along with an adhesive disc on its undersurface give the organism an unmistakable appearance **(figure 25.22).** *Giardia lamblia* can exist in two forms: as a vegetative **trophozoite** or as a resting form called a **cyst.** The trophozoite is the actively feeding form, and it colonizes the upper part of the small intestine. Generally, only trophozoites are present in the feces when an infected person is having diarrhea. When trophozoites are carried slowly by intestinal contents toward the large intestine, they develop into cysts. *Giardia* cysts have thick walls composed of chitin, a tough, flexible, nitrogen-containing polysaccharide. The cyst wall protects the organism from harsh environmental conditions. Because it appeared to lack mitochondria, *G. lamblia* has been considered to be part of an ancient group that arose before eukaryotes acquired these organelles. More recent evidence appears to refute this idea because the protozoan contains functional remnants of mitochondria. ■ chitin, p. 71

FIGURE 25.22 ***Giardia lamblia*** **Trophozoites in Human Intestine**
Scanning electron micrograph.

Pathogenesis

The cyst form is responsible for infection because, unlike the trophozoite, it is resistant to stomach acid. Two trophozoites emerge from each cyst when it reaches the small intestine. Some of these trophozoites attach to the epithelium by their adhesive disc, whereas others move freely in the intestinal mucus using their flagella. Some may even migrate up the bile duct to the gallbladder and cause crampy pain or jaundice. The protozoa do not destroy the host epithelium, but in severe infestations they may entirely cover the epithelial surface. They interfere with the ability of the intestine to absorb nutrients and secrete digestive enzymes. The result is malnutrition, bulky feces containing fat, and excessive intestinal gas from bacterial digestion of unabsorbed food material. Some of the intestinal impairment is probably a side effect of the host immune system attacking the parasites.

Epidemiology

Transmission of *G. lamblia* is usually by the fecal-oral route, especially via fecally contaminated water. Known or suspected sources of *G. lamblia* include beavers, raccoons, muskrats, dogs, cats, and humans. A single human stool can contain 300 million *G. lamblia* cysts; only 10 cysts are required to establish infection. The cysts are infectious immediately after passage and can remain viable in cold water for more than 2 months. Usual levels of chlorination of municipal water supplies are ineffective against the cysts, and water filtration is necessary to remove them. Hikers who drink from streams, even in remote areas thought to contain safe water, are at risk of contracting giardiasis.

Although waterborne outbreaks are the most common, person-to-person contact can also transmit the disease. This mode of transmission is especially likely in day care centers where workers' hands become contaminated in the process of diaper changing. People who promiscuously engage in anal intercourse and fellatio are also prone to contracting the disease. Transmission by fecally contaminated food has also been reported. Good personal hygiene, especially handwashing, decreases the chance of passing on the infection. ■ day care centers, p. 461

Prevention and Treatment

Filtration of community water supplies is effective. For hikers, the best way to make drinking water safe from giardiasis is to boil it for 1 minute or use a portable filter of 1 micrometer pore size or smaller. Other methods, such as using a few drops per quart of water of household sodium hypochlorite bleach, tincture of iodine, or commercial water-purifying tablets, are time consuming and considerably less reliable. As in all chemical sterilization procedures, time and temperature are important. Only an hour may be necessary to treat warm water, but many hours are required to kill cysts in cold water. Several medicines, including quinacrine (Atabrine) and metronidazole (Flagyl), effectively treat giardiasis.

Table 25.13 describes the main features of giardiasis.

Cryptosporidiosis

Cryptosporidiosis, disease of the small intestine caused by the *Cryptosporidium* genus of protozoa, was recognized in animals in the early 1900s, but not identified as a threat to human beings until the AIDS epidemic struck in the early 1980s. Since then, it has become clear that cryptosporidiosis is a major hazard not only to those with immunodeficiency, but to the public at large. A 1993 waterborne outbreak of the disease in Milwaukee involved more than 403,000 people!

TABLE 25.13	Giardiasis
Symptoms	Mild illness: indigestion, flatulence, nausea; severe: vomiting, diarrhea, abdominal cramps, weight loss
Incubation period	6 to 20 days
Causative agent	*Giardia lamblia,* a flagellated pear-shaped protozoan with two nuclei
Pathogenesis	Ingested cysts survive stomach passage; trophozoites emerge from the cysts in the small intestine, where some attach to epithelium and others move freely; mucosal function is impaired by adherent protozoa and host immune response.
Epidemiology	Ingestion of fecally contaminated water; person-to-person, in day care centers.
Prevention and treatment	Boiling or disinfecting drinking water; filtration of community water supplies. Treatment: quinacrine hydrochloride (Atabrine) or metronidazole (Flagyl).

Symptoms

Cryptosporidiosis is characterized by fever, loss of appetite, nausea, crampy abdominal pain, and profuse watery diarrhea, beginning after an incubation period of 4 to 12 days. The symptoms generally last 10 to 14 days, but in people with immunodeficiency diseases, they can last for months and be life threatening.

Causative Agent

Cryptosporidium parvum, a coccidian member of the Apicomplexa, is the causative agent of cryptosporidiosis. *Cryptosporidium parvum* multiplies intracellularly in the small intestinal epithelium, its entire life cycle occurring in a single host. The oocyst stage is a tiny (4 to 6 μm), acid-fast sphere that contains four banana-shaped sporozoites **(figure 25.23).** ■ Apicomplexa, p. 273

Pathogenesis

Few details are known about the pathogenesis of cryptosporidiosis. Following ingestion, the digestive juices of the small intestine release the sporozoites from the oocysts. The sporozoites invade the epithelial cells of the small intestine, deforming the epithelium and intestinal villi and causing an inflammatory response beneath the cells. Secretion of water and electrolytes increases and absorption of nutrients decreases, but the mechanism is unknown. Cell-meditated immunity is important in controlling the infection.

Epidemiology

The oocysts of *C. parvum* are infectious when eliminated with the feces, and fewer than 30 of them are sufficient to cause an infection. Person-to-person spread readily occurs under conditions of poor sanitation. Infected individuals often have prolonged diarrhea and can continue to eliminate infectious cysts for 2 weeks or more after the diarrhea ceases. Moreover, the cysts can survive for long periods of time in food and water, are even more resistant to chlorine than *Giardia,* and are too small to be removed from drinking water by some filtration methods. Equally troubling, *C. parvum* has a wide host range, infecting domestic animals such as dogs, pigs, and cattle. These animals, as well as humans, can contaminate food and drinking water. Epidemics have arisen from drinking water, swimming pools, a water slide, a zoo fountain, day care centers, unpasteurized apple juice, and other food and drink. The organism is responsible for many cases of traveler's diarrhea.

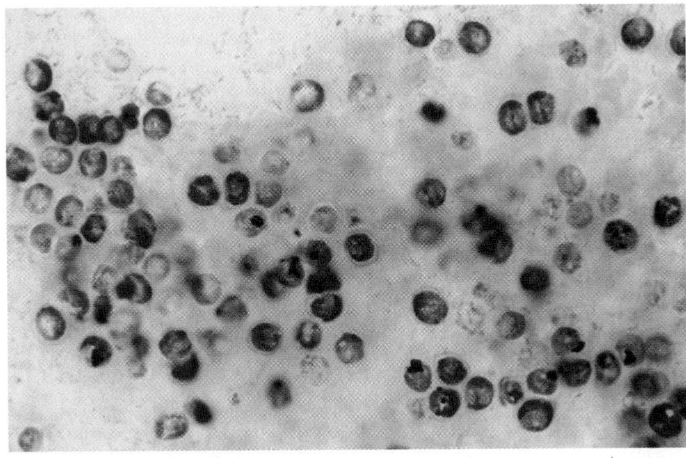

FIGURE 25.23 ***Oocysts of Cryptosporidium parvum*** Acid fast stain of feces. Person-to-person and waterborne transmission are common.

Prevention and Treatment

Effective measures for preventing cryptosporidiosis are limited to careful monitoring of municipal water supplies, with institution of filtration, ultraviolet radiation, and ozone treatments when necessary. Pasteurization of liquids for human consumption and sanitary disposal of human and animal feces are other important control measures. Food-handlers with diarrhea should not handle food until they are symptom-free, and all food-handlers should adhere to handwashing and other sanitary measures. Immunodeficient individuals are advised to avoid contact with animals, to boil or filter drinking water using a 1 μm or smaller pore size filter, and to avoid recreational water activities. No satisfactory treatment exists for cryptosporidiosis, although a combination of two antibiotics, paromomycin and azithromycin, has helped control the disease in some AIDS patients.

The main features of cryptosporidiosis are shown in **table 25.14.**

TABLE 25.14	Cryptosporidiosis
Symptoms	Fever, loss of appetite, nausea, crampy abdominal pain, watery diarrhea, usually lasting 10 to 14 days
Incubation period	Usually about 6 days (range, 4 to 12 days)
Causative agent	*Cryptosporidium parvum,* a coccidian member of the Apicomplexa. Its life cycle takes place entirely within the epithelial cells of the small intestine.
Pathogenesis	Following ingestion of oocysts, the intestinal juices cause the release of four banana-shaped sporozoites. The sporozoites invade the epithelial cells, causing deformity of the cells and the intestinal villi and inciting an inflammatory response. Secretion of water and electrolytes increases, and absorption of nutrients decreases. The process is brought under control largely by cell-mediated immunity.
Epidemiology	Oocysts of *C. parvum* are infectious when discharged with the feces. The infectious dose is small, and person-to-person spread readily occurs under unsanitary conditions. Infected individuals can discharge the oocysts for weeks after symptoms subside. The oocysts resist chlorination, and they pass through many municipal water filtration systems. The host range is wide, including cattle, pigs, and dogs.
Prevention and treatment	Pasteurization of beverages such as milk and apple juice, boiling of drinking water, or filtering it using a 1-μm or smaller pore size filter. Sanitary disposal of human and animal feces. No effective treatment exists.

Cyclosporiasis

Cyclosporiasis first came to medical attention in the late 1980s, with widely scattered epidemics of severe diarrhea. For several years, the causative organism was known only as a "cyanobacterium-like body." Later, the bodies were shown to be the oocysts of a coccidian protozoan, *Cyclospora cayetanensis.*
■ cyanobacteria, p. 258

Symptoms

Cyclosporiasis begins after an average incubation period of about 1 week, with fatigue, loss of appetite, slight fever, vomiting, and watery diarrhea, followed by weight loss. The diarrhea usually subsides in 3 to 4 days, but relapses occur for up to 4 weeks.

Causative Agent

Cyclospora cayetanensis is also a coccidian member of the Apicomplexa, and its oocysts are similar to those of *Cryptosporidium parvum,* except *C. cayetanensis* oocysts are larger, about 8 to 10 μm in diameter, and when passed in feces do not yet contain sporozoites. After days under favorable conditions outside the body, two sporocysts, each containing two sporozoites, develop within each oocyst, which is then infectious. Structural features place the protozoan within the *Cyclospora,* a genus described in 1881. Studies of its ribosomal RNA base sequence show that it is closely related to members of the genus *Eimeria.* This is an example of the turmoil arising from molecular biological findings as the newer science challenges the old on the naming of microorganisms.

Pathogenesis

Little is known about the pathogenesis of cyclosporiasis because laboratory animals are not susceptible to the disease. Scant information available from small intestinal biopsies of patients infected with *C. cayetanensis* confirms that both sexual and asexual forms of the protozoan are present in the intestinal epithelium.

Epidemiology

The epidemiology of *Cyclospora cayetanensis* differs in important ways from that of *Cryptosporidium parvum.* The oocysts of *C. cayetanensis* are immature when eliminated in the stool and are non-infectious, so that person-to-person spread does not occur. Most infections occur in spring and summer, probably because warm, moist conditions favor maturation of the *C. cayetanensis* oocysts. Travelers to tropical areas are more likely than others to become infected. So far, no animal source has been identified. Fresh produce, especially imported raspberries, has been implicated in a number of epidemics, but the source of the contaminating organisms is unknown. In most instances, the produce has been imported from a tropical region. The protozoa have been identified in natural waters, but a human source could not be ruled out. Outbreaks have occurred in Canada, the United States, Nepal, Peru, Haiti, and various other countries of South and Central America, Southeast Asia, and Europe.

Prevention and Treatment

No preventive measures are available except boiling or filtering drinking water and thoroughly washing produce such as berries and leafy vegetables during an epidemic. Cotrimoxazole (trimethoprim plus sulfamethoxazole) effectively treats most cases of cyclosporiasis.

The main features of cyclosporiasis are presented in **table 25.15.**

Amebiasis

Disease resulting from *Entamoeba histolytica* infection is called amebiasis. In the United States, cases occur mainly among male homosexuals who have many sexual partners, in poverty-stricken areas, and among migrant farm workers. Although usually a mild disease, worldwide it causes about 30,000 deaths per year, most of which occur in Mexico, parts of South America, Asia, and Africa. Life-threatening disease occurs in these areas because of the presence of virulent *E. histolytica* strains and crowded, unsanitary living conditions.

Symptoms

Amebiasis is commonly asymptomatic, but symptoms ranging from chronic, mild diarrhea lasting months or years, to acute dysentery and death, can occur.

Causative Agent

The causative organism, *Entamoeba histolytica* (see **figure 25.24** inset), is a member of the Sarcodina, ranging from about 20 to

TABLE 25.15	Cyclosporiasis
Symptoms	Fatigue, loss of appetite, vomiting, watery diarrhea, and weight loss. Symptoms improve in 3 to 4 days, but relapses can occur for up to a month.
Incubation period	Usually about 1 week (range, 1 to 12 days)
Causative agent	*Cyclospora cayetanensis,* a coccidian
Pathogenesis	Biopsies of human intestine show sexual and asexual stages in the intestinal epithelium. Laboratory animals are not susceptible to the infection.
Epidemiology	The oocysts of *C. cayetanensis* are not infectious when discharged in the feces, and so person-to-person spread does not occur. Travelers to tropical countries are at risk of infection. Produce, especially raspberries, imported from tropical Central America, has been implicated in most North American outbreaks.
Prevention and treatment	Use of boiled or filtered drinking water is advised in the tropics. Thorough washing of imported berries and leafy vegetables. Most victims are effectively treated with trimethoprim-sulfamethoxazole (trade names Bactrim, Septra).

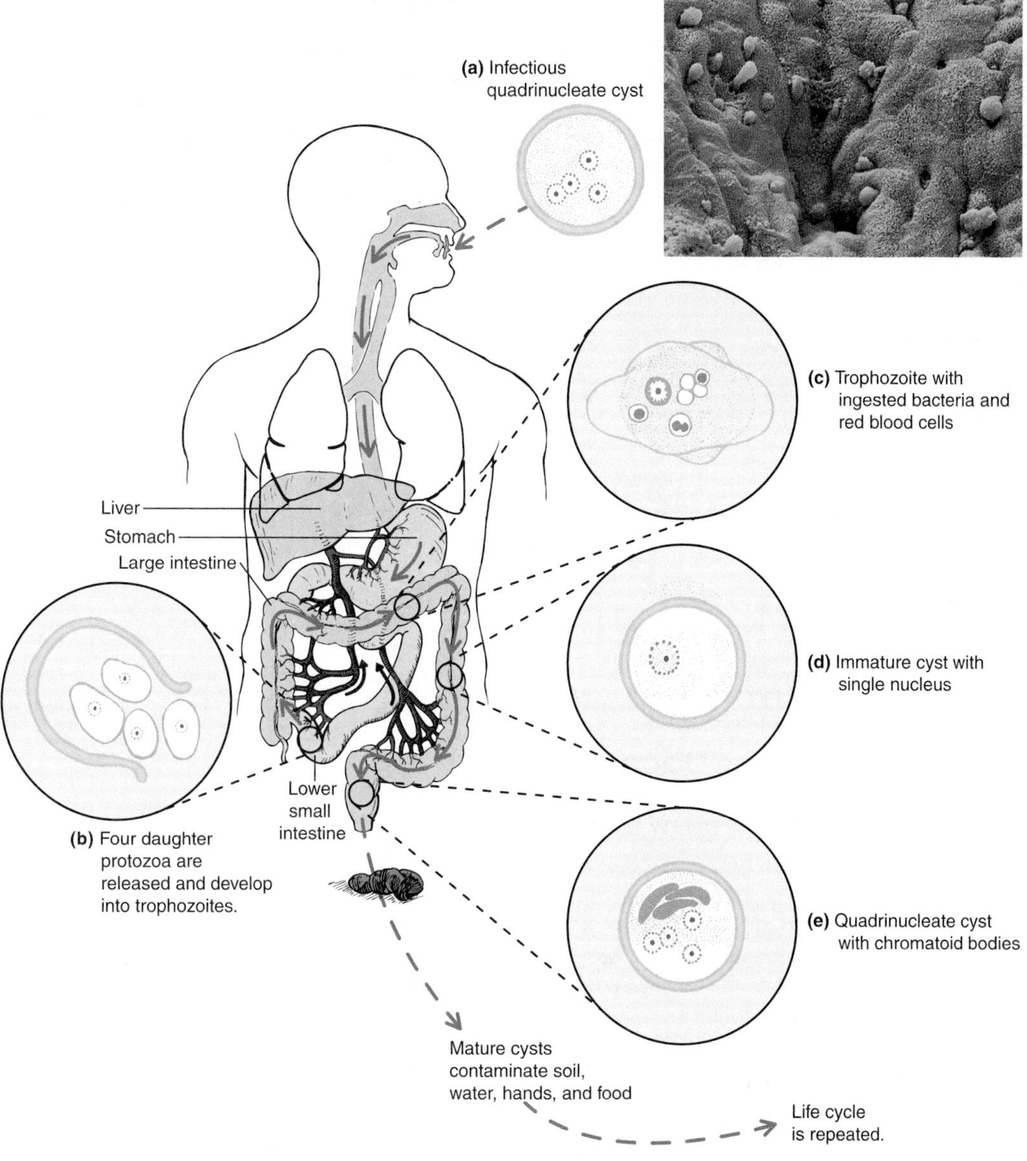

FIGURE 25.24 Life Cycle of *Entamoeba histolytica* Quadrinucleate cysts enter the mouth **(a)** and pass through the stomach to the lower small intestine **(b)** Four daughter protozoa are released from each cyst and develop into trophozoites **(c),** the feeding form **(d).** Dehydration in the large intestine stimulates progressive stages of cyst development **(e).** Mature cysts are passed in the feces to contaminate soil, water, hands and food. Trophozoites that burrow into intestinal blood vessels can be carried to the liver or other organs, causing abscesses. The electron micrograph shows amebic cysts and trophozoites on bowel mucosa.

40 μm in diameter. Like *Giardia,* it has a cyst form with a chitin-containing cyst wall. Immature cysts have a single nucleus. As the cyst matures, the nucleus divides twice to form a cyst with four nuclei—the infectious form for the next host. The life cycle of *E. histolytica* is shown in figure 25.24.

Pathogenesis

Ingested quadrinucleate cysts of *E. histolytica* survive passage through the stomach. The organisms are released from their cysts in the small intestine, whereupon the cytoplasm and nuclei divide, yielding eight trophozoites. Upon reaching the lower intestine, these trophozoites begin feeding on mucus and intestinal bacteria. Many, but not all, strains produce a cytotoxic enzyme that kills intestinal epithelium on contact, allowing the organisms to penetrate the lining cells and enter deeper tissues of the intestinal wall. Sometimes, they penetrate into blood vessels and are carried to the liver or other body organs. Multiplication of the organisms and tissue destruction in the intestine and in other body tissues can result in amebic abscesses. The irritating effect of the amebas on the cells lining the intestine causes intestinal cramps and diarrhea. Due to intestinal ulceration, the diarrheal fluid is often bloody, and the condition is referred to as **amebic dysentery.**

Epidemiology

Amebiasis is distributed worldwide, but it is more common in tropical areas where sanitation is poor. Humans constitute the only important reservoir. Transmission is fecal-oral. The disease is mainly associated with poverty, male homosexuality, and migrant workers.

Prevention and Treatment

Prevention of amebiasis depends on sanitary measures and avoiding fecal contamination of drinking water. Metronidazole, paromomycin, and other medications are available for treatment.

Table 25.16 gives the main features of amebiasis.

TABLE 25.16 Amebiasis

Symptoms	Diarrhea, abdominal pain, blood in feces
Incubation period	2 days to several months
Causative agent	*Entamoeba histolytica;* a protozoan of the group Sarcodina
Pathogenesis	Ingested cysts liberate trophozoites in the small intestine; upon reaching the large intestine, trophozoites feed on mucus and cells lining the intestine; enzymes are produced that allow penetration of the intestinal epithelium, sometimes the intestinal wall and blood vessels, and thence to the liver and other organs, resulting in abscesses.
Epidemiology	Ingestion of fecally contaminated food or water; disease associated with poverty, homosexual men, and migrant workers.
Prevention and treatment	Good sanitation and personal hygiene. Treatment: metronidazole, paromomycin.

MICROCHECK 25.7

Giardiasis, caused by the flagellated protozoan *Giardia lamblia,* is a common waterborne disease, but it can also be transmitted person-to-person. Cryptosporidiosis is another waterborne disease that is also transmissible person-to-person, but it is caused by a coccidian member of the Apicomplexa that multiplies within intestinal epithelial cells. There is no satisfactory treatment. Cyclosporiasis is a similar disease, but it is not transmissible person-to-person and it can be effectively treated. Amebiasis is caused by an ameba that ulcerates the large intestinal epithelium, resulting in dysentery.

- ✓ Explain why person-to-person spread does not occur in cyclosporiasis.
- ✓ Why might cryptosporidiosis be a threat to AIDS patients?
- ✓ Would you expect an individual with giardiasis who has diarrhea to be more likely to transmit the disease than an individual with giardiasis who does not have diarrhea? Explain.

FUTURE CHALLENGES

Defeating Diarrhea

Development of better preventive and treatment techniques for digestive tract diseases has an urgency arising from massive food and beverage production and distribution methods. For example, in 1994, an estimated 224,000 people became ill because a tanker truck used to transport ice cream mix had previously carried liquid eggs. One day's production from a ground beef factory can yield hundreds of thousands of pounds of hamburgers, which are soon sent to many parts of this or other countries. The challenge is to better educate the producers and transporters, to develop guidelines to help them avoid contamination, to develop fast and accurate ways of identifying pathogens in food, and to utilize newly approved methods such as meat irradiation.

Other challenges include:

- *Exploiting the power of molecular biology techniques to produce effective vaccines against hepatitis C virus and bacterial pathogens.* There is special need for bacterial vaccines that can be administered by mouth and evoke long-lasting mucosal immunity.
- *Exploring the influence of global warming on digestive tract diseases.* A study of Peruvian children by Johns Hopkins University scientists found an 8% increase in clinic visits for diarrhea with each 1°C increase in temperature from the normal.
- *Exploring new prevention and treatment options.* Scientists at the University of Florida have developed a genetically engineered *Streptococcus mutans* that does not produce lactic acid but readily displaces wild strains of the dental decay-causing bacterium. Researchers at the University of Alberta have custom-designed a molecule that binds circulating Shiga toxin, potentially preventing hemolytic uremic syndrome. Others work on finding an effective therapy for cryptosporidiosis.

SUMMARY

25.1 Anatomy and Physiology (figures 25.1, 25.2)

The digestive tract is composed of the mouth and teeth, saliva-producing glands, esophagus, stomach, small intestine, liver, pancreas, and large intestine. It is a major route for pathogens to enter the body.

The Mouth

The cheeks and tongue move food so it can be ground by the teeth to make it more easily digested. Teeth are composed of **enamel**—a protective, hard, outer layer—and **dentin**—a softer, more easily penetrated substance. The **gingival crevice** is a space between the tooth and the gum.

Salivary Glands

Salivary glands are located under the tongue, and laterally in the floor of the mouth. Two other glands, the parotids, are located in the face below the ears. These glands produce saliva, saturated with calcium, and contain lysozyme, amylase, lactoferrin, and IgA antibodies.

The Esophagus

The esophagus is a muscular walled tube that conveys food and liquids to the stomach by **peristalsis.**

The Stomach

The stomach is a distensible sack with a muscular wall and a lining epithelium that produces hydrochloric acid and the digestive enzyme, **pepsin.** The epithelium is protected by a thick layer of mucus. Emptying of the stomach is regulated by a valve called the **pylorus,** under the control of nerves and hormones.

The Small Intestine

Villi and **microvilli** markedly increase the surface area of the intestinal lining, which both secretes digestive enzymes and absorbs fluids and nutrients. The lining cells are continuously shed and replaced, which helps protect against microbial poisons.

The Pancreas

The pancreas produces **insulin,** a regulator of blood glucose levels, directly into the bloodstream. The pancreas also produces a large amount of alkaline fluid that contains digestive enzymes, conveyed via a tube into the upper small intestine.

The Liver

The liver produces a fluid, **bile,** that is stored in the gallbladder, ultimately flowing through a tube into the upper small intestine. Bile contains a yellowish pigment, **bile pigment,** a breakdown product of hemoglobin that the liver removes from the blood. Bile helps in the digestion of fats and oils, and the absorption of fat-soluble vitamins, A, D, E, and K. Other functions of the liver include inactivating bloodstream poisons and chemically altering medications.

The Large Intestine

The large intestine absorbs any remaining nutrients and water from the intestinal contents, leaving the residual material, feces, for excretion.

25.2 Normal Microbiota

The Mouth

The species of bacteria that inhabit the mouth colonize different locations depending on their ability to attach to specific receptors. **Dental plaque** consists of enormous quantities of bacteria of various species attached to teeth or each other (figure 25.3). The presence of teeth allows for colonization by anaerobic bacteria.

The Intestines

The normal fasting stomach is devoid of microorganisms. Microorganisms make up about one-third of the weight of feces. The biochemical activities of microorganisms in the large intestine include synthesis of vitamins, degradation of indigestible substances, competitive inhibition of pathogens, chemical alteration of medications, and production of carcinogens.

Upper Digestive System Infections

25.3 Bacterial Diseases of the Upper Digestive System

Tooth Decay (Dental Caries) (table 25.1)

Dental caries is caused mainly by *Streptococcus mutans* involved in formation of extracellular glucans from dietary sucrose. Penetration of the calcium phosphate tooth structure depends on acid production by **cariogenic dental plaque.** *S. mutans* is not inhibited by plaque acidity and stores fermentable intracellular polysaccharide (figure 25.4). Control of dental caries depends mainly on supplying fluoride and restricting dietary sucrose. Dental sealants fill the pits and crevices in children's teeth.

Periodontal Disease (table 25.1)

Periodontal disease is caused by an inflammatory response to the plaque bacteria at the gum line; it is mainly responsible for tooth loss in older people (figure 25.5).

Trench Mouth (table 25.1)

Trench mouth, or **acute necrotizing ulcerative gingivitis (ANUG),** can occur at any age in association with poor mouth care (figure 25.6).

Helicobacter pylori Gastritis (table 25.2)

H. pylori predisposes the stomach and the uppermost part of the duodenum to peptic ulcers (figures 25.7, 25.8). Treatment with antimicrobial medications can cure the infection and prevent peptic ulcer recurrence.

25.4 Viral Diseases of the Upper Digestive System

Herpes Simplex (Cold Sores or Fever Blisters) (table 25.3, figure 25.9)

Herpes simplex is caused by an enveloped DNA virus. Infected cells show intranuclear inclusion bodies. HSV persists as a latent infection inside sensory nerves; active disease occurs when the body is stressed.

Mumps (table 25.4, figure 25.10)

Mumps is caused by an enveloped RNA virus that infects not only the parotid glands, but the meninges, testicles, and other body tissues. Mumps virus generally causes more severe disease in persons beyond the age of puberty; it can be prevented using an attenuated vaccine (figure 25.11).

Lower Digestive System Infections

25.5 Bacterial Diseases of the Lower Digestive System

Cholera (table 25.5)

Cholera is a severe form of diarrhea caused by a toxin of *Vibrio cholerae* that acts on the small intestinal epithelium (figures 25.12, 25.13).

Shigellosis (table 25.6)

Shigellosis is caused by species of *Shigella,* common causes of **dysentery** because they invade the epithelium of the large intestine (figure 25.14). One *Shigella* species produces **Shiga toxin,** which causes the **hemolytic uremic syndrome.**

Escherichia coli Gastroenteritis (tables 25.7, 25.8)

Virulence factors often depend on plasmids, which can transfer virulence to other enteric bacteria. Some *E. coli* strains, such as O157:H7, originating from wild and domestic animals, can cause the hemolytic uremic syndrome.

Salmonellosis (table 25.9)

Strains of *Salmonella* that originate from animals usually cause **gastroenteritis.** The organisms are often foodborne, commonly with eggs and poultry. Typhoid fever is caused by *Salmonella* Typhi, which only infects humans. The disease is characterized by high fever, headache, and abdominal pain. Untreated, it has a high mortality rate. An oral attenuated vaccine helps prevent the disease.

Campylobacteriosis (table 25.10)

Campylobacter jejuni is the most common bacterial cause of diarrhea in the United States; it usually originates from domestic animals (figure 25.15).

25.6 Viral Diseases of the Lower Digestive System

Rotaviral Gastroenteritis (table 25.11)

Rotaviral gastroenteritis is the main diarrheal illness of infants and young children, but also can involve adults, as in traveler's diarrhea. Rotaviruses are segmented RNA viruses of the reovirus family (figure 25.16).

Norovirus Gastroenteritis (table 25.11)

Norovirus gastroenteritis accounts for almost half the cases of viral gastroenteritis in the United States. Noroviruses are small RNA viruses of the calcivirus family (figure 25.17).

Hepatitis A (table 25.12, figures 25.18, 25.21)

Hepatitis A is usually mild or asymptomatic in children; some cases are prolonged, with weakness, fatigue, and jaundice. Hepatitis A virus (HAV) is a picornavirus spread by fecal contamination of hands, food, or water. An injection of gamma globulin gives temporary protection from the disease; an inactivated vaccine is available actively to immunize against the disease.

Hepatitis B (table 25.12, figure 25.21)

Hepatitis B virus (HBV) is a hepadnavirus spread by blood, blood products, semen, and from mother to baby (figures 25.19, 25.20). Asymptomatic carriers are common and can unknowingly transmit the disease. Chronic infection is common and can lead to scarring of the liver (cirrhosis) and liver cancer.

Hepatitis C (table 25.12, figure 25.21)

Hepatitis C virus (HCV) is a flavivirus transmitted mainly by blood; approximately 60% of cases are acquired from needle sharing by injected drug abusers. Hepatitis C is asymptomatic in over 60% of acute infections; more than 80% of infections become chronic.

25.7 Protozoan Diseases of the Lower Digestive System

Giardiasis (table 25.13)

Transmission of *Giardia lamblia,* a mastigophoran, is usually via drinking water contaminated by feces. It is a common cause of traveler's diarrhea (figure 25.22). Its cysts survive in chlorinated water and must be removed by filtration.

Cryptosporidiosis (table 25.14)

The life cycle of *Cryptosporidium parvum,* a coccidium, a member of the Apicomplexa, takes place in the small intestinal epithelium. Its oocysts are infectious, resist chlorination, and are too small to be removed by many filters (figure 25.23). It is a cause of many water- and foodborne epidemics, and traveler's diarrhea.

Cyclosporiasis (table 25.15)

Transmission of *Cyclospora cayetanensis,* a coccidium, is fecal-oral, via water or produce such as berries; it causes traveler's diarrhea. Oocysts are not infectious when passed in feces, and thus there is no person-to-person spread; no hosts other than humans are known.

Amebiasis (table 25.16)

Entamoeba histolytica, a member of the Sarcodina group of protozoa, is an important cause of dysentery; often chronic; infection can spread to the liver and other organs (figure 25.24). The organism exists in ameba and cyst forms; the quadrinucleate cyst is infectious.

REVIEW QUESTIONS

Short Answer

1. Name a disease caused by *Clostridium difficile.*
2. Name three characteristics of *Streptococcus mutans* that contribute to its ability to cause dental caries.
3. Give the pathogenesis of periodontal disease.
4. Give three examples of stressful events that can reactivate herpes simplex.
5. At what stage of maturation in boys is mumps likely to be complicated by swelling of the testicles?
6. What is the hemolytic uremic syndrome? Name an organism that can cause it.
7. What is the usual source of *Campylobacter jejuni* infections?
8. How can shigellas move from one host cell to another even though they are non-motile?
9. Name four different groups of *Escherichia coli* based on their pathogenic mechanisms.
10. Contrast the transmission of hepatitis A and hepatitis B.
11. Name two kinds of hepatitis that can be prevented by vaccines.
12. Contrast the cause and epidemiology of giardiasis and amebiasis.

Multiple Choice

1. The following are all true of intestinal bacteria, *except*
   a) they produce vitamins.
   b) they can produce carcinogens.
   c) they are mostly aerobes.
   d) they produce gas from indigestible substances in foods.
   e) they include potential pathogens.
2. All of the following attributes of *Streptococcus mutans* are important in tooth decay, *except*
   a) it attaches specifically to tooth pellicle.
   b) it can grow at pH below 5.
   c) it produces lactic acid.
   d) it synthesizes glucan.
   e) it stores fermentable polysaccharide.
3. All of the following are true of *Helicobacter pylori, except*
   a) it is a helical bacterium with sheathed flagella.
   b) it has not been cultivated *in vitro.*
   c) it produces a powerful urease.
   d) it causes stomach infections that last for years.
   e) it has an important role in the causation of stomach ulcers.
4. All of the following are probably important to the cholera-causing ability of *Vibrio cholerae, except*
   a) it attaches firmly to small intestinal epithelium.
   b) it produces cholera toxin.
   c) lysogenic conversion.
   d) acid resistance.
   e) it survives in the sea in association with zooplankton.
5. Which of the following statements concerning *Salmonella* Typhi is *false?*
   a) It is commonly acquired from domestic animals.
   b) It can colonize the gallbladder for years.
   c) It is highly resistant to killing by bile.
   d) It can destroy Peyer's patches.
   e) It causes typhoid fever.
6. Which statement about rotaviral gastroenteritis is *false?*
   a) The name of the causative agent was suggested by its appearance.
   b) Most of the 600,000 deaths occurring worldwide from this disease are due to dehydration.
   c) Most cases of the disease occur in infants and children.
   d) The causative agent infects mainly the stomach.
   e) The disease is transmitted by the fecal-oral route.
7. Which of the following statements about noroviruses is *false?*
   a) They cause almost half the cases of viral gastroenteritis in the United States.
   b) They can be responsible for epidemics of gastroenteritis.
   c) They generally produce an illness lasting 1 to 2 weeks.
   d) Similar viruses are widespread among marine animals.
   e) They typically cause disease in children and adults rather than infants.

8. Which of the following statements about hepatitis is *false?*
   a) Both RNA and DNA viruses can cause hepatitis.
   b) Some kinds of hepatitis can be prevented by vaccines.
   c) More than half of the new hepatitis C cases are the result of injected-drug abuse.
   d) Lifelong carriers of hepatitis A are common.
   e) At least six different viruses can cause hepatitis.
9. Which of the following statements about hepatitis B virus is *false?*
   a) Replication involves reverse transcriptase.
   b) Infected persons may have large numbers of non-infectious viral particles circulating in their bloodstream.
   c) In the United States, the incidence of infection has been steadily increasing over the last few years.
   d) Asymptomatic infections can last for years.
   e) Infection can result in cirrhosis.
10. Choose the most accurate statement about cryptosporidiosis.
   a) Waterborne transmission is unlikely.
   b) The host range of the causative agent is narrow.
   c) It is prevented by chlorination of drinking water.
   d) Person-to-person spread does not occur.
   e) The life cycle of the causative agent occurs within small intestinal epithelial cells.

Applications

1. One reason given by Peruvian officials for not chlorinating their water supply is that chlorine can react with substances in water or in the intestine to produce carcinogens. How do you assess the relative risks of chlorinating or not chlorinating drinking water?
2. A medical scientist is designing a research program to determine the effectiveness of hepatitis B vaccine in preventing liver cell cancer. Because liver cell cancer probably has multiple causes, how would you measure the success of an anticancer vaccination program?

Critical Thinking

1. What might the lack of a brown color of feces indicate?
2. Mutant strains of *Helicobacter pylori* that lack the ability to produce urease fail to cause infection when they are swallowed. Infection occurs, however, if a tube is used to introduce them directly into the layer of mucus that overlies the stomach epithelium. What does this imply about the role of urease in the bacterium's pathogenicity?

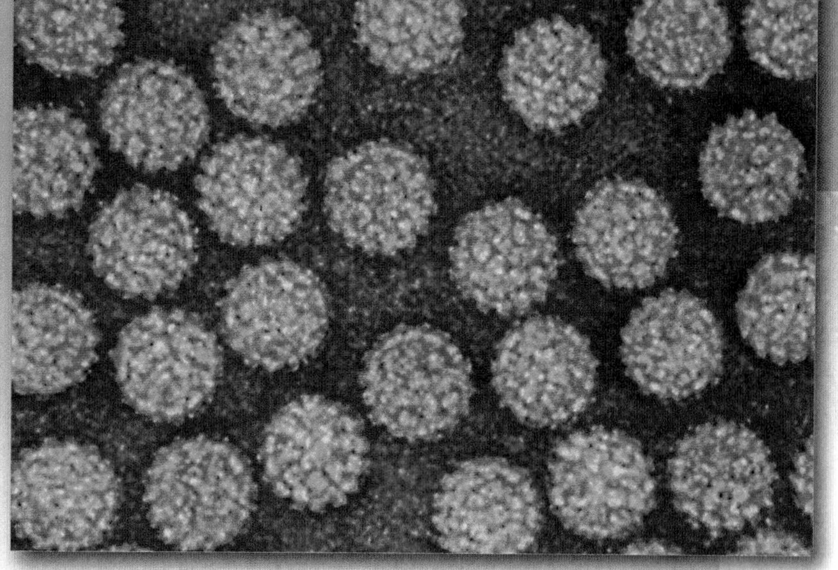

Color-enhanced electron micrograph of a human papillomavirus.

26

Genitourinary Infections

A Glimpse of History

Although some historians argue that syphilis was transported to Europe from the New World by Columbus's crew, others find convincing evidence, including biblical references, that syphilis existed in the Old World for many years before Columbus returned. History and literature record that kings, queens, statesmen, and heroes degenerated into madness or mental incompetence or were otherwise seriously disabled as a result of the later stages of syphilis. Henry VIII (King of England, 1509–1547), Ivan the Terrible (Czar of Russia, 1547–1584), Catherine the Great (Empress of Russia, 1762–1796), Merriwether Lewis (Lewis and Clark Expedition, 1803–1808), and Benito Mussolini (Premier of Italy, 1883–1945) are a few of the famous people who probably suffered from syphilis.

Syphilis was first named the "French pox" or the "Neapolitan disease" because it was believed to have come from France or Italy. In 1530, Girolamo Fracastoro, an Italian physician who suggested the germ theory long before the discovery of microorganisms, wrote a poem about a shepherd named Syphilis who had ulcerating sores covering his body. The description matched the symptoms of syphilis, and from that time on, the disease was known by the shepherd's name. Initially, the method of transmission was unknown, but gradually it became generally recognized that the disease was sexually transmitted. The symptoms of syphilis were very severe in the early years of the epidemic, often causing death within a few months. Treatment, consisting of doses of mercury, and guaiacum, the resin from a tropical American tree of the genus *Guaiacum,* had little beneficial effect. As the decades went by, mutations and natural selection resulted in more resistant hosts and a less-virulent microbe, and the disease evolved into the chronic illness known today.

In 1905, Fritz Schaudinn, a German protozoologist, examined some fluid from a syphilitic sore and saw a faintly visible organism "twisting, drilling back and forward, hardly different from the dim nothingness in which it swam." When Schaudinn used dark-field illumination, he was able to see the organism much more clearly. It appeared very thin and pale, similar to a corkscrew without a handle. The spirochetes appeared in specimens from other cases of syphilis, and Schaudinn later succeeded in staining them. By this time, he felt certain he had discovered the cause of syphilis, but the organism could not be cultivated on laboratory media. Schaudinn named the organisms *Spirochaeta pallida,* "the pale spirochete." This organism is now called *Treponema pallidum,* from treponema, "a turning thread."

Infections of the reproductive and urinary tracts are very common, often uncomfortable or quietly destructive, and sometimes tragic in their consequences. Urinary infections lead the list of infections acquired in hospitals and are the chief source of fatal, nosocomial, bacterial bloodstream invasions. Nosocomial uterine infections—**puerperal fever,** once a common cause of maternal deaths from childbirth—still require strict medical vigilance to be prevented. Much of this chapter will be devoted to sexually transmitted diseases (STDs).

The United States leads the industrialized nations in the reported incidence of STDs. About 1 million unintended pregnancies occur among the estimated 12 million college students each year. Only about 15% of the students have never engaged in sexual intercourse, and almost 35% of the remaining students have had six or more lifetime sexual partners. About 20% of women students say they have been forced to have sexual intercourse against their will. Approximately 70% of sexually active students report that they or their partner rarely or never use a condom. These data suggest a lack of knowledge or concern about the high risk of contracting and transmitting an STD. ■ puerperal fever, p. 450

KEY TERMS

Catheter A flexible tube, usually of plastic or rubber, inserted into the bladder or other body space, usually in order to drain it or instill medication.

Chancre Sore resulting from an ulcerating infection.

Cystitis Inflammation of the urinary bladder.

Fallopian Tube The tubes that convey ova from the ovaries to the uterus.

Ophthalmia Neonatorum Infection of the eyes of newborn babies contracted during passage through the birth canal; often caused by the gonococcus or *Chlamydia trachomatis.*

Papilloma A kind of tumor characterized by nipplelike projections of tissue; warts are an example.

Pelvic Inflammatory Disease (PID) Infection of the fallopian tubes.

Puerperal Fever Childbed fever; infection of the uterus following childbirth, commonly caused by *Streptococcus pyogenes.*

Pyelonephritis Infection of the kidneys.

Semen Fluid composed of sperm from the testicles and prostate gland secretions.

Toxic Shock Collapse of the blood pressure due to a circulating toxin.

Trachoma Potentially serious chronic eye disease caused by certain strains of *Chlamydia trachomatis.*

Urethra The tube draining the bladder.

26.1 Anatomy and Physiology

Focus Points

- List ways the urinary system protects itself from infection.
- Discuss the potential consequences of fallopian tube infections.

The reproductive and urinary systems are often considered together, and they are referred to as the *genitourinary system* because of their close proximity to one another. Also, both systems are commonly affected by the same pathogens. The genitourinary system is one of the portals of entry for pathogens.

The Urinary System

The urinary system consists of the kidneys, the ureters, the bladder, and the urethra **(figure 26.1).** The kidneys act as a specialized filtering system to cleanse the blood of many waste materials, selectively reabsorbing substances that can be reused. Waste materials are excreted in the urine, which is usually acidic because of excretion of excess hydrogen ions from foods and metabolism. Consistently alkaline urine suggests infection with a urease-producing bacterium that converts the urea in urine to ammonia. Antimicrobial medications are commonly excreted in the urine and reach concentrations higher than those in the bloodstream. Each kidney is drained by a ureter, which connects it with the urinary bladder. The bladder acts as a holding tank. Once filled, it empties through the **urethra.** Infections of the urinary tract occur far more frequently in women than in men, because the female urethra is short (about 4 cm, or 1.5 inches, compared with 20 cm, or 8 inches, in the male) and is adjacent to openings of the genital and intestinal tracts. Special groups of muscles near the urethra keep the system closed most of the time and help prevent infection. The downward flow of urine also helps clean the system by flushing out microorganisms before they have a chance to multiply and cause infection.

The urinary tract is protected from infection by a number of mechanisms besides its anatomy. Normal urine contains antimicrobial substances such as organic acids and small quantities of antibodies. During urinary tract infections, larger quantities of specific antibodies can be found in the urine. Antibody-forming lymphoid cells in the infected kidneys or bladder wall produce protective antibodies locally at the site where they are needed. In addition, during infection an inflammatory response occurs in which phagocytes enter the bladder and are of the utmost importance in engulfing and destroying the invading microorganisms.

■ phagocytes, p. 358

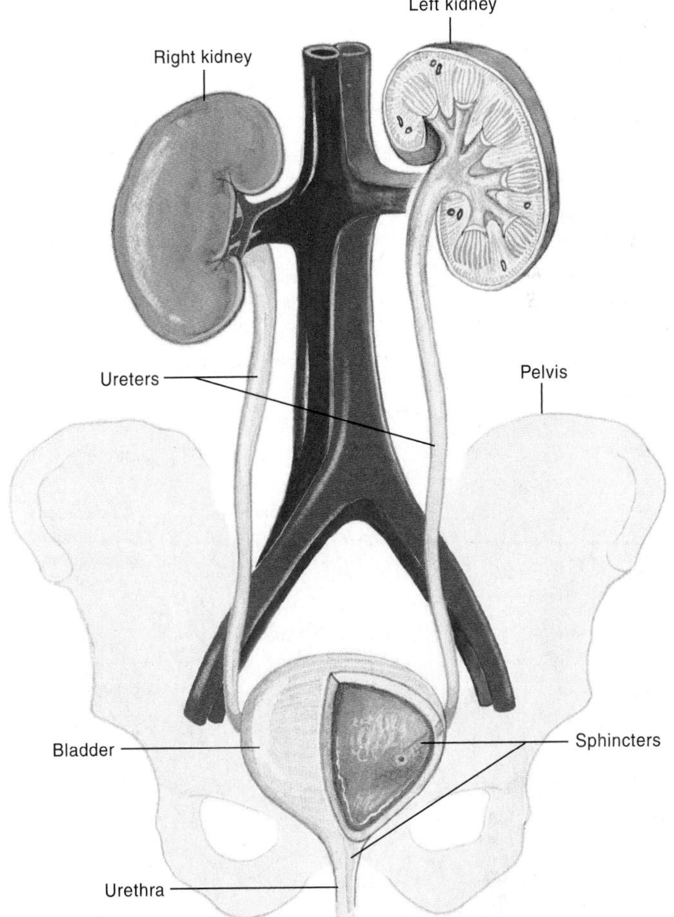

Figure 26.1 Anatomy of the Urinary System Urine flows from the kidneys, down the ureters, and into the bladder, which empties through the urethra. Sphincter muscles help to prevent microorganisms from ascending.

The Genital System

The anatomy of the female and male genital systems is shown in **figure 26.2.** In women during the childbearing years, an egg, or ovum, is expelled from one of the two ovaries each month and swept into the adjacent fallopian tube. Fertilization normally takes place in the fallopian tube; ciliated epithelium of the tube then moves the fertilized ovum to the uterus, where it implants itself in the epithelial lining. If fertilization does not occur, the epithelial lining of the uterus sloughs off, producing a menstrual period. Infection of the **fallopian tubes** can cause scarring and destruction of the ciliated epithelium, so that ova are not moved efficiently to the uterus. The fallopian tubes are open on both ends, providing a pathway into the abdominal cavity where the liver and other structures can be infected. The **uterine cervix** is a common site of sexually transmitted infections, and a place where cancer can develop. Except during menstruation, the cervical opening is tiny and filled with mucus. The vagina is a portal of entry for a number of infectious organisms that can advance to the uterus and fallopian tubes at the time of menstruation. The vaginal opening is framed on each side by two **labia,** or lips. These plus the **clitoris** and entryway of the vagina constitute the female external genitalia, or **vulva.**

In men, the paired reproductive organs, the **testes** (testicles), exist outside the abdominal cavity in the scrotum. Sperm from each testis collect in a tightly coiled tubule called the **epididymis** and are conveyed by a long tube called the vas deferens, which enters the abdomen in the groin to join the prostate gland. The sperm and secretions of the prostate gland compose the **semen.** The urinary and reproductive systems join at the prostate gland. The prostate can be infected by urinary or sexually transmitted pathogens. In older men it often enlarges and hinders the flow of urine, thus fostering urinary infection.

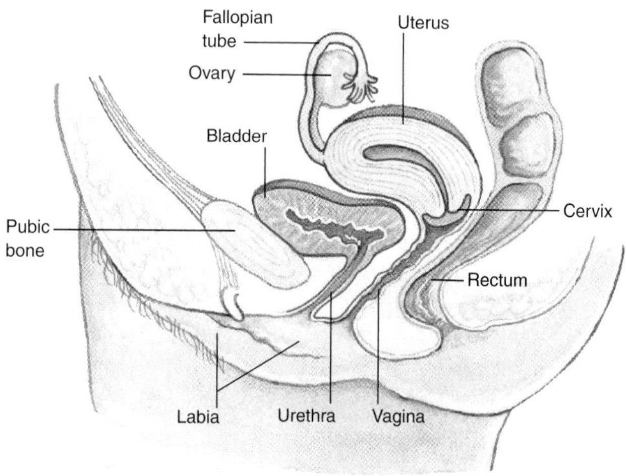

(a)

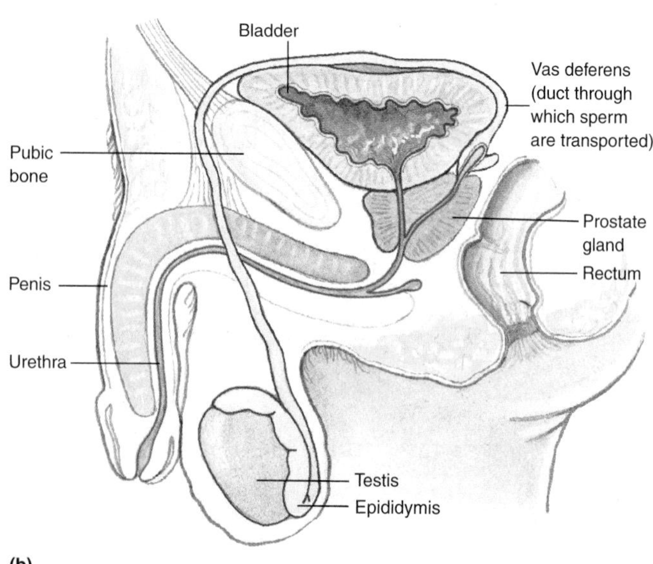

(b)

Figure 26.2 Anatomy of the Genital System (a) Female and **(b)** male.

MICROCHECK 26.1

The genitourinary system is one of the portals of entry for pathogens. Many antimicrobial medications are excreted in the urine in concentrations higher than in the blood. The flushing action of urination is a key defense mechanism against bladder infections. The fallopian tubes can provide a passageway for pathogens to enter the abdominal cavity. The uterine cervix is a common site of infection by sexually transmitted pathogens. Prostate enlargement predisposes men to urinary infection.

✓ Name two kinds of cells that help defend the bladder against infection.

✓ Would ammonia make the urine basic? How?

26.2 Normal Microbiota of the Genitourinary System

Focus Points

- Outline the effects of estrogen on the vaginal epithelium.
- List the parts of the genitourinary system that harbor normal microbiota.

Normally, the urine and urinary tract above the entrance to the bladder are essentially free of microorganisms; the lower urethra, however, has a normal resident microbiota. Species of *Lactobacillus, Staphylococcus* (coagulase-negative), *Corynebacterium, Haemophilus, Streptococcus,* and *Bacteroides* are common inhabitants.

The normal microbiota of the genital tract of women is influenced by the action of estrogen hormones on the epithelial cells of the vaginal mucosa. When estrogens are present, glycogen is deposited in these cells. The glycogen is converted to lactic acid by lactobacilli, resulting in an acidic pH that inhibits the growth of many potential pathogens. Lactobacilli may also release hydrogen peroxide, a powerful inhibitor of some anaerobic bacteria, as a by-product of metabolism. Thus, the normal microbiota and resistance to infection of the female genital tract vary considerably with the person's hormonal status. For example, prepubertal girls, having low estrogen levels, are much more susceptible to vaginal infections with *Streptococcus pyogenes* and *Neisseria gonorrhoeae* than are women in their childbearing years. ■ lactobacilli, p. 255

MICROCHECK 26.2

Normally the urine and urinary tract are free of microorganisms above the entrance to the bladder, but the lower urethra hosts a number of genera of bacteria. During the childbearing years, a woman's hormones are important in the vagina's resistance to infection, because estrogen promotes growth of lactobacilli and acidic conditions.

- ✓ Which parts of the genitourinary system are normally sterile?
- ✓ Name two species of pathogenic bacteria to which the vagina is especially susceptible during childhood.
- ✓ What changes might occur in the vagina if lactobacilli were eliminated?

26.3 Urinary System Infections

Focus Points

- Describe the factors that predispose the urinary tract to infections.
- Outline the pathogenesis and epidemiology of leptospirosis.

Urinary infections can involve the urethra, the bladder, or the kidneys, alone or in combination. They account for about 7 million visits to the doctor's office each year in the United States. Any situation in which the urine does not flow naturally increases the chance of infection. After anaesthesia and major surgery, for example, the reflex ability to void urine is often inhibited for a time, and urine accumulates and distends the elastic bladder. Even being too busy to empty the full bladder may predispose to infection. **Catheterization** of the bladder (meaning inserting a tube, **catheter,** into the bladder) is another common cause of infection. Most cases of urinary infection, however, occur in otherwise healthy young women without impaired urinary flow.

Pathogens infecting other body sites can spread to infect the urinary system. For example, in typhoid fever, *Salmonella* Typhi organisms disseminate throughout the body, infect the kidneys, and are excreted in the urine starting about 1 to 2 weeks after infection. ■ typhoid fever, pp. 601, 602

Bacterial Cystitis

Cystitis, meaning inflammation of the bladder, is the most common type of urinary track infection. Bacterial cystitis is common among otherwise healthy women, and is also a common nosocomial infection. ■ health-care related infections, p. 462

Symptoms

The onset of bacterial cystitis is typically abrupt, with a burning pain on urination, an urgent sensation to urinate, and frequent voiding in small amounts. The urine is usually cloudy due to the presence of leukocytes, often smells bad, and sometimes has a pale-red color due to bleeding. Tenderness may be present in the area above the pubic bone due to the underlying inflamed bladder. Some cases, however, are asymptomatic, especially among children and the elderly. Others develop sudden elevation in temperature, chills, vomiting, back pain, and tenderness overlying the kidneys, indicating a potentially serious complication—a kidney infection called **pyelonephritis.**

Causative Agents

Bladder infections usually originate from the normal intestinal microbiota. Specific strains of *Escherichia coli* cause most cases of bacterial cystitis, accounting for 80% to 90% of cystitis cases in women during the reproductive years and about 70% of all bladder infections. Other enterobacteria such as *Klebsiella* and *Proteus* species account for 5% to 10%, and the Gram-positive species, *Staphylococcus saprophyticus,* accounts for another 5% to 10% of infections in young women. Hospitalized patients are more likely than outpatients to have bladder infections with bacteria such as the Gram-negative rods *Serratia marcescens* and *Pseudomonas aeruginosa* and the Gram-positive coccus *Enterococcus faecalis,* which are resistant to antibacterial medications. Persons with long-standing bladder catheters are often chronically infected with multiple species of intestinal bacteria.

Pathogenesis

Generally, the causative agents of cystitis reach the bladder by ascending from the urethra. The process is aided by motility of the organisms. Urine is a good growth medium for many species of bacteria. Strains of *E. coli* that infect the urinary system possess pili that attach specifically to receptors on the cells that line the bladder. Experimental evidence indicates that attachment is followed by death and sloughing of the superficial layer of epithelium. The bacteria then penetrate the newly exposed cells by endocytosis. Intracellular bacteria encase themselves in a polysaccharide matrix in a kind of **biofilm.** Bacteria ascend the ureters and cause pyelonephritis in many cases of cystitis. Repeated episodes of pyelonephritis cause scarring and shrinkage of the kidneys and are an important cause of kidney failure. ■ pili, pp. 66, 398 ■ endocytosis, p. 73

Epidemiology

About 30% of women develop cystitis at some time during their life. Factors involved in urinary infections in women include:

- A relatively short urethra. The position of the urethra makes it subject to fecal contamination and colonization with potentially pathogenic intestinal bacteria. From there, the bacteria need only travel a few centimeters to access the bladder.
- Sexual intercourse. About one-third of urinary tract infections in sexually active women are associated with sexual intercourse. The massaging effect of sexual intercourse on the urethra propels bacteria from the urethra into the urinary bladder. Many women develop their first bladder infections following their first sexual intercourse—a condition referred to as "honeymoon cystitis."
- Use of a diaphram to avoid pregnancy. The ring of the diaphragm compresses the urethra and impedes the flow of urine, increasing by two to three times the risk of urinary infection.

Urinary infections are unusual in men until about age 50, when enlargement of the prostate gland compresses the urethra and makes it difficult to completely empty the bladder.

Medical conditions may require a bladder catheter to be inserted for periods ranging from several days to months. Unfortunately, this allows bacteria to reach the bladder and establish urinary tract infections both via the catheter lumen and the mucus between the wall of the catheter and the urethra. Pathogens establish a biofilm on the catheter, making it difficult or impossible to kill them with antibacterial medications. The risk of infection increases about 5% each day the catheter remains in place. In the United States, about 500,000 hospitalized patients develop bladder infections each year, mostly after catheterization.

Paraplegics, individuals with paralysis of the lower half of the body, are almost always afflicted with urinary infections. Because they lack bladder control, paraplegics are unable to void normally and require a catheter indefinitely to carry their urine to a container.

Prevention and Treatment

General measures for preventing urinary infections include taking enough fluid to ensure voiding at least four or five times daily, voiding immediately after sexual intercourse, and wiping from front to back after defecation to minimize fecal contamination of the vagina and urethra. Preventing recurrent infections may require taking a dose of antibacterial medication immediately before or after sexual intercourse, or taking daily a small dose of an antibiotic that is concentrated in the urine. Scientific evidence is equivocal regarding the value of cranberry juice in preventing or treating bacterial cystitis. Treatment of cystitis is usually easily carried out with a few days of an antimicrobial medication to which the causative bacterium is susceptible. Pyelonephritis usually requires prolonged treatment and often hospitalization.

The main features of bacterial cystitis are presented in **table 26.1.**

TABLE 26.1 Bacterial Cystitis

Symptoms	Abrupt onset, burning pain on urination, urgency, frequency, foul smell, red-colored urine; with pyelonephritis, fever, chills, back pain, and vomiting
Incubation period	Usually 1 to 3 days
Causative agents	Most due to *Escherichia coli;* other enterobacteria, *Staphylococcus saprophyticus* cause some cases; nosocomial infections with antibiotic-resistant strains of *Pseudomonas, Serratia,* and *Enterococcus* genera
Pathogenesis	Usually, bacteria ascend the urethra, enter the bladder, and attach by pili to receptors on urinary tract epithelium. Sloughing of cells and an inflammatory response ensue. Spread to the kidneys can occur via the ureters, causing pyelonephritis and potential kidney failure.
Epidemiology	Bacterial cystitis is common in women, promoted by a relatively short urethra, use of a diaphragm, and sexual intercourse. Middle-aged men are prone to infection because enlargement of the prostate gland partially obstructs their urethra. Placement of a bladder catheter commonly results in infection.
Prevention and treatment	Taking sufficient fluid to void urine at least four to five times daily, wiping from front to back. Single dose of antimicrobial medication with sexual intercourse may help prevent bacterial cystitis in women. Short-term antimicrobial therapy usually sufficient. Longer treatment for pyelonephritis.

Leptospirosis

In leptospirosis, the causative organism infects the urinary system after entering the body through a mucous membrane or wound. It is then carried to the urinary system by the bloodstream rather than by ascending from the urethra. The disease is probably the most widespread of all the zoonoses, but most cases are mild, recover without treatment, and remain undiagnosed. Epidemics have occurred in soldiers serving in tropical areas and in athletes competing in triathlons. Only one or two fatal cases occur in the United States each year and it has not been a reportable disease since 1994. Leptospirosis occurs around the world in all types of climates, although it is more common in the tropics. ■ zoonoses, p. 453

Symptoms

Many infections are asymptomatic. Symptomatic cases of leptospirosis typically begin after an incubation period averaging 10 days, with the abrupt onset of headache, spiking fever, chills, and severe muscle pain. The most characteristic feature of this phase of the disease is the development of redness of the eyes due to dilation of small blood vessels. In mild cases, which are the most common, symptoms usually subside within a week and all signs and symptoms of illness are gone in a month or less. Severe cases follow a similar course, except after 1 to 3 days of feeling well, symptoms recur accompanied by bleeding from various sites, confusion, and evidence of severe heart, brain, liver, and kidney damage. This biphasic course of the illness is characteristic of most cases that come to medical attention.

Causative Agent

Leptospirosis is caused by a slender spirochete with hooked ends, *Leptospira interrogans* (**figure 26.3**). Although some of the more

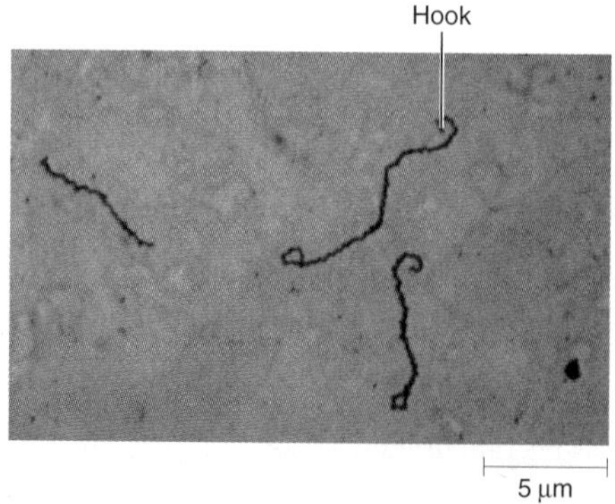

Figure 26.3 ***Leptospira interrogans,*** **the Cause of Leptospirosis** Note the hooked ends of this spirochete.

CASE PRESENTATION

The patient was a 32-year-old married woman complaining of 1 week of burning pain on urination, and frequent voiding of small amounts of bloody urine. About 8 days earlier, she completed 3 days of trimethoprim-sulfamethoxazole therapy for similar symptoms. Tests at that time showed that her urine was infected with *Escherichia coli,* resistant only to amoxicillin. When the symptoms returned, she began drinking 12 ounces of cranberry juice three times daily but had only partial relief. She denied having chills, fever, back pain, nausea, or vomiting. She was approximately 12 weeks pregnant. ■ **trimethoprim, p. 479**

Her medical history revealed that she had suffered two or three similar episodes of urinary symptoms every year for a number of years. Sometimes the symptoms would go away when she forced herself to take extra fluid, but at other times the symptoms would persist and she would obtain medical evaluation and treatment with an antibacterial medication. On one occasion several years before the present illness, chills, fever, back pain, nausea, and vomiting accompanied her symptoms and she was hospitalized for a "kidney infection." There was no history suggesting any underlying disease such as diabetes, cancer, or immunodeficiency.

On examination the patient appeared well, with no obvious distress. Her temperature was normal, as was the remainder of the physical examination.

The results of her laboratory tests included normal leukocyte count and kidney function tests. Microscopic examination of her urine showed numerous red and white blood cells. A Gram-stained smear of the uncentrifuged urine showed numerous rod-shaped, Gram-negative bacteria and polymorphonuclear neutrophils. Culture of the urine revealed more than 100,000 colonies of *Escherichia coli* per ml. The bacterium was resistant to amoxicillin, but sensitive to the other antibacterials useful for treating urinary infections. ■ **polymorphonuclear neutrophils, p. 350**

1. What is the diagnosis?
2. What is the treatment?
3. What is the prognosis?
4. What future preventive measures might be undertaken?

Discussion

1. This woman's symptoms and signs clearly lead to a diagnosis of bacterial cystitis, but there are a number of clues in the presentation of this case that point to a significant complication. First, most patients with uncomplicated bacterial cystitis are cured by 3 days of an antibacterial medication to which the causative bacterium is susceptible. Her symptoms recurred only 1 day after completing her medication. Second, her symptoms had been present for a full week before she sought medical evaluation. Third, she was pregnant. Fourth, she gave a past history of being hospitalized for pyelonephritis.

   These clues make it highly likely that she has a condition called subclinical pyelonephritis, in which her bladder infection has spread to her kidneys but has not yet produced the symptoms of pyelonephritis. As many as 30% of patients with cystitis have subclinical pyelonephritis, depending on risk factors such as the clues mentioned in this case. Although contraindicated in this patient because of her pregnancy, the kidney infection can be demonstrated with a scintogram, an image of the kidneys produced following injection into the bloodstream of a tiny amount of radioactive material, which is removed from the blood by the kidneys and excreted.
2. The patient can be treated with trimethoprim-sulfamethoxazole, the same medication that was given before. Because infection in the kidneys takes much longer to cure than bladder infections, however, the treatment must be continued for 2 weeks or longer. This medication diffuses well into vaginal secretions and can eliminate *E. coli* colonization of the vagina, which may be present. Although this drug is safe to use early in pregnancy, it cannot be used late in pregnancy because it can worsen jaundice in the newborn. A urine culture was done 1 week after completion of treatment to be sure that the infection was truly gone. ■ **newborn jaundice, p. 419**
3. The outlook is good for a full recovery without any permanent damage to the kidneys. The patient was advised, however, that repeated future infections of the kidneys could lead to kidney failure if not treated promptly.
4. For the future, this patient should employ all the usual methods for preventing urinary infections, including taking enough fluid to ensure voiding urine at least four or five times daily, avoiding delays in emptying the bladder, not using a diaphragm for contraception, urinating promptly after intercourse, and taking a preventive antimicrobial medication. No vaccine for this infection has been approved for use in the United States.

than 200 antigenic types have been given different species names in the past, probably all belong to this single species. Most strains can be cultivated using cell-free media. ■ **spirochetes, p. 269**

Pathogenesis

The organisms enter the body through mucous membranes and breaks in the skin. No lesion develops at the site of entry, but the organisms multiply and spread throughout the body by way of the bloodstream, penetrating all tissues, including the eyes and the brain. Severe pain is characteristic of this first (septicemic) phase, and it may lead to unnecessary surgery for suspected appendicitis or gallbladder infection. A lack of inflammatory changes or tissue damage is a striking feature. Within a week, the immune response destroys the organisms present in most tissues, although they continue to multiply in the kidneys. The victim then characteristically enjoys 1 or more days of improvement before symptoms recur. The cause of the recurrent symptoms is presumed to be due to an immune response, and this second phase of the illness is often called the "immune phase." This phase is characterized by injury to the cells that line the lumen of tiny blood vessels, causing clotting and impaired blood flow in tissues throughout the body. This accounts for most of the serious effects of the illness, including kidney failure, the main cause of fatalities.

Epidemiology

Leptospira interrogans infects numerous species of wild and domestic animals, usually causing little or no apparent illness, but ranging to highly fatal, epidemic disease. Characteristically, the organisms are excreted in the animal's urine, which provides the principal mode of transmission to other hosts. Spots on the ground where urine has been deposited can remain a source of infection for as long as 2 weeks, while *Leptospira* in mud or water can survive for several weeks. Warm summer temperatures and neutral or slightly alkaline, moist conditions promote survival of the bacteria. Swimming and getting splashed with urine-contaminated water account for many cases. Infected humans usually excrete the organisms for up to a few weeks, but sometimes excretion can continue for many months. Rodents often excrete *Leptospira* for a lifetime.

Prevention and Treatment

Because of the ubiquity of the causative organism, there are few effective, preventive measures other than to avoid animal urine.

Maintaining general sanitary conditions is helpful in the care of domestic animals raised for food. Multivalent vaccines—those that contain a number of different serotypes of *L. interrogans*—are available for preventing the disease in domestic animals, but they do not consistently prevent the carrier state. In an epidemic situation, small doses of a tetracycline antibiotic can prevent the disease in exposed individuals. A number of antibiotics are effective for treating the disease, but only if started during the first 4 days of the illness. As in a number of other microbial diseases, the onset of treatment is often followed in 4 to 6 hours by a transitory worsening of symptoms due to massive release of antigens from organisms lysed by the antimicrobial medication.

The main features of leptospirosis are shown in **table 26.2.**

MICROCHECK 26.3

Situations that interfere with the normal flow of urine predispose to urinary system infections. Bladder infections are common, especially in women, caused by bowel bacteria ascending from the urethra. Pyelonephritis is a feared complication. In leptospirosis, a widespread zoonosis spread by urine, the kidneys are infected from the bloodstream. Usually mild, the disease can be a severe biphasic illness characterized by tissue invasion and pain in the first phase and tissue destruction in the second.

- ✓ What organism is the leading cause of bladder infections in otherwise healthy women? Where does this organism come from?
- ✓ How do people become infected with *Leptospira interrogans*?
- ✓ How might insertion of a catheter foster bladder infection?

TABLE 26.2 Leptospirosis

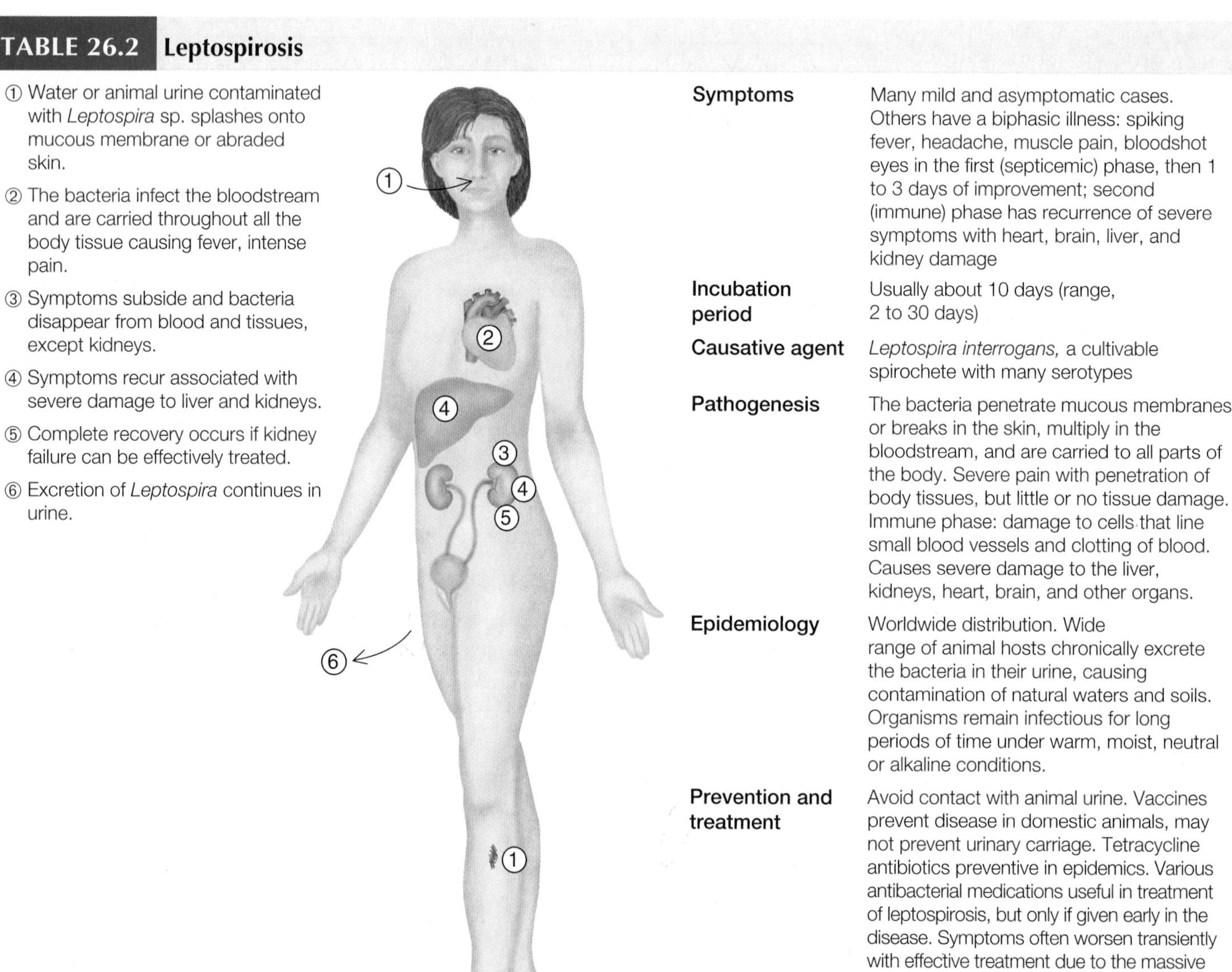

① Water or animal urine contaminated with *Leptospira* sp. splashes onto mucous membrane or abraded skin.
② The bacteria infect the bloodstream and are carried throughout all the body tissue causing fever, intense pain.
③ Symptoms subside and bacteria disappear from blood and tissues, except kidneys.
④ Symptoms recur associated with severe damage to liver and kidneys.
⑤ Complete recovery occurs if kidney failure can be effectively treated.
⑥ Excretion of *Leptospira* continues in urine.
1
1
2
3
4
4
5
6

Symptoms	Many mild and asymptomatic cases. Others have a biphasic illness: spiking fever, headache, muscle pain, bloodshot eyes in the first (septicemic) phase, then 1 to 3 days of improvement; second (immune) phase has recurrence of severe symptoms with heart, brain, liver, and kidney damage
Incubation period	Usually about 10 days (range, 2 to 30 days)
Causative agent	*Leptospira interrogans,* a cultivable spirochete with many serotypes
Pathogenesis	The bacteria penetrate mucous membranes or breaks in the skin, multiply in the bloodstream, and are carried to all parts of the body. Severe pain with penetration of body tissues, but little or no tissue damage. Immune phase: damage to cells that line small blood vessels and clotting of blood. Causes severe damage to the liver, kidneys, heart, brain, and other organs.
Epidemiology	Worldwide distribution. Wide range of animal hosts chronically excrete the bacteria in their urine, causing contamination of natural waters and soils. Organisms remain infectious for long periods of time under warm, moist, neutral or alkaline conditions.
Prevention and treatment	Avoid contact with animal urine. Vaccines prevent disease in domestic animals, may not prevent urinary carriage. Tetracycline antibiotics preventive in epidemics. Various antibacterial medications useful in treatment of leptospirosis, but only if given early in the disease. Symptoms often worsen transiently with effective treatment due to the massive release of leptospiral antigens.

26.4

Genital System Diseases

Focus Points

- List the distinctive characteristics of three microbial genital system diseases.
- Outline prevention of staphylococcal toxic shock syndrome.

The genital tract is the portal of entry for numerous infectious diseases, both sexually and non-sexually transmitted. This section discusses some examples of genital system diseases that are not generally transmitted sexually. Puerperal ("childbed") fever, which can still occur occasionally following childbirth, is an example of such a genital system infection. Other serious infections are associated with menstruation and spontaneous or induced abortions. Organisms from dirt and dust and from a woman's own fecal bacteria can attack the traumatized uterus. A much-feared bacterium, *Clostridium perfringens,* can cause uterine gas gangrene and has been responsible for many fatalities following abortions induced under unclean conditions. Also, the normal vagina can be colonized by various pathogens, producing symptoms that range from annoying to life threatening. ■ puerperal fever, p. 450 ■ *Clostridium perfringens*, p. 569

Bacterial Vaginosis

In the United States, bacterial vaginosis is the most common vaginal disease of women in their childbearing years. It is termed *vaginosis* rather than *vaginitis* because inflammatory changes are absent. Pregnant women with this condition have a sevenfold increase in the risk of having a premature baby or other complications. Nevertheless, the risk is small (less than 10%).

Symptoms

Bacterial vaginosis is characterized by a thin, grayish-white, slightly bubbly vaginal discharge that has a characteristic pungent, "fishy" odor. For many, the odor is more distressing than the vaginal discharge, which is often slight; half of those with the disease are asymptomatic.

Causative Agent

The cause or causes of bacterial vaginosis are unknown. A marked decrease in vaginal lactobacilli is a constant feature, suggesting that conditions supressing lactobacilli, or promoting other microbiota, play a causative role. *Gardnerella vaginalis*, a small, aerotolerant bacterium with a Gram-positive type of cell wall, is commonly present in large numbers, as are strictly anaerobic bacteria of the *Mobiluncus* and *Prevotella* genera, *Mycoplasma* sp., and anaerobic streptococci. Pure cultures of these species do not consistently produce bacterial vaginosis when healthy people are voluntarily inoculated, although the discharge from people with the condition can produce the disease. Each of these species of bacteria can occur in vaginal secretions of women without the disease, although in much smaller numbers.

Pathogenesis

The key changes in this disease are a decrease in the normal acidity of the vagina, a marked derangement of the normal vaginal flora, and a substantial increase in the numbers of **clue cells** (**figure 26.4**). Clue cells are epithelial cells that have sloughed off the vaginal wall and are covered with masses of bacteria. The pH of the vaginal secretions is elevated above the normal of 4.5. No inflammation occurs unless there is another, concurrent vaginal infection. The strong, fishy odor is due to metabolic products of the anaerobes, including the amines putrescine and cadaverine.

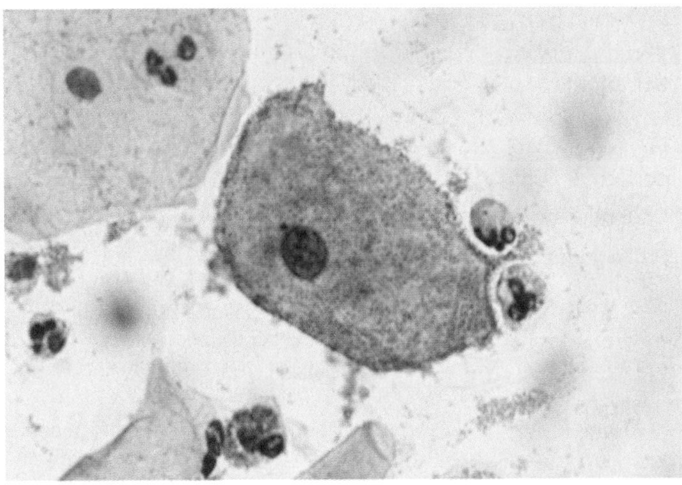

Figure 26.4 Clue Cell in an Individual with Bacterial Vaginosis The cells in the photograph are epithelial cells that have sloughed from the vaginal wall, one of which, the clue cell, is a dark color because it is completely covered with adherent anaerobes.

Epidemiology

The cause of bacterial vaginosis is not known, so its epidemiology is not well understood. The disease is most common among sexually active women and sometimes occurs in children who have been sexually abused. Sexual promiscuity, a new sex partner, douching, and using an intrauterine device (IUD) also increase risk of the disease. Virgins rarely get the disease. Proof is lacking, however, that bacterial vaginosis is a sexually transmitted infection.

Prevention and Treatment

There are no proven measures to prevent bacterial vaginosis. Studies on the use of yogurt by mouth or vaginally to restore vaginal lactobacilli have given conflicting results. Treatment of the male sex partners of patients with recurrent disease does not prevent recurrences. Most cases respond promptly to treatment with metronidazole, a medication active against anaerobes. The main features of bacterial vaginosis are summarized in **table 26.3.**

Vulvovaginal Candidiasis

Vulvovaginal candidiasis is the second most common cause of vaginal symptoms after bacterial vaginosis. As with bacterial vaginosis, vulvovaginal candidiasis appears to follow a disruption of normal microbiota. As the name indicates, the infection often involves not only the vagina, but the woman's vulva, or external genitalia, as well.

Symptoms

The most common symptoms of vulvovaginal candidiasis are itching and burning that can be intense and unremitting. Typically, the vaginal discharge is scanty and whitish, often occurring in curdlike clumps. The involved area is usually red and somewhat swollen.

TABLE 26.3	Bacterial Vaginosis
Symptoms	Gray-white vaginal discharge and unpleasant fishy odor
Incubation period	Unknown
Causative agent	Unknown
Pathogenesis	Uncertain. Marked distortion of the normal microbiota. Increased sloughing of vaginal epithelium in the absence of inflammation. Odor due to metabolic products of anaerobic bacteria. Association with complications of pregnancy, including premature births.
Epidemiology	Associated with many sexual partners or a new partner, but can occur in the absence of sexual intercourse. Probably not a sexually transmitted disease.
Prevention and treatment	No proven preventive measures. Treatment with metronidazole is effective.

Causative Agent

Vulvovaginal candidiasis is caused by *Candida albicans* **(figure 26.5),** a yeast that is part of the normal microbiota of the vagina in about 35% of women. Because *C. albicans* is a fungus, it has a eukaryotic cell structure. ■ yeast, p. 289

Pathogenesis

Normally, vaginal colonization by *C. albicans* causes no symptoms. The large numbers of the normal vaginal lactobacilli keep these fungi in check, probably because they compete for nutrients. When this balance is upset by intensive antibacterial treatment, *C. albicans* multiplies without restraint, causing an inflammatory response. The symptoms of vulvovaginitis occur within about 10 days. Other predisposing factors to *Candida* infection are late pregnancy, the use of oral contraceptives, and poorly controlled diabetes with glucose excretion in the urine. However, most patients with vulvovaginal candidiasis have no known predisposing factors.

Epidemiology

The disease does not spread from person to person and is generally not transmitted sexually. Treatment with antibacterial medications poses an increased risk of the disease, which increases with the duration of treatment.

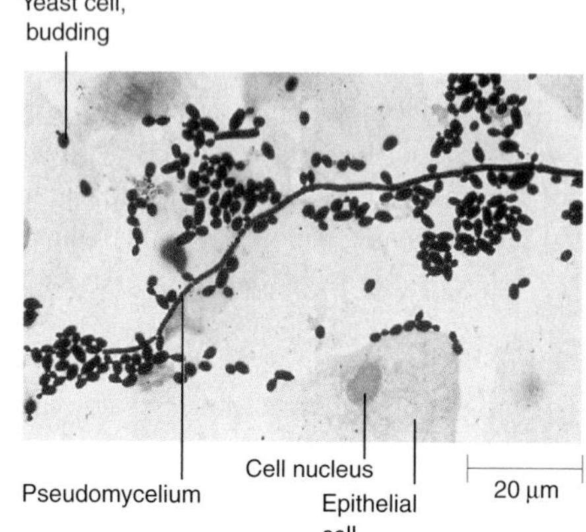

Figure 26.5 ***Candida albicans*** **in the Vaginal Discharge of a Woman with Vulvovaginal Candidiasis**

Prevention and Treatment

Prevention depends on minimizing the use and duration of antibacterial medications, and on effective treatment of underlying conditions such as diabetes. Intra-vaginal treatment of *C. albicans* infections with antifungal medicines such as nystatin or clotrimazole is usually effective. Fluconazole (Diflucan) given by mouth is generally safe and effective, although some recipients have annoying side effects such as headache and nausea, and it can interact with other medications or cause rare, serious reactions. Self diagnosis and treatment with over-the-counter medications is often in error and can contribute to development of treatment-resistant organisms. Simultaneous treatment of any male sex partners is rarely helpful, except when a *Candida* infection of the penis is present.

The main features of vulvovaginal candidiasis are presented in **table 26.4.**

Staphylococcal Toxic Shock Syndrome

Toxic shock syndrome was described in the late 1970s in several children with staphylococcal infections. In 1980, it became epidemic in young, healthy, menstruating women in association with use of a brand of high-absorbency tampon that has since been removed from the market **(figure 26.6).** The term *toxic shock syndrome* was used to represent the symptoms and signs of the illness. Now that we know its cause, it is appropriately called *staphylococcal toxic shock syndrome.* This was not a new disease, but one that emerged in a new form and became much more common as a result of changes in technology and human behavior.

Symptoms

Staphylococcal toxic shock syndrome is characterized by sudden onset of high temperature, headache, muscle aches, bloodshot eyes, vomiting, diarrhea, a sunburn-like rash, and confusion. Without treatment, dropping blood pressure that can lead to kidney failure and death may result. Typically, the skin peels about a week after the onset.

TABLE 26.4	Vulvovaginal Candidiasis
Symptoms	Itching, burning, thick white vaginal discharge, redness and swelling
Incubation period	Usually unknown. Generally 3 to 10 days when associated with antibacterial medications
Causative agent	*Candida albicans,* a yeast
Pathogenesis	Inflammatory response to overgrowth of the yeast, which is often present among the normal microbiota. Associated with antibacterial therapy, use of oral contraceptives, pregnancy, and uncontrolled diabetes, but most cases have no identifiable predisposing factor.
Epidemiology	Not contagious. Usually not sexually transmitted.
Prevention and treatment	No proven preventive measures. Intravaginal antifungal medications such as clotrimazole usually effective.

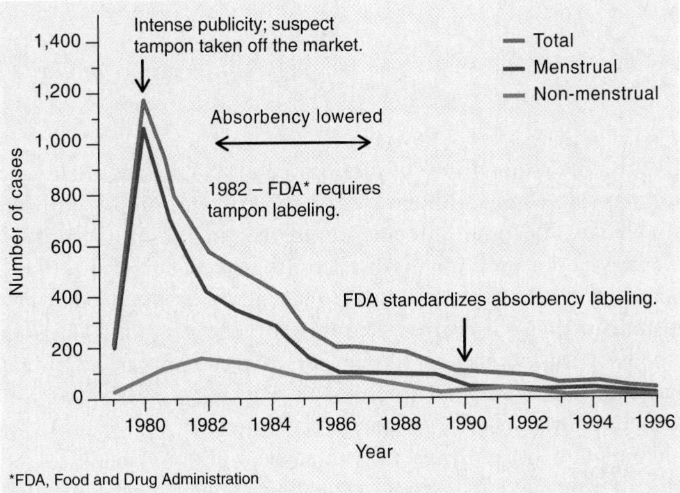

Figure 26.6 Staphylococcal Toxic Shock Syndrome, United States A sharp drop in cases occurred when a brand of high-absorbency tampon was taken off the market. Since 1996, only six or fewer total cases have generally been reported per 100,000 population each year. Non-menstrual cases (men, women, and children) have sometimes exceeded the number of cases associated with menstruation (almost all of them tampon users).

Causative Agent

Staphylococcal toxic shock syndrome is caused by strains of *Staphylococcus aureus* that produce one or more of a particular kind of exotoxin. The involved toxins are all superantigens. Toxic shock syndrome toxin-1 (TSST-1) accounts for about 75% of the cases; the remainder are due to other staphylococcal exotoxins. ■ *Staphylococcus aureus*, p. 535 ■ exotoxins, p. 403 ■ superantigens, p. 406 ■ staphylococcal enterotoxins, p. 763

Pathogenesis

Tampon-associated toxic shock usually begins 2 to 3 days after the start of menstruation and use of vaginal tampons. Generally, it is not an infection because the causative staphylococci grow in the menstrual fluid-soaked tampon and rarely spread throughout the body. Staphylococcal toxic shock results from absorption into the bloodstream of one or more of the toxins from the site where the organisms are growing, a situation analogous to staphylococcal scalded skin syndrome. Tampons that abrade the vaginal wall probably promote toxin absorption. Toxic shock syndrome toxin-1 and the other responsible toxins, being superantigens, cause a massive release of cytokines, which in turn cause a drop in blood pressure and kidney failure—the most dangerous aspect of the illness. ■ staphylococcal scalded skin syndrome, p. 537

Epidemiology

Staphylococcal toxic shock syndrome can occur after any infection with *Staphylococcus aureus* strains that produce one of the responsible toxins. It does not spread from person to person. The syndrome can occur after infection of surgical wounds, infections associated with childbirth, and other types of staphylococcal infections. Using tampons increases the risk of staphylococcal toxic shock, and the higher-absorbency tampons may pose a greater risk. Use of intravaginal contraceptive sponges also increases risk. Most of the menstruation-associated cases occur in women under age 30, probably because younger women are less likely to have protective antibody against the toxin. Recovery from the disease does not consistently give rise to immunity. In fact, about 30% of those who recover will suffer a recurrence of the disease, although it is usually milder than the original illness. Except for the tampons responsible for the 1980s epidemic, which were removed from the market, the type of absorbent fiber is probably not an important factor in the current incidence of the disease. Since 1990, there has been a slow, steady decline in the incidence of staphylococcal toxic shock syndrome, now estimated to be only one to two cases per 100,000 menstruating women per year. In recent years, the prevalence of non-menstrual toxic shock has been increasing and is generally more prevalent than menstruation-related cases. Close monitoring continues, however, because of the introduction of prolonged-use tampons and those of different fiber composition.

Prevention and Treatment

More than 1,100 cases were reported 1980; only 100 to 200 cases have been reported each year since 1994. The improved figures result largely from women better understanding how to use vaginal tampons, and the fact that they no longer use certain highly absorbent types that promoted development of the disease. The incidence of toxic shock syndrome associated with the use of menstrual tampons can be minimized in these ways:

- Wash hands thoroughly before and after inserting a tampon.
- Use tampons with the lowest absorbency that is practical. Since 1990, the law requires that absorbency be stated on the package, ranging from less than 6 to 15 ml.
- Change tampons about every 6 hours and use a pad instead of a tampon while sleeping. Tampons for overnight use introduced in the last few years, however, appear so far to be safe.
- Avoid trauma to the vagina when inserting tampons.
- Know and understand the symptoms of staphylococcal toxic shock syndrome, and remove any tampon immediately if symptoms occur.
- Do not use tampons if you have had staphylococcal toxic shock syndrome previously.

Staphylococcal toxic shock syndrome can be effectively treated with an antibacterial medication active against the infecting *S. aureus* strain, intravenous fluid, and other measures to prevent shock and kidney damage. Most people recover fully in 2 to 3 weeks. With treatment the mortality rate is less than 3%.

Table 26.5 describes the main features of staphylococcal toxic shock syndrome.

MICROCHECK 26.4

A marked derangement of the normal vaginal flora characterizes bacterial vaginosis, the most prevalent vaginal disease of women in their childbearing years. Vulvovaginal candidiasis often occurs as a result of antibacterial therapy suppressing normal vaginal flora, but many other cases arise for unknown reasons. Staphylococcal toxic shock syndrome is caused by certain strains of *Staphylococcus aureus* whose exotoxins are absorbed into the bloodstream, causing the massive release of cytokines responsible for shock.

✓ Why was the incidence of staphylococcal toxic shock syndrome higher in menstruating women than in other people during the early 1980s?

✓ Why is puerperal fever not regarded as a sexually transmitted disease? What spreads the disease?

TABLE 26.5 Staphylococcal Toxic Shock Syndrome

Symptoms	Fever, vomiting, diarrhea, muscle aches, low blood pressure, and a rash that peels
Incubation period	3 to 7 days
Causative agent	*Staphylococcus aureus,* certain toxin-producing strains
Pathogenesis	Toxin (TSST-1 and others) produced by certain strains of *S. aureus;* toxins are superantigens, causing cytokine release and drop in blood pressure.
Epidemiology	Associated with certain high-absorbency tampons, leaving tampons in place for long periods of time, and abrasion of the vagina from tampon use. Also as a result of infection by certain toxin-producing *S. aureus* strains in other parts of the body.
Prevention and treatment	Awareness of symptoms. Prompt treatment of *S. aureus* infections; frequent change of tampons by menstruating women. Antimicrobial medication effective against the causative *S. aureus* strain; intravenous fluids.

26.5

Sexually Transmitted Diseases: Scope of the Problem

Focus Points

- List three behaviors that increase the risk of acquiring a sexually transmitted disease.
- List three measures that help prevent sexually transmitted diseases.

Despite the expenditure of approximately $8 billion each year for sexually transmitted disease (STD) control, 15 million Americans become infected annually, including 3 million teenagers. An individual who acquires an STD is apt to have acquired others without knowing it, and he or she needs to be tested for that possibility. The risk of acquiring an STD rises steeply with the number of partners with whom an individual has unprotected sexual intercourse, and with the numbers of sexual partners of those partners **(figure 26.7).** The use of alcohol and other drugs that release inhibitions adds to the risk. **Table 26.6** lists some often-overlooked symptoms that can indicate the possibility of a person having a sexually transmitted disease. These warrant clinical evaluation even if they go away without treatment, especially if they appear within a few weeks of sexual intercourse with a new partner. A number of STDs can cause few or no symptoms and yet produce serious effects and be transmissible to other people.

Simple measures, although generally not popular, are highly effective in preventing STDs. A small but increasing percentage of students are abstaining from sexual intercourse, or conducting a monogamous relationship with a non-infected person. For others, proper and consistent use of latex or polyurethane condoms, while not an absolute guarantee of protection, markedly reduces the risk of acquiring an STD.

The list of diseases that can be transmitted sexually is very long. Shigellosis, giardiasis, scabies, viral hepatitis, and many others that have non-sexual modes of transmission cannot be overlooked. The epidemiology can be obscure, such as when a drug

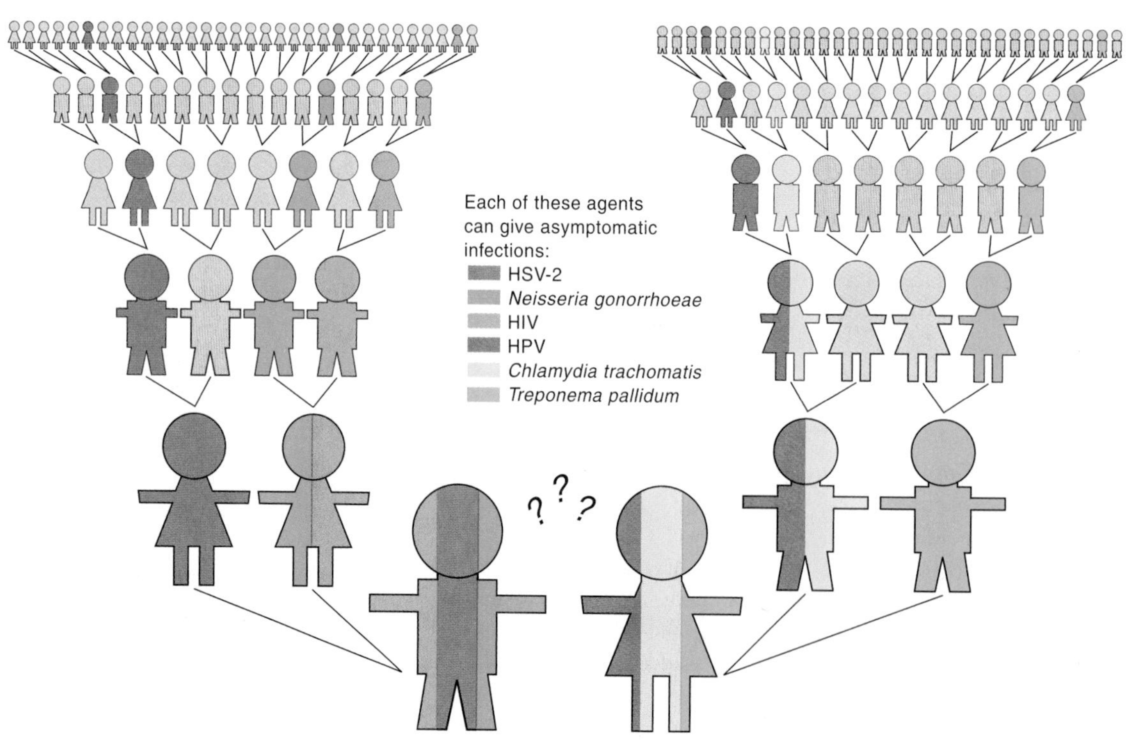

Figure 26.7 The Possible Risk of Acquiring a Sexually Transmitted Disease in Two Individuals Contemplating Unprotected Sexual Intercourse Each partner had two previous sexual partners, and each of these partners had two previous partners, and so on. This risk of contracting an STD rises with the number of sexual partners. (HSV-2—herpes simplex virus, type 2; HIV—human immunodeficiency virus; HPV—human papillomavirus.)

TABLE 26.6 Symptoms that Suggest STD

1. Abnormal discharge from the vagina or penis
2. Pain or burning sensation with urination
3. Sore or blister, painful or not, on the genitals or nearby; swellings in the groin
4. Abnormal vaginal bleeding or unusually severe menstrual cramps
5. Itching in the vaginal or rectal area
6. Pain in the lower abdomen in women; pain during sexual intercourse
7. Skin rash or mouth lesions

abuser contracts hepatitis B from sharing a contaminated needle. The disease can smolder unrecognized for years, and then is transmitted to another person by semen during sexual intercourse. ■ **shigellosis, p. 598** ■ **giardiasis, p. 609** ■ **hepatitis, pp. 605–609**

Table 26.7 lists some common sexually transmitted diseases discussed in subsequent sections of this chapter.

MICROCHECK 26.5

An individual who acquires an STD may well have acquired others without knowing it. The chance of a person acquiring an STD increases steeply with the number of his or her sexual partners, and with the number of sexual partners of those partners. Symptoms of an STD can sometimes easily be overlooked. Simple measures are highly effective for preventing STDs.

✓ Name two diseases that have both sexual and non-sexual modes of transmission.

✓ Why might an individual with an STD need to be checked for other STDs even though he or she has no symptoms of any others?

26.6 Bacterial STDs

Focus Points

- Compare and contrast gonorrhea and chlamydial genital system infections.
- Compare and contrast syphilis and chancroid.

Most of the bacteria that cause sexually transmitted diseases survive poorly in the environment. Because of this, transmission from one person to another usually requires intimate physical contact but is highly unlikely to occur via a handshake or contact with a contaminated toilet seat.

Gonorrhea

Gonorrhea is not a new problem; it was common among World War I recruits. More than 1 million cases per year were reported during the late 1970s, now down to an average around 342,250 per year. Today's concerns are its continued prevalence and its slowly increasing resistance to antibacterial treatment.

Symptoms

The incubation period of gonorrhea is generally only 2 to 5 days. Asymptomatic infections can occur in both sexes. In men, gonorrhea is characterized by urethritis, with pain during urination, and a thick, pus-containing discharge from the penis **(figure 26.8).** These symptoms are dramatic and unpleasant, and they usually inspire prompt treatment. In women, the usual symptoms of painful urination and vaginal discharge tend to be mild, and they may be overlooked. Therefore, women become unknowing carriers of gonorrhea.

TABLE 26.7 Common Sexually Transmitted Diseases

Disease	Cause	Comment
Bacterial		
Gonorrhea ("clap")	*Neisseria gonorrhoeae*	Average reported cases per year—340,000. True incidence much higher.
Chlamydial infections	*Chlamydia trachomatis*	Average reported cases per year—800,000. True incidence much higher.
Syphilis	*Treponema pallidum*	Average reported cases (primary and secondary) per year—6,600.
Chancroid	*Haemophilis ducreyi*	Average reported cases per year—50. True incidence much higher.
Viral		
Genital herpes simplex	Herpes simplex virus (HSV)	Not reportable. Estimated 45 million Americans infected; about 85% HSV, type 2.
Papillomavirus infections	Human papillomavirus (HPV)	Not reportable. Estimated 40 million Americans infected.
AIDS	Human immunodeficiency virus (HIV)	Average reported cases per year—40,000.
Protozoal		
Trichomoniasis ("trich")	*Trichomona vaginalis*	Not reportable. Estimated 5 million Americans infected per year.

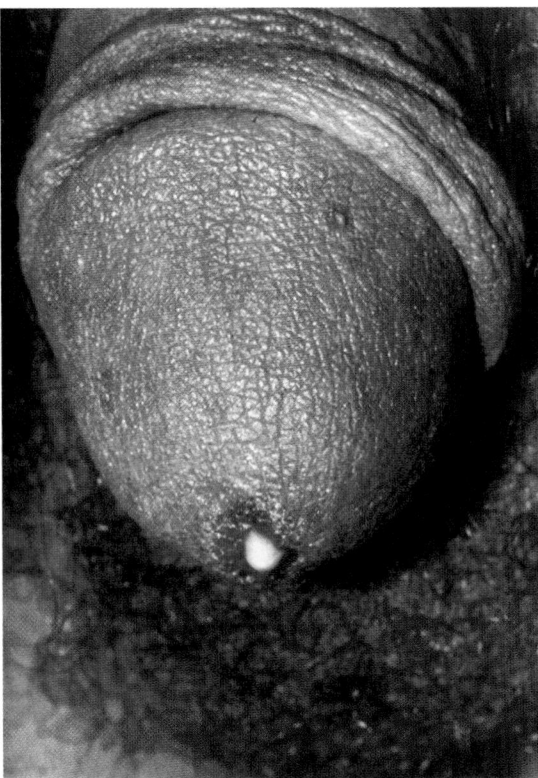

Figure 26.8 Urethral Discharge in a Man with Gonorrhea Gonorrhea in men is usually, but not always, highly symptomatic.

Causative Agent

Gonorrhea is caused by *Neisseria gonorrhoeae,* the gonococcus—a Gram-negative diplococcus that can be cultivated on chocolate agar medium. The organisms are typically found on and within the leukocytes in urethral pus **(figure 26.9).** Gonococci are parasites of humans only, preferring to live on the mucous membranes of their host. Most strains are susceptible to cold and drying and, hence, do not survive well outside the host. For this reason, gonorrhea is transmitted primarily by direct contact. Because the bacteria mainly live in the genital tract, this contact is almost always sexual. An increasing percentage of strains contain R plasmids that render them resistant to antibiotics such as penicillin and tetracycline. ■ **chocolate agar, p. 96** ■ **R plasmids, p. 205**

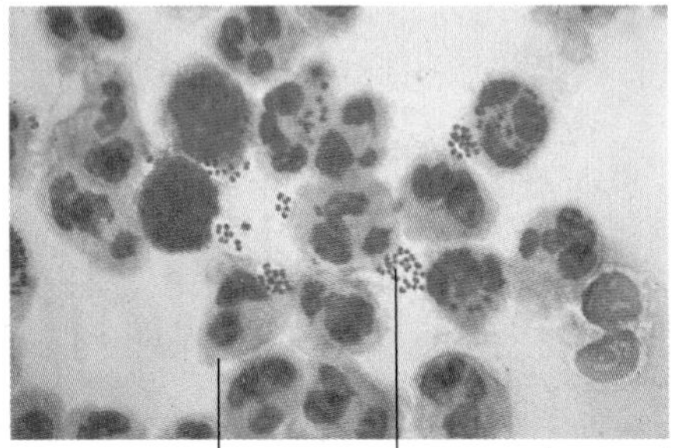

Figure 26.9 Stained Smear of Pus from the Urethra Showing *Neisseria gonorrhoeae* The bacteria are the tiny, Gram-negative, bean-shaped diplococci next to the neutrophils.

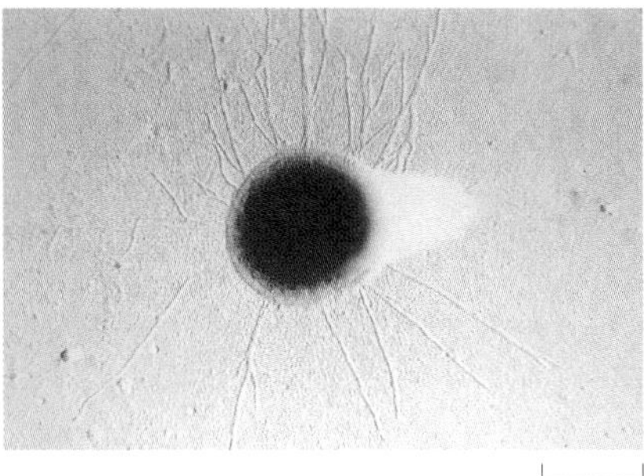

Figure 26.10 Electron Micrograph of a Single Coccus of *Neisseria gonorrhoeae* Notice the many long pili.

Pathogenesis

Gonococci selectively attach to certain non-ciliated epithelial cells of the body, notably those of the urethra, uterine cervix, pharynx, and conjunctiva. Infection begins when pili (fimbriae) **(figure 26.10)** attach specifically to receptors on host cells. Pili and certain other surface proteins involved in attachment can either be expressed or not, an example of phase variation. Also, a single strain of gonococcus can express many different kinds of pili by antigenic variation, brought about by chromosomal rearrangements within the pili genes. This variation explains why cultures from different body sites in an infected individual and those from his or her sex partner may yield *N. gonorrhoeae* with different types of pili. The ability of gonococci to express different surface antigens allows it to attach to many different kinds of cell receptors. Also, phase and antigenic variation allow the organism to escape the effects of antibody formed against the bacterium. So far, the large variety of *N. gonorrhoeae* surface antigens has defeated efforts to develop a vaccine for prevention of gonorrhea. Also, virtually all pathogenic *N. gonorrhoeae* produce an enzyme that destroys IgA antibody found on mucosal surfaces. ■ **phase and antigenic variation, p. 181**

Outer membrane proteins called Opa represent another important pathogenicity-associated property of *N. gonorrhoeae.* Some Opa proteins attach specifically to receptors on CD4 T lymphocytes, thereby preventing activation and proliferation of the lymphocytes and impairing the immune response to the infection.

Untreated disease in men can lead to complications. Inflammatory reaction to the infection can cause scar tissue formation that partially obstructs the urethra, predisposing the man to urinary tract infections. The infection may spread to the prostate gland and testes, producing hard-to-treat prostatic abscesses and **orchitis** (inflammation of the testicle). Sterility can result when scar tissue blocks the tubes that carry the sperm, or if testicular tissue is destroyed by the infection.

Gonorrhea follows a different course in women. Besides infecting the urethra, gonococci thrive in the cervix and fallopian tubes, as well as in glands in the vaginal wall and other areas of the genital tract. Some 15% to 30% of untreated infections progress upward through the uterus into the fallopian tubes, causing **pelvic inflammatory disease (PID).** Occasionally the infection exits the fallopian tube and passes into the abdominal cavity, where it attacks the surface of the liver or other abdominal organ. It is not clear how *N. gonorrhoeae* (which is non-motile) can traverse the uterus to reach the fallopian tubes. One possibility is that the organisms hitch a ride on sperm, to which the bacteria are known to attach. Scar tissue formed as a result of the infection in the fallopian tubes can block normal passage of the ova through the tubes, causing sterility. Scarring of a fallopian tube can also lead to a dangerous complication—**ectopic pregnancy**—in which the embryo develops in the fallopian tube or even in the abdominal cavity outside the uterus. Ectopic pregnancy can lead to life-threatening internal hemorrhaging.

Infrequently, *Neisseria gonorrhoeae* can also cause **disseminated gonococcal infection (DGI).** Curiously, this complication is not usually preceded by urogenital symptoms. Disseminated gonococcal infections are characterized by one or more of the following: arthritis caused by growth of gonococci within the joint spaces, fever, and rash. Any joint may be affected, especially the larger ones. The heart valves can also be infected and destroyed in DGI.

Ophthalmia neonatorum is a destructive *N. gonorrhoeae* infection of the eyes of newborn babies whose mothers have symptomatic or asymptomatic gonorrhea. The bacteria are transmitted during the infant's passage through the infected birth canal. This form of gonococcal infection is now unusual in the United States because laws require that an antibiotic ointment such as 0.5% erythromycin be placed directly into the eyes of all newborn infants. This treatment must be given within 1 hour of birth. Some mothers who feel certain that they do not have gonorrhea have challenged these laws, but the long-lasting and asymptomatic nature of gonorrhea in many women makes it risky to omit prophylactic treatment of a baby's eyes.

Epidemiology

Gonorrhea is among the most prevalent of the sexually transmitted diseases. In the United States, its incidence is the highest of any reportable bacterial disease other than *Chlamydia* infection. A steady rise in reported cases of gonorrhea from about 350,000 in 1966 to more than 1 million cases in 1976 occurred despite the availability of effective treatment. From 1976 through 1980, more than 1 million cases were reported each year, but since then the incidence has progressively declined, probably due largely to increased condom use from fear of AIDS.

Factors that influence the incidence of gonorrhea include:

- Birth control pills. About 35% of sexually active American college and university students say that they or their partner employs a contraceptive pill. Oral contraceptives without use of a condom offer no protection against sexually transmitted diseases and may increase susceptibility to them. Use of oral contraceptives leads to migration of gonorrhea-susceptible epithelial cells from the cervical lumen onto more exposed areas of the outer cervix. Oral contraceptives also tend to increase both the pH and the moisture content of the vagina, favoring infection with gonococci and other agents of sexually transmitted diseases. Besides being more vulnerable to gonorrhea, women taking oral contraceptives are also more likely to develop serious complications from the disease.
- Carriers. Carriers of gonococci, both male and female, can unknowingly transmit these bacteria over months or even years.
- Lack of immunity. There is little or no immunity following recovery from the disease. Individuals can contract gonorrhea repeatedly.

Prevention and Treatment

Prevention depends on abstinence, monogamous relationships, consistent use of condoms, and prompt identification and treatment of sexual contacts. No vaccine is available. Resistance of *N. gonorrhoeae* to penicillins, tetracyclines, and sulfonamides is now widespread. Fluoroquinolones were formerly recommended but resistance has risen progressively over the past decade reaching 30% or higher in some groups of men who have sex with men. Only a single class of antibiotics, the cephalosporins, is currently recommended for the initial treatment of gonorrhea. ■ fluoroquinolones, p. 478 ■ cephalosporins, p. 475

Table 26.8 describes the main features of the disease.

Chlamydial Genital System Infections

Most college men with symptoms suggesting gonorrhea actually have other infections. From 25% to 40% of these students are infected with *Chlamydia trachomatis*, a common cause of sexually transmitted disease of men and women. Chlamydial infections mimic gonorrhea in several ways, including production of urethritis, and testicle and fallopian tube damage. Like *N. gonorrhoeae*, *C. trachomatis* attaches to sperm, and some have speculated that its rapid ascent to the fallopian tubes occurs by hitching a ride on sperm. The main importance of the infection is that it can produce pelvic inflammatory disease (PID) in women, damaging the fallopian tubes and promoting sterility or ectopic pregnancy. Most alarming, tubal damage can occur without symptoms. The infection in men can produce sterility by infecting the epididymis. Chlamydial genital infections have only been reportable nationally since 1995, and they substantially exceed the numbers of reported cases of gonorrhea. ■ *Chlamydia,* p. 274

Symptoms

The symptoms of *C. trachomatis* infection generally appear 7 to 14 days after exposure. In men, the main symptom is a thin, gray-white discharge from the penis, sometimes with painful testes. Women most commonly develop an increased vaginal discharge, sometimes accompanied by painful urination, abnormal vaginal bleeding, and upper or lower abdominal pain. Many infections of men and women are asymptomatic.

Causative Agent

Chlamydia trachomatis is a spherical, obligate intracellular bacterium. Inclusion bodies containing a glycogen-like material form at the site of bacterial replication in the cytoplasm of host cells. These inclusions stain deeply with iodine and can provide a rapid way of identifying *C. trachomatis* infections. There are a number of different antigenic types of *C. trachomatis*, and different types may cause different diseases. Approximately eight types are responsible for most *C. trachomatis* sexually transmitted disease. Three other types cause **lymphogranuloma venereum,** a rare STD in the United States, in which lymph nodes in the groin swell up and drain pus; after years, gross swelling of

TABLE 26.8 Gonorrhea

① Eyes of adults and children are susceptible to the gonococcus; serious infections leading to loss of vision are likely in newborns.

② Organisms carried by the bloodstream infect the heart valves and joints.

③ The outer covering of the liver is infected when gonococci enter the abdominal cavity from infected fallopian tubes.

④ Prostatic gonococcal abscesses may be difficult to eliminate.

⑤ Infection of the fallopian tubes results in scarring, which can lead to sterility or ectopic pregnancy.

⑥ The cervix is the usual site of primary infection in women.

⑦ Urethral scarring from gonococcal infection can predispose to urinary infections by other organisms.

⑧ Scarring of testicular tubules can cause sterility.

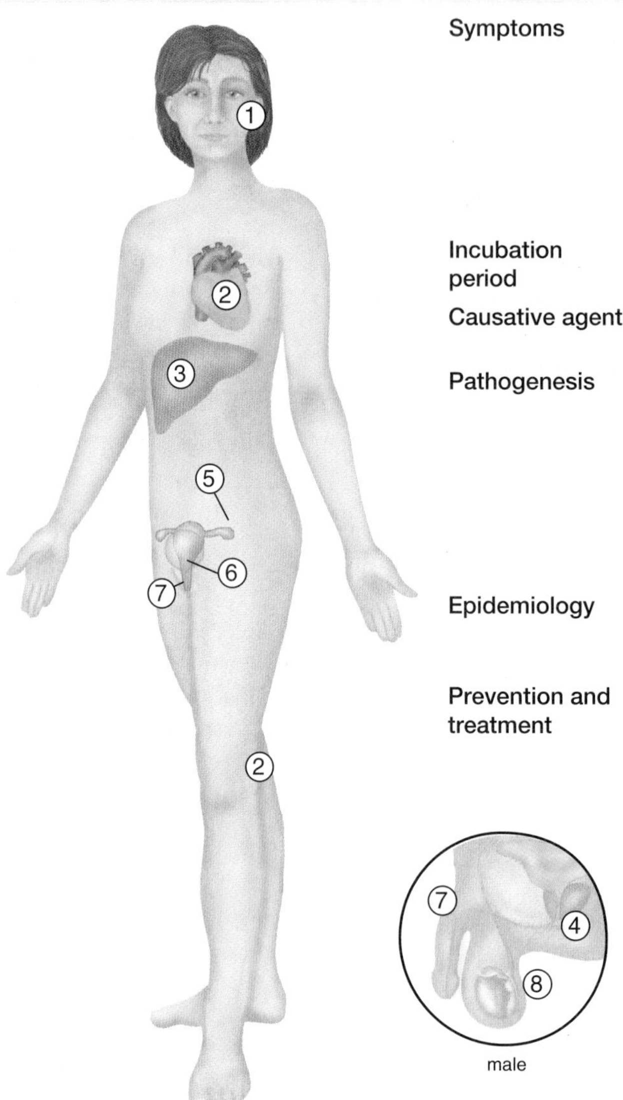

Symptoms	Men: no symptoms, pain on urination, discharge; with complication impaired urinary flow, sterility, or arthritis. Women: no symptoms or pain on urination, discharge, fever, pelvic pain, sterility, ectopic pregnancy, arthritis can occur
Incubation period	2 to 5 days
Causative agent	*Neisseria gonorrhoeae,* a Gram-negative diplococcus
Pathogenesis	Organisms attach to certain non-ciliated epithelial cells by pili; phase and antigenic variation in surface proteins allows attachment to different host cells and escape from immune mechanisms. Inflammation, scarring; can spread by bloodstream.
Epidemiology	Transmitted by sexual contact. Asymptomatic carriers. No immunity.
Prevention and treatment	Abstinence, monogamous relationships, condoms, early treatment of sexual contacts. Treatment: intramuscular ceftriaxone.

the genitalia can occur. Four other types cause the chronic eye disease **trachoma,** an important preventable cause of blindness worldwide. ■ glycogen, p. 32

Pathogenesis

The infectious form of *C. trachomatis,* called an **elementary body,** attaches specifically to receptors on the surface of the host epithelial cell, thereby inducing the cell to take in the bacterium by endocytosis. In the endocytic vacuole, the bacterium enlarges and becomes a non-infectious form called a **reticulate body.** The reticulate body divides repeatedly by binary fission, resulting in numerous elementary bodies. These are released from the host cell and infect nearby cells. The infected cells release cytokines that provoke an intense inflammatory reaction. Much of the tissue damage results from the cell-mediated immune response. The infection usually involves the urethra in both men and women. In men, it spreads to the tubules that collect sperm from the testicles, causing acute pain and swelling. In women, the infection commonly involves the cervix, making it bleed easily, often with sexual intercourse. The uterus is commonly involved, causing pain and bleeding. From the uterus, the infection spreads to the fallopian tubes **(figure 26.11),** causing PID, and like gonorrhea, it can exit the fallopian tubes and infect the surface of the liver. Subsequent scar tissue formation is responsible for many of the serious consequences of the disease such as sterility and ectopic pregnancy. ■ endocytosis, p. 73

Epidemiology

Chlamydial genital infections lead all reportable bacterial infectious diseases. In contrast to gonorrhea, the number of reported chlamydial infections has tended to rise each year, probably a reflection of increased awareness of the disease and better diagnostic tests. Only a fraction of the total cases, estimated at 4 million per year in the United States, is reported by physicians to public health departments. More than 14% of sexually active high school and college women were asymptomatic carriers of

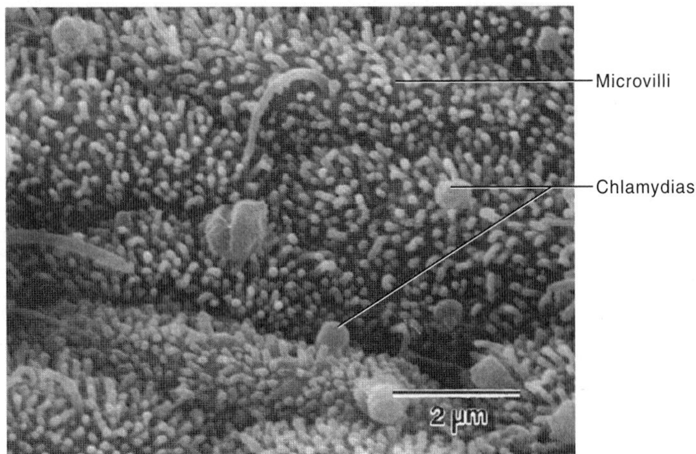

Figure 26.11 Scanning Electron Micrograph of *Chlamydia trachomatis* The bacteria are attached to fallopian tube mucosa.

C. trachomatis in one study. Non-sexual transmission of this agent also occurs, for example, in non-chlorinated swimming pools. Newborn babies of infected mothers often develop chlamydial ophthalmia neonatorum, also called **inclusion conjunctivitis,** and pneumonia from *C. trachomatis* infection contracted during passage through the birth canal.

Prevention and Treatment

Abstinence, monogamous relationships, and condoms properly used can prevent transmission of the disease. All sexually active women are advised to get tested for *Chlamydia* each year, or twice yearly if they have multiple partners or if their partner has multiple partners. Several antibiotics offer effective treatment and prevent serious complications if the disease is diagnosed and treated promptly. Azithromycin can be given as a single dose, whereas tetracyclines and erythromycin are less expensive alternatives. The sexual partner is treated at the same time. Chlamydial ophthalmia neonatorum is treated with oral erythromycin.

The main features of chlamydial genital infections are summarized in **table 26.9.**

TABLE 26.9 Chlamydial Genital System Infections

Symptoms	Men: thin, gray-white penile discharge, painful testes. Women: vaginal discharge, vaginal bleeding, lower or upper abdominal pain
Incubation period	Usually 7 to 14 days
Causative agent	*Chlamydia trachomatis,* an obligate intracellular bacterium, certain serotypes
Pathogenesis	Elementary body attaches to specific receptors on the epithelial cell, causing endocytosis; transforms to reticulate body in the endocytic vacuole; repeated replication by binary fission and differentiation into elementary bodies; rupture of the vacuole and release of elementary bodies to infect adjacent cells; release of cytokines results in inflammatory response; cellular immune response against infection causes extensive damage; scar tissue formation responsible for ectopic pregnancy and sterility.
Epidemiology	The leading reportable bacterial infection in the United States. Large numbers of asymptomatic men and women carriers. Non-sexual transmission can occur in non-chlorinated swimming pools.
Prevention and treatment	Abstinence, monogamous relationship, condom use. Test sexually active men and women at least once yearly to rule out asymptomatic infection. Treatment: azithromycin, single dose; other antibacterial medications.

Syphilis

During the first half of the twentieth century, syphilis was a major cause of mental illness and blindness, and a significant contributor to the incidence of heart disease and stroke. Syphilis was common in the United States during World War II, present in about 5% of military recruits, but by the mid-1950s, it was almost eradicated. This was accomplished by aggressively locating syphilis cases and their sexual contacts and treating them with penicillin, which became generally available after the war. A number of factors, however, conspired to cause a resurgence of the disease. Inner-city poverty, prostitution, and drug use were linked to a high incidence of syphilis, which exceeded 100 new cases per 100,000 population in at least seven cities in 1990. Since then, renewed efforts in education, case finding, and treatment have caused a dramatic drop in new cases. The presence of the AIDS epidemic added urgency to syphilis control efforts, because syphilis, like other STDs that cause genital sores, promotes the spread of AIDS by increasing the risk of HIV infection. By the end of 1998, the syphilis rate was down to 2.6 per 100,000, the lowest level since reporting began in 1941, surpassing the national health objective for the year 2000 of four or fewer cases per 100,000 population. Unfortunately rates have tended to rise since then and there has been a worrisome increase in the disease among women.

Symptoms

Syphilis occurs in so many forms that it is easily confused with other diseases and is often called "the great imitator." Generally, its manifestations occur in three clinical stages. The characteristic feature of **primary syphilis** occurs about 3 weeks after infection and consists of a painless, red ulcer with a hard rim called a hard **chancre** (pronounced "shanker") that appears at the site of infection **(figure 26.12).** The local lymph nodes enlarge. Primary syphilis often goes unnoticed in women and homosexual men because the painless chancre is hidden from view in the anus or vagina. After 2 to 10 weeks or longer, the manifestations of **secondary syphilis** usually appear, including runny nose and watery eyes, aches and pains, sore throat, a rash that includes the palms and soles, and whitish patches on the mucous membranes. After a latent period that can last for many years, manifestations of **tertiary syphilis** occur and typically include mental illness, blindness, stroke, and other nervous system disorders.

PERSPECTIVE 26.1

The Demise of Syphilis?

For 40 years, 1932 to 1972, the U.S. Public Health Service conducted a study, now known as the Tuskegee Syphilis Experiment, of 399 poor Alabama black men with advanced syphilis. For roughly the first half of the study there was no effective therapy for syphilis, and the study was meant to assess the natural progression of the disease in black men. The men were not told their diagnosis nor educated about the nature of their condition and its transmission. In a blatant union of bad science and raw racism, the study was allowed to continue without giving treatment, even after effective therapy became available. Twenty-eight of the subjects died of the disease, and another hundred died from its complications. Forty wives became infected, and there were 19 children born with congenital syphilis. Nothing of scientific value was learned, and many people to this day remain bitter and distrustful of the U.S. Public Health Service. In 1997, President Clinton, speaking for the government, formally apologized to the surviving eight subjects.

Fortunately, in more recent years, major governmental efforts to conquer the disease have met with some success. In the year 2006, only 9,756 cases of primary and secondary syphilis were reported, down from 50,223 in 1990. The Centers for Disease Control and Prevention (CDC) has even considered the possibility that transmission of the disease could be eliminated in the United States.

Syphilis is a tempting target for eradication because it has no animal reservoir, infectious cases can generally be treated with a single injection of penicillin, and the incubation period is long enough so that sexual contacts of an infected person can be found and treated before they spread the disease any further. Moreover, instead of being widely endemic, most syphilis cases are now concentrated in a relatively few areas and certain populations. Also, new techniques are now available for rapid diagnosis, treatment, and epidemiological tracing of the disease. Finally, control of syphilis is closely linked to AIDS control efforts because the risk of HIV transmission is six times as great if either sex partner has syphilis.

One population targeted for treatment is persons detained in county and city jails—representing more than 500,000 individuals at any one time—with a known high incidence of STDs including syphilis. Screening these people for syphilis and getting them treated has proved frustrating because almost half of them are discharged within 48 hours—too soon to get the results of blood tests and arrange for medical evaluation and treatment. After release from jail, these individuals are notoriously hard to find and are often unwilling or unable to reveal the names and locations of their sexual contacts. Studies carried out in Illinois during 1996 showed that by using a screening blood test for syphilis that gave results in 15 minutes, and having a physician's assistant available for examination and treatment, almost 80% of jail inmates found to have syphilis got treatment the same day. Unfortunately, in the past, few jailers have had the interest or resources to focus on STD control to this extent.

Causative Agent

Syphilis is caused by *Treponema pallidum,* an extremely slender, motile spirochete with tightly wound coils. It ranges in length up to 20 μm. Dark-field microscopy is used to observe *T. pallidum* **(figure 26.13)** and its characteristic slow rotational and flexing motions. The organism is difficult to study because it cannot be cultivated *in vitro* and must be grown in the testicles of laboratory rabbits. Although the organism can be maintained *in vitro* and its metabolism studied, it undergoes little or no multiplication. Like most strains of gonococci, *T. pallidum* is killed by drying and chilling and is therefore transmitted almost exclusively by sexual or oral contact. Normally, *T. pallidum* is only a parasite of humans. We now know the sequence of nucleotides in its genome, which should give better understanding of its virulence, why it cannot be cultivated *in vitro,* its evolution, and how to make an effective vaccine. ■ dark-field microscopy, p. 43

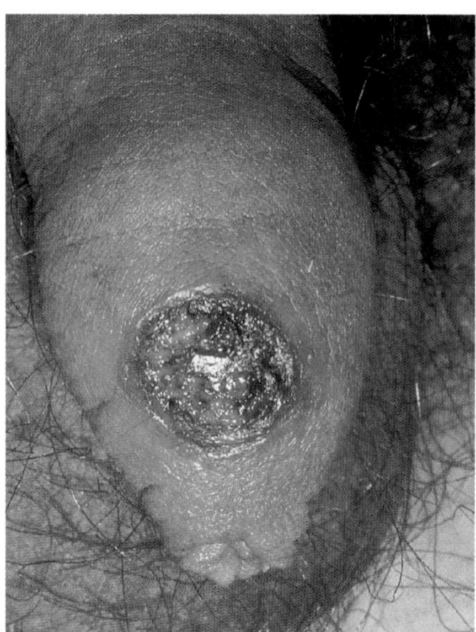

Figure 26.12 Syphilitic Chancre on the Foreskin of an Uncircumcised Man This is the site where *Treponema pallidum* entered the man's body.

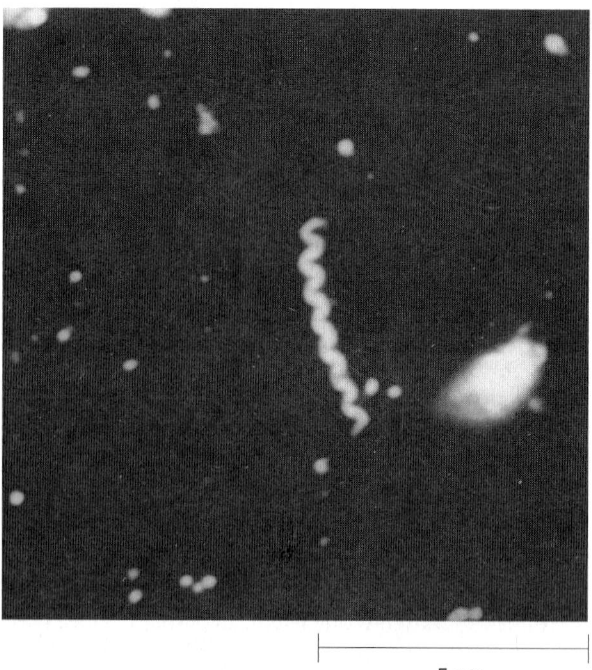

Figure 26.13 Appearance of *Treponema pallidum* with Dark-Field Illumination This technique readily detects the organism in skin and mucous membrane lesions of syphilis.

Pathogenesis

Like the spirochetes that cause leptospirosis, *T. pallidum* readily penetrates mucous membranes and abraded skin. The infectious dose is very low—less than 100 organisms. In primary syphilis, *T. pallidum* grows and multiplies in a localized area of the genitalia, spreading from there to the lymph nodes and bloodstream. The hard chancre represents an intense inflammatory response to the bacterial invasion. Sometimes no chancre develops, only a pimple small enough to go unnoticed. Examination of a drop of fluid squeezed from the chancre reveals that it is teeming with infectious *T. pallidum.* Whether or not treatment is given, the chancre disappears within 2 to 6 weeks, and the victim may mistakenly believe that recovery from the disease has occurred. The organisms resist destruction by the body's defenses via an unknown mechanism, however, and progression of the disease can continue for years.

Many of the manifestations of secondary syphilis are due to immune complexes that form as specific antibodies bind to circulating *T. pallidum.* By this time, the spirochetes have spread throughout the body, and infectious lesions occur on the skin and mucous membranes in various locations, especially in the mouth **(figure 26.14).** Syphilis can be transmitted by kissing during this stage. The secondary stage lasts for weeks to months, sometimes as long as 1 year, and then gradually subsides. About 50% of untreated cases never progress past the secondary stage. After a latent period of from 5 to 20 years or even longer, however, some people with the disease develop tertiary syphilis. ■ **immune complexes, p. 420**

Tertiary syphilis (the third stage of syphilis) represents a hypersensitivity reaction to small numbers of *T. pallidum* that grow and persist in the tissues. In this stage, the patient is no longer infectious. The remaining organisms may be present in almost any part of the body, and the symptoms of tertiary syphilis depend on where the hypersensitivity reactions occur. If they occur in the skin, bones, or other areas not vital to existence, the disease is not life threatening. If, however, they occur within the walls of a major blood vessel such as the aorta, the vessel may become weakened and even rupture, resulting in death. Hypersensitivity reactions to *T. pallidum* in the eyes cause blindness; central nervous system involvement most commonly manifests itself as a stroke. A granulomatous necrotizing mass called a **gumma (figure 26.15),** analogous to the tubercle of tuberculosis, can involve any part of the body. A characteristic pattern of symptoms and signs called **general paresis** develops an average of 20 years after infection. Typical findings include personality change, emotional instability, delusions, hallucinations, memory loss, impaired judgment, abnormalities of the pupils of the eye, and speech defects. ■ **tubercle, p. 515**

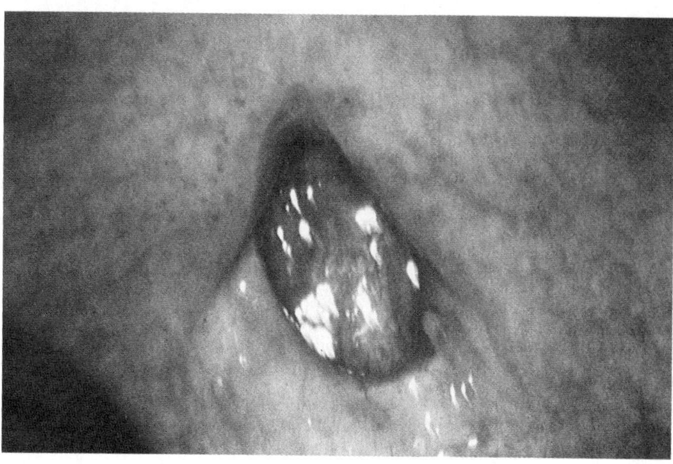

Figure 26.15 A Gumma of Tertiary Syphilis Gummas consist of an inflammatory mass which can perforate, as in this example in the roof of the mouth. Gummas can occur anywhere in the body.

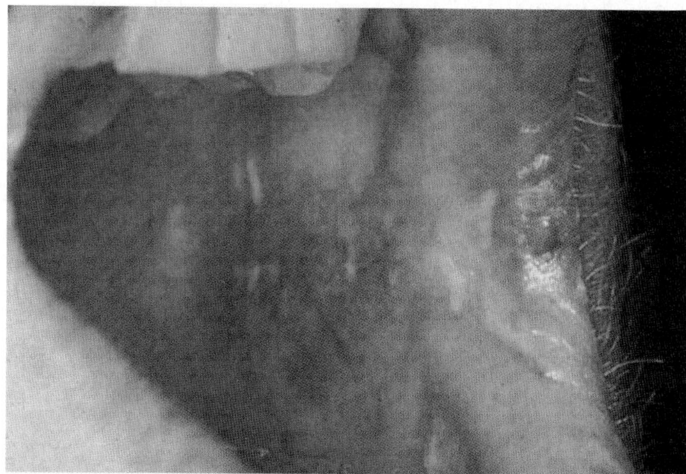

Figure 26.14 Appearance of Mucous Patches in the Mouth of an Individual with Secondary Syphilis These lesions are swarming with *Treponema pallidum* and are highly infectious.

The main characteristics of the three stages of syphilis are summarized in **table 26.10.**

Congenital Syphilis During pregnancy, *T. pallidum* readily crosses the placenta and infects the fetus. This can occur at any stage of pregnancy, but damage to the fetus does not generally occur until the fourth month. Therefore, if the mother's syphilis is diagnosed and treated before the fourth month of pregnancy, the fetus will also be treated and will not develop the disease. Without treatment, the risk to the fetus depends partly on the stage of the mother's infection. Three-fourths or more of the fetuses become infected if the mother has primary or early secondary syphilis; risk decreases with the duration of the mother's infection but is still significant into the latent period. Fetal infections can occur in the absence of any symptoms of syphilis in the mother. About two out

TABLE 26.10 Stages of Syphilis

Stage of Disease	Main Characteristics	Infectious?
Primary	Firm, painless ulcer (hard chancre) at site of infection; lymph node enlargement	Yes
Secondary	Rash, aches, and pains; mucous membrane lesions	Yes
Tertiary	Gummas; damage to large blood vessels, eyes, nervous system; insanity	No

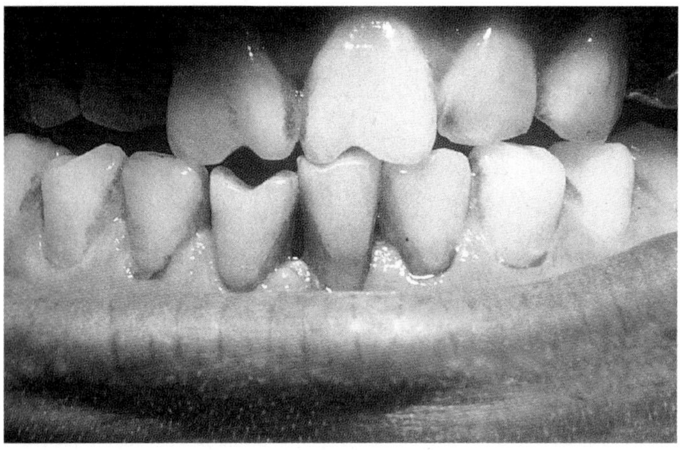

Figure 26.16 Hutchinson Teeth Notice the notched, deformed incisors, a late manifestation of congenital syphilis.

of every five infected fetuses are lost through miscarriage or stillbirth. The remainder are born with congenital syphilis. Although these infants frequently appear normal at the time of birth, some develop secondary syphilis, which is often fatal, within a few weeks. Others develop characteristic deformities of their face, teeth **(figure 26.16),** and other body parts later in childhood.

Epidemiology

There is no animal reservoir. Syphilis is usually transmitted by sexual intercourse, but infection can be contracted from kissing. Elimination of transmission in the United States is within reach, but depends on identifying and treating cases within high-risk groups such as prostitutes, promiscuous homosexual men, and jail inmates, and guarding against reintroduction of the disease from countries where it is still rampant. Effective screening to detect infected individuals can easily be accomplished using a blood test.

Prevention and Treatment

No vaccine against syphilis is currently available. Abstinance, monogamous relationships, condoms, and other safer sex practices decrease the risk of contracting the disease. Prompt identification and treatment of sexual contacts are important in limiting spread of the disease. Primary and secondary syphilis are effectively treated with an antibiotic such as penicillin. Treatment must be continued for a longer period for tertiary syphilis, however, probably because most of the organisms are not actively multiplying.

Table 26.11 summarizes the main features of syphilis.

Chancroid

Chancroid is another bacterial sexually transmitted disease that showed a marked increase in frequency, reaching 5,000 reported cases in 1988, followed by a progressive decline to the lowest frequency in decades. Although widespread in the United States, it is not commonly reported. It is another STD with genital sores that promote the spread of HIV, the cause of AIDS.

Symptoms

Chancroid is characterized by a single or multiple genital sores called **soft chancres (figure 26.17a),** which are painful, unlike the hard chancre of syphilis. The lymph nodes in the groin enlarge (figure 26.17b); they become tender, and sometimes pus-filled.

TABLE 26.11 Syphilis

① *Treponema pallidum* enters the body through a microscopic abrasion or mucous membrane, usually genitalia, mouth, or rectum.

② A chancre develops at site of entry.

③ Organisms multiply locally and spread throughout the body by the bloodstream.

④ Infectious mucous patches and skin rashes of secondary syphilis appear. A fetus will become infected, resulting in miscarriage or a live-born infant with congenital syphilis.

⑤ An asymptomatic latent period occurs. *T. pallidum* disappears from blood, skin, and mucous membranes.

⑥ After months or years, symptoms of tertiary syphilis appear:
- heart and great vessel defects
- gummas
- strokes
- eye abnormalities
- general paresis
- insanity.

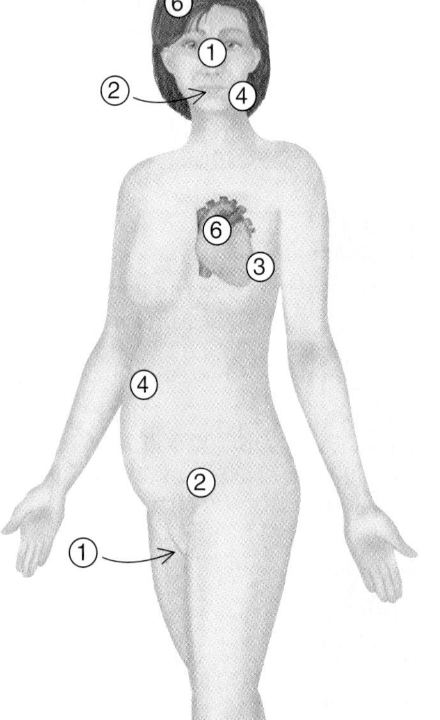

Symptoms	Chancre, fever, rash, stroke, nervous system deterioration; can imitate many other diseases
Incubation period	10 to 90 days
Causative agent	*Treponema pallidum,* a non-culturable spirochete
Pathogenesis	Primary lesion, or chancre, appears at site of inoculation, heals after 2 to 6 weeks; *T. pallidum* invades the blood vessel system and is carried throughout the body, causing fever, rash, mucous membrane lesions; damage to brain, arteries, and peripheral nerves appears years later.
Epidemiology	Sexual contact with infected partner; kissing; transplacental passage.
Prevention and treatment	Monogamous relationships, use of condoms, treatment of sexual contacts, reporting cases. Treatment: penicillin.

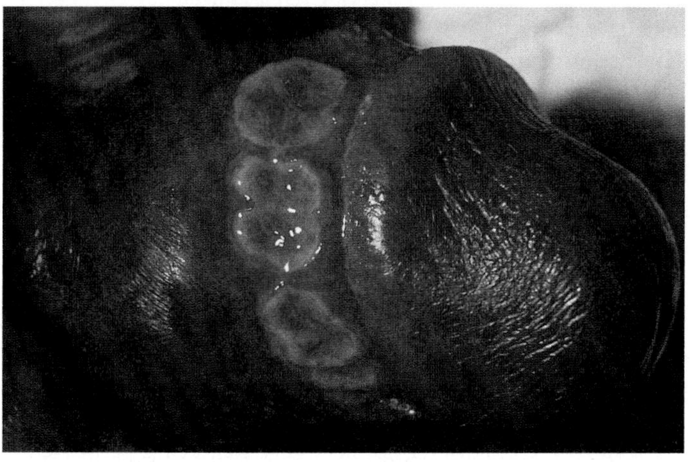

(a)

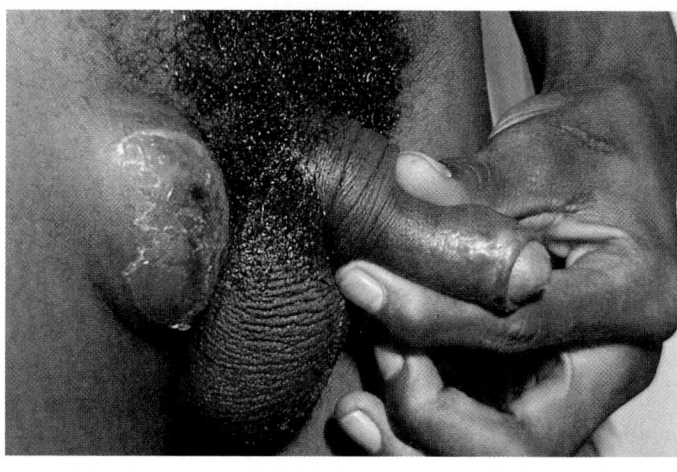

(b)

Figure 26.17 Chancroid (a) Lesions of the penis ("soft chancres"). These ulcerations are soft and painful, in contrast to the chancres of syphilis. **(b)** Swollen groin lymph nodes. The nodes are tender, often pus-filled, and may break open and drain.

Causative Agent

The causative organism of chancroid is *Haemophilus ducreyi,* a small, pleomorphic, Gram-negative rod that requires X-factor, a blood component called hematin, for growth. ■ *Haemophilus,* p. 273

Pathogenesis

Knowledge of the pathogenesis of chancroid is incomplete. Typically, a small pimple appears, presumably at the site of entry of the organisms. After a few days, this ulcerates and gradually enlarges, reaching an inch or more in diameter, often joining other lesions. Organisms that reach the lymph nodes incite an intense inflammatory response, with congregation of neutrophils, release of proteolytic enzymes, and liquefaction of the lymphoid tissue. The pus-filled nodes enlarge, stretch the skin, and rupture through to the surface, discharging their contents. ■ neutrophils, p. 350

Epidemiology

Epidemics in American cities are associated with prostitution; in some tropical countries, chancroid is second only to gonorrhea in prevalence of the sexually transmitted diseases.

Prevention and Treatment

Abstinence from sexual intercourse, monogamous relationships, and use of condoms help prevent chancroid. The disease usually responds well to treatment with the antibiotics erythromycin, azithromycin, or ceftriaxone. Some strains, however, possess R plasmids that code for resistance to multiple antibacterial medications. Effectiveness of therapy is also sharply reduced if the patient has AIDS. As in a number of bacterial infections, body defenses and antibacterial medications work together to defeat the disease.

Table 26.12 gives the main features of chancroid.

MICROCHECK 26.6

Transmission of bacterial STDs usually requires direct human-to-human contact. Unsuspected sexually transmitted infections of pregnant women pose a serious threat to fetuses and newborn babies. Even asymptomatic infections can cause genital tract damage and be transmissible to other people.

- ✓ Can a baby have congenital syphilis without its mother ever having had symptoms of syphilis? Explain.
- ✓ What, if any, is the relationship between chancroid and development of AIDS?
- ✓ Why should scarring of a fallopian tube raise the risks of an ectopic pregnancy?

TABLE 26.12	**Chancroid**
Symptoms	One or more painful, gradually enlarging, soft chancres on or near the genitalia; large, tender regional lymph nodes
Incubation period	3 to 10 days
Causative agent	*Haemophilus ducreyi,* a small, pleomorphic, Gram-negative rod requiring X-factor for growth
Pathogenesis	A small pimple appears first, which ulcerates and gradually enlarges; multiple lesions may coalesce; lymph nodes enlarge, liquefy, and may discharge to the skin surface.
Epidemiology	Sexual transmission. Common in prostitutes; fosters the spread of AIDS.
Prevention and treatment	Abstinence from sex, monogamous relationships; avoidance of sexually promiscuous partners; proper use of condoms. Treatment: several antibacterial medications effective. Resistance can be a problem.

26.7

Viral STDs

Focus Points

- Discuss the distinctive characteristics of genital herpes simplex and human papilloma virus infection.
- Outline the relationship between HIV infection and AIDS.

Viral sexually transmitted diseases are probably as common or more common than the bacterial diseases, but they are incurable. Their effects can be severe and long-lasting. Genital herpes simplex can cause recurrent symptoms for years, papillomavirus infections can lead to cancer, and untreated HIV infections usually result in AIDS.

Genital Herpes Simplex

Genital herpes simplex is among the top three or four most common sexually transmitted diseases. An estimated 45 million Americans are infected with the causative virus and about 500,000 new cases occur each year.

Symptoms

Symptoms of genital herpes simplex begin 2 to 20 days (usually about a week) after exposure, with genital itching, burning, and in women, often severe pain. Infection of the urethra may imitate the symptoms of a bladder infection. Clusters of small, red bumps appear on the genitalia, which then become blisters surrounded by redness **(figure 26.18).** The blisters break in 3 to 5 days, leaving an ulcerated area. The ulcers slowly dry and become crusted and then heal without a scar. Symptoms are the worst in the first 1 to 2 weeks and disappear within 3 weeks. They recur in an irregular pattern; about 75% of patients will have at least one recurrence within a year, and the average is about four. The recurrent symptoms are usually not quite as severe as those of the first episode, and recurrences generally decrease in frequency with time. Some individuals may have recurrences for life.

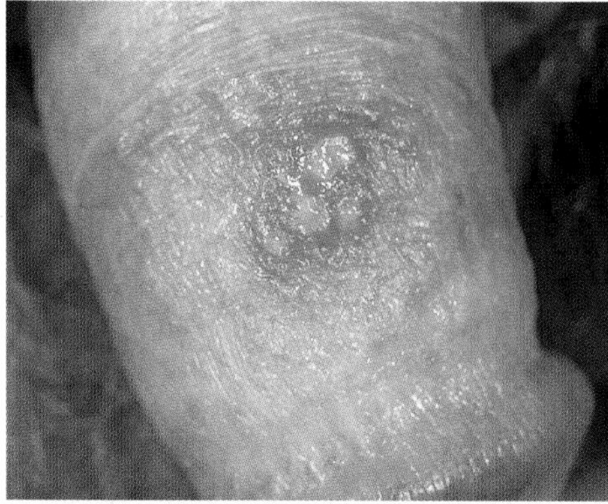

Figure 26.18 Genital Herpes Simplex on the Shaft of the Penis The lesions typically begin as a group of small vesicles surrounded by redness. Within 3 to 5 days, they break, leaving the small superficial ulcers as shown. The lesions can be very painful, especially in women.

Causative Agents

The causative agent of genital herpes is usually herpes simplex virus type 2 (HSV-2), a medium-sized, enveloped, double-stranded DNA virus. On electron micrographs it looks like herpes simplex virus type 1 (HSV-1), the cause of "cold sores" ("fever blisters," herpes simplex labialis), and the genomes of the two viruses are about 50% homologous. This disease, like cold sores, often recurs because the virus becomes latent. Either virus can infect the mouth and the genitalia, but the type 2 virus causes more severe genital lesions whose frequency of recurrence is greater than that of the type 1 virus. During initial infection and for as long as 1 month after, the virus is found in genital secretions. During recurrence, the virus is usually present in large numbers for less than a week. ■ HSV-1, p. 592

Pathogenesis

Lesions begin with infection of a group of epithelial cells that lyse following viral replication, creating small, fluid-filled blisters (vesicles) containing large numbers of infectious virions. Rupture of the vesicles produces painful ulcerations. Latency of the disease is incompletely understood. The viral DNA exists within nerve cells in a circular, non-infectious form during times when there are no symptoms. In this state, only a single gene is expressed and it encodes for a small segment of RNA called a microRNA (miRNA). HSV miRNA is probably responsible for latency of the virus. At times, however, the entire viral chromosome can be transcribed and complete infectious virions replicated. These reinfect the area supplied by the nerve and cause a recurrence. The mechanisms by which the latent infection is maintained or reactivated are not known in detail, but they probably depend on cellular immunity. ■ latent infections, p. 330 ■ miRNA, p. 180

Genital herpes can pose a serious risk to newborn babies. If the mother has a primary infection near the time of delivery, the baby has about a one in three risk of acquiring the infection. The baby often dies from overwhelming infection or is permanently disabled by it. To prevent contact with the infected birth canal, these babies must be delivered by cesarian section. If the mother has recurrent disease, the risk to the baby is very low, presumably because transplacental anti-HSV antibody from the mother protects the baby, but physicians will often do a cesarian section to minimize the risk.

Epidemiology

There are no animal reservoirs. The virus can survive for short periods on fomites or in bathwater, but non-sexual transmission is rare. Sexual transmission of HSV is most likely to occur during the first few days of symptomatic disease, but it can happen in the absence of symptoms or signs of the disease. Once infected, an individual poses a life-long risk of transmitting the virus to another person. Herpes simplex, like other ulcerating genital diseases, promotes the spread of AIDS by increasing the risk of HIV infection.

Prevention and Treatment

Avoidance of sexual intercourse during active symptoms and the use of condoms with a spermicide reduce, but do not eliminate, the chance of transmission. Spermicidal jellies and creams can inactivate herpes simplex viruses. There is no cure for genital herpes, although medications such as acyclovir and famciclovir can decrease the severity of the first attack and the incidence of recurrences.

The main features of genital herpes simplex are presented in **table 26.13.**

Papillomavirus STDs: Genital Warts and Cervical Cancer

Sexually transmitted human papillomaviruses (HPVs) are among the most common of the sexually transmitted disease agents, infecting an estimated 40 million Americans. Although they cannot be cultivated in the laboratory, their presence in human tissues can be studied by using nucleic acid probes. Some HPVs are responsible for **papillomas**—warty growths of the external and internal genitalia; others cause non-warty lesions of mucosal surfaces such as the uterine cervix and are a major factor in the development of cervical cancer. Invasive cervical cancer strikes about 12,500 Americans each year and kills more than 4,500 of them. ■ nucleic acid probes, p. 227

TABLE 26.13 Genital Herpes Simplex

Symptoms	Itching, burning pain at the site of infection, painful urination, tiny blisters with underlying redness. The blisters break, leaving a painful superficial ulcer, which heals without scarring. Recurrences are common.
Incubation period	Usually 1 week (range, 2 to 20 days)
Causative agent	Usually herpes simplex virus, type 2. The cold sore virus, herpes simplex type 1, can also be responsible. Herpesviruses are enveloped and contain double-stranded DNA.
Pathogenesis	Lysis of infected epithelial cells results in fluid-filled blisters containing infectious virions. Rupture of these vesicles causes a painful ulceration. The acute infection is controlled by body defenses; genome persists within nerve cells in a non-infectious form beyond the reach of body defenses. Replication of infectious virions can occur and cause recurrent symptoms in the area supplied by the nerve. Newborn infants can contract fatal generalized herpetic infection if their mother has a primary infection at the time of delivery.
Epidemiology	No animal reservoirs. Transmission by sexual intercourse, oral-genital contact. Transmission risk greatest first few days of active disease. Transmission can occur in the absence of symptoms. Herpes simplex increases the risk of contracting HIV.
Prevention and treatment	Abstinence, monogamy, and condoms help prevent transmission. Medications help prevent recurrences, shorten duration of symptoms. No cure.

Symptoms

Warts are the most easily recognized manifestation of infection, with about 750,000 new cases in the United States each year. They often appear on the head or shaft of the penis **(figure 26.19),** at the vaginal opening, or around the anus. Occasionally, the warts become inflamed or bleed. These warts rarely, if ever, become cancerous but they can have a large emotional impact because of the embarrassment of having such an obvious STD. Important precancerous lesions of the cervix are generally asymptomatic and can only be detected by vaginal examination.

Causative Agents

Human papillomaviruses are small, non-enveloped, double-stranded DNA viruses of the papovavirus family. There are more than 100 types of HPV based on DNA homology, and at least 30 of them are transmitted by sexual contact. Approximately 15 types are strongly associated with cancer of the cervix and other cancers. The different papillomavirus types infect different kinds of epithelium. For example, HPVs that cause the common warts of the hands and plantar warts of the feet generally do not infect the genitalia. The HPVs that cause genital warts may infect the cervix, but they do not generally predispose to cancer.

Pathogenesis

HPVs are thought to enter and infect the deeper layers of epithelium through microscopic abrasions. A latent infection, without

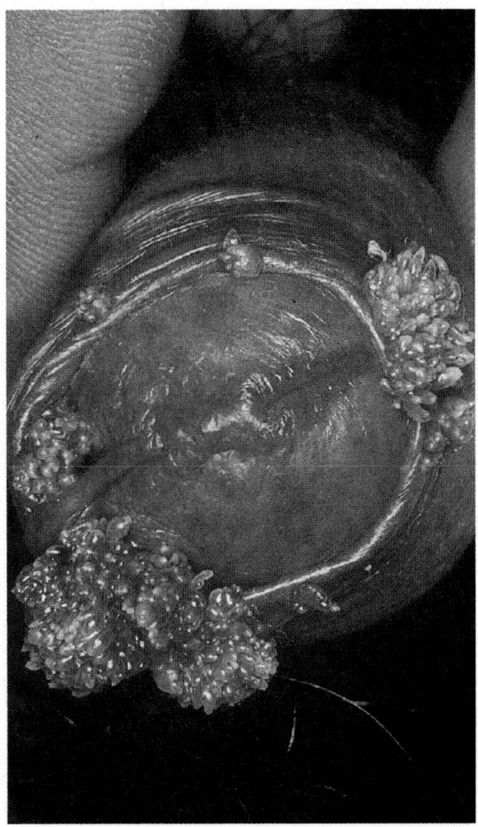

Figure 26.19 Genital Warts on the Penis, a Manifestation of Human Papillomavirus Infection

warts or microscopic lesions, is the most common result of infection. Latent infection can last for years, but studies in college-age women reveal that about 90% of HPV infections are eliminated by body defenses within 2 years. Viral replication and release are intermittent and require that the deeper layer of epithelial cells begins the normal process of changing to surface layer cells. Different kinds of lesions develop depending on the virus type and the location of the infection. Lesions can be flat, raised or unraised, cauliflower-like, or hidden within the epithelium.

The mechanism giving rise to warts is unknown. The viral genome exists in infected cells as extrachromosomal, closed DNA circles. Warts usually appear about 3 months after infection (range, 3 weeks to 8 months). Following removal or destruction of warts, HPV persists in surrounding normal-appearing epithelium and can give rise to additional warts. Warts can occasionally partly obstruct the urethra or, if very large, the birth canal. Newborn infants can become infected with HPV at birth and develop warts that obstruct their respiratory tract, a serious condition that occurs in less than one out of every 100,000 births.

Worldwide, cancer of the cervix is second in frequency only to breast cancer among the malignant tumors of women. Most of these cancers are associated with certain HPV types. Cancer-associated HPV types differ from wart-causing HPV types in that the former can integrate into the chromosome of a host cell and code for a protein that permits excessive cell growth. These oncogenic types of HPV tend to cause infections that persist longer than other HPV types, and to cause precancerous lesions. A cancer-associated HPV is present in almost 100% of cervical cancers, but only a small percentage of infections by these viruses result in cancer. This indicates that other unknown factors must be present for cancer to develop. Cervical cancer-associated HPVs have also been found in vaginal cancers, cancers of the penis, and in anal cancers of men and women who engage in receptive anal intercourse. ■ oncogenes, p. 333

Epidemiology

HPVs are readily spread by sexual intercourse; a single sexual exposure to an infected person transfers the infection 60% of the time. Asymptomatic individuals infected with HPV can transfer the virus to others. Forty-six percent of women university students seeking routine gynecological examinations were found to be infected with HPV in one study. HPV infection is the most common reason for an abnormal Papanicolaou (Pap) test in teenage women (see next section). A history of having multiple sex partners is the most important risk factor for acquiring genital HPV infection. Warts can develop in the mouth from HPV infection transferred by oral sex.

Prevention and Treatment

HPV-infected individuals are advised to use condoms to help decrease the chance of transmission. Women should have a **Papanicolaou smear** every 12 months. The "Pap test" consists of removing cells from the cervix with a small brush or spatula, staining them on a microscope slide, and examining them for any abnormal cells **(figure 26.20).** Cervical cancer is generally preceded by precancerous lesions—areas of abnormal cell growth that can be detected by the Papanicolaou test. The abnormal growth can be removed, thereby preventing development of a cancer. Currently available vaccines are designed to protect against infection by HPV types 16 and 18, which are responsible for about 70% of cancers, and types 6 and 11, which are responsible for more than 90% of genital warts. The vaccine is given in a three-dose series, and is recommended for girls and young women ages 9 to 26. A vaccine will probably soon be approved for men. Warts on the skin can be removed by laser treatment or freezing with liquid nitrogen, but these measures do not cure the infection. Imiquimod (Aldara Cream), an immune modifier, can be effective for anal and genital warts.

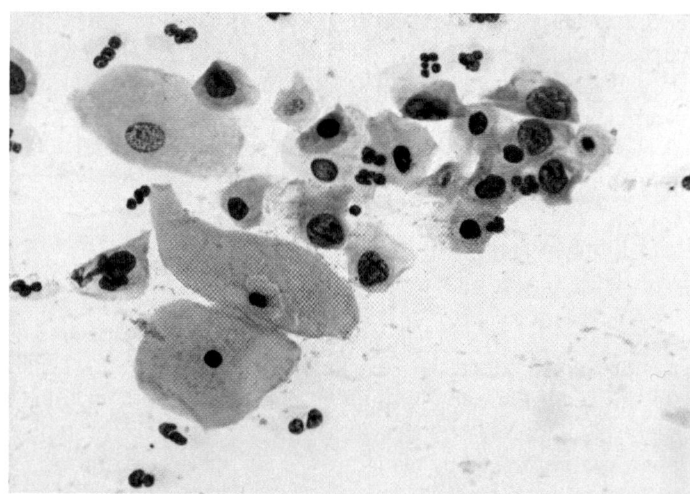

Figure 26.20 Abnormal Papanicolaou Smear The pink and blue objects are squamous epithelial cells; abnormalities include doubling of the nuclei and a clear area around them. Most abnormal smears in young women are due to human papillomavirus infection, that when persistent are an important factor in the development of cancer of the cervix.

Table 26.14 summarizes some important features of papillomavirus STDs.

AIDS

AIDS, acquired immunodeficiency syndrome, unquestionably the most important sexually transmitted disease of the twentieth century, is covered briefly here and more fully in chapter 29. The AIDS epidemic was first recognized in the United States in 1981. Over the first decade and a half of the epidemic, 500,000 Americans developed AIDS and 300,000 died of the disease. With major advances in prevention and treatment of the disease, and expenditures of huge amounts of money, the epidemic shows signs of leveling off. Nevertheless, in 2007, an estimated 33.2 million people around the world were living with human immunodeficiency virus (HIV) infection, the causative agent of AIDS, and more than 2 million died of AIDS. There were an estimated 1.7 million new infections in sub-Saharan Africa alone. The advances in prevention and treatment of AIDS are not uniformly available to many of those around the world who suffer because of the disease. ■ AIDS, p. 698

Symptoms

AIDS is the end stage of what is more properly called **HIV disease.** Six days to 6 weeks after contracting the virus, some individuals develop a fever, head and muscle aches, enlarged lymph nodes, and a rash. These symptoms are often mild, attributed to the flu, and go

TABLE 26.14 Papillomavirus STDs

Symptoms	Many have no symptoms. Warts of the external and internal genitalia the most common symptom.
Incubation period	Usually 3 months (range, 3 weeks to 8 months)
Causative agents	Human papillomaviruses, many types, small, non-enveloped, double-stranded DNA viruses of the papovavirus family. Different types infect different tissues and produce different lesions.
Pathogenesis	Virus enters epithelium through abrasions, infects deep layer of epithelium; establishes latency; cycles of replication occur when host cell begins maturation; cancer-associated viral types can integrate into the host cell chromosome and can cause precancerous lesions.
Epidemiology	Asymptomatic individuals can transmit the disease; 60% transmission with a single sexual contact; multiple sex partners the greatest risk factor; warts can be transmitted to the mouth with oral sex, and to newborn babies.
Prevention and treatment	Latex condoms advised to minimize transmission and avoidance of sexual contact with those having multiple sex partners. Pap tests at least yearly for sexually active women. Wart removal by multiple techniques, does not cure the infection. Imiquimod useful in treating multiple warts about anus and external genitalia.

away by themselves. Many individuals have no symptoms at all. The typical case of HIV disease then smolders unnoticed for almost 10 years before immunodeficiency develops, as evidenced by certain malignancies and unusual microbial infections that often attack the lungs, intestines, skin, eyes, or the central nervous system.

Causative Agents

Most AIDS cases are caused by human immunodeficiency virus type 1 (HIV-1), an enveloped, single-stranded, RNA virus of the retrovirus family. Each virion contains reverse transcriptase and two copies of the viral genome. In parts of western Africa, another retrovirus, HIV-2, is a common cause of AIDS; it is rarely a cause in the United States. ■ retroviruses, pp. 323, 332

Pathogenesis

HIV can attack a variety of types of cells in the human body, but most critically a subset of lymphocytes—the helper T (T_H) cells. These cells possess the CD4 surface protein to which the virus attaches, and they are referred to as $CD4^+$ cells. In order for the virus to enter the cell, it must also attach to certain cytokine receptors on the cell surface. Following entry of the virus, a DNA copy is made using reverse transcriptase. This copy integrates itself into the cell's genome and hides there, inaccessible to antiviral chemotherapy. If the cell becomes activated, however, the virus leaves the cell genome, replicates, and destroys the cell, releasing virions to infect additional cells. Despite the body's ability to replace hundreds of billions of $CD4^+$ lymphocytes, the number of these lymphocytes slowly declines over many months or years. Infection of macrophages contributes to the decline in cellular immunity. Macrophages, also vitally important to the immune system, are also $CD4^+$ cells. Although they are generally not killed, their function is impaired both by the viral infection and the lack of interaction with T_H lymphocytes. Eventually, the immune system becomes so impaired it can no longer respond to infections or cancers. ■ helper T cells, p. 368 ■ macrophages, p. 350

Epidemiology

HIV is present in blood, semen, and vaginal secretions in symptomatic and asymptomatic infections. HIV disease spreads mainly by sexual intercourse, sharing hypodermic needles, or from mother to newborn. The virus is not highly contagious, and most transmissions could be halted by simple changes in human behavior.

Prevention and Treatment

One of the most important developments in the prevention of AIDS has been the finding that transmission from mother to newborn can be interrupted by chemotherapy in about two-thirds of the cases. Supplying injected-drug abusers with sterile needles and syringes in exchange for used ones decreases transmissions of the virus without encouraging addiction. Specifically designed educational programs targeting certain groups of people have also shown short term success in decreasing risky sexual practices. Lastly, identification and treatment of other STDs decreases the risk of contracting HIV disease, as does consistent use of latex condoms.

Treatment is designed to block replication of HIV, but it does not affect viral nucleic acid already integrated in the host genome. Four kinds of medications are currently available—(1) those that block reverse transcriptase activity; (2) those that act against viral protease; (3) those that block viral integrase; and (4) those that interfere with attachment to host cells. The first three block enzymes essential at different stages of viral replication. The fourth type is represented by a drug that interferes with HIVattachment to the cytokine receptor CCR5. Viral mutants resistant to a single medication, however, quickly arise. To help prevent the selection of resistant variants, a "cocktail" of several medications, each acting in a unique way is given. In many cases this therapy can clear the patient's blood of detectable viral nucleic acid (but does not affect HIV integrated in the genome of cells), halt the progress of the disease, and even allow partial recovery of immune function. Therapy is very expensive and the medications often have serious side effects, two features that preclude use of the regimen in many of the world's HIV disease sufferers. While not curative, chemotherapy can markedly prolong and improve the quality of life. ■ HIV replication, p. 703

Each person has a role to play in controlling the AIDS pandemic by studying the disease and doing his or her part in preventing spread from one person to another. About 25% of the estimated 900,000 people in the United States infected with HIV are unaware of their infection. Individuals who, since the 1980s, have engaged in injected drug use or risky sexual behavior, and anyone having had unprotected sexual intercourse with them, should consider obtaining a blood or saliva test to rule out HIV infection. Persons who learn that their HIV test is positive can receive optimum treatment sooner and also prevent transmission of the virus to others.

Table 26.15 presents the main features of HIV disease and AIDS.

TABLE 26.15	HIV Disease and AIDS
Symptoms	No symptoms, or "flu"-like symptoms early in the illness; an asymptomatic period typically lasting years; symptoms of lung, intestine, skin, eyes, brain, and other infections, and certain cancers
Incubation period	About 6 days to 6 weeks for "flu"-like symptoms; many months or years for cancers and unusual infections
Causative agents	Generally human immunodeficiency virus, type 1 (HIV-1)
Pathogenesis	The virus infects CD4+ lymphocytes and macrophages, thereby slowly destroying the ability of the immune system to fight infections and cancers.
Epidemiology	HIV present in blood, semen, and vaginal secretions in symptomatic and asymptomatic infections; spread usually by sexual intercourse, sharing of needles by injected-drug abusers, and from mother to infant at childbirth. Other STDs foster transmission.
Prevention and treatment	Abstinence from sexual intercourse and drug abuse; monogamy; consistent use of latex condoms; avoidance of sexual contact with injected-drug abusers, those with multiple partners or history of STDs. Anti-HIV medication for expectant mothers and their newborn infants. Treatment: reverse transcriptase and protease inhibitors in combination.

MICROCHECK 26.7

Viral STDs are at least as common as bacterial STDs, but so far they are incurable. Genital herpes simplex is widespread and, like most other STDs, can be transmitted in the absence of symptoms. Human papillomaviruses cause genital warts; some play an important role in cancer of the cervix and probably cancers of the vagina, penis, and rectum. HIV disease generally ends in AIDS, but treatment can markedly improve longevity and quality of life.

- ✓ What are the possible consequences of sexually transmitted papillomavirus infections for men, women, and babies?
- ✓ What kinds of sex partners present a high risk of transmitting human immunodeficiency virus?
- ✓ Why should a person be concerned about genital herpes simplex—is it not just a cold sore on the genitals? Explain.

26.8 Protozoal STDs

Focus Points

- List the distinctive characteristics of the organism that causes trichomoniasis.
- Outline the epidemiology of trichomoniasis.

A number of protozoan diseases are frequently transmitted sexually even though they do not infect the genital system. For example, giardiasis and cryptosporidiosis are intestinal diseases transmitted by the fecal-oral route in individuals who engage in oral-genital and -anal contact as part of sexual activity. By contrast, trichomoniasis is a sexually transmitted protozoan disease that involves the genital system. ■ protozoa, p. 285 ■ cryptosporidiosis, p. 610

"Trich" (Trichomoniasis)

Trichomoniasis is caused by *Trichomonas vaginalis*. It ranks third after bacterial vaginosis and candidiasis among the diseases that commonly cause vaginal symptoms. Infected men usually are asymptomatic carriers of the organism. An estimated 5 million Americans contract this STD each year. An African study indicated that trichomoniasis can cause a two to three fold increase in HIV transmission.

Symptoms

Most symptomatic *T. vaginalis* infections occur in women and are characterized by itching of the vulva and inner thighs; itching and burning of the vagina; and a frothy, sometimes malodorous, yellowish-green vaginal discharge, at times accompanied by burning discomfort with urination. Although most infected men are free of symptoms, some have penile discharge, burning pain with urination, painful testes, or a tender prostate gland.

Causative Agent

Trichomonas vaginalis is a protozoan measuring about 10 by 30 µm. It has four anterior flagella and a posterior flagellum attached to an undulating membrane **(figure 26.21).** It also has a slender, posteriorly protruding, rigid rod called an axostyle. *Trichomonas vaginalis*

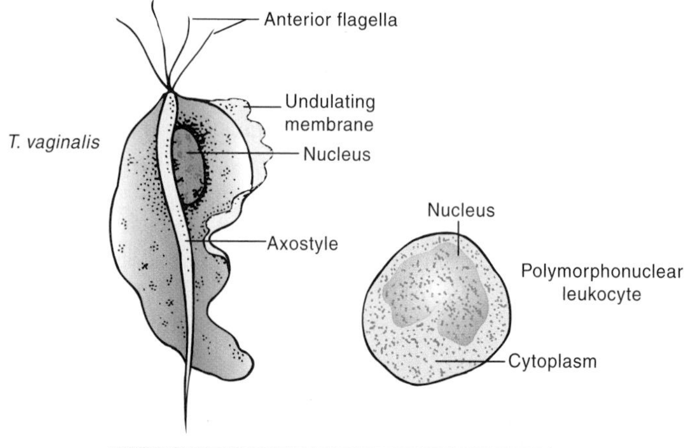

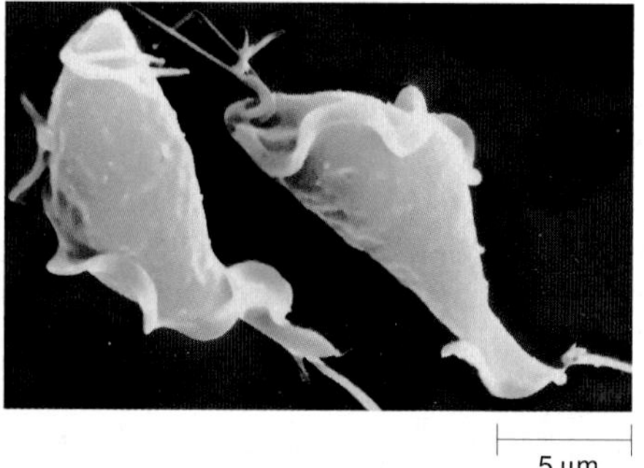

Figure 26.21 *Trichomonas vaginalis*, a Common Sexually Transmitted Cause of Vaginitis Diagram shows the relative sizes of the protozoan and a polymorphonuclear leukocyte. Although it is a flagellated protozoan, *T. vaginalis* often exhibits pseudopodia.

has an unmistakable, jerky motility on microscopic examination of the vaginal drainage. Unlike most other pathogenic protozoa, *T. vaginalis* lacks a cyst form to aid its survival in the environment away from the host's body. The organism is a eukaryote, but it lacks mitochondria. *Trichomonas vaginalis* has interesting cytoplasmic organelles called hydrogenosomes. Enzymes within these double membrane-bounded organelles remove the carboxyl group (COOH) from pyruvate and transfer electrons to hydrogen ions, thereby producing hydrogen gas. ■ mitochondria, p. 76

Pathogenesis

The pathogenesis of trichomoniasis is not fully understood. The vulva and walls of the vagina are diffusely red and slightly swollen. Often there are scattered pinpoint hemorrhages. These changes have been attributed to mechanical trauma by the moving protozoa, but toxins or exoenzymes have not been ruled out. The frothy discharge is probably due to gas produced by the organisms.

Epidemiology

Trichomonas vaginalis is distributed worldwide as a human parasite and has no other reservoirs. Because it lacks a cyst form, it is easily killed by drying. Therefore, transmission is usually by sexual contact. The organism can survive for a time on moist objects such as towels and bathtubs, so that it can occasionally be transmitted non-sexually. Nevertheless, *T. vaginalis* infections in children should at least raise the question of sexual abuse and possible exposure to other sexually transmitted diseases. Newborn infants can contract the infection from infected mothers at birth. There is a high percentage of asymptomatic carriers, especially among men, and this fosters transmission of the disease. Infection rates are highest in men and women with multiple sex partners.

Prevention and Treatment

Transmission is prevented by abstinence, monogamy, and the use of condoms. Most strains of *T. vaginalis* respond quickly to metronidazole treatment, but a few are resistant.

Table 26.16 gives the main features of trichomoniasis.

TABLE 26.16 Trichomoniasis

Symptoms	Women: itching, burning, swelling, vaginal redness, frothy, sometimes malodorous, yellow-green discharge, and burning on urination. Men: discharge from penis, burning on urination, painful testes, tender prostate. Many women, most men asymptomatic
Incubation period	4 to 20 days
Causative agent	*Trichomonas vaginalis,* a protozoan with four anterior flagella, and a posterior flagellum attached to an undulating membrane; a rigid rodlike structure called an axostyle protrudes posteriorly; unmistakable jerky motility; no mitochondria; hydrogenosomes are present
Pathogenesis	Unexplained. Inflammatory changes and pinpoint hemorrhages suggest mechanical trauma from the motile organisms.
Epidemiology	Worldwide distribution; asymptomatic carriers foster spread; easily killed by drying due to lack of cyst form, transmission by intimate contact; high rate of infection with multiple sex partners. Newborn infants of infected mothers can acquire the infection at birth.
Prevention and treatment	Abstinence, monogamy, and consistent use of condoms prevent the disease. Treatment: metronidazole.

MICROCHECK 26.8

Trichomoniasis is a protozoan STD that involves the genital tract (other protozoan STDs involve the intestine). The causative agent is a peculiar, flagellated protozoan that does not form cysts and lacks mitochondria. Organelles called hydrogenosomes produce hydrogen gas from pyruvate.

✓ What significance does the lack of a cyst form of *Trichomonas vaginalis* have on the epidemiology of trichomoniasis?

✓ Why is there a strong relationship between being "easily killed by drying" and "transmission usually by sexual contact"?

FUTURE CHALLENGES

Getting Control of Sexually Transmitted Diseases

Few problems are as complicated as getting control of sexually transmitted diseases because of the psychological, cultural, religious, and economic factors that are involved—factors that vary from one population and culture to another. Gaining control means focusing on diagnosis, interruption of transmission, education, and treatment. The challenge is to develop innovative approaches in each of these areas and to apply them worldwide on a sustained basis.

In the area of diagnosis, the challenge is to use molecular biological techniques to identify and produce specific antigens and antibodies that can be used to quickly identify causative agents and determine the extent of their spread. The development of nucleic acid probes to identify genes coding for resistance to therapeutic agents assists in choosing prompt treatment.

Interruption of transmission by the consistent use of condoms can be highly effective, but in practice their use presents a number of problems. When infected college students are asked why they did not use a condom, they often state that "it takes away the romance," "it decreases pleasurable sensation," or "that's for sissies." Also, many individuals are allergic to latex, and the alternatives, perhaps except for polyurethane, are unreliable for preventing disease transmission. Worldwide, as many as one out of three condoms is defective, and the poorest countries cannot even afford them. The challenge is to develop cheaper, more reliable, and more acceptable barrier methods, perhaps using new space-age polymers or antimicrobial vaginal gels.

Effective vaccines could block the spread of STDs, but their development has been extremely problematic and remains a long-term goal. Sequencing the genomes of the causative agents is helping to identify appropriate antigens to include in vaccines, but using them to stimulate a protective immune response is a continuing challenge.

Educational efforts focusing on groups at high risk for STDs have had mixed success. Education can be a useful tool for gaining control of STDs, and the challenge is to better understand the reasons for its failures.

SUMMARY

26.1 Anatomy and Physiology

The genitourinary system is one of the portals of entry for pathogens to invade the body.

The Urinary System (figure 26.1)
The kidneys, ureters, bladder, and urethra compose the urinary system. Frequent, complete emptying is an important bladder defense mechanism against infection. Infections occur more frequently in women than in men because of the shortness of the female urethra and its closeness to openings of the genital and intestinal tracts.

The Genital System (figure 26.2)
The fallopian tube, which is open on both ends, provides a passageway for infection to enter the abdominal cavity. The uterine cervix is a frequent site of infection and a place where cancer can develop. The vagina is a portal of entry for a number of infections. In men, the prostate can enlarge and partially obstruct urinary flow.

26.2 Normal Microbiota of the Genitourinary System

The distal urethra is inhabited by various microorganisms, sometimes including potential pathogens. The normal vaginal microbiota is influenced by estrogen hormones, which cause deposition of glycogen in the cells lining the vagina. Vaginal lactobacilli release lactic acid and hydrogen peroxide when they metabolize the glycogen, making the vagina more resistant to colonization by pathogens.

26.3 Urinary System Infections

Any condition that impairs normal bladder emptying increases the risk of infection. The urinary system usually becomes infected by organisms ascending from the urethra, but it can also be infected from the bloodstream.

Bacterial Cystitis (table 26.1)
Urine is a nutritious medium for many bacteria. Kidney infection—**pyelonephritis**—may complicate untreated bladder infection when pathogens ascend through the ureters and involve the kidneys. Most urinary tract infections in healthy people are caused by *Escherichia coli* or other enterobacteria from the person's own normal intestinal flora. Nosocomial urinary infections are common and caused by *Pseudomonas aeruginosa, Serratia marcescens,* and *Enterococcus faecalis*—organisms that are generally resistant to many antibiotics.

Leptospirosis (table 26.2)
In leptospirosis, the urinary system is infected by *Leptospira interrogans* from the bloodstream (figure 26.3). This biphasic illness causes fever, bloodshot eyes, and pain in the septicemic phase. Improvement occurs, then recurrent symptoms emerge, with damage to multiple organs during the immune phase. Many species of animals are chronically infected with *L. interrogans* and excrete the organism in their urine.

26.4 Genital System Diseases

Bacterial Vaginosis (table 26.3, figure 26.4)
Bacterial vaginosis is the most common cause of vaginal symptoms, including a gray-white discharge from the vagina and a pungent fishy odor; there is no inflammation. The causative agent or agents are unknown.

Vulvovaginal Candidiasis (table 26.4)
Vulvovaginal candidiasis is second among causes of vaginal disorders. Symptoms include itching, burning, vulvar redness and swelling, and a thick, white discharge. The causative agent is a yeast, *Candida albicans,* that is commonly part of the normal vaginal flora (figure 26.5). Antibacterial treatment, uncontrolled diabetes, and oral contraceptives are predisposing factors, but in most cases no such factor can be identified.

Staphylococcal Toxic Shock Syndrome (table 26.5)
First described in children, it became widely known with a 1980 epidemic in menstruating women who used a certain kind of vaginal tampon that has since removed from the market (figure 26.6). Symptoms include sudden fever, headache, muscle aches, bloodshot eyes, vomiting, diarrhea, a sunburn-like rash that later peels, and confusion. The blood pressure drops, and without treatment, kidney failure and death may occur.

26.5 Sexually Transmitted Diseases: Scope of the Problem (tables 26.6, 26.7)

In the United States, 15 million new sexually transmitted infections occur each year, including 3 million in teenagers. Simple measures exist for controlling STDs: abstinence from sexual intercourse, a monogamous relationship with an uninfected person, and consistent use of latex or polyurethane condoms (figure 26.7).

26.6 Bacterial STDs

Most of the bacteria that cause STDs survive poorly in the environment and require intimate contact for transmission.

Gonorrhea (table 26.8)
Gonorrhea, caused by *Neisseria gonorrhoeae,* has been generally declining in incidence, but it is still one of the most commonly reported bacterial diseases (figures 26.9, 26.10). Men usually develop painful urination and thick pus draining from the urethra; women may have similar symptoms, but they tend to be milder and are often overlooked (figure 26.8). Expression of different surface antigens allows attachment of *N. gonorrhoeae* to different types of cells, and frustrates development of a vaccine. Inflammatory reaction to the infection causes scarring, which can partially obstruct the urethra or cause sterility in men and women.

Chlamydial Genital System Infections (table 26.9, figure 26.11)
Chlamydial genital infections, caused by *Chlamydia trachomatis,* are reported more often than any other bacterial disease. Symptoms and complications of chlamydial infections are very similar to those of gonorrhea, but milder; asymptomatic infections are common and readily transmitted. Sexually active individuals should be tested at least once a year so that transmission can be halted and complications prevented.

Syphilis (tables 26.10, 26.11)
Syphilis, caused by *Treponema pallidum,* is called "the great imitator" because its many manifestations can resemble other disease (figure 26.13). Primary syphilis is noted by a painless, firm ulceration called a **hard chancre;** the organisms multiply and spread throughout the body (figure 26.12). In secondary syphilis, skin and mucous membranes show lesions that teem with the organisms; a latent period of months or years separates secondary from tertiary syphilis (figure 26.14). Tertiary syphilis is not contagious. It is manifest mainly by damage to the eyes and the cardiovascular and central nervous systems. An inflammatory, necrotizing mass called a **gumma** can involve any part of the body (figure 26.15). Syphilis in pregnant women can spread across the placenta to involve the fetus, resulting in congenital syphilis (figure 26.16).

Chancroid (table 26.12)
Chancroid is another widespread bacterial STD, but it is not commonly reported because of difficulties in making a bacterial diagnosis. Epidemics in the United States have been associated with prostitution. Caused by *Haemophilus ducreyi*, chancroid is characterized by single or multiple

soft, tender genital ulcers, and enlarged, painful groin lymph nodes (figure 26.17). Some strains of *H. ducreyi* possess R factors, making treatment more difficult.

26.7 Viral STDs

Viral STDs are at least as common as bacterial STDs, but they are not yet curable.

Genital Herpes Simplex (table 26.13, figure 26.18)

It is a very common disease, important because of the discomfort and emotional trauma it causes, its potential for causing death in newborn infants, its association with cancer of the cervix, and the increased risk it poses of transmitting HIV infection and AIDS. Symptoms may include a group of vesicles with itching, burning, or painful sensations; these break, leaving a tender, superficial ulcer. Local lymph nodes enlarge. Many infections have few or no symptoms; some have painful recurrences. The virus establishes a latent infection in sensory nerves and cannot be cured. It can be transmitted in the absence of symptoms, but the risk is greatest when lesions are present.

Papillomavirus STDs: Genital Warts and Cervical Cancer (table 26.14)

Papillomavirus STDs are probably more prevalent than any other kind of STD; HPVs are the main cause of abnormal Pap smears in young women (figure 26.20). They are manifest in two ways—as warts on or near the genitalia (figure 26.19) and as precancerous lesions. The latter are asymptomatic and can only be detected by medical examination. The causative agents, human papillomaviruses, are small DNA viruses that have not been cultivated in the laboratory.

AIDS (table 26.15)

AIDS is the end stage of disease caused by human immunodeficiency virus (HIV). Worldwide, millions of new HIV infections and deaths from AIDS occur each year. Advances in treatment, available to HIV disease victims in some countries, are not available to many of the world's HIV victims. HIV disease is usually first manifest as a "flu"-like illness that develops about 6 days to 6 weeks after an individual contracts the virus. An asymptomatic interval follows that typically lasts almost 10 years, during which the immune system is slowly and progressively destroyed. Unusual cancers and infectious diseases then herald the onset of AIDS. No vaccine or medical cure is yet available, but spread of infection can be slowed significantly by consistent use of condoms and employment of sterile needles by injected-drug abusers. A marked reduction in mother-to-newborn transmission can be achieved with medication.

26.8 Protozoal STDs

Intestinal protozoan disease such as giardiasis and cryptosporidiosis are transmitted by the fecal-oral route in those individuals who engage in oral-genital and -anal contact as part of sexual activity.

"Trich" (Trichomoniasis) (table 26.16)

Symptoms include itching, burning, swelling, and redness of the vagina, with frothy, sometimes smelly, yellow-green discharge, and burning on urination. Men may have discharge from the penis, burning on urination, sometimes accompanied by painful testes and a tender prostate gland. Many women and most men are asymptomatic. The causative agent is *Trichomonas vaginalis,* a protozoan with four anterior flagella, and a posterior flagellum attached to an undulating membrane (figure 26.21).

REVIEW QUESTIONS

Short Answer

1. Name two substances released by lactobacilli that help protect the vagina from potential pathogens.
2. List four things that predispose to the development of infection of the urinary bladder.
3. Name two genera of bacteria that infect the kidneys from the bloodstream.
4. What possible danger lurks in a spot on the ground where an animal has urinated 1 week earlier?
5. What is a clue cell?
6. What is ophthalmia neonatorum?
7. List three diseases caused by different antigenic types of *Chlamydia trachomatis.*
8. Which bacterium leads all others as a cause of reportable infectious disease?
9. Why is dark-field microscopy used to view *Treponema pallidum?*
10. Give two ways in which the chancre of chancroid differs from the chancre of syphilis.
11. What is the relationship between AIDS and HIV disease?
12. What is the reservoir of *Trichomonas vaginalis?*

Multiple Choice

1. All of the following are true of bacterial cystitis, *except*
   a) about one-third of all women will have it at some time during their life.
   b) catheterization of the bladder markedly increases the risk of contracting the disease.
   c) individuals who have a bladder catheter in place indefinitely risk bladder infections with multiple species of intestinal bacteria at the same time.
   d) bladder infections occur as often in men as they do in women.
   e) bladder infections can be asymptomatic.
2. Choose the one correct statement about leptospirosis.
   a) Humans are the only reservoir.
   b) Most infections produce severe symptoms.
   c) Transmission is by the fecal-oral route.
   d) It can lead to unnecessary abdominal surgery.
   e) Effective vaccine is generally available for preventing human disease.
3. Which one of the following statements about bacterial vaginosis is *false?*
   a) It is the most common vaginal disease in women of childbearing age.
   b) In pregnant women it is associated with a seven-fold increased risk of obstetrical complications.
   c) Inflammation of the vagina is a constant feature of the disease.
   d) The vaginal microbiota shows a marked decrease in lactobacilli and a marked increase in anaerobic bacteria.
   e) The cause is unknown.
4. Pick the one *false* statement about vulvovaginal candidiasis.
   a) It often involves the external genitalia.

b) It is readily transmitted by sexual intercourse.
c) It is caused by a yeast present among the normal vaginal microbiota in about one-third of healthy women.
d) It is associated with prolonged antibiotic use.
e) It involves increased risk late in pregnancy.

5. All of the following statements about staphylococcal toxic shock are true, *except*
   a) it can quickly lead to kidney failure.
   b) the causative organism usually does not enter the bloodstream.
   c) it occurs only in vaginal tampon users.
   d) almost one-third of victims of the disease will suffer a recurrence sometime after recovery.
   e) person-to-person spread does not occur.
6. Which of the following statements about gonorrhea is *false?*
   a) The incubation period is only a few days.
   b) Disseminated gonococcal infection (DGI) is almost invariably preceded by prominent urogenital symptoms.
   c) DGI can result in arthritis of the knee.
   d) Phase variation helps the causative organism evade the immune response.
   e) Pelvic inflammatory disease (PID) is common in untreated women.
7. Which one of these statements about chlamydial genital infections is *false?*
   a) The incubation period is usually shorter than in gonorrhea.
   b) Infected cells develop inclusion bodies.
   c) Pelvic inflammatory disease (PID) can be complicated by infection of the surface of the liver.
   d) Tissue damage largely results from cell-mediated immunity.
   e) Fallopian tube damage can occur in the absence of symptoms.
8. Which symptom is *least* likely to occur as a result of tertiary syphilis?
   a) Gummas
   b) White patches on mucous membranes
   c) Emotional instability
   d) Stroke
   e) Blindness
9. During the first 15 years of the AIDS epidemic, approximately how many Americans died of the disease?
   a) 10,000
   b) 50,000
   c) 300,000
   d) 5 million
   e) 50 million
10. All of the following are true of "trich" (trichomoniasis), *except*
    a) it can cause burning pain on urination and painful testes in men.
    b) it occurs worldwide.
    c) asymptomatic carriers are rare.
    d) transmission can be prevented by proper use of condoms.
    e) individuals with multiple sex partners are at high risk of contracting the disease.

Applications

1. Religious restrictions of a small North African community are preventing a World Health Organization project from reducing the incidence of gonorrhea. The community will not permit the testing of females for the disease. They can be treated, however, if they show outward evidence of the disease. Only males are allowed to participate fully in the project, with testing for the disease and treatment. The village elders argue that eradicating the disease from males would eventually remove it from the population. What would be the impact of these restrictions on the success of the project?
2. Former president Ronald Reagan once commented at a press conference that the best way to combat the spread of AIDS in the United States was to prohibit everyone from having sexual contact for 5 years. What would be the success of such a program if it were possible to carry it out?

Critical Thinking

1. The middle curve of figure 26.6 shows the occurrence of staphylococcal toxic shock syndrome in menstruating women from 1979 to 1996. What aspect of these data argue that high-absorbency tampons were not the only cause of staphylococcal toxic shock syndrome associated with menstruation?
2. In early attempts to identify and isolate the cause of syphilis, various bacteria in the discharge from syphilitic lesions in experimental animals were isolated in pure culture. None of them, however, would cause the disease when used in attempts to infect healthy animals. Why was it considered a critical step to have the cultivated bacteria reproduce the disease in the healthy animals?

TEM of West Nile virus, a virus introduced into the United States in 1999.

Nervous System Infections

A Glimpse of History

Today it is hard to appreciate the fear and loathing formerly attached to leprosy. In the Greek translation of the Bible from Hebrew, a variety of disfiguring skin diseases, including leprosy, are covered by the word *lepros,* meaning "scaly," but having the connotation of being disgustingly filthy, outcast, condemned by God for sin. In the Bible, Moses calls lepers "unclean" and proclaims that they must live away from others. In the Middle Ages, lepers wore bells to warn others of their presence, and attended a symbolic burial of themselves before being sent away.

Gerhard Henrik Armauer Hansen (1841–1912) was a burly Norwegian physician who lived in the southern part of that country in the bustling seaport of Bergen. He was a man of many interests, ranging from music, religion, and polar exploration to marine biology. A vigorous trade of goods and ideas with other countries was going on at the time, especially with Germany and England and, like many of the city's inhabitants, Dr. Hansen was fluent in several languages. He was well aware of the scientific advances of the day, including Darwin's studies, and Pasteur's proof that bacteria could cause fermentation. His Norwegian contemporaries included such luminaries as his friend, Edvard Grieg, the famous composer, and Henrik Ibsen, a world-renowned playright.

When he was 32 years old, Hansen decided to go into medical research, and was named assistant to Dr. Daniel C. Danielson, a leading authority on leprosy. Danielson had developed the hypothesis that leprosy was a disease of the blood that once acquired, was hereditary. He had discarded the idea that the disease was contagious as a "peasant superstition." To "dispose of this foolish notion," he injected himself with material from leprosy patients. Fortunately for him and others who had done the same thing, no disease developed, because effective treatment of leprosy did not become available for another three-quarters of a century. Luckily, Danielson and Hansen had great mutual respect, because Hansen, in a meticulous series of 58 studies over a number of years, disproved Danielson's hypothesis and showed that a unique bacillus was associated with the disease in every leprosy patient he studied. His findings, finally reported in 1873, almost a decade before Koch's proof of the cause of tuberculosis, linked for the first time a specific bacterium to a disease. His studies forever changed the way people looked at leprosy, and helped overturn a viewpoint held by many people, one that had existed for many centuries.

In the United States, even during the first half of the twentieth century, persons diagnosed with leprosy risked having their houses burned to destroy contagion. Their names were changed to avoid embarrassing their family, and they were whisked off to a leprosarium such as the one at Carville, Louisiana, surrounded by a 12-foot fence topped with barbed wire. They were separated from spouses and children, and denied the right to marry or to vote. Those who attempted to escape were captured and brought back in handcuffs. The story of the Carville leprosarium is told by one who did escape, Betty Martin, in her wonderful book, *Miracle at Carville*. The leprosarium was finally closed and converted to a military-style academy for high school dropouts in 1999.

Because the English word *leprosy* carries these innuendoes from earlier days, many people prefer to use the term *Hansen's disease,* so named after the discoverer of the causative organism.

Infections of the nervous system are apt to be serious because they threaten a person's ability to move, to feel, and to think normally. Just consider poliomyelitis, in which nerve cells are destroyed, leaving the victim with a paralyzed arm or leg, or unable to breathe without mechanical assistance. In Hansen's disease, nerve damage can result in complete loss of fingers or toes, or deformity of the face. Infections of the brain or its covering membranes can render a child deaf or mentally retarded. Fortunately, nervous system infections are uncommon. Examples given in this chapter, however, reveal a multiplicity of causes, including bacteria, viruses, fungi, and even some protozoa.

KEY TERMS

Arbovirus Arthropod-borne virus. One of a large group of RNA viruses carried by insects and mites that act as biological vectors.

Axon The long, thin extension of a nerve cell.

Blood-Brain Barrier Property of the central nervous system blood vessels that restricts passage of infectious agents and certain molecules (such as medications) into the brain and spinal cord.

Botulism A paralytic illness caused by a toxin produced by *Clostridium botulinum.*

Creutzfeld-Jakob Disease A rare, degenerative brain disease of humans caused by prions.

Ganglion (pl., **ganglia**) Small body near the spinal column representing a bulge in a peripheral nerve at the site where the sensory nerve cells are located.

Meninges Membranes covering the brain and spinal cord.

Meningoencephalitis Inflammation of the brain and its covering membranes.

Negri Body Viral inclusion body characteristic of rabies.

Neurotoxin A toxin that damages the nervous system.

Parasitemia Parasites (such as protozoan pathogens) circulating in the bloodstream.

Petechia (Petechiae) Small, purplish spot on the skin or mucous membrane caused by a hemorrhage.

Spongiform encephalopathy Central nervous disease in which the brain has the appearance of a sponge; caused by prions.

Viremia Viruses circulating in the bloodstream.

27.1 Anatomy and Physiology

Focus Points

- List the main components of the nervous system.
- Describe the routes a pathogen can take to enter the central nervous system.

The brain and spinal cord make up the **central nervous system (CNS).** Both are enclosed by bone—the brain by the skull and the spinal cord by the vertebral column **(figure 27.1).** The network of nerves throughout the body is called the **peripheral nervous system.** All nerves are made up of cells with very long, thin extensions called **axons,** which transmit electrical impulses. The peripheral nervous system is connected with the CNS by bundles of axons that penetrate the protective bony covering. Nerves can be damaged if the bones are infected at sites of nerve penetration. **Motor nerves** carry messages from the CNS to different parts of the body and cause them to act; **sensory nerves** transmit sensations like heat, pain, light, and sound from the periphery to the central nervous system. The sensory nerve cells are located in small bodies called **ganglia** (singular, **ganglion**) located near the vertebral column; motor nerve cells are located in the central nervous system. Nerves can sometimes regenerate if severed or damaged, but they cannot be repaired if the nerve cells are killed, as occurs, for example, in poliomyelitis. Some viruses and toxins can move through the body within the cytoplasm of the axons, and some herpesviruses can remain latent in nerve cells for many years.

The brain is a very complex structure—distinct parts have different functions. This means that if an individual has symptoms from a brain abscess, for example, physicians will know in which part of the brain the abscess is located. Generalized inflammation or infection of the brain is termed **encephalitis.** Deep inside the brain are four cavities called **ventricles** that are filled with a clear fluid called **cerebrospinal fluid (CSF).** This fluid is continually produced by structures in the wall of ventricles, and it flows out through a small opening at the base of the brain. It then spreads over the surface of the brain and spinal cord, and is reabsorbed into the bloodstream at specialized sites between the skull and the brain **(figure 27.2a).**

Three membranes called **meninges** cover the surface of the brain and spinal cord. Inflammation or infection of these membranes is called **meningitis.** The outer membrane, or **dura,** is tough and fibrous and adheres closely to the skull and vertebrae. It provides a barrier to the spread of infection from bones surrounding the central nervous system. The two inner membranes, the **arachnoid** and **pia,** are separated by a space through which the cerebrospinal fluid flows. Blood vessels and nerves that pass through the meninges and cerebrospinal fluid can become damaged by meningitis. Sometimes both the meninges and the brain

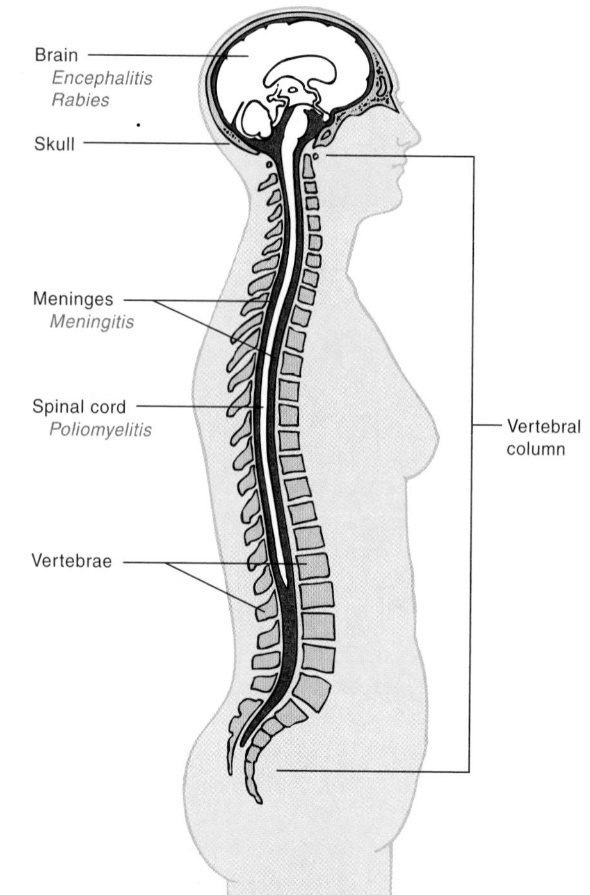

FIGURE 27.1 The Central Nervous System Terms in color indicate diseases associated with that area.

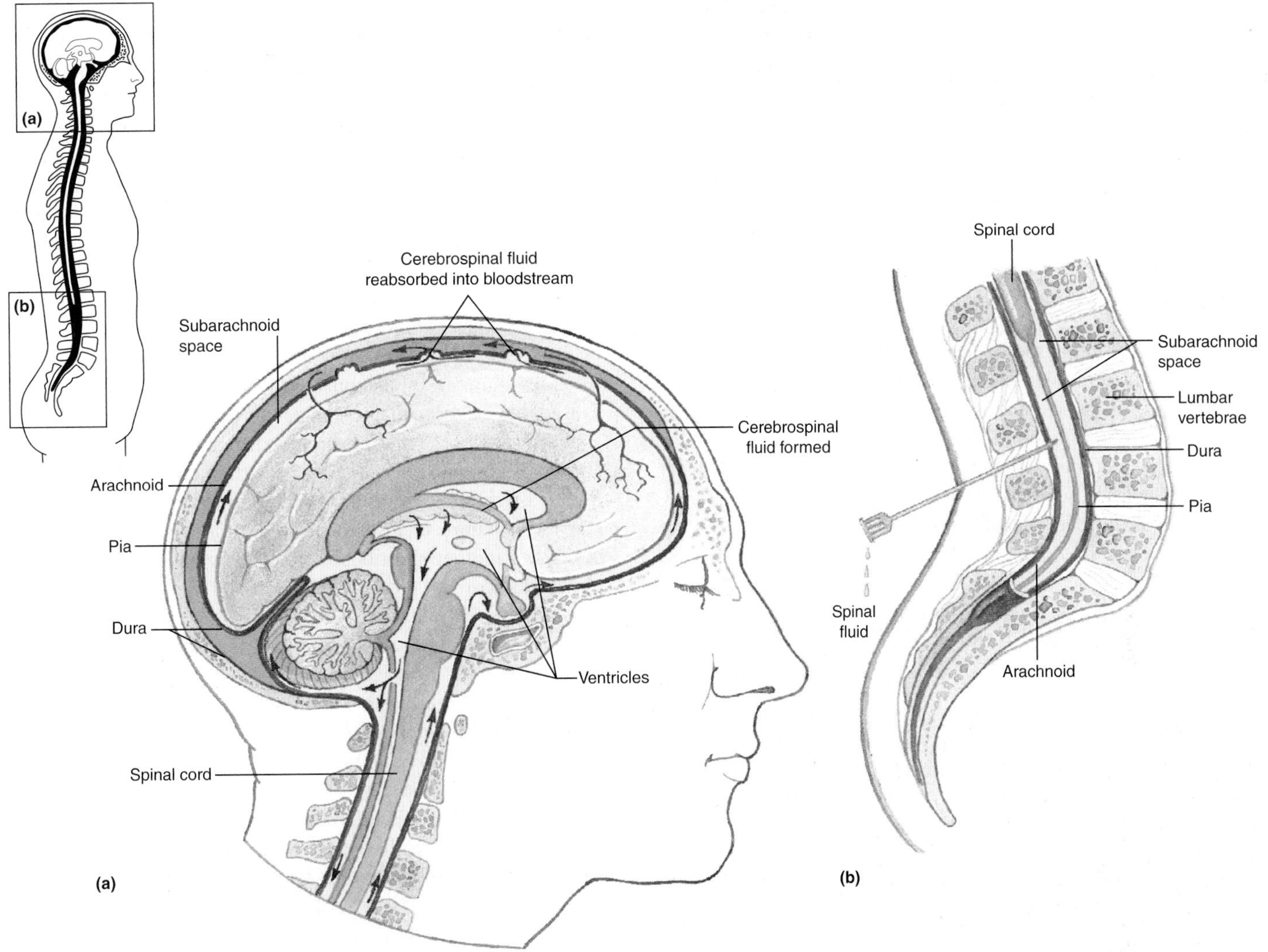

FIGURE 27.2 Cerebrospinal Fluid (a) The fluid is formed in the ventricles, flows out over the surface of the brain and spinal cord, and is reabsorbed by structures at the top of the brain. **(b)** A sample of cerebrospinal fluid can be withdrawn from the subarachnoid space.

are involved by an infection—a condition called **meningoencephalitis.**

To determine if a person has meningitis, a needle can be safely inserted between the vertebrae in the small of the back and a sample of cerebrospinal fluid withdrawn and examined. In this anatomical region, the spinal cord has tapered to only a threadlike structure (see figure 27.2b). The causative agent of a central nervous system infection can thus be identified microscopically and cultivated on laboratory media. Once the cause of the disease is known, it can usually be treated effectively with an appropriate antimicrobial medication.

Pathways to the Central Nervous System

Because the nervous system lies entirely within body tissues, it has no normal flora. It is sterile, and no viruses or microorganisms are normally found there. Also, its well-protected environment prevents infectious agents from getting to it readily. Most microorganisms that infect the nervous system infect other parts of the body much more frequently. The routes that pathogenic microorganisms and viruses take to reach the brain and spinal cord include the bloodstream, the nerves, and extensions from bone.

The Bloodstream

The bloodstream is the chief source of CNS infections, although it is very difficult for infectious agents to cross from it to the brain. This barrier to the passage of harmful agents is called the **blood-brain barrier.** It depends on special cells lining the central nervous system capillaries. These cells, only a single layer thick, are so close to each other that except for essential nutrients that are actively taken up, most molecules cannot pass through them. Consequently, many medications, including penicillin, cannot cross the barrier unless their concentrations in the blood are very high; antimicrobial medications must be able to cross the barrier in order to treat CNS infections. The blood-brain barrier generally prevents pathogens from entering nervous tissue except in rare cases when a high concentration of the infectious agent circulates for a long time in the bloodstream.

Infections of the face above the level of the mouth can involve veins that communicate through bone with veins on the brain surface. These communicating veins can provide a direct avenue for spread of infection to the central nervous system. A good rule, therefore, is to never squeeze pimples situated on the upper part of the face.

The Nerves

Some pathogenic agents penetrate the CNS by traveling up the nerves, probably within the cytoplasm of the axons. Examples of such agents are the rabies and herpes simplex viruses and the tetanus toxin. ■ rabies, p. 663 ■ herpes simplex, p. 592 ■ tetanus, p. 566

Extensions from Bone

To reach the CNS, microorganisms must penetrate not only the protecting bone of the skull or spine, but also the tough membrane that surrounds all bones, and the dura. Only rarely do infections in the bone surrounding the CNS succeed in reaching the brain or spinal cord. Sometimes, however, they extend through bone from the sinuses, mastoids, or middle ear. Skull fractures commonly produce non-healing injuries that predispose a person to recurring infection of the CNS. ■ mastoids, p. 496 ■ middle ear, p. 505

MICROCHECK 27.1

The brain and spinal cord make up the central nervous system (CNS). The peripheral nervous system includes sensory nerves, which convey sensations to the CNS, and motor nerves, which carry messages from the CNS to parts of the body, causing those parts to act. The cerebrospinal fluid fills cavities within the brain and flows out over the brain and spinal cord. Inflammation of the meninges is called meningitis. Infection can reach the CNS via the bloodstream, via the nerves, or by extension from bone. Antimicrobial medicines are effective in central nervous system infections only if they can cross the blood-brain barrier.

- ✓ Describe the flow of cerebrospinal fluid in the central nervous system.
- ✓ Why is it uncommon that infectious agents pass from the bloodstream to the CNS?
- ✓ Why can an infection in the brain's ventricles usually be detected in spinal fluid obtained from the lower back?

27.2

Bacterial Nervous System Infections

Focus Points

- Outline the pathogenesis of meningococcal meningitis.
- Give reasons why listeriosis is often contracted from contaminated food or drink.
- Explain why botulism generally is not an infectious disease.

Bacteria can infect the brain and spinal cord, causing abscesses. Or they can infect the peripheral nerves, as in Hansen's disease. More commonly, bacteria infect the meninges and cerebrospinal fluid, causing meningitis.

Meningitis provokes an intense inflammatory response, with a marked accumulation of white blood cells in the cerebrospinal fluid and swelling of brain tissue. The incidence of the disease is, on average, fewer than 10 cases per 100,000 population per year. A large number of bacterial species can infect the meninges. In many cases, these organisms are carried in the upper respiratory tract of healthy people, are transmitted by inhalation, and only produce meningitis in a small percentage of people who are infected. Organisms in this category include three important causes of bacterial meningitis—*Haemophilus influenzae, Neisseria meningitidis,* and *Streptococcus pneumoniae*. Age plays an important role in susceptibility to these bacteria. *Streptococcus pneumoniae, H. influenzae,* and *N. meningitidis* are uncommon causes of meningitis in newborn babies, because most mothers have antibodies against them. The mother's antibodies cross the placenta and protect her baby. ■ *Streptococcus pneumoniae*, p. 509

Meningitis in infants during the first month of life is usually caused by bacteria that colonize the mother's birth canal. For example, newborn babies become infected during labor and delivery by Gram-negative rods such as certain encapsulated strains of *E. coli* originating from the mother's intestinal tract. The reason that the newborn is susceptible to Gram-negative rod infections is that its mother's antibodies to the organisms are of the IgM class, which do not cross the placenta and therefore cannot confer immunity to the infant. ■ IgM antibody, p. 374

Streptococcus agalactiae, a Lancefield group B streptococcus, which colonizes the vagina in 15% to 40% of pregnant women, exceeds *E. coli* as a cause of meningitis in newborn infants. Babies also acquire this infection from the mother's genital tract around the time of birth. The American Academy of Pediatrics recommends that cultures selective for group B streptococci be obtained from the vagina and rectum late in pregnancy. Women with positive cultures can then be treated with an appropriate antibacterial medication shortly before or during labor. Evidence indicates that by using this screening procedure, doctors can decrease the incidence of serious group B streptococcal disease by more than 75%. ■ Lancefield grouping, p. 495

Of the six antigenic types of *Haemophilus influenzae,* labeled **a** through **f,** type **b** is responsible for most cases of serious disease. Starting in the late 1980s, children in the United States have been routinely immunized using a T-cell-dependent antigen consisting of *H. influenzae* b polysaccharide joined covalently to a bacterial protein such as diphtheria toxoid. The result of using these **conjugate vaccines** has been a dramatic decline in meningitis and other serious infections caused by *H. influenzae* type b **(figure 27.3).** The decline in meningitis due to *H. influenzae* has resulted in a marked overall decrease in bacterial meningitis and a shift in the peak incidence to older age groups. ■ T-dependent antigen, p. 377 ■ conjugate vaccines, pp. 383, 435

Streptococcus pneumoniae **(figure 27.4)** is the leading cause of meningitis in adults. The organism is a prominent cause of otitis media, sinusitis, and pneumonia—conditions that often precede pneumococcal meningitis. *Neisseria meningitidis,* the meningococcus, differs from the other causes in that it is often responsible for epidemics of meningitis. ■ otitis media, p. 504

Meningococcal Meningitis

Meningococcal meningitis occurs most commonly in children aged 6 to 11 months, but it also occurs frequently in older children and adults. It is greatly feared because it can sometimes

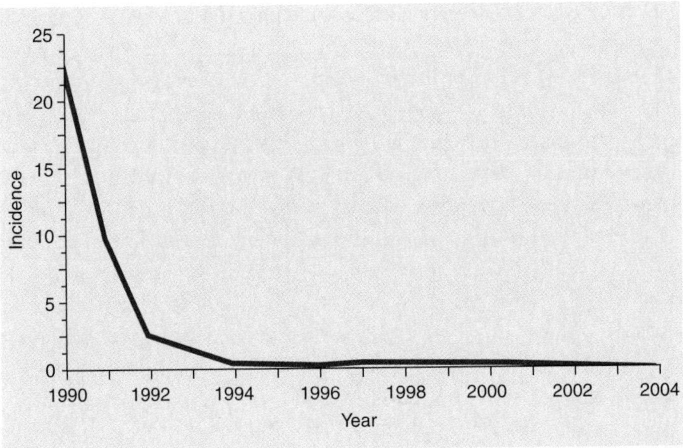

FIGURE 27.3 Rate of Serious *Haemophilus influenzae* Disease per 100,000 Children Less than Age Five, United States, 1990 Through 2004 Before the availability of conjugate vaccines in late 1987, *H. influenzae* type b was the most common cause of bacterial meningitis in preschool children.
(Source: CDC, Vaccine Information Statements.)

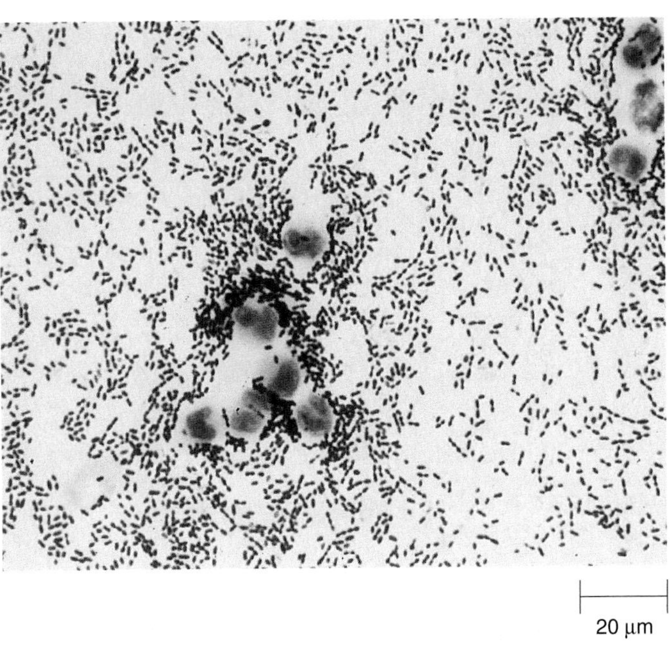

FIGURE 27.4 *Streptococcus pneumoniae* (Pneumococci) in the Spinal Fluid of a Person with Pneumococcal Meningitis This Gram-stained smear shows large numbers of the Gram-positive diplococci; the large red objects are polymorphonuclear leukocytes that have entered the spinal fluid in response to the infection.

progress to death within a few hours, but most patients respond well to treatment and recover without permanent nervous system damage.

Symptoms

The first symptoms of bacterial meningitis are similar regardless of the causative agent. They usually begin with a mild cold; followed by the sudden onset of a severe, throbbing headache; fever; pain and stiffness of the neck and back; nausea; and vomiting. Deafness and alterations in consciousness progressing to coma

CASE PRESENTATION

The patient was a 31-month-old girl admitted to the hospital because of fever, headache, drowsiness, and vomiting. She had been previously well until 12 hours before admission, when she developed a runny nose, malaise, and loss of appetite.

Her birth and development were normal. There was no history of head trauma. Her routine immunizations had been neglected.

On examination, her temperature was 40°C (104°F), her neck was stiff, and she did not respond to verbal commands.

Her white blood cell count was elevated and showed a marked increase in the percentage of polymorphonuclear neutrophils (PMNs). Her blood sugar was in the normal range.

A spinal tap was performed, yielding cloudy cerebrospinal fluid under increased pressure. The fluid contained 18,000 white blood cells per microliter (normally, there are few or none), a markedly elevated protein, and a markedly low glucose. Gram stain of the fluid showed many tiny, Gram-negative coccobacilli, most of which were outside the white blood cells.

1. What is the diagnosis and what is the causative agent?
2. What is the prognosis in this case?
3. What age group is most susceptible to this illness?
4. Compare the pathogenesis of this disease in children and adults.

Discussion

1. The patient had bacterial meningitis caused by ***Haemophilus influenzae,*** serotype b. The fact that she had been healthy prior to her illness makes it highly unlikely that other serotypes could be responsible.
2. With treatment, the fatality rate is approximately 5%. Formerly, ampicillin was effective in most cases, but beginning in 1974, an increasing number of strains possessed a plasmid coding for β-lactamase. So far, however, these strains have been susceptible to the newer cephalosporin-type antibiotics. Unfortunately, about one-third of those who are treated and recover from the infection are left with permanent damage to the nervous system, such as deafness or paralysis of facial nerves. Prompt diagnosis and correct choice of antibacterial treatment minimize the chance of permanent damage.
3. The peak incidence of this disease is in the age range of 6 to 18 months, corresponding to the time when protective levels of transplacental antibody are lacking and adaptive active immunity has not yet developed. Since 1987, vaccines consisting of type b capsular antigen conjugated with a protein, such as the outer membrane protein of *Neisseria meningitidis* or diphtheria toxin, have been used to immunize infants and thus eliminate the immunity gap. As a result, meningitis caused by *H. influenzae* type b is now rare. It is important to know, however, that strains other than type b can sometimes cause meningitis. So far, such strains are uncommon, although they represent an increasing percentage as type b strains decline in importance.
4. *Haemophilus influenzae* type b strains are referred to as "invasive" strains because they establish infection of the upper respiratory tract, pass the epithelium, and enter the lymphatic vessels and bloodstream. In this way, they gain access to the general circulation and are carried to the central nervous system. Most strains of *H. influenzae,* unlike type b, are non-invasive. Although they can cause infections of respiratory epithelium, they usually do not enter the circulation. Most adults are immune to type b strains, but some adults develop meningitis from non-invasive *H. influenzae* strains that gain access to the nervous system because of a skull fracture or by direct extension to the meninges from an infected sinus or middle ear.

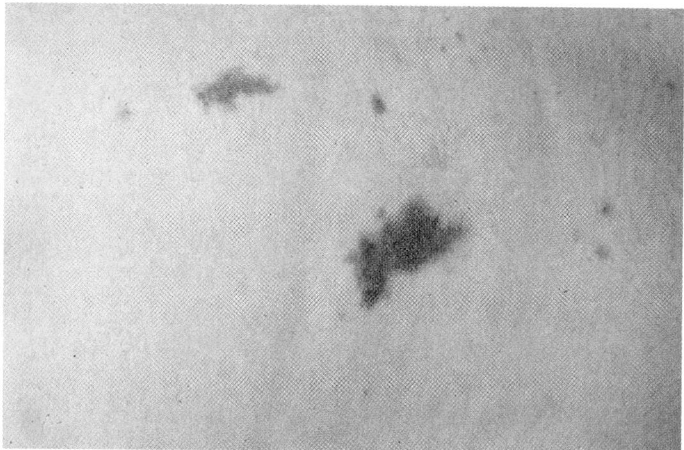

FIGURE 27.5 Petechiae of Meningococcal Disease *Neisseria meningitidis* is present in these skin lesions.

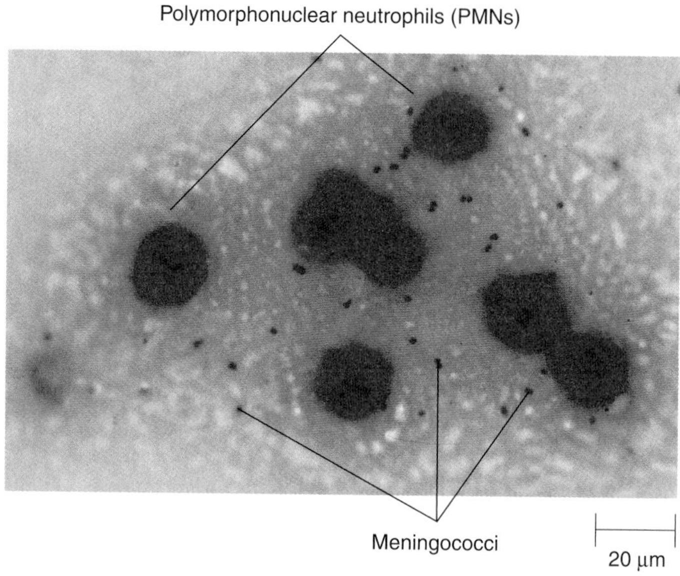

FIGURE 27.6 Meningococcal Meningitis Stained smear of spinal fluid showing polymorphonuclear neutrophils and meningococci. This case is somewhat unusual because so many of the meningococci can be seen outside the leukocytes.

may develop. Especially likely in meningococcal meningitis, purplish spots called **petechiae,** resulting from small hemorrhages, may appear on the skin **(figure 27.5).** The infected person can develop endotoxic shock and die within 24 hours. Usually, though, progression of the illness is slower, allowing time for effective treatment. ■ endotoxic shock, p. 407

Causative Agent

The causative organism of meningococcal meningitis is the Gram-negative diplococcus, *Neisseria meningitidis.* Although there are 13 antigenic groups of *N. meningitidis,* most serious infections are due to A, B, C, Y, and W135. The process of identifying the different strains, called **serogrouping,** employs serum containing specific antibody, obtained from the blood of laboratory animals that have been immunized with known strains of meningococci.

Pathogenesis

Individuals acquire the infection by inhaling airborne droplets from the respiratory tract of another person. The meningococci attach to the mucous membrane by pili and multiply. They are then taken in by the respiratory tract epithelial cells, pass through them, and invade the bloodstream. The blood carries the organisms to the meninges and cerebrospinal fluid. There they multiply faster than they can be engulfed and destroyed by the polymorphonuclear neutrophils (PMNs) that enter the fluid in large numbers in response to the infection **(figure 27.6).** The bacteria and leukocytes degrade glucose and thereby cause a marked drop in the glucose concentration normally present in the cerebrospinal fluid. The inflammatory response, with its formation of pus and clots, may cause brain swelling and **infarcts**—death of tissue resulting from loss of blood supply. It can also lead to obstruction of the normal outflow of cerebrospinal fluid, causing the brain to be squeezed against the skull by the buildup of internal pressure. The infection can damage the nerves of hearing or vision or the motor nerves, producing paralysis. Moreover, *N. meningitidis* circulating in the bloodstream releases endotoxin, causing a drop in blood pressure that can lead to **shock.** Shock is a state in which there is not enough blood pressure to circulate the blood adequately to meet the oxygen requirements of vital body tissues such as the kidneys and central nervous system. The smaller blood vessels of the skin are also damaged by the circulating meningococci, causing the petechiae. ■ pili, pp. 66, 398 ■ endotoxin, pp. 61, 407

The organism's endotoxin also increases the body's sensitivity to repeated endotoxin exposure. This strikingly increased sensitivity is probably responsible for the rapidly fatal outcome in some cases of meningococcal meningitis. The tendency of meningococci to autolyse—to rupture spontaneously—may enhance their release of endotoxin.

Epidemiology

The reasons are unknown why *N. meningitidis,* unlike *Haemophilus influenzae* and *Streptococcus pneumoniae,* is prone to cause epidemics of meningitis. Meningococcal meningitis can spread rapidly in crowded and stressed populations such as military recruits **(figure 27.7),** and outbreaks also occur in college dormitories. More typically, however, it appears at widely separate locations and occurs throughout the year. Humans are the only source of the infection, and transmission can occur with exposure to a person with the disease or an asymptomatic carrier of *N. meningitidis.* The majority of meningococcal infections are limited to the nose and throat and pass unnoticed. If a selective medium is used to suppress the growth of other nasopharyngeal microbiota, meningococci can be recovered from the throats of 5% to 15% of apparently healthy people.

Prevention and Treatment

A vaccine composed of purified A, C, Y, and W135 capsular polysaccharides has been available for many years to control epidemics caused by these groups and to immunize people at high risk for the disease. A conjugate vaccine for the same four serogroups was approved in 2005 for ages 11 to 55 years. It is now recommended for routine administration to all preadolescents and those entering high school, and to college freshmen who will live in dormatories.

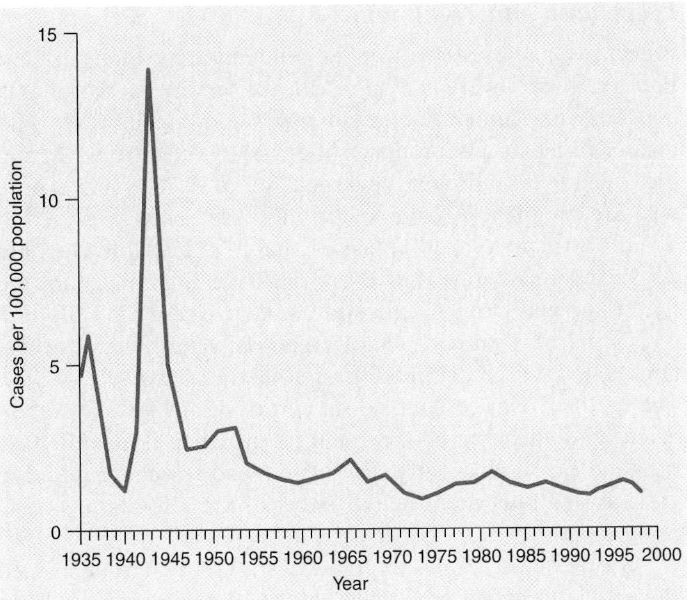

FIGURE 27.7 Meningococcal Disease in the United States, 1935 to 1998 Notice the high incidence during World War II, when epidemics occurred among military recruits crowded in barracks. Outbreaks also occur in college dormitories.

Group B strains possess a poorly immunogenic polysaccharide and, as yet, no vaccine is available for their control. Mass prophylaxis with an appropriate antibacterial medication, however, can be useful in controlling epidemics in small, confined groups such as people in jails, nursing homes, and schools. Because meningococcal strains from patients with active disease might be more virulent than other strains, people intimately exposed to cases of meningococcal disease are routinely given prophylactic treatment with the antibiotic rifampin. Meningococcal meningitis can usually be cured with penicillin or ceftriaxone unless severe brain injury or shock is present. The mortality rate is less than 10% in treated cases.

Table 27.1 gives the main features of this disease.

Listeriosis

Meningitis is a common manifestation of listeriosis, a foodborne disease caused by *Listeria monocytogenes.* Although this organism generally causes only a small percentage of meningitis cases in the United States, epidemics sometimes occur.
■ foodborne illness, p. 762

Symptoms

The vast majority of healthy people have asymptomatic infections or trivial symptoms. Symptomatic infections are usually characterized by fever and muscle aches, and sometimes nausea or diarrhea. About three-quarters of the cases coming to medical attention have meningitis, heralded by fever, headache, stiff neck, and vomiting, that may or may not be preceded by other symptoms. Pregnant women who become infected often miscarry, or deliver terminally ill premature or full-term infants.

Causative Agent

Listeria monocytogenes is a motile, non-spore-forming, facultatively anaerobic, Gram-positive rod that can grow at 4°C.

TABLE 27.1 Meningococcal Meningitis

① *Neisseria meningitidis* inhaled, infects upper airways.

② Bacteria enter the bloodstream and are circulated throughout the body.

③ The bacteria lodge in the skin and cause petechiae.

④ Bacteria on the meninges causes meningitis.

⑤ Lysing bacteria in the circulation release endotoxin, producing shock.

⑥ Inflammatory response in meninges can damage nerves of hearing causing deafness and obstruct the flow of cerebrospinal fluid causing increased pressure inside the brain.

⑦ Bacteria exit with respiratory secretions.

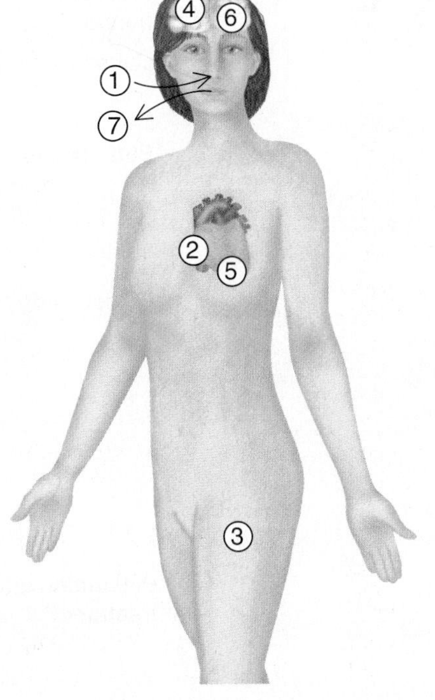

Symptoms	Mild cold followed by headache, fever, pain, stiff neck and back, vomiting, petechiae
Incubation period	1 to 7 days
Causative agent	*Neisseria meningitidis,* the meningococcus; a Gram-negative diplococcus
Pathogenesis	Meningococci adhere by pili, colonize upper respiratory tract, enter bloodstream; carried to meninges and spinal fluid; inflammatory response obstructs normal outflow of fluid; increased pressure caused by obstructed flow impairs brain function; damage to motor nerves produces paralysis; endotoxin release causes shock.
Epidemiology	Close contact with a case or carrier; inhalation of infectious droplets; crowding and fatigue predispose to the disease.
Prevention and treatment	Conjugate vaccine against serogroups A, C, W135, and Y used to immunize ages 11–55 years; rifampin given to those exposed. Penicillin, ceftriaxone, for treatment.

Pathogenesis

The mode of entry in isolated cases is usually obscure, but it is generally via the gastrointestinal tract during epidemics. Gastrointestinal symptoms may or may not occur, but the bacteria promptly penetrate the intestinal mucosa and enter the bloodstream. The resulting bacteremia is usually associated with fever and muscle aches, and it is the source of meningeal infection. In pregnant women, *L. monocytogenes* crosses the placenta and produces widespread abscesses in tissues of the fetus. Babies infected at the time of childbirth usually develop meningitis after an incubation period of 1 to 4 weeks.

Epidemiology

Listeria monocytogenes is widespread in natural waters and vegetation, and it can be carried in the intestines of asymptomatic animals and humans. It occurs on all continents except Antarctica. Pregnant women, the elderly, and those with underlying illnesses such as immunodeficiency, diabetes, cancer, and liver disease are especially susceptible to listeriosis. Epidemic disease has resulted from *L. monocytogenes* contamination of foods including coleslaw, non-pasteurized milk, pork tongue in jelly, some soft cheeses, and hot dogs. The fact that the organisms can grow in commercially prepared food at refrigerator temperatures has resulted in thousands of infections originating from a single food-processing plant.

Prevention and Treatment

Preventive measures reflect the epidemiology of listeriosis. Poultry, pork, beef, and other meats should be thoroughly cooked. Contamination of countertops and kitchenwares with uncooked meats and produce should be avoided. Raw vegetables need to be thoroughly washed before eating them. Those who are pregnant or have underlying diseases are advised to avoid Mexican-style soft cheeses, and to reheat cold cuts, hot dogs, and refrigerated leftovers before eating. In 2006, the U.S. Food and Drug Administration approved a food additive consisting of a number of different of bacteriophage strains that lyse food-contaminating strains of *L. monocytogenes.* The additive can be sprayed on a variety of meats normally eaten without being cooked, and its presence is recorded on the food label. Hopefully, use of this bacteriophage additive will substantially increase the safety of these foods. ■ **bacteriophages, p. 303**

So far, most strains of *L. monocytogenes* have remained susceptible to antibacterial medications such as penicillin. Even though the disease is often mild in pregnant women, prompt diagnosis and treatment are important to protect the fetus.

Some of the main features of listeriosis are presented in **table 27.2.** The main causes of acute bacterial meningitis are compared in **table 27.3.**

TABLE 27.2 Listeriosis

① Causative organism *Listeria monocytogenes* is ingested with food such as Mexican cheese, soft cheeses, non-pasteurized milk, hot dogs, or coleslaw.

② The bacteria rapidly penetrate the intestinal epithelium and establish bacteremia, especially in pregnant women, the elderly, and the immunodeficient.

③ In pregnant women, circulating *L. monocytogenes* crosses the placenta and fatally infects the fetus or bacteria transmitted to the baby at birth cause meningitis in one to four weeks. The mother usually does not have a serious illness.

④ In older people and those with underlying diseases *L. monocytogenes* attacks brain and meninges, causes meningitis, brain abscesses.

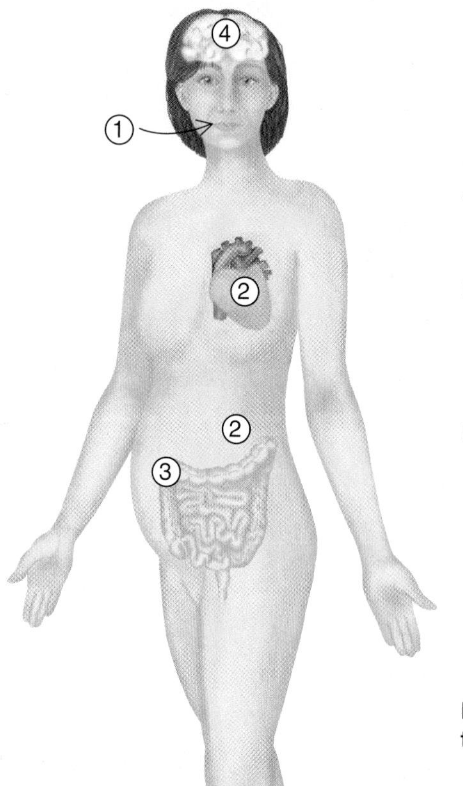

Symptoms	Fever and muscle aches, with or without gastrointestinal symptoms; headache and stiff neck mark the onset of meningitis
Incubation period	A few days to 2 to 3 months; in newborn babies, 1 to 4 weeks
Causative agent	*Listeria monocytogenes,* a non-spore-forming Gram-positive rod able to grow at 4°C
Pathogenesis	Ingested *L. monocytogenes* penetrate the intestinal epithelium and enter the bloodstream; the resulting bacteremia seeds the meninges, causing meningitis.
Epidemiology	Epidemics from contaminated soft cheeses, non-pasteurized milk, coleslaw, hot dogs. Pregnant women get bacteremia, fetal infection, miscarriage. Infants contract infection at birth, develop meningitis in the first month of life. Elderly and those with immunodeficiency, cancer, and diabetes also at high risk.
Prevention and treatment	Care in handling, cooking of raw meats; thorough washing of vegetables; reheating of cold cuts, hot dogs, and refrigerated leftovers. Antibacterial medications such as penicillin, given promptly, are effective treatment.

TABLE 27.3 Main Causes of Acute Bacterial Meningitis Compared

Agent	Characteristics	Source	Vaccine ?	Usual Victims
H. influenzae	Tiny, Gram-negative coccobacilli	Human respiratory system	Yes, type b	Young children
N. meningitidis	Gram-negative diplococci	Human respiratory system	Yes, four types	Mostly ages 2–20 years; can cause epidemics
S. pneumoniae	Encapsulated, Gram-positive diplococci	Human respiratory system	Yes, multiple types	Mostly late teens and adults
S. agalactiae	Gram-positive cocci in chains	Bowel, vagina	No	Mostly neonates; others with underlying diseases
E. coli	Gram-negative rods; usually a specific encapsulated type	Bowel	No	Mostly neonates
L. monocytogenes	Gram-positive motile rods; multiply at refrigerator temperatures	Environment; contaminated cheeses, coldcuts, other foods	No	Pregnant women, neonates; elderly

Hansen's Disease (Leprosy)

Although Hansen's disease, also known as leprosy, is now a relatively minor problem in the Western world, it was once common in Europe and America. Like tuberculosis, the disease began to recede for unknown reasons even before an effective treatment was discovered. Today, it is chiefly a problem in tropical and economically underdeveloped countries, with an estimated worldwide occurrence of about 600,000 active cases. Most cases are concentrated in India, Nepal, Myanmar, Madagascar, Mozambique, and Brazil. In recent years, reported new cases in the United States have averaged about 150 per year; most were acquired outside the country.

Symptoms

Hansen's disease begins gradually, usually with the onset of increased or decreased sensation in certain areas of skin. These areas typically have either increased or decreased pigmentation. Later, they enlarge and thicken, losing their hair, sweating ability, and all sensation. The nerves of the arms and legs become visibly enlarged with accompanying pain, later changing to numbness, muscle wasting, ulceration, and loss of fingers or toes **(figure 27.8).** In some patients the skin lesions are not sharply defined, but slowly enlarge and spread. Changes are most obvious in the face, with thickening of the nose and ears and deep wrinkling of the facial skin. Collapse of the supporting structure of the nose occurs with accompanying congestion and bleeding.

Causative Agent

Mycobacterium leprae **(figure 27.9)** is identical in appearance to *M. tuberculosis.* It is aerobic, acid-fast, rod-shaped, and typically stains in a beaded manner. Despite many attempts, *M. leprae* has not been grown in the absence of living cells. In the 1960s, it was successfully cultivated to a limited degree in the footpads of laboratory mice, and in 1971, in a major advance, Dr. Eleanor E. Storrs and colleagues demonstrated that the organisms can be grown in armadillos. Later, mangabey monkeys were also found to be susceptible. *Mycobacterium leprae* grows very slowly, with a generation time of about 12 days. A clone bank of the genome of *M. leprae* has been made in *E. coli,* thus making large quantities of the organism's antigens available for study. ■ acid-fast stain, p. 50 ■ *Mycobacterium,* p. 261 ■ cloning, p. 215

Pathogenesis

The earliest detectable finding in human infection with *M. leprae* is generally the invasion of the small nerves of the skin demonstrated by biopsy of skin lesions. Indeed, *M. leprae* is the only known human pathogen that preferentially attacks the peripheral nerves. The bacterium also initially grows very well within macrophages, before cellular immunity develops. The

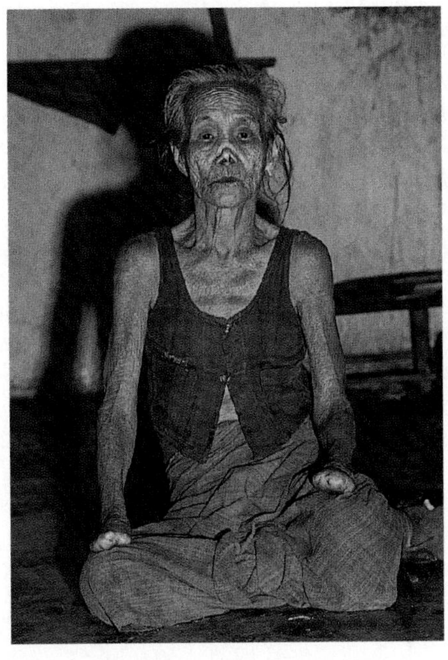

FIGURE 27.8 A Person with Leprosy Notice the absence of her fingers and sunken nose as a result of the disease.

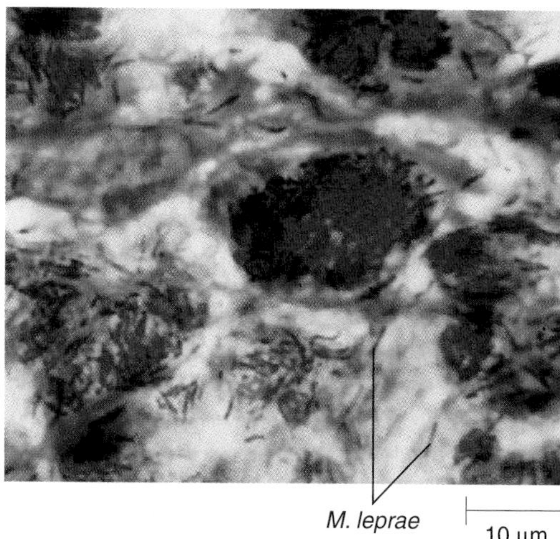

FIGURE 27.9 ***Mycobacterium leprae* in a Biopsy Specimen** The red areas are dense masses composed of millions of the bacteria, a typical finding in lepromatous leprosy.

course of the infection depends on the immune response of the host. In most cases, cellular immunity develops against the invading bacteria. Activated macrophages limit the growth of *M. leprae,* and therefore the bacteria do not become numerous. The attack of immune cells against chronically infected nerve cells can progressively damage the nerves, however, which leads to disabling deformity, resorption of bone, and skin ulceration. The disease often spontaneously stops progressing, and thereafter the nerve damage, although permanent, does not worsen. This limited type of Hansen's disease in which cellular immunity suppresses proliferation of the bacilli is called **tuberculoid leprosy.** People with tuberculoid leprosy rarely, if ever, transmit the disease to others. In many cases of tuberculoid leprosy, however, the immune system is gradually overwhelmed by the growth of *M. leprae,* leading in multiple stages to the more serious lepromatous form of the disease discussed in detail next. These stages are distinguished largely by the quantity of the bacteria in the patient's tissues. Early treatment is important to prevent the disease from progressing.
■ macrophage, pp. 350, 358 ■ cellular immunity, p. 367

When cellular immunity to *M. leprae* fails to develop or is suppressed, unrestricted growth of *M. leprae* first occurs in the cooler tissues of the body, notably in skin macrophages and peripheral nerves, and later throughout the body. This relatively uncommon form of Hansen's disease is called **lepromatous leprosy.** The tissues and mucous membranes contain billions of *M. leprae,* but there is almost no inflammatory response to them. The mucus of the nose and throat is loaded with the bacteria, which can readily be transmitted to others. Lymphocytes are present in the lesions but there is little or no evidence of macrophage activation.

In lepromatous leprosy, cell-mediated immunity to *M. leprae* is absent, although immunity and delayed hypersensitivity to other infectious agents are usually normal. Normal immune function against *M. leprae* tends to return when the disease is controlled by medication.

The very long generation time of this bacterium most likely accounts for the lengthy incubation period of Hansen's disease, usually about 3 years (range, 3 months to 20 years).
■ generation time, p. 84

Epidemiology

Transmission of *M. leprae* is by direct human-to-human contact. The source of the organisms is mainly nasal secretions of a lepromatous case, which transport *M. leprae* to mucous membranes or skin abrasions of another individual. Although leprosy bacilli readily infect people exposed to a case of lepromatous leprosy, the disease develops in only a tiny minority, being controlled by body defenses in the rest. Hansen's disease is no longer a much-feared contagious disease, and the morbid days when lepers were forced to carry a bell or horn to warn others of their presence are fortunately past.

Natural infections with *M. leprae* occur in wild nine-banded armadillos and in mangabey monkeys. An epidemiological study appears to exclude armadillos as an important source of human leprosy, but occasional transmissions to humans have not been ruled out.

Prevention and Treatment

No proven vaccine to control leprosy is yet available. Tuberculoid leprosy can be arrested by treatment with the antimicrobials dapsone and rifampin administered in combination for 6 months.

Lepromatous leprosy is generally treated for a minimum of 2 years, with a third drug, clofazimine, included in the treatment regimen. As in tuberculosis, multiple drug therapy is required to minimize the development of strains resistant to the antimicrobials.

Some features of Hansen's disease are summarized in **table 27.4.**

TABLE 27.4 Hansen's Disease (Leprosy)

Symptoms	Skin lesions that lack sensation, deformed face, loss of fingers or toes
Incubation period	3 months to 20 years; usually 3 years
Causative agent	*Mycobacterium leprae,* an acid-fast, nonculturable rod
Pathogenesis	Invasion of small nerves of skin; multiplication in macrophages; course of disease depends on immune response of host; activated macrophages limit growth of bacterium; attack of immune cells against infected nerve cells produces nerve damage, leading to deformity; in lepromatous leprosy, lymphocytes fail to react to the bacteria, allowing unrestrained growth of *M. leprae.*
Epidemiology	Direct contact with *M. leprae* from mucous membrane secretions.
Prevention and treatment	No vaccine. Treatment: dapsone plus rifampin for months or years; clofazimine added for lepromatous disease.

Botulism

Botulism, classically a severe form of food poisoning, is not a nervous system infection. It is considered here because its principal symptom is paralysis, but it is also discussed in chapter 32, which covers foodborne illness. It is one of the most feared of all diseases because it strikes without warning and is often fatal. The name "botulism" comes from *botulus,* meaning "sausage," chosen because some of the earliest recognized cases occurred in people who had eaten contaminated sausage. The majority of cases worldwide are **foodborne botulism,** which individuals contract by eating food in which the organism has grown and released a poisonous substance. A small percentage of cases occur when the organisms colonize the intestine or a wound, however, causing **intestinal botulism** and **wound botulism,** respectively. Botulism does not involve the central nervous system. ■ foodborne botulism, p. 763

Symptoms

Symptoms usually begin 12 to 36 hours after ingestion of improperly canned, non-acidic food. Most cases begin with dizziness, dry mouth, and blurred or double vision, indicating eye muscle weakness. Abdominal symptoms, including pain, nausea, vomiting, and diarrhea or constipation, are present in 30% to more than 90% of cases, depending on the strain of the causative organism. Progressive paralysis then ensues, generally involving all voluntary muscles, but respiratory paralysis is the most common cause of death. Paralysis distinguishes botulism from most other forms of food poisoning

Causative Agent

Botulism is caused by the strictly anaerobic, Gram-positive, spore-forming, rod-shaped bacterium *Clostridium botulinum.* Its endospores **(figure 27.10),** generally resist boiling for hours, but they are killed by autoclaving. Different strains of *C. botulinum* vary markedly in their biochemical activity, but all produce an exotoxin called **botulinum toxin,** which causes paralysis. Eight antigenically distinct types of botulinum toxin are synthesized by different strains of *C. botulinum.* Types A, B, and E are responsible for most human cases, whereas other types such as C and D affect only birds and other animals. In strains C and D, toxin production results from lysogenic conversion. Botulinum toxin is released in an inactive form upon lysis of the bacterial cells, and it is then activated by proteolytic enzymes. ■ exotoxin, p. 403 ■ lysogenic conversion, p. 310

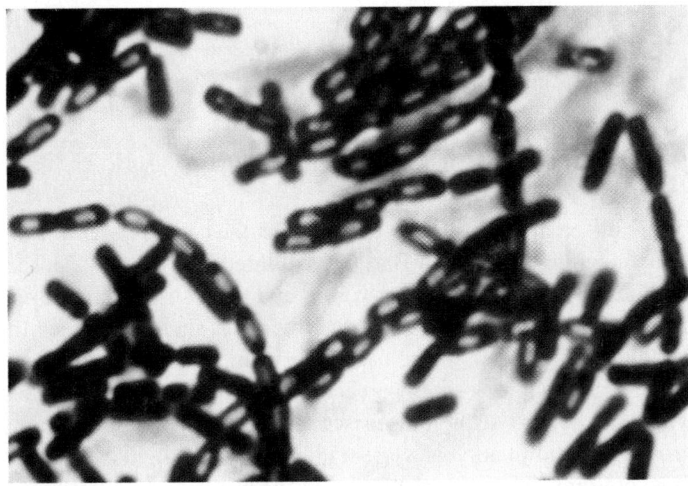

FIGURE 27.10 ***Clostridium botulinum*** Notice the spores. These spores are not reliably killed by the temperature of boiling water.

Pathogenesis

Like other clostridia, *C. botulinum* produces endospores that are generally highly resistant to heat. These endospores can persist in foods such as vegetables, fruit, meat, seafood, and cheese despite cooking and some home-canning processes. The spores germinate if the environment is favorable—such as having a suitable nutrient, anaerobic conditions, a pH above 4.5, and a temperature above 39°F (4°C). Growth of the bacteria results in the release of botulinum toxin into the food. When a person eats the food, the toxin resists digestion by stomach enzymes and acid, is absorbed by the small intestine, and can circulate in the bloodstream for 3 weeks or more. Botulinum toxin is a **neurotoxin**—it acts against the nervous system and is one of the most powerful poisons known. A few milligrams of the toxin would be sufficient to kill the entire population of a large city. Indeed, cases of botulism have resulted from a person eating a single, contaminated string bean, and from licking a finger in unknowingly tasting a food contaminated with botulinum toxin.

Circulating botulinum toxin attaches to motor nerves, blocking transmission of nerve signals to the muscles, and thereby producing paralysis. Normally, nerve signals are transmitted by a type of chemical known as a **neurotransmitter,** present in the nerve in tiny vesicles. The neurotoxin probably acts by preventing the vesicles from attaching to the cytoplasmic membrane of the nerve cell, thereby preventing discharge of the neurotransmitter by exocytosis. Like a number of other bacterial exotoxins, botulinum toxin is composed of two portions—A and B. The B portion attaches to specific receptors on motor nerve endings, and the A portion enters the nerve cell. The A portion then becomes an active peptidase enzyme that degrades the protein on the vesicle surface responsible for attachment to the cytoplasmic membrane. ■ A-B toxins, p. 405

Intestinal botulism occurs occasionally when *C. botulinum* colonizes the intestine, especially in infants 6 months of age or less, and produces a mild form of the disease characterized by constipation followed by generalized paralysis that can range from mild lethargy to respiratory insufficiency. In infants, *C. botulinum* and its toxin are often demonstrable in their feces, but the toxin levels are too low to be detectable in their blood. Most recover without receiving antitoxin treatment, although respiratory support and tube feeding may be needed until the organisms are replaced by normal intestinal microbiota. These infections arise from ingestion of *C. botulinum* spores, which are commonly present in dust and contaminate foods such as honey. Indeed, ingestion of honey has been implicated in 10% to 30% of infant botulism cases. Intestinal botulism also occurs in adults, particularly in immunodeficient patients whose normal intestinal microbiota has been suppressed by antibiotic treatment.

Clostridium botulinum can also colonize dirty wounds, especially those containing dead tissue. The bacteria do not invade, but they can multiply in dead tissue. Botulinum toxin then diffuses into the bloodstream. Many cases have been due to wounds caused by abuse of injected drugs.

Botulinum toxin can be used to treat people with certain chronic, spastic conditions. Minute amounts of the toxin injected into the area of spasm give prolonged relief of symptoms. In the United States, only one company is licensed to market botulinum toxin for human use (Botox); severe paralysis has resulted when other sources were used for cosmetic treatments. ■ Botox, p. 764

Epidemiology

Clostridium botulinum is widely distributed in soils and aquatic sediments around the world. Foodborne botulism, formerly the most common and most severe of the three types of botulism, was first recognized in the late eighteenth century. In the early part of the twentieth century, outbreaks of foodborne botulism were common in the United States, often traceable to commercially canned foods. Strict controls were then placed on commercial canners to ensure adequate sterilizing methods. Since then, outbreaks caused by commercially canned foods have been infrequent. Today, in the United States, intestinal botulism is much more common than foodborne botulism.

Prevention and Treatment

Prevention of botulism depends on proper sterilization and sealing of food at the time of canning. Fortunately, the toxin is heat-labile, and it is recommended that home-canned, low-acid foods be heated to 100°C for 15 minutes just prior to serving to ensure that they are safe to eat. One cannot rely on a spoiled smell, taste, or appearance to detect contamination because such changes are not always present. The disease is treated by the intravenous administration of antitoxin as soon as possible after the diagnosis is made. The antitoxin, however, only neutralizes toxin circulating in the bloodstream, and the nerves already affected by it recover slowly, over weeks or months. Enemas, gastric washing, and surgical removal of dirt and dead tissue from infected wounds help to remove any unabsorbed toxin. Artificial respiration with a mechanical ventilator may be required for prolonged periods.

The main features of botulism are presented in **table 27.5.**

MICROCHECK 27.2

Bacterial meningitis is uncommon. Most cases occur in children under 5 years of age. Organisms from the mother's birth canal are responsible in infants less than 1 month old. Bacteria commonly carried in the upper respiratory tract cause most other cases. Meningococcal meningitis sometimes occurs in epidemics and occasionally causes death within 24 hours. Meningitis is a common complication of listeriosis, caused by a Gram-positive rod that can multiply at refrigerator temperatures and is usually foodborne. Hansen's disease (leprosy) is caused by an acid-fast, unculturable bacterium; the disease exists in two forms, depending on the immune status of the victim. Botulism is a toxin-mediated disease characterized by generalized paralysis. It is not a nervous system infection.

- ✓ Why are most newborn babies unlikely to contract meningococcal, pneumococcal, or *Haemophilus* meningitis?
- ✓ Describe the only bacterial pathogen that preferentially invades peripheral nerves.
- ✓ Why is such a high percentage of infant botulism cases associated with honey?

TABLE 27.5 Botulism

Symptoms	Blurred or double vision, weakness, nausea, vomiting, diarrhea; generalized paralysis and respiratory insufficiency
Incubation period	Usually 12 to 36 hours
Causative agent	*Clostridium botulinum,* an anaerobic, Gram-positive, spore-forming, rod-shaped bacterium
Pathogenesis	*Clostridium botulinum* endospores germinate in food and release neurotoxin. Toxin is ingested, survives stomach acid and enzymes, is absorbed by the small intestine, and is carried by the bloodstream to motor nerves; toxin acts by blocking the transmission of nerve signals to the muscles, producing paralysis; *C. botulinum* can also colonize intestine or wounds, and cause generalized weakness or paralysis.
Epidemiology	Ingestion of contaminated, often home-canned, non-acid food that was not heated enough to kill *C. botulinum* spores. Spores widespread in soil, aquatic sediments, and dust. Can result in colonization of the intestine of adults and infants with deficiencies in normal microbiota, and wounds containing dirt and dead tissue, including those caused by injected-drug abuse.
Prevention and treatment	Education in proper home-canning methods; heating food to boiling for 15 minutes just prior to serving. Treatment: enemas and stomach washing to remove toxin, cleaning infected wounds of dirt and dead tissue, intravenous administration of antitoxin, and artificial respiration.

27.3 Viral Diseases of the Nervous System

Focus Points

- Name the main causative agents for viral meningitis.
- Describe the difference between sporadic and epidemic viral encephalitis.
- Explain why the biggest impact of poliomyelitis in the 1950s occurred in countries with good sanitation.
- Outline the steps in the pathogenesis of rabies.

Many different kinds of viruses can infect the central nervous system, including the Epstein-Barr virus of infectious mononucleosis; the mumps, rubeola, varicella-zoster, and herpes simplex viruses; and, more commonly, human enteroviruses and the viruses of certain zoonoses. In most cases, nervous system involvement occurs in only a very small percentage of people infected with the viruses. The next section discusses four kinds of illness resulting from viral central nervous system infections: meningitis, encephalitis, poliomyelitis, and rabies.

Viral Meningitis

Viral meningitis is much more common than bacterial meningitis. It is usually a mild disease that does not require specific treatment, and patients generally recover in 7 to 10 days. Viruses are responsible for most cases of "aseptic meningitis," meaning those with symptoms and signs of meningitis but with negative tests for bacterial pathogens.

Symptoms

The onset of viral meningitis is typically abrupt, with fever and severe headache above or behind the eyes, sensitivity of the eyes to light, and a stiff neck with increased pain on bending the head forward. Nausea and vomiting are common. In addition, depending on the causative agent, there may be a sore throat, chest pain with inspiration, swollen parotid glands, or a skin rash. ■ parotid glands, p. 582

Causative Agents

Small, non-enveloped RNA viruses—members of the enterovirus subgroup of picornaviruses—are responsible for at least half of the cases of viral meningitis. Of these, the most common offenders are coxsackie viruses, which can cause throat or chest pain, and echoviruses, which can cause a rash. The mumps virus, formerly common, is now an infrequent cause because of widespread immunization against the disease. ■ picornaviruses, p. 324 ■ mumps, p. 593

Pathogenesis

Enteroviruses characteristically infect the throat and intestinal epithelium and lymphoid tissue, and then seed the bloodstream. The **viremia**—viruses circulating in the bloodstream—results in meningeal infection and sometimes rashes or chest infection. The inflammatory response in the meninges differs from bacterial meningitis in that fewer cells usually enter the cerebrospinal fluid, and a high percentage of them are mononuclear rather than polymorphonuclear neutrophils (PMNs). Typically, the cerebrospinal fluid glucose remains normal.

Epidemiology

Enteroviruses are relatively stable in the environment, and they can sometimes even survive in chlorinated swimming pools. Infected individuals, including those with asymptomatic infections, often eliminate the viruses in their feces for weeks. Enteroviral meningitis is transmitted by the fecal-oral route, and the peak incidence occurs in the late summer and early fall in temperate climates. Mumps virus is transmitted by the respiratory route, and mumps meningitis is generally most common in the fall and winter months.

Prevention and Treatment

Handwashing and avoidance of crowded swimming pools are reasonable preventive measures when cases of aseptic meningitis are present in the community. There are no vaccines against coxsackie and echoviruses. Mumps virus disease can be prevented by immunization. No specific treatment is available.

The main features of viral meningitis are presented in **table 27.6.**

TABLE 27.6	Viral Meningitis
Symptoms	Abrupt onset, fever, severe headache, stiff neck, often vomiting; sometimes sore throat, large parotid glands, rash, or chest pain
Incubation period	Usually 1 to 2 weeks for enteroviruses, 2 to 4 weeks for mumps
Causative agents	Most cases: small non-enveloped RNA enteroviruses of the picornavirus family, usually coxsackie or echoviruses. Mumps virus common in unimmunized populations
Pathogenesis	Viremia from primary infection seeds the meninges. Fewer leukocytes enter cerebrospinal fluid than with bacterial infections, and many are mononuclear, usually no decrease in CSF glucose.
Epidemiology	Enteroviruses transmitted by the fecal-oral route, mumps by respiratory secretions and saliva. Enteroviruses transmission mainly summer and early fall; mumps in fall and winter.
Prevention and treatment	Handwashing, avoiding crowded swimming pools during enterovirus epidemics; mumps vaccine for mumps prevention. No specific treatment.

Viral Encephalitis

Whereas viral meningitis is usually a benign illness, viral encephalitis is much more likely to cause death or permanent disability. Viral encephalitis can be sporadic—there are generally a few widely scattered cases occurring all the time; or it can be epidemic—a number of cases appear in a limited period of time in a given geographical area.

Sporadic encephalitis is usually due to herpes simplex virus, but other viruses, including those that cause mumps, measles, and infectious mononucleosis, can occasionally cause encephalitis. Two mechanisms have been proposed to explain herpes simplex encephalitis. Most occurrences are thought to be due to activation of latent virus—the result of primary infection in the past. In about one-fourth of the cases, however, the encephalitis strain differs from latent virus in the person's ganglia, suggesting that the infection causing the encephalitis is newly acquired; the virus probably entering the CNS via the nerves of smell. Most people recover from the disease but are left with permanent impairment, such as epilepsy, paralysis, deafness, or difficulty thinking. The yearly incidence of herpes simplex encephalitis is estimated to be two to three cases per million people. The disease occurs throughout the year, mostly affecting those under age 30 and the elderly. Cases in newborn infants are usually acquired from the mother's birth canal and are due to herpes simplex type 2, but more than 95% of cases in children and adults are caused by herpes simplex type 1. No proven preventive measures for herpes simplex encephalitis exist except for the newborn, for whom cesarian section can be performed to prevent exposure of the baby to a type 2 herpes simplex infection. Medication given promptly can shorten the illness and improve the outcome. ■ herpes simplex virus, pp. 592, 638

Epidemic viral encephalitis is a bigger problem because of the numbers of people that can be affected. Although its symptoms are quite similar to those of sporadic encephalitis, the causative agents, pathogenesis, epidemiology, prevention, and treatment are different.

Symptoms

The onset is usually abrupt, with fever, headache, vomiting, and one or more nervous system abnormalities such as disorientation, localized paralysis, deafness, seizures, or coma.

Causative Agents

Epidemic viral encephalitis is usually caused by **arboviruses** (arthropod-borne viruses), a diverse group of viruses transmitted by insects, mites, or ticks. The four leading causes of epidemic encephalitis in the United States are all arboviruses transmitted by mosquitoes. The West Nile virus, also mosquito-borne, was introduced into New York from the Middle East in the summer of 1999. Migrating birds soon spread the virus to 20 eastern states from Maine to Florida, and then across the country **(figure 27.11)**.

The arboviruses are enveloped, single-stranded RNA viruses that are classified in different families:

- LaCrosse encephalitis virus, bunyavirus family;
- St. Louis and West Nile encephalitis viruses, flavivirus family; and the
- Eastern and western equine encephalitis viruses, togavirus family. ■ mosquitoes, p. 296

Pathogenesis

Knowledge of the pathogenesis of arboviral encephalitis is incomplete. The viruses multiply at the site of a mosquito bite and in local lymph nodes, producing viremia. Levels of viremia are very low and transitory, yet the ratio of overt encephalitis to mild or inapparent infection can be 1:25 or greater. It is not known how these viruses cross the blood-brain barrier so readily. The viruses replicate in nerve cells and cause extensive destruction of brain tissue in severe cases. The process is halted with the appearance of neutralizing antibody. Mortality ranges from about 2% with LaCrosse, to 35% to 50% with eastern equine encephalitis. Disabilities occur in those who recover, ranging from 5% to more than 50% of cases, depending largely on the kind of virus and the age of the patient—the very young and the elderly suffer the most. The kinds of disability include emotional instability, mental retardation, epilepsy, blindness, deafness, and paralysis of one side of the body.

Epidemiology

During epidemic viral encephalitis, only a minority of those infected develop encephalitis. Others develop viral meningitis, fever and headache only, or no symptoms at all. These diseases are all zoonoses maintained in nature in birds or rodents; humans are an accidental host. In the eastern half of the United States, LaCrosse encephalitis virus usually causes most of the reported encephalitis cases. In its natural cycle **(figure 27.12)**, the LaCrosse virus infects *Aedes* mosquitoes, which pass it directly from one mosquito to another in semen. It can survive the winter in the mosquito's eggs. These mosquitoes feed on and infect squirrels and chipmunks, which markedly amplify the amount of virus and spread it to uninfected female mosquitoes feeding on the rodents' blood. These mosquitoes are forest dwellers that normally breed in water-containing cavities in hardwood trees or in discarded tires.

The natural cycle of other encephalitis-causing viruses involves wild birds rather than rodents, and involves different mosquitoes with various habitats and feeding habits.

There is still much to be learned about the ecological factors responsible for the spread of these viruses to humans, but the proximity of humans to their natural cycles, rainfall, winter and summer temperatures, numbers and range of amplifying hosts, and predators could all be important.

Prevention and Treatment

As equine encephalitis viruses spread from their natural hosts, they generally infect horses 1 or 2 weeks before the first human cases appear. Cases in horses provide a warning to increase protection against mosquitoes. Sentinel chickens serve the same function. They are stationed in cages with free access to mosquitoes, and their blood

FIGURE 27.11 Areas Reporting West Nile Virus Infection United States, 2007 (as of November 13, 2007)

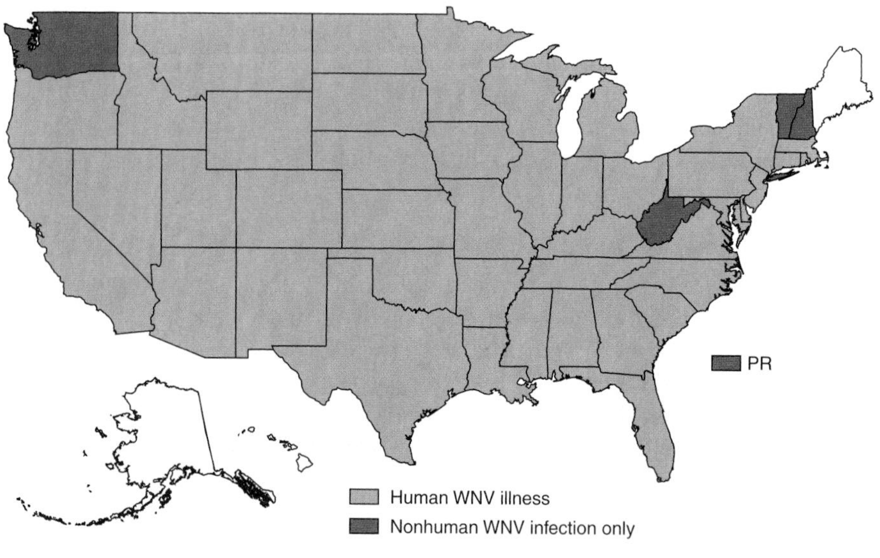

PR

Human WNV illness

Nonhuman WNV infection only

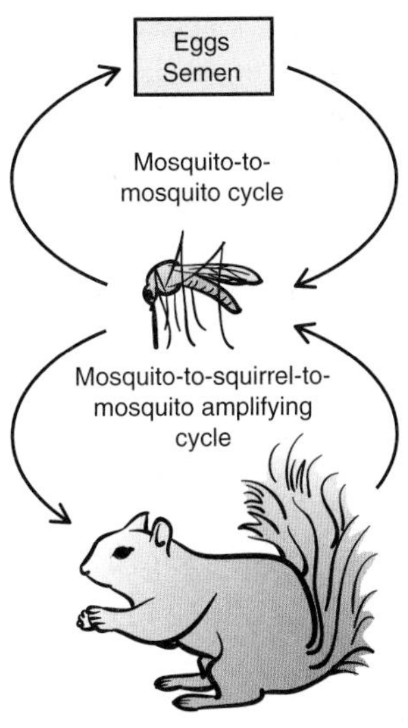

FIGURE 27.12 LaCrosse Encephalitis Virus, Natural Cycles The virus is maintained both by mosquito-to-mosquito transmission through semen and eggs, and by mosquito-to-squirrel (or chipmunk)-to-mosquito transmission. Infection of the rodent markedly increases the quantity of virus and the potential to infect many mosquitoes.

is tested periodically for evidence of arbovirus infection. A positive test would trigger an encephalitis alert, as for St. Louis encephalitis in Florida: "Avoid outdoor activity during evening and night, the peak hours of biting for the *Culex* mosquito vector. If outdoors, wear long sleeves and pants. Make sure windows and porches are properly screened. Use insect repellents and insecticides."

A vaccine against eastern equine encephalitis is approved for use in horses and has also been used to protect emus—a large, domesticated, meat-producing fowl that is highly susceptible to this virus. There is no proven antiviral therapy for arbovirus encephalitis.

The main features of epidemic viral encephalitis are presented in **table 27.7.**

Infantile Paralysis, Polio (Poliomyelitis)

The characteristic feature of poliomyelitis is selective destruction of motor nerve cells, usually of the spinal cord, resulting in permanent paralysis of one particular group of muscles, such as those of an arm or leg. The two individuals most responsible for control of this terrifying disease will not see its imminent elimination from the world—Albert Sabin died in 1992; Jonas Salk in 1995. The two men were bitter rivals, both of whom expected, but did not receive, the Nobel Prize.

Symptoms

Poliomyelitis usually begins with the symptoms of meningitis: headache, fever, stiff neck, and nausea. In addition, pain and spasm of some muscles generally occur, later followed by paralysis. Over the ensuing weeks and months, muscles shrink and bones do not develop normally in the affected area **(figure 27.13).** In more severe cases, the muscles of respiration are paralyzed, and the victim requires an artificial respirator—a machine to pump air in and out of the lungs. Some recovery of function is the rule if the person survives the acute stage of the illness. The nerves of sensation, touch, pain, and temperature are not affected.

Causative Agent

Poliomyelitis is caused by three types of polioviruses, designated 1, 2, and 3, that are distinguished by using antisera. These small,

TABLE 27.7 Epidemic Viral Encephalitis

① Infected mosquito introduces encephalitis virus.

② Virus multiplies locally, establishes brief low-level viremia.

③ Virus crosses blood-brain barrier and preferentially attacks the brain.

④ Destruction of brain tissue causes death or permanent disabilities such as emotional instability, mental retardation, paralysis of face, arm, leg.

⑤ Due to brief viremia, there is no exit for the virus, thus humans are the final host.

Symptoms	Abrupt onset, fever, headache, vomiting, disorientation, paralysis, seizures, deafness, coma
Incubation period	First symptoms within a few days; encephalitic symptoms often within the first week
Causative agent	Usually caused by one of four arboviruses, LaCrosse, St. Louis, western equine, or eastern equine
Pathogenesis	Replication of virus at the site of the mosquito bite, further replication in lymph nodes, then viremia that seeds brain tissue. Nerve cells in the brain invaded, destroyed. Process halted by neutralizing antibody.
Epidemiology	Viruses transmitted to humans from birds or rodents by mosquitoes.
Prevention and treatment	Chicken sentinels to warn of arbovirus epidemics. Insecticides and other anti-mosquito preventive measures. No accepted treatment for arboviral encephalitis.

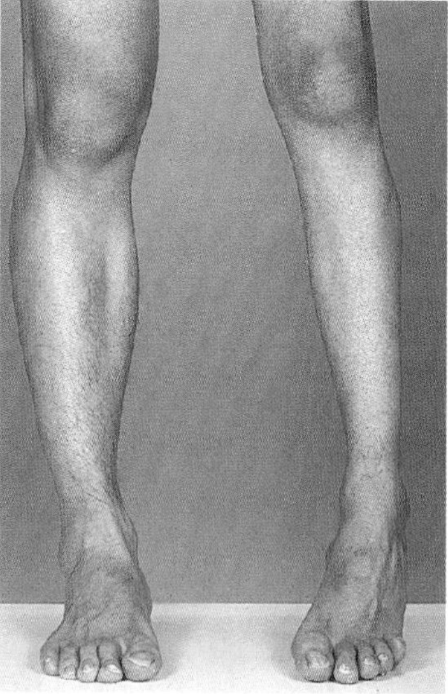

FIGURE 27.13 Atrophy of the Left Leg Due to Poliomyelitis

non-enveloped, single-stranded RNA viruses are members of the enterovirus subgroup of the picornavirus family. They can be grown *in vitro* in cell cultures, where they cause cell destruction. With low concentrations of virus, the areas of cell destruction, termed **plaques,** are separated from one another and are readily seen with the unaided eye **(figure 27.14).** ■ plaques, p. 337

Pathogenesis

Polioviruses enter the body orally, infect the throat and intestinal tract, and then invade the bloodstream. In most people, the immune system conquers the infection and recovery is complete. Only in a small percentage of people does the virus enter the nervous system and attack motor nerves. The viruses can infect a cell only if the surface of that cell possesses specific receptors to which the virus can attach. This specificity of receptor sites helps explain the fact that polioviruses selectively infect motor nerve cells of the brain and spinal cord, while sparing many other kinds of cells. The infected cell is destroyed when the mature virus is released.

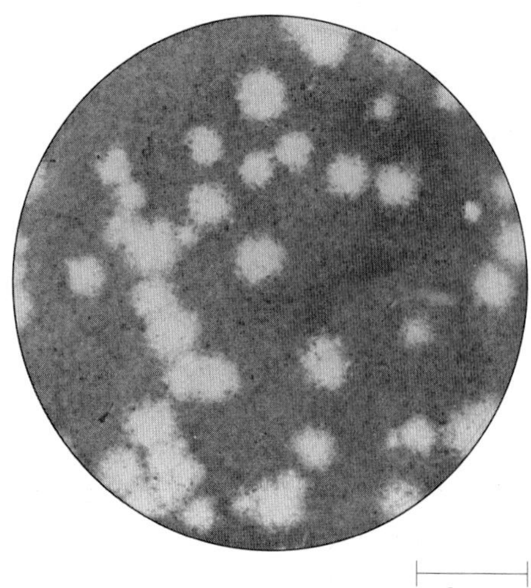

FIGURE 27.14 Plaques Produced by Poliomyelitis Virus in a Cell Culture Monolayer Each plaque represents an area of the monolayer that has been destroyed by replicating poliovirus.

Even though poliomyelitis transmission no longer occurs in the United States, there are about 300,000 survivors of the disease, most of whom recovered years ago. Some of them develop muscle pain, increased weakness, and muscle degeneration 15 to 50 years after they had acute poliomyelitis. This condition, called the **post-polio syndrome,** sometimes involves muscles not obviously affected by the original illness. The late progression of muscle weakness is not due to recurrent multiplication of polio viruses. During recovery from acute poliomyelitis, surviving nerve cells branch out to take over the functions of the killed nerve cells. The late appearance of symptoms is probably due to the death of these nerve cells that have been doing double duty for so many years.

Epidemiology

Because only a small percentage of infected individuals develop nervous system involvement, even one case of poliomyelitis indicates that the virus is widespread in the community.

Poliomyelitis has had its greatest impact in countries when the level of sanitation is improved. In areas where sanitation is poor, the polioviruses are widespread, transmitted by the fecal-oral route. Very few people in areas with poor sanitation escape childhood without becoming infected and developing immunity to the disease. Therefore, most babies have received antibodies against polio from their mothers, the antibodies having crossed the placenta and entered the circulation of the fetus. Newborn infants in these nations thus are partially protected against nervous system invasion by poliomyelitis virus for as long as their mothers' antibodies persist in the babies' bodies—usually about 2 or 3 months. During this time, because of exposure to the poliovirus through crowding and unsanitary conditions, infants are likely to develop mild infections of the throat and intestine, thereby achieving lifelong immunity with little chance of paralysis.

In contrast, in areas with efficient sanitation, the poliomyelitis virus sometimes cannot spread to enough susceptible people to sustain itself, and it disappears from the community. When it is reintroduced, people of all ages may lack antibody and be susceptible, and a high incidence of paralysis will result. This situation occurred in the United States in the 1950s, resulting in many cases of respiratory paralysis **(figure 27.15)** and death. With most people now routinely immunized against the disease, however, and the likelihood of imported disease rapidly waning, this scenario should no longer occur.

Prevention and Treatment

Like other enteroviruses, polioviruses are quite stable under natural conditions, often surviving in swimming pools, but are inactivated by pasteurization and properly chlorinated drinking water. Control of poliomyelitis using vaccines represents one of the greatest success stories in the battle against infectious diseases **(figure 27.16).** Ironically, all cases of paralytic polio acquired in the United States since 1980 were caused by Sabin's oral, attenuated polio vaccine, which was introduced in 1961. These cases arise because in rare instances the vaccine strain can mutate and become virulent. This small risk of developing paralytic poliomyelitis from the vaccine virus, approximately one case per 2.4 million doses given, led to discontinuing routine use of the live vaccine in the United States in mid-1999. However, despite its drawback, the Sabin oral vaccine remains key to controlling polio in areas of the world where

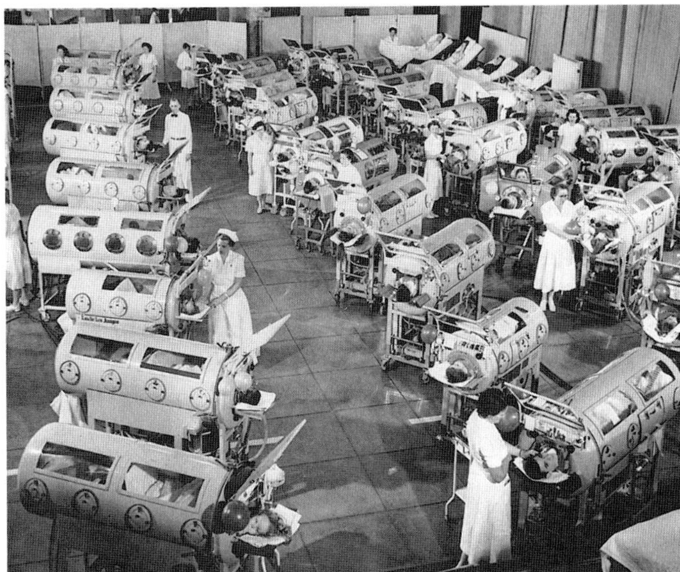

FIGURE 27.15 The Horror of Poliomyelitis These tanklike respirators ("iron lungs") were used during the 1950s epidemics of poliomyelitis to keep alive people whose respiratory muscles were paralyzed by the disease. Now we can hope to see polio forever banished from the earth!

transmission of the wild virus still occurs. It produces better local immunity in the throat and intestine, can spread from person to person, does not require an injection, provides herd immunity, and is less expensive. In the 15 years following 1988, when the World Health Organization resolved to eradicate poliomyelitis, the number of countries with endemic disease was reduced from 125 to only six. ■ **herd immunity, p. 456** ■ **campaign to eliminate poliomyelitis, p. 436**

Poliomyelitis is summarized in **table 27.8.**

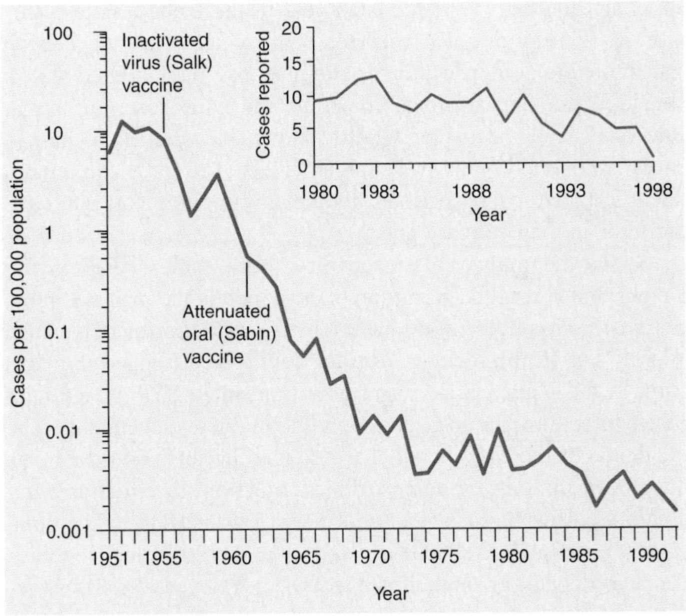

FIGURE 27.16 Incidence of Poliomyelitis in the United States, 1951 to 1998 Ironically, since 1980, all cases acquired in the United States have been caused by the oral attenuated (Sabin) vaccine. Endemic poliomyelitis was eliminated from the United States by 1980, and the entire Western Hemisphere by 1991.

TABLE 27.8	Poliomyelitis
Symptoms	Headache, fever, stiff neck, nausea, pain, muscle spasm, followed by paralysis
Incubation period	7 to 14 days
Causative agents	Polioviruses 1, 2, 3, members of the picornavirus family
Pathogenesis	Virus infects the throat and intestine, circulates via the bloodstream, and enters some motor nerve cells of the brain or spinal cord; infected nerve cells lyse upon release of mature virus.
Epidemiology	Spreads by the fecal-oral route; asymptomatic and nonparalytic cases common.
Prevention and treatment	Prevented by injecting Salk's inactivated vaccine, or by Sabin's orally administered attenuated vaccine in areas of epidemic or endemic disease. Treatment: artificial ventilation for respiratory paralysis; physical therapy and rehabilitation.

Rabies

In the United States, immunization of dogs and cats against rabies has practically eliminated them as a source of human disease. The rabies virus remains rampant among wildlife, however—a constant threat to non-immunized domestic animals and humans. Many questions remain about the pathogenesis of rabies, and no effective treatment exists for the disease.

Symptoms

Rabies is one of the most feared of all diseases because its terrifying symptoms almost invariably end with death. Like many other viral diseases, it begins with fever, head and muscle aches, sore throat, fatigue, and nausea. The characteristic symptom that strongly suggests rabies is a tingling or twitching sensation at the site of viral entry, usually an animal bite. These early symptoms generally begin 1 to 2 months after viral entry and progress rapidly to symptoms of encephalitis, agitation, confusion, hallucinations, seizures, and increased sensitivity to light, sound, and touch. The body temperature then rises steeply, and increased salivation combined with difficulty swallowing result in frothing at the mouth. Hydrophobia—painful spasm of the throat and respiratory muscles provoked by swallowing or even upon seeing liquids—occurs in half the cases. Coma develops, and about 50% of patients die within 4 days of the first appearance of symptoms, the rest soon after.

Causative Agent

The cause of rabies is the rabies virus **(figure 27.17a),** a member of the rhabdovirus family. This virus has a striking bullet shape, is enveloped, and contains single-stranded, negative-sense RNA. It buds from the surface of infected cells.

Pathogenesis

The principal mode of transmission of rabies to humans is via the saliva of a rabid animal (figure 27.17b) introduced into

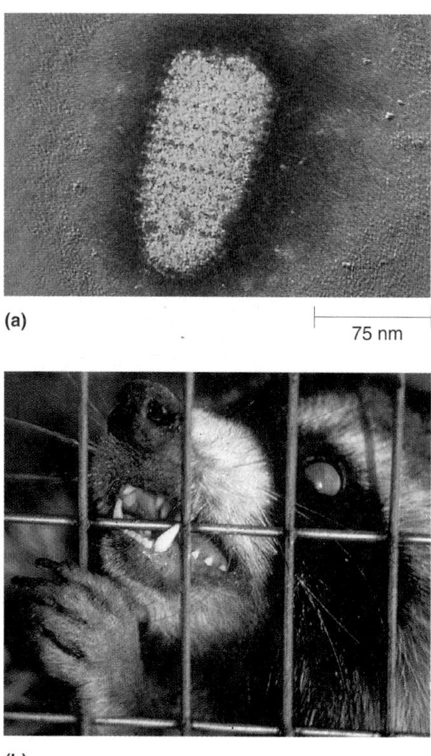

FIGURE 27.17 Rabies Virus (a) Color-enhanced transmission electron micrograph of rabies virus. Notice the bullet shape. **(b)** Rabid raccoon caged in a Virginia animal shelter. Rabies virus is usually transmitted in the saliva of a biting animal.

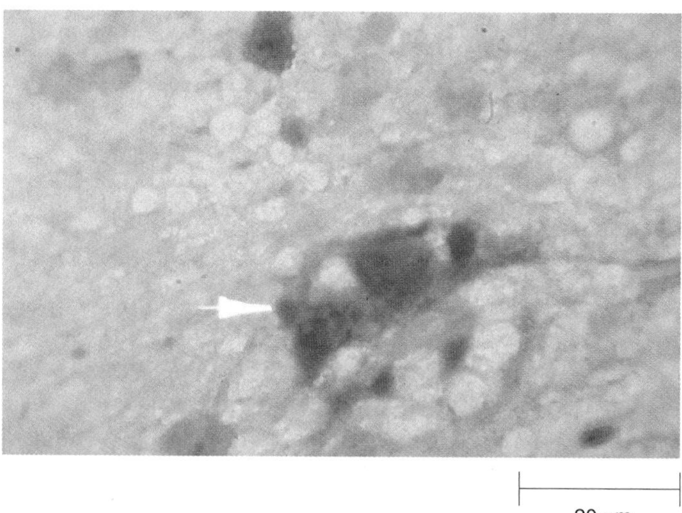

FIGURE 27.18 Stained Smear of Brain Tissue from a Rabid Dog The arrows point to one of several Negri bodies within the triangular-shaped nerve cell. The Negri bodies represent the sites of rabies virus replication.

bite wounds or abrasions of the skin. Individuals can also contract rabies by inhaling aerosols containing the virus, such as from bat feces. Only a few details of the events that follow introduction of the virus into the body are known. During the incubation period of the disease, the virus multiplies in muscle cells and probably other cells at the site of infection. Knoblike projections on the viral surface attach to receptors in the region where the nerve joins the muscle. At some point, the virus enters an axon and is carried along the nerve by the normal flow of the axon's cytoplasm, eventually reaching the brain. The long incubation period, usually 1 to 2 months but sometimes exceeding a year, is partly determined by the length of the journey to the brain. Patients with head wounds into which the virus is introduced tend to have a shorter incubation period than those with extremity wounds. Severe wounds and those with a large amount of introduced rabies virus also generally result in short incubation periods. The virus then multiplies extensively in brain tissue, causing the symptoms of encephalitis. Characteristic inclusion bodies, called **Negri bodies (figure 27.18),** form at the sites of viral replication in the brain, but the cells are not lysed. The virus spreads outward from the brain via the nerves to various body tissues, notably the salivary glands, eyes, and fatty tissue under the skin, as well as to the heart and other vital organs. The presence of the virus in the eyes is of some practical significance, because victims can be diagnosed before their death by identifying rabies in stained smears made from the surface of their eyes. Moreover, several cases of rabies have occurred in individuals who received corneal transplants from donors who died of an atypical form of rabies that escaped diagnosis (the cornea is the clear middle part of the eye in front of the lens).

Epidemiology

Rabies is widespread in wild animals; about 5,000 wild animal cases are generally reported each year in the United States. This represents an enormous reservoir from which infection can be transmitted to domestic animals and humans. In the United States, skunks, raccoons, and bats constitute the chief reservoir hosts. Raccoons lead the list of wildlife cases, but almost all human cases are due to contact with infected bats. The virus can remain latent in bats for long periods, and healthy-looking bats can have the virus in their salivary glands.

Until the mid-twentieth century, most rabies cases in the United States resulted from dog bites, which is the primary mode of transmission in the non-industrialized world. Fortunately, since World War II, the incidence of dog rabies has dropped dramatically in the United States, and now more than 85% of reported rabies infections are in wild, as opposed to domestic, animals. The incidence of rabies in people has declined because dogs and cats have been immunized against rabies infection, in effect creating a partial barrier to the spread of the rabies virus from wild animal reservoirs to humans. Because pets vary in their tolerance to rabies vaccines depending on their age and species, several kinds of vaccines are available.

About three-quarters of dogs that develop rabies excrete rabies virus in their saliva, and about one-third of these begin excreting it 1 to 3 days before they get sick. Therefore, when a person is bitten

PERSPECTIVE 27.1

A Rabies Survivor!

In October 2004, a 15-year-old girl was admitted to the hospital with typical signs and symptoms of rabies. A month earlier, while in church, she picked up a bat that had fallen to the floor. The bat bit her left index finger and she carried it outside and released it. The bite wound, which was small and showed a small amount of bleeding, was cleaned with hydrogen peroxide, and healed uneventfully.

Because most rabies deaths result from respiratory failure, a tube was placed in the patient's trachea and hooked to a respirator. She was given medication to put her into a coma to rest her nervous system, and she was given the antiviral medication ribavirin, according to an experimental protocol. After 7 days, the coma-inducing medication was reduced and the girl was allowed to wake up. A little more than a month after her illness began, her breathing tube was removed, she regained speech, solved mathematics problems, and walked with assistance. Slow, steady improvement continued after she returned home.

Survival after the onset of rabies symptoms is known to have occurred in only five individuals prior to this case, and all five had received anti-rabies vaccine either before they were exposed to rabies or before they showed signs or symptoms of the disease. All but one of them suffered persisting neurological damage, unlike the present patient, who made a complete recovery. Over the years, many treatments have been tried without benefit, including hyperimmune human anti-rabies serum, anti-rabies vaccines, and interferon. The current patient did not receive any of these.

What is to be learned from the survival of this patient? Perhaps her survival had nothing to do with her treatment, but was due to her own immune system fighting a small infectious dose or a rabies viral strain of low virulence. Is there some clue in her survival that will perhaps lead to the first effective rabies treatment in the 3,000 years the disease has been recognized? So far, attempts to cure other patients using the treatment used in this case have been unsuccessful.

by an unvaccinated, apparently healthy dog, the animal should be confined for 10 days to see if symptoms of rabies appear. Some dogs become irritable and hyperactive with the onset of rabies, produce excessive saliva, and attack people, animals, and inanimate objects. Perhaps more common is the "dumb" form of rabies, in which an infected dog simply stops eating, becomes inactive, and suffers paralysis of throat and leg muscles. Obviously, one should not be tempted to try to remove a suspected foreign body from the throat of a sick, choking, non-vaccinated dog! The reported number of rabies cases in humans now generally ranges from zero to four per year in the United States. Only about one-fourth of the cases have a history of animal bites or exposure to sick animals. Some of them possibly contracted the infection by inhaling dust contaminated by rabies virus. The long incubation period, however, reportedly up to 6 years, and the patients' illness make the animal bite history unreliable.

Prevention and Treatment

A person who has been bitten by an animal should wash the wound immediately and thoroughly with soap and water and then apply an antiseptic. In people bitten by dogs having rabies virus in their saliva, the risk of developing rabies is about 30%. Louis Pasteur discovered that this risk can be lowered considerably by administering rabies vaccine as soon as possible after exposure to the virus. Presumably, the vaccine provokes a better immune response than the natural infection, neutralizing free virus and killing infected cells during the long incubation period before the virus enters the nerves. Pasteur's vaccine was made from dried spinal cords of rabies-infected rabbits. Unfortunately, the rabbit nervous system tissue in the vaccine sometimes stimulated an immune response against the patient's own brain, causing **allergic encephalitis.** Current vaccines are essentially devoid of nervous tissue and the risk of serious side effects is very low.

If there is a reasonable possibility that the animal is rabid, the bitten individual should then receive a series of five injections of vaccine intramuscularly. Anti-rabies antibody is also injected at the wound site and intramuscularly to provide passive immunity. The anti-rabies antibody is obtained from humans who have been immunized against rabies. In the United States, about 30,000 people annually receive rabies vaccine to prevent rabies after having been bitten by suspected rabid animals. ■ passive immunity, p. 432

There is no effective treatment for rabies, and only six people are known to have ever recovered from the disease (see **Perspective 27.1**).

Some features of rabies are summarized in **table 27.9.**

TABLE 27.9 Rabies

Symptoms	Fever, headache, nausea, vomiting, sore throat, cough at onset; later, spasms of the muscles of mouth and throat, coma, and death
Incubation period	Usually 30 to 60 days; sometimes many months or years
Causative agent	Rabies virus, single-stranded RNA, rhabdovirus family; has an unusual bullet shape
Pathogenesis	During incubation period, virus multiplies at site of bite, then travels via nerves to the central nervous system; here it multiplies and spreads outward via multiple nerves to infect heart and other organs.
Epidemiology	Bite of rabid animal, usually a bat. Inhalation is another possible mode.
Prevention and treatment	Avoid suspect animals; immunize pets. Effective post-exposure measures: immediately wash wound with soap and water and apply antiseptic; inject rabies vaccine and human rabies antiserum as soon as possible. No effective treatment once symptoms begin.

MICROCHECK 27.3

Many different kinds of viruses can attack the nervous system, but they generally do so in only a small percentage of the people they infect. At least half of viral meningitis cases are caused by the enterovirus subgroup of picornaviruses, which are generally spread by the fecal-oral route. Viral meningitis is usually benign, but viral encephalitis often causes permanent disability. Arboviruses maintained in nature in a mosquito-bird or mosquito-rodent cycle are a leading cause of epidemic viral encephalitis. Sporadic viral encephalitis is usually due to herpes simplex virus. Poliomyelitis, characterized by paralysis of one or more muscle groups, is caused by three other enteroviruses. Rabies is a widespread zoonosis, almost uniformly fatal for humans, usually transmitted by animal bites.

- ✓ Name one cause of viral meningitis preventable by vaccination.
- ✓ Why is rabies now rare in humans when it is still so common in wildlife?
- ✓ Why is the Sabin vaccine against poliomyelitis the preferred vaccine to use in countries where polio is still prevalent?

27.4

Fungal Diseases of the Nervous System

Focus Point

- Discuss the main steps in the development of cryptococcal meningoencephalitis.

Fungi rarely invade the central nervous system of healthy people, but patients with cancer, diabetes, and AIDS, as well as those receiving immunosuppressive medications, risk serious disease. Common fungi from the soil and decaying vegetation sometimes infect the nose and sinuses of these patients; from there, they penetrate to the brain and cause the patient's death. These infections are dangerous because they are difficult to treat with antimicrobial medication. Cryptococcal meningoencephalitis differs somewhat from this general pattern—about half the cases occur in otherwise healthy people, and many cases are cured by medication.

Cryptococcal Meningoencephalitis

Cryptococcal meningoencephalitis was an uncommon disease until the onset of the AIDS epidemic. Now 2 to 4 cases occur among every 1,000 AIDS patients, as opposed to 0.2 to 0.9 cases per 100,000 in the general population. The disease is among the top four life-threatening infectious complications in AIDS.

Symptoms

In apparently healthy people, symptoms of cryptococcal meningoencephalitis develop very gradually in most cases and generally consist of difficulty in thinking, dizziness, intermittent headache, and slight or no fever. After weeks or months of slow progression of these symptoms, vomiting, weight loss, paralysis, seizures, and coma may appear. In people with immunodeficiency, the disease generally progresses much faster; without treatment death can occur in as little as 2 weeks.

Causative Agent

Cryptococcal meningoencephalitis is an infection of the meninges and brain by the encapsulated yeast, *Cryptococcus neoformans*, a member of the basidiomycetes. Different antigenic varieties labeled by capital letters are distinguished using antisera; varieties A and D are opportunistic pathogens, whereas varieties B and C can attack otherwise healthy people. The latter two antigenic varieties are also known as *C. neoformans* var *gattii*. In nature, *C. neoformans* probably can produce tiny spores (basidiospores), but this has been difficult to study in the laboratory. As seen in infected material from patients, the organism is a small, spherical yeast generally 3 to 7 μm in diameter surrounded by a large capsule **(figure 27.19).** ■ basidiomycetes, p. 288

Pathogenesis

Presumably, basidiospores of *C. neoformans* become airborne and enter the human body by inhalation. Infection is established in the lung, usually producing mild or no symptoms. Body defenses of healthy people can often eliminate the infection, but phagocytic killing is slow and inefficient. In some cases, particularly in immunocompromised individuals, the organisms continue to multiply, enter the bloodstream, and are then distributed throughout the body. The capsule is essential to pathogenicity; non-encapsulated strains do not cause disease. Capsular material inhibits phagocytosis of the organisms despite their being coated with opsonin from the complement system. Migration of leukocytes to the site of infection is impaired. Capsular material diffuses from the organisms and can be detected in spinal fluid and urine, aiding diagnosis. Progressive infection and dissemination are likely to occur when a person's cellular immunity is impaired, as in AIDS and certain cancers. Meningoencephalitis is the most common infection outside of the lung, but organisms spread by the bloodstream can also infect skin, bones, and other body tissues. In meningoencephalitis, the organisms typically cause thickening of the meninges, sometimes impeding the flow of cerebrospinal fluid, thereby increasing the pressure within the brain. They also

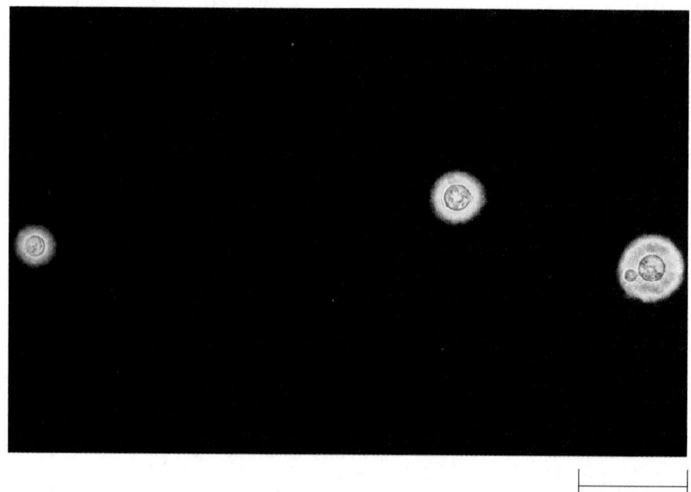

FIGURE 27.19 *Cryptococcus neoformans* in the Spinal Fluid of an Individual with Cryptococcal Meningoencephalitis India ink has been added to the fluid to outline the organism's capsule.

invade the brain tissue, producing multiple abscesses. The sites of infection often show little inflammatory response. ■ opsonins, pp. 356, 373

Epidemiology

Cryptococcus neoformans is distributed worldwide in soil and vegetation, but infections have been linked only to sites of well-decomposed pigeon droppings (varieties A and D), and to flowering of certain trees (B variety of *C. neoformans*). For every case of cryptococcal meningoencephalitis, millions of people are infected by the organism without harm. Symptomatic infection is often, but not always, the first indication of AIDS. Person-to-person transmission of the disease does not occur.

Prevention and Treatment

There is no vaccine or other preventive measure available. Treatment with the antibiotic amphotericin B is often effective, particularly if given concurrently with flucytosine (5-fluorocytosine) followed by the oral medicine, fluconazole. Amphotericin B must be given intravenously and the dose carefully regulated to minimize the toxic effects of the antibiotic, mainly against the kidneys. Because amphotericin B does not reliably cross the blood-brain barrier, it is often necessary to administer it through a plastic tube inserted through the skull into a lateral ventricle of the brain. Except in AIDS patients, treatment is successful in about 70% of cases. AIDS patients respond poorly to treatment, most likely because they lack T-cell-dependent killing of *C. neoformans* that normally assists the action of the antifungal medications. Unless their T-cell function can be restored by treatment, AIDS patients are rarely cured of their infection and must remain on antifungal medication for life. ■ antifungal medicines, p. 488

The main features of cryptococcal meningoencephalitis are summarized in **table 27.10.**

TABLE 27.10 Cryptococcal Meningoencephalitis

Symptoms	Headache, vomiting, confusion, and weight loss; slight or no fever; symptoms may progress to seizures, paralysis, coma, and death
Incubation period	Widely variable, few to many weeks
Causative agent	*Cryptococcus neoformans*, an encapsulated yeast
Pathogenesis	Infection starts in lung; encapsulated organisms multiply, enter bloodstream, and are carried to various parts of the body; phagocytosis inhibited and opsonins neutralized; meninges and adjacent brain tissue become infected.
Epidemiology	Inhalation of dust containing dried pigeon droppings contaminated with the fungus; other sources; most people resistant to the disease.
Prevention and treatment	No preventive measures. Treatment; amphotericin B with flucytosine or itraconazole.

MICROCHECK 27.4

Fungi rarely invade the CNS of healthy people, but even common soil fungi are a serious threat to individuals with diabetes, cancer, or immunodeficiency. Cryptococcal meningoencephalitis occurs rarely in healthy people, but it is among the leading life-threatening complications in AIDS. The causative organism is a small yeast with a large capsule, often found in soil contaminated with pigeon droppings.

✓ What underlying condition should a physician immediately suspect when confronted with a patient with cryptococcal meningoencephalitis?

✓ Why might it be a good idea for persons with immunodeficiency to avoid soil contaminated with decomposed pigeon droppings?

✓ Why is it so difficult to cure AIDS victims of cryptococcal meningoencephalitis?

27.5 Protozoan Diseases of the Nervous System

Focus Point

■ Outline the pathogenesis of African sleeping sickness.

Only a few species of protozoa are important central nervous system pathogens for humans. One interesting example, *Naegleria fowleri,* causes **primary amebic meningoencephalitis** after penetrating the skull along the nerves responsible for the sense of smell. The protozoan occurs worldwide, but less than 200 cases have been reported, almost all rapidly fatal. The disease is usually acquired when individuals swim in or are splashed with warm, fresh water, with or without chlorine. Apparently it can also be contracted by inhaling dust laden with cysts of the protozoa. For every case of *Naegleria* meningoencephalitis, many millions of people are exposed to the organism without harm. *Naegleria fowleri* can exist in three forms: ameba, flagellate, and cyst (see figure 12.7). It is one of only a few free-living protozoa pathogenic for humans. In contrast, African sleeping sickness is a quite different protozoan disease. ■ protozoa, p. 285

African Sleeping Sickness

African sleeping sickness, also known as **African trypanosomiasis,** is transmitted by its biological vector—the daytime-biting tsetse fly. The disease is important because it can be contracted by residents and visitors in a wide area across the middle of the African continent. The causative protozoan has an interesting way of protecting itself against the host's immune system. ■ biological vector, p. 455

Symptoms

The first symptoms of African trypanosomiasis appear within a week after a person is bitten by an infected tsetse fly. A tender nodule develops at the site of the bite. The regional lymph nodes might enlarge, but symptoms may all disappear spontaneously. Weeks to several years later, recurrent fevers develop that can continue for months or years. Involvement of the central nervous system is marked by gradual loss of interest in everything, decreased activity, and indifference to food. The eyelids droop, the individual falls asleep while eating or even standing, the speech becomes slurred, coma develops, and eventually death ensues.

Causative Agent

African sleeping sickness is caused by the flagellated protozoan, *Trypanosoma brucei.* These organisms are slender, and have a wavy, undulating membrane and an anteriorly protruding flagellum **(figure 27.20).** There are two subspecies that are morphologically identical—*T. brucei rhodesiense* and *T. brucei gambiense.* The *rhodesiense* subspecies occurs mainly in the cattle-raising areas of East Africa, whereas the *gambiense* subspecies occurs mainly in forested areas of Central and West Africa. Both are transmitted by tsetse flies, a group of biting insects of the genus *Glossina.*

Pathogenesis

When an infected tsetse fly bites its victim, the protozoan enters the bite wound in the fly's saliva. The organism multiplies at the skin site and within a few weeks enters the lymphatics and blood circulation. The patient responds with fever and production of antibody against the protozoa, and symptoms improve. Within about a week and at roughly weekly intervals thereafter, however, there are recurrent increases in the number of parasites in the blood. Each of these bursts of increased **parasitemia,** meaning parasites in the circulating blood, coincides with the appearance of a new glycoprotein on the surface of the trypanosomes. More than a thousand genes, each coding for a different surface glycoprotein, are present in the protozoan chromosome. Only one of these genes is activated at a time, and the patient's immune system must respond to each gene product with production of a new specificity of antibody. The recurrent cycles of parasitemia and antibody production continue until the patient is treated or dies.

In *T. brucei rhodesiense* infections, the disease tends to progress rapidly—the heart and brain are invaded within 6 weeks of infection. Irritability, personality changes, and mental dullness result from brain involvement, but the patient usually dies from heart failure within 6 months. With *T. brucei gambiense,* progression of infection is much slower—years may pass before death occurs, often from secondary infection. Much of the damage to the host is due to immune complexes formed when antibody reacts with complement and high levels of protozoan antigen. ■ immune complexes, p. 420

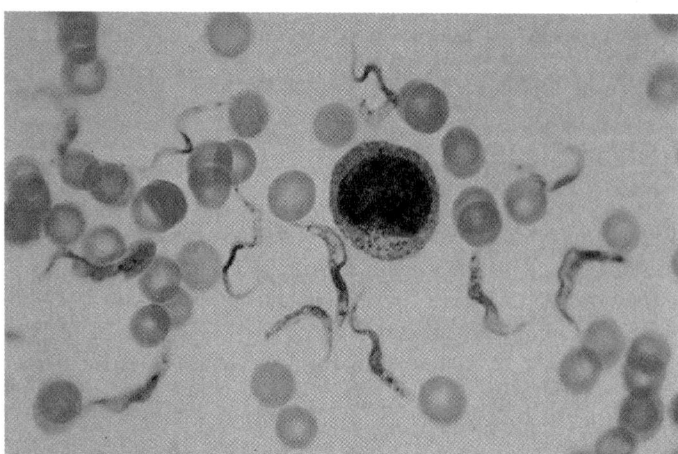

FIGURE 27.20 ***Trypanosoma brucei* in a Blood Smear** Notice the slender protozoa among the blood cells of an individual with African sleeping sickness (African trypanosomiasis)

Epidemiology

African sleeping sickness occurs on the African continent within about 15° of the equator, with 10,000 to 20,000 new cases each year, including a number of American tourists. The occurrence of the disease is determined by the distribution of the tsetse fly vectors. The severe Rhodesian form of the disease is a zoonosis and the main reservoirs are wild animals; for the milder Gambian form, humans are the main reservoir and human-to-human transmission is more common than animal-to-human. Bites of the infected tsetse fly transmit the disease, but less than 5% of the flies are infected. ■ reservoir, p. 453

Prevention and Treatment

Preventive measures directed against tsetse fly vectors include use of insect repellents and wearing protective clothing to prevent bites, use of traps containing bait and insecticides, and clearing of brush that provides breeding habitats for the flies. Populations can be screened for *T. brucei* infection by examining blood specimens. Treatment of infected people helps reduce the protozoan's reservoir. A single, intramuscular injection of the medication pentamidine prevents the Gambian form of the illness for a number of months, although it will not necessarily prevent an infection that could progress at a later time. As with other eukaryotic pathogens, treatment is problematic because of toxic side effects of the available medications. Suramin can be used if the disease has not progressed to involve the central nervous system; melarsoprol and eflornithine cross the blood-brain barrier and can be used when the central nervous system is involved. ■ antiprotozoan medications, p. 490

The main characteristics of African sleeping sickness are presented in **table 27.11.**

TABLE 27.11 African Sleeping Sickness

Symptoms	Tender nodule at site of tsetse fly bite; fever, enlargement of lymph nodes; later, involvement of the central nervous system, uncontrollable sleepiness, headache, poor concentration, unsteadiness, coma, death
Incubation period	Weeks to several years
Causative agent	*Trypanosoma brucei,* a flagellated protozoan
Pathogenesis	The protozoa multiply at site of a tsetse fly bite, then enter blood and lymphatic circulation; as new cycles of parasites are released, their surface protein changes and the body is required to respond with a new antibody.
Epidemiology	Bites of infected tsetse flies transmit the trypanosomes through fly saliva; wild animal reservoir for *T. brucei rhodesiense.*
Prevention and treatment	Protective clothing, insecticides, clearing of brush where flies breed, pentamidine. Treatment: suramin; when central nervous system is involved, melarsoprol or eflornithine.

MICROCHECK 27.5

Only on rare occasions can free-living amebas such as *Naegleria fowleri* cause meningoencephalitis in human beings. Residents and visitors to a wide swath of tropical Africa, however, are at risk of contracting African sleeping sickness, caused by the flagellated protozoan parasite *Trypanosoma brucei,* and transmitted by a biting insect, the tsetse fly. These protozoa can circulate in the bloodstream for extended periods by changing their surface antigens to escape the host's antibodies. Eventually they are able to penetrate the CNS, causing indifference, sleepiness, coma, and death.

✓ How likely is it that a person who swims in warm, fresh water will contract primary amebic meningoencephalitis?

✓ How can one explain repeated, abrupt increases in *T. brucei* in the blood of African sleeping sickness victims?

27.6 Transmissible Spongiform Encephalopathies

Focus Point

- List the ways in which the transmissible spongiform encephalopathies differ from the infectious diseases discussed earlier in this chapter.

A mysterious group of chronic, degenerative brain diseases involves wild and domestic animals and humans. Mink, elk, mule deer, sheep, goats, and cattle are among the animals that can be afflicted **(figure 27.21).** Microscopically, brain tissue affected by these diseases has a spongy appearance due to the loss of nerve cells and other changes—this is why these diseases are referred to as **spongiform encephalopathies.** Brain and other tissues from affected animals can transmit the disease to normal animals of the same, and sometimes different, species. Some of these encephalopathies can be transmitted to small laboratory animals such as mice, allowing scientific studies. These diseases assumed new prominence when, in 1996, a previously unrecognized spongiform encephalopathy in humans, variant **Creutzfeld-Jakob disease,** appeared during a mad cow disease epidemic in the United Kingdom. ■ mad cow disease, p. 320

FIGURE 27.21 A Captive Elk With Chronic Wasting Disease, A Kind of Transmissible Spongiform Encephalopathy Afflicted animals develop incoordination and weight loss, ending in death.

Transmissible Spongiform Encephalopathy in Humans

Transmissible spongiform encephalopathy is rare in humans, occurring in only 0.5 to 1 case per million people. Most cases occur sporadically as Creutzfeld-Jakob disease, although there are other forms of the disease that run in families. Another form, **kuru,** is associated with cannibalism, as formerly practiced by some New Guinea natives.

Symptoms

Early symptoms include vague behavioral changes, anxiety, insomnia and fatigue, which progress over weeks or months to the hallmarks of the disease, muscle jerks, lack of coordination, and dementia. Individuals suffer deteriorating intellectual function, impaired judgment, memory loss, and loss of normal muscle function, ending in death, generally within a year.

Causative Agent

The causative agents of these diseases are called *pr*oteinaceous *i*nfectious *p*articles, or **prions (PrP).** Prions appear to be a misfolded form of a normal cellular protein (PrP^c). ■ prions, p. 341

The main characteristics of prions are:

- They increase in quantity during the incubation period of the disease.
- They resist inactivation by ultraviolet and ionizing radiation.
- They resist inactivation by formaldehyde and heat.
- They are not readily destroyed by proteases.
- They are not destroyed by nucleases.
- They are much smaller than the smallest virus.
- They are composed of protein coded by a normal cellular gene but modified after transcription.

Pathogenesis

In the normal course of infection, multiplication (see figure 14.21) first occurs in the spleen and other lymphoid tissues. Immune cells and proinflammatory cytokines are required for prion multiplication at these sites. The prions are then transported to the central nervous system by B lymphocytes, and probably by other means. There they aggregate in insoluble masses (plaques) outside the nerve cells. Malfunction and nerve cell death probably result from abnormal PrP replacing normal PrP^c protein anchored to the cell surface. But not all prions act the same. They differ in host range, incubation period, and the areas of the nervous system that are attacked.

Epidemiology

Creutzfeld-Jakob disease generally occurs in individuals older than 45 years. It can be transmitted experimentally to chimpanzees, and from human to human with corneal transplants and contaminated surgical instruments.

Spongiform encephalopathy of sheep, scrapie, has been known for more than two centuries, without any evidence that it is directly transmissible to humans. Current evidence indicates that cattle in the United Kingdom became infected with scrapie prions in protein food supplements made from sheep carcasses, giving rise to "mad cow disease." Based on laboratory and epidemiological evidence, the cattle prions can be transmitted to humans and cause a variant Creutzfeld-Jakob disease marked by differences in symptoms, brain pathology, and age of onset. The median age of individuals with the variant disease is only 28 years.

Prevention and Treatment

Prions are highly resistant to disinfectants including formaldehyde. They can be inactivated by autoclaving in 1N sodium hydroxide. There is no treatment for the spongiform encephalopathies, and they are invariably fatal.

The main features of transmissible spongiform encephalopathies are presented in **table 27.12.**

MICROCHECK 27.6

Transmissible spongiform encephalopathies are degenerative nervous system diseases that occur in a variety of wild and domestic animals. They are rare in humans, but the appearance of a new form of the disease during an epidemic of mad cow disease in the United Kingdom strongly suggests the possibility that epidemic disease in humans could originate from animals, transmitted via ingested meat. These diseases appear to be caused by prions—infectious agents consisting only of protein and highly resistant to inactivation by heat, radiation, and disinfectants.

✓ By what route might scrapie prions from sheep infect cattle?

✓ If you were a famous American eye surgeon about to transplant a cornea to restore the vision of a woman, would you rather the donor be under 35 or over 45 years of age?

TABLE 27.12 Transmissible Spongiform Encephalopathy

Symptoms	Behavioral changes, anxiety, insomnia, fatigue, progressing over weeks or months to muscle jerks, lack of coordination, dementia
Incubation period	Usually many years
Causative agents	Proteinaceous infectious particles known as prions; lack nucleic acids; identical amino acid sequence to a normal protein, but folded differently, and relatively resistant to proteases; resistant to heat, radiation and disinfectants
Pathogenesis	Prions replicate in dendritic cells in various parts of the body by converting normal protein to more prions; transmission to the brain; aggregation into masses outside the nerve cells; cell malfunction and death.
Epidemiology	Human-to-human transmission by corneal transplantation and by contaminated surgical instruments; probable transmission of cattle prions to humans by eating contaminated beef; sporadic Creutzfeld-Jakob disease in those over age 45 years; median age of variant Creutzfeld-Jakob cases only 28 years.
Prevention and treatment	Prions are inactivated by autoclaving in concentrated sodium hydroxide; no treatment, invariably fatal.

FUTURE CHALLENGES

Eradicate Polio: Then What?

It is hoped that poliomyelitis will be among the next ancient scourges to follow smallpox down the road to eradication. However, for a variety of political and scientific reasons, the causative virus of smallpox still exists many years after the last naturally acquired case of the disease. Will it be any easier to rid the world of the polioviruses after the last case of poliomyelitis occurs?

Dramatic progress toward polio eradication has mainly been accomplished using "immunization days," when the entire population in a given area is immunized using attenuated (Sabin) vaccine, often with one or more follow-up days to immunize individuals missed earlier. The vaccine viruses then disappear from the population within a few months. Although a small percentage of the population suffers paralytic illness from the vaccine, the population is protected from virulent polioviruses until new generations are born. If there were no polioviruses left in the world, there would be no need to immunize the new generations.

Potential sources of virulent polioviruses include wild viruses circulating in remote populations, individuals with mild or atypical illness, long-term carriers (especially those with immunodeficiency, who excrete the viruses for months or years), and laboratory freezers around the world that are known to harbor the viruses, as well as frozen specimens such as feces that could harbor them. It is even possible to synthesize a poliovirus in the laboratory. Lastly, vaccine viruses can change genetically and acquire full virulence, as occurred in the Hispaniola polio epidemic of 2000–2001.

The challenge is to have a continuous, reliable, global polio surveillance system, and maintain immunizations and strategically located stockpiles of vaccine during what is likely to be a long time after the last case of paralytic polio occurs.

SUMMARY

27.1 Anatomy and Physiology (figure 27.1)

The brain and spinal cord make up the central nervous system (CNS); the peripheral nervous system is composed of **motor nerves** and **sensory nerves. Cerebrospinal fluid** is produced by structures in cavities inside the brain and flows out over the brain and spinal cord (figure 27.2). **Meninges** are the membranes that cover the surface of the brain and spinal cord.

Pathways to the Central Nervous System

Infectious agents can reach the CNS by way of the bloodstream when they penetrate the **blood-brain barrier,** via the cytoplasm of nerve cell **axons,** and by direct extension through bone.

27.2 Bacterial Nervous System Infections

Bacteria can infect the brain, spinal cord, and peripheral nerves, but more commonly they infect the meninges and cerebrospinal fluid, causing **meningitis** (figure 27.4). Bacterial meningitis is uncommon; formerly most victims were children, but childhood immunization has decreased the incidence in these age groups. In most but not all of the victims, the causative bacterium is one commonly found among the normal upper respiratory flora of healthy people. *Haemophilus influenzae,* a tiny, pleomorphic, Gram-negative rod, once the leading cause of childhood bacterial meningitis, is now mostly controlled by a vaccine (figure 27.3).

Meningococcal Meningitis (table 27.1)

Meningococcal meningitis is greatly feared because it can result in shock and death, and it can occur in both childhood and adult epidemics. Symptoms are similar to other forms of meningitis: cold symptoms followed by abrupt onset of fever, severe headache, pain and stiffness of the neck and back, nausea, and vomiting. Small hemorrhages into the skin (figure 27.5), deafness, and coma can occur. The bacterium *Neisseria meningitidis* causes a massive PMN response (figure 27.6); metabolic activity of the leukocytes and bacteria consumes glucose normally present in the cerebrospinal fluid; shock results from the release of endotoxin into the bloodstream.

Listeriosis (table 27.2)

Listeriosis is caused by *Listeria monocytogenes,* a non-spore-forming, Gram-positive rod; it is a foodborne illness often manifest as meningitis in newborn infants and others. The bacterium is widespread, commonly contaminates foods such as non-pasteurized milk, cold cuts, and soft cheeses, and can grow in the refrigerator. The bacteria readily penetrate the intestinal epithelium, enter the bloodstream, and infect the meninges.

Hansen's Disease (Leprosy) (table 27.4, figure 27.8)

Hansen's disease is characterized by invasion of peripheral nerves by the acid-fast bacillus *Mycobacterium leprae,* which has not been cultivated *in vitro* (figure 27.9). The disease occurs in two main forms—**tuberculoid** and **lepromatous,** depending on the immune status of the patient.

Botulism (table 27.5)

Botulism is not a nervous system infection, but is an often-fatal type of food poisoning that causes severe generalized paralysis. The causative bacterium, *Clostridium botulinum,* is an anaerobic, Gram-positive rod that forms heat-resistant spores (figure 27.10). Spores that survive canning or other heat treatment of foods germinate, and the bacteria multiply, releasing a powerful toxin into the food. Because of strict controls on food processing, intestinal botulism is now the most common form of the disease in the United States. Relatively mild symptoms result from *C. botulinum* colonization of the intestine of infants and some adults. Wound botulism, caused when *C. botulinum* colonizes dirty wounds containing dead tissue, is rare.

27.3 Viral Diseases of the Nervous System

Most viral nervous system infections are caused by human enteroviruses or by the viruses of certain zoonoses. Many common viruses of human beings can occasionally infect the nervous system, including those that cause infectious mononucleosis, mumps, measles, chickenpox, and herpes simplex ("cold sores," genital herpes).

Viral Meningitis (table 27.6)

Viral meningitis is much more common than bacterial meningitis. It is generally a mild disease for which there is no specific treatment.

Viral Encephalitis (table 27.7)

Viral encephalitis has a high fatality rate and often leaves survivors with permanent disabilities. It can be sporadic or epidemic. Herpes simplex virus is the most important cause of sporadic encephalitis; epidemic encephalitis is usually caused by **arboviruses.** LaCrosse encephalitis virus, maintained in *Aedes* mosquitoes and squirrels and chipmunks, is usually the most frequently reported (figure 27.12).

Infantile Paralysis, Polio (Poliomyelitis) (table 27.8, figure 27.16)

Destruction of motor nerve cells of the brain and spinal cord leads to paralysis, muscle wasting, and failure of normal bone development (figures 27.13, 27.15). **Post-polio syndrome** occurs years after poliomyelitis, and it is probably caused by the death of nerve cells that had taken over for those killed by poliomyelitis virus.

Rabies (table 27.9)

Rabies is a widespread zoonosis transmitted to humans mainly through the bite of an infected animal (figure 27.17). Once symptoms appear in an infected human being, the disease is almost uniformly fatal. Because of the long incubation period, prompt immunization with inactivated vaccine begun after a rabid animal bite is effective in preventing the disease. Passive immunization given at the same time increases the protection.

27.4 Fungal Diseases of the Nervous System

Fungi rarely invade the nervous system of healthy people, but they can be a threat to the life of individuals with underlying diseases such as diabetes, cancer, and immunodeficiency. Treatment of these infections is usually very difficult.

Cryptococcal Meningoencephalitis (table 27.10)

Infection originates in the lung after a person inhales dust laden with spores of *Cryptococcus neoformans,* an encapsulated yeast, which resists phagocytosis because of its large capsule (figure 27.19). Some varieties of this organism are associated with soil contaminated with decomposed pigeon droppings.

27.5 Protozoan Diseases of the Nervous System

Only a few free-living protozoa infect the human nervous system. Certain parasitic protozoa are a much more widespread threat.

African Sleeping Sickness (table 27.11)

African sleeping sickness is a major health problem in a wide area across equatorial Africa. In its late stages, it is marked by indifference, sleepiness, coma, and death. The disease is caused by *Trypanosoma brucei* (figure 27.20), a flagellated protozoan that lives in the blood of its victims and in its biological vector, the tsetse fly. During infection, the organism shows bursts of growth, each appearing with different surface proteins. The organism has more than a thousand genes coding for these antigens, but only expresses one at a time. Each antigen requires that the body respond with a new antibody.

27.6 Transmissible Spongiform Encephalopathies

The spongiform encephalopathies are a group of rare diseases characterized by a spongelike appearance of brain tissue caused by loss of nerve cells. They afflict a variety of wild and domestic animals as well as humans (figure 27.21). Mad cow disease is an example. Brain and other tissues from afflicted animals can transmit the disease to the same, or sometimes other, species. A new form of the disease appeared in humans during an epidemic of mad cow disease in the United Kingdom.

Transmissible Spongiform Encephalopathies in Humans (table 27.12)

The main symptoms are progressive loss of muscle control, and dementia. The diseases are caused by prions—abnormal proteins that are resistant to heat, radiation, and disinfectants. They appear to be identical to a normal protein except they are folded differently. Prions convert the normal protein into copies of themselves. In the brain they aggregate in masses and kill nerve cells. There is no apparent inflammatory or immune response. The most common form of the disease in humans, Creutzfeld-Jakob disease, occurs in only about one in a million people, generally in those over 45 years of age. It can be transmitted from one person to another by corneal transplants and by contaminated instruments. The variant of this disease that appeared during the mad cow disease epidemic affected much younger people. Evidence indicates that scrapie prions, from protein supplements made from sheep carcasses, infected cattle; humans then became infected by eating beef. Prions can be inactivated by autoclaving them in strong alkali. So far, there is no treatment for these diseases and they are invariably fatal.

REVIEW QUESTIONS

Short Answer

1. How can the incidence of meningitis due to Lancefield group B streptococci be reduced?
2. Name and describe the organism that is the leading cause of bacterial meningitis in adults.
3. What cell wall component of *Neisseria meningitidis* is probably responsible for the shock and death that sometimes occur with infections by this bacterium?
4. What measures can be undertaken to prevent meningococcal meningitis?
5. Why is listeriosis so important to pregnant women even though it usually causes them few symptoms?
6. Describe the differences between tuberculoid and lepromatous leprosy.
7. What is the difference between sporadic encephalitis and epidemic encephalitis? Name one cause of each.
8. Give two ways in which viral meningitis usually differs from bacterial meningitis.
9. Give an example of an enterovirus.
10. What are arboviruses? Give an example.
11. Why does poliomyelitis virus attack motor but not sensory neurons?
12. Can a dog infected with rabies virus transmit the disease while appearing well? Explain.

Multiple Choice

1. Which is the best way to prevent meningococcal meningitis in individuals intimately exposed to the disease?
   a) Vaccinate them against *Neisseria meningitidis.*
   b) Treat them with the antibiotic rifampin.
   c) Culture their throat and hospitalize them for observation.
   d) Withdraw a sample of spinal fluid and begin antibacterial treatment if the cell count is high and the glucose is low.
   e) Have them return to their usual activities, but seek medical evaluation if symptoms of meningitis occur.
2. Which of these statements concerning the causative agent of listeriosis is *false?*
   a) It can cause meningitis during the first month of life.
   b) It is a Gram-positive rod that can grow in refrigerated food.
   c) It is usually transmitted by the respiratory route.
   d) Infection commonly results in bacteremia.
   e) It is widespread in natural waters and vegetation.
3. Of the following statements about Hansen's disease, which one is *false?*
   a) It was once common in the United States.
   b) An early symptom is loss of sensation, sweating, and hair in a localized patch of skin.
   c) The incubation period is usually less than 1 month.
   d) Treatment should include more than one antimicrobial medication given at the same time.
   e) The form the disease takes depends on the immune status of the victim.
4. Pick the one *false* statement about foodborne botulism.
   a) It is not a central nervous system infection.
   b) Pathogenicity of some strains of the causative agent is the result of lysogenic conversion.
   c) Food can taste normal but still cause botulism.
   d) Treatment is based on choosing the correct antibiotic.
   e) Control of the disease depends largely on proper food-canning techniques.
5. Which of the following statements about enteroviral meningitis is *true?*
   a) Vaccines are generally available to protect against the disease.
   b) The main symptom is muscle paralysis.
   c) Transmission is usually by the fecal-oral route.
   d) The causative agents do not survive well in the environment.
   e) Recovery is rarely complete.
6. Choose the one *false* statement about arboviral encephalitis.
   a) It is likely to occur in epidemics.
   b) Mosquitoes can be an important vector.

c) Epilepsy, paralysis, and thinking difficulties are among the possible sequels to the disease.
d) Use of sentinel chickens helps warn about the disease.
e) In the United States, the disease is primarily a zoonosis involving cattle.

7. Which one of the following statements about poliomyelitis is *false?*
a) The nerves of sensation are usually involved.
b) It can be caused by any of three specific enteroviruses.
c) Only a small fraction of those infected will develop the disease.
d) The disease is transmitted via the fecal-oral route.
e) A postpolio syndrome can develop years after recovery from the original illness.

8. Choose the one *false* statement about rabies.
a) It is widespread in wildlife.
b) Vaccines are available to protect humans and domestic animals.
c) Hydrophobia is a characteristic symptom.
d) Animals that do not bite do not get the disease.
e) The incubation period is often a month or more.

9. Which statement concerning cryptococcal meningoencephalitis is *true?*
a) It is caused by a yeast with a large capsule.
b) It is a disease of pigeons transmissible to humans.
c) Typically it attacks the meninges, but spares the brain.
d) Person-to-person transmission commonly occurs.
e) Cell-mediated immunity is of little importance in the body's struggle against the disease.

10. Which of the following statements about African sleeping sickness is *true?*
a) It is transmitted by a species of night-time-biting mosquito.
b) It is a threat to visitors to tropical Africa.
c) The onset of sleepiness is usually within 2 weeks of contracting the disease.
d) It is caused by free-living protozoa.
e) Distribution of the disease is determined mainly by the distribution of standing water.

Applications

1. An outbreak of viral meningitis in a small eastern city was linked epidemiologically to a group who swam a non-chlorinated pool in an abandoned quarry outside of town. What might public health officials surmise about the probable cause of the outbreak?
2. Two microbiologists are trying to write a textbook, but they cannot agree where to place the discussion of botulism. One favored the chapter on nervous system infections, whereas the other insisted on the chapter covering digestive system infections. Which microbiologist is correct?

Critical Thinking

1. A pathologist stated that it was much easier to determine the causative agent of meningitis than of an infection of the skin or intestine. Is her statement valid? Why or why not?
2. Why is it important to learn about rabies when only a few cases occur in the entire United States each year?

SEM of the tip of the mouthparts of a female mosquito.

28

Blood and Lymphatic Infections

A Glimpse of History

Alexandre Emile John Yersin (1863–1943) is one of the most interesting, though relatively unknown, contributors to the conquest of infectious diseases. He was born in 1863 in the Swiss village of Lavaux, where his family lived in a gunpowder factory. His father, director of the factory and a self-taught insect expert, died unexpectedly 3 weeks before Yersin's birth. Alexandre developed an interest in science when, as a young boy, he discovered in an attic trunk the microscope and dissecting instruments of the father he never knew. A local public health physician befriended the family and influenced Yersin to study medicine, and at age 20 he began pre-med studies in Lausanne. Later, while a medical student in Paris, Yersin volunteered to help at the Pasteur Institute, and in his fourth year of medical training, he was hired by Emile Roux, a coworker of Pasteur. Yersin became increasingly recognized for his work at the Institute, but he was bored with research, and he did not want to practice medicine because he felt it was wrong for physicians to make a living from sickness.

He abruptly left the Pasteur Institute and was employed as a physician on a ship sailing from Marseilles to Saigon. Most of Indochina was under French control and was largely unexplored and undeveloped. Yersin became enchanted with the land and its peoples, exploring crocodile-infested rivers in a dugout canoe and making expeditions on elephant into jungles where tigers roared. For the rest of Yersin's life, Vietnam was his home. His many contributions to the region included the introduction of improved breeds of cattle, cultivation of rubber and quinine-producing trees, and the establishment of a medical school.

Yersin also studied the diseases that affected the Vietnamese people. One of them, plague, was a horrifying recurring problem. The disease had killed millions in Europe in medieval times and had struck France as recently as 1720. This great killer, an incurable disease, came and went unpredictably, its cause and transmission completely unknown.

In 1894, Yersin went to study a Hong Kong outbreak of plague because the facilities for studying the disease were better in Hong Kong than in Vietnam. Unfortunately for Yersin, Shibasaburo Kitasato, the famous colleague of Robert Koch and Emil von Behring, had arrived 3 days earlier with a large team of Japanese scientists. British authorities had given them access to patients and laboratory facilities. Yersin did not speak English, so it was difficult for him to communicate with the authorities. Moreover, although both he and the Japanese scientists spoke fluent German, the Japanese team treated him coolly, perhaps as a reflection of the intense rivalry between Pasteur and Koch. Yersin was reduced to setting up a laboratory in a bamboo shack. He bribed British sailors responsible for disposing of the bodies of plague victims to allow him to get samples of material from their swollen, pus-filled lymph nodes (called *buboes*).

A week after his arrival in Hong Kong in the summer of 1894, Yersin reported to the British authorities his discovery of a bacillus of characteristic appearance, invariably present in the buboes of plague victims. The bacterium could be cultivated, and it caused plaguelike disease when injected into rats. Later, he showed that the disease was transmitted from one rat to another. Yersin's plague bacillus, now known as *Yersinia pestis,* was used to make an antiplague vaccine, and in 1896, antiserum prepared against the organism provided the world's first cure of a patient with plague. Not long afterward, another Pasteur Institute scientist proved that the rat flea was crucial in transmission. ■ **Robert Koch, p. 83** ■ **Shibasaburo Kitasato, p. 559**

Kitasato, working with blood from the hearts of plague victims, announced shortly after his arrival in Hong Kong that he had isolated the bacterium that caused plague. Although subsequently it proved to be only a laboratory contaminant, Kitasato was named co-discoverer of the plague bacillus because of his great prestige.

KEY TERMS

Aneurysm Localized, abnormal enlargement of a blood vessel resulting from disease of the blood vessel wall.

Bubo An enlarged, tender lymph node characteristic of plague and some venereal diseases.

Disseminated Intravascular Coagulation (DIC) Condition in which clots form in small blood vessels throughout the body, leading to failure of vital organs.

Embolus (pl., **Emboli**) An abnormal object such as an air bubble or blood clot that enters the blood circulation.

Endocarditis Inflammation of the heart valves or lining of the heart chambers.

Lymphangitis Inflammation of lymphatic vessels.

Necrosis Death of tissue.

Osteomyelitis Inflammation of the bone marrow and adjacent bone.

Petechiae Small, purple spots on the skin and mucous membranes caused by hemorrhage from small blood vessels.

Pneumonic Referring to the lung.

Septicemia Acute illness caused by infectious agents or their products circulating in the bloodstream: "blood poisoning."

Septic Shock An array of effects that results from infection of the bloodstream or circulating endotoxin; includes fever, drop in blood pressure, and disseminated intravascular coagulation.

Subacute Bacterial Endocarditis A chronic illness caused by infection of the heart lining or valves, generally caused by bacteria of low virulence.

Zygote A diploid cell formed by the fusion of two haploid cells (gametes) during sexual reproduction.

The circulation of blood and lymph fluids supplies nutrients and oxygen to the body's tissue cells and carries away the cells' waste products. The circulatory system also heats and cools body tissues to maintain an optimum temperature. Infection of the system can have devastating effects because infectious agents become **systemic,** meaning they can be carried to all parts of the body, producing disease in one or more vital organs or causing the circulatory system itself to stop functioning. Thus, even a small scratch can cause considerable harm if it results in a pathogen entering the bloodstream. The circulation of an agent in the bloodstream is given a name ending in *-emia* that specifies the nature of the agent, as with **bacteremia, viremia,** and **fungemia.** In many cases, there are no symptoms associated with the circulating agent. When illness results from a circulating agent or its toxins, the condition is referred to as **septicemia,** or blood poisoning. When, as a result of septicemia, the blood pressure falls to such low levels that blood flow to vital organs is insufficient to maintain their functioning, the condition is called **septic shock.**

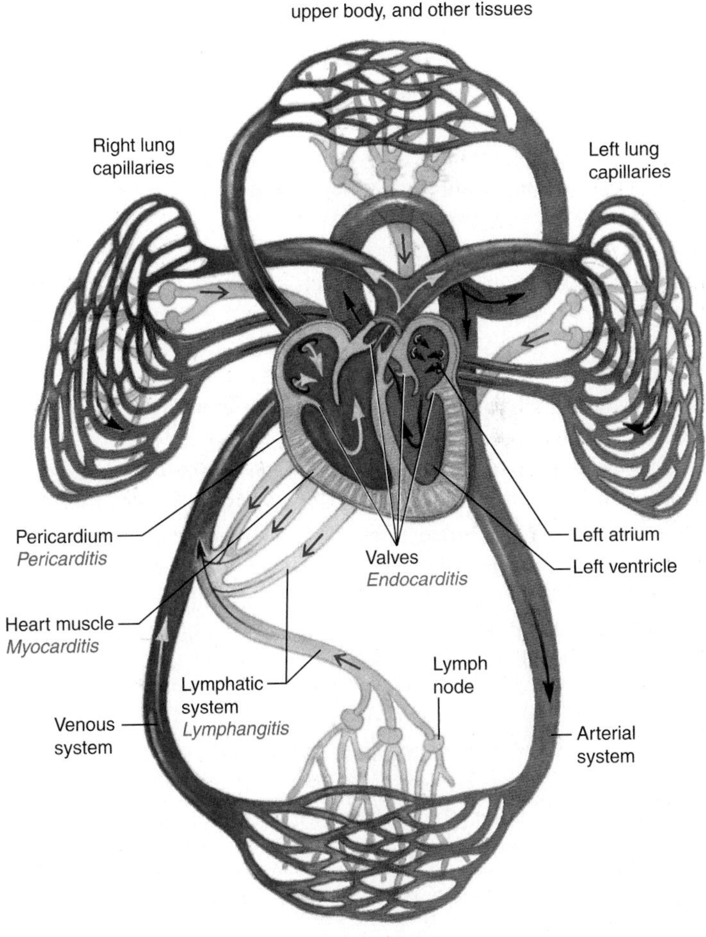

FIGURE 28.1 The Blood and Lymphatic Systems Disease conditions are indicated in red type. For simplicity, the spleen is not shown. It is a fist-sized, blood-filled lymphoid organ located high in the left side of the abdomen, behind the stomach.

28.1 Anatomy and Physiology

Focus Points

- Outline the function of valves in the circulatory system.
- List the functions of the lymphatic system.

The Heart

The heart, a muscular double pump enclosed in a fibrous sac called the **pericardium,** supplies the force that moves the blood. As shown schematically in **figure 28.1,** the heart is divided into right and left sides separated by a septum—a wall of tissue through which blood

PERSPECTIVE 28.1

Arteriosclerosis: The Infection Hypothesis

Arteriosclerosis, the main cause of heart attacks and strokes, is characterized by lipid-rich deposits that develop in arteries and can impair or stop the flow of blood.

In 1988, researchers in Helsinki, Finland, reported that patients with coronary artery disease—meaning arteriosclerosis of the arteries that supply the heart muscle—commonly had antibodies against *Chlamydia pneumoniae*. This organism is a tiny, obligate, intracellular bacterium that is responsible for a variety of upper and lower respiratory infections including sinusitis and pneumonia. The infections are common, and most people have been infected by *C. pneumoniae* by the time they reach adulthood. What was intriguing about the antibody studies was that the higher the titer of antibody, the greater the risk of coronary disease. Then, in 1996, viable *C. pneumoniae* were shown to be present in many arteriosclerotic lesions, and other studies conclusively established that there was an association between the bacterium and arteriosclerotic lesions of both heart and brain arteries.

Does *C. pneumoniae* cause arteriosclerosis, does it worsen the effects of arteriosclerosis, or does it merely exist harmlessly in the lesions? How can the hypothesis that *C. pneumoniae* contributes to heart attacks and strokes be tested? Some studies of coronary patients treated with antibiotics effective against the bacterium have suggested a beneficial effect, and others have not. It now appears that many inflammatory conditions, not just *C. pneumoniae* infections, increase the risk of heart attacks and strokes. The risk correlates with the level of acute-phase proteins arising from the release of pro-inflammatory cytokines. Undoubtedly there will be more to come in determining the relationship, if any, between arteriosclerosis and acute-phase proteins. ■ **acute-phase proteins, p. 360**

normally does not pass after birth. The right and left sides of the heart are both divided into two chambers—the **atrium,** which receives blood, and the **ventricle,** which discharges it. Blood from the right ventricle flows through the lungs and into the atrium on the left side of the heart. From the atrium, the blood passes into the left ventricle and is then pumped through the aorta to the arteries and capillaries that supply the tissues of the body. Blood that exits the capillaries is returned to the right atrium by the veins. The heart valves are situated at the entrance and exit of each ventricle and ensure that the blood flows in only one direction. Although not common, infections of the heart valves, muscle, and pericardium can be disastrous because they affect the vital function of blood circulation.

Arteries

Arteries have thick muscular walls to withstand the high pressure of the arterial system. The blood in arteries is bright red, the color of oxygenated hemoglobin. Infection of the arteries is unusual, although the aorta can be dangerously weakened by infection in syphilis. Arteries are subject to arteriosclerosis—"hardening of the arteries"—a process that begins in teenage years or before and progresses through life, ultimately putting individuals at risk for heart attacks and strokes. A possible role of bacteria and viruses in arteriosclerosis has been considered for some years. ■ **syphilis, p. 633**

Veins

Blood becomes depleted of O_2 in the capillaries, causing it to assume the dark color characteristic of venous blood. Because pressure in the veins is low, one-way valves help keep the blood flowing in the right direction. Veins are easily compressed, so the action of muscles aids the flow of venous blood.

Thus, as the blood flows around the circuit, it alternately passes through the lungs and through the tissue capillaries. During each circuit, a portion of the blood passes through organs such as the spleen, liver, and lymph nodes, all of which, like the lung, contain phagocytic cells of the mononuclear phagocyte system. These phagocytes help cleanse the blood of foreign material, including infectious agents, as the blood passes through these tissues.

Lymphatics (Lymphatic Vessels)

The system of lymphatic vessels (see figure 16.2) begins in tissues as tiny tubes that resemble blood capillaries but differ from them in having closed ("blind") ends and in being somewhat larger. Lymph, an almost colorless fluid, is conveyed within these lymphatic vessels. It originates from plasma, the non-cellular portion of the blood, that has oozed through the walls of the blood capillaries to become the interstitial fluid that surrounds tissue cells. This fluid bathes and nourishes the tissue cells and then enters the lymphatics. Unlike the blood capillaries, the readily permeable lymphatic vessels take up foreign material such as invading microbes and their products, including toxins and other antigens. The tiny lymphatic capillaries join progressively larger lymphatic vessels. Many one-way valves in the lymphatic vessels keep the flow of lymph moving away from the lymphatic capillaries. Both contraction of the vessel walls and compression by the movements of the body's muscles force the lymph fluid along. ■ **lymphatic vessels, p. 370**

At many points in the system, lymphatic vessels drain into small, bean-shaped bodies called lymph nodes. These nodes are constructed so that foreign materials such as bacteria are trapped in them; the nodes also contain phagocytic cells and antibody-producing cells. Generally, lymph flows out of the nodes through vessels that eventually unite into one large tube (thoracic duct), which then discharges into a large vein (subclavian vein) behind the left collarbone, and thus back into the main blood circulation. ■ **lymphoid system, p. 369** ■ **phagocytic cells, p. 358**

When a hand or a foot is infected, a visible red streak may spread up the limb from the infection site **(figure 28.2).** This streak represents the course of lymphatic vessels that have become inflamed in response to the infectious agent. This condition is called **lymphangitis.** It may stop abruptly at a swollen and tender lymph node, only to continue later to yet another lymph node. This pause in the progression of lymphangitis demonstrates the ability, even though sometimes temporary, of the lymph nodes to clear the lymph of an infectious agent.

Blood and lymph carry infection-fighting leukocytes and such antimicrobial proteins as antibodies, complement, lysozyme, β-lysin,

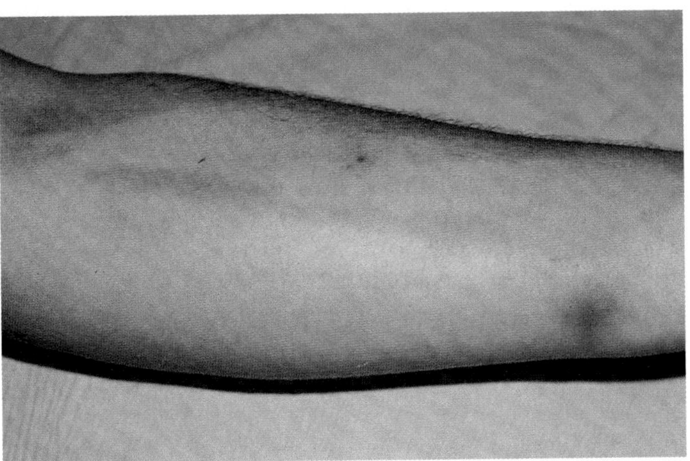

FIGURE 28.2 Lymphangitis Notice the red streak extending up the arm. The streak represents an inflamed lymphatic vessel.

and interferon. An inflammatory response may cause lymph and blood to clot in vessels that are close to areas of infection or antibody-antigen reactions—one of the ways the body has to localize an infection from the rest of the body. As with the nervous system, the blood and lymphatic systems lie deep within the body and are normally sterile. ■ antimicrobial substances, p. 348 ■ inflammatory response, p. 360

Spleen

The spleen is a fist-sized organ situated high up on the left side of the abdominal cavity, behind the stomach. It is composed of two kinds of tissue—one consists of multiple blood-filled passageways, and the other, of lymphoid tissue. One of its functions is to cleanse the blood, much as the lymph nodes cleanse the lymph. Large numbers of phagocytes located in the spleen remove aging or damaged erythrocytes, bacteria, and other foreign materials from the blood. Another function, carried on by the lymphoid tissue, is to provide an immune response to microbial invaders. A third function is to produce new blood cells in rare situations where the bone marrow is unable to meet the demand. The spleen enlarges in a number of infectious diseases, such as infectious mononucleosis and malaria, in which its immune and filtering functions are challenged.

MICROCHECK 28.1

The heart can be thought of as two pumps that work together to move the blood around a circular course. The main function of the lymphatic system is to cleanse interstitial fluid and to return it to the bloodstream. Systemic infections threaten the transportation of oxygen and nutrients to body tissues and removal of waste products. The circulatory system can expose all of the body's tissues to infectious agents and their toxins.

✓ What are the functions of the spleen?

✓ Which side of the heart—right or left—do bacteria in infected lymph reach first? Where do they go from there?

28.2 Bacterial Diseases of the Blood Vascular System

Focus Points

- Outline the pathogenesis of subacute bacterial endocarditis.
- List the effects of endotoxemia.

Bacterial infections of the vascular system can be rapidly fatal, or they can smolder for months, causing a gradual decline in health. They are not common, but they are always dangerous. Usually the bacteria are carried into the bloodstream by the flow of lymph from an area of infection in the tissues. Some pathogens multiply in the bloodstream, and they may colonize and form biofilms on structures such as the heart valves or abnormalities in the heart or major arteries. ■ biofilm, p. 85

Endocarditis is the term used for infections of the heart valves or the inner, blood-bathed surfaces of the heart, and it can be acute or subacute. **Acute bacterial endocarditis** starts abruptly with fever, and usually an infection such as pneumonia is present somewhere else in the body or there is evidence of injected-drug abuse. Virulent species such as *Staphylococcus aureus* and *Streptococcus pneumoniae* are usually the cause, and they infect both normal and abnormal heart valves. They can often produce a rapidly progressive disease, often with valve destruction and formation of abscesses in the heart muscle, leading to heart failure. By contrast, **subacute bacterial endocarditis** is usually caused by organisms with little virulence, and it has a much more protracted course. ■ *Staphylococcus aureus*, p. 535 ■ *Streptococcus pneumoniae*, p. 509

Septicemia is caused by both Gram-negative and Gram-positive bacteria, as well as other infectious agents.

Subacute Bacterial Endocarditis

Subacute bacterial endocarditis (SBE) is usually localized to one of the valves on the left side of the heart. It commonly occurs on valves that are deformed as a result of a birth defect, rheumatic fever, or some other disease. ■ subacute bacterial endocarditis, p. 677 ■ rheumatic fever, p. 501

Symptoms

People with subacute bacterial endocarditis usually suffer from marked fatigue and slight fever. They typically become ill gradually and slowly lose energy over a period of weeks or months. They may abruptly develop a stroke.

Causative Agent

The causative organisms of SBE are usually members of the normal bacterial microbiota of the mouth or skin, notably α-hemolytic viridans streptococci and *Staphylococcus epidermidis*. The infecting organisms are usually shed from the infected heart valve into the circulation and can be found by culturing samples of blood drawn from an arm vein. In 5% to 15% of cases, however, culturing the blood from an arm vein fails to yield the causative bacteria. This is especially likely to occur when the infection is on the right side of the heart, and blood from the infected site must pass through the lung, which is richly supplied with phagocytes, before entering the left side of the heart, traversing the capillaries, and reaching the arm vein. Also, some bacteria that cause endocarditis, such as *Coxiella burnetii,* the cause of Q fever, cannot be cultivated on cell-free media. At least one, *Tropheryma whippelii,* the cause of Whipple's disease, can be identified at the time of valve surgery by using the polymerase chain reaction and a nucleic acid probe. Cultivation of the bacterium requires a special medium. ■ Q fever, p. 274 ■ Whipple's disease, p. 243 ■ polymerase chain reaction, p. 223 ■ nucleic acid probes, p. 227

Pathogenesis

The bacteria that cause SBE can gain entrance to the bloodstream during dental procedures, toothbrushing, or other trauma. In an abnormal heart, a thin blood clot may form in areas where there is turbulent blood flow around a deformed valve or other defect. This clot traps circulating organisms; these multiply and can form a biofilm that makes them inaccessible to phagocytic killing, and tends to protect them from antimicrobial treatment. People with bacterial endocarditis often have high levels of antibodies, which are of little value in eliminating the bacteria and may even be harmful because they make the bacteria clump together and adhere to the clot. As the organisms multiply, more clot is deposited around them, gradually building up a fragile mass. Bacteria continually wash off the mass into the circulation, and pieces of infected clot (septic **emboli**) can break off. If large enough, these clots can block important blood vessels and lead to death of the tissue (infarction) supplied by the vessel, as occurs in a stroke. They can also cause a vessel to weaken and balloon out, forming an **aneurysm.** Circulating immune complexes may lodge in the skin, eyes, and other body structures. In the kidney, they produce a kind of glomerulonephritis by their inflammatory effects. Even though the organisms normally have little invasive ability, great masses of them growing in the heart are sometimes able to burrow into heart tissue to produce abscesses or damage valve tissue, resulting in a leaky valve. This is a good example of how pathogenicity depends on both host factors and the virulence of the microorganism. ■ immune complexes, p. 420 ■ virulence, p. 394

Epidemiology

In recent years, viridans streptococci have accounted for a smaller percentage of SBE cases than previously; an increasing percentage of cases are caused by more antibiotic-resistant organisms. This may be an unintended effect of giving antibiotic treatment before dental procedures to patients with prominent heart murmurs. A heart murmur is an abnormal sound the physician hears when listening to the heart, often indicative of a deformed valve or other structural abnormality. Also, more SBE cases occur in injected-drug abusers, in hospitalized patients who have plastic intravenous catheters for long periods, and in those with artificial heart valves. These people are usually infected with *S. epidermidis* or a wide variety of species other than viridans streptococci.

Prevention and Treatment

No scientifically proven preventive methods are available. In people with known or suspected defects in their heart valves, however, the accepted practice is to give an antibacterial medication shortly before dental or other bacteremia-causing procedures. The medication is chosen according to the expected species of bacterium and its likely susceptibility to antimicrobials. To prevent nosocomial SBE, rigorous attention to sterile technique when inserting plastic intravenous catheters, changing the catheters to new locations every several days, and discontinuing their use as soon as possible probably helps prevent colonization of the catheters and consequent bacteremia that could lead to heart valve infection. For treatment, antibacterial medications are chosen according to the susceptibility of the causative organism. Only bactericidal medications are effective, and usually two or more antimicrobials are used together. Prolonged treatment such as penicillin and gentamicin over 1 or more months is usually required. Infection of foreign material such as an artificial heart valve is almost invariably associated with biofilm formation and the object must often be replaced to effect a cure. Sometimes, it is necessary to perform surgery to remove the mass of infected clot or to drain abscesses. ■ nosocomial infections, p. 462

The main characteristics of SBE are presented in **table 28.1.**

Gram-Negative Septicemia

Septicemia is a common nosocomial illness, with an estimated 400,000 cases occurring in the United States each year. Approximately 30% of the cases are caused by Gram-negative bacteria. Endotoxin release by Gram-negative bacteria can lead to shock and death. ■ endotoxic shock, p. 407

Symptoms

The symptoms of septicemia include violent shaking, chills, and fever, often accompanied by anxiety and rapid breathing. If

TABLE 28.1	Subacute Bacterial Endocarditis
Symptoms	Fever, loss of energy over a period of weeks or months; sometimes, a stroke
Incubation period	Poorly defined, usually weeks
Causative agents	Usually oral α-hemolytic viridans streptococci or *Staphylococcus epidermidis*
Pathogenesis	Normal microbiota gain entrance to bloodstream through dental procedures, other trauma; in an abnormal heart, turbulent blood flow causes formation of a thin clot that traps circulating organisms; a biofilm forms, makes them inaccessible to phagocytic killing; pieces of clot break off, block important blood vessels, leading to tissue death.
Epidemiology	Persons at risk are mainly those with hearts that have congenital defects or are damaged by disease such as rheumatic fever; situations that cause bacteremia.
Prevention and treatment	Administration of an antibiotic immediately prior to anticipated bacteremia, such as before dental work. Treatment: Bactericidal antibiotics given together, such as penicillin and gentamicin.

septic shock develops, urine output drops, respiration and pulse become more rapid, and the arms and legs become cool and dusky-colored.

Causative Agents

Probably because they possess endotoxin, Gram-negative bacteria are more likely to cause fatal septicemias than are other infectious agents. Shock is common, and despite treatment, only about half of all people afflicted with this kind of infection survive. Cultures of blood from these patients usually reveal facultative anaerobes such as *Escherichia coli.* Among the aerobic, Gram-negative, rod-shaped bacteria encountered in septicemia are organisms commonly found in the natural environment, such as *Pseudomonas aeruginosa.* Some of the Gram-negative organisms that cause septicemia are anaerobes. For example, *Bacteroides* sp., which make up a sizable percentage of the normal microbiota of the large intestine and the upper respiratory tract, cause many septicemia cases. ■ enterobacteria, p. 262 ■ *Pseudomonas*, p. 261

Pathogenesis

Septicemia almost always originates from an infection somewhere in the body other than the bloodstream—a kidney infection, for example. Alterations in normal body defenses as the result of medical treatments, such as surgery, placement of catheters, and medications that interfere with the immune response, may allow microorganisms that normally have little invasive ability to infect the blood. Endotoxin is released from the outermost cell walls of Gram-negative bacteria growing in a localized infection or in the bloodstream. Unfortunately, antibiotics that act against the bacterial cell wall can also enhance the release of endotoxin from the organisms. These antibiotics are typically used in treating Gram-negative bacterial infections. ■ endotoxin, p. 407

The response of the body cells to endotoxin is appropriate for localizing Gram-negative bacterial infections in tissues and killing the invaders. When localization fails and endotoxin enters the circulation, however, it causes the nearly simultaneous triggering of macrophages throughout the body and a cascade of harmful events **(figure 28.3).** Although macrophages normally are of central importance in body defense, they also play a key role in septic shock.

The interaction of endotoxin with toll-like receptors in macrophage cell membranes causes the cell to synthesize and release pro-inflammatory cytokines. For example, tumor necrosis factor (TNF) is released from macrophages within minutes of exposure to endotoxin. TNF has diverse effects, one of which is a change in the setting of the body's thermostat, causing the temperature to rise. TNF also causes circulating polymorphonuclear neutrophils (PMNs) to adhere to capillary walls, leading to large accumulations of these cells in tissues such as the lung, which have large populations of macrophages. Experimentally, antibody against TNF gives substantial protection against endotoxic shock but its use in humans has been disappointing. ■ cytokines, p. 353 ■ toll-like receptors, p. 355

Interleukin-1 (IL-1) is another cytokine released from macrophages. Besides acting with TNF to cause fever and the release of leukocytes from bone marrow, it has many other effects. One potentially harmful action is to cause the release of proteolytic enzymes from neutrophils.

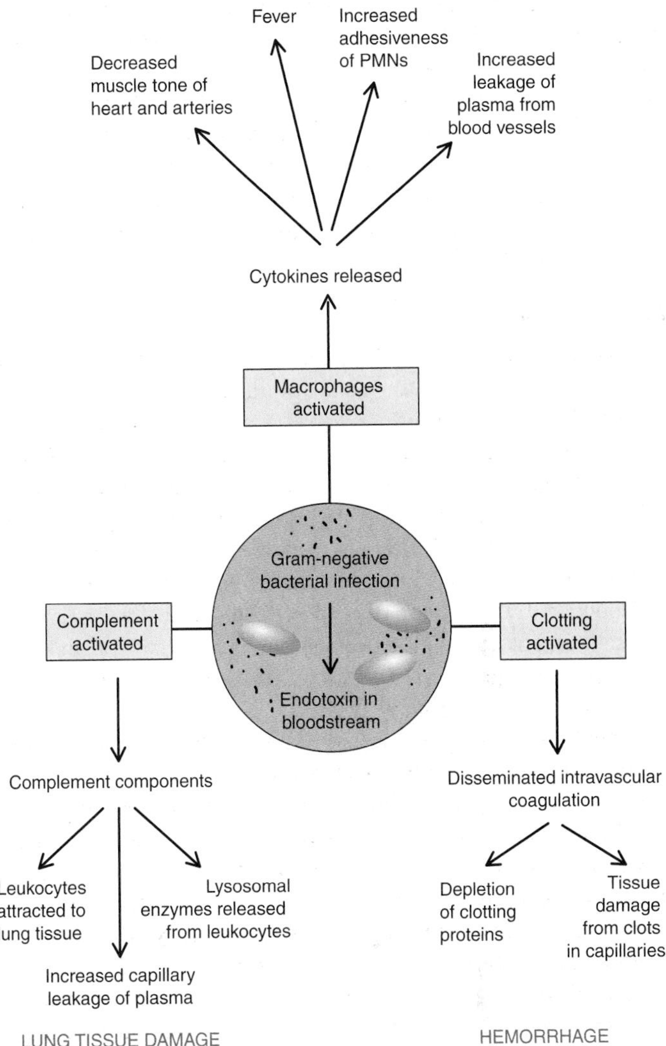

FIGURE 28.3 Events in Gram-Negative Septicemia

Macrophages also synthesize and secrete complement system proteins. Components of activated complement attract leukocytes and cause them to release tissue-damaging lysosomal enzymes. Activated complement also causes capillaries to leak excessive amounts of plasma. ■ complement system, p. 355

The circulating proteins responsible for blood clotting are also activated by endotoxin. Activation causes small clots to form, plugging capillaries, cutting off blood supply, and causing tissue **necrosis.** Paradoxically, this condition, called **disseminated intravascular coagulation (DIC),** is often accompanied by hemorrhage, bleeding. This is because not enough clotting proteins are left to stop blood from flowing out of damaged blood vessels.

These harmful effects of **endotoxemia**—endotoxin circulating in the bloodstream—are made worse by the hypotension (low blood pressure) that usually is present. The hypotension is caused by decreased muscular tone of the heart and blood vessel walls, and the low blood volume that results from leakage of plasma out of the blood vessels. Current evidence indicates that the release of cytokines and their effects on the heart and blood vessels play a key role in causing hypotension in Gram-negative septicemia. Shock results when the blood pressure falls so low that vital organs are no longer supplied with adequate amounts of blood to maintain their function.

Although multiple organs are affected by endotoxemia, the lung is particularly vulnerable to serious, irreversible damage because of its high concentrations of tissue macrophages. Cytokines released from these macrophages in response to endotoxin attract leukocytes, which in turn discharge tissue-destroying enzymes. The resulting damage to lung tissue often results in death despite successful cure of the individual's infection and correction of shock.

Epidemiology

Gram-negative septicemia is mainly a nosocomial disease, reflecting the high incidence of Gram-negative bacteremia in hospitalized patients with impaired host defenses. Patients with cancers and other malignancies, diabetes, and organ transplants are particularly vulnerable. There is a general trend toward an increasing incidence of the disease that relates to longer life span, antibiotic suppression of normal microbiota, immunosuppressive medications, and medical equipment where biofilms readily develop, such as respirators, and catheters placed in blood vessels and the urinary system.

Prevention and Treatment

Prevention of septicemia depends largely on the prompt identification and effective treatment of localized infections, particularly in people whose host defenses are impaired. Also, conditions such as bedsores and pyelonephritis, which commonly lead to septicemia in patients with cancer and diabetes, can usually be prevented. For treatment, antimicrobial medications directed against the causative organism are given. Measures are taken to correct shock and poor oxygenation. Despite these treatments, the mortality rate remains high, generally 30% to 50%, partly because most of the patients have serious underlying conditions. Treatment with monoclonal antibody directed against endotoxin or TNF has shown some benefit when given early in the illness before shock develops. Drotrecogin alpha, another medication being evaluated, is activated human protein C produced by recombinant technology. Protein C normally acts to oppose the effects of cytokines such as TNF and IL-1. Evaluation of these kinds of treatments is difficult because many of the patients have underlying life-threatening illnesses in addition to their septicemia. ■ **pyelonephritis, p. 621** ■ **monoclonal antibodies, p. 439**

MICROCHECK 28.2

Bacteria of low virulence can cause serious, even fatal, infections in individuals whose only defect is a structural abnormality of the heart. A systemic infection represents failure of the body's mechanisms for keeping infections localized to one area. The inflammatory response, although vitally important in localizing infections, can be life-threatening if generalized. People die of Gram-negative septicemia despite antibacterial therapy because of the endotoxin-mediated release of cytokines in the lung and bloodstream.

- ✓ What is a "systemic" infection?
- ✓ What is the difference between bacteremia and septicemia?
- ✓ Why might clots on the heart valves make microorganisms inaccessible to phagocytic killing?

28.3 Bacterial Diseases of the Lymph Nodes and Spleen

Focus Points

- Compare and contrast the epidemiology of tularemia and brucellosis.
- Give reasons why plague is listed as a category A bioterrorism disease.

Enlargement of the lymph nodes and spleen is a prominent feature of diseases that involve the mononuclear phagocyte system. Three examples of these diseases—tularemia, brucellosis, and plague—are discussed next. All three are now uncommon human diseases in the United States, but they represent a constant threat because of their widespread existence in animals. Understanding these diseases is the best protection against contracting them. ■ **mononuclear phagocyte system, p. 350**

"Rabbit Fever" (Tularemia)

Tularemia is widespread among wild animals in the United States, involving species as diverse as rabbits, muskrats, and bobcats. Many human cases are acquired when people are skinning animals that appear to be free of disease. The causative organism enters through unnoticed scratches or by penetration of a mucous membrane. The disease can also be acquired from the bites of flies and ticks and by inhalation of the causative organism. It is potentially a disease transmitted by bioterrorism.

Symptoms

Tularemia is characterized by development of a skin ulceration and enlargement of the regional lymph nodes 2 to 5 days after a person is bitten by a tick or insect or handles a wild animal. The usual symptoms of fever, chills, and achiness that occur in many other infectious diseases are also present in tularemia. When contracted by inhalation, there is also a dry cough and pain beneath the sternum (breastbone) due to enlarged lymph nodes. Symptoms usually clear in 1 to 4 weeks, but sometimes they last for months.

Causative Agent

Tularemia is caused by *Francisella tularensis,* a non-motile, aerobic, Gram-negative rod that derives its name from Edward Francis, an American physician who studied tularemia in the early 1900s, and from Tulare County, California, where it was first studied. The organism is unrelated to other common human pathogens and is unusual in that it requires a special medium enriched with the amino acid cysteine in order to grow.

Pathogenesis

Typically, *F. tularensis* causes a steep-walled ulcer where it enters the skin **(figure 28.4).** Lymphatic vessels draining the area carry the organisms to the regional lymph nodes. These nodes then become large and tender, and they may become filled with pus and

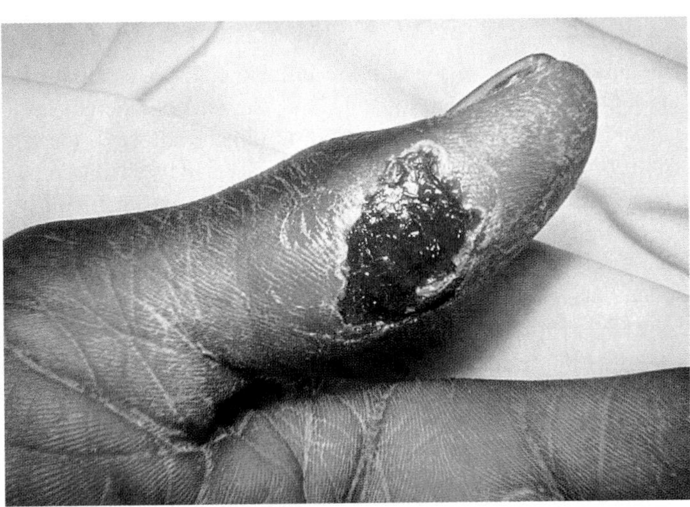

FIGURE 28.4 Tularemic Ulcer of the Thumb This kind of lesion is generally acquired through skinning rabbits or other wild game. Tularemia can also be contracted from insect or tick bites, and by inhalation.

drain spontaneously. Later, the organisms spread to other parts of the body via the lymphatics and blood vessels. Pneumonia, which occurs in 10% to 15% of the cases, occurs when the organisms infect the lung from the bloodstream or by inhalation. Without treatment, tularemic pneumonia has a mortality rate as high as 30%. *Francisella tularensis* is ingested by phagocytic cells and grows within them. This may explain why tularemia persists in some people despite the high titers of antibody in their blood. Cell-mediated immunity is responsible for ridding the host of this infection, as it is with other pathogens that can live intracellularly. Both delayed hypersensitivity and serum antibodies quickly arise during infection, so that even without treatment over 90% of infected people survive. ■ **cell-mediated immunity, p. 367** ■ **delayed hypersensitivity, p. 422**

Epidemiology

Tularemia has been contracted from the bite of a pet hampster. It occurs among wild animals in many areas of the Northern Hemisphere, including all the states of the United States except Hawaii. In the eastern United States, human infections usually occur in the winter months, as a result of people skinning rabbits. Hence, the common name, "rabbit fever." Hunters and trappers in various parts of the country, however, have contracted the disease from muskrats, beavers, squirrels, deer, and other wild animals. The animals are generally free of illness. In the West, infections mostly result from the bites of infected ticks and deer flies and, thus, usually occur during the summer. Generally, 150 to 250 cases of tularemia are reported each year from counties across the United States **(figure 28.5).** Especially in Europe, epidemics of inhalation tularemia have occurred from dust arising from mowing lawns or rodent-infested buildings. Because of its very low infecting dose, it has long been included among biological warfare agents, and in the United States *F. tularensis* is listed among the category A (highest risk) agents of biological terrorism.

Prevention and Treatment

Rubber gloves and goggles or face shields are advisable for people skinning wild animals; remember that the bacteria can enter the body via mucous membranes. Insect repellants and protective clothing help guard against insect and arachnid vectors. It is a good practice to inspect routinely for ticks after exposure to the out-of-doors and to remove them carefully. A vaccine is available for laboratory workers, veterinarians, trappers, game wardens, and others at high risk for infection. Most cases of tularemia are effectively treated with ciprofloxacin or gentamicin.

The main features of this disease are summarized in **table 28.2.**

"Undulant Fever" (Brucellosis, "Bang's Disease")

Only about 150 cases of human brucellosis are reported each year in the United States. Between 10 and 20 times that number of cases, however, go unreported annually. Brucellosis is often called "undulant fever," or "Bang's disease," after Frederik Bang (1848–1932), a Danish veterinary professor who discovered the cause of cattle brucellosis.

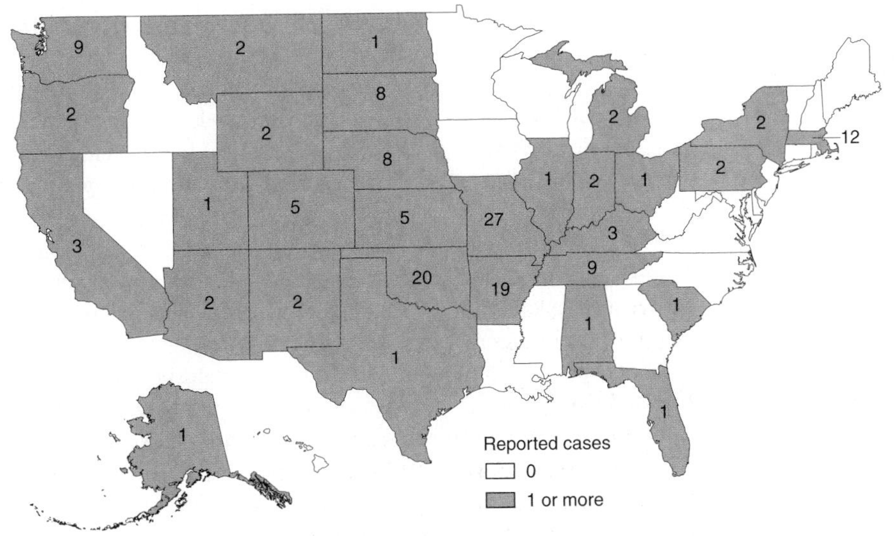

FIGURE 28.5 Reported Cases of Tularemia, United States, 2005 About 60% of the cases were reported from Arkansas, Massachusetts (Martha's Vineyard), Missouri, and Oklahoma.

TABLE 28.2	"Rabbit Fever" (Tularemia)
Symptoms	Ulcer at site of entry, enlarged lymph nodes in area, fever, chills, achiness
Incubation period	1 to 10 days; usually 2 to 5 days
Causative agent	*Francisella tularensis,* an aerobic, Gram-negative rod
Pathogenesis	Organisms are ingested by phagocytic cells, grow within these cells, and then spread throughout body.
Epidemiology	Present among wildlife in most states of the United States. Risk mainly to hunters, trappers, game wardens, and others who handle wildlife. Mucous membrane or broken skin penetration of the organism, as with skinning rabbits, for example; bite of infected insect or tick. Occasionally, inhalation.
Prevention and treatment	Vaccination for high-risk individuals; avoiding bites of insects and ticks; wearing rubber gloves, goggles, when skinning rabbits; taking safety precautions when working with organisms in laboratory. Treatment: gentamicin or ciprofloxacin.

Symptoms

The onset of brucellosis is usually gradual, and the symptoms are vague. Typically, patients complain of mild fever, sweating, weakness, aches and pains, enlarged lymph nodes, and weight loss. The recurrence in some cases of fevers over weeks or months gave rise to the alternative name, "undulant fever." Even without treatment, most cases recover within 2 months, and only 15% will be symptomatic for more than 3 months.

Causative Agent

Brucella sp. are small, aerobic, nonmotile, Gram-negative rods with complex nutritional requirements. Four varieties of the genus *Brucella* cause brucellosis in humans. DNA studies show that all members of the genus fall into a single species, *Brucella melitensis,* but traditionally, the different varieties were assigned species names depending largely on their preferred host: *B. abortus* invades cattle; *B. canis,* dogs; *B. melitensis,* goats; and *B. suis,* pigs. The distinctions between the various strains are mainly useful epidemiologically and generally have little pathogenic significance.

Pathogenesis

As with tularemia, the organisms responsible for brucellosis penetrate mucous membranes or breaks in the skin and are disseminated via the lymphatic and blood vessels to the heart, kidneys, and other parts of the body. The spleen enlarges in response to the infection. Like *Francisella tularensis, Brucella* sp. are not only resistant to phagocytic killing but also can grow intracellularly in phagocytes, where they are inaccessible to antibody and some antibiotics. Mortality, generally due to endocarditis, is about 2%. Bone infection—**osteomyelitis**—is the most frequent serious complication.

Epidemiology

Brucellosis is typically a chronic infection of domestic animals involving the mammary glands and the uterus, thereby contaminating milk and causing abortions in the affected animals. Abortion is not a feature of human disease. Sixty percent of the cases of brucellosis occur in workers in the meat-packing industry; less than 10% arise from ingesting raw milk or other unpasteurized dairy products. Worldwide, brucellosis is a major problem in animals used for food, causing yearly losses of many millions of dollars. In the United States, hunters have acquired infections from elk, moose, bison, caribou, and reindeer. About 20% of the Yellowstone Park bison herd is infected, and more than 1,000 have been killed when they wander outside the park because of fear that they will transmit the disease to cattle. *B. melitensis* is listed as a category B (intermediate risk to national security) bioterrorism agent.

Prevention and Treatment

The most important control measures against brucellosis are pasteurization of dairy products and inspection of domestic animals for evidence of the disease. The use of goggles or face shield and rubber gloves helps protect veterinarians, butchers, and slaughterhouse workers; remember, the bacterium can penetrate mucous membranes. An attenuated vaccine effectively controls the disease in domestic animals. Treatment using tetracycline with rifampin for 6 weeks is usually effective. ■ pasteurization, p. 112

Table 28.3 gives the main features of brucellosis.

"Black Death" (Plague)

Plague, once known as the "black death," was responsible for the death of approximately one-fourth of the population of Europe between 1346 and 1350. Crowded conditions in the cities and a large rat population undoubtedly played major roles in the spread of the disease. Plague is a potential bioterrorism disease, listed as category A.

Symptoms

The symptoms of plague develop abruptly 1 to 6 days after an individual is bitten by an infected flea. The person characteristically develops markedly enlarged and tender lymph nodes called **buboes**—hence the name **bubonic plague**—in the region that receives lymph drainage from the area of the flea bite. High fever, shock, delirium, and patchy bleeding under the skin quickly develop. Some victims develop cough and bloody sputum indicating lung involvement; these cases are called **pneumonic plague.**

Causative Agent

Plague is caused by *Yersinia pestis,* which, like other enterbacteria, is a facultatively anaerobic, Gram-negative rod. It is non-motile and grows best at 28°C. The organism resembles a safety pin in stained preparations of material taken from infected lymph nodes because the ends of the bacterium stain more intensely than does the middle **(figure 28.6).** Extensive study has revealed some of the complex mechanisms by which *Y. pestis* achieves its impressive virulence. The bacterium has plasmids that code for virulence factors. One plasmid codes

TABLE 28.3 "Undulant Fever" (Brucellosis, "Bang's Disease")

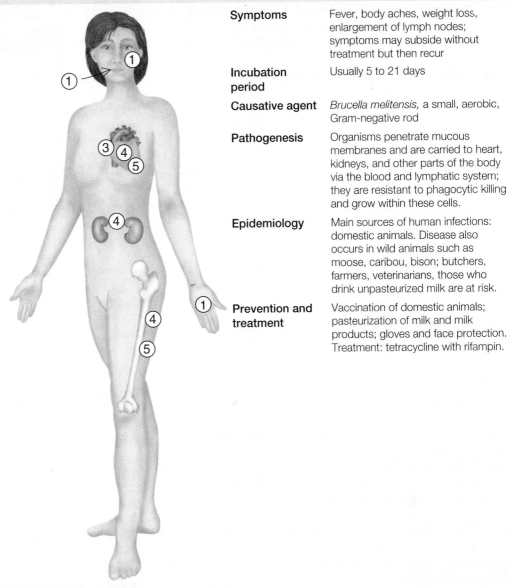

① Causative organism *Brucella melitensis* enters the body through mucous membranes, skin abrasions or ingestion of unpasteurized milk.

② The bacteria are taken up by phagocytes but resist digestion and grow within cells.

③ The bacteria enter the lymphatics and bloodstream and are carried throughout the body.

④ Infection is established in other body tissues, such as the heart valves, kidneys, and bones.

⑤ Osteomyelitis is the most common serious complication. Most deaths are due to endocarditis.

Symptoms	Fever, body aches, weight loss, enlargement of lymph nodes; symptoms may subside without treatment but then recur
Incubation period	Usually 5 to 21 days
Causative agent	*Brucella melitensis,* a small, aerobic, Gram-negative rod
Pathogenesis	Organisms penetrate mucous membranes and are carried to heart, kidneys, and other parts of the body via the blood and lymphatic system; they are resistant to phagocytic killing and grow within these cells.
Epidemiology	Main sources of human infections: domestic animals. Disease also occurs in wild animals such as moose, caribou, bison; butchers, farmers, veterinarians, those who drink unpasteurized milk are at risk.
Prevention and treatment	Vaccination of domestic animals; pasteurization of milk and milk products; gloves and face protection. Treatment: tetracycline with rifampin.

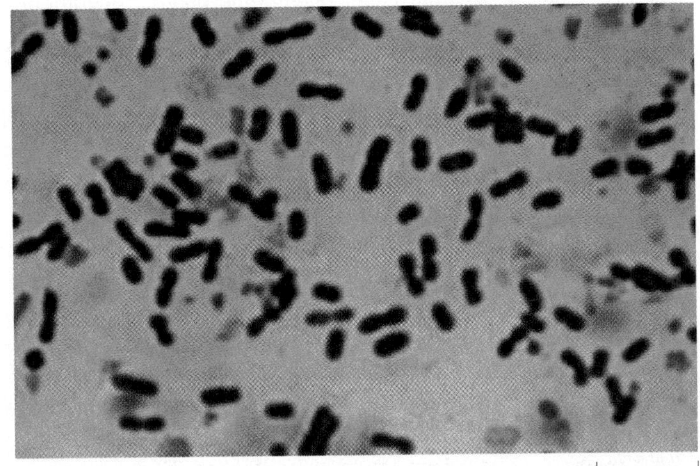

FIGURE 28.6 ***Yersinia pestis*** Each pole of the organism is stained intensely by certain dyes, producing a safety pin appearance.

for a virulence factor called **Pla,** an important protease that causes blood clots to dissolve by activating a substance in the body called plasminogen activator. Another activity of the Pla protease is that it destroys the C3b and C5a components of the complement system. ■ plasmids, pp. 68, 205

Another plasmid codes for a group of proteins that interfere with phagocytosis. These proteins are referred to as **Yops,** for *Y*ersinia *o*uter membrane *p*rotein*s*. The bacterium loses its virulence if this plasmid is lost. Yops are injected into host cells by a type III secretion system. They are responsible for several kinds of actions, the sum of which is to interfere with the phagocytes that normally would kill *Y. pestis* and initiate an immune response to the bacterium. A plasmid-coded protein, V (for virulence), regulates the type III secretion system. ■ secretion systems, p. 398

Yet another plasmid codes for an important antigen, **F1.** This protein becomes part of an antiphagocytic capsule and is an important component of a plague vaccine. The stimulus for capsule production is the relatively increased temperature of a mammalian host (37°C for humans) compared with that of a flea (about 26°C). The genes for some other *Y. pestis* virulence factors reside on the bacterial chromosome. These encode resistance to the lytic action of activated complement, mechanisms for storing hemin and using it as an iron source, and production of a pilus adhesion (PsaA). These genes are probably activated as a result of intracellular growth. ■ **iron requirement, p. 93** ■ **adhesins, p. 398**

A summary of these virulence factors is presented in **table 28.4.**

Pathogenesis

Masses of *Y. pestis* partially obstruct the digestive tract of infected rat fleas **(figure 28.7).** Consequently, not only is the flea ravenously hungry, causing it to bite repeatedly, but it also regurgitates infected material into the bite wounds. Chronically infected fleas do not have their digestive tract obstructed but discharge *Y. pestis* in their feces; the organisms can then be introduced into human tissue when a person scratches the flea bite.

The *Y. pestis* protease Pla is essential for the spread of the organisms from the site of entry of the bacteria by clearing the lymphatics and capillaries of clots. The organisms are carried to the regional lymph nodes, where they are taken up by macrophages. The bacteria are not killed by the macrophages and instead multiply and produce F1 capsular material and other virulence factors in response to the intracellular growth conditions. The macrophages die and release the bacteria, which now are encapsulated and express Yops, pilus adhesin, complement resistance, and heme storage. After several days, an acute inflammatory reaction develops in the nodes, producing enlargement and marked tenderness. The lymph nodes become necrotic, allowing large numbers of virulent *Y. pestis* to spill into the bloodstream. This stage of the disease is called **septicemic plague,** and endotoxin release results in shock and disseminated intravascular coagulation (DIC). Infection of the lung from the bloodstream occurs in 10% to 20% of the cases, resulting in pneumonic plague. Organisms transmitted to another person from a case of pneumonic plague are already fully virulent and, therefore, are especially dangerous. The dark hemorrhages into the skin from DIC and the dusky color of skin and mucous membranes probably inspired the name "black death" for the plague.

The mortality rate for persons with untreated bubonic plague is between 50% and 80%. Untreated pneumonic plague progresses rapidly and is nearly always fatal within a few days.

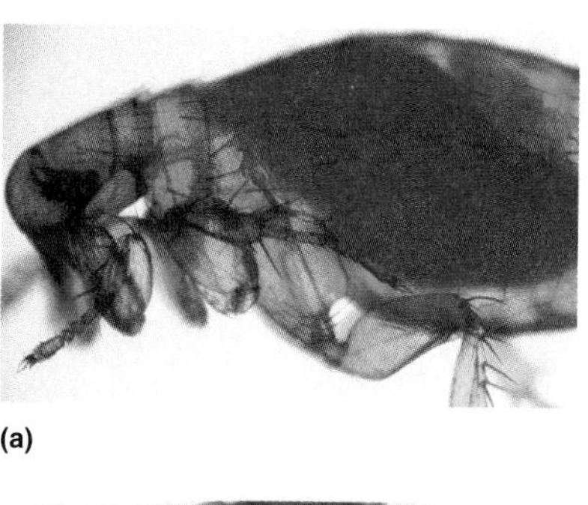

(a)

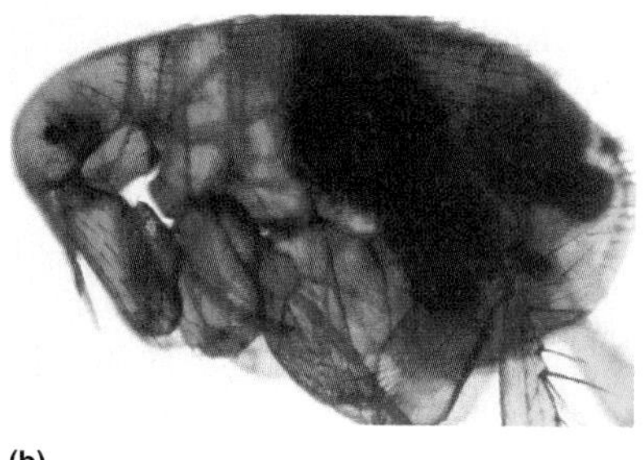

(b)

FIGURE 28.7 Fleas Following a Blood Meal (a) Healthy flea; **(b)** flea with obstruction due to *Yersinia pestis* infection.

Epidemiology

Plague is endemic in rodent populations of all continents except Australia. In the United States, the disease is mostly confined to wild rodents in about 15 states in the western half of the country. Prior to 1974, only a few cases of plague were reported each year. Over the last few decades, however, the number of cases has generally been higher, averaging about 15 reported cases per year, as towns and cities expand into the countryside. Prairie dogs, rock squirrels, and their fleas constitute the main reservoirs, but rats, rabbits, dogs, and cats are potential hosts. Hundreds of species of fleas can transmit plague, and the fleas can remain infectious for a year or more in abandoned rodent burrows. Epidemics in humans, initiated by infected rodent fleas, can spread from one person to another by household fleas, as well as by respiratory droplets produced by coughing patients with pneumonic plague. Pneumonic plague is the most dangerous because *Y. pestis* is fully virulent at the time of transmission. ■ **endemic disease, p. 451** ■ **reservoir, p. 453**

TABLE 28.4 Virulence Factors of *Yersinia pestis*

Factor	Coded by	Action
Pla (protease)	9.5-kbp plasmid	Activates plasminogen activator; destroys C3b, C5a
Yops (proteins)	72-kbp plasmid	Interferes with phagocytosis and the immune response by differing mechanisms
V antigen	Plasmid	Controls type III secretion system
F1	110-kbp plasmid	Forms antiphagocytic capsule at 37°C
PsaA (adhesin)	Chromosome	Role in attachment to host cells
Complement resistance	Chromosome	Protects against lysis by activated complement
Iron acquisition	Chromosome	Traps hemin and other iron-containing substances; stores iron compounds intracellularly

Prevention and Treatment

Plague epidemics can be prevented by rat control measures such as proper garbage disposal, constructing rat-proof buildings, installing guards on the ropes that moor ocean-going ships to keep rats from entering, and rat extermination programs. The latter must be combined with the use of insecticides to prevent the escape of infected fleas from dead rats. The killed whole cell vaccine used for many years is no longer available commercially. It did not provide protection against pneumonic plague, the primary concern in the event of biological warfare. New, genetically engineered vaccines containing the F1 and V antigens are being evaluated under a joint agreement among the United States, Canada, and the United Kingdom. The antibiotic tetracycline can be given as a preventive for someone exposed to plague, and is useful in controlling epidemics because of its immediate effect. Treatment with gentamicin, ciprofloxacin, or doxycycline is effective, especially if given within 24 hours of the onset of symptoms of the disease. The main features of plague are presented in **table 28.5.**

MICROCHECK 28.3

Generally, tularemia is a bacteremic disease of wild animals transmitted to humans by exposure to their blood or tissues, or by biting insects. Brucellosis is most commonly a disease of domestic animals transmitted to humans when individuals handle the animals' flesh or drink unpasteurized milk. Plague is endemic in rodents in the western United States and is transmitted to humans by flea bites. Unlike tularemia and brucellosis, plague has the potential to spread rapidly from human to human by coughing, with a high fatality rate.

- ✓ Workers in what industry are especially likely to contract brucellosis? How can they protect themselves from the disease?
- ✓ How would growth within phagocytes protect *Francisella tularensis* from destruction by antibodies?
- ✓ How would crowded conditions in cities favor spread of plague?

TABLE 28.5 "Black Death" (Plague)

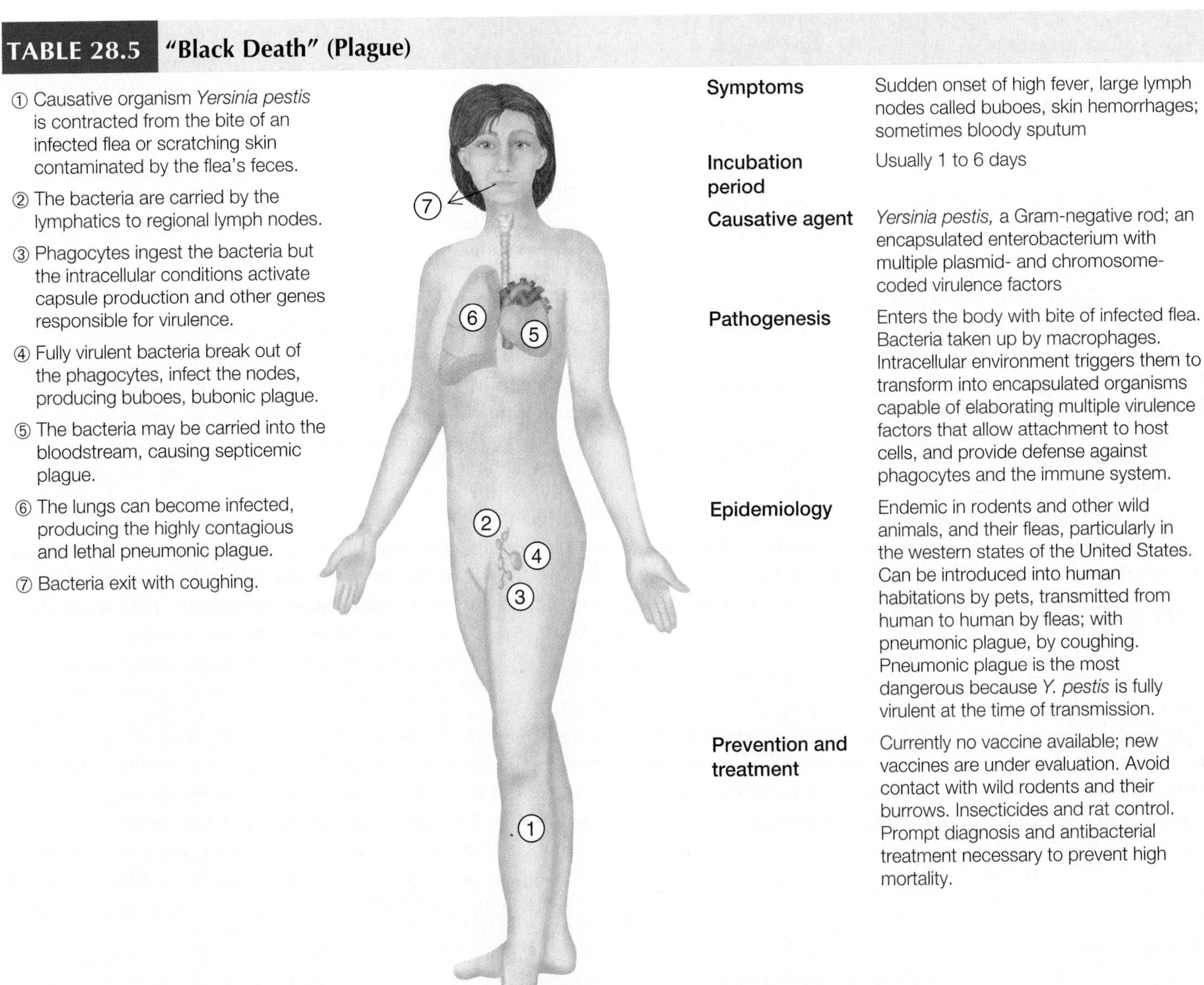
① Causative organism *Yersinia pestis* is contracted from the bite of an infected flea or scratching skin contaminated by the flea's feces.
② The bacteria are carried by the lymphatics to regional lymph nodes.
③ Phagocytes ingest the bacteria but the intracellular conditions activate capsule production and other genes responsible for virulence.
④ Fully virulent bacteria break out of the phagocytes, infect the nodes, producing buboes, bubonic plague.
⑤ The bacteria may be carried into the bloodstream, causing septicemic plague.
⑥ The lungs can become infected, producing the highly contagious and lethal pneumonic plague.
⑦ Bacteria exit with coughing.
7
6
5
2
4
3
1

Symptoms	Sudden onset of high fever, large lymph nodes called buboes, skin hemorrhages; sometimes bloody sputum
Incubation period	Usually 1 to 6 days
Causative agent	*Yersinia pestis,* a Gram-negative rod; an encapsulated enterobacterium with multiple plasmid- and chromosome-coded virulence factors
Pathogenesis	Enters the body with bite of infected flea. Bacteria taken up by macrophages. Intracellular environment triggers them to transform into encapsulated organisms capable of elaborating multiple virulence factors that allow attachment to host cells, and provide defense against phagocytes and the immune system.
Epidemiology	Endemic in rodents and other wild animals, and their fleas, particularly in the western states of the United States. Can be introduced into human habitations by pets, transmitted from human to human by fleas; with pneumonic plague, by coughing. Pneumonic plague is the most dangerous because *Y. pestis* is fully virulent at the time of transmission.
Prevention and treatment	Currently no vaccine available; new vaccines are under evaluation. Avoid contact with wild rodents and their burrows. Insecticides and rat control. Prompt diagnosis and antibacterial treatment necessary to prevent high mortality.

28.4

Viral Diseases of the Lymphoid and Blood Vascular Systems

Focus Points

- Outline the pathogenesis of infectious mononucleosis.
- Describe how infectious mononucleosis is transmitted.
- Explain why yellow fever has not been eradicated despite the availability of an effective vaccine.

A number of viral illnesses mainly involve the lymphoid system and blood vessels. Two notable examples discussed in chapter 29 are the human immunodeficiency virus (HIV), the cause of AIDS; and cytomegalovirus, which causes lymph node enlargement in healthy adults and more serious disease in infants. Infectious mononucleosis has similar symptoms, but it is a much more familiar disease. Yellow fever more prominently involves the circulatory system. ■ human immunodeficiency virus disease, p. 699 ■ cytomegalovirus disease, p. 715

"Kissing Disease" (Infectious Mononucleosis, "Mono")

Infectious mononucleosis ("mono") is a disease familiar to many students because of its high incidence among people between the ages of 15 and 24 years. The term *mononucleosis* refers to the fact that people afflicted with this condition have an increased number of mononuclear leukocytes in their blood. ■ leukocytes, p. 350

Symptoms

Typically, symptoms of infectious mononucleosis appear after a long incubation period, usually 30 to 60 days. They consist of fever, a sore throat covered with pus, marked fatigue, and enlargement of the spleen and lymph nodes. In most cases, the fever and sore throat are gone in about 2 weeks; the enlarged lymph nodes in 3. Persons can usually return to school or work within 4 weeks, but some suffer severe exhaustion and difficulty concentrating that prohibits return to normal activities for months.

Causative Agent

Infectious mononucleosis is caused by the Epstein-Barr (EB) virus, named after its discoverers, M. A. Epstein and Y. M. Barr. It is a double-stranded DNA virus of the herpesvirus family, and although identical in appearance to the other known herpesviruses that cause human disease, it is not closely related to any of them. This interesting virus was discovered in the early 1960s when it was isolated from **Burkitt's lymphoma,** a malignant tumor derived from B lymphocytes, common in parts of Africa. Subsequent studies showed that EB virus is the cause of infectious mononucleosis. ■ B lymphocytes, p. 352

Pathogenesis

Primary infection with EB virus is analogous to throat infections with the herpes simplex virus in that both viruses initially infect the mouth and throat and then become latent in another cell type. The probable sequence of events is shown in **figure 28.8.** Following replication in epithelium of the mouth, saliva-producing glands, and throat, the Epstein-Barr virus is carried by the lymphatics to the lymph nodes. There it infects B lymphocytes, which have specific surface receptors for the virus. During the illness, up to 20% of the circulating B lymphocytes are infected with the virus. The B-cell infection can be (1) productive, in which the virus replicates and kills the B cell; or (2) non-productive, in which the virus establishes a latent infection, existing as either extrachromosomal circular DNA or integrated into the host cell chromosome at random sites. For most of these cells, the infection is non-productive, but profound changes in the cells result from the infection. The virus activates the B cells, causing multiple clones of B lymphocytes to proliferate and produce immunoglobulin. The infected cells are also "immortal," meaning that they can reproduce indefinitely in laboratory cultures.

The T lymphocytes respond actively to the infection and destroy B cells with productive infections because these B cells display viral antigens on their surfaces. The abnormal-appearing lymphocytes characteristically seen in smears of the patient's blood are effector cytotoxic T cells responding to the infected B cells **(figure 28.9).**

The proliferating lymphocytes are responsible for the large numbers of mononuclear cells that give the disease its name. Their numbers and appearance sometimes mistakenly suggest the diagnosis of leukemia, which is disproved when the patient spontaneously recovers. In many cases, a consequence of B-cell infection is the appearance of an IgM antibody that will react with an antigen on the red blood cells of certain animal species, notably sheep, horses, and oxen. This kind of antibody arising against antigens of another animal is called a **heterophile antibody.** It generally has no pathologic significance, but its presence helps in diagnosing infectious mononucleosis. The antibody does not react specifically with EB virus. Enlargement of the lymph nodes and spleen reflects the active replication of lymphocytes. Hemorrhage from rupture of the enlarged spleen is the chief cause of the rare deaths due to infectious mononucleosis. It is most likely to occur within 3 to 4 weeks of the onset of illness. Patients are instructed to avoid exertion and contact sports, but almost half of the cases of splenic rupture occur in the absence of trauma.

The possibility that EB virus could play a role in causing certain malignant tumors has been intensely investigated. The malignancies most closely related to EB virus infection—Burkitt's lymphoma and nasopharyngeal carcinoma—cluster dramatically in certain populations. This suggests that factors other than simple EB viral infection must be present for these malignancies to develop. EB virus-associated Burkitt's lymphoma cases occur mainly in children in East Africa and New Guinea, whereas nasopharyngeal carcinoma is common in Southeast China. The EB virus genome is detectable in 90% to 100% of these two malignancies. Infectious mononucleosis has not been shown to increase risk of malignant tumors in otherwise healthy American college students. Evidence suggests that EB virus may be a factor, however, in some malignancies in patients with immunodeficiency from AIDS or organ transplantation.

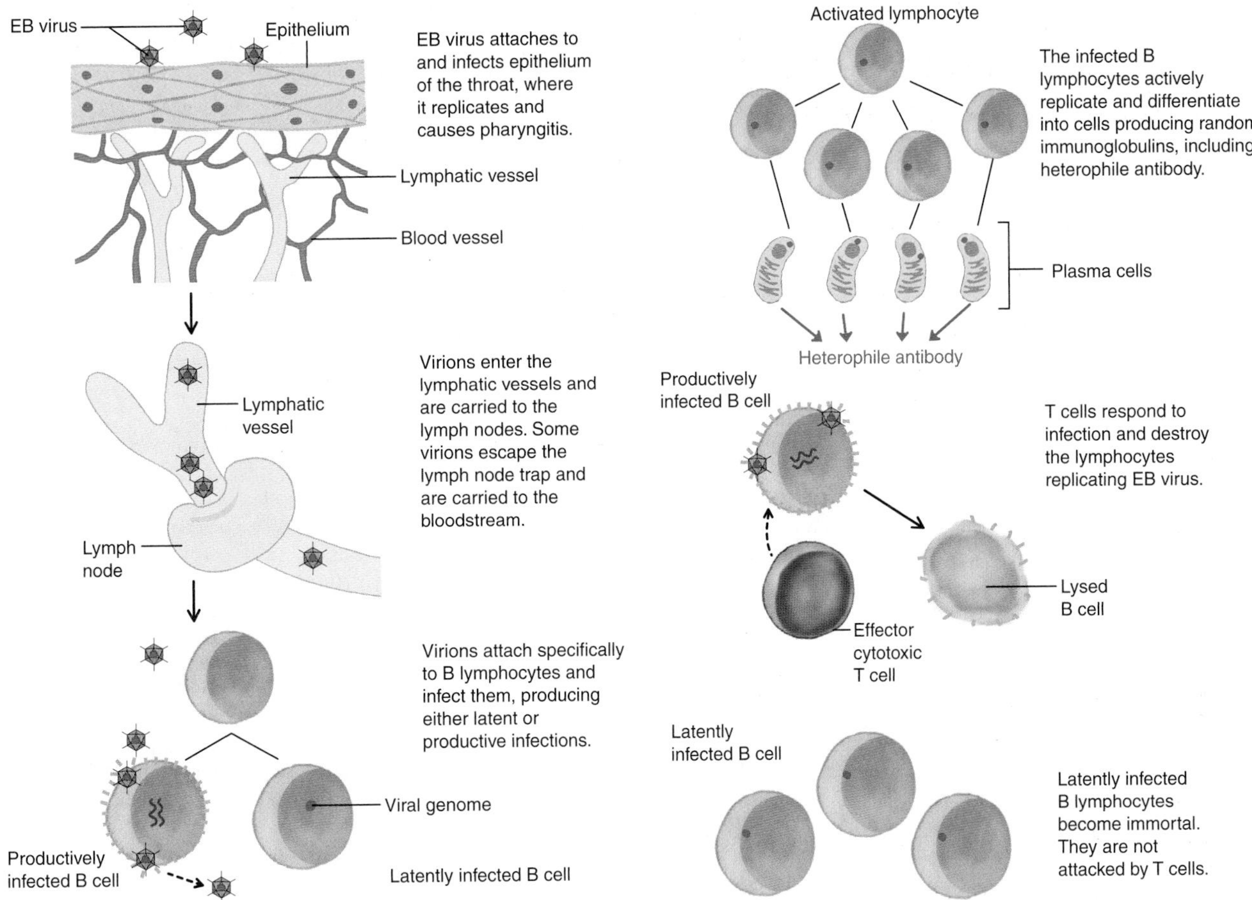

FIGURE 28.8 Pathogenesis of Infectious Mononucleosis

Epidemiology

The EB virus is distributed worldwide and infects individuals in crowded, economically disadvantaged groups at an early age without producing significant illness. In such populations, the characteristic infectious mononucleosis syndrome is quite rare. More affluent populations such as students entering college in the United States often have escaped past infection with the agent and lack immunity to EB virus. In any year, an estimated 1.5 million college students will either have had the disease or will get it during the college year. Infectious mononucleosis occurs almost exclusively in adolescents and adults who lack antibody to the virus. Even in the age group of 15 to

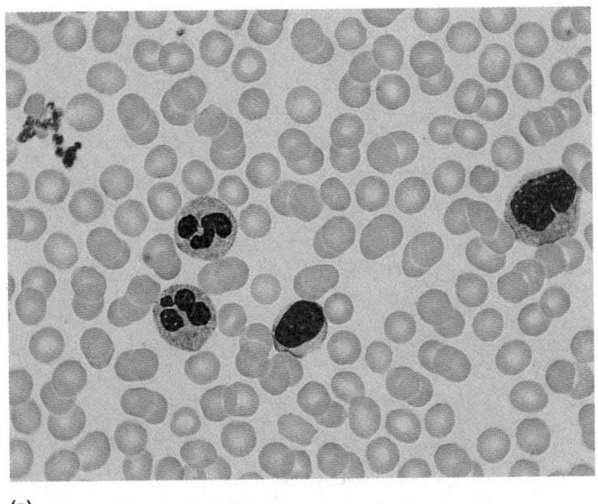
(a)

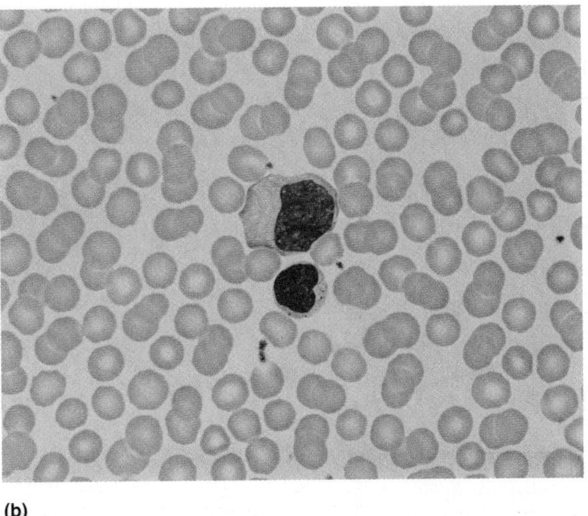
(b)

FIGURE 28.9 Normal and Infectious Mononucleosis Blood Smears Photomicrographs of **(a)** a normal blood smear showing two polymorphonuclear neutrophils (PMNs), a lymphocyte and a monocyte and **(b)** an infectious mononucleosis blood smear showing an abnormal lymphocyte.

CASE PRESENTATION

The patients were two American boys, 7 and 9 years old, living in Thailand, who almost simultaneously developed irritated eyes and slightly runny noses, progressing to fever, headache, and severe muscle pain. It hurt them to move their eyes, and they refused to walk because of pain in their legs. The symptoms subsided after a couple of days, only to recur at lesser intensity. No treatment was given, and they were completely back to normal within a week. Their illness was diagnosed as **dengue** (pronounced DEN-gay), also known as "breakbone fever."

1. What causes dengue and where does it occur?
2. Why is dengue important?
3. What is known about the pathogenesis of dengue?
4. What can be done to prevent dengue?

Discussion

1. Dengue is caused by any of four closely related flaviviruses—dengue 1, 2, 3, and 4. The disease occurs in large areas of the tropics and subtropics around the world. In the Americas, the disease is transmitted mainly by *Aedes aegypti* mosquitoes, which have staged a comeback after being almost eradicated in the 1960s. This vector now occurs year-round along the Gulf of Mexico. Serious dengue epidemics occurred in Brazil and Cuba in 2002.
2. The two individuals presented here had a mild form of the disease that characteristically involves newcomers to an area where the disease commonly occurs. In endemic areas, the disease is characterized by fever, headache, muscle aches, rash, nausea, and vomiting. In a small percentage of cases, however, the effects of dengue infection are much more serious, resulting in dengue hemorrhagic fever. This form of the disease is characterized by bleeding and leakage of fluid from the capillaries. An important result is that the blood pressure drops and the blood thickens. With expert treatment, the mortality is about 1% to 2%. Dengue shock syndrome is another potential development. It is characterized by profound shock and disseminated intravascular coagulation (DIC) and has a mortality above 40%.
3. About 90% of the cases of hemorrhagic fever or shock occur in subjects who have previously been infected by a dengue virus and have antibody to it. The remaining cases occur largely in infants who still have transplacentally acquired maternal antibody against dengue viruses. These antibodies, whether from earlier infection or from the mother, attach to the infecting dengue virus strain and thereby promote its uptake by macrophages. The virus, instead of being killed, reproduces in the macrophage. This results in the death of the macrophage and release of chemicals that cause leaky capillaries and shock. T lymphocytes may also play a role in this process in older children and adults. Evidence also exists that the virus causes a depression of the bone marrow, accounting for the very low white blood cell and platelet counts seen in dengue.
4. Scientists at Mahidol University in Bangkok and other research centers around the world are working on vaccines designed to bring dengue under control. At present, control efforts are largely directed at killing mosquitoes and their larvae and eliminating water containers around houses where the mosquitoes breed. Scientists at Colorado State University have successfully genetically engineered mosquitoes to render their cells incapable of reproducing dengue virus. This was accomplished by using another virus to introduce an antisense segment of the dengue genome into the mosquito cells. This was a dramatic accomplishment, but years of work remain before it can play a role in the control of dengue.

24 years, however, only about half of the EB virus infections produce infectious mononucleosis; the remainder develop few or no symptoms.

EB virus is present in the saliva for up to 18 months after infectious mononucleosis, and thereafter it occurs intermittently for life. Continuous salivary shedding of virus is common in people with AIDS or other immunodeficiencies. Mouth-to-mouth kissing is an important mode of transmission of infectious mononucleosis in young adults, giving rise to the name "kissing disease." The donor of the virus is usually asymptomatic and may have been infected in the past without developing symptoms of infectious mononucleosis. By middle age, most people have demonstrable antibody to the virus, indicating past infection. There is no animal reservoir.

Prevention and Treatment

Prevention is aided by avoiding the saliva of another person, as well as objects such as toothbrushes or drinking glasses possibly contaminated with another's saliva. No vaccine for infectious mononucleosis is available. Antiviral medications such as acyclovir and famciclovir inhibit productive infection by the virus and are of value in rare serious cases; however, they have no activity against the latent infection. Cortisone-like medication is sometimes useful in relieving airway obstruction from swollen tissue.

Table 28.6 gives the main features of infectious mononucleosis.

TABLE 28.6 "Kissing Disease" (Infectious Mononucleosis, "Mono")

Symptoms	Fatigue, fever, sore throat, and enlargement of lymph nodes
Incubation period	Usually 1 to 2 months
Causative agent	Epstein-Barr (EB) virus, a DNA virus of the herpesvirus family
Pathogenesis	Productive infection of epithelial cells of throat and salivary ducts; latent infection of B lymphocytes; activation of B and T lymphocytes; hemorrhage from enlarged spleen is a rare but serious complication.
Epidemiology	Spread by saliva; lifelong recurrent shedding of virus into saliva of asymptomatic, latently infected individuals.

Yellow Fever

Yellow fever was first recognized with an epidemic in the Yucatan of Mexico, in 1648, probably introduced there from Africa. One of the worst outbreaks of yellow fever in the twentieth century occurred in Ethiopia in the 1960s, producing 100,000 cases and 30,000 deaths. In 1989, an epidemic of yellow fever occurred in

Bolivia among poor people who moved into the jungle to try to make a living growing coca. There have been no outbreaks in the United States since 1905, but the vector mosquito has reappeared in the Southeast, raising the possibility of outbreaks of the disease if the virus is again introduced.

Symptoms

The symptoms of yellow fever can range from very mild to severe. Symptoms of mild disease, the most common form, may be only fever and a slight headache lasting a day or two. Patients suffering severe disease, however, may experience a high fever, nausea, bleeding from the nose and into the skin, "black vomit" (from gastrointestinal bleeding), and jaundice (hence, the name *yellow fever*). The mortality rate of severe yellow fever cases can reach 50% or more. The reasons for the wide variation in severity of symptoms are unknown, but probably have more to do with the size of the infecting dose and the status of human host defenses than with differences among strains of the causative virus.

Causative Agent

Yellow fever is caused by an enveloped, single-stranded, positive-sense RNA arbovirus of the flavivirus family. The virus multiplies in species of mosquitoes, apparently without harming them, and the mosquitoes transmit the infection to humans.

Pathogenesis

The yellow fever virus is introduced into humans by the bite of an *Aedes* mosquito, the biological vector. It multiplies, enters the bloodstream, and is carried to the liver and other parts of the body. Viral liver damage results in jaundice and decreased production of clotting proteins, and injury to small blood vessels produces **petechiae**—tiny hemorrhages—throughout the body. The virus affects the circulatory system by directly damaging the heart muscle, by causing bleeding from blood vessel injury in various tissues, and by causing disseminated intravascular coagulation (DIC). Kidney failure is a common consequence of loss of circulating blood and low blood pressure. ■ **biological vector, p. 455** ■ **disseminated intravascular coagulation, p. 679**

Epidemiology

The reservoir of the virus is mainly infected mosquitoes and primates living in the tropical jungles of Central and South America and in Africa **(figure 28.10)**. Periodically, the disease spreads from the jungle reservoir to urban areas, where it is transmitted to humans by *Aedes* mosquitoes.

Prevention and Treatment

In urban areas, control of yellow fever is achieved by spraying insecticides and eliminating the breeding sites of its principal vector, *Aedes aegypti*. In the jungle, control of yellow fever is almost impossible because the mosquito vectors live in the forest canopy and transmit the disease among canopy-dwelling monkeys. A highly effective attenuated vaccine is available to immunize people who might become exposed, including foreign

(a)

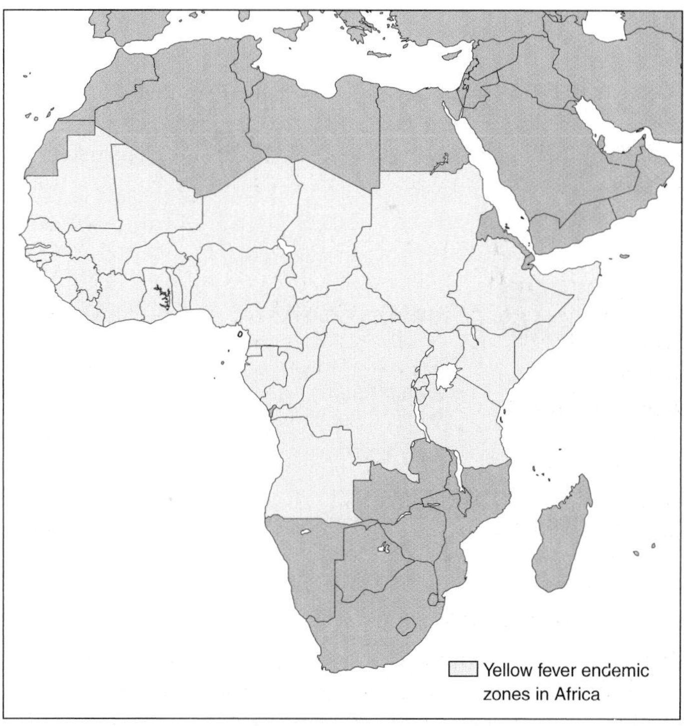

(b)

FIGURE 28.10 Distribution of Endemic Yellow Fever (a) South America **(b)** Africa.
(Source: CDC Health Information for International Travel 2008.)

PERSPECTIVE 28.2

Walter Reed and Yellow Fever

Walter Reed was born in Virginia in 1851. After receiving medical degrees, he entered the Army Medical Corps. He was appointed professor of bacteriology at the Army Medical College in 1893, and in 1900, Reed was named president of the Yellow Fever Commission of the U.S. Army. Although improved sanitation had effectively controlled other diseases, the incidence of yellow fever remained high and its cause and transmission were unknown.

Reed suspected that an insect was the vector of yellow fever. As early as 1881, Dr. Carlos Finlay of Havana, Cuba, had suggested that a mosquito might be the carrier of this disease, but he was unable to prove his hypothesis. Beginning with mosquitoes raised from eggs, Reed and the others in his commission proceeded with a series of experiments. First, the laboratory-raised mosquitoes were allowed to feed on patients with yellow fever and, second, on members of the commission.

After initial failures to transmit the disease by mosquitoes, one member of the commission, Dr. James Carroll, came down with classic yellow fever 3 days after an experimental mosquito bite. Dr. Jesse Lazear, another commission member, noted that the yellow fever patient on which the mosquito had originally fed was in his second day of the disease and that 12 days had elapsed before it bit Carroll. This timing was critical for the transmission of the disease because, as we now know, the mosquito can only contract the infection during a brief period when the patient is viremic. Then, the mosquito can only transmit the infection to another individual after the virus has replicated to a high level in the mosquito.

Later, Lazear was bitten by a mosquito while working in a hospital yellow fever ward. He developed yellow fever and died. He was the only yellow fever fatality among commission members, although other volunteers lost their lives before the mysteries of this potentially fatal disease were resolved. The commission had a mosquito-proof testing facility constructed, called Camp Lazear in honor of their deceased colleague, to house volunteers for their experiments.

As a result of their studies, Dr. Reed and his colleagues made the following conclusions: (1) mosquitoes are the vectors of the disease; (2) an interval of about 12 days must elapse between the time the mosquito ingests the blood of an infected person and the time it can transmit the disease to an uninfected person; (3) yellow fever can be transmitted from a person acutely ill with the disease to a person who has never had the disease by injecting a small amount of the ill person's blood; (4) yellow fever is not transmitted by soiled linens, clothing, or other items that have come into contact with infected persons; and (5) yellow fever is caused by an infectious agent so small that it passes through a filter that excludes bacteria.

Armed with these findings, Major William C. Gorgas, the chief sanitary officer for Havana, instituted mosquito control measures that resulted in a dramatic reduction in yellow fever cases. Later, Gorgas used mosquito control in the Panama Canal Zone to allow construction of the Panama Canal. Previously, French workers had tried to build the canal but had failed because of heavy losses from yellow fever and malaria.

travelers to the endemic areas. There is no proven antiviral treatment. ■ attenuated vaccine, p. 433

The main features of yellow fever are presented in **table 28.7.**

TABLE 28.7 Yellow Fever

Symptoms	Often only headache and fever. Severe cases characterized by high fever, jaundice, black vomit, and hemorrhages into the skin
Incubation period	Usually 3 to 6 days
Causative agent	Yellow fever virus, an enveloped, single-stranded RNA virus of the flavivirus family
Pathogenesis	Virus multiplies locally at site of introduction by an infected mosquito; spreads to the liver and throughout the body by the bloodstream. Virus destroys liver cells, causing jaundice and decreased production of blood-clotting proteins. Hemorrhages and decreased strength of the heart result in circulatory failure and kidney failure.
Epidemiology	Virus persists in forest primates and the mosquitoes that feed on them, in Africa and Central and South America; human epidemics occur when the virus infects household mosquitoes that feed on humans.
Prevention and treatment	There is a highly effective attenuated viral vaccine. No proven antiviral therapy is available.

MICROCHECK 28.4

Infectious mononucleosis occurs worldwide and is transmitted from person to person by saliva. Curiously, the causative virus is associated with different malignancies, but only in certain geographic areas or in AIDS patients. Yellow fever virus infects monkeys that live in tropical forests of Africa and South and Central America, and the mosquitoes that bite them. Transmission of the virus among human beings can occur wherever there is a human-biting mosquito species that is susceptible to the virus.

- ✓ What characteristic changes occur in the blood of patients with infectious mononucleosis?
- ✓ What is the name of the tumor from which EB virus was first isolated?
- ✓ Why does it take more than a week before a mosquito just infected with yellow fever virus can transmit the disease?

28.5 Protozoan Diseases

Focus Points

- Outline the *Plasmodium vivax* life cycle.
- Discuss why malaria has been so difficult to control.

Protozoa infect the blood vascular and lymphatic systems of millions of people worldwide. One example, discussed in an earlier chapter, is *Trypanosoma brucei,* fundamentally a bloodstream parasite of African animals and cause of African sleeping sickness in humans. Another trypanosome, *T. cruzi,* causes chagas disease (American trypanosomiasis), often manifest as a chronic heart infection. Protozoans of the genus *Leishmania* cause visceral leishmaniasis, with enormous splenic enlargement, and a problem for American troops involved in war in the Middle East. (There is more information about leishmaniasis at the Online Learning Center www.mhhe.com/nester6.) Malaria, another protozoan disease, is much more widespread and is a leading cause of morbidity and mortality, meaning illness and death, worldwide. An estimated 300 to 500 million people become ill with malaria each year, and a child dies of the disease every 40 seconds. ■ protozoa, p. 285 ■ African sleeping sickness, p. 667

Malaria

Malaria is an ancient scourge, as evidenced by early Chinese and Hindu writings. During the fourth century B.C., the Greeks noticed its association with exposure to swamps and began drainage projects to control the disease. The Italians gave the disease its name, *malaria,* which means "bad air," in the seventeenth century. In early times, malaria ranged as far north as Siberia and as far south as Argentina. In 1902, Ronald Ross received a Nobel Prize for demonstrating the life cycle of the protozoan cause of malaria.

Malaria is the most common serious infectious disease worldwide. In 1955, the World Health Organization (WHO) began a program for the worldwide elimination of malaria. Initially there was great success, as WHO employed insecticides such as DDT against the mosquito vector, detected infected patients by obtaining blood smears, and provided treatment for those who were infected. Fifty-two nations undertook control programs and, by 1960, 10 of them had eradicated the disease. Unfortunately, strains of *Anopheles* mosquitoes resistant to insecticides began to appear, and in cooperation with bureaucracy and complacency, malaria began a rapid resurgence. In 1976, the World Health Organization acknowledged that the eradication program was a failure. Today, over 300 million people are infected with malaria annually worldwide and it causes more than 1 million deaths per year. More people are dying of the disease than when the eradication programs first began. ■ DDT, p. 749

Symptoms

The first symptoms of malaria are "flu"-like, with fever, headache, and pain in the joints and muscles. These symptoms generally begin about 2 weeks after the bite of an infected mosquito, but in some cases they can begin many weeks afterward. After 2 or 3 weeks of these symptoms, the pattern changes abruptly and symptoms tend to fall into three phases highly suggestive of malaria: (1) The patient feels cold and develops shaking chills that can last for as much as an hour (cold phase): (2) following the chills, the temperature begins to rise steeply, often reaching 40°C (104°F) or more (hot phase); and (3) after a number of hours of fever, the patient's temperature falls, and drenching sweating occurs (wet phase). These abrupt changes in symptoms are referred to as **paroxysms**—a sudden intensification of symptoms. Except for fatigue, the patient feels well until 24 or 48 hours later, depending on the causative species, when the next paroxysm occurs.

Causative Agent

Human malaria is caused by protozoa of the genus *Plasmodium.* Four species are involved—*P. vivax, P. falciparum, P. malariae,* and *P. ovale.* These species differ in microscopic appearance and, in some instances, life cycle, type of disease produced, severity, and treatment. In recent years, the majority of returning travelers and immigrants diagnosed with malaria in the United States have been infected with *P. vivax,* but up to 30% have been infected with the more dangerous species, *P. falciparum.*

The *Plasmodium* life cycle is complex **(figure 28.11),** with a number of different forms that differ in microscopic appearance and antigenicity. In the human host, **sporozoites**—the infectious form injected by the mosquito—are carried by the bloodstream to the liver, where they infect liver cells. In these cells, each parasite enlarges and subdivides, producing thousands of **merozoites.** These merozoites are then released into the bloodstream and establish the cycle involving the erythrocytes. The parasite grows and divides in the erythrocytes of the host. The earliest form resembles a ring, with a large, pale food vacuole in the central area, and the nucleus and cytoplasm pushed to the periphery. This develops into a larger, motile **trophozoite,** which goes on to subdivide, producing a **schizont.** The infected erythrocyte then breaks open, and the offspring of the division, also called merozoites, are released into the plasma. The merozoites then enter new erythrocytes and multiply, repeating the cycle. Some merozoites that enter erythrocytes develop into **gametocytes,** which are specialized sexual forms. These are different from the other circulating plasmodia in both their appearance and susceptibility to antimalarial medicines. These sexual forms do not rupture the red blood cells. They cannot develop further in the human host and are not important in causing the symptoms of malaria. They are, however, infectious for certain species of *Anopheles* mosquitoes and are thus ultimately responsible for the transmission of malaria from one person to another.

When a mosquito dines on a person's blood (see figure 28.11), it digests the erythrocytes and liberates the gametocytes. Shortly after entering the intestine of the mosquito and stimulated by the drop in temperature, the male and female gametocytes change in form to become gametes. The male gametocyte transforms into about a half dozen tiny, whiplike gametes that swim about until they unite with the female gamete in much the same way as the sperm and ovum unite in higher animals. The resulting **zygote** transforms into a motile form that burrows into the wall of the midgut of the mosquito and forms a cyst. The cyst enlarges as the diploid nucleus undergoes meiosis, dividing asexually into numerous offspring. The cyst then ruptures into the body cavity of the mosquito, and the released parasites, called sporozoites, find their way to the mosquito's salivary glands and saliva, from which they may be injected into a new human host.

Pathogenesis

The characteristic feature of malaria—recurrent paroxysms of fever followed by feeling healthy again—results from the erythrocytic cycle of growth and release of merozoites. Interestingly, the infections in all the millions of different red blood cells become nearly synchronous. Thus, cell rupture and release of daughter protozoa occur at roughly the same time for all infected cells, and each release causes a fever. For *P. malariae,* the growth cycle

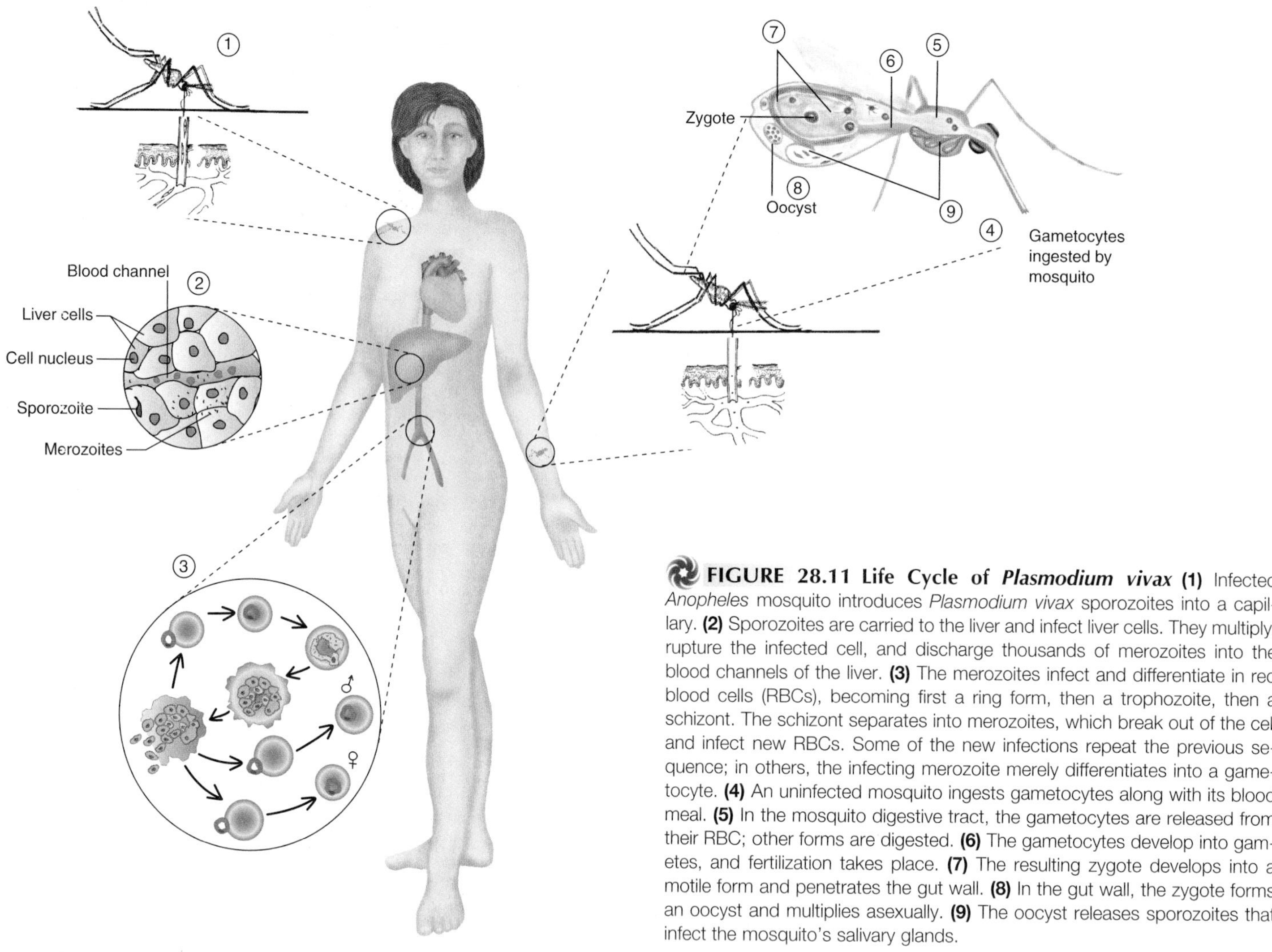

FIGURE 28.11 Life Cycle of *Plasmodium vivax* **(1)** Infected *Anopheles* mosquito introduces *Plasmodium vivax* sporozoites into a capillary. **(2)** Sporozoites are carried to the liver and infect liver cells. They multiply, rupture the infected cell, and discharge thousands of merozoites into the blood channels of the liver. **(3)** The merozoites infect and differentiate in red blood cells (RBCs), becoming first a ring form, then a trophozoite, then a schizont. The schizont separates into merozoites, which break out of the cell and infect new RBCs. Some of the new infections repeat the previous sequence; in others, the infecting merozoite merely differentiates into a gametocyte. **(4)** An uninfected mosquito ingests gametocytes along with its blood meal. **(5)** In the mosquito digestive tract, the gametocytes are released from their RBC; other forms are digested. **(6)** The gametocytes develop into gametes, and fertilization takes place. **(7)** The resulting zygote develops into a motile form and penetrates the gut wall. **(8)** In the gut wall, the zygote forms an oocyst and multiplies asexually. **(9)** The oocyst releases sporozoites that infect the mosquito's salivary glands.

takes 72 hours, so that fever recurs every third day. For the other species, fevers generally occur every other day.

Infections by *P. falciparum* tend to be very severe, probably because all erythrocytes are susceptible to infection, whereas other *Plasmodium* species infect only young or old erythrocytes. Thus, very high levels of parasitemia—parasites in the bloodstream—can develop with *P. falciparum* infections. The infected red blood cells become rigid, in contrast to normal red cells, which are flexible. Also, they adhere to each other and to the walls of capillaries. These tiny blood vessels therefore become plugged, and the affected tissue becomes deprived of oxygen as a result. Involvement of the brain, or **cerebral malaria,** is particularly devastating, but almost any organ can be severely affected.

Plasmodium vivax and *P. ovale* malaria often relapse after treatment of the blood infection because treatment-resistant forms of the organisms **(hypnozoites)** continue to reside in the liver in a dormant state. Months or even years later, hypnozoites can begin multiplying in an **exoerythrocytic stage** (*exo-* means "outside of," *-erythrocytic* refers to red blood cells). The exoerythrocytic stage can initiate new erythrocytic cycles of infection after the earlier bloodstream infection has been cured.

The spleen characteristically enlarges in malaria to cope with the large amount of foreign material and abnormal red blood cells, which it removes from the circulation. Malaria is the most common cause of splenic rupture, which can occur with or without trauma.

Especially with *P. falciparum,* the parasites cause anemia by destroying red blood cells and converting the iron in hemoglobin to a form not readily recycled by the body. The large amount of foreign material in the bloodstream strongly stimulates the immune system. In some cases, the overworked immune system fails and immunodeficiency results.

Those who live continuously in areas where malaria is endemic develop some immunity from repeated infections, and this protects them from the lethal effects of the disease. This immunity crosses the placenta and gives partial protection to the newborn. The greatest risk of death from malaria is to children over 6 months of age as this immunity wanes. Currently, worldwide, a child dies of malaria about every 40 seconds. Others at high risk of death are women in their first pregnancy, and individuals who move into an endemic area.

Epidemiology

Malaria was once common in both temperate and tropical areas of the world, and endemic malaria was only eliminated from the continental United States in the late 1940s. Today, malaria is predominantly a disease of warm climates, but 41% of the world's population lives in endemic areas. Certain human-biting mosquito species of the genus *Anopheles* are biological vectors of malaria. Because suitable vectors are abundant in North America, the potential exists for the spread of malaria whenever it is introduced. Infected mosquitoes and humans constitute the reservoir for malaria. Besides mosquitoes, malaria can be transmitted by blood transfusions or the sharing of syringes among drug users. Malaria contracted in this manner is easier to treat because it involves only red blood cells and not the liver—only sporozoites from mosquitoes can infect the liver. Some people of black African heritage are genetically resistant to *P. vivax* malaria because their red blood cells lack the receptors for the parasite. Also, some genetically determined blood diseases such as sickle cell anemia have survived over the eons, despite their negative effect on health, because they provide partial protection against malaria. ■ biological vector, p. 455

Prevention and Treatment

Treatment of malaria is complicated by the fact that different stages in the life cycle of the parasite respond to different medications. Chloroquine, and the newer, chemically related mefloquine, are effective against the erythrocytic stages of sensitive strains, but will not cure the liver infection with *P. vivax* or kill the gametocytes of *P. falciparum.* Primaquine or a newer derivative, tafenoquine, is generally effective against the hypnozoites and the *P. falciparum* gametocytes.

In 1998, a new initiative called Roll Back Malaria was begun, linking the World Health Organization, the United Nations Childrens Fund (UNICEF), the United Nations Development Program, and the World Bank in the fight against malaria. The program's goal is to reduce malaria deaths by 50% by the year 2010, and again by 2015. The initial focus was on detailed mapping of malarious areas using satellite imagery and climate information, and documenting the level of malaria treatment and prevention at the village level. The aim was to organize and fund a sustained effort to improve access to medical care, strengthen local health facilities, and promote the delivery of medications and insecticide impregnated mosquito netting. There was a better understanding that malaria control must be part of overall economic development because it is difficult for one to move forward without the other.

Unfortunately, the rate of malaria continued to increase under this initiative largely because chloroquine, the main antimalarial medication used for many years, is no longer effective in many areas of the world. Resistance has also quickly developed to newer medications such as mefloquine and sulfadoxine/ pyrimethamine **(figure 28.12).** By 2004, it was generally agreed that for malaria to be controlled, it is necessary to turn to more effective, although much more costly, medications. Derivatives of artemisinin, the active ingredient of an ancient Chinese herbal medicine, are now used in combination with another medication to minimize the risk of developing resistance. With better funding, these combinations, called ACTs (artemisinin-based combination therapies), are now widely available and are helping to reverse the resurgence of malaria. Other highly active antimalarial medications are being evaluated, and a number of potential vaccines are being tested. The most promising vaccine, so far, gives only 30% protection to children. It consists of a sporozoan coat protein attached to a hepatitis B vaccine protein, and is administered with an adjuvant. Judicious indoor spraying of DDT, the use of insecticide-impregnated bed nets, and elimination of mosquito breeding areas are also being emphasized. Adequate funding of malaria control is a continuing problem.

Table 28.8 gives the main features of malaria.

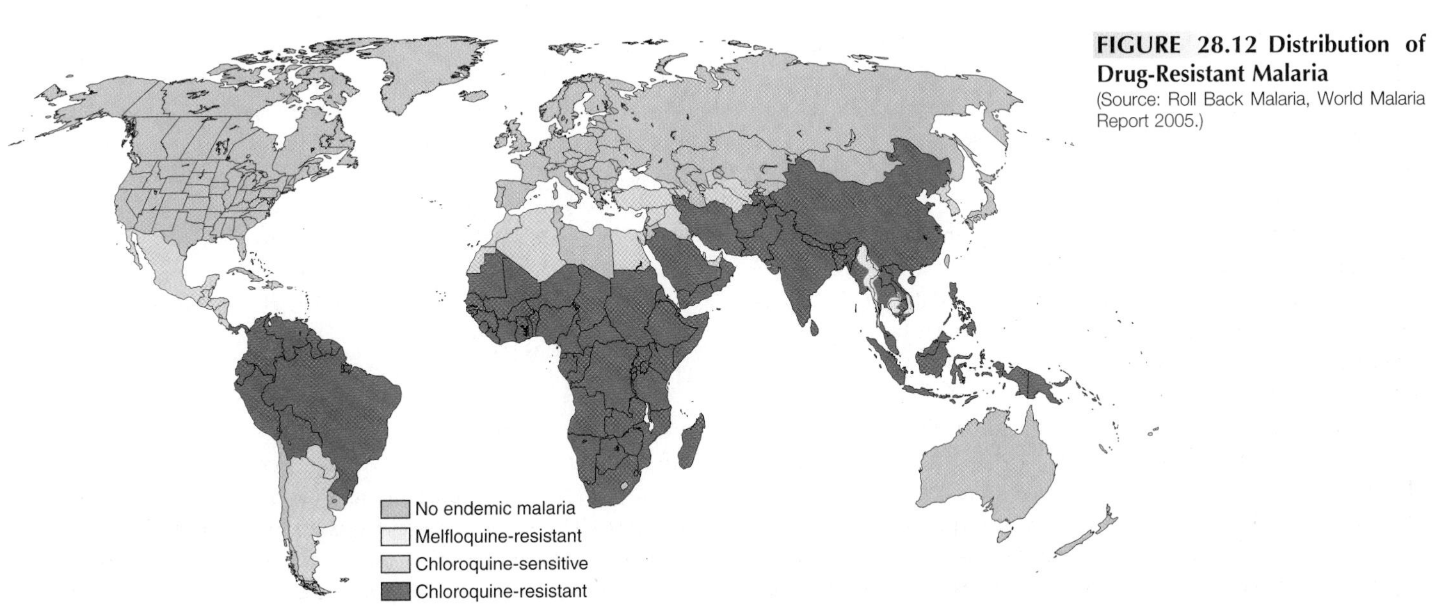

FIGURE 28.12 Distribution of Drug-Resistant Malaria
(Source: Roll Back Malaria, World Malaria Report 2005.)

TABLE 28.8 Malaria

Symptoms	Recurrent bouts of violent chills and fever alternating with feeling healthy
Incubation period	Varies with species; 6 to 37 days
Causative agent	Four species of protozoa of the genus *Plasmodium*
Pathogenesis	Cell rupture, release of protozoa causes fever; infected red blood cells adhere to each other and to walls of capillaries; in the case of *falciparum* malaria; vessels plug up, depriving tissue of oxygen; spleen enlarges in response to removing large amount of foreign material and many abnormal blood cells from the circulation.
Epidemiology	Transmitted from person to person by bite of infected anopheline mosquito. Some individuals genetically resistant to infection.
Prevention and treatment	For prevention, weekly doses of chloroquine if in chloroquine-sensitive malarial areas; doxycycline, mefloquine, or other alternative if in chloroquine-resistant areas; after leaving, primaquine is given for liver stage; ACTs or other medicines for resistant strains; eradication of mosquito vectors; mosquito netting impregnated with insecticide; vaccines under development. Treatment: usually ACTs; other medicines if sensitivity known.

MICROCHECK 28.5

There are a number of different stages in the life cycles of the protozoa that cause malaria, and they have differing microscopic appearance, antigenicity, and susceptibility to antimalarial medications. Malaria is a major hindrance to economic improvement in many countries. It is a leading serious, infectious disease worldwide, and unless sustained, well-funded control measures are undertaken, it is expected to advance into new geographic areas.

✓ Why is malaria contracted from a blood transfusion easier to treat than if contracted from a mosquito?

✓ Why did insecticide-resistant mosquitoes begin to appear?

✓ Are sporozoites diploid or haploid?

FUTURE CHALLENGES

Rethinking Malaria Control

Previous attempts at malaria control have relied heavily on the use of DDT, a non-biodegradable insecticide. Use of this substance was banned in the United States in 1972 because of its accumulation in the environment and its damaging effects on fowl, notably the peregrine falcon and the bald eagle. Many other concerns have been raised about its possible carcinogenicity and potential effects on human fetuses. Although DDT is not generally used in agriculture, more than 20 countries still use it for malaria control, and more would use it if it were affordable. Biodegradable insecticides are available, but they are more expensive than DDT, and not without toxicity. With children dying of malaria every minute worldwide, the options for rapid relief from malaria all seem bad, and a choice has to be made of which ones are the least bad. Insecticides may be necessary short term, but other options, including education, economic development, vaccines, biological control of the *Anopheles* vectors, and mosquito habitat alteration, may prove more important in achieving sustained relief from the disease.

The challenge for the future is to better understand the ecology of malaria as it applies in each location, with the aim of minimizing insecticide use and discovering new options for maintaining long-term control.

SUMMARY

28.1 Anatomy and Physiology (figure 28.1)

The Heart

The left side of the heart receives oxygenated blood from the lungs and pumps it into thick-walled arteries; the right side of the heart receives blood depleted of oxygen from the veins and pumps it through the lungs.

Arteries

Arteries have thick, muscular walls. People begin to develop arteriosclerosis in the teenage years or before. The possible role of bacteria and viruses in the formation of arteriosclerotic lesions is under investigation.

Veins

The blood in veins is a dark color because it is depleted of O_2. The pressure is much lower than in arteries. Blood flow is aided by contracting muscles compressing the veins. Direction of flow is maintained by one-way valves.

Lymphatics

Lymphatics—lymph vessels—are blind-ended tubes that take up fluid that leaks from capillaries. They also take up bacteria, which are normally trapped by lymph nodes distributed along the course of the lymphatics.

Spleen

The spleen filters unwanted material such as bacteria and damaged erythrocytes from the arterial blood. It becomes enlarged in diseases such as infectious mononucleosis and malaria.

28.2 Bacterial Diseases of the Blood Vascular System

Bacteria circulating in the bloodstream can colonize the inside of the heart, and they can cause collapse of the circulatory system and death. Infections of the heart valves and lining of the heart are called **endocarditis;** illness resulting from circulating pathogens is called **septicemia. Acute bacterial endocarditis** is caused when virulent bacteria enter the bloodstream from a focus of infection elsewhere in the body, or when contaminated material is injected by drug abusers; normal heart valves are commonly infected and destroyed.

Subacute Bacterial Endocarditis (table 28.1)

Subacute bacterial endocarditis (SBE) is commonly caused by organisms of little virulence, including oral streptococci and *Staphylococcus epidermidis,* and infection usually begins on structural abnormalities of the heart. Biofilm formation may complicate treatment.

Gram-Negative Septicemia (figure 28.3)

Gram-negative septicemia is commonly a nosocomial illness; many afflicted individuals have serious underlying illnesses such as cancer and diabetes. Shock precipitated by release of endotoxin from the bacteria is a common complication; it is the result of an exaggerated response to endotoxin in which massive release of cytokines occurs in the circulation.

28.3 Bacterial Diseases of the Lymph Nodes and Spleen

Tularemia, brucellosis, and plague involve the mononuclear-phagocyte system and are characterized by enlargement of the lymph nodes and spleen. The causative organisms grow within phagocytes, protected from antibody.

"Rabbit Fever" (Tularemia) (table 28.2, figure 28.4)

Tularemia is usually transmitted from wild animals to humans by exposure to the animals' blood or via insects and ticks. The cause is the Gram-negative aerobe *Francisella tularensis,* which is found throughout the United States except Hawaii (figure 28.5).

"Undulant Fever" (Brucellosis, "Bang's disease") (table 28.3)

Brucellosis, caused by *Brucella melitensis,* is usually acquired from cattle or other domestic animals, less often from wild animals such as elk, moose, and bison. Hunters, butchers, and those who drink unpasteurized milk or milk products are at increased risk for the disease. The organisms can infect via mucous membranes and minor skin injuries.

"Black Death" (Plague) (table 28.5)

Plague, once pandemic, now persists endemically in rodent populations, including those in many western states of the United States. It is caused by *Yersinia pestis,* an enterobacterium with many virulence factors, chromosomally or plasmid coded, that interfere with phagocytosis and immunity (table 28.4, figure 28.6). **Bubonic plague** is transmitted to humans by fleas (figure 28.7). **Pneumonic plague** is transmitted from person to person. Untreated, bubonic plague mortality is 50% to 80%; pneumonic plague, almost 100%.

28.4 Viral Diseases of the Lymphoid and Blood Vascular Systems

"Kissing Disease" (Infectious Mononucleosis, "Mono") (table 28.6)

The incidence is high in 15- to 24-year-olds; most individuals with the disease become lifelong carriers, capable of transmitting the disease by their saliva. The causative agent is Epstein-Barr virus, which establishes a lifelong latent infection of B lymphocytes; the disease is called mononucleosis because victims have an increased number of mononuclear cells in their blood (figure 28.9). The disease is confirmed by the presence of a **heterophile antibody.**

Yellow Fever (table 28.7)

Yellow fever is a zoonosis of mosquitoes and monkeys that exists mainly in tropical jungles; it can become epidemic in humans where a suitable *Aedes* mosquito vector is present (figure 28.10). The disease involves the heart and blood vessels throughout the body and is characterized by fever, jaundice, and hemorrhaging. A highly effective attenuated vaccine is available for preventing the disease.

28.5 Protozoan Diseases

African sleeping sickness is present over much of tropical Africa; of the protozoan blood and lymphatic diseases, however, malaria is the most widespread.

Malaria (table 28.8)

Malaria is caused by four species of *Plasmodium* and is transmitted from person to person by the bite of certain species of *Anopheles* mosquitoes, its biological vector. Now confined mainly to impoverished warm regions of the world, malaria survives despite massive eradication programs; it is the most widespread of all serious infectious diseases (figure 28.12). The life cycle is complex; different forms of the organism invade different body cells and have different susceptibility to anti-malarial medication (figure 28.11). Replication of the organism inside red blood cells results in the almost simultaneous rupture of the infected cells and release of the progeny protozoa to infect new red cells.

REVIEW QUESTIONS

Short Answer

1. What is the significance of immune complex formation in SBE?
2. What is disseminated intravascular coagulation (DIC)?
3. What activities of humans are likely to expose them to tularemia?
4. Why is brucellosis a threat to big game hunters?
5. Why might the *Yersinia pestis* bacteria from a patient with pneumonic plague be more dangerous than the bacteria from fleas?
6. Why might rodent burrows be a source of plague months after they are abandoned?
7. What type of leukocytes does EB virus infect and immortalize?
8. Name a viral disease that can be complicated by disseminated intravascular coagulation.
9. Travelers to and from which areas of the world should have certificates of yellow fever vaccination?
10. Which microorganism causes the most dangerous form of malaria?

Multiple Choice

1. Which of the following infection fighters are found in lymph?
   a) Leukocytes
   b) Antibodies
   c) Complement
   d) Interferon
   e) All of the above
2. All the following are true of the spleen, *except*
   a) it is located low on the right side of the abdomen.
   b) it cleanses the blood of foreign material and damaged cells.
   c) it provides an immune response to circulating pathogens.
   d) it can help produce new blood cells.
   e) it enlarges in a number of infectious diseases.
3. Which one of the following statements about SBE is *false?*
   a) It is generally a chronic illness characterized by fatigue and slight fever.
   b) Kidney involvement can be one complication.
   c) Infection occurs exclusively on the left side of the heart.
   d) Injected-drug abuse can be responsible for the disease.
   e) It can result in aneurysm formation.
4. Choose the one *true* statement about Gram-negative bacterial septicemia.
   a) It is a rare nosocomial disease.
   b) The output of urine increases if shock develops.
   c) It can only be caused by anaerobic bacteria.
   d) An antibiotic that kills the causative organism can be depended on to cure the disease.
   e) Lung damage is an important cause of death.
5. Which of these statements about tularemia is *false?*
   a) It can be contracted from muskrats and bobcats.
   b) Biting insects and ticks can transmit the disease.
   c) The causative organism is closely related to *E. coli.*
   d) A steep-walled ulcer at the site of entry of the bacteria and enlargement of nearby lymph nodes is characteristic.
   e) Without treatment, 9 out of 10 people can be expected to survive.
6. All of the following statements about brucellosis are true, *except*
   a) fevers that come and go over a long period of time gave it the name "undulant fever."
   b) the causative agent can infect via mucous membranes.
   c) the causative agent is readily killed by phagocytes.
   d) the disease in cattle is characterized by chronic infection of the mammary glands and uterus.
   e) butchers are advised to wear goggles or a face shield to help protect against the disease.
7. Pick out the one *false* statement about *Yersinia pestis.*
   a) Growth conditions inside human phagocytes activate virulence genes.
   b) The bacterium can multiply in the flea digestive system.
   c) The code for YOPs exists on the bacterial chromosome.
   d) The organism resembles a safety pin in certain stained preparations.
   e) It was responsible for the "black death" in Europe during the 1300s.
8. Which of the following statements about yellow fever is *false?*
   a) There is no animal reservoir.
   b) The name "yellow" comes from the fact that many victims have jaundice.
   c) Certain mosquitoes are biological hosts for the causative agent.
   d) Outbreaks of the disease could occur in the United States because a suitable vector is present.
   e) An attenuated vaccine is widely used to prevent the disease.
9. The malarial form infectious for mosquitoes is called a
   a) gametocyte.
   b) trophozoite.
   c) sporozoite.
   d) schizont
   e) merozoite
10. Choose the one *true* statement concerning malaria.
    a) Transmission cannot occur in temperate climates
    b) Transmission usually occurs with the bite of a male *Anopheles* mosquito.
    c) The disease is currently well controlled in tropical Africa.
    d) *P. falciparum* infects only old erythrocytes and therefore causes milder disease than other *Plasmodium* species.
    e) The characteristic recurrent fevers are associated with release of merozoites from erythrocytes.

Applications

1. Some years ago, dentists and doctors began noticing an association between subacute bacterial endocarditis and prior dental work, and they began advising that an antibiotic be administered at the time of dental procedures to those with known or suspected heart defects. What was the rationale for this advice?
2. Several children attending a New Mexico school developed a serious illness characterized by high fever and enlarged, tender lymph nodes. Their physicians diagnosed their illnesses as plague, but were mystified because the children all lived and played in town and denied seeing any rodents. Health officials checked and found no evidence of rodents at their homes or school. What other investigations should they carry out to try to establish the source of the disease?
3. A health worker in Honduras is concerned about a potential outbreak of yellow fever in his town. A laborer from a jungle area known to be endemic for the disease had come to the town 2 weeks earlier to work and subsequently developed yellow fever. Several coworkers reported getting mosquito bites while working with him. Why is it important that the health worker determine how long it is since the workers were bitten by the mosquitoes?

Critical Thinking

1. The finding that there is an association between *Chlamydia pneumoniae* infection and arteriosclerotic lesions raised hopes that new methods to combat arteriosclerosis could be developed. An investigator reviewing this research, however, stated that even a perfect correlation between infection and lesion formation would not prove that infection causes arteriosclerosis. Moreover, even showing that therapeutic antibiotics could prevent infection and lesion formation would not be definitive proof. Is the investigator justified in making this argument? Why or why not?
2. Distinguish between EB virus infection and infectious mononucleosis.
3. Even though genetically engineered mosquitoes might be developed that do not allow the reproduction of malaria protozoa, these mosquitoes would have little, if any, immediate effect on the spread of the disease. Why should this be so? What would have to happen for these mosquitoes to significantly affect the spread of malaria?

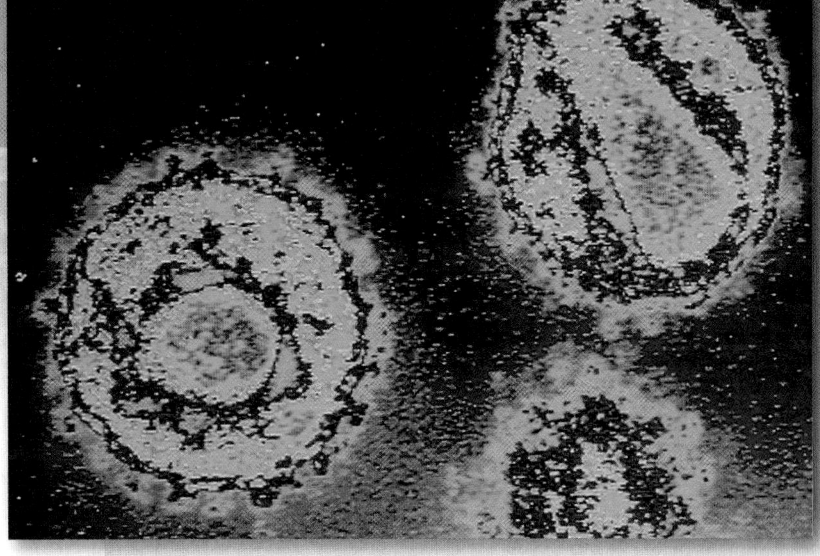

Color-enhanced TEM of human immunodeficiency virus (HIV).

HIV Disease and Complications of Immunodeficiency

A Glimpse of History

In 1981, five reports described an illness in previously healthy, young, homosexual men characterized by unusual infections, certain malignant tumors, and immunodeficiency. This illness came to be known as AIDS, an acronym for *a*cquired *i*mmuno*d*eficiency *s*yndrome. Initially, there was wild speculation about the cause of AIDS, but by 1982, the Centers for Disease Control had convincing epidemiological evidence that AIDS was caused by a new infectious agent. Scientists around the world scrambled to identify it.

Fortunately, the scientists were able to build on some important scientific advances of the 1960s and 1970s—specifically, the discovery of subsets of lymphocytes, the functions of T cells, and the role of cytokines. In 1975, the Nobel Prize was awarded to H. Temin and D. Baltimore for discovering reverse transcriptase. Also, in early 1978, Dr. Robert Gallo's laboratory at the National Cancer Institute discovered the first human retrovirus, human T-lymphotrophic virus (HTLV). Dr. Gallo developed a technique for cultivating retroviruses in normal lymphocytes activated with interleukin-2 (IL-2). ■ **lymphocytes, p. 352** ■ **cytokines, p. 353** ■ **reverse transcriptase, p. 332** ■ **interleukins, p. 353** ■ **retroviruses, p. 332**

In January 1983, an important scientific breakthrough occurred in the laboratory of Dr. Luc Montagnier at the famous Pasteur Institute. Using techniques developed earlier by Gallo for cultivating HTLV, Montagnier recovered a new virus from a patient with lymphadenopathy syndrome, an AIDS-associated condition. Because the IL-2 activated lymphocyte cultures died out quickly, he was only able to obtain small quantities of the new virus, named LAV, an acronym for *l*ymph*a*denopathy *v*irus, but enough to use as antigen in a blood test that showed that AIDS patients were infected with the virus. He sent a sample of this virus to Gallo at the National Cancer Institute, and the two exchanged material from other patients. He also made application to the U.S. Patent Office for a patent on his blood test.

Meanwhile, Gallo's laboratory began recovering a virus from AIDS patients, and reported the finding in the same issue of the journal *Science* in which Montagnier reported his finding. A number of these viral isolates were introduced together in cell cultures to see if a strain of the virus could replicate in the cells. One did replicate well, enabling Gallo to obtain large quantities of the virus. Gallo named the virus HTLV-III because it resembled human T-lymphotrophic viruses discovered earlier, but it soon became clear that the new virus was a lentivirus. Using "HTLV-III," Gallo perfected a blood test for AIDS and, like Montagnier, applied for a patent. These conflicting patent claims caused a bitter scientific and legal battle that raged for years. Under pressure from President Reagan, a settlement was negotiated in 1987 wherein Gallo and Montagnier were named codiscoverers of the test. Eighty percent of the royalties were to go to an AIDS foundation, the remainder to be divided equally between the National Institutes of Health (NIH) and the Pasteur Institute.

The settlement did not end the conflict. Genetic analysis showed that HTLV-III was actually Montagnier's LAV, introduced into Gallo's cultures as described earlier. Later, the Pasteur Institute suffered its own embarrassment, acknowledging that due to a laboratory mixup of its own, LAV actually came from a different patient than stated in its publications. In 1994, the earlier settlement was modified, NIH admitting that HTLV-III and LAV were the same and giving the French side a larger share of the royalties.

The American share of the royalties has generated many millions of dollars for the NIH over the years, but the value of the blood test far exceeded any monetary figure. It helped prove that HTLV-III/LAV—renamed HIV for *h*uman *i*mmunodeficiency *v*irus—caused AIDS. It also showed that many infected people were asymptomatic, that they could transmit the virus, and that the epidemic was far more extensive than previously suspected. These findings helped generate support for controlling the disease. By 1985, the blood test was generally available for routine testing of donated blood, thus markedly improving the safety of blood transfusions and blood products.

KEY TERMS

Carcinoma A malignant tumor that arises from epithelium.

Clade A group of organisms or viruses that have the same ancestor; subtype of a virus such as human immunodeficiency virus (HIV) defined by similar amino acid sequences of their envelope proteins.

Hemophiliac An individual with an inherited bleeding disorder caused by failure of the blood to clot normally.

Integrase Enzyme of human immunodeficiency virus that integrates its linear DNA form into the host chromosome.

Lymphoma A malignant tumor that arises from lymphatic tissue.

Metastasize To spread to another area of the body, as with tumors or infections.

Nucleoside A compound composed of a purine or pyrimidine base covalently bound to ribose or deoxyribose.

Protease Enzyme that degrades protein; the protease that is encoded by HIV is the target of several anti-HIV medications.

Reverse Transcriptase Enzyme that synthesizes double-stranded DNA complementary to an RNA template; target of a number of anti-HIV medications.

Sarcoma A malignant tumor that arises from connective tissue, bone, cartilage, or muscle.

Therapeutic Vaccine A vaccine administered after infection occurs with the hope of activating immunity in time to prevent or ameliorate a disease; effective in preventing rabies.

AIDS

At the end of the second decade after its recognition in 1981, the disease AIDS had killed more American citizens than the Korean and Vietnam wars combined. By 1994, it had become the leading cause of death among those 25 to 44 years of age in the United States. The total number of individuals alive and infected with HIV exceeded 850,000 and the numbers were increasing by more than 40,000 each year. Initially introduced into a population of homosexual men, the virus increasingly infected heterosexuals, especially among economically disadvantaged, racial and ethnic minorities.

Worldwide, at the end of the second decade of the epidemic, an estimated 40 million people were living with HIV infection and more people had died of AIDS than were killed by the "black death" of Europe in the Middle Ages. It became the number one killer in Africa south of the Sahara, surpassing malaria, and the disease spread rapidly into India and China. Millions of children lost one or both parents to the disease. ■ black death, p. 682

Billions of dollars have been spent trying to control the AIDS pandemic, with the result that new HIV infections have declined **(figure 29.1).** Nevertheless, AIDS is far from conquered.

FIGURE 29.1 The Global HIV/AIDS Epidemic at the End of 2007 An estimated 2.5 million people (figure in red) were newly infected during the year. Altogether, 33.2 million people (figures in black) were living with HIV/AIDS. The arrows indicate an increase (↑) or decrease (↓) in the numbers of cases as compared with 2001 using an updated statistical method over the 7-year period.

(Source: UNAIDS and WHO '07 AIDS Epidemic update.)

29.1 Human Immunodeficiency Virus (HIV) Infection and AIDS

Focus Points

- Outline the pathogenesis of HIV disease.
- Explain the importance of the three main modes of transmission of HIV.

Microorganisms such as *Pneumocystis jirovecii* (formerly *Pneumocystis carinii*) are common, but of such low virulence that they only cause disease in individuals with immunodeficiency disorders. When such a disease appears, it strongly suggests that the victim is immunodeficient. In the United States in 1981, a number of cases of *P. jirovecii* pneumonia in previously healthy homosexual men first led to the recognition of AIDS. The acronym **AIDS,** for "acquired immunodeficiency syndrome," was first used in 1982 by the Centers for Disease Control for diseases that were at least moderately predictive of a defect in cellular immunity in subjects with no apparent cause for low resistance to the diseases. These "AIDS-defining conditions" **(table 29.1)** were useful in studying the AIDS epidemic before its cause was known, and still alert physicians to the possibility of AIDS. Immunodeficiency, however, is the end stage of a disease with many

TABLE 29.1 "AIDS-Defining Conditions"

Cancer of the uterine cervix, invasive	Kaposi's sarcoma
Candidiasis involving the esophagus, trachea, bronchi, or lungs	Lymphomas, such as Burkitt's, or arising in the brain
Coccidioidomycosis, of tissues other than the lung	Mycobacterial diseases, including tuberculosis
Cryptococcosis, of tissues other than the lung	Pneumocystosis (pneumonia due to *Pneumocystis jirovecii*)
Cryptosporidiosis of duration greater than 1 month	Pneumonias occurring repeatedly
Cytomegalovirus disease of the retina with vision loss or other involvement outside liver, spleen, or lymph nodes	Progressive multifocal leukoencephalopathy (a brain disease caused by the JC polyomavirus)
Encephalopathy (brain involvement with HIV)	*Salmonella* infection of the bloodstream, recurrent
Herpes simplex virus causing ulcerations lasting 1 months or longer or involving the esophagus, bronchi, or lungs	Toxoplasmosis of the brain
Histoplasmosis of tissues other than the lung	Wasting syndrome (weight loss of more than 10% due to HIV); also known as **slim disease**
Isosporiasis (a protozoan disease of the intestine) of more than 1 month's duration	

other manifestations. AIDS is therefore more appropriately called **HIV disease,** after its causative agent, human immunodeficiency virus (HIV). In this chapter, the term *HIV disease* implies that replication of the virus is causing symptoms or demonstrable damage to the body, whereas *HIV infection* means only that the virus has entered the body and is replicating, whether or not disease has occurred. The term *AIDS* refers to the end stage of HIV disease characterized by unusual tumors and immunodeficiency. ■ immunodeficiency disorders, p. 426

HIV Disease

Almost everyone who becomes infected with HIV develops HIV disease, marked by slow destruction of their immune system and eventually ending in AIDS.

Symptoms

The first symptoms of HIV disease appear after an incubation period of 6 days to 6 weeks and usually consist of fever, headache, sore throat, muscle aches, enlarged lymph nodes, and a generalized rash. Some people develop central nervous system symptoms ranging from moodiness and confusion to seizures and paralysis. These symptoms constitute the **acute retroviral syndrome (ARS),** and they typically subside within 6 weeks. Many HIV infections are asymptomatic, however, or the symptoms are mild and attributed to the "flu."

Following the acute illness, if any, there is an asymptomatic period that typically lasts for years, even though the disease advances in the infected person and can be transmitted to others. The asymptomatic period may end with persistent enlargement of the person's lymph nodes, a condition known as **lymphadenopathy syndrome (LAS).** In some cases, symptoms heralding immunodeficiency include fever, weight loss, fatigue, and diarrhea. These symptoms are referred to as the **AIDS-related complex (ARC),** and indicate a poor prognosis. The first symptoms in many cases are those due to tumors or opportunistic infections resulting from severe immunodeficiency. As one might expect, symptoms at this stage of the disease vary widely according to the kind of infection. For example, a frequently encountered symptom is a fuzzy white patch on the tongue—**hairy leukoplakia** (**figure 29.2**)—a result of latent Epstein-Barr virus (EBV) reactivation. Severe skin rashes, cough and chest pain, stiff neck, confusion, visual problems, and diarrhea are manifestations of other infections. ■ Epstein-Barr virus, p. 686

Causative Agent

In the United States, and in most other parts of the world, AIDS is usually caused by human immunodeficiency virus, type 1 (HIV-1), a single-stranded RNA virus of the retrovirus family. There are many kinds of retroviruses that naturally infect hosts as diverse as fish and humans. HIV-1 belongs to the lentivirus subgroup of the retroviruses, which characteristically infect mononuclear phagocytes.

HIV-1 viruses can be classified into subtypes based on nucleic acid sequences. In the M (for "major") group of HIV-1 viruses, 10 subtypes—**clades**—have been described, designated A through J. Members of each clade are closely related, as determined by sequencing their genomes. Infection by two different subtypes can give rise to "hybrids" that have properties of both. Other HIV-1

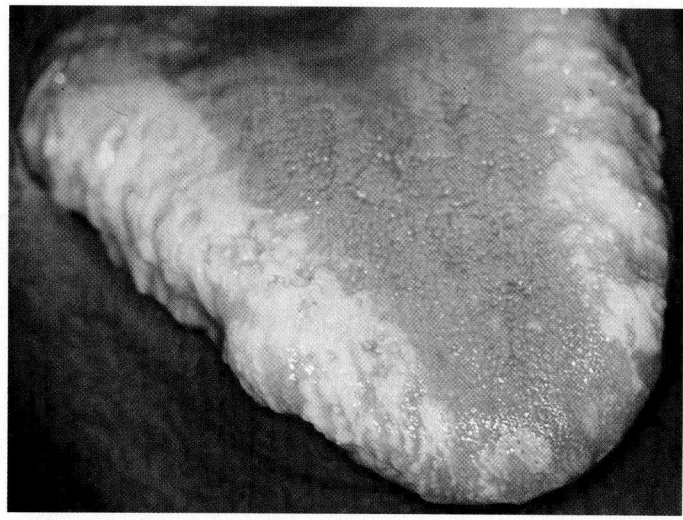

FIGURE 29.2 Hairy Leukoplakia This AIDS-related condition is probably due to activation of latent Epstein-Barr virus infection.

groups such as O (for "outlier") and N differ from M group viruses. The importance of being able to recognize all of these different strains of HIV-1 viruses is to aid epidemiological studies and to ensure that the tests for HIV reliably detect infection.

Human immunodeficiency viruses, type 2 (HIV-2), are lentiviruses similar in structure to HIV-1, but antigenically distinct, their genomes differing from HIV-1 by more than 55%. They are a prominent cause of AIDS in parts of West Africa and India, and have appeared in the United States and other countries. HIV-2 transmission has generally been less efficient than that of HIV-1, and disease progression is slower. Otherwise, the biology of HIV-2 is quite similar to that of HIV-1. For the remainder of this chapter, we will use the term HIV to indicate HIV-1.

The structure of HIV is shown schematically in **figure 29.3a.** The locations of its important antigens are indicated. The numerous knobs projecting from the surface of the virion represent SU

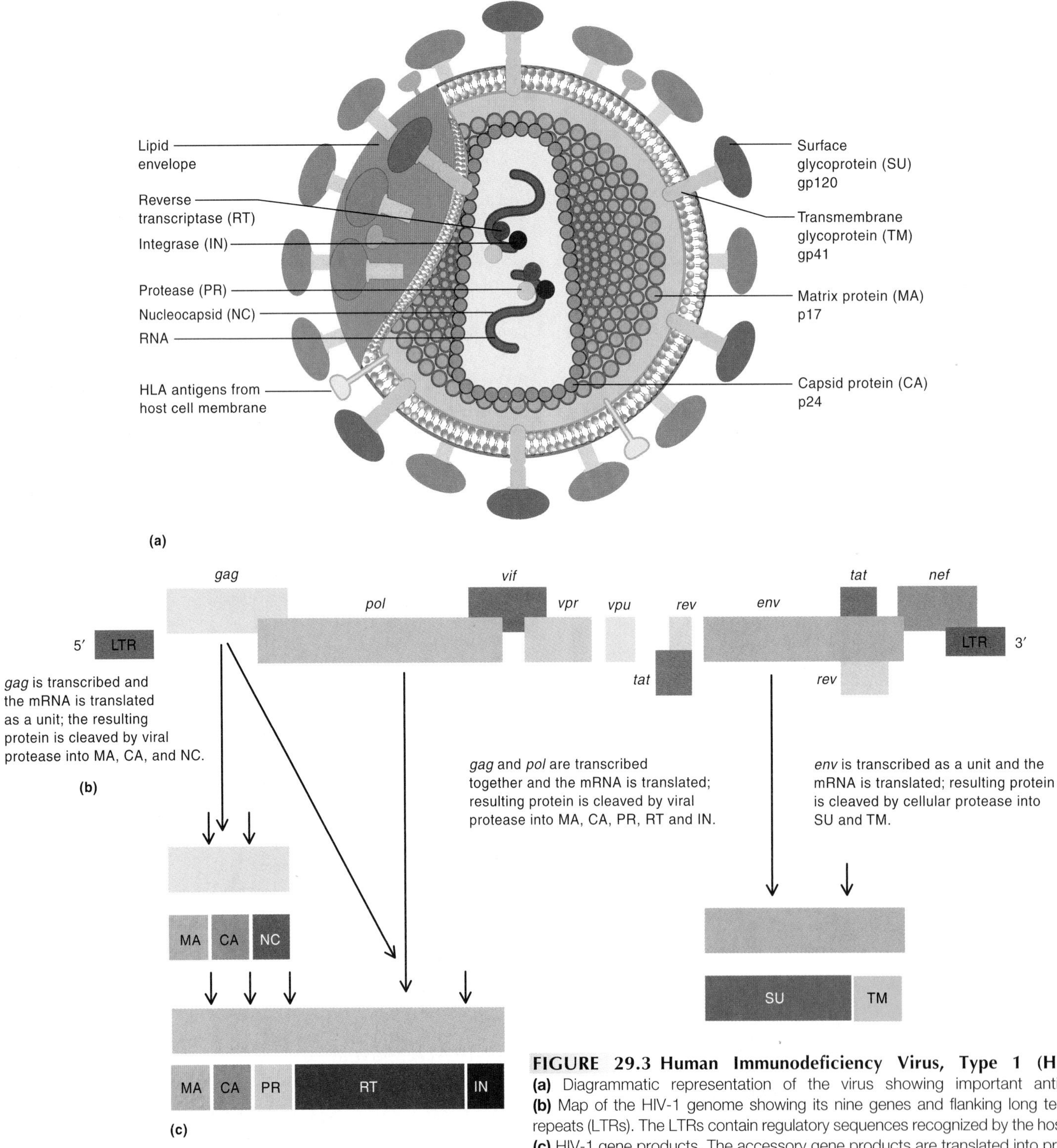

FIGURE 29.3 Human Immunodeficiency Virus, Type 1 (HIV-1) **(a)** Diagrammatic representation of the virus showing important antigens. **(b)** Map of the HIV-1 genome showing its nine genes and flanking long terminal repeats (LTRs). The LTRs contain regulatory sequences recognized by the host cell. **(c)** HIV-1 gene products. The accessory gene products are translated into proteins of final size, while *gag* and *pol* products must be cleaved by viral protease, and *env* products by host cell protease. The small arrows indicate the sites of cleavage.

(for "surface") antigen, which is partly responsible for attachment of the virus to the host cell. The "gp120" indicates that SU is a glycoprotein and the relative weight and size of its molecule is 120. The TM (for "transmembrane") antigen is also a glycoprotein (gp41). It traverses the viral envelope, is closely associated with SU and plays a role in entry of the virus into the host cell. The MA (for matrix protein) is a protein, p17, located inside the viral envelope. It helps to maintain viral structure and transports the viral genome to the host cell nucleus for assembly of new virions. The core of the virion is composed of the p24 CA (for capsid) antigen, two copies of the single-stranded RNA viral genome, and various other protein antigens, including nucleocapsid (NC) and three important viral enzymes—reverse transcriptase (RT), protease (PR), and integrase (IN).

The HIV genome is shown schematically in figure 29.3b. The first two genes, *gag* (from "group antigen") and *pol* (from "polymerase"), can be translated as a unit from the full-length viral messenger RNA (figure 29.3c). The resulting protein is split into functional segments by viral **protease.** This enzyme releases itself from the large protein and enzymatically splits the remaining protein into functional enzymes and structural units. The final yield is three enzymes: protease, **reverse transcriptase,** and **integrase,** and a number of other proteins, including p24 (CA) capsid and p17 (MA) matrix. The virus, however, has a greater need for structural proteins MA, CA, and NC than for the enzymes PR, RT, and IN, and so *gag* is usually expressed by itself, only rarely joining *pol* through a frameshift. Reverse transcriptase, protease, and integrase are the targets of the medicines currently available for treating AIDS. The *env* gene is translated from a spliced messenger RNA, yielding a precursor protein that is processed by host cell enzymes to give the gp120 (SU) surface glycoprotein and the gp41 (TM) transmembrane glycoprotein. ■ RNA splicing, p. 175

There are six additional genes, known as accessory genes, all translated from spliced messenger RNAs. These genes—*tat, rev, nef, vif, vpr,* and *vpu*—code for the proteins Tat, Rev, Nef, Vif, Vpr, and Vpu. These gene products interact with host cell proteins in complex ways to regulate HIV replication and release from the cell.

Pathogenesis

HIV virions enter the body and attach to and infect certain types of cells. Which cells are affected and their response to infection vary with the particular viral strain and the type of host cell. Some of the cell types and the consequences of infection are shown in **figure 29.4.** Infection of intestinal epithelium is partly responsible for the chronic diarrhea and weight loss, and infection of brain cells for HIV dementia. Infected macrophages and other antigen-presenting cells are a continuing source of infectious virus and show impairment of chemotaxis, phagocytosis, and antigen presentation. They and the T_H (helper T) lymphocytes are exceedingly important targets of HIV because of their central role in the body's specific immune response. The cytokines of T_H lymphocytes regulate cytotoxic action of T_C cells, immunoglobulin production by B cells, and chemotaxis of antigen-presenting cells such as macrophages.

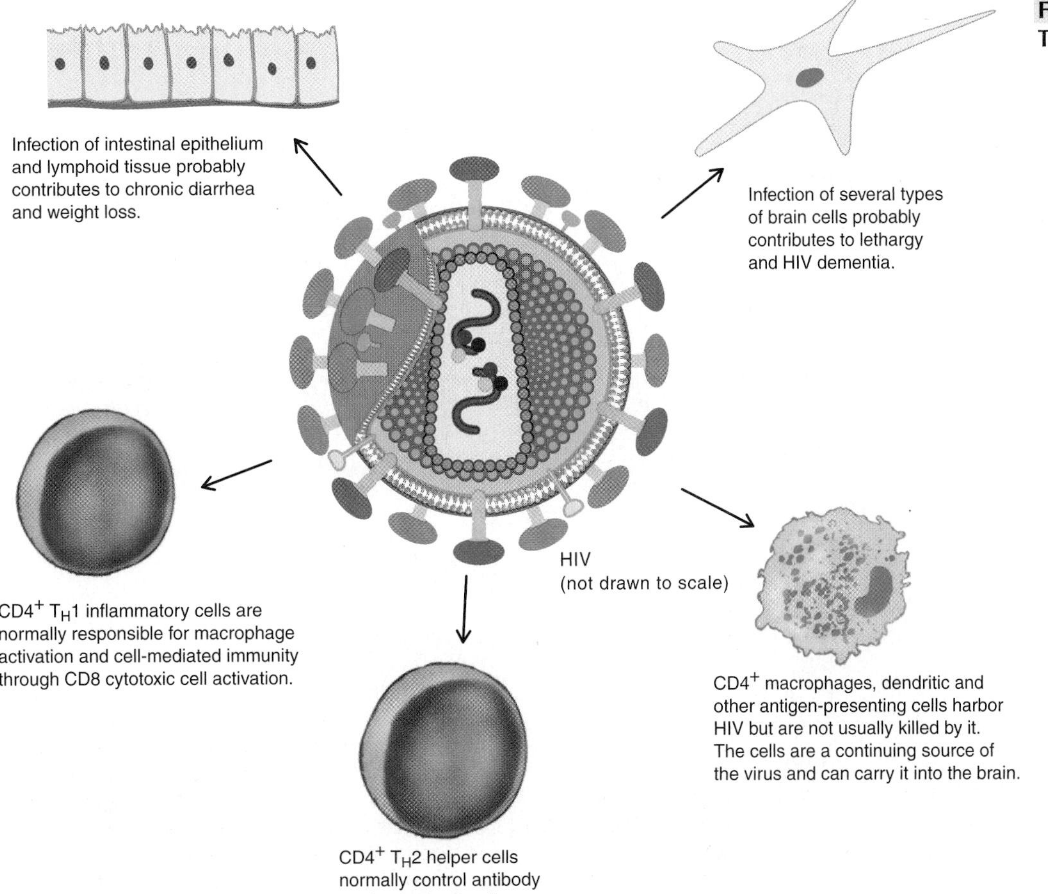

Infection of intestinal epithelium and lymphoid tissue probably contributes to chronic diarrhea and weight loss.
Infection of several types of brain cells probably contributes to lethargy and HIV dementia.
HIV (not drawn to scale)
CD4+ TH1 inflammatory cells are normally responsible for macrophage activation and cell-mediated immunity through CD8 cytotoxic cell activation.
CD4+ macrophages, dendritic and other antigen-presenting cells harbor HIV but are not usually killed by it. The cells are a continuing source of the virus and can carry it into the brain.
CD4+ TH2 helper cells normally control antibody production by B cells.

FIGURE 29.4 Some of HIV's Cellular Targets

HIV must attach to and enter the body's cells to establish infection. Like most other cells susceptible to HIV, T_H lymphocytes and macrophages are $CD4^+$—they possess the CD4 surface antigen. HIV attaches to CD4 by its surface gp120 (SU) antigen. The presence of the CD4 antigen, however, is by itself not sufficient to cause entry of HIV. In addition, there must be a host cell co-receptor specific for HIV in order for viral entry to occur **(figure 29.5).** The co-receptor differs for different cell types. CXCR-4, a chemokine receptor, is an HIV co-receptor. CCR5, a chemokine receptor on macrophages, is also an HIV co-receptor.

The nature of HIV attachment and entry into the host cell has been a mystery subject to intense study because antibodies and other substances designed to block SU and CD4 have not been very successful in preventing infection. SU is known to be flexible, irregular in shape, and divided into two parts connected by peptide strands. The main binding sites for CD4 and the co-receptor are only exposed after initial contact of the virus with CD4. Penetration into the host cell probably involves gp41 (TM).

Once HIV enters the host cell, its reverse transcriptase makes a DNA copy of the viral RNA genome. A complementary DNA strand is then added, and the ends of the resulting double-stranded DNA segment are joined non-covalently. The resulting circular DNA is then moved to the nucleus and is inserted into the host cell chromosome by the viral integrase (IN) enzyme. The viral segment thus becomes a provirus, a step necessary for efficient replication of HIV. No specific location is required for the inser-

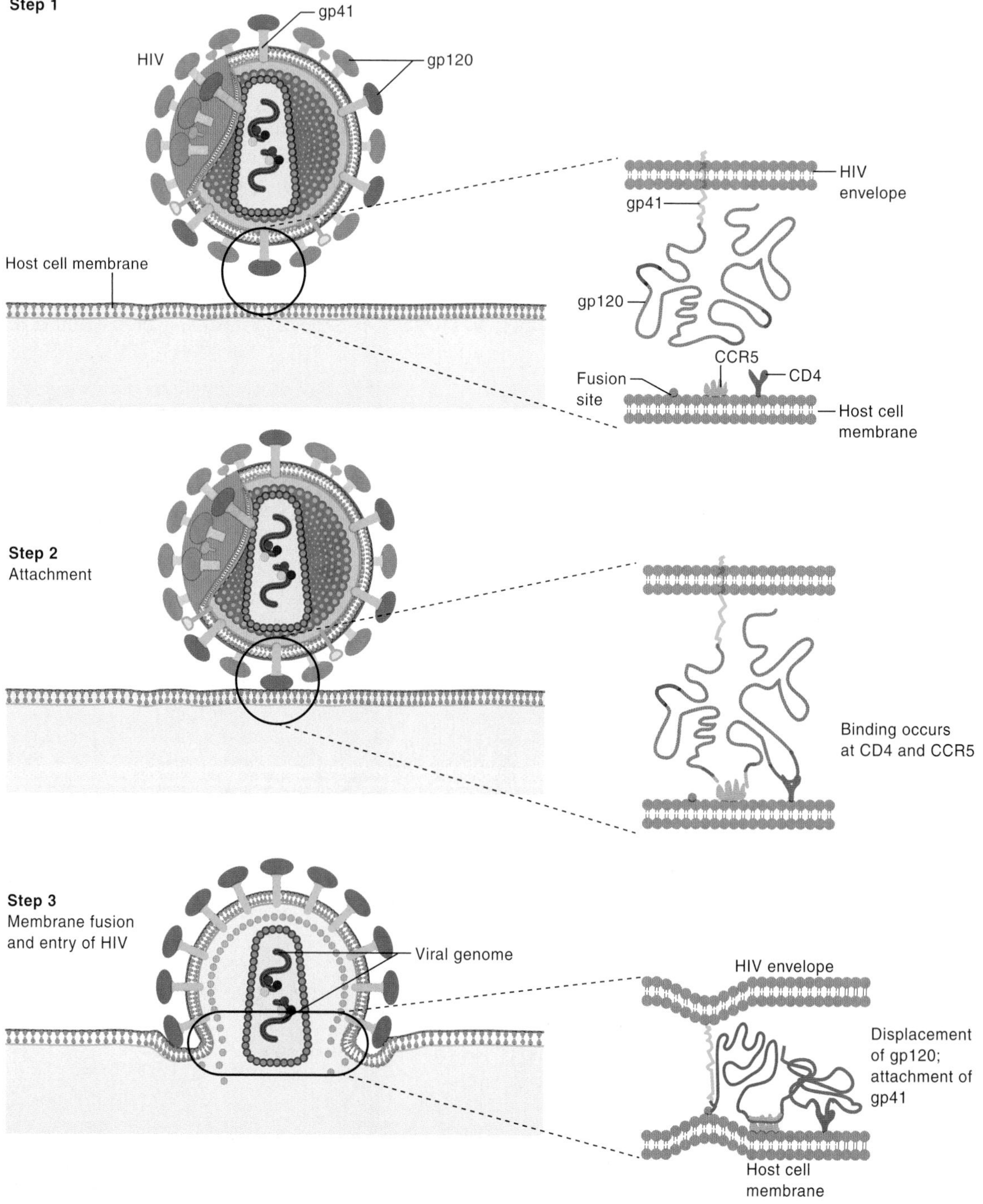

FIGURE 29.5 Attachment and Entry of HIV into a Host Cell, Schematic Representation **Step 1:** Virion in close proximity to the cell membrane; blowup showing gp120 protein and sites of reaction with host cell receptors. **Step 2:** Initial contact of gp120 (SU) is with CD4. Attachment to a chemokine receptor such as CCR5 must occur before membrane fusion and entry of the viral genome can take place. **Step 3:** Membrane fusion probably is mediated by gp41 (TM).

tion into the host chromosome. Following integration, spliced mRNA segments appear, mainly transcribed from the regulatory genes: *tat, rev,* and *nef,* and are translated into proteins that are involved in HIV gene expression. High levels of Tat are associated with replication of infectious virions. Nef protein, despite the name derivation from negative factor, may cause increased virion production and other effects that increase pathogenicity, depending on the viral strain and type of cell infected. In the case of T_H lymphocytes, replication of mature virions results in lysis of the cells and release of infectious HIV. Macrophages, on the other hand, usually release infectious virus over long periods of time without death of the cells. The life cycle of HIV is shown schematically in **figure 29.6,** but in an infected person, at any one time, the vast majority of infected cells show neither lysis nor latency. Instead, most of the HIV-infected cells show accumulations of various viral products that impair the normal functioning of the cells. ■ provirus, p. 332

HIV is genetically highly variable. The variant viruses show differences in their preferred host cell, rates of replication, response to host immunity, and other characteristics. The variability is due to a high rate of "error" when reverse transcriptase copies the viral genome. Some regions of the genome are highly conserved, meaning they do not change much from one strain of virus to another. Other regions are highly variable. This is reflected for example in the glycoprotein gp120 (SU), which has conserved and variable regions **(figure 29.7).** The V3 variable region is important because it plays a role in the virulence of HIV. This antigenic variability of HIV enormously complicates the task of developing an effective vaccine against the virus. ■ reverse transcriptase, p. 332

Destruction of immune system T_H cells by HIV can occur via multiple mechanisms:

- *Lysis following HIV replication.* Lysis of activated helper $CD4^+$ T lymphocytes during replication of infectious virus is one mechanism, but by itself it cannot account for the devastation that results from HIV disease.
- *Attack by HIV-specific cytotoxic $CD8^+$ T lymphocytes.* The earliest detectable immune response to HIV infection involves HIV-specific $CD8^+$ T lymphocytes, which attack and lyse infected cells.
- *Natural killer (NK) cells.* NK cells probably also play a role in cell destruction. ■ natural killer cells, p. 384
- *Antibody-dependent cellular cytotoxicity.* Humoral antibody may also play a role through antibody-dependent cytotoxicity. ■ antibody-dependent cellular cytotoxicity, p. 372

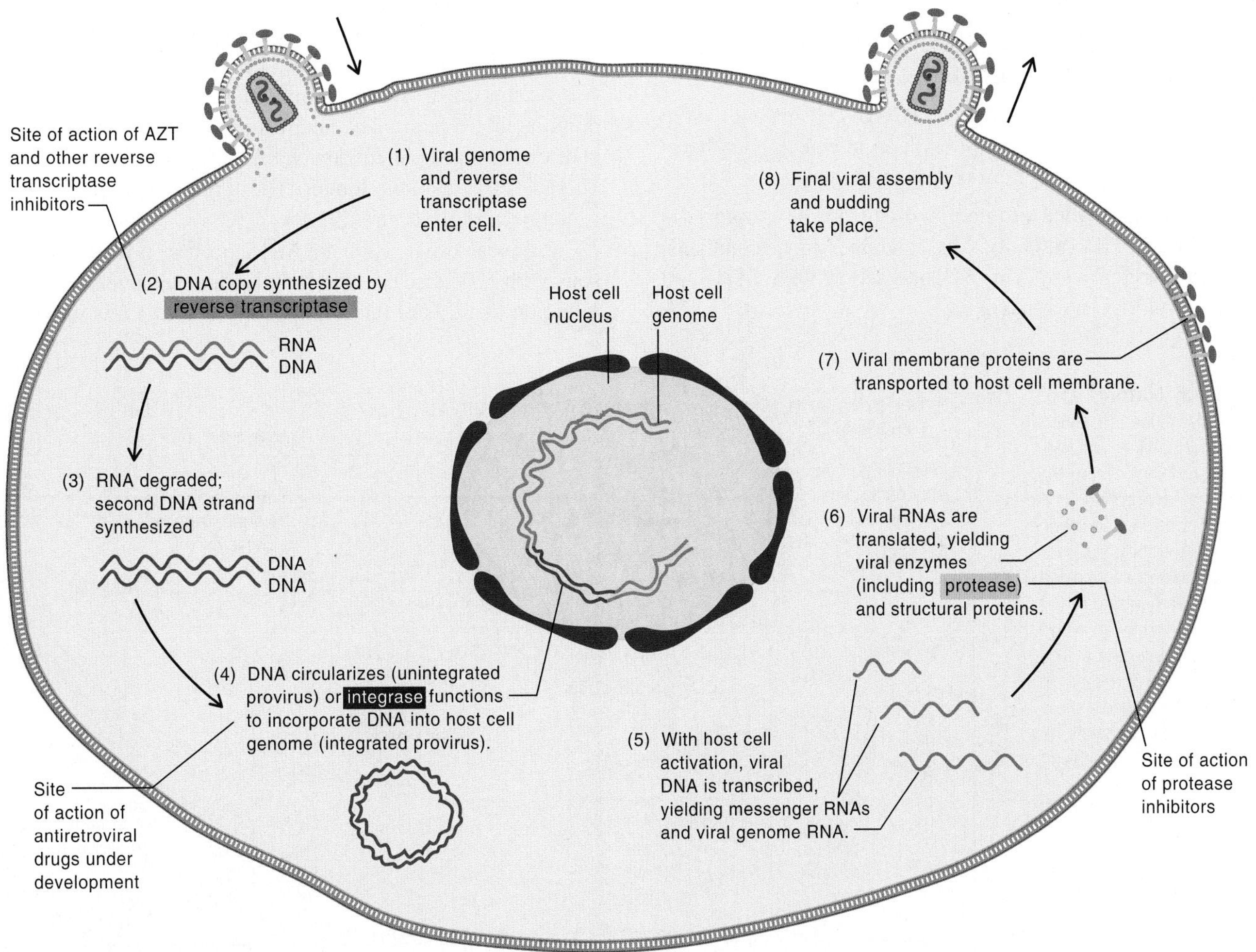

FIGURE 29.6 The Steps in HIV Replication In step 4, the enzyme, integrase, mediates integration of the provirus into the host cell genome. This enzyme is potentially a target of new anti-HIV medications. In step 7, the accessory gene product Vpr and MA transport viral proteins to the cell membrane.

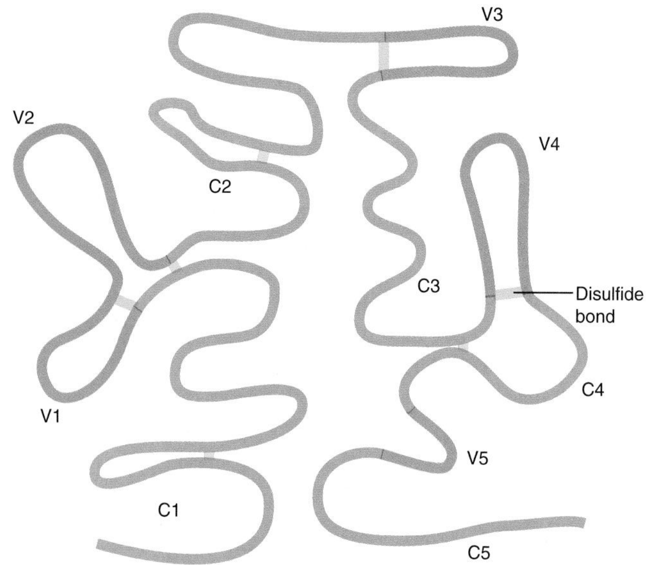

FIGURE 29.7 Diagram of the SU Glycoprotein There are five "constant" segments (blue) that are conserved in different HIV strains. The segments in orange, labeled V1 to V5, are "variable" and differ from strain to strain, frustrating attempts to develop a vaccine. Note the V3 loop, implicated in pathogenicity.

- *Autoimmune process.* Autoimmunity could be responsible because HIV envelope proteins show some homology with MHC class II molecules. ■ major histocompatibility complex, class II (MHC-II), p. 380
- *Fusion of infected and uninfected cells.* In cell fusion, a large number of uninfected cells fuse with an infected cell, whereupon the resulting syncytium is destroyed.
- *Apoptosis.* Also called programmed cell death, apoptosis is accelerated in HIV infections by a number of mechanisms, including induction by Tat, SU interaction with CD4, and certain cytokines. ■ apoptosis, p. 362

Accumulation of viral products such as viral RNA and unintegrated viral DNA inside the cytoplasm of the infected cell can also result in its death.

An acute retroviral syndrome (ARS) begins in 50% to 70% of HIV-infected subjects after an incubation period of 6 days to 6 weeks, with sore throat, fever, muscle and headaches, enlarged lymph nodes, and a rash. These symptoms generally last 1 to 4 weeks and disappear without treatment. Whether or not symptoms occur, the concentration of HIV in the blood rises to high levels as the newly infected cells release their progeny virus **(figure 29.8).** The marked viremia usually subsides as the supply of uninfected cells dwindles and anti-HIV $CD8^+$ cytotoxic T cells appear. The $CD4^+$ T-cell level falls initially, then slowly rises but does not reach the preinfection level. Following the acute episode, the HIV-infected individual generally enters an asymptomatic period ranging from months to many years. Regardless of the lack of symptoms, a silent struggle goes on between HIV and the immune system for the rest of the person's life. Infectious virus, a threat to others, continues to be present in the blood and body secretions.

In the typical case, as seen in approximately 80% of HIV-infected people, the immune system slowly loses ground to HIV even though the body can normally replace over a billion $CD4^+$ cells per day. The peripheral blood $CD4^+$ count (normally about 1,000 cells per microliter) steadily falls at a rate of roughly 50 cells per microliter per year.

Symptoms of AIDS usually appear when $CD4^+$ counts fall below 200 cells per microliter. Levels of infectious virus again rise dramatically. At this point in the illness, the pathology is dominated by malignant neoplasms or opportunistic infections. Half of untreated patients reach this stage of the illness within 9 to 10 years; the rest take longer. The typical course of HIV disease is summarized in figure 29.8.

Atypical progression to AIDS occurs in about 10% of persons with HIV disease who have high virus levels with the acute infection that do not fall dramatically within a few months. These

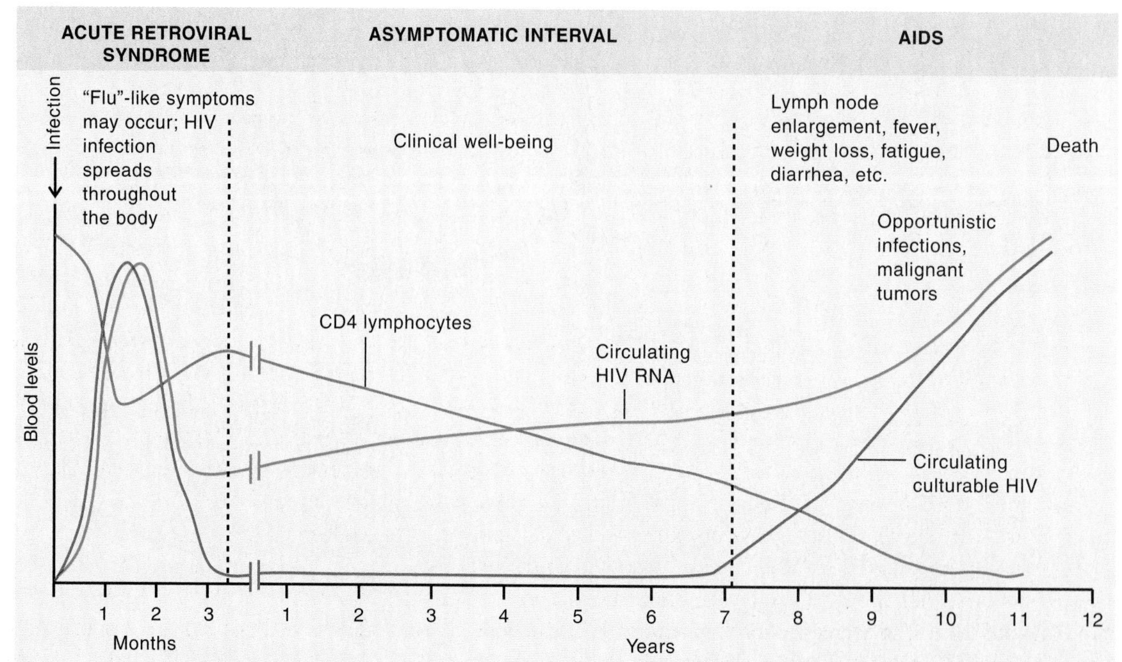

FIGURE 29.8 Natural History of HIV Disease Blood levels of infectious virus are very high at the beginning, during the acute retroviral syndrome, and at the end of the disease when AIDS ensues. Antibody tests for diagnosing the disease are often negative in the early stage of the disease even though infected people are highly infectious. The disease steadily progresses in the absence of symptoms, as shown by the rising levels of plasma viral RNA and falling $CD4^+$ cell count.

PERSPECTIVE 29.1

Origin of AIDS-Causing Viruses

Where did the AIDS-causing viruses come from? Genetic evidence indicates that the HIV-1 virus mutated to its present form fairly recently, between 50 and 150 years ago. Although it first appeared in the United States in the 1970s, serological evidence indicates that it was present in Africa in only a few individuals in the 1950s.

Viruses similar to HIV exist in a number of wild and domestic animals, including cats, dairy cattle, and monkeys. A virus closely related to HIV-1 has been found in animals belonging to a single West African subspecies of chimpanzees. The virus does not appear to harm the chimpanzees, suggesting that they have been living together for eons. Another virus, closely related to HIV-2, is present in a species of large monkey, the sooty mangabey. A likely theory is that the AIDS-causing viruses "jumped" to humans from these simian relatives, presumably through contact with their blood. Indeed, in Africa, chimps and mangabeys are often killed for food, exposing humans to their blood, and incidentally, driving some species nearly to extinction. Also, a number of humans were intentionally injected with chimpanzee and mangabey blood in the course of research on malaria carried out between 1922 and 1955.

Genetic comparisons of a large number of simian lentiviruses and AIDS-causing viruses from humans support the idea that the simian viruses can jump to humans. In the case of HIV-1, a jump to humans probably occurred only once, between 1910 and 1950. It is unlikely that HIV was transferred to humans by inadequately sterilized Salk polio vaccine grown in simian kidney cell cultures. There is no credible evidence for another popular theory that the viruses resulted from botched biological warfare experiments by Russia or the United States. It is possible that AIDS-causing viruses have existed for many years in people living in isolated African villages, perhaps even for centuries. According to this idea, population increases and migration to big, crowded cities allowed the viruses to spread rapidly, becoming more virulent in the process.

The answer to the question about the origins of AIDS-causing viruses might never be known precisely, but the question is intriguing and may lead to better understanding of the emergence of new infectious diseases.

subjects progress rapidly to AIDS within a few years. Another 5% to 10% of HIV-infected individuals show no fall in $CD4^+$ cells. They maintain high levels of anti-HIV antibody and HIV-specific $CD8^+$ cytotoxic T cells. Estimates suggest that, with these cases and the more slowly progressing typical cases, 10% to 17% of HIV-infected persons will not have progressed to AIDS 20 years after becoming infected with HIV.

Epidemiology

Sexual intercourse with multiple partners without the use of condoms is a major factor in the spread of HIV disease. In the United States, initially, promiscuous homosexual men were the hardest hit by the epidemic. An estimated 1.4% to 10% of American men are homosexuals. An additional number of men are bisexual, meaning they have sexual intercourse with both men and women. A survey done before the arrival of AIDS found that 33% to 40% of homosexual men had more than 500 lifetime sexual partners. By 1984, two-thirds of one group of homosexual men in San Francisco were infected with HIV and almost one-third had developed AIDS. In the 1-year period ending in mid-1999 there were more than 22,000 new AIDS cases reported among men who had sex with men, and these accounted for an estimated 45% of new HIV infections over the same period. A newer study carried out between 2001–2006 indicated that men who have sex with men continue to lead all other groups in the incidence of HIV infection and AIDS.

Heterosexual spread of HIV is increasing and promises to become the dominant mode of transmission, as it is in many African countries. Sex with multiple partners; sex with a person who has had multiple partners; being the receptive partner, especially receptive anal sex; traumatic sex; and any irritation or inflammatory process as from another sexually transmitted disease—all increase the risk of sexual HIV transmission. Contrary to popular notion, the disease can be contracted by the insertive partner during vaginal or rectal intercourse with an infected person. Saliva is very unlikely to transmit the disease, but there is evidence that oral-genital contact may be risky.

The next most important mode of transmission of HIV is through blood and blood products. Individuals infected with HIV who donated blood before a screening test became available in 1985 unknowingly infected thousands of transfusion recipients. One of the products from pooled donated blood, clotting factor VIII, was used to treat bleeding episodes among **hemophiliacs.** By 1984, over half of the hemophiliacs in the United States and 10% to 20% of their sexual partners were HIV positive. Fortunately, the risk of HIV transmission by factor VIII was eliminated in 1992 when recombinant factor VIII was licensed. However, transmission by blood is still a major factor in the HIV disease pandemic because of needle sharing by those who abuse injected drugs. In the United States, the population of abusers of injected drugs is estimated to be 1.5 million. Many of them, both men and women, support their drug habits with prostitution, thereby furthering the spread of HIV. Because HIV is acquired and spread through sexual intercourse and through the sharing of hypodermic needles, drug abusers and their sexual partners have been a big factor in spreading the virus. A study of the first 640,000 AIDS cases revealed that more than one-third were directly or indirectly related to injected-drug abuse.

The third most important mode of HIV spread is from mother to infant. Women represent an increasing percentage of the total AIDS cases as heterosexual spread of HIV increases **(figure 29.9).** An estimated estimated 70% of new HIV infections in women in the 1-year period ending in mid-1999 were acquired heterosexually. If untreated, about 1 out of 10 pregnant HIV-positive women will miscarry, and of live-born babies, 15% to 40% will develop AIDS. Remarkably, however, in a study of 219 newborn babies positive for HIV by culture or PCR, almost 3% appeared to clear their infection without any treatment. Breast feeding carries a significant risk of mother-infant transmission, especially if the mother has high levels of HIV in her blood as a result of an acute HIV infection.

Prevention and Treatment

There is no approved vaccine against HIV infection, and current clinical vaccine trials do not look promising. Many people with HIV

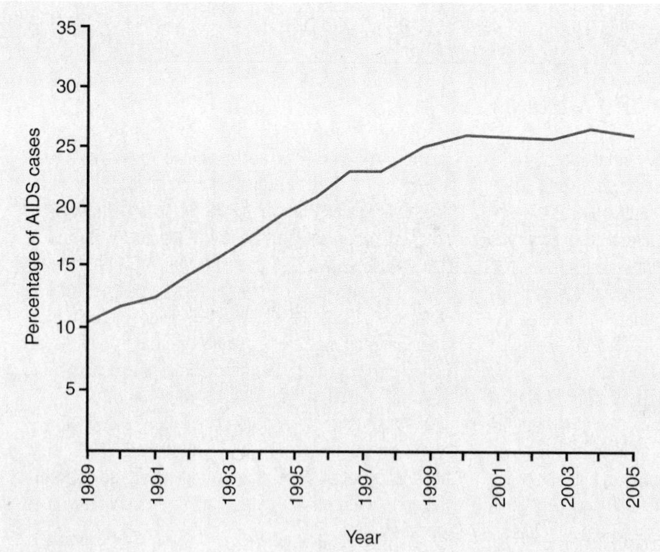

FIGURE 29.9 After steadily rising during the 1990s, the percentage of AIDS cases in women appears to be leveling off.

disease do not know they are infected, and it is advisable to consider all blood as potentially containing the virus. Infectious HIV persists in samples of blood plasma for at least 1 week after they are taken from AIDS patients. HIV on objects and surfaces contaminated by body fluids is easily inactivated by commercially available, high-level disinfectants and heat at 56°C or more for 30 minutes. Freshly opened household sodium hypochlorite 5.25% bleach, diluted 1:10, is a cheap and effective disinfectant for general use. Virus present in dried blood or pus, however, may be difficult to inactivate.

Knowing how HIV is transmitted is a powerful weapon against the AIDS epidemic. This weapon can be far more effective than any vaccine or treatment now on the horizon. HIV is not highly contagious, and the risk of contracting and spreading it can be eliminated or markedly reduced by assuming behaviors that prevent transmission of the virus **(table 29.2).**

Some groups of homosexual men lowered the incidence of new HIV infections from 10% to 20% annually to only 1% to 2% annually by using condoms and avoiding practices that favor HIV transmission. The improvement in infection rates has not always been sustained, however.

Since 1985, the risk of acquiring HIV from blood transfusions has been lowered dramatically by screening potential donors for HIV risk factors and testing their blood for antibody to HIV. The risk is now estimated to be less than 1 in 400,000 transfused units. Newer tests for HIV antibody and nucleic acid give positive results within 1 month of infection in most but not all cases. Screening tests used for infection have also markedly reduced the risk of HIV transmission from artificial insemination and organ transplantation.

Education about how HIV is transmitted can be a very effective tool in helping to bring the worldwide epidemic of HIV disease under control. Education of schoolchildren has been shown effective in decreasing risky sexual behavior among teenagers. Videotapes and written material developed for one group of people, however, can be completely ineffective and even offensive to another group.

TABLE 29.2 Behaviors that Help Control the AIDS Epidemic

1. Not engaging in sexual intercourse.
2. Staying in a mutually monogamous relationship.
3. Avoiding sexual intercourse with persons at risk for HIV infection (see table 29.3).
4. Avoiding trauma to the genitalia and rectum. Small breaks in the skin and mucous membranes allow HIV to infect.
5. Not engaging in sexual intercourse when sores from herpes simplex or other causes are present. They represent sites where HIV can infect.
6. Not engaging in anal intercourse. Receptive anal intercourse carries a high risk of HIV transmission.
7. Using latex condoms from beginning to end of sexual intercourse. Polyurethrane condoms are a reasonable alternative for those allergic to latex. Condoms made from other materials are not reliable for disease prevention, nor are those marketed in many African and other countries outside the United States. Oil-based lubricants are not compatible with latex. Condoms for women are available.
8. Postponing pregnancy indefinitely if you are a woman infected with HIV. If you are not sure of your HIV status, blood tests to rule out HIV disease before considering pregnancy.
9. Using extreme care to avoid needles, razors, toothbrushes, etc., that could be contaminated with someone else's blood.

All persons unsure of their HIV status and especially those at increased risk of HIV disease **(table 29.3)** are advised to get tested for HIV. With this knowledge, those with HIV disease can receive prompt preventive treatment for the infections and cancers that complicate the disease. Antiretroviral treatment of HIV disease shows promise of preventing the complications of immunodeficiency for decades, and it may also reduce transmissibility. Also, HIV-positive individuals who know their status are able to do their part in preventing transmission. Federally approved home test kits coupled with counseling are reliable and commercially available. ■ **sexually transmitted diseases, p. 628**

TABLE 29.3 Persons at Increased Risk for HIV Disease

1. Injected-drug abusers who have shared needles.
2. Persons who received blood transfusions or pooled blood products between 1978 and 1985.
3. Sexually promiscuous men and women, especially prostitutes, drug abusers, and homosexual and bisexual men.
4. People with history of hepatitis B, syphilis, gonorrhea, or other sexually transmitted diseases that may be markers for unprotected sexual intercourse with multiple partners.
5. People who have had blood or sexual exposure to any of the people listed.

HIV transmission from infected mother to newborn can safely be prevented in two-thirds of cases by administering zidovudine (AZT) to the mother during pregnancy and to the newborn infant for 12 weeks. Newer, more potent combinations of AZT and other antiretroviral therapies are widely employed and may be more effective, but their long-term safety is still being evaluated. Elective cesarian section significantly reduces the risk of HIV transmission to the newborn baby, perhaps by avoiding HIV-containing fluids in the birth canal. Also, medications are effective for preventing AIDS acquired from HIV-contaminated instruments, including bloody hypodermic needles.

Hundreds of needle and syringe exchange programs, operating in at least 30 states and the District of Columbia, help prevent the spread of HIV and other blood-borne diseases among injected-drug abusers and to their sexual partners and children. In these programs, sterile syringes and needles are exchanged for used ones. Programs also provide drug rehabilitation efforts and education about use of condoms and other safer sexual practices.

Better prevention and treatment of opportunistic infections and better antiviral therapy against HIV have significantly lengthened the asymptomatic stage of the disease and prolonged life once AIDS develops **(figure 29.10).** Advances in antiviral treatment mainly result from the development of new medications with different modes of antiviral action. Use of these medications in "cocktails," such as combinations of reverse transcriptase and protease inhibitors, is referred to as **HAART**—"highly active antiretroviral therapy." Other important advances include the combining of medications in a single dosage form, and developing "once-a-day therapy." The effectiveness of HAART probably stems from the fact that each of the medications act on the replicating virus at different parts of the virus life cycle. Despite the high mutation rate of HIV, it is much less likely that any one mutant could develop resistance to all the medications at the same time.

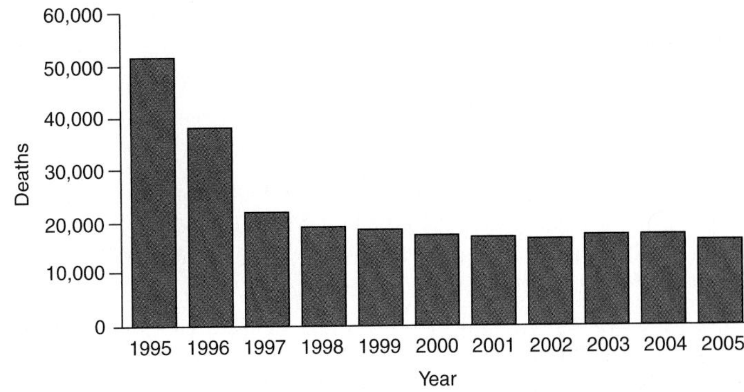

FIGURE 29.10 Deaths Due to AIDS, United States, 1995 to 2005
Notice the marked drop in deaths after HAART became available in late 1995. (Source: CDC HIV/AIDS Surveillance Report v.17 revised June 2007; earlier volumes.)

CASE PRESENTATION

The patient was a young woman from West Africa presenting with the complaint of generalized enlargement of her lymph nodes. She had recently moved to the United States. She had no history of drug abuse or of receiving blood transfusions. She had had three sex partners in her life. Her single pregnancy the year before her arrival was delivered by emergency cesarian section. She and the baby's father had routine tests for HIV at that time, and both tests were negative. The baby and the father remained well. Ten years before her evaluation, she was treated for a fever by scarification deliberately making superficial cuts in the skin, and this was repeated for an unrelated complaint 4 years before her evaluation. The native healer performing the scarification used a razor blade, the sterility of which is unknown to the patient.

Her initial test results included non-diagnostic lymph node biopsies and a negative HIV test (enzyme-linked immunosorbent assay, ELISA). A repeat HIV test some months later was weakly positive by ELISA, and the confirmatory Western blot showed only questionable reactions of the patient's serum with gp41 and two other HIV antigens. A test for HIV-2 was negative. A $CD4^+$ T lymphocyte cell count was very low. A test to detect HIV-1 using the polymerase chain reaction was negative.

1. Could this patient have HIV disease? If so, how could the negative tests be explained?
2. Does the history give any possible ways in which she could have contracted HIV disease?
3. How could the diagnosis be established?
4. Could her baby and the baby's father have the disease?
5. Does this case suggest that any changes should be made in the way HIV disease is diagnosed?

Discussion

1. This patient could well have had HIV disease/AIDS because of her persistent, generalized lymphadenopathy and very low $CD4^+$ cell count. Her HIV strain might have differed enough from the pandemic group M HIV-1 strains so that the usual laboratory tests did not detect antibody to it.
2. The woman could have contracted an AIDS-causing virus during sexual intercourse, or from a blood-contaminated razor blade used for scarification. Contracting such a virus from unsterile instruments during her emergency cesarian is a possibility, but this seems less likely because of the short time span before presenting with severe immunodeficiency.
3. The Centers for Disease Control and Prevention have a global surveillance system designed to detect cases like this because they raise the possibility of new or rare AIDS-causing viruses being introduced into a population. In the present case, samples of the patient's blood were examined by the CDC and she was shown to be infected with a group O HIV-1 virus. The patient's serum reacted with peptides specific to group O strains, and the virus was isolated from her blood. Nucleic acid sequences of the *env, gag,* and *pol* genes matched those of previously isolated group O strains. Group O strains of HIV-1 were first found in Cameroon, a country on the West African coast, where they account for 6% of HIV infections.
4. Despite the absence of symptoms, both her baby and the baby's father could have the disease. The baby's risk of infection was reduced but not eliminated by the emergency cesarian section, and infection could have occurred subsequently if it were breast fed. The mother's very low $CD4^+$ cell count suggests a high level of viremia and, therefore, increased risk of transmitting the disease. She could have infected the baby's father during sexual intercourse, or he could have infected her.
5. This patient was the first case of group O HIV disease identified in the United States. Studies indicate that previous introductions, if any, were not accompanied by spread of the virus. Nevertheless, federal agencies worked with manufacturers of HIV tests to increase sensitivity of the tests to group O strains.

Source: Centers for Disease Control and Prevention. 1996. *Morbidity and Mortality Weekly Report* 45(6): 122.

Presently available medications fall into four groups: inhibitors of reverse transcriptase, inhibitors of viral protease, inhibitors of viral integrase, and a fusion inhibitor. Other types of medications being developed are fusion inhibitors to prevent viral entry into host cells, and medications that interfere with other polyproteins. Medications that interfere with reverse transcriptase fall into two categories: **nucleoside** reverse transcriptase inhibitors (NRTIs), and non-nucleoside reverse transcriptase inhibitors (NNRTIs). Zidovudine (AZT), stavudine (D4T), and lamivudine (3TC) are among the half dozen NRTIs in wide use, and a number of others are undergoing clinical trials. All are nucleoside analogs, but they differ chemically and in their site of action on the enzyme. These substances owe their effectiveness to their resemblance to the normal purine and pyrimidine building blocks of nucleic acids. During nucleic acid synthesis, the viral reverse transcriptase enzyme incorporates the medication molecule into the growing DNA chain, thereby blocking completion of the DNA strand. This process for AZT is illustrated in **figure 29.11.**
■ nucleosides, figure 2.22, p. 33

The non-nucleoside inhibitors of reverse transcriptase—nevirapine, efavirenz, and delavirdine—are among those in wide use, and at least five newer NNRTIs are in clinical trials. Generally, the NNRTIs are not nucleoside analogs, and they act in a completely different manner from NRTIs, by binding tightly to reverse transcriptase in an example of non-competitive inhibition.
■ non-competitive inhibition, p. 138

Protease inhibitors were a major addition to the anti-HIV arsenal. The first protease inhibitor, saquinavir, was approved for therapy in December 1995. Currently, there are at least six protease inhibitors approved for use, and more in clinical trials. Unlike the reverse transcriptase inhibitors, the protease inhibitors act late in HIV replication to prevent packaging of viral proteins in the virion.

The introduction of integrase inhibitors, with raltegravir in 2007, represented another important advance in treatment options that is very welcome in view of increasing resistance to older agents.

Fusion inhibitors are a useful addition to therapy with the three types of enzyme inhibitors just mentioned. They act by interfering with viral attachment to the host cell. The first one to be approved, enfuvirtide (Fuzeon), in 2003, acts by blocking gp41attachment; maraviroc, approved in 2007, blocks the CCR5 co-receptor.

HAART does not cure AIDS. Viremia becomes undetectable in only about half the cases, and even then, the viremia returns if the medications are discontinued. While it stops production of virions, it does not eliminate HIV provirus hidden in host cell genomes. In successful HAART, production of infectious virus is largely halted, many fewer $CD4^+$ lymphocytes are killed, and so the $CD4^+$ cell count rises. When the medications are stopped, persistently infected cells release infectious virions that can infect a new population of $CD4^+$ cells, resulting in a quick rise in viremia. Many authorities now feel that the secret to controlling HIV disease lies ultimately in minimizing or eliminating body cells containing the hidden HIV genomes. These authorities are advocating HAART for the acute retroviral syndrome to prevent the buildup in the body of cells containing HIV provirus.

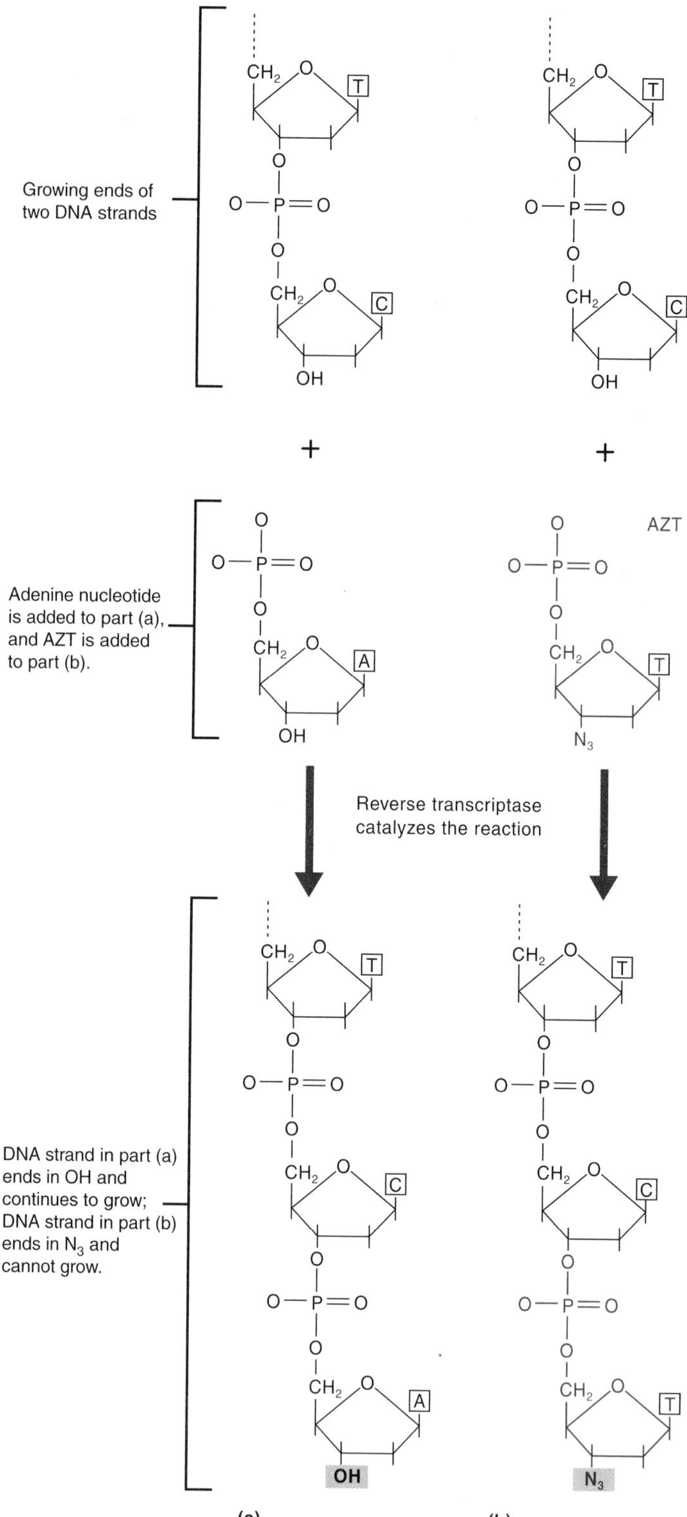

FIGURE 29.11 Mode of Action of Zidovudine (AZT) (a) Normal elongation process of DNA in which reverse transcriptase catalyzes the reaction between the OH group on the chain with the phosphate group of the nucleotide being added. **(b)** Reverse transcriptase catalyzes the reaction of zidovudine with the growing DNA chain. Since zidovudine lacks the reactive OH group, no further additions to the chain can occur, and DNA synthesis is halted. T = thymine, C = cytosine, A = adenine. (It should be noted that zidovudine is a nucleoside. It becomes phosphorylated after entry into the cell.)

Many strains of HIV fail to respond to HAART because they have become resistant to the medications during past treatment. These resistant strains of HIV are transmissible to other persons. Toxic effects of the medications are another limitation of anti-HIV therapy. For example, AZT can cause anemia, low white blood cell count, vomiting, fatigue, headache, and muscle and liver damage. Painful peripheral nerve injury, inflammation of the pancreas, rash, mouth and esophagus ulceration, and fever are side effects of other NRTIs. Also, indinavir may be responsible for kidney stone formation, and ritonavir, another protease inhibitor, causes nausea and diarrhea. Diabetes mellitus is another potential side effect of protease inhibitors.

Finally, a big limitation of the use of anti-HIV medications is their cost. The cost of a combination of these medications can easily exceed $1,000 per month for either an asymptomatic or a symptomatic individual and is expected to be required for life. While not considered excessive for a serious chronic disease in the United States, the cost of medications puts them out of reach for many of the world's HIV victims. Negotiated price decreases and the early 2005 approval of three generic drugs by the U.S. Food and Drug Administration have made treatment available to many more HIV victims worldwide.

The main features of HIV disease are presented in **table 29.4.**

HIV Vaccine Prospects

There are no approved vaccines for preventing HIV disease. Development of potential HIV vaccines began soon after the discovery of the causative agent. In theory, a vaccine could be used in either of two ways. One, "preventive vaccine," would be to immunize uninfected individuals against the disease. The other **"therapeutic vaccine,"** would be to boost the immunity of those already infected with HIV before they become severely immunodeficient. The latter approach would be similar to the post-exposure treatment of rabies infections. A successful vaccine must induce both mucosal and bloodstream immunity, because HIV disease is primarily sexually transmitted. The vaccine must also get around the problem of HIV antigenic variability and stimulate cellular and humoral responses against virulence determinants. Moreover, the vaccine has to be safe. An attenuated agent must not be capable of becoming a disease-causing strain, and it must not be oncogenic, meaning cancer-causing. Additionally, the vaccine must not stimulate an autoimmune response, and it must not cause production of "enhancing antibodies" that could aid the passage of HIV into the body's cells. The vaccine should induce neutralizing antibodies against cell-free virions and also prevent direct spread of HIV from a cell to neighboring cells. ■ rabies, p. 663

Despite enormous difficulties, HIV vaccine research continues to offer the best hope for eventually controlling a worldwide epidemic that every year produces tens of thousands of new HIV infections in the United States alone. Findings that have encouraged vaccine researchers include failure of some African prostitutes to develop AIDS although repeatedly exposed to HIV. Also, HIV-infected pregnant women that have high titers of HIV-neutralizing antibodies are less likely to transmit the infection to their offspring than women with low antibody titers. Finally, infection by some strains of HIV apparently progresses to AIDS very slowly.

TABLE 29.4 HIV Disease

Symptoms	Over half develop fever, sore throat, head and muscle aches, rash, enlarged lymph nodes early in the infection. After an asymptomatic period symptoms from unusual malignant tumors, pneumonia, meningoencephalitis, diarrhea, etc.
Incubation period	Usually 6 days to 6 weeks for acute symptoms; immunodeficiency symptoms within 10 years in half the infections (10% within 5 years and 90% within 17 years)
Causative agent	Human immunodeficiency virus, type 1 (HIV-1), many subtypes and strains. HIV-2 mainly in West Africa.
Pathogenesis	HIV infects various body cells, notably those vital to specific immunity, CD4$^+$ T lymphocytes and antibody-presenting cells. T cells killed, numbers slowly decline until the immune system can no longer resist infections or development of tumors.
Epidemiology	Three main routes of transmission of HIV: intimate sexual contact, via transfer of blood or blood products, and from mother to child around the time of childbirth. Risk of transmission by breast milk, and possibly by oral-genital contact.
Prevention and treatment	No vaccine yet available. Medications and vaccines can prevent many of the infections that can complicate HIV disease. Anti-HIV medications and cesarian section decrease mother-to-newborn transmission. Effective preventive measures: sex education of schoolchildren, needle exchange programs for drug addicts, use of condoms. Treatment: HAART therapy, consisting of combinations of several anti-HIV medications, effective for many AIDS victims, and delays progression of HIV disease to AIDS. Not a cure, and too expensive for most of the world's HIV disease victims.

Candidate vaccines undergo an extensive evaluation for safety and immunogenicity in experimental animals before they are tested in humans. Some of the many substances in this preclinical stage of development are inactivated HIV, various viruses and bacteria carrying parts of the HIV genome, HIV peptides and proteins, HIV proteins with adjuvants, and naked DNA containing HIV genes. ■ DNA-based vaccines, p. 437

Vaccine trials in humans have been undertaken for at least 10 experimental vaccines. Vaccine evaluations in humans progress from phase I (testing safety and ability to provoke an immune response), to phase II (determining optimum dose, and characteristics and duration of immune response), to phase III (testing effectiveness in preventing HIV infection or AIDS). Substances that have entered human trials include recombinant gp120 and gp160; vaccinia virus containing HIV envelope genes; canary poxvirus containing HIV genes *env, gag,* and *pol;* a synthetic lipopeptide; a DNA vaccine containing HIV envelope genes; a *Salmonella* typhi strain engineered to express HIV gp120; and purified p17 (MA). Phase III trial employing gp120 from subtypes B and E gave negative results as did a later trial of prime-boost vaccine. The "prime-boost" technique employs an initial injection of a virus such as genetically engineered vaccinia containing HIV surface and core genes, followed later by an injection of recombinant viral surface antigen, gp120. The first injection gives primarily a cytotoxic T-lymphocyte response, and the second injection boosts the production of neutralizing antibodies. ■ poxviruses, p. 327 ■ vaccinia, pp. 324, 554, 709

The potential of synthetic peptides in immunization against HIV is also being explored. The amino acid sequence of the surface proteins of HIV is known; thus, peptides that exactly mimic immunologically important segments such as the V3 loop (see figure 29.7) can be synthesized. ■ peptide, p. 28 ■ peptide vaccines, p. 437

MICROCHECK 29.1

The signs and symptoms of persons with AIDS are mainly due to the opportunistic infections and tumors that complicate HIV disease. HIV is not highly contagious, and the AIDS worldwide epidemic could be stopped by changes in human behavior. Highly active antiretroviral therapy has given many patients with AIDS miraculous improvement.

✓ What is hairy leukoplakia?

✓ How would the ability to recognize differing strains of HIV-1 aid in epidemiological studies?

✓ If AIDS was present in Africa in the 1950s, why did it not appear in the United States until the 1970s?

29.2 Malignant Tumors That Complicate Acquired Immunodeficiencies

Focus Points

- Explain how HIV disease might increase the risk of malignant tumors.
- Outline the relationship between Kaposi's sarcoma and human herpesvirus-8.

Certain malignant tumors are associated with HIV disease, organ transplantation, and other acquired immunodeficiency states. Most of these malignancies fall into one of only three types: Kaposi's **sarcoma, lymphomas,** and **carcinomas** arising from anal or cervical epithelium. They tend to **metastasize,** meaning jump to new areas, and are difficult to treat. Evidence indicates that viruses are a factor in their causation. A popular theory is that certain viral antigens, perhaps with the aid of cytokines, cause rapid multiplication of a host cell. A mutation to malignancy then occurs among the rapidly dividing cells, the result of insertion of viral DNA into their genome or other carcinogen. Finally, the malignant cell escapes detection and destruction by immune surveillance because of defective cellular immunity, and it is able to multiply without restraint.

Kaposi's Sarcoma

Kaposi's sarcoma **(figure 29.12)** is an unusual tumor arising from blood or lymphatic vessels in multiple locations. Formerly, it was seen rarely and afflicted mainly older men of Mediterranean and Eastern European origin. It also occurs among all age groups in certain parts of tropical Africa. It is not associated with immunodeficiency in any of these varieties.

Coincident with the spread of HIV, the tumor began to appear in young men with HIV disease and its incidence rose dramatically. Among a group of never-married San Francisco men, the incidence of Kaposi's sarcoma was 2,000 times as high in 1984 as it was in the period before HIV became widespread. The tumor was so common among AIDS patients that it became an AIDS-defining condition, even though it generally appeared before the development of severe immunodeficiency. One of the many unsolved mysteries about this tumor is that its incidence among AIDS patients has fallen dramatically since the adoption of condom use and safer sex practices.

Besides HIV disease, Kaposi's sarcoma is often associated with the use of medications to suppress the immune system in organ transplant patients. Indeed, the incidence of Kaposi's sarcoma is more than 400 times as high in transplant patients as in the general public. Another peculiarity of the tumor is that it can sometimes disappear if a patient's organ graft is rejected, suggesting a strong influence for cellular immunity.

In 1994, scientists at Columbia University detected a previously unknown herpesvirus in Kaposi's sarcomas, named Kaposi's sarcoma-associated herpesvirus (KSHV), or human herpesvirus-8 (HHV-8). The virus can be detected in essentially all cases of Kaposi's sarcoma, whether associated with immunodeficiency or not. It is also strongly associated with certain malignant tumors, including multiple myeloma. Serological evidence of infection is present in about 25% of the healthy adult U.S. population. HHV-8 has been detected in the saliva of patients with HIV disease and Kaposi's sarcoma, and in semen of healthy men. Although there is still much to be learned about the role HHV-8 plays in Kaposi's

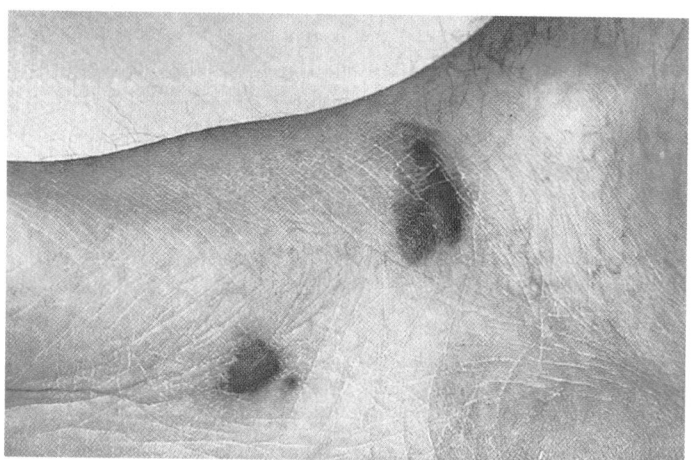

FIGURE 29.12 Kaposi's Sarcoma Two lesions on a person's foot are shown.

sarcoma, the virus appears to be necessary for formation of the tumor, but additional factors must be present for the tumor to develop. ■ multiple myeloma, p. 427

In Kaposi's sarcoma, HHV-8 infects the endothelial cells that line blood and lymphatic vessels and persists there mostly in a latent form; only a small percentage of the infected cells evidence a lytic infection at any one time. Presence of the virus is associated with two dramatic changes that result in tumor formation: (1) The cells assume a spindle shape and proliferate; and (2) extensive formation of new blood vessels occurs. These changes are probably due to release from the host cells of the cytokines normally responsible for stimulating cell growth and new vessel formation. This release of cytokines appears to be due largely to a gene product of latent HHV-8.

The changes typically seen in cancers and other malignant tumors are usually not present—namely, proliferation of a single clone of microscopically abnormal cells with abnormal chromosomes, and origin in a single location with later distant metastases. When malignant changes occur in Kaposi's sarcoma, as they do occasionally, the abnormal clone of cells does not contain the HHV-8 genome, showing that a factor other than the virus is responsible for the transformation.

Only a fraction of individuals infected with both HHV-8 and HIV-1 develop Kaposi's sarcoma, but that fraction is enormously greater than that seen in people infected only with HHV-8. Why is HHV-8 infection so much more likely to result in Kaposi's sarcoma in patients with HIV-1 disease? Scientists around the world, hard at work on this question, have discovered some clues. Tat, the protein product of the HIV *tat* gene, is released from infected T lymphocytes and enhances proliferation of endothelial cells. The Tat of HIV-2 does not have the same effect, and there is no increase in Kaposi's sarcoma in patients with HIV-2 disease. Many of the molecular details of the interaction of the two viruses are now known and undoubtedly will lead to new treatment options and better understanding of how malignancy develops.

B-Lymphocytic Tumors of the Brain

Lymphomas are a group of malignant tumors that arise from lymphoid cells. Most of these tumors arise from B lymphocytes, although T-lymphocytic lymphomas also occur. B-cell lymphomas are 60 to 100 times as common in AIDS patients as in the general public, and 60 times as frequent in individuals with organ transplants. Intense, sustained replication of lymphoid cells is a constant feature of HIV. As stated at the beginning of this chapter, HIV was first isolated from a patient with the lymphadenopathy syndrome, a condition in which the lymph nodes are markedly enlarged for 3 months or more, now known to be a prelude to AIDS. The lymph node enlargement reflects proliferation of lymphoid cells in response to high-level, unregulated cytokine release and to the HIV gp41 (TM) antigen. Also in HIV disease, sustained replication of T cells occurs as billions of new cells are produced each day to replace those destroyed by HIV. In contrast to Kaposi's sarcoma, with B-lymphocytic tumors there is no epidemiologic evidence of involvement of a sexually transmitted infection.

Epstein-Barr virus (EBV), however, plays a role in many of the B-cell lymphomas associated with AIDS. EBV is present in essentially all cases that involve the brain. Lymphomas rarely arise in the brain except in AIDS patients. EBV is also present in all the B-cell lymphomas that occur in transplant patients. The EBV-related tumor, Burkitt's lymphoma, is at least 1,000 times as frequent among AIDS patients as in the general public, representing about one-fourth of all the lymphomas associated with AIDS. EBV probably plays an indirect role in B-cell lymphoma formation rather than being the direct cause. HIV infection causes activation of latent EBV infection, with release of the virus to infect new B lymphocytes. This, in turn, causes polyclonal B-cell proliferation and increased life span. Malignant B-cell clones are thought to arise from this population of rapidly dividing cells. ■ Epstein-Barr virus, Burkitt's lymphoma, p. 686

Cervical and Anal Carcinoma

Carcinoma (cancer) of the uterine cervix in women and carcinoma of the anus in women and homosexual men are strongly associated with human papillomaviruses (HPV) types 16 and 18. The cells involved in these cancers are epithelial cells and, therefore, differ from those in Kaposi's sarcomas and lymphomas. HPV is transmitted during sexual activity and infects the cervical and anal epithelium, appearing to cause increased replication of the cells by blocking expression of a cellular gene responsible for controlling cell growth. In HIV disease, organ transplantation, and other immunodeficient conditions, HPV replication increases with the decline of the host's cellular immunity. Precancerous changes can be demonstrated in the anal cells of HIV-positive homosexual men twice as frequently as in homosexual men who are HIV negative, and six times as frequently among HIV-positive homosexual men with low $CD4^+$ T-cell counts as in HIV-positive homosexual men with high $CD4^+$ T-cell counts. Interestingly, even before the arrival of HIV, the incidence of anal carcinoma in homosexual men exceeded the incidence of cervical carcinoma in women. One important implication of these findings is that women who engage in anal intercourse should be screened for precancerous lesions of the cervix and anus, and homosexual men for lesions of the anus, at least twice yearly if they are HIV positive. ■ human papillomavirus infections, p. 638

MICROCHECK 29.2

Certain DNA viruses are strongly associated with development of malignant tumors in patients with HIV disease. These viruses are not sufficient by themselves to cause malignancy, but require the presence of other conditions. The tumors all arise in a setting of increased cell proliferation caused in part by the viruses. Many of these tumors are preventable by practicing safer sex practices and screening for premalignant lesions.

- ✓ What HIV gene product appears to play a role in development of Kaposi's sarcoma?
- ✓ What member of the herpesvirus family is associated with almost all of the B-cell lymphomas of the brain in AIDS patients?
- ✓ What measure is important to prevent cervical and anal carcinomas in persons with HIV disease?
- ✓ Would rapidly dividing cells favor mutation?

29.3

Infectious Complications of Acquired Immunodeficiency

Focus Points

- Describe the significance of *Pneumocystis jirovecii* infection as it relates to HIV disease.
- Outline the epidemiology of toxoplasmosis.
- Discuss the relationship between CMV disease and HIV disease.
- Outline the pathogenesis of *Mycobacterium avium* complex infection in individuals with immunodeficiency.

Immunodeficient individuals are susceptible to the same infectious diseases as other people. In addition, infections that pose no threat to immunologically normal people can cause severe, even fatal disease in those with impaired immune systems. Prevention of infectious diseases is key to the care of these people. **Table 29.5** shows some immunizations that are advisable; they should be given as early as possible in HIV disease. Except for measles-mumps-rubella, attenuated vaccines are generally unsafe to use in HIV disease.

Bacteria, viruses, fungi, protozoa, and even parasitic worms such as *Strongyloides stercoralis* can be life threatening. The geographical area may determine which infectious complications are most important. For example, the mold *Penicillium marneffei* is the third most common opportunistic agent among AIDS patients in Thailand but is almost unheard of elsewhere.

This section will present some examples of infectious diseases common in patients with immunodeficiency.

Pneumocystosis

Pneumocystosis, a severe, infectious lung disease, was recognized just after World War II in Europe when it caused deaths among hospitalized, malnourished, premature infants. Subsequently, scattered cases were recognized among immunodeficient patients until the onset of the AIDS epidemic, when the incidence soared. By 1995, almost 128,000 cases had been reported to the Centers for Disease Control and Prevention. Even today it remains the leading opportunistic infection in AIDS patients.

TABLE 29.5 Immunizations in Individuals with HIV Disease

Advised	Not Advised
Measles-mumps-rubella	Oral poliomyelitis
Pneumococcal polysaccharide	Oral typhoid
Influenza	BCG (tuberculosis vaccine)
Hepatitis A	Varicella-zoster
Hepatitis B	Yellow fever

Symptoms

The symptoms of pneumocystosis typically begin slowly, with gradually increasing shortness of breath and rapid breathing. Fever is usually slight or absent, and only about half of the patients have a cough, which is non-productive. As the disease progresses, a dusky coloration of the skin and mucous membranes appears and gradually worsens—a reflection of poor oxygenation of the blood, which can become fatal.

Causative Agent

Pneumocystosis is caused by *Pneumocystis jirovecii* **(figure 29.13)**, a tiny fungus belonging to the phylum Ascomycota, class Archiascomycetes. Formerly the organism was classified as *P.carinii* and it is still widely known by that name. It differs from many fungi in the chemical make-up of its cell wall, which makes it resistant to medications often used against fungal pathogens. The organism has not reliably been cultivated *in vitro*. ■ ascomycetes, p. 288

Pathogenesis

The small spores of *P. jirovecii,* measure only 1 to 3 μm and are easily inhaled into lung tissue. In experimental infections, the spores attach to the alveolar walls, and the alveoli fill with fluid, mononuclear cells, and masses of *P. jirovecii* cells in various stages of development. Later, the alveolar walls become thickened and scarred, preventing the free passage of oxygen.

Epidemiology

Various species of *Pneumocystis* are widespread among animals, including dogs, cats, horses, and rodents, persisting in their lungs as a latent infection. Although identical in appearance to *P. jirovecii,* they are genetically distinct and play no role in human disease. Most children are infected with *P. jirovecii* by age 2 and a half. The infection is asymptomatic and is generally

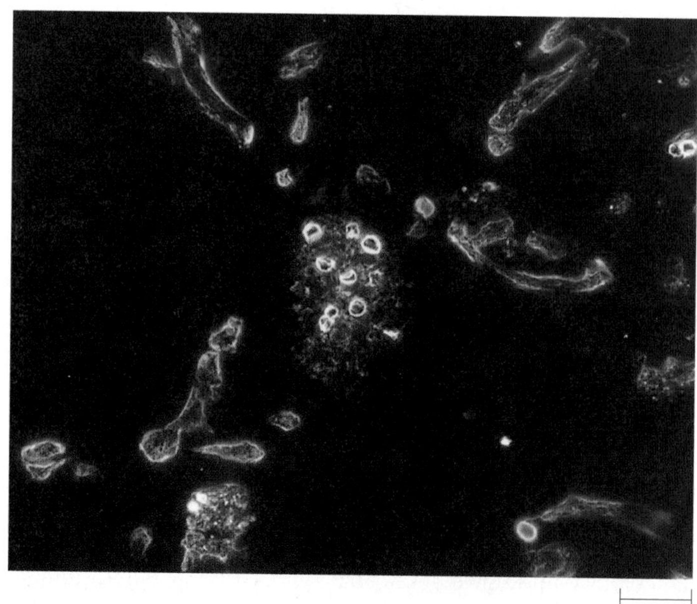

FIGURE 29.13 Fluorescent Antibody Stain of *Pneumocystis jirovecii* The yellow circles are *P. jirovecii* cysts.

eliminated within a year. The source and transmission of human infections are unknown. Most cases of pneumocystosis occur in persons with immunodeficiency, but it is uncertain whether their disease is caused by activation of latent infection, or infection newly acquired from inhalation of airborne spores. Epidemics among hospitalized malnourished infants and elderly nursing home residents suggest airborne spread, and *P. jirovecii* has been detected in indoor and outdoor air by using the polymerase chain reaction.

Prevention and Treatment

Pneumocystosis used to occur in about four-fifths of AIDS patients and was the leading cause of death. The disease is now largely prevented by starting regular doses of a medication such as trimethoprim-sulfamethoxazole to persons with HIV disease as soon as their CD4$^+$ T-cell count falls below 200 per μl, or if other hallmarks of immunodeficiency appear, such as **thrush,** a *Candida* infection of the mouth and throat.

Trimethoprim-sulfamethoxazole is also among the best medications for treating pneumocystosis and, along with oxygen and other measures, can reduce mortality from nearly 100% to about 30%. Alternative medications are available for treating those who cannot tolerate trimethoprim-sulfamethoxazole because of its side effects—mainly rash, nausea, and fever. For unknown reasons, people with HIV disease are more likely to develop these side effects than others. After treatment for pneumocystosis, individuals with HIV disease must receive preventive medication indefinitely, or until they have a sustained rise in CD4$^+$ T-cell count to above 200 cells/μl.

The main features of pneumocystosis are presented in **table 29.6.**

TABLE 29.6 Pneumocystosis

Symptoms	Gradual onset, shortness of breath, rapid breathing, non-productive cough, slight or absent fever, dusky color of skin and mucous membranes
Incubation period	4 to 8 weeks
Causative agent	*Pneumocystis jirovecii,* a tiny fungus related to the ascomycetes
Pathogenesis	Pneumocystosis can result from reactivation of latent infection or be newly acquired. Spores of *P. jirovecii* escape body defenses, enter the lungs with inspired air, attach to alveolar walls, multiply. Alveoli fill with fluid, macrophages, and *P. jirovecii.* The walls thicken, impairing oxygen exchange.
Epidemiology	*Pneumocystis carinii* widespread in domestic and wild animals as a latent lung infection, but the source of animal and human infections is unknown. Most humans become infected in early childhood. Disease arises in individuals with immunodeficiency; epidemics can occur in hospitalized premature infants and elderly nursing home residents.
Prevention and treatment	Formerly leading cause of death in those with AIDS, now usually prevented by medication (e.g., trimethoprim-sulfamethoxazole) as soon as the CD4$^+$ lymphocyte count drops to 200 cells per μl. Same medication is used for treatment; alternatives available. Medication is continued for life, or until the CD4$^+$ cell count rises and remains above 200 as a result of HAART or other treatment of the underlying immunodeficiency.

Toxoplasmosis

Toxoplasmosis is a protozoan disease that rarely develops among healthy people but can be a serious problem for those with malignant tumors, recipients of organ transplantation, the unborn child, and people with HIV disease.

Symptoms

The disease presents itself differently in the three main categories of patients: (1) the immunologically normal, (2) unborn children, and (3) those with immunodeficiency.

Toxoplasmosis can be acquired by immunocompetent individuals who eat raw or undercooked meat or who are exposed to cat feces. Most infections are asymptomatic, but 10% to 20% develop symptoms similar to those of infectious mononucleosis. They usually consist of sore throat, fever, enlarged lymph nodes and spleen, and sometimes a rash. These symptoms subside over weeks or months and do not require treatment. Rarely, a severe, even life-threatening illness develops, due to involvement of the heart or central nervous system. ■ infectious mononucleosis, p. 686

Toxoplasmosis of the unborn results from almost half of maternal infections that occur during pregnancy. Fetal toxoplasmosis during the first trimester of pregnancy is the least common but most severe, often resulting in miscarriage or stillbirth. Babies born live may have severe birth defects including small or enlarged heads, and their lungs and liver may also be damaged by the disease. Later, these babies can develop seizures or manifest mental retardation. Almost two-thirds of fetal toxoplasmosis cases occur during the last trimester of pregnancy, however, and the effects on the fetus are usually less severe. Most of these infants appear normal at birth, although later in life, **retinitis**—infection of the retina, the light-sensitive part of the eye—can be a big problem for them. This manifests itself as recurrent episodes of pain, sensitivity to light, and blurred vision, usually involving only one eye. Less common late consequences of congenital toxoplasmosis include mental retardation and epilepsy.

Toxoplasmosis that complicates immunodeficiency is commonly life threatening. Brain involvement in the form of encephalitis occurs in more than half the cases, manifested by confusion, weakness, impaired coordination, seizures, stiff neck, paralysis, and coma. Involvement of the brain, heart, and other organs often results in death.

Causative Agent

Toxoplasmosis is caused by *Toxoplasma gondii,* a tiny (3 by 7 μm), banana-shaped protozoan **(figure 29.14a)** that has a worldwide distribution and infects most warm-blooded animals, including household pets, pigs, sheep, cows, rodents, and birds. Its species name derives from the fact that the protozoan was first

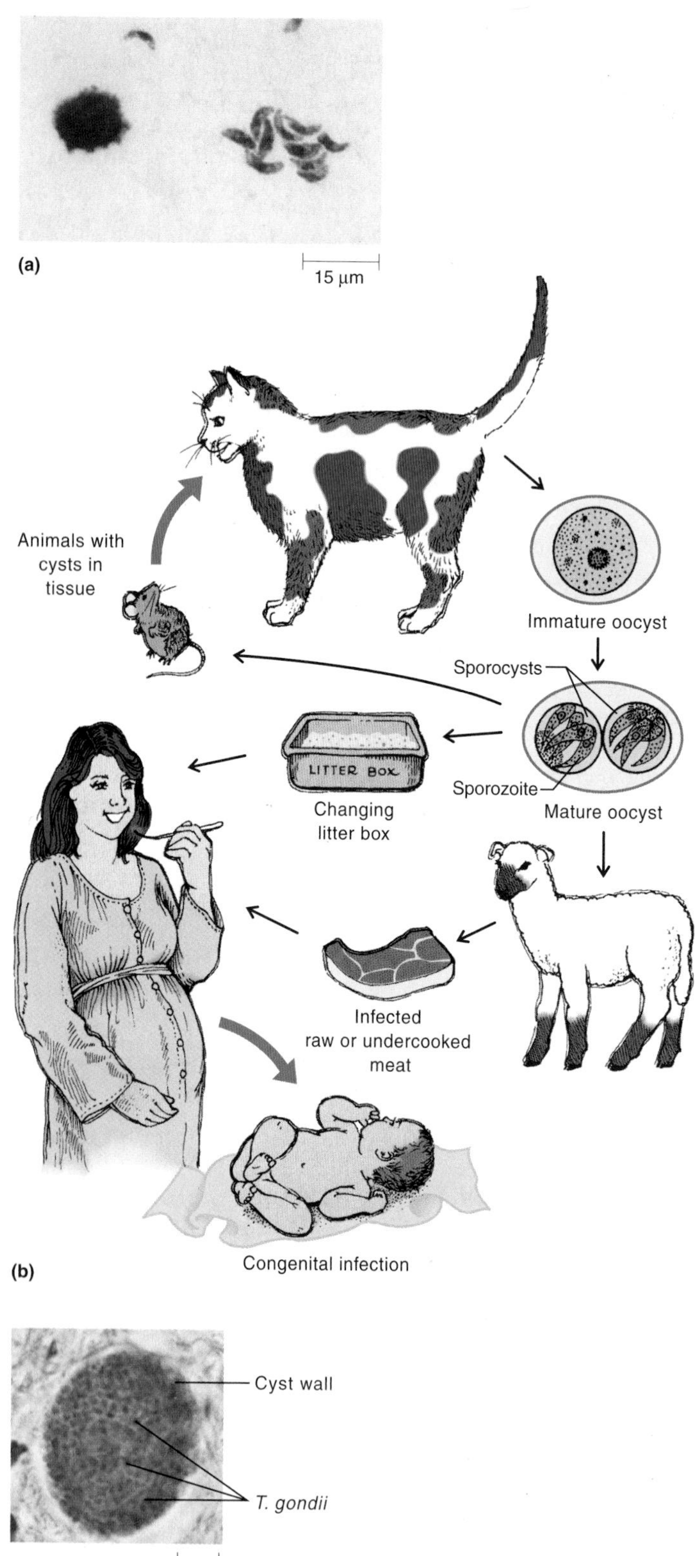

FIGURE 29.14 ***Toxoplasma gondii*** **(a)** Invasive forms. **(b)** Life cycle. Oocysts from cat feces and cysts from raw or inadequately cooked meat can infect humans and many other animals. **(c)** Cyst in tissue.

discovered in the gondi, an African desert rodent. The life cycle of *T. gondii* is shown in figure 29.14b. The definitive host, the one in which sexual reproduction occurs, is the cat or other feline (ocelot, puma, bobcat, Bengal tigers, and so on). The organism reproduces in the cat's intestinal lining, resulting in numerous progeny, some of which spread throughout the body and others that differentiate into male and female gametes (sexual forms). Union of gametes results in the formation of a thick-walled, oval structure (oocyst) about 12 μm in diameter that is shed in the cat's feces. Millions of the oocysts are shed each day, generally over a period of 1 to 3 weeks. In the soil, the oocysts undergo further development over 1 to 5 days into an infectious form containing two sporocysts, each with four sporozoites. These can remain viable for up to a year, contaminating soil and water and, secondarily, hands and food. In general, the cats recover from the acute infection and do not shed oocysts again. The oocysts are infectious for cats and other animals, including humans.

When non-feline animals ingest the oocysts, usually from contaminated food or water, the sporozoites emerge from the oocysts and invade the cells of the small intestine, especially its lower part, the ileum, but there is no sexual cycle. The intestinal infection spreads by the lymphatics and blood vessels throughout the tissues of the host, infecting cells of the heart, brain, and muscles. As host immunity develops, multiplication slows, and a tough, fibrous capsule forms, surrounding large numbers of a smaller form of *T. gondii.* These capsules packed with organisms are called cysts (figure 29.14c), and they remain viable for months or years. The life cycle is completed when a cat becomes infected by eating an animal with cysts in its tissues.

Pathogenesis

The organisms enter the body by ingestion of oocysts or inadequately cooked meat containing tissue cysts. *Toxoplasma gondii* is infectious for any kind of warm-blooded animal cell except non-nucleated red blood cells. Entry into the host cell is aided by an enzyme produced by the organism, which alters the host cell membrane. Proliferation of *T. gondii* in cells of the host causes destruction of the cells. Unless the infecting dose of the organism is very high, this process is normally brought under control by the immune response of the host, tissue cysts develop, and infected humans usually show few, if any, symptoms. In patients with immunodeficiency, however, infection can be widespread and uncontrolled, producing many areas of tissue necrosis. The disease process can result from a newly acquired infection, or declining immunity can allow reactivation of latent infection with escape of *T. gondii* from a person's tissue cysts.

Epidemiology

Toxoplasma gondii is distributed worldwide, but it is less common in cold and in hot, dry climates. Human infection is widespread. Serological surveys show the infection rate increases with age, ranging from 10% to 67% among those over age 50. Most infections are acquired when a person ingests oocysts that contaminate fingers, food, or drink or inhales contaminated dust in an enclosed space such as a barn. Young, homeless cats that commonly eat rodents and birds are likely sources of the organism. Small epidemics have occurred from drinking water contaminated with oocysts. Gardening in areas frequented by cats can result in *T. gondii* contamination of hands or vegetables. Eating rare meat poses a definite risk. Tissue cysts are present in about 25% of pork, in 10% of lamb, and less commonly in beef and chicken.

Prevention and Treatment

General measures for preventing *T. gondii* infection include washing hands after handling raw meat, coming in contact with soil, and changing cat litter. Litterboxes should be changed frequently, before oocysts have a chance to mature. Meat, especially lamb, pork, and venison, should be cooked until the pink color is lost from its interior. Fruits and vegetables should be washed before eating. Cats should not be allowed to hunt birds and rodents, nor should they be fed undercooked or raw meat. Children's sandboxes should be kept tightly covered when not in use. These measures are especially important for pregnant women and persons with immunodeficiency.

HIV-infected patients and those about to receive immunosuppressant medications are tested for antibody to *T. gondii.* A positive test indicates they have latent infection. Therefore, those with positive tests are given prophylactic trimethoprim-sulfamethoxazole if they have fewer than 100 $CD4^+$ T cells per microliter. Treatment of toxoplasmosis employs related medications—pyrimethamine with sulfadiazine. Alternatives are available.

The main features of toxoplasmosis are presented in **table 29.7.**

TABLE 29.7 Toxoplasmosis

Symptoms	In healthy individuals: sore throat, fever, enlarged lymph nodes, rash; with fetal infections: miscarriage, stillbirth, birth defects, epilepsy, mental retardation, retinitis; in immunodeficient individuals: confusion, poor coordination, weakness, paralysis, seizures, coma
Incubation period	Usually indeterminate
Causative agent	*Toxoplasma gondii,* a protozoan infectious for most warm-blooded animals. Sexual reproduction occurs in the intestinal epithelium of cats, the definitive hosts. Infected cats discharge oocysts with their feces. Ingested organisms are released from the oocysts, multiply rapidly, spread throughout the body. As immunity develops, infected cells become filled with the organisms, resulting in tissue cysts, which remain viable and infectious for the lifetime of the animal.
Pathogenesis	Organisms penetrate host cells causing necrosis. With development of immunity cell destruction stops, tissue cysts develop. Most healthy individuals have few or no symptoms unless the numbers of ingested organisms is very large. Organisms released from tissue cysts if immunity becomes impaired.
Epidemiology	Occurs worldwide, less common in cold or dry locations. Infection acquired by ingesting oocysts from cat feces, or eating inadequately cooked meat.
Prevention and treatment	Prevention: avoiding foods potentially contaminated with oocysts from cat feces, and not consuming inadequately cooked meat. Trimethoprim-sulfamethoxazole is given to immunodeficient persons with $CD4^+$ T lymphocyte counts below 100 cells per μl if they have antibodies to *T. gondii* indicating latent infection. Treatment: pyrimethamine with sulfadiazine, or alternative medication.

Cytomegalovirus Disease

Cytomegalovirus (CMV) is a member of the herpesvirus family, which includes herpes simplex virus, Epstein-Barr virus, and varicella-zoster virus, any of which can cause troublesome symptoms in patients with immunodeficiency. CMV, like other herpesviruses, is commonly acquired early in life and then remains latent. With impairment of the immune system, the infection activates and can cause severe symptoms.

Symptoms

Symptoms of cytomegalovirus disease follow a pattern similar to that of toxoplasmosis. Acute infections in immunocompetent individuals are usually without symptoms, but adolescents and young adults sometimes develop illness that resembles infectious mononucleosis, with fever, fatigue, and enlarged lymph nodes and spleen for weeks or months.

Severe damage can occur to the fetus if the mother develops an acute infection during pregnancy. This condition is known as **congenital cytomegalic inclusion disease** and is characterized by jaundice, large liver, anemia, eye inflammation, and birth defects. The vast majority of infected infants appear normal at birth, but 5% to 25% manifest hearing loss, mental retardation, or other abnormalities later in life.

Blindness is one of the most feared complications of cytomegalovirus disease in immunodeficient individuals. Other symptoms include fever, loss of appetite, painful joints and muscles, rapid and difficult breathing, ulcerations of the gastrointestinal tract with bleeding, lethargy, paralysis, dementia, coma, and death.

Causative Agent

Human cytomegalovirus is an enveloped, double-stranded DNA virus that looks like other herpesviruses on electron micrographs but has a larger genome. Its name (*cyto* for "cell" and *megalo* for "large") derives from the fact that cells infected by the virus are two or more times the size of uninfected cells. Infected cells show a large, intranuclear inclusion body surrounded by a clear halo, inspiring its description as an "owl's eye" **(figure 29.15).** The envelope is acquired as the virion buds from the Golgi apparatus membrane. ■ Golgi apparatus, p. 78

There are many different strains of the virus detected by the patterns of endonuclease digests, and antigenic differences also occur. Like other herpesviruses, CMV can cause lysis of the infected cell or become latent and subject to later reactivation. ■ genomic typing, p. 243

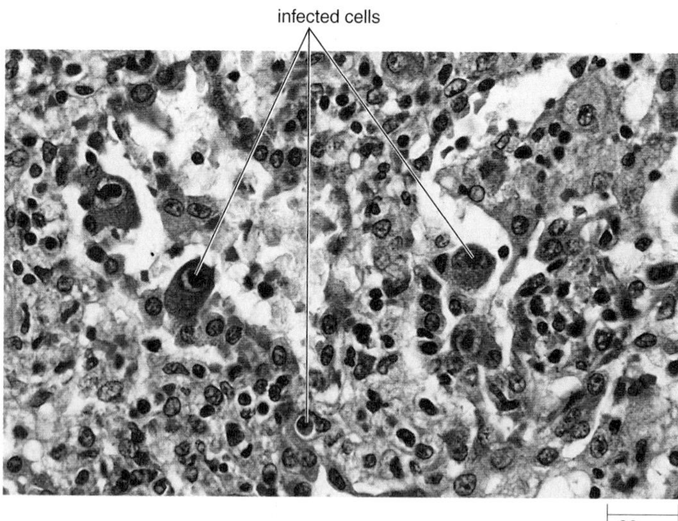

FIGURE 29.15 Cytomegalovirus Infection Infected cells are two to four times normal size; their nuclei contain large inclusions and are surrounded by a clear zone suggesting an "owl's eye" appearance.

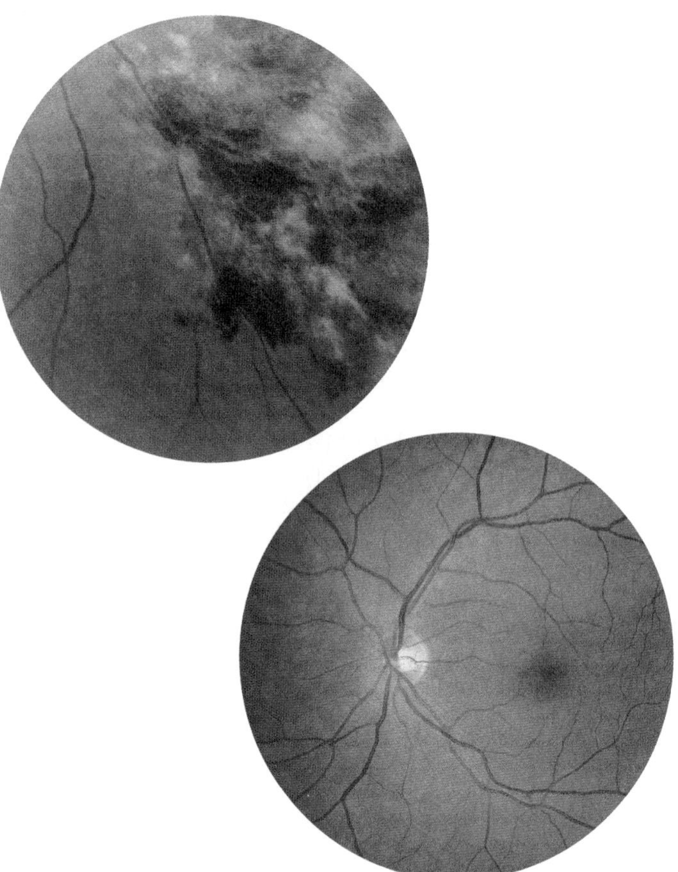

FIGURE 29.16 Cytomegalovirus (CMV) Retinitis, a Common Cause of Blindness in Persons with AIDS Photograph of a CMV-infected retina as it appears when viewed through the eye's pupil using an ophthalmoscope.

Pathogenesis

In cytomegalovirus disease, a wide variety of tissues, including eye **(figure 29.16),** central nervous system, lung, and liver, are susceptible to infection. Once cell entry has occurred, the viral genome can exist in a latent, non-infectious form, in a slowly replicating form, or as a fully productive infection. Control of viral gene expression depends partly on the type of cell infected. Monocytes allow low levels of infectious virus production. The CMV genome is ordinarily quiescent in T and B lymphocytes, expressing some viral genes but not producing viral DNA. Integration of viral DNA into the host cell genome probably does not occur. If CMV-infected T cells are also infected with HIV, productive CMV infection occurs. Fully productive infection of a wide variety of tissues occurs during acute infections, causing tissue necrosis. Transplanted organs and blood transfusions can transmit the disease, indicating that virus-producing cells are present or that non-producing cells in the blood or organ start producing infectious CMV under conditions of immune suppression. Cellular immunity probably plays a role in suppressing production of infectious virus as well as in lysis of infected cells. In cells infected with CMV, however, transfer of MHC molecules to the cell surface is impaired and, therefore, CMV antigens are not recognized as being "foreign." CMV infection is associated with an increase in $CD8^+$ cells and a decrease in $CD4^+$ cells, thus enhancing the effect of HIV infection. Latent infections activate with AIDS, organ transplants, and other immunodeficient states. Also, immunodeficient subjects are highly susceptible to newly acquired infection. ■ **MHC molecules, p. 380**

Epidemiology

CMV is found worldwide. One U.S. study found that more than 50% of adults, ages 18 to 25 years, and more than 80% of people over age 35 had been infected. Infection is lifelong. Infants born with CMV infection and those who acquire it shortly after birth excrete the virus in their saliva and urine for months or years. Virus is found in saliva, semen, and cervical secretions in the absence of symptoms, and sexual intercourse is a common mode of transmission in young adults. Up to 15% of pregnant women secrete the virus, and 1% of newborn infants have CMV in their urine. Breast milk, blood, and tissue transplants may contain CMV and can be responsible for transmission. Almost all prostitutes and promiscuous gay men are infected with CMV. CMV spreads readily in day care centers.

Prevention and Treatment

There is no approved vaccine for preventing cytomegalovirus disease. The use of condoms is effective in decreasing the risk of sexual transmission. People with HIV disease or other immunodeficiency syndromes or those who lack CMV antibody should avoid contact with day care centers if possible and, in any case, should wash their hands if exposed to saliva, urine, or feces. Tissue and blood donors can be screened for antibody to CMV. Those who have anti-CMV antibody are assumed to be infected and should not donate to those lacking antibody to CMV. The anti-herpesviral medication ganciclovir, given orally, halves the incidence of CMV retinitis in HIV disease patients with low $CD4^+$ cell counts and positive tests for CMV antibody. A ganciclovir implant designed to release the medication into the eye over a long period also delays the progression to blindness.

Combination drug therapy with the antiviral drugs ganciclovir and foscarnet can reduce the severity of CMV disease. They inhibit CMV DNA polymerase, the enzyme responsible for assembling viral DNA, at different sites. Both medications have serious side effects, mainly bone marrow suppression with ganciclovir and kidney impairment with foscarnet. ■ DNA polymerase, p. 192

The main features of cytomegalovirus disease are presented in **table 29.8.**

Mycobacterial Diseases

Initial exposure to *Mycobacterium tuberculosis,* the cause of tuberculosis, usually causes an asymptomatic infection that is controlled by the immune system and becomes latent. As might be expected, defects in cellular immunity are associated with reactivation of latent tuberculosis, which commonly results in unrestrained disease in AIDS and other immunodeficiencies. *Mycobacterium tuberculosis* is not the only mycobacterium that causes disease in immunodeficient people. Disease has been reported due to *M. kansasii, M. scrofulaceum, M. xenopi, M. szulgai, M. ganavense, M. haemophilum, M. celatum,* BCG vaccine, and others. This section discusses disease caused by organisms of *Mycobacterium avium* complex, which, next to *M. tuberculosis,* is the most common mycobacterial opportunists that complicate immunodeficiency diseases.

TABLE 29.8 Cytomegalovirus Disease

Symptoms	Symptoms rare in immunocompetent individuals, but sometimes an infectious mononucleosis-like illness develops. Infection of the mother during pregnancy can result in disease of the newborn. Immunocompromised individuals may experience blindness, lethargy, dementia, coma and brain damage
Incubation period	In immunocompetent adults, 20 to 60 days
Causative agent	Cytomegalovirus, a member of the herpesvirus family
Pathogenesis	Many tissues susceptible to infection, damage, especially eyes, brain, and liver. CMV latent infection can reactivate, produce infectious virions, tissue necrosis. $CD4^+$ T lymphocyte count is depressed and thus can enhance HIV disease.
Epidemiology	Common worldwide; lifelong infection. More than 50% of 18- to 25-year-olds are infected. Infants with congenital infections and those infected shortly after birth shed the virus for months or years. Body fluids, including breast milk, blood, urine, semen, and vaginal secretions, can transmit the disease. Almost all prostitutes and promiscuous homosexual men are infected with CMV.
Prevention and treatment	No vaccine available. Condoms decrease transmission. CMV-negative immunodeficient persons advised to avoid day care centers, wash their hands following contact with bodily fluids of infants. Blood and tissue transplants are tested for CMV before being given to CMV-negative individuals. The antiviral medication ganciclovir is considered for immunodeficient individuals who have antibody to CMV and whose $CD4^+$ T lymphocyte count falls below 50 cells per μl. Treatment: ganciclovir plus foscarnet.

Symptoms

The vast majority of *Mycobacterium avium* complex (MAC) infections in immunologically normal people are asymptomatic. Elderly people, especially those with underlying lung disease from smoking and those with alcoholism, develop a chronic cough productive of sputum and sometimes lesions in the lungs resembling tuberculosis. Children sometimes develop chronic enlargement of lymph nodes on one side of their neck, easily treated by surgical removal of the affected nodes. Patients with AIDS and other severe immunodeficiencies have slowly progressing symptoms ranging from chronic cough productive of sputum to fever, drenching sweats, marked weight loss, abdominal pain, and diarrhea.

Causative Agent

Mycobacterium avium complex is a group of mycobacteria consisting of two closely related species—*M. avium* and *M. intracellulare.* More than two dozen strains of these acid-fast rods fall into the MAC, distinguishable by serological tests, optimum growth temperature, and host range. Their growth rate is almost as slow as that of *M. tuberculosis,* but they are easily distinguished by using biochemical tests and nucleic acid probes. ■ tuberculosis, p. 514

Pathogenesis

MAC organisms enter the body via the lungs and the gastrointestinal tract. They are phagocytized by macrophages, but resist destruction because they inhibit acid production in the phagosome. Surviving organisms multiply within phagocytes and are carried by the bloodstream to all parts of the body. When cellular immunity is intact, most organisms are destroyed, and disease is localized. With profound immunodeficiency, the disease spreads throughout the body. In these patients, there is persistent bacteremia, occasionally with counts as high as 1 million per milliliter of blood. The small intestine contains macrophages packed with the bacteria, and infected tissues **(figure 29.17)** may resemble leprosy, sometimes containing more than 100 billion of the acid-fast bacteria per gram of tissue. Despite the presence of enormous numbers of the bacteria, there is little or no inflammatory reaction, and the clinical effect is a slow decline in the patient's well-being rather than a quickly lethal effect. One cause of deterioration may be activation of HIV replication when HIV-infected macrophages are infected with MAC. ■ leprosy, p. 655

Epidemiology

MAC organisms are widespread in natural surroundings and have been found in food, water, soil, and dust. In the United States, they are most prevalent in the Southeast, parts of the Pacific Coast, and the North Central region. Some strains are important pathogens of chickens and pigs. In AIDS patients, they are the most common bacterial cause of generalized infection. It is not known whether infection in AIDS patients is mostly newly acquired or a reactivation of latent disease. Most infections are from environmental sources rather than from person-to-person spread.

Prevention and Treatment

No effective measures are available to prevent exposure to MAC organisms. For HIV patients whose $CD4^+$ count is below 50 cells/μl, the antibacterial medication clarithromycin is recommended to help prevent MAC disease. If MAC bacteremia develops, patients

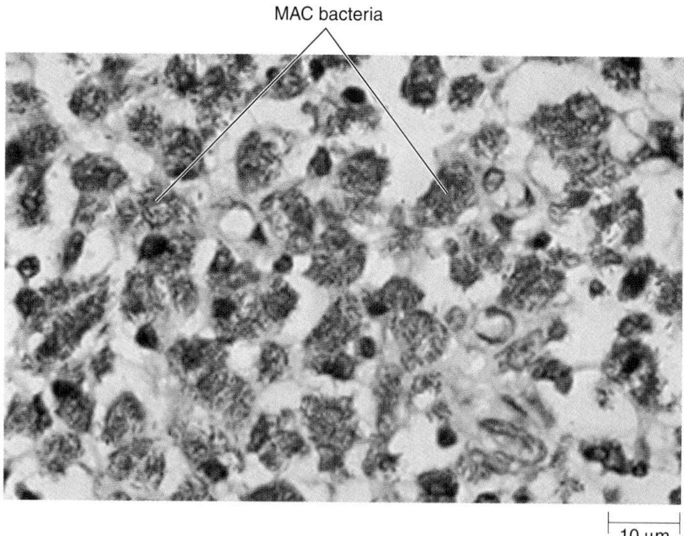

FIGURE 29.17 MAC Infection Acid-fast stain showing massive numbers of MAC bacteria infecting cells.

must be given two or more medications together, such as clarithromycin plus ethambutal.

The main features of MAC disease are presented in **table 29.9.**

MICROCHECK 29.3

Infectious diseases are a major threat to individuals with immunodeficiency. They get the same infectious diseases that afflict healthy people and, in addition, they are much more susceptible to reactivation of latent infections and organisms of low virulence that are generally harmless to others. Prevention of infectious diseases in AIDS patients is key to improving the quality and duration of their lives.

- ✓ Bacteria, fungi, protozoa, and viruses can all cause life-threatening diseases in AIDS patients. Name a fifth kind of infectious agent that can do the same.
- ✓ The disease pneumocystosis mainly involves which part of the body?
- ✓ Why should raw or uncooked meat not be eaten?

TABLE 29.9 MAC Disease

Symptoms	Usually asymptomatic in healthy people. Children can develop chronic localized enlargement of lymph nodes. Can cause chronic cough similar to tuberculosis in the elderly. Cough, fever, sweating, marked weight loss, abdominal pain, and diarrhea in those with severe immunodeficiencies.
Incubation period	Usually indeterminate
Causative agents	A group of mycobacterial strains in the species *M. avium* and *M. intracellulare*
Pathogenesis	MAC bacteria enter the body via lungs or gastrointestinal tract, are taken up by macrophages. In immunocompetent individuals, most are destroyed and infection controlled by cellular immunity. In severe immunodeficiency the bacteria multiply without restraint, are carried by macrophages throughout the body, and grow to enormous numbers in tissues but cause little or no inflammatory reaction.
Epidemiology	MAC organisms are widespread in food, water, soil, and dust. MAC disease of immunodeficient persons could result from environmental sources, or activation of latent infection.
Prevention and treatment	No proven measures to prevent exposure to MAC bacteria. The antibiotic clarithromycin is used to prevent MAC disease in severe immunodeficiency. If MAC bacteremia develops, two or more medications, such as clarithromycin plus ethambutal, are effective.

FUTURE CHALLENGES

AIDS and Poverty

In the United States and other economically advanced countries, antiviral medications have contributed significantly to the lessening of mother-child transmission of HIV, to delaying progression of HIV disease to AIDS, and to improving the quality and duration of life of AIDS sufferers. Newer drugs have also decreased transmission of HIV somewhat, thereby slowing the progress of the epidemic. The cost in the United States, while high, amounts to less than 1% of the country's total healthcare expenditure.

Antiviral medications have provided little benefit to most of the world's AIDS sufferers because they cannot afford them. Only an estimated 440,000 of the 5.5 million people needing anti-HIV medications are actually receiving them. Although HIV disease is spreading rapidly in India and China—two countries that together contain more than one-third of the world's population—and in the Caribbean, Southeast Asia, and Eastern Europe, the situation is the most desperate in African countries south of the Sahara desert. Only 10% of the world's population, but two-thirds of all those infected with HIV, lives in this area. About 75% of the region's young people (ages 15 to 24) infected with HIV are girls or young women. Vitally important research is underway to develop a procedure that would give young women an option within their control for preventing HIV infection.

The humanitarian, social, and economic implications of uncontrolled HIV disease are enormous. Unimaginable suffering; large numbers of starving, uneducated street children; loss of productive work years; spread of other infectious diseases; breakdown of public health; and setbacks for economic development are possible consequences that can affect neighboring countries and the rest of the world.

The challenge is urgent: to find effective HIV control measures, especially for the world's poor.

SUMMARY

AIDS

Since **AIDS** was first recognized in 1981 in sexually promiscuous homosexual men, it has claimed millions of lives in the United States, and is a leading cause of death in people 25 to 44 years of age. Worldwide, over 30 million people are infected with an AIDS-causing virus (figure 29.1); it is estimated that a new infection occurs about every 6 seconds, and a person dies of AIDS every 5 minutes. In Africa south of the Sahara, AIDS is now the number one cause of death, surpassing malaria; it is spreading rapidly in India and into China.

29.1 Human Immunodeficiency Virus (HIV) Infection and AIDS

"AIDS-defining conditions," such as unusual tumors or serious infections by agents that normally have little virulence, usually reflect immunodeficiency and were especially useful "markers" in early studies of the AIDS epidemic (table 29.1).

HIV Disease (table 29.4, figure 29.8)

Acquired immunodeficiency syndrome (AIDS) is a late manifestation of **human immunodeficiency virus (HIV) disease.** "Flu"-like or infectious mononucleosis-like symptoms representing the **acute retrovirus syndrome** often occur 6 days to 6 weeks after infection by HIV; these symptoms subside without treatment. Despite the absence of symptoms, HIV disease progresses and can be transmitted to others for years. The asymptomatic period ends with the appearance of certain tumors, or the onset of immunodeficiency (AIDS) as marked by **LAV, ARC,** and opportunistic or reactivated infections. Human immunodeficiency virus, type 1 (HIV-1), possesses single-stranded RNA. It belongs to the lentivirus genus of the retrovirus family, and is the main cause of AIDS (figure 29.3). The disease is spread mainly by sexual intercourse, by blood or contaminated hypodermic needles, and from mother to fetus or newborn (tables 29.2, 29.3). Different kinds of cells can be infected by HIV, notably two kinds vital to the immune system—helper T (T_H) lymphocytes and macrophages (figures 29.4, 29.5). Inside the host cell, a DNA copy of the viral genome is made through the action of reverse transcriptase, a complementary DNA strand is made, and the resultant double-stranded DNA is inserted into the host genome as a provirus by viral integrase (IN) (figure 29.6). Antimicrobial medications are used to prevent and treat opportunistic infections, and combinations of antiretroviral medications **(HAART)** are used to treat HIV disease (figure 29.11).

HIV Vaccine Prospects

No approved vaccine is currently available, but a number of prospects have moved from animal to human trials.

29.2 Malignant Tumors that Complicate Acquired Immunodeficiencies

Kaposi's Sarcoma (figure 29.12)

Kaposi's sarcoma is a tumor that arises from blood or lymphatic vessels. The incidence of the tumor is markedly increased among immunodeficient individuals. Infection by human herpesvirus-8 appears to be required for the tumor to develop.

B-Lymphocytic Tumors of the Brain

Lymphomas are malignant tumors that arise from lymphoid cells. Both B and T lymphocytes can give rise to lymphomas in individuals with immunodeficiencies, but B-cell lymphomas are more common. In contrast to non-immunodeficient cases, B-cell tumors often arise in the brain. Strong evidence exists that Epstein-Barr virus plays a causative role in these tumors.

Cervical and Anal Carcinoma

There is an increased rate of anal, genital, and cervical carcinoma in people with HIV disease. These tumors arise from squamous epithelial cells, thus differing from Kaposi's sarcoma and lymphomas. These cancers are strongly associated with human papillomaviruses (HPV), transmitted by sexual intercourse.

29.3 Infectious Complications of Acquired Immunodeficiency

Infections that occur in healthy individuals also occur and produce more severe disease in those with immunodeficiency. Latent infections such as those by *Mycobacterium tuberculosis* and herpesviruses commonly activate, and organisms rarely capable of causing disease in healthy individuals can be life threatening.

Pneumocystosis (table 29.6)

Before effective preventive regimens were developed, the disease was one of the most common causes of death among AIDS sufferers. Symptoms develop slowly, with gradually increasing shortness of breath and rapid breathing; patients can die from lack of oxygen. The causative agent, *Pneumocystis jirovecii,* a tiny fungus, is widespread in humans and other animals, generally causing asymptomatic infections that become latent.

Toxoplasmosis (table 29.7)

Toxoplasmosis is rare among healthy people but can be a serious problem for those with cancer, organ transplants, and HIV disease. The disease can also be congenital. Although infection is common in healthy people, only about 10% develop symptoms of sore throat, fever, and enlarged lymph nodes and spleen, sometimes with a rash. Symptoms usually disappear without treatment. With infections early in pregnancy, miscarriage can occur due to toxoplasmosis or babies can be born with birth defects or lung and liver damage. Infections later in pregnancy are usually milder but can result in epilepsy, mental retardation, or recurrent **retinitis** in the child. More than half of the cases with toxoplasmosis and AIDS develop encephalitis, and death often occurs because of involvement of the brain, heart, and other organs. *Toxoplasma gondii* is a tiny, banana-shaped protozoan that undergoes sexual reproduction in the intestinal epithelium of cats but can infect humans and many other vertebrates. Oocysts discharged in the feces of acutely infected cats become infectious in soil and contaminate food, water, and fingers (figure 29.14). *T. gondii* is present worldwide, and most people become infected from oocysts. Epidemics have resulted from contaminated drinking water. Eating rare meat can also cause infection.

Cytomegalovirus Disease (table 29.8)

Cytomegalovirus is a common cause of impaired vision in people with AIDS (figure 29.16). Normal individuals generally have asymptomatic infections; fetuses may develop **cytomegalic inclusion disease** or appear normal at birth but show mental retardation or hearing loss later; immunodeficient individuals develop fever, gastrointestinal bleeding, mental dullness, and blindness. Cytomegalovirus (CMV) is an enveloped, double-stranded DNA virus; infected cells are enlarged and have an "owl's eye" appearance (figure 29.15). The virus can exist in a latent form, a slowly replicating form, or a fully replicating form. Co-infection with HIV results in fully productive infection and tissue death. The virus occurs worldwide in breast milk, semen, and cervical secretions, and in saliva and urine of infected infants. No vaccine is available. Condoms decrease the risk of sexual transmission. The antiviral medication ganciclovir can be given to prevent CMV retinitis.

Mycobacterial Diseases (table 29.9)

Mycobacterium tuberculosis and *Mycobacterium avium* complex (MAC) organisms are most commonly responsible for infection. Normal people usually get asymptomatic or mild infections with MAC organisms, but immunodeficient patients may have fever, drenching sweats, severe weight loss, diarrhea, and abdominal pain. MAC organisms enter the body via the lungs and gastrointestinal tract, are taken up by macrophages but resist destruction, and are carried to all parts of the body. In immunodeficiency, MAC organisms multiply without restriction, producing massive numbers of the organisms in blood, intestinal epithelium, and other tissues (figure 29.17). No generally effective measures are available for preventing exposure to MAC bacteria. Prophylactic medication is advised for severely immunodeficient patients, but it can fail to prevent infection.

REVIEW QUESTIONS

Short Answer

1. What is the main symptom of patients with lymphadenopathy syndrome (LAS)?
2. Which cells of the immune system are prime targets of HIV?
3. What role do asymptomatic people with HIV disease play in the epidemiology of AIDS?
4. Why might the infant son of a hemophiliac man develop AIDS when the son's parents were strictly monogamous non-abusers of drugs?
5. Give two reasons it is a good idea to know whether you are infected with HIV.
6. What are four requirements of an acceptable HIV vaccine?
7. What is a phase III vaccine trial?
8. What are the three main types of malignant tumors that complicate HIV disease?
9. How do physicians prevent pneumocystosis in AIDS patients?
10. In AIDS patients with toxoplasmosis, which part of the body is affected in more than half the cases?
11. Name a feared complication of cytomegalovirus infection in AIDS patients.
12. Where in an AIDS patient's surroundings might MAC organisms be found?

Multiple Choice

1. About how many people died of AIDS in the United States in the first 10 years after the introduction of HAART?
   a) 4,000,000 b) 4,100 c) 42,000
   d) 250,000 e) 120,000
2. All of the following symptoms are characteristic of the AIDS related complex (ARC), *except*
   a) fever. b) fatigue. c) diarrhea.
   d) blindness. e) weight loss.
3. Which one of the following is true of Kaposi's sarcoma?
   a) HHV-8 is necessary for development of the tumor.
   b) HIV-1 is necessary for development of the tumor.
   c) Both HHV-8 and HIV-1 are necessary for development of the tumor.
   d) HHV-8 alone is sufficient for development of the tumor.
   e) Both HHV-8 and HIV-1 together are sufficient for the tumor to develop.
4. All of the following are HIV accessory genes, *except*
   a) *tat.* b) *env.* *vpr.* d) *rev.* e) *vpu.*
5. When was AIDS first recognized as representing a new disease?
   a) 1973 b) 1959 c) 1981 d) 1989 e) 1999
6. All of the following are AIDS-defining conditions, *except*
   a) influenza.
   b) herpes simplex of the esophagus.
   c) *Pneumocystis jirovecii* pneumonia.
   d) invasive cancer of the uterine cervix.
   e) Kaposi's sarcoma.
7. Which of the following types of cells can be infected by HIV?
   a) T_H cells
   b) Intestinal epithelium
   c) Antigen-presenting cells
   d) Brain cells
   e) All of the above
8. All of the following are HIV antigens, *except*
   a) CD4. b) TM. c) RT. d) MA. e) CA.
9. Which of the following is a cause of T_H cell death in HIV disease?
   a) Replication of HIV lyses the cell.
   b) Infected cells are destroyed by cytotoxic T cells (T_C).
   c) Infected cells are attacked by natural killer cells.
   d) Cells are killed by fusion and syncytium formation.
   e) All of the above.
10. Highly active antiretroviral therapy (HAART) is less than ideal because
    a) it does not eliminate latent HIV infection.
    b) its cost is too great for 90% of AIDS sufferers.
    c) it often has severe side effects.
    d) some HIV strains are resistant to it.
    e) All of the above.

Applications

1. An epidemiologist from the Centers for Disease Control and Prevention was presenting a report on the status of AIDS to a congressional committee. In concluding her remarks, she noted that from an epidemiological perspective it was more important to focus on HIV infection than on AIDS, and urged that the Congress consider redirecting funding of AIDS research to reflect this fact. What was the rationale for her request?
2. A historian researching the influence of society on the spread of communicable disease began to speculate on what it would be like if AIDS had appeared at a different time. What differences might one expect, for example, if AIDS had appeared in 1928 instead of 1978?
3. A newly emerging virus lethal for all felines is rapidly killing off household cats and related zoo animals. The CDC urgently appeals for funds to develop a vaccine against the virus, but a scientific adviser to Congress states that it would be very expensive and may not be possible. Moreover, she states that getting rid of cats would have the side benefit of ridding the world of toxoplasmosis. Is she correct? If so, how long would it take?

Critical Thinking

1. Vaccines have effectively prevented many viral diseases—witness smallpox and poliomyelitis. Attempts over many years to develop an effective vaccine against HIV disease and AIDS, however, have so far met with little success. Why might this be so?
2. Why is reverse transcriptase needed in order for HIV to become a provirus?

Farming relies on the activities of microorganisms.

30 Microbial Ecology

A Glimpse of History

For centuries, farmers have understood they could not continue growing the same crop on the same piece of land year after year without reducing the crop's yield. They knew that allowing a field to lie unplanted for one or more seasons enables it to recover its productivity. The wild plants that grow on the field for a year or two appear to rejuvenate the soil. It was not until the late nineteenth century that scientists began to discover why this was so. They isolated soil microorganisms that could take nitrogen from the air and transform it into forms plants could use. This process is called nitrogen fixation and the nitrogen is said to be "fixed." Although many scientists have worked for years to understand how microorganisms fix nitrogen, one scientist from the Netherlands, Martinus Willem Beijerinck, stands out as an early contributor in these studies.

In 1887 Beijerinck reported on the properties of the root nodule bacterium, which he called *Bacterium radicola.* (The genus name of this bacterium was later changed to *Rhizobium.*) He showed in 1890 that root nodules were formed when *B. radicola* was incubated with legume seedlings. Russian microbiologist Sergei Winogradsky then showed that the bacterium formed a symbiotic relationship with the roots of the legumes. We now know that the bacterial cells in the nodules fix nitrogen.

Beijerinck also made major contributions to other areas of microbiology. He worked on yeasts, plant viruses such as tobacco mosaic virus, and plant galls. *Beijerinckia,* a group of Gram-negative aerobic rods, is named for him. The genera *Azotobacter* and *Beijerinckia* include bacteria that can fix nitrogen under aerobic conditions in the absence of plants.

Beijerinck was described as a "keen observer," a person who was able "to fuse results of remarkable observations with a profound and extensive knowledge of biology and the underlying sciences." This ability was undoubtedly partly responsible for the great success of his work. ■

Microbes cycle nutrients, maintain fertile soil, and decompose wastes and other pollutants. Without microbial activities, life on earth could not survive. People would quickly become buried by the tons of wastes they generate, and nutrients would be depleted, halting growth and reproduction. In view of the crucial functions microorganisms perform, it would seem we should know a great deal about the diverse microbial species that inhabit our surroundings. Quite the opposite is true, however, as less than a mere 1% have been successfully grown in culture.

Even if all microorganisms could be cultivated in the laboratory, the information gained might not accurately reflect their role in the environment. In the laboratory, organisms are grown as pure cultures under controlled conditions that ensure the optimal growth. In nature, however, organisms generally grow as members of heterogeneous communities in poorly defined and often changing conditions. Nutrients are normally in short supply, limiting microbial growth. Thus, with respect to environmental microbiology, results obtained in the artificial setting of the laboratory, although useful, must be interpreted with caution. ■ **bacterial growth in nature, p. 85**

Chapters 4 and 11 discussed bacterial growth in nature, and microbial diversity. This chapter will expand on some of those concepts, focusing on activities of microorganisms that make them essential in the biosphere.

KEY TERMS

Consumers Organisms that consume organic material and rely on the activities of primary producers.

Decomposers Organisms that digest the remains of primary producers and consumers.

Denitrification Reduction of nitrate to gaseous nitrogen by anaerobic respiration.

Eutrophic An environment that is nutrient rich, supporting the excessive growth of algae and other organisms.

Hydrothermal Vents Undersea geysers that spew out mineral-laden hot water.

Hypoxic An environment very low in dissolved oxygen.

Microbial Mat A type of microbial community characterized by distinct layers of different groups of microbes that together make up a thick, dense, highly organized structure.

Nitrification Oxidation of ammonia (NH_3) to nitrate (NO_3^-).

Nitrogen Fixation Conversion of nitrogen gas to ammonia.

Oligotrophic A nutrient-poor environment.

Primary Producers Organisms that convert CO_2 into organic compounds, sustaining other life forms.

Rhizosphere Zone around plant roots containing organic materials exuded by the roots.

30.1 Principles of Microbial Ecology

Focus Points

- Describe the roles of primary producers, consumers, and decomposers.
- Describe how some microbes are able to grow in low-nutrient environments and adapt to environmental changes.
- Compare and contrast microbial competition and antagonism.
- Describe the structural organization of a microbial mat.
- Describe FISH and DDGE, focusing on how each enables researchers to better understand complex microbial communities.

Ecology is the study of the relationships of organisms, plant and animal, to each other and to their environment. Likewise, **microbial ecology** is the study of the relationships of microorganisms to each other and to their environment. Living organisms interact with one another in symbiotic relationships, such as commensalism, mutualism, and parasitism, described in chapter 17. Organisms in a given area, the **community,** interact with each other and the non-living environment, forming an ecological system, or **ecosystem.** Major ecosystems include the oceans, rivers and lakes, deserts, marshes, grasslands, forests, and tundra. Each ecosystem possesses a certain spectrum of organisms and characteristic physical conditions. The region of the earth inhabited by living organisms is called the **biosphere.** Within the biosphere, ecosystems vary both in **biodiversity** (number and variety of species present and their evenness of distribution) and **biomass** (the weight of all organisms present). ■ symbiotic relationships, p. 392

Microorganisms play a major role in most ecosystems, and many ecosystems host microbes unique to themselves. The role an organism plays in a particular ecosystem is called its **ecological niche.** The environment immediately surrounding an individual microorganism, the **microenvironment,** is most relevant to that cell, but because microbes are so small the microenvironment is difficult to identify and measure. The more readily measured gross environment, or **macroenvironment,** may be very different from the microenvironment. Consider a bacterial cell living within a biofilm (see figure 4.3); growth of aerobic organisms in the biofilm can deplete oxygen and create microzones that will support the survival of obligate anaerobes. Fermenters can produce organic acids that may then be metabolized by other organisms in the film. Various growth factors can be transferred directly within the microenvironment of the biofilm. Thus, certain microorganisms that might seem unexpected in a given macroenvironment actually thrive there because of microenvironments. ■ biofilm, p. 85 ■ growth factor, p. 95

Nutrient Acquisition

Organisms are categorized according to their **trophic level,** or their source of food, which is intimately related to the cycling of nutrients. There are three general trophic levels: primary producers, consumers, and decomposers **(figure 30.1):**

- **Primary producers** are autotrophs; they convert carbon dioxide into organic materials. Producers include both photoautotrophs, which use sunlight for energy, and chemolithoautotrophs, which oxidize inorganic chemicals for energy. Primary producers serve as a food source for consumers and decomposers. ■ photoautotroph, p. 95 ■ chemolithoautotroph, p. 95
- **Consumers** are heterotrophs. Because they utilize organic materials, they rely on the activities of primary producers. Herbivores, which eat plants or algae, are **primary consumers.** Carnivores that eat herbivores are **secondary consumers;** carnivores that eat other carnivores are **tertiary consumers.** This chain of consumption is called a **food chain.** Interacting food chains are called a **food web.**
- **Decomposers** are heterotrophs that digest the remains of primary producers and consumers. The fresh or partially decomposed organic matter used as a food source, including carcasses, excreta, and plant litter, is called **detritus.** Decomposers specialize in digesting complex materials such as cellulose, converting them into small molecules that can more readily be reused by other organisms. The complete breakdown of organic molecules into inorganic molecules such as ammonia, sulfates, phosphates, and carbon dioxide is called **mineralization.** Microorganisms, particularly bacteria and fungi, play a major role in decomposition processes owing to their ubiquity and unique metabolic capabilities.

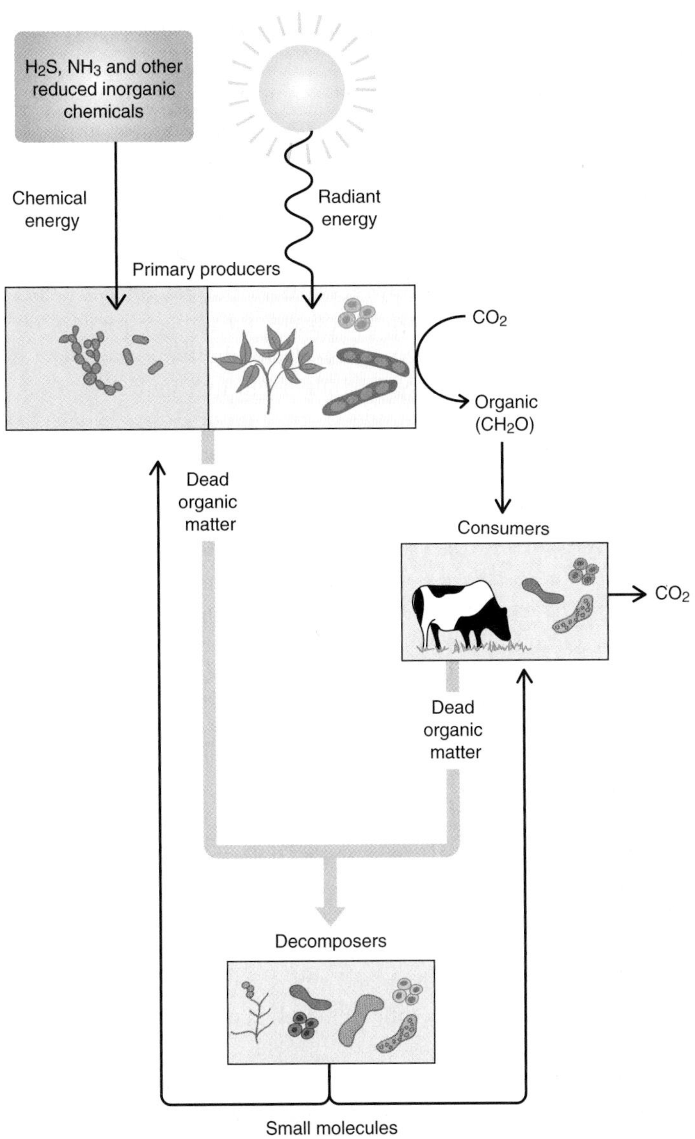

FIGURE 30.1 Relationship Among Primary Producers, Consumers, and Decomposers in an Ecosystem

Bacteria in Low-Nutrient Environments

Because low-nutrient environments are common in nature, microorganisms capable of growth in dilute aqueous solutions are also common. Most microbial growth in these environments is in biofilms, and the organisms are shed from the film into the aqueous solution. These organisms are by no means restricted to lakes, rivers, and streams. Indeed, microorganisms even grow in distilled-water reservoirs such as those found in research laboratories and pulmonary mist therapy units used in hospitals. In these environments, the microbes can extract trace amounts of nutrients absorbed by the water from the air or adsorbed onto the biofilm. Although the organisms grow slowly, they can reach concentrations as high as 10^7 per milliliter. This cell concentration is not high enough to result in a cloudy solution, so the growth usually goes unnoticed. This can have serious consequences for the health of hospitalized patients and for the success of laboratory experiments that depend on water purity. Organisms that grow in dilute environments contain highly efficient transport systems for moving nutrients inside the cell. Other mechanisms that bacteria use to thrive in dilute aquatic environments are described in chapter 11.

■ transport systems, p. 58 ■ thriving in aquatic environments, p. 266

Microbial Competition and Antagonism

Perhaps nowhere in the living world is competition more fierce and the results of competition more quickly evident than among microorganisms. The ability of an organism to compete successfully for a habitat is generally related to the rate at which the organism multiplies, as well as to its ability to withstand adverse environmental conditions. Because bacteria multiply logarithmically, any small differences in their generation times will result in a very large difference in the total number of cells of each species after a relatively short time **(figure 30.2).**

Antagonism among groups of organisms also helps determine the make-up of a community. In the soil, for example, some microbes resort to a type of chemical warfare, producing antimicrobial compounds. **Bacteriocins,** proteins produced by bacteria that kill closely related strains, are an example of antagonistic chemicals that play an important role in microbial ecosystems, promoting biodiversity through competition. It is tempting to speculate that antibiotics produced by *Streptomyces* species share a similar function, but their natural role is still poorly understood.

■ the genus *Streptomyces*, p. 265

Microorganisms and Environmental Changes

Environmental changes often result in alterations in a community. Those organisms that have adapted to live several inches beneath the surface of an untilled field will probably not be well suited

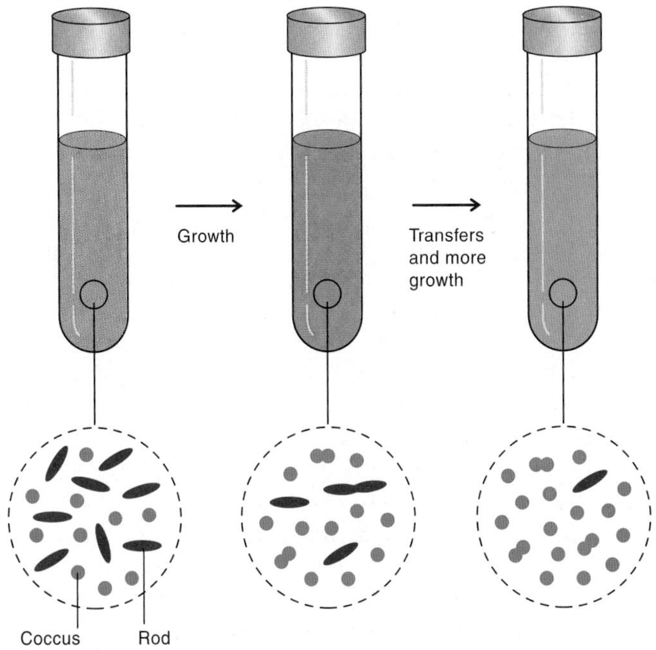

FIGURE 30.2 Competition The bacterium that multiplies faster yields the larger population.

to growth in that field if it is plowed, fertilized, and irrigated. In addition to external sources of environmental change, the growth and metabolism of organisms themselves can alter the environment dramatically. Nutrients may become depleted, and a variety of waste products, many of which are toxic, may accumulate.

In some environments, the changing conditions bring about a highly ordered and predictable succession of bacterial species. An example of such a microbial succession occurs in unpasteurized milk, which usually contains various species of microbes derived mainly from the immediate environment around the cow. Initially, the dominant organism is the bacterium *Lactococcus lactis,* which breaks down the milk sugar lactose, forming lactic acid as an end product **(figure 30.3).** This inhibits most other organisms in the milk, and eventually enough acid is produced to inhibit *L. lactis.* The acid sours the milk and also curdles it, a result of denaturation of the milk proteins. Members of the genus *Lactobacillus* can multiply in this highly acidic environment. These metabolize any remaining sugar, forming more acid until their growth is also inhibited. Yeasts and molds, which grow very well in this highly acidic environment, then become the dominant group and convert the lactic acid into non-acidic products. Because most of the sugar has already been used, the streptococci and lactobacilli cannot resume multiplication. Milk protein (casein) is still available and can be utilized for energy by bacteria of the spore-forming genus *Bacillus,* and some other bacteria, all of which secrete proteolytic enzymes that digest the protein. This breakdown of protein, known as **putrefaction,** yields a completely clear and very odorous product. The milk thus goes through a succession of changes with time, first souring and finally putrefying.

Microbial Communities

Microorganisms most often grow as biofilms attached to solid substrates or at air-water interfaces. General aspects of biofilms were described in detail in chapter 4. In this section, we focus on a specific type of biofilm—a microbial mat. ■ biofilms, p. 85

A **microbial mat** is a thick, dense, highly organized structure composed of distinct layers. Frequently they are green, pink, and black, which indicate the growth of different groups of microorganisms **(figure 30.4).** The green layer is the uppermost and is typically composed of various species of cyanobacteria. The color is due to the photosynthetic pigments of these microbes. The pink layer directly below consists of purple sulfur bacteria. The light-harvesting pigments of these anoxygenic phototrophs can use wavelengths of light not collected by the cyanobacteria. The black layer at the bottom results from iron molecules reacting with hydrogen sulfide produced by a group of bacteria called sulfate-reducers. These obligate anaerobes oxidize the organic compounds produced by the photosynthetic bacteria growing in the upper layers of the mat, using sulfate as a terminal electron acceptor. ■ cyanobacteria, p. 258 ■ photosynthetic pigments, p. 152 ■ purple sulfur bacteria, p. 256 ■ sulfate-reducers, p. 254 ■ terminal electron acceptor, p. 131

Although microbial mats can be found in a variety of areas, those near hot springs in Yellowstone National Park are some of the most intensively studied. The mats in these extreme areas are undisturbed by grazing eukaryotic organisms and, consequently, they provide an important model for the study of microbial interactions.

Studying Microbial Ecology

Because so few microorganisms can be successfully cultivated in the laboratory, investigating only those that have been isolated often does not portray an accurate picture of what actually occurs in nature. Molecular techniques are now complementing the traditional methods such as culture and microscopy, enabling researchers to better understand complex microbial communities.

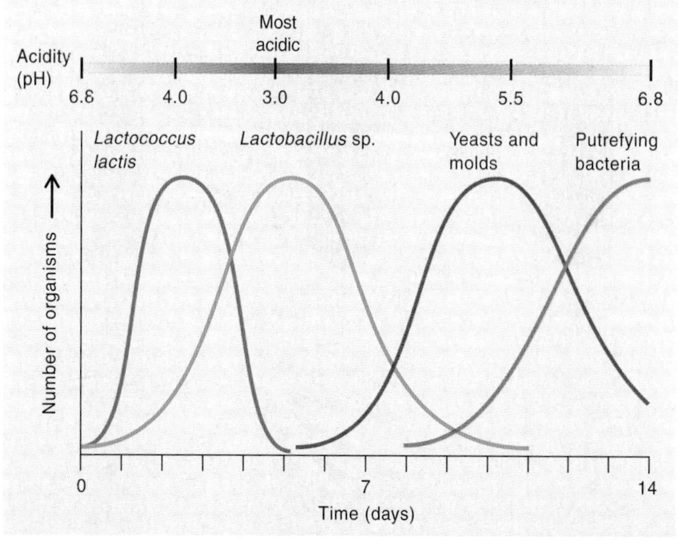

FIGURE 30.3 Growth of Microbial Populations in Unpasteurized Raw Milk at Room Temperature Production of acid causes souring and encourages growth of fungi. Eventually bacteria digest the proteins, causing putrefaction.

FIGURE 30.4 A Microbial Mat A microbial mat is a thick, dense, highly organized structure composed of distinct layers of different groups of microorganisms.

Microscopic methods can now be used to examine the composition of microbial populations. For example, certain dyes are made fluorescent by metabolic activities carried out only by living cells, and therefore can be used to observe only those cells that are viable (see figure 3.19a). A different technique, **fluorescence *in situ* hybridization (FISH),** uses nucleic acid probes that are labeled with a fluorescent molecule to observe only cells that contain specific nucleotide sequences (see figure 9.20). Confocal scanning laser microscopes enable researchers to observe sectional views of a three-dimensional specimen such as a biofilm (see figure 3.8). ■ fluorescent dyes, p. 51 ■ FISH, p. 228 ■ confocal scanning laser microscope, p. 46

Polymerase chain reaction (PCR) can be used to detect certain organisms and assess population characteristics. To detect a specific organism, primers are selected that amplify only DNA unique to that organism (see figure 9.14). To study the composition of a population, total 16S rRNA gene segments can be amplified. Individual fragments can then be cloned and studied. Alternatively, the set of amplified sequences can be separated and examined using a technique called **denaturing gradient gel electrophoresis (DGGE).** This procedure gradually denatures double-stranded nucleic acid during gel electrophoresis and, as a consequence, separates fragments of similiar size according to their melting point, which is related to the nucleotide sequence. Using DGGE, a mixture of 16S rRNA fragments with different sequences will resolve into a distinct pattern of bands. PCR and DDGE studies have confirmed that standard culture techniques can be poor indicators of the composition of natural microbial populations. Based on these molecular techniques that can show the relative abundance of specific nucleotide sequences in the sample, the species that predominate in laboratory culture often represent only a minute portion of the total population. ■ polymerase chain reaction, p. 223 ■ gel electrophoresis, p. 214

Genomics is also advancing the study of microbial ecology because sequence information gleaned from one species can be applied to others. For example, researchers found that variations of a gene coding for bacterial rhodopsin, a light-sensitive pigment that provides a mechanism for harvesting the energy of sunlight, are widespread in marine bacteria. This gene provides bacteria with a mechanism for phototrophy that does not require chlorophyll and might be an important mechanism for energy accumulation in ocean environments. Scientists are also using **metagenomics,** the study of total genomes in a sample, to study microbial populations in their natural environment. ■ genomics, p. 181 ■ bacterial rhodopsin, p. 275 ■ metagenomics, p. 221

MICROCHECK 30.1

Microorganisms play a major role in most ecosystems. Organisms are categorized as primary producers, consumers, or decomposers. Competition among microorganisms in a habitat can be intense. Microbial mats have distinct layers. Molecular techniques are enabling researchers to better understand natural microbial communities.

- ✓ What are the roles of primary producers, consumers, and decomposers?
- ✓ What microorganisms live within the green, pink, and black layers of a microbial mat?
- ✓ How could FISH (fluorescence *in situ* hybridization) be used to determine the relative proportions of *Archaea* and *Bacteria* in a population?

30.2 Aquatic Habitats

Focus Point

- Compare and contrast the habitats provided by marine, freshwater and specialized aquatic environments.

Oceans, covering more than 70% of the earth's surface, are the most abundant aquatic habitat, representing about 95% of the global water. They compose the marine environment. Lakes and rivers, the freshwater environments, represent only a small fraction of the total water.

Deep lakes and oceans have characteristic zones that influence the distribution of microbial populations. The uppermost layer, where sufficient light penetrates, supports the growth of photosynthetic microorganisms, including algae and cyanobacteria. The organic material synthesized by these primary producers gradually descends and is then metabolized by heterotrophs. In **oligotrophic** waters, meaning nutrient poor, the growth of photosynthetic organisms and other autotrophs is limited by the lack of certain inorganic nutrients, particularly phosphate, nitrate, and iron. When waters are **eutrophic,** or nutrient rich, photosynthetic organisms flourish, sometimes forming a visible layer on the surface **(figure 30.5).** In turn, the photosynthetic activities of

FIGURE 30.5 Eutrophication in a Polluted Stream Photosynthetic organisms flourish in the nutrient-rich water. The organic compounds they produce permit luxuriant growth of heterotrophs in lower layers.

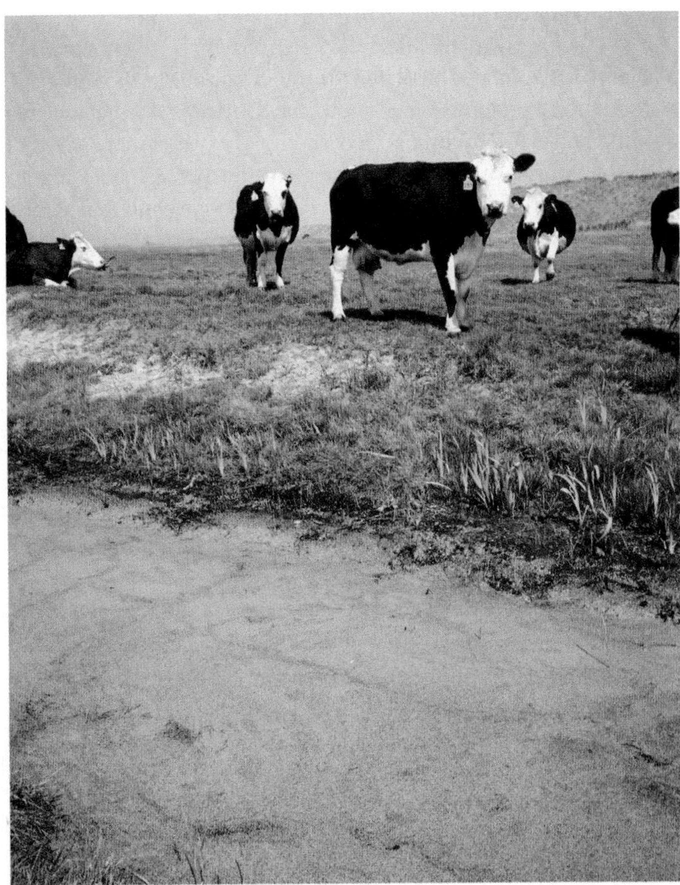

these organisms produce organic compounds that permit luxuriant growth of heterotrophs in lower layers. The heterotrophs consume dissolved O_2 as they metabolize the organic material. Because O_2 consumption can outpace the slow rate of diffusion of atmospheric O_2 into the waters, the environment can become very low in dissolved oxygen, or **hypoxic.** Lack of sufficient O_2 leads to the death of resident fish and other aquatic animals.

Marine Environments

Marine environments range from the deep sea, where nutrients are scarce, to the shallower coastal regions, where nutrients may be abundant due to runoff from the land. Seawater contains about 3.5% salt, compared with about 0.05% for fresh water. Consequently, it supports the growth of halophilic organisms, which prefer or require high salt concentrations, and halotolerant ones. Temperatures may vary widely at the surface, depending upon locale and other factors, but decrease with depth until reaching about 2°C in the deeper waters; an exception is the areas around hydrothermal vents, which will be described later. ■ hydrothermal vents, p. 732

Ocean waters are typically oligotrophic, limiting the growth of microorganisms. The little organic material produced by photosynthetic organisms is consumed as it descends, so that only scant amounts reach the sediments below. Even in the deep sea, marine water is O_2-saturated due to mixing associated with tides, currents, and wind action.

The ecology of inshore areas is not as stable as the deep sea, and can be dramatically affected by nutrient-rich runoff. An unfortunate example is a region devoid of fish and other marine life—a **dead zone**—that forms in the Gulf of Mexico as well as other areas **(figure 30.6).** The Mississippi River, carrying nutrients accumulated as it runs through agricultural, industrial, and urbanized regions, feeds into the Gulf. As a consequence of this nutrient-enrichment, populations of algae and cyanobacteria flourish in the Gulf waters in the spring and summer when sunlight is also plentiful. Heterotrophic microbes then metabolize the organic compounds synthesized by these primary producers, consuming dissolved O_2 in the process. This causes a large region in the Gulf, sometimes in excess of 7,000 square miles, to become hypoxic. Animals in the area either flee or die. Enrichment of coastal waters also contributes to blooms of toxic algae. ■ toxic algae, p. 284

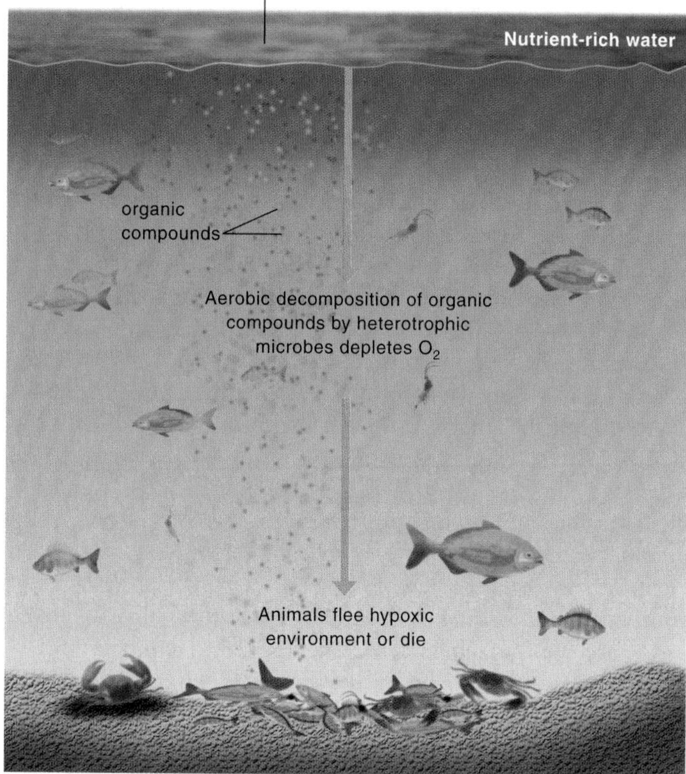
Algae and cyanobacteria flourish, using photosynthesis to produce organic compounds.
Nutrient-rich water
organic compounds
Aerobic decomposition of organic compounds by heterotrophic microbes depletes O_2
Animals flee hypoxic environment or die

FIGURE 30.6 Dead Zone Formation

Freshwater Environments

As with marine environments, the types and relative numbers of microbes inhabiting fresh waters depend on multiple factors including light, concentration of dissolved O_2 and nutrients, and temperature.

Oligotrophic lakes in temperate climates may have anaerobic layers due to **thermal stratification** resulting from seasonal temperature changes. During the summer months, the surface water warms. This decreases the density of the water, causing it to form a distinct layer that does not mix with the cooler, denser water below. The upper layer, called the **epilimnion,** is generally oxygen-rich due to the activities of photosynthetic organisms. In contrast, the lower layer, the **hypolimnion,** may be anaerobic due to the consumption of O_2 by heterotrophs. Separating these two layers is the **thermocline,** a zone of rapid temperature change. As the weather cools, the waters mix, providing O_2 to the deep water.

Rapidly moving waters, such as rivers and streams, are very different from lakes. They are usually shallow and turbulent, facilitating O_2 circulation, so they are generally aerobic. Light may penetrate to their bottom, making photosynthesis possible. Sheathed bacteria such as *Sphaerotilus* and *Leptothrix* species commonly adhere to rocks and other solid structures, enabling the microbes to remain stationary, utilizing nutrients that flow by. ■ sheathed bacteria, p. 266

Specialized Aquatic Environments

Specialized aquatic environments include salt lakes, such as the Great Salt Lake in Utah, which have no outlets. Water in these lakes evaporates, leaving concentrations of salt much higher than that in seawater. Extreme halophiles thrive in this environment. Other specialized habitats include iron springs that contain large quantities of ferrous ions; these springs are habitats for species of *Gallionella* and *Sphaerotilus.* Sulfur springs support the growth of both photosynthetic and non-photosynthetic sulfur bacteria. There are many other aquatic environments, both natural and human-made, ranging from groundwater to stagnant ponds and swimming pools to drainage ditches, each offering its own opportunity for bacterial growth. ■ extreme halophiles, p. 275

MICROCHECK 30.2

Aquatic habitats include marine, freshwater, and specialized environments. Nutrient-rich waters can become hypoxic. Lakes often exhibit thermal stratification during summer months.

- ✓ Compare the salt content of seawater with that of fresh water and salt lakes.
- ✓ Explain how nutrient-rich runoff can cause waters to become hypoxic.
- ✓ Why would the nutrient content of a body of water be more homogeneous than that of a terrestrial environment?

30.3 Terrestrial Habitats

Focus Point

- Describe soil as a microbial habitat.

Although microbes can adhere to and grow on a variety of objects on land, the focus in this section is soil—a critical component of terrestrial ecosystems. Extreme terrestrial habitats, such as volcanic vents and fissures, and some of the extremophiles that inhabit them are described in chapter 11. ■ archaea that thrive in extreme conditions, p. 275

Human interest in the microbiology of soil stems partly from the ability of microbes to synthesize a variety of useful chemicals. For example, over 500 different antibiotic substances are produced by *Streptomyces* species, at least 50 of which have useful applications in medicine, agriculture, and industry. The pharmaceutical industry has tested many thousands of soil microorganisms in search of those that produce useful antibiotics. In addition, soil microbes are being investigated for their ability to degrade toxic chemicals, an application of environmental microbiology called bioremediation, which will be discussed in chapter 31. Probably in no other habitat can one hope to find a greater range of biosynthetic and biodegradative capabilities than are represented in the soil. ■ bioremediation, p. 749

Characteristics of Soil

Soil is composed of pulverized rock, decaying organic material, air, and water. It teems with life, including bacteria, fungi, algae, protozoa, worms, insects, and plant roots. Soil communities may contain more than 4,000 different species per gram of soil. The top 6 inches of fertile soil may harbor more than 2 tons of bacteria and fungi per acre! Soil represents an environment that can fluctuate abruptly and dramatically. Heavy rains, for example, can cause a soil to rapidly become waterlogged. Trees dropping their leaves can suddenly enrich the soil with organic nutrients. Farmers and gardeners can rapidly change the nutrient mix by applying fertilizers.

Soil forms as rock weathers. Water, temperature changes, wind-blown particles, and other physical forces gradually lead to cracking and fragmentation of the rock. Photosynthetic organisms including algae, mosses, and lichens growing on the surfaces of rocks synthesize organic compounds. Various chemoorganoheterotrophic bacteria and fungi then use these compounds as carbon and energy sources. Their metabolism results in the production of acids and other chemicals, which gradually decompose the rocks. As soil slowly forms, some plants begin to grow. When these die and decay, the residual organic material functions as a sponge, retaining water and thus enabling more plants to grow. Over time, more organic compounds accumulate, forming a slowly degrading complex polymeric substance called **humus.** ■ lichens, p. 292 ■ chemoorganoheterotroph, p. 95

The texture of the soil influences its porosity, which in turn impacts the amount of air exchange and how much water can flow through. Finely textured soils, such as clay soils, are more apt to become waterlogged and anaerobic. In contrast, sandy soils that dry quickly allow water to pass through and are generally aerobic.

Microorganisms in Soil

The density and composition of the microbiota of the soil are dramatically affected by environmental conditions. Wet soils, for example, are unfavorable for aerobic microbes because the spaces in the soil fill up with water, thus diminishing the amount of air in the soil. When the water content of soil drops to a very low level, as during a drought or in a desert environment, the metabolic activity and number of soil microorganisms decrease. Many species of soil organisms produce survival forms such as endospores and cysts that are resistant to drying. Other environmental influences that impact soil microbes include acidity, temperature, and nutrient supply. For example, acidity suppresses the growth of bacteria, allowing fungi to thrive with less competition for nutrients. This is why mushrooms often appear in a lawn fertilized with an acid-producing fertilizer such as ammonium chloride.

Prokaryotes are the most numerous soil inhabitants. Their physiological diversity allows them to colonize all types of soil. In general, Gram-positive bacteria are more abundant in soils than Gram-negative bacteria. Among the most common Gram-positive bacteria are members of the genus *Bacillus.* These form endospores, enabling them to survive prolonged periods of adverse conditions such as drought or extreme heat. *Streptomyces* species produce conidia, which are dessication-resistant structures. They also produce metabolites called **geosmins,** which give soil its characteristic musty odor. As discussed earlier, *Streptomyces* species produce many medically useful antibiotics. Some of the other bacteria adapted to thrive in terrestrial environments, including myxobacteria and species of *Clostridium, Azotobacter, Agrobacterium,* and *Rhizobium,* were discussed in chapter 11. ■ endospore-formers, p. 264 ■ the genus *Streptomyces,* p. 265 ■ thriving in terrestrial environments, p. 263

While prokaryotes are the most numerous in soil, the biomass of fungi is much greater. Because most fungi are aerobes, they are usually found in the top 10 cm of soil. The soil fungi are crucial in decomposing plant matter—degrading and using complex macromolecules such as lignin, the major component of cell walls of woody plants, and cellulose. Some soil fungi are free-living, and others live in symbiotic relationships. The latter include **mycorrhizae,** fungi growing in a symbiotic relationship with certain plant roots. ■ mycorrhizae, p. 733

In addition to bacteria and fungi, various algae and protozoa are found in most soils. Because algae depend on sunlight for energy, they mostly live on or near the soil surface. Most protozoa require oxygen, so they too are found near the surface. They may be found in higher numbers in areas where microbes on which they feed are plentiful.

The Rhizosphere

The **rhizosphere** is the zone of soil that adheres to plant roots. The root cells excrete organic molecules including sugars, amino acids, and vitamins. The enriched soil in this zone fosters the growth of microorganisms. As a result, the concentration of microbes, particularly Gram-negative bacteria, is generally much higher than that of the surrounding soil. Particular bacterial species appear to preferentially interact with certain plants. For example, the rhizosphere of certain grasses is enriched with *Azospirillum* species, which fix nitrogen. ■ nitrogen fixation, p. 730

MICROCHECK 30.3

The density and composition of the soil are dramatically affected by environmental conditions. The concentration of microbes in the rhizosphere is generally much higher than that of the surrounding soil.

✓ Why are wet soils unfavorable for aerobic organisms?

✓ What is the significance of the rhizosphere?

✓ How can the biomass of fungi in soil be greater considering that bacteria are more numerous?

30.4 Biogeochemical Cycling and Energy Flow

Focus Points

- Diagram the carbon, nitrogen, sulfur, and phosphorus cycles, and describe some of the important microbial contributors to these cycles.
- Compare and contrast energy cycling in environments with sunlight versus those far removed from sunlight.

Biogeochemical cycles are the cyclical paths that elements take as they flow through living (biotic) and non-living (abiotic) components of ecosystems. These cycles are important because a fixed and limited amount of the elements that make up living cells exists on the earth and in the atmosphere. Thus, in order for an ecosystem to sustain its characteristic life forms, elements must continuously be recycled. For example, if the organic carbon that animals use as an energy source and exhale as carbon dioxide (CO_2) were not eventually converted back to an organic form, we would run out of organic carbon to build cells. The carbon and nitrogen cycles are particularly important because they involve stable gaseous forms (carbon dioxide and nitrogen gas), which enter the atmosphere and thus have global impacts.

While elements continually cycle in an ecosystem, energy does not. Instead, energy must be continually added to an ecosystem, fueling the activities required for life.

Understanding the cycling of nutrients and the flow of energy is becoming increasingly important as the burgeoning human population impacts the environment in a major way. For example, industrial processes that convert nitrogen gas (N_2) into ammonia-containing fertilizers have increased food production substantially, but they have also altered the nitrogen cycle by increasing the amount of fixed nitrogen, such as ammonium and nitrate, in the environment. Pollution of lakes and coastal areas with these nutrients has far-reaching effects, including depletion of dissolved O_2, which leads to the death of aquatic animals. It also decreases the biodiversity in terrestrial ecosystems. Excavation and burning of coal, oil, and other carbon-rich fossil fuels provides energy for our daily activities, but releases additional CO_2 and other carbon-containing gases into the atmosphere. Fossil fuels, the ancient remains of partially decomposed plants and animals, are nutrient reservoirs that are unavailable without human intervention, and therefore would not normally participate in biogeochemical cycles. The increase of carbon-containing gases in the atmosphere raises global temperatures because the gases absorb infrared radiation and reflect it back to earth.

When studying biogeochemical cycles, it is helpful to bear in mind the role of a given element in a particular organism's metabolism. Elements are used for three general purposes:

- **Biomass production.** As an example, all organisms require nitrogen to produce amino acids. Plants and many prokaryotes assimilate nitrogen by incorporating ammonia (NH_3) to synthesize the amino acid glutamate (see figure 6.28a). Some prepare for this step using the process of assimilatory nitrate reduction, which converts nitrate (NO_3^-) to ammonia. Once glutamate has been synthesized, the amino group can then be transferred to other carbon compounds in order to produce the necessary amino acids. Animals cannot incorporate ammonia and instead require amino acids in their diet. Some prokaryotes can reduce atmospheric nitrogen to form ammonia—the process of nitrogen fixation. The ammonia can then be incorporated into cellular material. ■ amino acid synthesis, p. 156
- **Energy source.** For example, reduced carbon compounds such as sugars, lipids, and amino acids are used as energy sources by chemoorganotrophs. Chemolithotrophs can use reduced inorganic molecules such as hydrogen sulfide (H_2S), ammonia (NH_3), and hydrogen gas (H_2) (see table 4.5). ■ energy source, p. 130
- **Terminal electron acceptor.** In aerobic conditions, O_2 is used as a terminal electron acceptor. In anaerobic conditions, some prokaryotes can use nitrate (NO_3), nitrite (NO_2), sulfate (SO_4), or carbon dioxide (CO_2) as a terminal electron acceptor. ■ terminal electron acceptor, p. 131

Carbon Cycle

All organisms are composed of organic molecules including proteins, lipids, and carbohydrates. Consumers eat plants as well as other consumers to acquire organic carbon for building biomass and as a source of energy. Decomposers use the remains of primary producers and consumers for the same purposes. As the organic carbon is degraded, respiration and some fermentations release CO_2, which must then be converted back to an organic form to complete the cycle **(figure 30.7).**

A fundamental aspect of the carbon cycle is carbon fixation, the defining characteristic of primary producers. These organisms all convert CO_2 into an organic form. The mechanisms used are described in chapter 6. Without the activities of the primary producers, no other organisms, including humans, could exist. We depend on them to generate the organic carbon we require.
■ **carbon fixation, p. 154**

The organic carbon travels through the food chain as primary producers are eaten by primary consumers, which are then eaten by secondary consumers. Through these events, one form of biomass is transformed into another. Not all of the organic material consumed is employed to create biomass, however; some is used as an energy source, generating carbon dioxide as a product.

As plants lose their leaves, and as members of the food web die, various decomposers degrade the resulting detritus, using it both as an energy source and to create biomass. The type of organic material helps dictate which species are involved in the degradation. For example, a wide variety of organisms utilize the more readily decomposable organic substances such as sugars, amino acids, and proteins. Bacteria, which usually multiply rapidly, generally play the dominant role in the decomposition of animal flesh. In contrast, only certain fungi can break down lignin, a major component of wood **(figure 30.8).** Aerobic conditions are required for this degradation, which is why water-saturated wood in anaerobic conditions such as a marsh resists decay.

The supply of oxygen has a profound influence on the carbon cycle. Not only does oxygen allow the degradation of certain compounds such as lignin, it also helps determine the types of carbon-containing gases produced. During the aerobic decomposition of organic matter, a great deal of carbon dioxide is produced through aerobic respiration. When the level of oxygen is low, however, as is the case in wet rice-paddy soil, marshes, swamps, and manure piles, the degradation is incomplete, generating some CO_2 and a variety of other products. Some of the CO_2 produced is used by the methanogens, anaerobic members of the Domain *Archaea.* These prokaryotes gain energy by oxidizing hydrogen gas, using carbon dioxide as a terminal electron acceptor, generating methane (CH_4). Methane that enters the atmosphere is oxidized by ultraviolet light and chemical ions to carbon monoxide (CO) and carbon dioxide.
■ **methanogens, p. 254**

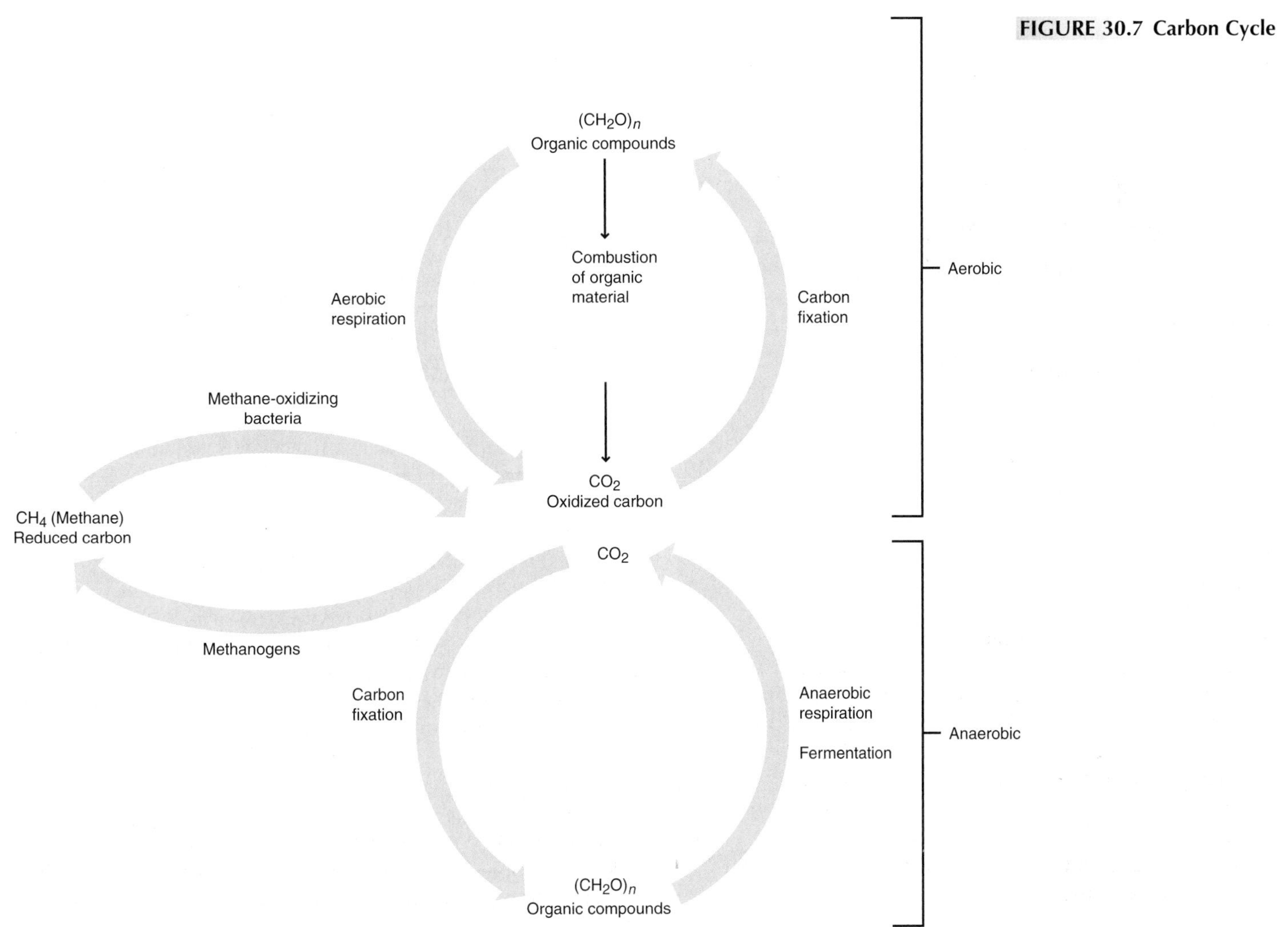

FIGURE 30.7 Carbon Cycle

FIGURE 30.8 Wood-Degrading Fungus Growing on a Dead Tree These fungi thrive in wet conditions. They digest lignin, the major cell wall component of woody plants, and consequently degrade the wood.

Nitrogen Cycle

Nitrogen is an essential constituent of proteins and nucleic acids. As consumers ingest plants and animals to fill their carbon and energy needs, they also obtain their required nitrogen, using it solely to build biomass. Prokaryotes, as a group, are far more diverse in their use of nitrogen-containing compounds. Some use oxidized nitrogen compounds such as nitrate (NO_3^-) and nitrite (NO_2^-) as a terminal electron acceptor; others used reduced nitrogen compounds such as ammonium (NH_4^+) as an energy source. These metabolic activities represent essential steps in the nitrogen cycle **(figure 30.9).**

Nitrogen Fixation

Nitrogen fixation is the process in which nitrogen gas (N_2) is reduced to form ammonia (NH_3), which can then be incorporated into cellular material. The process, which is mediated by the enzyme complex **nitrogenase,** requires a tremendous expenditure of energy. At least 16 molecules of ATP must be expended for every molecule of nitrogen fixed, because N_2 has a very stable triple bond. Although the atmosphere consists of approximately 79% N_2, relatively few organisms, all of which are prokaryotes, can reduce this gaseous form of the element. Thus, just as animals, including humans, depend on other organisms to fix carbon, they rely on prokaryotes to convert atmospheric nitrogen to a form they can assimilate to create biomass.

Nitrogen-fixing prokaryotes, or **diazotrophs,** may be free-living, or they may live in symbiotic association with higher organisms, particularly certain plants. Those that form symbiotic relationships will be discussed later. Among the free-living examples are members of the genus *Azotobacter.* These heterotrophic, aerobic, Gram-negative rods may be the chief suppliers of fixed nitrogen in grasslands and other similar ecosystems that lack plants with nitrogen-fixing symbionts. Certain cyanobacteria species are also diazotrophs, enabling these photosynthetic organisms to use both nitrogen and carbon from the atmosphere. The dominant free-living, anaerobic, nitrogen-fixing organisms of soil are certain members of the genus *Clostridium,* which are distributed widely in nature. ■ symbiotic nitrogen-fixers and plants, p. 734 ■ the genus *Azotobacter*, p. 264 ■ nitrogen-fixing cyanobacteria, p. 258 ■ the genus *Clostridium,* p. 254

Energy-expensive chemical processes developed to fix nitrogen are widely used to make fertilizers. These synthetic compounds are playing an increasingly larger role in the nitrogen cycle. In fact, fixed nitrogen sources associated with human intervention, including fertilizer production and planting crops that foster the growth of symbiotic nitrogen-fixers, now appear to surpass natural biological nitrogen fixation.

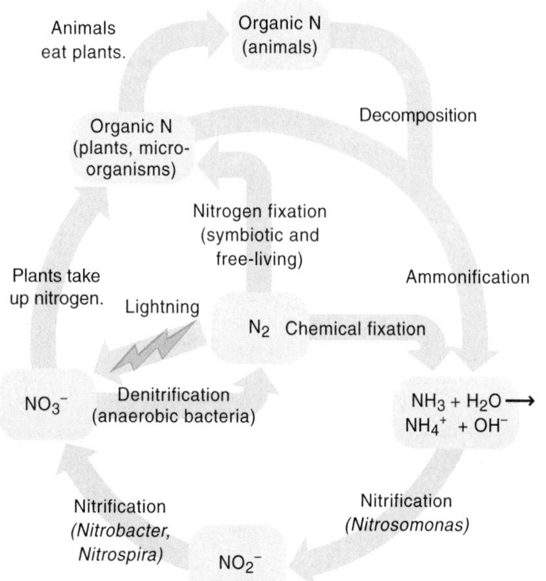

FIGURE 30.9 Nitrogen Cycle

Ammonification

Ammonification is the decomposition process that converts organic nitrogen into ammonia (NH_3). In neutral environments, ammonium (NH_4^+), a positively charged ion that adheres to negatively charged particles, is formed; in alkaline environments, such as heavily limed soil, the gaseous ammonia may enter the atmosphere.

A wide variety of organisms, including aerobic and anaerobic bacteria as well as fungi, can degrade proteins, which are among the most prevalent nitrogen-containing organic compounds. The microbes do this initially through the action of extracellular proteolytic enzymes that break down proteins into short peptides or amino acids. After transport of the breakdown products into the cell, the amino groups are removed, releasing ammonium. The decomposer will assimilate much of this compound to create biomass. Some, however, will be released into the environment, where it can then be assimilated by other organisms such as plants. ■ deamination, p. 150

Nitrification

Nitrification is the process that oxidizes ammonium (NH_4^+) to nitrate (NO_3^-). A group of bacteria known collectively as **nitrifiers** do this in a cooperative two-step process, using ammonium and an intermediate, nitrite (NO_2^-), as energy sources. They are obligate aerobes, using molecular oxygen as a terminal electron acceptor. Consequently, nitrification does not occur in waterlogged soils or in anaerobic regions of aquatic environments. ■ nitrifiers, p. 260

Nitrification has some important consequences with respect to agricultural practices and pollution. Farmers often apply ammonium-containing compounds to soils as a source of nitrogen for plants. The ammonium is retained by soils, because its positive charge enables it to adhere to negatively charged soil particles. Nitrification converts the ammonium to nitrate, a form of nitrogen more readily used by plants, but rapidly leached from soil by rainwater. To impede nitrification, certain chemicals can be added to ammonium-fertilized soils. Another negative aspect of ammonium oxidation is that nitrite can accumulate in soil if insufficient numbers of nitrite oxidizers are present. If the nitrite leaches into groundwater, it might contaminate wells used for drinking water. Nitrite is toxic because it combines with hemoglobin of the blood, reducing blood's O_2-carrying capacity. Even nitrate, which in itself is not very toxic, can be dangerous if high levels contaminate groundwater. When ingested, it can be converted to nitrite by intestinal bacteria that use it as a terminal electron acceptor.

Denitrification

Denitrification is the process that converts nitrate (NO_3^-) to gaseous nitrogen. Nitrate (NO_3^-) represents fully oxidized nitrogen. Some *Pseudomonas* species and a variety of other bacteria can use nitrate as a terminal electron acceptor when molecular oxygen is not available; this is the process of anaerobic respiration. The nitrate is reduced to gaseous forms of nitrogen such as nitrous oxide (N_2O) and molecular nitrogen (N_2). Release of these gases to the atmosphere represents a loss of nitrogen from an ecosystem. In addition, nitrous oxide contributes to global warming. ■ anaerobic respiration, pp. 133, 145

Under anaerobic conditions in wet soils, denitrifying bacteria can use the oxidized nitrogen compounds of expensive fertilizers, resulting in the release of gaseous nitrogen to the atmosphere and consequent economic loss to the farmer. In some areas, this process may represent 80% of nitrogen lost from fertilized soil. Denitrification is not always undesirable, however. The process can be actively fostered in certain steps of wastewater treatment as a means to remove nitrate. This compound could otherwise act as a fertilizer in the waters to which the wastewater is discharged, thereby promoting algal growth. ■ microbiology of wastewater treatment, p. 739

Anammox

Certain bacteria oxidize ammonium under anaerobic conditions, using nitrate as a terminal electron acceptor. This reaction, called **anammox** (for anoxic ammonia oxidation), forms N_2 and might provide an economical means of removing nitrogen compounds during wastewater treatment.

Sulfur Cycle

Sulfur occurs in all living matter, chiefly as a component of the amino acids methionine and cysteine. Like the nitrogen cycle, key steps of the sulfur cycle depend on the activities of prokaryotes **(figure 30.10)**. Some prokaryotes use the reduced form of sulfur, hydrogen sulfide (H_2S), and elemental sulfur (S^0) as energy sources or electron donors; others use oxidized sulfur compounds such as sulfate (SO_4^{-2}) as terminal electron acceptors.

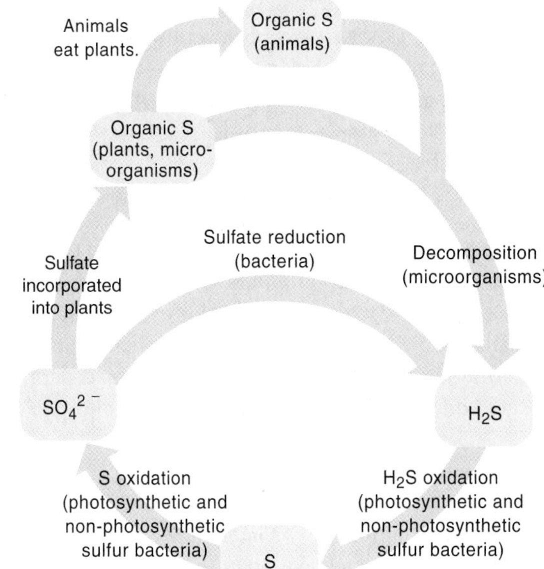

FIGURE 30.10 Sulfur Cycle

Most plants and microorganisms assimilate sulfur as sulfate (SO_4^{-2}), reducing it to form biomass. Like nitrogen, organic sulfur is present chiefly as a part of proteins. These organic compounds are first degraded into their constituent amino acids by proteolytic enzymes secreted by a wide variety of microorganisms. Decomposition of the sulfur-containing amino acids releases hydrogen sulfide, a gas.

Sulfur Oxidation

Hydrogen sulfide (H_2S) and elemental sulfur (S^0) can both serve as an energy source for certain chemolithotrophs. Sulfur-oxidizing prokaryotes, including *Beggiatoa, Thiothrix,* and *Thiobacillus* species, oxidize these molecules to sulfate (SO_4^{-2}). Certain bacteria in anaerobic marine environments can oxidize elemental sulfur, using nitrate as a terminal electron acceptor. As discussed in chapter 11, these organisms, including *Thioploca* species and the largest known bacterium, *Thiomargarita namibiensis,* have unusual mechanisms to cope with the fact that their energy source and terminal electron acceptor are found in two different environments. ■ sulfur-oxidizing bacteria, p. 259 ■ sulfur-oxidizing, nitrate-reducing marine bacteria, p. 270

Hydrogen sulfide and elemental sulfur are oxidized anaerobically by photosynthetic green and purple sulfur bacteria. These bacteria use sunlight for energy, but require reduced molecules as a source of electrons to generate reducing power. Like the chemolithotrophs that use hydrogen sulfide and elemental sulfur, the photosynthetic sulfur oxidizers produce sulfate. ■ green sulfur bacteria, p. 257 ■ purple sulfur bacteria, p. 256

Sulfur Reduction

Under anaerobic conditions, sulfate generated by the sulfur-oxidizers can then be used as a terminal electron acceptor by certain organisms. The sulfur- and sulfate-reducing bacteria and archaea use sulfate in the process of anaerobic respiration, reducing it to hydrogen sulfide (H_2S). In addition to its unpleasant odor, the H_2S is a problem because it reacts with metals, resulting in corrosion. ■ **sulfur- and sulfate-reducing bacteria, p. 254**

Phosphorus Cycle and Other Cycles

Phosphorus is a component of several critical biological compounds including nucleic acids, phospholipids, and ATP. Most plants and microorganisms readily take up phosphorus as orthophosphate (PO_4^{-3}), assimilating it into biomass. From there, the phosphorus is passed along the food web. When plants and animals die, decomposers convert organic phosphate back to inorganic phosphate.

In many aquatic habitats, growth of algae and cyanobacteria, the primary producers, is limited by low concentrations of phosphorus. Addition of phosphates from sources such as agricultural runoff, phosphate-containing detergents, and wastewater can result in eutrophication.

Other important elements, including iron, calcium, zinc, manganese, cobalt, and mercury, are also recycled by microorganisms. Many prokaryotes contain plasmids coding for enzymes that carry out oxidation of metallic ions.

Energy Sources for Ecosystems

All chemotrophs, including animals, harvest the energy trapped in chemical bonds to generate ATP. This energy cannot be totally recycled, however, since a portion is always lost as heat when bonds are broken. Thus, energy is continually lost from biological systems. To compensate for this outflow, energy must be added to ecosystems.

Photosynthesis, carried out by chlorophyll-containing plants and microorganisms, converts radiant energy (sunlight) to chemical energy in the form of organic compounds, which can be used by chemoorganotrophs. The requirement for radiant energy has traditionally been used to explain why life is not equally abundant everywhere. The discovery of different types of communities far removed from sunlight, including those near hydrothermal vents and within rocks, however, has dramatically altered these ideas. These communities rely on chemolithoautotrophs, which harvest the energy of reduced inorganic compounds and use it to form organic compounds. ■ **chemolithoautotroph, pp. 95, 150**

A number of **hydrothermal vents** have been discovered, some thousands of meters below the ocean surface. These vents form when water seeps into cracks in the ocean floor and becomes heated by the molten rock, finally spewing out in the form of mineral-laden undersea geysers. The hydrogen sulfide discharged supports thriving deep-sea communities, oases in the otherwise desolate ocean floor **(figure 30.11).** Large numbers of sulfur-oxidizing chemolithoautotrophs (bacteria and archaea) are found in and around the vents. Many are free-living but some live in symbiotic association with the large tube worms and clams that inhabit the areas. The chemolithoautotrophs harvest energy from oxidation of hydrogen sulfide, and they fix CO_2, providing the animals with both a carbon and energy source.

(a)

(b)

FIGURE 30.11 Hydrothermal Vent Community (a) This diverse community is supported by the metabolic activities of chemolithoautotrophs. **(b)** Water escaping from the vent is rich in minerals and dissolved gases, including hydrogen sulfide.

In 1994, microbial populations were reported living almost 3 km under the ground. Then, in 1995, scientists found large populations of microbes thriving in iron-rich volcanic rocks from nearly a thousand meters below the surface of the Columbia River (see Perspective 4.1). These organisms gain energy from hydrogen (H_2) produced in the subsurface. It has been estimated that if (and it is a big "if" at this point) most similar rocks contain microbes, there could be as much as 2×10^{14} tons of underground microorganisms—equivalent to a layer 1.5 meters thick over the entire land surface of the earth!

MICROCHECK 30.4

Recycling of elements occurs as organisms incorporate them to produce biomass, oxidize reduced forms as energy sources, and reduce oxidized forms as terminal electron acceptors. Carbon fixation utilizes atmospheric CO_2 to produce organic material; the CO_2 is regenerated during respiration and some fermentations. Prokaryotes are essential for several steps of the nitrogen cycle including nitrogen fixation, nitrification, denitrification, and anammox. Prokaryotes are also essential for several steps of the sulfur cycle, including sulfur reduction and sulfate oxidation.

- ✓ What is a diazotroph?
- ✓ Why do farmers try to impede nitrification?
- ✓ Although chemoautotrophs serve as the primary producers near hydrothermal vents, animals in that environment still ultimately depend on the photosynthetic activities of plants and cyanobacteria. Why?

30.5

Mutualistic Relationships between Microorganisms and Eukaryotes

Focus Point

- Describe the mutalistic relationships between fungi and plant roots, symbiotic nitrogen-fixers and plants, and microorganisms and herbivores.

As described in chapter 19, mutualism is a symbiotic association in which both partners benefit. A variety of other ecologically important symbiotic relationships exist, but mutualistic relationships highlight the vital role of microorganisms to life on this planet. This section will describe three important mutualistic relationships between microbes and eukaryotes—mycorrhizae and plants, symbiotic nitrogen-fixers and plants, and microorganisms and herbivores. ■ symbiotic relationships, p. 392

Mycorrhizae

Mycorrhizae are fungi growing in symbiotic relationships with plant roots **(figure 30.12).** They enhance the competitiveness of plants by helping them take up phosphorus and other substances from the soil. In turn, the fungi gain nutrients for their own growth from root secretions. It is estimated that over 85% of vascular plants (plants with specialized water and food conducting tissues) have mycorrhizae.

There are two common types of mycorrhizal relationships:

- **Endomycorrhizae.** The fungi penetrate root cells, growing as coils or tight, bushlike masses within the cells. These are by far the most common mycorrhizal relationships, and are found in association with most herbaceous plants. Relatively

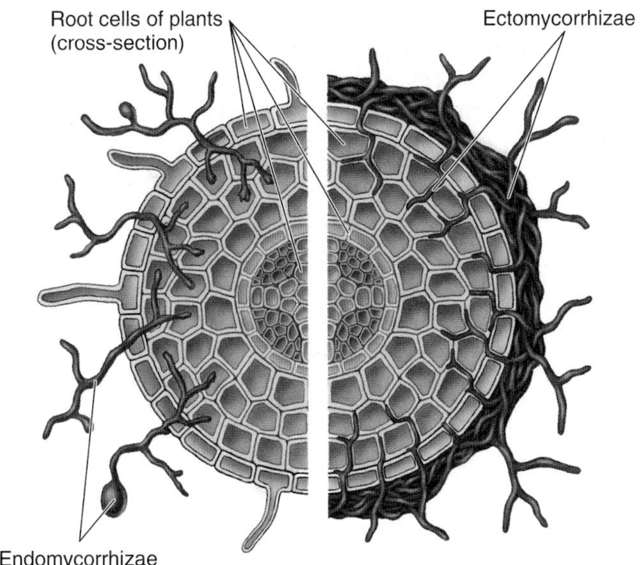

(a) Diagram of Mycorrhizae

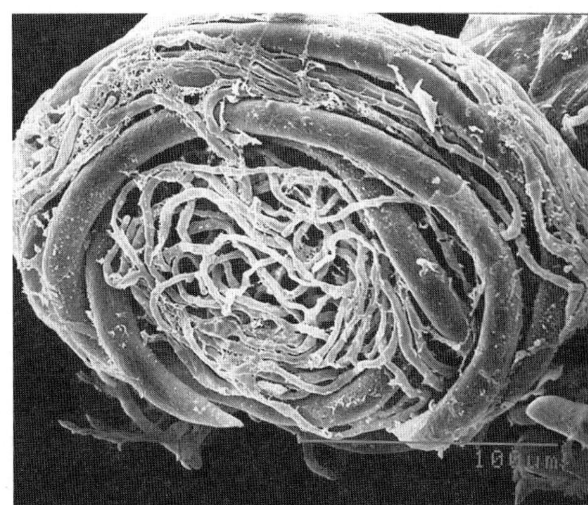

(b) Endomycorrhiza

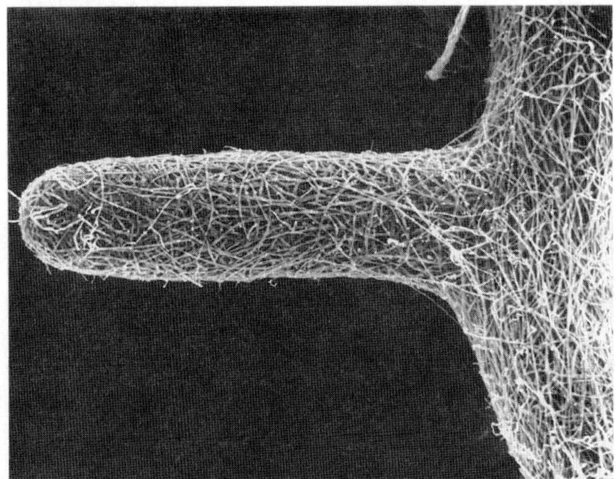
(c) Ectomycorrhiza

FIGURE 30.12 Mycorrhizae (a) Diagram illustrating the two types of mycorrhizae. **(b)** In an endomycorrhizal relationship, the fungi penetrate root cells growing within the cells. **(c)** In an ectomycorrhizal relationship, the fungi grow around the plant root cells, forming a fungal sheath around the root.

few species of fungi are involved, perhaps only 100 or so, and most appear to be obligate symbionts. The relationship for some plants is also obligate; for example, most orchid seeds will not germinate without the activities of a fungal partner.

- **Ectomycorrhizae.** The fungi grow around the plant cells, forming a fungal sheath around the root. These fungi mainly associate with certain trees, including conifers, beeches, and oaks. Over 5,000 species of fungi are involved in ectomycorrhizal relationships, but many are restricted to a single type of plant. Chanterelles and truffles are examples of commercially valuable ectomycorrhizal fungi.

Symbiotic Nitrogen-Fixers and Plants

Although free-living bacteria are potentially capable of adding a considerable amount of fixed nitrogen to the soil, symbiotic nitrogen-fixing organisms are far more significant in benefiting plant growth and crop production. They are important in both terrestrial and aquatic habitats.

Members of several genera, including *Rhizobium, Bradyrhizobium, Sinorhizobium,* and *Azorhizobium,* collectively referred to as rhizobia, are the most agriculturally important symbiotic nitrogen-fixing bacteria. These organisms associate with leguminous plants including alfalfa, clover, peas, beans, and peanuts. The input of soil nitrogen from the microbial symbionts of these plants may be roughly 10 times the annual rate of nitrogen fixation attainable by non-symbiotic organisms in a natural ecosystem. To encourage plant growth, people often plant the seeds of certain legumes together with inocula of the appropriate symbionts.

The symbiotic association between rhizobia and plants involves chemical communication between the partners. First, root exudates of the leguminous plant attract the appropriate bacterial species, which then attach to and colonize the root cells **(figure 30.13).** Substances secreted by the root cells cause the bacteria to produce chemical signals called **nod factors,** which in turn, induce the root hairs to curl. The bacteria then invade the root hair, multiplying and moving into the root cells by means of an **infection thread,** a tube produced by the plant cells in response to infection, through which the bacteria invade. Once inside the root, the bacterial cells change in appearance, forming a cell type called a **bacterioid.** Repeated division of both root cells and bacteroids leads to the development of root nodules (figure 30.13b). The plant cells synthesize a special oxygen carrier called **leghemoglobin,** which binds to O_2 and regulates its concentration in the nodule. This protects the O_2-sensitive nitrogenase by keeping free O_2 at a low level. The bacterioids in the nodule fix nitrogen, releasing ammonia that then diffuses into the root cell.

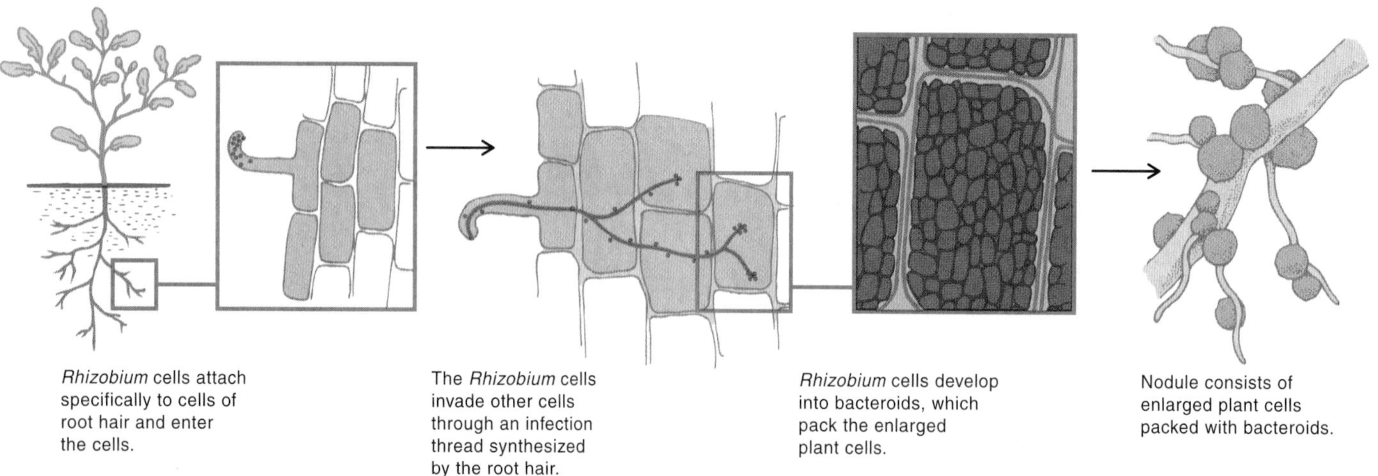

Rhizobium cells attach specifically to cells of root hair and enter the cells.
The *Rhizobium* cells invade other cells through an infection thread synthesized by the root hair.
Rhizobium cells develop into bacteroids, which pack the enlarged plant cells.
Nodule consists of enlarged plant cells packed with bacteroids.
(a)

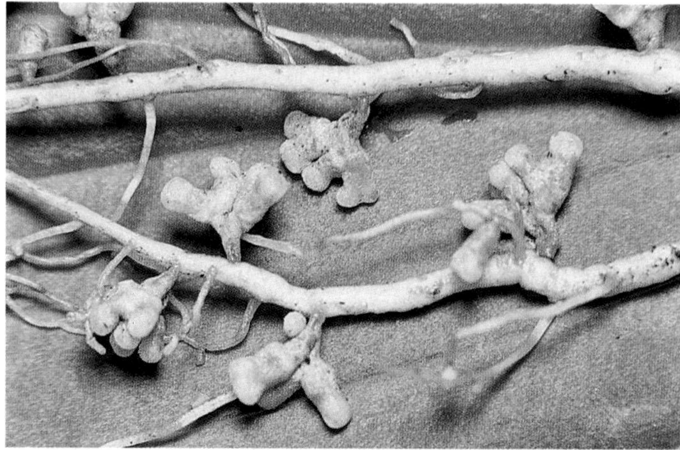

FIGURE 30.13 Symbiotic Nitrogen Fixation (a) The major steps leading to the formation of a root nodule in leguminous plants by *Rhizobium* species. **(b)** Root nodules.

There it is assimilated into amino acids for use by the plant. In return, the bacteria receive nutrients from the plant.

Although the relationship between the plant and bacterium is not obligate, it offers a distinct competitive advantage to both partners. The rhizobia do not fix nitrogen in soils lacking leguminous plants, and they compete poorly with other environmental organisms, slowly disappearing from soils in which leguminous plants are not grown. Likewise, leguminous plants compete poorly against other plants in heavily fertilized soils.

Several genera of non-leguminous trees, including alder and gingko, possess nitrogen-fixing root nodules at some stages of their life cycle. The bacteria involved in the symbiosis are members of the genus *Frankia.*

In aquatic environments, the most significant nitrogen-fixers are cyanobacteria. They are especially important in flooded soils such as rice paddies. In fact, rice has been cultivated successfully for centuries without the addition of nitrogen-containing fertilizer because of the symbiotic relationship between the cyanobacterium *Anabaena azollae* and the aquatic fern *Azolla.* The bacterium grows in specialized sacs in the leaves of the fern, providing nitrogen to the fern. Before planting rice, the farmer allows the flooded rice paddy to overgrow with *Azolla* ferns. Then, as the rice grows, it eventually crowds out the ferns. As the ferns die and decompose, their nitrogen is released into the water.

Microorganisms and Herbivores

Another mutualistic relationship occurs between microbes and certain herbivores. In order to subsist on grass and other plant material, herbivores such as cattle and horses rely on a community of microbes that inhabit a specialized digestive compartment. The microbes digest cellulose and hemicellulose, two of the major components of plant material, releasing compounds that can then serve as a nutrient source for the animal. In ruminants such as cattle, sheep, and deer, the digestive compartment, called the **rumen,** precedes the true stomach; in non-ruminant herbivores such as horses and rabbits, the **cecum,** which lies between the small intestine and the large intestine, serves a similar purpose.

■ cellulose, p. 32

The rumen functions as an anaerobic fermentation vessel to which nutrients in the form of plant materials are intermittently added. In some cases the animal produces over 150 liters of saliva per day. In addition to providing water, the saliva also contains bicarbonate, a buffer, which helps maintain the pH. A remarkably complex variety of microorganisms degrade the ingested material, liberating sugars that are then fermented, producing various organic acids. Each milliliter of rumen content contains approximately 10^{10} bacteria, 10^6 protozoa, and 10^3 fungi. Of the over 200 species identified in the rumen, no single one accounts for more than 3% of the total flora. The organic acids released during fermentation are absorbed by the cells that line the rumen, providing the animal with an energy and nutrient source. Copious quantities of gas are produced as a result of fermentation, which are discharged when the animal belches.

The contents of the rumen then enter another compartment (the omasum), eventually reaching the acidic true stomach (abomasum). There, more organic acids are absorbed. In addition, lysozyme is secreted, enabling the animal to lyse and then digest members of the microbial population, providing even more nutrients. A critical aspect of the anatomy of ruminants is that the microbial population gets the first opportunity to use the nutrients that are ingested. The animal then uses the end products of microbial metabolism as well as the microbial cells themselves.

The cecum of non-ruminant herbivores serves a similar function as a rumen, but because it follows the stomach, there is no anatomical mechanism to utilize the microbes as a food source. The benefit of this arrangement, however, is that the animal can digest and absorb readily available nutrients without competition from microbes.

MICROCHECK 30.5

Mycorrhizal fungi gain nutrients from plant root secretions while helping plants take up substances from soil. Symbiotic nitrogen-fixers provide plants with a source of usable nitrogen while being provided with an exclusive habitat. Microorganisms in the rumen and cecum of herbivores digest cellulose and hemicellulose, enabling the animal to subsist on plant material.

- ✓ Describe the differences between endomycorrhizae and ectomycorrhizae.
- ✓ Describe the differences between a rumen and a cecum.
- ✓ Gardeners sometimes plant clover between productive growing seasons. Why would this practice be beneficial?

SUMMARY

30.1 Principles of Microbial Ecology

Within the biosphere, ecosystems vary in their biodiversity and biomass. The microenvironment immediately surrounding a microorganism is most relevant to its survival and growth.

Nutrient Acquisition

Primary producers convert carbon dioxide into organic material; **consumers** utilize the organic materials, directly or indirectly, produced by plants; **decomposers** digest the remains of primary producers and consumers (figure 30.1).

Bacteria in Low-Nutrient Environments

Microorganisms capable of growing in dilute aqueous solutions are common in nature; often they grow in biofilms.

Microbial Competition and Antagonism

Microorganisms in the environment vie for the same limited pool of nutrients. A species can competitively exclude others, or produce compounds that inhibit others.

Microorganisms and Environmental Changes

External and internal sources of environmental fluctuations are common. As a result, different species can become dominant (figure 30.3).

Microbial Communities

A **microbial mat** is a thick, dense, highly organized biofilm composed of distinct layers of different groups of microbes, often green, pink, and black (figure 30.4).

Studying Microbial Ecology

Microbial ecology has been difficult to study because so few environmental prokaryotes can be successfully grown in the laboratory. Molecular techniques, including **fluorescence *in situ* hybridization (FISH),** polymerase chain reaction (PCR), **denaturing gradient gel electrophoresis (DGGE),** and DNA sequencing are enabling researchers to better understand complex microbial communities.

30.2 Aquatic Habitats

Oligotrophic waters are nutrient poor; **eutrophic** waters are nutrient rich (figure 30.5). Excessive growth of aerobic heterotrophs may cause an aquatic environment to become **hypoxic,** resulting in the death of fish and other aquatic animals.

Marine Environments

Ocean waters are generally oligotrophic and aerobic, but inshore areas can be dramatically affected by nutrient-rich runoff (figure 30.6).

Freshwater Environments

Oligotrophic lakes may have anaerobic layers due to **thermal stratification.** Shallow, turbulent streams are generally aerobic.

Specialized Aquatic Environments

Salt lakes and mineral-rich springs support the growth of microbes specifically adapted to thrive in these specialized environments.

30.3 Terrestrial Habitats

Characteristics of Soil

Soil represents an environment that can fluctuate abruptly and dramatically.

Microorganisms in Soil

The environmental conditions affect the density and composition of the flora of the soil.

The Rhizosphere

The concentration of microbes in the **rhizosphere** is generally much higher than that of the surrounding soil.

30.4 Biogeochemical Cycling and Energy Flow

Organisms use elements to produce biomass, as sources of energy, and as terminal electron acceptors.

Carbon Cycle (figure 30.7)

One of the fundamental aspects of the carbon cycle is carbon fixation. As consumers and decomposers degrade organic material, respiration and some fermentations release CO_2.

Nitrogen Cycle (figure 30.9)

The steps of the nitrogen cycle include **nitrogen fixation, ammonification, nitrification, denitrification,** and **anammox.**

Sulfur Cycle (figure 30.10)

Certain steps of the sulfur cycle—sulfur reduction and sulfate oxidation—depend on the activities of prokaryotes.

Phosphorus Cycle and Other Cycles

Most plants and microorganisms take up orthophosphate, assimilating it into biomass. Iron, calcium, zinc, manganese, cobalt, and mercury are recycled by microorganisms.

Energy Sources for Ecosystems

Photosynthetic organisms convert radiant energy to chemical bond energy in the form of organic compounds. Chemolithoautotrophs harvest energy from reduced inorganic chemicals (figure 30.11).

30.5 Mutualistic Relationships between Microorganisms and Eukaryotes

Mycorrhizae

Mycorrhizae help plants take up phosphorus and other substances from soil; in turn the fungal partners gain nutrients for their own growth (figure 30.12). **Endomycorrhizal** fungi penetrate root cells; **ectomycorrhizal** fungi grow around root cells.

Symbiotic Nitrogen-Fixers and Plants

Rhizobia fix nitrogen in nodules of leguminous plants (figure 30.13). *Frankia* species fix nitrogen in nodules of alder and gingko. A species of cyanobacteria fixes nitrogen in specialized sacs in the leaves of the *Azolla* fern.

Microorganisms and Herbivores

In order to subsist on grass and other plant material, herbivores rely on a community of microbes that inhabit a specialized digestive compartment, either a **rumen** or a **cecum.**

REVIEW QUESTIONS

Short Answer

1. Describe why a microbial mat has green, pink, and black layers.
2. Why do lakes in temperate regions stratify during the summer months?
3. Why is there a high concentration of microbes in the rhizosphere?
4. What dictates whether a form of an element is suitable for use as an energy source versus a terminal electron acceptor?
5. Why does wood resting at the bottom of a bog resist decay?
6. What is the importance of nitrogen fixation?
7. Describe the relationship between ammonia oxidizers and nitrite oxidizers.
8. How do hydrothermal vents support a thriving community of microbes, clams, and tube worms?
9. Give examples of free-living and symbiotic nitrogen-fixing microorganisms. Are these prokaryotic or eukaryotic?
10. Describe the steps that lead to the formation of the symbiotic relationship between rhizobia and leguminous plants.

Multiple Choice

1. Cyanobacteria are
   a) primary producers. b) consumers.
   c) herbivores. d) decomposers.
   e) more than one of the above.
2. Which of the following is *false?*
   a) Culture techniques provide an accurate way of determining the predominant members of a microbial community such as a biofilm.
   b) Fluorescence *in situ* hybridization (FISH) can be used to distinguish subsets of prokaryotes that contain a specific nucleotide sequence.
   c) Polymerase chain reaction (PCR) can be used to distinguish subsets of prokaryotes based on their 16S rRNA sequences.

d) Denaturing gradient gel electrophoresis (DGGE) can be used to separate PCR products.
e) Studying the genome of one organism can give insights into the characteristics of another.

3. The decomposition of organic matter
   a) is carried out by only a few bacterial species.
   b) produces oxygen.
   c) involves all the biogeochemical cycles discussed.
   d) involves primarily photosynthesis.
   e) is largely symbiotic.
4. Which of the following pairs that relate to aquatic environments does not match?
   a) Oligotrophic—nutrient poor
   b) Hypoxic—oxygen poor
   c) Hypolimnion—lower layer
   d) Epilimnion—oxygen poor
   e) Eutrophic—nutrient rich
5. Which of the following pairs that relate to terrestrial environments does not match?
   a) Soil—minimal biodiversity
   b) *Bacillus*—endospores
   c) *Streptomyces*—geosmin production
   d) Fungi—lignin degradation
   e) Rhizosphere—soil that adheres to plant root
6. Atmospheric nitrogen can be used
   a) directly by all living organisms.
   b) only by aerobic bacteria.
   c) only by anaerobic bacteria.
   d) in symbiotic relationships between rhizobia and plants.
   e) in photosynthesis.
7. Which process converts ammonium (NH_4^+) into nitrate (NO_3^-)?
   a) Nitrogen fixation b) Ammonification c) Nitrification
   d) Denitrification e) Anammox
8. Energy for ecosystems can come from
   a) sunlight via photosynthesis.
   b) oxidation of reduced inorganic chemicals by chemoautotrophs.
   c) both A and B.
9. Mycorrhizae represent associations between plant roots and microorganisms that
   a) are antagonistic.
   b) help plants take up phosphorus and other nutrients from soil.
   c) involve algae in the association with plant roots.
   d) form nodules on the plant's leaves.
   e) lead to the production of antibiotics.
10. In symbiotic nitrogen fixation by rhizobia and legumes
   a) the amount of nitrogen fixed is much greater than by non-symbiotic organisms.
   b) neither the bacteria nor the legume can exist independently.
   c) the bacteria enter the leaves of the legume.
   d) the bacteria operate independently of the legume.

Applications

1. A farmer who was growing soybeans, a type of legume, saw an Internet site advertising an agricultural product for safely killing soil bacteria. The ad claimed that soil bacteria were responsible for most crop losses. The farmer called the agricultural extension office at a local university for advice. Explain what the extension office crop adviser most likely told the farmer about the usefulness of the product.
2. Recent reports suggest that human activities, such as the generous use of nitrogen fertilizers, have doubled the rate at which elemental nitrogen is fixed, raising concerns of environmental overload of nitrogen. What problems could arise from too much fixed nitrogen, and what could be done about this situation?

Critical Thinking

1. Each colony growing on an agar plate arises from a single cell (see photo). Colonies growing close together are much smaller than those that are well separated. Why would this be so?
2. An entrepreneur found an economically feasible way of collecting large amounts of sulfur from underwater hot vents in the Pacific Ocean. The sulfur will be harvested from the microorganisms found in the vent areas. A group of ecologists argued that the project would destroy the fragile ecosystem by depleting it of usable sulfur. The entrepreneur argued that the environment would not be harmed because the vents produce an unlimited source of sulfur for the clams and tube worms in the area. Explain who is correct.

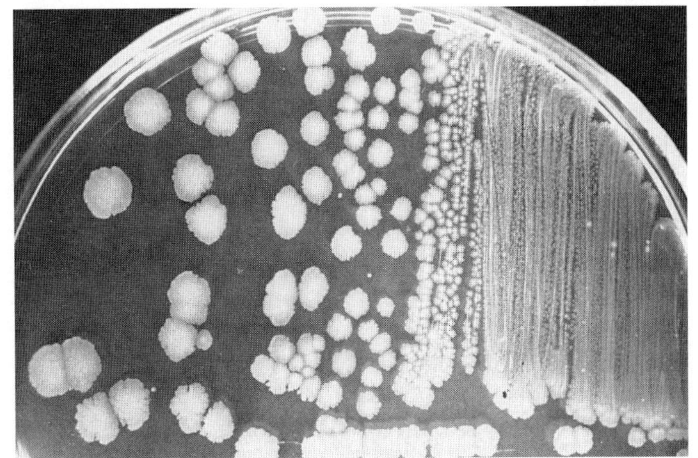

A pristine mountain lake.

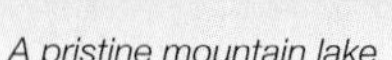

31

Environmental Microbiology:

Treatment of Water, Wastes, and Polluted Habitats

A Glimpse of History

Delivering fresh water to urban areas and removing human wastes have been practiced at least since Roman times. The ruins of aqueducts that delivered fresh water long distances can be seen today in many parts of Europe. Ridding cities of human wastes has been more difficult, and the sewers that were used until the mid-nineteenth century were not much more than large, open cesspools.

Long before the discovery of the microbial world, it was recognized that some diseases are associated with water supplies. As early as 330 B.C., Alexander the Great had his armies boil their drinking water, a habit that probably contributed to his huge successes. Certainly, many battles have been lost over the years as a result of waterborne diseases that decimated the combatants. Years before the cholera-causing *Vibrio cholerae* was identified, it was obvious that cholera epidemics were associated with drinking water. The desire for clean, clear water led to the use of a sand filtration system in London and elsewhere in the early nineteenth century. Late in that century, Robert Koch showed that not only did this kind of filtration yield clear water, it also removed more than 98% of bacteria from the water.

As early as the 1840s, Edwin Chadwick, an English activist, championed a new idea on how wastes could be removed. His idea was to construct a system of narrow, smooth ceramic pipes through which water could be flushed along with solid waste materials. This system would carry the waste materials away from the inhabited part of the city to a distant collection site. There, he hoped to collect the waste materials and turn them into fertilizer to sell to farmers. The system he envisioned required the installation of new water and sewer pipes along with pumps to deliver water under pressure to houses. With the water under pressure and smooth narrow pipes, the system could be kept well flushed.

In 1848, with the threat of a cholera epidemic imminent, the Board of Health in England instituted widespread reforms and began installing a sewage system along the lines envisioned by Chadwick. New York City did not establish its Board of Health and a proper, sewage disposal system until 1866, again in response to a threatened cholera epidemic. By the end of the nineteenth century, most large European and U.S. cities had established water-sewer systems to deliver safe drinking water and remove and treat waste materials. Cholera in the industrialized nations of Europe and North America virtually disappeared.

Most people living in developed countries take for granted that their tap water is safe to drink, their wastes will reliably disappear into sewers or landfills for proper disposal, and **pollutants,** substances that are harmful or injurious, will not accumulate in the environment. They seldom consider the role that microbes play in these essential aspects of modern life.

Microorganisms are important in the treatment of water, waste, and polluted environments for two very distinct reasons. First, we benefit from the fact that microbes are the ultimate recyclers, playing an essential role in the decomposition of our wastes. At the same time, pathogenic microorganisms and viruses must be eliminated from sewage before it is discharged, and removed from drinking water before it is deemed **potable,** or safe for human consumption. Recreational waters such as swimming pools, water parks, lakes, rivers, and shorelines are also monitored to ensure they do not contain or accumulate harmful levels of certain pathogens.

Treatment of water, waste, and polluted habitats is a formidable challenge, particularly in densely populated areas. Consider that every day the average American uses about 150 gallons of water, and produces 120 gallons of wastewater and 5 pounds of trash. This means that a city with only 1 million inhabitants is faced with the disposal of approximately 44 billion gallons of wastewater and a million tons of trash each year!

KEY TERMS

Advanced Treatment Any physical, chemical, or biological purification process beyond secondary treatment of wastewater.

Anaerobic Digestion Process that uses anaerobic microbes to degrade the sludge obtained during wastewater treatment.

Biochemical Oxygen Demand (BOD) The amount of O_2 required for the microbial decomposition of organic matter in a sample.

Bioremediation Process that uses microorganisms to degrade harmful chemicals.

Effluent The liquid portion of treated wastewater.

Indicator Organisms Microbes commonly found in the intestinal tract whose presence in other environments suggests fecal contamination.

Primary Treatment A physical wastewater treatment process designed to remove materials that will settle out of sewage.

Sanitary Landfill A site used for disposal of non-hazardous solid wastes in a manner that minimizes damage to human health and the environment.

Secondary Treatment A biological wastewater treatment process in which microbial growth is actively encouraged, allowing the microbes to convert most of the suspended solids to inorganic compounds and removable cell mass.

Septic Tank A large tank used for individual wastewater treatment systems.

Sludge The solid portion of wastewater that settles to the bottom of sedimentation tanks during primary and secondary treatment.

Wastewater (Sewage) Material that flows from household plumbing systems; municipal wastewater also include business and industrial wastes and storm water runoff.

31.1 Microbiology of Wastewater Treatment

Focus Points

- Describe the concept of BOD.
- Compare and contrast primary treatment, secondary treatment, advanced treatment, and anaerobic digestion.
- Describe how a septic tank functions.

Wastewater, or **sewage,** is composed of all the material that flows from household plumbing systems, including washing and bathing water and toilet wastes. Municipal wastewater also includes business and industrial wastes. In many cities, storm water runoff that flows into street drains enters the system as well.

The most obvious reason that wastewater must be treated before discharge is that pathogenic microbes can be transmitted in feces, including those that cause diarrheal diseases and hepatitis. If untreated sewage is released into a river or lake that is then used as a source of drinking water, disease can easily spread. In a similar manner, if marine waters become contaminated with untreated sewage, consumption of the shellfish grown there can result in disease. Shellfish are filter feeders and they concentrate microbes from the waters in which they live.

A less obvious problem is the impact that the high nutrient content of wastewater has on the receiving water. When any nutrient-rich substance is added to an aqueous environment, microorganisms quickly utilize the compounds as energy sources, employing metabolic pathways such as glycolysis and the TCA cycle (see figure 6.8). As a result, microbes that use aerobic respiration consume available O_2 in the water, using it as a terminal electron acceptor (see figure 6.18). The amount of dissolved O_2 in lakes and rivers is limited and can easily be depleted during the microbial breakdown of nutrients. Fish and other aquatic animals in the environment die because they require O_2 for respiration (see figure 30.6). Thus, effective treatment of wastewater must decrease the level of organic compounds substantially, in addition to eliminating pathogens, toxic materials, and other pollutants. ■ aerobic respiration, pp. 133, 144

Biochemical Oxygen Demand (BOD)

An important goal of wastewater treatment is to lessen the environmental impact of sewage by reducing the **biochemical oxygen demand (BOD),** the amount of O_2 required for the microbial decomposition of organic matter in a given sample. The BOD is roughly proportional to the amount of degradable organic material present in a sample. To determine the BOD, the O_2 level in a well-aerated sample of microbe-containing test water is first measured. The sample is then incubated in a sealed container in the dark under standard conditions of time and temperature, usually 5 days at 20°C. The O_2 level is then determined again. The difference between the dissolved O_2 at the beginning of the test and at the end reflects the BOD of the sample. In many cases the sample must be diluted first in order to accurately determine the BOD. High BOD values indicate that large amounts of degradable materials were present in the test water, resulting in correspondingly large amounts of O_2 being used during its biological degradation. The BOD of raw sewage is approximately 300 to 400 mg/liter, which could easily deplete the dissolved O_2 in the receiving water. The dissolved O_2 content of natural waters is generally 5 to 10 mg/liter.

Municipal Wastewater Treatment Methods

Large-scale wastewater treatment plants in the United States use a series of two processes—primary and secondary treatment—as mandated by the 1972 Federal Water Pollution Control Act, now known as the Clean Water Act. Once treated, the liquid portion, or **effluent,** can be discharged into a body of water, the **receiving water.** The solid portion, or **sludge,** is further treated in an **anaerobic digester.**

Primary Treatment

Primary treatment is a physical process designed to remove materials that will settle out, removing approximately 50% of the solids and 25% of the BOD. Raw sewage is first passed through a series of screens to remove large objects such as sticks, rags, and trash **(figure 31.1).** Skimmers then remove scum and other floating materials. The sewage is allowed to settle in a sedimentation tank, facilitating removal of the solids. Once the settling period

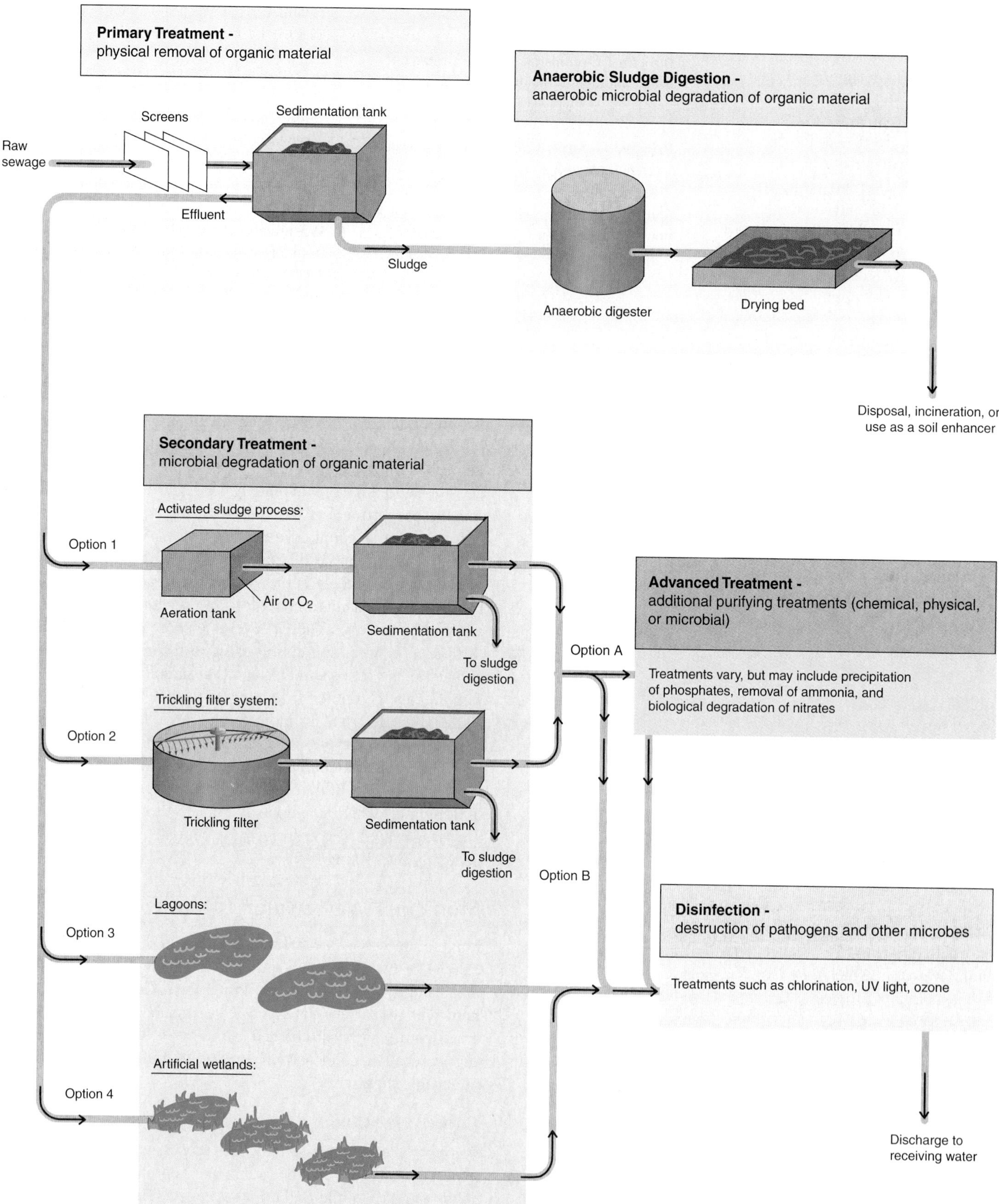

FIGURE 31.1 Municipal Wastewater Treatment The processes consist of primary treatment, secondary treatment, advanced treatment (optional), disinfection, and anaerobic sludge digestion.

is complete, the sludge is removed and the primary treated wastewater that remains is sent for secondary treatment.

Secondary Treatment

Secondary treatment is chiefly a biological process designed to convert most of the suspended solids in sewage to inorganic compounds and cell mass that can then be removed, eliminating as much as 95% of the BOD. Microbial growth is actively encouraged during secondary treatment, allowing aerobic organisms to oxidize the biologically degradable organic material to CO_2 and H_2O. Note that because secondary treatment relies on the metabolic activities of microorganisms, the processes could be devastated if too much toxic industrial wastes or hazardous household materials were dumped into sewage systems, killing the microbial population. Methods used for secondary treatment of wastewater include:

- **Activated sludge process.** This common system employs mixed aerobic microbes that are adapted to utilize the nutrients available in sewage and grow as suspended biofilms, or **flocs.** Although the organisms are often naturally present in sewage, large numbers are inoculated into the wastes by introducing a small portion of leftover sludge from the previous load of treated wastes. An abundance of O_2 is supplied by mixing the sewage in an aerator. As the microbes proliferate, the organic matter is converted into both biomass and waste products such as CO_2. Following the aeration, the wastewater is again sent to a sedimentation tank. There, most of the flocs settle and the resulting sludge is removed; a portion of this sludge is introduced to a new load of wastewater to act as an inoculum. A complication of the activated sludge process occurs when filamentous bacteria such as *Thiothrix* species overgrow in the wastewater during treatment, creating a buoyant mass that does not settle. This problem, **bulking,** interferes with the separation of the solid sludge from the liquid effluent. ■ biofilm, p. 85 ■ *Thiothrix,* p. 259
- **Trickling filter system.** This method is frequently used for smaller wastewater treatment plants. A rotating arm sprays sewage over a bed of plastic or coarse gravel and rocks, the surfaces of which become coated with a biofilm of microbes that aerobically degrade the organic materials **(figure 31.2).** The film consists of a heterogeneous mix of bacteria, fungi, algae, protozoa, and nematodes. The rate of sewage flow can be adjusted so that waste materials are maximally degraded.
- **Lagoons.** The wastewater is channeled into shallow ponds, or lagoons, where it remains for several days to a month or more, depending on the design of the lagoon. Algae and cyanobacteria that grow at the surface provide O_2, enabling aerobic organisms in the ponds to degrade the sewage.
- **Artificial wetlands.** These employ the same principles as lagoons, but their more advanced designs not only offer a means to treat wastewater, but also provide a habitat for birds and other wildlife **(figure 31.3).** For example, the wastewater treatment processes in Arcata, California, use a series of marshes that now attract a variety of shorebirds and serve as a wildlife sanctuary.

Advanced Treatment

As water supplies are becoming scarce, and in order to comply with discharge standards designed to protect surface groundwaters, many communities are finding **advanced treatment** necessary. This encompasses any purification process beyond secondary treatment; it may involve physical, chemical, or biological processes, or any combination of these. Advanced treatment is expensive, however, and has not been common in the past.

Often, advanced treatment is designed to remove ammonia, nitrates, and phosphates—compounds that foster the growth of algae

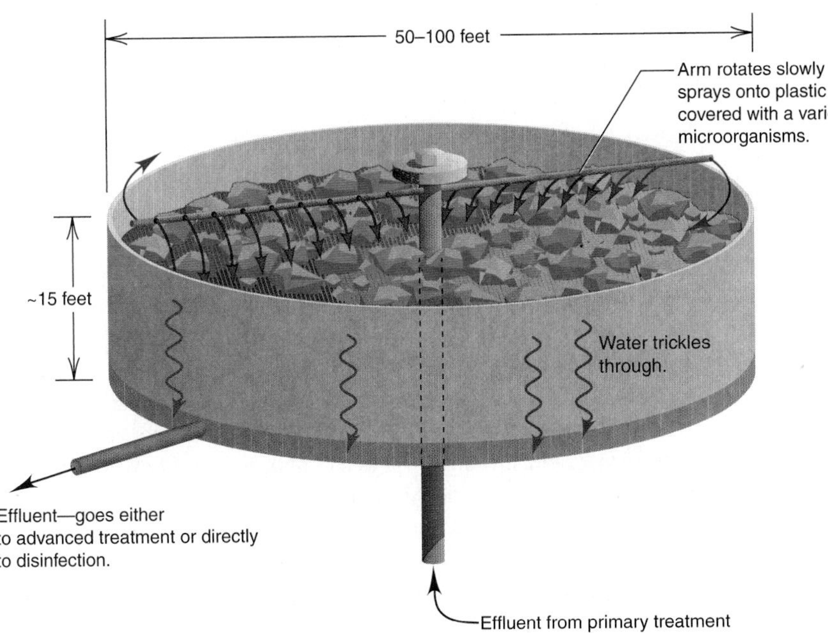
50–100 feet
Arm rotates slowly as effluent sprays onto plastic or rocks covered with a variety of microorganisms.
~15 feet
Water trickles through.
Effluent—goes either to advanced treatment or directly to disinfection.
Effluent from primary treatment

FIGURE 31.2 Trickling Filter Wastewater is channeled into the revolving arm, and then trickles through holes in the bottom of the arm onto plastic or a gravel and rock bed. The surfaces are coated with a biofilm of microbes that aerobically degrade the organic materials as they trickle through the bed. The effluent has a greatly reduced BOD.

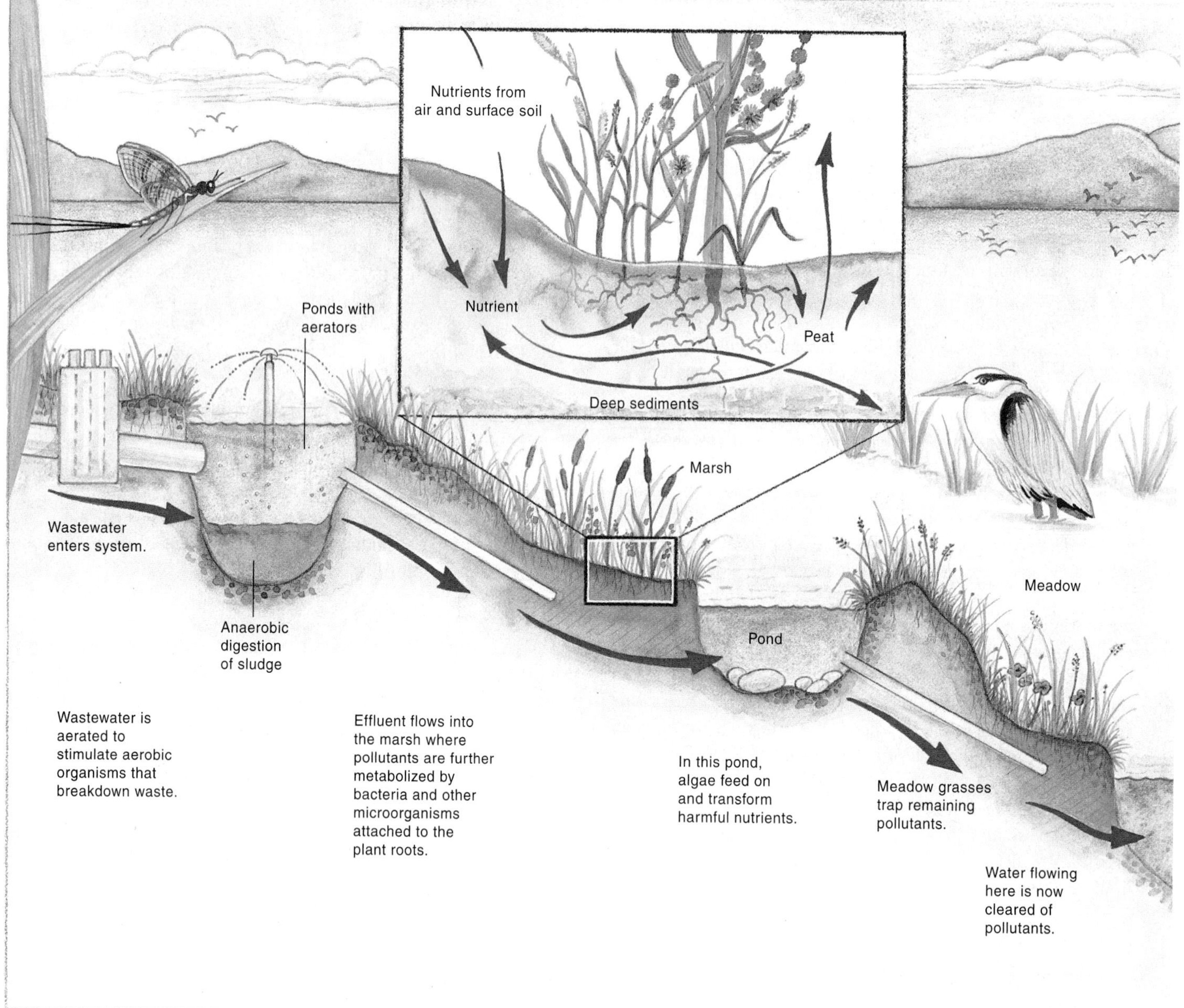

FIGURE 31.3 Artificial Wetland

and cyanobacteria in the receiving waters. The concentrations of these nutrients, which are very low in unpolluted waters, normally limit the growth of the photosynthetic organisms. Consequently, if the nutrients are added, photosynthetic organisms proliferate, often leading to a surface scum composed of buoyant masses of cells (see figure 11.7). In addition, the photosynthetic organisms provide a source of carbon for other microbes, increasing the BOD and, consequently, threatening other forms of aquatic life.

Ammonia is removed by a process called ammonia stripping, which liberates gaseous ammonia from the water. Nitrates can be removed by exploiting the activities of denitrifying bacteria. These organisms use nitrate as a terminal electron acceptor during anaerobic respiration, forming nitrogen gas. This gas is inert, non-toxic and easily removed. Phosphates are removed using chemicals that combine with phosphates, causing them to precipitate. ■ **denitrification, p. 731**

Disinfection

Before discharge into the receiving water, the effluent is disinfected with chlorine, ozone, or UV light to decrease the numbers of microorganisms and viruses. If chlorine is used, the disinfected water can then be dechlorinated to avoid releasing excessive amounts of the toxic chemical into the environment. ■ **chlorine, p. 119** ■ **ozone, p. 120** ■ **UV light, p. 116**

Anaerobic Digestion

Within the anaerobic digester, microorganisms act on the solids removed during the sedimentation steps of primary

PERSPECTIVE 31.1

Now They're Cooking with Gas

In rural areas of China, fuel for cooking and heating is often in short supply; as a consequence, the hay that should go for animal feed must be used for fuel. To alleviate this problem, many of the farmers build methane-producing tanks on their farms **(figure 1).** Near the farmhouse is an underground cement tank connected to the latrine and pigpen. Human and animal wastes along with water and some other organic materials such as straw are added to the tanks. As the natural process of fermentation occurs, methane gas (CH_4) is produced. This gas rises to the top of the tank and is connected to the house with a hose. Enough gas is produced to provide lights and cooking fuel for the farm family. By producing its own gas, a family also saves money because it does not need to buy coal for cooking.

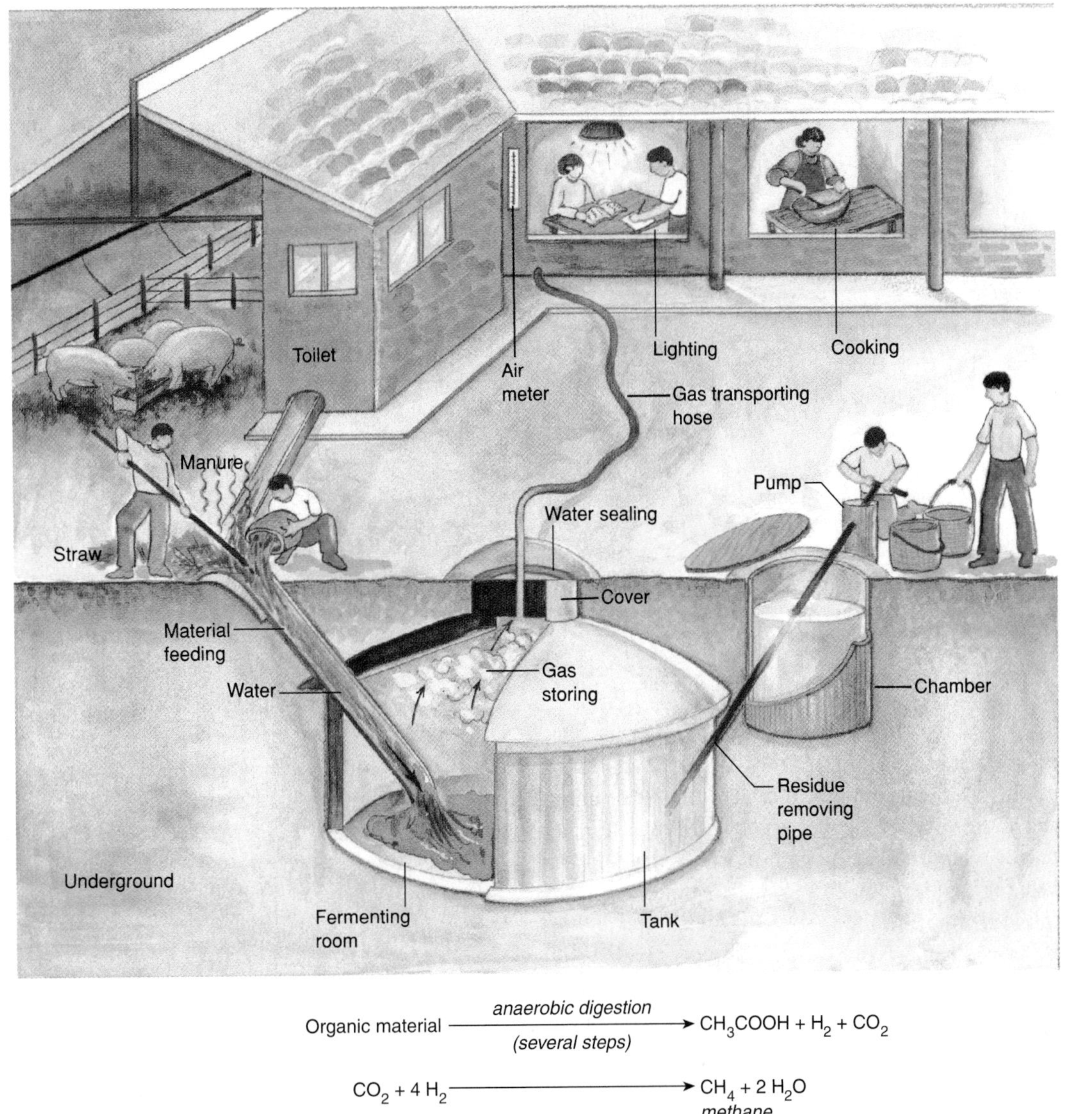

$$\text{Organic material} \xrightarrow[\text{(several steps)}]{\text{anaerobic digestion}} CH_3COOH + H_2 + CO_2$$

$$CO_2 + 4\,H_2 \longrightarrow \underset{\text{methane}}{CH_4} + 2\,H_2O$$

$$CH_3COOH \longrightarrow CH_4 + CO_2$$

and secondary treatment. Various anaerobic populations act sequentially, ultimately converting much of the organic material to methane:

- Organic compounds ⟶ organic acids, CO_2, H_2
- Organic acids ⟶ acetate, CO_2, H_2
- Acetate, CO_2, H_2 ⟶ methane (CH_4)

Many wastewater treatment plants are equipped to use the methane generated, thereby avoiding the cost of other sources of energy to run their equipment.

After anaerobic digestion, water is removed from the remaining sludge, generating a nutrient-rich product called **stabilized sludge.** This can be incinerated or disposed of in landfills, but may also be used to improve soils and promote plant growth. The sludge generated by the city of Milwaukee, Wisconsin, is used to produce Milorganite®, a fertilizer for lawns, gardens, golf courses, and playfields. An increasing number of wastewater treatment facilities are finding similar ways to recycle their treated sludge. Concerns exist, however, about heavy metals and other pollutants that can sometimes be concentrated in the product.

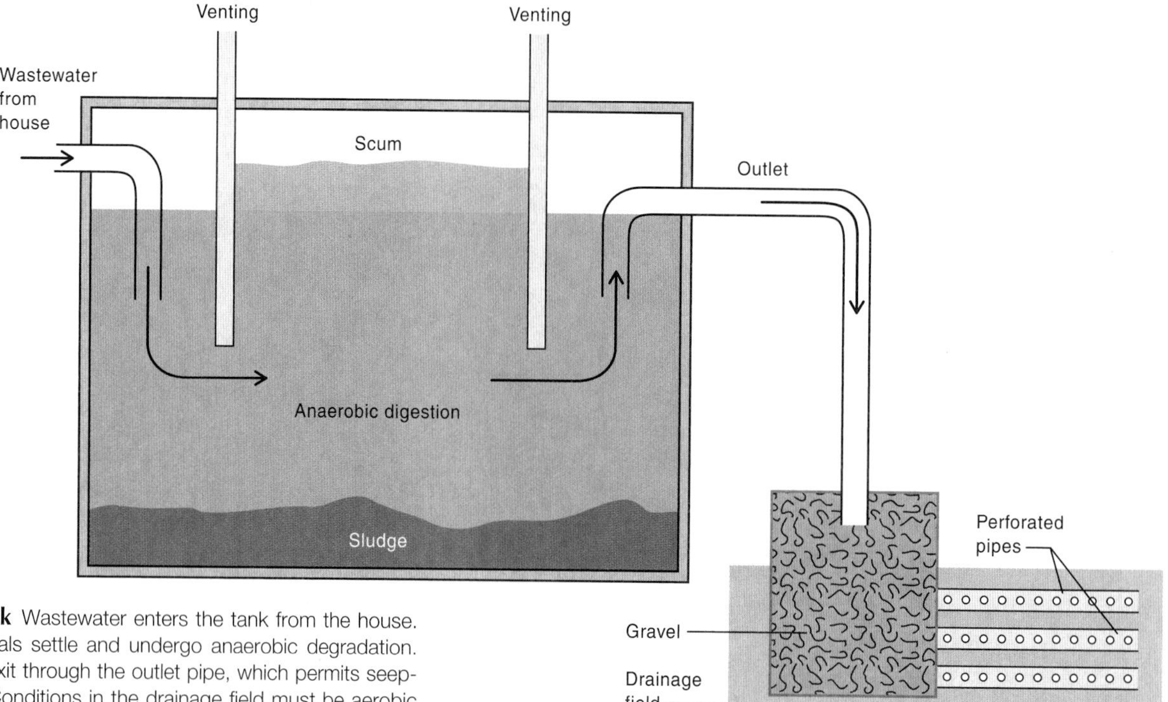

FIGURE 31.4 Septic Tank Wastewater enters the tank from the house. Within the tank, solid materials settle and undergo anaerobic degradation. Materials that do not settle exit through the outlet pipe, which permits seepage into the drainage field. Conditions in the drainage field must be aerobic so that the materials remaining can be degraded by the activities of aerobic microorganisms. If the drainage area is not properly designed, contaminated materials readily enter adjacent waters.

Individual Wastewater Treatment Systems

Rural dwellings customarily rely on **septic tanks** for wastewater disposal. In theory, the septic tank makes sense; in practice, however, it often does not work correctly. Wastewater is collected in a large tank in which much of the solid material settles and is degraded by anaerobic microorganisms **(figure 31.4).** The fluid overflow from the tank has a high BOD and must be passed through a drainage field of sand and gravel designed to allow oxidation of the organic material in the same manner described for the trickling filter. The process, however, depends on adequate aeration and sufficient action by aerobic organisms in the drainage field. Unfortunately, certain conditions, such as a clay soil under a drainage field, can prevent adequate drainage, allowing anaerobic conditions to develop. Toxic materials can inhibit microbial activity in the drainage field. Drainage from a septic tank may contain pathogens; therefore, the tank must never be allowed to drain where it can contaminate water supplies.

MICROCHECK 31.1

Primary treatment is a physical process that removes material that will settle out. Secondary treatment is chiefly a biological process that converts most of the suspended material into inorganic compounds and microbial biomass. Advanced treatment is often designed to remove ammonia, phosphates, and nitrates. Anaerobic digestion converts much of the organic matter to methane. Septic tanks are individual wastewater treatment systems.

✓ What does the term BOD mean? What is its significance?

✓ What is the advantage of removing phosphates and nitrates from wastewater?

✓ Why would the processes of secondary treatment preclude denitrification?

31.2 Drinking Water Treatment and Testing

Focus Points

- Describe how water is typically treated to make it safe for drinking.
- Describe the importance of indicator organisms, and explain how total coliforms are detected.

Large cities generally obtain their drinking water from **surface waters** such as lakes or rivers. Because surface water may serve as the receiving water for another city's wastewater effluent, drinking water treatment is intimately connected to wastewater treatment. The quality of the surface water is also affected by the characteristics of the **watershed**—the land over which water flows into the river or lake. Even pristine rivers are likely contaminated with feces of animals that inhabit the watershed.

Smaller communities often use **groundwater,** pumped from a well, as a source of drinking water **(figure 31.5).** This water occurs in **aquifers**—water-containing layers of rock, sand, and gravel—that is replenished as water from rain and other sources seeps through the soil. Because aquifers are not directly exposed to rain, animals, and the atmosphere, they are somewhat protected from contamination. However, poorly located or maintained septic tanks and sewer lines, as well as sludge or other fertilizers, can lead to groundwater contamination.

Public water systems in the United States are regulated under the Safe Drinking Water Act of 1974, amended in 1986 and 1996.

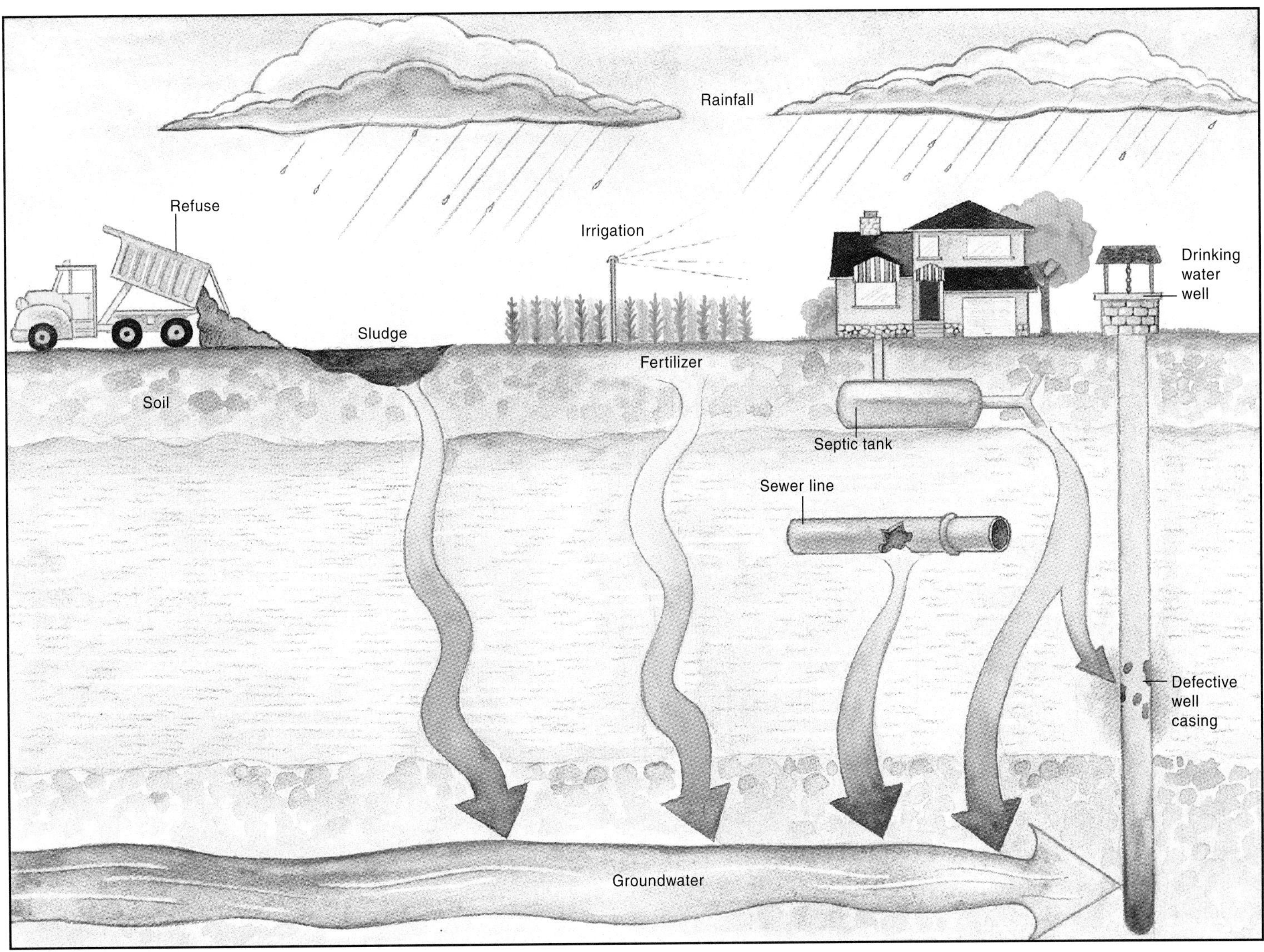

FIGURE 31.5 Groundwater The water in the aquifer is replenished as water from various sources seeps through the soil. Poorly located or maintained septic tanks and sewer lines, as well as sludge or other fertilizers, can lead to groundwater contamination.

This gives the Environmental Protection Agency (EPA) the authority to set drinking water standards in order to control the level of contaminants in drinking water. Standards are modified in response to new concerns; for example, regulations now govern the maximum levels of *Cryptosporidium* oocysts, *Giardia* cysts, and enteric viruses in drinking water. ■ ***Cryptosporidium*, p. 610** ■ ***Giardia*, p. 609**

Water Treatment Processes

The treatment of metropolitan water supplies is designed to eliminate pathogenic microbes as well as harmful chemicals **(figure 31.6).** First, water flows into a reservoir and is allowed to stand long enough for the particulate matter to settle. The water is then transferred to a tank where it is mixed with a flocculent chemical, such as aluminum potassium phosphate, or alum. This causes materials still suspended in the liquid to coagulate, forming aggregates that slowly sink to the bottom. As the clumps settle, they remove unwanted materials from the water, including some bacteria and viruses.

Following the flocculation, the water is filtered, often through a thick bed of sand and gravel, to remove various microorganisms including bacteria and protozoan cysts and oocysts. Organic chemicals that may be harmful or impart undesirable tastes and odors can be removed by additional filtration through an activated charcoal filter, which adsorbs dissolved chemicals. Not only does filtration physically remove various particles, but microorganisms growing in biofilms on the filter materials use carbon from the water as it passes. This lowers the organic carbon content of the water, resulting in less microbial growth in pipes delivering the water. ■ **filtration, p. 115**

Finally, the water is treated with chlorine or other disinfectants to kill harmful bacteria, protozoa, and viruses that might remain. A concern with using chlorine, however, is that some of the disinfection by-products might be carcinogenic. In response to this concern, ultraviolet irradiation and ozone are increasingly being used as alternatives, but a small amount of chlorine must still be added to prevent problems associated with post-treatment contamination. Note that disinfection of waters with a high organic content requires more chlorine because organic compounds consume free

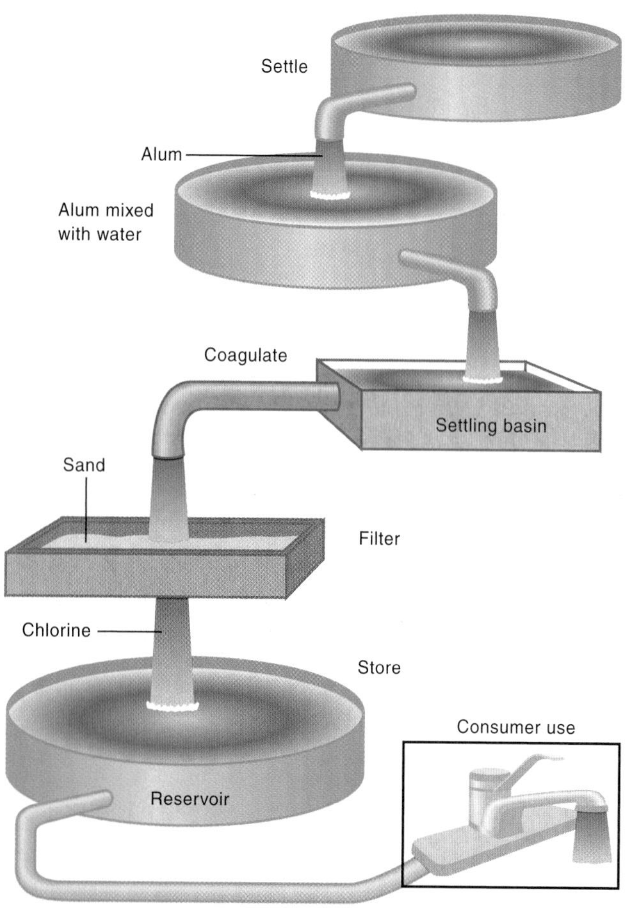
Settle
Alum
Alum mixed with water
Coagulate
Settling basin
Sand
Filter
Chlorine
Store
Consumer use
Reservoir
Step 1 Water is held in a series of reservoirs, where large materials sediment. Aluminum potassium sulfate (alum) may be added to cause flocculation of organic matter, which then settles out.
Step 2 The water is then filtered through beds of sand, which removes almost all of the bacteria. Filtration through activated charcoal may be used to remove toxic or objectionable organic materials.
Step 3 Finally, the water is disinfected to kill any pathogens that might remain. Chlorine, ultraviolet irradiation or ozone may be used for disinfection. Some cities also add fluoride to protect against dental cavities.

FIGURE 31.6 Steps in the Treatment of Municipal Water Supplies

chlorine. ■ chlorine, p. 119 ■ disinfection by-products, p. 110 ■ ultraviolet irradiation, p. 116 ■ ozone, p. 120

Water Testing

A primary concern regarding the safety of drinking water is the possibility that it might be contaminated with any of a wide variety of intestinal pathogens, such as those discussed in chapter 25. It is not feasible to test for all of the pathogens, however, so **indicator organisms** function as surrogates. These microbes are routinely found in feces, survive longer than intestinal pathogens, and are relatively easy to detect and enumerate. The most common group of bacteria used as indicator organisms in the United States is **total coliforms**—lactose-fermenting members of the family *Enterobacteriaceae,* including *E. coli.* The group is functionally defined as facultatively anaerobic, Gram-negative, rod-shaped, non-spore-forming bacteria that ferment lactose, forming acid and gas within 48 hours at 35°C. Although total coliforms are routinely present in the intestinal contents of warm-blooded animals, certain species can also thrive in soils and on plant material. Thus, the presence of these organisms does not necessarily imply fecal pollution. To compensate for this shortcoming, **fecal coliforms,** a subset of total coliforms more likely to be of intestinal origin, are also used as indicator organisms. The most common fecal coliform is *E. coli.* Note that although some strains of *E. coli* can cause intestinal disease, the organism is used in water testing merely to indicate fecal pollution.

Methods used to detect total coliforms in a water sample include:

- **ONPG/MUG test.** A water sample is added to a minimal medium containing ONPG (*o*-nitrophenyl-β-D-galactopyranoside) and MUG (4-methylumbelliferyl-β-D-glucuronide). Lactose-fermenting bacteria hydrolyze ONPG, yielding a yellow-colored compound; thus, all coliforms turn the medium yellow **(figure 31.7).** *E. coli* produces an enzyme that hydrolyzes MUG, generating a fluorescent compound. Because of ONPG and MUG, a sample can be assayed simultaneously for the presence of both total coliforms and *E. coli.*
- **Presence/absence test.** A 100-ml water sample is added to a lactose-containing broth that contains a vial to trap gas. If gas is produced, the broth is then tested to confirm the presence of coliforms.
- **The most probable number (MPN) method.** This statistical assay of cell numbers employs successive dilution of a water sample in broth similar to that used in the presence/absence test. Positive tubes from the MPN test (see figure 4.20) are further tested to confirm that the tubes actually contain coliforms.
- **Membrane filtration.** A water sample is passed through a filter that retains bacteria (see figure 4.19), concentrating the bacteria from a known volume of water. The filter is then placed on a lactose-containing selective and differential agar medium. ■ selective media, p. 96 ■ differential media, p. 97

The **total coliform rule** establishes a maximum number of positive samples (100 ml) permitted. That maximum relates to the number of samples routinely collected by the water system. The number collected ranges from 1 to 480 samples per month, depending on the size of the population served by the system. Systems that collect at least 40 samples per month are in violation if more than 5% are total coliform-positive in a month. Systems that collect fewer samples are in violation if more than one sample tests positive per month. If a sample tests positive, repeat samples within 24 hours are mandated. Total coliform-positive samples are also tested for either fecal coliforms or *E. coli.* If a water system exceeds the monthly total coliform limit, the system must notify the state and the public. Notification is also required if either of two sequential samples that test positive for total coliforms also test positive for fecal coliforms or *E. coli.*

Because of limitations of total coliform and fecal coliform assays in predicting contamination with protozoan cysts and oocysts, alternatives are being explored. Other microbes that can be used as indicators of fecal pollution include enterococci, some *Clostridium* species, and certain types of bacteriophages.

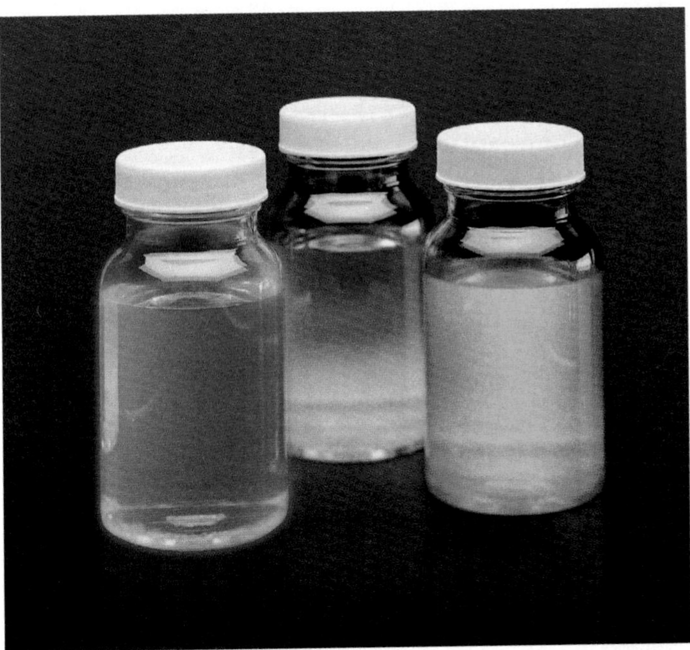

FIGURE 31.7 ONPG/MUG Test Coliforms hydrolyze ONPG, yielding a yellow-colored compound. *E. coli* hydrolyzes MUG, generating a blue fluorescent compound.

MICROCHECK 31.2

Drinking water may be obtained from surface water or groundwater. Treatment of drinking water is designed to eliminate pathogens and harmful chemicals. In the United States, total and fecal coliforms are the most commonly used indicator organisms.

✓ What is an aquifer?

✓ Describe two methods of water testing.

✓ Which would be more likely to cause illness—a water sample that tested positive for coliforms or one that tested positive for *E. coli* O157:H7?

31.3 Microbiology of Solid Waste Treatment

Focus Point

- Compare and contrast sanitary landfills and composting programs.

In addition to ridding our environment of wastes in water, we must dispose of the solid wastes (garbage) generated each day. Eliminating these has become an increasingly complex problem.

Sanitary Landfills for Solid Waste Disposal

Sanitary landfills are widely used to dispose of non-hazardous solid wastes in a manner that minimizes damage to human health and the environment. Before sanitary landfills were developed, solid wastes were often piled up on the ground in open-burning dumps, attracting insects and rodents and causing aesthetic and public health problems.

Federal standards dictate that sanitary landfills must be located away from wetlands, earthquake-prone faults, flood plains, or other sensitive areas. The excavated site is lined with plastic sheets or a special membrane atop a thick layer of clay, which minimizes leaching of contaminates from the wastes into the surrounding area. Next comes a layer of sand with drainage pipes; above this, the wastes are compacted and covered with a layer of soil every day. When a landfill is full, it is covered with soil and plants, and can be used for recreation and eventually as a site for construction. Methane and other gases are vented, and the methane is burned or recovered for use.

There are several disadvantages to this type of waste management. First, only a limited number of sites are available for use near urban and suburban areas. Second, the organic content of landfills anaerobically decomposes very slowly, over a period of at least 50 years. During this time, the methane gas produced must be removed. If buildings are constructed before the methane is removed, disastrous gas explosions can occur. Pollutants such as heavy metals and pesticides can leak from landfill sites into the underground aquifers. It is very difficult to purify these aquifers once they have become contaminated.

Sanitary landfills have traditionally been a low-cost method of handling large quantities of solid waste. Because of increased costs and decreased availability of land, however, many cities are looking for ways to decrease the amount of solid waste dumped in landfills. In some cities, the fees charged to people for garbage collection are based on the size of the container collected. The smaller the can, the lower the cost. This is intended to raise people's awareness of how much solid waste they are generating as well as offer an incentive to recycle. Programs to recycle paper, plastics, glass, and metal are being implemented in many cities and counties with great success. Through these programs, landfill areas can be expected to be available for a longer period of time.

Municipal and Backyard Composting—Alternative to Landfills

Composting is the natural decomposition of organic solid material. Municipal and home composting programs are becoming popular in many areas and are succeeding in reducing the amount of organic waste added to landfills.

Backyard composting involves mixing garden debris with kitchen organic waste, excluding meats and fats **(figure 31.8)**. If proper amounts of warmth, water, and air are provided to the mix, the bulk of the waste is reduced by two-thirds in a matter of months. The black organic material generated by composting can then be used to improve the soil in garden beds.

A backyard compost pile usually starts with a supply of organic material such as leaves and grass clippings, as well as kitchen wastes. Often, some soil and water are added. Within a few days, microbial metabolism causes the inside of the pile to heat up. At 55°C to 66°C, pathogens are killed but thermophilic organisms are not affected. If the pile is frequently aerated, which

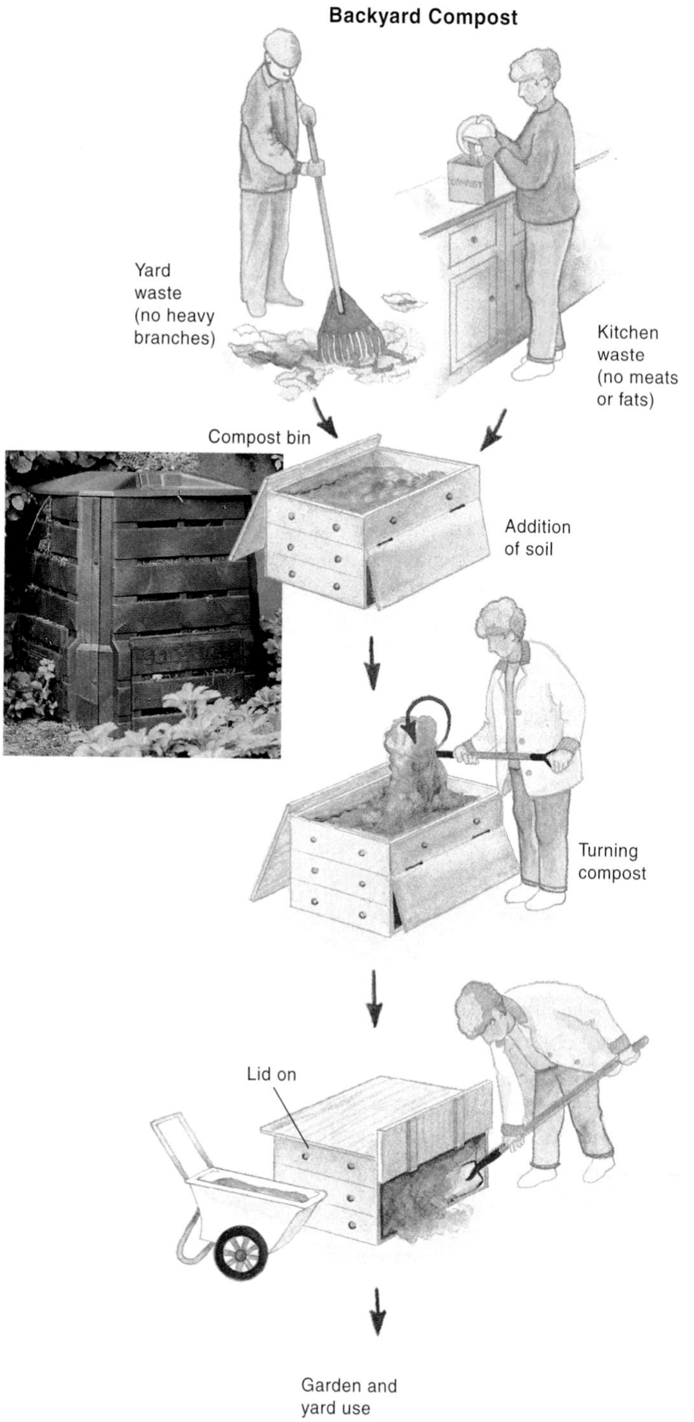

FIGURE 31.8 Backyard Compost Heap

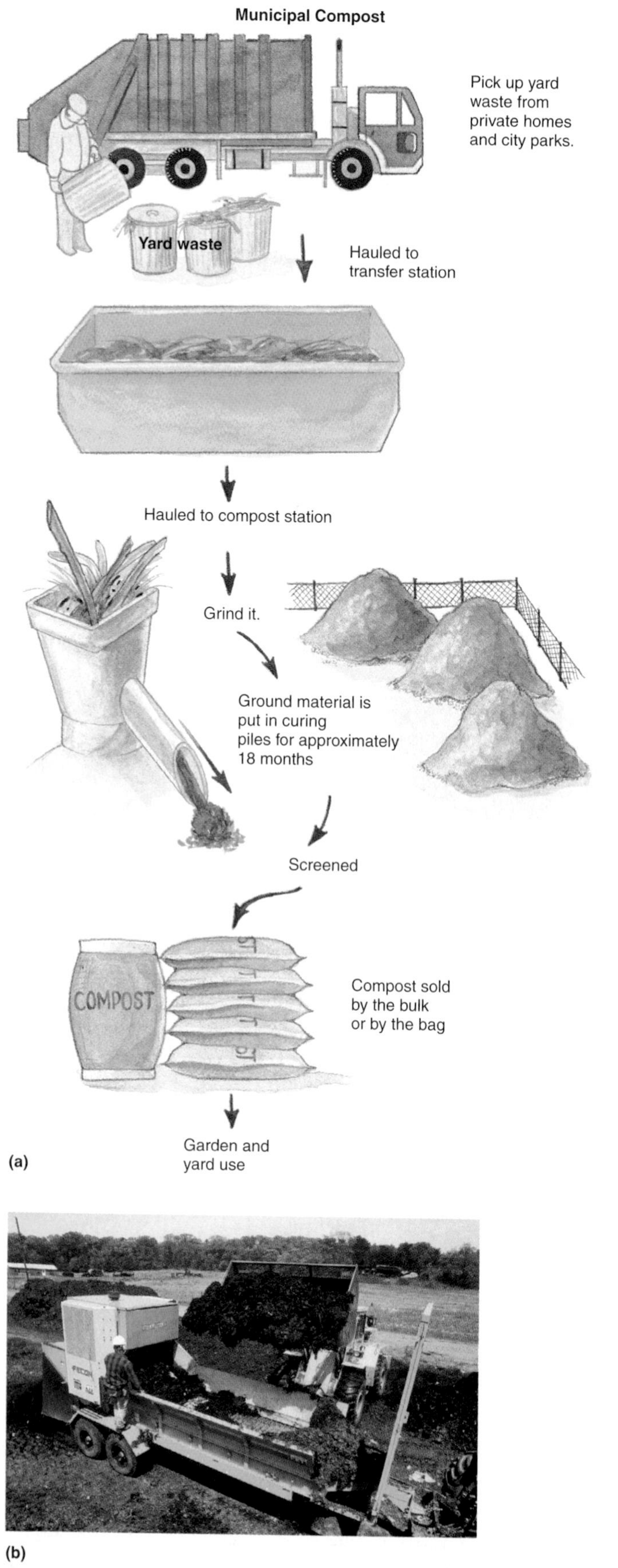

(b)

FIGURE 31.9 Municipal Composting (a) Process, **(b)** Composting station.

can be done by physically stirring and turning it, and it is kept moist, the composting can be completed in as little as 6 weeks.

■ **thermophile, p. 91**

Composting on a large scale offers cities a way to reduce the amount of garbage sent to their landfills. In some cities, yard wastes are collected separately from the main garbage. These wastes are then composted and used in various ways, including improving soils in city parks. Special machinery is used to compost on a large scale, and the composting can be accomplished in a very short time **(figure 31.9).**

MICROCHECK 31.3

Composting makes use of the decomposing ability of natural organisms. It dramatically reduces the need for large landfills to dispose of solid waste.

✓ What are the advantages and disadvantages of landfills?
✓ List the steps in successful composting.
✓ Why would soil and water be added to a compost pile?

31.4

Microbiology of Bioremediation

Focus Point

- Describe how two bioremediation strategies remove pollutants.

Bioremediation is the use of microorganisms to degrade or detoxify pollutants in a given environment. It may involve the use of specific organisms introduced into the polluted environment or, more commonly, it may take advantage of organisms already present, possibly adding nutrients to encourage their growth.

Pollutants

Pollutants from domestic and industrial wastes have often been dumped into the environment as a matter of convenience. Fortunately, most organic compounds of natural origin can be degraded by one or more species of soil or aquatic organisms under appropriate conditions for microbial growth. As oil spills dramatically demonstrate, however, some natural materials can cause devastating effects before they are degraded. Synthetic compounds are more likely to be degradable if they have a chemical composition similar to that of naturally occurring substances. **Xenobiotics,** synthetic compounds totally different from any that occur in nature, often persist for long periods of time. This is because microorganisms are unlikely to have enzymes necessary for degrading foreign substances; such an enzyme would not give them a competitive advantage in a natural situation.

Relatively slight molecular changes markedly alter the biodegradability of a compound. Perhaps the best-studied example involves the herbicides 2,4-dichlorophenoxyacetic acid (2,4-D) and 2,4,5-trichlorophenoxyacetic acid (2,4,5-T). The only difference between these two compounds is the additional chlorine atom on the latter. When 2,4-D is applied to the soil, it disappears within a period of several weeks, as a result of its degradation by microbes in the soil. When 2,4,5-T is applied, however, it is often still present more than a year later **(figure 31.10).** Its persistence is apparently due to the additional chlorine atom, which blocks the enzyme that makes the initial attack on 2,4-D.

Most herbicides and insecticides not only are toxic to their target, but also have deleterious effects on fish, birds, and other animals. For example, the pesticide DDT accumulates in the fat of predatory birds through biological magnification **(table 31.1).** Small amounts of the pollutant that contaminate water are concentrated in minute plankton, which are eaten by minnows, accumulating even more in the fish. When large birds eat the fish,

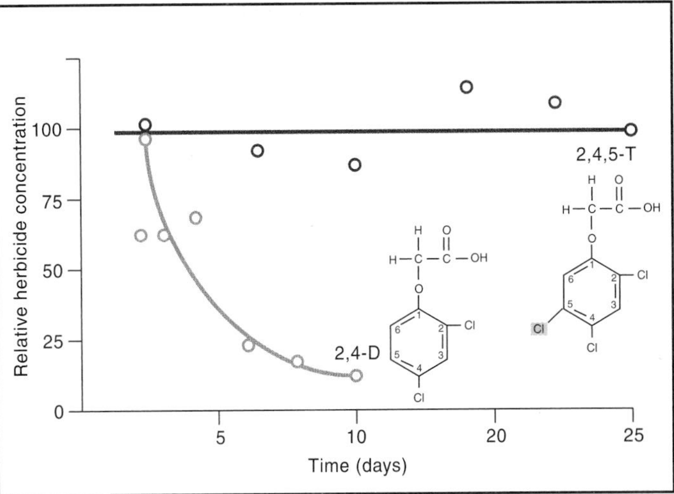

FIGURE 31.10 Comparison of the Rates of Disappearance of Two Structurally Related Herbicides, 2,4-D, and 2,4,5-T The addition of a third chlorine atom in the 2,4,5-T molecule blocks the enzyme that degrades the substance, so the compound remains in the environment.

the amount of chemical in tissues is tremendously magnified. The continuing ingestion of DDT, which accumulates in fat, results in an ever-greater concentration of the DDT as it passes upward through the food chain. DDT interferes with the reproductive process of birds, leading to the production of fragile eggs, which break before the young can hatch. Although banned in the United States, DDT is still used in other countries, particularly to control mosquitoes that transmit malaria.

Means of Bioremediation

Many factors influence the degradation rate of pollutants. As a general rule, any practice that favors multiplication of microorganisms will increase the rate of degradation. Thus, providing adequate nutrients, maintaining the pH near neutrality, raising the temperature, and providing an optimal amount of moisture are all likely to promote pollutant degradation. ■ **environmental factors that influence microbial growth, p. 90** ■ **nutritional factors that influence microbial growth, p. 93**

There are two general bioremediation strategies—biostimulation and bioaugmentation. **Biostimulation** enhances growth of indigenous microbes in a contaminated site by providing additional nutrients. Petroleum-degrading bacteria are naturally present in seawater, but they degrade oil at a very slow rate because the low levels of certain nutrients, including nitrogen

TABLE 31.1 Biological Magnification of DDT

Parts per Million DDT	Source
0.00005	Water
0.04	Plankton
0.23	Minnow
3.57	Heron
22.8	Merganser (fish-eating duck)

and phosphorus, limit their growth. To enhance bioremediation of oil spills, a fertilizer containing these nutrients, and which adheres to oil, was developed. When this fertilizer is applied to an oil spill, microbial growth is stimulated, leading to at least a threefold increase in the speed of degradation by the bacteria. **(figure 31.11). Bioaugmentation** relies on activities of microorganisms added to the contaminated material, complementing the resident population. The activated sludge process used during secondary treatment of wastewater is a form of bioaugmentation. A great deal of research is underway to develop microbial strains suited for bioaugmentation. One example, *Burkholderia (Pseudomonas) cepacia,* is capable of growth on 2,4,5-T and has been used successfully to remove this chemical from soil samples in the laboratory. However, microbes that thrive under laboratory conditions may not compete well in natural habitats.

■ activated sludge processes, p. 741

Successful bioremediation may also involve controlling metabolic processes by manipulating the availability of O_2 and specific growth substrates. For example, anaerobic degradation of trichloroethylene (TCE), a solvent used to clean metal parts, results in the accumulation of vinyl chloride, a compound more toxic than TCE. Aerobic conditions are important to prevent this buildup. Some pollutants are degraded only when specific substrates are made available to the microbes. This phenomenon, called **co-metabolism,** occurs because the enzyme produced by the microbe to degrade the additional substrate degrades the pollutant as well. As an example, the enzymes produced by some microbes to degrade methane also degrade TCE. In this case, adding methane enhances the degradation of TCE.

Bioremediation may be done either *in situ* ("in place") or off-site. *In situ* bioremediation generally relies on biostimulation and is less disruptive. Oxygen (O_2) can be added to contaminated groundwater and soil either by injecting hydrogen peroxide, which rapidly decomposes to liberate O_2 and water, or pumping air into soil. Off-site processes may be performed using a **bioreactor,** a large tank designed to accelerate microbial processes. Both nutrients and oxygen may be added to facilitate microbial growth and metabolism, while the slurry is agitated to ensure that the microbes remain in contact with the contaminants. A slower process involves mounding the contaminated soil over a layer that traps seeping chemicals. To provide O_2, the soil can be turned occasionally or air forced through.

FIGURE 31.11 Oil Spill Bioremediation Bioremediation was used to clean the shoreline contaminated by oil spilled from the *Exxon Valdez.* The growth of indigenous oil-degrading bacteria was stimulated by addition of nutrients. Comparing the uncleaned rocks on the left to the rocks cleaned by bioremediation on right shows the dramatic efficiency of bioremediation.

MICROCHECK 31.4

Bioremediation uses microorganisms to degrade pollutants. Biostimulation and bioaugmentation are two methods employed.

- ✓ Give three reasons why pollutants can persist in the environment.
- ✓ What is meant by biological magnification?
- ✓ How would adding a third chlorine atom to 2,4-D (see figure 31.10) block the enzyme attack? (Hint: Explain in terms of enzyme and substrate structure.)

FUTURE CHALLENGES

Better Identification of Pathogens in Water and Wastes

One of the most important challenges in the field of water and waste treatment is the development of new and better methods to detect waterborne contaminants in both drinking water and environmental water samples. This would make it possible to follow the occurrence and persistence of pathogens in water supplies with greater accuracy, and aid in better reporting of waterborne illnesses. Methods being developed, but not yet perfected for use in this field, include the polymerase chain reaction (PCR) and a variety of fluorescence techniques and radioactivity labeling methods.

Cysts of *Giardia* and oocysts of *Cryptosporidium* have been detected by amplifying specific regions of their DNA by PCR. In fact, using this method, it is possible to detect a single cyst of *Giardia* and to distinguish between species that are pathogenic for humans and those that are not. But problems arise in using these techniques with environmental samples that contain substances which inhibit the reaction. In addition, PCR detects DNA from dead organisms as well as living. Studies are needed to make these techniques feasible for use in water testing.

Viruses can also be detected by PCR. The viruses are concentrated by filtration onto membranes. Many different viruses can be detected simultaneously by combining gene probes from various groups of viruses. A problem is that viruses inactivated by disinfection procedures are still detected by PCR. To overcome this, viruses can be put into cell cultures to allow them to replicate, indicating that they are not inactivated, and the PCR is then performed on the infected cell cultures.

SUMMARY

31.1 Microbiology of Wastewater Treatment

Biochemical Oxygen Demand (BOD)

An important goal of wastewater treatment is the reduction of the **biochemical oxygen demand (BOD).**

Municipal Wastewater Treatment Methods (figures 31.1, 31.2, 31.3)

Primary treatment is a physical process designed to remove materials that sediment out. **Secondary treatment** is chiefly a biological process designed to convert most of the suspended solids to inorganic compounds and microbial biomass, removing most of the BOD. **Advanced treatment** is often designed to remove ammonia, nitrates, and phosphates. **Biosolids** that result from anaerobic digestion of **sludge** can be used to improve soils and promote plant growth.

Individual Wastewater Treatment Systems (figure 31.4)

Rural dwellings customarily rely on septic tanks for sewage disposal.

31.2 Drinking Water Treatment and Testing

Water Treatment Processes (figure 31.6)

Metropolitan water supplies are treated to remove particulate and suspended matter, various microorganisms, and organic chemicals. Chlorine or other disinfectants are then used to kill harmful microbes.

Water Testing

Total coliforms and fecal coliforms are used as indicator organisms; their presence suggests the possible presence of pathogens.

31.3 Microbiology of Solid Waste Treatment

Sanitary Landfills for Solid Waste Disposal

Landfills are used to dispose of solid wastes near towns and cities.

Municipal and Backyard Composting—Alternative to Landfills (figures 31.8, 31.9)

Composting reduces the amount of garbage sent to landfills.

31.4 Microbiology of Bioremediation

Pollutants

Synthetic compounds are more likely to be biodegradable if they have a chemical composition similar to that of naturally occurring compounds (figure 31.10).

Means of Bioremediation

Biostimulation can be used to increase the effectiveness of oil degradation by naturally occurring bacteria (figure 31.11).

REVIEW QUESTIONS

Short Answer

1. Describe how the BOD of a water sample is determined.
2. Which step of wastewater treatment removes most of the BOD?
3. Compare and contrast the activated sludge process and the trickling filter system used in secondary treatment of wastewater.
4. Why is it beneficial to remove nitrates and phosphates in wastewater?
5. How does a septic tank system work?
6. What is an aquifer?
7. Why do water-testing procedures look for coliforms rather than pathogens?
8. How does the ONPG/MUG test allow a sample to be assayed simultaneously for the presence of both coliforms and *E. coli?*
9. What aspect of 2,4,5-T makes it more likely to persist in the environment than 2,4-D?
10. Describe the use of bioremediation in the cleanup of oil spills.

Multiple Choice

1. A marked decrease in BOD during secondary treatment indicates
   a) lack of oxidation during treatment.
   b) effective aerobic decomposition during treatment.
   c) effective anaerobic decomposition during treatment.
   d) removal of all pathogenic bacteria.
   e) removal of all toxic chemicals.
2. Advanced treatment is often designed to remove
   a) BOD. b) nitrates and phosphates. c) bacteria.
   d) protozoa. e) methane.
3. Which of the following pairs does *not* match?
   a) Potable water—presence of pathogens
   b) High BOD—high organic content
   c) Stabilized sludge—fertilizer
   d) Primary treatment—removal of material that settles
   e) Bulking—growth of filamentous bacteria
4. Which of the following is *false?*
   a) Bulking interferes with trickling filter systems.
   b) Artificial wetlands provide a habitat for wildlife.
   c) Removal of nitrates by microorganisms requires anaerobic conditions.
   d) Methane is a by-product of anaerobic digestion.
5. Which of the following pairs does *not* match?
   a) Surface water—watershed
   b) Groundwater—aquifer
   c) Sand and gravel filters—removes organic chemicals
   d) Alum—causes suspended material to coagulate
   e) Disinfection—chlorine, ozone, or ultraviolet light
6. Septic tanks should be placed
   a) as close to the well as possible.
   b) at least 500 feet from the house.
   c) under the house.
   d) in deep clay soil.
   e) where the overflow cannot contaminate any water supply.

7. Which of the following statements about coliform testing methods is *true?*
   a) All determine the number of *E. coli* present in a sample.
   b) The MPN procedure precisely indicates the concentration of coliforms.
   c) The media employed test for the ability to ferment lactose.
   d) A positive test indicates that pathogens are definitely present in the sample.
   e) All coliforms hydrolyze ONPG and MUG.
8. Landfills are often used to dispose of
   a) household wastewater. b) commercial wastewater.
   c) solid wastes. d) petroleum wastes.
   e) wastewater effluent.
9. Backyard composting is an excellent way to dispose of
   a) cooking fats. b) garden debris. c) spoiled meats.
   d) insecticides. e) cleaning supplies.
10. Synthetic compounds are most likely to be biodegradable if they
    a) are totally different from anything found in nature.
    b) have three chlorine atoms per molecule.
    c) are plastics.
    d) are present in very large amounts.
    e) are chemically similar to naturally occurring substances.

Applications

1. A developer is interested in building vacation homes on 150 acres of oceanfront property. A priority is to retain as much natural beauty of the area as possible. Safe and effective sewage treatment must be part of the plan. What advantages and disadvantages of each of the following options proposed by an environmental consultant must the developer consider before selecting one?
   a. Individual septic tanks for each home
   b. Trickling filter central treatment system
   c. Artificial wetlands
2. A public health official is investigating waterborne diseases in Illinois. She notes that over half of the actual cases of waterborne diseases originating from drinking water were caused by *Giardia lamblia.* Other data showed that most cases of gastroenteritis attributed to exposure to recreational waters were caused by *Cryptosporidium parvum.* What does this suggest about controlling waterborne diseases?

Critical Thinking

1. Why is oil not degraded when in a natural habitat underground, yet susceptible to bioremediation in an oil spill?
2. The accompanying figure shows the effects of different treatments of drinking water on the incidence of typhoid fever in Philadelphia, 1890–1935. If filtration of drinking water caused such a dramatic decrease in the disease incidence, was it necessary to introduce chlorination a few years later? Why or why not?

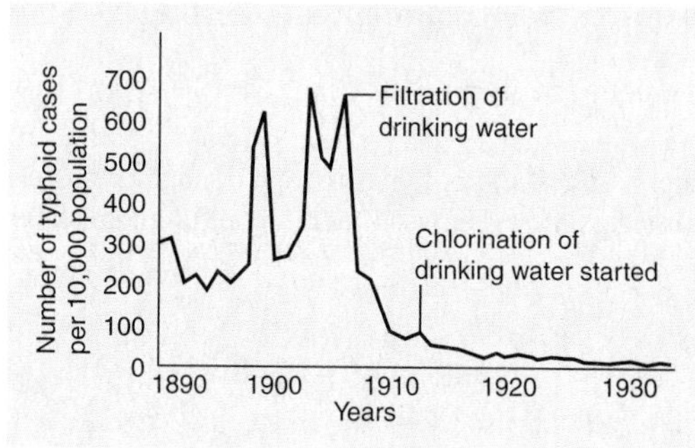

32

Refrigeration is a method of food preservation.

Food Microbiology

A Glimpse of History

Pasteurization of milk and milk products is so common today it is hard to fathom that only during the twentieth century have federal, state, and local laws mandated it for all milk products. Although most large cities have required milk to be pasteurized since 1900, as late as 1930 most milk sold in rural areas was not routinely pasteurized. As a result, human diseases such as brucellosis and tuberculosis were fairly common. ■ **brucellosis, p. 681** ■ **tuberculosis, p. 514**

Alice Catherine Evans, the first woman president of the Society of American Bacteriology (now the American Society for Microbiology), helped establish the connection between unpasteurized milk and brucellosis in humans. A graduate of both Cornell University and the University of Wisconsin, Evans worked for the U.S. Department of Agriculture, seeking out the sources of microbial contamination of dairy products. In 1917, Evans reported that cases of human brucellosis were related to the finding of *Brucella abortus* in cows' milk. Her conclusion that *B. abortus* could be transmitted from cows to humans through milk conflicted with the prevailing view of a number of prominent scientists, including Robert Koch. In 1900, Koch had declared that bovine tuberculosis and brucellosis could not be transmitted to humans. As a result, at least 30 years elapsed before many scientists and dairy workers would accept the increasing evidence that diseases could be transmitted from cows to humans through the milk supply.

In the 1920s, dairy herds were inspected and vaccinated for tuberculosis. Herds that passed inspection and were vaccinated were called certified herds, and their milk could be sold commercially without pasteurization. In the late 1930s, after a number of the children of dairy workers had died of brucellosis even though they had drunk only certified milk, the problem of milk-borne disease was finally acknowledged. Today, milk is routinely pasteurized, and only very small amounts of unpasteurized milk are sold in the United States.

When you prepare a meal, you also invite a host of microorganisms to dinner. Practically all the food we purchase or grow—fruit, vegetable, meat, or dairy product—harbors a variety of microorganisms. This is not surprising when one considers that bacteria and fungi are ubiquitous and are especially plentiful in soil and around animals. From the microbial perspective, food can be viewed as a fertile ecosystem in which these organisms vie for the nutrients. The successful microorganisms are able to multiply and predominate.

Microorganisms on foods are not necessarily undesirable. Sometimes, their growth results in a more pleasant taste or texture. Foods that have been intentionally altered during production by carefully controlling the activity of bacteria, yeasts, or molds are called **fermented (figure 32.1).** For example, food manufacturers

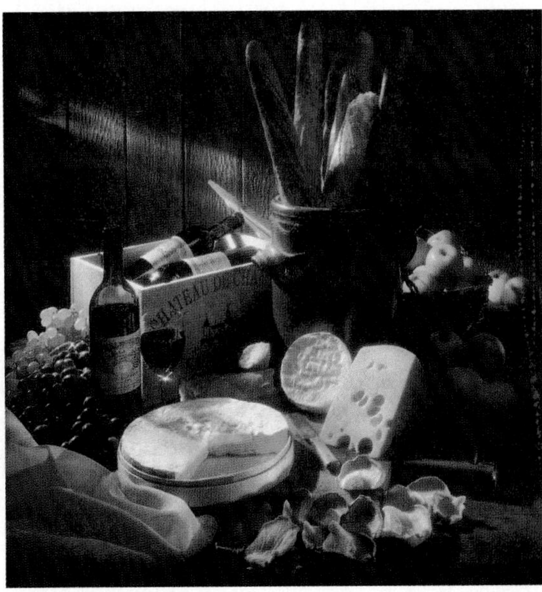

Figure 32.1 Fermented Foods Fermented foods have been intentionally altered in their production by carefully controlling the growth and activity of microorganisms.

KEY TERMS

Fermented Foods Foods that have been intentionally altered during production by encouraging the activity of bacteria, yeasts, or molds.

Food Preservation Increasing the shelf life of foods by preventing the growth and concurrent metabolic activities of microorganisms.

Food Spoilage Biochemical changes in foods that are perceived as undesirable.

Foodborne Infection An illness that results from consuming microbes in a food product; ingested microbes colonize the host and cause disease.

Foodborne Intoxication An illness that results from the consumption of an exotoxin produced by a microorganism growing in a food product.

Water Activity (a_w) The relative amount of water available to microorganisms; pure water has an a_w of 1.0.

purposely encourage some microorganisms to flourish in milk in order to produce foods such as sour cream and cheese. Alcoholic beverages, such as beer and wine, and many Asian food condiments, such as soy sauce and miso, also rely on microbial metabolism for their production. Strictly speaking, the term *fermentation* is used to describe only those metabolic activities that utilize pyruvate or another organic compound as an electron acceptor, with the result that alcohols and acids are produced. Food scientists, however, use the term more generally, to encompass any desirable change that a microorganism imparts to food. ■ fermentation, p. 147

Biochemical changes in foods, when perceived as undesirable, are called **spoilage (figure 32.2).** The processes that cause spoilage are often the same ones involved in fermentation of foods. In fact, a food product considered by one cultural population to be fermented may be considered spoiled by another. Sour milk and moldy bread are examples of foods considered spoiled as a result of microbial growth. The souring of milk, however, involves microbial processes analogous to those that cause the agreeable acidic flavor of sour cream, and the mold growing on bread may be related to the one that causes the blue veining in Gorgonzola cheese.

Growth of pathogens such as *Clostridium botulinum, Staphylococcus aureus, Salmonella* species, and *E. coli* O157: H7 can result in foodborne illness but generally does not result in perceptible changes in quality of a food. Depending on the type of pathogen, the illness may result from consuming either the living organisms or the toxins they have produced during growth.

Limiting microbial growth can preserve the quality of foods and prevent foodborne illnesses. Foods can be canned, pasteurized, or irradiated to eliminate or decrease the numbers of microorganisms. Alternatively, the multiplication of microorganisms can be suppressed by storing food at cold temperatures, or by adding growth-inhibiting ingredients, called **preservatives.** The end products of some fermentation processes can preserve food by inhibiting the growth of many undesirable microorganisms.

32.1 Factors Influencing the Growth of Microorganisms in Foods

Focus Point

- Describe four intrinsic factors and two extrinsic factors that influence the growth of microorganisms in foods.

Microorganisms have a competitive advantage in those foods with characteristics that enable them to multiply most rapidly. Understanding the factors that influence microbial growth is essential to maintaining food quality when producing fermented foods or prolonging the shelf life of perishable foods.

The conditions naturally present in the food, such as moisture, acidity, and nutrients, are called **intrinsic factors.** Environmental conditions, such as the temperature and atmosphere of storage, are called **extrinsic factors.** All of these factors combine to determine which microorganisms can grow in a particular food product and

(a)

(b)

(c)

FIGURE 32.2 Examples of Spoiled Food **(a)** Apples infected by the fungus *Venturia inaequalis*, cause of apple scab. **(b)** Orange affected by a *Penicillium* mold. **(c)** Ear of corn and a lemon after several weeks in the refrigerator.

the rate of that growth. For example, bacteria are well suited for growth in the environment found in fresh meat and other moist, pH-neutral, nutrient-rich foods. Yeasts and molds can also grow under these conditions, but the more rapid increase of bacteria overwhelms these competitors. When conditions such as lack of moisture or high acidity restrict the growth of bacteria, fungi predominate despite their relatively slow growth.

Intrinsic Factors

The growth of microorganisms in a food is influenced by the inherent characteristics of that food. In general, microbes multiply rapidly in moist, nutritionally rich, pH-neutral foods.

Water Availability

Food products vary in terms of how much water is accessible to microorganisms. Fresh meats and milk, for example, have ample water and support growth of many microbes. Bread, nuts, and dried foods, on the other hand, provide a relatively arid environment. Some sugar-rich foods, such as jams and jellies, are seemingly moist, but most of that water is chemically interacting with the sugar, making it unavailable for use by microbes. Highly salted foods, for similar reasons, have little available moisture.

The term **water activity** (a_w) is used to designate the amount of water available in foods. By definition, pure water has an a_w of 1.0. Most fresh foods have an a_w above 0.98, whereas ham has an a_w of 0.91, jam has an a_w of 0.85, and some cakes have an a_w of 0.70.

Most bacteria require an a_w above 0.90 for growth, which explains why fresh, moist foods spoil more quickly than dried, sugary, or salted foods. Fungi can grow at an a_w as low as 0.80, which is why forgotten bread, cheese, jam, and dried foods often become moldy. *Staphylococcus* species, which are adapted to grow on the dry, salty surfaces of human skin, can grow at an a_w of 0.86, lower than the minimum required by most common spoilage bacteria. *Staphylococcus* species normally do not compete well with other bacteria, but on salty products such as ham and other cured meats, they can multiply with little competition. Ham is a common vehicle for *S. aureus* food poisoning. ■ *Staphylococcus aureus* foodborne illness, p. 763

pH

The pH of a food is important in determining which organisms can survive and thrive on it. Many species of bacteria, including most pathogens, are inhibited by acidic conditions and cannot grow at a pH below 4.5. An exception is the lactic acid bacteria, which can grow at a pH as low as 3.5 and are used in the production of fermented foods such as yogurt and sauerkraut. This group of bacteria produces lactic acid as a result of fermentative metabolism. Although they are useful in food production, their growth and accompanying acid production are also prime causes of spoilage of unpasteurized milk and other foods. Fungi can grow at a lower pH than most spoilage bacteria, leading to some acidic foods eventually becoming moldy. For example, the pH of lemons is approximately 2.2, which inhibits the growth of bacteria, including the lactic acid group, but some fungi can grow. ■ pH, p. 24 ■ lactic acid bacteria, p. 255 ■ pasteurization, p. 112

The pH of a food product can also determine whether toxins can be produced. For instance, *Clostridium botulinum*, the causative agent of botulism, does not grow or produce toxin below pH 4.5, and so it is not considered a danger in highly acidic foods. This is why the canning process for acidic fruits and pickles is less stringent than that for foods with a higher pH. However, some newer varieties of tomatoes are less acidic than older types, and require the addition of acid if they are to be safely canned using these less stringent procedures.

Nutrients

The nutrients present in a food determine the kinds of organisms that can grow in it. An organism requiring a particular vitamin cannot grow in a food lacking that vitamin. A microbe capable of synthesizing that vitamin, however, can grow if other conditions are favorable. Members of the genus *Pseudomonas* often spoil foods because they can synthesize essential nutrients and can multiply in various environments, including refrigeration.

Biological Barriers

Rinds, shells, and other coverings help protect foods from invasion by microorganisms. Eggs, for example, retain their quality much longer with intact shells. Whole lemons keep longer than slices. Even so, microorganisms will eventually break down these coverings and cause spoilage.

Antimicrobial Chemicals

Some foods contain natural antimicrobial chemicals that help prevent spoilage. Egg white, for instance, is rich in lysozyme. If lysozyme-susceptible bacteria breach the protective shell of an egg, they are destroyed by lysozyme before they can cause spoilage. Other examples of naturally occurring antimicrobial chemicals are benzoic acid in cranberries and allicin in garlic. ■ lysozyme, p. 62

Extrinsic Factors

The extent of microbial growth varies greatly depending on the conditions under which a food is stored. Microorganisms multiply rapidly in warm, oxygen-rich environments such as the surface of meat stored at room temperature.

Storage Temperature

The temperature of storage affects the rate of growth of microorganisms in food. Below its freezing point, water becomes crystalline and inaccessible, effectively halting microbial growth. At low temperatures above freezing, many enzymatic reactions are either very slow or non-existent, with the result that some microorganisms are unable to grow. Those that can do so at a reduced rate. Microorganisms that grow on refrigerated foods are most likely psychrophiles or psychrotrophs such as some members of the genus *Pseudomonas*. ■ psychrophiles, p. 91 ■ psychrotroph, p. 91

Atmosphere

The presence or absence of O_2 affects the type of microbial population able to grow in food. For example, members of the genus *Pseudomonas* are obligate aerobes, and consequently they cannot grow in foods stored under conditions that exclude all of their required oxygen. Excluding O_2 from a food, however, may enable the growth of other bacteria, including the obligate anaerobe *Clostridium botulinum*. A case of botulism was traced to the consumption of a thick, homemade stew that had been slowly cooked and then left at room temperature overnight. The cooking process did not destroy the endospores of *C. botulinum* and had driven off the O_2, thereby creating anaerobic conditions in which the organism thrived and produced toxin. ■ oxygen requirements, p. 92

MICROCHECK 32.1

Intrinsic factors (such as available moisture, pH, and the presence of antimicrobial chemicals) and extrinsic factors (including storage temperature and atmosphere) influence the type of microorganisms that grow and predominate in a food product.

✓ Why is *Staphylococcus aureus* more likely to be found in high numbers on ham than on fresh meat?

✓ Which is more important to refrigerate: homemade stew or bread? Why?

✓ Why would the cooking process create anaerobic conditions?

32.2 Microorganisms in Food and Beverage Production

Focus Points

- Compare and contrast the production of cheese, yogurt, and acidophilus milk.
- Compare and contrast the production of pickled vegetables and fermented meat products.
- Compare and contrast the production of wine, beer, distilled spirits, and vinegar.
- Describe the production of soy sauce.

Yogurt, cheese, pickled vegetables, and other fermented foods are not only perceived as pleasant tasting, but the acids produced as a by-product of microbial metabolism inhibit the growth of many spoilage organisms as well as foodborne pathogens. Thus, fermentation historically has been, and continues to be today, an important method of food preservation, particularly when modern conveniences such as refrigeration are lacking.

Lactic Acid Fermentations by the Lactic Acid Bacteria

The tart taste of yogurt, pickles, sharp cheeses, some sausages, and other foods is due to the production of lactic acid by one or more members of a group of bacteria known as the lactic acid bacteria **(table 32.1)**. This group of organisms, including members of the genera *Lactobacillus, Lactococcus, Streptococcus, Leuconostoc,* and *Pediococcus,* are obligate fermenters that characteristically produce lactic acid as an end product of their metabolism. Some also produce flavorful and aromatic compounds that contribute to the overall quality of fermented foods. ■ lactic acid bacteria, p. 255 ■ obligate fermenters, p. 92

Table 32.1 Foods Produced Using Lactic Acid Bacteria

Food	Characteristics
Milk Products	
Cheese (unripened)	Employs a starter culture usually containing *Lactococcus cremoris* and *L. lactis*
Cheese (ripened)	Employs rennin and a starter culture containing *Lactococcus cremoris* and *L. lactis*, ripened for weeks to years; other bacteria and/or fungi may be added to enhance flavor development
Yogurt	Employs a starter culture containing *Streptococcus thermophilus* and *Lactobacillus delbrueckii* subspecies *bulgaricus*
Sweet acidophilus milk	*Lactobacillus acidophilus* added for purported health benefits
Vegetables	
Sauerkraut	Cabbage; succession of naturally occurring bacteria including *Leuconostoc mesenteroides, Lactobacillus brevis*, and *Lactobacillus plantarum*
Pickles	Cucumbers; naturally occurring bacteria
Poi	Taro root; naturally occurring bacteria; Hawaii
Olives	Green olives
Kimchee	Cabbage and other vegetables; Korea
Meats	
Dry and semidry sausages	Employs a starter culture containing species of *Lactobacillus* and *Pediococcus*, meat is stuffed into casings, incubated, heated, and then dried

Cheese, Yogurt, and Other Fermented Milk Products

In a cow's udder, milk is sterile, but it rapidly becomes contaminated with a variety of microorganisms during milking and handling. Various species of lactic acid bacteria are inevitably introduced because they commonly reside on the udder. If the milk is not refrigerated, these bacteria readily ferment lactose, the predominant sugar in milk, producing lactic acid. The removal of the primary carbohydrate as a nutrient source, combined with the production of lactic acid, inhibits the growth of many other microorganisms in the milk. Aesthetic features of the milk change as well because lactic acid lowers the pH, which in turn causes the milk proteins to coagulate or curdle, and sours the flavor. ■ milk spoilage, p. 724

Today, with high quality control standards and the use of pasteurized milk, the commercial production of fermented milk products does not rely on naturally introduced lactic acid bacteria. Instead, **starter cultures** containing one or more strains of lactic acid bacteria are added to the milk. These strains are carefully selected to produce the most desirable flavors and textures. Precious starter cultures must be carefully maintained and protected against contamination, particularly by bacteriophages, which can damage or destroy them. ■ bacteriophage, p. 303

Cheese Cheese can be made from the milk of a wide variety of animals, but most common cheeses are made with cow's milk. Sheep or goat's milk can be used to give characteristic flavors. Cheeses are classified as very hard, hard, semisoft, and soft, according to their percentage of water.

Cottage cheese is one of the simplest cheeses to make. Pasteurized milk is inoculated with a starter culture, usually containing *Lactococcus cremoris* and *L. lactis,* and then incubated until fermentation products cause the proteins in milk to coagulate. The coagulated proteins, or **curd,** are heated and cut into small pieces to facilitate drainage and removal of the liquid waste portion, or **whey.** Unlike most cheeses that undergo further microbial processes called **ripening** or **curing,** cottage cheese is unripened.

The initial steps in the production of ripened cheese are the same as those of cottage cheese, except the enzyme **rennin** is added to the fermenting milk to hasten protein coagulation **(figure 32.3).** Rennin is an enzyme naturally found in the stomach of young calves, where it aids in digestion of the mother's milk. Today, rennin is commercially produced using genetically engineered microorganisms.

After the whey is removed, the curds are salted and pressed, and then they are shaped into the traditional forms of cheese, usually bricks or wheels. The cheese is then ripened to encourage characteristic changes in texture and flavor. Depending on the type of cheese, ripening can take from several weeks to several years. Changes imparted during this time are generally due to the metabolic activities of naturally occurring or starter lactic acid bacteria. Longer ripening gives rise to more acidic, sharper cheeses. Some cheeses are inoculated with other bacteria or fungi that impart characteristics particular to the kind of cheese. For example, the bacterium *Propionibacterium shermanii* ripens Swiss cheese and gives it the characteristic holes, or eyes, and a nutty flavor. This bacterium ferments organic compounds to produce propionic acid and CO_2. The CO_2 gas causes the holes in the cheese, while the propionic acid imparts the typical flavor. Propionic acid also inhibits spoilage organisms. Roquefort, Gorgonzola, and Stilton cheeses are ripened by the fungus *Penicillium roquefortii.* Growth of the fungus along cracks in the cheese gives these cheeses the distinctive bluish-green veins. Brie and Camembert are ripened by a white fungus such as *P. candidum* or *P. camemberti* inoculated on the surface of the cheese. The mycelia of the fungus produce enzymes that alter texture and flavor as they gradually work their way into the cheese. Limburger cheese is made in a similar manner, but with the bacterium *Brevibacterium linens.* ■ mycelium, p. 290

Lactic acid production and rennin activity cause the milk proteins to coagulate. The coagulated mixture is cut to facilitate the separation of the solid curd and liquid whey.

The curd is heated and cut into small pieces. The liquid whey is removed by draining.

Curds are salted and pressed into blocks or wheels for aging.

Figure 32.3 Commercial Production of Cheese Why does a longer ripening process give rise to a sharper cheese?

Yogurt To produce yogurt, pasteurized milk is concentrated slightly by evaporation and then inoculated with a starter culture containing *Streptococcus thermophilus* and *Lactobacillus delbrueckii* subspecies *bulgaricus.* The mixture is incubated at 40°C to 45°C for several hours, during which time these thermophilic bacteria grow rapidly and produce lactic acid and other end products, such as acetaldehyde, that contribute to the flavor. Carefully controlled incubation conditions favoring the balanced growth of the two species ensure the proper levels of acid and flavor compounds.

Acidophilus Milk Traditional acidophilus milk is the product of fermentation by *Lactobacillus acidophilus.* The more readily available **sweet acidophilus milk** retains the flavor of fresh milk because it is not fermented. Instead, a culture of *L. acidophilus* is added immediately before packaging. The bacteria are simply included for their purported health benefits. Some evidence suggests they may aid in the digestion of lactose as well as prevent and reduce the severity of some diarrheal illnesses, but the role they play in the complex interactions of the human intestinal tract is not clear. Unlike most lactic acid bacteria used as starter cultures, *L. acidophilus* can potentially colonize the intestinal tract.

Pickled Vegetables

Another fermentation process known as **pickling** originated as a way to preserve vegetables such as cucumbers and cabbage. Today, pickled products such as sauerkraut (cabbage), pickles (cucumbers), and olives are valued for their flavor. Unlike the fermentation of milk products that rely on the use of starter cultures in the manufacturing process, fermentation of most vegetables utilizes naturally occurring lactic acid bacteria residing on the vegetables.

One of the most well-studied natural fermentations is the production of sauerkraut. The cabbage is first shredded and layered with salt. The layers are firmly packed to provide an anaerobic environment. The salt draws water and nutrients from the cabbage, creating a brine that inhibits the growth of many microbes but permits the growth of the naturally occurring lactic acid bacteria. Under the correct conditions, natural successions of lactic acid bacteria grow. These bacteria—*Leuconostoc mesenteroides, Lactobacillus brevis,* and *Lactobacillus plantarum*—produce lactic acid, which lowers the pH, further inhibiting undesired microbes. The lactic acid and other end products of the fermentation give sauerkraut its characteristic tangy taste. When the desired flavor has been attained, usually after 2 to 4 weeks at room temperature, the sauerkraut is often canned. Similar processes are used to make some pickles, olives, and other vegetable products.

Fermented Meat Products

Traditionally, fermented meat products, such as salami, pepperoni, and summer sausage, were produced by enabling the small numbers of lactic acid bacteria naturally present to multiply to the point of dominance. Relying on the natural fermentation of meat is inherently risky, however, because the incubation conditions used to initiate fermentation can potentially support the growth and toxin production of pathogens such as *Staphylococcus aureus* and *Clostridium botulinum.* The development and use of reliable starter cultures assure rapid production of lactic acid, inhibiting the growth of pathogens and enhancing flavor development. Starter cultures used by U.S. sausage-makers typically contain species of *Lactobacillus* and/or *Pediococcus,* depending on the type of sausage.

To make fermented sausages, meat is ground and combined with a starter culture and other ingredients including sugar, salt, and nitrite. The sugar serves as a substrate for fermentation, because meat does not naturally contain enough fermentable carbohydrate to produce sufficient amounts of lactic acid. Salt and nitrite contribute to the flavor of sausage; they also inhibit the growth of spoilage microorganisms and, most importantly, *C. botulinum.* After thorough blending, the mixture is stuffed into a casing and incubated from one to several days. The product can then be smoked or otherwise heated to kill bacteria. Finally, it is dried.

Alcoholic Fermentations by Yeast

Some yeasts, such as members of the genus *Saccharomyces,* ferment simple sugars to produce ethanol and carbon dioxide. These yeasts are used to make a variety of alcoholic beverages as well as vinegar and bread **(table 32.2).**

Table 32.2 Foods and Beverages Produced Using Alcoholic Fermentation by Yeast

Product	Characteristics
Alcoholic Beverages	
Wine	Sugars in grape juice are fermented by *Saccharomyces cerevisiae.*
Sake	Amylase from mold (*Aspergillus oryzae*) converts the starch in rice to sugar, which is then fermented by *S. cerevisiae.*
Beer	Enzymes in germinated barley convert starches of barley and other grains to sugar, which is then fermented by *S. cerevisiae.*
Distilled spirits	Sugars, or starches that are converted to sugars, are fermented by *S. cerevisiae*; distillation purifies the alcohol.
Vinegar	Alcohol produced by fermentation is oxidized to acetic acid by species of *Gluconobacter* or *Acetobacter.*
Breads	*S. cerevisiae* ferments sugar; expansion of CO_2 causes the bread to rise; alcohol is lost to evaporation during baking.

Wine

Wine is the product of the alcoholic fermentation of naturally occurring sugars in the juices of fruit, most commonly grapes. One of the most important variables in the quality of wine is the variety and quality of grapes used. The growing conditions and ripeness as well as other factors affect the grapes' content of sugar, acids, and various organic compounds, which in turn critically influences the final product. Commercially, wine is made by crushing carefully selected grapes in a machine that removes the stems and collects the resulting solids and juices, or **must (figure 32.4).** For red wine, the entire must, including skins and pulp, of red grapes is put into the fermentation vat. The red color and complex flavors contributed by the **tannins** of red wines are derived from components of the grape skin and seeds. The solids are removed during fermentation by a wine press that separates them from the partially fermented juice once the desired amount of color and tannins have been extracted. For the production of white wines, the solids are removed immediately and only the clear juice of white, or occasionally red, grapes is fermented. Rose wines obtain their light pink color from the entire crushed red grape fermented for about 1 day, after which the juice is removed and fermented alone.

The fermentation must be carefully controlled to ensure that desired reactions occur. Sulfur dioxide (SO_2) is generally added to inhibit the growth of the natural microbial population of the grape, especially acetic acid bacteria. These bacteria can convert alcohol to acetic acid (vinegar) and are the most prevalent cause of wine spoilage.

The fermentation process is initiated by the addition of specially selected strains of *Saccharomyces cerevisiae.* These strains are more resistant to the antimicrobial action of SO_2 and produce a higher alcohol content than naturally occurring yeasts.

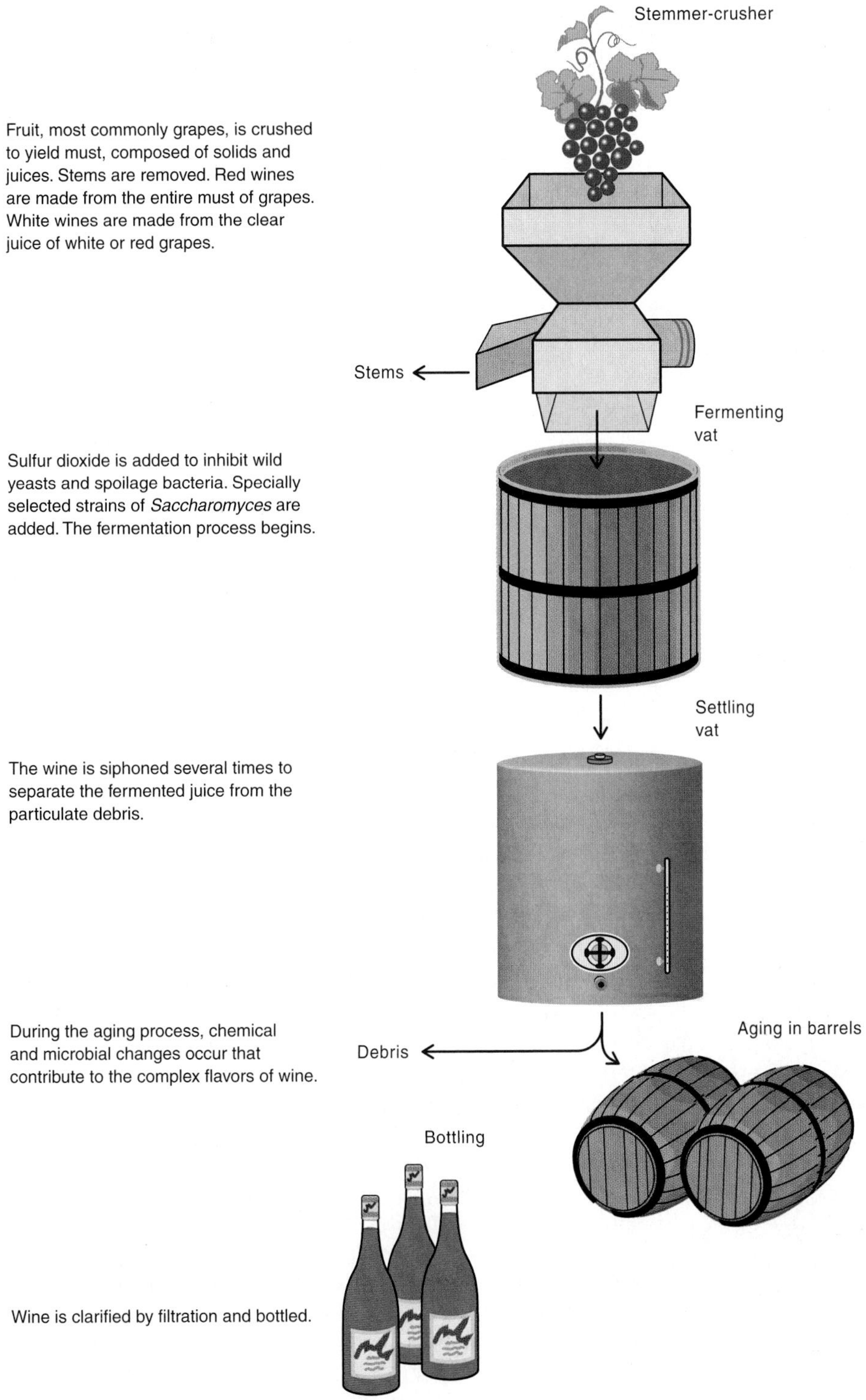

Stemmer-crusher

Fruit, most commonly grapes, is crushed to yield must, composed of solids and juices. Stems are removed. Red wines are made from the entire must of grapes. White wines are made from the clear juice of white or red grapes.

Stems

Fermenting vat

Sulfur dioxide is added to inhibit wild yeasts and spoilage bacteria. Specially selected strains of *Saccharomyces* are added. The fermentation process begins.

Settling vat

The wine is siphoned several times to separate the fermented juice from the particulate debris.

Aging in barrels

During the aging process, chemical and microbial changes occur that contribute to the complex flavors of wine.

Debris

Bottling

Wine is clarified by filtration and bottled.

Figure 32.4 Commercial Production of Wine

Fermentation is carried out at a carefully controlled temperature, which varies with the type of wine, for a period ranging from a few days to several weeks. During fermentation most of the sugar is converted to ethanol and CO_2, generally resulting in a final alcohol content of less than 14%. Dry wines result from the complete fermentation of the sugar, whereas sweet wines contain residual sugar.

In addition to the alcoholic fermentation of the grape sugars, a distinctly different type of fermentation, called **malolactic fermentation,** can occur during wine production. Lactic acid bacteria, primarily species of *Leuconostoc*, convert malic acid to the less acidic lactic acid. Red wines made from grapes grown in cool regions tend to have high levels of malic acid, and their flavor is mellowed by this fermentation.

After fermentation, the wine is siphoned several times to remove the clear juice from the sediment of yeast and particulate debris. Most red wines and some white wines are then aged in oak barrels, contributing to the complexity of the flavor. Wine is clarified by filtration and bottled. The CO_2 produced during fermentation is usually released before the wine is bottled, resulting in a "still" (noncarbonated) wine. Other processes are used to prepare carbonated wines such as champagne.

The Japanese wine sake depends on several microbial fermentation reactions. First, cooked rice is inoculated with the fungus *Aspergillus oryzae*. The fungus produces the enzyme amylase, which degrades the rice starch to sugar. Then, a strain of *Saccharomyces cerevisiae* is added to convert the sugar to alcohol and CO_2. Lactic acid bacteria add to the flavor by producing lactic acid and other fermentation end products.

Beer

Beer production is a multistep process designed to break down the starches of grains such as barley to produce simple sugars, which are then fermented by yeast. Yeasts alone cannot convert grain to alcohol because they lack the enzymes that degrade starch, the primary carbohydrate of grain. Sprouted or germinated barley, however, known as **malted barley** or **malt,** naturally contains these and other important enzymes.

Dried, roasted malt is ground, mixed with **adjuncts** (starches, sugars, or whole grains such as rice, corn, or sorghum), and then soaked in warm water in a process called **mashing (figure 32.5).** During this process, enzymes of the malt act on the starches, converting them to fermentable sugars. The final characteristics of the beer, such as color, flavor, and foam, are derived entirely from compounds in the roasted malt. The adjuncts simply serve as readily available, less expensive sources of carbohydrates for alcohol production.

After mashing, the residual solids, or **spent grains,** are removed to yield the sugary liquid called **wort. Hops,** the flowers of the vinelike hop plant, are added to the wort to impart a

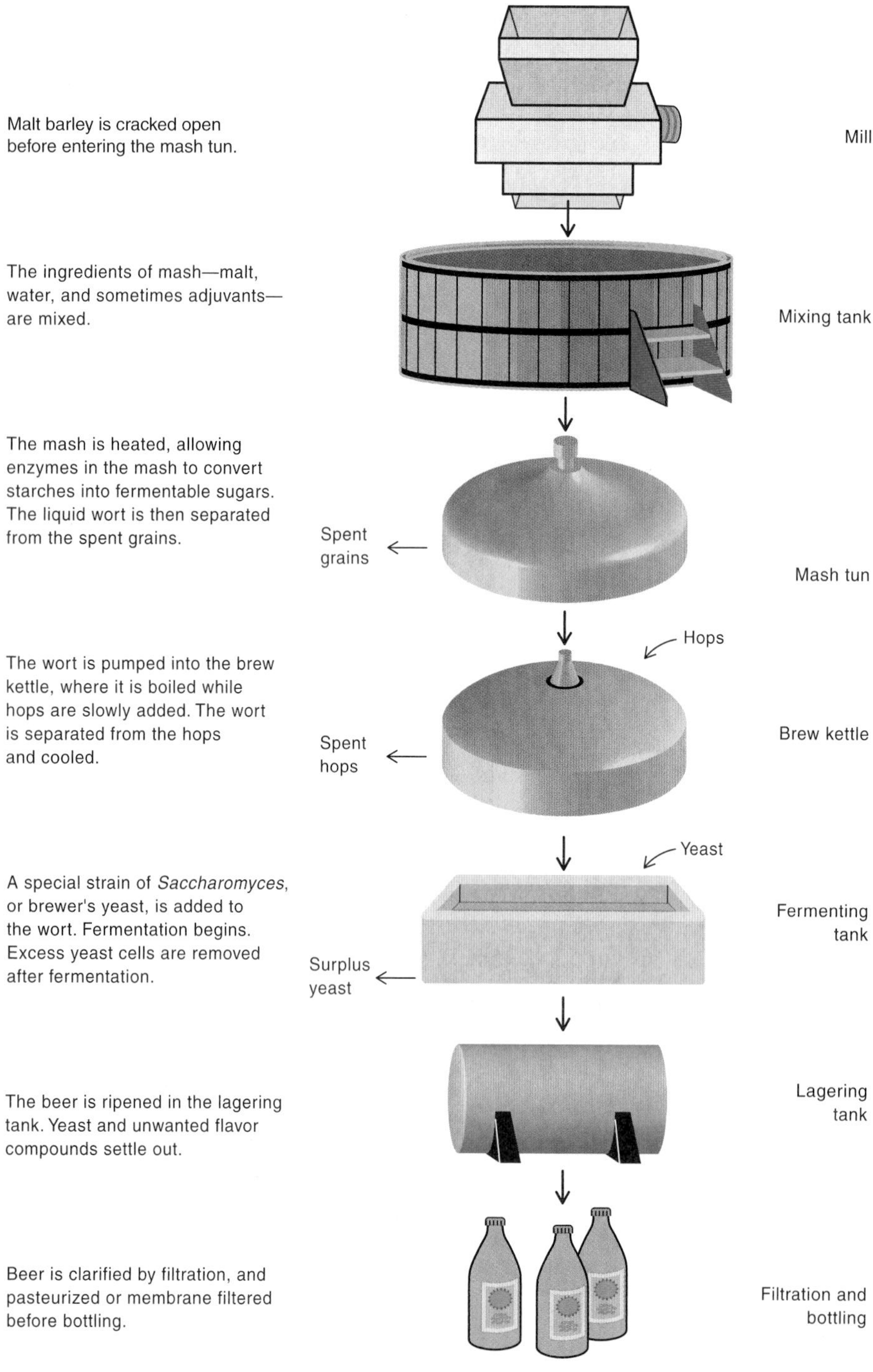

Malt barley is cracked open before entering the mash tun.
Mill
The ingredients of mash—malt, water, and sometimes adjuvants—are mixed.
Mixing tank
The mash is heated, allowing enzymes in the mash to convert starches into fermentable sugars. The liquid wort is then separated from the spent grains.
Spent grains
Mash tun
Hops
The wort is pumped into the brew kettle, where it is boiled while hops are slowly added. The wort is separated from the hops and cooled.
Spent hops
Brew kettle
Yeast
A special strain of Saccharomyces, or brewer's yeast, is added to the wort. Fermentation begins. Excess yeast cells are removed after fermentation.
Surplus yeast
Fermenting tank
The beer is ripened in the lagering tank. Yeast and unwanted flavor compounds settle out.
Lagering tank
Beer is clarified by filtration, and pasteurized or membrane filtered before bottling.
Filtration and bottling

Figure 32.5 Commercial Production of Beer What is the purpose of the membrane filtering in the last step?

desirable bitter flavor to the beer and contribute antibacterial substances. The mixture is boiled to extract the flavor components of hops, concentrate the wort, inactivate enzymes, kill most microorganisms, and precipitate proteins, facilitating their removal. The wort is then centrifuged to remove the solids, including hops and precipitated proteins, and cooled before being transferred to the fermentation tank.

Special strains of **brewer's yeasts** (as opposed to baker's yeasts) are commonly used in beer-making. **Bottom yeasts,** such as a *Saccharomyces cerevisiae* strain referred to as *S. carlsbergensis,* tend to form clumps that sink to the bottom of the fermentation vat. These yeasts ferment best at temperatures between 6°C and 12°C and usually take 8 to 14 days to complete fermentation. In contrast, **top-fermenting yeasts**—certain strains of *Saccharomyces cerevisiae*—are distributed throughout the wort but are carried to the top of the vat by the rising CO_2. They ferment at higher temperatures (14°C to 23°C) and over a shorter period (5 to 7 days). Most American beers are lagers produced by the bottom yeasts. Ales, porters, and stouts are made using top-fermenting yeasts.

The fermentation process generates beer with an alcohol content ranging from 3.4% to 6%. Most of the yeasts settle out following fermentation and are removed. These yeasts can be sold as flavor and dietary supplements. The beer is then aged, during which time residual, unwanted flavor compounds are metabolized by remaining yeast cells or settle out. Cask-conditioned beer undergoes a second fermentation, which generates CO_2 in the cask. Other beers must be carbonated to replace the CO_2 that escapes during fermentation. After aging, beer is clarified by filtration, microorganisms are removed or killed using membrane filtration or pasteurization, and the product is packaged.

Distilled Spirits

The manufacturing of distilled spirits such as scotch, whiskey, and gin is initially similar to that of beer, except the wort is not boiled. Consequently, degradation of starch by the enzymes in the wort continues during the fermentation. When the fermentation is complete, the ethanol is collected by distillation.

Different types of spirits are made in different ways. For example, rum is made by fermenting sugar cane or molasses. Malt scotch whiskey is the product of the fermentation of barley that is then aged for several years in oak sherry casks. The wood and the residual sherry contribute both flavor and color to the whiskey as it ages. Lactic acid bacteria are used to produce lactic acid in grain mash for making sour-mash whiskey. The yeast *S. cerevisiae* subsequently ferments the sour mash to form alcohol. The distilled spirit tequila is traditionally made from the fermentation of juices from the agave plant using the bacterium *Zymomonas mobilis*. This bacterium ferments sugars to ethanol and CO_2 via a pathway similar to the yeast alcoholic fermentation pathway.

Vinegar

Vinegar, an aqueous solution of at least 4% acetic acid, is the product of the oxidation of ethanol by the acetic acid bacteria—

Acetobacter and *Gluconobacter* species. Acetic acid bacteria are strictly aerobic, Gram-negative rods, characterized by their ability to carry out a number of oxidations. They can tolerate high concentrations of acid as they oxidize alcohol to acetic acid.

Alcohol is commercially converted to vinegar using processes that provide readily available oxygen to hasten the oxidation reaction. The **vinegar generator** sprays alcohol onto loosely packed wood shavings that harbor a biofilm of acetic acid bacteria. As the alcohol trickles through the bacteria-coated shavings, it is oxidized to acetic acid. In principle, the vinegar generator operates much like the trickling filter used in wastewater treatment by providing a large surface area for aerobic metabolism. The **submerged culture reactor** is an enclosed system that continuously pumps small air bubbles into alcohol that has been inoculated with acetic acid bacteria. ■ trickling filter, p. 741 ■ biofilm, p. 85

Bread

Yeast bread rises through the action of **baker's yeast**—strains of *Saccharomyces cerevisiae* carefully selected for the commercial baking industry. The CO_2 produced during fermentation causes the bread to rise, producing the spongy texture characteristic of yeast breads. The alcohol evaporates during baking.

Yeast bread is made from a mixture of flour, sugar, salt, milk or water, yeast, and sometimes butter or oil **(figure 32.6).** Packaged baker's yeast that can be reconstituted in warm water is readily available as pressed cakes or dried granules. An excess of yeast is added to enable adequate production of CO_2 in a time period too short to permit multiplication of spoilage bacteria.

Sourdough bread is made with a combination of yeast and lactic acid bacteria. Lactic acid is produced as well as alcohol and CO_2, giving the bread its sour flavor.

Changes Due to Mold Growth

Molds contribute to the flavor and texture of some cheeses, as already discussed. In addition, many traditional dishes and condiments used throughout the world are produced by encouraging the growth of molds on food **(table 32.3).** Successions of naturally occurring microorganisms are often involved. The microbiological and chemical aspects of many of these traditional foods have not been extensively studied.

Soy Sauce

Soy sauce is made by inoculating equal parts of cooked soybeans and roasted cracked wheat with a culture of either *Aspergillus oryzae* or *A. sojae*. The mixture, called **koji,** is allowed to stand for several days, during which time carbohydrates and proteins in the soybeans are broken down, producing a yellow-green liquid containing fermentable sugars, peptides, and amino acids. After this initial step, the mixture is put into a large container with an 18% NaCl solution, or brine. Salt-tolerant microorganisms then grow and impart changes over an extended period. Microorganisms involved in this stage of fermentation include lactobacilli, pediococci, and yeasts. After the brine mixture is allowed to ferment for 8 to 12 months, the liquid soy sauce is removed. The residual solids are used as animal feed.

Ingredients (mix)

Ingredients of bread include yeast, milk or water, oil, flour, salt, and sugar.

Knead

The dough is thoroughly mixed and kneaded.

Rise (fermentation)

The dough is put in a bowl and allowed to rise. During this time, the yeast produces ethanol and CO_2. The dough rises to approximately double the original volume. The gas generated creates air pockets in the dough, producing the texture we see in the finished bread.

Following the period of rising, the dough is shaped into loaves and then goes through a second rising. The bread dough is then baked.

Baking

Note the holes caused by the production of CO_2 and the evaporation of ethanol in the finished loaf of bread.

Figure 32.6 Bread Production

Table 32.3 Foods Produced Using Molds

Food	Characteristics
Soy sauce	Koji is produced by inoculating soybeans and cracked wheat with a starter culture of *Aspergillus oryzae* or *A. sojae*; the mixture is then added to a brine and incubated for many months.
Tempeh	Soybeans are fermented by lactic acid bacteria and then inoculated with a species of the mold *Rhizopus*; Indonesia.
Miso	Rice, soybeans, or barley are inoculated with *Aspergillus oryzae*; Asia.
Cheeses	
Roquefort, Gorgonzola, and Stilton	Curd is inoculated with *Penicillium roquefortii*.
Brie and Camembert	Wheels of cheese are inoculated with selected species of *Penicillium*.

MICROCHECK 32.2

Lactic acid bacteria are used to produce a variety of foods including cheese, yogurt, sauerkraut, and some sausages. Yeast is used to produce alcoholic beverages and breads. Some cheeses and many traditional foods owe their characteristics to changes caused by molds.

✓ Describe how the metabolism of lactic acid bacteria differs from that of most other microorganisms that can grow aerobically.

✓ How does the use of starter cultures improve the safety of fermented meat products?

✓ How could cottage cheese be produced without bacteria?

32.3 Food Spoilage

Focus Point

- Distinguish between fermented foods and spoiled foods.

Food spoilage encompasses any undesirable changes in food. Spoilage microorganisms produce metabolites that have repugnant tastes and odors. While these are aesthetically disagreeable, they are generally not harmful. This is not surprising when microbial growth requirements are considered. Most human pathogens grow best at temperatures near 37°C, whereas most foods are usually stored at temperatures well below the normal body temperature. Similarly, the nutrients available in fruits, vegetables, and other foods are generally not suitable for the optimum growth of human pathogens. As a result, the non-pathogens can easily outgrow the pathogens when competing for the same nutrients. Spoiled foods are considered unsafe to eat, however, because high numbers of spoilage organisms indicate that foodborne pathogens may be present as well.

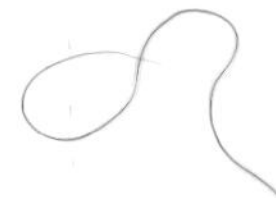

Common Spoilage Bacteria

Numerous types of bacteria are important in food spoilage. *Pseudomonas* species can metabolize a wide variety of compounds, and grow on and spoil many different kinds of foods, including meats and vegetables. Psychrophilic species are notorious for spoiling refrigerated foods. Members of the genus *Erwinia* produce enzymes that degrade pectin, and so they commonly cause soft rot of fruits and vegetables. *Acetobacter* species transform ethanol to acetic acid, the principal acid of vinegar. Although this property is very beneficial to commercial producers of vinegar, it presents a great problem to wine producers. Milk products are sometimes spoiled by members of the genus *Alcaligenes* that form a glycocalyx, causing strings of slime, or "ropiness," in raw milk. The lactic acid bacteria, including species of *Streptococcus, Leuconostoc,* and *Lactobacillus,* all produce lactic acid. Anyone who has unexpectedly consumed sour milk knows that this can be disagreeable. Members of the genera *Bacillus* and *Clostridium* are particularly troublesome causes of food spoilage because their heat-resistant endospores survive cooking and, in some cases, canning. *Bacillus coagulans* and *B. stearothermophilus* spoil some canned foods. ■ glycocalyx, p. 64

Common Spoilage Fungi

A wide variety of fungi, including species of *Rhizopus, Alternaria, Penicillium, Aspergillus,* and *Botrytis,* spoil foods. Because fungi grow readily in acidic as well as low-moisture environments, fruits and breads are more likely to be spoiled by fungi than by bacteria. *Aspergillus flavus* grows on peanuts and other grains, producing **aflatoxin,** a potent carcinogen monitored by the Food and Drug Administration.

MICROCHECK 32.3

The metabolites of microbes can spoil foods by imparting undesirable flavors, odors, and textures.

✓ What characteristics of *Pseudomonas* species enable them to spoil such a wide variety of foods?

✓ Why do fungi most commonly spoil breads and fruits?

32.4 Foodborne Illness

Focus Point

- Distinguish between foodborne intoxication and foodborne infection, and give two examples of each.

Foodborne illness, commonly referred to as **food poisoning,** occurs when a pathogen, or a toxin it produced, is consumed in a food product. Food production is carefully regulated in the United States to prevent foodborne illness. Federal, state, and local agencies cooperate in inspections to help enforce protective laws.In spite of strict controls, millions of cases of food poisoning are estimated to occur each year from foods prepared either commercially, at home, or in institutions such as hospitals and schools. The vast majority of these cases could have been prevented with proper storage, sanitation, and preparation. **Table 32.4** lists some bacteria that cause foodborne illness in the United States.

To more accurately determine the burden of foodborne illness in the United States, a government program called **FoodNet** (Foodborne Disease Active Surveillance Network) now collects data on laboratory-confirmed cases of diarrheal illness in 10 states, covering approximately 15% of the population (www.cdc.gov/foodnet). By gaining a better understanding of the epidemiology of foodborne diseases, these diseases can hopefully be prevented more easily.

Foodborne Intoxication

Foodborne intoxication is an illness that results from the consumption of an exotoxin produced by a microorganism growing in a food product. When such a food product is ingested, it is the toxin that causes illness, not the living organisms **(figure 32.7).** *Staphylococcus aureus* and *Clostridium botulinum* are two examples of organisms that cause foodborne intoxication. ■ exotoxin, p. 403

Staphylococcus aureus

Many strains of *Staphylococcus aureus* produce a toxin that, when ingested, causes nausea and vomiting. Although *S. aureus* does not compete well with most spoilage organisms, it thrives in moist, rich foods in which other organisms have been killed or their growth inhibited. For example, *S. aureus* can grow with little competition on unrefrigerated salty products such as ham (a_w of 0.91). Creamy pastries and starchy salads stored at room temperature also offer ideal conditions for the growth of this pathogen, because the cooking of the ingredients kills most other competing organisms. The source of *S. aureus* is usually a human carrier who has not followed adequate hygiene procedures, such as handwashing, before preparing the food. If the organism is inoculated into a food that can support its growth, and the food is left at room temperature for several hours, *S. aureus* can grow and produce the toxin. Unlike most exotoxins, *S. aureus* toxin is heat-stable, so that cooking the food will not destroy it. ■ *Staphylococcus aureus*, p. 535

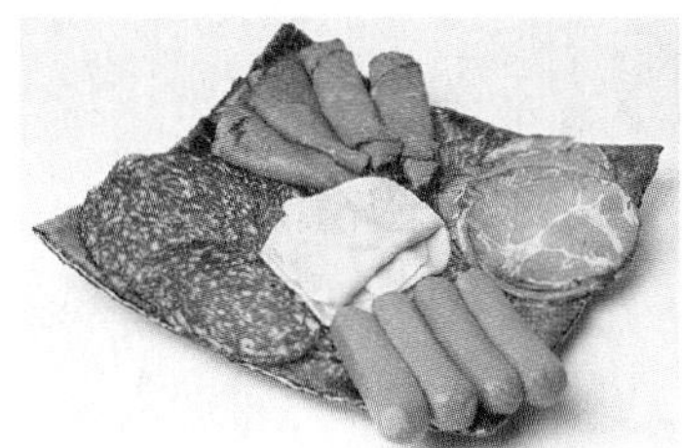

- Most bacteria that normally compete with *Staphylococcus aureus* are either killed by cooking or inhibited by high salt conditions.
- Food handler inoculates *S. aureus* onto food.
- *S. aureus* grows and produces toxin when food is allowed to slowly cool or is stored at room temperature.
- A person ingests the toxin-containing food. Symptoms of staph food poisoning, including nausea, abdominal cramping, and vomiting, begin after 4 to 6 hours.

- *Clostridium botulinum* endospores, common in soil and marine sediments, contaminate many different foods.
- Endospores survive inadequate canning processes. Canned foods are anaerobic.
- Surviving *C. botulinum* endospores germinate, grow, and produce toxin in low-acid canned foods.
- A person ingests the toxin-containing food. Symptoms of botulism, including weakness, double vision, and progressive inability to speak, swallow, and breathe, begin in 12 to 36 hours.

FIGURE 32.7 Typical Events Leading to *Staphylococcus aureus* and *Clostridium botulinum* Foodborne Intoxications

Botulism

Botulism is a paralytic disease caused by ingestion of a neurotoxin produced by the anaerobic, spore-forming, Gram-positive

Table 32.4 Common Foodborne Illnesses

Organism	Symptoms	Foods Commonly Implicated
Intoxication		
Clostridium botulinum	Weakness; double vision; progressive inability to speak, swallow, and breathe	Low-acid canned foods such as vegetables and meats
Staphylococcus aureus	Nausea, vomiting, abdominal cramping	Cured meats, creamy salads, cream-filled pastries
Infection		
Campylobacter species	Diarrhea, fever, abdominal pain, nausea, headache	Poultry, raw milk
Clostridium perfringens	Intense abdominal cramps, watery diarrhea	Meats, meat products
Escherichia coli O157:H7	Severe abdominal pain, bloody diarrhea; may progress to hemolytic uremic syndrome	Ground beef, raw vegetables, unpasteurized juices
Listeria monocytogenes	Influenza-like symptoms, fever, may progress to septicemia, meningitis	Raw milk, cheese, meats, raw vegetables
Salmonella species	Nausea, vomiting, abdominal cramps, diarrhea, fever	Poultry, eggs, milk, meat
Shigella species	Abdominal cramps; diarrhea with blood, pus, or mucus; fever; vomiting	Salads, raw vegetables
Vibrio parahaemolyticus	Diarrhea, abdominal cramps, nausea, vomiting, headache, fever, chills	Fish and shellfish

PERSPECTIVE 32.1

Botox for Beauty and Pain Relief

The exotoxin produced by *Clostridium botulinum* is one of the most powerful poisons known. These anaerobic soil bacteria are sporeformers found naturally on many foods. They survive usual cooking methods and inadequate canning procedures. Should conditions become anaerobic in the food, toxin can be produced, and if ingested it causes botulism. A few milligrams of this exotoxin is sufficient to kill the entire population of a large city. The botulinum toxin blocks transmission of acetylcholine nerve signals to the muscles, resulting in paralysis and often in death. This is a very dangerous toxin. Yet surprisingly, in recent years the toxin, known as botox, has become useful in treating various conditions.

Botox is often used as a cosmetic treatment to remove facial lines, such as frown lines. Extremely dilute botox is injected directly into the area, paralyzing the muscles that are causing the frown or other lines. While the lines are erased, the effect is temporary and the treatment must be repeated after several months.

Botox is also used to relieve a number of very painful and disabling conditions involving muscle contractions such as dystonia (severe muscle cramping). For example, cervical dystonia is a painful disease in which muscles in the neck and shoulders contract involuntarily, causing jerky movements, muscle pain, and tremors. Injections of botox directly into the affected areas give relief for 3 to 4 months, after which the treatment can be repeated. Another example is Parkinson's disease, a condition in which certain nerve cells are lost, resulting in tremor, impaired movement, and in some cases, dystonia. Botox injected into the affected muscles can give dramatic, although temporary, relief. So, in spite of its powerful and dangerous properties, botulinum toxin, when used with great care, can be a useful therapeutic agent.

rod *Clostridium botulinum*. Unfortunately, growth of the organism and production of the deadly toxin may not result in any noticeable changes in the taste or appearance of the food. Canning processes for low-acid foods are specifically designed to destroy the endospores of this organism. If the endospores are not destroyed due to processing errors, or they are introduced post-processing, they can germinate in these foods. The resulting vegetative cells can then grow and produce toxin. Such errors are rare in commercially canned foods, and most cases of botulism are due to processing errors in home-canned foods. As an added safety measure, low-acid, home-canned foods should be boiled for at least 15 minutes immediately before consumption. The toxin is heat-labile (sensitive), and so the heat treatment will destroy any toxin that may have been produced. Cans that are damaged should be discarded, because they might have small holes through which *C. botulinum* could enter. Likewise, bulging cans indicate gas production, which could indicate microbial growth, and should be discarded. ■ **botulism, p. 657** ■ **neurotoxin, p. 405**

Foodborne Infection

Unlike foodborne intoxication, **foodborne infection** requires the consumption of living organisms. The symptoms of the illness, which usually do not appear for at least 1 day after eating the contaminated food, usually include diarrhea. Thorough cooking of food immediately before consumption will kill the organisms, thereby preventing infection. *Escherichia coli* O157:H7 and *Salmonella* and *Campylobacter* species are examples of organisms that cause foodborne infection **(figure 32.8).**

Salmonella and Campylobacter

Salmonella and *Campylobacter* are two genera commonly associated with poultry products such as chicken, turkey, and eggs. Inadequate cooking of these products can result in foodborne infection. **Cross-contamination** of other foods can result in the transfer of pathogens to those foods. For example, if a cutting board on which raw chicken was cut is then immediately used to cut up vegetables for a salad, the salad can become contaminated with *Salmonella* or *Campylobacter* species. ■ ***Salmonella*, p. 601** ■ ***Campylobacter*, p. 602**

Escherichia coli O157:H7

Escherichia coli O157:H7 causes bloody diarrhea that sometimes develops into hemolytic uremic syndrome (HUS), a life-threatening condition. The bacterium is commonly found in the intestinal tract of healthy cattle and other livestock, where it is subsequently shed in their feces. Because of this, meats can easily become contaminated. In 2007, a multistate, foodborne outbreak of *E. coli* O157:

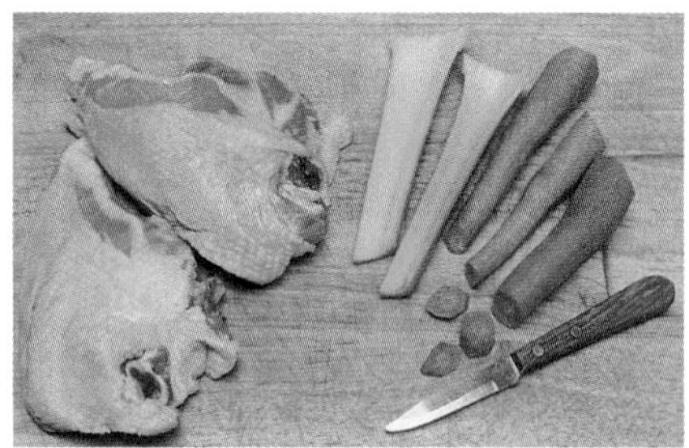

- Incomplete cooking fails to kill all pathogens. Surviving *Salmonella* and/or *Campylobacter* can multiply as food is cooled slowly or stored at room temperature.
- Live organisms are ingested. They multiply in the intestinal tract and cause disease. Symptoms include diarrhea, abdominal pain, and nausea.

- Incomplete cooking fails to kill all pathogens. Even low numbers of surviving *E. coli* O157:H7 can cause illness.
- Live organisms are ingested. They multiply in the intestinal tract and cause disease. Symptoms include severe abdominal pain and bloody diarrhea.

FIGURE 32.8 Typical Events Leading to *Salmonella, Campylobacter,* and *E. coli* O157:H7 Foodborne Infection

H7 was linked to frozen ground beef patties, resulting in a recall of 21.7 million pounds of the product. Ground meats such as these are a particularly troublesome source of infection. While the initial contamination by the bacterium normally occurs on the surface and is easily destroyed by searing the exterior of meats such as steaks, grinding the meat distributes the bacterial cells throughout the product. Prevention then involves more thorough cooking so that enough heat reaches the center to kill all *E. coli* cells. Outbreaks have also been linked to unpasteurized milk and produce that was contaminated with animal manure. ■ hemolytic uremic syndrome, p. 598

MICROCHECK 32.4

Foodborne intoxication results from the consumption of toxins produced by organisms growing in a food. Foodborne infection results from consumption of living organisms.

✓ How does cooking a home-canned food immediately prior to consumption prevent botulism?

✓ Which foodborne pathogen can cause hemolytic uremic syndrome?

✓ Why would a large number of competing microorganisms in a food sample result in lack of sensitivity of culture methods for detecting pathogens?

32.5 Food Preservation

Focus Point

- Describe the methods used to preserve foods.

Preventing the growth and concurrent metabolic activities of microorganisms that cause spoilage and foodborne illness preserves the quality of food. Some methods of **food preservation,** such as drying and salting, have been known throughout the ages, whereas others have been discovered or developed more recently. The major methods of preserving foods—high-temperature treatment, low-temperature storage, addition of antimicrobial chemicals and irradiation—are briefly summarized here and described in more detail in chapter 5.

- **Canning.** The canning process destroys all spoilage and pathogenic organisms capable of growth at normal storage temperatures. Low-acid foods are processed using steam under pressure (autoclaving) in order to reach temperatures high enough to destroy the endospores of *Clostridium botulinum.* Acidic foods are not subjected to as stringent heating conditions because *Clostridium botulinum* cannot grow and produce toxin in those foods. ■ canning, p. 114
- **Pasteurization.** Heating foods under controlled conditions at high temperatures for short periods of time destroys non-spore-forming pathogens and reduces the numbers of spoilage organisms without significantly altering the flavor. ■ pasteurization, p. 112
- **Cooking.** Cooking, like pasteurization, can destroy non-spore-forming organisms. Cooking obviously alters the characteristics of food, however. Heat distribution may be uneven, resulting in survival of organisms in inadequately heated regions.
- **Refrigeration.** Refrigeration preserves food by slowing the growth rate of microbes. Many organisms, including most pathogens, are unable to multiply at low temperatures. ■ temperature requirements, p. 90
- **Freezing.** Freezing stops microbial growth because water in the form of ice is unavailable for biological reactions. Some of the microbial cells will be killed by damage caused by ice crystals, but those remaining can grow and spoil food once it is thawed.
- **Drying/reducing the a_w.** Drying foods or adding high concentrations of sugars or salts inhibits microbial growth by decreasing the available moisture. Eventually, however, molds may grow. ■ drying food, p. 122
- **Lowering the pH.** Lowering the pH, either by adding acids or encouraging fermentation by lactic acid bacteria, inhibits a wide range of spoilage organisms and pathogens.
- **Adding antimicrobial chemicals.** Organic acids such as propionic acid, benzoic acid, and sorbic acid are naturally occurring antimicrobial chemicals that are added to a variety of foods to inhibit fungal growth. Nitrates are added to cured meats to inhibit growth of *Clostridium botulinum* and other organisms. Wine, fruit juices, and other products are preserved by the addition of sulfur dioxide. ■ chemical preservatives, p. 122
- **Irradiation.** Gamma irradiation destroys microorganisms without significantly altering the flavor of foods such as spices and meats. ■ irradiation, p. 116

MICROCHECK 32.5

Food spoilage can be eliminated or delayed by destroying microorganisms or altering conditions to inhibit their growth.

✓ Why does the canning process for low-acid foods require higher temperatures than the temperature required in the canning process for acidic foods?

✓ Why are nitrates added to cured meats?

✓ Microorganisms are often grouped according to their optimum growth temperatures. Which of these groups is most likely to spoil refrigerated foods?

FUTURE CHALLENGES

Using Microorganisms to Nourish the World

The world's steadily increasing population mandates the efficient use of finite natural resources and the development of new protein supplies to nourish that growing populace. One potential solution that addresses both of these needs is to cultivate microorganisms as a protein source, employing industrial by-products currently considered wastes as the growth medium.

The term **single-cell protein,** or **SCP,** was coined in the 1960s to describe the use of unicellular organisms such as yeasts and bacteria as a protein source. Today, the term generally encompasses the use of multicellular microorganisms as well, and might more accurately be called *microbial biomass.* While whole cells can be consumed, a more palatable alternative is to extract proteins from those cells and use them to form textured protein products.

Yeasts are considered the most promising, large-scale source of single-cell protein. They multiply rapidly, are larger than bacteria, and are more readily acceptable as a potential food. The most suitable type of yeast depends on the growth medium employed. The various genera of yeasts utilize differing carbohydrate sources and conditions for growth. For example *Kluyveromyces marxianus* can be grown on whey, a by-product of cheese-making; *Saccharomyces cerevisiae* can grow on molasses, a by-product of the sugar industry; and *Candida utilis* can multiply on cellulose-containing by-products of the pulp and paper industry. Most of these wastes must be supplemented with a nitrogen source, as well as various vitamins and minerals, in order to support the growth of yeasts or other microorganisms.

To a lesser extent, the use of bacteria as SCP is also being explored. The cyanobacterium *Spirulina maxima* can be cultivated in alkaline lakes and then harvested and dried. This requires adequate sunlight and warmth, making large-scale, year-round production practical in only some parts of the world.

One of the chief concerns regarding the consumption of microorganisms as a protein source is their high concentration of nucleic acid, primarily RNA. Yeast and bacteria contain as much as eight times more RNA per gram as does meat. High levels of nucleic acid in the diet cause an increase in uric acid in the blood, which can lead to gout and kidney stones. Chemical and enzymatic methods are being developed to decrease the nucleic acid in SCP without altering the nutritional value of the protein.

SUMMARY

32.1 Factors Influencing the Growth of Microorganisms in Foods

Intrinsic Factors

Bacteria require a high a_w. Fungi can grow in foods that have an a_w too low to support the growth of bacteria. Many species of bacteria, including most pathogens, are inhibited by acidic conditions. The nutritional content of a food determines the kinds of organisms that can grow in it. Rinds, shells, and other coverings aid in protecting some foods from the invasion of microorganisms. Some foods contain natural antimicrobial chemicals that may help prevent spoilage.

Extrinsic Factors

Low temperatures halt or inhibit the growth of most foodborne microorganisms. Psychrophiles and psychrotrophs, however, grow at refrigeration temperatures. The presence or absence of O_2 affects the type of microbial population able to grow in a food.

32.2 Microorganisms in Food and Beverage Production

Not only are fermented foods perceived as pleasant tasting, the acids produced inhibit the growth of many spoilage organisms as well as foodborne pathogens.

Lactic Acid Fermentations by the Lactic Acid Bacteria (table 32.1)

The tart taste of yogurt, pickles, sharp cheese, and some sausages is due to the metabolic products of the lactic acid bacteria. Starter cultures, and sometimes rennin, are added to pasteurized milk to make cheese. Other bacteria or fungi are sometimes added to ripened cheese to impart characteristic flavors or textures (figure 32.3). Fermentation of vegetables, **pickling,** relies on naturally occurring lactic acid bacteria. Commercial sausage production utilizes starter cultures to rapidly decrease the pH and prevent the growth of pathogens.

Alcoholic Fermentations by Yeast (table 32.2)

Wine is the product of the alcoholic fermentation of naturally occurring sugars in the juices of fruit by specially selected strains of *Saccharomyces cerevisiae* (figure 32.4). Beer production is a multistep process designed to break down the starches of grains such as barley to produce simple sugars, which can then serve as a substrate for alcoholic fermentation by yeast (figure 32.5). Distilled spirits are produced using distillation to collect the alcohol generated during fermentation. Vinegar is the product of the oxidation of alcohol by the acetic acid bacteria. In bread-making, the CO_2 produced by yeast causes bread to rise, and the alcohol is lost to evaporation (figure 32.6).

Changes Due to Mold Growth (table 32.3)

Some cheeses and other foods are produced by encouraging the growth of molds on foods. Soy sauce is made by allowing species of *Aspergillus* to degrade a mixture of soybeans and wheat, which is then fermented in brine.

32.3 Food Spoilage

Food spoilage is most often due to the metabolic activities of microorganisms as they grow and utilize the nutrients in the food.

Common Spoilage Bacteria

Pseudomonas, Erwinia, Acetobacter, Alcaligenes, lactic acid bacteria, and endosporeformers are important causes of food spoilage.

Common Spoilage Fungi

Fungi grow readily in acidic as well as low-moisture environments; therefore, fruits and breads are more likely to be spoiled by fungi than by bacteria.

32.4 Foodborne Illness (table 32.4)

Foodborne Intoxication

Foodborne intoxication results from consuming a toxin produced by a microorganism growing in a food product (figure 32.7). Many strains of *Staphylococcus aureus* produce a toxin that, when ingested, causes nausea and vomiting. Botulism is caused by ingestion of a neurotoxin produced by the anaerobic, spore-forming, Gram-positive rod *Clostridium* *botulinum*. As an added safety measure, low-acid home-canned foods should be boiled at least 15 minutes immediately before consumption to ensure destruction of potential botulinum toxin.

Foodborne Infection

Foodborne infection requires the consumption of living organisms (figure 32.8). Thorough cooking of food immediately before consumption will kill bacteria, thereby preventing foodborne infection. *Salmonella* and *Campylobacter* species are commonly associated with poultry products. Some outbreaks of *E. coli* O157:H7 have been traced to undercooked contaminated hamburger patties and produce contaminated with manure.

32.5 Food Preservation

Food spoilage can be eliminated or delayed by destroying microorganisms or altering conditions to inhibit their growth. Methods used to preserve foods include canning, pasteurization, cooking, freezing, refrigeration, reducing the a_w, lowering the pH, adding antimicrobial chemicals, and irradiation.

REVIEW QUESTIONS

Short Answer

1. What is the purpose of rennin in cheese-making?
2. What causes the bluish-green veins to form in blue cheese?
3. What causes the holes to form in Swiss cheese?
4. What is the difference between traditional acidophilus milk and sweet acidophilus milk?
5. What is the purpose of the mashing step in beer-making?
6. Explain how *Alcaligenes* species cause "ropiness" in raw milk.
7. Explain the significance of *Aspergillus flavus* growing in grain products.
8. Explain the typical sequence of events that lead to botulism.
9. Explain the typical sequence of events that lead to staphylococcal food poisoning.
10. How does canning differ from pasteurization?

Multiple Choice

1. Benzoic acid is an antimicrobial chemical naturally found in which of the following foods?
   a) Apples b) Cranberries c) Eggs
   d) Milk e) Yogurt
2. The a_w of a food product reflects which of the following?
   a) Acidity of the food
   b) Presence of antimicrobial constituents such as lysozyme
   c) Amount of water available
   d) Storage atmosphere
   e) Nutrient content
3. What is a generally minimum pH for growth and toxin production by *Clostridium botulinum* and other foodborne pathogens?
   a) 8.5 b) 7.0 c) 6.5 d) 4.5 e) 2.0
4. Which of the following organisms cause foodborne intoxication?
   a) *E. coli* O157:H7 b) *Campylobacter* species
   c) *Lactobacillus* species d) *Salmonella* species
   e) *Staphylococcus aureus*
5. Which group of organisms most commonly spoils breads, fruits, and dried foods?
   a) *Acetobacter* b) Fungi
   c) Lactic acid bacteria d) *Pseudomonas*
   e) *Saccharomyces*
6. Canned pickles require less stringent heat processing than canned beans, because pickles
   a) contain fewer nutrients.
   b) are more acidic.
   c) have a lower a_w.
   d) contain antimicrobial chemicals.
   e) are less likely to be contaminated with endospores.
7. Which of the following genera is used in bread, wine, and beer production?
   a) *Lactobacillus* b) *Pseudomonas*
   c) *Saccharomyces* d) *Streptococcus*
   e) *Staphylococcus*
8. Most spoilage bacteria cannot grow below an a_w of
   a) 0.3. b) 0.5. c) 0.7. d) 0.9. e) 1.0.
9. Which of the following is often added to wine to inhibit growth of the natural microbial population of grapes?
   a) Benzoic acid b) Lactic acid c) Carbon dioxide
   d) Sulfur dioxide e) Oxygen
10. In the brewing process, the sugar and nutrient extract obtained by soaking germinated grain in warm water is called
    a) baker's yeast. b) hops. c) malt.
    d) must. e) wort.

Applications

1. A small cheese-manufacturing company in Wisconsin is looking for ways to reduce the costs of disposing of whey, a cheese by-product. As a food microbiologist working at the company, what would you suggest that the company do with the thousands of liters of whey being produced per month so the company can actually profit from it?
2. A microbiologist is troubleshooting a batch of home-brewed ale that did not ferment properly. She noticed that the alcohol content was only 2%, well below the desired level. Microscopic examination showed numerous yeast cells. Chemical analysis showed low levels of sugar, high levels of CO_2, and large amounts of protein in the liquid. What did the microbiologist conclude as the probable cause of the beer not coming out properly?

Critical Thinking

1. It has been argued that the nature of the growth of fungi in Roquefort cheese, indicated by the appearance of bluish-green veins, is evidence that these fungi require oxygen for growth. How does this evidence lead to the conclusion?
2. In the production of sauerkraut, a natural succession of lactic acid bacteria is observed growing in the product. What causes the succession? What does this tell you about the optimal growth conditions of the different species of lactic acid bacteria?

Microbial Mathematics

Because prokaryotes are very tiny and can multiply to very large numbers of cells in short time periods, convenient and simple ways are used to designate their numbers without resorting to many zeros before or after the number. In the study of microbiology, it is important to gain an understanding of the metric system, which is used in scientific measurements.

The basic unit of measure is the meter, which is equal to about 39 inches. All other units are fractions of a meter:

1 decimeter is one tenth = 0.1 meter
1 centimeter is one hundredth = 0.01 meter
1 millimeter is one thousandth = 0.001 meter

Because prokaryotes are much smaller than a millimeter, even smaller units of measure are used. A millionth of a meter is a micrometer = 0.000001 meter, and is abbreviated μm. This is the most frequently used size measurement in microbiology, since bacteria are in this size range. For comparison, a human hair is about 75 μm wide.

Since it is inconvenient to write so many zeros in front of the 1, an easier way of denoting the same number is through the use of superscript, or exponential, numbers (exponents). One hundred dollars can be written 10^2 dollars. The 10 is called the base number and the 2 is the exponent. Conversely, one hundredth of a dollar is 10^{-2} dollars; thus the exponent is negative. The base most commonly used in biology is 10 (which is designated as $\log_{10}$). The above information can be summarized as follows:

1 millimeter = 1 mm = 0.001 meter = 10^{-3} meter
1 micrometer = 1 μm = 0.000001 meter = 10^{-6} meter
1 nanometer = 1 nm = 0.000000001 meter = 10^{-9} meter

The same prefix designations can be used for weights. The basic unit of weight is the gram, abbreviated g. Approximately 450 grams are in a pound.

1 milligram = 1 mg = 0.001 gram = 10^{-3} g
1 microgram = 1 μg = 0.000001 gram = 10^{-6} g
1 nanogram = 1 ng = 0.000000001 gram = 10^{-9} g
1 picogram = 1 pg = 0.000000000001 gram = 10^{-12} g

Note that the number of zeros before the 1 is one less than the exponent.

The value of the number is obtained by multiplying the base by itself the number of times indicated by the exponent.

Thus, $10^1 = 10 \times 1 = 10$
$10^2 = 10 \times 10 = 100$
$10^3 = 10 \times 10 \times 10 = 1{,}000$

When the exponent is negative, the base and exponent are divided into 1.

For example, $10^{-2} = 1/10 \times 1/10 = 1/100 = 0.01$

When multiplying numbers having exponents to the same base, the exponents are added.

For example, $10^3 \times 10^2 = 10^5$ (not 10^6)

When dividing numbers having exponents to the same base, the exponents are subtracted.

For example, $10^5 \div 10^2 = 10^3$

In both cases, only if the bases are the same can the exponents be added or subtracted.

Microbial Teminology

Singular and Plurals

Singular	Plural	Singular	Plural
alg*a*	alg*ae*	fung*us*	fung*i*
ameb*a*	ameb*ae*	hydroly*sis*	hydroly*ses*
bacill*us*	bacill*i*	hyph*a*	hyph*ae*
bacteri*um*	bacteri*a*	inocul*um*	inocul*a*
cili*um*	cili*a*	medi*um*	medi*a*
clostridi*um*	clostridi*a*	mucos*a*	mucos*ae*
cocc*us*	cocc*i*	myceli*um*	myceli*a*
conidi*um*	conidi*a*	mycos*is*	mycos*es*
dat*um*	dat*a*	phyl*um*	phyl*a*
diagnos*is*	diagnos*es*	pil*us*	pil*i*
fimbri*a*	fimbri*ae*	nucle*us*	nucle*i*
flagell*um*	flagell*a*	sept*um*	sept*a*
foc*us*	foc*i*	synthes*is*	synthes*es*

Meanings of Prefixes and Suffixes

Prefix or Suffix	Meaning	Example
a-, an-	not, without	avirulent (lacking virulence), anaerobic (without air)
aer-	air	aerobic
anti-	against	antiseptic
-ase	enzyme	penicillinase
chlor-	green	chlorophyll
-chrom-	color	metachromatic (staining differently with the same dye)
-cide	causing death	germicide
co-, com-, con-	together	coenzyme
-cyan-	blue	pyocyanin (a blue bacterial pigment)
de-	down, from	dehydrate (remove water)
-dem-	people, district	epidemic
endo-	within	endospore (spore within a cell)
-enter-	intestine	enteritis (inflammation of the intestine)
epi-	upon	epidermis
erythro-	red	erythocyte (red blood cell)
eu-	well, normal	eukaryotic (true nucleus)
exo-	outside	exoenzyme (enzyme that acts outside the cell that produced it)
extra-	outside of	extracellular
flav-	yellow	flavoprotein (a protein containing a yellow enzyme)
-gen	produce, originate	antigen (a substance that induces a production of antibodies)

Prefix or Suffix	Meaning	Example
glyc-	sweet	glycemia (the presence of sugar in the blood)
hetero-	other	heterotroph (organism that obtains carbon from organic compounds)
homo-	common, same	homologous (similar in structure or origin)
hydr-	water	dehydrate
hyper-	excessive, above	hypersensitive
hypo-	under	hypotonic (having low osmotic pressure)
iso-	same, equal	isotonic (having the same osmotic pressure)
-itis	inflammation	appendicitis, meningitis
leuko-	white	leukocyte (white blood cell)
ly-, -lys, -lyt-	loosen; dissolve	bacteriolysis (dissolution of bacteria)
meso-	middle	mesophilic (preferring moderate temperatures)
meta-	changed	metachromatic (staining differently with the same dye)
micro-	small; one-millionth part	microscopic
milli-	one-thousandth part	millimeter (10^{-3} meter)
mito-	thread	mitochondrion (small, rod-shaped or granular organelle)
mono-	single	monotrichous (having a single flagellum)
multi-	many	multinuclear (having many nuclei)
myc-	fungus	mycotic (caused by fungus)
myx-	mucus	myxomycete (slime mold)
-oid	resembling	lymphoid (resembling lymphocytes)
-ose	a sugar	lactose (milk sugar)
-osis	disease of	coccidioidomycosis (disease caused by *Coccidioides*)
pan-	all	pandemic (widespread epidemic)
para-	beside	parasite (an organism that feeds in and at expense of the host)
patho-	disease	pathogenic (producing disease)
peri-	around	peritrichous (having flagella on all sides)
-phag-	eat	phagocyte (a cell that ingests other cell substances)
-phil	like, having affinity for	eosinophilic (staining with the dye eosin)
-phot-	light	photosynthesis
-phyll	leaf	chlorophyll (green leaf pigment)
pleo-	more	pleomorphic (occurring in more than one form)
poly-	many	polymorphonuclear (having a many-shaped nucleus)
post	after	postnatal (after birth)
pyo-	pus	pyogenic (producing pus)
-sta-	stop	bacteriostatic (inhibiting bacterial multiplication)
sym-, syn-	together	symbiosis (life together)
thermo-	heat	thermophilic (liking heat)
trans-	through, across	transfusion
-trich-	hair	monotrichous (having a single flagellum)
-troph	nourishment	autotroph (organism that obtains carbon from CO_2)
tox-	poison	toxin
zym-	ferment	enzyme

APPENDIX III

Pronunciation Key for Bacterial, Fungal, Protozoan, and Viral Names

A

Acetobacter (a-see′-toe-back-ter)
Achromobacter (a-krome′-oh-back-ter)
Acinetobacter calcoaceticus (a-sin-et′-oh-back-ter kal-koh-ah-see′-ti-kus)
Actinomyces israelii (ak-tin-oh-my′-seez iz-ray′-lee-ee)
Actinomycetes (ak-tin-oh-my′-seats)
Adenovirus (ad′-eh-no-vi-rus)
Agrobacterium tumefaciens (ag-rho-bak-teer′-ee-um too-meh-faysh′-ee-enz)
Alcaligenes (al-ka-li′-jen-ease)
Amoeba (ah-mee′-bah)
Arbovirus (are′-bow-vi-rus)
Aspergillus niger (ass-per-jill′-us nye′-jer)
Aspergillus oryzae (ass-per-jill′-us or-eye′-zee)
Azolla (aye-zol′-lah)
Azotobacter (ay-zoh′-toe-back-ter)

B

Bacillus anthracis (bah-sill′-us an-thra′-siss)
Bacillus cereus (bah-sill′-us seer′-ee-us)
Bacillus coagulans (bah-sill′-us coh-ag′-you-lans)
Bacillus fastidiosus (bah-sill′-us fas-tid-ee-oh′-sus)
Bacillus stearothermophilus (bah-sill′-us steer-oh-ther-maw′-fill-us)
Bacillus subtilis (bah-sill′-us sut′-ill-us)
Bacillus thuringiensis (bah-sill′-us thur′-in-jee-en-sis)
Bacteroides (back′-ter-oid′-eez)
Baculovirus (back′-you-low-vi-rus)
Beggiatoa (beg-gee-ah-toe′-ah)
Beijerinckia (by-yer-ink′-ee-ah)
Bordetella pertussis (bor-deh-tell′-ah per-tuss′-iss)
Borrelia burgdorferi (bor-real′-ee-ah berg-dor′-fir-ee)
Bradyrhizobium (bray-dee-rye-zoe′-bee-um)
Branhamella (bran-ham-el′-lah)
Brucella abortus (bru-sell′-ah ah-bore′-tus)

C

Campylobacter jejuni (kam′-peh-low-back-ter je-june′-ee)
Candida albicans (kan′-did-ah al′-bi-kanz)
Caulobacter (caw′-loh-back-ter)
Ceratocystis ulmi (see′-rah-toe-sis-tis ul′-mee)
Chlamydia trachomatis (klah-mid′-ee-ah trah-ko-ma′-tiss)
Claviceps purpurea (kla′-vi-seps purr-purr′-ee-ah)
Clostridium acetobutylicum (kloss-trid′-ee-um a-seat-tow-bu-till′-i-kum)
Clostridium botulinum (kloss-trid′-ee-um bot-you-line′-um)
Clostridium difficile (kloss-trid′-ee-um dif′-fi-seal)
Clostridium perfringens (kloss-trid′-ee-um per-frin′-gens)
Clostridium tetani (kloss-trid′-ee-um tet′-an-ee)
Coccidioides immitis (cock-sid-ee-oid′-eez im′-mi-tiss)
Coronavirus (kor-oh′-nah-vi-rus)
Corynebacterium diphtheriae (koh-ryne′-nee-bak-teer-ee-um dif-theer′-ee-ee)
Coxsackievirus (cock-sack-ee′-vi-rus)
Cryptococcus neoformans (krip-toe-cock′-us knee-oh-for′-manz)
Cytophaga (sigh-taw′-fa-ga)

D

Desulfovibrio (dee-sul-foh-vib′-ree-oh)

E

Eikenella corrodens (eye-keh-nell′-ah kor-roh′-denz)
Entamoeba histolytica (en-ta-mee′-bah his-toh-lit′-ik-ah)
Enterobacter (en′-ter-oh-back-ter)
Enterococcus faecalis (en′-ter-oh-kock′-us fee-ka′-liss)
Enterovirus (en′-ter-oh-vi-rus)
Epidermophyton (eh-pee-der′-moh-fy-ton)
Escherichia coli (esh-er-ee′-she-ah koh′-lee)

F

Filobasidiella neoformans (fee-loh-bah-si-dee-ell′-ah knee-oh-for′-manz)
Flavivirus (flay′-vih-vi-rus)
Flavobacterium (flay-vo-back-teer′-ee-um)
Francisella tularensis (fran-siss-sell′-ah tu-lah-ren′-siss)
Frankia (frank′-ee-ah)
Fusobacterium (fu′-zoh-back-teer-ee-um)

G

Gallionella (gal-ee-oh-nell′-ah)
Gardnerella vaginalis (gard-nee-rel′-lah va-jin-al′-is)
Giardia intestinalis (jee-are′-dee-ah in-test′-tin-al-is)
Giardia lamblia (jee-are′-dee-ah lamb′-lee-ah)
Gluconobacter (glue-kon-oh-back′-ter)
Gonyaulax (gon-ee-ow′-lax)
Gymnodinium breve (jim-no-din′-i-um brev-eh)

H

Haemophilus influenzae (hee-moff′-ill-us in-flew-en′-zee)
Helicobacter pylori (he′-lih-koh-back-ter pie-lore′-ee)
Hepadnavirus (hep-ad′-nah-vi-rus)
Hepatitis virus (hep-ah-ti′-tis vi-rus)
Herpes simplex (her′-peas sim′-plex)
Herpes zoster (her′-peas zoh′-ster)
Histoplasma capsulatum (his-toh-plaz′-mah cap-su-lah′-tum)
Hyphomicrobium (high-foh-my-krow′-bee-um)

I

Influenza virus (in-flew-en′-za vi-rus)

K

Klebsiella pneumoniae (kleb-see-ell′-ah new-moan′-ee-ee)

L

Lactobacillus brevis (lack-toe-ba-sil′-lus bre′-vis)
Lactobacillus bulgaricus (lack-toe-ba-sil′-lus bull-gair′-i-kus)
Lactobacillus casei (lack-toe-ba-sil′-us kay′-see-ee)
Lactobacillus plantarum (lack-toe-ba-sil′-us plan-tar′-um)
Lactobacillus thermophilus (lack-toe-ba-sil′-us ther-mo′-fil-us)
Lactococcus lactis (lack-toe-kock′-us lak′-tiss)
Legionella pneumophila (lee-jon-ell′-ah new-moh′-fill-ah)
Leptospira interrogans (lep-toe-spire′-ah in-ter-roh′-ganz)
Leuconostoc citrovorum (lew-kow-nos′-tok sit-ro-vor′-um)
Listeria monocytogenes (lis-tear′-ee-ah mon′-oh-sigh-to-jen′-eze)

M

Malassezia (mal-as-seez′-e-ah)
Methanobacterium (me-than′-oh-bak-teer-ee-um)
Methanococcus (me-than-oh-ko′-kus)
Microsporum (my-kroh-spore′-um)
Mobiluncus (moh-bi-lun′-kus)
Moraxella catarrhalis (more-ax-ell′-ah kah-tah-rah′-liss)
Moraxella lacunata (more-ax-ell′-ah lak-u-nah′-tah)
Mucor (mu′-kor)
Mycobacterium leprae (my-koh-bak-teer′-ee-um lep-ree)
Mycobacterium tuberculosis (my-koh-bak-teer′-ee-um too-ber-kew-loh′-siss)
Mycoplasma pneumoniae (my-koh-plaz′-mah new-moan′-ee-ee)

N

Neisseria gonorrhoeae (nye-seer′-ee-ah gahn-oh-ree′-ee)
Neisseria meningitidis (nye-seer′-ee-ah men-in-jit′-id-iss)
Neurospora sitophila (new-rah′-spor-ah sit-oh-phil′-ah)

O

Orthomyxovirus (or-thoe-mix′-oh-vi-rus)
Oscillatoria (os-sil-la-tor′-ee-ah)

P

Papillomavirus (pap-il-oh′-ma-vi-rus)
Parainfluenza virus (par-ah-in-flew-en′-zah vi-rus)
Paramecium (pair′-ah-mee-see-um)
Paramyxovirus (par-ah-mix′-oh-vi-rus)
Parvovirus (par′-vo-vi-rus)
Pasteurella multocida (pass-ture-ell′-ah mul-toe-sid′-ah)
Pediococcus soyae (ped-ih-oh-ko′-kus soy′-ee)
Penicillium camemberti (pen-eh-sill′-ee-um cam-em-bare′-tee)
Penicillium roqueforti (pen-eh-sill′-ee-um rok-e-for′-tee)
Peptostreptococcus (pep′-to-strep-to-ko-kus)
Phytophythora infestans (fy′-toe-fy-thor-ah in-fes′-tanz)
Picornavirus (pi-kor′-na-vi-rus)
Plasmodium falciparum (plaz-moh′-dee-um fall-sip′-air-um)
Plasmodium malariae (plaz-moh′-dee-um ma-lair′-ee-ee)
Plasmodium ovale (plaz-moh′-dee-um oh-vah′-lee)
Plasmodium vivax (plaz-moh′-dee-um vye′-vax)
Pneumocystis carinii (new-mo-sis′-tis car′-i′-nee-ee)
Poliovirus (poe′-lee-oh-vi-rus)
Polyoma virus (po-lee-oh′-mah vi-rus)
Propionibacterium acnes (proh-pee-ah-nee-bak-teer′-ee-um ak′-neez)
Propionibacterium shermanii (proh-pee-ah-nee-bak-teer′-ee-um sher-man′-ee-ee)
Proteus mirabilis (proh′-tee-us mee-rab′-il-us)
Pseudomonas aeruginosa (sue-dough-moan′-ass aye-rue-gin-o′-sa)

R

Rabies virus (ray′-bees vi-rus)
Retrovirus (re′-trow-vi-rus)
Rhabdovirus (rab′-doh-vi-rus)
Rhinovirus (rye′-no-vi-rus)
Rhizobium (rye-zoh′-bee-um)
Rhizopus nigricans (rise′-oh-pus nye′-gri-kanz)
Rhizopus stolon (rise′-oh-pus stoh′-lon)
Rhodococcus (roh-doh-koh′-kus)
Rickettsia rickettsii (rik-kett′-see-ah rik-kett′-see-ee)
Rotavirus (row′-tah-vi-rus)
Rubella virus (rue-bell′-ah vi-rus)
Rubeola virus (rue-bee-oh′-la vi-rus)

S

Saccharomyces carlsbergensis (sack-ah-row-my′-sees karls-berg-en′-siss)
Saccharomyces cerevisiae (sack-ah-row-my′-sees sara-vis′-ee-ee)
Saccharomyces rouxii (sack-ah-row-my′-sees roos′-ee-ee)
Salmonella enteritidis (sall-moh-nell′-ah en-ter-it′-id-iss)
Salmonella typhi (sall-moh-nell′-ah tye′-fee)
Salmonella typhimurium (sall-moh-nell′-ah tye-fe-mur′-ee-um)
Serratia marcescens (ser-ray′-sha mar-sess-sens)
Shigella dysenteriae (shig-ell′-ah diss-en-tair′-ee-ee)
Spirillum minus (spy-rill′-um my′-nus)
Sporothrix schenckii (spore′-oh-thrix shenk-ee-ee)
Staphylococcus aureus (staff-ill-oh-kok′-us aw′-ree-us)
Staphylococcus epidermidis (staff-ill-oh-kok′-us epi-der′-mid-iss)
Streptobacillus moniliformis (strep-tow-bah-sill′-us mon-ill-i-form′-is)
Streptococcus agalactiae (strep-toe-kock′-us a-ga-lac′-tee-ee)
Streptococcus cremoris (strep-toe-kock′-us kre-more′-iss)
Streptococcus mutans (strep-toe-kock′-us mew′-tanz)

Streptococcus pneumoniae (strep-toe-kock′-us new-moan′-ee-ee)
Streptococcus pyogenes (strep-toe-kock′-us pie-ah-gen-ease)
Streptococcus salivarius (strep-toe-kock′-us sal-ih-vair′-ee-us)
Streptococcus sanguis (strep-toe-kock′-us san′-gwis)
Streptococcus thermophilis (strep-toe-kock′-us ther-moh′-fill-us)
Streptomyces griseus (strep-toe-my′-seez gree′-see-us)

T

Thiobacillus (thigh-oh-bah-sill′-us)
Torulopsis (tore-you-lop′-siss)
Treponema pallidum (tre-poh-nee′-mah pal′-ih-dum)
Trichomonas vaginalis (trick-oh-moan′-as vag-in-al′-iss)
Trichophyton (trick-oh-phye′-ton)
Trypanosoma brucei (tri-pan′-oh-soh-mah bru′-see-ee)

V

Varicella-zoster virus (var-ih-sell′-ah zoh′-ster vi-rus)
Veillonella (veye-yon-ell′-ah)
Vibrio anguillarum (vib′-ree-oh an-gwil-air′-um)
Vibrio cholerae (vib′-ree-oh kahl′-er-ee)

Y

Yersinia enterocolitica (yer-sin′-ee-ah en-ter-oh-koh-lih′-tih-kah)
Yersinia pestis (yer-sin′-ee-ah pess′-tiss)

APPENDIX IV

Metabolic Pathways

FIGURE IV.1 **The Entner-Doudoroff Pathway**

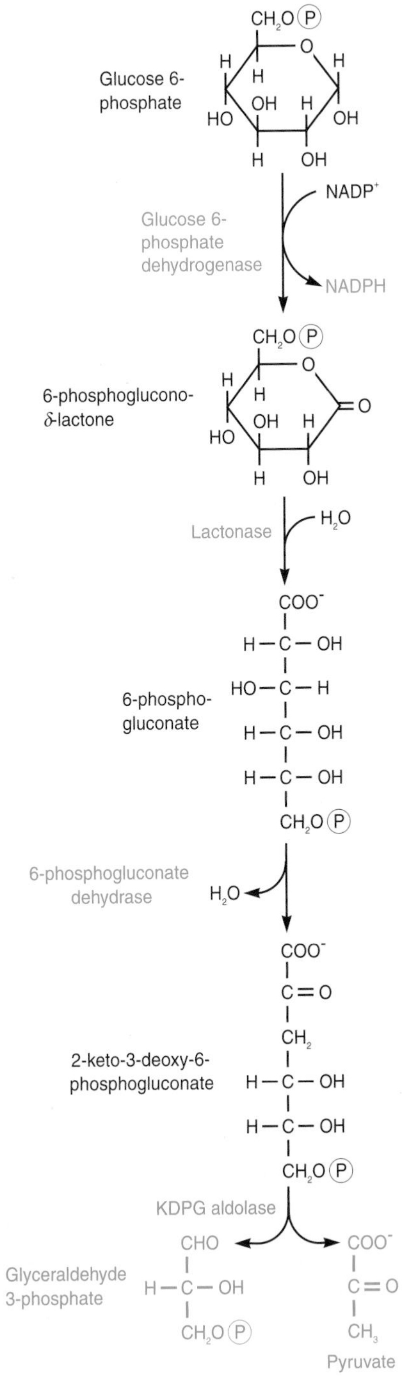

Glucose 6-phosphate

Glucose 6-phosphate dehydrogenase ($NADP^+$ → NADPH)

6-phosphoglucono-δ-lactone

Lactonase (H_2O)

6-phospho-gluconate

6-phosphogluconate dehydrase (H_2O)

2-keto-3-deoxy-6-phosphogluconate

KDPG aldolase

Glyceraldehyde 3-phosphate

Pyruvate

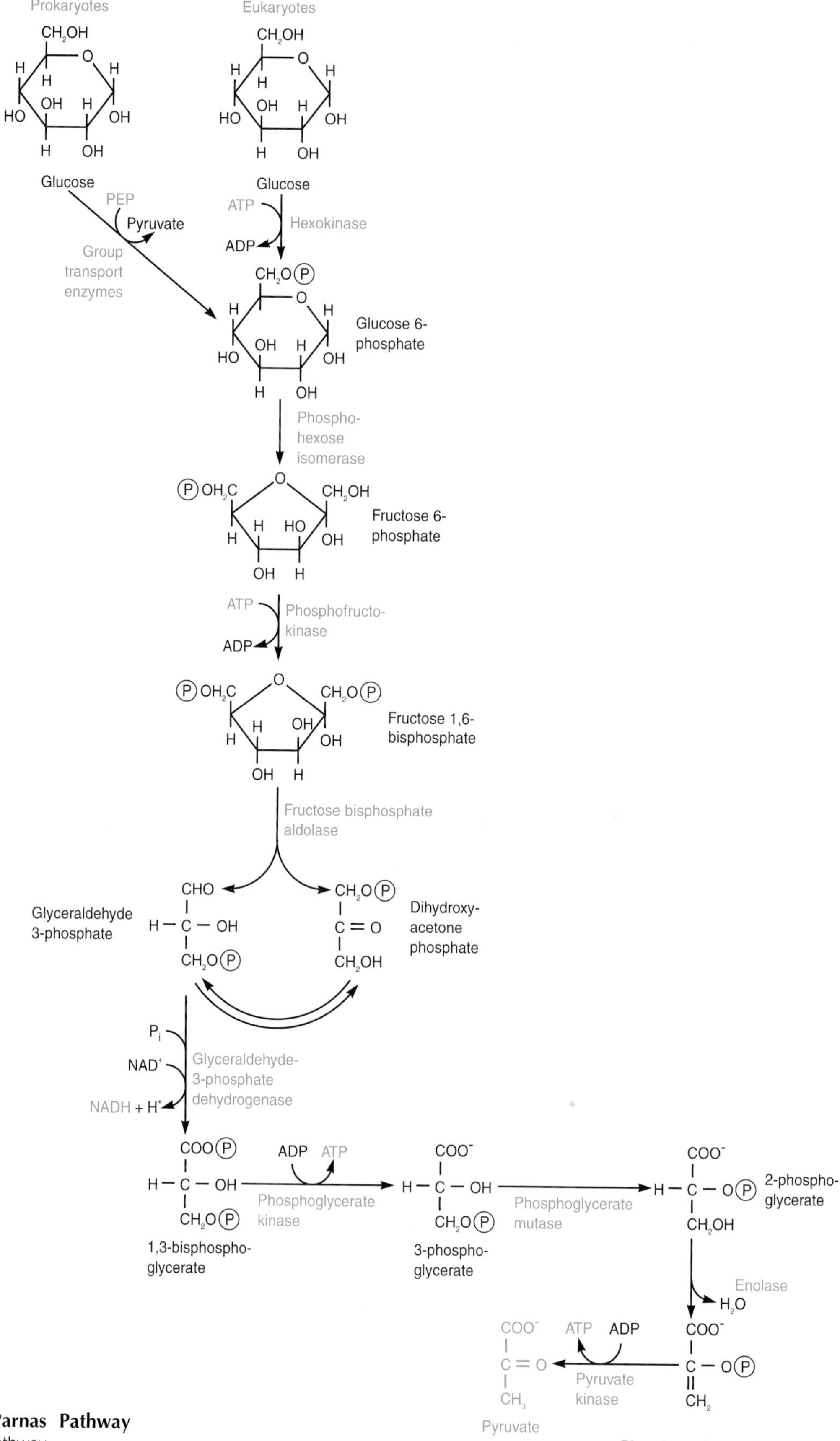

FIGURE IV.2 The Embden-Meyerhof-Parnas Pathway
Commonly called glycolysis or the glycolytic pathway.

FIGURE IV.3 The Pentose Phosphate Pathway

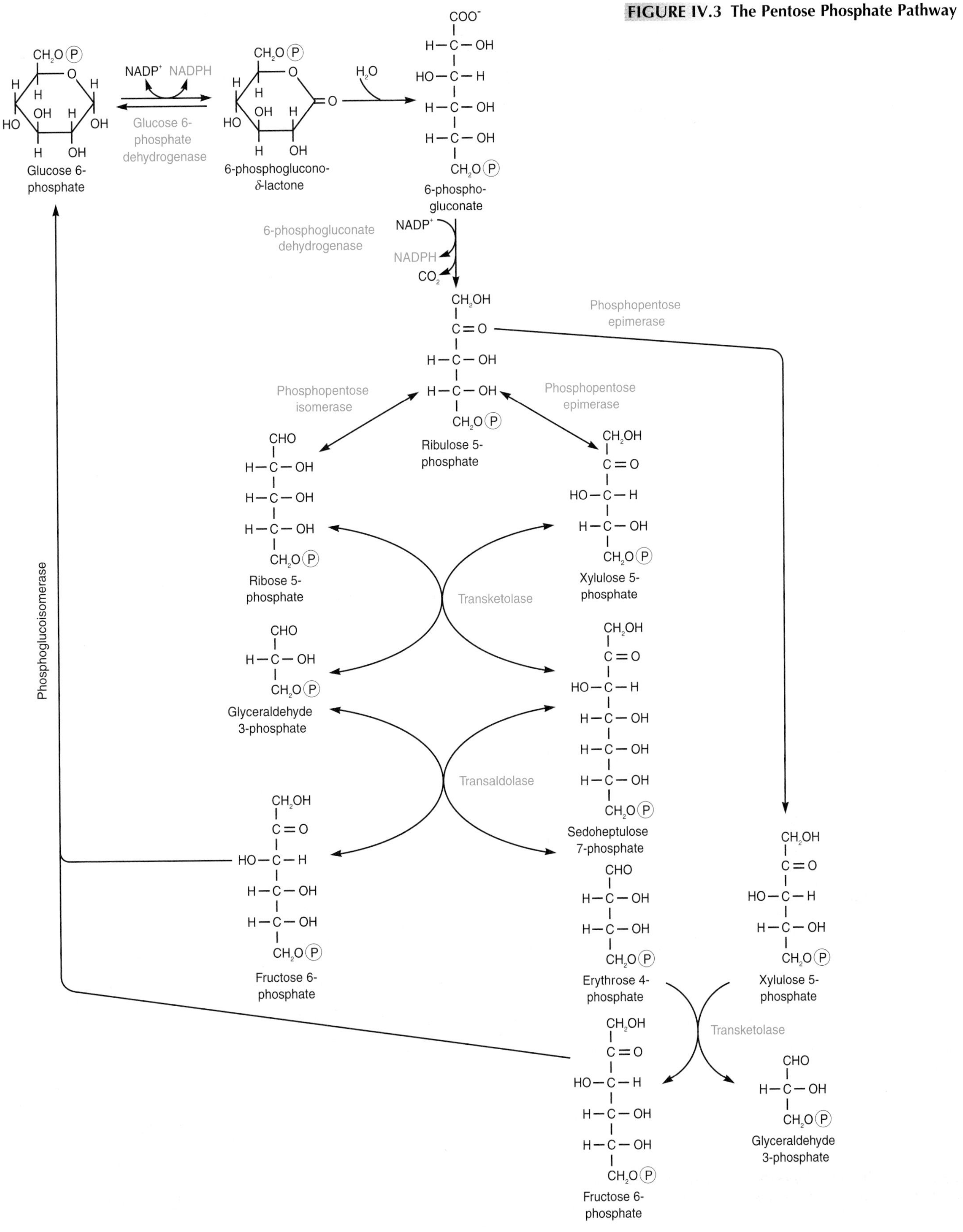

Glucose 6-phosphate

NADP$^+$ NADPH

Glucose 6-phosphate dehydrogenase

6-phosphoglucono-δ-lactone

H_2O

6-phosphogluconate

6-phosphogluconate dehydrogenase

NADP$^+$ NADPH CO_2

Ribulose 5-phosphate

Phosphopentose isomerase

Phosphopentose epimerase

Phosphopentose epimerase

Ribose 5-phosphate

Xylulose 5-phosphate

Transketolase

Glyceraldehyde 3-phosphate

Sedoheptulose 7-phosphate

Transaldolase

Fructose 6-phosphate

Erythrose 4-phosphate

Xylulose 5-phosphate

Transketolase

Fructose 6-phosphate

Glyceraldehyde 3-phosphate

Phosphoglucoisomerase

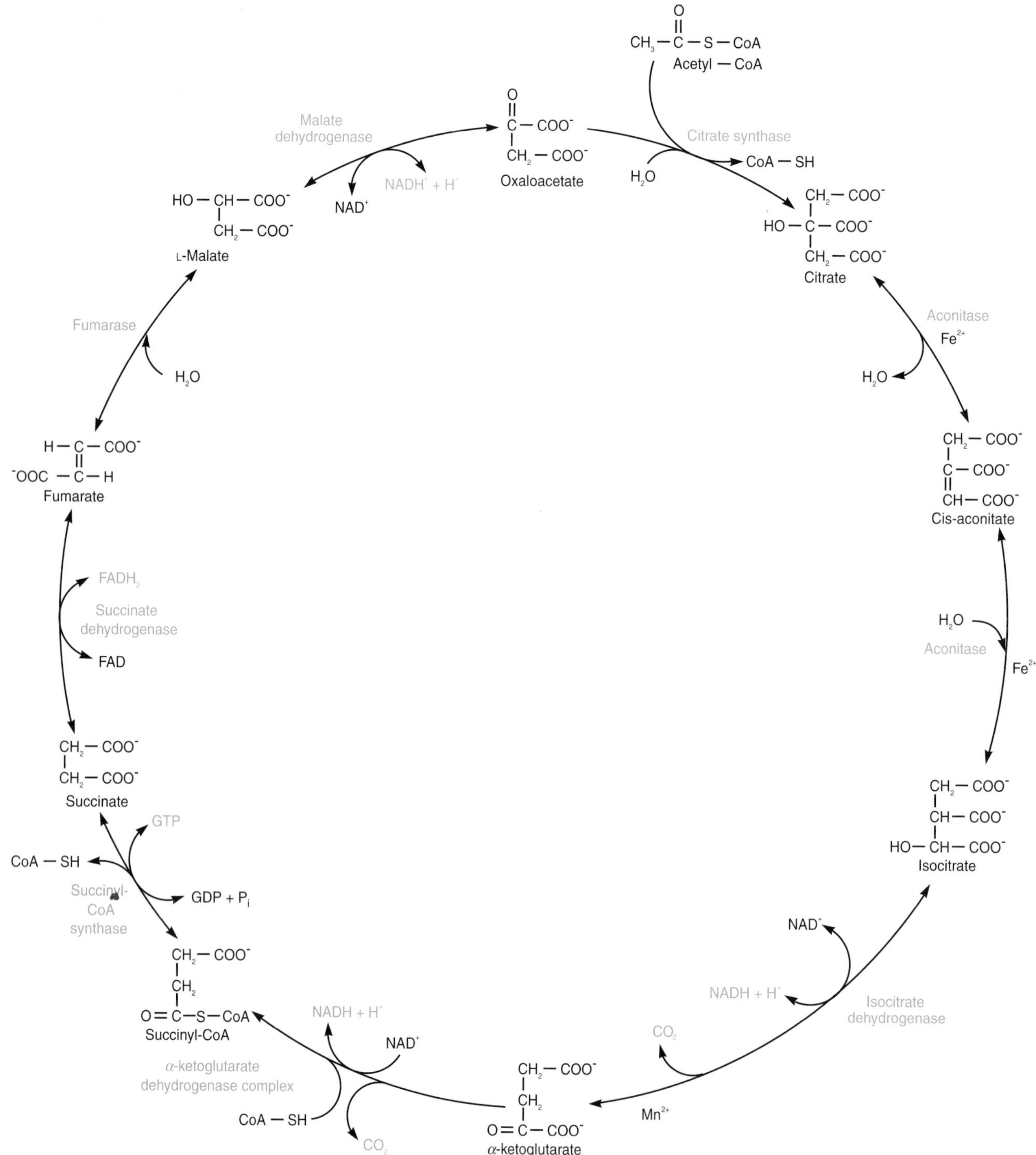

FIGURE IV.4 The Tricarboxylic Acid Cycle (TCA Cycle) Also referred to as the Krebs cycle or the citric acid cycle.

APPENDIX V

Answers to Multiple Choice Questions

Chapter 1

1. D
2. A
3. B
4. C
5. B
6. B
7. C
8. A
9. E
10. A

Chapter 2

1. C
2. A
3. E
4. A
5. B
6. C
7. B
8. D
9. E
10. A

Chapter 3

1. A
2. A
3. E
4. B
5. B
6. C
7. E
8. C
9. D
10. A

Chapter 4

1. B
2. B
3. C
4. B
5. D
6. B
7. E
8. A
9. D
10. B

Chapter 5

1. C
2. D
3. A
4. B
5. B
6. C
7. E
8. D
9. C
10. A

Chapter 6

1. D
2. C
3. D
4. E
5. B
6. A
7. D
8. D
9. D
10. E

Chapter 7

1. B
2. A
3. A
4. A
5. C
6. B
7. B
8. C
9. D
10. D

Chapter 8

1. C
2. A
3. D
4. A
5. B
6. D
7. A
8. B
9. A
10. B

Chapter 9

1. C
2. C
3. E
4. B
5. B
6. C
7. B
8. B
9. B
10. D

Chapter 10

1. E
2. B
3. E
4. B
5. D
6. B
7. C
8. D
9. B
10. E

Chapter 11

1. E
2. B
3. E
4. B
5. D
6. A
7. D
8. B
9. C
10. D

Chapter 12

1. C
2. A
3. B
4. A, C
5. A
6. B
7. A, B, E
8. A
9. C
10. A, B

Chapter 13

1. C
2. B
3. E
4. A
5. C
6. E
7. D
8. C
9. A
10. A

Chapter 14

1. D
2. A
3. B
4. B
5. D
6. B
7. A
8. D
9. E
10. A

Chapter 15

1. B
2. E
3. A
4. B
5. A
6. E
7. A
8. C
9. D
10. C

Chapter 16

1. E
2. E
3. D
4. D
5. B
6. D
7. D
8. A
9. B
10. D

Chapter 17

1. C
2. A
3. D
4. D
5. A
6. C
7. C
8. D
9. D
10. A

Chapter 18

1. A
2. E
3. A
4. D
5. E
6. C
7. A
8. C
9. A
10. A

Chapter 19

1. D
2. C
3. E
4. D
5. A
6. D
7. C
8. A
9. C
10. C

Chapter 20

1. A
2. B
3. A
4. B
5. E
6. B
7. B
8. E
9. C
10. E

Chapter 21

1. D
2. C
3. A
4. B
5. A
6. B
7. D
8. D
9. A
10. C

Chapter 22

1. E
2. C
3. E
4. A
5. B
6. A
7. A
8. C
9. D
10. D

Chapter 23

1. E
2. E
3. A
4. E
5. B
6. A
7. C
8. A
9. D
10. D

Chapter 24

1. B
2. E
3. C
4. A
5. A
6. C
7. D
8. C
9. E
10. B

Chapter 25

1. C
2. A
3. B
4. D
5. A
6. D
7. C
8. D
9. C
10. E

Chapter 26

1. D
2. D
3. C
4. B
5. C
6. B
7. A
8. B
9. C
10. C

Chapter 27

1. B
2. C
3. C
4. D
5. C
6. E
7. A
8. D
9. A
10. B

Chapter 28

1. E
2. A
3. C
4. E
5. C
6. C
7. C
8. A
9. A
10. E

Chapter 29

1. D
2. D
3. A
4. B
5. C
6. A
7. E
8. A
9. E
10. E

Chapter 30

1. A
2. A
3. C
4. D
5. A
6. D
7. C
8. C
9. B
10. A

Chapter 31

1. B
2. B
3. A
4. A
5. C
6. E
7. C
8. C
9. B
10. E

Chapter 32

1. B
2. C
3. D
4. E
5. B
6. B
7. C
8. D
9. D
10. E

GLOSSARY

ABC transport systems A type of active transport system that requires ATP as an energy source (*A*TP *B*inding-*C*assette).
abscess A localized collection of pus within a tissue.
A-B toxin Exotoxin composed of an active subunit (A-subunit) and a binding subunit (B-subunit).
acellular Not composed of cells, therefore not living.
acid-fast staining A procedure used to stain certain microorganisms, particularly members of the genus *Mycobacterium*, that do not readily take up dyes.
acidic amino acids Amino acids with more carboxyl ($—COOH$) than amino ($—NH_2$) groups.
acidophiles Organisms that grow optimally at a pH below 5.5.
acquired resistance Development of antimicrobial resistance in a previously sensitive organism; occurs through spontaneous mutation or acquisition of new genetic information.
actin Protein that makes up actin filaments of eukaryotic cells; it can rapidly assemble and subsequently disassemble to cause motion.
actin filaments Cytoskeletal structures of eukaryotic cells that enable the cell cytoplasm to move.
actinomycetes Filamentous bacteria; many are valuable in the production of antibiotics.
activated macrophages Macrophages stimulated by cytokines to enlarge and become metabolically active, with greatly increased capability to kill and degrade intracellular organisms and materials.
activated sludge method A method of sewage treatment in which wastes are degraded by complex populations of aerobic microorganisms.
activated T cell T cell activated by exposure to antigen in conjunction with required accessory signals.
activation energy Initial energy required to break a chemical bond.
activator-binding site Sequence of DNA that precedes an ineffective promoter; binding of an activator to this site enhances the ability of RNA polymerase to initiate transcription at that promoter.
active immunity Protective immunity produced by an individual in response to an antigenic stimulus.
active site Site on an enzyme molecule to which substrate binds; also known as the catalytic site.
active transport Energy-consuming process by which molecules are carried across cell boundaries; can accumulate compounds against a concentration gradient.
acute infections Infections in which the symptoms and signs have a rapid onset and are usually severe, often with fever, but short-lived.
acute inflammation Short-term inflammatory response, marked by a prevalence of neutrophils.
acylated homoserine lactone (AHL) Small signal molecules that can move freely in and out of cells; provides cells with a mechanism of assessing cell density (quorum sensing).
adaptive immunity Protection provided by host defenses that develop throughout life; involves B cells and T cells.
ADCC (antibody-dependent cellular cytotoxicity) Killing of target cells by macrophages, granulocytes, or natural killer cells that contact the target via their Fc receptors binding to Fc of antibodies attached to the target.
adenosine diphosphate (ADP) The acceptor of free energy in a cell; that energy is used to add an inorganic phosphate (P_i) to ADP, generating ATP.
adenosine triphosphate (ATP) The energy currency of a cell, serving as the ready and immediate donor of free energy.
adherence A necessary first step in colonization and infection, in which the pathogen attaches to host cells to avoid being removed from the body.
adhesin Component of a microorganism that is used to bind to surfaces.
adhesion molecule Molecule on the surface of a cell that allows that cell to adhere to other cells.
adjuvant Substance that increases the immune response to antigen.
ADP Abbreviation for adenosine diphosphate.
adsorption Attachment of one substance to the surface of another.
aerobic respiration Metabolic process in which electrons are transferred from the electron transport chain to molecular oxygen (O_2).
aerosol Material dispersed into the air as a fine mist.
aerotaxis Movement toward or away from molecular oxygen.
aerotolerant anaerobes Organisms that can grow in the presence of O_2 but never use it as a terminal electron acceptor; also called obligate fermenters.
affinity maturation The "fine-tuning" of the fit of an antibody molecule for an antigen; it is due to mutations that occur as activated B cells multiply.
aflatoxin Potent toxin made by *Aspergillus flavus*.
agar Polysaccharide extracted from marine algae; used to solidify microbiological media.
agarose Highly purified form of agar used in gel electrophoresis.
agar slant Microbiological medium that has been solidified with agar and stored in a tube that was held at a shallow angle as the medium solidified, creating a larger surface area.
agglutination Clumping together of cells or particles.
AIDS Acquired immunodeficiency syndrome.
AIDS-related complex (ARC) A group of symptoms–fever, fatigue, diarrhea, and weight loss–that herald the onset of AIDS.
alga (pl. algae) A primitive photosynthetic eukaryotic organism.
alkalophiles Organisms that grow optimally at a pH above 8.5.
alkylating agent Chemical that adds alkyl groups (short chains of carbon atoms) to purines and pyrimidines; promotes mutations.
alkyl group Short chain or single carbon atom such as a methyl group ($—CH_3$).
allele One form of a gene.
allergen Antigen that causes an allergy.
allergic rhinitis Hay fever; sneezing, runny nose, teary eyes resulting from exposure of a sensitized person to inhaled antigen; an IgE-mediated allergic reaction.
allergy Hypersensitivity, especially of the IgE-mediated type.
allograft Organ or tissue graft transplanted between genetically nonidentical members of the same species.
allosteric enzyme An enzyme that contains a site—the allosteric site—to which a small molecule can bind and change enzyme activity.
alpha (α) hemolysis Type of hemolysis observed on blood agar, characterized by a zone of greenish clearing around the colonies.
alternative pathway Pathway of complement activation initiated by the binding of a complement protein (C3b) to cell surfaces.
amalgam Mixture of mercury with other metals to form a paste that hardens; used to fill cavities in teeth.
amino acids Subunits of a protein molecule.
aminoglycosides Group of antimicrobial medications; they interfere with protein synthesis.
amino terminal (or N terminal) The end of the protein molecule that has an unbonded $—NH_2$ group.
ammonification The decomposition process that converts organic nitrogen into ammonia (NH_3).
amphibolic pathways Metabolic pathways that play roles in both catabolism and anabolism.
amylases Enzymes that digest starches.
anabolism Cellular processes that use the energy stored in ATP to synthesize and assemble subunits such as amino acids; synonymous with biosynthesis.
anaerobic Without molecular oxygen (O_2).
anaerobic respiration Metabolic process in which electrons are transferred from the electron transport chain to a terminal electron acceptor other than O_2.
analytical study An epidemiological study done to identify specific risk factors associated with developing a certain disease.
anamnestic response (See *secondary response.*)
anaphylaxis Allergic reaction caused by IgE; generalized hypersensitive reaction to an allergen that can cause a profound drop in blood pressure.
anaplerotic reactions Chemical reactions that bypass certain steps of the central metabolic pathways; they are used to replenish some of the intermediates drawn off for biosynthesis.
anion Negatively charged ion.
anneal Form a double-stranded duplex from two complementary strands of DNA.

anoxic Devoid of O_2.
anoxygenic phototrophs Photosynthetic bacteria that use hydrogen sulfide or organic compounds rather than water as a source of electrons for reducing power; they do not generate O_2.
antagonistic In antimicrobial therapy, a combination of antimicrobial medications in which the action of one interferes with the action of the other.
antenna complex Complex in photosynthetic organisms composed of hundreds of light-gathering pigments; acts as a funnel, capturing light energy and transferring it to reaction-center chlorophyll.
antibacterial drug Chemical used to treat bacterial infections.
antibiogram Antibiotic susceptibility pattern; used to distinguish between different bacterial strains.
antibiotic Chemical produced by certain molds and bacteria that kills or inhibits the growth of other microorganisms.
antibiotic-associated colitis Intestinal disease caused by overgrowth of toxin-producing strains of *Clostridium difficile;* typically, occurs when a person is taking antimicrobial medications.
antibody Immunoglobulin protein produced by the body in response to a substance; it reacts specifically with that substance.
anticodon Sequence of three nucleotides in a tRNA molecule that is complementary to a particular codon in mRNA.
antigen Molecule that reacts specifically with an antibody or immune lymphocyte.
antigen-antibody complex Linked group of antibodies bound to antigen.
antigen-binding sites Regions at the ends of the two arms of an antibody molecule that recognize a specific antigen; two identical antigen-binding sites are on each monomer of antibody.
antigen-presenting cells (APCs) Cells such as B cells, macrophages, and dendritic cells that can present exogenous antigen to helper T cells.
antigenic determinant Part of an antigen molecule that binds the specific antibody; an epitope.
antigenic drift Slight changes that occur in the antigens of a virus; specific antibodies made in response to the antigen before the change occurred are only partially protective.
antigenic shift Major changes that can occur in the antigens of a virus.
antigenic variation Routine alteration by a microorganism in the characteristics of certain of its surface proteins.
antigen presentation Process in which cells display antigen in the groove of MHC molecules for inspection by T cells.
antimicrobial drug Chemical used to treat microbial infections; also called an antimicrobial.
antiparallel Term used to describe opposing orientations of the two strands of DNA in the double helix; one strand is oriented in the 5′ to 3′ direction and its complement is oriented in the 3′ to 5′ direction.
antisense strand Complement to the sense (or plus) strand of RNA; also called the minus (–) strand.
antiseptic A disinfectant that is non-toxic enough to be used on skin.
antiserum A preparation of serum containing protective antibodies.
antitoxin An antibody preparation that protects against a given toxin.
aplastic anemia Potentially lethal condition in which the body is unable to make blood cells.
apoptosis Programmed cell death.
arbovirus Arthropod-borne virus. One of a large group of RNA viruses carried by insects and mites that act as biological vectors.
Archaea One of the two domains of prokaryotes; many archaea grow in extreme environments.
arteriosclerosis Condition characterized by thickening and loss of elasticity of the walls of arteries; "hardening of the arteries."
arthropod Classification grouping of invertebrate animals that includes insects, ticks, lice, and mites.
Arthus reaction Hypersensitivity reaction caused by immune complexes and neutrophils.
artificial wetland method Method of sewage treatment in which sewage is channeled into successive ponds where both aerobic and anaerobic degradation occurs.
artificially acquired immunity Active or passive immunity acquired through artificial means such as vaccination or administration of immune serum globulin.
aseptic Free of microorganisms and viruses; sterile.
aseptic technique Use of specific methods and sterile materials to exclude contaminating microorganisms from an environment.
asexual Reproduction not preceded by the union of cells or exchange of DNA.
A-site (or **aminoacyl site**) Site on the ribosome to which tRNAs enter to donate their amino acid; acceptor site.
asthma Immediate respiratory allergy resulting from mediator release from mast cells in the lower airways.
astrobiology Study of life in the universe.
atomic force microscope Type of scanning probe microscope that has a tip mounted so it can bend in response to the slightest force between the tip and the sample.
ATP Abbreviation for adenosine triphosphate.
ATP synthase Protein complex that harvests the energy of a proton motive force to synthesize ATP.
attack rate Proportional number of cases developing in a population exposed to an infectious agent.
attenuated vaccine Vaccine composed of a weakened form of a disease-causing microorganism or virus that is generally unable to cause disease; the vaccine strain is able to replicate inside its host.
autoantibodies Antibodies that bind to "self" proteins.
autoclave Device employing steam under pressure to sterilize materials that are stable to heat and moisture.
autoimmune disease Disease produced as a result of an immune reaction against one's own tissues.
autolyze To spontaneously disintegrate as a result of enzymes within the cell.
autoradiography The use of film to detect a radioactive molecule.
autotroph Organism that uses CO_2 as its main source of carbon.
auxotroph A microorganism that requires an organic growth factor.
avirulent Lacking disease-causing attributes.
a_w Abbreviation for water activity.
axial filaments Characteristic structure of motility found in spirochetes.
axon The long thin extension of a nerve cell.
azoles Large family of chemically synthesized medications, some of which have antifungal activity.

---

bacillus (pl. **bacilli**) Cylindrical-shaped bacterium; also referred to as a rod.
bacitracin Antimicrobial medication that inhibits cell wall biosynthesis by interfering with the transport of peptidoglycan precursors across the cytoplasmic membrane.
bacteremia Bacterial cells circulating in the bloodstream.
Bacteria One of the two domains of prokaryotes; all medically important prokaryotes are in the domain *Bacteria*.
bactericidal Able to kill bacteria.
bacteriochlorophyll Type of chlorophyll used by purple and green bacteria; absorbs wavelengths of light that penetrate to greater depths and are not used by other photosynthetic organisms.
bacteriocins Proteins made by bacteria that kill certain other bacteria.
bacteriophage A virus that infects bacteria; often abbreviated to phage.
bacteriorhodopsin Pigment of some prokaryotes that absorbs energy from sunlight and uses it to expel protons from the cell, generating a proton gradient.
bacteriostatic Able to inhibit the growth of bacteria.
balanced pathogenicity Host parasite relationship in which the parasite persists in the host while causing minimal harm.
basal body Structure that anchors the flagellum to the cell wall and cytoplasmic membrane.
base Refers to the purine or pyrimidine ring structure found in nucleic acids.
base analog Compound that resembles a purine or pyrimidine base closely enough to be incorporated into a nucleotide in place of a natural base; the resulting molecule can be incorporated into DNA.
base-pairing The hydrogen bonding of adenine (A) to thymine (T) and cytosine (C) to guanine (G); occurs between two complementary strands of DNA.
basement membrane Thin layer of fibrous material that underlies epithelial cells.
basic amino acids Amino acids with more basic ($—NH_3^+$) groups than acid ($—COO^-$) groups.
basophil Leukocyte with large dark-staining granules that contain histamine and other mediators of inflammation; receptors on cell surfaces bind monomers of IgE.
B-cell receptor Membrane-bound derivative of the antibody that a B cell is programmed to make; it enables the B cell to recognize a specific antigen.
B cells Lymphocytes programmed to produce antibody molecules.
beta- (β) hemolysis Type of hemolysis observed on blood agar that is characterized by a clear zone around a colony.
beta- (β) lactam drugs Group of antimicrobial medications that inhibit peptidoglycan synthesis and have a shared chemical structure called a β-lactam ring.
bilayer membrane (or **unit membrane**) Double layer of phospholipid molecules that forms the major structure of the cytoplasmic (plasma) membrane.

bile Yellow-colored fluid produced by the liver that aids in the absorption of nutrients from the intestine.

binary fission Asexual process of reproduction in which one cell divides to form daughter cells.

binding protein Protein that functions in the ABC transport system; resides immediately outside of the cytoplasmic membrane to deliver a given molecule to a specific transport complex within the membrane.

binomial system System of naming each species of organism with two Latin words that indicate the genus and species.

biochemical oxygen demand (BOD) Measure of the amount of biologically degradable organic material in water.

biocide Compound such as a disinfectant that is toxic to many forms of life, including microorganisms.

biodiversity Diversity in the number of species inhabiting an ecosystem and their evenness of distribution.

biofilm Polysaccharide-encased community of microorganisms.

bioinformatics Developing and using computer technology to store, retrieve, and analyze nucleotide sequence data.

bioleaching Conversion of metals to a soluble form due to the metabolic oxidation of insoluble metal sulfides by microorganisms.

biological vector Organism that acts as a host for a pathogen before it is transmitted to another organism; the pathogen can multiply to high numbers within it.

bioluminescence Biological production of light.

biomass Total weight of all organisms in any particular environment.

bioremediation Process that uses microorganisms to degrade harmful chemicals.

biosphere The sum of all the regions of the earth where life exists.

biotechnology The use of microbiological and biochemical techniques to solve practical problems and produce more useful products.

biotype A group of strains that have a characteristic biochemical pattern different from other strains; also called a biovar.

blood agar Type of rich agar medium that contains red blood cells and can be used to detect hemolysis.

blood-brain barrier Property of the central nervous system blood vessels that restricts passage of infectious agents and certain molecules (such as medications) into the brain and spinal cord.

blunt end The type of DNA ends generated by a restriction enzyme that cuts directly in the middle of the recognition sequence.

BOD Abbreviation for biochemical oxygen demand.

boil Painful localized collection of pus within the skin and subcutaneous tissue; a furuncle.

bonds Forces that hold atoms or molecules together.

bone marrow Soft material that fills bone cavities and contains stem cells for all blood cells.

botulinum toxin Toxin produced by *Clostridium botulinum* that can cause a fatal paralysis in people who consume it.

bright-field microscope Type of light microscope that illuminates the field of view evenly.

brine Salty water; used to cure fish and meats.

broad host range plasmid A plasmid that can replicate in a wide variety of unrelated bacteria.

broad-spectrum antimicrobials Antimicrobials that inhibit or kill a wide range of microorganisms, often including both Gram-positive and Gram-negative bacteria.

Bt-toxin Protein crystal naturally produced by *Bacillus thuringiensis* as it forms endospores; toxic to insect larvae that consume it.

bubo Enlarged, tender lymph node characteristic of plague and some venereal diseases.

bubonic plague Form of plague that typically develops when *Yersinia pestis* is injected via the bite of an infected flea.

budding Asexual reproductive technique that involves a pushing out of a part of the parent cell that eventually gives rise to a new daughter cell. Also, a process by which some viruses are released from host cells.

buffer Substance in a solution that acts to prevent changes in pH.

bulking Overgrowth of filamentous microorganisms in sewage at treatment facilities; interferes with the separation of the solid sludge from the liquid effluent.

burst size Number of newly formed virus particles released from a single cell.

Calvin cycle Metabolic pathway used by many autotrophs to incorporate CO_2 into an organic form; also called the Calvin-Benson cycle.

cAMP Abbreviation for cyclic AMP.

cancer Abnormally growing cells that can spread from their site of origin; malignant tumors.

candidiasis Fungal diseases caused by *Candida albicans*.

candle jar Closed jar in which a lit candle converts some of the O_2 in air to CO_2 and water vapor; used to cultivate capnophiles.

CAP Abbreviation for cyclic AMP-activating protein.

cap Methylated guanine derivative added to the 5′ end of eukaryotic mRNA before transcription is complete.

capnophiles Organisms that require increased concentrations of CO_2 (5% to 10%) and approximately 15% O_2.

capsid Protein coat that surrounds the nucleic acid of a virus.

capsule A distinct thick gelatinous material that surrounds some types of microorganisms; sometimes correlated with an organism's ability to cause disease.

carbapenems Group of antimicrobial medications that interferes with peptidoglycan synthesis; very resistant to inactivation by β-lactamases.

carbohydrate Compounds containing principally carbon, hydrogen, and oxygen atoms in a ratio of 1:2:1.

carbon fixation Process of converting inorganic carbon (CO_2) to an organic form.

carboxyl terminal (or **C terminal**) The end of the protein molecule that has an unbonded —COOH group.

carbuncle Painful infection of the skin and subcutaneous tissues; manifests as a cluster of boils.

carcinogen A chemical or radiation that causes cancer.

cariogenic Causing dental caries, tooth decay.

carotenoids Accessory pigment found in a wide variety of photosynthetic organisms that increases the efficiency of light capture by absorbing wavelengths of light not absorbed by chlorophylls; mammals can use one of the carotenoids (beta-carotene) as a source of vitamin A.

carrier (1) Type of protein found in cell membranes that transports certain compounds across the membrane; may also be called a permease or transporter protein. (2) A human or other animal that harbors a pathogen without noticeable ill effects.

carrier cell A virus-infected cell that extrudes virions as it multiplies.

carrier state State of infection in which the agent can be detected in body fluids without causing disease symptoms.

cascade In biology, a series of reactions that, once started, continues to the final step by each step triggering the next in a special order; activation of the complement cascade is an example.

caseous necrosis Type of localized tissue death resulting in a cheeselike consistency, characteristic of tuberculosis and certain other chronic infectious diseases.

catabolism Cellular processes that harvest the energy released during the breakdown of compounds such as glucose and use that energy to synthesize ATP, the energy currency of all cells.

catabolite Product of catabolism.

catabolite repression The mechanism by which cells decrease the expression of genes that encode certain degradative enzymes in the presence of a compound such as glucose.

catalase Enzyme that breaks down hydrogen peroxide (H_2O_2) to produce water (H_2O) and oxygen gas (O_2).

catalyst Substance that speeds up the rate of a chemical reaction without being altered or depleted in the process.

cations Positively charged ions.

CD markers Abbreviation for cluster of differentiation markers.

CD4 lymphocytes T lymphocytes bearing the CD4 markers; helper T cells are CD4 cells.

CD8 lymphocytes T lymphocytes bearing the CD8 markers; cytotoxic T cells are CD8 cells.

cDNA DNA obtained by using reverse transcriptase to synthesize DNA from an RNA template *in vitro;* lacks introns that characterize eukaryotic DNA.

cell culture (or **tissue culture**) Cultivation of animal or plant cells in the laboratory.

cell envelope The layers surrounding the contents of the cell; includes the cytoplasmic membrane, cell wall, and capsule (if present).

cell wall Rigid barrier that surrounds a cell, keeping the contents from bursting out; in prokaryotes, peptidoglycan provides rigidity to the cell wall.

cell-mediated immunity (CMI) (See *cellular immunity.*)

cellular immunity (also called cell-mediated immunity or CMI) The immune response mediated by T lymphocytes (T cells).

cellulose Polymer of glucose subunits; principal structural component of plant cell walls.

central metabolic pathways Glycolysis, the TCA cycle, and the pentose phosphate pathway.

cephalosporins Group of antimicrobial medications that interfere with peptidoglycan synthesis.

cestode Tapeworm.

chain terminator A dideoxynucleotide; when this molecule is incorporated into a growing strand of DNA, no additional nucleotides can be added, and elongation of the strand ceases.

challenge In immunology, to give an antigen to provoke an immunologic response in a subject previously sensitized to the antigen.

chancre Sore resulting from an ulcerating infection; the "hard chancre" of primary syphilis is typically firm and painless.
chaperones Proteins that help other proteins fold properly.
chemical bond Force that holds atoms together to form molecules.
chemically defined media Bacteriological media composed of ingredients of fixed chemical composition; generally used for specific experiments when nutrients must be precisely controlled.
chemiosmotic gradient Accumulation of protons on one side of a membrane due to expulsion of protons by the electron transport chain; used by prokaryotes to power the synthesis of ATP, fuel certain transport processes, and drive the rotation of flagella; also called the proton motive force.
chemiosmotic theory The theory that a proton gradient is formed by the electron transport chain and is then used to power the synthesis of ATP.
chemoautotrophs Organisms that use chemicals as a source of energy and CO_2 as the major source of carbon.
chemoheterotrophs Organisms that use chemicals as a source of energy, and organic compounds as a source of carbon.
chemokine Cytokine important in chemotaxis of immune cells.
chemolithoautotrophs Organisms that obtain energy by degrading reduced inorganic compounds such as hydrogen gas (H_2), and use CO_2 as a source of carbon.
chemolithotrophs Organisms that obtain energy by degrading reduced inorganic chemicals such as hydrogen gas (H_2); in general, chemolithotrophs are chemolithoautotrophs.
chemoorganoheterotrophs Organisms that obtain both energy and carbon from organic compounds.
chemoorganotrophs Organisms that obtain energy by degrading organic compounds such as glucose; in general, chemoorganotrophs are chemoorganoheterotrophs.
chemostat Device used to grow bacteria in the laboratory that allows nutrients to be added and waste products to be removed continuously.
chemotaxis Directed movement of an organism in response to a certain chemical in the environment.
chemotherapeutic agent Chemical used as a therapeutic medication to treat a disease.
chemotrophs Organisms that obtain energy by degrading chemical compounds.
chickenpox Disease caused by the herpesvirus, varicella.
chloramphenicol Antimicrobial medication that interferes with protein synthesis.
chlorophylls The primary light-absorbing pigments used in photosynthesis.
chloroplasts Organelles in photosynthetic eukaryotic cells that harvest the energy of sunlight and use it to synthesize ATP, which is then used to fuel the synthesis of organic compounds.
chlorosomes Structures in which the accessory pigments in green bacteria are located.
chocolate agar Type of agar medium that contains red blood cells heated under controlled conditions to lyse them, releasing their nutrients; used to culture fastidious bacteria.
cholesterol Sterol found in animal cell membranes; provides rigidity to eukaryotic membranes.
chromatin Complex of histones and DNA that make up the chromosomes of eukaryotic cells.
chromosome Array of genes responsible for the determination and transmission of hereditary characteristics.
chronic infections Infections that develop slowly and persist for months or years.
chronic inflammation Long-term inflammatory response, marked by the prevalence of macrophages, giant cells, and granulomas.
cilium (pl. **cilia**) Short, projecting hairlike organelle of locomotion, similar in function to a flagellum.
circulative transmission Transmission of viruses to plants by insects within which the virus circulates but does not multiply.
cirrhosis A chronic progressive condition affecting the liver characterized by formation of strands of scar tissue; various causes, but commonly due to alcoholism.
citric acid cycle Metabolic pathway that incorporates acetyl-CoA, ultimately generating CO_2 and reducing power, also known as the tricarboxylic acid (TCA) cycle and the Krebs cycle.
clade Subtype of a virus such as human immunodeficiency virus (HIV), defined by similar amino acid sequences of their envelope proteins.
class In classification, a collection of similar orders; a collection of several classes makes up a phylum.
class switching The process that allows a B cell to change the antibody class it is programmed to make; through class switching plasma cells that descend from the B cell can make antibodies other than IgM.
classical pathway Pathway of complement activation initiated by antigen-antibody complexes.
classification Process of arranging organisms into similar or related groups, primarily to provide easy identification and study.
clonal selection and expansion Selection and activation of a lymphocyte by interaction of antigen and specific antigen receptor on the lymphocyte surface, causing the lymphocyte to proliferate.
clonal deletion Elimination of lymphocytes that have an antigen receptor that binds to normal host molecules.
clone Group of cells derived from a single cell.
closed system Batch system (such as a tube or flask of broth, or an agar plate) used for growing microorganisms; nutrients are not replenished and wastes are not removed.
clusters of differentiation (CD) markers Molecules on the surface of T cells and other white blood cells that are used by scientists to distinguish subsets of cells.
CMI Abbreviation for cell-mediated immunity.
CO_2 fixation Process of converting inorganic carbon (CO_2) to an organic form.
coagulase Non-enzymatic product synthesized by *Staphylococcus aureus* that clots plasma.
coccus (pl. **cocci**) Spherical-shaped bacterial cell.
codon Set of three nucleotides.
coenzyme Non-protein organic compound that assists some enzymes, acting as a loosely bound carrier of small molecules or electrons.
cofactor Non-protein component required for the activity of some enzymes.
cohesive ends Single-stranded overhangs generated when DNA is digested with a restriction enzyme that cuts asymmetrically within the recognition sequence; sticky ends.
cohort group Population with a known exposure to a specific risk factor that is followed over time in a prospective study.
coliforms (See *total coliform.*)
colonization Establishment of a site of reproduction of microbes on a material, animal, or person without necessarily resulting in tissue invasion or damage.
colony Population of bacterial cells arising from a single cell.
colony blotting Technique that uses a probe to detect a given DNA sequence in colonies growing on an agar plate.
colony-forming unit A unit that gives rise to a single colony; may be a single cell or multiple cells attached to one another.
colony-stimulating factors A group of cytokines that direct the formation of the various types of blood cells from stem cells.
combination therapy Administration of two or more antimicrobial medications simultaneously to prevent growth of mutants that might be resistant to one of the antimicrobials.
commensalism Relationship between two organisms in which one partner benefits from the association and the other is unaffected.
commercially sterile Free of all microorganisms capable of growing under normal storage conditions; the endospores of some thermophiles may remain.
common-source epidemic Outbreak of disease due to contaminated food, water, or other single source of infectious agent.
communicable diseases Diseases that are spread from an infected animal or person to another animal or person.
community All of the living organisms in a given area.
competent Physiological condition in which a bacterial cell is capable of taking up DNA.
competitive inhibition Type of enzyme inhibition that occurs when the inhibitor competes with the normal substrate for binding to the active site.
complement system Series of serum proteins involved with innate immunity; complement system proteins can be rapidly activated, contributing to protective outcomes including inflammation, lysis of foreign cells, and opsonization.
complementary Describes bases in nucleic acid which hydrogen bond to one another; A (adenine) is complementary to T (thymine), and G (guanine) is complementary to C (cytosine).
complex medium Medium for growing bacteria that has some ingredients of variable chemical composition.
compound microscope Microscope that employs multiple magnifying lenses; the lenses in combination visually enlarge an object by a factor equal to the product of each lens's magnification.
condenser lens Lens used to focus the illumination of a microscope; positioned between the light source and the specimen and does not affect the magnification.
confocal scanning laser microscope Type of microscope that focuses a laser beam to illuminate a given point on one vertical plane of a specimen; after successive regions and planes have been scanned, a computer can construct a three-dimensional image of a thick structure.
congenital A condition existing from the time of birth.

conidia Asexual spores borne on hyphae; produced by fungi and bacteria of the genus *Streptomyces*.

conjugate vaccine A vaccine composed of a polysaccharide antigen covalently attached to a large protein molecule; this type of vaccine converts what would be a T-independent antigen into a T-dependent antigen.

conjugation Mechanism of gene transfer in bacteria that involves cell-to-cell contact.

conjugative plasmid Plasmid that carries the genes for sex pili and can transfer copies of itself to other bacteria during conjugation.

constant region The part of an antibody molecule that does not vary in amino acid sequences among molecules of the same immunoglobulin class.

co-stimulatory molecules Surface proteins expressed by antigen presenting cells (APCs) when the cell senses molecules that signify an invading microbe or tissue damage; they facilitate activation of naive T cells that recognize antigen presented by dendritic cells.

constitutive enzyme An enzyme that is constantly synthesized by a cell.

contact dermatitis A T-cell-mediated inflammation of the skin occurring in sensitized individuals as a result of contact with the particular antigen; a form of delayed hypersensitivity.

contagious diseases Diseases that are spread from one host to another very readily.

continuous culture Method used to maintain cells in a state of uninterrupted growth by continuously adding nutrients and removing waste products; a type of open system.

convalescence Period of recuperation and recovery from an illness.

convergent evolution Process of evolution when two genetically different organisms develop similar environmental adaptations.

corepressor Molecule that binds to an inactive repressor and, as a consequence, enables it to function as a repressor.

cortex Layer of the endospore that helps maintain the core in a dehydrated state, thereby protecting it from the effects of heat.

counterstain In a differential staining procedure, the stain applied to impart a contrasting color to bacteria that do not retain the primary stain.

covalent bond Strong chemical bond formed by the sharing of electrons between atoms.

critical instruments Medical instruments such as needles and scalpels that come into direct contact with body tissue.

cross-contamination Transfer of pathogens from one item to another.

cross-sectional study Study that surveys a range of people to determine the prevalence of characteristics including disease, risk factors associated with disease, or previous exposure to a disease-causing agent.

croup Acute obstruction of the larynx occurring mainly in infants and young children, often resulting from respiratory syncytial or other viral infections.

crown gall tumor A tumor on a plant caused by *Agrobacterium tumefaciens*.

CSF Abbreviation for colony-stimulating factor.

curd Coagulated milk proteins, produced during cheese-making.

cyanobacteria Group of Gram-negative oxygenic phototrophs genetically related to chloroplasts.

cyclic AMP-activating protein (CAP) Protein that binds to cAMP to promote gene transcription.

cyclic photophosphorylation Type of photophosphorylation in which electrons are returned directly to the chlorophyll; used to synthesize ATP without generating reducing power.

cyst Dormant resting protozoan cell characterized by a thickened cell wall.

cysticercus (pl. **cysticerci**) Cystlike larval form of tapeworms.

cystitis Inflammation of the urinary bladder.

cytochromes Proteins that carry electrons, usually as members of electron transport chains.

cytokines Low molecular weight regulatory protein made by cells that affect the behavior of other cells; cytokines attach to specific cytokine receptors and are essential for communication between cells.

cytokine receptor Type of surface receptor that binds a chemokine.

cytopathic effect Observable change in a cell *in vitro* produced by viral action such as lysis of the cell.

cytoplasm Viscous fluid within a cell.

cytoplasmic membrane Thin, lipid bilayer that surrounds the cytoplasm and defines the boundary of a cell.

cytoskeleton Dynamic filamentous network that provides structure and shape to eukaryotic cells.

cytotoxic Kills cells.

cytotoxic T cells Type of lymphocyte programmed to destroy corrupt "self" cells.

cytotoxin Toxin that damages a variety of different cell types.

---

dark-field microscope Type of microscope that directs light toward the specimen at an angle, so that only light scattered by the specimen enters the objective lens; materials in the specimen stand out as bright objects against a dark background.

dark reactions Process of carbon fixation in photosynthetic organisms; the ATP used to drive the process is obtained in the light reactions; the dark reactions are called the light-independent reactions.

dark repair Enzymes of DNA repair that do not depend on visible light.

death phase Stage in which the number of viable bacterial cells in a population decreases at an exponential rate.

decarboxylation Removal of carbon dioxide from a chemical.

decimal reduction time Time required for 90% of the organisms to be killed under specific conditions; D value.

decontamination Treatment to reduce the number of pathogens to a level considered safe.

defensins Short antimicrobial peptides produced naturally by a variety of eukaryotic cells to fight infections.

degerm Treatment used to decrease the number of microbes in an area, usually skin.

degranulation Release of mediators from granules in the cell, as histamine is released from mast cells.

dehydration synthesis Chemical reaction in which H_2O is removed with the result that two molecules are joined together.

dehydrogenation Oxidation reaction in which both an electron and an accompanying proton are removed.

delayed hypersensitivity Hypersensitivity caused by cytokines released from sensitized T lymphocytes; reactions occur within 48 to 72 hours after exposure of a sensitized individual to antigen.

denaturation (1) Disruption of the three-dimensional structure of a protein molecule. (2) The separation of the complementary strands of DNA.

denaturing gradient gel electrophoresis (DGGE) A procedure that gradually denatures double-stranded nucleic acid during gel electrophoresis and, as a consequence, separates similar-sized fragments according to their melting point, which is related to the nucelotide sequence.

dendritic cells Antigen-presenting cells that play an essential role in activation of naive T cells.

denitrification Bacterial conversion of nitrate to gaseous nitrogen by anaerobic respiration.

dental plaque A biofilm on teeth.

deoxyribonucleic acid (DNA) Macromolecule in the cell that carries the genetic information.

deoxyribose A 5-carbon sugar molecule found in DNA.

depth filter Type of filter with complex, tortuous passages that allow the suspending fluid to pass through while retaining microorganisms.

dermatophytes Fungi that live on the skin and can be responsible for disease of the hair, nails, and skin.

dermis The layer of skin that underlies the epidermis.

descriptive study Type of study that seeks to characterize a disease outbreak by determining the characteristics of the persons involved and the place and time of the outbreak.

dessication Dehydration.

detritus Fresh or partially degraded organic matter used as a food source by decomposers.

diapedesis Movement of leukocytes from blood vessels into tissues in response to a chemotactic stimulus during inflammation.

diatomaceous earth Sedimentary soil composed largely of the skeletons of diatoms; contains large amounts of silicon.

diauxic growth Two-step growth frequency observed when bacteria are growing in media containing two carbon sources.

diazotroph Organism that can fix nitrogen.

dichotomous key Flowchart of tests used for identifying an organism; each test gives either a positive or negative result.

dideoxy chain termination method A technique used to determine the nucleotide sequence of a strand of DNA; in the procedure, a small amount of chain terminator (a dideoxynucleotide) is added to an *in vitro* synthesis reaction.

dideoxynucleotide (ddNTP) Nucleotide that lacks the 3′ OH group, the portion required for the addition of subsequent nucleotides during DNA synthesis.

differential media Culture media that contain certain ingredients such as sugars in combination with pH indicators; used to distinguish among organisms based on their metabolic traits.

differential staining Type of staining procedure used to distinguish one group of bacteria from another by taking advantage of the fact that certain bacteria have distinctly different chemical structures in some of their components.

differentiate In cell development, a change in a cell associated with the acquisition of distinct morphological and functional properties.

diffusion Movement of substances from a region of high concentration to a region of low concentration.

digesting Treating with an enzyme such as a restriction enzyme, thereby generating degradation products, such as restriction fragments.

diluent Sterile solution used to make dilutions.

dimorphic Able to assume two forms, as the yeast and mold forms of pathogenic fungi.

diphtheroids Gram-positive cells that are club-shaped and arranged to form V-shapes and palisades; resembles the typical microscopic morphology of *Corynebacterium diphtheriae*.

diplococci Cocci that typically occur in pairs.

directly observed therapy Method used to ensure that patients comply with their antimicrobial therapy; health care workers routinely visit patients in the community and watch them take their medications.

direct microscopic count Method of determining the number of microbial cells in a measured volume of liquid by counting them microscopically using special glass slides.

direct selection Technique of selecting mutants by plating organisms on a medium on which the desired mutants but not the parent will grow.

disaccharide Carbohydrate molecule consisting of two monosaccharide molecules.

disease Process resulting in tissue damage or alteration of function, producing body changes noticeable by physical examination or laboratory tests.

disinfectant A chemical used to destroy many microorganisms and viruses.

disinfection Process of reducing or eliminating pathogenic microorganisms or viruses in or on a material so that they are no longer a hazard.

disinfection by-product (DBPs) Compounds formed when chlorine or other disinfectants react with naturally-occurring chemicals in water.

disseminate To spread.

disseminated intravascular coagulation Devastating condition in which clots form in small blood vessels, leading to failure of vital organs.

division Taxonomic rank that groups similar classes; also called a phylum. A collection of similar divisions makes up a kingdom.

DNA Abbreviation for deoxyribonucleic acid.

DNA-based vaccine Vaccine composed of segments of naked DNA from infectious organisms that can be introduced directly into muscle tissue; the host tissue expresses the DNA for a short time, producing the microbial antigens encoded by the DNA. Vaccines of this type are still in developmental stages.

DNA cloning Procedure by which DNA is inserted into a replicon such as a plasmid or bacteriophage, which is then introduced into cells where the replicon can replicate. (See *gene cloning*.)

DNA fingerprinting The use of characteristic patterns in the nucleotide sequence of DNA to match a specimen to a probable source.

DNA gyrase Enzyme that helps relieve the tension in DNA caused by the unwinding of the two strands of the DNA helix.

DNA library Collection of cloned molecules that together encompass the entire genome of an organism of interest; each clone can be viewed as one "book" of the total genetic information of the organism of interest.

DNA ligase Enzyme that forms covalent bonds between adjacent fragments of DNA.

DNA-mediated transformation Process of gene transfer in which DNA is transferred as a "naked" molecule.

DNA microarray (See *microarray*.)

DNA polymerases Enzymes that synthesize DNA; they use one strand as a template to generate the complementary strand.

DNA probe A piece of DNA, labeled in some manner, that is used to identify the presence of a certain sequence DNA by hybridizing it to its complement.

DNA replication Duplication of a DNA molecule.

DNA sequencing Determining the sequence (order) of nucleotide bases in a strand of DNA.

domain (1) Level of taxonomic classification above the kingdom level; there are three domains—*Bacteria*, *Archaea*, and *Eucarya*. (2) Distinct globular regions that characterize immunoglobulin molecules.

donor Refers to the cell that donates DNA in DNA transfer.

double-blind Type of study where neither the physicians nor the patients know who is receiving the actual treatment.

doubling time Time it takes for the number of cells in a population to double; the generation time.

downstream Direction toward the 3′ end of an RNA molecule or the analogous (+) strand of DNA.

droplet transmission Transmission of infectious agents through inhalation of respiratory droplets.

Durham tube Small inverted tube placed in a broth of sugar-containing media that is used to detect gas production by a microorganism.

D value Abbreviation for the decimal reduction time.

dysentery Condition characterized by crampy abdominal pain and bloody diarrhea.

---

eclipse period Time during which viruses exist within the host cell separated into their protein and nucleic acid components.

ecological niche The role that an organism plays in a particular ecosystem.

ecosystem An environment and the organisms that inhabit it.

eczema Condition characterized by a blistery skin rash, with weeping of fluid and formation of crusts, usually due to an allergy.

edema Swelling of tissues caused by accumulation of fluid.

edible vaccine Vaccine created by transferring genes encoding key antigens from infectious agents into plants. Vaccines of this type are still in developmental stages.

effector Any molecule that causes an effect.

effector T cell A descendent of an activated T cell that has become armed with the ability to produce specific cytokines and other substances, endowing the cell with specific protective attributes.

electrochemical gradient A separation of charged ions across the membrane.

electron Negatively charged component of an atom that orbits the nucleus of the atom.

electron microscope Microscope that uses electrons instead of light and can magnify images in excess of 100,000×.

electron transport chain Series of electron carriers that transfer electrons from donors such as NADH to acceptors such as oxygen, ejecting protons in the process.

electrophoresis Technique that uses an electric current to separate either DNA fragments or proteins.

electroporation Process of treating cells with an electric current to introduce DNA into them.

element A substance that is composed of a single type of atom.

elementary body Small dense-appearing infectious form of *Chlamydia* species that is released upon death and rupture of the host cell.

ELISA Abbreviation for enzyme-linked immunosorbent assay.

Embden-Meyerhof pathway Metabolic pathway that oxidizes glucose to pyruvate, generating ATP and reducing power; also known as glycolysis and the glycolytic pathway.

emerging diseases Diseases that have increased in incidence in the past two decades.

encephalitis Inflammation of the brain.

endemic Constantly present in a population.

endergonic Chemical reaction that requires a net input of energy because the products have more free energy than the starting compounds.

endocarditis Inflammation of the heart valves or lining of the heart chambers.

endocytosis Process through which cells take up particles by enclosing them in a vesicle pinched off from the cell membrane.

endogenous antigen An antigen produced within a given host cell.

endogenous pyrogen Fever-inducing substance (such as cytokines) made by the body.

endonuclease An enzyme that cleaves bonds internally in the backbone of DNA.

endoplasmic reticulum (ER) Organelle of eukaryotes where macromolecules destined for the external environment of other organelles are synthesized.

endosome Vesicle formed when a cell takes up material from the surrounding environment using the process of endocytosis.

endospore A kind of resting cell, characteristic of a limited number of bacterial species; highly resistant to heat, radiation, and disinfectants.

endosymbiont Microorganism that resides within another cell, providing a benefit to the host cell.

endosymbiont theory Theory that the ancestors of mitochondria and chloroplasts were bacteria that had been residing within other cells in a mutually beneficial partnership.

endothelial cell Cell type that lines the blood and lymph vessels.

endotoxic shock Septic shock that occurs as a result of endotoxin (lipopolysaccharide) circulating in the bloodstream.

endotoxin Lipopolysaccharide, a toxic component of the outer membrane of Gram-negative cells that can elicit symptoms such as fever and shock; lipid A is the molecule responsible for the toxic effects of endotoxin.

end product inhibition Inhibition of gene activity by the end product of a biosynthetic pathway.

energy The capacity to do work.

energy source Compound a cell oxidizes to harvest energy; also called an electron donor.

enrichment culture Culture method that provides conditions to enhance the growth of one particular organism in a mixed population.

enterics A common name for members of the family *Enterobacteriaceae*.

enterobacteria (See *enterics*.)

enterotoxin Poisonous substance, usually of bacterial origin, that causes diarrhea and vomiting.

Entner-Doudoroff pathway Pathway that converts glucose to pyruvate and glyceraldehyde-

3-phosphate by producing 6-phosphogluconate and then dehydrating it.

enveloped viruses Viruses that have a lipid bilayer surrounding their nucleocapsid.

enzyme A protein that functions as a catalyst.

enzyme-linked immunosorbent assay (ELISA) Technique used for detecting and quantifying specific antigens or antibodies by using an antibody labeled with an enzyme.

enzyme-substrate complex Transient form that occurs in an enzyme-mediated reaction, as the enzyme converts a substrate into a product.

eosinophil A type of white blood cell; thought to be primarily important in expelling parasitic worms from the body.

EPA Abbreviation for Environmental Protection Agency, a federal agency.

epidemic A disease or other occurrence whose incidence is higher than expected within a region or population.

epidemiology The study of factors influencing the frequency and distribution of diseases.

epidermis The outermost layer of skin.

epithelial cell Cell type that lines the surfaces of the body.

epitope Region of an antigen recognized by antibodies and antigen receptors on lymphocytes.

ergosterol Sterol found in fungal cell membranes; the target of many antifungal drugs.

ergot Poisonous substance produced by the fungus that causes rye smut.

erythrocytes Red blood cells.

E-site (or **exit site**) Site on the ribosome from which tRNAs exit after donating their amino acid to the adjacent tRNA.

ester bond Covalent bond formed between a —COOH group and an —OH group with the removal of H_2O.

ethambutol Antimycobacterial drug that inhibits enzymes required for synthesis of mycobacterial cell wall components.

ethidium bromide Mutagenic dye that binds to nucleic acid by intercalating between the bases; ethidium bromide-stained DNA is fluorescent when viewed with UV light.

Eucarya Name of the domain comprising eukaryotic organisms.

eukaryote Organism composed of one or more eukaryotic cells.

eukaryotic cell Complex cell type differing from a prokaryotic cell mainly in having a nuclear membrane surrounding its chromosomes.

eutrophic A nutrient-rich environment supporting the excessive growth of algae and other autotrophs.

evolutionary chronometer A molecule such as rRNA that can be used to measure the time elapsed since two organisms diverged from a common ancestor.

exanthem A skin rash.

excision repair Mechanism of DNA repair in which a fragment of single-stranded DNA containing mismatched bases is removed.

exergonic Describes a chemical reaction that releases energy because the starting compounds have more free energy than the products.

exfoliatin A bacterial toxin that causes sloughing of the outer epidermis.

exocytosis Process by which eukaryotic cells expel material; membrane-bound vesicles inside the cell fuse with the plasma membrane, releasing their contents to the external medium.

exoenzyme Enzyme that acts outside the cell that produces it.

exoerythrocytic Occurring outside the red blood cells, as the developmental cycle in malaria that occurs in the liver.

exogenous antigen An antigen that originated outside a given host cell.

exogenous pyrogen Fever-inducing substance (such as bacterial endotoxin) made from an external source.

exons Portions of eukaryotic genes that are expressed; interrupted by introns.

exotoxin Soluble poisonous protein substance released by a microorganism.

experimental study Type of study done to assess the effectiveness of measures to prevent or treat disease.

exponential phase Stage of growth of a bacterial culture in which cells are multiplying exponentially; log phase.

external node Point on a phylogenetic tree that represents a named species that still exists.

external transmission (or **temporary transmission**) Refers to transmission of viruses to plants by insects in which the virus is associated with the external mouthparts of the insect.

extrachromosomal DNA in a cell that is not part of the chromosome.

extremophiles Organisms that live under extremes of temperature, pH, or other environmental conditions.

extrinsic factors In food microbiology, environmental conditions, such as the temperature and atmosphere, that influence the rate of microbial growth.

Fab (fragment antigen-binding) region Portion of an antibody molecule that binds to the antigen.

facilitated diffusion Transport process that enables movement of impermeable compounds from one side of the membrane to the other by exploiting a concentration gradient; does not require expenditure of energy by the cell.

facultative Flexible with respect to growth conditions; for example, able to live with or without O_2.

facultative anaerobe Organism that grows best in the presence of oxygen (O_2), but can grow in its absence.

FAD Abbreviation for flavin adenine dinucleotide, an electron carrier.

fallopian tube The tubes that convey ova from the ovaries to the uterus.

FAME Stands for fatty acid methyl ester; a component of a technique that identifies bacteria based on their cellular fatty acid composition.

family Taxonomic group between order and genus.

fasciitis Inflammation of the fascia, bands of fibrous tissue that underlie the skin and surround muscle and body organs.

fastidious Exacting; refers to organisms that require growth factors.

fatty acid A molecule consisting of long chains of carbon atoms bonded to hydrogen atoms with an acidic group (—COOH) at one end.

F^- cell Recipient bacterial cell in conjugation.

F^+ cell Donor bacterial cell in conjugation, transfers the F plasmid.

F plasmid (See *fertility plasmid.*)

Fc region Crystallizable end of the constant region of an immunoglobulin molecule; responsible for binding to Fc receptors on cells, for initiating the classical pathway of complement activation, and for other biological functions.

fecal coliforms Thermotolerant coliform bacteria.

fecal-oral transmission Transmitting organisms that colonize the intestine by ingesting fecally contaminated material.

feedback inhibition Inhibition of the first enzyme of a biosynthetic pathway by the end product of that pathway; also called allosteric or end product inhibition.

feeding tolerance Lack of immune response to a specific antigen resulting from introducing the antigen orally.

fermentation Metabolic process in which the final electron acceptor is an organic compound such as pyruvate or a derivative.

fertility plasmid (or **F plasmid**) Plasmid found in donor cells of *E. coli* that codes for the F or sex pilus and makes the cell F^+.

fever An increase in internal body temperature to 37.8°C or higher.

fibronectin Glycoprotein occurring on the surface of cells and also in a circulating form that adheres tightly to medical devices; certain pathogens attach to it to initiate colonization.

filterable viruses The old terminology for viruses.

fimbria (pl. **fimbriae**) Type of pilus that enables cells to attach to a specific surface.

first-line antimicrobials In antimycobacterial drug therapy, the antimicrobials that are preferred because they are most effective as well as least toxic.

first-line defenses The barriers that separate and shield the interior of the body from the surrounding environment.

flagellin Protein subunits that make up the filament of flagella.

flagellum (pl. **flagella**) (1) In prokaryotic cells, a long protein appendage composed of subunits of flagellin that provides a mechanism of motility. (2) In eukaryotic cells, a long whiplike appendage composed of microtubules in a 9+2 arrangement that provides a mechanism of locomotion.

flavin adenine dinucleotide (FAD) A derivative of the vitamin riboflavin that functions as an electron carrier.

flavoprotein A flavin-containing electron carrier that functions in the electron transport chain.

flow cytometer Instrument that counts cells in a suspension by measuring the scattering of light by individual cells as they pass by a laser.

fluid mosaic model Model that describes the dynamic nature of the cytoplasmic membrane.

fluke Short, non-segmented, bilaterally symmetrical flatworm.

fluorescence-activated cell sorter (FACS) Machine that sorts fluorescent-labeled cells in a mixture by passing single cells in a stream past photodetectors.

fluorescence *in situ* hybridization (FISH) A procedure that uses a fluorescently labeled probe to detect specific nucleotide sequences within intact cells affixed to a microscope slide.

fluorescence microscope Special type of microscope used to observe cells that have been stained or tagged with fluorescent dyes.

fluoroquinolones Group of antimicrobial drugs that interferes with nucleic acid synthesis.

fomites Inanimate objects such as books, tools, or towels that can act as transmitters of pathogenic microorganisms or viruses.

foodborne intoxication Disease resulting from ingestion of food that contains a toxin produced by a microorganism.

foraminifera Protozoa that have silicon or calcium in their cell walls.

forespore Portion of the endospore formed during the process of sporulation that will ultimately become the core of the endospore.

fowl cholera Worldwide septicemic illness of wild and domestic fowl, caused by *Pasteurella multocida;* focus of the discovery by Pasteur that an attenuated organism could be used as a vaccine.

fragmentation Form of asexual reproduction in which a filament composed of a string of cells breaks apart, forming multiple reproductive units.

frameshift mutation Mutation resulting from the addition or deletion of a number of nucleotides not divisible by three.

free energy Amount of energy that can be gained by breaking the bonds of a compound; does not include the energy that is always lost as heat.

freeze-etching Process used to prepare specimens for transmission electron microscopy that allows viewing of the shape of underlying regions within structures of a cell.

fruiting body With respect to myxobacteria, a complex aggregate of cells, visible to the naked eye, produced when nutrients or water are depleted.

fungemia Fungi circulating in the bloodstream.

fungicide Kills fungi; used to describe the effects of some antimicrobial chemicals.

fungistatic Able to inhibit the growth of fungi.

fungus (pl. **fungi**) A non-photosynthetic eukaryotic heterotroph.

furuncle A boil; a localized skin infection that penetrates into the subcutaneous tissue, usually caused by *Staphylococcus aureus.*

GALT Abbreviation for gut-associated lymphoid tissue.

gametes Haploid cells that fuse with other gametes to form the diploid zygote in sexual reproduction.

gamma globulin Portion of blood serum proteins that contains IgG.

ganglion (pl. **ganglia**) Small body near the spinal column representing a bulge in a peripheral nerve at the site where the sensory nerve cells are located.

gas chromatography Technique of separating and identifying gaseous components of a substance.

gastroenteritis Acute inflammation of the stomach and intestines; often applied to the syndrome of nausea, vomiting, diarrhea, and abdominal pain.

gas vesicles Small rigid compartments produced by some aquatic bacteria that provide buoyancy to the cell; gases, but not water, flow freely into the vesicles, thereby decreasing the density of the cell.

G + C content Percentage of guanine plus cytosine in double-stranded DNA; also called the GC content.

gel electrophoresis Technique that uses electric current to separate either DNA fragments or proteins according to size by drawing them through a slab of gel.

gene The functional unit of a genome.

gene cloning Procedure by which genes are inserted into a replicon such as a plasmid or bacteriophage, which is then introduced into cells where the replicon can replicate.

gene fusion The joining of two genes.

gene library Sum total of all of the genes of an organism that have been inserted into cloning vectors.

generalized transducing phage Bacteriophage that is capable of transferring any part of the bacterial chromosome from one cell to another. (By contrast, a specialized transducing phage transfers only specific parts of the genome.)

generalized transduction Transfer of any bacterial gene to another bacterium by a phage.

general paresis Group of symptoms arising from nervous system damage, usually occurring 10 to 20 years after contracting syphilis; often manifest by emotional instability, memory loss, hallucinations, abnormalities of the eyes, and paralysis.

general secretory pathway Primary mechanism bacterial cells use to secrete proteins; proteins destined for secretion are recognized by their characteristic sequence of amino acids that make up the amino terminal end.

generation time Time it takes for the number of cells in a population to double; doubling time.

genetic engineering Process of deliberately altering an organism's genetic information by changing its nucleic acid sequences.

genetic reassortment Exchange of genetic information following two different segmented viruses infecting the same cell.

genetic recombination The joining together of genes from different organisms.

genetics The study of the function and transfer of genes.

genome Complete set of genetic information in a cell.

genome mining Searching genomic databases; for example, companies might search genomic databases to locate ORFs that may encode proteins of medical value.

genomic island Large segment of DNA that has been acquired from another species through horizontal gene transfer. Examples include pathogenicity islands and antibiotic resistant islands.

genomics Study and analysis of the nucleotide sequence of DNA.

genotype The sequence of nucleotides in the DNA of an organism.

genus (pl. **genera**) Category of related organisms, usually containing several species. The first name of an organism in the Binomial System of Nomenclature.

germicide Agent that kills microorganisms and inactivates viruses.

germination Sum total of the biochemical and morphological changes that an endospore or other resting cell undergoes before becoming a vegetative cell.

giant cell Very large cell with many nuclei, formed by the fusion of many macrophages during a chronic inflammatory response; found in granulomas.

gingivitis Inflammation of the gums.

global control The simultaneous regulation of numerous unrelated genes.

glucans Polysaccharides composed of repeating subunits of glucose; involved in formation of dental plaque.

glucose-salts Type of chemically defined medium that contains only glucose and certain inorganic salts; supports the growth of *E. coli.*

glycan chain High molecular weight linear polymer of alternating subunits of *N*-acetylglucosamine and *N*-acetylmuramic acid that serves as the backbone of the peptidoglycan molecule.

glycocalyx Gel-like layer that surrounds some cells and generally functions as a mechanism of either protection or attachment.

glycogen Polysaccharide composed of glucose molecules.

glycolipids Lipids that have various sugars attached.

glycolysis Metabolic pathway that oxidizes glucose to pyruvate, generating ATP and reducing power; also called the Embden-Meyerhoff pathway and the glycolytic pathway.

glycolytic pathway Glycolysis.

glycoproteins Proteins with covalently bonded sugar molecules.

glycosylase An enzyme that removes oxidized guanine from DNA by breaking a bond between deoxyribose and the oxidized guanine.

goblet cells Mucus-secreting epithelial cells.

Golgi apparatus Series of membrane-bound flattened sacs within eukaryotic cells that serve as the site where macromolecules synthesized in the endoplasmic reticulum are modified before they are transported to other destinations.

Gram-negative bacteria Bacteria that lose the crystal violet in the Gram stain procedure and therefore stain pink; the cell wall of these organisms is composed of a thin layer of peptidoglycan surrounded by an outer membrane.

Gram-positive bacteria Bacteria that retain the crystal violet stain in the Gram stain procedure and therefore stain purple; the cell wall of these organisms is composed of a thick layer of peptidoglycan.

Gram stain Staining technique that divides bacteria into one of two groups, Gram-positive or Gram-negative, on the basis of color; among bacteria, the staining reaction correlates well with cell wall structure.

granulation tissue New tissue formed during healing of an injury, consisting of small, red, translucent nodules containing abundant blood vessels.

granulocytes White blood cells characterized by the presence of prominent granules; basophil granules stain dark with basophilic dyes, eosinophils stain bright red with eosinophilic dyes, and neutrophils do not take up either stain.

granuloma Found in a chronic inflammatory response, collections of lymphocytes and stages of macrophages; an attempt by the body to wall off and contain persistent organisms and antigens.

griseofulvin Antifungal medication that appears to interfere with the action of tubulin, a necessary factor in nuclear division.

group translocation Type of transport process that chemically alters a molecule during its passage through the cytoplasmic membrane.

growth curve Growth pattern observed when cells are grown in a closed system; consists of five stages—lag phase, log phase (or exponential phase), stationary phase, death phase, and the phase of prolonged decline.

growth factors Compounds that a particular bacterium cannot synthesize and therefore must be included in a medium which supports the growth of that organism.

gumma Localized area of chronic inflammation and necrosis in tertiary syphilis, often manifest as a swelling.

HAART Highly active antiretroviral therapy; a cocktail of medications that act at different sites during replication of human immunodeficiency virus.

hairy leukoplakia Whitish patch, usually appearing on the tongue of individuals with severe immunodeficiency, thought to be caused by reactivation of latent Epstein-Barr virus (EBV) infection.

half-life Time it takes for one-half of the original number of molecules of a compound to be eliminated or degraded.

halophile Organism that prefers or requires a high salt (NaCl) medium to grow.

haploid Containing only a single set of genes.

hapten Substance that can combine with specific antibodies but which cannot incite the production of those antibodies unless it is attached to a large carrier molecule.

haustoria Specialized hypha of parasitic fungi that can penetrate plant or animal cell walls.

heavy chain The two higher molecular weight polypeptide chains that make up an antibody molecule; the type of heavy chain dictates the class of antibody molecule.

helicase Enzyme that unwinds the DNA helix ahead of the replication fork.

helminth A parasitic worm.

helper T cells Type of lymphocyte programmed to activate B cells and macrophages, and assist other aspects of adaptive immunity.

hemagglutination Clumping of red blood cells.

hemagglutination inhibition Immunological test used to detect antibodies against certain viruses which naturally cause red blood cells to agglutinate; antibodies that bind the virus inhibit the usual agglutination.

hemagglutinin A protein important in the virulence of the influenza virus.

hematopoietic stem cells Bone marrow cells that give rise to all blood cells.

hemolytic disease of the newborn (HDN) Disease of the fetus or newborn caused by transplacental passage of maternal antibodies against the baby's red blood cells, resulting in red cell destruction; usually anti-Rhesus (Rh) antibodies are involved and the disease is called Rh disease; also called erythroblastosis fetalis.

hemolytic uremic syndrome (HUS) Serious condition characterized by red cell breakdown and kidney failure; a sequel to infection by certain Shiga toxin-producing strains of *Shigella dysenteriae* and *Escherichia coli*.

hepatitis Inflammation of the liver; various causes, but most commonly the result of viral infection, particularly by the hepatitis viruses.

hepatitis B virus An enveloped DNA virus with an unusual mode of replication involving reverse transcriptase; cause of hepatitis B.

herd immunity Phenomenon that occurs when a critical concentration of immune hosts prevents the spread of an infectious agent.

hermaphroditic Having both male and female reproductive structures in the same organism.

herpes zoster Another name for shingles; a disease that results from reactivation of the herpesvirus causing chickenpox.

heterocyst Specialized non-photosynthetic cells of cyanobacteria within which nitrogen fixation occurs.

heterophile antibody Antibody that reacts with the red blood cells of another animal.

heterotroph Organism that obtains carbon from an organic compound such as glucose.

Hfr cells (high frequency of recombination cells) Rare cells in the F^+ population that can transfer their chromosome to an F^- cell.

high-copy-number plasmid Plasmid whose numbers in the cell range from 50 to 500.

high efficiency particulate air (HEPA) filters Special filters that remove from air nearly all particles, including microorganisms, that have a diameter greater than 0.3 μm.

high-energy phosphate bond Bond that joins a phosphate group to a molecule and releases a relatively high amount of energy when hydrolyzed; denoted by the symbol ~.

high-level disinfectant Chemical used to destroy all viruses and vegetative cells, but not endospores.

high-temperature-short-time (HTST) method Most common pasteurization protocol; using this method, milk is pasteurized by holding it at 72°C for 15 seconds.

histamine A substance found in basophil and mast cell granules that upon release can cause dilation and increased permeability of blood vessel walls and other effects; a mediator of inflammation.

HIV disease The illness caused by human immunodeficiency virus, marked by gradual impairment of the immune system, ending in AIDS.

HLA Abbreviation for human leukocyte antigen.

homologous With respect to DNA, stretches that have similar or identical nucleotide sequences and probably encode similar characteristics.

homologous recombination Genetic recombination between stretches of similar or identical nucleotide sequences.

homoserine lactone (HSL) Freely diffusible molecule used by certain types of bacteria to sense the density of cells within their population.

hook Curved flagellar structure that connects the filament of the flagellum to the cell surface.

hops Flowers of the vinelike hop plant; they are added to wort to impart a desirable bitter flavor to beer and contribute antibacterial substances.

horizontal gene transfer Transmission of DNA from one bacterium to another through conjugation, DNA-mediated transformation, or transduction; also called lateral gene transfer.

horizontal transmission Transfer of a pathogen from one person to another through contact, ingestion of food or water, or via a living agent such as an insect.

host Organism on or in which smaller organisms or viruses live, feed, and reproduce.

host cell In immunology, one of the body's own cells.

host range The range of animals or cell types that a pathogen can infect.

HSV-1 (herpes simplex virus-1) Member of the herpes family of viruses that causes cold sores and other types of infection.

HSV-2 (herpes simplex virus-2) Member of the herpes family of viruses; principal cause of genital herpes.

HTST Abbreviation for high-temperature-short-time pasteurization.

human leukocyte antigen (HLA) Human MHC molecules.

humoral immune response Antibody response.

hybridization The annealing of two complementary strands of DNA from different sources to create a hybrid double-stranded molecule.

hybridoma Cell made by fusing a lymphocyte, such as an antibody-producing B cell, with a cancer cell.

hydrogenation Reduction reaction in which an electron and an accompanying proton is added to a molecule.

hydrogen bond Weak attraction between a positively-charged hydrogen atom of one compound and a negatively charged atom of another compound; the charges of the two atoms are due to polar covalent bonds.

hydrolysis Chemical reaction in which a molecule is broken down as H_2O is added.

hydrophilic Water loving; soluble in water.

hydrophobic bonds Weak bonds formed between molecules as a result of their mutual repulsion of water molecules.

hyperimmune globulin Immunoglobulin prepared from the sera of donors with large amounts of antibodies to certain diseases, such as tetanus; used to prevent or treat the disease.

hypersensitivity Also termed allergy; heightened immune response to antigen.

hyperthermophiles Organisms that have an optimum growth temperature between 70°C and 110°C.

hypervariable regions Small areas in the Fab portion of the immunoglobulin light and heavy polypeptide chains that bind the antigenic epitope.

hypha (pl. **hyphae**) Threadlike structure that characterizes the growth of most fungi and some bacteria such as members of the genus *Streptomyces*.

hyposensitization (or **desensitization**) Form of therapy for immediate IgE-mediated allergies in which extremely small but increasing amounts of antigen are injected regularly over a period of months, directing the response from IgE to IgG.

hypoxic Deficient in oxygen.

ID_{50} The number of organisms that, when administered, will cause infection in approximately 50% of hosts.

identification In taxonomy, the process of characterizing an isolate in order to determine the group (taxon) to which it belongs.

IFN Abbreviation for interferons.

IgA proteases Enzymes that degrade IgA; they may have other roles as well.

illness Period of time during which symptoms and signs of disease occur.

immune complex Complex of antigen and antibody bound together, often with some complement system components included.

immune serum globulin Immunoglobulin G portion of pooled plasma from many donors, containing a wide variety of antibodies; used to provide passive protection.

immunity Protection against infectious agents and other substances.

immunoassay Tests using immunological reagents such as antigens and antibodies.

immunocompromised A host with weaknesses or defects in the innate or adaptive defenses.

immunodeficiency Inability to produce a normal immune response to antigen.

immunodiffusion tests Precipitation reactions carried out in agarose or other gels.
immunoelectrophoresis Technique for separating proteins by subjecting the mixture to an electric current followed by diffusion and precipitation in gels using antibodies against the separated proteins.
immunofluorescence Technique used to identify particular antigens microscopically in cells by the binding of a fluorescent antibody to the antigen.
immunogen Antigen that induces an immune response.
immunoglobulin Glycoprotein molecules that react specifically with the substance that induced their formation; antibodies.
immunological tolerance (See *tolerance*.)
immunology The study of immunity, or protection against infectious and other agents, and conditions arising from the mechanisms involved in immunity, such as hypersensitivities.
immunosuppression Non-specific suppression of adaptive immune responses.
immunotherapy Techniques used to modify the immune system action for a favorable effect.
inactivated vaccine Vaccine composed of killed bacteria, inactivated virus, or fractions of the agent; the agent in the vaccine is unable to replicate.
inapparent (or **subclinical**) **infections** Infections in which symptoms do not occur or are mild enough to go unnoticed.
incidence rate Number of new cases of a disease within a specific time period in a given population.
inclusion body Microscopically visible structure within a cell representing the site at which an infecting virus replicates; can occur within the nucleus or the cytoplasm.
incubation period Interval between entrance of a pathogen into a susceptible host and the onset of illness caused by that pathogen.
index case First identified case of a disease in an epidemic.
indirect contact Means of transmitting infectious disease via fomites.
indirect selection In microbial genetics, a technique for isolating mutants and identifying organisms unable to grow on a medium on which the parents do grow; often involves replica plating.
induced mutation Mutation that results from the organism being treated with an agent that alters its DNA.
inducer Substance that activates transcription of certain genes.
inducible enzyme Enzyme synthesized only under certain environmental conditions.
induction Process by which a prophage is excised from the host cell DNA; activation of gene transcription.
infection Growth and multiplication of a parasitic organism or virus in or on the body of the host with or without the production of disease.
infectious disease Disease caused by a microbial or viral infection.
infectious dose Number of microorganisms or viruses sufficient to establish an infection; often expressed as ID_{50} in which 50% of the hosts are infected.
inflammation Innate response to injury characterized by swelling, heat, redness, and pain in the affected area.
initiation complex Complex of a 30S ribosomal subunit, a tRNA that carries f-Met, and elongation factors that comes together at a start codon on mRNA and begins the process of translation.
innate immunity Immunity that is not affected by prior contact with the infectious agent and is not mediated by lymphocytes.
innate resistance Resistance of an organism to an antimicrobial medication due to the inherent characteristics of that type organism; also called intrinsic resistance.
inner membrane (1) In prokaryotic cells, the cytoplasmic membrane of Gram-negative bacteria. (2) In eukaryotic cells, the membrane on the interior side of an organelle that has a double membrane.
inorganic A compound that contains no C—H bonds.
insert DNA that is (or will be) joined to a vector to create a recombinant DNA molecule.
insertion mutation Mutation resulting from the integration of a transposon into a gene.
insertion sequence (IS) Short piece of DNA that has the ability to move from one site on a DNA molecule to another; simplest type of transposable element.
insulin-dependent diabetes mellitus (IDDM) Diabetes caused by autoimmune destruction of pancreatic cells by cytotoxic T cells.
intercalating agents Agents that insert themselves between two nucleotides in opposite strands of a DNA double helix.
interference microscope Type of light microscope that employs special optical devices to cause the specimen to appear as a three-dimensional image; an example is the Nomarski differential interference contrast microscope.
interferons Cytokines that induce cells to resist viral replication.
interleukins Cytokines produced by leukocytes.
intermediate fibers Component of the eukaryotic cell cytoskeleton.
intermediate-level disinfectant Type of chemical used to destroy all vegetative bacteria including mycobacteria, fungi, and most, but not all, viruses.
internal node Branch point on a phylogenetic tree that represents an ancestor to modern organisms.
intranuclear inclusion body Structure found within the nucleus of cells infected with certain viruses such as the cytomegalovirus.
intrinsic resistance (See *innate resistance*.)
intron Part of the eukaryotic chromosome that does not code for a protein; removed from the RNA transcript before the mRNA is translated.
inverted repeat Sequence of nucleotides on one strand of DNA that is identical to DNA on another strand when both are read in the same direction, that is, 5′ to 3′; associated with transposable elements.
in vitro In a test tube or other container as opposed to inside a living plant or animal.
in vivo Inside a living plant or animal as opposed to a test tube or other container.
ion Positively or negatively charged atom or molecule.
ionic bond Bond formed by the attraction of positively charged atoms or molecules to negatively charged ones.
IS Abbreviation for insertion sequence.
isomer Molecule with the same number and types of atoms as another but differing in its structure.
isotope Form of an element that differs in atomic weight from the form most common in nature.

Jarisch-Herxheimer reaction Abrupt but transitory worsening of symptoms after starting effective antibacterial treatment, thought to be caused by substances released by the death of the bacteria.

---

keratin A water-repelling protein found in hair, nails, and the outermost cells of the epidermis.
kinetic energy Energy of motion.
kingdom Taxonomic rank that groups several phyla or divisions; a collection of similar kingdoms makes up a domain.
Kirby-Bauer disc diffusion test Procedure used to determine whether a bacterium is susceptible to concentrations of an antimicrobial compound usually present in the bloodstream of an individual receiving the antimicrobial.
Koch's Postulates The criteria used to determine the cause of an infectious disease by culturing the agent and reproducing the disease.
Koplik spots Lesions of the oral cavity caused by measles virus that resemble a grain of salt on a red base.
Krebs cycle Metabolic pathway that incorporates acetyl-CoA, and generates ATP (or GTP), CO_2, and reducing power; also called the tricarboxylic acid (TCA) cycle and the citric acid cycle.

---

labeled Tagged with a detectable marker such as a radioactive isotope, a fluorescent dye, or an enzyme.
***lac* operon** Operon that encodes the proteins required for the degradation of lactose; it has served as one of the most important models for studying gene regulation.
lactic acid bacteria Group of Gram-positive bacteria that generate lactic acid as a major end product of their fermentative metabolism.
lactoferrin Iron-binding protein found in leukocytes, saliva, mucus, milk, and other substances; helps defend the body by depriving microorganisms of iron.
lactose Disaccharide consisting of one molecule of glucose and one of galactose.
***lacZ′* gene** Gene used to visually determine whether or not a vector contains a fragment inserted into a multiple cloning site.
lagging strand In DNA replication, the strand that is synthesized as a series of fragments.
lagooning Sewage treatment method in which sewage is channeled into shallow lagoons, during which time it is degraded by anaerobic and/or aerobic organisms.
lag phase Stage in the growth of a bacterial culture characterized by extensive macromolecule and ATP synthesis but no increase in the number of viable cells.
laminar flow hood Biological safety cabinet in which laboratory personnel work with potentially dangerous airborne pathogens; a continuous flow of incoming and outgoing air is filtered through HEPA filters to contain microorganisms within the cabinet.
Lancefield grouping Classification of β-hemolytic streptococci based on serological identification.
latent infection Infection in which the infectious agent is present but not active.
latent state The state of a phage when its DNA is integrated into the genome of the host.
lateral gene transfer (See *horizontal gene transfer*.)

leading strand In DNA replication, the DNA strand that is synthesized as a continuous fragment.
leaky Refers to a mutation in which the mutant gene codes for a protein that is partially functional.
lecithin Component of mammalian cell membranes; attacked by the α-toxin of *Clostridium perfringens* and other lecithinases.
lectin pathway Pathway of complement system activation initiated by binding of mannan-binding lectins to microbial cell walls.
leghemoglobin Protein synthesized by leguminous plants that carries O_2 within a *Rhizobium*-harboring nodule.
lethal dose (LD) Concentration of an infectious agent that causes death; often expressed as LD_{50}, the concentration of substances in which 50% of the hosts are killed by the agent.
leukemia Cancer of the leukocytes (white blood cells).
leukocidins Substances that kill white blood cells.
leukocytes White blood cells.
leukotrienes Substances active in inflammation, leading to chemotaxis and increased vascular permeability; produced by mast cells, basophils, and macrophages.
L-forms Bacterial variants that have lost the ability to synthesize the peptidoglycan portion of their cell wall.
lichen Organism composed of a fungus in a symbiotic association with either a green alga or a cyanobacterium.
ligand A specific molecule that binds to a given receptor.
light chains In an antibody, the two lighter molecular weight polypeptide chains that make up the molecule.
light-dependent reactions Processes used by phototrophs to harvest energy from sunlight; the energy-gathering component of photosynthesis.
light-independent reactions Stage of photosynthesis in which the ATP generated in the light-dependent reactions is used to fix carbon; also called dark reactions.
light microscope Microscope that uses visible light to observe objects.
light reactions (See *light-dependent reactions.*)
light repair Process by which bacteria repair UV damage to their DNA only in the presence of light.
lincosamides Group of antimicrobials that interferes with protein synthesis.
lipid One of a diverse group of organic substances all of which are relatively insoluble in water, but soluble in alcohol, ether, chloroform, or other fat solvents.
lipid A Portion of lipopolysaccharide (LPS) that anchors the molecule in the lipid bilayer of the outer membrane of Gram-negative cells; it plays an important role in the body's ability to recognize the presence of invading bacteria, but is also responsible for the toxic effects of LPS.
lipopolysaccharide (LPS) Molecule formed by bonding of lipid to polysaccharide; a part of the outer membrane of Gram-negative bacteria.
lipoprotein Macromolecule formed by the bonding of lipid to protein.
lipoteichoic acids Component of the Gram-positive cell wall that is linked to the cytoplasmic membrane.
localized infections Infections limited to one site in or on the body, as a furuncle.
locus Designated location on a chromosome.
log phase Stage of growth of a bacterial culture in which the cells are multiplying exponentially.
low-copy-number plasmid Plasmid whose numbers in the cell are one or two copies.
low-level disinfectants Type of chemical used to destroy fungi, enveloped viruses, and vegetative bacteria except mycobacteria.
LPS Abbreviation for lipopolysaccharide.
luciferase Enzyme that catalyzes the chemical reactions that produce bioluminescence.
lymph Clear yellow liquid that flows within lymphatic vessels; generally contains lymphocytes and may contain globules of fat.
lymphadenopathy syndrome (LAS) Marked generalized enlargement of lymph nodes that often occurs at the end of the period of clinical well-being in HIV disease.
lymphangitis Inflammation of lymphatic vessels.
lymphatic vessels Vessels that carry lymph, which is collected from the fluid that bathes the body's tissues.
lymphocyte Small, round, or oval white blood cell with a large nucleus and a small amount of cytoplasm; involved in adaptive immunity.
lymphoid tissues and organs Collections of lymphocytes and related cells involved in immune responses.
lymphokines Cytokines secreted by lymphocytes.
lysate Remains of cells and virions that are released after lysis of cells.
lyse To burst.
lysogenic conversion Modification of the properties of a cell resulting from expression of phage DNA integrated into a bacterial chromosome.
lysogens Bacteria that carry a prophage integrated into their chromosome.
lysosome Membrane-bound structure in eukaryotic cells that contains powerful degradative enzymes.
lysozyme Enzyme that degrades the peptidoglycan layer of the bacterial cell wall.

MacConkey agar Type of selective and differential bacteriological medium used to isolate certain Gram-negative rods such as those that typically reside in the intestine.
macroenvironment Overall environment in which an organism lives; opposed to microenvironment.
macrolides Group of antimicrobial medications that interfere with protein synthesis.
macromolecule Very large molecule usually composed of repeating subunits.
macrophages Differentiated phagocytes of the mononuclear phagocyte system that can engulf and destroy microorganisms and other extraneous materials, function as antigen-presenting cells, and carry out ADCC (antibody-dependent cellular cytotoxicity).
magnetotaxis Movement by bacterial cells containing magnetite crystals in response to a magnetic field.
major histocompatibility complex (MHC) Cluster of genes coding for key cell surface proteins important in antigen presentation.
malaise Vague feeling of uneasiness or discomfort.
malignant tumor Abnormal growth of cells no longer under normal control that have the potential to spread to other parts of the body.
MALT Abbreviation for mucosal-associated lymphoid tissue.
mannan-binding lectins (MBLs) Pattern-recognition molecules the body uses to detect polymers of mannose, which are typically found on microbial but not mammalian cells.
mast cells Granule-containing tissue cells similar in appearance and function to the basophils of the blood, with receptors for the Fc portion of IgE; important in the inflammatory response and immediate allergic reactions.
maturation The stage in viral replication in which the various components of the virion assemble to form a whole virion; also called assembly.
MBC Abbreviation for minimum bactericidal concentration.
M cells Specialized epithelial cells lying over Peyer's patches that collect material in the intestine and transfer it to the lymphoid tissues beneath.
mechanical vector Organism such as a fly that physically moves contaminated material from one location to another.
mechanisms of pathogenicity Methods that pathogens use to evade host defenses and cause damage to the host.
medium (pl. **media**) Any material used for growing organisms.
megakaryocyte The large blood cell from which platelets arise.
meiosis Process in eukaryotic cells by which the chromosome number is reduced from diploid (2*N*) to haploid (1*N*).
melting Denaturating of double-stranded DNA.
membrane attack complex (MAC) Complex of certain components of the complement system that forms pores in the cell membrane, resulting in lysis of the cell.
membrane-damaging toxin Toxin that disrupts plasma membranes of eukaryotic cells.
membrane filtration A technique used to determine the number of bacterial cells in a liquid sample that has a relatively low number of organisms; concentrates bacteria by filtration before they are plated.
membrane proteins Specialized proteins embedded in the membrane bilayer; some function as receptors and others function as transport proteins.
memory cells Lymphocytes that persist in the body after an immune response to an antigen; upon subsequent exposure to the same antigen, they must differentiate and usually proliferate to become effector cells.
memory response (See *secondary response.*)
meninges Membranes covering the brain and spinal cord.
meningitis Inflammation of the meninges.
merozoite Stage in the life cycle of certain protozoa, such as the malaria-causing *Plasmodium* species.
mesophiles Bacteria that grow most rapidly at temperatures between 20°C and 45°C.
messenger RNA (mRNA) Single-stranded RNA that binds to ribosomes and directs the synthesis of protein.
metabolism Sum total of all the enzymatic chemical reactions in a cell.
metabolite Any product of metabolism.
metachromatic granules Polyphosphate granules found in the cytoplasm of some bacteria that appear as different colors when stained with a basic dye.
methanogens Group of *Archaea* that obtain energy by oxidizing hydrogen gas, using CO_2 as a terminal electron acceptor, thereby generating methane (CH_4).
MHC Abbreviation for major histocompatibility complex.

MHC molecules Molecules on the surface of cells used to present antigen to T cells; the MHC molecules, encoded by the major histocompatibility complex (MHC), are also involved in the immunological rejection of transplanted tissue and organs.

MIC Abbreviation for minimum inhibitory concentration.

microaerophiles Organisms that require small amounts of oxygen (2% to 10%) for growth, but are inhibited by higher concentrations.

microarray A solid support that contains a fixed pattern of numerous different single-stranded nucleic acid fragments of known sequences.

microenvironment Environment immediately surrounding an individual microorganism.

microtiter plate A small tray containing numerous wells, usually 96.

microtubules Cytoskeleton structures of a eukaryotic cell that form mitotic spindles, cilia, and flagella; long hollow cylinders composed of tubulin.

microvillus (pl. **microvilli**) Tiny cylindrical process from luminal surfaces of cells such as those lining the intestine; increases surface area of the cell.

mineralization Complete breakdown of organic molecules to inorganic molecules.

minimum bactericidal concentration (MBC) Lowest concentration of a specific antimicrobial medication that kills 99.9% of cells in a culture of a given strain of bacteria.

minimum inhibitory concentration (MIC) Lowest concentration of a specific antimicrobial medication that prevents the growth of a given strain of bacteria *in vitro.*

minus (−) strand (1) The DNA strand used as a template for RNA synthesis. (2) The complement to the plus (or sense) strand of RNA. Also called the antisense strand.

miracidium First larval form of a fluke, hatching from the ovum as a ciliated organism.

mismatch repair Repair mechanism in which a repair enzyme recognizes improperly hydrogen-bonded bases and excises a short stretch of nucleotides containing these bases.

mitochondrion Organelle in eukaryotic cells in which the majority of ATP synthesis occurs.

mitogen Substance that induces mitosis; causes proliferation of cells.

mitosis Nuclear division process in eukaryotic cells that ensures the daughter cells receive the same number of chromosomes as the original parent.

MMWR Abbreviation for *Morbidity and Mortality Weekly Report*, published by the Centers for Disease Control and Prevention (CDC).

mobilize (plasmids) To prepare DNA for transfer by conjugation.

mold A filamentous fungus.

mole Amount of a chemical in grams that contains 6.023×10^{23} molecules; it is equal to the molecular weight of the chemical, or the sum of the atomic weights of all the atoms in a molecule of that chemical.

molecular postulates The criteria used to study a microbe's virulence factors by using genetic and other molecular techniques.

molecular weight Relative weight of an atom or molecule based on a scale in which the H atom with one proton is assigned the weight of 1.0.

molecule Chemical consisting of two or more atoms held together by chemical bonds.

monobactams Group of antimicrobial medications that interferes with peptidoglycan synthesis; very resistant to β-lactamases.

monocistronic RNA transcript that carries the gene.

monoclonal antibodies Antibody molecules with a single specificity produced *in vitro* by lymphocytes that have been fused with a type of malignant myeloma cell.

monocytes Mononuclear phagocytes of the blood; part of the mononuclear phagocyte system of professional phagocytes.

monomer Subunit of a polymer.

mononuclear phagocyte system (MPS) System of mononuclear cells (monocytes and macrophages) that are scattered throughout the body and highly efficient at phagocytosis.

monosaccharide A sugar; a simple carbohydrate generally having the formula $C_nH_{2n}O_n$, where n can vary in number from three to eight.

morbidity Illness; most often expressed as the rate of illness in a given population at risk.

mordant Substance that increases the affinity of cellular components for a dye.

morphology Form or shape of a particular organism or structure.

mortality Death; most often expressed as a rate of death in a given population at risk.

most probable number (MPN) method Statistical estimate of cell numbers based on the theory of probability; a sample is successively diluted to determine the point at which subsequent dilutions receive no cells.

M protein A protein found in the cell walls of Group A streptococci that is associated with virulence.

mRNA Messenger RNA.

mucociliary escalator Moving layer of mucus and cilia lining the respiratory tract that traps bacteria and other particles and moves them into the throat.

mucosa (See *mucous membrane.*)

mucosal-associated lymphoid tissue (MALT) Lymphoid tissue present in the mucosa of the respiratory, gastrointestinal, and genitourinary tracts.

mucous membrane Epithelial barrier that is coated with mucus.

multiple-cloning site Small sequence of DNA that contains several unique restriction enzyme recognition sites into which foreign DNA can be cloned.

mushroom Filamentous multicelled fungus with macroscopic fruiting bodies.

mutagen Any agent that increases the frequency at which DNA is altered (mutated).

mutant Organism that has a changed nucleotide sequence or arrangement of nucleotides in its DNA.

mutation A change in the nucleotide sequence of a cell's DNA, which is then passed on to daughter cells.

mutualism A symbiotic association in which both partners benefit.

myasthenia gravis Autoimmune disease characterized by muscle weakness, caused by autoantibodies.

mycelium (pl. **mycelia**) Tangled, matlike mass of fungal hyphae.

mycology The study of fungi.

mycorrhiza Symbiotic relationship between certain fungi and the roots of plants.

mycosis (pl. **mycoses**) Disease caused by a fungus.

myxobacteria Group of Gram-negative bacteria that congegrate to form complex structures called fruiting bodies.

***N*-acetylglucosamine (NAG)** One of the two alternating subunits of the glycan chains that make up peptidoglycan.

NADP/NADPH Abbreviations for the oxidized/reduced forms of nicotinamide adenine dinucleotide phosphate, an electron carrier.

NAD/NADH Abbreviations for the oxidized/reduced forms of nicotinamide adenine dinucleotide, an electron carrier.

narrow host range plasmid Plasmid that only replicates in one or a few closely related species of bacteria.

narrow-spectrum antimicrobials Antimicrobial medications that inhibit or kill a limited range of bacteria.

National Molecular Subtyping Network for Foodborne Disease Surveillance (See *PulseNet.*)

natural killer cell (See *NK cell.*)

natural selection Selection by the environment of those cells best able to grow in that environment.

naturally acquired immunity Active or passive immunity acquired through natural means such as exposure to a disease-causing agent, breastfeeding, or transfer of IgG to a fetus in *utero.*

necrotic Dead; refers to dead cells or tissues in contact with living cells, as necrotic tissue in wounds.

negative staining Staining technique that employs an acidic dye to stain the background against which colorless cells can be seen.

negative (−) strand (See *minus (−) strand.*)

Negri body Viral inclusion body characteristic of rabies.

nematodes Roundworms.

neurotoxin Toxin that damages the nervous system.

neurotransmitter Any of a group of substances released from the terminations of nerve cells when they are stimulated; they cross to the adjacent cell and cause it to be excited or inhibited.

neutron Uncharged component of an atom found in the nucleus.

neutrophiles Organisms that can live and multiply within the range of pH 5 (acidic) to pH 8 (basic) and have a pH optimum near neutral (pH 7).

neutrophils (See *polymorphonuclear neutrophils.*)

nitrification Conversion of ammonia (NH_3) to nitrate (NO_3^-).

nitrifiers Group of Gram-negative bacteria that obtain energy by oxidizing inorganic nitrogen compounds such as ammonia or nitrate.

nitrogen fixation Conversion of nitrogen gas to ammonia.

NK cell Large granular, non-T, non-B lymphocyte that can kill cells to which antibody has bound and cells that do not bear MHC class I molecules on the surface.

Nomarski differential interference contrast microscope Type of microscope that has a device for separating light into two beams that pass through the specimen and then recombine; light waves are out of phase when they recombine, resulting in the three-dimensional appearance of material in the specimen.

nomenclature System of assigning names to organisms; a component of taxonomy.

non-communicable diseases Disease that cannot be transmitted from one individual to another.

non-competitive inhibition Type of enzyme inhibition that results from a molecule binding to the enzyme at a site other than the active site.

non-critical instruments Medical instruments and surfaces such as stethoscopes and countertops that only come into contact with unbroken skin.

non-cyclic photophosphorylation Type of photophosphorylation in which high-energy electrons are drawn off to generate reducing power; electrons must still be returned to chlorophyll, but they must come from a source such as water.

non-polar covalent bond Bond formed by sharing electrons between atoms that have equal attraction for the electrons.

nonsense mutation A mutation which generates a stop codon, resulting in a shortened protein.

normal microbiota (or **normal flora**) That group of microorganisms that colonizes the body surfaces but does not usually cause disease.

nosocomial infection Infection acquired during hospitalization.

notifiable diseases Group of diseases that are reported to the CDC by individual states; typically these diseases are of relatively high incidence or otherwise a potential danger to public health.

nuclear envelope Double membrane that separates the nucleus from the cytoplasm in eukaryotic cells.

nucleic acids hybridization (See *hybridization.*)

nucleic acids Ribonucleic acid (RNA) and deoxyribonucleic acid (DNA).

nucleocapsid Viral nucleic acid and its protein coat.

nucleoid Region of a prokaryotic cell containing the DNA.

nucleolus Region within the nucleus where ribosomal RNAs are synthesized.

nucleosome Unit of the chromatin of eukaryotic cells that consists of a complex of histones around which the linear DNA wraps twice.

nucleotides Basic subunits of ribonucleic or deoxyribonucleic acid consisting of a purine or pyrimidine covalently bonded to ribose or deoxyribose, which is covalently bound to a phosphate molecule.

nucleus Membrane-bound organelle in a eukaryotic cell that contains chromosomes and the nucleolus.

numerical taxonomy Method of classification based on the phenotypes of prokaryotes; determines the relatedness of different organisms based on the percentage of characteristics that two groups have in common.

O antigen Antigenic polysaccharide portion of lipopolysaccharide, the molecule that makes up the outer leaflet of the outer membrane of Gram-negative bacteria.

objective lens Lens of a compound microscope that is closest to the specimen.

obligate aerobes Organisms that require oxygen for growth.

obligate anaerobes Organisms that cannot multiply if O_2 is present; they are often killed by traces of O_2 because of its toxic derivatives.

obligate fermenters Organisms that can grow in the presence of O_2 but never use it as a terminal electron acceptor; also called aerotolerant anaerobes.

obligate intracellular parasites Organisms that grow only inside living cells.

occlusion bodies Masses of viruses inside or outside cells.

ocular lens Lens of a compound microscope that is closest to the eye.

oil A liquid fat.

Okazaki fragment Nucleic acid fragment synthesized as a result of the discontinuous replication of the lagging strand of DNA.

oligonucleotide Short chain of nucleotides.

oligosaccharide Short chain of monosaccharide subunits joined together by covalent bonds; shorter than a polysaccharide.

oligotrophic environment A nutrient-poor environment.

oligotrophs Organisms that can grow in a nutrient-poor environment.

oncogene Gene whose activity is involved in turning a normal cell into a cancer cell.

open reading frame (ORF) Stretch of DNA, generally longer than 300 base pairs, that has a reading frame beginning with a start codon and ending with a stop codon; it suggests that the region encodes a protein.

open system Method used to maintain cells in a state of continuous growth by continuously adding nutrients and removing waste products; also called a continuous culture.

operator Region located immediately downstream of a promoter to which a repressor can bind; binding of the repressor to the operator effectively prevents RNA polymerase from progressing past that region and blocks transcription.

operon Group of linked genes whose expression is controlled as a single unit.

opine Unusual amino acid derivative; a portion of the Ti plasmid of *Agrobacterium tumefaciens* is transferred to plant cells and directs the plant cells to synthesize this compound.

opportunistic pathogen Organism that causes disease only in hosts with impaired defense mechanisms or when introduced into an unusual location; also called an opportunist.

opsonization Enhanced phagocytosis, usually caused by coating of the particle to be ingested with either antibody or complement system components.

opthalmia neonatorum Eye infection of newborns usually caused by *Neisseria gonorrhoeae* or *Chlamydia trachomatis*, acquired from infected mothers during the birth process.

optical isomer (or **stereoisomer**) Mirror image of a compound.

optimal proportion Relative proportions of antigen and antibody at which both are fully incorporated in a precipitate.

optimum growth temperature Temperature at which a microorganism multiplies most rapidly.

oral tolerance Decreased reactivity of immune cells resulting from feeding an antigen.

order Taxonomic classification between class and family.

organ A structure composed of different tissues coordinated to perform a specific function.

organelle A cell structure that performs a specific function.

organic A compound in which a carbon atom is covalently bonded to a hydrogen atom.

origin of replication Distinct region of a DNA molecule at which replication is initiated.

origin of transfer Short stretch of nucleotides, a part of which is transferred first when a plasmid is transferred to a recipient cell; necessary for plasmid transfer.

osmosis Movement of water across a membrane from a dilute solution to a more concentrated solution.

osmotic pressure Pressure exerted by water on a membrane due to a difference in the concentration of molecules on each side of the membrane.

O-specific polysaccharide side chain The portion of LPS that is directed away from the membrane, at the end opposite of lipid A; because its composition varies, it can sometimes be used to identify species or strains.

outbreak Cluster of cases occurring during a brief time interval and affecting a specific population; may herald the onset of an epidemic.

outer membrane (1) In prokaryotic cells, the unique lipid bilayer of Gram-negative cells that surrounds the peptidoglycan layer. (2) In eukaryotic cells, the membrane on the cytoplasmic side of organelles that have double membranes.

oxazolidinones Group of antimicrobial drugs that interferes with protein synthesis.

oxidase test Rapid biochemical test used to detect the activity of cytochrome *c* oxidase.

oxidation Removal of an electron.

oxidation-reduction reactions Chemical reactions in which one or more electrons is transferred from one molecule to another; the compound that loses electrons becomes oxidized and the chemical that gains electrons becomes reduced.

oxidative phosphorylation Synthesis of ATP from ADP and inorganic phosphate using the energy of a proton motive force, which is generated by harvesting chemical energy.

oxygenic photosynthesis Photosynthetic reaction that releases O_2.

oxygenic phototrophs Phototrophic organisms that produce O_2.

---

palindrome Two stretches of DNA on opposite strands that are identical when oriented in the same direction, that is, 5′ to 3′ or 3′ to 5′.

pandemic A worldwide epidemic.

para-aminobenzoic acid (PABA) Intermediate in the pathway for folic acid synthesis in bacteria; sulfa drugs have a similar structure to PABA.

parasitism Association in which one organism, the parasite, benefits at the expense of the other organism, the host.

parent strain Refers to the original strain of a bacterium used in an experiment; term is often used in place of wild-type strain.

passive diffusion Process in which molecules flow freely into and out of a cell so that the concentration of any particular molecule is the same on the inside as it is on the outside of the cell.

passive immunity Protective immunity resulting from the transfer of antibody-containing serum produced by other individuals or animals.

pasteurization Process of heating food or other substances under controlled conditions of time and temperature to kill pathogens and reduce the total number of microorganisms without damaging the substance.

pathogen Organism or virus causing a disease.

pathogenesis Process by which disease develops.

pathogenicity islands Stretches of DNA in bacteria that code for virulence factors and appear to have been acquired from other bacteria.

pattern recognition Method by which the innate immune system recognizes invading microbes; the system uses receptors and other molecules that bind lipopolysaccharide, peptidoglycan, and other molecular patterns associated with microbes.

peliosis hepatis Serious condition characterized by formation of blood-filled cysts in the liver, caused by *Bartonella henselae*; usually a complication of AIDS or other severe immunodeficiency.

penicillin Antibiotic that interferes with the synthesis of the peptidoglycan portion of bacterial cell walls.

penicillin-binding proteins (PBPs) Target of β-lactam antimicrobial drugs; their role in bacteria is peptidoglycan synthesis.

penicillin enrichment Method for increasing the relative proportion of auxotrophic mutants in a population by killing off the growing prototrophic cells with penicillin.

pentamer Polymer composed of five monomeric structural units.

pentose phosphate pathway Metabolic pathway that initiates the degradation of glucose, generating reducing power in the form of NADPH, and two precursor metabolites.

peptide bond Covalent bond formed between the —COOH group of one amino acid and the —NH_2 group of another amino acid; characteristic of proteins.

peptide interbridge Component of the peptidoglycan layer of Gram-positive bacteria; the short chain of amino acids that links the peptide side chains of adjacent *N*-acetylmuramic acid molecules.

peptide vaccine Vaccine composed of key antigenic peptides from disease-causing microbes. Vaccines of this type are still in developmental stages.

peptidyl site (or **P-site**) Site on the ribosome to which the tRNA that temporarily carries the elongating amino acid chain resides.

peptidoglycan Macromolecule found only in bacteria that provides rigidity to the bacterial cell wall. The basic structure of peptidoglycan is an alternating series of two major subunits *N*-acetylmuramic acid (NAM) and *N*-acetylglucosamine (NAG); chains of these alternating subunits, are cross-linked by peptide chains.

peptone Common component of bacteriological media; consists of proteins originating from any of a variety of sources that have been hydrolyzed to amino acids and short peptides by treatment with enzymes, acids, or alkali.

perforin Molecule produced by T_C cells and NK cells to destroy target cells.

periplasm (or **periplasmic gel**) Gel that fills the region between the outer membrane and the cytoplasmic membrane in Gram-negative bacteria.

peristalsis The rhythmic contractions of the intestinal tract that propel food and liquid.

peritrichous flagella Distribution of flagella over the entire surface of a cell.

peroxidase enzymes Enzymes found in neutrophil granules, saliva, and milk that together with hydrogen peroxide and halide ions make up an effective antimicrobial system.

persistent Refers to infection in which the causative agent remains in the body for long periods of time, often without causing symptoms of disease.

petechia (pl. **petechiae**) Small purplish spot on the skin or mucous membrane caused by hemorrhage.

Petri dish Two-part dish of glass or plastic often used to contain medium solidified with agar, on which bacteria are grown.

Peyer's patches Collections of lymphoid cells in the gastrointestinal tract; part of the mucosal-associated lymphoid tissue (MALT).

pH Scale of 0 to 14 that expresses the acidity or alkalinity of a solution.

phage Shortened term for bacteriophage.

phage induction Process by which phage DNA is excised from bacterial chromosomal DNA.

phagocytes Cells that specialize in engulfing and digesting microbes and cell debris.

phagocytosis (v. **phagocytize**) The process by which certain cells ingest particulate matter by surrounding and enveloping those materials, bringing them into the cell in a membrane-bound vesicle.

phagolysosome Membrane-bound vacuole generated when a phagosome fuses with a lysosome.

phagosome Membrane-bound vacuole that contains the material engulfed by a phagocyte.

phase-contrast microscope Type of light microscope that employs special optical devices to amplify the difference in the refractive index of a cell and the surrounding medium, increasing the contrast of the image.

phase variation The reversible and random alteration of expression of certain bacterial structures such as fimbriae by switching on and off the genes that encode those structures.

phenotype The properties of a cell determined by the expression of the genotype.

phospholipase Membrane-damaging toxin that enzymatically removes the polar head group on phospholipids.

phospholipid Lipid that has a phosphate molecule as part of its structure.

phosphotransferase system Type of group translocation in which the transported molecule is phosphorylated as it passes through the cytoplasmic membrane.

photoautotrophs Organisms that use light as the energy source and CO_2 as the major carbon source.

photoheterotrophs Organisms that use light as the energy source and organic compounds as the carbon source.

photooxidation Chemical reaction occurring as a result of absorption of light energy in the presence of oxygen.

photophosphorylation Synthesis of ATP from ADP and inorganic phosphate using the energy of a proton motive force, which is generated by harvesting radiant energy.

photoreactivation (or **light repair**) Breakage of the covalent bonds joining thymine dimers in the light, thereby restoring the DNA to its original state.

photosynthesis Reactions used to harvest the energy of light to synthesize ATP, which is then used to power carbon fixation.

photosystems Protein complexes within which chlorophyll and other light-gathering pigments are organized; located in special photosynthetic membranes.

phototaxis Directed movement in response to variations in light.

phototrophs Organisms that use light as a source of energy.

phycobiliproteins Light-harvesting pigments of cyanobacteria; they absorb energy from wavelengths of light that are not well absorbed by chlorophyll.

phylogenetic tree Type of diagram that depicts the evolutionary heritage of organisms.

phylogeny Evolutionary relatedness of organisms.

phylum (pl. **phyla**) Collection of similar classes; a collection of similar phyla makes up a kingdom; a phylum may also be called a division.

phytoplankton Floating and swimming algae and photosynthetic prokaryotic organisms of lakes and oceans.

pilus (pl. **pili**) Hairlike appendages on many Gram-negative bacteria that function in conjugation and for attachment.

pinocytosis Process by which eukaryotic cells take in liquid and small particles from the surrounding environment by internalizing and pinching off small pieces of their own membrane, bringing along a small volume of liquid and any material attached to the membrane.

plankton Primarily microscopic organisms floating freely in most waters.

plaque (1) Clear area in a monolayer of cells. (2) In dentistry, a polysaccharide-encased community of bacteria (a biofilm) that adheres to a tooth surface.

plasma Fluid portion of non-clotted blood.

plasma cell End cell of the B-cell series, fully differentiated to produce and secrete large amounts of antibody.

plasma membrane Semipermeable membrane that surrounds the cytoplasm in a cell; cytoplasmic membrane.

plasmid Small extrachromosomal circular DNA molecule that replicates independently of the chromosome; often codes for antibiotic resistance.

plasmolysis Process in which water diffuses out of a cell, causing the cytoplasm to dehydrate and shrink from the cell wall.

plate count Method used to determine the number of viable cells in a specimen by determining the number of colonies that arise when the specimen is added to an agar medium.

platelets (or **thrombocytes**) Small cell fragments in the blood that are essential for blood clotting; arise from large bone marrow cells called megakaryocytes.

pleomorphic Bacteria that characteristically vary in shape.

pleurisy Inflammation of the pleura, membranes that line the lung and chest cavity; often marked by a sharp pain associated with breathing.

plus (+) strand (1) The DNA strand that is complementary to the strand used as a template for RNA synthesis. (2) Of the two RNA molecules that can theoretically be transcribed from double-stranded DNA, the one that can be translated to make a protein; also called the sense strand.

PMN Abbreviation for polymorphonuclear neutrophil.

pneumonia Inflammation of the lungs accompanied by filling of the air sacs with fluids such as pus and blood.

pneumonic plague Disease that develops when *Yersinia pestis* infects the lungs.

point mutation Mutation in which only a single base pair is involved.

polar covalent bond Bond formed by sharing electrons between atoms that have unequal attraction for the electrons.

polarity (1) The degree of affinity that an atom has for electrons; this results in positive or negative charges on atoms in a molecule. (2) The 5′ to 3′ directionality of a nucleic acid fragment. (3) The transfer of DNA from one cell to another in an exclusive direction, for example from an F^+ cell to an F^- cell.

poly A tail Series of approximately 200 adenine derivatives that are added to the 3′ end of an mRNA transcript in eukaryotic cells; thought to stabilize the transcript and enhance translation.

polycistronic An mRNA molecule that carries more than one gene.

polygenic (1) A trait encoded by more than one location on the chromosome. (2) An mRNA molecule that carries more than one gene.

polymer Large molecules formed by the joining together of repeating small molecules (subunits).

polymerase chain reaction (PCR) Method used to create millions of copies of a given region of DNA in only a matter of hours.

polymorphic Having different distinct forms.

polymorphonuclear neutrophils (PMNs) Type of phagocytic cell; the nuclei of these cells are segmented and composed of several lobes.

polymyxin B Type of antimicrobial medication that damages cytoplasmic membranes.

polypeptide Chain of amino acids joined by peptide bonds; also called a protein.

polyribosome Assembly of multiple ribosomes attached to a single mRNA molecule; also called a polysome.

polysaccharide Long chains of monosaccharide subunits.

polysaccharide vaccine Vaccine composed of polysaccharides, which make up the capsule of certain organisms.

polyunsaturated fatty acid Fatty acid that contains numerous double bonds.

porins Proteins in the outer membrane of Gram-negative bacteria that form channels through which small molecules can pass.

portal of entry Place of entry of microorganisms into the host.

portal of exit Place where infectious agents leave the host to find a new host.

positive selection Process that permits only those T cells that recognize MHC molecules to some extent to develop further.

positive (+) strand (See *plus (+) strand.*)

potential energy Stored energy; it can exist in a variety of forms including chemical bonds, a rock on the top of a hill, and water behind a dam.

pour-plate method Method of inoculating an agar medium with bacterial cells while the agar is liquid and then pouring it into a Petri dish, where the agar hardens; the colonies grow both on the surface and within the medium.

precipitation reaction Reaction of antibody with soluble antigen to form an insoluble substance.

precursor metabolites Metabolic intermediates of catabolic pathways that can be used in anabolic pathways.

preservation The process of inhibiting the growth of microorganisms in products to delay spoilage.

prevalence Total number cases, both old and new, in a given population at risk at a point in time.

primary culture Cells taken and grown directly from the tissues of an animal.

primary immune response Immune response that occurs upon first exposure to an antigen.

primary infection Infection in a previously healthy individual, such as measles in a child who has not had measles before.

primary lymphoid organs Organs in which lymphocytes mature; the thymus and bone marrow.

primary metabolites Compounds synthesized by a cell during the log phase.

primary pathogen Microbe able to cause disease in an otherwise healthy individual.

primary producers Organisms that convert CO_2 into organic compounds; by doing so, they sustain other life forms, including humans.

primary response The response that marks the adaptive immune system's first encounter with a particular antigen.

primary stain First dye applied in a multistep differential staining procedures; generally stains all cells.

primary structure Refers to the sequence of amino acids in a protein.

primary treatment In treatment of wastewater, a physical process designed to remove materials that will settle out.

primase Enzyme that synthesizes small fragments of RNA to serve as primers for DNA synthesis during DNA replication.

primer RNA molecule that initiates the synthesis of DNA.

prion Infectious protein that has no nucleic acid.

probe In nucleic acid hybridization, a single-stranded piece of nucleic acid that has been tagged with a detectable marker; it is used to identify homologous nucleic acid by allowing it to hybridize to its complement.

productive infection Virus infection in which more virions are produced.

proglottid One of the segments that make up most of the body of a tapeworm.

pro-inflammatory cytokines Any of a group of cytokines that contribute to the inflammatory response.

prokaryote Single-celled organism that does not contain a membrane-bound nucleus.

prokaryotic cell Cell characterized by lack of a nuclear membrane and thus no true nucleus.

promoter Nucleotide sequence to which RNA polymerase binds to initiate transcription.

propagated epidemic Outbreak of disease in which the infectious agent is transmitted to others, resulting in steadily increasing numbers of people becoming ill.

prophage Latent form of a temperate phage whose DNA has been inserted into the host's DNA.

prophylaxis Prevention of disease.

prospective study Study that looks ahead to see if the risk factors identified by a retrospective study predict a tendency to develop the disease.

prosthecate bacteria Group of gram-negative bacteria that have extensions projecting from the cells, thereby increasing their surface area.

protease Enzyme that degrades protein; the protease encoded by HIV is the target of several anti-HIV medications.

protein Macromolecule consisting of amino acid subunits.

protein A Protein produced by *Staphylococcus aureus* that inhibits phagocytosis of the organism by binding to the Fc portion of antibodies.

protein subunit vaccine Vaccine composed of key protein antigens or antigenic fragments of an infectious agent, rather than whole cells or viruses.

protist Designation for eukaryotic organisms other than plants, animals, and fungi; may be unicellular or multicellular.

proton Positively charged component of an atom.

proton motive force Form of energy generated by the electron transport chain, which expels protons to create a chemiosmotic gradient.

proton pump Complex of electron carriers in the electron transport chain that ejects protons from the cell.

proto-oncogene A type of gene involved in tumor formation; it codes for a protein that activates transcription.

prototroph Organism that has no organic growth requirements other than a source of carbon and energy.

protozoa Group of single-celled eukaryotic organisms.

provirus Latent form of a virus in which the viral DNA is incorporated into the chromosome of the host.

pseudomembranous colitis Disease of the colon caused by *Clostridium difficile* in which patches called pseudomembranes, composed of dead epithelium, inflammatory cells and clotted blood, form on the intestinal lining.

pseudopods Transient armlike extensions formed by phagocytes and protozoa; they surround and enclose extracellular material, including bacteria, during the process of phagocytosis.

P-site (or **peptidyl site**) Site on the ribosome where the tRNA that temporarily carries the elongating amino acid chain resides.

psychrophile Microorganism that grows best between −5°C and 15°C.

psychrotroph Organism that has an optimum temperature between 20°C and 30°C.

puerperal fever Childbed fever; infection of the uterus following childbirth, commonly caused by *Streptococcus pyogenes.*

pulsed-field gel electrophoresis Type of gel electrophoresis that is used to separate very large fragments of DNA.

PulseNet Surveillance network established by the Centers for Disease Control and Prevention to facilitate the tracking of foodborne disease outbreaks; catalogues the RFLPs of certain pathogenic organisms. Also called the National Molecular Subtyping Network for Foodborne Disease Surveillance.

pure culture A population of organisms descended from a single cell.

purine Component of RNA and DNA; the two major purines are adenine and guanine.

pus Thick, opaque, often yellowish material that forms at the site of infection, made up of dead neutrophils and tissue debris.

putrefaction Digestion of proteins by enzymes to yield foul-smelling products.

pyelonephritis Infection of the kidneys.

pyoderma Any skin disease characterized by production of pus.

pyogenic Pus-producing.

pyrimidine Component of RNA and DNA; the three major pyrimidines are thymine, cytosine, and uracil.

pyrogens Fever-inducing substances.

pyruvate End product of glycolysis; a precursor metabolite used in the synthesis of certain amino acids.

quaternary ammonium compounds Cationic (positively charged) detergents that are non-toxic enough to be used to disinfect food preparation surfaces; also called quats.
quaternary structure Level of structure of a protein molecule consisting of several polypeptide chains.
quinone A lipid-soluble electron carrier that functions in the electron transport chain.
quorum sensing Communication between bacterial cells by means of small molecules, permitting the cells to sense the density of cells.

---

radial immunodiffusion test Quantitative antigen-antibody precipitation in gel test in which one reactant is distributed throughout the gel and the other reactant diffuses into the gel, producing a ring of precipitation.
rDNA DNA that encodes ribosomal RNA (rRNA).
reactive oxygen A form of oxygen that is highly toxic to cells because it damages DNA.
reading frames Grouping of a stretch of nucleotides into sequential triplets; an mRNA molecule has three reading frames, but only one is typically used in translation.
receptor Type of membrane protein that binds to specific molecules in the environment, providing a mechanism for the cell to sense and adjust to its surroundings.
receptor-mediated endocytosis Type of pinocytosis that allows cells to internalize extracellular ligands that bind to the cell's receptors.
recognition sequence The DNA sequence recognized by a particular restriction enzyme.
recombinant DNA molecule DNA molecule created by joining DNA from two different sources *in vitro;* a vector-insert chimera is a recombinant DNA molecule.
recombinant vaccines Subunit vaccines produced by genetic engineering.
redox reactions Transfer of electrons from one compound to another; one compound becomes reduced and the other becomes oxidized.
reducing agents Compounds that readily donate electrons to another compound, thereby reducing the other compound.
reducing power Reduced electron carriers such as NADH, NADPH, and $FADH_2$; their bonds contain a form of usable energy.
reduction Process of adding electrons to a molecule.
refraction Bending of light rays that occurs when light passes from one medium to another.
regulatory protein Protein that binds to DNA, either blocking or enhancing the function of RNA polymerase.
regulatory T cells Type of lymphocyte that helps control the immune response.
regulon Set of related genes that are transcribed as separate units but are controlled by the same regulatory protein.
replica plating Technique for the simultaneous transfer of organisms in separated colonies from one medium to another medium.
replication fork In DNA synthesis, the site at which the double helix is being unwound to expose the single strands that can function as templates.
replicon Piece of DNA that is capable of replicating; contains an origin of replication.
reporter gene Gene that has a detectable phenotype and can be fused to a gene of interest, providing a mechanism by which to monitor the expression of the gene of interest.
repressible enzyme Enzyme whose synthesis can be turned off by certain conditions.
repressor Protein that binds to the operator site and prevents transcription.
reservoir Source of a disease-causing organism.
resident microbiota Normal microbiota that typically inhabit body sites for extended periods.
resistance plasmid (or **R plasmid**) Plasmid that carries genetic information for resistance to one or more antimicrobial medications and heavy metals.
resolve To clearly separate.
respiration Process that involves transfer of electrons stripped from a chemical energy source to an electron transport chain, generating a proton motive force that is then used to synthesize ATP.
respire To use the processes of respiration.
response regulator In a two-component regulatory system, the protein that binds to DNA.
restriction enzyme Type of enzyme that recognizes and cleaves a specific sequence of DNA.
restriction fragment length polymorphism (RFLP) Pattern of fragment sizes obtained by digesting DNA with one or more restriction enzymes.
restriction fragments Fragments generated when DNA is cut with restriction enzymes.
reticulate body Fragile, replicating, non-infectious intracellular form of *Chlamydia* species.
retrospective study Type of study done following a disease outbreak; compares the actions and events surrounding clinical cases with those of controls.
retroviruses Group of viruses that carry their genetic information as single-stranded RNA; they have the enzyme reverse transcriptase, which forms a DNA copy that is then integrated into the host cell chromosome.
reverse transcriptase Enzyme that synthesizes double-stranded DNA complementary to an RNA template.
reversion Process by which a second mutation corrects a defect caused by an earlier mutation.
Reye's syndrome Often fatal condition characterized by vomiting, coma, and brain and liver damage, mostly occurring in children treated with aspirin for influenza or chickenpox.
RFLP Abbreviation for restriction fragment length polymorphism.
rheumatoid arthritis Severe crippling autoimmune disease of the joints, caused by cytokines from inflammatory T_H1 cells and immune complexes.
rhizosphere Zone around plant roots containing organic materials exuded by the roots.
rhuMab (recombinant human monoclonal antibody) Monoclonal antibody derived from a laboratory animal in which part of the molecule has been replaced with the human equivalent.
ribonucleic acid (RNA) Macromolecules in a cell that play a role in converting the information coded by the DNA into amino acid sequences in protein.
ribose A 5-carbon sugar found in RNA.
ribosomal RNA (rRNA) Type of RNA present in ribosomes; the nucleotide sequences of these are increasingly being used to classify and, in some cases, identify microorganisms.
ribosome Structure that facilitates the joining of amino acids during the process of translation; composed of protein and ribosomal RNA.
ribosome-binding site Sequence of nucleotides in bacterial mRNA to which a ribosome binds; the first time the codon for methionine (AUG) appears after that site, translation generally starts.
ribotyping Technique used to distinguish among related strains; detects RFLPs in ribosomal RNA genes.
ribozymes RNA molecules that have a catalytic function.
rifamycins Group of antimicrobial medications that block transcription.
risk factors Specific conditions associated with high frequencies of disease.
RNA Abbreviation for ribonucleic acid.
RNA polymerase Enzyme that catalyzes the synthesis of RNA using a DNA template.
RNases Enzymes that degrade RNA.
rod Cylindrical-shaped bacterium; also called a bacillus.
rolling circle replication Mechanism of DNA replication in which a single strand of DNA is synthesized.
rough endoplasmic reticulum Organelle where proteins destined for locations other than the cytoplasm are synthesized.
roundworm A helminth that has a cylindrical, tapered body; a nematode.
R plasmids Plasmids that encode resistance to one or more antimicrobial medications and heavy metals.
rRNA Ribosomal RNA.
RTF Abbreviation for resistance transfer factor.
rubisco Enzyme that initiates the Calvin cycle by joining CO_2 to the 5-carbon compound ribulose-1,5-bisphosphate.

---

salinity Amount of salt in a solution.
SALT Skin-associated lymphoid tissues.
sanitization Process of substantially reducing the microbial populations on objects to achieve acceptably safe public health levels.
saprophyte Organism that takes in nutrients from dead and decaying matter.
saturated Refers to a fatty acid that contains no double bonds.
scanning electron microscope (SEM) Type of electron microscope that scans a beam of electrons back and forth over the surface of a specimen; used for observing surface details, but not internal structures of cells.
scavenger receptor Receptor on phagocytes that facilitates the engulfment of various materials that have charged molecules on their surface.
schizogony Process of multiple fission in which the nucleus divides a number of times before individual daughter cells are produced.
schizont Multinucleate stage in the development of certain protozoa, such as the ones that cause malaria.
scolex Attachment organ of a tapeworm, the head end.
scrapie Common name for a neurological disease of sheep; it is caused by a prion.
sebum Oily secretion of the sebaceous glands of the skin.
secondary response (or **memory response**) Enhanced immune response that occurs upon second or subsequent exposure to specific antigen, caused by the rapid activation of long-lived memory cells; anamnestic response.
secondary infection Infection that occurs along with or immediately following another infection, usually as a result of the first infection.
secondary lymphoid organs Peripheral lymphoid organs where mature lymphocytes function in immune responses; they include the adenoids,

tonsils, spleen, appendix, and lymph nodes, among others.

secondary metabolites Metabolic products synthesized during late-log and stationary phase.

secondary structure Refers to the arrangement of amino acids in a protein; the two major arrangements are helices and sheets.

secondary treatment In treatment of wastewater, a biological process designed to convert most of the suspended solids to microbial mass and inorganic compounds.

segmented virus Virus that has a genome consisting of multiple different nucleic acid fragments.

selectable marker Gene that encodes a selectable phenotype such as antibiotic resistance.

selective enrichment Method of increasing the relative proportion of one particular species in a broth culture by including a selective agent that inhibits the growth of other species.

selectively permeable Material that allows some but not other molecules to pass through freely.

selective medium Culture medium that inhibits the growth of certain microorganisms and therefore favors the growth of desired microorganisms.

selective toxicity Causing greater harm to a pathogen than to the host.

self-assembly Spontaneous formation of a complex structure from its component molecules without the aid of enzymes.

self-transmissible plasmid Plasmid that codes for all of the information necessary for its own transfer.

semiconservative replication Nucleic acid replication that results in each of the two double-stranded molecules containing one of the original strands (the template strand) and one newly synthesized strand.

semicritical instruments Medical instruments such as endoscopes that come into contact with mucous membranes, but do not penetrate body tissues.

semipermeable Describes material that allows the passage of some but not other molecules.

sense strand Of the two RNA molecules that can theoretically be transcribed from double-stranded DNA, the one that can be translated to make a protein; also called the plus (+) strand.

sensitization Prior immunization; allergic reactions to an antigen occur only in sensitized individuals who have been exposed to that particular antigen.

sepsis A bloodstream infection.

septic shock An array of effects including fever, drop in blood pressure, and disseminated intravascular coagulation, that results from infection of the bloodstream or circulating endotoxin.

septic tank A large tank used for individual sewage treatment systems; wastes are collected in the tank and degraded by anaerobic organisms, with the organic compounds in the resulting fluid degraded in a drainage field by aerobic organisms.

septicemia Acute illness caused by infectious agents or their products circulating in the bloodstream; blood poisoning.

serial dilutions Series of dilutions, usually twofold or tenfold, used to determine the titer or concentration of a substance in solution.

seroconversion Change from serum without specific antibodies to serum positive for specific antibodies.

serogroup Microorganisms within a species that are the same antigenically as determined by specific antisera.

serology The study of *in vitro* antibody-antigen reactions.

serotype A group of strains that have a characteristic antigenic structure that differs from other strains; also called a serovar.

serum Fluid portion of blood that remains after blood clots.

sex pilus Thin protein appendage required for attachment of two bacteria prior to DNA transfer by conjugation. Also called F pilus.

shake tube Tube of agar medium that has been uniformly inoculated with a bacterial culture in order to determine the oxygen requirement of that organism.

shingles (or **herpes zoster)** Condition resulting from the reactivation of the varicella-zoster virus.

shock Condition with multiple causes characterized by low blood pressure and circulation of the blood inadequate to sustain normal function of vital organs; septic shock results from growth of microorganisms in the body; toxic shock results from a circulating exotoxin.

siderophore Iron-binding substance produced by bacteria to scavenge iron.

sigma (σ) factor Component of RNA polymerase that recognizes and binds to the promoter.

signal sequence Characteristic series of hydrophobic amino acids at the amino terminal end of a protein destined for secretion; functions as a tag, directing transport of the protein through the membrane.

signal transduction Process that transmits information from outside of a cell to the inside, allowing that cell to respond to changing environmental conditions.

signature sequences Characteristic sequences in the genes encoding ribosomal RNA that can be used to classify or identify certain organisms.

signs Effects of a disease observed by examining the patient.

similarity coefficient Numerical value that can be used to classify prokaryotes based on their phenotypic characteristics.

simple diffusion Movement of molecules or ions in solution from a region of high concentration to a region of low concentration; does not involve transport proteins.

simple staining Staining technique that employs a basic dye to impart color to cells.

single-cell protein (SCP) Use of microorganisms such as yeast and bacteria as a protein source.

site-specific recombination Mechanism by which a piece of DNA becomes part of a larger piece of DNA; involves identical sequences on each piece of DNA.

size standard In gel electrophoresis of DNA, a series of DNA fragments of known sizes added to a lane of the gel to be used as a basis for later size comparison.

skin-associated lymphoid tissue (SALT) Secondary lymphoid tissue consisting of collections of lymphoid cells under the skin.

slime layer Type of glycocalyx that is diffuse and irregular.

slime mold Terrestrial organism that is similar to the fungi but not related genetically.

slow infection An infection that takes a long period of time before symptoms appear.

smear In a staining procedure, the film obtained by placing a drop of a liquid containing a microbe on a glass microscope slide and allowing it to air dry.

smooth endoplasmic reticulum Organelle of eukaryotic cells that is the site of lipid synthesis and degradation and calcium ion storage.

solute Dissolved molecule in a solution.

SOS repair Complex, inducible repair process used to repair highly damaged DNA.

specialized transduction Transfer of only specific bacterial genes by phage from one bacterium to another.

species Group of related isolates or strains; the lowest basic unit of taxonomy.

spikes (or **attachment proteins)** Structures on the outside of the virion that bind to host cell receptors.

spirillum (pl. **spirilla)** Curved rod long enough to form spirals.

spirochetes Long helical bacteria that have a flexible cell wall and an axial filament.

splicing Process that removes introns from eukaryotic precursor RNA to generate mRNA.

spoilage Biochemical changes in foods that are perceived as undesirable.

spontaneous generation Discredited theory that organisms can arise from non-living matter.

spontaneous mutation Mutation that occurs naturally without the addition of mutagenic agents.

spore Type of differentiated, specialized cell formed by certain organisms; includes some types of dormant cells that are resistant to adverse conditions and the reproductive structures formed by fungi.

sporozoite Elongated infectious form of certain protozoa; for example, in malaria, the form entering the body from a mosquito bite, infectious for liver cells.

sporulation In bacteria, a complex, highly ordered sequence of morphological changes during which a bacterial vegetative cell produces a specialized cell greatly resistant to environmental adversity; eukaryotes can also undergo sporulation.

spread plate Technique used to cultivate bacteria by uniformly spreading a suspension of cells onto the surface of an agar plate.

sputum Material coughed from the lungs.

start codon Codon at which translation is initiated; in prokaryotes, typically the first AUG after a ribosome-binding site.

starter cultures Strains of microorganisms added to a food to initiate the fermentation process.

stationary phase Stage of growth of a culture in which the number of viable cells remains constant.

stereoisomer (or **optical isomer)** Mirror image of a compound.

sterilant A chemical used to destroy all microorganisms and viruses in a product, rendering it sterile.

sterile Completely free of all microorganisms and viruses; an absolute term.

sterilization The process of destroying or removing all microorganisms and viruses through physical or chemical means.

steroid Type of lipid with a specific four-membered ring structure.

stock culture Culture stored for use as an inoculum in later procedures.

stop codon Codon that does not code for an amino acid and is not recognized by a tRNA; signals the end of the polypeptide chain.

strain Population of cells descended from a single cell.

streak plate Simplest and most commonly used technique for isolating bacteria; a series of successive streak patterns is used to sequentially dilute an inoculum on the surface of an agar plate.

streptococcal pyrogenic exotoxins (SPEs) Family of genetically related toxins produced by certain strains of *Streptococcus pyogenes*, responsible for scarlet fever, toxic shock, and "flesh-eating" necrotizing fasciitis.

stromatolite Coral-like mat of filamentous microorganisms.

structural isomers Molecules that contain the same elements but in different arrangements that are not mirror images; structural isomers have different names and properties.

subclinical Infection or disease with no apparent symptoms.

substrate (1) Substance on which an enzyme acts to form products. (2) Surface on which an organism will grow.

substrate-level phosphorylation Transfer of the high-energy phosphate from a phosphorylated compound to ADP to form ATP.

sucrose Disaccharide consisting of a molecule of glucose bonded to fructose; common table sugar.

sugar-phosphate backbone Series of alternating sugar and phosphates moieties of a DNA molecule.

sulfa drugs Group of antimicrobial drugs that inhibit folic acid synthesis.

sulfanilamide Antimicrobial drug that inhibits folic acid synthesis; one of the sulfa drugs.

sulfate-reducers Group of obligate anaerobes that use sulfate (SO_4^{2-}) as a terminal electron acceptor, producing hydrogen sulfide as an end product.

sulfur-oxidizing bacteria Group of Gram-negative bacteria that obtain energy by oxidizing elemental sulfur and reduced sulfur compounds, thereby generating sulfuric acid.

S unit Unit of measurement that expresses the sedimentation rate of a compound; reflects the mass and density of the compound; "S" stands for Svedberg.

superantigens Molecules that stimulate T lymphocytes by binding to MHC class II molecules and to part of the T-cell receptor distinct from the antigen-binding site, resulting in activation of many T cells, overproduction of cytokines, severe reactions, and sometimes fatal shock.

superficial mycoses Fungal infections that affect the hair, skin, or nails.

superoxide (O_2^-) Toxic derivative of O_2.

superoxide dismutase Enzyme that degrades superoxide to produce hydrogen peroxide.

surface receptors Proteins in the membrane of a cell to which certain signal molecules bind; they enable the inner workings of the cell to sense and respond to signals outside of the cell.

swarmer cells Motile cells of sheathed bacteria that disperse to new locations.

symbiosis The living together of two dissimilar organisms or symbionts.

symptoms Effects of a disease experienced by the patient.

syncytium (pl. **syncytia**) Multinucleate body formed by the fusion of cells.

synergistic Describes the acting together of agents to produce an effect greater than the sum of the effects of the individual agents.

synergistic infection An infection in which two or more species of pathogens act together to produce an effect greater than the sum of effects if each pathogen were acting alone.

synthetic medium Medium in which the chemical composition and quantity of every component is known.

systemic infection Infection in which the infectious agent spreads throughout the body.

systemic mycoses Fungal infections that affect the tissues deep in the body.

---

tapeworm A helminth that has a segmented, ribbon-shaped body; a cestode.

***Taq* polymerase** Heat-stable DNA polymerase of the thermophilic bacterium *Thermus aquaticus*.

target cell In immunology, cell that is the direct recipient of a T cell's effector functions.

target DNA In the PCR procedure, the region to be amplified.

taxa Groups into which organisms are classified.

taxonomy The science that studies organisms in order to arrange them into groups; those organisms with similar properties are grouped together and separated from those that are different. Taxonomy encompasses identification, classification, and nomenclature.

T_C cells Effector form of a cytotoxic T cell; it induces apoptosis in infected or cancerous "self" cells.

T-cell receptor Molecule on a T cell that enables the T cell to recognize a specific antigen.

T cells Lymphocytes that mature in the thymus; they are responsible for cellular immune responses and function as helper cells in the antibody response.

T-dependent antigens Antigens that evoke an antibody response only with the participation of T_H cells.

T-DNA Portion of the Ti plasmid of *Agrobacterium tumefaciens* that is transferred into a plant cell.

teichoic acids Component of the Gram-positive cell wall, composed of chains of a common subunit, either ribitol-phosphate or glycerol-phosphate, to which various sugars and D-alanine are usually attached.

temperate phage Bacteriophage that can either become integrated into the host cell DNA as a prophage or replicate outside the host chromosome leading to cell lysis.

template Strand of nucleic acid that a polymerase uses to synthesize a complementary strand.

terminal electron acceptor Chemical that is ultimately reduced as a consequence of chemotrophic metabolism.

tertiary structure Level of structure of a protein described by its three-dimensional nature; two major shapes exist, globular and fibrous.

tertiary treatment In treatment of wastewater, any purification process beyond secondary treatment, generally designed to remove nitrates and phosphates.

tetracyclines Group of antimicrobial medications that interfere with protein synthesis.

T_H cell Effector form of a helper T cell; it activates B cells and macrophages, and releases cytokines that stimulate other aspects of the immune system.

therapeutic index Ratio of minimum toxic dose to minimum effective dose of a medication.

thermophile Organism with an optimum growth temperature between 45°C and 70°C.

thrush Infection of the mouth by *Candida albicans*.

thylakoids Membrane-bound disclike structures within the stroma of chloroplasts; they contain chlorophyll.

thymine dimer Two adjacent thymine molecules on the same strand of DNA joined together through covalent bonds.

thymus Primary lymphoid organ, located in the upper chest, in which T lymphocytes mature.

T-independent antigens Antigens that can activate B cells without the assistance of a T_H cell.

Ti plasmid (See *tumor-inducing plasmid*.)

tissue culture Culture of plant or animal cells that grows in an enriched medium outside the plant or animal.

titer Measure of the concentration of a substance in solution; for example, the amount of a specific antibody in serum, usually measured as the highest dilution of serum that will test positive for antibody.

tolerance Specific unresponsiveness of the adaptive immune system that reflects its ability to ignore any given molecule, such as a normal cellular protein.

toll-like receptors (TLRs) A group of surface receptors that recognize specific compounds unique to microbes, enabling the cell to sense the presence of invading microbes and then alert other components of the host's defenses.

total coliforms Facultative, non-spore-forming, Gram-negative rods that ferment lactose, producing acid and gas within 48 hours at 35°C; because most typically reside in the intestine, they are used as indicators of fecal pollution.

toxemia Circulation of toxins in the bloodstream.

toxin Poisonous chemical substance.

toxoid Modified form of a toxin that is no longer toxic but is able to stimulate the production of antibodies that will neutralize the toxin.

trace elements Elements that are required in very minute amounts by all cells; they include cobalt, zinc, copper, molybdenum, and manganese.

trachoma Potentially serious chronic eye disease caused by certain strains of *Chlamydia trachomatis*.

transamination Transfer of an amino group from an amino acid to a recipient compound, converting the recipient compound to an amino acid.

transcript Fragment of RNA, synthesized using one of the two strands of DNA as a template.

transcription Process of transferring genetic information coded in DNA into messenger RNA (mRNA).

transcytosis Transport of a substance across a border made up of cells; a cell takes up the substance from one side of the border and then releases it to the other side.

transduction Mechanism of gene transfer between bacteria in which bacterial DNA is transferred inside a phage.

transfer RNA (tRNA) Type of RNA that delivers the appropriate amino acid to the ribosome during translation.

transferrin An iron-binding protein found in blood and tissue fluids.

transformed cells Bacterial, animal, or plant cells containing inheritable changes.

transfusion reaction Reaction characterized by fever, low blood pressure, pain, nausea, and vomiting, resulting from the transfusion of immunologically incompatible blood.

transgenic Plants and animals into which new DNA has been introduced.

transient expression Expression of a gene for a short period of time. Often refers to expression of genes introduced into cells and not integrated into the genome.

transient microbiota Microorganisms that are only temporary residents of body sites.

transition step Step in metabolism that links glycolysis to the TCA cycle; converts pyruvate to acetyl-CoA.

translation Process by which genetic information in the messenger RNA directs the order of amino acids in protein.

translocation Advancement of a ribosome a distance of one codon during translation.

transmissible spongiform encephalopathies Group of fatal neurodegenerative diseases of humans and animals in which brain tissue develops spongelike holes.

transmission electron microscope (TEM) Type of microscope that directs a beam of electrons at a specimen; used to observe fine details of cell structure.

transport protein Type of protein found in cell membranes that functions in the transport of certain compounds across the membrane; may be called a permease or a carrier.

transport systems Mechanisms used to transport small molecules across the cytoplasmic membrane.

transposable element (or **transposon**) Gene that moves from one DNA molecule to another within the same cell or from one site on a DNA molecule to another site on the same molecule.

transposition Movement of a piece of DNA from one site in a molecule to another site in the same cell.

trematodes Flatworms known as flukes.

tricarboxylic acid (TCA) cycle Metabolic pathway that incorporates acetyl-CoA, ultimately generating ATP (or GTP), CO_2, and reducing power, also called the Krebs cycle and the citric acid cycle.

trichomes Filamentous multicellular associations of cyanobacteria that may or may not be enclosed within a sheath.

trickling filter method Treatment method for small sewage plants in which a rotating arm sprays sewage onto a bed of rocks coated with a biofilm of organisms that aerobically degrades the wastes.

triglyceride Molecule consisting of three molecules of the same or different fatty acids bonded to glycerol.

trimethoprim Antimicrobial medication that interferes with folic acid synthesis.

tRNA Transfer RNA.

trophozoite Vegetative feeding form of some protozoa.

tubercle Granuloma formed in tuberculosis.

tumble Rolling motion of a motile cell that is caused by an abrupt change in the direction of rotation of flagella.

tumor-inducing (Ti) plasmid Plasmid of *Agrobacterium tumefaciens* that enables the organism to cause tumors in plants; a derivative of the plasmid is used as a vector by scientists to introduce DNA into plants.

tumor necrosis factors (TNFs) A group of cytokines that play an important role in the inflammatory response and other aspects of immunity.

turbidity Cloudiness; the turbidity of a bacterial suspension is proportional to the number of cells in that suspension.

tyndallization Repeated cycles of heating and incubation to kill spore-forming bacteria.

type III secretions system Mechanism by which bacterial pathogens transfer gene products directly into host cells.

---

ubiquity Widespread prevalence.

ultra-high-temperature (UHT) method A method that uses heat to render a product free of all microorganisms that can grow under normal storage conditions.

ultraviolet (UV) light Electromagnetic radiation with wavelengths between 175 and 350 nm; invisible.

uncoating In virology, the separation of the protein coat from the nucleic acid of the virion.

unsaturated Refers to a fatty acid with one or more double bonds.

upstream Direction toward the 5′ end of either an RNA molecule or the analogous (+) strand of DNA.

urea A waste product of protein catabolism by the body's cells; present in various body fluids, notably urine.

urticaria Hives; an allergic skin reaction characterized by the formation of itchy red swellings.

UV Abbreviation for ultraviolet light.

---

vaccine Preparation of attenuated or inactivated microorganisms or viruses or their components used to immunize a person or animal against a particular disease.

vancomycin Antimicrobial medication that interferes with peptidoglycan synthesis.

variable region The portion of an antibody molecule that contains the antigen-binding sites; tremendous variation exists between the amino acid sequences of variable regions in different antibody molecules.

vector (1) In molecular biology, a piece of DNA that acts as a carrier of a cloned fragment of DNA. (2) In epidemiology, any living organism that can carry a disease-causing microbe; most commonly arthropods such as mosquitoes and ticks.

vegetative cell Typical, actively multiplying cell.

vehicle Inanimate carrier of an infectious agent from one host to another.

vertical transmission Transfer of a pathogen from a pregnant woman to the fetus, or from a mother to her infant during childbirth.

vibrio (pl. **vibrios**) Short, curved rod-shaped bacterial cell.

villus (pl. **villi**) Narrow protrusion from a membrane such as the intestinal lining.

viremia Viruses circulating in the bloodstream.

virion Viral particle in its inert extracellular form.

viroid Piece of RNA that does not have a protein coat but does replicate within living cells.

virulence Relative ability of a pathogen to overcome body defenses and cause disease.

virulence determinants Arsenal of mechanisms of pathogenicity of a given microbe.

virus Acellular or non-living agent composed of nucleic acid surrounded by a protein coat.

vitamin One of a group of organic compounds found in small quantities in natural foodstuffs that are necessary for the growth and reproduction of an organism; usually converted into coenzymes.

volutin Storage form of phosphate found inside certain bacterial cells; because granules of volutin exhibit characteristic staining with the dye methylene blue, they are called metachromatic granules.

---

water activity (a_w) Quantitative measure of the water available.

water molds Non-photosynthetic members of the heterokons; similar to the fungi but not related genetically.

Western blotting Procedure that uses labeled antibody molecules to detect specific proteins.

whey Liquid portion that remains after milk proteins coagulate during cheese-making.

wide host range plasmid Plasmid that can replicate in unrelated bacteria.

wild type Form of an organism that is isolated from nature.

---

yeast Unicellular fungus.

---

zone of inhibition Region around a chemical saturated disc where bacteria are unable to grow due to adverse effects of the compound in the disc.

zoonosis (pl. **zoonoses**) Disease of animals that can be transmitted to humans.

zooplankton Floating and swimming small animals and protozoa found in marine environments, usually in association with the phytoplankton.

zygote Diploid cell formed by the sexual fusion of two haploid cells.

CREDITS

Chapter 1 Opener: © Science VU/Visuals Unlimited; **1.3:** © Bettmann/Corbis; **1.5:** © Kathy Talaro/Visuals Unlimited; **1.6:** © Cabisco/Visuals Unlimited; **1.7a:** © Dr. John D. Cunningham/Visuals Unlimited; **1.7b:** © C. Shih & R. Kessel/Visuals Unlimited; **1.8:** © Manfred Kage/Peter Arnold; **1.9a:** © K.G. Murti/Visuals Unlimited; **1.9b:** © Thomas Broker/Phototake; **1.9c:** © K.G. Murti/Visuals Unlimited; **1.10:** © U.S. Department of Agriculture/Dr. Diemer; **1.11:** © Stanley B. Prusiner/Visuals Unlimited; **Perspective 1.1 (Figure 1):** Courtesy of Milton P. Gordon and Syed Tanveer Haider; **Perspective 1.1 (Figure 2):** Courtesy of Dr. Heide N. Schulz/Max Planck Institute for Marine Microbiology; **Perspective 1.1 (Figure 3):** Courtesy of Reinhard Rachel and Harald Huber, University of Regensburg, Germany.

Chapter 2 Opener: © Scott Camazine/Photo Researchers, Inc.; **Perspective 2.1 (Figure 1a):** © Mark Antman/The Image Works; **Perspective (Figure 1b):** © SIU/Peter Arnold

Chapter 3 Opener: © Terry Beveridge/Biological Photo Service; **3.1:** Courtesy of Leica, Inc., Deerfield, FL; **3.2 (both):** © W. A. Jensen; **3.3a:** © Richard Megna/Fundamental Photographs; **3.4:** © Wim van Egmond/Visuals Unlimited; **3.5:** © P. Johnson/Photo Researchers, Inc.; **3.6:** © T. E. Adams/Visuals Unlimited; **Table 3.1 (Bright field):** From S. T. Williams, M.E. Sharpe and J.G. Holt (Eds.), Bergey's Manual of Systematic Bacteriology, Vol. 4. © 1989 Williams and Wilkins Co., Baltimore. Micrograph from T. Cross, University of Bradford, Bradford, U.K.; **Table 3.1b (Phase -Contrast):** © Wim van Egmond/Visuals Unlimited; **Table 3.1 (Interference):** © P. Johnson/Photo Researchers, Inc.; **Table 3.1 (Dark-field):** © T. E. Adams/Visuals Unlimited; **Table 3.1 (Fluorescence):** © Evans Roberts; **Table 3.1 (Confocal):** Courtesy of Michael W. Davidson/National High Magnetic Field Lab; **Table 3.1 (Transmission):** © R. Kessel & C. Shih/Visuals Unlimited; **Table 3.1 (Scanning):** © David M. Phillips/Visuals Unlimited; **Table 3.1 (Atomic Force Microscope):** © Torunn Berge/Photo Researchers, Inc.; **3.7:** © Evans Roberts; **3.8 (both):** Courtesy of Michael W. Davidson/National High Magnetic Field Lab; **3.10a:** © R. Kessel & C. Shih/Visuals Unlimited; **3.10b:** © H. Aldrich/Visuals Unlimited; **3.11:** © David M. Phillips/Visuals Unlimited; **3.12:** © Torunn Berge/Photo Researchers, Inc.; **3.14b:** © Leon J. Le beau/Biological Photo Service; **3.15:** Public Health Image Library/Centers for Disease Control, Photo by Dr. George P. Kubica; **3.16:** © Dr Gladden Willis/Visuals Unlimited/Getty; **3.17:** © Jack M. Bostrack/Visuals Unlimited; **3.18:** © E. Chan/Visuals Unlimited; **3.19a:** Courtesy of Molecular Probes, Eugene, OR; **3.19b:** © Richard L. Moore/Biological Photo Service; **3.19c:** © Evans Roberts; **3.20a, b:** © David M. Phillips/Visuals Unlimited; **3.20c:** © Dennis Kunkel/Phototake; **3.20d:** © Veronika Burmeister/Visuals Unlimited; **3.20e, f:** © David M. Phillips/Visuals Unlimited; **3.21a:** Courtesy of Walther Stoeckenius; **3.21b:** Courtesy of James T. Staley; **3.22a (top):** © George Musil/Visuals Unlimited; **3.22a (bottom):** © David M. Phillips/Visuals Unlimited; **3.22b:** © R. Kessel & C. Shih/Visuals Unlimited; **3.22c:** © Oliver Mecks/Photo Researchers, Inc; **3.23b:** Courtesy of L. Santo, H. Hohl, and H. Frank, Journal of Bacteriology 99:824, 1969. American Society for Microbiology; **3.31:** Courtesy of Dale C. Birdsell; **Table 3.5:** © Leon J. Lebeau/Biological Photo Service; **3.33c, 3.34d:** © Terry Beveridge, University of Guelph; **3.35:** Courtesy of Dr. Edwin S. Boatman; **3.36a:** Courtesy of K.J. Cheng and J. W. Costerton; **3.36b:** Courtesy of A. Progulske and S.C. Holt, Journal of Bacteriology, 143:1003-1018, 1980; **3.37a:** © Fred Hossler/Visuals Unlimited; **3.37b:** © Science VU/Visuals Unlimited; **3.40:** © D. Blackwill and D. Maratea/Visuals Unlimited; **3.41a:** Courtesy of Dr. Charles Brinton, Jr.; **3.41b:** © U.S. Department of Agriculture/Harley W. Moon; **3.42a:** © CNRI/SPL/Photo Researchers, Inc.; **3.42b:** © Dr. Gopal Murti/SPL/Photo Researchers; **3.44:** Courtesy of Dr. Edwin S. Boatman; **3.45:** Courtesy of J.F.M. Hoeniger, J. Bacteriol. 96:1835, 1968. American Society for Microbiology; **3.47c:** Courtesy of Thomas Fritsche; **3.52b:** © Garry T. Cole/Biological Photo Service; **3.53b:** © Keith Porter/Photo Researchers, Inc.; **3.54:** © George Chapman/Visuals Unlimited/Getty; **3.55:** © R. Bolendar & D. Fawcett/Visuals Unlimited; **3.56:** Courtesy of Charles J. Flickinger

Chapter 4 Opener: © Michael Klein/VISUM/The Image Works; **4.2:** © Dr. Dennis Kunkel/Visuals Unlimited; **4.3:** Courtesy of the Center for Biofilm Engineering, Montana State University-Bozeman, Image by P. Stoodley; **4.4:** © Michael Gabridge/Visuals Unlimited; **4.5:** © Fred Hossler/Visuals Unlimited; **4.11a:** © Christine Case/Visuals Unlimited; **4.11b:** © L. M. Pope and D. R. Grote/Biological Photo Service; **4.12:** © Dr. Elmer Koneman/Visuals Unlimited; **4.13:** © Becton, Dickson and Company; **4.14:** Courtesy of Thermoforma, Forma Scientific Division; **4.19a:** © Dennis Kunkel/Phototake; **4.19b:** © Kathy Talaro/Visuals Unlimited; **4.21:** © Richard Megna/Fundamental Photographs

Chapter 5 Opener: © Southern Illinois University/Photo Researchers, Inc.; **5.1a:** © Yoav Levy/Phototake; **5.1c, d:** © Rosenfeld Images/LTD/SPL/Photo Researchers, Inc.; **5.1d:** © Bob Daemmrich/The Image Works; **5.1e:** © Nieto/Herrican/Photo Researchers, Inc.; **5.4 (both):** © Evans Roberts; **5.6:** © Pall/Visuals Unlimited; **5.8:** © Dennis Kunkel/Phototake

Chapter 6 Opener: © Brock May/Photo Researchers, Inc.; **6.2:** © Farrell Grehan/Photo Researchers, Inc.; **6.3 (top):** © Photodisc Vol. Series 74, photo by Robert Glusie; **6.3 (bottom):** © Jane Burton/Bruce Coleman, Inc; **6.9b, c:** From Voet: Biochemistry, 1/e, 0471617695, 1990, John Wiley & Sons; **6.21 (acetone), 6.21 (wine, beer):** © Brian Moeskau/McGraw-Hill; **6.21 (acetic acid, lactic acid):** © Dennis Strete/Fundamental Photographs; **6.21 (cheese):** © Photodisc/McGraw-Hill; **6.21 (yogurt, dairy, pickle):** © Brian Moeskau/McGraw-Hill; **6.21 (Formic acid, Ethyl alcohol):** © Dennis Strete/Fundamental Photographs

Chapter 7 Opener: © D. Struthers/Toney Stone Images/Getty; **7.4:** From J. Cairns, "The Chromosome of E. coli" in Cold Spring Harbor Symposia on Quantitative Biology, 77. © 1963 by Cold Spring Harbor Laboratory Press

Chapter 8 Opener: © Dr. Gopal Murti/SPL/Custom Medical Stock Photo; **8.6:** © Matt Meadows/Peter Arnold, Inc; **8.22:** Courtesy of C. Brinton, Jr. and J. Carnahan

Chapter 9 Opener: © Jean Claude Revy/Phototake; **9.2b:** © Richard T. Nowitz/Photo Researchers, Inc.; **9.5:** Courtesy of Pamela Silver and Jason Karana, Harvard Medical School, Dana Farber Cancer Institute. Photo provided by Howell Martin; **9.9:** Reprinted with permission from Edvotek, Inc.; **9.13:** © Josh Westrich/zefa/Corbis; **9.20:** Dr. R. Rachel, and Prof. Dr. K.O. Stetter, University of Regensburg, Lehrstuhl fuer Mikrobiologie,

Regensburg, Germany.; **9.21:** Courtesy of Rosetta Inpharmatics © 2002 Rosetta Inpharmatics, Inc. All rights reserved.

Chapter 10 Opener: © Meckes/Ottawa/Photo Researchers, Inc.; **10.2a:** © George Wilder/Visuals Unlimited; **10.2b:** Courtesy of Dr. Thomas R. Fritsche, M.D., Ph.D., Clinical Microbiology Division, University of Washington, Seattle; **10.3a:** © E. Koneman/Visuals Unlimited; **10.3b:** © Biophoto Associates/Photo Researchers, Inc.; **10.4a:** © Denise Anderson; **10.4b,c:** © Dennis Strete/Fundamental Photographs; **10.6a:** Courtesy of bioMerieux, Inc.; **10.6b:** Courtesy of Becton, Dickinson and Company; **10.9b:** Courtesy of Patricia Exekiel Moor; **10.10c, 10.11:** © Evans Roberts

Chapter 11 Opener, 11.1a: © F. Widdel/Visuals Unlimited; **11.1b:** © Ralph Robinson/Visuals Unlimited; **11.2a:** © Thomas Tottleben/ Tottleben Scientific Company; **11.2b:** © David M. Phillips/Visuals Unlimited; **11.3:** © John Walsh/Photo Researchers, Inc.; **11.4a:** Courtesy of Dr. Heinrich Kaltwasser. From ASM News 53(2): Cover, 187.; **11.4b:** From M. P. Starr et al (Eds.), The Prokaryotes. Springer-Verlag.; **11.5:** From J. G. Holt (Ed.), The Shorter Bergey's Manual of Determinative Bacteriology, 8/e, 1977. Williams and Wilkins Co., Baltimore; **11.6a:** Courtesy of Isao Inouye, University of Tsukuba, Mark A. Shneegurt, Wichita State University and Cyanosite (www.cyanosite.bio.purdue.edu/ index.html); **11.6b:** © M. I. Walker/Photo Researchers, Inc.; **11.7:** Courtesy of Wayne Carmichael, Wright State University, Mark A. Schneegurt, Wichita State University and Cyanosite (www.cyanosite.bio. purdue.edu/index.html); **11.8:** Courtesy of Roger Burks, University of California Riverside, Mark A. Schneegurt, Wichita State University and Cyanosite (www.cyanosite.bio.purdue.edu/index.html); **11.9a:** Courtesy of J. T. Staley and J. P. Dalmasso; **11.9b:** Courtesy of Dr. James Staley; **11.10:** © John D. Cunningham/Visuals Unlimited; **11.11:** © Dennis Kunkel/Phototake; **11.12:** © Evans Roberts; **11.13:** © CNRI/Phototake; **11.15a:** © Eric Grave/Photo Researchers, Inc.; **11.15b:** © Soad Tabaqchali/ Visuals Unlimited; **11.16 (both):** Courtesy of Harold L. Sadoff; **11.17a:** Courtesy of Mary F. Lampe; **11.17b:** © Patricia L. Grilione/Phototake; **11.18:** From S. T. Williams, M.E. Sharpe and J.G. Holt (Eds.), Bergey's Manual of Systematic Bacteriology, Vol. 4. © 1989 Williams and Wilkins Co., Baltimore. Micrograph from T. Cross, University of Bradford, Bradford, U.K.; **11.19, 11.20a:** © John D. Cunningham/Visuals Unlimited; **11.20b:** From N. R. Krieg and J. G. Holt (Eds.), Bergey's Manual of Systematic Bacteriology, Vol. 1, 1984. Williams and Wilkins Co., Baltimore; **11.21, 11.22a:** Courtesy of J. T. Staley and J. P. Dalmasso; **11.23a:** From J.T. Staley, M.P. Bryant, N. Pfennig, and J.G. Holt (Eds.) Bergey's Manual of Systematic Bacteriology, Vol. 3. © 1989 Williams and Wilkins Co., Baltimore; **11.24a:** © Alfred Pasieka/Peter Arnold, Inc.; **11.25b: 11.25a:** Courtesy of Amy Cheng Vollmer, Swarthmore College; **11.25b:** © Kenneth Lucas/Biological Photo Service; **11.26:** Courtesy of Betsy L. Williams; **11.27:** © Michael Gabridge/Visuals Unlimited; **11.28:** © Alfred Pasieka/Peter Arnold, Inc.; **11.29:** Courtesy of R. R. Friis, J. Bacteriol. 110:706, 1972. American Society for Microbiology; **11.30:** © Wayne P. Armstrong, Life Sciences Department, Palomar College; **11.31:** Dr. R. Rachel, and Prof. Dr. K.O. Stetter, University of Regensburg, Lehrstuhl fuer Mikrobiologie, Regensburg, Germany.; **11.32:** US Geological Survey, Photo by David E. Wieprecht

Chapter 12 Opener: © John D. Cunningham/Visuals Unlimited; **12.2a:** © Cabisco/Visuals Unlimited; **12.2b:** © John D. Cunningham/Visuals Unlimited; **12.3:** © Martha Nester; **12.5:** © David M. Phillips/Visuals Unlimited; **12.6:** © Arthur M. Siegelman/Visuals Unlimited; **12.7 (all):** Courtesy of Fritz Schoenknecht; **12.9a:** © E. Chan/Visuals Unlimited; **12.9b:** © Inga Spence/Tom Stack & Associates; **12.9c:** © Bob Pool/Tom Stack & Associates; **12.10:** © A.B. Dowsette/SPL/Photo Researchers; **12.13a:** © Martha Nester; **12.13b:** © Bruce Iverson; **12.14b, 12.15 (all):** © Martha Nester; **12.16b:** © Carolina Biological Supply/Phototake; **12.18, 12.20, 12.21:** Courtesy of Sam Eng and Fritz Schoenknecht; **Future Chalanges, p. 299:** Centers for Disease Control

Chapter 13 Opener: © Oliver Mecks/Photo Researchers, Inc; **13.1a:** Courtesy of Harold Fisher, University of Rhode Island and Robley Williams, University of California Berkeley; **13.1b:** © Dennis Kunkel/Phototake; **13.1c:** © Thomas Broker/Phototake; **13.8:** Courtesy of Harold Fisher; **13.12a:** © Prof. L. Caro/SPL/Photo Researchers; **13.12b:** Courtesy of Dr. Rafael Martinez, University of California, Los Angeles; **13.12c:** © Oliver Mecks/Photo Researchers, Inc.

Chapter 14 Opener: © Scott Camazine/Photo Researchers, Inc.; **14.1a:** © CDC/Science Source/Photo Researchers; **14.1b:** © K. G. Murti/Visuals Unlimited; **14.1c:** Courtesy of Dr. Frederick A. Murphy, Centers for Disease Control and Prevention; **14.6b:** © Dr. Dennis Kunkel/Visuals Unlimited; **14.12:** © Michael Gabridge/Custom Medical Stock; **14.13 (all):** Courtesy of M. Cooney; **14.14:** © Terry Hazen/Visuals Unlimited; **DROP:** © E. Chan/ Visuals Unlimited; **14.15:** Courtesy of Fred Williams, National Exposure Research Laboratory/U.S. EPA; **14.16b:** Courtesy of M. Cooney; **14.17a:** © Dean Glawe/Biological Photo Service; **14.17b:** © Bruce Iverson; **14.17c:** © Dean Glawe/Biological Photo Service; **14.18:** © Science VU/Visuals Unlimited; **14.19:** © Barbara J. Miller/Biological Photo Service; **14.20:** © BSIP/Photo Researchers

Chapter 15 Opener: © David M. Phillips/Visuals Unlimited; **15.8b:** Reprinted from Wendell F. Rosse et al., "Immune Lysis of Normal Human and Parozysmal Nocturnal Hemogloginuria (PNH) Red Blood Cells," Journal of Experimental Medicine, 123:969, 1996. Rockefeller University Press; **15.10:** © Meckes/Ottawa/SPL/Photo Researchers, Inc.

Chapter 16 Opener: © Lennart Nilsson/Boehringer Ingelheim International GMBH; **16.11a:** Courtesy of Dr. James W. Vardiman; **16.11b:** © NIBSCS/ SPL/Photo Researchers; **16.11c:** © D. Fawcett/Visuals Unlimited

Chapter 17 Opener: © Kevin Noran/Tony Stone/Getty; **17.5:** Courtesy of Chihiro Sasakawa, University of Tokyo; **17.6:** Courtesy of Mark A. Jepson, from Trends in Microbiology (98)v6:359-365, "Studying M cells and their role in infection"; M.A. Jepson and M.A. Clark, Elsevier Press; **17.8:** Courtesy of Pascale F. Cossart, From B. Finlay and P. F. Cossart, Science 276:720, April 4, 1997

Chapter 18 Opener: © Mario Tama/Getty Images; **18.2:** © STU/Custom Medical Stock; **18.6:** © Bart's Medical Library/Phototake; **18.7e:** © Public Health Image Library; **18.8:** © Kenneth E. Greer/Visuals Unlimited; **18.9:** © SIU/Visuals Unlimited

Chapter 19 Opener, 19.1 (top left): © SPL/Photo Researchers; **19.1 (top right):** © Dag Sundberg/Photographer's Choice/Getty Images; **19.1 (bottom left):** © BSIP/Phototake; **19.1 (bottom right):** © SPL/Photo Researchers; **19.3:** © Jean Claude Revy - ISM/Phototake; **19.5b:** © Dr. Dan Schadler/ Oglethorpe University; **19.9:** © Science VU/Wellcome/Visuals Unlimited; **19.10:** © Centers for Disease Control; **19.12:** © Bob Coyle/McGraw-Hill; **19.13:** Courtesy of Genelabs Diagnostics

Chapter 20 Opener: © Gianni Tortoli/Photo Researchers; **20.1a (top):** © Photodisc/McGraw-Hill; **20.1a (bottom):** © Corbis Digital Stock CD/ McGraw-Hill; **20.1b (hands):** © Photodisc/McGraw-Hill; **20.1b (tick):** Courtesy of University of Nebraska/Department of Entomology; **20.1b (phone):** © Photodisc/McGraw-Hill; **20.1b (sneeze):** © Kent Wood/Photo Researchers; **20.1b (foods):** © Photodisc/McGraw-Hill; **20.3 (both):** © Joan K. Mann; **20.9:** © WHO, Geneva Switzerland; **20.11:** © B. Seitz/ Photo Researchers

Chapter 21 Opener: © SPL/Photo Researchers; **21.5 (both):** Reprinted with permission of Robert L. Perkins from A. S. Klainer and R. L. Perkins, Journal of Infectious Diseases, 122:4, 1970 © University of Chicago Press; **21.10:** © Fred Marsik/Visuals Unlimited; **21.11:** Courtesy bioMerieux, St. Louis, MO; **21.12 (both):** AB Biodisk

Chapter 22 Opener: © Mediscan/Visuals Unlimited; **22.2, 22.3, 22.5:** © Evans Roberts; **22.7:** © SPL/Photo Researchers; **22.9:** © A.B. Dowsette/SPL/Photo Researchers; **22.10:** © LeBeau/Custom Medical Stock Photo; **22.11 (both):** © St. Francis Hospital/Joan K. Mann; **22.12a:** © Evans Roberts; **22.12b:** © Larry Jensen/Visuals Unlimited; **22.13:** Courtesy of J.B. Baseman; from P.C. Hu, A.M. Collier, and J.B. Baseman "J. Exp. Med." 145:1328, 1977. Rockefeller University Press.; **22.14:** Courtesy of Dr. Kenneth Bromberg; **22.17:** © Biophoto Associates/Photo Researchers, Inc.; **22.18a:** © John D. Cunningham/Visuals Unlimited; **22.18b:** © CNRI/Phototake; **22.19:** © SPL/Photo Researchers; **22.20:** © Science VU/Visuals Unlimited; **22.24a:** Centers for Disease Control; **22.24b:** © E. Chan/Visuals Unlimited; **22.26 (both):** © Evans Roberts

Chapter 23 Opener: © SPL/Photo Researchers; **23.1:** Courtesy of Karen A. Holbrook; **23.2a, b:** © Carroll H. Weiss/Camera M.D. Studios; **23.2c:** Courtesy of Glenn D. Roberts, Mayo Clinic; **23.3:** © SPL/Photo Researchers; **23.4:** Courtesy of National Institute Slide Bank/The Welcome Center for Medical Sciences; **23.5:** © SPL/Photo Researchers; **23.6:** © Evans Roberts; **23.7:** © Ken Greer/Visuals Unlimited; **23.8:** Courtesy of Rocky Mountain Laboratory, Hamilton, Montana, CDC; **23.10:** © Ed Reschke/Peter Arnold, Inc.; **23.11:** © SPL/Photo Researchers; **23.12:** © Charles W. Stratton/Visuals Unlimited; **23.14:** © C. C. Duncan/Medical Images, Inc.; **23.15a:** © CDC/Dr. Libero Ajello; **23.15b:** © Dr. G. W. Willis/Visuals Unlimited; **23.16:** © John D. Cunningham/Visuals Unlimited; **23.17:** © Carroll H. Weiss/Camera M.D. Studios; **23.18:** © Oliver Mecks/Photo Researchers, Inc; **23.19:** © Lowell Georgia/Photo Researchers, Inc; **23.20:** © Dr. P. Marazzi/SPL/Photo Researchers, Inc; **23.21:** © Carroll H. Weiss/Camera M.D. Studios; **23.23 (both):** © CDC; **23.24:** © CNRI/SPL/Photo Researchers, Inc.; **23.25a:** © Dr. P. Marazzi/SPL/Photo Researchers, Inc; **23.25b:** © Evans Roberts; **23.26a:** © Dr. P. Marazzi/SPL/Photo Researchers, Inc; **23.26b:** © A.M. Siegelman/Visuals Unlimited

Chapter 24 Opener: © Moore/Custom Medical Stock Photo; **24.3:** © Michael English/Medical Images, Inc.; **24.4:** © Evans Roberts; **24.5:** © Kenneth E. Greer/Visuals Unlimited; **24.6a:** © Leonard Morse, M.D./Medical Images, Inc.; **24.6b:** © Carroll H. Weiss/Camera M.D. Studios; **24.6c:** © A.M. Siegelman/Visuals Unlimited; **24.7:** Centers for Disease Control; **24.8:** © A.M. Siegelman/Visuals Unlimited; **24.10, 24.11:** Courtesy of National Institute Slide Bank/The Welcome Center for Medical Sciences; **24.12:** Courtesy of Dr. John Slack; **24.13:** © Kenneth E. Greer/Visuals Unlimited; **24.14:** © Leonard Morse, M.D./Medical Images, Inc.

Chapter 25 Opener: © Garry Watson/Photo Researchers, Inc; **25.3:** © Fred Hossler/Visuals Unlimited; **25.5 (both), 6a:** From S. Schluger, R.A. Yaodels and R.C. Page, "Periodontal Disease," 1977. © Lea & Febiger, Philadelphia; **25.6b:** © Evans Roberts; **25.9a:** © Dr P. Marazzi/Photo Researchers, Inc.; **25.9b:** © Frederick C. Skvara, M.D.; **25.10:** © Science VU/Visuals Unlimited; **25.12:** © Veronika Burmeister/Visuals Unlimited; **25.14:** Courtesy of Philippe J. Sansonette, M.D., Professeur Institut Pasteur; **25.15:** © David M. Rollins/Visuals Unlimited; **25.16:** © Moredun Animal Health Ltd./Photo Researchers, Inc.; **25.17:** © Science Source/Photo Researchers, Inc.; **25.22:** © P.M. Motta & F. M. Magliocca/SPL/Photo Researchers, Inc.; **25.23:** © Michael Abbey/Photo Researchers, Inc.; **25.24a:** © Oliver Meckers/Photo Researchers, Inc.

Chapter 26 Opener: © Kwangshin Kim/Photo Researchers, Inc.; **26.3:** © Science VU/Fred Marsik/Visuals Unlimited; **26.4:** © Custom Medical Stock Photo; **26.5:** Centers for Disease Control; **26.8:** © Biological Photo Service/Terraphotographics; **26.9:** © Biophoto Associates/Photo Researchers, Inc.; **26.10:** Courtesy of Stanley Falkow; **26.11:** Courtesy of Morris D. Cooper, Ph.D., Professor of Medical Microbiology, Southern Illinois University School of Medicine, Springfield, IL; **26.12:** Courtesy of National Institute Slide Bank/The Welcome Center for Medical Sciences; **26.13:** Courtesy of S. Eng and F. Schoenknecht; **26.14:** Courtesy of Thomas Diepgen, University Heidelberg; **26.15:** © Carroll H. Weiss/Camera M.D. Studios; **26.16:** © L. Winograd/Biological Photo Service; **26.17a:** © Carroll H. Weiss/Camera M.D. Studios; **26.17b:** © Harvey Blank, M.D./Camera M.D. Studios; **26.18:** Courtesy of National Institute Slide Bank/The Welcome Center for Medical Sciences; **26.19:** © Biophoto Associates/Photo Researchers, Inc.; **26.20:** © Dr. F. C. Skvara/Peter Arnold, Inc; **26.21:** © David M. Phillips/Visuals Unlimited

Chapter 27 Opener: © Dr. Linda Stannard, UCT/Science Photo Library/Photo Researchers; **27.4:** © John Durham/SPL/Photo Researchers, Inc.; **27.5:** © Kenneth Greer/Visuals Unlimited; **27.6:** © Mary K. Stallone/The Medical File Inc.; **27.8:** © Palo Koch/Photo Researchers, Inc; **27.9:** © John D. Cunningham/Visuals Unlimited; **27.10:** © Science VU/Dr. Soad Tabaqchali/Visuals Unlimited; **27.13:** Courtesy of National Institute Slide Bank/The Welcome Center for Medical Sciences; **27.14:** © Evans Roberts; **27.15:** © Bettmann/Corbis; **27.17a:** © Tektoff-BM/CNRI/SPL/Photo Researchers, Inc.; **27.17b:** © Tobey Sanford; **27.18:** © Evans Roberts; **27.19:** © Mary K. Stallone/The Medical File Inc.; **27.20:** © Cabisco/Visuals Unlimited; **27.21:** © Dr. Terry Kreeger/Visuals Unlimited

Chapter 28 Opener: © Dr. Tony Brain/SPL/Photo Researchers, Inc; **28.2:** © Kenneth Greer/Visuals Unlimited; **28.4:** © Science Source/Photo Researchers, Inc; **28.6:** © A.M. Siegelman/Visuals Unlimited; **28.7 (both):** Courtesy of Joe Hinnebusch, From Hinnebusch, B.J., Perry, R.D., and Schwan, T.G., Science 273:367, 19 July 1996.; **28.9a:** © Ed Reschke/Peter Arnold, Inc.; **28.9b:** © Jacquin Carrillo-Farga/Science Source/Photo Researchers, Inc.

Chapter 29 Opener: © Manfred Kage/Peter Arnold, Inc.; **29.2:** Courtesy of Stuart Fischman, D.M.D; **29.12:** © St. Mary's Hospital Medical School/SPL/Photo Researchers, Inc.; **29.13:** © Cecil H. Fox/Photo Researchers; **29.14a:** © E. Koneman/Visuals Unlimited; **29.14c:** Courtesy of Dr. Thomas R. Fritsche, M.D., Ph.D., Clinical Microbiology Division, University of Washington, Seattle; **29.15:** © Dr. F.C. Skvara/Peter Arnold, Inc.; **29.16 (top):** © ISM/Phototake; **29.16 (bottom):** © Paula Inhad/Custom Medical Stock Photo; **29.17:** Courtesy of Dr. Edwin P. Ewuing, Jr./Centers for Disease Control

Chapter 30 Opener: © Grant Heilman Photography; **30.4:** From Y. Cohen and E. Rosenberg, "Microbial Mats," Fig. 1a, pg. 4, 1989, American Society for Microbiology; **30.5:** © SPL/Photo Researchers; **30.8:** © Nancy Pearsall; **30.11a:** © WHOI-D. Foster/Visuals Unlimited; **30.12b:** © Stephan Imhof/Philipps-Universität; **30.12c:** © R.L. Peterson/Biological Photo Service; **30.13b:** © John D. Cunningham/Visuals Unlimited; **p. 737:** Courtesy of J.P. Dalmasso

Chapter 31 Opener: © Myrleen Cate/PhotoEdit; **31.7:** Courtesy of IDEXX Laboratories, Inc.; **31.8:** © Alan L. Detrick/Photo Researchers, Inc.; **31.9:** © Phil Degginger/Color-Pic, Inc.; **31.11 (both):** © Science VU/Visuals Unlimited

Chapter 32 Opener: © Michael Newman/PhotoEdit; **32.1:** © Paul Webster/Tony Stone/Getty; **32.2 (all):** © Martha Nester; **32.3 (all):** Courtesy of the Wisconsin Milk Marketing Board; **32.6 (all):** © Martha Nester; **32.7 (left):** © Rachael Epstein/PhotoEdit; **32.7 (right):** © William J. Weber/Visuals Unlimited; **p. 764 (both):** © Felicia Martinez/PhotoEdit

INDEX

Page numbers followed by *t* and *f* indicate tables and figures, respectively.

B

C

D

G

H

J

K

L

N

O

P

Q

T

U

W

X

Y

Z

Nobel Prizes Awarded for Research in Microbiology or Immunology

Year	Winners	Research in Microbiology	Research in Immunology
1901	Emil Adolph von **Behring**		Developed serum treatment, especially in diphtheria
1902	Sir Ronald **Ross**	Discovered the life cycle of the malaria parasite in humans and mosquitoes	
1905	Robert **Koch**	Developed scientific methods in bacteriology and proved the cause of tuberculosis	
1907	Charles Louis Alphonse **Laveran**	Showed that protozoa cause some infectious diseases	
1908	Ilya Ilyich **Mechnikov** Paul **Ehrlich**		Studied immune reactions and phagocytic cells
1912	Alexis **Carrel**		Developed organ and blood vessel transplantation
1913	Charles Robert **Richet**		Discovered and treated anaphylactic shock
1919	Jules **Bordet**		Made fundamental discoveries in immunity
1927	Julius **Wagner-Jauregg**	Used malaria to treat late-stage syphilis	
1928	C. J. H. **Nicolle**	Made fundamental discoveries on typhus	
1930	Karl **Landsteiner**		Discovered the ABO human blood groups
1939	Gerhard **Domagk**	Discovered sulfa drugs	
1945	Sir Alexander **Fleming** Sir E. B. **Chain** Lord H. W. **Florey**	Discovered and developed penicillin	
1951	Max **Theiler**		Developed vaccine for yellow fever
1952	Selman A. **Waksman**	Discovered streptomycin	
1954	John F. **Enders** T. H. **Weller** F. C. **Robbins**	First to grow poliovirus in cell cultures, making a polio vaccine possible	
1957	Daniel **Bovet**		Developed antihistamines and synthetic curare
1958	George W. **Beadle** E. L. **Tatum** Joshua **Lederberg**	Discovered genetic recombination in *Escherichia coli* and proved the relationship of genes to enzymes	
1959	Severo **Ochoa** Arthur **Kornberg**	Made fundamental discoveries on the synthesis of DNA and RNA	
1960	F. M. **Burnet** Peter B. **Medawar**		Discovered the basis of acquired immunological tolerance
1962	F. H. C. **Crick** J. D. **Watson** H. F. **Wilkins**	Elucidated the molecular structure of DNA	
1965	Francois **Jacob** Andre **Lwoff** Jacques **Monod**	Made fundamental discoveries on gene regulation in bacteria	
1966	Charles B. **Huggins** Francis Peyton **Rous**		Studied the role of hormones in causing cancer; demonstrated that viruses can cause cancer in animals